MICROCOMPUTER THEORY AND SERVICING

FOURTH EDITION

Stuart M. Asser, PE

Queensborough Community College of the City University of New York

Vincent J. Stigliano, PE

Maui Community College of the University of Hawaii

Richard F. Bahrenburg

Northrop Grumman Corporation

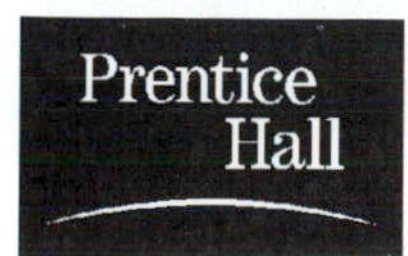

Upper Saddle River, New Jersey
Columbus, Ohio

Library of Congress Cataloging in Publication Data

Asser, Stuart.
Microcomputer theory and servicing / Stuart M. Asser, Vincent J. Stigliano, Richard F. Bahrenburg.—4th ed.
p. cm.
ISBN 0-13-010955-X
1. Microcomputers—Maintenance and repair. I. Stigliano, Vincent. II. Bahrenburg, Richard. III. Title.

TK7887.A88 2001
621.39′16—dc21 00-029847

Vice President and Publisher: Dave Garza
Editor in Chief: Stephen Helba
Assistant Vice President and Publisher: Charles E. Stewart, Jr.
Production Editor: Alexandrina Benedicto Wolf
Production Coordination: York Production Services
Design Coordinator: Robin G. Chukes
Cover Designer: Brian Huber
Cover Image: PhotoDisc
Production Manager: Matthew Ottenweller
Marketing Manager: Barbara Rose

This book was set in Times Roman by York Graphic Services, Inc. It was printed and bound by R.R. Donnelley & Sons Company. The cover was printed by Phoenix Color Corp.

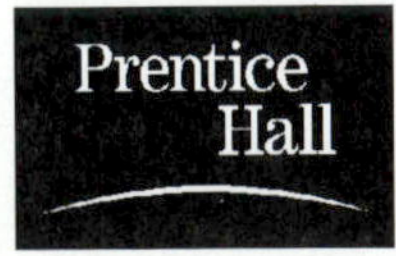

10 9 8 7 6 5 4 3 2 1
ISBN 0-13-010955-X

For Our Families

Diane, Jason, Rebecca
Jeane, Joey, Robby
Maureen, Rachel, Daniel

I Tuoi Genitori E I Tuoi Maestri
Non Si Possono Mai Ringraziare Abbastanza

(You can never thank your family, friends, and teachers enough)

PREFACE

The computer service industry has been growing in geometric proportions for many years. That growth would not have been possible without the bright young minds that have emerged from our schools. The next several years will place an even greater burden on our educational institutions as the need for computer specialists, engineers, technicians, programmers, service personnel, and salespeople continues to grow.

Over the past 50 years, remarkable changes have taken place in digital computers. The amount of computational power that once required a roomful of vacuum tubes can now be found in small calculators. Complex calculations that would once have taken days to perform can now be done in seconds. Yet with all these changes, certain fundamental design concepts have remained unchanged. The basic organization and operating design of ENIAC (one of the first digital computer systems) is still used in the supercomputers of today; that is, data bits are retrieved from memory, brought to the central processing unit (CPU) to be operated on, and returned to memory for storage. What has changed is the size and speed of the modern-day computer system.

Not so long ago, computers were huge pieces of equipment. They were hidden in antiseptic rooms and were operated by gurus who seemed to speak a different language. The advent of the microprocessor made computer systems available to everyone. Secretaries replaced their typewriters with computers and plastic disks containing word-processing programs. Accountants put away their ledger pads in favor of more efficient electronic spreadsheets. Managers are relying less on educated guesses; instead, they are basing their decisions on information gained from the computer's ability to do hundreds of complex calculations almost instantly. Globally, bankers, merchandisers, police, firefighters, and defense and government personnel all use computers. Today, we can't register our car at the motor vehicle bureau or land an airplane at an airport without a computer. Meanwhile, out of the arcades came the video game, introducing millions of American families to the computer as entertainment. Probably the most significant advancement in the past few years is the Internet. Although the Internet has been around since 1969, the Internet we know today as the World Wide Web has existed only since 1993. Remarkably, today there are more microprocessors in the world than people.

Due to the widespread use of computers, it has become a serious matter whenever one fails to work. Quick, efficient, reliable, and competent repair is necessary. Most people and businesses cannot afford to have their computer inoperative for extended periods of time. Thus, computer service has become one of the largest segments of the electronics industry today.

Microcomputer Theory and Servicing, Fourth Edition, is intended for a one- or two-semester course at two- and four-year colleges and technical institutes. It can also be used for high school electronics programs and industry-based training programs, or as a reference for practicing technicians and engineers. The only prerequisite knowledge is a basic understanding of electrical circuit theory and electronic devices.

This edition is designed to take the student from the introductory concepts of microprocessor operation to the analysis and servicing of the complete microcomputer system. The authors are all seasoned computer-service professionals. The emphasis is on the need-to-know information, practical applications, and the fault symptoms and analysis required for computer repair.

The text is organized into four parts to allow flexibility in application to different courses and/or different levels. Part I, Essentials of Digital Logic, presents the fundamental concepts of the real-world considerations of digital logic circuits. Part II, Basics of Microprocessors, introduces the microprocessor and its support devices. The Intel microprocessor family is used to present a balanced hardware/software approach to this subject. Part III, Computer Systems and Peripherals, exposes the theory of operation and servicing of various components that together form the microcomputer system. The IBM-compatible PC is used as a representative model of a modern computer system. Part IV, Installation and Service, describes the skills and tools necessary for the setup, installation, and servicing of a microcomputer system. Customer relations as well as diagnostics and troubleshooting are emphasized.

The reading level has been kept as simple as possible. A list of key terms is presented at the beginning of each chapter. When a key term is first introduced, it is defined and printed in boldface type. Illustrated examples are presented using a step-by-step method that includes the use of standard units where applicable. The illustrated examples and homework exercises become progressively more challenging as the use of practical examples and circuit values are introduced. Each chapter also contains a crossword puzzle exercise and a section titled Tech Tips and Troubleshooting—T^3. The crossword puzzle incorporates the key terms found in the chapter and is intended to provide an interesting and entertaining way of building the student's technical vocabulary. The Tech Tips and

Troubleshooting—T^3 section offers realistic examples of what, and why, things go wrong with a computer system and how to fix common problems. It also serves as a chapter summary.

The text follows a consistent format and pedagogy. Each chapter contains key terms, an introduction, sequentially numbered equations, illustrated examples, Tech Tips and Troubleshooting—T^3, homework exercises, and a crossword puzzle. Many illustrated examples, charts, diagrams, photographs, and tables are used to present the material.

Parts II, III, and IV of the fourth edition have been completely revised to reflect the rapid changes that have taken place over the past five years in personal computer systems. Chapter 9, which includes an introduction to computer programming, has been revised and expanded into two chapters. Chapter 9 provides a solid introduction to assembly language with an expanded coverage of addressing modes. Chapter 10 is a completely new chapter that guides the student through the steps required to create executable assembly language programs. Microsoft's MASM, A86 (a popular shareware assembler), and DEBUG are presented in detail. Chapter 11 now includes expanded coverage of memory and I/O address decoding as well as a section on building a basic, single-board (a PC board is available) 8088 microprocessor system. The building project is intended to give students practical, hands-on experience that incorporates design, building, and programming. It uses a stage-by-stage approach and is easily expandable. Chapter 12 has been completely rewritten to include coverage of all of the popular microprocessors, chipsets, bus structures, motherboards, and memory. Chapter 13 has been completely revised and updated to include coverage of all of the new storage devices and their interfaces. Chapter 14 has also been completely revised and updated with expanded coverage of peripherals such as video monitors, printers, and modems. An entirely new Chapter 15 provides an introduction to PC networking, including theory, setup, and servicing. Chapter 16 presents a completely new and updated coverage of theory, setup, installation, and troubleshooting. Chapter 17 builds on the previous chapter to cover diagnostic tools and troubleshooting. Both chapters have been completely rewritten and geared to today's PCs. Finally, a section on A+ Certification has been added to Chapter 18.

Acknowledgments

The development and production of this type of textbook requires the energy and dedication of many individuals. The authors feel fortunate to have had the opportunity to work with the many talented professionals at Prentice Hall, particularly Charles Stewart, Kate Linsner, and Alex Wolf, and Kirsten Kauffman of York Production Services.

We also acknowledge the following reviewers for sharing their knowledge and expertise with us in this project: Nasser Hedayat, Valencia Community College; Jeremy Smith, Thompson Institute; and Donald Wade, Nassau Community College.

Every author and teacher must learn his or her craft. Our learning began at Queensborough Community College of the City University of New York. We would like to express our deepest gratitude and respect to our colleagues and teachers at Queensborough, particularly Professors Joseph B. Aidala (in memoriam), Robert L. Boylestad, Leon Katz, Gabriel Kousourou, Edward Leff, Louis Nashelsky, and Peter Stark for teaching, guiding, and inspiring us.

Finally, the authors especially thank their wives, Diane Asser, Jeanne Stigliano, and Maureen Bahrenburg, for their patience, encouragement, and love during the writing of this book.

Stuart M. Asser
Vincent J. Stigliano
Richard J. Bahrenburg

CONTENTS

MICROCOMPUTER THEORY AND SERVICING

PART

ESSENTIALS OF DIGITAL LOGIC

I

The modern computer age dawned in 1946 with the completion of the Electronic Numerical Integrator Analyzer and Computer (ENIAC) by scientists and engineers at the University of Pennsylvania. The first electronic computer consisted of over 18,000 vacuum tubes and filled a very large room. Today the same processing power is available in a hand-held calculator.

While ENIAC is considered the grandparent of the modern computer, the history of computers actually began many years before ENIAC, with the invention of the abacus. The abacus was invented in Babylonia around 3000 B.C. It was made up of sliding beads on a frame and was used to perform arithmetic. Over 4000 years later, the next scientific milestone was the invention of a multiplication scheme by John Napier in 1617. Napier constructed a lattice of rods that could perform multiplication. Napier's "bones," as they became known, were essentially a movable column of rods. Between 1623 and 1674 a number of mechanical calculators were invented. Among the more famous of these were the mechanical calculators invented by the French mathematician Blaise Pascal in 1645 and the slide rule, invented by William Oughtred and Richard Delamain in 1630.

Whether or not they actually worked, Charles Babbage's ideas and designs for a mechanical calculator, known as a differential engine and analytical engine, were the inspiration for the modern day computer. These designs were hampered by the limitations of technical machining of the early 1800s.

The next significant event in computer history was the publication by George Boole in 1847 of *The Mathematical Analysis of Logic*. Boole's work led to the mathematics of computers known as Boolean algebra.

In 1890 Herman Hollerith invented punch cards and tabulating equipment for the U.S. census. Later, Hollerith formed his own company called Computing–Tabulating–Recording Corporation (CTR). In 1914 Thomas Watson, Sr. joined CTR, which became IBM in 1924.

During the next 20 years, a great deal of research and development was done on various types of electromechanical calculators. Also during this time, Lee DeForrest developed the triode vacuum tube, W.H. Eccles and F.W. Jordan developed the flip-flop, and Claude Shannon showed the feasibility of switching circuits in electronic applications. These events culminated in the first fully functional electronic calculator, ENIAC, in 1946.

ENIAC was 8 ft high and 80 ft long and weighed 30 tons. Its 18,000 vacuum tubes, 70,000 resistors, 10,000 capacitors, 1500 relays, and 6000 manual switches consumed approximately 175 kW of power. This enormous machine was probably less powerful than your hand-held calculator. However, it demonstrated the feasibility of the electronic computer and led to the development of the modern computer.

Part I, "Essentials of Digital Logic," presents the fundamental concepts and skills of digital logic circuits, which are the basis of today's computers.

CHAPTER 1

NUMBER SYSTEMS AND CODES

KEY TERMS

Alphanumeric Codes
ASCII
Binary
Binary-coded Decimal (BCD)
Binary Point
Bit
Code
Complementing
Even Parity
Hexadecimal
LSB (Least Significant Bit)
MSB (Most Significant Bit)
Octal
Odd Parity
1s Complement
Parity Bit
2s Complement

1.0 INTRODUCTION—THE ONE/ZERO CONCEPT

The base 10, or decimal, number system is familiar to all of us. It is based on the 10 fingers of the hands. We learn to use the decimal number system at a very early age. Therefore, it has become a natural method of counting for us. Why then must we use a different number system in today's computers? The first computing machines were mechanical and operated on the base 10 number system. For example, the abacus, Babbage's engine, and the many mechanical calculators that followed were all designed for the base 10 number system. The first electronic computing machines were an outgrowth of telephone switching circuits. Switches are inherently binary in nature. They are either open or closed. This led to the use of the binary number system to describe switching circuits.

The **binary,** or base 2, number system has only two digits, zero (0) and one (1). This number system is very easy to reproduce electrically. The *on/off* states of a switch can be represented by the binary digit 1 and the binary digit 0. A light can be *on* or *off.* Voltage can be *high* or *low.* Current can *flow* or *not flow.* Magnetism can be *north* or *south.* Questions can be answered *yes* or *no.* As you can see, these examples are all binary in nature.

1.1 THE DECIMAL NUMBER SYSTEM

The decimal, or base 10, number system has 10 digits, 0 through 9. It is known as a *positional number system.* The value of any number depends on the position of the digits. For example, the number 747 can be represented as

7 hundreds	or	7×100	or	7×10^2
4 tens	or	4×10	or	4×10^1
7 units	or	7×1	or	7×10^0

From this example we can see that all sevens are not equal. Beginning at the leftmost digit, the first seven occupies the hundreds, or 10^2, position. It is called the *most significant digit* (MSD). Continuing to the right, the 4 occupies the tens, or 10^1, position. The last digit, 7, occupies the units, or 10^0, position. It is called the *least significant digit* (LSD). The total value of the number is based on the position of the digits and can be computed as follows:

$$7 \times 10^2 + 4 \times 10^1 + 7 \times 10^0 = 747$$

$$\text{or} \quad 700 + 40 + 7 = 747$$

From this example we can see that the value of each position is a power of the base 10. The power or exponents can be either *positive, negative,* or *zero,* as shown in Table 1.1.

TABLE 1.1
Powers of 10

$10^0 = 1$	
$10^1 = 10$	$10^{-1} = 1/10 = 0.1$
$10^2 = 100$	$10^{-2} = 1/100 = 0.01$
$10^3 = 1{,}000$	$10^{-3} = 1/1{,}000 = 0.001$
$10^4 = 10{,}000$	$10^{-4} = 1/10{,}000 = 0.0001$
$10^5 = 100{,}000$	$10^{-5} = 1/100{,}000 = 0.00001$
$10^6 = 1{,}000{,}000$	$10^{-6} = 1/1{,}000{,}000 = 0.000001$

Let's consider another example, 1492.76. In this case we have a *decimal,* or *radix,* point, which is used to separate a number into its whole and fractional parts. In this example, the number 1492.76 can be represented as

1 thousand	or	1×1000	or	1×10^3
4 hundreds	or	4×100	or	4×10^2
9 tens	or	9×10	or	9×10^1
2 units	or	2×1	or	2×10^0
7 tenths	or	7×0.1	or	7×10^{-1}
6 hundredths	or	6×0.01	or	6×10^{-2}

The value of the number can be computed as follows:

$$1 \times 10^3 + 4 \times 10^2 + 9 \times 10^1 + 2 \times 10^0 + 7 \times 10^{-1} + 6 \times 10^{-2} = 1492.76$$

$$\text{or} \quad 1000 + 400 + 90 + 2 + 0.7 + 0.06 = 1492.76$$

From these examples we can conclude that the value of any number can be expressed as *the sum of the products of each digit multiplied by its positional value.* Figure 1.1 illustrates the positional value of the base 10 number system expressed in scientific notation.

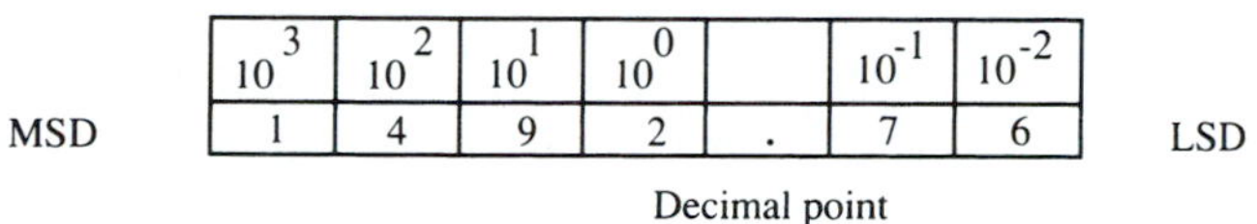

FIGURE 1.1
Base 10 number system

1.2 THE BINARY NUMBER SYSTEM

The binary, or base 2, number system has only two digits, 0 and 1. It is also a positional number system. In this case the positions are weighted as powers of the base 2. Table 1.2 illustrates the positional values for the binary or base 2 number system.

TABLE 1.2
Powers of 2

	$2^0 = 1$
$2^1 = 2$	$2^{-1} = 1/2 = 0.5$
$2^2 = 4$	$2^{-2} = 1/4 = 0.25$
$2^3 = 8$	$2^{-3} = 1/8 = 0.125$
$2^4 = 16$	$2^{-4} = 1/16 = 0.0625$
$2^5 = 32$	$2^{-5} = 1/32 = 0.03125$
$2^6 = 64$	$2^{-6} = 1/64 = 0.015625$
$2^7 = 128$	$2^{-7} = 1/128 = 0.0078125$
$2^8 = 256$	$2^{-8} = 1/256 = 0.00390625$

The radix point is called the **binary point** in the base 2 number system. Each binary digit is called a **bit** (binary digit). The leftmost digit or bit is referred to as the **most significant bit (MSB).** The rightmost digit or bit is called the **least significant bit (LSB),** as shown in Figure 1.2.

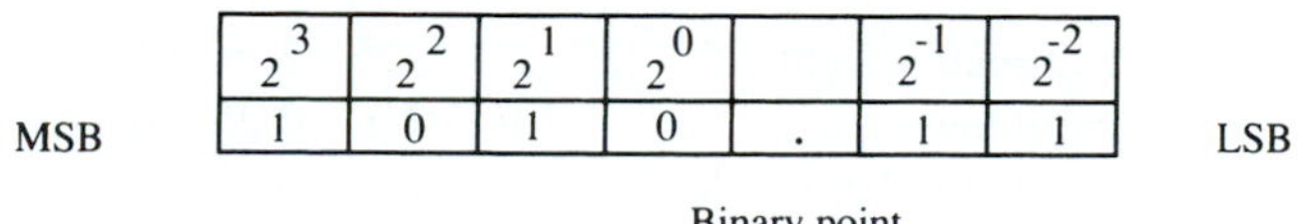

FIGURE 1.2
Binary number system

For example, the binary number $(1010.11)_2$ can be represented as

$$\begin{aligned}
1 \times 2^3 &= 1 \times 8 &&= (8)_{10} \\
0 \times 2^2 &= 0 \times 4 &&= (0)_{10} \\
1 \times 2^1 &= 1 \times 2 &&= (2)_{10} \\
0 \times 2^0 &= 0 \times 1 &&= (0)_{10} \\
1 \times 2^{-1} &= 1 \times 0.5 &&= (0.5)_{10} \\
1 \times 2^{-2} &= 1 \times 0.25 &&= (0.25)_{10}
\end{aligned}$$

The equivalent decimal value can be computed as follows:

$$\begin{aligned}
&1 \times 2^3 + 0 \times 2^2 + 1 \times 2^1 + 0 \times 2^0 + 1 \times 2^{-1} + 1 \times 2^{-2} \\
&= 1 \times 8 + 0 \times 4 + 1 \times 2 + 0 \times 1 + 1 \times 0.5 + 1 \times 0.25 \\
&= 8 + 0 + 2 + 0 + 0.5 + 0.25 = (10.75)_{10}
\end{aligned}$$

Note the use of the parentheses and subscript to signify the base of the number.

1.3 BINARY-TO-DECIMAL CONVERSION

As we have just seen, numbers can be converted from binary to decimal by computing the sum of the products of each digit multiplied by its positional value. Let's consider the following examples.

EXAMPLE 1.1 Convert the binary number $(11001)_2$ to decimal.

Solution

Positional weight	2^4	2^3	2^2	2^1	2^0
Positional value	16	8	4	2	1
Binary number	1	1	0	0	1

$$\begin{aligned}
&1 \times 2^4 + 1 \times 2^3 + 0 \times 2^2 + 0 \times 2^1 + 1 \times 2^0 \\
&= 1 \times 16 + 1 \times 8 + 0 \times 4 + 0 \times 2 + 1 \times 1 \\
&= 16 + 8 + 0 + 0 + 1 = (25)_{10}
\end{aligned}$$

EXAMPLE 1.2 Convert the binary number $(101.101)_2$ to decimal.

Solution

Positional weight	2^2	2^1	2^0	2^{-1}	2^{-2}	2^{-3}
Positional value	4	2	1	0.5	0.25	0.125
Binary number	1	0	1.	1	0	1

$$\begin{aligned}
&1 \times 2^2 + 0 \times 2^1 + 1 \times 2^0 + 1 \times 2^{-1} + 0 \times 2^{-2} + 1 \times 2^{-3} \\
&= 1 \times 4 + 0 \times 2 + 1 \times 1 + 1 \times 0.5 + 0 \times 0.25 + 1 \times 0.125 \\
&= 4 + 0 + 1 + 0.5 + 0 + 0.125 = (5.625)_{10}
\end{aligned}$$

Notice that in the preceding examples the process could have been simplified by leaving out all the zero terms. Furthermore, starting at the units position (2^0), each position to the left is multiplied by 2 (the base number) to obtain its positional value. Each position to the right is divided by 2 to obtain its positional value.

1.4 DECIMAL-TO-BINARY CONVERSION

The method of converting whole numbers from one base system to another is known as *repeated division.* The decimal number is divided by the base number to obtain a quotient plus a remainder. The remainder is then recorded as part of the binary number. The quotient is then divided by the base again to produce a second quotient and a second remainder. The second remainder becomes the second binary digit. The process is then continued until the quotient becomes zero. For example, in this case, since we are converting to binary we shall divide by 2. Note that because we are dividing by 2, the remainder can only be a 0 or a 1. The binary result is obtained by writing the first remainder as the LSB and the last remainder as the MSB.

EXAMPLE 1.3

Convert the decimal number $(29)_{10}$ to binary.

Solution

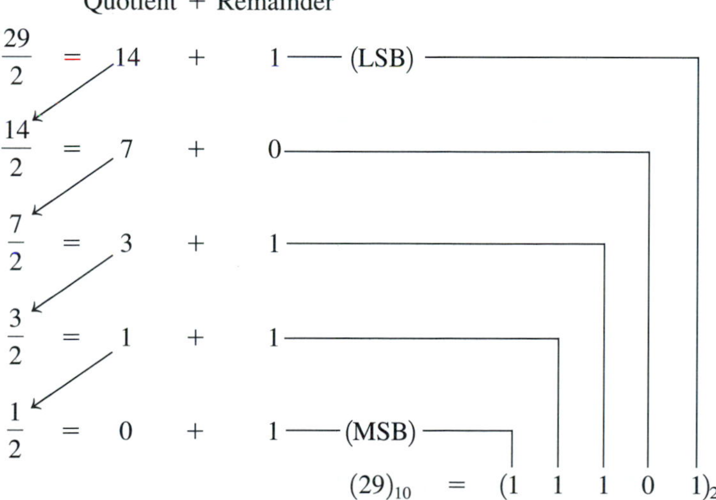

The method of converting fractional decimal numbers to binary is known as *repeated multiplication.* The fraction is multiplied by the base with the carry becoming the binary number. In this process, the first carry becomes the MSB and the last carry becomes the LSB.

EXAMPLE 1.4 Convert the fractional decimal number $(0.625)_{10}$ to binary.

Solution

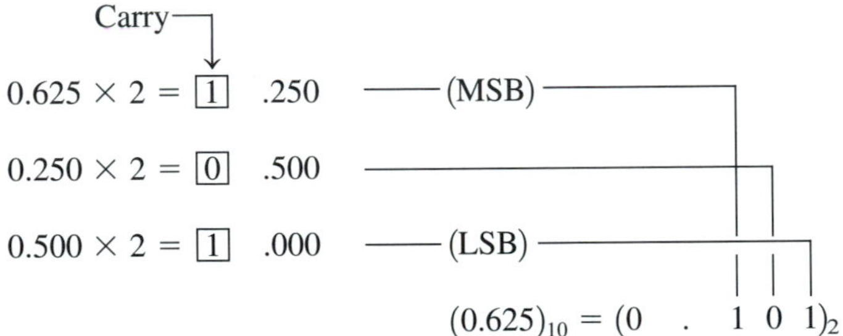

In summary, to convert whole numbers from decimal to binary, the following relationship can be used:

$$\frac{\text{Decimal number}}{\text{Base number}} = \text{quotient} + \text{remainder}$$

The remainders become the binary number. To convert fractional numbers from decimal to binary, the following relationship can be used:

$$\text{Fractional decimal number} \times \text{base number} = \text{carry} + \text{product}$$

The carries become the binary number.

1.5 THE OCTAL NUMBER SYSTEM

The **octal,** or base-8, number system has eight digits. They are represented by the symbols

0, 1, 2, 3, 4, 5, 6, 7

As with the binary and decimal number systems, the octal number system is a positional number system. Figure 1.3 illustrates the positional nature of the octal number system.

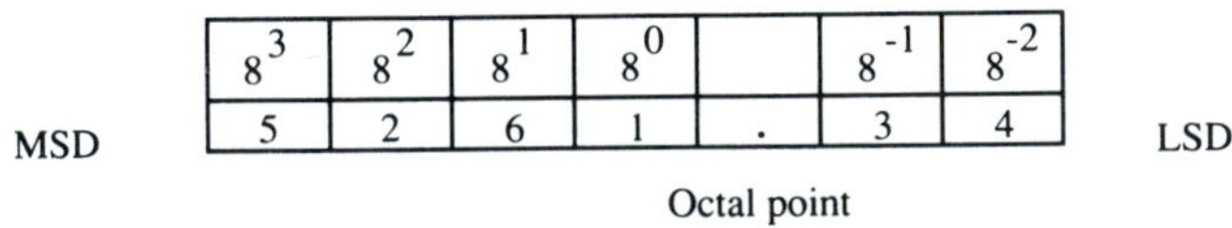

FIGURE 1.3
Octal number system

Table 1.3 illustrates the relationships between decimal, binary and octal numbers.

TABLE 1.3
Decimal, binary, and octal numbers

Decimal	Octal	Binary	Decimal	Octal	Binary
0	0	0000	9	11	1001
1	1	0001	10	12	1010
2	2	0010	11	13	1011
3	3	0011	12	14	1100
4	4	0100	13	15	1101
5	5	0101	14	16	1110
6	6	0110	15	17	1111
7	7	0111	16	20	10000
8	10	1000	17	21	10001

Note that the counting in all three systems is somewhat similar. All three number systems start at zero. In each case, the counts are incremented by 1. However, when we reach 2 or higher in the binary system (8 or higher in the octal system) we make the same move as is taken when we reach 10 in the decimal system. That is, we place a 1 in the next higher position and a 0 in the original position. Thus binary or octal counting is analogous to decimal counting with a carry to the next higher position each time a 2 or an 8 is reached instead of each time a 10 is reached.

1.6 OCTAL CONVERSIONS

Octal to Decimal Conversion

The procedure for converting an octal number to its equivalent decimal number is to *multiply each octal digit by its positional value and obtain the sum of the products.* Since the octal number system is a base-8 system, each position has a weight equal to a power of 8.

EXAMPLE 1.5 Convert the octal number $(137)_8$ to decimal.

Solution

Positional weight	8^2		8^1		8^0	
Positional value	64		8		1	
Octal number	1		3		7^1	
	1×8^2	+	3×8^1	+	7×8^0	
=	1×64	+	3×8	+	7×1	
=	64	+	24	+	7	$= (95)_{10}$

EXAMPLE 1.6 Convert the octal number $(24.1)_8$ to decimal.

Solution

Positional weight	8^1	8^0	8^{-1}
Positional value	8	1	0.125
Octal number	2	4.	1

$$\begin{aligned} & 2 \times 8^1 + 4 \times 8^0 + 1 \times 8^{-1} \\ = \; & 2 \times 8 + 4 \times 1 + 1 \times 0.125 \\ = \; & 16 + 4 + 0.125 = (20.125)_{10} \end{aligned}$$

Decimal-to-Octal Conversion

The method of converting a decimal number to an octal number is similar to the repeated-division method used to convert decimal numbers to binary numbers. The main difference is that in the case of decimal to octal conversions we *divide by the base number* 8.

EXAMPLE 1.7 Convert the decimal number $(149)_{10}$ to octal.

Solution

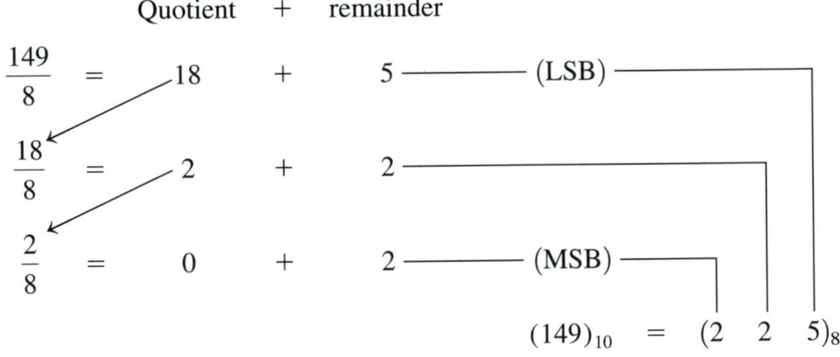

EXAMPLE 1.8 Convert the decimal number $(60.3)_{10}$ to octal.

Solution

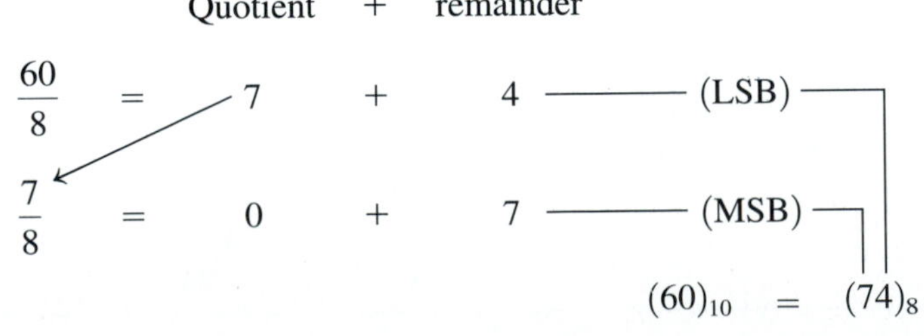

$$
\begin{aligned}
&\text{Carry} \downarrow \\
0.3 \times 8 &= \boxed{2}\ .4 \text{ (MSB)} \\
0.4 \times 8 &= \boxed{3}\ .2 \\
0.2 \times 8 &= \boxed{1}\ .6 \text{ (LSB)}
\end{aligned}
$$

$$(0.3)_{10} = (0.231)_8$$

Note: Because this is a repeating decimal, we have stopped the process after three iterations for three-place accuracy. The final answer is

$$(60.3)_{10} = (74.231)_8$$

Octal-to-Binary Conversion

Octal numbers are easily converted to the binary number system. In fact, octal numbers are used as a convenient way to express binary numbers. One octal number replaces *three binary digits,* as shown in Table 1.4.

TABLE 1.4
Octal/binary digits

Octal Digit	Binary Equivalent
0	000
1	001
2	010
3	011
4	100
5	101
6	110
7	111

To convert octal numbers to binary numbers, simply *replace each octal digit with its equivalent 3-bit binary number.*

EXAMPLE 1.9 Convert the following octal numbers to binary

a. $(64)_8$ b. $(156)_8$
c. $(37)_8$ d. $(714.12)_8$

Solution

a. 6 → 110, 4 → 100

$$(64)_8 = (110100)_2$$

b. 1 → 001, 5 → 101, 6 → 110

$$(156)_8 = (001101110)_2$$

c. 3 → 011, 7 → 111

$(37)_8 = (011111)_2$

d. 7 → 111, 1 → 001, 4 → 100, . → ., 1 → 001, 2 → 010

$(714.12)_8 = (111001100.001010)_2$

Binary-to-Octal Conversion

To convert a binary number to its octal equivalent, simply group the binary bits by three, starting at the radix point. If necessary, you may have to add *zeros* to complete the groups.

EXAMPLE 1.10 Convert the following binary numbers to octal.

a. $(101011)_2$ b. $(11111)_2$

Solution

a. 101 → 5, 011 → 3

$(101011)_2 = (53)_8$

b. 011 → 3, 111 → 7

$(11111)_2 = (37)_8$

Note the leading zero that was added in (b) to complete the group.

1.7 THE HEXADECIMAL NUMBER SYSTEM

Modern computers use another number system called the **hexadecimal** number system. The hexadecimal number system is a positional base 16 system. It uses the digits 0 through 9 and the letters A through F as its symbols, as shown in Table 1.5.

TABLE 1.5
Hexadecimal number system

Hexadecimal	Decimal	Binary
0	0	0000
1	1	0001
2	2	0010
3	3	0011
4	4	0100
5	5	0101
6	6	0110
7	7	0111
8	8	1000
9	9	1001
A	10	1010
B	11	1011
C	12	1100
D	13	1101
E	14	1110
F	15	1111

As with previous number systems, the value of a hexadecimal number is determined by the position of the digits. Table 1.6 illustrates the positional value of base-16 digits.

TABLE 1.6
Powers of 16

	$16^0 = 1$
$16^1 = 16$	$16^{-1} = 1/16 = 0.0625$
$16^2 = 256$	$16^{-2} = 1/256 = 0.00390625$
$16^3 = 4096$	$16^{-3} = 1/4096 = 0.000244140625$
$16^4 = 65536$	$16^{-4} = 1/65{,}536 = 0.000015258789$

1.8 HEXADECIMAL CONVERSIONS

Hexadecimal-to-Decimal Conversion

Hexadecimal numbers can be converted to decimal by computing the *sum of the products of each digit times its positional value,* as shown in the following examples.

EXAMPLE 1.11

Convert the following hexadecimal numbers to decimal.

a. $(5B)_{16}$ b. $(2AF)_{16}$ c. $(1C5)_{16}$

Solution

a.

Positional weight	16^1	16^0
Positional value	16	1
Hexadecimal number	5	B

$$\begin{aligned} & 5 \times 16^1 + B \times 16^0 \\ = \; & 5 \times 16 + 11 \times 1 \\ = \; & 80 + 11 = (91)_{10} \end{aligned}$$

b.

Positional weight	16^2	16^1	16^0
Positional value	256	16	1
Hexadecimal number	2	A	F

$$\begin{aligned} & 2 \times 16^2 + A \times 16^1 + F \times 16^0 \\ = \; & 2 \times 256 + 10 \times 16 + 15 \times 1 \\ = \; & 512 + 160 + 15 = (687)_{10} \end{aligned}$$

c.

Positional weight	16^2	16^1	16^0
Positional value	256	16	1
Hexadecimal number	1	C	5

$$\begin{aligned} & 1 \times 16^2 + C \times 16^1 + 5 \times 16^0 \\ = \; & 1 \times 256 + 12 \times 16 + 5 \times 1 \\ = \; & 256 + 192 + 5 = (453)_{10} \end{aligned}$$

Decimal-to-Hexadecimal Conversion

To convert decimal numbers to hexadecimal numbers, simply apply the repeated-division method, as shown in the following example.

EXAMPLE 1.12

Convert the following decimal numbers to hexadecimal.

a. $(93)_{10}$ b. $(142)_{10}$ c. $(1776)_{10}$

Solution

a. Quotient + remainder

$$\frac{93}{16} = 5 + 13, \text{ or } D \quad \text{(LSB)}$$

$$\frac{5}{16} = 0 + 5 \quad \text{(MSB)}$$

$$(93)_{10} = (5\ D)_{16}$$

b. Quotient + remainder

$$\frac{142}{16} = 8 + 14, \text{ or } E \quad \text{(LSB)}$$

$$\frac{8}{16} = 0 + 8 \quad \text{(MSB)}$$

$$(142)_{10} = (8\ E)_{16}$$

c. Quotient + remainder

$$\frac{1776}{16} = 111 + 0 \quad \text{(LSB)}$$

$$\frac{111}{16} = 6 + 15 \text{ or } F$$

$$\frac{6}{16} = 0 + 6 \quad \text{(MSB)}$$

$$(1776)_{10} = (6\ F\ 0)_{16}$$

Hexadecimal-to-Binary Conversion

As with octal numbers, hexadecimal numbers are easily converted to binary numbers. In fact, hexadecimal numbers are also a convenient way of expressing binary numbers. Referring to Table 1.5, to convert hexadecimal numbers to binary numbers, simply *replace each hexadecimal digit with its 4-bit binary equivalent.*

EXAMPLE 1.13 Convert the following hexadecimal numbers to binary.

a. $(37)_{16}$ b. $(C4)_{16}$ c. $(DF6)_{16}$

Solution

a.

3	7
↓	↓
0011	0111

$$(37)_{16} = (00110111)_2$$

b.

C	4
↓	↓
1100	0100

$$(C4)_{16} = (11000100)_2$$

c.

D	F	6
↓	↓	↓
1101	1111	0110

$$(DF6)_{16} = (110111110110)_2$$

Binary-to-Hexadecimal Conversion

To convert a binary number to its hexadecimal equivalent, simply *group the binary bits by four, starting at the radix point.* If necessary, add zeros to complete the groups.

EXAMPLE 1.14 Convert the following binary numbers to hexadecimal.

a. $(10101110)_2$ b. $(1110101)_2$ c. $(1111101000010000)_2$

Solution

a.

1010	1110
↓	↓
A	E

$$(10101110)_2 = (AE)_{16}$$

b.

0111	0101
↓	↓
7	5

$$(1110101)_2 = (75)_{16}$$

Note the leading zero that was added to complete the most significant group.

c.

1111	1010	0001	0000
↓	↓	↓	↓
F	A	1	0

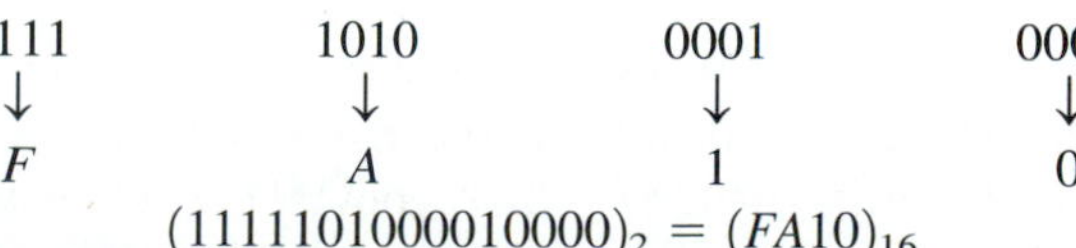

$$(1111101000010000)_2 = (FA10)_{16}$$

1.9 BINARY-CODED DECIMAL (BCD) CODE

A **code** is a group of symbols used to represent information. Sometimes codes are used as a means of keeping information secret, such as security codes in military applications. Sometimes codes are used for universal understanding, such as the international symbols used on road signs and automobile control switches. A code that is used in many digital applications is the **binary-coded decimal (BCD),** or 8421, code. It is used to specify the decimal numbers 0 through 9 in a binary form. Each decimal digit is represented by a 4-*bit binary number, or code,* as shown in Table 1.7.

TABLE 1.7
BCD code

Decimal	BCD
0	0000
1	0001
2	0010
3	0011
4	0100
5	0101
6	0110
7	0111
8	1000
9	1001

Note that although the BCD code uses 4 binary bits, only the decimal digits 0 through 9 are valid. The main advantage of BCD coding is that it can be read and recognized quickly and easily by both computers and people. This is demonstrated by the ease of conversion between BCD code and decimal numbers.

EXAMPLE 1.15 Convert the following decimal numbers to BCD.

a. $(37)_{10}$ b. $(1492)_{10}$ c. $(8506)_{10}$

Solution

a.

Decimal	3	7
	↓	↓
BCD	0011	0111

$(37)_{10} = (0011\ \ 0111)_{BCD}$

b.

Decimal	1	4	9	2
	↓	↓	↓	↓
BCD	0001	0100	1001	0010

$(1492)_{10} = (0001\ \ 0100\ \ 1001\ \ 0010)_{BCD}$

c.

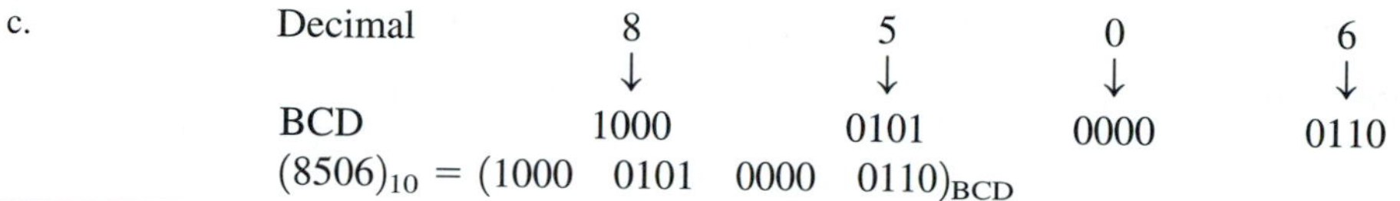

Decimal	8	5	0	6
	↓	↓	↓	↓
BCD	1000	0101	0000	0110

$(8506)_{10} = (1000\ \ 0101\ \ 0000\ \ 0110)_{BCD}$

Conversions from BCD to decimal are similar to conversions from binary to hexadecimal. *Starting at the radix point, simply group the BCD digits by fours.*

EXAMPLE 1.16

Convert the following BCD numbers to decimal.

a. $(00010110)_{BCD}$ b. $(10011000.0111)_{BCD}$ c. $(0110100100010000)_{BCD}$

Solution

a.

BCD	0001	0110
	↓	↓
Decimal	1	6

$(00010110)_{BCD} = (16)_{10}$

b.

BCD	1001	1000	.	0111
	↓	↓		↓
Decimal	9	8	.	7

$(10011000.0111)_{BCD} = (98.7)_{10}$

c.

BCD	0110	1001	0001	0000
	↓	↓	↓	↓
Decimal	6	9	1	0

$(0110100100010000)_{BCD} = (6910)_{10}$

1.10 ALPHANUMERIC CODES

Codes have also been developed to represent letters as well as numbers and special symbols. These codes are called **alphanumeric codes.** The **American Standard Code for Information Interchange (ASCII)** is a 7-bit code that is commonly used in today's computer systems. Since ASCII is a 7-bit code, there are 2^7, or 128, possible coding combinations. Each letter of the alphabet (upper- and lowercase) as well as the decimal digits 0 through 9 and special characters are represented by a unique code, as shown in Table 1.8.

Conversion to ASCII and other alphanumeric codes is best accomplished by looking up values in a table. For each alphanumeric character you wish to convert, an ASCII equivalent code can be found in Table 1.8. For example, the uppercase letter *A* is equivalent to 1000001, or 41H in ASCII code.

TABLE 1.8
ASCII code

Column Bits { B_5, B_6, B_7 } → Column bits

ASCII Code
Column bits ($B_7B_6B_5$) Row bits ($B_4B_3B_2B_1$)
$B_7B_6B_5B_4B_3B_2B_1$
A = 1000001
or
A = 41(HEX)

	Bits(row)						000	001	010	011	100	101	110	111	Column bits
ROW (HEX)	B_4	B_3	B_2	B_1			0	1	2	3	4	5	6	7	Column (HEX)
							↓	↓	↓	↓	↓	↓	↓	↓	Control characters
0	0	0	0	0	⟶		NUL	DLE	SP	0	@	P	\	p	NUL = Null DLE = Data link escape
1	0	0	0	1	⟶		SOH	DC1	!	1	A	Q	a	q	SOH = Start of heading DL1 = Device control 1
2	0	0	1	0	⟶		STX	DC2	"	2	B	R	b	r	STX = Start of text DC2 = Device control 2
3	0	0	1	1	⟶		ETX	DC3	#	3	C	S	c	s	ETX = End of text DC3 = Device control 3
4	0	1	0	0	⟶		EOT	DC4	$	4	D	T	d	t	EOT = End of transmission DC4 = Device control 4
5	0	1	0	1	⟶		EQN	NAK	%	5	E	U	e	u	ENQ = Enquiry NAK = Negative acknowledge
6	0	1	1	0	⟶		ACK	SYN	&	6	F	V	f	v	ACK = Acknowledge SYN = Syn idle
7	0	1	1	1	⟶		BEL	ETB	'	7	G	W	g	w	BEL = Bell ETB = End of transmission block
8	1	0	0	0	⟶		BS	CAN	(	8	H	X	h	x	BS = Backspace CAN = Cancel
9	1	0	0	1	⟶		HT	EM	)	9	I	Y	i	y	HT = Horizontal tab EM = End of medium
A	1	0	1	0	⟶		LF	SUB	*	:	J	Z	j	z	LF = Line feed SUB = Substitute
B	1	0	1	1	⟶		VT	ESC	+	;	K	[	k	{	VT = Vertical tab ESC = Escape
C	1	1	0	0	⟶		FF	FS	,	<	L	\	l	¦	FF = Form feed FS = File separator
D	1	1	0	1	⟶		CR	GS	-	=	M	]	m	}	CR = Carriage return GS = Group separator
E	1	1	1	0	⟶		SO	RS	.	>	N	^	n	~	SO = Shift out RS = Record separator
F	1	1	1	1	⟶		SI	US	/	?	O	–	o	DEL	SOH = Shift in US = Unit separator

(HEX)

EXAMPLE 1.17

Convert the letters JOE to ASCII.

Solution

Alphanumeric	J	O	E
ASCII	1001010	1001111	1000101
Hex	4*A*	4*F*	45

1.11 PARITY

Another type of code, which is used for error detection and correction, is known as *parity.* Two types of parity codes exist: **odd parity** and **even parity.** In both cases an extra bit known as the **parity bit** is attached to the code of information. The parity bit may be either a 0 or a 1, depending on the type of parity to be used. In *odd parity* the total number of 1s in the complete coded word (including the parity bit) must be an *odd* number. For example, suppose we wish to send the ASCII uppercase character A to a printer using odd parity. The ASCII code for uppercase A is

100001

This code sequence has two 1s. Therefore, we must add a parity bit of 1 to obtain odd parity. This makes the total number of 1s equal to three, which is, of course, odd. The odd-parity-coded ASCII A is then

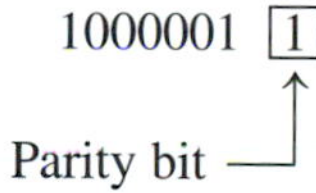

It should be noted that the parity bit may be added as the LSB or the MSB, depending upon the system design. In this case it is shown as the LSB.

In *even parity* the total number of 1s in the complete coded word (including the parity bit) must be an *even* number. For example, suppose we wish to send the ASCII uppercase character A to a printer using even parity. The ASCII code for uppercase A is

1000001

This code sequence has two 1s. Therefore, we must add a parity bit of 0 to obtain even parity. This makes the total number of 1s equal to two, which is even. The even-parity-coded ASCII A is then

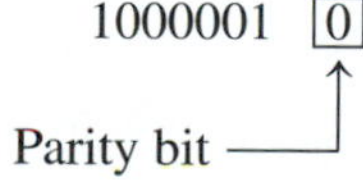

EXAMPLE 1.18

Determine the odd parity bit for the following letters.

a. J b. E c. Z

Solution

a.

ASCII J	1001010
Number of 1s	three
Odd parity bit	0
Odd parity code	1001010[0]
	parity bit ↑

b.

ASCII E	1000101
Number of 1s	three
Odd parity bit	0
Odd parity code	1000101[0]
	parity bit ↑

c.

ASCII Z	1011010
Number of 1s	four
Odd parity bit	1
Odd parity code	1011010[1]
	parity bit ↑

EXAMPLE 1.19

Determine the even parity bit for the following letters.

a. J b. E c. Z

Solution

a.

ASCII J	1001010
Number of 1s	three
Even parity bit	1
Even parity code	1001010[1]
	parity bit ↑

b.

ASCII E	1000101
Numbers of 1s	three
Even parity bit	1
Even parity code	1000101[1]
	parity bit ↑

c.

ASCII Z	1011010
Number of 1s	four
Even parity bit	0
Even parity code	1011010[0]
	parity bit ↑

1.12 BINARY ARITHMETIC

Addition

Binary arithmetic is similar to decimal arithmetic. Bits or digits are added one at a time. The sum is written down and the carry, if any, is added to the next pair of bits. The rules for binary addition are

$$\begin{aligned} 0 + 0 &= 0 \\ 0 + 1 &= 1 \\ 1 + 0 &= 1 \\ 1 + 1 &= 10 \end{aligned}$$

(the 1 of 10 ← Carry)

EXAMPLE 1.20 Add the following binary numbers.

$$\text{a.}\ \begin{array}{r} 101 \\ +010 \\ \hline \end{array} \qquad \text{b.}\ \begin{array}{r} 101 \\ +011 \\ \hline \end{array} \qquad \text{c.}\ \begin{array}{r} 1011 \\ +1100 \\ \hline \end{array}$$

Solution

a.
$$\begin{array}{r} 101 \\ +010 \\ \hline 111 \end{array} \qquad \text{Check:} \quad \begin{array}{r} (101)_2 = (5)_{10} \\ (010)_2 = (2)_{10} \\ (111)_2 = (7)_{10} \end{array} \qquad \begin{array}{r} (5)_{10} \\ +(2)_{10} \\ \hline (7)_{10} \end{array}$$

b.
$$\begin{array}{r} 111 \leftarrow \text{Carry} \\ 101 \\ +\ 011 \\ \hline 1000 \end{array} \qquad \text{Check:} \quad \begin{array}{r} (101)_2 = (5)_{10} \\ (011)_2 = (3)_{10} \\ (1000)_2 = (8)_{10} \end{array} \qquad \begin{array}{r} (5)_{10} \\ +(3)_{10} \\ \hline (8)_{10} \end{array}$$

c.
$$\begin{array}{r} 1 \leftarrow \text{Carry} \\ 1011 \\ +\ 1100 \\ \hline 10111 \end{array} \qquad \text{Check:} \quad \begin{array}{r} (1011)_2 = (11)_{10} \\ (1100)_2 = (12)_{10} \\ (10111)_2 = (23)_{10} \end{array} \qquad \begin{array}{r} (11)_{10} \\ +(12)_{10} \\ \hline (23)_{10} \end{array}$$

Subtraction

Subtraction can be considered a special case of addition. That is, the opposite, or negative, of the number to be subtracted can be added. For example,

$$7 - 2 = 7 + (-2) = 5$$

The rules for binary subtraction are

$$\begin{aligned} 0 - 0 &= 0 \\ 1 - 0 &= 1 \\ 1 - 1 &= 0 \\ 10 - 1 &= 1 \end{aligned}$$

(Borrow → the 1 of 10)

Note that when subtracting numbers, sometimes it is necessary to *borrow* from the next higher order position.

EXAMPLE 1.21 Subtract the following binary numbers.

a. 101 − 011   b. 1101 − 0110   c. 1110 − 0101

Solution

a. When 1 is borrowed, a 0 is left → 0 1 (above the crossed-out 1)

Borrowing 1 from next position results in 10 − 1 for this position (10 − 1 = 1)

```
   01
   ɫ01
 − 011
 ─────
   010
```

0 − 0 = 0 → 010 ← 1 − 1 = 0

Check $(101)_2 = (5)_{10}$ $(5)_{10}$
$(011)_2 = (3)_{10}$ $-(3)_{10}$
$(010)_2 = (2)_{10}$ $(2)_{10}$

b.
```
  0¹01
  ɫɫ01
− 0110
──────
  0111
```
Check: $(1101)_2 = (13)_{10}$ $(13)_{10}$
$(0110)_2 = (6)_{10}$ $-(6)_{10}$
$(0111)_2 = (7)_{10}$ $(7)_{10}$

c.
```
    0₁
  11ɫ0
− 0101
──────
  1001
```
Check: $(1110)_2 = (14)_{10}$ $(14)_{10}$
$(0101)_2 = (5)_{10}$ $-(5)_{10}$
$(1001)_2 = (9)_{10}$ $(9)_{10}$

Multiplication

Multiplication of binary numbers is relatively easy. It is performed by *multiplying, shifting left, and adding.* The *multiplicand* is multiplied by the *multiplier* to obtain *partial products.* The partial products are then added to obtain the *final product.* The rules for binary multiplication are:

$$0 \times 0 = 0$$
$$0 \times 1 = 0$$
$$1 \times 0 = 0$$
$$1 \times 1 = 1$$

EXAMPLE 1.22 Multiply the following binary numbers.

a. 10 ×11   b. 101 ×101   c. 1101 ×110

Solution

a.

		Check:	
Multiplicand	10	$(10)_2 = (2)_{10}$	$(2)_{10}$
Multiplier	× 11	$(11)_2 = (3)_{10}$	× $(3)_{10}$
Partial Products	10	$(110)_2 = (6)_{10}$	$(6)_{10}$
	10		
Final Product	110		

b.

```
    101
  × 101
    101
   000
  101
  11001
```

Check: $(101)_2 = (5)_{10}$, $(101)_2 = (5)_{10}$, $(11001)_2 = (25)_{10}$

$(5)_{10} \times (5)_{10} = (25)_{10}$

c.

```
    1101
   × 101
    0000
   1101
  1101
 1001110
```

Check: $(1101)_2 = (13)_{10}$, $(110)_2 = (6)_{10}$, $(1001110)_2 = (78)_{10}$

$(13)_{10} \times (6)_{10} = (78)_{10}$

Division

As with multiplication, binary division is simpler than decimal division. The procedure involves dividing the dividend by the divisor to obtain the quotient:

$$\text{Divisor}\overline{)\text{dividend}}\quad\text{quotient}$$

It should be noted that the quotient must be either a zero or a one. That is, the divisor can only go into the dividend either *one time or zero times.* Division is, in effect, a case of repeated subtraction.

EXAMPLE 1.23 Divide the following binary numbers.

a. $101\overline{)1010}$ b. $11\overline{)11110}$ c. $100\overline{)11000}$

Solution

a.

```
       10
  101)1010
      101
       00
```

Check: $(1010)_2 = (10)_{10}$, $(101)_2 = (5)_{10}$, $(10)_2 = (2)_{10}$

$5\overline{)10}$ = 2

b.

```
        1010
   11)11110
      11
       11
       11
       00
```

Check: $(11110)_2 = (30)_{10}$
$(11)_2 = (3)_{10}$
$(1010)_2 = (10)_{10}$

```
  10
3)30
```

c.

```
         110
  100)11000
       100
        100
        100
         00
```

Check: $(11000)_2 = (24)_{10}$
$(100)_2 = (4)_{10}$
$(110)_2 = (6)_{10}$

```
   6
4)24
```

1.13 SUBTRACTION BY COMPLEMENTS

Most computer systems perform subtractions using a method known as **complementing.** The complement of a number is found by *negating* the number, which is the process of changing a 1 to a 0 or a 0 to a 1. This is often referred to as the **1s complement.** For example, the 1s complement of the binary number 10110 is

Original binary number	10110
1s complement	01001

The **2s complement** of a binary number is found by *adding* 1 to the 1s complement. For example, the 2s complement of the binary number 10110 is found as follows:

Original binary number	10110
1s complement	01001
Add 1	+ 1
2s complement	01010

Subtraction may be performed using the complementing method by first obtaining the 2s complement of the original number to be subtracted (subtrahend). Next, the 2s complement of the subtrahend is added to the minuend. The final carry is always ignored, as follows:

101	minuend
− 011	subtrahend
010	difference

Subtrahend	011
1s complement	100
Add 1	+ 1
2s complement	101

101	minuend
+ 101	2s complement
[1]010	difference

Carryout ignored ↑ (the boxed 1)

Answer $= (010)_2 = (2)_{10}$

EXAMPLE 1.24

Perform the following binary subtractions using the 2s complement method.

a.	111	b.	1101	c.	1001
	− 110		− 1001		− 1100

Solution

a.

Subtrahend	110	111	minuend
1s complement	001	+ 010	2s complement
Add 1	+ 1	[1] 001	answer
2s complement	010	↑	carryout ignored

Answer = $(001)_2 = (1)_{10}$

b.

Subtrahend	1001	1101	minuend
1s complement	0110	+ 0111	2s complement
Add 1	+ 1	[1] 0100	answer
2s complement	0111	↑	carryout ignored

Answer = $(0100)_2 = (4)_{10}$

c. Note that in this example we are subtracting a larger number from a smaller number. This results in a negative answer.

Subtrahend	1100	1001	minuend
1s complement	0011	+ 0100	2s complement
Add 1	+ 1	1101	result
2s complement	0100		

The result 1101 is a negative number in 2s complement form. We must realize this when we begin the problem. To obtain the final answer, the 2s complement of the result must be determined and the sign changed.

Result	1101
1s complement	0010
Add 1	+ 1
2s complement	0011
Final answer	−0011

Check:

$$(1001)_2 = (9)_{10} \qquad (9)_{10}$$
$$(1100)_2 = (12)_{10} \qquad -(12)_{10}$$
$$-(0011)_2 = -(3)_{10} \qquad -(3)_{10}$$

EXERCISES

1.1 For most of human existence the __________ number system has been used.

1.2 Evaluate the following:
a. 2^5 b. 2^8
c. 2^{10} d. 2^{16}

1.3 Evaluate the following:
a. 2^{-1} b. 2^{-4}
c. 2^{-8} d. 2^{-10}

1.4 Convert the following binary numbers to decimal:
a. 1101 b. 1111
c. 1001 d. 0010

1.5 Convert the following binary numbers to decimal:
a. 01110001 b. 00010001
c. 11110101 d. 00101111

1.6 Convert the following binary numbers to decimal:
a. 0.01 b. 0.0101
c. 0.1101 d. 0.00100

1.7 Convert the following binary numbers to decimal:
a. 1101.0101 b. 10111.0111
c. 1000001.1111 d. 0101.11001

1.8 Convert the following decimal numbers to binary:
a. 19 b. 35
c. 47 d. 69

1.9 Convert the following decimal numbers to binary:
a. 109 b. 128
c. 250 d. 257

1.10 Convert the following decimal numbers to binary:
a. 0.50 b. 0.75
c. 0.625 d. 0.128

1.11 Convert the following decimal numbers to binary:
a. 14.12 b. 39.56
c. 125.7 d. 110.0625

1.12 Convert the following octal numbers to decimal:
a. 14 b. 65
c. 112 d. 347

1.13 Convert the following octal numbers to decimal:
a. 1247 b. 7777
c. 6053 d. 5432

1.14 Convert the following octal numbers to decimal:
a. 0.3 b. 0.12
c. 0.23 d. 14.53

1.15 Convert the following decimal numbers to octal:
a. 27 b. 153
c. 942 d. 787

1.16 Convert the following decimal numbers to octal (three places):
a. 0.4 b. 0.35
c. 0.125 d. 67.55

1.17 Convert the following octal numbers to binary:
a. 7 b. 6
c. 62 d. 111

1.18 Convert the following octal numbers to binary:

a. 347 b. 605
c. 1234 d. 7071

1.19 Convert the following octal numbers to binary:

a. 0.12 b. 15.03
c. 167.531 d. 6754.1302

1.20 Convert the following binary numbers to octal:

a. 101111 b. 110000
c. 110010 d. 001010

1.21 Convert the following binary numbers to octal:

a. 111111111 b. 1100010
c. 10101010 d. 110010.111010

1.22 Evaluate the following:

a. 16^0 b. 16^2
c. 16^3 d. 16^5

1.23 Convert the following hexadecimal numbers to decimal:

a. *B*5 b. *FF*
c. 10 d. 29

1.24 Convert the following hexadecimal numbers to decimal:

a. *FA*2 b. *C*5*E*
c. 11*D*3 d. 0*BC*4

1.25 Convert the following decimal numbers to hexadecimal:

a. 39 b. 78
c. 256 d. 1000

1.26 Convert the following decimal numbers to hexadecimal:

a. 1787 b. 1492
c. 2048 d. 4096

1.27 Convert the following hexadecimal numbers to binary:

a. *A*4 b. *FD*
c. 11 d. 1*C*1

1.28 Convert the following hexadecimal numbers to binary:

a. *ABCD* b. *FE*15
c. *F*00*E* d. 04051

1.29 Convert the following binary numbers to hexadecimal:

a. 001100110011 b. 1010.1111
c. 100101101000 d. 001001111100

1.30 Convert the following binary numbers to hexadecimal:

a. 1111110 b. 1000011101
c. 00100001101 d. 11111101110010000111

1.31 Convert the following decimal numbers to BCD:

a. 49 b. 96
c. 201 d. 537

1.32 Convert the following decimal numbers to BCD:

a. 8964 b. 09238
c. 1987.176 d. 1066.1860

1.33 Convert the following BCD numbers to decimal if possible:

a. 00010110 b. 10101111
c. 10010001 d. 001001110011

1.34 Convert the following BCD numbers to decimal if possible:

a. 10000111100 b. 1110110
c. 1000010101 d. 0111001100011000

1.35 Write the following characters and symbols in ASCII code:
a. R b. r
c. 9 d. *

1.36 Write the following characters and symbols in ASCII code:
a. BECCA b. ROBBY
c. 1579 d. STOP

1.37 Decode the following from ASCII:
a. 0110000 b. 1001111
c. 1011010 d. 1000001

1.38 Decode the following from ASCII:
a. 0000000 1001010
b. 0100100 0110101
c. 1010011 1010100 1010101
d. 1010010 1001001 1000011 1001011

1.39 State whether the following are examples of even or odd parity:
a. 1001101 b. 10000000
c. 11111111 d. 10101000

1.40 Write the following ASCII characters with even parity:
a. 12 b. DIANE
c. JEANE d. 5%

1.41 Write the following ASCII characters with odd parity:
a. WORK b. RUN
c. 3 + 2 = 5 d. MENU #2

1.42 Find the following binary sums.

a. $\begin{array}{r} 101 \\ +\ 100 \\ \hline \end{array}$ b. $\begin{array}{r} 1001 \\ +\ 110 \\ \hline \end{array}$

c. $\begin{array}{r} 1010 \\ +\ 0111 \\ \hline \end{array}$ d. $\begin{array}{r} 1100 \\ +\ 1001 \\ \hline \end{array}$

1.43 Find the following binary differences.

a. $\begin{array}{r} 111 \\ -\ 100 \\ \hline \end{array}$ b. $\begin{array}{r} 1010 \\ -\ 0111 \\ \hline \end{array}$

c. $\begin{array}{r} 1000 \\ -\ 0011 \\ \hline \end{array}$ d. $\begin{array}{r} 1100 \\ -\ 101 \\ \hline \end{array}$

1.44 Find the following binary products.

a. $\begin{array}{r} 110 \\ \times\ 10 \\ \hline \end{array}$ b. $\begin{array}{r} 101 \\ \times\ 101 \\ \hline \end{array}$

c. $\begin{array}{r} 1101 \\ \times\ 0111 \\ \hline \end{array}$ d. $\begin{array}{r} 10010 \\ \times\ 1010 \\ \hline \end{array}$

1.45 Divide the following binary numbers:
a. $11\overline{)1101}$ b. $101\overline{)10110111}$
c. $100\overline{)111000}$ d. $110\overline{)10010111}$

1.46 Find the following binary differences using the 2s complement method:

a. $\begin{array}{r} 1110 \\ -\ 1010 \\ \hline \end{array}$ b. $\begin{array}{r} 10011 \\ -\ 1101 \\ \hline \end{array}$

c. $\begin{array}{r} 1101 \\ -\ 1110 \\ \hline \end{array}$ d. $\begin{array}{r} 0111 \\ -\ 1001 \\ \hline \end{array}$

1.47 Crossword Puzzle

ACROSS

3. In even parity the total number of _____ must be an even number.
7. A code is a group of _____ that is used to represent information.
9. The number into which you are dividing.
10. An 8421 code that is used in many digital applications.
11. Hexadecimal base number.
12. Codes used to represent letters as well as numbers and special symbols.
15. The number to be subtracted from.
16. The _____ of 0000 is 1111.
17. Modern computers use another number system called the _____ number system.
22. The decimal number divided by the base number equals the quotient plus the _____.
23. A binary digit can be represented by a high or low _____.
25. In ASCII 1001010/1001111/1000101/1011001 means _____.
27. The American Standard Code for Information Interchange.
28. The octal number system is a base _____ system.
29. The number by which you are dividing.
30. 1000001 is an example of _____ parity code for the ASCII letter A.

DOWN

1. To convert hexadecimal numbers to binary, replace each digit with its _____ bit binary equivalent.
2. The rightmost digit or bit (base two) is referred to as the _____.
4. In a _____ bit code, there are 128 possible coding combinations.
5. The decimal point and the binary point are also called the _____ point.
6. The binary and base 10 number systems are known as _____ number systems.
7. The number you are subtracting.
8. Babbage's engine was an example of a _____ calculator.
13. Converting fractional decimal numbers to binary uses a process known as repeated _____.
14. The answer in a division problem.
18. The leftmost digit or bit (base two) is referred to as the _____.
19. Converting whole numbers from base 10 to another base system is known as _____ division.
20. Each _____ digit is represented by a 4-bit binary number in BCD.
21. One octal number replaces _____ binary digits (bits).
24. In binary arithmetic 1 + 1 = sum of 0 and a _____ of 1.
26. ASCII code for 1001010 1000001 1010011 1001111 1001110.

1 2 3 4 5 6 7 8 9 10 11 12 13 14 15 16 17 18 19 20 21 22 23 24 25 26 27 28 29 30

CHAPTER 2

FUNDAMENTALS OF DIGITAL LOGIC

KEY TERMS

Amplitude
AND
AND Gate
Boolean Algebra
Bubble
Chip
Comparator
Complementary Metal-Oxide Semiconductor (CMOS)
Dual-In-Line Package (DIP)
Duty Cycle
EXCLUSIVE-OR (XOR)
EXCLUSIVE-NOR (XNOR)
Fall Time
Float
Frequency
Integrated Circuit (IC)
Inverter
Leading Edge
Logic Families
Logic Gates
Logic Probe
Logic Pulses
NAND
Negative-Going Pulse
Negative Logic System
NOR
NOT
OR
OR Gate
Period
Positive-Going Pulse
Positive Logic System
Pulse Train
Pulse Width
Rise Time
Trailing Edge
Transistor–Transistor Logic (TTL)
Tristate
Truth Table

2.0 INTRODUCTION

Logic gates are essentially a combination of switches. These switches can be mechanical, electromechanical, or electronic in nature. The position, or *state,* of the switches can be used to represent information. Switches are inherently binary in nature, that is, they can be either open or closed and can be used to represent the binary digits 0 or 1, as shown in Figure 2.1.

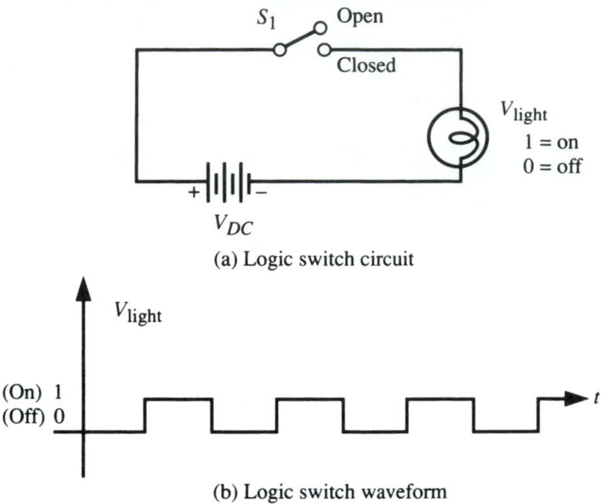

FIGURE 2.1
Logic switch circuit

Figure 2.1(a) shows a simple logic switch circuit. When switch S_1 is open, no current will flow through the circuit. Therefore, the light will be off. This can be represented as a logic level 0. When switch S_1 is closed, the circuit is complete, and current will flow from the source V_{DC} to the light. The light will then be on. This can be represented as a logic level 1. If the switch is repeatedly opened and closed, the light will blink on and off. This is illustrated by the waveform in Figure 2.1(b).

2.1 LOGIC LEVELS AND WAVEFORMS

In general, the binary digits 0 and 1 can be easily represented by any two different voltage levels. If the higher voltage is used to represent a logic level 1, a **positive logic system** is said to exist. If the higher voltage is used to represent a logic level 0, a **negative logic system** is said to exist.

Positive Logic System	**Negative Logic System**
High voltage level = 1	High voltage level = 0
Low voltage level = 0	Low voltage level = 1

Today most digital circuits use a nominal voltage level of 0 V_{DC} to represent a logic level zero and a nominal voltage level of +5 V_{DC} to represent a logic level one. Systems of this nature are positive logic systems. The word *nominal* is used because 0 V_{DC} and +5 V_{DC} represent ideal levels referenced to a particular power supply. Due to the voltage drops across resistors and transistors inside digital logic circuits, these levels are usually not attained. This detail is fully explained later in the text when we look at how these circuits work.

Repeated logic level transitions from either 0 to 1 or 1 to 0 are often referred to as **logic pulses.** A series of pulses is sometimes called a **pulse train,** or *waveform,* as shown in Figure 2.2(a). The time to complete one cycle of the waveform is called the **period** (T) and is measured in seconds. The **frequency** (f), or rate at which the waveform repeats itself, is measured in *pulses per second* (PPS). The relationship between period and frequency is given by the formula

$$T = \frac{1}{f} \quad \text{or} \quad f = \frac{1}{T} \tag{2.1}$$

The ratio between the **pulse width** (t_w) and the total **period** (T) is defined as the **duty cycle.** The duty cycle is used to express the percentage of time that a pulse or circuit is on, as shown in Figure 2.2(b). The duty cycle can be calculated as follows:

$$\text{Duty cycle} = \frac{t_w}{T} \times 100\% \tag{2.2}$$

Note that the duty cycle is usually presented as a percentage and is sometimes called the percent duty cycle.

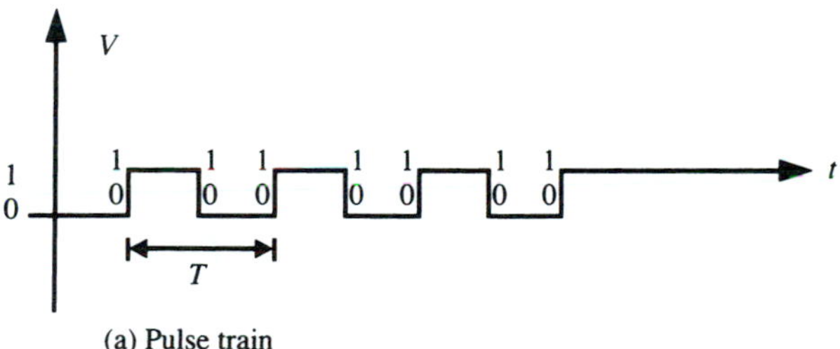

(a) Pulse train

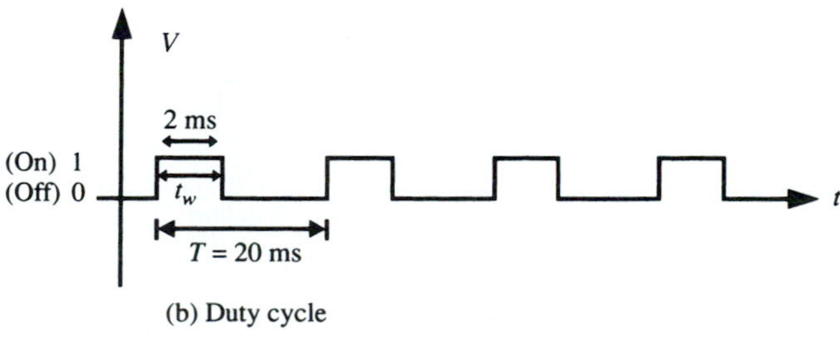

(b) Duty cycle

FIGURE 2.2
Pulse waveforms

EXAMPLE 2.1 Determine the duty cycle for the waveform shown in Figure 2.2.

Solution

$$\begin{aligned}\text{Duty cycle} &= \frac{t_w}{T} \times 100\% \\ &= \frac{2\text{ ms}}{20\text{ ms}} \times 100\% \\ &= 0.1 \times 100\% \\ &= 10\%\end{aligned}$$

Pulse waveform, as the name implies, consist of a string of pulses. Let's examine the concept of a digital pulse in greater detail. Figure 2.3(a) represents a string of ideal pulses. Note that the pulses can be either positive going or negative going. A **positive-**

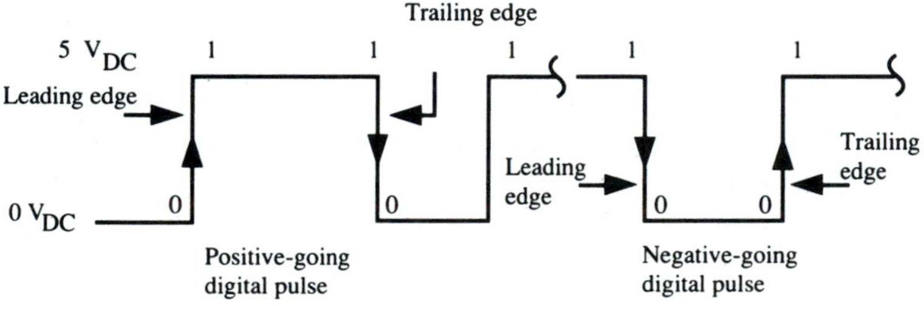

(a) Ideal pulses

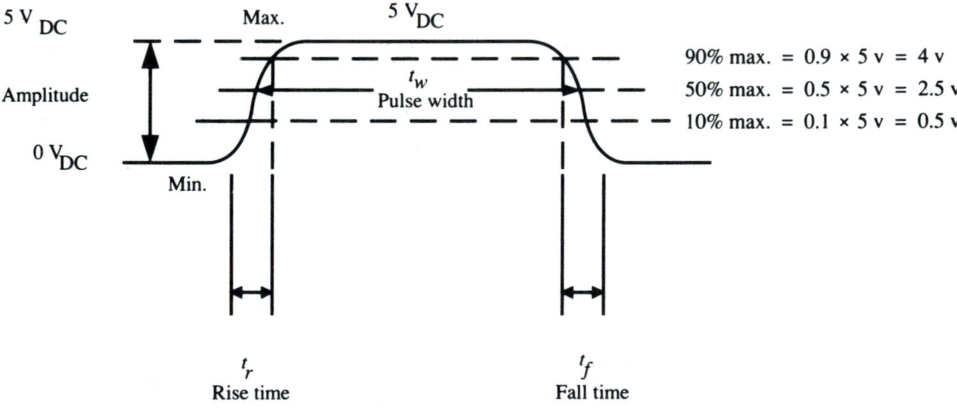

(b) Non-ideal pulse

FIGURE 2.3
Digital pulse waveform

going pulse is one that *starts* at a logic level 0, goes to a logic level 1, and then *returns* to a logic level 0. A **negative-going pulse** is one that *starts* at a logic level 1, goes to a logic level 0, and then *returns* to a logic level 1. In both cases, the first transition is called the **leading edge** of the pulse, and the second transition is called the **trailing edge** of the pulse.

In the real world nothing is ideal. Pulses cannot go from one logic level to another instantaneously, as shown in Figure 2.3(b). Therefore, it is important for us to understand how a real or nonideal pulse is defined.

The **amplitude** can be defined as the difference between the maximum and minimum values of the pulse waveform voltage. The **rise time** (t_r) is defined as the time it takes the pulse to rise from 10% of the amplitude to 90% of the amplitude. This allows a convenient method for making measurements with an oscilloscope. The **fall time** (t_f) is defined as the time it takes the pulse to fall from 90% of the amplitude to 10% of the amplitude. The actual pulse width (t_w) is measured between the 50% point of the leading edge of the pulse and the 50% point of the trailing edge of the pulse.

2.2 BOOLEAN EXPRESSIONS AND TRUTH TABLES

In about the year 1850, George Boole, an English mathematician, developed the basic laws and theorems that became known as **Boolean algebra.** Boole's work was an outgrowth of a topic in philosophy known as logic. Logic can be used to break down complex problems into simple questions that can be answered as either true or false. The binary nature of logic was recognized by Claude Shannon of MIT in 1938. Shannon applied Boolean algebra to relay-logic switching circuits as a means to analyze these circuits.

From Boolean algebra come the three logical functions that form the basis of all digital and computer circuits. These logical functions are called the *AND* function, the *OR* function, and the *NOT* function. These functions can be expressed mathematically using Boolean algebra or in tabular form using a **truth table.** The input and output variables are usually represented as letters such as *A*, *B*, and *C* or *X*, *Y*, and *Z*. The logic state of these variables is represented by the binary numbers 0 and 1.

The AND Function

The **AND** function can be thought of as a series circuit containing two or more logic switches, as shown in Figure 2.4(a). The logic indicator L1 will be on only when logic switch *A and* logic switch *B* are both closed. If we examine the circuit in Figure 2.4(a) further, we can see that logic switch *A* and logic switch *B* each have two possible logic states (open or closed). The logic states can be represented in binary form using a 0 for *open* and 1 for *closed.* Furthermore, the logic indicator L1 can also have two possible binary logic states, 0 for *off* and 1 for *on.* The truth table is used to illustrate all the possible combinations of input and output conditions that can exist in a logic circuit, as shown in Figure 2.4(b). The Boolean expression used to represent *this* AND function is written as

$$A \cdot B = \text{L1}.$$

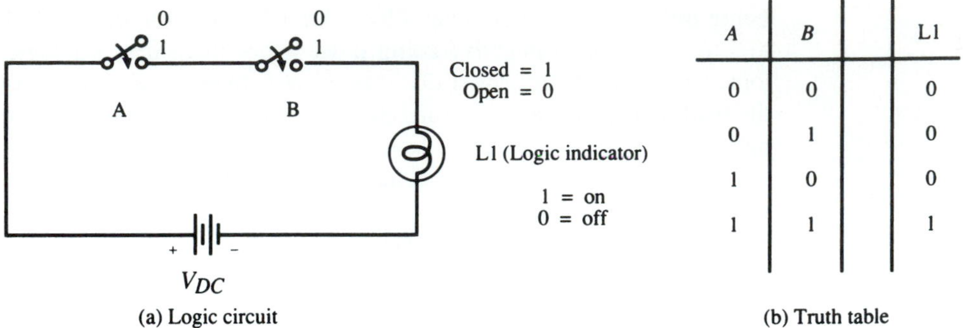

A	B	L1
0	0	0
0	1	0
1	0	0
1	1	1

(a) Logic circuit

(b) Truth table

$A \cdot B = \text{L1}$
(c) Boolean expression

FIGURE 2.4
AND function

This expression can be read as "*A and B* equals L1." It gives us a clear and precise mathematical method of describing the behavior of the circuit. It should be noted that the dot (·), or AND symbol, is often deleted. When this is done, the expression is written as

$$AB = \text{L1}.$$

The OR Function

The **OR** function can be thought of as a parallel circuit containing two or more logic switches, as shown in Figure 2.5(a). Here the logic indicator L1 will be on whenever ei-

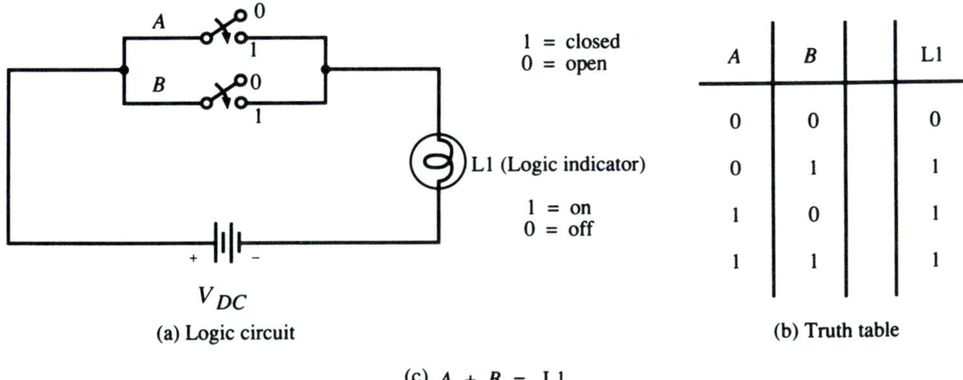

A	B	L1
0	0	0
0	1	1
1	0	1
1	1	1

(a) Logic circuit

(b) Truth table

(c) $A + B = \text{L1}$
Boolean expression

FIGURE 2.5
OR function

ther logic switch *A or* logic switch *B or* both are closed. As previously described with the AND function, each logic switch can have two possible logic states (open or closed, which correspond to 0 and 1). The truth table shown in Figure 2.5(b) is again used to illustrate all the possible combinations of input and output conditions that can exist. It should be noted that the number of possible logic combinations is determined by the number of inputs. The number of logic combinations can be calculated as follows:

$$\text{Number of logic combinations} = 2^n$$

where n is the number of logic inputs. For example, if there are three inputs, A, B, and C, the number of possible logic combinations is calculated as follows:

$$\text{Number of logic combinations} = 2^n = 2^3 = 8$$

The Boolean expression used to represent the OR function in Figure 2.5 is

$$A + B = \text{L1}$$

This expression can be read as "*A or B* equals L1." It is used to describe the behavior of the circuit in Figure 2.5(a).

The NOT Function

The **NOT** function can be thought of as an *inverter,* or negation circuit, as shown in Figure 2.6(a). The logic indicator L1 will be on whenever the logic switch *A* is open. If logic switch *A* is open, current will flow through the logic indicator L1, which will cause the indicator to light. If logic switch *A* is closed, L1 will be *shorted,* and the

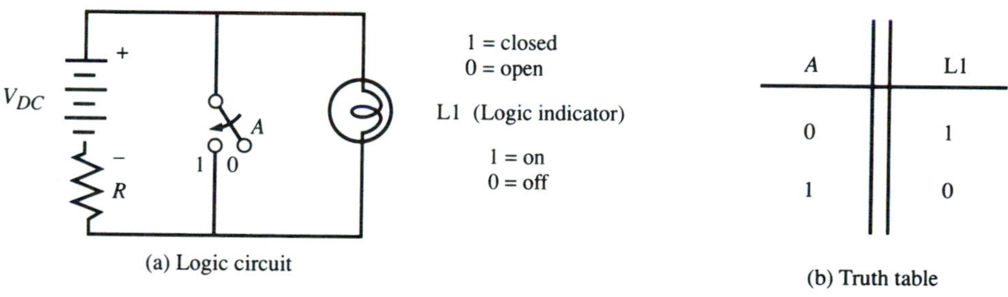

A	L1
0	1
1	0

(b) Truth table

$$\overline{A} = \text{L1}$$

(c) Boolean expression

FIGURE 2.6
NOT function

current will flow through the logic switch back to the power source. Remember, electricity will take the path of least resistance. Note that the resistor is shown here to protect the power source when logic switch *A* is closed. Once again the truth table of Figure 2.6(b) is used to describe the circuit operation. Unlike the AND and OR functions, the NOT function can have only one input. The Boolean expression used to represent this NOT function is written as

$$\overline{A} = \text{L1}.$$

This expression is read as "not *A* equals L1." That is, when logic switch *A* is open (0), the logic indicator L1 is on (1). Conversely, when logic switch *A* is closed (1), the logic indicator L1 is off (0).

2.3 INTEGRATED CIRCUITS

The electronic circuits or devices that perform the Boolean logic functions are called **logic gates.** The logic gates are made up of circuits containing transistors, diodes, resistors, and capacitors. These components are usually integrated onto a single device known as an **integrated circuit (IC),** or **chip.** The IC itself is rather small and fragile. To facilitate handling, ICs are usually packaged in various sizes, shapes, and styles. One of the most common styles is the **dual-in-line package,** or **DIP,** as shown in Figure 2.7. Notice that pin 1 is on the left-hand side of the end with the U-shaped notch, when looking down from the top of the DIP. Some DIP manufacturers use a dot instead of a notch to indicate the location of pin 1. Referring to Figure 2.7, the pin immediately to the right of pin 1 is pin 2. The pin numbers continue sequentially around the DIP, as shown in Figure 2.7.

Digital integrated circuit devices are manufactured using different techniques. These manufacturing techniques are classified by **logic families.** Integrated circuits contained within a logic family usually have similar electrical characteristics. For example, devices within a logic family all operate at the same power supply voltages and are designed to be connected together. The two most popular logic families in use today are the TTL and CMOS families.

Transistor–transistor logic (TTL) circuits are integrated circuits designed around the basic bipolar junction transistor (BJT). All TTL devices operate from a single $+5\ V_{DC}$ power source. TTL devices are identified by a part number, which is composed of five segments:

S N	7 4	L S	0 8	N
Manufacturer	Series	Subseries	Device type	Package

The first two letters, SN, indicate the manufacturer. In this case, SN stands for Semiconductor Network Program, which is manufactured by Texas Instruments. The next two numbers, 74, represent the series: 74 is the code for a standard TTL commercial device; 54 represents a high-reliability military version. The next letters, LS, define the subseries. LS represents a low-power, high-speed device known as a low-power Schottky. The absence of a letter would represent a standard TTL device. The next digits, 08, define the number and type of logic function to be performed. In this case, 08 defines a device containing

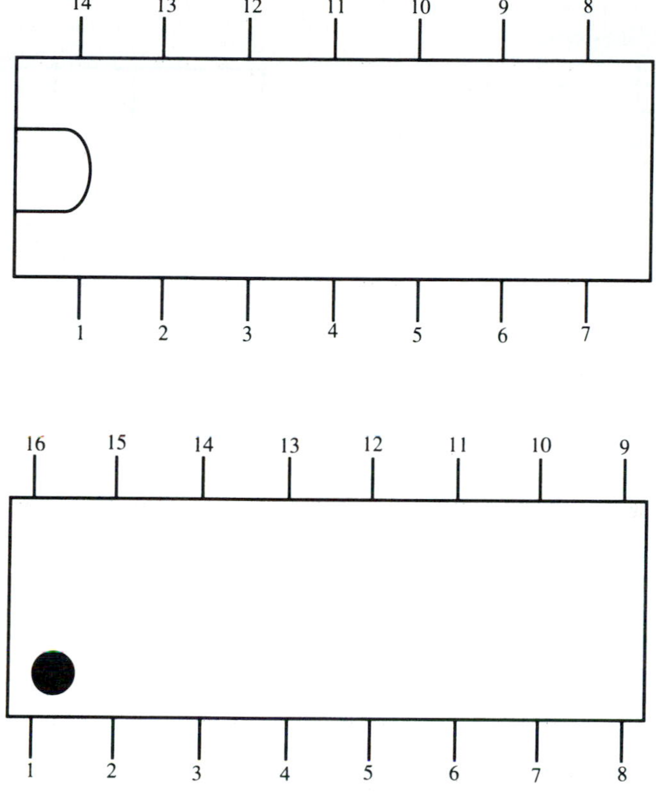

FIGURE 2.7
DIP

four 2-input AND gates. The last letter, N, indicates the package type. Here the N describes a plastic DIP device.

Complementary metal-oxide semiconductor (CMOS) devices are integrated circuits designed around the field-effect transistor (FET). The power supply voltage for a digital CMOS device can range anywhere between +3 and +18 V_{DC}. CMOS devices consume much less power than comparable TTL devices. However, CMOS devices tend to be slower than comparable TTL devices. The identification process for a CMOS device is similar to that of the TTL device. The internal workings of these and other types of logic families are covered in greater detail in Chapter 6.

2.4 BASIC LOGIC GATES

The three basic digital logic circuits are the *AND gate,* the *OR gate,* and the *inverter.* These three electronic circuits or gates form the basis of all digital and computer circuits by performing the Boolean AND, OR, and NOT functions electronically.

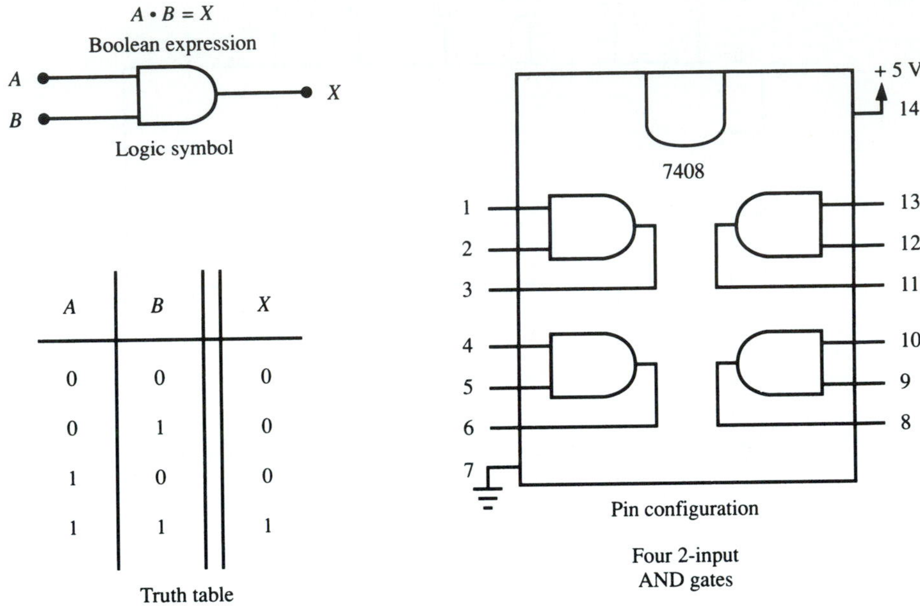

FIGURE 2.8
AND gate

The AND Gate

The electronic circuit that performs the AND function is called the **AND gate.** Figure 2.8 illustrates the Boolean expression, logic symbol, truth table, and pin configuration for a basic 2-input AND gate. From the truth table we can see that the output will be high, or a logic 1, *only* when all of the inputs are high, or at a logic 1. In *all* other cases, the output is low, or a logic 0. This is true regardless of the number of inputs the device has. Table 2.1 summarizes a number of available TTL and CMOS AND gates.

TABLE 2.1
TTL and CMOS AND gates

TTL Series	CMOS Series	Description
7408	4081	Four 2-input AND gates
7411	4073	Three 3-input AND gates
7421	4082	Two 4-input AND gates

EXAMPLE 2.2

Input waveforms *A* and *B* are applied to an AND gate as shown in Figure 2.9. Determine the output waveform *X*.

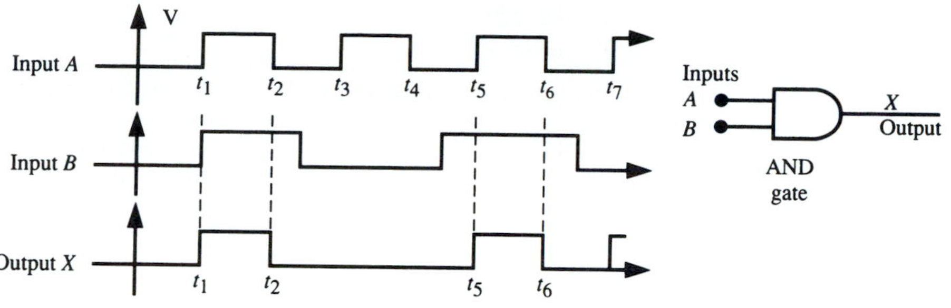

FIGURE 2.9

Solution

The output waveform X is high whenever input A and input B are both high.

The OR Gate

The electronic circuit that performs the OR function is called the **OR gate.** Figure 2.10 illustrates the Boolean expression, logic symbol, truth table, and pin configuration for a basic 2-input OR gate. From the truth table we can see that the output is high whenever *any* single input or both inputs are high. The output is low *only* when both inputs are low. This is true regardless of the number of inputs the device has. Table 2.2 summarizes a number of available TTL and CMOS OR gates.

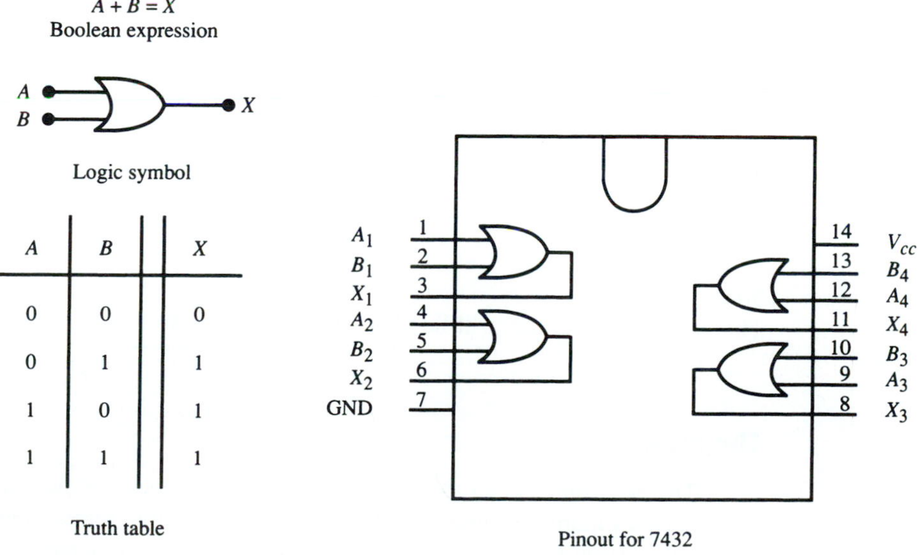

A	B	X
0	0	0
0	1	1
1	0	1
1	1	1

FIGURE 2.10
OR gate

TABLE 2.2
TTL and CMOS OR gates

TTL Series	CMOS Series	Description
7432	4071	Four 2-input OR gates
—	4075	Three 3-input OR gates
—	4072	Two 4-input OR gates

EXAMPLE 2.3 Input waveforms *A* and *B* are applied to an OR gate as shown in Figure 2.11. Determine the output waveform *X*.

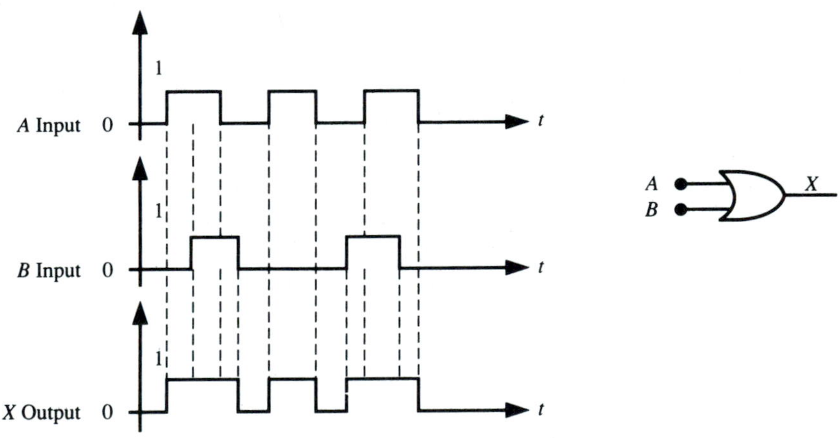

FIGURE 2.11

Solution

The output waveform *X* will be high whenever input *A* or input *B* or both are high.

The Inverter

The electronic circuit that performs the NOT, or negate, function is better known as the **inverter.** Figure 2.12 illustrates the Boolean expression, logic symbol, truth table, and pin configuration for the basic inverter. Note that the inverter has only one input. From the truth table we can see that the output is *always* the opposite logic state of the input. For example, if the input is high, the output is low. Table 2.3 summarizes a number of available TTL and CMOS inverters.

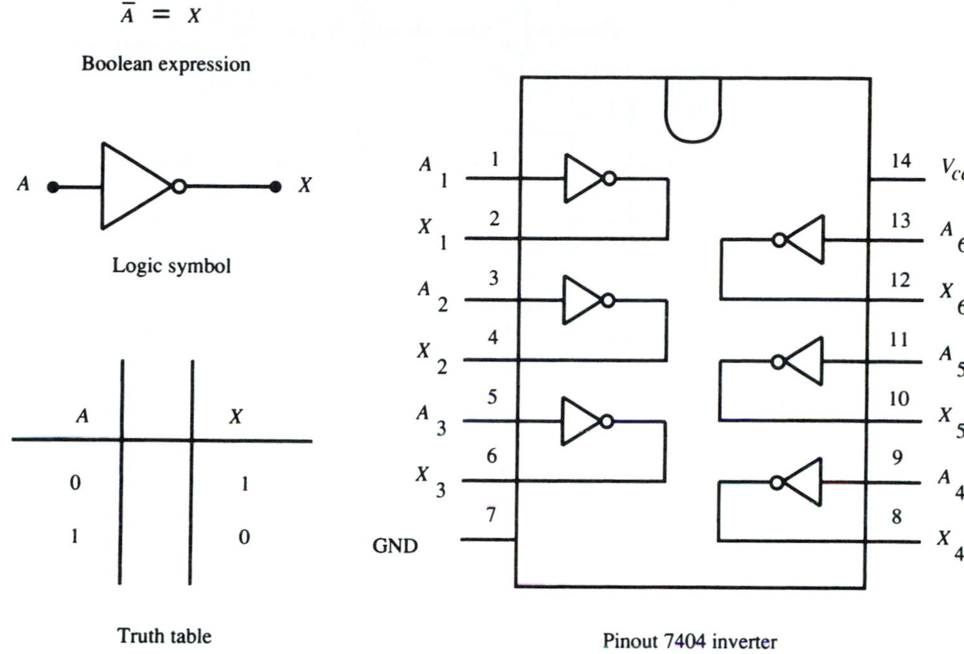

FIGURE 2.12
Inverter

TABLE 2.3
TTL and CMOS inverters

TTL Series	CMOS Series	Description
7404	4049	Hex inverter
7414	4585	Hex inverter

EXAMPLE 2.4

Input waveform *A* is applied to the inverter circuit shown in Figure 2.13. Determine the output waveform *X*.

Solution

The output of inverter 1 is the *opposite* of the input waveform *A*. It is then applied to the input of inverter 2. The output of inverter 2 is the *opposite* of its input and therefore the same as the original waveform *A*.

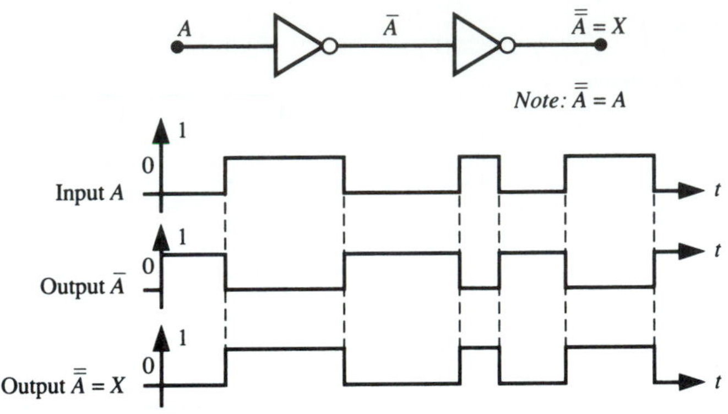

FIGURE 2.13

2.5 NAND/NOR LOGIC

All digital logic circuits are formed using combinations of the three basic logic functions. Many of the most commonly used combinations are available preassembled into one package. Two of the most popular combinations are the *NAND* and *NOR* gates.

The NAND Gate

The **NAND** logic function is equivalent to an AND function whose output is inverted, as shown in Figure 2.14(a). The name NAND comes from NOT-AND. This means that the NAND function is the *opposite,* or complement, of the AND function. This is represented by the normally closed switching circuit of Figure 2.14(b). Referring to the circuit of Figure 2.14(b), we can see that the logic indicator, lamp L1, will be lit, or on, whenever either switch is closed. Lamp L1 will be off only when both switches are open. The truth table in Figure 2.14(c) summarizes the operation of the NAND function. *A* and *B* represent the possible input conditions. *X* represents the intermediate output, or AND function. L1 represents the final output of the NOT-AND or NAND function. Note that L1 is the exact opposite of *X.*

The device that performs the NAND function is called the *NAND gate.* Figure 2.15 illustrates the Boolean expression, logic symbol, truth table, and pin configuration for a basic 2-input NAND gate.

The Boolean expression for the NAND gate describes its operation mathematically. The NAND expression is composed of the combination of the AND expression $(A \cdot B)$ and the NOT expression symbolized by the NOT bar ($^{-}$) over the *entire* expression. The logic symbol is also a combination of the AND symbol and the inverter symbol. The small circle on the output of the AND symbol is known as a **bubble** and is used to represent the inverter, or NOT function. From the truth table of Figure 2.15 we can see that the *only*

(a) NAND NOT-AND

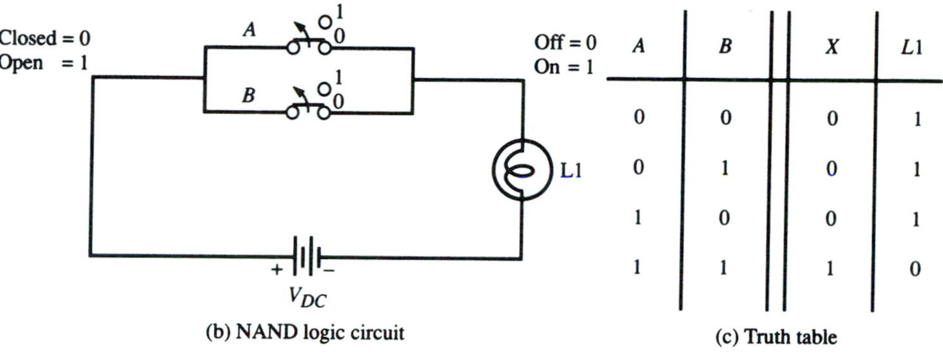

(b) NAND logic circuit

A	B	X	L1
0	0	0	1
0	1	0	1
1	0	0	1
1	1	1	0

(c) Truth table

FIGURE 2.14
NAND function

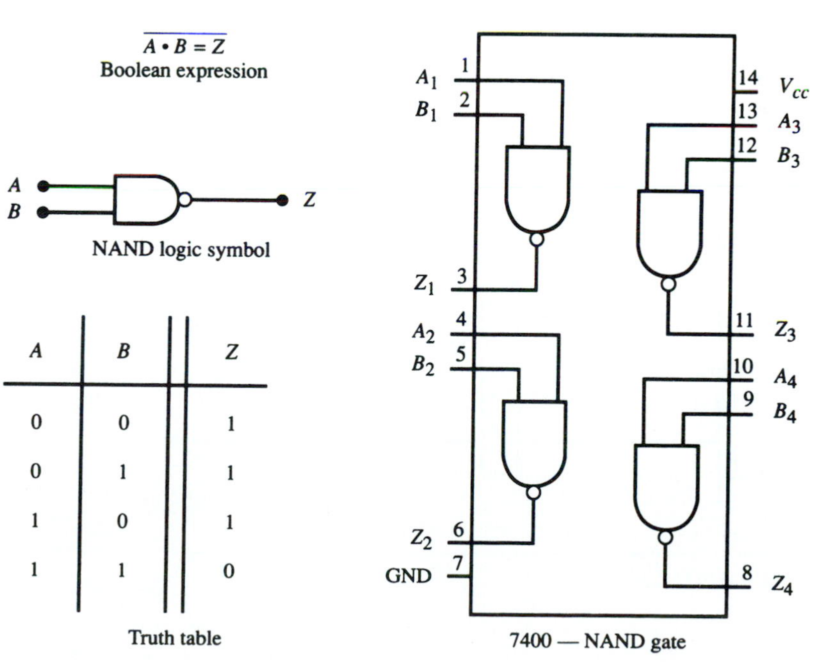

A	B	Z
0	0	1
0	1	1
1	0	1
1	1	0

Truth table

FIGURE 2.15
NAND gate

time that the output of the NAND gate is low (logic 0) is when *both* of the inputs are high (logic 1). This is true regardless of the number of inputs.

Table 2.4 summarizes a number of available TTL and CMOS NAND gates.

TABLE 2.4
TTL and CMOS NAND gates

TTL Series	CMOS Series	Description
7400	4011	Four 2-input NAND gates
7410	4023	Two 3-input NAND gates
7420	4012	Two 4-input NAND gates
7430	4068	One 8-input NAND gate
74132	—	One 12-input NAND gate

EXAMPLE 2.5

Input waveforms *A*, *B*, and *C* are applied to a NAND gate as shown in Figure 2.16. Determine the output waveform *Z*.

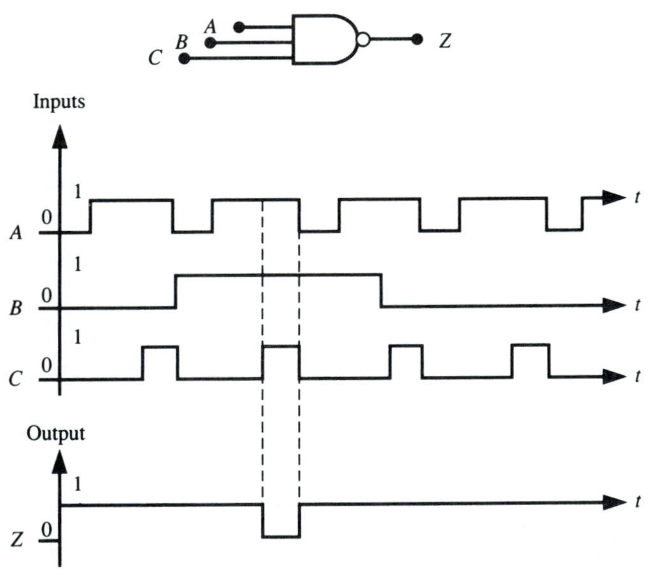

FIGURE 2.16

Solution

The output waveform *Z* is low only when all points (*A*, *B*, and *C*) are high.

The NOR Gate

The **NOR** logic function is equivalent to an OR function whose output is inverted, as shown in Figure 2.17(a). The name NOR comes from NOT-OR. This means that the NOR function is the *opposite,* or complement, of the OR function. This function is represented by the normally closed switching circuit of Figure 2.17(b). Referring to the circuit of Figure 2.17(b), we can see that the logic indicator, lamp L1, is lit *only* when both switches are closed. L1 is off whenever *either* switch is open. The truth table in Figure 2.17(c) summarizes the operation of the NOR function. A and B represent the possible input conditions. X represents the intermediate output, or OR function. L1 represents the final output of the NOT-OR, or NOR, function. Note that L1 is the exact opposite of X.

The device that performs the NOR function is called the *NOR gate.* Figure 2.18 illustrates the Boolean expression, logic symbol, truth table, and pin configuration for a basic 2-input NOR gate.

The Boolean expression for the NOR gate describes its operation mathematically. The NOR expression is composed of the combination of the OR expression $(A + B)$ and the NOT expression symbolized by the NOT bar ($\bar{}$) over the entire expression. The logic symbol is also a combination of the OR symbol and the inverter symbol. The bubble on the output of the OR symbol is again used to represent the inverter, or NOT function. From the truth table of Figure 2.18(c), we can see that the *only* time the output of the

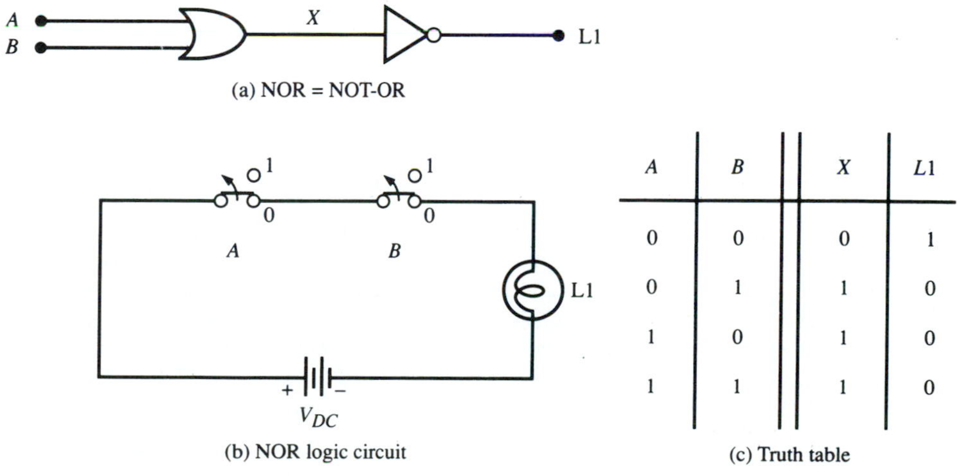

(a) NOR = NOT-OR

(b) NOR logic circuit

A	B	X	L1
0	0	0	1
0	1	1	0
1	0	1	0
1	1	1	0

(c) Truth table

FIGURE 2.17
NOR function

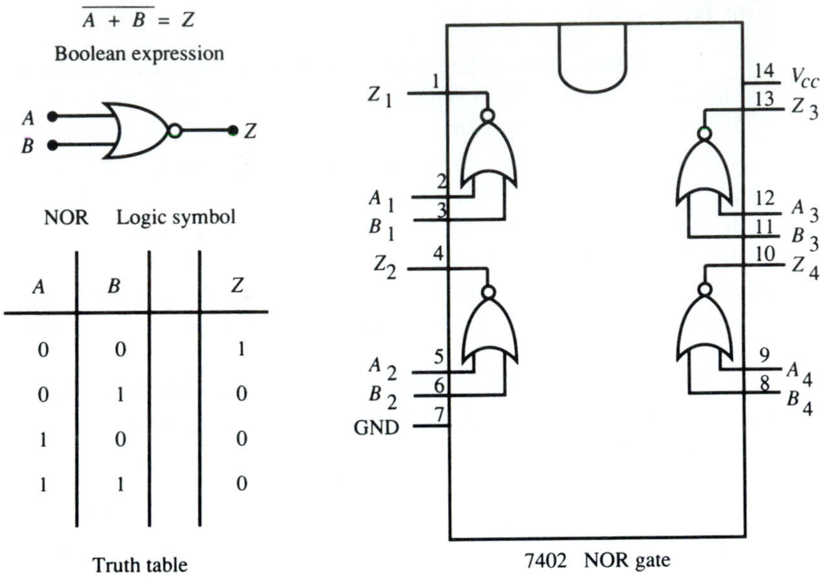

A	B	Z
0	0	1
0	1	0
1	0	0
1	1	0

FIGURE 2.18
NOR gate

NOR gate is high (logic 1) is when *both* of the inputs are low (logic 0). This is true regardless of the number of inputs. Table 2.5 summarizes a number of available TTL and CMOS NOR gates.

TABLE 2.5
TTL and CMOS NOR gates

TTL Series	CMOS Series	Description
7402	4001	Four 2-input NOR gates
7427	4025	Three 3-input NOR gates
—	4002	Two 4-input NOR gates
74260	—	Two 5-input NOR gates

EXAMPLE 2.6

Input waveforms A and B are applied to a NOR gate as shown in Figure 2.19 (p. 49). Determine the output waveform Z.

Solution

The output Z is high *only* when both inputs are low.

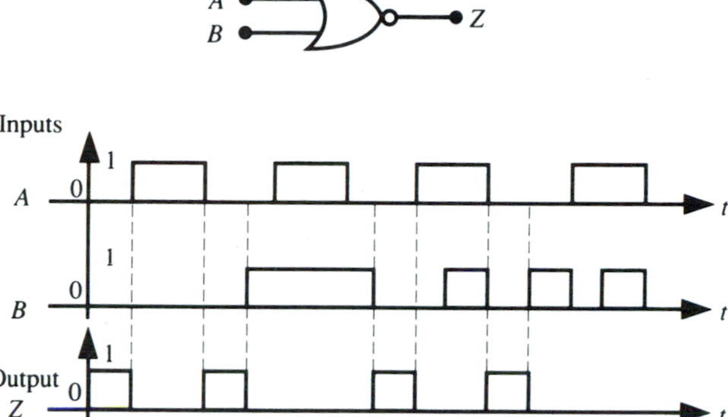

FIGURE 2.19

2.6 EXCLUSIVE-OR/EXCLUSIVE-NOR LOGIC

Two special logic functions that are used in microcomputer circuits are the *EXCLUSIVE-OR* and *EXCLUSIVE-NOR* functions. We will use these functions to illustrate how logic gates can be combined to form higher-order logic functions.

EXCLUSIVE-OR

The **EXCLUSIVE-OR (XOR)** logic function produces a high output (logic 1) whenever the two inputs are at *opposite* levels. Therefore, it can be used to *compare* two input signals. If the two input signals are the *same,* the output is low (logic 0). If the two inputs signals are *different,* the output is high (logic 1). Figure 2.20 illustrates the Boolean expression, logic symbol, truth table, and pin configuration for a two-input EXCLUSIVE-OR gate. The symbol $\oplus$ is used to represent the EXCLUSIVE-OR operation. Also note the modified OR symbol used to represent the EXCLUSIVE-OR gate in Figure 2.20(b). From the truth table we can see that the EXCLUSIVE-OR gate is a special type of 2-input OR gate that determines when two input signals are different.

The circuit of Figure 2.21 represents a combination of basic logic gates that perform the EXCLUSIVE-OR function. The input to inverter 1 is A. Therefore, the output is $\overline{A}$. The input to inverter 2 is B. Therefore, its output is $\overline{B}$. The inputs to AND gate 3 are $\overline{A}$ and B. Therefore, its output X is equal to $\overline{A} \cdot B$. The inputs to AND gate 4 are A and $\overline{B}$. Therefore, its output Y is equal to $A \cdot \overline{B}$. Finally, the outputs to OR gate 5 are X and Y, where

$$X = \overline{A} \cdot B$$
$$Y = A \cdot \overline{B}$$

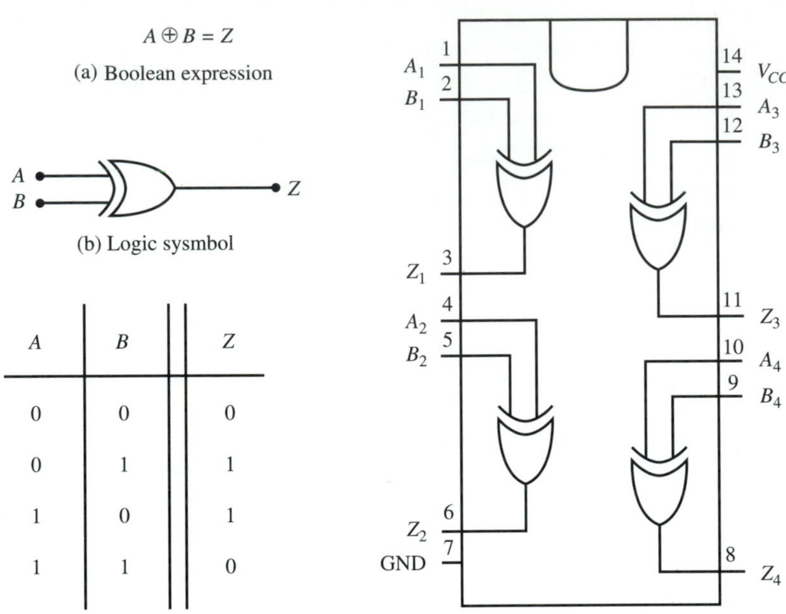

A	B	Z
0	0	0
0	1	1
1	0	1
1	1	0

(c) Truth table

FIGURE 2.20
EXCLUSIVE-OR

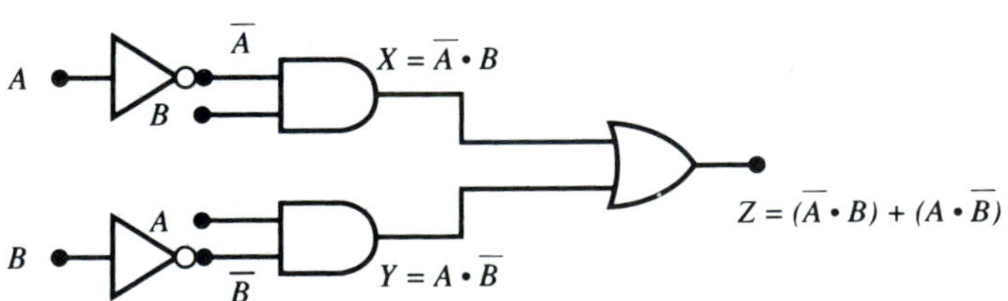

Inputs				Outputs		
A	B	$\overline{A}$	$\overline{B}$	X	Y	Z
0	0	1	1	0	0	0
0	1	1	0	1	0	1
1	0	0	1	0	1	1
1	1	0	0	0	0	0

(c) Truth table

$$Z = (\overline{A} \cdot B) + (A \cdot \overline{B}) = A \oplus B$$

FIGURE 2.21
EXCLUSIVE-OR function

Therefore, the output of OR gate 5, Z, is equal to

$$Z = (\overline{A} \cdot B) + (A \cdot \overline{B})$$

Note the use of the *parentheses* to clarify the order of operations of the circuit. By comparing the truth table of Figures 2.20 and 2.21 we can see that the circuit of Figure 2.21 performs the EXCLUSIVE-OR function. Therefore, we can conclude that

$$Z = (\overline{A} \cdot B) + (A \cdot \overline{B}) = A \oplus B$$

$$\overline{A \oplus B} = Z$$

(a) Boolean expression

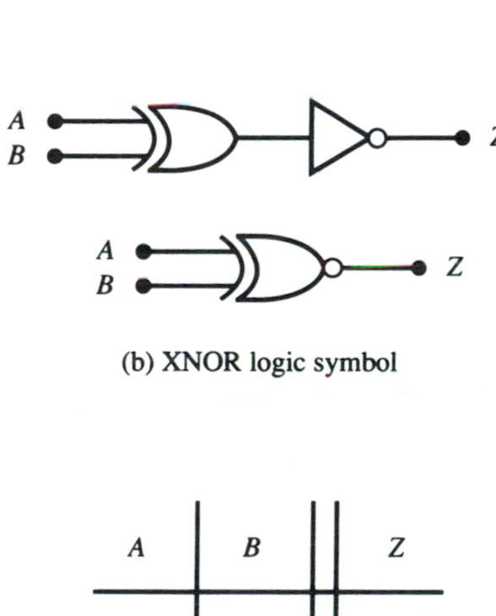

(b) XNOR logic symbol

A	B	Z
0	0	1
0	1	0
1	0	0
1	1	1

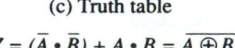

(c) Truth table

$$Z = (\overline{A} \cdot \overline{B}) + A \cdot B = \overline{A \oplus B}$$

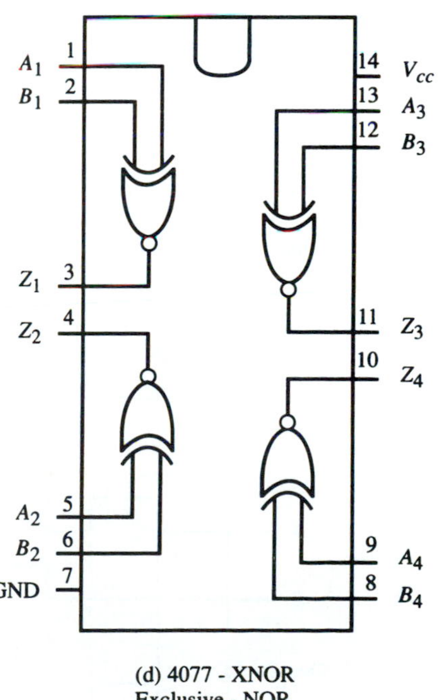

(d) 4077 - XNOR
Exclusive - NOR

FIGURE 2.22
EXCLUSIVE-NOR

EXCLUSIVE-NOR

The **EXCLUSIVE-NOR** (XNOR) gate is a special type of 2-input NOR gate. It is used to determine when the inputs are the same. This is sometimes referred to as an *equity checker,* or **comparator.** Figure 2.22 illustrates the EXCLUSIVE-NOR function. The Boolean expression for the EXCLUSIVE-NOR gate is written as

$$Z = \overline{A \oplus B}$$

Note that the EXCLUSIVE-NOR function can be constructed from an EXCLUSIVE-OR gate with an inverted output, as shown in Figure 2.22(b). From the truth table we can see that the output of an EXCLUSIVE-NOR gate is high (logic 1) whenever the inputs are the same. The output is low (logic 0) whenever the inputs are different.

Table 2.6 summarizes a number of available TTL and CMOS EXCLUSIVE-OR and EXCLUSIVE-NOR gates.

TABLE 2.6
TTL and CMOS EXCLUSIVE-OR/EXCLUSIVE-NOR gates

TTL Series	CMOS Series	Description
7486	4030	Four 2-input EXCLUSIVE-OR gates
—	4077	Four 2-input EXCLUSIVE-NOR gates

EXAMPLE 2.7 Input waveforms A and B are applied to an EXCLUSIVE-OR gate. Determine the output waveforms X and Y for the circuit of Figure 2.23.

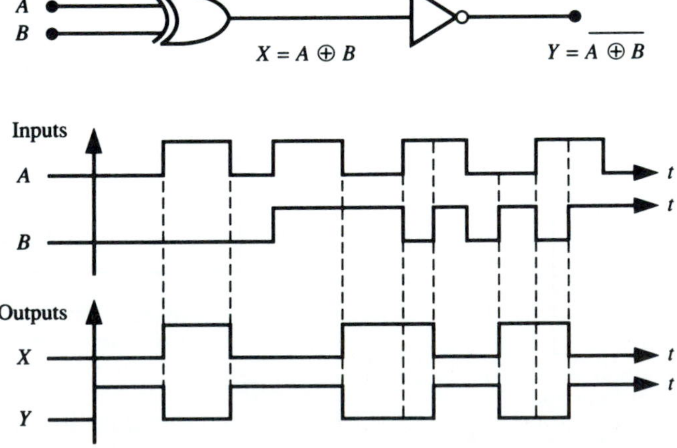

FIGURE 2.23

Solution

The output waveform X is high whenever inputs A and B are at different logic states. The output waveform Y is the opposite of waveform X.

2.7 SUMMARY OF LOGIC GATES

It is important to understand thoroughly the operation of all the basic logic gates. Figure 2.24 summarizes the Boolean expressions, logic symbols, and truth tables of all the gates presented up to this point. The information in Figure 2.24 should be memorized and completely understood because it is the foundation to all that is to follow.

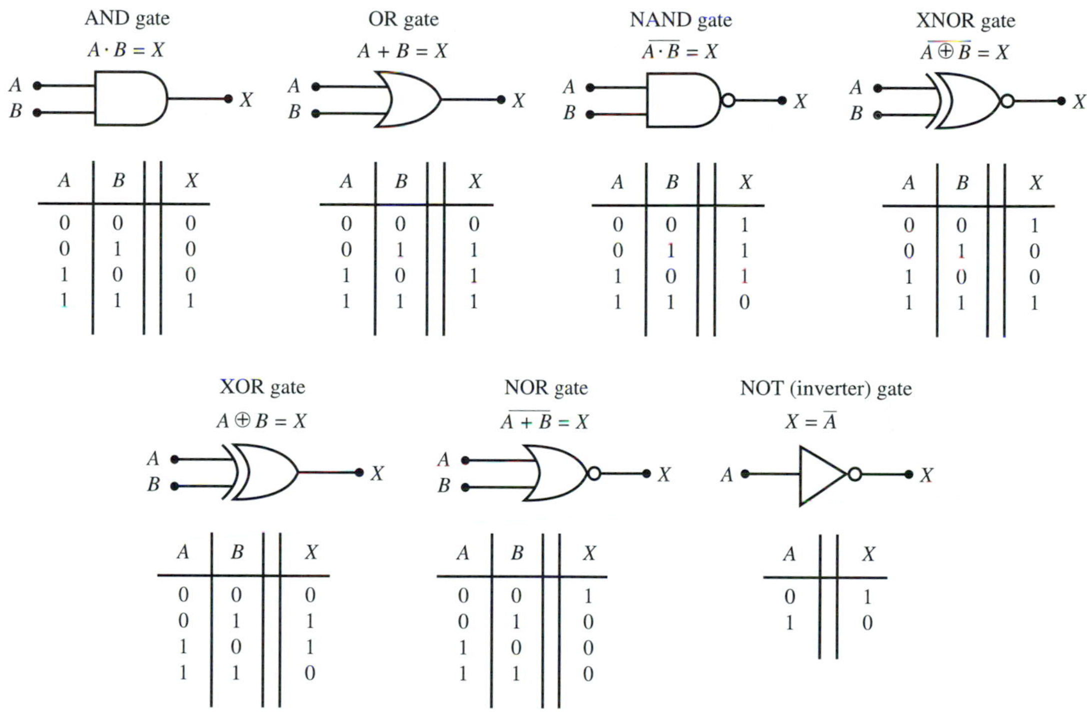

AND gate

A	B	X
0	0	0
0	1	0
1	0	0
1	1	1

OR gate

A	B	X
0	0	0
0	1	1
1	0	1
1	1	1

NAND gate

A	B	X
0	0	1
0	1	1
1	0	1
1	1	0

XNOR gate

A	B	X
0	0	1
0	1	0
1	0	0
1	1	1

XOR gate

A	B	X
0	0	0
0	1	1
1	0	1
1	1	0

NOR gate

A	B	X
0	0	1
0	1	0
1	0	0
1	1	0

NOT (inverter) gate

A	X
0	1
1	0

FIGURE 2.24
Summary of logic gates

2.8 TECH TIPS AND TROUBLESHOOTING—T^3

The **logic probe** is a tool that you will find extremely useful for troubleshooting digital circuits. Figure 2.25 shows an example of a typical logic probe. The logic probe is a very handy instrument for determining the logic state of a device. Although logic probes may vary in their features and capabilities, they are all quite similar.

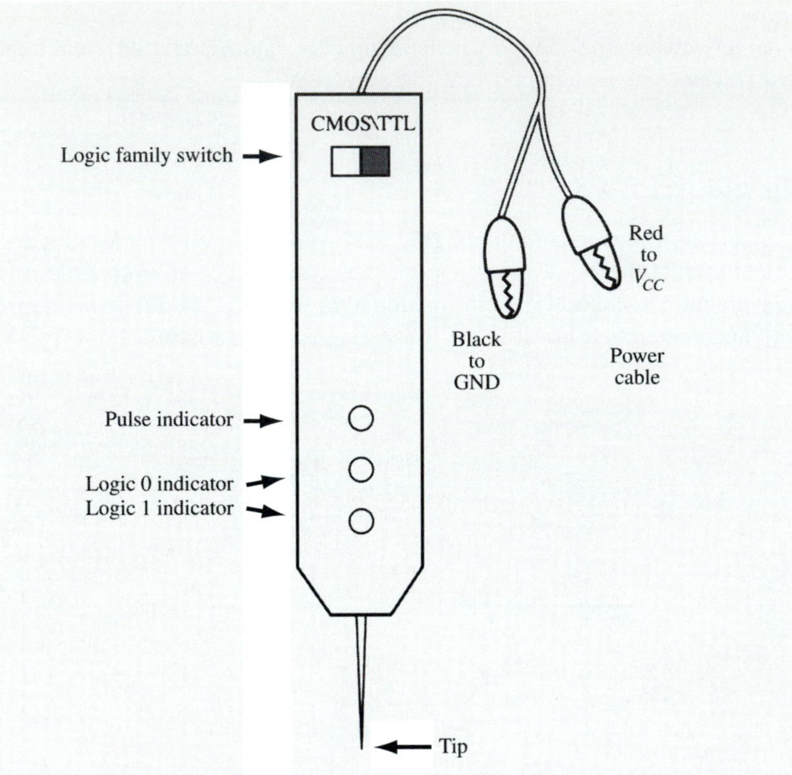

FIGURE 2.25
Logic probe

From our discussion on logic devices, we have seen that there are different types of logic families. Each logic family operates at different voltages and has different input and output characteristics. Because of these differences, logic probes are designed to operate with a particular logic family. Some logic probes have a switch that enables operation with two or more logic families by simply changing the position of the switch.

Most logic probes are powered by the system under test. Power is obtained by connecting a cable from the logic probe to the system power supply (V_{CC}) and ground.

At the bottom of the logic probe is a naillike tip. The tip is used to *probe,* or test, the state of a logic device. When the tip is placed on an IC pin or test point in a circuit, the logic probe indicates the current logic state of that point.

Most probes have visual indicators to signify a logic state. For example, a red light may indicate a logic 1 condition, and a green light may be used to indicate a logic 0 condition. If no light is lit, it usually indicates an open-circuit, or **tristate,** condition. This is sometimes referred to as a **float.** Some logic probes have a third light that pulses to indicate that a constant pulsing is occurring. Others have an internal 1-bit memory built in that is very useful in determining transient pulses.

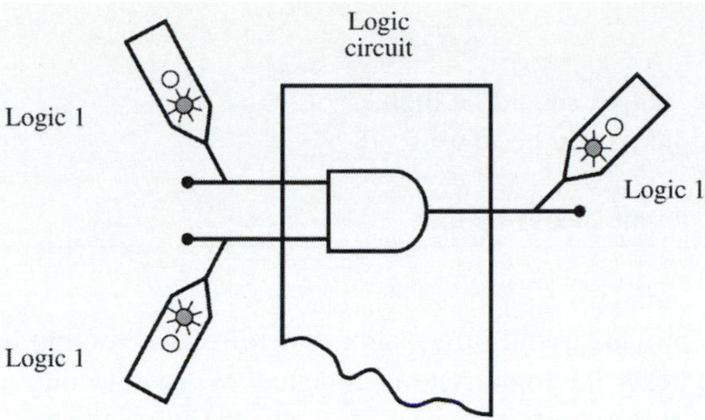

FIGURE 2.26
Use of logic probe

Let's consider the standard AND gate in Figure 2.26 to examine a simple use of the logic probe. Recall that the output of an AND gate is high (1) only when all the inputs are high (1). By probing the inputs we can verify that both inputs are high (1). If both inputs are high (1), we conclude that the output should be high (1). By probing the output, this can be verified. If the output is high (1), we can assume that the gate appears to be operating correctly. To be sure, we would have to test all the possible combinations of input and output conditions dynamically. If the probed output is *not* high (1), we can conclude that either the gate is bad or that something else in the circuit is causing a problem. For example, a short circuit to ground or a second gate that is loading down the output of this gate may be causing the problem.

EXAMPLE 2.8

Identify each logic gate in Figure 2.27 as either good or possibly bad.

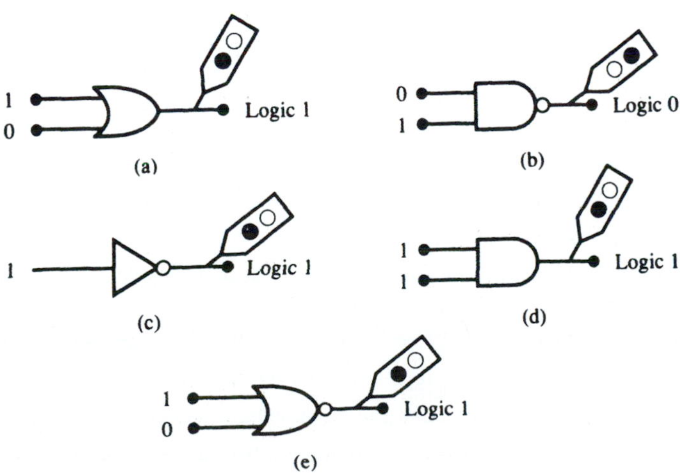

FIGURE 2.27

Solution

a. Good OR gate.
b. Bad NAND gate; output should be high.
c. Bad inverter; output should be low.
d. Good AND gate.
e. Bad NOR gate; output should be low.

Logic probes provide useful information when troubleshooting because they allow you to determine quickly the logic state of a signal. When selecting a logic probe, you must consider what features you will need. Usually the more features, the more expensive the logic probe will be. However, having a feature may save you a lot of money in time and effort. As we study and examine more complex digital circuits, you will appreciate this device even more.

EXERCISES

2.1 What is a logic gate?

2.2 Describe the types of devices that can be used to make a logic gate.

2.3 If the greater of two voltages is used to represent a logic 1, this is called __________.

2.4 If the greater of two voltages is used to represent a logic 0, this is called __________.

2.5 Most systems today use __________ V_{DC} to represent a logic 1 and __________ V_{DC} to represent a logic 0.

2.6 Find the frequency of a waveform that has a period of 250 μs.

2.7 If the pulse width of a waveform is 75 μs, and its period is 250 μs, calculate its duty cycle.

2.8 Explain the difference between a positive-going pulse and a negative-going pulse.

2.9 Draw an ideal pulse. Show the leading edge and trailing edge.

2.10 Draw a real, or nonideal, pulse showing t_r, t_f, and t_w.

2.11 Identify and explain the three basic logic functions.

2.12 Boolean algebra was first applied to what type of circuits?

2.13 Describe a truth table.

2.14 Identify the logic function that can be thought of as a series circuit containing two or more logic switches.

2.15 Identify the logic function represented by the Boolean expression $A \cdot B \cdot C = \text{L1}$.

2.16 Identify the logic function represented by the Boolean expression $A + B + C = \text{L1}$.

2.17 Identify the logic function that can be thought of as logic switches in parallel.

2.18 Calculate the number of logic combinations that are possible from a 5-input logic circuit.

2.19 Using logic switches, draw the logic circuits to perform the following functions:

a. AND
b. OR
c. NOT

2.20 Explain a DIP. Draw one of the most common styles.

2.21 Identify the two most common logic families.

2.22 Which logic family can use a DC power supply voltage of +3 V to +18 V?

2.23 Identify the function of the following devices:

a. 7408 b. 7411
c. 7432 d. 7404
e. 4081 f. 4082
g. 4071 h. 4049

2.24 Give the logic symbol, truth table, and Boolean expression for the following:

a. AND gate b. OR gate
c. Inverter

2.25 If four inverters are connected in series, determine the output of the last inverter when the input to the first inverter is a logic 1.

2.26 Design a three-input NAND gate using only AND gates, OR gates, and inverters.

2.27 Design a three-input NOR gate using only AND gates, OR gates, and inverters.

2.28 Give the logic symbol, truth table, and Boolean expression for the following:

a. NAND gate b. NOR gate
c. XOR gate d. XNOR gate

2.29 Using a combination of the basic logic gates, draw the following logic circuits and verify their operation using the truth table.

a. XOR b. XNOR

2.30 Identify the functions of the following devices:

a. 7486 b. 7427
c. 7402 d. 7400
e. 4011 f. 4001
g. 4030 h. 4070

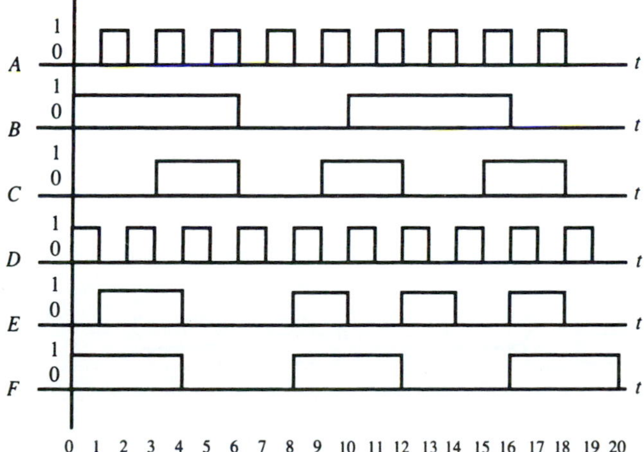

FIGURE 2.28

2.31 Explain the operation and use of a logic probe.

2.32 In each of the following diagrams, determine the output waveform Z using the input waveforms in Figure 2.28 (p. 57). Use graph paper.

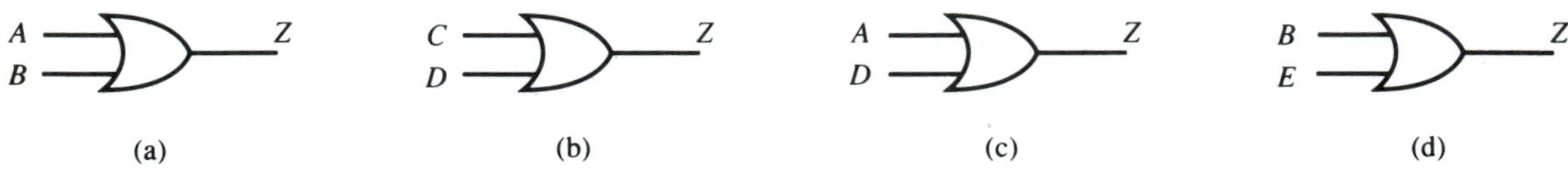

FIGURE 2.29

2.33

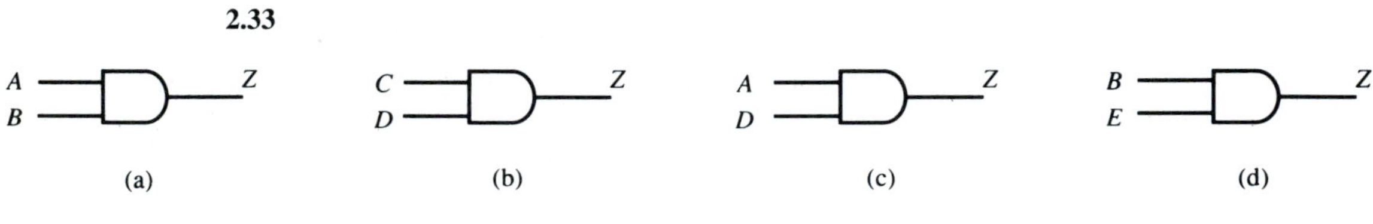

FIGURE 2.30

2.34

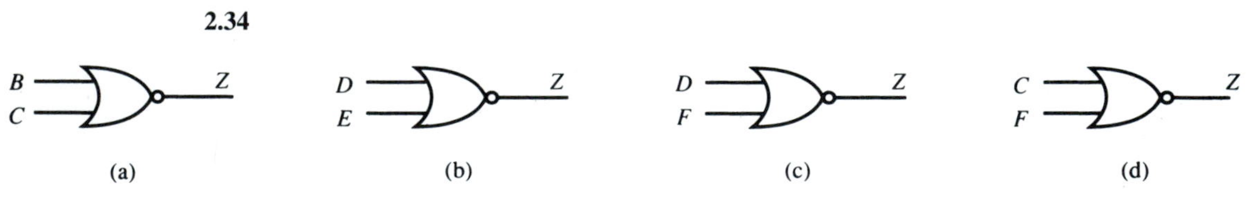

FIGURE 2.31

2.35

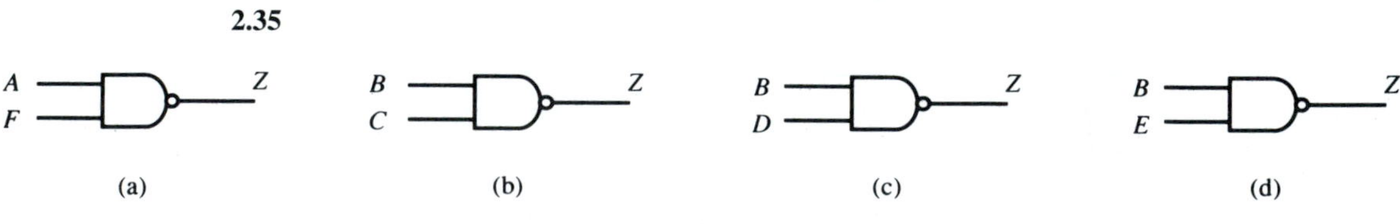

FIGURE 2.32

2.36

A Z
(a)
B Z
(b)
C Z
(c)
F Z
(d)

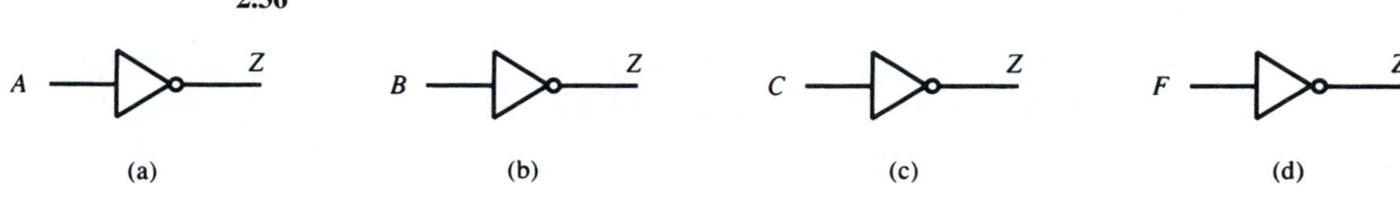

FIGURE 2.33

2.37

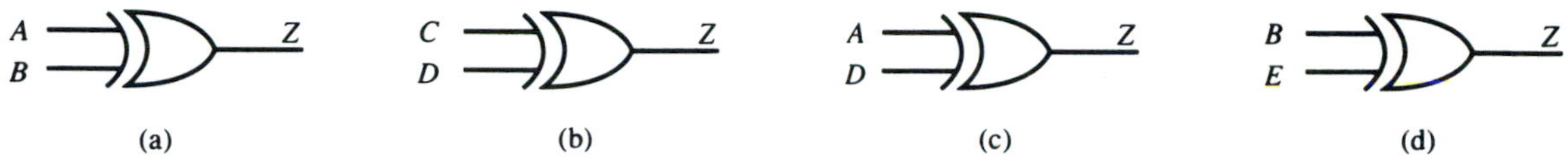

A B Z
(a)
C D Z
(b)
A D Z
(c)
B E Z
(d)

FIGURE 2.34

2.38

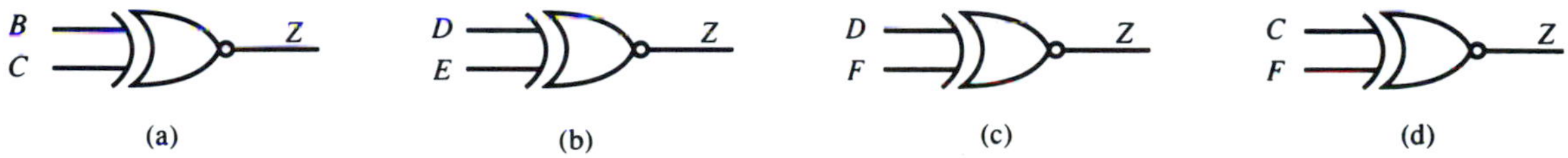

B C Z
(a)
D E Z
(b)
D F Z
(c)
C F Z
(d)

FIGURE 2.35

2.39

A B C Z
D E F Z
A C F Z

FIGURE 2.36

2.40

A B C Z
D E F Z
A C F Z

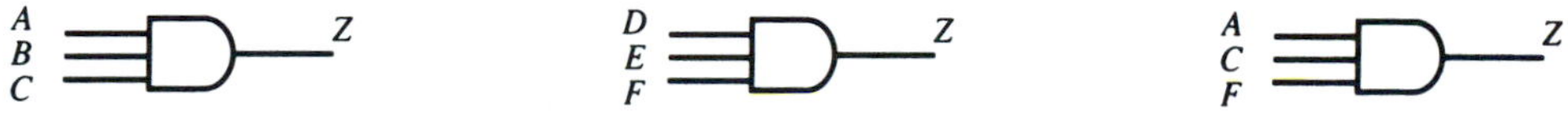

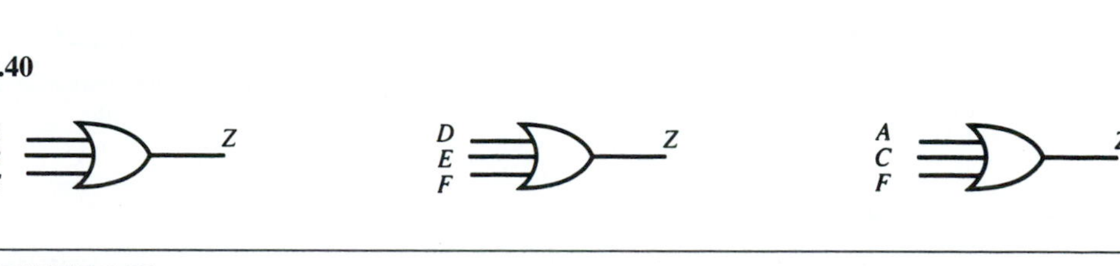

FIGURE 2.37

2.41

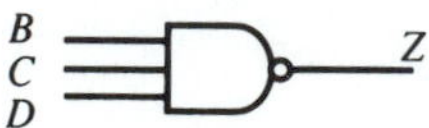

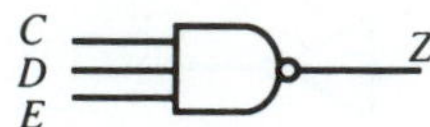

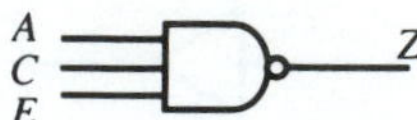

FIGURE 2.38

2.42

FIGURE 2.39

2.43

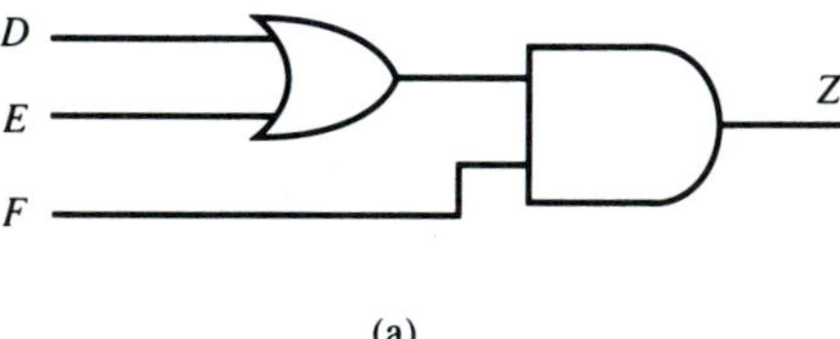

(a)

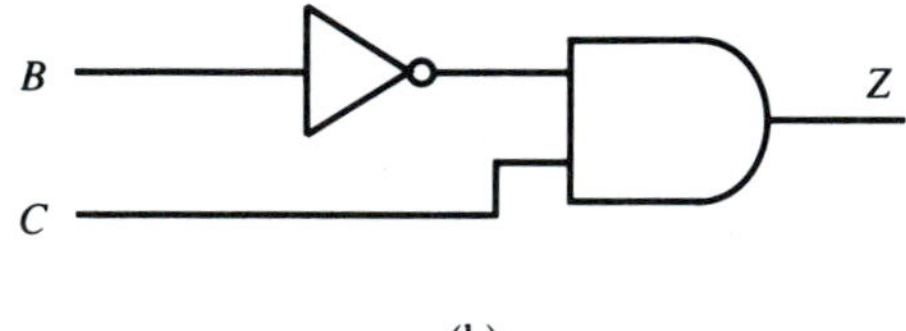

(b)

FIGURE 2.40

2.44

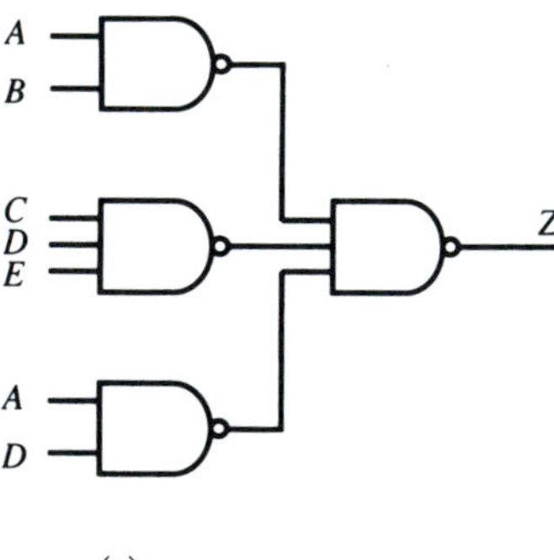

(a)

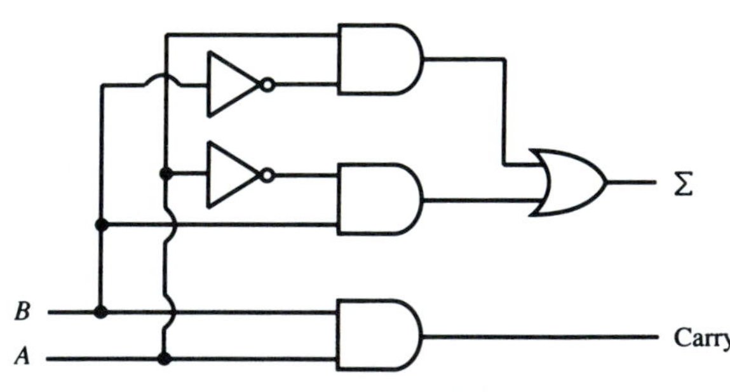

Determine the waveforms for Σ and Carry

(b)

FIGURE 2.41

2.45 Crossword Puzzle

ACROSS

1. 4071 gate.
8. A series of pulses.
10. The logic function that is equivalent to an AND function whose output is inverted.
12. An IC packaging style.
13. The time to complete one cycle of a waveform.
16. The first transition of a waveform pulse is called the _____ edge.
17. Devices that perform the Boolean logic function are called logic _____.
22. The difference between the maximum and minimum values of a pulse waveform.
24. The logic function that is equivalent to an OR function whose output is inverted.
25. A small circle that is used to represent the inverter function.
26. An open switch can represent a logic level _____.
28. _____ circuit (IC).
29. Logic functions can be expressed by mathematics, logic symbols, or _____ _____.

DOWN

2. _____ time is the time it takes to go from 10% to 90% of the amplitude of a pulse.
3. Logic function whose output will be a logic 1 if its inputs are different.
4. SN7408N gate.
5. The ratio between pulse width and period.
6. Logic switches are inherently _____ in nature.
7. Manufacturing techniques are classified by logic _____.
9. 74-series.
11. George Boole's work with logic formed the basic laws and theorems of _____ algebra.
14. The second transition of a waveform pulse is the _____ edge.
15. A closed switch can represent the logic level _____.
18. _____ time is the time it takes the pulse to go from 90% to 10% of its amplitude.
19. Measured between the 50% points of the pulse.
20. A handy instrument for determining the logic state of a device is called a logic _____.
21. The electronic circuit that performs the NOT function.
23. Logic level transition.
27. $Z = \overline{A}$ is an example of the _____ function.

1 2 3 4 5 6 7 8 9 10 11 12 13 14 15 16 17 18 19 20 21 22 23 24 25 26 27 28 29

CHAPTER 3

COMBINATION LOGIC

KEY TERMS

Active-State Logic
Associative Law
Boolean Algebra Theorems
Commutative Law
DeMorgan's Theorem
Distributive Law
Don't-Care Terms
Float
Full-Adder Circuit
Half-Adder Circuit
Identity Law
Karnaugh Map
Logic Pulser
Nashelsky's Theorem
Redundancy Law
Sum-of-the-Products (SOP)
Universal Logic Gates

3.0 INTRODUCTION

In the previous chapter we learned about the three basic logic operations: AND, OR, and NOT. We saw how these can be described mathematically using Boolean expressions. We also combined these three basic operations to form more complex functions (NAND, NOR, XOR, XNOR). In this chapter we build upon this basic knowledge to develop complex digital circuits using more sophisticated techniques. The techniques you will learn in this chapter can be used to design the simplest and the most complex digital circuits.

3.1 DEVELOPING LOGIC CIRCUITS FROM BOOLEAN EXPRESSIONS

Boolean expressions are used to describe the operation of logic circuits. For Boolean expressions to be of practical use to the technician, they must be converted into circuits that can be constructed and tested accurately. The first step in the process is to draw a logic diagram based on the Boolean expression.

For example, suppose we are given a Boolean expression that describes the operation of an alarm system. A bell is to ring under certain constraints described by the Boolean expression

$$\text{BELL} = A\overline{B} + \overline{A}B + C$$

Step 1. Identify the input and output variables as shown in Figure 3.1(a). It should be noted that logic diagrams are usually drawn with the inputs on the left and the outputs on the right. This is similar to the way we read, left to right, top to bottom.

Step 2. Break the logic expression down into parts or terms that can be implemented using simple logic gates. One method of breaking down a logic expression is to place parentheses around each AND function or single input variable.

$$\text{BELL} = \underset{\text{term 1}}{(A\overline{B})} + \underset{\text{term 2}}{(\overline{A}B)} + \underset{\text{term 3}}{(C)}$$

Step 3. Implement the first logic term $(A\overline{B})$. In this case it can be implemented using an AND gate and an inverter, as shown in Figure 3.1(b).

Step 4. Repeat the process with the second term $(\overline{A}B)$ and add it to the diagram, as shown in Figure 3.1(c).

Step 5. Combine the three terms using a single 3-input OR gate. Notice that the single input term (C) can be directly connected as an input to the OR gate. Connect the output of the OR gate to BELL, the output variable in the original Boolean expression.

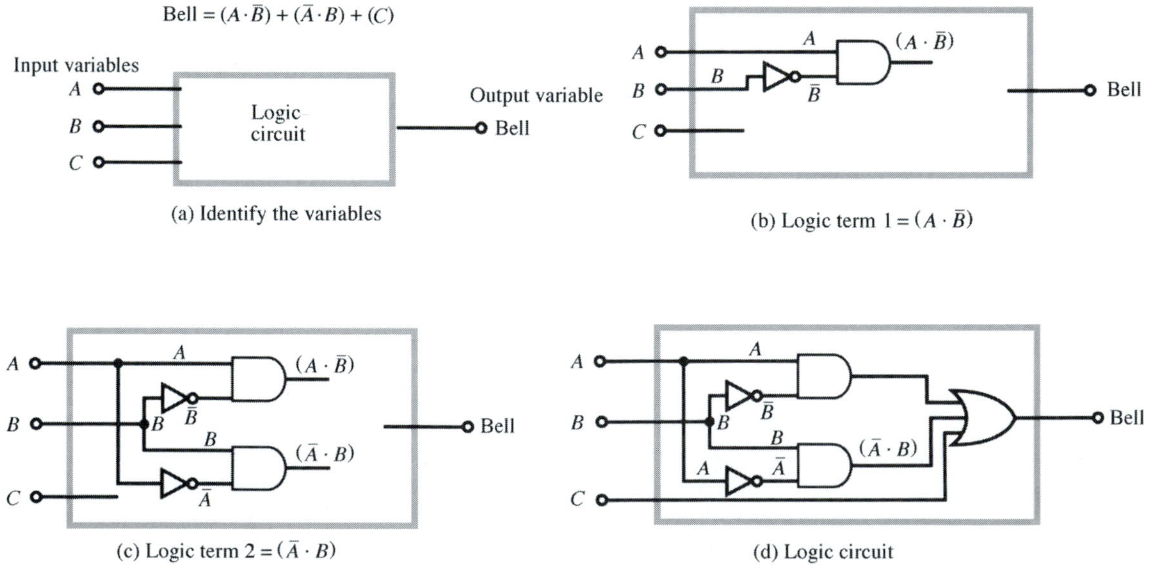

FIGURE 3.1
Developing a logic circuit from a Boolean expression

EXAMPLE 3.1

Draw the logic diagram that implements the Boolean expression

$$\text{CLOCK} = A \cdot B + \overline{A \cdot B}$$

Solution

See Figure 3.2. The expression can be broken down into two terms (AB) and $(\overline{AB})$. Note that the second term $(\overline{AB})$ can be implemented using a single NAND gate.

$\text{Clock} = (A \cdot B) + (\overline{A \cdot B})$

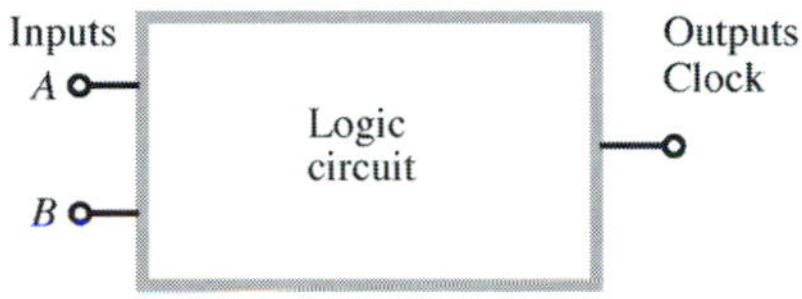

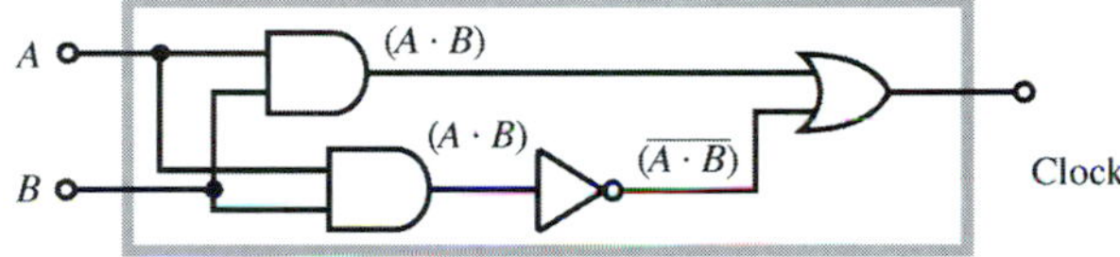

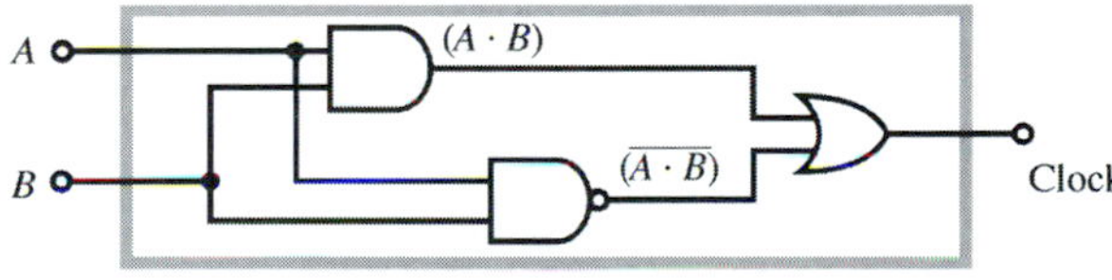

FIGURE 3.2
Logic circuit using a NAND gate

3.2 DEVELOPING BOOLEAN EXPRESSIONS FROM LOGIC CIRCUITS

The Boolean expression of a logic circuit can easily be obtained by simply writing the output expression for each logic gate. One method of doing this is to start at the inputs and proceed gate by gate to the outputs. For example, consider the logic diagram of Figure 3.3.

The input variables are *A*, *B*, and *C*. The output variable is labeled *Z*. Input variables *A* and *B* are connected to OR gate 1. The output of OR gate 1 can be written as

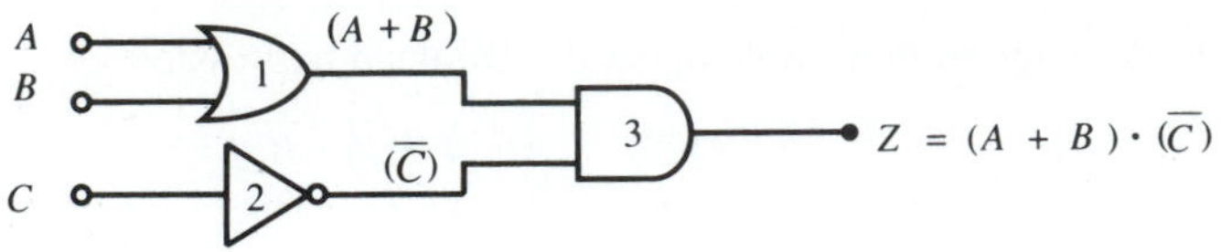

FIGURE 3.3
Boolean expression from logic circuit

$(A + B)$. Input variable C is connected to inverter 2. The output of inverter 2 is then written as $(\overline{C})$. The inputs to AND gate 3 are the outputs of OR gate 1 and inverter 2. The output of AND gate 3 is then written as $(A + B) \cdot (\overline{C})$. Since this is the final output of the circuit, it is equal to the output variable Z. The Boolean expression that describes the total circuit operation is then written as

$$Z = (A + B) \cdot (\overline{C})$$

Note that the use of the parentheses to indicate the individual logic gate outputs. Parentheses are not always required; however, they provide a convenient way of keeping things simple.

EXAMPLE 3.2 Obtain the Boolean expression for the logic circuit in Figure 3.4.

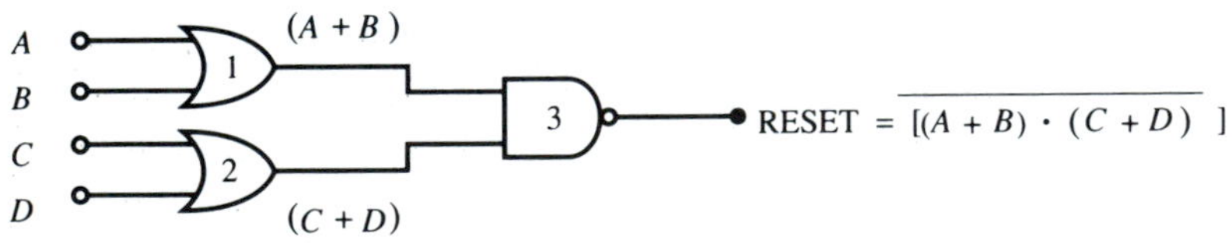

FIGURE 3.4

Solution

The input variables are A, B, C, and D. The output of OR gate 1 is $(A + B)$. The output of OR gate 2 is $(C + D)$. The outputs of OR gate 1 and OR gate 2 become the input to NAND gate 3. The output of NAND gate 3 is obtained by ANDing the two input expressions and inverting the entire result. Thus the final result is $\text{RESET} = \overline{[(A + B) \cdot (C + D)]}$.

Note the use of the brackets to contain the total expression that is to be inverted by the NOT bar. Note also that in this example the output variable is called RESET, which describes the function of the output. Using the name of a function for a variable is commonly done for clarity.

3.3 BOOLEAN ALGEBRA THEOREMS

Boolean expressions can be combined, manipulated, and simplified through a process known as Boolean algebra. As with any type of mathematics, certain rules, laws, and procedures must be followed. These rules, laws, and procedures can be grouped under the general heading **Boolean algebra theorems.** Boolean theorems can be used to combine, manipulate, and simplify logic circuits.

Basic Theorems

Three basic theorems describe the fundamental logic operations of the OR, AND, and NOT functions (see Figure 3.5).

Boolean Algebra Laws

The next group of theorems involve fundamental mathematical laws. They describe basic algebraic operations as well as the order in which these operations can be performed. See Figure 3.6 (p. 69).

Theorem 4 is called the **identity law.** It defines the operation of a gate when the inputs are the same. If the inputs are the same, they can be considered to be connected together. The output of a basic gate is then the same as the input.

Theorem 5 is called the **commutative law.** It states that it does not matter in which order an AND or OR operation is performed. It is similar to saying that it does not matter whether we add $5 + 6$ or $6 + 5$ to obtain 11. The same can be said for multiplication. We can multiply 5×6 or 6×5 to obtain 30. Note that the OR operation is similar to addition, and the AND operation is similar to multiplication.

Theorem 6 is called the **associative law.** It states that variables can be grouped in any way that we choose, as indicated by the placement of parentheses in algebraic and Boolean expressions.

Theorem 7 is called the **distributive law.** It states that expressions can be expanded by multiplying. The distributive law works the same way in Boolean algebra as it does in algebra. For example,

$$\begin{aligned} A(B + C) &= AB + AC \\ (A + B)(A + B) &= AA + AB + AB + BB \\ &= A^2 + 2AB + B^2 \end{aligned}$$

Another important aspect of the distributive law is that it may be used backward. That is, it implies that factoring must be allowed. Factoring in Boolean algebra works just as in algebra. Common terms or variables can be taken out of the expression. For example,

$$\begin{aligned} ABC + CDE &= C(AB + DE) \\ ABC + BCE &= BC(A + E) \end{aligned}$$

Theorem 1: **Basic OR operations**

(a) $A + 0 = A$ — inputs A, 0; output [A]

(b) $A + 1 = 1$ — inputs A, 1; output [1]

(c) $A + \overline{A} = 1$ — inputs A, $\overline{A}$; output [1]

Theorem 2: **Basic AND operations**

(a) $A \cdot 0 = 0$ — inputs A, 0; output [0]

(b) $A \cdot 1 = A$ — inputs A, 1; output [A]

(c) $A \cdot \overline{A} = 0$ — inputs A, $\overline{A}$; output [0]

Theorem 3: **Basic negation (NOT) operations**

(a) $(\overline{A}) = \overline{A}$ — input A; output [$\overline{A}$]

(b) $\overline{(\overline{A})} = A$ — A, $(\overline{A})$, $\overline{(\overline{A})} = A$

(c) $\overline{\overline{(\overline{A})}} = \overline{A}$

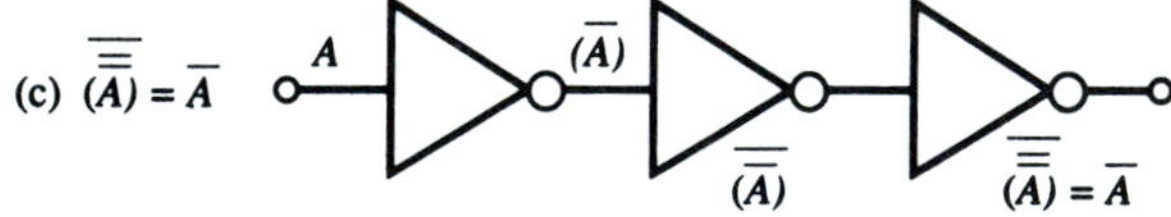

FIGURE 3.5

Theorem 4: Identity law

(a) $A + A = A$

(b) $A \cdot A = A$

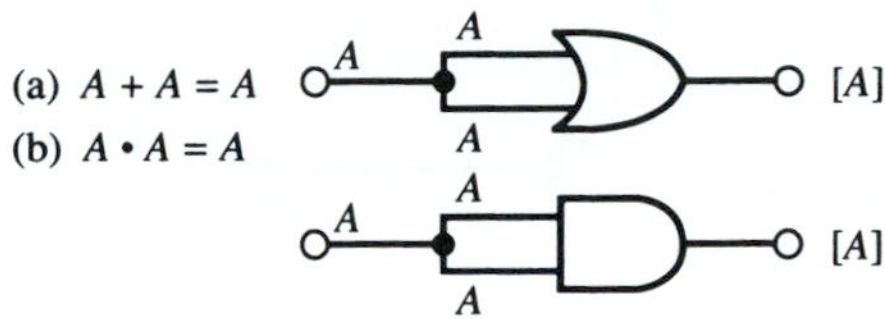

Theorem 5: Commutative law

(a) $A + B = B + A$

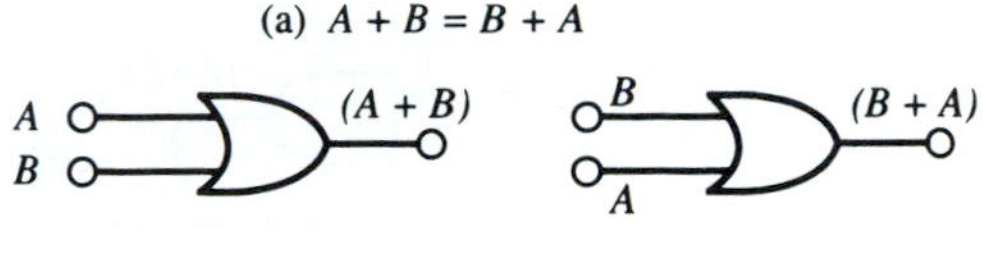

(b) $A \cdot B = B \cdot A$

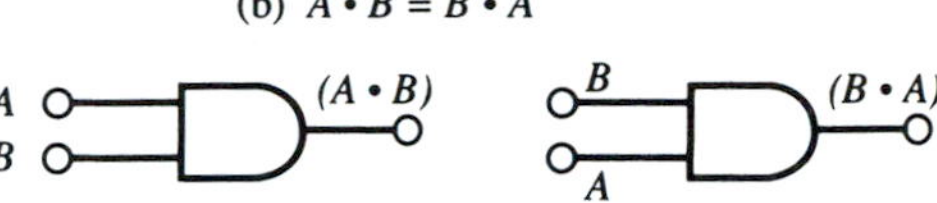

Theorem 6: Associative law

(a) $A + (B + C) = (A + B) + C$

B
C
A
(B + C)
[A + (B + C)]

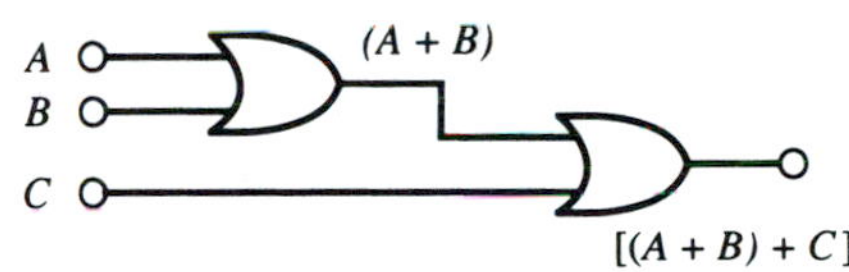

(b) $A \cdot (B \cdot C) = (A \cdot B) \cdot C$

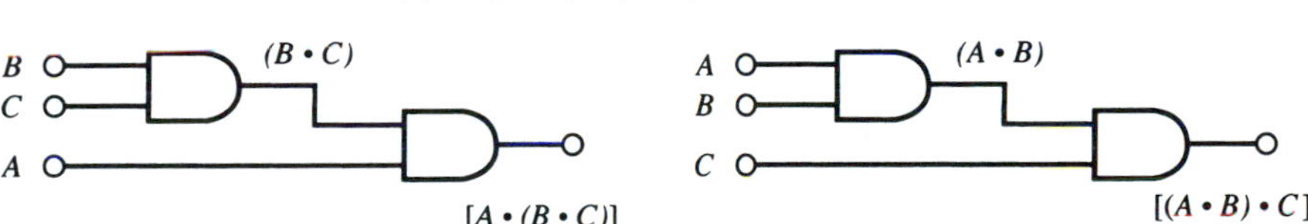

Theorem 7: Distributive law

(a) $A \cdot (B + C) = (A \cdot B) + (A \cdot C)$

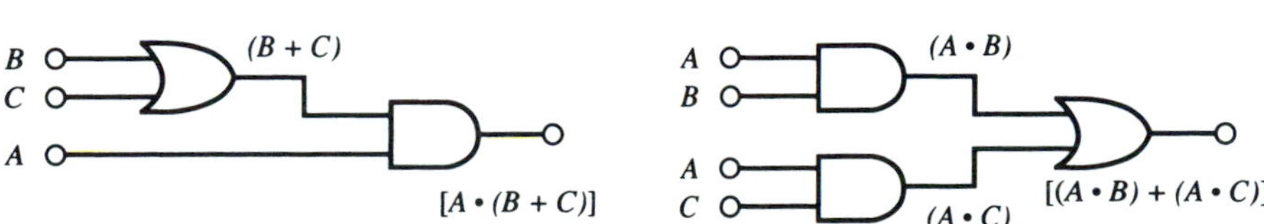

(b) $A + (B \cdot C) = (A + B) \cdot (A + C)$

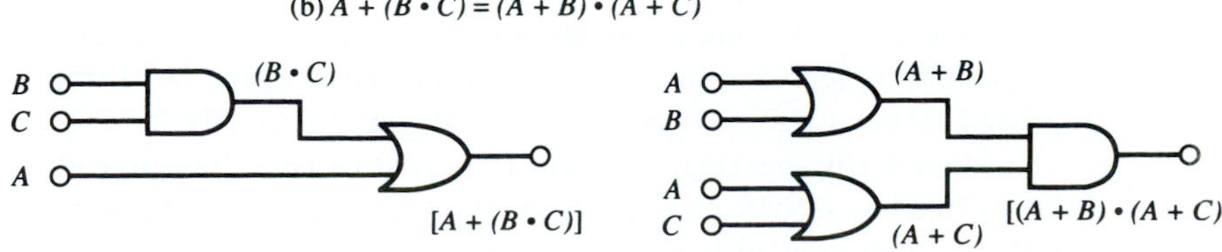

FIGURE 3.6

Theorem 8: Redundancy law

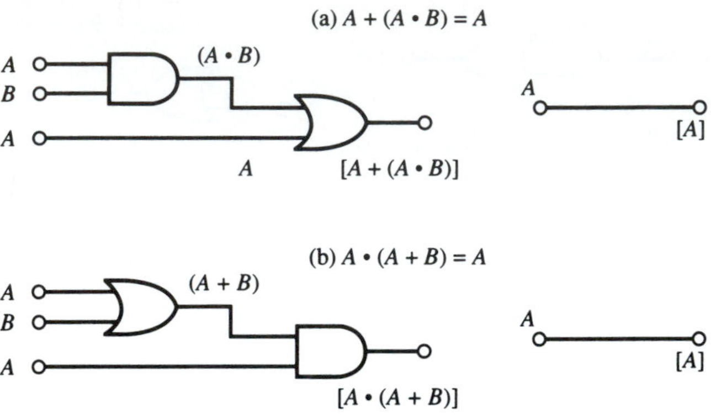

FIGURE 3.6 (continued)

Theorem 8 is called the **redundancy law.** It is really a special case of the identity law. It states that under certain conditions the second term of an expression may be redundant and ignored.

Advanced Theorems

There are a number of Boolean relationships that are not as obvious as the ones that have already been described. We refer to these as advanced theorems. These theorems are very useful for simplifying Boolean expressions. A simple way to prove their validity is by the use of the truth table. See Figure 3.7 (p. 71).

Theorem 9 is sometimes known as **Nashelsky's theorem.** These theorems were named after Dr. Louis Nashelsky, one of the pioneers of digital technology. Nashelsky's theorems can also be considered a special case of the redundancy law:

Theorem 9a: $A + (\overline{A} \cdot B) = A + B$

Theorem 9b: $A \cdot (\overline{A} + B) = A \cdot B$

At a first glance, Nashelsky's theorems are not obvious. However, they can be proved using Boolean algebra. Figure 3.7 illustrates a more practical proof using a truth table.

Theorem 10 is known as **DeMorgan's theorem.** It was developed by the famous nineteenth-century mathematician Augustus DeMorgan. DeMorgan's theorem can be proved by the use of the truth table. See Figure 3.8 (p. 72).

Practically speaking, Theorem 10a says that an AND gate with inverted inputs may be replaced by a NOR gate. Theorem 10b says that a NAND gate may be replaced by an OR gate with inverted inputs (and vice versa).

One way to remember and use DeMorgan's theorem is to remember this rule: Break the bar and change the sign. Careful inspection of DeMorgan's theorem will indicate the validity of this rule.

Advanced Theorems
Theorem 9 : Nashelsky's theorems

(a) $A + (\bar{A} \cdot B) = A + B$

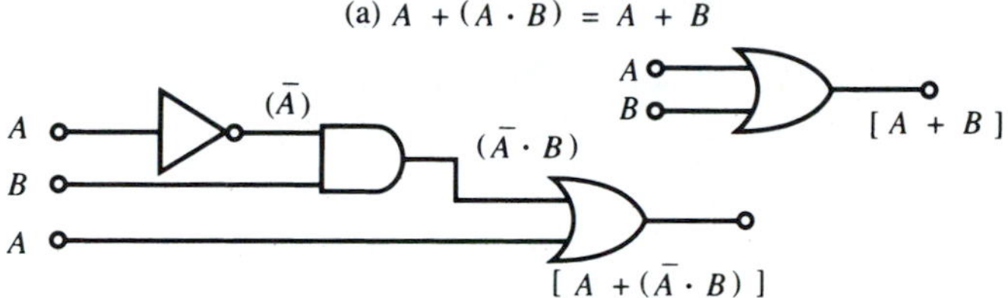

A	$\bar{A}$	B	$(\bar{A} \cdot B)$	$[A + (\bar{A} \cdot B)]$	$[A + B]$
0	1	0	0	0 + 0 = 0	0 + 0 = 0
0	1	1	1	0 + 1 = 1	0 + 1 = 1
1	0	0	0	1 + 0 = 1	1 + 0 = 1
1	0	1	0	1 + 0 = 1	1 + 1 = 1

Truth table

(b) $A \cdot (\bar{A} + B) = A \cdot B$

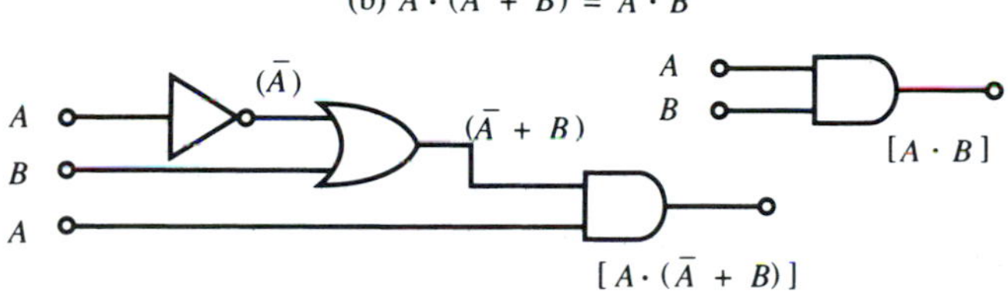

A	$\bar{A}$	B	$(\bar{A} + B)$	$[A \cdot (\bar{A} + B)]$	$[A \cdot B]$
0	1	0	1 + 0 = 1	0 · 1 = 0	0 · 0 = 0
0	1	1	1 + 1 = 1	0 · 1 = 0	0 · 1 = 0
1	0	0	0 + 0 = 0	1 · 0 = 0	1 · 0 = 0
1	0	1	0 + 1 = 1	1 · 1 = 1	1 · 1 = 1

Truth table

FIGURE 3.7

$$\text{Theorem 10a:} \quad \overline{(A + B)} = \bar{A} \cdot \bar{B}$$

$$\text{Theorem 10b:} \quad \overline{(A \cdot B)} = \bar{A} + \bar{B}$$

Breaking the NOT bar and changing the logic operation results in an equivalent expression. For example, consider the following Boolean equation:

$$Z = \overline{[(\bar{A}B) + (A\bar{B})]}$$

Theorem 10 : De Morgan's theorems

(a) $(\overline{A + B}) = \bar{A} \cdot \bar{B}$

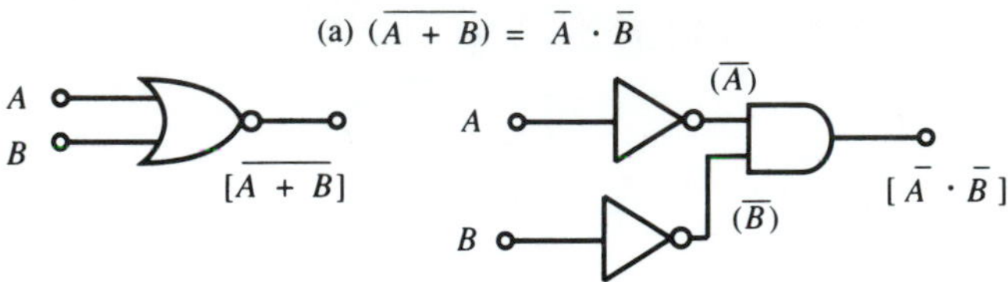

A	B	$\bar{A}$	$\bar{B}$	$[\overline{A + B}]$	$[\bar{A} \cdot \bar{B}]$
0	0	1	1	$[\overline{0 + 0}] = 1$	$[1 \cdot 1] = 1$
0	1	1	0	$[\overline{0 + 1}] = 0$	$[1 \cdot 0] = 0$
1	0	0	1	$[\overline{1 + 0}] = 0$	$[0 \cdot 1] = 0$
1	1	0	0	$[\overline{1 + 1}] = 0$	$[0 \cdot 0] = 0$

Truth table

(b) $(\overline{A \cdot B}) = \bar{A} + \bar{B}$

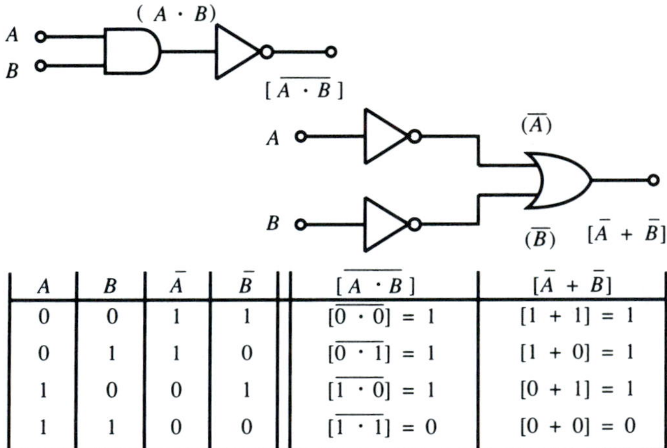

A	B	$\bar{A}$	$\bar{B}$	$[\overline{A \cdot B}]$	$[\bar{A} + \bar{B}]$
0	0	1	1	$[\overline{0 \cdot 0}] = 1$	$[1 + 1] = 1$
0	1	1	0	$[\overline{0 \cdot 1}] = 1$	$[1 + 0] = 1$
1	0	0	1	$[\overline{1 \cdot 0}] = 1$	$[0 + 1] = 1$
1	1	0	0	$[\overline{1 \cdot 1}] = 0$	$[0 + 0] = 0$

Truth table

FIGURE 3.8

By applying the rule for Theorem 10a (break the bar and change the sign), we get the equivalent expression

$$Z = [\overline{(\bar{A}B)} \cdot \overline{(A\bar{B})}]$$

Applying Theorem 10b in the same manner gives the result

$$Z = [(\bar{\bar{A}} + \bar{B}) \cdot (\bar{A} + \bar{\bar{B}})]$$

Applying Theorem 3b, the basic negation theorem ($\bar{\bar{A}} = A$) yields

$$Z = [(A + \bar{B}) \cdot (\bar{A} + B)]$$

Next, applying the distributive law, Theorem 7a, we have

$$Z = [(A \cdot \overline{A}) + (A \cdot B) + (\overline{B} \cdot \overline{A}) + (\overline{B} \cdot B)]$$

Finally, applying Theorem 2c, the basic AND operation ($A \cdot \overline{A} = 0$), we have

$$\begin{aligned} Z &= [0 + (A \cdot B) + (\overline{B} \cdot \overline{A}) + 0] \\ &= A \cdot B + \overline{B} \cdot \overline{A} \end{aligned}$$

or by applying Theorem 5b, the commutative law ($A \cdot B = B \cdot A$), we may write the expression as

$$Z = AB + \overline{A}\overline{B}$$

Other applications of the use of Boolean theorems to simplify Boolean expressions are given in the following examples.

EXAMPLE 3.3 Simplify the following logic expression:

$$Z = (X + \overline{Y}) \cdot (\overline{X}Y + X)$$

Solution

$Z = (X + \overline{Y}) \cdot (\overline{X}Y + X)$	
$= X\overline{X}Y + XX + \overline{Y}\overline{X}Y + \overline{Y}X$	Theorem 7 (distributive law)
$= X\overline{X}Y + X + \overline{Y}\overline{X}Y + \overline{Y}X$	Theorem 4b (identity law: $X \cdot X = X$)
$Z = 0 + X + 0 + \overline{Y}X$	Theorem 2c (basic AND operation: $X \cdot \overline{X} = 0$, $Y \cdot \overline{Y} = 0$)
$= X + \overline{Y}X$	
$= X(1 + \overline{Y})$	Theorem 7 (distributive law—factoring)
$= X(1)$	Theorem 1b (basic OR operation: $1 + \overline{Y} = 1$)
$= X$	Theorem 2b (basic AND operation: $X \cdot 1 = X$)

EXAMPLE 3.4 Simplify the logic expression

$$Z = (\overline{Y} + X) \cdot (\overline{X} + Y) + \overline{(\overline{W} + Y + \overline{X})}$$

Solution

$Z = (\overline{Y} + X) \cdot (\overline{X} + Y) + \overline{(\overline{W} + Y + \overline{X})}$	
$= \overline{Y}\overline{X} + \overline{Y}Y + X\overline{X} + XY + \overline{(\overline{W} + Y + \overline{X})}$	Theorem 7 (distributive law)
$= \overline{Y}\overline{X} + XY + \overline{(\overline{W} + Y + \overline{X})}$	Theorem 2c (basic AND operation: $\overline{Y} \cdot Y = 0$, $\overline{X} \cdot X = 0$)
$Z = \overline{Y}\overline{X} + XY + \overline{\overline{W}} \cdot \overline{Y} \cdot \overline{\overline{X}}$	Theorem 10a (DeMorgan's theorem: $\overline{(\overline{W} + Y + \overline{X})} = \overline{\overline{W}} \cdot \overline{Y} \cdot \overline{\overline{X}}$)

$= \overline{Y}\overline{X} + XY + (W \cdot \overline{Y} \cdot X)$ Theorem 3b (basic negation operation: $\overline{\overline{W}} = W, \overline{\overline{W}} = X$)

$= \overline{Y}\overline{X} + XY + X\overline{Y}W$ Theorem 5b (commutative law)

$= \overline{Y}\overline{X} + X(Y + \overline{Y}W)$ Theorem 7 (distributive law)

$= \overline{Y}\overline{X} + X(Y + W)$ Theorem 9a (Nashelsky's theorem: $Y + \overline{Y}W = Y + W$)

$= \overline{Y}\overline{X} + XY + XW$ Theorem 7 (distributive law)

EXAMPLE 3.5

Simplify the logic expression

$$\text{BELL} = WX + \overline{WX}Y + W$$

Solution

$\text{BELL} = WX + \overline{WX}Y + W$

$= WX + (\overline{W} + \overline{X})Y + W$ Theorem 6b (DeMorgan's theorem)

$= WX + \overline{W}Y + \overline{X}Y + W$ Theorem 7 (distributive law)

$= WX + W + \overline{W}Y + \overline{X}Y$ Theorem 5a (commutative law)

$= W(X + 1) + \overline{W}Y + \overline{X}Y$ Theorem 7 (distributive law—factoring)

$= W(1) + \overline{W}Y + \overline{X}Y$ Theorem 1b (basic OR operation: $X + 1 = 1$)

$= W + \overline{W}Y + \overline{X}Y$ Theorem 2b (basic AND operation: $W \cdot 1 = W$)

$= W + Y + \overline{X}Y$ Theorem 9a (Nashelsky's theorem: $W + \overline{W}Y = W + Y$)

$= W + Y(1 + \overline{X})$ Theorem 7 (distributive law—factoring)

$= W + Y(1)$ Theorem 1b (basic OR operation: $1 + \overline{X} = 1$)

$= W + Y$ Theorem 2b (basic AND operation: $Y \cdot 1 = Y$)

3.4 UNIVERSAL LOGIC GATES

The NAND and NOR gates are known as **universal logic gates** because they can be used to perform all the other logic functions. This feature of universality can be helpful to the technician because it may eliminate the need to carry many different types of ICs and give greater flexibility in utilizing existing devices for modification of a system. Figure 3.9 illustrates how a NAND or NOR gate can be used to implement the basic logic functions. Any circuit may be implemented using only NAND or NOR gates by replacing the logic function with its equivalent universal logic gate.

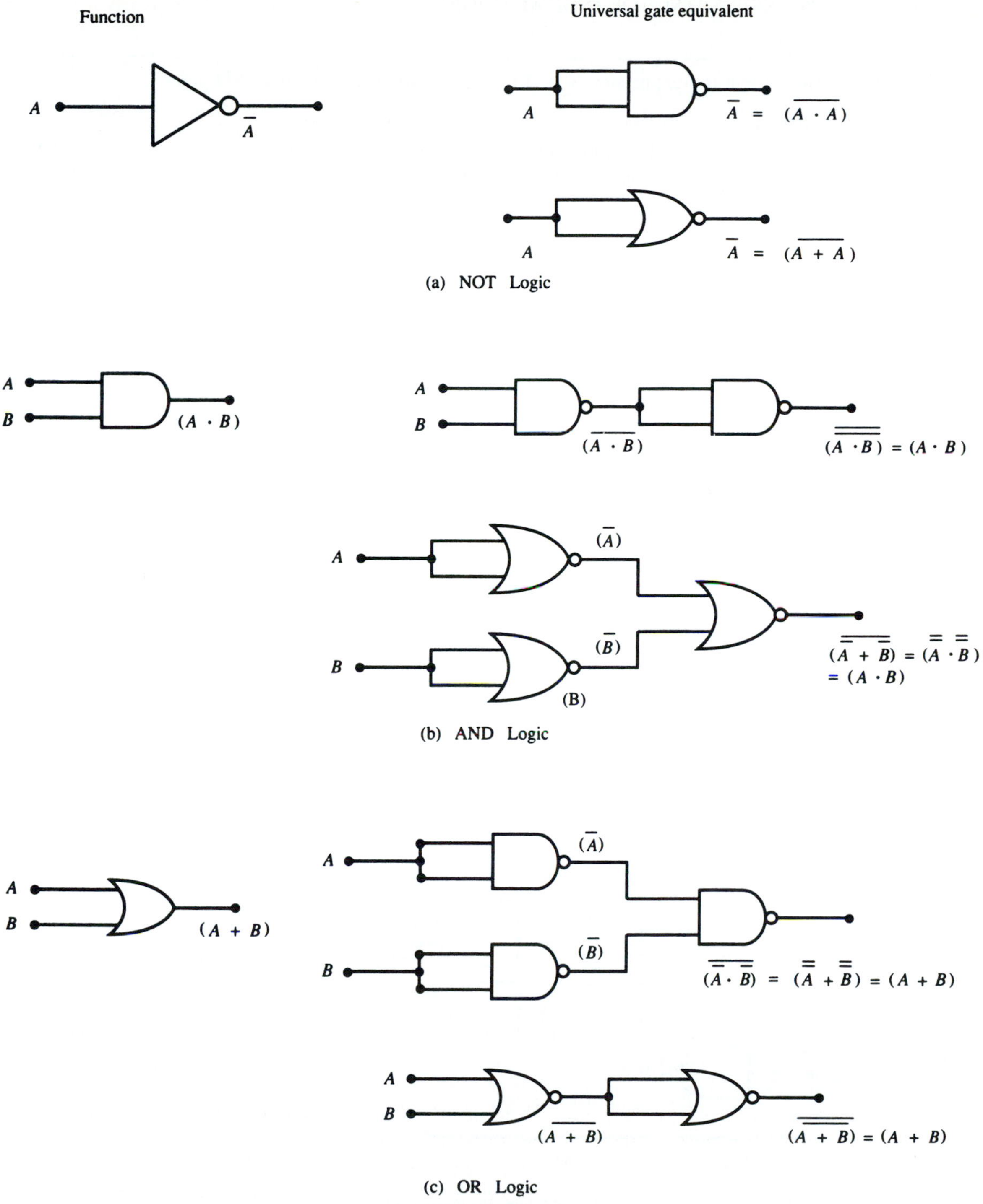

Function
Universal gate equivalent
A
Ā
A
Ā = (A · A)‾
A
Ā = (A + A)‾
(a) NOT Logic
A
B
(A · B)
A
B
(A · B)‾
(A · B)‾‾ = (A · B)
A
(Ā)
B
(B̄)
(B)
(Ā + B̄)‾ = (A̿ · B̿)
= (A · B)
(b) AND Logic
A
B
(A + B)
A
(Ā)
B
(B̄)
(Ā · B̄)‾ = (A̿ + B̿) = (A + B)
A
B
(A + B)‾
(A + B)‾‾ = (A + B)
(c) OR Logic

FIGURE 3.9
Universal gate equivalent

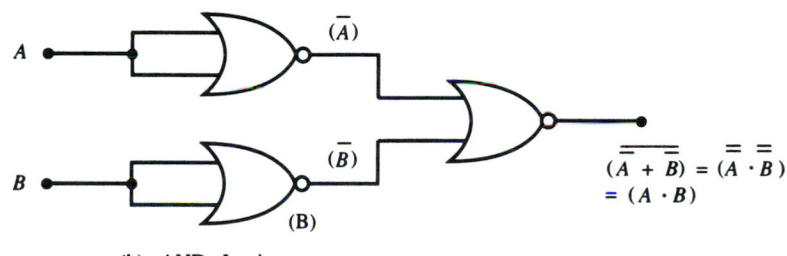

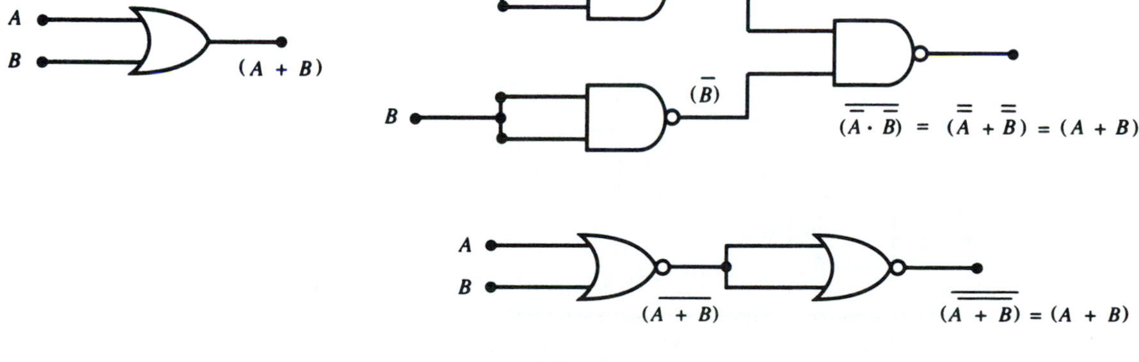

EXAMPLE 3.6

Implement the expression $Z = AB + AC + B\overline{C}$ using only NAND gates.

Solution

First, implement the expression using AND, OR, and inverter gates (see Figure 3.10a). Next, replace each logic function with its equivalent universal gate (see Figure 3.10b). Finally, realizing that two inversions result in the original function (Theorem 3b, the basic NOT operation), reduce the circuit as shown in Figure 3.10(c) (p. 77).

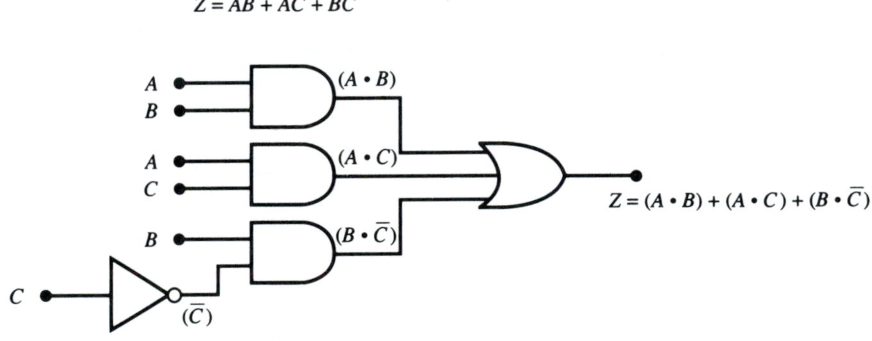

(a) Logic circuit using AND, OR, NOT gates

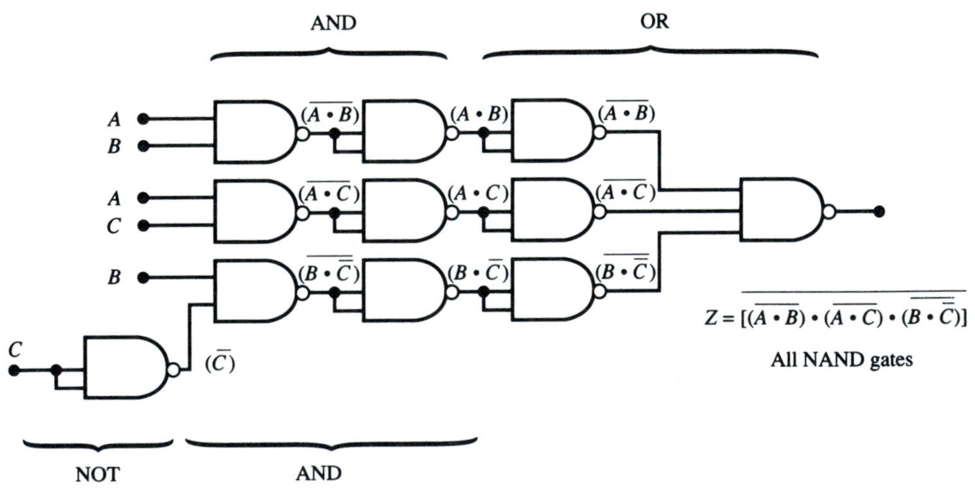

(b) Logic circuit using only NAND gates

FIGURE 3.10

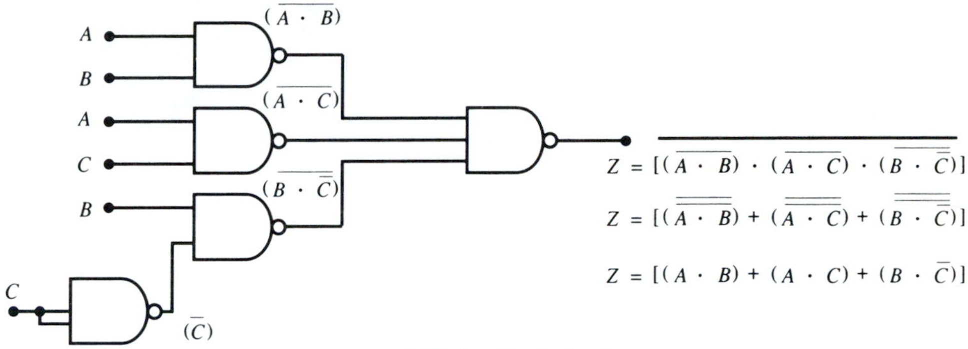

$$Z = \overline{[(\overline{A \cdot B}) \cdot (\overline{A \cdot C}) \cdot (\overline{B \cdot \overline{C}})]}$$

$$Z = [(\overline{\overline{A \cdot B}}) + (\overline{\overline{A \cdot C}}) + (\overline{\overline{B \cdot \overline{C}}})]$$

$$Z = [(A \cdot B) + (A \cdot C) + (B \cdot \overline{C})]$$

(c) Reduced logic circuit

FIGURE 3.10 (continued)

Example 3.6 could have been performed mathematically. Recalling from Theorem 3b (the basic NOT operation) that a double negation does not change the expression, we can conclude

$$Z = (AB) + (AC) + (B\overline{C})$$
$$= \overline{\overline{[(AB) + (AC) + (B\overline{C})]}}$$

Applying DeMorgan's theorem, we break the lower bar and change the sign of the logic operation as follows:

$$Z = \overline{[(\overline{AB}) \cdot (\overline{AC}) \cdot (\overline{B\overline{C}})]}$$

Realizing that each term in parentheses is a Boolean expression for a NAND gate, we can construct the logic circuit of Figure 3.10(c). Note that the original expression required *three different types* of ICs to implement and therefore would require a minimum of three ICs. The NAND-only implementation is not really any simpler. It can, however, be implemented using two ICs. This results in a savings due to reduced cost and PC board space (real estate). It also results in higher reliability because fewer components are used.

3.5 ACTIVE-STATE LOGIC

There is a second set of logic symbols that is used to indicate logic functions, called **active-state logic** symbols. Active-state logic attempts to describe the logic state of a gate (0 or 1) that is performing a task. Figure 3.11 illustrates the logic symbols that are used in the active-state logic representation of a circuit.

Active-state logic can be derived using DeMorgan's theorem. Active-state logic tells a technician to expect that a logic level 1 (active high) or a logic level 0 (active low) is going to cause something to happen. For example, suppose that the output of an AND

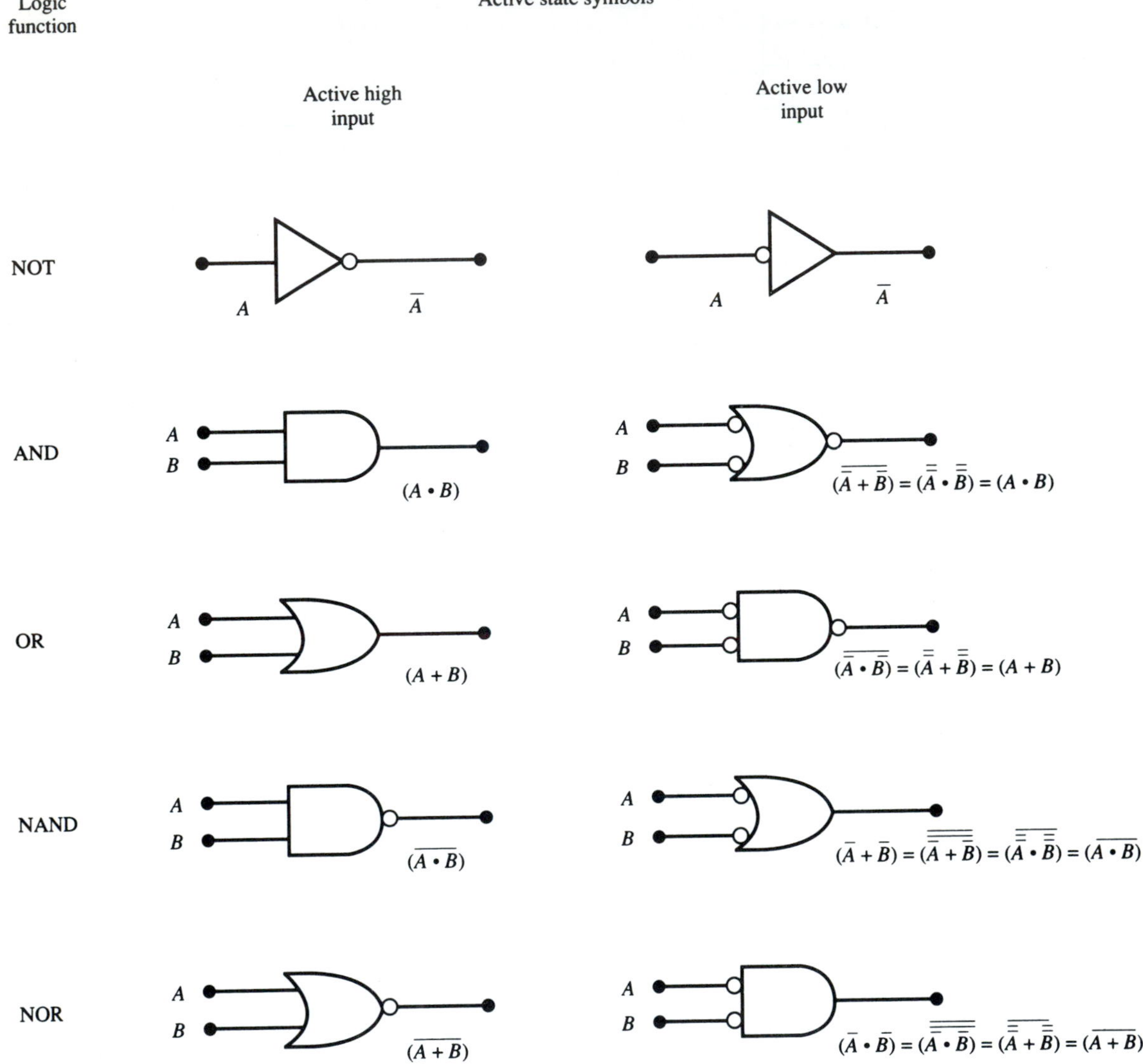

FIGURE 3.11
Active-state logic symbols

gate is going to be used to light a light. In standard notation we cannot predict whether a logic level 1 or a logic level 0 will cause the light to light. Using active-state symbols, we can represent the circuit in either of the two ways shown in Figure 3.12.

Figure 3.12(a) says that the lamp will light whenever 1 AND 1 appears on the input to the AND gate, causing the output to be active high. Figure 3.12(b) says that the lamp will light whenever 0 OR 0 appears on the input to the AND gate, causing the output of the AND gate to be active low. Note that both symbols in Figure 3.12 indicate an AND gate. Note also that in active-state logic, the true function of the gate's use is described as well as the active-logic state.

Referring to Figure 3.11, we see that a standard symbol can be converted to an active-state symbol by applying DeMorgan's theorem. The rules for implementing active-state symbols are as follows:

1. Invert all inputs and outputs as indicated by the addition or deletion of bubbles.
2. Change AND symbols to OR symbols and OR symbols to AND symbols.
3. Leave inverter symbols unchanged but use bubbles to indicate the active state.

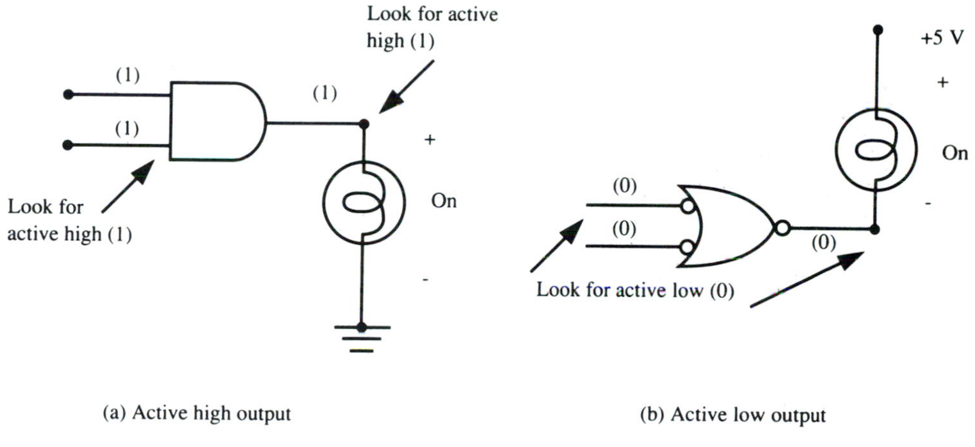

FIGURE 3.12
Active-state AND gate

EXAMPLE 3.7 Redraw the circuit of Figure 3.13 using active-state symbols to indicate that an active high output will light a light.

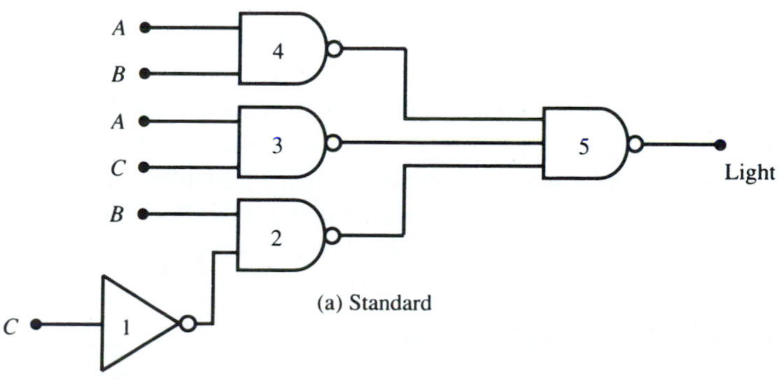

FIGURE 3.13

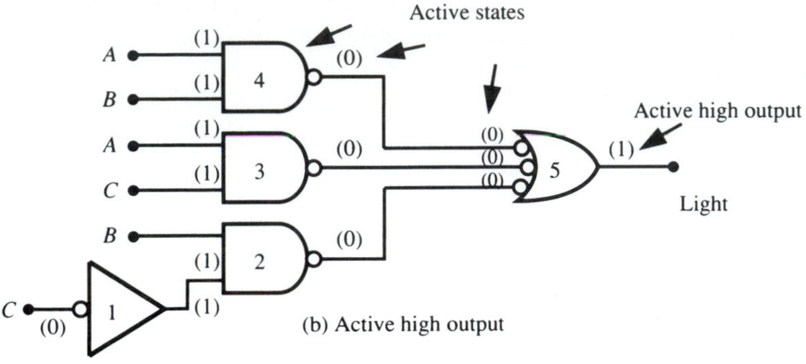

FIGURE 3.13 (continued)

Solution

Beginning at the output, we see that NAND gate 5 has to be redrawn. Thus we change the AND symbol to an OR symbol and invert all the inputs and outputs. Analyzing the output of NAND gate 4, we see that it already yields the correct active state 0 (required for the input to NAND gate 5) and does not have to be redrawn. NAND gate 3 and NAND gate 2 are also correct. Analyzing the output of inverter 1, we see that it does not indicate the correct active-state output (a 1, which is the required input to NAND gate 2). Therefore, it is redrawn to indicate the correct active state.

3.6 KARNAUGH MAPS

A **Karnaugh map** is a systematic graphical method for simplifying Boolean expressions. The Boolean algebra method for simplifying expressions can be cumbersome and difficult to use. Furthermore, it is difficult to determine how much a logic expression can be simplified. Karnaugh maps provide us with a technique that overcomes this limitation of Boolean algebra. They tell whether a Boolean expression *can be simplified at all* and provide a simplified method for *minimizing* the expression.

A Karnaugh map is a graphical representation of the truth table. It uses rows and columns to form boxes that represent each term or line on a truth table. Therefore, there are 2^n boxes in a Karnaugh map, where n represents the number of input variables in the logic expression. Figure 3.14 shows an example of 2-, 3-, and 4-variable Karnaugh maps. Although logic expressions involving more than 4 variables can be solved using Karnaugh maps, this method becomes impractical at that point. Other methods, such as the *Quine–McCluskey method,* can be used to solve logic expressions with any number of variables.

Mapping Boolean Expressions

To map a Boolean expression, it must be in the **sum-of-the-products (SOP)** form. This means that each term in the expression must be an AND function, and all the terms are ORed together. The following are some examples of Boolean expressions in SOP form:

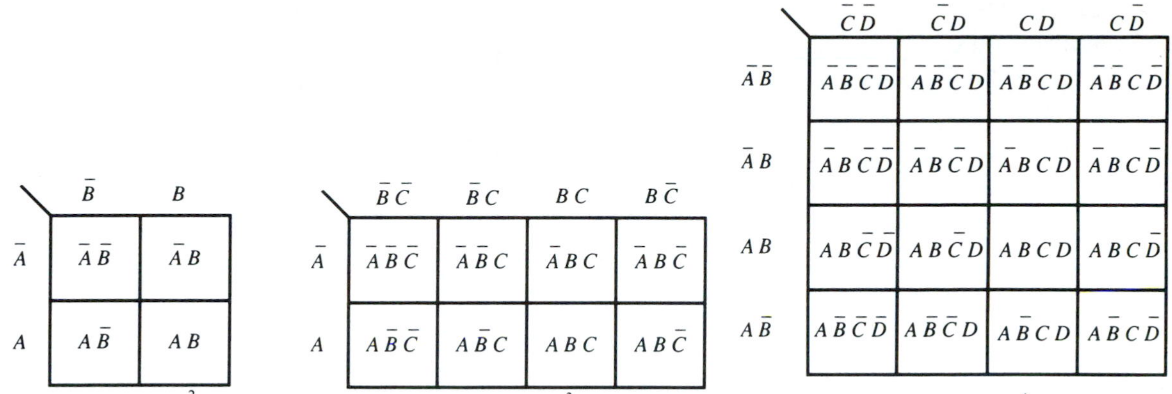

	$\overline{B}$	B
$\overline{A}$	$\overline{A}\,\overline{B}$	$\overline{A}B$
A	$A\overline{B}$	AB

(a) 2-Variable map (2^2 = 4 boxes)

	$\overline{B}\,\overline{C}$	$\overline{B}C$	BC	$B\overline{C}$
$\overline{A}$	$\overline{A}\,\overline{B}\,\overline{C}$	$\overline{A}\,\overline{B}C$	$\overline{A}BC$	$\overline{A}B\overline{C}$
A	$A\overline{B}\,\overline{C}$	$A\overline{B}C$	ABC	$AB\overline{C}$

(b) 3-Variable map (2^3 = 8 boxes)

	$\overline{C}\,\overline{D}$	$\overline{C}D$	CD	$C\overline{D}$
$\overline{A}\,\overline{B}$	$\overline{A}\,\overline{B}\,\overline{C}\,\overline{D}$	$\overline{A}\,\overline{B}\,\overline{C}D$	$\overline{A}\,\overline{B}CD$	$\overline{A}\,\overline{B}C\overline{D}$
$\overline{A}B$	$\overline{A}B\overline{C}\,\overline{D}$	$\overline{A}B\overline{C}D$	$\overline{A}BCD$	$\overline{A}BC\overline{D}$
AB	$AB\overline{C}\,\overline{D}$	$AB\overline{C}D$	$ABCD$	$ABC\overline{D}$
$A\overline{B}$	$A\overline{B}\,\overline{C}\,\overline{D}$	$A\overline{B}\,\overline{C}D$	$A\overline{B}CD$	$A\overline{B}C\overline{D}$

(c) 4-Variable map (2^4 = 16 boxes)

FIGURE 3.14
Karnaugh maps

$$Z = AB + \overline{A}\,\overline{B}$$
$$W = A\overline{B}C + ABC + \overline{A}BC + \overline{A}\,\overline{B}C$$
$$R = WXYZ + \overline{W}\,\overline{X}\,\overline{Y}\,\overline{Z} + \overline{W}X\overline{Y}\,\overline{Z} + \overline{W}X\overline{Y}Z + \overline{W}XYZ$$

The mapping procedure is as follows:

1. Write the expression in SOP form.
2. Determine the number of input or independent variables, and select the appropriate Karnaugh map. For example, a 2-variable expression would use the map shown in Figure 3.14(a).
3. Plot each term of the expression in the Karnaugh map by placing a 1 in the box corresponding to the term.

Figure 3.15 illustrates how to plot the three preceding SOP-form expressions on the appropriate Karnaugh map.

Grouping for Simplification

The first step in the simplification process is to group adjacent boxes that have been filled in with 1s. The *larger* the grouping, the *simpler* the expression will be. Adjacent boxes can be *vertical, horizontal,* or *looped around the ends.* Identify each grouping according to the following rules by drawing a loop around the grouping:

1. The adjacent grouping must be combined in groups of 1, 2, 4, 8, 16, and so forth.
2. Each grouping should be as large as possible.
3. Every 1 on the map must be included in at least one group.
4. The same 1 may be used in two or more overlapping groupings.
5. The map can be considered closed, so that the end boxes are considered adjacent (top and bottom or left and right).

Figure 3.16 (p. 83) illustrates a number of examples of groupings.

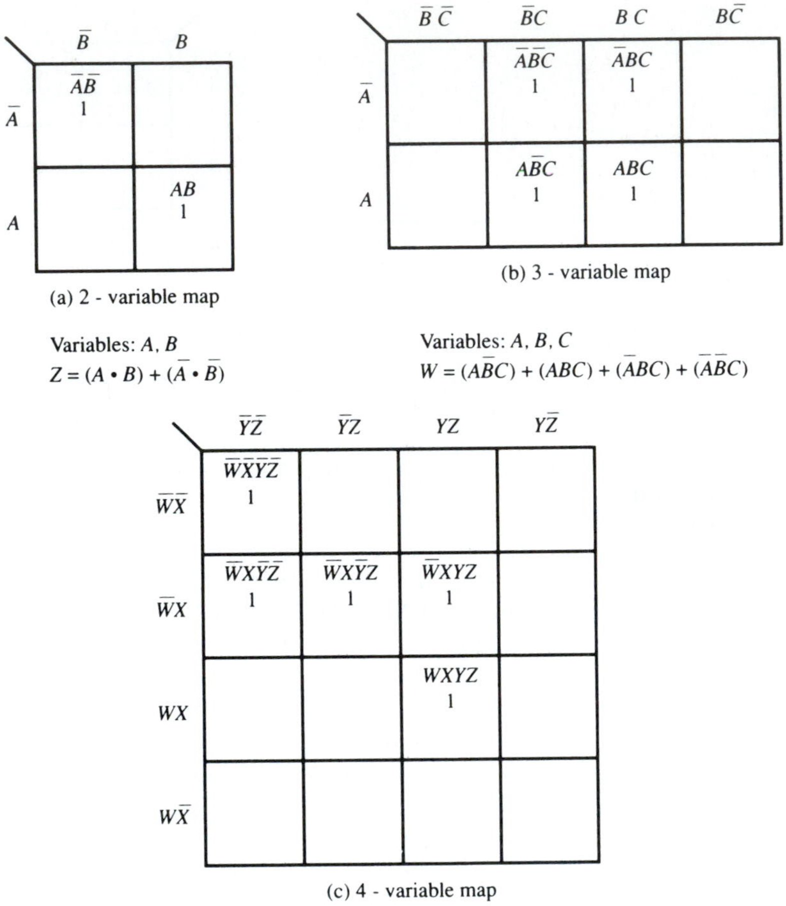

	$\bar{B}$	B
$\bar{A}$	$\bar{A}\bar{B}$ 1	
A		AB 1

(a) 2 - variable map

Variables: A, B
$Z = (A \bullet B) + (\bar{A} \bullet \bar{B})$

	$\bar{B}\,\bar{C}$	$\bar{B}C$	$B\,C$	$B\bar{C}$
$\bar{A}$		$\bar{A}\bar{B}C$ 1	$\bar{A}BC$ 1	
A		$A\bar{B}C$ 1	ABC 1	

(b) 3 - variable map

Variables: A, B, C
$W = (A\bar{B}C) + (ABC) + (\bar{A}BC) + (\bar{A}\bar{B}C)$

	$\bar{Y}\bar{Z}$	$\bar{Y}Z$	YZ	$Y\bar{Z}$
$\bar{W}\bar{X}$	$\bar{W}\bar{X}\bar{Y}\bar{Z}$ 1			
$\bar{W}X$	$\bar{W}X\bar{Y}\bar{Z}$ 1	$\bar{W}X\bar{Y}Z$ 1	$\bar{W}XYZ$ 1	
WX			$WXYZ$ 1	
$W\bar{X}$				

(c) 4 - variable map

$R = (WXYZ) + (\bar{W}\bar{X}\bar{Y}\bar{Z}) + (\bar{W}X\bar{Y}\bar{X}) + (\bar{W}X\bar{Y}Z) + (\bar{W}XYZ)$

FIGURE 3.15
Karnaugh map plotting

Simplifying the Expression

Each grouping will result in a term of the final simplified Boolean expression according to the following rules:

1. The grouping will reduce to an AND term made up of only the common variables.
2. The final simplified Boolean expression is made up of each grouping's terms ORed together.

More specifically, the rules for *two-variable maps* are as follows:

1. Any single grouping results in a 2-variable term made from the row and the column.

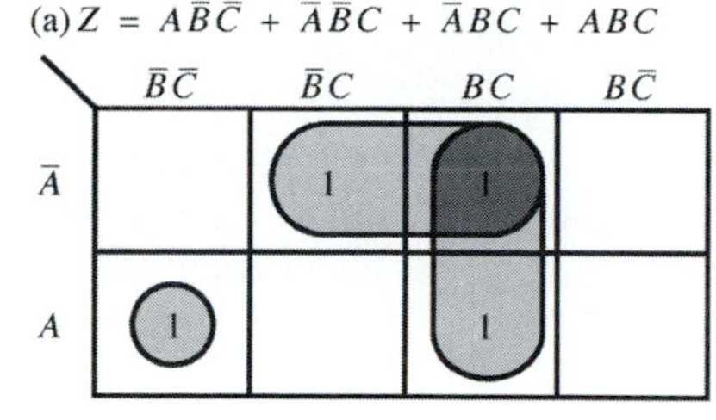

(b) $W = \overline{A}\overline{B}\overline{C}\overline{D} + A\overline{B}\overline{C}\overline{D} + \overline{A}B\overline{C}D + \overline{A}\overline{B}CD$
$+ \overline{A}\overline{B}C\overline{D} + ABCD + A\overline{B}CD + A\overline{B}C\overline{D}$

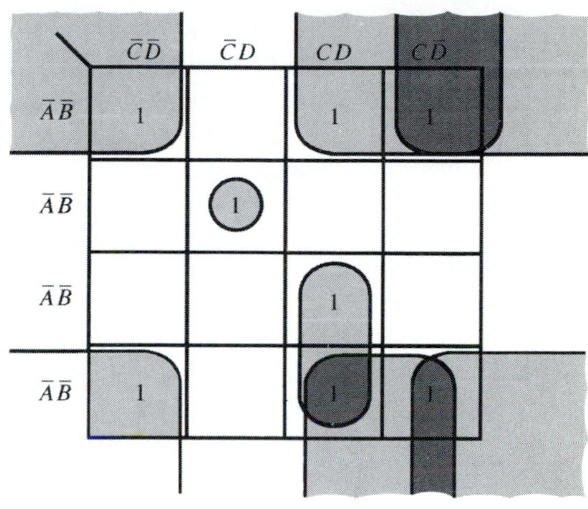

FIGURE 3.16
Groupings of terms

2. Any two adjacent 1s can be combined to form a single-variable term, which is the variable that is common to both.
3. The final simplified Boolean expression is made up of each grouping's terms ORed together.

EXAMPLE 3.8

Simplify the Boolean equation

$$Z = \overline{A}\overline{B} + \overline{A}B + AB$$

Solution

See Figure 3.17.

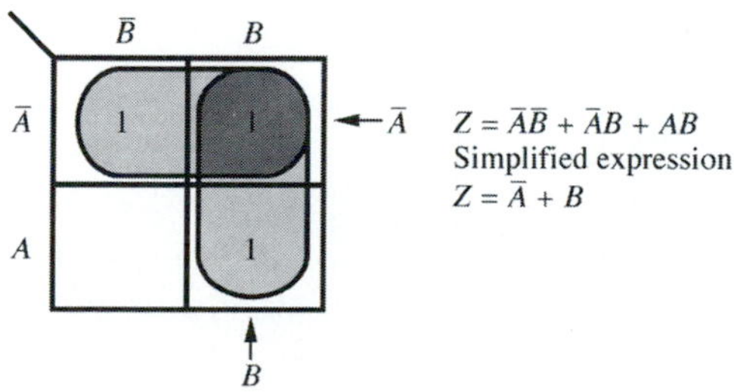

FIGURE 3.17

Specific rules for *3-variable maps* are as follows:

1. Any single grouping results in a 3-variable term.
2. Any two adjacent 1s can be combined to form a 2-variable term.
3. Any four adjacent 1s can be combined to form a single-variable term.
4. The final simplified Boolean expression is made up of each grouping's terms ORed together.

EXAMPLE 3.9 Simplify the Boolean equation

$$W = A\overline{B}C + ABC + \overline{A}BC + \overline{A}\overline{B}C$$

Solution
See Figure 3.18.

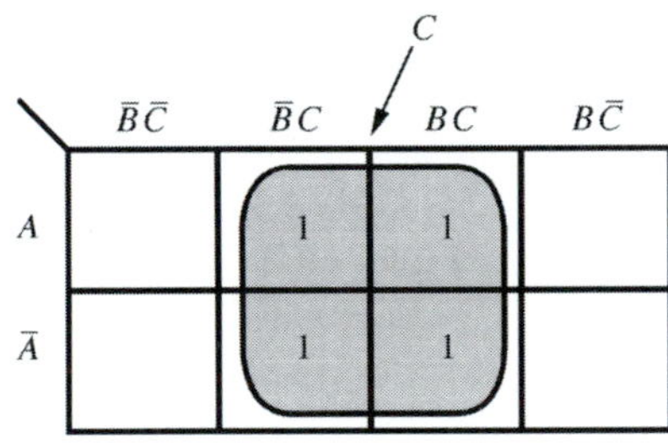

FIGURE 3.18

EXAMPLE 3.10 Simplify the Boolean equation

$$Z = A\overline{B}\overline{C} + \overline{A}\overline{B}C + \overline{A}BC + ABC$$

Solution
See Figure 3.19.

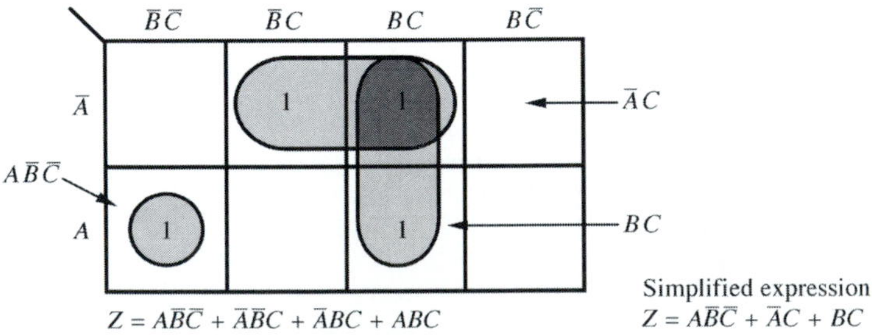

FIGURE 3.19

Specific rules for *4-variable maps* are as follows:

1. Any single grouping results in a 4-variable term.
2. Any two adjacent 1s can be combined to form a 3-variable term.
3. Any four adjacent 1s can be combined to form a 2-variable term.
4. Any eight adjacent 1s can be combined to form a single-variable term.
5. The final simplified Boolean expression is made up of each grouping's terms ORed together.

EXAMPLE 3.11

Simplify the Boolean expression

$$X = ABCD + \overline{A}\,\overline{B}\,\overline{C}D + \overline{A}B\overline{C}\,\overline{D} + \overline{A}B\overline{C}D + \overline{A}BCD$$

Solution

See Figure 3.20.

Note: Joining the rightmost grouping (BCD) to the other two would result in the addition of another term. This process should be avoided.

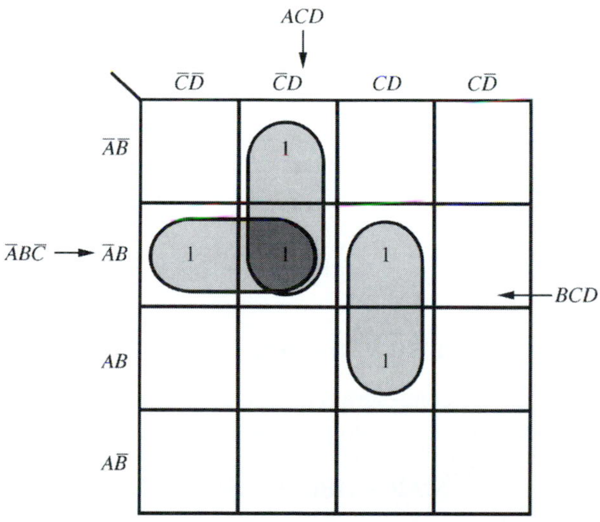

$X = ABCD + \overline{A}\,\overline{B}\,\overline{C}D + \overline{A}B\overline{C}\,\overline{D} + \overline{A}B\overline{C}D + \overline{A}BCD$

Simplified expression

$X = \overline{A}B\overline{C} + \overline{A}\,\overline{C}D + BCD$

FIGURE 3.20

EXAMPLE 3.12 Simplify the Boolean expression

$$Y = \overline{A}\,\overline{B}\,\overline{C}\,\overline{D} + \overline{A}\,\overline{B}\,\overline{C}D + \overline{A}\,\overline{B}C\overline{D} + \overline{A}B\overline{C}\,\overline{D} + \overline{A}BC\overline{D} + AB\overline{C}\,\overline{D} + ABC\overline{D} +$$
$$ABC\overline{D} + A\overline{B}\,\overline{C}\,\overline{D} + A\overline{B}\,\overline{C}D + A\overline{B}C\overline{D}$$

Solution

See Figure 3.21.

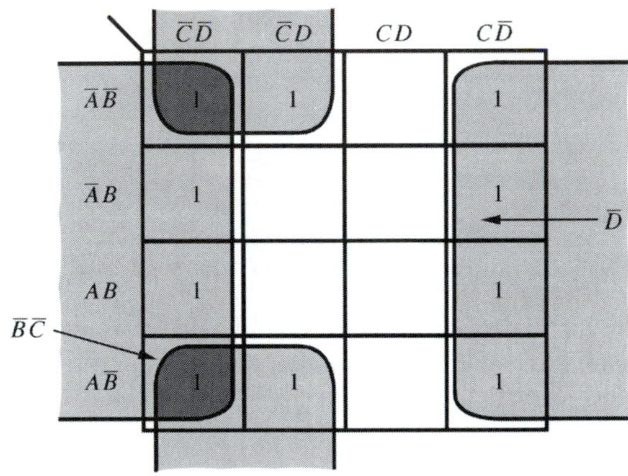

$$Y = \overline{A}\,\overline{B}\,\overline{C}\,\overline{D} + \overline{A}\,\overline{B}\,\overline{C}D + \overline{A}\,\overline{B}C\overline{D} + \overline{A}B\overline{C}\,\overline{D} + \overline{A}BC\overline{D} + AB\overline{C}\,\overline{D}$$
$$+ ABC\overline{D} + A\overline{B}\,\overline{C}\,\overline{D} + A\overline{B}\,\overline{C}D + A\overline{B}C\overline{D}$$

Simplified expression

$$Y = \overline{D} + \overline{B}\,\overline{C}$$

FIGURE 3.21

Mapping Terms with Missing Variables

In some cases you may come across a logic expression in which each term does not contain all the input variables. For example, consider the following logic expression:

$$\text{MAP} = ABC + \overline{A}\,\overline{B}\,\overline{C} + \overline{A}C$$

Notice that the last term, $\overline{A}C$, does not contain the input variable B. This signifies that $\overline{A}C$ must be true (1) regardless of the logic state of B. Therefore, the term $\overline{A}C$ can be expressed as two terms encompassing all the combinations of B as follows:

$$\overline{A}C = \overline{A}CB + \overline{A}C\overline{B}$$

Figure 3.22 shows a mathematical and graphical interpretation of this technique.

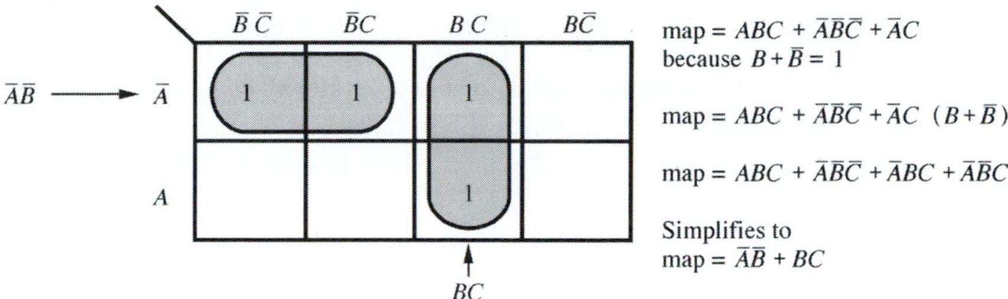

FIGURE 3.22

EXAMPLE 3.13 Simplify the following Boolean expression using a Karnaugh map:

$$Z = AB\overline{C}\,\overline{D} + ABD + A\overline{B} + \overline{A}C + ABC\overline{D}$$

Solution

See Figure 3.23.

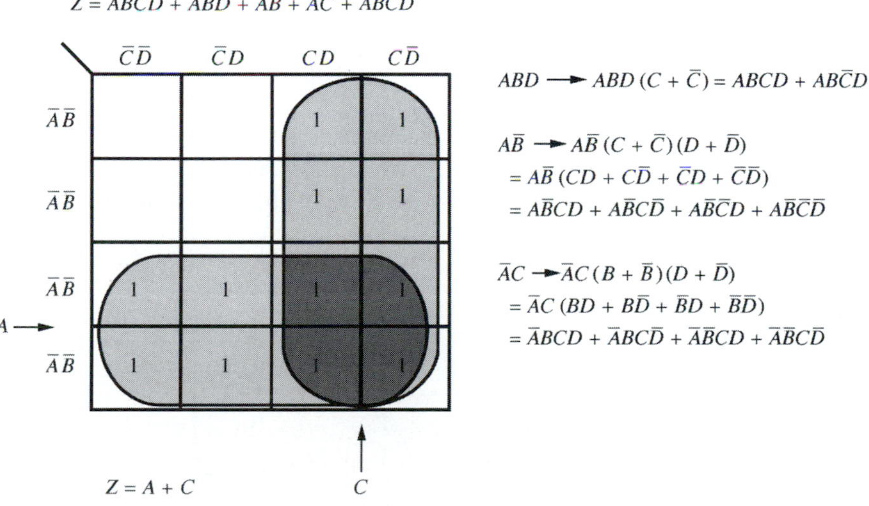

FIGURE 3.23

3.7 DESIGN BASICS

The design of digital circuits is best accomplished through the use of systematic methods. A systematic method avoids the pitfalls of trial-and-error, or *seat-of-the-pants,* techniques. The classical systematic method for digital design involves these steps:

1. Stating the problem.
2. Developing a truth table.
3. Obtaining a Boolean expression.
4. Simplifying a Boolean expression.
5. Implementing the expression.

To demonstrate the use of the systematic method, let's consider the following.

EXAMPLE 3.14

Mr. Dunnigan wishes to design an alarm circuit to protect his expensive car. Two switches are to be used to monitor the opening and closing of the left and right doors. A third switch is to be used to turn the alarm on and off.

Solution

1. *State the problem.*

   Given: Let A = left-door switch
   B = right-door switch
   S = control switch
   ALARM = Output signal to sound the siren

   Assume: Logic 1 = door open, control switch on, siren on
   Logic 0 = door closed, control switch off, siren off

   Constraints: The alarm is to sound whenever the control switch is *on* AND either OR *both* doors are open.

2. *Develop a truth table.* There are three input variables, which we have defined as *A, B,* and *S.* Therefore, there are 2^3, or 8, possible combinations of input conditions that can exist. The output variable is called ALARM. The truth table then is

A	*B*	*S*	ALARM
0	0	0	
0	0	1	
0	1	0	
0	1	1	
1	0	0	
1	0	1	
1	1	0	
1	1	1	

Recalling from the statement of the problem that the alarm is to sound (1) whenever $S = 1$ AND either A OR B OR both equal 1, the truth table can be completed as follows:

A	B	S	ALARM	
0	0	0	0	
0	0	1	0	
0	1	0	0	
0	1	1	1	$\overline{A} \cdot B \cdot S$
1	0	0	0	
1	0	1	1	$A \cdot \overline{B} \cdot S$
1	1	0	0	
1	1	1	1	$A \cdot B \cdot S$

3. *Obtain a Boolean expression.* Referring to the truth table, we see that there are three considerations for which the alarm will sound.
   a. $A = 0$ AND $B = 1$ AND $S = 1$
      (This results in the Boolean term $\overline{A} \cdot B \cdot S$.)
   b. $A = 1$ AND $B = 0$ AND $S = 1$, giving $A \cdot \overline{B} \cdot S$
   c. $A = 1$ AND $B = 1$ AND $S = 1$, giving $A \cdot B \cdot S$

Realizing that the alarm is to sound whenever condition (a) OR condition (b) OR condition (c) results, we obtain the Boolean expression

$$\text{ALARM} = \overline{A}BS + A\overline{B}S + ABS$$

Recall that this procedure is known as the SOP method.

4. *Simplify the Boolean expression.* To simplify the Boolean expression, we utilize the Karnaugh map method (see Figure 3.24).

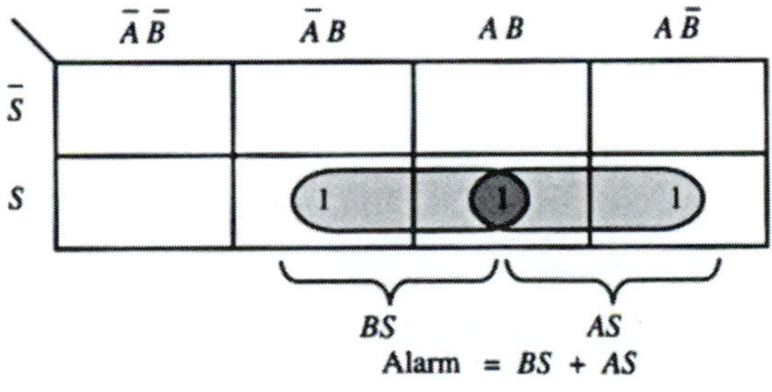

FIGURE 3.24

5. *Implement the expression.* The logic diagram for the simplified Boolean expression results in two AND gates and an OR gate, as shown in Figure 3.25.

B
S
A
S
Alarm

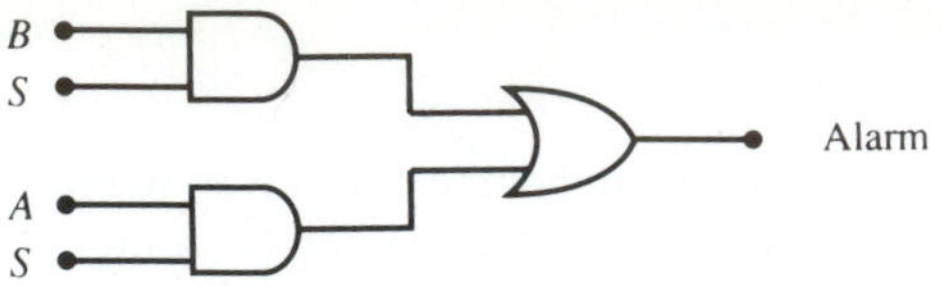

FIGURE 3.25

EXAMPLE 3.15 Design a logic circuit to add two binary digits (A, B). This is sometimes referred to as a **half-adder circuit.**

Solution

When two binary digits are added, two output conditions exist, the *sum* output and the *carry* output, as shown in Figure 3.26. The truth table of Figure 3.26(b) illustrates the

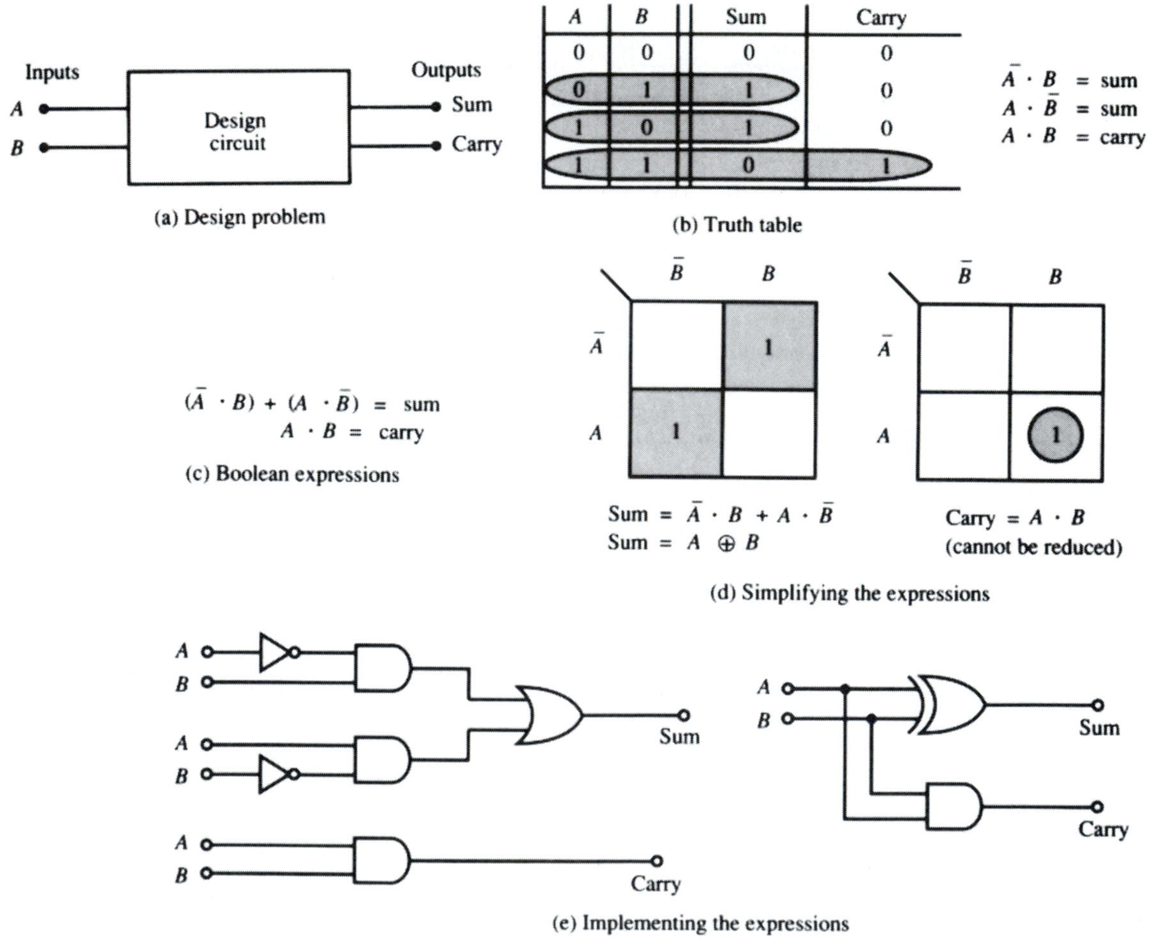

Inputs
A
B
Design circuit
Outputs
Sum
Carry
(a) Design problem
A B Sum Carry
0 0 0 0
0 1 1 0
1 0 1 0
1 1 0 1
$\bar{A} \cdot B$ = sum
$A \cdot \bar{B}$ = sum
$A \cdot B$ = carry
(b) Truth table
$(\bar{A} \cdot B) + (A \cdot \bar{B})$ = sum
$A \cdot B$ = carry
(c) Boolean expressions
$\bar{B}$ B
$\bar{A}$ 1
A 1
Sum = $\bar{A} \cdot B + A \cdot \bar{B}$
Sum = $A \oplus B$
$\bar{B}$ B
$\bar{A}$
A 1
Carry = $A \cdot B$
(cannot be reduced)
(d) Simplifying the expressions
A
B
A
B
Sum
A
B
Carry
A
B
Sum
Carry
(e) Implementing the expressions

FIGURE 3.26
Design example

adding process. Using the SOP method, we obtain Boolean expressions for the sum and carry outputs. The Karnaugh map for each expression is shown in Figure 3.26(d). Note that since there are no adjacent 1s, the expression appears to be irreducible. However, the checkerboard pattern indicates the possibility of an EXCLUSIVE-OR reduction ($A \oplus B = \bar{A}B + A\bar{B}$). Since this is not always obvious to all designers, two implementations are presented in Figure 3.26(e).

EXAMPLE 3.16

Design a **full-adder** logic circuit that consists of two binary digits and a carry-in bit from a previous adder circuit.

Solution

See Figure 3.27.

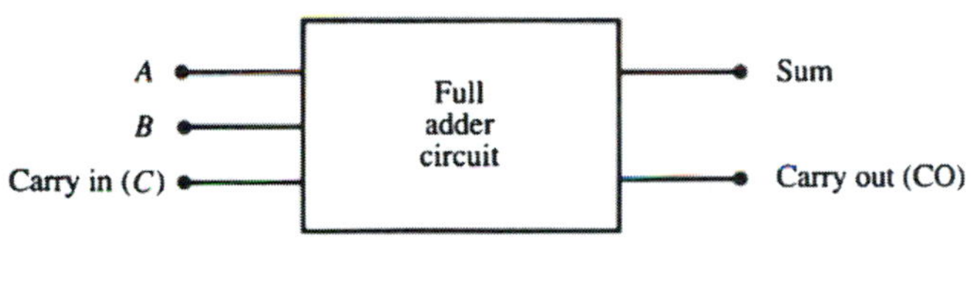

(a) Design problem

A	B	C	Sum	CO
0	0	0	0	0
0	0	1	1	0
0	1	0	1	0
0	1	1	0	1
1	0	0	1	0
1	0	1	0	1
1	1	0	0	1
1	1	1	1	1

(b) Truth table

$$\text{Sum} = \bar{A}\bar{B}C + \bar{A}B\bar{C} + A\bar{B}\bar{C} + ABC$$

$$\text{CO} = \bar{A}BC + A\bar{B}C + AB\bar{C} + ABC$$

(c) Boolean expressions

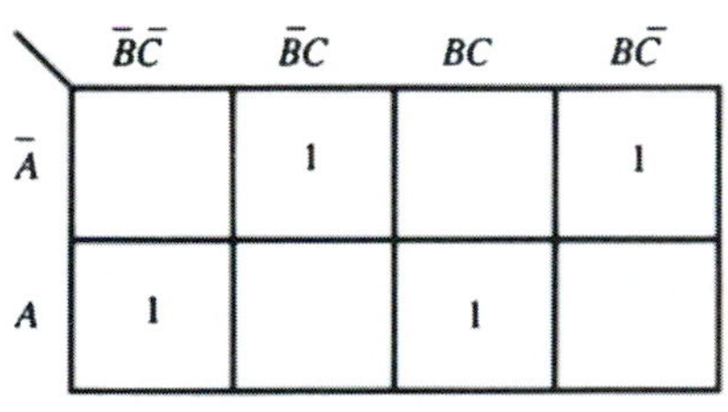

Sum has no adjacent pairs

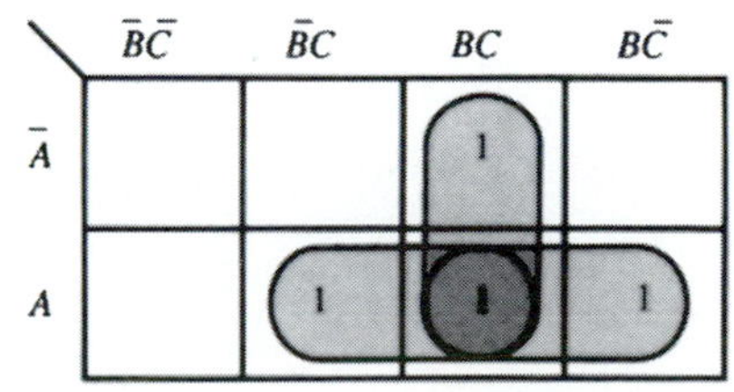

CO = $AC + BC + AB$

(d) Simplifying the expression

FIGURE 3.27
Design example

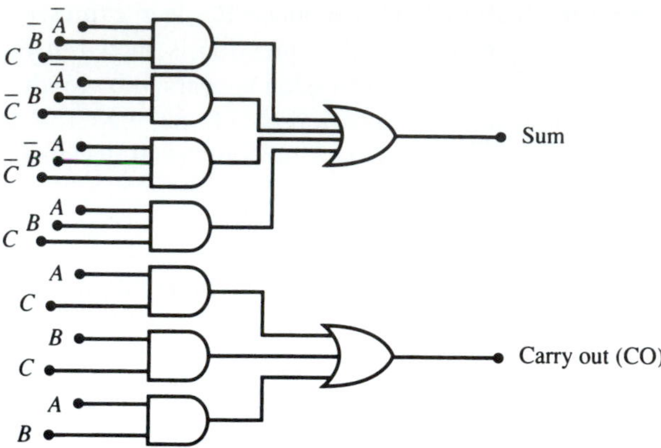

(e) Implementing the expression

FIGURE 3.27 (continued)
Design example

Note that the Karnaugh map for the SUM expression has no adjacent 1s; therefore, it cannot be simplified by classical Karnaugh map techniques. Normally when there are no adjacent 1s in the Karnaugh map, the Boolean expression is not reducible. However, the one exception is the *checkerboard pattern* obtained for the SUM expression Karnaugh map. Whenever you see this pattern, it indicates that an EXCLUSIVE-OR or EXCLUSIVE-NOR reduction is possible using Boolean algebra techniques. In this case,

$$\text{SUM} = \overline{A}\,\overline{B}C + \overline{A}B\overline{C} + A\overline{B}\,\overline{C} + ABC$$

Recalling the XOR relationship $B \oplus C = \overline{B}C + B\overline{C}$ and the XNOR relationship $\overline{B \oplus C} = \overline{B}\,\overline{C} + BC$, rearranging, and factoring gives

$$\text{SUM} = \overline{A}\,(\overline{B}C + B\overline{C}) + A(\overline{B}\,\overline{C} + BC)$$

Therefore,

$$\text{SUM} = \overline{A}(B \oplus C) + A(\overline{B \oplus C})$$

Now if we let $X = B \oplus C$, then by substitution

$$\text{SUM} = \overline{A}(X) + A(\overline{X})$$

This is an EXCLUSIVE-OR of A and X; therefore,

$$\text{SUM} = A \oplus X$$

or

$$\text{SUM} = A \oplus (B \oplus C)$$

Another circuit for implementing the full adder is then as shown in Figure 3.28.

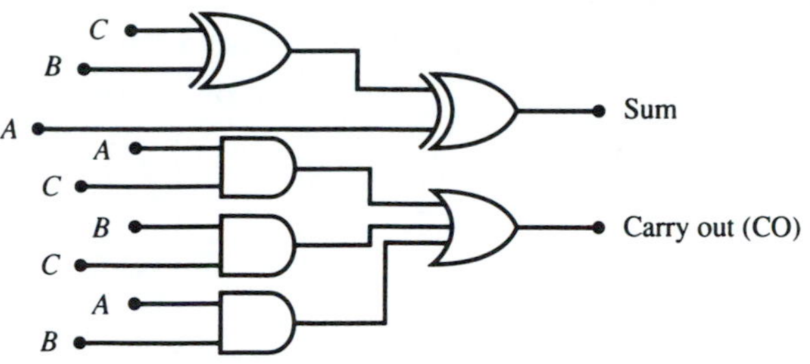

FIGURE 3.28
Full adder using EXCLUSIVE-OR circuits

Don't-Care Terms

In some design problems we may find combinations of input variables that cannot occur. These combinations are referred to as **don't-care terms.** Don't-care terms can be used to simplify further the solutions of a design problem. For example, suppose we wish to design a logic circuit that is to control a 7-segment decimal display, as shown in Figure 3.29.

Since this is to be a decimal display, we must design a circuit that will display only the digits 0 through 9. Therefore, all other input variable combinations can be considered to be don't-care terms, as shown in the truth table of Figure 3.30(a). The solution to this problem requires seven Karnaugh maps, one map for each of the seven output segments *a* through *g*. For demonstration purposes, we design the logic circuit for the *a* segment only. You will have an opportunity to design the circuits for the remaining outputs in the exercises at the end of the chapter. The *a*-segment logic expression and Karnaugh map reduction are shown in Figure 3.30(b). The implemented circuit is shown in Figure 3.30(c).

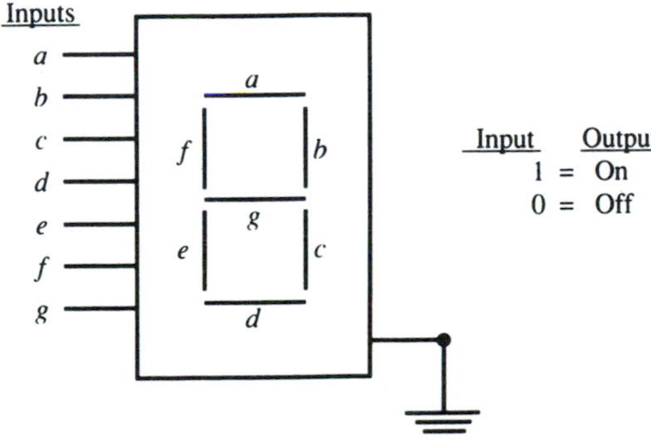

FIGURE 3.29
Seven-segment LED display

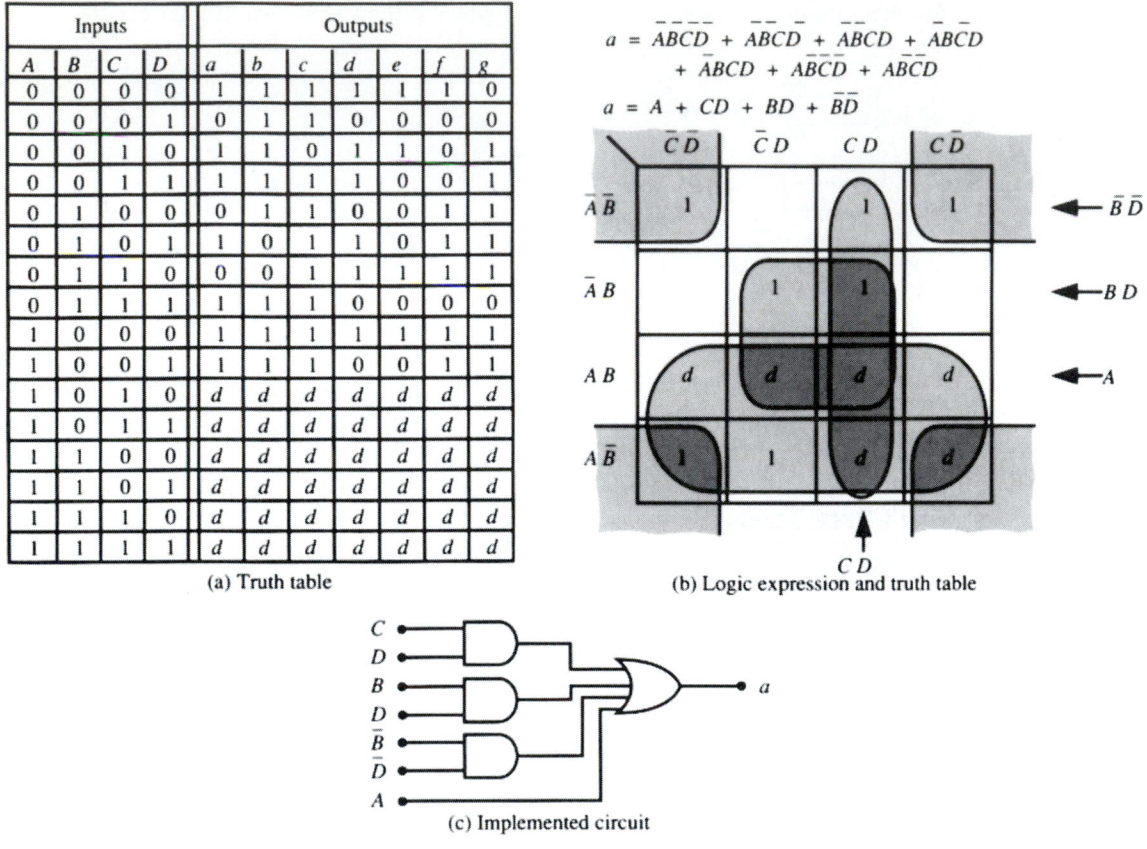

Inputs				Outputs						
A	B	C	D	a	b	c	d	e	f	g
0	0	0	0	1	1	1	1	1	1	0
0	0	0	1	0	1	1	0	0	0	0
0	0	1	0	1	1	0	1	1	0	1
0	0	1	1	1	1	1	1	0	0	1
0	1	0	0	0	1	1	0	0	1	1
0	1	0	1	1	0	1	1	0	1	1
0	1	1	0	0	0	1	1	1	1	1
0	1	1	1	1	1	1	0	0	0	0
1	0	0	0	1	1	1	1	1	1	1
1	0	0	1	1	1	1	0	0	1	1
1	0	1	0	d	d	d	d	d	d	d
1	0	1	1	d	d	d	d	d	d	d
1	1	0	0	d	d	d	d	d	d	d
1	1	0	1	d	d	d	d	d	d	d
1	1	1	0	d	d	d	d	d	d	d
1	1	1	1	d	d	d	d	d	d	d

(a) Truth table

$$a = \overline{A}\,\overline{B}\,\overline{C}\,\overline{D} + \overline{A}\,\overline{B}C\overline{D} + \overline{A}\,\overline{B}CD + \overline{A}B\overline{C}D + \overline{A}BCD + A\overline{B}\,\overline{C}\,\overline{D} + A\overline{B}\,\overline{C}D$$

$$a = A + CD + BD + \overline{B}\,\overline{D}$$

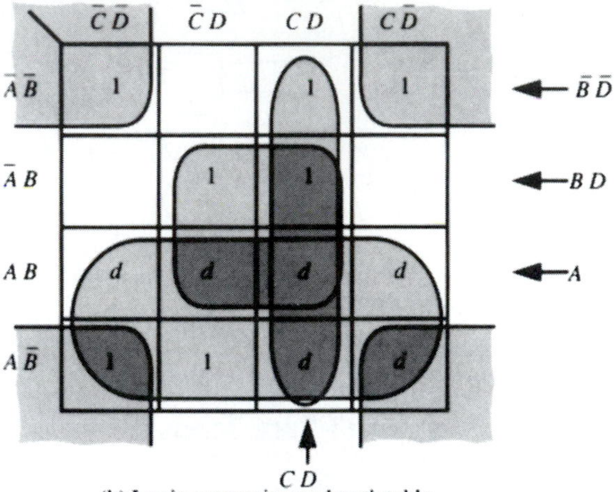

(b) Logic expression and truth table

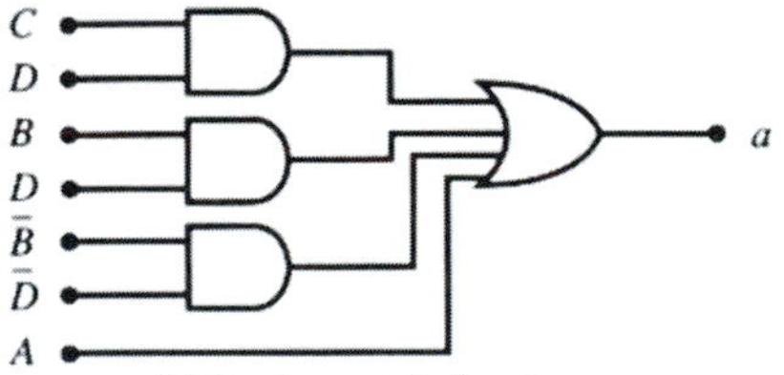

(c) Implemented circuit

FIGURE 3.30
Design example 7-segment decimal display

3.8 TECH TIPS AND TROUBLESHOOTING—T³

The logic probe can be used to verify proper circuit operations. The proper operation of a logic circuit can be determined by probing the input and output pins of a circuit. In general there are four types of problems to look for:

1. No V_{CC} or ground
2. Defective gate
3. Open connections or wires
4. Shorted connections or wires

No V_{CC} or Ground

When troubleshooting any device, the first step is to verify that power is applied to the circuit. Without power an IC cannot function; therefore, the first step is to check for V_{CC}

and ground connections. This test can easily be performed using the logic probe. V_{CC} will be indicated as a logic 1 and ground as a logic 0.

Defective Gate

Figure 3.31 illustrates a method that can be used to detect a defective gate. Beginning at the output of gate 3, we use the logic probe to determine the logic state of Z. In this case we find it is *always* a logic 0. Next we probe the two-input pins of gate 3. We find the top to be pulsing and the bottom pin to be a logic 0. Since gate 3 is an OR gate, we realize that any 1 on an input pin should force the output to be high. Therefore, the output should be pulsing. Since the logic probe indicated that the output of gate 3 was always a 0, we should suspect that gate 3 is defective.

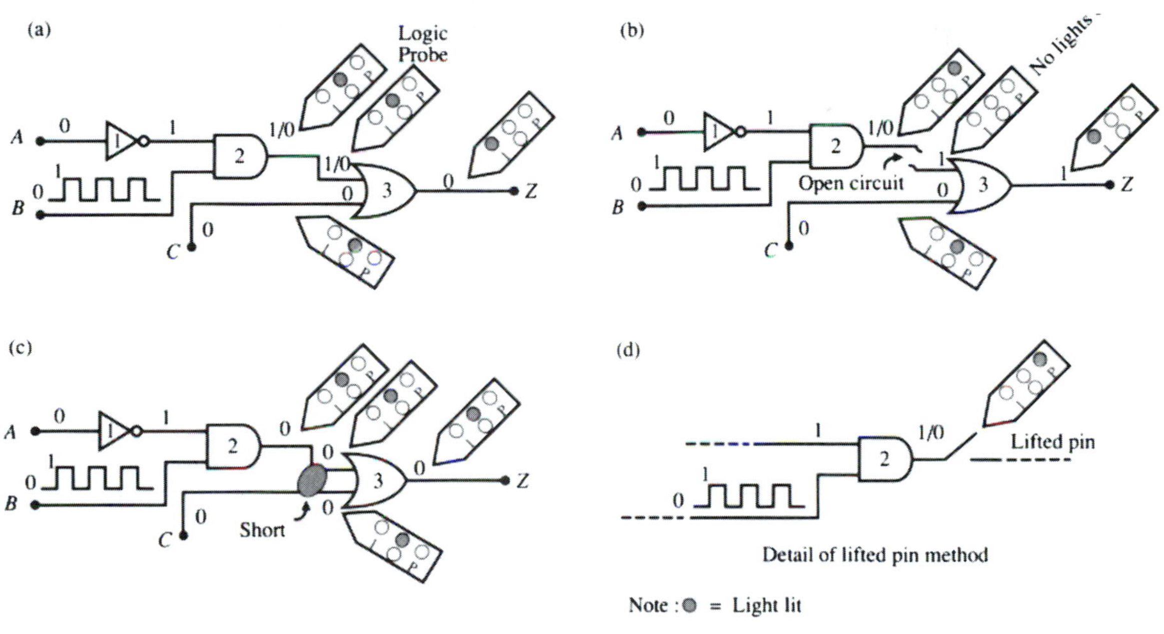

FIGURE 3.31
Logic probe troubleshooting

Open Connections or Wires

Referring to Figure 3.31, suppose that our logic probing indicates no pulses on the top input to gate 3. Furthermore, suppose that by moving the logic probe back to the output of gate 2, we detect a pulsing condition. This indicates an open or broken connection between the output of gate 2 and the input of gate 3. It is important to note the logic probe indication at the top input to gate 3. No lights are lit. This condition is sometimes referred to as a **float.** For TTL gates, floats are always interpreted *by a logic gate as a logic level* 1.

Shorted Connections or Wires

Referring to Figure 3.31(c), we note a logic level 0 on both inputs to gate 3. Probing the output of gate 2 also indicates a logic level 0. Probing the inputs to gate 2 indicates a logic level 1 on the top input and pulsing on the bottom input. This might cause us to think that gate 2 is defective. If after replacing gate 2, the condition remains the same, we need to look for another cause of the problem. One technique we may try is to lift the output pin of gate 2 out of the socket. If we lift the output pin out of the socket, we should observe pulsing on the logic probe, as shown in Figure 3.31(d). But what if probing the inputs to gate 3 still indicates a logic level 0 on both input pins? This is an indication of a shorted connection at the input to gate 3, since the top pin to gate 3 has been forced to a float (logic 1).

It should be noted that ICs are not always socketed; therefore, it may be necessary to cut a pin in order to force a float condition. Although this is not a preferred procedure, it can be performed as a test. The IC can then be either repaired by soldering or replaced.

Another technique is to use a device known as a **logic pulser.** This device functions as a signal-substitution source that overrides the current logic state at the input to a gate. Therefore, this device can also be used to test the output of a logic gate, as shown in Figure 3.32.

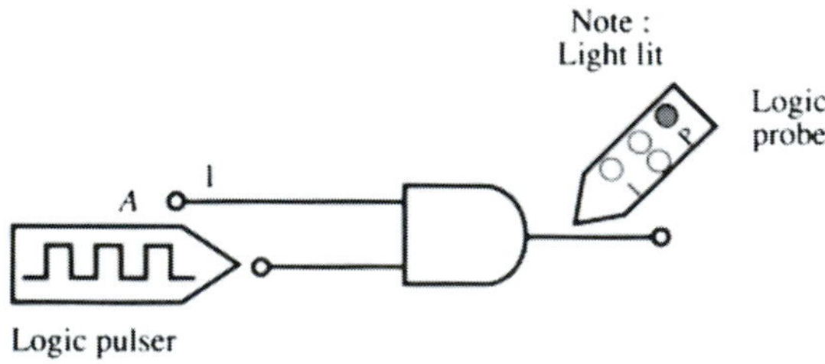

FIGURE 3.32
Logic pulser troubleshooting

EXERCISES

3.1 List the steps used to draw the logic diagram for the expression ALARM $= \overline{XY} + XY + W$.

3.2 Draw the logic diagram for the expression ALARM $= \overline{XY} + XY + W$.

3.3 Given the logic expression, $Z = (X + Y) \cdot (X + W)$.
   a. Simplify the expression, stating the theorems being applied.
   b. Draw the logic diagram for the simplified expression.
   c. Using truth tables, prove that your simplified expression is correct.

3.4 Repeat Exercise 3 for the expression ON $= \overline{A}B + AB + \overline{A}\,\overline{B}$.

3.5 Repeat Exercise 3 for the expression GO $= \overline{A}\,\overline{B}C + A\overline{B} + AB\overline{C}$.

3.6 Repeat Exercise 3 for the expression $Z = \overline{B}(A + C) + C(\overline{A} + B) + AC$.

3.7 Repeat Exercise 3 for the expression ENABLE $= (A + \overline{C}) \cdot (\overline{A} + ABC)$.

3.8 Using a truth table, prove $\overline{(X + Y + Z)} = \overline{X} \cdot \overline{Y} \cdot \overline{Z}$.

3.9 Using a truth table, prove $\overline{(X \cdot Y \cdot Z)} = \overline{X} + \overline{Y} + \overline{Z}$.

3.10 Using DeMorgan's theorem, simplify the expression

$$C = \overline{[(\overline{AB}) + (AB)]}.$$

3.11 Simplify the expression BELL $= XYZ + X\overline{Y}Z + XY\overline{Z}$.

3.12 Simplify the expression HOT $= (\overline{A} + B) \cdot (A + \overline{B})$.

3.13 Simplify the expression READY $= \overline{(A + \overline{B}) + C}$.

3.14 Simplify the expression XCK $= \overline{\overline{(A + \overline{B})} + \overline{CD}}$.

3.15 Simplify the expression $MX = RJ + J(D + J) + R(J + D)$.

3.16 How many boxes will there be in a Karnaugh map of a 4-variable Boolean equation?

3.17 List the steps for mapping an expression.

3.18 Use a Karnaugh map to simplify the expression $X = \overline{JR} + \overline{J}R$.

3.19 Use a Karnaugh map to simplify the expression

$$X = AB\overline{C} + \overline{A}\,\overline{B}\,\overline{C} + \overline{A}\,\overline{B}C + A\overline{B}C.$$

3.20 Given the expression BELL $= \overline{X}Y\overline{Z} + XY\overline{Z} + \overline{X}\,\overline{Y}Z + X\overline{Y}Z$.

a. Simplify using a Karnaugh map.
b. Verify your simplified expression using a truth table.
c. Implement the simplified expression.

3.21 Repeat Exercise 20 for the expression

$$\text{OFF} = \overline{A}\,\overline{B}\,\overline{C}\,\overline{D} + \overline{A}\,\overline{B}C\overline{D} + CD + A\overline{B}C\overline{D}.$$

3.22 List the steps used in the classical, systematic method for digital design.

3.23 Design a logic circuit to detect whenever any two of three switches are closed.

3.24 Ms. Jerome wishes to design an alarm circuit for her four-door car. An on/off switch is not required.

a. List your design method.
b. Using your method, design the alarm.
c. If possible, wire up your design and test the circuit with a logic probe.

3.25 Design a logic circuit that will activate a light whenever the output of a 3-bit counter is 4 or less (use A as the LSB and C as the MSB).

3.26 Design a logic circuit that will activate a high whenever the output of a 3-bit counter is 5 or greater.

3.27 Complete the design of the 7-segment display shown in Figure 3.27.

3.28 Design a logic circuit for a comparator circuit that produces a logic 1 output whenever bits A and B are the same *and* bits C and D are the same.

3.29 Design a logic circuit that will produce an output whenever any three of four input switches are the same.

3.30 A candy machine releases candy whenever 15¢ or more is deposited. If more than 15¢ is deposited, no change is returned. Assuming the machine accepts nickels and dimes only, design the logic circuit for the candy machine shown in Figure 3.33.

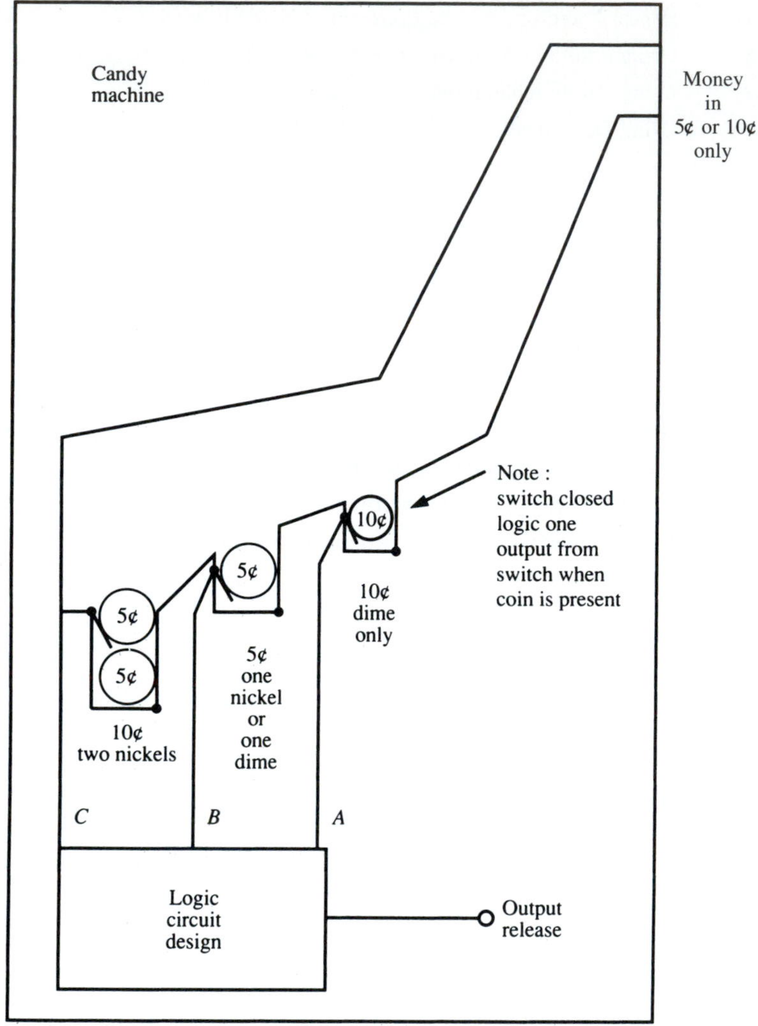

FIGURE 3.33

3.31 Discuss the steps you would follow if you were to test a digital circuit using a logic probe.

3.32 How can you verify V_{CC} and ground with a logic probe?

3.33 How can you test for open connections in a digital circuit?

3.34 How can you test for shorted connections in a digital circuit?

3.35 Crossword Puzzle

ACROSS

2. Combinations of input variables that cannot occur are referred to as _____ terms.
4. When testing circuits, there are _____ general types of problems to look for.
5. _____ boxes can be vertical, horizontal, or looped around the end.
9. Active-state NAND function.
10. When troubleshooting a device, the first step is to check if _____ is applied to the circuit.
11. Boolean law that states that variables can be grouped in any way that we choose.
15. $A \cdot B$.
18. Boolean law that defines the operation of a gate when the inputs are the same.
19. Logic signal substitution source.
21. Logic 1.
22. NAND and NOR gates are sometimes referred to as _____ logic gates.
23. Four-variable map grouping that results in a single variable term.
25. Boolean law that states that the second term of an expression may be ignored.

DOWN

1. When using a logic probe, a ground is indicated as a logic _____.
3. An exclusive OR reduction can be indicated on a Karnaugh map by a _____ pattern.
4. For TTL gates, a _____ is always interpreted as a logic level 1.
6. Alternate set of logic symbols.
7. Karnaugh map form of a Boolean expression.
8. Gates with incorrect output and correct inputs are suspected as being _____.
12. Boolean law that states that it does not matter in which order an AND operation is performed.
13. A systematic graphical method for simplifying Boolean expressions.
14. Factoring in Boolean algebra is an example of this law.
16. Three-variable map grouping that results in a three-variable term.
17. Break the bar and change the sign.
20. Each grouping on a Karnaugh map should be as _____ as possible.
24. 1010.

1 2 3 4 5 6 7 8 9 10 11 12 13 14 15 16 17 18 19 20 21 22 23 24 25

CHAPTER 4

SEQUENTIAL LOGIC

KEY TERMS

Armed
Arming the Latch
Astable
Asynchronous
Asynchronous Counter
Bistable
Clear
D *Flip-Flop*
D *Latch*
Decoding
Edge-triggered Flip-Flops
Enabling the Latch
Flip-Flops
Free-running Multivibrator
Frequency Divider
Hold Time
Infamous Decode Spike
JK *Flip-Flop*
JK *Master-Slave Flip-Flop*
Latch Flip-Flop
Medium Scale Integrated Circuits (MSI)
Modulus (MOD)
Monostable
Multivibrator Circuit
One-shot
Oscillator
Preset
Propagation Delay Time
Race Condition
Ripple Counter
Setup Time
Switch-contact Bounce
Synchronous Counter
Toggling

4.0 INTRODUCTION

In the previous chapter we studied combinations of logic gates. These types of circuits exhibited no memory; the output was always dependent on the input. In this chapter, we study a type of circuit that has memory and can store or hold its output state. These circuits are generally classified as **multivibrator circuits** and are often referred to as **flip-flops.** A flip-flop is a device that has two stable output states. That is, the output can be either a logic 1 or a logic 0. These output states are stored or stable even if the inputs are removed. Because of this feature, flip-flops are referred to as **bistable** (two states) devices that exhibit memory.

Flip-flops can be classified into two general types: *latch flip-flops* and *edge-triggered flip-flops.* In a **latch flip-flop,** the output is controlled by the logic *level* of the input. In an **edge-triggered flip-flop,** the output is controlled by the leading or trailing *edge* of an input pulse (Figure 4.1).

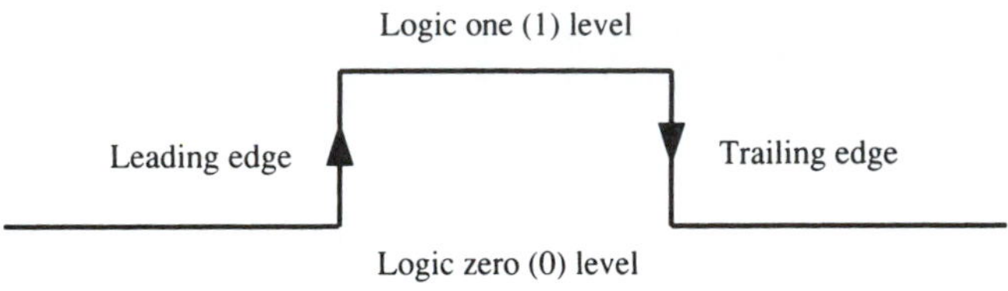

FIGURE 4.1
Trigger pulse

4.1 LATCH FLIP-FLOPS

The basic flip-flop is shown in Figure 4.2(a). The inputs are commonly labeled S (SET) and R (RESET). Most flip-flops contain two outputs, which are generally called Q and $\overline{Q}$. $\overline{Q}$ is the inverted output of Q. $\overline{Q}$ is always at the logic state opposite that of the logic state of Q. Therefore, if Q is a logic 0, $\overline{Q}$ will be a logic 1. If Q is a logic 1, $\overline{Q}$ will be a logic 0. This can be thought of as a seesaw. That is, when one side is high, the other side must be low.

Recall that with two input variables, four possible combinations of input conditions can exist. The input combinations and the resulting output states are described in the truth table of Figure 4.2(b).

To explain the operation of the flip-flop, consider the circuit of Figure 4.2(c). Two NOR gates are wired so that their outputs are fed back to the other's input. We begin by considering the first condition of the truth table. Both inputs S and R are at a logic 0, as shown in Figure 4.3(a). Under this condition the outputs remain in whatever state they were in previously. The reason for this should become clear in a few moments. Assume for now that the Q output is at a logic level 0 and the $\overline{Q}$ output is a logic level 1. Next consider the condition where the S input goes to a logic level 1 and the R input remains at a logic level 0. Under this condition the inputs to NOR gate 2 are 1 and 0, as shown in Figure 4.3(b). This will force the output of NOR gate 2 to become a 0. The output of NOR gate 2 is fed back to the input of NOR gate 1. Thus the inputs to NOR gate 1 are 0 and 0, as shown in Figure 4.3(b). The output of NOR gate 1 will then be forced to a logic level 1 and fed back to the input of NOR gate 2. The inputs to NOR gate 2 are now 1 and 1. Therefore, the output of NOR gate 2 will remain at a logic level 0. Thus, the Q output becomes a logic level 1, and the $\overline{Q}$ output becomes a logic level 0. This is sometimes referred to as the *SET* condition ($Q = 1$) as shown in the stable state in Figure 4.3(c).

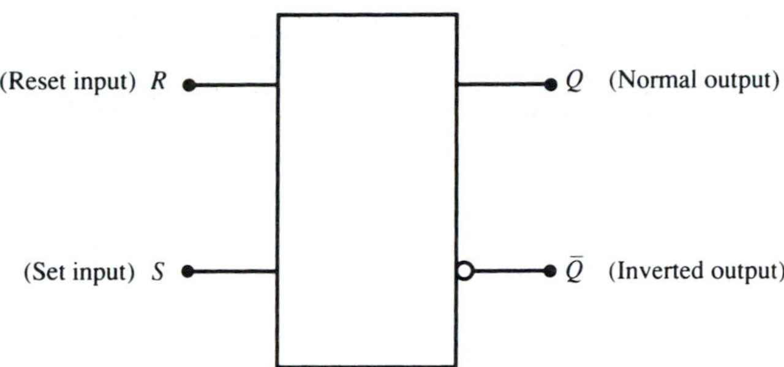

(a) Basic R-S latch flip-flop (NOR gates) symbol

R	S	Q	$\bar{Q}$	Action
0	0	Q	$\bar{Q}$	No change (Q, $\bar{Q}$ stay the same)
0	1	1	0	Set
1	0	0	1	Reset (clear)
1	1	-	-	Illegal

(b) Truth table

Active high input
(c) R-S latch flip-flop

FIGURE 4.2
***RS* latch flip-flop**

Notice that if the S input returns to a logic level 0, causing both inputs S and R to be at a logic level 0, the output remains unchanged. The inputs to NOR gate 2 are now 0 and 1, as shown in Figure 4.3(d), so the output of NOR gate 2 remains at a logic 0. The inputs to NOR gate 1 are then 0 and 0, which causes the output to remain at a logic 1. Thus the RS input condition 0–0 causes the outputs to remain the same as they previously were (no change).

All possible output conditions can be summarized as follows:

1. $S = 0, R = 0$: Q and $\overline{Q}$ remain in whatever state they were in previously.
2. $S = 1, R = 0$: Q flips to a 1; $\overline{Q}$ flops to a 0. This is called the *SET* condition.
3. $S = 0, R = 1$: Q flips to a 0; $\overline{Q}$ flops to a 1. This is called the *RESET* condition.
4. $S = 1, R = 1$: This condition tries to force Q and $\overline{Q}$ to a logic 0 at the same time. This is an illegal condition, since Boolean algebra Q and $\overline{Q}$ must always be in the opposite state. Therefore, although this condition can exist, it should be avoided.

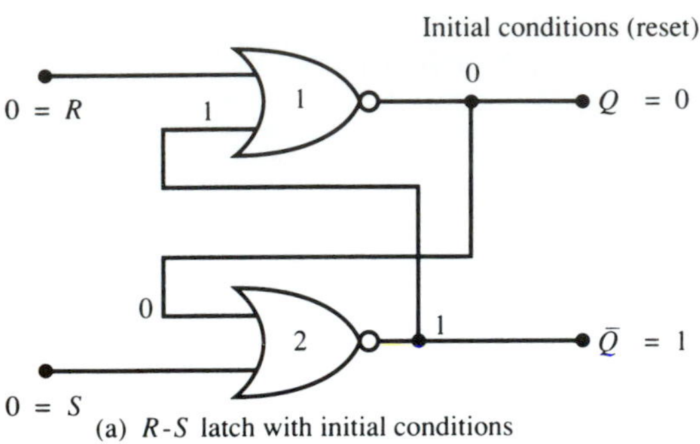

(a) R-S latch with initial conditions

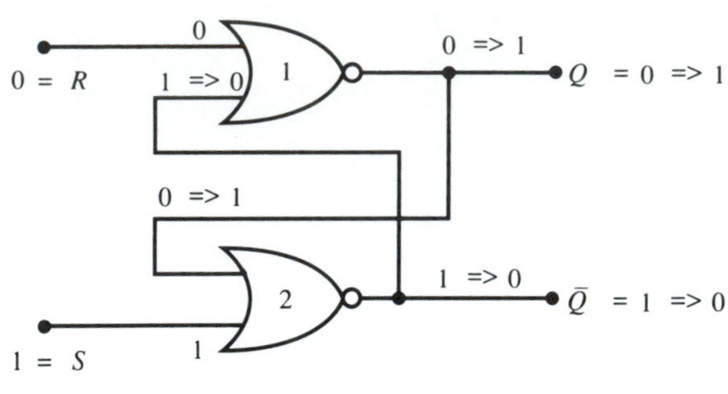

(b) Set input conditions, $S = 1, R = 0$ with Q feeding back to the input to gate 2 and the $\overline{Q}$ feeding back to the input to gate 1

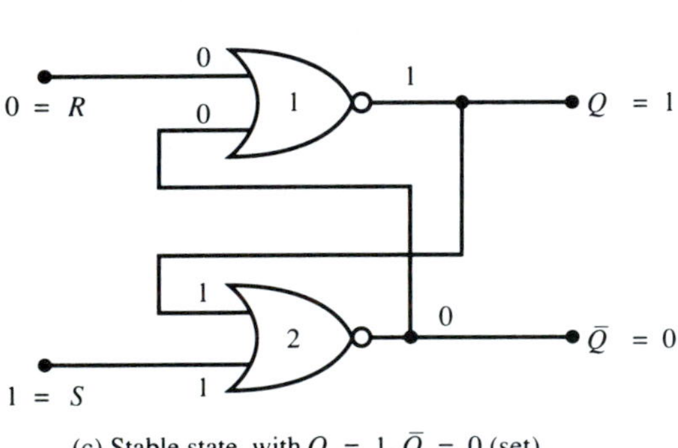

(c) Stable state, with $Q = 1, \overline{Q} = 0$ (set)

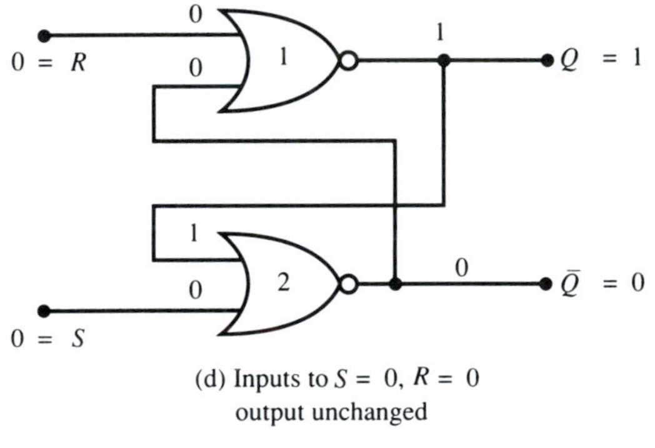

(d) Inputs to $S = 0, R = 0$ output unchanged

FIGURE 4.3
***RS* latch flip-flop explained**

NAND Gate RS Latch

Two NAND gates can be connected in the same manner as the NOR gates to create an active low input *RS* latch, as illustrated in Figure 4.4(a). The SET and RESET inputs are labeled $\overline{S}$ and $\overline{R}$, respectively. The outputs are labeled Q and $\overline{Q}$. All the possible input and output combinations are described in the truth table of Figure 4.4(b). Figure 4.4(c) shows the conventional and equivalent active-state logic diagram representation of the NAND gate *RS* latch. Generally speaking, a low on the SET input $\overline{S}$ will force the Q output to a logic level 1 and the $\overline{Q}$ output to a logic level 0. A low on the RESET input $\overline{R}$ will force the $\overline{Q}$ output to a logic level 1 and the Q output to a logic level 0. To summarize:

1. $\overline{S} = 1$, $\overline{R} = 1$: Q and $\overline{Q}$ remain in whatever state they were in previously.
2. $\overline{S} = 0$, $\overline{R} = 1$: Q flips to a 1; $\overline{Q}$ flops to a 0. This is called the SET condition.
3. $\overline{S} = 1$, $\overline{R} = 0$: Q flips to a 0; $\overline{Q}$ flops to a 1. This is called the RESET condition.

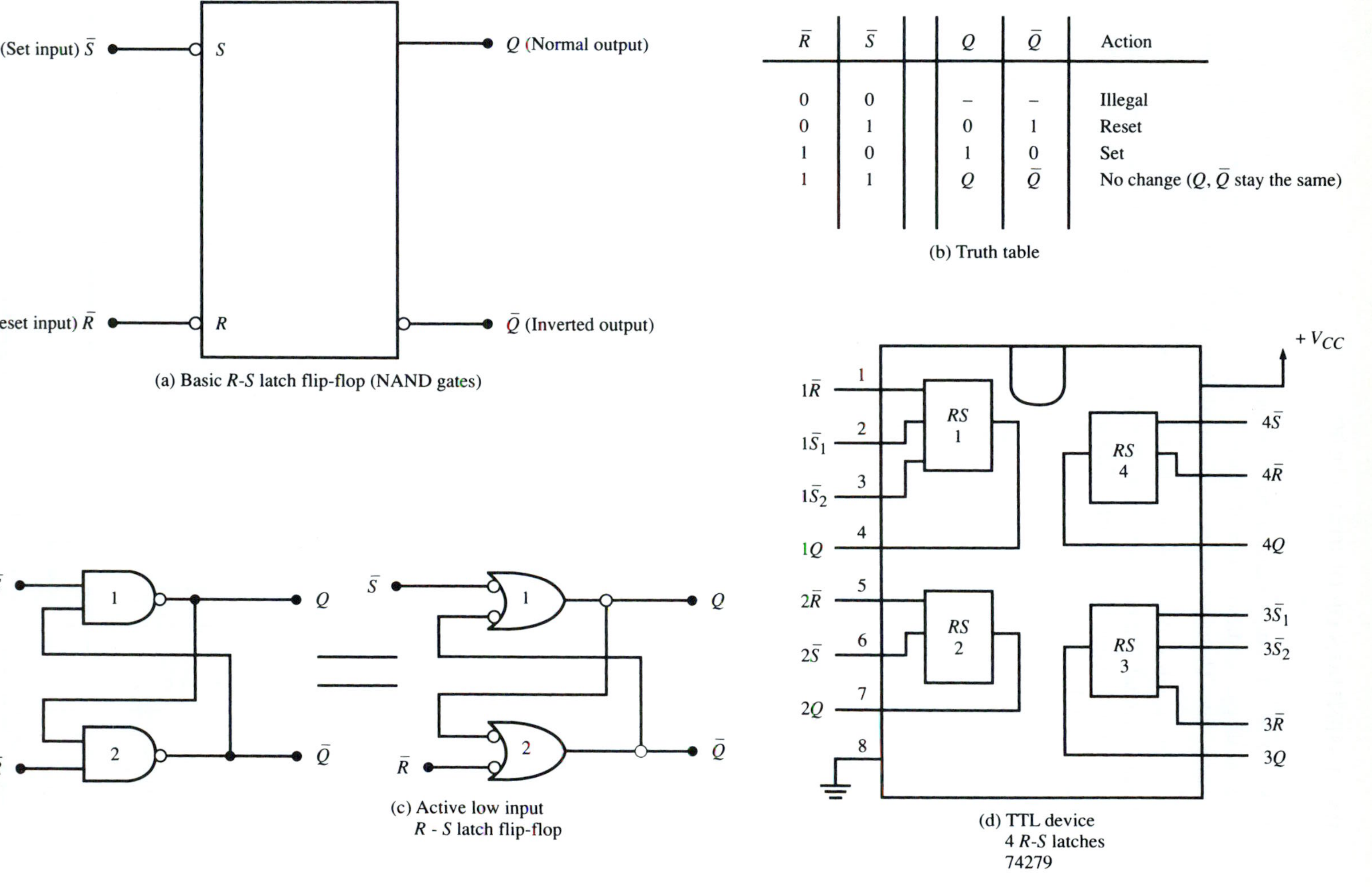

$\bar{R}$	$\bar{S}$	Q	$\bar{Q}$	Action
0	0	–	–	Illegal
0	1	0	1	Reset
1	0	1	0	Set
1	1	Q	$\bar{Q}$	No change (Q, $\bar{Q}$ stay the same)

(b) Truth table

FIGURE 4.4
***RS* latch using NAND gates**

4. $\overline{S} = 0$, $\overline{R} = 0$: This condition tries to force Q and $\overline{Q}$ to a logic 1 at the same time. This is an illegal condition, since in Boolean algebra Q and $\overline{Q}$ must always be in the opposite state. Therefore, although this condition can exist, it should always be avoided.

A device that contains four *RS* NAND gate latches in one package is the TTL 74279 integrated circuit. Figure 4.4(d) shows the pin configuration for this device. All the integrated wiring is provided internally. This device simplifies the use of the *RS* NAND gate latch in digital logic circuits.

EXAMPLE 4.1

For the input conditions shown in Figure 4.5(b), determine the output waveform for the Q output of the NAND gate *RS* latch of Figure 4.5(a). Assume Q is initially high.

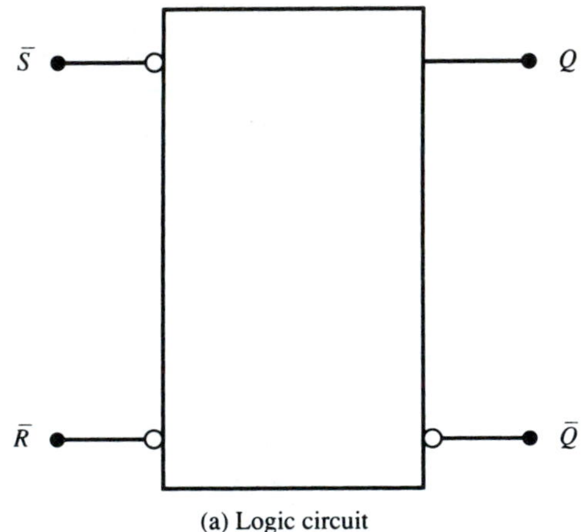

(a) Logic circuit

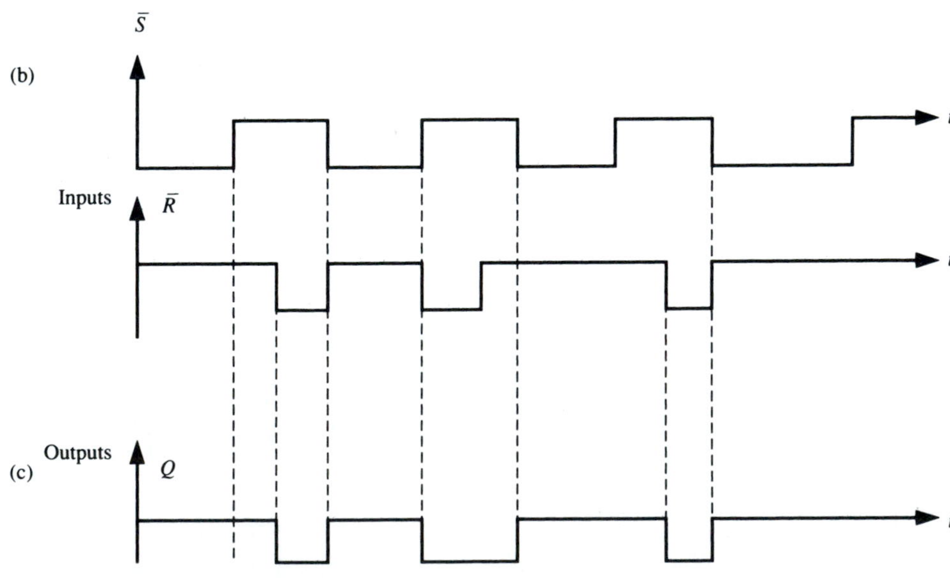

FIGURE 4.5

Solution
The Q output waveform is shown in Figure 4.5(c). Whenever the $\overline{S}$ input is low, the Q output will be high, and the $\overline{Q}$ output will be low. Whenever the $\overline{R}$ input is low, the Q output will be high, and the Q output will be low. Whenever both $\overline{S}$ and $\overline{R}$ are high, the Q output will remain in its previous state (no change).

4.2 LATCH APPLICATIONS

The basic *RS* latch forms the foundation for many types of logic circuit applications. A number of examples are discussed next.

Switch-Contact Debounce

One common application of the *RS* latch circuit is for switch-contact debouncing. When a switch is thrown from one position to another (i.e., *on* to *off*), the contacts will actually *make* and *break* several times. This phenomenon is referred to as **switch-contact bounce** and is illustrated in the waveform of Figure 4.6(b) (p. 108). The effect is the occurrence of many transitions from a logic level 0 to a logic level 1 and from a logic level 1 to a logic level 0. Many transitions on the input of a logic gate or circuit can result in unwanted transitions on the output. The switch-debounce circuit of Figure 4.6(a) filters out any unwanted switch-bouncing pulses. The switch is initially shown in the S = UP position. The SET input to the *RS* latch is then a logic 0. The RESET input to the *RS* latch is then at a logic level 1. This causes the Q to be at a logic level 1, as shown in the waveform of Figure 4.6(c). When the switch is thrown to the S = DOWN position, the RESET input immediately goes to a logic 0. But as the contacts of the switch break and make (bounce), the RESET input alternates between a logic 0 and a logic 1, as shown in Figure 4.6(b). The Q output makes the transition to a logic 0 as soon as the RESET input goes to a logic 0. The Q output will remain stable at a logic 0 during the switch-bounce period because when the switch contacts bounce to a logic level 1, both inputs SET and RESET are at a logic level 1. Therefore, the Q output will remain stable or unchanged at a logic level 0.

Gated Latches

Sometimes we wish to control when a latch will be allowed to change state. This is sometimes referred to as **enabling the latch,** or **arming the latch.** The circuit of Figure 4.7 (p. 109) illustrates a method of controlling when a latch will be allowed to change state. When the C input is low (logic 0), the output of NAND gate 1 and 2 is high (logic 1). Recall from the truth table of an *RS* latch that this condition results in a *no change* output. Placing a high on the C input will enable, or arm, the *RS* latch and allow it to function as previously described by its truth table. It should be noted that the inputs S and R are inverted by NAND gates 1 and 2.

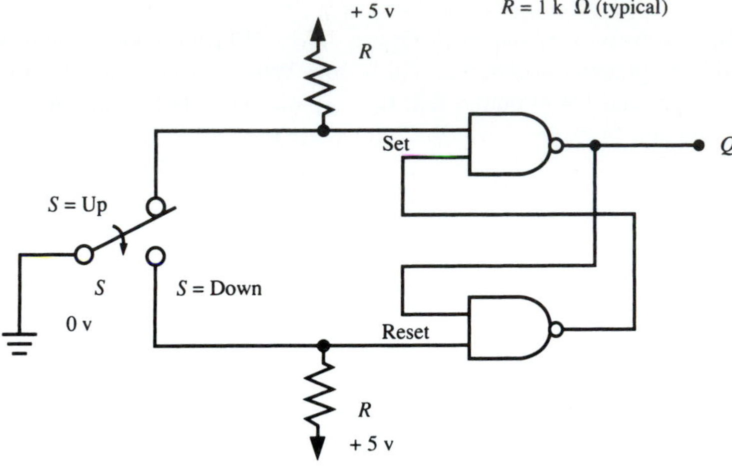

(a) Debounce circuit

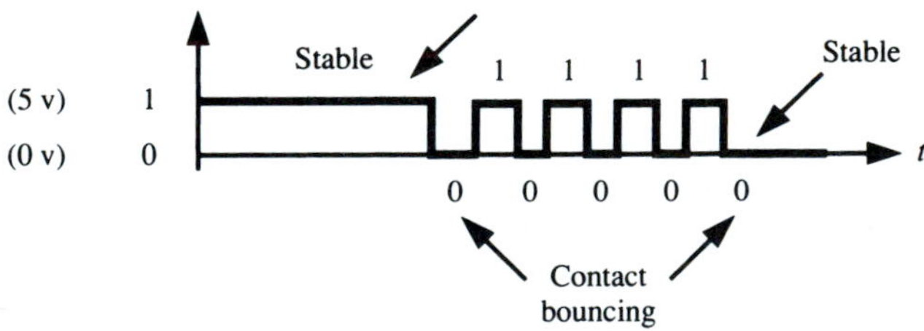

(b) Reset input waveform

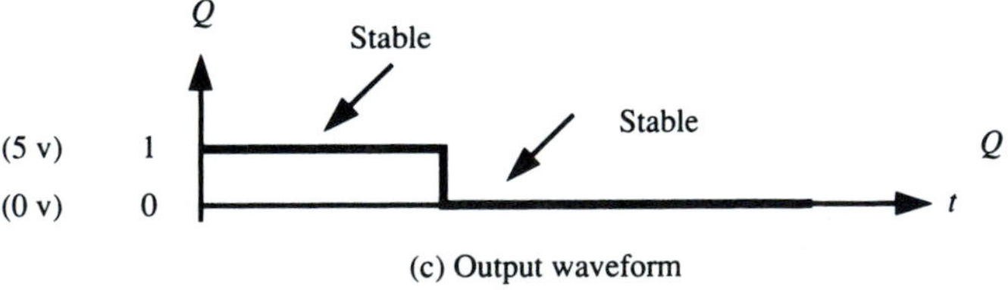

(c) Output waveform

FIGURE 4.6
Switch-debounce circuit

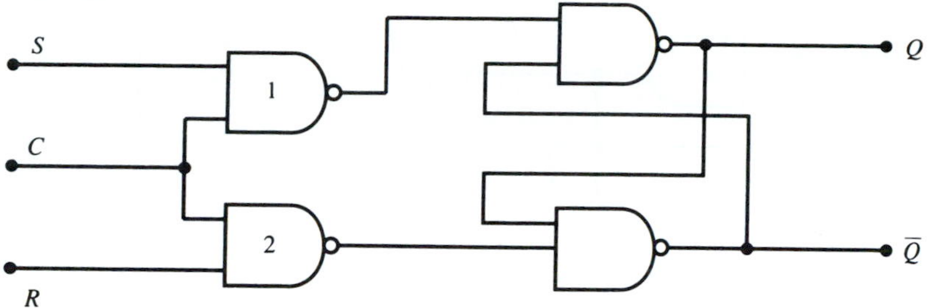

(a) Gated latch logic circuit

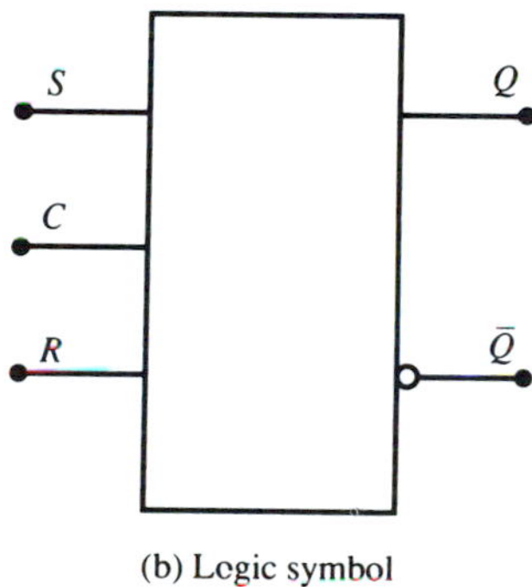

(b) Logic symbol

FIGURE 4.7
Gated latch

D Latch

Latches can have only one input. Consider the circuit of Figure 4.8. By using an inverter, we can create both inputs for the basic *RS* latch. When the *D* input is high, the *S* input is high and the *R* input is low. When the *D* input is low, the *S* input is low and the *R* input is high. This circuit is referred to as a ***D* latch.** It performs like the *RS* latch. Note from the truth table of Figure 4.8(c) that with only one input there are only two possible input conditions. Figure 4.8(d) shows the pin configuration for a 7475 *D*-type latch. This device contains four *D*-type latches. Note the enable inputs. These perform the gated latch functions previously described and illustrated in the truth table of Figure 4.8(e). Also note that the latch of Figure 4.8(a) has active low inputs. Therefore, when the *D* input is a logic 0, the *Q* output is a logic 1. This is sometimes referred to as an *inverting latch.* The 7475 *D*-type latch of Figure 4.8(d) is a noninverting latch. That is, when the *D* input is a logic 0, the Q output is a logic 0. This is illustrated in the truth tables of Figure 4.8(c) and 4.8(e).

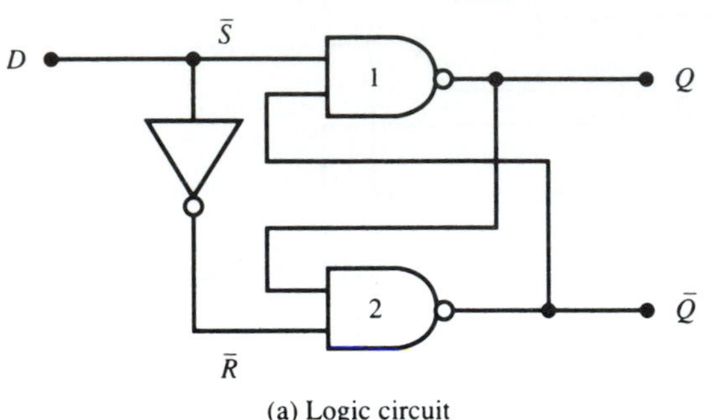

(a) Logic circuit

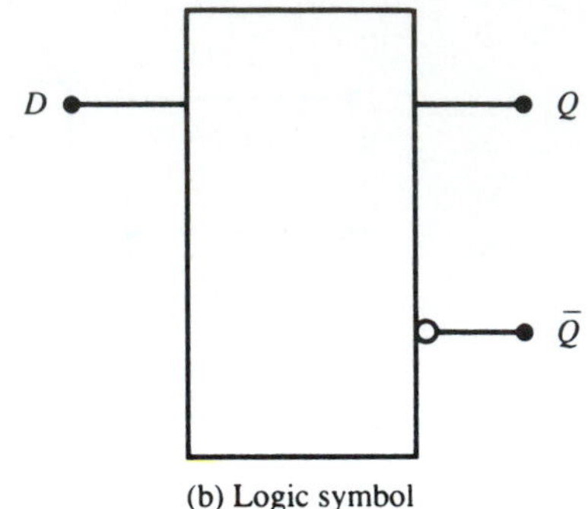

(b) Logic symbol

Input D	Output Q
0	1
1	0

(c) Truth table

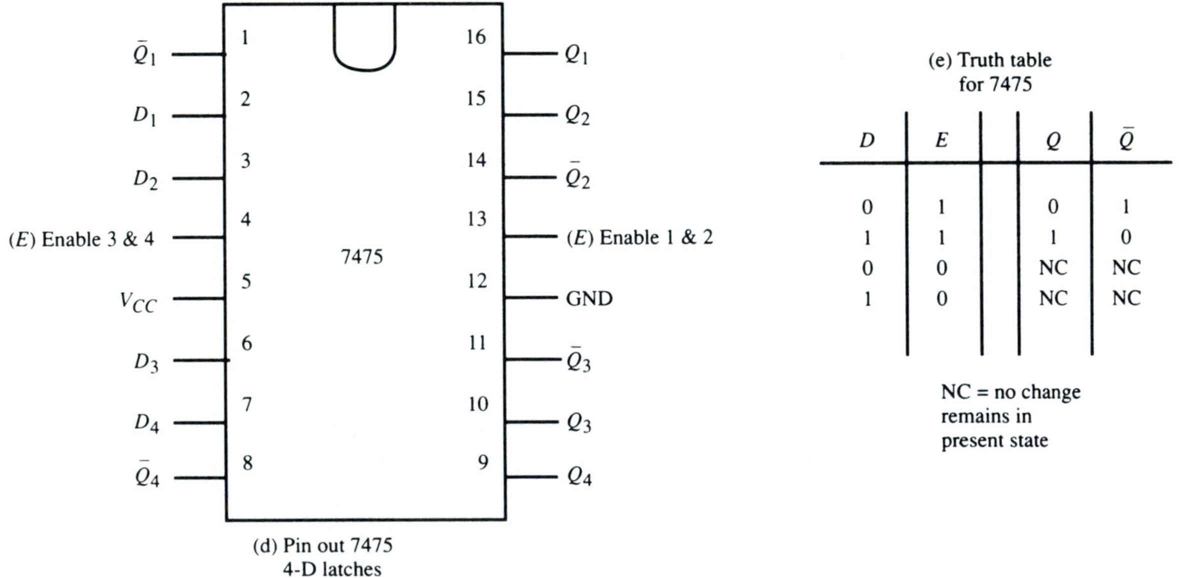

(d) Pin out 7475
4-D latches

(e) Truth table
for 7475

D	E		Q	$\bar{Q}$
0	1		0	1
1	1		1	0
0	0		NC	NC
1	0		NC	NC

NC = no change
remains in
present state

FIGURE 4.8
***D* latch**

4.3 EDGE-TRIGGERED FLIP-FLOPS

Edge-triggered flip-flops are flip-flops whose output is controlled by the leading or trailing edge of a gated clock pulse. That is, the output can change state only on an edge of a clock pulse. The output logic (Q) is still determined by the input logic levels. The two most popular types of edge-triggered flip-flops are the D flip-flop and the JK flip-flop.

D Flip-Flop

The ***D* flip-flop** is a single-input edge-triggered or edge-controlled flip-flop. On the positive-going, or leading, edge of a clock pulse, the input data D is transferred to the Q output, as shown in Figure 4.9(a). Note the use of the *chevron,* or small triangle, on the clock input to indicate an edge-triggered device. Functionally, the D flip-flop can be thought of as performing according to the circuit diagram of Figure 4.9(b). Notice that it is a combination of the gated RS latch and the D inverter circuits previously described. Table 4.1 lists a number of common TTL and CMOS D-type flip-flops. Note that although

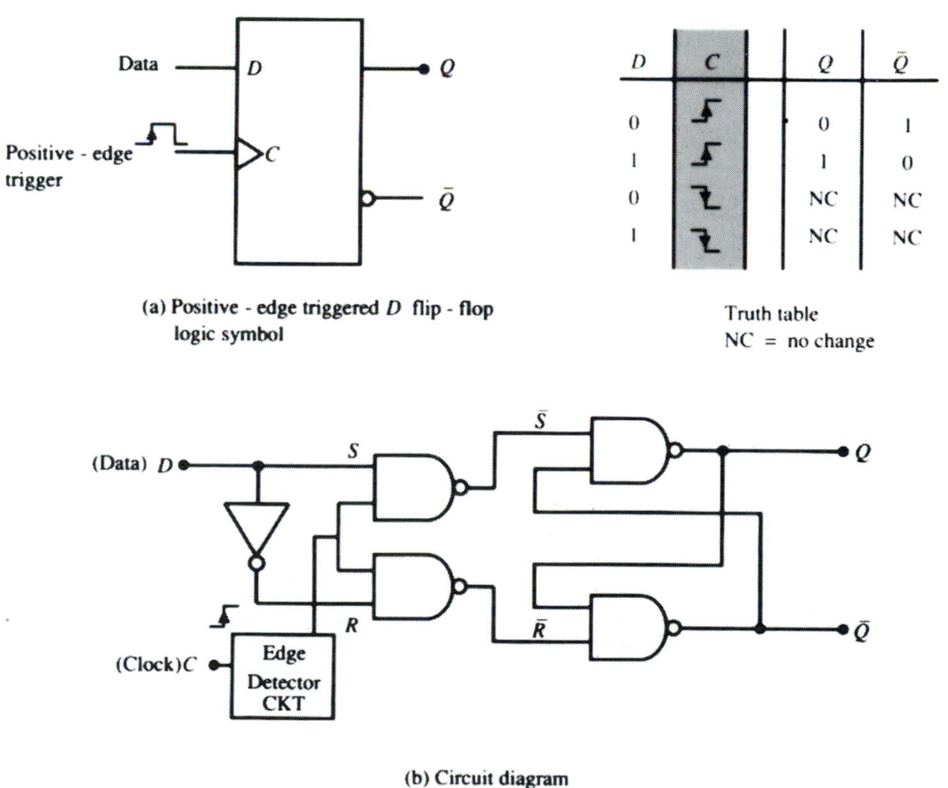

FIGURE 4.9
Edge-triggered *D* flip-flop

D-type flip-flops can be either positive-going edge-triggered or negative-going edge-triggered, most *D* flip-flops use positive edge triggering.

TABLE 4.1
D-type flip-flops

TTL	CMOS	Description
7474	4013	Two *D* flip-flops
74175	40175	Four *D* flip-flops
74174	40174	Six *D* flip-flops

EXAMPLE 4.2 Determine the *Q* output waveform for the *D* flip-flop with the *D* and clock input waveforms shown in Figure 4.10. Assume *Q* is initially low.

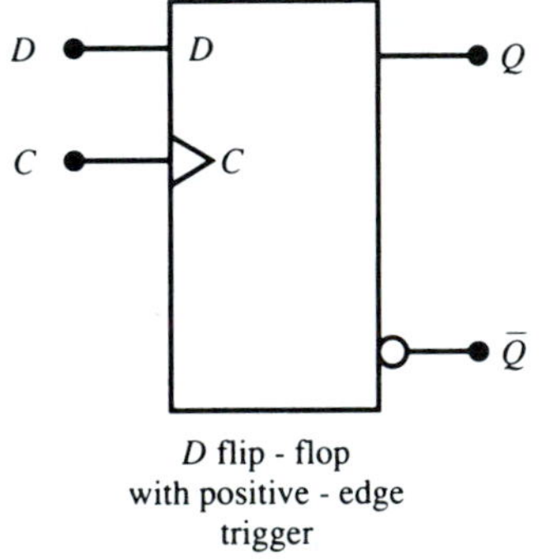

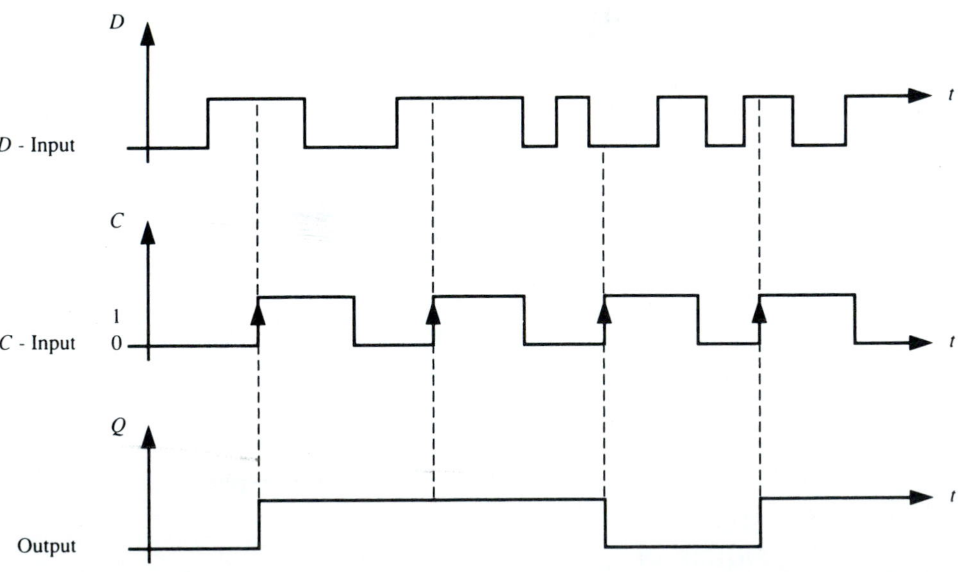

FIGURE 4.10

Solution

The D input is transferred to the Q output on the leading edge of each clock pulse. Note that no change can occur at any other time.

JK Flip-Flops

The ***JK* flip-flop** is a two-input edge-triggered or edge-controlled flip-flop. *JK* flip-flops can be triggered on either the positive- or negative-going edge of a clock pulse. However, negative-going edge-triggered *JK* flip-flops tend to be more commonly used in industry today. Since *JK* flip-flops have two inputs, there are four possible combinations of input conditions. The truth table of Figure 4.11 defines the output for each input combination. If both inputs are low, the output remains in its previous state (no change). When both inputs are different, the contents of the J input will be transferred to the Q output, and the contents of the K input will be transferred to the $\overline{Q}$ output. When both inputs are high, the outputs change to their opposite states. This is called **toggling.** The functional operation of a negative edge-triggered *JK* flip-flop is shown in Figure 4.11(b). Notice that it is a gated *RS* latch with the outputs fed back to the inputs in a cross-coupled manner. The *JK* flip-flop is probably the most widely used of all the different types of flip-flops in use today.

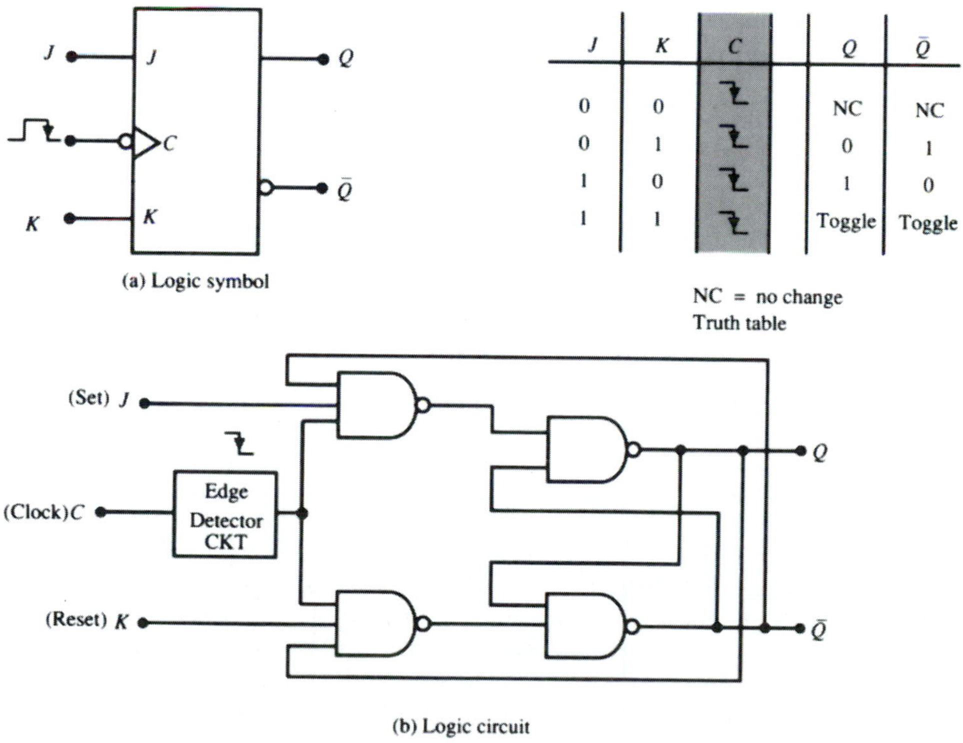

J	K	C	Q	$\overline{Q}$
0	0	↓	NC	NC
0	1	↓	0	1
1	0	↓	1	0
1	1	↓	Toggle	Toggle

NC = no change
Truth table

FIGURE 4.11
Basic *JK* flip-flop

EXAMPLE 4.3

Determine the Q output waveform for the JK flip-flop with the J, K, and clock inputs shown in Figure 4.12. Assume Q is initially high.

Solution

At the time of the first negative-going edge of the clock pulse, J is high and K is high. The Q output is therefore toggled to its opposite state. In this case it goes low. On the second negative-going edge of the clock pulse, J is low and K is low. Therefore, Q does

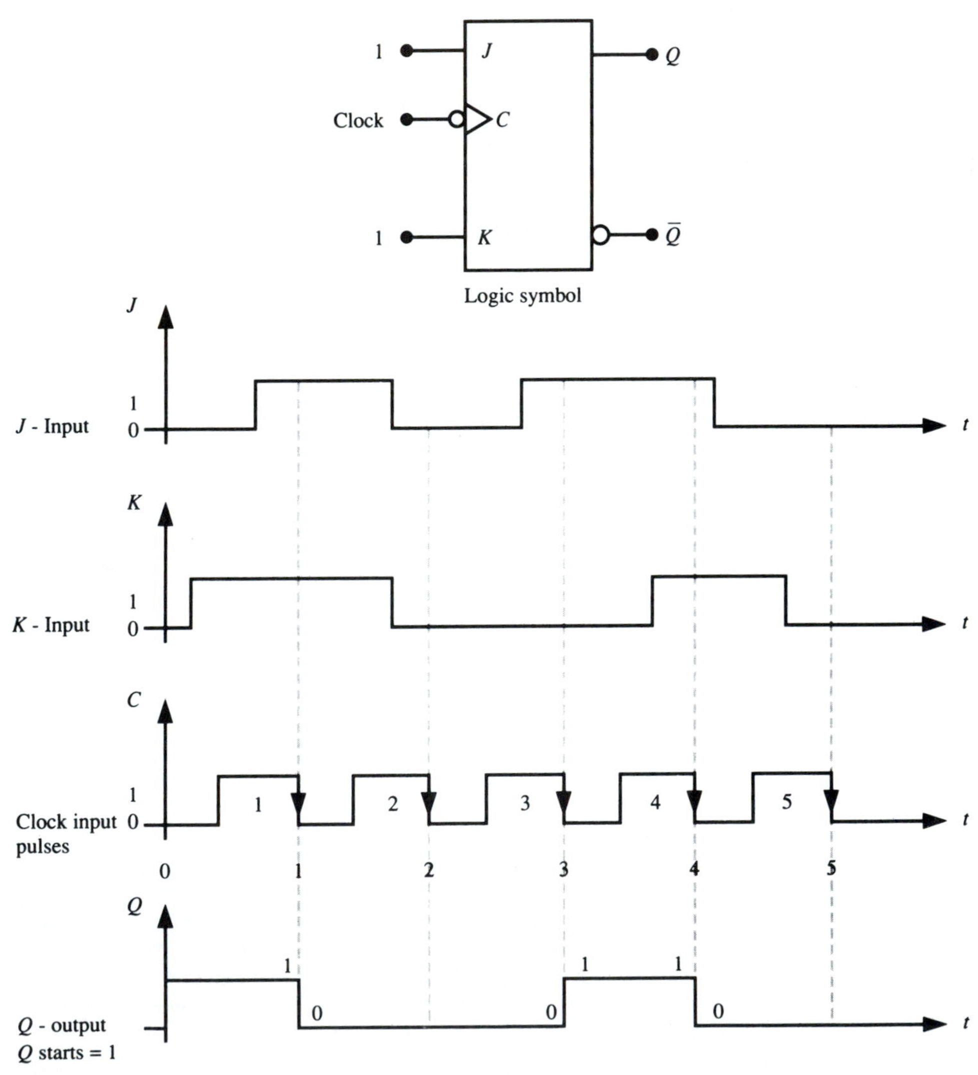

FIGURE 4.12

not change. On the third clock pulse, *J* is high and *K* is low. Therefore, *Q* goes high. On the fourth clock pulse, *J* and *K* are both high. Therefore, *Q* toggles low. On the fifth clock pulse, *J* and *K* are both low, which results in no change of *Q*.

JK Master-Slave Flip-Flops

A practical problem with negative edge-triggered flip-flops occurs when the input changes at the same time that the clock trigger pulse occurs. This problem arises because the inputs are not clearly defined, or *stable,* at the time of triggering, as shown in Figure 4.13. This is sometimes referred to as a **race condition.**

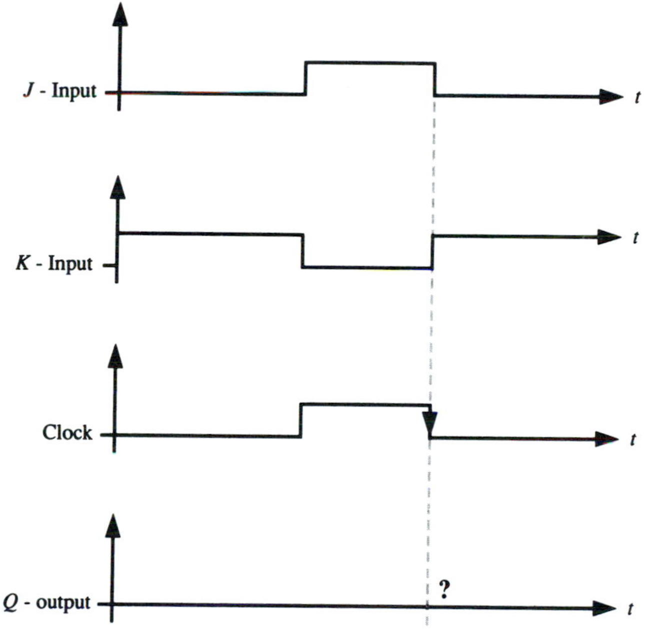

FIGURE 4.13
Race condition

The ***JK* master-slave flip-flop** of Figure 4.14 offers a solution for this type of situation. Consider the diagram of Figure 4.14. The *J* and *K* inputs are latched into the *master JK* flip-flop on the positive-going edge of the clock pulse. The *JK* inputs are then transferred to the output *slave JK* flip-flop on the negative-going edge of the clock pulse. The advantage here is that since the master has stored the inputs on the positive-going edge of the clock pulse, it is not necessary for the *J* and *K* inputs to remain stable once the negative-going edge of the clock pulse transitions.

The *JK* master-slave flip-flop is actually two *JK* flip-flops configured into one device. Figure 4.14(c) illustrates the operation of the *JK* master-slave flip-flop. When the clock is a low, the outputs of NAND gates 1 and 2 are forced high. This results in a

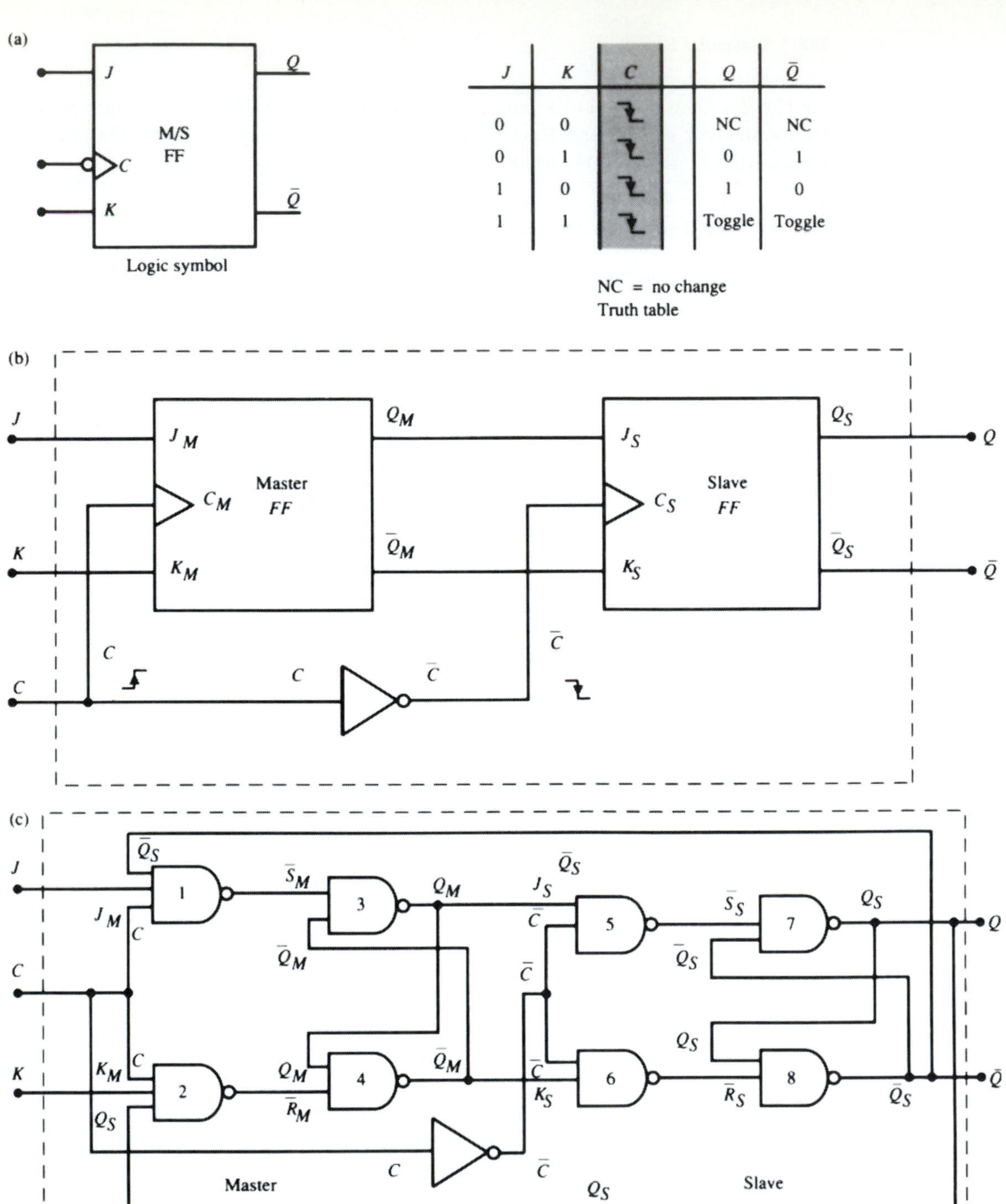

FIGURE 4.14
Master-slave *JK* flip-flop

no-change condition for the master *RS* latch. When the clock transitions to a high, the *J* and *K* inputs will determine the state of the master *RS* latch. The slave *RS* latch is controlled, however, by the inverted clock signal. Therefore, when the clock *C* is high, $\overline{C}$ is low, and the inputs to NAND gates 7 and 8 are forced high. This is the no-change condition for the slave flip-flop. When the clock *C* transitions from a high to a low, the slave clock $\overline{C}$ goes high. The *JK* inputs to the slave, J_S and K_S, are then free to control the slave. Thus the data bits are transferred to the *Q* and $\overline{Q}$ outputs. The key control element is the clock inverter. It disables the slave when the master is enabled and vice versa. It is important to realize that the input data must remain stable during the time that the clock signal is high because NAND gate 1 and NAND gate 2 are both enabled during this time.

Flip-Flop Timing

As we can see from our previous discussion, timing can be a very important consideration when dealing with flip-flops. Before using flip-flops, a technician must investigate a number of practical timing characteristics, which can be obtained from the manufacturer's specifications or data sheet (Figure 4.15, p. 118). The most important of these are setup time, hold time, and propagation delay.

The **setup time** is the minimum time that the input data must be stable *before* the clock transition occurs. This is referred to as t_s on the manufacturer's data sheet. The setup time of a TTL flip-flop is typically 20 ns. It is typically 0 ns for a TTL master-slave flip-flop.

The **hold time** is the minimum time that the input must remain stable *after* the clock transition occurs. The hold time requirement for a TTL flip-flop is typically very short (5 ns). It is referred to as t_H on the data sheet.

The **propagation delay time** refers to the time required for the *output* of a flip-flop to actually change state. The propagation delay time from a low to a high transition may be different than the propagation delay from a high to a low. Typical TTL flip-flop propagation delays are 15 ns for a low to a high change (t_{PLH}) and 25 ns for a high to a low change (t_{PHL}).

It is important to note that setup time, hold time, and propagation delay time are all measured using the midpoint, or 50% point, of the pulse edge as a reference.

The propagation delay of a device can be used to demonstrate how edge triggering is accomplished in a flip-flop. The propagation delay of inverter 2 in Figure 4.16 (p. 119) causes a sharp spike in the output of the NAND gate because the inverted pulse arrives at the input to the NAND gate a few nanoseconds after the actual pulse. This circuit demonstrates a method of accomplishing negative edge triggering as shown by the timing diagram of Figure 4.16. Deleting inverter 1 results in a positive edge-triggering circuit.

Asynchronous Inputs

Asynchronous means not synchronous or not clocked. Asynchronous inputs are inputs that act directly on a flip-flop regardless of the state of the clock and *JK* inputs (synchronous inputs). The asynchronous inputs of a flip-flop are labeled **preset** (PS) and **clear** (CL). The preset input *sets* the flip-flop, that is, the *Q* output is set to a 1 and the $\overline{Q}$ becomes a 0. The clear input *resets* the flip-flop and causes the *Q* output to become a 0 and

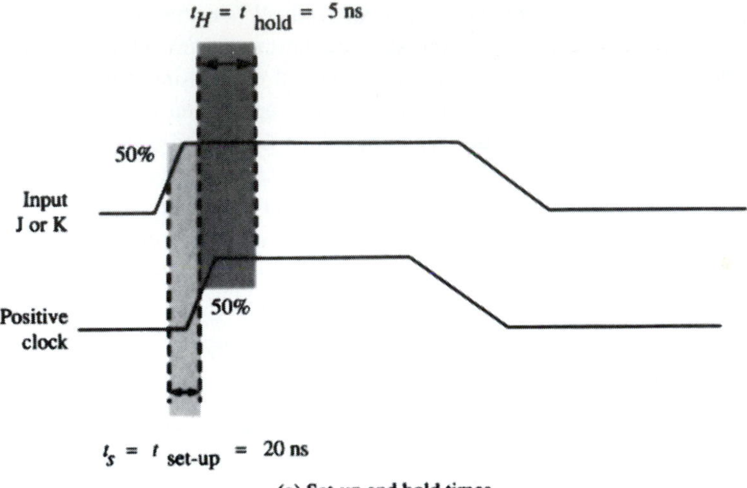

(a) Set-up and hold times

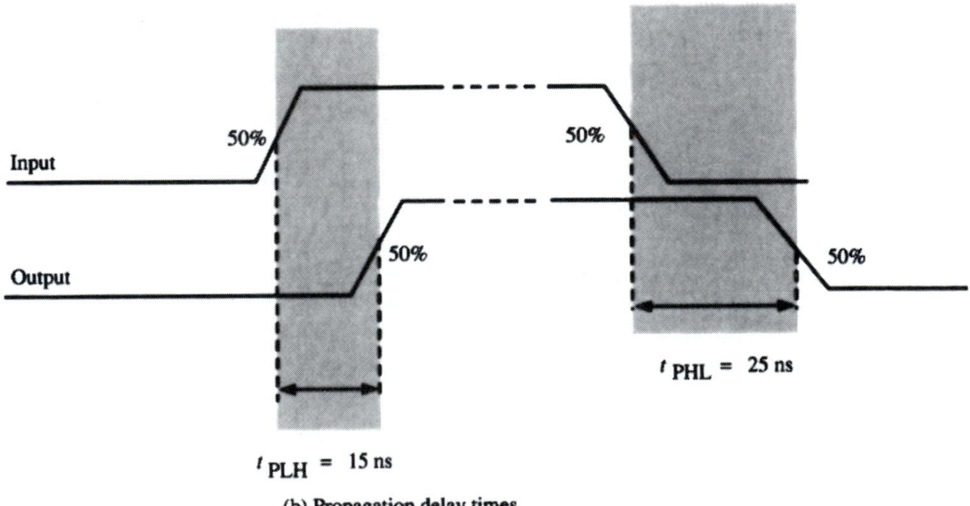

(b) Propagation delay times

FIGURE 4.15
Flip-flop timing

the $\overline{Q}$ output to be a 1. Both asynchronous inputs are active low and therefore are usually referred to as $\overline{PS}$ and $\overline{CL}$. When both $\overline{PS}$ and $\overline{CL}$ are high, the flip-flop is ready for synchronous, or clocked, operation. This is referred to as the **armed** condition. The conditions $\overline{PS} = 0$ and $\overline{CL} = 0$ should be avoided. This condition would force Q and $\overline{Q}$ to be a logic 1 at the same time. This is an illegal condition, since in Boolean algebra Q and $\overline{Q}$ must always be in the opposite state. Furthermore, since it is impossible for $\overline{PS}$ and $\overline{CL}$ to go high (release) at exactly the same time, it is not possible to predict reliably the final new output state.

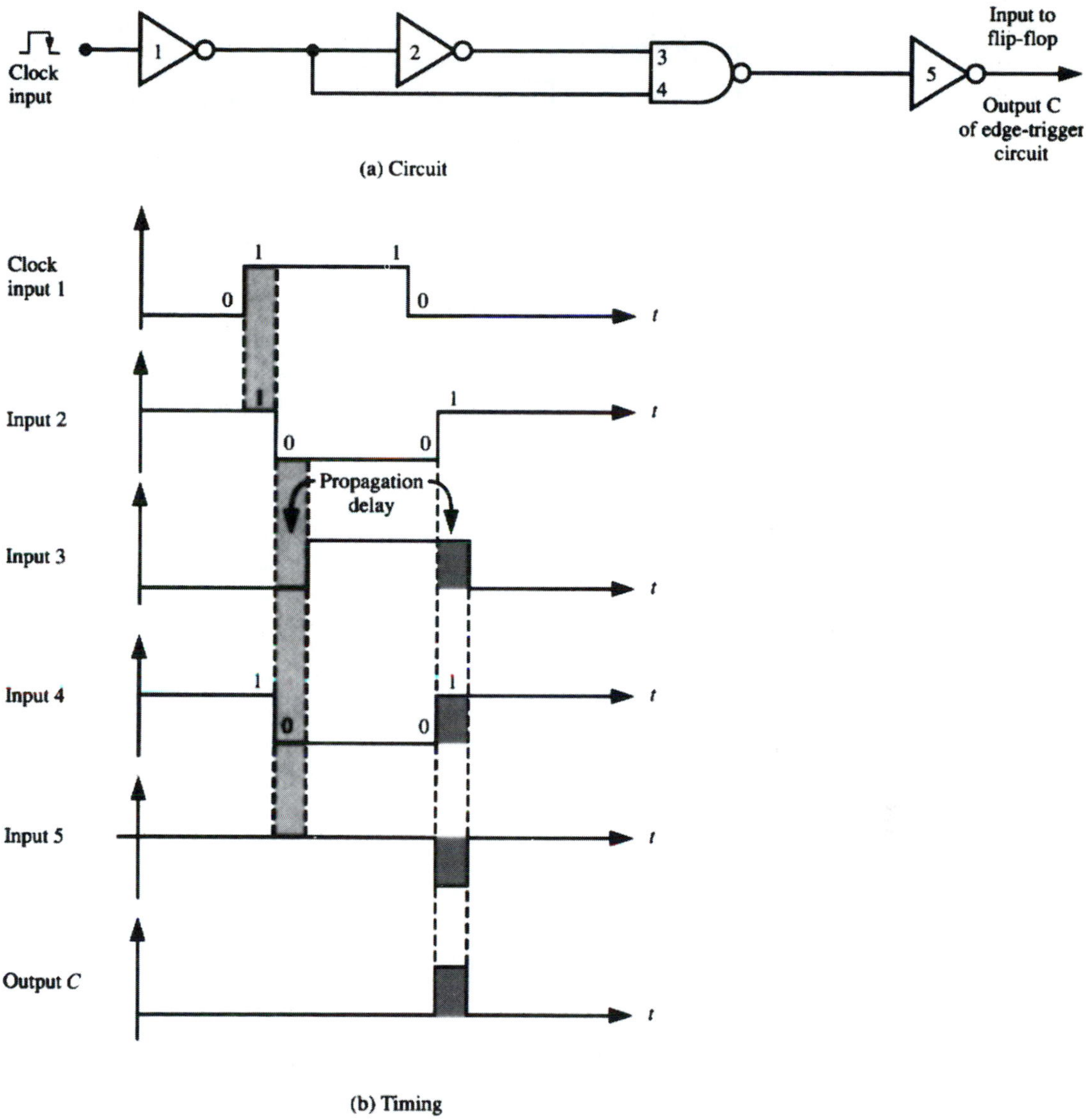

(a) Circuit

(b) Timing

FIGURE 4.16
Edge-triggering circuit

It is important to note that the asynchronous inputs *always* take precedence over the synchronous. Figure 4.17 shows the symbol for a flip-flop with asynchronous inputs, the truth table for these asynchronous inputs, and their functional operation.

Asynchronous inputs are often used to initialize flip-flops after the power to a circuit is first turned on. Asynchronous inputs are available on all types of flip-flops. Table 4.2 summarizes a number of TTL and CMOS *JK* flip-flops.

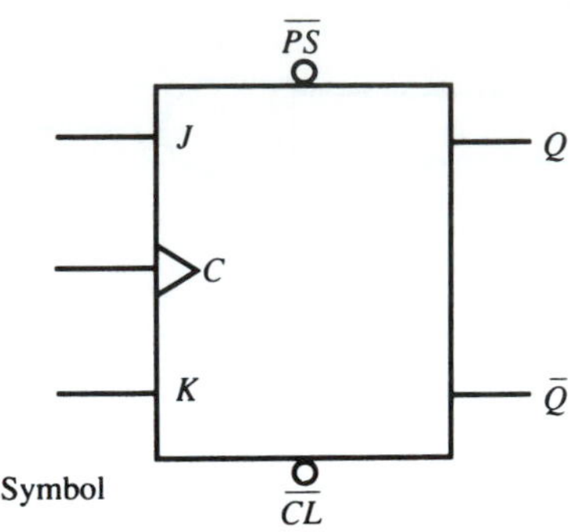

(a) J - K flip-flop with asynchronous preset and clear inputs

Asynchronous inputs

	$\overline{PS}$	$\overline{CL}$	Q	$\bar{Q}$
	0	0	—	—
(Set)	0	1	1	0
(Reset)	1	0	0	1
(Armed)	1	1	NC	NC

NC = no change
Truth table

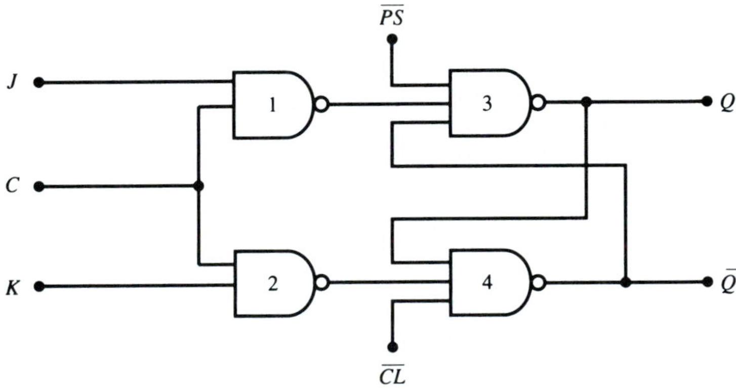

(b) J - K flip-flop with $\overline{PS}$ and $\overline{CL}$

FIGURE 4.17
Asynchronous inputs

TABLE 4.2
JK flip-flops

TTL	CMOS	Description
7473	74C73	*JK* master-slave with clear
74107	74C107	*JK* with clear
7476	74C76	*JK* master-slave with preset and clear
7478	74C78	*JK* with preset, common clock, and common clear
74109	74C109	*JK* with preset and clear
74112	74C112	*JK* with preset and clear
74113	74C113	*JK* with preset

4.4 COUNTERS

Flip-flops can be used as the basic building blocks for digital counters, which are circuits that can be used to count a number of input pulses. These pulses can be from an input clock frequency or any other digital timing event. Counter circuits are possible because flip-flops have an inherent memory capability. There are two basic types of counter circuits: *asynchronous* and *synchronous.*

Asynchronous Counters

An **asynchronous counter** is a counter in which all the flip-flops do not change state at the same time. The output of one flip-flop is connected to and used as the input to the next flip-flop, as shown in Figure 4.18. For this reason, asynchronous counters are also known as **ripple counters.**

The flip-flops shown in Figure 4.18 are negative edge-triggered *JK* flip-flops. Both the *J* and *K* inputs are connected to a logic level 1. Recall from the truth table of a *JK* flip-flop that whenever both inputs *J* and *K* are at a logic level 1, the flip-flop will toggle. That is, the flip-flop will change its state from whatever state it was in to its opposite state.

Let's assume that all three flip-flops are initially reset. This means that the Q outputs are all at a logic level 0. The input clock waveform is being fed into the first *JK* flip-flop (FF_A). Triggering occurs only on the negative-going edge, that is, only on a 1 to 0, or high to a low, transition, as indicated in Figure 4.18. On the negative-going edge of the first clock pulse (time 1), and Q_A output goes from a logic 0 to a logic 1. Q_B and Q_C do not change state. On the second clock pulse (time 2), the Q_A output toggles again. This time it goes from a logic 1 to a logic 0. This is a negative-going edge, or trigger, to flip-flop B (FF_B.) Thus the Q_B output toggles from a logic 0 to a logic 1. Note that Q_C does not change state. On the third clock pulse Q_A toggles from a logic 0 to a logic 1. Since this is a positive-going transition, it has no effect on Q_B or Q_C. On the fourth clock pulse Q_A toggles from a logic 1 to a logic 0. This negative transition causes Q_B to toggle from a logic 1 to a logic 0. The negative transition of Q_B in turn causes Q_C to toggle from a logic 0 to a logic 1. The count process continues as illustrated in Figure 4.18(b) until all three outputs, Q_A, Q_B, and Q_C, are at a logic 1. The next clock pulse (time 8) resets the counter by causing Q_A, Q_B, and Q_C to toggle from a logic 1 to a logic 0. Figure 4.18(c) summarizes the output conditions of the flip-flops for each clock pulse. This table also illustrates the counting ability of the circuit.

The number of counts obtained before the circuit resets is known as the **modulus (MOD)** of the counter. The number of counts can be calculated using equation (4.1)

$$\text{MOD \#} = 2^N \tag{4.1}$$

where N is the number of flip-flop stages. The highest number that a counter circuit can count to is obtained using equation (4.2).

$$\text{Maximum count} = 2^N - 1 \tag{4.2}$$

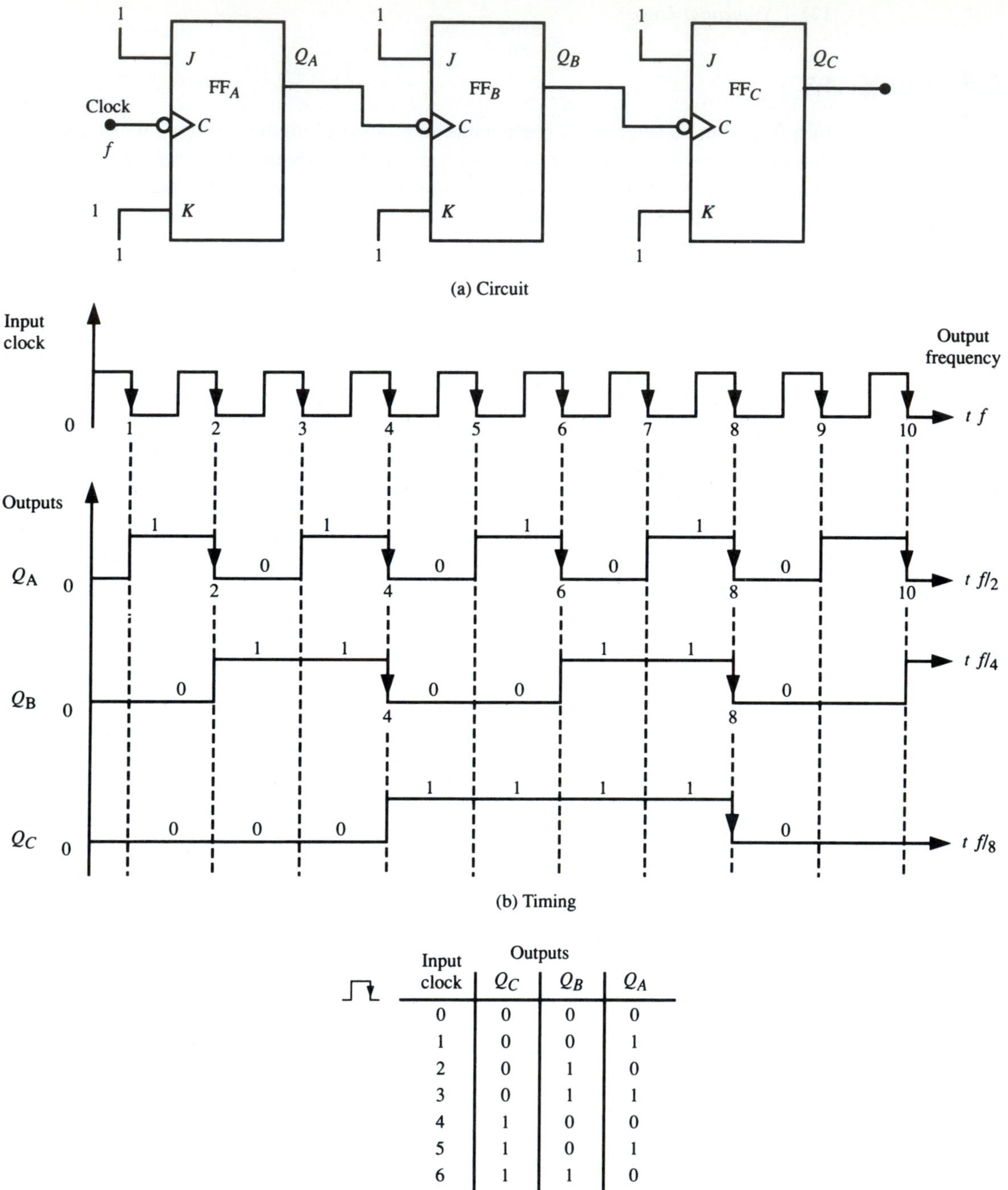

Input clock	Q_C	Q_B	Q_A
0	0	0	0
1	0	0	1
2	0	1	0
3	0	1	1
4	1	0	0
5	1	0	1
6	1	1	0
7	1	1	1
8	0	0	0

(c) Truth table

FIGURE 4.18
Asynchronous up counter

Note that the output frequency of flip-flop A in Figure 4.18 is exactly half of the input clock frequency. The output frequency of flip-flop B is exactly half of its input frequency from Q_A, or one-fourth of the input clock frequency. The output frequency of flip-flop C is exactly one-half of its input frequency from Q_B, or one-eighth of the input clock frequency. Therefore, counters can be used as **frequency dividers.** Equation (4.3) expresses the relationship between the input and output frequency of a counter.

$$f_{out} = \frac{f_{in}}{2^N} \tag{4.3}$$

$$= \frac{f_{in}}{\text{MOD \#}}$$

EXAMPLE 4.4

For the circuit of Figure 4.18(a) determine the modulus number, the highest count, and the output frequency Q_C if the input clock rate is 8 megahertz.

Solution

a.

$$\begin{aligned} \text{MOD \#} &= 2^N \\ &= 2^3 \\ &= 8 \end{aligned}$$

This is a MOD 8 counter.

b.

$$\begin{aligned} \text{Maximum count} &= 2^N - 1 \\ &= 2^3 - 1 \\ &= 8 - 1 \\ &= 7 \end{aligned}$$

The maximum counter is 7.

c.

$$\begin{aligned} f_{out} &= \frac{f_{in}}{\text{MOD \#}} \\ &= \frac{8 \times 10^6\text{Hz}}{8} \\ &= 1 \times 10^6\text{Hz} \\ &= 1\text{MHz} \end{aligned}$$

The output frequency at $Q_C = 1\text{MHz}$.

Asynchronous Down Counters

Ripple counters can be made to count down by using the $\overline{Q}$ output as the trigger for the following flip-flops. Figure 4.19 shows the circuit configuration for an asynchronous

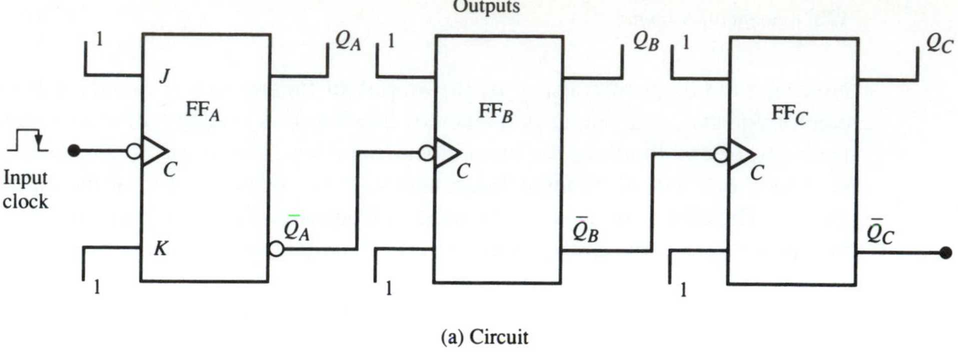

(a) Circuit

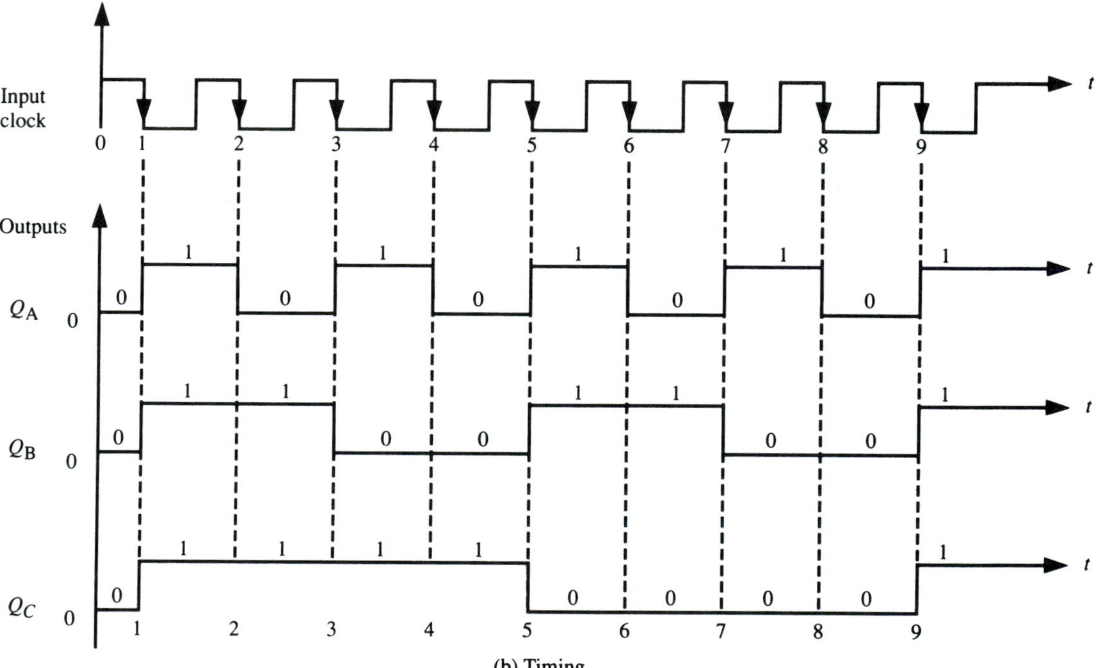

(b) Timing

Clock	Outputs Q_C	Q_B	Q_A	Count
0	0	0	0	0
1	1	1	1	7
2	1	1	0	6
3	1	0	1	5
4	1	0	0	4
5	0	1	1	3
6	0	1	0	2
7	0	0	1	1
8	0	0	0	0

(c) Truth table

FIGURE 4.19
Asynchronous down counters

down counter. Note that the output is still taken from the Q_A, Q_B, and Q_C outputs of the flip-flops, and the triggering is accomplished using $\overline{Q}_A$, $\overline{Q}_B$, and $\overline{Q}_C$.

If we begin with all three flip-flops reset, then $Q_A = 0$, $Q_B = 0$, Q_C, $\overline{Q}_A = 1$, $\overline{Q}_B = 1$, and $\overline{Q}_C = 1$. On the negative-going edge of the first clock pulse, Q_A goes to a logic 1; therefore, $\overline{Q}_A$ goes from a logic 1 to a logic 0. This causes Q_B to toggle, which in turn causes Q_C to toggle. The resulting count is then $Q_A = 1$, $Q_B = 1$, and $Q_C = 1$. The truth table and timing diagram of Figure 4.19 illustrate the entire counting sequence. It should be noted that the inverting bubble on the output of the flip-flop and the inverting bubble on the clock input can be considered to cancel each other out. Therefore, the triggering action seen on the Q_B and Q_C outputs appears to be occurring on the positive-going edge of the input clock cycle.

Synchronous Counters

A **synchronous counter** is a counter in which all the flip-flops are clocked at the same time. Unlike asynchronous ripple counters, where the output of one flip-flop is used to clock the next flip-flop, synchronous counters trigger all the flip-flops at the same time.

Figure 4.20 illustrates the operation of a simple 2-bit synchronous counter. The first flip-flop (FF_A) is set up to toggle on each negative-going edge of the input clock pulses by connecting both the J and K inputs to a logic 1. The J and K inputs of FF_B are tied together and connected to the output of Q_A. Recall that when J and K are both logic 0, the flip-flop does not change state. When J and K are both logic 1, the flip-flop toggles.

Let's begin our analysis by assuming that both flip-flops are initially reset. On the negative-going edge of the first input clock pulse, FF_A toggles to a logic 1. At the exact time of triggering, the J and K input of FF_B is at a logic 0; therefore, the Q_B output remains a logic 0. On the second clock pulse, FF_A toggles to a logic 0. At this exact time of triggering the J and K input of FF_B is at a logic 1; therefore, the Q_B output toggles to a logic 1. On the third clock pulse, FF_A toggles to a logic 1. At the exact time of triggering the J and K input of FF_B is at a logic 0; therefore, the Q_B output does not change and remains at a logic 1. On the fourth clock pulse, FF_A again toggles to a logic 0. At the exact time of triggering, the J and K input of FF_B is at a logic 1; therefore, it toggles to a logic 0. It should be noted that the synchronous counter takes advantage of the propagation delay (t_p) of the flip-flop, as illustrated in the timing diagram of Figure 4.20(b).

The primary advantage of the synchronous counter is speed, since all the flip-flops are triggered at the same time. There are, however, some practical considerations. Each individual flip-flop should be matched in speed. Using a high-speed flip-flop for FF_A and a standard flip-flop for FF_B could be disastrous. Synchronous counters are also more difficult to design than asynchronous counters. Both of these practical considerations are easily overcome through the use of MSI counters.

MSI Counters

Since many digital circuits require the use of counters, manufacturers have prepackaged counters into ICs. These devices tend to be more complex and require more internal components; they are therefore classified as **medium-scale integrated circuits (MSI).**

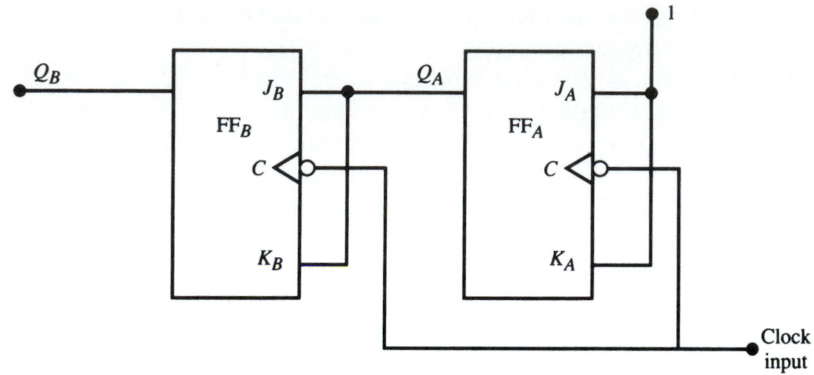

(a) Synchronous binary counter

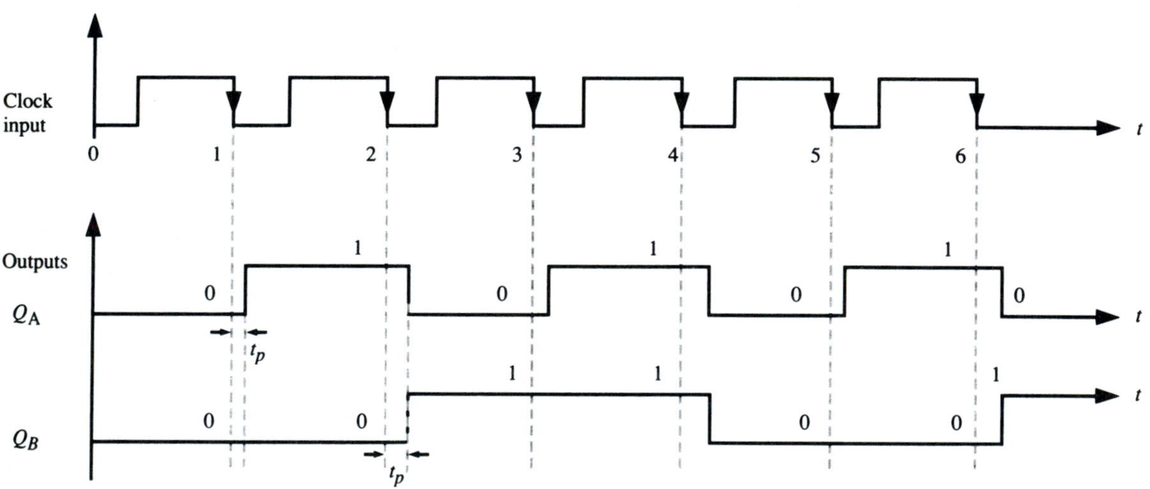

(b) Timing diagrams

After clock	Q_B	Q_A
0	0	0
1	0	1
2	1	0
3	1	1
4	0	0
5	0	1
6	1	0

(c) Truth (count) table

FIGURE 4.20
Synchronous counter

7493A MSI Counter

The 7493A is a 4-bit asynchronous binary counter. It consists of a single flip-flop and a group of three flip-flops connected together as an asynchronous counter, as shown in Figure 4.21. This is done for flexibility. The group of three flip-flops can be used as a 3-bit (MOD 8) asynchronous counter. The single flip-flop can be used as a MOD 2 counter or a divide-by-2 frequency divider. If the Q_A output of the single flip-flop is connected to the clock B (CKB) input of the three-flip-flop group, the device will be configured as a MOD 16 asynchronous counter. Q_A will be the LSB and Q_D the MSB. The reset input R_0 (1) and R_0 (2) can be used to clear or reset the counter upon initialization. They can also be used to change the MOD number of the counter. For example, suppose we wish to use the 7493A as a decade counter. By examination of the truth table or count table we note that the binary number for a counter of 10 is $(1010)_2$. Further examination and analysis reveals

Q_D	Q_C	Q_B	Q_A
1	0	1	0

Since this output sequence occurs only on the tenth counter, we can connect the Q_D output to the R_0 (1) input and the Q_B output to the R_0 (2) input to form a MOD 10 counter. The operation of the device is such that on the tenth input pulse Q_D and Q_B are both logic 1s. This causes the reset NAND gate to go low and clear all four flip-flops. The counter is then reset and never gets to a count above 10. Many other count sequences are possible through the use of connections between the two reset inputs R_0 (1) and R_0 (2) and the outputs Q_A, Q_B, Q_C, and Q_D.

74193 MSI Up/Down Counter

Figure 4.22 (p. 129) shows the pin configuration and organization for a 74193 up/down counter. This device can provide 4-bit synchronous counting operation in either direction. If the input clock pulses are applied to pin 5, the device will count up. If the input clock pulses are applied to pin 4, the device will count down. Furthermore, this device is *presetable.* That is, it can be initialized to any count by applying a binary number to the inputs *A*, *B*, *C*, and *D* and applying a logic 0 pulse to the $\overline{\text{LOAD}}$ pin. The $\overline{\text{BO}}$ and $\overline{\text{CO}}$ outputs provide a borrow-out and carry-out pulse for cascading the operation to other counters.

MOD Counter Design

As we noted with MSI counters, MOD counters can be made to count to any value. Recall that the maximum count of a MOD counter is equal to 2^N, where N represents the number of flip-flops. To change the MOD number of a counter, a method known as **decoding** can be used.

For example, suppose we wish to design a MOD 5 counter (0–4 counter). Realizing that two flip-flops can achieve a maximum count of only four, we implement a three-flip-flop asynchronous counting circuit. With three flip-flops the maximum count will be

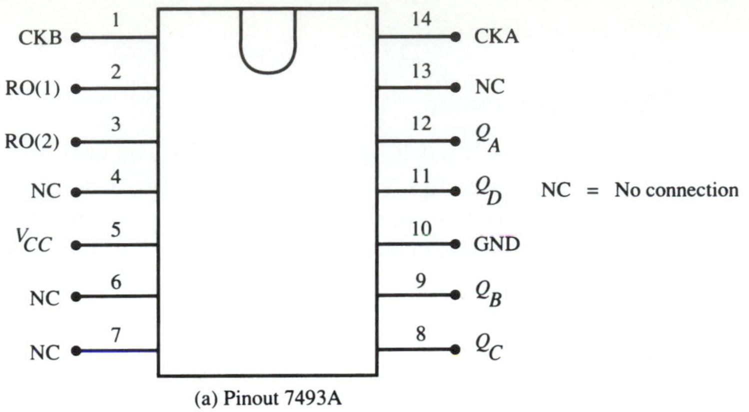

(a) Pinout 7493A

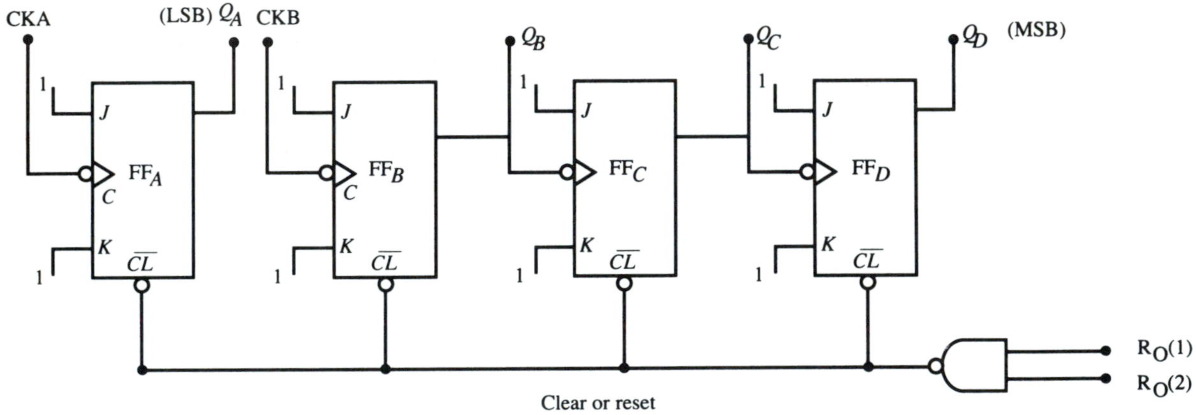

(b) Logic circuit for 7493A

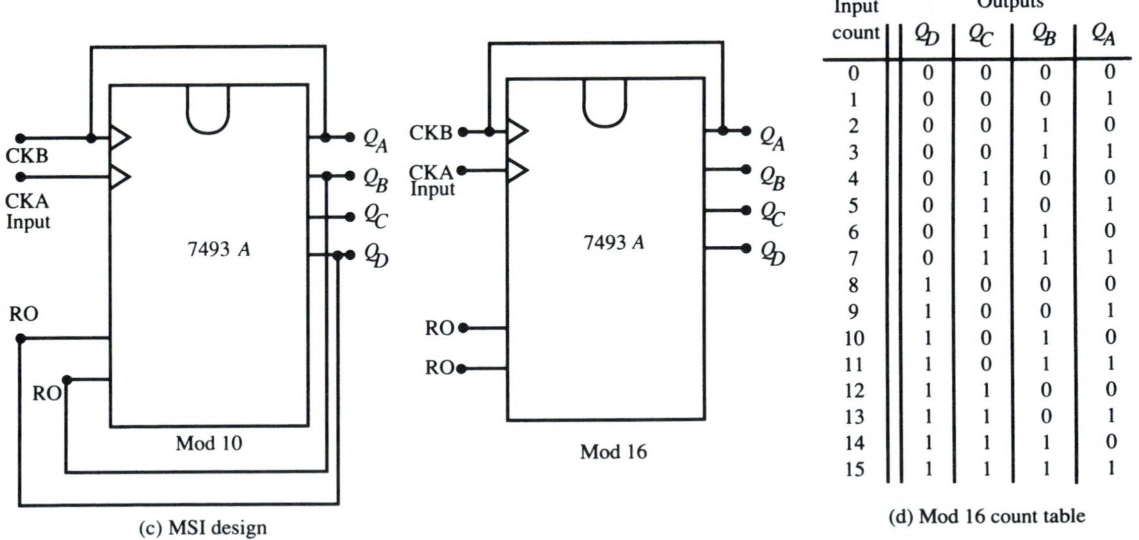

(c) MSI design

Input count	Outputs Q_D	Q_C	Q_B	Q_A
0	0	0	0	0
1	0	0	0	1
2	0	0	1	0
3	0	0	1	1
4	0	1	0	0
5	0	1	0	1
6	0	1	1	0
7	0	1	1	1
8	1	0	0	0
9	1	0	0	1
10	1	0	1	0
11	1	0	1	1
12	1	1	0	0
13	1	1	0	1
14	1	1	1	0
15	1	1	1	1

(d) Mod 16 count table

FIGURE 4.21
7493 MSI counter

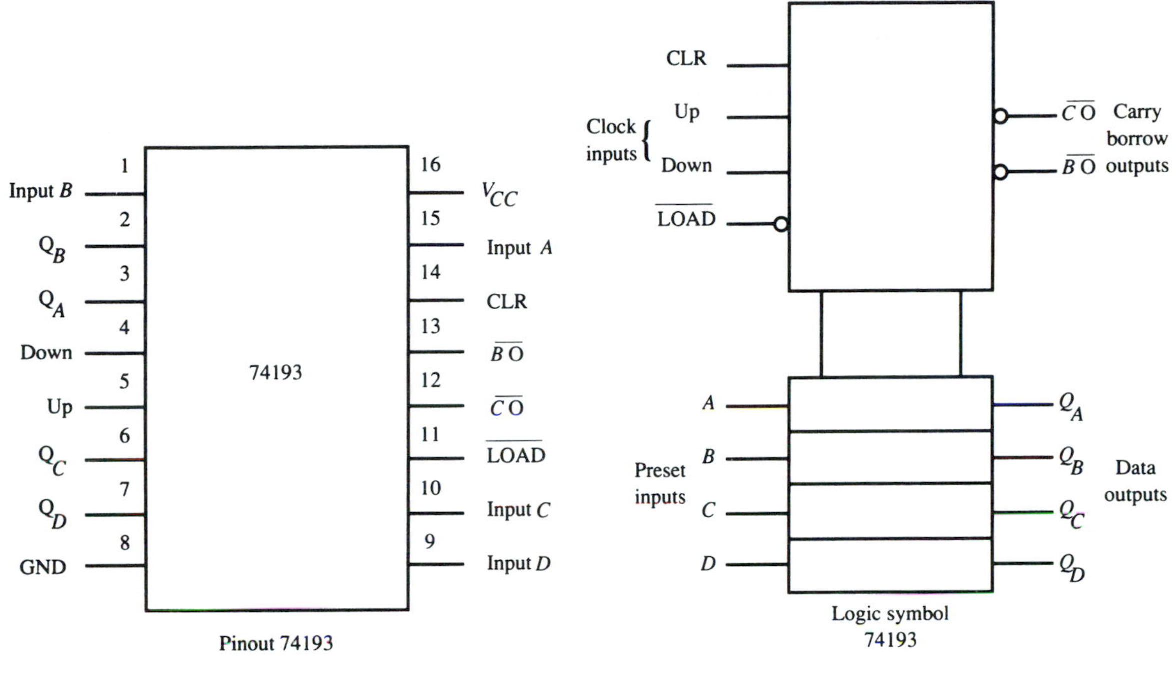

FIGURE 4.22
74193 MSI up/down counter

2^3, or 8. A simple method of making a MOD 5 counter is shown in Figure 4.23. The binary number for 5 is 101. By decoding the binary number 101 as $Q_C\overline{Q}_BQ_A$ and connecting these outputs to a NAND gate, we can control the maximum count to 5. It should be noted that the $\overline{Q}_B$ connection is not needed; however, it systematizes the design process and therefore makes it easier. On the fifth count all three inputs to the NAND gate are at logic 1. The output of the NAND gate goes low and resets all three flip-flops.

The one disadvantage of this design is that we must hit the fifth count in order to reset the counter. Therefore, for a brief moment, the counter goes to a count of 101 before being reset. This problem is illustrated in the timing diagram of Figure 4.23 by the spike in the Q_A output at the fifth count. This spike is sometimes referred to as the **infamous decode spike.** Decode spikes are too short in time to be seen on a display or LED; however, they can cause problems if the outputs are being used to control some other timing function. One solution to this problem is to use a synchronous counter. Synchronous counters trigger all the flip-flops at the same time and thereby eliminate decode spiking.

Recall that MOD counters can also be used as frequency dividers. The output frequency equals the input frequency divided by the MOD number. However, note that in the MOD 5 counter of Figure 4.23 the output frequency is not symmetrical. The Q_C output is low for four clock cycles and high for one clock cycle. Therefore, the duty cycle in this case is $\frac{1}{5} \times 100\%$, or 20%.

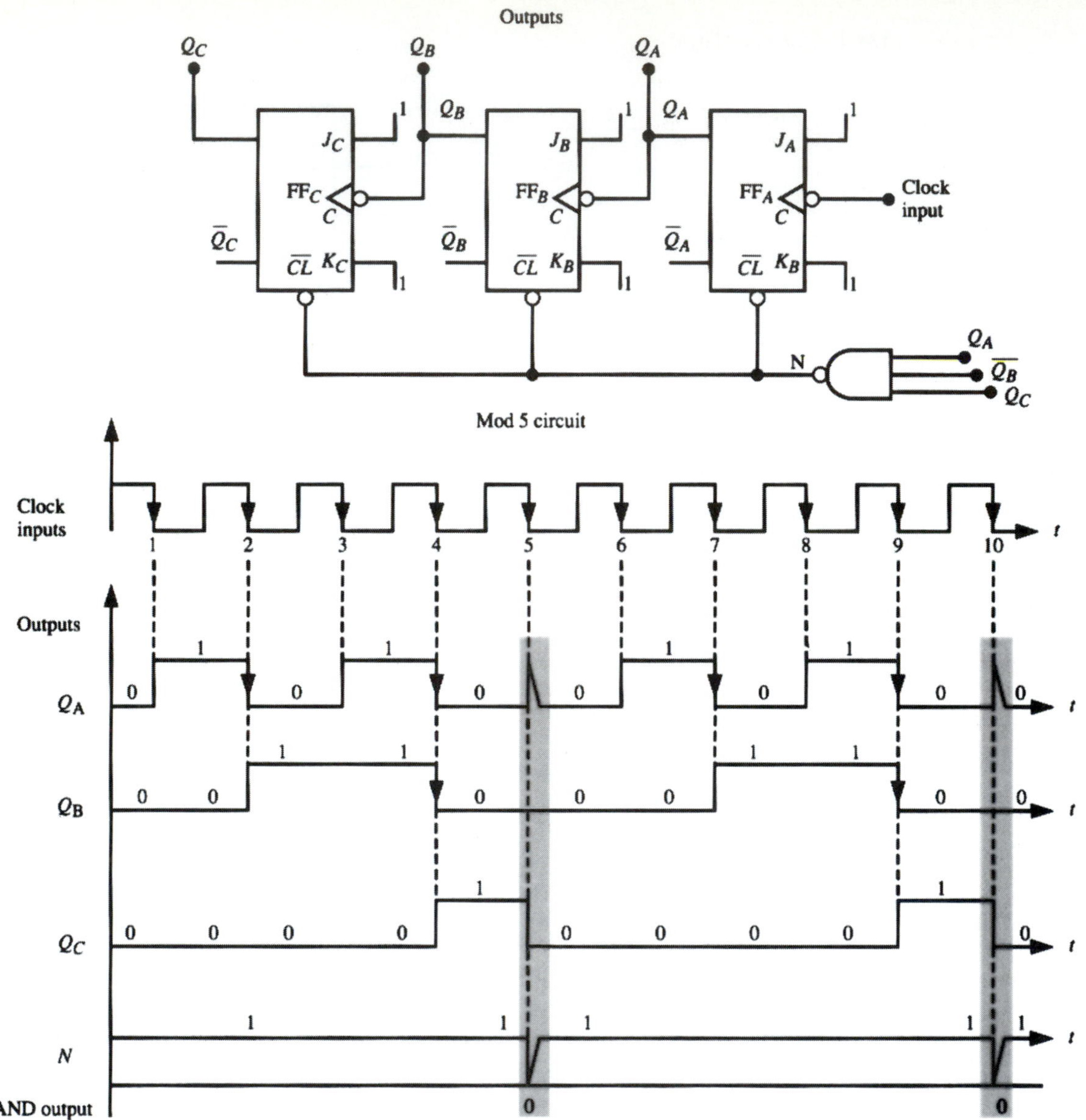

Timing diagram for mod 5 counter

Count Clock input	N	Q_C	Q_B	Q_A
0	High	0	0	0
1	High	0	0	1
2	High	0	1	0
3	High	0	1	1
4	High	1	0	0
5	Spike	0	0	0

FIGURE 4.23
MOD 5 counter

4.5 MONOSTABLE AND ASTABLE MULTIVIBRATORS

In the beginning of this chapter we stated that the flip-flop is also known as a bistable multivibrator, that is, it has two stable states. There are two other common types of multivibrator circuits; the **monostable,** or one-stable-state, multivibrator and the **astable** multivibrator, which has no stable states (i.e., an oscillator).

Monostable Multivibrators

The monostable multivibrator, or **one-shot** as it is often called, has only one stable state. The device produces an output pulse for each input pulse. The output pulse width can be controlled by the use of an *RC* network, as shown in Figure 4.24(a). Figure 4.24(b) illustrates the operation of a one-shot using NAND gates.

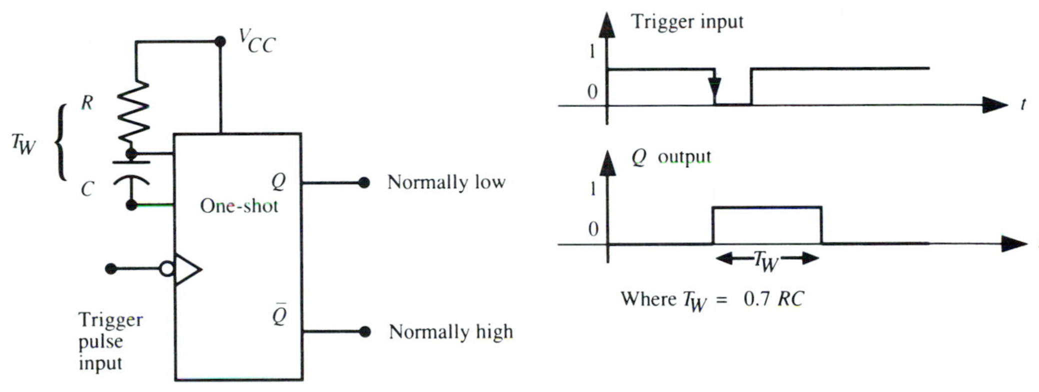

(a) Basic one-shot

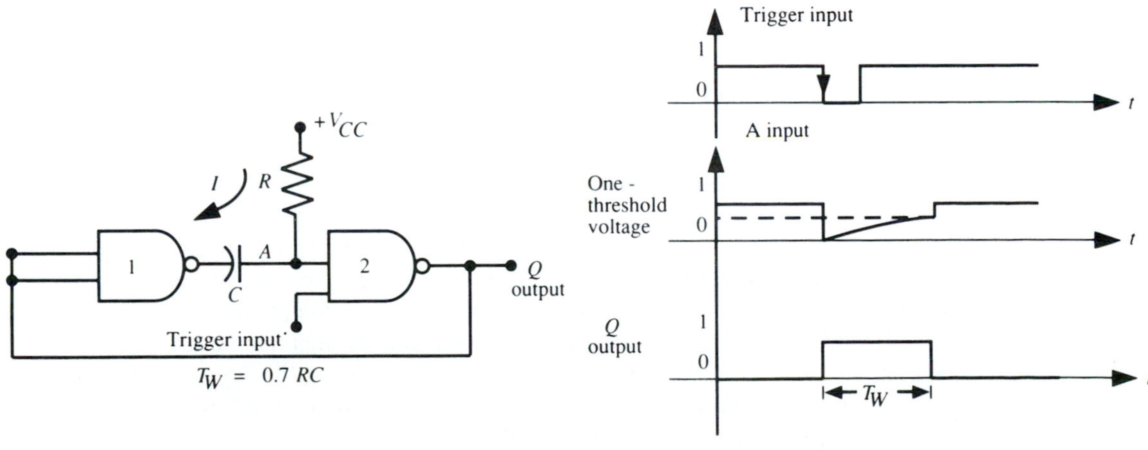

(b) Discrete one-shot

FIGURE 4.24
One-shot

In the stable state the Q output is low due to a normally high trigger input and V_{CC} (no current through R) on the inputs to NAND gate 2. When the trigger input goes low for a brief moment, the output of NAND gate 2 is forced high immediately. The output of NAND gate 1, which is functioning as an inverter, then goes low and creates a path for current to flow from V_{CC} through R and C to ground. The capacitor, C, initially looks like a short circuit and begins to charge. Thus the voltage at point A immediately drops to 0 and then begins to rise exponentially. When the voltage at point A reaches the logic 1 threshold voltage, the Q output returns to a logic 0. This assumes that the trigger input has previously returned to a logic 1. The pulse width of the output waveform can be calculated using the relationship

$$T_w = 0.7RC \tag{4.4}$$

because it takes less than one time constant to achieve a voltage level that equals the logic level 1 threshold voltage.

The 74121 is a typical TTL monostable multivibrator, or one-shot device. Figure 4.25(a) (p. 133) illustrates the functional operation and pin configuration for this device. Inputs A_1 and A_2 can be used to provide output pulses for negative-edge-going input trigger pulses. Input B is used for positive-edge-going trigger circuits. An external resistor and capacitor can be added to control the width of the output pulse.

The 74121 is known as a *nonretriggerable* one-shot. That is, if the output pulse intervals exceed the input pulse intervals, the additional input pulses will be ignored as shown in Figure 4.25(b).

A *retriggerable* one-shot will result in an extension of the output pulse for each input pulse. Such a device is illustrated in Figure 4.26(a) (p. 134). The 74122 retriggerable one-shot is similar to the 74121. This device has two negative-edge input triggers, A_1 and A_2, and two positive-edge input triggers, B_1 and B_2. It also has a clear line for added control. Figure 4.26(b) illustrates the effect of retriggering on this device. If retriggering occurs before the output pulse *times out,* the result is an extension of the output pulse.

Astable Multivibrators

The astable multivibrator has *no* stable states. For this reason it is often referred to as an **oscillator,** or **free-running multivibrator.** Figure 4.27(a) (p. 137) illustrates the operation of an astable multivibrator that is made up of NAND gates. The operation is similar to that of the monostable multivibrator of Figure 4.24(b) (p. 131) except that two capacitators are used, so both NAND gates can never enter the stable state. Astable multivibrators are often used to provide master timing, or *clock signals,* in digital circuits.

The 555 timer is an example of a very versatile IC that can be used as an astable multivibrator as well as for many other applications. Figure 4.27(c) (p. 137) illustrates the pin configuration and use of a 555 timer as an astable multivibrator. The output frequency is controlled by the external components R_1, R_2, and C. The output frequency can be calculated by

$$f = \frac{1.443}{(R_1 + 2R_2)C} \tag{4.5}$$

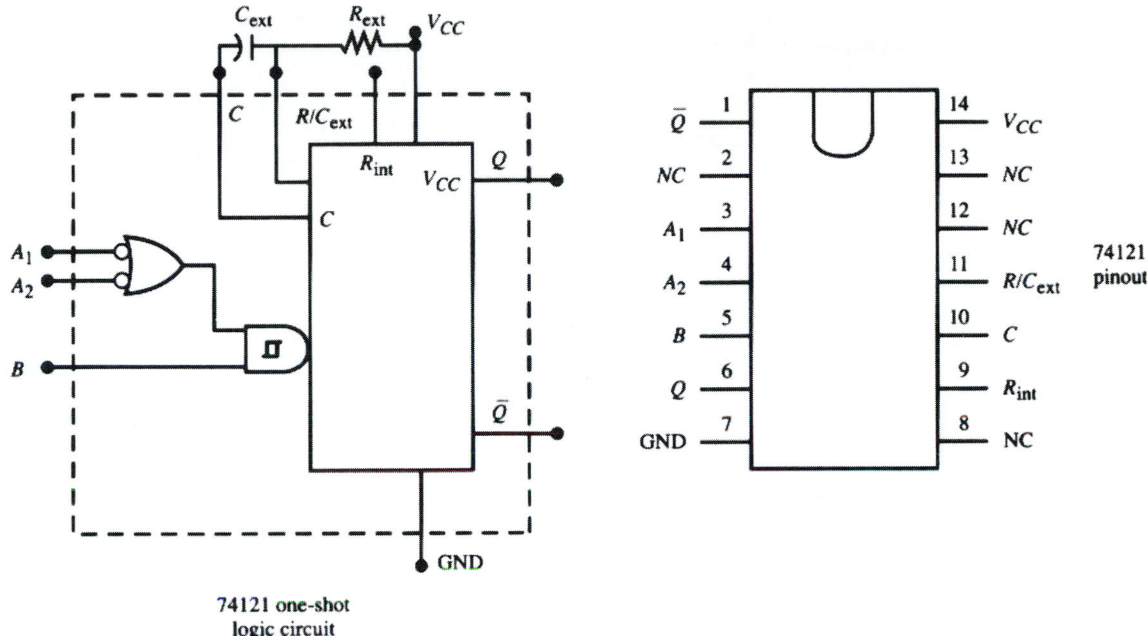

(a) 74121 non-retriggerable one-shot

Positive trigger input

t

Q output

T_W

Trigger input

Q output

T_W

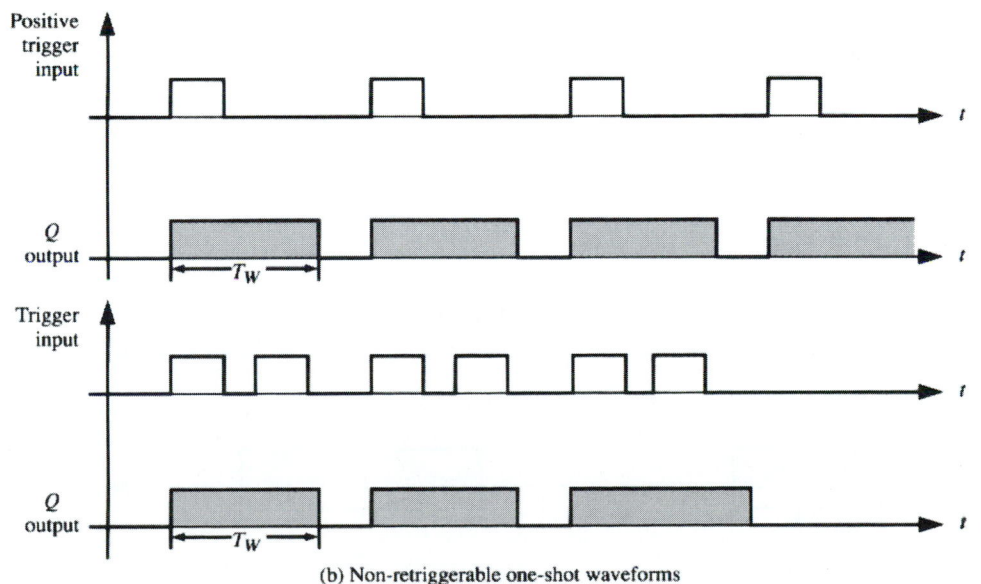

(b) Non-retriggerable one-shot waveforms

FIGURE 4.25
Nonretriggerable one-shot

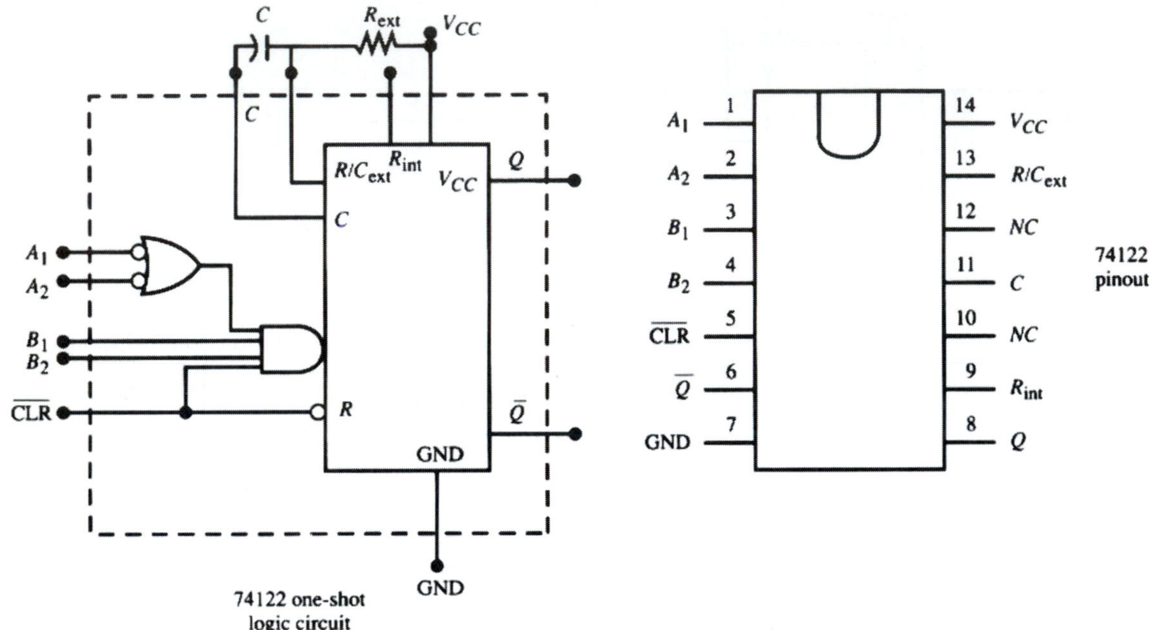

(a) 74122 retriggerable one-shot

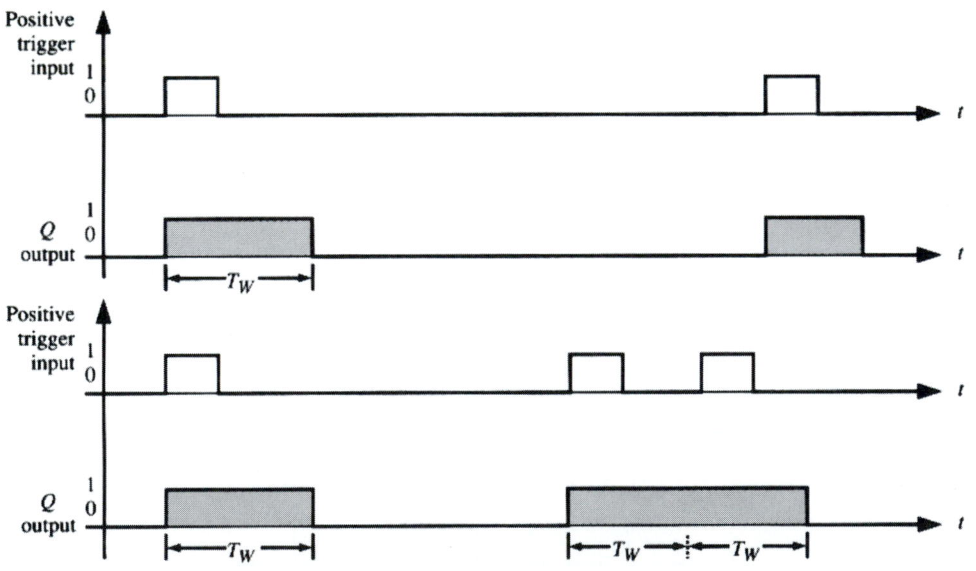

(b) Retriggerable one-shot waveforms

FIGURE 4.26
Retriggerable one-shot

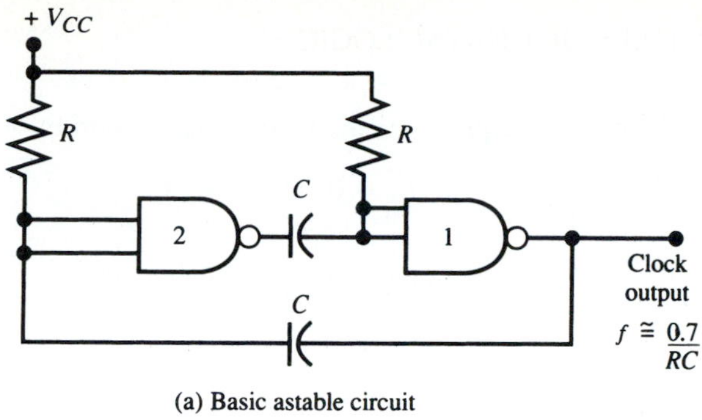

(a) Basic astable circuit

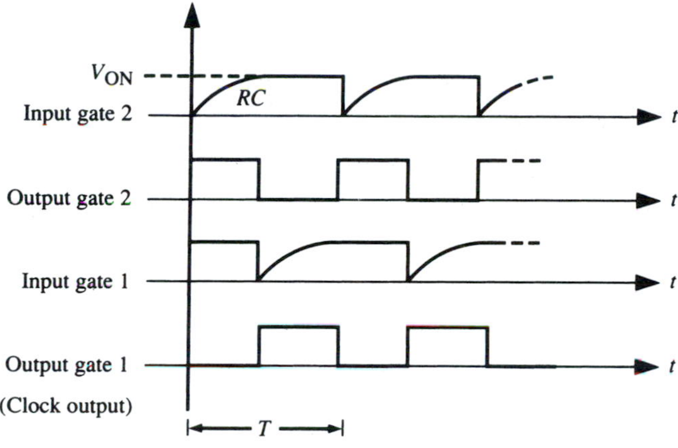

(b) Astable (clock) waveforms

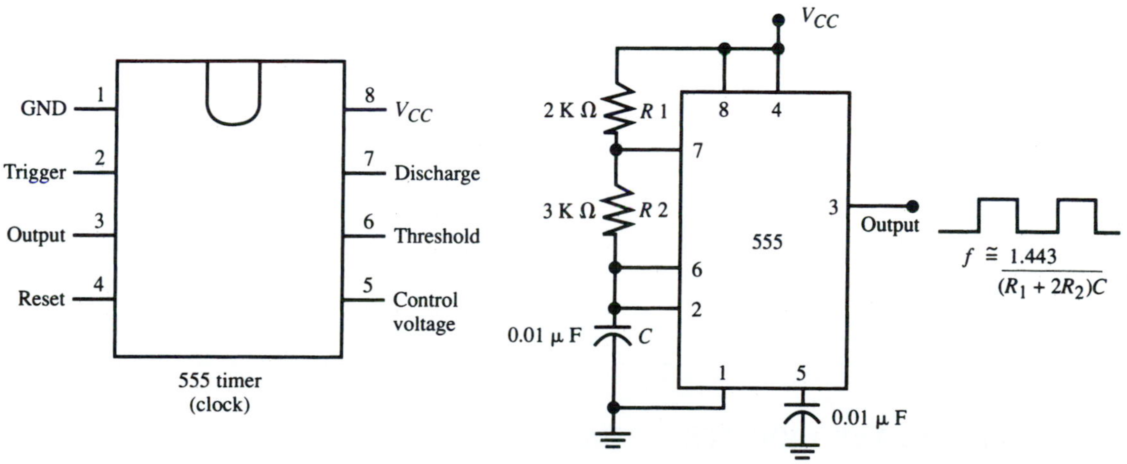

(c) 555 *IC* timer (astable) circuit

FIGURE 4.27
Astable multivibrator

The time that the output frequency is high can be calculated by

$$t_H = 0.7(R_1 + R_2)C \tag{4.6}$$

The time that the output frequency is low can be calculated by

$$t_L = 0.7R_2C \tag{4.7}$$

The output-frequency active-high duty cycle can be calculated by

$$\%\text{ duty cycle} = \frac{R_1 + R_2}{R_1 + 2R_2} \times 100\% \tag{4.8}$$

Thus when R_2 is very large with respect to R_1, the duty cycle approaches 50%. For practical considerations, the value of $R_1 + 2R_2$ should always be between 1 kΩ and 10 MΩ. The value of the capacitor, C, may be any practical value greater than 0.01 μF. The maximum practical output frequency is about 200 kHz.

EXAMPLE 4.5

Calculate the frequency and duty cycle for the 555 astable multivibrator circuit shown in Figure 4.27(b).

Solution

$$\begin{aligned} f &= \frac{1.443}{(R_1 + 2R_2)C} \\ &= \frac{1.443}{(2 \times 10^3 + 2 \times 3 \times 10^3)0.01 \times 10^{-6}} \\ &= 18 \times 10^3\,\text{Hz} \\ &= 18\,\text{KHz} \end{aligned}$$

$$\begin{aligned} \%\text{ duty cycle} &= \frac{R_1 + R_2}{R_1 + 2R_2} \times 100 \\ &= \frac{2 \times 10^3 + 3 \times 10^3}{2 \times 10^3 + 2 \times 3 \times 10^3} \times 100 \\ &= \frac{5 \times 10^3}{8 \times 10^3} \times 10 \\ &= 62.5\% \end{aligned}$$

4.6 TECH TIPS AND TROUBLESHOOTING—T[3]

The techniques used to troubleshoot sequential circuits are basically the same as the techniques used to troubleshoot combination logic circuits. The primary differences involve the increase in the number of functions that must be considered due to timing considerations. The circuit illustrated in Figure 4.28 is an asynchronous ripple counter. The Q output of each flip-flop is connected to a light-emitting diode (LED) to indicate the current

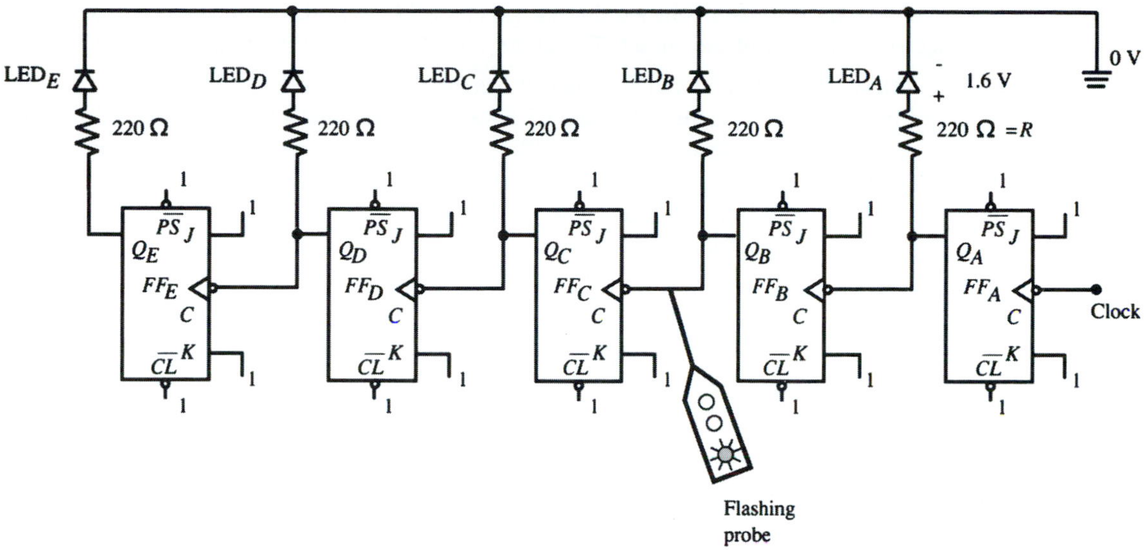

FIGURE 4.28
Tech tips and troubleshooting

binary count. The 220-Ω resistors are being used to limit the current through the LEDs. Assuming the LEDs are rated at 16 mA with a forward voltage drop of 1.6 V, we can calculate the value of R as follows:

$$R = \frac{V_{CC} - V_f}{I}$$
$$= \frac{5 - 1.6}{0.016}$$
$$= 212.5\Omega$$

Therefore, we will use a 220-Ω resistor.

In practice the logic 1 Q output voltage will not equal 5 V but is typically about 4 V due to the voltage drop across internal resistors in the IC. The output voltage may drop even lower if we are driving more than one IC because of increased current draw.

Let's assume for a moment that LED C, LED D, and LED E are not flashing. Investigation reveals that LED A is flashing at half the clock frequency, and LED B is flashing at half the rate of LED A. Therefore, we should conclude that FF_A and FF_B are functioning properly.

We begin our troubleshooting by logic probing the input clock at FF_C. If this point is pulsing, it indicates that the connection between FF_B and FF_C is not open or shorted. Probing the Q_C output of FF_C should indicate the same result as the LED C display. If this is the case, we know that the Q_C output and the input to the LED anode and resistor are all connected.

We have now established that the inputs to FF_C and the outputs from FF_C are connected to the rest of the circuit. Therefore, our next step will be to investigate FF_C itself.

1. Verify proper V_{CC} and GND connections and voltage levels.
2. Check that the PRESET ($\overline{PS}$) and CLEAR ($\overline{CL}$) are always high. If either of these signals ever goes low, the flip-flop will be set or reset, respectively, Note that transient spikes on these lines could cause intermittent errors.
3. Check that the J and K inputs are both high (the toggling mode). If they are not both high, the FF will not count properly.

If all these checks reveal no problems, we might suspect that FF_C is defective. Before changing the flip-flop, we should consider one other possible cause of the problem. The output wires or PC board connections from Q_C to FF_D could be shorted to ground. This can be checked by lifting or cutting the Q_C output pin of the flip-flop. Placing the logic probe on the connection with the pin removed should indicate a float (no light lit on logic probe). If the connection is shorted, a logic 0 will be indicated on the logic probe. If no shorts are found, replace the FF_C.

EXERCISES

4.1 Design an *RS* flip-flop using
a. NAND gates b. NOR gates

4.2 Using the input waveforms for *R* and *S* shown in Figure 4.29(a), draw the output waveforms for *Q* and $\overline{Q}$ produced by an *RS* flip-flop made from NOR gates.

4.3 Using the input waveforms for $\overline{R}$ and $\overline{S}$ of Figure 4.29(b), draw the output waveforms for *Q* and $\overline{Q}$ produced by an *RS* flip-flop made from NAND gates.

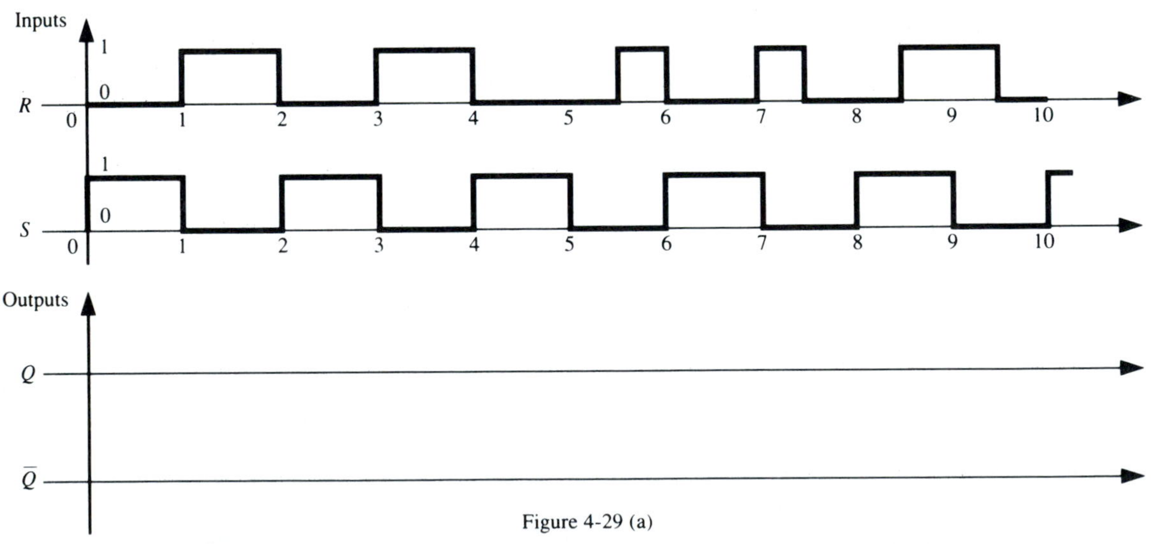

Figure 4-29 (a)

FIGURE 4.29

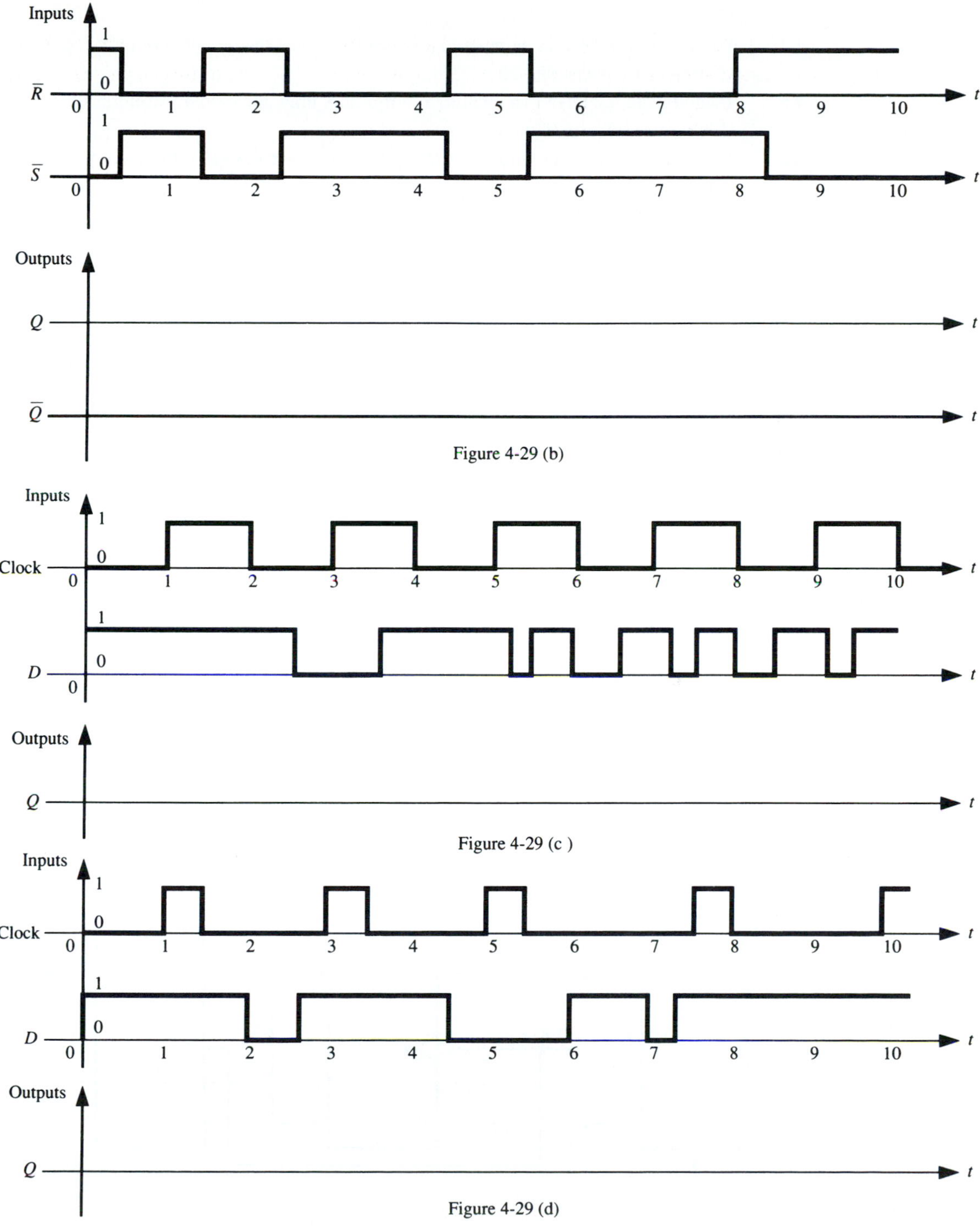

Figure 4-29 (b)

Figure 4-29 (c)

Figure 4-29 (d)

FIGURE 4.29 continued

4.4 Using Figure 4.29(c) as an input to a *D* flip-flop, draw the output waveform for *Q*. (Flip-flop is initially set.)

4.5 Using Figure 4.29(d) as an input to a *D* flip-flop, draw the output waveform for *Q*. (Flip-flop is initially reset.)

4.6 Draw the output waveform for the *D* flip-flop of Figure 4.30(a) if the clock input frequency is 10 kHz.

4.7 Draw the truth table for the *JK* flip-flop in Figure 4.30(b).

4.8 For Figure 4.30(b), if $J = 1$, $K = 1$, and the clock input is connected to a 10-kHz square wave with a 50% duty cycle, draw the output waveform.

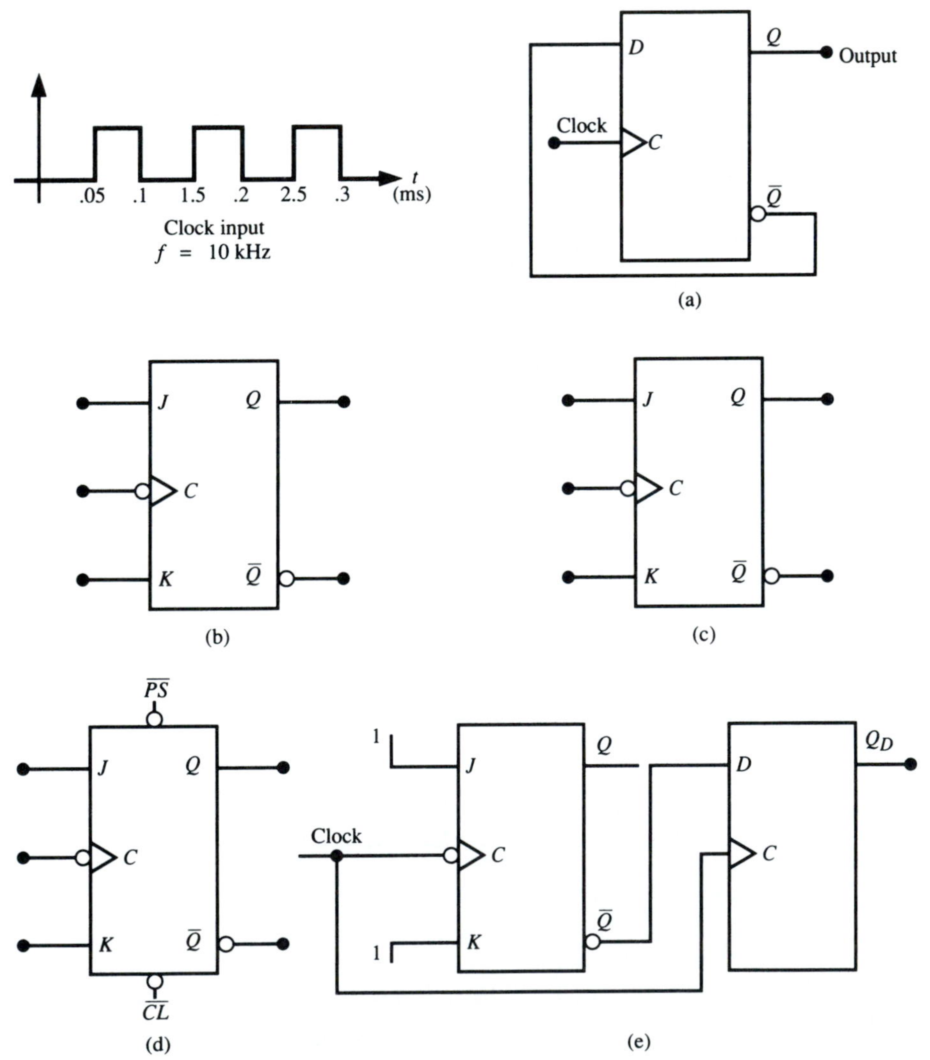

FIGURE 4.30

4.9 Draw the truth table for the *JK* flip-flop in Figure 4.30(c).

4.10 For Figure 4.30(c), if $J = 1$, $K = 1$, and the clock input is connected to a 10-kHz square wave with a 50% duty cycle, draw the output waveform.

4.11 Determine the output frequency for the *JK* flip-flop of Figure 4.30(c) when the input clock frequency is 10 kHz and $J = 1$ and $K = 1$.

4.12 Explain the operation of the $\overline{PS}$ and $\overline{CL}$ inputs of the *JK* flip-flop in Figure 4.30(d).

4.13 Draw the truth table for the *JK* flip-flop in Figure 4.30(d).

4.14 Draw the Q output waveform for the *JK* flip-flop in Figure 4.30(b) if the J, K, and clock inputs from Figure 4.31 are applied.

4.15 Draw the Q output waveform for the *JK* flip-flop of Figure 4.30(c) if the J, K, and clock inputs from Figure 4.31 are applied. (Flip-flop is initially reset.)

4.16 Draw the Q output waveform for the *JK* flip-flop in Figure 4.30(d) if the J, K, $\overline{PS}$, $\overline{CL}$, and clock inputs from Figure 4.31 are applied.

4.17 Draw the output waveforms for Q and Q_D outputs in Figure 4.30(e) if the clock input is a 10-kHz, 50% duty cycle waveform.

4.18 Draw the diagram of a *JK* master-slave flip-flop with a negative edge trigger.

4.19 Define the following terms for flip-flop devices.

a. Setup time
b. Propagation delay time
c. Asynchronous inputs
d. Race condition

4.20 Design a 4-bit ripple counter using negative-edge-triggered *JK* flip-flops. Summarize the output conditions for each flip-flop in a count table. Determine the MOD number and maximum count for the counter.

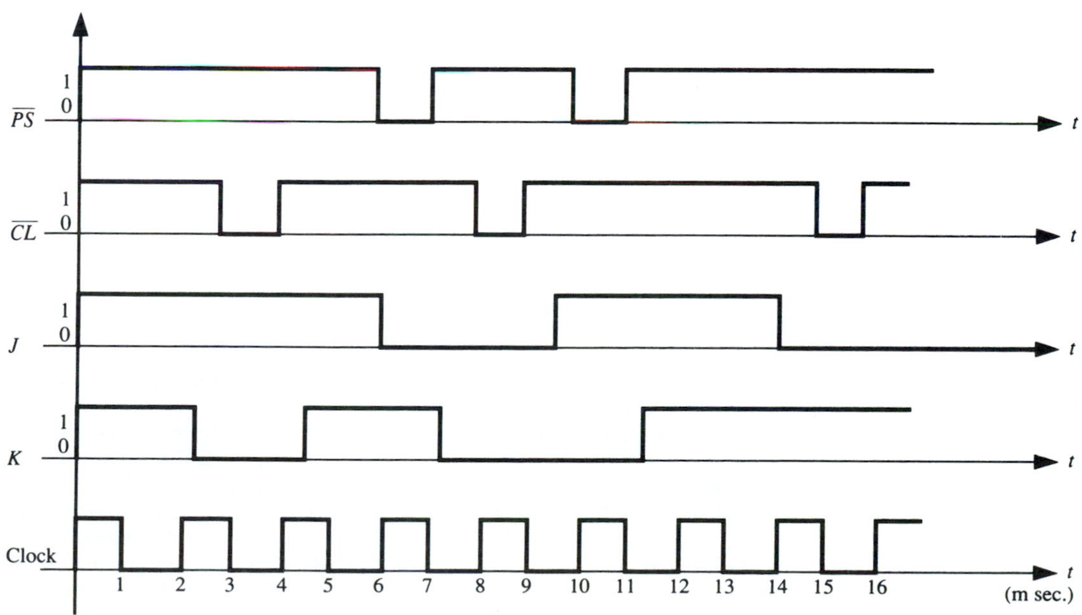

FIGURE 4.31

4.21 Determine the output frequency for an 8-bit ripple up counter if the output is taken from the last flip-flop, and the input frequency is 100 kHz.

4.22 Draw the diagram for a 3-bit down counter. Develop the count table for the circuit.

4.23 Draw the diagram for a 4-bit synchronous counter using *JK* flip-flops. Show the count table for the circuit.

4.24 What is the advantage of a synchronous counter over a ripple counter?

4.25 Draw the diagram of a MOD 8 counter using a TTL 7493A MSI device.

4.26 Draw the diagram of a 4-bit down counter using a TTL 74193 MSI device.

4.27 Explain the differences between the 74121 and 74122 one-shots.

4.28 Design and draw the circuit for a one-shot using 74121 and a 10-kΩ resistor to produce an output pulse with a period of $7\,\mu$s.

4.29 If a 1-MHz clock is applied to the input of a 74122 with $R_{\text{EXT}} = 7.5\,\text{k}\Omega$ and $C = 0.001\,\mu\text{F}$, determine the output waveform for the Q and $\overline{Q}$ outputs.

4.30 What are the output frequency and percent duty cycle for an astable circuit using a 555 timer with $R_1 = R_2 = 1\,\text{k}\Omega$ and $C = 0.01\,\mu\text{F}$? Draw the waveform for the output of this device.

4.31 Crossword Puzzle

ACROSS

1. Ripple counter with $\overline{Q}$ output connected to trigger of following flip-flop.
4. Edge triggering of *D* flip-flops.
7. Time required for a flip-flop to change state.
8. In ripple counters the output of one flip-flop is used as the _____ to the next flip-flop.
10. A condition when the clock and input signal to a flip-flop occur at the same time.
11. Edge-triggered input symbol.
13. The preset and clear inputs.
15. A use of the clear input after the power to a circuit is first turned on.
16. When $Q = 1$ in a flip-flop.
20. A type of flip-flop.
23. In an active low *RS* latch, if $\overline{S} = 0$ and $\overline{R} = 1$, Q will be a logic _____.
24. An *RS* flip-flop condition when $S = 1$ and $R = 1$.
25. Time that the input data must be stable before the clock transition occurs.
26. Controlling when a latch will be allowed to change state is referred to as enabling, or _____.
27. Counter in which all the flip-flops are clocked at the same time.

DOWN

1. Infamous counter spikes that are too short to be seen on a display or LED.
2. A base *RS* flip-flop can be made from these gates.
3. This type of *JK* flip-flop is actually two flip-flops configured in one device.
5. An MSI counter that can be initialized to any count is _____.
6. A method of writing all the input combinations and resulting outputs.
9. Circuits that can be used to record a number of input pulses.
11. To reset a flip-flop.
12. Type of circuit that can store or hold an output state.
14. A gate in which any 0 on an input will result in a 1 output.
17. When both inputs to a *JK* flip-flop are high.
18. The number of counts before a counter resets is known as the _____ number.
19. The _____ input sets the flip-flop regardless of the state of the *J* input.
21. Asynchronous counter.
22. The TTL 74279 IC contains _____ RS NAND gate latches.

1 2 3 4 5 6 7 8 9 10 11 12 13 14 15 16 17 18 20 21 22 23 24 25 26 27

CHAPTER 5

DATA CONTROL CIRCUITS

KEY TERMS

Accuracy
Analog-to-Digital Converter
Buffers
Conversion Time
Data Bus
Data Distributor
Data Register
Data Selector
Decoding
Demultiplexer
Digital-to-Analog Converters
Encoding
Hystersis Loop
Interface
Interface Circuit
Johnson Counter
Multiphase Timing
Multiplexer
Offset Voltage
Parallel In–Parallel Out
Registers
Resolution
Ring Counter
Schmitt Trigger
Serial In–Serial Out
Settling Time
Shift Registers
Subsystems
Timing Devices
Transceiver
Tristate
Universal Asynchronous Receiver Transmitter (UART)
Wave-shaping Device

5.0 INTRODUCTION

In the previous chapter we studied methods for implementing combination and sequential logic devices. In this chapter we investigate devices and circuits that are used to control the flow of data. **Registers** are used to hold data temporarily and then move the data. **Timing** and **wave-shaping devices** are used to provide clocking and triggering synchronization. **Interface circuits,** which consist of decoders, encoders, multiplexers, demultiplexers, transceivers and buffers, and D/A and A/D converters, are used to interconnect different subsystems.

5.1 REGISTERS

Registers can be classified into two types: **data registers** and **shift registers.** These devices are used to hold, move, and manipulate data. They consist of a series of flip-flops that are used to *latch* or hold data, manipulate the data, and then pass the data on to the next device or circuit at the proper time.

Data Registers

In computer systems different groupings of circuits called **subsystems** operate at different speeds. Data registers are used to hold data temporarily and provide the correct timing for the movement of data.

The four flip-flops shown in Figure 5.1 form what is called a **parallel in–parallel out** data register. The data register is connected to a high-speed data bus. A **data bus** can be thought of as a highway where each lane represents a wire that is used to transmit a data bit. Each lane in the data bus is connected to its own flip-flop in the data register. The data register latches the data received from the data bus. The clock pulses control when the data bits are transferred into the register. The data bits are then available on the flip-flop outputs Q_0 through Q_3 until the next clock pulse transitions.

Figure 5.2 (p. 147) shows the pin configuration and logic diagram for a 74175 TTL data register. This device provides both the Q and $\overline{Q}$ outputs for each D flip-flop. The clear and clock inputs are used to control the operation of the device.

Shift Registers

Shift registers are used primarily to control the movement or shifting of data. They are often used in arithmetic and counting applications.

The four flip-flops shown in Figure 5.3 (p. 148) represent a basic **serial in–serial out** shift register. Let's begin by assuming that all the flip-flops are initially reset. On the first clock pulse, the first data bit is shifted into (to the right) the first flip-flop. Each clock pulse that follows shifts the data one flip-flop to the right. After four clock pulses, the initial data bit is lost. The timing diagram and truth table of Figure 5.3 illustrates the internal movement of data through the shift register.

Shift registers can be used to input data serially from a telephone line and output the data in a parallel fashion to a computer data bus for processing. Figure 5.4 (p. 149) illustrates a variety of shift register configurations.

Figure 5.5 (p. 150) shows the pin configuration and functional truth table for the TTL 74194 bidirectional universal shift register. Pins 3, 4, 5, and 6 are used for parallel data input. Pin 2 is used for serial shift-left data input, and pin 7 is used for serial shift-right data input. This device can be operated in any one of four mode states controlled by pin 9 (S_0) and pin 10 (S_1). When S_0 and S_1 are both high, parallel data bits are loaded. If

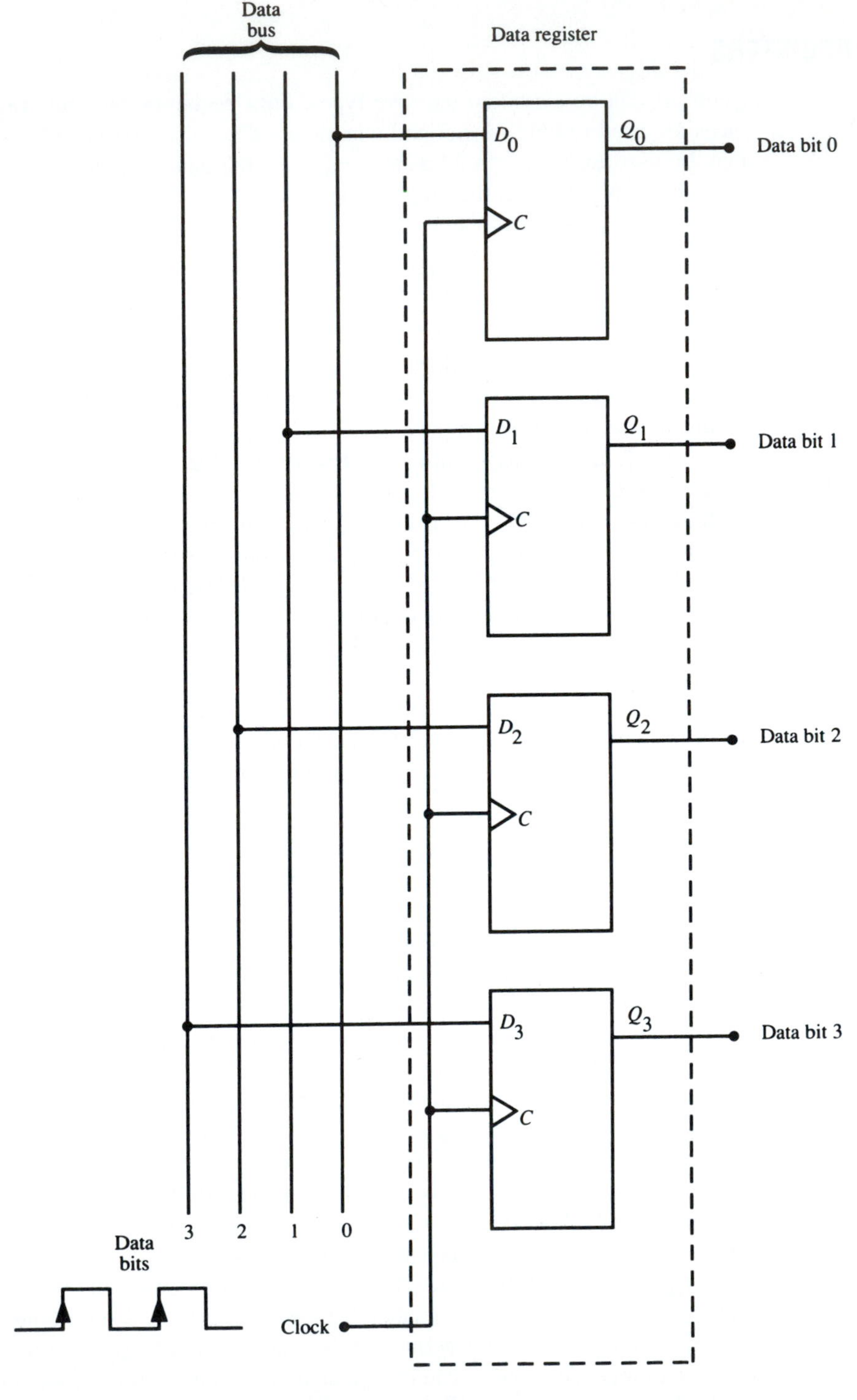

FIGURE 5.1
Data register

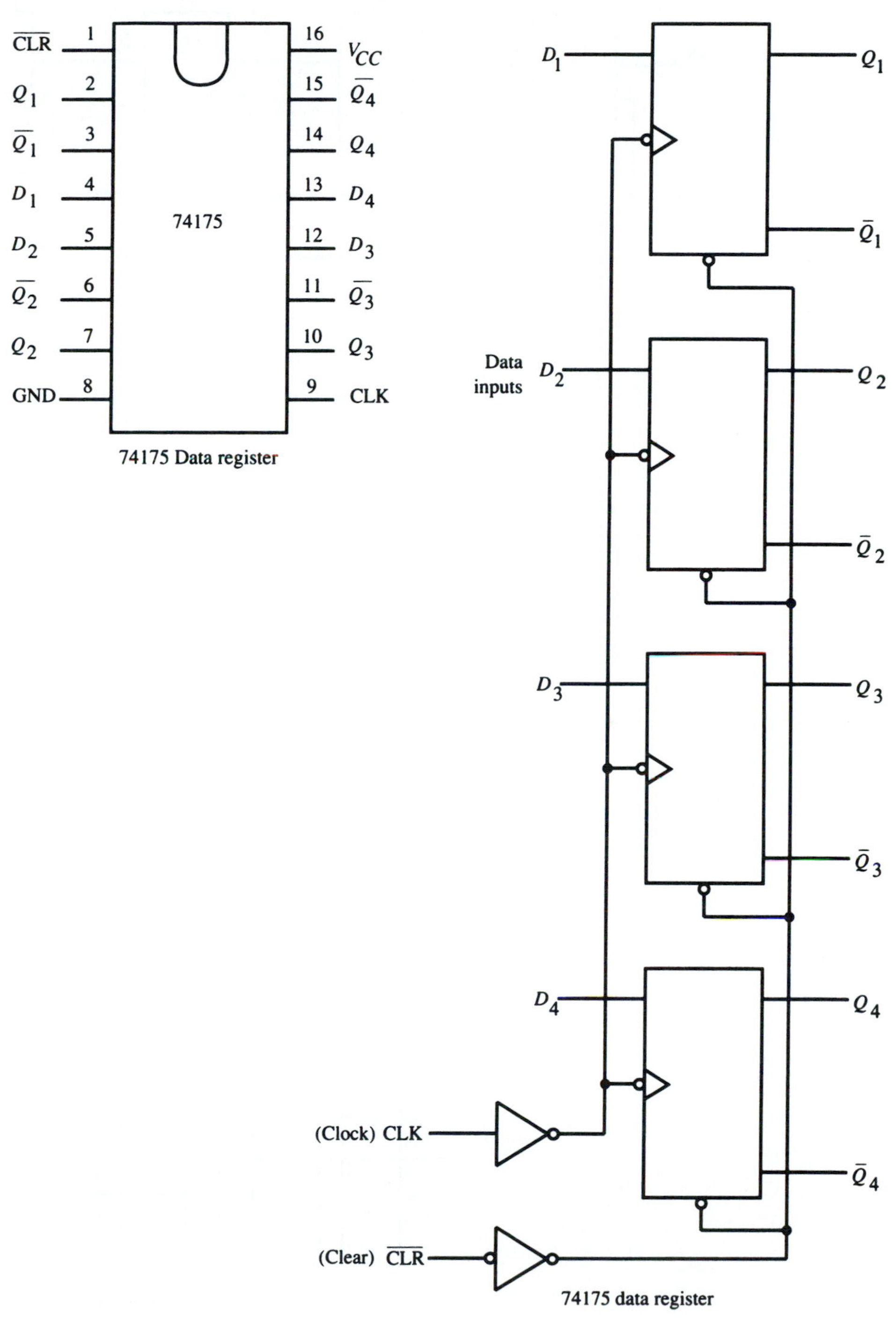

FIGURE 5.2
74175 data register

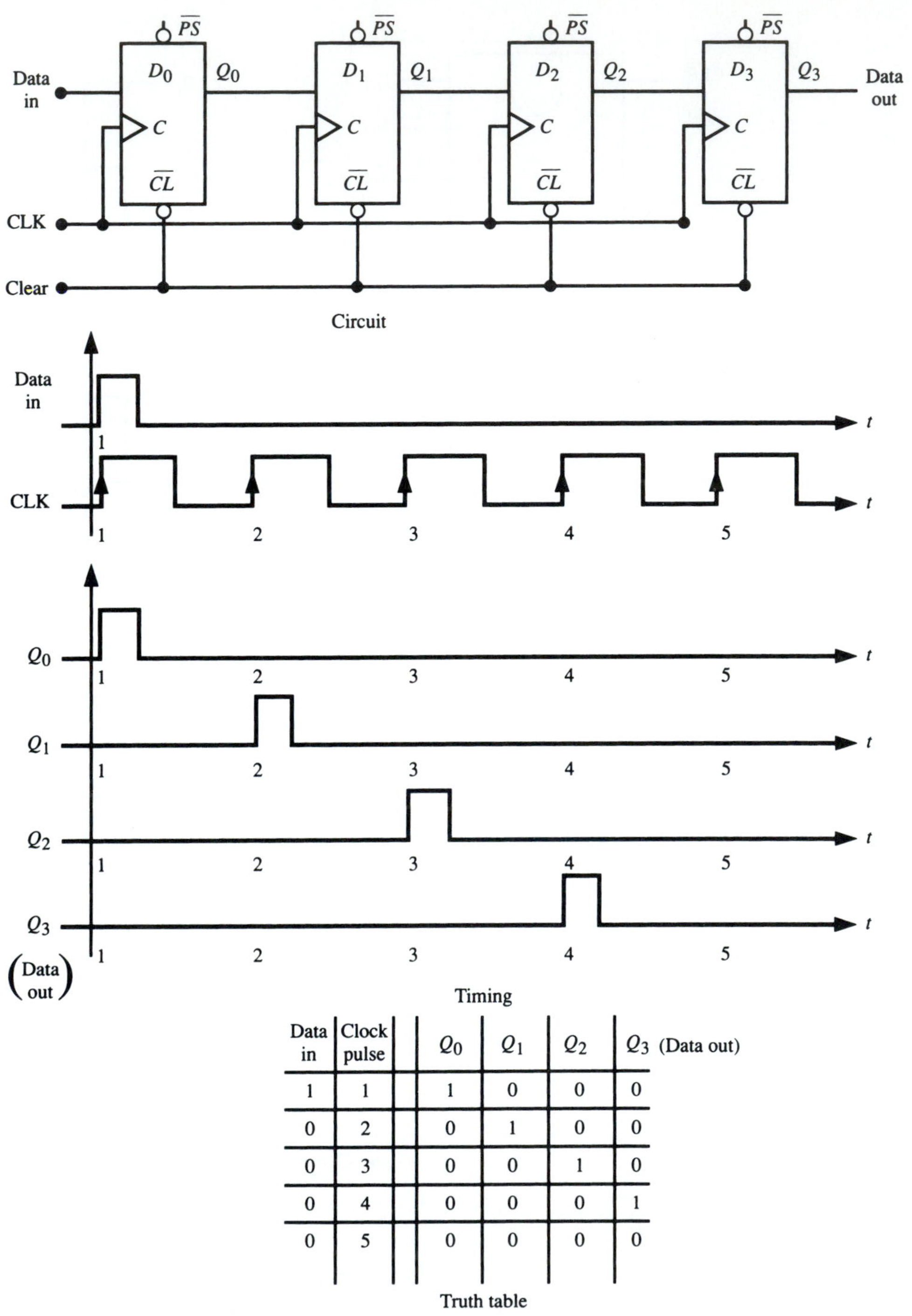

Data in	Clock pulse	Q_0	Q_1	Q_2	Q_3 (Data out)
1	1	1	0	0	0
0	2	0	1	0	0
0	3	0	0	1	0
0	4	0	0	0	1
0	5	0	0	0	0

FIGURE 5.3
Basic shift register

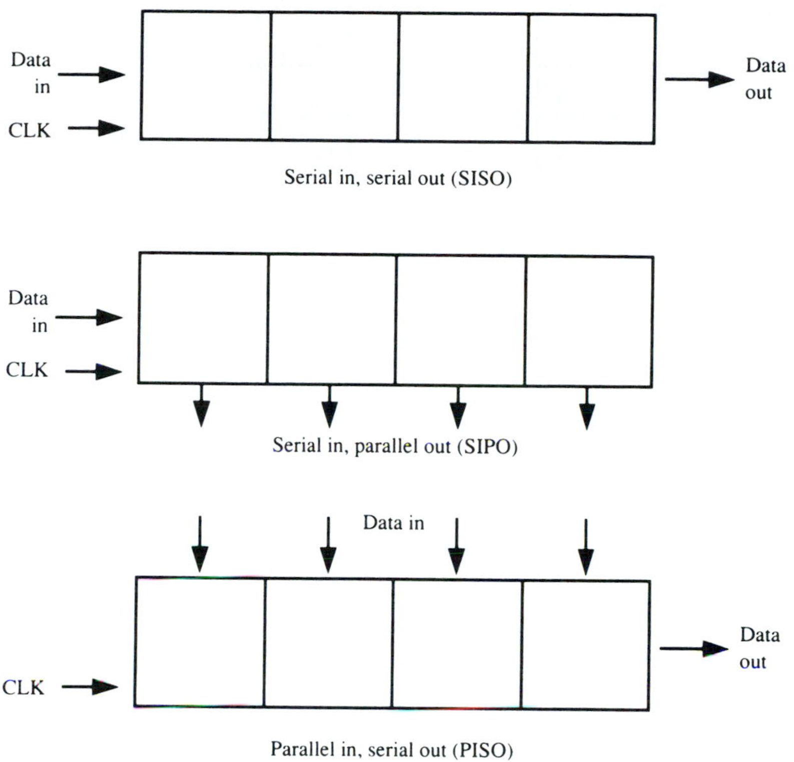

FIGURE 5.4
Shift register configuration

S_0 is high and S_1 is low, serial data bits are shifted right. If S_0 and S_1 are both low, the device enters a hold mode and functions as a latch. Table 5.1 summarizes the four modes of operation.

TABLE 5.1
74194 operation modes

S_1	S_0	Function
0	0	Latch or hold data
0	1	Serial shift right
1	0	Serial shift left
1	1	Parallel load

Table 5.2 summarizes a number of common TTL and CMOS shift registers.

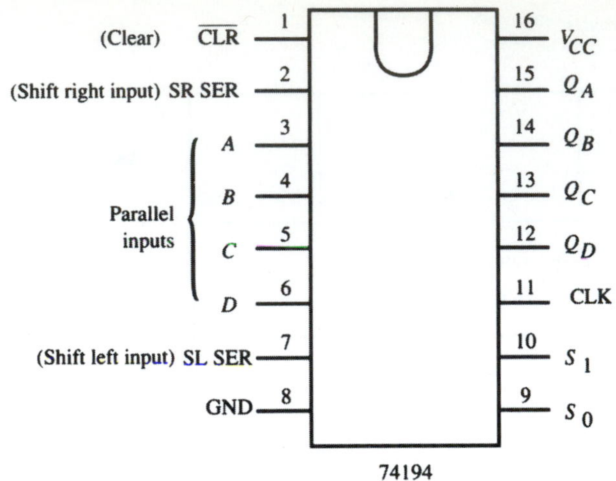

Control inputs			Clock	Serial inputs		Parallel inputs				Outputs			
$\overline{CLR}$	Mode S_1	Mode S_0	CLK	SR SER	SL SER	A	B	C	D	Q_A	Q_B	Q_C	Q_D
0	X	X	X	X	X	X	X	X	X	0	0	0	0
1	X	X	0	X	X	X	X	X	X	H	O	L	D
1	1	1	↑	X	X	a	b	c	d	a	b	c	d
1	0	1	↑	X	1	X	X	X	X	1	Q_{AN}	Q_{BN}	Q_{CN}
1	0	1	↑	X	0	X	X	X	X	0	Q_{AN}	Q_{BN}	Q_{CN}
1	1	0	↑	1	X	X	X	X	X	Q_{BN}	Q_{CN}	Q_{DN}	1
1	1	0	↑	0	X	X	X	X	X	Q_{BN}	Q_{CN}	Q_{DN}	0
1	0	0	X	X	X	X	X	X	X	H	O	L	D

Note : X = any input (0 or 1)
$Q_{AN}, Q_{BN}, Q_{CN}, Q_{DN}$ = previous state of Q_A, Q_B, Q_C, Q_D

FIGURE 5.5
74194 four-bit bidirectional universal shift register

TABLE 5.2
TTL and CMOS shift registers

TTL	CMOS	Description
7491	4006	SISO
7494	4014	PISO
74164	74C164	SIPO
—	4035	PIPO
74194	74C194	Universal

5.2 SHIFT REGISTER COUNTERS

A shift register can be used as a counter because each output state has a unique binary code. The counting sequence, however, does not follow the binary number system. For this reason, shift register counters are not used for arithmetic. Instead, they are used to develop **multiphase timing** waveforms. Each output state is a unique binary code or phase-shift output of the previous state. The two most common types of shift register counters are the *ring counter* and the *Johnson counter.*

Ring Counter

The **ring counter** is like a circulating shift register. It utilizes one flip-flop for each output state. Usually the first flip-flop's Q output is preset to a 1, and all the rest of the flip-flops are cleared upon initialization. The logic 1 is then made to circulate around the register. The Q output of the last flip-flop is fed back to the D input of the first flip-flop to form an endless ring or circle. The operation of the ring counter is illustrated in Figure 5.6.

The output frequency of a ring counter can be calculated using equation (5.1).

$$f_{\text{out}} = \frac{f_{\text{in}}}{N} \tag{5.1}$$

where f_{in} is the input clock frequency and N is the number of flip-flops. The duty cycle of the output waveform can be calculated using equation (5.2).

$$\%\ \text{duty cycle} = \frac{1}{N} \times 100 \tag{5.2}$$

It should be noted that ring counters are very *inefficient* frequency dividers because one flip-flop is required for each count state. Thus to divide the input frequency by 10 would require 10 flip-flops. By comparison, it would take only 4 flip-flops if we used an asynchronous ripple circuit. Another disadvantage of the ring counter is that it is *not self-starting.* The initial 1 must be preset into the ring counter before counting can begin.

Johnson Counters

A variation of the ring counter is called the **Johnson counter.** The Johnson counter is a standard ring counter except that the $\overline{Q}$ output of the last flip-flop is fed back to the input of the first flip-flop. For this reason, the Johnson counter is sometimes called a *twisted ring counter,* or a *switch tail counter.* The operation of the Johnson counter is illustrated in Figure 5.7. Note the unique output count state pattern.

Johnson counters are more efficient than standard ring counters. The output frequency can be calculated using equation (5.3).

$$f_{\text{out}} = \frac{f_{\text{in}}}{2N} \tag{5.3}$$

Another advantage of the Johnson counter is that the output waveform is symmetrical (50% duty cycle). Also, because of the feedback arrangement ($\overline{Q}$ to the input), the Johnson counter is self-starting.

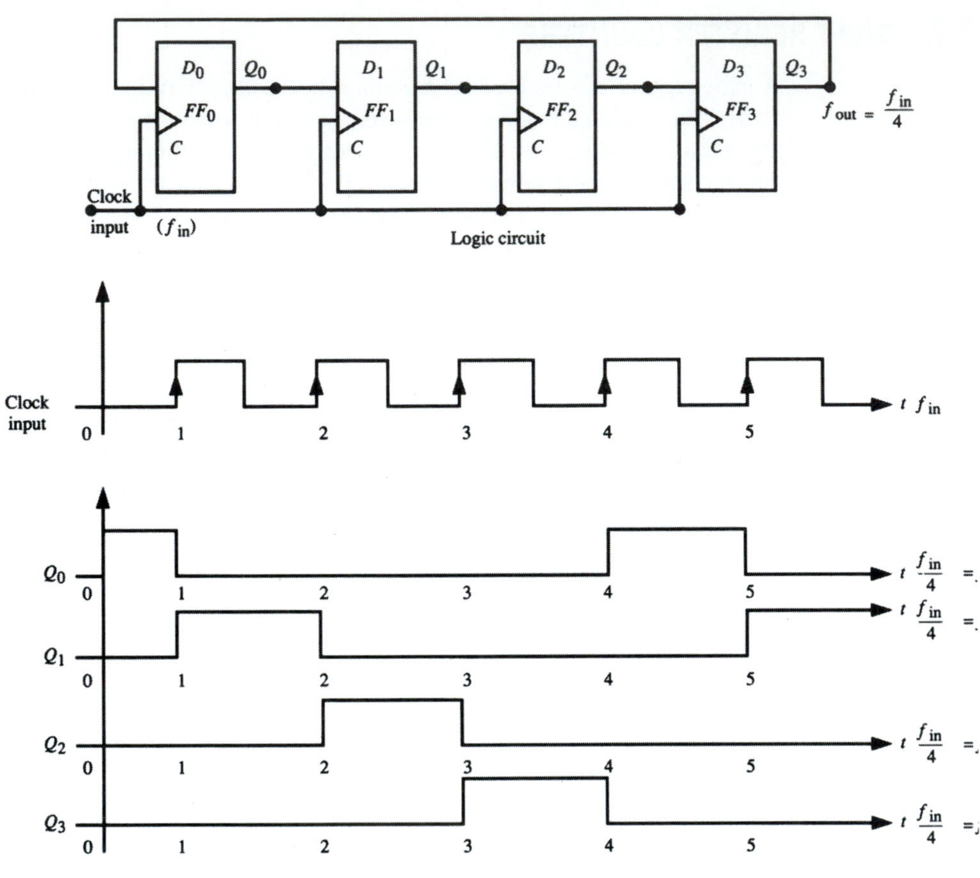

Timing diagram

Input clock	Q_0	Q_1	Q_2	Q_3
0	1	0	0	0
1	0	1	0	0
2	0	0	1	0
3	0	0	0	1
4	1	0	0	0
5	0	1	0	0

Count table

FIGURE 5.6
Four-stage ring counter

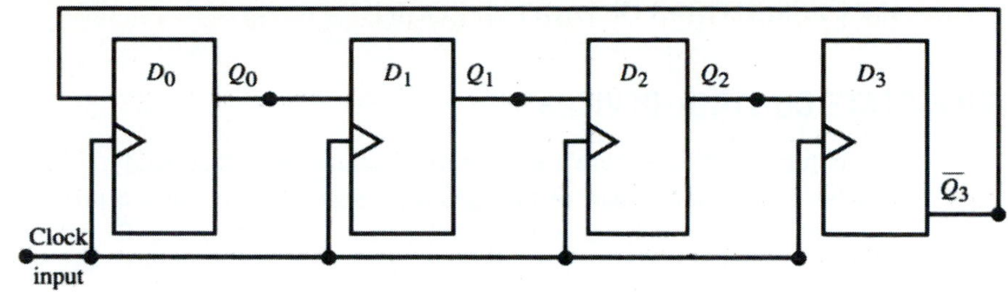

Johnson counter logic circuit

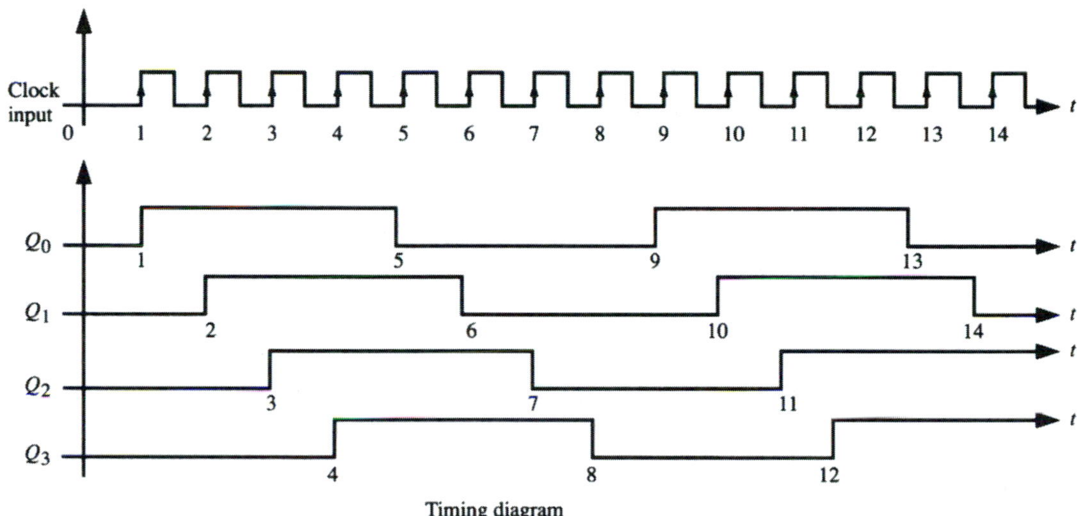

Timing diagram

Input clock	Q_0	Q_1	Q_2	Q_3
0	0	0	0	0
1	1	0	0	0
2	1	1	0	0
3	1	1	1	0
4	1	1	1	1
5	0	1	1	1
6	0	0	1	1
7	0	0	0	1
8	0	0	0	0

Count table

FIGURE 5.7
Four-stage Johnson counter

5.3 THREE-STATE CONTROL DEVICES

In a computer system there are often many subsystems that need to connect, or **interface,** to a bus or to another part of the system. Figure 5.8 illustrates a problem where a memory subsystem, a disk subsystem, and a tape subsystem are all connected to a microprocessor's data bus. Obviously, all three subsystems cannot be allowed to put data on the bus at the same time. For this reason, three-state, or **tristate,** logic devices were developed. They are used to interface multiple devices onto a common bus.

Three-state logic devices, unlike conventional logic devices, have three possible output logic states. These are:

Logic 0 state
Logic 1 state
Hi-*Z* state (high impedance state)

To explain the operation of a three-state device, we consider the inverter of Figure 5.9 (p. 156). This enable control line can be considered to be an extra input. When the enable line is active, the device functions normally. When the enable line is inactive, the device enters the high impedance (Hi-*Z*), or disconnected, state. Thus the device looks like an open circuit to the common bus. Enable lines can be either active high or active low. Figure 5.9 illustrates the operation of an active high and an active low three-state inverter for all possible combinations of input conditions. Most types of TTL and CMOS devices are available in three-state and conventional form. A small triangle (▽) added to the symbol is used to indicate a three-state device. Looking back at Figure 5.8 we can see the purpose and function of a three-state device. It is important to note that only one subsystem can be enabled to send data onto a bus at any given instant. Special control logic circuits must be designed to ensure that only one subsystem will be enabled at any time.

Buffers

Three-state **buffers** are devices that provide bus isolation and *driving power* in digital circuits. Buffers can be either inverting or noninverting. As their symbol implies, they can be thought of as *amplifiers.* Figure 5.10 (p. 157) illustrates the operation of the 74126 noninverting buffer.

Schmitt Trigger Devices

Many devices are available in **Schmitt trigger** versions. A Schmitt trigger is a circuit that switches states at threshold or trigger points. As soon as the input voltage reaches the trigger point, the output is caused to switch. The operation of a Schmitt trigger device is described in Figure 5.11 (p. 158). At the V_{TH} threshold level of the input signal, the output is switched high. At the V_{TL} threshold of the input signal, the output is switched low.

Most digital devices have specifications for pulse rise times. If the rise time is too long, the device will not operate properly. Schmitt trigger devices are often used as buffers to clean up distorted or slow-rising pulses. This is sometimes required when digital signals are transmitted long distances because the line capacitance tends to distort the sig-

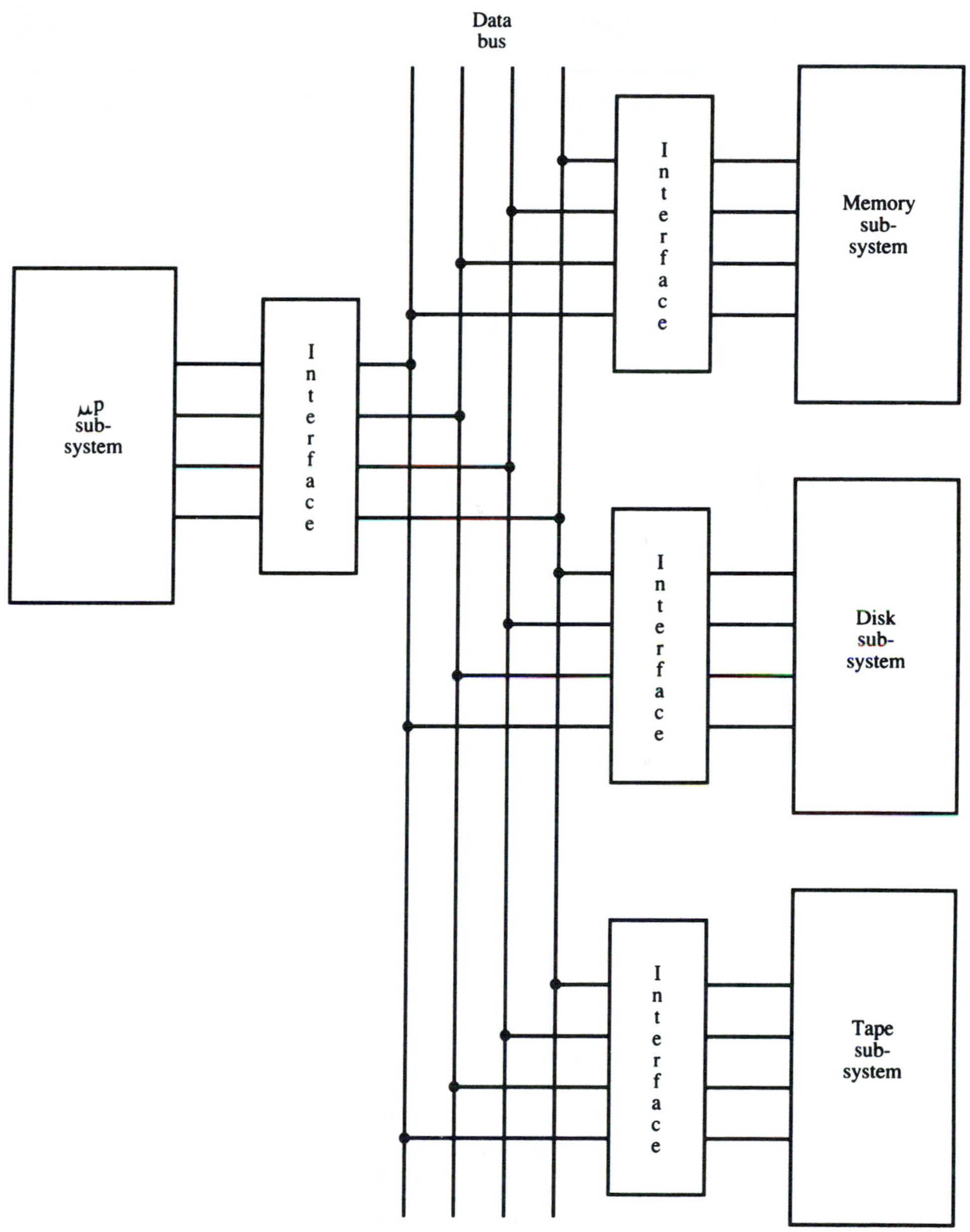

FIGURE 5.8
Microcomputer system

nal. Schmitt triggers can be used to restore and correct distorted signals. Line lengths for TTL devices usually should not exceed 12 in. between devices.

The symbol used to indicate a Schmitt trigger–type device is the **hystersis loop** (⊏⊐).

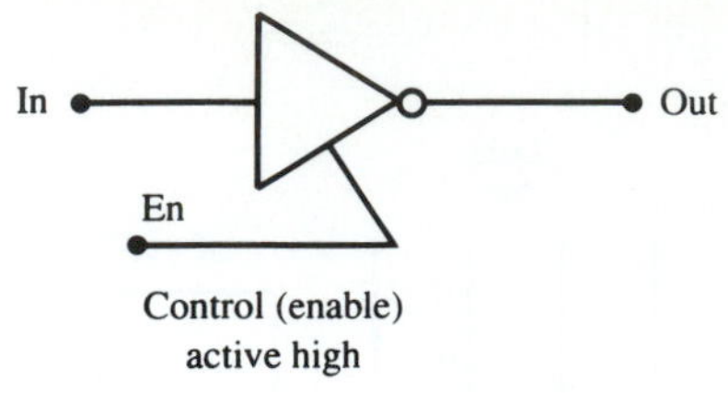

Control (enable)
active high

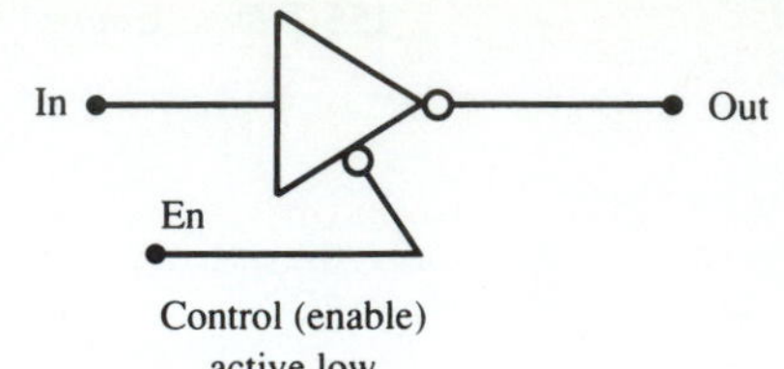

Control (enable)
active low

In	En	Out
0	0	Hi - Z
0	1	1
1	0	Hi - Z
1	1	0

Truth table
active high
enable

In	En	Out
0	0	1
0	1	Hi - Z
1	0	0
1	1	Hi - Z

Truth table
active low
enable

Hi - Z = High impedance or disconnected

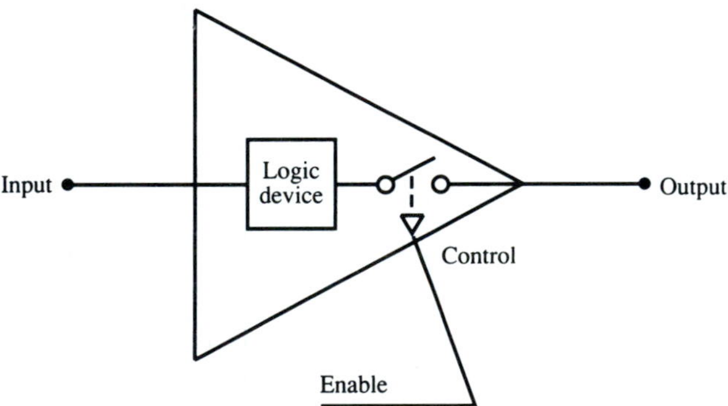

Block diagram of a three - state(tri - state)
device

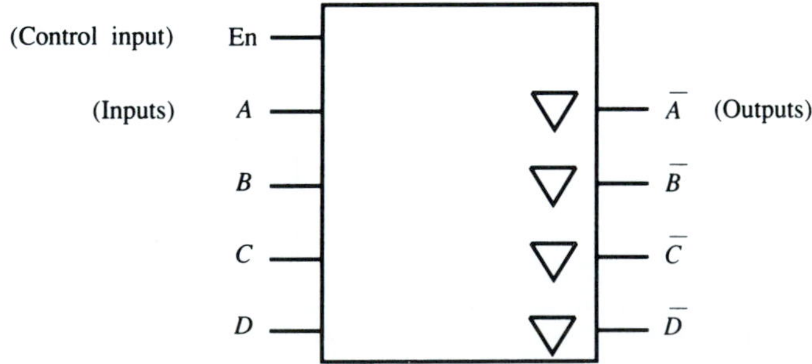

Four three - state inverters

FIGURE 5.9
Three-state logic

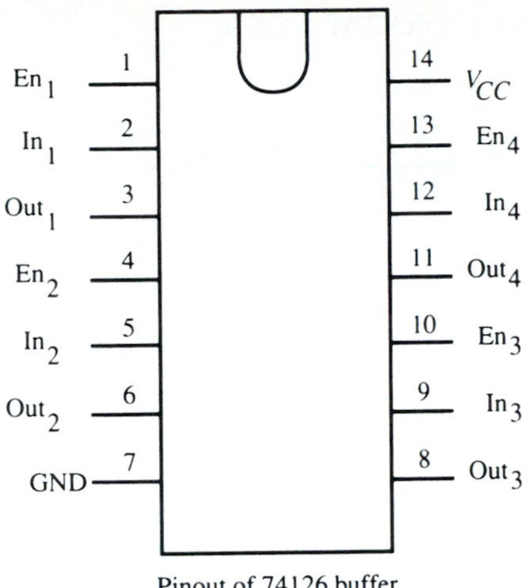

Pinout of 74126 buffer

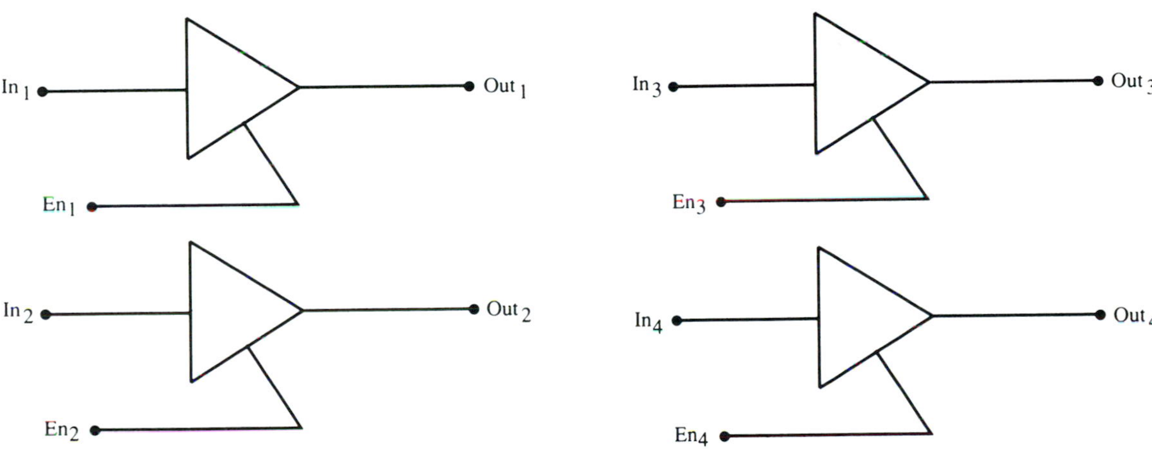

Logic circuit

In	En	Out
0	0	Hi - Z
0	1	0
1	0	Hi - Z
1	1	1

FIGURE 5.10
Four-bit noninverting buffer

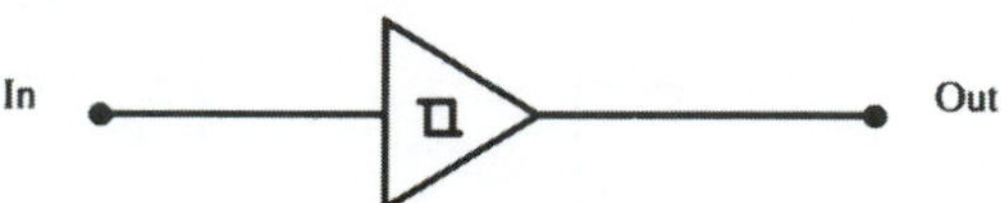

Logic symbol
Non - inverting Schmitt trigger

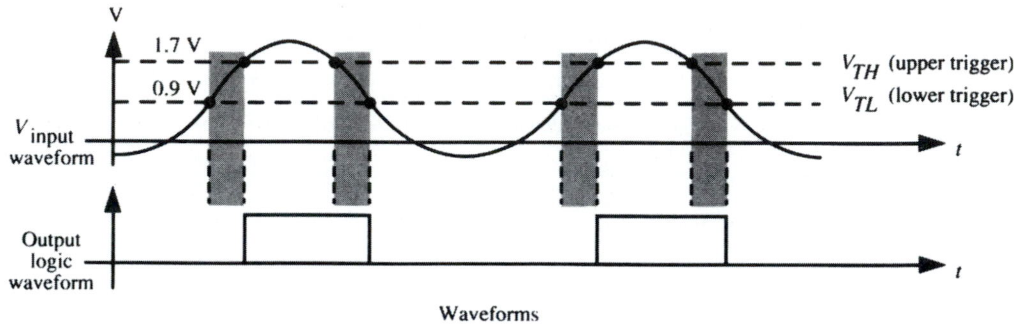

Waveforms

V_{TH} = upper trigger (threshold) level
V_{TL} = lower trigger (threshold) level

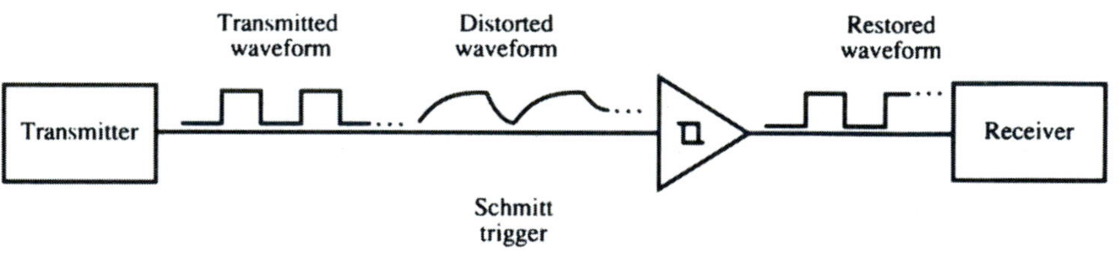

FIGURE 5.11
Schmitt trigger

Transceivers

When buffers are used to send data, they can be thought of as *transmitters.* When buffers are used to receive data, they can be thought of as *receivers.* In either case, these devices are *unidirectional.* A **transceiver** is a *bidirectional* device that can transmit or receive data in *either direction* but not at the same time. Figure 5.12 describes the operation of a 74245 octal bus transceiver. This device has Schmitt trigger inputs and three-state outputs. The direction of data flow is controlled by pin 1. When pin 1 is low, data travel from *B* to *A*. When pin 1 is high, data travel from *A* to *B*. Three-state output enable is controlled by pin 19. When pin 19 is low, the device is enabled. When pin 19 is high, the device is disconnected (Hi-Z). Table 5.3 (p. 160) lists a number of common TTL and CMOS interface devices.

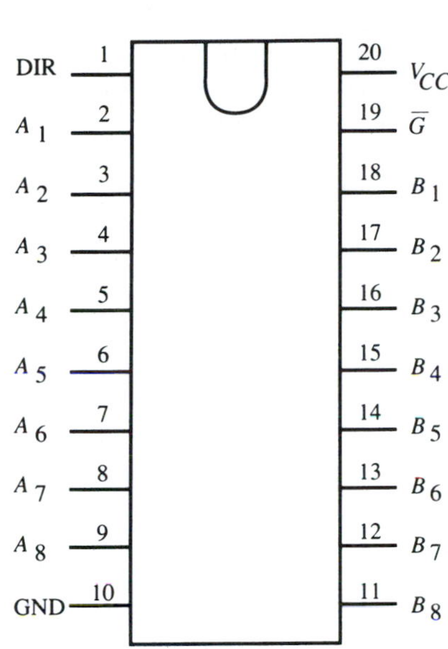

Pinout 74245

Enable $\overline{G}$	Direction Control DIR	Operation
0	0	Data from B to A
0	1	Data from A to B
1	X	Isolation

DIR, $\overline{G}$, A_1–A_8, B_1–B_8

Logic circuit 74245

FIGURE 5.12
Octal bus transceiver with three-state outputs

TABLE 5.3
TTL and CMOS interface devices

TTL	CMOS	Description
7414	40106	Hex inverter Schmitt trigger
74132	4093	4 two-input NAND Schmitt triggers
74125	74C125	4 noninverting tristate buffers
74368	74C368	6 inverting tristate buffers
74245	74C245	8 noninverting transceivers

5.4 DATA CONVERTERS

Data converters are used to change data from one format to another. For example, sometimes it is necessary to change BCD data to decimal, analog data to digital, parallel data to serial, and so forth. In this section we study the types of devices that are used to perform these tasks.

Multiplexers

A **multiplexer,** or **data selector,** is a device that is used to select one of many input signals and connect that input to the output of the device. It can be thought of as a single-pole, multiposition rotary switch, as shown in Figure 5.13(a). By manually rotating the switch position, we can control or select which input will be sent to the output.

The electronic multiplexer functions in a similar manner except that selection and switching are performed electronically instead of manually. Figure 5.13(b) illustrates the operation of a 4-line digital multiplexer. The input lines are labeled D_0, D_1, D_2, and D_3, and the output line is labeled D_{out}. The select lines S_0 and S_1 are used to determine which input line will be connected to the output line. This is performed in a binary manner with S_0 as the LSB and S_1 the MSB. The operation and truth table are given in Figure 5.13(c).

The 74150 and the 74C150 are typical examples of TTL and CMOS data selectors or multiplexers. Figure 5.14 (p. 162) shows the pin configuration and truth table data for the 74150 TTL device. This device selects or multiplexes 1 of 16 lines of data. The data input lines are labeled D_0 through D_{15}. The data output line is labeled $\overline{D_{out}}$. Note that the output is the *inversion* of the selected input. Data selection is controlled in a binary manner using the select lines S_0, S_1, S_2, and S_3. S_0 is the LSB and S_3 is the MSB. The strobe line $\overline{G}$ is used to enable or disable the device. When $\overline{G}$ is low, the device is enabled, and the selected input is switched or connected to the output. When $\overline{G}$ is high, the output pin, $\overline{D_{out}}$, is forced high.

Demultiplexers

The **demultiplexer,** or **data distributor,** performs the opposite task of the multiplexer. This device has a signal input and multiple outputs, as shown in Figure 5.15 (p. 163). It is

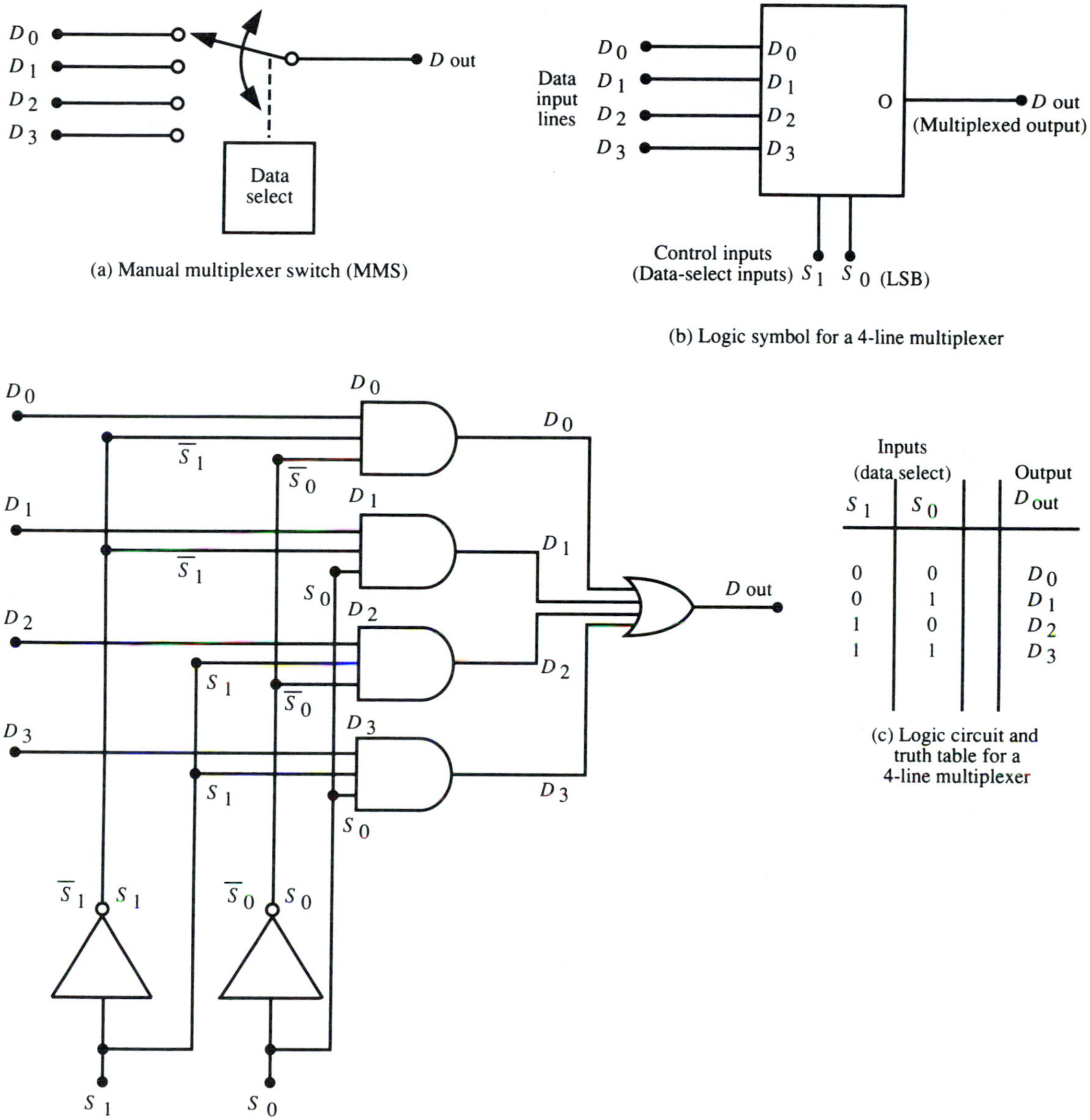

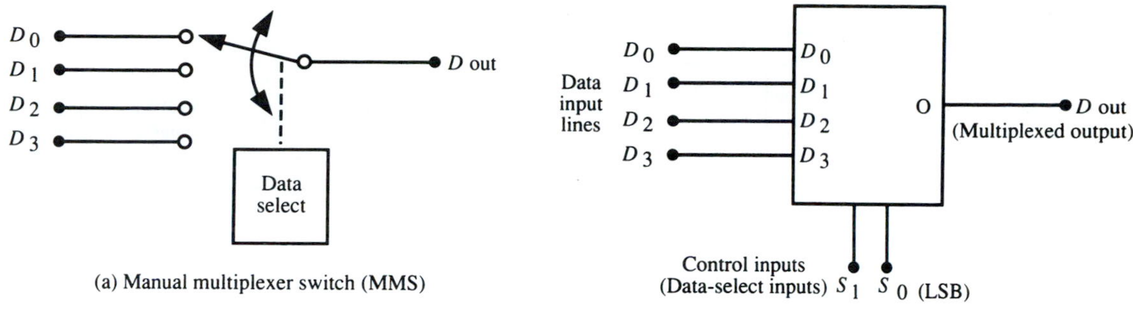

(a) Manual multiplexer switch (MMS)

(b) Logic symbol for a 4-line multiplexer

Inputs (data select) S_1	S_0	Output D_{out}
0	0	D_0
0	1	D_1
1	0	D_2
1	1	D_3

(c) Logic circuit and truth table for a 4-line multiplexer

FIGURE 5.13
Multiplexer

used to direct the input signal to the proper output path. In Figure 5.15(b) the input line is labeled D_{in}, and the output lines are labeled D_0, D_1, D_2, and D_3. Path selection is controlled by select lines S_0 and S_1 in a binary manner. S_0 is again defined as the LSB and S_1 is the MSB.

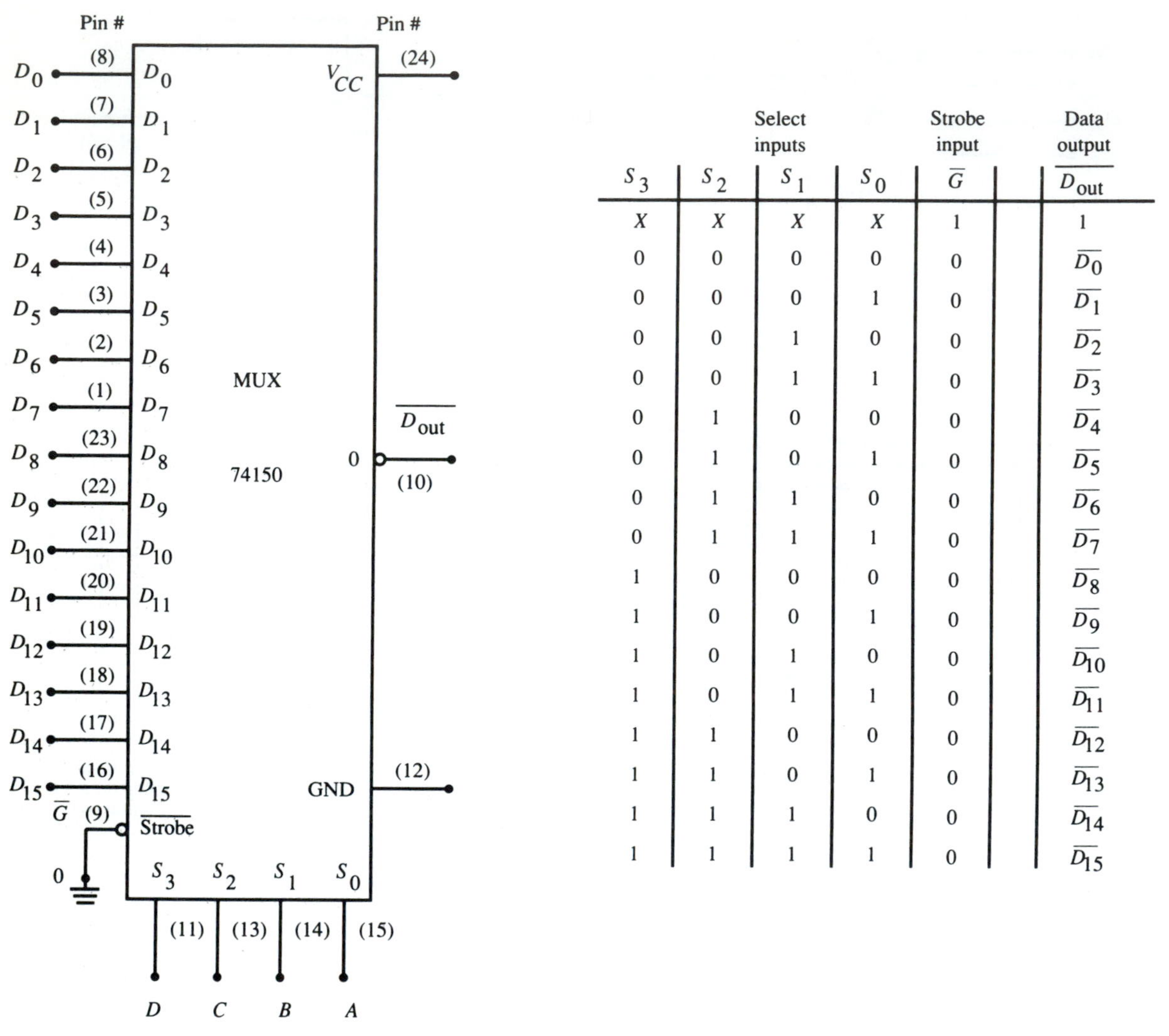

Select inputs				Strobe input	Data output
S_3	S_2	S_1	S_0	$\overline{G}$	$\overline{D_{out}}$
X	X	X	X	1	1
0	0	0	0	0	$\overline{D_0}$
0	0	0	1	0	$\overline{D_1}$
0	0	1	0	0	$\overline{D_2}$
0	0	1	1	0	$\overline{D_3}$
0	1	0	0	0	$\overline{D_4}$
0	1	0	1	0	$\overline{D_5}$
0	1	1	0	0	$\overline{D_6}$
0	1	1	1	0	$\overline{D_7}$
1	0	0	0	0	$\overline{D_8}$
1	0	0	1	0	$\overline{D_9}$
1	0	1	0	0	$\overline{D_{10}}$
1	0	1	1	0	$\overline{D_{11}}$
1	1	0	0	0	$\overline{D_{12}}$
1	1	0	1	0	$\overline{D_{13}}$
1	1	1	0	0	$\overline{D_{14}}$
1	1	1	1	0	$\overline{D_{15}}$

FIGURE 5.14
74150 Sixteen-line multiplexer (MUX)

The 74154 and the 74C154 are typical examples of TTL and CMOS demultiplexers. Figure 5.16 (p. 164) shows the pin configuration and truth table for the 74154 TTL device. This device has four binary-coded select lines labeled S_0, S_1, S_2, and S_3. S_0 is the LSB and S_3, the MSB. The strobe line $\overline{G}$ is used to enable and disable the device. When the device is enabled, the input data is switched by the binary-coded select lines to the appropriate output. This device is actually very flexible and can be used to function as a decoder. (Decoding is explained in the next section.)

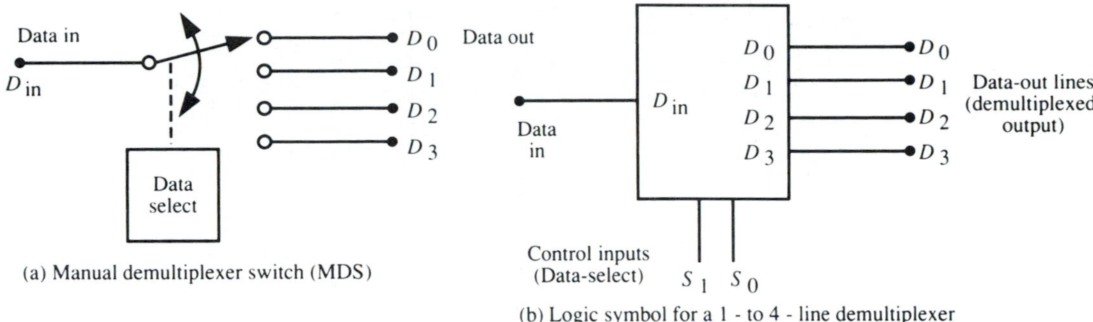

(a) Manual demultiplexer switch (MDS)

(b) Logic symbol for a 1 - to 4 - line demultiplexer

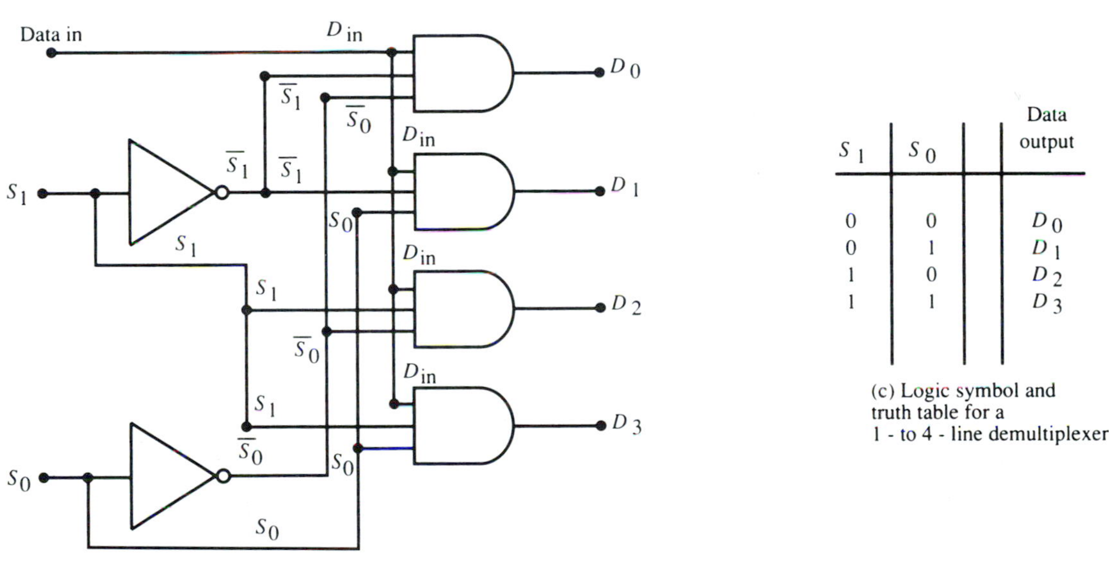

S_1	S_0		Data output
0	0		D_0
0	1		D_1
1	0		D_2
1	1		D_3

(c) Logic symbol and truth table for a 1 - to 4 - line demultiplexer

FIGURE 5.15
Demultiplexer

Decoders and Encoders

The process of converting from a *primary coding system* to a *secondary coding system* is called **encoding.** The process of converting from a *secondary coding system* back to the *primary coding system* is called **decoding.** Generally speaking, encoders and decoders perform the same function. That is, they are used to convert data from one format to another, as shown in Figure 5.17 (p. 165). Since decimal is most familiar to us, it can be considered our primary system. All other number systems and codes can be considered secondary systems. Therefore, the act of converting decimal numbers to binary can be considered encoding. The act of converting binary numbers to decimal numbers can be considered decoding. For example, a calculator keyboard would use an encoder to

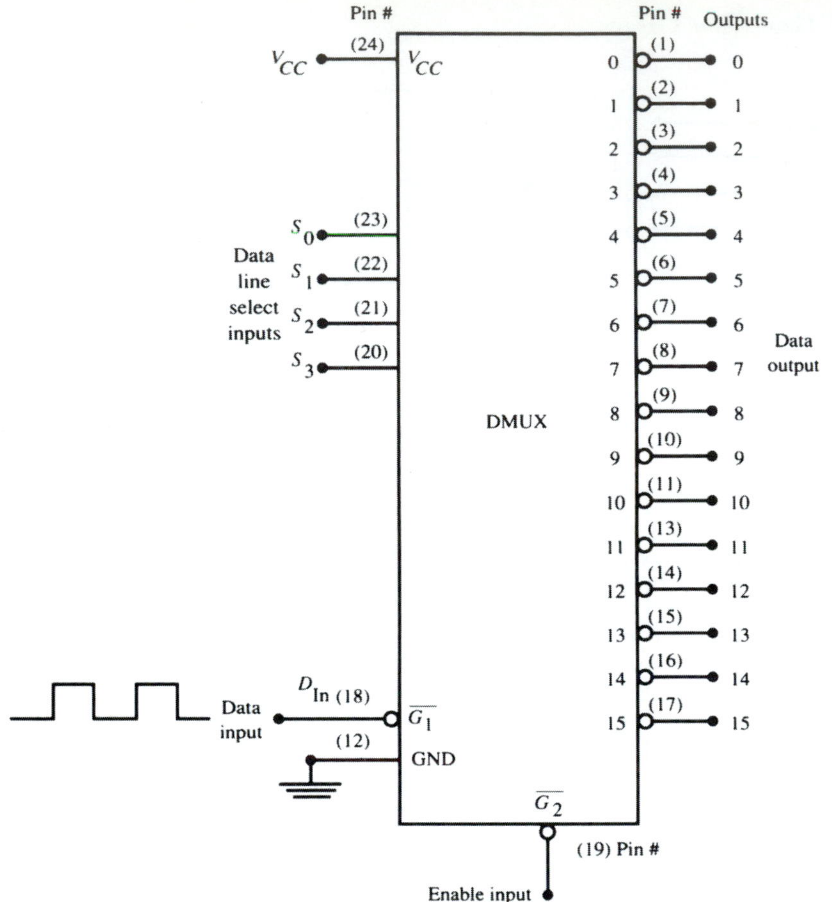

Data output lines

Data in ($\overline{G_1}$)	S_3	S_2	S_1	S_0	$\overline{G}$	0	1	2	3	4	5	6	7	8	9	10	11	12	13	14	15
D_{In} = **X**	0	0	0	0	0	X	1	1	1	1	1	1	1	1	1	1	1	1	1	1	1
X	0	0	0	1	0	1	X	1	1	1	1	1	1	1	1	1	1	1	1	1	1
X	0	0	1	0	0	1	1	X	1	1	1	1	1	1	1	1	1	1	1	1	1
X	0	0	1	1	0	1	1	1	X	1	1	1	1	1	1	1	1	1	1	1	1
X	0	1	0	0	0	1	1	1	1	X	1	1	1	1	1	1	1	1	1	1	1
X	0	1	0	1	0	1	1	1	1	1	X	1	1	1	1	1	1	1	1	1	1
X	0	1	1	0	0	1	1	1	1	1	1	X	1	1	1	1	1	1	1	1	1
X	0	1	1	1	0	1	1	1	1	1	1	1	X	1	1	1	1	1	1	1	1
X	1	0	0	0	0	1	1	1	1	1	1	1	1	X	1	1	1	1	1	1	1
X	1	0	0	1	0	1	1	1	1	1	1	1	1	1	X	1	1	1	1	1	1
X	1	0	1	0	0	1	1	1	1	1	1	1	1	1	1	X	1	1	1	1	1
X	1	0	1	1	0	1	1	1	1	1	1	1	1	1	1	1	X	1	1	1	1
X	1	1	0	0	0	1	1	1	1	1	1	1	1	1	1	1	1	X	1	1	1
X	1	1	0	1	0	1	1	1	1	1	1	1	1	1	1	1	1	1	X	1	1
X	1	1	1	0	0	1	1	1	1	1	1	1	1	1	1	1	1	1	1	X	1
X	1	1	1	1	0	1	1	1	1	1	1	1	1	1	1	1	1	1	1	1	X
X	*X*	*X*	*X*	*X*	1	1	1	1	1	1	1	1	1	1	1	1	1	1	1	1	1

FIGURE 5.16
74154 used as a 1-to-16 demultiplexer (DMUX)

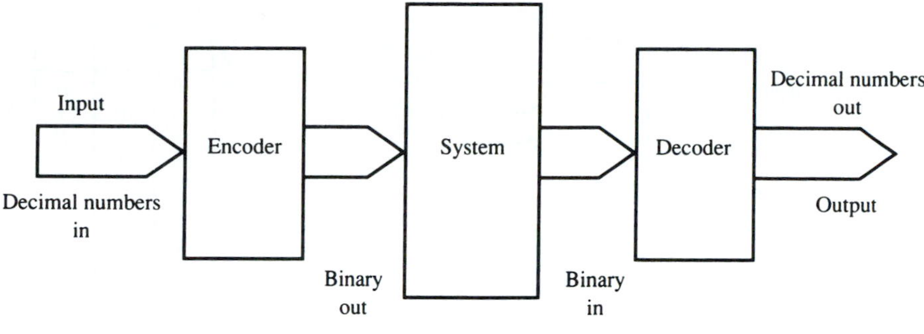

FIGURE 5.17
Encoder/decoder system

convert decimal numbers to BCD. A decimal LED display would require a decoder to convert BCD code to decimal.

The 74154 demultiplexer can also function as a decoder. By using the data input pin as a second enable input, the 4 select lines are decoded to 1 of the 16 output lines. When used in this manner, the 74154 is referred to as a 4-line to 16-line decoder.

The 7447 is an example of a BCD to 7-segment decoder/driver. Figure 5.18 illustrates a simplified application of how this device is used to control a 7-segment LED display. The BCD inputs to the 7447 are labeled A, B, C, and D, with A as the LSB and D as the MSB. The 7-segment outputs are labeled $\overline{a}$, $\overline{b}$, $\overline{c}$, $\overline{d}$, $\overline{e}$, $\overline{f}$ and $\overline{g}$. The truth table of Figure 5.18 illustrates the code conversions. The 7447 decodes the BCD inputs and activates the proper output segment to display decimal numbers. The device is also called a driver because it can provide enough current to light the segments of the LED. Resistors are used to limit the current to the level required by the display. Table 5.4 lists a number of common TTL and CMOS data converter devices.

TABLE 5.4
TTL and CMOS data converter devices

TTL	CMOS	Description
74157	74C157	Quad 2-line to 1-line multiplexer
74151	74C151	8-line to 1-line multiplexer
74150	74C150	16-line to 1-line multiplexer
74139	74C139	Dual 2-line to 4-line demultiplexer
74138	74C138	3-line to 8-line demultiplexer
74154	74C154	4-line to 16-line demultiplexer
7447	74C4511	BCD to 7-segment decoder
7445	74C42	BCD to decimal decoder

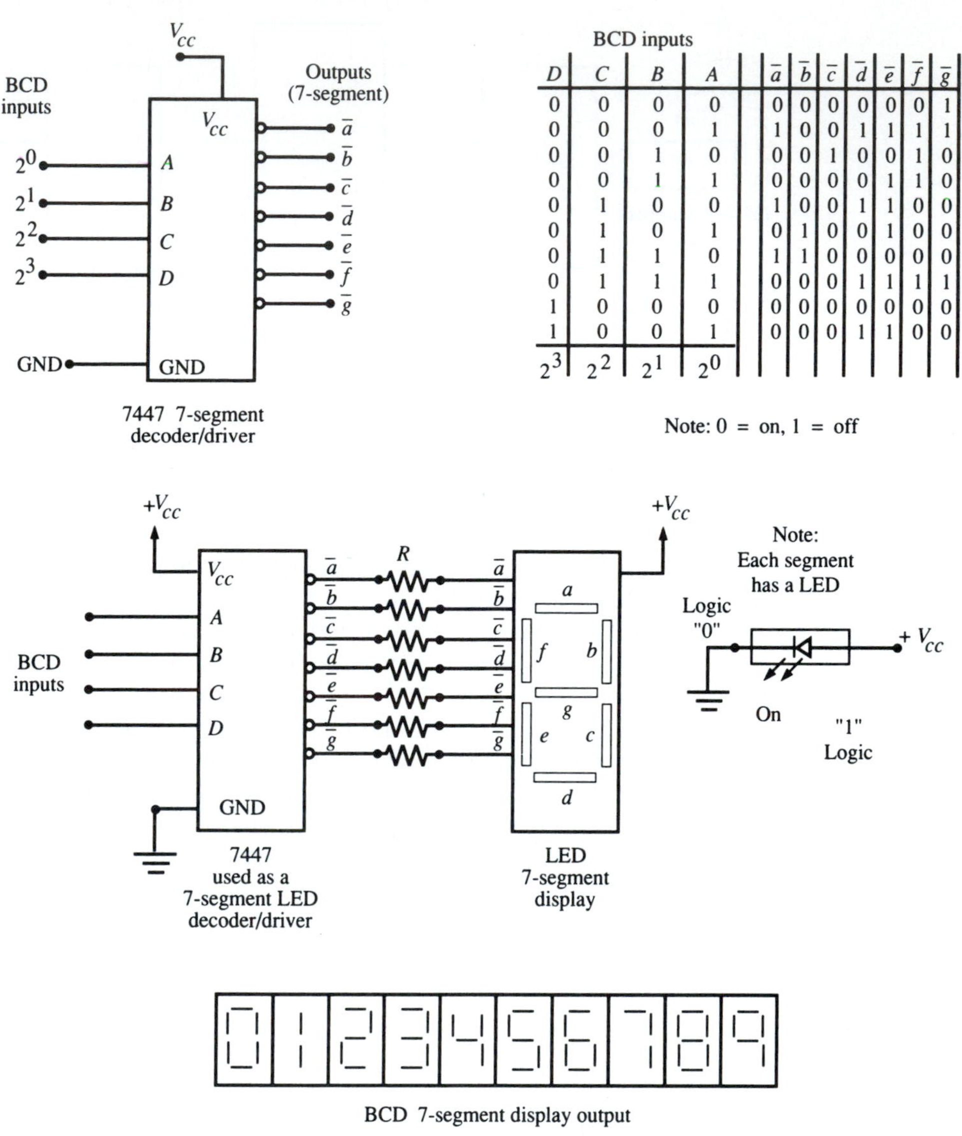

D	C	B	A	$\bar{a}$	$\bar{b}$	$\bar{c}$	$\bar{d}$	$\bar{e}$	$\bar{f}$	$\bar{g}$
0	0	0	0	0	0	0	0	0	0	1
0	0	0	1	1	0	0	1	1	1	1
0	0	1	0	0	0	1	0	0	1	0
0	0	1	1	0	0	0	0	1	1	0
0	1	0	0	1	0	0	1	1	0	0
0	1	0	1	0	1	0	0	1	0	0
0	1	1	0	1	1	0	0	0	0	0
0	1	1	1	0	0	0	1	1	1	1
1	0	0	0	0	0	0	0	0	0	0
1	0	0	1	0	0	0	1	1	0	0
2^3	2^2	2^1	2^0							

FIGURE 5.18
7447 seven-segment decoder/driver

5.5 DIGITAL-TO-ANALOG CONVERTERS

One of the simplest **digital-to-analog converters** (DAC) is formed using resistors that double in value. This DAC is sometimes referred to as a *binary-weighted ladder DAC* circuit, as shown in Figure 5.19(a). Each resistor is twice the value of the preceding resistor, just as the weight of each binary bit position is twice the weight of the previous bit

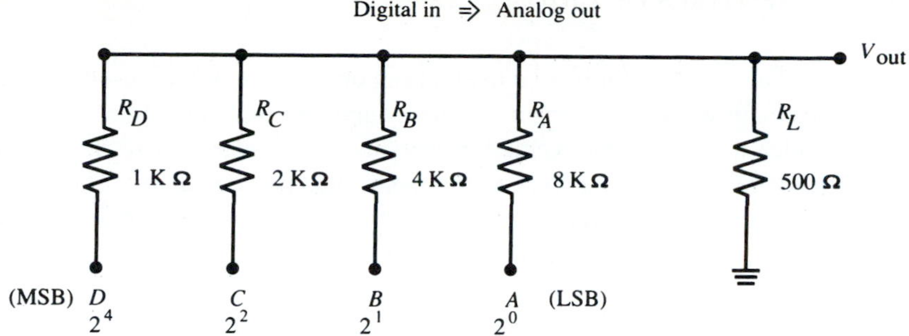

(a) 4-bit weighted ladder DAC

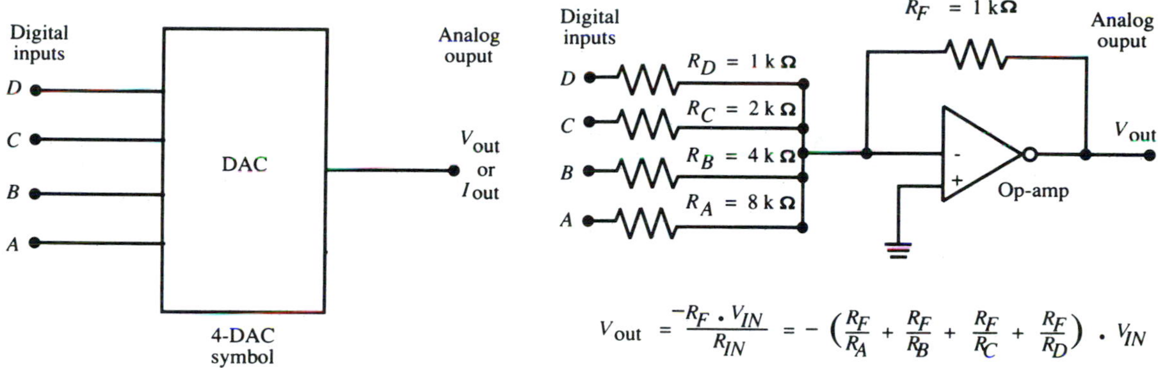

(b) 4-bit DAC symbol and circuit

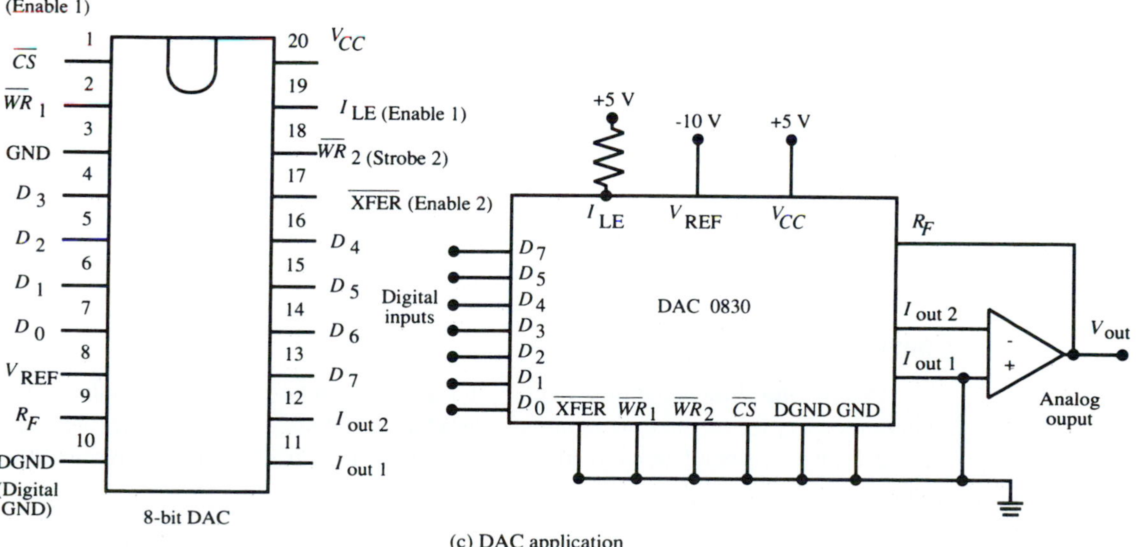

(c) DAC application

FIGURE 5.19
DAC

position. The load resistor is used to create an output voltage. It must be equal to or smaller than the smallest resistor to ensure accuracy and minimize the effect on the summing current. Notice that the largest value becomes the LSB because it results in the lowest current flow. The DAC is basically a current-controlled parallel circuit. The current is summed at the common tie point and fed through resistor R_L to create the analog output voltage. The greater the current, the greater the output voltage. Figure 5.19(b) illustrates a binary-weighted DAC using an op-amp summing amplifier.

Today most IC manufacturers provide DAC devices in a single package. Figure 5.19(c) illustrates a typical integrated circuit DAC. Because these devices usually operate on a wide voltage range, they are compatible with both TTL and CMOS devices. V_{REF} is the maximum analog output voltage. $\overline{WR_1}$, $\overline{WR_2}$, $\overline{CS}$, I_{LE}, and $\overline{XFER}$ can be used to interface the DAC to a microprocessor. The function of the remaining pins is indicated on the application diagram. When selecting a DAC, it is important to understand DAC terminology in order to evaluate the specifications.

Resolution

Resolution is a measure of how fine the output voltage steps or increments will be. It is usually determined by the number of input bits or as the reciprocal of the number of input combinations expressed as a percentage.

$$\text{Resolution} = \frac{1}{2^n} \times 100\% \qquad \textbf{(5.4)}$$

where n equals the number of input bits. For example, an 8-bit DAC will have a resolution of 0.39% and is said to have a finer or a smaller resolution than a 4-bit DAC (6.25%).

Accuracy

Accuracy is a measure of the difference between the expected voltage output and the actual voltage output. It is usually expressed as a percentage of the maximum deviation between the expected and the actual output voltage.

$$\text{Accuracy} = \frac{V_{\text{exp}} - V_{\text{act}}}{V_{\text{exp}}} \times 100 \qquad \textbf{(5.5)}$$

Conversion Time

Conversion time is the time required for a DAC to produce an output voltage. It is sometimes specified as the time required to produce a full-scale output voltage when the input is changed from all zeros to all ones. Conversion time is also referred to as **settling time.**

Offset Voltage

Offset voltage is the DAC output voltage when the binary inputs are all zeros.

5.6 ANALOG-TO-DIGITAL CONVERTERS

The basic op-amp comparator can be used to form a simple **analog-to-digital converter** (ADC). The comparator functions as a switch. When the input voltage in Figure 5.20(a) is equal to or greater than the reference voltage, the comparator will produce an output. When the input voltage is less than the reference voltage, the comparator output will be zero. In Figure 5.20(b) the reference voltage is divided by a series of equal resistors. Thus the reference voltage on comparator 1 is 1 V, on comparator 2 is 2 V, and so on. The input voltage is then applied to each comparator, and the resulting output is indicated by the truth table. A code converter can be used to convert the comparator truth table output to a true binary form. Figure 5.20(c) illustrates the use of an integrated circuit ADC.

5.7 UNIVERSAL ASYNCHRONOUS RECEIVER TRANSMITTER (UART)

The **universal asynchronous receiver transmitter (UART)** is a type of serial-to-parallel or parallel-to-serial data converter. UARTs can be used to connect serial-type data devices to parallel-type data systems, like the microprocessor, as shown in Figure 5.21(a) (p. 171). The receiver and transmitter can operate simultaneously. The transmitter section accepts parallel binary data and converts it to a serial asynchronous output. The receiver section accepts serial asynchronous binary data and converts it to a parallel output.

The serial data format for a typical UART is shown in Figure 5.21(b). Control bits are provided to separate each parallel data word length in the serial format as well as to define the word length and type of parity being used. The low-transition start bit indicates the beginning of data. The bits that follow up to the word length define the converted parallel data. The next bit is the parity bit, which is followed by stop bits indicating the end of the data word. The stop bits may be 1, 1½, or 2 bits long.

5.8 TECH TIPS AND TROUBLESHOOTING—T^3

The tech tips used to troubleshoot data control circuits are essentially the same as the techniques described in the chapters on combination and sequential logic. In the previous chapters we analyzed troubleshooting techniques applied to individual devices and circuits. We now consider troubleshooting an entire system. Figure 5.22 (p. 172) is the schematic diagram for a 12-h digital clock.

When troubleshooting any system, it is important to understand the basic theory of operation. Starting at the power source:

1. The 120 V_{AC}, 60 Hz is stepped down to 6.3 V_{AC}, full-wave bridge-rectified, filtered and regulated by the LM 309 to produce the +5 V_{DC} required for V_{CC}.
2. The 60-Hz reference signal provides the basic timing for the clock. It is limited to the 5-V TTL level by the 620-Ω resistor and the 1N751 zener diode.
3. The 74121 one-shot, U_1, provides wave shaping by producing sharp, clean pulses for each input pulse from the 60-Hz reference.
4. U_2 is a basic 7490 decade counter wired to provide a divide-by-10 output. U_3 is a 7490 wired to produce a divide-by-6 output. The result is a 1-Hz, or one pulse per second, signal that is applied to U_6.

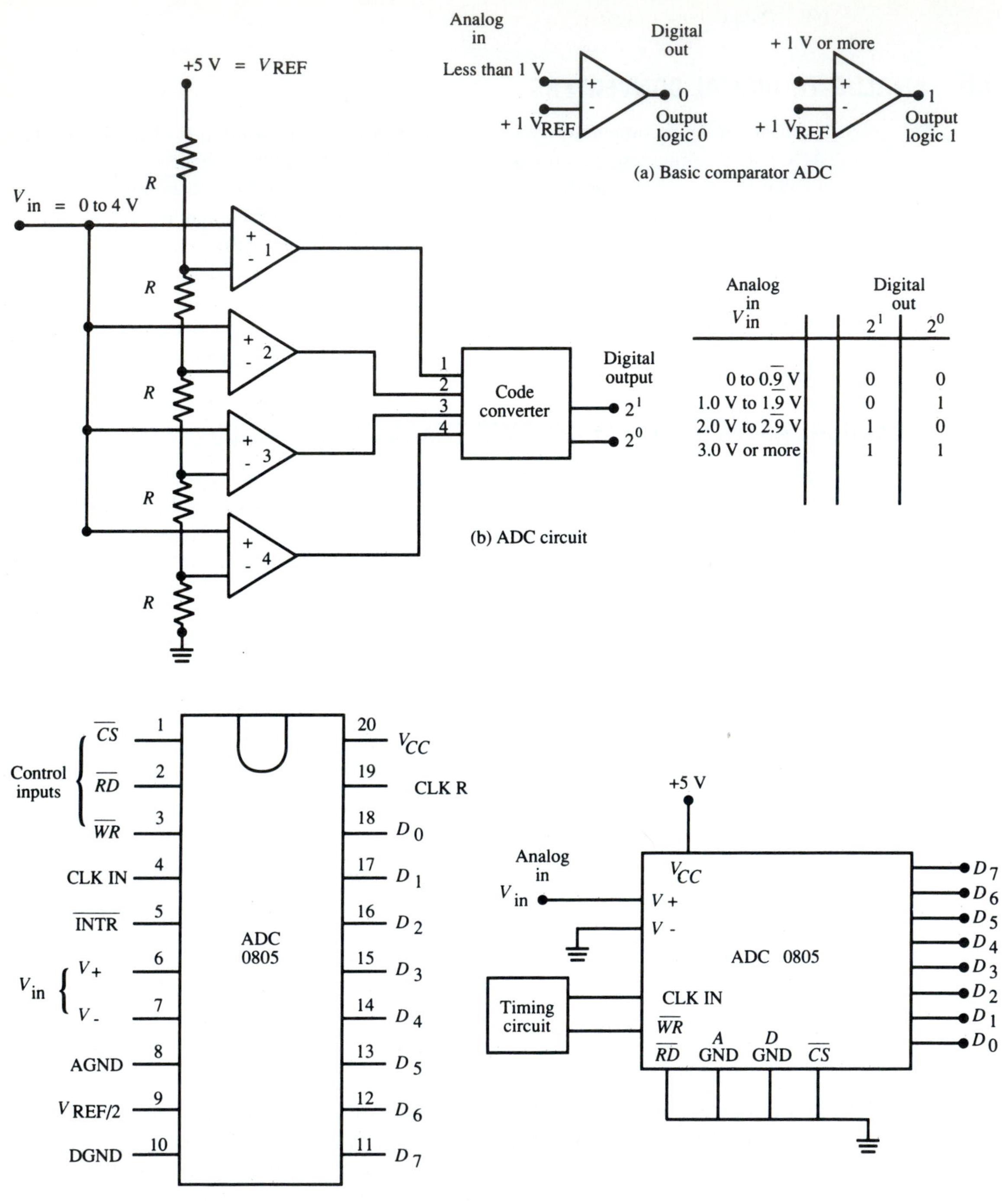

FIGURE 5.20
ADC

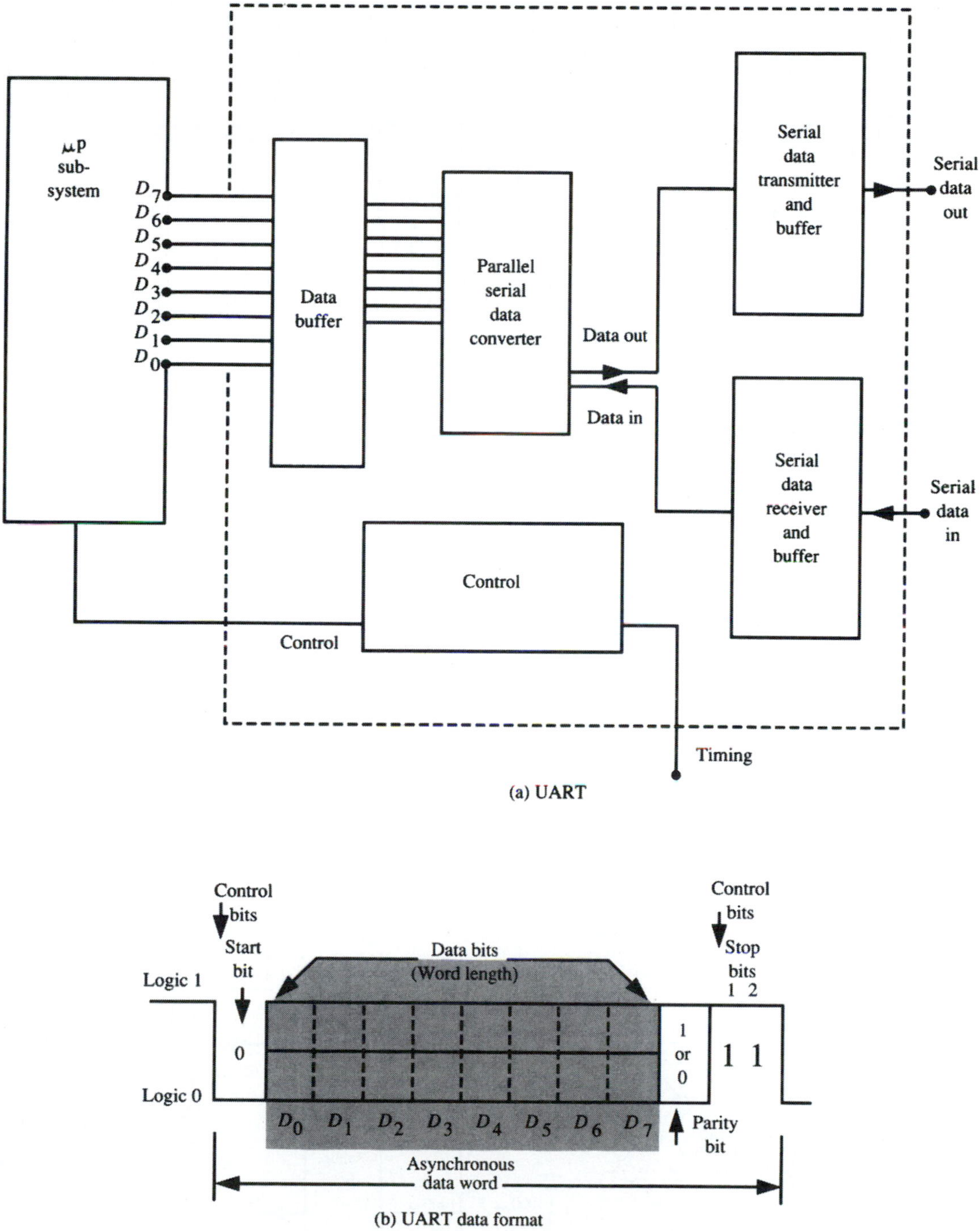

μp
sub-
system
D_7
D_6
D_5
D_4
D_3
D_2
D_1
D_0
Data buffer
Parallel serial data converter
Data out
Data in
Serial data transmitter and buffer
Serial data out
Serial data receiver and buffer
Serial data in
Control
Control
Timing

(a) UART

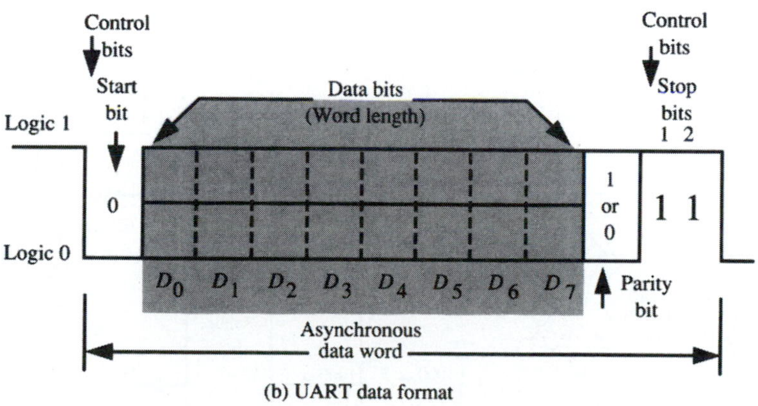

(b) UART data format

FIGURE 5.21
UART

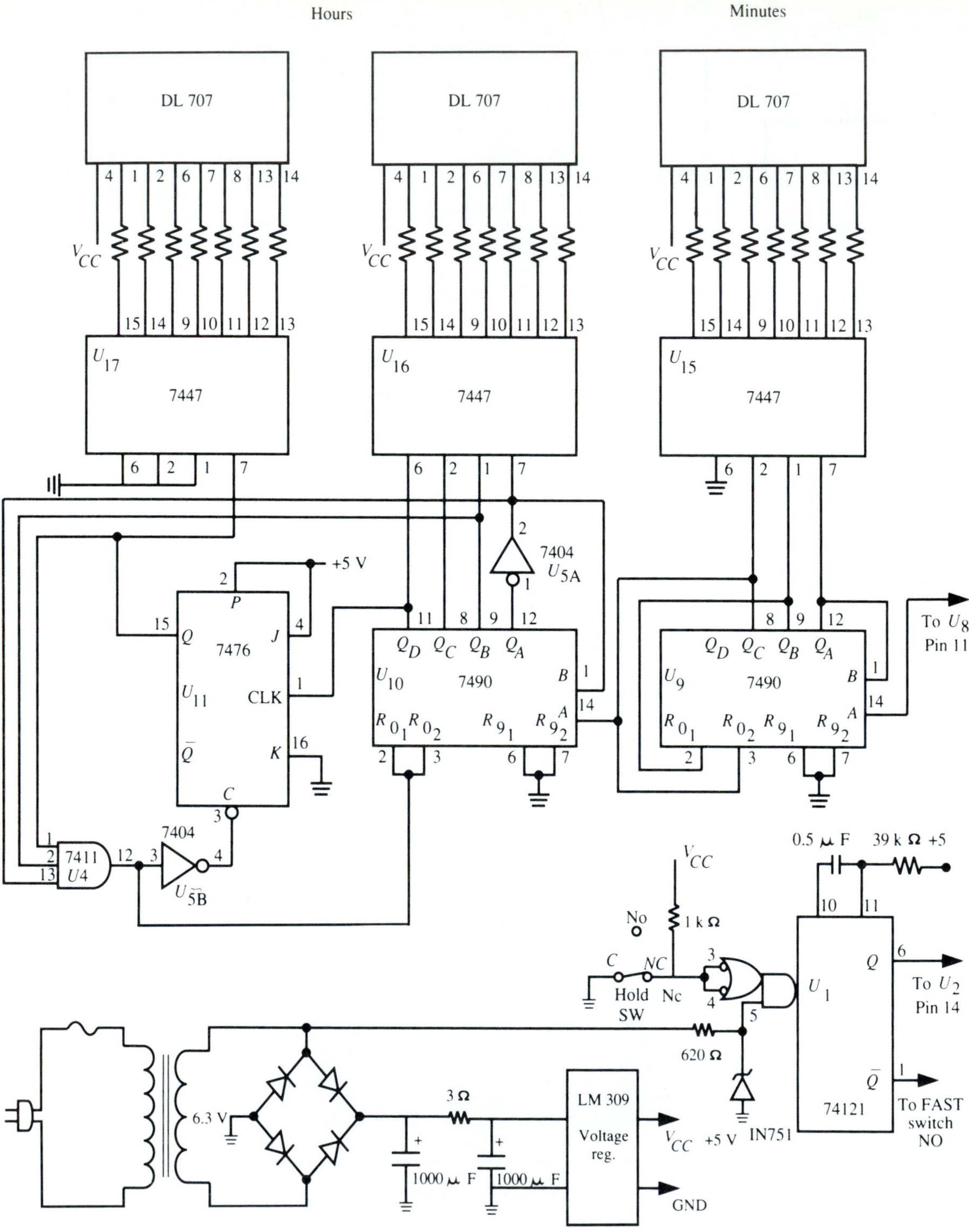

FIGURE 5.22
Twelve-hour clock

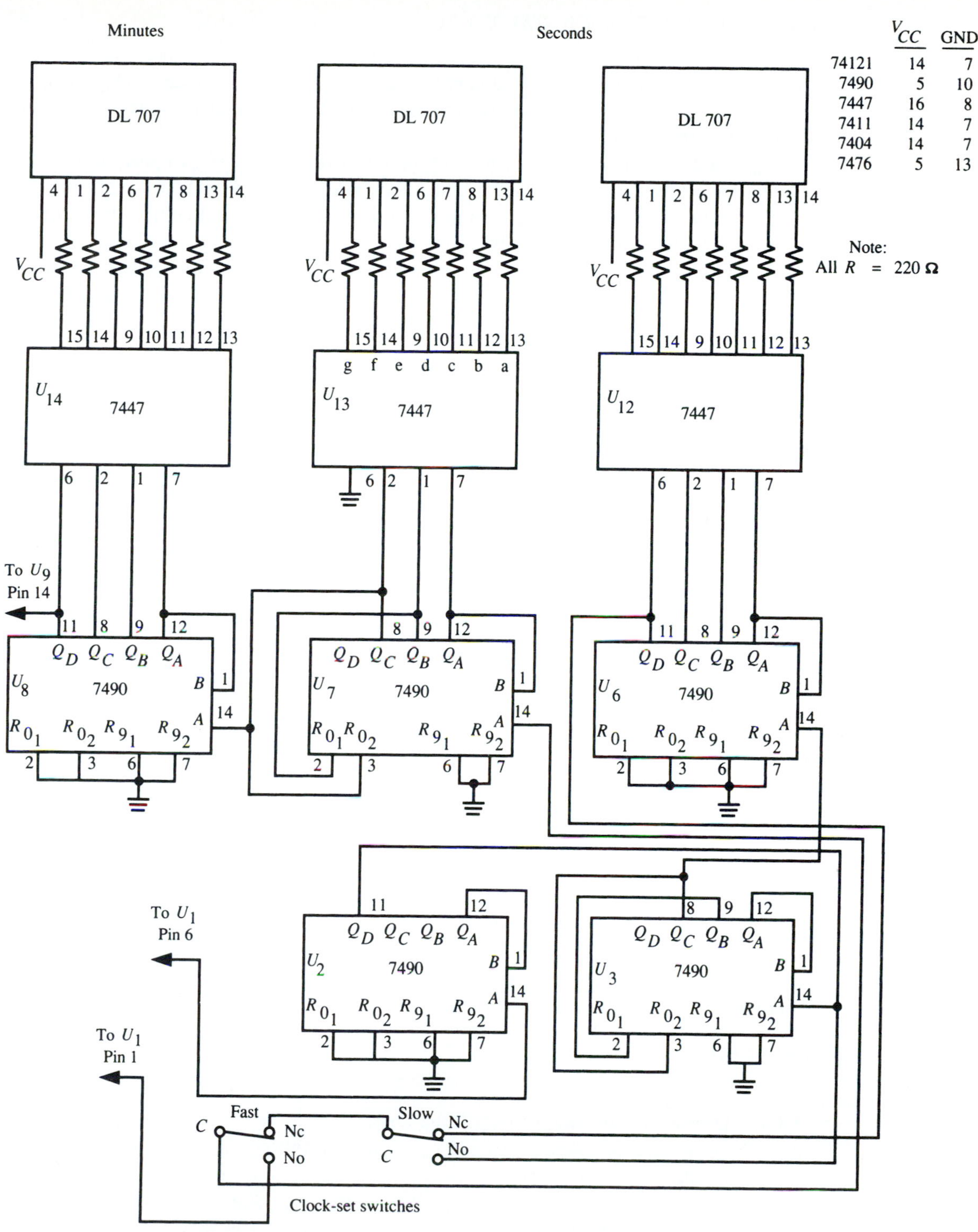

FIGURE 5.22
continued

5. U_6 and U_8 are divide-by-10 counters to provide the units and minutes, respectively. U_7 and U_9 are the divde-by-6 counters that provide the tens of seconds and tens of minutes, respectively.
6. U_{10} is a divide-by-10 counter whose Q_A output is inverted by U_{5A} to provide the unit hours reset. Recall that at 12:59:59, the clock must reset to 01:00:00.
7. U_{11} provides the tens of hours control. We do not need an MSI counter here, since in a 12-h clock this digit is only a 0 or a 1.
8. U_4 and U_{5B} provide decode logic to reset the clock. At 12:59:59 the minutes and seconds will automatically reset to zero. The hours must be reset to 01 h. U_4 decodes 13 o'clock and resets the hours to 01 o'clock. U_{5B} provides the correct logic to reset the flip-flop U_{11}.
9. U_{12} through U_{17} are 7-segment decoder/drivers that are used to control the common anode displays.
10. Switches S_1, S_2, and S_3 are used to set the time of a clock.

Consider the case where the unit minutes display is not functioning properly. We see that the seconds are operating normally but the minutes and hours are not changing or keeping time correctly. This should direct us to believe that the problem is in the minute section of the clock. Begin by *probing* the input to U_8 (pin 14) with a logic probe or oscilloscope. Detecting pulses at this point verifies that we are receiving a signal from the previous stage, U_7. Probing pin 12 of U_9 (Q_A) also indicates that pulses are present. Probing pins 9, 8, and 11, the Q_B, Q_C, and Q_D outputs of U_8, we find no pulse activity present. This indicates either a break or short in the connection between pin 12 and pin 1 or that U_8 is defective. Pin 14 is the input to the Q_A flip-flop. Pin 1 is the input to the Q_B, Q_C, and Q_D asynchronous counter section of the device.

Let's consider another problem. Suppose this time that all the displays appear to be functioning properly except for the minutes. This should direct us to either U_{14} or the unit minutes display itself. U_8 is probably all right, since the succeeding displays appear to be working. We can verify that U_8 is working by probing pins 12, 9, 8, and 11 and noting pulse activity. To check U_{14}, the decoder driver, we probe pins 13, 12, 11, 10, 9, 14, and 15 to look for pulse activity. No pulses indicate that U_{14} is defective. Pulses indicate that the display is defective. Other possible problems could be a bad V_{CC} or ground connection or an open or short on any of the lines between U_{14} and the display.

Understanding how a system works is a key element in the troubleshooting of a system. The techniques used will give you the experience and knowledge to troubleshoot circuits you are not familiar with in the future.

EXERCISES

5.1 Draw a 4-bit data register using *D* flip-flops.

5.2 Design an interface for a 4-bit data bus using a 74175 TTL data register to provide parallel-in, parallel-out data.

5.3 Draw a 5-bit serial in–serial out shift register. Develop a truth table for the circuit operation. Begin with all flip-flops reset and continue for six clock pulses.

5.4 Draw the logic diagram for a serial shift-left shift register using the 74194 universal shift register.

5.5 Draw the logic diagram, timing diagram, and count table for a three-stage ring counter.

5.6 Calculate the output frequency of a three-stage ring counter if the input frequency is 1 MHz.

5.7 Calculate the duty cycle of the output waveform of a 3-stage ring counter.

5.8 Draw the logic diagram, timing diagram, and count table of a 3-stage Johnson counter.

5.9 Calculate the output frequency of a 3-stage Johnson counter if the input frequency is 5 MHz.

5.10 Draw the logic diagram and truth table for a noninverting, active high–enable tristate buffer.

5.11 Draw the logic diagram and truth table for an inverting, active low–enable tristate device.

5.12 Draw the logic diagram and truth table for a 74126 buffer used to interface a 4-bit data bus to a 4-bit output device.

5.13 Draw the logic symbol and input/output waveforms for an inverting Schmitt trigger device.

5.14 Draw the wiring diagram for a 16-bit transceiver using 74245 devices. Have the devices set to receive data. Show the proper connections for all pins.

5.15 Draw the output waveforms for D_{out} if the input waveforms of Figure 5.23 are applied to the multiplexer of Figure 5.13(c).

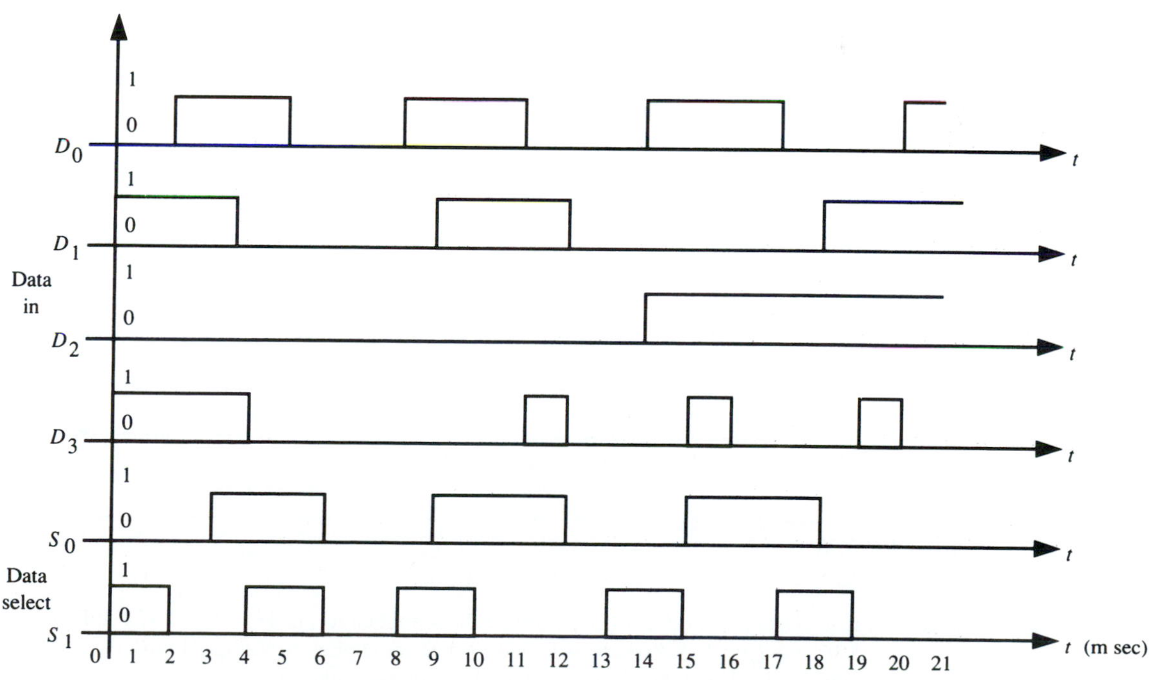

FIGURE 5.23

5.16 Draw the output waveforms D_0, D_1, D_2, and D_3 if the waveforms of Figure 5.24 are applied to the D multiplexer of Figure 5.15(c).

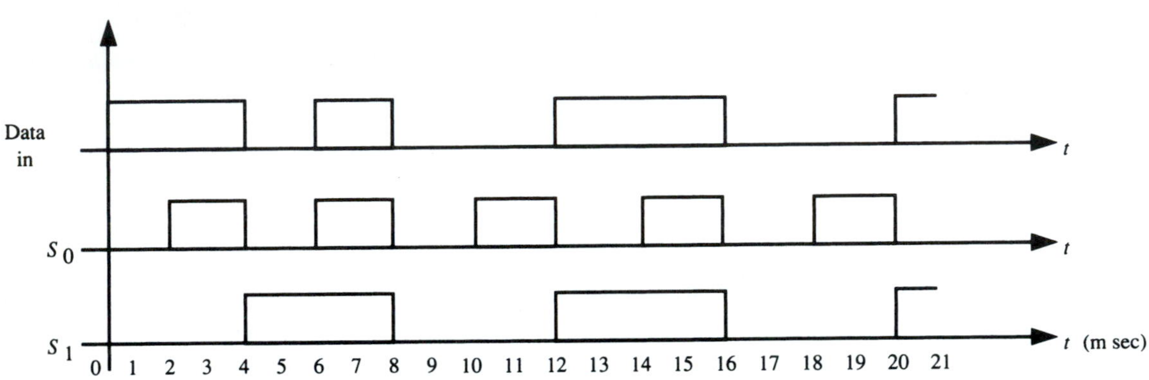

FIGURE 5.24

5.17 List the output that will be seen for each second of LED 7-segment display of Figure 5.18(b) if the waveform of Figure 5.25 is applied to the inputs of the 7447 of Figure 5.18(b).

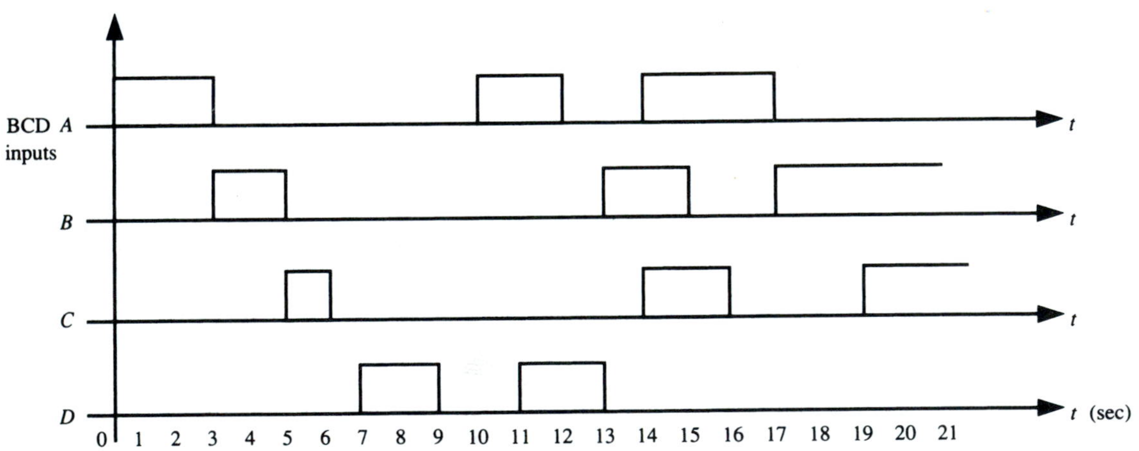

FIGURE 5.25

5.18 If the Data in waveform from Figure 5.24 is applied to pin 18 of the 74154 in Figure 5.16 and $S_0 = 1$, $S_1 = 1$, $S_2 = 0$, and $S_3 = 0$ are applied to the select pins, draw the output waveform that will result at the Data output line 3.

5.19 Determine the resolution of

a. A 10-bit DAC

b. A 12-bit DAC

5.20 Determine the logic output for Figure 5.20(b) if 2.532 V is applied to V_{in}.

5.21 For the asynchronous data word 000000011111, which is being applied to a UART, identify the function and content of each bit or group of bits.

5.22 Explain the operation of the seconds portion of the 12-h clock of Figure 5.22.

5.23 If the tens of seconds display in the clock of Figure 5.22 is not counting, explain what you should test and why.

5.24 The hours display in the clock of Figure 5.22 does not reset to 01:00:00 after counting to 12:59:59 but instead resets to 00:00:00. Explain what you would test and why.

5.25 Crossword Puzzle

ACROSS

5. Converting form a primary coding system to a secondary coding system.
7. Circuits used to interconnect different subsystems together.
8. Signal into a gate.
9. Used to hold data temporarily.
10. A data _____ can be thought of as a highway where each wire represents a lane.
12. Circuit which switches states at threshold or trigger points.
14. A device used to connect serial type data devices to parallel type data devices.
16. Twisted ring counter.
18. A type of register.
23. A device used to convert a digital signal to an analog signal.
25. DAC voltage specification.
26. When buffers are used to accept data they are called _____.
27. A device which selects one of many input signals.

DOWN

1. A DAC time specification.
2. Measure of the difference between the expected output voltage and the actual output voltage.
3. _____ and wave-shaping devices are used to provide clocking and triggering synchronization.
4. SISO shift register.
6. DAC output voltage when the binary inputs are zero.
11. Term for a register used to hold data.
13. Term for buffers that are being used to send data.
15. Bidirectional device.
17. Unit of resistance.
19. Symbol used to indicate a tristate device.
20. Changes data from one format to another.
21. Converting from a secondary coding system back to the primary coding system.
22. Term for the 74194 shift register.
24. Type of amplifier that provides isolation.
25. Circulating shift register.

1 2 3 4 5 6 7 8 9 10 11 12 13 14 15 16 17 18 19 20 21 22 23 24 25 26 27

CHAPTER 6

TROUBLESHOOTING LOGIC FAMILIES

KEY TERMS

Current Sinking
Current Sourcing
Decoupling Capacitors
Fan-out
Input Diode Anomaly
MOSFET
Noise Immunity
Noise Margin
Open Collector
Power Dissipation
Propagation Delay Time
Schottky Barrier Diode
Unit Load
Wired-AND

6.0 INTRODUCTION

When troubleshooting computer systems, the most important single component you will deal with is the IC. In this chapter we study the specifications, characteristics, and circuits that make up ICs. Recall that ICs are grouped by common characteristics called logic families. The two most common logic families in use today are the TTL and CMOS logic families. Devices within the same logic family are designed to be connected or interfaced directly.

In this chapter we examine each of the important electrical characteristics of ICs. We also examine how to connect or interface ICs from different logic families. The electrical characteristics of ICs can be found on manufacturer's data sheets, which are often grouped together by logic family in a manufacturer's data book. Learning to read and understand IC specifications can be very helpful when troubleshooting digital and computer systems because you do not have to memorize all the facts about a specific IC. Furthermore, being able to read and interpret IC data sheets will be helpful when you encounter a new IC for the first time.

6.1 IC SPECIFICATIONS

The typical IC data sheet can be broken down into three parts:

1. Summary
2. Schematic diagram
3. Electrical characteristics

The *summary part* contains information describing the functional type and part number of the device. It tells us the operating temperature range, the truth or function table, logic symbol or diagram, Boolean output equation, and packaging and pin configuration for the device. Figure 6.1 shows the summary part of a data sheet for a TTL quadruple 2-input positive logic NAND gate (74LS00). Referring to Figure 6.1(a) we note that the 74LS00 data sheets tell us the following:

1. The device contains four independent 2-input NAND gates.
2. The operating temperature is between 0°C and 70°C.
3. From the truth table, the output will be low (L) *only* when both inputs are high (H).
4. The logic diagram shows the symbol for a 2-input NAND gate with A and B as the input and Y as the output.
5. The Boolean expression is given as $Y = \overline{A \cdot B}$, or $Y = \overline{A} + \overline{B}$.
6. The pin configuration and packaging styles that are available from this manufacturer are shown.

The *schematic diagram part* of the data sheet shows the internal circuit diagram of the device. Figure 6.1(b) shows the 74LS00 schematic. Notice that four different schematics are shown, one for each available subseries of the device. For example, the first schematic is for the standard version, 00, and the low-power version, L00. The second schematic is for the high-speed version, H00. The third is for the low-power Schottky version, LS00, and the fourth schematic is for the Schottky version, S00. We discuss the operation of these circuits later in the chapter. Also given in the schematic diagram part of the data sheet are the absolute maximum ratings for V_{CC}, input voltage, operating temperature, and storage temperature.

The *electrical characteristics part* of the data sheet is divided into three subsections: recommended operating conditions, electrical characteristics, and switching characteristics. Figure 6.1(c) shows the electrical characteristics part of the 74LS00 data sheet. The most useful characteristics are as follows:

- V_{CC}—*supply voltage:* Typically the nominal value is $+5\,V_{DC}$ with a range of $+4.75\,V_{DC}$ to $+5.25\,V_{DC}$.
- V_{IH}—*high-level input voltage:* The minimum input voltage needed to be recognized as a logic 1 by the device. Typically $V_{IH} = 2.0\ V_{DC}$.
- V_{IL}—*low-level input voltage:* The maximum input voltage allowed to be recognized as a logic 0 by the device. Typically $V_{IL} = 0.8\ V_{DC}$.
- I_{OH}—*high-level output current:* The maximum output current when the device is in the logic 1 output state. Typically $I_{OH} = -0.4$ mA (minus sign indicates current is leaving the device).
- I_{OL}—*low-level output current:* The maximum output current when the device is in the logic 0 output state. Typically $I_{OL} = 8$ mA.
- V_{OH}—*high-level output voltage:* The logic 1 output voltage. It defines the lowest voltage that the device will output in the logic 1 state. Typically $V_{OH} = 3.4\ V_{DC}$; minimum $V_{OH} = 2.7\ V_{DC}$.
- V_{OL}—*low-level output voltage:* The logic 0 output voltage. It defines the highest voltage that the device will output in the logic 0 state. Typically $V_{OL} = 0.35\ V_{DC}$; maximum $V_{OL} = 0.5\ V_{DC}$.

REVISED DECEMBER 1983

- **Package Options Include Both Plastic and Ceramic Chip Carriers in Addition to Plastic and Ceramic DIPs**
- **Dependable Texas Instruments Quality and Reliability**

description

These devices contain four independent 2-input NAND gates.

The SN5400, SN54H00, SN54L00, and SN54LS00, and SN54S00 are characterized for operation over the full military temperature range of −55°C to 125°C. The SN7400, SN74H00, SN74LS00, and SN74S00 are characterized for operation from 0°C to 70°C.

FUNCTION TABLE (each gate)

INPUTS		OUTPUT
A	B	Y
H	H	L
L	X	H
X	L	H

logic diagram (each gate)

A, B → Y

positive logic

$$Y = \overline{A \cdot B} \text{ or } Y = \overline{A} + \overline{B}$$

SN5400, SN54H00, SN54L00 . . . J PACKAGE
SN54LS00, SN54S00 . . . J OR W PACKAGE
SN7400, SN74H00 . . . J OR N PACKAGE
SN74LS00, SN74S00 . . . D, J OR N PACKAGE
(TOP VIEW)

Pin	Name	Pin	Name
1	1A	14	V_{CC}
2	1B	13	4B
3	1Y	12	4A
4	2A	11	4Y
5	2B	10	3B
6	2Y	9	3A
7	GND	8	3Y

SN5400, SN54H00 . . . W PACKAGE
(TOP VIEW)

Pin	Name	Pin	Name
1	1A	14	4Y
2	1B	13	4B
3	1Y	12	4A
4	V_{CC}	11	GND
5	2Y	10	3B
6	2A	9	3A
7	2B	8	3Y

SN54LS00, SN54S00 . . . FK PACKAGE
SN74LS00, SN74S00 . . . FN PACKAGE
(TOP VIEW)

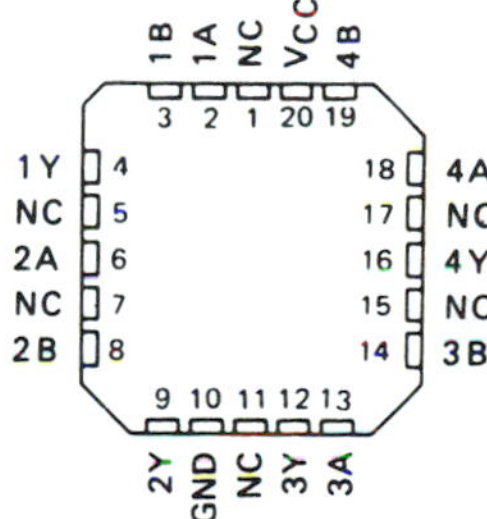

NC – No internal connection

3 TTL DEVICES

FIGURE 6.1a
74LS00 data sheet (pp. 181–183 reprinted by permission of Texas Instruments Inc.)

schematics (each gate)

'00, 'L00

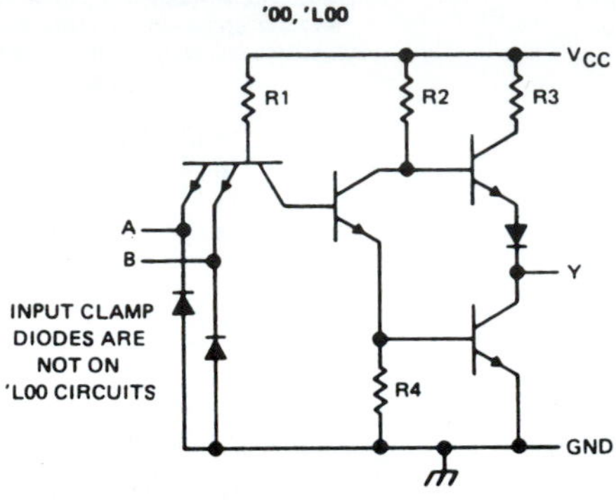

CIRCUIT	R1	R2	R3	R4
'00	4 kΩ	1.6 kΩ	130 Ω	1 kΩ
'L00	40 kΩ	20 kΩ	500 Ω	12 kΩ

'H00

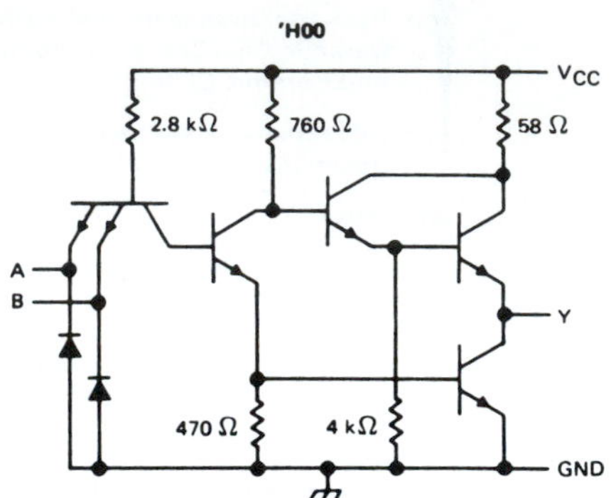

'LS00

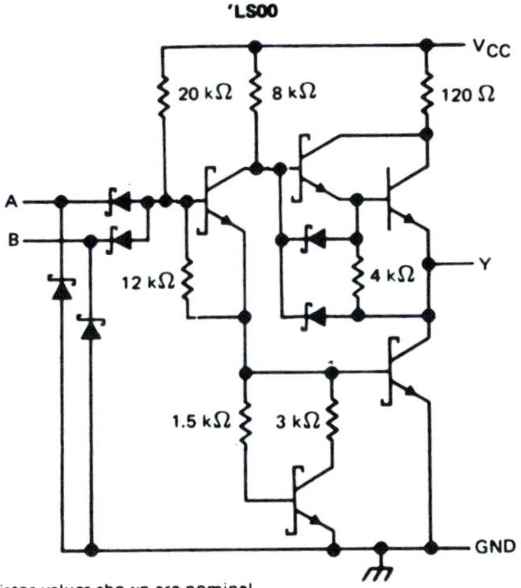

'S00

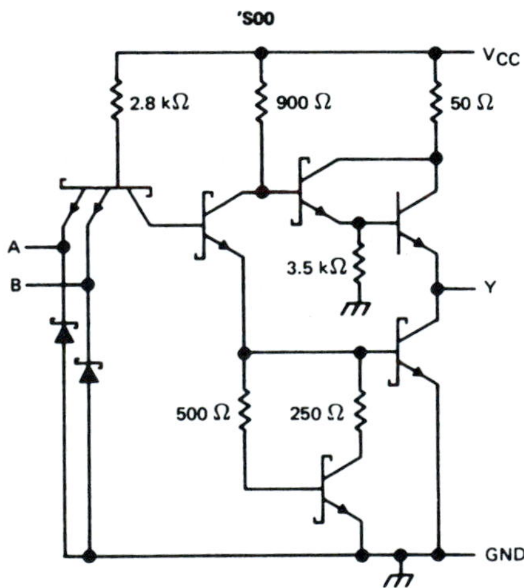

Resistor values shown are nominal.

absolute maximum ratings over operating free-air temperature range (unless otherwise noted)

Supply voltage, V_{CC} (see Note 1) '00, 'H00, 'LS00, 'S00 . . . . . . 7 V
'L00 . . . . . . 8 V
Input voltage: '00, 'H00, 'L00, 'S00 . . . . . . 5.5 V
'LS00 . . . . . . 7 V
Operating free-air temperature range: SN54' . . . . . . − 55°C to 125°C
SN74' . . . . . . 0°C to 70°C
Storage temperature range . . . . . . − 65°C to 150°C

NOTE 1: Voltage values are with respect to network ground terminal.

FIGURE 6.1b

recommended operating conditions

		SN54LS00			SN74LS00			UNIT
		MIN	NOM	MAX	MIN	NOM	MAX	
V_{CC}	Supply voltage	4.5	5	5.5	4.75	5	5.25	V
V_{IH}	High-level input voltage	2			2			V
V_{IL}	Low-level input voltage			0.7			0.8	V
I_{OH}	High-level output current			– 0.4			– 0.4	mA
I_{OL}	Low-level output current			4			8	mA
T_A	Operating free-air temperature	– 55		125	0		70	°C

electrical characteristics over recommended operating free-air temperature range (unless otherwise noted)

PARAMETER	TEST CONDITIONS †	SN54LS00			SN74LS00			UNIT
		MIN	TYP‡	MAX	MIN	TYP‡	MAX	
V_{IK}	V_{CC} = MIN, I_I = – 18 mA			– 1.5			– 1.5	V
V_{OH}	V_{CC} = MIN, V_{IL} = MAX, I_{OH} = – 0.4 mA	2.5	3.4		2.7	3.4		V
V_{OL}	V_{CC} = MIN, V_{IH} = 2 V, I_{OL} = 4 mA		0.25	0.4		0.25	0.4	V
	V_{CC} = MIN, V_{IH} = 2 V, I_{OL} = 8 mA					0.35	0.5	
I_I	V_{CC} = MAX, V_I = 7 V			0.1			0.1	mA
I_{IH}	V_{CC} = MAX, V_I = 2.7 V			20			20	μA
I_{IL}	V_{CC} = MAX, V_I = 0.4 V			– 0.4			– 0.4	mA
I_{OS}§	V_{CC} = MAX	– 20		– 100	– 20		– 100	mA
I_{CCH}	V_{CC} = MAX, V_I = 0 V		0.8	1.6		0.8	1.6	mA
I_{CCL}	V_{CC} = MAX, V_I = 4.5 V		2.4	4.4		2.4	4.4	mA

† For conditions shown as MIN or MAX, use the appropriate value specified under recommended operating conditions.
‡ All typical values are at V_{CC} = 5 V, T_A = 25°C
§ Not more than one output should be shorted at a time, and the duration of the short-circuit should not exceed one second.

switching characteristics, V_{CC} = 5 V, T_A = 25°C (see note 2)

PARAMETER	FROM (INPUT)	TO (OUTPUT)	TEST CONDITIONS	MIN	TYP	MAX	UNIT
t_{PLH}	A or B	Y	R_L = 2 kΩ, C_L = 15 pF		9	15	ns
t_{PHL}					10	15	ns

NOTE 2: See General Information Section for load circuits and voltage waveforms.

FIGURE 6.1c

- I_{IH}—*high-level input current:* The logic 1 input current that will flow into a device's input. Typically $I_{IH} = 20\ \mu A$ maximum.
- I_{IL}—*low-level input current:* The logic 0 input current that will flow through a device's input. Typically $I_{IL} = -0.4$ mA.
- I_{CCH}—*supply current high:* The supply current when the output is a logic 1 (high). The total current supplied to the device from V_{CC}. Typically $I_{CCH} = 0.8$ mA; maximum $I_{CCH} = 1.6$ mA.
- I_{CCL}—*supply current low:* The supply current when the output is a logic 0 (low). The total current supplied to the device from V_{CC}. Typically $I_{CCL} = 2.4$ mA; maximum $I_{CCL} = 4.4$ mA.
- t_{PLH}—*propagation low to high:* The time required for the device to switch or change its output state from a logic 0 to a logic 1 (low to high). Typically $t_{PLH} = 9$ ns; maximum $t_{PLH} = 15$ ns.
- t_{PHL}—*propagation high to low:* The time required for the device to switch or change its output state from a logic 1 to a logic 0 (high to low). Typically $t_{PHL} = 10$ ns; maximum $t_{PHL} = 15$ ns. The parameters t_{PLH} and t_{PHL} are also known as the **propagation delay times** of the device.

Note that all the preceding typical, minimum, and maximum specifications are from the 74LS00 data sheet shown in Figure 6.1.

6.2 DATA SHEET ANALYSIS

Three useful electrical characteristics can be derived from the data sheet by calculation, power dissipation, noise immunity, and fan-out.

Power Dissipation

Power dissipation is the power consumed by the device. In general, power dissipation can be calculated by equation (6.1).

$$P_D = V_{CC} \times I_{CC} \tag{6.1}$$

where V_{CC} is the voltage applied to the power supply pin on the device and I_{CC} is the current flowing through that pin. Note that for TTL logic family devices the power pin is usually labeled V_{CC}, whereas for CMOS logic family devices the power pin is usually labeled V_{DD}.

As previously described, I_{CC} varies depending on the logic state of the device (i.e., I_{CCH}, I_{CCL}). Therefore, it is often useful to consider the average power dissipation by calculating an average I_{CC} as follows:

$$I_{CC(\text{avg})} = \frac{I_{CCH} + I_{CCL}}{2} \tag{6.2}$$

and

$$P_{D(\text{avg})} = V_{CC} \times I_{CC(\text{avg})} \quad \textbf{(6.3)}$$

Noise Immunity

Noise immunity is a measure of a circuit's ability to handle noise voltage on its inputs. Noise voltage is an unwanted electrical signal that can be picked up by circuit wiring leading into a device. The noise voltage can be added to or subtracted from the intended input voltage, causing a false or incorrect logic signal state to be interpreted by the device, as shown in Figure 6.2.

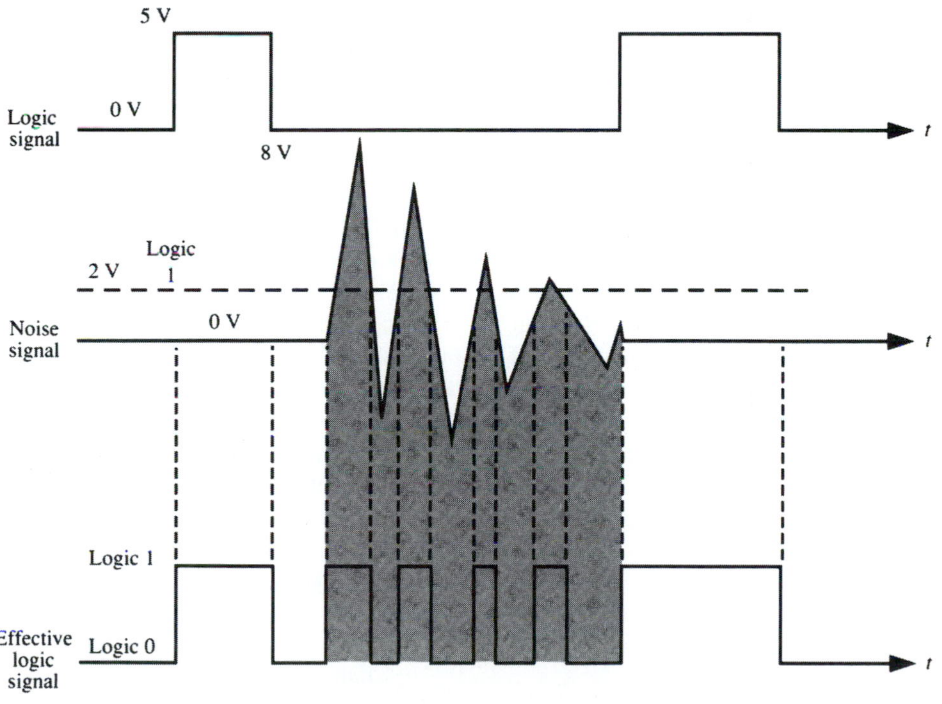

FIGURE 6.2
Noise immunity

Noise immunity calculations are referred to as the **noise margin** of a device. Output voltages greater than $V_{OH(\text{min})}$ will be interpreted by the device as a logic 1. Input voltages greater than $V_{IH(\text{min})}$ will be interpreted by the device as a logic 1. The high-state noise margin becomes

$$V_{NMH} = V_{OH(\text{min})} - V_{IH(\text{min})} \quad \textbf{(6.4)}$$

For the 74LS00 data sheet of Figure 6.1 the high-state noise margin is calculated as follows.

$$\begin{aligned} V_{NMH} &= V_{OH(\text{min})} - V_{IH(\text{min})} \\ &= 2.4 - 2.0 \\ &= 0.4\,\text{V} \end{aligned}$$

Input voltages less than $V_{IL(\text{max})}$ will be interpreted as a logic 0. Output voltages less than $V_{OL(\text{max})}$ will be interpreted as a logic 0. Thus the low-state noise margin becomes

$$V_{NML} = V_{IL(\text{max})} - V_{OL(\text{max})} \tag{6.5}$$

For the 74LS00 data sheet of Figure 6.1 the low-state noise margin is calculated as follows.

$$\begin{aligned} V_{NML} &= V_{IL(\text{max})} - V_{OL(\text{max})} \\ &= 0.8 - 0.4 \\ &= 0.4\text{ V} \end{aligned}$$

Fan-Out

Fan-out is defined as the maximum number of device inputs of the same logic family that a device output can drive in parallel. In Figure 6.3 the output of the 74LS00 (gate 0) is being used to drive 20 74LS00 (10 devices, gates 1–10) inputs. For most TTL logic family devices this number is 10 inputs or more.

The calculation of fan-out requires the analysis of two parameters—the current available from the output device and the current required by the device input. Recall that the output current available depends on the logic state of the device (I_{OH}, I_{OL}). This is also true for device input current requirements (I_{IH}, I_{IL}). Therefore, two calculations for fan-out are required, one for the high state and one for the low state. The actual device fan-out is said to be the worst case or lowest condition. Specifically,

$$\text{Fan-out}_{\text{high}} = \frac{I_{OH}}{I_{IH}} \tag{6.6}$$

$$\text{Fan-out}_{\text{low}} = \frac{I_{OL}}{I_{IL}} \tag{6.7}$$

For the 74LS00 data sheet of Figure 6.1, fan-out is calculated as follows:

$$\begin{aligned} \text{Fan-out}_{\text{high}} &= \frac{I_{OH}}{I_{IH}} \\ &= \frac{0.4\text{ mA}}{20\ \mu\text{A}} \\ &= \frac{400\ \mu\text{A}}{20\ \mu\text{A}} \\ &= 20 \end{aligned}$$

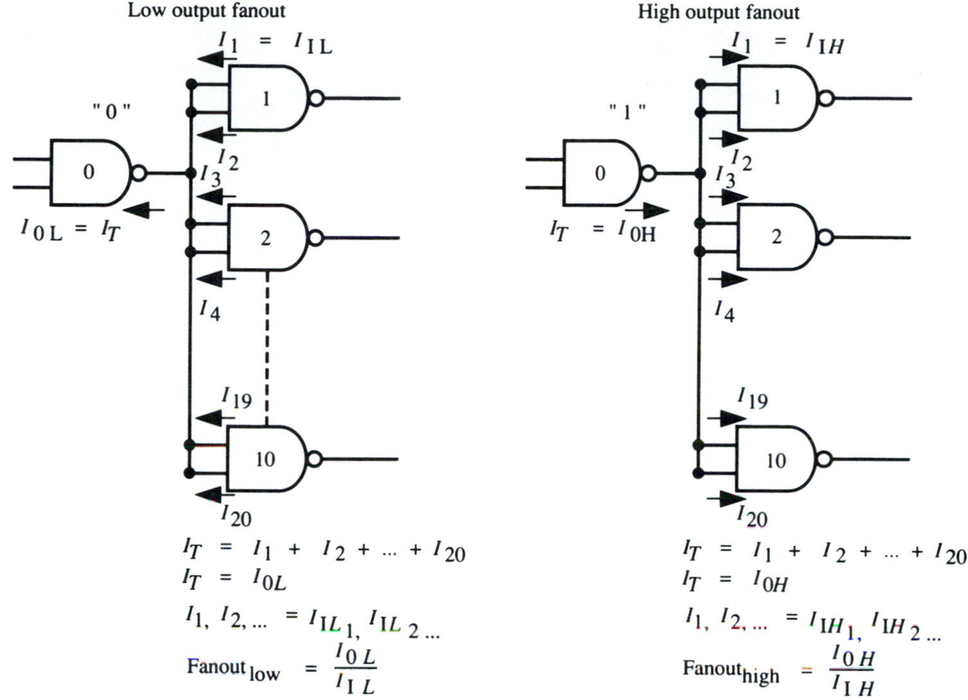

FIGURE 6.3
Fan-out

$$\text{Fan-out}_{low} = \frac{I_{OL}}{I_{IL}} = \frac{8\text{ mA}}{0.4\text{ mA}} = 20$$

Note that the fan-out for a device is the worst case, or lowest condition. In this case, for the 74LS00, fan-out$_{high}$ and fan-out$_{low}$ turn out to be the same value. This will not always be the case with other devices. It should also be noted that fan-out is sometimes referred to as **unit load** calculations.

Current Sourcing and Current Sinking

When the output of a gate is high and it is supplying current to the input of another gate or gates, this condition is known as **current sourcing.** Current sourcing can then be defined as I_{OH} as shown in Figure 6.4(a). In Figure 6.4(a) the output of NAND gate 1 is high and supplying current to the input of NAND gate 2, where ground is supplied externally.

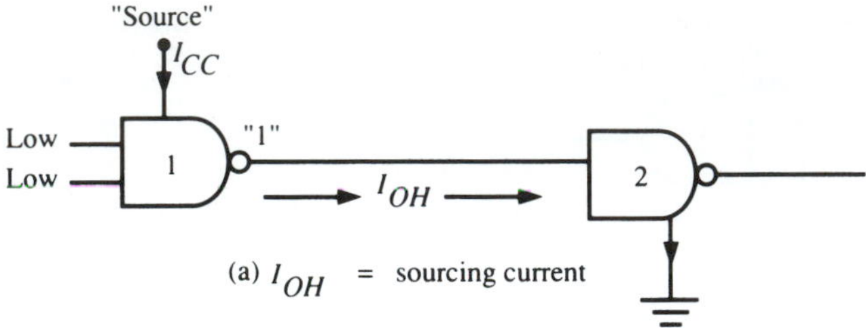

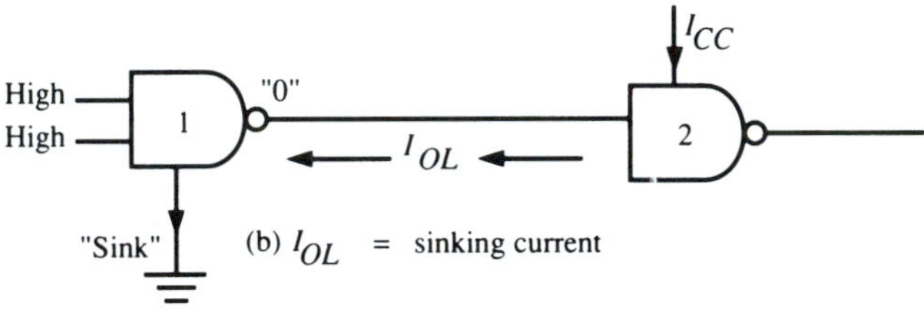

FIGURE 6.4
Sourcing/sinking

When the output of a gate is low and current is flowing from the input of another gate or gates into the output, this condition is known as **current sinking.** Current sinking is then a measure of how much current can enter a device output when it is in the logic 0 state and not have the output voltage rise above a specified limit. Current sinking can then be defined as I_{OL}, as shown in Figure 6.4(b). In Figure 6.4(b) the output of NAND gate 1 is low (providing ground internally), and NAND gate 1 must sink current from the input of NAND gate 2.

6.3 TTL CIRCUIT ANALYSIS

The basic bipolar transistor is used as an inverter and functions as a transistor switch. For the *NPN* transistor of Figure 6.5, when the base is positive with respect to the emitter, the transistor is said to be switched on. Current then flows through the transistor to ground. The collector is effectively at ground potential. Thus when the input to the transistor base is a logic 1 (+5 V), the output taken off the collector is approximately 0 V, or logic 0. When the input to the base is a logic 0, (0 V) the transistor is switched off. No current is allowed to flow through the transistor. Thus there is no voltage drop across resistor R_C, and the output voltage at the collector is approximately 5 V, or a logic 1. Thus the basic bipolar transistor can be made to function as an inverter.

The diode arrangement of Figure 6.6(a) (p. 190) functions as an AND gate. Recall that when a diode is forward biased (anode positive with respect to cathode), the diode

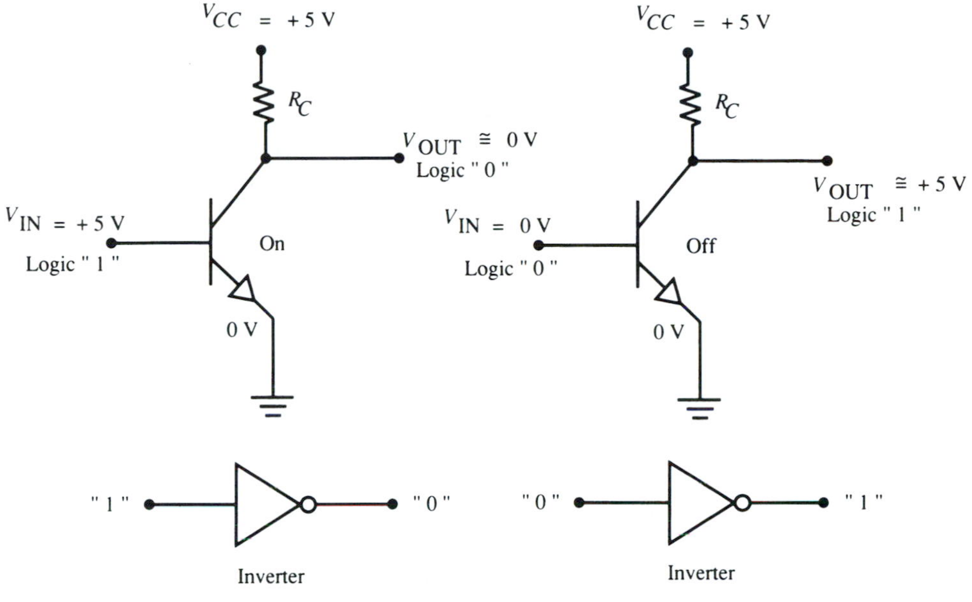

FIGURE 6.5
NOT gate (inverter)

acts like a closed switch. When a diode is reverse biased (anode negative with respect to cathode), the diode acts like an open switch. For the circuit of Figure 6.6(a), whenever a logic 0 (0 V, or ground) is applied to either input *A* or *B*, the respective diode is forward biased and made to conduct. The output voltage for a logic 0 input condition is 0 V, or a logic 0. Whenever a logic 1 (+5 V) is applied to either input *A* or *B*, the respective diode is reverse biased. If a logic 1 is applied to both inputs simultaneously, both diodes are reverse biased. No current flows through resistor *R*. Thus there is no voltage drop across resistor *R*, and the resulting output voltage is approximately V_{CC} or +5 V (logic 1). The truth table and voltage table of Figure 6.6(a) show all the possible combinations of input and resulting output conditions. This shows clearly that the circuit functions as an AND gate.

The diode arrangement of Figure 6.6(b) functions as an OR gate. Whenever a logic 1 (+5 V) is applied to either input *A* or *B*, the respective diode conducts. The output voltage is approximately 5 V (logic 1). Whenever a logic 0 (0 V or ground) is applied to either input *A* or *B*, the respective diode is switched off. The output voltage is 0 V (logic 0) because there is no input voltage. The truth table and voltage table of Figure 6.6(b) show all the possible combinations of input and resulting output conditions. This shows clearly that the circuit functions as an OR gate.

It should be noted that a NAND gate can be constructed by combining the AND gate and inverter circuit as shown in Figure 6.7(a) (p. 191). A NOR gate can be constructed by combining an OR gate and an inverter circuit as shown in Figure 6.7(b).

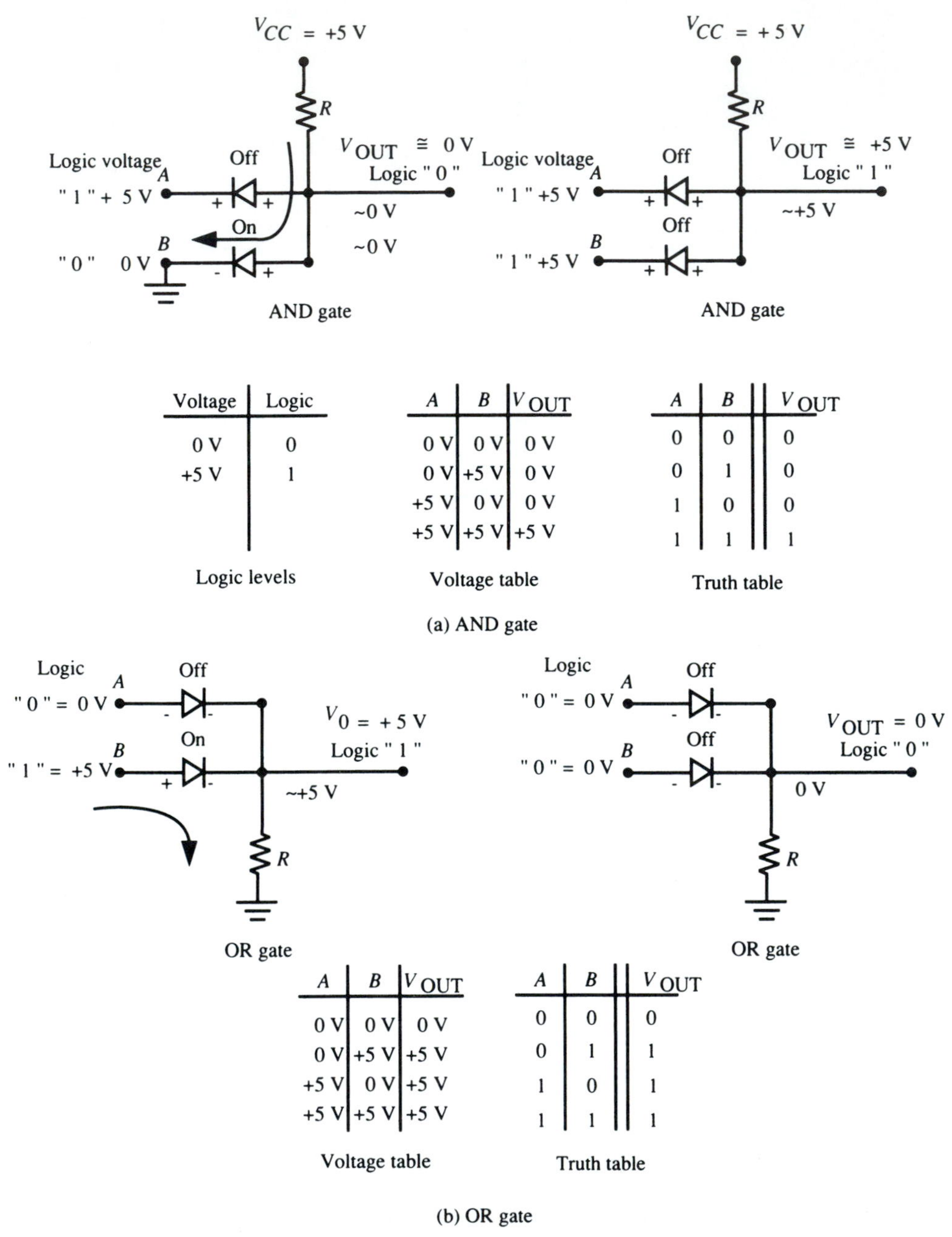

Voltage	Logic
0 V	0
+5 V	1

Logic levels

A	B	V_{OUT}
0 V	0 V	0 V
0 V	+5 V	0 V
+5 V	0 V	0 V
+5 V	+5 V	+5 V

Voltage table

A	B	V_{OUT}
0	0	0
0	1	0
1	0	0
1	1	1

Truth table

(a) AND gate

A	B	V_{OUT}
0 V	0 V	0 V
0 V	+5 V	+5 V
+5 V	0 V	+5 V
+5 V	+5 V	+5 V

Voltage table

A	B	V_{OUT}
0	0	0
0	1	1
1	0	1
1	1	1

Truth table

(b) OR gate

FIGURE 6.6
Diode logic gates

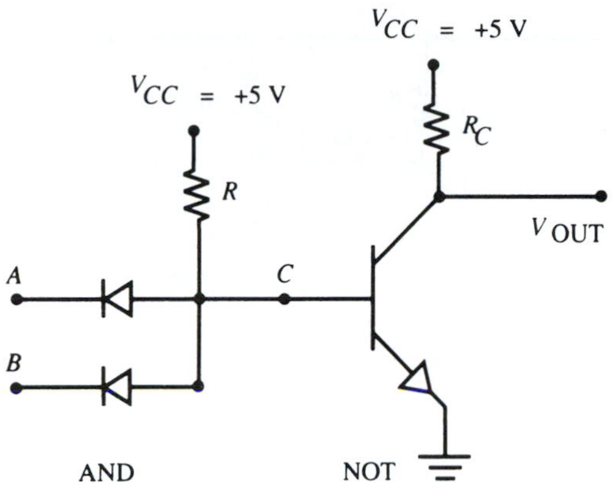

A	B	C	V_{OUT}
0 V	0 V	0 V	+5 V = 1
0 V	+5 V	0 V	+5 V = 1
+5 V	0 V	0 V	+5 V = 1
+5 V	+5 V	+5 V	0 V = 0

Truth/voltage table

(a) NAND gate

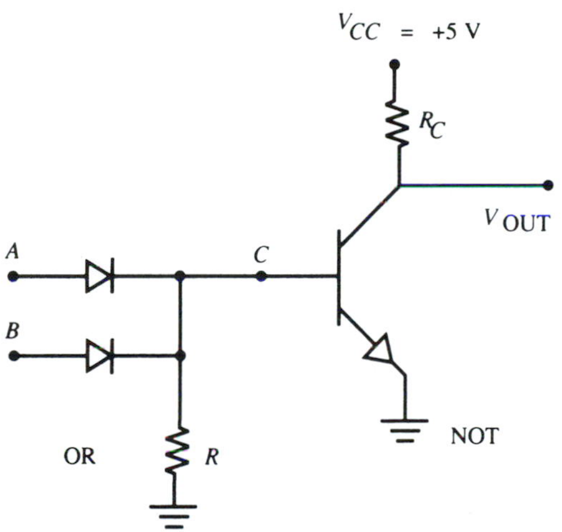

A	B	C	V_{OUT}
0 V	0 V	0 V	+5 V = 1
0 V	+5 V	+5 V	0 V = 0
+5 V	0 V	+5 V	0 V = 0
+5 V	+5 V	+5 V	0 V = 0

Truth/voltage table

(b) NOR gate

FIGURE 6.7
NAND/NOR gate

TTL NAND Gate

The circuit of Figure 6.8(a) is a typical TTL family NAND gate. The circuit operation is not as difficult as it might first appear; however, it is important to remember the switching rules for diodes and transistors. Input transistor Q_1 has two emitters. This may appear confusing, but by recalling that a transistor is analogous to multiple diodes, we can rely on the equivalent circuit of Figure 6.8(b) to explain the circuit operation. Whenever either input A or B is a logic 0 (0 V or ground), the respective base–emitter diode of Q_1 is switched on. Thus points C and D are at ground potential, or at approximately 0 V.

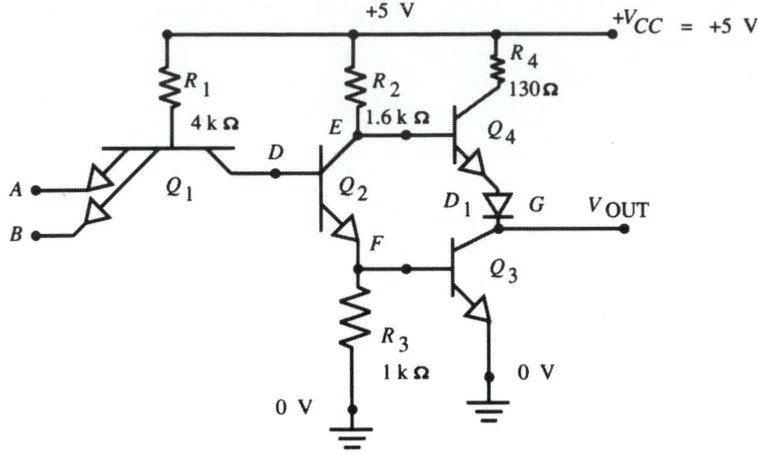

(a) Basic TTL NAND gate circuit

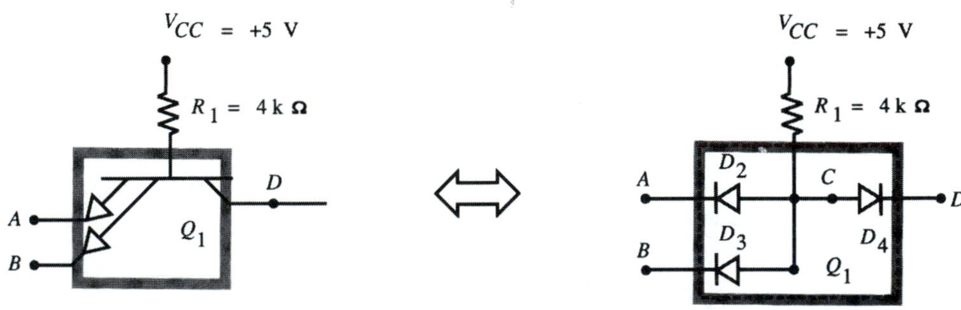

(b) Q_1 equivalent circuit

A	B	Q_1	D	Q_2	E	F	Q_3	Q_4	G	V_0
0 V	0 V	On	0 V	Off	+5 V	0 V	Off	On	+3.6 V	1
0 V	+5 V	On	0 V	Off	+5 V	0 V	Off	On	+3.6 V	1
+5 V	0 V	On	0 V	Off	+5 V	0 V	Off	On	+3.6 V	1
+5 V	+5 V	Off*	1.4 V	On	0.7 V	0.7 V	On	Off	~0 V	0

* Current will flow through D_4 to Q_2.

(c) Truth/voltage table

FIGURE 6.8
Basic TTL NAND gate

Since the base of Q_2 is at 0 V, Q_2 is switched off. When Q_2 is switched off, there is no base current for Q_3, and Q_3 is switched off. Also when Q_2 is off, current from V_{CC} flows through R_2 to the base of Q_4, which turns Q_4 on. With Q_4 on and Q_3 off, current from V_{CC} also flows through R_4, Q_4, and D_1 to the output. The resulting output voltage, after subtracting the voltage drop across R_4, Q_4, and D_1, is approximately 3.6 V, which is still a logic 1.

Whenever both inputs A and B are at a logic 1 (+5 V), both base–emitter junctions of Q_1 are reverse biased, and Q_1 is switched off. With Q_1 off, current can flow through R_1 and D_4 into the base of Q_2. This turns Q_2 on. Now current flows from V_{CC} through R_2, Q_2, and R_3 to ground. The Q_2 emitter current also flows into the base of Q_3 and switches Q_3 on. At the same time, the voltage drop across R_2 is large enough to reduce Q_2's collector voltage to a level that is not enough to switch Q_4 on. Thus Q_4 is switched off when Q_3 is switched on. When Q_3 is switched on, the output is effectively at ground potential, or a logic 0.

The need for D_1 now becomes apparent. It helps to assure that Q_4 will remain off when Q_3 is on. With D_1 in the circuit, the base voltage required to turn on Q_4 is approximately 1.4 V. The voltage table of Figure 6.8(c) indicates the circuit operation for all possible input conditions.

Open-Collector TTL

When R_4, Q_4, and D_1 are removed from the basic TTL circuit, the resulting TTL device is known as an **open-collector** device. Open-collector devices are useful when high current is necessary. They are also useful because the outputs of open-collector devices can be connected through a common pull-up resistor, which is typically about 2 kΩ. Standard TTL device *outputs* cannot be connected together unless an open-collector-type device is used. When open-collector devices are connected together as in Figure 6.9(a), the result is referred to as a **wired-AND** operation. Thus the only time the output can be high is when both NAND gate outputs are high. If either NAND gate's output is low, the circuit output will be low, which effectively creates an AND operation without the use of an AND gate. Figure 6.9(b) shows the circuit of a TTL open-collector NAND gate. The circuit operation is explained by the voltage table of Figure 6.9(b).

Low-Power TTL (74Lxx)

Low-power TTL devices are essentially the same as standard TTL devices except that the resistor values are increased. The increased resistor values result in less current flow and less power consumption. The larger resistor values do, however, increase the propagation delay of the device. This increase is due to increased internal transistor switching times in the device.

High-Speed TTL (74Hxx)

High-speed TTL devices are similar to standard TTL devices except that the resistor values are lower. Also, Q_3 is replaced by a Darlington transistor configuration. The result is significantly faster switching speeds but increased power consumption. This change occurs because lower resistor values result in the higher current flows required for faster switching operation.

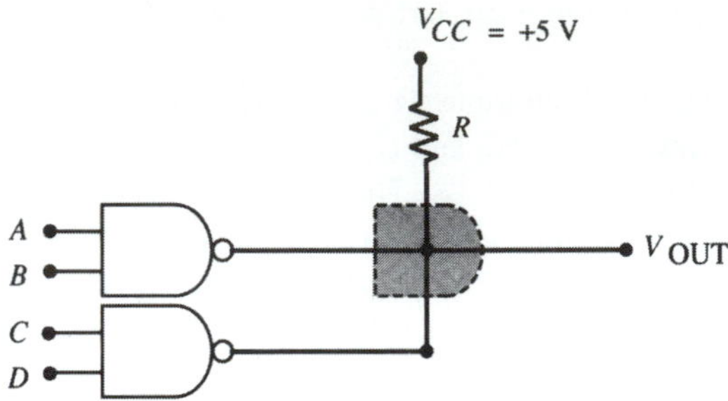

(a) Open-collector wired-AND

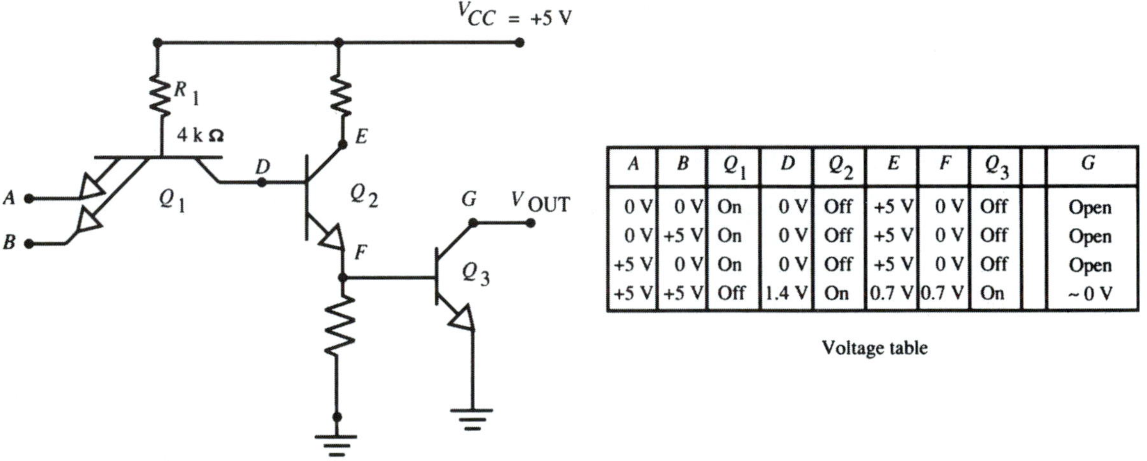

A	B	Q_1	D	Q_2	E	F	Q_3		G
0 V	0 V	On	0 V	Off	+5 V	0 V	Off		Open
0 V	+5 V	On	0 V	Off	+5 V	0 V	Off		Open
+5 V	0 V	On	0 V	Off	+5 V	0 V	Off		Open
+5 V	+5 V	Off	1.4 V	On	0.7 V	0.7 V	On		~0 V

Voltage table

(b) Basic TTL open-collector NAND gate circuit

FIGURE 6.9
Open-collector TTL

Schottky TTL (74Sxx)

Schottky TTL devices are similar to standard TTL devices except that the transistors are replaced with Schottky-clamped transistors. These devices are approximately two times faster than H-series devices. The increased speed is accomplished by not allowing the transistors to go as deeply into saturation, which increases transistor-switching speed. To limit the level of saturation, a diode is connected between the base and collector as shown in Figure 6.10. This diode is known as a **Schottky barrier diode** (SBD). The SBD has a forward voltage drop of approximately 0.25 V. When the SBD becomes forward biased, it conducts some of the base current away from the base, which decreases the saturation current. Note the use of a slightly different symbol to indicate a Schottky transistor. As with H-series devices, Schottky devices generally replace Q_3 with a Darlington transistor configuration to help to increase switching speed.

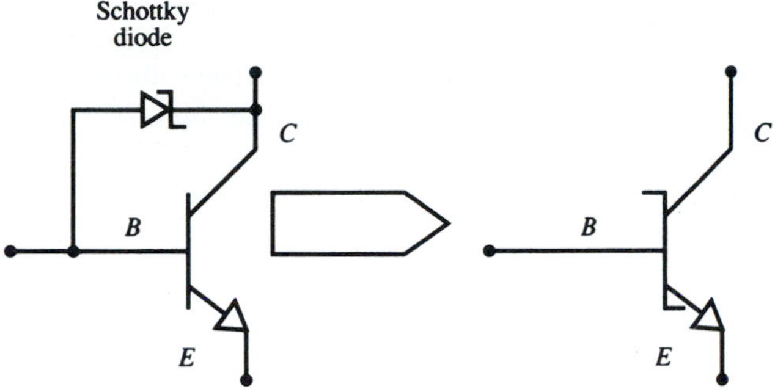

FIGURE 6.10
Schottky transistor

Low-Power Schottky TTL (74LSxx)

This series is a low-power version of the Schottky series. It uses higher resistor values to reduce power consumption. The higher resistor values do, however, decrease transistor switching speeds and increase propagation delay. Since the speed of the LS series device is comparable with that of the standard TTL device at far less power consumption, it has become the series of choice today.

TTL Series Comparison

Table 6.1 gives a comparison of the propagation delay times and power consumption for a typical TTL series device. These characteristics vary from one type of device to another and from one manufacturer to another.

TABLE 6.1
TTL series comparison

	Standard	L	H	S	LS
Propagation delay	10 ns	35 ns	6 ns	3 ns	10 ns
Power consumption	10 mW	1 mW	20 mW	20 mW	2 mW

6.4 CMOS CIRCUIT ANALYSIS

As previously mentioned, CMOS stands for complementary metal-oxide semiconductor. Generally speaking, CMOS family logic devices have slower propagation delays but require far less power to operate than bipolar TTL devices. CMOS devices also exhibit a greater supply voltage range, have higher fan-out capability, and have a greater noise immunity than their TTL counterparts.

Propagation Delay

The propagation delay for a typical CMOS device is about 150 ns. The propagation delay can be decreased by raising the supply voltage; however, the power consumption increases with supply voltage. The 74HC high-speed CMOS series is much faster than the standard 4000 CMOS series because silicon is used instead of aluminum in the device. Furthermore, the 74HC series is pin-for-pin compatible with the LS TTL family.

Power Dissipation

In the steady-state condition, a CMOS family device consumes virtually no power. During switching, the transistors are on for only a brief period of time, which results in only a small amount of current flow. Typical steady-state power consumption for the 74HC series at 5 V is less than 3 nW per gate. The power consumption increases with increased supply voltage and frequency. It has been found that the power consumption of a CMOS device equals that of a TTL low-power Schottky device at approximately 500 kHz.

Supply Voltage

CMOS devices can operate at supply voltages between 3 and 18 V_{DC}. This is a much greater range than that of TTL family devices, which are fixed around 5 V_{DC} ± 0.25 V. The low power consumption and wide supply voltage range make CMOS family devices perfect for battery-operated equipment.

Fan-Out

Fan-out for CMOS family devices is ideally infinite due to its high input impedance ($10^{12}\,\Omega$ typical). Practical values of fan-out are greater than 50.

Noise Immunity

The noise margin for CMOS family devices is superior to that of TTL family devices. For a supply voltage of 5 V, both the high noise margin (V_{NMH}) and the low noise margin (V_{NML}) are approximately 1.5 V.

CMOS Inverter

Figure 6.11 illustrates a basic CMOS inverter circuit. It is composed of a complementary pair of **MOSFET**s (metal-oxide semiconductor field-effect transistor). One MOSFET is a *P*-channel device, and the other MOSFET is an *N*-channel device. The two input gates (*G*) are connected together to form the input to the device. The two output drains (*D*) are connected together to form the device output. The source of the *P*-channel MOSFET is connected to the supply voltage common or ground.

MOSFET devices are nearly ideal switches. When a MOSFET device is off, the drain-to-source impedance is approximately $10^{12}\,\Omega$ and is equivalent to an open switch. When the MOSFET device is on, the drain-to-source impedance is only a few ohms and is equivalent to a closed switch.

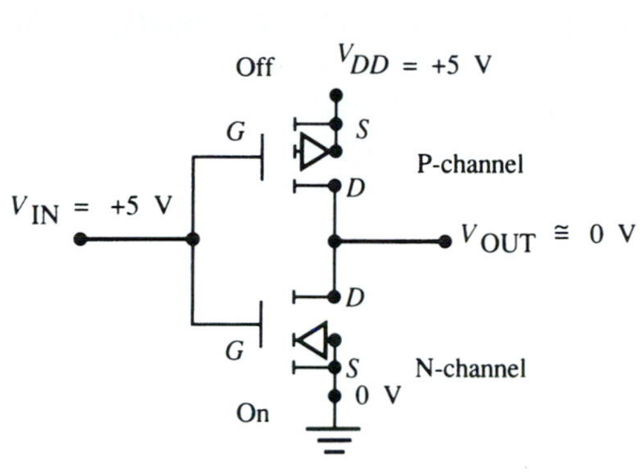

CMOS inverter circuit

V_{IN}	V_{OUT}
+5 V	~0 V

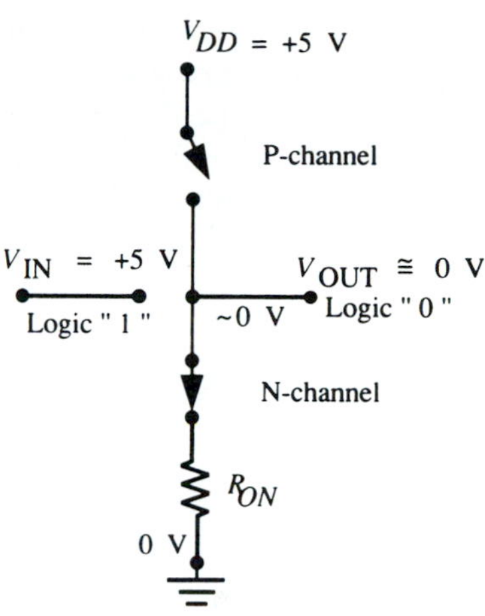

CMOS inverter equivalent

Logic in	Logic out
1	0

(a)

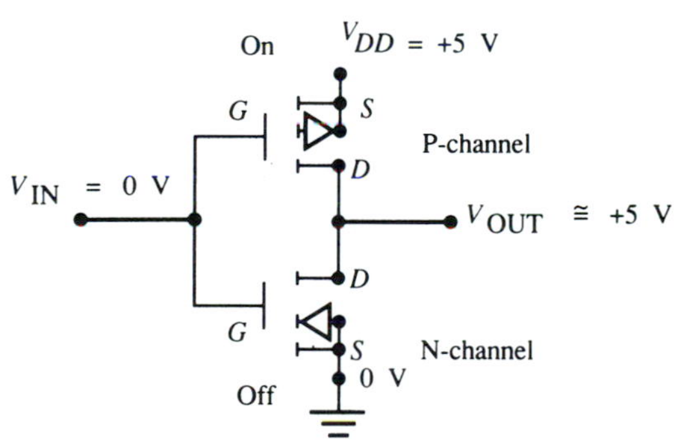

CMOS inverter circuit

V_{IN}	V_{OUT}
0 V	~+5 V

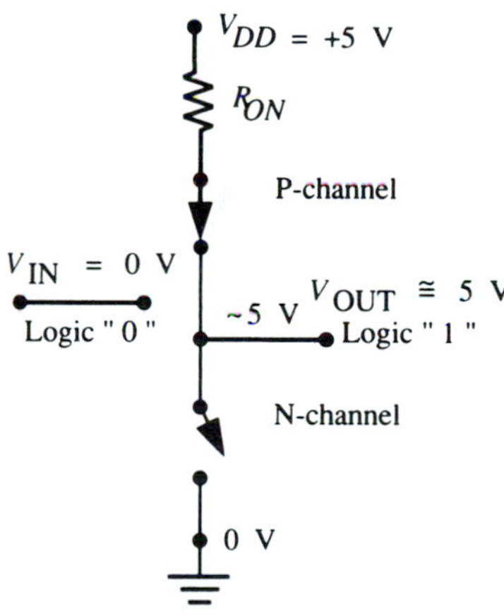

CMOS inverter equivalent

Logic in	Logic out
0	1

(b)

FIGURE 6.11
CMOS inverter

The inverter circuit of Figure 6.11 functions like a push-pull amplifier. When the gate voltage of the inverter is positive (logic 1), the *N*-channel MOSFET is turned on and the *P*-channel MOSFET is turned off (Figure 6.11a). This produces a logic 0 output (approximately 0 V), since the *N*-channel MOSFET switches the output to ground potential. When the gate voltage of the inverter is 0 V (logic 0), the *P*-channel MOSFET is turned on and the *N*-channel MOSFET is turned off (Figure 6.11b). This produces a logic 1 output (approximately V_{DD}), since the *P*-channel MOSFET switches V_{DD} to the output.

CMOS Logic Gates

The *N*- and the *P*-channel MOSFETs form the basic building blocks for all CMOS family devices. Recall that a logic 1 input will turn on an *N*-channel MOSFET, and a logic 0 will turn on a *P*-channel MOSFET. Referring to the diagram of Figure 6.12, we can see that connecting two *N*-channel MOSFETs in series with two *P*-channel MOSFETs in parallel will result in a NAND gate (Figure 6.12a). Connecting two *P*-channel MOSFETs with two *N*-channel MOSFETs in parallel will result in a NOR gate (Figure 6.12b).

Logic Family Comparison

Table 6.2 is a comparison of TTL and CMOS logic family characteristics.

TABLE 6.2
Logic family comparison

	TTL		CMOS	
	7400	74LS	4000	74HC
Supply voltage	4.75–5.25 V	4.75–5.25 V	3–18 V	3–6 V
Supply current	4 mA	0.8 mA	150 μA	150 μA
Propagation delay	15 ns	15 ns	250 ns	15 ns
Noise immunity	0.4 V	0.4 V	1.5 V	1 V
Fan-out	10	10	50	50

6.5 INTERFACING LOGIC FAMILIES

Sometimes it is necessary to connect logic circuits from one logic family to logic circuits of another logic family. For example, we may wish to connect TTL devices to CMOS devices or vice versa.

Interfacing TTL Devices to CMOS Devices

A comparison of the TTL output characteristics and the CMOS input characteristics reveals that the output currents of the TTL family are more than adequate to supply the input current requirements of CMOS family devices. It would seem then that there would be no problem connecting the output of a TTL device to the input of a CMOS device, provided that the CMOS device was operating at a V_{DD} of 5 V. The problem, however, is that V_{OH} of the TTL device is very close to V_{IH} of the CMOS device. If the V_{OH}

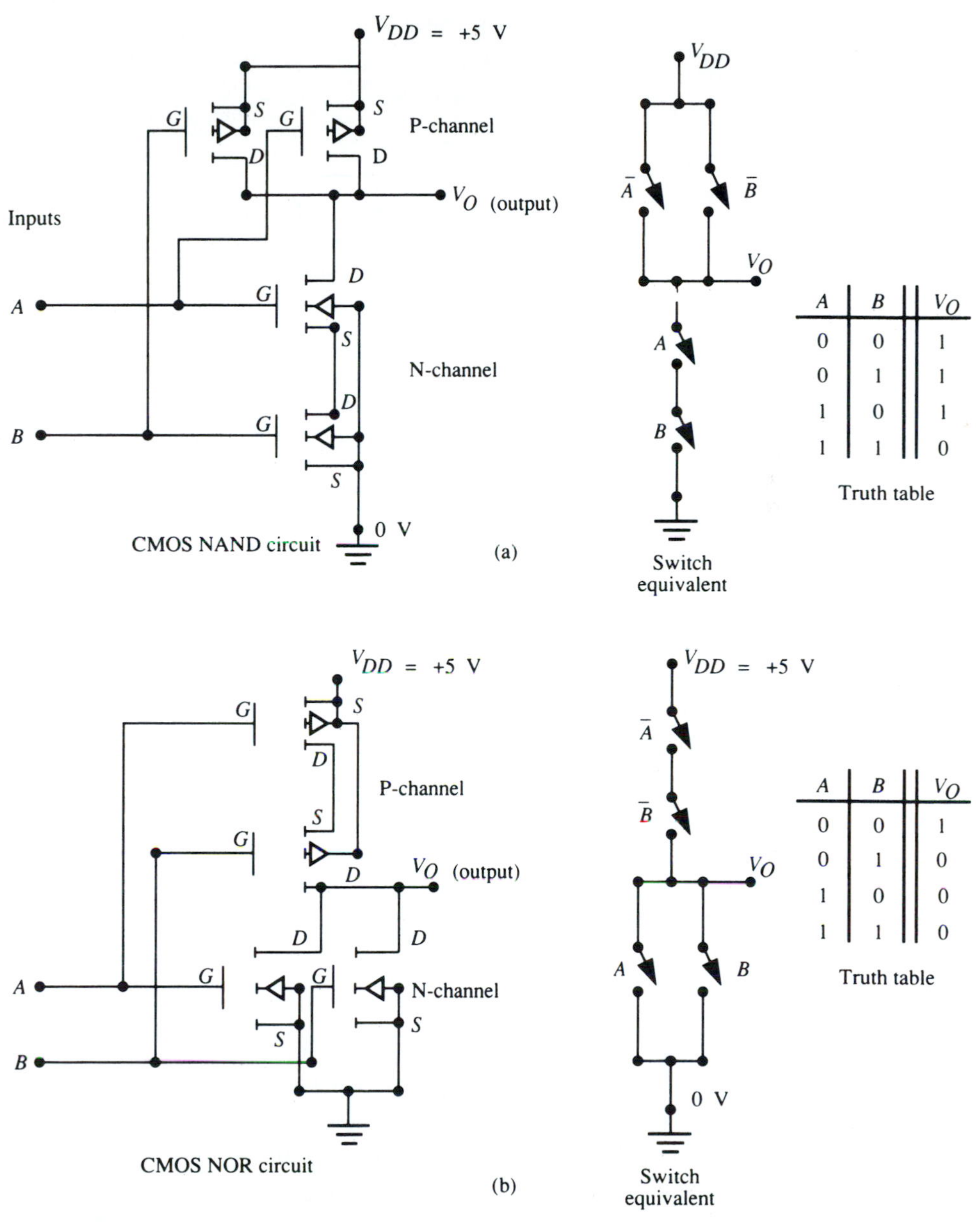

FIGURE 6.12
CMOS NAND/NOR gates

of a TTL device were to fall below the V_{IH} requirement of the CMOS device, unpredictable operation would result. The typical solution is to connect the two logic families together using a pull-up resistor, as shown in Figure 6.13(a). Since the input impedance of a CMOS device is so high, the pull-up resistor will ensure that V_{IH} is approximately 5 V when the

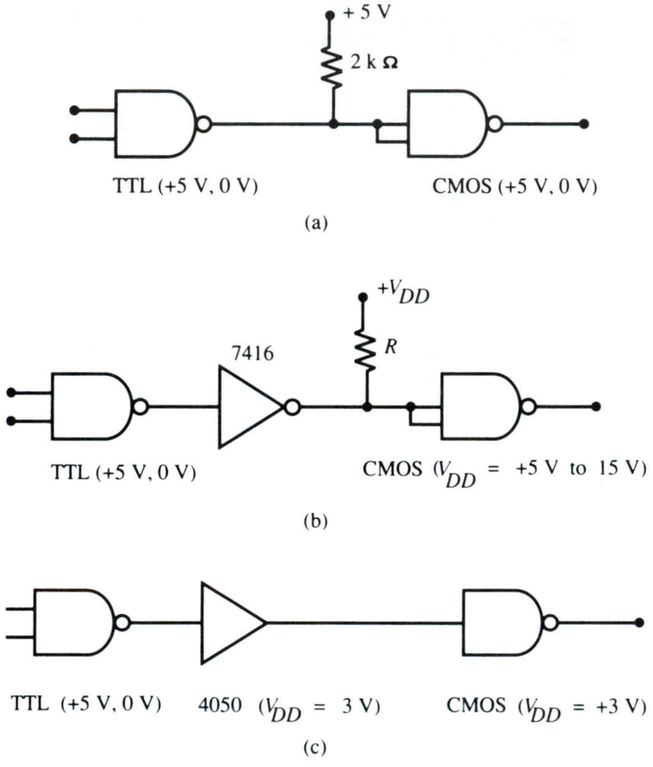

FIGURE 6.13
TTL-to-CMOS circuits

TTL device is in the high or logic 1 state. When the TTL device is in the low or logic 0 state, the output voltage must not exceed 1.5 V, which is the maximum value of V_{IL} for CMOS. The typical value for the pull-up resistor when operating at 5 V is 2kΩ. *R* can be calculated for other operating voltages using the following equations:

$$R_{\text{min}} = \frac{V_{DD} - V_{OL(\text{max})}}{I_{OL(\text{max})}} \tag{6.8}$$

$$R_{\text{max}} = \frac{V_{CC} - V_{IH(\text{min})}}{I_{IH}} \tag{6.9}$$

If the supply for the CMOS device is higher than 5 V, the simplest solution is to use an open-collector TTL device to drive the CMOS device. If an open-collector TTL device is used, the same technique of using a pull-up resistor can be used. The pull-up resistor must be connected to V_{DD}, as shown in Figure 6.13(b).

Another interfacing technique is to use another CMOS IC. The CMOS family 4050 allows an input voltage of up to 20 V regardless of the supply voltage, V_{DD}. This means that the CMOS device can be operating at less than or greater than the standard TTL V_{CC} voltage of 5 V. This technique is illustrated in Figure 6.13(c).

Interfacing CMOS Devices to TTL Devices

When CMOS devices are connected to TTL devices, the primary concern is with the output parameters of the CMOS device and the input parameters of the TTL device. The problem that arises is with I_{OL} of the CMOS device. What happens is that the CMOS device output cannot sink enough current for the input of the TTL device in the logic 0, or low, state. The standard solution is to use a buffer, as shown in Figure 6.14. This technique is similar to the interfacing technique just described when the TTL device and the CMOS device are operating at different voltages.

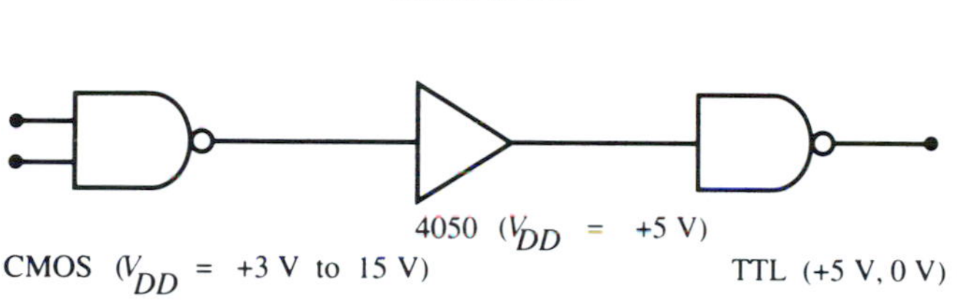

FIGURE 6.14
CMOS-to-TTL circuit

6.6 TECH TIPS AND TROUBLESHOOTING—T³

The unused inputs to a TTL or CMOS device should always be connected to something. This is especially important when dealing with CMOS devices. As unused input on a TTL device will be perceived by the device as a logic 1 applied to its input. Whereas unused TTL inputs that are left open become more susceptible to EMI and stray electronic interference, they generally can provide reliable operation. Unused inputs of CMOS devices should be connected to either V_{DD} or ground. Open inputs on CMOS devices can lead to unpredictable operation and even oscillation. Furthermore, the unused inputs on a CMOS device can increase the power consumption of the device and even destroy it. For this reason it is important to connect all unused inputs, including those on an unused gate, to either V_{DD} or ground.

CMOS devices have a very high input impedance because MOS transistors are in effect an open circuit ($10^{12}\Omega$). This fact makes CMOS devices highly susceptible to damage from static electricity. For example, the static electricity generated from walking across a carpet is usually more than enough to destroy a CMOS device. To help solve this problem, most CMOS devices have input circuit protection built into the device in the form

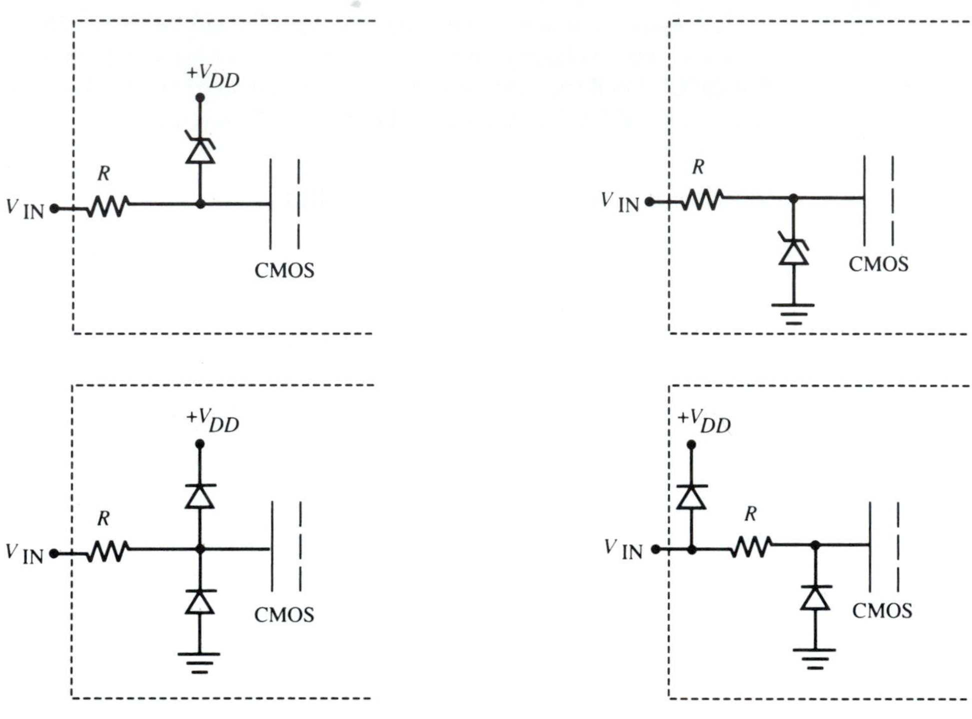

FIGURE 6.15
CMOS input-protection circuits

of input diodes. Figure 6.15 shows a number of CMOS input-protection techniques. The specific technique used varies from one manufacturer to another and by type and function of the device.

TTL devices also use input diodes for protection. The input diodes on TTL devices are used to prevent ringing on the negative-going transitions. Prevention occurs by limiting the maximum negative voltage that can appear on the input to a TTL device to −0.5 V. Figure 6.16 illustrates a TTL device with input diode protection. If the input voltage should ring negative in excess of −0.5 V, the diode begins to conduct, clamping the input.

Since input diodes are the first components seen by the input signals, they are prone to overload damage. If an input diode opens, the device may still function normally. If an input diode shorts, the device will not work at all. The problem that occurs is that often when an input diode is blown, it becomes a high-resistance short circuit because it is a small diode with a small junction. The silicon semiconductor ends up shorting the junction, and the silicon creates a short of about 30 to 200 Ω. The device may still function, but unreliably. Depending on device threshold voltages and loading, enough current may be available for proper operation some of the time. This phenomenon is sometimes known

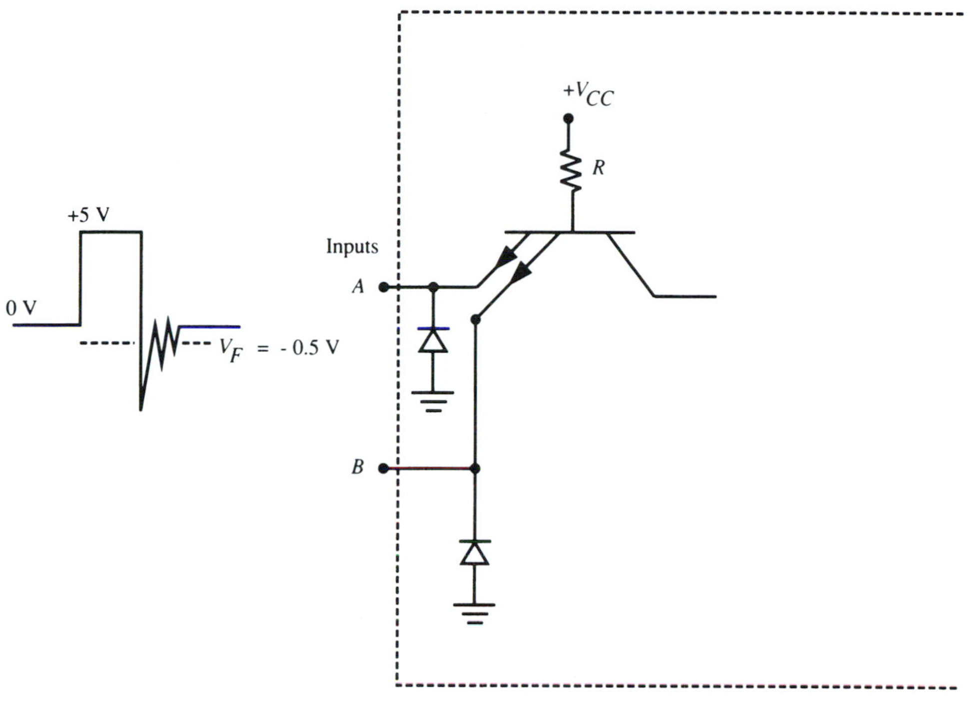

FIGURE 6.16
TTL input protection

as the **input diode anomaly.** If the device is checked out statically, it will still appear to function properly as defined by the truth table. However, dynamic operation in a circuit may be unreliable and cause intermittent errors. For this reason it is important to check the input diodes in a device. Simply use a DMM with a diode scale and note the proper forward voltage drop. When the black lead of the DMM is placed on the input pin and the red lead to ground, the DMM should indicate about 500 mV. Reversing the leads will indicate an overload (OL) condition. It should be noted that analog meters usually do not have enough voltage to perform this test.

One final consideration when dealing with TTL devices is that of *decoupling.* **Decoupling capacitors** are used to eliminate unwanted *transients,* or *spikes,* on power supply lines. These spikes are created when the device output switches states. To remove these unwanted transients, decoupling capacitors are installed between V_{CC} and ground. Usually a large capacitor, 0.1 to 1 μF, is installed at the V_{CC} input pin of an edge connector on a printed circuit board. Smaller decoupling capacitors, between 0.01 and 0.1 μF are also installed between V_{CC} and ground for every one to three ICs. These capacitors are installed as close as possible to each IC, with the lead length of the capacitors being made as short as possible.

EXERCISES

6.1 Describe the package types and pin outs that are available for the 7400 device shown in Figure 6.1.

6.2 Referring to Figure 6.1, determine the operating temperature for
a. The 7400
b. The 5400

6.3 From the 74LS00 device specifications of Figure 6.1, determine the absolute maximum ratings for
a. V_{CC}
b. Input voltage
c. Storage temperature

6.4 From the 74LS00 device specifications of Figure 6.1, determine the maximum recommended operating conditions for
a. V_{CC}
b. V_{IL}
c. I_{OH}
d. I_{OL}
e. T_P (max)

6.5 From the 74LS00 device specifications of Figure 6.1, determine the minimum recommended operating conditions for
a. V_{CC}
b. V_{IH}
c. V_{OH}

6.6 What is the average power dissipated by a 74LS00 device?

6.7 Determine the noise margins V_{NMH} and V_{NML} for
a. 54LS00
b. 74LS76 (see appendix C)

6.8 Determine the fan-out for the 54LS00 and the 74LS76 devices

6.9 Determine the sourcing current for the 74LS00 and the 74LS76 devices.

6.10 Determine the sinking current for the 74LS00 and the 74LS76 devices.

6.11 Referring to Figure 6.17, determine the approximate voltage levels and logic states at B_1, E_1, C_1, V_{out_1}, V_{in_1}, B_2, E_2, C_2, and V_{out_2}.

6.12 Draw the schematic diagram, voltage table, and truth table for a 3-input diode AND gate. Complete the tables.

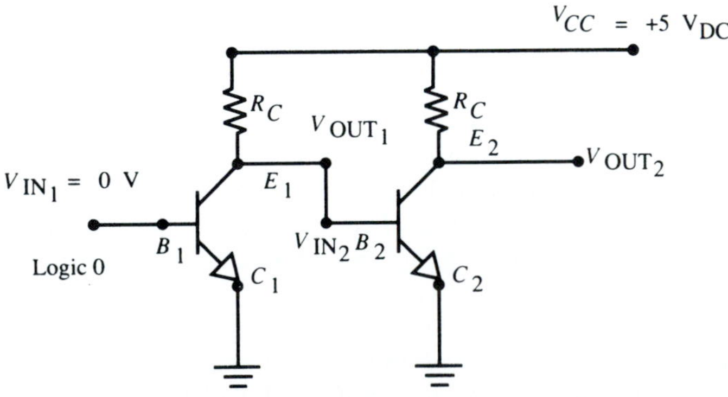

FIGURE 6.17

6.13 Draw the schematic diagram, voltage table, and truth table for a 3-input diode OR gate. Complete the tables.

6.14 Referring to Figure 6.18, determine the approximate voltage levels and logic states at V_{out_1}, V_{out_2}, and V_{out_3}.

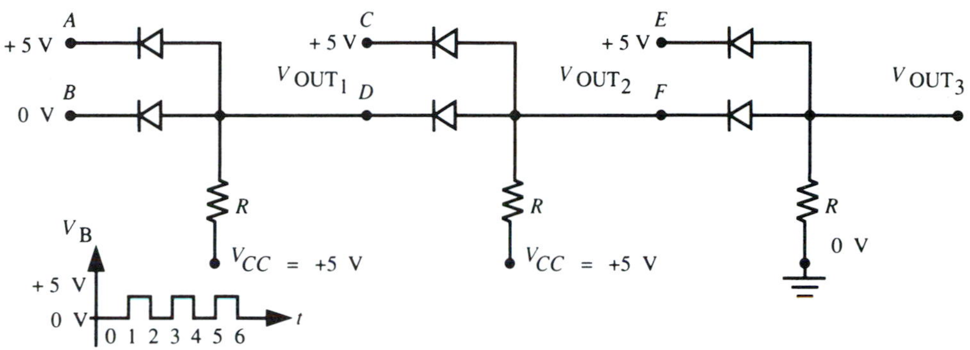

FIGURE 6.18

6.15 For Figure 6.18, if input *B* is changed to waveform V_B (shown), draw the waveform seen at V_{out_3}.

6.16 Draw the schematic diagram, voltage table, and truth table for a 3-input diode-transistor NAND gate. Complete the tables.

6.17 Draw the schematic diagram, voltage table, and truth table for a 3-input diode-transistor NOR gate. Complete the tables.

6.18 For Figure 6.8, if $A = +5\,V$ and $B = +5\,V$, determine which transistors will be on and which transistors will be off. Redraw this circuit replacing each on transistor with a closed switch and each off transistor with an open switch.

6.19 Explain where open-collector TTL devices are useful.

6.20 Compare the advantages and disadvantages of the TTL 74Lxx, 74Hxx, 74Sxx, and 74LSxx devices.

6.21 What effect does increasing the supply voltage have on power dissipation and propagation delay in CMOS devices?

6.22 Compare the fan-out and noise immunity of CMOS and TTL devices.

6.23 Referring to Figure 6.19, determine the approximate voltage levels and logic levels at V_{in}, *A*, *B*, and V_{out}.

6.24 Referring to Figure 6.19, if V_{in} is changed to waveform V_1 (shown), draw the waveform seen at V_{in}, *A*, *B*, and V_{out}.

6.25 Discuss the primary concerns when interfacing TTL to CMOS devices.

6.26 Discuss the primary concerns when interfacing CMOS to TTL devices.

6.27 Explain what should be done with unused inputs on CMOS devices and why.

6.28 Explain what should be done with unused inputs on TTL devices and why.

6.29 How are the inputs to TTL and CMOS devices protected?

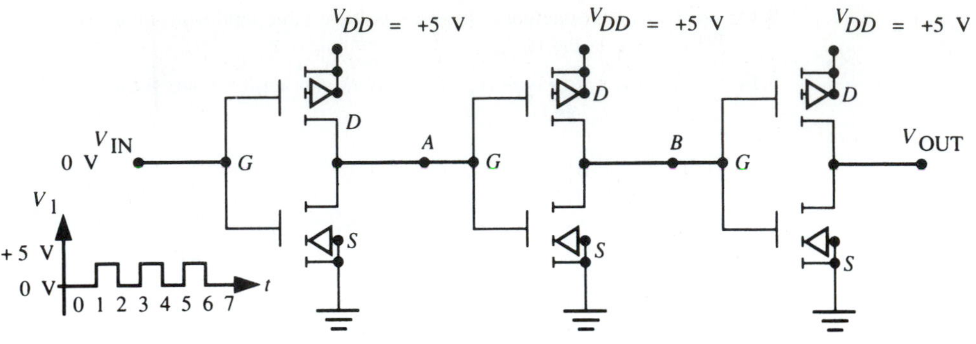

FIGURE 6.19

6.30 What is the input diode anomaly? Explain how input diodes can be tested.

6.31 Discuss the use of decoupling capacitors on TTL devices.

6.32 Crossword Puzzle

ACROSS

8. IC.
9. Maximum number of device inputs a device output can drive.
12. A resistor used to connect a device input to V_{CC}.
13. High-level output current.
15. IC wattage specification.
16. When logic 0 current enters a device output.
19. CMOS devices use a complementary pair of _____ transistors.
20. In CMOS devices power consumption will increase with increased supply voltage and _____.
21. High-level input voltage.
22. Term used to connect different logic families together.
23. Maximum CMOS supply voltage.

DOWN

1. Where electrical characteristics of an IC are found.
2. Standard TTL device with lower resistor values.
3. IC switching time.
4. Transistor transistor logic.
5. A device that when forward biased acts like a closed switch.
6. Low-level input voltage.
7. Circuit's ability to handle noise.
10. Logic 1 output of a device supplying current.
11. TTL wave-shaping device that uses special diodes.
14. Capacitors used to eliminate unwanted transients on power supply lines.
17. Low-level output current.
18. Basic bipolar transistor is used to perform this logic function.

1
2
3
4
5
6
7
8
9
10
11
12
13
14
15
16
17
18
19
20
21
22
23

CHAPTER 7

MEMORY

KEY TERMS

Address
Address Decoding
Address Multiplexing
Application-specific Integrated Circuits (ASICs)
Bit Organized
Bit Parity
Blank PLD
Cell
Check Sums
Column-address Selector (CAS)
Combinational PALS
Cyclic Redundancy Check (CRC)
Dynamic RAM (DRAM)
Erasable PROM (EPROM)
Fuse Map
Maskable ROM
Memory Cell
Memory Location
Nonvolatile Memory
Product Line
Product Term
Programmable Array Logic (PAL)
Programmable Logic Array (PLA)
Programmable Logic Device (PLD)
Programmable ROM (PROM)
PROM Burner
Random Access
Random-access Memory (RAM)
Read Cycle
Read-only Memory (ROM)
Read Operation
Refreshed
Registered PALS
Row-address Selector (RAS)
Sequential PALS
Static RAM (SRAM)
Sum Line
Word Organized
Word Parity
Write Cycle
Write Operation
Volatile Memory

7.0 INTRODUCTION

Since the early 1970s, ICs or semiconductor memory have been the most widely used type of *primary memory* found in microcomputers. The simplest form of computer memory is the basic flip-flop. A flip-flop is called a **memory cell** when it represents a single storage bit (0 or 1). In most microcomputers 8 to 64 cells are connected together to form a memory byte or memory word. Each memory byte or word has a unique location in memory called an **address.** This address tells the computer where to find the memory cells and can be likened to the address of your house. Memory is a place, therefore, where data bits (0 or 1) can be stored and then later retrieved when the computer needs it. The process by which a computer stores data into memory is called *writing.* Therefore we say that a computer is in a **write cycle** or performing a **write operation** when it is storing data into memory. The process by which a computer retrieves data from memory is called a **read cycle** or **read operation** as illustrated in Figure 7.1.

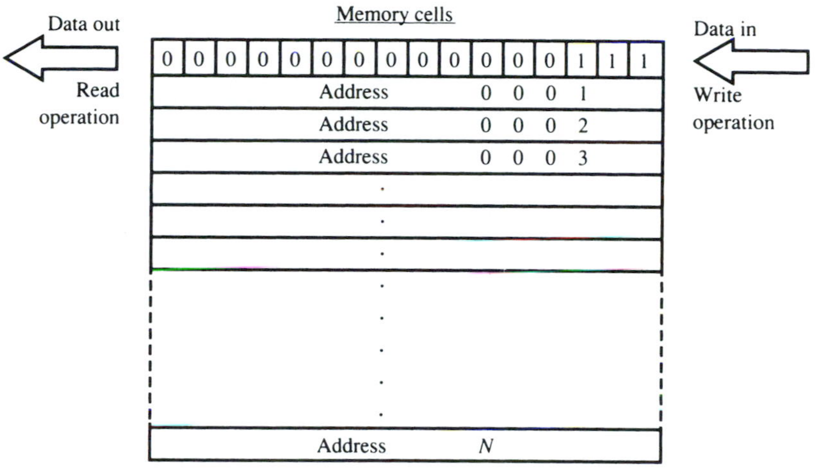

FIGURE 7.1
Basic memory

7.1 MEMORY FUNDAMENTALS

Memory can be classified into two general types, **ROM** and **RAM.** ROM stands for **read-only memory,** and RAM stands for **random-access memory.** ROM generally contains permanently stored data that cannot be changed. It can be read but not written into. ROM is always available to the computer and is not lost when the power is turned off. For this reason, it is known as **nonvolatile memory.** RAM, on the other hand, is memory that can be read from or written to. Thus a computer can store or save data and later retrieve that data. The data can also be changed at any time. RAM memory is **volatile memory,** that is, it is lost or erased whenever the power is switched off. RAM memory can be compared with your notebook. You write notes in your notebook, read your notes, and sometimes even change your notes. On the other hand, ROM memory is like your textbook. Generally the information in your textbook may be read but not written or changed.

The term **random access** means that any memory location can be enabled or addressed in any sequence. That is, information does not have to be read or written to sequentially. Actually, both RAM and ROM memory are random access. The major difference is that RAM is memory that can be read and written to. A better name for RAM might be read/write memory.

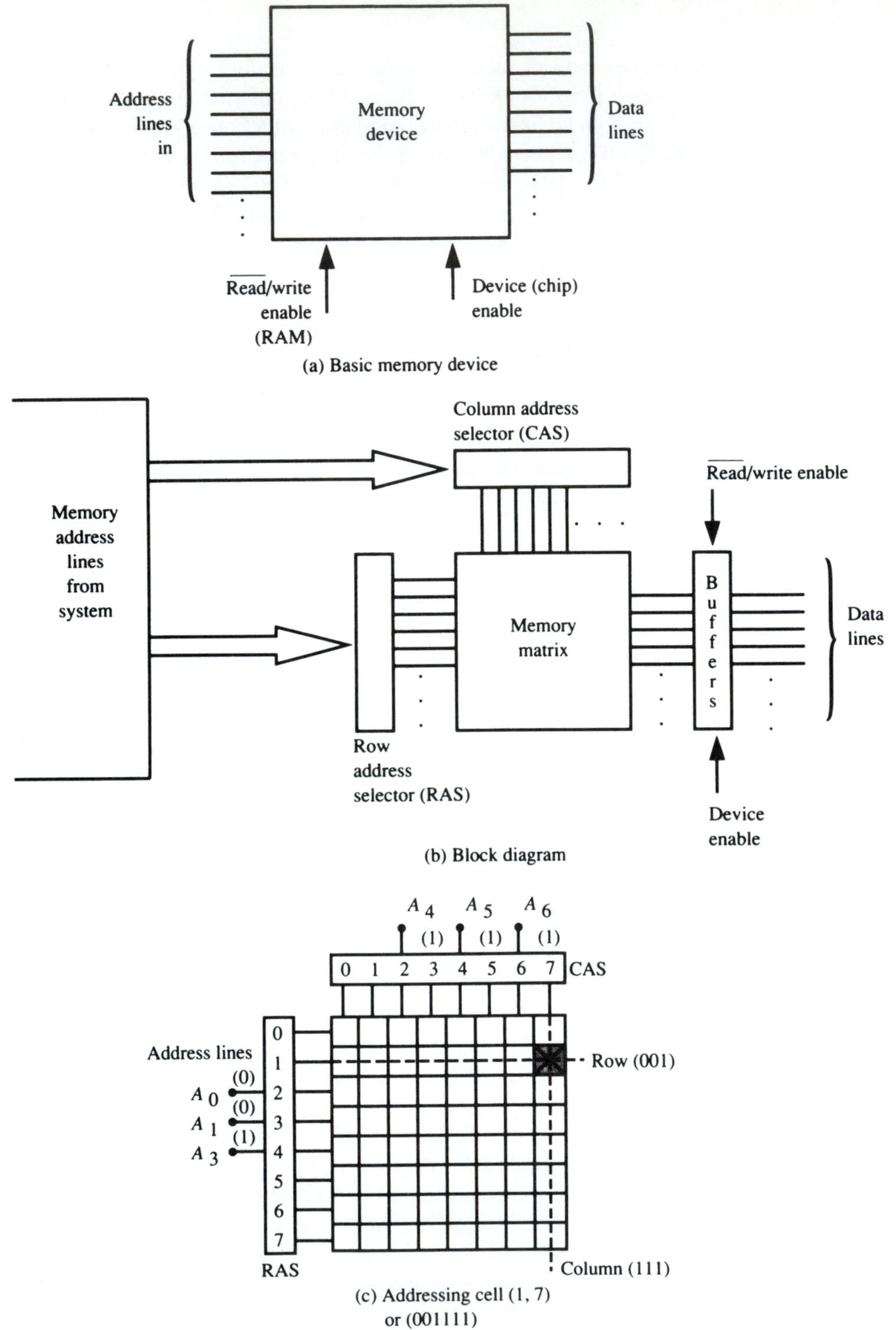

FIGURE 7.2
Basic memory device

Memory Structure

Figure 7.2 illustrates a basic block diagram of a memory device. A memory device is connected to the rest of a microcomputer system through its address and data lines. It is controlled by its device-enable line and, in the case of RAM, its $\overline{\text{Read}}$/write control line.

Memory devices are generally organized in a matrix format. The matrix is organized into rows and columns to define an individual memory location or address as shown in Figure 7.2(c). This matrix can be thought of as a postal address scheme where the rows represent the street addresses and the columns represent the house numbers. Each box in the memory matrix is called a **cell,** or **memory location.** A memory cell or location is selected by the address selection or decode circuitry. This circuitry is broken up into two parts: the **row-address selector (RAS)** and the **column-address selector (CAS).** The memory address lines define the RAS and the CAS locations. The device-enable control line is used to turn on the memory output buffers from the tristate condition. The $\overline{\text{Read}}$/write control defines the type of operation or cycle being performed.

Memory is either **bit organized** or **word organized.** A bit-organized memory device can store a single bit for each address location. Thus for a bit-organized device, each box in the matrix of Figure 7.2(c) represents one binary digit. A word-organized memory device selects a group of memory cells at the same time for each address location. Thus for a word-organized device, each box in the matrix of Figure 7.2(c) represents a group of binary digits. Each group of cells is usually a byte (8 bits) or a word (16 bits) in length. Regardless of the type of organization being used, the maximum number of memory locations is determined by the number of address lines, using the following relationship:

$$\text{Maximum number of memory locations} = 2^N \tag{7.1}$$

where N represents the number of address lines.

Table 7.1 shows the relationship between the number of address lines and the number of memory locations. Note the use of the standard notation, which rounds the actual

TABLE 7.1
Address lines/memory locations

Address Lines	Memory Locations	Standard Notation
10	1,024	1K
11	2,048	2K
12	4,096	4K
13	8,192	8K
14	16,384	16K
15	32,768	32K
16	65,536	64K
17	131,072	128K
18	262,144	256K
19	524,288	512K
20	1,048,576	1M
24	16,777,216	16M
32	4,294,467,296	4G

maximum number of memory locations. For example, 64K of memory actually has 65,536 memory locations.

7.2 ROM

As previously stated, ROMs are referred to as nonvolatile memory because the information stored in a ROM is not lost or destroyed when the power is turned off. ROMs can be classified into three general types. A **maskable ROM** is a ROM that is programmed with information or data by the manufacturer. Once programmed, these data bits cannot be altered or changed. A **programmable ROM,** or **PROM,** is a device that can be programmed by an individual user through the use of specialized equipment called a **PROM burner.** Once programmed, the data in a PROM, like a ROM, cannot be altered or changed. An **erasable PROM,** or **EPROM,** is a type of ROM that can be programmed by an individual user but whose data may be erased or changed with the use of specialized equipment. Erasing is accomplished with ultraviolet light. ROMs are used in applications where the data bits are of a permanent nature—for example, code converters, character generators, fixed computer program instructions, fixed constants, and data tables like those used for the trigonometric functions of sine, cosine, and tangent.

Figure 7.3 illustrates a masked ROM configured from a diode array matrix. The ROM is programmed through the use of diodes. A diode is connected between an input address line and an output data line whenever a logic 1 is desired. For example, if a logic 1 (+5 V) is applied to one of the input address row lines, the corresponding diode, D_1, will be forward biased, resulting in a logic 1 output on the data line. If we wish the output to be a logic 0, we simply omit the diode. Therefore, if we apply a logic 1 to input address line A_0 and a logic 0 to all other input address lines, the output data on lines D_3, D_2, D_1, and D_0 will be 0101, respectively. The truth table of Figure 7.3 summarizes the output data patterns for various input addresses of the diode matrix ROM shown in Figure 7.3(a).

A PROM can be constructed using bipolar transistors (BJTs) or from enhancement field-effect transistors (MOSFETs) as shown in Figure 7.4 (p. 214). Programming is accomplished through fuses. If the fuse is connected, a logic 1 results. If the fuse is blown, a logic 0 results. To program the PROM, a device known as a PROM programmer or PROM burner is used. Once the desired address is selected, a large amount of current is used to blow out the desired fuses. This is why the data bits are permanently stored and cannot be changed or altered in this type of device.

The erasable PROM or EPROM uses enhancement MOSFET devices. Storage or nonstorage is accomplished by trapping charges at the gate of the enhancement MOSFET. Erasing is performed by exposing the MOSFET to intense ultraviolet light.

Although ROMs may vary in size and shape, some terminology has become standard. ROMs may be classified by the number of storage locations and the number of output data bits. For example, a 256 × 4 ROM will have 256 storage locations of 4 bits each. A 1024 × 8 ROM has 1024 storage locations of 8 bits each. A 512 × 1 ROM has 512 single-bit storage locations, and so forth.

Figure 7.5 (p. 215) shows the block diagram and pin configuration for the Intel 2716, 2048 × 8 EPROM. The address lines are decoded to select the desired output data. Tristate output buffers are used to enable the device. Programming is accomplished by activating the $\overline{\text{PD}}$/PGM pin.

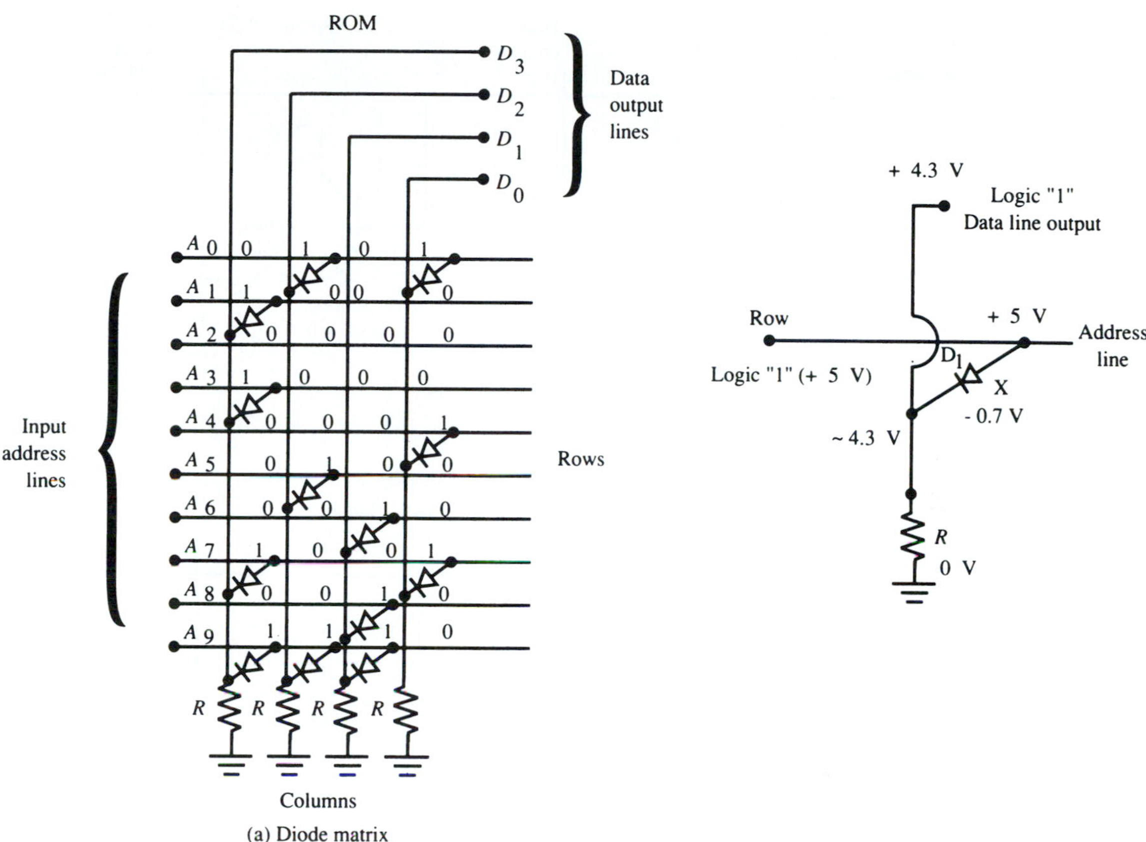

(a) Diode matrix

Input address										Output data			
A_9	A_8	A_7	A_6	A_5	A_4	A_3	A_2	A_1	A_0	D_3	D_2	D_1	D_0
0	0	0	0	0	0	0	0	0	1	0	1	0	1
0	0	0	0	0	0	0	0	1	0	1	0	0	0
0	0	0	0	0	0	0	1	0	0	0	0	0	0
0	0	0	0	0	0	1	0	0	0	1	0	0	0
0	0	0	0	0	1	0	0	0	0	0	0	0	1
0	0	0	0	1	0	0	0	0	0	0	1	0	0
0	0	0	1	0	0	0	0	0	0	0	0	1	0
0	0	1	0	0	0	0	0	0	0	1	0	0	1
0	1	0	0	0	0	0	0	0	0	0	0	1	0
1	0	0	0	0	0	0	0	0	0	1	1	1	0

(b) Truth table

FIGURE 7.3
Masked ROM from diode matrix

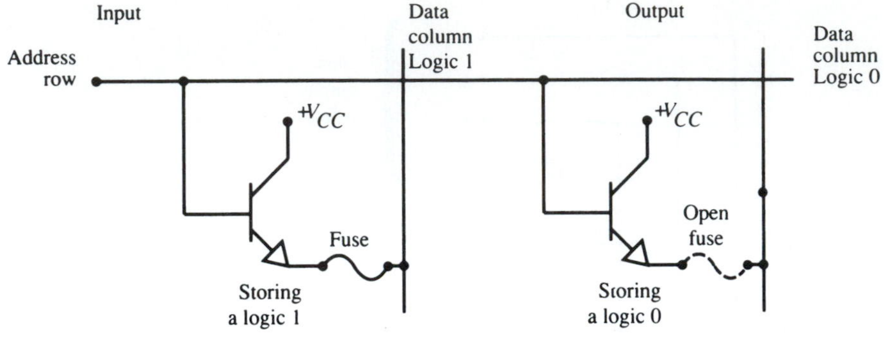

(a) Bipolar cells

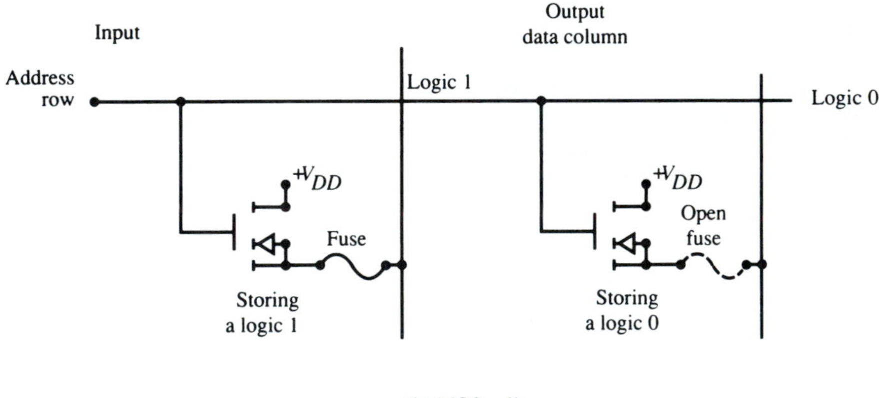

(b) MOS cells

FIGURE 7.4
PROM cells

7.3 RAM

RAM, or read/write memory, is a type of volatile memory from which data can be read and into which it can be written. Any memory cell of RAM memory can be selected in any sequence, and the data contents, unlike ROM memory, can be changed at any time. RAMs can be classified as either static or dynamic. A **static RAM (SRAM)** holds data as long as the power is applied to the device. It is essentially an array of flip-flops in which each flip-flop represents 1 bit or one memory cell, as shown in Figure 7.6(a) (p. 216). A **dynamic RAM (DRAM)** is a type of RAM in which the data must be periodically recharged, or **refreshed.** Refreshing is accomplished by performing a repetitive read or write operation or a special refresh operation. The added cost and complexity of the DRAM are more than offset by the higher cell density, which lowers the per-bit cost of the device. DRAMs are made up from MOSFET devices, which appear to act like capacitors, as shown in Figure 7.6(b). If the capacitor is not recharged periodically, the data

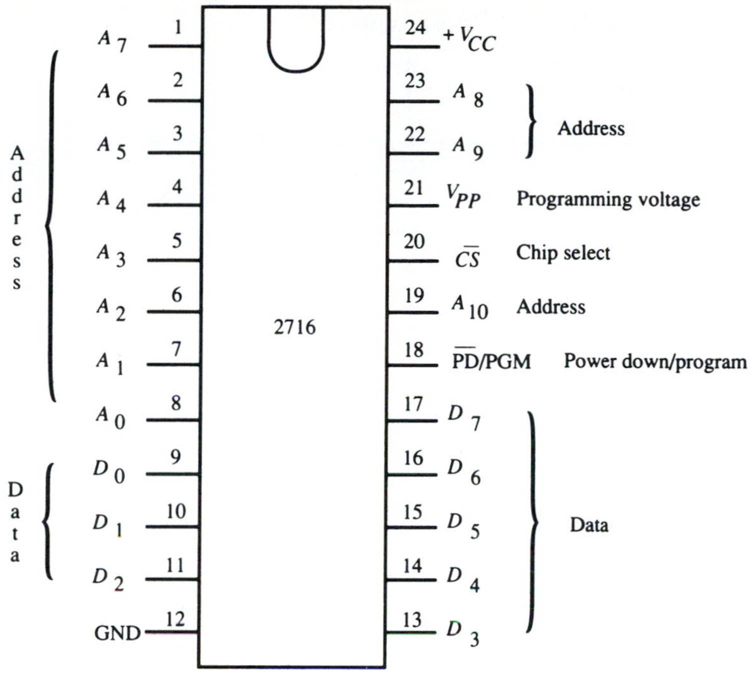

(a) 2716 Prom pinout

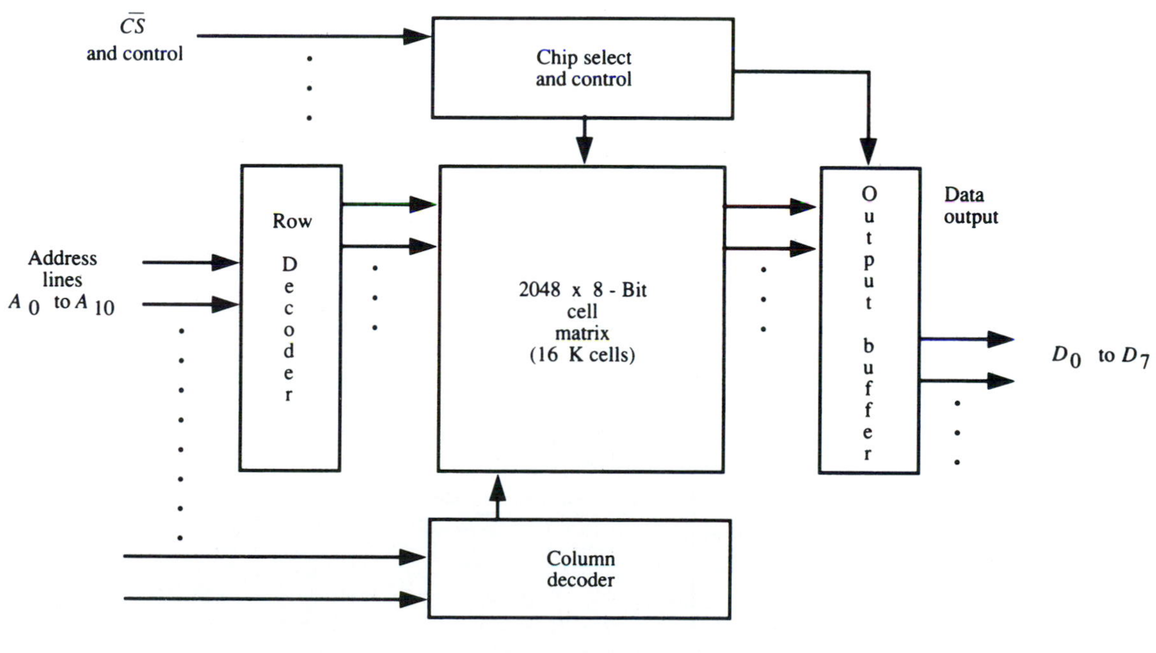

(b) 2716 prom block diagram

FIGURE 7.5
2716, 2048 × 8 EPROM

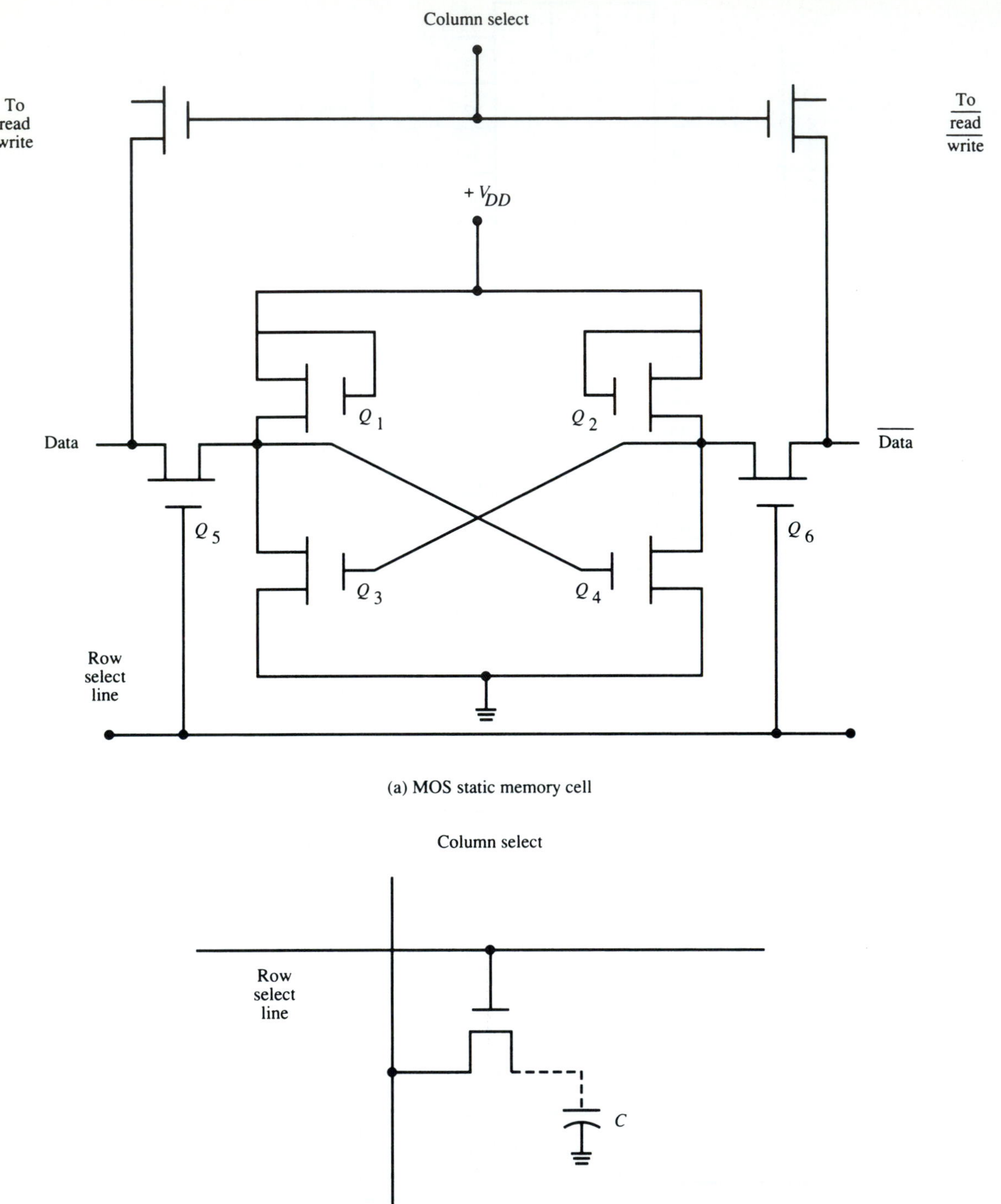

FIGURE 7.6
RAM memory cells

will be lost due to leakage. DRAMs require refreshing approximately every 2 to 4 ms. Refreshing is accompanied by reenergizing the capacitor to store a logic 1 and by maintaining the discharge of the capacitor to store a logic 0.

SRAM

Figure 7.7 illustrates the functional diagram, pin configurations, and truth table for a TMS 4016 SRAM. The TMS 4016 is a MOS device organized in a 2048 × 8-bit memory cell array. Pin 20, $\overline{G}$, controls the output buffer enable line. If $\overline{G}$ is high, the output is in the tristate. If $\overline{G}$ is low, the output is connected to the system. Pin 18, $\overline{S}$, controls the chip-select line. If $\overline{S}$ is high, the device is disabled and cannot read or write data. If $\overline{S}$ is low, the device is selected and can then be used to read or write data. Pin 21, $\overline{W}$, controls the read/write operation. If $\overline{W}$ is high, the device is in the read mode. If $\overline{W}$ is low, the device is in the write mode. A_0 through A_{10} are the address lines that define an individual memory cell in the array. D_1 through D_8 are the memory cell data input or output lines. The truth table illustrates the operation of the device. For example, to write or store data into the device, $\overline{W}$ must be low and $\overline{S}$ must be low. The data bits on the data lines are then written or stored in the address defined on the address lines. Note that during a write operation, the logic state of $\overline{G}$, the output control, is not important.

DRAM

Simplicity, cost, and high data density make the DRAM the most popular type of memory device used in microcomputer systems today. Since each memory cell is essentially a single MOSFET, large numbers of memory cells can be cheaply packaged into a single device. The drawbacks are generally slower speed and the need to refresh. Since the internal capacitance of the MOSFET is great enough to make it appear that a small capacitor (a few picofarads) exists in the MOSFET, data can be stored as charge or no charge in the capacitor. No charge or a logic 0 can be stored indefinitely. A logic 1 or a charged capacitor must be refreshed, or recharged, at least once every 2 ms, or the capacitor will lose its charge and the data.

Figure 7.8 (p. 219) shows the pin configuration for the TMS 4116 DRAM. The TMS 4116 is a MOS device organized in a 16K × 1-bit array. Recall from Table 7.1 that 16K of memory is actually 16,384 memory locations. This would require 14 address bits or 14 address lines, since 2^{14} equals 16,384. To save address lines and reduce the number of pins required on the IC, most DRAMs employ a technique known as **address multiplexing.** Instead of 14 address lines, only 7 address lines are provided. During a read or write operation, the 7 address lines first contain the row information and then the column information. This is controlled by using the $\overline{\text{RAS}}$ and $\overline{\text{CAS}}$ control lines, as shown in Figure 7.9 (p. 220). Whenever the $\overline{\text{RAS}}$ line is low, the information on the address line will be held or latched in the *row-address latch.* Whenever the $\overline{\text{CAS}}$ line is low, the information on the address line will be held or latched in the *column-address latch.* The $\overline{\text{RAS}}$ and $\overline{\text{CAS}}$ control lines must never be allowed to both be low at the same time, or confusion within the device may result. The write enable line, $\overline{\text{WE}}$, determines the read/write mode. When $\overline{\text{WE}}$ is low, data on the D_{in} line will be written into the selected address. When the $\overline{\text{WE}}$ line is high, data from the selected address will appear on the D_{out} line.

Refreshing is accomplished by reading data, writing data, or a separate refresh operation. Basically, the refresh control circuitry must sequentially select each row of mem-

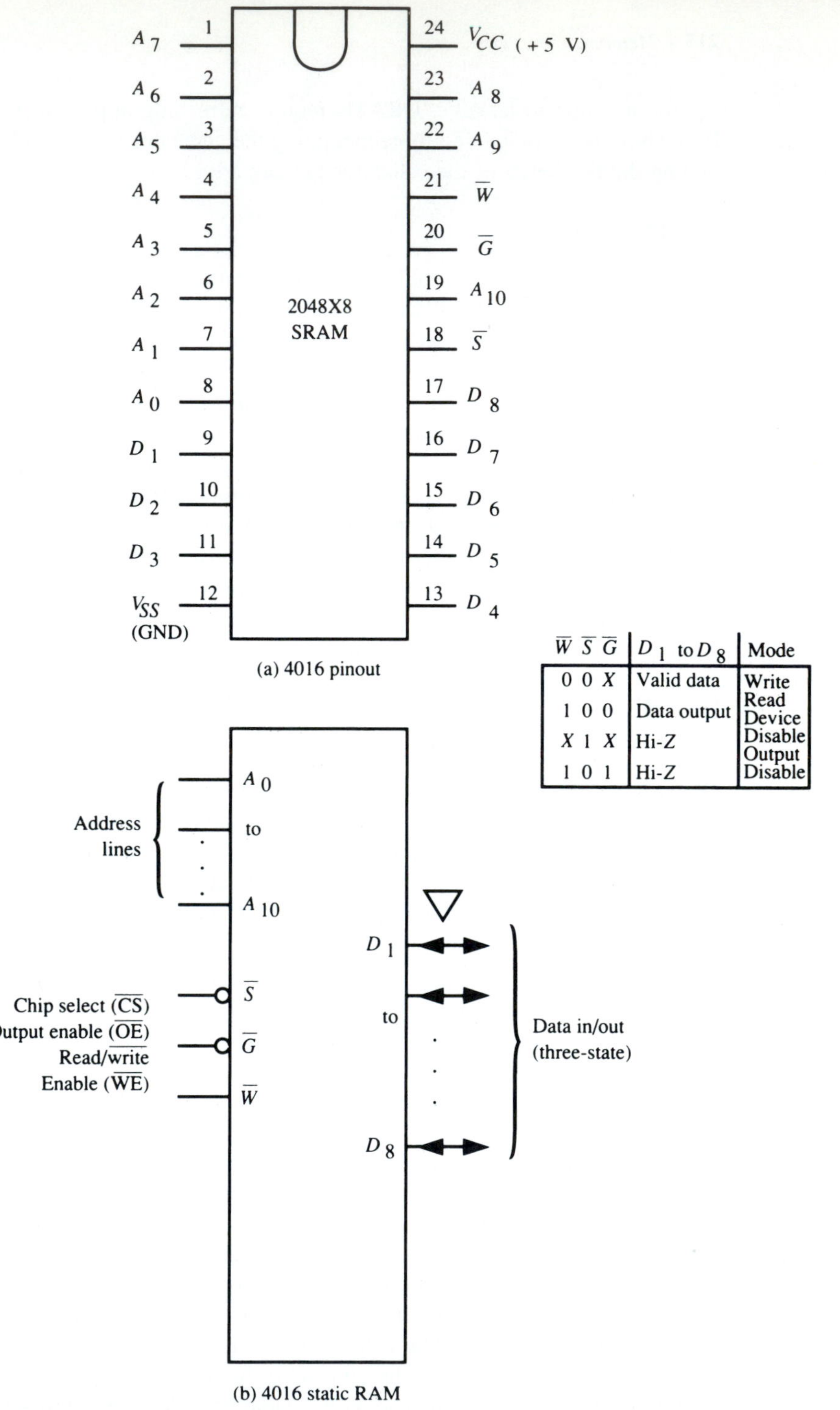

$\overline{W}$ $\overline{S}$ $\overline{G}$	D_1 to D_8	Mode
0 0 X	Valid data	Write
1 0 0	Data output	Read
X 1 X	Hi-Z	Device Disable
1 0 1	Hi-Z	Output Disable

FIGURE 7.7
TMS 4016 static MOS RAM

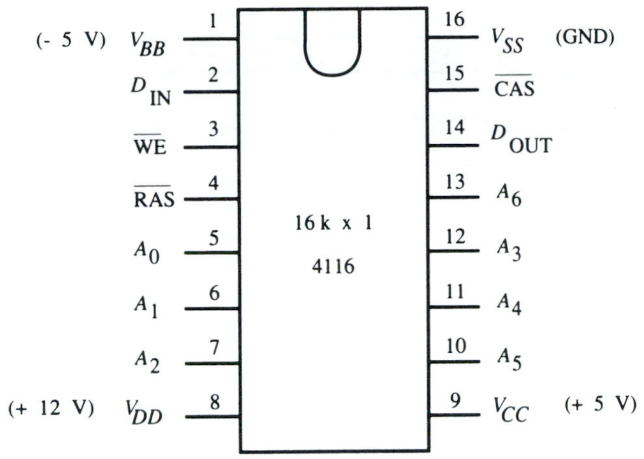

FIGURE 7.8
4116 DRAM pinout

ory cells, one row at a time, until all rows have been refreshed. This is called *burst mode refreshing*. During this time, data cannot be read from or written into the device until the entire refresh operation is complete. Another technique refreshes each row in discrete cycles. This is called *single-cycle refresh*. In Figure 7.10(a) (p. 221) the refresh control circuitry causes the multiplexer to switch from the address line to the refresh row address counter. Once the refresh row address counter is selected, the counter supplies the row address information (RA_0–RA_6) to step sequentially through each row of the memory array. When the refresh counter reaches its final count, which is the last row in the memory cell array, the refresh cycle is ended. The multiplexer is switched back to the address lines, and the memory cell array is then available to the system. Figure 7.10(b) illustrates the refresh cycle timing. The control signal initiates the refresh cycle by selecting the refresh row address counter. The $\overline{\text{RAS}}$ signal causes each selected row to be refreshed. The $\overline{\text{CAS}}$ signal is kept high to disable the output buffers.

7.4 MEMORY EXPANSION

We have seen that memory arrays can be organized in many ways, for example, 2048 × 8, 4096 × 4, and 16,384 × 1. Many microcomputer systems organize their data in 8-, 16-, and even 32-bit formats. A typical problem that arises is how to organize a certain size memory system. For example, suppose we want a 16K × 8-bit memory. We could use either eight 2048 × 8 memory chips or eight 16,384 × 1 memory chips. Let's first consider the case where we wish to make a 16K × 8-bit memory system out of eight 16,384 × 1-bit memory chips. The TMS 4116, which was previously discussed, is a 16,384 × 1-bit memory chip that can be used for this application. Essentially, all that is required for this type of memory expansion is to connect each memory chip in parallel, as illustrated in Figure 7.11 (p. 222). Notice that all the address lines and control lines

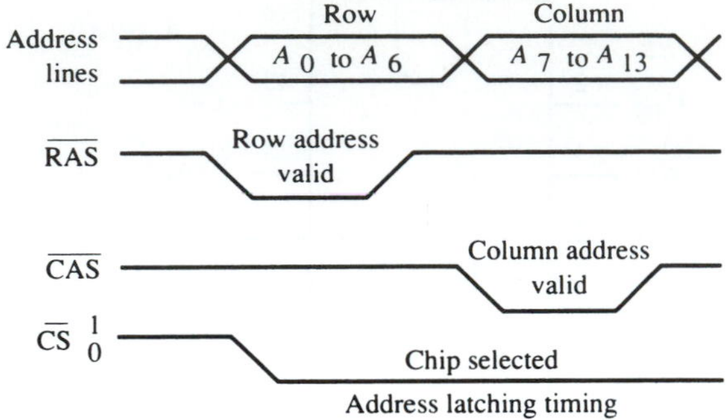

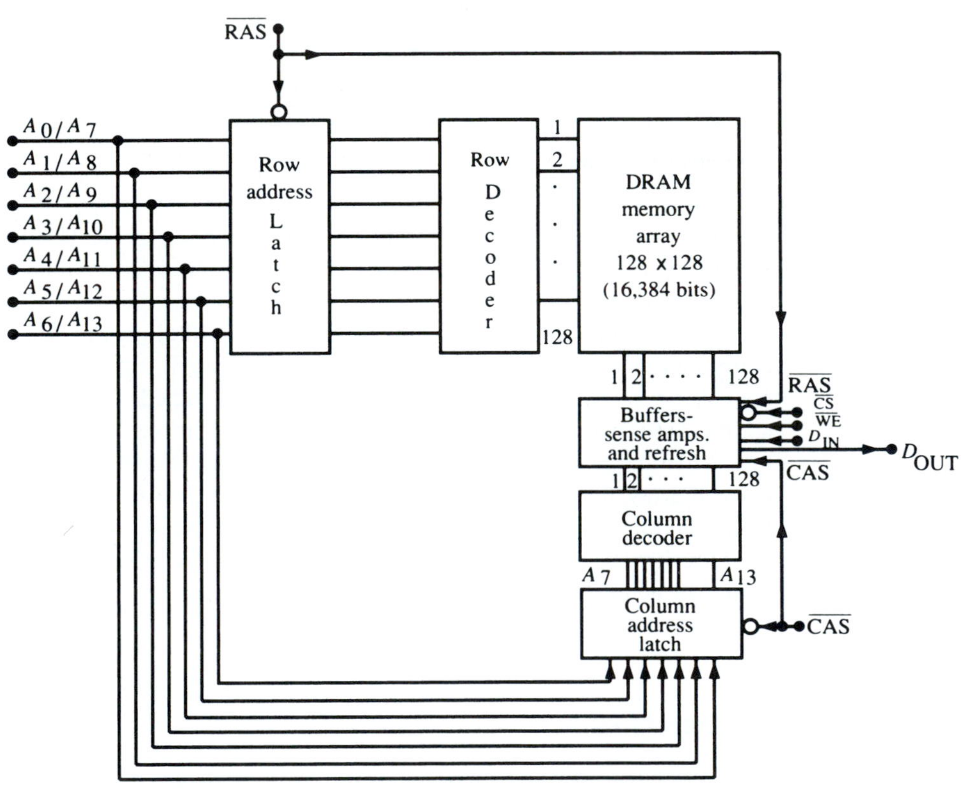

FIGURE 7.9
DRAM block diagram

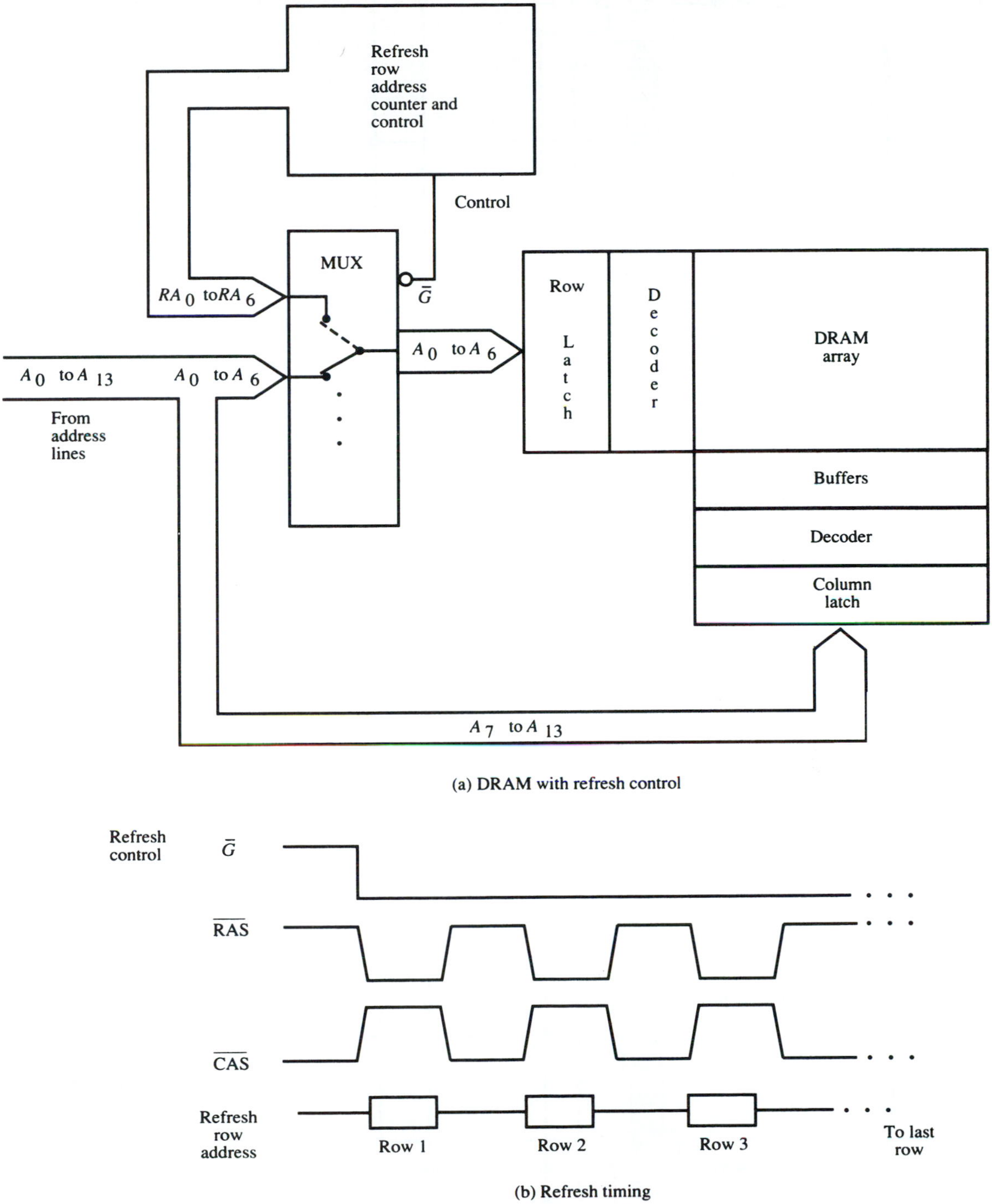

Refresh row address counter and control

Control

MUX

$\bar{G}$

RA_0 to RA_6

A_0 to A_{13}

A_0 to A_6

From address lines

A_0 to A_6

Row Latch

Decoder

DRAM array

Buffers

Decoder

Column latch

A_7 to A_{13}

(a) DRAM with refresh control

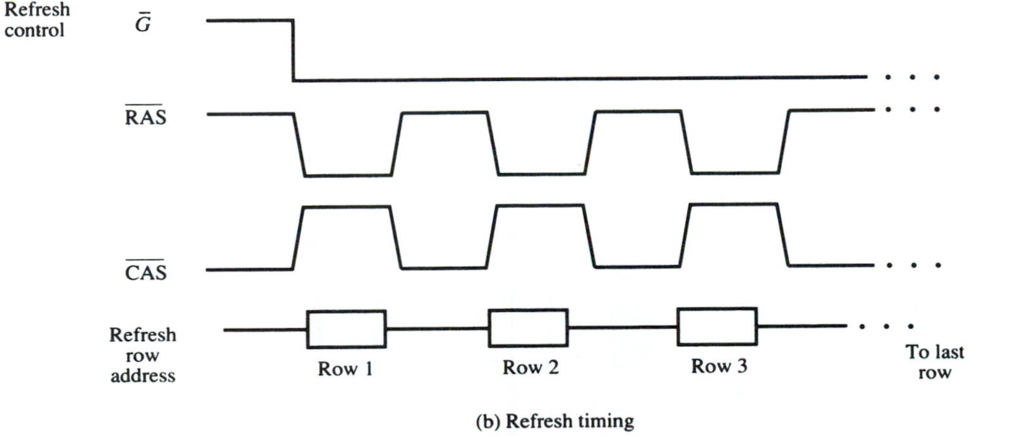

(b) Refresh timing

FIGURE 7.10
DRAM refresh

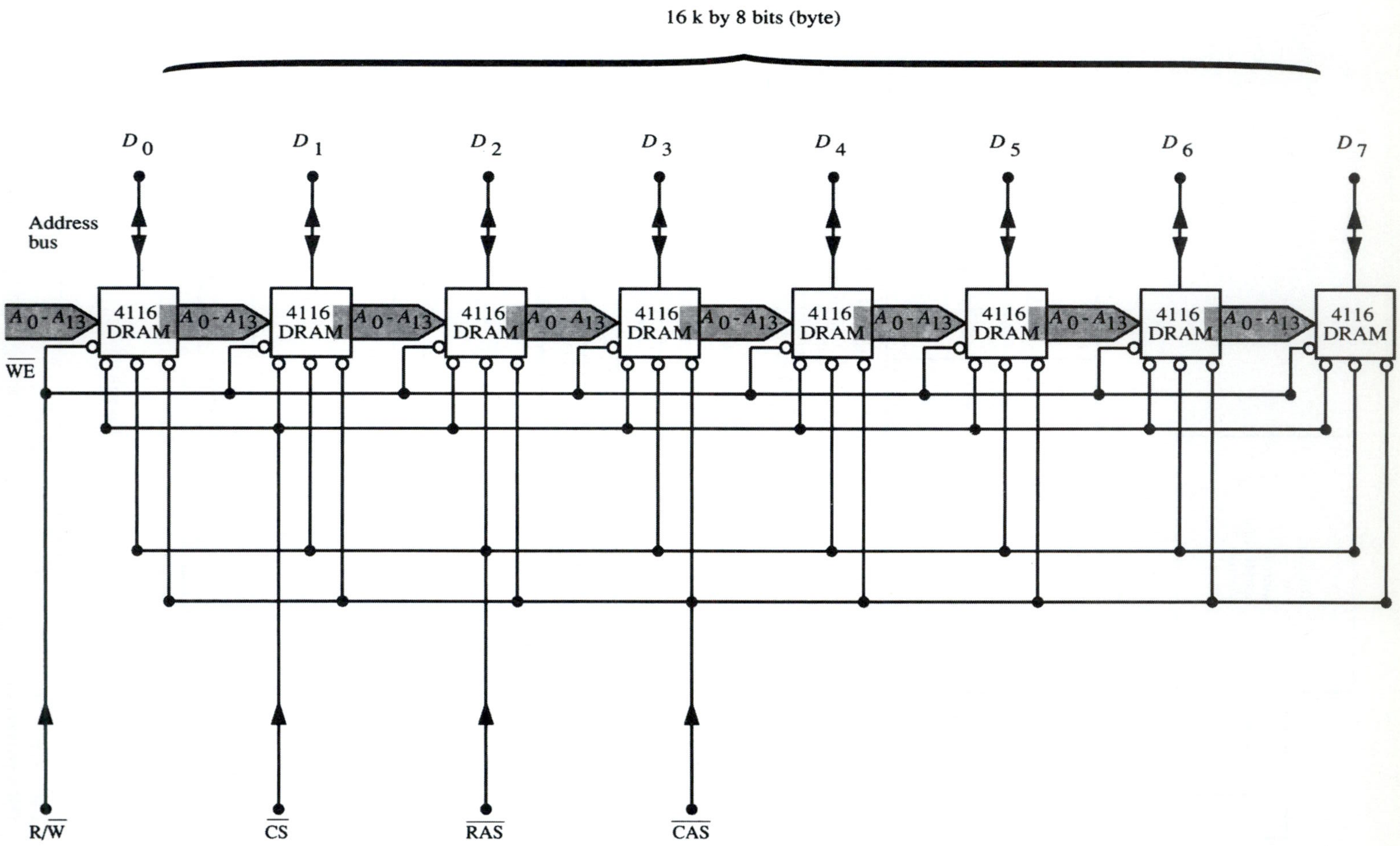

FIGURE 7.11
16K × 8-bit DRAM memory expansion using 4116 DRAMS (16K × 1 bit each)

are connected and parallel. If we are in the read mode, all eight memory chips will be in the read mode. If we are in the write mode, all eight memory chips will be in the write mode. The data lines are used to form the 8-bit-wide, or byte-wide, memory system.

To illustrate how to use eight 2048 × 8 memory chips to make a 16K × 8-bit memory system, we first consider a number of techniques that are commonly used to expand memory. We begin by expanding to a 4K × 8 system, then to an 8K × 8 system, and finally to a 16K × 8-bit memory system.

The 4K × 8-bit memory system is shown in Figure 7.12. It consists of two TMS 4016 static RAMs. Each 4016 is organized in a 2048 × 8-bit array. The 4016 has 11

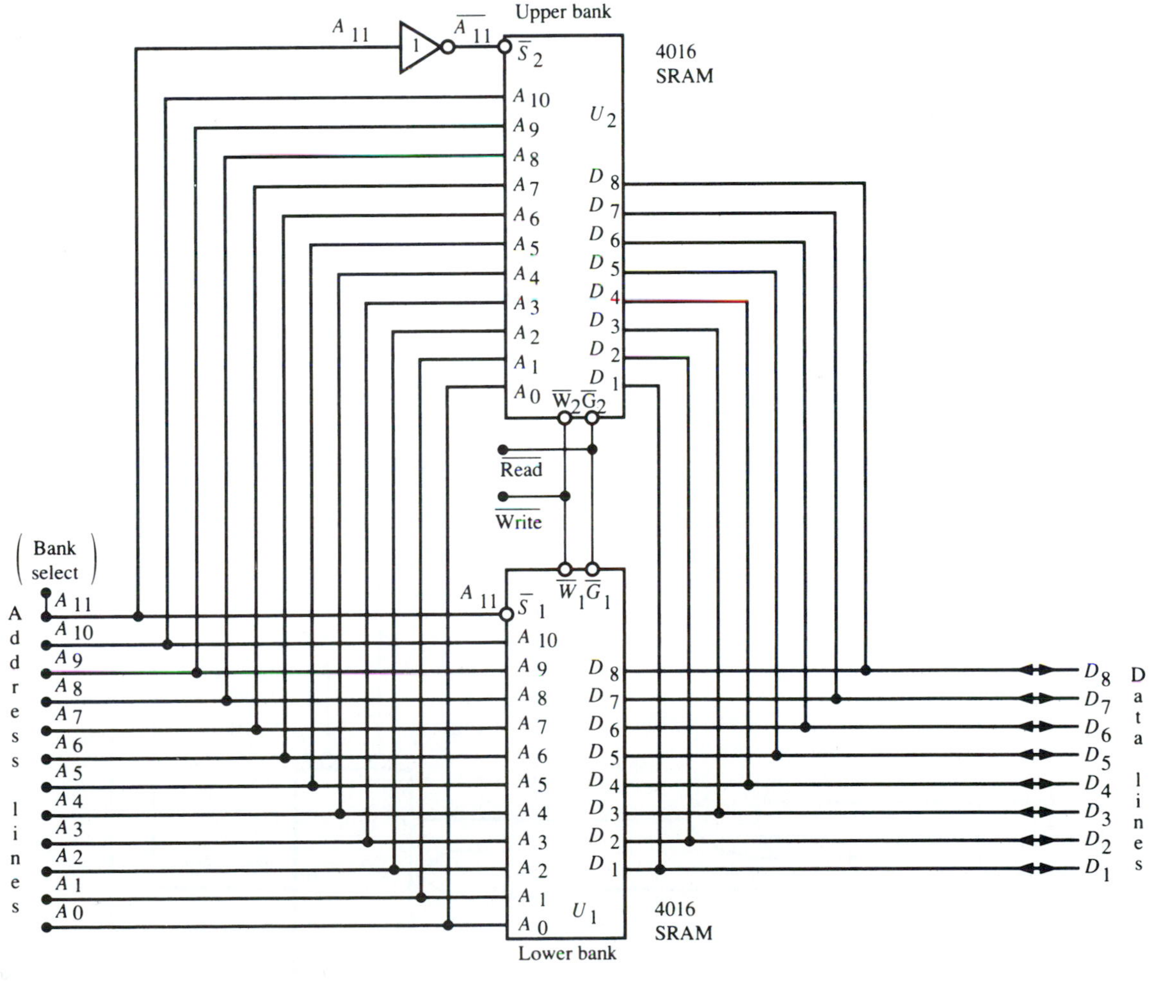

FIGURE 7.12
4K × 8-bit (byte) memory system using 4016 SRAMs of 2K × 8 bits

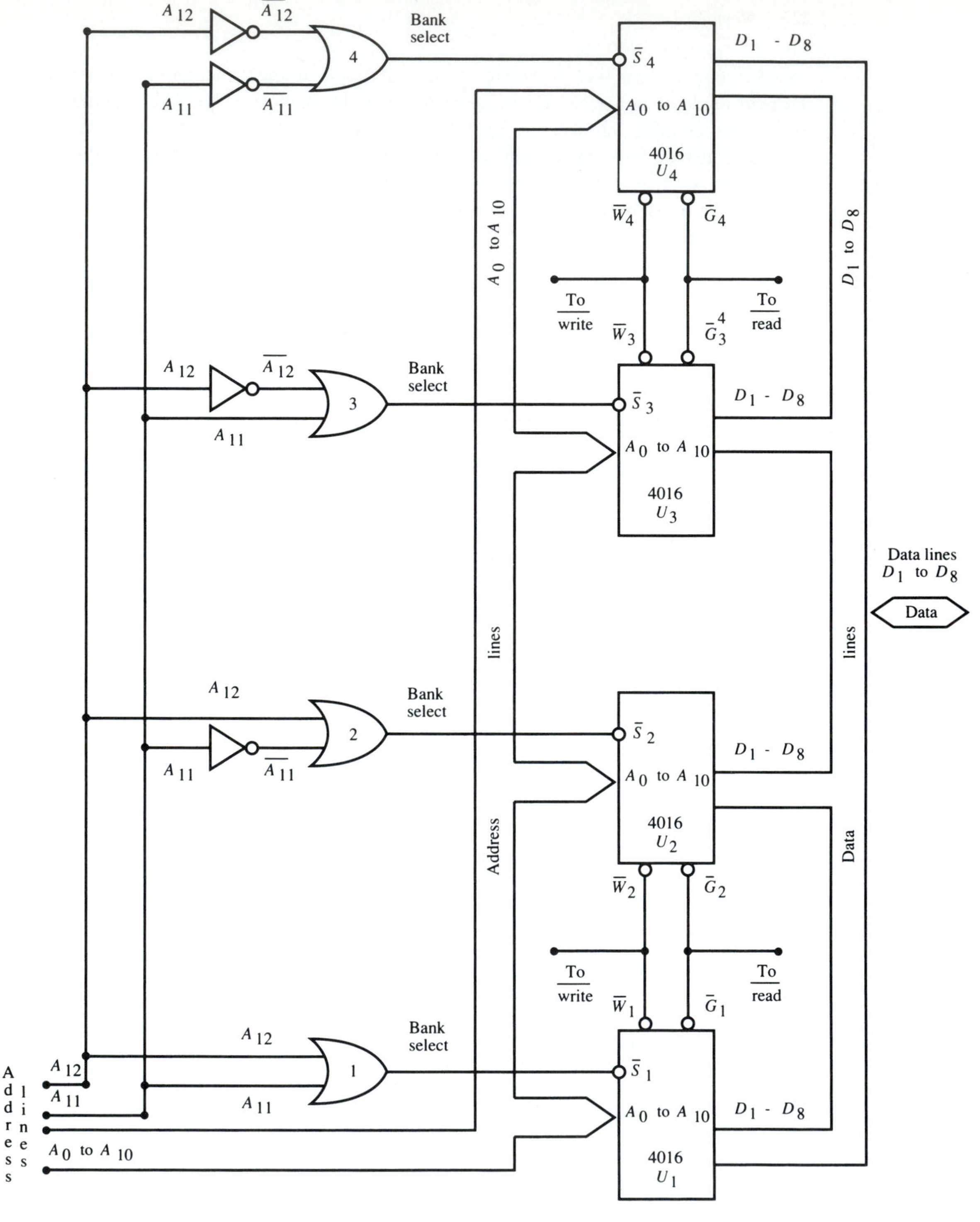

FIGURE 7.13
8K × 8-bit (byte) memory system

address lines, A_0 through A_{10}. This is what you would expect for a 2K memory chip, since 2^{11} equals 2048 memory locations. To address 4K, or 4096, memory locations, an additional address line is required ($2^{12} = 4096$). This additional line is used to select which memory chip we wish to use. Therefore, A_{11} is connected to the chip-select pin. To distinguish between the lower 2048 memory locations (0–2047) and the upper 2048 memory locations (2048–4095), an inverter is used. For all memory addresses between 0 and 2047, A_{11} will be a logic 0, selecting the lower bank. For all memory addresses between 2048 and 4095, A_{11} will be a logic 1, selecting through the inverter the higher bank. As we can see, A_{11} is used as a bank selector. Using the higher-order address lines for bank selection is a common technique. It is sometimes referred to as **address decoding.**

The 8K × 8-bit memory expansion system is shown in Figure 7.13. It consists of four TMS 4016 2K × 8-bit static RAMS. To address 8K, or 8192, memory locations requires 13 address lines (A_0 through A_{12}), since 2^{13} equals 8192 memory locations. Each TMS 4016 has 11 address lines, A_0 through A_{10}. The additional address lines, A_{11} and A_{12}, are used to select the memory chip we wish to use as follows:

- U_1 defines memory locations $(0000\text{–}2047)_{10}$ or $(0000\text{–}07FF)_{16}$.
- U_2 defines memory locations $(2048\text{–}4095)_{10}$ or $(0800\text{–}0FFF)_{16}$.
- U_3 defines memory locations $(4096\text{–}6143)_{10}$ or $(1000\text{–}17FF)_{16}$.
- U_4 defines memory locations $(6144\text{–}8191)_{10}$ or $(1800\text{–}1FFF)_{16}$.

Chip selection is controlled by the gating arrangement illustrated in Figure 7.13. For all memory addresses between 0000 and 2047, both A_{11} and A_{12} will be low. Referring to Figure 7.13, we can see that if A_{11} and A_{12} are both low, the output of OR gate 1 will be low, selecting only memory chip U_1. For all memory addresses between 2048 and 4095, A_{11} will be high, but A_{12} will be low. The combination of A_{11} high and A_{12} low will result in only memory chip U_2 being selected. For all memory addresses between 4096 and 6143, A_{11} will be low and A_{12} will be high, resulting in the selection of only U_3. Finally, for all memory addresses between 6144 and 8191, A_{11} and A_{12} will be high, resulting in the selection of only U_4.

The technique of using a gating arrangement to perform address decoding in memory-expansion systems would become cumbersome for larger memory systems. The 16K × 8-bit memory system of Figure 7.14 employs a slightly different technique. Eight TMS 4016 2K × 8-bit static RAMS are used to make the 16K × 8-bit memory system. Fourteen address lines, A_0 through A_{13}, are now required to define 16,384 memory locations, since 2^{14} equals 16,384. The additional three lines, A_{11}, A_{12}, and A_{13}, are used for chip selection. Address decoding or memory chip selection is accomplished using a 74138, three- to eight-line decoder. In Figure 7.14 address lines A_0 through A_{10} are connected to each TMS 4016. Address lines A_{11}, A_{12}, and A_{13} are connected to inputs *A*, *B*, and *C* of the 74138, respectively. The 74138 outputs 0 through 7 are connected to memory banks U_1 through U_8, respectively. As memory addresses are defined, the 74138 decodes and selects the proper memory bank. For example, if A_{11}, A_{12}, and A_{13} are all low, memory bank U_1 is selected. For larger systems, additional decoders can be added.

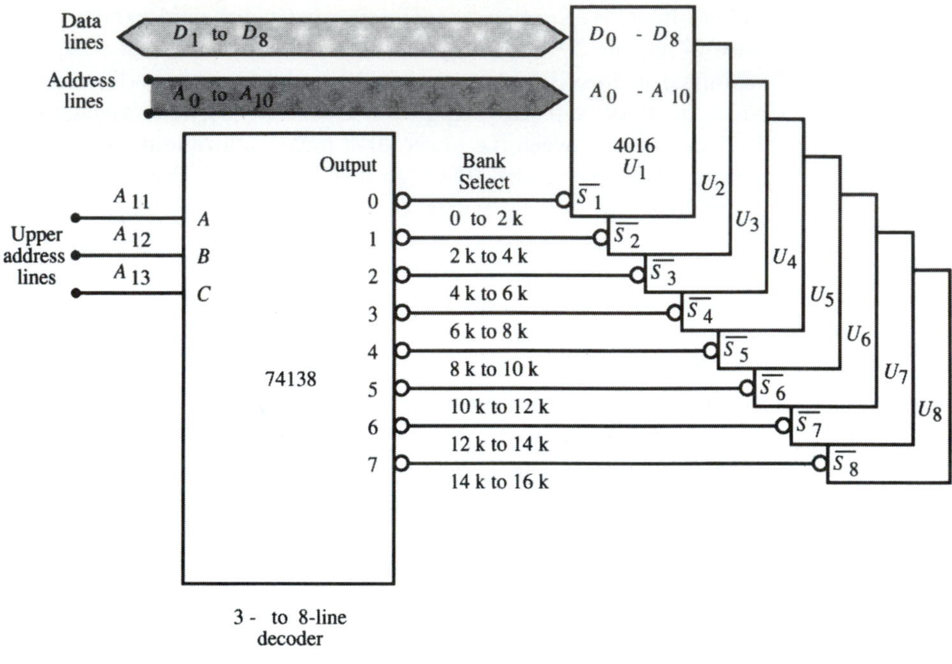

FIGURE 7.14
16K × 8-bit (byte) memory system

7.5 PROGRAMMABLE LOGIC DEVICES

Another type of memory device that has the characteristics of not only storing data but of also performing logical functions is called a **programmable logic device (PLD).** The PLD is an IC that contains numerous logic gates. These gates can be combined by programming to produce any logical function. A PLD can be used to replace a large number of logic devices. Instead of using individual logic gates, the manufacturer programs a PLD to perform the same function. Thus only one IC is required instead of many. This reduces cost and increases system reliability. For example, the circuit of Figure 7.15(a) can be replaced by the single PLD of Figure 7.15(b). Instead of wiring the circuit of Figure 7.15(a), the designer can program the PLD to perform the same function.

If you look at circuit boards used in recently manufactured computers, such as the IBM PS/2 series, and compare them with earlier ones, you will see that many of the individual gates have been replaced by PLDs. PLDs enable the manufacturer to save a lot of valuable PC board space and also provide more complex functions. This allows more features for the same given PC board area. For example, newer microcomputers have been able to provide all their graphic functions using PLDs. Design corrections or changes can be made by simply programming a new PLD and replacing the old one.

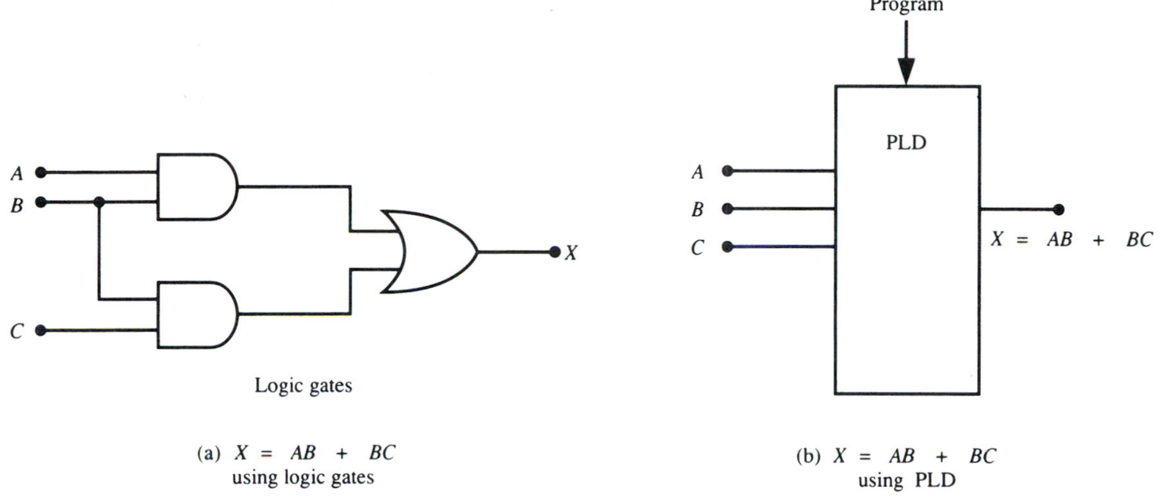

(a) $X = AB + BC$ using logic gates

(b) $X = AB + BC$ using PLD

FIGURE 7.15
PLD

The most popular PLD is the PROM. Other PLDs are reprogrammable, such as the ultraviolet erasable (EPROM) and the electrically erasable (EEPROM) PLD. With these types of PLDs, the logic function can be erased and a new one programmed in, thus eliminating the need to replace the chip.

7.6 PLD INTERNAL STRUCTURE

A **product line** is an input line connected to an *AND gate.* The inputs to the PLD can be programmed such that they will be either connected or not connected to the product line. All the inputs that are connected are ANDed together. The result is called the **product term.** For example, if the product term $X \cdot Y$ is needed, lines X and Y remain connected, but line Z does not, as shown in Figure 7.16(a). If the term $X \cdot Z$ is desired, we connect lines X and Z only as shown in Figure 7.16(b).

A **sum line** is a line connected by an *OR gate* to the output. For example, if the expression $(X \cdot Y) + Z$ is desired, the product line from AND gate 1 and the product line from AND gate 3 will be connected to the sum line, as shown in Figure 7.16(c). Note that the product line connections must also be made.

A PLD that is not programmed is referred to as a **blank PLD.** Each connection is made by a fuse. When the PLD is programmed, the fuses of the lines that are not to be connected are blown open. Internally, the PLD consists of a matrix of product lines and sum lines that is called a **fuse map.** The blowable fuses are used to connect the product and sum lines to the output. All the inputs to a PLD are connected through a buffer. This provides both inverted and noninverted inputs, as shown in Figure 7.17 (p. 229).

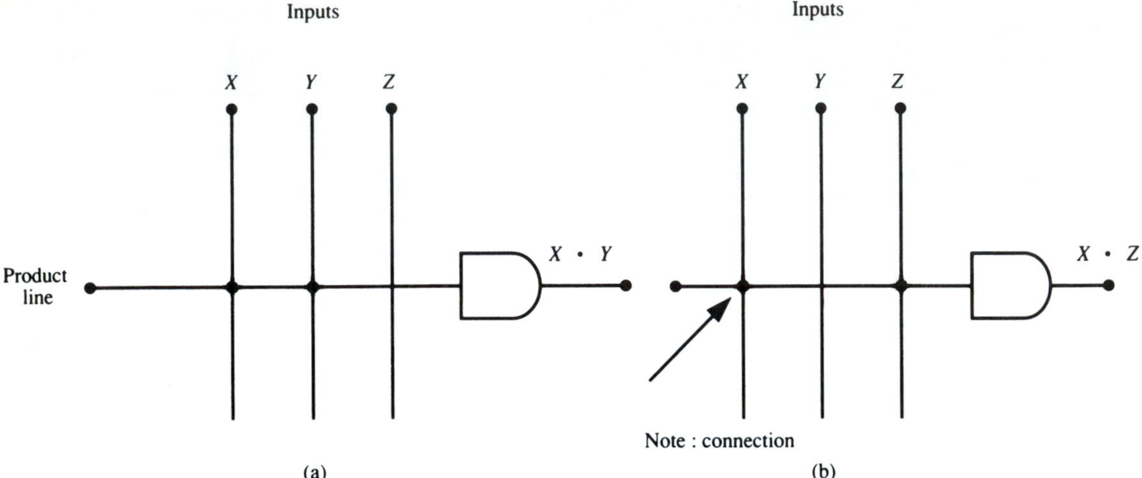

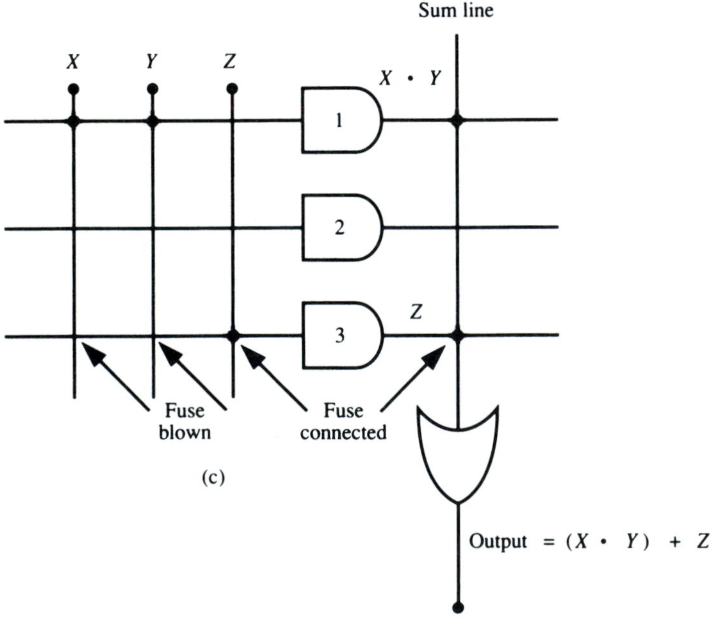

FIGURE 7.16
Fuse map

Some PLDs are completely programmable, whereas others have one level of gates fixed or permanently connected, which is the basic difference among different types of PLDs.

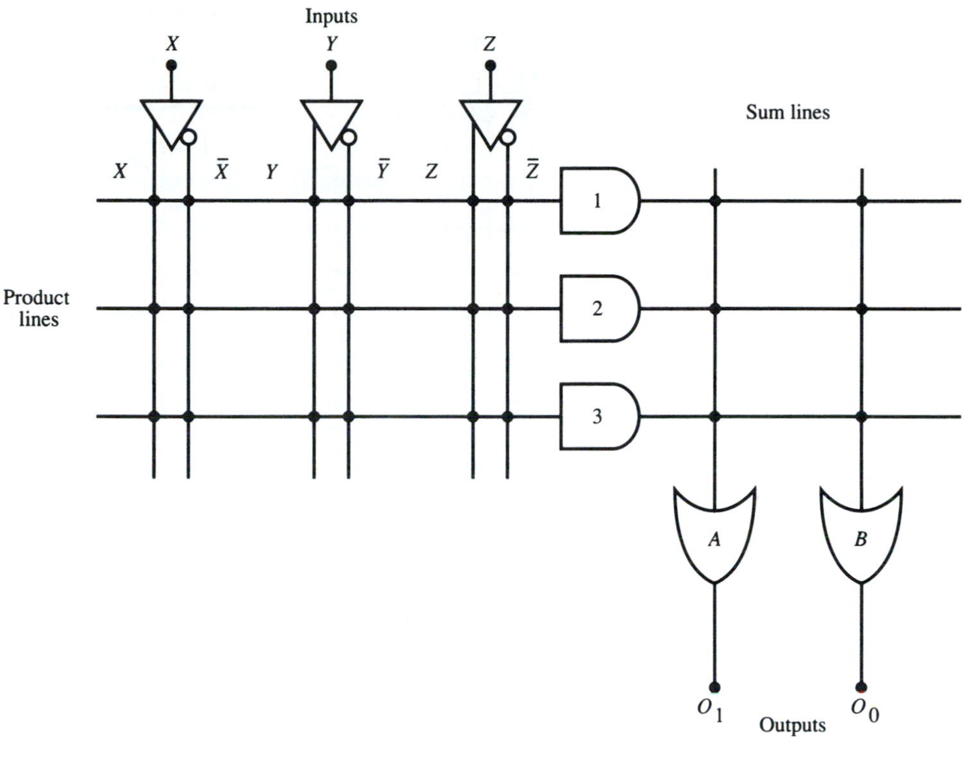

FIGURE 7.17
PLD with buffered inputs

For example, a PROM is a PLD with fixed AND gates and programmable OR gates. The array of AND gates form the PROM address decoder, as shown in Figure 7.18. For a given address input, its product term from the AND gate will be high, whereas all the others will be low. For example, in Figure 7.18, if address inputs I_0, I_1, I_2. and I_3 are all high (logic 1), product line 15 through AND gate 15 will be high, and all others will be low. This will select or decode address 15. If we wish to store 1010 at this memory location, we must program the OR array to store this number. The number 1010 is stored by programming the OR array to blow the fuses of the sum lines *A* and *C* and leave connected the fuse connections of the sum lines *B* and *D*.

A **programmable logic array (PLA)** is a PLD that has both its AND gates and OR gates programmable. This provides complete control over both the product and sum lines. A **programmable array logic (PAL)** is a PLD that has programmable AND gates and fixed OR gates. Table 7.2 summarizes the internal structures of the different types of PLDs.

It should be noted that PLAs have certain disadvantages compared with PALs and PROMs. First, a given signal has to pass through two programmable arrays, which increases propagation delays. Second, PLA programming equations are generally more cumbersome, making them more difficult to program. PROMs and PALs are generally very easy to program, thus making them more popular.

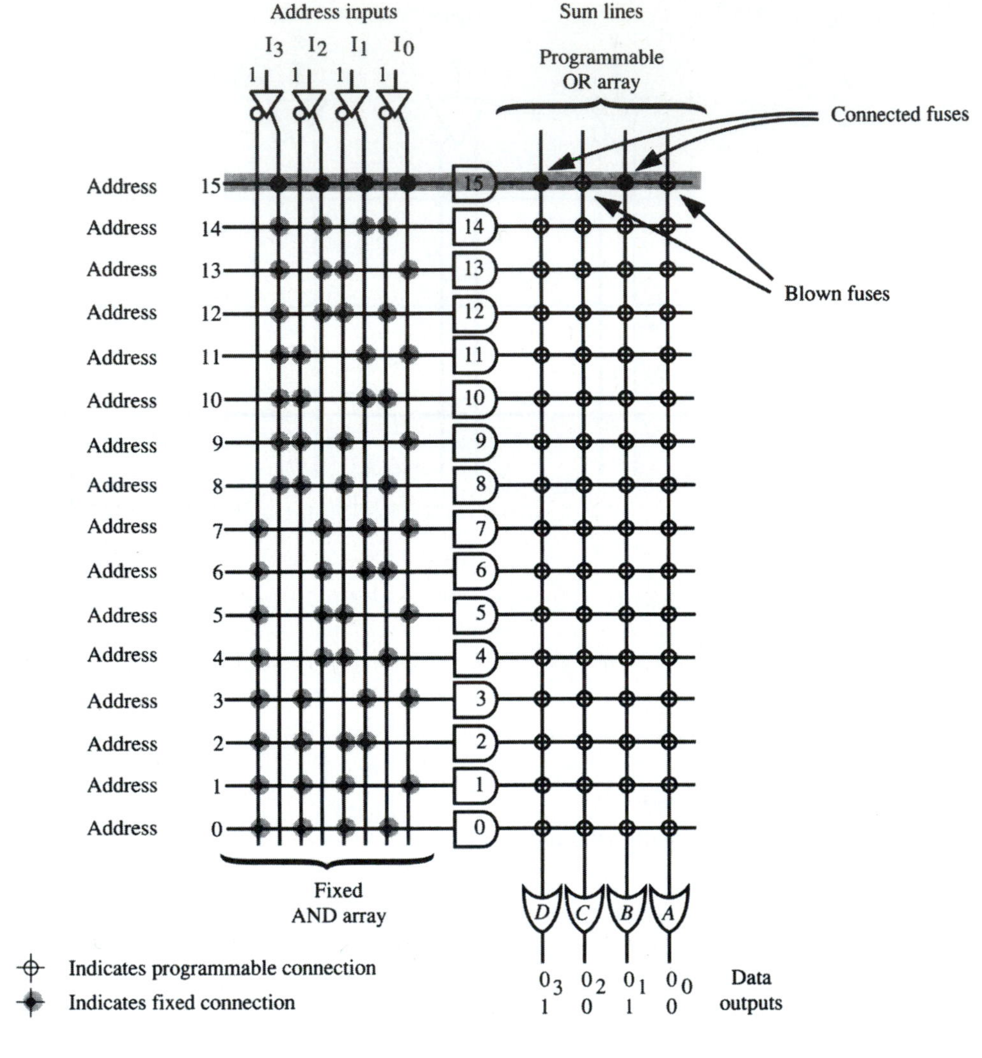

FIGURE 7.18
PROM array structure

TABLE 7.2
PLD types

	PROM	PAL	PLA
AND	Fixed	Programmable	Programmable
OR	Programmable	Fixed	Programmable

7.7 PALs

PALs can be divided into two general types. **Combinational PALs** contain only logic gates; **sequential PALs** contain both logic gates and flip-flops.

Combinational PALs

Recall that in a PAL the AND gates are programmable, whereas the OR gates are fixed. Figure 7.19 illustrates the logic diagram of a PAL with three inputs, one output, and three product lines. The number of programmable fuses is then 3 inputs times 2 (inverted and noninverted) times 3 product lines, or 18 (3 × 2 × 3) fuses. The PAL shown in Figure 7.19 is programmed for the Boolean equation

$$O_1 = (I_1 \cdot I_2 \cdot \overline{I_3}) + (\overline{I_1} \cdot \overline{I_2})$$

Note that product line 3 will always result in a logic 0 output because all the fuses have been left connected, resulting in $I \cdot \overline{I} = 0$. If all the fuses were blown, the product term would result in a logic 1 output.

As you can see, PALs are easily programmed from Boolean equations in the sum-of-the-product (SOP) form. Each product term is made by one product line. All the product lines for a given output are then ORed to produce the SOP term.

The first step to be performed in programming a PAL is to write the Boolean equation in SOP form from the truth table. The number of product lines is limited, so the Boolean equation must be optimized. This can be done using the classical Karnaugh map method. Computer programs are now available to program the PAL and optimize the Boolean equation for you.

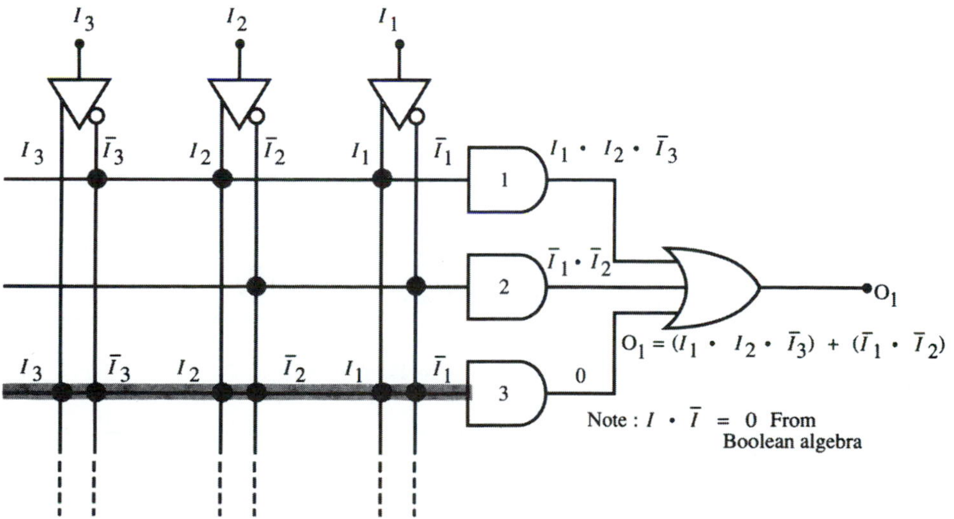

FIGURE 7.19
PAL

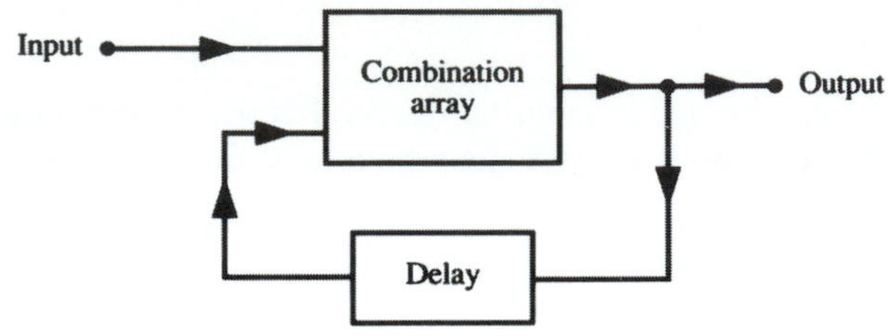

(a) Sequential logic

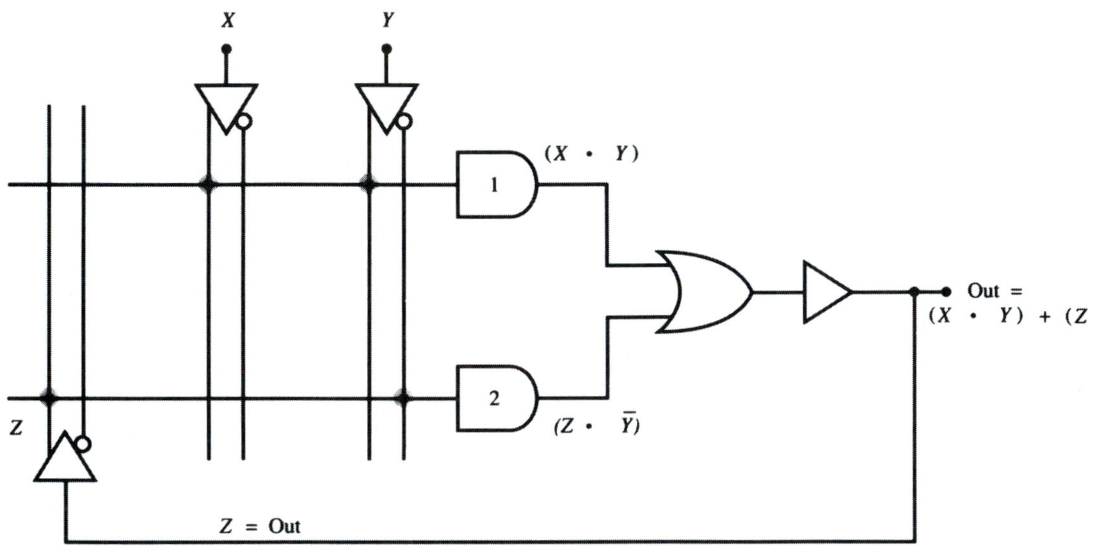

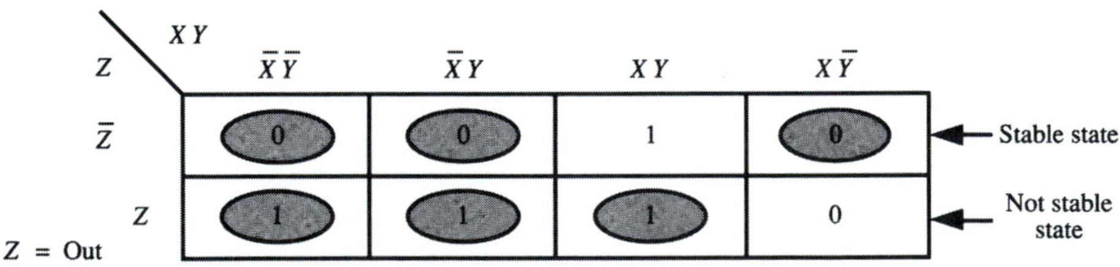

(b) Feedback PAL

FIGURE 7.20
Sequential PAL

Sequential PALs

Sequential PALs employ feedback, as shown in Figure 7.20(a). That is, the output is fed back to the input, which can provide the ability to implement asynchronous sequential logic on a PAL. In this case the inputs plus the present state of the output are used to generate the next output state. The PAL becomes stable when the present state and the next state are the same. In Figure 7.20(b) we have programmed a PAL to implement the function

$$\text{OUT} = (X \cdot Y) + (\overline{Y} \cdot Z)$$

where Z represents the feedback output of the previous state. The Karnaugh map shows the resulting outputs, with the stable states circled.

Some PALs provide the ability to implement synchronous sequential logic. They are called **registered PALs.** In a registered PAL some of the outputs are terminated by D flip-flops. Once the product terms have been added, the sum is connected to the D input of the flip-flop. The Q output becomes the output of the PAL, and the $\overline{Q}$ output is used for the feedback path, as shown in Figure 7.21. The clock is the same for all flip-flops; it is usually on pin 1 of the chip. Here again, the outputs are a function not only of the inputs but also of the present state of the flip-flops.

PAL Identification

PALs are identified by a numbering system that defines the number of input lines, the number of output lines, and the type of PAL. For example, a 14H4 PAL is a PAL with 14 inputs and 4 outputs and is active high. A 14L4 is similar except that it is active low. A 16R4 is a PAL with 16 inputs and 4 registered outputs (feedback paths). Figure 7.22 shows the logic diagram for a 16R4 PAL.

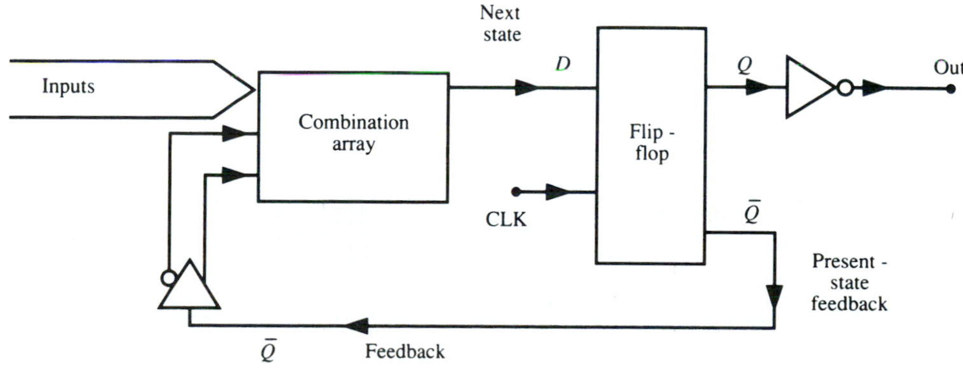

FIGURE 7.21
Registered PAL

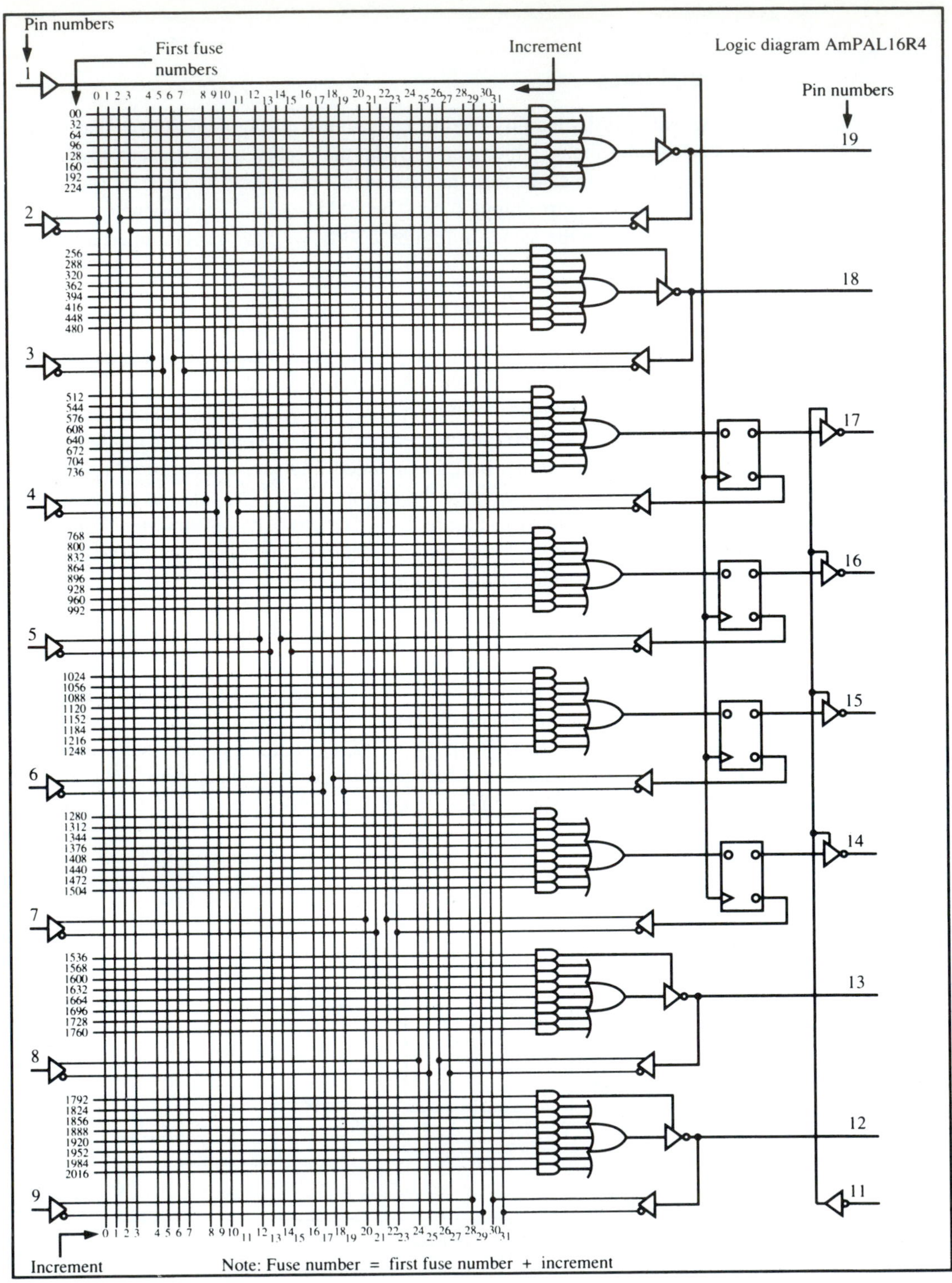

FIGURE 7.22
(Reprinted by permission of Advanced Micro Devices)

7.8 PAL DESIGN EXAMPLE

To illustrate the use of a PAL, we will design a 4-to-1 multiplexer using a PAL. Figure 7.23(a) illustrates the design problem. There are four input lines, I_0 to I_3. Since there are

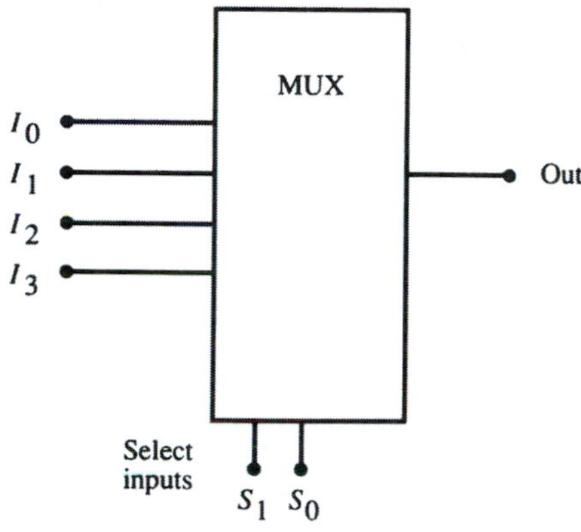

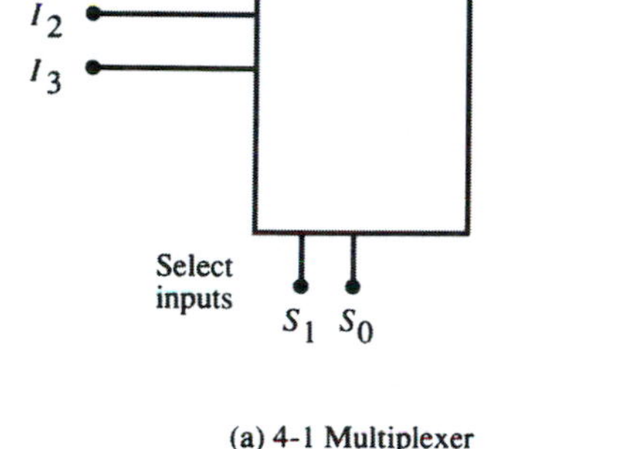

I	S_1	S_0	Output
I_0	0	0	I_0
I_1	0	1	I_1
I_2	1	0	I_2
I_3	1	1	I_3

$$\text{Out} = I_0 = (\overline{S_1} \cdot \overline{S_0} \cdot I_0)$$
$$\text{Out} = I_1 = (\overline{S_1} \cdot S_0 \cdot I_1)$$
$$\text{Out} = I_2 = (S_1 \cdot \overline{S_0} \cdot I_2)$$
$$\text{Out} = I_3 = (S_1 \cdot S_0 \cdot I_3)$$

(a) 4-1 Multiplexer

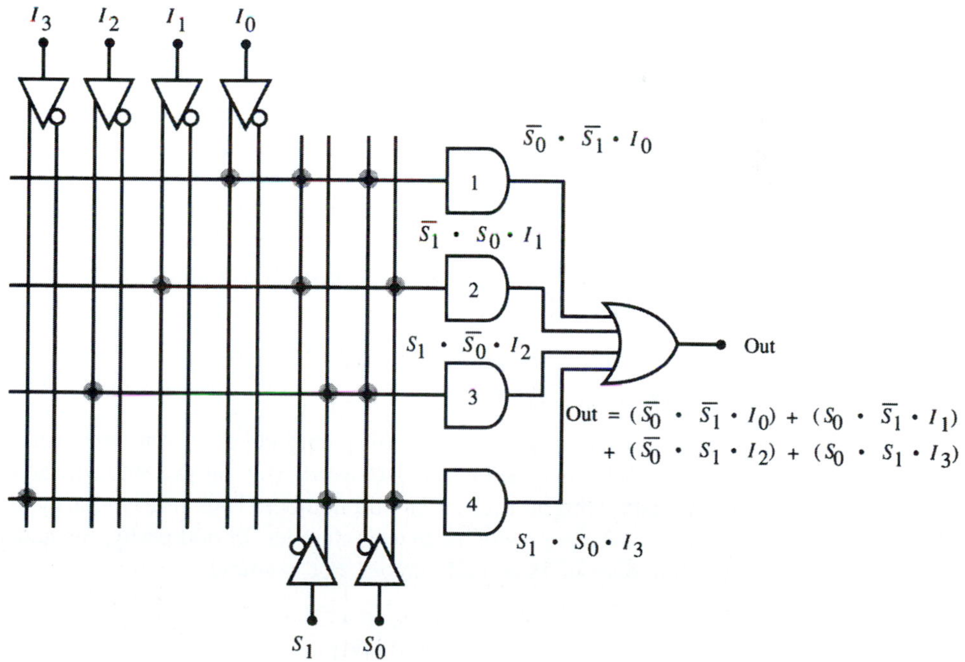

(b) Circuit

FIGURE 7.23
PAL programmed for multiplexer

four inputs that we wish to switch to the output, two control lines ($2^2 = 4$) are required. The control lines are labeled S_0 and S_1. The control lines define which input will be switched to the output. The truth table defines the desired operation of the PAL. The Boolean equation for the output is then

$$\text{OUT} = (\overline{S_0} \cdot \overline{S_1} \cdot I_0) + (S_0 \cdot \overline{S_1} \cdot I_1) + (\overline{S_0} \cdot S_1 \cdot I_2) + (S_0 \cdot S_1 \cdot I_3)$$

The programmed PAL is shown in Figure 7.23(b).

7.9 APPLICATION-SPECIFIC INTEGRATED CIRCUITS (ASICs)

Today, many microcomputers, such as the IBM PS/2 series and the Apple Macintosh II, have replaced many of the standard SSI and MSI integrated circuits with **application-specific integrated circuits (ASICs).** The ASIC can be thought of as a further advancement or enhancement of the PLD.

ASICs and ICs were designed for many of the same reasons, including increased reliability, smaller PC board area, less power consumption, simplified design, and improved speed. Designers are now using ASICs to replace hundreds of standard MSI chips. Furthermore, ASICs make it much more difficult to copy, or *clone,* circuits that utilize them. ASICs are designed using advanced *computer-aided engineering* (CAE) workstations. They represent one of the fastest growing segments of the IC industry today.

7.10 TECH TIPS AND TROUBLESHOOTING—T^3

Most microcomputer memory systems use some method of *validating* the data bits that are stored in its memory. Validating means that the data bits in the memory system are checked and even corrected when they are received (written) or transmitted (read). Thus, *lost* or *stuck bits* will result in a system failure. Memory failures are one of the most common problems with microcomputer systems today. Many different techniques are used to detect memory errors, the most common being bit parity, word parity, check sums, and cyclic redundancy checks.

Bit Parity

There are two basic types of **bit parity** checks, *even parity* and *odd parity.* In both cases an extra bit is added to the data word. This bit is used to make the total number of 1s in the data word an even or an odd number. Thus in even parity, the added parity bit makes the total numbers of 1s an even number. In odd parity, the added parity bit makes the total number of 1s an odd number. For example,

Data Word
1100 0001
Even parity
1100 0001 1 ← Parity bit
Odd Parity
1100 0001 0 ← Parity bit

It should be noted that bit parity methods can detect only a single bit error. If two bits in a data word were to change at the same time, the parity bit *might not* detect the error.

Word Parity

A variation of the bit parity technique is **word parity.** In this technique a parity bit is added to each data word (row) and also for each vertical data position (column). For example, consider the following words using odd parity:

Data Word	OP	
1101 0001	1	
1110 0100	1	
1100 0101	1	
1110 0010	1	
1111 0000	1	
1111 0010	0	
1111 0011	1	
1001 1100	1	
1000 0000	0	← Parity word

Eight data words are shown with an additional odd parity bit. The parity word is calculated by determining odd parity for each vertical position (column). If a single-bit error occurs, it can be detected and corrected automatically by the computer. For example, if during transmission word five develops an error in the fourth column from the left, the computer can detect and correct this problem, as illustrated.

Data Word	OP	
110 1 0001	1	
111 0 0100	1	
110 0 0101	1	
111 0 0010	1	
111 0 0000	1	← Parity error detected
111 1 0010	0	
111 1 0011	1	
100 1 1100	1	
100 0 0000	0	← Parity word
↑ Parity error detected		

Check Sums

Most ROMs and PROMs in microcomputer systems today employ a technique known as **check sums** for error detection. In this technique the last location of the ROM's or PROM's memory is reserved for a check sum word. The check sum word is a word that produces a sum of zero when added to the other data words in memory. For example, let's consider an 8-bit ROM with eight memory locations.

Location	Data Word	Hex Equivalent	
0	1101 0101	D5	
1	1100 0000	C0	
2	1110 0010	E2	
3	1100 1000	C8	
4	1100 0101	C5	
5	1101 0000	D0	
6	1110 0010	E2	
7	0100 1010	4A	Check sum word
	110 0000 0000	6 00	Sum
Overflow →	↑		

Memory location 7 contains the check sum word. Note that the sum of all the data words in memory including the check sum word is zero if we neglect the overflow bits. The check sum word is obtained by adding memory locations 0 through 6 and subtracting the result from the next higher order word (1 0000 0000B or 100H). The result becomes the check sum word.

Cyclic Redundancy Checks

The **cyclic redundancy check** (CRC) system can be used to detect double errors. This technique is commonly used in data communications such as those used in microcomputer disk drive subsystems. CRC calculations are performed using multisectional shift registers that are fed into EXCLUSIVE-OR gates, as shown in Figure 7.24(a). Data bits are fed into the multisectional shift register to produce *block check characters* (BCCs). The BCC of the transmitting device is compared with the BCC of the receiving device using an EXCLUSIVE-OR gate, as shown in Figure 7.24(b). As long as the output of the EXCLUSIVE-OR gate is zero, no error exists. If the output of the EXCLUSIVE-OR gate is a 1, this signals the computer that an error has occurred. For a 16-bit system the CRC polynomial can be calculated using the algorithm

$$\mathrm{CRC}_{16} = X^{16} + X^{15} + X^{2} + 1$$

As each data bit is received through the data in line, it is cycled around through EXCLUSIVE-OR gate 1. The output of EXCLUSIVE-OR gate 1 is fed back into shift register A and is simultaneously fed back into shift registers B and C through EXCLUSIVE-OR gates 2 and 3, respectively, to produce a unique BCC word. This technique is duplicated at the transmitter and the receiver. The results are compared through another EXCLUSIVE-OR gate to perform the CRC. The algorithm is unique, so it is highly unlikely that a combination of errors will produce the same BCC at the receiver and transmitter CRC circuits. The most important property of CRC is that, due to the feedback elements in the circuit, the contents of the shift register depend upon the past manipulation of its states.

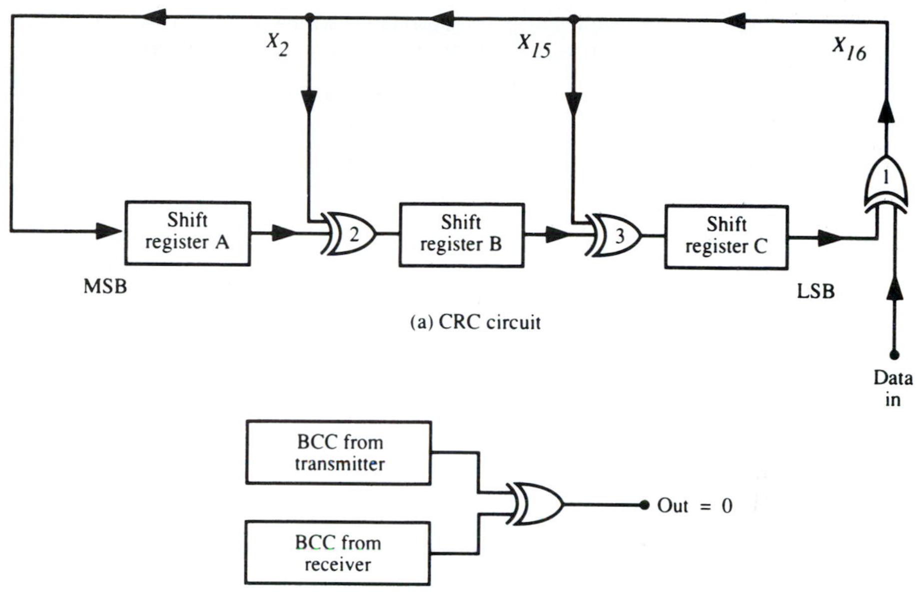

FIGURE 7.24
Cyclic redundancy check

EXERCISES

7.1 Define the following terms:

a. Memory cell
b. Memory address
c. Memory read cycle
d. Memory write cycle

7.2 Explain how memory devices are generally organized or structured.

7.3 Compare volatile and nonvolatile memory.

7.4 Discuss the difference between bit-organized memory and word-organized memory.

7.5 The TTL 7489 RAM is a word-organized device with four address lines. Determine how many memory locations it contains.

7.6 List and describe the different classifications of ROMs.

7.7 List three applications of ROMs.

7.8 Redraw Figure 7.3(a) (for addresses A_0 to A_3 only). Use the following data: $A_0 = 1111$, $A_1 = 1100$, $A_2 = 0000$, $A_3 = 0101$.

7.9 Redraw Figure 7.4(a) so that the data 1100 will be stored in the BJT PROM.

7.10 Redraw Figure 7.4(b) so that the data 0101 will be stored in the MOSFET PROM.

7.11 Explain the differences between static and dynamic RAM.

7.12 Discuss how refreshing is accomplished.

7.13 a. Draw a diagram showing the wiring needed to write the data word 11111111 into a 4016 static RAM at memory address 0000000000.

b. How would the diagram change if the data just stored was to be read out?

7.14 List the advantages and disadvantages of DRAMs.

7.15 Explain how to address all 16K of memory in a 4116 DRAM with only seven address pins.

7.16 What is the function of the $\overline{\text{RAS}}$ and the $\overline{\text{CAS}}$ pins on a DRAM device?

7.17 Draw a diagram for a 16K $\times$ 4-bit memory using 4116 DRAMs.

7.18 In Figure 7.12 what logic levels should be applied to store the data word $(4)_{16}$ in memory location $(3)_{16}$?

7.19 In Figure 7.13 what logic levels should be applied to read the contents of memory location $(1FFF)_{16}$?

7.20 In Figure 7.14 what logic levels must be applied to select bank 1 (U_1)?

7.21 Explain the use of a PLD.

7.22 Referring to Figure 7.16, redraw the fuse map to implement the following Boolean expression: OUT = $(X \cdot Y) + (X \cdot Z)$

7.23 Referring to Figure 7.17, redraw the fuse map for the following Boolean expression:

$$\text{OUT} = (\overline{\text{X}} \cdot \text{Z}) + (\overline{\text{Y}} \cdot \overline{\text{Z}}) + (\text{X} \cdot \text{Y} \cdot \text{Z})$$

7.24 Discuss the differences between a PROM, PAL, and PLA.

7.25 Redraw Figure 7.19 to obtain $O_1 = 1 + (\overline{I_1} \cdot I_2 \cdot \overline{I_3}) + 0$

7.26 Show how the PAL in Figure 7.22 can be used to implement the function OUT = $(X \cdot Y) + (Z \cdot \overline{Y})$ from Figure 7.20.

7.27 Design a 2-to-1 multiplexer using a PAL.

7.28 Add the correct parity bits to obtain odd and even parity for the following data words.

a. 11001111 b. 01010100

7.29 Explain the meaning of the term "parity word." Explain how it is used to detect and correct errors.

7.30 Explain the meaning of the terms:

a. Check sum b. CRC

7.31 Crossword Puzzle

ACROSS

1. Programmable logic array.
3. A check system that can be used to detect double errors.
5. PLD input line connected to an AND gate.
8. Type of PAL that contains both logic gate and flip-flops.
10. The use of higher-order address lines for memory bank selection.
15. Type of RAM that does not require refreshing.
17. Type of memory that loses its data when the power is turned off.
20. Programmable logic device.
21. Device used to program 19 Down.
23. PLD input connected by an OR gate to the output.
24. Type of parity.
25. Same as 17 Across.

DOWN

1. PLD with programmable AND gates and fixed OR gates.
2. Matrix of product lines and sum lines.
4. PALs that contain only logic gates.
6. Application-specific integrated circuits.
7. A type of parity that is computed vertically.
9. Opposite of 24 Across.
11. DRAM operation.
12. Type of RAM.
13. Read-only memory.
14. Type of memory that is programmed with information by the manufacturer.
16. ROM and PROM error detection technique.
18. Method of validating memory data.
19. Programmable ROM.
22. Erased using ultraviolet light.

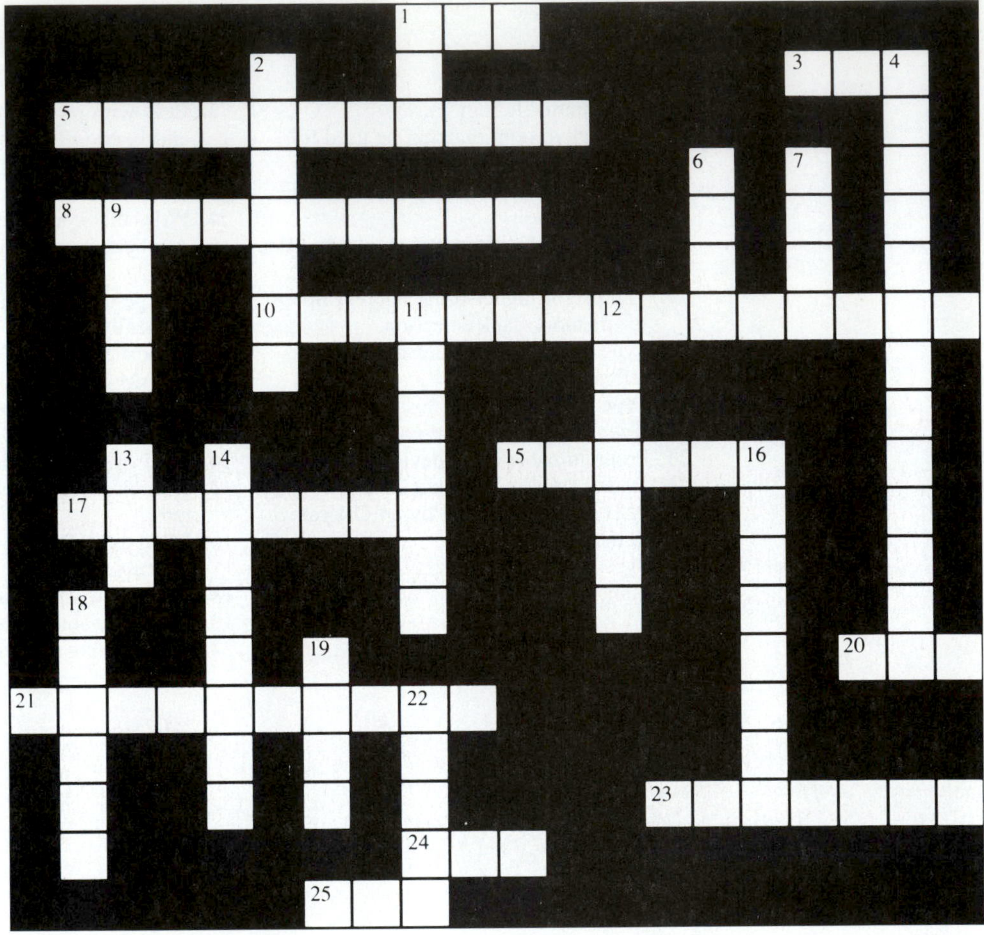
1
2
3
4
5
6
7
8
9
10
11
12
13
14
15
16
17
18
19
20
21
22
23
24
25

PART

BASICS OF MICROPROCESSORS

II

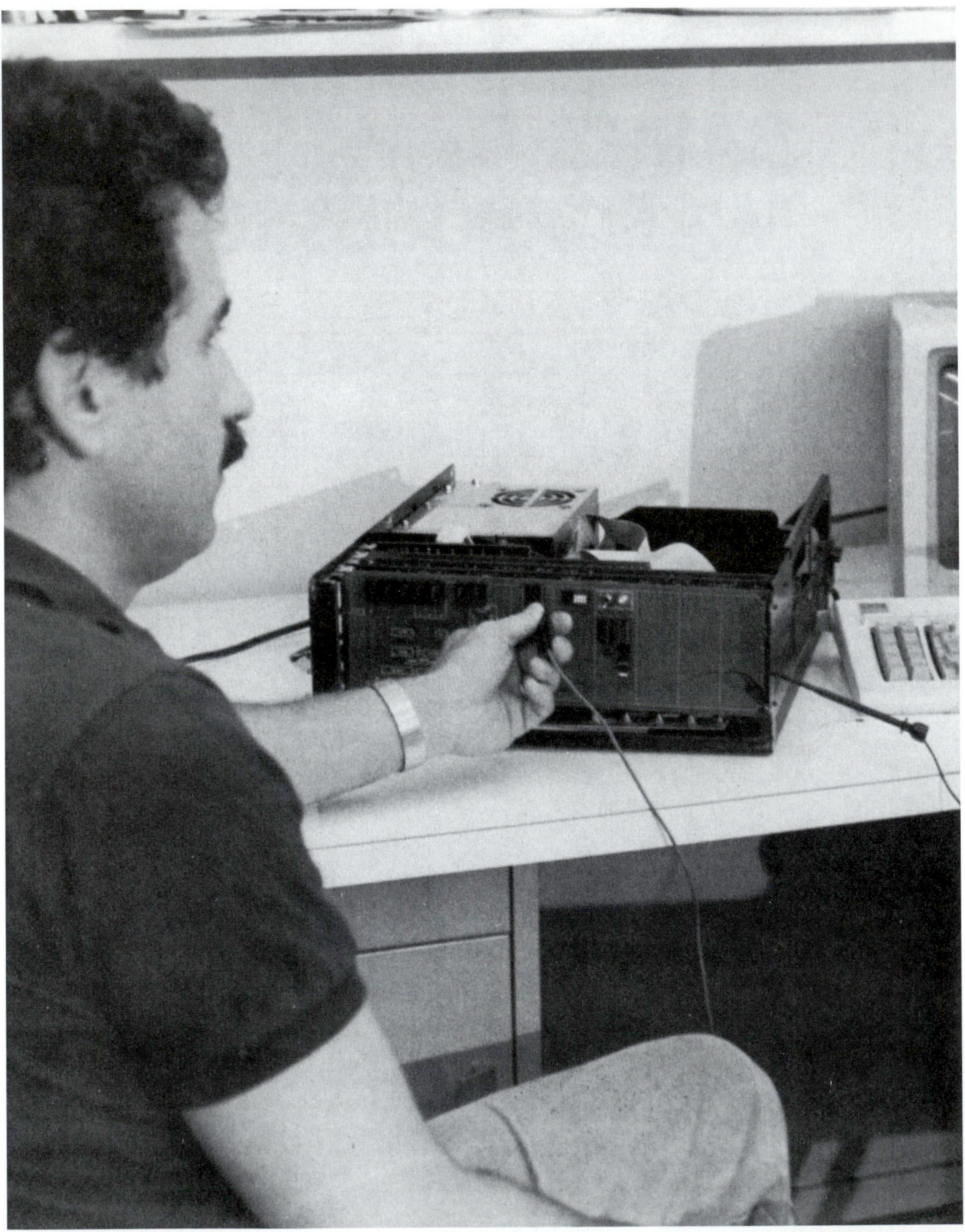

PART

BASICS OF MICROPROCESSORS

II

In a world inundated with electronic gadgets, words like computer and microprocessor have become household words. Computer software and hardware terminology, jargon, and acronyms are heard every day. Words like bits, bytes, RAM, and ROM are becoming commonplace in our language. We have discussed the meaning of many of these terms in Part I, but let's briefly review a few definitions before going on.

A **bit** is a single binary digit. A **nibble** is a collection of 4 binary digits or 4 bits. A **byte** is 8 binary digits, which is equal to 8 bits or 2 nibbles. A **word** equals 16 binary digits, 2 bytes, or 4 nibbles. A **bus** is a group of common wires in which signals travel. **RAM** stands for random-access memory, which is memory that can be written to or read from. **ROM,** or read-only memory, is similar to RAM but can only be read from. **I/O** stands for input or output, for example, keyboards and printers.

Since the beginning of digital computers, the basic architecture has not changed much. All digital computers can be considered to be made up of five parts as shown in Figure A:

1. A central processing unit (CPU)
2. Memory unit or local storage (RAM and ROM)
3. Input units
4. Output units
5. Mass storage units such as disk drives and tape drives

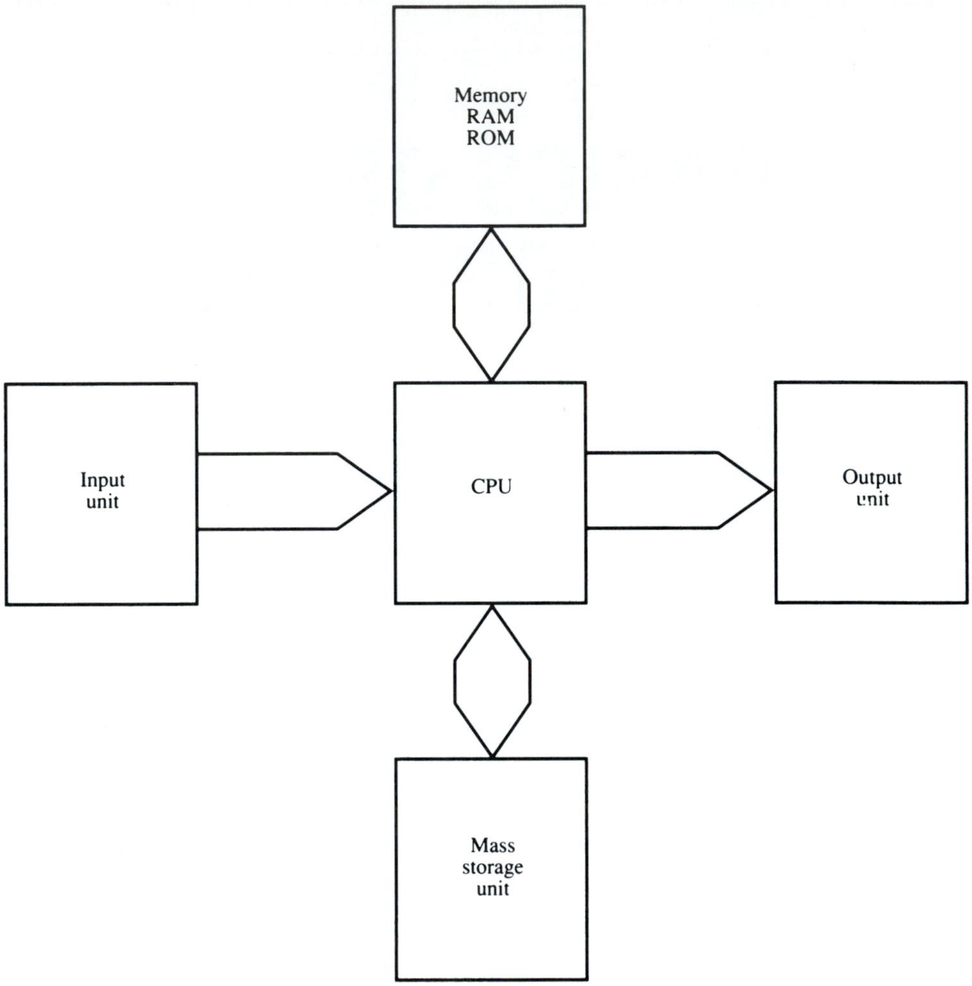

FIGURE A
Digital computer block diagram

The big difference in the evolution of computers has been their size and speed. As we have seen, computer systems are getting smaller and faster every year. Early computer systems of the 1950s, 1960s, and even the 1970s required air conditioned rooms and many kilowatts of electric power to operate them. Today, comparable computing power sits on a deck, plugged into a standard wall outlet.

There are two major contributors to this change in computing power. First are the advancements in IC design and technology. Low-power Schottky transistor–transistor logic (LSTTL), metal-oxide semiconductor (MOS), and, for very low power, complementary metal-oxide semiconductor (CMOS) as well as large-scale integration (LSI) and very large-scale integration (VLSI) have all helped to reduce the size and power requirements as well as to increase the speed of digital computers.

The second major contributor is memory technology. Early computers used memories made up of arrays of magnetic cores that looked like tiny donuts. Today semiconductor memories have revolutionized the digital computer. Large amounts of memory are now available in a single IC. Semiconductor memory devices have also increased operating speeds and reduced size, power, and even cost requirements.

While these technological changes were taking place something unique happened: The concept of a microprocessor was introduced. A microprocessor is a large-scale IC that contains many of the logic circuits used in a computer—for example, the ALU, registers, and control circuits.

One of the first microprocessors was the INTEL 4004. The INTEL 4004 was a 4-bit, simple microprocessor designed for controller applications. Everything was fine until somebody decided to make it into a smart calculator; then the microcomputer revolution took place. Like everything else, the microcomputer has continued to evole. The 4-bit microprocessor became 8 bits, and then 16 bits, and now some are 32 bits wide. The early, simple software, *instruction sets,* became very sophisticated and complex. Combine the new microprocessor with high-density memory and VLSI circuits and what was a supercomputer or mainframe computer can now be on your desk or your lap. Furthermore, it may even operate on batteries.

In this section we study the basic theory of microprocessors. Since complete textbooks are devoted to the study of microprocessors, our main objective will be to introduce only the principles that are required to be understood for the servicing of microprocessor-based systems. Although we concentrate only on what the technician needs to know for servicing, this section provides a solid foundation for future study in this area.

CHAPTER 8

FUNDAMENTALS OF MICROPROCESSORS

KEY TERMS

Address Bus
Arithmetic Logic Unit (ALU)
Base Register
Bidirectional
Bus-control Unit
Bus Cycle
Bus Interface Unit (BIU)
Code Segment (CS)
Control Bus
Data Bus
Data Registers
Data Segment (DS)
Execution Unit (EU)
Extra Segment (ES)
Fetch
FIFO
Flag Register
Index Registers
Instruction Pointer
Instruction Queue
Logical Address
Memory-mapped I/O
Multiplex
Peripheral Devices
Physical Address
Pointer Registers
Read Operation
Stack Segment (SS)
Write Operation

8.0 INTRODUCTION

The microprocessor, or microprocessor unit (MPU) as it is often known, is the brain of the computer system. The MPU is one of the most useful and flexible of all digital integrated circuits. It is found in an increasing number of practical applications, from digital clocks, microwave ovens, washing machines, and robots to temperature control devices in buildings, electronic fuel injection in automobiles, aerospace vehicles, and of course the general-purpose computer itself.

There are a number of microprocessors available today. In this text we use the Intel 8086/8088 series of microprocessors as a practical example. The Intel 8086/8088 microprocessor series is one of the most popular and powerful microprocessors in use today.

8.1 INTERNAL ARCHITECTURE

The term *architecture,* as used in microprocessor circuits, describes the functional components that make up the MPU and the interaction between them. These include the temporary storage devices known as registers, which are used to hold data, instructions, and status information. There are also devices to perform arithmetic, such as addition and subtraction, and the logic operations such as AND and OR. Control devices are used to control the flow of information through the MPU.

The basic 8086/8088 microprocessor consists of two sections known as the **execution unit (EU)** and the **bus interface unit (BIU),** as shown in Figure 8.1. The EU performs all the arithmetic and logic operations. The BIU obtains, or *fetches,* the instructions and/or data from memory. These instructions are used to control and operate the MPU.

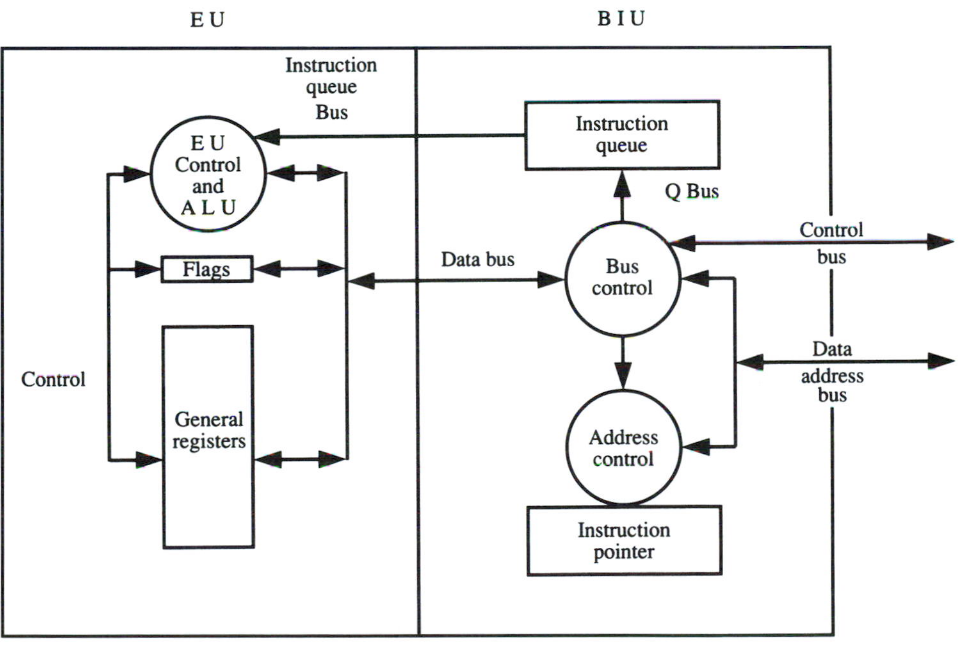

FIGURE 8.1
8088/8086 Microprocessor

Execution Unit (EU)

The EU is where the actual processing of data takes place inside the 8086/8088 MPU. It is here that the **arithmetic logic unit (ALU)** is located, along with the registers used to manipulate data and store intermediate results. The EU accepts instructions and data that have been fetched by the BIU and then processes the information. Data processed in the EU can be transmitted to memory or **peripheral devices** through the BIU. It should be noted that the EU has no direct connection to the outside world and relies solely on the BIU to feed it with instructions and data as indicated in Figure 8.2.

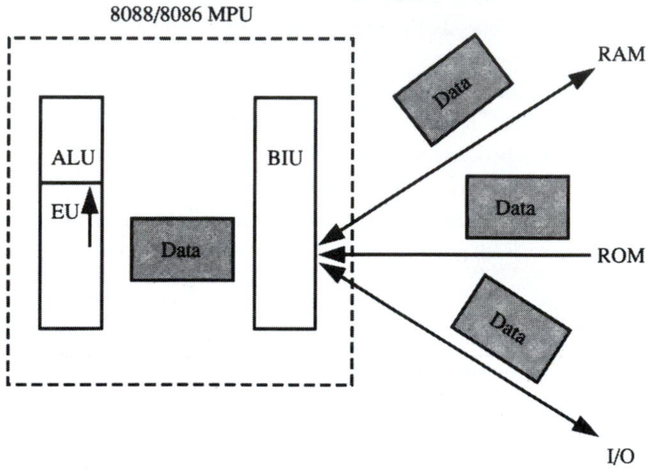

FIGURE 8.2
BIU/EU data feed

Bus Interface Unit (BIU)

The BIU is made up of the *address generation and bus-control unit, the instruction queue,* and the *instruction pointer.* It has the task of making sure that the bus is used to its fullest capacity in order to speed up operations. This function is carried out in two ways. First, by fetching the instructions before they are needed by the execution unit and storing them in the instruction queue, the 8086/8088 MPU is able to increase computing speed. Second, by taking care of all bus-control functions, the EU is free to concentrate on processing data and carrying out the instructions. The instruction pointer contains the location or address of the next instruction to be executed.

8.2 ADDRESS BUS

An address is a unique location in memory. It is like a mailbox in the post office, where each mailbox has its own unique number to identify its location. An address is necessary because there may be two people named Jones on the same street. An address gives a way of determining which person we want to contact.

In the Intel 8086/8088 the address is determined by a 20-bit number. This gives us 2^{20} possible address locations, or 1,048,576 bytes of memory.

An **address bus** is made up of 20 wires, or conductors, labeled A_0 through A_{19}, with A_0 as the LSB and A_{19} as the MSB, as shown in Figure 8.3. It is used to locate or find information in memory. It is also used to define a location in memory where information is to be stored. The address bus is sometimes used to identify which I/O port is used for input/output operations. For example, if the MPU were being used in an automobile to control the lights, power windows, and air conditioner vents, the address bus would define which light, power window, or air conditioner vent to control.

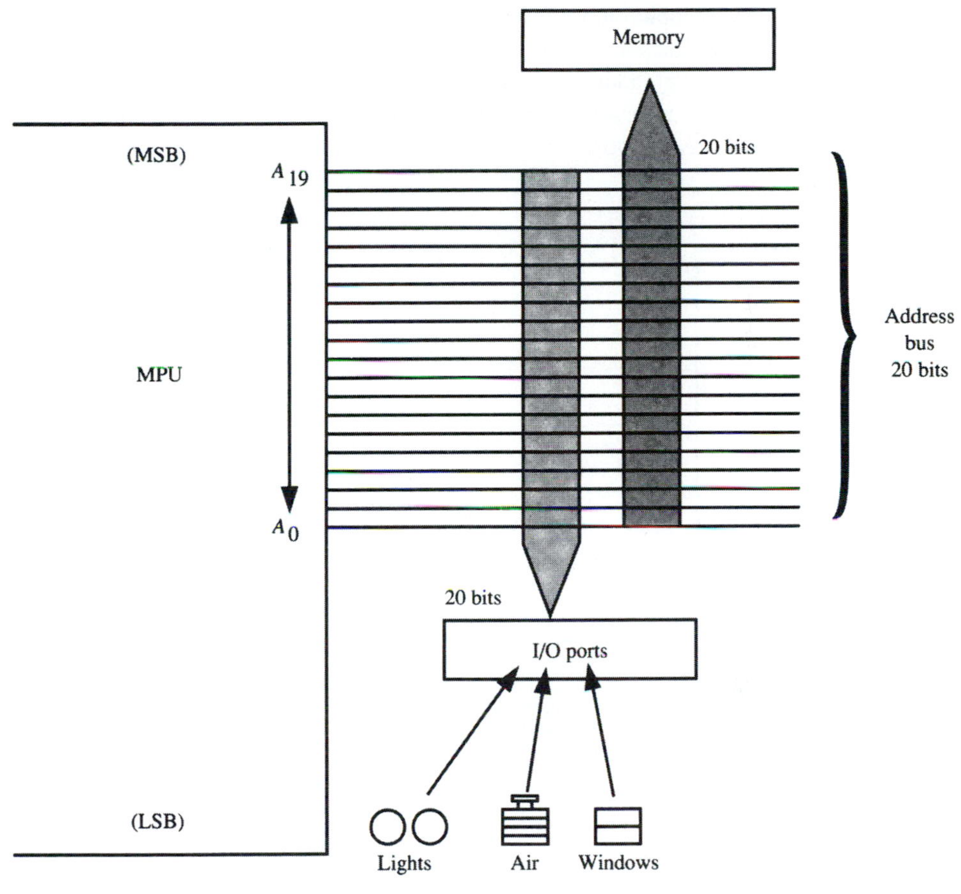

FIGURE 8.3
Address bus

8.3 DATA BUS

A **data bus** is used to move information (data and instructions) from the MPU to memory and other devices. This is sometimes referred to as a **write operation.** The data bus is also used to receive information into the MPU. This is called a **read operation.** Because the data bus receives and transmits information, it is known as a **bidirectional** bus. However, it cannot receive and transmit data at the same time.

The Intel 8086 has a 16-bit data bus labeled D_0 to D_{15}, where D_0 is the LSB and D_{15} is the MSB, as shown in Figure 8.4(a). The 8088 uses an 8-bit data bus labeled D_0

(LSB) to D_7 (MSB), as shown in Figure 8.4(b). This is the most significant difference between the 8088 MPU and the 8086 MPU.

Both the 8088 and 8086 microprocessors **multiplex** the address and data buses. Multiplexing is the process of using the same wires or pins to do different things at different times. When acting as a data bus, the signal lines carry read/write information for memory or input/output information for I/O devices. When acting as an address bus, the same signal lines are used to locate information. Figure 8.5 illustrates the multiplexing of the 8086/8088 address and data bus. This is analogous to using a single set of railroad tracks to carry a passenger train and a freight train over the same bridge. Both may use the tracks to cross the bridge in either direction but not at the same time.

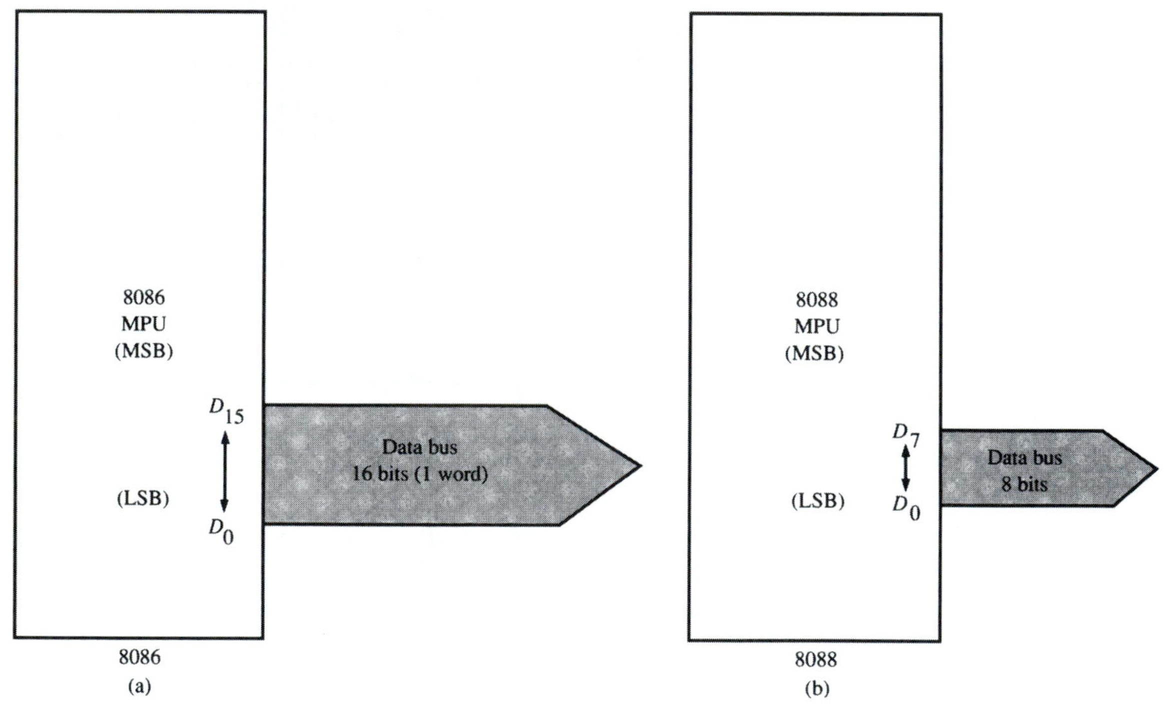

FIGURE 8.4
Data bus 8086/8088

8.4 CONTROL LINES

The 8086/8088 control lines determine how and when an operation is to be performed. For example, the control lines carry signals that determine whether the information on the bus is data or an address location. Figure 8.6 (p. 254) shows the pinout of the 8088 microprocessor with all its control lines. The control lines are sometimes referred to as the

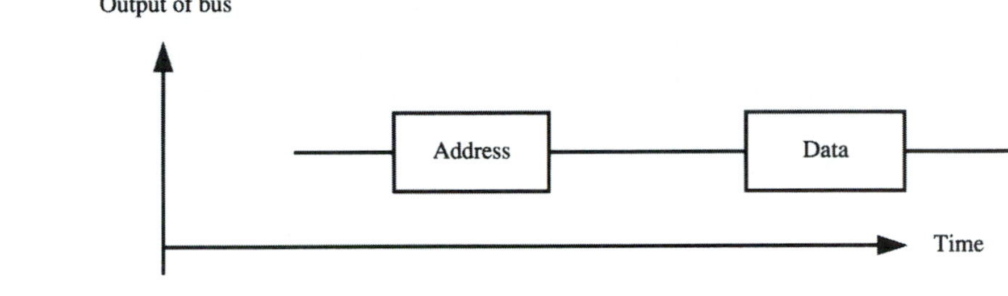

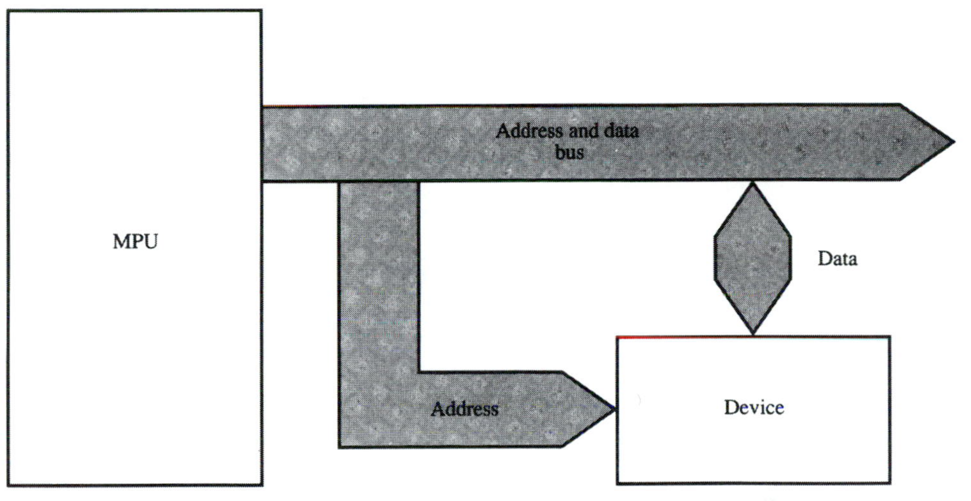

FIGURE 8.5
Multiplexed address and data bus

control bus. The pin configuration for the 8088 MPU is summarized and can be grouped functionally as follows:

1. Power lines
2. Address/status lines
3. Data lines
4. Data/address control lines
5. Interrupt control lines
6. Operation control lines

1. *Power lines.* Pins 1 and 20 are used for power supply ground connections. Pin 40 is connected to +5 V, which is also called $+V_{CC}$.
2. *Address/status lines.* Pins 2 through 16 and pins 35 through 39 are used for the address bus. Pins 35 through 38 are also used by multiplexing to provide information or status about the MPU. These status signals are labeled S_3, S_4, S_5, and S_6, as shown.

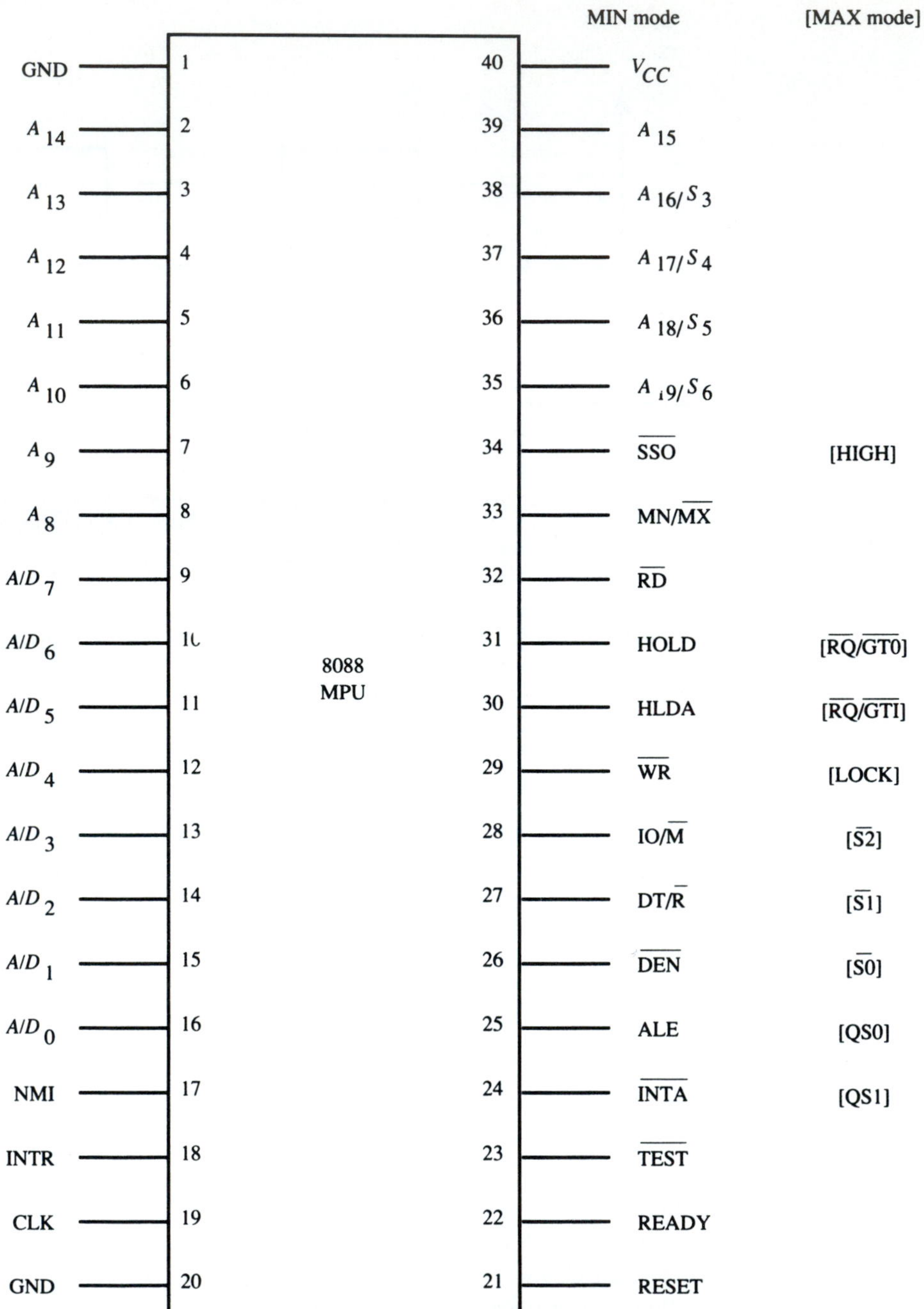

FIGURE 8.6
Pinout 8088 MPU

3. *Data lines.* Pins 9 through 16 are also used by multiplexing with the address signals for the data bus. These pins are labeled D_7 through D_0, respectively.
4. *Data/address-control lines.* Pins 25 through 29, and 32 and 34 provide control and status of the movement of information on the data and address buses.

- *Pin 25 ALE (address latch enable).* An active high output signal that indicates that a *valid* or stable address word is on the address bus.
- *Pin 26 $\overline{DEN}$ (data enable).* A tristate active low output signal that determines whether the data buffer is enabled or disabled. A logic level 0 turns on the data bus.
- *Pin 27 DT/$\overline{R}$ (data transmit receive).* A tristate output signal used to control the direction of data flow. A logic level 1 indicates data bits are being transmitted *from* the MPU. A logic level 0 indicates that data bits are being received *into* the MPU.
- *Pin 28 IO/$\overline{M}$ (input/output/memory).* This tristate output signal determines if the address bus is connected to memory or I/O. A logic level 1 indicates that the I/O is being addressed. A logic level 0 indicates that memory is being addressed, or *selected.*
- *Pin 29 $\overline{WR}$ (write).* A tristate output signal that indicates that the MPU has put valid and stable data on the data bus. The active logic level 0 can be used as a latch pulse, or *strobe pulse,* to indicate that valid data bits are on the bus.
- *Pin 32 $\overline{RD}$ (read).* A tristate active low output signal that indicates that the MPU is ready to *read* data from the data bus.
- *Pin 34 $\overline{SSO}$ (status line* 0). A tristate active low output signal that when combined with IO/$\overline{M}$ and DT/$\overline{R}$ provides bus cycle information, as shown in Table 8.1.

TABLE 8.1
Bus cycle status

IO/$\overline{M}$	DT/$\overline{R}$	SSO	Bus Cycle Function
0	0	0	Instruction fetch
0	0	1	Read memory
0	1	0	Write memory
0	1	1	Passive
1	0	0	Interrupt acknowledge
1	0	1	Read I/O port
1	1	0	Write I/O port
1	1	1	Halt

5. *Interrupt-control lines.* Pins 17, 18, 21, 23, and 24 allow the software or devices in the system to stop or interrupt the MPU's operation.

- *Pin* 17 *NMI (Nonmaskable interrupt).* An active high input signal that interrupts MPU processing. This signal cannot be ignored, or *masked,* by the MPU.
- *Pin* 18 *INTR (Interrupt request).* An active high input signal that is used to interrupt MPU processing. This interrupt is maskable and can be made to be ignored by the MPU through software control.

- *Pin* 24 $\overline{INTA}$ *(Interrupt acknowledge).* An active low output signal that indicates that an interrupt has been received and accepted by the MPU. This can be referred to as *servicing the interrupt.*
- *Pin* 21 *RESET (restart).* An active high input signal used externally to start or reset MPU activity.
- *Pin* 23 $\overline{TEST}$ *(test interrupt).* An active low input signal that is tested by the software WAIT instruction. If $\overline{\text{TEST}}$ is a logic level 1, the MPU will wait or interrupt the program execution until $\overline{\text{TEST}}$ is a logic level 0.

6. *Operation-control lines.* The remaining pins, 19, 22, 30, 31 and 33, as well as pins 21 and 23, are used to manage and run the execution of the MPU.

- *Pin* 19 *CLK (clock input).* The master timing signal for the MPU, which synchronizes all operations.
- *Pin* 22 *READY (data transfer ready).* An active high input signal that provides a means for the memory and I/O devices to tell the MPU that they are ready for data transfer.
- *Pin* 31 *HOLD (hold request).* An active high input signal that provides a way for a device to request access to the system data/address bus.
- *Pin* 30 *HLDA (hold acknowledge).* This active high output signal indicates that the MPU has received or accepted a hold request. This causes the address, data, and control buses to go into the high-impedance tristate.
- *Pin* 33 $MN/\overline{MX}$ *(minimum/maximum mode).* An input signal that determines whether the MPU is in a single or multiprocessor mode. A logic level 1 selects a single MPU, or *minimum* system mode. A logic level 0 selects a multiprocessor, or *maximum* system mode.

Table 8.2 summarizes the 8088 I/O pin connections and includes a description of signal names with their associated functions.

8.5 INSIDE THE EU

The EU is made up of two parts known as the *ALU* and the *general registers.* It is here that instructions are received, decoded, and executed from the instruction queue portion of the BIU. The instructions are taken from the top of the instruction queue on a first-in, first-out, or FIFO, basis.

ALU

The *ALU* is the calculator part of the execution unit. It consists of electronic circuitry that performs arithmetic operations or logical operations on the binary-represented electrical signals. The control system for the execution unit can also be thought of as part of the ALU. It provides a path for the flow of instructions into the ALU, the general registers, and the flag register.

The **flag register** is a special register associated with the ALU. The flag register is used to store information about the conditions of the operation of the ALU, as shown in

TABLE 8.2
8088 Input/output pin configuration

Label	Pins	Description	Function
GND	1, 20	Power ground	Power
V_{CC}	40	$+5\ V_{DC}$	Power
A/D_0–A/D_7	16–9	Address/data lines	Address/data
A_8–A_{14}	8–2	Address lines	Address
A_{15}	39	Address line	Address
A_{16}/S_3–A_{19}/S_6	38–35	Address/status lines	Address/status
ALE	25	Address latch enable	Address control
$\overline{\text{DEN}}$	26	Data enable	Data control
DT/$\overline{\text{R}}$	27	Data transmit/receive	Data control
IO/$\overline{\text{M}}$	28	Input output/memory	Address control
$\overline{\text{WR}}$	29	Write data	Data control
$\overline{\text{RD}}$	32	Read data	Data control
$\overline{\text{SSO}}$	34	Status line 0	Address/data control
NMI	17	Nonmaskable interrupt	Interrupt control
INTR	18	Interrupt request	Interrupt control
RESET	21	Restart	Interrupt/operation control
$\overline{\text{TEST}}$	23	Alternate form interrupt	Interrupt/operation control
$\overline{\text{INTA}}$	24	Interrupt acknowledge	Interrupt control
CLK	19	Clock input	Operation control
READY	22	Data transfer ready	Operation control
HLDA	30	Hold acknowledge	Operation control
HOLD	31	Hold request	Operation control
MN/$\overline{\text{MX}}$	33	Minimum/maximum mode	Operation control

Figure 8.7. For example, if two binary numbers are multiplied and the resulting product is too large to be stored in the assigned location, a flag is *set* in the flag register to tell the MPU of this condition.

MSB 15	14	13	12	11	10	9	8	7	6	5	4	3	2	1	LSB 0
				OF	DF	IF	TF	SF	ZF		AF		PF		CF

FIGURE 8.7
Flag register

The 8086/8088 has nine flags to record processor status information and control operations. Six flags are status flags—AF, CF, OF, SF, PF, and ZF. The remaining three flags are control flags—DF, IF, and TF. Table 8.3 presents a flag summary and highlights key concerns. Each flag is next discussed in detail.

TABLE 8.3
Flag summary

Status Flags	Description
AF (auxiliary flag)	Indicates if the instruction generated a carry out of the 4 LSBs.
CF (carry flag)	Indicates if the instruction generated a carry out of the MSB.
OF (overflow flag)	Indicates if the instruction generated a signed result that is out of range.
SF (sign flag)	Indicates if the instruction generated a negative result.
PF (parity flag)	Indicates if the instruction generated a result having an even number of 1s.
ZF (zero flag)	Indicates if the instruction generated a zero result.
DF (direction flag)	Controls the direction of the string manipulation instructions.
IF (interrupt-enable flag)	Enables or disables external interrupts.
TF (trap flag)	Puts the processor into a single-step mode for program debugging.

- *AF (auxiliary flag).* If this flag is set, there has been a carry out or a borrow of the 4 least significant bits. This flag is used during decimal arithmetic instructions.
- *CF (carry flag).* If this flag is set, there has been a carry out or overflow of the most significant bit. It is used by instructions that add and subtract multibyte numbers.
- *OF (overflow flag).* If this flag is set, an arithmetic overflow has occurred; that is, a significant digit has been lost because the size of the result exceeded the capacity of its destination location.
- *SF (sign flag).* Since negative binary numbers are represented in the 8086/8088 in standard 2s complement notation, SF indicates the sign of the result (0 = positive, 1 = negative).
- *PF (parity flag).* If this flag is set, the result has even parity, an even number of 1s. This flag can be used to check for transmission errors.
- *ZF (zero flag).* If this flag is set, the result of the operation is 0.
- *DF (direction flag).* Setting DF causes string instructions to auto-decrement (count down); that is, to process strings from the high address to the low address, or from right to left. Clearing DF causes string instructions to auto-increment (count up), or process strings from left to right.
- *IF (interrupt-enable flag).* Setting IF allows the MPU to recognize external (maskable) interrupt requests. Clearing IF disables these interrupts. IF has no effect on either non-maskable external or internally generated interrupts.
- *TF (trap flag).* Setting TF puts the processor into single-step mode for debugging. In this mode the MPU automatically generates an internal interrupt after each instruction, allowing a program to be inspected as it executes instruction by instruction.

General Registers

The general registers consists of a set of data registers, which are used to hold intermediate results, and the pointer and index registers, which are used to locate information within a specified portion of memory.

The **data registers** are 16-bit registers labeled AX, BX, CX, and DX and are split into upper and lower halves of 8 bits, or 1 byte, each. The *H* represents the high-order or most-significant byte and the *L* represents the low-order or least-significant byte, as shown in Figure 8.8(a). The halves of each of these registers may be used separately as two 8-bit storage areas or combined to form one 16-bit (*one word*) storage area. The *H and L* group, as the general registers are sometimes called, can be used in most arithmetic and logic operations to hold or *accumulate* data.

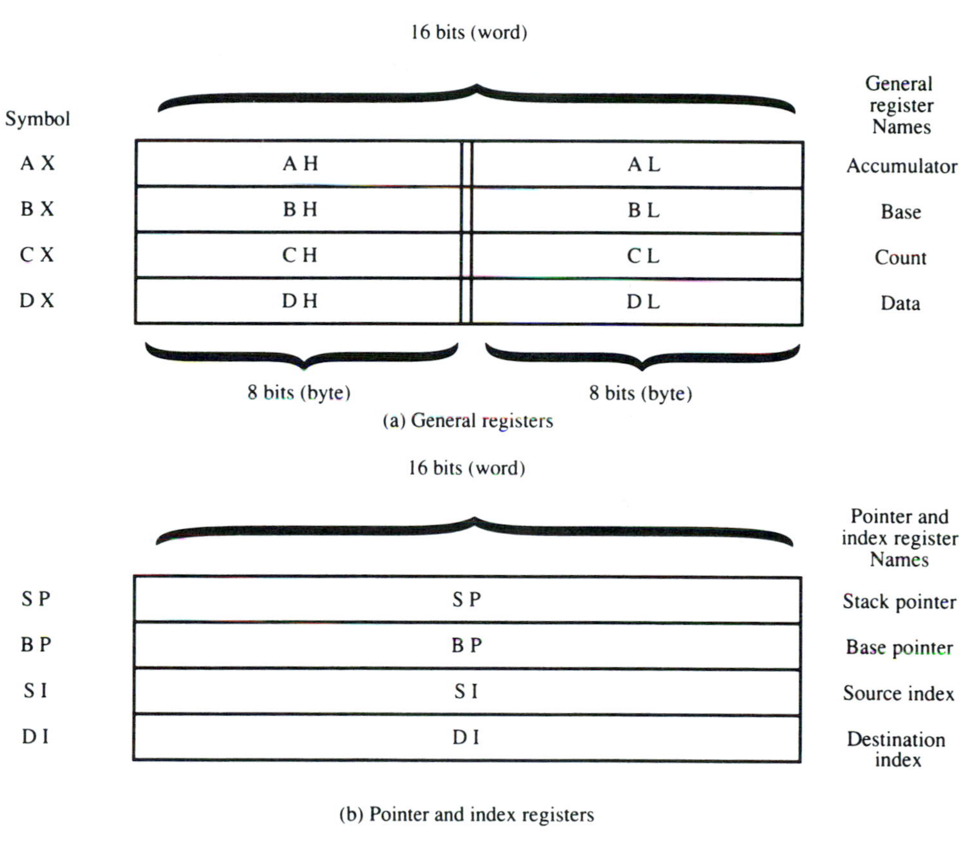

FIGURE 8.8
General and pointer/index registers

Pointer and Index Registers

The **pointer registers** are labeled SP and BP, and the **index registers** are labeled SI and DI. These registers can also be used in most arithmetic and logic operations and can be considered under the general register heading (see Figure 8.8b); however, the pointer and index registers are usually used to point to or index to an address in memory. When used in this manner, these registers are address registers that designate a specific location in memory that may be frequently used by the program. The addresses contained in these registers can be combined with information from the BIU to physically locate the data in memory.

Table 8.4 summarizes all the registers that are part of the execution unit and includes some of their special functions and uses.

TABLE 8.4
Execution unit registers

Registers	Description	Special Functions
AX	Accumulator	Word arithmetic; I/O
AL	Accumulator (low byte)	Byte arithmetic; I/O
AH	Accumulator (high byte)	Byte arithmetic; I/O
BX	Base	Data transfer; memory address
BL	Base (low byte)	Byte transfer
BH	Base (high byte)	Byte transfer
CX	Count	String operations; loops
CL	Count (low byte)	Shifts; rotates
CH	Count (high byte)	Shifts; rotates
DX	Data	Indirect I/O
DL	Data (low byte)	Byte-wide I/O
DH	Data (high byte)	Byte-wide I/O
SP	Stack pointer	Stack operations
BP	Base pointer	Base register
SI	Source index	String source; index register
DI	Destination index	String destination; index register

8.6 INSIDE THE BIU

The BIU is the portion of the microprocessor that directly accesses or interfaces with the rest of the computer system. The BIU can be thought of as three functional blocks: *bus control, instruction queue,* and *address control,* as shown in Figure 8.9(a). The BIU is responsible for *prefetching* instructions to fill the instruction queue. It also sends and receives system control signals to and from other devices. Its final responsibility is to act as the interface to memory and the I/O ports.

Bus Control

The **bus-control unit** performs the bus operations for the MPU. It fetches and transmits instructions, data, and control signals between the MPU and the other devices of the system. For example, it is used to determine the direction that data is flowing on the data bus through the *data transmit/receive* control line (DT/$\overline{\text{R}}$).

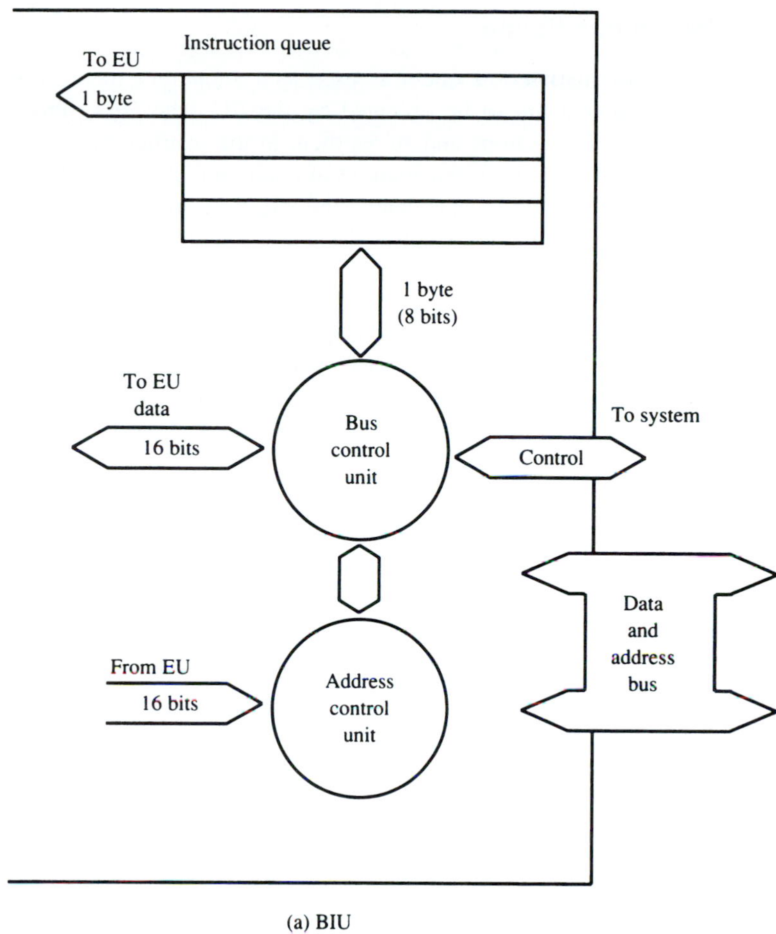

(a) BIU

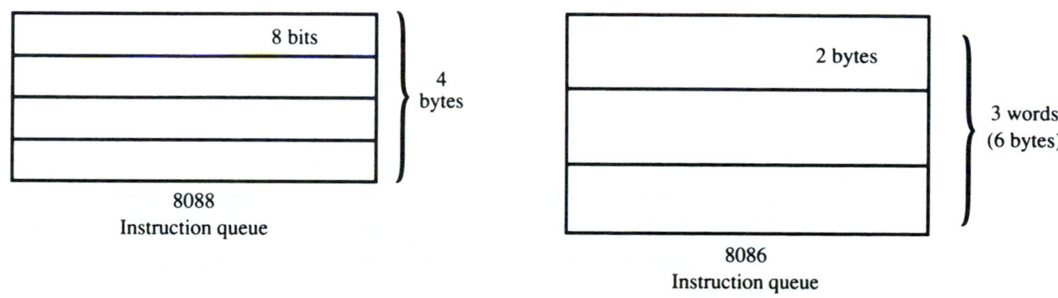

(b) Instruction queues

FIGURE 8.9
Inside the BIU

Instruction Queue

The **instruction queue** is used as a temporary memory storage area for data instructions that are to be executed by the EU. The BIU, through the bus-control unit, prefetches instructions and stores them in the instruction queue. This allows the execution unit to perform its calculations at maximum efficiency. Because the BIU and the EU essentially operate independently, the BIU concentrates on loading instructions into the instruction queue. This usually takes more time to do than the calculations performed by the execution unit. In effect, the BIU and the EU work in parallel.

The instruction queue is a *first-in, first out* **(FIFO)** memory. This means that the first instruction loaded into the instruction queue by the bus control unit will be the first instruction to be used by the ALU. It should be noted that the 8088 MPU has a *4-byte-wide* instruction queue. Remember that the 8088 uses a byte-wide, or 8-bit, data bus. Therefore, there can be up to 4 bytes, or 2 words, of data in the instruction queue at any time. The 8086 MPU has a 3-word (16-bit) instruction queue. Its data bus is 16 bits wide; therefore, it brings in information 1 word at a time. This is another important difference between the two microprocessors, as shown in Figure 8.9(b).

Address Control

The address-control unit is used to generate the 20-bit memory address that gives the *physical* or actual location of the data or instruction in memory. This unit is composed of the *instruction pointer,* the *segment registers,* and the *address generator,* as shown in Figure 8.10.

Instruction Pointer The **instruction pointer (IP)** is a 16-bit register that is used to *point* to, or tell the MPU, the instruction to execute next. Therefore, the instruction pointer is used to control the *sequence* in which the program is executed. Each time the execution unit accepts an instruction, the instruction pointer is *incremented* to point to the next instruction in the program.

Segment Registers There are four segment registers. They are the **code segment (CS),** the **data segment (DS),** the **stack segment (SS),** and the **extra segment (ES).** These registers are used to define a *logical memory space,* or memory segment that is set aside for a particular function.

The CS register points to the current code segment. Instructions are fetched from this segment. The DS register points to the current data segment. Porgram variables and data are held in this area. The SS register points to the current stack segment. Stack operations are performed on locations in the SS segment. The ES register points to the current extra segment, which is also used for data storage. Each of the segment registers can be up to 64 kilobytes long. Each segment is made up of an uninterrupted section of memory locations. Each segment can be addressed separately using the *base address* that is contained in its segment register. The base address is the starting address for that segment. For example, if the code segment register has the address 200H in it, then the code segment instructions start at the *logical address* 200H. See Figure 8.11 (p. 264).

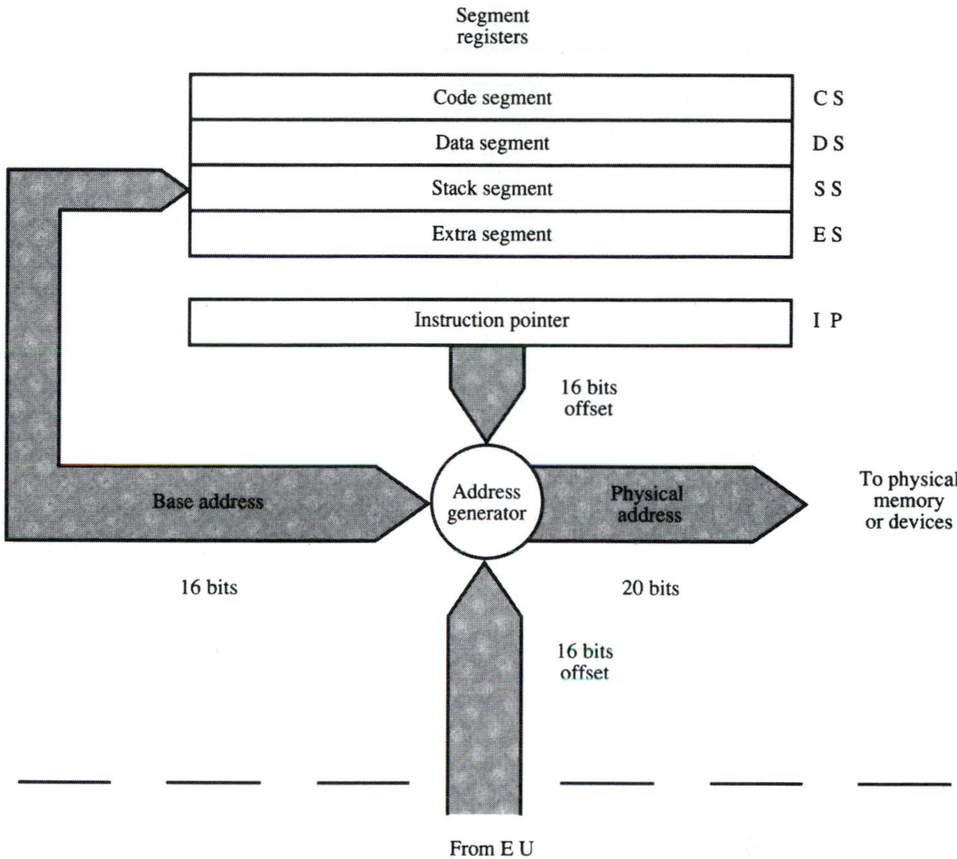

FIGURE 8.10
Address control

Address Generator The address-generator unit is used with the segment registers to generate the 20-bit physical address required to identify all the possible memory addresses. The 20 address lines give a maximum *physical memory* size of 2^{20} address locations, or 1,048,576 bytes of memory. This creates a problem, since all the registers in the MPU are only 16 bits wide. Figure 8.12 (p. 265) shows how a 20-bit physical address is generated from a 16-bit segment register base address and a 16-bit segment register offset address. The physical address is obtained by shifting the segment base value four bit positions (one hex position) and adding the offset or logical address of the segment.

We have introduced some new terms here. The **base register** contains the starting address of a segment in memory. The **logical address** is the address of a piece of information within the 64-kilobyte block of the memory segment. This logical address is used to obtain the offset. This is done as shown in Figure 8.12. The segment base address is

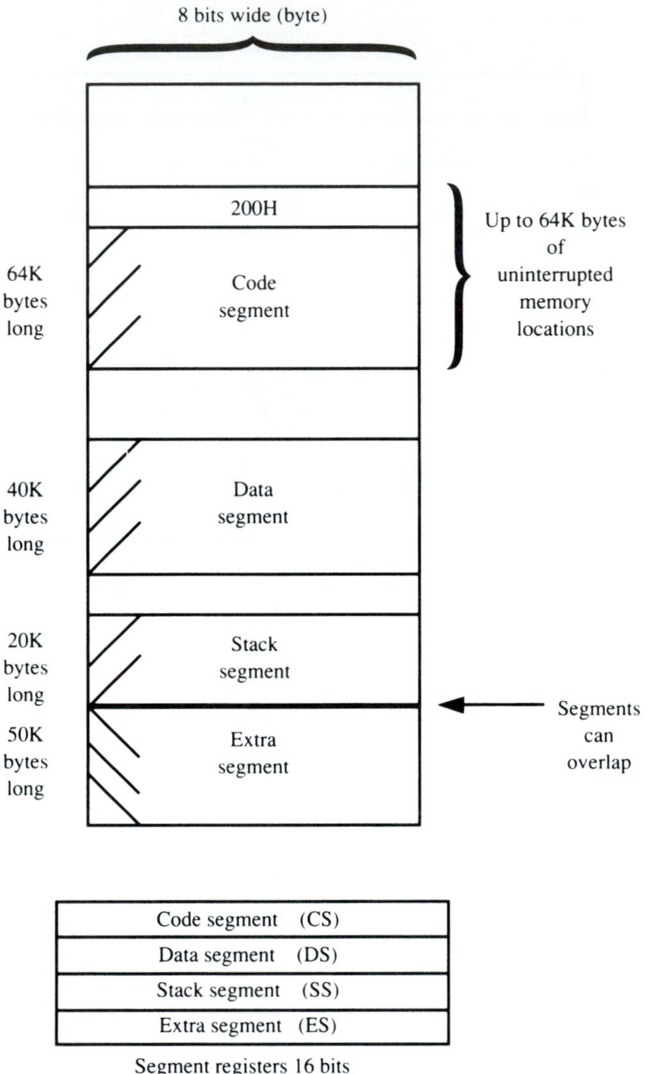

FIGURE 8.11
Segmentation of memory

shifted left 4 bit positions (one hex position) to create a 20-bit address number. The logical address is used as an offset and is added to the shifted segment base address to obtain the **physical address,** which is a uniquely addressable memory location.

Since this idea tends to be somewhat confusing, we work through several examples.

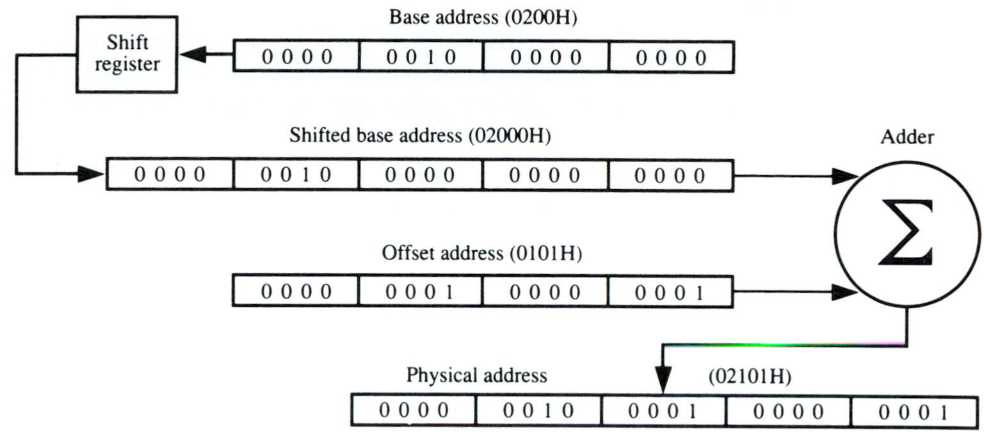

FIGURE 8.12
To produce a 20-bit physical address

EXAMPLE 8.1

The data segment base address is FF00H. The logical offset address is 0321H. Find the physical address of memory.

Solution

FF00H	Base address (DS)
0321H	Offset
FF000H	Shifted-left base address
+ 0321H	Added offset (logical)
FF321H	Physical address of memory

EXAMPLE 8.2

The stack segment base address is 1111H. The stack pointer register contains 0293H. Find the physical address of memory.

Solution

1111H	Stack segment base address
0293H	Offset (SP)
11110H	Shifted-left base address
+ 0293H	Added offset
113A3H	Physical address of memory

EXAMPLE 8.3

The instruction pointer points to 001FH. This is the next instruction to be executed. When we try to execute this instruction, we seem to have a memory problem. We suspect that there may be a problem in a memory device or circuit. Where do we look?

Solution

First we must obtain the base address from the active segment register. In this case, we find that the code segment is the register of interest and contains the value 0200H. We can now calculate the physical address and then examine the contents of that location in memory.

0200H	Code segment base address
001FH	Instruction pointer offset
02000H	Shifted-left base address
+ 001FH	Added offset
0201FH	Physical address of memory

We should examine the contents of memory location 0201FH.

Table 8.5 illustrates the segment registers, the offset source registers, and the type of operation being performed.

TABLE 8.5
Segment register operation

Segment	Offset Source	Operation
CS	IP	Instruction fetch
DS	BX, SI, DI	Data; string source
SS	SP	Stack
ES	DI	String destination

8.7 MEMORY

As we previously stated, memory can be thought of as a number of mailboxes, or storage locations. Each location must have its own unique address to identify it.

In a microcomputer system the memory locations are used to store data and instructions that tell the MPU what it is supposed to do. For the microprocessor to use the information stored in memory, it must **fetch,** or read, the information from memory. The information is read into the BIU of the microprocessor over the data bus. Once the data bits are processed, the results can be *stored* or written back into memory for use at a later time. This is called *writing to memory.* Again, the information flows back into memory on the data bus. Therefore, the data bus must be *bidirectional.* Figure 8.13 shows the relationship between the MPU and the memory subsystem. Notice that the address bus and the necessary control signals are also represented in Figure 8.13. Recall that the lower-order address signals are *multiplexed* with the data signals.

The operations of the memory control signals are as follows:

1. *ALE (address latch enable).* The falling edge of the ALE signal is used to latch or hold the memory address. The memory address is held, so that reading and writing operations can be performed, until the ALE signal returns high.

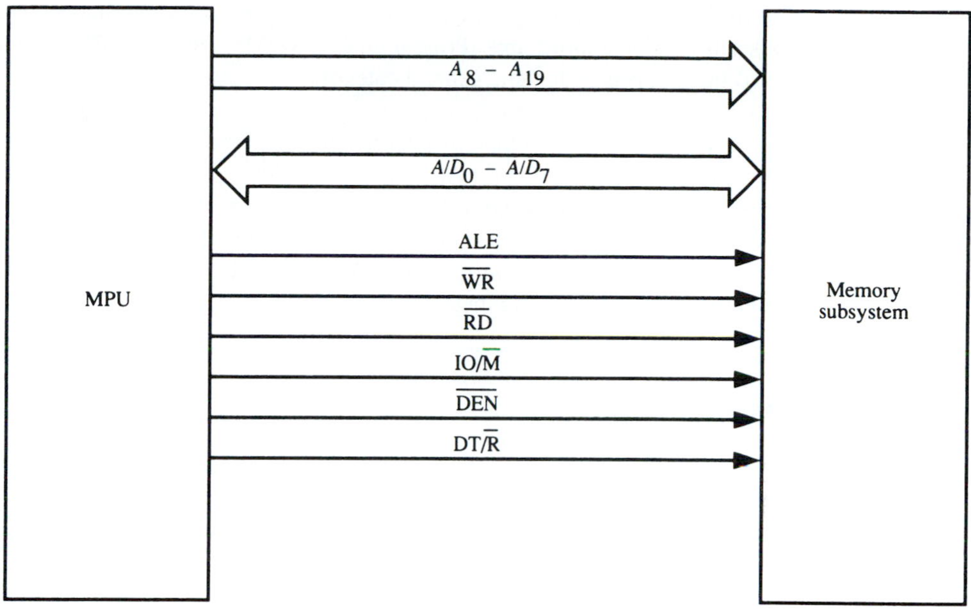

FIGURE 8.13
MPU's relationship memory subsystem

2. *$\overline{WR}$ (write)/$\overline{RD}$ (read).* Once the address has been latched, the MPU activates either the $\overline{WR}$ or the $\overline{RD}$ signal. This activates a memory location or a device. These tristate signals are active low and cannot both be turned on simultaneously. Circuitry inside the MPU ensures that the $\overline{WR}$ and $\overline{RD}$ signals cannot be low at the same time.
3. *$IO/\overline{M}$ (input/output/memory).* When this signal is low, a memory address is being selected. When this signal is high, an input or output port is being selected. This signal is necessary because the same address may be used for a memory location or for an I/O port.
4. *$\overline{DEN}$ (data enable).* Due to bus timing problems most systems connect to the data bus through a buffer circuit. The $\overline{DEN}$ signal is a tristate active-low signal that is used to turn on or *enable* the buffer.
5. *$DT/\overline{R}$ (data transmit/receive).* This tristate output signal is high when data bits are being transmitted by the MPU. It is low when the MPU is capable of receiving data.

RAM and ROM

Memory can be divided into two general types: ROM and RAM.

ROM ROM contains permanent information that can only be read. Start-up or *bootstrap* programs are stored in ROM. Other examples of information that may be stored in ROM include test diagnostics, language interpreters (BASIC), character generators, and speech synthesis programs. Programmable ROMs, such as PROMs and EPROMS, can also be grouped under this general category.

RAM RAM is nonpermanent memory that can be read or written to. It acts as temporary storage within the microcomputer for either program instructions or data. RAM is broken down into two general types called *static* and *dynamic.* Static RAM retains the information as long as power is not removed. Dynamic RAM requires the system constantly to *refresh,* or rewrite, the information. As previously discussed in Chapter 7, each type of memory has its own advantages and applications.

Memory Maps

A memory map is a diagram that represents all the occupied or used locations in memory. The memory map is used to identify the locations and their purposes. The memory map shows the locations or addresses, usually as a hexadecimal number, of ROM, RAM, port locations, interrupt vectors, program segments, and the like. Figure 8.14 shows an example of a memory map.

A method of treating a peripheral device as a memory location is known as **memory-mapped I/O.** For example, if a technician designs a microcomputer to be used as a fire alarm controller, the technician may physically connect the output of the fire sensors to specific memory addresses. These inputs are now memory mapped; this allows the programmer to access the data through these memory locations.

The memory map is used as an aid in partitioning the available memory. It shows the locations of all memory devices and segments, which is useful because some memory locations are dedicated for special purposes. For example, the bootstrap or start-up ROM may be located in memory addresses E6F2H through E728H. Some memory locations are reserved or set aside by the microprocessor manufacturer to allow for compatibility with future designs.

8.8 TIMING

All microprocessors use an oscillator to generate a master frequency to synchronize or time operations. For the 8086/8088 microprocessor the oscillator frequency, or *clock frequency,* is typically 5 MHz. The period of one clock cycle is then equal to

$$\begin{aligned} T &= \frac{1}{F} \\ &= \frac{1}{5 \times 10^6 \text{ H}} \\ &= 0.2 \times 10^{-6} \text{ s} \\ &= 200 \text{ ns} \end{aligned}$$

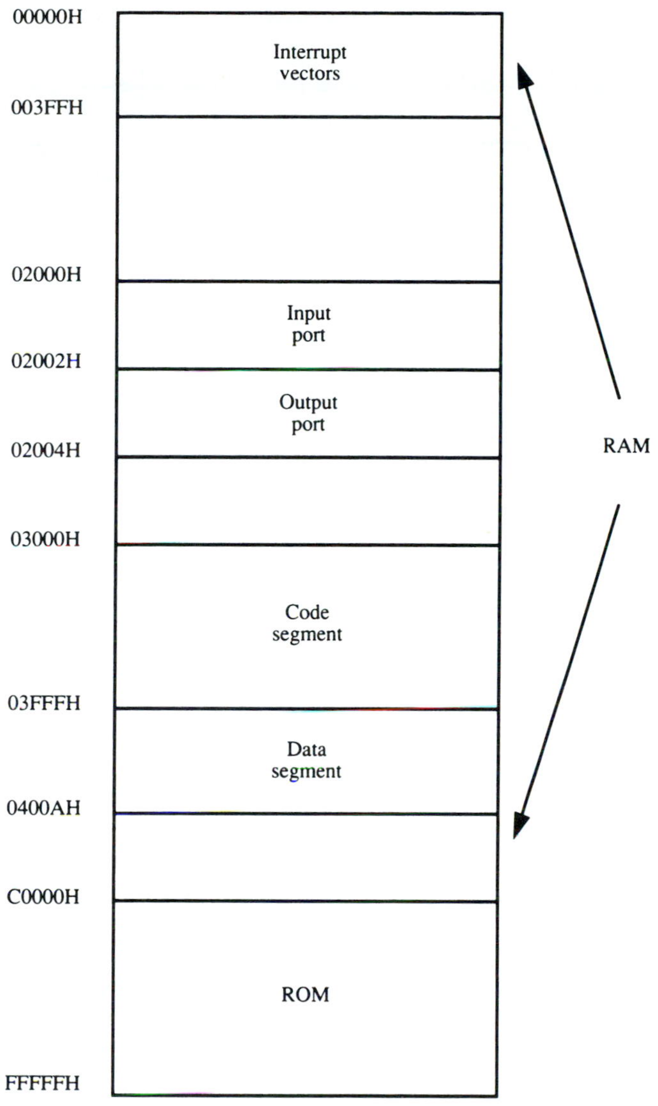

FIGURE 8.14
Memory map

The 8086/8088 operates in time periods called **bus cycles.** Each bus cycle requires *4 clock cycles* to complete. Therefore, in this example, the bus cycle is completed every 800 ns. A typical bus cycle is shown in Figure 8.15(a).

The two major bus cycles are the read bus cycle and the write bus cycle. The read bus cycle is activated when the MPU is *reading* information from memory or an I/O device. During the read bus cycle, there are normally four clock cycles. T_1, T_2, T_3, and T_4.

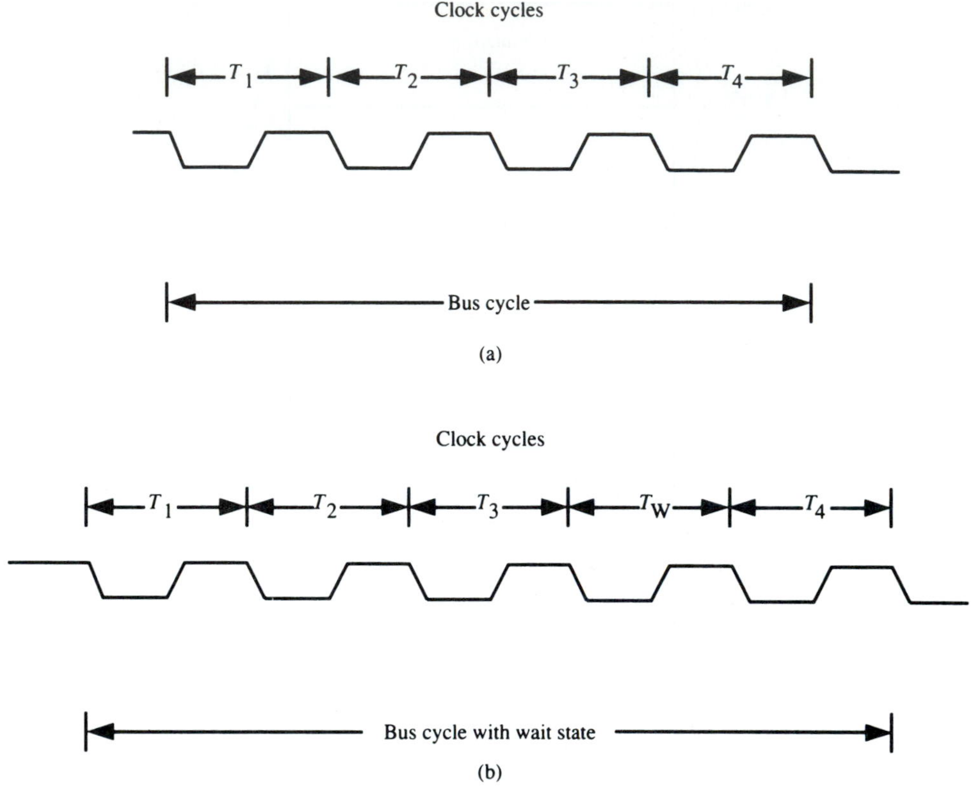

FIGURE 8.15
Timing-bus cycle

However, if the device outputting data to the MPU needs more time to send that data, a *wait state* (T_W) is initiated by placing extra clock cycles (T_W's) beween clock cycle T_3 and T_4, as shown in Figure 8.15(b).

Read Bus Cycle

Figure 8.16 shows an example of the typical read timing for the 8088 MPU.

T_1 During the first clock cycle the address/data bus is used to output the address of a memory or I/O location. Also outputted during the first clock cycle are control signals ALE, DT/$\overline{\text{R}}$, and IO/$\overline{\text{M}}$ by the 8088 MPU. At the end of T_1, ALE goes low and the address on the bus is latched.

T_2 At the beginning of clock cycle T_2 the multiplexed bus lines switch to their alternate function. The lower-order bus lines (A/D_0–A/D_7) go to their high-impedance state. The higher-order bus lines (A_{16}/S_3–A_{19}/S_6) go to their status output state. Bus lines A_8–A_{15}

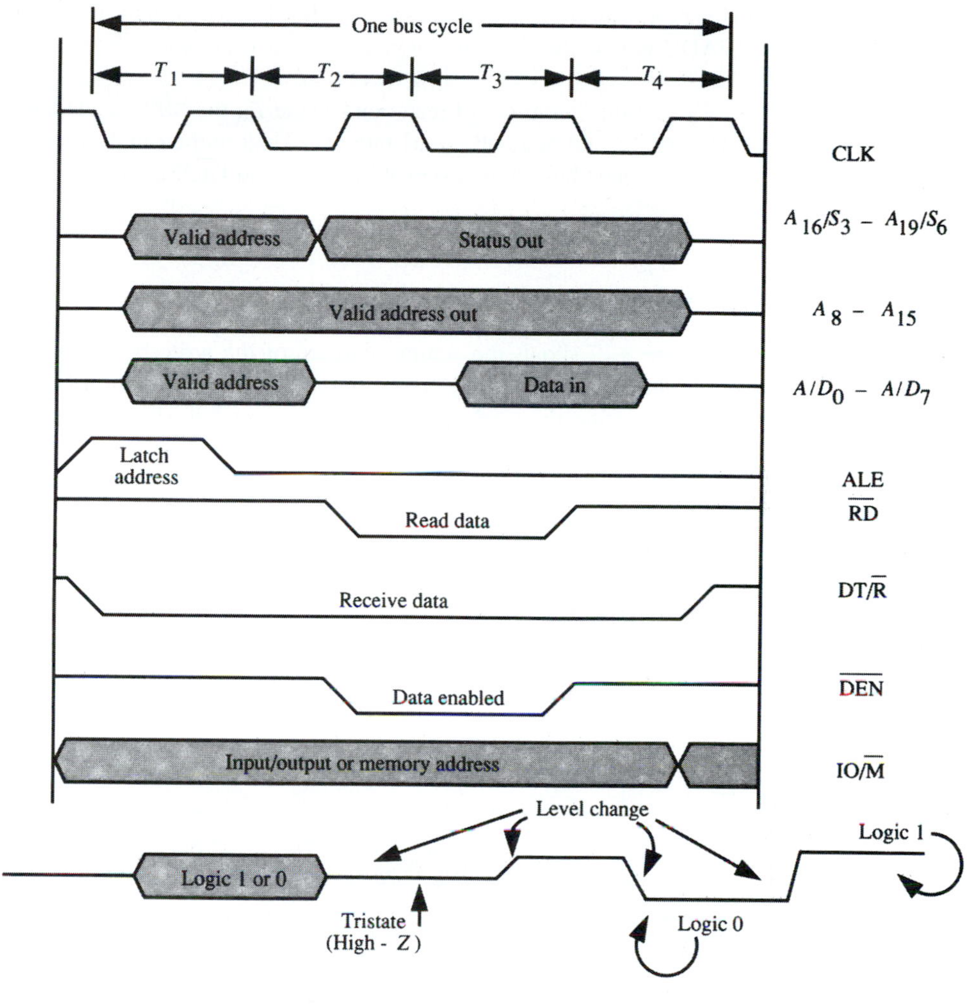

FIGURE 8.16
Read bus cycle

do not change and are still outputting address information. Since A/D_0–A/D_7 and A_{16}/S_3–A_{19}/S_6 are no longer outputting address information, ALE is now low, indicating that the address is no longer valid on all the address lines. Note that A/D_0–A/D_7 stay in the high-impedance state for one clock cycle. This gives the data-sending device time to respond to the data request. During the read clock cycle T_2 the MPU outputs active-low signals for control lines $\overline{\text{RD}}$ and $\overline{\text{DEN}}$. The $\overline{\text{RD}}$ signal causes the memory or I/O device to output data. The $\overline{\text{DEN}}$ signal is used to activate the data bus buffers, allowing data to be sent onto the bus to the MPU.

T_3 During read clock cycle T_3 the memory or I/O device is putting data onto the bus. This cycle provides additional time for the data to become stable on the bus. The

MPU samples the READY pin during T_3. If READY is high, the next clock cycle is T_4. If READY is low, the next clock cycle is T_W, a wait state.

T_4 At the beginning of read clock cycle T_4, the MPU reads the data present on the data bus. At the end of T_4 all the tristate lines float to their high-impedance state in preparation for the next bus cycle. Control lines $\overline{RD}$ and $\overline{DEN}$ go high, signaling the end of the read bus cycle.

Write Bus Cycle

Figure 8.17 shows the timing diagram of the write bus cycle. This can be seen to be similar to the read bus cycle. The differences are that the one clock cycle delay on lines A/D_0–A/D_7 during the read cycle is no longer needed, since the microprocessor is writ-

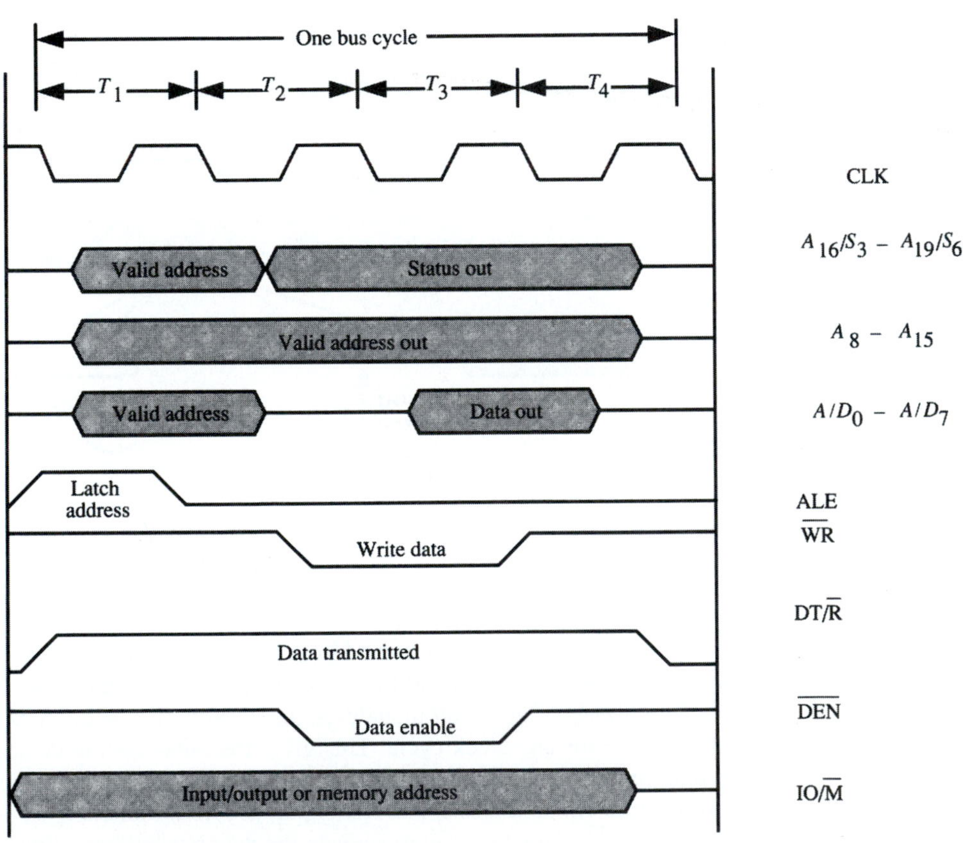

FIGURE 8.17
Write bus cycle

ing to memory or to an I/O device. Therefore, valid data bits are available after the beginning of clock cycle T_2. Note that just before the start of the write bus cycle the $DT/\overline{R}$ signal is latched high to signify that the MPU will be transmitting data. This signal remains high until the end of the write bus cycle. The $\overline{WR}$ line goes low after clock time T_2 to signal a write operation.

8.9 FETCH-EXECUTE CYCLE

We have looked at the read and write control timing. Now let's briefly investigate instruction timing. The microprocessor has two primary functions: fetch and execute. First it must fetch or read the program instruction or data. As we have seen, this can take one or more bus cycles. Once it has fetched the necessary program instructions and data through the BIU, the microprocessor's next step is to execute the instructions. The EU receives the instruction from the instruction queue and executes it. Some instructions may take 2 clock cycles to execute, whereas others may require as many as 100 clock cycles to execute. In older microprocessors this left the bus idle while the MPU was executing a long instruction, as shown in Figure 8.18. However, since the 8086/8088 MPU is broken up into two functional units, the BIU and the EU, it avoids much of the idle time required by older microprocessors. It does this by having the BIU prefetch instructions and place them into the instruction queue and data registers while the EU is executing the program instructions. Therefore, while the bus is busy during a read cycle, the EU can be executing a previous instruction. When the bus is busy during a write cycle, the EU can be executing another instruction. This greatly increases the effective speed of the entire system.

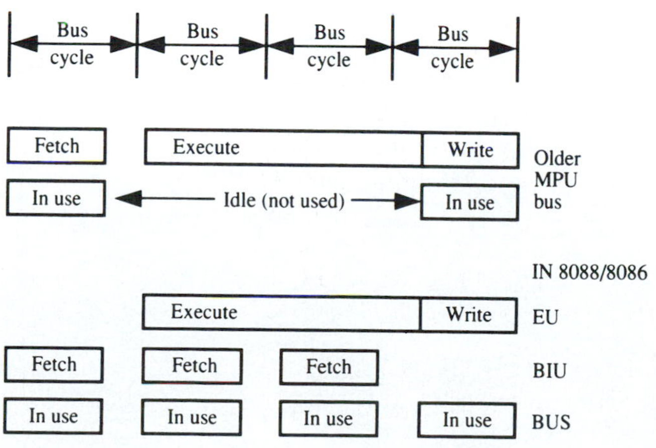

FIGURE 8.18
Fetch-execute cycle

8.10 TECH TIPS AND TROUBLESHOOTING—T³

MPU-related problems can be very difficult to troubleshoot. Due to the complex bus structure, the precise timing cycle relationships, and the software interactions and operation, the traditional types of test equipment will not be able to test the MPU adequately. Simple relationships between the software, hardware, and logic signals do not generally exist. Therefore, it is difficult to diagnose and localize MPU failures. Fortunately, MPU failures are rare. This is partially due to the fact that the MPU is in the center of the system. Other ICs that interface the control lines and bus signals to the rest of the system act as buffers to the MPU and provide the MPU with a certain amount of protection.

However, when an MPU failure is suspected, the first thing to do is visually inspect the IC. Look for burn marks on the top of the chip and on all the pins. Check for broken pins and connections. Clean away any dirt, grease, or corrosion. Touch the top of the MPU chip and feel its temperature. Although MPU chips will run hotter than other ICs, they should not be so hot that you cannot touch them.

If the visual inspection or operating temperature of the MPU leads you to believe that the MPU is defective, one technique is to *swap* the MPU, as shown in Figure 8.19.

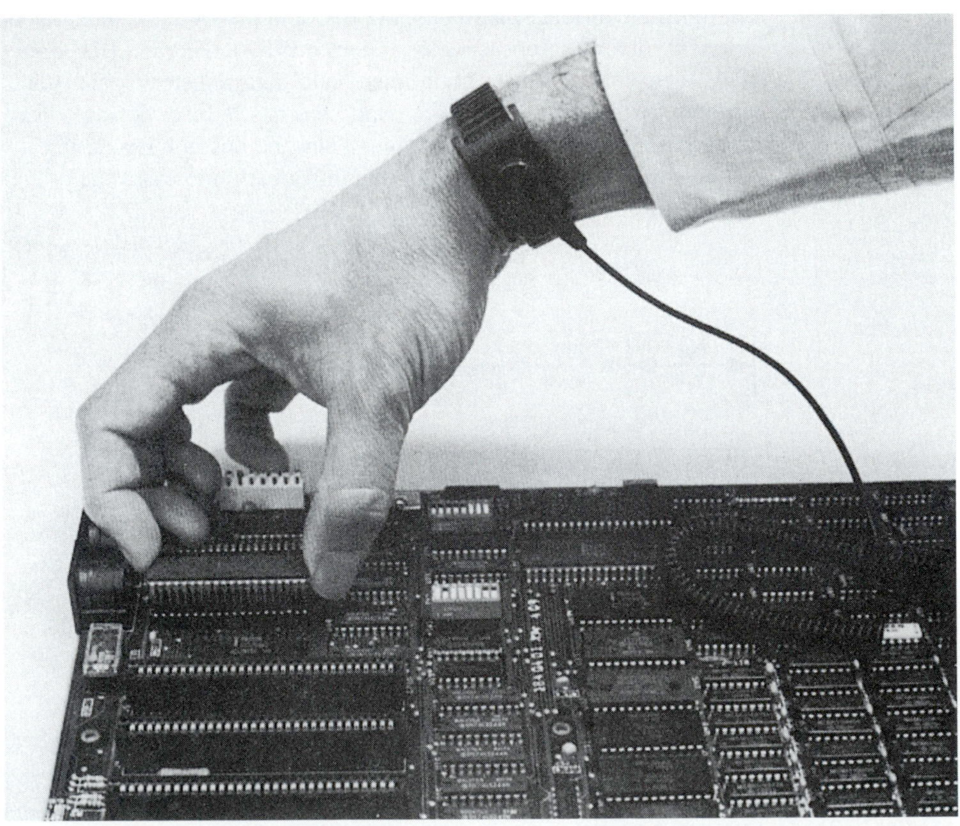

FIGURE 8.19

If the system works with the new MPU but does not work with the old one, then the old MPU is considered defective. Often MPU devices are installed in sockets for easy replacement. If this is the case, a swap will be very easy to do and can save a great deal of time and effort. However, a great deal of caution is required. MPU devices are very easily damaged by static electricity. Even if the MPU is not destroyed, static electricity can cause *intermittent* problems and shorten the life of the device. Therefore, always remember to ground yourself before you touch the MPU. If possible, use antistatic table mats and wrist straps. Also make certain that all the pins of the MPU are correctly inserted into the socket.

While the MPU chip rarely fails, several related items can cause problems. MPU input control lines can *hang* the processor or stop its operation. Bus lines can be stuck high or low. Master clock operation can be incorrect. Bus *contention* conflicts can occur when two or more devices attempt to drive a bus line at the same time. The defective devices can cause constant interrupts that will slow or stop MPU operation.

Without the use of sophisticated test equipment, these types of problems can be difficult to identify. However, you can use an oscilloscope or logic probe to provide a general indication of what is going on by performing what are called *gross signal checks.*

1. Check for V_{CC} and ground signals from the power supply. Verify voltage levels, ripple, and noise.
2. Check for clock signals. Verify rise and fall times as well as frequency.
3. Check for the presence of signals on the address and data buses. There should be signal activity. Verify that these signals are not stuck high or low at all times. Remember that some lines may not show activity because the program does not call for any. For example, the high-order address lines may remain in the zero state because the program and the data reside in a lower part of memory. Therefore, you may have to run a special program that will exercise all the lines.
4. Check for control bus activity. Remember that the MPU outputs control and status information. Signals like ALE, $\overline{\text{DEN}}$, DT/$\overline{\text{R}}$, IO/$\overline{\text{M}}$, $\overline{\text{WR}}$, $\overline{\text{RD}}$, and $\overline{\text{SSO}}$ should be changing logic levels constantly during MPU operation.
5. Check the interrupt control lines. A constant interrupt can halt or drastically slow down MPU operation. Furthermore, a constant interrupt can cause address, data, and control lines to become inactive. For example, a constant high on the RESET line will cause the MPU to remain reset and stop it from beginning operation. A constant low on the $\overline{\text{TEST}}$ line will generate continual wait states.
6. If general activity is present, check for decoded outputs such as chip selects or output enables.

By understanding the operation and function of the different signal lines discussed earlier in the chapter, a technician can probe MPU pins to determine the health of the system. If signals are detectable and look good, the circuitry is probably good. If signals appear inactive or stuck high or low, then the possibility of a fault exists in this area.

Parts 3 and 4 of this text are specifically concerned with the theory, operation, and servicing of the microcomputer system. A more detailed explanation of servicing and troubleshooting procedures is found there.

EXERCISES

8.1 List six practical uses for a microprocessor.

8.2 Name the functional parts of the 8088 MPU.

8.3 Name the 16-bit data registers.

8.4 Identify the pointer and index registers.

8.5 Discuss the major differences between the 8088 MPU and the 8086 MPU.

8.6 Which functional unit is responsible for interfacing the 8088 MPU to the rest of the system?

8.7 What are the functional parts of the EU? the BIU?

8.8 Name the signal lines used to determine a memory location.

8.9 Which signal lines are multiplexed? What is their function?

8.10 Which microprocessor bus is bidirectional?

8.11 What is the addressable memory size of a microprocessor with
a. 8 address lines b. 16 address lines
c. 20 address lines

8.12 Following is a list of 8088 signal lines and functional groupings. For each signal line, choose the letter that identifies its functional grouping. (Letters may be used more than once.) Then, briefly, describe its use and operation.

Signal Lines	*Functional Groups*
1. ALE	A. Power supply line
2. HLDA	B. Address/status line
3. NMI	C. Data line
5. $\overline{\text{WR}}$	D. Data/address control line
6. $\overline{\text{INTA}}$	E. Interrupt control line
7. $\overline{\text{SSO}}$	F. Operation control line
8. A_{15}	
9. V_{CC}	
10. DT/$\overline{\text{R}}$	

8.13 For each of the following, identify which signal or signals a technician should monitor.
a. To determine if a memory write cycle is being performed.
b. To determine the direction of data flow.
c. To determine whether the output from a microprocessor is going to memory or an I/O port.
d. To determine if a valid address is on the bus.
e. To determine if the MPU is being interrupted.
f. To determine if the MPU will accept the data transfer.
g. To determine if the MPU is in multiprocessor mode.

8.14 Explain the function and operation of the instruction queue.

8.15 Describe the function of the flag register.

8.16 What flags are used to indicate the status of the MPU?

8.17 List the names and describe the functions of the general registers.

8.18 What is the use of the stack pointer? The instruction pointer?

8.19 Identify the segment register that is used to determine the starting location of a program's instructions.

8.20 What is the difference between a logical address and a physical address?

8.21 Explain the differences between static and dynamic RAM.

8.22 What is meant by memory-mapped I/O?

8.23 Draw a memory map that positions ROM at location E6F2H through E728H, interrupts at 0000H through 00FFH, RAM at 1000H through E000H.

8.24 Draw the timing diagram for a read bus cycle.

8.25 Draw the timing diagram for a read bus cycle with a wait state.

8.26 Determine whether either of the following statements is true or false.

a. A bootstrap program will usually be found in RAM.
b. An I/O device sends an active-high signal to the MPU to signify that it is ready for a data transfer.
c. The $\overline{RD}$ and $\overline{WR}$ control lines cannot be active at the same time.
d. During clock cycle T_2 the MPU outputs the address of a memory location.
e. During clock cycle T_3 the MPU reads the data on the data bus.

8.27 Explain the advantage of the 8088/8086 microprocessor over older microprocessors with relation to the fetch and execute cycles.

8.28 RMB associates designs a turbo microcomputer system that uses an 8-MHz clock frequency. Calculate the time for one clock cycle. Calculate the time for one bus cycle with two wait states.

8.29 The base address is F100H. The offset is 0200H. Determine the physical address of memory.

8.30 The instruction pointer contains 0777H. The code segment register contains 0200H. The data segment register contains 0900H. The stack segment register contains 1F00H. Determine the physical address for the first program instruction.

8.31 The physical address of a memory location is 421FH. The base address from the segment register is 0200H. Find the offset.

8.32 If it is suspected that the MPU may be faulty, what procedures and precautions should be taken to correct the problem?

8.33 Crossword Puzzle

ACROSS

4. Sharing a bus.
5. Write.
6. Base register.
8. A term that describes the functional components of an integrated circuit.
10. 8088 queue width.
11. String destination.
12. Read.
15. Processing part of the 8088 MPU.
17. Count.
22. Connects to memory.
23. Five volts.
26. Wait state.
29. Valid address-control signal.
30. A signal used to stop the MPU.
31. Negate.
33. Computer chip.
37. Calculator part of the MPU.
38. MPU operational information.
40. 200 ns.
41. Indicates zero results.
42. Pin 22.

DOWN

1. Instruction storage prior to processing.
2. String source.
3. *AX*, *BX*, *CX*, *DX*.
5. Sixteen bits.
7. Next instruction to be executed.
8. Memory location.
9. Uses instruction pointer.
10. Four clock cycles.
13. Group of common signals or wires.
14. Four bits.
16. Interrupt enable.
18. Five megahertz.
19. Information.
20. Turns on a buffer.
21. 8086 data lines.
24. Name of a bus.
25. Binary digit.
27. EXTRA, CODE, DATA, STACK.
28. An instruction cycle.
29. Accumulator.
32. Address lines.
34. Overflow.
35. Register and segment.
36. Memory.
38. Indicates negative results.
39. Stack operations.

1 2 3 4 5 6 7 8 9 10 11 12 13 14 15 16 17 18 19 20 21 22 23 24 25 26 27 28 29 30 31 32 33 34 35 36 37 38 39 40 41 42

CHAPTER 9

INTRODUCTION TO PROGRAMMING

KEY TERMS

Addressing Mode
Argument Field
Assembling
Assembly Language Programming
Byte Swapping
Comment Field
Comments
Compilers
Computer Program
DEBUG
Destination Operand
Directive Field
Editor
EDLIN
Flowcharting
Footer
Header
Immediate Value
Label Field
Linker Program
Looping
Machine Language
MACRO Assembler (MASM)
Masking Off Bits
Name Field
Object Program
Op-code
Operands
Poll
Source Operand
Source Program
Stack
Subroutine

9.0 INTRODUCTION

The microprocessor and the microcomputer system cannot be operated without a sequence of instructions known as a **computer program.** This program tells the computer what to do, how to do it, and when it should be done. The language that a computer can directly understand is called its **machine language.** Each microprocessor has its own machine language that is made up of a series of ones and zeros. These ones and zeros tell the system exactly what to do. For example, the machine language instruction to ADD a number may be written as

Function	*Machine Language*
ADD	00000011

Clearly, a program made up of many machine language instructions would be tedious and error prone. For this reason, so called *higher-level languages* such as Fortran, COBOL, Pascal, and BASIC were developed. These languages use English-like words for commands that make the programming task easier for people. Most high-level languages use **compilers** to translate instruction statements into machine language (1s and 0s) as shown in Figure 9.1. The compiler takes in instructions and turns them into machine language. The microprocessor can now understand these machine language instructions and execute them. However, the translation process increases processing time because of the added steps.

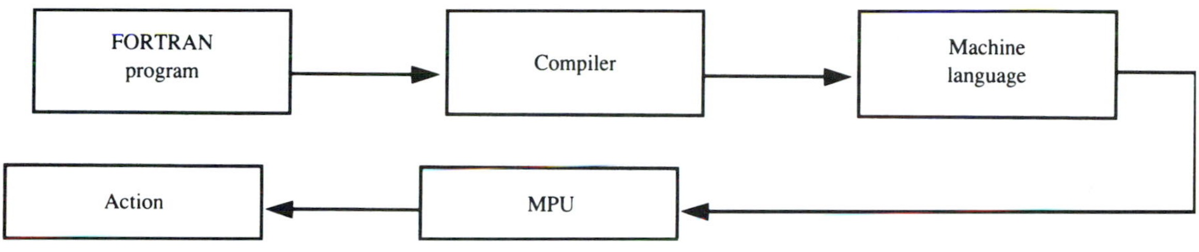

FIGURE 9.1
Converting a high-level language program

A compromise between the people-oriented, high-level languages and the computer-oriented, machine-level language is called **assembly language programming.** Assembly language does not require compiling, thus speeding up execution time and allowing faster access to peripherals. Instructions are assembled into machine code executed directly by the CPU.

The assembly language program is also known as a **source program.** The act of converting the source program into the machine language that the computer understands is called **assembling** the program. This machine language output is also known as an **object program.** The entire process is summarized in Figure 9.2

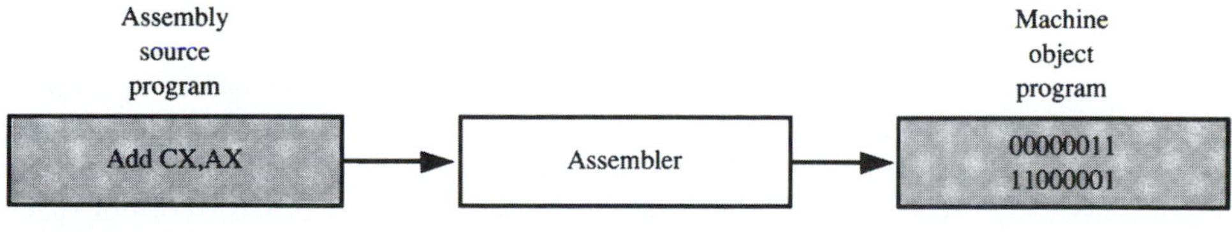

FIGURE 9.2
Assembling a program

It should be noted that compiling and assembling are very similar in nature. However, assembly language code is usually directly related to the resulting machine language code and therefore more efficient. Although compiled languages are easier to use, they usually require more machine language steps and are therefore less efficient.

9.1 ASSEMBLY LANGUAGE PROGRAMMING

Learning to write 8086/8088 assembly language programs is similar to learning to operate an electronic calculator. All data values must be entered and each operation, such as addition or subtraction, must be performed separately. Although assembly language programs and high-level language programs look different, the same procedures are used to develop each of them. These procedures are summarized as follows:

1. *Look and think:* What is the problem to be solved? What are the givens? What is to be found? This may require you to draw a flowchart or block diagram of how the program works.
2. *Develop an assembly language program* that uses the givens to solve for the desired output.
3. *Enter the program* into the computer.
4. *Assemble the program:* Use the assembler to translate the program into machine language. If the assembler finds errors, correct them and reassemble the program.
5. *Run the program.*
6. *Verify the results:* Check to see that the results are what you expected. If the results are incorrect, find the error and correct it.

Instruction Format

Each assembly language instruction in a source program is composed of up to four fields, as shown:

Label	*Op-code*	*Operands*	*Comments*
DATA:	MOV	CX,AX	;load CX with AX

1. The **label field** assigns a symbolic name to an assembly language instruction. This is sometimes called a *tag.* It lets other instructions reference the label by name rather than by a numeric location in memory. The first character in a label field must be a letter and the last character of a label field must be a colon. All letters A–Z, numerals 0–9, an underscore (—), and an at sign (@) can be used in label fields. Spaces are not allowed. If a label field is not used, a space character or tab (series of spaces) is used in its place.
2. The **op-code** is a mnemonic shorthand for an 8086/8088 microprocessor instruction. For example, MOV is the mnemonic for a move instruction. The beginning of the op-code field follows the colon of the label field or a space character. The end of the op-code field is determined by a space character or tab (series of spaces). See Appendix B for a complete listing of the 8086/8088 op-code instructions.
3. **Operands** tell the MPU where to find the data to be operated on. For example, in the instruction

   ```
   MOV CX,AX
   ```

   the operands are CX and AX. They tell the MPU to copy or move the contents of the AX register into the CX register. Some instructions use no operands, so the operand field is left blank. Other instructions use one or two operands. If there are two operands, they must be separated by a comma. When there are two operands, the first is called the **destination operand,** and the second is called the **source operand.**

```
MOV CX,AX
```

Destination operand = CX
Source operand = AX

The end of the operand field is determined by a semicolon if a comment is to be used. If *no* comment is to be used, the end of the operand field is determined by a carriage return.

4. **Comments** are used to describe a statement in the program to make a program easier to understand. Comments become very useful in longer programs. They will help you to remember why you did something a certain way. They will also help others to understand what you are doing. Comments must begin with a semicolon (;) and end with a carriage return. If a comment runs past the end of a line, it may be continued on the next line as long as the line begins with a semicolon.

Programming Model

The programming model of the 8088 in Figure 9.3 shows the 8088 as a collection of 8- and 16-bit registers. These registers can be made to operate using the software instructions that are incorporated as part of the 8088 design. The program instructions have to do with the control of the registers and the digital data path that are physically contained inside the 8088.

Figure 9.3 shows each register as a rectangular box. The name of each register is shown as large capital letters to the left of the model. Furthermore, the 8-bit halves of the general registers are shown by the H and the L above the general register set, indicating the high and low byte paths of each register. The programming model is an abstract of the register set and represents only the registers that we may use in a program.

Addressing Modes

The way in which the location of an operand is determined is called the **addressing mode.** How an operand is addressed in a program depends on the type and location of the data. There are three general types of addressing modes:

1. Immediate addressing modes
2. Register addressing modes
3. Memory addressing modes

Immediate Addressing Modes The immediate addressing mode uses a number or a constant as its source operand. For example, in the instruction

```
MOV AX,0001H
```

the immediate value 0001H is moved into the AX register. The equivalent machine language instruction is written as

```
10111 000 0000000000000001
```

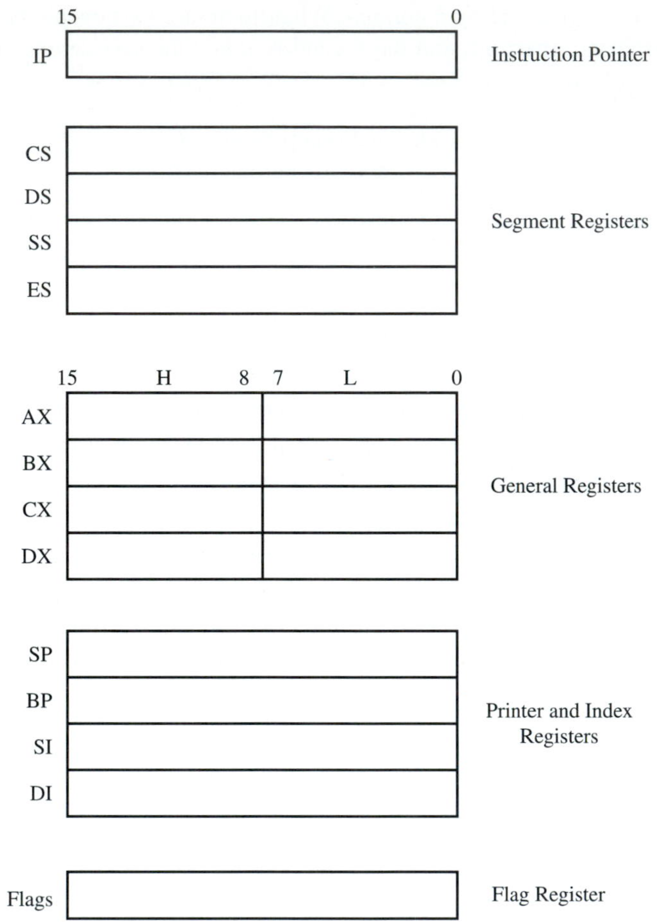

FIGURE 9.3
Programming model

The first 5 bits, 10111, tell the MPU to perform an immediate word (16 bits) move. The next 3 bits, 000, define the destination operand as the AX register. The final 16 bits define the immediate value 0001H, which in binary is equal to 0000 0000 0000 0001B.

Figure 9.4 illustrates the operation of the MOV AX, 0001H instruction. Notice that the immediate operand 0001 is stored in the code segment of memory in the two bytes immediately following the op-code of the instruction and is part of the instruction. Referring to Figure 9.4, we see that the code segment register and the instruction pointer combine to define the physical memory address (01000H) of the instruction to be fetched. The byte of data, B8 (10111000), defines the immediate MOV instruction to the AX register. The MPU knows that it must perform two more fetches to obtain the immediate value to be placed into the AX register. The next two memory locations must contain this data. Notice that the least significant byte is in memory location 01001 and the most significant byte is in memory location 01002. When combined, they form the number 0001H.

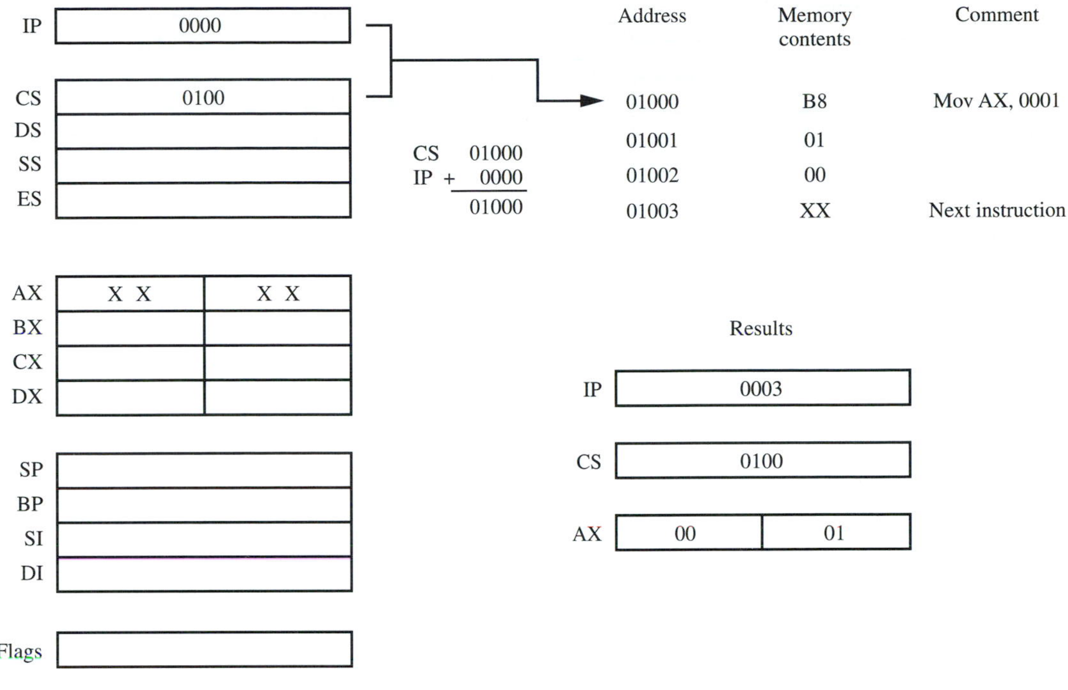

FIGURE 9.4
Immediate addressing mode

Intel processors always put the low order byte in the lower memory address. This is known as **byte swapping.** After the instruction is executed, the instruction pointer will be set at 0003H, the code segment register will be at 0100H, and the AX register will contain 0001H as shown in Figure 9.4.

The immediate addressing mode can have a register or a memory location as its destination operand. If the destination is a register, the instruction will take approximately 4 clock cycles to execute. If the destination is a memory address, the instruction will take approximately 10 clock cycles to execute.

Register Addressing Modes In the register addressing mode both the source and destination operands of the instruction are in registers. Because this mode operates entirely within the MPU it is very fast. A typical instruction in this mode takes approximately 2 clock cycles to execute. For example, in the instruction

```
MOV AX,CX
```

the contents of the CX register are copied or moved from the AX register.

Referring to Figure 9.5, we once again see that the code segment and the instruction pointer combine to define the physical memory address (01000H) of the instruction to be fetched. The memory contents of address 01000H is 89, which defines a 16-bit register MOV instruction into the AX register. C8 defines the CX register as the source

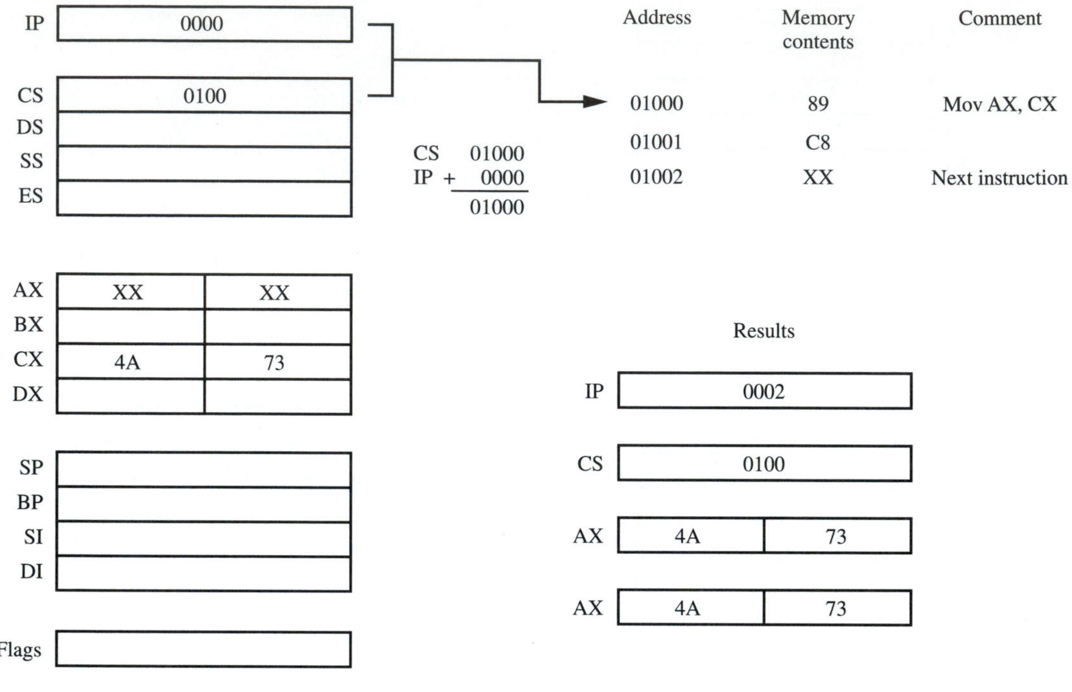

FIGURE 9.5
Register addressing mode

operand. After the instruction is executed, the instruction pointer will be at 0102H and the AX and CX registers will both contain 4A73H, which is the original data in the CX register.

Memory Addressing Modes The memory addressing modes use either the source or destination operand as a memory address or label. There are two general types of memory addressing modes:

1. Direct addressing
2. Indirect addressing

Direct Addressing Mode In the direct addressing mode, either the source or the destination operand is a specific memory location defined by the address number or a label. For example, in the instruction

```
MOV AX,MEM1
```

the contents of the memory address labeled MEM1 is copied or moved into the AX register.

Direct addressing mode differs from immediate addressing in that the locations following the op-code hold an effective memory address instead of data. The effective address is combined with the contents of the data segment register to produce the physical address of the operand in memory.

The instruction MOV AX, MEM1 is an example of direct memory addressing. The contents of the memory location labeled MEM1 are moved into the AX register. The physical address of MEM1 is computed by combining the offset stored in the two memory locations that follow the instruction with the contents of the data segment register. Thus the physical address of the data to be moved into the AX register is 06789H (04000 + 2789). Once again, notice the byte swapping of MEM1 (2789). Memory location 01001 contains 89, and memory location 01002 contains 27. It is also important to note that, since we are moving into the AX register, two bytes are actually moved. The byte, AA, in memory location 6789 is moved into the AL register and the byte, 90, in memory location 678A is moved into the AH register. This is another example of byte swapping.

After the instruction is executed, the instruction pointer is at 01003, the code segment register remains at 01000, the data segment register remains at 0400, and the AX register contains 90AA. Figure 9.6 illustrates the execution of the MOV AX, MEM1

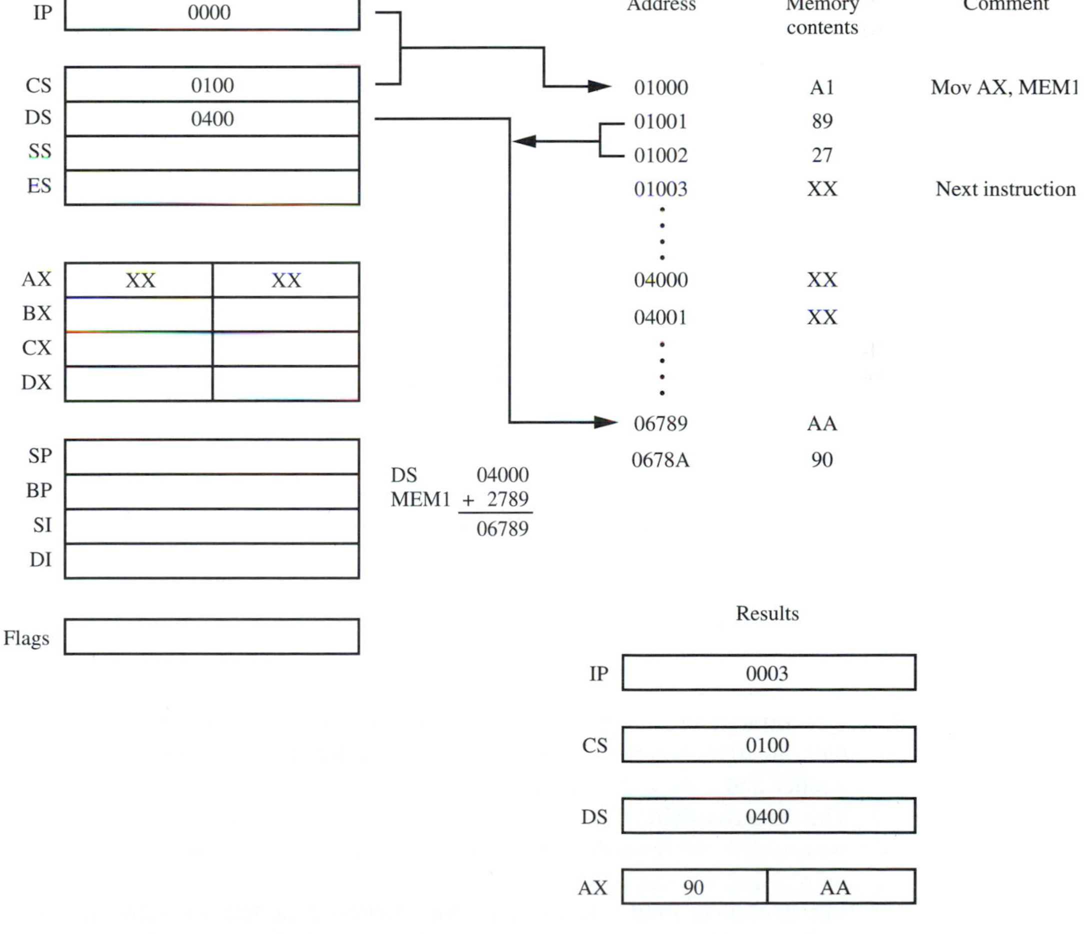

FIGURE 9.6
Direct memory addressing mode

instruction. Notice that the instruction is fetched by combining the instruction pointer with the code segment as in the previous examples.

Indirect Addressing Modes In the indirect addressing mode, the memory address is not directly given. A register is used to indicate the address where the data can be found. Therefore, the register acts as an indirect address to locate the data. For example, in the instruction

```
MOV [BX],CX
```

the source of the data is the CX register. The destination, where the data are to be placed or copied to, is the *address* pointed to by the BX register. The brackets ([ ]) around BX indicate that the BX register contains an *address* and not a numeric value. There are many variations of indirect addressing modes that are available to the 8086/8088 assembly language programmer. However, the basic concept in all the indirect addressing modes remains the same.

Figure 9.7 illustrates the operation of the MOV [BX],CX instruction. The physical address of the destination is computed by combining the data segment with the effective address in the BX register. After execution, the instruction pointer is at 01003, the code segment register remains at 0100, the data segment register remains at 0400, the BX register still contains the effective address 2789 and the CX register still contains the data 55AA. The results are displayed in physical memory address 06789, which contains AAH, and physical memory address 0678A, which contains 55H.

9.2 ASSEMBLER DIRECTIVES

Assembler directives are instructions to the assembler program and are not executed by the MPU. The programs that you write will include assembler directives. However, the assembler directives are not translated into machine language. Assembler directives are used to pass information to the assembler such as where to begin the program (ORG), to reserve a space in memory for data (RS), or to indicate to the assembler that there are no more instructions to assemble (END). The assembler directives look very much like assembly language instructions. Each assembler directive consists of four fields.

Name	Directive	Argument	Comments
Pi	EQU	3.14	;define value of PI

1. The **name field** is similar to the label field in the instruction statement. Some assembler directives require a name field; others do not. The name field always begins with a letter and ends with a space.
2. The **directive field** is a mnemonic shorthand for the assembler directive operation. This is similar to the op-code in the instruction statement. It begins with a space and ends with a space or carriage return.
3. The **argument field** contains a memory address or a numeric value that is used with and determined by the directive. It also begins with a space and ends with a semicolon or carriage return.

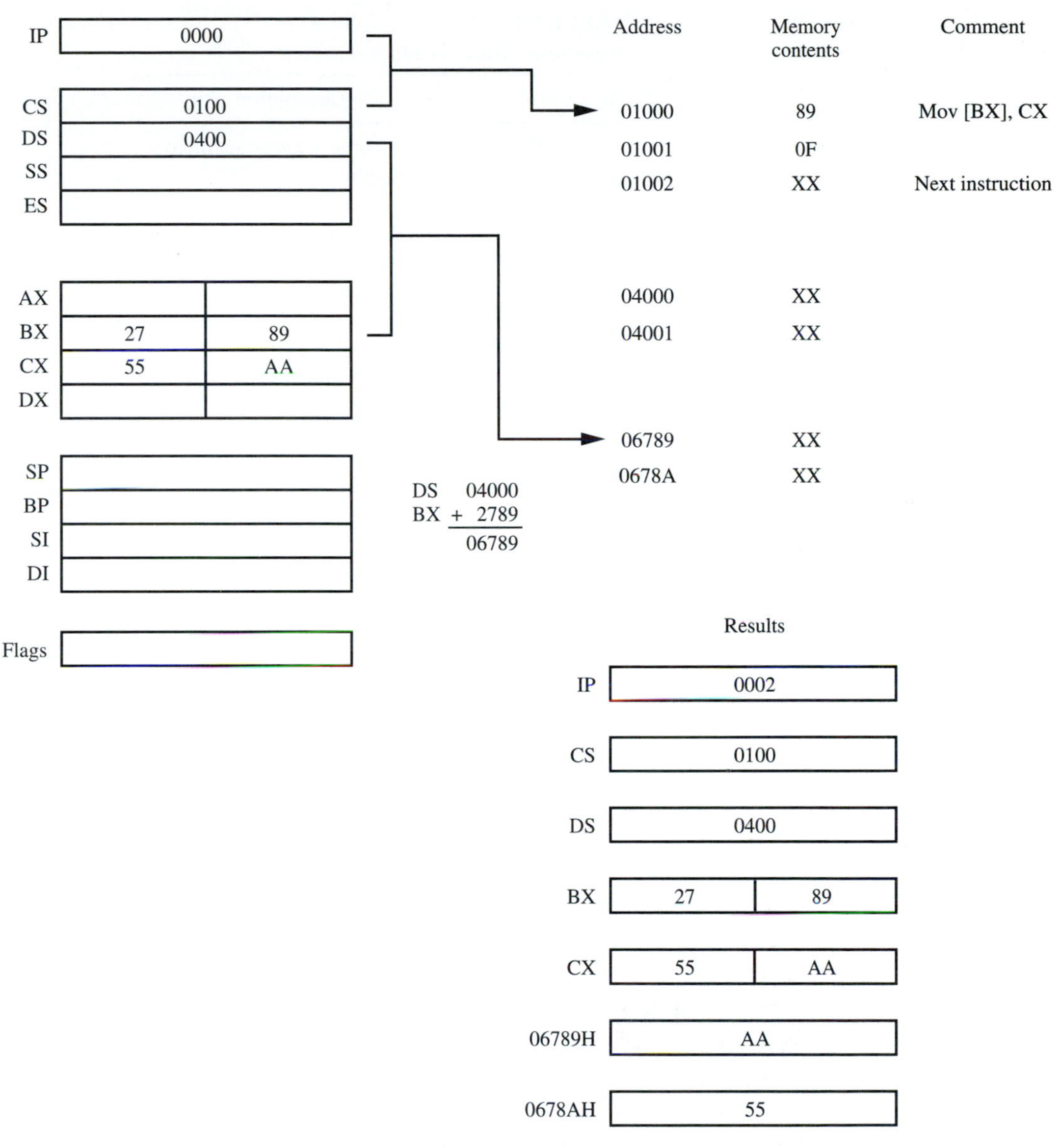

FIGURE 9.7
Indirect addressing mode

4. The **comment field** is used exactly like the comment field in the instruction statement. It begins with a semicolon and ends with a carriage return.

Although there are many assembler directives, for our purposes we will limit our discussion to the directives shown in Table 9.1

TABLE 9.1
Assembler directives summary

Directive	Description
ASSUME	Assume register segment
ORG	Originate
DB	Define byte
DW	Define word
DUP	Duplicate
EQU	Equate
PROC	Procedure
ENDP	End procedure
END	Program end

ASSUME The assume assembler directive tells the assembler which segment registers the program is using. Otherwise, the assembler will not know the difference between the code segment and the data segment, and so forth—for example,

```
ASSUME CS:CODE, DS:DATA
```

ORG The originate directive is used to set the instruction point to the starting address of the program in memory—for example,

```
ORG 0100H
```

This program begins at memory location 100H.

DB The define byte directive can be used to allocate, define, and name a byte space in memory—for example,

```
DATA DB 90H
```

In this case the name DATA will be associated with memory location that contains the value 90H.

DW The define word directive is used to allocate, define, and name two consecutive locations in memory.

DUP The duplicate directive is used as the operand of the defined byte or defined word directive to produce a block of data. It is often used to establish the stack—for example,

```
DB 100 DUP (0) ;100 bytes initially all 0
DB 100 DUP (?) ;100 bytes not initialized
```

EQU The equate directive is used to associate a name with a value or another symbol—for example,

```
PI EQU 3.14
```

In this case the variable name PI is set equal to the value 3.14.

PROC The procedure directive PROC is used to break up an assembly language program into blocks. It helps to modularize a program into smaller pieces. Only the simplest program and task can be written without being broken down into smaller blocks. The PROC directive is often used to define a subroutine, or portion of a program that is performed many times—for example,

```
DELAY PROC NEAR
DELAY PROC FAR
```

NEAR is used if the procedure is performed in one segment. FAR is used if the procedure extends into two or more segments.

ENDP The end procedure directive ENDP is used to terminate a procedure in an assembly language program—for example,

```
DELAY    PROC    NEAR    ;    begin delay procedure
           .
           .
           .
DELAY    ENDP            ;    end of delay procedure
```

END The end directive is used to tell the assembler that there are no more instructions left to assemble.

9.3 8086/8088 INSTRUCTION SET

The 8086/8088 instruction set is made up of over 100 basic assembly language instruction mnemonics that tell the MPU what to do. Each of the basic instructions can have many variations. For example, there are 28 different variations of the basic move instruction (MOV). The complete instruction set can be found in Appendix B. In this section we look at some of the more useful instructions from a technician's point of view. For clarity the instructions we examine in this section are classified into four groups:

1. Data-transfer instructions
2. Arithmetic instructions
3. Logical instructions
4. Program-control instructions

Data-Transfer Instructions

MOV The fundamental data-transfer instruction is the move (MOV) instruction. This instruction can be used to transfer bytes (8 bits) or words (16 bits) of data. The general format of the MOV instruction is

MOV destination, source

The MOV instruction can be used to transfer data between two registers, as shown in Figure 9.8.

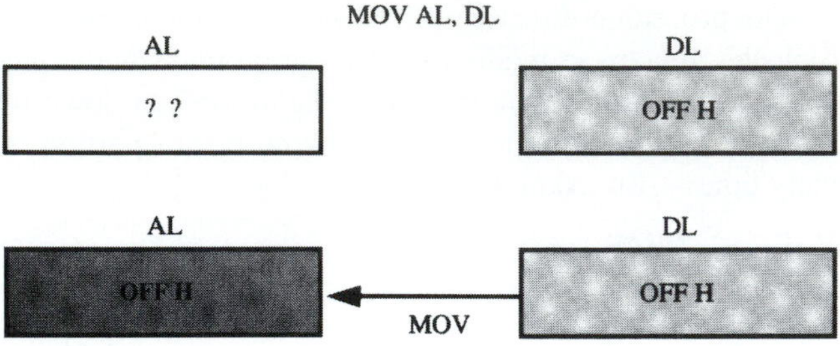

FIGURE 9.8
The MOV instruction register to register

Here the contents of the DL register (lower byte of the data register) contain the value of 0FFH. The AL register contains an unknown value or a value to be discarded. After the MOV instruction is executed, the contents of the DL register (0FFH) are also found in the AL register. Furthermore, the value 0FFH remains in the DL register after the move is completed. It should be noted that the leading zero is used here to differentiate or indicate that FFH is a constant value as opposed to a label. This is necessary to avoid confusion between interpreting hexadecimal numbers and label names. In assembly language programming a constant value is often referred to as an **immediate value.** Table 9.2 summarizes the various destinations and sources that are used with the move instruction.

TABLE 9.2
MOV instruction formats

	Destination	Source	Example
1	Memory	Register	MOV 100H,AX
2	Register	Memory	MOV AX,MEM1
3	Register	Register	MOV AX,BX
4	Register	Immediate	MOV AX,OFFFFH

The first example in Table 9.2 MOVes the contents of the AX register into memory location 100H. In the second example the contents of a memory location pointed to or named by the label MEM1 is loaded into the AX register. In the third example the contents of the BX register are loaded into the AX register. Finally, in the fourth example the constant value FFFFH is loaded into the AX register.

LEA Load effective address (LEA) can be considered another type of move instruction. The LEA instruction is used to load a register with the *address* given. It is *not* used to move the contents of the given address. The general format of the LEA instruction is

LEA destination register, source label

The destination operand must be a register, and the source operand must be a label. For example, in the instruction

```
LEA AX,MEM1
```

the AX register is loaded with the *address* of MEM1 and *not* the data contained in MEM1.

Before introducing the next instructions, we examine the concept of a **stack.** Think of the image of a stack of cafeteria trays in a holding device. When a new tray is placed on top of the stack of trays, it *pushes* all the trays beneath it down one level. When the top tray is removed from the stack all the trays *pop* up one level. The last tray placed on the stack was the first tray removed. This is called a last-in, first-out (LIFO) operation and is illustrated in Figure 9.9.

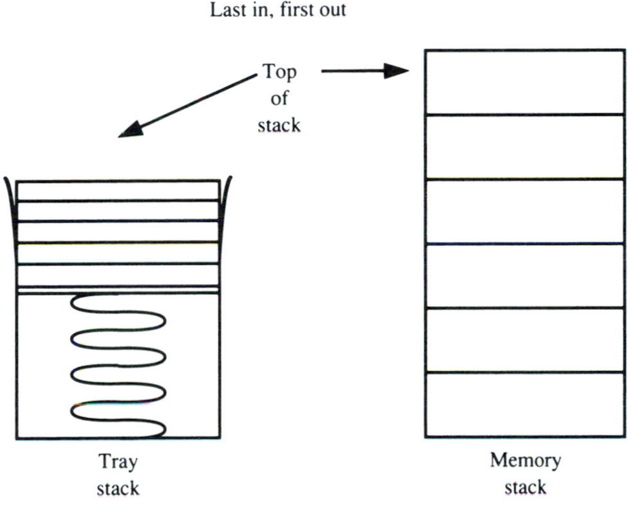

FIGURE 9.9
Stack operation

PUSH/POP The stack register is a convenient place to temporarily deposit data and memory operands from a program. For example, a program might want to save the contents of the AX register while it puts the AX register to some other use. The PUSH and POP instructions can be used to accomplish this task. The general formats of the PUSH and POP instructions are

PUSH source
POP destination

The source and destination operand can be a register or a memory address—for example,

```
PUSH AX    ;Save AX on top of the stack
POP AX     ;Retrieve AX from the top of the stack
```

PUSH/POPF The contents of the flag register may be stored or retrieved from the memory stack using the PUSHF and POPF instructions—for example,

```
PUSHF ;pushes contents of flag register onto the stack
POPF  ;retrieves contents of the flag register from the
      ;stack
```

The physical memory address of the stack is obtained by combining the contents of the stack segment register with the contents of the stack pointer register (SS:SP). The initial values of the stack segment and the stack pointer in a program define the top of the stack. The 8088 pushes data onto the stack one word at a time. Each time the contents of a register are to be pushed onto the top of the stack, the stack pointer is *first* automatically decremented by two, and then contents of the register are written into the stack memory. This causes the stack to grow down in memory.

Figure 9.10 illustrates the operation of the instruction PUSH AX. The stack segment is initially 1000H, and the stack pointer is initially 2000H. This defines the physical address

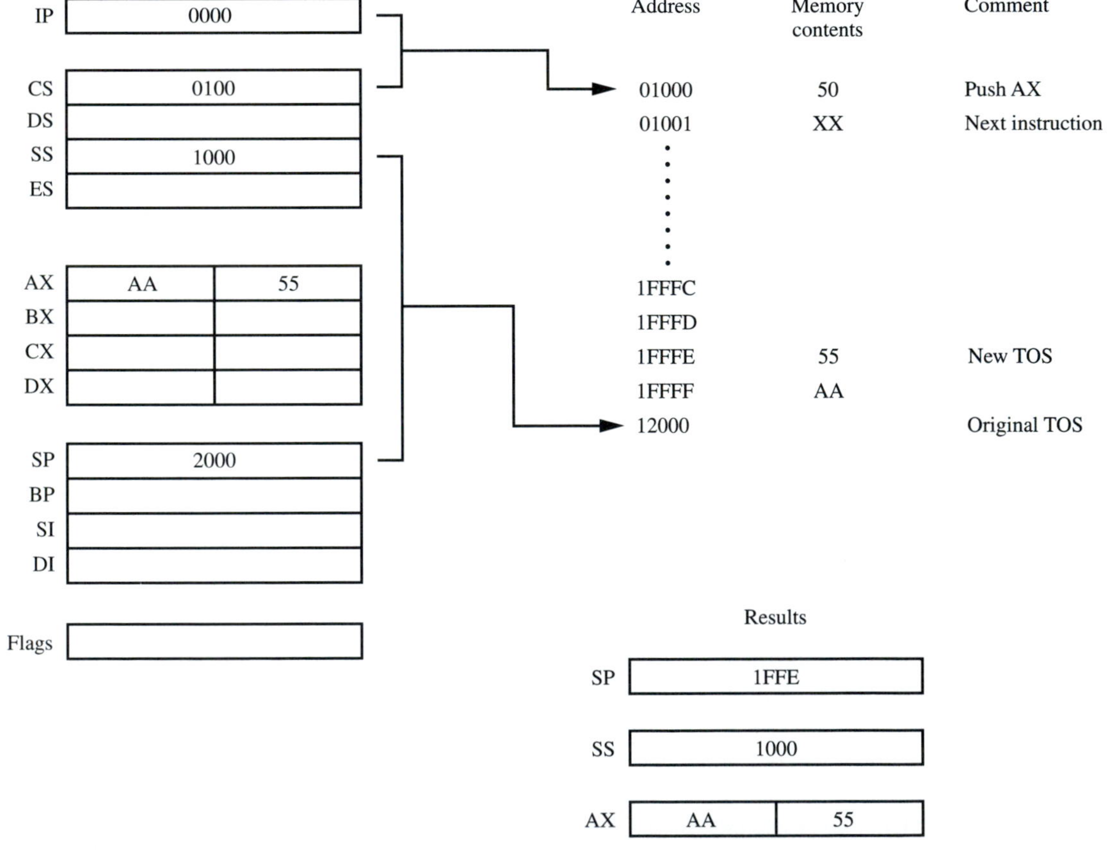

FIGURE 9.10
Pushing onto the stack

of the top of the stack at 12000H (segment register shifted left + pointer register). Executing the instruction, PUSH AX caused the stack pointer to first be decremented by two bytes to 1FFEH and then the contents of the AX register to be written into physical memory address 01FFF and 01FFE. Note that the high-order byte is written into the higher memory address and the low-order byte is written into the lower memory address (byte swapping).

When a value is popped from the top of the stack, the reverse sequence occurs. The physical memory address, defined by the stack segment register and the stack pointer register, points to the location of the last value pushed onto the stack. The contents of memory are first popped off the stack and put into the register defined by the POP instruction. The stack pointer is then automatically decremented by two. Figure 9.11 illustrates the operation of the POP AX instruction.

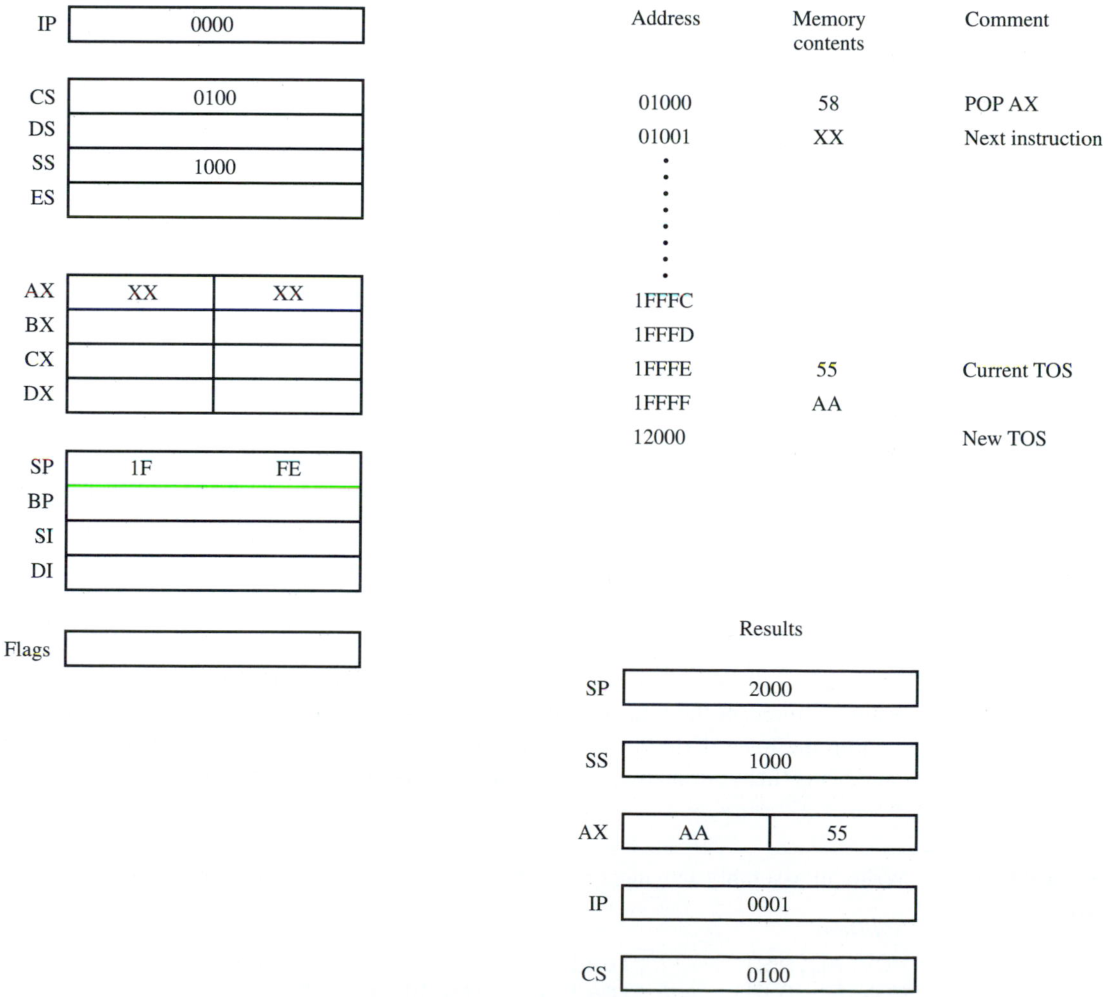

IP	0000	
CS	0100	
DS		
SS	1000	
ES		
AX	XX	XX
BX		
CX		
DX		
SP	1F	FE
BP		
SI		
DI		
Flags		

Address	Memory contents	Comment
01000	58	POP AX
01001	XX	Next instruction
⋮		
1FFFC		
1FFFD		
1FFFE	55	Current TOS
1FFFF	AA	
12000		New TOS

Results

SP	2000	
SS	1000	
AX	AA	55
IP	0001	
CS	0100	

FIGURE 9.11
Popping off the stack

Arithmetic Instructions

The arithmetic instructions cover the four basic mathematical operations of addition, subtraction, multiplication, and division.

ADD/SUB The general formats of the add and subtract instructions are

ADD destination, source
SUB destination, source

Table 9.3 summarizes the various types of destination and source operands that may be used with the addition and subtraction instruction.

TABLE 9.3
ADD/SUB instruction formats

Destination	Source
Register	Register
Register	Memory
Memory	Register
Memory	Immediate (constant)
Register	Immediate (constant)

In all cases the result of an addition or subtraction will be found in the destination operand. For example, in the instruction

```
ADD AX,BX ;add BX to AX
```

the contents of the BX register are added to the contents of the AX register. The results are stored in the AX register.

In the instruction

```
SUB CL;AL ;subtract AL from CL
```

the contents of the AL register are subtracted from the contents of the CL register. The result is stored in the CL, or destination, register. Note that in this example we are subtracting the lower byte of the accumulator register from the lower byte of the count register. Thus there are many possible variations of the ADD and SUB instructions.

EXAMPLE 9.1

Write an assembly language program to add 5H plus 3H using the AL and BL registers.

Solution

```
MOV AL,05H      ;load 5H into AL register
MOV BL,03H      ;load 3H into BL register
ADD AL,BL       ;add BL to AL, result in AL = 8H
MOV 100H,AL     ;move result from AL into memory
                ;location 100H
```

EXAMPLE 9.2

Write an assembly language program to solve

$$ANSWER = 5H + 3H - 2H$$

Solution

```
MOV AL,05H        ;load 5H into AL register
MOV BL,03H        ;load 3H into BL register
ADD AL,BL         ;add BL to AL, results in AL = 8H
MOV BL, 02H       ;move 2H into BL register
SUB AL,BL         ;subtract BL from AL
                  ;results in AL = 6H
MOV ANSWER,AL     ;move results to memory address
                  ;labeled ANSWER
```

Note that in the solutions to the previous examples the assembler directives have been omitted. This has been done for clarity. Complete programs that include assembler directives are shown later in this chapter.

MUL/DIV The general formats of the multiply (MUL) and divide (DIV) instructions are

MUL source multiplier
DIV source divisor

When the multiply command is used, the multiplicand must be moved into the AX or AL register. The source multiplier can be moved into any other register or memory location. Examples of multiply instructions are

```
MUL BX          ; multiplier in BX register
MUL MEM1        ; multiplier in memory address
                ; labeled MEM1
```

When two bytes are multiplied together, the result, or product, is stored in the AX register. Note that in Figure 9.12 the multiplication of the two single-byte numbers results in a 16-bit, or 1-word, product in the AX register.

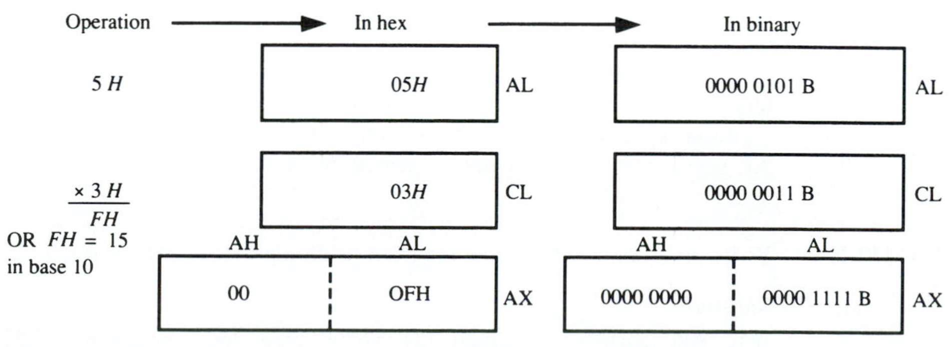

FIGURE 9.12
Using the MUL instruction

EXAMPLE 9.3

Write an assembly language program to multiply 5H by 3H using the CL register.

Solution

```
MOV AL,05H      ; move multiplicand 5H
                ; into the AL register
MOV CL,03H      ; move multiplier 3H
                ; into the CL register
MUL CL          ; multiply 3H by 5H
                ; store product in AX register
MOV MEM1,AL     ; move product (0FH) from AL register
                ; into memory location labeled
                ; MEM1
```

When two 16-bit words are multiplied, the multiplicand must be moved into the AX register. The multiplier can be in any other register or 16-bit memory location. The results or product will be a 32-bit, or *double-word,* number and will be found in the DX and AX registers. The higher-order word will be in the DX register, and the lower-order word will be in the AX register.

EXAMPLE 9.4

Write an assembly language program to multiply 3A62H by 2B14H.

Solution

```
MOV AX,3A62H   ; move multiplicand 3A62H into AX register
MOV CX,0B214H  ; move multiplier B214H into CX register
MUL CX         ; multiply AX by CX, product = 289C63A8H
MOV MEM1,AX    ; move low-order word (63A8) into MEM1
MOV MEM2,DX    ; move high-order word (289C) into MEM2
```

Note again the leading zero in the second move instruction.

```
MOV CX,0B214H
```

This leading zero is used to tell the assembler that *B*214H is a number and not a label. Recall that labels begin with a letter and therefore can be confused with some hexadecimal numbers.

The divide command is fundamentally similar to the multiply command. In byte number division, the divisor is a byte that can be a register or memory location. The dividend is a word located in the AX register. The results or quotient will be found in the AL register with the remainder located in the AH register as shown in Figure 9.13.

EXAMPLE 9.5

Write an assembly language program to divide 6H by 3H using the CL register.

Solution

```
MOV AX,0006H  ; move 6H into the AX register (dividend)
MOV CL,03H    ; move 3H into the CL register (divisor)
DIV CL        ; divide AX by CL
```

$$\frac{6H}{3H} = 2H + 0 \text{ (Remainder)}$$

$$\frac{\text{Dividend}}{\text{Divisor}} = \text{Quotient} + \text{Remainder}$$

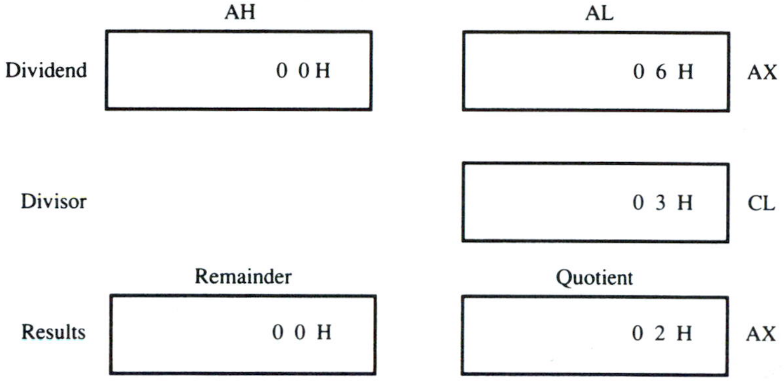

	AH	AL	
Dividend	0 0 H	0 6 H	AX
Divisor		0 3 H	CL
	Remainder	Quotient	
Results	0 0 H	0 2 H	AX

FIGURE 9.13
Divide operation

```
                ; quotient = 02H remainder = 00H
MOV MEM1,AL     ; move quotient into MEM1 (02H)
MOV MEM2,AH     ; move remainder into MEM2 (00H)
```

Note that 6H was entered as 0006H in order to fill the entire AX register. This was done to clear the high-order byte of the AX register of any erroneous data.

In word number division, the divisor is a word that can be a register or memory location. The dividend is a double word that is located in the DX and AX registers. The DX register will hold the high-order word, and the AX register will hold the low-order word. The result or quotient will be found in the AX register with the remainder located in the DX register.

EXAMPLE 9.6

Write an assembly language program to divide 1A034H by 1002H using the BX register.

Solution

```
MOV AX,0A034H    ; move low-order word
                 ; into AX register (dividend)
MOV DX,0001H     ; move high-order word
                 ; into DX register (dividend)
MOV BX,1002H     ; move divisor into BX register
DIV BX           ; divide DX AX by BX
MOV MEM1,AX      ; move quotient (1AH) from AX to MEM1
MOV MEM2,DX      ; move remainder (00H) from DX to MEM2
```

INC/DEC The increment (INC) command adds one to the operand. The decrement (DEC) command subtracts one from the operand. These commands are very useful for counting operations. The general format of the INC and DEC commands is

INC source
DEC source

The source operand may be a register or a memory address. The source may be a 16-bit word or an 8-bit byte—for example,

```
INC AX      ; add one to AX register
INC AL      ; add one to AL register
DEC CX      ; subtract one from CX register
DEC CL      ; subtract one from CL register
DEC MEM1    ; subtract one from MEM1
```

As with many instructions, the increment and decrement instructions affect status bits in the flag register. For example, consider the case below where the AL register, which initially contains 05, is continually decremented. The zero flag bit and the sign flag bit are initially reset (0). This is because the AL register is not zero or negative. By decrementing the AL register a number of times, both the zero flag and the sign flag will eventually be set (1), although not at the same time.

Instruction	AL Results	ZF	SF
DEC AL	04	0	0
DEC AL	03	0	0
DEC AL	02	0	0
DEC AL	01	0	0
DEC AL	00	1	0
DEC AL	*FF*	0	1

Note that *FF* is actually −1 in 2s complement form.

EXAMPLE 9.7 Write an assembly language program to add 1 + 2 + 3 + 4 using the increment operand.

Solution

```
MOV AL,01H     ; move 1H into AL register
MOV BL,02H     ; move 2H into BL register
ADD AL,BL      ; add BL to AL (01H + 02H = 03H)
INC BL         ; add one to BL to obtain 03H
ADD AL,BL      ; add BL to AL (03H + 03H = 06H)
INC BL         ; add one to BL to obtain 04H
ADD AL,BL      ; add BL to AL (06H + 04H = 0AH)
MOV MEM1,AL    ; save sum in MEM1
```

Logical Instructions

Logical instructions include the Boolean operations NOT, AND, and OR. The NOT instruction inverts all the bits in a word or byte operand. The AND/OR instructions perform the Boolean AND/OR operations on each pair of bits in the source and destination operands. These instructions may be used with word or byte operands.

NOT The general format of the NOT instruction is

NOT source

where the source operand may be a 16-bit word or an 8-bit byte. The source operand may be a register or a memory location: for example,

```
NOT AX
NOT BL
NOT MEM1
```

To illustrate the use of the NOT instruction, consider the following program:

```
MOV BL,00110011B        ; move binary number
                        ; into BL register
NOT BL                  ; negate BL
MOV MEM1,BL             ; save results in MEM1
```

The contents of the BL register were originally 00110011B. After the NOT operations are performed, the contents of the BL register are 11001100B.

$$BL = 00110011B$$
$$\overline{BL} = 11001100B$$

AND/OR The general format of the AND/OR instruction is

AND destination, source
OR destination, source

AND/OR performs the Boolean operation between the source and destination operands. The results are stored in the destination operand. For example, in the instruction

```
AND AL,BL
```

the contents of the BL register are *ANDed* with the contents of the AL register. The results are stored in the AX register. If the number in the AL register is 00001101B and the number in the BL register is 00110011B, the result in the AL register after the AND operation is performed is

	AL	00001101B
	BL	00110011B
Results	AL	00000001B

Table 9.4 summarizes the various types of destination and source operands that may be used with AND and OR instructions.

TABLE 9.4
AND/OR instruction formats

Destination	Source
Register	Register
Register	Memory
Memory	Register
Memory	Immediate (constant)
Register	Immediate (constant)

EXAMPLE 9.8

Write an assembly language program to implement the logic function of Figure 9.14 sixteen times.

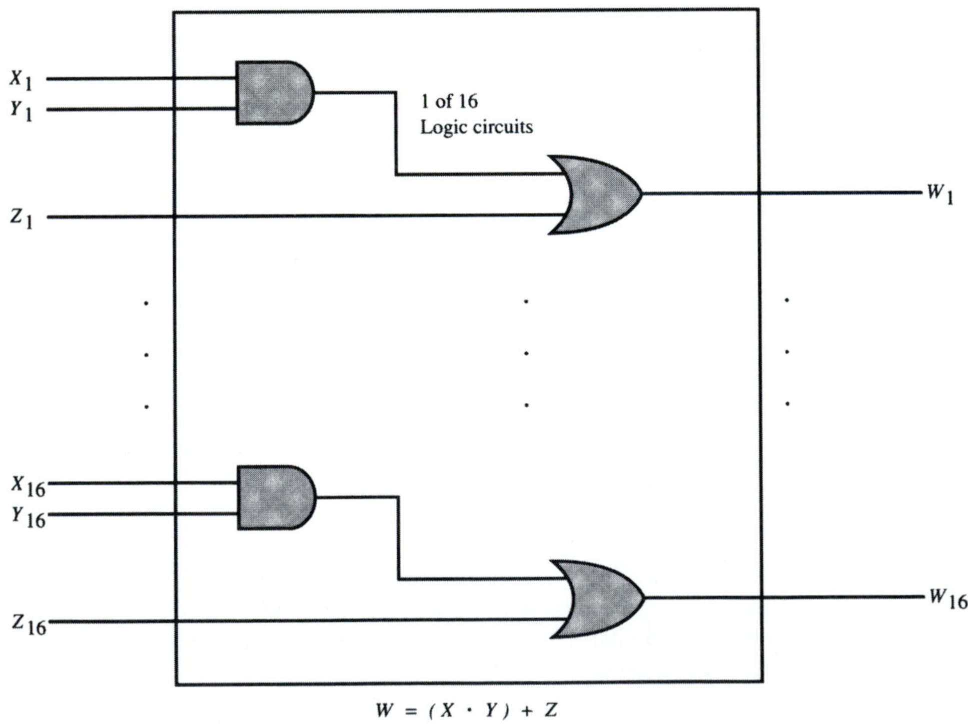

FIGURE 9.14
Logic diagram

Solution

```
MOV AX,X       ; move X into AX
MOV BX,Y       ; move Y into BX
MOV CX,Z       ; move Z into CX
AND AX,BX      ; (X · Y)
OR AX,CX       ; (X · Y) + Z
MOV W,AX       ; store results in W = (X · Y) + Z
```

Note that the label variables X, Y, and Z can represent 16 bits of information. Thus the program can perform the work of many logic AND and OR gates.

A technician wishes to use the program in Example 9.8 to solve for the output of W_3 only. By adding one additional AND instruction, the technician can set all the outputs except W_3 to zero. This process is known as **masking off bits** and is as follows:

```
MOV AX,X                   ; move X into AX
MOV BX,Y                   ; move Y into BX
MOV CX,Z                   ; move Z into CX
AND AX,BX                  ; (X · Y)
OR AX,CX                   ; (X · Y) + Z
AND AX,0000000000000100B ; mask off bits
MOV W3,AX
```

The results of the OR operation found in the AX register are:

	1010 1111 1001 1100B
Masked with (AND)	0000 0000 0000 0100B
Results in (W)	0000 0000 0000 0100B

Program-Control Instructions

Up to this point we have examined the instructions that perform a specific task once. From this we might conclude that if we need to perform a specific operation more than once in a program, we must duplicate the entire sequence of instructions each time we need to perform the operation. Duplicating a sequence of instructions many times in a program would be frustrating and time-consuming. Program-control instructions help to eliminate this duplication.

JMP When you read a set of instructions and reach a direction like "JUMP to STEP 10," you have come upon a jump instruction. The jump (JMP) instruction makes the MPU take its next instruction from someplace other than the next consecutive memory location. Jump instructions can be *unconditional* or *conditional*. An unconditional jump instruction is one that is always taken. Whenever it occurs in a program, it is executed immediately. A conditional jump requires the MPU to make a decision based on the contents of the flag register. The general format of the jump instruction is

JMP destination

The destination can be a label, memory address, or 16-bit register. A direct jump might be a label. An indirect jump is to a register or memory location that *points* to the target destination. For example, consider the following:

```
         MOV BX,0001H    ; load BX with 1H
         SUB AX,AX       ; initialize AX to zero
REPEAT:  ADD AX,BX       ; add BX to AX
         JMP REPEAT      ; count by repeated addition
```

In this program the AX register will be incremented by 1 continuously until the computer is turned off. There is no condition to stop the program.

Conditional jump instructions can be used to control the program. They allow us to set a condition that causes the MPU to make a decision on whether to jump or not. For example, in the following program the JNZ (jump not zero) instruction is used to control the number of times a mathematical operation is performed

```
         SUB AX,AX         ; initialize AX to zero
         MOV BX,0007H      ; load 7H in BX
         MOV CX,0003H      ; load 3H in CX
MULT:    ADD AX,BX         ; add BX to AX
         DEC CX            ; decrement CX by 1
         JNZ MULT          ; jump if not zero to MULT
         MOV PROD,AX       ; store results in PROD
```

This program can be used to do multiplication by repeated addition. We begin by initializing the AX register to zero. We could have used the instruction

```
MOV AX,0000H
```

to accomplish the same operation. Next we load the numbers we wish to multiply together into separate registers. The

```
ADD AX,BX
```

instruction will be performed until the contents of the CX register is zero. Each time the ADD instruction is performed, we will decrement the CX register by one (DEC CX) and test to see if the result is zero (JNZ). The JNZ instruction looks at the zero flag (ZF) in the flag register. If the ZF flag is reset (ZF = 0), we jump to the label MULT. If the ZF flat is set (ZF = 1), we skip to the next instruction, which in this case saves the answer. The zero flag will be set by the MPU anytime that an arithmetic operation results in a zero.

There are many types and variations of conditional and unconditional jump instructions. A complete list is located in Appendix B.

LOOP The loop instruction (LOOP) is a special form of the jump instruction with a built-in count capability. Like the jump instruction, it is used to repeat an operation or a sequence of instructions. This is commonly referred to as **looping.** Loop instructions can be conditional or unconditional and are always used with the CX register. The general format of the loop instruction is

LOOP destination label

The LOOP instruction uses the CX or count register to determine the number of times the loop is to be performed. When the LOOP instruction is used, a value, or number, is first placed in the CX register. Every time the LOOP instruction is executed, the CX register is automatically decremented by 1. Next, the CX register is tested to see if it is zero. If it is zero, the loop is terminated and the next instruction performed. If it is not zero, the program will loop to the destination label.

There are also conditional LOOP instructions that use the count register and the flag register to determine when to loop. For example,

```
LOOPZ - loop while zero (ZF = 1 CX = 0)
LOOPNZ - loop while not zero (ZF = 0 CX = 0)
```

To illustrate the use of the LOOP instruction we will rewrite the previous program example (multiplication by repeated addition) using the LOOP instruction.

```
        SUB AX,AX       ; initialize AX to zero
        MOV BX,0007H    ; load 7H in BX
        MOV CX,0003H    ; load 3H in CX
MULT:   ADD AX,BX       ; add BX to AX
        LOOP MULT       ; repeat addition
        MOV PROD,AX     ; store results in PROD
```

9.4 PROGRAM STRUCTURE

As programs get longer and more complex, the need for organization becomes more important. One method of organization is to develop an outline. The outline can be used to break the problem down into smaller parts. These parts must be ordered or sequenced to

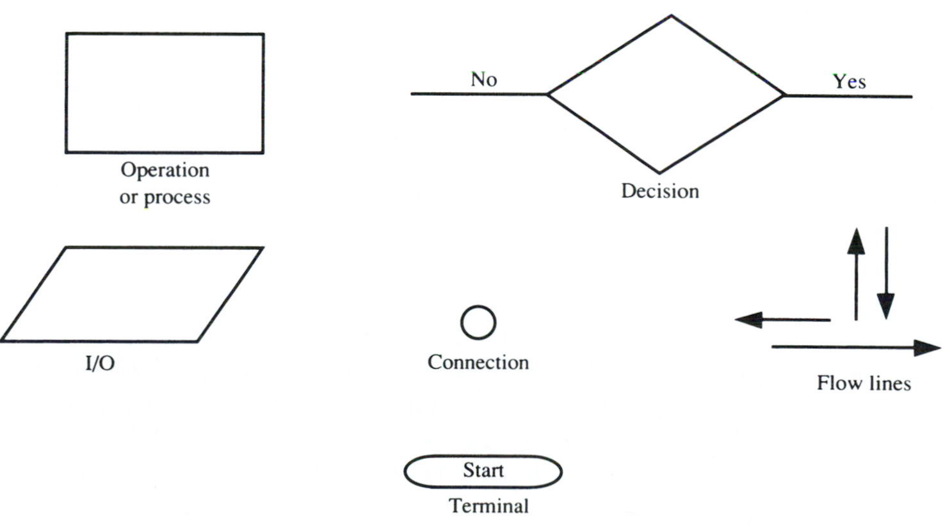

FIGURE 9.15
Flowchart symbols

allow the problem to be solved correctly. A method of outlining or mapping a problem is known as **flowcharting.** The flowchart is a block diagram that uses standard symbols to represent operations. Figure 9.15 shows some of the basic flowcharting symbols.

Let's create a flowchart and an assembly language program to control an alarm in a truck weighing station. If a truck exceeds a certain weight, we wish to sound an alarm. If a truck is below the weight limit, we keep the alarm off and allow the truck to pass. Figure 9.16 describes a solution to this problem using a flowchart.

We begin by defining the data variables. The variable DATA_IN will be used to read in the truck weight. Note the use of the "__" in the variable name. This is done to separate the words DATA and IN without using a space. Remember that spaces are illegal in labels and variables. Programmers often use this technique when defining variables and label names for the purpose of clarity. The variable DATA_OUT will be used to control the alarm. An output code of 00H will keep the alarm off, and an output code of *FF*H will turn the alarm on. The output code could be sent to a digital-to-analog converter circuit to control an audible alarm. Finally, the variable MAX_WT will be used to define the maximum allowable truck weight, which has been set equal to 10 tons or 0*A*H in this problem.

The next box in our flowchart initializes the alarm output to be off. We then read in the weight of a truck (DATA_IN). If the truck's weight is *less than or equal to* the maximum allowable weight, we hold the alarm off and go on to read the next truck's weight. If the truck's weight *exceeds* the maximum allowable weight, we sound the alarm by outputting the *FF*H code. The program then returns to read in the next truck's weight. If the truck's weight is less than or equal to the maximum allowable weight, the alarm will be turned off. Otherwise, the alarm will continue to sound.

Figure 9.17 (p. 308) illustrates an assembly language program that can be used to implement the solution to our truck weighing station problem. Note the use of the program segments to separate the data portion from the code portion of the program. Furthermore, in this example we have presented the use of assembler directives to illustrate their use in an assembly language program. The following points should be noted:

1. The program begins with a comment, which is used as a title for this program.
2. The use of the semicolon followed by the star border is purely aesthetic and is used for highlighting portions of the program.
3. The assembler directive DATA SEGMENT defines the start of the data segment.
4. The assembler directive ORG 0100H starts the data segment at logical address 0100H.
5. The variables DATA_IN, DATA_OUT, and MAX_WT are as described before.
6. The assembler directive DATA ENDS signifies the end of the data segment.
7. The assembler directive CODE SEGMENT defines the start of the code segment.
8. The assembler directive ASSUME CS:CODE, DS:DATA is used to tell the assembler that the program will have two segments (CODE and DATA).
9. The assembler directive ORG 0200H starts the code segment at logical address 0200H.
10. The code segment instructions were also described above.
11. The assembler directive CODE ENDS signifies the end of the code segment.
12. The assembler directive END signifies the end of the program.

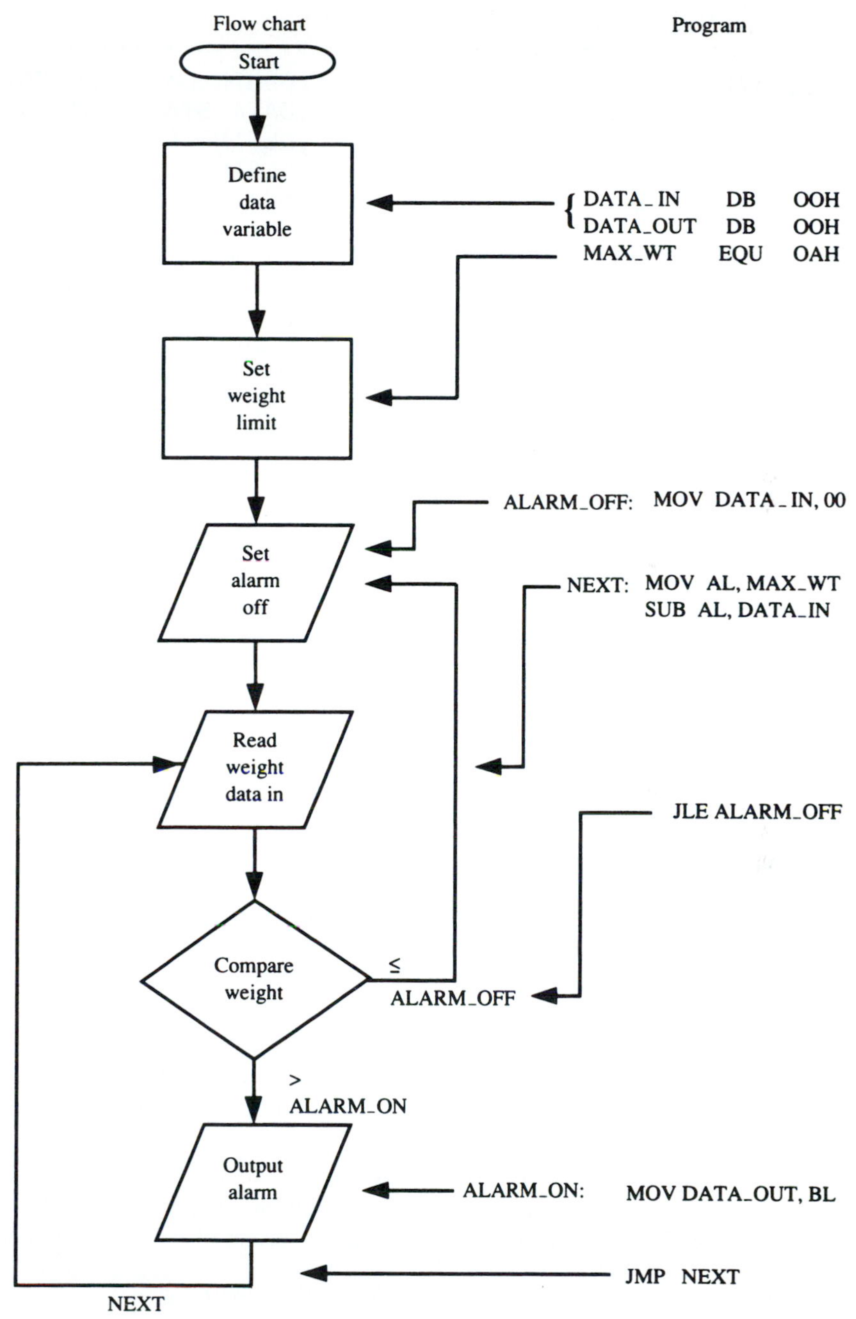

FIGURE 9.16
Flowchart—Weight station program

```
;WEIGHT STATION ALARM PROGRAM
;*************************************************************
DATA SEGMENT                                ;BEGIN DATA SEGMENT
                ORG 0100H                   ;DATA STARTS AT 0100H
DATA_IN         DB  00H                     ;DEFINE & INITIALIZE TO 00H
DATA_OUT        DB  00H                     ;DEFINE & INITIALIZE TO 00H
MAX_WT          EQU 0AH                     ;SET MAXIMUM WEIGHT LIMIT
DATA ENDS                                   ;END OF DATA SEGMENT
;*************************************************************
CODE SEGMENT                                ;BEGIN CODE SEGMENT
ASSUME CS:CODE,DS:DATA                      ;DEFINE SEGMENTS
                ORG 0200H                   ;CODE STARTS AT 0200H
ALARM_OFF:      MOV DATA_OUT,00H            ;TURN ALARM OFF, OUTPUT 00H
NEXT:           MOV AL,MAX_WT               ;LOAD AL WITH MAX WGT LIMIT
                SUB AL,DATA_IN              ;COMPARE TRUCK WEIGHT TO
                                            ;MAXIMUM WEIGHT LIMIT
                JLE ALARM_OFF               ;JUMP TO ALARM_OFF IF
                                            ;DATA_IN < MAX_WT
ALARM_ON:       MOV DATA_OUT,0FFH           ;TURN ALARM ON, OUTPUT 0FFH
                JMP NEXT                    ;READ NEXT TRUCK WEIGHT
CODE ENDS                                   ;END OF CODE SEGMENT
END                                         ;END OF PROGRAM
```

FIGURE 9.17
Weight station alarm program

9.5 SUBROUTINES

A **subroutine** is a portion of a program that is used to perform a particular task. Usually subroutines are used to perform a task that needs to be repeated many times in the program. For example, a program that computes the square root of a number can be used as a subroutine in another program. Any time we wish to calculate the square root of a number we can call upon our subroutine without having to rewrite the instructions each time, which is the advantage of a subroutine. This also saves memory space. Thus if we need to calculate the hypotenuse of a right triangle 1000 times, the square root subroutine will be very helpful.

The instruction that allows us to go to a subroutine is the CALL instruction. The instruction that allows us to return to the main program once the subroutine is completed is the RET (return) instruction. The general format of the CALL and RET instruction is

```
CALL    destination
RET
```

The CALL destination can be a label, an address, or a 16-bit register. The RET instruction usually does not have an operand.

When the microprocessor executes a CALL instruction, the address of the next main program instruction must be saved. If this was not done, the MPU would not know where to return in the main program. This is performed in coordination with the stack. The CALL instruction pushes the current contents of the instruction pointer onto the stack. Next the MPU puts the memory address of the subroutine into the instruction pointer. Thus the stack is used to save the return address of the main program, and the instruction pointer is used to point to the beginning address of the subroutine. When the subroutine is completed, the return address is popped back out of the stack and into the instruction pointer, and the program is continued. To summarize:

1. The CALL instruction is read and decoded by the MPU.
2. The instruction pointer is now incremented to point to the *next* instruction in the main program.
3. The contents of the instruction pointer are now pushed onto the stack.
4. The instruction pointer is now loaded with the starting address of the subroutine.
5. The subroutine is executed.
6. The RET instruction, which must be the last instruction in the subroutine, pops the return address of the main program from the stack.
7. The return address is loaded into the instruction pointer, and the main program continues.

To demonstrate the use of the subroutine, let us now consider the previous weight station alarm program. Suppose our weight station actually has three scales. We wish to use our alarm program to **poll,** or check, the weight of each scale. Due to the speed of the microprocessor we can read the weight of the first scale and check to see if we should sound the alarm. Next, we read the weight of the second scale and check to see if we should sound the alarm. Finally, we read the weight of the third scale and again determine if we should sound the alarm. The entire process is repeated over and over again. Figure 9.18 illustrates the flowchart that will be used to create the assembly language program. Note that the original weight station alarm program has been modified to be used as a subroutine. Figure 9.19 (p. 311) shows an assembly language program that can be used to solve this problem.

Let's analyze the program in detail.

1. Since we will be using the CALL instruction, a stack will be required to save the contents of the instruction pointer when we jump into our subroutine. The stack segment instruction DB 20 DUP (?) establishes the stack area in memory.
2. The data segment defines the variables that are used in the program. Furthermore, it should be noted that the DATA_IN1, DATA_IN2, and DATA_IN3 are memory-mapped I/O addresses, as shown in Figure 9.20 (p. 312), that connect the scales directly to memory addresses 0100H, 0101H, and 0102H, respectively.
3. DATA_OUT is also memory mapped and connects the alarm directly to memory address 0103H.
4. The instruction MAX_WT EQU 0AH is used to set the maximum allowable truck weight equal to 10 tons.
5. The code segment of the program begins at 0200H (ORG 0200H).
6. The instruction MOV BL,DATA_IN1 moves the truck weight at scale 1 into the BL register.
7. The next instruction, CALL WEIGHT, causes the program to jump to the subroutine labeled WEIGHT.

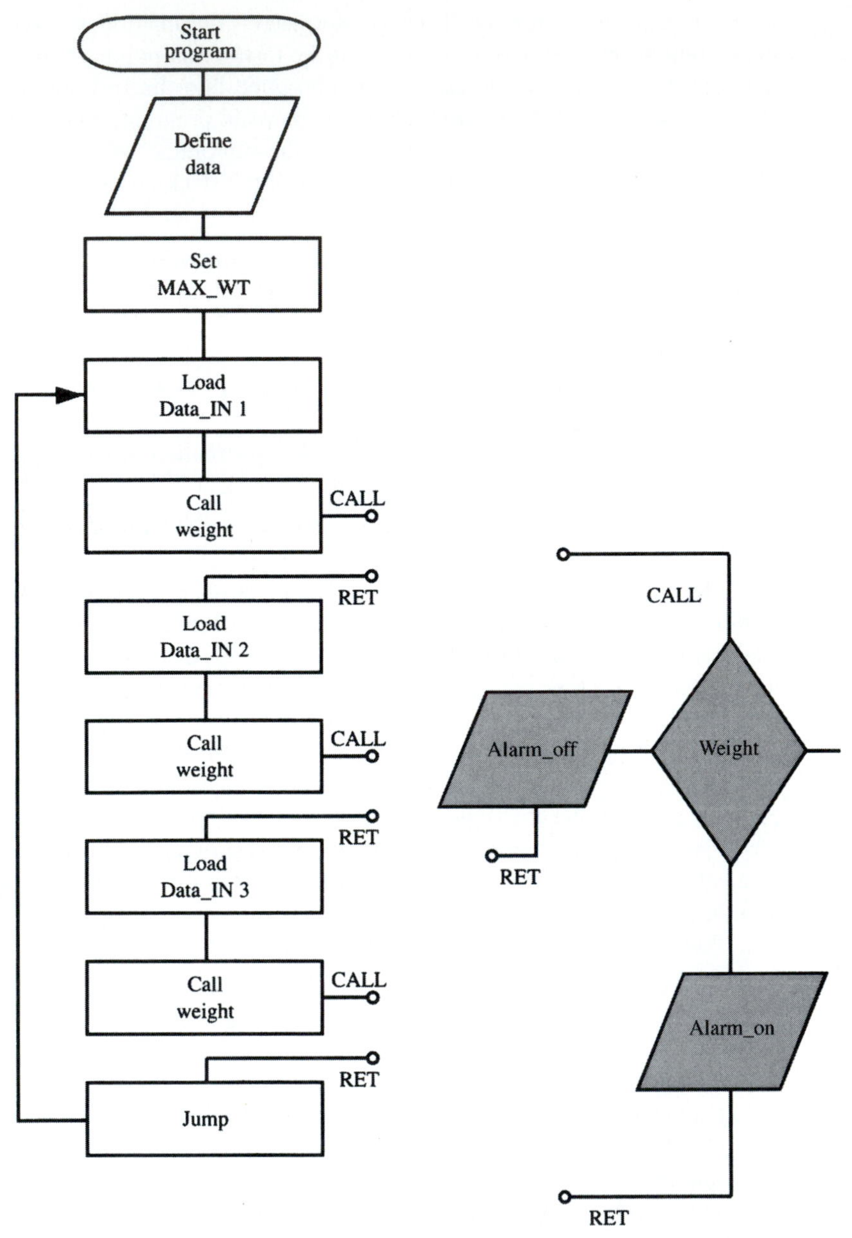

FIGURE 9.18
Flowchart—Three-scale weight station program

```
;THREE SCALE WEIGHT STATION ALARM PROGRAM
;*************************************************************
STACK SEGMENT STACK                          ;BEGIN STACK SEGMENT
              DB   20 DUP (?)                ;STACK SIZE = 20 BYTES
STACK ENDS                                   ;END OF STACK SEGMENT
;*************************************************************
DATA SEGMENT                                 ;BEGIN DATA SEGMENT
              ORG   0100H                    ;DATA STARTS AT 0100M
DATA_IN1      DB    00H                      ;DEFINE & INITIALIZE TO 00H
DATA_IN2      DB    00H                      ;DEFINE & INITIALIZE TO 00H
DATA_IN3      DB    00H                      ;DEFINE & INITIALIZE TO 00H
DATA_OUT      DB    00H                      ;DEFINE & INITIALIZE TO 00H
MAX_WT        EQU   0AH                      ;SET MAXIMUM WEIGHT LIMIT
DATA ENDS                                    ;END OF DATA SEGMENT
;*************************************************************
CODE SEGMENT                                 ;BEGIN CODE SEGMENT
ASSUME CS:CODE,DS:DATA,SS:STACK              ;DEFINE SEGMENTS
              ORG  0200H                     ;CODE STARTS AT 0200H
START:        MOV  BL,DATA_IN1               ;READ SCALE 1
              CALL WEIGHT                    ;WEIGHT ALARM SUBROUTINE
              MOV  BL,DATA_IN2               ;READ SCALE 2
              CALL WEIGHT                    ;WEIGHT ALARM SUBROUTINE
              MOV  BL,DATA_IN3               ;READ SCALE 3
              CALL WEIGHT                    ;WEIGHT ALARM SUBROUTINE
              JMP  START                     ;RESTART PROCESS
WEIGHT        PROC NEAR                      ;BEGIN SUBROUTINE
ALARM_OFF:    MOV  DATA_OUT,00H              ;TURN ALARM OFF, OUTPUT 00H
              MOV  AL,MAX_WT                 ;LOAD AL WITH MAX WGT LIMIT
              SUB  AL,BL                     ;COMPARE TRUCK WEIGHT TO
                                             ;MAXIMUM WEIGHT LIMIT
              JLE  RETURN                    ;JUMP TO RETURN IF
                                             ;DATA_IN < MAX_WT
ALARM_ON:     MOV   DATA_OUT,0FFH            ;TURN ALARM ON, OUTPUT 0FFH
RETURN:       RET                            ;READ NEXT TRUCK WEIGHT
WEIGHT        ENDP                           ;END OF WEIGHT PROCEDURE
                                             ;SUBROUTINE
CODE ENDS                                    ;END OF CODE SEGMENT
END                                          ;END OF PROGRAM
```

FIGURE 9.19
Three-scale weight station alarm program

8. The instruction WEIGHT PROC NEAR defines the beginning of a subroutine procedure within the code segment labeled WEIGHT.
9. The program continues as previously described to determine whether or not to sound the alarm until the RET instruction is reached.

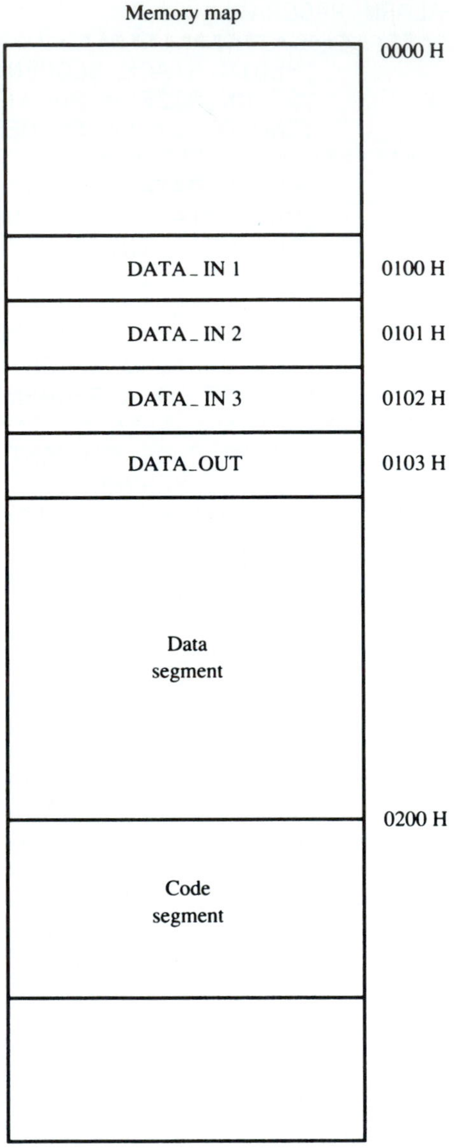

FIGURE 9.20
Memory map

10. The ENDP directive tells the assembler that the procedure is complete.
11. RET returns us to the main program, causing the second scale to be read (DATA_IN2).
12. This process repeats itself for DATA_IN3 (the third scale).
13. The JMP START instruction restarts the entire process.

9.6 TECH TIPS AND TROUBLESHOOTING—T[3]

DEBUG is an essential *utility* program for assembly language programs. DEBUG comes with IBM PC DOS and allows us to examine internal registers to determine what is going on in a program. Rarely will programs work perfectly the first time they are executed. DEBUG will help clarify what is going on because:

1. It allows you to set break points and halt the program's execution.
2. It allows you to single step, or trace through the program one instruction at a time.

To use DEBUG, type the following at the DOS prompt:

```
A>DEBUG <ENTER>
```

The computer responds

```
A>DEBUG
-_
```

Now type in the following program

```
-A 0100 <ENTER>
101D:0100 MOV AL,22 <ENTER>
101D:0102 MOV BL,05 <ENTER>
101D:0104 ADD AL,BL <ENTER>
101D:0106 INT 20 <ENTER>
101D:0108 <ENTER>
```

When you are finished entering the program, press the ENTER key. Note that DEBUG expects everything to be in hex. It will not accept the H at the end of a number.

This program will simply add the two numbers 22H and 5H. The INT 20 command is a PC DOS command that is used to terminate a program. It is similar to END. The A 0100 command tells DEBUG to assemble the program starting at memory address 0100. Before executing this program, type U0100 0106 <ENTER> to list the program. The computer will respond with

```
-U0100 0106
101D:0100 B022    MOV    AL,22
101D:0102 B305    MOV    BL,05
101D:0104 00D8    ADD    AL,BL
101D:0106 CD20    INT    20
-
```

If you had just typed the command U, the computer would have listed the next 15 lines following the current position of the instruction pointer. Now type R <ENTER> to examine the registers. The computer will respond with

```
-R
AX=0000   BX=0000 CX=0000 DX=0000 SP=FFEE BR=0000 SI=0000 DI=0000
DS=101D   ES=101D SS=101D CS=101D IP=0100 NV UP EI PL NZ NA PO NC
101D:0100 B022    MOV AL,22
```

Next type T <ENTER> to trace through the program one instruction at a time. The computer will respond with

```
-T
AX=0022   BX=0000 CX=0000 DX=0000 SP=FFEE BP=0000 SI=0000 DI=0000
DS=101D   ES=101D SS=101D CS=101D IP=0102 NV UP EI PL NZ NA PO NC
101D:0102 B305    MOV BL,05
```

Next type T <ENTER> two more times, and the computer will respond with

```
-T
AX=0022   BX=0005 CX=0000 DX=0000 SP=FEEE BP=0000 SI=0000 DI=0000
DS=101D   ES=101D SS=101D CS=101D IP=0104 NV UP EI PL NZ NA RO NC
101D:0104 00D8    ADD AL,BL
-T

AX=0027   BX=0005 CX=0000 DX=0000 SP=FFEE BP=0000 SI=0000 DI=0000
DS=101D   ES=101D SS=101D CS=101D IP=0106 NV UP EI PL NZ NA PE NC
101D:0106 CD20    INT 20

-
```

Note the contents of the registers each time you type the T command. Furthermore, notice that the instruction pointer is updated each time to point to the next instruction to be executed. Typing T one more time will execute the INT 20 instruction. If we do this, we will have to step through many lines of instructions that reside in DOS and perform the INT 20 command. This command terminates the program and returns us to DOS. Instead, we will type G <ENTER> for GO to execute the program. The computer responds

```
G
Program terminated normally
-
```

You can now examine the registers again to see the results.

DEBUG has many other commands, features, and uses. For example, you can examine and change the contents of any register by typing R followed by the register name (RAX). You can enter or change the data in any location using the E or ENTER command (E0101 55). To exit DEBUG type Q for QUIT. You will find that DEBUG is a valuable aid in troubleshooting your programs. You may want to consult the IBM DOS manual or some other text to learn more about DEBUG.

EXERCISES

9.1 Give an example of a high-level computer programming language.

9.2 High-level programming languages use __________ to translate instructions into machine language.

9.3 List the procedures used to develop an assembly language program.

9.4 The way in which the location of an operand is determined is called the __________.

9.5 Explain the meaning of MOV CX,DX.

9.6 List the three general types of addressing modes.

9.7 Explain the meaning of MOV AX,[BX].

9.8 What is the EQU directive used for?

9.9 What is the general format of a data-transfer instruction?

9.10 Give an example of a data-transfer instruction using
a. Register addressing mode b. Memory addressing mode
c. Immediate addressing mode

9.11 List four data-transfer instructions and give an example for each.

9.12 Give an example of an arithmetic instruction using
a. Register addressing mode b. Memory addressing mode
c. Immediate addressing mode

9.13 List four different types of arithmetic instructions and give an example for each.

9.14 What is the difference between an assembler directive and an instruction?

9.15 What will the results of the instruction NOT AX, be if the AX register contains FFFFH?

9.16 Define the term *masking.*

9.17 What is the general format of the OR instruction? Give an example of its use.

9.18 What is the difference between a conditional and an unconditional program control instruction? Give two examples of each.

9.19 What is the difference between a JUMP and a LOOP instruction? Give an example of each.

9.20 Define a subroutine.

9.21 Write a program to add 4H to 3H using the AX and BX registers. Load the results into a memory address labeled SUM.

9.22 Write a program to compute the equation $Z = (2X + 5Y)/W$ where $X = 4$, $Y = 3$, and $W = 2$.

9.23 Write a program to compute $Z = (A \cdot B) + \overline{C}$ using MEMA for the value of A, MEMB for the value of B, and MEMC for the value of C. Store the results of Z in the memory location labeled STORE.

9.24 Write a program using a subroutine to compute the value of X, $2X$, $3X$, and X^2 for integer values of X between 1 and 10. Draw a flowchart for the program and explain what the program is doing by using comments. Be sure to include required assembler directives.

9.25 Write a program to sound an alarm when the temperature of a refrigerator rises above 50°F. Draw a flowchart for the program and explain what the program is doing by using comments. Be sure to include the required assembler directives.

9.26 Crossword Puzzle

ACROSS

3. Translates into machine language.
5. A special form of the jump instruction with a built-in count capability.
7. Mnemonic field.
8. A logical instruction used to complement a binary number.
9. An instruction used to tell the MPU to copy the contents of a register or memory location.
12. A place where information is stored.
13. Also called a machine language program.
17. MOV.
18. Define byte directive.
19. A language that uses English-like words for commands.
21. Arithmetic instruction.
23. Immediate, register, and memory __________ modes.
24. The field used to contain a memory address or a numeric value used with the directive.
25. A procedure contained in one segment.
29. When a register is used to indicate the address where the data can be found.
30. MASM.
32. The last instruction in a procedure.
33. Creates an executable program.

DOWN

1. A type of move instruction used to load an effective address.
2. The instruction to remove data from the stack.
4. What the computer directly understands.
6. A list of instructions that tells the MPU what to do.
8. Field similar to label field.
10. A constant source value.
11. Also called an assembly language program.
14. A field used to describe an instruction in a program.
15. Tells MPU where the data are found.
16. Repeatedly checking status.
20. Field assigned symbolic name.
22. Assembler instruction.
26. Subroutine instruction.
27. Intersegment procedure instruction.
28. Procedure instruction.
29. Adds one to the operand.
31. The last instruction.

1 2 3 4 5 6 7 8 9 10 11 12 13 14 15 16 17 18 19 20 21 22 23 24 25 26 27 28 29 30 31 32 33

CHAPTER 10

CREATING AND DEBUGGING PROGRAMS

KEY TERMS

A86	*Footer*
Assembler	*Header*
DEBUG	*Linker*
Dump	*MASM*
EDIT	*NotePad*
Editor	*Relocatable*
EDLIN	*Trace*

10.0 INTRODUCTION

The assembler converts assembly language instructions into machine language instructions. Recall that it reads the source code instructions and outputs object code instructions. The final step is to link or merge the object code with the operating system of the computer in order to create a fully executable program. In the IBM PC, a program called **MACRO Assembler (MASM)** performs all of these tasks for us.

Let's assume that we have created a flow chart and worked out the logic of an assembly language program. We have written the program down on paper and are ready to see it work. The next step is to input our source code into the computer. This task is usually performed using a program called an **editor.**

10.1 THE EDITOR

There are many screen editors that we may use to develop our source code file. We may even use a word processor providing that it outputs a standard ASCII file—most do. The original text editor that came with the IBM disk operating system (DOS) was called **EDLIN** (LINe EDitor program). Previous editions of this book describe the use of EDLIN in detail. EDLIN was replaced with a full-screen editor program in DOS 5.0 called **EDIT.** EDIT functions very much like a word processor in the DOS environment and was developed with QBASIC. EDIT is still available with Windows 98. Another alternative for an editor is **Notepad,** which comes with all versions of Windows. Notepad is an ASCII editor, which will work well with files up to 64 KB in size. Word processors like Word for Windows can be used as an assembly language editor, but the Save as file format must be changed from the default .DOC format to the .TXT format. Note that EDIT and Notepad only operate in ASCII and will by default save files automatically as .TXT files. It is important to note that although most ASCII editors will name files with a .TXT extension, most assemblers require the extension .ASM. For this reason, it is a good idea to rename the file to utilize the extension .ASM. For example, if the original source code developed using an editor is named PROGRAM.TXT, simply rename it to PROGRAM.ASM. Depending on the editor, this can be done when originally saving the program, by using the REName command at the DOS prompt, or by selecting Rename from the file menu in Windows.

To start the DOS EDIT program, simply type EDIT at the DOS prompt as follows:

```
C:\>EDIT
```

The EDIT window shown in Figure 10.1 will appear.

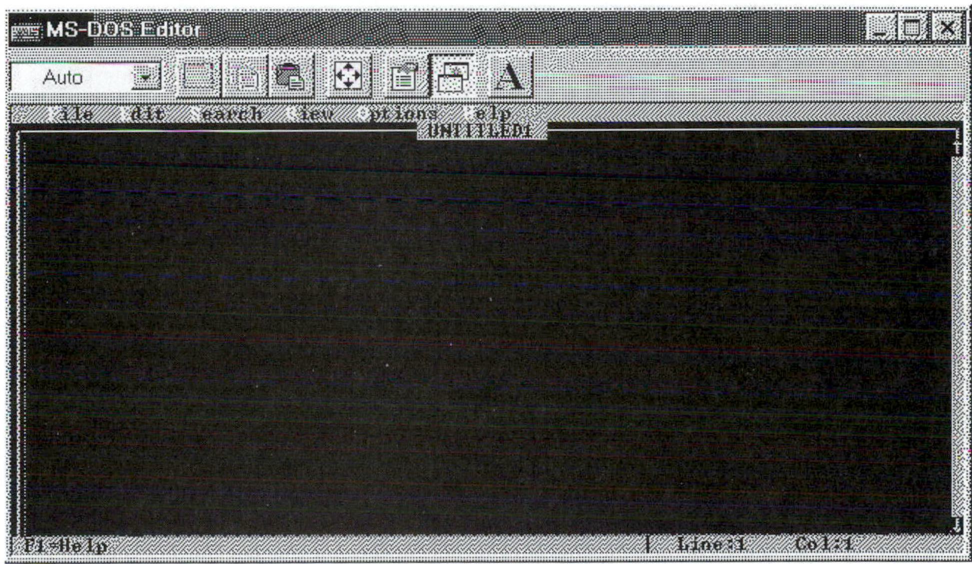

FIGURE 10.1
MSDOS editor

Press the Escape key to remove the welcome screen and you can begin to enter your source code. Note that either the mouse or the Alt key can be used to access the menu choices.

In the Windows environment, Notepad provides a convenient alternative. In Windows 3.X Notepad is located in the Accessories program group, or in Windows 95/98, select Start/Programs/Accessories/Notepad. Notepad is a Windows ASCII word processor that is simple to use. However, the filename will have to be renamed after it is saved, to work with most assemblers. Even saving a Notepad file with the .ASM extension will generally not work with most assemblers. The file may need to be renamed using the .ASM extension again anyway. Figure 10.2 illustrates the Notepad editing environment.

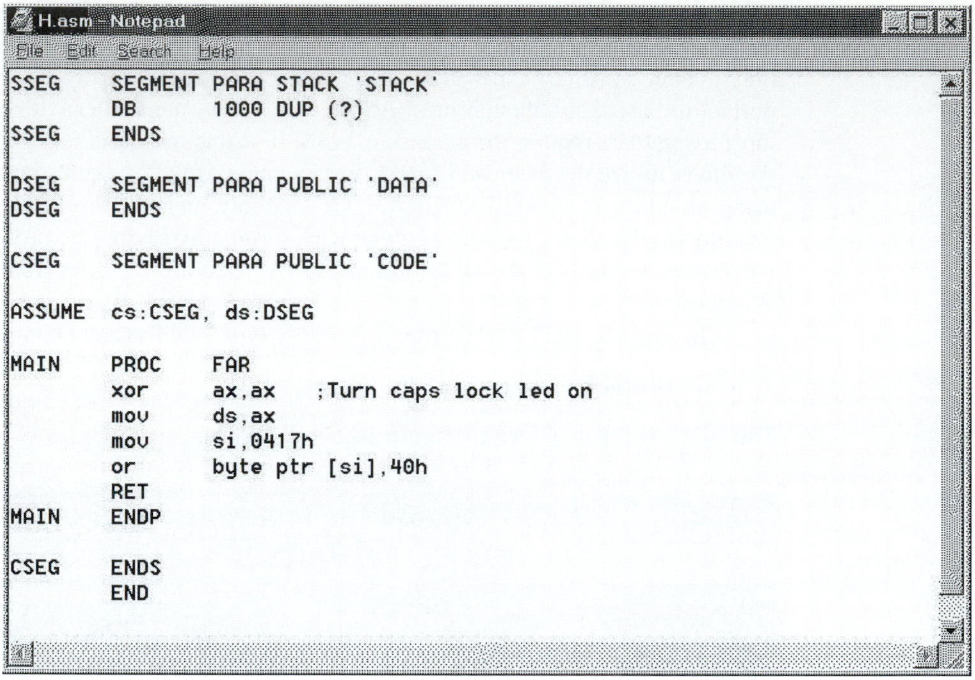

H.asm - Notepad

File Edit Search Help

```
SSEG    SEGMENT PARA STACK 'STACK'
        DB      1000 DUP (?)
SSEG    ENDS

DSEG    SEGMENT PARA PUBLIC 'DATA'
DSEG    ENDS

CSEG    SEGMENT PARA PUBLIC 'CODE'

ASSUME  cs:CSEG, ds:DSEG

MAIN    PROC    FAR
        xor     ax,ax    ;Turn caps lock led on
        mov     ds,ax
        mov     si,0417h
        or      byte ptr [si],40h
        RET
MAIN    ENDP

CSEG    ENDS
        END
```

FIGURE 10.2
Notepad

10.2 THE ASSEMBLER

The assembler lets you write programs at the level the microprocessor understands but doesn't force you to memorize a set of numeric codes. Instead, you write instructions as English-like abbreviations and then use an assembler program to convert the abbreviations to their numeric equivalents. Some popular assembler programs on the market today are Microsoft's Macro Assembler (MASM), A86 (available as shareware), and DEBUG, which comes with all versions of DOS and Windows.

Let us assume that a source code program called ROTATE.ASM has been created using an editor and is ready to be assembled. To assemble a program using MASM, we proceed as follows: At the DOS prompt type

```
C:\>MASM
```

The computer should respond

```
Source filename [.ASM]:_
```

Type the name of the file to be assembled, which in this case is ROTATE, and then press the Enter key.

```
Source filename [.ASM]:ROTATE <ENTER>
```

MASM will respond with three more questions. Answer each of them by typing the filename ROTATE and pressing the Enter key.

```
Source filename [.ASM]:ROTATE
Object filename [ROTATE.OBJ]:ROTATE
Source listing [NUL.LST]:ROTATE
Cross reference [NUL.CRF]:ROTATE
```

If there are no errors, MASM will assemble the program and respond

```
Warning          Severe
Errors           Errors
0                0
```

If there were any errors in the program, the program did not assemble. Correct the errors using the assembler, and reassemble the program. The .LST file can be used to locate the errors as well as to view the assembled machine code. It should be noted that MASM expects to see a stack segment in the source code file. If no stack segment exists, MASM will report a warning error. This error is not critical and will not halt assembly.

10.3 THE LINKER

After we have obtained an error-free assembled program, we are ready to convert the object file into a PC DOS executable file. Note that this step is only required for programs that must run under the DOS operating system or any operating system that requires **relocatable** files. This is because DOS can store a program at any convenient place in memory, which frees you from having to tell DOS where to put the program in memory. However, to use this feature you must convert the assembled program (.OBJ file) to a form that can be moved around (the computer term "relocatable"). This involves using a program called **LINK.** To call the linker program, type

```
C:\>LINK
```

The linker will respond with the message

```
Object Modules [.OBJ]:_
```

Since the filename extension .OBJ was created by the assembler, type

```
Object Modules [.OBJ]: ROTATE <ENTER>
```

The linker will respond with three more questions, which may be responded to by pressing the Enter key each time.

```
Object Modules [.OBJ]: ROTATE <ENTER>
Run File [ROTATE.EXE]:  <ENTER>
List File [NULL.MAP]:  <ENTER>
Libraries [.LIB]:  <ENTER>
```

The linker has now created and saved the executable program file on the disk drive using the filename ROTATE.EXE. You can verify this by typing the DOS command DIR.

The .EXE file can be easily converted to a .COM file by using the DOS program EXE2BIN.COM. A .COM file only uses one segment, and thus no stack segment can exist. COM files are considerably smaller in size than .EXE files. To convert ROTATE.EXE to ROTATE.COM, type the following at the DOS prompt

```
C:\>EXE2BIN  ROTATE.EXE   ROTATE.COM
```

Typing DIR will verify the conversion and also the reduction in file size.

10.4 CREATING ASSEMBLY LANGUAGE PROGRAMS

As a first example, enter the program shown in Figure 10.3 and save it using the name HMASM.ASM. If you used Notepad as your editor in Windows 95 or Windows 98, the file will be saved as HMASM.ASM.TXT, so you will have to rename the file back to HMASM.ASM. The program doesn't do very much. It simply performs some moves and additions and prints the ASCII character H on the screen. It does, however, demonstrate all of the steps necessary to create and execute an assembly language program.

The program actually starts with the first move instruction, MOV AL, 41H. Note that the lines, which precede the move instruction, are called assembler directives and are required to assemble and link the program using MASM. They will be required on most

```
SSEG      SEGMENT PARA STACK 'STACK'
          DB      1000 DUP (?)
SSEG      ENDS

DSEG      SEGMENT PARA PUBLIC 'DATA'
DSEG      ENDS

CSEG      SEGMENT PARA PUBLIC 'CODE'

ASSUME    cs:CSEG, ds:DSEG

MAIN      PROC     FAR
          push     ds
          sub      ax,ax
          mov      al,41h
          mov      bl,07h
          add      al,bl
          mov      dl,al
          mov      ah,02h
          int      21H
          RET
MAIN      ENDP

CSEG      ENDS
          END
```

FIGURE 10.3
Display H program (HMASM.ASM)

programs, so we refer to them as a **header.** Included in the header are three instructions for proper termination, which are only required on DOS systems. Furthermore, the two instructions (MOV AX, DSEG, and MOV DS, AX) are only required for .EXE files. Recall that in a .COM file, the program code, data, and stack (segments) are all stored in one 64K block of memory. In a .EXE file, the code, data, and stack may be stored in different segments, thus the requirement to define the segment locations. Until you get more familiar with assembly language programming, you may wish to use the header instructions with all programs that you write. The next few lines simply define an ASCII character to display on the screen. Recall that hex 41 is equivalent to an uppercase A. In this case, hex 48 defines an uppercase H character. (Note that MOV DL,48H would have accomplished the same thing.) INT 21H is a DOS command that when executed with 02H in the AH register, will display the character defined in the DL register on the screen. INT 21H is a command that can only be used on DOS-based microprocessor systems because it is actually a small program that is embedded in the BIOS of the PC. The remaining lines (beginning with the RET instruction) are required by MASM and the linker to terminate the program and return a DOS prompt to the user. We refer to them as a **footer.** If you do not terminate a program, the 8088 will continue executing whatever else is in memory.

If you are new to assembly language programming, it may be best to simply use the model header and footer described in "Tech Tips and Troubleshooting—T^3" in this chapter. You may be adding a few unnecessary lines of code to your program, but it will make getting started easier. The next steps involve the actual assembly and linkage of the program HMASM.ASM

```
Source filename [.ASM]:HMASM
Object filename [ROTATE.OBJ]:HMASM
Source listing [NUL.LST]:HMASM
Cross reference [NUL.CRF]:HMASM
```

If there are no errors, MASM will assemble the program and respond

```
Warning          Severe
Errors           Errors
0                0
```

Next we utilize the linker

```
Object Modules [.OBJ]: HMASM <ENTER>
Run File [HMASM.EXE]:  <ENTER>
List File [NULL.MAP]:  <ENTER>
Libraries [.LIB]:  <ENTER>
```

The linker has now created and saved the executable program file HMASM.EXE. To execute the program, simply type the name of the program after the DOS prompt. Note that in Windows 95 or Windows 98, you may have to exit to a DOS prompt for the program to execute properly.

The HMASM program could have been assembled using the A86 shareware assembler in a similar manner. To illustrate this procedure, open the file HMASM.ASM using Notepad (or your editor) and resave it using the filename HA86.ASM. At the DOS prompt type

```
C:\>A86  HA86.ASM
```

If there are no errors, the computer will generate the file HA86.COM directly without the need of linking, and output the following to let you know that the assembly is complete.

```
A86 Macro Assembler
Source:
HA86.ASM
Object:HA86.COM
Symbols:HA86.SYM
```

The A86 assembler is much simpler to use, but it only generates .COM files and not .EXE files. COM files, however, will run very nicely on the PC and will normally be smaller in size. In this case, because of the stack segment, you'll notice that MASM generates more efficient code. Redoing the preceding examples without the stack segment will illustrate the size advantage of a .COM file over a .EXE file. Furthermore, .EXE files (without stack segments) can be converted to .COM files by utilizing the DOS program EXE2BIN.COM as follows:

```
C:\>EXE2BIN HMASM.EXE HMASM.COM
```

10.5 PROGRAMMING THE KEYBOARD LEDS

Physical memory address 00417H contains one byte of information, which controls eight keyboard functions as shown in Figure 10.4. The upper four bits of this byte may be changed by the user; the lower four bits must remain undisturbed. When the PC is started, the BIOS loads physical memory address 00417H with its initial conditions. The program CAPSON.EXE can be used to turn on the Caps Lock LED and cause all typing to be in uppercase letters.

The source code for the development of CAPSON.EXE is shown in Figure 10.5. The first two lines load the data segment register with 0000H; the next line loads the SI

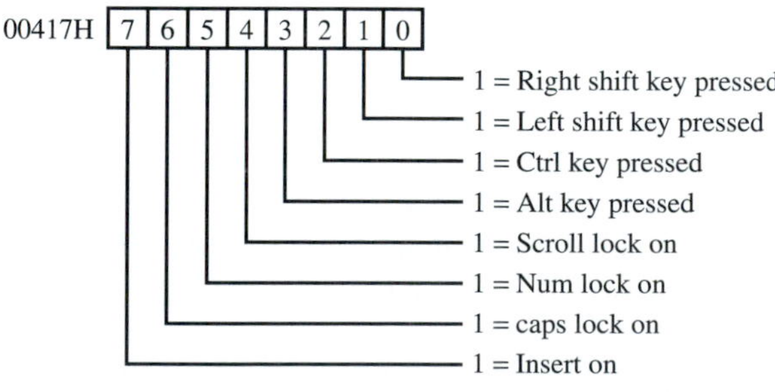

FIGURE 10.4
Keyboard status byte

```
SSEG        SEGMENT  PARA  STACK  'STACK'
            DB       1000  DUP  (?)
SSEG        ENDS

DSEG        SEGMENT  PARA  PUBLIC  'DATA'
DSEG        ENDS

CSEG        SEGMENT  PARA  PUBLIC  'CODE'

ASSUME      cs:CSEG, ds:DSEG

MAIN        PROC     FAR
            xor      ax,ax          ;Turn caps lock led on
            mov      ds,ax
            mov      si, 0417h
            or       byte ptr [si], 40h
            RET
MAIN        ENDP

CSEG        ENDS
            END
```

FIGURE 10.5
CAPSON.EXE

register with offset 0417H to develop access to physical memory address 00417H. Finally, the OR statement forces a 1 into bit six of the physical memory address pointed to by the DS:SI registers to turn on the Caps Lock LED.

Once the source code file is created, you can use MASM and LINK to create the program CAPSON.EXE. The A86 assembler can also be used to create a .COM version of the program, but requires that the OR statement be written as

```
OR [byte ptr SI],40H
```

Figure 10.6 illustrates a program that uses subroutines to blink the three keyboard LEDs in sequence. The Scroll Lock LED is lit by ORing 10H with the keyboard status byte. The Num Lock LED is turned on by ORing 20H with the keyboard status byte, and the Caps Lock LED is lit by ORing 40H with the keyboard status byte. The LEDs are turned off by ANDing 0*EF*H, 0*DF*H, and 0*BF*H respectively with the status byte. By ANDing with 1, the original contents of the status byte remain unchanged. Note that there are other ways of accomplishing the same result. INT 16 is a DOS function call that tests the keyboard to see if a key is being pressed. It is being used here to exit the program. INT 16 will set the zero flag if no character is in the keyboard buffer. It will reset the zero flag if a character is in the keyboard buffer.

```
SEGMENT DATA word public
DATA    ENDS

SEGMENT  CODE byte public

ASSUME     cs:CODE, ds:DATA

           PUBLIC MAIN

DELAY      EQU    40000

MAIN       PROC   NEAR

Start:     call   offcap    ;begin by turning off all 3 leds
           call   numoff
           call   offscr
BLINK:     NOP
Repeat:    call   oncaps    ;turn on cap led
           call   pause     ;delay
           jnz    Close     ;jmp if key depressed
           call   offcap    ;turn off cap led
           call   numon     ;turn on num led
           call   pause     ;delay
           jnz    Close     ;jmp if key depressed
           call   numoff    ;turn off num led
           call   onscrl    ;turn on scroll led
           call   pause     ;delay
           jnz    Close     ;jmp if key depressed
           call   offscr    ;turn off scroll led
           jmp    Repeat    ;do again until a key is pressed
Close:     call   offcap    ;turn all leds off on exit
           call   numoff
           call   offscr
           ret
MAIN       ENDP

ONCAPS:    xor    ax,ax     ;Turn caps lock led on
           mov    ds,ax
           mov    si, 0417h
           or     [byte ptr si], 40h
           ret

OFFCAP:    xor    ax,ax     ;Turns caps lock led off
           mov    ds,ax
           mov    si, 0417h
           and    [byte ptr si], 0bfh
           ret
```

FIGURE 10.6
KEYBOARD.ASM

```
NUMON:      xor     ax,ax     ;Turn num lock led on
            mov     ds,ax
            mov     si, 0417h
            or      [byte ptr si], 20h
            ret

NUMOFF:     xor     ax,ax     ;Turn num lock led off
            mov     ds,ax
            mov     si, 0417h
            and     [byte ptr si], 0dfh
            ret

OFFSCR:     xor     ax,ax     ;Turn scroll lock led off
            mov     ds,ax
            mov     si,0417h
            and     [byte ptr si], 0efh
            ret

ONSCRL:     xor     ax,ax     ;Turn scroll lock led on
            mov     ds,ax
            mov     si, 0417h
            or      [byte ptr si], 10h
            ret

PAUSE:      mov     cx,DELAY ;Pause according to the equate DELAY
@@10:       mov     ah, 01h
            int     16h       ;Test for key depress
            jnz     Quit      ;Quit if key depressed
            loop    @@10
QUIT:       ret

CODE        ENDS
            END
```

FIGURE 10.6 (continued)

10.6 DEBUG

The **DEBUG** utility, which comes with all versions of DOS and Windows 95/98, can be extremely useful. It is almost a must for assembly language programmers and can also provide insight into the operation of a program and the system at the bit level. It has several nice features, including the ability to display and change any of the registers in the 8088 microprocessor, start and stop program execution at any time, change the program, and look at diskettes, sector by sector. DEBUG works at the machine code level, but it also has the ability to disassemble machine code and assemble instructions directly into machine code. This section is intended to show some examples of how to use the various

DEBUG commands. Note that the segment registers will most likely *not* be exactly the same as what is shown but that the offset values will be the same. This is normal, because we do not know exactly where DOS will load the Debug program. The command syntax is described in Table 10.1 (p. 329).

Starting Debug

The procedure for starting DEBUG is simple. From a DOS prompt, simply type debug as follows:

```
A>DEBUG <ENTER>
```

DEBUG is not especially user friendly and responds with a hyphen (**-**) prompt indicating that debug is ready to accept a command.

Starting DEBUG with a File

There are two ways to start DEBUG with a file. Both ways produce the same results, and either can be used. To start DEBUG with the file PROG1.COM, we proceed as follows:

Method I

```
C:\>debug prog1.com <ENTER>
```

Method II

```
C:\>debug <ENTER>
-n prog1.com
```

With either method, you will get the DEBUG prompt of a hyphen (**-**). DEBUG has loaded your program and is ready to run. If you have problems, you can enter the command **Q** (Quit) any time you have the DEBUG prompt (-). This should return you to the DOS prompt.

Display Commands

Register Command The first thing we will look at are the registers, using the **R** command. If you type in an R with no parameters, the registers should be displayed as shown:

```
-R
AX=0000  BX=0000  CX=0446  DX=0000  SP=FFFE  BP=0000  SI=0000  DI=0000
DS=6897  ES=6897  SS=6897  CS=6897  IP=0100  NV UP DI PL NZ NA PE NC
6897:0100 E96B01        JMP     026E
```

If you opened DEBUG with a file, the CX register contains the length of the file (0446H). If the file is larger than 64K, BX will contain the high order of the size. This is very important to remember when using the Write command, as this is the size of the file to be written. Remember, once the file is in memory, DEBUG has no idea how large the file is,

TABLE 10.1
Debug commands

Command	Syntax	Function
Register	R [Register Name]	Examine or modify the contents of one or all registers
Quit	Q	Exit Debug and return to DOS
Dump	D [Starting Address] [Ending Address]	Display the contents of memory to the display
Enter	E [Address] [List]	Examine or change the contents of memory locations
Fill	F [Starting Address] [Ending Address] [List]	Fill a block of memory with a specified value
Move	M [Starting Address] [Ending Address] [Destination Address]	Move a block of data from a source location to a destination location
Compare	C [Starting Address] [Ending Address] [Destination Address]	Compare two blocks of data and display the locations that differ
Search	S [Starting Address] [Ending Address] [List]	Search a block of data and display all locations that match data in list
Input	I [Address]	Display the contents from a specified I/O port address
Output	O [Address]	Output data to a specified I/O port address
Hexadecimal	H [Number1] [Number2]	Calculate the sum and difference of the two numbers
Unassemble	U [Starting Address] [Ending Address]	Unassemble machine code into assembler instructions
Write	W [Starting Address] [Drive] [Starting Sector] [# Sectors]	Save to a file on a disk
Load	L [Starting Address] [Drive] [Starting Sector] [# Sectors]	Load memory with the contents of a file on a diskette
Name	N [File Name]	Define the name of a file
Assemble	A [Starting Address]	Assemble instructions into machine code
Trace	T = [Starting Address] [Number]	Execute one or more instructions and display register contents with each execution
Proceed	P = [Starting Address] [Number]	Proceed to execute a specified number of instructions
Go	G = [Starting Address] [Breakpoint Address]	Execute instructions up to the breakpoint

or if you may have added to it. The amount of data to be written will be taken from the BX and CX registers. Note that if DEBUG is opened without a file the CX register will contain 0000H.

If we want to change the contents of one of the registers, we enter R and the register name. Let's place 1234 (hexadecimal) in the AX register:

-RAX	R and AX register
AX 0000	DEBUG responds with register and contents
:1234	: is the prompt for entering new contents. We respond 1234
-	DEBUG is waiting for the next command.

Now if we display the registers again, by typing an *R,* we see the following:

-R

```
AX=1234  BX=0000  CX=0446  DX=0000  SP=FFFE  BP=0000  SI=0000  DI=0000
DS=6897  ES=6897  SS=6897  CS=6897  IP=0100  NV UP DI PL NZ NA PE NC
6897:0100 E96B01        JMP     026E
```

Note that nothing has changed, with the exception of the AX register. The new value has been placed in it, as we requested. One additional note, the Register command can only be used for 16-bit registers (AX, BX, etc.). It cannot change the 8-bit registers (AH, AL, BH, etc.). To change just AH, for instance, you must enter the data in the AX register, with your new AH and the old AL values.

Dump Command One of the other main features of DEBUG is the ability to display areas of storage. Unless you are very good at reading 8088 machine language, the **Dump** command is mostly used to display data (text, flags, etc.). To display code, the Unassemble command which is described later in this section, is a better choice. If we enter the Dump command at this time, DEBUG will default to the start of the program. It uses the DS register as its default, and, since this is a .COM file, begins at DS:0100. It will by default display 80H (128D) bytes of data, or the length you specify. The next execution of the Dump command will display the following 80H bytes, and so on. For example, the first execution of D will display DS:0100 for 80H bytes, the next one DS:0180 for 80H bytes, etc. Of course, absolute segment and segment register overrides can be used, but only hex numbers can be used for the offset. For example, D DS:BX is invalid.

With our program loaded, if we enter the Dump command, we will see this:

-D

```
6897:0100  E9 6B 01 50 52 4F 47 31-2E 43 4F 4D 43 6F 70 79  ik.PROG1.COMCopy
6897:0110  72 69 67 68 74 20 28 43-29 20 31 39 39 38 53 69  right (C) 1998Si
6897:0120  6D 6F 6E 20 41 2E 20 47-72 65 65 6E 65 72 50 75  mon A. GreenerPu
6897:0130  62 6C 69 63 20 64 6F 6D-61 69 6E 20 73 6F 66 74  blic domain soft
6897:0140  77 61 72 65 00 00 00 00-00 00 00 00 00 00 00 00  ware............
6897:0150  00 00 00 00 00 00 00 00-00 24 00 00 00 00 00 00  .........$......
6897:0160  00 00 00 00 00 00 00 00-00 00 00 00 00 00 00 00  ................
6897:0170  00 00 00 00 00 00 00 00-00 00 00 00 44 4F 53 20  ............DOS
```

Notice that the output from the Dump command is divided into three parts. On the left, we have the address of the first byte on the line (6897:0100 or physical address 68A70).

This is in the format **Segment:Offset.** Next comes the hex data at that location. DEBUG will always start the second line at a 16 byte boundary. For example, if you entered D 109, you would get 7 bytes of information on the first line (109–10*F*), and the second line would start at 110. The last line of data would have the remaining 9 bytes of data, so 80H bytes are still displayed. The third area is the ASCII representation of the data. Only the standard ASCII character set is displayed. Special characters for the IBMPC are not displayed. Periods (.) are shown in their place. This makes it much easier to search for plain text.

Dump can be used to display up to 64 KB of data, with one restriction: It cannot cross a segment boundary; that is, D 0100 *F*000 is valid (display DS:0100 to DS:*F*000), but D 9000 8000 is not (9000H + 8000H = 11000H and crosses a 64K segment boundary). If at any time you want to terminate the display of data, Ctrl-Break will stop the display and return you to the DEBUG prompt.

Search Command Search is used to find the occurrence of a specific byte or series of bytes within a segment. The address parameters are the same as for the Dump command, so we will not duplicate them here. However, we also need the data to be searched for. This data can be entered as either hexadecimal or character data. Hexadecimal data is entered as bytes, with a space or a comma as a separator. Character data is enclosed by single or double quotes. Hexadecimal and character data can be mixed in the same request: S 0 100 12 34 'abc' 56 is valid, and requests a search from DS:0000 through DS:0100 for the sequence of 12H 34H a b c and 56H, in that order. Uppercase characters differ from lowercase characters, and a match will not be found if the case does not match. For instance, 'ABC' is not the same as 'abc' or 'Abc' or any other combination of upper- and lowercase characters. However, 'ABC' is identical to "ABC", since single and double quotes may be used as separators.

As an example let's search for the string 'COM'. Here's what would happen (assuming the program PROG1.COM was loaded):

```
-S 0 FFFF 'COM'
6897:0109
-
```

Again, the actual segment would be different in your system, but the offset should be the same. Thus the occurrence of COM occurred at offset 0109. Note that if we wanted to find every place we did an INT 21H (machine code for INT is CD), we would do the following:

```
-S 0 FFFF CD 21
6897:0050
6897:0274
6897:027F
6897:028B
6897:02AD
6897:02B4
6897:0332
6897:0345
6897:034C
```

```
6897:043A
6897:0467
6897:047A
6897:0513
6897:0526
6897:0537
6897:0544
-
```

The display indicates that DEBUG has found the hex data CD 21 at 16 offset locations. This does not mean that all these addresses are INT 21s, only that that data was there. It could be (and most likely is) an instruction, but it could also be an address, the last part of a JMP instruction, and so on. You will have to manually inspect the code to make sure it is an INT 21 instruction.

Compare Command Along the same lines of Dump and Search commands, we have the Compare command. Compare will take two blocks of memory and compare them, byte for byte. If the two addresses do not contain the same information, both addresses are displayed, with their respective data bytes. As an example, we will compare DS:0100 with DS:0200 for a length of 8 addresses.

-C 0100 0107 0200

```
6897:0100  E9  65  6897:0200
6897:0101  6B  70  6897:0201
6897:0102  01  74  6897:0202
6897:0103  50  65  6897:0203
6897:0104  52  6D  6897:0204
6897:0105  4F  62  6897:0205
6897:0106  47  65  6897:0206
6897:0107  31  72  6897:0207
```

Since there were no comparisons, we got an output for each byte. If we had gotten a match on any of the bytes, DEBUG would have skipped that byte. If all of the locations requested matched, DEBUG would have simply responded with another prompt. No other message is displayed. This is useful for comparing two blocks of data from a file or a program with the BIOS ROM.

Unassemble Command For program debugging, one of the main commands you will use is the Unassemble command. This command will take machine code and convert it to instructions. Addressing is the same as for previous commands with one exception. Since we are now working with code (the previous commands are mainly for data), the default register is the CS register. In a .COM program, this makes very little difference, unless you reset the DS register yourself. However, in a .EXE file, it can make a lot of difference, because the CS and DS registers are usually set to different values.

Unassembling data can lead to some interesting results. For example suppose, CS:IP is set to 6897:0100. If we look at the program, we see a JMP as the first instruction, followed by data. If we just enter U, we will start at CS:IP (6897:0100) and start unassembling data. What we will get is a good instruction, followed by more or less nonsense—for instance,

-U

```
6897:0100  E96B01   JMP  026E
6897:0103  43       INC  BX
6897:0104  4C       DEC  SP
6897:0105  4F       DEC  DI
6897:0106  43       INC  BX
6897:0107  4B       DEC  BX
```

and so on, through 6897:011*D*. We know the INC BX, DEC SP, and so forth are not valid instructions because of the jump instruction, but DEBUG doesn't, so we do have to look at the code. After working with DEBUG a little, you will be able to spot code versus data with the Unassemble command. For now, understand that the first instruction will take us to CS:026*E* and we can start to unassemble code again from there.

If we unassemble CS:026*E*, we will find something which looks a little more like what we might expect.

-U 026E

```
6897:026E  8D167802  LEA  DX,[0278]
6897:0272  B409      MOV  AH,09
6897:0274  CD21      INT  21
6897:0276  EB05      JMP  027D
6897:0278  1B5B32    SBB  BX, [BP+DI+32]
6897:027B  4A        DEC  DX
6897:027C  24B4      AND  AL,B4
6897:027E  30CD      XOR  CH,CL
6897:0280  213C      AND  [SI],DI
6897:0282  027D0A    ADD  BH,[DI+0A]
6897:0285  8D167C01  LEA  DX,[017C]
6897:0289  B409      MOV  AH,09
6897:028B  CD21      INT  21
6897:028D  CD20      INT  20
```

The first few instructions look fine, but after the JMP 027*D*, things start to look a little funny. Also, note that there is no instruction starting at 027*D*. We have instructions at 027*C* and 027*E*, but not 027*D*. This is again because DEBUG doesn't know data from instructions. At 027*C*, we should (and do) have the end of our data, but this also translates into a valid AND instruction, so DEBUG will treat it as such. If we wanted the actual instruction at 027*D*, we could enter U 027*D* to get it, but from here, we don't know what it is. What we are trying to illustrate is that DEBUG will do whatever you tell it. If you tell it to Unassemble data, it will do so to the best of its ability. So, you have to make sure you have instructions where you think you do.

Data Entry Commands

Enter Command The Enter command is used to place bytes of data in memory. It has two modes: Display/Modify and Replace. The difference is in where the data is specified, in the Enter command itself, or after the prompt.

If you enter E address, alone, you are in display/modify mode. DEBUG will prompt you one byte at a time, displaying the current byte followed by a period. At this time, you have the option of entering one or two hexadecimal characters. If you hit the space bar, DEBUG will *not* modify the current byte, and go on to the next byte of data. If you go too far, the hyphen (-) will back up one byte each time it is pressed.

```
-E 103
6897:0103  50.41  52.42  4F.43  47.     31.45
6897:0108  2E.46  43.40  53.-
6897:0109  40.47  53.
```

In this example, we entered E 103. DEBUG responded with the address and the information at that byte (50). We entered the 41 and DEBUG automatically showed the next byte of data (52). Again, we entered 42, debug came back. The next byte was 4*F*; we changed it to 43. At 106, 47 was fine with us, so we just hit the space bar. DEBUG did not change the data, and went on to the following byte. After entering 40 at location 109, we found we had entered a bad value. The hyphen key was pressed, and DEBUG backed up one byte, displaying the address and current contents. Note that it has changed from the original value (43) to the value we typed in (40). We then type in the correct value 47 and terminate by pressing the Enter key.

As you can see, this can be very awkward, especially where large amounts of data are concerned. Also, if you need ASCII data, you have to look up each character and enter its hex value. That's where the Replace mode of operation comes in handy. Where the Display/Modify mode is handy for changing a few bytes at various offsets, the Replace mode is better for changing several bytes of information at one time. Data can be entered in hexadecimal or character format, and multiple bytes can be entered at one time without waiting for the prompt. If you wanted to store the characters 'My name' followed by a hexadecimal 00 starting at location 103, you would enter

```
-E 103 'My name' 0
```

As in the Search command, data can be entered in character (in quotes) or hexadecimal forms and can be mixed in the same command. This is the most useful way of entering large amounts of data into memory. The dump command can be used to display the results.

Fill Command The Fill command is useful for storing a lot of data of the same value. It differs from the Enter command in that the list will be repeated until the requested amount of memory is filled. If the list is longer than the amount of memory to be filled, the extra items are ignored. Like the Enter command, it will take hexadecimal or character data. Unlike the Enter command, though, large amounts of data can be stored without specifying every character. As an example, to clear 32K (8000H) of memory to 00H, you only need to enter

```
-F 0  8000 0
```

which translates into Fill, starting at DS:0000 for a length of 32K (8000) with 00H. If the data were entered as '1234', (note the quotes) the memory would be filled with the repeating string ‘123412341234’, and so on. Usually, it is better to enter small amounts of data with the Enter command, because an error in the length parameter of the Fill command can destroy a lot of work. The Enter command, however, will only change the number of bytes actually entered, minimizing the effects of a parameter error.

Move Command The Move command does just what it says. It moves data around inside memory. It takes bytes from the starting address and moves them to the ending address. If you need to add an instruction into a program, Move can be used to make room for the instruction. Beware, though. Any data or labels referenced after the move will not be in the same place. Move can be used to save a part of the program in free memory while you play with the program, and then restore it at a later time. It can also be used to copy ROM BIOS into memory, where it can be written to a file or played with to your heart’s content. You can then change things around in BIOS without having to worry about programming a ROM.

```
-M 100  200 ES:100
```

This will move the data from DS:0100 to DS:0200 to the address pointed to by ES:0100. Later, if we want to restore the data, we can say

```
-M ES:100  200 100
```

which will move the data back to its starting point. Unless the data has been changed while at the temporary location (ES:0100), we will restore the data to its original state.

Assemble Command The Assemble command will take assembly language instructions and convert them to machine code directly. Some of the things it can’t do, however, are reference labels, set equates, use macros, or anything else that cannot be translated to a value. Data locations have to be referenced by the physical memory address, segment registers, if different from the defaults, must be specified, and RET instructions must specify the type (NEAR or FAR) of return to be used. Also, if an instruction references data but not registers (i.e., Mov [278],5), the Byte ptr or Word ptr overrides must be specified to define whether a byte or word is to be moved. There is one other restriction: To tell DEBUG the difference between moving 1234H into AX and moving the data from location 1234 into AX, the latter is coded as Mov AX,[1234], where the brackets indicate the reference is an addressed location. A short comparison between MASM and DEBUG instructions follows.

MASM	DEBUG	Comments
Mov AX,1234	Mov AX,1234	Place 1234 into AX
Mov AX,[1234]	Mov AX,[1234]	Contents of address 1234 to AX
Mov AX,CS:1234	CS:Mov AX,[1234]	Move from offset of CS.
Mov Byte ptr SI,[1234]	Mov Byte ptr SI,[1234]	Move byte string
Mov Word ptr SI,[1234]	Mov Word ptr SI,[1234]	Move word string
Ret	Ret	Near return
Ret	Retf	Far return

Let's try a very simple routine to clear the screen.

-A 0100

```
6897:0100 mov ax,0600
6897:0103 mov cx,0
6897:0106 mov dx,184F
6897:0109 mov bh,07
6897:010B int 10
6897:010D int 20
6897:010F

-
```

We are using the BIOS interrupt 10H program, which is the video interrupt. (If you would like more information on the interrupt, there is a very good description in the IBM DOS Technical Reference Manual.) We need to call BIOS with AX = 0600, BH = 07, CX = 0, and DX = 184*F*H. First we had to load the registers, which is done in the first four instructions. The instruction at offset 6897:010*B* actually calls the BIOS. The INT 20 instruction at offset 010*D* is for safety only. We really don't need it, but with it in, the program will stop automatically. Without the INT 20, and if we did not stop, DEBUG would try and execute whatever occurs at 010*F*. If this happens to be a valid program (unlikely), we would just execute the program. Usually, though, we will find it to be invalid, and will probably hang the system, requiring a Ctrl-Alt-Del (maybe) or a power-off and on again (usually). So, be careful and double check your work!

Now, we need to execute the program. To do this, enter the Go command, a G followed by the Enter key. If you have entered the program correctly, the screen will clear and you will get a message "Program terminated normally" (more on the Go command later).

I/O Commands

Name Command The Name command has just one purpose: specifying the name of a file that DEBUG is going to load or write. It does nothing to change memory or execute a program, but does prepare a file control block for DEBUG to work with. If you are going to load a program, you can specify any parameters on the same line, just as in DOS. One difference is, the extension *must* be specified. DEBUG will load or write any file, but the full file name must be entered.

-n chkdsk.com /f

This statement prepares DEBUG for loading the program CHKDSK.COM, passing the /f switch to the program. When the Load command is executed, DEBUG will load CHKDSK.COM and set up the parameter list (/f) in the program's input area.

Load Command The Load command has two formats. The first will load a program that has been specified by the Name command into storage, set the various registers, and prepare for execution. Any program parameters in the Name command will be set into the program segment prefix (PSP), and the program will be ready to run. If the file is a .HEX file, it is assumed to have valid hexadecimal characters representing memory values, two hexadecimal characters per byte. Files are loaded starting at CS:0100 or at the address specified in the command. For .COM. .HEX, and .EXE files, the program will be loaded, the registers set, and CS:IP set to the first instruction in the program. For other files, the registers are undetermined, but basically, the segment registers are set to the segment of the PSP (100H bytes before the code is actually loaded), and BX and CX are set to the file length. Other registers are undetermined. For example,

```
-n prog1.com
-L
```

This sequence will load PROG1.COM into memory and set IP to the entry point of 0100, and CX will contain 0446, the hexadecimal size of the file. The program is now ready to run.

The second form of the Load command does not use the Name command. It is used to load absolute sectors from a disk into memory. The sector count starts with the first sector of track 0 and continuing to the end of the track. The next sector is track 0, second side (if double sided), and continues to the end of that track. Then, back to the first side, track 1, and so on, until the end of the disk. Up to 80H (128D) sectors can be loaded at one time. To use this form, you must specify starting address, drive (0 = A, 1 = B, etc.), starting sector, and number of sectors to load. For example,

```
-L 100 0 10 20
```

This instruction tells DEBUG to load, starting at DS:0100, from drive A, sector 10H for 20H sectors. DEBUG can sometimes be used this way to recover part of the information on a damaged sector. If you get an error, check the memory location for that data. Often times, part of the data has been transferred before the error occurs and the remainder (especially for text files) can be manually entered. Also, repetitive retries will sometimes get the information into memory. This can then be rewritten on the same diskette (see the "Write Command" section) or copied to the same sector on another diskette. In this way, the data on a damaged disk can sometimes be recovered.

Write Command The Write command is very similar to the Load command. Both have two modes of operation, and both will operate on files or absolute sectors. As you have probably guessed, the Write command is the opposite of the Load command. Since all the parameters are the same, we will not cover the syntax in detail. However, when using the file mode of the Write command, the amount of data to be written is specified in BX and CX, with BX containing the high-order file size. The start address can be specified or is defaulted to CS:0100. Also, files with an extension of .EXE or .HEX cannot be written out, and an error message to that effect will be displayed. If you do need to change a .EXE or .HEX file, simply rename and load it, make your changes, save it, and name it back to its original filename. Extensions like .COM and .ASM are valid. Note you must be very careful when using the Write command because you can overwrite existing files on a disk. Furthermore if you try to Write on the hard drive you could easily overwrite a piece of the operating system. This use of the Write command will work, but we will demonstrate a safer use. Let us use the clear screen program previously described as an example.

First assemble the program as follows:

-A 0100

```
6897:0100 mov ax,0600
6897:0103 mov cx,0
6897:0106 mov dx,184F
6897:0109 mov bh,07
6897:010B int 10
6897:010D int 20
6897:010F
```

Next we name the program as follows:

```
-N a:clear.com
```

Unassembling the program would also show that the program contains 15 bytes of data but you can also see this from the assembly. Now we must set up the BX and CX registers to Write 15 bytes as follows:

```
-RBX
BX xxxx
:0000

-RCX
CX xxxx
:000F
```

Finally we issue the Write command.

```
-W 0100
```

Once completed we can see the filename by using the DIR A: command from a DOS prompt. We can also run the program and see that the screen is cleared. To reload the program, start DEBUG and issue the following command sequence.

```
-N A:clear.com
-L 0100
-U 0100
```

Input Command

```
-I port address
```

```
-I 3f8
-7d
```

This is the input port address for the first asynchronous adapter. Your data may be different, because it depends on the current status of the port. It indicates the data in the register at the time it was read was 7*D*H. Depending on the port, this data may change, as the ports are not controlled by the PC.

Output Command As you can probably guess, the Output command is the reverse of the Input command. Use the Output command to send a single byte of data to a port.

Note that certain ports can cause the system to hang (especially those dealing with system interrupts and the keyboard), so be careful with what you send where!

```
-O 3FC 1
-
```

Port 3*FC*H is the modem control register for the first asynchronous port. Sending a 01H to this port turns on the DTR (data terminal ready) bit. A 00H will turn all the bits off. If you have a modem with indicator LEDs, you can watch the light flash as you turn the bit on and off.

Execution Commands

Go Command The Go command is used to start program execution. It can be used to start the execution at any point in the program, and optionally stop at any of 10 points (breakpoints) in the program. If no breakpoints are set (or the breakpoints are not executed), program execution continues until termination, in which case the message "Program terminated normally" is sent. If a breakpoint is executed, program execution stops, the current registers are displayed, and the DEBUG prompt is displayed. Any of the DEBUG commands can be executed, including the Go command to continue execution. Note that the Go command *cannot* be terminated by Ctrl-Break. This is one of the few commands that cannot be interrupted while executing.

-G=100

The Go command without breakpoints starts program execution at the address (in this case CS:0100) in the command. The equal sign before the address is required (without the equal sign, the address is taken as a breakpoint). If no starting address is specified, program execution starts at CS:IP. In this case, since no breakpoints are specified, PROG1. COM will continue execution until the program terminates. At this time, you will get the message "Program terminated normally." Note that, after the termination message, the program should be reloaded before being executed again. Any memory alterations made during program execution (storing data, etc.) will not be restored unless the program is reloaded.

-G=100 276 47C 528

This version of the Go command will start the program and set breakpoints at CS:276, CS:47C, and CS:528.

Some notes about breakpoints. The execution stops just before the instruction is executed. Setting a breakpoint at the current instruction address will not execute any instructions. DEBUG will set the breakpoint first, then try to execute the instruction, causing another breakpoint. Also, the breakpoints use Interrupt 3 to stop execution. DEBUG intercepts interrupt 3 to stop the program execution and display the registers. Finally, breakpoints are not saved between Go commands. Any breakpoints you want will have to be reset with each Go command.

Trace Command Along the same lines as Go is the **Trace** command. The difference is that, whereas Go executes a whole block of code at one time, the Trace command executes instructions one at a time, displaying the registers after each instruction. Like the Go instruction, execution can be started at any address. The start address again must be preceded by an equal sign. However, the Trace command also has a parameter to indicate how many instructions are to be executed.

-T=100 5

This Trace command will start at CS:100 and execute five instructions. Without the address, execution will start at the current CS:IP value and continue for five instructions. T alone will execute one instruction.

When using Trace to follow a program, it is best to go around calls to DOS and interrupts, because some of the routines involved can be lengthy. Also, DOS cannot be traced, and doing so has a tendency to hang the system. Therefore, trace to the call or interrupt and go to the next address after the call or interrupt or use the Proceed command.

Proceed Command The Proceed command is very similar to the Trace command. Both execute instructions and display register contents. The difference is that the Proceed command will execute loop, interrupt, call, and repeated string instructions to completion. For example consider the following lines of code:

```
6897:0100 MOV AH,02
6897:0102 MOV DL,07
6897:0104 INT 20
```

Executing the Proceed command (**P**) three times would complete the program (and result in the message "Program terminated normally"), whereas Trace would display each line of code that makes up the INT 20 function call. As another example:

```
6897:0100 MOV CX,0005
6897:0103 LOOP 103
```

Only two Proceed commands are required to complete this code, but five Trace commands are necessary to accomplish the same thing. These examples should indicate that although the Trace command can supply more detail, the Proceed command is often more convenient.

Arithmetic Commands

Hexarithmetic Command The Hexarithmetic command is handy for adding and subtracting hexadecimal numbers. It has just two parameters: the two numbers to be added and subtracted. DEBUG's response is the sum and difference of the numbers. The numbers can be one to four hexadecimal digits long. The addition and subtraction are unsigned, and no carry or borrow is shown beyond the fourth (high-order) digit.

```
-H 5 6
000B FFFF

-h 5678 1234
68AC 4444
-
```

In the first example, we are adding 0005 and 0006. The sum is 000*B*, the difference is −1. However, since there is no carry, we get *FFFF*. In the second example, the sum of 5678 and 1234 is 68*AC*, and the difference is 4444.

If you give it a chance, DEBUG can be a very useful tool for the 8088 microprocessor. It is almost a requirement for debugging assembly language programs, because no nice error messages are produced at run time. DEBUG works at the base machine level, so you need some experience to use it effectively, but with practice, it will be your most useful assembly language debugging tool.

10.7 TECH TIPS AND TROUBLESHOOTING—T³

Getting started with assembly language programming can be a frustrating experience. Different assemblers, linkers, and operating systems can all have different requirements. Some programs require different segments, and some do not. Furthermore, programs designed to run on the IBM PC and compatible systems have specific requirements for proper termination and return to the DOS prompt. To aid the beginner, the general model presented in Figure 10.7 should solve most programming problems. You may be adding a few extra lines of unnecessary code, but at least the program will assemble, link, and terminate properly.

```
STACK SEGMENT STACK
                DB        256 DUP (?)
STACK ENDS
;**********************************************************************
DATA SEGMENT
          (ADD DATA DEFINITIONS HERE)
DATA ENDS
;**********************************************************************
CODE SEGMENT
ASSUME CS:CODE,DS:DATA,SS:STACK
MAIN            PROC      FAR
                PUSH      DS
                SUB       AX,AX
       (INSERT MAIN PROGRAM LOGIC HERE)
                RET
       (INSERT SUBROUTINES HERE)
MAIN         ENDP
CODE ENDS
END
```

FIGURE 10.7
MASM general format for assembly language programming

Start by using the same header and footer section for all programs. As your knowledge of assembly language programming increases, you should be able to correctly delete any unnecessary lines of code or add other directives that may be required for special circumstances.

EXERCISES

10.1 What is the function of the assembler?

10.2 Name three PC application programs that can be used as assembly language editors.

10.3 What file format must assembly language source code files be saved in?

10.4 Name two PC assembler programs.

10.5 What is the function of the linker?

10.6 What DOS utility can be used to convert a .COM file to an .EXE file?

10.7 Which file format will create smaller file sizes, .COM or .EXE?

10.8 List the three general steps that must be performed to create an assembly language program for the PC.

10.9 What symbol is used for the DEBUG prompt?

10.10 Explain the difference between the DEBUG Trace and Proceed commands.

10.11 What DEBUG command will display the contents of the CX register?

10.12 What DEBUG command will allow you to change the contents of the BX register?

10.13 What is the keyboard command to halt DEBUG when you lose control?

10.14 Explain how the DEBUG Compare command reports a mismatch between files.

10.15 Crossword Puzzle

ACROSS

1. Instructions to the assembler.
4. Converts abbreviations into machine code.
5. Number system with ones and zeroes
7. Segment required for Microsoft assembler.
8. Windows ASCII editor.
10. Addressing mode.
12. Instruction to send data to a port.
13. Segment for program instructions.
14. Converts an object file into an executable file.
16. Single-step DEBUG command.
17. Program that runs in one segment.

DOWN

1. To convert to machine code to assembly language code.
2. Portion of a program that performs a specific task.
3. Instruction to read data from a port.
6. DEBUG command that displays memory contents.
9. An ordered set of instructions.
11. Executable file.
15. Microsoft assembler.

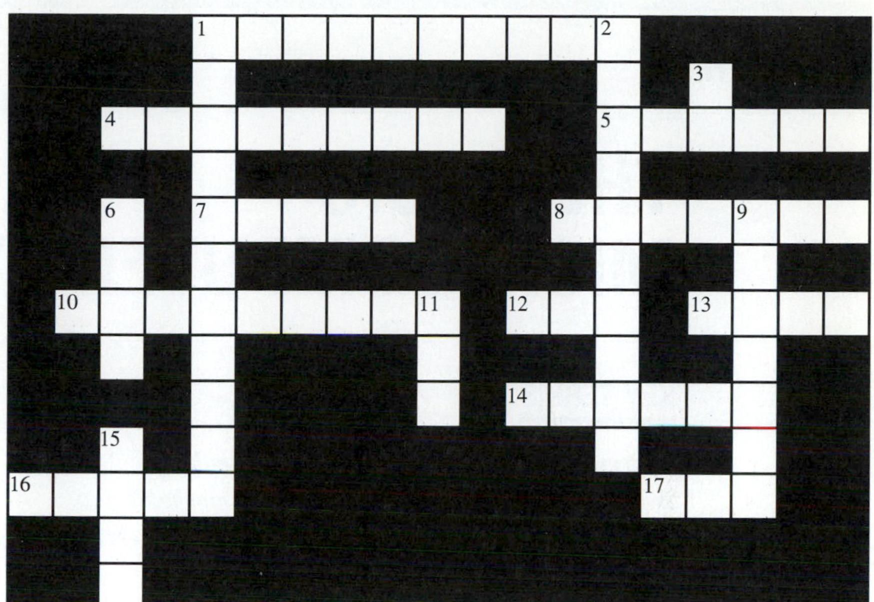
1
2
3
4
5
6
7
8
9
10
11
12
13
14
15
16
17

CHAPTER 11

THE BASIC MICROCOMPUTER SYSTEM

KEY TERMS

Address Decoder
Byte High Enable
Command Register
Direct Memory Access (DMA)
Full Address Decoding
Grant
Interrupt
Interrupt Service Routine
Maskable
Nonmaskable
Partial Address Decoding
Ports
Programmable Peripheral Interface (PPI)
Pushing
Request
Vectoring

11.0 INTRODUCTION

Microcomputer systems can control and perform many tasks. Design applications may vary, but all basic microcomputer systems consist of four subsystems, as shown in Figure 11.1.

1. *Microprocessor (MPU) subsystem.* This is the *brain* and *heart* of any microcomputer system. It performs all the logical instructions, sequences, and arithmetic computations. It also includes all the basic system timing.
2. *Memory subsystem.* This area is used to store both instructions and data for the basic microcomputer system. It consists of two types of memories: memories that never forget or change (ROM) and memories that can be changed (RAM).
3. *Input/output subsystem.* This section handles the communication between *peripheral devices* such as CRT displays, keyboards, and disk drives. It allows us to talk to the outside world.
4. *Bus subsystem.* This network of paths routes information between the MPU, memory, and I/O subsystems. It provides a highway for data, address, and control signals to flow.

These components, in conjunction with various support devices, form the basis for a microcomputer system. Note that the 8086 or 8088 microprocessor by itself is not a

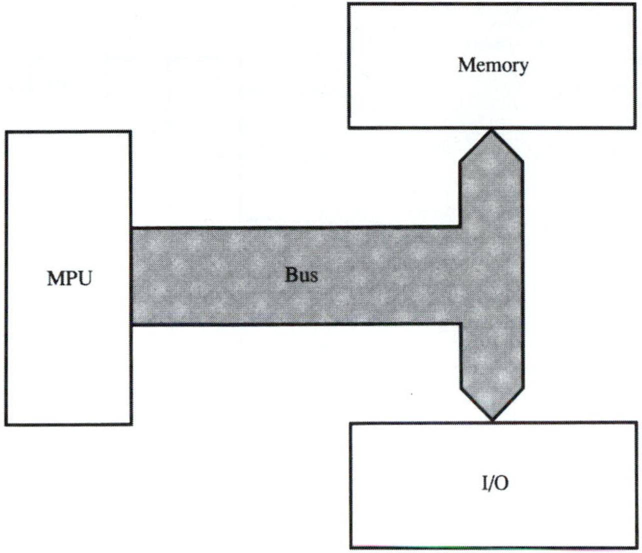

FIGURE 11.1
Microcomputer subsystems

microcomputer. The MPU does not contain any memory or I/O ports; therefore the MPU can think but by itself cannot remember or communicate.

11.1 THE MICROPROCESSOR (MPU) SUBSYSTEM

The microprocessor was introduced in Chapter 8 using the Intel 8088 microprocessor. In this chapter we study the microprocessor and how it works and interfaces with its support chips. We begin, however, by investigating the Intel 8086 microprocessor. This is basically a more powerful version of the 8088 and was used in the IBM Series 2 Model 30 personal computer.

The 8086 Pinout

Recall that the major difference between the 8086 and the 8088 microprocessors is the size of the data bus. The 8086 has a 16-bit data bus, whereas the 8088 has an 8-bit data bus. Figure 11.2 illustrates the basic pin configuration of both microprocessors. From this diagram we can see that the 8086 pin configuration is almost the same as the 8088 in maximum mode. The major differences are these:

1. A/D_0–A/D_{15} of the 8086 are the multiplexed address/data lines. This allows for a 16-bit or full-word data bus. The 8088 multiplexes only A/D_0–A/D_7 for an 8-bit or byte-wide data bus.

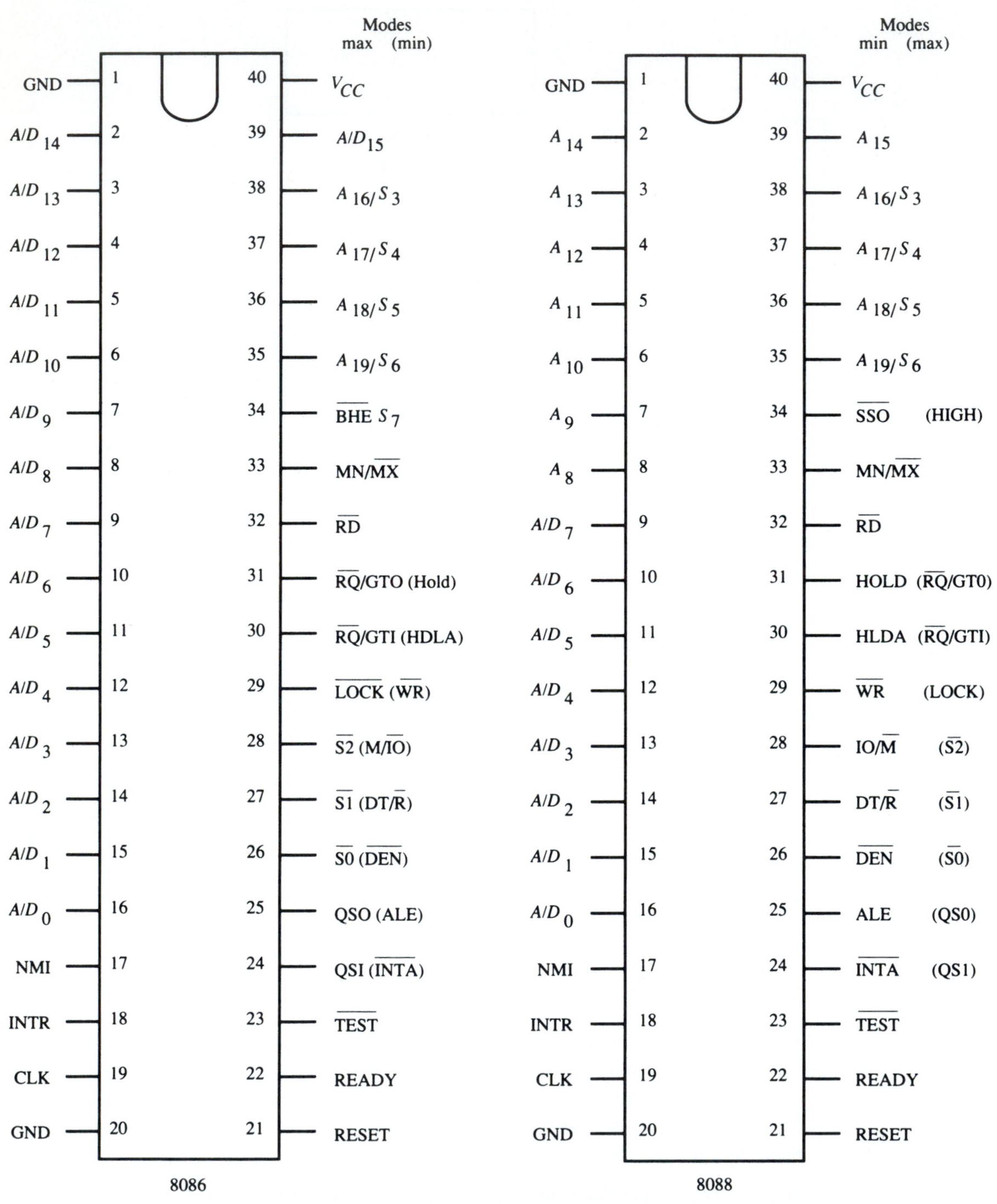

FIGURE 11.2
8088/8086 MPU pinout

2. M/$\overline{IO}$ (pin 28) of the 8086 functions opposite to that of the 8088. When memory is addressed, this signal is *active high.* When I/O information is on the address bus, this signal is *active low.* Recall that this is the opposite of the 8088's operation.
3. $\overline{BHE}/S_7$ (pin 34) is used on the 8086 to tell the memory circuits whether or not to access the 8 higher-order bits (D_{16}–D_8) on the data bus. This signal, **byte high enable,** is a tristate active low output signal. It is available during clock cycle T_1. During clock cycles T_2, T_3, and T_4, status information (S_7) is available on this pin. The $\overline{BHE}$ signal is not used on the 8088, since it has only an 8-bit bus. Instead, the 8088 uses this pin to provide the status information $\overline{SSO}$.

Maximum Mode Pins

For operation with a coprocessor, the 8086 and the 8088 MPU must operate in maximum mode. This is achieved by applying a logic level 0 to pin 33 MN/$\overline{MX}$. In maximum mode pins 24–31 take on an alternate function.

QS_1 (pin 24) and QS_0 (pin 25) tell us status information about the instruction queue. Recall that the instruction queue for the 8088 is 4 bytes long; for the 8086, it is 6 bytes long. Table 11.1 describes the function of the queue status pins.

TABLE 11.1
Queue status

QS_1	QS_0	Queue Status
0	0	No operation
0	1	Indicates first byte of op-code from queue
1	0	Indicates queue is empty
1	1	Indicates subsequent byte from queue

$\overline{S_0}$ (pin 26), $\overline{S_1}$ (pin 27), and $\overline{S_2}$ (pin 28) are used for bus-control status information. This status is used by the 8288 bus-controller chip to generate all memory and I/O control signals. Table 11.2 describes the bus-control functions of status lines $\overline{S_0}$, $\overline{S_1}$, and $\overline{S_2}$.

TABLE 11.2
Bus-control functions

$\overline{S_2}$	$\overline{S_1}$	$\overline{S_0}$	Control Function
0	0	0	Interrupt acknowledge
0	0	1	Read I/O port
0	1	0	Write to I/O port
0	1	1	Halt
1	0	0	Code access
1	0	1	Read memory
1	1	0	Write to memory
1	1	1	Passive

$\overline{\text{LOCK}}$ (pin 29) is a tristate active low signal that prohibits coprocessors from gaining control of the bus. It remains active for instructions prefixed by LOCK.

$\overline{\text{RQ}}/\overline{\text{GT}}_1$ (pin 30) and $\overline{\text{RQ}}/\overline{\text{GT}}_0$ (pin 31) are used with coprocessors to **request** and **grant** bus control to the coprocessor. These are time-multiplexed bidirectional active low signals. Thus the coprocessor requests bus control during one clock cycle and the master MPU grants bus control during another clock cycle.

MPU Timing

All microprocessors require timing signals to synchronize operations. The 8284A clock generator chip provides the basic timing requirements for the 8086/8088 microprocessor. Furthermore, it provides the basic timing for the entire microcomputer system, as shown in Figure 11.3. It is interesting to note that the 8284A and the 82C84A are not exact replacements.

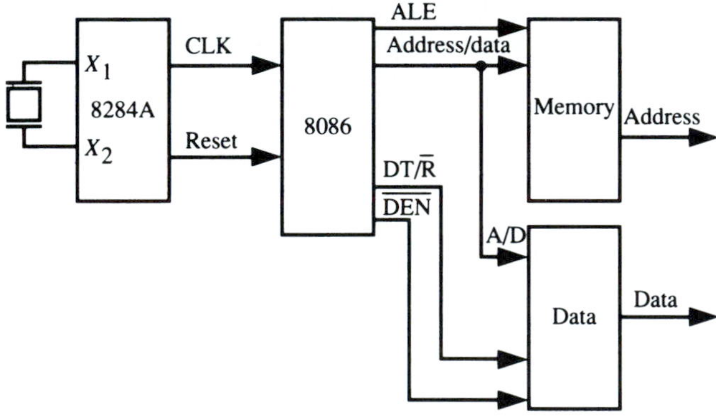

FIGURE 11.3
Basic timing circuit for system

The 8284A is an 18-pin integrated circuit and is illustrated in Figure 11.4. We now discuss the functions of each of the signal pins.

1. V_{CC} (pin 18). Power supply input pin, which is equal to $+5\ V_{DC}$.
2. GND (pin 9). Power supply ground.
3. X_1 (pin 17), X_2 (pin 16). External crystal input pins. These are used to provide the source input frequency for the clock generator. On the IBM PC this fundamental frequency is 14.31818 MHz.
4. OSC (pin 12). The oscillator output signal is a TTL-level signal at the same frequency as the crystal. It is used as an input to other clock generator chips in the microcomputer system.
5. CLK (pin 8). The clock signal is the input timing signal for the MPU. Its frequency is one-third of the crystal frequency with a 33% duty cycle, as shown in Figure 11.4 (one-third on, two-thirds off).

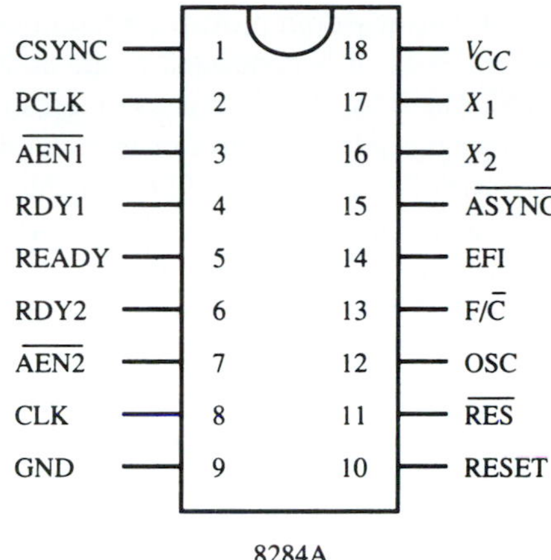

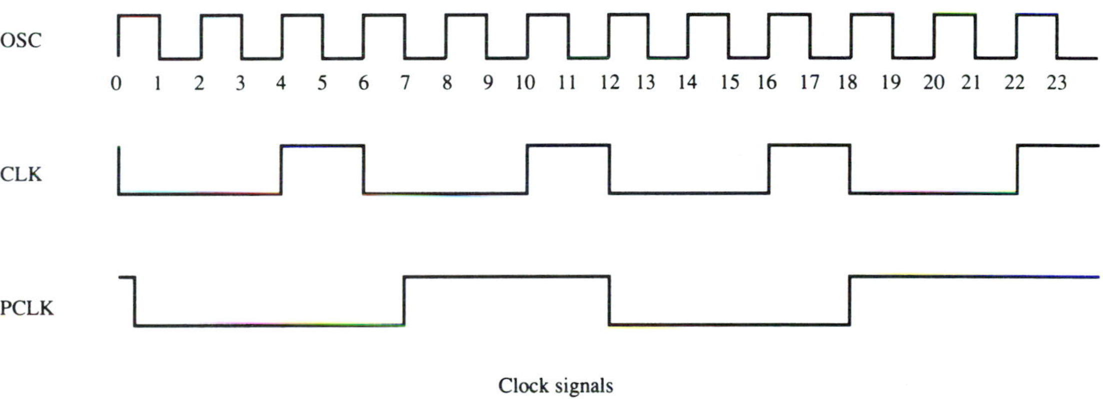

FIGURE 11.4
8284A clock generator and waveforms

6. PCLK (pin 2). The peripheral clock signal provides timing for various peripheral devices in the system. Its frequency is one-half of CLK, or one-sixth of the crystal, with a 50% duty cycle, as shown in Figure 11.4.
7. $F/\overline{C}$ (pin 13). The frequency/crystal input determines if the input signal is a crystal or an external clock frequency. A logic level 0 defines a crystal input to X_1 and X_2 (pins 17 and 16). A logic level 1 defines an external input to EFI (pin 14).
8. EFI (pin 14). The external frequency input pin is used as a timing input signal from a source other than a crystal (i.e., another 8284A).
9. CSYNC (pin 1). The clock synchronization pin is used to synchronize multiple 8284A chips to one clock frequency. When a crystal is used, this input pin is tied low.
10. $\overline{\text{RES}}$ (pin 11). The system reset pin is used to provide a power-on reset pulse. An *RC* circuit is used to provide a pulse that lasts for at least four clock cycles.

11. RESET (pin 10). This reset pin provides a reset output that synchronizes the $\overline{\text{RES}}$ input to the system clock. It is connected to the 8086/8088 RESET input pin.
12. READY (pin 5). The ready pin is an active high signal that tells the MPU that an I/O device or memory is ready to receive or transmit data.
13. $\overline{\text{AEN1}}$ (pin 3), RDY1 (pin 4). The address enable 1 input signal and the ready 1 input signal are gated together to control the READY output line, as shown in Figure 11.5. They are used to create WAIT states.
14. $\overline{\text{AEN2}}$ (pin 7), RDY 2 (pin 6). This is a second set of ready inputs that are ORed with $\overline{\text{AEN1}}$ and RDY1 to control the READY output line, as shown in Figure 11.5.
15. $\overline{\text{ASYNC}}$ (pin 15). The ready synchronization select pin defines the type of ready input being applied to the 8284A. For devices that are normally not ready, $\overline{\text{ASYNC}}$ is high. For devices that are normally ready, $\overline{\text{ASYNC}}$ is kept low.

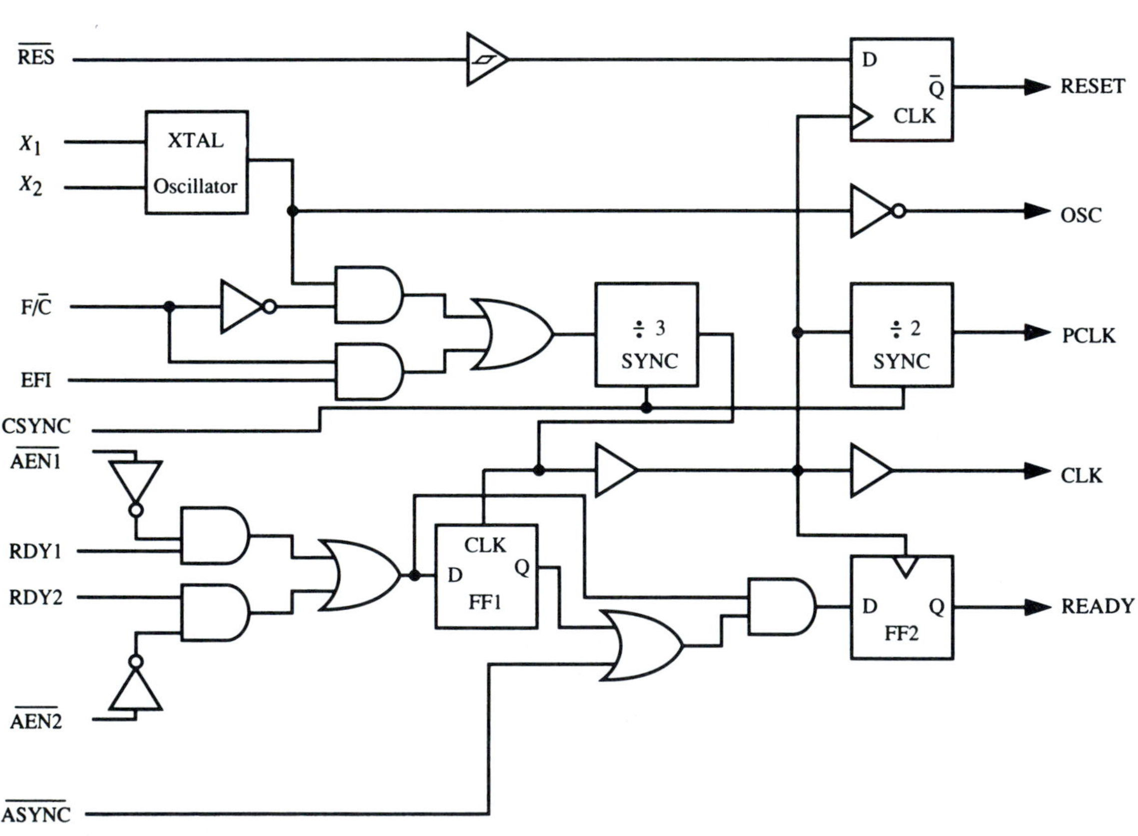

FIGURE 11.5
8284A block diagram

11.2 THE BUS SUBSYSTEM

The bus subsystem is centered around the 8288 bus-controller chip. The 8288 supplies I/O and memory read and write control lines to the other subsystems. Also included in the bus-control system are the address bus, latch/buffers (74LS373), and the data bus transceivers (74LS245). These devices buffer and interface the MPU subsystem to the memory and I/O subsystems, as shown in Figure 11.6.

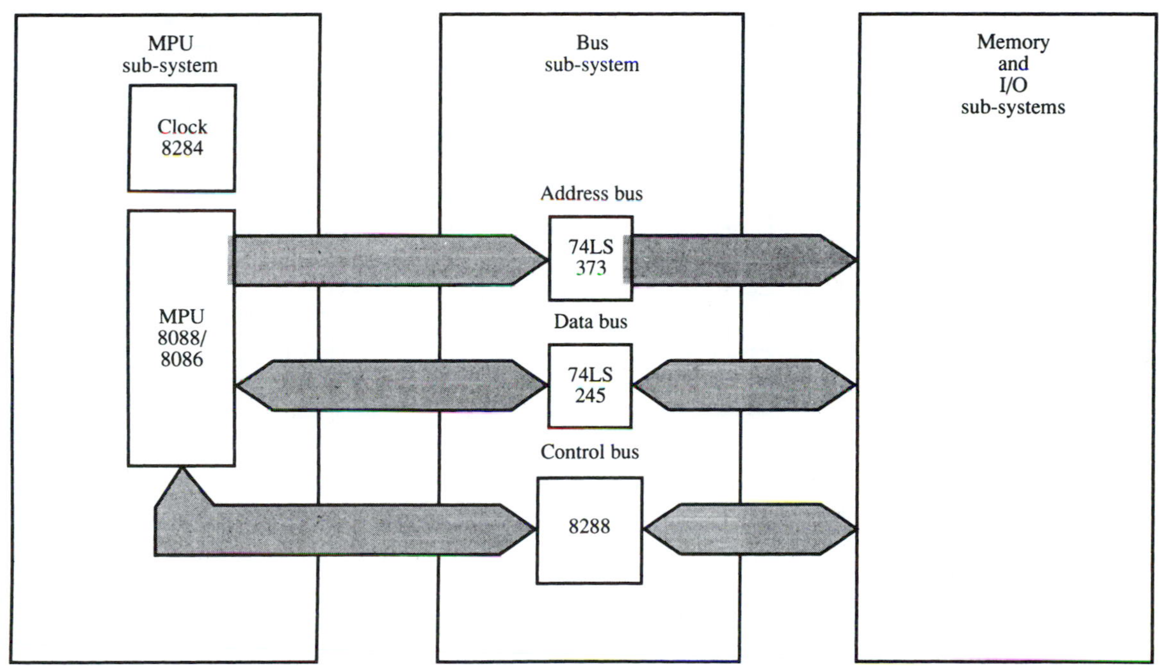

FIGURE 11.6
Maximum mode bus subsystem

The 8288 Bus Controller

When the MPU subsystem is operated in maximum mode, the 8288 bus controller *must* be used to provide the control signals that were *eliminated* by setting the MPU in *maximum mode.* This was previously described in Figure 11.2. It should be noted that the IBM PC and compatibles are operated in maximum mode to allow for the use of an optional 8087 math coprocessor.

The block diagram and pin configuration for the 8288 bus controller are shown in Figure 11.7. The status decoder block is used to decode status lines $\overline{S_0}$, $\overline{S_1}$, and $\overline{S_2}$, as was described previously in Table 11.2. The control logic block is used to determine what command the 8288 is to issue. The command signal generator block issues the output command. The output commands are as follows on page 353:

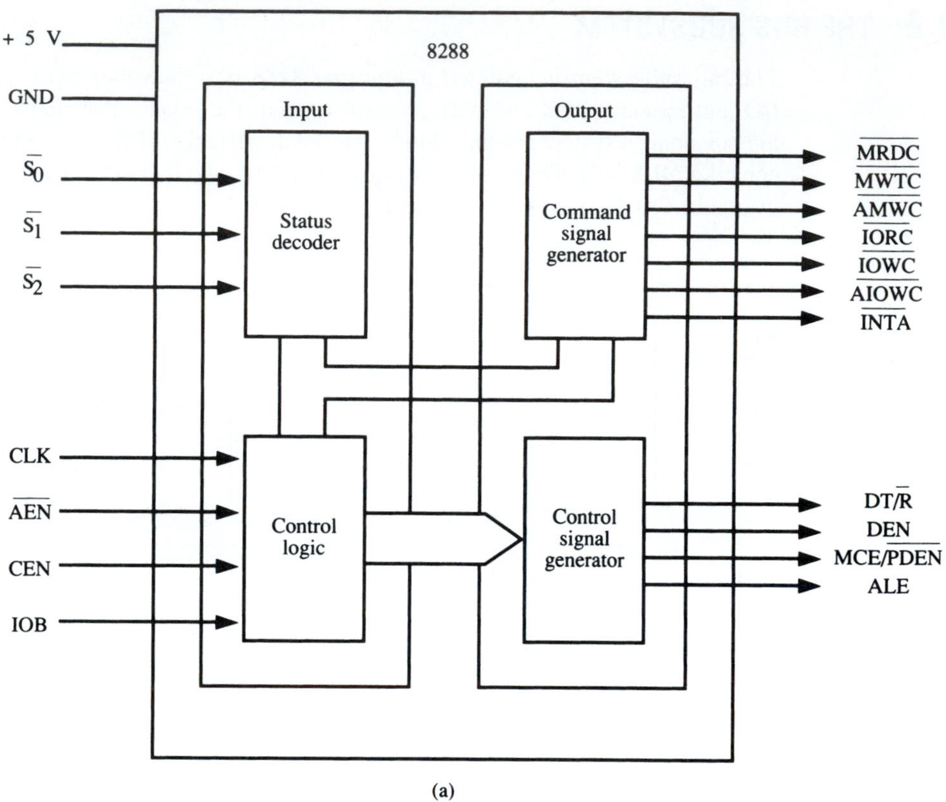

(a)

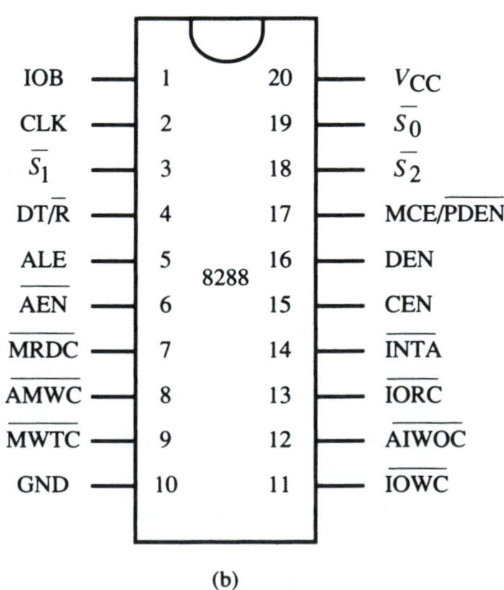

(b)

FIGURE 11.7

$\overline{\text{MRDC}}$ Memory read command
$\overline{\text{MWTC}}$ Memory write command
$\overline{\text{IORC}}$ I/O read command
$\overline{\text{IOWC}}$ I/O write command
$\overline{\text{AMWC}}$ Advanced memory write command
$\overline{\text{AIOWC}}$ Advanced I/O write command
$\overline{\text{INTA}}$ Interrupt acknowledge

The control signal generator block interfaces to the address latch/buffers, the data transceivers, and the interrupt-control signals. The output-control signals are as follows:

DT/$\overline{\text{R}}$ Data transmit/receive
DEN Data enable
MCE/$\overline{\text{PDEN}}$ Master cascade enable/peripheral data enable
ALE Address latch enable

The signal functions of the 8288 bus controller are as follows:

1. V_{CC} (pin 20). Power supply input voltage (+5 V_{DC}).
2. GND (pin 10). Power supply ground.
3. $\overline{S_0}$, $\overline{S_1}$, $\overline{S_2}$ (pins 19, 3, and 18, respectively). Status input signals from the MPU subsystem. These signals are decoded by the 8288 to generate the bus control signals as previously described in Table 11.2.
4. CLK (pin 2). Input clock from the 8284A. This signal provides bus-control timing.
5. $\overline{\text{AEN}}$ (pin 6). Address enable tells the 8288 to issue the memory-control signals.
6. CEN (pin 15). Command enable controls all the command outputs as well as control outputs DEN and $\overline{\text{PDEN}}$. When CEN is high, the appropriate command/control signals are activated. When CEN is low, these signals are disabled.
7. IOB (pin 1). Input/output bus mode signal. When this input signal is high, the 8288 is in the I/O bus mode. When it is low, the 8288 operates in the system bus mode. The I/O mode is used when separate buses are available for I/O and memory. The system bus mode isused when a single bus is shared for I/O and memory.
8. $\overline{\text{MRDC}}$ (pin 7). The memory read command signal is an active low output. It tells the memory to put data onto the data bus for the MPU to read. This is called a memory read operation.
9. $\overline{\text{MWTC}}$ (pin 9). The memory write command signal is an active low output. It tells the memory that the data on the data bus is to be written or recorded into memory. This is called a memory write operation.
10. $\overline{\text{AMWC}}$ (pin 8). The advance memory write command signal is an active low output. It is used to provide an early or advance notice to memory that a write operation is going to be performed.
11. $\overline{\text{IORC}}$ (pin 13). The I/O read command signal is similar to the $\overline{\text{MRDC}}$ signal except that it applies to an I/O device. It tells the I/O device that the MPU wants to read data from the device.
12. $\overline{\text{IOWC}}$ (pin 11). The I/O write command signal tells an I/O device that the MPU wants to write data to the I/O device. This is called an I/O write operation.
13. $\overline{\text{AIOWC}}$ (pin 12). The advance I/O write command signal is similar to the AMWC signal except it applies to an I/O device. It provides an early, or advanced, notice to the I/O device that a write operation is going to be performed.

14. $\overline{\text{INTA}}$ (pin 14). The interrupt acknowledge signal is used to tell an interrupting device that the MPU has accepted or acknowledged the interrupt request.
15. DT/$\overline{\text{R}}$ (pin 4). The data transmit/receive signal is used to control the direction of data through a transceiver. A high indicates that the MPU is transmitting data. A low indicates that the MPU is receiving data.
16. DEN (pin 16). The data enable signal is an active high output signal. It is used to turn on or enable the data transceivers.
17. MCE/$\overline{\text{PDEN}}$ (pin 17). Master cascade enable (MCE), with IOB low during an interrupt operation, signals the interrupt controller that a cascaded address is to be read. It combines the interrupt address (vector) with the address bus to create the service address. *Peripheral* data enable ($\overline{\text{PDEN}}$), together with IOB, is used to enable the I/O bus transceivers.
18. ALE (pin 6). The address latch enable signal is an active high output. It is used to strobe an address into the address latch circuitry.

The 74LS373 Latch

The 74LS373 is a tristate output 8-bit latch. It is specifically designed for high-capacitance or low-impedance loads. It contains eight *D*-type latches in which the *Q* outputs will follow the data (*D*) inputs. Figure 11.8 shows the logic diagram, pin configuration, and truth table for the 74LS373.

The output control pin ($\overline{\text{OC}}$) determines whether the latches are in the high-impedance state or attached to the bus as indicated in the truth table (Figure 11.8). $\overline{\text{OC}}$ does not affect the internal operation of the latches. Data can be retained and new data entered even while the outputs are in the high-impedance or floating state.

The latch enable input pin (*G*), when high, allows the *Q* outputs to *follow* the data (*D*) inputs. When the latch enable goes low, the *Q* outputs will be latched to the state that the data (*D*) was set up to.

In an 8086 MPU microprocessor system the 74LS373 is often used to latch and buffer the address lines from the multiplexed buses. Recall that the address information is on the bus for only a short time. The latch, controlled by the ALE (address latch enable) signal, is used to capture the address information.

The 74LS245 Transceiver

The 74LS245 is an 8-bit bidirectional bus transceiver. This tristate device is used to buffer and control the direction of data flow on the data bus. Figure 11.9 shows the logic diagram, pin configuration, and truth table for this device.

When the enable pin $\overline{G}$ is high, the device enters the high-impedance state. When the enable pin $\overline{G}$ is low, the data flows in the direction determined by the DIR pin. If DIR is low, data flows from the *B* inputs to the *A* outputs. If DIR is high, data flows from the *A* inputs to the *B* outputs.

In an 8086 MPU microcomputer system the 74LS245 is often used to control the direction of data flow on the data bus. When the DIR pin is tied to the DT/$\overline{\text{R}}$ (data transmit/receive) signal line, it is used to determine whether data is to be transmitted or received by the MPU. The enable input $\overline{G}$ is tied to the DEN (data enable) signal to control the output connection to the data bus.

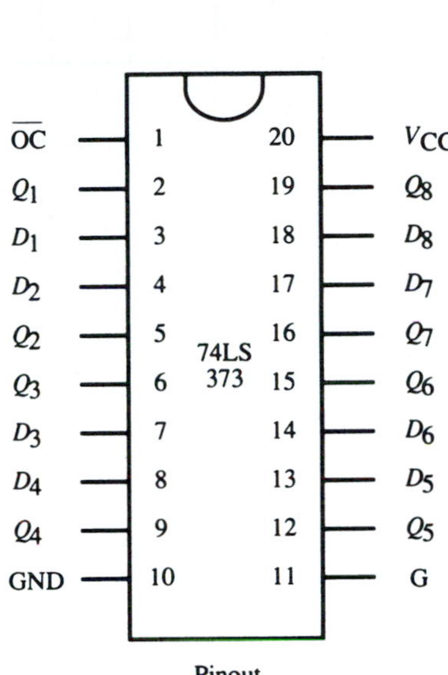

$\overline{OC}$ Output control	G Enable Latch	D Input	Q Output
L	H	H	H
L	H	L	L
L	L	X	No change
H	X	X	Hi-Z

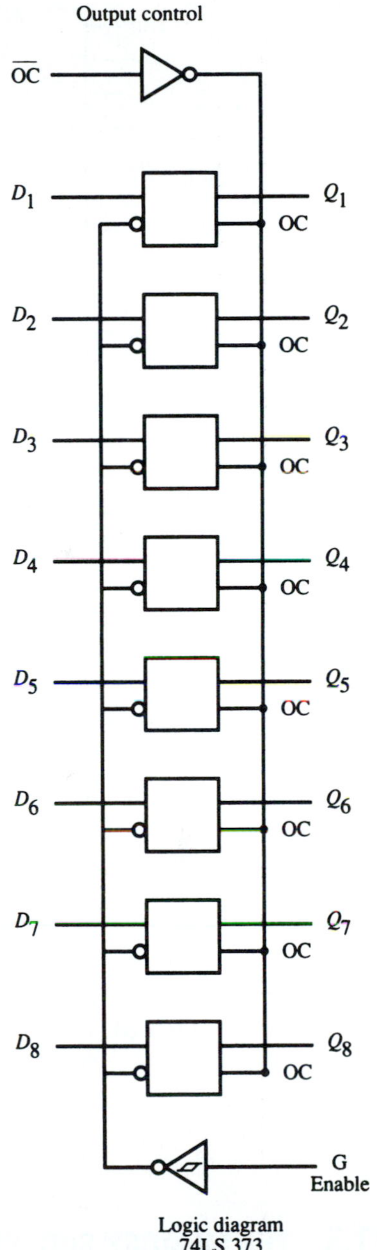

FIGURE 11.8
74LS373 *D*-type latch (8-bit)

Signal	Pin	Pin	Signal
DIR	1	20	V_{cc}
A_1	2	19	$\overline{G}$
A_2	3	18	B_1
A_3	4	17	B_2
A_4	5	16	B_3
A_5	6	15	B_4
A_6	7	14	B_5
A_7	8	13	B_6
A_8	9	12	B_7
GND	10	11	B_8

Pinout

$\overline{G}$	DIR	OUT
L	L	*B-A*
L	H	*A-B*
H	X	Hi-Z

Truth Table

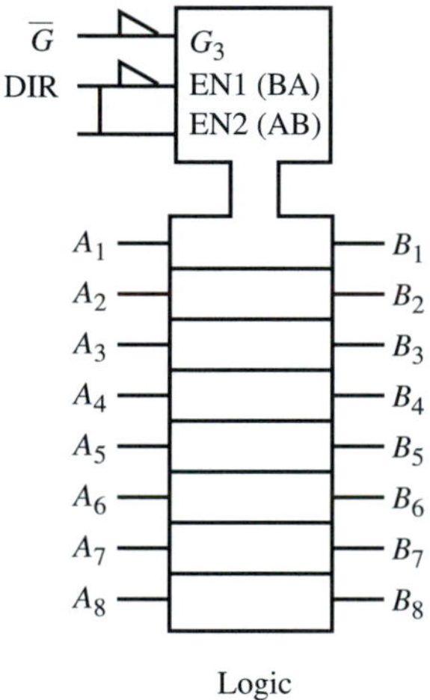

Logic

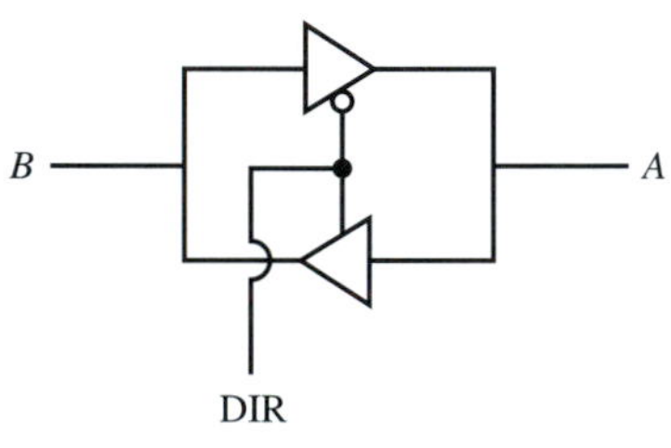

Functional model

FIGURE 11.9
74LS245 bus transceiver (8-bit)

11.3 THE MEMORY SUBSYSTEM

The memory subsystem is centered around the microcomputer system's RAM and ROM devices. Since the amount of addressable memory is large, a method of selecting the desired memory address is necessary. The selection method is known as **address decoding.** In more sophisticated systems it is desirable to have a method that allows direct and automatic access to the memory without utilizing the MPU. This frees up the MPU for other tasks, thereby speeding up the entire system operation. This process is called **direct memory access (DMA)** and requires a special, dedicated device like the 8237 DMA controller.

RAM and ROM

Memory devices were discussed in detail in Chapter 7. Recall that there are two basic types of memory devices. ROM is like a textbook. The information is preprinted and meant to be read only. The ROM is used to store information that the microcomputer system always needs to operate. ROMs are referred to as nonvolatile, since the information is not destroyed or lost when the power is turned off. RAM is like a notebook. The information is meant to be written to or read from. RAM is used in a microcomputer system for temporary storage of information. RAMs are volatile, that is, their information is lost or destroyed when the power is turned off. As was discussed in Chapter 7, there are several different types of RAM and ROM devices. RAM can be either static or dynamic. ROM can be preprogrammed by the manufacturer or field programmed by the user (i.e., PROMs, EROMs, and EEROMs).

Address Decoding

The **address decoder** is used to interface the MPU subsystem to the memory subsystem. The MPU address bus has 20 address lines, but most memory devices do not. Therefore, some circuitry is required to handle the mismatch in the number of address lines. This circuitry is called the address decoder. Address decoding facilitates memory expansion and efficiency.

The simplest address decoder circuit utilizes a single gate as shown in Figure 11.10. Since the ROM chip select is active low, we utilize a NAND gate to perform the decode.

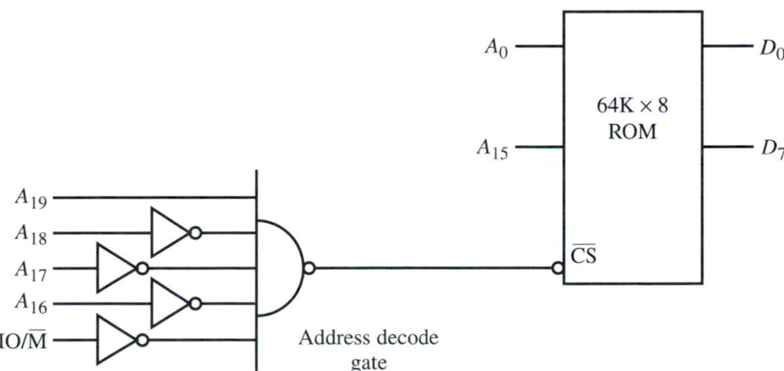

A_{19}	A_{18}	A_{17}	A_{16}	A_{15}	A_{14}	A_{13}	A_{12}	A_{11}	A_{10}	A_9	A_8	A_7	A_6	A_5	A_4	A_3	A_2	A_1	A_0	
1	0	0	0	X	X	X	X	X	X	X	X	X	X	X	X	X	X	X	X	
1	0	0	0	0	0	0	0	0	0	0	0	0	0	0	0	0	0	0	0	Lowest
1	0	0	0	1	1	1	1	1	1	1	1	1	1	1	1	1	1	1	1	Highest

Address range 80000 to 8FFFF

FIGURE 11.10
NAND gate decoding

When all of the inputs to the NAND gate are high, the output will be low, which will enable the ROM. When performing a memory operation, the signal $IO/\overline{M}$ will be low. The inverter converts the low to a high and insures that the ROM will only be activated on memory operations. To determine the active memory range, write the required bits followed by all zeros and then all ones. All zeros represents the lowest address in the range and all ones represents the highest address in the range.

Figure 11.11 illustrates another example of NAND gate decoding. Both illustrations are examples of **partial address decoding.** Partial address decoding involves the use of only some, but not all, of the available address lines. Partial address decoding results in a range of addresses that will select the device. If all of the address lines are utilized, only one unique address will select the device. When all the address lines are utilized, it is referred to as **full address decoding.** Full address decoding helps to avoid conflicts and address overlaps but is more expensive to implement.

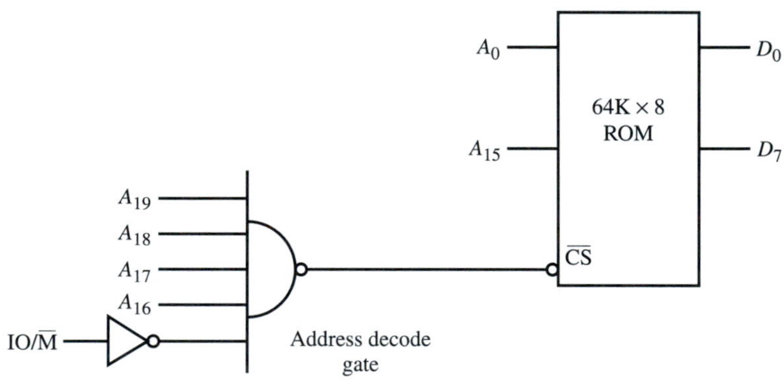

A_{19}	A_{18}	A_{17}	A_{16}	A_{15}	A_{14}	A_{13}	A_{12}	A_{11}	A_{10}	A_9	A_8	A_7	A_6	A_5	A_4	A_3	A_2	A_1	A_0	
1	1	1	1	X	X	X	X	X	X	X	X	X	X	X	X	X	X	X	X	
1	1	1	1	0	0	0	0	0	0	0	0	0	0	0	0	0	0	0	0	Lowest
1	1	1	1	1	1	1	1	1	1	1	1	1	1	1	1	1	1	1	1	Highest

Address range F0000 to FFFFF

FIGURE 11.11
NAND gate address decoder

The 74LS138, a one-of-eight line decoder, is a commonly used device for address decoding. This device is illustrated in Figure 11.12. Selection lines A, B, and C are used to decode the desired output. Since there are three select lines, there are 2^3 combinations, or eight possible outputs (0–7). E_1, E_2, and E_3 are chip enable inputs. E_1 and E_2 are active low, and E_3 is active high. When the device is disabled, the output pins are all high, as indicated in the truth table of Figure 11.12. When the device is enabled, only the selected output will go low.

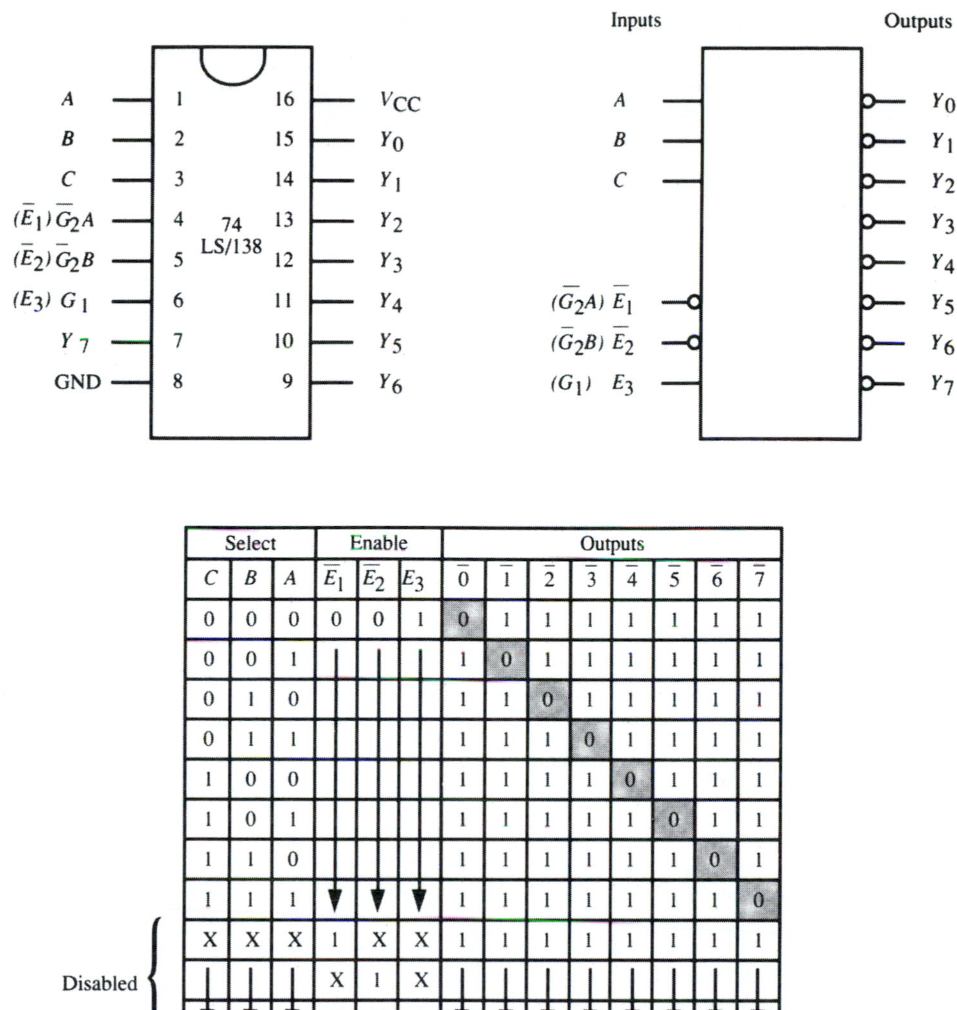

Select			Enable			Outputs							
C	B	A	$\overline{E}_1$	$\overline{E}_2$	E_3	$\overline{0}$	$\overline{1}$	$\overline{2}$	$\overline{3}$	$\overline{4}$	$\overline{5}$	$\overline{6}$	$\overline{7}$
0	0	0	0	0	1	0	1	1	1	1	1	1	1
0	0	1	↓	↓	↓	1	0	1	1	1	1	1	1
0	1	0				1	1	0	1	1	1	1	1
0	1	1				1	1	1	0	1	1	1	1
1	0	0				1	1	1	1	0	1	1	1
1	0	1				1	1	1	1	1	0	1	1
1	1	0				1	1	1	1	1	1	0	1
1	1	1	↓	↓	↓	1	1	1	1	1	1	1	0
X	X	X	1	X	X	1	1	1	1	1	1	1	1
↓	↓	↓	X	1	X	↓	↓	↓	↓	↓	↓	↓	↓
			X	X	0								

Note : X = don't care

FIGURE 11.12
74LS 138 8-output line decoder

Figure 11.13 illustrates a typical ROM decoding circuit. The ROM decoding circuit is controlled by the 74LS138 IC. Since ROM is usually located at the high end of the memory map, address lines A_{19}, A_{18}, A_{17}, and A_{16} (the highest-order lines) are used to enable the decoder device (74LS138). The 74LS138 is enabled for all addresses above *F*0000. This occurs when A_{19}, A_{18}, A_{17}, and A_{16} are all logic level ones ($1111 = F$). M/$\overline{\text{IO}}$ is gated with the high-order address lines to define a memory operation. This signal is high for memory operations and low for I/O operations. The next three address lines, A_{15}, A_{14}, and A_{13}, are used to determine which ROM device to select. Each of the 74LS138 outputs is

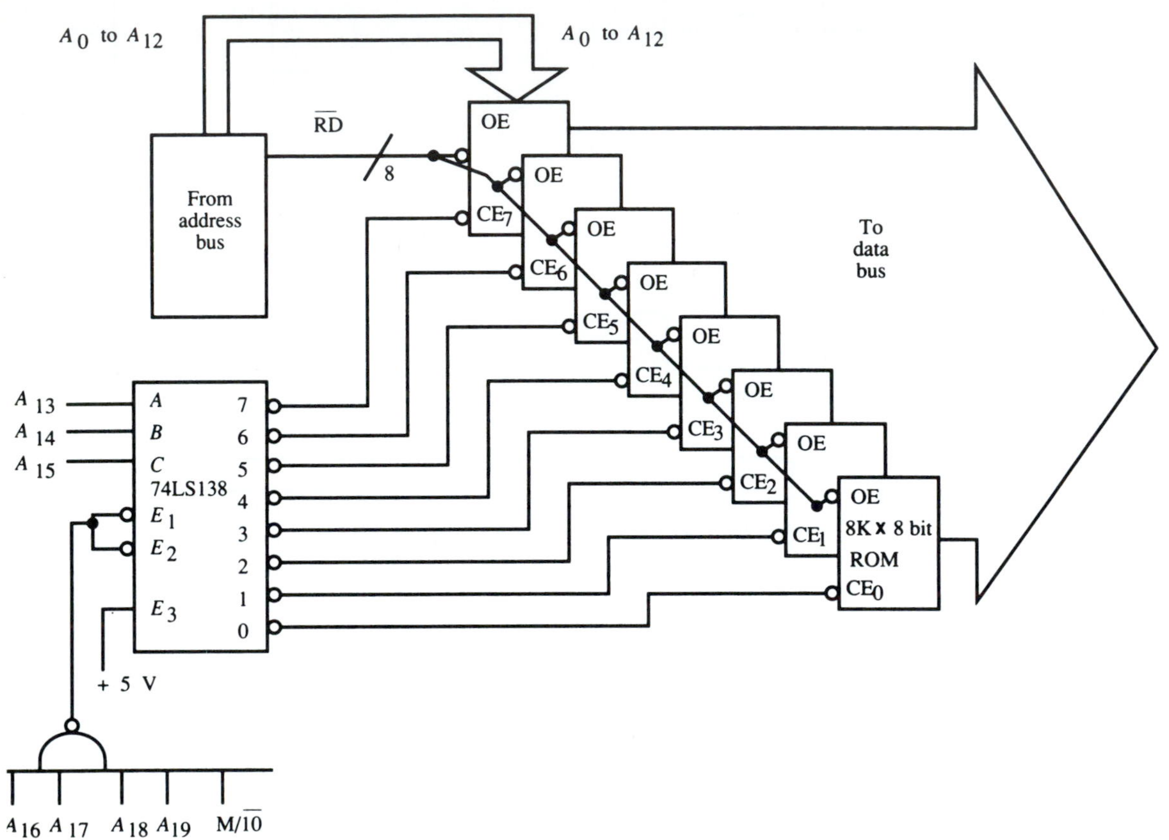

FIGURE 11.13
ROM decode circuit

connected to only one ROM chip enable input (CE). Thus only one ROM device is enabled at any time. Address lines A_{12} through A_0 from the bus subsystem are connected to each individual ROM device. These lines determine a unique address in the selected ROM device. The data bits are outputted on the data lines when $\overline{RD}$ goes low. $\overline{RD}$ is connected to the output enable (OE) pin of each ROM device.

Figure 11.14 illustrates how the actual ROM decoding is performed. The 4 high-order bits (A_{19}, A_{18}, A_{17}, A_{16}) are used to enable the 74LS138, which defines a ROM memory select. The next 3 bits (A_{15}, A_{14}, A_{13}) define which ROM is to be selected, in this case, ROM 0. The 13 low-order bits (A_{12}–A_0) define the unique address of interest on ROM 0. They also define the memory range of ROM 0 as *F*0000 to *F*1*FFF*. The memory range of ROM 1 would be *F*2000 to *F*3*FFF*. Listed in Table 11.3 are the memory ranges for each of the eight decoded ROMs.

A_{19}	A_{18}	A_{17}	A_{16}	A_{15}	A_{14}	A_{13}	A_{12}	A_{11}	A_{10}	A_9	A_8	A_7	A_6	A_5	A_4	A_3	A_2	A_1	A_0
1	1	1	1	0	0	0	X	X	X	X	X	X	X	X	X	X	X	X	X

ROM memory select — ROM device select — ROM address

ROM 0 range

A_{19}	A_{18}	A_{17}	A_{16}	A_{15}	A_{14}	A_{13}	A_{12}	A_{11}	A_{10}	A_9	A_8	A_7	A_6	A_5	A_4	A_3	A_2	A_1	A_0	
1	1	1	1	0	0	0	X	X	X	X	X	X	X	X	X	X	X	X	X	
1	1	1	1	0	0	0	0	0	0	0	0	0	0	0	0	0	0	0	0	Lowest
1	1	1	1	0	0	0	1	1	1	1	1	1	1	1	1	1	1	1	1	Highest

Address range F0000 to F1FFF

FIGURE 11.14
A ROM address decode

DMA

DMA is a type of I/O technique in which data can be transferred between the microcomputer memory and an external device without utilizing the MPU. DMA is typically used to transfer blocks of data between the memory subsystem and an external device. A DMA read operation transfers data from the memory to an external device. A DMA write operation transfers data from an external device to memory.

Since the main purpose of the DMA operation is to transfer data between external devices and memory without involving the MPU, another device is required. This device is called a DMA controller. The DMA controller must be capable of performing read and write operations in the same manner as the MPU. Therefore, the DMA controller is

TABLE 11.3
ROM address decode ranges

ROM	Lowest Address	Highest Address
0	*F*0000	*F*1*FFF*
1	*F*2000	*F*3*FFF*
2	*F*4000	*F*5*FFF*
3	*F*6000	*F*7*FFF*
4	*F*8000	*F*9*FFF*
5	*FA*000	*FBFFF*
6	*FC*000	*FDFFF*
7	*FE*000	*FFFFF*

actually a special-purpose microprocessor whose only task is to perform high-speed data transfers between memory and an external device.

The INTEL 8237 DMA controller is a 40-pin programmable device that is compatible with the 8086/8088 microprocessor. The 8237 has four independent DMA channels, as shown in Figure 11.15. This means that one 8237 can provide DMA transfers to several external devices—for example, a cassette recorder, a floppy disk drive, a Winchester disk drive, and some other external hardware circuitry. Data transfers begin with the DMA request lines $DREQ_0$–$DREQ_3$. $DREQ_0$ has the highest priority, and $DREQ_3$ has the lowest priority. Handshaking with the DREQ are the acknowledge lines $DACK_0$–

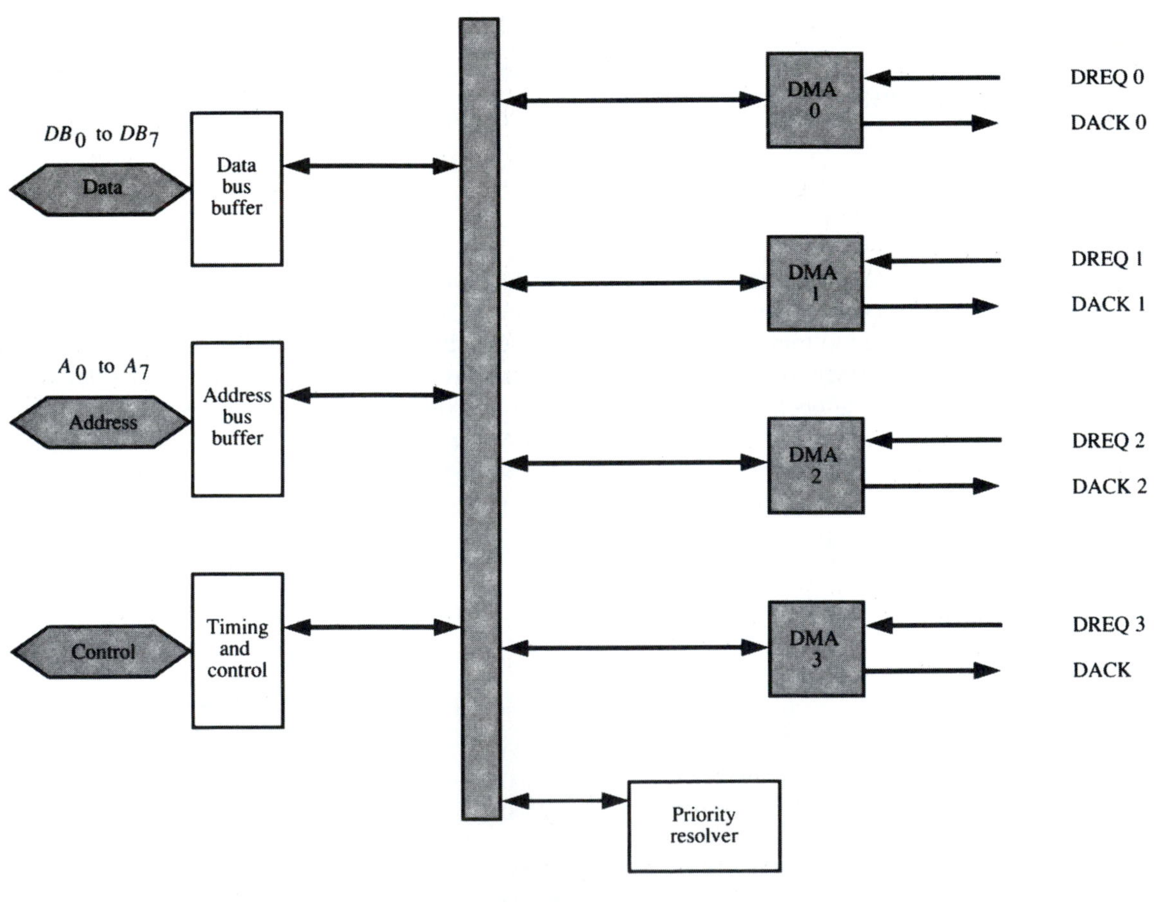

FIGURE 11.15
8237A DMA controller simplified block diagram

$DACK_3$. As their names imply, these lines are used to *acknowledge* a DMA channel request. The 8237 uses the 8086/8088 HOLD signal to take over the system bus. After being initialized by the MPU, the 8237 takes control of the bus in order to perform the DMA operation. Data bits are then transferred between a peripheral or external device without involving the microprocessor. Before discussing the DMA operation in detail, let's look at the basic pin configuration of the 8237, as shown in Figure 11.16.

- CLK (pin 12). Clock input controls the interval operations of the 8237 and may be operated at frequencies up to 10 MHz.
- $\overline{\text{CS}}$ (pin 11). The chip select signal is an active low input used to select the 8237 as an I/O device. This allows the MPU to communicate on the data bus.
- RESET (pin 13). This is an active high input signal that clears command, status, request, and temporary registers.

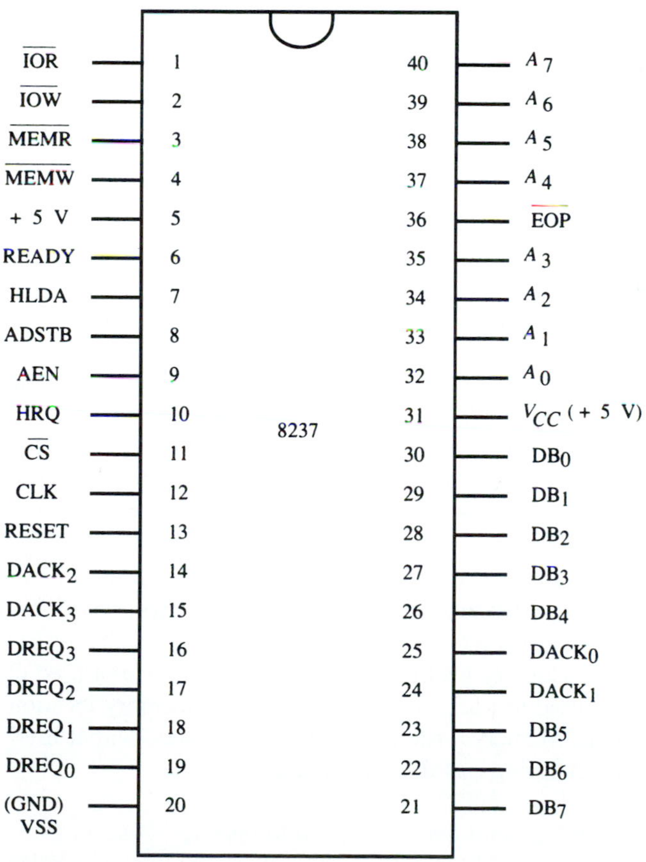

FIGURE 11.16
8237A DMA controller pinout

- READY (pin 6). This is an active high input signal used to accommodate the read and write lines with slow memories or I/O peripheral devices.
- HLDA (pin 7). Hold acknowledge is an active high input signal from the MPU that indicates that the MPU has given up control of its buses.
- $DREQ_0$–$DREQ_3$ (pins 19–16). The DMA request lines are asynchronous channel inputs used by external devices to obtain DMA operations. These signals are programmable to be active high or active low. The RESET pin initializes $DREQ_0$–$DREQ_3$ as active high.
- DB_0–DB_7 (pins 30–26 and 23–21). The data bus lines are tristate, bidirectional signals connected to the system data bus. During DMA operations these lines are multiplexed to output the most significant address bits A_{15}–A_8.
- $\overline{IOR}$ (pin 1). I/O read is an active low, bidirectional, tristate line used during the idle cycle when the MPU is waiting for a DMA request. It is used to read the control registers. In the DMA, or active, cycle, it lets the 8237 access data from external devices during a DMA write transfer.
- $\overline{IOW}$ (pin 2). I/O write is an active low, tristate, bidirectional line. In the idle cycle it is used by the MPU to load information into the 8237. In the DMA, or active, cycle, it is an output control signal used by the 8237 to load data to external devices during a DMA transfer.
- $\overline{EOP}$ (pin 36). End of process is an active low, bidirectional signal. Placing a low on this line terminates DMA service. When used as an output, this signal is used to interrupt the processor to signal the end of the DMA operation.
- A_0–A_3 (pins 32–35). The four least significant address lines are bidirectional, tristate signals. During the idle cycle they are used as inputs. During the DMA, or active, cycle, they are used as outputs.
- A_4–A_7 (pins 37–40). The four higher-order address lines are tristate outputs and provide 4 bits of address information. They are enabled only during a DMA service.
- HRQ (pin 10). Hold request asks the MPU for control of the system's buses.
- $DACK_0$–$DACK_3$ (pins 24, 25, 14, 15). DMA acknowledge is used to notify external devices that a DMA cycle has been granted to one of the devices.
- AEN (pin 9). Address enable is an active high output signal. It is used to enable an address latch that is connected to the data lines D_0–D_7. Recall that these lines are multiplexed to provide the most significant address byte A_{15}–A_8.
- ADSTB (pin 8). The active high address strobe signal is used to strobe the upper address byte into an external latch. AEN enables the latch. ADSTB strobes the address data into the latch.
- $\overline{MEMR}$ (pin 3). Memory read is an active low tristate output used during a DMA read operation. It is used to access data from a selected memory location.
- $\overline{MEMW}$ (pin 4). Memory write is an active low tristate output used to write data to a selected location during a DMA write operation.

The 8237 DMA controller is designed to operate in two major cycles, the idle and active cycles. When the 8237 is in the idle cycle, no external device is requesting a DMA transfer. In this cycle the 8237 samples the DREQ lines every clock cycle to determine if any of the four channels are requesting service. When one of the DMA request lines $DREQ_0$–$DREQ_3$ becomes active, the 8237 will enter the active cycle. When the 8237 is in the idle cycle and a channel requests a DMA transfer, the 8237 outputs an HRQ (hold request) to the MPU and enters the active cycle, as shown in Figure 11.17. It is in this cycle that the DMA transfer actually takes place. When ready, the MPU responds to the 8237 with an HLDA (hold acknowledge), indicating that it has released control of the buses by entering the tristate. The 8237 then responds to the external device requesting the DMA with a DACK (DMA acknowledge), indicating the start of the active cycle. Figure 11.18 illustrates the basic block diagram of the DMA interface. The 8237 interfaces to all four subsystems to perform memory data transfers for the MPU.

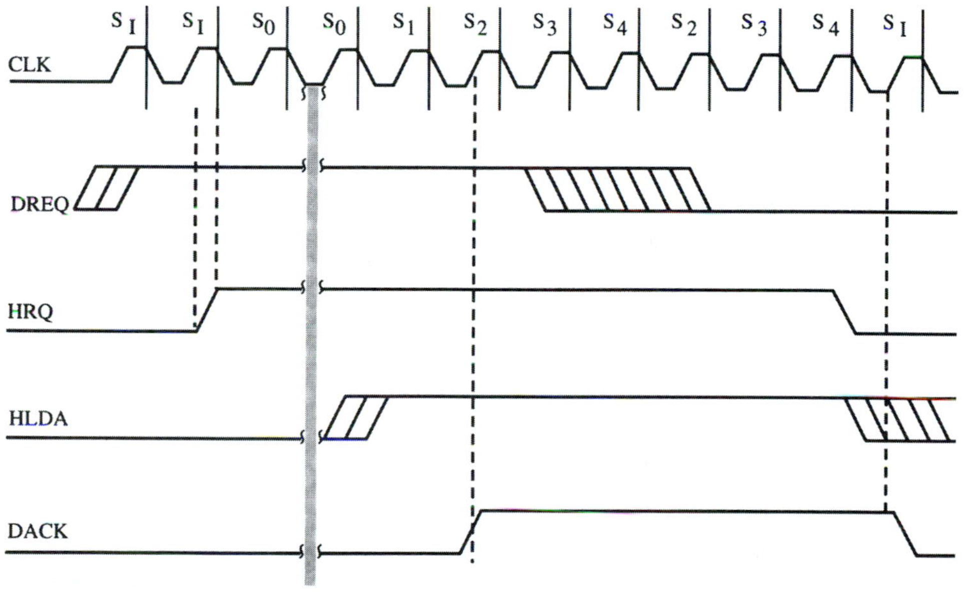

FIGURE 11.17
DMA timing diagram

When the 8086/8088 MPU is in the maximum mode it uses the $\overline{RQ}/\overline{GT}$ (request/grant) signal line to interface with the 8237. These two signals come from a single bidirectional pin and function similar to the HOLD and HLDA in the minimum mode. The 8086/8088 contains two request/grant pins ($\overline{RQ_0}/\overline{GT_0}$, $RQ_1/\overline{GT_1}$), which allow interfacing for up to two DMA controllers or coprocessors, or a maximum of eight external devices. $\overline{RQ}$ (request) is like the HOLD signal. $\overline{GT}$ (grant) is like the HLDA (hold acknowledge) signal. The request/grant pin is used for the request/grant/release cycle.

First, the DMA controller requests a HOLD from the MPU in response to a DREQ (DMA request) from an external device. Next, when the MPU recognizes the request and is ready, it outputs a grant signal on the same signal line back to the DMA controller, as shown in Figure 11.19. The DMA controller can now use the buses to transfer data between memory and an external device. The DMA controller now signals the requesting

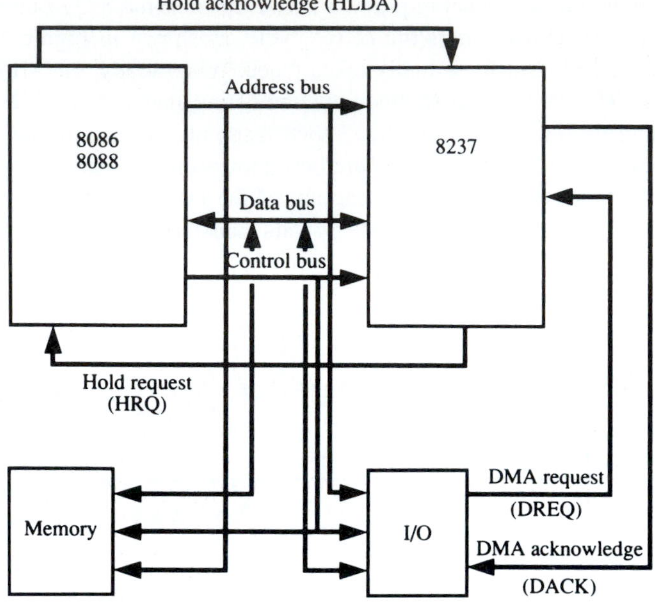

FIGURE 11.18
DMA interface basic block diagram

external device with a DACK (DMA acknowledge) signal. This indicates the start of a DMA transfer. When the DMA controller completes the data transfer, it outputs an EOP (end of process), which indicates the completion of the DMA transfer. It now returns bus control to the MPU.

The 8237 in its active cycle can perform a DMA transfer in one of four modes, *single mode, block mode, demand mode,* and *cascade mode.*

1. *Single transfer mode.* In this mode the 8237 is programmed to transfer one byte of data each time the request is active.

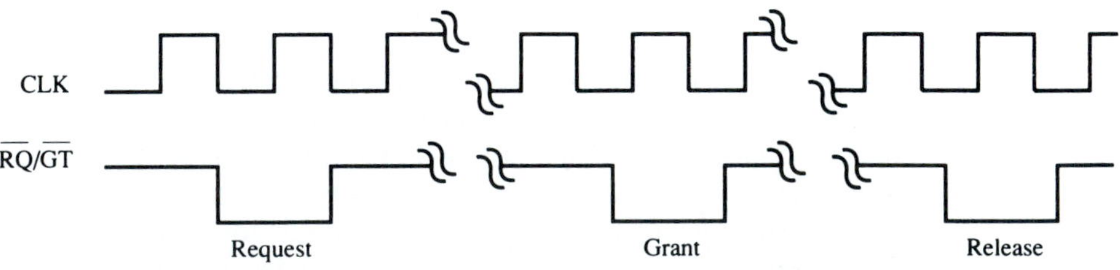

FIGURE 11.19
$\overline{RQ}/\overline{GT}$ timing

2. *Block transfer mode.* In this mode the 8237, once activated by DREQ, continues to make transfers until a block of data is transferred. The 8237 is programmed with the starting address of the data and the number of bytes to be transferred. The transfers continue until the word count register in the 8237 reaches its final count or an EOP is received. The number of bytes transferred will always be one more than the initial number loaded into the word count register.
3. *Demand transfer mode.* In this mode the 8237 is programmed to continue making transfers until an EOP is received or until a DREQ from an external device goes inactive. In this mode one byte of data is transferred for each demand (DREQ) received.
4. *Cascade transfer mode.* This mode is used to cascade more than one 8237 together for system expansion.

The 8237 has a number of internal registers. These registers are used to control the DMA cycle operation.

1. *Current address register.* Each channel has a 16-bit current address register. This register holds the address of the data to be transferred. It is programmed to be *incremented* or *decremented* after each transfer.
2. *Current word register.* Each channel has a 16-bit word count register. This register determines the number of transfers to be performed. It is incremented after each byte transfer.
3. *Base address register.* This 16-bit register stores the original or starting address of the current address register.
4. *Base word register.* This 16-bit register stores the original value of the current word register.
5. *Command register.* This 8-bit register is used to program and control the 8237. It is used to initialize the device.
6. *Mode register.* Each channel has a 6-bit mode register to define its mode of operation (single, block, demand, cascade).
7. *Request register.* This 4-bit register is used to request a DMA transfer by software.
8. *Mask register.* Each channel has a mask register bit that is used to disable incoming DREQ signals.
9. *Temporary register.* This is an 8-bit register used to hold data during a memory-to-memory transfer.
10. *Status register.* This is an 8-bit register used for the microprocessor to read the present status of the 8237.

To use the DMA feature the programmer must first access the internal registers of the 8237. The **command register** must be loaded with a command word to define the initial conditions, mode of operation, and type of operation. Next, depending on the mode and type of operation, the internal registers must be loaded with information defining the starting address of the data in memory, the number of bytes to transfer, where the data is to go, the DMA channel number, and so forth. Finally, the DMA channel's request signal must be enabled. Initialization and programming of the 8237 is accomplished by writing control words to the 8237 using the I/O OUT instruction, which is discussed in "I/O Instructions" in Section 11.4. The OUT instruction is used to send control words to the internal registers. Each bit in the control word defines the setup of an operation. The definition of each bit for each internal register can be found in the manufacturer's 8237 data sheet.

11.4 THE I/O SUBSYSTEM

The I/O subsystem is responsible for the movement of data between the basic microcomputer system and the peripheral or external devices connected to it. It performs the same functions as a seaport or airport for a city. Data bits are moved *in* or *out* of the I/O subsystem in the same way as people and goods are moved *in* and *out* of the seaport or airport. The I/O subsystem exchanges data with peripheral devices through interface circuitry known as **ports.** The peripheral device is physically connected to the port. The port is physically connected to the interface control circuitry, as shown in Figure 11.20.

Once the peripheral device is connected to the port, it needs a method of accessing the MPU. This method is called an **interrupt.** An interrupt is used to cause a temporary halt in the execution of a program. The MPU responds to the interrupt with an **interrupt service routine,** which is a short program or subroutine that instructs the MPU on how to handle the interrupt.

There are two basic types of interrupts, **maskable** and **nonmaskable** interrupts. A nonmaskable interrupt requires an immediate response by the MPU. It is usually used for serious circumstances such as power failure. A maskable interrupt is an interrupt that the

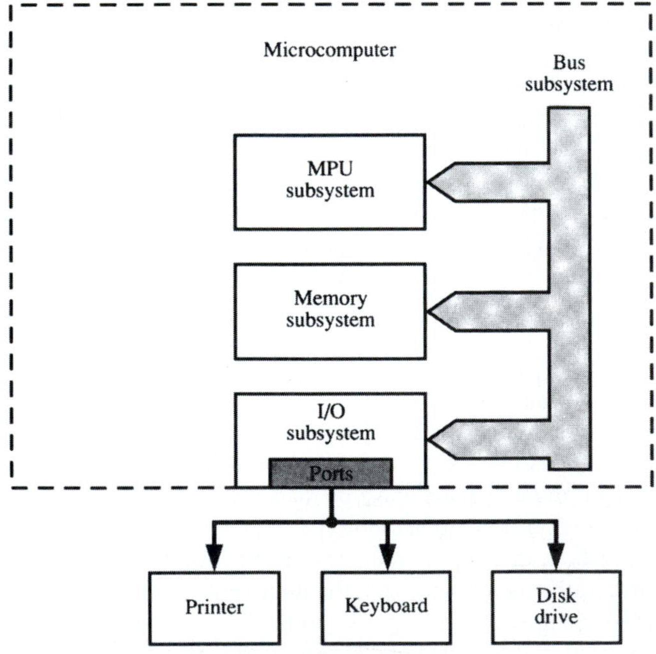

FIGURE 11.20
Microcomputer with I/O subsystem

MPU can ignore depending upon some predetermined condition defined by the status flag register. Interrupts can be generated by both hardware and software. Interrupts are also prioritized to allow for the case when more than one interrupt needs to be serviced at the same time. For example, a *power fail* interrupt has a higher priority than a *printer out-of-paper interrupt.*

I/O Instructions

Before discussing the I/O interface control devices, we need to study two additional 8086/8088 programming instructions, the IN instruction and the OUT instruction. The IN instruction is used to *input* data *to* the MPU *from* a peripheral or external device *through* the I/O port. The OUT instruction is used to *output* data *from* the MPU *to* a peripheral or external device *through* the I/O port. The general format of the IN and OUT instructions is

```
IN      destination, source
OUT     destination, source
```

The IN instruction operand *must be* the AX or AL register. The source operand can be a constant value between 0 and 255 (fixed port addressing) or a variable that can access up to 64K of port locations (variable port addressing). The address used in variable port addressing *must be* contained in the DX register. Two examples of fixed port addressing instructions are

```
1.              IN     AH,0244H
2. PORT__IN     EQU    0F8H
                IN     AX,PORT__IN
```

Two examples of variable port addressing instructions are

```
1. PORT__IN     EQU    0F37AH
                MOV    DX,PORT__IN
                IN     AX,DX
2.              MOV    DX,0F37AH
                IN     AX,DX
```

The OUT instruction is the opposite of the IN instruction. The source operand *must be* the AX or AL register. The destination operand can be a constant value between 0 and 255 (fixed port addressing) or a variable that can access up to 64K of port locations (variable port addressing). The address used in variable port addressing *must be* contained in the DX register. For example,

```
1.              OUT    0F3H,AL
2. PORT__OUT    EQU    0FF7FH
                MOV    DX,PORT__OUT
                OUT    DX,AX
```

In the second example the data in the AX register is outputted through the port defined or pointed to by the DX register. In this case the DX register is pointing to port number *FF7F*H.

I/O Address Decoding

As with memory devices, I/O devices need to have an address. The address is sometimes referred to as a device number or I/O port. Decoding an I/O address is very much the same as decoding a memory address. Figure 11.21 illustrates an example of full I/O

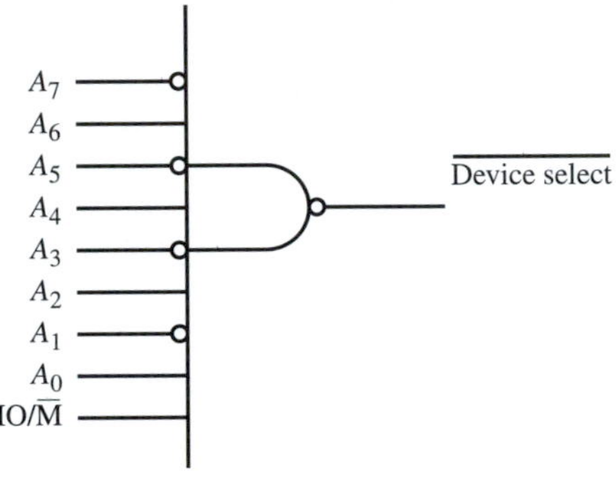

A_7	A_6	A_5	A_4	A_3	A_2	A_1	A_0
0	1	0	1	0	1	0	1
	5				5		

FIGURE 11.21
Full I/O address decoding

decoding on a system with eight address lines. When using the IN or the OUT instruction, the microprocessor will output a 1 on IO/$\overline{\text{M}}$, indicating an I/O instruction. Outputting address 55H on the address bus will cause all of the inputs to the NAND gate to be high. Thus the output of the NAND gate will be low. Figure 11.22 illustrates an example of partial I/O decoding. Any I/O instruction to a device number in the address range *C*000 to *CFFF* will cause the output of the decode date to become active low and select the device.

A technique known as multiple port address decoding is illustrated in Figure 11.23. By utilizing a single 74138 decoding IC, eight different device addresses can be selected. This eliminates eight different decode circuits. The decode NAND gate, the read signal, and the IO/$\overline{\text{M}}$ signal combine to enable the 74138 address lines A_2, A_1, and A_0 select a unique device. Finally, Figure 11.24 on p. 372 illustrates a technique of using the same device number for an input and output port without causing a conflict. The distinction is made by using the read and write lines. The NAND gate again performs the address de-

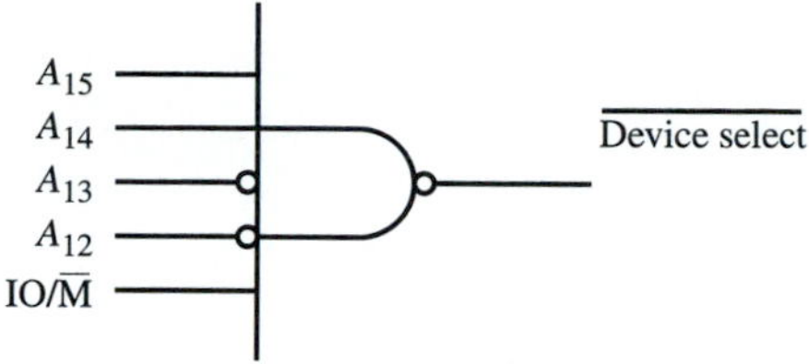

A_{15}	A_{14}	A_{13}	A_{12}	A_{11}	A_{10}	A_9	A_8	A_7	A_6	A_5	A_4	A_3	A_2	A_1	A_0	
1	1	0	0	X	X	X	X	X	X	X	X	X	X	X	X	
1	1	0	0	0	0	0	0	0	0	0	0	0	0	0	0	Lowest
1	1	0	0	1	1	1	1	1	1	1	1	1	1	1	1	Highest

Device select address C000 to CFFF

FIGURE 11.22
Partial I/O device decoding

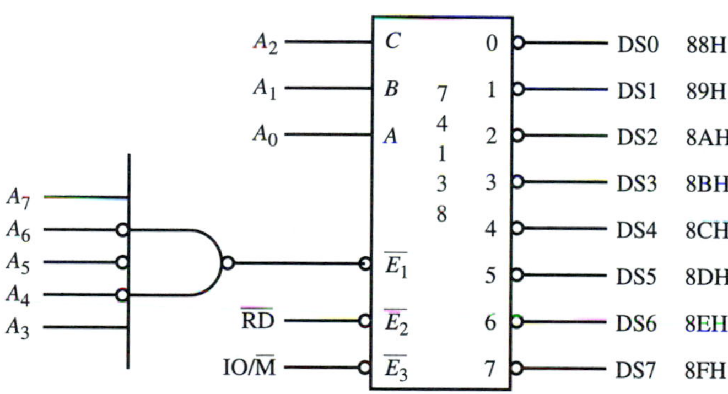

A_7	A_6	A_5	A_4	A_3	A_2	A_1	A_0		
1	0	0	0	1	X	X	X		
1	0	0	0	1	0	0	0	Lowest	88H
1	0	0	0	1	1	1	1	Highest	8FH

88 Activates DS0
8F Activates DS7

FIGURE 11.23
Multiple port address decoding

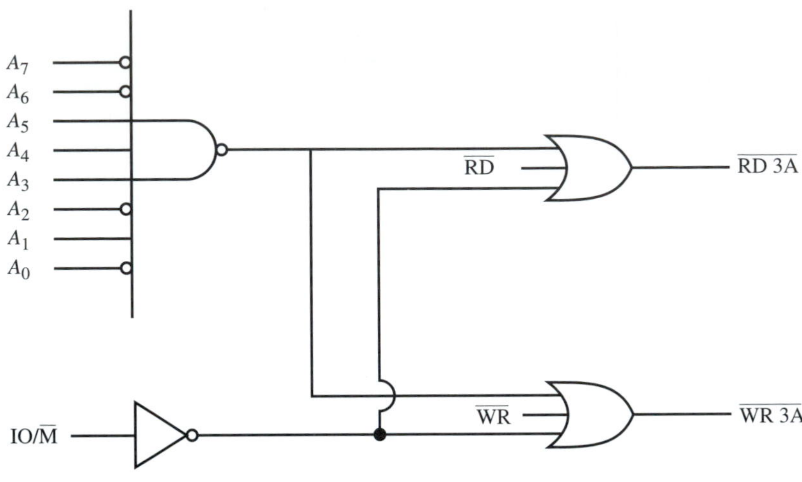

A_7	A_6	A_5	A_4	A_3	A_2	A_1	A_0
0	0	1	1	1	0	1	0
	3				A		

FIGURE 11.24
One address–two device control

coding but selection is controlled by the read and write lines. Note that the IN instruction will activate both the I/O line and the read line only. The OUT instruction will activate both the I/O and the write line only. The circuit works because the microprocessor will never activate both the read and the write lines at the same time. Note that all zeros on the input to either OR gate will enable the device.

The 8255A Programmable Peripheral Interface (PPI)

The 8255A is a **programmable peripheral interface (PPI)** device that connects peripheral devices to the microcomputer system. It is compatible with the 8086/8088 microprocessor and is designed for the implementation of parallel I/O ports to the microcomputer system without the need for additional external circuitry in most cases. The 8255A provides a very flexible parallel interface that is software controlled.

Figure 11.25 gives the functional block diagram of the 8255A PPI. The MPU side of the 8255A PPI includes the 8-bit bidirectional data bus buffers (D_0–D_7) and the read/write control logic signals ($\overline{RD}$, $\overline{WR}$, A_1, A_0, RESET, $\overline{CS}$). The I/O side of the 8255A PPI is represented by port A (I/O signals PA_0–PA_7), port B (I/O signals PO_0–PB_7), and

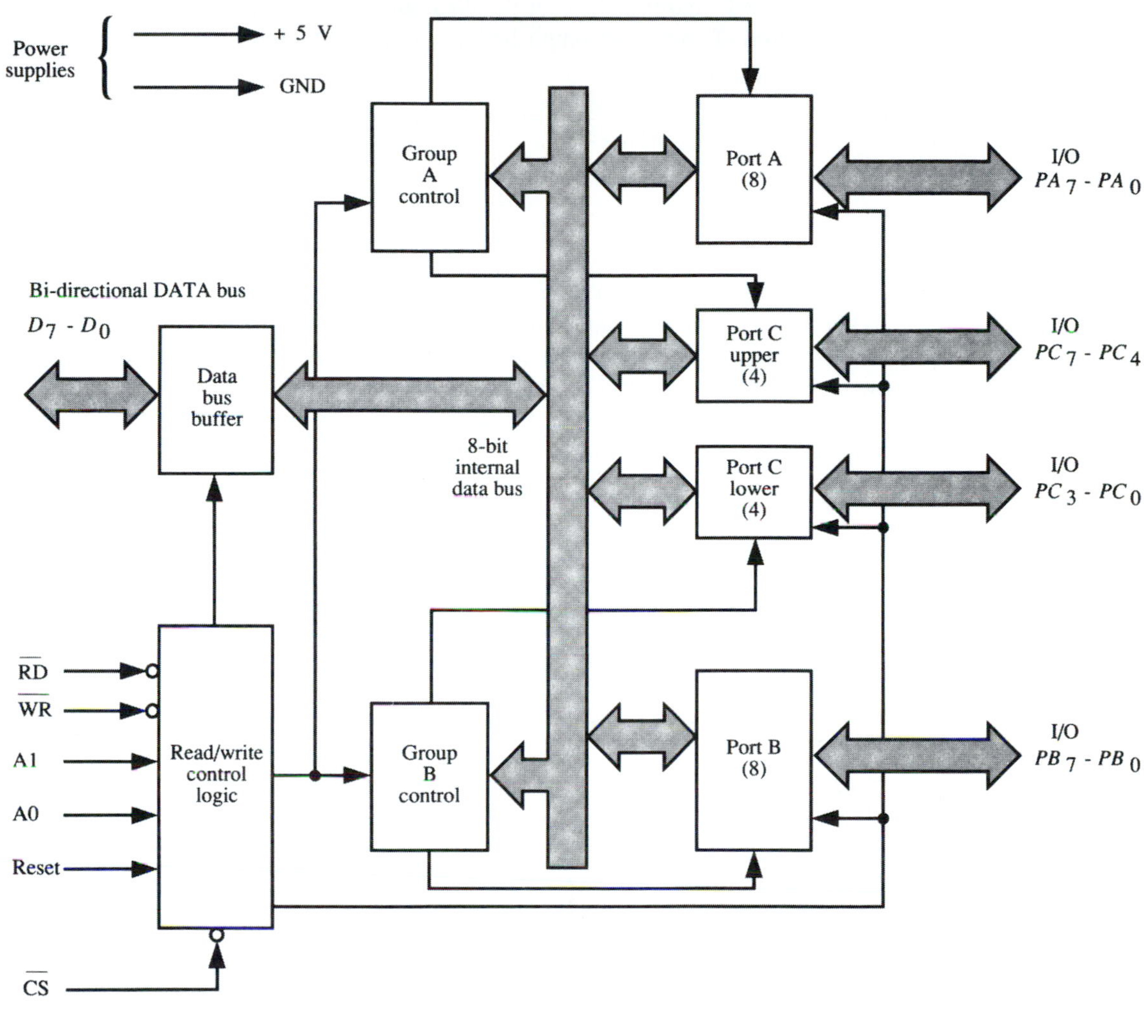

FIGURE 11.25
8255 PPI

port *C* (I/O signals PC_0–PC_7). Ports *A* and *B* are 8-bit bidirectional ports. Port *C* is divided into two 4-bit (nibble) ports. The upper 4 bits are defined as PC_7–PC_4, and the lower 4 bits are defined as PC_3–PC_0. These signal lines are used to transfer data, commands, and status information between the MPU, the 8255A PPI, and the peripheral devices.

Timing of the data transfers to the PPI is controlled by the read ($\overline{RD}$) and write ($\overline{WR}$) control signals. These signals allow the MPU to read from the PPI or write to the PPI. When the read ($\overline{RD}$) signal is active low, the MPU reads data or status information from the PPI over the data bus. When the write ($\overline{WR}$) signal is active low, the MPU writes data or control words into the PPI over the data bus.

The selection of ports is accomplished using input signal lines A_0 and A_1 as follows:

A_1	A_0	Port
0	0	*A*
0	1	*B*
1	0	*C*
1	1	Control

The 8255A is configured for operation by software control. The control register is mapped as an I/O port. Accessing the mapped address allows the user to write a control word into the control register. The control word *initializes* the 8255A by defining its configuration and how it is going to be used. For example, it defines which ports are going to be used. It also defines whether the ports are being used to input or output data. Figure 11.26 illustrates the meaning of the control word bits.

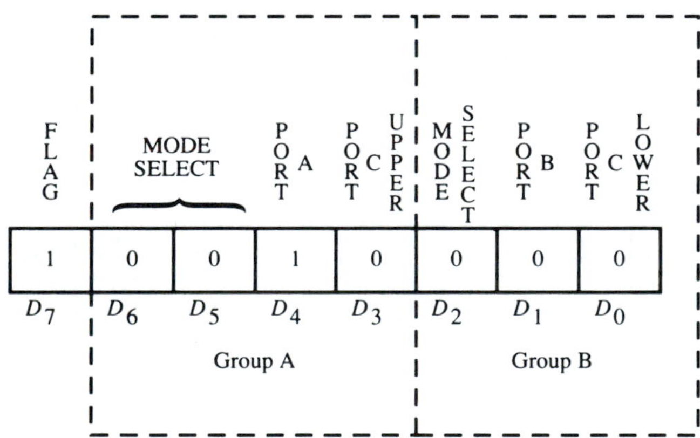

FIGURE 11.26
8255A control registers

Bit D_0 defines whether the port *C* lower nibble is being used as an input or an output. D_1 defines whether port *B* is being used as an input or an output, and so on. A logic level 1 indicates that the port is being used as an input port. A logic level 0 indicates that the port is being used as an output port. Bits D_2, D_5, and D_6 are used to define the mode of operation for the device. D_2 defines the mode of operation for the port *B* and port *C* lower nibbles. D_5 and D_6 define the mode of operation for the port *A* and *C* upper nibbles. Basically there are three modes of operation:

Mode 0: Basic I/O
Mode 1: Strobed I/O
Mode 2: Bidirectional operation

In mode 0 the 8255A functions as a latch—that is, the data once outputted by the MPU remains in the 8255A until it is changed. In mode 1 the data remains in the 8255A for a brief period of time. It must be strobed into an external latch if it needs to be saved. Mode 2 is an advanced mode that is used for bidirectional operation. Note that the port *B* and port *C* lower nibbles can be operated only in mode 0 or mode 1 because there is only one data bit to define the mode of operation. The last control word bit, D_7, is the mode–set flag bit. It must be set to a logic level 1 whenever the mode of operation is to be changed.

In Figure 11.26 the control word defines the operation of the 8255A. Bit D_7 is set, indicating that we are initializing the control register (mode flag set). D_6, D_5, and D_2 are a logic level 0, indicating that all ports are being used for basic I/O operations (mode 0). D_4 is set, defining port *A* as an input port. D_3, D_1, and D_0 are all logic level 0, defining ports *B* and *C* (upper and lower) as output ports. Note that in assembly language programming it is customary to write constants in hexadecimal. Grouping the bits by four, we convert the control word from a binary to a hexadecimal number. In this case, it is equal to 90H.

The following assembly language program illustrates how to initialize and use the 8255A PPI.

```
DATA  SEGMENT
                ORG 100H
CTRLR           EQU 0FFFFH   ;CONTROL REGISTER ADDRESS
INIT            EQU 80H      ;CONTROL WORD
PA              EQU 0FFF8H   ;PORT A ADDRESS
DATA 1          EQU AAH      ;1/0 PATTERN
DATA ENDS
CODE SEGMENT
ASSUME CS:CODE, DS:DATA
                MOV AL,INIT  ;LOAD CONTROL WORD
                MOV DX,CTRLR ;LOAD CONTROL REGISTER ADDRESS
                OUT DX,AL    ;OUTPUT CONTROL WORD
                MOV AL,DATA1 ;LOAD 1/0 PATTERN
                MOV DX,PA    ;LOAD PORT A ADDRESS
                OUT DX,AL    ;OUTPUT DATA
CODE ENDS
END
```

The data segment defines the values that will be used to initialize and control the 8255A. 0*FFFF*H is the mapped address of the control register. 0*FFF*8H is the address of port *A*. 80H is used to initialize all ports as mode 0 operation output ports. Recall that 80H equals 10000000B and corresponds to D_7–D_0, respectively. The data to be outputted is a value *AAH*, which results in a pattern of alternating 1s and 0s. *AAH* equals 10101010B. The code segment begins by moving the initialization value 80H into the AL register. Next, the control register address 0*FFFF*H is moved into the DX register. The OUT instruction outputs the initialization value to the address pointed to by the DX register. In this case it is the address of the 8255A control register. We then move the data into the AL register and the address of port *A* into the DX register. The next OUT instruction sends the data to the address pointed to by the DX register. In this case the data bits are sent to the 8255A port *A*, which was initialized as a mode 0 output port.

Before completing our discussion of the 8255A, let's take a brief look at the 8255A signal lines. Figure 11.27 illustrates the 8255A pin configuration and defines the functions of each signal. Since these signals have all been discussed, we do not dwell on them any further.

The 8259A Programmable Interrupt Controller (PIC)

In most microcomputer systems there is a method that allows an I/O device to gain the attention of the MPU. We have already seen that the 8086/8088 microprocessor has several signals that are dedicated to the *interrupt* function. When the MPU receives an interrupt request (INTR or NMI), it performs a sequence of steps that allows it to respond or service the interrupt request. It first completes the current process instruction and then determines if it is going to acknowledge the interrupt request. In the case of the nonmaskable interrupt (NMI) request, it *must* acknowledge and service the request. In the case of the interrupt request (INTR), the MPU first checks the flag register to see if the interrupt has been turned off or masked. It then services the interrupt if necessary. When the MPU services the interrupt, it first must save the contents of the instruction pointer and the affected registers. It does this by **pushing** the contents of the instruction pointer and the pertinent registers onto the memory stack. It then services the interrupt by looking up the location of the *interrupt service routine,* which is stored in ROM. This is called **vectoring.** The interrupt service routine is a program that defines the steps to be taken to service a particular type of interrupt.

The concept of interrupt is familiar to all of us. When a teacher is lecturing to a class and a student wishes to ask a question, the student must interrupt the teacher. The teacher must determine whether to acknowledge the question (interrupt) at the present time. If the question is to be answered, the teacher must first note where the interrupt occurred in order to return to that point to continue the lecture. The teacher can then answer the question or service the interrupt.

If two students were to ask a question at the same time, the teacher would have to determine which student to acknowledge first. Likewise, the microcomputer system must have a method of handling multiple interrupts. The 8259A Programmable Interrupt Controller (PIC) is a device designed for this purpose. It simplifies the implementation of the interrupt interfacing for the 8086/8088 microprocessor.

Figure 11.28 illustrates the basic block diagram and pin configuration for the 8259A PIC. The 8259A PIC is a software-controlled device that can be used to interface eight interrupt signals to the MPU. Eight additional 8259As may be *cascaded* or connected together to handle a maximum of 64 interrupts. The 8259A signals perform the following functions:

1. V_{CC} (pin 28). Power supply, $+5\ V_{DC}$.
2. GND (pin 14). Power supply ground.
3. $\overline{CS}$ (pin 1). Chip select is an active low input signal. It is used to enable the 8259A.
4. $\overline{WR}$ (pin 2). Write is an active low input signal. This signal, in conjunction with $\overline{CS}$, enables the 8259A to accept command words from the MPU.

Pin	Name	
34 to 27	$D_0 - D_7$,	Data bus
4 to 1	$PA_0 - PA_3$,	Port A
40 to 37	$PA_4 - PA_7$,	Port A
18 to 25	$PB_0 - PB_7$,	Port B
14 to 17	$PC_0 - PC_3$,	Port C (lower)
13 to 10	$PC_4 - PC_7$,	Port C (upper)
25	RESET,	Sets all ports to input
6	$\overline{\text{CS}}$,	Chip select
5	$\overline{\text{RD}}$,	Read input
36	$\overline{\text{WR}}$,	Write input
9,8	$A_0 - A_1$,	Port address select
26	V_{CC},	+ 5 Volts power in
7	GND,	0 Volts power in

(a)

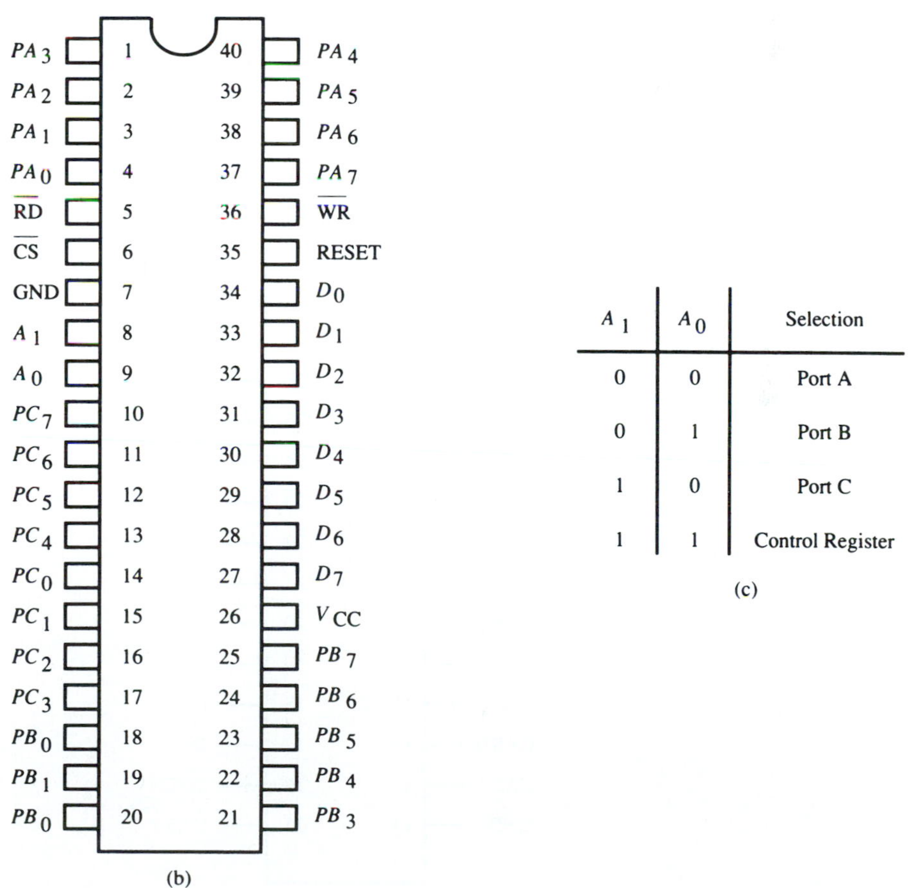

A_1	A_0	Selection
0	0	Port A
0	1	Port B
1	0	Port C
1	1	Control Register

(c)

FIGURE 11.27
8255A pin function

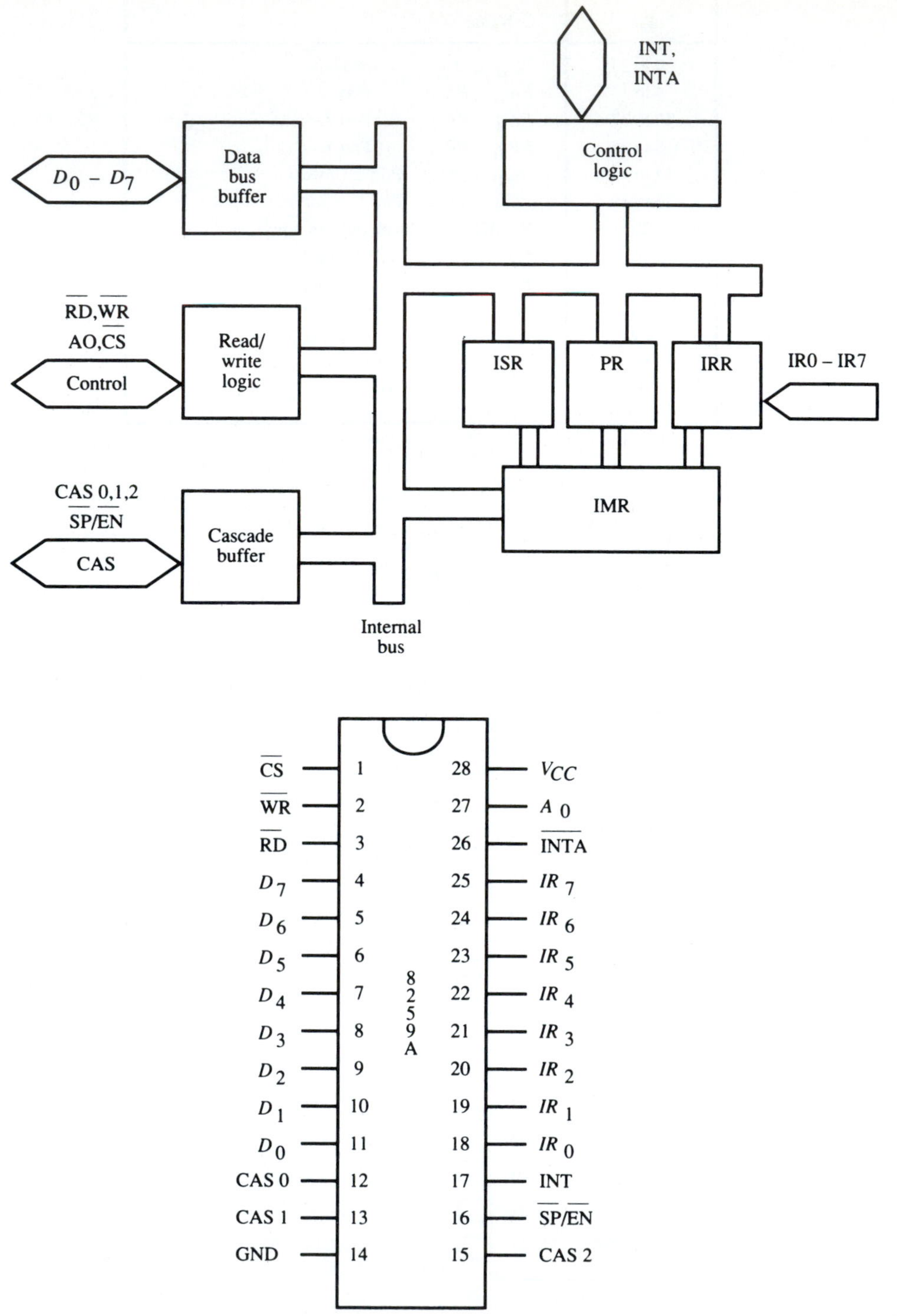

FIGURE 11.28
8259A Programmable Interrupt Controller

5. $\overline{RD}$ (pin 3). Read is an active low input signal. This signal in conjunction with $\overline{CS}$ enables the 8259A to release status information onto the data bus for the MPU.
6. CAS_0–CAS_2 (pins 12, 13, 15). The cascade lines form a control bus for use with multiple 8259A system designs. They are the outputs of the master 8259A and the inputs for the slave 8259As.
7. $\overline{SP}/\overline{EN}$ (pin 16). Slave program enable buffer is a dual function pin. It is used as an output to control buffer transceivers in the buffer mode (EN). It is used as an input to designate cascading 8259As in the SP mode.
8. INT (pin 17). Interrupt is an active high output that is used to interrupt the MPU.
9. IR_0–IR_7 (pins 18–25). Interrupt requests are asynchronous active high inputs used to signal an interrupt request from a system device.
10. $\overline{INTA}$ (pin 26). Interrupt acknowledge is an active low input signal from the MPU that is used to acknowledge that the MPU is going to service the interrupt.
11. A_0 (pin 27). The A_0 address line in conjunction with the chip select, write, and read signals is used to select different command words for the 8259A.
12. D_7–D_0 (pins 4–11). The bidirectional data bus signals are used for transfer, control, status, and interrupt vector information.

The internal architecture of the 8259A is made up of eight basic functional blocks, as shown in Figure 11.28. The data bus buffers interface the data bus to the internal bus of the 8259A. This 8-bit bidirectional tristate buffer is enabled by the read/write logic block. The read/write logic block provides for direction, timing, and source or destination for the data transferred through the data bus buffer block. Control signals are provided by the $\overline{RD}$, $\overline{WR}$, A_0, and $\overline{CS}$ control input signals. The interrupt request register (IRR) stores the status of the interrupt request inputs. The inservice register (ISR) stores the interrupt level that is presently being serviced. The priority resolver (PR) determines which of the active interrupt inputs has the highest priority. The interrupt mask register (IMR) is used to enable or mask out individual interrupt request inputs. The control logic block uses the information provided by the IRR, ISR, and PR to control the INT (interrupt) output signal that requests an interrupt to the MPU. It also handles the $\overline{INTA}$ (interrupt acknowledge) input signal from the MPU. The cascade buffer/comparator block provides an interface for the master and slave 8259As during cascaded operation.

The 8253 Programmable Interval Timer

The 8253 is a programmable interval timer/counter device. Its main use is to allow the programmer to create accurate time delays under software control. It can be used to perform *timing/counter* functions such as variable rate generation, real-time clock generation, digital one-shot control, event counting, frequency generation, and motor control.

The 8253 consists of three independent 16-bit programmable counters. Each 16-bit counter is a presetable digital down counter. They can operate in either binary or BCD count modes. The block diagram and pin configuration of the 8253 timer/counter is shown in Figure 11.29. The 8253 can be divided into three sections. The *input section* contains the data bus buffers, the read/write control logic, and the control word register. The *internal bus section* is used to transfer data and control signals between the input section

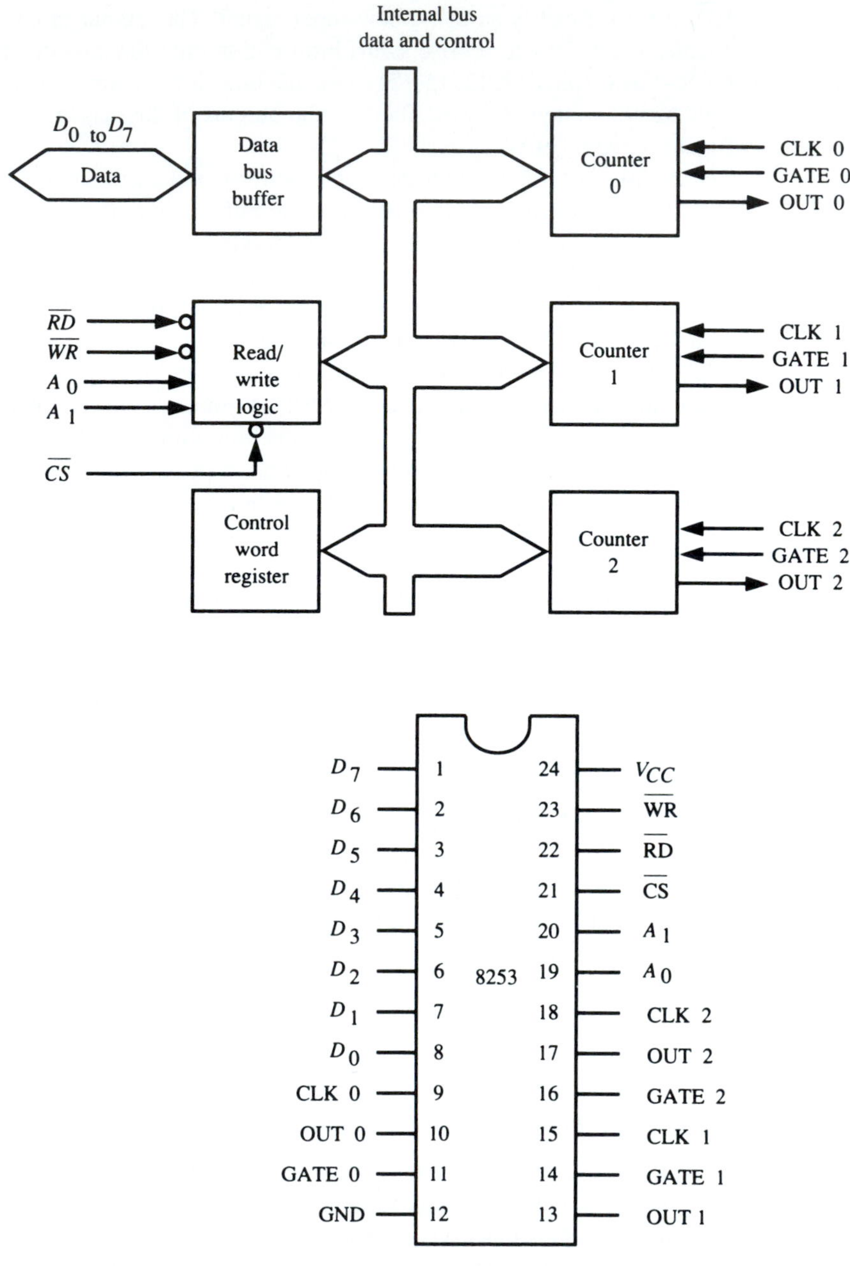

FIGURE 11.29
8253 programmable interval timer block diagram

and the output section. The *output section* consists of three independent 16-bit down counters.

The 8253 signal lines perform the following functions:

1. V_{CC} (pin 24). Power supply input +5 V_{DC}.
2. GND (pin 12). Power supply ground.
3. D_7–D_0 (pins 1–8). Data bits D_7 to D_0 are tristate bidirectional signals that are connected to the data bus buffers. Data can be transmitted or received by the buffer by using the 8086/8088 OUT or IN instruction. These signal lines are used to set the 8253's mode of operation, set the initial value of the counters, and to read the data in the counters.
4. $\overline{\text{RD}}$ (pin 22). Read is an active low input signal that is used to tell the 8253 that the MPU is inputting data.
5. $\overline{\text{WR}}$ (pin 23). Write is an active low input signal that is used to tell the 8253 that the MPU is outputting data.
6. A_0/A_1 (pins 19, 20). These address bus signals are used to select one of the three counters or the control word register as shown next.

A_1	A_0	Function
0	0	Select counter 0
0	1	Select counter 1
1	0	Select counter 2
1	1	Select control word register

7. $\overline{\text{CS}}$ (pin 21). Chip select is an active low input signal that enables the read/write control logic section. Thus it activates the $\overline{\text{RD}}$, $\overline{\text{WR}}$, A_0, and A_1 input signals.
8. CLK_0, CLK_1, CLK_2 (pins 9, 14, 18). The clock input signals are timing inputs to the respective counters.
9. OUT_0, OUT_1, OUT_2 (pins 10, 13, 17). The OUT signals are the individual outputs of the respective counters.
10. GATE_0, GATE_1, GATE_2 (pins 11, 14, 16). The GATE signals are active high inputs that are used to enable the respective counters.

The control word register is used to program and initialize the 8253. When A_0 and A_1 are both at logic level 1, the control register is selected. It is then able to receive information from the data bus buffers. This information is stored in the control register and is used to select the operational mode of each counter. Each counter of the 8253 is individually programmed by writing a control word into the control word register. The format of the control word is as follows:

D_7	D_6	D_5	D_4	D_3	D_2	D_1	D_0
SC_1	SC_0	RL_1	RL_0	M_2	M_1	M_0	BCD

The control word bits are defined as follows.

SC: Select Counter

D_7 SC_1	D_6 SC_0	Function
0	0	Select counter 0
0	1	Select counter 1
1	0	Select counter 2
1	1	Illegal

RL: Read/Load

D_5 RL_1	D_4 RL_0	Function
0	0	Counter latching operation
0	1	Read/load least significant byte only
1	0	Read/load most significant byte only
1	1	Read/load least significant byte first, then most significant byte

M: Mode

D_3 M_2	D_2 M_1	D_1 M_0	Function
0	0	0	Mode 0
0	0	1	Mode 1
X	1	0	Mode 2
X	1	1	Mode 3
1	0	0	Mode 4
1	0	1	Mode 5

Note: X indicates a don't-care state.
Mode 0: Interrupt on terminal count; output goes high when final count is reached.
Mode 1: Programmable one-shot
Mode 2: Rate generator
Mode 3: Square-wave rate generator
Mode 4: Software-triggered strobe
Mode 5: Hardware-triggered strobe

11.5 BUILDING A BASIC 8088 MICROPROCESSOR SYSTEM

A microprocessor system can be complex and difficult to understand. However, the basic system is not as complicated as you might think. Using a stage-by-stage approach, a working microprocessor system can be developed, built, and tested in an organized and controlled manner. The system shown in Figure 11.30 on pages 384 and 385 is an ideal project for students wishing to get some hands-on experience. The figure also demonstrates many of the concepts presented in this text.

The Timing Stage

All microprocessors require timing signals to synchronize operations. These timing signals are often provided by an external oscillator circuit. The 8284A clock generator chip provides the basic timing requirements for the 8088/8086 microprocessor. It also provides the reset logic for the entire microcomputer system. Figure 11.31 (p. 386) illustrates the basic 8284A clock circuit.

The 8088 has simple and easy clock requirements. The clock has to be asymmetrical, with a 33% duty cycle. The rise and fall times of the pulses can be no greater than 10 ns. Since the 8088 registers are made from dynamic RAM devices, they have to be constantly refreshed. This requirement means that the minimum clock speed for any 8088 microprocessor is 2 MHz. The maximum clock rate depends on the version of the chip: 5 MHz (8088), 8 MHz (8088-2), or 10 MHz (8088-1).

Fortunately, the 8284A clock generator chip generates clock pulses that are perfectly tailored to the 8088 microprocessor. Basically, all that is required is an external oscillator or crystal input. The 8284A will provide three output frequencies for the system. The OSC (pin 12) output signal is a TTL-level signal at the same frequency as the input oscillator or crystal. The CLK (pin 8) output signal is one-third of the input frequency or crystal with a 33% duty cycle. It is used as the basic input timing signal for the 8088 MPU. The PCLK (pin 2) output signal is one-half of the CLK, or one-sixth of the crystal frequency, with a 50% duty cycle. It is used to provide timing signals for the various peripheral devices in the system. Figure 11.4 illustrates the basic timing relationships for the 8284A clock generator chip. (Refer to Section 11.1, where the 8284A is discussed in detail.) After building the circuit of Figure 11.31, you can verify the waveforms, frequency, period and duty cycle for the output signals at pin 12 (OSC), pin 8 (CLK), and pin 2 (PCLK) to determine if the oscillator is working correctly. Furthermore, pressing the reset switch should cause a reset pulse on the reset output at pin 10.

When the power is first applied to the system, the $\overline{RES}$ pin must be held low for at least 4 clock cycles. The R_4, C_2 combination ensures the delay. Diode D_1 provides a safe path for capacitor C_2 to discharge to ground when power is removed.

CPU Bus Demultiplexing

Our building of the 8088 circuitry will demonstrate the basic operation of any microprocessor-based circuit. It will also show you what to consider when you are doing microprocessor design and troubleshooting.

The 8088 has 20 address lines labeled A_0 through A_{19}. Twenty address lines result in a maximum of 2^{20} addresses, or 1,048,576 addresses. The first hexadecimal address is $(00000)_{16}$ and the highest is $(FFFFF)_{16}$. Whenever the 8088 receives a reset pulse, it begins its operation at address $(FFFF0)_{16}$. This starting address is known as the *reset vector.* The 8088 begins its operation by outputting address $(FFFF0)_{16}$ on the address lines and initiating a read cycle. The contents of memory address $(FFFF0)_{16}$ will be read and interpreted by the microprocessor. Generally, a jump instruction is stored in memory address $(FFFF0)_{16}$. The jump instruction is used to point to the starting address of the first instruction of a program. When this program is a start-up program, it is often referred to as a *boot* program.

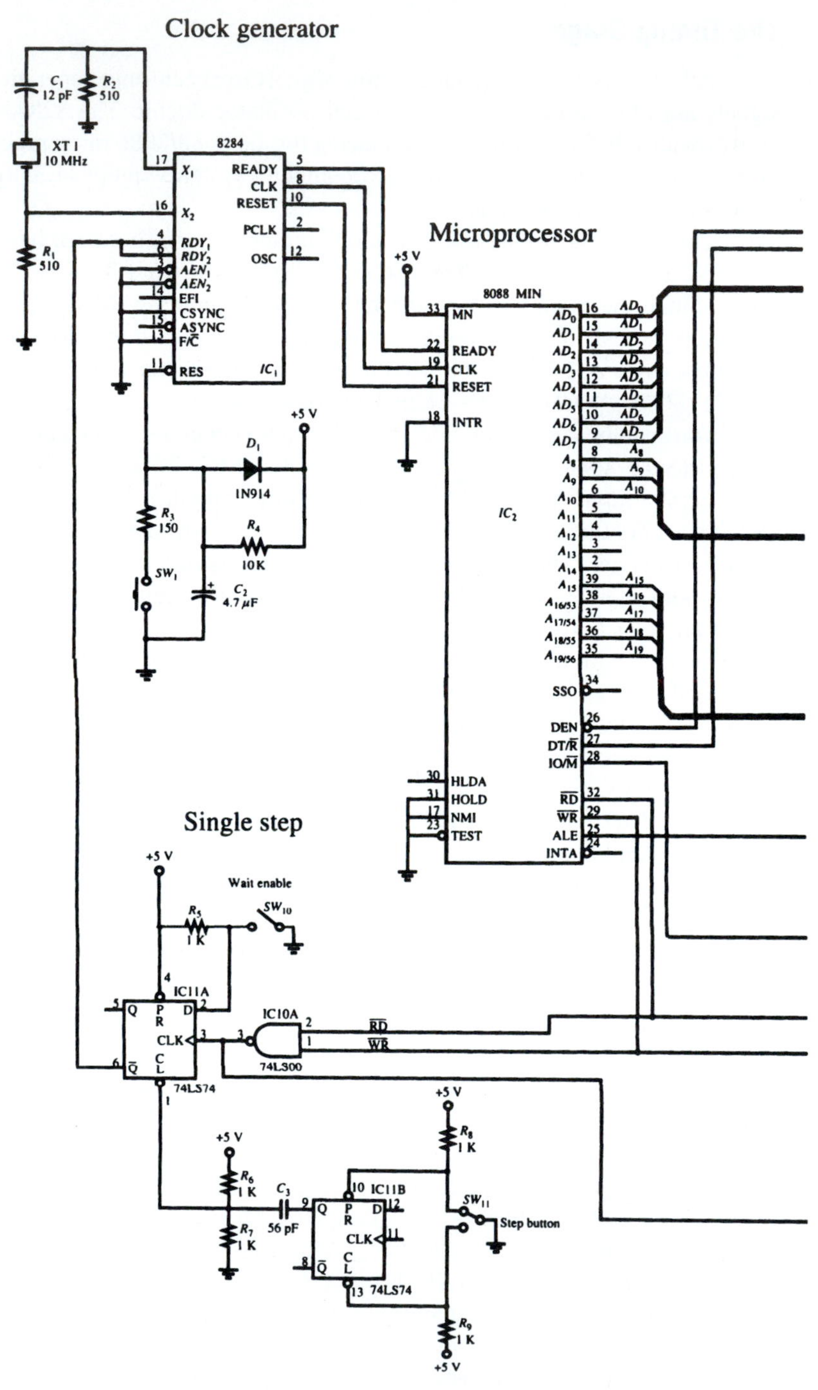

FIGURE 11.30
Microprocessor project

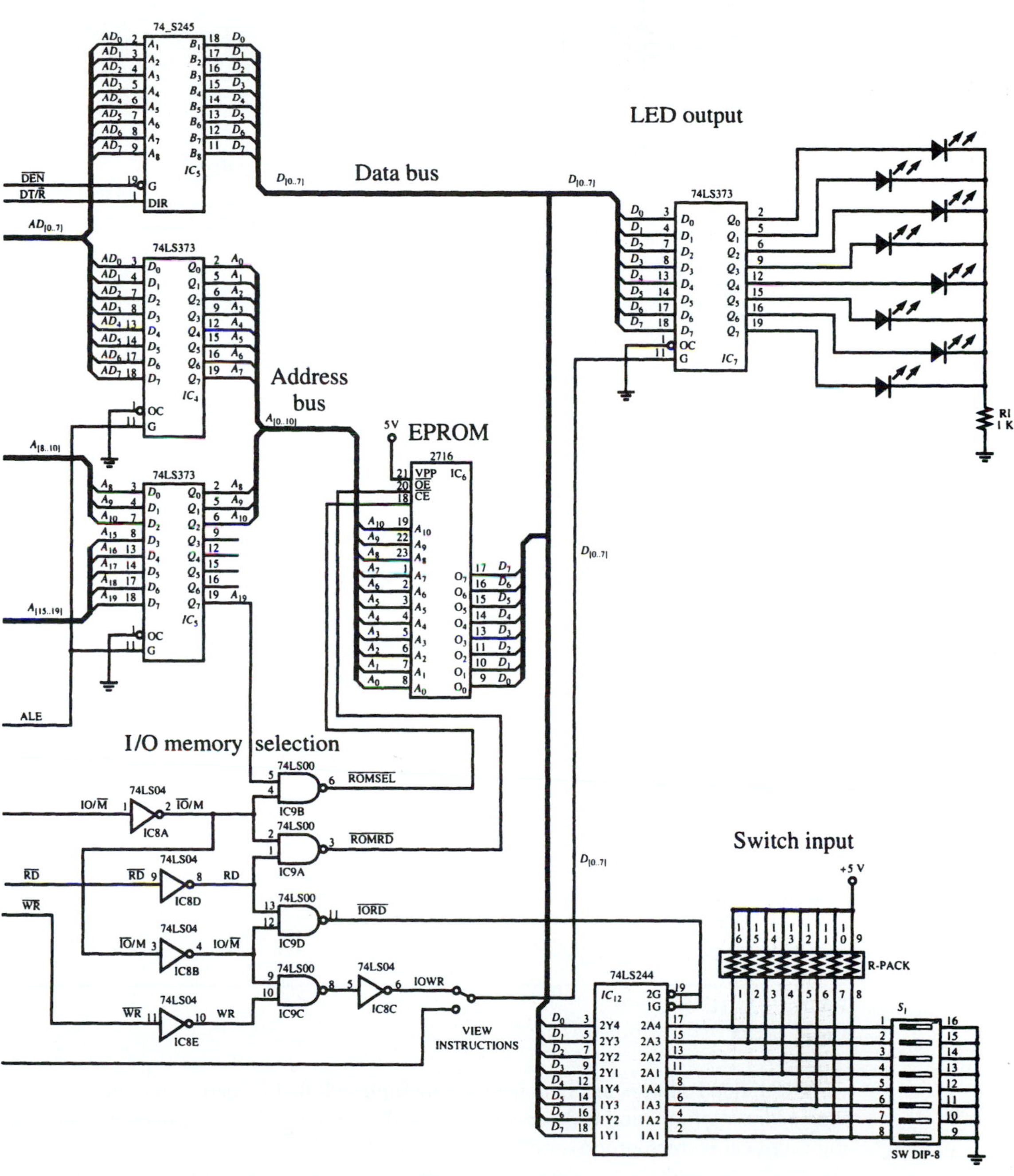

FIGURE 11.30 (continued)

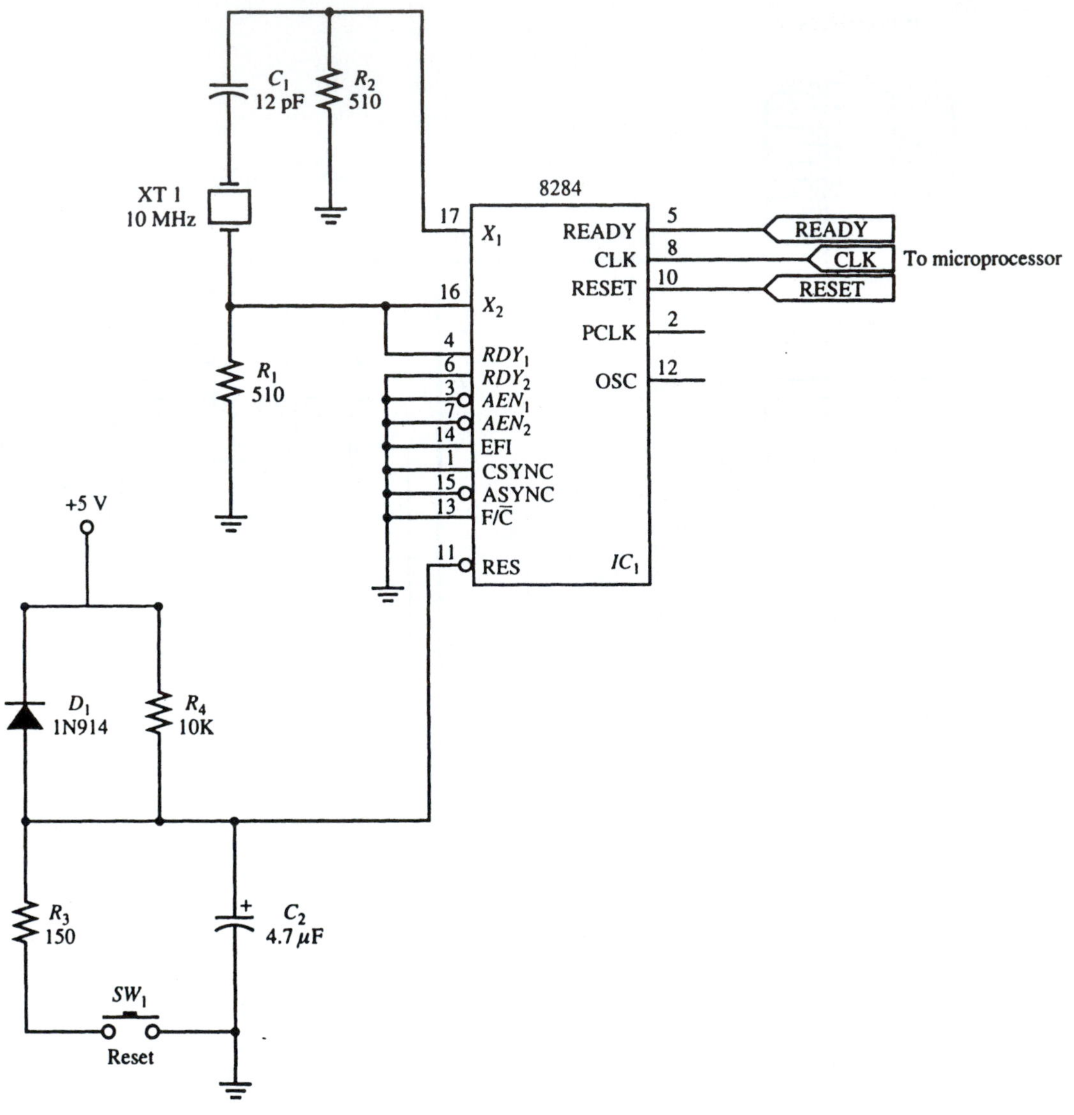

FIGURE 11.31
Clock circuit

The 8088 address and data lines are multiplexed; that is, part of the time they provide address information and part of the time they provide data information. Thus, timing is critical in microprocessor circuitry. Control lines $\overline{\text{DEN}}$, DT/$\overline{\text{R}}$, and ALE provide the control for the address and data bus demultiplexing. The data enable line, $\overline{\text{DEN}}$, is high when the microprocessor is outputting address information. It is low when the microprocessor expects to receive or transmit data information. The data transmit and receive line, DT/$\overline{\text{R}}$, is high when the microprocessor is transmitting data and low when it expects to receive data. The address latch enable line, ALE, is used to tell the system that the microprocessor has placed valid address information on the bus.

Figure 11.32 illustrates stage 2 of the project. Be certain to connect V_{CC} and ground to each IC. Note that the 8088 microprocessor has two ground pins. Both 8088 ground pins must be connected to ground. Also be certain to connect a separate 0.1-μF decoupling capacitor from V_{CC} to ground on every IC. Note that because only 11 address lines (A_0 to A_{10}) are used in the system, the starting address or reset vector for the system will be $(7F0)_{16}$. Using an oscilloscope, the reset vector can be verified. Disconnect the reset wire on pin 21 of the microprocessor and connect the TTL level output of an oscillator to pin 21 (the reset pin). Set the oscillator for a frequency of 50,000 Hz. This will cause the microprocessor to be reset 50,000 times per second. Use the oscilloscope to display ALE (pin 25) on channel one and each address line (A_{19}–A_0) on channel two, one line at a time. Be certain the oscilloscope is set to trigger on channel one, ALE. Record the logic levels of each address line on the microprocessor at the trailing edge of the first ALE pulse.

Another good test of the circuit is to verify the address cycle timing. Disconnect the oscillator and reconnect the reset line to pin 21 of the 8088. Set up the oscilloscope to display the CLK signal (pin 19 on the 8088) on channel one and ALE (pin 25 of the 8088) on channel two. Be certain to trigger the oscilloscope on channel one (CLK). Depress the reset button several times until the ALE signal appears. In general, a bus cycle will take four clock pulses. Thus, you should see one ALE pulse on every fourth CLK pulse. You can also set up the oscilloscope to display ALE (pin 25) on channel one and $\overline{\text{RD}}$ (pin 32) on channel two. Be certain to trigger on channel one (ALE).

Memory I/O Decode Logic

The decode circuitry of Figure 11.33 ensures that only one device is enabled on the data bus at a time. This enabling is performed by controlling the tristate chip enable pins on every memory and I/O device. For example, the 2716 EPROM is controlled by using the $\overline{\text{ROMSEL}}$ line in our decode circuit.

Memory

The 2716 EPROM provides the basic memory for the system. Using a standard EPROM programmer, the following program will enable you to sense a switch input and light a light.

Address	Machine Code	Assembler Code
07F0	E410	IN AL,10
07F2	E610	OUT 10,AL
07F4	EBFA	JMP 07F0

Note that both the switch input circuit and LED output circuit are not decoded to any particular address. Thus, any I/O address can be used. In this program, I/O address 10 is used. Using the single-step circuit, you should be able to verify and display the hex instruction codes on the LEDs. Figure 11.34 illustrates the memory writing.

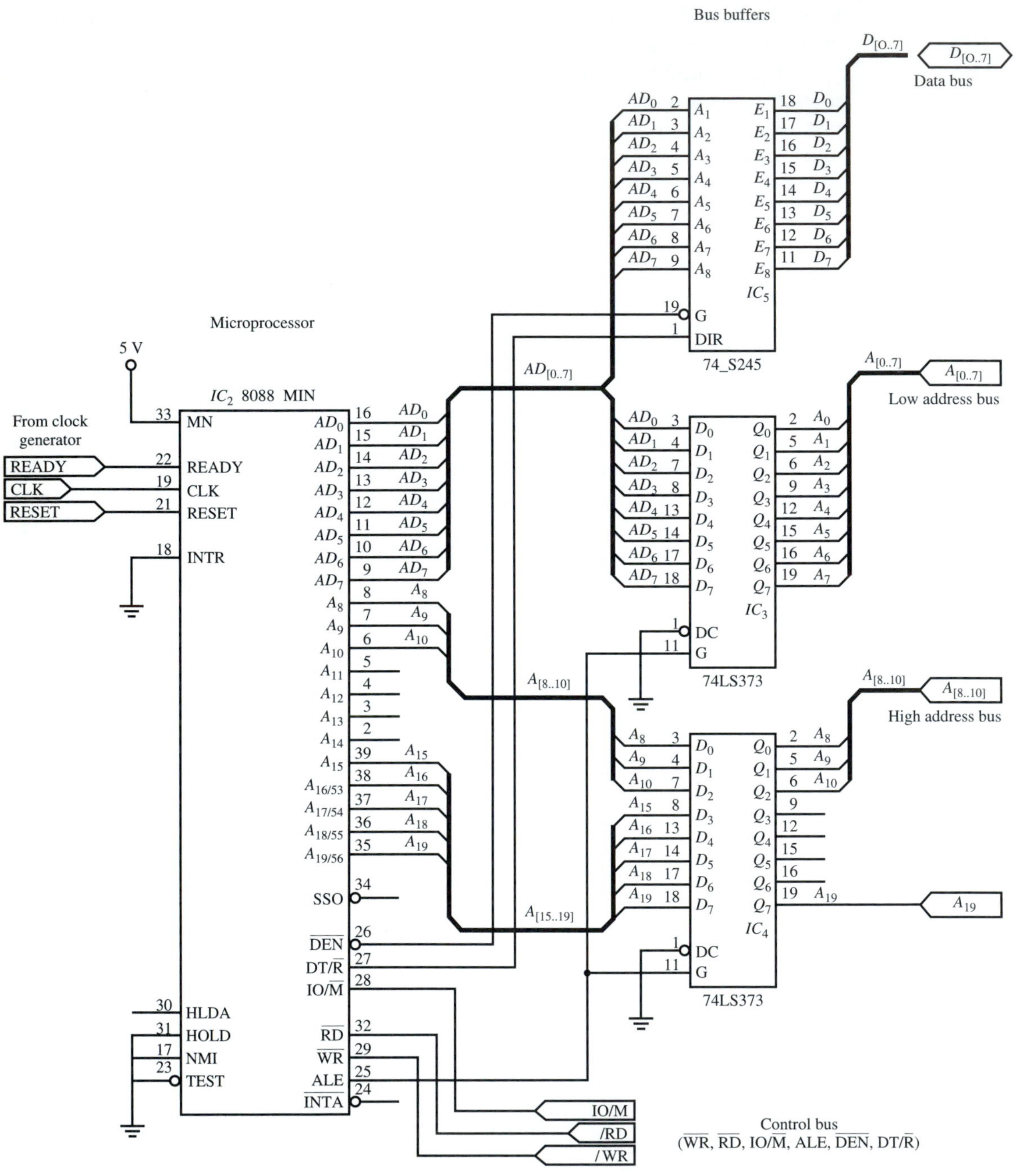

FIGURE 11.32
Microprocessor bus demultiplexing

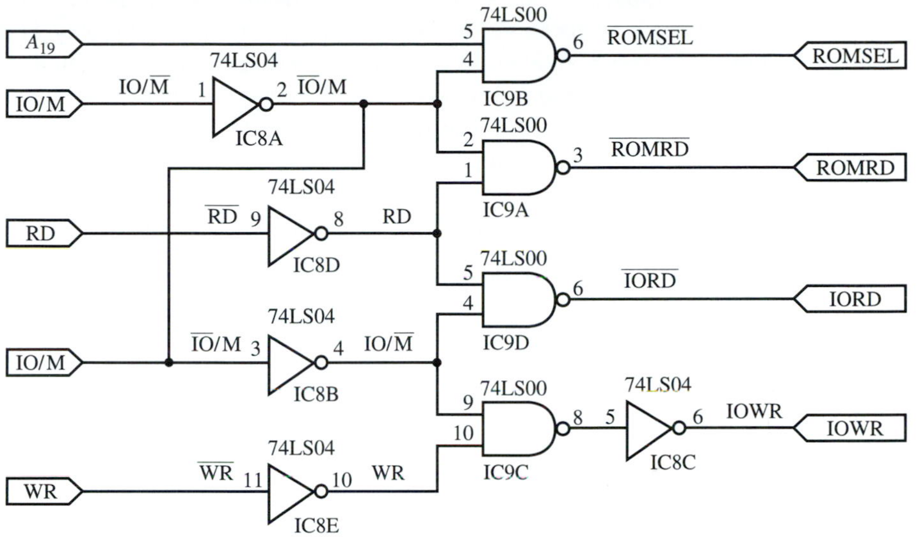

FIGURE 11.33
Memory I/O decode logic

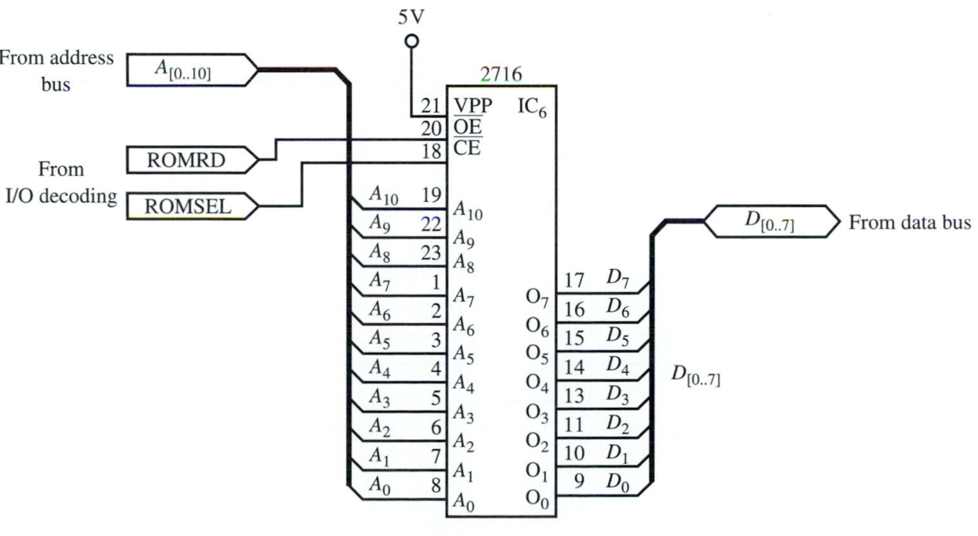

FIGURE 11.34
EPROM memory circuit

Another classical type of program involves rotating the lights of the display. Since this program is a little longer, the program starts with a jump to address 0100H instruction. The actual program starts at location 0100H. Recall that the 8088 starts executing instructions from location *FFFF*0H. Since we have wired up only 11 address lines, our start-up address is 07*F*0H. In either case, there are only 16 locations left before the top of memory. The jump instruction to lower memory increases the amount of space available to develop programs. The following program illustrates the jump on start-up technique and will rotate the lights on the display.

Address	Machine Code	Assembler Code
07F0	EA000100F0	JMP F000:0100
0100	B001	MOV AL,01
0102	E610	OUT 10,AL
0104	BB0400	MOV BX,0004
0107	B9FFFF	MOV CX,FFFF
010A	E2FE	LOOP 010A
010C	4B	DEC BX
010D	83FB00	CMP BX,+00
0110	75F5	JNZ 0107
0112	D0C0	ROL AL,1
0114	EBEC	JMP 0102

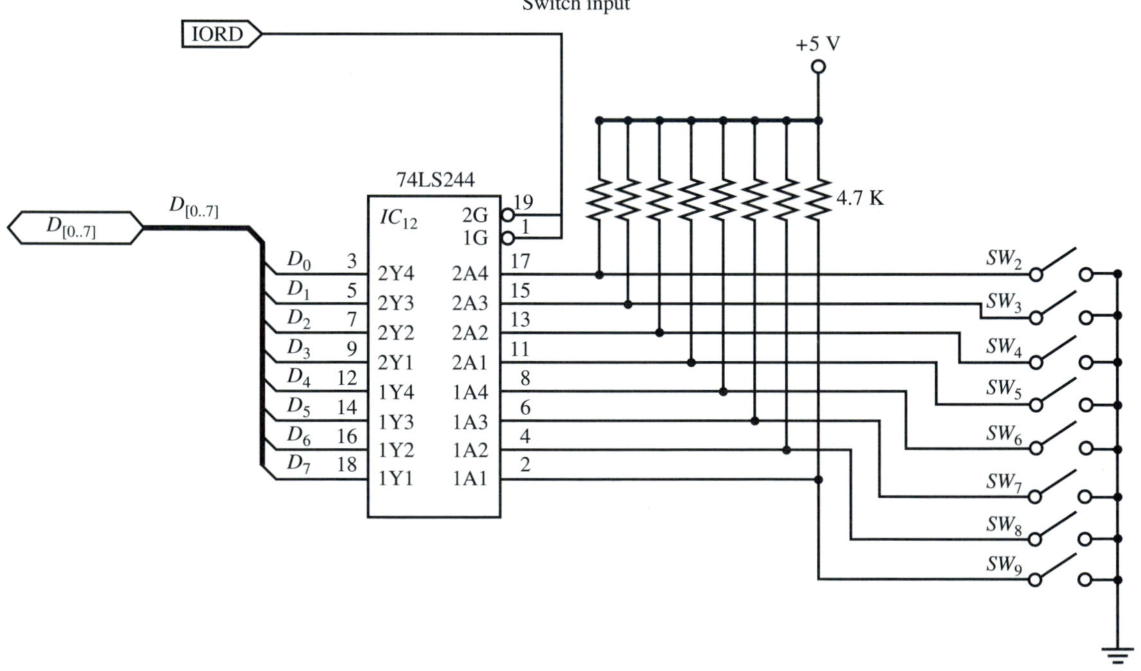

FIGURE 11.35
Switch input port

Note that most of the program in the example involves a nested delay loop. This nested delay loop is required to keep the LEDs lit long enough for you to see the light. Remember the microprocessor in this project is running and executing instructions at 3.3 MHz (10 MHz/3).

Input Stage

Figure 11.35 illustrates the switch input wiring. A 74LS244 provides simple input buffering so that the switch inputs are applied to the data bus at only the proper times. Tristate control is provided using pins 1 and 19. As with all of the circuits, be certain to include V_{CC} and ground connections. (They are not shown here for the sake of simplicity.)

Output Stage

Figure 11.36 illustrates the LED output wiring. A 74LS373 tristate latch provides buffering and control. A seven-segment hexadecimal display and driver can be added in parallel with the LEDs to make it easier to read the output display.

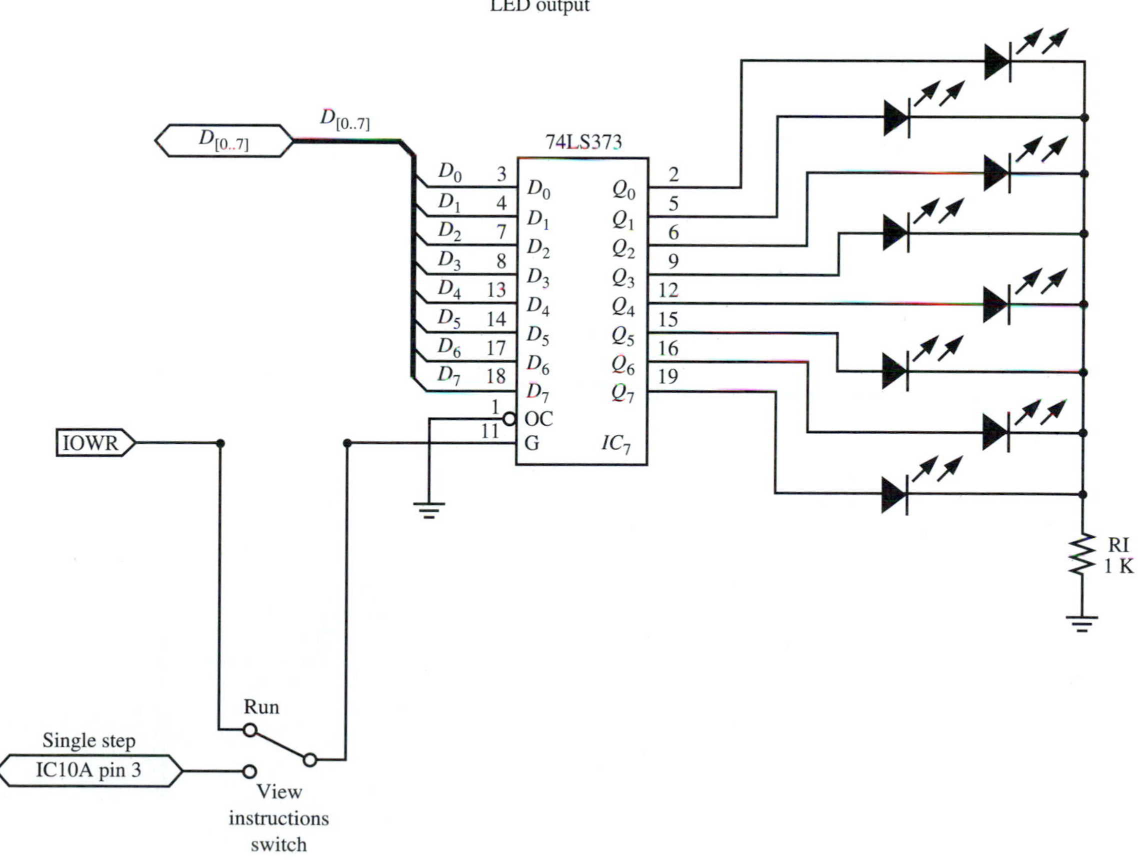

FIGURE 11.36
LED output port

Single-Step Stage

The single-step circuit provides a simple way of controlling the circuit to execute and display one instruction at a time. In single-step mode, the system is put into a wait state. Depressing the single-step push button releases the wait state for one read or write operation. The $\overline{RD}$ or $\overline{WR}$ signal puts the system back into the wait state. Figure 11.37 illustrates the wiring for the single-step circuit.

Sample Program

The following program example, SINE.ASM, was created using Notepad as an editor and assembled using the A86 assembler. It illustrates the process of using the A86 as-

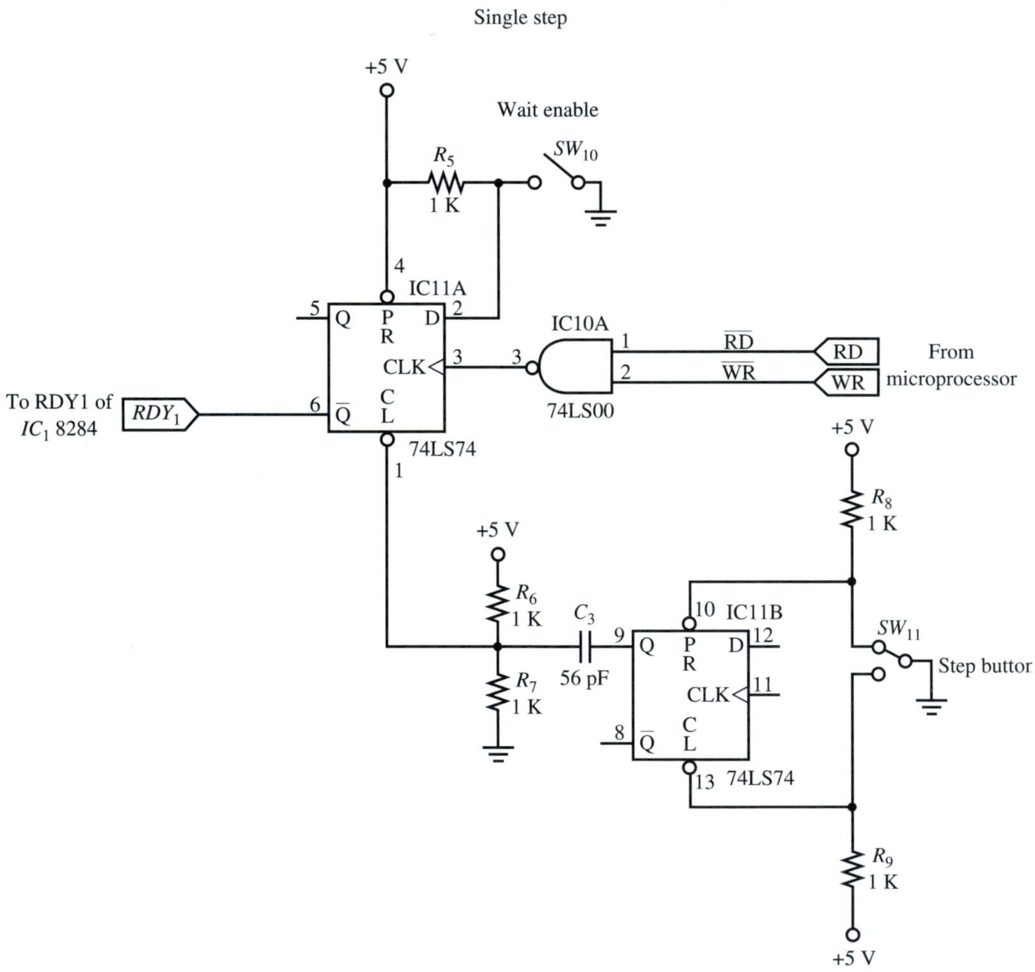

FIGURE 11.37
Single-step circuit

sembler instead of DEBUG to assemble a program for the project. It uses a lookup table technique to display the sine of an angle. The angle is entered using the switches and the sine of the angle is displayed in BCD format on the LEDs.

```
        ORG  0000h
SINES   DB   00,02,03,05,07,09   ;0–5
        DB   10,12,14,16,17      ;6–10
        DB   19,21,23,24,26      ;11–15
        DB   28,29,31,33,34      ;16–20
        DB   36,37,39,41,42      ;21–25
        DB   44,45,47,48,50      ;26–30
        DB   52,53,54,56,57      ;31–35
        DB   59,60,62,63,64      ;36–40
        DB   66,67,68,69,71      ;41–45
        DB   72,73,74,75,77      ;46–50
        DB   78,79,80,81,82      ;51–55
        DB   83,84,85,86,87      ;56–60
        DB   87,88,89,90,91      ;61–65
        DB   91,92,93,93,94      ;66–70
        DB   95,95,96,96,97      ;71–75
        DB   97,97,98,98,98      ;76–80
        DB   99,99,99,99,99      ;81–85
        DB   99,99,99,99         ;86–89
;
;This is a lookup table procedure.
;
START:  SUB AX,AX                    ;ZERO REGISTERS
        SUB BX,BX
        MOV DX,F000
        MOV DS,DX
        IN  AL,10                    ;GET ANGLE
        CMP AL,90
        JB  SIN
        NEG AX
        ADD AX,0180
SIN:    MOV SI,AX
        MOV AL,SINES[SI]
DISPL:  OUT 10,AL                    ;SEND TO LEDs
        JMP START
;
;BOOT SECTION
        ORG 7F0                      ;RESET VECTOR
        DB  0EA,00,00,00,0F0
        ORG 7FF                      ;LAST MEMORY BYTE
        DB  0F0
```

Assembling the program using the following command creates this listing:

```
C:\>A86 /L SINE.ASM
```

Once assembled, the BIN file can be read into a PROM burner or the op-codes on the left side of the listing can be entered into the PROM burner manually. Starting on the left side of the listing, the PROM memory address is indicated followed by the data which is to be stored in the address. Beginning at line 23 the data represents the actual op-codes for the instructions.

```
A86 V4.02 assembly of sine.BIN                              1999-11-28 18:06

Source: sine.asm                                                 Page 1

  Begin Source: sine.asm

  1                                        ORG     0000h
  2 0000    00 02 03 05 07 09   SINES      DB   00,02,03,05,07,09 ;0-5
  3 0006    0A 0C 0E 10 11                 DB   10,12,14,16,17    ;6-10
  4 000B    13 15 17 18 1A                 DB   19,21,23,24,26    ;11-15
  5 0010    1C 1D 1F 21 22                 DB   28,29,31,33,34    ;16-20
  6 0015    24 25 27 29 2A                 DB   36,37,39,41,42    ;21-25
  7 001A    2C 2D 2F 30 32                 DB   44,45,47,48,50    ;26-30
  8 001F    34 35 36 38 39                 DB   52,53,54,56,57    ;31-35
  9 0024    3B 3C 3E 3F 40                 DB   59,60,62,63,64    ;36-40
 10 0029    42 43 44 45 47                 DB   66,67,68,69,71    ;41-45
 11 002E    48 49 4A 4B 4D                 DB   72,73,74,75,77    ;46-50
 12 0033    4E 4F 50 51 52                 DB   78,79,80,81,82    ;51-55
 13 0038    53 54 55 56 57                 DB   83,84,85,86,87    ;56-60
 14 003D    57 58 59 5A 5B                 DB   87,88,89,90,91    ;61-65
 15 0042    5B 5C 5D 5D 5E                 DB   91,92,93,93,94    ;66-70
 16 0047    5F 5F 60 60 61                 DB   95,95,96,96,97    ;71-75
 17 004C    61 61 62 62 62                 DB   97,97,98,98,98    ;76-80
 18 0051    63 63 63 63 63                 DB   99,99,99,99,99    ;81-85
 19 0056    63 63 63 63                    DB   99,99,99,99       ;86-89
 20                                                               ;
 21                                   ;This is a lookup table procedure.
 22                                                               ;
 23 005A    2B C0           START:   SUB      AX,AX          ;ZERO REGISTERS
 24 005C    2B DB                      SUB    BX,BX
 25 005E    E4 0A                      IN     AL,10          ;GET ANGLE
 26 0060    3C 5A                      CMP    AL,90
 27 0062    72 05                      JB     SIN
 28 0064    F7 D8                      NEG    AX
 29 0066    05 80 01                   ADD    AX,0180
 30 0069    8B F0                      SIN:   MOV      SI,AX
 31 006B    8A 04                      MOV    AL,SINES[SI]
 32 006D    E6 0A                      DISPL: OUT     10,AL ;SEND TO LEDs
 33 006F    EB E9                      JMP    START
 34                                                          ;BOOT SECTION
 35                                    ORG    7F0            ;RESET VECTOR
 36 07F0    EA 00 00 00 F0             DB     0EA,00,00,00,0F0
 37                                    ORG    7FF            ;Last Byte
 38 07FF    F0                         DB     0F0
```

```
Symbols:

  : 006D   DISPL
  : 0000   SINES
  : 0069   SIN
  : 005A   START
```

11.6 TECH TIPS AND TROUBLESHOOTING—T³

Troubleshooting the MPU support devices can be an involved process. Besides being highly complex, many of the support devices are software dependent. They have many different modes of operation, which can hinder fault isolation. The basic problem is that the output has no simple relationship to the input. In a logic gate, the technician can easily determine what the output should be for a given input condition. The MPU support devices, on the other hand, are somewhat *intelligent*. The output can depend on the input, the mode of operation, how it was initialized, how it was programmed, and its present state. Thus it is very difficult for the technician to determine what the output should be for a given input condition. Although troubleshooting with conventional tools can be difficult, there are numerous types of microprocessor system testers and analyzers that can be used to aid the technician in the troubleshooting process. These testers are discussed in Chapter 17. They can automatically exercise, test, and analyze all possible modes of operation for intelligent programmable devices, a process that may not be possible by manual means. For example, one defective transistor in an LSI device could result in only one of 64 modes of operation not working properly.

There are, however, things that the technician can do using conventional test equipment and common sense to localize failures. If the problem can be localized, component swapping can save a great deal of investigative time and effort. If the system works with the new component, the old component can be considered defective.

1. Always begin with a visual inspection. Check for loose and broken wires or connectors. Look for burned or discolored components. Check switch settings and jumpers for proper initialization.
2. With the power off, reseat PC boards and cables. Humidity can cause corrosion on gold-plated contacts. Reseating can clean corrosion and remake connections. Cleaning the connections is better. A soft pencil eraser can be very useful for cleaning gold-plated contacts.
3. Check all power supply voltages and grounds. Is V_{CC} and ground present on each device? Is V_{CC} within tolerance and free of excessive noise and ripple?
4. Use the test equipment you have, together with common sense, to localize problems. Perform gross signal checks as described in the Chapter 8 Tech Tips. Combining an understanding of the functions of the signal lines with gross signal checks can be a good clue as to whether or not a device is functioning property. For example, a technician might use an oscilloscope to investigate the 8284A OSC, CLK, and PCK signals for proper operation. The 8288 bus controller's control, status, and bus signals can be checked for signal activity, and the like.

5. Sometimes it is possible to force a predictable output to occur. For example, placing a logic level 1 on an unused 8259A interrupt request line IR_0–IR_7 should force activity on the interrupt acknowledge ($\overline{INTA}$) line. Using a dual-channel oscilloscope, you can trigger on the interrupt request (IR_0–IR_7) and check for the interrupt (INT) and the interrupt acknowledge ($\overline{INTA}$) signals.
6. DEBUG can also be used as a diagnostic tool for the IBM PC. For example, the 8255A PPI can be checked by using DEBUG to output data to a port. The 8253 timer can be checked by using DEBUG to change a count register. In fact, many different types of checks can be performed using DEBUG. Let's see how. First install the DOS diskette containing DEBUG into drive A. Now type

```
A:\>DEBUG
```

The computer responds

```
A:\>DEBUG
—
```

On the IBM PC the speaker is connected to port *B* of the 8255 PPI. Outputting a logic level 1 on the two least significant bits will enable the speaker. Port *B* of the 8255A is I/O mapped as address 61H. Therefore, the following instruction should sound the speaker. Type

```
0 61 6F
```

The 0 in DEBUG is similar to the 8086/8088 OUT instruction. 61 is the address of the 8255's port *B*. 6*F* is used to output a logic level 1 on the two least significant data bits. Recall that 6*F* in hexadecimal is equal to 01101111 in binary. 6*F* is used so as not to disturb the functions of the other higher-order bits. To turn off the speaker, type

```
0 61 68
```

68 is normally the initial value of the port *B* register. You can examine this register by using the DEBUG input instruction I. To do this, type

```
I 61
```

The computer will respond

```
I 61
68
```

Here we are asking the computer for the contents of memory address 61 (the 8255A port *B* address). The computer tells us the data (in hexadecimal) in this address.

The 8253 timer can also be investigated. On the IBM PC the 8253 timer's counter 2 is memory mapped to address 42H. This is also used to determine the speaker output frequency. To change the tone of the speaker's output, we must first enable the speaker as we did before. Next, we must output data to the 8253 timer's counter 2 to change the tone. Recall that the 8253 timer's registers are all 16-bit registers. Now type

```
0 61 6F
```

to enable the speaker and

```
0 42 FF
0 42 FF
```

to change that tone. Two output instructions are required to generate 16 bits of data. The pitch of the tone can be changed by changing the data.

Successful results indicate that the devices are functioning properly. Obviously we have not checked every mode or register so we do not know for sure that the entire device is functioning properly. However, these simple tests give us an idea of what is going on. Tests can be created to check many of the support devices. Performing these checks on a working system will validate the test and the results.

EXERCISES

11.1 List the four subsystems found in all basic microcomputer systems.

11.2 Draw the block diagram of a basic microcomputer system.

11.3 Which subsystem is used to communicate between peripheral devices and the microcomputer?

11.4 Explain the function of the bus subsystem.

11.5 Compare the 8088 MPU to the 8086 MPU. What are the major differences?

11.6 When is maximum mode used?

11.7 Describe the function of the QS_1 and QS_0 pins.

11.8 If a write-to-memory operation is being performed, what is the state of each of the bus control status information pins?

11.9 Describe the function and operation of the 8284A chip.

11.10 A microcomputer system uses an 8284A clock generator chip. If a 15-MHz external crystal is connected to pins X_1 and X_2, describe the clock signal CLK output.

11.11 Describe the function and operation of the 8288 bus controller.

11.12 Draw the block diagram and pin configuration of the 8288 bus-controller chip.

11.13 Name the output control signals of the 8288 that interface to the address latch/buffers, data transceivers and interrupt control signals.

11.14 Explain how a 74LS373 8-bit latch might be used in an 8086 microcomputer system.

11.15 How is the 74LS245 8-bit transceiver used in an 8086 microcomputer system?

11.16 Explain how address decoding is used to interface the MPU and memory subsystems.

11.17 Draw a typical ROM decoding circuit using the 74LS138 decoder device.

11.18 What is DMA typically used for?

11.19 Draw the block diagram and pin configuration of the 8237 DMA controller chip.

11.20 Explain how a data transfer begins in a system that uses the 8237 DMA controller chip.

11.21 Describe the two major cycles of the 8237 DMA controller chip.

11.22 What is the function of the mode register in the 8237 DMA controller chip?

11.23 List the steps used to program the 8237 controller chip.

11.24 Design a microcomputer system using a DMA controller interfaced to all four subsystems.

11.25 What are the responsibilities of the I/O subsystem?

11.26 When the IN instruction uses a source operand with a constant value between 0 and 255, what type of addressing is being used?

11.27 The IN and OUT instructions must use the __________ or __________ registers.

11.28 To provide a very flexible parallel interface that is software controlled, __________ is used.

11.29 What signals control the timing of the data transfers to the 8255 PPI?

11.30 What signals are used by the MPU to receive interrupt requests?

11.31 What must be saved when the MPU services an interrupt?

11.32 Explain the meaning of the term "vectoring."

11.33 What can be used as a method of handling multiple interrupts in a microcomputer system?

11.34 Draw the block diagram pin and configuration of the 8259 PICA.

11.35 Describe the function of the eight basic functional blocks of the 8259A PIC.

11.36 What is the main use of the 8253 PIT?

11.37 Draw the block diagram pin configuration of the 8253 PIT.

11.38 How are the three counters of the 8253 PIT selected?

11.39 List and explain the five modes of operation of the 8253 PIT.

11.40 Explain why it is difficult to troubleshoot MPU support devices.

11.41 Explain how DEBUG can be used as a diagnostic tool for the IBM PC.

11.42 How can the 8253 PIT be checked for proper operation?

11.43 What is the function of the 8284A integrated circuit?

11.44 What is the purpose of diode D_1 in Figure 11.31?

11.45 What is the reset vector for the 8088 microprocessor?

11.46 Identify the address, data, and control bus signals in the project of Figure 11.30.

11.47 Crossword Puzzle

ACROSS

7. An 8284A output signal with the same frequency as the crystal.
8. Power supply reference level.
10. 74LS373.
13. Subsystem used to store data.
14. Instructions to programmable devices to control their operation.
18. To send data.
19. When the 74LS245 _____ pin is high, the transceiver will enter the high impedance state.
22. See 24 Across.
24. Placing a low on this line (22 across) will _____ a DMA transfer.
25. Causes a temporary halt in the execution of a program.
27. Memory that does forget.
28. A software diagnostic utility program.
29. A _____ interrupt requires an immediate response by the MPU.
30. A 64K block of memory.
32. Multiple 8259A system design.
33. An I/O instruction.

DOWN

1. Memory that never forgets.
2. An 8284A signal that is one-third of the crystal frequency.
3. A signal used by the 8237 to indicate the start of a DMA transfer.
4. The 8253 consists of three independent 16-bit programmable _____.
5. Line used by the 8237 to ask the MPU for control of the system's buses.
6. The control word _____ the 8255A by defining how it is going to be used.
9. Subsystem used to route information between other subsystems.
11. Data enable.
12. To send data into memory.
15. To send information to an external device.
16. 8253 function.
17. Address _____.
20. Address latch enable.
21. Circuitry that connects an external device to the microcomputer.
23. Pin configuration.
26. Opposite of serial.
31. Subsystem that provides basic system timing.

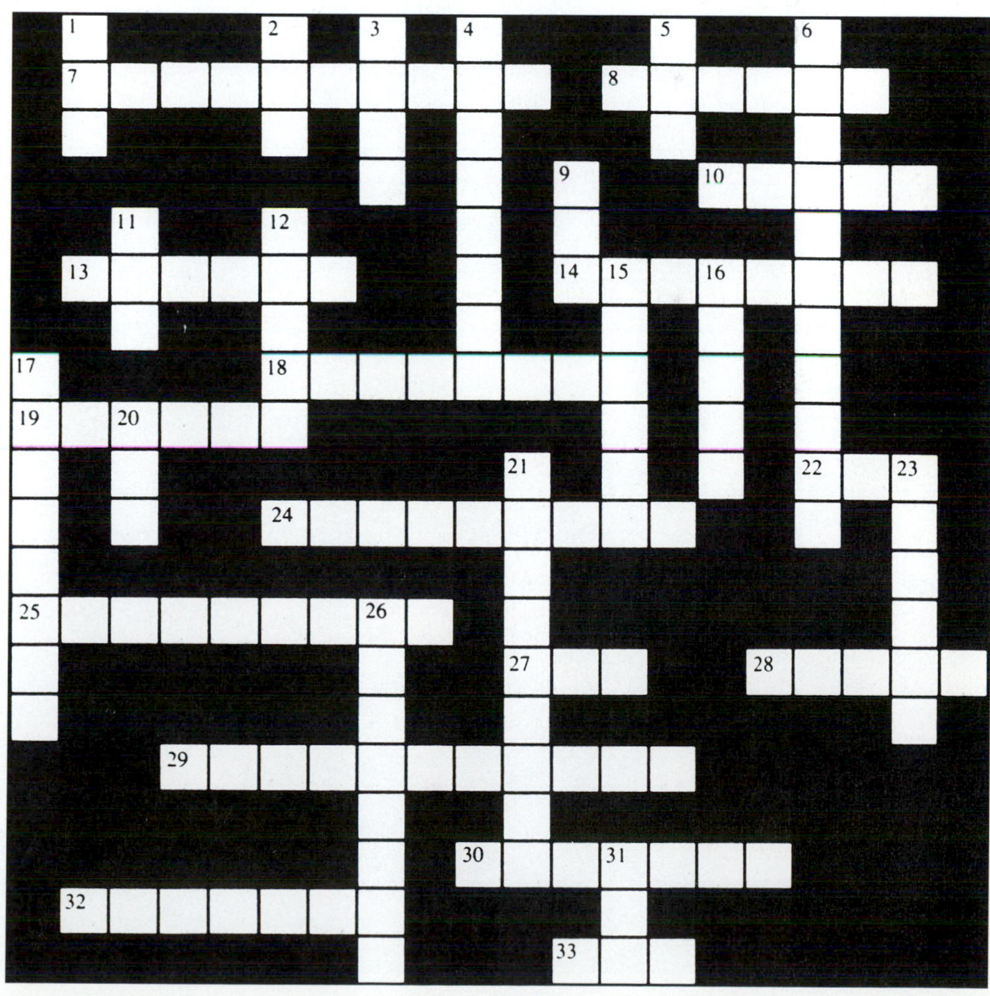

PART

COMPUTER SYSTEMS AND PERIPHERALS

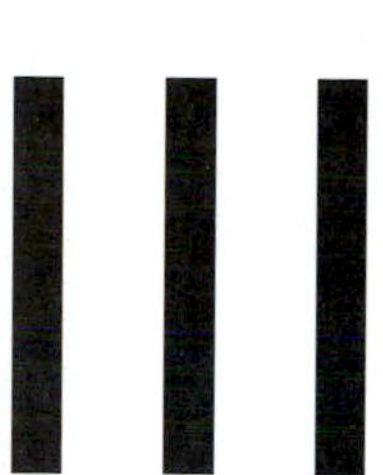

Up to this point, we have learned about the fundamentals of digital logic theory. We have also learned about the microprocessor and some of the related LSI devices. In this part, we discuss all the elements of digital logic from a systems approach as it relates to the computer. Since all digital computers are similar in their basic designs, we speak as generally as possible. However, to demonstrate principles we use the generic personal computer, known as the PC, for examples. The PC is based upon the original IBM PC introduced in 1980 by IBM Corporation. Over the years, the name has taken many forms, from the IBM PC to the IBM PC/XT and IBM PC/AT. Because of the open design, many manufacturers created similar systems. These systems were called *clones* or *PC compatibles.* One of the first manufacturers to mass-produce a clone was Compaq Computer Corporation. In fact, in the early days of clones it had been said that the Compaq was more compatible than the original IBM PC because of extra features and portability. Today you will rarely hear the term "clone"; the Windows/Intel system is simply called the PC. Another common term for the PC is the Wintel computer, because of Windows running on an Intel-based microprocessor.

Part 3 will discuss PC computer systems with enough detail to give the reader a good understanding of how things work inside the system unit. We will discuss the power supply, microprocessor evolution, motherboards, and memory. We also study in detail memory and display peripherals, magnetic recording theory, digital video display theory, printers, and laser printers. The topics are discussed from an analytical and service point of view rather than from a design viewpoint. This more detailed approach to the study of digital circuitry is necessary if a technican expects to troubleshoot a complex computer system. Two levels of troubleshooting will be presented. The first level is the process of diagnosing and replacing defective PC boards and subassemblies. The second level involves the troubleshooting of PC boards and subassemblies to the component level. The second level is much more difficult and requires a more detailed knowledge of digital circuitry.

Before we begin, let's briefly review a few terms. The digital computer system consists of a central processing unit, memory, input, and output devices. The central processing unit consists of the arithmetic unit and the control unit. Together they control the operation of the entire system. The memory section is used to store binary information. When information is stored in a computer's memory, it is referred to as a file. Recently memory devices have become faster and smaller in size. The input and output devices are used to get information into and out of the computer. These are devices like keyboards, display terminals, card readers, magnetic tape systems, magnetic disk systems, modems, scanners, printers, and analog-to-digital and digital-to-analog converters. These devices are referred to as peripherals. The circuit that connects or interfaces these devices to the central processor is called a peripheral controller.

Understanding computer jargon or computer terminology is very important. Often, simple terms can be misleading and can cloud the understanding of concepts. Without an understanding of the language used in the computer industry, it will be difficult to communicate with other persons in the field. Furthermore, since all computers are essentially the same, a competent technician can easily begin the troubleshooting and repair of larger computer systems. Serious troubleshooting on any type of computer system requires knowledge and skill.

Displays

Display devices can vary from one system configuration to another; however, most systems today come configured with high-resolution cathode-ray tube (CRT) monitors. Other, more expensive, display technologies include LCD or liquid crystal display (common on notebook computers), high-definition color television, and gas plasma displays. The last is the most expensive of all.

Most displays use a 15-pin D-shell VGA male connector, which connects to the video card in the back of the case. Older displays (CGA and EGA) use 9-pin D-shell male connectors. High-end displays sometimes use five BNC connectors. In almost all cases, power is supplied by the same 3-pin 120-V_{AC} cable that is used for the case.

Keyboards

Various types of keyboards are in use on PCs today, and several keyboard standards currently exist. The AT style of keyboard was popular in computers that followed the design of IBM's AT computers. This remained common in system configurations until 1996 or so. At that point, the IBM PS/2 style of keyboard became popular as the ATX form factor was introduced into the marketplace. Some keyboards take a different approach. Universal serial bus (USB) keyboards do away with the keyboard controller chip, relying instead on the USB controller for I/O. These keyboards are hot pluggable, making switching a faulty keyboard on the fly safer. Infrared (IR) keyboards do away with the cord attaching to the computer. This allows for increased freedom of movement as long as line of sight with the system is maintained. Another recent trend has been the introduction of special function and programmable keys for use with the Windows operating system.

Mouse

Like keyboards, mouse technologies also have a number of different iterations. The most common mouse system up until 1996 was the serial mouse, which used a serial port connection to the computer for communication. When the keyboard switched from the AT to PS/2 style, so did the mouse. Instead of using a serial bus, the mouse used a PS/2 port for communication, freeing up a serial port and an interrupt request (IRQ) for other uses. Mice also come in USB and IR formats. Some mice even come with a number of extra buttons and or wheels for increased versatility. The latest innovation is an optically scanning mouse that has no moving mechanical parts.

Floppy Disk Drive

The floppy, or flexible, disk is the most common means of temporary data storage and data exchange. The floppy disk drive works very much like a tape recorder. Information is recorded on an oxide-coated mylar disk. However, the read/write head moves in or out to access different areas of the disk, whereas the head in a tape recorder remains fixed.

The most common type of floppy disk is the 3.5-in. high-density 1.44 MB disk. Older types of floppy technology include 5.25-in. and 8-in. floppy disks and drives. The 3.5-in. floppy replaced all earlier floppy disks in IBM as well as Macintosh machines, but is currently being phased out by newer removable media technology. These include the Zip drive and the backward compatible LS-120 drives (which also reads 3.5-in. floppies). Until a replacement is universal, however, the floppy drive will still be an important component of the system.

Hard Drive

Computers can only work with data that is held in memory. Memory is very fast, but the most commonly used memory is also volatile, which means that once power is removed, the information it holds is lost. For long-term storage, some other medium was needed. The floppy drive sufficed at first when the operating system (OS) fit on a single floppy, but ever since the PC was introduced, the OS, applications, and data files have taken up more and more space. The hard drive has dominated as the medium for long-term storage within the system. Because of this close relationship to memory, hard disk space is often confused with memory, which is better applied to system memory. Modern multitasking operating systems actually use the hard drive as virtual memory, whereby system memory is swapped in and out of hard drive space.

All modern systems contain at least one hard drive, since the size has been reduced and cost per megabyte (MB) has dropped substantially in recent years. To give you some perspective here, in the early 1980s a 300 MB hard drive was the size of a washing machine, used ten 14-in. platters and cost tens of thousands of dollars. Now a typical PC hard drive has a 3.5-in. form factor, with at least 6 GB of storage capacity and costs under $200. Fortunately, this has coincided with the widespread use of the Internet to distribute files and the common practice of storing large multimedia files on hard disks.

With increasingly high densities of data being stored on hard disk media, the most common physical hard disk size is the 3.5-in. drive, though 5.25-in. hard drives are not uncommon. In storage size, hards disks can range from less than 1 GB to greater than 40 GB. Undoubtedly in the future this will seem like a small amount.

CD-ROM Drive

Before hard disk space became less expensive, compact disk–read-only memory (CD-ROM) represented a huge amount of storage space, 650 MB. Originally, this was a read-only medium, so it was not useful for backup but perfect for distribution of software. CDs are relatively inexpensive to create and far more durable than floppy or hard drives.

CD-ROMs use an optical medium adopted from the larger WORM (write once read many) optical storage system. This technology was applied to the music industry and then applied to the PC industry. CD-ROM drives are similar to hard disk drives in physical format and use the same connectors; however, due to the larger size of the CD medium, CD-ROM drives are only manufactured in the 5.25-in. size. CD-ROM drives can play your favorite music CDs and have an additional connector, a four-wire round cable used to transmit the sound data on music CDs to a sound card or sound processor on the systems board.

CHAPTER 12

PC ARCHITECTURE

KEY TERMS

ATX
BIOS
Boot Strap Loader
Chipset
CMOS
Duty Cycle
Flags
Foldback
Form Factor
Front Side Bus (FSB)
Industry Standard Architecture (ISA)
L1 Cache
L2 Cache
Linear Pass Power Supply
North Bridge
Peripheral Computer Interface (PCI)
Personal Computer Memory Card (PCMCIA)
Pin Grid Array (PGA)
POST
RDRAM
ROM
RAM
SECC
Slot 1
SMP
Socket 7
South Bridge
Switching Regulator Power Supply
VESA Local (VL) Bus
Zero Insertion Force (ZIF)

12.0 INTRODUCTION

All computer systems, whether they are small personal computers, large business servers, or workstations, are similar in nature. The variations occur in the size and number of peripherals connected to the system. The typical microcomputer system has a central processing unit (CPU), one or more disk drives, a keyboard and mouse, sound, a printer, a modem, and a display as shown in Figure 12.1. For a system to be functional, several parts must be connected or configured together. This chapter discusses the makeup and variations of these parts.

12.1 POWER SUPPLIES, COOLING, AND CASE

The power supply in almost all current systems is the **switching regulator** type, as opposed to the **linear pass** type (Figure 12.2). In many of the older microcomputers the power supply was the linear type, which required a large and heavy step-down transformer to convert 115 V_{AC} to low-voltage AC for the rectifiers and regulators. One big disadvantage

FIGURE 12.1
Microcomputer

of the linear type is the amount of heat that it generates. The output pass transistor acts like a variable resistor controlling the output voltage. The excess voltage is not passed to the output but is dissipated as heat.

In the older S-100 microcomputer systems the power was stepped down, rectified, filtered, and sent down the bus as unregulated DC or what is called *raw* DC. The regulation was done *locally,* on the boards themselves. Each board had a small local regulator as opposed to one large regulator. One big advantage of the linear type is ease of servicing, since no hazardous voltages are present at the regulators.

The switching regulator is entirely different. It is a much smaller, lighter, and far more efficient power supply. Instead of wasting the unused power, it switches only the power needed to the output.

Think of a conventional home lamp dimmer; it works in much the same way as the switching power supply. When the knob is rotated to dim the lamp, the power is switched on and off at a high rate of speed. When the lamp is dim, the *on time* is shorter than the *off time.* As the knob is rotated to make the lamp brighter, the *on time* becomes longer in proportion to the *off time* (Figure 12.3 on p. 408). The lamp dimmer can carry as much as 600 W. Imagine the size of a 600-W rheostat; it would be about 12 in. in diameter and very heavy.

The big disadvantage of a switching supply is that it is a bit tricky to troubleshoot and repair. Basically the switching supply is a voltage-controlled switch (see Figure 12.4 on p. 409). The primary power (raw AC, right from the wall) is rectified and filtered as high-voltage DC (1). It is then switched to a high rate of speed, approximately 20 to 40 kHz, and fed to the primary side of a step-down transformer (2). The step-down transformer is only a fraction of the size of a comparable 60-Hz unit, thus relieving the size and weight problem. The secondary side of the transformer is rectified and filtered and

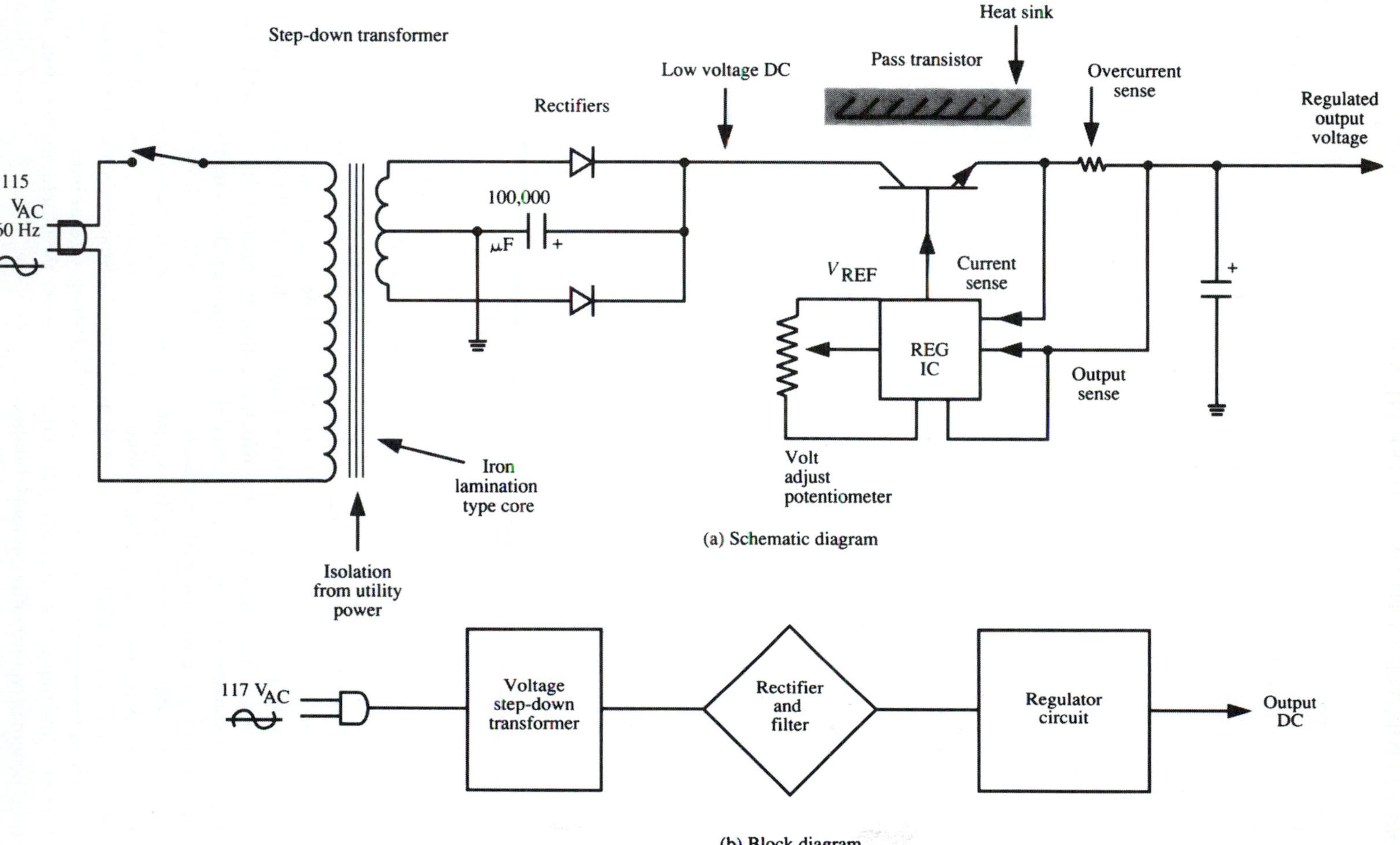

FIGURE 12.2
The linear power supply

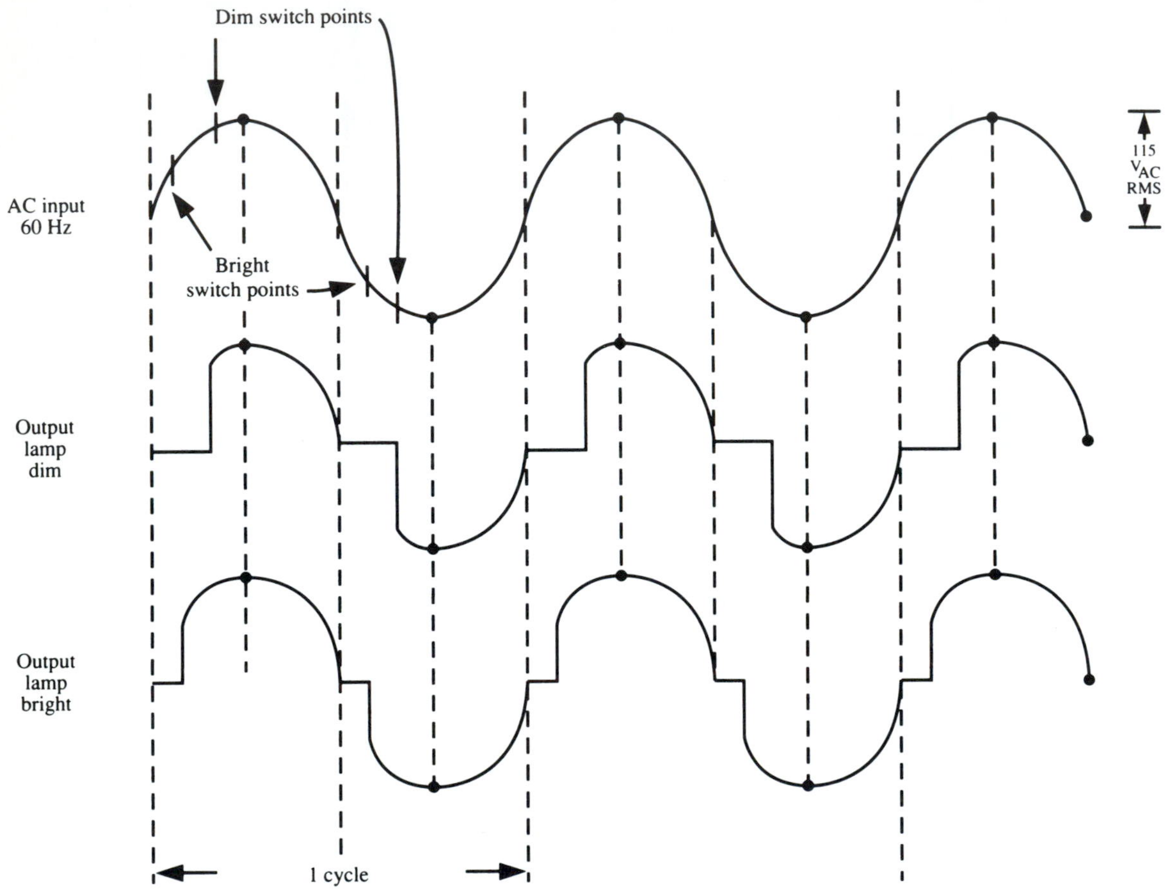

FIGURE 12.3
Lamp dimmer power output

then sent to the output of the power supply (3). A sample of this output is sent back to the switch to control the output voltage (4).

Most switching power supplies regulate their output using a method called **pulse-width modulation.** The power switch, which feeds the primary side of the step-down transformer, is driven by a pulse-width-modulated oscillator (5). In Figure 12.5 (p. 411), when the ratio of on time to off time, which is called the **duty cycle,** is at 50%, then the maximum amount of energy will be passed through the step-down transformer. As the on time or duty cycle is decreased, less energy will be passed through the transformer.

The width, or on time, of this oscillator is controlled by the voltage fed back from the secondary rectifier output, thus forming a closed-loop regulator. As we can see in Figure 12.5, the pulse width to the power switch is inversely proportional to the output voltage. When the output voltage drops, the switch is no longer, resulting in more energy delivered to the transformer and a higher output voltage. As the output voltage rises, the on time becomes shorter until the loop stabilizes.

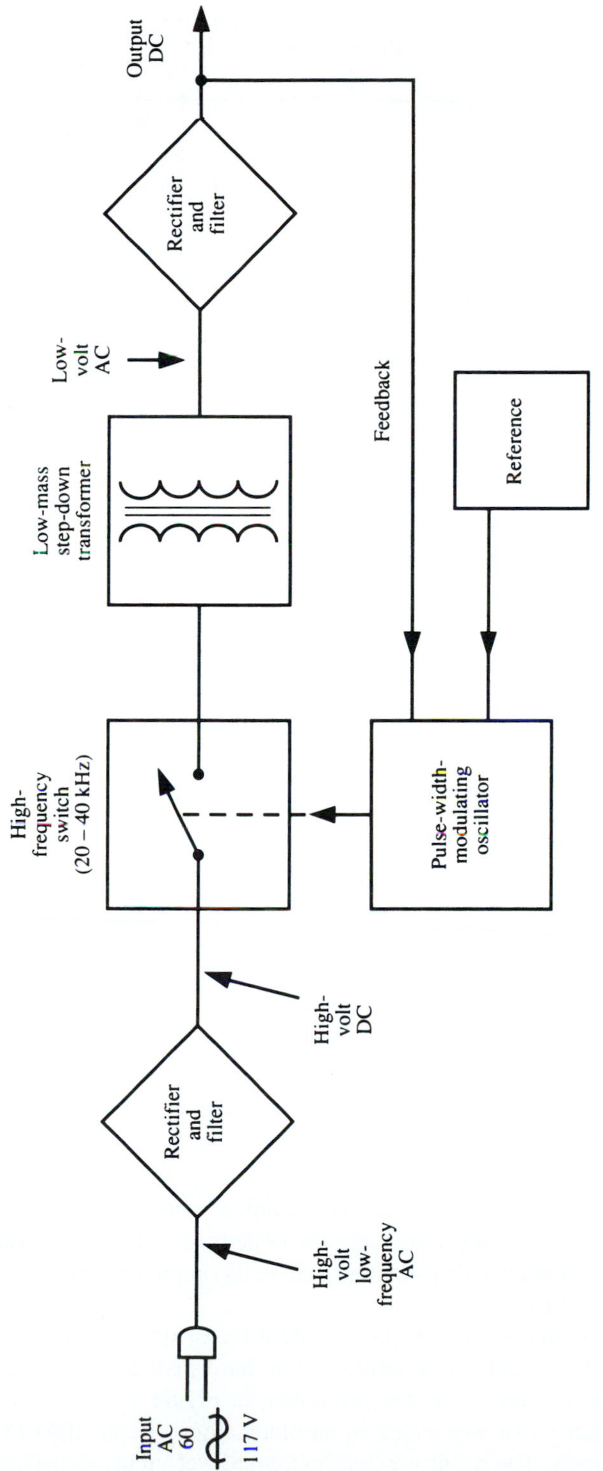

(a) Block diagram of a power switching supply

FIGURE 12.4
Switching power supply

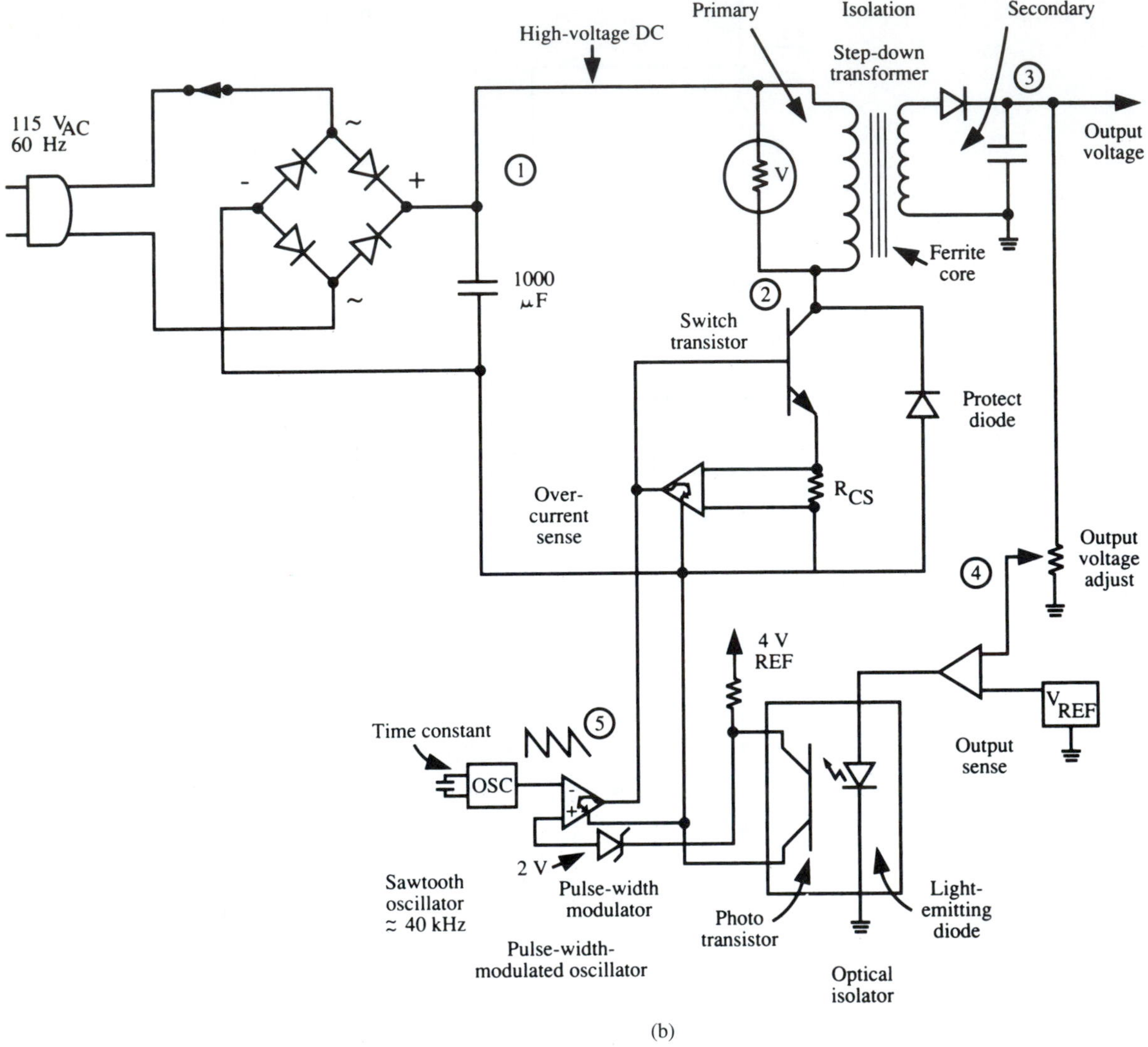

FIGURE 12.4 (continued)

Power supplies vary from one system unit to another, depending upon the number and type of peripherals mounted inside the system unit. The power rating is given as total power output in watts rather than the individual outputs in amps. Table 12.1 lists power supplies in the IBM PC family.

The power supply rating is an important factor when configuring a new system or upgrading an old system. For example, some hard disk drives require 5 A peak on the +12-V line just to spin up or start, as shown in Figure 12.6 (p. 412). There have been cases where a hard disk was added to an older, floppy-based IBM PC with the original 65-W power supply. The result was intermittent power-on problems and premature power supply failures.

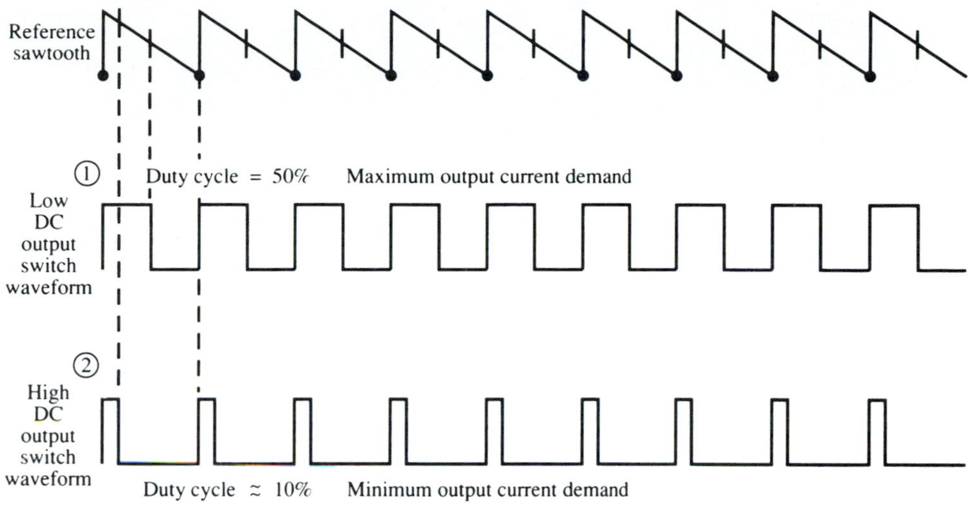

FIGURE 12.5
Switching power supply waveforms

TABLE 12.1
IBM PC power supplies

Type	Power Rating
IBM PC Model 5150	65 W
IBM PCXT Model 5160	135 W
IBM PCAT Model 5170	200 W

In Table 12.2 we must add the current draws of the system board, the display adapter, the floppy adapter, the floppy disk drives, and so forth to select the proper power supply. As a rule of thumb, always allow 25% to 50% margin of safety. For example, if 100 W of power is required, then use a 125- to 150-W power supply.

TABLE 12.2
IBM PC Board current usage

Component	Current Draw
System board	+5 V @ 2 A
Display controller	+5 V @ 1 A
Floppy controller	+5 V @ 0.5 A
Floppy drive	+5 V @ 0.5 A, + 12 V @ 1 A
Hard disk controller	+5 V @ 1 A
Hard disk drive	+5 V @ 0.75 A, + 12 V @ 5 A peak

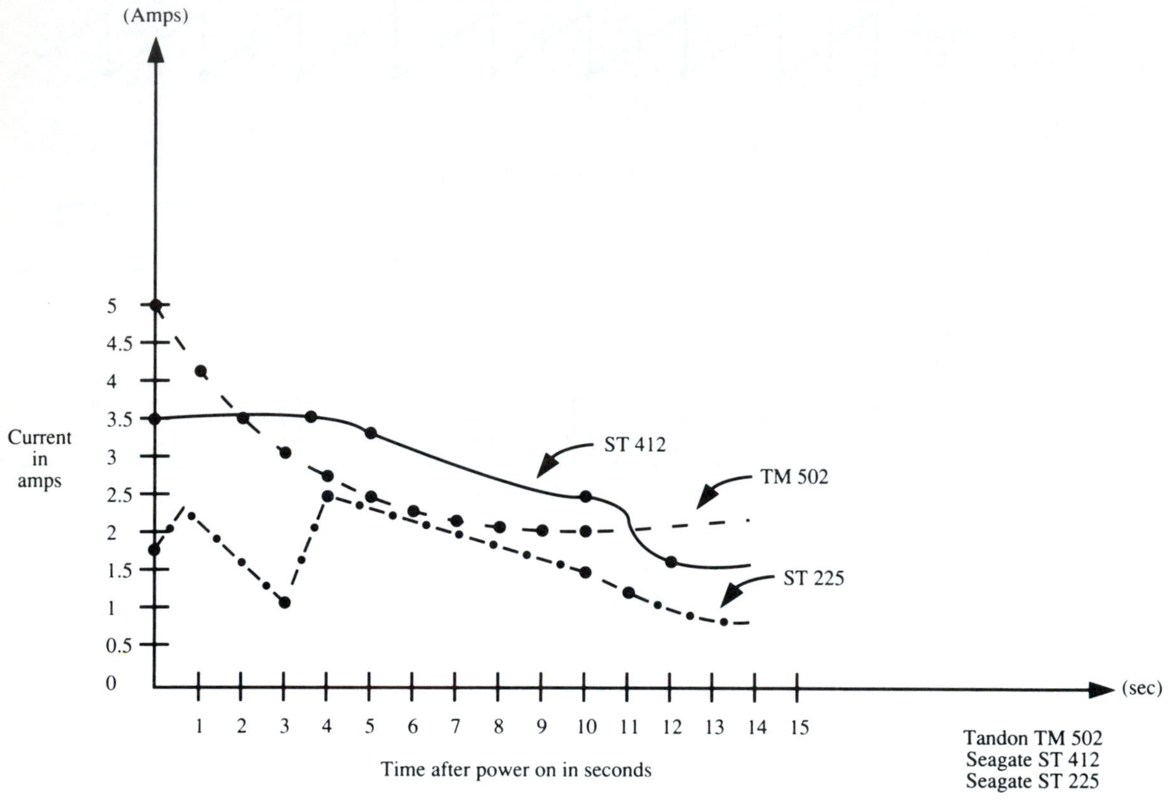

FIGURE 12.6
Hard disk power consumption

The ATX Power Supply and Case

The ATX power supply shown in Figure 12.7 is found in almost 100% of the computers being produced today. The ATX specification maintains the seven-slot configuration of the baby AT. A typical configuration has three ISA slots, three PCI slots, and one shared slot. The ATX form factor is essentially the next generation of the baby AT form factor, and incorporates several practical improvements. The orientation of the board is rotated 90° within the chassis, which allows for the following improvements:

- Relocated processor, which allows for all full-length expansion slots, and removal of the processor without removal of cards
- Single power connector, instead of the two (P8 and P9) connectors in the AT power system
- Relocation of power supply closer to CPU
- Side venting of power supply
- More I/O interfacing incorporated onto the motherboard
- Relocation of SIMM slots to allow for easier access to memory

FIGURE 12.7
ATX power supply

Figure 12.8 shows the ATX chassis layout.

Relocation of the Processor

Relocating the CPU resulted in two improvements. First, there is now enough space for full-length peripheral or adapter interface cards in the expansion slots. While this may not sound especially beneficial, it is a big advantage when installing either an older or cus-

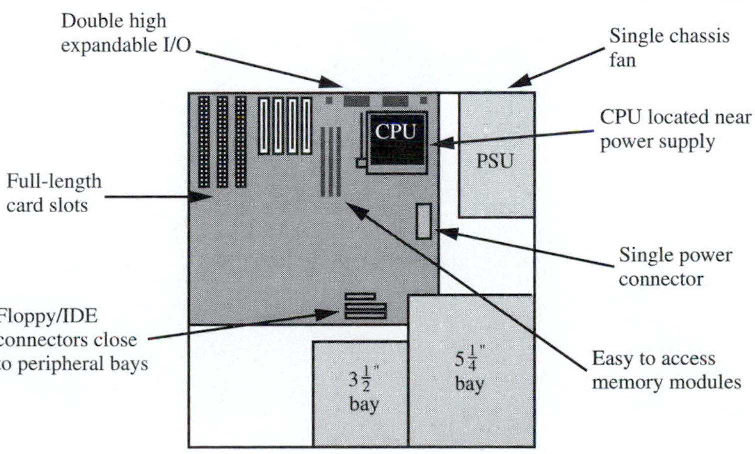

FIGURE 12.8
ATX chassis layout

tom made (perhaps wire wrapped by hand) expansion card. You simply may need the extra length. Second, if you have ever tried to upgrade or repair the processor in a non-ATX chassis, you know that you practically have to disassemble the computer to do so. In the ATX chassis you have clear access to the CPU.

Single Power Connector The ATX board uses a single 20-pin power connector, which incorporates ±5-V, ±12-V, 3.3-V, and soft power signals. The 5-v and 12-v signals have been standard for years, but the 3.3-V and soft power signals are not. With the evolution of processor technology comes the need for lower voltage requirements, hence the 3.3-V output. In addition, the 3.3-V signal was added to accommodate the expected transition to 3.3-V PCI peripheral cards. The soft power signal was added to take advantage of the ability to shut down the computer from software. Figure 12.9 shows both the AT and ATX power connectors. Figure 12.10 illustrates the ATX power connector pinout.

ATX also allows for an optional power connector. Although this is primarily a manufacturer's issue, it could provide benefits to the end user when implemented. It also provides options for fan cooling and monitoring, and fan speed control, which can be used during applications that have lower power requirements. The optional connector provides a 3.3-V sense line and an IEEE 1394 power source. Figure 12.11 illustrates the ATX optional power connector.

FIGURE 12.9
AT and ATX power connectors

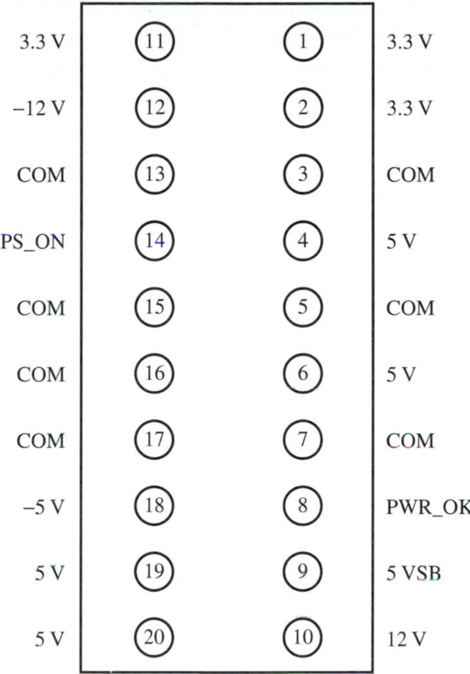

FIGURE 12.10
ATX power supply connector pinout

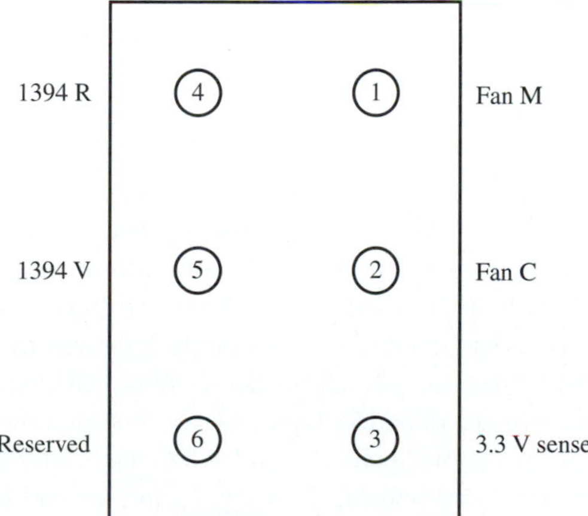

FIGURE 12.11
Optional ATX power connector

Location and Venting of Power Supply The ATX power supply differs from its predecessors in two ways. One, it is located directly next to the CPU, and two, it is side venting, as opposed to venting out the back of the chassis. According to the ATX specification v2.03, this allows for airflow over the CPU, thus eliminating the need for a CPU heat sink or fan. Some manufacturers configure computer systems this way. They use a heat sink on the CPU with a cowling going from the power supply fan to the CPU, and no CPU fan. The risk here is that if the power supply fan malfunctions, not only can the CPU overheat, but all of the other components in the system can overheat as well.

More I/O on the Motherboard The advantage here is obvious. It is cheaper to put I/O on the motherboard than to install separate add-in cards. This is particularly true with the increased sophistication of the audio and video needed for today's multimedia applications. Eliminating some cabling also reduces cost. In addition, the elimination of cables can reduce the amount of EMI emissions and interference, and therefore improve system performance and speed.

Relocation of SIMM Slots Those of you who have had to disassemble a computer to get to a processor have almost certainly had to do so in order to get to memory. Adding a few more megabytes should not have to entail major surgery. The ATX style chassis allows for clear access to all SIMM memory slots, which makes the process of adding or upgrading memory easier, faster, and less expensive.

12.2 MOTHERBOARDS

The microprocessor is often called the brain of a computer system, and the main system board, commonly known as the motherboard, can correspondingly be titled the central nervous system. Its importance becomes more obvious when one realizes that all major components such as the microprocessor, memory, long-term storage, and all I/O devices use the motherboard as their intermediate.

Motherboards can be categorized in several ways. They are usually classified by the microprocessor they support, chipset, and form factor. **Chipsets** function as the traffic controllers of data (including address and control information) throughout the motherboard. The chipset will dictate which type of microprocessor, what types of memory, what types of expansion buses, and which bus speeds (among other things) a motherboard can support. Since chipsets are so closely linked with the microprocessor, the introduction of a new microprocessor family is often accompanied by the introduction of a new chipset.

The original motherboards in the IBM PC had five 8-bit I/O slots. The IBM PC/XT had eight 8-bit I/O slots. The next generation of motherboards was the IBM PC/AT. The PC/AT motherboard had two 8-bit and six 16-bit (ISA) slots. All three of these motherboards were built from discrete combination logic. Motherboards now are built up from chipsets primarily in two form factors: baby AT and ATX. The older baby AT form factor is a compact rectangle usually measuring 9 in. by 13 in., around half the size of the original AT, and is commonly used in desktop (horizontal) cases. This motherboard, known as the socket-7 motherboard, uses a zero insertion force, or ZIF, microprocessor socket. The socket-7 motherboard can accommodate any Pentium class processor including chips from AMD and Cyrix. The Pentium Pro uses a similar motherboard except it uses the socket-8 standard. The placement of the processor usually blocks some of the expansion slots restricting the use of full-length cards. The introduction of the Pentium II has given rise to a new microprocessor connector known as slot 1.

As discussed earlier, better placement of the power supply and integration of I/O connectors onto ATX motherboards make the interior of ATX cases easier to work on and more easily upgradable. Air intake and output slots built into the case along with the cleaner layout help to fight the growing problem of overheating caused by high-temperature components such as the processor, graphics card, and hard drives.

Many settings on a motherboard either are changed by modifying settings in the **basic input/output system (BIOS)** during the boot-up process, or by jumpers and dip switches on the motherboard itself. DIP switches are more convenient and have rocker switches for a two-position setting. Some newer motherboards use a jumperless setup whereby all settings can be changed in the BIOS. This is discussed in detail in Chapter 16.

Cabling and Connectors

The motherboard contains numerous types of connectors. Among these are the MPU connector, memory sockets, adapter card slots, disk drive connectors, external I/O connectors, and the power connector. Devices should never be inserted or removed from connectors when the system is powered on because this may damage the device, the motherboard, or both. The only possible exception is when using USB or firewire devices, which can be removed or inserted with the power on.

Two MPU connector types are presently in use: socket and slot connectors. Intel's earlier Celerons, Pentium II, and Pentium III use the **slot 1** connector (the Xeon, a multiprocessor version of the Pentium II uses the similar slot 2). The slot 1 is a short brown slot about 5 in. long and is segmented by a small bar about two-thirds of the way along its length for orientation purposes. It is easer to orient these processors correctly by facing the heat sink/fan toward the front of the case. Figure 12.12 shows a typical motherboard with an Intel Celeron processor installed and a slot 1 connector.

The microprocessor socket form is cheaper to manufacture than the slot form, and later versions of Intel's Celeron return to the socket form. Microprocessors can be easily removed by unlocking a lever built into the **zero insertion force (ZIF)** socket. The microprocessor chip is a square package similar to the original Pentium processor with a pin grid array (PGA) style shape. The Celeron uses a different motherboard socket known as Socket 370, which is not pin compatible with the socket 7.

There are usually four memory slots at the front of the motherboard near the MPU connector. They are commonly dark brown with white levers on each side and about 6 in. long. The 168-pin **dual in-line memory modules (DIMMs)** are plugged in perpendicular to the motherboard with white retainer clips fastened after insertion. Figure 12.13 shows a typical DIMM memory socket.

In older systems, 72-pin **single in-line memory modules (SIMMs)** may be more common. A 72-pin SIMM is notched on the left side so that it can only be installed in one direction. To remove a SIMM, the retainers at either side are disengaged and the SIMM is rotated about 45° forward and lifted out. Installation is just the reverse. Figure 12.14 illustrates the SIMM installation and removal process.

FIGURE 12.12
Motherboard with Intel Celeron processor

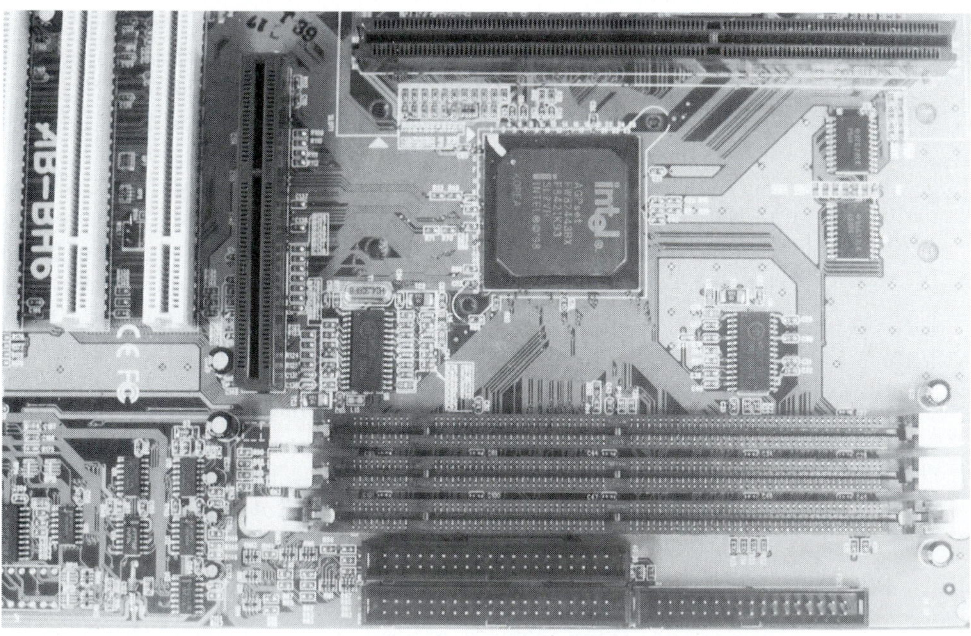

FIGURE 12.13
DIMM memory sockets

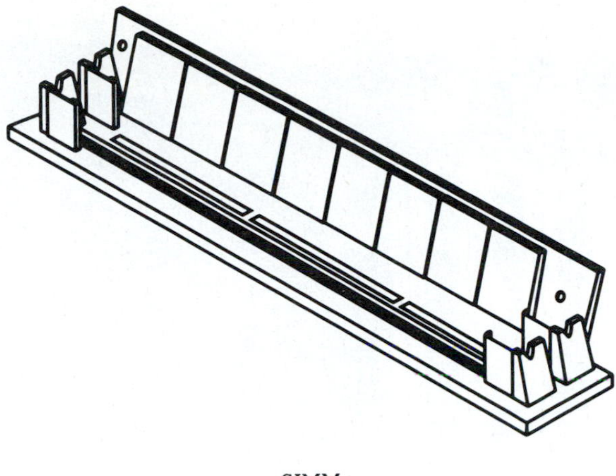

FIGURE 12.14
SIMM installation and removal

Whereas populated circuit boards in general are vulnerable to electrostatic damage, memory modules are especially vulnerable. Always use a grounded antistatic wrist strap or some other electrostatic sensitive device (ESD) protection tool when working with memory modules.

The adapter card slots can take up the majority of motherboard real estate and commonly includes the **advanced graphics port (AGP)**, **peripheral component interconnect (PCI),** and **industry standard architecture (ISA)** slots.

There is only one AGP slot on a motherboard, and it is usually brown, about 3 in. long, and the closest slot to the microprocessor connector. It is even smaller than the PCI slot and offset from the PCI slots. The AGP slot is used exclusively for video cards.

PCI slots are usually molded in white plastic, about 4 in. long, and are shorter than the two-segmented black ISA slots. ISA slots are about 6 in. long, and the last ISA slot is placed almost flush with the bottom edge of the board. Most ISA expansion cards only use the first segment of the slot. There is generally one shared ISA/PCI slot. Figure 12.15 shows a typical motherboard with the different types of expansion slot connectors.

Most motherboards have two IDE drive connectors located near the floppy drive connector. Orientation of these connectors is important. Some connectors have a keyed collar around the pin headers to aid orientation; others do not. If this is the case, look on the circuit board near the pin headers for the number 1. Pin 1 corresponds to the red ribbon running along one side of the IDE or floppy cable. Each IDE connector has 40 pins arranged in two rows; the floppy connector is slightly shorter, with 34 pins in two rows. One of the 40 IDE pins is unused, however, and in some systems this pin is missing. The cable for such a system aids orientation by placing a plug in the unused pin position. Be careful not to bend any of these pins when inserting or removing the cable. Each IDE

FIGURE 12.15
Expansion slot connectors

FIGURE 12.16
IDE motherboard connectors

cable can accept two IDE devices. The floppy cable can also accept two floppy drives. Figure 12.16 shows the IDE interface connectors on the motherboard.

The ATX specification groups all the external connectors in a double row, just behind the MPU connector including the keyboard and mouse PS/2 (also called mini-DIN) connectors, two USB connectors, one parallel (or printer) port and two serial connectors. Each of these connectors is keyed for a certain orientation, so if a device does not plug in correctly the first time, check to make sure of the orientation. The two six-pin PS/2 female connectors are identical, and the general convention is to have the keyboard connector on the row closest to the PCB, with the mouse connector on the upper row. The USB connectors are short white rectangles about 3/4 in. long. The parallel port is a female DB25 connector usually designated LPT1 because it is most often used for printers. The parallel port is also used to connect scanners and external storage devices. The serial ports are identical male DB9 connectors and are called RS-232 connectors. AT motherboards may have one male DB9 and one male DB25 connector for COM1 and COM2 respectively. The serial port closer to the USB ports is COM1; the other port is COM2.

Older AT-style motherboards may not have USB connectors, and the keyboard connector may be of the larger AT or DIN style. Adapters can be used to change the keyboard to the appropriate connector. Figure 12.17 shows the external I/O block connectors.

FIGURE 12.17
External I/O block connectors

There is usually one white 20-pin ATX power connector, which is keyed to protect against accidentally inserting the connector the wrong way. Never connect a power connector with the power on. Always remove the 120-V_{AC} power cord from the wall first. Figure 12.18 shows a typical ATX motherboard power connector.

Motherboards have miscellaneous pin headers for LED connectors, speaker, fans, and other optional functions, usually combined in one block. The only connectors that are required on an ATX system are the power switch and reset connectors; however, the other functions can be useful for troubleshooting a system.

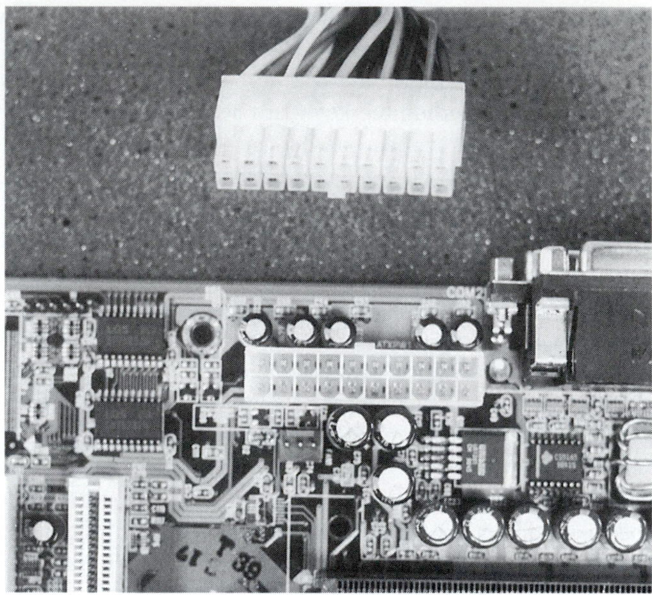

FIGURE 12.18
ATX motherboard power connector

12.3 THE MOTHERBOARD CHIPSETS

Soon after the introduction of the IBM PC/AT, many of the combination logic chips were combined into what is known as application-specific integrated circuits, or ASICs. This made good sense because of the open design used in the original PC. Anyone designing a motherboard would use the same basic layout and components. Thus it made sense to place all the common components inside a common IC or ASIC and use it for many different boards. This simplified the manufacture, reduced the size, and greatly reduced the price of a motherboard. As mentioned earlier, the chipset of a motherboard acts as a traffic controller for data, address, and control information between all major subsystems. It controls the several different timing functions that are necessary for operation and coordinates bus transfers. The chipsets are also very specific in what devices and technology they can support, and usually support a limited range of devices. This is why when some new technology—such as faster memory, faster processors, or new processor instruction sets—becomes available, a new chipset is usually introduced to support it.

In the block diagram of Figure 12.19, the chipset is composed of two main components. These are the host bridge, also known as the north bridge, and the PCI to ISA bridge, also known as the south bridge. As we can see, the processor connects only to the north bridge, as opposed to the traditional system design, in which the processor is on the system bus. The north bridge acts like a router or central nervous system and directs data to and from system busses. The north bridge is associated with the microprocessor, mem-

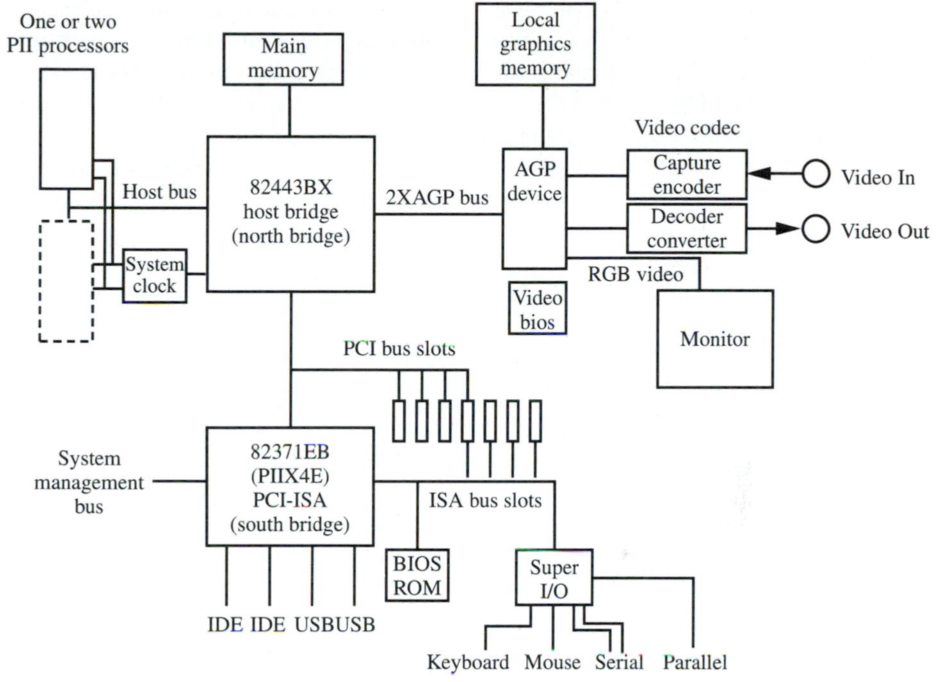

FIGURE 12.19
Typical chipset block diagram

ory subsystem, PCI bus, and AGP bus. It becomes obvious that the AGP was created to optimize the movement of graphical data between the MPU, system memory, and AGP video card memory. In this scenario, the processor can access the PCI bus and the AGP device may access main memory. The south bridge provides the legacy ISA (AT) bus services, two advanced busmaster IDE interfaces, and two USB ports. This design removes the slower low-priority devices out of the way of the higher-performance devices. The south bridge was first introduced in the Pentium motherboard to increase data transfers from the hard disk subsystem. With the exception of the PCI IDE header, the south bridge mostly handles lower-bandwidth data transfers. It is usually aided by another I/O chip known as the super I/O.

In the early days of personal computers, the motherboard would be loaded with over 100 chips, including the clock generators, memory controllers, DMA and IRQ controllers, CMOS RAM, keyboard controller, serial and parallel port controllers and buffers, CPU itself, and numerous other things. Intel made many of these components, but other manufacturers also made some. To produce a motherboard, one would need to obtain all of these separate components. This was very costly for motherboard manufacturers.

In 1986, Chips and Technologies produced the 82C206 component for the Intel 80286 processor. This chip combined many of the separate motherboard functions onto a single chip. Overnight, a motherboard became easier and less expensive to produce, and the idea of a chipset was born. Almost instantly, the company became a huge success because manufacturers were scrambling to use this chipset on their motherboards. Soon after, many IC manufacturers were producing PC chipsets. Some example chipsets are Suntac, Opti, VIA, Micronics, AMD, and of course Intel. Some manufacturers make single-chip solutions for embedded applications.

As mentioned earlier, the chipset becomes the traffic controller for data, address, and control information between all the major subsystems. It controls the several different timings necessary for operation and coordinates bus transfers. As new technologies emerged, chipsets, like processors or memory, had to cope with the changes. This generally meant consolidating more functions onto the chipset and had the result of lowering prices.

The first significant chipset that Intel introduced after the Pentium family chipset was released was the 440LX chipset. It added support for the new AGP graphic card interface and was designed exclusively for the Pentium II family of processors, but still ran the **front side bus (FSB)** at 66 MHz. The FSB is the data channel or processor to system bus interface. It is used to transmit data between the processor and system memory. This is critical for performance. With the revolutionary 440BX chipset, Intel officially introduced a 100-MHz FSB for the first time. In conjunction with the 100-MHz FSB, the BX chip also introduced support for PC100 SDRAM. Memory up to that point was run at a maximum clock speed of 66 MHz. The BX chipset also allowed the use of **serial presence detect (SPD),** whereby the chipset can query a chip on the memory module to automatically set memory timings instead of manually setting the timings in the BIOS.

Intel 430LX

The Intel 430LX chipset was introduced alongside Intel's first rendition of the Pentium processor, which came in 60 and 66 MHz. This chip set offered no built-in IDE support. All disk functions are provided by external circuitry. These chips operated at 5 V_{DC} and were used in socket-4 motherboards. The 430LX chipset could access a maximum of 192 MB of RAM.

Intel 430NX

The Intel 430NX chipset (code name Neptune) was introduced with the second generation of Pentium processors (75–133 MHz), and operated at 3.3 V, which is supplied by a voltage regulator installed onto the motherboard. The hallmark of this chipset was the first support for dual processors in a personal computer system. The 430 NX chipset could support up to 512 MB of RAM.

Intel 430FX

The Intel 430FX chipset (code name Triton) was introduced as a substitution for the NX. It implemented support for EDO (extended data out) RAM, higher-speed cache memory (pipeline burst), and IDE bus mastering using the PIIX2 controller. However, it lacked support for multiple processors and parity error checking in memory. The 430FX chipset could support a maximum of 128 MB of RAM.

Intel 430HX

The Intel 430HX was created as a replacement to the NX and FX chipsets. It included all of the support for EDO memory, pipeline burst cache, and bus mastering. It also reimplemented support for dual processors and parity error checking. In addition, it offered support for ECC (error correcting code) RAM, 512 MB of RAM, USB support, and PCI 2.1.

Intel 430VX

The Intel 430VX chipset was the first to support SDRAM, up to 128 MB. However, it lacked support for parity or ECC memory. This chip set included the PIIX3 busmaster controller supporting IDE PIO mode 4. In addition, the VX chipset is compliant with PCI 2.1. It supported the Pentium 75- to 200-MHz processors.

Intel 430TX

The Intel 430TX chipset was the last of the Pentium chipsets. The TX chipset used the PIIX4 controller, which added support for Ultra ATA/DMA IDE hard drives, and a low-power mode for mobile usage. However, it didn't support ECC or parity memory and was limited to only 128 MB of RAM. It offered support for the Pentium 75- to 233-MHz. MMX processors.

The Pentium II spawned a new generation of chipsets. These chipsets were based around a two-chip design. Older Pentium chipsets contained anywhere from two to four chips in the chipset. The two chips that were used in the Pentium II chipsets were the host/bridge controller and the PCI to ISA/IDE accelerator. They are also known respectively as the north bridge and south bridge chips.

An example of the north bridge chip is the Intel 82443BX, found in the BX chipset. The north bridge is responsible for the memory and AGP subsystem. It routes all data to and from the processor and any PCI or AGP expansion slot, main memory, and the south bridge chip. The south bridge is responsible for data traffic to the USB ports, IDE buses, CMOS and RTC, the ISA slots, keyboard, mouse, serial ports, parallel port, floppy drives, and the flash BIOS ROM. An example of this chip is the Intel PIIX4E, also found in the BX chipset.

Intel 440FX

The Intel 440FX was originally designed for the Intel Pentium Pro processor. However, when the Pentium II was ready for market, Intel designers had not finished their work on the LX chipset, which was supposed to be released concurrently with the Pentium II. Instead, they ported the FX chipset to the Pentium II. The FX chipset was similar to the Pentium HX chipset. It had support for USB, busmaster IDE, full parity, and ECC, and could address up to 1 GB of EDO RAM.

Intel 440LX

The Intel 440LX was designed around the new PII processor and totally replaced the FX chipset in all markets upon its release. It was the first chipset that could take advantage of the new PII design. It included support for the new AGP (accelerated graphics port) slot, as well as USB, PCI 2.1, UDMA IDE, and 66-MHz SDRAM, and dual-processor support. This was the first chipset to relinquish support of EDO RAM in favor of the faster SDRAM technology.

Intel 440EX

The Intel 440EX chipset was released concurrently with the new Pentium Celeron processors. It lacked many of the advanced features that were standard in the LX chipset such as ECC or parity memory and dual-processor support. Like the Celeron processor, it was designed for the low-cost market. It never became popular with consumers or OEMs.

Intel 440BX

The Intel 440BX chipset was a vast improvement over the LX design. It was the first to support 100-MHz RAM and was developed with the new 350- to 450-MHz processors in mind, but it also included support for 66-MHz RAM. Another feature is the ACPI, or advanced configuration and power interface, specification. This was released as the first mobile chipset to support the Pentium II architecture.

Intel 440MX

The Intel 440MX chipset was an extension of the venerable BX chipset. It included optimizations for the Celeron processor, advanced power management, modem and audio emulation in the CPU, and combined the 82443BX and PIIX4 into a single chip.

Intel 440ZX

The Intel 440ZX is designed solely for the entry-level computer. It is built upon the BX design, but it incorporates the entire chipset onto a single chip, therefore decreasing the size requirements on the motherboard. The ZX chipset also includes support for ATA66 IDE access, but does not support socket 370. It only supports slot-1 designs.

Intel 440ZX66

The Intel 440ZX66 is a compliment to the original ZX chipset. It adds a greater value to low-end computers because it is designed around the socket 370 instead of the slot 1. This makes it ideal for embedded or single-board computer and controller applications.

Intel 810

The Intel 810 design incorporates a number of functions that would normally be found as expansion options, such as audio, video, and modem. This is the first chipset to incorporate all the functions of a basic multimedia PC into the chipset. It is truly designed for low-price applications. It builds on the 440BX design but features design improvements such as double data rate PCI (PCI 2.2), integrated video, audio, modem support, an improved AGP bus, ATA66, and better system manageability.

Intel 810E

The Intel 810E chipset builds on the design of the 810. It offers 133-MHz system memory support, whereas the 810 only offers 100-MHz support. It also provides enhanced support for the Pentium III and Celeron processors, and digital video out for connection with new flat panel displays. This chipset is designed for the performance desktop market.

Intel 450GX

The Intel 450GX chipset is designed to support the Intel Xeon processor. It has the same basic design as the BX chipset except it can support twice the amount of RAM, 2 GB, compared to the BX's 1 GB. It includes support for the AGP bus and ECC/parity memory. The GX chipset is primarily used in the graphics workstation market.

Intel 450NX

The Intel 450NX chipset was designed around the capabilities of the Xeon processors. It allows up to 4-way **SMP (symmetric multiprocessing)** without the addition of support chips and 16-way SMP with additional support. It allows for up to 8 GB of ECC or parity 60-ns or 50-ns, 3.3-V EDO RAM. It has no support for the AGP bus or 100-MHz SDRAM. The 450NX chipset was designed for the enterprise server market. The design of the 450NX makes it very scalable and versatile in the high-performance business market.

Intel 840

The Intel 840 chipset replaces all previous chipsets in the performance desktop, workstation, and server markets. It supports both the Pentium III and the Pentium Xeon processors. It lacks support for the Celeron processor.

The 840 is a complete redesign of the chipset architecture intended for multiprocessor systems. It does away with the ISA slot and incorporates many new design features. The 840 includes support for AGP 4X, AGP Pro, and USB (v2). This chipset supports separate 32-bit and 64-bit PCI buses as well as a 133-MHz system bus. Also supported are PC400 and PC800 RDRAM (discussed later in this chapter), digital signing and security protocols, and a prefetch cache. This allows for highly efficient data flow.

Intel 820

The Intel 820 chipset is the latest release, optimized for the Pentium III processor. It is a scaled down version of the 840. The 820 chipset is intended for one- to two-processor systems. This allows for a lower chip count, down to three as opposed to six on the 840 set. The 820 chipset also provides better buffering and a deeper pipeline.

12.4 MAIN SYSTEM TIMING

The MPU clock speed is actually a product of two numbers, the MPU clock multiplier and the FSB. For example the 450-MHz Pentium II has a clock multiplier of 4.5 and Pentium II systems run at a FSB of 100 MHz, so 4.5×100 MHz $=$ 450 MHz.

Similarly, the FSB is used to determine the speed of system memory, the AGP, and the PCI bus. On a 440BX chipset Pentium II system, memory will have a clock speed equal to $1 \times$ FSB, or 100 MHz. The AGP clock will run at $2/3 \times$ FSB, or 66 MHz, and the PCI bus will run at $1/3 \times$ FSB, or 33 MHz. In a Celeron system, the FSB is normally 66 MHz. In these systems, the memory will have a clock speed of 66 MHz, the AGP clock will run at $1 \times$ FSB or 66 MHz, and the PCI bus will run at $2/3 \times$ FSB, or 33 MHz. Therefore, the chipset provides different multipliers for different FSBs to keep various components running within specified speeds.

The chipset itself does not provide the clock timing, however. A clock generator chip generates the actual clock signals. Different clock generator chips are designed for specific chipsets.

Level 1 (L1) cache is very high-speed memory inside the processor; **L2 cache** is memory intermediate between the L1 cache and system memory in both speed and manufacturing costs. Normally the L1 cache runs at the MPU clock speed. The L2 clock speed depends on the MPU architecture and can run at the FSB (Pentium, K6), at half the MPU clock speed (Pentium II), or at the full MPU clock speed (Pentium Pro, Xeon). The production cost of fabricating high-speed cache is usually the limiting factor for the design of L2 cache clock speed.

12.5 PC MICROPROCESSORS

The 80X86 Processor Evolution

Over the past ten years the Intel microprocessor has undergone many changes and enhancements, including second sourcing from AMD Corporation. All AT class computers use one of the Intel 80286, 80386-SX, 80386-DX, 80486-SX, or 80486-DX microprocessors. This new line of microprocessors includes

- 80286 in speeds of 6, 8, 10, 12, and 16 MHz
- 80386-SX in speeds of 16, 20, and 25 MHz
- 80386-DX in speeds of 16, 20, 25, and 33 MHz
- AMD 80386-DX at a speed of 40 MHz
- 80486-SX at a speed of 25 MHz
- 80486-DX in speeds of 25, 33, 50, 75, and 100 MHz
- Pentium in speeds of 60, 75, 90, 100, 133, 166, and 200 MHz

The IBM PC and PC/XT are centered around the 8088 microprocessor. The **AT class** systems are centered around the 80286, 80386, and 80486 microprocessors. This was done to improve performance, increase processing speed, and increase memory capacity. It is important to note that the processors are all **upward compatible;** that is, anything that will run on an 8088 will run the 8086, 80286, 80386, or 80486. The opposite, of course, is not the case. Programs developed for the 80486 may not work on lower-level processors.

The major differences among the processors revolve around bus size, speed, and memory capacity. For example, the 8088 is an 8-bit processor, the 80286 is a 16-bit processor, and the 80386 and 80486 are 32-bit processors.

The 80286 Processor

The 80286 microprocessor is a high-speed (6, 8, 10, 12, and 16 MHz) 16-bit microprocessor. The 80286 can directly address up to 16 MB of RAM and up to 1 GB of virtual memory. The 80286 operates in two modes: real address mode and protected mode. In the **real address mode,** the 80286 can execute any programs written for the 8086 and 8088. In the real address mode, programs being run on an 8-MHz 80286 will run up to five times faster than on a 5-MHz 8086. In the **protected mode,** the 80286 can still execute 8086 programs but will operate using memory protection and allow virtual memory addressing.

The 80286 microprocessor consists of four basic sections: the instruction unit, execution unit, address unit, and bus interface unit. Figure 12.20 illustrates the interaction of the four basic sections in a block diagram format.

The **bus interface unit** performs all the operations for the MPU that require bus operations. These include generating the command signals for the address and data bus. The address and data command signals allow the interfacing of the MPU to external memory or to input and output devices (buffers, flip-flops, keyboards, etc.). The bus unit also controls how the MPU operates with other processors on the same address bus (such as math coprocessors). It also looks ahead and gets the next instruction to be executed from memory. Instructions are stored in the instruction queue to be used by other units in the 80286.

The **instruction unit** gets (fetches) the next instruction to be executed from the instruction queue. The instruction is then decoded into **microcode** (the 1s and 0s used by the MPU) and passed on to the execution unit.

The **execution unit** takes the decoded instruction from the instruction unit queue and performs the action called out by the decoded instruction. The execution unit also uses the bus unit to transfer data into and from memory.

The **address unit** manages and protects the system memory for the MPU. The management that it performs is keeping the sections of memory that are in use by the MPU in consecutive (contiguous) blocks. Protection involves keeping any sections of memory (data) from being erased or corrupted unintentionally.

The 80286 is configured into a 68-pin, high-density package, as shown in Figure 12.21. Of the 68 pins two are +5 V (V_{CC}), three are ground (V_{SS}), and six are no connection. The signal lines are as follows:

1. The CLK (clock) line provides the fundamental timing used by the chip.
2. The D_0–D_{15} (data bus) lines provide the bidirectional, general-purpose path for data signals between the 80286 and other devices.

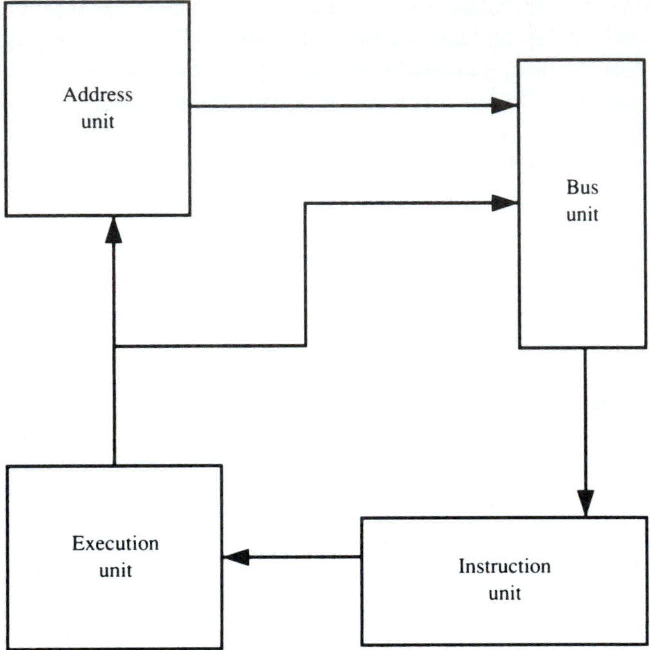

FIGURE 12.20
80286 block diagram

3. The A_0–A_{23} (address bus) lines provide the connections among the 80286, physical memory, and I/O devices.
4. The $\overline{\text{BHE}}$ (bus high enable) line indicates the transfer of data on the upper byte of the data bus (D_8–D_{15}).
5. The $\overline{S_0}$, $\overline{S_1}$ (bus cycle status) lines indicate when the bus cycle starts and, along with several other signals (M/$\overline{\text{IO}}$, COD/$\overline{\text{INTA}}$), indicate what type of bus cycle is being executed.
6. The M/$\overline{\text{IO}}$ (memory–I/O select) line indicates whether a memory or I/O cycle is being performed.
7. The COD/$\overline{\text{INTA}}$ (code/interrupt acknowledge) signal distinguishes between instruction fetch cycles from memory read cycles and acknowledge signals from interrupt cycles.
8. The $\overline{\text{LOCK}}$ (bus lock) signal distinguishes between locked and unlocked bus cycles.
9. The $\overline{\text{READY}}$ (bus ready) signal indicates that the bus cycle is complete.
10. The HOLD (bus hold request) input signal indicates a request for bus control by a separate device.
11. The HLDA (bus hold acknowledge) output signal indicates that the 80286 has given control of the bus to another device.
12. The INTR (interrupt request) signal requests that the 80286 stop its present operation and service an interrupt, which can be masked by software.

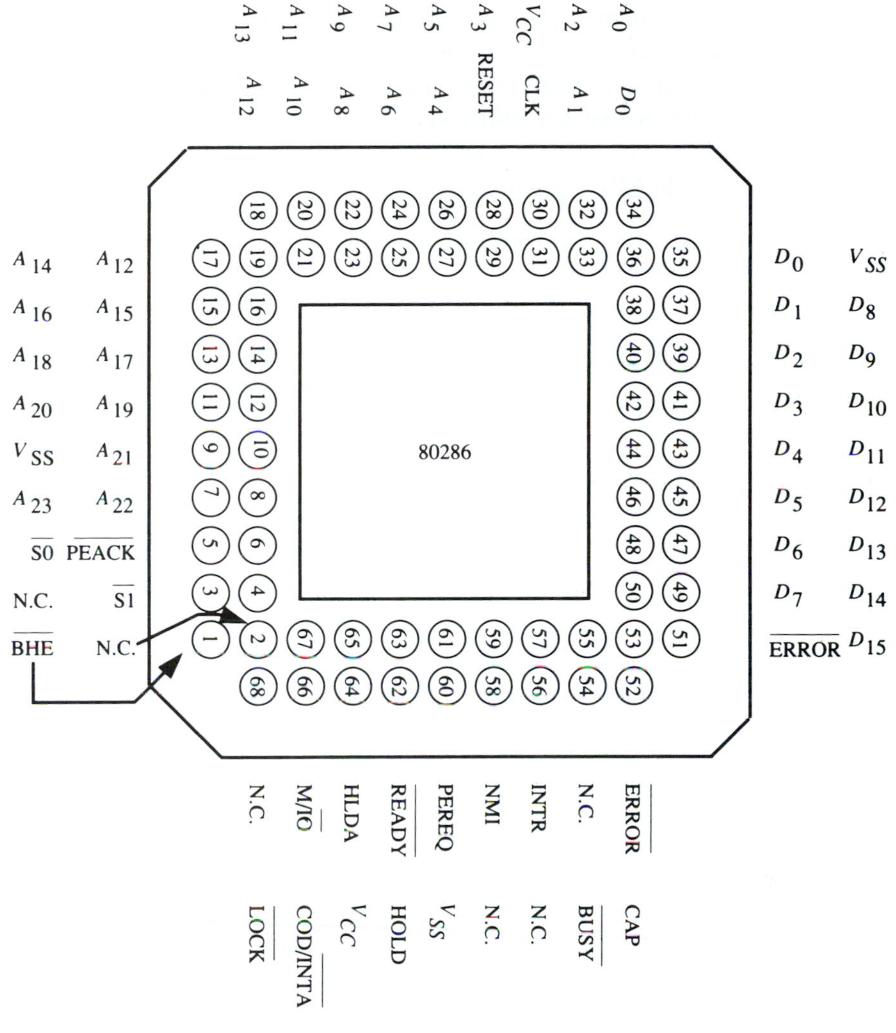

FIGURE 12.21
80286 pinout

13. The NMI (nonmaskable interrupt request) input signal indicates a request for interrupt service, which cannot be masked by software.
14. The PEREQ (coprocessor request) input signal indicates a coprocessor request for a data operand to be transferred to or from the 80286.
15. The PEACK (coprocessor acknowledge) signal tells the 80286 when the operand is being transferred.
16. The BUSY (coprocessor busy) input signal indicates that the coprocessor is executing and is unable to respond to another request.

17. The $\overline{\text{ERROR}}$ (coprocessor error) input signal indicates that the coprocessor has experienced an error in the previous instructions.
18. The RESET (reset) input signal suspends any operation in progress by placing the 80286 in a known reset state.

The 80386 Microprocessor

The 80386 microprocessor is a high-speed, 32-bit microprocessor. The 80386 consists of 275,000 transistors to perform all the functions of an advanced 32-bit computer. The 80386 can operate software that was written for the 8086 and the 80286, in addition to those programs that were written specifically for the 80386.

The Intel 80386 is a full 32-bit microprocessor, with 32-bit internal registers and a 32-bit data bus. It also has 32 memory address lines. The 80386 can be thought of as two 8086s combined into one package with additional support devices included. With its 32-bit address bus, it can directly access 4 GB ($2^{32} \approx 4.29 \times 10^9$) of physical memory. To put this into perspective, 4 GB is more than 4000 times as much memory as the IBM PC with an 8088. This is more than 250 times the maximum memory capability of an 80286-based computer.

The 80386 can operate in three different modes: *real, protected,* and *virtual.* In the **real mode,** the 80386 operates as an 8086 with the same segments and 1 MB of memory. In this mode, only one program can run at a time. In the protected mode, programs use the same segments and offset registers as in the real mode. In the **protected mode,** the segment register is not a *real* address. The base address for each segment is 32 bits wide and can be anywhere within the 4 GB of addressable memory. The offset addresses are also 32 bits wide. The upper 14 bits of the segment registers are used to look up a base address in a **descriptor table.** A descriptor table is a part of memory containing the physical address. This allows for over 16,000 different segments ($2^{14} = 16{,}384$), each of which can be 4 GB long (4 GB = 2^{32}). Thus, the total amount of **virtual memory** is 64×10^{12}, or 64 TB ($2^{14} \times 2^{32}$). The 80386 also uses on-chip memory management and multitasking. This allows for more than one program to be run at any given time. In the **virtual mode,** the 80386 has an addressing feature that is halfway between the 8086 and the fully protected 80386 mode. In this mode, the 80386 mimics the 1 MB of addressing space of the 8086. Unlike the real mode, there can be more than one virtual mode operating at one time. Each of these **virtual environments** can have its own DOS application. Figure 12.22 illustrates the layout of the 80386 microprocessor chip. Included are the real estate areas that coincide with the six functional areas of the chip.

The 80386 consists of six sections, which perform all the actions of the microprocessor. These sections are the bus interface unit, code prefetch unit, instruction decode unit, execution unit, segmentation unit, and paging unit. See Figure 12.23.

The **bus interface unit,** as its name implies, provides the interface between the 80386 and the outside world. It accepts requests for the internal transfer of information (data and addresses) and instructions. The interface unit also assigns levels of priority to these various requests. At the same time the bus interface unit also generates and processes all the information necessary to make up the bus cycles (read and write signals, data acknowledges, address strobes, etc.).

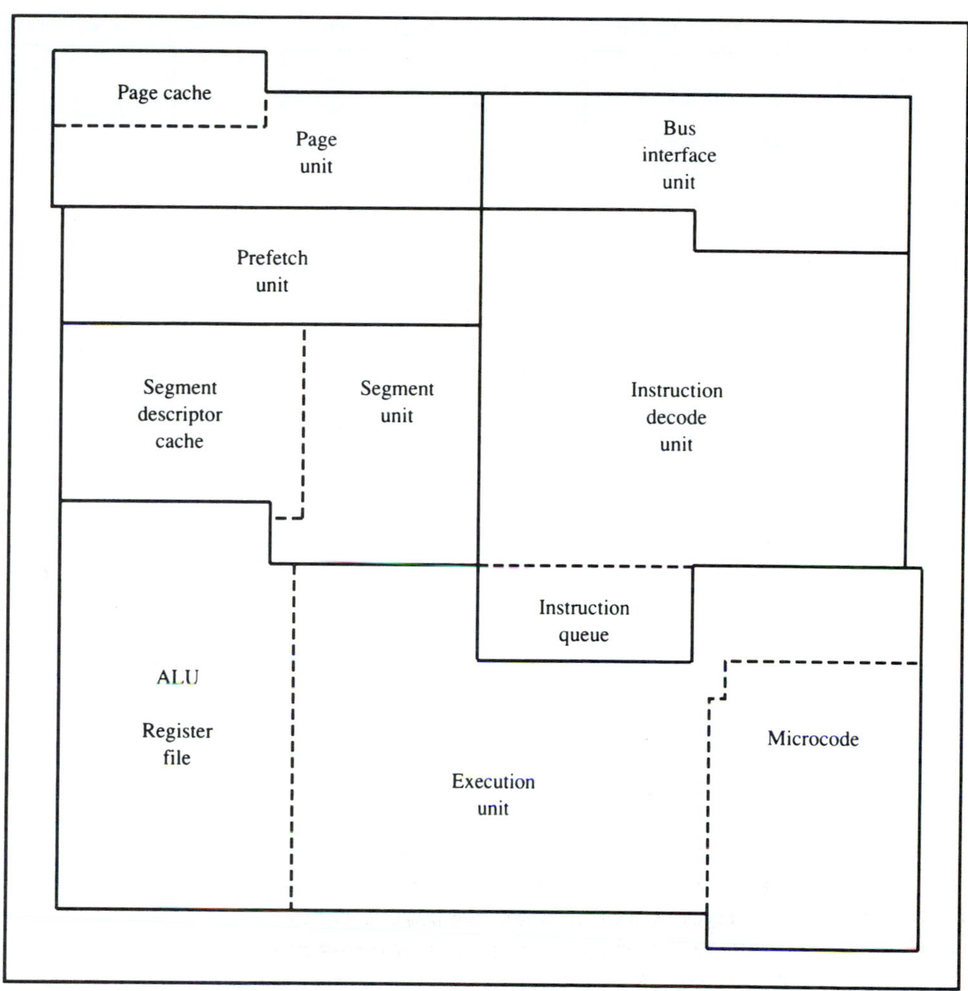

FIGURE 12.22
80386 microprocessor chip

The **code prefetch unit** performs a look-ahead function for the 80386. This means that after the information for the instruction that is presently being executed is loaded and is being processed, the prefetch unit looks ahead to the next instruction and loads it into the 16-bit instruction queue. Because this action is not a direct execution of a program that is currently running, the bus interface unit gives the prefetch requests a low priority, which means that the prefetch unit must await until the bus interface unit is not busy performing any instructions.

The **instruction decode unit** takes the instructions left in the instruction queue by the prefetch unit and translates them. The instructions written in 80386 operation codes

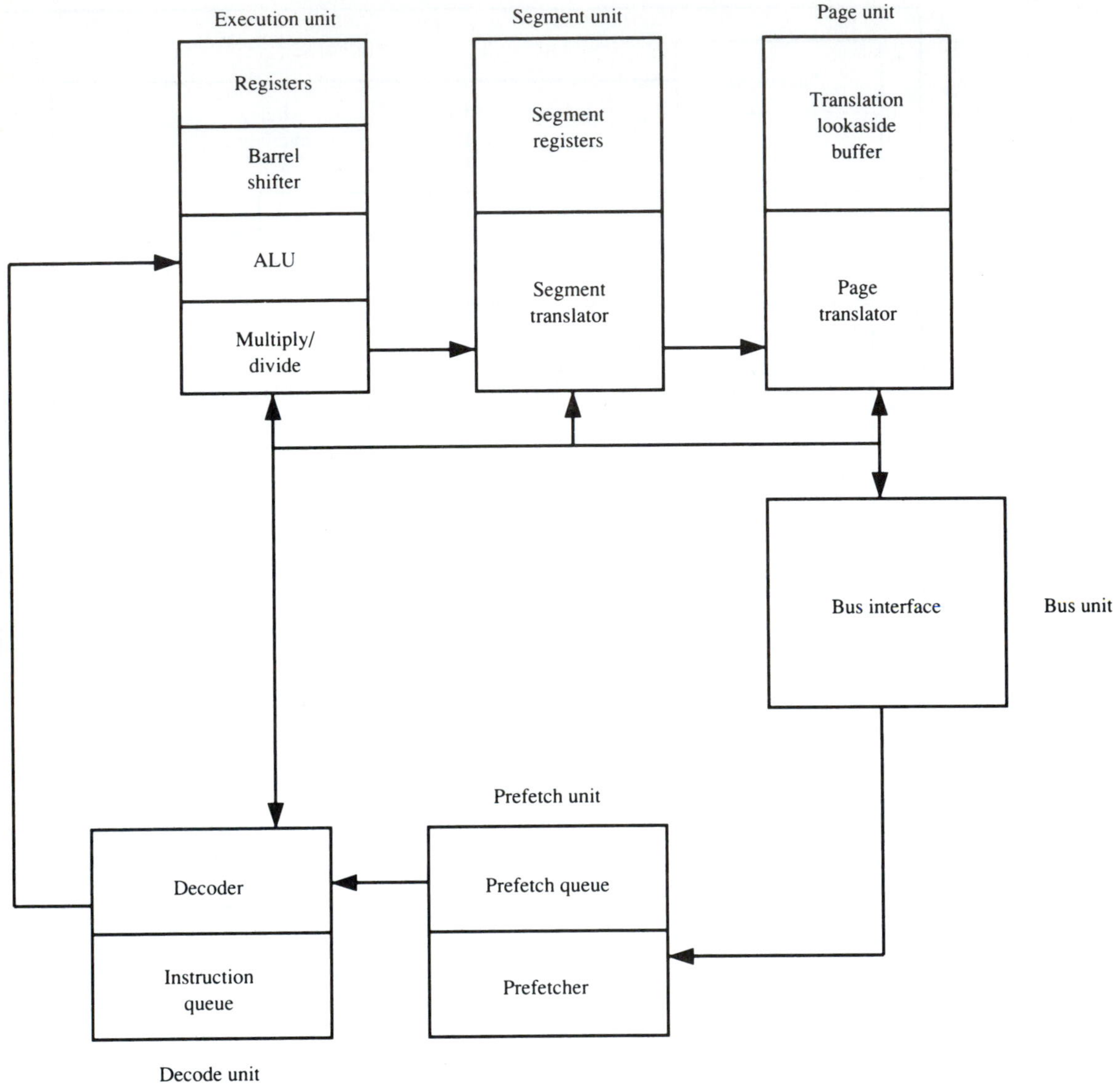

FIGURE 12.23
80386 block diagram

(op-codes) are translated down even further by the decode unit from the hexadecimal codes used by the programmer to microcode (1s and 0s that the logic circuits in the microprocessor can understand).

The **execution unit** of the 80386 takes the instructions decoded from the instruction queue and communicates them to all other units that are required to execute the instruction. The execution unit contains three subunits. These units are the **data unit,** which contains the **arithmetic logic unit (ALU);** the **control unit;** and the **protection test unit.** The data unit performs the operations requested by the control unit. The ALU in the data unit performs all the arithmetic operations. The control unit contains the microcode (taken from the instruction decode unit) and the circuitry that performs calculations to locate the proper addresses for the execution of the microcode. The protection unit makes sure that the code being executed does not attempt to violate the movement between segments (blocks) of code.

The **segmentation unit** translates the logical addresses in the code being run into addresses that can be used by the execution unit. A segment of memory is a block of memory, usually 64 KB long, used to store data or addressses separately. This type of memory is referred to as *flat memory.* It provides no protection to the variables in a program because all the information resides in the same block or area of memory.

The **paging unit** in the 80386 translates the addresses generated by the segmentation or prefetch unit into physical addresses. The paging unit also sends these physical addresses to the bus interface unit for use in executing code.

The package of the 80386 is a 132-**pin grid array (PGA)** with a 14 by 14 matrix formed by three rows of pins all around, as shown in Figure 12.24. Twenty of these pins are used for power (V_{CC}), 21 are used for ground (V_{SS}), and eight pins are left unused. The remaining 83 pins are for carrying information and timing signals. These signals are as follows:

1. The CLK2 (clock) provides fundamental timing that is divided by two for use by the chip.
2. The D_0–D_{31} (data bus) lines provide the bidirectional, general-purpose path for data signals between the 80386 and other devices.
3. The A_0–A_{31} (address bus) lines provides the tristate path to physical memory and I/O port addresses.
4. The $\overline{\text{BE}}_0$–$\overline{\text{BE}}_3$ (byte enable) outputs indicate which bytes of the 32-bit data bus are being used for the current transfer.
5. W/$\overline{\text{R}}$ (write or read indication) is the output signal that distinguishes between write and read cycles.
6. D/$\overline{\text{C}}$ (data-control indicator) is the output signal that distinguishes between data and control cycles.
7. M/$\overline{\text{IO}}$ (memory–I/O indication) is the output signal that distinguishes between memory and I/O cycles.
8. $\overline{\text{LOCK}}$ (bus lock indication) is the output signal that distinguishes between locked and unlocked bus cycles.
9. $\overline{\text{ADS}}$ (address status) is the tristate output signal indicating that a valid address is on the bus.
10. $\overline{\text{READY}}$ (transfer acknowledge) is the input signal that indicates a bus cycle is complete.
11. $\overline{\text{NA}}$ (next address request) is the input signal used to request the use of address pipelining.
12. $\overline{\text{BS16}}$ (bus size 16) is the signal used to indicate if 16- or 32-bit data transfers are to be used.
13. HOLD (bus hold request) is the input signal that indicates a request for bus control for a separate device.

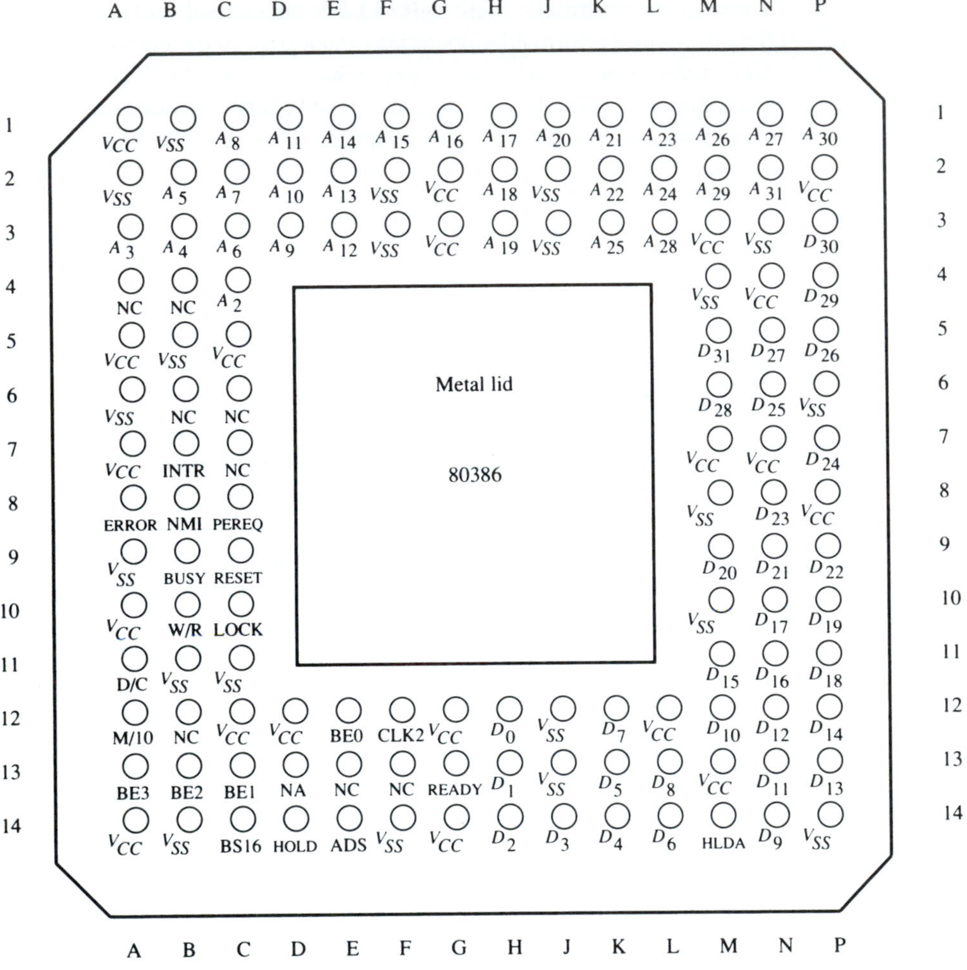

FIGURE 12.24
80386 pinout

14. HLDA (bus hold acknowledge) is the output signal indicating the 80386 has given control of the bus to another device.
15. PEREQ (coprocessor request) is the input signal indicating a coprocessor request for a data operand to be transferred to or from the 80386.
16. BUSY (coprocessor busy) is the input signal indicating the coprocessor is executing and is unable to respond to another request.
17. ERROR (coprocessor error) is the input signal indicating that the coprocessor has experienced an error in the previous instructions.
18. INTR (maskable interrupt request) is the input signal indicating a request for interrupt service, which can be masked by the flag register, IF, bit.

19. NMI (nonmaskable interrupt) is the input signal indicating a request for interrupt service, which cannot be masked by software.
20. RESET (reset) is the input signal used to initialize and suspend any operation in progress by placing the 80386 in a known reset state.

The **80386-SX** processor is the same internally as the 80386. It is designed to replace the 80286, although it is *not* pin for pin compatible. The bus lines are designed for a 16-bit bus design like the 80286. In fact, some clone boards can accept either the 80286 or 80386-SX processor. Basically, the 80386-SX can run all 80386 applications such as multitasking on Windows 3.1, but without the higher costs of the 80386-DX processor and the 32-bit memory system. Of course, the overall speed and performance will not be as good as the 80386-DX processor.

The 80486 Processor

The 80486 is the latest enhancement to the Intel processor line. The 80486 is functionally the same as the 80386, with the following changes:

1. The numeric coprocessor is included on the chip.
2. An internal 8 KB of cache memory is included on the chip, which greatly increases the instruction execution speed.
3. A reduced instruction operation time coding scheme reduces the amount of time required to execute many of the instructions to just one clock cycle.
4. A 100-MHz version is available from Intel, which makes this processor extremely fast.

Figure 12.25 shows a diagram of the 80486 pinout. The signals are defined as follows:

1. The CLK (clock) input provides the fundamental timing for the 80486. It is either 25, 33, or 50 MHz, depending on the version of the processor.
2. The D_0–D_{31} (data bus) data lines provide the bidirectional, general-purpose path for data signals between the 80486 and the other devices.
3. The A_0–A_{31} (address bus) address lines provide the tristate path to physical memory or I/O port addresses.
4. The $\overline{BE_0}$–$\overline{BE_3}$ (byte enable) output signals are used to indicate which bytes of the 32-bit data are being used when fewer than 32 bits of data are being transferred.
5. W/$\overline{R}$ (write or read indicator) is the output signal that distinguishes between write and read cycles.
6. D/$\overline{C}$ (data or control indicator) is the output signal that distinguishes between data and control cycles.
7. M/$\overline{IO}$ (memory–I/O indication) is the output signal that distinguishes between memory and I/O cycles.
8. $\overline{LOCK}$ (bus lock indicator) is the output signal that distinguishes between locked and unlocked bus cycles.
9. $\overline{PLOCK}$ (pseudo lock) is the output signal that indicates that more than one bus cycle is required to perform the current operation.
10. $\overline{ADS}$ (address status) is the tristate output signal that indicates that a valid address is on the bus.

	1	2	3	4	5	6	7	8	9	10	11	12	13	14	15	16	17
S	A27	A26	A23	NC	A14	VSS	A12	VSS	VSS	VSS	VSS	VSS	A10	VSS	A6	A4	ADS#
R	A28	A25	VCC	VSS	A18	VCC	A15	VCC	VCC	VCC	VCC	A11	A8	VCC	A3	BLAST#	NC
Q	A31	VSS	A17	A19	A21	A24	A22	A20	A16	A13	A9	A5	A7	A2	BREQ	PLOCK#	PCHK#
P	D0	A29	A30												HLDA	VCC	VSS
N	D2	D1	DP0												LOCK#	M/IO#	W/R#
M	VSS	VCC	D4												D/C#	VCC	VSS
L	VSS	D6	D7												PWT	VCC	VSS
K	VSS	VCC	D14						80486						BE0#	VCC	VSS
J	VCC	D5	D16												BE2#	BE1#	PCD
H	VSS	D3	DP2												BRDY#	VCC	VSS
G	VSS	VCC	D12												NC	VCC	VSS
F	DP1	D8	D15												KEN#	RDY#	BE3#
E	VSS	VCC	D10												HOLD	VCC	VSS
D	D9	D13	D17												A20M#	BS8#	BOFF#
C	D11	D18	CLK	VCC	VCC	D27	D26	D28	D30	NC	NC	NC	NC	FERR#	FLUSH#	RESET	BS16#
B	D19	D21	VSS	VSS	VSS	D25	VCC	D31	VCC	NC	VCC	NC	NC	NC	NWI	NC	EADS#
A	D20	D22	NC	D23	DP3	D24	VSS	D29	VSS	NC	VSS	NC	NC	NC	IGNNE#	INTR	AHOLD

FIGURE 12.25
80486 pinout

11. $\overline{\text{READY}}$ (transfer acknowledge) is the input signal that indicates a bus cycle is complete.
12. $\overline{\text{BRDY}}$ (burst ready) performs the same function as the $\overline{\text{READY}}$ input when a burst cycle is being performed.
13. $\overline{\text{BLAST}}$ (last burst) is an output signal that indicates the next $\overline{\text{BRDY}}$ signal will complete the burst bus cycle.
14. DP_0–DP_3 (data parity generation/detection) are used to generate or detect even parity on the data bus.
15. $\overline{\text{PCHK}}$ (parity check) is an output signal used to indicate that a parity error has occurred.
16. RESET (reset) is the input signal used to initialize the processor and suspend any operation.
17. INTR (maskable interrupt request) is an input signal indicating a request for service interrupt, which can be masked by the flag register interrupt enable bit.

18. NMI (nonmaskable interrupt) is the input signal indicating a request for interrupt service, which cannot be masked by software.
19. BREQ (bus request) is an output signal that indicates an internal bus request has been generated.
20. HOLD (hold) is an input used to request a DMA action that causes all address, data, and control lines to enter the tristate condition.
21. HLDA (hold acknowledge) is an output that indicates the acknowledgment of the hold request.
22. $\overline{\text{BOFF}}$ (back off) is an input signal that forces all buses to enter the tristate condition during the next clock style.
23. $\overline{\text{AHOLD}}$ (address hold) is an input signal that allows another MPU to access the address bus.
24. EADS (external address) is an input signal that indicates an external address on the address bus is to be used to perform an internal cache cycle.
25. $\overline{\text{KEN}}$ (cache enable) is an input signal used to indicate whether the current bus cycle is cacheable.
26. $\overline{\text{FLUSH}}$ (flush) is an input signal used to erase all the internal 8 KB of cache memory.
27. PWT (page write) is an output signal that reflects the state of the PWT attribute bit in the page table or page directory entry.
28. PCD (page cache) is an output signal that reflects the state of the PCD attribute bit in the page table or page directory entry.
29. $\overline{\text{A20M}}$ (address bit 20 mark) is a signal used to tell the 80486 to wrap its address around.
30. $\overline{\text{BS8}}$ (bus size 8) is an input signal used to indicate that 8-bit data transfers are being used.
31. $\overline{\text{BS16}}$ (bus size 16) is an input signal used to indicate if 16- or 32-bit data transfers are being used.
32. $\overline{\text{FERR}}$ (floating point error) is an output signal that indicates the 80387 math coprocessor has detected an error.
33. $\overline{\text{IGNNE}}$ (ignore numeric error) is an input signal that tells the 80486 to ignore a floating point error.

As previously mentioned, the 80486 is almost identical to the 80386 and the 80387 packaged together. There are eight 32-bit general-purpose registers, six segment registers that are used to form addresses along with one of the index registers or pointers, an instruction pointer register, and the flag register. See Figure 12.26.

The 80486 contains all of the instructions of the 80386 and the 80387 math coprocessor plus six additional new instructions as described in Table 12.3.

The 80486 also includes 8 KB of internal cache memory. The cache memory is organized into four 2 KB sections. Whenever data is written to the external bus or memory, it is also written into cache memory. This is known as a **write through operation.** The WBINVD instruction erases data just written into cache memory.

Cache memory significantly improves system performance because it is faster than having to go back to the bus to read the contents of memory through a memory read cycle. Because the 80486 is so fast, it would constantly be waiting for information from

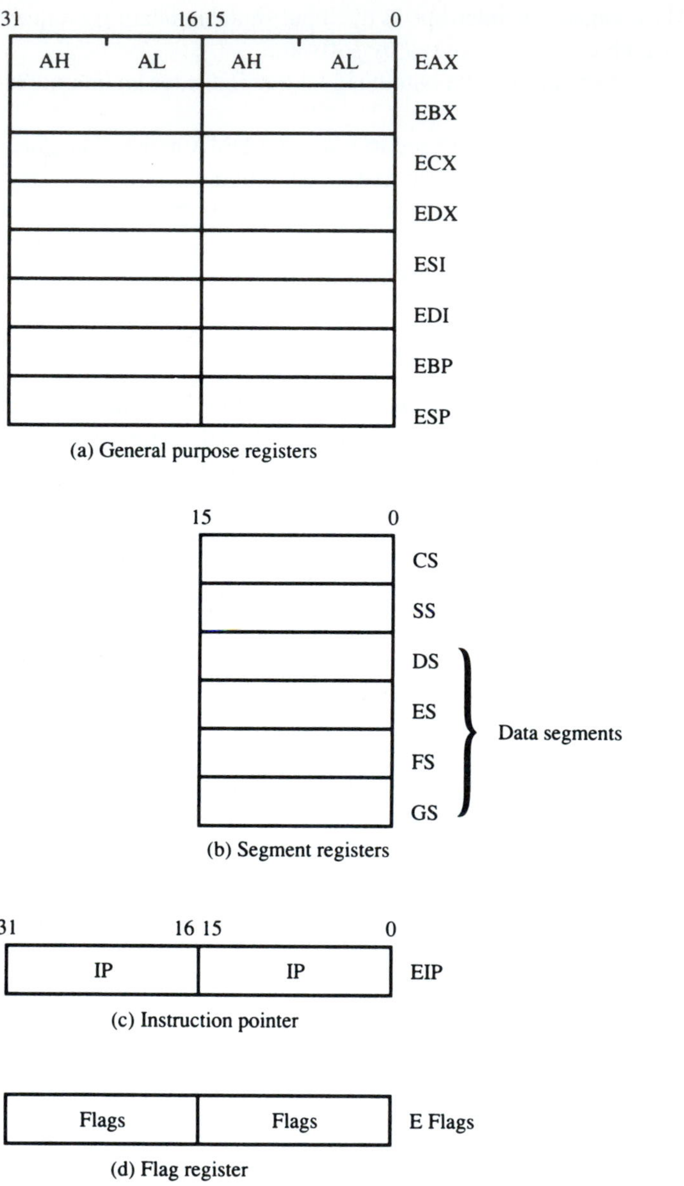

FIGURE 12.26
80486 registers

TABLE 12.3
80486 new instructions

Instruction	Description
XADD	Swaps low-order and high-order bytes on 16-bit data and also performs addition.
CMPXCHG	Reorders and compares 16-bit data.
BSWAP	Swaps the order of the 4 bytes in a 32-bit register.
INVD	Erases internal cache memory.
WBIVD	Erases internal cache memory after writing to memory.
INVLPG	Invalidates a page table entry.

memory to be transferred. Figure 12.27 is a block diagram of the 80486 MPU. It is functionally similar to the 80386 with a 80387 math coprocessor.

The Pentium Processor

The fifth generation of the Intel 8086/8088 microprocessor line is called the **Pentium,** or P5, processor. The processor is not called the 80586 because Intel discovered it could not trademark a number and wanted to prevent other manufacturers from using the same name for any clone processors that were developed.

The Pentium P5 has a 64-bit data bus and a 32-bit address bus, and operates at speeds from 60 MHz to 200 MHz. The internal register size, however, is only 32 bits wide. Since the data bus is 64 bits and the internal registers are only 32 bits, two completely separate pipelines, called the U and V pipes, were created. Thus, in many cases, two instructions can be processed simultaneously. Intel calls this **instruction pairing** or **superscaling** technology. However, not all instructions can be processed simultaneously because some are data dependent. For example, in the following program, the second instruction MOV EAX, 1500 cannot be executed until the first instruction ADD EAX, EBX is completed or the incorrect result will end up in address 1500. Thus, the result of instruction 2 is dependent on instruction 1.

```
ADD  EAX,EBX;  add EBX to EAX
MOV   EAX,1500;   move EAX to address 1500
```

Often the instructions are not data dependent, however, and can be processed simultaneously, as shown in the following programming example.

```
ADD  EAX,EBX;  add EBX to EAX
DEC  DI;     decrement   the   pointer
```

In this example, instruction 2 is not data dependent on instruction 1 and therefore will be processed simultaneously.

In addition to being able to process two instructions simultaneously, most instructions can be executed in one or two clock cycles. Thus, current software will run much faster on a Pentium. Figure 12.28 shows a simplified block diagram of the internal architecture of the Pentium P5 processor.

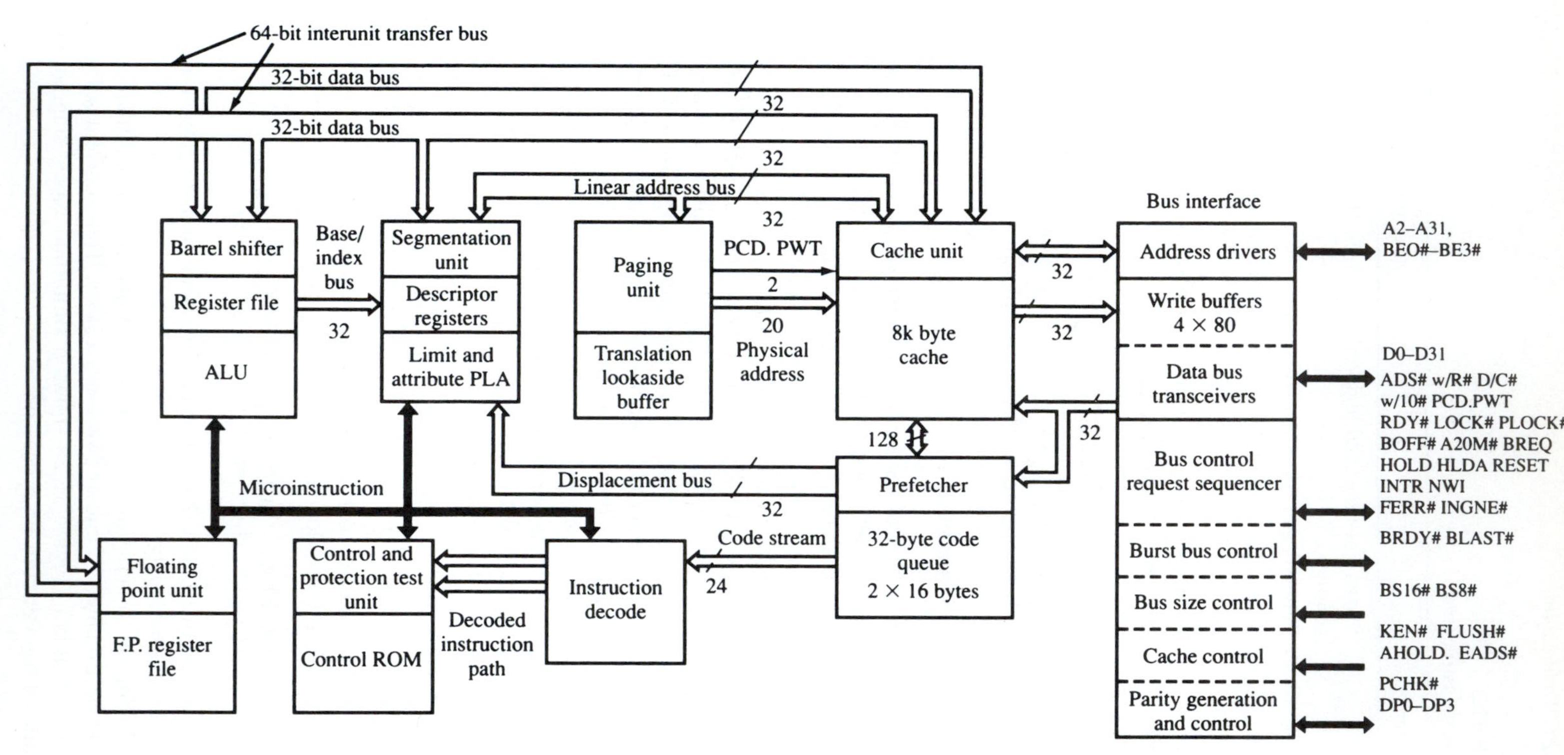

FIGURE 12.27
80486 block diagram

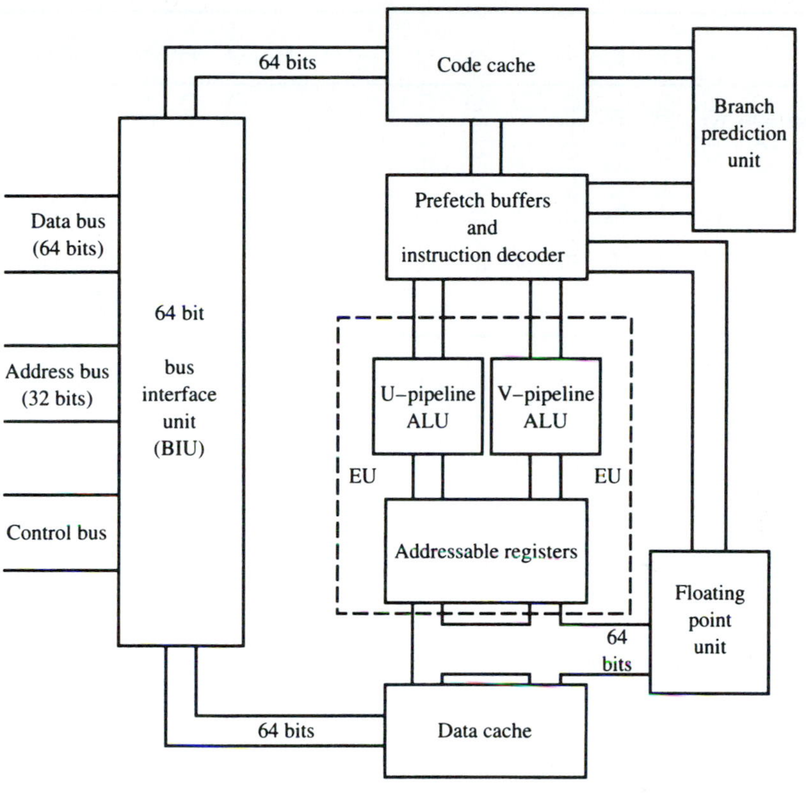

FIGURE 12.28
Pentium simplified block diagram

The original Pentium P5 processors were available in 60- and 66-MHz speeds. This design used an 0.8-micron process to create a 273-pin grid array (PGA) form factor. The chip ran on 5 V and consumed 3.2 amps, or 16 W, of power. To reduce the heat and power consumption, the Pentium was redesigned using a 0.6-micron process. The new design resulted in a 320-pin staggered pin grid array (SPGA) form factor (the P5 actually uses only 296 of these pins) and operates at 3.3 V. The new Pentiums are available in 90-MHz and 100-MHz models, with a 75-MHz model intended for laptops. The 3.3-V design reduced power consumption and heat. The 100-MHz Pentium requires about 3.25 amps at 3.3 V, which equates to about 10.725 W. The 90-MHz model requires only 2.5 amps (9.735 W), and the 75-MHz model uses less than 7 W of power. Finally, to increase the speed even further, a 133-MHz, a 166-MHz, and a 200-MHz version consuming approximately 12, 14, 16 W, respectively, were released. Figure 12.29 illustrates the pinout for the 273-pin PGA form factor, and Figure 12.30 shows the 320-pin SPGA layout.

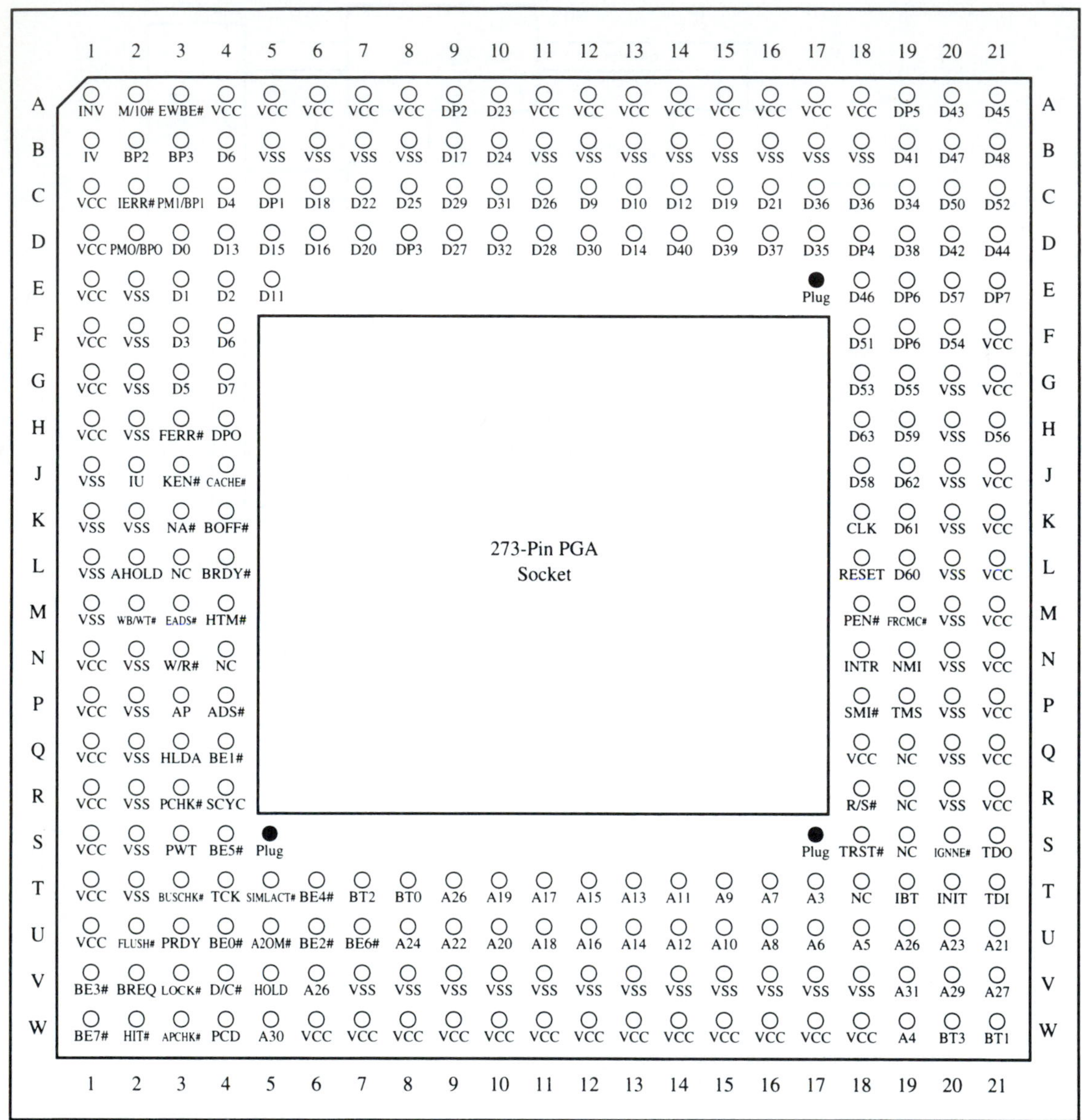

FIGURE 12.29
Pentium 273-pin PGA form factor

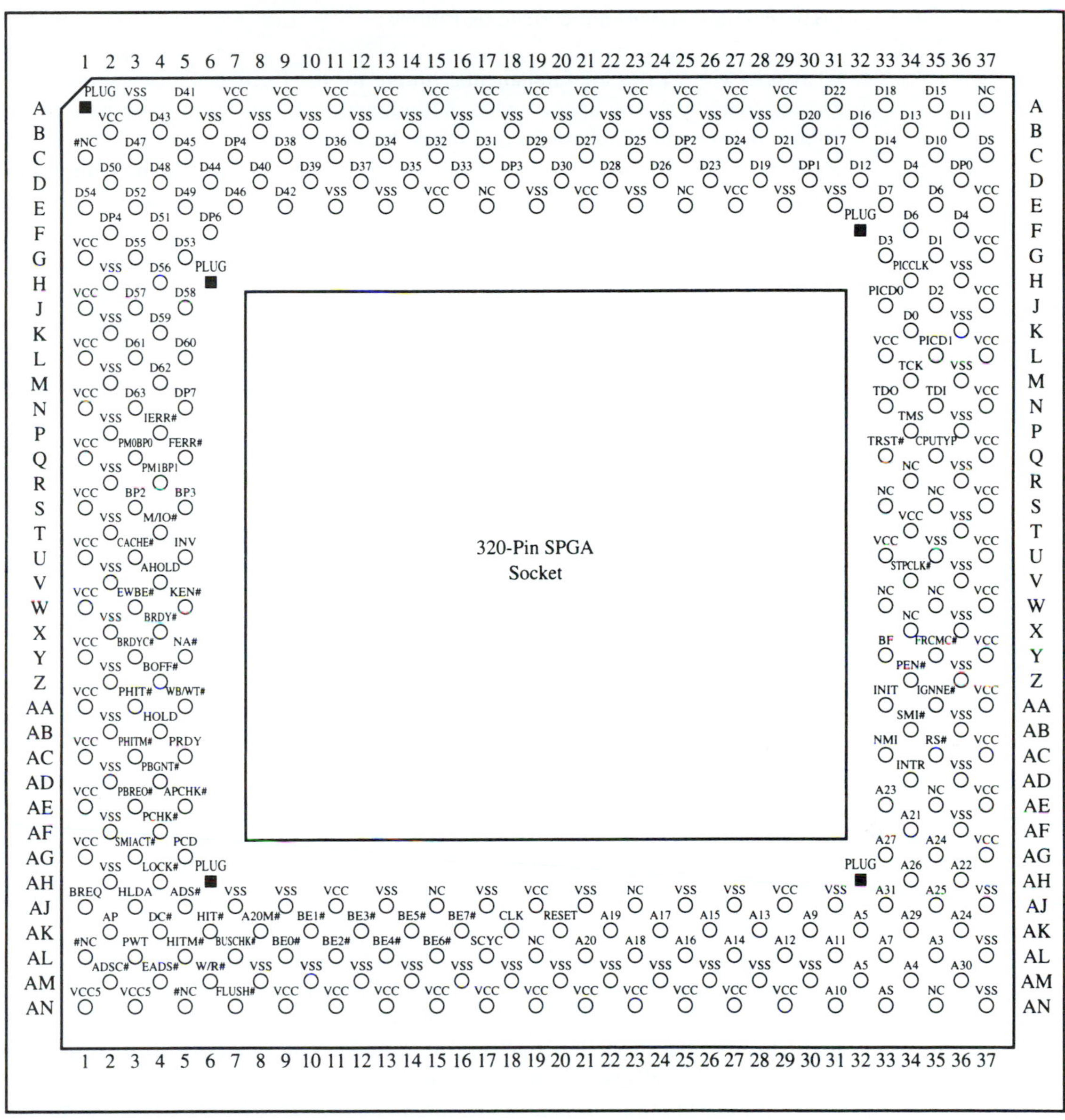

FIGURE 12.30
Pentium 320-pin SPGA form factor

The Pentium signals are defined as follows:

1. CLK (clock) provides the fundamental timing for the Pentium P5 processor.
2. D_0–D_{63} (data bus) data lines provide the bidirectional, general-purpose path for data signals between the Pentium P4 and other devices.
3. A_3–A_{31} (address bus) address lines together with BE_0–BE_7 provide the path to physical memory or I/O port addresses. Note that A_0, A_1, and A_2 are internally encoded to select any or all of the 8 byte enable (BE_0–BE_7) pins.
4. BE_0–BE_7 (byte enable) outputs are used with the address lines to provide the physical memory and I/O port addresses.
5. W/$\overline{\text{R}}$ (write or read indicator) is the output signal that distinguishes between write and read cycles.
6. D/$\overline{\text{C}}$ (data or control indicator) is the output signal that distinguishes between data and control cycles.
7. M/$\overline{\text{IO}}$ (memory or I/O indicator) is the output signal that distinguishes between memory and I/O cycles.
8. $\overline{\text{LOCK}}$ (bus lock indicator) is the output signal that distinguishes between locked and unlocked bus cycles. It is used to tell the system that the bus cannot be relinquished.
9. $\overline{\text{ADS}}$ (address status) is the output signal that indicates a valid address is on the bus.
10. $\overline{\text{A20M}}$ (address 20 mask) is the input signal used to emulate the 1 MB address limitation of the 8086/8088 microprocessor. When this signal is low, the Pentium masks physical address bit 20 before performing a bus operation.
11. $\overline{\text{BRDY}}$ (burst ready) is an output signal that indicates there is valid data on the data bus in response to a read request or a write request.
12. DP_0–DP_7 (data parity) is used to generate or detect even parity on the data bus. There is one parity pin for each byte (8 bits) of the data bus. Thus, DP_0 applies to D_0–D_7 and DP_7 applies to D_{56}–D_{63}.
13. $\overline{\text{PCHK}}$ (parity check) is an output signal used to indicate that a parity error has occurred.
14. AP (address parity) is a bidirectional, even parity address pin for the address bus.
15. $\overline{\text{APCHK}}$ (address parity check) is an output signal that becomes a logic 1 whenever the Pentium detects an address parity error.
16. $\overline{\text{IERR}}$ (internal error) is an output pin that indicates an internal parity error.
17. $\overline{\text{PEN}}$ (parity enable) is an input signal used to determine if a processor check exception should be taken if a parity error is detected.
18. RESET (reset) is an input signal used to initialize the processor to its reset vector (*FFFFFFF*0H) and suspend any operation.
19. INIT (initialization) is an input signal that initializes the processor in the same manner as reset, except that the internal caches and floating-point registers retain the values they had prior to the INIT.
20. INTR (maskable interrupt request) is an input signal that indicates a request for a service interrupt that can be masked off by the flag register interrupt enable bit.
21. NMI (nonmaskable interrupt) is an input signal indicating a request for interrupt service, which cannot be masked off by software.
22. BREQ (bus request) is an output signal that indicates that an internal bus request has been generated.
23. $\overline{\text{BUSCHK}}$ (bus check) is an input signal that tells the Pentium the bus cycle was unsuccessful.

24. HOLD (hold) is an input signal used to request a DMA action that causes all address, data, and control lines to enter their tristate condition and send an HLDA signal.
25. HLDA (hold acknowledge) is an output signal that indicates the acknowledgement of a hold request (HOLD).
26. $\overline{\text{BOFF}}$ (back off) is an input signal that forces all buses to enter the tristate condition during the next clock cycle.
27. AHOLD (address hold) is an input signal that tells the Pentium to hold the address on the address bus for the next clock cycle.
28. $\overline{\text{EADS}}$ (external address) is an input signal that indicates the external address, which is on the address bus, is to be used to perform an internal cache cycle.
29. $\overline{\text{EWBE}}$ (external write buffer empty) is an input signal that indicates a write cycle is pending in the external system.
30. $\overline{\text{KEN}}$ (cache enable) is an input signal used to indicate whether the current bus cycle is cacheable.
31. $\overline{\text{CACHE}}$ (cache) indicates cacheability of the current cycle.
32. $\overline{\text{FLUSH}}$ (flush cache) is an input signal used to erase all the internal 8 KB of cache memory.
33. PCD (page cache) is an output signal that reflects the state of the PCD attribute bit in the page table or page directory entry.
34. INV (invalidation request) is an input signal that determines the cache line state after an inquire.
35. WB/$\overline{\text{WT}}$ (write back/write through) selects the operation for the data cache.
36. SCYC (split cycle) is an output signal that indicates a misaligned locked transfer is on the bus.
37. PWT (page write) is an output signal that reflects the state of the PWT attribute bit in the page table or page directory entry.
38. $\overline{\text{FERR}}$ (flooding point error) is an output signal that indicates the numeric coprocessor has detected an error.
39. $\overline{\text{IGNNE}}$ (ignore numeric error) is an input signal that tells the Pentium to ignore a floating point error.
40. BP/PM (breakpoint and performance monitoring) is a type of status indication. The breakpoint pins, BP_0–BP_3, indicate a breakpoint match with debug registers DR_0–DR_3. The performance monitoring pins, PM_0 and PM_1, indicate the setting of BP_0 and BP_1 in the debug mode control register for breakpoint or performance monitoring.
41. FRMC (functional redundancy check) is used to poll the Pentium to determine whether it is configured in the master or checker mode.
42. BT (branch instruction) is an output signal that indicates the execution of a branch instruction.
43. BT_0–BT_3 (branch trace) are output signals that indicate a target linear address and the default operand size.
44. $\overline{\text{HIT}}$ (hit) shows that, in the inquire mode, the internal cache contains valid data.
45. HITM (hit modified) shows that, in the inquire mode, the address on the address bus is in the modified state in the data cache.
46. IU (U pipeline) is an output signal that shows the U pipeline instruction is complete.
47. IV (V pipeline) is an output signal that shows the V pipeline instruction is complete.
48. $\overline{\text{NA}}$ (next address) is an input signal used to indicate that the external memory system is ready to accept a new bus cycle.

49. PRDY (probe ready) is an output signal used to indicate that the probe mode has been entered for debugging.
50. $\overline{\text{R/S}}$ (request/service) is an input signal used with the debugging port to cause an interrupt.
51. $\overline{\text{SMI}}$ (system management interrupt) is an input signal that causes a management mode interrupt.
52. $\overline{\text{SMIACT}}$ (system management interrupt active) is an output signal that shows the processor is in the management mode.
53. TCK (test ability clock) is an input signal that provides a clocking function for the process boundary scan.
54. TDI (test data in) is an input signal used for receiving serial test data and instructions for the boundary scan and probe mode test logic.
55. TDO (test data out) is an output signal used for sending serial test data and instructions for the boundary scan and probe mode test logic.
56. TMS (test mode select) is an input signal that selects the boundary scan control logic.
57. TRST (test reset) is an input signal that resets the test mode.

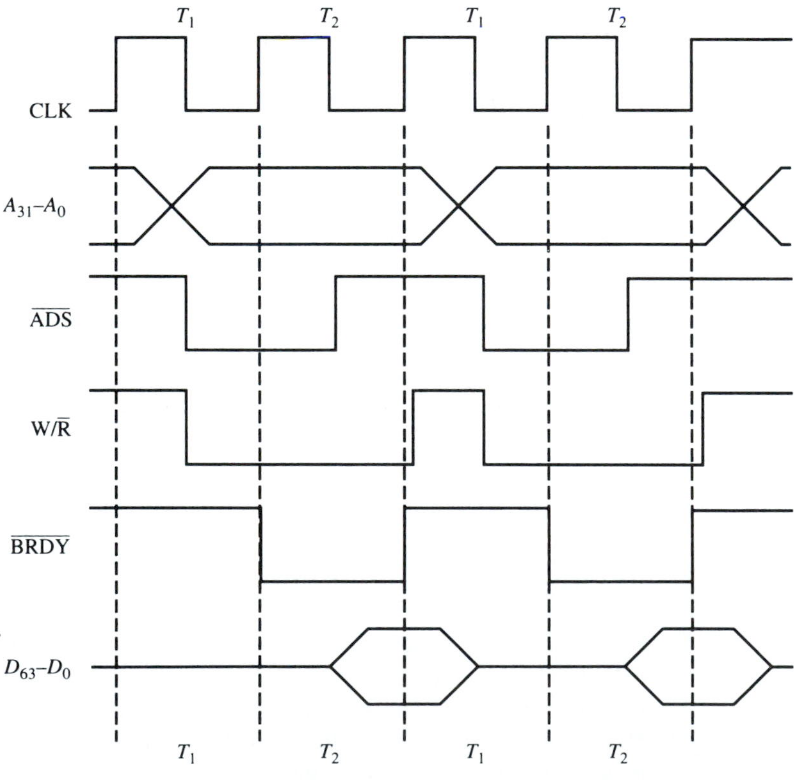

FIGURE 12.31
Pentium nonpipelined read cycle timing

The Pentium supports a number of different bus cycles. The basic nonpipelined memory read cycle consists of two clock pulses. Refer to Figure 12.31. The Pentium starts the cycle by placing address information, A_{31}–A_0, on the bus during clock time T_1 and activating $\overline{ADS}$ to indicate that a valid address is on the bus. The read/write line is also placed in the appropriate mode at this time. Synchronized with clock time T_2, the $\overline{BRDY}$ signal goes active low, indicating that valid data can be read at the end of the clock cycle. If $\overline{BRDY}$ is not active, wait states are generated, as illustrated in Figure 12.32, until the $\overline{BRDY}$ signal goes active.

The Pentium P5 microprocessor has a new 32-bit flag register, as shown in Figure 12.33. Four new flags are provided to control the new Pentium features. The new flags and their functions are described in Table 12.4. The Pentium contains all the instructions of the 80486, plus six additional new instructions, as described in Table 12.5.

Near the end of the Pentium production cycle, Intel was able to improve manufacturing capability enough to produce an improved Pentium variant, the Pentium MMX. The Pentium MMX (also designated P55C) is pin compatible with the Pentium (or P54C) and uses the same socket 7, but introduced several new features. The MMX refers to 57 **multimedia extensions (MMX)** in the Pentium instruction set. Although these extensions

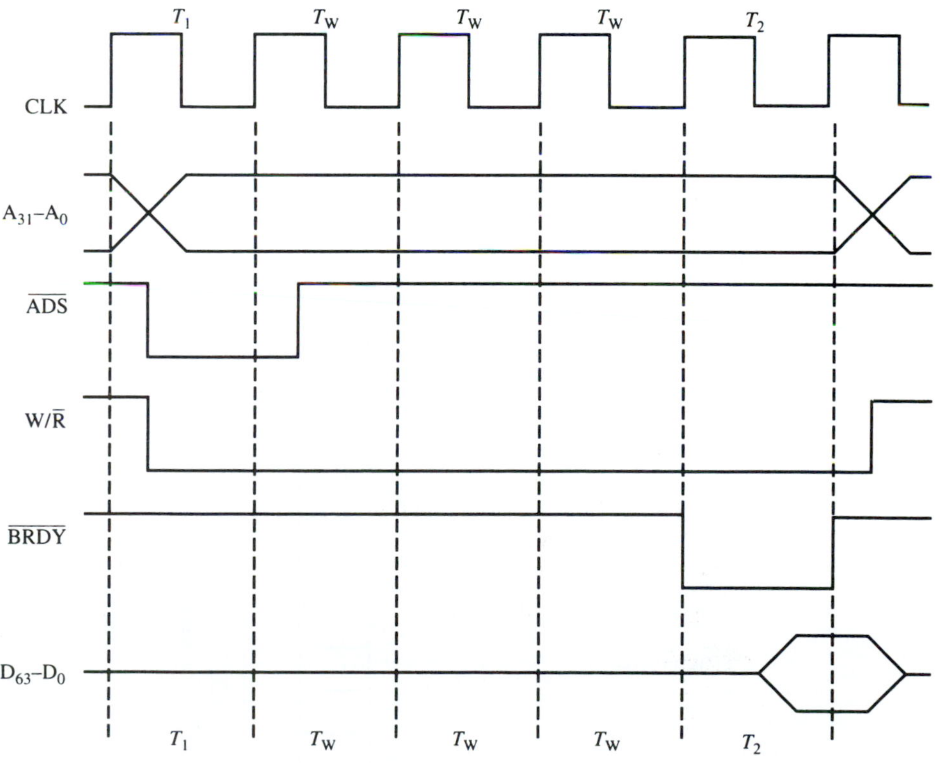

FIGURE 12.32
Pentium nonpipelined read cycle with wait states

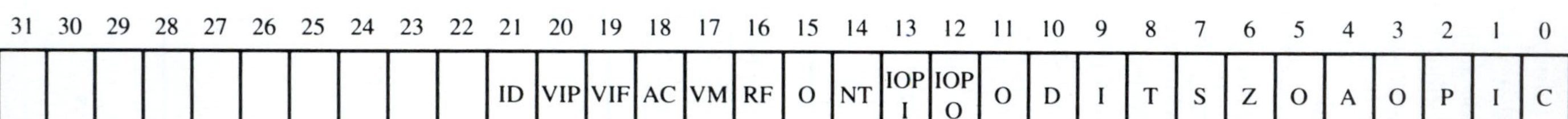

FIGURE 12.33
Pentium flag register

TABLE 12.4
New flag register bits

Flag	Bit	Function
ID	21	Identification
VIP	20	Virtual interrupt pending
VIF	19	Virtual interrupt
AC	18	Alignment check

TABLE 12.5
New Pentium instructions

Instruction	Description
CMPX-CHG8B	Compare and exchange 8 bytes
CPUID	Return the CPU identification code
RDTSC	Intel proprietary
RDMSR	Read model-specific register
WRMSR	Write model-specific register
RSM	Return from system interrupt

never found much support from software developers, MMX support is included in every Intel processor since the Pentium MMX. More important than the MMX instructions, the L1 cache was increased from 16 KB to 32 KB, which significantly improved performance. The Pentium MMX is also the first Intel MPU to feature a split voltage design. Besides the normal 3.3 V supplied to the MPU, the Pentium MMX requires a second, core voltage setting of 2.8 V. This was partly due to an improved fabrication process using a much smaller die size of 0.35 microns (compared to a Pentium's 0.8-micron die size). All subsequent Intel processors require a separate core voltage. The exact voltage varies between processor lines. Even within processor lines, core voltage settings may vary between the differently clocked versions. When installing or replacing a microprocessor, make sure that the correct core voltage is known, or damage may result. Core voltage settings are changed for each motherboard, but generally are modified using jumpers, dip switches, or within the system board BIOS. Whenever possible, consult the motherboard manual.

Intel's true successor to the Pentium processor was the Pentium Pro. It increased the transistor count to 5.5 million, but the greatest improvement was that the L2 cache was incorporated into the die of the microprocessor itself. This meant that the L2 cache could run at the same clock speed as the processor. Increasing the processor clock speeds to between 150 and 200 MHz resulted in an increased L2 cache speed of up to three times the Pentium's, whose L2 cache ran at 66 MHz. In addition, the L2 cache is increased to up to 2 MBs. The Pentium Pro architecture was also optimized for 32-bit code, though it remained backward compatible with 16-bit code. The Pentium Pro was also the first of the Pentium family designed to be used in high-end dual-processor systems. The Pentium Pro is not pin compatible with the Pentium but instead uses a socket-8 connector to accommodate the Pro's 387-pin grid array. The 440FX chipset is used exclusively for Pentium Pro system boards.

The manufacturing difficulties of producing high-speed cache as well as the integrated nature of the L2 cache (building it into the chip) combined to raise the Pentium Pro's price well beyond the range of the casual user. One reason is that if either the MPU or cache component was found to be faulty during testing, the entire expensive package had to be discarded. Pentium Pros were used in the server market and virtually nowhere else. At best, the Pentium Pro could be considered a transitional or niche product.

Pentium II

The Pentium II represents a substantial improvement over the previous Pentium CPU family. New manufacturing techniques allow much faster chips with less power consumption. The Pentium II incorporates a dual-bus architecture, which includes the system bus and the cache bus. Using the dual-bus architecture, performance is greatly enhanced because the processor will run code and data from cache on the cache bus, thus leaving the system bus free for other tasks. In addition, the tightly coupled cache bus runs at half of the processor speed, which is faster than the system bus speed.

The system bus speed was also substantially increased in the Pentium II. Multimedia extensions allow for increased gaming and graphics performance. Intel also presented improvements in cache and chipset technology with the Pentium II. The Pentium II also incorporates some features of the Pentium Pro in its design. The Pentium II uses a signaling method referred to as AGTL+, or assisted gunning transistor logic. As shown in Figure 12.34, AGTL is very different from the standard TTL and modified 3.3-V TTL used on past processors. The 0 logic level is similar to standard TTL logic of 0.8 V_{DC}. However, the 1 logic level is quite different and only rises to 1.5 V_{DC}. Another difference is that the signal lines are terminated by resistors. This logic requires very strict land runs and cannot tolerate excessive ringing. Ringing occurs when a signal switches very fast and overshoots or undershoots the driven level. Older technologies use clamping diodes to V_{CC} and ground to dampen the ringing. Clamping diodes are used here to provide ESD (static transients) protection only. It is for this reason that no wires or test points should be attached to the host bus on the motherboard. Doing so may damage the processor and

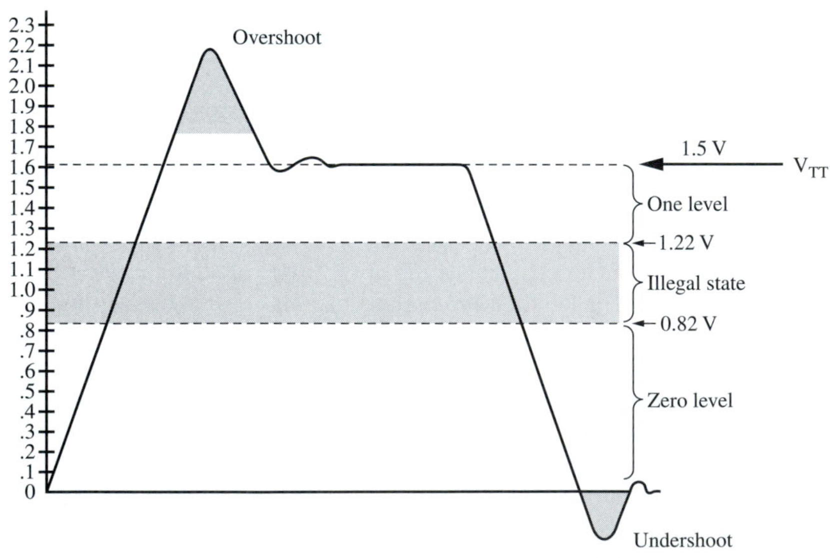

FIGURE 12.34
AGTL signal levels

host bus bridge chip. Anyone wishing to learn troubleshooting by inserting faults and probing should use an older technology such as an old PC or PC/AT.

Termination resistors are connected to each end of the host bus as shown in Figure 12.35 and are used to control and dampen the signals similar to high-speed networks. The termination resistors are connected to a reference supply called V_{TT}. The AGTL+ signaling method is an extension of the GTL method used by the Pentium Pro. (GTL and AGTL logic are compatible and can be used on the same system bus.) AGTL+ inputs use differential receivers that require a signal called V_{ref}. It is used to determine whether a signal is a logical 1 or 0. Terminating resistors are required at each end of the signal trace.

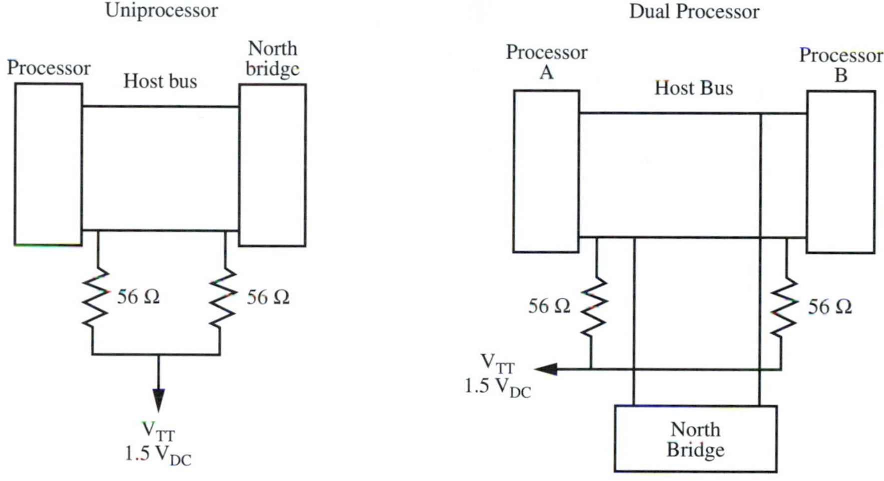

FIGURE 12.35
AGTL termination

With all these changes, Intel developed radically new packaging for the Pentium II processor as shown in Figure 12.36. All of these things helped to propel the Pentium II past the Pentium when it was introduced. Some new or improved features of the Pentium II include

- *MMX:* Intel incorporates its MMX technology in the Pentium II processor. MMX is essentially a new instruction set (57 new instructions), which allows for increased performance in a range of applications. These include full-screen, full-motion video, and advanced graphics.
- *Manufacturing:* With the introduction of the Pentium II 333-MHz CPU, Intel utilizes the 0.25-micron manufacturing process. This allows for greater core frequencies and at the same time, decreased power consumption. Earlier versions of the Pentium II used a 0.35-micron process.

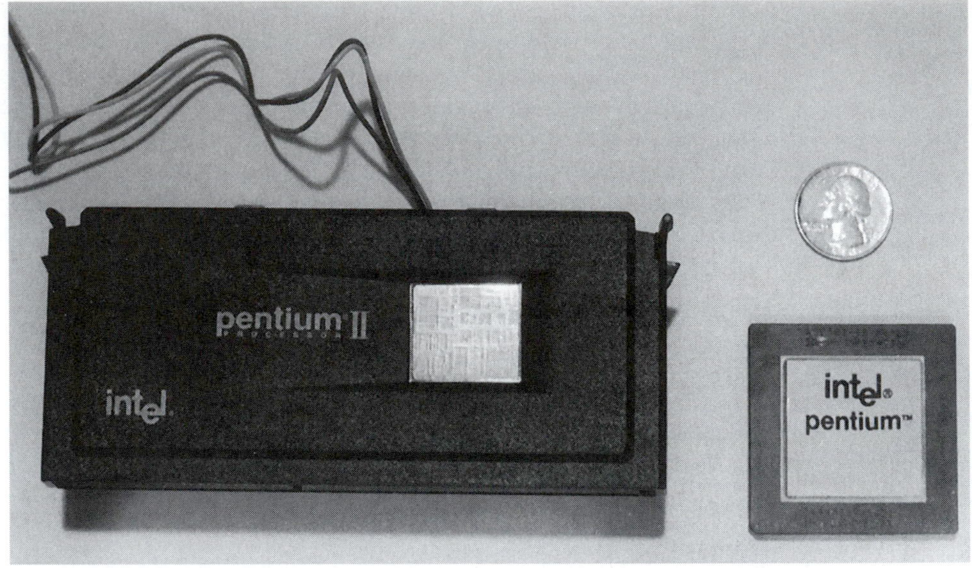

FIGURE 12.36
Intel Pentium II vs. Pentium

- *Packaging:* Significant improvements in packaging came about with the Pentium II processor. The Pentium II is contained in a package called **single edge contact cartridge,** or **SECC.** In this package, the components, CPU and cache, are surface mounted on a substrate and enclosed within a cover. This allows the L2 cache to be contained within the processor module (cache is discussed in detail later in this chapter). The first Pentium II processors used an SECC package with an extended thermal plate that allows for a heat sink attachment. Later versions of the Pentium II used SECC2 packaging, without this plate. These processors have a different core technology that allows the thermal solutions to be placed directly into those cores. Both of the SECC packages utilize an edge finger connection called the SC242 connector. The connector transmits signals in the same way as a peripheral card does. Figure 12.36 shows the SECC versus the original Pentium packages.
- *Cache:* The Pentium II processor contains 32 KB of L1 cache and 512 KB of L2 cache. It has a dedicated L2 cache bus. As discussed earlier, this configuration is called dual independent bus architecture, or DIB. It provides greater bandwidth and performance. The cache bus does not connect to the system bus and is invisible to other system components. The L2 cache consists of two types of memory, Tag RAM and BSRAM, or burst pipelined synchronous static ram. They are contained in a multiple die that is more readily available and less expensive than the die used in the Pentium Pro. Both of the cache components receive clocked data directly from the processor core.
- *Bus Frequency:* With the introduction of the Pentium II 350, Intel increased the bus speed, or front side bus, of the Pentium II from 66 MHz to 100 MHz. This provides a

substantial increase in bandwidth and peak performance. The processor, the frequency synthesizer, and the chipset determine the bus speed. The operating speed is controlled by bus signal BCLK. All system components must operate at the same frequency. This applies to multiple processor systems as well. The Pentium II does support multiple processors, but Intel does not validate the mixing of different frequencies.

- *Chipset:* With the PENTIUM II 350, Intel also introduced the 440BX chipset. The BX was the first chipset to support the 100-MHz front side bus. Remember that the BX was not the first AGP chipset. There was a 440LX AGP chipset that did not have 100-MHz support. Neither did the 440FX PCI chipset. This combination of hardware (Pentium II 350 and BX) lead to breakout performance for video operation.

Celeron

To cater to the lower-end market, Intel introduced the lower-cost Celeron (shown in Figure 12.37), which runs at the older 66-MHz FSB. The Celeron does not have the SECC packaging, which actually makes it easier to cool the MPU and cache chips. Cooling is a significant factor for processors, since they produce considerable amounts of heat, which can damage the MPU when unchecked. The first two Celerons, the 233- and 266-MHz versions, had no L2 cache. This deficiency readily showed in benchmark tests, and Intel soon abandoned that scheme. All future Celerons (designated Celeron A), contain a small L2 cache of 128 KB, but in compensation, run at the full MPU clock speed. This allows them to perform at a level almost indistinguishable from that of an equivalently clocked Pentium II in common applications. This results in the awkward position of choosing between a 400-MHz Celeron with a 400-MHz, 128 KB L2 cache and a 400 Pentium II with a 200-MHz, 512 KB L2 cache, (ignoring the considerable price difference). So which is faster, 128 KB of L2 cache running at full MPU speed or 512 KB of L2 cache running at half the MPU speed? It all depends on the size of the code the MPU is processing. If the code is smaller than 128 KB, the Celeron has the advantage. If the code is larger than 128 KB, the code will run faster on the Pentium II/III. The different FSB speeds also need to be taken into account, but benchmarks show negligible differences in most applications.

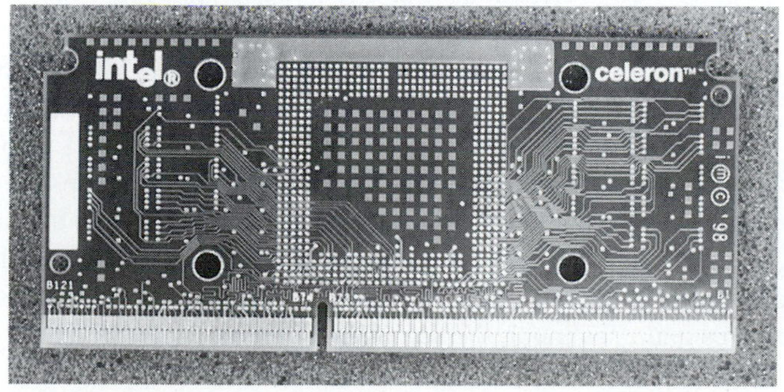

FIGURE 12.37
The Intel Celeron MPU

As manufacturing processes improved, Intel returned the Celeron to a cheaper-to-manufacture socket package. Intel stopped producing slot-1 Celerons and exclusively manufactures socket-370 Celerons at the higher processor speeds. However, several manufacturers produce socket 370 to slot 1 converter cards, so it is not necessary to change motherboards to take advantage of the faster Celerons.

Pentium III

The latest addition to the Intel microprocessor line is the Pentium III, which includes the Xeon version of the chip. In the Xeon, the transistor count increases to 9.5 million. Think of it as the Pentium Pro version of the Pentium III. Just like the Pentium Pro, the Xeon's L2 cache runs at full processor speed, but the Xeon is available with 512 KB, 1 MB, and 2 MB of L2 cache. Moreover, just like the Pentium Pro, the cost of the Xeon is prohibitive except for high-end server or workstation systems. The Xeon is also physically much larger than the Pentium II, even though it incorporates a fabrication die shrink down to 0.18 microns and utilizes the slot-2 packaging design. The Xeon chipset, the 450NX, natively supports four Xeons. With a special controller, up to eight Xeons can be used in a single system.

The Pentium III L2 cache size is the same as the Pentium II and still runs at half the processor speed, but the L1 cache increases to 64 KB. There have been some substantial improvements over the Pentium II. Chief among them may be the 133-MHz bus support. (There are versions of the chip that come with a 100-MHz bus, so make sure you check.) The bus is a multitransaction system bus, meaning that the bus can handle and process multiple requests simultaneously. Couple a 733-MHz P3 CPU with the recently released 820 chipset and populate the motherboard with direct RDRAM, and you have a serious machine. Another improvement lies in the cache. One version of the chip offers something called advanced transfer cache. This includes an on-die, full-speed 256K L2 cache with ECC. This means that the L2 cache runs at the speed of the processor. In the case of the Pentium III, the L2 cache will run at speeds up to 733 MHz. Intel also offers a version of the Pentium III that incorporates 512K of discreet, off-die, half-speed L2 cache. The Pentium II offered only off-die, half-speed cache. Intel also incorporates internet streaming SIMD (single instruction multiple data) extensions in the Pentium III. This consists of 70 new instructions that allow higher resolution and higher quality audio. It improves the performance of streaming video and audio, as well as 3D applications. Some other instructions include SIMD for FPU, and additional SIMD-integer and cachability control instructions. These SIMD extensions are available on all Pentium III chips. The Pentium III uses advanced system buffering, utilizing four writeback buffers, six fill buffers, and eight bus queue entries. Each Pentium III processor contains a unique processor serial number. The Pentium III also has many features in common with its predecessor. It utilizes Intel's MMX technology, dual independent bus architecture, and dynamic execution micro architecture. In addition it shares many of the same power management features and uses much the same signaling as the Pentium II.

AMD K5

The AMD K5 processor was introduced to compete directly with the Pentium processor. It was marketed as a best in its class processor because design enhancements over the

Pentium allowed the K5 to run faster at a given clock speed. Unfortunately, this confused manufacturers and customers, and the K5 never became very popular, even though its design was technically better than that of the Intel Pentium.

The K5 was offered in six different speed ratings, or P-ratings: PR75, PR90, PR100, PR120, PR133, and PR166. A number of the K5 processors ran at different speeds than the Pentium rating that they were given. An AMD K5 PR133 was equivalent in speed to the Pentium 133 even though it only ran at 100-MHz. Table 12.6 shows the operating frequency of the K5 family in comparison with its P-rating.

TABLE 12.6
AMD K5 speed settings

P-Rating	Actual CPU Core Speed
PR75	75
PR90	90
PR100	100
PR120	90
PR133	100
PR166	117

AMD K6

The AMD K6 processor was poised to compete against the Intel Pentium II design. Physically, it runs in the same socket 7 as the Pentium. This has limited many aspects of its performance, resulting in a typically slower CPU when compared to similar speed PIIs. The AMD K6-II and III, however, were designed to overcome some of the socket limitations. These processors are comparable in speed to the Pentium II, but were offered at a significantly lower price.

With the K6 processor AMD found a pretty comfortable niche in the low-end to midrange PC market. However, the shortcomings of the K6 prevented it from ever entering the server or workstation environment.

The K6-II introduced the 100-MHz memory bus onto the socket-7 market. This allowed the K6-II to perform notably better than the earlier K6. The K6-II also introduced AMD's version of MMX, called 3Dnow. 3Dnow allowed AMD to capture a larger share of the computer gaming industry, but it was not enough to let AMD compete in the all-important business arena.

The K6-III was AMD's response to the Pentium III. It features 256 KB of full-speed, on-die L2 cache. This was another significant performance improvement for AMD, and the K6-III became the chip of choice in the computer gaming sector. Businesses were still leery though, and the K6-III was never quite accepted over the Intel brand. Table 12.7 lists the K6 CPU and speed types.

AMD K7

The AMD K7, known as the Athlon, is a revolution in chip design. In many ways, it looks like the Pentium II processor, but that's where the similarities end. The K7 proces-

TABLE 12.7
AMD K6 CPU and speed types

CPU	Bus Speed	Multiplier	L2 Cache
K6 166	66	2.5	N/A
K6 200	66	3.0	N/A
K6 233	66	3.5	N/A
K6 266	66	4.0	N/A
K6 300	66	4.5	N/A
K6-II 300	100	3	N/A
K6-II 350	100	3.5	N/A
K6-II 400	100	4	N/A
K6-II 450	100	4.5	N/A
K6-III 400	100	4	256 KB full speed
K6-III 450	100	4.5	256 KB full speed

sor is the next generation of processor technology and has been dubbed the first 786. It is fully Intel compatible, including the MMX extensions. The K7 has put tremendous pressure on Intel, and AMD is looking to take over the number 1 spot in the high-end CPU market. With the recent introduction of the Intel Pentium III 733-MHz and 820 chipset, the war is on.

The Instruction Decoder The job of a CPU is to execute software. Before that can happen, the CPU needs to decode the machine language instructions that are fed to it into operations that it can execute. The decoded instructions are called ops (short for operation), and numerous ops may be required for a single complex instruction. A simple instruction, on the other hand, only requires a single op.

The Pentium III instruction decoder has three parallel units for decoding incoming instructions into ops. These are known as complex, simple, and simple. In other words, the Intel processor can simultaneously decode one complex instruction and two simple instructions. If two complex instructions were in need of decoding, the second instruction would have to wait until the instruction decoder is done with the first complex instruction. The K7, on the other hand, has three parallel complex instruction decoder units. It can decode any combination of simple and complex instructions in parallel. This adds to the design complexity of the K7, but it also adds considerably to its performance.

The Instruction Control Unit Once an instruction is decoded, it is stored temporarily in the instruction control unit and is then routed to either the integer unit schedulers or the floating-point unit schedulers. The K7 can store up to 72 ops while they are waiting to be dispatched to the schedulers. The PIII, on the other hand, can only store up to 20 ops in this stage.

Execution Ports The primary job of a microprocessor is to execute the instructions given to it. The Pentium III has a total of 12 parallel execution ports, which means it can execute 12 separate instructions per clock cycle. However, 7 of these ports are dedicated to internal functions, leaving only 5 available to process user code. The K7 processor has 9 ports available, almost twice as many as the Pentium III. This is definitely a significant performance advantage in that it allows the K7 to execute almost twice as many ops in the same amount of time as the PIII.

The Execution Pipeline If you can imagine an automobile manufacturing plant, you can understand the concept of a pipeline. When automobiles were first produced, a set team of people would work on the car from start til finish. This process was relatively slow and tedious. Modern automobile manufacturing facilities use a different approach, the assembly line. In this analogy, the assembly line is the pipeline. The old way of producing cars would be analogous to a single-stage pipeline; the assembly line is a multi-stage pipeline.

For a short pipeline with only a few stages, each stage takes a relatively long time and cannot be fed at a very high rate. A long pipeline with many stages, on the other hand, can be fed at a high rate. The more stages you have in your pipeline, the longer it will take to correct the error. A short pipeline, on the other hand, can correct errors much more rapidly. For design simplicity, Intel chose to make the pipelines in the PIII long and run the risk of having a few errors now and then. AMD, on the other hand, shortened their pipelines and added additional branch prediction tables. This makes the processor far more accurate, and therefore it would have to reprocess less often. This is another significant performance point.

The FPU The FPU or floating-point unit has always been a weak spot for AMD processors; that is, until the K7 processor. Previous AMD processor renditions had an unpipelined FPU. This made the AMD FPU significantly slower than the Intel FPU. With the K7, AMD included not just one, but three fully pipelined and parallel FPUs, each one with its own separate port. The PIII also has three separate FPUs, but they are all behind just one port, making simultaneous access impossible. Separating the FPUs into three separate ports effectively more than doubles the FPU speed of the AMD chip over that of the PIII.

The L2 Cache The cache subsystem of the K7 processor is very versatile. It can run without any L2 cache or with as much as 8 MB. It can run the cache at one-fourth, one-third, one-half, two-thirds, or even at the same speed as the processor. Currently, PIII CPUs only allow for 512K of cache running at half the processor clock speed. The K7 can offer a far wider range of processor configurations. It could offer a CPU with 8 MB of full-speed cache to manufacturers in the server and workstation markets, with the same CPU core and with no cache or a small amount of half-speed cache, to manufacturers providing systems for the home computing market.

The L1 Cache The K7 really shines in this arena. AMD has included no less than 64 KB of instruction and data cache. This is four times the amount that comes with a PIII. The L1 cache feeds directly into the CPU core and runs at the same speed as the CPU core, for limited latency or wait time. This makes it very expensive to produce, but a small increase in L1 cache can have big payoffs in the end. AMD quadrupled the PIII specification, giving it another performance advantage.

The System Bus Current system board technology has the front side bus running at 100 MHz. AMD and DEC (the designer of the EV6 protocol that AMD is using) have continued this with one slight change. The bus actually transmits on the rising and falling edges of the clock. This gives the AMD system bus an effective speed of 200 MHz, double that of the current design. AMD also claims that the bus architecture is scalable up to 200 MHz, or an equivalent of up to 400 MHz.

The K7 system bus also has an advantage in the multiprocessing arena. Current Intel multiprocessing systems have to share the bus between each of the processors. The EV6 chipset dedicated a separate path in and out of the chipset for each processor. This can get quite complicated at high processor counts, but it is far superior to sharing the bus.

The Chipset The AMD K7 processor is a departure from the Pentium clones that AMD has been making for the past number of years. So different in fact, it requires its own chipset, the first designed and built by AMD themselves. It is called the 750 Irongate. It is basically like the Intel 440BX chipset. It comes with a north bridge that communicates with the CPU via DEC's EV6 Bus protocol, as well as the main memory and AGP bus. It also comes with a south bridge that has most of the functions of the Intel PIIX4E chip. Future versions of the chipset are expected to have support for DDR-SDRAM, PC133, and even RDRAM.

12.6 SYSTEM MEMORY

Memory is essential to the operation of the computer. System memory is the workspace for the computers microprocessor (MPU). It is a temporary storage area where the data and the programs the MPU is working on reside. It is considered temporary because data will remain there only as long as the computer has electrical power. Once the power is shut off, all the data stored in memory are erased. This is why it is very important to save your work often.

There are actually several different types of memory. For most people, when they think of memory, they think of **random access memory (RAM).** It is by far the most well known type of memory, however, there is another form of memory known as **ROM, or read only memory.** RAM and ROM are the two basic categories of memory. There are many different types of ROM, and even more so, of RAM.

ROM, or read-only memory, is a type of memory that can permanently store information. It is termed read-only because it is difficult or impossible to write to it after the first information is entered. It can be referred to as nonvolatile memory because the data it holds remains, even after the machine has been powered off. ROM chips are the perfect place for holding the system's start-up instructions because you wouldn't want the instructions that boot the system to disappear every time the machine is powered down. Actually, the ROM on a computer motherboard generally contains four programs:

- The **POST,** or Power-on-self-test program. This ensures that the system is working properly on start-up. It is responsible for those annoying system beeps when the video card or keyboard has a problem. The POST can produce several different error messages that let the user know what is wrong with the system. These error codes may come in the form of messages that appear on the screen or sequences of beeps. The error codes are usually printed in the motherboard manual or are generally available for download. Appendix E details a list of typical error codes.

- The **CMOS,** or complementary metal-oxide semiconductor, setup program. This program allows a user to set system settings, boot options, security preferences, and so on.
- The BSL, or boot strap loader. This program scans the floppy and hard drive for bootable program code. If found, it turns the system over to that operating system.
- The **BIOS,** or basic input/output system. The BIOS is basically a set of drivers for running your system hardware, especially hardware that is part of the boot process. It contains basic drivers for your hard drive, floppy drive, and so on. The BIOS also serves as the interface between the computer hardware and the operating system. It controls the boot sequence and hardware settings, and is responsible for displaying the startup screen and summary display. The BIOS reads in its initial settings from the CMOS memory and provides a program to edit those settings, called appropriately enough the BIOS CMOS setup program.

Peripheral adapter cards such as video cards, SCSI (small computer systems interface) cards, and some network adapters also have built-in ROMs. These contain the basic instructions for operating the card. These instructions are executed on start-up so that the device will become operational without the support of an operating system. You wouldn't want to load your video card drivers from disk every time you booted your computer, because the screen would remain blank until the drivers are loaded. Most drivers that are stored on ROM are very basic drivers. Most operating systems will load larger, more advanced drivers over these basic drivers after the initial boot process is complete. These more advanced drivers allow the system to use additional features, or make more efficient use of the device.

There are four different types of ROM chips:

- ROM (read-only memory)
- PROM (programmable ROM)
- EPROM (erasable PROM)
- EEPROM (electrically erasable PROM, also known as flash ROM).

The original ROM chips were manufactured with the 1s and 0s set into the chip die. The program code in the chip was impossible to change without replacing the chip, and only the manufacturer could program the chips. These types of chips are no longer being manufactured and used.

PROM chips allow the user to program the chip with a special programming device called a PROM burner. The chip comes preloaded with all 1s, and the job of the PROM burner was to change these 1s into 0s in accordance with the program code. They can be programmed only once and never erased. They are typically found in imbedded devices, such as the electronic control modules in your car. This makes it possible to alter such things as spark advance, fuel delivery, idle speed, and much more by replacing one of these PROMs.

The EPROM is an erasable version of the PROM. It is functionally the same as PROM, except that it can be erased and reprogrammed. The clear quartz window in the center of the chip easily identifies the EPROM chip. This quartz window is what allows the EPROM to be erased. EPROM chips are erased by exposing the chip to an intense wavelength of ultraviolet radiation, effectively resetting the chip. Once the chip is reset, you are free to reprogram it.

EEPROM is the newest type of ROM. It is characterized by its ability to be erased and reprogrammed directly in the circuit in which it is installed. With EPROM technologies, one must remove the chip from the circuit to erase and reprogram it. EEPROM allows even the most inexperienced of computer users to upgrade the system ROMs. All one needs is the right software, which is usually supplied by the system manufacturer. Flash ROMs are also used to give additional capabilities to existing hardware. A prime example of this is updating your modem from X2 to V.90.

One bad thing about ROM chips is that they are relatively slow by nature. Most ROM chips have access times of around 150 ns. This is very slow compared to the 10-ns access time of RAM today.

RAM is so called because you can randomly and quickly access, read, and write to any location in a memory module. When we talk about a computer's memory, we are generally referring to the RAM in the system. Just like ROM technology, there are several different types of RAM. The different types of RAM tend to fall under two broad categories:

- DRAM (dynamic RAM)—one transistor and one capacitor per cell
- SRAM (static RAM)—four transistors per cell, flip-flop design

DRAM is the type of memory used as main memory in a computer system. One of the main advantages of DRAM technology is that it is very dense. This means a lot of memory can be packed into a very small physical space. This makes DRAM inexpensive compared to SRAM.

The memory cells in DRAM are tiny capacitors (actually the internal capacitance of FET devices) that represent bits, and tiny transistors that read the charge of the capacitors. If the capacitor is positively charged, then a logical 1 is represented. No charge represents a logical 0. Because of the design of DRAM, the memory cells must be constantly refreshed (recharged) or else the memory capacitors will drain and data will be lost.

One disadvantage of DRAM is that it is slow compared to the MPU. As processor performance improved, DRAM technology also had to improve. Therefore, there are several different forms of DRAM:

- Standard DRAM (the original high density RAM)
- FPM DRAM (fast page mode DRAM)
- EDO DRAM (extended data out DRAM)
- BEDO DRAM (burst EDO DRAM)
- SDRAM (synchronous DRAM)
- RDRAM (rambus DRAM)
- DDR SDRAM (double data rate SDRAM)

Over the years, system memory has come packaged in many different shapes and sizes. At first, it came packaged as individual DIP chips that installed into sockets onto the computer's motherboard. As time went on, memory manufactures and computer makers needed an easier way to install memory. They began putting several of these DIP chips onto a single printed circuit board (PCB) that could easily be installed onto the system board. This is the state of memory packaging today, with a few additional improvements.

The first commercially available computer systems had a fixed amount of memory installed in them. Soon afterward, however, technological improvements allowed users to add more memory than what was originally installed at the factory. With the advent of the 286 processor, users could install up to 16 MB of memory. At that time, most PC computers were coming with 1 MB or maybe 2 MB of memory for high-end systems. So, instead of installing all 16 MB of memory, system builders installed sockets for memory upgrades. Users could install 16 to 32 or more individual chips. Each one would have to be correctly installed, or the system could be damaged. This was not very easy for the user, and most didn't attempt it. They simply brought their machine to a computer store to have it upgraded.

When the 386 processor hit the market, industry manufacturers needed a better way for users to upgrade their systems, so they invented the **single in-line memory module (SIMM).** The SIMM was, in essence, a collection of the earlier DIP chips installed onto a single PCB. This PCB was designed to fit into a slot very easily. Users could finally upgrade their systems conveniently and safely. Instead of installing 16+ individual chips, they could simply install two of these SIMMs. This technology also allowed system designers to easily offer a wider array of memory configurations.

The SIMM was so simple and easy that we are still using variations of it today. Nothing better has yet appeared. However, over the years there have been many changes in memory technology, and the SIMM has had to undergo various transformations to cope. This has meant adding many more pins to the SIMM and socket for the additional data and address lines. As with any change, there is confusion about the different types of memory modules.

The first widespread SIMM was called the 30-pin SIMM because it had 30 contacts for address line and data transfer. These were 8-bit SIMMs; that is, they could transfer data 8 bits at a time. This was fine for the 386 and 486 processors, but not for the 586 or Pentium generation because the Pentium transfers data in and out at a rate of 64 bits. Thus, you would need eight of these 30-pin SIMMs to fill this demand. The 386 and 486 have 32-bit data buses, so only four of these SIMMs were required. Something else was needed to meet the demand of the Pentium.

The second widespread version of the SIMM was the 72-pin SIMM. This is a 32-bit module because it transfers data in and out 32 bits at a time. This was perfect for the 486 because you would only need one of these to fill its 32-bit demand. The 72-pin SIMM was also acceptable, but not perfect, for the Pentium generation. To fill the Pentium's 64-bit demand, users would have to install two identical 72-pin SIMMs at a time. This was acceptable for awhile, but increased speed requirements mandated continued improvements in memory packaging and design. The most common forms of memory for the 72-pin SIMM were fast page RAM and EDO RAM.

When processor speeds started growing leaps and bounds beyond the capabilities of memory, memory manufacturers were pressured into designing a new, fast, yet cheap solution. SDRAM was born. The 168-pin DIMM or dual in-line memory module became the next standard. It is dual because it supports two separate 64-bit banks of memory, one bank on each side of the module. This allowed memory manufacturers to easily produce different sizes of modules. For instance, a 64-MB DIMM may have chips installed on only one side of the DIMM, whereas a 128 MB DIMM would have chips on both sides. Also, because they are each 64-bit processors, they are independent of each other and can be installed individually into Pentium and higher systems with little or no confusion.

There have been many variations of each of the different memory types. System manufacturers generally use standard configurations, but over the years many different proprietary types of memory have emerged. Some manufacturers, in an attempt to require users to purchase upgrade memory from them alone, change certain aspects of the memory module. They may resize it slightly or package it in a completely different fashion. Some have resorted to changing the basic operating voltage of the module, and any attempt to upgrade with standard parts could damage either the part or the motherboard.

Another common variant was parity memory. Parity memory is a very simple addition to the memory module. Parity is a form of error detection in memory. The memory controller would add up all of the bits in a byte that is to be stored in memory. If it was even, the parity bit would be assigned a 0, and if it was odd the parity bit would be assigned a 1. When the memory controller read a memory location, it would again add up all of the bits and compare them to the parity bit. If the parity checked, then the memory contents would be passed on; if not, the memory would be discarded and refetched. To implement parity, memory manufacturers had to install another bit of memory for every 8 bits. This meant that an 8-bit, 30-pin SIMM became a 9-bit, 30-pin SIMM with parity. A 32-bit, 72-pin SIMM became a 36-bit, 72-pin SIMM with parity, and a 64-bit, 168-pin DIMM became a 72-bit, 168-pin DIMM with parity. Parity became fairly popular in critical-application computers, but it was never accepted in the mainstream market. Because of this extra bit, parity memory was more expensive and was produced in smaller quantities, driving the price up further and removing it as a viable mass market product.

Yet another common variant on the SIMM/DIMM scheme is that of ECC memory. ECC, or error correcting code, is closely related to parity memory. However ECC memory allows for correction of errors on the fly, something that parity memory cannot do. ECC memory was very expensive for awhile because it used even more bits than parity memory: 32-bit ECC memory actually uses 39 bits, 7 for ECC. However, 64-bit ECC memory only uses 8 extra bits for ECC, the same as parity. This makes ECC a viable replacement for parity memory in recent servers, workstations, and critical-application computers.

Installing memory into your system is quite easy. All of the different styles of RAM modules are different sizes and would be very difficult to install into the wrong slot. For instance, 30-pin SIMMs are about 3.5 in. long, 72-pin SIMMS are about 4 in. long, and 168-pin DIMMs are about 6 in. in length. Mistaking one for the other would be very difficult, and installing the wrong memory in the wrong socket is almost impossible. Figure 12.38 shows all three types of memory module respectively.

It is extremely important to observe the direction in which the memory module is to be installed. Each generation has a key mechanism that helps prevent misinstallation, but these can be easily defeated. Another consideration is to validate your memory voltage and to make sure that the memory you are installing is generally compatible with your machine. There is also no reason to use expensive ECC memory if the motherboard doesn't have an ECC memory controller and logic. Memory installation is discussed in detail in Part 4, "Installation and Service."

Table 12.8 details the evolution of the memory technology used in today's PC systems.

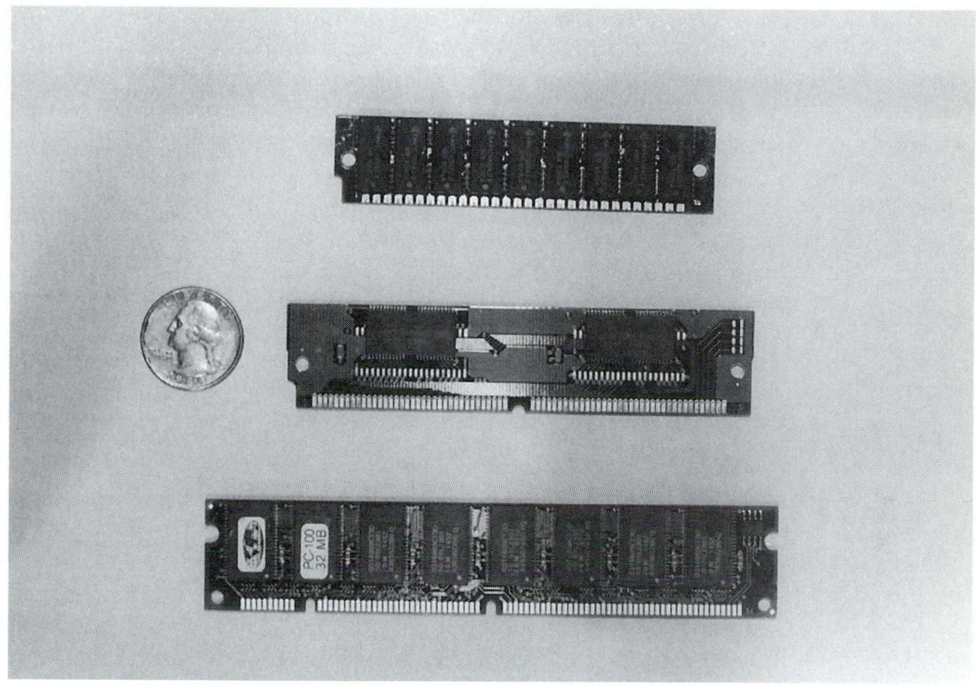

FIGURE 12.38
Memory modules

TABLE 12.8
Memory evolution

MPU Type	486	Pentium	Pentium Pro	Pentium II	Pentium III	Pentium Xeon
Year	1993	1997	1997	1999	1999	1999
Top MPU Speed	100 MHz	233 MHz	200 MHz	450 MHz	600 MHz	600 MHz
L1 Cache Speed	10 ns (100 MHz)	4 ns (233 MHz)	5 ns (200 MHz)	2.2 ns (450 MHz)	1.6 ns (600 MHz)	1.6 ns (600 MHz)
L2 Cache Speed	30 ns (33 MHz)	15 ns (66 MHz)	5 ns (200 MHz)	4.5 ns (225 MHz)	3 ns (300 MHz)	1.6 ns (600 MHz)
FSB Speed	33 MHz	66 MHz	66 MHz	100 MHz	100 MHz	100 MHz
Memory Speed	60 ns (16 MHz)	60 ns (16 MHz)	60 ns (16 MHz)	10 ns (100 MHz)	10 ns (100 MHz)	10 ns (100 MHz)

Once you determine the type of module to use, you must determine the speed and access cycle for your motherboard. The speed is displayed on the module tag or on the individual ICs on the module. A designation of -8 (dash 8) is 80 ns on 30-pin and 72-pin SIMMs. Next, determine the access cycle of the memory. SIMM memory is available in standard, fast page, and EDO cycles. DIMM memory is available in synchronous 66-MHz, PC-100, PC-133, and direct RDRAM cycles.

On a typical 486 system using standard DRAM, five clock cycles are required to access the memory. Sequential cycles will be 5-5-5-5 for the first and next three reads. This means it takes one processor clock cycle to start the transfer and four wait states until the data can be read. The data access sequence on a standard SIMM is as follows:

1. RAS low—row address bits are presented on the address lines.
2. CAS low—column address bits presented on the address lines.
3. RAS high—data are valid 80 ns later.
4. CAS high—output is off, ready for next cycle.

An improvement known as fast page mode has been made to the DRAM used in the standard SIMM. Fast page mode allows holding RAS low while changing only the column address bit, thus increasing access speeds by eliminating the RAS cycle. On a typical fast page ram system, five clock cycles are required to access the first address but sequential cycles will be only three processor cycles or 5-3-3-3. The access cycle for a fast page mode SIMM is as follows:

1. RAS low—row address bits are presented to the address lines.
2. CAS low—column address bits are presented on the address lines.
3. RAS high—data are valid 80 ns later.
4. CAS high—output is off, ready for next cycle.
5. Repeat CAS low/high for the entire ROW address.

The EDO module is a further improvement and uses a modified cycle, keeping the output valid after releasing CAS. This allows for an overlapped address in the read cycle, increasing the read performance around 20 to 30% over fast page mode. In a typical EDO system, five clock cycles are required to access the first address. Sequential cycles will be only two processor cycles, or 5-2-2-2. The speed increase is because the CAS line can be toggled one cycle sooner.

Burst EDO DRAM is basically a faster flavor of EDO RAM. It has special burst features that allow it to transfer more data than EDO RAM. However, it was produced in small quantities and had very limited support. It was quickly overshadowed by the superior SDRAM technology.

SDRAM is the newest type of main memory used in production Pentium systems and stands for synchronous DRAM. In SDRAM the access cycle is synchronous to the processor clock. SDRAM is not rated in access time as were previous memory types, but rather in megahertz for bus speed. SDRAM uses a burst counter on the chip, allowing the column portion of the address to increment very quickly. The memory controller supplies the location and size of the memory block needed, and the SDRAM supplies the CPU with the bits. The SDRAM chip uses an on board clock to synchronize the memory chip's clock to the system clock. This speeds up the subsequent reads tremendously, resulting in an access cycle of 5-1-1-1. Note that there are no wait states during subsequent read cycles.

The next evolution of SDRAM was 100-MHz SDRAM, also known as PC100 SDRAM. However, it was not until the introduction of the Intel 440BX chipset that the full advantage of PC100 was seen. The 440BX was the first chipset to run the system bus at 100 MHz. When you join the 440BX chipset with a Pentium II that has a 100-MHz bus, then you see the full advantage of PC100 SDRAM. The Pentium II 350-MHz was the first Intel CPU to incorporate the 100-MHz system bus. Depending on the manufacturer of the particular motherboard in a given system, you may or may not have that ideal combination of the Pentium II 350 with the BX chipset.

Next came the introduction of PC133 SDRAM. This memory was introduced while the industry waited for direct RDRAM (which will be discussed shortly). With speeds reportedly up to 1.6 gigabytes, twice that of PC100, PC133 SDRAM represents a significant advancement in memory technology. As always, there must be a supporting chipset to achieve optimum performance. At this point in the evolution of the memory model, we see another battle between Intel and other CPU manufacturers. With the arrival of direct RDRAM continually delayed, Intel's competitors began to produce chips that supported PC133 and Intel was left behind. Eventually Intel did introduce supporting chipsets. A further improvement on PC133 is VC133, or virtual channel SDRAM. This memory incorporates a small amount of very fast static RAM, which acts as cache for the SDRAM. There is yet another twist on the SDRAM memory model, and that is DDR DRAM, or double data rate DRAM. This memory transfers data on both the rising and falling edges of the clock cycle, thereby theoretically doubling the performance. DDR SDRAM, or double data rate SDRAM, is very similar to SDRAM—in fact it is the next evolutionary step in DRAM technology. It achieves faster data transfers by doubling the amount of information sent every clock cycle. Instead of making one transfer every clock cycle, as in SDRAM, DDR SDRAM makes two transfers every clock cycle. It transfers once on the rising of the clock signal, and once on the falling of the clock signal, effectively doubling the transfer rate. This memory technology will work at 100–133 MHz at 7–8 ns.

Direct RDRAM

Direct Rambus dynamic random access memory, or **direct RDRAM** is a new memory standard developed by Rambus Inc. in partnership with Intel. From the point of view of memory architecture, this truly is revolutionary. It operates in a completely different way than its predecessors. It utilizes bus mastering via the Rambus Channel Master, with a data pathway called the Rambus Channel, which is between memory devices, or Rambus Channel Slaves. Simply put, direct RDRAM performs concurrent memory transfers with no waiting in line. Here's how it works: The direct Rambus Channel consists of a memory controller and a number of RDRAM chips linked via a common bus. This bus is two bytes wide and carries high-speed signals up to 800 MHz. This signaling method is called Rambus Signaling Logic. Currently a single channel can transfer data at 1.6 gigabytes. The real advantage is that multiple channels can be implemented in parallel. The current generation of RDRAM can use four channels, for a data throughput of 6.4 gigabytes. RDRAM is packaged in modules very similar in size and shape to an SDRAM, and is called a RIMM, or Rambus in-line memory module. The RIMM uses a different connector and has a sheath covering the chips which acts as a heat sink. RDRAM also lowers the voltage from 3.3 V to 2.5 V, and actually lowers the internal power consumption to just over half a watt. RDRAM not only decreases power consumption, but also signal interference. Figure 12.39 compares SDRAM and RDRAM throughput.

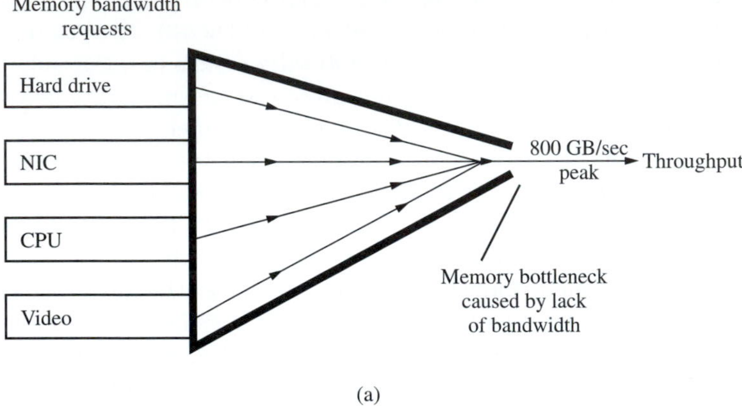

(a)

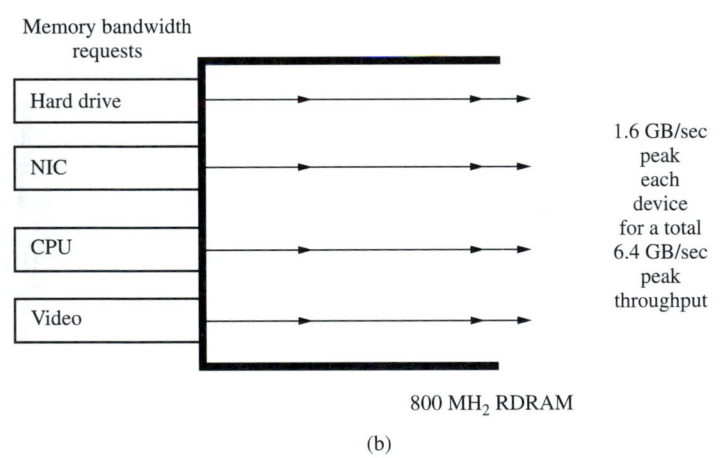

(b)

FIGURE 12.39
SDRAM vs. RDRAM throughput

Now we must discuss the inevitable downside of this radical new technology. First and foremost is the issue of cost. RDRAM technology can increase system cost by $600 to $700. As supply and acceptance increase, the cost will probably come down. Another factor that comes into play involves the memory manufacturers themselves. Manufacturers are displaying caution in using this technology in fear that some new and cheaper technology will emerge. Another potential pitfall for the technology is the heat factor. Even given the substantial reductions in power achieved, because of the high density of the components, large amounts of heat will be generated. Perhaps the biggest problem for RDRAM is that it may not offer a significant performance increase with current chipset and CPU technology. There are test results that seem to show a real lack of performance enhancement when using RDRAM with the current chipsets. However, when Intel releases a new chipset that supports RDRAM, significant performance increases should result. The new chipset will be called the Camino 820 and will be the first to have full support for RDRAM. Part of the problem is that RDRAM has outgrown current .25-micron chip technology. Intel is working on a new .18-micron technology in its Coppermine processor. Also, AMD is working on RDRAM support in its ATHLON chipset and hopes to realize a 200-MHz bus speed. The bottom line here is that the hardware support is not currently available for RDRAM.

Until recently, DRAM technology was limited to 16 MHz. When processors were slower than 16 MHz, they could easily read data out of main memory without having to wait. Once MPUs crossed that 16-MHz barrier, every time a MPU wanted something from memory it would effectively slow to 16 MHz while it waited for the data. What was needed was something between the fast MPU and the slow main memory bank that would buffer this speed difference. SRAM, or static RAM, is that something. Static RAM is radically different from DRAM technologies. It does not use a capacitor and therefore never needs to be refreshed. This makes SRAM much, much faster than DRAM. In fact, SRAM is fully capable of keeping pace with the fastest of modern processors. This means that SRAM can currently run at 600 MHz or more, as compared to the slower 100 MHz of DRAM.

The problem with SRAM is that it is big and expensive. Because of its design, it takes up about 30 times the space as DRAM, and it is also 30 times as expensive. Two megabytes of SRAM take up the same space and cost the same as 64 MB of DRAM.

The design of SRAM is fairly simple in that 4 to 20 transistors (or flip-flops) represent a single bit. The SRAM community has settled on an economically efficient version of SRAM that uses 6 transistors to represent a single bit.

To prevent the unwanted situation of a MPU reading data directly from system memory, chip designers have employed a two-stage caching structure; L1, or level one, cache, and L2, or level two, cache. The L1 cache is also called the internal cache because it is built directly into the MPU chip. Because of this, the L1 cache always runs at the full processor speed. This also makes it very expensive, and its use has been limited to 32 to 64 KB of data for the recent processors. Older processors had even less. L2 cache is called external cache because it generally resides outside the MPU chip. Originally, this meant that the cache was installed directly on the motherboard. Newer processors have incorporated the L2 cache into the MPU package, but it still remains external to the MPU chip itself. Because the L2 cache was external, it usually ran at the slower system clock rather than at the MPU clock. This meant that the L2 cache ran at 66 MHz on 486 or Pentium systems. This was still a lot faster than the 16 MHz of DRAM. In Pentium II machines, the cache was taken off the motherboard and put on the MPU package. This allowed Intel to make the cache as fast as possible. Instead of running at the system bus speed, they speed it up to half of the MPU clock speed. For Pentium II systems, this meant that a 400-MHz MPU had a 400-MHz L1 cache, a 200-MHz L2 cache, and 100-MHz system memory. Compare this to an older 200-MHz Pentium system with a 200-MHz L1 cache, 66-MHz L2 cache, and 16-MHz system memory.

The next step in MPU cache design was to incorporate the L2 cache into the MPU die itself. The first chip to do this was the Pentium Pro. The full-speed L2 cache significantly improved performance, and the chip quickly became the choice MPU for server and graphic workstation manufacturers. The next MPU to have full-speed L2 cache was the Intel Xeon MPU, which comes standard with 1 MB to 2 MB of L2 cache running at the same speed as the MPU. Because of this, the Pentium Pro and Pentium Xeon chips are very expensive to produce and therefore come at a price premium to the end user. A Pentium Pro with 1 MB of L2 cache retailed for $2500 until the equally expensive Pentium Xeon chips took over the high-end market.

12.7 DMA

In modern PC systems, several system resources are in limited supply when configuring devices. When two devices are configured to use the same resource, conflicts arise that can contribute to system instability, or complete system crashes. Even when free resources are available, some devices that are only designed to use certain resources may cause conflicts. Although this situation has greatly improved with the advent of plug and play (PNP) BIOS and devices, resource conflicts are still a common problem. The two most common resources that the technician must configure are **direct memory access (DMA)** channels and **interrupt requests (IRQs).**

The DMA Controller

DMA is a faster and more efficient way than *programmed I/O* to move data between I/O devices and memory. During programmed I/O the processor must look for data from a device, read it, and then write it out to memory. This procedure can take a considerable time to execute. DMA, on the other hand, links a device directly to memory without the use of the processor, thus freeing it up.

The DMA circuit is a bit complex. It is centered around an Intel 8237A-5 advanced direct memory access controller chip. This chip was discussed earlier in the book, so you may want to review it before going on.

Since the actual DMA transfer works independently of the processor, the DMA circuitry must supply the memory address as well as the read and write control lines for the bus. Think of the DMA function as a separate processor. In the IBM PC the longest transfer is 65,536 bytes of data and cannot cross into the next segment of memory. This limitation is due to the 16-bit DMA address registers. Recall that 16 bits limits you to 65,536.

Before we allow an I/O device to grab the bus and run, we must set some rules concerning where in memory the data will go to or come from and how much data can be transferred. Imagine what might happen to a program contained in RAM if we do not do this first. Some device might transfer its data right over our program, destroying it.

The DMA process is quite simple and requires that we program the base address and the base word count registers. We must also program the extended address register, if any. At this point we may tell the applicable I/O device that it is safe to perform DMA transfers by setting its DMA enable circuit. When the device has finished the transfer, it can inform the processor by causing an interrupt or setting a bit in its status register for the processor to poll or check.

DMA channels provide a faster and more efficient way to transfer data, by allowing devices to directly access memory without using the MPU. The benefit is twofold:

The device no longer has to wait for the MPU to service its request, and the MPU is not burdened with data transfer requests. A good example of this is seen when DMA bus mastering is available for hard disks. During data transfers from the disk, processor usage drops, from a sizable percentage of the MPU's time without DMA support to often less than 10% with bus mastering enabled. This can be important for PC systems, where slow disk throughput is sometimes the bottleneck for overall performance.

There are only eight available DMA settings, 0 through 7, though not very many devices require a DMA setting. Table 12.9 gives some of the more commonly used DMAs.

TABLE 12.9
Common DMA settings

DMA #	Device
0	
1	Sound Blaster sound card
2	Floppy disk controller
3	
4	DMA controller
5	
6	
7	

12.8 INTERRUPTS

Using an **interrupt-driven** I/O system is a faster and more efficient way of handling multiple I/O devices than polling. In a **polled I/O** system the processor must ask each I/O device that is on the system if it has any data to transfer. The processor must also ask each I/O device for its status. This process can waste considerable time. In an interrupt-driven system, the I/O device **flags** the processor when it has a request for data or status to the processor.

Visualize an office of five workers and one boss. The boss gives all five workers a task to perform. In a polled system the boss must continuously ask all five workers about their status on the assigned task to see if they are done or having a problem. As we can see, this can tie up the boss's time. As an alternative and more efficient system, we assign the workers their tasks and instruct the workers to interrupt the boss only when they have a problem or are finished with the task. This frees up the boss's time and allows the boss to perform other tasks and get more tasks ready for the workers. In a **prioritized interrupt** system the workers who are performing critical tasks always get the boss's attention first.

Almost all modern computer systems use interrupt-driven I/O schemes. In addition to the address lines and I/O control lines, they have interrupt request and interrupt acknowledge lines. Some systems use **daisy-chained** (in series) lines and some use separate lines. Others, like the PDP-11 minicomputers, use daisy-chained and separate interrupt lines together.

Since it may be possible for more than one I/O device to interrupt the processor at the same time, a priority system must be set up using hardware or software, or a combination of the two. In a daisy-chained system the I/O device electrically closest to the processor receives the interrupt acknowledge first. If that I/O device does not request the interrupt, it passes the acknowledge signal to the next device on the bus. When the requesting device receives the acknowledge signal, it then directs the processor to the program that can handle the request. When separate interrupt lines are used, the incoming interrupts are fed to a circuit known as a **priority encoder.** This circuit passes only the highest-priority interrupt to the processor and blocks the rest until the priority interrupt has been completed.

In a software-prioritized system, the processor must first poll the I/O devices to ask which one has requested the interrupt. The first device polled gets serviced first, establishing the order of priority. The address of the program to service that interrupt is implied by the polling routine. As an alternative to polling, **vectors** are used. An **interrupt vector** is an address generated by the interrupt controller. These vectors direct the processor to the proper interrupt service routine rather than having the processor figure out the address first.

In most PC systems 16 IRQs, 0 through 15, are available for devices to use. This may have seemed more than sufficient when AT personal computer systems were originally designed, but it is not uncommon to see systems with every IRQ used, or even with one IRQ shared between two devices (with certain devices this is a workable configuration). This is further complicated by the fact that certain IRQs are historically reserved for specific devices. Table 12.10 gives some of the more commonly used IRQs with their respective devices. If a system is short of available IRQs, it is possible to disable certain

TABLE 12.10
Common IRQ settings

IRQ #	Device
0	System timer
1	Keyboard
2	Cascade to second interrupt controller
3	COM2/COM4 (modem)
4	COM1/COM3 (serial mouse)
5	Sound Blaster sound card (ISA card)
6	Floppy disk controller
7	Printer port (LPT1)
8	System C/MOS real-time clock
9	Open for use, mapped to IRQ 2
10	Open for use
11	Open for use
12	PS/2 mouse
13	Numeric coprocessor
14	Primary IDE controller
15	Secondary IDE controller

devices in the BIOS (such as unused COM ports) to free up an IRQ. It also becomes important to see if there are any IRQs available before installing a new device, and which IRQs that device is capable of using. Some older modems and sound cards are only able to use one or two different IRQs and it may be necessary to move an existing device's IRQ to accommodate an inflexible older device. PNP devices will generally accept a much wider range of IRQs.

12.9 THE BUS SUBSYSTEM

The ISA Bus

The 16-bit IBM PC/AT system bus has come to be known as the **Industry Standard Architecture (ISA) bus.** This is by far the most popular bus for MS-DOS computers to date. Almost all adapter or controller boards designed for the 8-bit bus will work in the 16-bit ISA bus. Any 8-bit board will work in any (8- or 16-bit) adapter slot as long as the board will physically fit in the chassis. However, 16-bit boards will work only in a 16-bit slot. The ISA bus is easily identified by the second 36-pin connector just in front of the 62-pin PC bus connector, as shown in Figure 12.40.

Many new bus signals have been added along with the additional eight data bits, including four new DMA channels, six new interrupt requests, 24-bit addressing, a bus-master input, and two 16-bit bus selects. Table 12.11 shows the ISA bus signals with the new AT signals labeled accordingly.

DMA channels 5 through 7 are 16-bit transfer channels and can cross the 64-kb boundary to 128 kb because the address output lines of the 8237 DMA controller are shifted up 1 bit (bit 0 is always a 0). Thus, 16-bit transfers are always on an even boundary.

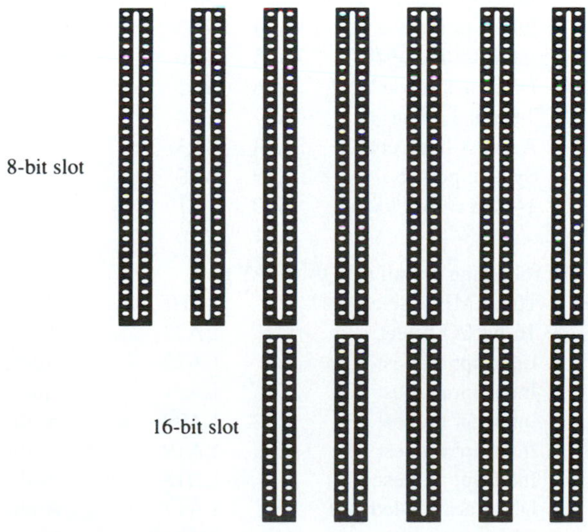

FIGURE 12.40
ISA bus connectors

TABLE 12.11
ISA bus connector pins

Left Side			Right Side		
Pin	Signal	Use	Pin	Signal	Use
B1	GROUND		A1	I/O CHCK	Parity errors
B2	RESET DRV	System reset	A2	SD7	System data bit 7
B3	+5 V	System power	A3	SD6	System data bit 6
B4	IR9	Interrupt 9 (new)	A4	SD5	System data bit 5
B4	−5 V	System power	A5	SD4	System data bit 4
B6	DRQ2	DMA request 2	A6	SD3	System data bit 3
B7	−12 V	System power	A7	SD2	System data bit 2
B8	OWS	Zero wait state (new)	A8	SD1	System data bit 1
B9	+12 V	System power	A9	SD0	System data bit 0
B10	GROUND		A10	I/O CHRDY	General wait state
B11	SMEMW	Memory write signal	A11	AEN	Address enable
B12	SMEMR	Memory read signal	A12	SA19	System address bit 19
B13	IOW	I/O write signal	A13	SA18	System address bit 18
B14	IOR	I/O read signal	A14	SA17	System address bit 17
B15	DACK3	DMA acknowledge 3	A15	SA16	System address bit 16
B16	DRQ3	DMA request 3	A16	SA15	System address bit 15
B17	DACK1	DMA acknowledge 1	A17	SA14	System address bit 14
B18	DRQ1	DMA request 1	A18	SA13	System address bit 13
B19	REFRESH	Memory refresh	A19	SA12	System address bit 12
B20	CLK	System clock	A20	SA11	System address bit 11
B21	IR7	Interrupt request 7	A21	SA10	System address bit 10
B22	IR6	Interrupt request 6	A22	SA9	System address bit 9
B23	IR5	Interrupt request 5	A23	SA8	System address bit 8
B24	IR4	Interrupt request 4	A24	SA7	System address bit 7
B25	IR3	Interrupt request 3	A25	SA6	System address bit 6
B26	DACK2	DMA acknowledge 2	A26	SA5	System address bit 5
B27	T/C	Terminal count	A27	SA4	System address bit 4
B28	BALE	Address latch enable	A28	SA3	System address bit 3
B29	+5 V	System power	A29	SA2	System address bit 2
B30	OSC	14.318 color clock	A30	SA1	System address bit 1
B31	GROUND		A31	SA0	System address bit 0
The following are all new to the AT bus:					
D1	MEM CS16	16-bit MEM select	C1	SBHE	Bus high enable
D2	I/O CS16	16-bit I/O select	C2	LA23	Address bus bit 23
D3	IRQ10	Interrupt request 10	C3	LA22	Address bus bit 22
D4	IRQ11	Interrupt request 11	C4	LA21	Address bus bit 21
D5	IRQ12	Interrupt request 12	C5	LA20	Address bus bit 20
D6	IRQ15	Interrupt request 15	C6	LA19	Address bus bit 19
D7	IRQ14	Interrupt request 14	C7	LA18	Address bus bit 18
D8	DACK10	DMA acknowledge 0	C8	LA17	Address bus bit 17
D9	DRQ0	DMA request 0	C9	MEMR	Memory read
D10	DACK5	DMA acknowlege 5	C10	MEMW	Memory write
D11	DRQ5	DMA request 5	C11	SD08	System data bit 8

(continued)

TABLE 12.11 (continued)

Left Side			Right Side		
Pin	Signal	Use	Pin	Signal	Use
D12	$\overline{\text{DACK6}}$	DMA acknowledge 6	C12	SD09	System data bit 9
D13	DRQ6	DMA request 6	C13	SD10	System data bit 10
D14	$\overline{\text{DACK7}}$	DMA acknowledge 7	C14	SD11	System data bit 11
D15	DRQ7	DMA request 7	C15	SD12	System data bit 12
D16	+5 V	System power	C16	SD13	System data bit 13
D17	$\overline{\text{MASTER}}$	Bus master control	C17	SD14	System data bit 14
D18	GROUND		D18	SD15	System data bit 15

In the PC/AT-type computer, DMA channel 0 has been freed for 8-bit transfers. In the original 8-bit PC, DMA channel 0 was for memory refresh. DMA channel 4 is reserved to cascade the original four DMA channels 0–3.

Seven more interrupt request lines have been added. IRQ8 through IRQ15 are now cascaded through IRQ2. The hard disk interrupt has been moved to IRQ14. IRQ9 replaces IRQ2 and is redirected to IRQ2's vector, thus resulting in a total of 15 IRQ lines.

Sixteen-bit transfers are facilitated through the use of the $\overline{\text{MEM CS16}}$ and $\overline{\text{I/O CS16}}$ signals. When these signals are high, normal 8-bit transfers occur in any slot. When these signals go low, 16-bit transfers may be performed as soon as the card is selected. The $\overline{\text{I/O CS16}}$ signal is for I/O operations, and the $\overline{\text{MEM CS16}}$ signal is for external memory operations.

A new signal $\overline{\text{MASTER}}$ is provided to allow an adapter or second processor to obtain control of the system bus. When this signal is pulled low, the system processor will pause and release control of the system bus.

ISA Bus System Setup

One vast improvement over the older PC system is the elimination of the system board switches. The setup information is now stored in a battery-powered RAM circuit contained in an on-board real-time clock chip. This chip is the Motorola MC 146818 real-time clock plus RAM, which contains a clock/calender circuit and 50 bytes of **CMOS RAM.** By supplying power to the chip with a battery, the time of day and all setup information can be maintained even when the system power is turned off. A block diagram of the MC14618 IC is shown in Figure 12.41.

The system setup information is loaded into the CMOS RAM by running a special program supplied by the computer manufacturer. An example of the PHOENIX BIOS setup program is shown in Figure 12.42. As you can see in the figure, much of the guesswork has been eliminated by a user-friendly, menu-driven screen complete with help screens. By pressing the F1 key, help for the selected field is displayed.

In most of the latest computers, this setup program is contained in the BIOS ROM and accessed by pressing the Del or other key during self-test. One such example of this is the AMI (American Megatrends Inc.) BIOS, which is self-contained and menu-driven for ease of use. Figure 12.43 shows the AMI BIOS setup menu.

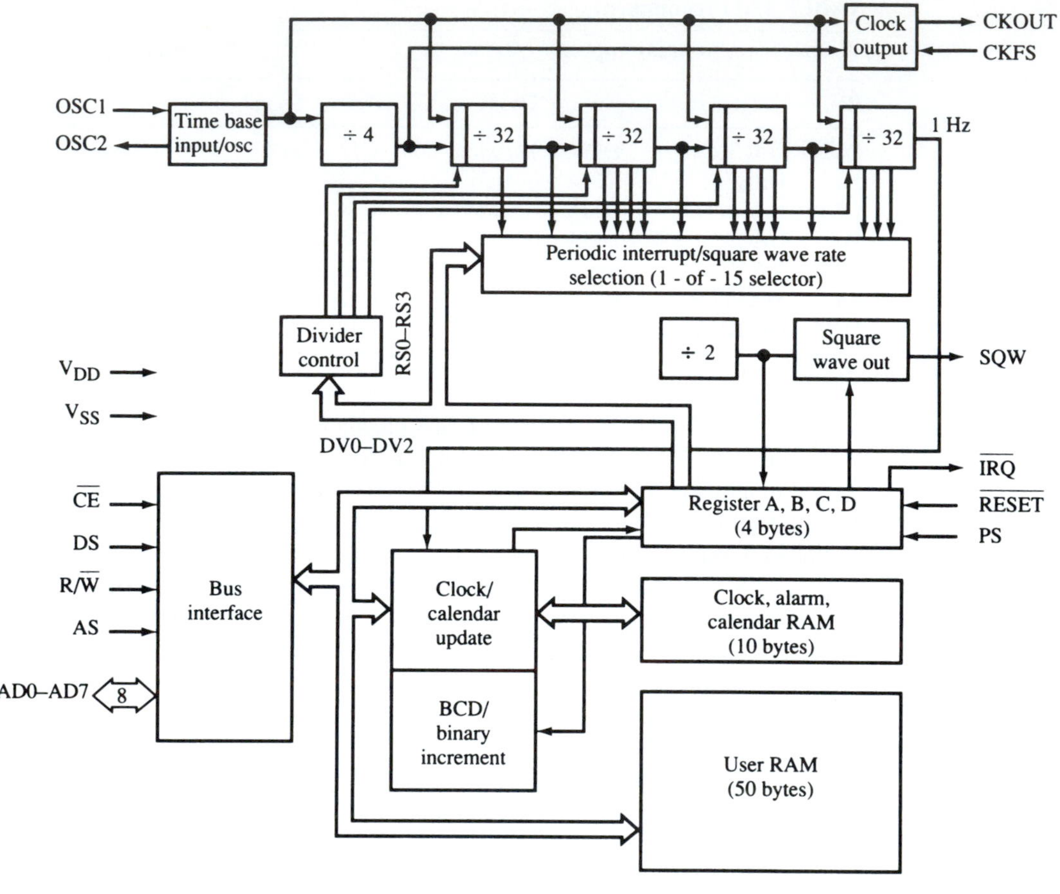

FIGURE 12.41
MC146818 block diagram

Local Bus

The development of the local bus was driven by the emerging dominance of the graphical user interface (GUI) over the traditional DOS-based operating system. When Microsoft first introduced Windows to compete with the Apple GUI environment, it was a poor imitation and was not terribly successful. As the program evolved and improved, however, it gained widespread acceptance. By the time Windows 3.1 was released, the combination of "IBM" hardware and Microsoft's Windows/DOS environment was clearly the dominant force in the PC industry. The number of available Windows applications grew very quickly, but PC operation abated. Of particular concern was the time it would take for screen writes or rewrites in a typical Windows application. The solution to this problem was to operate the video functions at a speed equal to or close to the CPU. This was particularly attractive because the operating speed of the CPU was increasing prac-

```
Phoenix Technologies Ltd.
System Configuration Setup V4.0

Time: 13:14:51
Date: Sun Feb 02, 1992

Diskette A:          5.25 Inch, 1.2 MB
Diskette B:          3.5 Inch, 1.44 MB
Hard Disk C:         Type 47
Hard Disk D:         Not Installed
Base Memory:         640 KB
Extended Memory:     3072 KB
Display:             EGA/VGA
Keyboard:            Installed
CPU Speed:           LOW

Coprocessor:         Not Installed

Up and Down Arrow to select entries
Left and Right Arrow to change entries
F1 to get help on current entry
F10 to exit Setup
Esc to reboot the system
```

FIGURE 12.42
Phoenix BIOS setup screen

tically every month. To speed up video transfers, a local bus technology was developed. The earliest designs actually consisted of video circuitry on the motherboard as opposed to board-mounted slots. As the local bus proved its value, other peripherals requiring a greater data throughput used this technology. The obvious first choice was the hard drive, which of course required an additional expansion slot for the controller card. At this point, there were several proprietary local bus designs on the market and compatibility became a problem. The need for compatibility led to the formation of the Video Electronics Standards Association (VESA) and the **VESA local bus** (VL bus), formally announced on August 28, 1992. To maintain compatibility with current ISA bus computers, the VESA local bus connectors are mounted behind the ISA connectors as viewed from the rear. See Figure 12.44.

The VL bus was designed around the Intel 486 chip and very closely duplicates its signals, with a few exceptions. Some signals were designed more broadly to allow for compatibility with older 386 chips. Those 486 signals that are not relevant to the expan-

```
CMOS SETUP (C)  Copyright 1985-1990, American Megatrends Inc.,

Date (mn/date/year) : Sat, Feb 01 1992     Base memory size   : 640 KB
Time (hour/min/sec) : 11 : 31 : 26         Ext. memory size   : 3072 KB
Floppy drive A:     : 1.2  MB, 5¼"         Numeric processor  : Not Installed
Floppy drive B:     : 1.44 MB, 3½"
                                            Cyln  Head  WPcom  LZone  Sect  Size
Hard disk C:  type  : 47 = USER TYPE        980   10    65535  980    17    81 MB
Hard disk D:  type  : Not Installed
Primary display     : VGA or EGA
Keyboard            : Installed

Scratch RAM option  : 1

Month  : Jan, Feb,.....Dec
Date   : 01, 02, 03,...31
Year   1901, 1902,.....2099

ESC = Exit,←↓→↑ = Select, PgUp/PgDn = Modify
```

Sun	Mon	Tue	Wed	Thu	Fri	Sat
26	27	28	29	30	31	1
2	3	4	5	6	7	8
9	10	11	12	13	14	15
16	17	18	19	20	21	22
23	24	25	26	27	28	29
1	2	3	4	5	6	7

FIGURE 12.43
AMI BIOS setup screen

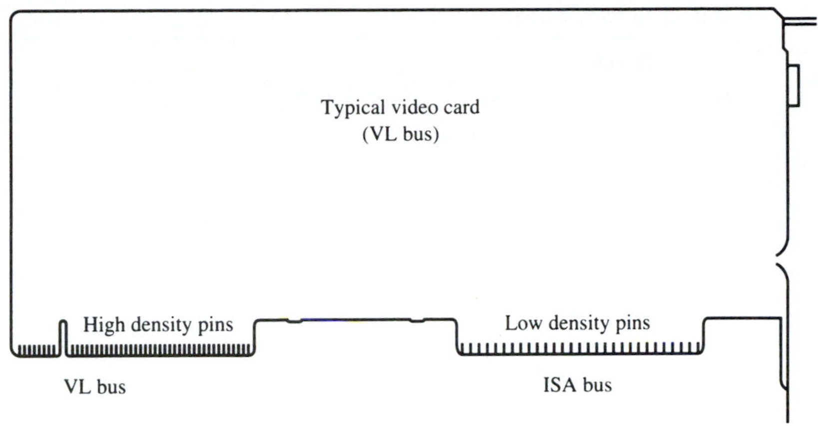

FIGURE 12.44
VESA local bus

sion bus and those that are redundant are not carried through to the local bus. In addition to 486-derived signals, the VL bus adds several signals of its own. Four static signal pins are provided to determine the microprocessor of the system. The VL bus peak transfer rate occurs during the burst mode. The VL burst mode follows that of the 486 chip: a single address cycle followed by four data cycles. At speeds above 33 MHz, at least one wait state is added per read or write operation. Below 33 MHz, 1 wait state is added per read operation, none for a write operation. The addition of one wait state reduces nonburst throughput by 50%; therefore, a 66 Mbps transfer becomes a 44 Mbps transfer. In this same scenario, burst transfer speed is cut in half. In addition, individual VL devices may add still more wait states, which further decreases throughput. Table 12.12 describes the VESA local bus signal connections.

TABLE 12.12
VESA local bus pinout

B Side			A Side	
64 Bit	32 Bit	Pin Number	32 Bit	64 Bit
	DAT00	01	DAT01	
	DAT02	02	DAT03	
	DAT04	03	GND	
	DAT06	04	DAT05	
	DAT08	05	DAT07	
	GND	06	DAT09	
	DAT10	07	DAT11	
	DAT12	08	DAT13	
	VCC	09	DAT15	
	DAT14	10	GND	
	DAT16	11	DAT17	

TABLE 12.12 (continued)

B Side			A Side	
64 Bit	32 Bit	Pin Number	32 Bit	64 Bit
	DAT18	12	VCC	
	DAT20	13	DAT19	
	GND	14	DAT21	
	DAT22	15	DAT23	
	DAT24	16	DAT25	
	DAT26	17	GND	
	DAT28	18	DAT27	
	DAT30	19	DAT29	
DAT63	ADR31	21	ADR30	DAT62
	GND	22	ADR28	DAT60
DAT61	ADR29	23	ADR26	DAT58
DAT59	ADR27	24	GND	
DAT57	ADR25	25	ADR24	DAT56
DAT55	ADR23	26	ADR22	DAT54
DAT53	ADR21	27	VCC	
DAT51	ADR19	28	ADR20	DAT52
	GND	29	ADR18	DAT50
DAT49	ADR17	30	ADR16	DAT48
DAT47	ADR15	31	ADR14	DAT46
	VCC	32	ADR12	DAT44
DAT45	ADR13	33	ADR10	DAT42
DAT43	ADR11	34	ADR08	DAT40
DAT41	ADR09	35	GND	
DAT39	ADR07	36	ADR06	DAT38
DAT37	ADR05	37	ADR04	DAT36
	GND	38	WBACK#	
DAT35	ADR03	39	BEO#	BE4
DAT35	ADR02	40	VCC	
LBS64#	NC	41	BE1#	BE5#
	RESET	42	BE2#	BE6#
	D/C#	43	GND	
DAT33	M/10#	44	BE3#	BE7#
DAT32	W/R#	45	ADS#	
		KEY		
		KEY		
	RDYRTN#	48	LRDY#	
	GND	49	LDEV X#	
	IRQ9	50	LREQ X#	
	BRDY#	51	GND	
	BLAST#	52	LGNT X#	
	ID0	53	VCC	
	ID1	54	ID2	
	GND	55	ID3	
	LCLK	56	ID4	ACK64#
	VCC	57	NC	
	LBS16#	58	LEADS#	

PCI Bus

Peripheral component interface (PCI) is fast beginning to eclipse the VL bus as the preferred method of peripheral connection. Although PCI is itself a type of local bus, it is quite different in many ways. One significant difference in PCI is the decoupling of the processor and main memory subsystems from the standard expansion bus (ISA, MCA, EISA), as shown in Figure 12.45.

Connection between the subsystems of the processor's main memory and the PCI bus is achieved via the PCI bridge. The PCI bridge and the CPU can operate in parallel provided that the CPU is not addressing a PCI unit at the same time. This convenience allows for communication between two PCI units while the CPU is addressing main memory to perform a task. Another significant difference in PCI is that it is microprocessor independent, meaning that it is not tied only to the Intel family of microprocessors. Another very important difference between VL bus and PCI is the PCI burst mode. In PCI burst mode, the PCI bridge combines incoming signals to form a data burst of unlimited size for transfer. Contrast this with the single-transfer mode of the older buses, or the four-

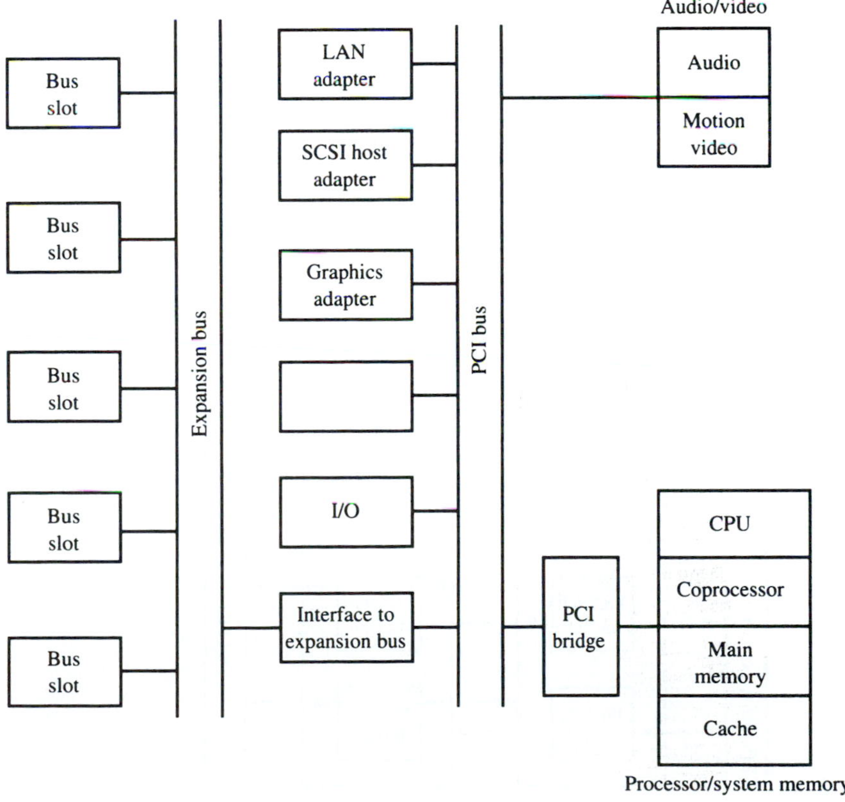

FIGURE 12.45
PCI bus

cycle maximum burst transfer of the VL bus, and you have a significant performance increase and bottleneck decrease. In addition, PCI provides configuration address space. A configuration area of 256 bytes is provided for every PCI unit (and each separate function of a multifunction unit); therefore, there are 64 registers for each 32 bits. A 64-byte fixed header is located at the top and is predefined for any unit with this structure. The remaining 192 bytes are used differently, depending on the type of PCI unit. In addition PCI bus arbitration is handled separately for each access. Thus, a busmaster cannot hold up the bus between accesses.

As discussed earlier, in the PCI configuration there is a decoupling of the main processor and memory subsystems from the bus via the PCI bridge. An example layout of the PCI bridge is shown in Figure 12.46. The PCI bridge is far more intuitive than the ISA, EISA, or MCA bus controllers. It directs CPU access to the PCI unit, to provide optimal access. As with SCSI, the unit requesting data is called the initiator, and the unit receiving data is called the target. The PCI bridge acts as a buffer between the initiator and the target, synchronizing the two units. The bridge will receive and direct TRDY (target ready), and IRDY (initiator ready) signals, which is how the PCI bridge creates the PCI burst.

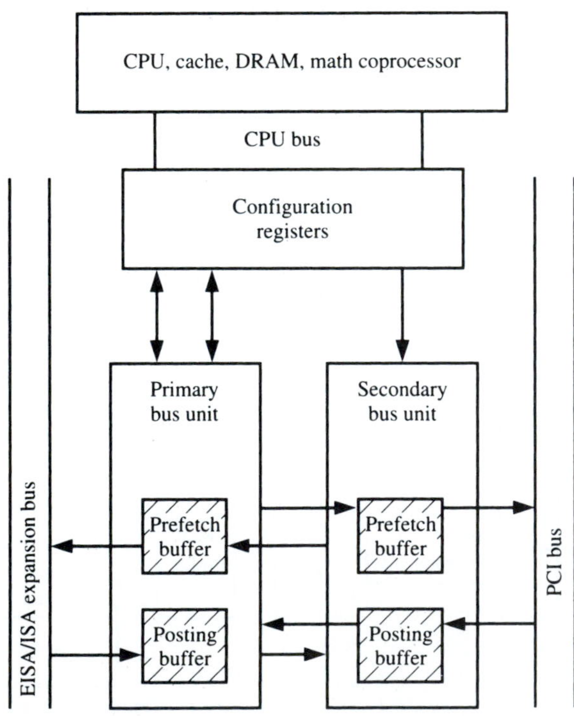

FIGURE 12.46
PCI bridge

Currently PCI recognizes 12 types of access:

- INTA sequence (0000)
- Special cycle (0001)
- I/O read access (0010)
- I/O write access (0011)
- Memory read access (0110)
- Memory write access (0111)
- Configuration read address (1010)
- Configuration write address (1011)
- Memory multiple read address (1100)
- Dual addressing cycle (1101)
- Line memory read access (1110)
- Memory write access with invalidation (1111)

Every PCI data transfer begins with an address phase. During this phase, the address/data pins AD_x transfer the address and the C/BE_x pins transfer the instruction code. One or more data phases follow, with the same data/address pins AD_x transferring data and the C/BE_x pins transferring the byte enable signals. A transfer then begins with activation of the FRAME signal. The target indicates to the bridge that it is ready with an active TRDY signal, and the initiator indicates the same with an active IRDY signal. An active TRDY during a write access indicates that the target can accept data from lines AD_{31}–AD_0 in a 32-bit transaction, or from lines AD_{63}–AD_0 during a 64-bit transaction. An active TRDY during a read access indicates that addressed data has been sent along the same lines. The PCI bridge acts as a traffic cop between the various signals, or between two PCI devices. This is how the bridge performs bursting, by combining single signals into multiple bursts. For example, if the PCI bus were to pass on data faster than the CPU could supply it, the PCI bridge would deactivate the IRDY signal to let the target know that the transfer is not yet finished. The initiator ends a transfer by deactivating the FRAME signal. The target may also end a transfer by activating the STOP signal, which leads to a target abort. The bridge also performs prefetching and writing posting. During a read prefetch, the PCI bridge reads data very quickly from the target and stores it in a prefetch buffer, later passing it on to the CPU. If the prefetch buffer is full because data is not being collected quickly enough, the bridge will deactivate IRDY to inform the target not to deliver any more data. During write posting, the CPU writes data to a buffer (perhaps faster than the PCI bus can pass it on). The posting buffer then passes the data to the CPU later.

The PCI bus uses two control signals for bus arbitration, request (REQ), and grant (GNT). Each busmaster has its own REQ and GNT signals, which are interpreted by central arbitration logic. The PCI spec does not mandate that a certain model be used for arbitration, but it does require that a PCI busmaster activate the REQ signal to indicate its request for the PCI bus. In addition, the arbitration logic must activate the GNT signal for the master making the request to gain access to the bus. The busmaster must then begin the transfer within 16 clock cycles or a time overrun error will occur. Because there is no mandated model, some programming must be done by the system designer.

As previously mentioned, a configuration area of 256 bytes is provided for each PCI unit, with a fixed header of 64 bytes, and the remaining 192 bytes is used as needed by individual PCI devices. See Figure 12.47. The configuration software for the device will determine the use of the 192-byte segment. The header is divided into a 16-byte section (which remains the same for all PCI devices), with the remaining 48 bytes used by individual devices in the prescribed fashion.

The legal values for unit IDs of PCI devices are between 0000H and *FFFE*H, with the value *FFFF*H indicating that a PCI device is not installed. Therefore, the start-up routine of the PCI BIOS can identify all installed PCI devices. This situation is similar to the way SCSI device IDs are assigned and recognized. The first 16 bytes of the CAS header (00H to 0*F*H) are the same for all PCI units, with the remaining 48 bytes of the header varying according to individual devices. The layout is differentiated by the header entry (offset 0*E*H). The PCI specification currently defines only a single header type (00H). The most significant header bit, 7, indicates if the unit is multifunctional (bit 7 = 1) or single function (bit 7 = 0). The class code identifies the type of PCI device, for example, a network adapter, multimedia unit, or video controller. The class code is divided into three 1 byte sections. The most significant byte (offset 0*B*H) indicates the basic class code, the middle byte (offset 0*A*H) indicates the subclass code, and the least significant byte (offset 09H) provides a programming interface for the unit.

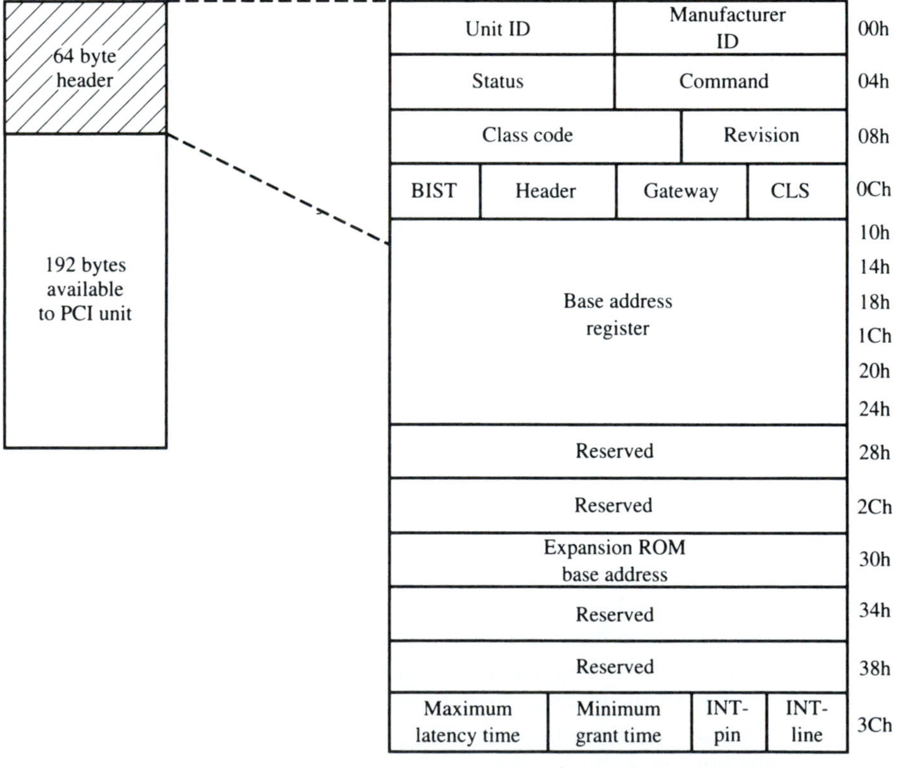

FIGURE 12.47
PCI configuration address space

PCI Bus Setup

PCI is as close to *plug and play* as you can get. PCI bus motherboards contain a PCI BIOS addition for automatic configuration. The BIOS ROM is usually located at *E*000:0000, so beware when setting up high-memory managers. Most motherboards allow the PCI bus to be configured as shown in Figure 12.48. In the figure, all four PCI slots can be set up to assign the IRQ lines automatically or they may be set to any open IRQ. That's about all there is to setting up PCI cards, and using the default configuration is usually all that is required.

The PCMCIA Bus

The explosion in the use of portable computing has fueled the advances in PCMCIA technology. **The Personal Computer Memory Card International Association (PCMCIA)** was established in 1989 to formulate a set of standards for credit card size memory modules. In 1990 PCMCIA specification release 1.0 designated standards for Type I PCMCIA memory cards. The Type I specification for sockets is 3.3 mm high with a 68-pin pinout. In September 1991, the specification was revised to include standards for I/O cards, modems, LAN cards, mass storage, and other peripherals. Release 2.0 included a specification for 5-mm Type II cards. The latest release, 2.01, addresses certain problems and includes a specification for the Type III 10.5-mm card, which is used mostly for hard drives. PCMCIA considered and rejected a Type IV 16-mm slot; however, Toshiba adopted it for their T4500 notebook. Manufacturers were hoping to take advantage of the larger slot size to accommodate an RJ-11 jack to hook up a modem, rather than using a *dongle* to attach them. Also, hard drive manufacturers wanted the taller slot to allow for larger hard drive capacities. In December 1991, PCMCIA signed an agreement with the Japan Electronic Industry Development Association (JEIDA), and release 2.0 was adopted as the JEIDA 4.1 specification, thereby creating a global standard.

Progress has been made in the crucial area of compatibility, but PCMCIA manufacturers have given a lot of hype to the plug and play aspect of PCMCIA without reaching the promised land! In fact, the entire concept of plug and play in the computer industry may really mean plug it in and then play with it for a while to get it to work! There are still software drivers to be installed, and the sequence in which you do this may be important. Also, the application software that you intend to use may not work with PCMCIA hardware. Card and socket services software can be different, although a standard is supposed to eliminate these problems.

Unlike ISA, EISA, microchannel, PCI, and local bus, PCMCIA connectors do not connect to one of the system board buses directly. PCMCIA resembles more closely a network stack with a hierarchical structure, as shown in Figure 12.49 (p. 487). The PCMCIA bus is connected to the system bus through hardware known as the PCMCIA controller or chip. To date there are a half dozen or so PCMCIA controllers on the market, including ISA boards providing PCMCIA services to a desktop computer. Since the PCMCIA interface is a standard unto itself, it becomes platform independent. By using a host-specific PCMCIA controller, you can use available PCMCIA cards on any type of computer, including Apple Macintosh and Sun Microsystems. It is also possible to have more than one PCMCIA controller per system.

```
                    ROM PCI/ISA BIOS (2A5IAE12)
                     PCI & ONBOARD I/O SETUP
                      AWARD SOFTWARE, INC.

Slot 1 Using INT#    : AUTO    Onboard FDC Controller    : Enabled
Slot 2 Using INT#    : AUTO    Onboard Serial Port 1     : COM1
Slot 3 Using INT#    : AUTO    Onboard Serial Port 2     : COM2
Slot 4 Using INT#    : AUTO    COM3 & COM4 Address       : 338H,238H
                               Onboard Parallel Port     : 278H
1st Available IRQ    : 10      Parallel Port Mode        : Normal
2nd Available IRQ    : 11
3rd Available IRQ    : 9       IDE HDD Block Mode        : Enabled
4th Available IRQ    : 5       IDE 32-bit Transfer Mode  : Enabled
                               Onboard CMD IDE Mode 3    : Enabled
PCI IDE IRQ Map To   : ISA

                               ESC : Quit         ↑↓→ : Select Item
                               F1  : Help         PU/PD/+/- : Modify
                               F5  : Old Values   (Shift)F2 : Color
                               F6  : Load BIOS Defaults
                               F7  : Load Setup Defaults
```

FIGURE 12.48
PCI bus BIOS configuration

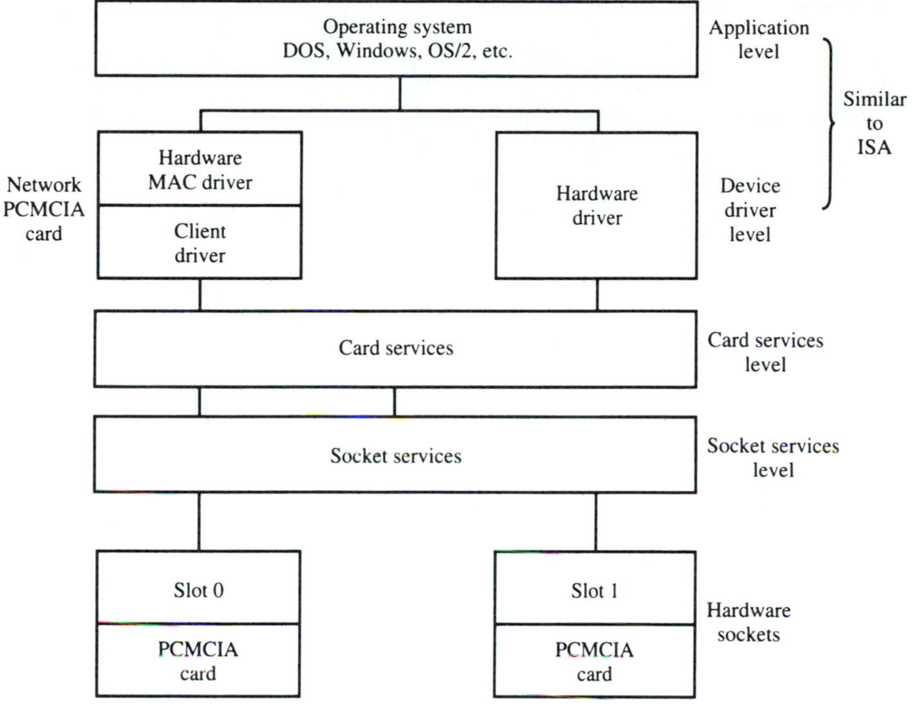

FIGURE 12.49
PCMCIA interface stack

The current PCMCIA standard allows the same types of devices found on traditional PC buses and includes I/O devices such as serial ports and modems, ATA (IDE) devices, SCSI controllers, and memory cards like those that are found in a traditional PC. A PCMCIA memory device is allowed a maximum of 64 MB of common memory and an additional 64 MB of attribute memory (for a total of 128 MB of memory) compared to the maximum of 16 MB on the 24-bit ISA bus. One of the advantages of PCMCIA is that the cards can be removed and inserted while the system is powered up and running. Don't try that on an internal system bus. Obviously the system should not be accessing the device while a PCMCIA device is removed.

The PCMCIA bus connection is somewhat similar to the ISA bus because it is a 16-bit, parallel, bidirectional bus, as shown in Figure 12.50; however, it differs in the control lines and allows for 26 bit addresses. There are no DRQ or IRQ signals, but these lines are generated by the PCMCIA controller back to the host bus. Table 12.13 lists the PCMCIA signal name, pinout, and associated function for memory cards only. Table 12.14 (p. 490) illustrates the I/O card changes.

The PCMCIA sockets, or slots, are controlled by a chip called a host adapter. The host adapter responds to requests from the PC card, handles interrupts, and sometimes provides power management. In addition, the host adapter may contain address buffers to handle hot swapping. Card and socket services (CSS) software provides the interface be-

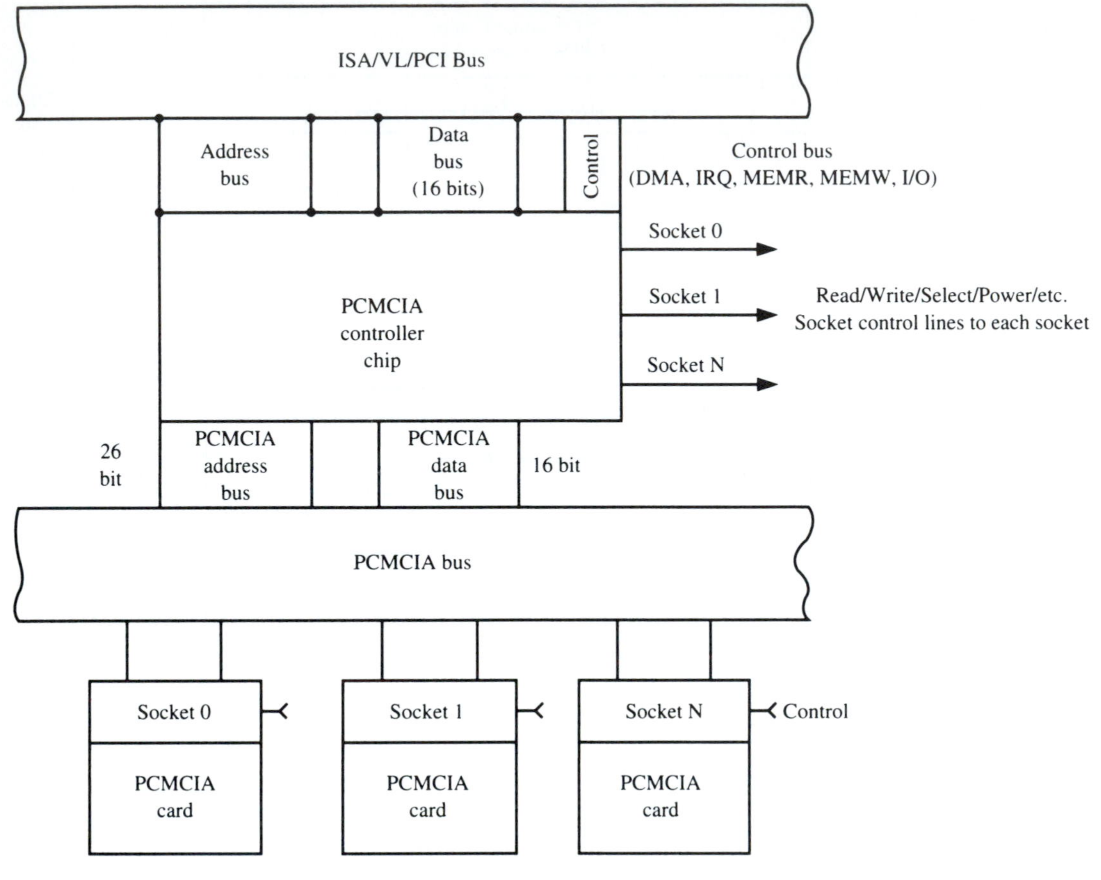

FIGURE 12.50
PCMCIA hardware

tween the host adapter chip and the drivers for PC cards. Socket services are BIOS level software that interface directly with the host adapter and with card services. Card services manages the communication between the driver (the client) and socket services. Card and socket services work jointly to identify the number of PCMCIA sockets in the computer, assign system resources, manage communications, and detect removal and insertion of PC cards. Some computer manufacturers will provide CSS software that is installed as a BIOS extension. Card and socket services may also come in the form of device drivers that are loaded through the CONFIG.SYS file. Today most if not all new notebook computers come with the CSS software. Most PC cards today also come with CSS software as well. If your computer does not come with CSS software, you will have to install a third-party CSS product such as Card Soft by System Soft.

TABLE 12.13
PCMCIA connector pinout (memory card only)

Pin	Signal	Function	Pin	Signal	Function
1	GND	Ground	35	GND	Ground
2	D_3	Data bit 3	36	CD_1	Card detect
3	D_4	Data bit 4	37	D_{11}	Data bit 11
4	D_5	Data bit 5	38	D_{12}	Data bit 12
5	D_6	Data bit 6	39	D_{13}	Data bit 13
6	D_7	Data bit 7	40	D_{14}	Data bit 14
7	CE_1	Card enable	41	D_{15}	Data bit 15
8	A_{10}	Address bit 10	42	CE_2	Card enable
9	OE	Output enable	43	RFSH	Refresh
10	A_{11}	Address bit 11	44	RFU	Reserved
11	A_9	Address bit 9	45	RFU	Reserved
12	A_8	Address bit 8	46	A_{17}	Address bit 17
13	A_{13}	Address bit 13	47	A_{18}	Address bit 18
14	A_{14}	Address bit 14	48	A_{19}	Address bit 19
15	WE/PGM	Write enable	49	A_{20}	Address bit 20
16	RDY/BSY	Ready/busy	50	A_{21}	Address bit 21
17	$\mathrm{V_{CC}}$	Power	51	V_{CC}	Power
18	Vpp1	Programming supply voltage 1	52	V_{pp2}	Programming supply voltage 2
19	A_{16}	Address bit 16	53	A_{22}	Address bit 22
20	A_{15}	Address bit 15	54	A_{23}	Address bit 23
21	A_{12}	Address bit 12	55	A_{24}	Address bit 24
22	A_7	Address bit 7	56	A_{25}	Address bit 25
23	A_6	Address bit 6	57	RFU	Reserved
24	A_5	Address bit 5	58	RESET	Card reset
25	A_4	Address bit 4	59	WAIT	Extend bus cycle
26	A_3	Address bit 3	60	RFU	Reserved
27	A_2	Address bit 2	61	REG	Register select
28	A_1	Address bit 1	62	BVD_2	Battery voltage detect 2
29	A_0	Address bit 0	63	BVD_1	Battery voltage detect 1
30	D_0	Data bit 0	64	D_8	Data bit 8
31	D_1	Data bit 1	65	D_9	Data bit 9
32	D_2	Data bit 2	66	D_{10}	Data bit 10
33	WP	Write protect	67	CD_2	Card detect
34	GND	Ground	68	GND	Ground

12.10 TECH TIPS AND TROUBLESHOOTING—T^3

One of the most common failures of a microcomputer system is the power supply. Since the power supply is connected to the common utility power, it is actually in parallel with the inductive loads of appliance motors and other unforeseen transients; therefore, it is subject to an unstable environment. Furthermore, power supplies are usually designed to

TABLE 12.14
PCMCIA connector pinout changes for I/O Cards*

Pin	Signal	Function
16	IREQ	Interrupt request
33	IOIS16	I/O port is 16 bit
44	IORD	I/O read
45	IOWR	I/O write
60	INPACK	Input port acknowledge
62	SPKR	Audio digital waveform
63	STSCHG	Card status changed

*All pins are the same with the exception of pins 16, 33, 44, 45, 60, 62, and 63.

run at their rated maximum power output in order to keep their physical size and weight to a minimum. Thus power supplies naturally tend to run warm. While a warm operating temperature is normal and poses no danger to the unit, excessive heat buildup due to overloading or poor cooling or ventilation will cause electronic components to fail prematurely.

The problem of power line transients is easily solved by using commercial **line conditioners** or **surge protectors.** These devices are inexpensive and can protect the entire system. Heat buildup problems can usually be solved by using good common sense. Be sure that the air flow passages are clean and not blocked. If an air filter exists, replace it periodically. Consider the location of the system. Try not to place it near a hot appliance such as an oven or near a radiator.

If a power supply failure is suspected, consider the following steps:

1. Check the condition of the line cord. Often the cord breaks from flexing at the ends.
2. Check the fuse if one exists. Sometimes fuses are hidden internally in power supplies.
3. Check the condition of the output connectors. Excessive humidity will cause corrosion to build up over time, causing an open circuit. This is a common problem and is solved by cleaning the connectors. Simply removing and reconnecting the connector a few times will clean it. A soft pencil eraser will also do a good job on edge connectors. Be careful not to rub too hard, because it is possible to remove some of the gold plating.
4. Measure the output voltages. Remember that an overload or short on the system board will cause the power supply to shut down. This is known as **foldback** current limiting.
5. Check the output for ripple. Remember that a VOM will average excessive ripple, so an oscilliscope must be used.
6. When bench testing switching power supplies, be aware that some load is required, or no output will be available. A 5-Ω load across the +5-V output is usually sufficient.

EXERCISES

12.1 What types of power supplies are found in microcomputers?

12.2 Draw the block diagram of a linear power supply.

12.3 Draw the block diagram of a switching power supply.

12.4 Compare the linear and switching power supplies.

12.5 Define BIOS.

12.6 What does the BIOS do?

12.7 What is DMA used for?

12.8 Compare an interrupt-driven system with a polled system.

12.9 How many bits are there on the data bus for the 80486, 80386, 80286, and 8088 microprocessors?

12.10 Identify the four basic sections of the 80286 microprocessor.

12.11 Identify the three memory modes of the 80386 microprocessor.

12.12 What is the ISA bus and where is it found?

12.13 Will an 8-bit ISA video card work in a 16-bit slot?

12.14 How many IRQ lines are there in an ISA bus AT class computer system?

12.15 Which IRQ line is used for the hard disk in the ISA bus AT class computer system?

12.16 What happened to the IRQ2 line in the ISA bus AT class computer?

12.17 Where is the setup information saved on an AT class computer system?

12.18 Why isn't the setup information lost when the system power is turned off?

12.19 Name the different bus technologies used on personal computers.

12.20 What is meant by the term *card and socket services*?

12.21 What Pentium signal indicates that the data on the data bus is ready to be read?

12.22 How is the FSB speed related to the MPU speed?

12.23 What does the chipset do?

12.24 What is the difference between L1 and L2 cache?

12.25 What is the purpose of the AGP connector?

12.26 Why is DRAM not used for the cache?

12.27 Before picking up a memory module, what should a technician do?

12.28 What is the purpose of the IRQs?

12.29 What is the DMA setting for a sound card?

12.30 Where is PCMICA used?

12.31 Crossword Puzzle

ACROSS

4. An I/O device.
5. DMA is faster and more efficient than _____ I/O system.
8. The system unit is also known as _____.
10. One additional hardware interrupt that is not for general I/O use.
11. _____ unit allows the user to add more peripherals than the system unit can handle.
14. Group of wires.
15. Similar to a floppy disk, but ten times faster.
17. Disadvantage of linear power supply.
20. Signal applied to the input RDY1 on the 8284A chip in the PC.
21. Dynamic RAM requirement.
24. _____ checking is a means of checking memory for correct data.
26. The 8088 in the IBM PC is in the _____ mode.
27. The peripheral clock signal.
30. Some interrupt systems make use of series of _____ chained interrupt request lines.
31. 64K of memory.

DOWN

1. The big disadvantage of a switching power supply is that it is tricky to _____.
2. Column address strobe NOT, active low.
3. Input device.
4. The main RAM in the PC uses _____ RAM chips.
6. The MFM hard disk uses two flat _____ cables.
7. _____ bus links the MPU, coprocessor, interrupt controller and the bus controller.
9. A type of power supply.
12. IBM PC model.
13. Interrupt-driven system improves this method.
16. 8087 coprocessor function.
18. Memory chips, memory bus, memory parity, and refresh circuitry are all part of the _____.
19. Row-address strobe NOT, active low.
22. Most PCs today use a _____ regulator power supply.
23. System clocks, wait circuitry, and master system reset are all part of the main system _____.
25. The 8237 chip, page RAM, and control logic are part of the _____ controller.
28. Connectors are _____ to prevent being reversed.
29. Type of parity.

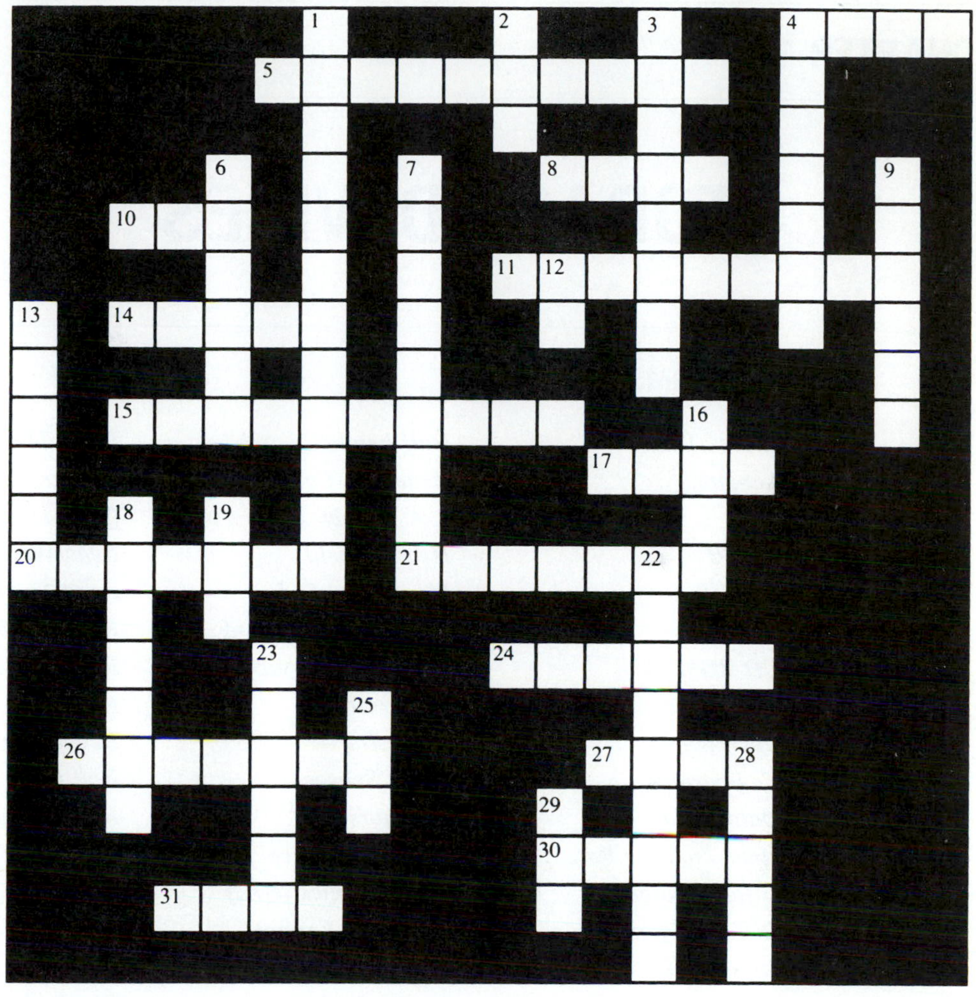
1
2
3
4
5
6
7
8
9
10
11
12
13
14
15
16
17
18
19
20
21
22
23
24
25
26
27
28
29
30
31

CHAPTER 13

STORAGE DEVICES

KEY TERMS

Access Time
Actuator
ATA
ATAPI
Bit Cell
CDR
CD-ROM
CDRW
Coercivity
Comparator
Data Field
Data Transfer Rate
Differentiator
Differentiator Droop
Directory
Door Open Sensor
Double Density
DVD
Encoding
Enhanced Small Device Interface (ESDI)
External Storage
Fields
File Allocation Table
Floppy Disk
Formatting
Frequency Modulation
Frequency Shift Keying
Full-height
Half-height
Head and Carriage Assembly
Header Field
Head Load Pad
Head Load Solenoid
High-level Format
Hysteresis
Index Burst
Index Sensor
Initiators
Integrated Drive Electronics (IDE)
Low-level Format
LS-120
Magnetized
Magnetic Floppy Disks
Magnetic Flux Reversal
Magnetic Recording Tape
Master
Modified Frequency Modulation (MFM)
Overlay
Partition
Permeable
Phase Distortion
Phase-locked Loop
Postamble Field
Preamble Field
Primary Channel
Quad Density
Read/Write Head
Read Amplifier
Read Chain
Recalibrate
Recording Head
Retentivity
Roll Off
Run Length Limited
Runout
Secondary Channel
Sector
Seek Time
Single-density Recording
Slave
SCSI
Sync Bytes
Targets
Tone Encoding
Track
Track 00 Sensor
Wobble
Write-protect Notch
Write-protect Sensor
Zip Drive

13.0 INTRODUCTION

One of the most important components of any computer system is **external storage.** An external storage device is required for initial program loading and for mass data storage and retrieval. In the early days of computing, programs were loaded from paper tapes. Paper tapes are long strips of paper in which holes are punched to represent the 1 state, as shown in Figure 13.1. The paper tapes were punched and then read by a *teletype* reader/printer device. The teletype also served as the local console terminal. One major drawback of paper tape is that it is somewhat clumsy to handle and represents a low-den-

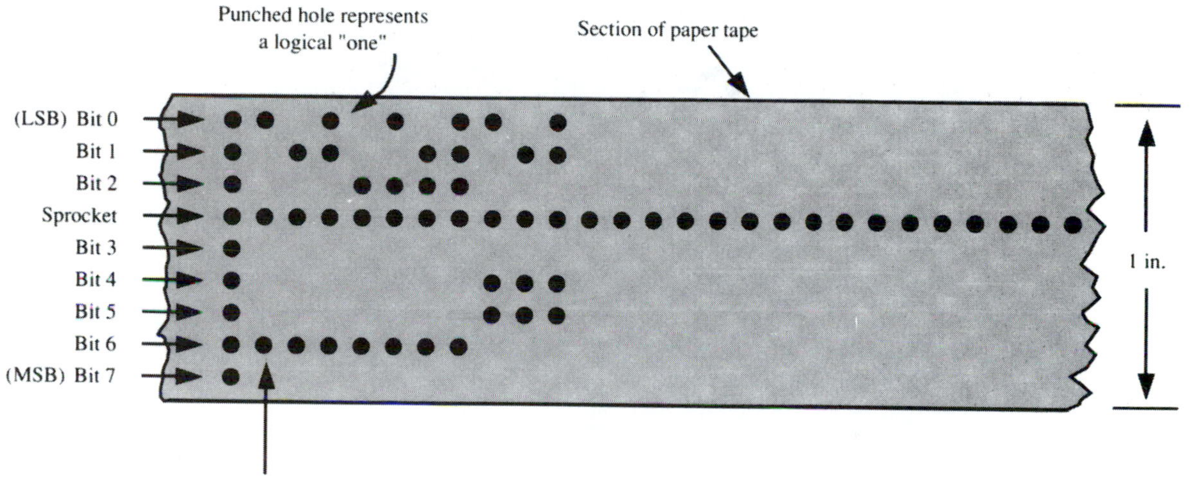

ASCII character

A	B	C	D	E	F	G	1	2	3
Binary code									
1	0	1	0	1	0	1	1	0	1
0	1	1	0	0	1	1	0	1	1
0	0	0	1	1	1	1	0	0	0
0	0	0	0	0	0	0	0	0	0
0	0	0	0	0	0	0	1	1	1
0	0	0	0	0	0	0	1	1	1
1	1	1	1	1	1	1	0	0	0
0	0	0	0	0	0	0	0	0	0

Note: On some paper tape readers, the sprocket hole is used for the data strobe signal.

FIGURE 13.1
Eight-level paper tape

sity storage. That is, the number of 1s and 0s or data per inch, is relatively small. Another disadvantage is that once the paper tape is punched, it cannot be reused. If an error is made, the paper tape must be discarded.

Some earlier computer system used *punched cards,* which are often referred to as IBM cards, instead of paper tapes for their mass storage. Punched cards are similar to paper tape in that a punched hole represents a 1 state, as shown in Figure 13.2.

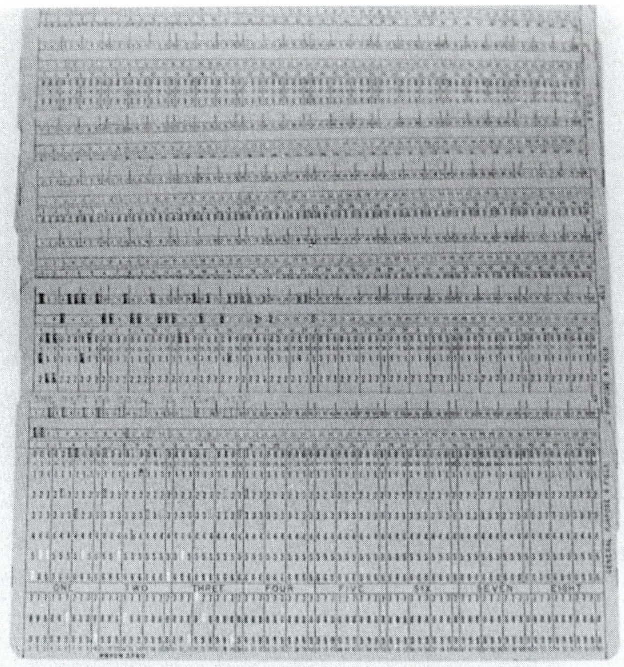

FIGURE 13.2
Punched card

As an alternative to the paper tape and punched cards, **magnetic recording tape,** which was already being used by the audio industry, was applied to recording computer data. Instead of a hole punched to represent a 1 state, a **magnetic flux reversal** is recorded on the tape, as shown in Figure 13.3. This technology was followed by *magnetic drum memories, magnetic disk memories,* and **magnetic floppy disks.** The latest innovation is the *optical disk memory,* or *compact disk.*

13.1 MAGNETIC RECORDING FUNDAMENTALS

In magnetic recording a magnetic charge is stored on a permeable medium. This permeable medium must be a ferrous-type material that exhibits some magnetic retentivity. **Permeable** means that the applied magnetic flux is able to penetrate and flow through the medium easily, whereas **retentivity** means that the medium will hold the flux or become

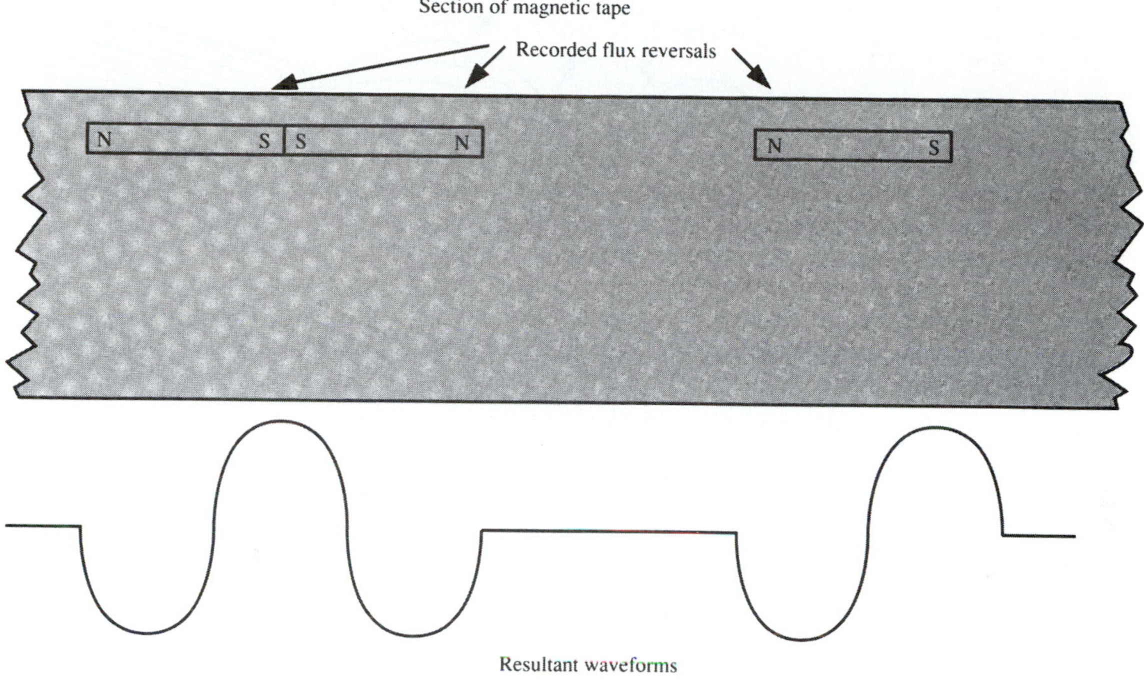

FIGURE 13.3
Recorded magnetic flux reversals

magnetized after the applied energy is removed. Different ferrous and ferrite materials exhibit varying levels of permeability and retentivity.

To demonstrate this effect, take a common household screwdriver and a permanent magnet. While holding the magnet in one hand, rap the end of the screwdriver blade onto one pole of the magnet. Repeat this several times on the same pole. The screwdriver will become magnetized. The end of the blade will be the opposite pole of the magnet you were using, as shown in Figure 13.4. The screwdriver can be neutralized by placing the blade in an alternating magnetic field such as a standard tape head demagnetizer, illustrated in Figure 13.5.

The ease with which a material can become magnetized is known as its **coercivity.** A material that is known to have a high coercivity will not require as much flux density to become magnetized as will a material of low coercivity.

One of the most commonly used materials for magnetic recording is ferric oxide. This material was chosen because it demonstrates good permeability, high coercivity, and excellent retentivity. It is also low in cost and readily available.

Magnetic Tape

One of the oldest and most common types of magnetic media is magnetic recording tape. Magnetic recording tape is defined in Webster's dictionary as "a thin plastic ribbon coated with a suspension of ferromagnetic iron oxide particles used as a storage

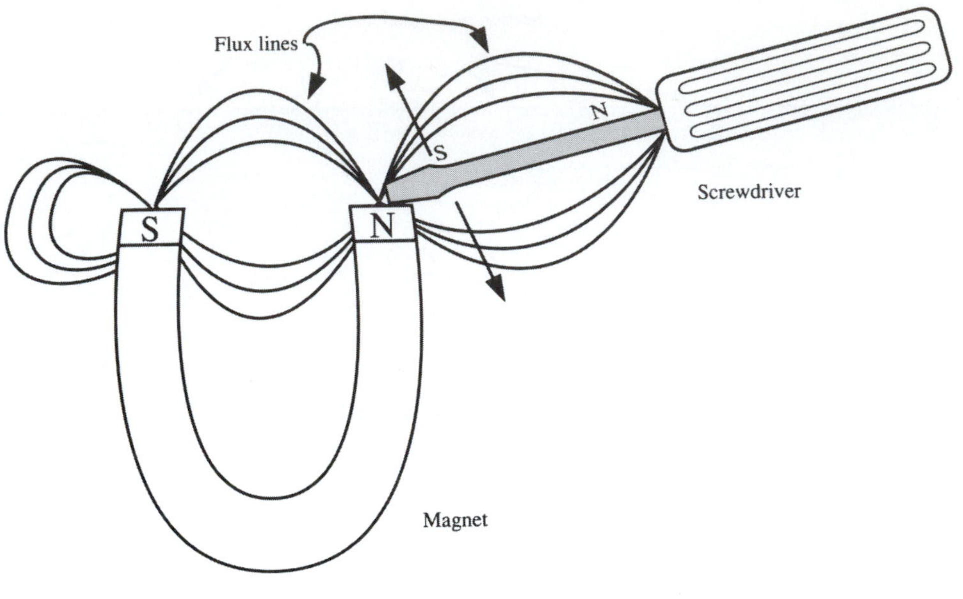

FIGURE 13.4
Magnetizing a screwdriver

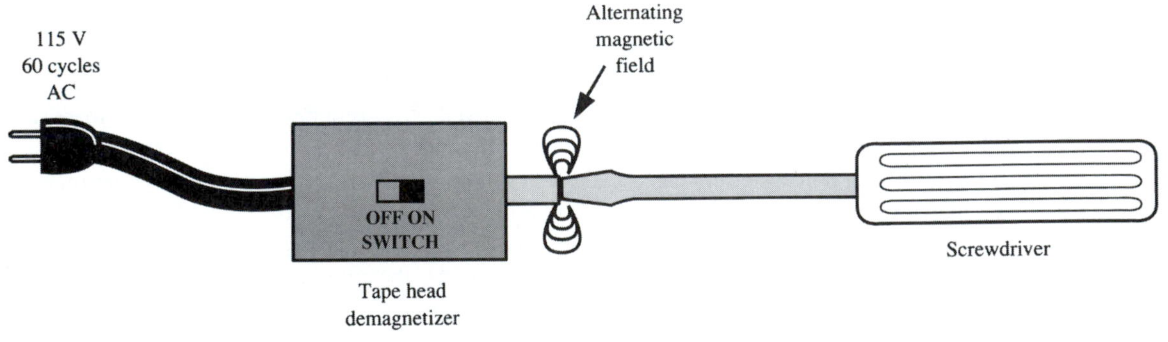

FIGURE 13.5
Erasing the field

medium for electrical signals, as from sound, video material, digital computer data, etc." The most common type of magnetic recording tape is the standard *Phillips cassette* found in almost every home. Many of the earlier hobby microcomputers and some modern low-cost home computers use this cassette as their storage medium. Figure 13.6 illustrates the construction of magnetic tape.

Information is recorded on the magnetic tape by a **recording head.** As the tape is recording, it is pulled across the head by a *tape transport* mechanism, as shown in Figure 13.7.

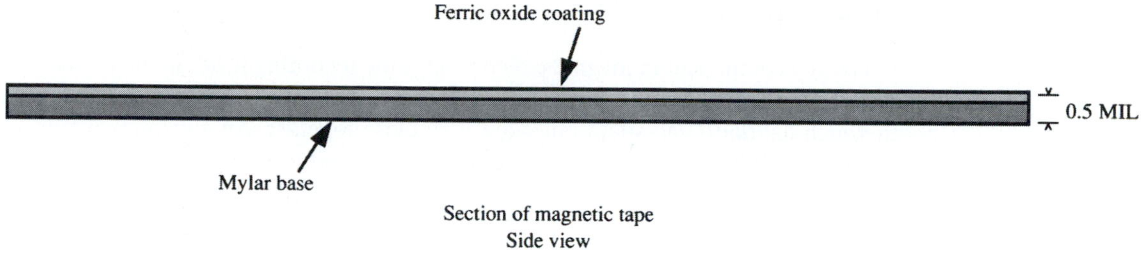

FIGURE 13.6
Magnetic tape construction

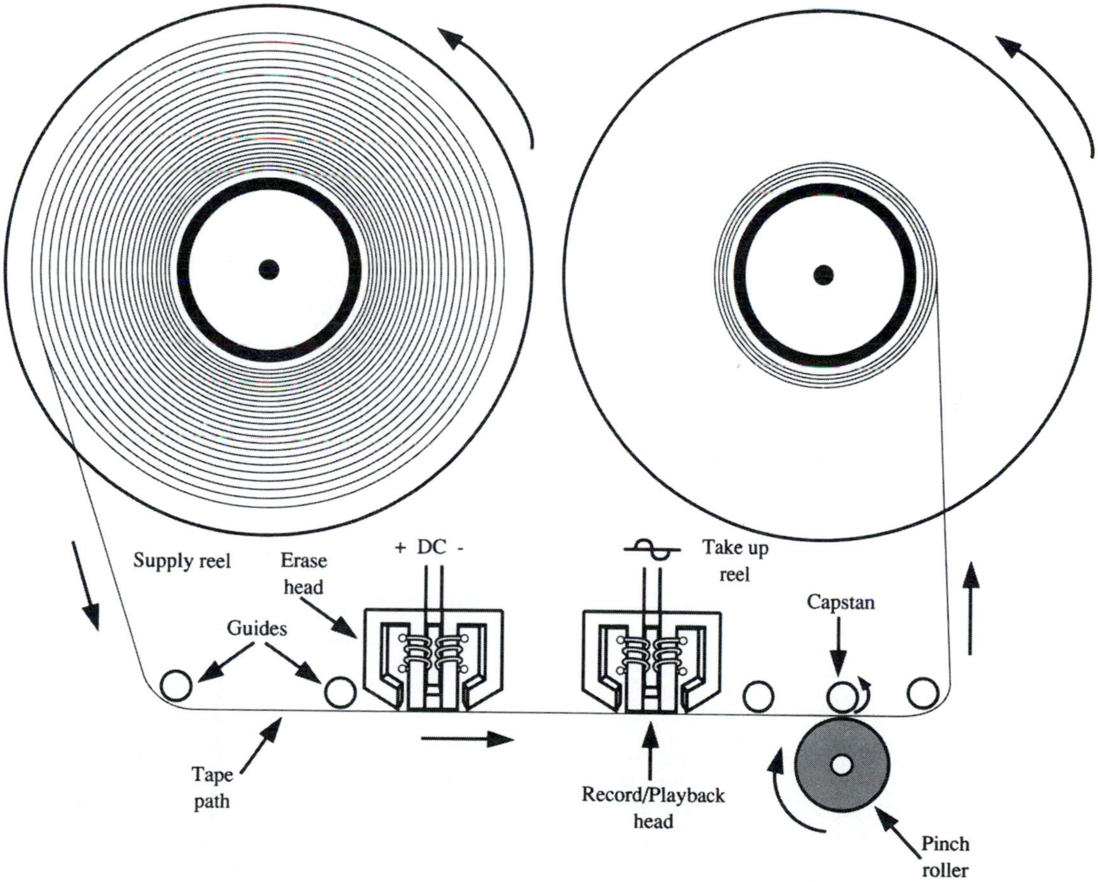

FIGURE 13.7
Tape transport

Read/Write Heads

The key component in magnetic recording is the recording head. In the digital world it is also known as the **read/write head.** The read/write head is basically an electromagnet in which the north and south poles are very close together. An electromagnet is simply a core of soft iron with a coil of wire wrapped around it. When an electric current is applied to the coil, the core becomes magnetized as shown in Figure 13.8. The core remains magnetized only as long as the electric current is applied. The magnetic field will collapse when the current is removed. In fact, if an external magnetic field is applied and

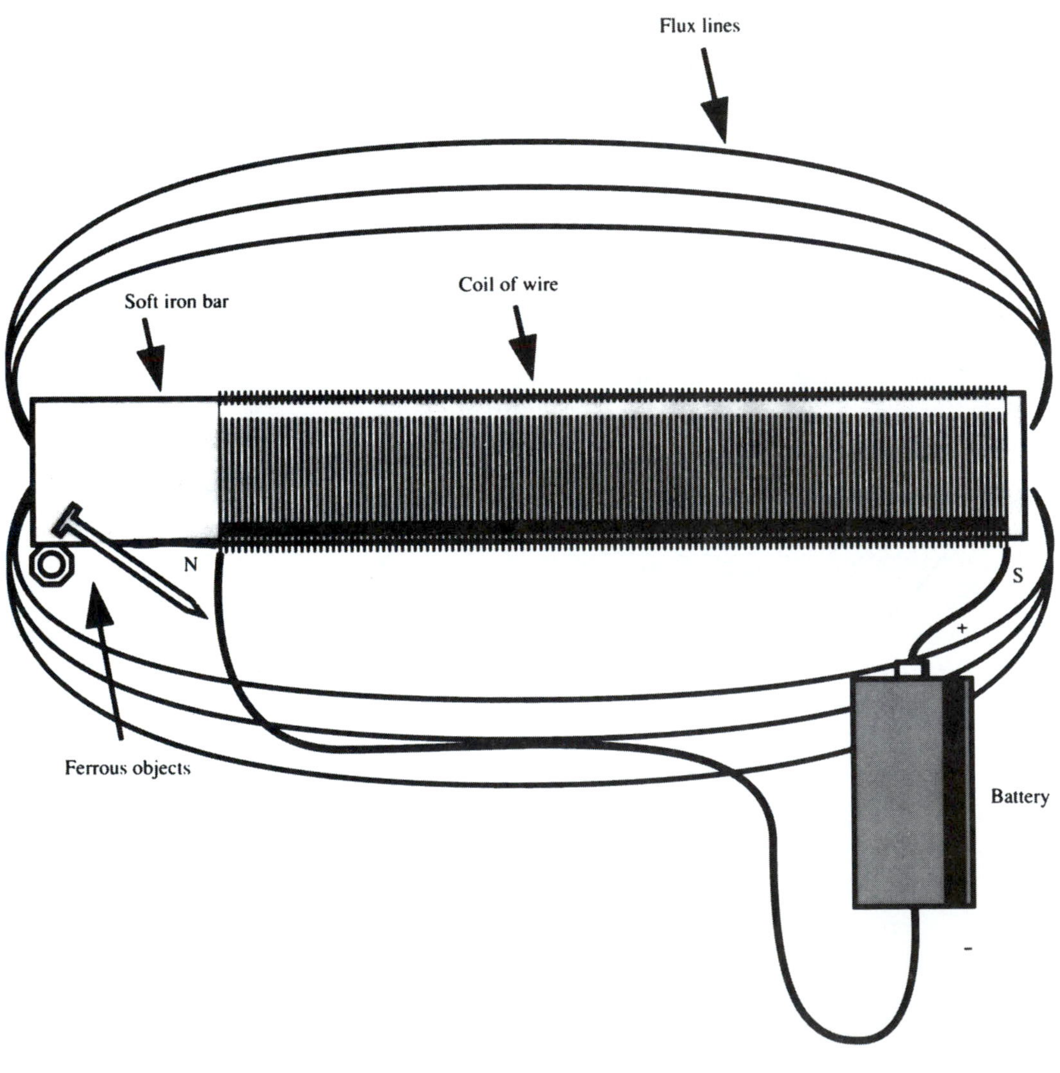

FIGURE 13.8
The electromagnet

then removed, an electrical field will be induced in the coil, thus forming a small generator. Thus the same head that is used to record the information can be used to read the information back. Recall that to generate a voltage, we need only a magnetic field, motion, and a conductor. If the core is made up of hard iron or hard steel, as was the screwdriver, some of the magnetic field remains due to retentivity.

The read/write head is simply a modified version of the electromagnet, as shown in Figure 13.9. In the magnetic recording process, the tape is pulled across the read/write

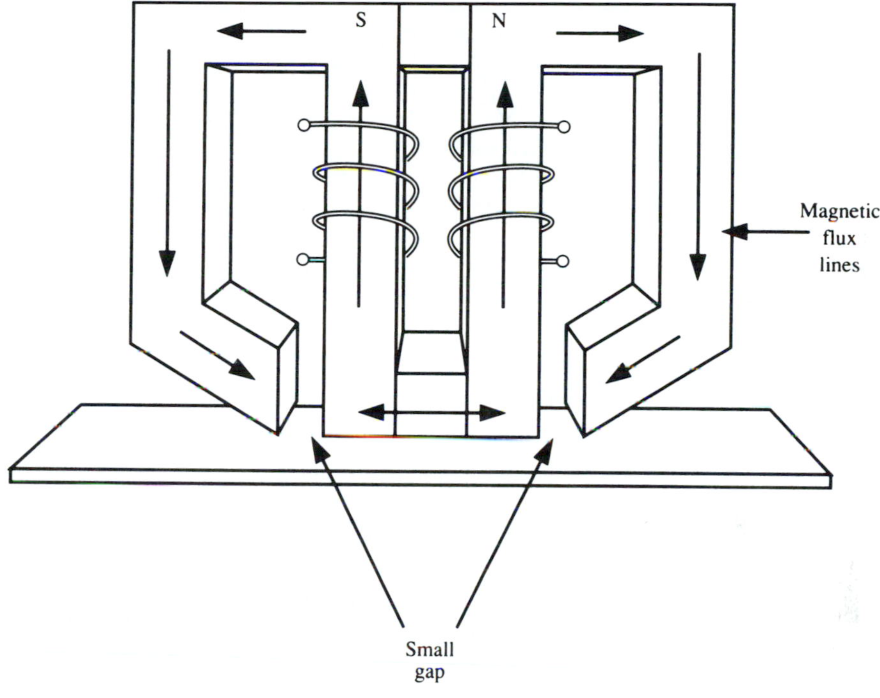

FIGURE 13.9
The read/write head

head. As the tape is moving, magnetic flux reversals can be recorded by switching an electrical current across the coil in the head, as shown in Figure 13.10). When the tape is rewound and pulled across the head again, without current applied to the coil, small voltage changes will be present at the coil when the previously recorded flux reversals pass under the gap between the poles of the head. Because these voltages are so small, they must be amplified before they can be used. This is done by electronic circuitry known as the **read amplifier.**

Read Amplifier Circuits

The read amplifier circuitry, which may also be called the **read chain,** differs greatly between analog and digital recording systems. In the analog system the head is connected to a high-gain preamplifier, which is followed by an equalizer circuit to boost the high-

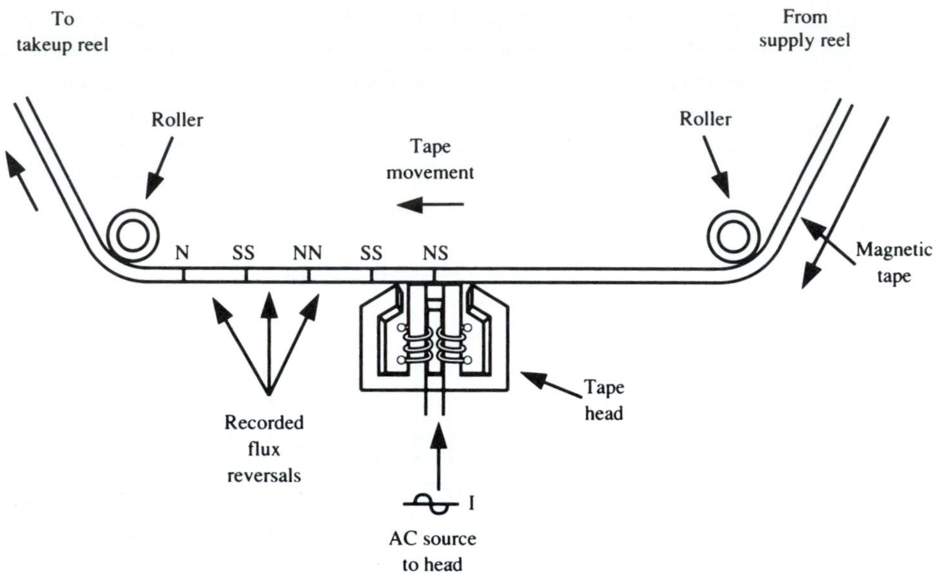

FIGURE 13.10
Magnetic tape recording

frequency and cut the low-frequency signals. This is necessary because high-frequency loss is characteristic of magnetic tape. The equalizer is followed by more stages of amplification and fed to the output of the tape deck, as shown in Figure 13.11.

When magnetic recording is used in digital applications, it is not possible to record and then play back DC levels on magnetic tape. It is not possible for the same reason that DC cannot pass through a transformer; an *alternating* field must be set up. The flux re-

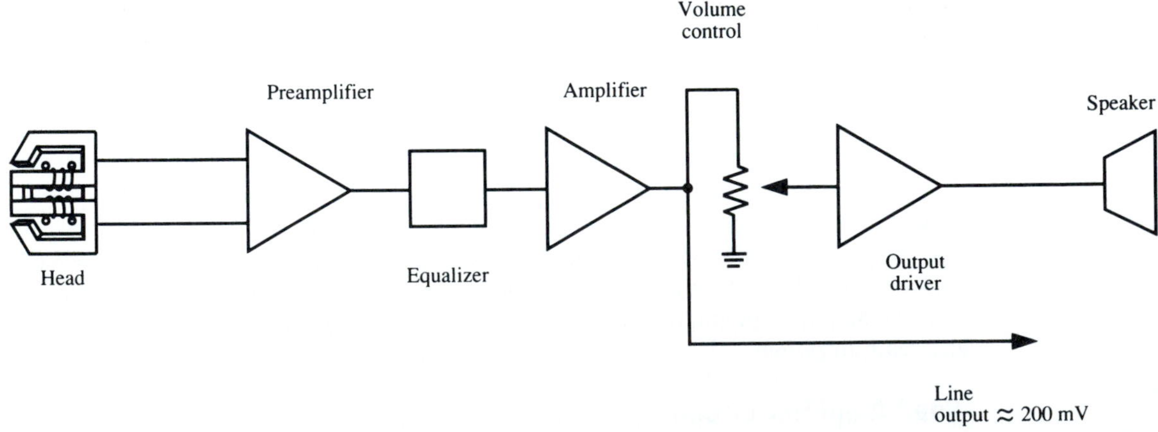

FIGURE 13.11
The analog read chain

versal is recorded on the medium. When a DC current is present on the head coil, the oxide particles are all magnetized in the same direction. This has the effect of *erasing* the tape. When the tape is rewound and played back, there is no voltage present at the head coil.

The digital read chain is designed to operate at a fixed set of frequencies; therefore, no equalization is required. A **differentiator** circuit is used instead, as shown in Figure 13.12. The differentiator is required to restore the square waveform of the original digital signal, as shown in Figure 13.13. Remember that the DC level will not record, and the playback signal will **roll off** until another flux reversal is recorded, as shown in Figure 13.14.

At high frequencies, when the flux reversals are at or near the high-frequency saturation of the medium, the roll-off is almost gone, and the playback signal resembles a sine wave. The problem of roll-off causes **phase distortion** and noise to be introduced into the playback signal. With the addition of the differentiator circuit, these problems are reduced because the circuit restores the recorded DC levels. As we can see in Figure 13.15, the output of the differentiator begins to droop as the input voltage begins to roll off. This is known as **differentiator droop.** The problem of differentiator droop is virtually eliminated with the addition of a **comparator** at the output of the differentiator, as shown in Figure 13.16. As long as the droop is not excessive and does not come close to the comparator threshold, the output will be immune to the droop and any noise that may be present on the signal.

The output of the comparator is followed by a noise-filtering circuit. Some comparators use a digital filter and a bidirectional one-shot. The one-shot triggers on both the positive and negative transitions of the recovered signal. The output is a pulse for every flux reversal read back. This output is referred to as *raw data.*

The Write Amplifier

The *write amplifier* in the digital recording system is essentially a current switch, as shown in Figure 13.17. When the write enable signal goes true, the write current is turned on to the write driver transistors. When a data bit is present at the input, the current across the head switches polarity. In digital recording, most systems drive the medium into saturation, so no erase head is required.

13.2 DIGITAL MAGNETIC RECORDING

Before going on, let's stop and recall what we have just learned about magnetics and the magnetic recording process and apply it to the digital world. Is it possible to record digital data bits directly onto the medium? Recall that it is not possible to record DC onto the medium. As long as the bits are changing states, we can record the pattern. If the data stream is 1 1 1 1 or 0 0 0 0 . . . , the tape will become DC erased where the bits did not change states. For this reason we must change the data stream into an alternating pattern. This is known as **encoding** the data, and numerous encoding methods are used today. The simplest is **tone encoding,** which is used with voice-grade equipment. The most common encoding methods used in microcomputers are frequency modulation, modified frequency modulation, group code recording, and frequency shift keying.

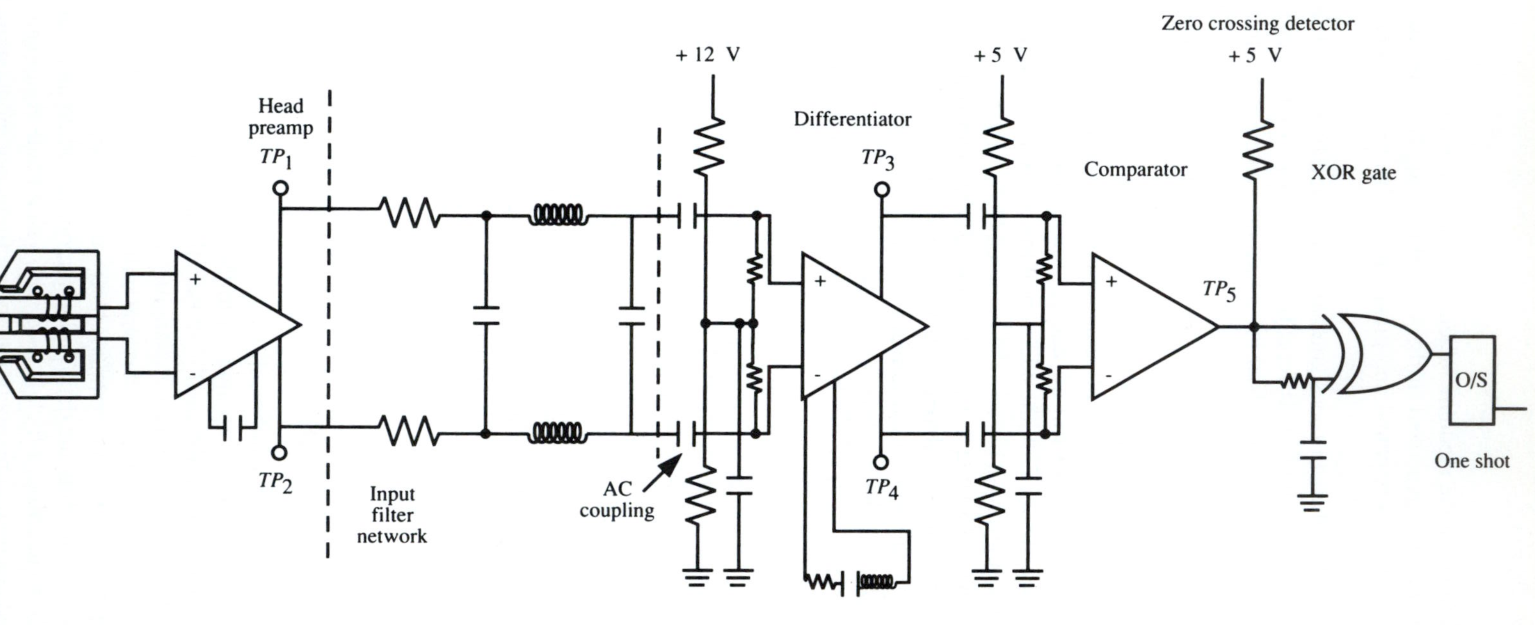

FIGURE 13.12
The digital read chain

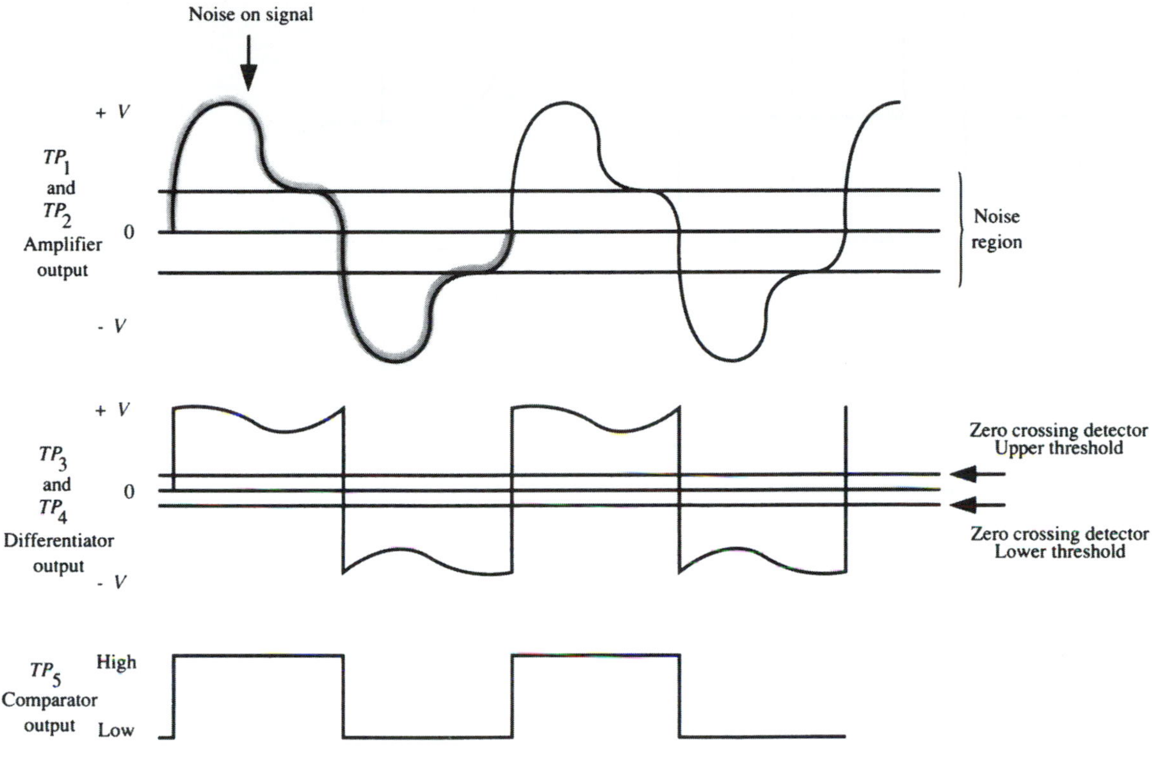

FIGURE 13.13
Read waveforms

Encoding/Decoding Techniques

Frequency shift keying (**FSK**) is used primarily for voice-grade devices such as audio cassette records. It is also used for telephone modems. This method encodes the data stream into a series of audio tones, as shown in Figure 13.18.

One of the original cassette standards was called the *Kansas City standard,* in which a 0 state was represented by a 2400-Hz tone and a 1 state was a 1200-Hz tone. This method was extremely speed- and noise-tolerant but very slow. The data-transfer rate was 300 bits per second, or 30 bytes per second. The IBM PC uses a similar method but transfers the data at about 1500 bits per second. In the IBM PC the cassette data is encoded and decoded by the software in the BIOS ROM. The hardware for the cassette interface in the IBM PC is very simple. It uses a relay to control the cassette motor and a comparator to square up the playback data.

Frequency Modulation

Frequency modulation (**FM**) was one of the earlier methods used by floppy disk systems. This method is also referred to as **single-density recording.** FM is similar to frequency shift keying in that the frequency is shifted to represent the state of the data

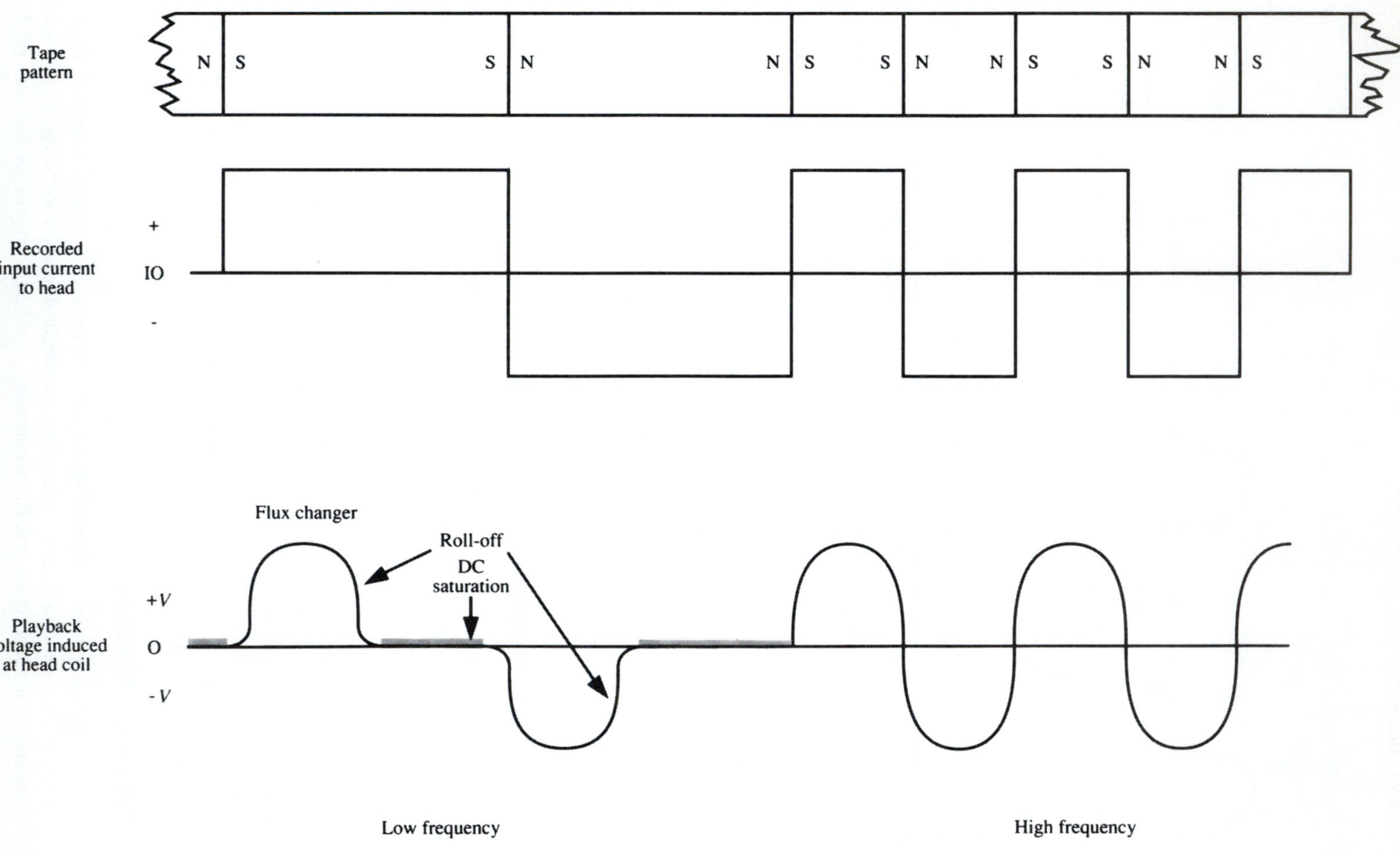

FIGURE 13.14
Recorded roll-off

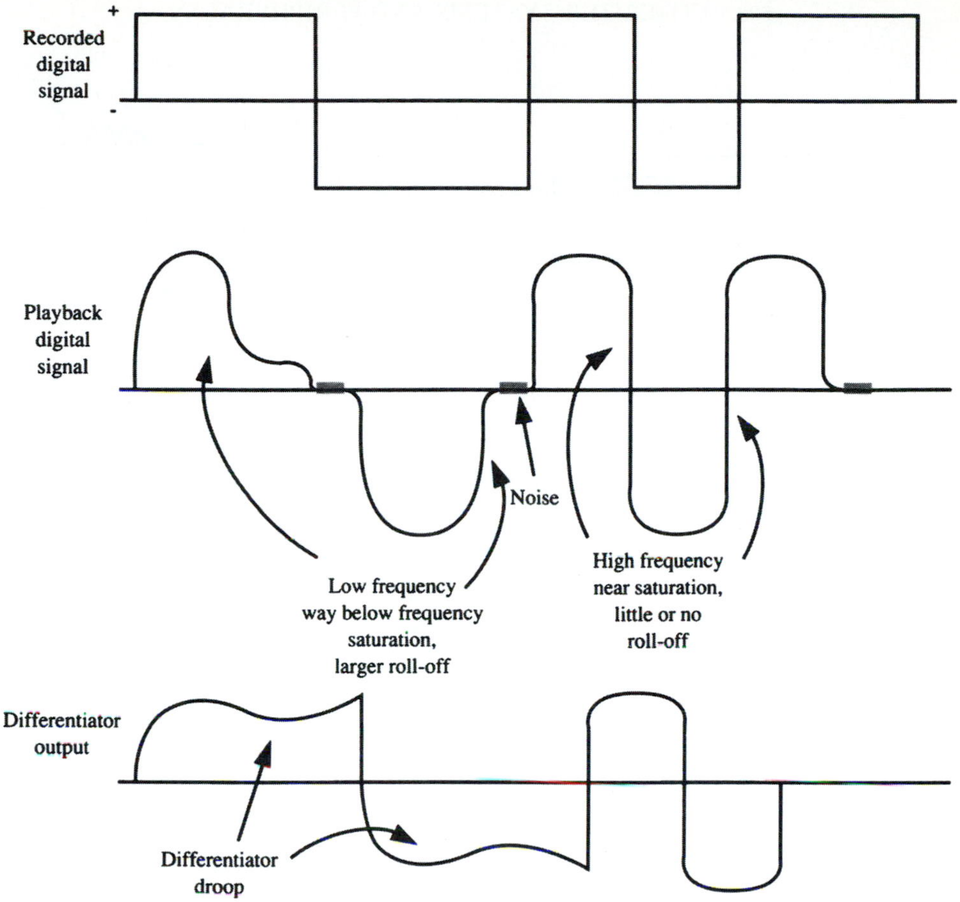

+
Recorded
digital
signal
-
Playback
digital
signal
Noise
Low frequency
way below frequency
saturation,
larger roll-off
High frequency
near saturation,
little or no
roll-off
Differentiator
output
Differentiator
droop

FIGURE 13.15
Differentiator droop

Differentiator
output
Noise
Comparator
switching
threshold
Comparator
output
0
+ V
- 0-Baseline

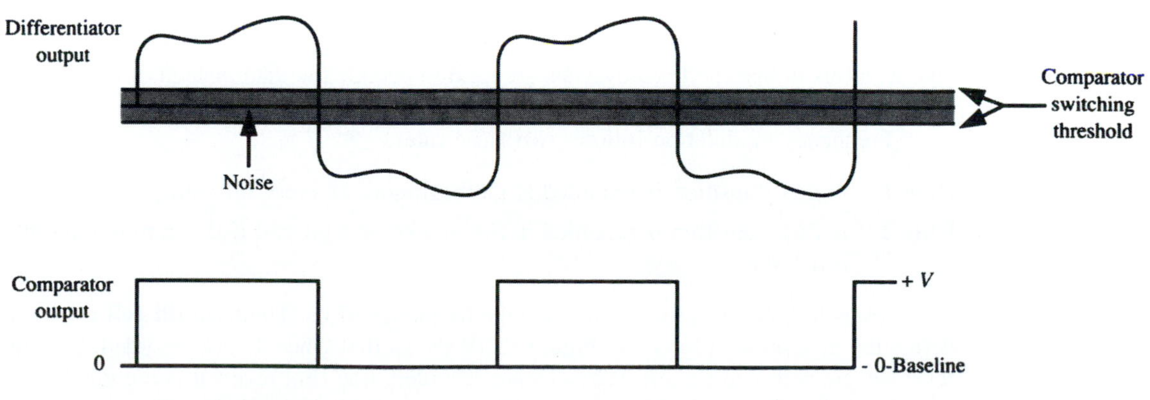

FIGURE 13.16
DC restoration

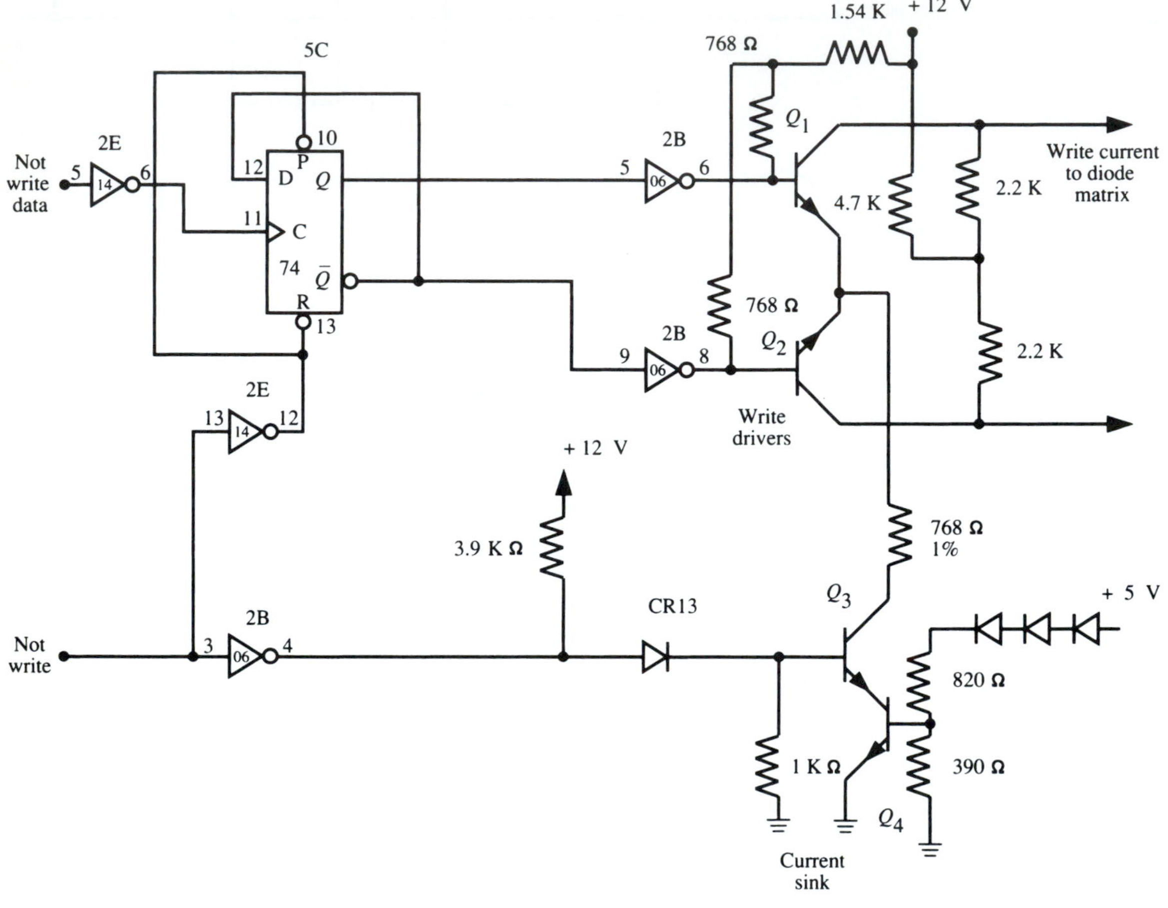

FIGURE 13.17
Write current circuit

bits. It differs in that the flux reversals are used to decode the data instead of a frequency discriminator.

Frequency modulation follows two basic rules:

Rule 1 A flux transition is recorded at the beginning of every bit cell.

Rule 2 A flux transition is recorded in the middle of a bit cell if the current data bit is a 1 state.

This means that 1s are recorded at *twice* the frequency of 0s. The term **bit cell** is used to define the time allotted for 1 bit. Figure 13.19 shows that when 1s are recorded, two flux reversals are recorded per bit. When 0s are recorded, one flux reversal is recorded.

Decoding this method is simple and requires only timing circuitry. This method can also be decoded by software timing loops if the processor is fast enough. This method of

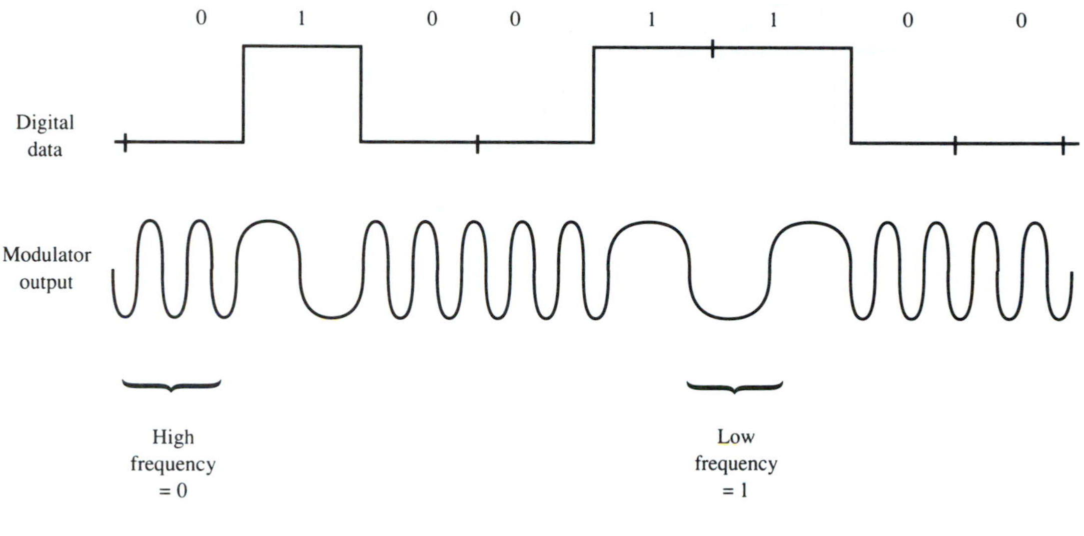

FIGURE 13.18
Frequency shift keying

encoding is known as *self-clocking,* since there is always one flux reversal per bit. Since the rules state that there must be a flux reversal at the beginning of a bit cell, we refer to the first reversal as the *clock.* If within a specified period of time another flux reversal is encountered, the bit must be a 1.

Modified Frequency Modulation

Modified frequency modulation (**MFM**) is a revised version of FM and is the most common encoding method used today. The major difference from FM encoding is that only *one* flux reversal is required for each bit cell regardless if the bit is a 1 or a 0. Recall that FM requires *two* flux reversals for a 1 state. For this reason, *twice* as many bits can be recorded using the MFM method rather than the FM method. Hence the term **double density** is used.

Modified frequency modulation follows two basic rules:

Rule 1 A flux reversal is written in the *middle* of the bit cell if the current bit is at a logic 1 state.

Rule 2 A flux reversal is recorded at the *beginning* of the bit cell if the current bit *and* the proceeding bits are at a logic 0 state.

There is one important factor to consider when using MFM. If the preceding bit was at a logic 1 state *and* the current bit is at a logic 0 state, *no* flux reversal is recorded for this bit cell. For this reason, the MFM method *is not* self-clocking; therefore, an external clock generator circuit must be used. The external clock generator circuit usually contains

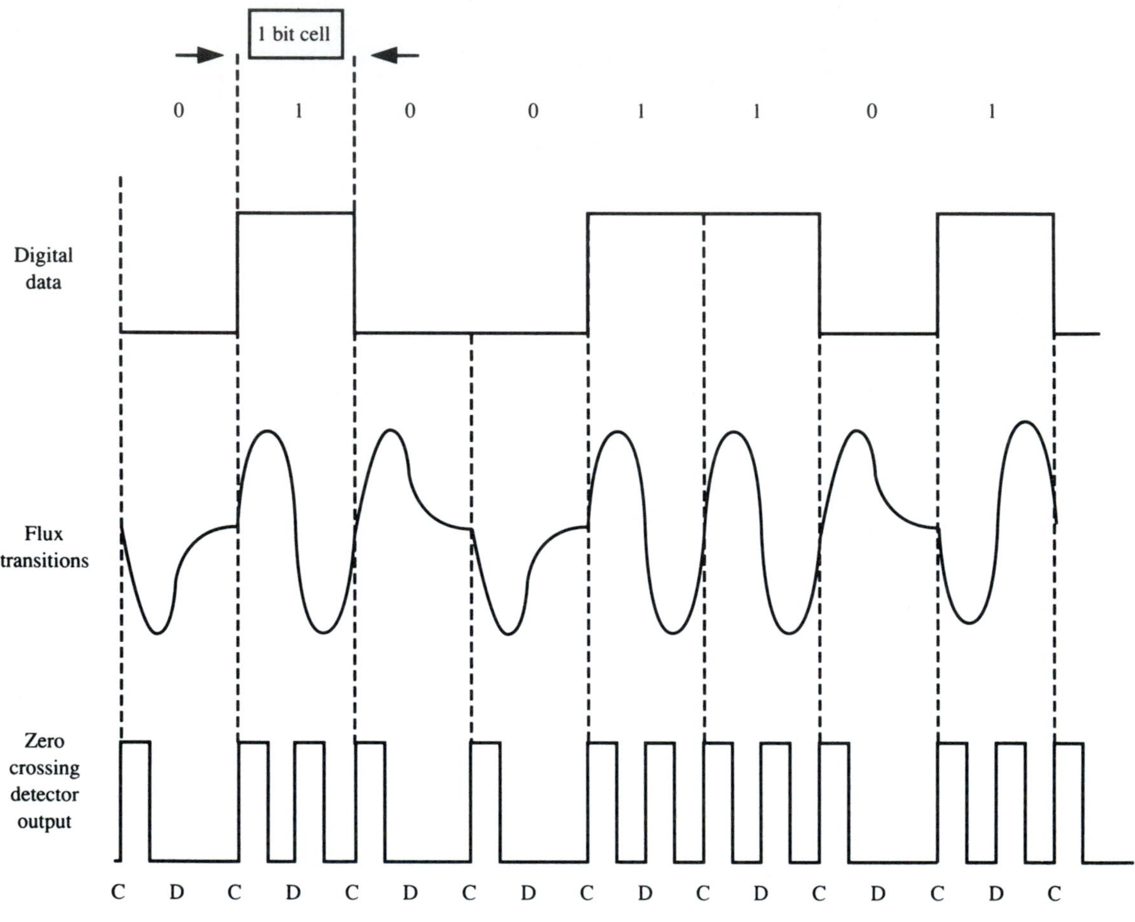

FIGURE 13.19
Frequency modulation (FM) encoding/decoding

a **phase-locked loop** that is synchronized during a steady data stream of all 0s. MFM is illustrated in Figure 13.20.

13.3 THE FLOPPY DISK SUBSYSTEM

Since the inception of the personal computer system, the floppy drive has served as an effective form of removable storage. Over the years, it has seen many permutations, but the underlying principles behind the floppy drive have remained the same.

The floppy drive was invented in 1967 by a team at IBM, lead by Alan Shugart. They designed a flexible medium system with a protective jacket and a cloth lining. The first of the **floppy disks** were 8 in. in diameter. These never really appeared in personal systems, but it was the beginning of the removable medium we call the floppy.

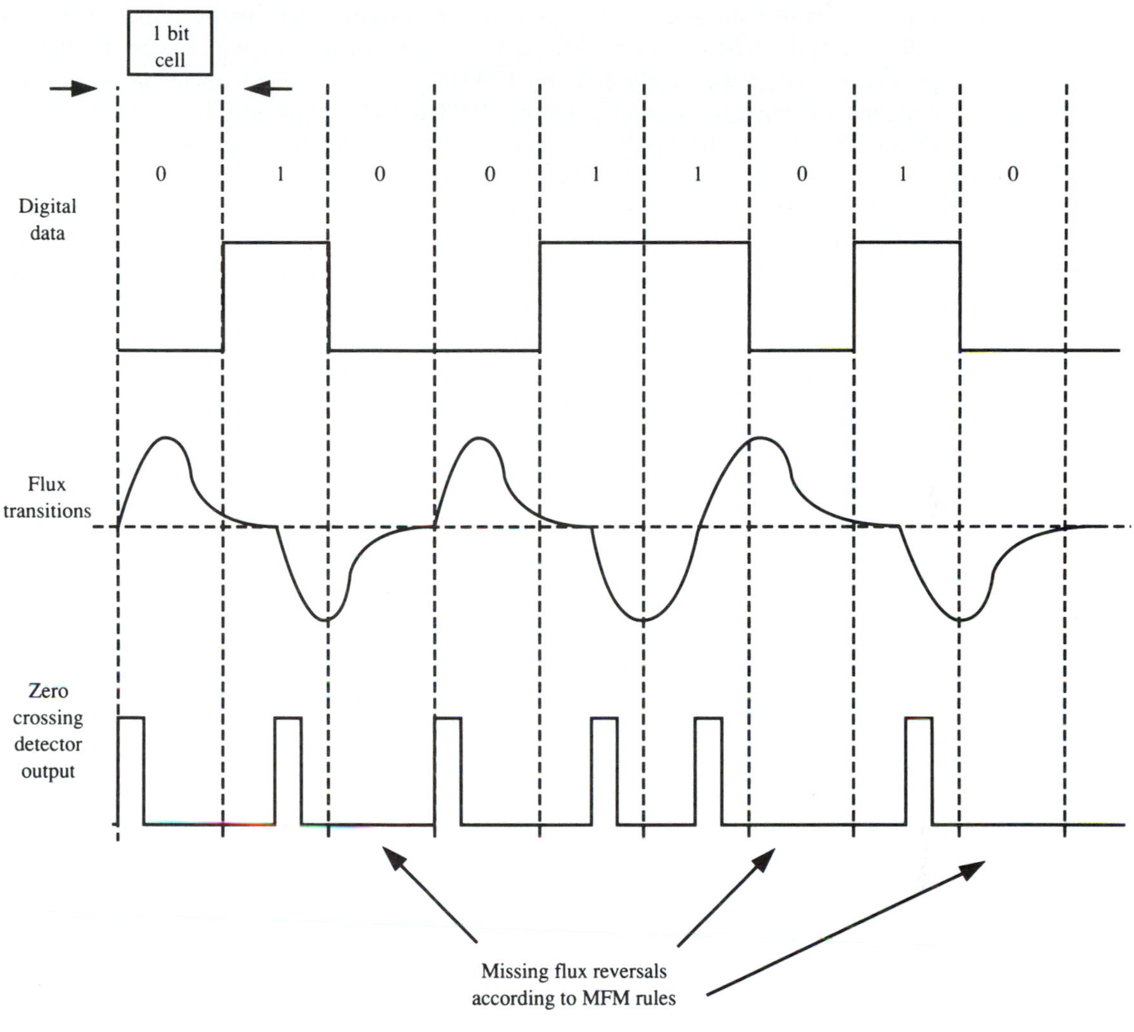

FIGURE 13.20
Modified frequency modulation

The first floppy disk drive system that became popular in personal systems was the 5.25-in. floppy. These were very similar to the 8-in. disks and drives, except that they were smaller and easier to manage. The 5.25-in. drives came in a couple of different formats. The first rendition was able to store 360 KB of information; the next and most popular format could store 1.2 MB of information. There were other, less common forms of the 5.25-in. floppy disk.

The next variety of floppy drives was the 3.5-in. disk and drive system. These were designed to be a little more rugged than the 5.25-in. and earlier floppy disks. They had a sturdy plastic shell and a metal slider that prevented the medium from being constantly exposed. These also came in a number of different formats. The first was the 720 KB disk, followed by 1.44 MB and 2.88 MB disks. The 1.44 MB variation was by far the most popular format of the floppy disk: By 1990 or so, it was commonplace in just about all computers being manufactured; and by 1995 it had completely displaced the 5.25-in. drives. Even today, the 3.5-in., 1.44 MB floppy disk and drive format is standard equipment in nearly all machines produced.

Only recently has the computer industry seen effective replacements to the floppy drive. Some are complete departures from the floppy disk system; others are backward compatible with floppy disks, meaning that they can read and write to standard floppy disks as well as their own media. The 100 MB Iomega zip drive is an example of a floppy replacement that is not compatible with the older floppy disks. Iomega departed from the floppy system in favor of their Bernoulli disk technology, which is totally incompatible. The Imation LS-120, on the other hand, can read and write to standard 1.44 MB floppies as well as its own 120 MB variety. The Imation drive has been designed to replace the 1.44 MB drive and to become the new industry standard; however, the Iomega zip drive's popularity has quickly made it the de facto standard.

Diskette Format

When a diskette is used for the first time, there is no information on it. The first thing we must do is to organize the diskette so we will know where to store data. We must also keep a record of where the data is stored so we can find it later on. This process of organizing the diskette is known as **formatting** the diskette.

To illustrate this concept more clearly, think of a file cabinet with many drawers in it. When the file cabinet is new, all the drawers are empty. If we start storing information in all the drawers, finding a specific piece of information will be a difficult task. Instead, we organize the file cabinet by labeling the drawers and putting labeled dividers in each drawer. In the front of the file cabinet we keep an index of all the information we have stored in the cabinet. Now the task of finding a specific piece of information is a simple one. All that is required is to look at the index, open the specified drawer, and go to the specified divider.

The diskette is organized in very much the same way. Think of each **track** as a drawer in the cabinet, and think of each **sector** as a divider in the drawer. A formatted diskette also contains a **directory,** which is like our index. The directory contains the name of the file and a pointer to the track and sector where the file begins. The diskette also contains a **file allocation table** (**FAT**), which keeps a record of which sectors are free for storing information and which are in use by existing files.

There are actually *two* levels of formatting involved here. The first is the physical track and sector format, which is like the physical file cabinet. The second level is the logical format, where the FAT and the root directory are written. Sometimes the actual operating system is also written on the disk at this time.

The IBM PC disk operating system (DOS) formats each track of the diskette into sectors containing 512 bytes per sector. This allows 18×512, or 9216, bytes to fit on each track. Since there are 80 tracks, we can store 737,280 bytes of information on each side. If we use both sides, then we can store a maximum of 1.44 MB on one diskette. In either case, 33 sectors are reserved for directory and FAT information. These 33×512, or 16,896 bytes are not available to the user. The maximum amount of user storage is 1,474,560 − 16,896, or 1,457,664 bytes. For convenience, the IBM PC diskette drives used today are often referred to as 1.44 MB drives. The physical diskette format is illustrated in Figure 13.21.

As we stated before, each track is divided into sectors. Each sector is divided into **fields,** known as the **preamble, header, data,** and **postamble fields.** The preamble consists of **sync bytes,** which are used by the floppy controller for data synchronization. The

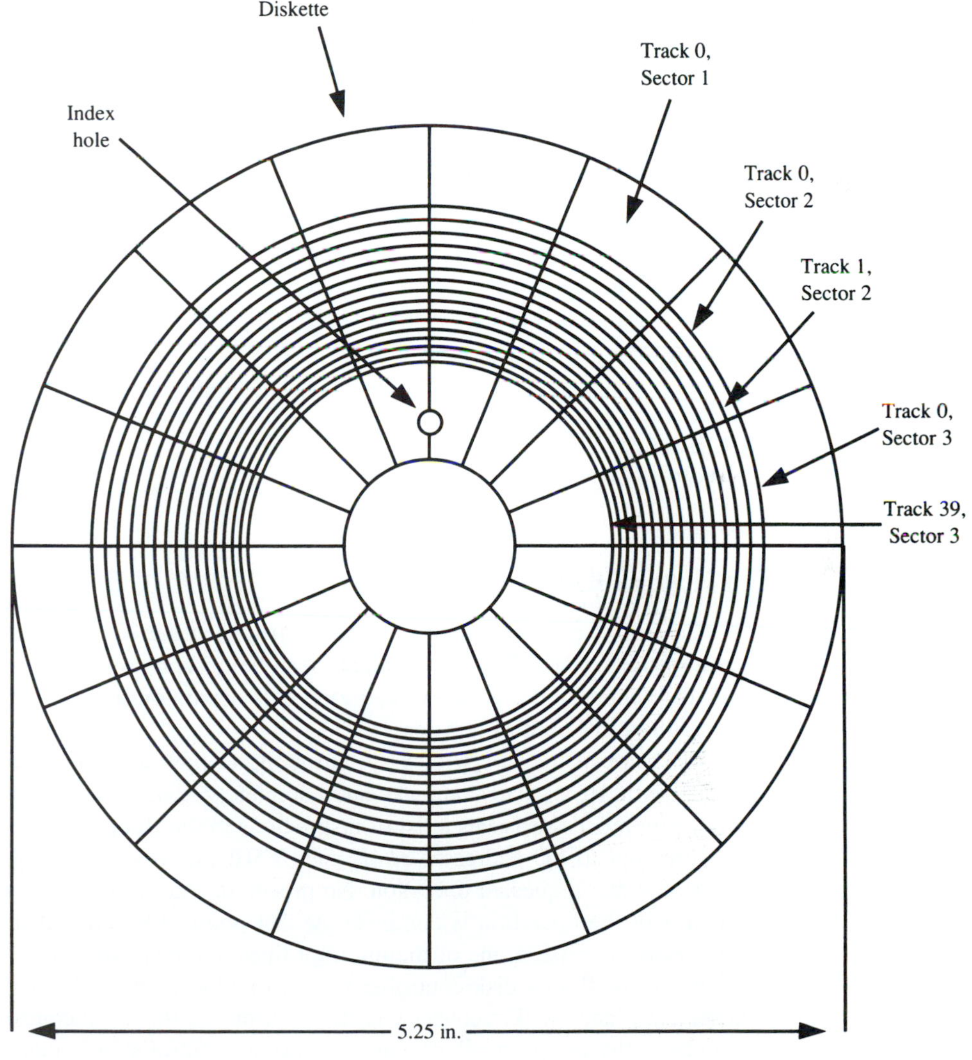

FIGURE 13.21
The formatted diskette

header contains the ID bytes, which are the track, side, sector, and sector size numbers that the floppy disk controller uses to find where to store or read data. The data field is where the actual data is written to or read from. The postamble is a small gap at the end of the sector, as shown in Figure 13.22.

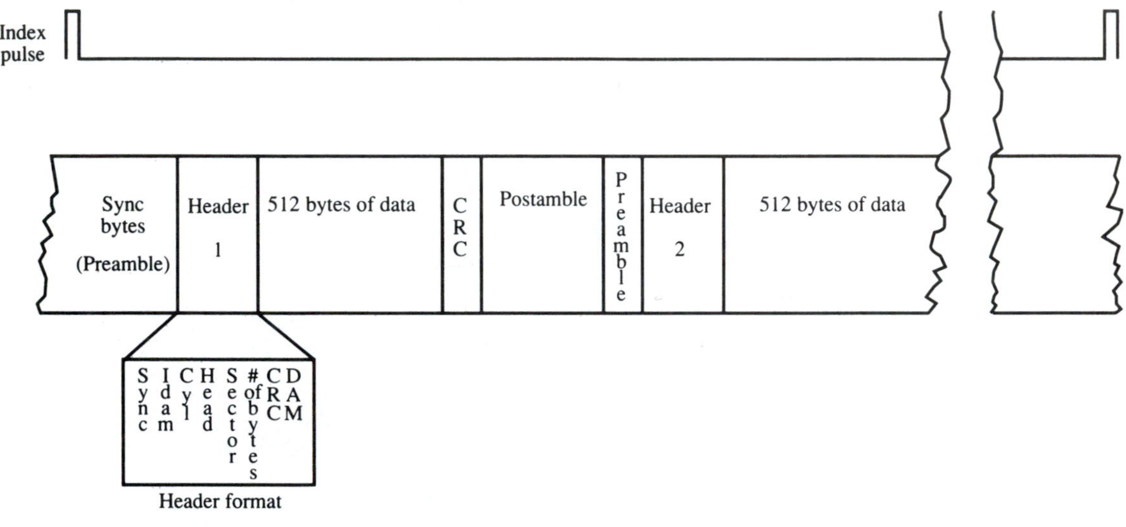

FIGURE 13.22
Typical track format

The Floppy Adapter

The floppy adapter contains all the electronics required to interface the floppy with the rest of the system. The floppy adapter can be broken down into four basic parts: the floppy disk controller, a phase-locked loop data separator, a write precompensation circuit, and the drive control register.

The heart of the floppy disk system is the floppy disk controller. The floppy disk controller is itself a small computer and completely controls the floppy disk drive. The main system processor issues a list of instructions to the floppy disk controller by way of I/O instructions to I/O addresses 3*F*4H and 3*F*5H. The floppy controller then takes over and performs the requested operation. No processor intervention is required unless there is an error or the operation is complete. At that point, the floppy disk controller signals the processor by raising one of the interrupt lines, typically IRQ 6.

When the floppy disk controller receives a read command from the processor, the data separator and clock recovery circuit synchronizes the drive controller to the all zero preamble on the floppy medium. Once the floppy controller is synchronized, it can read information from the data portion of the track. The controller will remain synchronized until the end of the read operation.

The write precompensation circuit shifts the write data pulse early or late by 250 ns as commanded by the floppy disk controller. This circuitry was added to improve the data reliability due to the high degree of peak shift common to the MFM method. This is more apparent on the inner tracks because of the tight bit density, where the frequency is at or near the maximum frequency response of the medium.

The drive control register is an 8-bit write-only port. The address of this port is 3*F*2H. This register is used to select one of four drives and to control the motors in all four drives. This register is also used to control the DMA and interrupt enables from the adapter circuitry.

Floppy Disk Drive Operation

The operation of the floppy disk drive is fairly simple. The drive motor spins the disk at 300 rpm, and the drive head and carriage assembly moves in and out along the surface of the disk. The heads can move about 1 in. and write as many as 80 tracks on both sides of the disk. Since there are tracks on both sides of the disk, they are collectively called cylinders. The drives can record at 500,000 flux reversals per second, which is equivalent to 500,000 bits per second in the MFM format. Disks may be single-sided or double-sided; however the double-sided disks are the primary standard in use today. The floppy disk drive consists of the following major assemblies:

1. The main motor assembly rotates the disk at 300 rpm.
2. The head and carriage assembly positions the read/write head over the desired track for reading or writing the data.
3. The write-protect sensor reports the presence of a write-protected disk.
4. The electronics board interfaces all major subsystems to the floppy controller.

Main Motor Assembly

The disk drive uses a "pancake" brushless-type motor that directly couples to the spindle shaft. These drives are referred to as direct drive units. This design relieves two major problems found on older units. The first problem was belt slippage on belt-driven units, and caused the disk speed to vary. The second problem was brush wear on older motors. This resulted in poor contact to the commutator, introduced noise into the electronic circuitry, and caused motor speed variation.

This motor is a multiphase synchronous type requiring no brushes. The armature (rotor) is a permanent magnet that rotates in an alternating magnetic field (stator) as shown in Figure 13.23. This form of motor has proven to be more reliable and has better speed regulation accuracy over older designs.

The speed of the disk is very important to the read data. Recall that we are using FM techniques to encode the data, so any variation in speed causes a shift in frequency. If the diskette is not correctly centered, the diskette will wobble as it rotates, which is called runout. This causes amplitude variations in the read signal and may also cause cross talk from adjacent tracks due to the alignment error. This will distort the read signal and cause read errors during playback.

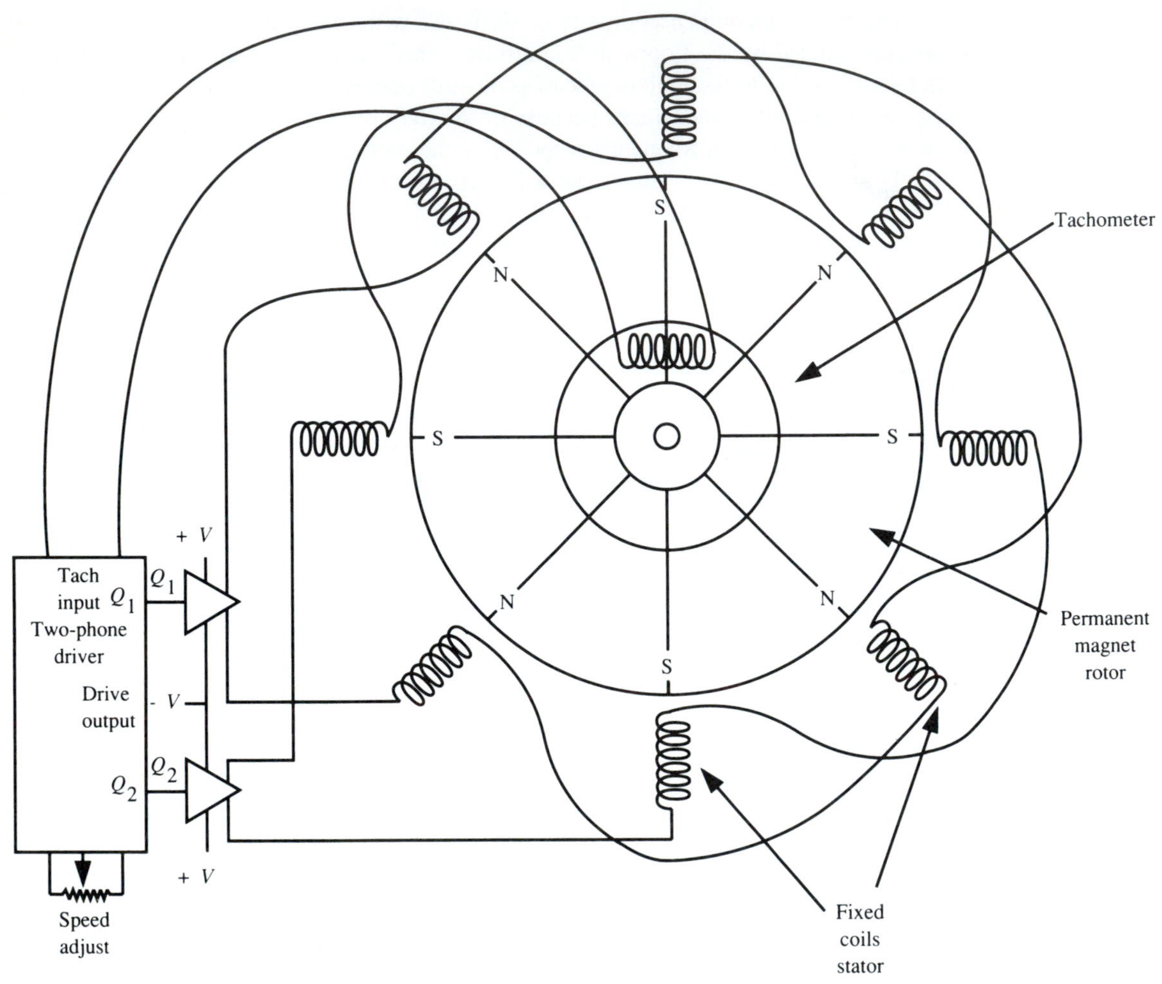

FIGURE 13.23
The direct-drive brushless motor

Head and Carriage Assembly

The head and carriage assembly is the most critical part of the floppy disk drive. This assembly is made up of the read/write head mounted on a sliding carriage. The carriage is driven by an electromechanical actuator, which is a multiphase stepper motor.

The read/write heads used on the 3.5-in. floppy drives are similar from one manufacturer to another. These heads are very much like the earlier example with one difference. The heads also contain erase windings and pole pieces. The pole pieces are mounted alongside both ends of the read/write gap, as shown in Figure 13.24. This design is known as tunnel erase. It trims the top and bottom fringes of the recorded flux reversals. It is not meant to be used as an erase head to erase the data fields. The resultant recorded pattern

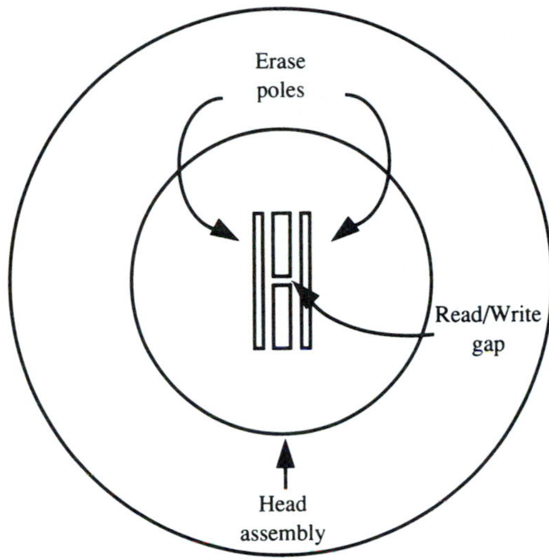

FIGURE 13.24
Erase poles in head

is shown in Figure 13.25. The tunnel erase head reduces track to track cross talk and minimizes the errors induced by minor runout problems on the disk or diskette drive.

Head 0 is the lower head, which is mounted rigidly onto the carriage assembly. Head 1 is mounted onto a spring-loaded flexure assembly. This allows the upper head to conform to the lower head and places even pressure on the diskette. Like the tape recorder, the diskette must be in physical contact with the read/write head during operation.

The actuator that drives the head carriage assembly is a stepper motor. The motor drives the head along a lead screw. The lead screw actuator works just like a screw and nut. The stepper motor shaft is coupled directly to the lead screw, which threads into the carriage assembly. As the stepper motor rotates, the carriage assembly is pulled toward or pushed away from the stepper motor, as shown in Figure 13.26.

Write-Protect Sensor

The **write-protect sensor** is a fairly simple device. There is a small tab in the corner of the diskette. This tab is called the write-protect tab and is used to signal the controller when a write-protected diskette is in the drive. The tab is sensed by the write-protect sensor on the disk drive. Most manufacturers use a photosensor arrangement, whereas others use a mechanical switch.

Electronics Board

The drive electronics board houses all the circuitry in the floppy drive and interfaces to the floppy controller. It has the floppy disk controller IC, the read/write amplifier, the motor driver circuitry, and the connections to the floppy disk bus and power bus.

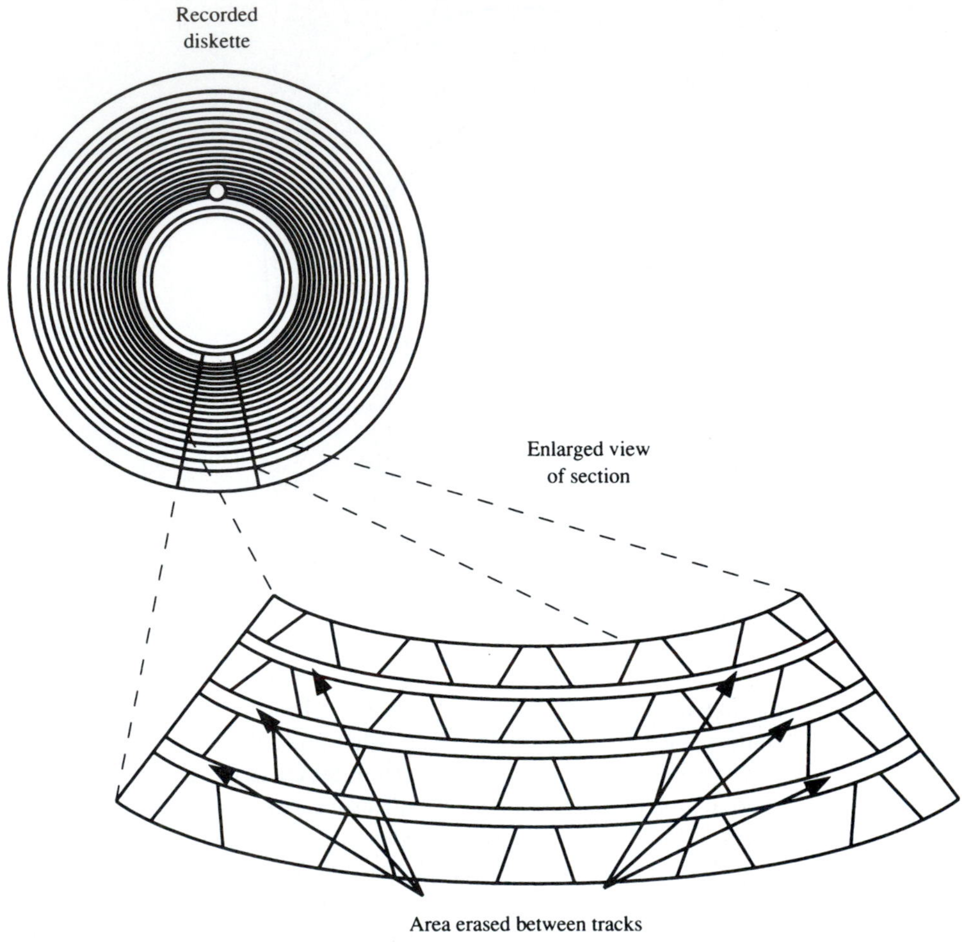

FIGURE 13.25
Tunnel erase (between tracks)

The drive control register is an 8-bit write-only port. The address of this I/O port is 3*F*2H. This register is used to select one of four drives and to control the motors in all four drives. This register is also used to control the DMA and interrupt enables from the adapter. A breakdown of these bits is shown below.

Bits 0 and 1 are the drive select bits, decoded as follows:

Bit	*1*	*0*	*Drive Select*
	0	0	0, drive A
	0	1	1, drive B
	1	0	2, drive C
	1	1	3, drive D

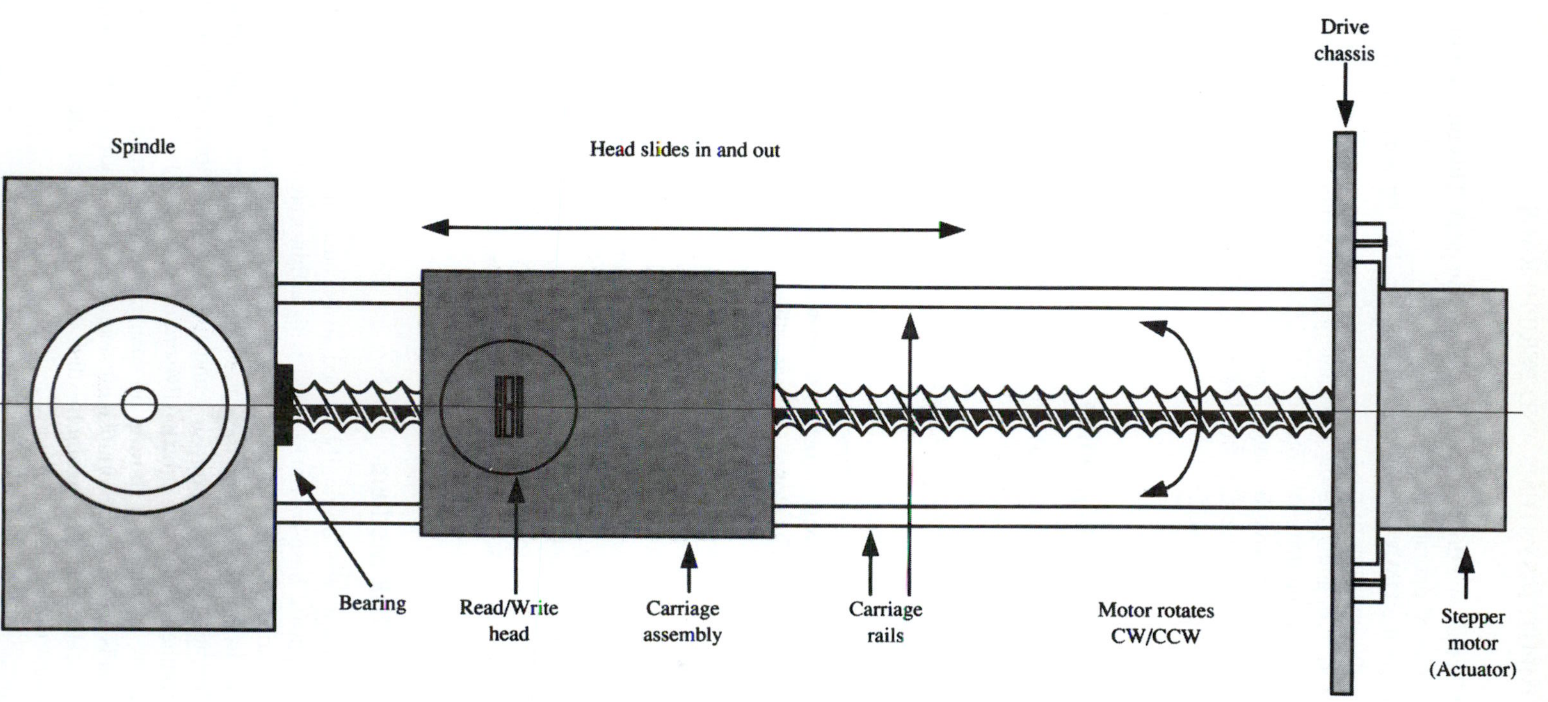

FIGURE 13.26
Lead screw–driven actuator

Bit 2 resets the floppy disk controller when clear. This bit must be set for the floppy disk controller to operate.

Bit 3 is the DMA and interrupt enable bit. This bit must be set to a 1 state to allow the DMA and interrupt to pass.

Bits 4, 5, 6, and 7 are the drive motor control bits and are defined as follows:

Bit 4 turns on drive A motor if set.

Bit 5 turns on drive B motor if set.

Bit 6 turns on drive C motor if set.

Bit 7 turns on drive D motor if set.

Installation and Configuration

The floppy drives of today are fairly straightforward to install and configure. For the most part, the 5.25-in. floppy has been completely replaced by the 3.5-in. version. In most machines produced today you will find one of these 3.5-in. drives. So if the floppy drive goes bad, you will most likely be replacing it with a 3.5-in. drive.

Personal computer systems use the drive letters A and B to designate the floppy drive. As you can see, a computer can have up to two floppy drives installed, one for each letter. The floppy drive is connected to the computer using a 34-pin IDC flat ribbon cable with a seven-wire twist. It is this seven-wire twist that tells the computer which drive is A and which is B. The floppy drive installed after the twist is given the A designation, and the drive installed before the twist is assigned B.

The 3.5-in. floppy drive is so common in computer systems today that since its inception it has had its own power connecter, aptly called the 3.5-in. floppy power connecter, or the mini power connecter due to its smaller size. This connecter is keyed, but it can be inserted incorrectly and could cause damage to the system.

Once the floppy is installed, the next step is to configure the system's CMOS settings. The settings for floppy drives are generally contained in the standard setup portion of the CMOS setup program. Most CMOSs today will have settings for the 5.25-in. 360K and 1.2M plus the 3.5-in. 720K, 1.44M, and 2.88M.

The Floppy Disk Medium

The floppy was originally named so because of its flexible medium. The disk is made of a plastic base (mylar) coated with a thin magnetic compound (ferrite, cobalt–ferrite, or barium–ferrite). This compound holds the information written to the disk.

This disk is then housed in a plastic casing to protect it from the environment. The earlier floppies (5.25-in.) were thin and flimsy, and left part of the medium exposed to the environment. They could be easily scratched by dust buildup. The newer floppy (3.5-in.) disks are housed in a more rigid casing, and they leave no part of the medium exposed. This is accomplished by placing a small spring-loaded metal sheath over the opening. The sheath only opens while the disk is being inserted into the floppy drive, keeping harmful dust out.

13.4 Hard Drive Subsystems

The hard drive used in the PC uses the flying head Winchester technology. The hard drive is quite similar to the floppy drive except that the medium is a rigid moving aluminum disk. Modern hard drives often have more than one disk or platter; often up to three or even five platters are used. The platter rotates at speeds of up to 12,000 revolutions per minute (rpm) as opposed to the 300 rpm for the floppy diskette. The read/write head actually flies over the disk surface on a cushion of air. Both the top and bottom surfaces of a platter are used to store data, so there is a read/write head for each surface. Since the head flies very close to the surface, *absolutely no airborne contamination* is allowed inside the head and disk assembly. The head and disk assembly is *sealed* from the outside environment with the exception of a pressure-equalizing filter. There is an air-scrubbing filter inside the head and disk assembly to capture any internal contaminations. If any contamination is present, it will build up on the leading edge of the head. This will disturb the flight altitude and may cause a loss of pressure or create a vacuum under the head. The loss of pressure will cause the head to crash on the surface and destroy the disk and head. Figure 13.27 illustrates the effects of contamination on a hard disk drive.

For the same reason, disk drives are susceptible to sharp mechanical shocks. Although today's drives are engineered to withstand up to 30 gravities (*G*) delivered within 2 ms, you should not expect the drive to absorb such abuse while powered up and working.

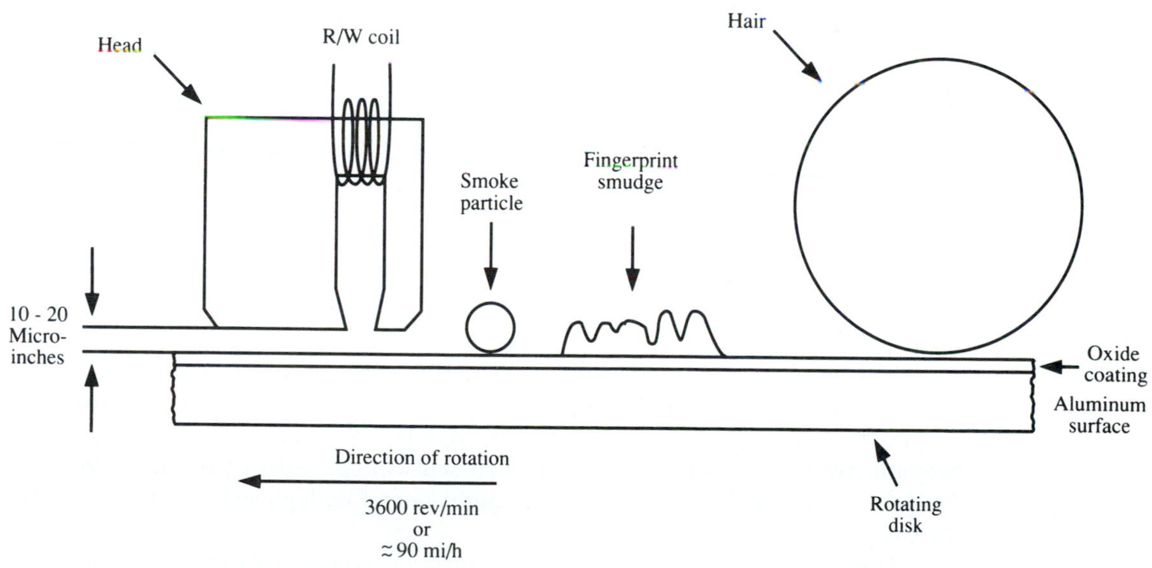

FIGURE 13.27
Hard disk contamination

Hard Disk Specifications

Disk drives are classified according to interface, speed, and capacity. There are several ways to measure speed: data transfer rate, access time, and seek time. The **seek time** is the amount of time it takes to position the head over a track. A more balanced measurement is the **average seek time,** or the time it takes to move the head to a randomly located request. The access time is the seek time plus the average amount of time it takes to move a sector of data under the head, also called latency. For example, a disk drive that rotates at 5400 rpm will complete one rotation in 11.11 ms. Using half of 11.11 ms as an average time for the data to rotate under the head results in an additional 5.55 ms. Thus, access time is equal to the seek time plus approximately 8 ms of latency. Today, average seek times of less than 9 ms and access times below 15 ms are common.

Capacity is a measure of how much data a formatted disk can hold. Capacity is linked to **areal density,** the number of bits that can be stored in a square inch on a platter, and the number of platters. The **data transfer rate** is a measure of how fast the disk drive can transfer data to the computer under ideal conditions. Increasing the number of platters a disk drive has will increase its capacity, but not necessarily its transfer rate. With the corresponding improvements in read/write heads, however, increasing the area density can increase the data transfer rate. With more data stored in the same amount of area, data will be presented to the read/write head at a faster rate than a drive with a lower area density. Increasing the rpm will also generally increase the data transfer rate for the same reason.

Another trend in disk drive manufacturing is to increase the buffer size of hard disks. Just as the system board uses system memory to cache data while waiting for slower subsystems, a hard drive uses a buffer to cache data while data is being read/written from the platter(s). While older drives contained as little as 128 KB of buffer, recent drives carry as much as 2 MB. Increasing the buffer size by itself may not increase performance, since reading/writing to the platter will always be slower than reading/writing to the buffer, and the buffer will be drained of data in any sustained read/write operation. With improvements in caching algorithms and for the transfer of small files, though, an increased buffer size can increase the temporary burst transfer rate of disk drives.

When evaluating a drive's specifications, the best indicators of performance are area density and rpm. Seek and access times are not usually obtained in a consistent manner, making it difficult to compare drives, and are not very good indicators of performance in the first place.

The physical size of the hard drive is almost universally standardized to fit into a 3.5-in. drive bay, though previously some drives were available for 5.25-in. bays. An adapter bracket is necessary to use a 3.5-in. drive in a 5.25-in. bay. Most drives are half height drives that measure about 1.6 in. high and will only take up one drive bay each. Some drives, called full-height drives, measure about 3 in. high and may take up to two drive bays.

One of the most commonly asked questions concerning hard drive installation is the discrepancy between advertised and actual disk capacity. This is due to a couple of caveats in the manufacturer's advertising figures. Manufacturers often define a megabyte as 1,000,000 bytes as opposed to the more conventional 1,048,576 Bytes. They also usually cite the unformatted capacity of the drive, which is very slightly larger than the formatted capacity. Both factors result in a smaller HD size than expected. This sometimes leads the technician to believe that the wrong drive was shipped or the wrong translation mode was used, when in fact there is no error.

Hard Drive Interface

The hard disk drive communicates with the rest of the PC through the adapter or controller board. Basically, the controller directs the drive's physical operations and performs functions such as telling the drive where to position its heads. The most commonly used types of hard disk controllers are the **integrated drive electronics (IDE)** interface and the **small computer systems interface (SCSI).** Some older, rarely used controllers include modified frequency modulation (MFM), run length limited (RLL), and enhanced small computer system interface (ESDI).

SCSI uses its own language to control the drive's heads and other operations. The SCSI controller translates the PC's commands to the particular language that the drive understands. IDE drives are designed to directly understand the PC's commands and translation is not necessary. As a result, the adapter for an IDE drive contains far less electronics than an adapter for the SCSI interface. Usually the IDE adapter is incorporated into the systems board, eliminating the need for an expansion adapter board. There are systems boards with an integrated SCSI adapter, though they are far less common than their IDE counterpart. Figure 13.28 illustrates some of the common hard disk drive cable and system operations.

MFM and RLL

The original technology to encode data onto a hard disk's magnetic surface was plain frequency modulation, or FM. FM was quickly replaced by a variation called **modified frequency modulation,** or **MFM.** As hard disk manufacturing matured and became more reliable, a more efficient encoding system that was originally used on mainframe computers came to the PC. It was called **run length limited,** or **RLL.** Actually, MFM and RLL hard disk drives are physically nearly identical. The only difference is in the method used to encode and store data. MFM drives use a 17-sectors-per-track data encoding format and transfers data at a rate of 5 Mbps. RLL drives use a 26-sectors-per-track format and transfers data at 7.5 Mbps. As a result, RLL drives are 50% faster and can hold 50% more data than MFM drives.

For example, the Seagate ST251 hard drive can store 40 MB of data using MFM encoding. The Seagate ST251 hard drive could store 60 MB of data using RLL encoding. This is not recommended because the ST251 is not RLL certified and may be unreliable with a RLL controller. Furthermore, it will void the manufacturer's warranty. If RLL encoding is used, a drive certified for RLL encoding should be used. The Seagate ST277R is the certified equivalent RLL drive for the ST251.

MFM and RLL drives are quickly becoming obsolete. Most manufacturers have stopped producing MFM and RLL drives. One advantage of the MFM and RLL standard is that they have become mature technologies. That is, they are the most reliable and least expensive of all the different types of interfaces available today. MFM encoding techniques are discussed in detail in Section 13.2.

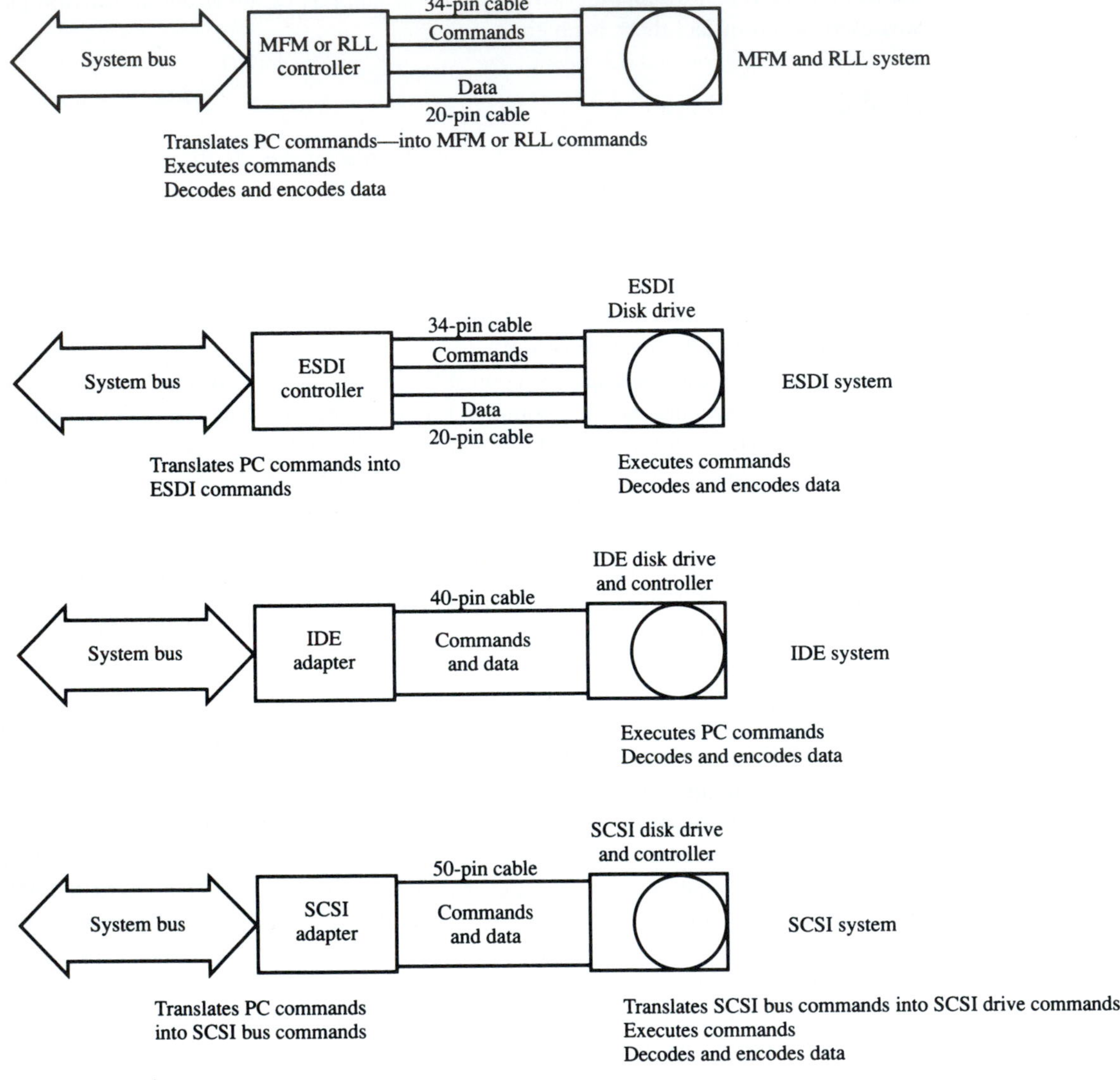

FIGURE 13.28
Hard disk system operation

ESDI

The next high-performance standard for the PC is the **enhanced small device interface (ESDI).** ESDI drives currently offer the highest data transfer rate of any of the popular interfaces. Almost all ESDI disks use RLL data encoding, but with a more efficient, high-level interface command language. ESDI drives translate commands on the controller but execute the commands and encode and decode the data on the disk drive electronics. MFM and RLL drives require the controller to translate commands, execute the commands, and encode and decode the data, thus slowing down the data transfer rate.

ESDI data transfer rates typically run about 10 Mbps. However, many manufacturers offer drives with transfer rates as high as 20 Mbps. The controller and the disk *must* be certified for the same transfer rate. Because of a more precise control over the head movement, ESDI drives can store more data than the largest MFM or RLL drives can. Thus, ESDI drives offer high reliability and high speed, and permit very large capacity drives to be used. However, they are more costly.

ESDI drives use the same type of cables as MFM or RLL drives, which makes it difficult to tell the drives apart. This can create problems when upgrading or replacing a drive. Even if the cables and controller match physically, there is no guarantee that a drive and controller will support the same interface.

IDE/ATA

The **integrated drive electronics (IDE)** interface is the most popular HD interface used in PCs today. The controller is built directly into the disk so that the drive can accept commands directly from the computer. Since most of the work is offloaded onto the drive controller, the interface adapter requires very little electronics and is usually incorporated directly into the motherboard. IDE drives connect to their adapters with a single 40-pin cable that incorporates both control and data signals. An IDE cable can have two or three 40-pin connectors, and all the connectors are identical, so either end can be plugged into the drive or the adapter.

In fact, because IDE drives have become so affordable and popular, most motherboards have two built-in IDE adapter connectors. Each adapter and cable represent one data pathway, or channel. Each channel can accept up to two IDE devices, with one device configured as **master** and one configured as **slave.** For most systems, this means that the system board can support up to four IDE devices. However, each adapter can only talk to one IDE device at a time, so the most efficient configuration is to have only one device on each channel. If only one device is attached to a channel, then that device must be configured as master. Drive master/slave status is changed using a set of jumpers on the front or back of the drive. Jumper values are usually written on the drive itself, but in the rare case that they are not, refer to either the documentation that came with the drive or to the manufacturer's website. Figure 13.29 illustrates the IDE drive jumper settings.

The IDE interface defines several data transfer protocols, or modes. Programmed I/O (PIO) mode is the basic protocol for transferring data over the IDE interface, and a higher-number mode denotes faster transfer rates. Direct memory access (DMA) mode improves upon the PIO mode by decreasing the dependence upon the MPU for data transfers. The fastest modes, PIO mode 2 or single-word DMA 2, allow a theoretical maximum transfer rate of 8.3 Mbps.

There is some confusion of terms, since the IDE interface is more correctly called the **AT attachment (ATA)** standard. These terms are used interchangeably and for all intents and purposes are identical. This is further complicated by the fact that most modern IDE devices are in fact **enhanced IDE (EIDE)** devices, a modification to the original IDE specifications. EIDE is in turn also more correctly called ATA-2 (also sometimes called

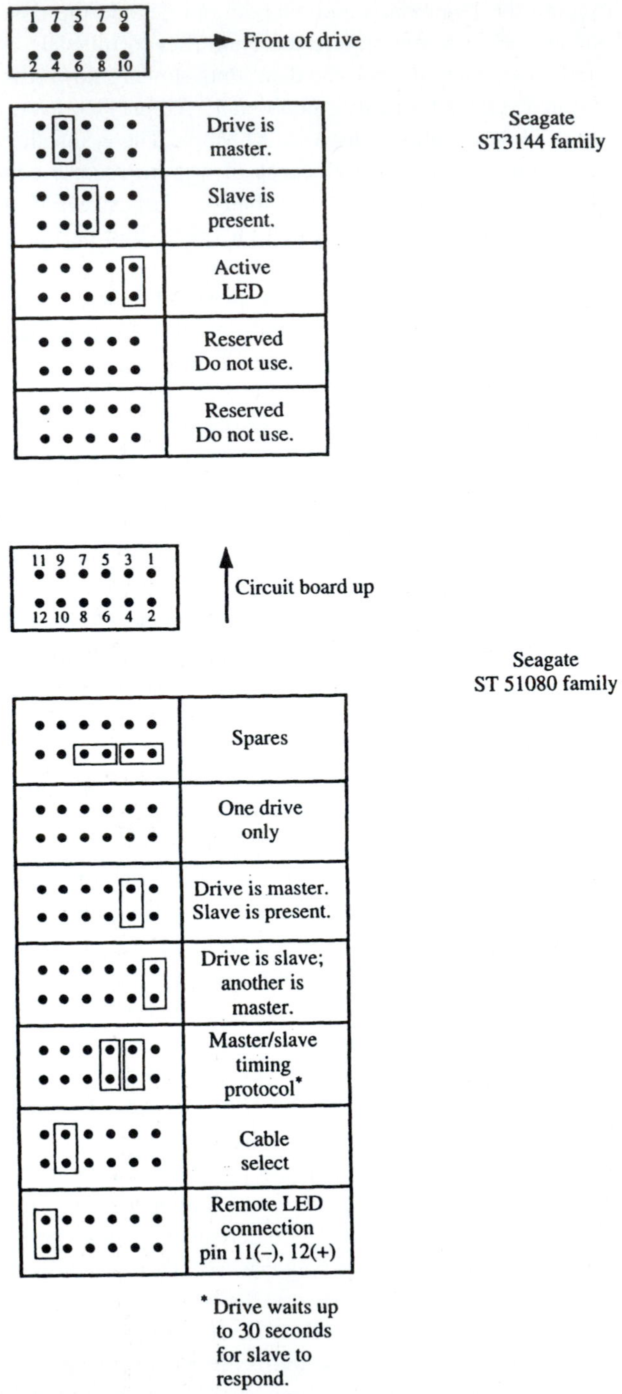

1 3 5 7 9
2 4 6 8 10
Front of drive

Seagate ST3144 family

Jumper	Function
	Drive is master.
	Slave is present.
	Active LED
	Reserved Do not use.
	Reserved Do not use.

11 9 7 5 3 1
12 10 8 6 4 2
Circuit board up

Seagate ST 51080 family

Jumper	Function
	Spares
	One drive only
	Drive is master. Slave is present.
	Drive is slave; another is master.
	Master/slave timing protocol*
	Cable select
	Remote LED connection pin 11(–), 12(+)

* Drive waits up to 30 seconds for slave to respond.

FIGURE 13.29
IDE drive select jumpers

fast ATA or fast ATA-2). In practical terms, they all mean roughly the same thing, and most modern hardware is backward compatible. This means that newer specification devices can be used with older ones, though they may not function up to their full potential.

EIDE/ATA-2

One improvement introduced by EIDE is protocol support for removable media such as CD-ROMs or tape drives, called ATA packet interface (ATAPI). EIDE also supports **block mode address,** which groups multiple read/write commands together for more efficient data transfers, and **large block address (LBA).** LBA is the most common translation mode used by the system BIOS to address a hard drive. It differs from earlier translation modes by abstracting the geometry of the disk (cylinders, tracks, and sectors) into a sequential numbering system. This helps circumvent some of the problems other translation modes had with larger drives. Virtually all BIOSs and hard drives manufactured today support LBA.

EIDE also allows faster and more efficient data transfer modes, PIO mode 3 (maximum 11.1 Mbps), multiword DMA 1 (13.3 Mbps), and PIO mode 4 or multi-DMA mode 2 (16.6 Mbps). All transfer rates are a possible theoretical maximum and refer to the interface burst transfer rate. For example, today's faster EIDE HDs average a 10 Mbps sustained transfer rate. However, this is a misleading figure because this is an average transfer rate. Because the outer edge of the platter spin faster than the inner edge, data can be written/read faster on the outside, or beginning of the drive, so a 10 Mbps sustained transfer rate may actually mean a 15 Mbps transfer rate at the outer edge of the disk, and 5 Mbps transfer rate at the inner edge. The maximum disk transfer rate may actually be pushing the limits of the interface. To reduce any bottleneck effects, the interface transfer rate should always exceed the disk internal transfer rates. This is even more important in SCSI systems, where multiple devices may be using the interface simultaneously. The internal disk transfer rate is practically the platter-to-buffer transfer rate and the interface transfer rate is the buffer-to-adapter transfer rate, since the buffer speed is so much faster than that of the data path.

Ultra ATA

The ATA-3 standard introduced reliability and power management capabilities, but no faster transfer modes. To keep the interface from limiting current drives while the ATA-4 standard is being finalized, an extension to the ATA-3 standard was created, called Ultra ATA or Ultra DMA. While it introduced a data integrity check in the form of **cyclical redundancy checking (CRC)** (see Chapter 7), the main improvement was an increased maximum interface burst transfer rate of 33 Mbps. This is accomplished by using the multi-DMA mode 2 2 (16.6 Mbps) transfer mode as a base, but using the falling as well as the rising edge of the clock signal to transfer data, much like the AGP or DDR SDRAM. This new specification is known as Ultra DMA/33 or Ultra ATA/33. The Ultra ATA/33 standard needs to be supported both by the hard drive and the adapter of the systems board to achieve maximum performance. The move to this standard occurred just in time, since current leading-edge hard drives can have an internal data transfer rate of 24 Mbps at their outer tracks, well beyond the limits of the ATA-2 interface. The goal was to improve the interface transfer rate such that it was not a limit to internal disk transfer rate, not to improve the internal transfer rates. That newer Ultra ATA/33 drives show a performance increase over ATA-2 drives can be atttributed more to advances in areal density, rpm, and the drive's electronics and firmware (similar to a BIOS for a device), than to the Ultra ATA/33 standard itself.

As Ultra ATA/33 drives have become almost universal, a newer Ultra ATA standard has been introduced, Ultra ATA/66. By increasing the number of conductors in the IDE cable from 40 to 80 lines while keeping the connector the same, Ultra ATA/66 increases the maximum burst transfer rate to 66 Mbps while remaining physically backward compatible. The additional conductor lines decrease signal cross stalk and allow more aggressive timing signals to be used. Since modern hard drives have yet to push the boundaries of the Ultra ATA/33 standard, Ultra ATA/66 can best be viewed as preparation for future drive performance increases.

Since it doesn't cost manufacturers much more to support the newer standard, almost all new hard drives support Ultra ATA/66. Ultra ATA/66 drives are also designed to function with Ultra ATA/33 (or even ATA-2) adapters. However, there have been a few cases where the motherboard BIOS attempts to run an Ultra ATA/66 drive at the 66 Mbps interface without actually supporting the Ultra ATA/66 standard. For these situations, manufacturers provide utilities to force the drive into Ultra ATA/33 compatibility until such issues are resolved. Also remember that a 80-wire 40-pin Ultra ATA/66 cable must be used with Ultra ATA/66 drives for the 66 Mbps interface standard to work.

SCSI

The history of the SCSI bus begins as far back as the 1960s with the IBM 360 mainframe. The 360 had bytewide parallel I/O bus to do block transfers to and from peripherals. This bus, called the selector channel in the earliest 360 models, could address only one logical device at a time. Later, this bus became the block multiplexer channel and could address several peripherals at once. It was also known as the OEM channel and was the preferred way for third-party peripherals to interface with IBM equipment. The OEM channel became so popular with the federal government that it made the channel the Federal Government Processing Standard 60. Other computer manufacturers sued, claiming that this gave IBM an unfair advantage. The suit was lost; however, enough pressure was exerted to keep ANSI from establishing OEM as a standard. ANSI *did* want to establish a nonproprietary parallel I/O bus. In the early 1980s the ANSI X3T9.3 committee began working on a bus called the intelligent peripheral interface, or IPI. IPI was similar to OEM because the CPU was the bus master, and it had similar transition states. However, IPI had the capability of performing 16-bit transfers, whereas OEM could accomplish only 8-bit transfers. During this period, Shugart was also working on a parallel I/O bus of their own called the Shugart Associates System Interface, or SASI. Shugart intended SASI to be a low-cost, peer-to-peer interface. ANSI was eventually asked to make SASI a standard and was faced with a small dilemma: Another committee at ANSI was working on a bus of its own. Rather than compete with yet another bus, SASI proponents agreed to work with the X3T9.2 committee and the new bus was called SCSI. The X3T9.3 committee completed work on the SCSI specification in 1984, and in 1986 it was approved and published. SCSI gained widespread acceptance, far more than IPI, which today is used primarily in mainframes.

The SCSI bus transfers data via a logical block command known as a command descriptor block. In the ANSI spec X3.131, the data transfer is described as follows:

> The command definitions assume a data structure providing the appearance at the interface of a contiguous set of logical blocks of a fixed or explicitly defined data length. . . . A single command may transfer one or more logical blocks of data.

To explain more simply what this means, let us take the example of a data block transfer from a disk drive. With another bus type, when the CPU issues a Find Sector ID command and the drive signals that the sector ID was found, the CPU would then need to issue a Read Sector or Write Sector command for the drive to know what to do. With the SCSI bus, the drive is given commands that contain both the address of the sector and instruction on what to do with that sector. Also, commands are issued by the host adapter and not the CPU, therefore keeping the CPU free to do other functions. This accomplishes a very fast channel turnaround time. We will go into greater detail on the SCSI bus later. SCSI implements true peer-to-peer communications; that is, it can accommodate connections between multiple CPUs and multiple peripherals. SCSI also allows several commands to peripherals at the same time. The host may issue a command and then disconnect, thus freecing the bus until the peripheral is ready to respond. Meanwhile the host can issue commands to other peripherals, or other hosts can use the bus. When a peripheral is finished with its command, it can reconnect to the host to transfer data or status information. In this way SCSI can do multiple I/O operations simultaneously. SCSI uses logical addressing as opposed to physical addressing. To use our earlier example of a disk drive transfer, this would mean that, rather than giving the physical address of the cylinder head and sector, the host would issue a command containing a logical address block.

The SCSI bus can support up to eight devices, which are either **initiators** that issue commands or **targets** that execute commands. Typically the host adapter operates as the initiator and the peripherals operate in target mode. SCSI controls bus arbitration, which allows one SCSI device to gain control of the bus to assume the role of target or initiator. Figure 13.30 illustrates several SCSI configurations.

The SCSI bus consists of eight data-bit signals (D_7-D_0), and a parity-bit signal to form a data bus. Each data bit is assigned an SCSI ID. DB_7 is the most significant bit and has the highest priority during the bus arbitration phase; DB_0 is the least significant bit and has the lowest priority during arbitration, as shown in Figure 13.31.

The SCSI bus architecture contains eight phases, or states. These phases in essence encompass the entire operation of the SCSI bus. That is, they do all the necessary functions to accomplish data transfer over the SCSI bus. The eight bus phases are

- Bus free
- Arbitration
- Selection
- Reselection
- Command*
- Data*
- Status*
- Message*

*Collectively, these four phases are known as the information transfer phase.

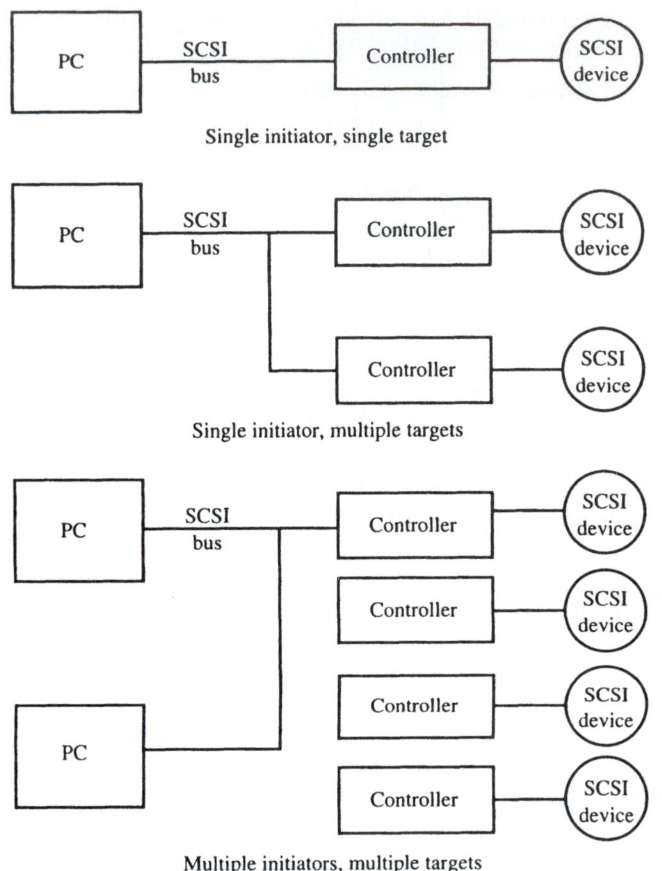

FIGURE 13.30
Sample SCSI configurations

Note that the SCSI bus may not be in more than one phase at a time. Table 13.1 explains the function of the eight bus phases.

SCSI transactions occur independently of the host computer; that is, SCSI devices can communicate with each other, arbitrate control of the bus, and perform functions, all without the intervention of the CPU. Of course, this will free up system resources to perform other operations and therefore improve overall system performance.

The SCSI hardware configuration is similar to that of ATA because the controller is built into the drive itself and what is required is a bus interface. Unlike ATA, however, the SCSI interface, called the *host adapter,* is much more complicated and actually contains a small microprocessor. The host adapter is itself an SCSI device with its own address. (SCSI addresses are determined at power-up, although the host adapter is generally assigned SCSI address 7.) Therefore, it can be connected to another host adapter in another PC via the 50-pin SCSI cable and can perform operations between the two PCs. For example, two host adapters in two PCs can be hooked up back to back, and the hard drives

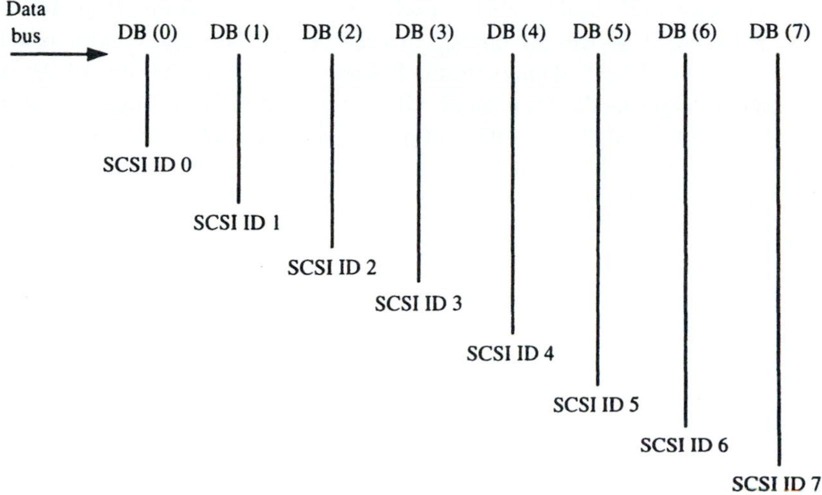

FIGURE 13.31
SCSI ID bits

TABLE 13.1
SCSI bus phases

Bus Phase	Function
Bus free	Indicates that no SCSI device is currently using the bus and that the bus is available
Arbitration	Allows one device to gain control of the bus so that the device may become either an initiator or a target
Selection	Allows an initiator to select a target to perform a particular target function (i.e., a read or write command)
Reselection	Allows a target to reconnect to an initiator to continue an operation previously begun by the initiator but suspended by the target
Command	Allows a target to request command information from the initiator
Data	Includes both data in and data out phases. The data in phase allows a target to request that data be sent to the initiator from the target. The data out phase allows a target to request that data be sent from the initiator to the target.
Status	Allows target to request that status information be sent from the target to the initiator
Message	Includes both the message in and message out phases. The message in phase allows a target to request that message(s) be sent to the initiator from the target. The message out phase allows a target to request that message(s) be sent from the initiator to the target.

in the two PCs will be able to exchange data. In effect this stimulates a peer-to-peer network. SCSI devices are connected via a 50-pin flat cable in what is called a *daisy chain.* Using SCSI host adapters in this fashion will require that one of the host adapters operate in target mode. Most SCSI adapters in a DOS PC operate only in initiator mode.

All SCSI devices attach in parallel with each other and the host adapter. Some SCSI devices will have two SCSI connectors and may designate one as IN and the other as OUT, giving the appearance of a series connection. The connectors are wired in parallel, and it does not matter which one you use. One important note is that *both* ends of an SCSI bus must be terminated using termination resistors. Some SCSI devices provide built-in termination that is switchable; therefore, check for termination before adding an external terminator. The cable and connectors are available in a variety of types, with most internal cabling using ribbon cable with 50-pin Berg (dual-in-line, push-on) style connectors. The original external SCSI cabling used Centronics style 50-pin connectors similar to a standard printer cable. This is sometimes called fat SCSI. Many newer devices use a smaller connector known as thin SCSI. The thin-style connector is used for both single-ended and differential SCSI, so check the type and terminators when using thin SCSI. Single-ended SCSI is limited to 6 meters, or appoximately 18 feet, which includes all internal cabling.

One SCSI controller or adapter card can support a maximum of eight SCSI devices, as shown in Figure 13.32. Many SCSI devices such as disks, printers, scanners, modems, CD ROMs, tape drives, and so on, are available. Thus, one SCSI controller can be used to control many different devices. Daisy chaining up to 64 SCSI devices is possible.

Since one controller operates many devices, SCSI devices can achieve greater speed performance when communicating to each other through the common SCSI controller. For this reason, SCSI has become very popular on network systems. Figure 13.33 illustrates several typical hard disk interface connections, including the SCSI connection.

The SCSI controller card, which is known as the host adapter, can attach to both internal and external devices at the same time. Before installing an SCSI adapter inside the PC, decide if you will be connecting internal, external, or both. If using only external devices, you must leave the internal SCSI terminator resistors installed on the adapter board. If using only internal devices inside the PC, you may leave the internal terminator resistors installed, or you may use an external terminator on the adapter external connector. Using an external terminator makes the job of adding external SCSI devices easier because you will not have to remove the PC cover later. If you are using both internal and external devices, you must remove the internal terminator resistors on the host adapter board. Proper SCSI termination is very important and one of the largest causes of problems during installation. Both ends of the SCSI chain must be terminated using the correct terminators. Only one device can provide the power to the terminators, and it is usually the host adapter. Since the host adapter provides some system power to the SCSI bus, the adapter contains a fuse to protect the system from shorts. Improper cabling, incorrect terminators, or a faulty cable can blow the fuse on the host adapter, thus causing a systemwide failure.

Next, you must set the board jumpers for the SCSI ID (usually 7), IRQ, DMA, parity, VL-BUS clock speed, and so on. An important thing to note here is that most DOS PC SCSI adapters contain a BIOS ROM. The BIOS ROM is required because the primary disk drive, drive C, may be attached to the SCSI adapter and could not boot DOS without the ROM. Be sure to check the address of the BIOS ROM and exclude the space from the EMM386 line in the CONFIG.SYS file. If the internal disk drives are all connected

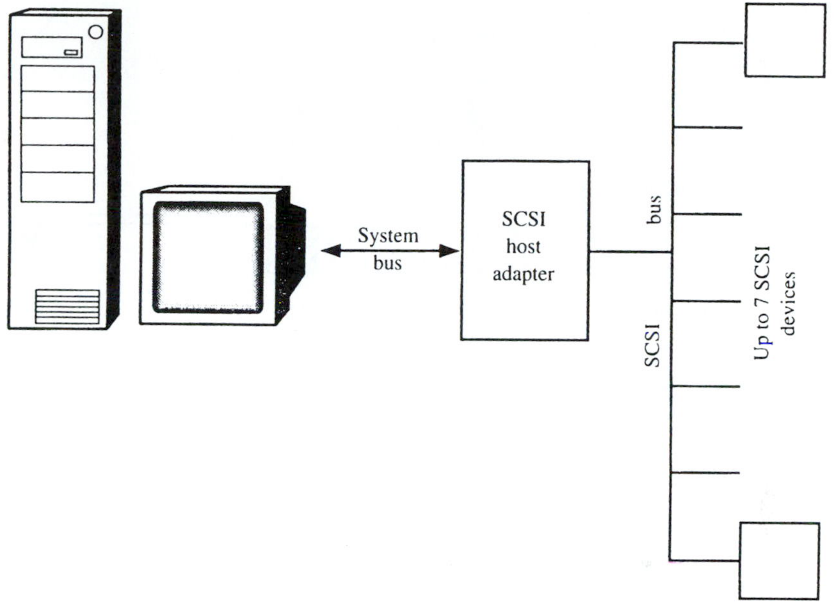

FIGURE 13.32
SCSI connections

to the SCSI host adapter, be sure to set the CMOS drive type to NONE because the BIOS ROM on the SCSI host adapter takes care of this for you.

Using the SCSI host adapter for only two internal disk drives requires no additional drivers other than the on-board BIOS. Most SCSI host adapters will handle DOS drives C and D automatically. If the SCSI host adapter provides advanced SCSI programming interface (ASPI) services, you should load them also. ASPI has become a standard for writing SCSI device drivers. Using ASPI services will improve the performance of any internal DOS disk drive. Any additional DOS device attached to the host adapter will require separate device drivers loaded at boot-up time as well as the ASPI services. The following is an example of a DOS PC setup for one hard disk, one CD-ROM disk, and a DTC 3270VL SCSI host adapter:

CONFIG.SYS

```
LASTDRIVE=Z
DEVICE=C:\WINDOWS\HIMEM.SYS
DEVICE=C:\DOS\EMM386.EXE NOEMS X=C800-CBFF I=CC00-EFFF
DEVICEHIGH=C:\ASPI3X70.SYS
DEVICEHIGH=C:\ADTC-CD.SYS /D:MSCD000
DEVICEHIGH=C:\DOS\SETVER.EXE
DOS=HIGH,UMB
FILES=30
DEVICEHIGH=C:\WINDOWS\IFSHLP.SYS
STACKS=9,256
```

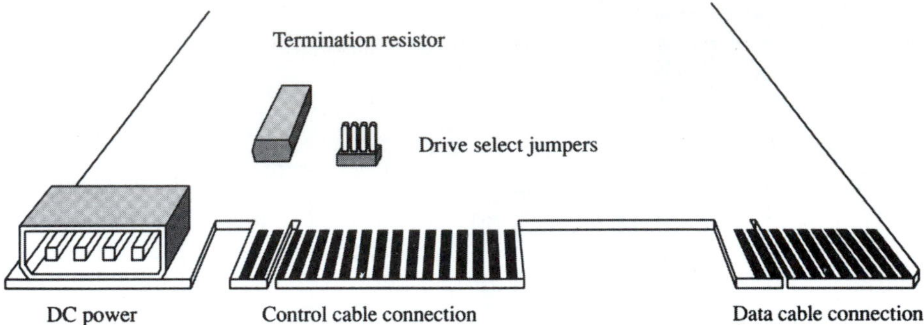

(a) MFM, RLL, ESDI

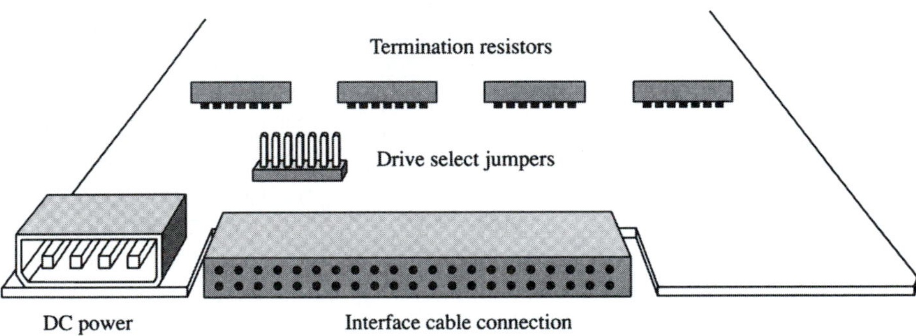

(b) SCSI

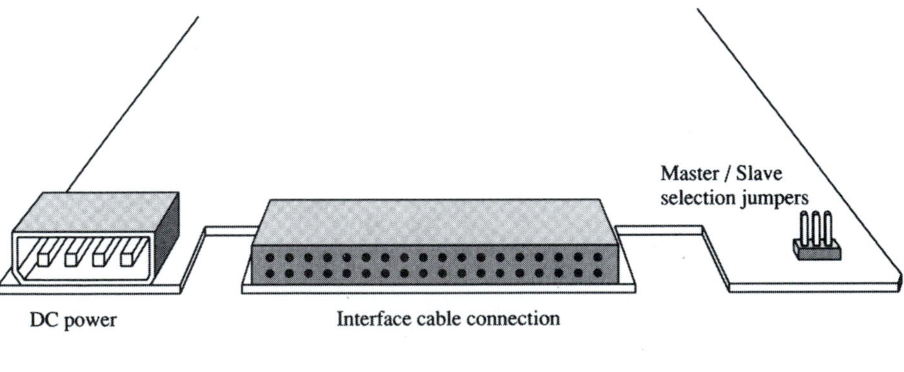

(c) IDE

FIGURE 13.33
Hard drive interface connections

AUTOEXEC.BAT

```
1h C:\BIN\MSCDEX.EXE /D:MSCD000 /M:10
C:\WINDOWS\NET START
1h C:\DOS\SMARTDRV.EXE /X 2048 128
@ECHO OFF
PROMPT $P$G
PATH C:\WINDOWS;C:\DOS;C:\QUICKCMD
SET TEMP=C:\DOS
```

Note in this example that the SCSI host adapter contains a BIOS ROM at address *C*800:0000 and the properly formatted EMM386 exclude statement. The default BIOS address was *D*800:0000, but it will place a hole in the high memory area. We moved the BIOS address to an address just above the VGA BIOS ROM to provide a more contiguous high memory area.

The following implemented SCSI standard is Ultra-2, which uses a 40-MHz clock speed for a maximum burst data transfer rates of 80 Mbps. It provides a cabling distance of up to 12 meters. Earlier forms of SCSIs use a single wire that ends in a terminator with a ground. Ultra-2 SCSI sends the signal over two wires with the data represented as the difference in voltage between the two wires. This allows support for longer cables.

The next SCSI standard is Ultra-3, which increases the maximum rate from 80 to 160 Mbps by being able to operate at the full clock rate rather than the half-clock rate of Ultra-2. Ultra-3 also includes CRC. The standard is also sometimes referred to as Ultra160/m. New disk drives supporting Ultra-3 will offer much faster data transfer rates.

Table 13.2 lists the different SCSI generations.

TABLE 13.2
SCSI generations

Technology Name	Maximum Cable Length (m)	Maximum Speed (Mbps)	Maximum # of Devices
SCSI-1	6	5	8
SCSI-2	6	5/10	8/16
Fast SCSI-2	3	10/20	8
Wide SCSI-2	3	20	16
Fast Wide SCSI-2	1.5	20	16
Ultra SCSI-3 8-bit	1.5	20	8
Ultra SCSI-3 16-bit	12	40	16
Ultra-2 SCSI	12	80	16
Ultra-3 SCSI	12	160	16

HD Installation and Configuring

A new HD needs to be recognized either by the systems board BIOS (basic input/output system) or by a **BIOS drive overlay.** The drive overlay acts as an interpreter between the HD and the systems board BIOS.

Some BIOSs have an 8-GB limit (older ones may have a 2-GB or even a 540-MB limit); however, most BIOSs can be updated by the user. Check the systems board manual to see if the systems board is equipped with a flashable BIOS EEROM. Systems board BIOS updates risk the chance of losing the BIOS (resulting in a nonfunctioning systems board), but may have other benefits such as updated options and support. It may also introduce new conflicts. Updating the BIOS is recommended for advanced users.

If the systems board BIOS doesn't support larger HDs, BIOS drive overlays are available with retail packages of HD or directly available from manufacturers' websites. Overlay works by modifying the master boot record (MBR), which is used by the BIOS when first booting. Installing or reinstalling Windows 9x normally overwrites the MBR, which will require using the overlay installation floppy to make the drive readable again. Drive overlay installation programs will usually walk the user through the installation process. The driver overlay does add a few seconds to boot-up time and requires an additional step to boot from a floppy, but doesn't noticeably decrease performance. Overlays may also cause some problems when using third-party partition utilities. For many of the preceding reasons, using a drive overlay should be the last resort for the technician.

Physical Installation

Up to four IDE devices can be installed on most personal computer systems today:

- Two IDE channels (primary & secondary) with one cable each.
- Two devices maximum per cable: 1 master (required) and 1 slave.
- Removable media drives normally need to be set as a slave if a HD is on the same channel.

When installing a hard disk drive, consider the following:

1. Power off the machine, but leave it plugged in.
2. Use an ESD wrist strap or other electrostatic protection device.
3. Set jumpers on drive to master or slave (some drives have a setting, "one drive only," which is the same as master).
4. When you need to move devices around to accommodate cable length and orientation, remember that the red ribbon on the cable corresponds to pin 1 on the device/motherboard.
5. If installing a 3.5-in. device in a 5.25-in. bay, you will need to use an adapter/brackets.
6. Be careful not to use long screws because they may enter the drive unit and damage the HD. Also, if the screw holes become stripped, it will be very difficult to remove or replace the unit.
7. Attach the 4-pin molex power connector. The power connector is keyed to fit in only one way, so do not try to force the connector in. If no power connectors are available and the power supply is sufficient, use a Y power splitter cable.

CMOS

Figures 13.34 and 13.35 illustrate typical CMOS hard disk setting. Begin by accessing the motherboard CMOS setup (press Delete, F2, Esc, or other key at boot-up):

```
                 ROM PCI/ISA BIOS (2A5IAE12)
                    CMOS SETUP UTILITY
                   AWARD SOFTWARE, INC.

Primary HDDs                CYLS.  HEADS  PRECOMP  LANDZONE  SECTORS  MODE
 Master :        (    Mb)

              Select Drive  Option (N=Skip) : N

OPTIONS     SIZE     CYLS.  HEADS  PRECOMP  LANDZONE  SECTORS  MODE
-------------------------------------------------------------------
  1(Y)      1083     2099    16     65535     2098      63    NORMAL
  2         1081      524    64         0     2098      63     LBA
  3         1082     1049    32     65535     2098      63    LARGE

                          ESC : Skip
```

FIGURE 13.34
CMOS setup screen

```
          ROM PCI/ISA BIOS (2A5IAE12)
             CMOS SETUP UTILITY
            AWARD SOFTWARE, INC.

Primary HDDs          CYLS. HEADS PRECOMP LANDZONE SECTORS MODE
 Master :     (   Mb)

          Select Drive  Option (N=Skip) : N

OPTIONS   SIZE   CYLS.  HEADS  PRECOMP  LANDZONE  SECTORS  MODE
----------------------------------------------------------------
 1(Y)     1083   2099    16     65535     2098      63    NORMAL
 2        1081    524    64         0     2098      63     LBA
 3        1082   1049    32     65535     2098      63    LARGE

                     ESC : Skip
```

FIGURE 13.35
LBA and large mode

1. Autodetect the new HD. If possible, choose the LBA translation mode, but refer to the drive's documentation first. If the HD is not detected or does not detect the full capacity, use the BIOS overlay utility. Alternately, refer to the documentation or contact the manufacturer for usable cylinder, head, and sector parameters.
2. Save settings and exit the CMOS setup (system will reboot).

Partitions

Partitions are a way of dividing up a single physical disk into several discrete portions. Partition sizes are limited by the FAT (file allocation table) version used. DOS/Windows allows a maximum of two partitions per HD: primary and extended.

- *Primary Partition.* The boot partition normally must be a primary partition and is assigned the lowest drive letter.
- *Extended Partition.* If the primary partition does not span the entire hard disk, the remaining space is assigned to an extended partition. The extended partition is not assigned a drive letter! Further "logical drives" are created within the extended partition. These logical drives are assigned drive letters.

FAT

The FAT is used by the OS (operating system) to find files on the disk. There are currently two versions in use by DOS, Windows 9x: FAT16 and FAT32.

- FAT16 sets a limit of 2-GB partitions, but can be recognized by Windows NT 4.0.
- FAT32 (with Windows 95 OSR2 and Windows 98) has a limit of 4-TB partitions.

There is a slight performance decrease compared to FAT16 (especially when using the SCANDISK or DEFRAG utilities).

Drive Letters

Drives are assigned letters in DOS and Windows according to the following scheme:

- First floppy = A; second floppy = B.
- ***Primary*** partition of master device on ***primary*** channel = C.
- ***Primary*** partition of slave device on ***primary*** channel = D.
- ***Primary*** partition of master device on ***secondary*** channel = E.
- ***Primary*** partition of slave device on ***secondary*** channel = F.
- Logical drives are assigned letters in the same priority scheme.
- CD-ROM drives and other removable peripherals (Zip drives, tape drives) = up to Z.

For example, a system has one floppy drive, two hard drives, and one CD-ROM drive. The floppy drive receives drive letter A. The first hard drive, the boot drive, is alone on the primary channel and is configured as master. It has a primary partition and one logical drive in an extended partition. The primary partition is assigned drive letter C. The logical drive letter assignment is deferred until all primary partitions receive a letter. Next, on the secondary channel are the second hard drive as master and the CD-ROM drive as slave. The second hard drive is configured as a single partition and is given the drive letter D. There are no more primary partitions, so the logical drive on the first hard drive is drive letter E. Last, the CD-ROM drive is given drive letter F.

Note that the first hard drive is given two letters out of sequence, C and E. This is not very intuitive, but becomes very important when adding a drive. If the technician is not careful, the wrong drive may be formatted when upgrading a system. Figure 13.36 illustrates the drive letter assignment of a typical system.

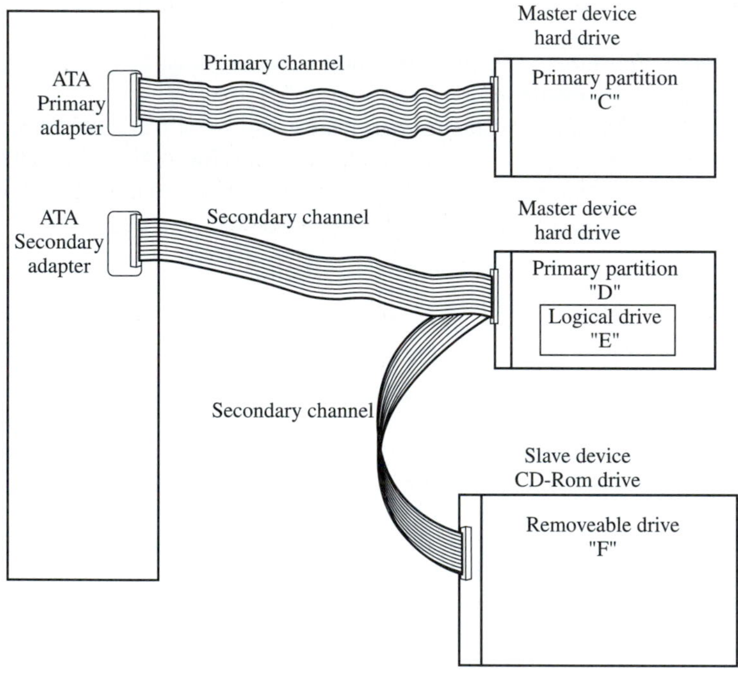

FIGURE 13.36
Drive letter assignment

FDISK

FDISK is the DOS and Windows partition configuring utility.

1. Boot to DOS either off of an existing bootable HD or from a bootable floppy disk.
2. In the directory where the FDISK utility is located, run FDISK.
3. If the FDISK is from a later version of Win9x, it will ask you if you want to enable large disk support (FAT32). Type N to use FAT16 (default) or Y to enable FAT32.
4. If there is more than one physical disk, use option 5 to choose the correct HD.
5. Use option 4 to verify any existing partition information.

6. Use option 1 to create a primary partition of the size you desire.
7. If this HD is to be the boot drive, set the primary partition to "active" with option 2.
8. If you did not use the entire HD for the primary partition, create an extended partition.
9. Create the desired number of logical drives within the extended partition.

If you want to use another OS on the same HD, leave some space on the extended partition unused. Figures 13.37 and 13.38 illustrate the use and menus of the FDISK utility.

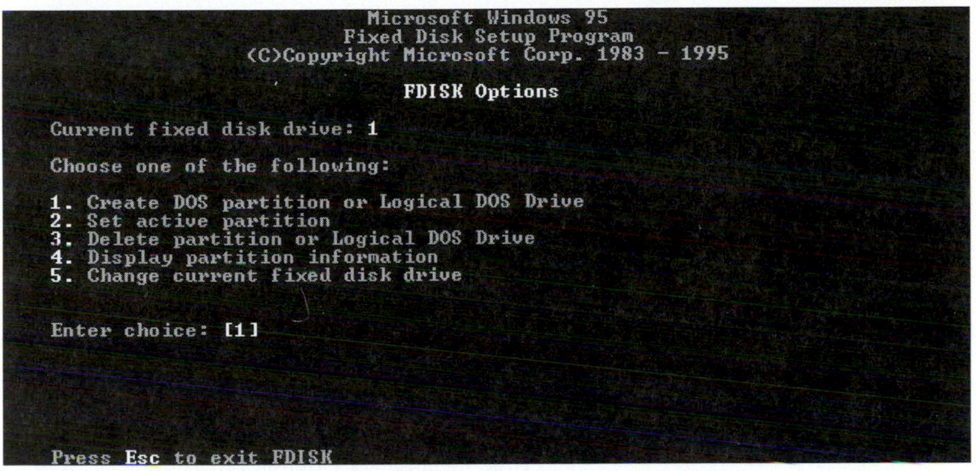

FIGURE 13.37
FDISK menu

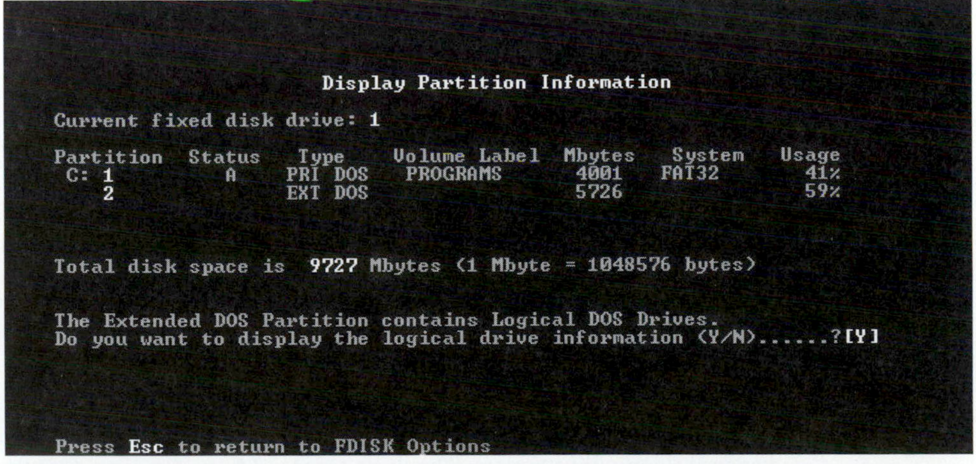

FIGURE 13.38
Display Partition screen

Formatting

Formatting creates a file system for OS use. The FORMAT command for DOS and Windows is FORMAT. Use FORMAT with the ? switch (FORMAT /?) for usage and syntax notes.

1. Format with FORMAT d: /s to copy system files over to the new disk (system files are necessary to make a HD bootable).
2. If more than one partition was created, each one must be formatted with FORMAT *x*:, where *x* is the letter of the drive. Only the boot drive (where the OS resides) needs to be formatted with the /s switch.
3. At the end of the format, the program will ask you for a drive label. Use descriptive labels to keep the HDs straight. For example, in a two-drive system, you may want to label the first drive "windows," and the second drive "data". The drive label can be changed later on with the LABEL command.

Hard Drive Manufacturer's Websites

Manufacturers generally keep information about their hard drives on their websites. This includes setup parameters, jumper settings, and disk utilities. Table 13.3 lists URLs of a number of popular hard disk manufactures.

TABLE 13.3
Hard Disk Manufacturer's Websites

Manufacturer	URL
Fujitsu	*http://www.fujitsu.com/*
IBM	*http://www.storage.ibm.com/storage/*
Maxtor	*http://www.maxtor.com/*
Quantum	*http://www.quantum.com/*
Seagate	*http://www.seagate.com/*
Western Digital	*http://www.westerndigital.com/*

13.5 REMOVABLE MEDIA DRIVES

Along with advances in hard drive technology has come growth in the removable storage area. Some of these technologies continue the evolution of the floppy drive, with the ability to backup data, while others are strictly read-only media.

Zip Drives

The **Zip drive** was introduced by Iomega and has gained significant popularity. It is based on Iomega's Bernoulli technology, making it completely incompatible with the standard floppy drive system. The drive accepts 3.5-in. Zip cartridges that are about twice as thick as floppy disk, and store 100 MB or 250 MB of information. The design of the Zip drive makes it quite fast as compared to the floppy drive. It has a 1 Mbps transfer rate and an access time of 29 ms.

The Zip drive comes in a number of flavors. It can be installed in your computer using SCSI or IDE, or it can reside outside the computer and connect using SCSI, USB, or a parallel port. It is this versatility that has added to the Zip's popularity. Most system BIOS's today come with support for the Zip drive built in. This means that you can boot off the Zip drive and therefore would not need to have a floppy disk drive; however, since the Zip cannot read floppy disks, you may still need one to load software or drivers.

LS-120 Drives

The LS-120 drive was introduced by the Imation (3M) corporation. It is designed around the industry standard 3.5-in. floppy disks. The LS-120 can read and write to 3.5-in. floppy disks as well as its own 120 MB media, and can do so five times faster than the standard floppy.

The LS-120 installs into your machine just like a standard floppy with the exception that it uses the IDE interface instead of the floppy interface. This allows it to transfer data much faster. Most system BIOSs today have built-in support for the LS-120, allowing you to boot off of one of these drives, and since they can read and write standard 3.5-in. floppy disks, they can totally replace the venerable floppy drive.

The LS-120 drive works by using a laser beam to track the movements of the read/write head and place it more accurately over the medium surface. This allows the LS-120 to write 83 times more data to a single LS-120 disk than the standard 1.44 MB 3.5-in. drives. The LS-120 actually stores the extra information in the spaces between tracks in the 3.5-in. disk standard.

CD-ROM Drives

Compact disk–read-only memory (CD-ROM) media can hold up to 650 MB, large enough to hold the installation files for large program suites or even operating systems. Also being relatively cheap to create and more durable than floppy disks, CD-ROMs have become the medium of choice for distributing software. Due in part to this fact, CD-ROM drives has become as common as floppy drives in most systems.

CD-ROM technology uses optical media, as opposed to magnetic media. To create a CD-ROM, a laser is used to write digitally encoded data onto a blank disk. A high-power laser diode is used to burn microscopic craters into the substrate of a CD-ROM. These craters are called **pits,** and the areas left unburned are called **lands.** This binary encoding can be read just like the magnetic flux reversals on magnetic media. To read from a CD-ROM, a lower-powered laser is focused on the series of pits and lands. A diode detector will see the light reflected by the lands, whereas the pits will scatter the light and not be detected. Unlike the tracks of a floppy disk, which are composed of concentric rings, the tracks of a CD-ROM are written in a spiral fashion that originates at the middle of the CD.

CD-ROM drives are often labeled IDE or EIDE to distinguish them from older proprietary models, but CD-ROM drives do not use the IDE or ATA command set. Instead, the CD-ROM drive and certain other removable drives use the **ATA packet interface (ATAPI).** The ATAPI command set resembles SCSI more than IDE, since the ATA command set is not sufficient for removable storage command structures. The reason for creating the hybrid command set is that the EIDE interface is so commonly used and low in cost that it made sense to adapt CD-ROM and tape devices to use the existing architecture. There are also SCSI CD-ROM drives available, though they are rarely used except in all SCSI systems.

CD-ROM drive speeds use the audio CD spinning speed as their reference, designated 1× with a data read rate of 150 Kbps. A 40× speed CD-ROM drive should therefore theoretically reach read rates of 12,000 Kbps, but in reality rarely reaches half as much. Just as the hard disk platter has higher transfer rates at its outer edge, the CD-ROM has a differential transfer rate. Early CD-ROM drives compensate for this by using a variable rpm and a **constant linear velocity (CLV).** As CD-ROM drive speeds increased beyond the 8× speed, CD-ROM drives return to the hard disk method of a constant rpm or **constant angular velocity (CAV).** Recent CD-ROM drives often use a combination of these two methods. Regardless, CD-ROM drives rarely live up to their speed rating. The more accurately rated CD-ROM drives give a range of speeds such as 12×/24×, expressing the physical constraints of a disk-based storage system.

Installation

CD-ROM drives only fit in 5.25-in. drive bays (excluding laptop models) and use the same IDE cable connector and power connector as hard drives. They are physically installed the same way as hard drives, though some drives may need to be configured as slaves even when they are the only drive on the channel. If the CD-ROM drive is to be used to listen to audio CDs, then an audio cable must be used to connect the CD-ROM drive to a sound card or integrated sound device. The audio connector on the drive is usually a standard four-pin MPC2 female connector or a two-pin digital SPDIF female connector. The connector on the sound card or sound device may be a four-pin MPC2 female connector, though four-pin Sound Blaster or other proprietary female connectors are still common. In all cases, the connector is keyed for the correct orientation. Figure 13.39 illustrates a typical MPC2 audio connector.

CD-ROM drives cannot be autodetected or configured in the BIOS setup, but the BIOS can recognize them, as it can hard drives, during boot-up. You will often see the CD-ROM drive's manufacturer information displayed after the BIOS power-on self-test

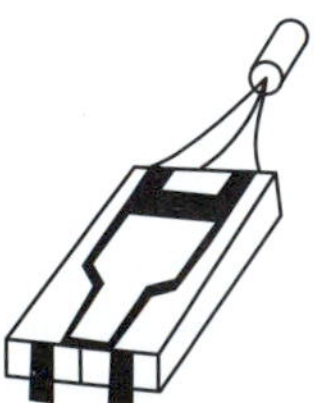

FIGURE 13.39
MPC2 audio connector

(POST), which the BIOS obtains from the drive's ROM. Also unlike hard drives, CD-ROM drives usually need a device driver to be recognized by an operating system. Windows 95/98 provides a **virtual device driver (VxD)** for CD-ROM drive support, but the real mode (DOS-compatible) device drivers are still needed to load the Windows OS from the CD in the first place. Retail drive packages may include an installation disk with the appropriate CD-ROM drive device driver. If possible, refer to the manufacturer for the correct way to modify a system's CONFIG.SYS and AUTOEXEC.BAT configuration start-up files to run the CD-ROM drive device driver.

Becoming increasingly popular as their prices have become more affordable are **CD recordable (CDR)** and **CD rewritable (CDRW)** drives. CDR drives use special CDR media either to make copies of other CDs or to write from hard disk to CD. CDR media can only be written to (all at once, in tracks, or in sessions) but cannot be erased. The writing, or burning, process requires a constant stream of data, and any pauses or buffer underruns can result in an unusable CD. CDRW drives improve upon CDR drives by using an additional medium, CDRW disks. CDRW media function just like CDR blank medium, but they can be erased after use. This makes a very durable backup media since optical media are more resistant to shock damage and not susceptible to the loss of electrical charge that makes magnetic media vulnerable over the long term. The almost universal inclusion of CD-ROM drives in PC systems make the CDR/CDRW media an attractive distribution and backup system, but even 650 MB may be too limited for some users. The reflective properties of CDR and CDRW media are different from commercially produced CDs, however, and earlier CD-ROM drives had difficulties reading them. Most CD-ROM drives now available should support both media.

DVD Drives

The **digital versatile disk (DVD)** continues the evolution of the CD-ROM with a much larger capacity ranging from 4.7 to 17 GB. The DVD uses the same size medium as the CD, but a laser with a smaller wavelength is used to encode the data. This allows the pits and lands to be more densely packed, with the tracks laid more closely together. Also, a more efficient data encoding and error correcting scheme is used compared to CDs. The different capacities result from the DVD specifications for single or dual layers, and for both sides of the disk to be used. The second layer does not double the capacity of a single layer DVD but adds an additional 3.8 GB. Most current DVD drives can only read a single side, so the DVD must be flipped for half of the data to be available on double-sided disks. Despite their differences, the DVD's physical appearance is identical to a CD-ROM. Table 13.4 lists various DVD capacities.

TABLE 13.4
DVD capacities

DVD Format	Capacity
Single side, single layer	4.7 GB
Single side, dual layer	8.5 GB
Double side, single layer	9.4 GB
Double side, dual layer	17 GB

The main push for DVD was for a medium that could hold an entire digitally encoded movie with theater quality digital sound. This explains the large capacities that the DVD is capable of holding. Because the data being streamed from the DVD is much more complex and larger than the relatively simple audio data of music CDs, the reference, or 1× speed, transfer rate is 1108 Kbps.

DVD drives are backward compatible with CD media, and most current models also support CDR and CDRW media. DVD drive specifications therefore list two speeds, the DVD read speed followed by the CD-ROM read speed. Current DVD drives reach 10×/40× speeds. Just as with the disk itself, a DVD drive's physical appearance is identical to a CD drive, the only difference being a DVD logo on the faceplate. In installation and configuration, DVD drives are the same as CD-ROM drives.

Tape Drives

With the ever-growing capacities of hard drives, one storage device remains as an economical and convenient backup system. With a hard drive size of 30 GB, it would take over 47 CDs or 313 100-MB Zip disks to completely back up the drive, and it would require the user to be present to change media until the backup was complete. Neither of these conditions is conducive to a regular backup program.

The importance of a carefully thought out and strictly followed backup program cannot be stressed enough. Although the operating system and applications can usually be rebuilt fairly quickly, the loss of data files and the downtime they represent to the user is often prohibitively expensive. The frequency and scope (full or incremental) of the backup will differ from situation to situation and will depend upon such factors as the frequency that the data changes, the value of the data, and the amount of resources available for backup.

Earlier tape drives interfaced with the system through the floppy drive connector or other proprietary adapters, but this was found to be too slow for today's needs. Almost all tape drives available now use the SCSI standard or the ATAPI/EIDE interface. There are several tape media formats, but the most common is the TRAVAN or QIC tape with a capacity of 8 to 25 GB per tape. Be aware that the advertised capacity of some tape systems assume compression is used with a 2:1 compression factor. This compression scale is optimistic, and the technician should base the required tape capacity using the uncompressed value and allowing for moderate growth of the target system. The installation of tape drives is the same as that of CD-ROM drives, because they use the same connectors and the same 5.25-in. bays.

13.6 TECH TIPS AND TROUBLESHOOTING—T^3

When trying to gauge capacity for a disk drive upgrade, a good rule of thumb is to at least double the current usage. Even when replacing a defective drive, it is not usually economical to replace a drive with one of similar capacity because drive capacity increases so rapidly. A replacement drive of the same capacity may cost little less, or even more than a larger-capacity drive.

Keep program files and data files separate by placing them on separate partitions. Install Windows and all program files on C, and use D for data files. Backing up the OS and the data separately allows more versatile restoration schemes. Defragment your HD often (whenever more than 10% fragmented) to keep performance from degrading. Low-level formatting is a process that normally only occurs at the factory, before a drive is shipped. Using a low-level format program on today's drives may result in total disk loss, so use one only as a last resort, with great caution.

EXERCISES

13.1 Name the low-density storage medium made up of a long strip of paper.

13.2 A material that will allow magnetic flux lines to penetrate and pass through easily is a ___________ ___________ material.

13.3 What is the term used to describe how well a material will hold a magnetic charge?

13.4 If a material can become magnetized and demagnetized easily, it is said to have a high ___________.

13.5 What is the most common material used for magnetic recording?

13.6 What is the key component for magnetic recording?

13.7 Name the device that will become magnetized only while an electric current is applied.

13.8 What is actually recorded on magnetic media?

13.9 Is it possible to record DC onto magnetic media?

13.10 The digital read chain will use a ___________ circuit to help restore the digital waveform.

13.11 When the playback signal begins to roll off, the differentiator will begin to ___________.

13.12 The process of converting digital data into a recordable pattern is known as ___________ the data.

13.13 For what does FSK stand?

13.14 Explain the two rules for frequency modulation.

13.15 Explain the advantage of modified frequency modulation.

13.16 Explain the two rules for modified frequency modulation.

13.17 True or false? There will be a flux reversal for every bit cell in modified frequency modulation.

13.18 Disk drive types can often be identified by the number of pins on the cable that connects to the drive. Identify the following drive types:

a. 34 pin
b. 40 pin
c. 50 pin

13.19 What two measurements of hard drive specifications are the best indicators of hard disk performance?

13.20 Why is the actual capacity of a hard drive often less than expected?

13.21 How many IDE devices can a typical systems board support?

13.22 What is the difference between the interface transfer rates and the disk transfer rates?

13.23 How many partitions and of what type does DOS/Windows 95/98 allow on a single drive?

13.24 A system has one floppy drive, two hard drives, and one Zip drive. The first drive is alone on the primary channel. On the secondary channel are the second hard drive and the Zip drive. The second hard drive has a primary partition and two logical drives in the extended partition. How should each device be configured (master/slave) and what drive letter would each device receive?

13.25 What switch option should be used with the DOS FORMAT command to make a drive bootable after formatting?

13.26 What is the transfer rate of the reference or 1× CD-ROM drive?

13.27 What is necessary to listen to music CDs on a PC CD-ROM drive if the system already has a sound card and speakers?

13.28 How is the DVD able to hold so much more data than a CD-ROM?

13.29 What is the capacity of a single-side, dual-layer DVD?

13.30 Why is the magnetic tape the logical choice for backing up data on today's hard drives?

13.31 Crossword Puzzle

ACROSS

4. The first thing that is done to a new diskette.
6. The motor that moves the head from track to track.
8. One recorded circle on the diskette.
10. One millionth of 1 s.
12. A measurement of speed.
13. MFM records in _____.
14. The alignment used to place the head over the track center.
15. A device that lets current flow in one direction only.
17. The notch on the side of the diskette.
19. The ID portion of a sector.
22. Modified frequency modulation.
23. The ability to hold a magnetic charge.
24. Stored data can be called _____.
27. The speed sensor in the motor.
28. Another word for wobble.

DOWN

1. File allocation table.
2. Instantaneous speed variation.
3. A diode that produces light.
4. A type of magnetic material.
5. The last part of a sector.
7. The first part of a sector.
9. A device that senses light.
11. Information on a diskette.
15. Where the file names are stored.
16. The ease with which magnetic flux lines will pass through a material.
18. Part of one track.
20. Frequency modulation.
21. The reference hole on the diskette.
25. Part of a disk drive that centers the diskette.
26. The part of a disk drive that reads and writes data.

1 2 3 4 5 6 7 8 9 10 11 12 13 14 15 16 17 18 19 20 21 22 23 24 25 26 27 28

CHAPTER 14

PERIPHERAL DEVICES

KEY TERMS

Anode
Asynchronous Serial Data
Baud Rate
Cathode
Cathode-ray Tube (CRT)
Character Generator
Color Graphics Adapter (CGA)
Composite Video
Convergence Yoke
Daisy Wheel Printer
Data Circuit Terminating Equipment (DCTE)
Data Terminating Equipment (DTE)
Deflection Yoke
Dot Clock
Dot-matrix Printer
Dot Pitch
Electrostatic Deflection
Enhanced Graphics Adapter (EGA)
Filament
Firewire
Full Duplex
Grids
Handshaking
Hard Copy
Half Duplex
Impact Printers
Interlaced Scanning
Ink Jet Printer
Keyboard Encoder
Laser Printer
Magnetic Deflection
Modem
Modulator-demodulator
Multimedia
Nonimpact Printers
Null Modem
Pixels
Raster
Reactance Scanning
Retrace
RS-232C
Start Bit
Stop Bit
Super VGA (SVGA)
Thermal Dot-matrix Printer
Triads
USB
Video Graphics Array (VGA)

14.0 INTRODUCTION

A computer is useless without input and output devices. These devices, also known as peripherals, include keyboards, displays, modems, and printers. In this chapter we discuss the theory of operation of keyboards and video displays. We also review modems and printers.

14.1 KEYBOARDS

The keyboard is the simplest input device on a computer system; it is merely a collection of momentary switches. The outputs of the key switches are fed to electronic circuitry known as the **keyboard encoder,** which converts them into binary-coded values. These values are then fed into the computer, which interprets the key pressed.

The design of computer keyboards has evolved over time and differs among manufacturers. On some of the earlier computers the keyboard output was many parallel wires connected from the key switches into the computer, as shown in Figure 14.1.

To reduce the number of wires required to interface the keyboard, the key switches can be arranged into a *matrix,* as shown in Figure 14.2. In the matrixed configuration the key switches are arranged as 4 columns by 4 rows of switches. The interface reads the keys by selecting one column at a time and then reading the row outputs. In Figure 14.2, a 16-key keypad was shown for simplicity. The matrixed configuration reduced the wire count from 16 to 8. If we apply the matrixed configuration to the typewriter-style keyboard consisting of 64 keys, the wire count can be reduced from 64 to only 16 by arranging the keys into 8 rows of 8 columns (Figure 14.3). Virtually all modern keyboards are set up this way.

Keyboard Encoding Methods

Most modern computers have the keyboard encoding circuitry built into the keyboard assembly. Some exceptions are low-cost home computers, which use the system processor to scan and encode the key switches. Some earlier encoding designs used a large diode matrix to encode the key switches, as shown in Figure 14.4.

Encoding the output of the matrixed keyboard requires more logic circuitry than the parallel keyboard because the key switches must be scanned by using a counter, as shown in Figure 14.5. Each column must be scanned for a pressed key. If a key is pressed, the counter stops when the column with the pressed key is selected. The value that is in the counter reflects the binary value of the key pressed.

With changing technology, many manufacturers put the keyboard encoding circuitry on a single large-scale IC. The method used by many manufacturers today is a single-chip microcomputer, which scans the key switches and encodes the output, as shown in Figure 14.6. The operation of the single-chip microcomputer is quite simple: All the column wires are connected to one port of the microcomputer, and the row wires are connected to a second port. A program simulates the hardware model to scan the keys and encode the key pressed to a binary value. Changing the encoded value for a key or keys requires only a program change and not a hardware change. At first glance this may seem like overkill, but the economy and flexibility of using a standard microcomputer for this application is very attractive. The encoded data may then be converted to bit-serial format by the single-chip microcomputer, which further reduces the number of conductors to just three or four wires, allowing a thin, flexible cable to be used.

The keyboard used on the IBM PC is controlled by an INTEL 8048 single-chip microcomputer, as shown in Figure 14.7. The key switches used in the IBM keyboard are not contact switches but capacitive switches. When a key is pressed, a conductive plastic plate falls down on the etched plates on the PC board, which changes the capacitance of the etched plates. The capacitance change is detected by the sense amplifier, chip Z_1.

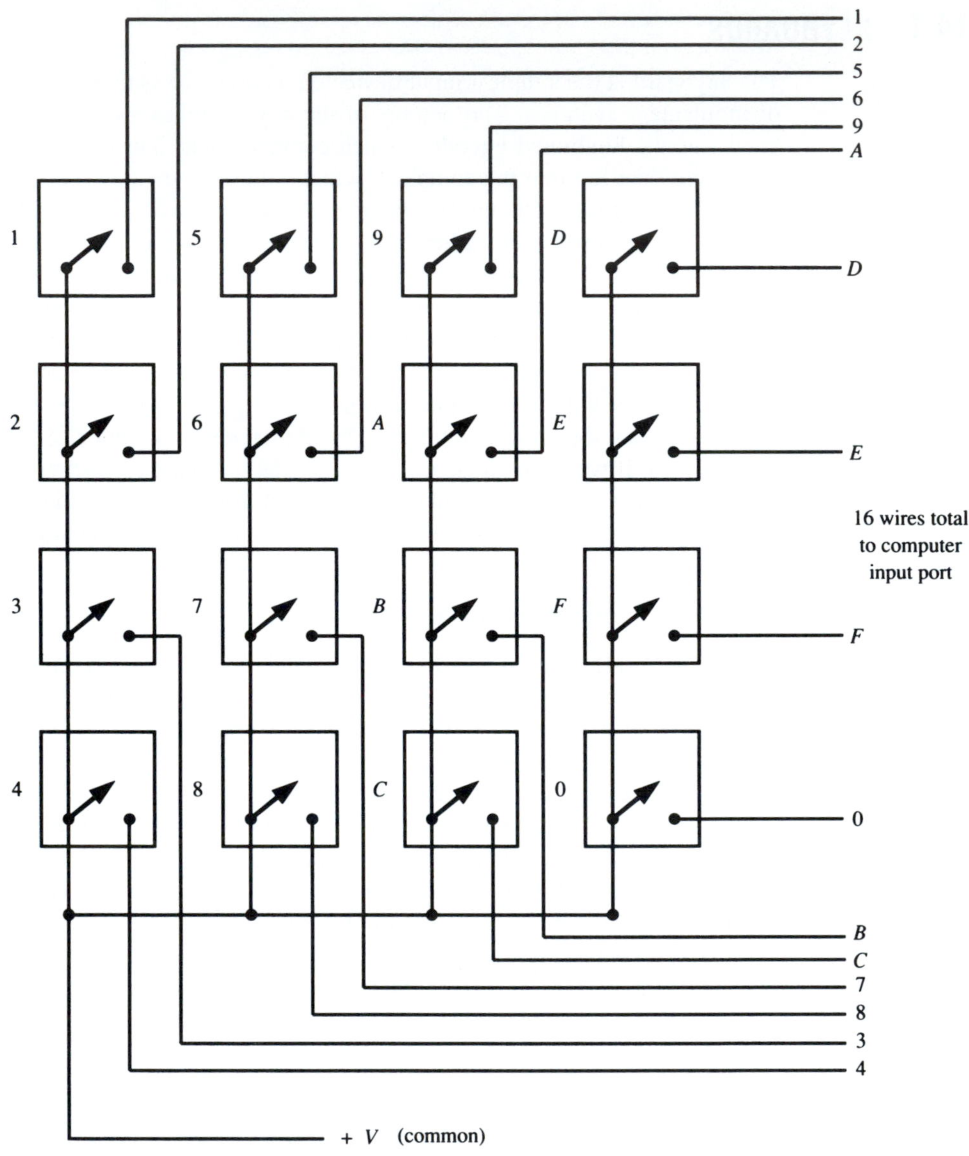

FIGURE 14.1
Simple output keypad

14.2 VIDEO DISPLAYS

Most modern computer systems use *raster scan video displays* as the primary user output device. The raster scan video display is very much like a conventional home television receiver. In fact, many low-cost home computers use a television as their output de-

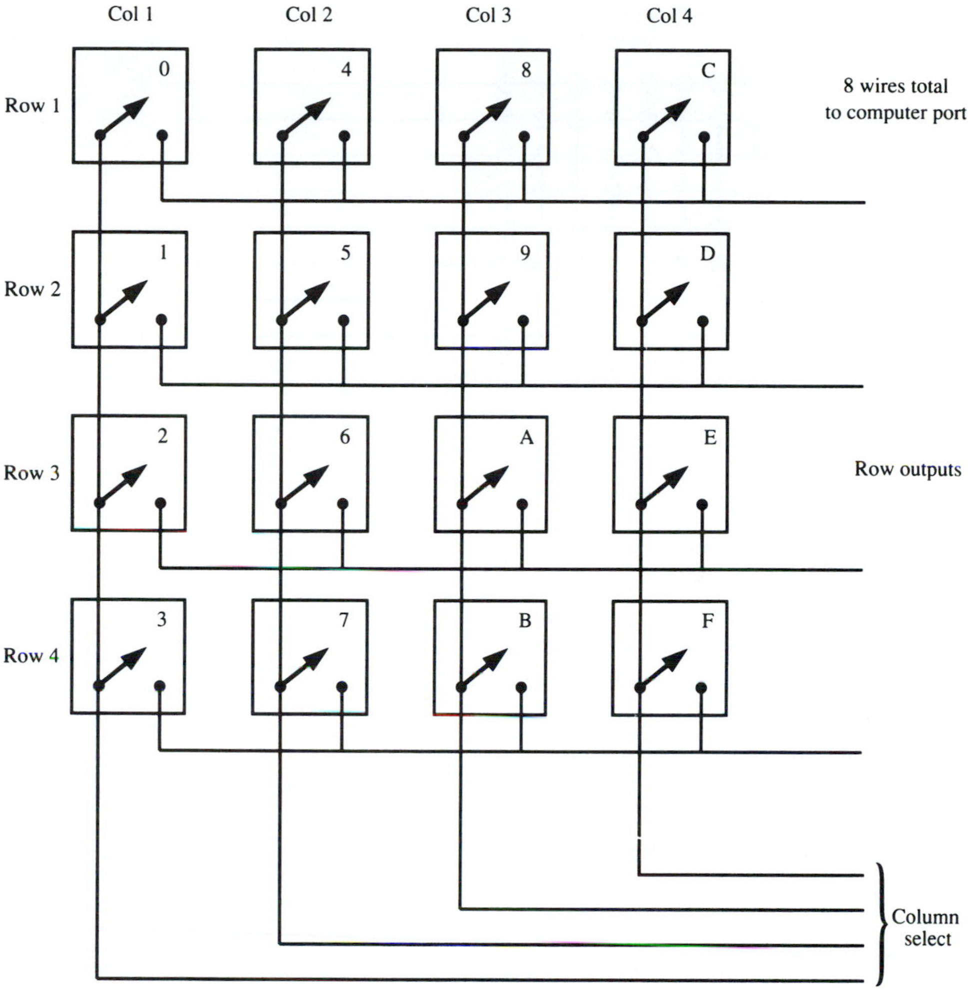

FIGURE 14.2
Matrix output keyboard

vice. The major drawback of using a television is the lack of resolution or fine detail, which is discussed later in this chapter.

The video display is a **cathode-ray tube (CRT).** An electron beam, which originates at the **cathode,** is scanned across the phosphor-coated face of the CRT by the deflection circuit. The intensity of the electron beam is controlled by voltage changes on the cathode with respect to grid 1, the control grid. The electron beam is pulled across the face of the CRT from left to right and top to bottom, as shown in Figure 14.8. The left-to-right motion, which is the horizontal rate, is much greater than the top-to-bottom motion, which is the vertical rate. The period of time during which the electron beam is returning to the top or left of the screen is known as the **retrace** time. During this time, the electron beam is

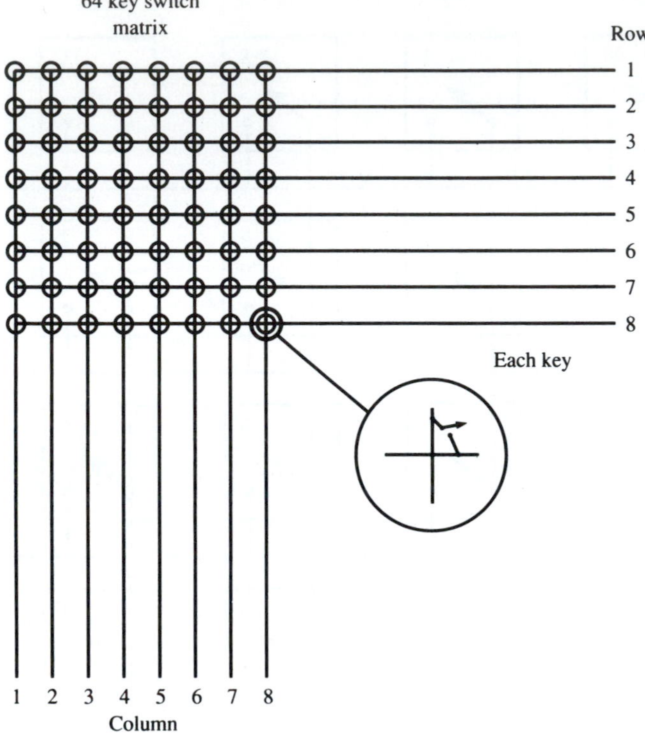

FIGURE 14.3
64-key keyboard

cut off, or blanked. The two rates for the horizontal and vertical deflection form a series of horizontal lines on the face of the CRT. This series of horizontal lines is known as the **raster.** The formula for obtaining the maximum number of lines in the raster is

$$\text{Max scan lines} = \frac{\text{horizontal rate}}{\text{vertical rate}}$$

The standard television scan rates are 15,750 Hz for the horizontal rate and 60 Hz for the vertical rate. When we divide the horizontal rate by the vertical rate, the result is 262.5:

$$\text{Scan lines} = \frac{15{,}750}{60} = 262.5$$

This means that the *maximum* number of vertical lines is 262.5. Since some lines must be allowed for overscan and retrace time, slightly fewer are available. For example, the IBM color graphics adapter uses the standard scan rates and allows 200 vertical dots, or **pixels.** It is possible to double the number of vertical lines by using a method known as **interlaced scanning,** in which the odd raster starts at the top left edge and the even raster at the top middle of the CRT, as shown in Figure 14.9. The even lines fall between the

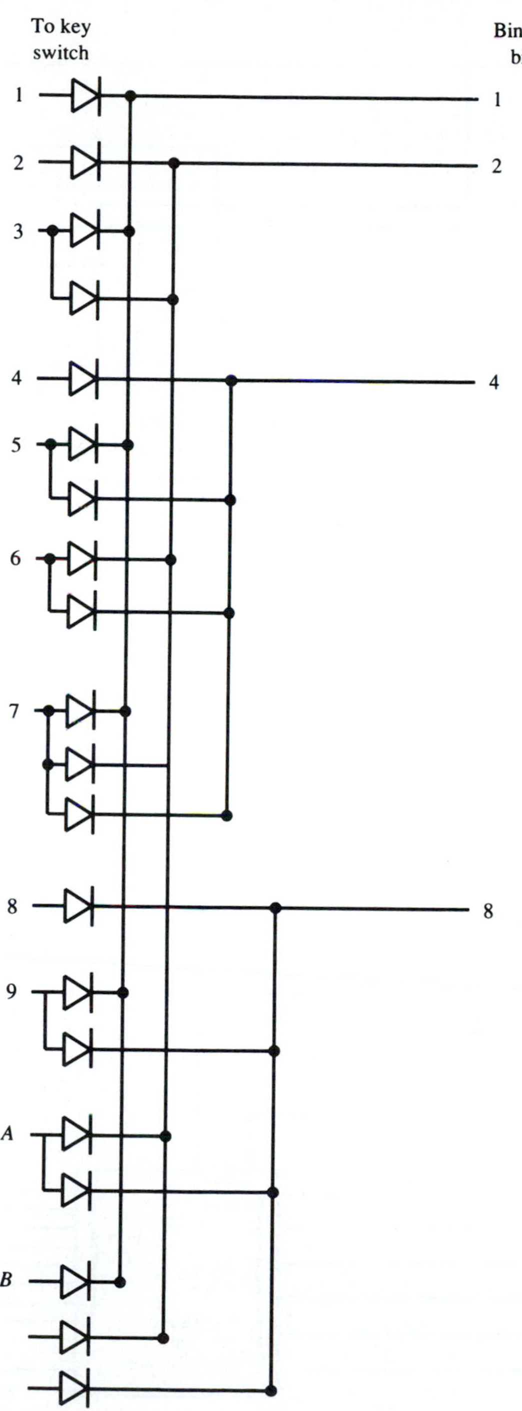

FIGURE 14.4
Diode encoder

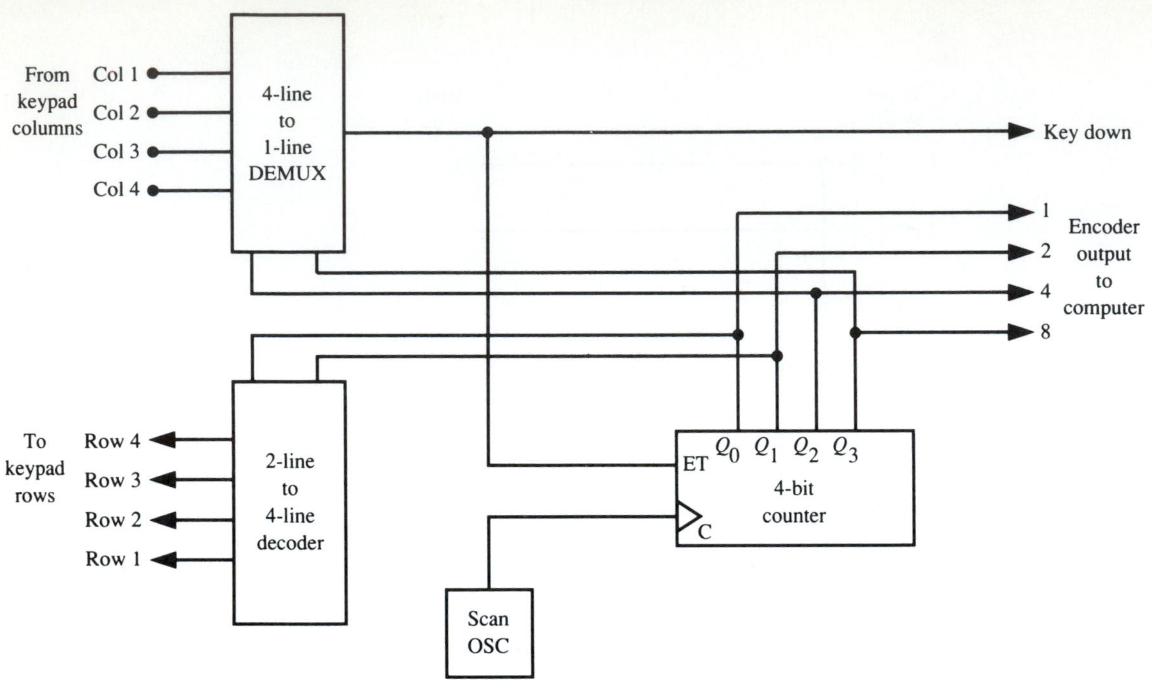

FIGURE 14.5
Scanning matrix decoder

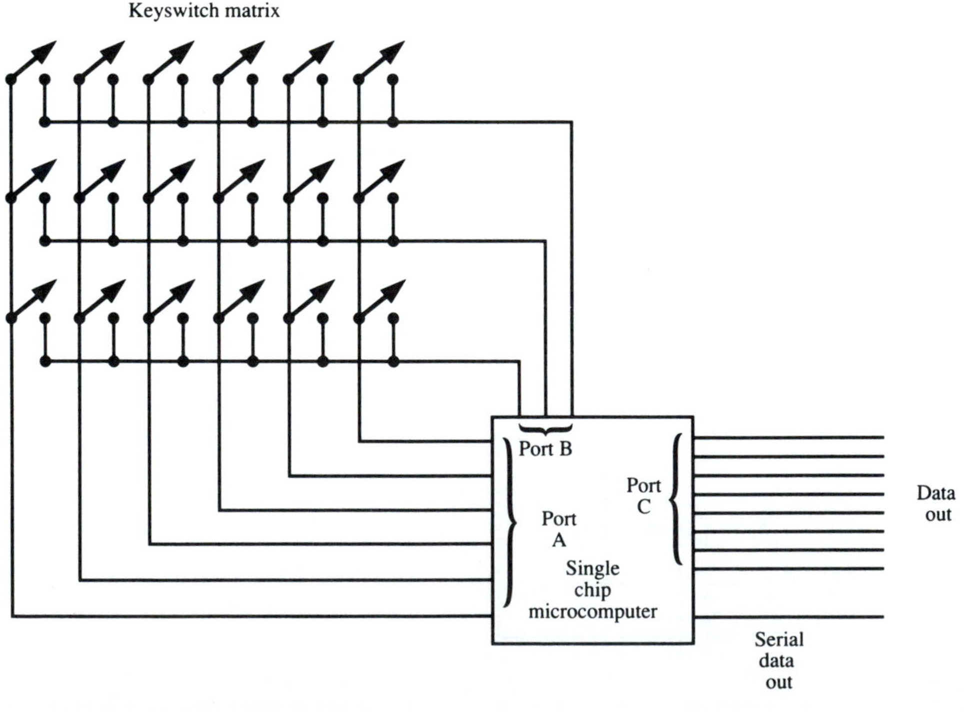

FIGURE 14.6
Microcomputer keyboard encoder

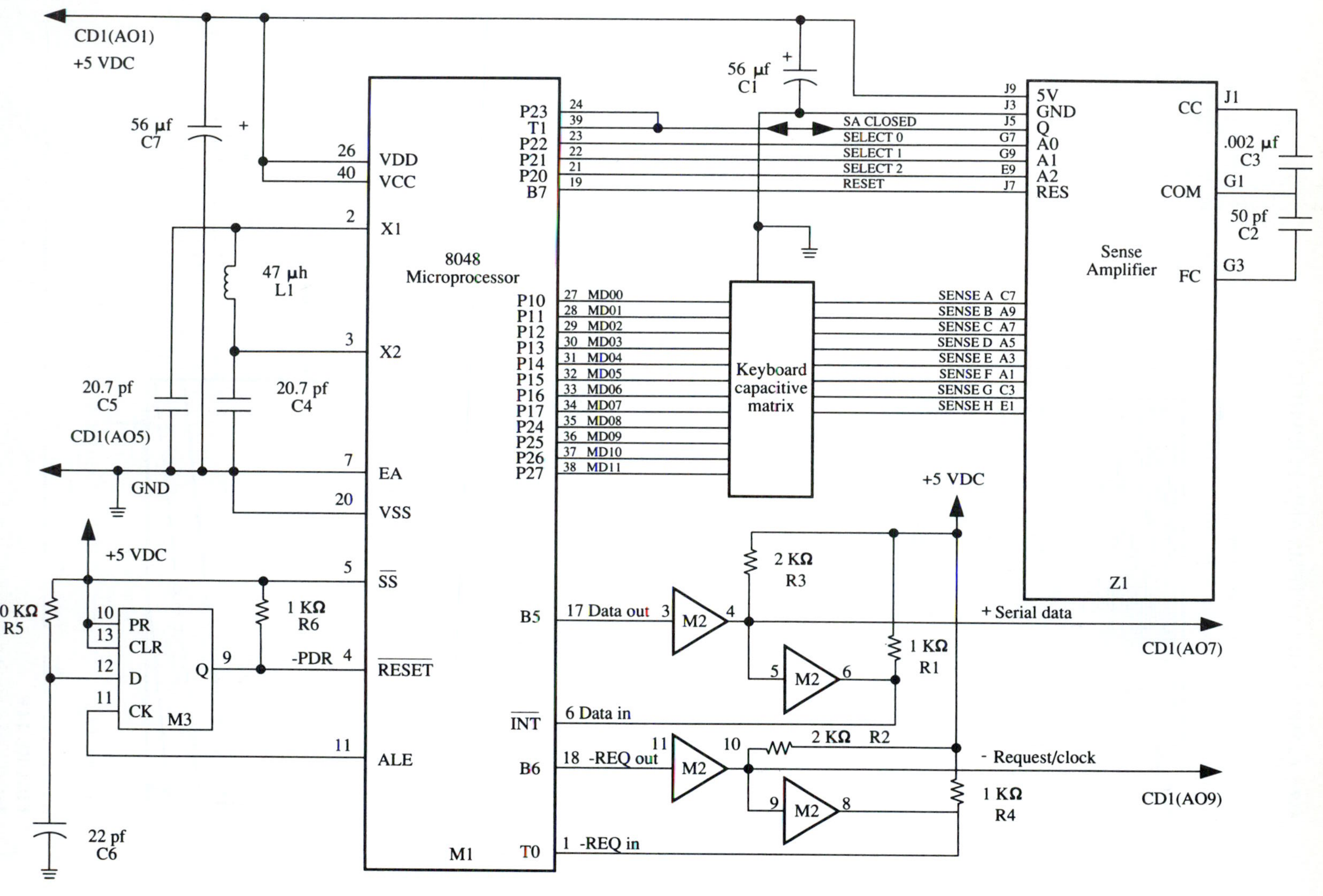

FIGURE 14.7
IBM keyboard (Courtesy of International Business Machines Corp.)

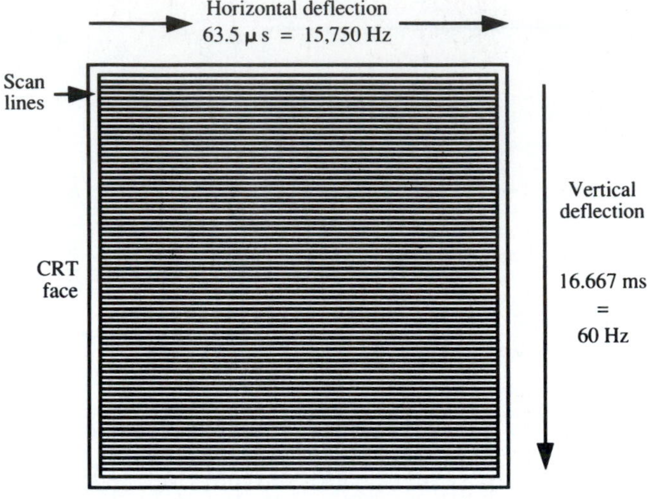

FIGURE 14.8
The raster display

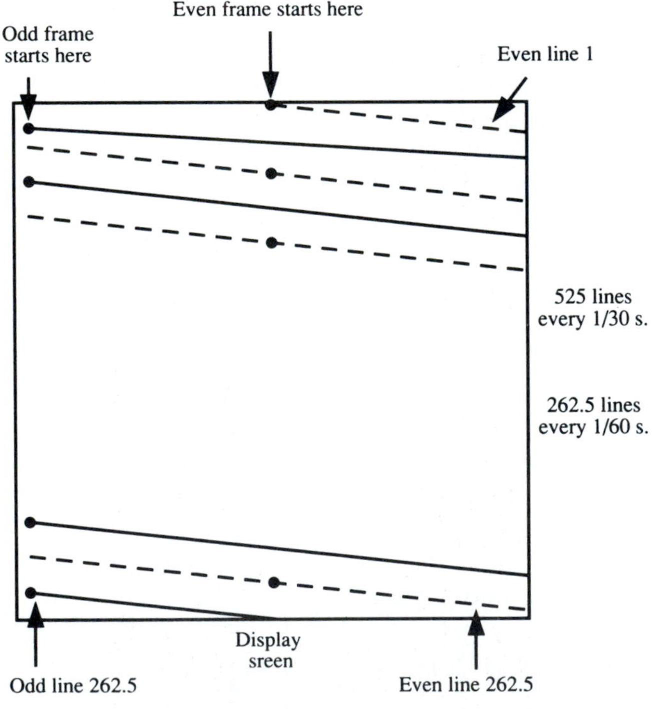

FIGURE 14.9
Interlaced scanning

odd lines, which results in a maximum total of 525 vertical lines. The interlaced scanning method is used in standard broadcast television but not in most computer display applications because the overall frame rate is reduced from 60 frames per second to 30 frames per second. As we discussed, the screen must be scanned *twice* to complete one frame. In broadcast video there are infinitely varying levels of gray in the picture, so differences between the odd and even lines are not noticeable on a line-by-line basis; however, in digitally generated video there are abrupt changes in the levels of intensity, which are perceived as flickering, or jumping, on the screen. This flickering is especially evident when text or fine graphic lines are displayed. Studies show that the human eye cannot perceive light variations greater than 45 cycles per second due to the eye's persistence of vision. The IBM monochrome display adapter uses a 350-line display by changing the horizontal rate to a frequency of 18,432 Hz and the vertical rate to 50 Hz. If we use the scan-line formula, we have

$$\text{Scan lines} = \frac{18{,}432}{50} = 368.64$$

maximum vertical lines. A CRT with a long-persistence phosphor is also used here because the frame rate is near the flicker rate perceivable by humans.

14.3 THE CRT

The CRT, which is commonly known as the **picture tube,** is a large glass envelope containing an electron gun. All air must be evacuated from the CRT in order to create a nearly perfect vacuum. A vacuum in the CRT is necessary for three reasons. First, if air molecules are present in the CRT, the electron beam scatters when it strikes them. Second, gases ionize when subjected to high voltage. Ionized gases are conductive and effectively short out the electron beam. Third, the filament burns up in the presence of oxygen. The basic elements of the conventional CRT are the filament, cathode, grids, and the anode, as shown in Figure 14.10.

When heated the cathode begins to emit electrons. It is heated by the **filament,** which is sometimes called the heater. The **grids** control the flow of the electrons from the cathode to the **anode,** which is the most positively charged element. Since electrons are negatively charged particles, they are attracted toward a positive potential and repelled away from a negative potential.

Electron beams are also affected by magnetic fields. Since current is flowing through the electron beam, a circular magnetic field is associated with it. When a magnetic field is placed through an electron beam, the beam is deflected within the externally induced magnetic field in the same way that two magnets attract or repeal each other. This phenomenon is known as **magnetic deflection.**

The *inside* face of the CRT is coated with phosphor, which is a material that exhibits luminescence (it glows) when excited by electrons (beta radiation) or other sources of radiation. When the electron beam from the cathode strikes the phosphor, it produces a bright dot of light, which is known as cathode-luminescence. The intensity of the dot of light is directly proportional to the intensity of the electron beam, which can be controlled by varying the voltage differential between the cathode and the control grid. When the control grid becomes more negative than the cathode, somewhat greater than −60 V, the electron beam is cut off. As the voltage at the control grid becomes less negative, some

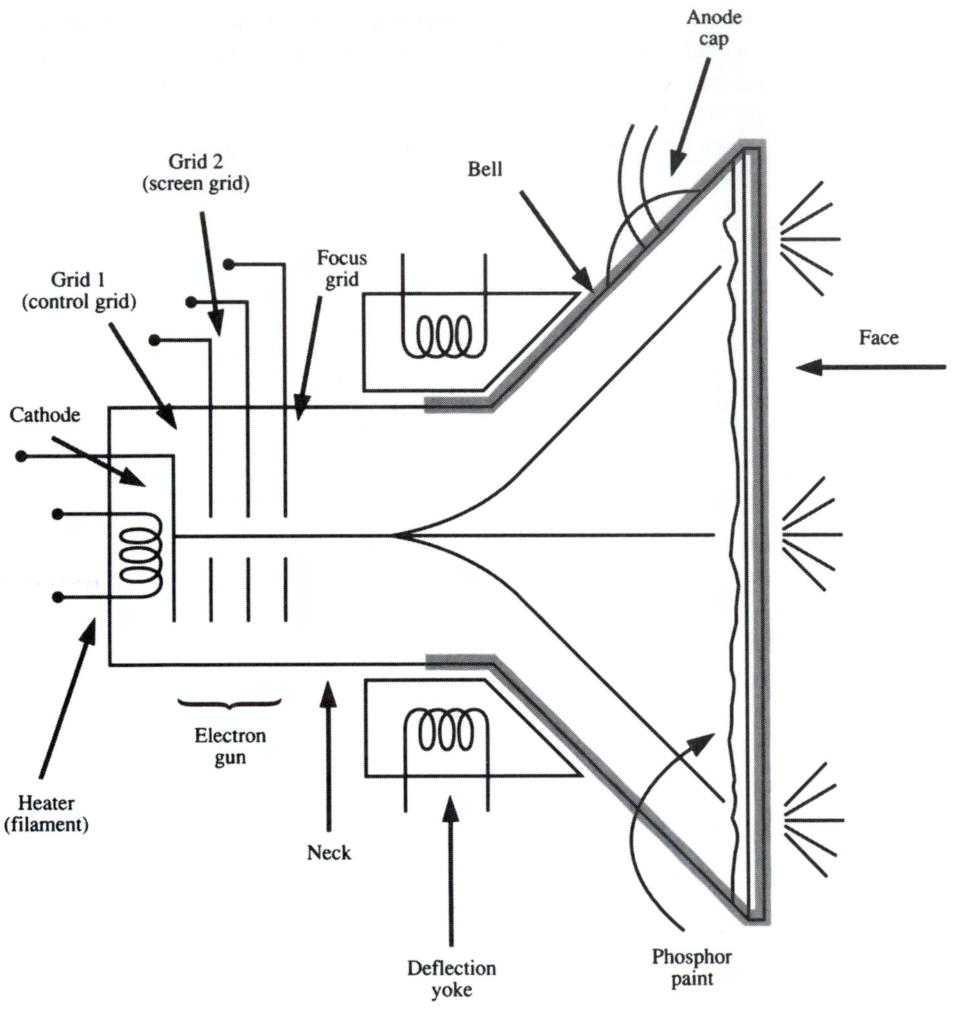

FIGURE 14.10
The CRT

of the electrons begin to pass through the control grid and strike the phosphor. The intensity may also be controlled by holding the control grid at a fixed voltage and by varying the voltage at the cathode. Most video display designs apply the video information to the cathode to create the image and control the overall brightness by varying the voltage at the control grid. The screen grid, or grid 2, is used to pull the electron beam toward the anode. Voltage changes at the screen grid also affect the brightness of the picture, but the voltage is usually set to a fixed value, at least +400 V for a monochrome CRT. The last grid is the focus grid, which focuses the electron beam to a pinpoint. The voltage present on the focus grid is usually adjustable from about −100 to +1000 V, depending on the design and type of CRT.

The last element is the anode. The inside surface of the "bell" is coated with conductive paint to form the anode. Since the electrons must travel a great distance from the cathode to the anode, high voltage must be present on the anode. The high voltage is usually 1000 V per inch of the face, or 12,000 V on a 12-in. CRT. Most color CRTs require 25,000 V regardless of the size of the face. A color CRT is similar to a monochrome CRT except that three electron guns are used, one for red, one for green, and one for blue, the primary colors of light. By mixing the three colors, any color is possible. The other main difference from the monochrome CRT is that a *shadow mask,* a metal plate with many tiny holes in it, is placed just before the face of the CRT. The face of the CRT is painted with many groups of red, green, and blue phosphor dots called **triads,** as shown in Figure 14.11. The shadow mask is set up in such a way that only the red electron beam hits the red dots, only the green beam hits the green dots, and only the blue beam hits the blue dots. Since the three electron beams do not originate from the same point, an additional

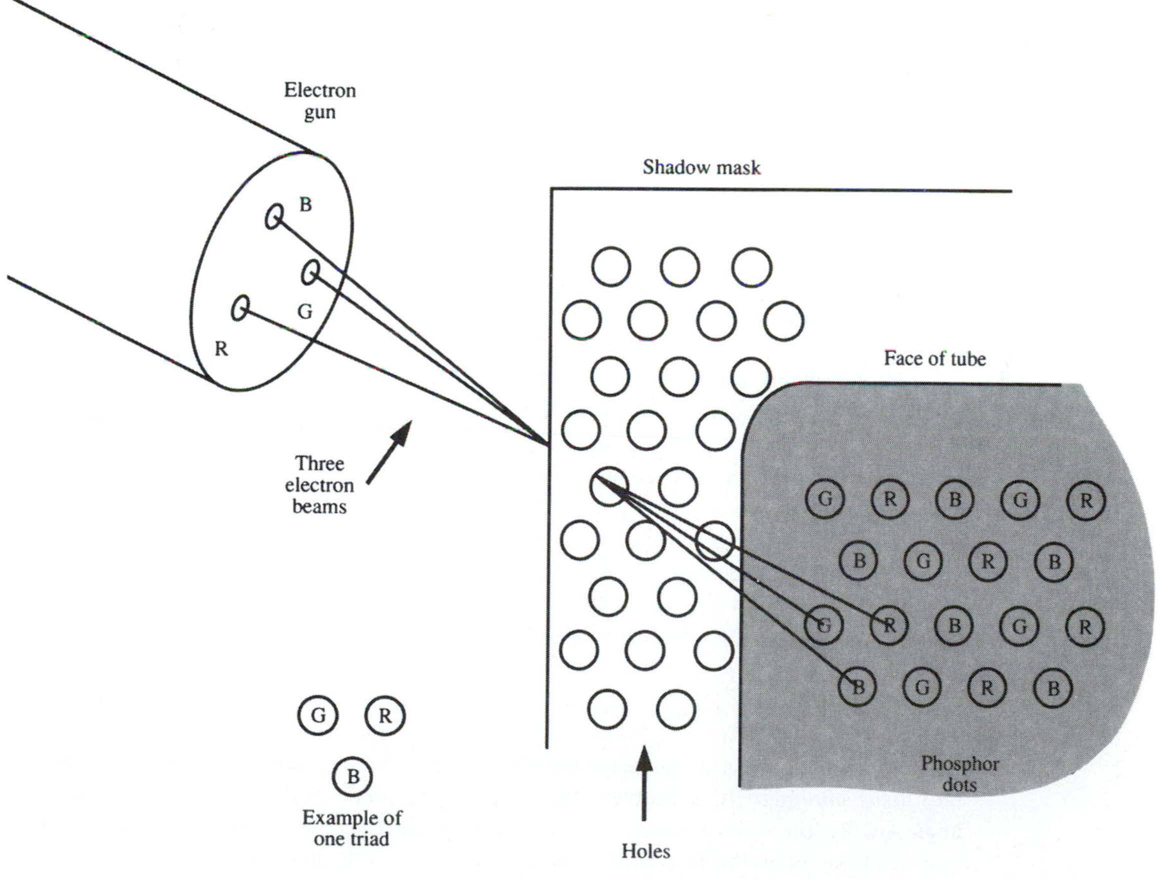

FIGURE 14.11
The color CRT

set of deflection magnets, called the **convergence yoke,** is required. This type of color CRT is called a delta gun CRT because the electron guns are arranged in delta form.

Many of the newer color CRT designs put all three electron beams into one common electron gun. Another difference in the new CRT design is that the phosphor is painted as groups of vertical stripes instead of dots. Directly behind the stripes is a slotted shadow mask instead of holes. One example of this design is the Sony Trinitron™ CRT. This single-gun design greatly reduces the number of convergence components required on the CRT and in some cases eliminates the need for the convergence yoke. Almost all the newer color monitors are designed this way and do not require periodic convergence adjustments.

High Voltage

Most video displays use the horizontal sweep circuit to generate the high voltage through a flyback transformer. In rare cases, on some very expensive monitors, a separate high-voltage power supply is used. As shown in Figure 14.12, the high-voltage circuit is a simple half-wave rectifier off the secondary of the flyback transformer.

No filter capacitor is required because the conductive coatings on the CRT act as a capacitor. It is for this reason that you must discharge the anode-to-chassis ground *before* removing or working around the CRT or high-voltage section. Discharging the anode to chassis ground is extremely dangerous and should be performed only by a qualified and experienced technician. *Failure to follow this precaution could result in a shock.* The high-voltage rectifier in the color display is a bit more complex because a tripler circuit is used to step the high voltage up to 25,000 V, as shown in Figure 14.13.

Another popular variation of the tripler is the multiple secondary flyback with integral diodes. In this configuration the secondaries are wired in series with diodes coupling the windings, as shown in Figure 14.14. This type of flyback is used in the IBM color monitor.

High-voltage failures are usually due to the absence of horizontal sweep or excessive loads on the flyback transformer. An additional winding is usually provided on the flyback to generate the grid biasing voltages, as shown in Figure 14.12. Diode D505 rectifies the 450-V screen grid voltage. Diode D504 rectifies the +55-V cathode bias voltage. Diode D503 rectifies the −170-V brightness control voltage, which is sent to the control grid from the brightness control circuit as shown in Figure 14.15.

Some monitor designs use the flyback output to generate *all* the low-voltage power supplies, as shown in Figure 14.14. This design puts additional loads on the horizontal scan circuit but has the advantage of reducing the cost of the monitor.

Dynamic Focus

Many video display monitors employ dynamic focus circuits in their design to correct *focus nonlinearity,* a problem that is most prevalent in the newer wide-deflection-angle CRTs. Since the face of the CRT is relatively flat, the distance from the electron gun to all points on the face is not constant. To correct for this, the focus voltage must be varied relative to the electron beam position on the face of the CRT. If the correction is not made, the image will appear a bit fuzzy on the edges of the CRT. The addition of the inductor and capacitor circuits on the horizontal output generates a parabolic waveform,

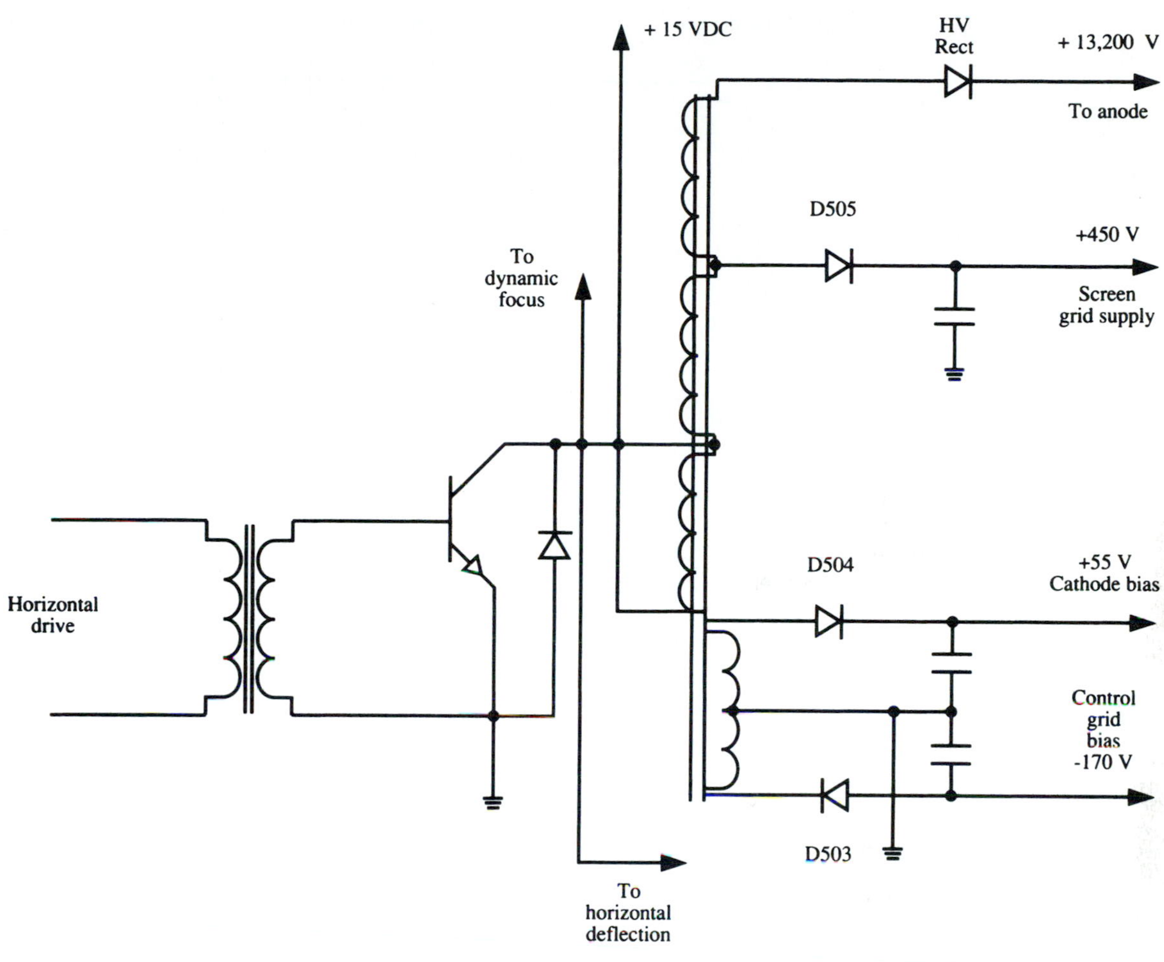

FIGURE 14.12
IBM monochrome display high voltage

as shown in Figure 14.16. This waveform is fed into the focus circuit to change the voltage "on the fly." The focus voltage on the edges is different from the voltage when the beam is in the center. These voltage changes cause the electron beam to remain in focus across the entire face of the CRT.

14.4 DEFLECTION

As was discussed earlier, the electron beam must be pulled across the face of the CRT in order to create a raster of horizontal lines. The position of the electron beam is controlled by the deflection yoke and the deflection circuits. There are in fact two types of deflection methods used in modern CRTs. The first is **electrostatic deflection,** which is used primarily in oscilloscopes. Some electrostatic deflection is rarely used in video display applications, it will not be discussed here. The second type is **magnetic deflection,** which

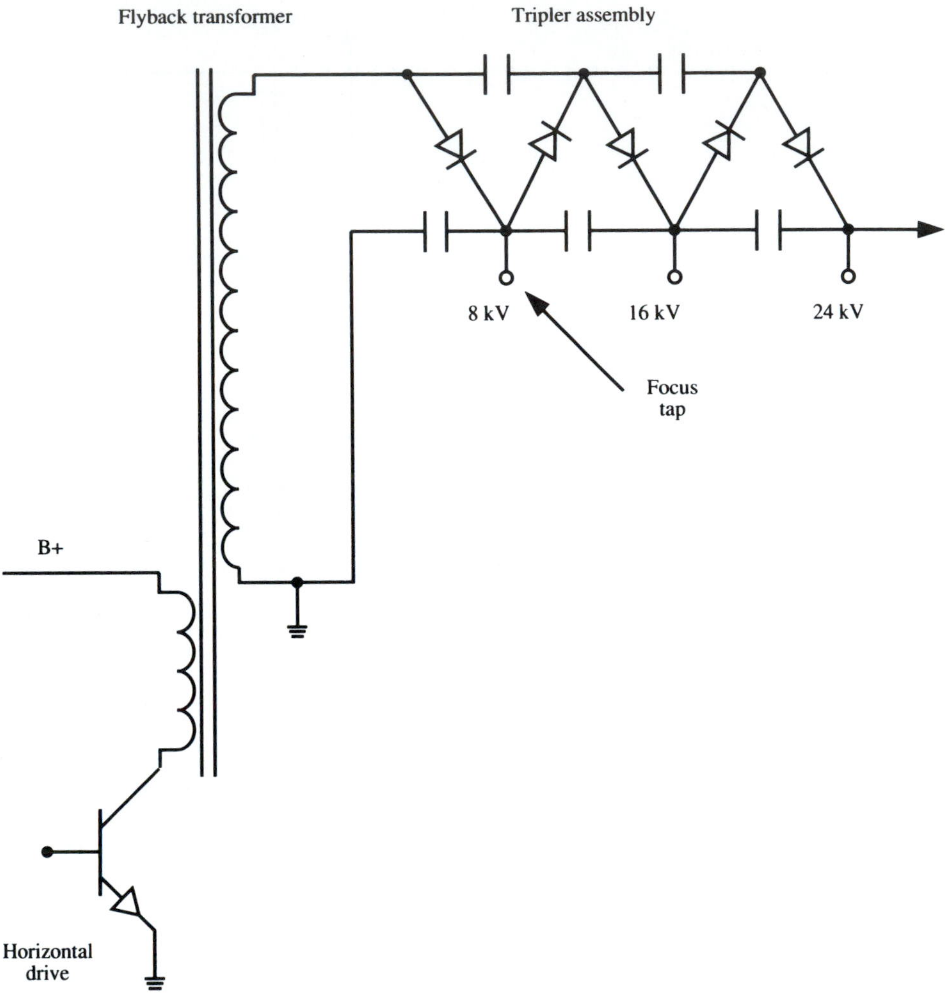

FIGURE 14.13
High-voltage tripler

is used in most video display applications. The main components of magnetic deflection are two sets of opposed electromagnets known as the **deflection yoke.** The deflection yoke is made up of four coils of wire mounted at 90° increments. The deflection yoke is slid over the neck and up against the bell of the CRT as shown in Figure 14.17.

Since the magnetic flux from a coil is at a right angle to it, the horizontal deflection coils are mounted above and below the neck. The vertical deflection coils are mounted on the right and left sides of the neck. In this arrangement when the vertical coils are energized, they pull the electron beam up or down. When the horizontal coils are energized, they pull the electron beam right or left. The deflection yoke coils are driven by the horizontal and vertical deflection circuits, which generate a current ramp waveform. The ramp

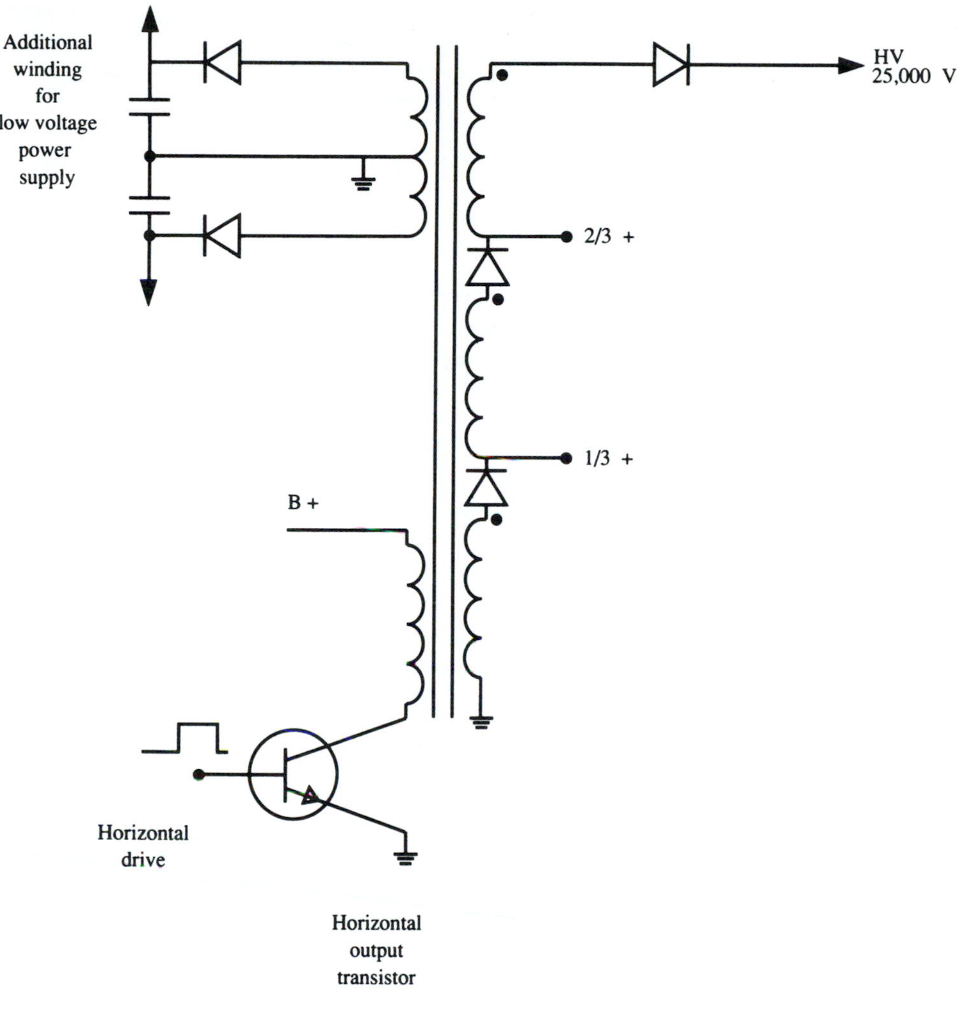

FIGURE 14.14
Multiple secondary flyback with integral diodes

pulls the electron beam at an even rate across the face and then back as fast as possible. When the waveform is viewed on an oscilloscope, it resembles a sawtooth, as shown in Figure 14.18.

Horizontal Deflection

The horizontal deflection section is typically made up of the horizontal oscillator, the horizontal output, the flyback transformer, and the deflection yoke. In most video display applications the output of the horizontal oscillator is a square wave and not a ramp. The square wave is then fed to the horizontal output, which is simply a switch. The com-

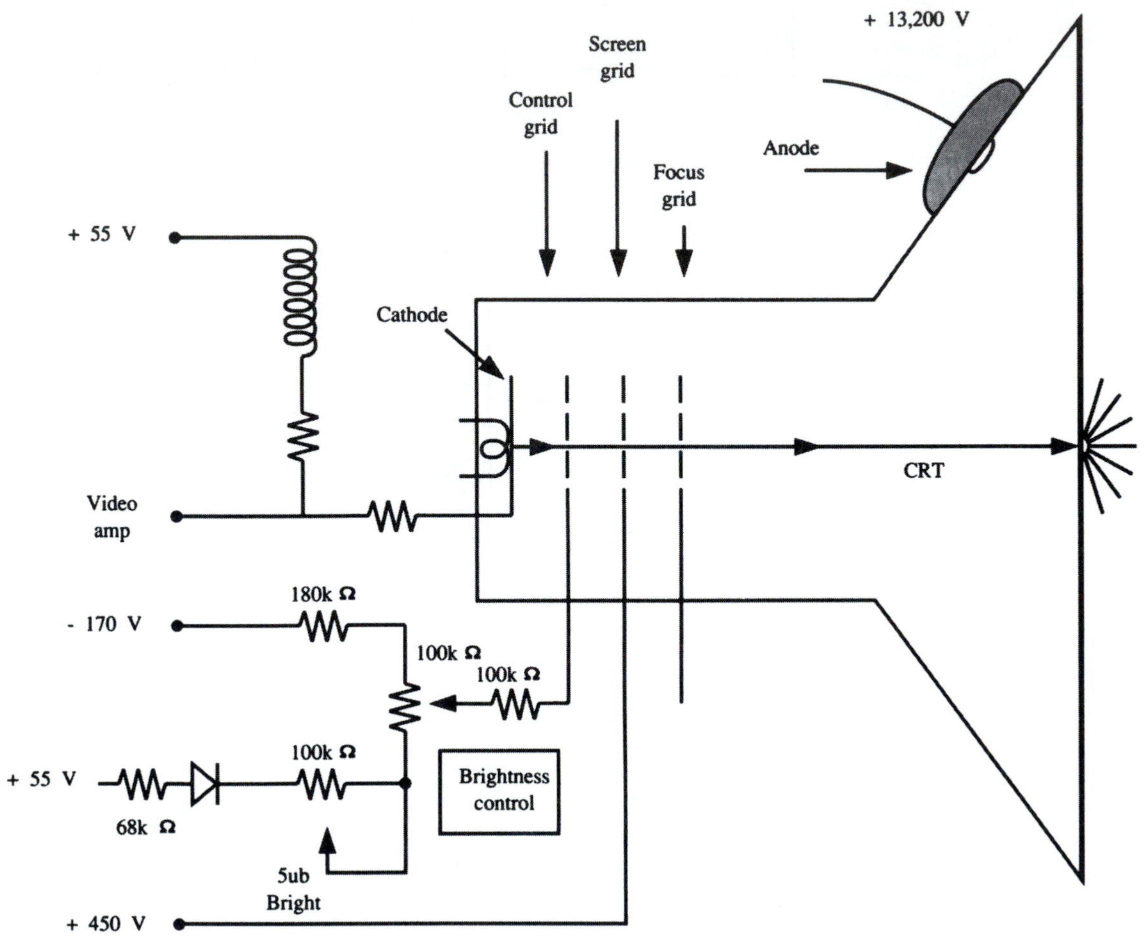

FIGURE 14.15
CRT and brightness control

bination of the flyback transformer, yoke coils, diodes, and capacitors forms the ramp for the deflection while also generating the high voltage through the flyback transformer. It is important to remember that the entire horizontal section is precisely tuned by the combination of *all* the components involved. *Any* component failure in the horizontal section may result in the loss of sweep and/or high voltage.

The horizontal deflection circuit used in the IBM monochrome display does not implement a free-running horizontal oscillator. The display adapter provides the horizontal drive signal directly to the horizontal driver circuit, so the IBM monochrome display screen is dark when the computer is not powered on. The entire horizontal deflection section consists of only three transistors, the flyback transformer, a coupling transformer, five capacitors, a diode, the deflection yoke, and three coils, as shown in Figure 14.19.

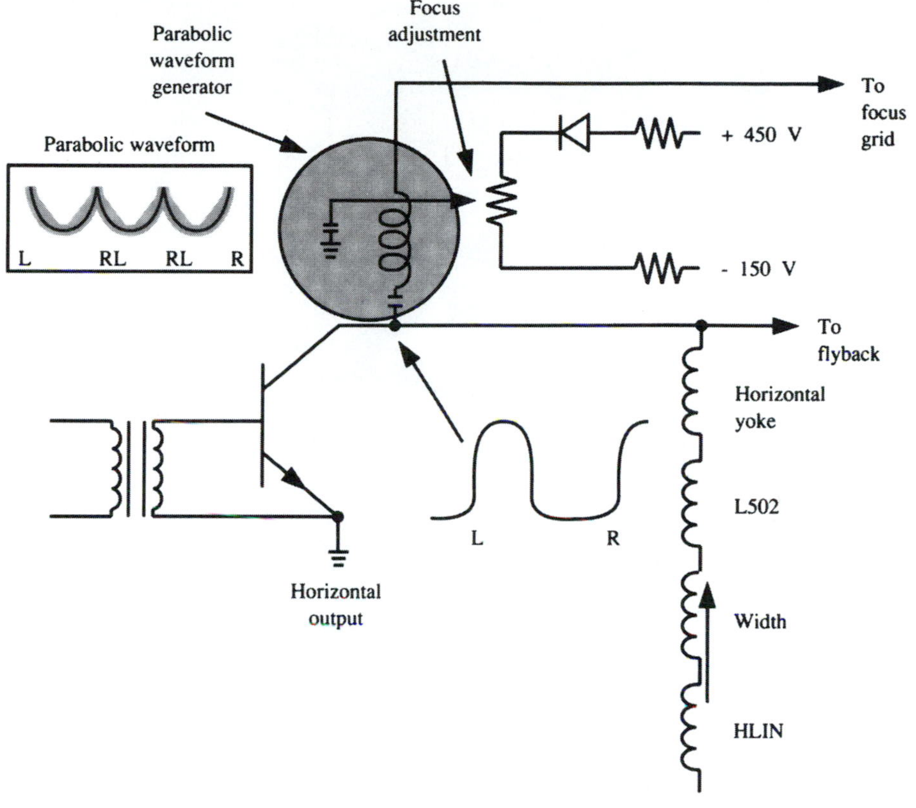

FIGURE 14.16
Dynamic focus circuit

The horizontal drive signal is fed to the base of the transistor TR22, which is configured as a saturated switch. The output of transistor TR22 is coupled through transformer T501 to the base of the horizontal output transistor TR23. The actual horizontal ramp is formed by the yoke inductance and the capacitor C505 while being controlled by the horizontal output transistor TR23 and damper diode D502. The operation of this circuit is illustrated in Figure 14.20. When the horizonal output transistor is first turned on, energy is stored in the flyback transformer primary. The moment that TR23 turns off, the energy is transferred to the secondary, creating the flyback pulse and high voltage. At this same time energy is stored in the yoke, creating the **retrace** interval. A short time later the yoke current reaches the peak level, and diode D502 begins to conduct, creating the first half of the trace. As the current reverses polarity, transistor TR23 begins conducting, completing the trace until retrace time, when the cycle repeats. Components D502 and C504 protect the output transistor from excessive voltage transients. Coils L502, L503, and L504 are used to tune the resonance of the yoke and control the shape of the current ramp waveform. Transistor TR21 and associated components are used to cut off the horizontal drive signal if the power supply voltage becomes excessive due to a failure of the power supply.

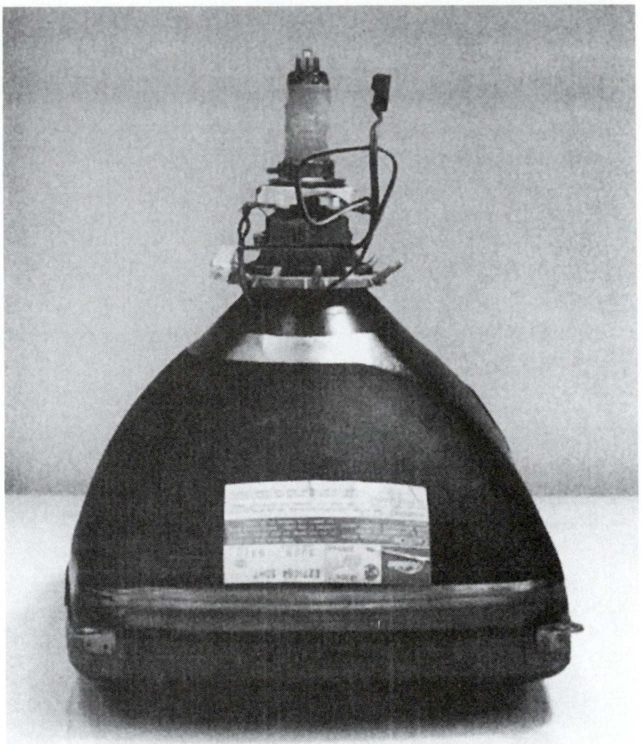

FIGURE 14.17
Picture of CRT and yoke

If this protection were not provided, the high voltage could possibly become high enough to allow some electron radiation to pass through the CRT, causing a radiation hazard to the user.

The scan method described is known as **reactance scanning** and is designed to operate at a fixed frequency. More sophisticated monitors that support variable scan rates employ more circuitry and usually have a separate yoke driver circuit. The IBM EGA and the new NEC MULTISYNC are such monitors, and they are far more expensive.

Vertical Deflection

The vertical deflection circuit is more complex than the horizontal deflection circuit because of the lower frequency used. Recall from the previous section that the scan frequency of the horizontal deflection is 15,750 Hz. However, the frequency of the vertical scan is 60 Hz, so the reactance scan method is not possible. A *linear* current ramp must be generated and applied to the vertical deflection yoke coils. A typical vertical deflection circuit is shown in Figure 14.21.

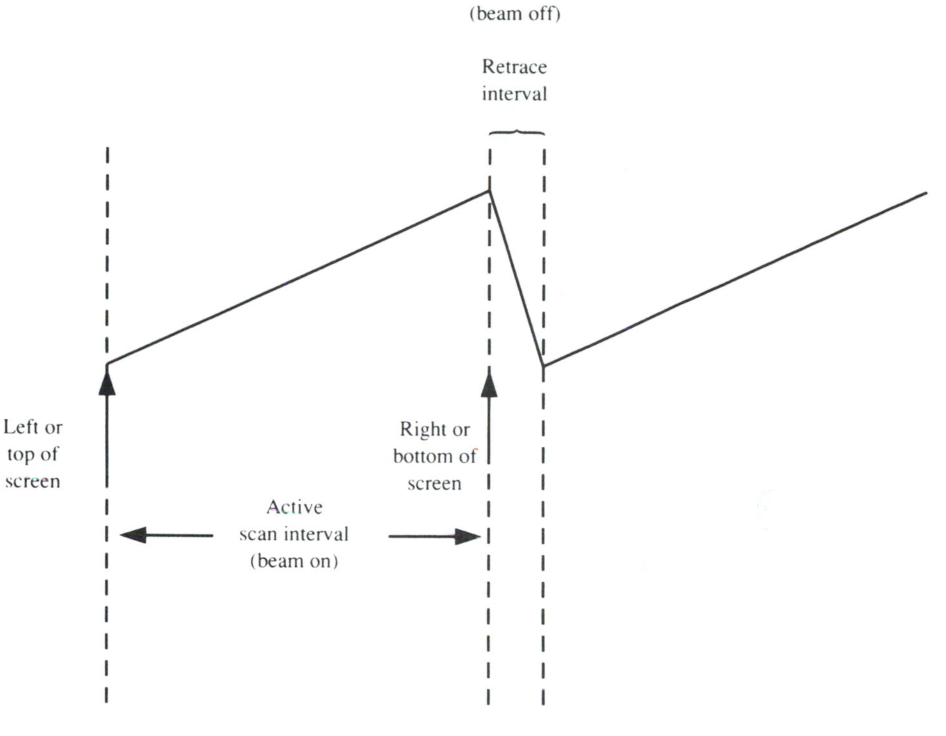

FIGURE 14.18
Sawtooth waveform

The exact design of the vertical deflection circuits used by different manufacturers varies greatly, but their basic operation is the same. As shown in Figure 14.21, the vertical deflection circuit consists of various stages, each with a specific function. The first stage is the vertical oscillator, which has a free-running frequency of 50 or 60 Hz. The vertical oscillator is synchronized to the incoming vertical sync signal from the video controller board. The next stage is the vertical driver, which forms a linear ramp from the output of the oscillator. The shape of the ramp may be changed by adjusting the vertical linearity control. The final stage is the vertical output, which provides the required current to drive the yoke coils directly. Most solid-state monitors use a two-transistor push-pull design similar to an audio amplifier. Failures in the vertical deflection circuit can easily be diagnosed by observing the picture. A single horizontal line in the middle of the screen is an indication of total vertical failure, also known as *vertical collapse.* In most cases this is caused by a failure of the vertical oscillator. Loss of the top or bottom of the picture is usually caused by a failure of the output stage transistors. Distorted or elongated pictures are caused by wave-shape problems, as shown in Figure 14.22. Many new monitors use a single IC for the vertical deflection circuit. The IBM monochrome monitor uses the TDA1170 IC for the vertical deflection circuit, as shown in Figure 14.23.

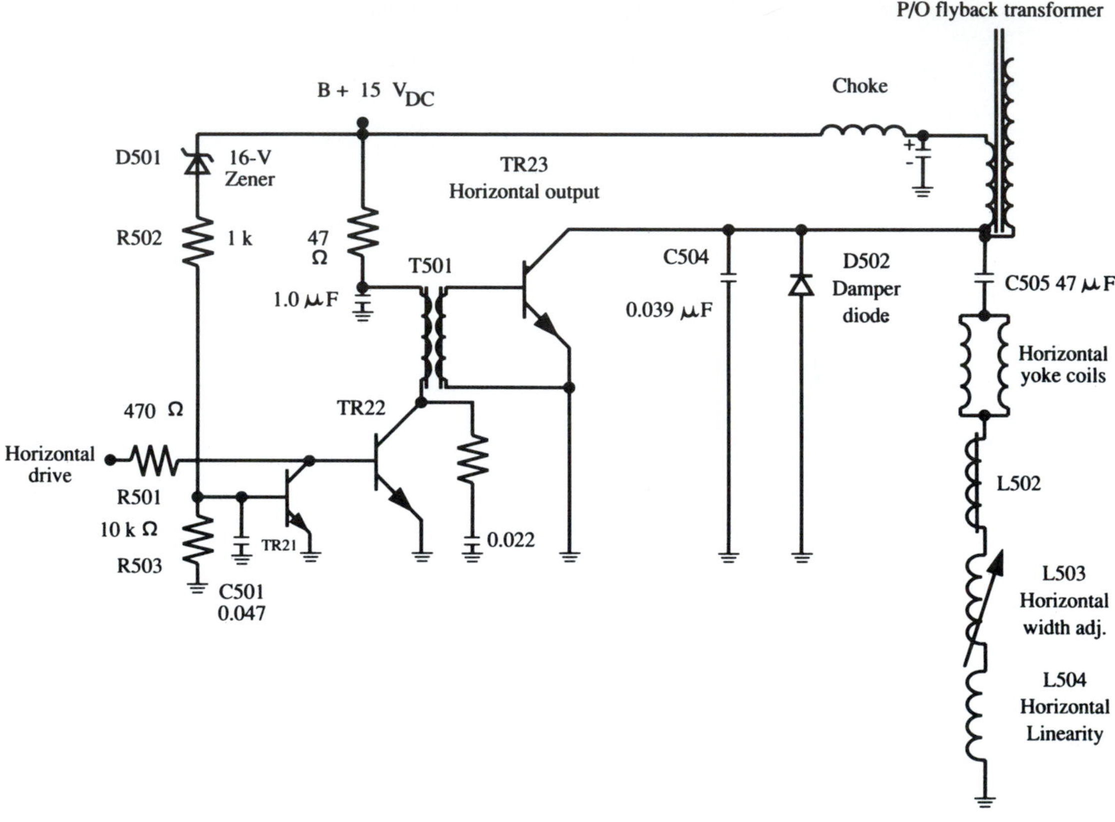

FIGURE 14.19
IBM monochrome monitor horizontal deflection

Synchronization Circuits

To create a usable picture on a raster scan video display, the vertical and horizontal scan circuits must be precisely synchronized to the incoming video signal. In the case of the composite monitor, the sync signal must be stripped from the video and then separated into the vertical and horizontal sync signals. This tripping of the sync from the video is done by circuitry known as the *sync separator.* The sync separator is discussed in the composite video section. Most video displays used on IBM and compatible computers require that the vertical and horizontal sync signals be supplied separately. Recall that the IBM monochrome display requires a horizontal drive signal, which differs from a conventional horizontal sync signal, as shown in Figure 14.24. It is for this reason that the IBM monochrome display *should not* be connected to the color graphics adapter.

It is a well-known fact that when an external signal is applied to an oscillator near the oscillator's frequency, the oscillator tends to synchronize itself with the induced signal. This can be used to synchronize the scan rates of the display with the incoming video signal. This method is used in the vertical scan system by passing the signal through an

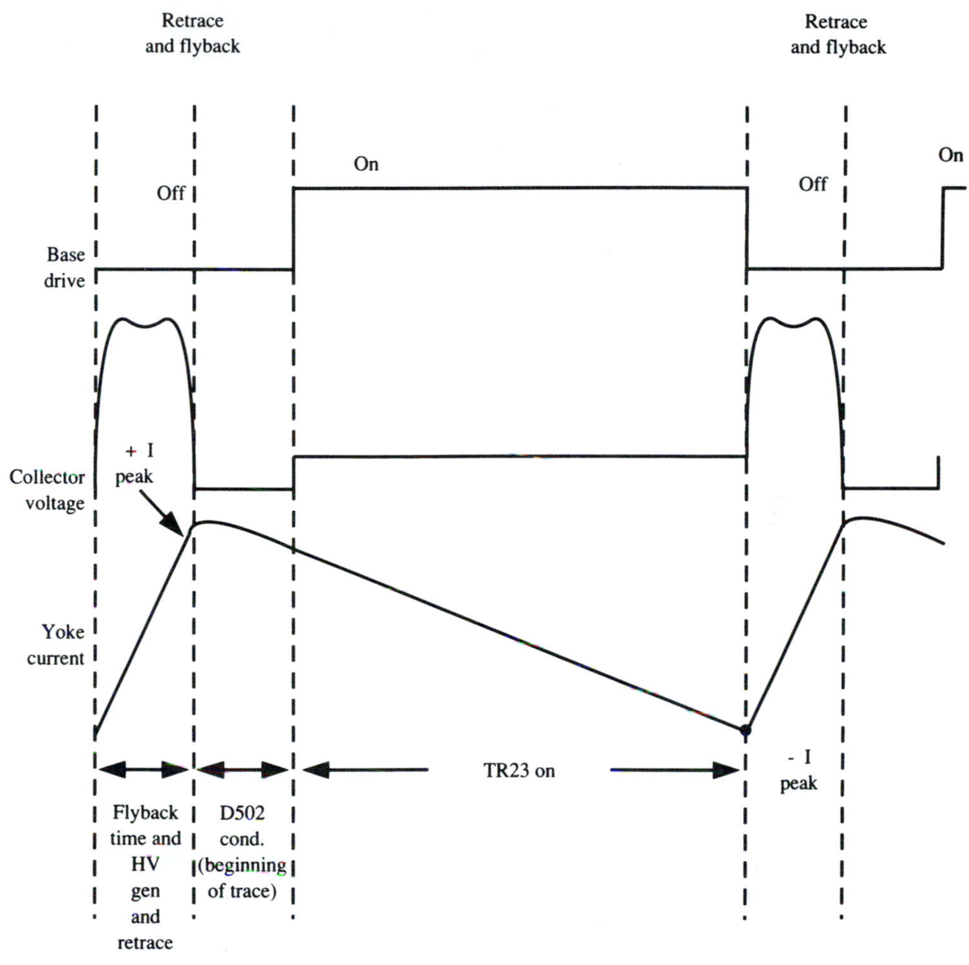

FIGURE 14.20
Horizontal sweep waveforms

integrator or low-pass filter into the vertical oscillator circuit. The integrator circuit is used to remove any noise or variations and to improve stability. In the case of the composite display the integrator also removes horizontal sync pulses.

The synchronization circuit required for the horizontal scan oscillator is more complex due to drift, phase, and noise problems. A phase-locked loop circuit is used here, which is often referred to as the horizontal automatic frequency control circuit. A typical horizontal oscillator is shown in Figure 14.25. As shown in Figure 14.25, the incoming sync pulse is split up into positive and negative pulses by transistor Q_1. A sample of the flyback pulse, which is the retrace time, is integrated into a ramp and fed to diodes D_1 and D_2. If the oscillator is running too slowly, the sync pulse arrives late and causes a negative voltage change at the junction of resistors R_4 and R_5. This negative change could cause the oscillator to speed up. If the oscillator is running too fast, the voltage change is

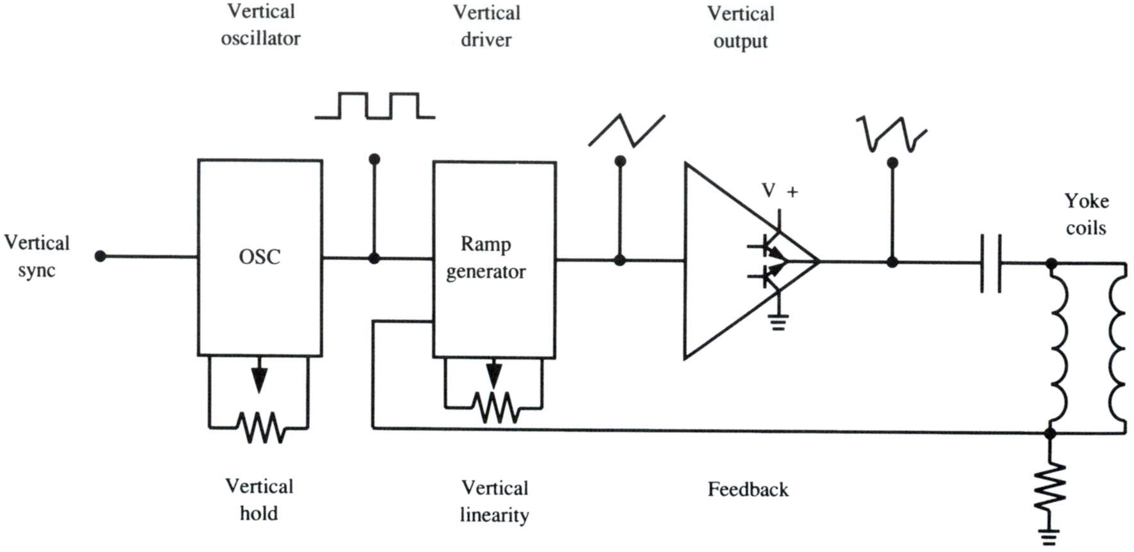

FIGURE 14.21
Vertical deflection circuit

positive and should cause the oscillator to slow down. An *antihunt* circuit is placed between the phase comparator and the voltage-controlled oscillator to prevent the loop from hunting or oscillating.

Many of the newer monitors are replacing the discrete components of the oscillator and synchronization circuits with IC chips. The IBM color monitor employs one IC with built-in synchronization circuitry for both the vertical and horizontal oscillators.

14.5 VIDEO AMPLIFIER

Recall the operation of the CRT and how the intensity of the beam is controlled. The brightness of the phosphor is controlled by voltage changes between the cathode and the control grid. Once a raster is achieved and is in synchronization with the video source, an image can be reproduced by applying video information to the cathode or control grid. In most video display applications the video information is applied to the cathode. A useful tip to remember is that the cathode wire on the CRT socket is *almost always yellow.* In the case of the color CRT there are three cathode wires, *yellow with a red or green or blue stripe* (one for each gun color). The required voltage swing to go from black (beam cutoff) to maximum brightness (beam near saturation) is around 45 V for a monochrome CRT and as high as 150 V for a color CRT. For this reason, the video signal must be amplified before it is sent to the cathode of the CRT. The circuitry used here is called the video amplifier and must have a *bandwidth* from DC to more than 15 MHz for good resolution. The video amplifier design is a simple one- or two-transistor design. It must have an external gain adjustment, which serves as the contrast control. Varying the gain of the

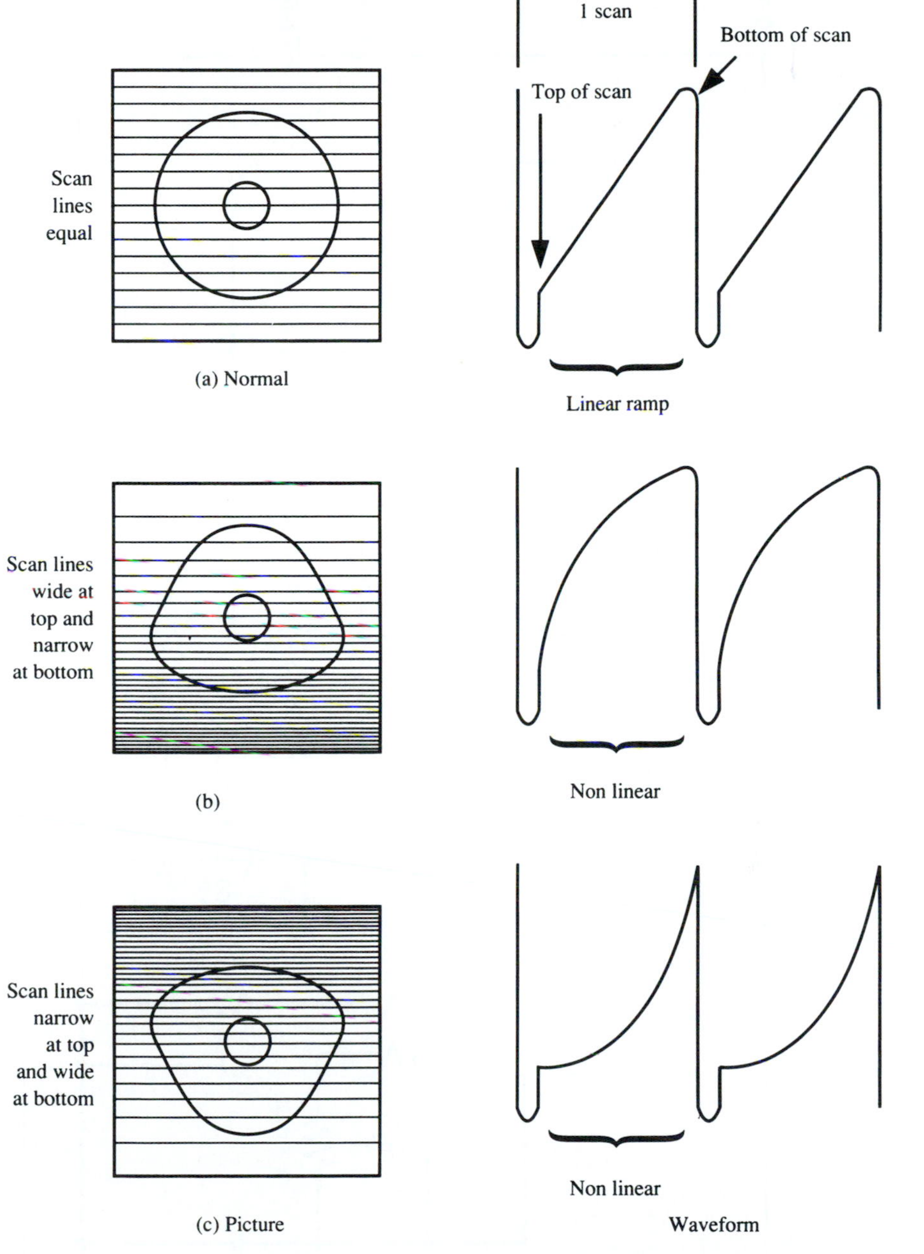

FIGURE 14.22
Distorted wave shapes

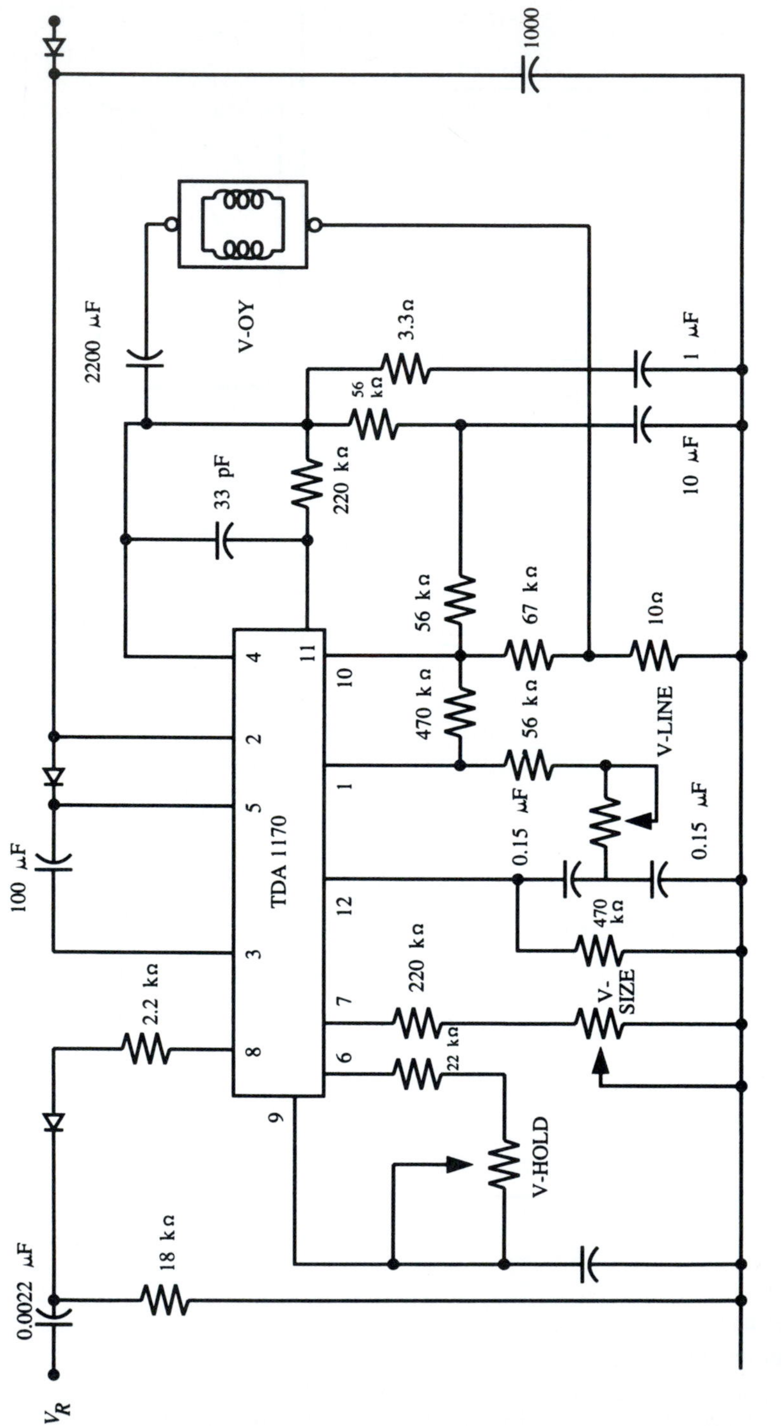

FIGURE 14.23
Vertical deflection

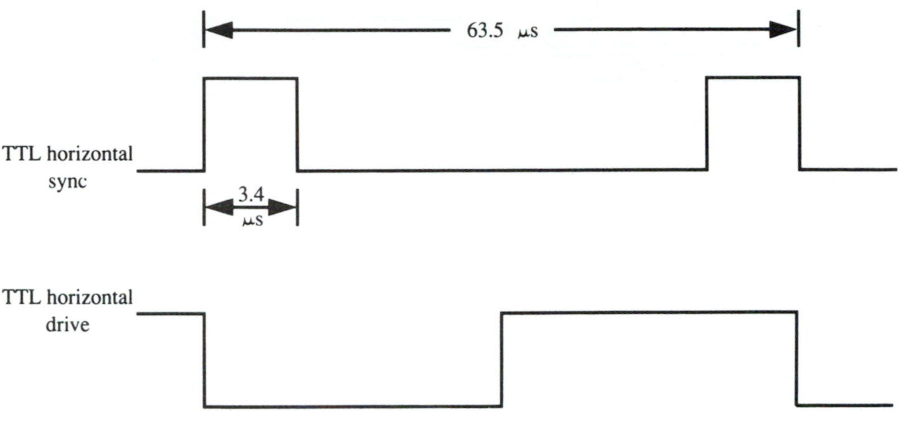

FIGURE 14.24
Horizontal sync and drive signals

video amplifier changes the sensitivity between white and black images and thus changes the overall contrast of the picture. This gain control is usually in the emitter circuit of a common-emitter amplifier, as shown in Figure 14.26. As can be seen in Figure 14.26, varying the contrast control changes the amount of bypass in the emitter circuit, thus changing the amount of gain of the amplifier.

Many computers supply the video information as TTL logic levels and also supply a separate intensity bit for video attributes. In the case of the TTL logic levels, a modified video circuit is used, as shown in Figure 14.27.

The video signal is buffered and sent directly into the video amplifier through a resistor divider. A second video signal called dual, which is the intensity bit, is buffered, and its amplitude is adjustable by the contrast control. This signal attenuates the supply voltage to the output of the video buffer, which adjusts the brightness of the intensified video.

14.6 COLOR VIDEO

The color video display monitor is actually *three* monitors in one package. The power supply, deflection, and high voltage are all shared in common, but there are three separate video channels. The three video channels are required to drive the three separate CRT cathodes, as shown in Figure 14.28. Figure 14.28 is a block diagram of the IBM color display. Each video amplifier operates very much the same as those in the monochrome video display. One important difference is that each video amplifier must have a separate gain adjustment. All three electron guns in the CRT may not exhibit the same sensitivity to the video levels, so they may not produce true gray across the entire brightness range. That is, when the CRT is set up for white at full brightness, dark gray may appear to have a color tint like dark blue. This is known as a **tracking error** in the CRT and is corrected by changing the gains of the individual video amplifiers to match the CRT sensitivity.

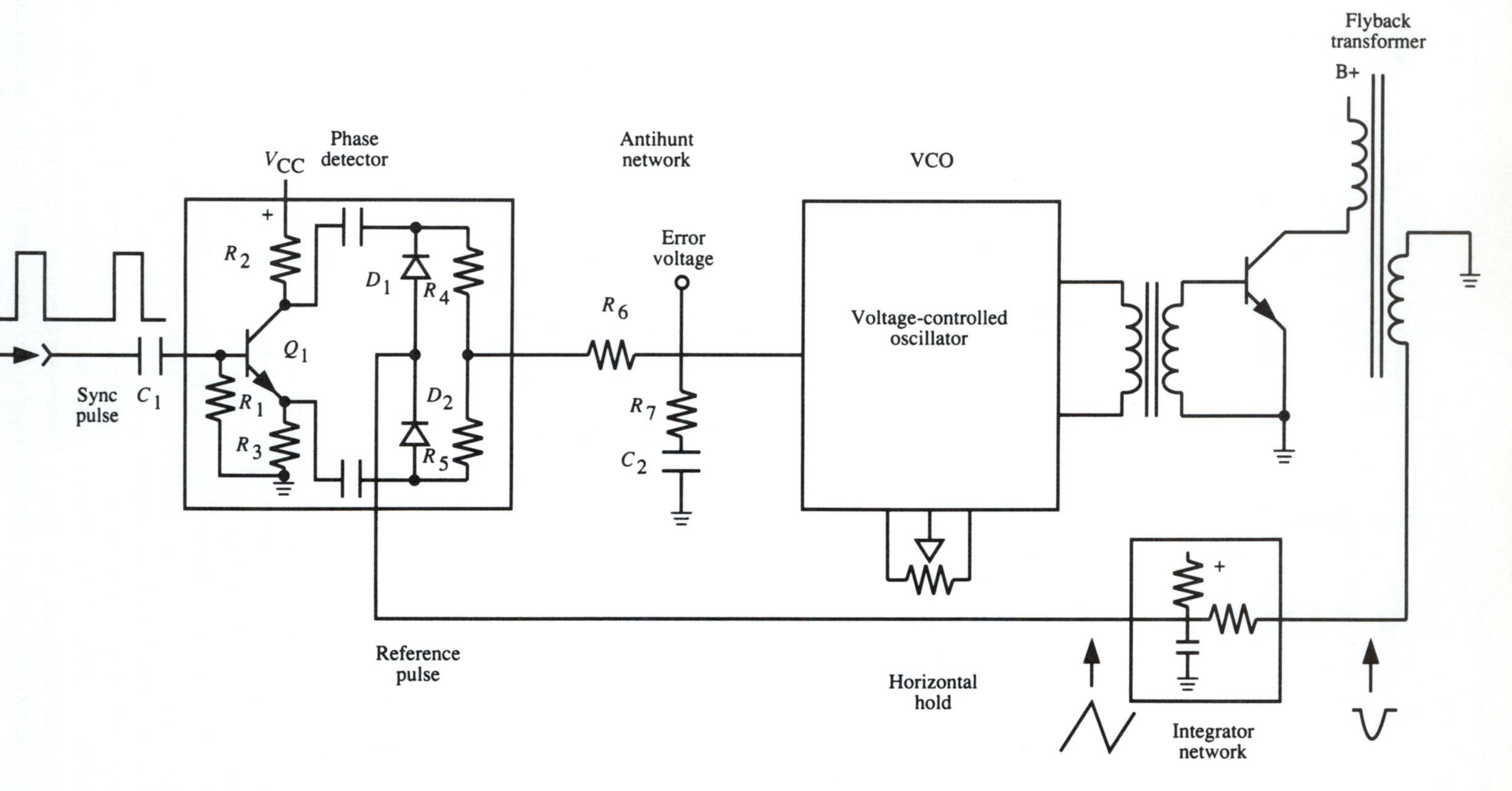

FIGURE 14.25
Typical horizontal phase-locked loop

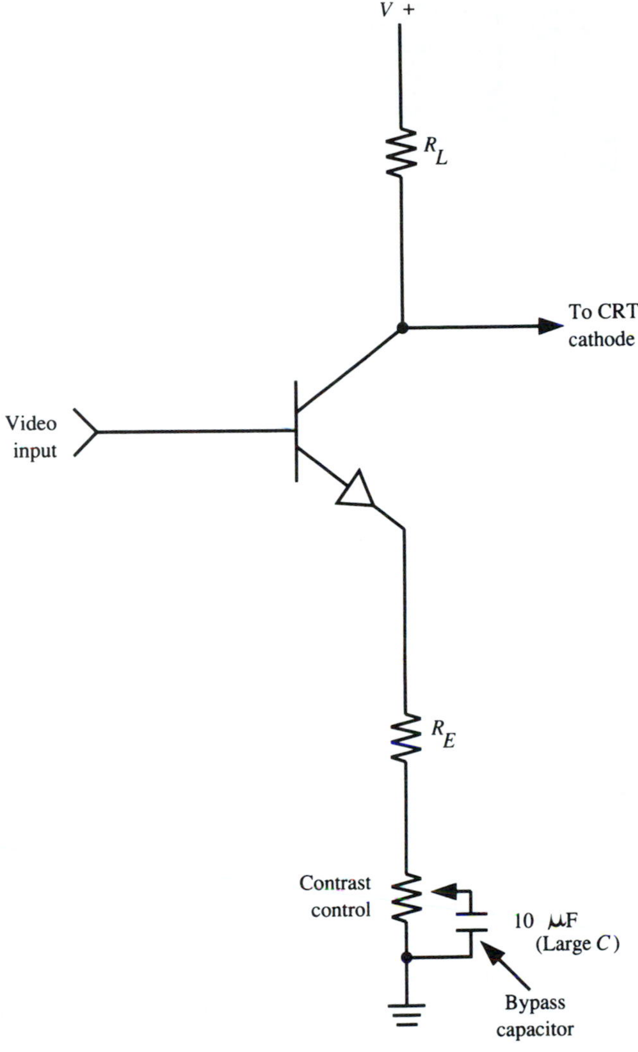

FIGURE 14.26
Typical video amplifier

Composite Video

Composite video is the term used when the synchronization, video information, and color subcarrier are all mixed on the same line. A typical waveform for an NTSC (National Television Standards Committee) video signal for one horizontal line is shown in Figure 14.29(a).

All voltages below the *black level* are considered synchronization information and are separated out and sent to the deflection circuits. These levels are also known as "blacker than black." The last portion of the horizontal sync pulse is called the **back porch** and is the location of the color burst signal. The color burst is the color subcarrier reference

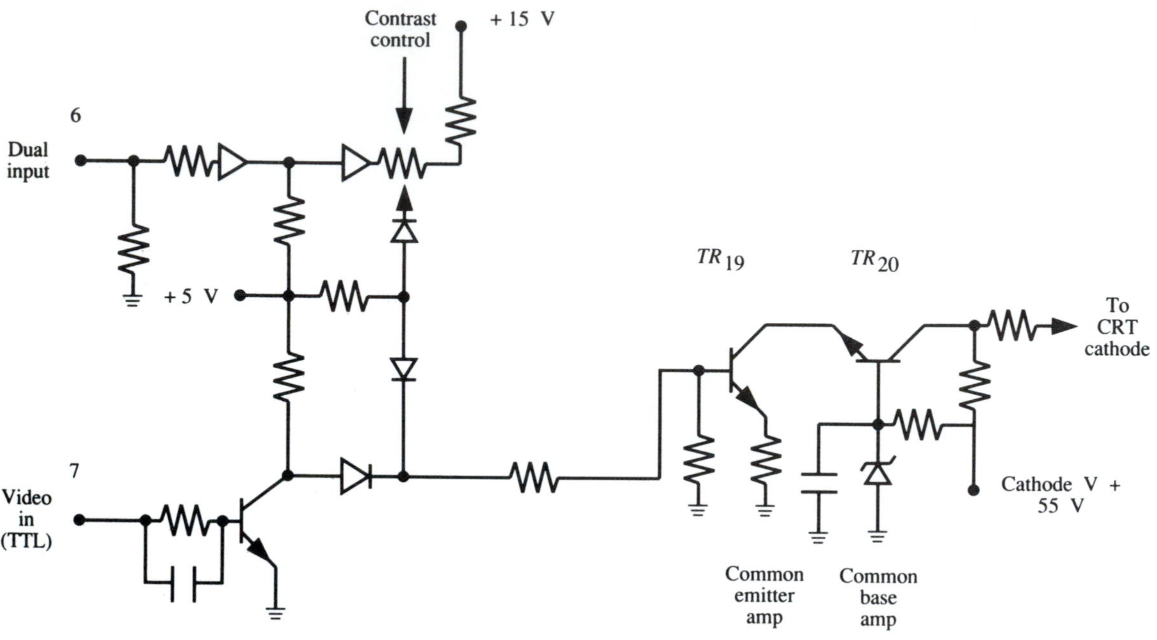

FIGURE 14.27
IBM monochrome video amplifier

frequency that is used by the color demodulator circuitry. Figure 14.29(b) shows the video waveform for one complete field of 262.5 lines. Note that the vertical sync period *is not* a solid sync level but is serrated. The serrations are necessary so that the horizontal deflection will not slip out of synchronization during vertical synchronization time. The color information is a *phase-modulated* subcarrier riding on the video information and is referred to as the chromanance signal. It is separated out of the composite video signal by a simple band-pass filter and sent to a three-output phase demodulator, as shown in Figure 14.30.

In the composite color display the incoming video information is processed in two ways. First, the video information is split up by some filters, stripped of the color subcarrier, and amplified as monochrome video. This is known as the luminance channel, or the Y signal. Second, the color information is filtered and amplified by a band-pass amplifier, which is known as the chromanance channel. The color information is phase demodulated by comparing the color signal phase with a reference signal phase provided by a crystal oscillator. The resultant output is presented as two difference signals called R − Y and B − Y. The R − Y signal is the red video minus the luminance signal, and the B − Y signal is the blue video minus the luminance signal. The R − Y, B − Y, and luminance (Y) are decoded by a summing matrix, which generates the red, green, and blue video signals. The reference oscillator is kept in phase synchronization with the burst signal by the burst gate. The burst gate allows the chromanance signal to sync the oscillator only during flyback time or retrace. The one major disadvantage of using a composite color

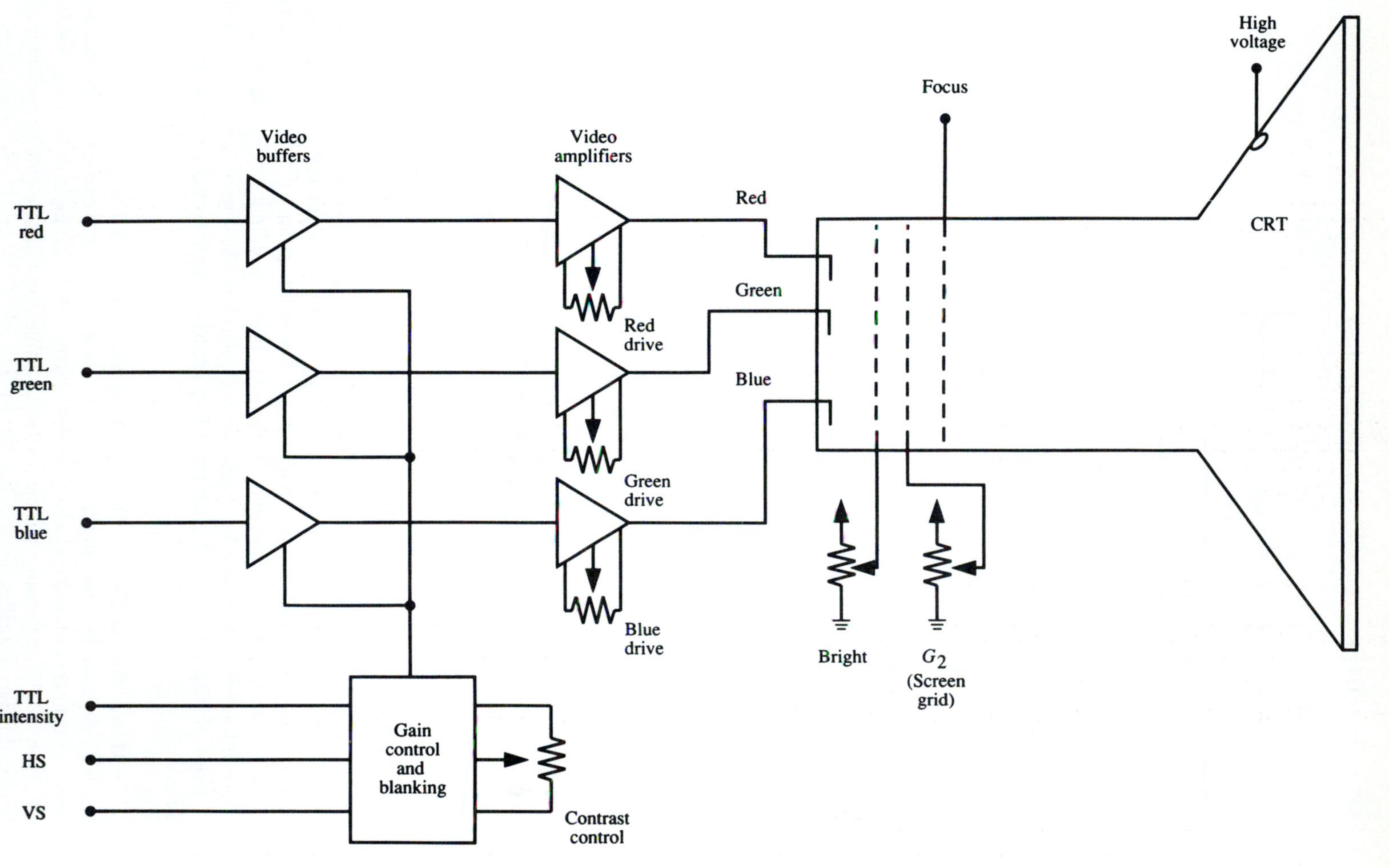

FIGURE 14.28
Color video amplifiers

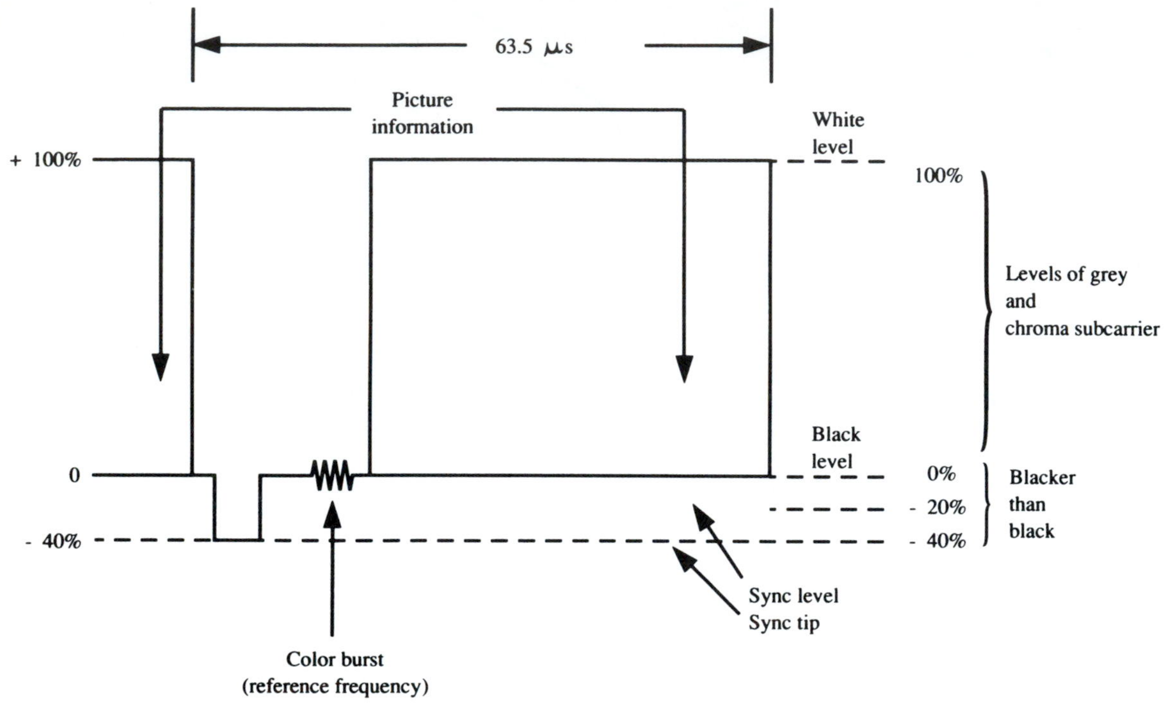

FIGURE 14.29a

video display is the reduced video bandwidth imposed by the chroma subcarrier, which results in a greatly reduced horizontal resolution. The reduced resolution limits the displayable text to 40 characters across the screen. This is why low-cost home computers generally display the 40-character line.

The Video Display Generator

Computer video displays are sometimes referred to as "glass teletypes" or "TV typewriters." They were so named because text information is displayed on a video screen rather than being typed onto a piece of paper. Up to this point we have discussed video display monitor principles and related electronic circuitry. In this section we discuss how video information is created by digital circuitry and how the output from a computer, as a series of binary codes, is transformed into readable English characters on the display screen.

Master Timing Chain The first and most important step in generating a video signal is to generate the appropriate timing. The timing circuitry must create the horizontal and vertical synchronization pulses if a raster scan video display is to be used. As shown in Figure 14.31, the timing can be created by using one high-frequency oscillator and a series of binary counters.

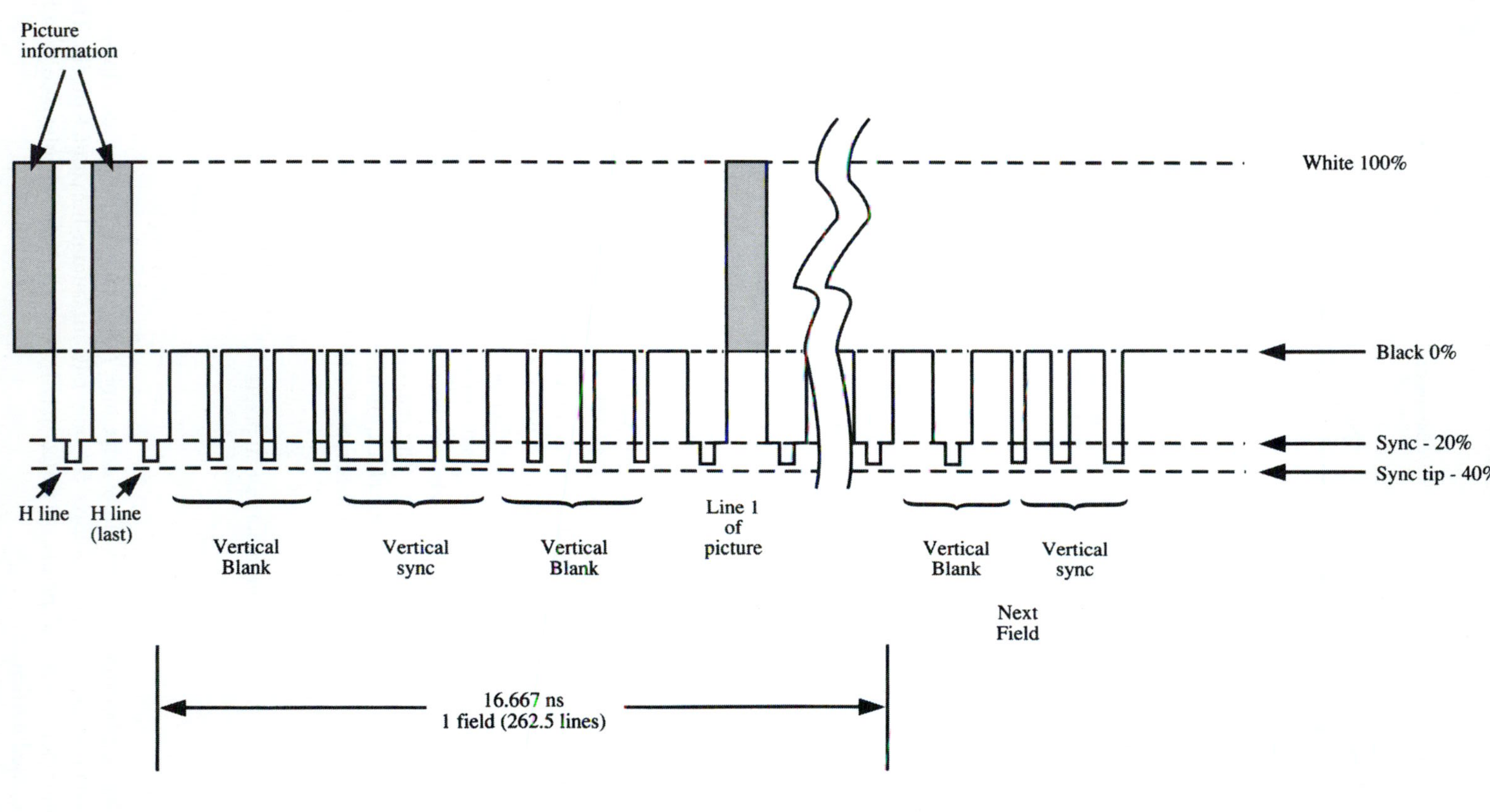

FIGURE 14.29b

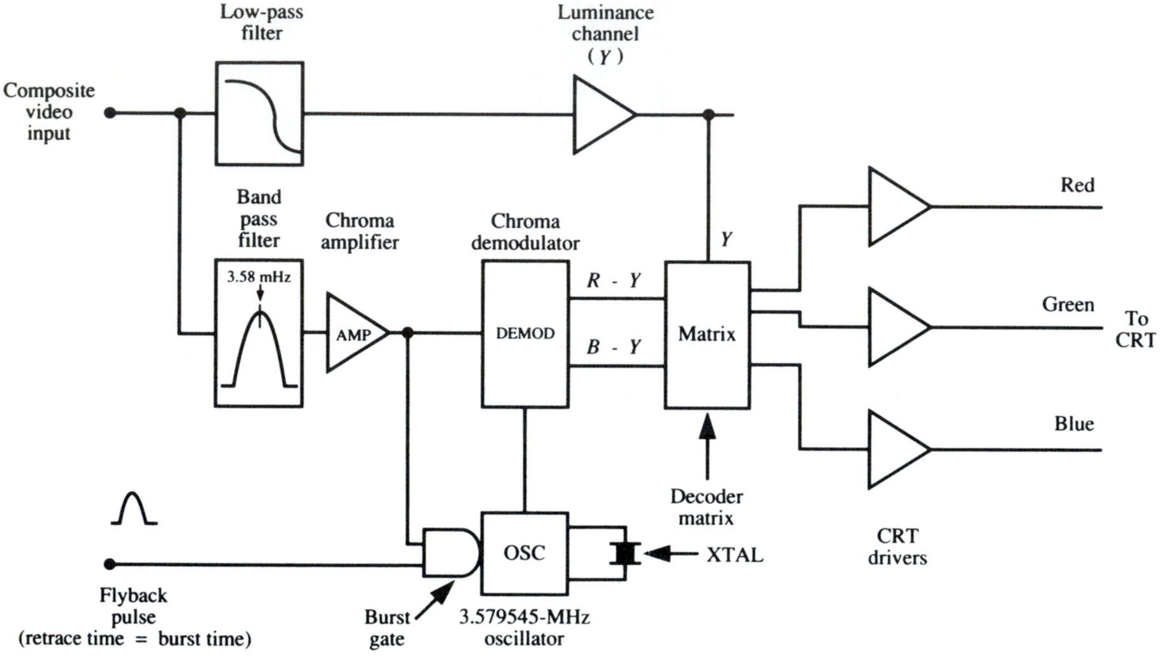

FIGURE 14.30
Composite color decoder

In Figure 14.31 a 14.318181-MHz clock frequency was used to accommodate the NTSC color subcarrier of 3.579545 MHz, which is the clock frequency divided by four. The horizontal sync time is obtained by using a counter that divides by 910, which yields a frequency of 15,734 Hz. In the case of color TV, the horizontal scan frequency is shifted slightly to accommodate the even divisor. The vertical sync is obtained by dividing the horizontal sync by 263, which yields a frequency of 59.82 Hz. The intermediate counter outputs can be used to generate the video memory addresses, which contain the picture element, or pixel, information. The maximum number of horizontal pixels is the maximum clock frequency, or **dot clock,** divided by the horizontal scan rate. The IBM color graphics adapter uses a dot clock rate of 14.318181 MHz. If we use the standard scan rate of 15,734 Hz, we get

$$\text{Maximum horizontal pixels} = \frac{\text{dot clock}}{f_{\text{HOR}}} = \frac{14.318181\,\text{MHz}}{15{,}734\,\text{Hz}} = 910\ \text{pixels}$$

Remember that we must allow some time for retrace and overscan. This is why the IBM color graphics adapter allows for only 640 maximum horizontal pixels. The maximum number of vertical pixels is limited only by the number of displayable scan lines, as discussed earlier.

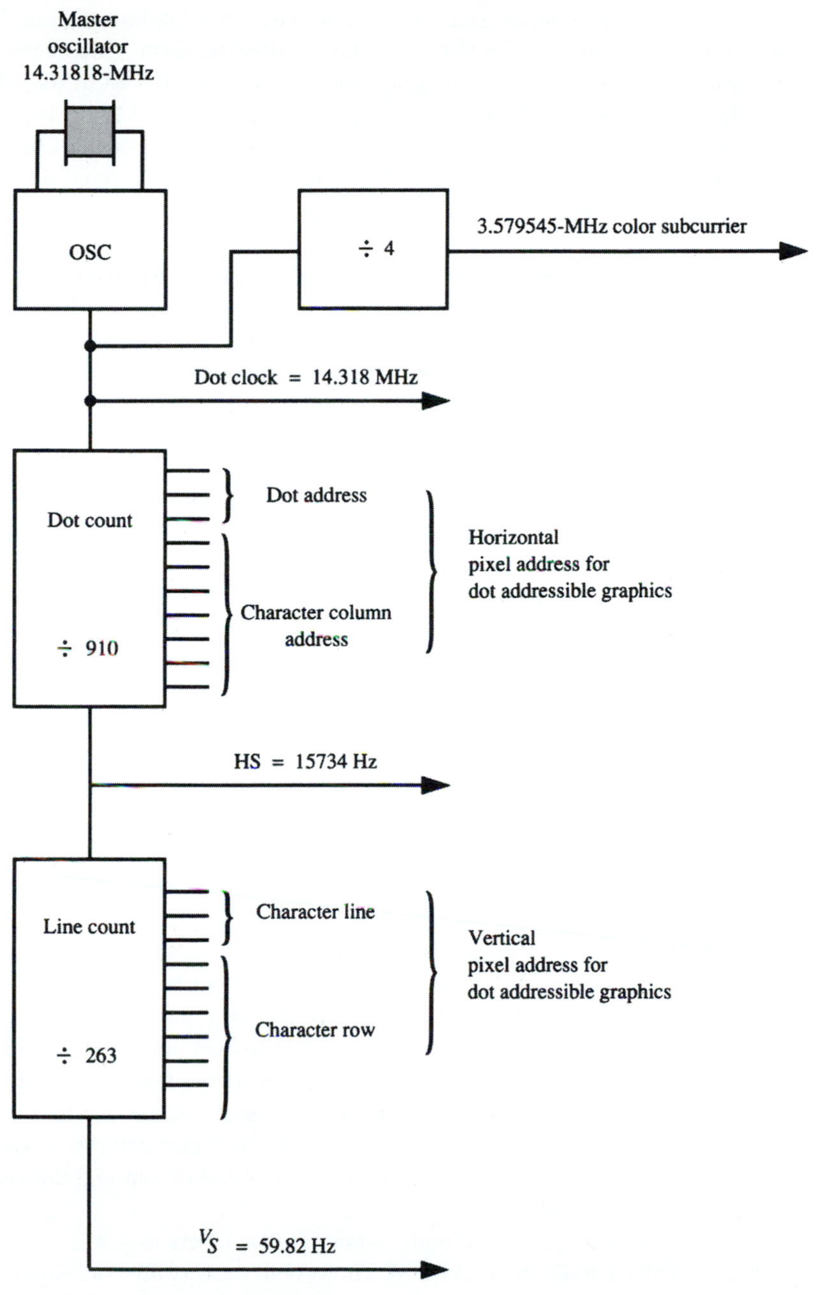

FIGURE 14.31
Example of a digital video timing chain

Recently the counter and timing logic have been implemented on LSI ICs by many different manufacturers. Some manufacturers offer the entire video generator with the exception of the video RAM (pixel memory) on a single IC. Both the IBM monochrome display adapter and the IBM color graphics adapter use the Motorola MC6845 CRT controller chip, as shown in Figure 14.32. The implementation of this chip not only reduces the component count of the adapter but greatly increases the flexibility of the adapter by allowing for software modification of the display format.

Creating an Image Once the appropriate timing has been generated and the display is scanning in synchronization, how is the image created? Recall that an image is created on the CRT by changing the intensity of the electron beam while the electron beam is scanning the face of the CRT. If we then connect a digital signal to the video input of the display, we can turn the beam on or off at will. When we apply a pattern like alternating 1s and 0s in time with the sync pulses, the resultant pattern should resemble a series of horizontal lines. The width and total number of lines will depend upon the rate, or dot clock, in which we are sending the alternating patterns, as shown in Figure 14.33. Of course, this simple bit pattern is useless except as a test pattern to adjust the display. A programmable pattern generator is necessary.

The programmable pattern area is known as the refresh buffer or video RAM and is simply a random access memory matrix. The video RAM data outputs are fed into a shift register, clocked by the dot clock, which converts the parallel RAM data into bit serial data (the pixels) one bit at a time. A bit counter must be provided to load the next group of pixels after the last pixel is shifted out. In the case of an 8-bit RAM, an 8-bit shift register is used and must be reloaded after each group of 8 pixels is shifted out to the display. The memory address lines from the video RAM are connected to the timing chain counter outputs. The first set of counter outputs, which is the byte counter, keeps track of the current pixel group, which is the horizontal pixel address. The second set of counter addresses keeps track of the current pixel line, which is the vertical pixel address, as shown in Figure 14.34.

When the memory matrix is cleared by the host computer, which must be dual ported into the RAM matrix, the video screen will appear to be blank. An image may then be created by writing logical 1s at the appropriate locations in RAM to illuminate the desired pixels on the display. This is fine and necessary for graphics but time-consuming and cumbersome for text-only displays. Since readable text is made up of a fixed set of characters, the patterns required to generate these characters can be stored in ROM. The character ROM that is preprogrammed with the character patterns is known as the **character generator.** It is placed between the video RAM output and the video shift register, as shown in Figure 14.35.

The character generator implementation shown here uses a 256- by 8- by 6-bit ROM, which is really a 2048- by 6-bit ROM. The system shown allows a 5-by-7 character mapped within a 6-by-8 box, as shown in Figure 14.36. We must allow for at least one dot space between the sides and bottoms of the characters so that they will blend together. Also, the memory addresses are rearranged slightly. The pixel counter counts to six and reloads the shift register after the sixth pixel is shifted out. The character column counter keeps track of the current displayed character column and supplies the current horizontal character address to the video RAM. The character line counter keeps track of which line within the character box is currently displayed and supplies a 3-bit address to the character

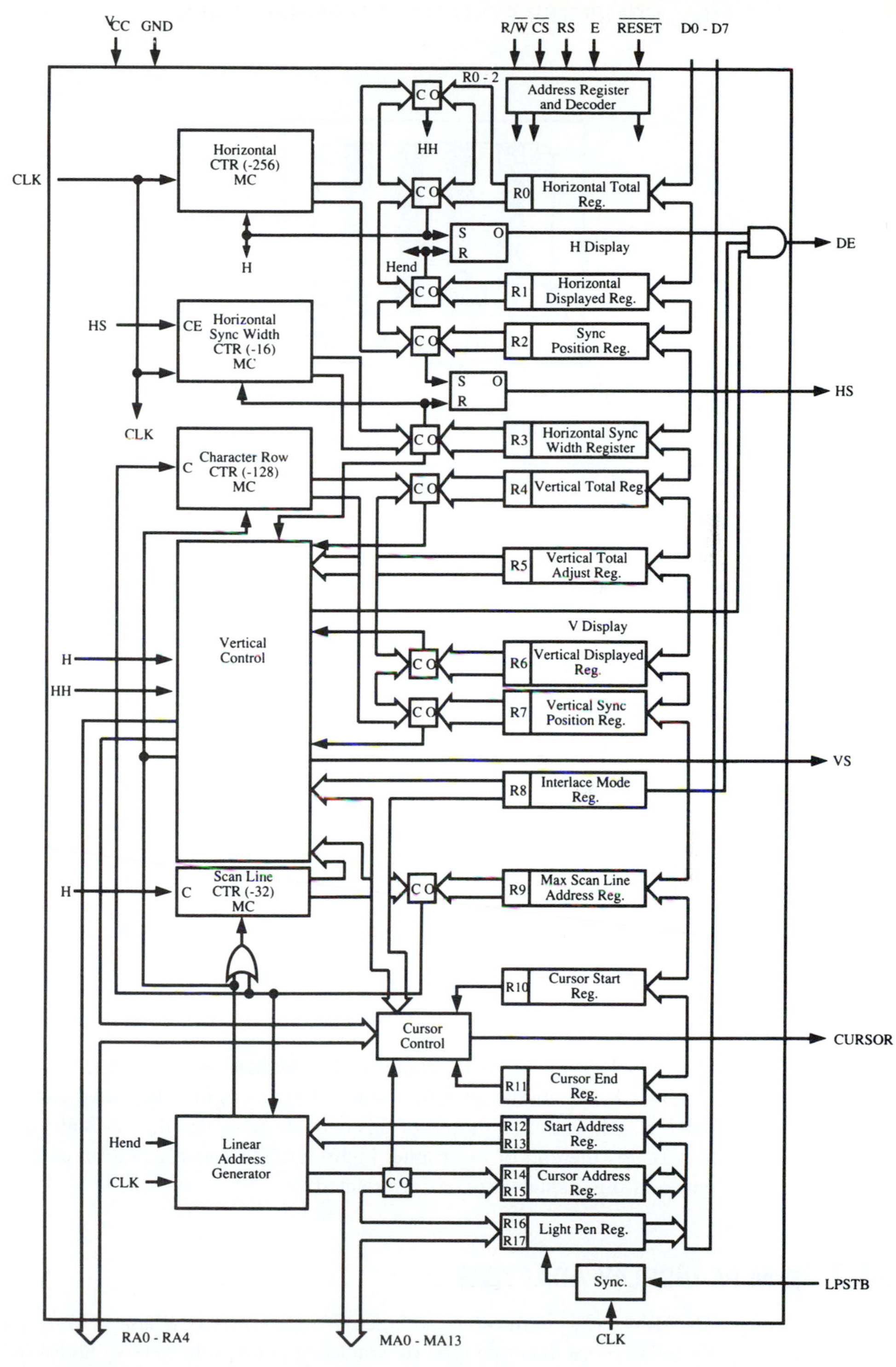

FIGURE 14.32
CRTC block diagram (Reprinted with permission of Motorola Inc.)

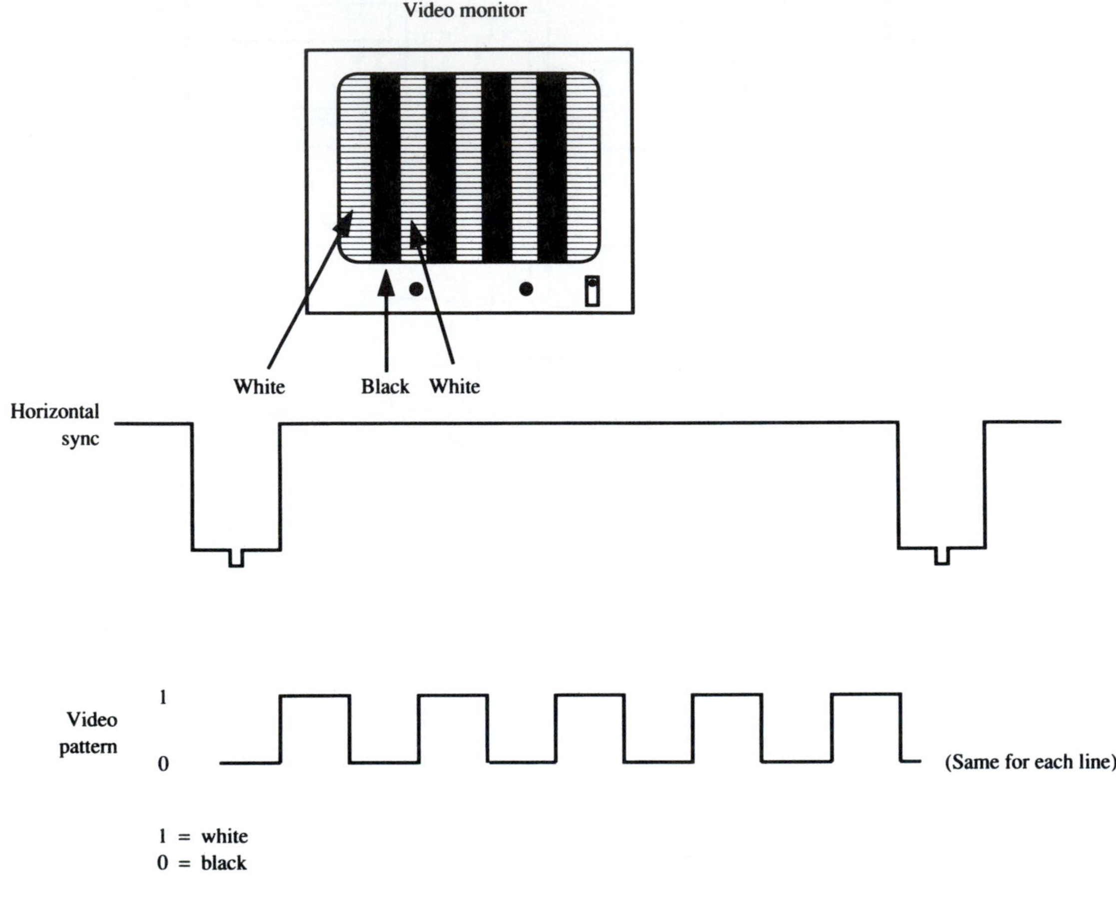

FIGURE 14.33
Simple video pattern

generator ROM. The final counter is the character row counter, which keeps track of the current row of characters displayed and also supplies the character row address to the video RAM. The final output of this counter generates the vertical synchronization, since it indicates the end of the frame. Figure 14.37 is an illustration of how the counter outputs and the displayed pixels are related.

14.7 IBM PC DISPLAY ADAPTERS

The IBM PC system offers a choice of video display adapters to accommodate the user. By using more than one type of adapter, it is possible to have multiple displays on a single system. For example, the user may wish to present a graphic drawing on one display while listing a set of instructions on the second display.

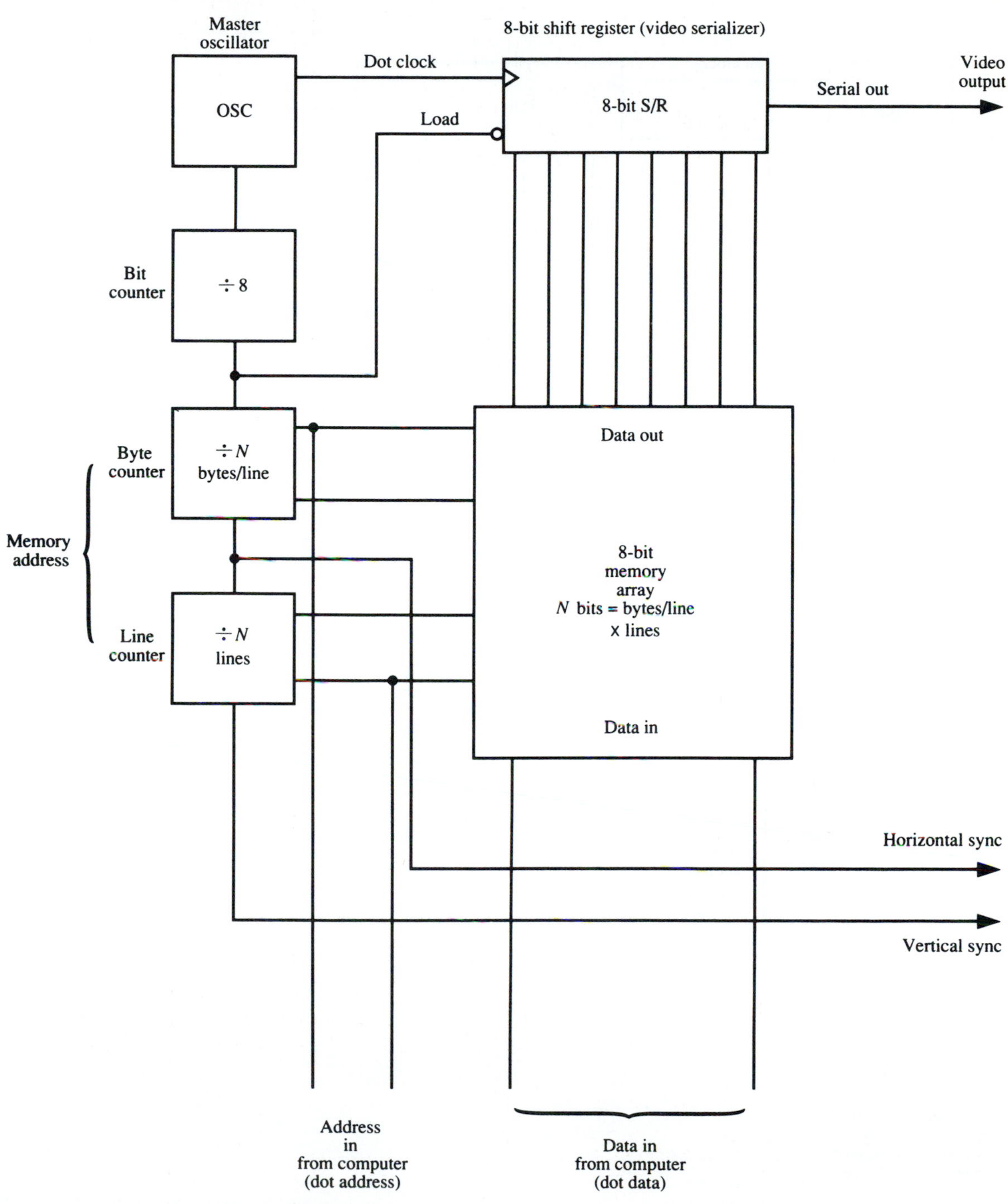

FIGURE 14.34
Memory-mapped video generator

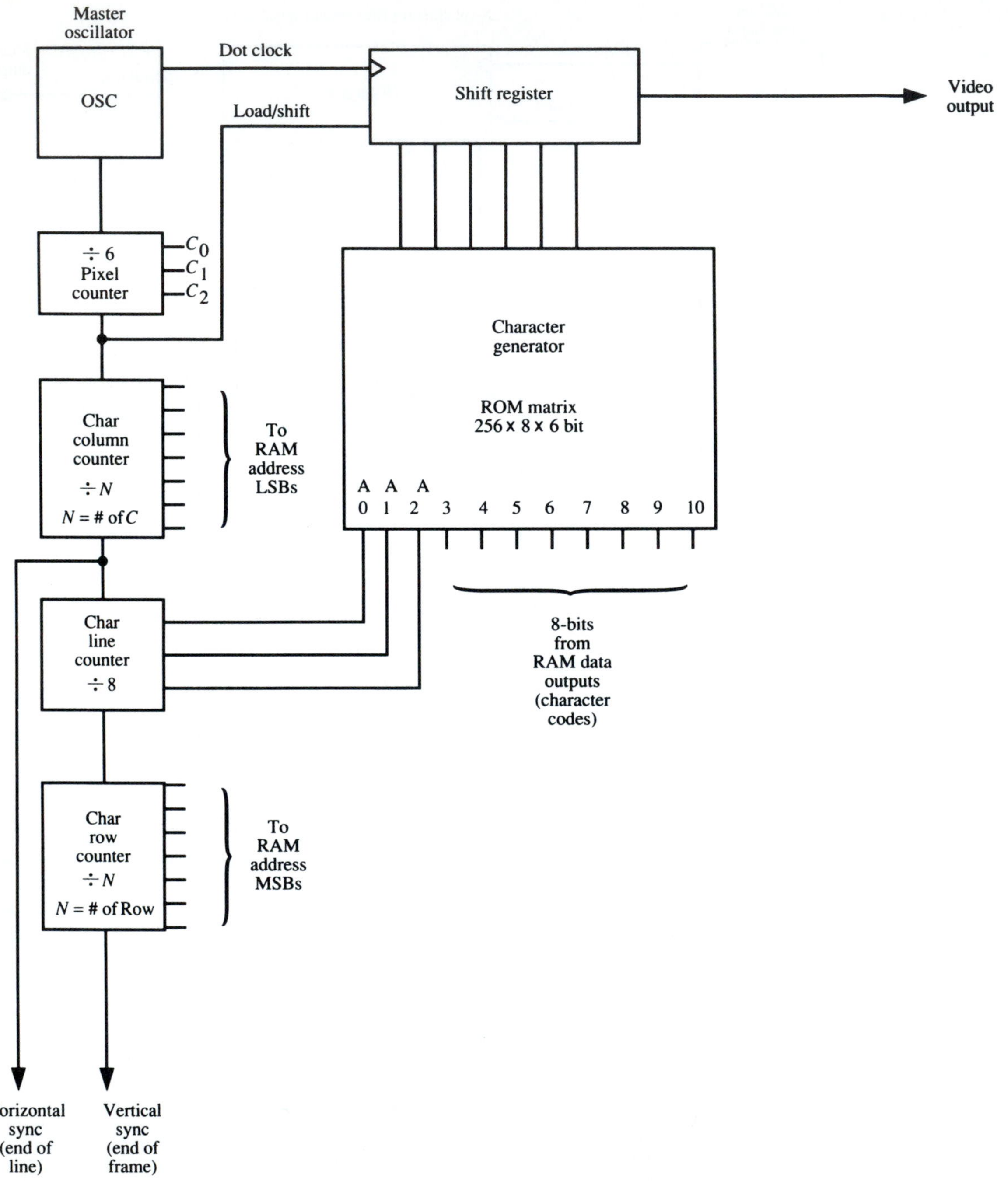

FIGURE 14.35
Character generator implementation

ROM address	ROM data
X X X 0 0 0	0 0 1 0 0 0
0 0 1	0 1 0 1 0 0
0 1 0	1 0 0 0 1 0
0 1 1	1 0 0 0 1 0
1 0 0	1 1 1 1 1 0
1 0 1	1 0 0 0 1 0
1 1 0	1 0 0 0 1 0
1 1 1	0 0 0 0 0 0

Character spacing

FIGURE 14.36
One ROM character

The IBM Monochrome Display Adapter

The IBM monochrome display adapter is designed for a text-only display. The text is displayed as 80 characters across by 25 rows down. The character format is presented in a 9-pixel-wide by 14-pixel-deep character box, which allows for a finely shaped and detailed character. The display adapter also supports three character attributes: inverse video, underline, and intensified video. This display adapter has 4096 bytes of video RAM and is memory mapped in the main system address from hex *B*000:0000 through hex *B* 000:0*FFF*. Address hex *B*000:0000 is the upper-left character position, with hex *B*000:0001 as the attribute byte for that character. The adapter control registers are I/O mapped at I/O addresses hex 03*B*0 to hex 03*BF*. This display adapter *must* be used with a TTL-level direct-drive monitor. Also, the monitor *must* support a horizontal scan rate of 18 kHz. The IBM monochrome display adapter also contains a centronics parallel printer port. A block diagram for the monochrome display adapter is shown in Figure 14.38.

The IBM Color Graphics Adapter

The IBM **color graphics adapter** is designed to support both text and color graphics display formats. The text mode displays the characters inside an 8-pixel-wide by 8-pixel-deep box and has two available modes. The first text mode is an 80-character by 25-line mode for direct-drive high-resolution display monitors, and the second mode is the 40-character by 25-line mode for television-type display monitors. The text attributes may be any one of 16 colors with any one of 8 background colors, blinking characters, and inverse characters on a character-by-character basis. This display adapter supports three graphics modes, as follows:

Mode 1: Low-resolution color graphics of 160 horizontal pixels by 100 vertical pixels in 16 colors. This mode is not supported in the IBM BIOS ROM and must be programmed without the use of ROM BIOS calls.

Mode 2: Medium-resolution color graphics of 320 horizontal pixels by 200 vertical pixels in two sets of 4 colors each. Only one color set may be used at a time, but the

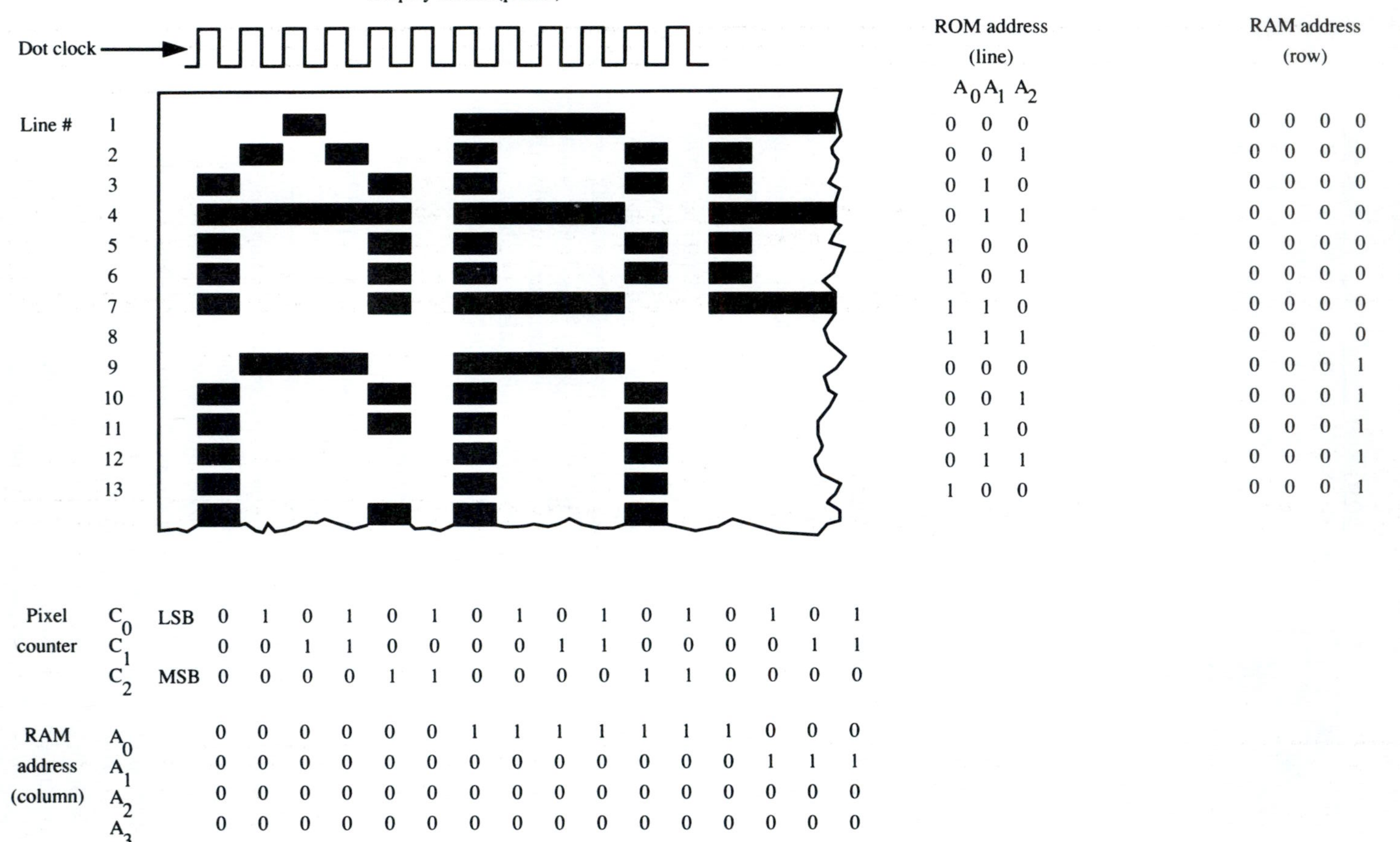

FIGURE 14.37
Character generator patterns and addresses

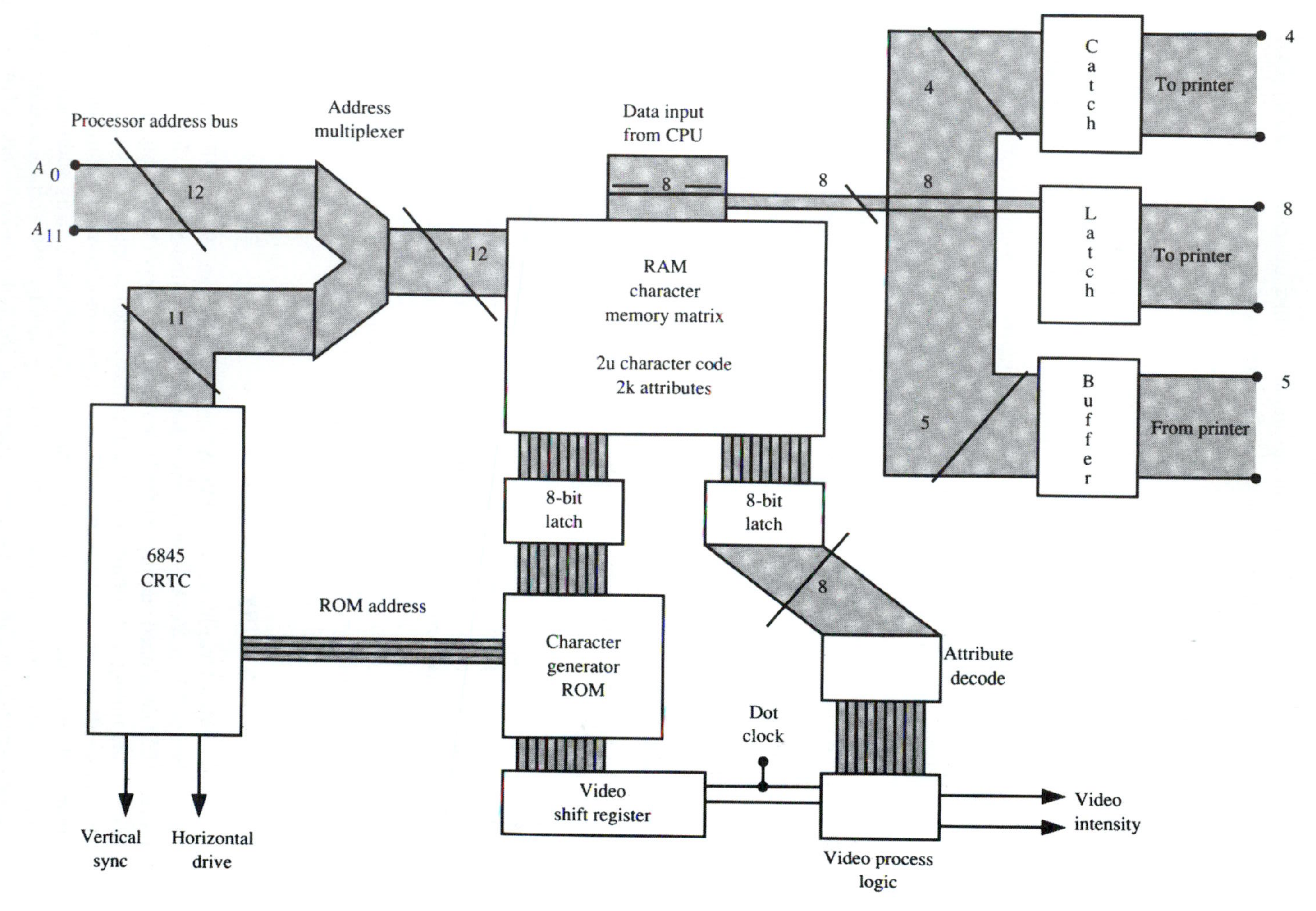

FIGURE 14.38
IBM monochrome display adapter

border may be any one of 16 colors. In this mode, *two* bits are required for *one* pixel. The combination of bits determines what color the pixel will appear, as follows:

Color Set 1 **Bits**		**Color Set 2**
0 0	Pixel is background color	Same
0 1	Pixel is green	Pixel is cyan
1 0	Pixel is red	Pixel is magenta
1 1	Pixel is brown	Pixel is white

Mode 3: High-resolution black-and-white graphics of 640 horizontal pixels by 200 vertical pixels. In this mode 1 bit per byte represents 1 pixel on the screen.

The color graphics adapter has 16,384 bytes of video RAM, which is memory mapped in the main system at addresses hex *B*800:0000 through hex *B*800:3*FFF*. In the 40-character text mode eight text pages are available. Four text pages are allowed in the 80-character mode. Since all three graphics modes require the entire memory range, only one graphics page is allowed. The control registers are I/O mapped at I/O addresses hex 03*D*0 through hex 03*DB*. The color graphics adapter has three video outputs, which are for an RGB direct-drive monitor, a composite video color monitor, and an RF modulator for use with a standard color television set. A separate light pen port is also provided to allow the use of a light pen.

Enhanced Graphics Adapter

The **Enhanced Graphics Adapter (EGA)** was introduced with the original IBM PC AT in 1984. It was the first high-resolution (for its time) color display standard for the IBM PC and compatibles. Basically, it provides the same text modes as CGA but uses a greater number of pixels for a clearer, more readable display. It also provides a selection of extended video modes, with resolutions up to 640 horizontal pixels by 350 vertical pixels. Graphics can be displayed in any of 16 colors from a palette of 64 colors or in any of two colors (monochrome mode). Most EGA adapters can support all of the CGA modes. The monitor interface is the same as it is the CGA. Today, EGA has been replaced with the VGA standard. The EGA interface is TTL digital, with red, green, blue, and an intensity signal. For this reason an EGA monitor can not be used with a VGA interface card unless it has an analog option switch.

Video Graphics Array

The **Video Graphics Array (VGA)** was developed in 1987 by IBM as an integral part of the PS/2 line. VGA is an acronym for the Application-Specific Integrated Circuit (ASIC) IBM developed and named the Video Graphics Array (VGA) chip. This chip contains all the timing and control logic used on past video adapters, thus allowing it to be placed directly on the motherboard. VGA is a higher-resolution standard, with a greater number of colors than past video standards. VGA has added many additional modes of operation, with resolutions up to 640 horizontal pixels by 480 vertical pixels. Graphics can be displayed in 256 colors from a palette of 262,144 (256K) colors. VGA is much more technically sophisticated than EGA in that it allows very flexible control over the displayed image but still maintains compatibility with past video adapters. The brightness of the screen can easily be scaled up or down, and the relative color balance can be accurately adjusted. Any adjustments to the colors can be corrected by the software. The timing and control circuitry are similar to the CGA and EGA circuitry and are based upon the 6845 CRT controller chip. Further enhancements include the amount of on-board memory, programmable color palette, and analog output. The analog video is created by feeding the output of the palette memory into a high-speed digital-to-analog converter or DAC as shown in Figure 14.39. VGA falls short in the lack of color depth and high-resolution modes. Also, VGA is limited to the ISA bus and uses paged memory. All on-board memory is mapped into memory addresses 0*xA*0000–0*xAFFFF* which is 64 KB per page. The CGA memory map of 0*xB*8000–0*xBFFFF* is supported.

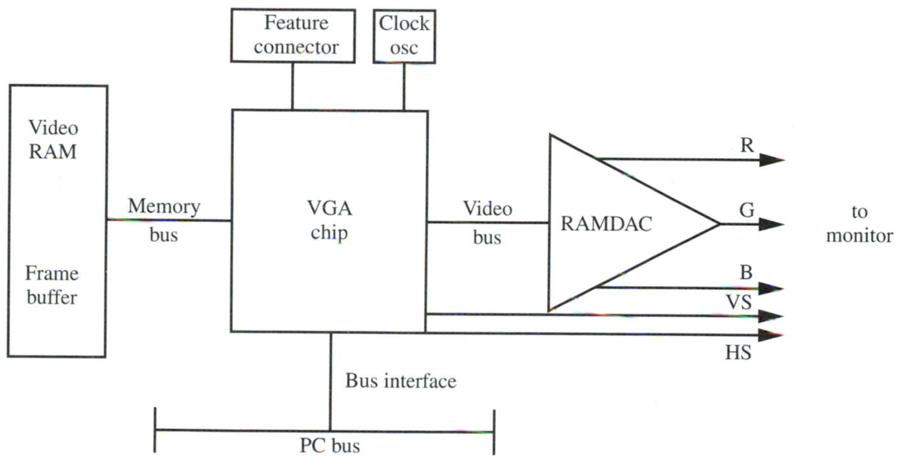

FIGURE 14.39
VGA video card

An important difference between VGA and EGA or CGA monitors is the video output method and connector used as shown in Figure 14.40. VGA adapters produce an analog or continuously variable signal for the monitor. As discussed earlier, EGA and CGA systems use the digital RGBI method. The analog signal with its greater capacity for information content permits the vast increase in the number of colors available to VGA. In fact, since the input is analog, an indefinite number of colors is possible. This is only limited to the pixel depth of the video card driving the monitor. VGA adapters are also made to interface with the 16-bit ISA bus. This doubles the video speed over the 8-bit EGA or CGA interface cards. The original VGA card used a maximum of eight bits per pixel, which yields a maximum of 256 colors on the screen. When compared to EGA or CGA, this may seem like a lot of capability, but it is not nearly enough for photo-realistic images. VGA provides an internal interface giving access to internal video pixel data, the pixel clock, blanking, and horizontal and vertical sync signals. This is known as the VGA feature connector and used to extend the capabilities of the video card. This feature is rarely used but is provided for compatibility.

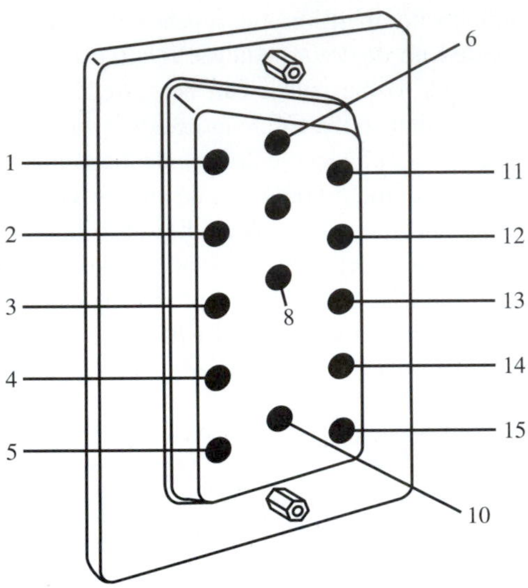

FIGURE 14.40
The VGA connector

IBM 8514/A, XGA The IBM 8514A, XGA standard was introduced shortly after VGA to add more functionality for high-end graphic and CAD applications. It provides for a 1024 by 768 resolution and interfaces to an IBM 8514 monitor. Although it is superior to standard VGA (640 × 480), the XGA standard is now obsolete and rarely used anymore.

VESA As soon as IBM released VGA in their PS/2 line of computers, everyone jumped on the bandwagon. No one would sell a computer without this wonderful colorful video. Of course, now the race was on to make each card a little better and incorporate more features. The ASIC manufacturers began producing VGA chipsets for generic card production. An organization was formed known as the Video Electronics Standards Association or VESA to keep some control over the chaos. Now everyone producing video products will follow the standards set forth by the VESA. The standards include monitor specifications, display resolutions, color depth, operating modes, additional feature connector functionality and the 486 local bus, also known as the VESA local bus. VESA also defined a set of rules for new video BIOS extensions, video memory, and the Super Video Graphics Array, SVGA.

Super VGA Adapter

The **super VGA** or **SVGA** is a much higher resolution standard than VGA. For example, resolution modes such as 1600 horizontal pixels by 1200 vertical pixels are possible. In addition, the pixel depth has been increased and programmable in steps of 8, 16,

24, and 32 bits per pixel. Modes using 24 or more bits are considered true color. This will produce photographic quality images. All this depends upon the video controller card and amount of on-board video RAM. Standard VGA adapters generally have 256K of video memory. Super VGA starts at 1 MB of video RAM and goes up to 64 MB of RAM depending on price.

The amount of memory required for a particular mode can be calculated by multiplying the desired pixel depth times the desired resolution and dividing by 8 to convert to bytes. For example, if you plan to use 16-bit color at a resolution of 640 × 480 (16 × 640 × 480/8 = 614,400), you will need a video card with at least 614,400 bytes of memory. Since memory generally comes in fixed amounts, you will probably require a video card with 1 MB of video RAM. If your card has 1 MB of video RAM you could consider operating at a higher video mode. For example 640 × 480 × 24/8 = (requires) 921,600 bytes or 800 × 600 × 16/8 = (requires) 960,000 bytes of video RAM. For photographic work, increasing the pixel depth is better than increasing the resolution. For CAD applications, increasing the resolution should be better.

The next big advance in SVGA is the addition of a linear frame buffer or LFB (the video RAM). Using a linear address allows the entire frame buffer to be treated as one large array. The memory is no longer paged and accessed in 64K blocks. Of course, the old modes are supported for backward compatibility. The LFB is mapped way up in the memory map. This greatly speeds up memory access and makes animation seem very smooth. SVGA video cards are available for the local, PCI, and AGP buses. This greatly enhances the video memory access speeds, aiding in smooth animation. Another improvement was the addition of a device known as a RAMDAC. This is a custom memory array used for the color palette and contains the video DAC. The RAMDAC provides a better data path to convert the digital color values to the analog output.

Dot Pitch

The spacing and number of dots or pixels on a screen will determine the degree of image detail or resolution. This spacing is called the **dot pitch** of the monitor. Dot pitch is usually measured in millimeters. There are four general classifications of monitors according to dot pitch.

Monitor Type	**Dot Pitch**
1. Standard television	0.62 mm or larger
2. Low resolution	0.41 to 0.61 mm
3. Medium resolution	0.30 to 0.40 mm
4. High resolution	0.29 mm or smaller

The smaller and closer the dots are, the greater the image detail that is possible. For example, a 12-in. diagonal monitor will typically have a horizontal display of 9.5 in. or approximately 240 mm. Characters are usually displayed in a 7 by 9 matrix (7 horizontal by 9 vertical) configuration within a 9-by-14 box as shown in Figure 14.41. The unused rows and columns of dots provide good separation between the characters. If each character requires 9 dots and 80 characters must be displayed in each horizontal line, a total

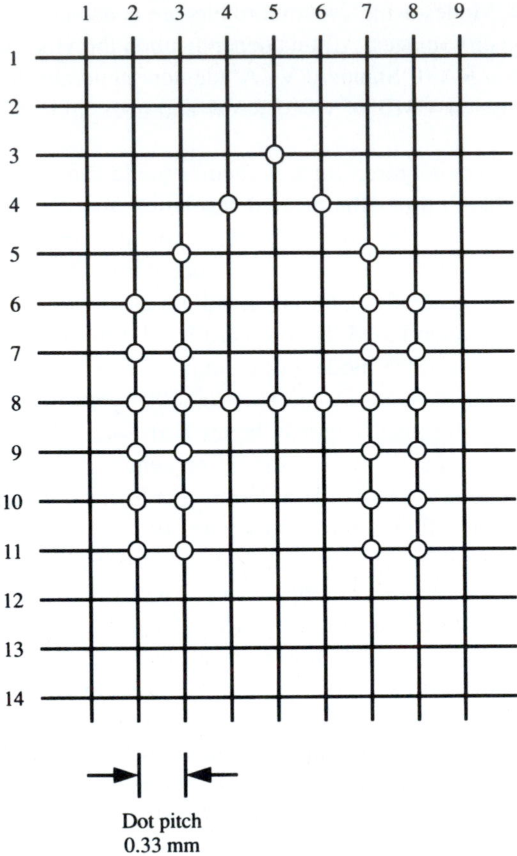

FIGURE 14.41
Dot pitch

of 720 dots are required for a 240-mm line. The required dot pitch can be calculated as follows:

$$\text{Dot pitch} = \frac{240\text{ mm}}{720\text{ dots}}$$

$$= 0.33\,\text{mm/dot}$$

SVGA Accelerators

Now that we have true colors and very high resolution, we have introduced a new problem. We must manipulate and move very large amounts of data, and in the case of animation, we must do it quickly. Placing the card on faster buses helps but places a tremendous load on the processor. Adding complex photographic images and texture mapping will quickly max out the processor, leaving power for nothing else. To help solve this problem some of the processing has been moved onto the video card. This relieves some of the burden from the processor, so it can perform other tasks. As video cards mature, more of the common video graphic and texture mapping functions are being added to the video card. In fact, now the video card can request the system bus and act like the processor, grabbing information to process and display.

AGP

The **Accelerated Graphics Port,** or **AGP,** is a high-performance interconnect aimed at 3D graphics applications. AGP is based on improvements and extensions to the PCI bus. AGP uses an entirely different type of connector as shown in Figure 14.42. This plugs into a dedicated AGP slot on the motherboard as shown in Figure 14.43. Unfortunately, AGP is not compatible physically or electrically with PCI video cards, so you will have to buy a new video card to take advantage of AGP.

FIGURE 14.42
The AGP video card

Today's 3D graphics applications require tremendous amounts of bandwidth. AGP addresses this by providing for the transfer of large amounts of data between the graphics controller and the system memory, thus eliminating the need to cache the data in local video memory. As we discussed in Chapter 12, there is now a data router, which directs data between AGP, PCI, memory, and the host bus. This removes traffic from the PCI bus, thus freeing the PCI bus up to perform other I/O functions. Remember that high-

FIGURE 14.43
The AGP slot on a motherboard

resolution photo-realistic image data is huge in size. All this data travels across the PCI or system bus and eats a lot of bandwidth. Add to this a fast UDMA hard disk and a 100-MHz PCI Ethernet card, and you really create a data path bottleneck. AGP relieves this bottleneck by providing a separate path for video data.

The separate AGP video data path provides the following advantages:

- *Dedicated high-speed bus.* By providing a dedicated high-speed bus directly between the graphics controller and the chipset, bandwidth-hungry video traffic is removed from the PCI bus. The PCI bus has a maximum bandwidth of 132 MBps, while AGP has a maximum of 533 MBps. You can imagine how quickly 100-MHz Ethernet adapters and Ultra DMA drives can gobble up those 132-MB. Conversely, consider how much better is it to have those 533 MB available for the bandwidth of 3D graphics applications.
- *Improved memory mapping.* AGP allows textures to be accessed from system memory rather than from local video memory. Segments of system memory can be earmarked by the OS for the exclusive use of the graphics controller. This memory is called AGP memory, or nonlocal video memory. The 440BX chipset performs address translations that allow for noncontiguous spaces in memory to be seen by the graphics controller as one contiguous page. This provides access to large data structures such as texture bitmaps as a single block of data or page. The chipset hardware, which performs this function, is called the GART, or graphics address remapping table. The GART is sim-

ply a table that performs address translation and returns an address to the graphics controller.

14.8 PRINTERS

So far in this chapter we have discussed the primary input and output devices, which are the keyboard and the CRT display, respectively. Sometimes it is necessary to produce the processed data as **hard copy,** or printed on a piece of paper. Since most modern computer systems use a CRT instead of a teletypewriter device, an additional peripheral device, a **printer,** is required. The printer is an electromechanical device that receives data in the form of binary-coded characters and prints the characters onto a piece of paper. New technology has had a tremendous effect on the evolution of the printer. Electronic and mechanical technological advances have transformed the huge mechanical machines into compact desktop units. There are many different types of printers available today. The type of printer chosen depends upon the user's specific needs, such as print speed, print quality, and economy. There are two major categories of printers: **impact printers** and **nonimpact printers.**

The impact printer forms the printed character by a mechanical contact with an ink-impregnated ribbon onto the paper. The many types of impact printers include dot-matrix, daisy wheel, drum, chain, and band printers. Impact printers have an electromagnet, known as a hammer, that supplies the mechanical energy to print the character.

A nonimpact printer forms the character without physically hitting an inked ribbon on the paper. The many forms of nonimpact printing include thermal transfer, ink jet, electrostatic, and laser printers. Nonimpact printers have the unique advantage of being extremely quiet.

Dot-Matrix Printers

The **dot-matrix printer** is the most widely used printer on microcomputers today. The characters are formed by printing a series of dots on the paper, similar to the method used by the display adapter, as shown in Figure 14.44.

Dot matrix printers are usually fast, but since the characters are made up of dots, the characters tend to be somewhat choppy. Some dot-matrix printers offer a "near-letter-quality" mode by printing more dots per character, but this slows down the print speed. The most widely used impact dot-matrix printers use a group of needles, or wires, connected to electromagnets. The needle-and-electromagnet assembly is known as the printhead. The needles are arranged in a group of 7 to 24 vertical dots, depending on the printer. The needles are often staggered to allow for the dots to overlap, which causes the dots to blend together for a smoother image, as shown in Figure 14.45.

The printhead is mounted on a motor-driven sliding carriage assembly. The older printer carriage drives were fixed-speed motors. The dot timing, which is the spacing between the dot groupings, was generated by an optical timing fence, which is a transparent strip containing a series of vertical stripes. Mounted below the printhead is a photosensor that senses these stripes and determines when the next group of dots are to be printed.

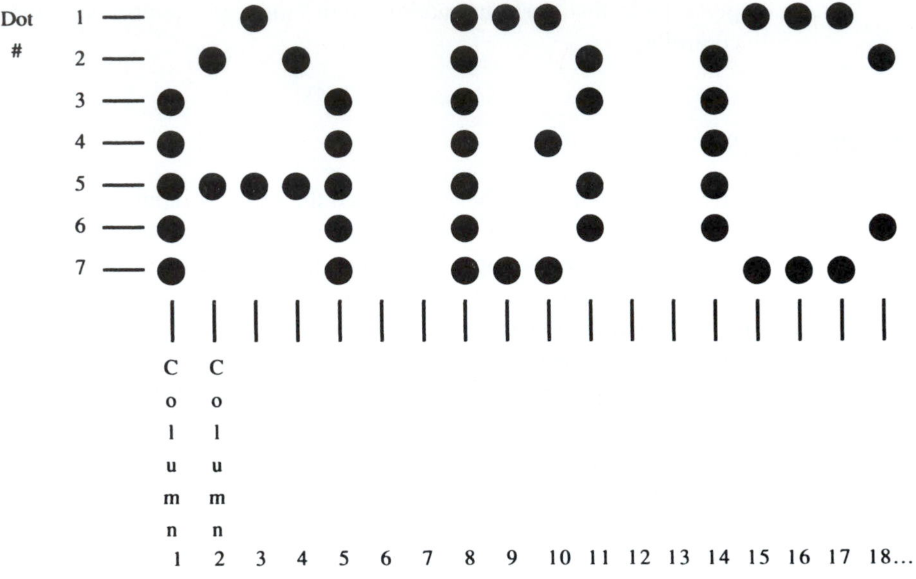

FIGURE 14.44
Dot-matrix characters

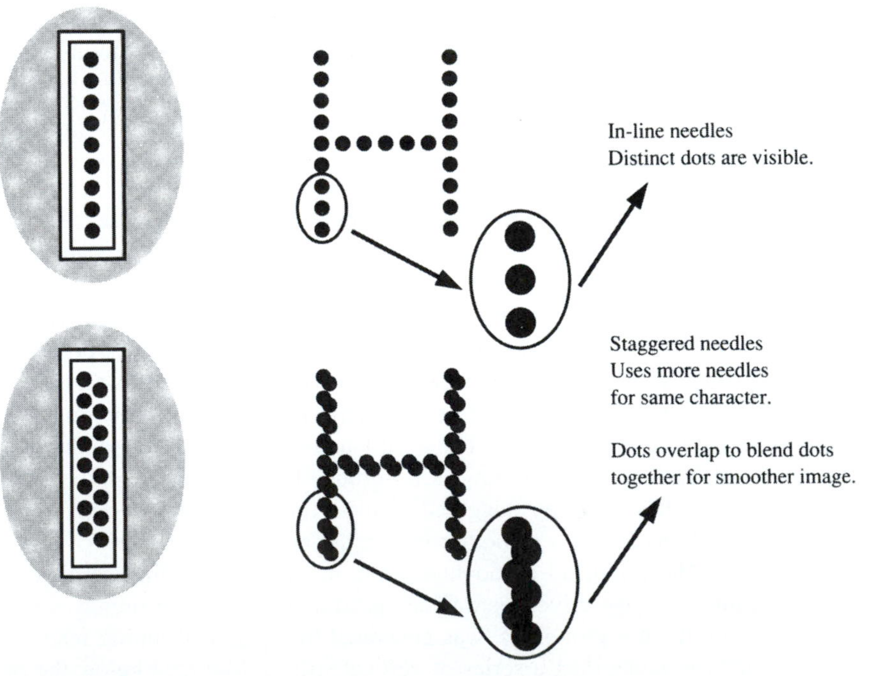

FIGURE 14.45
The dot-matrix impact printerhead

One variation of this design is the use of a slotted disk on the carriage motor shaft instead of the timing fence. In the newer printers the carriage drive system has been simplified with the use of a stepper motor. This allows for precise head positioning, so the head position sensor is not required. The paper feed mechanism is also stepper motor driven, which allows for precise paper positioning. Most newer printers have the ability to print graphics as well as text due to the precise positioning mechanisms.

Two variations of the dot-matrix printer are a single reciprocating hammer printer and the reciprocating "comb" printer. In the single reciprocating hammer printer the printhead consists of a single dot and electromagnet. The printhead reciprocates up and down the length of one character to form the vertical dot matrix while also sliding sideways on the carriage assembly. This is a simple and economical means of printing but has the disadvantage of being slow.

In the second variation a series of hammers is mounted on a comb of dots. The comb is mounted on a reciprocating shuttle assembly. The characters are formed by printing an entire line of dots at a time by firing the hammers while the shuttle is reciprocating horizontally. The paper is advanced a fraction of one character after each line is printed until all the lines have been printed to form the entire character.

The **thermal dot matrix printer** functions similarly, but the printhead is entirely different. Instead of electromechanical hammers and an inked ribbon to print dots on the paper, a group of heaters is used. The earlier thermal printers required the use of treated thermal paper, which changed color when heated. While these printers are usually less expensive than impact printers, the paper is far more expensive. If the paper is accidentally heated or left out in the sunlight, it turns dark, making the printed copy unreadable. As a solution to this problem the newer printers use a thermal ribbon. When the ribbon is heated against plain paper, the heated pigment transfers to the paper, forming an image on regular untreated paper.

Daisy Wheel Printers

Daisy wheel printers are impact printers. The entire printable character set is contained on a plastic wheel that resembles a daisy. Each petal of the flower has a different character on it. This method produces the best type quality, since the entire character is typed just as on a typewriter. Since the daisy wheel must start, spin, stop, and settle at the desired spot, these printers are much slower than dot-matrix printers. A motor rotates the daisy wheel to position the desired character between the hammer and the inked ribbon, as shown in Figure 14.46. The hammer is then fired, which drives the character type onto the ribbon and the paper, thus forming an image of that character on paper. The daisy wheel and hammer assembly is mounted on a sliding carriage similar to the dot-matrix printer. The one major disadvantage is that it is not possible to produce graphics on this type of printer.

Ink Jet Printers

The **ink jet printer** is another form of nonimpact matrix printer. The characters or images are formed by firing ink droplets at a high velocity onto the paper. The ink droplets are formed by pumping liquid ink into a vibrating chamber with a small orifice at the

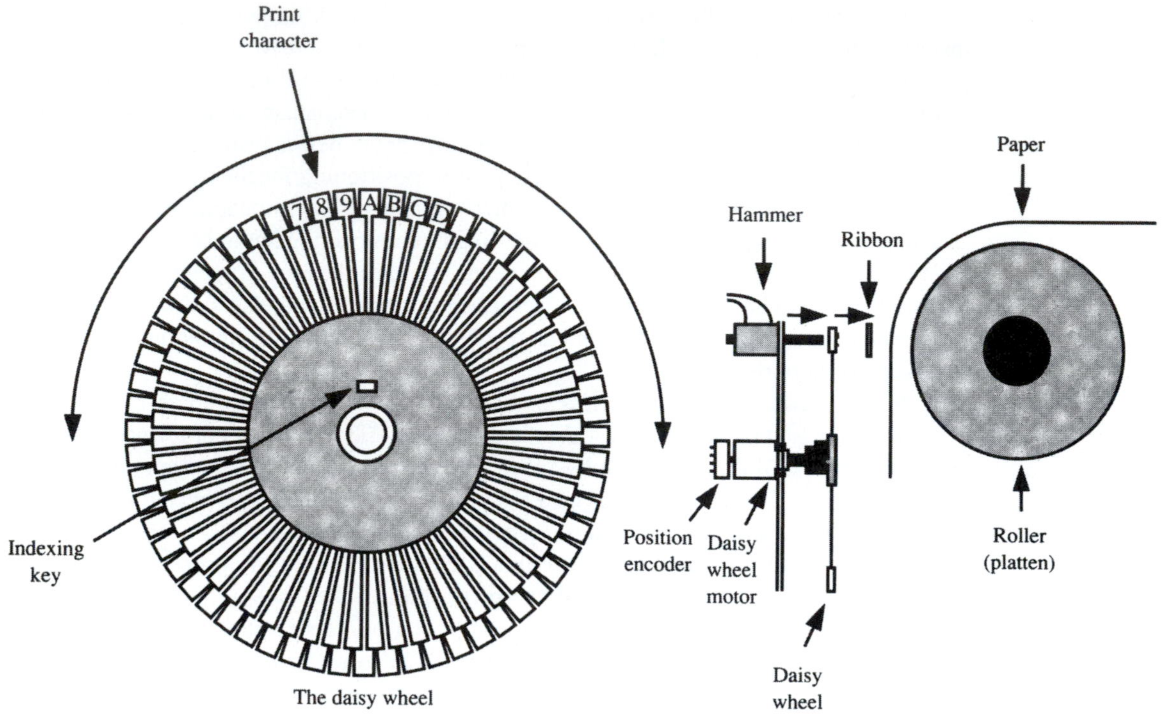

FIGURE 14.46
The daisy wheel printer

opposite end. The ink droplets are electrically charged as they pass through an electrostatic field generated by charging plates. The droplets are then deflected up or down by the deflection plates to create the vertical matrix while the printhead is sliding sideways to create the horizontal matrix, as shown in Figure 14.47.

A variation of this method uses several orifices and impulse fires the ink droplets without using a deflection system, similar to the impact matrix system. The carriage and paper feed mechanisms for the ink jet printer are similar to those used for the impact matrix printer.

The Laser Printer

Laser printing is another type of nonimpact printing. Laser printers have been around for quite some time. The first ones were very large, heavy, and expensive. Advances in solid-state electronics and materials have been applied to this technology. As a result laser printers are fast and quiet, and have become small, lightweight, and economical. Laser printers have become the most common type of printer used in offices, industry and the home office. The photosensitive and toner components have been placed inside one user-replaceable cartridge known as the toner cartridge. When the toner cartridge is replaced, you replace the photosensitive drum, two of the corona wires, the developer unit, and the

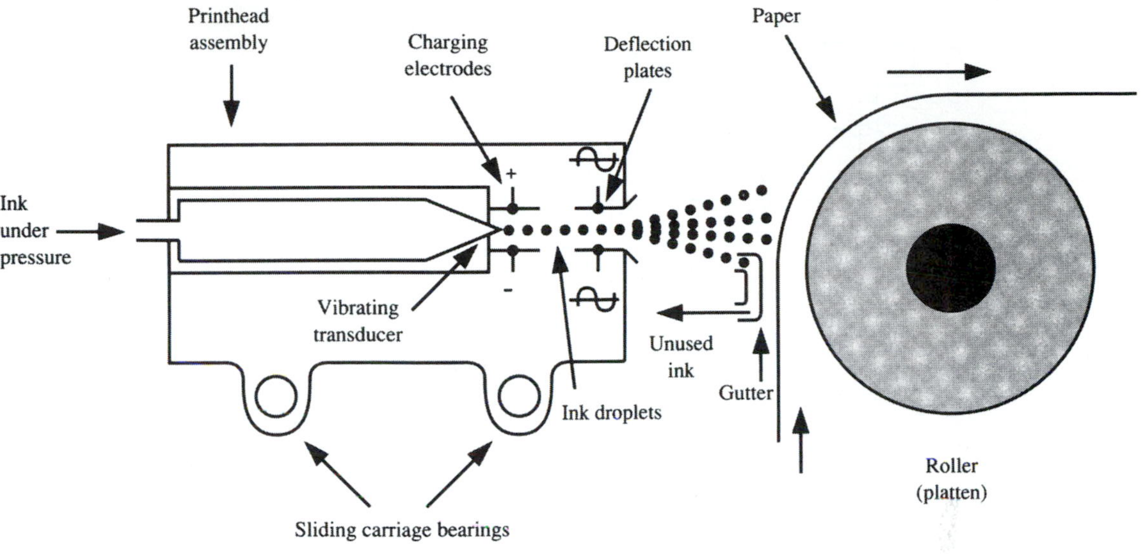

FIGURE 14.47
The ink jet printer

toner supply. Laser printers operate using a method known as transfer electrophotography imaging. They operate in the same fashion as a plain paper photocopy machine. In fact there are multipurpose office machines that serve as laser printers, copiers, image scanners, and fax machines. The difference between office copiers and laser printers is in the form of image exposure. As implied in the name, a laser printer uses a laser light source to generate the printed image. While the office copier uses a reflected image, laser printers contain a very sophisticated image-processing microprocessor to generate an image.

Electrophotography uses electrostatic charges to paint or transfer an image consisting of charged toner particles onto a reversed charged sheet of paper. Since opposite charges attract, the toner will be attracted to the paper. You may have noticed that a television screen or a computer monitor becomes dusty very quickly. This is because the high voltage potential behind the glass attracts dust particles out of the air onto the glass face. This is how a laser printer or office copier places an image onto the paper. In the case of a laser printer an image is created electronically by using a controlled laser source and a photosensitive drum.

Referring to Figure 14.48, the first step in the process requires an entire page of text and image to be loaded into the printer's memory from the computer. Once the page is loaded, the internal processor will create an image representing one entire page to be printed. Next the one-page image will be rasterized. Rasterizing the image is similar to the way a CRT creates an image on the face of the tube as discussed earlier in this chapter. Rasterizing is required because the image is placed on a photosensitive drum one line at a time using a scanning laser light beam. This is similar to the way an image is created by a CRT using a scanning electron beam.

The next step in the process is to create an electrostatic image from the raster memory onto a rotating photosensitive drum. Referring to Figure 14.48, the photosensitive drum

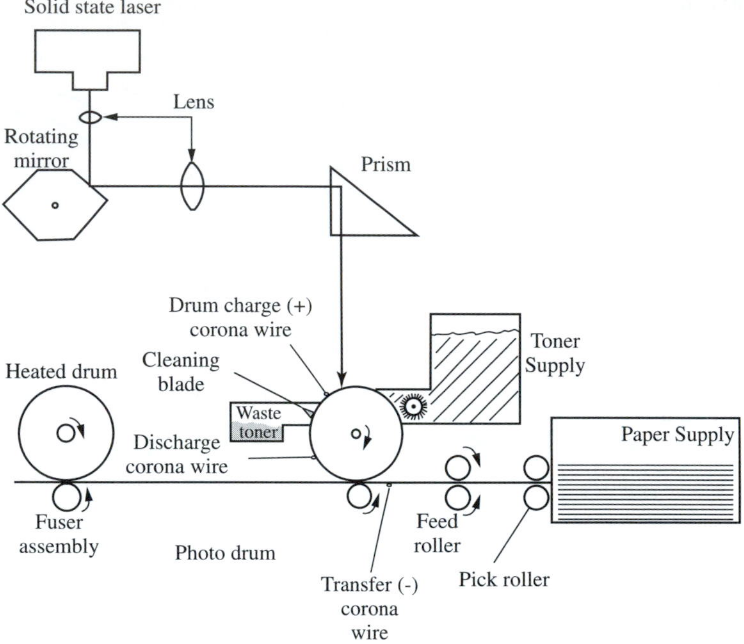

FIGURE 14.48
The laser printer

known as the photo drum is the heart of the laser printer and contains the latent image in the form of electrostatic charges. The photo drum is coated with a light-sensitive semiconductive material. This material is made up of selenium or cadmium disulfide (CdS). These materials exhibit a quality that makes them conductive when exposed to light. You can think of the photo drum as a large photocell. Surrounding the photo drum is a series of wires known as corona wires. The corona wires control the charge on the surface of the photo drum. The first corona wire is used to charge the photo drum surface. The second corona wire is used to transfer the image from the photo drum to the paper. The third corona wire is used to erase the charge on the photo drum, thus releasing any remaining toner for photo drum cleaning. The image is formed on the photo drum as follows: The first corona wire charges the photo drum. A scanning modulated laser beam forms an optical image one line at a time. This is accomplished by using a fixed solid-state laser source pointed at a rotating polygon mirror. The rotating mirror reflects the laser beam into a special flat lens to keep the beam focused on a prism. The prism bends the laser beam 90° down onto the surface of the photo drum. An image is written on the surface of the photo drum by scanning the drum with the controlled laser beam. When the beam is turned on, the part of the photo drum under the beam will discharge because the photo drum material will conduct the charge to ground but only at the point directly under the finely focused beam. Now the photo drum will contain a series of modulated image lines, similar to those in a CRT display.

The next part of the process is to apply a dry toner to the photo drum. The toner used in laser printers and copiers is a dry, very finely ground black material. In a laser printer the toner is mixed with powdered iron and is applied to the drum by a rotating magnet very close to the drum. This forms a sort of brush, which gently applies the toner to the drum. In older printers the powdered iron is added separately and is known as the developer. Where there is a difference in potential between the drum and toner, the toner will be attracted to the drum, thus forming an image of toner on the photo drum surface as shown in Figure 14.49.

The next step in the process is to transfer the image from the surface of the photo drum onto the paper. While the image is being written onto the drum, paper is fed through

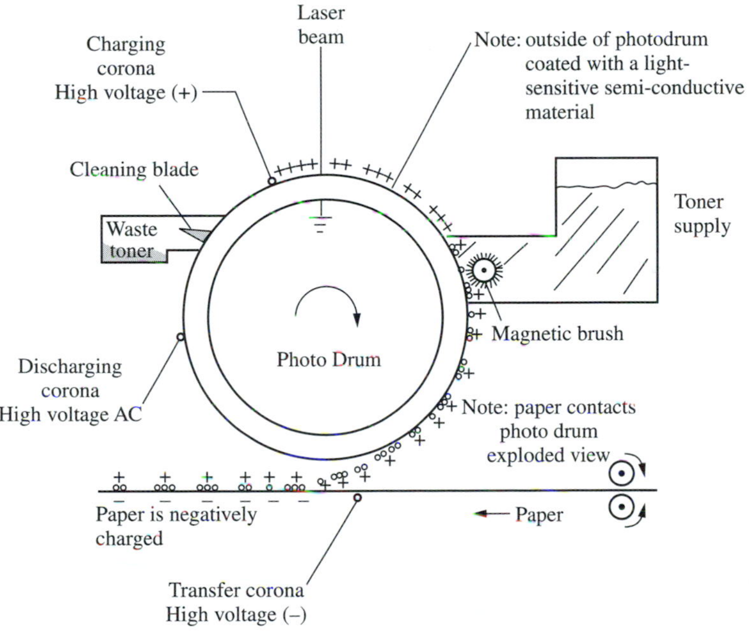

FIGURE 14.49
Toner application

the printer at the same speed that the drum is rotating. The paper feed is controlled by a solenoid-activated clutch. You will hear a slight click sound as soon as the printer feeds a sheet of paper to the photo drum. Keep this in mind when working on feed or jam problems. Now the second corona wire comes into play. As the paper nears the drum, it is charged with a potential opposite that of the drum. Again, opposites attract and the toner particles will jump from the drum surface onto the paper surface. Keep in mind that the toner particles are only lying on the surface of the paper held by a tiny charge. At this point the image can be easily wiped off the paper. As the drum continues to rotate it is discharged by the third corona wire. A soft Mylar cleaning blade then cleans the drum of any remaining toner.

The last step in the process is to get the toner affixed or fused to the paper permanently. This was a problem in the early days of copiers and laser printers. The first designs

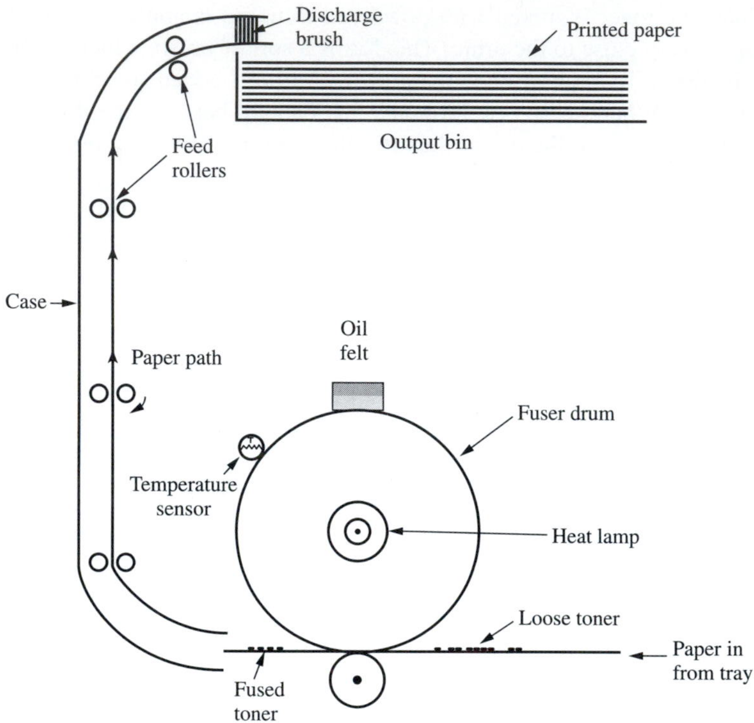

FIGURE 14.50
The fuser unit

used a liquid toner that was allowed to dry onto the paper, thus affixing it. This doesn't work well for high-speed and high-volume printing. Today, laser printers use a dry powdered toner and affix it by using pressure and high temperatures. It is literally melted into the paper. The part used to affix the image to the paper is known as the fuser unit. The fuser unit consists of a small soft rubber roller held tightly against a special rigid heated drum as illustrated in Figure 14.50. The fuser drum is coated with a porous material. The porous material is impregnated with a special silicone fusing oil to keep the toner from sticking to the fuser drum. There is also a wick saturated with the fusing oil rubbing on top of the fuser drum. The oil wick serves to clean and reapply the oil to the fuser drum continuously. A long glass heat lamp in the center of the fuser drum provides the heat. As the paper is ejected from the print path, you will notice a small series of wires gently brushing the paper. This brush is conductive and discharges any residual charge from the paper before stacking paper in the output bin.

14.9 INTERFACE STANDARDS

Since there are many different manufacturers of computers and computer peripherals, some form of standards is necessary. If interface standards were not used, it would be difficult to connect a printer from one manufacturer to a computer from another manufacturer. Most microcomputer systems support two types of peripheral interfaces, the centronics parallel interface and the asynchronous serial interface standards. There is another parallel printer standard known as the data products interface, but it is rarely used on microcomputers.

The Centronics Printer Interface

The centronics printer interface standard is used by the IBM PC and many other computers as well. This interface is an 8-bit parallel data interface with handshaking. The term **handshaking** means some form of status communication between devices. For example, when the printer is out of paper or is off-line, it must be able to tell the computer of this condition. The logic levels are defined as TTL logic levels, typically 0.80 V for a 0 state and 2.4 V for the 1 state. Not all printers use all the available status lines supplied by the IBM printer interface, but the strobe and busy lines must be used. The strobe line signals the printer that a character of data to be printed is currently on the data lines. The busy line is used to signal the computer when the printer is busy and cannot accept any data. The acknowledge line can be used when sending a stream of characters to accomplish the handshaking while filling the printer's internal buffer, if it exists. The rest of the status lines may or may not be used, depending on the printer. The IBM printer adapter card provides 12 output and 5 input lines. The first 8 bits are the character data and are decoded to I/O address 378 hex. The second group of 4 bits is used for printer control and is decoded to I/O address 37*A* hex.

The connector pinpoints are defined in Table 14.1.

TABLE 14.1
Centronics/IBM printer interface pinouts

Pin	Signal Name	Direction	Address	Data Bit
1	Strobe	To printer	37*A*	0
2	Data bit 0 (LSB)	To printer	378	0
3	Data bit 1	To printer	378	1
4	Data bit 2	To printer	378	2
5	Data bit 3	To printer	378	3
6	Data bit 4	To printer	378	4
7	Data bit 5	To printer	378	5
8	Data bit 6	To printer	378	6
9	Data bit 7 (MSB)	To printer	378	7
10	Acknowledge	From printer	379	6
11	Busy	From printer	379	7
12	PE (out of paper)	From printer	379	5
13	Select	From printer	379	4
14	Auto linefeed	To printer	37*A*	1
15	Error	From printer	379	3
16	Initialize printer	To printer	37*A*	2
17	Select input	To printer	37*A*	3
18	Ground	To pin 25 all grounds		
25	Ground			

The Asynchronous Serial Interface

Most computers provide at least one general-purpose communications port. This communications port may be used for almost any external peripheral device or as a link to another computer. The most universal and widely used method for data communications is known as the **asynchronous serial data** method. The term "serial data" means that the data bits are sent down the line one bit at a time, usually by loading a data byte into a shift register and clocking the data out in bit serial format. The term "asynchronous" means that no external clock or timing signals are required to accompany the data. The data is sent as a group of bits known as a **frame.** Each frame of data is preceded by one **start bit,** which is the first mark-to-space transition. The mark state is the TTL high state, and the space state is the TTL low state, *before any level conversion.* The start bit must be held for 1 bit time. Next, the data bits are sent, starting with the LSB. The data bits are followed with an optional parity bit. After the last bit is sent out, the line *must* return to the mark state for at least 1 bit time, which is known as the **stop bit(s).** A typical frame of data for an ASCII A with even parity is shown in Figure 14.51. Since the data is not synchronous, the stop bit may be any length as long as it is *at least* 1 bit time long. One exception is if the receiving equipment is expecting two stop bits, then two must be sent or a timing error known as a **framing error** will occur. The data bit field may be any length but is usually 7 or 8 bits long. Again, the sending and receiving equipment must be set to the same data length, or a framing error will occur.

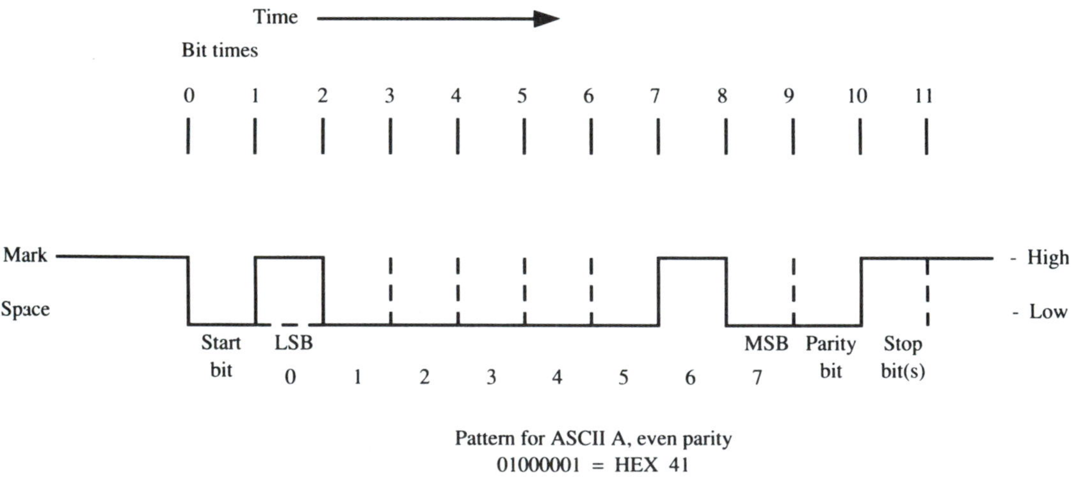

FIGURE 14.51
Asynchronous data frame

The rate at which the data bits are sent is known as the **baud rate.** A speed of 9600 baud means that 9600 bits per second are being transmitted. If the serial format is set for 8 bits, no parity, then 10 bit times are required for each character. This means that 960 characters are sent per second. There are many LSI IC chips on the market for serial data formatting, transmitting, and receiving. These ICs are often referred to as UARTs, which stands for universal asynchronous receiver transmitter. The latest chips are designed for microprocessor bus compatibility, and some contain built-in programmable baud rate generators. The IBM asynchronous communications adapter uses the INS8250 LSI chip and is completely programmable regarding speed, bits per character, parity, and stop bits.

Serial data is rarely interfaced at the TTL level but is usually converted to current loop or line driver levels. Most all serial interface lines today conform to the **RS-232C** standard. The logic levels are converted to the following values:

- *Marking level.* Logical true = −3 to −15 V, typically −12 V.
- *Spacing level.* Logical false = +3 to +15 V, typically +12 V.

Several status and modem control lines also accompany the transmit and receive data lines and are defined as follows:

- *Received data* (pin 2). Serial data into the computer or peripheral device.
- *Transmit data* (pin 3). Serial data from the computer or peripheral device.
- *Request to send* (pin 4). A status line from the computer that signals that the computer is ready to send data.
- *Clear to send* (pin 5). A status line to the computer that informs the computer when to stop or start sending data.
- *Data set ready* (pin 6). A status line, usually from a **modem** (a serial data transmission and receiving device), that signals the computer that the modem is ready to operate.
- *Carrier detect* (pin 8). A status line, usually from a modem, that signals the computer that a connection to a remote modem is established.
- *Data terminal ready* (pin 20). A status line from the computer or peripheral device that indicates that the device is on line or ready.
- *Ring indicator* (pin 22). A status line from a modem that indicates that the telephone line is ringing.

Two popular terms that are used in describing the connections on an RS-232 link are **DTE** and **DCE.** Without trying to confuse the issue, we will describe them briefly. DTE stands for **data terminating equipment,** and DCE stands for **data circuit terminating equipment.** The DTE is the computer or peripheral device, and the DCE is the modem or modem eliminator (crossover box) for direct connections without the use of modems. Most RS-232 interfaces use a 25-pin connector known as a DB-25-P or a DB-25-S connector. The P means pins (for male) and the S means sockets (for female). The pinouts for most DTE are shown in Table 14.2.

The interface and connections just discussed were originally designed to connect DTE, such as computers and peripheral devices, together through DCE, which are modems. There seems to be much confusion when connecting serial devices together, such as one computer to another computer or a serial printer to a computer through the serial port. It is obvious that it is not possible to wire the serial DTE devices together pin for pin. We must cross some of the pins or use a device called a ***null modem,*** which simply crosses pins 2–3, 3–2, 4–5, 5–4, 6–8–20, 20–8–6. In some applications—for example, the serial printer—the ready/busy status line from the printer must be wired to clear to send on the IBM PC for proper handshaking. In the worst case the receive data is simply wired to transmit data and signal ground is wired to signal ground. The rest of the status inputs

TABLE 14.2
RS-232 DTE pinout chart

Pin	Signal Name
1	Safety ground (chassis)
2	Received data
3	Transmit data
4	Request to send
5	Clear to send
6	Data set ready
7	Signal ground (common)
8	Data carrier detect
19	Reverse channel
20	Data terminal ready
22	Ring indicator
Additional Nonstandard Pin Usage	
9	+ Transmit current loop data
11	− Transmit current loop data
11	Ready/busy used by some printers
18	+ Receive current loop data
25	− Receive current loop data

Note: Be extremely careful when using nonstandard pins.

are wired to the data terminal ready line (forced true always). This usually works if all else fails.

Some serial devices use the X-ON/X-OFF protocol for handshaking, which eliminates the need for status line wiring. The X-ON/X-OFF protocol is accomplished by sending control codes back to the host computer. The X-OFF control character tells the computer to stop sending data while the printer or other device is busy. The X-ON code tells the computer to resume sending data.

A Word About Current Loop The current-loop communication method dates back to the days of the telegraph. This method was widely used in the 1960s and early 1970s when the teletype was commonly used as a computer terminal. In some cases the current-loop interface is still used today. There seems to be much confusion when connecting current-loop peripherals to computers. This confusion is easily eliminated when you understand how to connect to the interface. In the current-loop system there are two loops, one for the receiver and one for the transmitter. In the case of the printer, in which no status is returned, only one loop is required. Each loop must have an active end and a passive end. The active end supplies the current for the loop. There is no standard that determines which end of the loop is active or passive. The teletype, for example, has an active printer and a passive keyboard, as shown in Figure 14.52. The keyboard circuit is a simple rotary switch. The printer circuit contains the current source and the print mechanism solenoid. The IBM asynchronous adapter will support current-loop devices. The receive circuit at the adapter card is a passive circuit and is simply an input to an optical isolator. The transmit circuit at the adapter card is active and supplies approximately 30 mA of current.

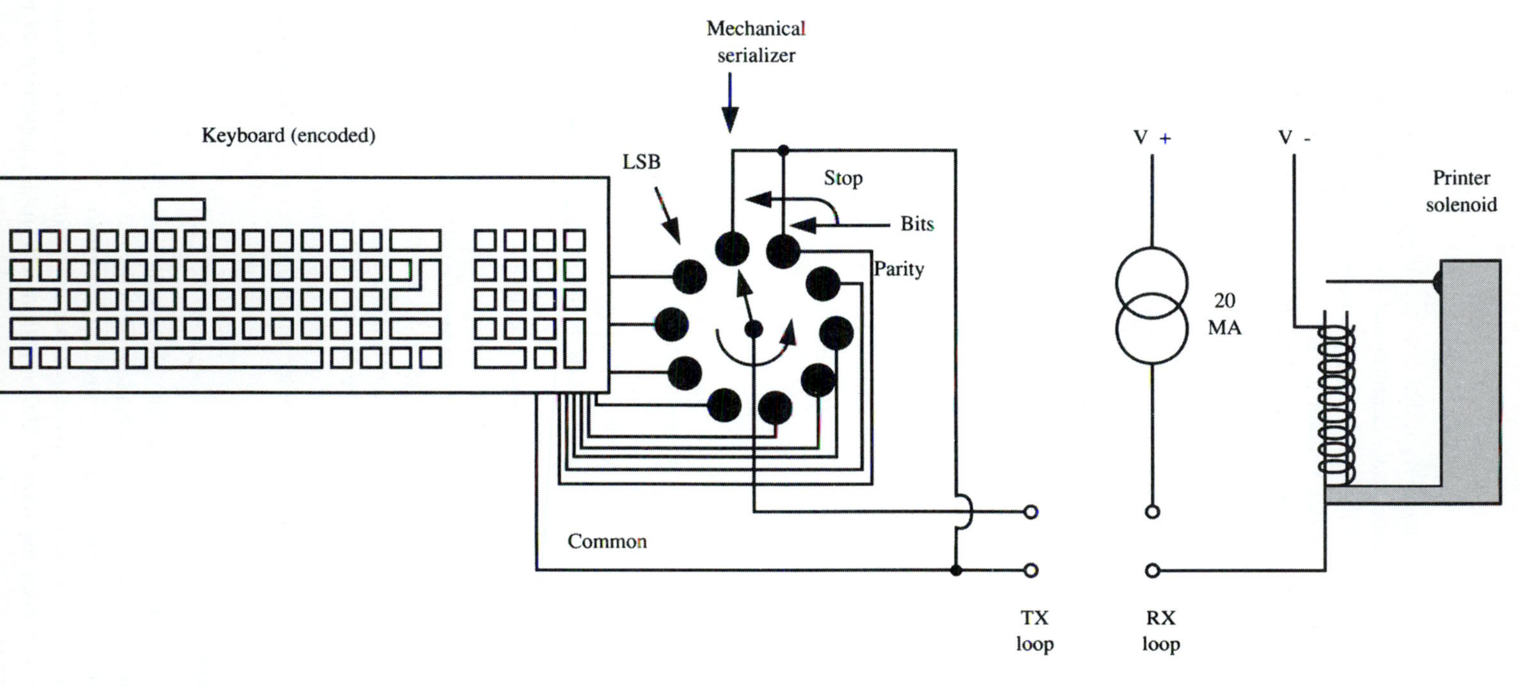

FIGURE 14.52
The teletype current loop

14.10 MODEMS

A modem is an electronic device that converts digital data into a series of audio tones for transmission over a voice-grade communication channel. This device will also convert audio tones received over the voice-grade channel back into digital data. The word **modem** is an acronym for modulator–demodulator. The modulator, similar to a radio transmitter, will take a data bit and convert, modulate, or mix it with a carrier signal. The demodulator side is similar to a radio receiver and decodes a carrier back into the original signal before the data bit it was modulated.

The term **full duplex** means that the modem is capable of sending data at the same time that it is receiving data. This is usually accomplished by using two sets of audio tones. The term **half duplex** means that the modem may only send or receive at one time. It is not capable of sending and receiving simultaneously due to bandwidth restrictions or other problems.

With the computer revolution came a need to share information. Originally, data was shared using removable magnetic medium and carried or mailed to another user. This is slow and clumsy. A few years ago, anyone using the common telephone system or POTS (plain ordinary telephone service), for digital communication had to do so by using an acoustic coupler. This was nothing more than a speaker and microphone. Thanks to anti-monopoly laws, competition, and common sense, these acoustic couplers are now only found in the Smithsonian. Modern modems are connected directly to the telephone line, coupled with a simple transformer to provide isolation. Modems, like the computers they are installed on, have gone through tremendous advances in technology. The first PC modems communicated at a speed of 300 baud or bits per second (bps), or roughly 30 characters per second. Today modems communicate at 56,000 bps (actually 53,333 if you get a real good connection). Furthermore the price of a modem has become less and less expensive. Not too long ago, in 1984, a Hayes Smart Modem 1200 (1200 bps) sold for about $700. Today you can get a 56K modem for $50. Until a few years ago, the theoretical maximum speed for data transmission was 33.6 kilobits per second (Kbps), known as Shannon's limit. Recent changes in the common carriers allow for 56 Kbps as long as your ISP (Internet service provider) has a digital connection to the telephone company.

PC modems come in two forms, internal and external. The internal modem, as the name implies, installs inside the system unit, usually on the ISA bus. The latest modems are available for the PCI bus. Internal modems also contain a COM (serial) port. Some of the cheaper modems only contain the line interface and use the computer to encode and decode the audio. This of course places an increased load on the processor, slowing other tasks. External modems attach to one of the existing COM (serial) ports on the rear of the system unit. These are the easiest to set up because no changes are made to the PC.

Modulation

Modulation can be defined as taking a constant medium and changing it in a way that it can be used to carry information. Perhaps the oldest form of modulation is the Indian smoke signal. Here a steady stream of smoke is interrupted (encoded) to carry information to a distant group. By observing the smoke patterns, a distant group can discriminate (decode) the intended information. The first application of modulation in the electronic era was the telegraph. In the telegraph, a DC current is interrupted with a sequence of long and short connections commonly called dashes and dots. The sender activates a switch known as a code key. When the switch is closed, a magnet will pull a bar, making a click sound on the distant side. By listening to the bar, the receiver can decode the dots from dashes. The code used is often known as Morse code. This technique was later applied to the wireless radio.

Amplitude Modulation (AM)

The first wireless system transmitted a modulated carrier wave. Instead of using a DC current, it worked on the principle of sending a carrier or not sending a carrier. The receiver would beep when the carrier was present. Again, the modulation method here was the duration of the carrier for dots and dashes. Later, voice was sent by modulating the carrier with a variable audio analog signal. The receiver decodes the modulated signal by simply using a diode detector fed to an audio amplifier. This regenerates the original audio signal and is known as amplitude modulation or AM as shown in Figure 14.53. Amplitude modulation is susceptible to noise and interference. Anyone listening to an AM radio has experienced this problem. This is why the frequency modulation, or FM, method was developed.

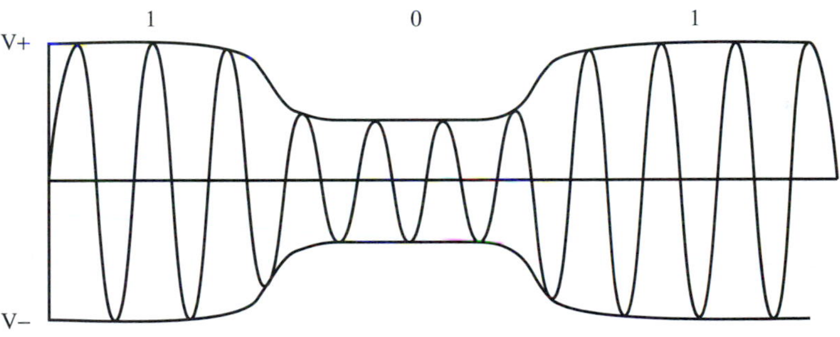

FIGURE 14.53
Amplitude modulation

Frequency Modulation (FM)

Frequency modulation uses a constant-carrier wave level but changes or modulates the frequency of the carrier wave shown in Figure 14.54. The signal is demodulated by using a frequency-sensitive circuit or frequency discriminator. Only changes in frequency are detected, and the level of the signal is ignored. Noise and interference affect the level of the signal but not the frequency. Anyone comparing an AM radio to a FM radio will immediately hear the difference and benefits of FM.

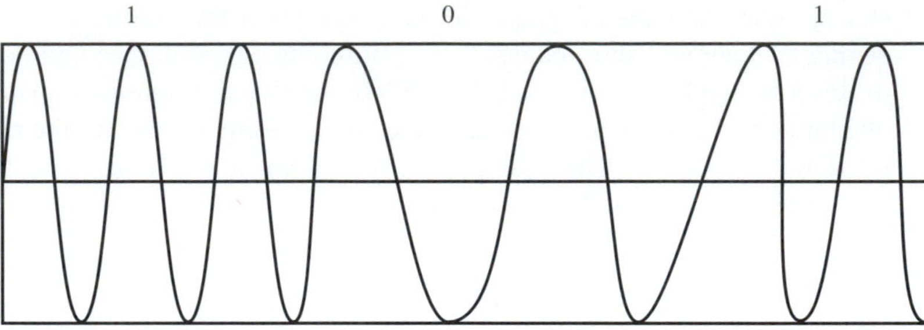

FIGURE 14.54
Frequency modulation

Phase Modulation

Another type of carrier wave modulation is known as phase modulation, or PM. In phase modulation, only the phase of the carrier is changed, as shown in Figure 14.55. The receiver maintains a local oscillator that is keyed to the originating oscillator. The phase of the sample signal is compared to the reference oscillator, thus decoding the signal. Phase modulation is used in the color television. In fact, a color television uses all three methods described earlier. AM is used for the video luminance, FM is used for the sound and PM is used to carry the color portion of the video.

A modified version of phase modulation known as quadrature modulation has been developed. It is still phase modulation, but the phase is controlled in increments of 90°, resulting in four quadrants of coding per wave. As shown in Figure 14.56, using all four

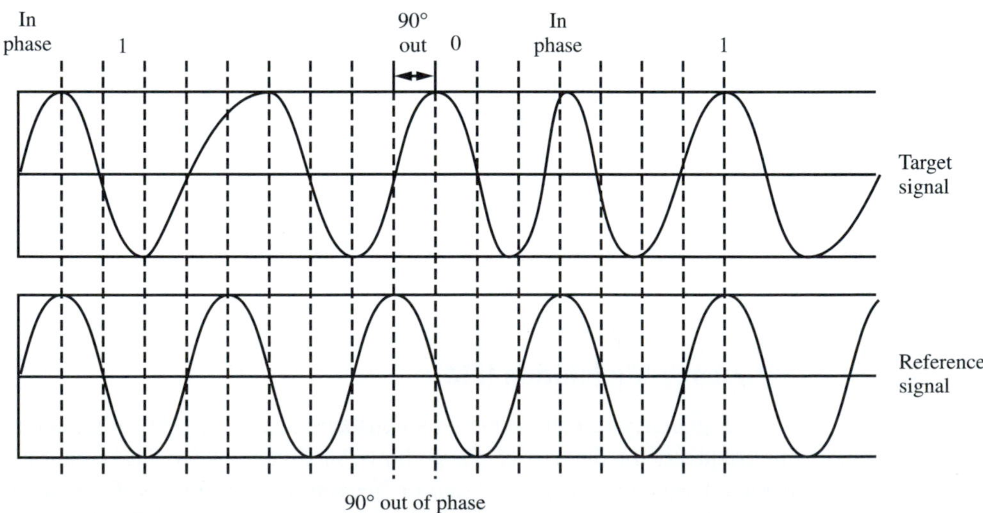

FIGURE 14.55
Phase modulation

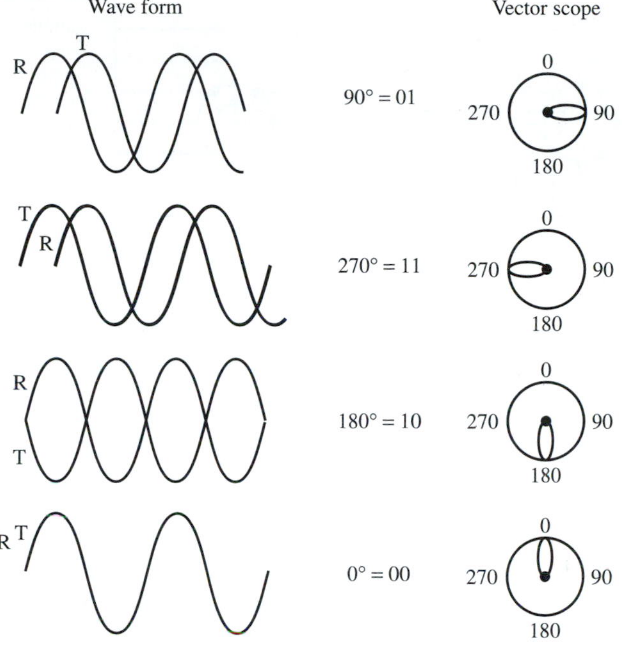

FIGURE 14.56
Quadrature modulation

states we can see that it is possible to encode two bits of information per baud. This results in what is called group coding. Instead of single bits being modulated, the bits are modulated in groups.

Transmission Medium

Before going into how the modem works and the modulation techniques used, we must first consider the transmission medium, the telephone. Anyone on hold listening to music can hear the poor audio quality. It is nothing like a radio and is missing all the bass. That is because the bandwidth is very limited, transmitting only enough to make human conversation clear. The bandwidth is limited to allow for many individual lines to be combined or multiplexed onto a single intermediate medium. Let's stop to think about this and use an example. You place a telephone call from New York to a friend in California. At first, you may think the telephone company connects your telephone to your friend's telephone directly. Now try to imagine how many wires would be needed for everyone to connect to everyone else. Today this would just not be practical. However in the very early days, this is exactly how it was done. Today the problem is solved by sharing a common line known as a trunk. Using a method known as time domain multiplexing, many lines can be placed on one wire (or fiber cable). A multiplexer is similar to a large rotary switch driven by a motor. Each incoming line is sampled and placed onto the common line in turn. There is a similar switch on the other end in synchronization with the sending switch to receive the call. Figure 14.57 illustrates this technique.

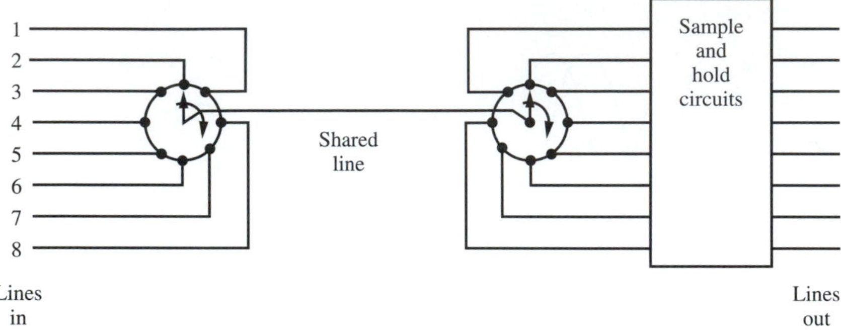

FIGURE 14.57
Multiplexing

The information is stored and sent to the destination. What the destination gets is a chopped and sampled version of the original signal. This is the limiting factor because you can only change states at or less than the sample rate. In fact, according to the Nyquist rate you must sample at two times the incoming frequency to reconstruct the original waveform. Also, most of the power required in audio is in the bass range, so to limit the power needed we simply eliminate the bass. Currently the telephone will only pass frequencies between 300 Hz and 3500 Hz. This is actually a boost from just a few years ago where the upper limit was 3000 Hz. Thus we have to get our information across a line that can't go any faster than 3500 cycles per second. Furthermore, that's not considering two way communication and error tolerance. Now one may ask, how does a 56,000-bps modem do it? The answer is, it's all in the modulation method and the type of connection to your ISP.

Baud vs. BPS

Often we hear the words "baud," "bps" (bits per second) and "bytes," or characters per second, used interchangeably. They are not the same. The term "baud" is named after famous French inventor J.M.E Baudot. In fact, he developed the first teletype code known as the Baudot code, which is a 5-level binary character code. The term "baud" is used to describe the number of carrier wave changes in one second, similar to cycles per second or Hertz. In baseband transmission, baud is equal to the bits per second. When applied to a modulated carrier wave it usually is not. This is so because it is possible to use a combination of modulation methods, which allows for more than one bit per wave change.

FSK

The earliest PC modems used frequency modulation and communicated at 300 bits per second, or bps. This isn't quite the same as an FM radio and is called frequency shift keying, or FSK. These modems used a simple decoder made up of two band-pass filters. The first band-pass filter is tuned to the one frequency, and the other band-pass filter is tuned to the zero frequency. Since these modems are capable of full duplex operation, they use two sets of frequencies and four band-pass filters. The caller is in originate mode, and the answering side is in the answer mode. The originate modem sends 2225 Hz for a one, and 2025 Hz for a zero, and listens at 1225 Hz for a one and 1025 Hz for a zero. This method is defined as the Bell 103 standard. The fastest possible communication speed using FM over a telephone line is 600 bps. Although these modems are slow, they are reliable because of the FM encoding method which is used.

Bell212A

The next advance for modems were the 1200-bps modems defined as Bell 212A. This standard uses DPSK or differential phase shift keying. This modulation method makes it possible to encode two bits per baud. Bell 212A uses a rate of 600 baud, which yields 1200 bps.

Multidimensional Modulation

The first major leap in modem modulation technology was applying multiple modulation techniques. The first to use multiple modulation was QAM, or quadrature amplitude modulation. QAM was first used on 2400-bps modems where 4 bits are encoded per baud. The first three bits are the phase portion, and the fourth bit is the level of the carrier. This method still used the base rate of 600 baud, but since 4 bits are encoded, you multiply the baud rate by the number of bits encoded which yields 2400 bps. When applied to a carrier rate of 2400 baud, you can achieve bit rates of 9600 bps. Using multiple modulation methods you must consider the modulation/encoding as a 2D matrix. If you plot all four quadrants on a graph and plot the bit possibilities as a function of amplitude and phase, it will look like a lattice, hence the term "lattice modulation," as shown in Figure 14.58.

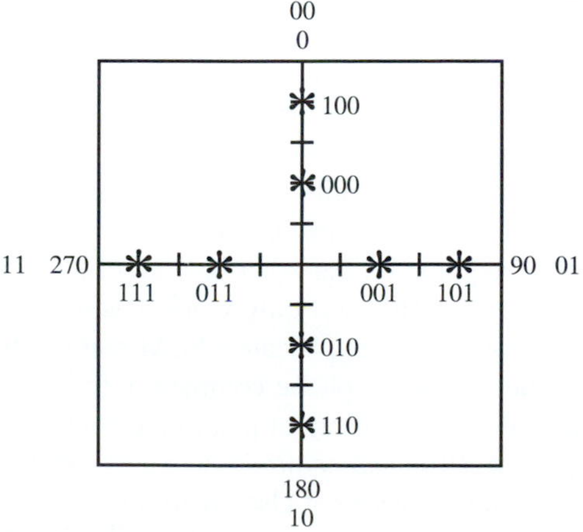

FIGURE 14.58
Multidimensional lattice modulation

Trellis-Coded Modulation (TCM)

TCM is similar to the method just discussed, but the data patterns are tighter (more resolution). The data bits are grouped so that the most frequently occurring patterns are sent at the high power levels and the least frequently occurring patterns are sent at a lower level. This reduces the possibilities of error. When combined with error calculation and correction logic (ECC), very high data rates, at or near the Shannon limit, can be attained. Using this method, the highest possible data rates are 33,600 bps when a perfect end-to-end POTS connection is available. In fact, this really does happen today because the internal infrastructure of the telephone network is digital. Only the terminating ends are analog.

Compression

Just like any other medium, digital data can be compressed. This has been applied to FAX transmissions for many years and is known as CCITT Group 3 Fax, which is based upon a method known as modified Huffman encoding. For example, a bitmap image may consist of a lot of redundant data. A single black line across the page may be thousands of pixels or bits of data of all zeros. Instead of sending all the zeros, why not send a zero followed by the number of zeros (black pixels) in the line? Using this method, you can compress the entire line down to four or five bytes. This is also known as run length logic, or RLL, encoding. Variations of this have been developed and applied to disk storage and modem transmission. Of course, the actual compression varies depending upon data redundancy. Images and text compress well, but binary data does not. Furthermore, there are two error correcting data standards for data compression: LAPM (link access procedure for modems) and MNP5 (microcom networking protocol).

PCM—Pulse-Coded Modulation

V.90 56K modems use an entirely different modulation method known as **pulse coded modulation,** or **PCM**. The V.90 modems are called 56K modems but in reality they can only go as fast as 53.333 Kbps due to modulation power limitations placed by the FCC. This was done to prevent cross talk problems with other lines. V.90 modems only receive using PCM but send using the more traditional 33.6=Kbps method.

PCM sends a pulse representing one entire byte or eight bits encoded as a 256-level pulse. Because only one analog conversion is made, the data are sampled and sent at 8000 samples per second. All that is required on the other side is a fast ADC or digital signal processor (DSP) to decode the pulse. This will only work if you are close to the central office (CO) of your local telephone company, because PCM can not tolerate noise.

The problem with V.90 lies in the telephone company infrastructure (i.e., the wires on the poles you see outside your house). Many of these date back to the 1950s and have been taped, patched, and spliced. Birds and squirrels nest and eat the insulation, exposing the conductors to rain and who knows what else. With all that, we try to send a low-level signal near or at the theoretical maximum allowed by physics. Most of the time it actually works, if it's not storming outside. Furthermore, there are physical limits such as distance to consider. For example, if you are 20 miles from the central office, 56K just won't work.

Modem Standards

The original modem standards were created by AT&T. Today, modems and their protocols are governed by the ITU, or International Telecommunications Union, which was formerly the CCITT. Currently the standards are referred to as V dot standards and are applied to both modulation and compression. Table 14.3 lists the current modem standards.

TABLE 14.3
Modem Standards

Standard	Bit Rate	Bits	Baud	Modulation	Notes
Bell 103	300	1	300	FSK	Obsolete
Bell 212A	1,200	2	600	DPSK	Quadrature modulation
V.17	14,400	6	2400	TCQAM	Fax standard
V.21	300	1	300	FSK	Not compatible with 103
V.22	1,200	2	600	PSK	Not compatible with 212A
V.22bis	2,400	4	600	QAM	International standard
V.23	1,200	2	600	PSK	1200 bps forward, 75 bps reverse
V.24	Any	6–9	Same	None	RS-232 serial definition
V.29	9,600	4	2400	QAM	Fax only, half duplex
V.32	9,600	4	2400	TCQAM	Includes negotiation and ECC
V.32bis	14,400	6	2400	TCQAM	Fast version of V.32, 6 bits/baud
V.34	28,800	8	3600	TCQAM	Added 2 bits, at maximum baud rate
V.34bis	33,600	9	3600	TCQAM	Added one more bit to V.34
V.42	None	None	None	None	LAPM protocol only
V.42bis	None	None	None	None	Data compression only
V.90	53,333	8	8000	PCM	One side digital to CO

Beyond Modems

Emerging technologies have given rise to alternatives to using modems, even for the average home PC user. Currently, alternatives like ISDN, DSL, cable TV, and satellite exist. These services are faster and more reliable than any POTS connection and are decreasing in cost everyday.

ISDN **ISDN,** or **integrated services digital network,** can completely replace your analog telephone service. ISDN is a completely digital circuit and places the CO inside your home. The ISDN modem is really not a modem, but is a small telephone switch. In fact, you can set up features like call forwarding and call conferencing just as if you had your own PBX. An ISDN modem will provide two analog RJ11 ports and a Digital RS-232 interface. Some are available for installation inside your PC just like a modem. When you get ISDN, you get a second line practically for free and it works fine with V.90. ISDN comes in two sizes, PRI (primary rate interface) and BRI (basic rate interface). The PRI is really a T-1 ISDN running at 1.544 MHz. The PRI is broken down into 24 BRIs running at 64 Kbps each. The BRI ISDN line consists of two 64-Kbps B (bearer) channels and one 16-Kbps D (delta) channel. The B channel carries voice and data. The D channel carries channel signaling information. Both B channels can be combined to achieve 128 Kbps or used separately at will. For example, you can call your ISP and begin downloading files while calling your friend on the other channel and talking about what you are downloading.

DSL—Digital Subscriber Line DSL is a purely digital service similar to ISDN, but no telephone services are provided. A DSL provider leases lines and space inside the telephone company CO to place their routers. Think of a DSL connection as a really long twisted-pair patch cable. You are literally on their network. There is a variation of DSL called **ADSL** for **asymmetric digital subscriber line.** In ADSL, you receive at a higher rate than you send. This is preferred in cases where you surf and download but rarely upload anything. The downside is distance and price. Your speed degrades greatly with distance, and the service is the most expensive of all home services.

Cable Modems Cable modems are not modems at all but broadband network interfaces. This topic is discussed further in Chapter 15. Cable modems may replace your telephone in the future or at least compete with local telephone companies. The performance is excellent. One big problem for cable companies is the major infrastructure improvements which are required to many of the existing cable TV areas. Current bandwidth is not adequate in some areas. In addition, some serious challenges await cable TV when high-definition television (HDTV) arrives.

Satellite This is one of the latest modem replacements today. In fact, you still need a modem but only to make requests. Using a small satellite dish (TV), you connect to a wireless broadband network, but the network is half duplex. You send over a traditional modem but receive from the satellite antenna. The broadband aspects of this technology are further discussed in Chapter 15.

14.11 UNIVERSAL SERIAL BUS—USB

The **universal serial bus (USB)** is an external bus and aims to both simplify and expand upon connecting devices to a computer. USB is not really a bus in the traditional sense, where devices connect to the same physical wire, but more like a star network topology. The bus originates at the computer called the **root hub** and is directly connected to another device or **expander hub.** USB devices are not connected to each other or daisy chained as SCSI devices are. All USB connections are device to hub, hub to hub or device to computer as shown in Figure 14.59.

USB devices are **hot swappable,** which means that you can add or remove devices while the computer is on without having to restart, and have the devices instantly accessible. USB only requires a single IRQ, and multiple devices can be daisy chained, up to 127 devices per root USB port. Most recent system boards include two USB ports at the rear of the computer case. Easily accessible USB devices can serve as hubs with several connectors. USB can also supply up to 2.5 W of power to each device reducing the tangle of power cords normally found behind the computer.

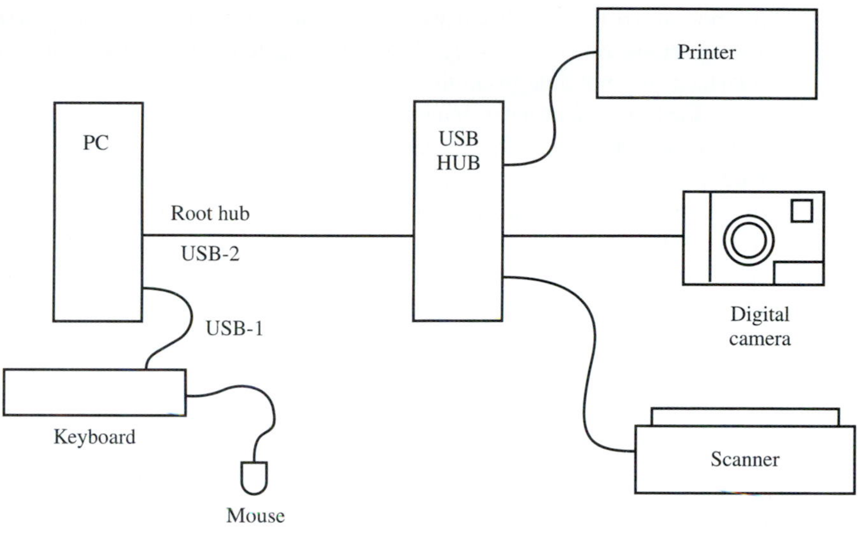

FIGURE 14.59
USB application

The current release of USB is version 1.1, which supports transfer speeds up to 12 Mbps or 1.5 MBps. The speed will throttle down to 1.5 Mbps for slower devices such as a printer. Clearly, this is not sufficient for motion video or disk storage. USB 1.1 is intended for scanners, digital cameras, extra serial and parallel ports, and equipment control. Under development is USB 2.0, which provides an enhanced command set and multiplies the speed by a factor of 40. This yields a transfer speed of 480 Mbps. Such a speed improvement makes USB usable for full-motion video, audio, DVD, and other high-bandwidth applications. USB 2.0 is backward compatible with USB 1.1 devices and the interconnect cables. Using a USB 1.1 hub will throttle devices on the 1.1 hub down to USB1.1 speeds. To fully implement USB connectivity, both the BIOS and OS (Win 98 or MacOS) must support USB in addition to USB devices. USB was not meant to replace EIDE or SCSI as a mass storage bus.

Firewire—IEEE 1394

Similar to USB, **IEEE 1394** is the IEEE designation for a high-performance serial bus. This serial bus, also known as **Firewire,** defines both a backplane physical layer and a point-to-point cable-connected virtual bus. The cable version supports data rates of 100, 200, and 400 Mbps across the cable medium supported in the current standard.

The primary application of the cable version is the integration of I/O connectivity at the motherboard of personal computers using a low-cost, scalable, high-speed serial interface. The 1394 standard also provides new services such as real-time I/O and live connect/disconnect capability for external devices, including disk drives, printers, scanners and cameras. This contrasts to parallel high-speed communications, such as SCSI, which are not suited to long distances and do not support live connect/disconnect, making reconfiguration time-consuming. The hot swap feature is convenient, as the current SCSI interfaces are not able to do this.

Factors driving next generation protocols such as IEEE 1394 include the need for reliability, durability, and universal interconnection. The IEEE 1394 serial bus is organized as if it were memory space interconnected between devices, or as if devices resided in slots on the motherboard. Device addressing is 64 bits wide, partitioned as 10 bits for network IDs, 6 bits for node IDs, and 48 bits for memory addresses. The result is the capability to address 1023 (2^{10}) networks of 63 (2^6) nodes, each with 281 terabytes (2^{48} bytes) of memory. Each bus entity is termed a "node," to be individually addressed, reset, and identified. Multiple nodes may physically reside in a single module, and multiple ports may reside in a single node.

Some key features of the Firewire topology are multimaster capabilities, live connect/disconnect (hot swapping) capability, and dynamic node address allocation as nodes are added to the serial chain. Another feature is that transmission speed is scalable from approximately 100 Mbps to 400 Mbps. Each node also acts as a repeater, allowing nodes to be chained together to form a tree topology. Due to the high speed of Firewire, the distance between each node or hop should not exceed 4.5 m, and the maximum number of hops in a chain is 16, for a total maximum end-to-end distance of 72 m. Cable distance between each node is limited primarily by signal attenuation. An inexpensive cable with 28-gauge signal pairs can be up to 4.5 m long. This gives an end-to-end distance of 72 to 224 m. An additional feature is the ability of transactions at different speeds to occur on a single cable (for example, some devices can communicate at 100 Mbps while others communicate at 200 Mbps and 400 Mbps). The features of Firewire will allow plugging in a computer expansion system as easily as plugging into AC power, providing communications on demand without having to shut down and reconfigure each time an I/O device is added or removed.

14.12 MULTIMEDIA

The term **multimedia** refers to a collection of hardware including a CD-ROM drive, sound card, PC speakers, video capture board to produce full-motion video, microphones, VCRs, digitalk cameras, and so on. Of course, we can't forget a huge array of software available on CD, ranging from mediocre to fantastic! For the purpose of our discussion on multimedia, we will include only the CD-ROM drive and sound card.

Perhaps the single most difficult element in multimedia is the sound card. Setting up a sound card in a PC can be difficult due to hardware conflicts. Which IRQ, what COM ports, what other installed devices may conflict with the card? A video setup can be less difficult, but troublesome nonetheless. But now, like a knight in shining armor riding a white horse, comes plug and play! Microsoft's Windows 95 boasts full plug and play capability. The caveat is that full plug and play capability must be on the list of supported devices! If you are unfortunate enough to have hardware that is not on the list, BEWARE! However, if your multimedia hardware is recognizable by Windows 95, it can be very easy, and in fact 95 may do the setup for you. Of course, the explosion of multimedia accounts for the advances in CD-ROM technology.

Advancements in CD-ROM technology have come fast and furious, with CD-ROM drives going from single speed to 10× speed in perhaps two years or less. CD-ROM drives come in proprietary, IDE, SCSI, parallel port, and PCMCIA versions, and internal and external versions, with capacities up to perhaps 800 to 900 MB per disk. Imagine the 2400 to 2700 360K floppy disks needed to store the information contained on just one CD! There are several types of CD-ROM interfaces in use today. Here is a partial list of the interfaces available:

- Panasonic
- Sony
- Mitsumi
- IDE
- SCSI
- PCMCIA
- Parallel port
- On the motherboard (usually IDE)
- On sound card (as in Soundblaster Pro and others)

Most of the original DOS PC CD-ROM drives were shipped with an interface card. Using the proprietary interface configuration requires careful planning for I/O addresses, and IRQ and DMA lines. Most proprietary CD-ROM interface cards use I/O addresses in the 300H to 340H range, as do most LAN cards, so keep in mind other devices that are using this address range. Many sound cards provide the proprietary interface built into the sound card, thus freeing up a slot in the PC. Be sure to configure the jumpers on the sound card CD-ROM section to match the CD-ROM drive you are installing. If you are installing the separate interface, be sure to disable the CD-ROM interface on the sound card to prevent conflicts. Generally, using the separate proprietary interface will provide better performance than using the interface on the sound card. The sound card's CD-ROM interface provides only PIO data transfers, while the proprietary interface provides both PIO and DMA transfers, thus yielding better performance.

The latest CD-ROM drives are supplied using an existing IDE interface. Using the IDE interface for installation is the simplest method because the I/O and IRQ settings are well defined. The CD-ROM drive can be daisy chained to the hard disk or connected to a second IDE interface. Most modern motherboards provide a dual IDE interface built into the motherboard. Installing an IDE CD-ROM is the same as adding a second IDE hard disk drive. Be sure to set the CD-ROM jumpers to SLAVE if the CD-ROM is connected to the primary hard disk drive. Also, check the primary hard disk drive jumpers for the slave present position. Refer to Chapter 12 for a description of the IDE jumpers and cabling. If the CD-ROM drive is to be connected to a second IDE interface with no other devices on the second interface, be sure to set the CD-ROM drive as the MASTER. Do not change the jumpers on the hard disk drive connected to the primary IDE interface.

After you have completed the hardware installation, you must load the CD-ROM disk driver software. CD-ROM drives are shipped with an install program that makes this task easier. If you must reconfigure the system, it may be simpler to edit the configuration line in CONFIG.SYS manually. The following is a sample CONFIG.SYS file showing the CD-ROM driver on the last line:

```
LASTDRIVE=Z
DEVICE=C:\WINDOWS\HIMEN.SYS
DEVICE=C:\DOS\EMM386.EXE NOEMS I=C800-E7FF X=E800-EFFF
DEVICEHIGH=C:\QUICKCMD\CMD640X.SYS /L
DEVICEHIGH=C:\DOS\SETVER.EXE
DOS=HIGH,UMB
FILES=30
DEVICEHIGH=C:\WINDOWS\IFSHLP.SYS
STACKS=9,256
DEVICEHIGH=C:\DEV\MTMCDAS.SYS /D:MSCD001 /P:340 /A:0
```

After the driver is added to CONFIG.SYS, you must add the Microsoft CD-ROM extensions to the AUTOEXEC.BAT file by adding the MSCDEX command as follows:

```
1h C:\BIN\MSCDEX.EXE /D:MSCD001 /M:10
C:\WINDOWS\NET START
C:\DOS\SMARTDRV.EXE /X 2048 128
@ECHO OFF
PROMPT $P$G
PATH C:\WINDOWS;C:/DOS;C:\QUICKCMD
SET TEMP=C:\DOS
```

The MSCDEX program is loaded in the first line of the file AUTOEXEC.BAT. Make sure the /D parameters match in both the CONFIG.SYS and AUTOEXEC.BAT.

There are dozens of sound cards available for the DOS PC. Most sound cards will emulate or support the AdLib, Sound Blaster, or ProAudio sound card standards. The most popular sound cards are

- Adlib, from Adlib
- Sound Galaxy, from Aztec
- Soundblaster, from Creative Labs
- Pro Audio Spectrum, from Mediavision

Since Creative Lab's Soundblaster is generally accepted as the de facto standard for multimedia, we will confine most of our discussion to that manufacturer. Most other sound cards will support the Soundblaster standard as well as their native mode. There are three very important things to keep in mind when setting up your sound card: the DMA channel, IRQ, and I/O address used by your card. These three items will probably be responsible for most of the trouble you will encounter in setting up your sound card. The original Soundblaster uses the following default settings:

- DMA channel 1
- IRQ 7 (newer Soundblasters use IRQ 5)
- I/O addresses 220H to 22*F*H

The Pro Audio Spectrum uses the following default settings:

- DMA 3 (DMA 5 will perform better in stereo due to 16-bit transfers over 16-bit DMA)
- I/O address 388H
- IRQ 2, 3, 5, 7, 10, 11, 12, and 15 (software selectable)

The default settings will generally present no problem for you. Depending on what peripherals you have installed, however, you may need to change the settings. Some peripherals that may present conflicts are scanners, network adapters, serial devices, CD-ROM drives, and SCSI adapters. Changing these settings is a simple matter of changing the jumpers on the corresponding jumper block on the card. In later versions of the sound cards, some of these settings are software switchable. It is always a good idea to keep a log of all your system settings. This includes not only the IRQ, DMA, and I/O addresses for all installed cards and peripherals, but the hard drive type and geometry, memory mapping, and serial and parallel port settings. This will save you an untold amount of grief if you have to reconfigure your system after a disaster.

The Soundblaster card does not have a selectable DMA channel. However, it can share its DMA channel with another device (provided that the device can use DMA channel 1). If the device presenting the conflict has a selectable DMA channel, it is preferable to change that. This is also the case with other conflicts: If you can change the settings on the conflicting cards, this is preferable to avoid reloading of software. Of course, it is possible that you will have to reload software for the other device as well! In the Soundblaster Pro, you do have the option of changing the DMA channel. The default setting is still 1. However, you also have 0 and 2 available. You may also share a DMA channel with another device, although the default setting is nonshare. You need to keep in mind that the DACK jumper must have the same setting as the DRQ jumper.

The Soundblaster card has 4 IRQ settings available to it: 2, 3, 5, and 7 (with 7 being the default setting). An example of a possible IRQ conflict is that some printer interfaces may require IRQ 7 and you will need to change the Soundblaster setting to IRQ 2, 3, or 5. Please keep in mind that if you change the default IRQ setting, you *must* reinstall the sound drivers with the changes! In the Soundblaster Pro, you have 4 IRQ lines available to you: 2, 5, 7 (default), and 10.

The factory default I/O address setting for the Soundblaster card is 220H. The available jumper selectable settings are 210H, 220H (default), 230H, 240H, 250H, and 260H. The Soundblaster Pro has only two I/O address options available. They are jumper selectable to 220H (default) and 240H.

14.13 TECH TIPS AND TROUBLESHOOTING—T^3

In this chapter we have covered video display theory in great detail. A thorough understanding of the video display and display generator is required if economical and timely repairs are to be made. Since the circuitry is complex—analog, digital, and critical timing are involved—troubleshooting even with the best test equipment will be difficult in some cases. As complex as all this may seem, if you have a good understanding of the theory of video display, the symptoms on the display screen will almost always tell you what is wrong. The most important thing to remember is: *Don't panic* and *don't start turning things or changing parts at random.* Keep in mind that you may be working around *high voltage, so be careful.*

At first the screen may not make any sense to you, so try to break the symptoms down to the related circuitry. For example, is it a sweep failure, or is the problem in the video circuit or the display generator card? First check for a raster by turning the contrast all the way down and the brightness up, which should result in an evenly lighted screen.

In the following sections we present some tips and clues to make the job of repairing computer peripheral devices a little easier.

Troubleshooting Keyboards

Most keyboard failures are due to stuck keys or debris lodged in the contacts. Check that *every* key has a "good feel" and returns up rapidly. If multiple characters show up when only one key is pushed, that indicates key bounce, which is a bad contact on that key. The contacts must be cleaned or the keyswitch replaced. If a group of keys are dead, one wire to the keyswitch matrix may be broken, or the keyboard scanner may have a dead line. This can be checked with an oscilloscope. Using the schematics and some common sense should get you to the problem fast.

Troubleshooting the Video Display

The most common failures in video monitors are in the deflection circuitry (sweep). Horizontal sweep failures usually result in no brightness on the screen because the high voltage is generated by the horizontal deflection circuitry. In some cases all the low-voltage DC power is generated by the horizontal deflection circuitry. The first thing to do when troubleshooting a "dead," or dark, screen is to check for the presence of high voltage. The recommended procedure is to use a high-voltage probe on the anode lead of the CRT. If the probe is not available, then simply hold an NE-2 neon bulb near the anode lead. If the bulb glows, some high voltage is present. If no high voltage exists, check the horizontal output for the flyback pulse using an oscilloscope. The flyback pulse should be a large pulse with a "dimple" in the top and have a smooth rise and fall. Any deformation of the pulse indicates an excessive load due to a faulty flyback transformer or component on the secondary of the flyback.

The most common failures here are usually the horizontal output transistor or the damper diode. These failures blow the power supply fuse. If no flyback pulse exists, check the base drive to the output transistor. Failure of the base drive indicates that the horizontal oscillator or driver is failing. Some monitors, like the IBM monochrome, have X-ray protection built in if the power supply voltage becomes too high. The protection works by grounding the input to the driver circuit, thus killing the high voltage.

A gassy CRT or shorted high-voltage rectifier will kill the high voltage. To check for this, remove the anode cap from the CRT. Power on the monitor and check for high voltage at the anode wire. Using an insulated-handle screwdriver, try to draw an arc off the anode wire. Here is a little tip: *high-voltage AC will arc to floating metal, but high-voltage DC will not arc to floating metal.* A heavy arc indicates a shortened high-voltage rectifier. Carefully replace the anode cap onto the CRT. If the high voltage is present but the screen is dark, check the bias voltages on the CRT socket. Recall that if the cathode is more positive than grid 1 by about 70 V or more, the beam will be cut off. If the bias voltages are within specification and the screen is still dark, a bad CRT is indicated. Conversely, if the screen is too bright, check the bias voltages on the CRT socket. An open pin inside the CRT will sometimes cause the brightness to run away.

Many solid-state monitors AC couple the horizontal yoke coil through a nonpolarized electrolytic capacitor. A failure of this capacitor is indicated by a single vertical line in the middle of the screen. Since the yoke requires approximately a 15-A peak current at a high frequency, a special high-current capacitor is required here. If this capacitor is replaced with a standard-type electrolytic capacitor, *it will explode after a few minutes of operation.* We have found that replacing the original electrolytic type with a standard foil or mica type proves far more reliable. This replacement is physically larger and may not fit in some cases. Remember, the high voltage situation can be extremely dangerous. *Do not attempt to work on a high-voltage problem until you have learned the proper safety precautions.*

Vertical deflection problems are a dead giveaway. A single horizontal line indicates no vertical sweep. In this case the vertical oscillator is dead or the yoke circuit is open. Most monitors use a push-pull circuit for the vertical output, similar to an audio amplifier. No top or bottom indicates one of the output transistors or its related circuitry is bad. If the picture is distorted, then a wave-shape problem exists, and there is probably a faulty capacitor in the vertical circuit. Small aluminum electrolytic capacitors are notorious for changing value, usually to open. If the problem is no video, simply follow the cathode wire and an oscilloscope back through the video amplifier to the source of the video until the failure is found.

Troubleshooting the Display Generator

Carefully observe the picture and try to isolate the failure to a major functional block. Measure the horizontal and vertical synchronization pulses. If the sync pulses are the correct timing values, the master timing must be correct. If the horizontal is good but the vertical is bad, the problem is in the countdown chain. If no sync pulses exist, check the master oscillator. If the picture is synchronized but the displayed image is bad, it will tell a story.

Command the computer to clear the screen. The computer writes all ASCII spaces to the display RAM. The code for a space is hex 20, 00100000 binary. Now let's say, for example, that the screen is full of 0s. The code for a zero is hex 30, 00110000 binary. It becomes obvious that 1 bit is "picked" in memory due to a bad RAM chip or buffer. If the computer displays only the top half of the characters and repeats them twice, the character ROM or an address line to the ROM must be bad.

Here is an experiment to try on the IBM PC with a color graphics card.

1. Type CLS; the screen should clear.
2. Type DEBUG (it must be on your DOS disk).
3. At the '-' prompt, type E B800:0 <Enter>.
4. Type 41 <Enter>: an A should appear on the top left of the screen.

What we just did was to write the hex code 41, which is an ASCII *A*, at memory address *B*800:0000, which is the first location of the video RAM. This is a simple way to test the RAM on the display generator.

Troubleshooting Printers

Handle printer problems in the same way as you would for a video display. Observe the printout sample for anomalies and determine if the problem is mechanical or electronic. Printing wrong characters, such as an A that should be a C, indicates electronic problems on both impact and nonimpact printers.

In the case of dot-matrix and daisy wheel printers, fuzzy, improperly spaced, or missing dots may indicate mechanical problems. Common problems are binding of the carriage, print head, or daisy wheel and worn ribbons. Dust or debris and lack of oil cause binding. Be very careful when using oil and carefully follow the manufacturer's recommendation. The most common problem with impact printers is the print heads. Another common problem is the driver transistors for print head coils and stepper motors. Don't be too hasty in replacing stepper motors because they rarely fail and are expensive. The cause for a stepper motor failure is usually shorted drive transistor. A shorted driver will keep the stepper motor winding energized, causing it to overheat and short out. A quick check for a stepper motor is to slowly rotate the shaft by hand. Each step should feel even without binding or skipping. If a winding is shorted you will feel a skip or uneven step while rotating the shaft.

In the case of laser printers, poor print quality is the most common problem. Laser printers must be kept clean. Light print is caused by low toner and is solved by adding toner or replacing the toner cartridge. The most common print quality problem is streaking or unwanted lines on the page. The photo drum is very fragile and easily scratched. A worn cleaning blade or badly scratched photo drum can cause long vertical lines. Toner stuck to the rollers or a chip on the photo drum can cause repeating unwanted patterns. A ghost of the previous page over the new image may be caused by a faulty discharge corona wire or worn cleaning blade. Worn rollers cause paper jams as well as a slipping paper feed clutch.

A special note on paper: All paper has a natural curl to it. Most paper manufacturers will indicate which side should face up to help relieve some jams. High humidity may cause excessive curling, resulting in frequent paper jams. Always try fresh paper before working on a paper jam problem.

Troubleshooting Networks

When a problem is reported on a thin Ethernet run, first check the resistance at any workstation tee connector. Remove the tee from the Ethernet card but leave on both wires, or leave the wire and terminator connected. Then check the resistance on the tee for 25 ohms. If the resistance is less than 25 ohms, there is a short in the line. If the resistance is 50 ohms, a terminator is missing or one side is open. If the resistance is greater than 50 ohms, the tee may be faulty, there may be a break on both sides, or terminators may be faulty and/or missing. This type of wiring is easily broken, and all BNC connectors should be checked for loose ends. If the problem is intermittent, check the resistance while wiggling the connectors. If the resistance changes, a poor connection is evident.

Troubleshooting Multimedia Systems

The largest problem when adding a sound card is conflicting DMA and IRQ settings. I/O addresses are rarely a problem since they are well known and not used by other devices. DMA 1 is the default DMA line for some hand scanners. IRQ 7 is the default LPT 1 IRQ line, but it is rarely used. Common problem symptoms are no sound or skipping and repeating sounds. The most likely causes of no sound are DMA and I/O conflicts. The most likely causes of skipping and repeating sounds are IRQ conflicts.

EXERCISES

14.1 Briefly describe a keyboard.

14.2 What does the encoder do?

14.3 What is the term used when the keys are wired as rows and columns?

14.4 What chip is used to encode the IBM keyboard?

14.5 What data format will reduce the wire count to a minimum?

14.6 What type of display is used on most computers?

14.7 What does CRT stand for?

14.8 Where does the electron beam originate?

14.9 How do we control the beam?

14.10 What is a raster?

14.11 The electron beam is __________ during retrace.

14.12 Show the formula to calculate the maximum number of scan lines.

14.13 If the horizontal rate is 15,750 and the vertical rate is 60, what is the maximum number of scan lines?

14.14 Explain why a vacuum is required in the CRT.

14.15 Name the four basic elements of the CRT.

14.16 Why do we need the heater?

14.17 True or false: The anode has a high negative potential on it.

14.18 How is the electron beam bent in most CRTs?

14.19 What is painted on the inside face of the CRT?

14.20 How do we control the brightness?

14.21 True or false: The screen grid is usually set to a fixed value.

14.22 What causes the screen to glow?

14.23 True or false: The high voltage comes from the power transformer.

14.24 What does the tripler do?

14.25 A __________ waveform is used to correct the focus nonlinearity.

14.26 The yoke is made up of __________.

14.27 A __________ waveform is used on the yoke.

14.28 True or false: The horizontal output generates a ramp.

14.29 When is the flyback pulse created?

14.30 The vertical yoke is driven by a __________ waveform.

14.31 What type of circuit is the horizontal oscillator?

14.32 What is the name for the signal that contains the image?

14.33 What is usually applied to the cathode?

14.34 What does the contrast control do?

14.35 Why do color monitors have separate gain controls?

14.36 For what does NTSC stand?

14.37 Draw a typical video waveform.

14.38 What is the color burst for?

14.39 How is composite color modulated?

14.40 What is the high-frequency signal that feeds the shift register called?

14.41 The digital image data is stored in the __________ __________.

14.42 The smallest part of a picture is called a __________.

14.43 What contains the patterns of characters?

14.44 The display RAM in the IBM PC is __________ mapped.

14.45 What is the address of the display RAM in the color graphics adapter?

14.46 What is the maximum horizontal resolution of the IBM color graphics adapter?

14.47 An impact printer uses an __________ __________ on paper.

14.48 Describe a dot-matrix character.

14.49 What is a thermal printer?

14.50 What is on the daisy wheel?

14.51 How does the ink jet printer work?

14.52 What is the photosensitive part of the laser printer?

14.53 The laser printer uses a __________ __________ to scan the drum.

14.54 The most popular parallel interface is the __________.

14.55 Define handshaking.

14.56 Describe one synchronous data frame.

14.57 What is the typical voltage level for RS232C?

14.58 What is current loop?

14.59 Crossword Puzzle

ACROSS

3. A method of doubling the vertical resolution.
7. The picture information is called ______.
12. If a material will give off light when bombarded by electrons, it must have a ______ characteristic.
14. When the sync and picture information are combined on the same line.
16. Left-to-right deflection.
17. Most monitors use ______ deflection.
18. The print method used when an inked ribbon strikes the paper.
20. The most positive element.
21. When the beam is returning to the left or top.
22. Converts digital data into audio tones.
23. A sample of the color reference signal.
27. A mechanical device for displaying computer data.
28. The inside face of the CRT is painted with ______.
29. A series of lines.
30. The most popular printer interface standard.

DOWN

1. Clear to send.
2. The high-voltage transformer.
4. The ______ grid will help to make the picture clear.
5. Converting keys into binary is done by the ______.
6. The origination of the electron beam.
8. The elements that control the electron beam.
9. What will gases do when subjected to high voltage?
10. Top-to-bottom deflection.
11. A simple input device.
13. The electron beam is bent by a means called ______.
14. Cathode-ray tube.
15. The shape of the vertical waveform.
16. The ______ will supply the mechanical energy to print a character.
19. Characters made up from a group of dots.
24. An LSI chip that converts data to serial form and back.
25. The serial false level.
26. The component that deflects the electron beam.

CHAPTER 15

NETWORKS

KEY TERMS

AUI
Baseband
Bridge
Broadband
Brouter
Bus Topology
Collision
Cross Talk
CSMA/CD
Data Link Layer
Datagram
DHCP
DNS
Ethernet
Fiber Optic
Gateway
Head End
Host File
Hub
JAM
LAN
MAC
Manchester Encoding
NDIS Model
Network Layer
ODI Model
OSI Model
Physical Layer
Presentation Layer
Repeater
Ring Topology
Router
RTZ
Session Layer
Star Topology
Subnet
TCP/IP
Thick Ethernet
Thin Ethernet
Token Ring
Topology
Transport Layer
WAN

15.0 INTRODUCTION

Networks—both large area networks, known as **LANs,** and wide area networks, known as **WANs**—have provided unparalleled advances in communication and productivity. Think of a typical day in your life: ATM transactions, banking, the Internet, transportation, reservation systems, and telephone calls are all controlled by networks. At first networks were used to connect large mainframe computers to what are known as communication controllers. Connecting a user terminal to a network required the use of a small computer known as a terminal server. Today, desktop or personal computers possess more than enough power to provide network connectivity and full desktop computing. In fact, a network interface card contains its own microcomputer to handle network traffic before data ever reaches the internal PC bus. Using client/server technology, large central computers need only provide database, file, and mail services while the desktop computer can provide user processing locally.

When we discuss networks, we must consider not just the physical wiring but all the hardware and software that make up the network. All components must operate in harmony, from the network software to the cabling, or chaos will be the result. In this chapter we will discuss the Ethernet interface in detail as it applies to the Windows/Intel (sometimes called Wintel) computer. Before we discuss network topologies and network protocols, let's discuss some of the fundamentals of networking. Any method of connecting two or more devices over some common medium constitutes a network. Connecting two or more computers in this way allows sharing of resources between attached, or networked, computers.

The most common applications of networking are printer and file sharing. Before networks were available, people needed a printer in order to print their work. If you did not have a printer, you had to save your work onto a removable medium such as a floppy diskette and carry it to another computer that had a printer in order to print your work. This was often jokingly referred to as sneaker net because people had to run between computers to print out their work. Furthermore, if a group of people were working on a common project, then they would have to swap floppies and merge the other person's work into their own. To relieve this problem and increase productivity, we began to connect computers together.

15.1 NETWORK SIGNALING

Once a piece of wire is placed between any two or more devices, we must overcome certain limitations and meet electrical requirements. As with any electronic communication method, there are always problems with interference and signal degradation. Since the information is digital, any disruption in transmission will result in scrambled and useless information. Two methods of placing digital information onto a wire are broadband and baseband signaling. To overcome the electrical limitations of great distances or harsh electrical environments, we now use optical signaling or fiber optic cable.

Broadband Signaling

Broadband signaling is similar to radio and television transmission in that information is modulated onto a common carrier wave. Many types of modulation can be used to represent data. One example would be AM, or amplitude modulation. A low carrier level would represent a logical 0 and a high carrier level would represent a logical 1. To improve noise immunity, one could use frequency modulation or phase shift keying for greater bandwidth. The greatest advantage of a broadband network is the use of cable TV wiring. Furthermore, multiple channels or networks can exist on the same wire, just as multiple TV channels share the same wire coming into our television. Unlike baseband signaling, broadband is more tolerant of poor wiring conditions. To allow a computer to send and receive on the same wire, we can use one frequency range for receiving and another for sending. In such a scenario an additional device known as a **head end** is required. As shown in Figure 15.1, the head end is a channel or frequency converter. The head end must be present on the same wire as the network device to shift the transmit frequency to the receive frequency.

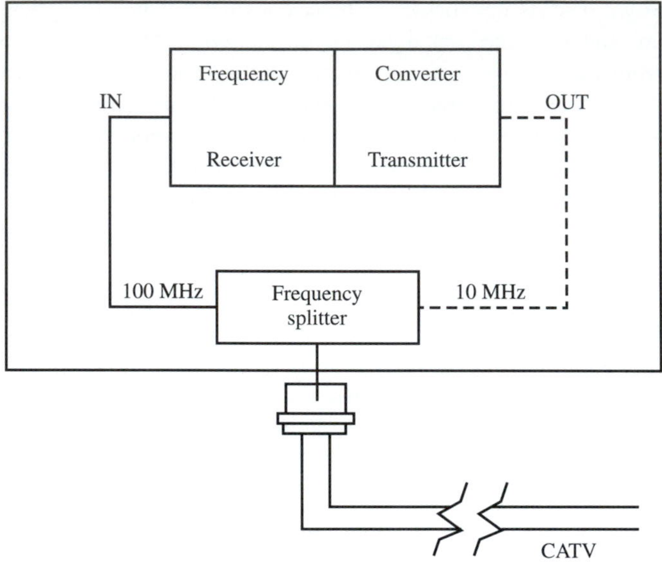

FIGURE 15.1
The head end

As shown in Figure 15.2, workstation 1 wants to send information to workstation 6. All workstations transmit at 100 MHz, but receive at 10 MHz. The head end will intercept the 100-MHz signal, convert it to 10 MHz, and continue the transmission. Each workstation receiving the signal will check for address information in the signal and process the information being sent to it. The disadvantages of broadband are bandwidth limitations and difficulty in troubleshooting. If great distances are to be covered, then bidirectional amplifiers must be added.

Baseband Signaling

No modulation is used in **baseband** signaling. Data levels are applied directly to the wire in the form of changing DC voltages. An example of this would be to represent a logical 1 as some voltage and a logical 0 as no voltage. Because the data bits are applied directly to the wire, high-quality wire and good connections are essential. Any noise or signal reflections, known as ringing, will scramble the data, as shown in Figure 15.3(a). This is why Ethernet requires terminators on both ends of a run of wire. The terminator is a 50-Ω resistor, which matches the electrical impedance of the wire. The terminating resistor will absorb the electrical signal and keep it from reflecting back to the sender as shown in Figure 15.3(b).

Optical Signaling

The latest advancement in communication medias is the use of transparent glass fibers. When light enters one end of a glass fiber it will pass though the fiber with very

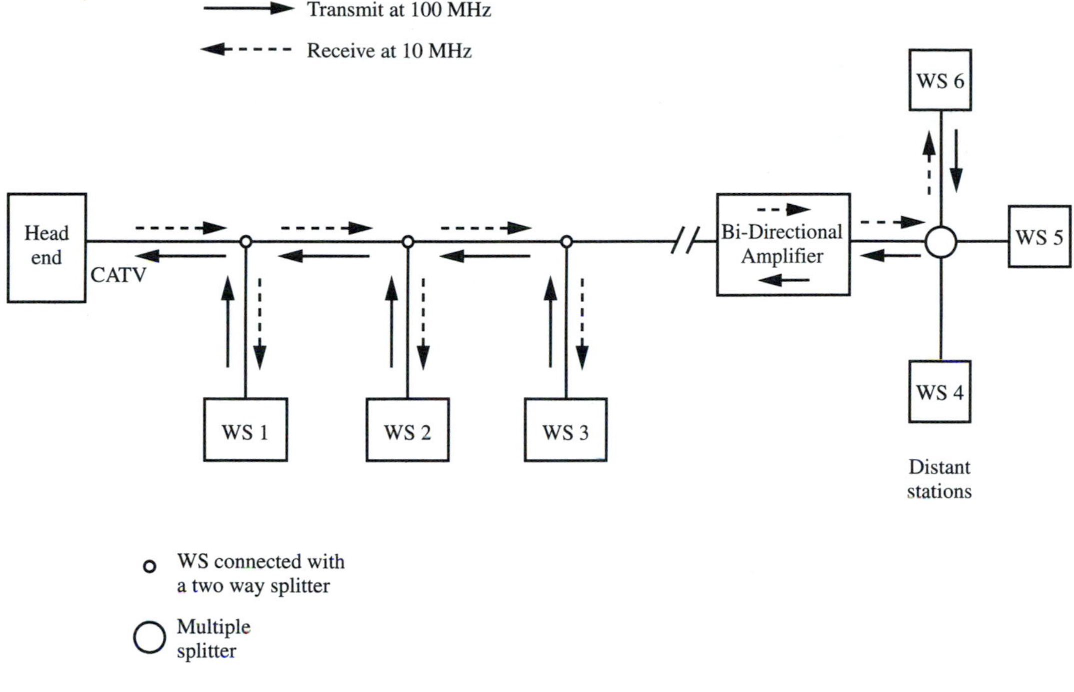

FIGURE 15.2
Broadband network

little loss in energy, similar to electricity flow in copper wire. Unlike electricity there are no induced electromagnetic fields. Outside electromagnetic interference will have no effect on the optical signal within the fiber. The conversion of electrical signals representing data is accomplished by using optical solid-state devices. The transmitting device is a light emitting diode, known as an LED. The receiving device used is a phototransistor.

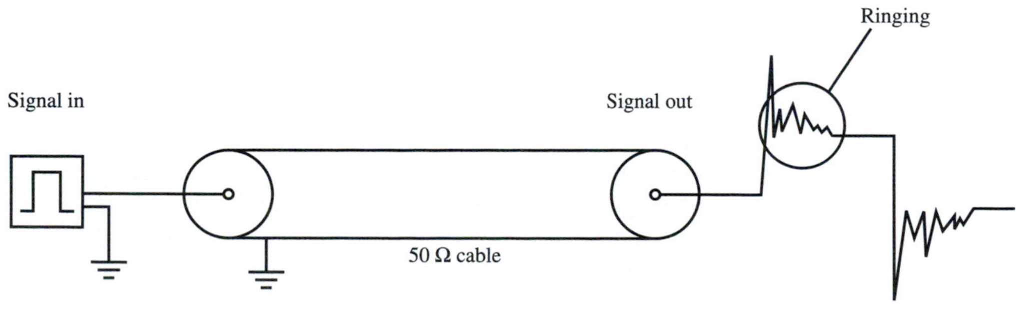

FIGURE 15.3a
Ethernet terminators

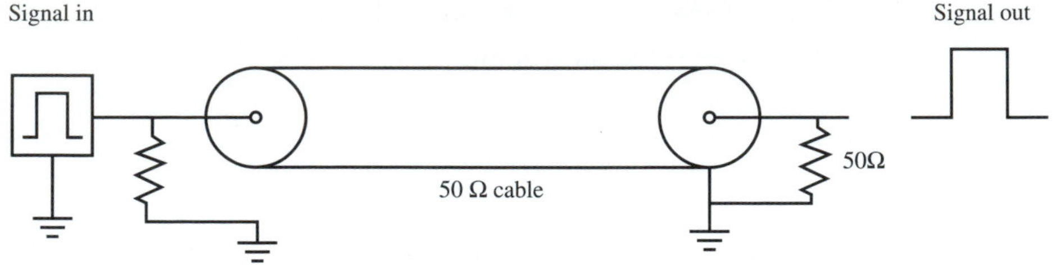

FIGURE 15.3b

When an LED is forward biased, it will emit light in the red or infrared range. When light enters a phototransistor, it will cause the transistor to begin conducting. As shown in Figure 15.4 an electrical impulse can be sent over an optical medium for great distances with very little distortion of the signal.

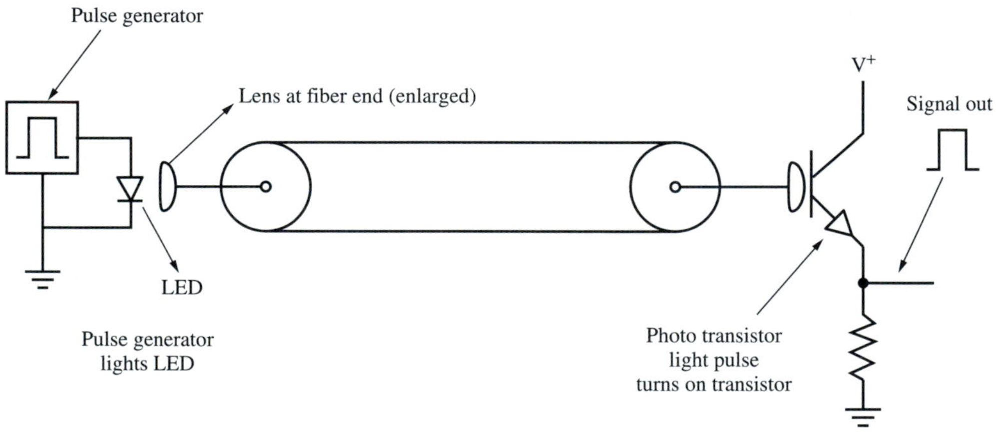

FIGURE 15.4
Fiber optic transmission

Converting digital signals to light impulses eliminates many problems. The greatest advantage of optical signaling over copper wire is that much greater distances can be covered without signal degradation and greater bandwidth is possible. Completely eliminated are the problems of outside electromagnetic interference. A disadvantage is that **fiber optic** cable is extremely fragile and difficult to work with. One must use extreme care when handling, splicing, terminating, or performing any other task with fiber optic cable. For these reasons, fiber optic installation can be very expensive. Figure 15.5 shows a fiber optic toolkit and connectors.

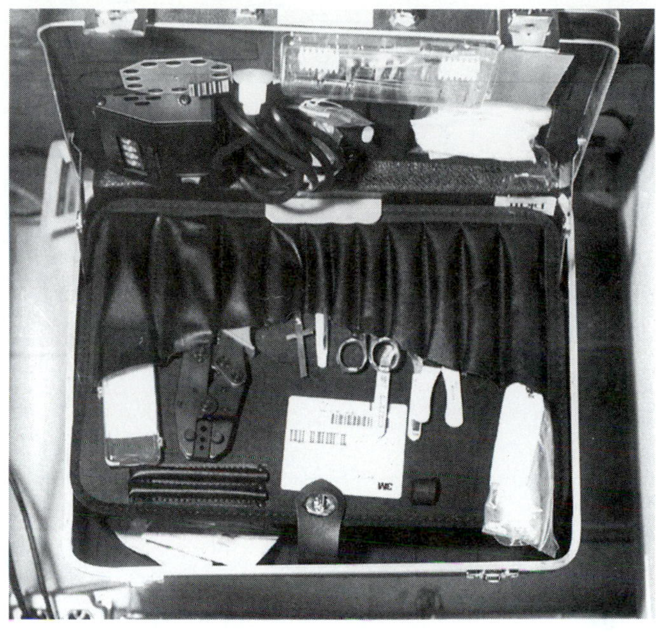

FIGURE 15.5
Fiber optic kit and connectors

15.2 NETWORK TOPOLOGIES

When we use the term **topology** in networks we mean the physical layout of the components and the logical connectivity between the devices. There are several network topologies in use today. The three most common are called bus, ring, and star topologies. A LAN may consist of one of these topologies, or use a combination of topologies.

Bus Topology

In the **bus topology,** workstations are connected in parallel. This is similar to the way in which devices inside the PC are connected, as shown in Figure 15.6. When devices on a network are connected using the bus topology, all devices receive data at the same time (or nearly the same time). It is up to the device to read the data and decide if the data is for it or not. If it is not, then no action is required. When a device wants to send data, the device must first determine if another device is using the network. Sensing an electrical level accomplishes this. If the network is free, then the device may send the data. Due to electrical delays, a device that wants to send data may not be aware that another device has just begun to send data. If this scenario occurs, both signals will collide on the wire. This condition is known as a **collision.** It is up to all devices on a bus topology network to sense and correct for this condition. The biggest problem with the bus topology is that if a break or short occurs anywhere in the line, all of the devices on the bus will fail to communicate.

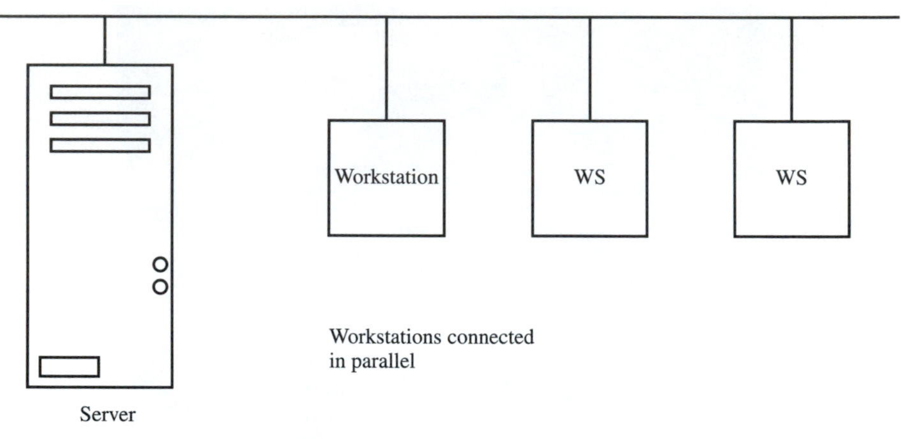

FIGURE 15.6
Bus topology

Ring Topology

In the **ring topology** network devices are connected in series to form a ring as shown in Figure 15.7. In a ring there is usually one master device controlling the ring, and each device will pass information to the next device in the ring. If any device is removed from a ring, the network will fail unless the device is bypassed. Access control of the network is usually controlled by a token. The station or device that has possession of the token can send data onto the next device in the ring.

Star Topology

In the **star topology** separate runs to a central point connect the workstations together. A device known as a **hub** connects the workstations to the rest of the network as shown in Figure 15.8. Hubs may be connected to other hubs (in order to connect network segments or to increase the number of nodes on a network), or to a bus by using a device known as a transceiver. When the physical star topology is used in Ethernet systems, it is really still an electrical bus where all stations receive data in parallel. When the physical star is used in token ring, the topology looks like a star but really is an electrical ring where each device gets network traffic in turn. The greatest advantage to the star topology is that when one run is broken or disconnected, the remainder of the network continues to function.

15.3 NETWORK TYPES

The network types most commonly used with PCs are Ethernet and token ring. There are also low-cost two-node networks utilizing serial or parallel ports. Artisoft made a product called Lantastic that provided a fully functioning two-node network over a serial port. Microsoft shipped a file-sharing mini-network included with DOS known as Interserver.

WS
1
T
WS 1 captures the token and sends a data frame to WS 2. The data frame contains the address of WS 2.
T
WS
2
WS
4
WS
3

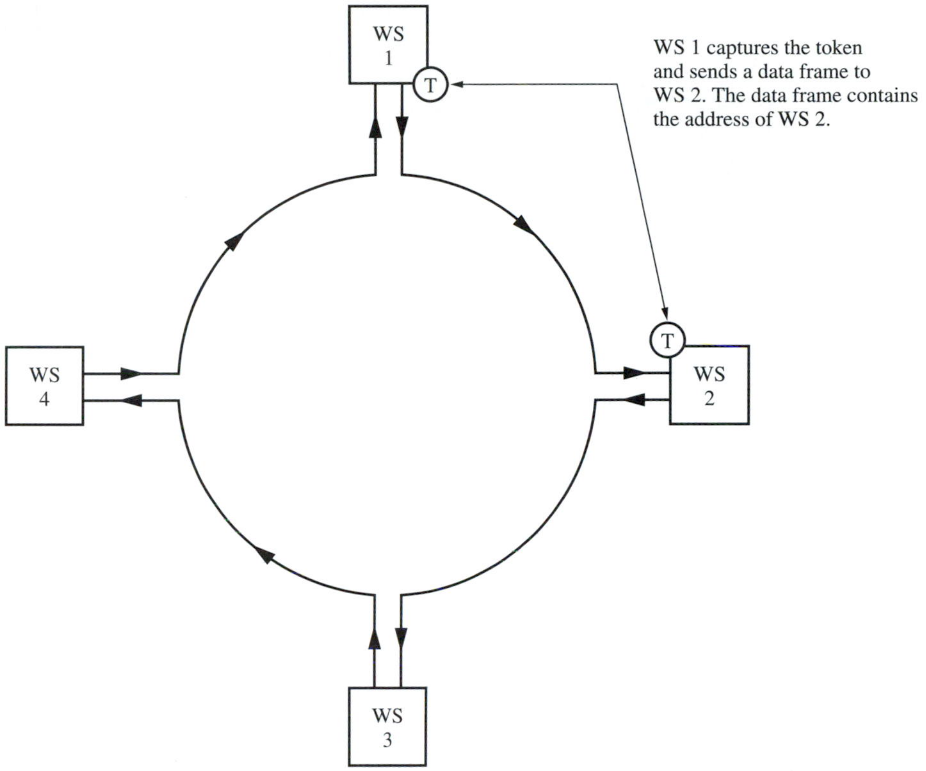

FIGURE 15.7
Ring topology

ARCNET was one of the first multinode low-cost personal computer networks and has become virtually obsolete. Token ring is found primarily on IBM mainframe systems. Ethernet has become the most popular network on small PC-based systems. Ethernet networks range in size from small two-node systems up to worldwide wide area networks. Ethernet has become the most common network in use today and has become very inexpensive. In fact, Ethernet network cards can be purchased for as little as $20 each. Microsoft Windows for Workgroups and Windows 95/98 provide all the software you need to get a small network up and running. Other operating systems such as Novell Netware, UNIX, LINUX, and Windows NT provide for large network control.

Token Ring

Token ring uses ring topology (wired as a star) and can be used with any protocol suite. Token ring can be joined to an Ethernet network by simply adding a bridge or router (but not a repeater) provided both networks are running the same protocol such as TCP/IP. Unlike Ethernet (or other bus topology), data on token ring travels in one direction, thus two conductor sets are used per port. The first conductor pair is called *ring in* and the

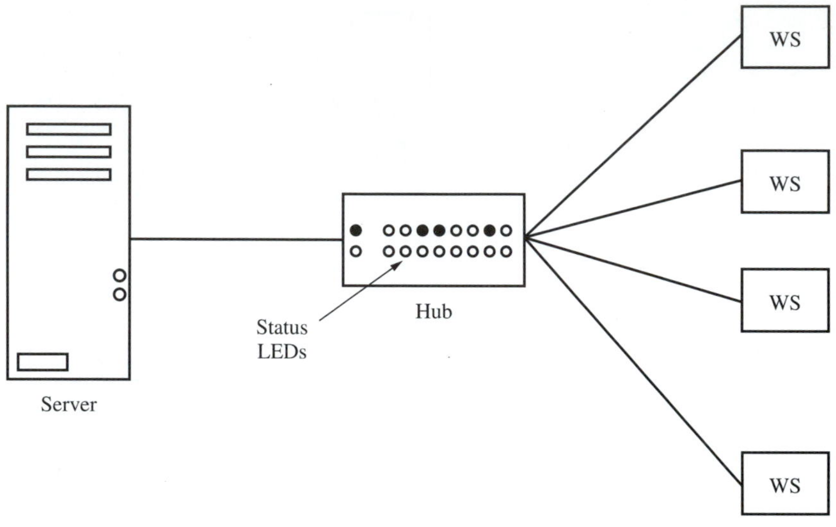

FIGURE 15.8
Star topology

second conductor pair is called *ring out*. Control of the interface is given to the interface in possession of the token. Upon start-up of the network, one of the interfaces will generate a free token. That token will be circulated forever or until someone picks it up. As shown in Figure 15.9, the terminals or stations will connect to a device known as a multi-

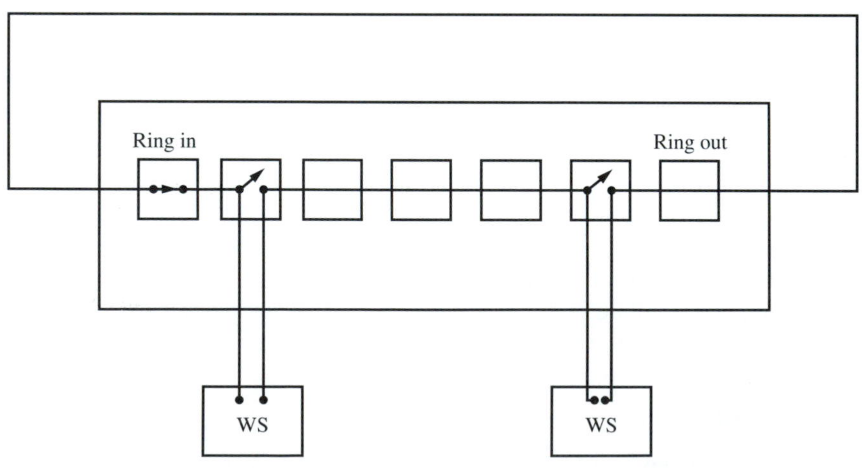

FIGURE 15.9
IBM token ring MAU

station access unit, or MAU. Behind each port in the MAU is a relay, which, when closed, completes the ring, indicating no terminal is connected. When a device wants to use the network, it waits for a free token. It then adds its information onto the network and sends it along with the token. Before sending information, the device will change the token status to busy as shown in Figure 15.7. The next device in the ring receives the packet with the token and checks the address. If the packet belongs to this device the interface will process the information, change the token status to free, strip the data, and pass the token along. If the information is destined for someone else on the network, the interface will simply send the entire packet of information along to the next station on the ring. This process continues as long as the network is running. A token ring network operates at 4 or 16 Mbps and uses baseband transmission. The IBM token ring conforms to the IEEE 802.5 standards. A popular variation of the token ring is the token bus. Token bus is described in the IEEE 802.4 specification and functions similar to token ring. Instead of the stations being connected in series they are connected in parallel on a bus similar to Ethernet. The token circulates from one station to another, forming a logical ring.

Ethernet

The origins of **Ethernet** can be traced back to the late 1960s, when the University of Hawaii was working on a wide area network known as ALOHA. This network contained the first carrier sense multiple access/collision detection **(CSMA/CD),** access method. This laid the foundation for the Ethernet system that is in use today.

The Xerox Corporation began work on Ethernet in 1972, and the new network type rapidly grew in popularity due to the success of the project. In the early 1980s Digital Equipment Corp (DEC), Intel, and Xerox joined together and released a standard for Ethernet known as Ethernet I. They continued to work together and released a new standard known as Ethernet II. In 1985 the IEEE released a standard from the project 802 committee known as the IEEE 802.3 standard for Ethernet. Modern Ethernet follows the 802.3 standard closely, but some minor variations do exist. Ethernet uses baseband signaling arranged in a bus topology. The most common signaling speeds are 10 MHz and 100 MHz. Basically Ethernet is a network architecture that provides a "best effort" data delivery service. Ethernet provides for error detection by way of CSMA/CD. CSMA/CD has no error correction. It is up to the network protocols or the application program to provide for error correction.

Ethernet Address

Every Ethernet network interface card, or NIC, has a unique six-byte address. The first three bytes of the address define the hardware manufacturer, and the last three bytes are the NIC card ID. Theoretically each and every network interface has a unique number. The standards committee, such as the IEEE, assigns the first three-byte portion of the ID, which is unique to each manufacturer. The manufacturer then assigns the last three bytes of the address. This hardware Ethernet address is known as the media access controller **(MAC)** address. Remember that the Ethernet or MAC address has nothing to do with TCP/IP Internet addresses, which are discussed later in this chapter.

Ethernet Access

Since Ethernet uses bus topology, every device attached to the network receives the same data packet. It is the responsibility of each individual device to determine if the packet is intended for its use. If the packet is not intended for the device, then the device must take no action. Before any device can send data it must "sniff" the network for the presence of activity, hence the CS (carrier sense) part of CSMA. If no activity is present at that moment, then the device will begin to send. During the sending process the device will also watch for an electrical anomaly known as data collision, hence the CD part of CSMA/CD. Collision detection works by monitoring the voltages you are sending and verifying that they make it onto the wire. If another device attempts to send data while you are sending, the voltage levels will shift, and this change in voltage signals a collision. Upon detecting a collision you must stop sending and send a special signal known as a **JAM** signal. The JAM signal notifies all devices that a collision has occurred and that the data are invalid. Also included in the Ethernet specification is a back-off algorithm. When a collision occurs, the algorithm will instruct each device involved in the transmission to wait for a period of time before sending the data again. If a collision occurs again, the length of the wait time is increased, which should break any deadlock that might occur.

Ethernet Encoding

Ethernet data bits are encoded using a method known as **Manchester encoding.** Manchester encoding states that a voltage transition must occur for every bit; thus every bit is self-clocking. In our example we will use a single-level DC voltage, also known as return to zero or **RTZ.** A 1 bit goes from a low level to a high level, and a 0 bit goes from a high level to a low level as shown in Figure 15.10.

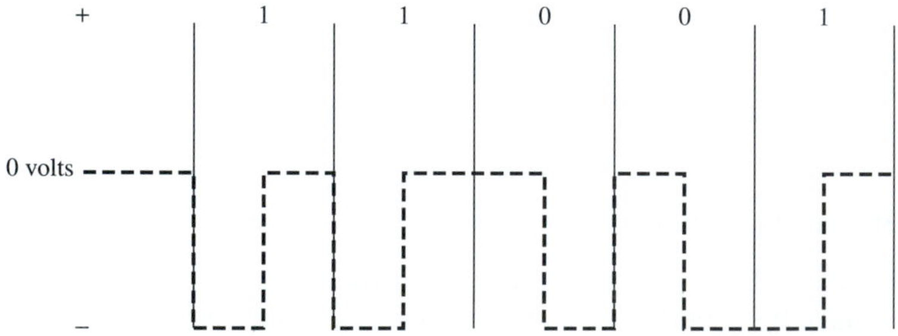
+
1
1
0
0
1
0 volts
–
1 bit goes from low to high; 0 bit goes from high to low.

FIGURE 15.10
Manchester encoding

Ethernet Cabling

Originally, Ethernet required the use of a special rigid cable known as the backbone into which taps and transceivers were inserted. A transceiver is connected to the workstation using an **AUI,** or attachment unit interface. This is a thick multiconductor cable with 15-pin connectors. This configuration has come to be known as **thick Ethernet.** When Ethernet workstations are connected using standard coax cable, it is called **thin Ethernet.** Thin Ethernet is sometimes called cheaper net because the cost is much lower than thick Ethernet. The drawback to using only thin Ethernet is that the maximum distance of the network will be less than if using thick Ethernet. Ethernet can operate over three electrical cable types; 10BASE2, 10BASE5, and 10BASET. The 10 stands for the signaling speed in megahertz, and BASE indicates baseband signaling. The 2 and 5 enumerate the length of the network segment in meters times 100. Thus, 10BASE2 cabling must not exceed 200 meters. The letter T is used to denote twisted-pair wiring. Twisted-pair wiring is more susceptible to electrical interference and cannot exceed 100 meters. Ethernet can also operate over fiber optic cable between hubs or repeaters. Ethernet implementations should not exceed the 5-4-3 rule per network. The 5-4-3 rule means 5 network segments, 4 repeaters maximum, and no more than 3 segments populated with users. An illustration of the 5-4-3 rule is shown in Figure 15.11.

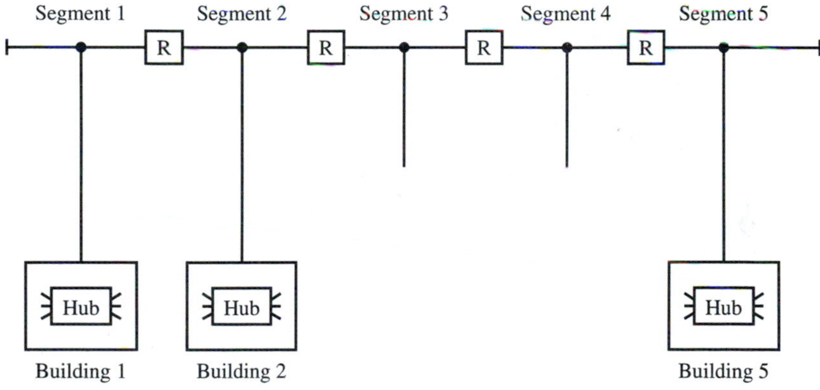

FIGURE 15.11
The 5-4-3 rule

Cable Definitions

10BASE5 stands for 10-MHz baseband signaling for a maximum of 500 m. The cable used is a special rigid type designed specifically for Ethernet. Connections are made by drilling through the cable jacket and outer conductor, and installing a connector known as a vampire tap. This is illustrated in Figure 15.12. An alternative tap is available that has a built-in barrel connector and is attached to the ends of the rigid cable as shown in Figure 15.13. These taps also contain the transceiver.

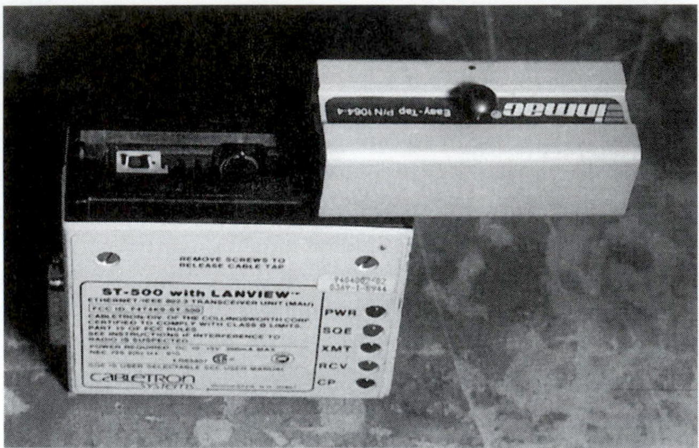

FIGURE 15.12
Vampire

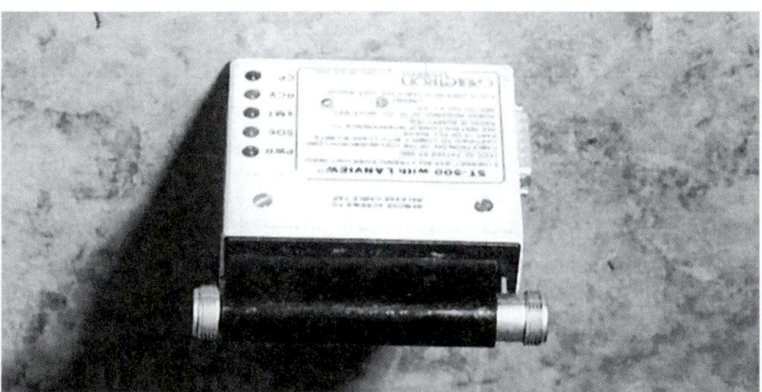

FIGURE 15.13
Barrel

Connections are made to the workstations by way of a special multiconductor cable, which carries power to the transceiver. This special cable is known as an attachment unit interface or AUI. AUI has also come to be known as thick net, and uses a 15 D-shell connector. This connector has also come to be known as a DIX connector as shown in Figure 15.14. DIX stands for the original DEC/Intel/Xerox standard. There can be a maximum of 100 taps or stations per 500-meter segment without using repeaters.

10BASE2 stands for 10-MHz baseband signaling for a maximum of 200 m (more precisely, it is 185 meters). The cable shown in Figure 15.15 is standard RG-58 a/u coaxial, which is also known as cheaper net for its low cost. There can be a maximum of 30 stations on each network segment before having to use repeaters. Each workstation is connected by the use of a standard BNC tee connector between the coaxial cable segments. If the station is the last one in the run, then a tee is still used, and a 50-Ω terminator is placed on the open end of the tee. A very important note, remember to *terminate!*

An important word about thin or coaxial Ethernet cable: It has been said that coaxial Ethernet cable is the same as cable TV wiring; however, this is not true. Cable TV wiring is 75-Ω broadband cable designed for RF transmission. Ethernet cable is designed for 10-MHz Ethernet operation and has an electrical impedance of 50-Ω. Also, cable TV

FIGURE 15.14
AUI/DIX connector

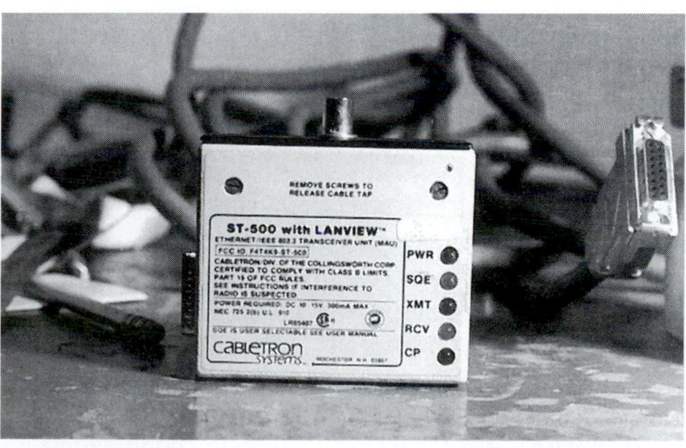

FIGURE 15.15
Cheaper Net

wire uses an aluminum shield for aluminum crimp-style F connectors, whereas 50-Ω Ethernet uses a silver-plated copper braid shielding, providing a superior connection when crimped. Mixing metals, particularly aluminum and copper, can cause corrosion, leading to poor connections. Avoid using cable TV wiring for Ethernet at all costs.

10BASET stands for 10-MHz baseband signaling over unshielded twisted pair cable. The maximum length for 10BASET is 100 meters per run. This is the newest Ethernet cabling method and is very different from other Ethernet layouts. The electrical topology is the bus type, but wired as a star, and can be referred to as a bus/star topology. Unshielded twisted pair or UTP is also used in the star topology and requires the use of a hub. One advantage of using UTP is that the wiring is far less expensive, and it is easier to handle than thick or coax cable. In addition, existing telephone wiring and punch down blocks may be used as long as the existing wire contains the same number of twists per foot as Ethernet UTP cable. Each run of 10BASET connects to a device known as a hub. The hub operates as a traffic cop by directing the traffic in the proper direction. The hub also acts as a repeater and repeats the signal to each node. Any node or connection to the hub can be connected or disconnected without affecting the rest of the network.

The original intent behind using unshielded twisted pair was to take advantage of preexisting telephone wire. In this way new network installations would be simplified because most buildings are prewired for telephones to every room. In older buildings this does not work out well because the wiring does not meet the requirements of category 3 and category 5 wiring. The EIA (Electronic Industries Association) has defined a set of standards governing unshielded twisted-pair wiring for buildings known as the EIA/TIA 568 commercial building standard. The standard is defined as follows:

Category 1 Ordinary voice-grade telephone wiring installed before 1983 with an unknown number of twists per foot.

Category 2 Unshielded twisted pair consisting of four pairs suitable for data up to 4 Mbps. This is similar to IBM type 3 cable.

Category 3 Unshielded twisted pair cable supporting Ethernet at a rate of 10 Mbps and token ring at 4 Mbps. This cable will have at least three twists per foot.

Category 4 Unshielded twisted-pair cable for token ring at 16 Mbps.

Category 5 Unshielded twisted-pair cable designed for 100-Mbps Ethernet. This type of cable is manufactured to a higher standard and is tightly twisted at around 12 twists per foot with a 100-Ω impedance. All modern wiring should only use this type of cable, since it is basically the same cost for any category wire. One hundred megahertz Ethernet is fast becoming the standard, and wiring with anything else would be impractical. In fact it has been shown that speeds greater than 600 Mbps per second are possible over category 5 cable at distances of 100 meters.

Twisted-Pair Wire What makes it possible to simply eliminate expensive coaxial cable and use twisted-pair wiring? What about electrical noise, interference, and cross talk? When electrical current flows through a piece of wire, a magnetic field is generated.

Conversely, anytime a magnetic field passes through a wire, an electrical current is induced in the wire. Therefore signals from one wire will be picked up by a nearby wire, a phenomenon known as **cross talk.** The reason twisted-pair Ethernet works is that it is based on differential transmission as shown in Figure 15.16.

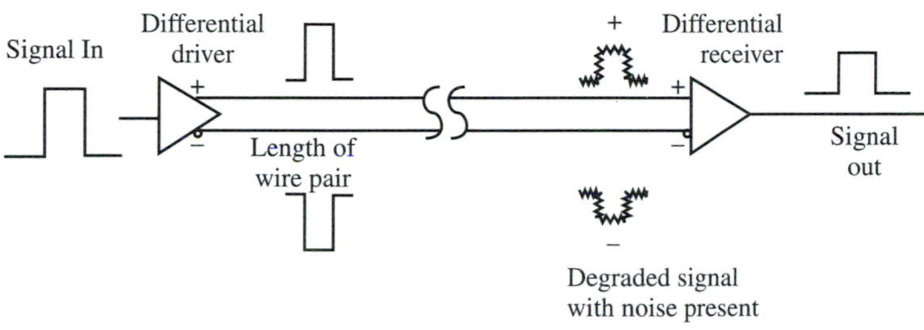

FIGURE 15.16
Differential Ethernet signaling

Because the signals are identical but switching in opposite directions, the resulting magnetic fields will cancel each other out. Also, the two conductors are twisted together, which helps to cancel the effect of stray electromagnetic fields. Furthermore, Ethernet receivers are designed to pass only differential signals and reject any common mode signals. If a stray signal were to be picked up by both conductors, the signal impulse would be the same polarity at both inputs to the receiver amplifier. A differential amplifier will only pass or amplify the difference presented at the inputs. Therefore a stray signal caused by interference or cross talk will have no effect on the output.

15.4 NETWORK PROTOCOLS

The set of programs used to set up the network connection in the workstation is organized into layers known as the network stack. Over the years, many attempts have been made to standardize the function of each layer and the overall network layout. One such standard has been achieved, and is known as the **OSI model.** OSI stands for the open system interconnect. The OSI model is organized into seven layers as shown in Figure 15.17. The original attempt of the OSI was to replace all other network models, but in actuality it has become a general reference model only. TCP/IP has become the standard today. Although most PC networking does not fit the OSI model exactly, the goal is to adapt to this model as closely as possible.

The lowest four layers provide the infrastructure for end-to-end data transfers over a network. The remaining three layers provide the application hooks to the transport. The lowest or first layer of the network stack is known as the **physical layer.** This is the actual interface hardware and cabling used for connectivity. The second layer is known as the **data link layer.** It is responsible for data transmission, framing and control of the network data. The third layer is known as the **network layer.** This layer is responsible for data transfer across the network independent of the media or network topology. The fourth

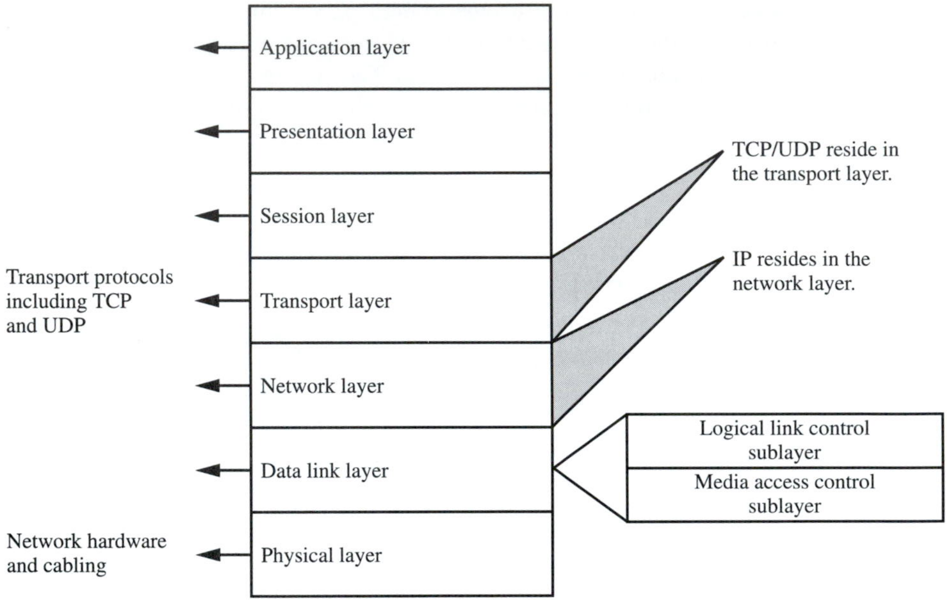

FIGURE 15.17
The OSI model

layer is known as the **transport layer.** This layer is responsible for the reliability and switching of the network data. The fifth layer is known as the **session layer.** Here more control is added to the network data. The sixth layer is known as the **presentation layer.** Here structure is added to the units of network data. The top or seventh layer is known as the **application layer.** This is where the application hooks are created between a user application and the network stack.

TCP/IP

The origin of transmission control protocol/Internet protocol or **TCP/IP** can be traced back to the Department of Defense (DoD) work done on ARPANET. TCP/IP is a set of network protocols developed by the earliest Internet users. TCP/IP is actually two layers, with TCP, which is the transport layer, sitting on top of IP, which is the network layer. The term actually incorporates elements of two different portions of the OSI model. Transmission control protocol, TCP, is one of the transport protocols available in the transport layer of the model. Internet protocol, IP, is found in the network layer of the OSI model as shown in Figure 15.17. As you can see, TCP/IP is equivalent to layers 3 and 4 of the OSI model. Taken as a whole, TCP/IP provides communication and data exchange between PCs and services such as routing and retransmission requests in the network environment.

In the earlier days of DOS-based PCs, TCP/IP was not used for LANs because of the additional overhead and memory space to decode and process information. This is no longer a factor on modern equipment. Perhaps the greatest service provided by TCP/IP is the access it provides to the Internet. TCP/IP has become the de facto standard for networking. Some other uses for TCP/IP are

- Network protocol for UNIX-based client/server systems
- WWW (World Wide Web) services (The Internet)
- One of the network protocols for Microsoft Windows 95/98, NT
- Mail services, SMTP (simple mail transfer protocol)
- File transfer, FTP (file transfer protocol)
- Virtual terminal service (Telnet)
- Network management, SNMP (simple network management protocol)

In a TCP/IP network each packet of information is called a **datagram.** There are two types of datagrams: TCP, for transmit control protocol, and UDP, for user datagram protocol. A TCP datagram contains information about the source address, destination, size, sequence, and error checking as shown in Figure 15.18(a). This is particularly important in a large network where reliable data exchange is critical. Conversely UDP has far less overhead, as shown in Figure 15.18(b). This is faster and more efficient when you are not faced with outside factors or use a small nonrouted network. UDP falls short in large multirouted networks, as packets sent may arrive out of sequence. When using TCP over a WAN or another very large network, packets may arrive out of sequence, or some may be missing. TCP provides a sequence number to check for this; UDP does not. TCP will

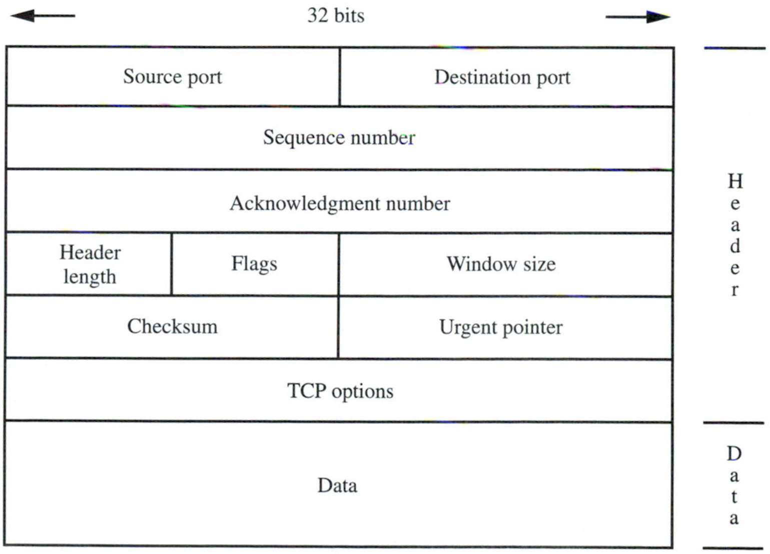

32 bits

Source port		Destination port	Header
Sequence number			Header
Acknowledgment number			Header
Header length	Flags	Window size	Header
Checksum		Urgent pointer	Header
TCP options			Header
Data			Data

(A) TCP datagram

FIGURE 15.18a
TCP/IP datagrams

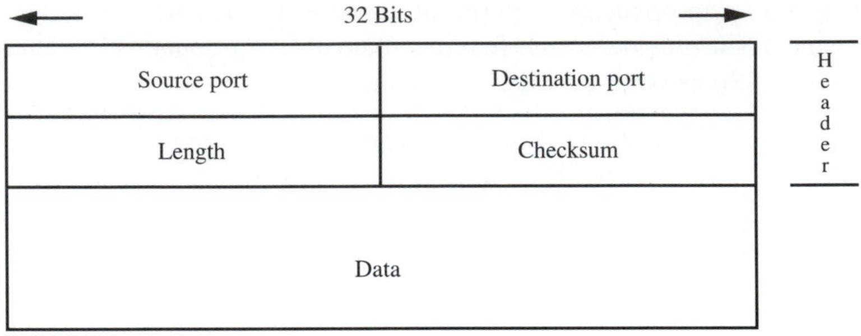

(B) UDP datagram

FIGURE 15.18b

assemble all of the packets of data received and sequence them in the correct order. UDP, on the other hand, will accept information as received without performing any "repair".

The IP protocol is the underlying network layer. All of the information across a TCP/IP network will be sent as IP packets. In fact the entire network is really an IP network. Included in the IP protocol are various control and information gathering packet types. The packet types are

- TCP (transmit control protocol)
- UDP (user datagram protocol)
- ICMP (internet control message protocol)
- IGMP (internet gateway message protocol)
- ARP (address resolution protocol)
- RIP (router information protocol)
- Broadcast: IP address 255.255.255.255 (special IP address to all devices)
- Multicast: IP address 224.*x*.*x*.*x* (special address to all devices listening to an address)

TCP/IP Addressing

Devices on a TCP/IP network are called hosts, and every device must have a unique address. A TCP/IP address is a 32-bit binary number represented in what is called dotted decimal notation. The address consists of four groups of three-digit numbers between 0 and 255, each separated by a period. An example of this would be 128.238.123.222. Each number group represents an eight-bit decimal value. The range of decimal values for each group is 0 to 255, because the highest value in a binary eight-bit number represented in decimal is 255. Every computer on the network (or the Internet) must have a unique IP address. Remember that the Ethernet address discussed earlier is not the same as the Internet address. If you change the network interface, you will have a different Ethernet address but you will retain the same I/P address because the TCP/IP protocol uses a process known as address resolution protocol, or ARP, to discover the hardware address of a host. TCP/IP addresses are broken down into what are known as network classes. The class type is defined by the most significant digits of the address. There are five classes of addresses; A through C are shown in Figure 15.19, and D and E are reserved for special and future use. The network class defines the number of computers or hosts that can be present on the network. On a TCP/IP network a host on one class of network can communicate with a host on a network with a different class type through a router. Classes only define how many hosts can be on a particular network. Each network class can be broken down into subnetworks as shown in Figure 15.19.

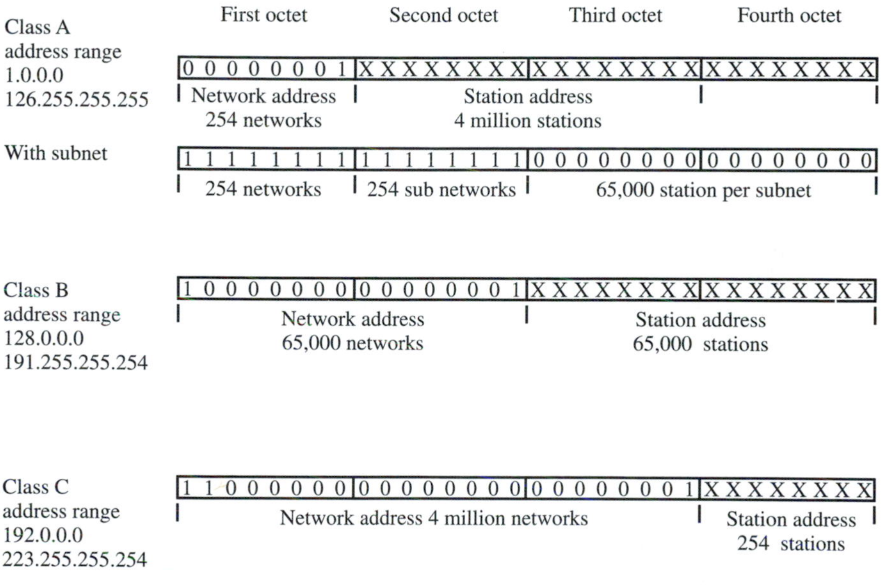

FIGURE 15.19
TCP/IP network classes

Dynamic Host Configuration Protocol (DHCP)

Address assignment is the responsibility of the network administrator. If an address already in use is chosen, obviously both devices will fail. Microsoft Windows 95/98/NT will display a message stating that a duplicate address is found, and will disable TCP/IP temporarily. TCP/IP networks provide an automatic method for IP address assignment known as dynamic host configuration protocol, or **DHCP.** Windows 95/98 and NT all default to DHCP. When using DHCP you need a server capable of acting as a DHCP provider. Microsoft Windows NT and most UNIX implementations provide this service. When a network device using DHCP is first started, it will broadcast a DHCP request message containing its hardware or MAC address. A DHCP server will listen for this special type of broadcast message and respond with an offer message. The requesting device will capture the message and respond with an accept message. The server will then add the address to its internal database and reply, completing the transaction. The server leases DHCP addresses for a specific period of time. The administrator may assign whatever amount of time is desired. DHCP can either assign a specific number of minutes, hours, or days in which to lease an address, or a date specifying the month and year of the lease expiration. There is no guarantee, however, that you will always get the same address unless the administrator sets the lease to never expire. In that case you will be assigned the same address unless you change the network card. DHCP listens for a broadcast message, so the addition of routers is not possible because broadcast messages are not routed. There is one exception to this: Microsoft Windows NT can be set to be a DHCP relay agent. If NT is the router to the subnet, then it will also relay DHCP information across the routes.

Subnetting

A TCP/IP network can be further divided into smaller networks known as **subnets.** There are many advantages to subnetting a large network. The greatest advantage is traffic reduction across the network. On a class B network you are allowed 65,534 hosts, because the host ID portion of the address is 16 bits. The actual decimal range for a 16-bit binary value is 0 to 65,535. This allows 65,536 combinations of values, but the lowest and highest addresses are reserved. For example, a class B network can contain 65,534 hosts. If all 65,534 hosts were on one network, then every host would see traffic from every other host. By manipulating a parameter known as the subnet mask you can define the scope of visibility of a smaller network or subnet. As an example, an address of 128.238.220.1 with a net mask of 255.255.0.0 can see any host within the 128.238.0.0 network.

The network mask uses a logical EXCLUSIVE-OR function of the network address bits and the mask bit. Changing the mask allows a large network to be broken down into smaller subnetworks. You might want to have a subnetwork of 254 hosts within the 65,534-host class B network. All you have to do is to change the network mask to 255.255.255.0. When this mask is applied, the network will have a subnetwork range of 254 devices or hosts. If the address is 128.238.220.10, it is part of the 128.238.220.0 network, and can only see hosts within that subnet. This means that if a computer wants to communicate with another computer outside its subnet, it must go through a router first. Subnetting is very useful in a class B network because it makes the network more manageable and network traffic can be isolated to local groups separated by routers. Furthermore any traffic within each subnet stays local unless the target address is outside the scope of the subnet. Broadcast traffic always stays local to the subnet, as do collisions or other network problems. Figure 15.20 illustrates the subnetting system.

Computer Names on a TCP/IP Network

Most people prefer to use names rather than numeric values, and TCP/IP makes provisions for this. It is easier to remember a machine name, (for example Hal), rather than 128.238.220.40. The first method of naming hosts on a TCP/IP network is by using a file called the Hosts file. The Hosts file is an ASCII text file describing the name and its IP address as shown in Figure 15.21.

A **Hosts file** is generally used with smaller networks and resides on each host connected to the network. The file is static and must be the same on all hosts in the network. If a computer on the network has its address changed or a host is added, then the host files on all computers must be updated. Most large networks, including the Internet, use a domain name server. This simplifies administration of host names, since they are con-

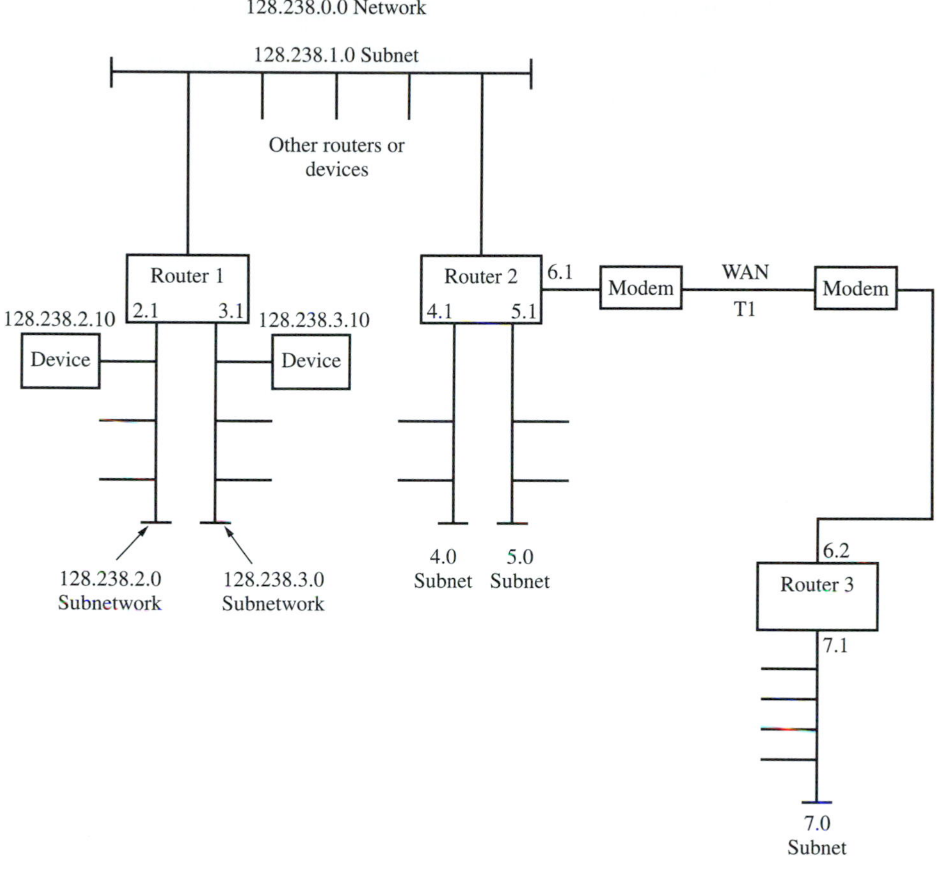

FIGURE 15.20
Subnetting

tained in one file within the domain name server. Each host using the network needs to know the address of at least one domain name server in order to translate a name into an address.

Domain Name Services

If your network is to be connected to the Internet, then domain name services **(DNS)** are a must. Remember that all IP addresses must be unique and are registered by an organization known as the INTERNIC. If your network will never be connected to the Internet, then you can pick any IP address you like, set up your network, and proceed. If you will be connecting to the Internet or World Wide Web, then you must register with the INTERNIC and obtain a range of TCP/IP addresses. Keep in mind that there are a finite number of addresses available, and the supply of numbers is running out. When you register you must pick a unique domain name. Domain names are hierarchical, with a root level at the top as shown in Figure 15.22.

```
# Copyright (c) 1993–1995 Microsoft Corp.
#
# This is a sample HOSTS file used by Microsoft TCP/IP for
# Windows NT.
#
# This file contains the mappings of IP addresses to host
# names.
# Each entry should be kept on an individual line.
# The IP address should be placed in the first column
# followed by the corresponding host name.
# The IP address and the host name should be separated by
# at least one space.
#
# Additionally, comments (such as these) may be inserted
# on individual lines or following the machine name denoted
# by a '#' symbol.
#
# For example:
#
#      102.54.94.97     rhino.acme.com         # source server
#       38.25.63.10     x.acme.com             # x client host
127.0.0.1        localhost
166.84.190.177   NTSERVER
166.84.190.178   CLIENT01
```

FIGURE 15.21
A sample Hosts file

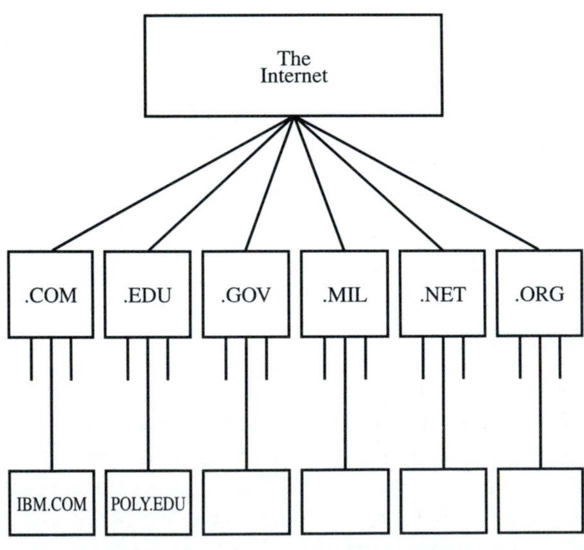

FIGURE 15.22
The Internet domain

In the United States six types of domains are available:

- .COM (commercial)
- .EDU (education)
- .GOV (government)
- .MIL (military)
- .NET (network providers)
- .ORG (not-for-profit organizations)

Examples would be ibm.com or poly.edu.

Microsoft has added an additional name resolution service known as WINS, or Windows Internet name service. This is even easier to use because it requires almost no administration. It is completely dynamic, and once a name server is set up, all that is required is to notify each workstation of the IP address of the WINS server. The WINS server "learns" the IP address of each Windows host as it uses the network and retains this information in a database.

15.5 NETWORK DEVICES

When multiple networks must be joined, the task may not be as simple as connecting a cable between them. When different network types are connected, you must use a device to translate the signal levels and data types so that the networks can communicate. These devices are known as repeaters, bridges, routers, and gateways. A repeater is the simplest device, with less functionality than the other three devices, and is basically an amplifier. The gateway is the most complex device because it must translate everything. These terms are often used incorrectly, perhaps because the devices perform similar tasks and the differences between them are less pronounced than they used to be.

Repeaters

Simply put, the function of a **repeater** is to take a weak signal and strengthen it. As you know, when any signal travels any length of wire or cable, it will deteriorate. If the wire or cable is long enough, the signal will eventually become unrecognizable. For example, in a coax network, the cable can run no more than 607 feet without major attenuation of the signal. In a twisted pair network the run is even shorter, 300 feet. Therefore if your network were to exceed these lengths, you would need to have repeaters between the segments in order to maintain signal integrity. A repeater functions at the physical layer of the OSI model, so it is the simplest of this type of network device. A repeater performs none of the higher functions that bridges, routers, and gateways perform. The repeater actually rebuilds a signal by first receiving the signal, discriminating it, amplifying or regenerating it, and finally sending it back out as shown in Figure 15.23.

Repeaters perform no translation therefore the packets being passed through them must conform to the same logical link control protocols. For example, a repeater will not allow communication between an Ethernet network and a token ring network. Also, repeaters do not provide filtering or traffic isolation. The consequence of this is that if a packet contains errors, these errors will be passed along from segment to segment, and this could result in delays in network traffic. Repeaters can, however, forward packets

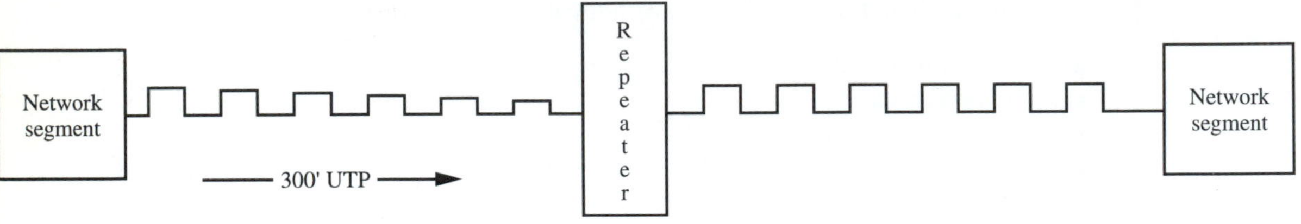

FIGURE 15.23
Signal regeneration

across different media types. For example a repeater may be capable (if it has the physical connectors) of sending data between coax segments through a fiber optic connection as shown in Figure 15.24. In fact one of the authors has this scenario at his university.

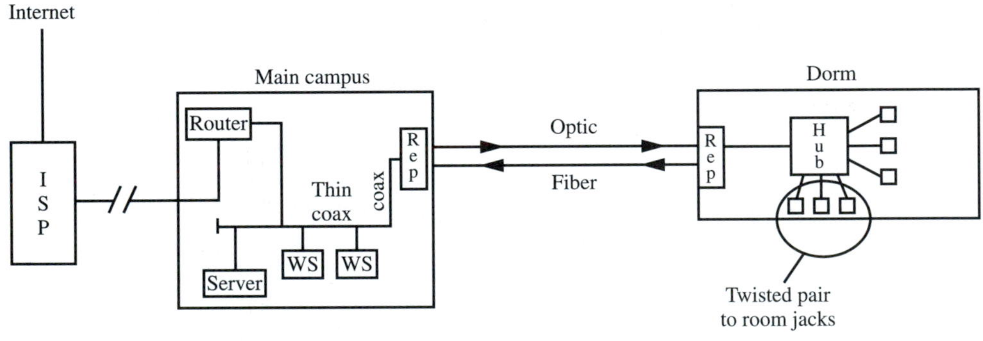

FIGURE 15.24
Coaxial to fiber-optic network repeater

Repeaters are a good way to expand small networks because they can link additional segments to the existing network. If network traffic is not very heavy or cost is a major concern, this expansion option may be preferable.

Bridges

Bridges are used when different networks with similar protocols are to be joined. In this case only moderate translation is required. An example of this would be a Windows workstation running a TCP/IP telnet session over Ethernet, connected to an IBM mainframe running TCP/IP over token ring as shown in Figure 15.25.

Bridges have certain functions in common with repeaters. They both can be used to expand a network by adding segments, and they can both link unlike media types. Bridges cannot distinguish between different protocols because they work at the data link layer and therefore do not have access to any information in the higher levels of the OSI model.

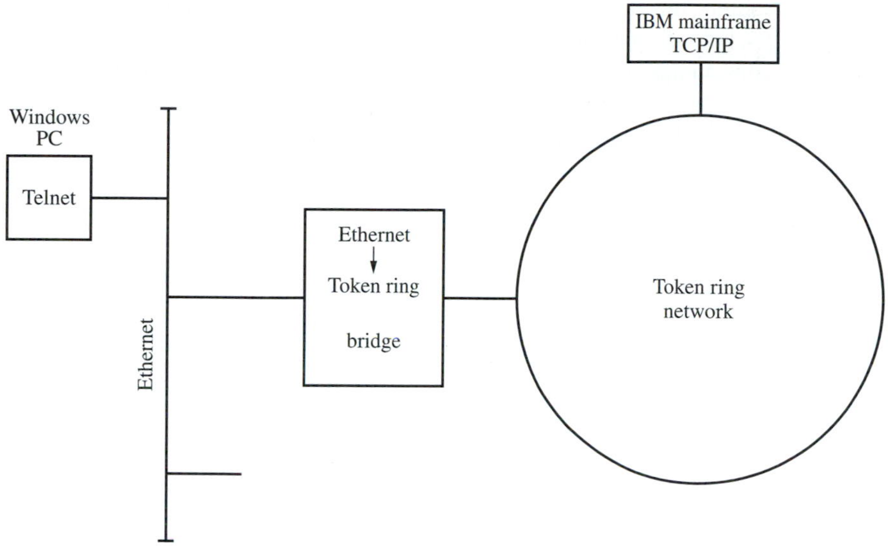

FIGURE 15.25
Ethernet to token ring bridge

Bridges will pass along any protocol and then it is the job of the individual PC or workstation to recognize a like protocol. Bridges work at the media access control sublayer of the data link layer. A bridge is somewhat intelligent in that it will assemble a database of addresses to assist it in forwarding data. The bridge will use the source address of individual packets that it encounters to build its address database. When a bridge receives a packet, it will compare the source address to its database and add it if it is not already contained therein. The bridge will then check the destination address of the packet and compare that to its database. If the destination address is the same as the source address, the packet will be discarded. This operation serves to reduce network traffic. If the destination address is not on the same segment as the source address, the packet will be forwarded. If the destination address is not in the database, then the bridge will forward the packet to all nodes.

Routers

Routers can of course perform some of the same functions as bridges, but they also can perform many more. Since routers function at a higher level of the OSI model, the network layer, they have access to information that none of the other devices do. Routers function in and between networks with different protocols and architectures. There are routers available that can accommodate multiple protocols on the same network. They provide filtering and isolation of network traffic, and can determine the most efficient path over which to transmit a packet. One way in which the router better manages traffic is by not passing broadcast traffic. This also avoids the problem of broadcast storms. Since routers function at a higher level of the OSI model, they have access to more complex packet information and can provide more efficient traffic management. Routers can also communicate with each other to share routing and status information. This allows a router to bypass specific network connections that may be experiencing difficulties and therefore further expedites traffic.

The Routing Table

The routing table contains the following information:

- All network addresses
- Paths between routers
- Connection options between networks
- Cost of data transmission (in terms of the number of "hops" required for a transmission)
- Host addresses (depending on the network protocol running)

Using the information in its routing table, the router chooses the most efficient route, determined by the available paths and cost of transmission. Routers communicate only with other routers and with local network adapters. They require specific addresses; therefore the routing table contains network numbers. Routers do not communicate with remote networks; however, they do communicate with the router managing the destination network. As a packet is passed from one router to another the source and destination addresses from the data link layer are stripped off and recreated. This is what enables the router to communicate between networks with differing protocols and architectures, and is a result of the fact that the router operates at a higher level of the OSI model (network layer). One disadvantage of the router is that because of its higher functionality it is generally slower than a bridge. Having said that, there are several ways in which routers handle network traffic more efficiently, thereby reducing the traffic and users' wait time.

Here are some ways in which routers provide better traffic management:

- Do not pass broadcast traffic (but can be set to pass multicast traffic)
- Avoid broadcast storms
- Do not route bad packets
- More sophisticated filtering and isolation
- Only pass packets with a known network address
- Listen to the network in order to determine the path that is least busy

Not all network protocols are routable. Routable ones include:

- Apple Talk, used by Apple Computer Inc.
- DEC-net, used by Digital Equipment Corp.
- IP, the Internet protocol
- IPX, developed and used by Novell Inc.

Protocols that are not routable include:

- Net BEUI, used by all Microsoft operating systems
- LAT, used by Digital Equipment Corp.

A router uses several methods to determine a data path. One way is by listening to the network to determine the least busy path. In addition routers can pass data along multiple data paths between LAN segments. Also, if there is a malfunctioning router anywhere along the way, it can be bypassed and the data can be sent to another router over an alternate route. A router also chooses a path by determining the number of hops between network segments.

Types of Routers

There are two types of routers, static and dynamic. There are a couple of important differences between the two; one is that the static router requires manual configuration of all routes. In the dynamic router only the first route requires manual configuration, then all additional routing is configured automatically. Another significant difference is that the route in the static device is hard coded and therefore is not necessarily the best route. The dynamic router, of course, can choose the best route based on the cost and amount of network traffic. There is a hybrid device called a **brouter,** which combines functions and features of both bridges and routers. For example, a brouter can route packets containing routable protocols, and can bridge nonroutable protocols.

Gateways

A gateway is the most highly functioning of these network components. In short, it allows networks that are completely dissimilar to communicate with each other. It allows communication between networks with different architectures, protocols, data formatting, and languages. It provides translation of data, signal levels, and network protocols. The gateway actually changes the format of the data so that it will be compatible with the application program on the receiving end of the transmission. Gateways are task specific, meaning they perform one function only—for example, an e-mail gateway. A gateway is usually a dedicated server on a network, and an important consideration is that it can use up a considerable amount of the server's available bandwidth because the gateway is performing resource-intensive tasks such as protocol conversion. Adequate CPU bandwidth and sufficient RAM should be allotted in order to assure peak performance of the gateway. When a gateway receives a data packet, it removes the protocol stack and translates the data to that of the protocol stack on the receiving network. A common application for a gateway is to connect a LAN using PCs to a mainframe or minicomputer environment. In this scenario a computer on the network would serve as the gateway. Application programs on the workstations allow them to access the mainframe (or mini) through the gateway computer and access all the resources of the mainframe. An example of this would be a Windows workstation running a TCP/IP telnet session over Ethernet, connected to an IBM mainframe running a 3270 session over token ring, as shown in Figure 15.26. Here signal levels, network protocol, and data translation must take place.

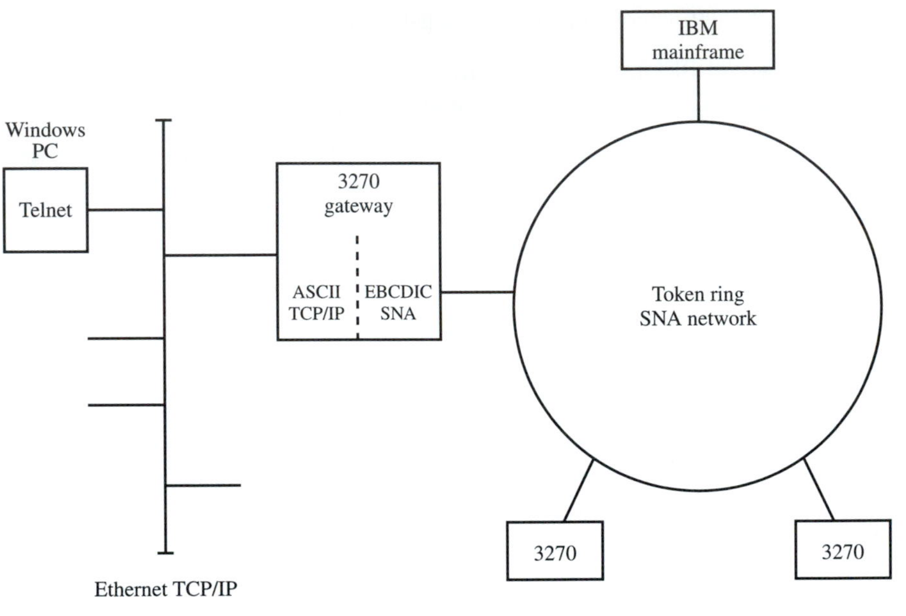

FIGURE 15.26
TCP/IP Telnet client to IBM 3270 gateway

15.6 PC NETWORK STACKS

Network stacks used in most personal computers are based upon either the NDIS or ODI standards. The **NDIS model,** which stands for network driver interface standard, is used by IBM, Microsoft, and many other network development companies. ODI stands for open data link interface. The **ODI model** is an open standard owned by Novell and is used with all Novell products. Some others have adopted the ODI standard, but most network venders currently use the NDIS model. Although Windows and Windows 95 default to the NDIS model, they support the ODI model and can be set up to run over the ODI stack.

The NDIS Model

The layout of the NDIS protocol stack is shown in Figure 15.27. The board driver, which is known as the MAC (media access control) driver, is usually loaded in the CONFIG.SYS file during boot-up. The file name for an NDIS MAC driver will have an extension of .DOS or .OS2—for example, 3C5X9.DOS—and is supplied with the interface card. The hardware configuration parameters are specified in a file called PROTOCOL.INI located in the network software installation directory. In the case of Windows for Workgroups the MAC driver is not loaded in the CONFIG.SYS file. Instead a configuration file is located in C:\WINDOWS\PROTOCOL.INI. Windows loads the MAC driver automatically. The next layer up is the program PROTMAN, which is the protocol manager. This is where one or more transport layers hook into the network layer.

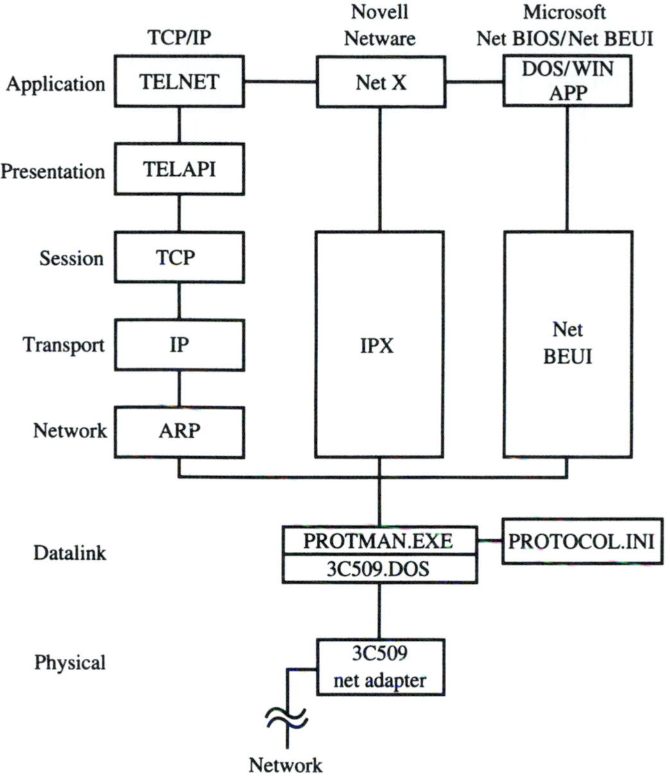

FIGURE 15.27
The NDIS Model

The ODI Model

The layout of the ODI protocol stack is shown in Figure 15.28. The network adapter driver resides in the MAC layer and is called the ODI driver. The ODI stack is loaded during the AUTOEXEC.BAT execution. ODI board drivers use the .COM file extension, for example 3C5X9.COM. If the driver is to be loaded in the CONFIG.SYS file, the board extension will be .SYS. The hardware configuration information is contained in a file called NET.CFG. The NET.CFG file is usually found either in the root directory or in C:\NET\NWCLIENT. The next layer in the ODI stack is the program LSL.COM, which is the link support layer. The function of LSL.COM is similar to the function of PROTMAN in the NDIS stack. One or more network transport stacks may be laid on top of the LSL. Novell provides an NDIS converter called ODINSUP.COM, which allows an NDIS protocol stack to be placed on the LSL.ODINSUP. This acts like an NDIS MAC driver but passes packets to the LSL rather than to an interface card. The LSL intern passes the NDIS packets to the interface card.

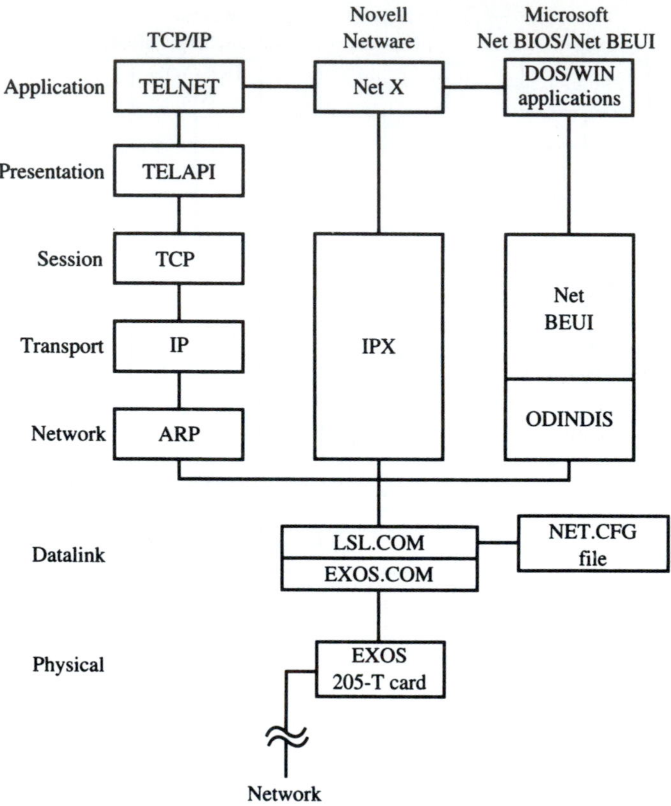

FIGURE 15.28
The ODI model

15.7 NETWORK HARDWARE AND CONFIGURATION

There are many network cards available for the desktop PC today. There are cards for specific network topology and type. There are network cards that are I/O mapped, some that are memory mapped, and some that use DMA access. The most commonly used network cards are I/O mapped and memory mapped. All network cards require some I/O space for initialization and at least one IRQ line. I/O-mapped network cards are the simplest to set up and require no high-memory space unless they contain a network boot ROM. Memory-mapped network cards usually require between 8 and 16 KB of space in high memory. Most can be placed anywhere in high memory between address 0*xA*0000 and 0*xEFFFF* but are most often placed between 0*xD*0000 and 0*xDFFFF*. If a network boot ROM is required, you will need an additional space in high memory for the boot ROM address.

Most of the older ISA bus network cards contain configuration jumpers. By moving the jumpers you can set up the memory address, I/O address, and IRQ line. Newer network cards are "soft programmable," meaning they are set up with a program that comes with the card. In this case there are no hardware jumpers to deal with. In addition to configuring the network card, you must also set up the network software configuration files—something to keep in mind when installing a network card is possible conflicts with other devices, or with memory. The rule is *think* before you install. It is helpful to have a map of the installed hardware on hand so you can set up the network card correctly and avoid conflicts.

Some network cards use I/O address 0*x*300 as their default, as do some CD-ROM interfaces. Address 0*x*300 was originally used as a prototype address by IBM and has become popular for many NIC interfaces, so there is a potential for conflict there. In addition, watch out for memory conflicts with other devices and high-memory managers, like EMM386. If you plan to use high-memory managers, be sure to exclude the memory area for the network card. For example, an EXOS 205-T card set for address 0*xC*8000 will require EMM386 to exclude address 0*xC*800 through 0*xCBFFF*. The correctly formatted statement in CONFIG.SYS is as follows:

```
DEVICE=EMM386.EXE NOEMS X=C800-CBFF
```

Next check the IRQ line used by the network card. Some default to IRQ 3, which is the IRQ line for COM2. Do not share an IRQ line between the network card and any other device. Use IRQ 5 or one of the higher IRQ lines available. In addition, watch out for the multimedia system IRQ requirements, because most sound cards use IRQ 5 or IRQ 7. If you are using Windows 95 with multimedia, use one of the higher IRQ lines. Careful planning will save hours of aggravation and possibly a service call to the site. After you set up the network card you must edit or create the network configuration files. Some network packages will do this for you by asking questions during the software installation. Figure 15.29 is an example of an ODI NET.CFG file for an EXOS 205-T Ethernet card.

The following is an example of the additions to the AUTOEXEC.BAT file:

```
C:\NET\NWCLIENT\LSL.COM
C:\NET\NWCLIENT\EXOS.COM
C:\NET\NWCLIENT\IPXODI.COM
```

```
Link Driver EXOS
      MEM   C8000
      PORT  310
      INT   5
      Frame ETHERNET_802.3
      Frame ETHERNET_802.2
      Frame ETHERNET_II
      Frame ETHERNET_SNAP
      Protocol IPX 0 ETHERNET_802.3
```

FIGURE 15.29
ODI NET.CFG configuration file

As discussed earlier, the NDIS stack uses the file PROTOCOL.INI to obtain the configuration information. Figure 15.30 is an example of the PROTOCOL.INI file for an NE2000 card.

```
Link Support
      Buffers 8 1500
      MemPool 4096

[NETBEUI$]
DriverName=NETBEUI$
Lanabase=1
sessions=10
ncbs=12
Bindings=MS2000$

[nwlink$]
DriverName=nwlink$
Frame_Type=0
cachesize=0
Bindings=MS2000$

[MS2000$]
DriverName=MS2000$
Interrupt=10
IOBase=0x300

[protman$]
priority=ndishlp$
DriverName=protman$

[ndishlp$]
DriverName=ndishlp$
Bindings=MS2000$

[data]
version=v4.00.950
netcards=MS2000$,*PNP80D6
```

FIGURE 15.30
The NDIS configuration file PROTOCOL.INI

15.8 TECH TIPS AND TROUBLESHOOTING—T^3

Troubleshooting a network can be a complex and tedious task. If the network covers a large geographical area, you will need the cooperation of all parties involved. First you must determine which network segment is failing. Then you must trace the failure to the specific component.

When working with a TCP/IP network you may use tools such as Ping and Trace-route. Ping sends a packet to the target address and waits for a response. Trace-route will use the ping method but shows the route taken, number of hops and length of time between each hop. If the ping fails then try the following procedure:

1. Ping the address of your own device. If the ping succeeds, proceed to step 2. If the ping fails, then the problem is in the device itself, and most likely the software setup.
2. Ping any nearby device. If the ping succeeds, proceed to step 3. If the ping fails, check the local network segment for missing, broken, or loose cables. If you are on a twisted-pair segment, check the link lights on both the network card and the local hub. If the problem persists, you may have a bad network card or an incorrectly set up protocol stack. You may also be using the wrong address or network mask. If the problem persists, ping from another station on the same network segment. If the second station fails, the problem is definitely on this network segment. If the ping succeeds, the problem is local to the faulty workstation. Go no further, and repair or replace this workstation.
3. Ping the next known functioning device after your local network segment. Consider a device on the other side of a repeater first. If the ping succeeds, proceed to step 4. If the ping fails, check the repeater. If a repeater fails, your local segment will function, but you will not be able to get outside. Repeat step 2 from any workstation on the other side of the repeater. If step 2 succeeds, then the repeater must be faulty. If step 2 fails, the problem is somewhere on this segment. Repair this segment before going on.
4. Ping the nearest router. If this step succeeds, proceed to step 5. If the ping fails, then ping a known device on each segment between you and the router according to steps 2 and 3. Repair any faulty segments along the way. If the problem persists and all segments pass, you have a faulty router.
5. Ping the other side of the router. If it is known, ping the address of the far side router interface; otherwise ping the nearest known device on that network. If the ping succeeds, proceed to step 6.
6. Continue steps 4 and 5 until you have found the bad segment. If the problem is beyond your control, for example your ISP (Internet service provider), then your problems are over. You must contact the responsible ISP and have them repair the problem. If the problem is on a WAN segment, then you may have to get the phone company or other WAN line provider involved. If the problem is at a distant location, then you must get assistance from someone else. Have your assistant repeat steps 1 through 5 on the segment you believe to be faulty until you have isolated the bad workstation or network segment.

When a problem is found on a thin Ethernet run, first check the resistance at any workstation tee connector. Remove the tee from the Ethernet card, leaving both wires, or wire and terminator, connected. Then check the resistance at the tee for 25 Ω. If the resistance is less than 25 Ω, there is a short in the line. If the resistance is 50 Ω, there is a missing terminator or one side is open. If the resistance is greater than 50 Ω, the tee may be faulty, or there is a break on both sides, or there may be faulty or missing terminators. This type of wiring is easily broken, and all BNC connectors should be checked for loose ends. If the problem is intermittent, check the resistance while wiggling the connectors. If the resistance changes, a poor connection is evident.

EXERCISES

15.1 Discuss the differences between broadband and baseband signaling.

15.2 What is the main advantage optical signal cabling?

15.3 Workstations in the bus topology are connected in ___________.

15.4 Workstations in the ring topology are connected in ___________.

15.5 In the star topology a ___________ is used to connect workstations to the network.

15.6 Hubs may connect to other hubs or to a bus network by the use of a ___________.

15.7 In a token ring network connectivity between workstations is provided by a ___________.

15.8 Ethernet uses ___________ signaling arranged in a BUS topology.

15.9 Ethernet provides error detection by the use of CSMA/CD, but has no ___________.

15.10 Error correction in Ethernet is provided by the ___________ or the ___________.

15.11 A ___________ notifies all workstations on an Ethernet network that a collision has occurred and that the data are invalid.

15.12 The Ethernet ___________ causes transmitting workstations to wait before retransmitting data.

15.13 Ethernet data bits are encoded using a method known as ___________, where a one bit goes from low to high, and a zero bit goes from high to low.

15.14 Explain the Ethernet 5-4-3 rule.

15.15 What are the types of cable utilized by Ethernet?

15.16 It is necessary to ___________ an Ethernet run with the use of a 50-Ω resistor.

15.17 What type of cable should never be used for Ethernet? Why?

15.18 Discuss the advantage of differential transmission over twisted pair in an Ethernet environment.

15.19 Which of the five categories of twisted-pair wire is the preferred wire today?

15.20 What does TCP/IP stand for, and where in the OSI model can it be found?

15.21 A TCP/IP address is a 32-bit binary number represented in a system known as ___________.

15.22 List the five classes in TCP/IP networking.

15.23 Devices on a TCP/IP network are called ___________.

15.24 Discuss use of the subnet mask in a TCP/IP network. Name some functions and create some specific examples of its use.

15.25 One way to name devices on a TCP/IP network, particularly a smaller one, is by use of a ___________.

15.26 A better way to name devices on a TCP/IP network, or on the Internet, is by using a ___________.

15.27 Microsoft has an additional name resolution service known as ___________.

15.28 The INTERNIC is the organization that registers a ___________ and assigns a range of ___________.

15.29 The basic function of a repeater is ___________.

15.30 What functions do repeaters perform?

15.31 Bridges function at the __________ layer of the OSI model.

15.32 Discuss how a bridge processes a packet it receives.

15.33 Routers, being the most complex of their type of network device, operate at the __________ layer of the OSI model.

15.34 Routers work on networks with different __________ and __________.

15.35 Routers provide __________ and __________ of network traffic.

15.36 Routers maintain a database of information called a __________.

15.37 Discuss the information contained in the routing table and its significance to the function of the network.

15.38 Routers can only communicate with __________ and __________.

15.39 Discuss the ways in which routers provide more efficient network traffic and management.

15.40 Discuss the differences between static and dynamic routers.

15.41 Discuss the application of a gateway in a network system.

15.42 Draw the OSI model with all layers in their proper order, and give a brief description of what each does.

15.43 The two network stacks used for DOS- and Windows-based networks are __________ and __________.

15.44 List some areas of conflict when setting up network cards.

15.45 Crossword Puzzle

ACROSS

1. Smaller TCP/IP networks.
3. Network signaling method that uses no modulation.
5. Protocol for the Internet.
6. Two signals on the bus simultaneously.
7. Network stack consisting of seven layers.
10. Network signaling method with the least signal degradation.
13. Network encompassing a smaller geographical area.
14. Network architecture that utilizes CSMA/CD.
18. Device used to pass on signals between network segments.
20. Rigid network cable.
21. Type of tap on coax cable.
23. Network topology where workstations are connected in a circle.
24. Network topology where workstations connect to a central point.

DOWN

2. Term that refers to the physical layout of a network.
3. Network signaling method that utilizes modulation on a common carrier wave.
4. Network topology where workstations are connected in parallel.
8. Organization that assigns and registers internet addresses.
9. Signals picked up by adjacent wires.
11. Network type.
12. Device on a network.
15. Runs of unshielded twisted-pair wire connect to a _____.
16. Device used to connect different networks with similar protocols.
17. Device used to connect completely dissimilar networks.
19. Data transmitted over a network.
22. Network troubleshooting command.

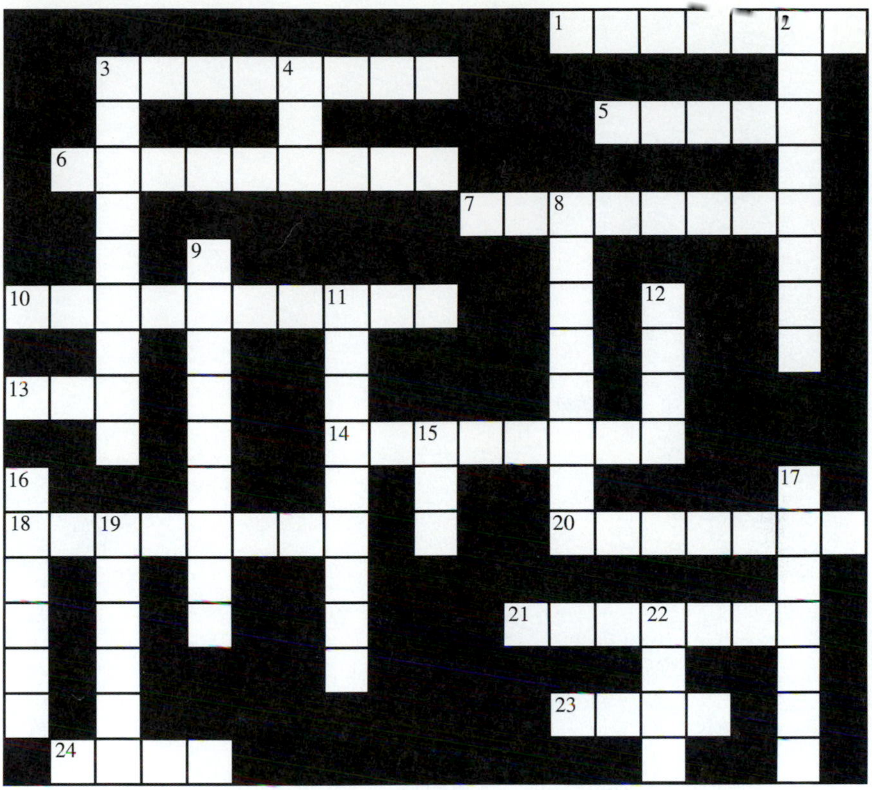
1
2
3
4
5
6
7
8
9
10
11
12
13
14
15
16
17
18
19
20
21
22
23
24

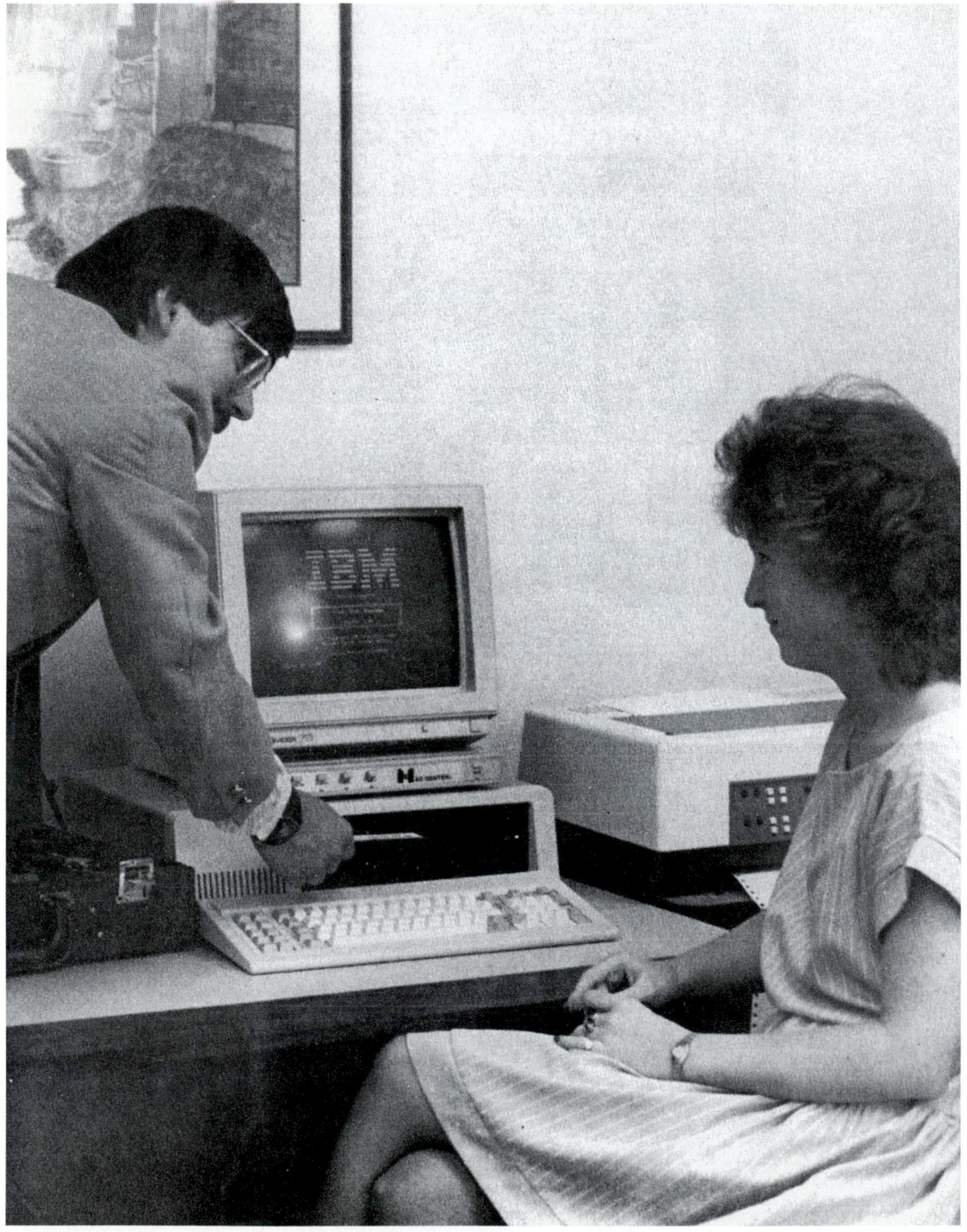
IBM

PART

INSTALLATION AND SERVICE

IV

The computer service industry holds an important and unique place in today's business world because it is critical when a business's computer fails. Quick, competent, and reliable servicing and repair are absolute necessities.

Most businesses cannot allow their computers to go down for any period of time. A decade ago, when microcomputers were first introduced into companies, they were used in only a limited fashion; therefore, it wasn't critical if the microcomputer was down for a few days. Today, with the growth of powerful microcomputers in the business world, even 1 day of down time is unacceptable. Imagine what would happen to a business if employee paychecks were not ready on payday, if invoices and billings were sent out late, or if proposals for new business opportunities could not be processed on time.

The traditional computer service process is a straightforward one. A well-trained computer service technician is sent to diagnose a problem on a computer system and to do whatever is necessary to repair the system. In general, this is the way most small business computers are repaired today.

The traditional service call has become increasingly expensive, due in part to the substantial and growing salaries of service technicians. While the cost to repair a microcomputer system is rising, the price of these systems keeps falling. These economic pressures have led to the use of a number of new servicing techniques. These techniques take advantage of improved technologies and are designed to lower service costs. The chapters in this section are intended to train the student to become the computer service technician that industry needs.

CHAPTER 16

SETUP, SERVICING, AND CUSTOMER RELATIONS

KEY TERMS

AUTOEXEC.BAT
BIOS
COMMAND.COM
CONFIG.SYS
DIMM
DMA
FAT (File Allocation Table)
FAT 32
FDISK
IDE
IRQ
Jumper
LBA Mode
MSCDEX
PNP (Plug and Play)
POST
SECC
SIMM
Slot 1
Socket 7
Standoff
UDMA

16.0 INTRODUCTION

In this chapter we discuss the initial setup and servicing of the Windows/Intel personal computer, also known as the Wintel computer. In the past, we have referred to this as the IBM PC. Today there are many variations to the design and layout that was originally developed by IBM, Intel, and Microsoft.

Since in many cases this work will be performed in the field, a good attitude and professional appearance are very important. The customer's opinion of how good your service company is will be based on the customer's impression of you. For this reason, a portion of this chapter is dedicated to customer relations.

All components used in PCs are made up of MOS field-effect transistors. These transistors are very sensitive to static electricity. Even using compressed air to clean dust can generate enough static electricity to destroy some MOS transistors. Good practice dictates the use of a conductive wrist strap while working on a computer as shown in Figure 16.1.

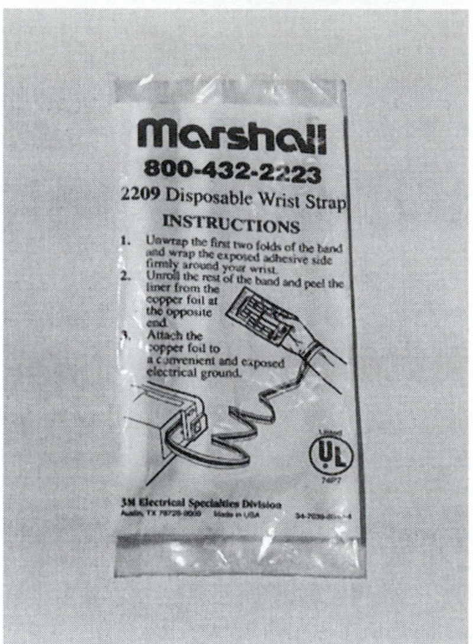

FIGURE 16.1
Anti-static wrist strap

16.1 CONFIGURATION

The original PC design made use of small switches known as DIP (dual in-line package) switches for the initial system configuration. Today you will not see many systems that use DIP switches. The introduction of the IBM PC/AT eliminated the need for many of the system configuration switches. In addition, the AT stores the system configuration on a nonvolatile RAM. The configuration RAM is part of a Motorola MC146818 real-time clock and CMOS RAM integrated circuit chip. A layout for the CMOS real-time clock plus RAM chip is shown in Figure 16.2.

The clock and RAM are kept alive when the system is powered off by a small battery mounted on the motherboard. Some manufactures have combined the battery and the CMOS IC into one package as shown in Figure 16.3.

The IBM PC, PC/XT, and PC/AT (80286 processor) had only the processor and memory on the motherboard. All peripheral controllers were on adapter boards that plugged into the motherboard. The adapter boards were configured separately, using jumpers or switches on each board. Some 386, most 486, and all Pentium motherboards contain serial, parallel, floppy, and IDE interfaces directly on the motherboard. The I/O address, IRQ, and DMA channel are configured using the BIOS setup program and stored in the CMOS RAM. The **BIOS** setup program is accessed during the power-on sequence by a key sequence during the POST sequence. Some common setup key sequences are F10, Ctrl-Alt-Esc, Ctrl-Alt-S, F1 or Del. Most will instruct you on how to start the configuration program during the POST sequence. All BIOS setup programs will automatically activate if a change is detected or an error exists. Almost all BIOS programs today contain

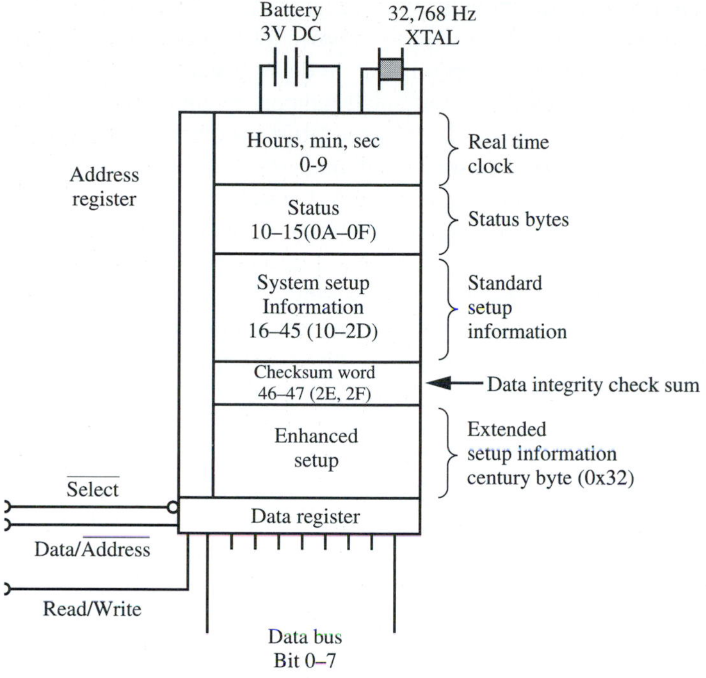

FIGURE 16.2
CMOS real-time clock plus RAM map

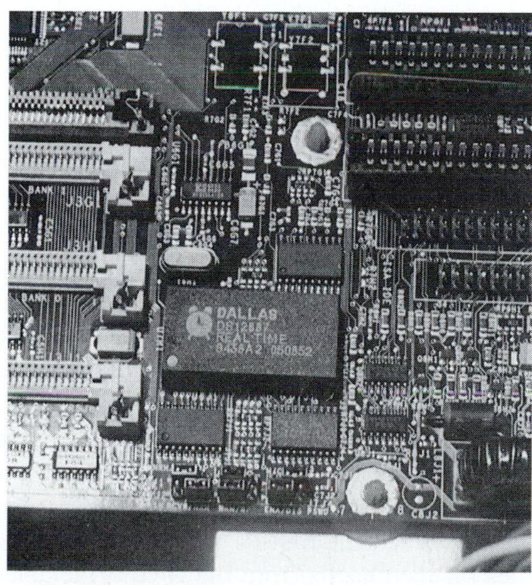

FIGURE 16.3
Real-time clock plus RAM and battery

enough intelligence to automatically configure most of the settings, but allow you to make changes if necessary or if desired to improve system operation and performance.

Some initial hardware configuration defaults cannot be assumed by the BIOS. This configuration is set by the use of small jumpers mounted on the motherboard. On some systems, a jumper is used to instruct the BIOS whether a color or monochrome monitor is present. On 386, 486, and some Pentium systems, jumpers are used to instruct the BIOS of the presence and amount of cache RAM installed.

The 486 and Pentium processors are available in many clock speeds. Most motherboards for these processors use jumpers to set the clock rate for the processor. Up to and including the 486, all processors use a single 5-volt power source. New transistor and manufacturing technologies were applied to the Pentium class processors. This allowed for lower voltages, thereby resulting in lower power dissipation and less heat. Also, since the voltage swings are lower, faster switching speeds and higher clock rates can be attained. Because of this, Pentium motherboards contain a internal voltage regulators and jumpers to set the processor voltage. Most Pentium motherboards comply with a manufacturing standard known as **Socket-7** (or socket 370 for some Pentium III and Celeron processors). Pentium motherboards generally use a zero insertion force (ZIF) microprocessor socket and can accommodate many different Pentium-type microprocessors. The higher-speed Pentium processors use a dual voltage supply, for example 3.3 V for the I/O and 2.8 V for the microprocessor core.

The introduction of the Pentium II simplified the process of adding or changing the processor. The processor voltage is controlled by the processor connector pins; the clock speed is detected by the BIOS. The microprocessor and cache memory are contained in a module called an **SECC** package, which stands for single-edge contact cartridge. The SECC module plugs into a special edge connector on a **Slot 1** motherboard. This is similar to an add-on board as shown in Figure 16.4.

16.2 SYSTEM CASE AND POWER SUPPLY

The first step in building a new system or configuring a prebuilt system is to prepare the case. The case consists of a main chassis and one or more covers. There are two major types of computer cases: the AT case and the ATX case. The AT case will accommodate all motherboards up to and including the socket-7 Pentium. The Pentium Pro (socket-8), socket-370, and all slot-1 processors must have the ATX case. The ATX case uses an entirely different power supply, which was discussed in detail in Chapter 12. The type of case is very important when considering an upgrade and will determine the type of motherboard you must use. For example, if you have an old 486 and you want to upgrade to a Pentium II, you will have to replace the case and power supply as well.

Today there are many sizes and styles of cases, some of which are desktop, slimline desktop, small tower, midtower and full tower. The full-tower case is generally placed on the floor next to the computer desk. The first PC design used a desktop case. On this style of case, the cover slides off the chassis from the front. The cover is fastened with five screws as shown in Figure 16.5. When working on this type of case, be careful to remove only the cover screws and not the power supply screws.

Next, if it is not already installed, is the power supply. There are many variations of power supply mountings today, but most will attach to the rear of the chassis as shown

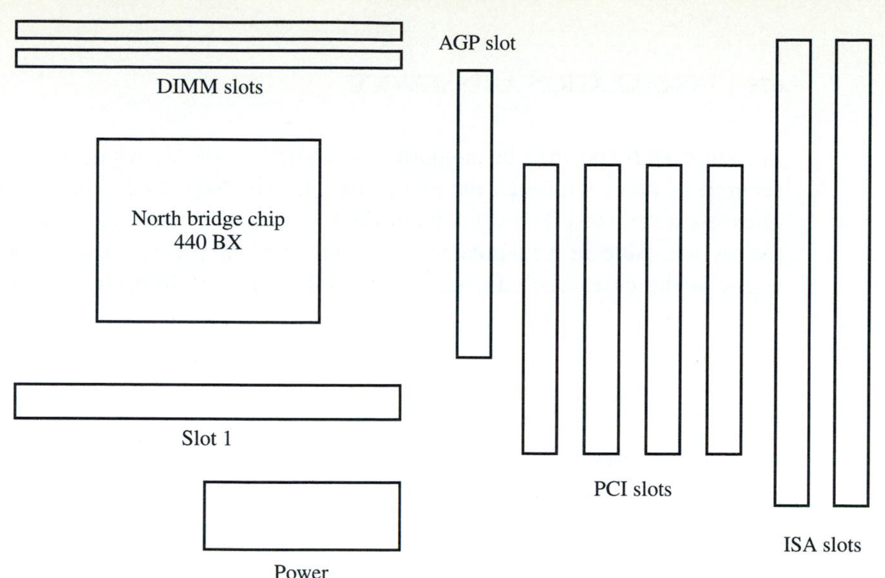

FIGURE 16.4
Slot 1 motherboard

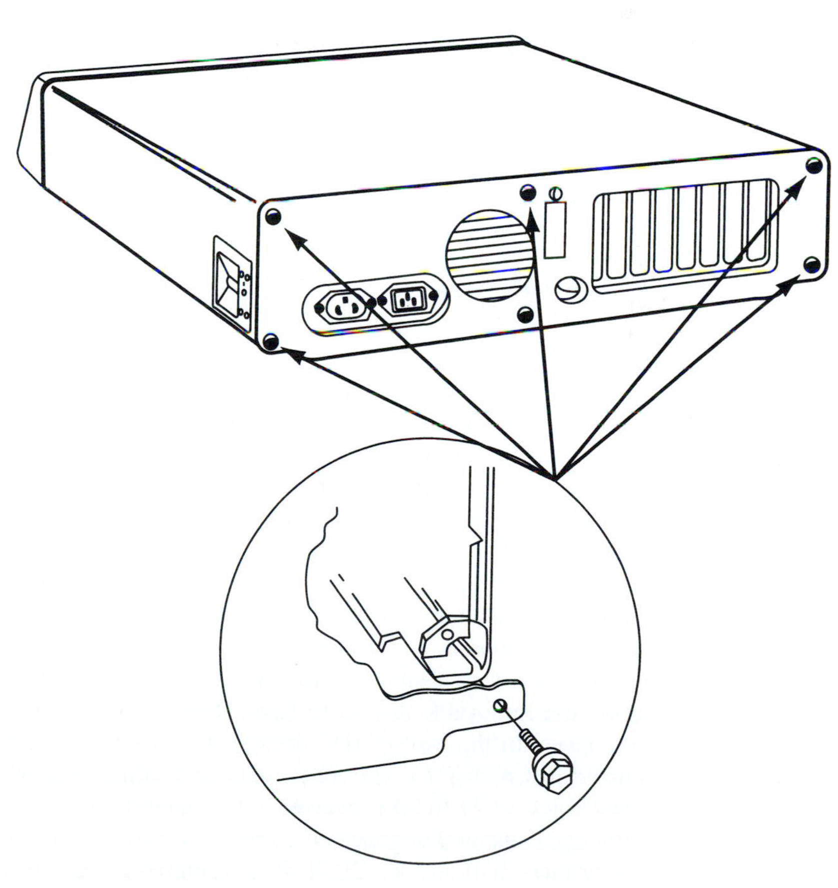

FIGURE 16.5
Cover removal

in Figure 16.6 and may be supported by a shelf inside the chassis. The power supply is secured by screws through the rear of the chassis. Newer desktop and tower power supplies use a remote power switch attached to the front of the chassis. Be careful to route the switch cable in a manner that will not chaff or pinch the wire by any sharp metal edges in the case. Such an occurrence will lead to a short circuit and a shock hazard.

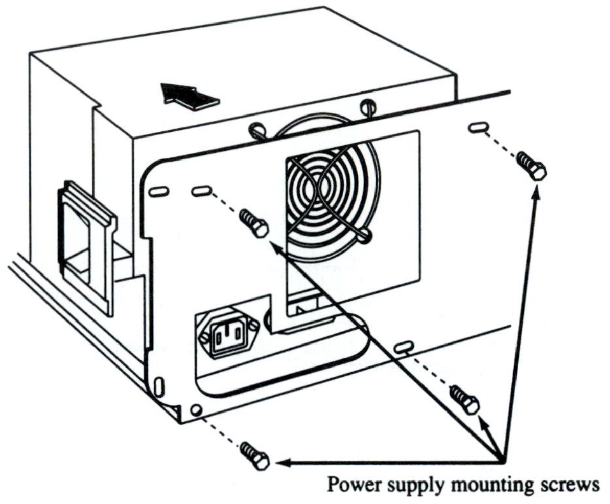

FIGURE 16.6
Power supply mounting

Next, consider installing any additional cooling fans into the case. When installing additional fans make sure to set the airflow in the same direction as the airflow of the power supply fan. Most power supply fans blow air out of the case to the rear, forcing cooling air to be pulled in from the front and across the internal components. It is important to install additional cooling fans because a fan fault can result in a destroyed system. Newer Pentium systems generate a great deal of heat. CPU cooling fans can easily become clogged, and you may have no indication of the failure until it is too late.

Next you may wish to install the disk drives into the case or wait until after the cables, motherboard, and adapter boards are installed. This will depend upon how much working room is inside the case and the ease of connecting the cables. Many computer cases use removable disk drive bays. Determine how the disk drives will connect to the interfaces. In the case of IDE devices, you need to set the master/slave jumpers as discussed in Chapter 13. The most common configuration is a single hard drive set as the master and a CD-ROM connected to the same cable set as a slave. This allows the use of only one cable and interface. The preferred method is to set the hard drive as a master on the primary IDE and the CD-ROM as a master on the secondary IDE. This will improve system performance and speed. In any case, don't forget to connect the drive power cables to the disk drives.

16.3 MOTHERBOARD SETUP

The motherboard in a new system should require no changes, because it is usually configured at the factory. If the system is being built or the processor replaced, then the motherboard settings must be carefully examined. An incorrect voltage setting can destroy the microprocessor. Since there are many different manufacturers and types of motherboards, we will give a general overview of the configuration. An important note here is to carefully review all of the manufacturer's documentation on the motherboard as well as the documentation on all other hardware in the system.

When installing a new motherboard be sure to install the supporting **standoffs** as shown in Figure 16.7. It is important that the board be properly supported and not allowed

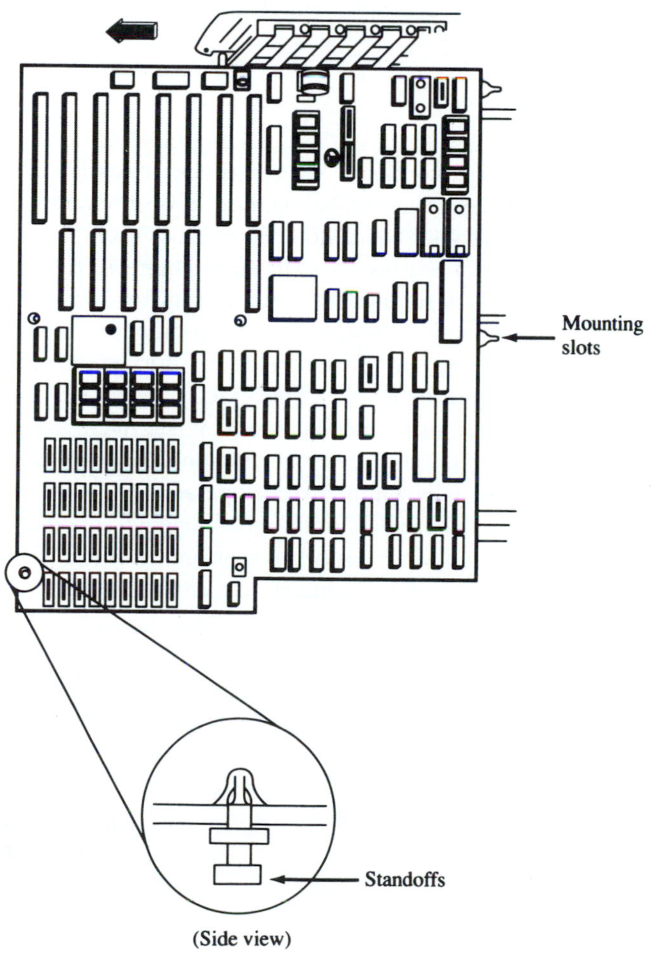

FIGURE 16.7
Plastic standoff

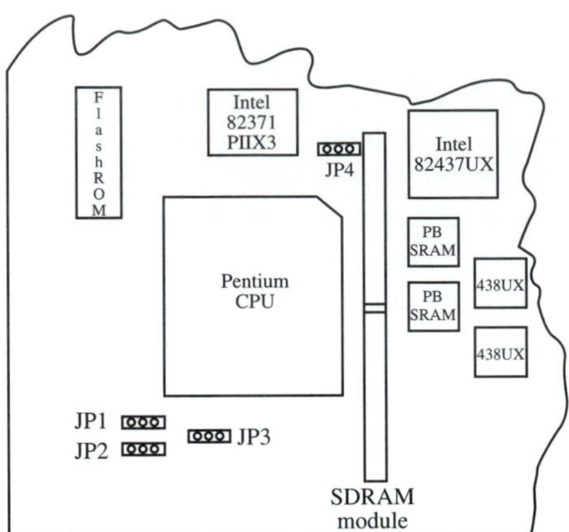

FIGURE 16.8
Voltage and clock jumpers

to flex. The motherboard is a multiple-layer board, and excessive flexing can cause the interconnections between layers to break, resulting in open circuits. Be very careful to provide support under the adapter board slots. Be certain to set the proper operating voltage and clock speed on the motherboard as shown in Figure 16.8. Included in Table 16.1 is a chart of voltages and clock settings for a Pentium **socket-7** motherboard. The processor identification will contain the voltage and speed specification as shown in Figure 16.9.

Some people set the jumpers for what is referred to as overclocking to make the system run faster. This is done by changing the clock jumpers to a faster than suggested processor speed. This is not a recommend practice. When microprocessors are manufactured, they are tested for speed and power dissipation. The chips are then stamped and sold at the fastest speed that falls within the rated parameters. Most microprocessors will run a step or two faster than rated under normal environmental conditions, but overclocking may cause the processor to draw too much power and overheat. Furthermore, running faster than specified may result in intermittent data errors and unreliable operation.

TABLE 16.1
Clock and voltage jumper settings

Processor	I/O Voltage	Core Voltage	Speed (MHz)	Comments
Intel P54C	3.3	Same	75–200	Older, non-MMX chip, single voltage
Intel P54C VRE	3.5	Same	75–200	Older, non-MMX chip, single voltage
Intel P55C MMX	3.3	2.8	150–233	Some require 2.9 core
AMD K5	Various	Same	75–166	See lid, single voltage
AMD K6	3.3	2.9/3.2	166–233	233 MHz requires 3.2 volts core
Cyrix 6x86	3.15–3.70	Same	90–200	See lid, single voltage
Cyrix 6x86L	3.3	2.8	200	Similar to P55C

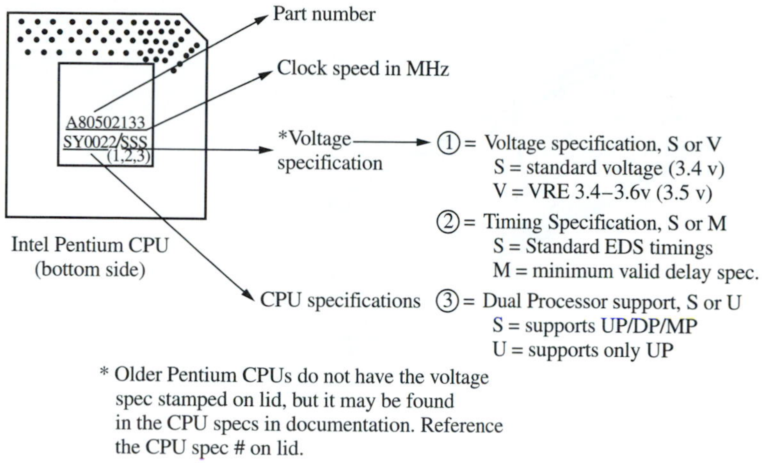

Part number
Clock speed in MHz
A80502133
SY0022/SSS
(1,2,3)
*Voltage specification
①= Voltage specification, S or V
S = standard voltage (3.4 v)
V = VRE 3.4–3.6v (3.5 v)
②= Timing Specification, S or M
S = Standard EDS timings
M = minimum valid delay spec.
Intel Pentium CPU (bottom side)
CPU specifications ③= Dual Processor support, S or U
S = supports UP/DP/MP
U = supports only UP

* Older Pentium CPUs do not have the voltage spec stamped on lid, but it may be found in the CPU specs in documentation. Reference the CPU spec # on lid.

FIGURE 16.9
Processor ID

16.4 MEMORY SETUP

When working with memory you must first determine the type of memory that is required by the motherboard. The first PCs used separate memory chips known as DIP, or dual in-line package, memory. These were only used on 8088 and 80286 computers. The 80386 and later computers used memory modules that contain the memory chips on a small circuit board. These modules are available in 30-pin SIMMs, 72-pin SIMMs, and 168-pin DIMMs as discussed in Chapter 12. Once you are sure that you have the correct type of memory for the motherboard, you can install the memory modules into the sockets on the motherboard.

If you are using DIP ICs, be certain that pin 1 of the IC is inserted into pin 1 of the socket. Installing the IC backward will destroy the IC. A small dot or notch indicates the location of pin 1 as shown in Figure 16.10. Be sure not to bend or break any pins. Also, make sure one of the pins does not miss the socket and bend under the IC.

If you are installing **SIMM** modules, be certain that pin 1 of the module is inserted into pin 1 of the socket. The SIMM is keyed by a small notch and therefore will not fit easily into the socket backward. If you force it, you may break the key and install the SIMM backward. This will destroy the SIMM if the power is turned on. Most SIMM sockets are the snap-in-place type. On this socket you simply align the pins with the SIMM at a slight angle and snap it into place as shown in Figure 16.10. To remove the SIMM, gently pull back on the locking arms and it will pop out. Be careful not to apply too much force or you will break the locks.

One thing to remember is the number of modules required for a complete memory word. Refer to the discussion on memory in Chapter 12 regarding the number of bits per SIMM module. Memory on the motherboard is installed in banks. One bank of memory depends upon the processor used. For example, a Pentium processor uses a 64-bit memory word. If you are using 72-pin SIMMs, you will need two SIMMs to complete one

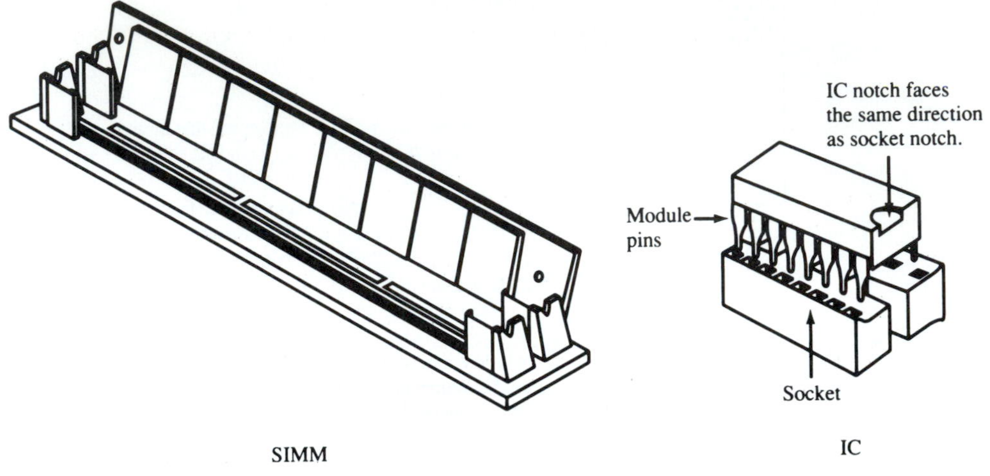
IC notch faces
the same direction
as socket notch.
Module pins
Socket
SIMM
IC

FIGURE 16.10
Inserting memory

bank, since one 72-pin SIMM contains only 32 bits (wide) of memory. Carefully note the location of the SIMM sockets, and install the memory in banks starting at bank zero.

If you are installing **DIMM** modules, follow the same precautions as those for installing SIMM modules. The difference here is that DIMM modules slide into dual-sided sockets, similar to the way a PCI bus adapter board is installed. The socket contains a lock/eject arm on both ends. The module is pin keyed, preventing incorrect installation. Avoid using excessive force when inserting the DIMM, because it should slide in easily. While inserting, snap the lock arms into place; the lock arm will aid the insertion process. To remove the DIMM, do not pull on the module. Gently pull the lock/eject arms outward and down from the DIMM module. The module should pop out of the socket. Never touch the contacts on any module, because finger oils and sweat can etch the contact plating.

Next, add the cache RAM chips if required. Depending upon the age and type of motherboard, the cache RAM will vary. Some motherboards use cache memory chips, which are soldered onto the motherboard and are not user replaceable. Older motherboards use DIP static memory chips. Newer motherboards use cache memory modules similar to a DIMM module. Be very careful when adding the DIP chips, and take care not to bend the pins under the socket. Once you complete the cache installation, you are ready to install the motherboard into the case.

When installing the motherboard into the case, make sure the supporting **standoffs** are properly placed. Some use threaded standoffs that screw directly into the case and the motherboard simply lies on top supported by small screws. Others use snap-in plastic standoffs that slide into place as shown in Figure 16.7. These will have one or two standoffs screwed in place to secure the board from sliding. Be very careful to observe the area around any metal standoff points. If any wire runs are close to a metal standoff or will contact the standoff or the screw, be certain to use a nonconductive washer between the standoff, the screw head, and the motherboard. Do not overtighten the screws. Use just enough force to snug the screws, because overtightening will distort and damage the motherboard. When using the slide-type standoff in place of the screw-type standoff, make sure that all the standoffs are correctly inserted into the slide slots. If one misses the slot in the case, it may place undue stress on the motherboard or cause it to twist, and will not provide any upward support.

16.5 CABLES AND CONNECTIONS

First, you must connect the power connectors to the motherboard. Prior to the ATX motherboard, the power connectors consisted of two separate but similar connectors (labeled P8 and P9). Do not confuse the motherboard power connectors with the drive power connectors. The drive power connectors are four-pin round connectors. The AT (socket-7) motherboard power connectors are six-pin flat connectors. If the two connectors are reversed, it will destroy the motherboard. As a rule, align the black (ground) wires on each connector so that they are adjacent, side by side. Table 16.2 is an illustration of the wire color used for the power supply. In the case of an ATX system, this is not a problem. The power is supplied to the motherboard using one keyed connector. This connector cannot be reversed, and the wire color is illustrated in Table 16.3.

TABLE 16.2
AT system power connectors

Connector	Pin	Signal	Wire Color
P8	1	Power good	Orange
	2	Not used	
	3	+12 V	Yellow
	4	−12 V	Blue
	5	Ground	Black
	6	Ground	Black
P9	1	Ground	Black
	2	Ground	Black
	3	−5 V	White
	4	+5 V	Red
	5	+5 V	Red
	6	+5 V	Red

At this point, you may wish to connect the power supply to the disk drives. The power connector is a four-pin plastic Molex-type connector and is keyed by a 45° chamfer on the topside. It is possible to force the connector in upside down if you push hard enough. Placing the connector upside down reverses the +5- and +12-volt power and will destroy the disk drive. The color code is shown in Table 16.4.

Next, you must connect any miscellaneous connectors to the motherboard as shown in Figure 16.11. These are usually for the speaker, external lights, and switches. Depending on the case you are using, you may have lights for power, hard disk activity, and turbo mode. Some cases provide reset, keyboard lock, and turbo mode switches. The turbo switch

TABLE 16.3
ATX system power connector

Pin	Signal	Wire Color	Pin	Signal	Wire Color
1	+3.3 V	Orange	11	+3.3 V +3.3 VSence	Orange Brown
2	+3.3 V	Orange	12	−12 V	Blue
3	COM	Black	13	COM	Black
4	+5 V	Red	14	PS ON	Green
5	COM	Black	15	COM	Black
6	+5 V	Red	16	COM	Black
7	COM	Black	17	COM	Black
8	PWR OK	Gray	18	−5 V	White
9	+5 VSB	Purple	19	+5 V	Red
10	+12 V	Yellow	20	+5 V	Red

TABLE 16.4
Disk drive power connector

Pin	Signal	Wire Color
1	+12 V	Yellow
2	Ground	Black
3	Ground	Black
4	+5 V	Red

is obsolete now and was provided to make the system run slower to accommodate old DOS games. The polarity of the connections for both the indicator lights and the switches is important. Incorrect polarity will *not* destroy anything but the light, or the switch may not operate properly.

Next, if the motherboard contains the disk interfaces, install the disk drive cables to the motherboard. Be careful to note the orientation of pin 1 on the cable. This is usually a red or black stripe on the ribbon cable as shown in Figure 16.12. The stripe, or pin 1, connects to pin 1 of the interface on the motherboard. Pin 1 is marked by a dot, arrow, or the numeral 1. The floppy disk cable is a 34-pin cable and has three connectors on it. Be sure to connect the long end to the motherboard as shown in Figure 16.12. The end past the twist in the cable is for floppy drive A.

Next, connect the IDE cables to the motherboard. The IDE cable is a 40-pin cable with two or three connectors on it. This is similar to the floppy drive cable but has no twist between the connectors. There are usually two IDE interfaces on the motherboard, labeled primary and secondary. The primary interface connects to the boot drive or disk drive C. Again, be careful to observe the location of pin 1 as shown in Figure 16.12.

Next, route the drive cables to the desired drive bay locations. The cables are kept neat by making 90°-folds along the desired path. Be careful to place the cables where they will not block cooling vents. Also, watch for the cables being cut by sharp metal edges on the chassis or crushed by inserting boards into the motherboard.

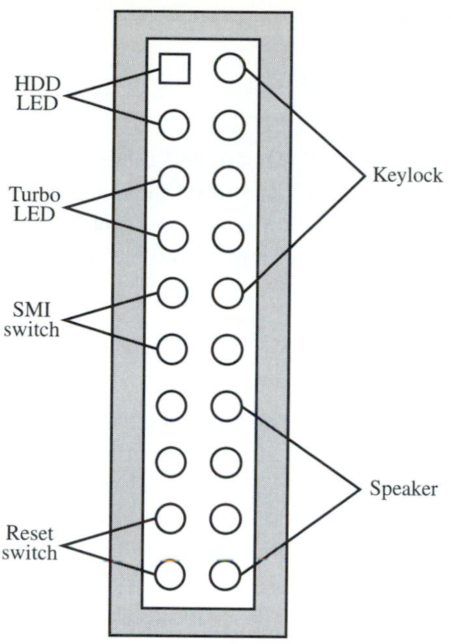

FIGURE 16.11
Motherboard lights and switches

FIGURE 16.12
Disk drive cables

16.6 ADAPTER BOARDS

Once you have completed installing the motherboard and cables, you are ready to add peripheral boards to the system. This is a good time to pause and think about what you are adding to the system. All adapter boards use some system resources. These resources are the **I/O addresses, IRQ lines,** and **DMA channels.** If you are installing older legacy ISA boards, you should set any I/O, IRQ, and DMA jumpers now. Review the I/O, IRQ, and DMA usage in the system to avoid conflicts before going on. When reviewing the available resources, don't forget to include any built-in devices on the motherboard. Table 16.5 shows a list of reserved and typical resources.

The adapter boards plug directly into the motherboard, which are secured to the chassis as shown in Figure 16.13. Before installing the board, be sure to remove the slot filler plate from the chassis and save the screw so you can reuse it to secure the board.

The first board to add is the video board. Depending upon your system, you may have some choices on the video board to use. If the system supports AGP, you should always try to use an AGP video board to take full advantage of the increased performance. However if you don't have an AGP video board, a PCI video board will also perform well. Stay away from an ISA video board unless you have no choice. Furthermore, if you're using a local bus system, then choose a local bus board over an ISA board. In all cases, when selecting a video board, choose a board that supports the linear frame buffer over the paged mode.

TABLE 16.5
Resource Map

Device	I/O Address	IRQ	DMA	Notes
Motherboard	00–0*xFF*	0, 1, 2, 8, 13	4	
Primary IDE controller	0*x*1*F*0–0*x*1*F*7	14		No DMA unless UDMA controller
Secondary IDE controller	0*x*170–0*x*17*f*	15		No DMA unless UDMA controller
Floppy controller	0*x*3*F*2	6	2	
PS/2 mouse	Uses keyboard control	12		
COM1	0*x*3*F*8–0*x*3*FF*	4		
COM2	0*x*2*F*8–0*x*2*FF*	3		
COM3	0*x*3*E*8–0*x*3*EF*	4		Can use IRQ 9(2)
COM4	0*x*2*E*8–0*x*2*EF*	3		Can use IRQ 9(2)
LPT1	0*x*378–0*x*37*B*	7		
Sound board	0*x*220–0*x*22*F* 0*x*330–0*x*33*F*	5	1, 5	DMA 5 for 16-bit audio
Game port	0*x*200–0*x*207			Typically on sound board
Video board	0*x*3B0–0*x*3*BB* 0*x*3*C*0–0*x*3*DF*	2, 5, 9, 10, 11		Typically IRQ 11
LAN board	0*x*300–0*x*31*F* Variable	5, 9, 10, 11		Typically IRQ 10

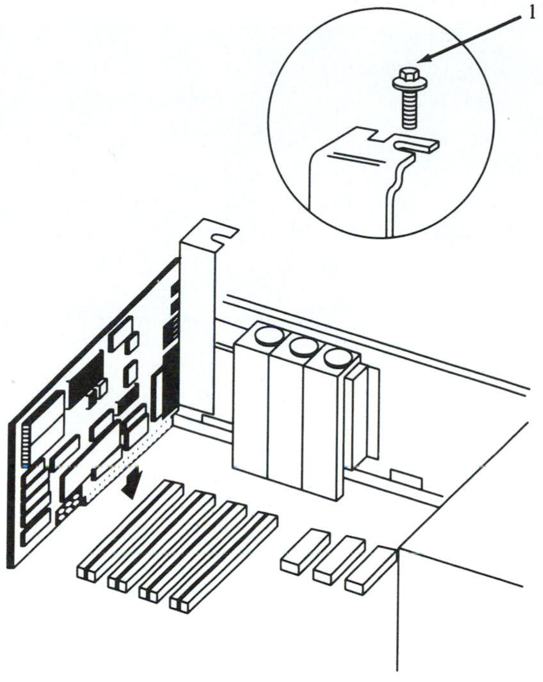

FIGURE 16.13
Adapter board installation

Little or no configuration is required before the video board is installed. Some exceptions are if memory or some other video option is added. Some examples are TV tuners, capture options, or 3D accelerators. The only jumper settings on older VGA video boards are the IRQ line being enabled or not. The reason for adding the video board first is that the system can be powered on and tested by POST. This is the minimum configuration when working on a totally dead or problematic system.

Next, install the audio board. Here there are many choices for audio boards and almost all support the Sound Blaster standard. Older audio boards are ISA boards and use small jumpers to set the configuration. Some will use a nonvolatile RAM, or NVR, to save settings. These audio boards are set up using a supplied set up program and can be run from a DOS boot disk. The latest audio boards are shipped as PCI boards, contain no jumpers and are plug and play. Some motherboards contain the audio hardware built in as a PCI board, which saves the slot for another board. Most audio boards also support a CD-ROM interface as shown in Figure 16.14.

The most popular CD-ROM interfaces are Mitsumi, Panasonic, Sony, and IDE. The IDE interface CD-ROM is the most common today, with the other three becoming obsolete. If the sound board has an IDE feature, it is best to disable it to avoid conflicts unless you absolutely need it, because the system configuration is easier without it and most motherboards already contain two IDE interfaces. The most common settings for the audio board are

FIGURE 16.14
Typical audio board

- I/O address–220 and 330
- IRQ 5 (most common for SB compatibility) or IRQ 2 or IRQ 7
- DMA–1 for 8-bit audio
- DMA–5 for 16-bit stereo audio

Try to avoid the use of IRQ 7, because Windows 95/98 and up use IRQ 7 for the printer port (LPT1). DOS and Windows 3.x do not care about the IRQ line for the printer.

Next, configure and install the internal modem. Most modems are ISA boards and install as an additional serial or COM port. Here is where many problems are encountered because of conflicts with internal COM ports or other hardware. Most AT and later motherboards contain at least one, but usually two internal COM ports. Multiple COM ports are allowed to share an interrupt line. COM1 and COM3 share IRQ 4. COM2 and COM4 share IRQ 3. It is important to avoid the use of a serial mouse on a COM port using the same IRQ as the modem. Most ISA modems will allow the selection of IRQ 2(9), IRQ 3, IRQ 4 and IRQ 5, but keep in mind the sound board or other hardware may be using IRQ 2 or IRQ 5. Most users will never need both of the serial interfaces or COM ports. If this is the case, you can disable the internal COM ports using the BIOS setup program. When disabling the internal COM ports, be sure the BIOS is plug and play, or Windows 95/98 won't reenable the ports. That will cause many problems later. A popular choice when using a serial mouse on a dual-port motherboard is to set the modem for COM3, IRQ 4, and plug the mouse into COM2. Another solution is to use an external modem connected to the internal COM ports.

Next, add the LAN board, also known as the NIC, for network interface card. Older LAN boards are configured using jumpers. Depending on the LAN board and network topology, you may have to set the medium type using a switch or jumpers. The most common example of this is an Ethernet board with an internal transceiver and an AUI connector. A switch is used to enable or disable the transceiver. Newer LAN boards are software programmable and are shipped with a special setup program. One problem with this is that if a conflict exists between the LAN board and something else in the system, the setup program will not be able to detect the LAN board and will report the LAN board does not exist. In this case you must remove the conflicting device temporarily and run the setup program to reconfigure the LAN board. If your system supports plug and play, this may be taken care of automatically.

Lastly add any other interface boards you may need such as SCSI, scanner, and video capture boards. Here there are too many to mention, but in some cases you may be forced to go back and reconfigure previous boards to allow for additional boards if they are inflexible on resource allocation.

16.7 BIOS SETUP

Once you have completed the assembly and cabling you are ready to begin the power-on configuration procedure. The first AT computers required the use of a bootable floppy diskette containing the setup programs. As larger-size BIOS ROMs became less expensive, it became possible to place the setup program within the BIOS ROM chip, thus eliminating the need for the special floppy diskette. Modern **BIOS** setup programs will attempt to automatically configure the system and then enter the setup dialog. The first option will be the standard setup screen, which has not changed much since the first AT computer. Additional options are available and are specific to the motherboard and chipset used. Since there are many different BIOS setup programs, we will discuss the general setup procedures that are common to them all.

The first procedure to be followed is the standard BIOS setup. Here you define the following:

- Date/time—current date and time; sets the hardware clock chip
- Primary display—EGA/VGA, CGA-80, CGA-40, or monochrome
- Keyboard installed—prevents an error if no keyboard is installed
- Floppy—type of floppy disk drives used
- Hard disk type and size (CHS/LARGE/LBA)

Most, if not all, of the standard setup will be automatically detected or set to a default but allow you to make changes. All of the settings given here have not changed much since the first AT computer except for the hard disk type.

Originally, the hard disk type was defined by a number in a BIOS drive table. For example, type 1 is a 10 MB drive using 306 cylinders, 4 heads, and 16 sectors per track. Today, a 10 MB hard disk drive might not even hold a single game. The number of hard drives grew too quickly for the drive type table to remain practical. The drive table is obsolete and is there for compatibility only. Almost all BIOS programs have a type 47 for user-defined parameters. Here you type in the actual drive geometry in cylinders, heads, and sectors per track, or C/H/S. Today, most BIOS programs will automatically detect the drive geometry and configure this section of the setup for you.

The introduction of hard disk drives larger than 528 MB has given rise to a totally new type of drive geometry known as **LBA mode.** This is because of a BIOS/DOS limitation on the geometry parameters. The maximum drive parameters were 1024 cylinders, 16 heads, and 63 (512 byte) sectors per track. If you do the math, $1024 \cdot 16 \cdot 63 \cdot 512 =$ 528,482,304 bytes (referred to as approximately 504 MB) total. If your BIOS only supports the C/H/S drive type then you can only access approximately 504 MB of hard disk space no matter how large the drive is. As a workaround to this problem, the drive manufacturers distributed a special program to override the BIOS limitation. The BIOS extender program modifies the boot sector and loads itself into the operating system. One thing to note here is that it works similar to a virus. Some BIOS extender programs will cause Windows 95/98 to warn that a possible virus is present. This limitation was solved by adding a new access mode known as logical block addressing, or **LBA mode.** In LBA mode, the drive is accessed in blocks starting at zero regardless of the actual geometry. Using this mode there is practically no BIOS limitation for the size of the hard drive. Do not confuse this with the DOS partition size limitation. The DOS partition limit is discussed later in the hard disk setup section of this chapter. Furthermore, this screen reports the amount of memory detected by the POST. If the amount of memory is not correct, you must find the problem before going on. A typical standard CMOS setup screen is shown in Figure 16.15.

The next setup option is usually the BIOS features. Some common BIOS features are as follows:

- *Virus warning.* This feature displays an alert whenever a write is attempted to the boot sector. Most viruses infect and propagate by adding code into the boot program. This way every time you boot the computer the virus program runs and becomes part of the operating system. Next, the virus replicates itself on every file that is opened, includ-

```
             ROM PCI/ISA BIOS (2A59CG0L)
                STANDARD CMOS SETUP
                AWARD SOFTWARE, INC.

Date (mm:dd:yy) : Sat, Dec 11 1999
Time (hh:mm:ss) : 22 : 35 : 14

HARD DISKS          TYPE  SIZE  CYLS  HEAD  PRECOMP  LANDZ  SECTOR  MODE

Primary Master    : Auto   0     0     0      0        0      0     AUTO
Primary Slave     : None   0     0     0      0        0      0     ——
Secondary Master  : None   0     0     0      0        0      0     ——
Secondary Slave   : None   0     0     0      0        0      0     ——

Drive A : 1.44M, 3.5 in.
Drive B : None                             Base Memory:   640K
Floppy 3 Mode Support : Disabled       Extended Memory: 15360K
                                          Other Memory:   384K
Video : EGA/VGA                        ----------------------
Halt On : All Errors                   Total Memory:   16384K

ESC : Quit        ↑ ↓ →      : Select Item      PU/PD/+/- : Modify
 F1 : Help        (Shift) F2 : Change Color
```

FIGURE 16.15
Standard CMOS setup screen

ing files saved on floppies and across networks. Imagine how fast a virus will propagate in a busy networked office.

- *Cache.* Cache internal/external enables or disables the cache memory on the motherboard and processor. In all cases, you want this enabled. If you are working on a problematic system with intermittent crashes, you will turn off the cache as an aid to diagnosing the problem.
- *Quick boot.* Quick boot bypasses the memory test part of POST to get the system started faster. Generally it is not a good idea to set this feature because the little memory testing that is normally done will now be disabled. Faulty memory is one of the largest problems of system crashes, so memory should be tested every time the system is powered on.
- *Boot sequence.* The boot sequence option specifies the device search order for the initial boot. This feature instructs the BIOS as to the order in which devices are searched for the operating system loader program. Most motherboards allow for floppy drive A, hard drive C, the CD-ROM, and from the network.
- *Swap floppy drive.* Swap floppy drive allows the user to change the letter designation of drive A and drive B. This is useful when two different-size floppy drives are present and you need to boot from different media. An example of this is if drive A is a 3.5-in. drive and drive B is a 5.25-in. drive but you need to boot from the older 5.25-in. diskette. If this feature were not available, you would have to swap the floppy drive cable connectors.
- *Security options.* Security options are used to lock the system until the correct keyboard sequence is typed. This replaces the hardware key lock on older systems by storing passwords in the CMOS RAM. Clues to the fact that the system is locked are flashing lights on the keyboard. We have witnessed a technician replacing a keyboard because he was sure the flashing lights indicated a bad keyboard. Using security passwords, you can also lock out other users from changing the BIOS settings. This is handy for keeping self-declared experts from attempting to hop up the performance. One warning here is not to forget the password. If you do, you will not be able to start or change the system settings. The only way out is to discharge the battery and reset the entire BIOS settings. Some motherboards provide a special jumper to clear the CMOS RAM. Clearing the CMOS will set the BIOS to the manufacturer's default.
- *Other BIOS features.* These are motherboard and manufacturer dependent options. The choices vary greatly between manufacturers and include default keyboard settings, initial floppy seek, and gate A-20 options. These are now set by the operating system and will have little effect unless you are running DOS or other older operating systems.

Next, you must set the BIOS chipset features. BIOS chipset features are setup parameters specific to the chipset logic used. You must be very careful when changing any of these settings. This should only be done by a qualified technician or a person who fully understands the hardware. Most BIOS programs will automatically set these features for you. Some common settings include the following:

- *Auto configuration.* Allows the user to manually configure the BIOS features.
- *DRAM timing.* Manually select the DRAM timing sequence.
- *Bus speed.* Some allow for setting of the bus clock speed.
- *Other features.* I/O wait states and other board specific options.

- *PCI configuration.* PCI motherboards require that the PCI bus and plug and play options be set. BIOS PnP/PCI configuration is used to enable the plug and play features. The first choice will be to select whether a PnP operating system is to be used. This is normally set to YES unless you are using a non-PnP operating system. Next, configure the PCI bus to legacy ISA bus mapping. This can be set to automatically configure the parameters or let the user define the parameters. In most cases, this should be set on automatic. Another feature is the internal PCI IDE interface to interrupt line mapping. This is normally set on automatic, which will map the primary IDE to IRQ 14 and the secondary IDE to IRQ 15. If required the IRQ mapping can be changed and remapped to a different IRQ if an unavoidable conflict exists.

The **plug and play** feature is only valid when an operating system that supports plug and play is installed. Devices that support the plug and play standard will present a unique identifier sequence when prompted. The operating system will look at the sequence and self-configure for the device's characteristics, including ranges for the I/O address, interrupt, and DMA settings. This allows for dynamic configuration while the operating system loads up. If you are using older ISA devices, the PnP BIOS has no way of knowing what resources the older device is using. If this is the case, you will have to instruct the BIOS about the resources used.

Controlling legacy ISA devices is done by setting the resources controlled to manual mode. Next, you will get a list of IRQ and DMA assignments. Setting the IRQ and DMA to legacy ISA or reserved will instruct the BIOS not to use those settings while dynamically configuring the system. Here you will also be able to reserve the high-memory area for devices with memory or ROM BIOS built in.

Some pitfalls to watch out for are leaving enough room for devices that are built in. Stick to well-known settings. Make sure the PnP device can be mapped around a legacy ISA device, or rejumper the ISA device around a PnP device.

- *BIOS integrated peripherals.* These devices are built onto the motherboard. This menu will vary greatly between motherboards, but many similarities do exist—for example, the IDE interface, floppy interface, printer interface, and serial port interface. In the IDE section you will enable or disable the internal IDE interfaces. If you are installing a special IDE controller, you may disable the internal interface. You may also disable the IDE interface if you are using only SCSI drives and you need the IRQ lines for some other device. The transfer mode of the IDE interface can also be set. PIO mode 4 is the fastest, and mode 0 is the slowest. In most cases, there is an automatic setting. Select automatic unless you are sure of the transfer mode of all drives on that interface. Later motherboards support a fast IDE standard known as the **UDMA,** or Ultra DMA. If the drives you are installing support UDMA, then select this option.

In the on-board floppy controller section you can enable or disable the on-board floppy controller. Leave this enabled unless you are installing a special controller. In the parallel port section you can change the I/O address and on later boards select the port mode. Newer motherboards allow for bidirectional and enhanced operation. This is useful when attaching parallel port devices such as tape drives, CD-ROMs, and scanners. In the serial port section you can disable or redefine the on-board serial ports. You may want to disable one or both serial ports if you are not using them and need the resources for other devices.

16.8 FIRST BOOT

Once you have completed the BIOS setup, you must prepare the system for an initial operating system load. Before an operating system can be loaded, one of the hard disk drives must be made ready to receive the operating system. The hard drive preparation will be discussed in the next section. Since no software exists on the system yet, we must load some preparation software from a removable or remote media. Here we will discuss the preparation of the first boot medium. This will also verify the basic functionality of the system.

The most common method is the floppy boot using Microsoft DOS. Most new systems are preloaded with the operating system. They also come with an emergency startup diskette containing a bare-bones DOS boot diskette. This is supplied in case the hard drive is replaced. The boot diskette must contain all the device drivers to enable the devices being loaded from and destination drives. For example, if a system is being loaded from a CD-ROM, you must load the CD-ROM drivers and DOS CD extensions, known as **MSCDEX.** If you are using a SCSI-based system, then you must also load the SCSI drivers. The following outlines a step-by-step process to create a boot load diskette.

First, create a bootable diskette from any DOS or Windows machine. To create this on a DOS machine, insert a blank diskette and type FORMAT A:/S at the DOS prompt. The /S option will place a boot sector and the command interpreter COMMAND.COM on the diskette. To create this diskette on a Windows 95 machine, insert a blank formatted diskette. Navigate to the floppy disk drive icon. Right click on the floppy and select Format. Check the Copy System Files Only button and click Start as shown in Figure 16.16

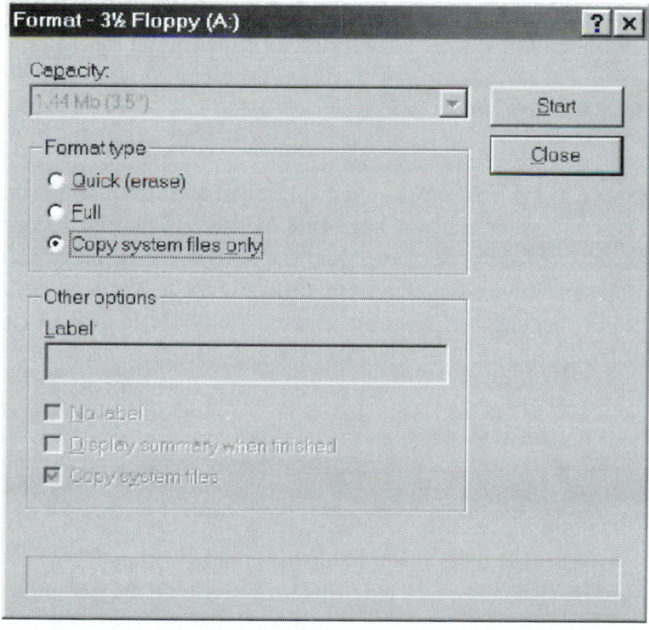

FIGURE 16.16
Creating a boot disk in Windows 95

```
device=HIMEM.SYS
DOS=HIGH,UMB
device=GSCDROM.SYS/D:MSCD001   ;This is the CDROM driver
```

FIGURE 16.17
Sample CONFIG.SYS file

Next, you must prepare the boot load configuration files. Using an editor, create the files **CONFIG.SYS** and **AUTOEXEC.BAT** as shown in Figures 16.17 and 16.18. Make sure to include the drivers (i.e., CD-ROM drivers) on the diskette to match the lines you create in the startup files. Next, create an AUTOEXEC.BAT file as shown in Figure 16.18. Make sure you place a copy of MSCDEX on the diskette. Once you complete this, you should copy at least the files shown in Table 16.6 onto the diskette for setting up and loading the hard disk.

```
SET BLASTER=A220 I5 D1 T4
SET SOUND16=C:\OPTi931
@ECHO OFF
SET PROMPT=$P$G
SET PATH=C:\
MSCDEX /D:MSCD001    ;The MSCDEX parameter must match the
                     ;CDROM driver
C:\OPTi931\sndtune
```

FIGURE 16.18
Sample AUTOEXEC.BAT file

Microsoft Windows 98 has a much improved startup disk utility. This will build a startup diskette containing a universal CD-ROM driver. The first thing that happens upon boot-up is that a RAM drive is created and the software is unpacked and loaded into the RAM drive. This is required because the software for a Windows 98 boot disk will not fit on a 1.44 MB diskette, so it is compressed. The boot loader will uncompress it and

TABLE 16.6
Boot disk files

File	Purpose
FDISK.EXE	Partitioning the disk
FORMAT.EXE	Formatting the disk
EDIT.EXE	To edit any changes on the startup files
SYS.COM	The boot load writer
HIMEM.SYS	The high-memory manager
MSCDEX	The CDFS file system extension

place it onto the RAM drive. Figure 16.19 shows the Windows 98 CONFIG.SYS file. You won't have to create this, it is here for reference only.

If the machine contains a network interface card, or NIC, and you have a fileserver, you might want to consider using the network boot method. This is the preferred method when building or reloading a large number of computers. Using this method, you place all the distribution files on a disk share and build a network boot diskette. All the diskette needs is DOS and a bare-bones network stack. All popular network stacks, such as NETBEUI, IPX, and TCP/IP, are supported. If you are running Windows NT as the file server, the NT server CD contains a program to build this boot disk. When you boot the diskette,

```
[menu]
menuitem=CD, Start computer with CD-ROM support.
menuitem=NOCD, Start computer without CD-ROM support.
menuitem=HELP, View the Help file.
menudefault=CD,30
menucolor=7,0

[CD]
device=himem.sys /testmem:off
device=oakcdrom.sys /D:mscd001
device=btdosm.sys
device=flashpt.sys
device=btcdrom.sys /D:mscd001
device=aspi2dos.sys
device=aspi8dos.sys
device=aspi4dos.sys
device=aspi8u2.sys
device=aspicd.sys /D:mscd001

[NOCD]
device=himem.sys /testmem:off

[HELP]
device=himem.sys /testmem:off

[COMMON]
files=10
buffers=10
dos=high,umb
stacks=9,256
devicehigh=ramdrive.sys /E 2048
lastdrive=z
```

FIGURE 16.19
Windows 98 CONFIG.SYS file

you log into the file server and issue a NET USE command. This will connect you to the share as shown in Figure 16.20.

```
path=a:\net
a:\net\net initialize
a:\net\netbind.com
a:\net\umb.com
a:\net\tcptsr.exe
a:\net\tinyrfc.exe
a:\net\nmtsr.exe
a:\net\emsbfr.exe
a:\net\net start
a:\net\net use z: \\longisland\ghost
z:
cd ghost1
ghost -ntil
```

FIGURE 16.20
Network boot

If you are setting up a new prebuilt system or have obtained a warranty-replacement hard disk, you may have a preloaded hard disk drive. If this is the case, simply set the drive up in the BIOS and turn on the system. The rest of the load is automatic and all you need to do is answer the questions.

16.9 HARD DISK SETUP

Unless the hard disk drive was provided as a preloaded replacement, nothing will be present on the disk. The first hard disk drives required low-level formatting. This was limited to older MFM, RLL, and ESDI hard drives that required a separate controller. If for some reason you continue to use one of these drives, you must run a low-level format program before going on.

As discussed in Chapter 12, **IDE** and SCSI drives actually contain the disk controller electronics on the disk drive unit. IDE disk drives are low-level formatted at the factory. They should not be low-level formatted by the user. The low-level format done at the factory uses very rigid testing parameters, and a bad sector map is stored on the medium. Most IDE drive manufactures provide a special test program to test and reallocate newly discovered bad sectors.

The next step is to partition the disk drive. Here is where you can define the physical disk drive into smaller logical drives or use it as one large drive. Partitioning the drive lays out the file system type and file allocation tables used by the target operating system. DOS, Windows 3.1, and original release of Windows 95 use a file system known as **FAT,** for file allocation table. The FAT uses variable cluster sizes ranging from 2 to 64 KB per cluster. The cluster size depends on the partition size. The entire partition has a limitation of 2 GB, which will use 64 KB per cluster. That means if you save a file containing one word or just a single byte on a 2 GB drive, the file will occupy one cluster or 64 KB of space on the drive.

Later releases of Windows 95 introduced a new file system known as **FAT32.** FAT32 uses a fixed cluster size of 4096 bytes and is a much more efficient method of using disk space on larger drives. Also, there is practically no limit to the partition size. The major problem with FAT and FAT32 is that there is no real file security across the network. The only security is password-protected shares, and no security exists on the local machine.

Microsoft introduced the NTFS file system for the Windows NT operating system. As with FAT32, there is no 2 GB limit, but NTFS provides file-level security to a user or group of users, similar to a UNIX system.

The initial partitioning is done by using a program known as **FDISK,** as shown in Figure 16.21. Using this program, you will define the partition type and sizes used. If you are using a later release of Windows 95 or Windows 98, the first step is to choose whether to use FAT32 or the original FAT16 file system. This is known as large disk support.

```
                 Microsoft Windows 98
               Fixed Disk Setup Program
        (C) Copyright Microsoft Corp. 1983-1998
                        FDISK

Your computer has a disk larger than 512 MB. This version of Windows
includes improved support for large disks, resulting in more efficient
use of disk space on large drives, and allowing disks over 2 GB to be
formatted as a single drive.

IMPORTANT:If you enable large disk support and create any new drives on
this disk, you will not be able to access the new drive(s) using other
operating systems, including some versions of Windows 95 and Windows NT,
as well as earlier versions of Windows and MS-DOS. In addition, disk
utilities that were not designed explicitly for the FAT32 file system
will not be able to work with this disk. If you need to access this disk
with other operating systems or older disk utilities, do not enable
large drive support.

Do you wish to enable large disk support (Y/N) . . . . . . . . . . ? [Y]
```

FIGURE 16.21
The initial FDISK screen

Next, you will get the FDISK options screen as shown in Figure 16.22. Here you will perform the partitioning tasks. Also provided in this screen is the ability to select a different physical drive. For now you will partition the first drive and set the partition to the active state. The system boots from the active drive partition. Once you select Create the Partition, you will have the ability to specify the partition size as shown in Figure 16.23. After you complete this task you can DOS format the drive for loading.

```
Microsoft Windows 98
Fixed Disk Setup Program
(C) Copyright Microsoft Corp. 1983-1998

                    FDISK options

 Current fixed disk drive: 1

 Choose one of the following:

 1. Create DOS partition or Logical DOS Drive
 2. Set active partition
 3. Delete partition or Logical DOS Drive
 4. Display partition information

Enter choice: [1]
Press Esc to exit FDISK
```

FIGURE 16.22
FDISK options screen

High-level format the hard disk drive by typing FORMAT C: /S at the DOS prompt. You will get a screen warning you of your actions. This is displayed to allow the user to cancel the operation. When formatting a disk containing data, all data will be erased. Actually, the data is not erased, but all reference to it is. This is because the file allocation table, root directory, and all sector pointers are cleared. The format process will read every

```
                  Create Primary DOS Partition


Current fixed disk drive: 1

Partition Status  Type  Volume Lable Mbytes System Usage
   C:1       A    PRI DOS                519    FAT16  100%


 Primary DOS Partition already exists.
 Press Esc to continue
```

FIGURE 16.23
Create the partition screen

sector, initialize the file allocation table, and make the disk bootable. Now the hard disk is ready for use. In fact, the hard disk now contains the bare-bones portion of DOS. Next, copy all files from the floppy diskette to the root directory of the hard drive. At this time you should reboot the computer without any floppy diskettes and allow DOS to load from the hard disk. You should have access to all devices including the CD-ROM drive on the system. In the next section we will break from the hardware discussion and present a brief discussion on DOS. Today you will rarely see DOS used in the workplace. DOS is a good diagnostic tool and still used on older and dedicated control computers.

16.10 INTRODUCTION TO DOS

The **Disk Operating System** (DOS) is a group of programs that manage the operation of a computer. These programs enable you to communicate with the computer's hardware. They control the operation and uses of the computer's disks, printers, keyboard, monitor, and other devices. DOS also organizes and keeps track of the information that has been stored on computer disks and provides utility programs to copy, delete, edit, move files, and so on.

A **file** is a group of related information that is stored in a highly organized manner. All programs, text, and data on a disk reside in files. A file on a disk can be compared to a file folder in a filing cabinet. DOS keeps track of all files using file names, which are kept in **directories** on the disk. Think of a directory as a drawer in the file cabinet. The file cabinet may have many drawers, and a disk may have many directories. These directories contain information about the size of a file, its location on the disk, and the dates that the file was created or updated.

At the beginning of each disk, DOS keeps track of the location of your files in an area called the **file allocation table (FAT).** The file allocation table is a type of index. It also allocates the free space on the disk to new files. Thus, the directories and file allocation table enable DOS to recognize and organize the files on a disk. The name of a file consists of the file name and the file name extension. The file name can be one to eight characters long. The extension can be one to three characters long. The general format for the file name is

```
File Name Extension
   ↓       ↓
  NAME.DTA
```

Notice that a period is used to separate the file name from the extension. File name extensions can be very useful for organizing your files. You can ask DOS to list all the files with a certain extension. For example, at the DOS prompt, typing DIR will list all of the files in the current directory. Typing DIR*.TXT will list only the files with the extension TXT. These techniques will be illustrated shortly. The asterisk (*) is called a global or wild card character. It indicates that any character can occupy that position or any of the remaining positions on the file name or extension.

The DOS prompt tells you that DOS is operational and ready to accept a command. It also tells you the current or working disk drive. Disk drives are labeled as single letters (beginning with the letter A). The DOS prompt when working with drive A is

```
A>
```

The DOS prompt may also display the current working directory. To work with a different disk drive, simply type the letter of the new drive you want, then type a colon, and press the enter key. For example, at the A> DOS prompt, type B: and press Enter. The process is

```
A>B:    <Enter>
B>
```

Directories

Recall that the directory and file allocation table form a type of index on a disk. The directory contains a list of file names, their size and location on the disk, and the dates they were created or updated. However, eventually the index will become cluttered or full. Therefore, DOS allows more than one directory (and even subdirectories) on a disk. Each directory can contain numerous files. Directories themselves are organized in a hierarchical structure as shown in Figure 16.24. The hierarchical directory structure can be thought of as a corporate organization chart. The creators of DOS thought of it as a "tree," except that the tree grows downward. Thus, the "root" is the highest level in the directory structure. The different levels of the subdirectories are the branches of the tree. The files within the directories and subdirectories are the leaves on the branches. The root directory is automatically created when you format a disk. The tree or directory structure grows as you create new directories for groups of files. The DOS TREE command will display the directory structure on the screen or on a printer. The backslash character (\) is used to identify the root directory.

You can create subdirectories using the make directory command (MD). For example, to create the subdirectory WORDPROC in Figure 16.24, type the following at the DOS prompt:

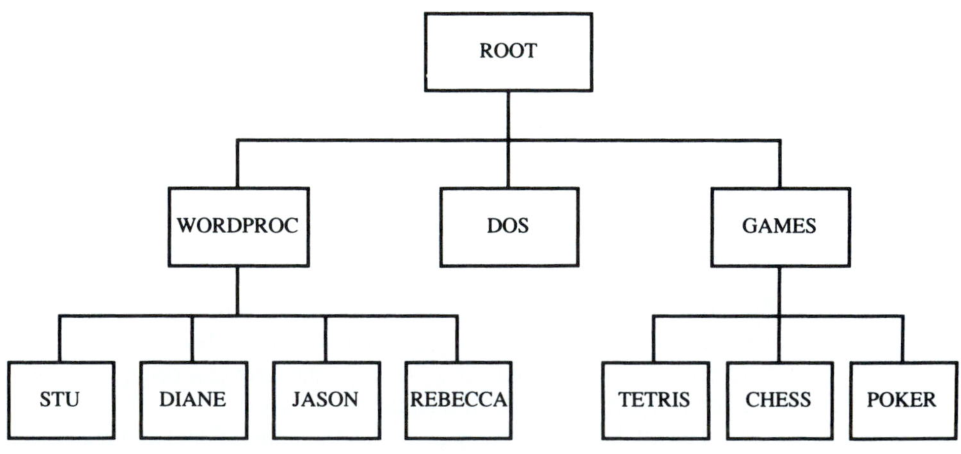

FIGURE 16.24
Directory structure

```
A>MD\WORDPROC
```

To create the next level subdirectory in Figure 16.24, type the following at the DOS prompt:

```
A>MD\WORDPROC\STU
```

You can change from your current working directory to another directory by using the change directory command (CD). For example, to change to the CHESS subdirectory, type the following at the DOS prompt:

```
A>CD\GAMES\CHESS
```

You can find out the name of the current directory that you are in by typing CD at the DOS prompt without any options. For example, typing

```
A>CD     <Enter>
```

will result in

```
A:\GAMES\CHESS>
```

Thus, typing CD displays your current drive plus the current working directory.

The directory command (DIR) displays a list of all files in the current directory. A number of useful options exist.

`A>DIR`	Will respond with a list of all files in the current directory on the A drive.
`A>DIR/W`	Will respond with a list of files in the current directory on the A drive displayed horizontally. (More files will fit on the screen, but less information is given.)
`A>DIR/P`	Will cause the screen to pause when the screen is full. Press any key to continue.
`A>DIR LETTER1.TXT`	Displays all of the directory information (name of file, size of file, date last edited) for the file LETTER1.TXT.
`A>DIR*.TXT`	Displays all files in the current directory with the extension TXT.

You can access any file in your current directory structure regardless of where you are or where the file is located by typing the **path.** The path is a sequence of directory names followed by a file name. Each name is separated by a backslash (\). For example, referring to Figure 16.24, the path for the file name STUD would be

```
A:\GAMES\POKER\STUD
```

The remove directory command (RD) will delete a directory from the tree. The directory to be removed must be empty except for what is referred to as the two hidden files. The hidden files are listed as . and . . , which are shorthand notations that represent the working directory and the parent directory (one level up from the working directory). Referring to Figure 16.24, to remove the TETRIS subdirectory, type the following from the root directory.

```
A>DEL\GAMES\TETRIS\*.*
```

Then from the root directory, type

```
A>RD\GAMES\TETRIS
```

In the previous example, the delete command (DEL), the path, and the wild card characters were all used to delete all of the files with any extension from the TETRIS subdirectory. Remember that you cannot be in the directory you want to delete.

Other DOS Commands

The check disk command (CHKDSK) will display a status report for a disk as well as report errors. By including the /F option, CHKDSK will also fix any of the errors it finds. For example, for a status report of the disk in drive B, type

```
A>CHKDSK B:
```

or

```
A>CHKDKS B:/F
```

Figure 16.25 shows a typical response. The response will vary depending on the version of DOS being used.

```
Volume DRAWDEMO  created 08-28-1991 2:28p

1457664  bytes total disk space
1439232  bytes in 4 user files
  18432  bytes available on disk

    512  bytes in each allocation unit
   2847  total allocation units on disk
     36  available allocation units on disk

655360   total bytes memory
630752   bytes free
```

FIGURE 16.25
CHKDSK response

If you had used the /F option, you would have been asked if you would like to recover any lost data sectors into data files. You would respond with Y or N.

The copy command (copy) is used to copy files from one disk to another or from one directory to another. The general format of the copy command is

```
COPY   <source file name>  <destination file name>
```

The file name should include the complete drive specification and path when necessary. For example,

`C>COPY A:FILE1.TXT B:`	Copies the file FILE1.TXT from the A drive to the B drive.
`C>COPY A:FILE1.TXT B:FILE2.TXT`	Copies the file FILE1.TXT from the A drive to the B drive and renames it FILE2.TXT.
`C>COPY A:*.* B:`	Copies all files from the A drive to the B drive.
`C>COPY B:*.* A:`	Copies all files from the B drive to the A drive.
`C>COPY A:*.TXT B:`	Copies all files with the extension TXT from the A drive to the B drive.
`A>COPY FILE1.TXT B:`	Copies the file FILE1.TXT from the A drive to the B drive.
`A>COPY *.* B:`	Copies all files from the A drive to the B drive.

The disk copy command (DISKCOPY) can also be used to copy files. The disk copy command copies all files from the source disk to the destination disk. DISKCOPY formats the destination disk as it copies. Therefore, any information on the destination disk will be erased and replaced with the information being copied from the source disk. For example,

`A>DISKCOPY A: B:`	Copies all data from drive A to drive B.
`A>DISKCOPY B: A:`	Copies all data from drive B to drive A.
`A>DISKCOPY A: A:`	Copies all data from drive A into memory, then prompts you to insert the destination disk into drive A and copies the information from memory onto the disk in drive A. (A number of iterations may be required.)

The rename command (REN) can be usd to change the name of a file without copying it first. For example,

`A>REN FILE1.TXT FILE1.OLD`	Changes the name of the file from FILE1.TXT to FILE1.OLD.
`A>REN TEXT.TXT NEWTEXT.TXT`	Changes the name of the file from TEXT.T\XT to NEWTEXT.TXT

The format command (FORMAT) initializes a new disk for use. It can also be used to erase all of the contents of a disk. For example,

`A>FORMAT A:`	Prepares the disk in drive A for use.
`A>FORMAT B:`	Prepares the disk in drive B for use.

Table 16.7 is a summary of many of the DOS commands. Not every command is available on every version of DOS. For a complete list of all of the DOS commands, consult the DOS manual for your version of DOS.

TABLE 16.7
Summary of DOS commands

ASSIGN	Changes drive letter assignments.
BACKUP	Backs up fixed disk files to a floppy diskette.
CD	Changes the current directory.
CHKDSK	Displays disk status and checks and fixes file structures.
COMP	Compares two files.
COPY	Copies files from one disk or directory to another.
DATE	Displays current date.
DEBUG	Assembles programs directly into memory.
DEL	Erases a file.
DIR	Displays disk directory.
DISKCOMP	Compares two diskettes.
DISKCOPY	Copies a diskette and formats the destination diskette if necessary.
EDLIN	Line-oriented text editor.
EDIT	Screen-oriented text editor.
ERASE	Erases a file.
EXE2BIN	Creates a COM file.
FDISK	Creates fixed disk logical partitions.
FIND	Locates text strings.
FORMAT	Formats or initializes a disk.
GRAPHICS	Enables the print screen key to print graphics.
HELP	Provides information about a DOS command.
JOIN	Joins a disk drive to a directory.
LABEL	Changes the volume label of a diskette.
MD	Creates a subdirectory.
MEM	Displays status of memory usage.
MODE	Sets display, printer, and asynchronous parameters.
MORE	Displays the contents of a file one screen at a time.
PATH	Defines the file search path.
PROMPT	Changes the DOS command prompt.
PRINT	Prints text files.
RECOVER	Recovers data from corrupted or damaged disks.
REN	Renames a file.
RESTORE	Restores a subdirectory.
RD	Removes a subdirectory.
SORT	Sorts lines in alphabetical order.
SUBST	Associates a path with a drive letter.
SYS	Copies system files to a diskette.
TIME	Displays current time.
TREE	Displays a disk's subdirectory structure.
TYPE	Displays an ASCII file on the screen.
VER	Displays the version of DOS.
VERIFY	Controls the verify after write feature.
VOL	Displays the volume label of a diskette.
XCOPY	Copies diskettes from one diskette or subdirectory to another.

16.11 FINAL OPERATING SYSTEM LOAD

In this section, we will discuss loading the advanced GUI, or graphical user interface, operating system, including Windows 95, Windows 98, and Windows NT. The most important thing is to have ready all the device driver diskettes for the installed hardware. If you do not have the driver handy for a device, leave it out of the system and add it later. It is much easier to add a driver than try to repair an incorrect installation. Do not guess the device type, it will only lead to countless hours of trying to repair the installation later. The only exception is in the case of a video board. If you do not have the correct drivers install it as standard VGA. You will be able to update that driver later.

Since all modern software is available on CD-ROMs, make certain the CD-ROM subsystem is operating correctly. Remember the discussion on creating the first boot diskette. There we created the start-up files and copied the CD-ROM drivers and the MSCDEX file onto the floppy diskette. Boot the diskette and observe the drive letter assigned to the CD-ROM. This is displayed when MSCDEX executes. Make sure you can read the CD by typing DIR <CD-ROM letter>. If you cannot access the CD-ROM, you will have to go back and repair the problem before going on.

If you are loading from a network boot diskette make sure you can access the entire distribution share. If space on your hard drive allows, copy the entire distribution directory to a directory on the hard drive before loading. Then, you can install the software from the hard drive. This is the most trouble-free and safest method of installing the operating system. The command to begin the installation or upgrade of Windows 95/98 is to simply type SETUP at the command prompt. If you copied the entire contents of the installation software to the hard drive as suggested, be certain you are in the correct directory. Otherwise the setup file will be on the CD-ROM, so don't forget the drive letter.

When you type SETUP and the installation begins, you will have one of two scenarios. The first is all goes well and in about 30 minutes to an hour, and a couple of reboots, the system functions normally. The second scenario will be hours of frustration because of an installation problem. The problem is that after the system boots first, it is running in real mode, which is the old DOS mode. In this mode it uses the segment/offset addressing mode and an entirely different set of device drivers than Windows uses in flat or virtual mode. Installation problems can occur because after the root code is loaded the system restarts. Next, Windows attempts to continue in its native or flat mode. Now one of the Windows drivers may not be compatible with your hardware, and the system can no longer access the CD-ROM or network interface. Another potential problem is that Windows enables plug and play, and may not be able to reconfigure the system correctly. If you are loading from one of these devices, the install continuation fails because it can no longer access files and you are left with a partially installed system.

As an alternative to this potential problem, copy the distribution files onto a directory on the hard drive first. Create a directory named WIN98 (or WIN95) and copy all the files from the CD-ROM WIN98 (or WIN95) directory into this directory. Then run SETUP from this directory. This has a few advantages. The first is the aforementioned problem. The second is that Windows loads faster from the hard drive than any other medium. The third is in changing or adding devices to the system. If the distribution is on the hard drive, you will not have to find the original Windows CD-ROM or network share.

Loading Windows NT is very different from loading other versions of Windows. First, Windows NT comes with three startup diskettes. These diskettes load up a bare-bones version of the NT kernel. Here NT creates a special boot sector, which allows for multiple OS loads. If you are to have Windows 95, some other OS, and Windows NT on the same partition, always load Windows 95 or any other operating system first. The NT loader will build a startup menu for those operating systems. Never attempt to load Windows 95 over a Windows NT system, because Windows 95 will overwrite the NT boot loader.

Another difference from other versions of Windows is that Windows NT can be loaded onto any partition or hard drive you may have in the system. The boot files are still placed on the C partition, but all other files can be somewhere else. You may also load Windows NT from a network share, just as we discussed for Windows 95/98. Instead of typing SETUP, you type WINNT -B and that will start the installation process. If you don't add the -B option, the setup program will create the three-diskette set also. If you are installing over the network, chances are you will never use these diskettes, so the option saves some time. You may also copy the distribution onto the hard drive, as was done for the Windows 95/98 installation. The difference here is the distribution files are contained in a directory named I386. Here we will create a directory on the hard drive named I386. Next, navigate to that directory and type WINNT -B. and the rest will be a guided installation process.

The NT installation is generally trouble-free as long as you are using hardware that is supported by Windows NT. A complete list of supported hardware is included with every release of Windows NT and is called the HCL, or hardware compatibility list.

One important thing to note here is that as of this writing, Windows NT 4.0 does not support plug and play. If your BIOS supports plug and play, disable it and configure the system manually. Recall that one of the BIOS setup options is PnP O/S Installed. Set this option to NO.

16.12 SERVICING

This section will address servicing from a field service perspective. Here we will discuss what to watch for and what to do when servicing a computer at a customer's site. The office environment usually consists of carpeted comfortable areas with lots of fabric and plastic. These are great static electricity generators, so be careful, and use a wrist strap to ground yourself. If your finger arcs to the motherboard, you will be in for more than you bargained for and the repair will be expensive. Completely disassembling a computer all over a user's desk is not recommended, because this is intrusive and unprofessional. If the system cannot be quickly repaired by replacing a board or drive, it should be returned to a depot repair facility for repair. Chapter 17, "Diagnostics and Troubleshooting," addresses much more detailed repair procedures. As an aid to diagnosing the problem, ask the customer for a brief history of the problem and any events that led up to the failure. Also, ask about any changes the customer or previous serviceperson made. Often the answer is, "I didn't do anything, it just stopped working," but someone actually deleted some important files, or tried to update software or add hardware. Be wary of the use of incorrect acronyms from the customer. One example of this is a customer who says, "The hard drive is bad" but really means the system unit. Now the technician shows up with a replacement hard drive, resulting in a wasted trip.

Never format or replace the customer's hard drive without checking with someone in authority and insisting upon a backup. If no backup is present, explain that the system will lose all stored data and let the customer make the decision. If the system is operational, insist that the customer perform a backup or if possible back up the system yourself to whatever medium is present. Even if the hard drive is beginning to fail, you may be able to save most of the customer's data somewhere else.

When servicing a computer system, you should also be aware of the startup sequences, lights, and sounds. For example as soon as you power on the system, the power light lights, the keyboard lights flash, the floppy drive makes a noise, and so on. By observing the system, you may be able to determine a problem even when nothing is displayed on the CRT. A classic example is a call for a dead system because the screen is dark. The problem may be as simple as the brightness is all the way down, the monitor is not on, or a cable has come loose.

The basic start-up operation is very similar on all PC-type computers. Upon power up the system initializes the motherboard hardware and quickly checks the RAM. Next, the video display is initialized, and the POST (power-on self-test) is run. During the POST, the results are displayed on the screen. POST errors are explained in the next section. Next, the BIOS will check the state of the internal battery, the CMOS check sum, and installed devices matching the CMOS device table. Any discrepancies found will cause the computer to pause and display a message to the user. The message is usually the discrepancy found along with a choice to continue loading or to enter the CMOS setup program. Next, the plug and play BIOS is run, the keyboard and floppy drive are initialized, and the system attempts to boot from the device list created during the BIOS setup process.

If any problems are found during the **POST,** an error code will be displayed on the screen. Early memory failures and video failures will sound a series of beep tones because the system is not able to display them. Table 16.8 is a list of displayable POST errors. A more detailed list can be found in Appendix E. The beep tones are unique to the BIOS manufacturer, so consult the manufacturer's, reference for these tones.

The early portion of the POST checks the motherboard hardware and a small portion of system RAM before the display is initialized. If a failure occurs here, there is no way of displaying the error on the screen. When the IBM PC/XT was introduced, the

TABLE 16.8
POST errors

Error Code	Cause	Possible solution
One long beep	Main RAM	Modules not seated; clean and reseat modules.
Repeating or series of beeps	No video detected	Board not seated correctly; reseat board.
		Board not seated correctly; reseat board.
03*xx*	Keyboard failure	Cable out, stuck keys, or plugged in wrong position; check mouse also.
04*xx*	Mono video	If you can see it, video RAM bad.
05*xx*	Color video	If you can see it, video RAM bad.
06*xx*	Floppy error	Cable off, something jammed inside.
17*xx*	Hard drive error	IDE cable on upside down, drive failure.

BIOS included checkpoints. The checkpoints are written out as a one-byte value to a special I/O address, which is 0*x*70 (address 70 Hex). You could monitor the progress of the POST and any errors by attaching a hex LED display or logic analyzer to trigger on that address. The PC/AT and later machines moved this special address to 0*x*80. If you wish to repair faulty motherboards, you can modify the tester project included in this book and decode that address.

After the POST completes, the system attempts to load the operating system. If the system fails to boot and no more information is displayed on the screen, check the status of the hard drive light. Also, don't forget to check for a diskette in the floppy drive. In this case, the system is attempting to boot a document or spreadsheet. If the hard drive light comes on as soon as the system is powered on and stays on, the interface cable may be installed upside down. As soon as the system attempts to load from the hard drive, the light will flash at least once. If the light comes on and stays on, you most likely have a bad drive, controller, or cable.

If more information is displayed on the screen, analyze this information. Now you must determine if the problem is hardware or software. A message similar to "Error Reading Drive C" is a serious error and usually indicates a bad or seriously corrupted hard drive. Other errors, including "Missing Operating System," "Command Interpreter Not Found," and "Incorrect DOS Version," indicate software errors and possibly that the boot sector or COMMAND.COM were overwritten or that essential files have been deleted. Solving this problem usually involves reinstalling the operating system. If the operating system loads partially or reports errors but works, examine the errors carefully. You may be able to repair the error by replacing any missing files. If the system loads up and runs but crashes during operation, you should suspect a hardware error. This will be discussed in detail in the next chapter.

16.13 CUSTOMER RELATIONS

Up to this point we have studied the theory and knowledge required to service microcomputers and digital systems. In industry this is known as **product training.** At this time it is appropriate to discuss **customer relations,** or servicing the customer. Product training provides the theory and knowledge of how products operate. Customer service training teaches methods that allow the technician to understand the customer's reaction to a situation. It should also provide the technician with methods to influence the customer and increase sales.

Interfacing with Customers

Effective customer service is difficult because each situation is different. Just as no two people are completely alike, no two customer situations are completely alike. Furthermore, an exact customer situation seldom repeats itself.

The customer service technician is viewed by the customer as a representative of his or her company. To many customers, the customer service technician *is* the company. Just being able to repair the machine is not enough. The customer must be served also. Good *customer perception* of your service is paramount. As anyone who has gone to the movies can attest, truth and reality are not always the same. For example, you can do a great job repairing a machine and leave the customer with a negative impression. Thus you have serviced the machine correctly but not the customer. The customer will perceive your service as poor. Interestingly, though you may do only an adequate job servicing the machine, a properly serviced customer will perceive you and your company as excellent. It is important to understand that although customer service is paramount, we still must repair the machine.

When interfacing with customers, there are two main factors to be considered. These are *appearance* and *communication.*

Appearance Your **appearance** is made up of several factors. The most important of these factors is a positive attitude. Proper attitude is projected through an appropriate dress and clothing, good hygiene and hair care, clear eye contact, and a firm handshake. A great sage once said that you can tell everything you need to know about a person by simply looking at his or her shoes. Look in the mirror and ask yourself, Is this a person you would want to service your valuable computer system? Professional apearance is the first step toward building customer trust.

Communication When interfacing with a customer, it is important to maintain a proper balance of **communication** between you and the customer. In order to maintain a proper balance of communication: (1) Avoid talking too much about yourself; (2) Avoid interrupting the customer when he or she is speaking; (3) Avoid contradictory discussion (don't say, "but . . . !"); (4) Avoid inflammatory statements; (5) Use words and phrases that make what you are saying easier for the customer to accept; (6) Use proper language; (7) Make an effort to create a good impression.

Remember, the customer is a human being like yourself. Customers want to be treated with understanding and respect. Furthermore, they are paying for the service. Good communication skills and a proper attitude will help to build a good customer perception of you and your company.

Handling Stressful Conditions

It is important to begin to handle stressful conditions before they occur by creating customer trust and understanding. You should always begin by greeting your customer by name. Learn as much as you can about the customer's business. Understand your customer's position, responsibilities, authority, and limitations. Always be prompt and prepared to work. Be certain to fulfill your promises. If you say you are going to do something (e.g., order a part), make sure you do it. Keep the customer well informed of the situation. Let the customer know what you are doing and how you are doing it. If possible, explain any options the customer might have. Before speaking, think about what you are going to say. Visualize the customer's reactions and tailor your responses to the individual customer. Building customer trust and understanding can temper stressful situations when they do occur.

Handling the Irate Customer When stressful conditions occur, it is important not to lose your composure. Try to defuse the situation and calm the customer. Sometimes, it is not what you say that is important but how you say it. Let the customer relieve some of the pressure by venting some of his or her frustrations. This is the wrong time to argue. When responding, express concern and display confidence. Show the customer that you are going to take charge of the situation. Develop your plan to correct the situation and explain it to the customer. Never blame the problem on a co-worker, supervisor or other organization in your company. For example, don't complain that your parts department is always out of the part you need. Try to assure the customer that you will do everything necessary to correct the problem as quickly as possible. This is probably the time to overreact in a positive way. Go the extra mile and get extra backup help, parts, and support. Show the customer you care.

Selling the Customer

Selling the customer requires a team approach. Remember that the customer has purchased the services of your entire company. You should always try to sell yourself and your entire company. Without sales, there are no jobs and no growth. Always look for opportunities to help customers with your services, but be careful to make your suggestions at an appropriate time. Sell the benefit to customers by letting them know what they will gain by using your company's services. Know your facts and give solid reasons for everything you suggest. Always express a positive attitude about your company's equipment and services. Be sure you pass on all leads about possible equipment purchases to the appropriate person in your company. Do not quote prices on new equipment—that is the function of the salesperson.

Remember that selling the customer starts and ends with good customer relations, which means creating customer trust, knowing the customers and their needs, and working as a team while keeping a positive attitude. Always explain what you are doing and your plan of attack. In one sense, solving a customer relations problem is like solving a digital circuit problem. The most important steps are to look and think. Look at the situation and analyze at first. Then think about your options and choose a plan to rectify the situation. Above all else, always think before you speak. Remember, illusion is as important as reality. The customer's perception of you is as important as your technical competence.

16.14 TECH TIPS AND TROUBLESHOOTING—T^3

Good customer perception is very important in field service. The field service technician is viewed by the customer as a representative of the company. For this reason, the field service technician must not only service the equipment but also serve the customer.

Serving the customer requires proper dress, proper attitude, courtesy, and preparation. When you arrive on a service call, you should be prepared to work, which means having the proper tools, test equipment, documentation, and spare parts with you. You should know whom you are going to see and you should have some idea of the problem. It is important not only to dress neatly and be clean but also to work in a neat and clean manner. Don't scatter your tools and parts all over the floor. Be careful in routing and dressing cables. Secure all connectors properly. Inspect and clean, if necessary, all connector contacts. Always remember to replace all screws and covers properly. Cover all switch settings and adjustments with a piece of tape to prevent accidental mishaps. Clean up your work area when you finish the job. Make sure you discuss the problem and what you did with the customer. If possible, instruct the customer on the preventive maintenance that could have avoided the problem, such as keeping air passages and filters clean, monitoring temperature and humidity settings, and taking general safety precautions. Above all, always be courteous. Consider yourself a guest in the customer's house and behave accordingly.

EXERCISES

16.1 In a dual-boot system, (Windows 95 and Windows NT), which OS should be loaded first, and why?

16.2 The drive access mode that broke the 528 MB DOS barrier is called __________.

16.3 Briefly describe the process of creating a boot disk in DOS and in Windows.

16.4 How would you jumper and cable an IDE drive and a CD-ROM drive in a system?

16.5 List several features or options contained in a typical BIOS program.

16.6 On a PCI motherboard with an integrated IDE controller, what IRQ settings should the primary and secondary IDE devices have?

16.7 Is it possible to disable the IDE interface on the motherboard? Why would you need to do so?

16.8 What are some possible hard drive error messages, and what do they mean?

16.9 Upon powering up the PC, what are the first three things that occur?

16.10 Briefly describe what happens during the POST.

16.11 What is the most common IRQ setting for Sound Blaster compatibility on a sound card?

16.12 What are the DMA settings for 8- and 16-bit audio on a typical sound card?

16.13 What IRQ does Windows 95 and up use for the printer?

16.14 Multiple COM ports can share IRQ lines. What are these shared combinations?

16.15 The Pentium II CPU plugs into what type of motherboard?

16.16 The red stripe on a drive cable denotes what?

16.17 How many pins are on the floppy cable connector?

16.18 How many connectors are on a typical floppy cable?

16.19 What does the longer end of the floppy cable connect to?

16.20 The end with the twist on a floppy cable connects to __________.

16.21 How many pins are on the IDE cable connector?

16.22 Does the IDE cable have a twist in it?

16.23 How many IDE interfaces are on a typical motherboard?

16.24 What is hard disk type 47 on the BIOS setup used for?

16.25 What are some of BIOS chipset features that may be changed by the user?

16.26 On a motherboard with an internal PCI IDE interface, the primary IDE will automatically be set to what IRQ?

16.27 In a system with full plug and play compatibility, what crucial settings will automatically be set by the operating system?

16.28 Will plug and play configure ISA devices?

16.29 What are the MS DOS CD-ROM extensions called?

16.30 Is it necessary to low-level format an IDE drive?

16.31 What is the file system in Windows NT that is similar to FAT32?

16.32 What is provided by NTFS that is not provided by FAT32?

16.33 Briefly describe what the format process (on a hard drive) does.

16.34 Does WIN NT 4.0 support plug and play?

16.35 A listing of hardware supported by NT is called what?

16.36 If the hard drive light comes on at power-up and stays on, what is the problem?

16.37 Crossword Puzzle

ACROSS

2. Type of processor.
3. Motherboard mounting.
7. Bypasses memory test.
9. ESD.
10. Pentium processor package.
11. Used on a floppy cable to denote drive A.
12. Older video standard.

DOWN

1. Initial startup diagnostic.
2. Commonly occurrs with IRQ settings.
4. Starting a computer.
5. Used to make motherboard system settings.
6. Preparing a disk for use.
8. System setup program.
9. Denotes pin 1.

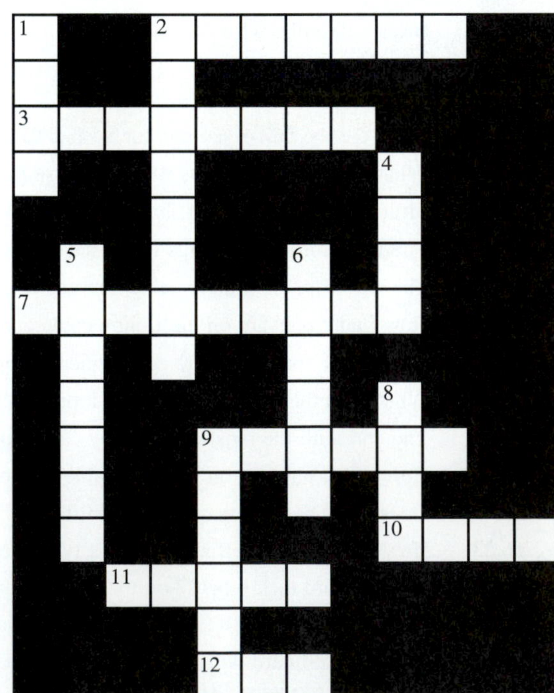

CHAPTER 17

DIAGNOSTICS AND TROUBLESHOOTING

KEY TERMS

Auto-Detect
Beep Codes
BIOS
Boot Failure
CMOS
DEBUG
Error Codes
FDISK
Norton Utilities
Ohmmeter
Partition
PING
POST
Safe Shutdown
TDR

17.0 INTRODUCTION

Chapter 16 discussed the initial setup, loading, and servicing of a typical PC-compatible computer. In this chapter, we discuss in detail good diagnostic and troubleshooting procedures. In many cases the process may be time-consuming and require partial disassembly of the computer. As always, exercise good service habits and use an antistatic wrist strap. Always keep your spare parts in antistatic bags and well packed to avoid shock and other hazards.

In past editions of this book we relied upon the IBM Advanced Diagnostics to test and determine a problem. Today, there are many excellent diagnostic programs available for PCs. First we will discuss the use of the **POST** diagnostic, because it is built into every PC. All but the earliest POST failures are displayed on the monitor or sounded as a series of beep tones. As discussed in the previous chapter, POST outputs its progress to I/O port 0*x*80. Very early POST failures provide no indication other than an output to port 0*x*80, and the system appears dead. You can modify the tester project in this book to decode I/O address 0*x*80 or purchase a commercially available POST display board. If the computer passes the POST, most failures can be diagnosed by observing the operating system load and then using the operating system to diagnose further.

It wasn't long ago that computers were diagnosed and repaired at the IC level. Faulty ICs on a controller board or on a motherboard were replaced and soldered back onto the board. Modern surface-mount technology, IC pin densities, and economics have made that practice outdated. Today you will diagnose a problem down to a defective board or plug in a new component and discard the faulty part.

17.1 INITIAL ASSESSMENT

Assume you have a encountered problem that is not a quick fix or operator error and must investigate the failure. Begin by discussing the problem and any events that led up to the failure with the user. The first thing to do is isolate which subsystem or assembly is failing. If not already done, connect a monitor, mouse, and keyboard. Assuming you are familiar with the operation of the system, power up the system, *stop, look,* and *listen.* A seasoned technician will also use their sense of smell. An overheating or burning component will give off a unique odor and often lead you right to the problem. Almost immediately you will see and hear clues to aid in diagnosing the problem. Failures can be broken down into the following categories:

1. *Totally dead system.* No lights, sounds, or beeps from the system unit
2. *Dead system.* Powers up and some signs of life or lights and sounds, but no beeps.
3. *System powers up.* Beeps, but the monitor shows nothing or garbage.
4. System powers up and initializes the monitor but fails the POST with a message.
5. Monitor displays BIOS greeting, passes the POST, but will not boot.
6. System begins to boot and hangs with one or more error messages.
7. System boots completely and crashes (blue screen of death).
8. System boots OK but has a peripheral problem.
9. Network problems.

17.2 TOTALLY DEAD SYSTEM

The most obvious problem would be a bad power supply. Often, you may notice that the power supply fan is not spinning. The power supply outputs many different voltages, so even if the fan is spinning the power supply may still be defective. However, when a power supply fan is not spinning, it is an indication that the power supply may be defective. Keep in mind that an ATX power supply uses a "soft start." A portion of the power supply is always on, providing a +5-V stand-by power supply. The main portion is electronically controlled by the motherboard. Switching regulator power supplies used in PCs are very high power and contain many safety features, including overvoltage and overcurrent protection. When an unsafe condition is detected, the power supply will shut down, preventing internal damage and the possibility of fire. No power is indicated by the absence of any lights, sounds and the fans not running. A voltmeter can be used to probe the power supply connector pins as shown in Figure 17.1.

Don't be fooled by a bad power cord or power strip. Just because the monitor powers up doesn't mean the outlet that the computer is plugged into is supplying power. In addition, a poor connection to the power supply may cause the power supply to go into a safe mode and shut down. A loose power cord will arc and spike the regulator, causing it to shut down. Turning the switch off and back on will keep the power supply in **safe shutdown** mode. The only way out of the shutdown condition is to remove the AC power

FIGURE 17.1
Probing the power supply

for at least a minute to allow the internal filter capacitors to fully discharge. The following is a series of steps for diagnosing power problems.

1. Remove the power cord from the power supply and check for 115 V_{AC}. Replace the power cord if no voltage is present.
2. Reconnect the power cord, verify that the 120/240 switch is in the correct position, and try the system again.
3. Remove the cover from the system and check all wires from the power supply. Look for chafing, scraping, or bare insulation and any possibility of a short to the case or chassis.

4a. *AT case, mechanical switch.* If an external power switch is used, check the wire and connection to the switch. Remove the power cord and check the switch using an ohmmeter.

4b. *ATX case, electronic power switch.* With the system connected to AC power, check pin 9 (purple wire) for the presence of $5V_{DC}$. If no $+5V_{DC}$ is present, replace the power supply. Next, jumper the motherboard power connector pin 14 (PS_ON) to pin 15 (COM). If the power comes on, the problem is on the motherboard. Sometimes the motherboard BIOS gets confused and will not power on. Try removing the CMOS battery to clear CMOS memory. Try the system again. If it still fails, check the power button and cable using an ohmmeter. If the button is good, replace the motherboard.

5. If the input power appears good, check for an overload condition. Remove the AC power and all power supply connectors, including those to the disk drives and motherboard. When checking an ATX power supply, don't forget to jumper pin 14 and pin 15 together.
6. Provide a load to the power supply by connecting a known good spare hard disk drive to one of the power connectors. This is because PC power supplies are switching power supplies, which require some load to operate.
7. Connect the AC power cord and turn on the power. Does the spare hard disk spin? If not, the power supply is faulty and must be replaced. If the hard disk spins up, you have a short or overload in the system.
8. Reconnect the power cables one at a time and check for system power until you isolate the device causing the overload. You may also wish to check for shorts using an ohmmeter.
9. If the overloading device is a disk drive, check for foreign objects or overlength screws causing a short to ground. Repair or replace the faulty component.
10. If the overloading device is the motherboard, remove the memory, processor, and all add-on cards, and check for an overload again by using an ohmmeter or by powering on the system. If the problem goes away, the problem is one of the devices you removed. Isolate and replace the faulty component.
11. If the problem persists, then you have a short on the motherboard. Check the motherboard mounting screw(s), and check for any foreign objects on the motherboard or in the card slots. If nothing is indicated, replace the motherboard.
12. In rare cases, all the components will power on but overload the power supply when all connected together. This is most likely a faulty power supply because it can't handle the load. Replace the power supply. Most power supplies manufactured today are more than adequate in capacity to power a system. However, older power supplies may not have the capacity to power a system and all of the peripheral components.

17.3 DEAD SYSTEM, APPEARS TO POWER UP

This is symptomatic of a dead processor, serious bus fault, or a very early POST failure. It may also be caused by a bad power supply or low AC power. The power supply provides a signal known as POWER GOOD (AT system) or PWR_OK on an ATX system. The POWER GOOD signal is used to hold the processor reset and prevent the system from operating until the power voltages are within rated limits. This signal is normally a TTL high for the system to operate. The following are steps to diagnose the problem.

1. *Power on the system.* Using a voltmeter, probe the power good signal on the motherboard power connector (pin- 8, gray wire on an ATX or the rearmost orange wire on an AT) for a voltage greater than 2.5 V_{DC}. If the signal is low, check the incoming AC voltage for greater than 95 V_{AC}. If the incoming power is good and the power good signal is low, then replace the power supply.
2. Probe the *red* wire for a voltage of $+5\ V_{DC} \pm 5\%$. If the voltage is out of tolerance, replace the power supply.

3. Probe the *yellow* wire for a voltage of $+12\ V_{DC} \pm 10\%$. If the voltage is out of tolerance, replace the power supply.
4. Probe the *blue* wire for a voltage of $-12\ V_{DC} \pm 10\%$. If the voltage is out of tolerance, replace the power supply.
5. Probe the *white* wire for a voltage of $-5V_{DC} \pm 15\%$. If the voltage is out of tolerance, replace the power supply.
6. *ATX only.* Probe the *orange* wire for a voltage of $+3.3\ V_{DC} \pm 4\%$. If the voltage is out of tolerance, replace the power supply.
7. Power is good but the system may be stuck on reset. Check the reset button for a stuck, pushed, or shorted condition. Sometimes the button is very small or recessed to prevent accidental activation. To verify a bad reset button, remove the reset-switch connector from the motherboard and try the system. You may also use an ohmmeter (with the AC line cord disconnected) to verify that the switch is normally open.
8. Verify that any IDE drives are spinning. If any IDE drives are not spinning, the system may be stuck reset or the IDE cables are on upside down. An indication of incorrectly installed IDE cables (upside down) is the access lights will be stuck on and the drives will not spin. This will also hang the system bus.
9. Power off the system and remove all add-in cards and all IDE cables from the motherboard. Power on the system. Look for any changes and listen for beeps. If there is no change, power down the system, remove all memory modules, and replace the memory with at least one known good bank. Power on the system. Look for any changes and listen for beeps. If there is no change, the motherboard or processor is bad. Use a POST display board to diagnose further or replace the motherboard and processor.
10. If step 9 was successful, then one or more peripherals are hanging the bus. Power down the system and remove all of the peripheral boards. One by one add the boards that were removed and power up the system until you isolate the faulty component. Replace the faulty component and retest the system.

17.4 SYSTEM POWERS UP, AND BEEPS, BUT NOTHING OR GARBAGE ON MONITOR

If the system powers up, but there is nothing or garbage displayed on the monitor, this is symptomatic of an early BIOS failure or faulty video board. There may be a rare case where the power supply is not regulating properly. You may wish to go back to Section 17.3 and test the DC voltages before going on. The first dozen or so POST steps are to initialize and test the motherboard chipset registers, initialize and quickly check the first 16K of memory, and set up the video board. Any failures up to this point cannot be displayed because the display controller is not programmed. Therefore, the BIOS will attempt to drive the system speaker, indicating the failure. One thing to keep in mind is that just to drive the speaker, much of the system board and bus must be operational. Early error codes are encoded as a series of short and long beeps or counts of beeps. By listening to the beeps, you can decipher the error codes. The sequence is unique to each BIOS vendor and BIOS version. The motherboard manual may list the codes, or you will have to contact the motherboard or BIOS manufacturer. The error codes and BIOS POST progress are also output to I/O port 0*x*80 and can be viewed using a logic analyzer or LED display connected to the ISA bus and decoded to I/O address 0*x*80. Here are some steps to follow to determine the cause of the failure.

1. Power off the system and remove all add-in cards except the video card and all IDE cables from the motherboard. Power on the system and observe the monitor for any BIOS error codes or messages. Then, one by one, add the components and power up the system until you isolate the faulty component. Replace the faulty component and retest the system.
2. Power off the system and remove the video board. Slowly and gently, using a soft pencil eraser, clean the contacts that insert into the motherboard. Do not use an ink eraser, as it will remove the gold plating. Remember to use a wrist strap and to be careful of static electricity. Reinsert the video board, power up the system, and check for any BIOS error codes or messages.
3. Power off the system and move the video board to a different slot. Power up the system and observe the monitor and check for any BIOS error codes or messages.
4. Power off the system and replace the video board with a known good spare. Power up the system, observe the monitor, and check for any BIOS error codes or messages.
5. Remove the system RAM modules. Slowly and gently, clean the RAM module contacts using a soft pencil eraser. Do not use an ink eraser, because it will remove the gold plating. Insert only enough RAM to complete one bank. Power up the system, observe the monitor, and check for any BIOS error codes or messages. Replace the remainder of the RAM and retest the system.
6. Replace the RAM modules with known good memory (preferred). If one is not available, use the second bank in the first bank position. Power up the system and observe the monitor and check for any BIOS error codes or messages.
7. If it is available, replace the processor chip or module. Power up the system and observe the monitor and check for any BIOS error codes or messages.
8. If it is available and replaceable, replace the BIOS chip. Many new motherboards use soldered-on flash BIOS chips. Power up the system, observe the monitor, and check for any BIOS error codes or messages.
9. If it is available and replaceable, replace the keyboard BIOS (8042 microcontroller) chip. Power up the system, observe the monitor, and check for any BIOS error codes or messages.
10. If you reach this step, you have a bigger problem and it is on the motherboard. You will need a logic analyzer or POST display to go further. The suggested action is to replace the motherboard.

The following is a partial list of common **beep codes** that are used by various motherboard and BIOS manufacturers.

AWARD BIOS

Beep Code	Meaning
Repeated single long beep	Missing or bad system RAM
1 long, 2 short, pause, 1 short	Missing or bad video board

AMI BIOS

Beep Code	Meaning
1 long, 2 short	Missing/bad video board
1, 2, or 3 beeps	Main RAM failure
4 beeps	PIT on motherboard bad (timer)
5 beeps	CPU failure
6 beeps	8042 keyboard/Gate A20 failure
7 beeps	CPU exception interrupt error
8 beeps	Video RAM failure
9 beeps	BIOS ROM bad
10 beeps	CMOS register failure

PHOENIX BIOS

Beep Code	Meaning
1–2–2–3 beeps	BIOS ROM bad
1–3–1–1 beeps	RAM refresh failure
1–3–1–3 beeps	8042 Gate A20 failure
1–3–4–1 beeps	Main RAM failure

17.5 SYSTEM FAILS THE POST WITH A MESSAGE ON THE MONITOR

If you get this far in the POST, the processor, the first 16K of RAM, and most of the motherboard is good. This is symptomatic of defective RAM above the first 16K or one or more defective peripherals. Older BIOS systems (286, 386, and some 486) present a four-digit **error code** followed by a brief description. Some examples of this type of error message are

0201: Memory error, followed by an address and data description
0601: Diskette controller failure
1701: Fixed disk controller failure

Refer to Appendix E for a complete list of **error codes.**

Most newer BIOS programs will display a more verbose message describing the problem followed by the message "Press F1 to enter setup," or the message, "Press Esc to continue." A comical message is "Keyboard bad or missing" along with "Press F1 to enter setup". If the keyboard is bad or missing, how are you supposed to press the F1 key? Someone must have a sense of humor. To diagnose further, follow the clues presented on the monitor and replace suspected hardware with known good spares. The following are some steps for working on this kind of a problem.

Memory Problems

1. Remove the system RAM modules. Slowly and gently, clean the contacts that insert into the motherboard, using a soft pencil eraser as shown in Figure 17.2. Do not use

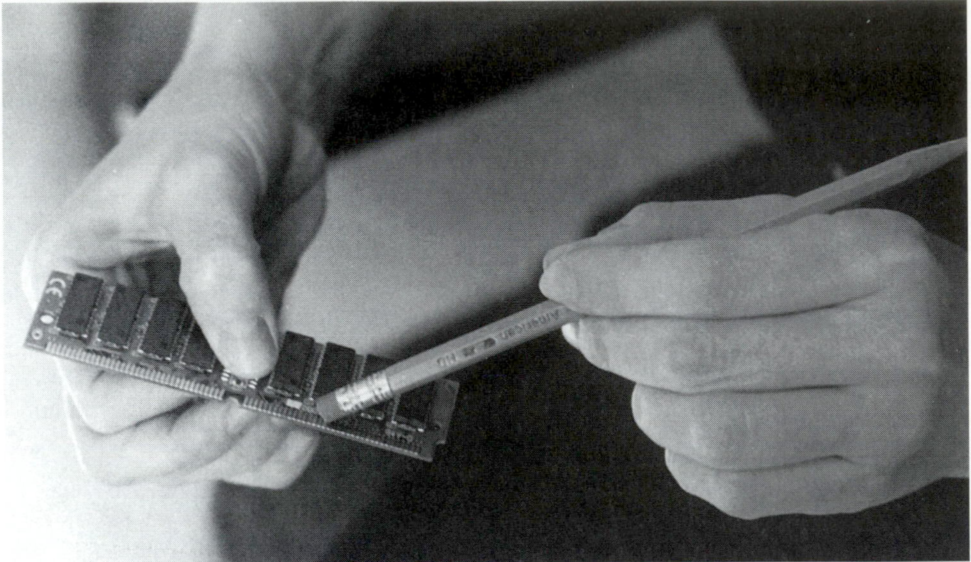

FIGURE 17.2
Cleaning memory module contacts

an ink eraser, because it will remove the gold plating. Insert only enough RAM to complete one bank. Power up the system and observe the monitor for any BIOS memory error messages. Next, replace the remainder of RAM and retest the system.

2. Replace the RAM modules with known good memory (preferred), or, if one is not available, use the second bank in the first bank position. Power up the system and observe the monitor for any BIOS memory error codes or messages. Then, replace the remainder of the RAM, replacing any bad modules with known good modules.
3. Possible bad motherboard. Verify the type of RAM modules used and run the BIOS setup program. Make sure the BIOS settings agree with the type of RAM used. Try using a slower RAM accesses method. Check to see if someone has attempted to overclock the motherboard and processor.
4. Power down the system. Remove everything from the motherboard except the first bank of RAM and the video board. If RAM still fails, replace the motherboard.
5. If the RAM checks out good, one by one reinsert everything removed and retest until you isolate the conflicting device.

Hard Disk Problems

1. As soon as you power up the system with the cover open, listen to the hard disk drive. You should hear the spindle spin up and settle out. If the drive does not spin, check the status of the hard disk activity light. The following is only applicable to IDE drives. If the light goes on immediately after power up and stays on, chances are one of the interface cables is on upside down. Try removing the interface cables and reapply power.

If the drive still doesn't spin, replace the hard disk drive. If the drive does spin, check the cable and any slave device on the cable. If you have a SCSI drive, it may have a spindle control jumper. This jumper is used to sequence the spin up in a multidrive system, thus reducing the initial load on the power supply. Make sure the drive jumpers are set correctly.

2. Next, you will hear the heads perform a short sequence of seeks. If you hear the heads repeatedly seek and, possibly, a light banging noise, the drive has suffered a head crash as shown in Figure 17.3. Sometimes if the crash was not very severe the drive may come ready after several retries but will emit a light scraping sound similar to roller skates on concrete.

FIGURE 17.3
Hard disk crash

3. If the hard disk drive spins up, seeks, and settles, it has performed and passed its own internal POST. This doesn't mean the drive is good, but it is able to spin up and find track 0. Enter the BIOS setup program and navigate to the hard disk setup portion. Almost all of the newer BIOS programs include an auto-detect option as shown in Figure 17.4. If the BIOS fails to detect the drive, try a different IDE port. Also check the MASTER/SLAVE jumpers.
4. If all else fails, try a known good spare drive and cable. If the spare fails, replace the motherboard, disk controller, or interface board.

```
ROM PCI/ISA BIOS (2A59CG0L)
CMOS SETUP UTILITY
AWARD SOFTWARE, INC.
```

STANDARD CMOS SETUP	PASSWORD SETTING
BIOS FEATURES SETUP	IDE HDD AUTO DETECTION
CHIPSET FEATURES SETUP	SAVE & EXIT SETUP
POWER MANAGEMENT SETUP	EXIT WITHOUT SAVING
PCI CONFIGURATION SETUP	
LOAD SETUP DEFAULTS	
ESC : Quit F10 : Save & Exit Setup	↑ ↓ → : Select Item (Shift) F2 : Change Color
Virus Protection, Boot Sequence . . .	

FIGURE 17.4
BIOS IDE auto-detect

Floppy Drive Problems

You can treat floppy drive failures the same way as hard disk failures. However, the floppy doesn't spin and seek the same way that a hard disk does. The BIOS will issue a seek, and it sounds like a buzzing noise. Most BIOS programs allow you to disable the floppy drive seek, so therefore you should check this before suspecting a bad floppy. It is common for the diskette protect shutter to come loose and fall off inside the floppy drive thus jamming the heads. Do not force the diskette in or out or try to pull a dislodged shutter out of the floppy drive. Doing so will permanently destroy the heads. You should remove and disassemble the drive to retrieve the shutter. In addition, if the shutter came off inside the floppy drive, chances are the small shutter spring has dislodged inside too. Do not use the drive until you have found the spring or are sure it is not inside. If the spring should short the $+5\ V_{DC}$ to the $+12\ V_{DC}$, you can destroy the entire system.

17.6 MONITOR DISPLAYS THE BIOS GREETING AND PASSES THE POST BUT WILL NOT BOOT

At this point, the POST has tested the processor, motherboard, memory, and status of known peripherals. A message similar to Figure 17.5 is shown, and the BIOS will attempt to load an operating system from the devices specified in the BIOS boot order as shown in Figure 17.6. Keep in mind that the BIOS messages vary greatly between BIOS manufacturers and year of publication. Remember to *stop, look,* and *listen* when the system

```
                 Award Software, Inc.
                 System Configurations
+------------------------------------------------------------------------------+
| CPU Type          : PENTIUM-S          Base memory          :     640K       |
| Co-Processor      : Installed          Extended Memory      :   15360K       |
| CPU Clock         : 133MHz             Cache Memory         :     256K       |
+------------------------------------------------------------------------------+
| Diskette Drive A : 1.44 M, 3.5 in.     Display Type         :  EGA/VGA       |
| Diskette Drive B : None                Serial Port(s)       :  3F8 2F8       |
| Pri. Master Disk : LBA, Mode 3, 545MB  Parallel Port(s)     :  378           |
| Pri. Slave Disk  : None                L2 Cache SRAM Type:  Pipeline         |
| Sec. Master Disk : None                                                      |
| Sec. Slave Disk  : None                                                      |
+------------------------------------------------------------------------------+
EDO DRAM : Installed
```

FIGURE 17.5
BIOS screen

first attempts to boot. Most systems are set to try the floppy, then try the hard disk. Newer systems have CD-ROM and network boot options. A common mistake is leaving a non-bootable floppy diskette in the diskette drive when the system is reset. When this occurs, a message similar to the following will be displayed:

```
              ROM PCI/ISA BIOS (2A59CG0L)
                 BIOS FEATURES SETUP
                AWARD SOFTWARE, INC.
+--------------------------------------------+---------------------------------+
| Virus Warning               : Disabled     | Video BIOS Shadow  : Enabled    |
| CPU Internal Cache          : Enabled      | C8000-CBFFF Shadow : Disabled   |
| External Cache              : Enabled      | CC000-CFFFF Shadow : Disabled   |
| Quick Power On Self Test    : Enabled      | D0000-D3FFF Shadow : Disabled   |
| Boot Sequence               : A, C         | D4000-D7FFF Shadow : Disabled   |
| Swap Floppy Drive           : Disabled     | D8000-DBFFF Shadow : Disabled   |
| Boot Up Floppy Seek         : Enabled      | DC000-DFFFF Shadow : Disabled   |
| Boot Up NumLock Status      : On           |                                 |
| Security Option             : Setup        |                                 |
| PCI/VGA Palette Snoop       : Disabled     |                                 |
|                                            +---------------------------------+
|                                            | ESC : Quit    ↑↓→ : Select Item |
|                                            | F1  : Help    PU/PD/+/- : Modify|
|                                            | F5  : Old Values (Shift) F2 : Color |
|                                            | F7  : Load Setup Defaults       |
+--------------------------------------------+---------------------------------+
```

FIGURE 17.6
BIOS boot order

```
Non-System disk or disk error
Replace and strike any key when ready
```

This situation can be avoided by changing the BIOS boot order from A, C to C, A. This will force the computer to boot the hard disk first. The downside to this is that when you need to boot a floppy you will have to reset the BIOS boot order first. Next, you must determine if the boot failure is caused by a hardware or software problem. By observing the hard disk (or floppy disk) and monitor activity you can get a general idea of how far the system has gotten through a boot process. If the process fails quickly or does not start, the problem is most likely a hardware problem. If the boot process runs for a while and then fails with messages, possibly with a deep blue screen, the problem is most likely software corruption or missing drivers.

Boot Problems

The following steps outline a procedure to isolate the cause of a boot failure.

1. As stated earlier, make sure nothing is in the floppy drive (unless you desire to boot from floppy) and reset the system. Immediately observe the system lights and listen for sounds. The usual sequence is the floppy light illuminates. Next, the floppy seeks, and last, the hard disk light flickers. If there is no activity on the hard disk and the light does not come on, or comes on and stays on, go back to the BIOS setup and check the hard disk type. If available, use the **auto-detect** feature as shown in Figure 17.7. If this fails, go back to Section 17.5 and troubleshoot the drive problem.

```
              ROM PCI/ISA BIOS (2A59CG0L)
                  CMOS SETUP UTILITY
                 AWARD SOFTWARE, INC.

HARD DISKS        TYPE  SIZE  CYLS  HEAD  PRECOMP LANDZ SECTOR MODE
---------------------------------------------------------------------
Primary Master :

        Select Primary Master Option (N=Skip) : N

  OPTIONS   SIZE   CYLS  HEAD  PRECOMP  LANDZ  SECTOR   MODE
  -----------------------------------------------------------
  2 (Y)     545     528   32        0   1056     63      LBA
  1         545    1057   16    65535   1056     63   NORMAL
  3         545     528   32    65535   1056     63    LARGE

Note: Some OSes (like SCO-UNIX) must use "NORMAL"  for installation
                         ESC : Skip
```

FIGURE 17.7
IDE auto detection screen

2. If the drive attempts to boot but fails with a message, the problem could be from many sources, including corrupted software or a virus, and you will need to gain clues from any error messages on the monitor. If the system appears to boot and read from the hard disk but hangs, insert a startup diskette or a bootable DOS diskette. Reset the system and allow it to come up. If the system will not boot from the floppy, go back and recheck all cables and setup parameters. If nothing works, you may have a bad motherboard. If the system boots from the floppy diskette, you should attempt to access the hard disk. If you cannot access the hard disk, proceed to step- 3. If you can, the boot sector may be corrupted. Using the distribution boot diskette attempt to reload the boot sector or reload the operating system.
3. From the boot diskette, run the **FDISK** program to determine if FDISK can recognize the hard disk. If FDISK cannot recognize the hard disk, go back to Section 17.5 and troubleshoot or replace the hard drive. Select the Display Partition Information option, and you should see a screen similar to Figure 17.8. Make sure partition C is set to A for active.

```
                    Display Partition Information

Current fixed disk drive:  1
 Partition   Status   Type    Volume Label    Mbytes    System    Usage
   C: 1         A    PRI DOS                    519      FAT16     100%


Total disk space is   519 Mbytes (1 Mbyte = 1048576 bytes)


Press Esc to continue
```

FIGURE 17.8
Partition screen

4. Once FDISK has properly recognized or set up the hard disk, you will have to reformat the disk. Keep in mind that doing this erases everything that was on the disk. If you are at this step, you could not see any data before, so be careful and only do this if you are sure that there is no other method of recovery. If formatting fails, replace the disk drive.
5. Finally you will need to reload all software or restore from backup.

17.7 SYSTEM BOOTS COMPLETELY AND CRASHES

When the system boots completely and then crashes, this is usually an intermittent problem and usually results in what is jokingly referred to as "the blue screen of death." The system boots up and runs for a while but will crash while running applications. Before going on, stop and consider the behavior of the failure, as it will play an essential role in failure diagnostics. If a particular application crashes the system at the same spot in the same way, you most likely have a bad or corrupted software application. This can be verified by reloading the suspect application or running other applications of equal complexity. If this is the case, you may want to try reinstalling the software.

There are a couple of additional possibilities for this kind of failure. The first is a virus such as the Trojan horse infecting the computer. You should have a good antivirus program and encourage the users to purchase one and to keep it updated. The second is attacks from computer hackers using programs know as "Win-nukes." Win-nuke programs will target a computer by sending a rapid series of bad network packets to a particular IP address. This is usually solved by applying the latest software patches to an operating system.

Most intermittent failures are caused by hardware. Many years of experience has proven that power supplies and anything on a connector is suspect. Modern solid-state components rarely fail, with digital components being the most reliable. The one exception to this rule is the RAM modules. Again, experience has shown random application crashes to be caused by memory failures. Often memory tests do not isolate memory problems. However, after changing the memory module, the crashing problem will be solved.

The first order of business is checking the power supply. Refer back to Section 17.3 and check the power supply voltages while accessing the hard disk. If the voltage begins to fluctuate beyond the specified limits, replace the power supply. Another indication that the power supply is having regulation problems is flickering on a VGA screen. After verifying the power supply and board and IC seating, the following steps will aid in diagnosing intermittent problems.

1. Check to see if someone has overclocked the system. This is a bad idea and causes unreliable operation and intermittent problems.
2. As shown in Figure 17.2, remove the memory modules and clean the contacts. Be very careful not to create static electricity. Check the speed and type of memory used, and never mix types within a bank. Generally, avoid mixing memory types at all. Also, try replacement of the memory.
3. Remove and reseat all the IO boards. Clean the contacts if there is any evidence of contamination.
4. Reseat the microprocessor in the socket.
5. Replace the disk drive IDE cable. Modern operating systems use the disk as virtual memory, swapping memory in and out as needed between the disk drive and memory subsystem. All it takes is a poor connection and eventually a bit or two will be dropped. This will result in a system crash.

17.8 SYSTEM BOOTS OK BUT HAS A PERIPHERAL PROBLEM

In our discussion, we will address the most common peripheral problems (keyboard, mouse, video, floppy, serial and parallel ports) one at a time.

Keyboards are cheap expendable items. No attempt should be made to repair them unless necessary. Even the best attempt may lead to sticky or repeating keys. This is not worth the frustration and repeated service call. Simply replace the keyboard with a new one.

Mice are also cheap, expendable items but are usually repaired by a simple cleaning. The problem is caused when contamination is picked up by the rolling ball and deposited onto the roller shafts. This causes sticky or jerky movement of the cursor. To solve these problems, simply remove the ball and using your fingernail or other soft tool remove any contamination from the shaft. You can also use compressed air to clean the mouse. Replace the ball and test the mouse operation. If that doesn't fix it, replace the mouse.

Video problems are diagnosed by observing the monitor. If the system appears to boot up with a blank or black screen, try another monitor. If you see the normal boot-up messages but the monitor blanks, cuts out, or appears scrambled when Windows starts, someone may have set the video mode beyond the monitor capability. To solve the problem in Windows 95/98, restart the system in safe mode and reset the monitor mode to standard VGA 640 by 480. After that, you can experiment with different resolutions. To solve the problem in Windows NT, restart the system in VGA mode. If the complaint is poor video due to jitter or unwanted artifacts on the screen, you may have a faulty video card. You may wish to try updated video drivers before replacing the board. If the complaint is poor focus, color, or brightness, you must consider that most monitors are based upon a cathode-ray tube, or CRT. The cathodes wear out over time and begin to lose electron emission. This results in the inability to drive the phosphorus illumination, and the beam goes out of focus. The three cathodes wear at different rates, which results in poor color balance. This can be corrected for awhile by adjusting the cathode gain controls and the screen grid voltage. However, this should never be attempted unless you are experienced with monitor repair because of the high voltage (40,000 volts) inside the monitor. In any event, this type of fix may be only a temporary solution because the problem will probably worsen over time.

The floppy diskette drive is another example of a cheap expendable item. Today you can replace the drive for as little as $20. If you have a failure, simply replace the drive, because it doesn't pay to repair them anymore. Sometimes in a large computer lab or company with many PCs you may find that half the floppy drives are aligned near one end of the specification and the other half are aligned at the other end of the specification. A little bit of wear may cause data stored on one floppy drive *not* to be read by another floppy drive. Both drives will read and write correctly to themselves but not to each other. Aligning the drive should solve the problem, but it may not be worth the time. Consider replacing one or both of the floppy drives.

Most serial and parallel ports are contained on the motherboard and rarely fail. Most problems are due to setup conflicts with other adapter boards or bad cables. If a failure does occur, replace the port. If the port is on the motherboard, remember to disable the motherboard, port before installing the replacement port board to avoid conflicts. This is usually done in BIOS or by jumper removal on the motherboard.

17.9 TROUBLESHOOTING NETWORKS

Troubleshooting network problems can be a difficult and time-consuming task. Since most networks have many connections and can cover a large geographical area, one should use the divide and conquer approach. A network problem will affect many users, and in a production environment, this is a disaster. Before you even touch a piece of wire, you should be able to isolate a problem down to a segment of the network. Work on a busy network must be handled with great care so as not to worsen the problem or cause busy working segments to be disconnected. Occasionally, there will be times when you must bring the network down to find a problem. When such a condition exists, notify all users, personally if necessary, before performing any work. Before proceeding, ascertain if the problem is local to just one machine, within a group, within the entire LAN, or outside of your control. One example may be that a user cannot access a website outside your network, but all internal networks are fine. Here the problem is with the ISP and can only be resolved by placing a call to the ISP. Another example is when a single user cannot access the network but everyone else can. Here the problem may be the PC, cable, or communicore. Always refer to your network topology diagram. If one does not exist, you should go over the network and make a sketch. It is very important to understand the electrical topology. For example, suppose you are working on a thin-wire Ethernet network consisting of two or three segments and a server. You discover the entire network is down because no one can access the server but the server is up and running. Thin-wire Ethernet is both electrically and logically the bus topology. When the network breaks, it is like those finicky Christmas tree lights—when one goes out they all go out. Here a short or open in the wire brings down the entire segment of network as shown in Figure 17.9. The same is true for a thick-wire Ethernet network.

Our discussion on troubleshooting a bus topology TCP/IP Ethernet network is based upon the example shown in Figure 17.9. This is a companywide network in a building

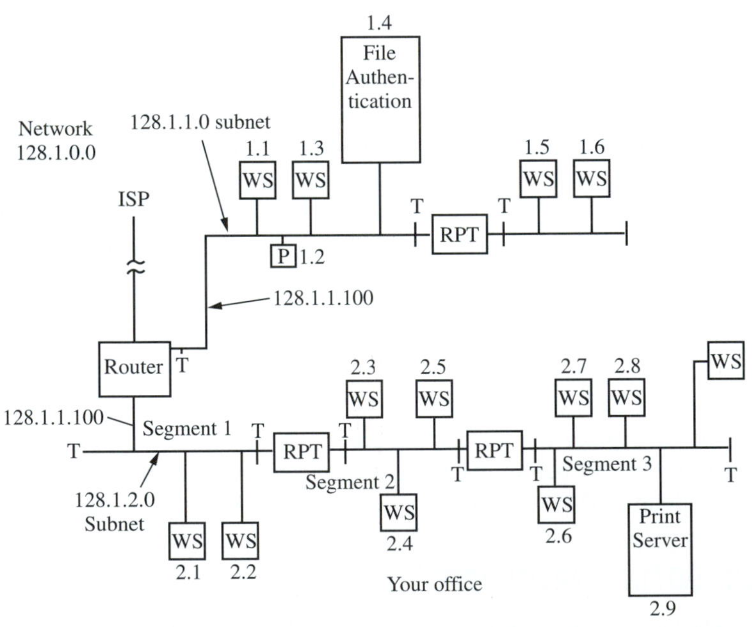

FIGURE 17.9
Bus network problem

with a couple of wings. Just before problems were reported, a construction company was called in to perform some building renovations. In this example, your office is on the 128.1.2.0 subnetwork and your station address is 128.1.2.4. A user on 128.1.2.7 places a call stating she cannot log into the file server. Soon after, the help desk begins receiving additional calls from around the company about similar problems. Now the problems seem to be everywhere on the network. Your first impression would be that the server has gone down, but careful analysis will reveal that assumption is false. The following sequence of steps will help to reveal the problem.

1. Attempt to contact the server's IP address by using the **PING** command as follows, PING 128.1.1.4. If the ping succeeds, the network between you and the server is fine. In our example, the ping will fail.
2. Attempt to contact a known powered-up workstation on the server segment—for example, 128.1.1.1. If the ping succeeds, there is a problem with the server. In our case the ping will fail, thus proving it is a network problem.
3. Attempt to contact your side of the router by issuing a ping to 128.1.2.100. If the ping fails, then you would attempt to contact a workstation on your segment. In our case, the ping to the router will succeed.
4. Attempt to contact the faulty side of the router by issuing a ping to 128.1.1.100. If the ping fails, a faulty router is evident. In our case, the ping will succeed.
5. You have now narrowed the problem down to the server segment of the network. There is no reason to bother anyone on the 128.1.2.0 network, because some may be using other services outside your network. You may also wish to ping outside to the ISP or World Wide Web just to verify operation.
6. Now is the time to touch the wire. With an **ohmmeter** and a LAN tester, proceed to the area containing the router. As shown in Figure 17.9, we can see that the server segment terminates at the router port and a repeater. Remove the router BNC- tee connector and measure the resistance at the exposed pin to the shell as shown in Figure 17.10. The resistance should be around 25 Ω. If the resistance is substantially lower, you have a short in the cable. If the resistance is around 50 Ω, the cable is cut or open. In our example the cable is open somewhere. Now comes the chore of finding the open, which may involve climbing in ceilings and closets.
7. Remove the LAN cable from the tee and connect it directly to a **time domain reflectometer (TDR)** as shown in Figure 17.11. Select the Wire Lengths option and observe the readout. It will give a good estimate of the distance between you and the open. If you don't have a TDR, simply proceed to the next station on the LAN segment and check the cable.
8. Reconnect the cable and tee to the router port, and proceed to the station closest to where you think the center of the faulty LAN segment is. Remove both ends of the cable from the tee. Using the ohmmeter, measure the resistance of both open ends of the cables by probing the center pin to the shell. One cable will be 50 Ω and the other will be open. If you are not sure which end is which, attach the good cable to one end of the tee and a 50-Ω terminator to the other end of the tee. Now connect the tee to the workstation.
9. Bring up the workstation and log in locally. Attempt to contact your side of the router. If this fails, attempt to contact the server. In our example this should succeed.

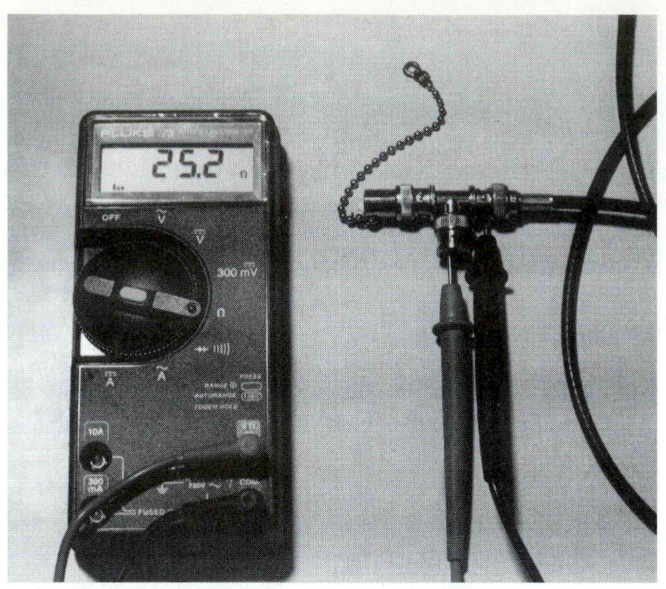

FIGURE 17.10
Measuring the resistance of the network segment

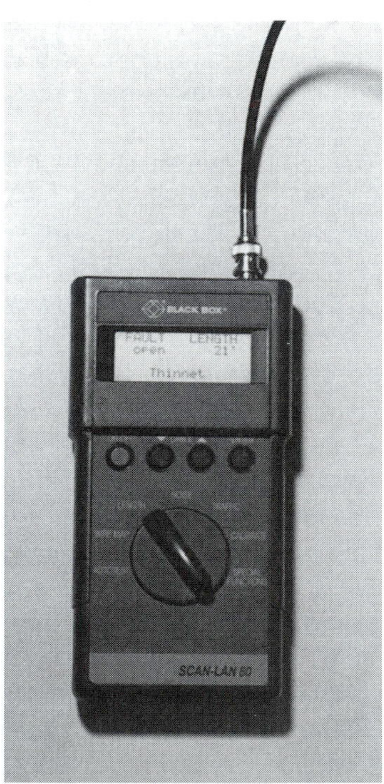

FIGURE 17.11
Measuring the length to a fault in the cable

10. Now you have determined the problem to be an open connection somewhere between you and the router. Repeat the process of divide and conquer until you isolate the wire in question.
11. Once you have found the fault, in this case a cut wire, repair the problem. If possible, avoid splicing the cable but replace the entire run. If you must splice due to a long length or difficult run, then crimp on new BNC connectors and use a barrel connector between the connectors.

Recall what we said earlier. Just before the problems were reported, a contractor was called in to perform some building renovations. With handy Saws-All in hand, they did indeed saw all, right through the network coax cable. Knowing this ahead of time would have saved a lot of time because it would be the first place we would look. As a general rule, always ask around if anything unusual was occurring just before the network went down. Another common problem on a thin-wire network is the BNC connectors where the wire is crimped. If these are poorly crimped or flexed a lot by moving the PC, the wire can pull out or short between the center conductor and braided shield. A close visual inspection should reveal a potential problem. Another test is to connect an ohmmeter to the center portion of the tee and observe the reading while wiggling the wires. Any deviation in the reading indicates a poor connection.

In the case of a wired star topology as in Ethernet over twisted pair, the network is a logical bus wired in a star configuration as shown in Figure 17.12. A short or open in one run will not affect the remainder of the network. This topology is far simpler to troubleshoot because you can quickly determine the faulty segment. The hub shown in Figure 17.13 has LEDs on every port, one to indicate a link (good connection) and one for traffic. Most network cards also have LEDs on them. Recall that twisted-pair cable uses two pairs of wires, one to receive and one to transmit. The LED indicates a good con-

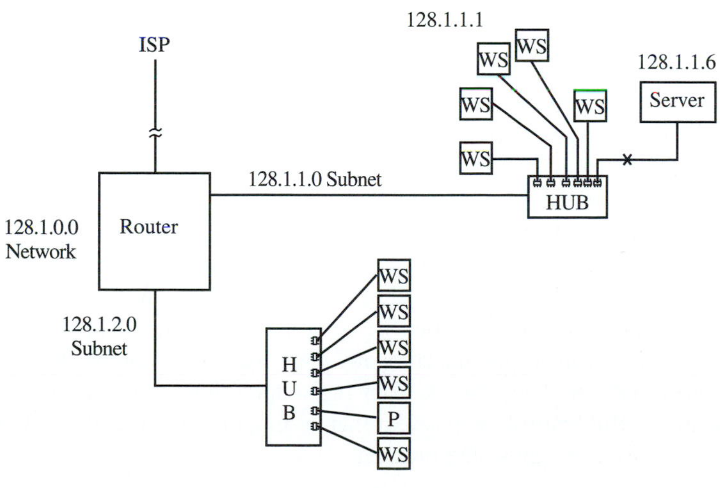

FIGURE 17.12
Ethernet star network

FIGURE 17.13
Network hub

nection to the receive pair only. In this example, you are working on a twisted-pair Ethernet network consisting of two segments and a server. Here you think the entire network is down because no one can access the server but the server is up and running. Remember, a twisted-pair network is logically bus topology wired as a star. When the network breaks, it only affects the broken wire run. Here a short or open in the wire does not bring down the entire segment of network.

Our discussion on troubleshooting a star topology, TCP/IP Ethernet network is based upon the example shown in Figure 17.12. This is a companywide network in a building with a couple of wings. Just before problems were reported, a construction company was called in to perform some building renovations. In this example, your office is on the 128.1.2.0 subnetwork and your station address is 128.1.2.6. A user on 128.1.1.1 places a call stating he cannot log into the file server. Soon after, the help desk begins receiving additional calls from around the company about similar problems. Now the problems seem to be everywhere on the network. Your first impression would be that the server has gone down, but careful analysis will reveal that assumption is false. The following sequence of steps will help to diagnose the problem.

1. Attempt to contact the server's IP address by using the PING command as follows, PING 128.1.1.6. If the ping succeeds, the network between you and the server is fine. In our example, the ping will fail.

2. Attempt to contact a known powered-up workstation on the server segment—for example, 128.1.1.1. If the ping succeeds, there is a problem with the server or the wire run to the server. In our case the ping will succeed, thus proving it is not the entire network.
3. Using a LAN or cable tester, proceed to the server and check the link LED on the network interface of the server. If the LED is not illuminated, check the wire using the LAN tester. If the LAN tester indicates a good link, you have a server or network interface problem. In our case, the link is bad.
4. Using a LAN or cable tester, proceed to the wiring closet or communicore containing the hub to which the server is connected. Since we disconnected the server, the link light should be out. Connect the wire going to the server into the cable tester and check the wire pairs as shown in Figure 17.14. You only need to worry about the wire pairs 1, 2 and 3, 6 for 10BASET and 100BASET. If the pairs check out, then you have a bad port on the hub, which can be simply resolved by moving the server to a spare port. You may wish to verify the port by connecting the LAN tester directly to the port using a spare patch cable. In our case, the server wire fails.
5. Connect the server wire directly to a time domain reflectometer (TDR). Select the Wire Lengths option and observe the readout. It will give a good estimate of the distance between you and the open.

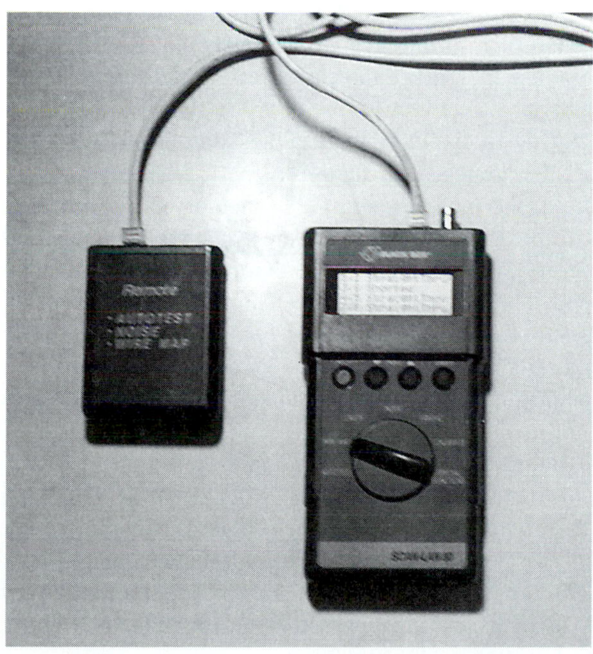

FIGURE 17.14
Twisted-pair LAN tester

6. Now you have determined the problem to be an open connection somewhere between the server and the hub. Using the fault information from the TDR if this is available, trace the wire run visually until you isolate the fault.
7. Once you have found the fault, in this case a chafed and shorted wire pair, repair the problem. If possible, avoid splicing the cable but replace the entire run.

17.10 ADDITIONAL DIAGNOSTIC TOOLS

A personal computer is a complex piece of equipment. In addition to its many subsystems, the operating system adds another level of complexity to any servicing problems.

When repairing personal computers, most people use a test and replace methodology. In other words, they troubleshoot the system until they find that one of the major components is defective and replace that component. Today component parts are so inexpensive that it no longer pays to repair most failures to the chip level. For example, in 1984, the IBM PC/XT cost around $6000 for a system with a 10 MB hard drive and 640K memory. The replacement motherboard cost around $700. As we can clearly see, spending a couple of hours to find and replace a $10 IC was worth the effort. Today a Pentium III 500-MHz computer costs around $1000 and contains a 6.4 GB hard drive and 64 MB of RAM. The motherboard costs only $85 (not including the processor) to replace. To make service worse, the motherboard contains very high-density surface-mounted ICs. Finding the bad IC is simpler, because you can guess at the part and be correct, but these are all but impossible to solder by hand. Besides, for $85, you can replace the motherboard with a newer BIOS, resulting in a virtually new machine. A motherboard failure today is a welcome excuse to upgrade the computer. A hard disk failure is an excuse to get a larger drive, although you will have to restore any data on it.

Several diagnostic tools aid in the troubleshooting process. Many companies sell general diagnostic software that you can use to check out the various parts of a computer system—for example, the motherboard, CPU, memory, I/O ports, and floppy and hard disk drives. You will find that some products are system independent, others require DOS, and still others operate under Windows. For the most part, diagnostic software products operate as advertised. They provide system benchmarks and point us in a troubleshooting direction. Diagnostics with an understanding of the inner workings of a personal computer and the operating system form the basis for correcting problems. Figures 17.15 and 17.16 show the opening menus for Micro-Scope and Troubleshooter diagnostics, respectively.

Floppy drive diagnostics are available to aid the test and alignment of floppy diskette drives. Using this tool will enable you keep all your equipment compatible and eliminate annoying reports that the hardware is causing problems. Figures 17.17, 17.18, and 17.19 illustrate the use of the Drive Probe floppy disk alignment utility. Another type of repair is data recovery. **Norton Utilities** from Symantec Corporation is very useful for many types of diagnostic analysis and preventive maintenance repairs. Figure 17.20 shows how to open the Norton Utilities.

When you install the Norton utilities onto your system, a new icon is added to the lower right corner of the Windows 95/98 screen, right next to the clock. This is an additional control panel for the Norton utilities. Simply double clicking on the icon will bring

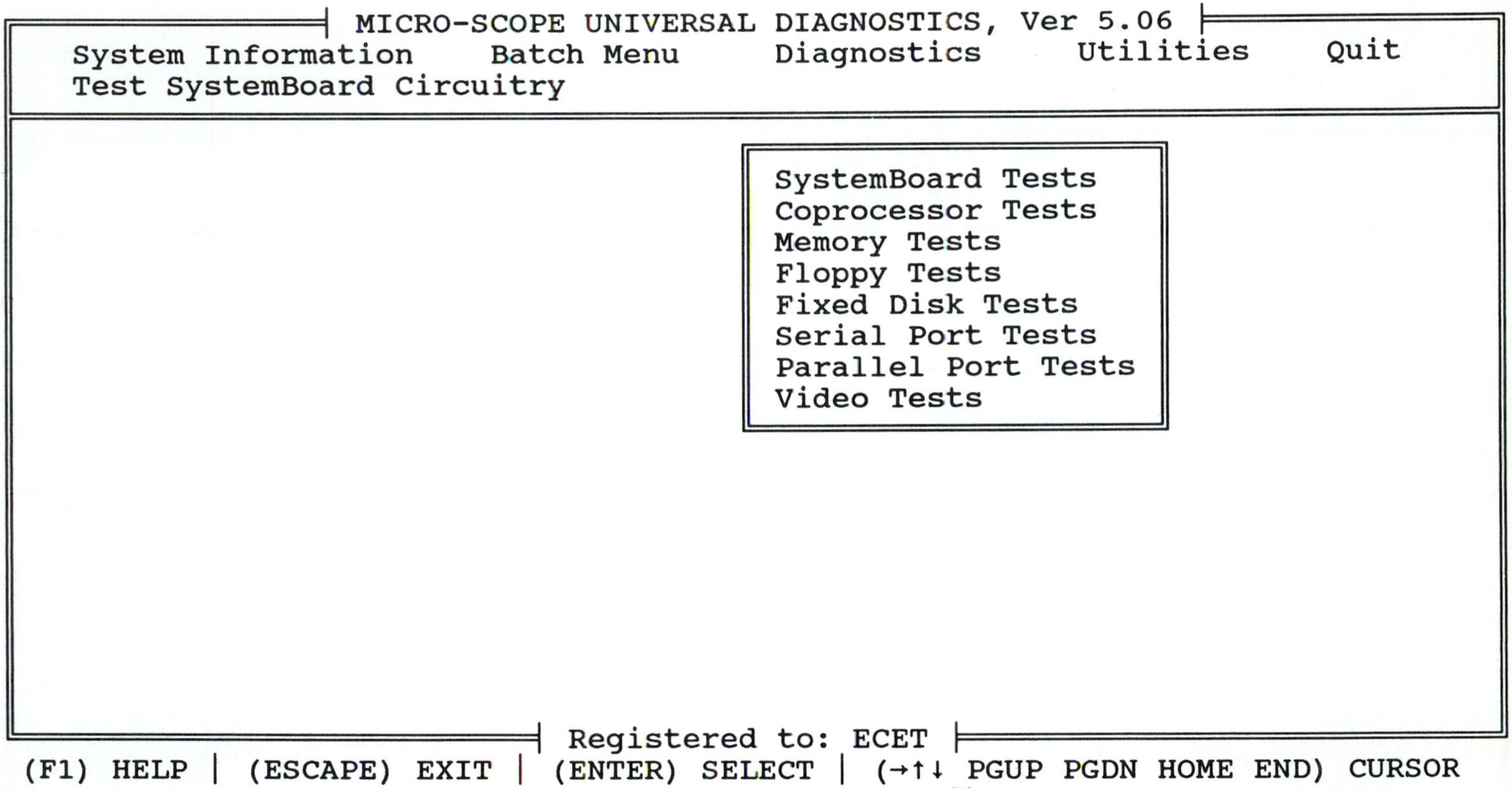

```
MICRO-SCOPE UNIVERSAL DIAGNOSTICS, Ver 5.06
System Information    Batch Menu    Diagnostics    Utilities    Quit
Test SystemBoard Circuitry

                                   SystemBoard Tests
                                   Coprocessor Tests
                                   Memory Tests
                                   Floppy Tests
                                   Fixed Disk Tests
                                   Serial Port Tests
                                   Parallel Port Tests
                                   Video Tests

Registered to: ECET
(F1) HELP | (ESCAPE) EXIT | (ENTER) SELECT | (→↑↓ PGUP PGDN HOME END) CURSOR
```

FIGURE 17.15
Micro-Scope diagnostics

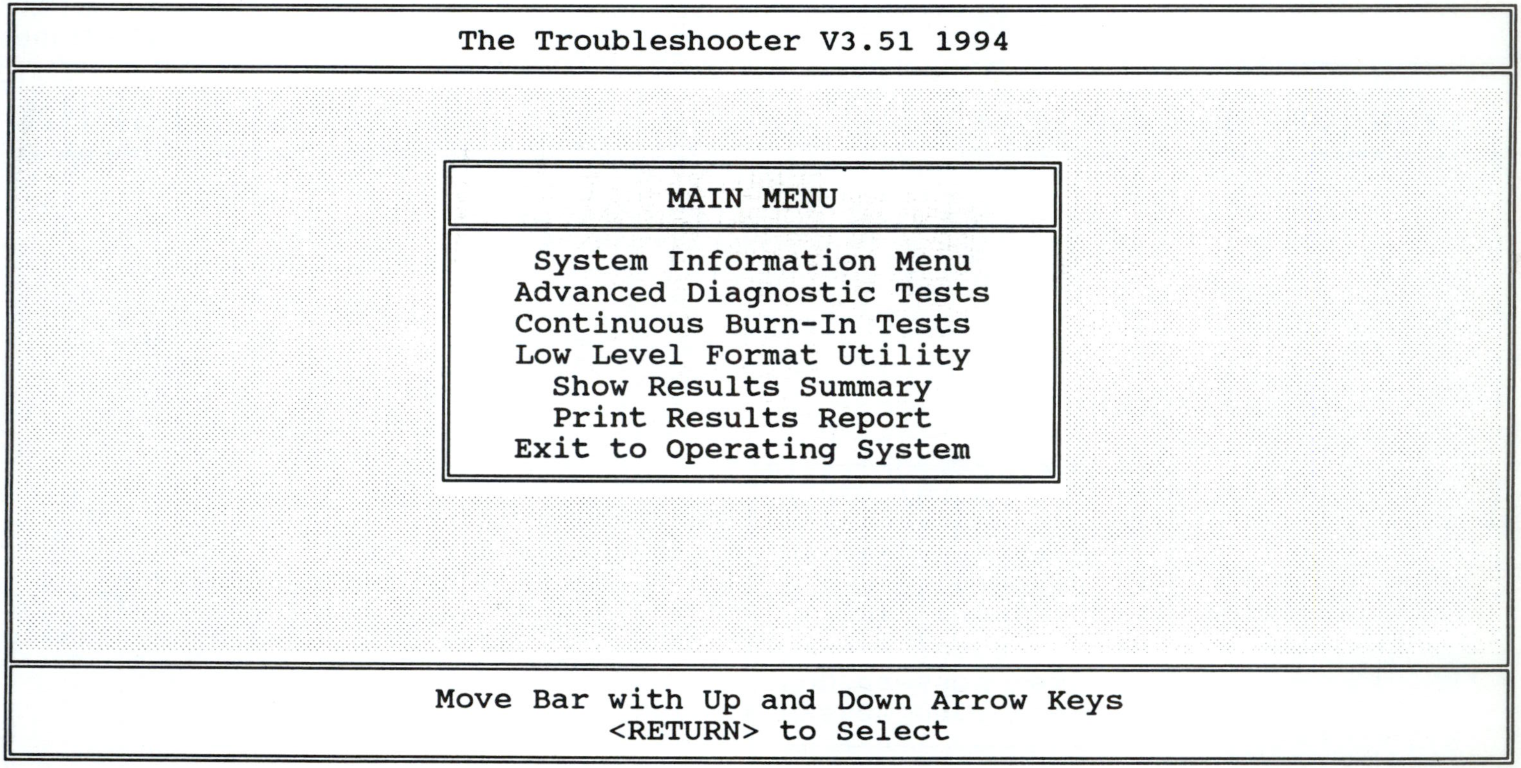

The Troubleshooter V3.51 1994

MAIN MENU

System Information Menu
Advanced Diagnostic Tests
Continuous Burn-In Tests
Low Level Format Utility
Show Results Summary
Print Results Report
Exit to Operating System

Move Bar with Up and Down Arrow Keys
<RETURN> to Select

FIGURE 17.16
Troubleshooter diagnostics

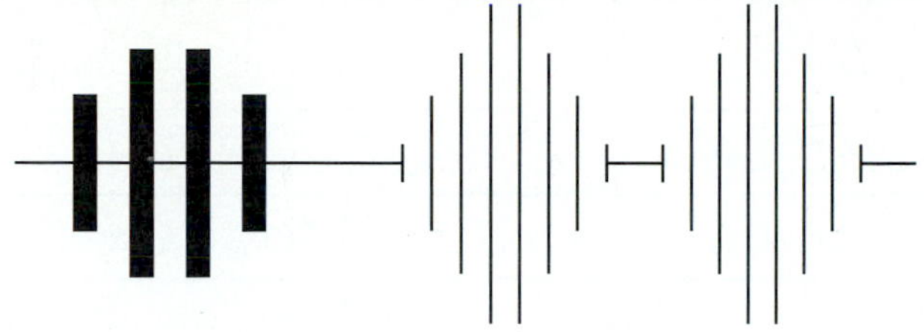

```
D r i v e   P r o b e                  Version 1.07a
A Product of
     Accurite Technologies Inc.
      48460 Lakeview Blvd
     Fremont, Ca 94538-6532
         (510) 668-4900

   © Copyright 1990-1996. All rights reserved.
   Drive Probe and Accurite Technologies are
    trademarks of Accurite Technologies Inc.
```

FIGURE 17.17
Drive Probe opening screen

```
  MAIN Menu    ↑↓Moves Menu Bar,  'Enter'-To Select,  'Exit' to End

 Main Menu
 Parameter Assignments
 Automatic Testing
 Manual Testing          Assign Parameters
 Read/Write Testing     Floppy drive type
 Utility Functions      Compute
 Save Configuration     Read an    Drive For Assignment
 Load Configuration     Step ra   1st physical drive 3 1/2 1.4 Mb
 Video Monitor          HRD rea   2nd physical drive - No Parameters -
 Exit                   Display   3rd physical drive - No Parameters -
                        Samples   4th physical drive - No Parameters -

 Drive 1 Selected as [ 3 1/2"  1.4Mb 300 RPM ] Location: Track 0 Head 0
```

FIGURE 17.18
Selecting the drive to test

AUTOMATIC Drive Test 'Esc'-For Previous Menu

Test	Track	Head 0 Data	Head 1 Data	Test Limits	Results	
Speed	NA	300 RPM / 199.8 mS		300 + 6 RPM	Pass	NA
Eccentricity	44	0 uI	NA	0 + 300 uI	Pass	NA
Radial	0	75% -400uI	87% -200 uI	50 - 100%	Pass	Pass
Radial	40	71% -500 uI	81% -300 uI	50 - 100%	Pass	Pass
Radial	79	52% -900 uI	58% -750 uI	50 - 100%	Pass	Pass
Azimuth	76	12 Min	-8 Min	0 + 30 Min	Pass	Pass
Index	0	595 uS	653 uS	400 + 600 uS	Pass	Pass
Index	79	554 uS	659 uS	400 + 600 uS	Pass	Pass
Hysteresis	40	50 uI	NA	0 + 250 uI	Pass	NA

UI = Micro-inches US = Microsecond MS = Millisecond
Min = Minutes NA = Not Applicable NT = Not Tested

Note: Radial is expressed as LOBE RATIO and OFFSET from track center line.
Auto Test Completed 'Esc' For Previous Menu

FIGURE 17.19
The test results

up the control panel as shown in Figure 17.21. The control panel displays common system functions and resources such as CPU usage, disk usage, fragmentation and other useful information about the system. It will also warn when resources are nearing limits or efficiency is impacted.

Included with Norton Utilities is the Norton System Information manager (shown in Figure 17.22) with a much more user friendly GUI panel than was available with older utilities. Figure 17.23 shows the results when the memory tab is selected. Figure 17.24 illustrates the disk optimizer known as Norton Speed Disk in action. This utility will defragment the disk drive, thus optimizing the performance of the system. Probably one of the best utilities ever created is the Norton Disk Doctor (NDD). If the drive begins to fail, you may be able to recover precious data before a total failure occurs using this utility. Norton Disk Doctor is part of Norton Utilities.

POST-code diagnostic boards are another diagnostic tool. These boards are plugged into one of the expansion slots on the PC motherboard. They display the POST codes that the computer generates during the booting process. Once you know a POST code, you can interpret its meaning by looking it up in a book supplied with the board by the manufacturer. Although POST-code boards are helpful, POST codes have never been standardized. The universal microprocessor tester (detailed in Chapter 18) was designed by the authors to obtain the same result. The universal microprocessor tester also offers single-step capability, which enables the user to step through the boot-up process one instruction at a time. This process can provide additional insight about the nature of a problem.

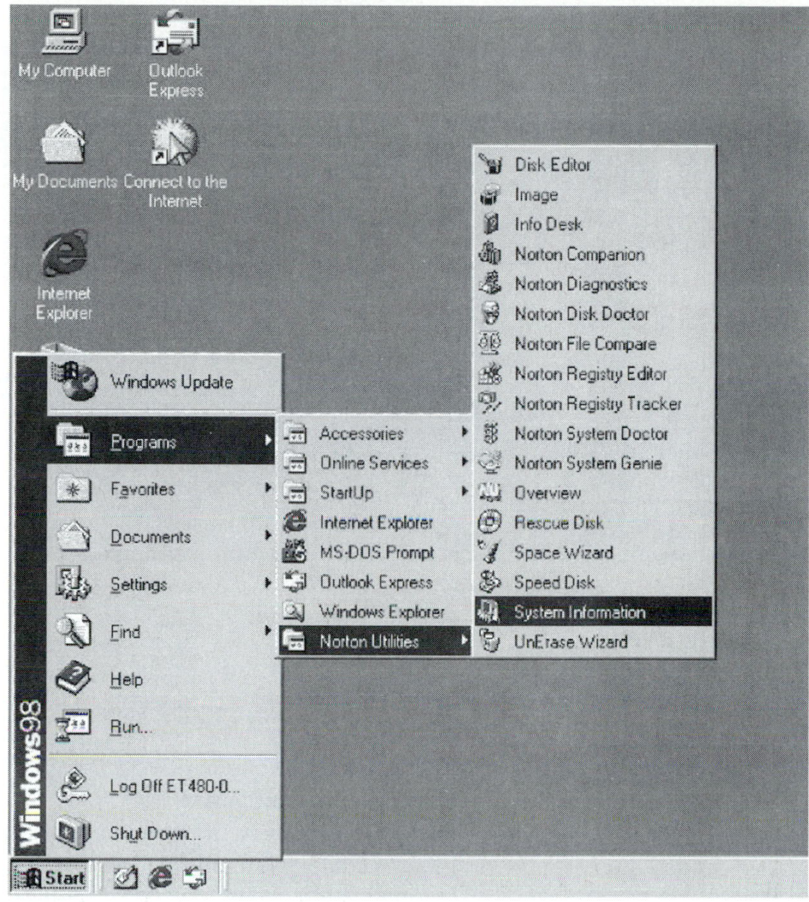

FIGURE 17.20
Norton Utilities startup

17.11 TROUBLESHOOTING CHECKLIST

The following troubleshooting checklist provides a general guide to the possible problems and related symptoms that a malfunctioning system can display.

System Problems

Symptom	**Checks**
System does not power up	System unit not plugged in
	Power cord loose or broken
	Wall socket faulty or circuit breaker open
	Short circuit in computer system causing power supply shutdown
	Power supply faulty

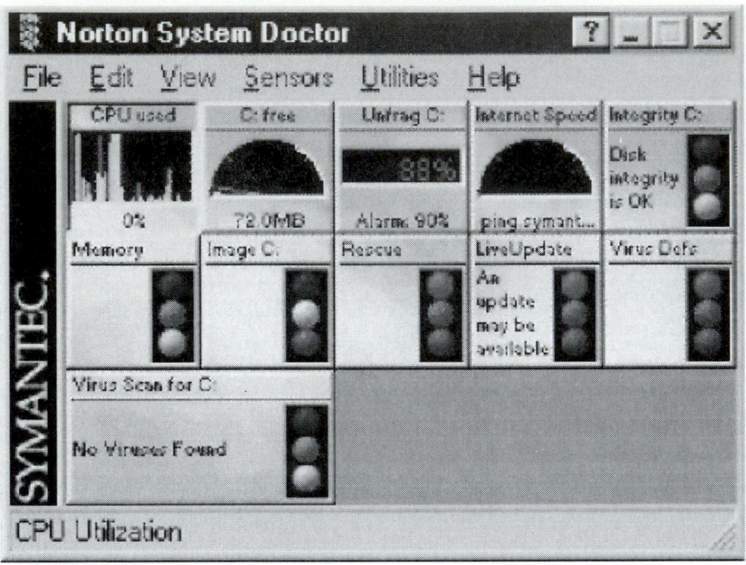

FIGURE 17.21
The Norton control panel

System does not boot	Nonbootable diskette in drive A
	Hard drive not initialized
	System files on hard drive missing or damaged
	Faulty CMOS setup information
	Incorrect motherboard dipswitch settings
	Faulty RAM memory chip (especially in first 64K of memory)
	Loose or defective disk drive control, or data or power cable
	Power supply loaded past wattage limit
	Expansion card not seated in slot
System freezes, crashes, or reboots spontaneously	Faulty RAM memory chip
	Faulty CACHE memory chip
	Loose motherboard power connector
	Expansion card not seated in slot
	Conflicting hardware assignments (DMA channel, IRQ, hardware port address)
	Incorrect or conflicting software configuration
	Memory resident program or driver conflict
Keyboard does not respond	Keyboard not firmly plugged into motherboard
	XT-style keyboard used with AT motherboard or AT-style keyboard used with XT motherboard

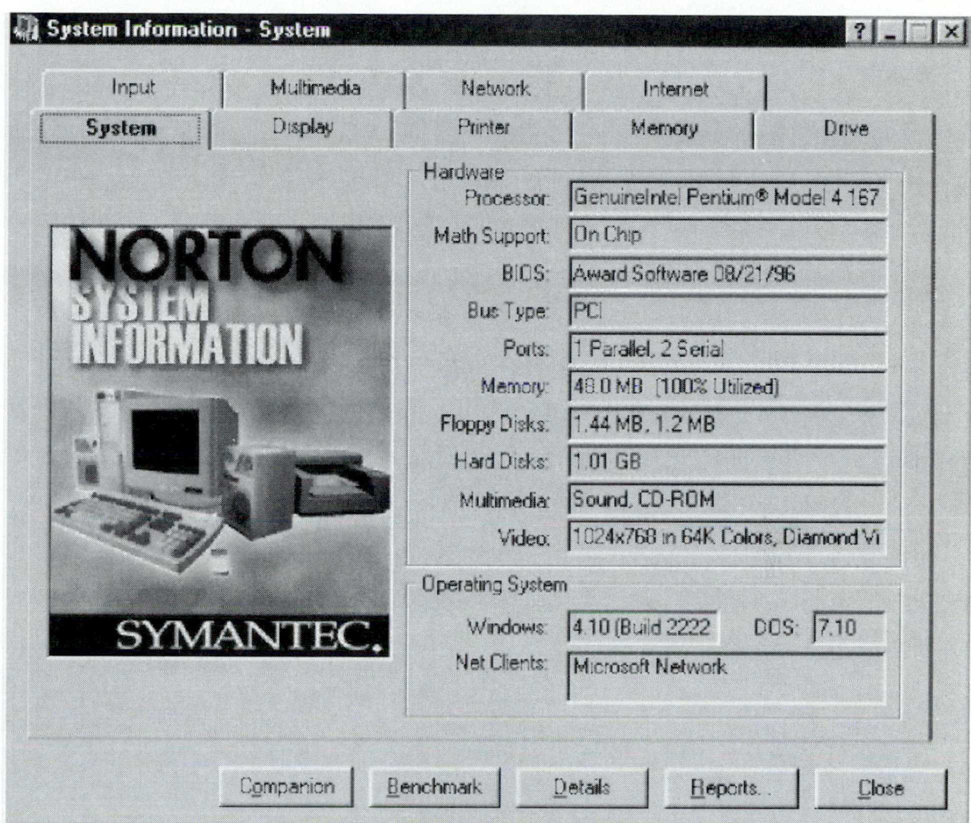

FIGURE 17.22
The System Information panel

	Switchable keyboard set to incorrect type of motherboard (XT/AT) Keyboard lock switch in lock position Keyboard lock switch wiring loose on motherboard or connector reversed
No sound from speaker	Speaker connection to motherboard loose or defective Speaker or wiring faulty Program does not make use of speaker
Real-time clock/calendar malfunction	No real-time clock/calendar installed (8088 system only) Clock support battery weak, dead, or missing Clock/calendar set incorrectly
Noise from system unit	Normal power supply noise (fan hum, air movement) Normal hard disk drive noise (hum or slight whine while running; chatter or faint squealing when heads move)

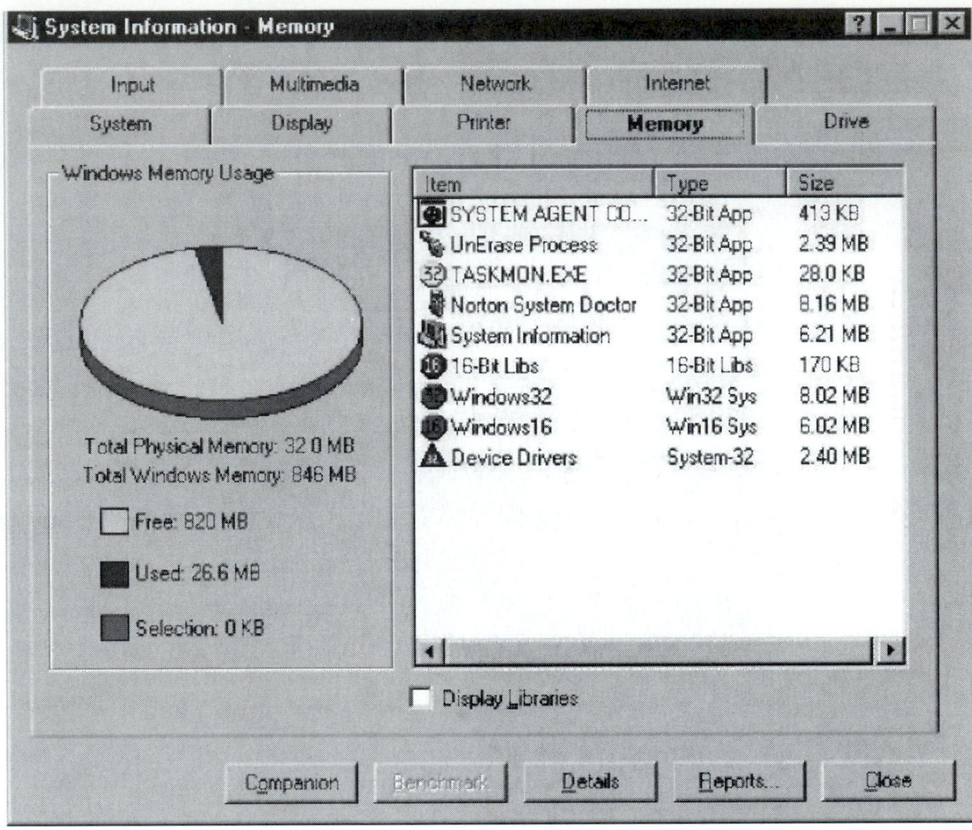

FIGURE 17.23
System Information memory tab

	Faulty cooling fan (loud hum, chatter, overheating smell) Faulty hard disk drive (loud or repeated chatter or thump; loud whine)

Video Problems

Symptom	**Checks**
No video display	Monitor not plugged in or not turned on Video cable not connected to video adapter, or connection to adapter or monitor loose Video cable faulty Incorrect monitor type for video adapter Motherboard color/mono select set incorrectly Incorrect video type specified in CMOS setup Video adapter, software, and/or monitor configured incorrectly

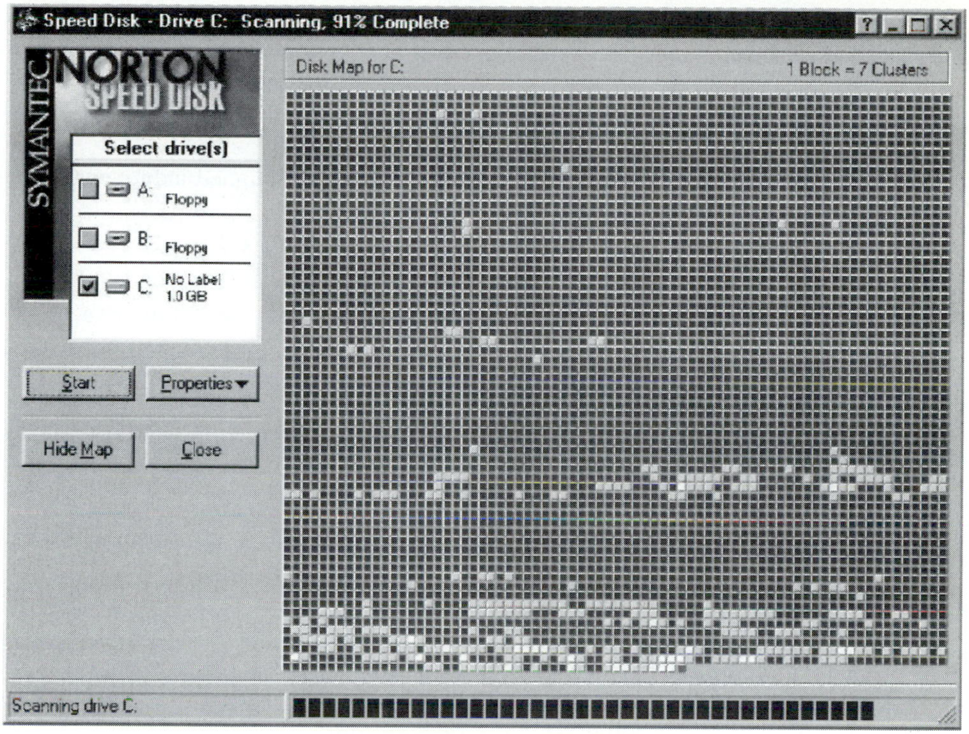

FIGURE 17.24
Speed Disk

	Incorrect video mode selected Monitor brightness or contrast turned down Screen blanking utility in operation
Faulty video display	Incorrect video mode selected Video adapter and/or software configured incorrectly Monitor vertical and/or horizontal hold and size controls adjusted incorrectly Monitor brightness and/or contrast controls adjusted incorrectly Insufficient video RAM Motherboard master oscillator trimmed incorrectly (on older CGA systems only)

Disk Drive Problems

Symptom	Checks
Read failures	No diskette in drive Diskette not firmly seated in drive Door not closed

	Diskette not formatted or initialized Bad boot sector Missing hidden files Corrupted FAT or directory table Head or speed alignment Defective or misaligned index sensor Controller failure Defective or loose cables
Write failures	No diskette in drive Diskette not firmly seated in drive Door not closed Diskette not formatted Diskette write protected Disk drive write protect sensor blocked, faulty, or misaligned Hard disk partition flagged "read only" Controller failure

Peripheral Problems

Symptom	**Checks**
Serial port not functioning	Expansion card, software, and/or serial device configured incorrectly Serial port not initialized (use DOS MODE command) Incorrect parmeters (baud rate, stop bits, parity) UART and/or driver ICs defective Serial cable defective or miswired Cable connection loose IRQ or port conflicts
Game port/joy stick not functioning	Incorrect type of joy stick Game port not enabled Multiple game ports on different adapters conflicting Joy stick cable not connected, or is loose or defective Software configured incorrectly
Printer fails to respond	Printer connected to wrong port or wrong port specified in software configuration Parallel port hardware configuration incorrect Parallel port IRQ disabled but required by software Printer not turned on Printer not on line or out of paper Cable loose or defective Printer buffer (hardware or software) configured incorrectly Windows Print Manager stalled
Mouse fails to operate	Mouse driver not loaded Mouse serial port not configured properly or initialized

Bus mouse card configured incorrectly
Mouse cable not plugged securely into serial port or bus card port
Mouse tracking ball dirty, sticking, or missing
Tracking surface too smooth, allowing ball to slide

17.12 TECH TIPS AND TROUBLESHOOTING—T^3

A good understanding of a typical system operation during start-up will aid in a quick diagnostic of a problem. Some examples are the normal sounds and the amount of time for start-up.

Listen to the power supply on a dead system. Here are some clues:

Allow the power to completely bleed off by removing the power cord for at least one minute. While listening to the power supply, attach the power cord and listen for any sounds. Do this again as soon as you hit the power switch. A ticking or whining sound is an indication of an overload or a power supply that has failed to regulate. You may also notice the power light flash and go out.

Listen carefully to the sound of the hard disk drive as soon as power is applied. Repeated seeks or unusual noise is evidence of a failure or impending failure. A common symptom is a low growl noise due to a failing spindle bearing. In most cases this is a nonfatal problem but is an indication of an impending failure. Furthermore, a scraping noise is evidence of a head crash.

Forgotten BIOS passwords can result in annoying problems. Consider the case where the BIOS password has been lost or forgotten. Here are a few things to try to gain access and use of the system.

1. If the system can boot to a floppy, run the DOS **DEBUG** utility and change the contents of the CMOS check sum word. This will clear the entire CMOS RAM chip, including the password, and allow you to reset the CMOS. The procedure is as follows: Start DEBUG and issue the following commands:

```
O 70 2E  (writes the CMOS RAM address of 0x2E, the address register)
O 71 00  (writes the CMOS RAM data of 0x00, the RAM at that address)
O 70 2F  (writes the CMOS RAM address of 0x2F, the address register)
O 71 00  (writes the CMOS RAM data of 0x00, the RAM at that address)
```

   Exit DEBUG and reset the system. The setup program will run automatically.
2. Most systems contain a CMOS clear jumper. Remove the case, set the jumper to clear, and power on the system. Power off the system and reset the jumper.
3. Remove the clock battery and allow to sit overnight to completely discharge the capacitors, thus erasing any memory.

EXERCISES

17.1 What is the POST, and what does the abbreviation stand for?

17.2 What port does the POST output its progress to?

17.3 In a totally dead system, where is the most logical place to start your diagnosis?

17.4 What occurs during the early part of the POST?

17.5 Why are errors found during early POST tests indicated by beeps?

17.6 True or false: The beep sequences are unique to each BIOS vendor and version.

17.7 Watch out for __________ when working with sensitive electronics.

17.8 It is important to always use __________ when working on computer boards.

17.9 How can you clean peripheral card connectors?

17.10 How can you clean memory module connectors?

17.11 Can you use an ink eraser to clean contacts? Why?

17.12 True or false: When troubleshooting memory, fill up all the banks and then troubleshoot.

17.13 What is the proper procedure to troubleshoot memory problems?

17.14 All newer BIOS have an option for identifying the hard drive, which can also help in diagnosing problems with the drive. What is this option in BIOS called?

17.15 What is FDISK used for?

17.16 If a monitor is scrambled or jittering when Windows boots, what should you check first?

17.17 If there is "garbage" on the monitor screen, that is, random characters out of place, what is the most likely problem?

17.18 Explain briefly how the cathode-ray tube fails.

17.19 If workstations cannot access the Internet, but all else seems to be fine on the network, where might the problem lie?

17.20 Explain how to use PING when troubleshooting a network. Give an example of its use.

17.21 Briefly discuss the process of troubleshooting a thin- or thick-wire Ethernet TCP/IP bus network.

17.22 Briefly discuss how to troubleshoot a twisted-pair Ethernet star network.

17.23 Crossword Puzzle

ACROSS

1. Start-up diagnostic.
3. Used for cooling.
5. Norton Speed Disk.
6. Displays error codes.
7. Use an ohmeter to check for this.
9. Primary IDE hard drive.
11. Holds system setup information.
12. Case style.
13. Error codes that cannot be displayed.
14. Type of hard disk.
15. Central connection for the nodes of a network.

DOWN

1. A network utility for diagnosing problems.
2. Memory package style.
3. Book from this device when you have a hard disk failure.
4. Type of processor.
7. Denotes pin 1 on ribbon cable.
8. Temporary memory.
10. BIOS manufacturer.
11. A fatal hard drive problem.

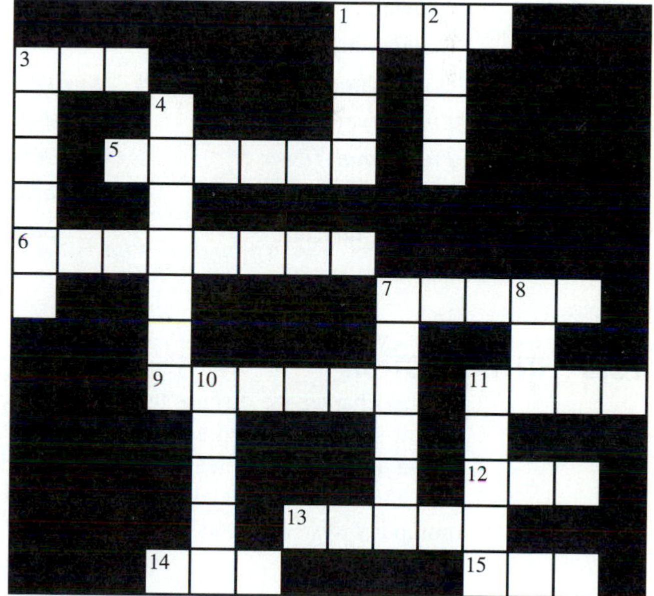

CHAPTER 18

TEST EQUIPMENT AND CERTIFICATION

KEY TERMS

A+ Certification
Aliasing
Checkpoints
Digital Storage Oscilloscopes
Emulative Tester
Functional Tester
In-circuit Tester
Logic Analyzer
Logic Probe
Logic Pulser
Power Line Monitor
PROM Burner
Undersampling
Universal Microcomputer Tester
Variable-persistence Oscilloscope
Volt Ohm Milliammeter (VOM)

18.0 INTRODUCTION

In this chapter we discuss the various types of test equipment used in the microcomputer repair shop. We begin with the simplest logic probes and meters, and work up to the latest in automatic testers. Included in this chapter is a universal microcomputer tester project. This is a project you can build and use to diagnose and repair most types of microcomputer systems in use today.

18.1 LOGIC PROBES AND PULSERS

Logic probes are simple logic level indicators (Figure 18.1). Most logic probes have three lamps, indicating the *high state, low state,* and *pulse.* The pulse indicator flashes whenever the logic level transitions from one valid level to the other valid level. The pulse indicator is required because it may be impossible to see a short or infrequently occurring pulse on the high and low indicators. Some more sophisticated logic probes may also include a separate gate input, which allows you to observe a logic level only when enabled by the gate input.

Using the logic probe is quite simple. Power to the probe is provided from the unit under test by connecting the probe power leads to the board. Be careful to observe the polarity of the power leads. When the probe tip is placed on an IC pin, the indicators

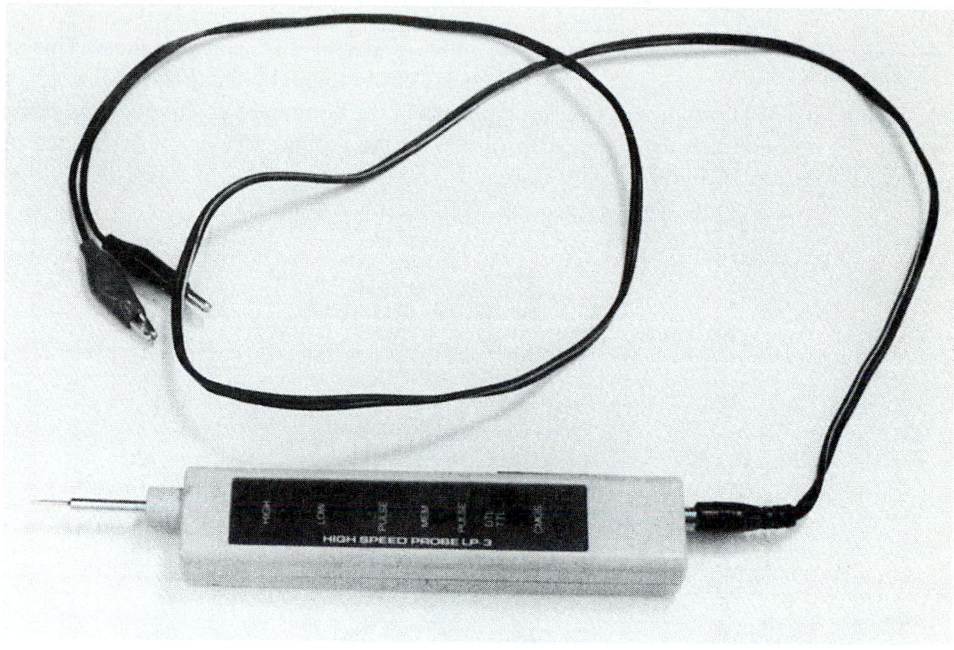

FIGURE 18.1
Digital logic probe

reflect the logic state of that pin. If no indicators are illuminated, an illegal logic state is present. When either the high or the low indicator is illuminated, the pin tested is at a valid stationary logic level. If the high and the pulse indicators are illuminated, the pin is mostly at the high state and pulsing low. If the low and the pulse indicators are illuminated, the pin is mostly at the low state and pulsing high. If all three—high, low, and pulse indicators—are illuminated, the pin is transitioning high and low at a ratio of approximately 50%. It is possible to approximate the high to low ratio, which is called the *duty cycle,* by observing the difference between the intensity of the high and low indicators.

The **logic pulser** is a device that will drive, or force, an IC pin to a desired logic level. This is a handy device when used in conjunction with the logic probe. A logic gate may be checked by driving an input pin with the logic pulser while observing the respective output pin with the logic probe.

18.2 METERS

No electronic repair shop is complete without at least a **volt ohm milliammeter (VOM).** The VOM is required for measuring the power supply output and for finding short and open circuits. If you decide to use a digital volt ohm milliammeter (DVM), the meter should include a diode check range. It is preferable to use a $3^1/_2$-digit DVM similar to the one shown in Figure 18.2. Recall the input diode anomaly discussed earlier. In many cases static discharge causes the input diodes of TTL gates to short. This short is usually approximately 30 to 200 Ω. This is easily found by using the VOM. A normally functioning TTL input will measure open when the ohmmeter positive lead is placed on the input pin and the negative lead is placed on the ground pin. A low resistance reading, such as 30 to 200 Ω, indicates a possible short. Always check if a resistance to ground or power is present on the suspect pin before replacing the chip.

FIGURE 18.2
Beckman digital volt milliammeter

18.3 OSCILLOSCOPES

As discussed throughout this text, an oscilloscope is required whenever waveform measurements are taken. The oscilloscope used should include at least two DC coupled vertical inputs. The horizontal time base should be the triggered sweep type, preferably with a delaying time base. The low-cost TV, service oscilloscope should not be used in this application due to its slow speed and lack of DC inputs.

There are many brands and types of oscilloscopes on the market, with a wide range in price and performance. An oscilloscope with a vertical bandwidth of 50 MHz is acceptable; however, a bandwidth of 100 MHz is preferred. Some manufacturers offer **variable-persistence oscilloscopes** for viewing single-shot events. The latest development in oscilloscopes is the **digital storage oscilloscope,** which converts the incoming analog waveform into digital data and stores the data in RAM. The RAM data is then displayed on the screen as the waveform. These digital scopes offer many advantages over the analog types, with the biggest advantage being the ability to capture single-shot events and store the waveforms for later viewing and comparison. The digital scope is not without a few disadvantages. The first is its relative high price, and the second is that the scope may miss high-frequency noise, or oscillation, which may occur during a single-shot or low-repetition event. If the noise, or oscillation, is at or above the sample rate of the scope, the noise will be missed or digitized improperly. This problem is known as **undersampling,** or **aliasing.**

Using the oscilloscope can be difficult for the novice technician. The first thing you must do is to familiarize yourself with all the controls. Most important, learn how to properly use the triggered sweep. As already discussed, many oscilloscopes include an automatic trigger mode. Do not depend on this mode all the time. Learn how to use the oscilloscope in the normal trigger mode. One example of using the oscilloscope to troubleshoot a microcomputer might be to check the BIOS ROM output pins. Since many devices are connected to the data bus, there will be a great deal of activity on the data bus lines. Simply observing the bus waveforms without setting the oscilloscope to trigger on a specific event is not adequate. To check the ROM output pins, connect the oscilloscope vertical input A to the ROM select pin. Set the trigger on the oscilloscope to the internal trigger mode from vertical channel A. Since most ROM select lines are active low, set the trigger slope to (−) or the negative edge. Set the vertical mode to display both channels in the chopped mode. Adjust both the vertical input gains to display 2 V/cm. Position the channel A trace to the upper portion of the display. Adjust the timebase and trigger level to display a stable negative (low) going pulse at least one-half the width of the display, as shown in Figure 18.3 for channel A. Connect vertical input channel B to one of the ROM data output lines. Channel B should display a waveform similar to the waveform shown in Figure 18.3. By observing the voltage levels on channel B, you can determine if the output drivers of the ROM are functioning properly. By observing the time delay from the point where the select line goes active to the point where the data output is at a valid level, you can measure the *access time* of the ROM chip. From this display you cannot determine if the bit is at the correct logic state; you can determine only that the output is at a correct logic voltage level. A voltage level between 0.8 V and 2.4 V indicates a bad ROM or a loaded output circuit. If, for example, some other device on the bus is active at the same time, a bus contention problem may exist. This will show up as an illegal voltage level on one of the data pins during the ROM select. Analyzing the output data patterns requires a multiple-channel display device with enough input channels to accommodate all the address and data lines simultaneously.

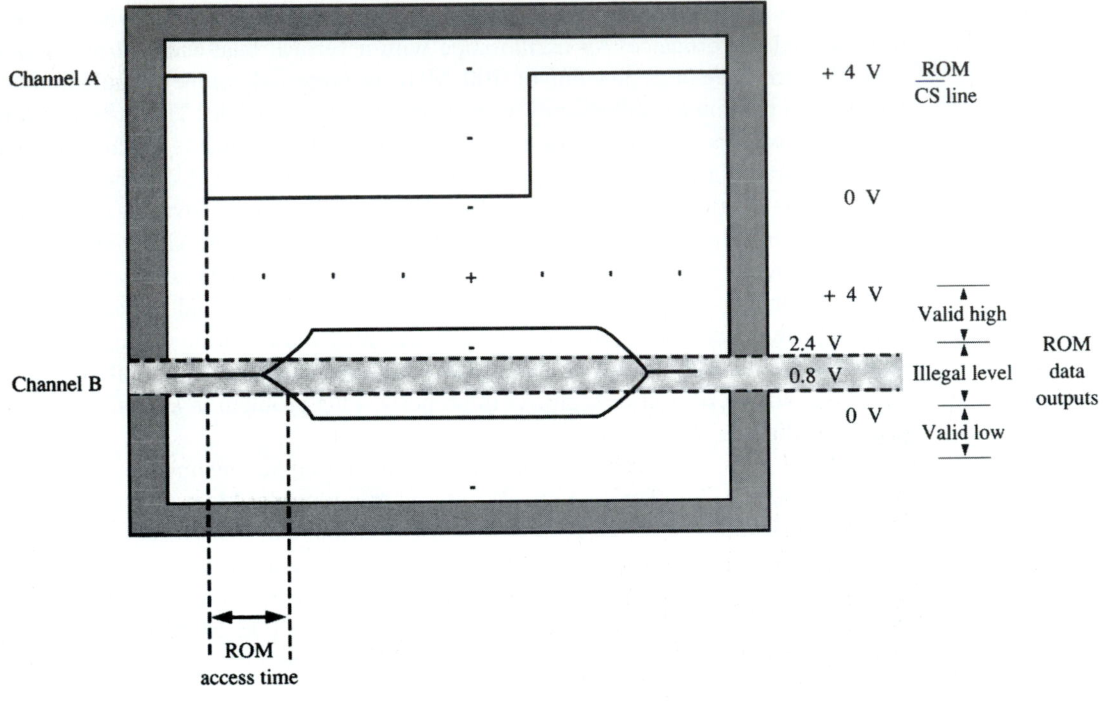

FIGURE 18.3
Oscilloscope

18.4 LOGIC ANALYZERS

The **logic analyzer** is a multichannel digital signal display device. Unlike the oscilloscope, the logic analyzer does not allow analog inputs. Logic analyzers are available with input channel numbers ranging from 8 up to 64, depending on price. Most logic analyzers will display the acquired data as timing waveforms, similar to an oscilloscope display, and as binary data in hexadecimal, octal, or binary format. The logic analyzer is well suited for analyzing bus activity. Using the logic analyzer, you will be able to observe the address bus and the data bus simultaneously. To perform this, connect some of the input channels to the address bus and connect the other input channels to the data bus. Set the logic analyzer to the external clock mode and connect the external clock input to the appropriate bus control line. In this case we will use the memory-select line. Many logic analyzers also include clock qualifier inputs, which allow the clock in only when the signals are valid. You must now set the logic analyzer to trigger on some desired memory address. By observing the bus activity, you will be able to follow the microprocessor instruction fetch sequence. You will also be able to observe RAM writes and reads. Many microprocessors use RAM when performing subroutine calls and returns. If a RAM error should occur during a subroutine, the microprocessor will "get lost," because the value of the calling routine's program counter is stored when the subroutine is run. When the subroutine is finished, the value of the calling routine's program counter is restored to the program counter from RAM. If the value is corrupted, the program counter will return to the wrong address, causing it to be lost. By observing this activity on the logic analyzer, you can determine the cause of the fault to the faulty bit. Consider the following MC6800 example:

```
ADDRESS DATA
   MAIN PROGRAM
0100 BD0200  Jump to subroutine at address 0200
0103 XXXXXX  Next instruction
01XX         Next instruction . . .

   SUBROUTINE
0200 B68002  Load accumulator a from port at 8002
0203 39      Return from subroutine, to caller
```

The logic analyzer should display the address and data, as follows:

```
0100  BD
0101  02
0102  00
XXFF  03     PROGRAM COUNTER LOW BYTE STORED TO STACK RAM
XXFE  01     PROGRAM COUNTER HIGH BYTE STORED TO STACK RAM
0200  B6
0201  80
0202  02
0203  39     RETURN INSTRUCTION
XXFE  01     PROGRAM COUNTER HIGH BYTE FROM STACK RAM
XXFF  03     PROGRAM COUNTER LOW BYTE FROM STACK RAM
0103  XX     NEXT INSTRUCTION IN MAIN PROGRAM
```

Now let's assume that a memory failure exists. In the next example we assume that memory bit 7, the MSB, is stuck high. The logic analyzer might display the following:

```
0100  BD
0101  02
0102  00
XXFF  03     STORED PROGRAM COUNTER LOW BYTE
XXFE  01     STORED PROGRAM COUNTER HIGH BYTE
0200  B6
0201  80
0202  02
0203  39     RETURN INSTRUCTION
XXFE  81     WRONG VALUE DUE TO PICKED BIT
XXFF  83     WRONG VALUE DUE TO PICKED BIT
8183  XX     JUNK INSTRUCTION DUE TO ILLEGAL ADDRESS
XXXX  XX     COULD BE ANY ADDRESS DEPENDING UPON JUNK
XXXX  XX     INSTRUCTION
             THE MICROPROCESSOR IS NOW LOST
```

As we can see in the preceding example, when the microprocessor executed the RETURN instruction, corrupted data were restored to the program counter (8183 instead of 0103). This is just one example of how the logic analyzer may be used to track down a bus problem.

Most logic analyzers are very flexible and are limited only by the user's imagination. Since the logic analyzer will also display the input channels as timing charts, many or all the bus-control lines may be viewed simultaneously and measured for correct timing. All the timing diagrams in Chapter 11 were obtained by using a logic analyzer connected to the IBM PC. Some of the latest logic analyzers include built-in disassemblers, which will translate the machine code into a format that humans can read. This feature will save the time of looking up the machine codes but will usually add to the cost of the logic analyzer.

18.5 PROM BURNERS

The name **PROM burner** is an acronym for a programmable read-only memory programmer. The PROM burner is usually a compact self-contained device designed to read in and program various types of programmable memory devices. There are many variations of the PROM programmer available on the market, including some in kit form. The newer versions of PROM programmers support various types of programmable logic devices (such as PLD, PAL, and PLA) by using adapter modules (Figure 18.4). Recent

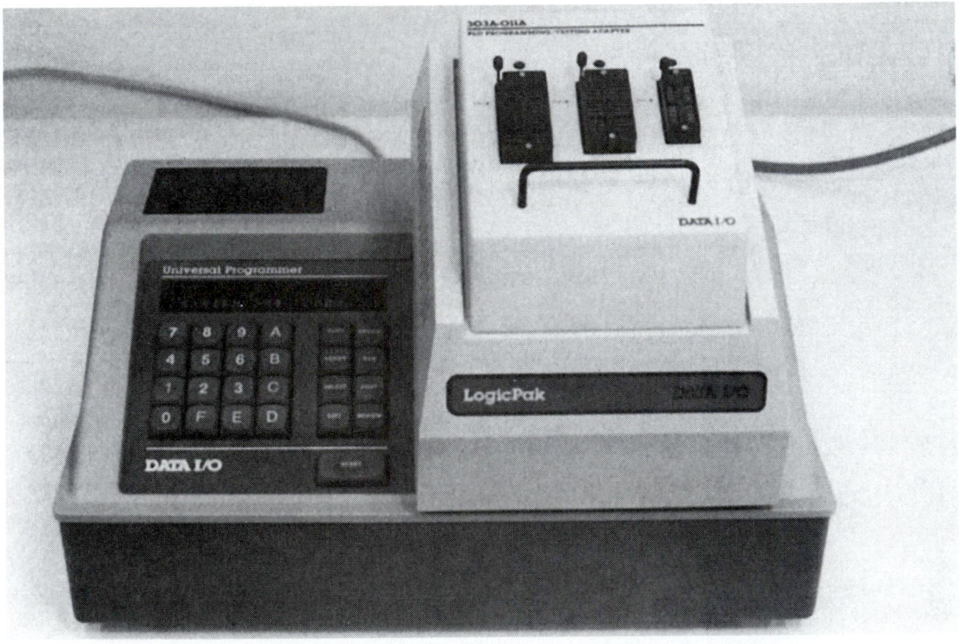

FIGURE 18.4
Data I/O PROM programmer

repair histories have shown that programmable devices have a high failure rate. In the repair shop, the PROM programmer can serve two functions. The first is the ability to program blank parts to use as replacement parts. The second is the ability to check quickly and verify a suspect part from a bad board.

18.6 POWER LINE MONITORS

The **power line monitor** is a recording voltmeter (Figure 18.5). This device is handy when analyzing the incoming AC power over a period of time. On occasion you will hear a complaint about a higher-than-normal failure rate or intermittent system crashes. In cases like this it makes sense to monitor the incoming power for unusual conditions. Excessive voltage drops, called *sags,* may cause the computer system to lock up or reset. Voltage spikes, called *surges,* may damage the power supply or sneak through the power supply and damage the system. Connecting the power line monitor to the supply power over the period of a few days to one week will give a good profile of the incoming power. If the results show excessive power variations, connect the computer to a different line or add

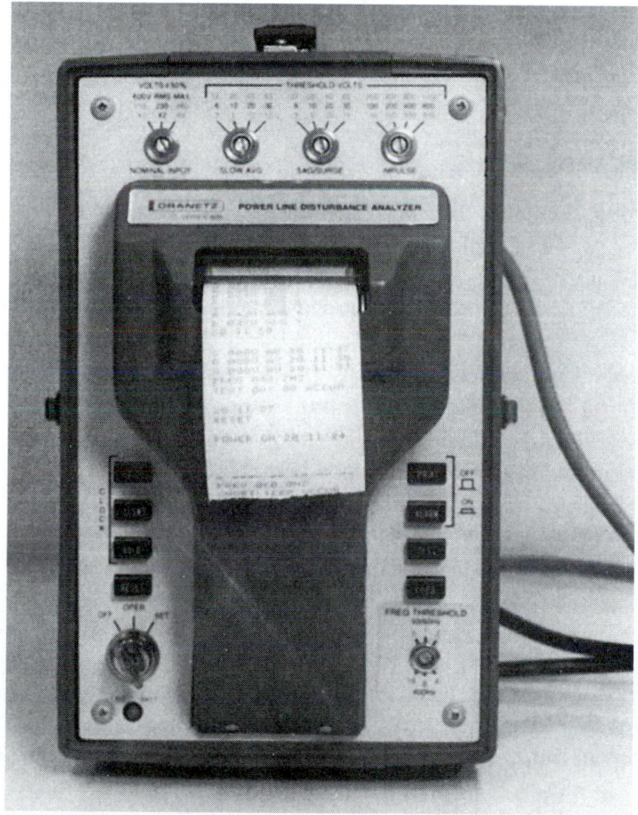

FIGURE 18.5
Dranetz power line monitor

a power line conditioner to the system. The worst offender to power line problems are inductive loads such as air conditioners and refrigerators.

18.8 THE UNIVERSAL MICROCOMPUTER TESTER PROJECT

Almost all the 1970 and earlier vintage minicomputers and large mainframe computers included various display panels. These display panels showed the status of various CPU buses and control lines. The display panels enabled the technician to diagnose the system by single-stepping the computer one instruction at a time and observing the status of the display panels. Even if the system were totally dead, the display panels would give some clue as to the problem. Most modern computers have eliminated these display panels. Today, when the system is dead, you do not have a clue as to the cause of the problem. What we offer here is an inexpensive alternative to the more expensive test equipment. This project was designed by the authors to enable you to observe the address bus and data bus in a way similar to that offered on the older computer systems. You will be able to select the tester to stop and display on I/O read, I/O write, memory read, or memory write operations or any combination thereof. The basic design of this project may be applied to almost any computer system. The project shown is designed for an XT-class system. Modify the design to decode I/O port 80 for newer systems.

Theory of Operation

The universal microprocessor tester project is basically an address and data line status display. We also include a wait state generator circuit, which performs a HALT function. The display will enable you to observe the bus activity visually. The wait state generator will allow you to stop and single-step the computer one instruction at a time.

The display is centered around seven TIL 311 hexadecimal LED display devices manufactured by Texas Instruments, Inc. These display devices include a data latch, decoder, and LED driver built in to the display, which greatly simplifies the construction of this project, as shown in Figure 18.6. The DATA inputs of the display are connected directly to the PC bus address and data lines, as shown in Figure 18.7.

The first five displays are for the 20 address bus lines. The last two displays are for the 8 data lines. The address and data bus information is displayed by selecting any one or any combination of read and write control lines. This is accomplished by using switches to connect the 4 inputs of a 4-input NAND gate to the 4 control lines IOR, IOW, MEMR, and MEMW. NAND gate IC2a is actually used as a negative-OR gate and drawn as such (Figure 18.7b). The output from IC2a is connected to NAND gate IC2b. NAND gate IC2b is used to block the display of refresh cycles by gating the select signal with the signal $DACK_0$, which is the refresh DMA cycle. The output of IC2b is connected to the display strobe input of all the displays, which causes the bus address and data to be latched on the displays.

The wait circuit consists of a *D*-type flip-flop, IC1a, which is clocked by the output of gate IC2a. The wait circuit is armed by flipping the wait enable switch up. When the selected display operation occurs, the output of IC2a triggers the flip-flop and causes the microprocessor to wait indefinitely by pulling the line IOCHRDY low.

FIGURE 18.6
TIL 311 hexadecimal displays

The single-step circuit consists of a double-pole, single-throw momentary switch connected to flop-flop IC1b used as a set/reset flip-flop. The flip-flop is required to *debounce,* or clean up, the noisy output of the momentary switch. The output of the single-step flip-flop is fed to a resistor-capacitor (*RC*) network, which is used to create a very narrow *low-going pulse.* This pulse is fed to the reset input of the wait flip-flop IC1a, which removes the wait condition when the button is depressed. This action will allow you to step to the next selected function.

Construction

The construction of this project is quite simple and requires no special skills except for some light soldering and wire-wrapping. Before beginning construction, review the physical layout of the pins on the components. Make certain you know the orientation of pin 1 of the display and other ICs. Keep in mind that when you are wiring the sockets, *you will be looking at the pins from the bottom view.* Normally, pin 1 is at the top left side of the IC. When you are wiring the sockets, pin 1 will be the top right side of the IC because you are looking at the bottom of the IC. We once observed a student lay out a printed

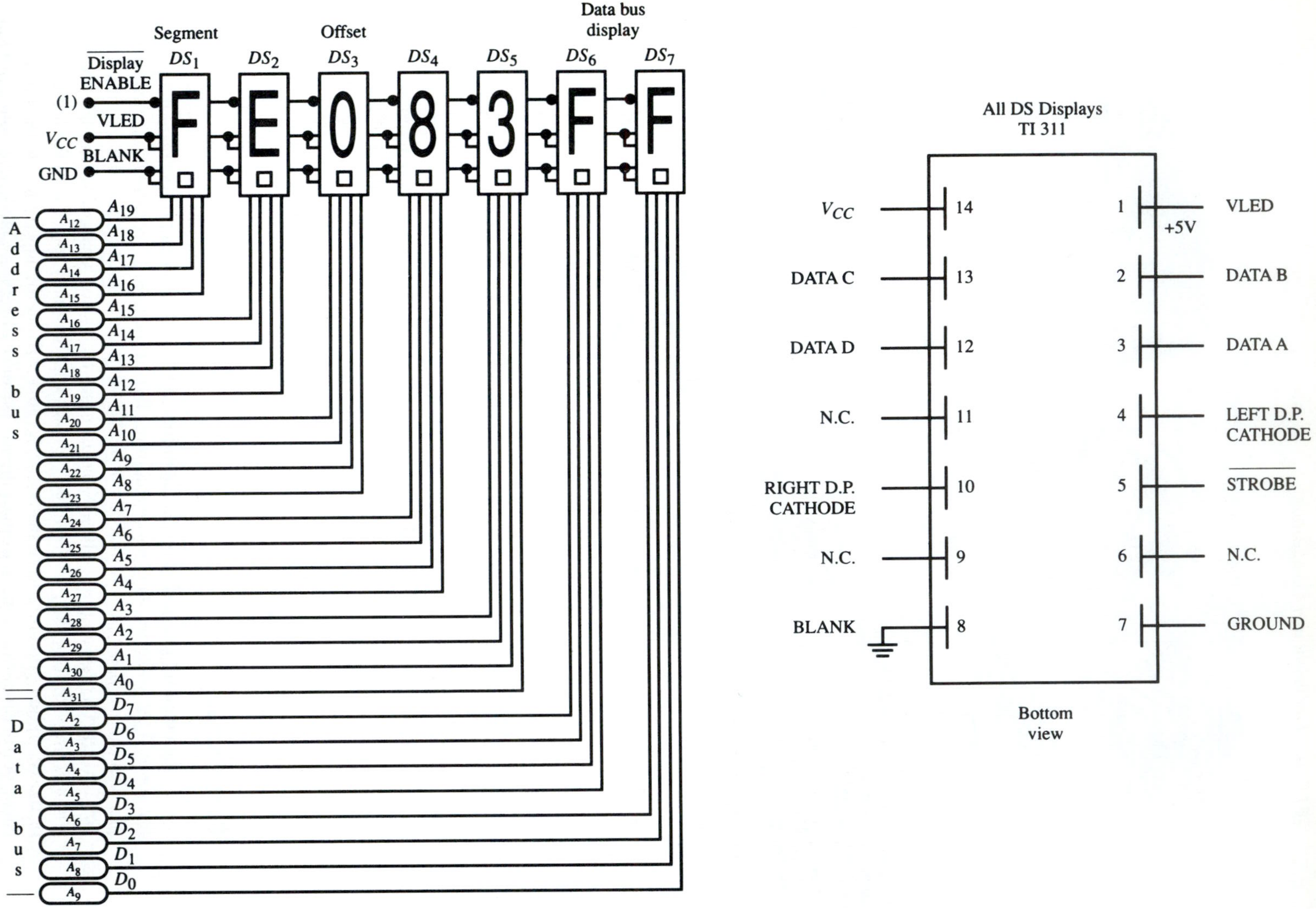

(a) Display circuitry

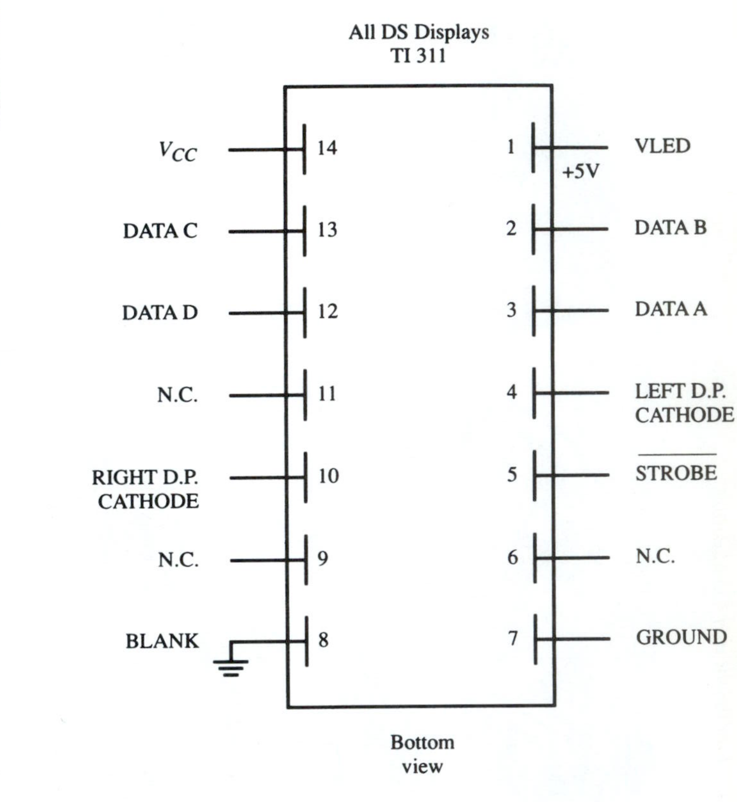

FIGURE 18.7a
Universal microcomputer tester project

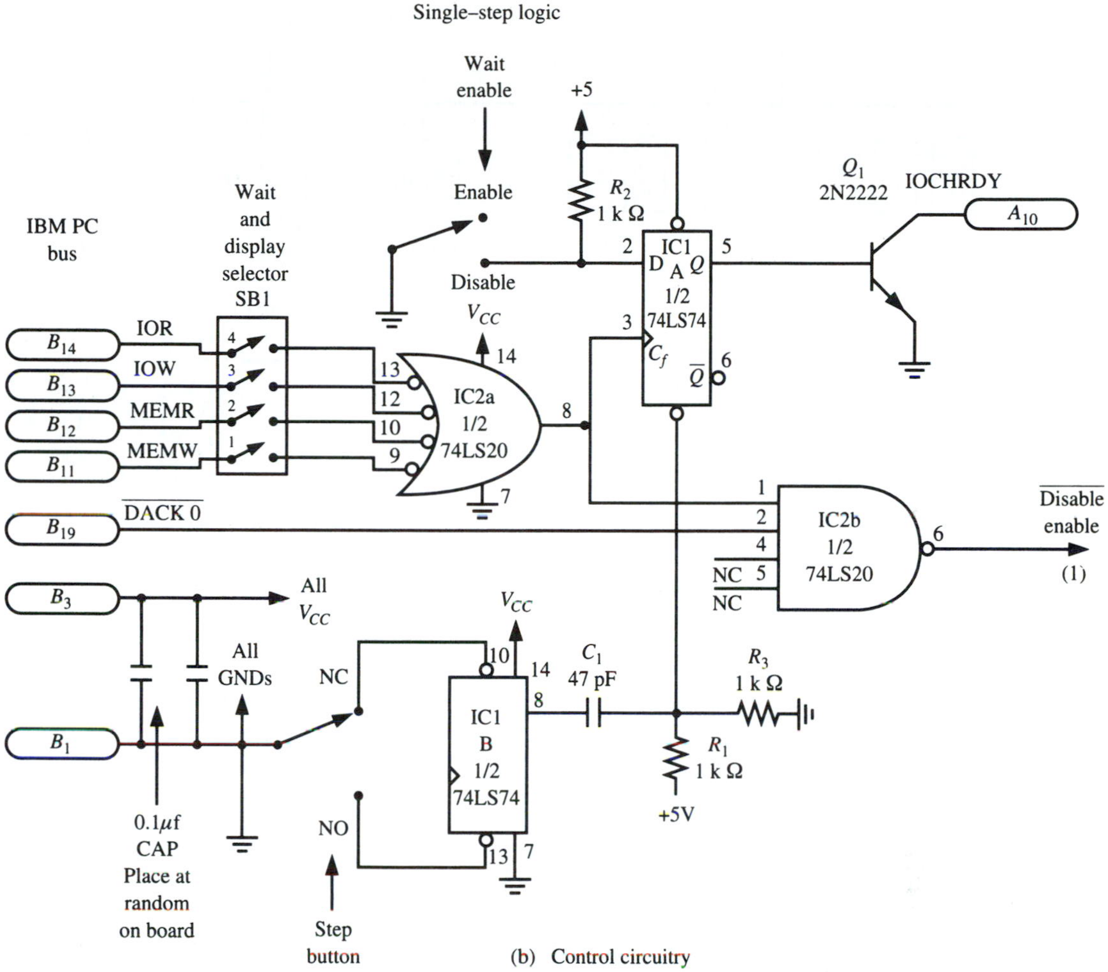

Single–step logic

(b) Control circuitry

FIGURE 18.7b

circuit board for a digital clock backward because he did not check the IC orientation. The result was that the project had to be scrapped and built over again, so don't let this happen to you. We constructed our prototype on a universal prototype board designed to fit into the IBM PC bus. These prototype boards are commercially available and provide easy access to the PC bus connections. The displays and switches should be mounted on the reverse side of the board, as shown in Figure 18.8. This should be done so that the display and switches are easily accessible when the board is placed in the last slot on the left side of the computer.

All the components, including the transistor and resistors, can be mounted on 14-pin wirewrap IC sockets. The placement of the components is not critical. Many of these universal wirewrap boards include *power* and *ground* buses, and care should be taken not to place the socket pins through these holes unless it is done intentionally. First, put the

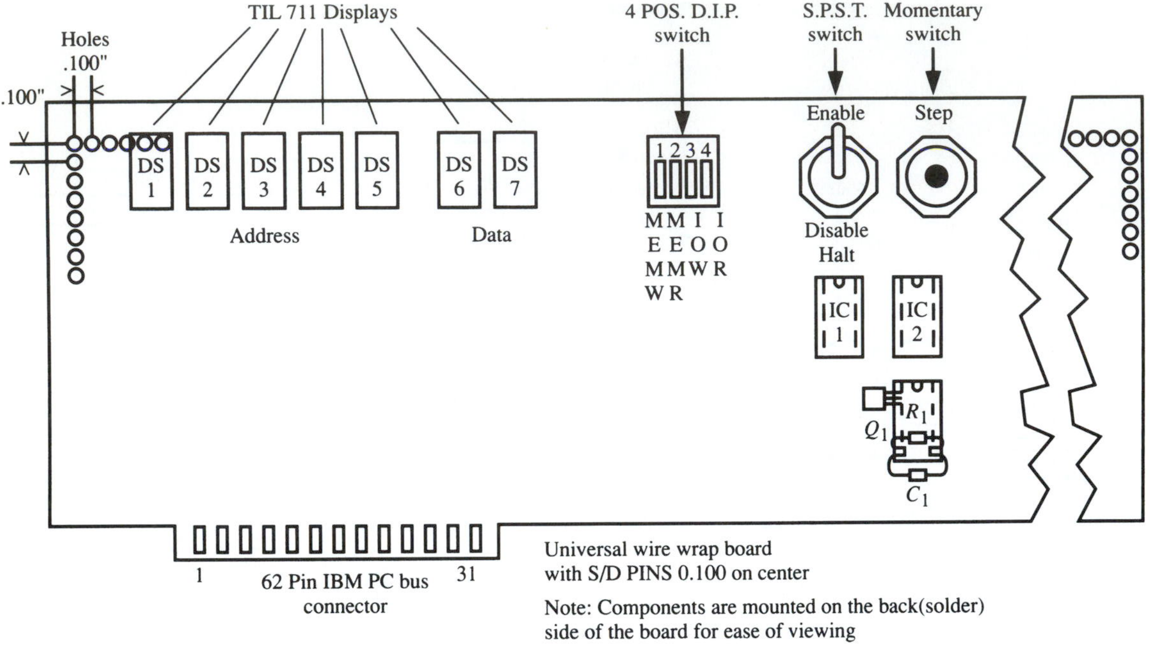

FIGURE 18.8
Universal microcomputer tester layout

wirewrap sockets into their respective places on the board. The sockets may be anchored by using cement or by soldering two opposite corner pins of the IC socket to the plated through holes (if equipped). Next, wire up the power and ground pins on all the sockets that require power and ground. Using an ohmmeter, check the wiring on the sockets and check for shorts between power and ground. This power check is an important step and should not be skipped. Next, drill larger holes to mount the switches. Remove the copper traces within 1/8 in. around the drilled holes. If this is not done, a short will exist between power and ground when the switch is installed. Install the switches and again check for any shorts between power and ground. Next, wire up the sockets according to the schematic and check *all* the wiring by using the ohmmeter. Next, install a few 0.1-μF capacitors at random locations on the board between the power and ground buses for signal decoupling. The number and placement of the capacitors are not critical. Install all the components into their respective sockets and verify their orientations. Last, prepare labels for the switches and install them in their proper places. These labels not only make the tester look nice, but they make it easy for you and other people to use the tester as well. The **universal microcomputer tester** is now ready for use.

A complete kit or assembled version of this project is available from:

RMB Associates Ltd.
51 Birchbrook Dr.
Smithtown, NY 11787

Operation

The operation of this tester is quite simple. First, remove the cover of any IBM PC or compatible *(always turn off the power first).* Next, install the tester into the end slot on the left side of the computer. If a board is already in this slot, move it to another slot. Set switch 2 of the switch block on the tester, labeled MEMR, to the on position. Set the wait enable switch to the enable position. What we just did was to set the tester to display and cause the computer to wait on any memory read operation. Power on the computer. The tester display should read

F F F F 0 E A

Under *no* condition should the display be dark! If the display is dark, *immediately* turn off the computer and check your work. This is an indication of no power to the display or possibly of a misplaced display blanking wire. If the digits appear fuzzy, the wait circuitry is miswired and the computer has not entered the wait state.

The RESET line, which is active during power-up, will cause the 8088 microprocessor to vector, or begin execution, at address *FFFF*0 hex. Since we set the tester to display and wait on a memory read operation, the display should show the address and data values of the first read from memory. The RESET vector address is *FFFF*0 hex, and the first instruction should be *EA* hex, a JUMP instruction. The next two bytes fetched will be the offset value placed into the program counter and the last two bytes are the segment value placed into the code segment register, CS. Recall that the Intel series microprocessors fetch the register values backward. That is, the LSB is fetched first. Each time you depress the step button on the tester, you should observe the next byte fetched along with its address. The sequence for the IBM PC with a BIOS ROM dated 10/27/82 from power on should look as follows,

F F F F 0 E A	The jump instruction
F F F F 1 5 B	The least significant offset byte
F F F F 2 E 0	The most significant offset byte
F F F F 3 0 0	The least significant segment byte
F F F F 4 F 0	The most significant segment byte
F F F F 5 3 1	Dummy read (execute cycle does prefetch)
F E 0 5 B F A	The first instruction of POST (CLI)
F E 0 5 C B 4	The next instruction
⋮	Subsequent instructions

You may continue to follow the POST execution by repeatedly depressing the step button. If the preceding sequence is correct, you can safely assume the system data bus is intact. If a bit on the data bus is stuck, the first few instructions will usually show the problem.

Next, power off the computer and set the wait enable switch to the disable position. Power on the computer. The display should appear fuzzy because the address and data lines are changing at a rate of 1 million times per second. The computer should come up normally. By observing the display you may notice some patterns, specifically on the high-order address display, where the changes occur less frequently. Perform the same test sequence for the MEMW, IOR, and IOW switches as we did for the MEMR switch to verify their operation. The displayed address and data values will vary, depending upon the current BIOS ROM revision installed into the computer. To use the tester properly, you must obtain the BIOS listing.

If the POST fails early in the sequence, the system will simply halt and give no clue to the problem. By using this tester with the wait enable switch in the disable position and the MEMR switch on, the address value displayed will be the address of the halt instruction plus a few locations. Recall that the 8088 microprocessor contains a prefetch queue and will fetch a few extra instructions while executing the current instruction. By referring to the BIOS listing, you can determine which test has failed by the address of the halt instruction. By analyzing the test that has failed, you will gain clues to accuse the faulty circuitry. The IBM PC/XT and the PC/AT have added **checkpoints** into the POST. The checkpoint is the currently executing test number output to an unused I/O port address. The port address on the IBM PC/XT is 60 hex, and on the PC/AT the port address is 80 hex. This may be observed by setting the tester to display IOW functions. If the computer is dead and the tester display shows address 00060 and data 01, the system failed at checkpoint 1.

With the wait enable switch set to disable, observing the display while selecting the various display functions should prove interesting. One example of this is observing the MEMW display function while running the IBM memory diagnostics. You will see the address value incrementing very rapidly while the data display shows a pattern of AA, 55, 01, 00. These are the actual patterns written to RAM during the test. The IOR and IOW displays will prove equally interesting.

If desired, you may add address decoding circuitry to the tester to only detect and display the checkpoint addresses. This feature will allow you to observe the progress of the POST test while the tests are running.

One important note on this tester and the IBM PC: This tester should not be used to single-step program routines in RAM or subroutines in ROM. You may, however, *display* any function at any time. When the tester causes the computer to wait, no RAM refresh cycles will occur. When we cause the system microprocessor to wait, we also cause the DMA and all other bus activity to wait. Any data stored in RAM will be lost, including any data in the stack RAM, which may be the calling routine's program counter. Recall the results when the stack memory is corrupted.

18.8 AUTOMATIC TEST EQUIPMENT

Troubleshooting modern microcomputers and microprocessor-based boards can be a tedious and time-consuming task. So far we have offered many ways of troubleshooting a microcomputer system manually. If you will be involved with production repairs where high volume and quick turnaround are required, then you will need a more efficient method of troubleshooting and repairing faulty boards and computers. Over the past decade a new breed of test equipment known as automatic test equipment has been developed. An automatic tester, put simply, is a collection of driver/sensor pins or wires controlled by a programmable computer. The computer, controlled by a set of instructions, will drive some of the pins while taking measurements on other pins. If we consider testing a two-input AND gate, we would require three pins, two to drive and one to sense. By driving the two inputs through all four possible logic combinations and monitoring the output, we can then compare the results to a software model. If the actual device and the model are the same, the device must be good. Any difference will indicate the device is faulty. Today's automatic testers are an elaboration on this simple principal. Currently there is a wide range of automatic testers, with price tags from $5000 to over $500,000. The lower-cost units are designed for the service shop, and the higher-cost units are intended for the large production factory. The test methods and implementation vary, depending upon the user's specific needs. For the remainder of this text we discuss three different types of automatic testers: in-circuit, functional, and emulative automatic testers.

In-Circuit Automatic Testers

The **in-circuit tester** to which we are referring is the "clip the chip" type (see Figure 18.9). This tester uses standard IC test clips and tests the board one chip at a time. There are two basic types of this tester, *software comparison* and *hardware comparison.* Software comparison testers use the *overdrive* method to stimulate the IC under test, which

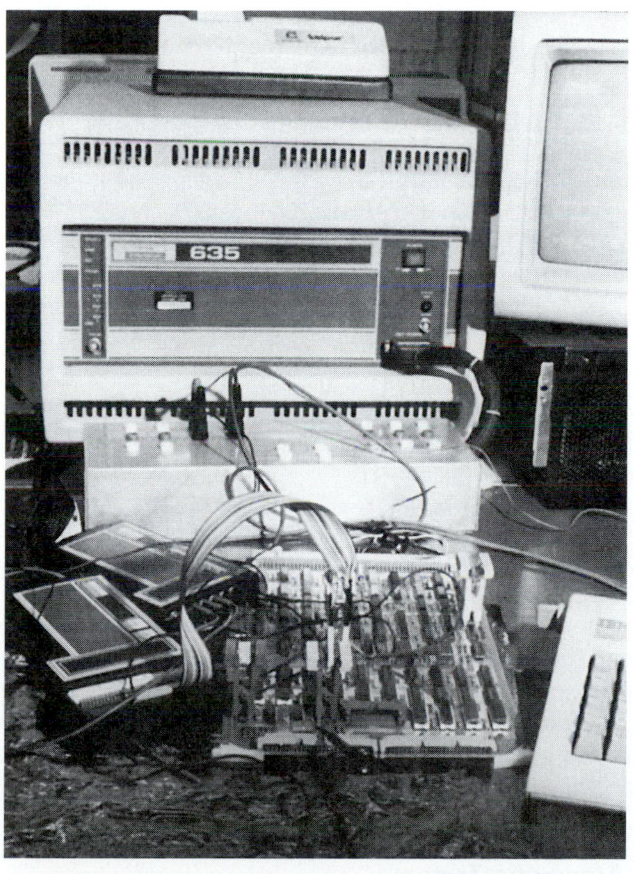

FIGURE 18.9
Factron in-circuit automatic tester

means that the tester can drive the test pins at high currents to force the input pins of the IC under test to the desired state, regardless of the preceding logic. At first, this method may seem dangerous and damaging to other components, but it has been proven that forcing IC output pins to the opposite logic state for short durations will not harm the IC. The input pins of the IC under test are driven to follow the software model truth table while the output pins are monitored and compared with the software model output truth table. If the output of the IC and the software model do not agree, the test will report a failure. The IC is flagged as faulty or the output pin may be loaded by some other component in the circuit. Recall the two-input AND gate example mentioned earlier. The truth table for this gate, where *A* and *B* are the inputs and *Y* is the output, is as follows:

A B Y	
0 0 0	NOT *A* and NOT *B* = False
0 1 0	NOT *A* and *B* = False
1 0 0	A and NOT *B* = False
1 1 1	*A* and *B* = True

The test sequence for this gate, written for a Factron/Slumberger 635 tester, looks as follows,

```
. (pin definitions and parameter block)
.
.
LOW A B     ;A AND B FALSE
TESTL Y     ;Y IS FALSE
HIGH B      ;B TRUE AND A FALSE
TESTL Y     ;Y IS FALSE
LOW B       ;B FALSE
HIGH A      ;AND A TRUE
TESTL Y     ;Y IS STILL FALSE
HIGH B      ;A AND B NOW TRUE
TESTH Y     ;Y IS NOW TRUE
CLEAR       ;END OF TEST
```

An error is reported by the word FAIL on the screen, and the failing statement will appear red when viewed in the debugger. The hardware comparison automatic tester with which we are familiar monitors only the pins of the IC under test. The board must be in the system and in the failing state. The tester contains a hardware library of components for comparison. The input pin signals are fed to the library device input pins inside the tester. The output signals of the device under test are fed to a comparitor circuit along with the output signals from the internal library device. If the IC under test and the library device do not match, the IC under test must be faulty or the output is loaded by the circuitry connected to the output of the IC.

Functional Automatic Testers

The **functional tester** is designed to test the entire board as opposed to testing one chip at a time (see Figure 18.10). Since this test method must emulate the host system,

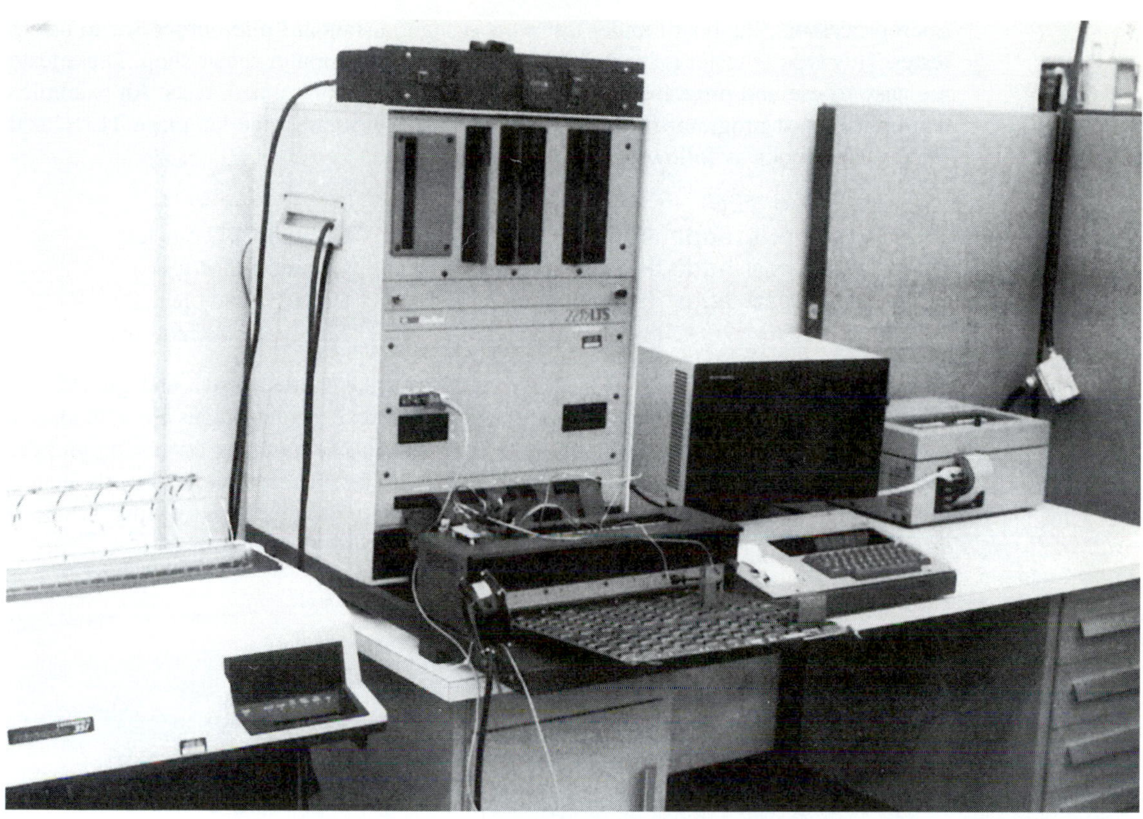

FIGURE 18.10
Gen Rad 2235 functional automatic tester

the entire board, including the artwork, is tested, from the board connectors inward. This is the best test method but requires far more program development time. The logical function of the board under test must be modeled in software along with the component interconnects, called the *node list,* or *image.* The modeling process takes considerable time to create. In some cases this test method cannot match the actual system timing, specifically where high-speed logic is concerned, which may become a major drawback. Using the guided probe fault isolation provided with the functional tester will prove to be the fastest repair method once the programs are developed. We have experienced cases where logic boards were tested and repaired in a matter of minutes.

Emulative Automatic Testers

The **emulative tester** is really a functional tester that uses a different test philosophy. The functional tester tests the board from the outside in. The emulative tester tests the board from the inside out. The tester is connected to the board's microprocessor chip socket. The tester then takes over the function of the microprocessor and controls the board under test. Since this type of tester is designed to test boards that contain off-the-shelf microprocessors, the board under test *must* contain a standard microprocessor to use this tester. This type of tester is ideal for the average microcomputer repair shop. These testers are easy to use and program, and some contain built-in, or canned, tests. An example of a simplified test program for the PC system board, written in the language TL/1 for the Fluke 9100, looks as follows:

```
program TEST_PC
  testbus addr A0000                     Canned CPU bus test
  testrom addr FE000 upto FFFFF          Canned ROM test
  testram addr 0000 upto 3FFFF           Canned RAM test
  :                                      More tests as required
```

The recent versions of this tester also contain external guided prove and guided fault-isolation software. One manufacturer has also included an asynchronous clock module for measuring asynchronous events and up to four 40-pin I/O modules for testing and driving various connectors and points on the board under test. These features greatly increase the flexibility of this tester up to and including in-circuit component testing, where the microprocessor may lack direct visibility to certain nodes. See Figure 18.11.

FIGURE 18.11
Fluke 9100 emulative automatic tester

18.9 A+ CERTIFICATION

The A+ certification exam is a multiple-choice timed test, officially known as the CompTIA Certified Computer Technician exam. This certification was developed by the Computer Technology Industry Association (CompTIA) to promote standards of excellence in computer technology. The members of CompTIA include almost 8000 companies in computer manufacturing, servicing, value-added reselling and training in the United States and Canada. There are over 60,000 A+ Certified Technicians worldwide. Many employers and customers see the A+ credentials as proof of professional achievement. Employers are increasingly requiring certification as a condition for employment or advancement.

The A+ exams are the result of an industry analysis of what a service technician with six months of work experience needs to know to be considered competent. Over 5000 A+ certified professionals validated the results of the analysis.

There are two exams, a core exam and software operating system (OS) exam (DOS/Windows). The core exam has 69 questions on microcomputers, displays (CRTs and LCDs), buses, CD-ROMs, printers, BIOS, modems, troubleshooting, basic networking, memory, and storage devices. The software OS exam has 70 questions. Each test is divided into technical areas or domains.

The Core exam consists of the following eight domains:

1.0 Installation, Configuration, and Upgrading
- Identify basic concepts, terms and functions, and normal operations of systems modules—for example, system board, MPU, power supply, BIOS, CMOS, memory modules, storage devices, monitor, and modem.
- Basic procedures for adding/removing field replaceable devices—for example, system board, MPU, power supply, I/O devices, memory, and storage devices.
- Identify and configure available IRQ, DMA, and I/O addresses for devices.
- Identify peripheral ports, connectors, and cables.
- Identify procedures for installing IDE/EIDE devices.
- Identify procedures for installing SCSI devices.
- Identify procedures for installing and configuring peripheral devices—for example, video cards, modems, monitors, and storage devices.
- Identify procedures for upgrading BIOS.
- Identify hardware methods for optimizing systems.

2.0 Diagnosing and Troubleshooting
- Identify and troubleshoot common problems with each module.
- Identify basic troubleshooting procedures and practices for solving customers' problems.

3.0 Safety and Preventive Maintenance
- Identify products and procedures used for preventive maintenance.
- Identify devices and procedures for protecting the environment.
- Identify hazards and safety procedures relating to lasers and high-voltage devices.
- Identify special EPA disposal procedures required for batteries, toner, solvents, and CRTs.
- Identify ESD procedures.

4.0 Motherboard, Processors, and Memory
- Identify characteristics and differences of popular MPU/CPU chips.

- Identify characteristics and categories of RAM.
- Identify the types, architecture, and components of popular motherboards.
- Identify the purpose of CMOS and how to update it.

5.0 Printers
- Identify basic printer concepts, components, operations, and field replaceable units.
- Identify printer connections and configurations.

6.0 Portable Systems
- Identify unique components and problems of portable systems—for example, battery life, LCDs, docking station, and cards.

7.0 Basic Networking
- Identify basic networking concepts and how networks work.
- Identify procedures for configuring and swapping network interface cards.
- Identify the effects of PC repairs on networks.

8.0 Customer Satisfaction
- Identify effective/ineffective behaviors contributing to customer satisfaction.

Note: The Customer Satisfaction domain is scored, but it doesn't impact the final score at this time.

The A+ DOS/Windows test module requirement includes the following five domains:

1.0 Function, Structure, Operation, and File Management
- Identify the operating system's functions, structure, and major system files.
- Identify ways to navigate the operating system.
- Identify basic concepts and procedures for creating viewing and managing files and directories, including procedures for changing attributes.
- Identify the procedures for basic disk management.

2.0 Memory Management
- Differentiate between types of memory.
- Identify typical memory conflict problems and how to optimize memory.

3.0 Installation, Configuration, and Upgrading
- Identify the procedures for installing DOS, Windows 3.x, and Windows 95, and for bringing the software to a basic operational level.
- Identify steps to perform an operating system upgrade.
- Identify the basic boot system sequences and alternative ways to boot the system software, including the steps to create an emergency disk.
- Identify procedures for loading software and device drivers.
- Identify the procedures for changing options, configuring, and using Windows.
- Identify the procedures for installing and running typical Windows applications.

4.0 Diagnosing and Troubleshooting
- Recognize and interpret the meaning of common error codes and start-up messages.
- Recognize Windows-specific printing problems and identify procedures for fixing printing problems.
- Recognize other common Windows problems.
- Identify concepts relating to viruses and virus types.

5.0 Networks
- Identify the networking capabilities of DOS and Windows.

- Identify concepts and procedures relating to the Internet and basic Internet setup procedures.

See the following list of sites for the latest test requirements.

CompTIA: www.comptia.org
Wave Technologies: www.wavetech.com/
Marcraft International: www.mic-inc.com/Aplus/
AFSMI: www.afsmi.org/aplus/

18.10 TECH TIPS AND TROUBLESHOOTING—T³

When using the oscilloscope to check bus timing, try to find a unique signal from which to trigger the sweep. It is impossible to tell what is happening on the bus by simply using the automatic trigger mode. When checking the RAM RAS and CAS timing, you should not trigger the sweep on the RAS line. Most dynamic RAM ICs are refreshed by causing RAS-only cycles, and the RAS to CAS timing will be difficult to observe. Try to find a signal unique to the RAM read function.

The BIOS ROM chips in the IBM PC are the masked program type. That is, the program code is placed on the ROM during the final metalization process. Most of these masked-type ROMs cannot be copied in the PROM programmer because the chip select pin must be toggled for every address change. Most PROM programmers do not do this; therefore, the ROMs cannot be read. There is a way in which these ROMs can be copied if your PROM programmer has a serial input port. Connect the serial port of the PC to the serial port of the PROM programmer. Write a small program, in any suitable language, to read the data from memory at address *FE*000 hex through *FFFFF* hex and send the data to the serial port. Before starting the program, command the PROM programmer to copy data from the serial port to the internal RAM. Once this is completed, a blank EPROM with the same pinout as the mask ROM may be programmed and installed in the PC under repair.

EXERCISES

18.1 A simple logic level indicating device is known as a ___________ ___________.

18.2 What three things will the device in Exercise 17.1 show? ___________, ___________, and ___________.

18.3 A device that will "pulse" a circuit is called a(n) ___________ ___________.

18.4 What should every electronic repair shop contain?

18.5 True or false: A DC-coupled scope should be used when checking digital logic.

18.6 What is the minimum bandwidth scope that should be used when testing a microcomputer system?

18.7 Briefly describe the digital storage oscilloscope.

18.8 Briefly describe the logic analyzer.

18.9 True or false: The logic analyzer can be used to analyze bus activity and bus timing.

18.10 What is a PROM burner?

18.11 Crossword Puzzle

ACROSS

4. Multichannel digital display device.
8. Power line recording voltmeter.
9. Light-emitting diode.
10. Voltage rises.
11. Voltage drops.
13. Used to perform waveform measurements.
15. Memory-write signal.
16. I/O read signal.
17. ATE that tests the entire PC board.
19. TIL 311.
21. Digital scope sampling-rate problem.

DOWN

1. Used to inject a signal into a gate.
2. Logic 1.
3. Memory-programming device.
5. Time to read data from a ROM.
6. An oscilloscope sweep control.
7. Clip-the-chip tester.
9. Logic-level-indicating tool.
12. Volt-ohmmeter.
14. ATE that tests from the inside out.
18. Logic gate.
20. Logic state.

1
2
3
4
5
6
7
8
9
10
11
12
13
14
15
16
17
18
19
20
21

APPENDIX A

BOOLEAN THEOREMS

Basic OR operation

$A + 0 = A$

$A + 1 = 1$

$A + \overline{A} = 1$

Basic AND operation

$A \cdot 0 = 0$

$A \cdot 1 = A$

$A \cdot \overline{A} = 0$

Basic NOT operation

$\overline{A} = \overline{A}$

$\overline{\overline{A}} = A$

$\overline{\overline{\overline{A}}} = \overline{A}$

Identity laws

$A + A = A$

$A \cdot A = A$

Commutative laws

$A + B = B + A$

$A \cdot B = B \cdot A$

Associative laws

$A + (B + C) = (A + B) + C$

$A \cdot (B \cdot C) = (A \cdot B) \cdot C$

Distributive law

$A \cdot (B + C) = (A \cdot B) + (A \cdot C)$

$A + (B \cdot C) = (A + B) \cdot (A + C)$

Redundancy law

$A + (A \cdot B) = A$

$A \cdot (A + B) = A$

Special theorems

$$\left.\begin{array}{r} A + (\overline{A} \cdot B) = A + B \\ A \cdot (\overline{A} + B) = A \cdot B \end{array}\right\} \text{Nashelsky}$$

$$\left.\begin{array}{r} \overline{(A + B)} = \overline{A} \cdot \overline{B} \\ \overline{(A \cdot B)} = \overline{A} + \overline{B} \end{array}\right\} \text{DeMorgan}$$

APPENDIX B

INTEL 8086/80888 INSTRUCTION SET

Effective address (EA) times.

Address	*Component*	*Clocks*
displacement	disp	6
base or index	BX, BP, DI, SI	5
displacement plus base or index	BX + disp, BP + disp, DI + disp, SI + disp	9
base plus index	BP + DI, BX + SI,	7
	BP + SI, BX + DI	8
displacement plus base plus index	BP + DI + disp, BX + SI + disp,	11
	BP + SI + disp, BX + DI + disp	12

The 8086/8088 instruction set.

AAA
0011 0111
Adjust for ASCII addition

Flags:	O	D	I	T	S	Z	A	P	C
	?				?	?	*	?	*

Operand	*Clocks*	*Transfers*	*Bytes*	*Example*
none	4	none	1	AAA

AAD
1101 0101 0000 1010
ASCII adjust for division

Flags:	O	D	I	T	S	Z	A	P	C
	?				*	*	?	*	?

Operand	*Clocks*	*Transfers*	*Bytes*	*Example*
none	60	none	2	AAD

AAM
1101 0110 0000 1010
ASCII adjust for multiplication

Flags:	O	D	I	T	S	Z	A	P	C
	?				*	*	?	*	?

Operand	*Clocks*	*Transfers*	*Bytes*	*Example*
none	83	none	2	AAM

AAS
0011 1111
ASCII adjust for subtraction

Flags:	0	D	I	T	S	Z	A	P	C
	?				?	?	*	?	*

Operand	*Clocks*	*Transfers*	*Bytes*	*Example*
none	4	none	1	AAS

ADC
0001 00dw oo rrr mmm dp-l dp-h
1000 00sw oo 010 mmm da-l da-h dp-l dp-h (immediate)
0001 010w da-l da-h (acc immediate)
Add with carry

Flags:	O	D	I	T	S	Z	A	P	C
	*				*	*	*	*	*

Operand	*Clocks*	*Transfers*	*Bytes*	*Example*
register, register	3	none	2	ADC AX,DX
register, memory	9(13) + EA	1	2–4	ADC CL,[SI]
memory, register	16(24) + EA	2	2–4	ADC DATA,CX
register, immediate	4	none	3–4	ADC AL,22H
memory, immediate	17(25) + EA	2	3–6	ADC DATA,[BX],3
accumulator, immediate	4	none	2–3	ADC AL,23H

(Continued)

ADD
0000 00dw oo rrr mmm dp-l dp-h
1000 00sw oo 000 mmm da-l da-h dp-l dp-h (immediate)
0001 010w da-l da-h (acc immediate)
Addition

Flags:	O	D	I	T	S	Z	A	P	C
	*				*	*	*	*	*

Operand	*Clocks*	*Transfers*	*Bytes*	*Example*
register, register	3	none	2	ADD AL,DX
register, memory	9(13) + EA	1	2–4	ADD CX,[SI]
memory, register	16(24) + EA	2	2–4	ADD DATA,CX
register, immediate	4	none	3–4	ADD CL,22H
memory, immediate	17(25) + EA	2	3–6	ADD DATA [BX],3
accumulator, immediate	4	none	2–3	ADD AX,2333H

AND
0010 00dw oo rrr mmm dp-l dp-h
1000 00sw oo 100 mmm da-l da-h dp-l dp-h (immediate)
0001 010w da-l da-h (acc immediate)
AND logical

Flags:	O	D	I	T	S	Z	A	P	C
	0				*	*	?	*	0

Operand	*Clocks*	*Transfers*	*Bytes*	*Example*
register, register	3	none	2	AND AL,BH
register, memory	9(13) + EA	1	2–4	AND CX,[SI]
memory, register	16(24) + EA	2	2–4	AND DATA,CX
register, immediate	4	none	3–4	AND CL,22H
memory, immediate	17(25) + EA	2	3–6	AND DATA [BX],3
accumulator, immediate	4	none	2–3	AND AX,2333H

CALL
1110 1000 dp-l dp-h intrasegment direct
1111 1111 oo 010 mmm intrasegment indirect
1001 1010 of-l of-h sg-l sg-h intersegment direct
1111 1111 oo 011 mmm intersegment indirect
CALL procedure

Flags:	O	D	I	T	S	Z	A	P	C

Operand	*Clocks*	*Transfers*	*Bytes*	*Example*
near	19(23)	1	3	CALL NEAR_PROC
far	28(36)	2	5	CALL FAR_PROC
memptr16	21(29) + EA	2	2–4	CALL TABLE[DI]
regptr16	16(24)	1	2	CALL BX
memptr32	37(57) + EA	4	2–4	CALL DATA[BX]

(Continued)

CBW
1001 1000
Convert byte to word

Flags: O D I T S Z A P C

Operand	*Clocks*	*Transfers*	*Bytes*	*Example*
none	2	none	1	CBW

CLC
1111 1000
Clear carry

Flags:	O	D	I	T	S	Z	A	P	C
									0

Operand	*Clocks*	*Transfers*	*Bytes*	*Example*
none	2	none	1	CLC

CLD
1111 1100
Clear direction flag

Flags:	O	D	I	T	S	Z	A	P	C
		0							

Operand	*Clocks*	*Transfers*	*Bytes*	*Example*
none	2	none	1	CLD

CLI
1111 1010
Clear interrupt-enable flag

Flags:	O	D	I	T	S	Z	A	P	C
			0						

Operand	*Clocks*	*Transfers*	*Bytes*	*Example*
none	2	none	1	CLI

CMC
1111 0101
Complement carry flag

Flags:	O	D	I	T	S	Z	A	P	C
									*

Operand	*Clocks*	*Transfers*	*Bytes*	*Example*
none	2	none	1	CMC

(Continued)

CMP
0011 10dw oo rrr mmm dp-l dp-h
1000 00sw oo 111 mmm da-l da-h dp-l dp-h (immediate)
0011 110w da-l da-h (acc immediate)
Compare Flags: O D I T S Z A P C
* * * * * *

Operand	*Clocks*	*Transfers*	*Bytes*	*Example*
register, register	3	none	2	CMP AX,DI
register, memory	9(13) + EA	none	2–4	CMP CX,[SI]
memory, register	9(13) + EA	none	3–4	CMP DATA,CX
register, immediate	4	none	3–4	CMP CL,22H
memory, immediate	10(14) + EA	none	3–6	CMP DATA,[BX],3
accumulator, immediate	4	none	2–3	CMP AX,2333H

CMPS
1010 011w
Compare strings Flags: O D I T S Z A P C
* * * * * *

Operand	*Clocks*	*Transfers*	*Bytes*	*Example*
dest,source	22(30)	2	1	CMPS DAT1,DAT2
repeated	9 + 22(30)/rep	2/rep	1	REP COMPS ID,DATA

CWD
1001 1001
Convert word to double word Flags: O D I T S Z A P C

Operand	*Clocks*	*Transfers*	*Bytes*	*Example*
none	5	none	1	CWD

DAA
0010 0111
Decimal adjust for addition Flags: O D I T S Z A P C
? * * * * *

Operand	*Clocks*	*Transfers*	*Bytes*	*Example*
none	4	none	1	DAA

(Continued)

DAS
0010 1111
Decimal adjust for subtraction

Flags: O D I T S Z A P C
? * * * * *

Operand	*Clocks*	*Transfers*	*Bytes*	*Example*
none	4	none	1	DAS

DEC
1111 111w oo 001 mmm dp-l dp-h
0100 1rrr
Decrement

Flags: O D I T S Z A P C
* * * * *

Operand	*Clocks*	*Transfers*	*Bytes*	*Example*
reg16	2	none	1	DEC CX
reg8	3	none	2	DEC DH
memory	15(23) + EA	2	2–4	DEC DATA[BP]

DIV
1111 011w oo 001 mmm dp-l dp-h
Divide

Flags: O D I T S Z A P C
? ? ? ? ? ?

Operand	*Clocks*	*Transfers*	*Bytes*	*Example*
reg8	80–90	none	2	DIV BH
reg16	144–162	none	2	DIV BP
mem8	(86–96) + EA	1	2–4	DIV DATA
mem16	(154–172) + EA	1	2–4	DIV NUMBER[DI]

ESC
1101 1xxx oo xxx mmm dp-l dp-h
Escape

Flags: O D I T S Z A P C

Operand	*Clocks*	*Transfers*	*Bytes*	*Example*
memory	8(12) + EA	1	2–4	FADD DATA
register	2	none	2	ESC 4,BL

HLT
1111 0100
Halt

Flags: O D I T S Z A P C

Operand	*Clocks*	*Transfers*	*Bytes*	*Example*
none	2	none	1	HLT

(Continued)

IDIV
1111 011w oo 111 mmm dp-l dp-h
Integer division

Flags: O D I T S Z A P C
? ? ? ? ? ?

Operand	Clocks	Transfers	Bytes	Example
reg8	101–112	none	2	IDIV CL
reg16	165–184	none	2	IDIV SI
mem8	(107–118) + EA	1	2–4	IDIV INFO
mem16	(175–194) + EA	1	2–4	INDIV MEM__DAT

IMUL
1111 011w oo 101 mmm dp-l dp-h
Integer multiplication

Flags: O D I T S Z A P C
* ? ? ? ? *

Operand	Clocks	Transfers	Bytes	Example
reg8	80–98	none	2	IMUL AL
reg16	128–154	none	2	IMUL AX
mem8	(86–104) + EA	1	2–4	IMUL BYTE__NUMB
mem16	(138–164) + EA	1	2–4	IMUL WORD__DATA

IN
1110 010w port fixed
1110 110w variable
Input byte or word

Flags: O D I T S Z A P C

Operand	Clocks	Transfers	Bytes	Example
acc,imm8	10(14)	1	2	IN AL,0FDH
acc,DX	8(12)	1	1	IN AL,DX

INC
1111 111w oo 000 mmm dp-l dp-h
0100 0rrr
Increment

Flags: O D I T S Z A P C
* * * * *

Operand	Clocks	Transfers	Bytes	Example
reg8	3	none	2	INC CL
reg16	2	none	1	INC DI
memory	15(23) + EA	2	2–4	INC DATA[BX]

(Continued)

INT
1100 1100 INT 3
1100 1101 type x INT x
Interrupt

Flags: O D I T S Z A P C

Operand	Clocks	Transfers	Bytes	Example
type 3	52(72)	5	1	INT 3
not type 3	51(71)	5	2	INT 55

INTO
1100 1110
Interrupt on overflow

Flags: O D I T S Z A P C

Operand	Clocks	Transfers	Bytes	Example
none	73 or 4	5 or 0	1	INTO

IRET
1100 1111
Interrupt return

Flags:	O	D	I	T	S	Z	A	P	C
	*	*	*	*	*	*	*	*	*

Operand	Clocks	Transfers	Bytes	Example
none	32(44)	3	1	IRET

JA/JNBE
0111 0111 disp
Jump above/not below or equal to

Flags: O D I T S Z A P C

Operand	Clocks	Transfers	Bytes	Example
short label	16/4	none	2	JA NEXT JNBE NEXT

JAE/JNB
0111 0011 disp
Jump above or equal to/not below

Flags: O D I T S Z A P C

Operand	Clocks	Transfers	Bytes	Example
short label	16/4	none	2	JAE DOWN JNB DOWN

(Continued)

JB/JNAE
0111 0010 disp
Jump below/not above or equal to — Flags: O D I T S Z A P C

Operand	Clocks	Transfers	Bytes	Example
short label	16/4	none	2	JB BELOW JNAE WHERE

JBE/JNA
0111 0110 disp
Jump below or equal to/not above — Flags: O D I T S Z A P C

Operand	Clocks	Transfers	Bytes	Example
short label	16/4	none	2	JBE LOOP JNA LABEL

JC
0111 0010 disp
Jump on carry — Flags: O D I T S Z A P C

Operand	Clocks	Transfers	Bytes	Example
short label	16/4	none	2	JC OUT

JCXZ
1110 0011 disp
Jump if CX is 0 — Flags: O D I T S Z A P C

Operand	Clocks	Transfers	Bytes	Example
short label	18/6	none	2	JCXZ FINISHED

JE/JZ
0111 0100 disp
Jump equal to/zero — Flags: O D I T S Z A P C

Operand	Clocks	Transfers	Bytes	Example
short label	16/4	none	2	JE END_IT JZ ONCE

(Continued)

JG/JNLE
0111 1111 disp
Jump greater/not less than or equal to   Flags: O D I T S Z A P C

Operand	*Clocks*	*Transfers*	*Bytes*	*Example*
short label	16/4	none	2	JG BIG JNLE NOT__SMALL

JGE/JNL
0111 1101 disp
Jump greater than or equal to/not less than   Flags: O D I T S Z A P C

Operand	*Clocks*	*Transfers*	*Bytes*	*Example*
short label	16/4	none	2	JGE UPWARD JNL LEFT

JL/JNGE
0111 1100 disp
Jump less than/not greater than or equal to   Flags: O D I T S Z A P C

Operand	*Clocks*	*Transfers*	*Bytes*	*Example*
short label	16/4	none	2	JL NEXT JNGE AGAIN

JLE/JNG
0111 1110 disp
Jump less than or equal to/not greater than   Flags: O D I T S Z A P C

Operand	*Clocks*	*Transfers*	*Bytes*	*Example*
short label	16/4	none	2	JLE NEXT

JMP

1110	1001	dp-l	dp-h					intrasegment direct
1110	1011	disp						intrasegment short
1111	1111	oo	100	mmm	dp-l	dp-h		intrasegment indirect
1110	1010	of-l			of-h	sg-l	sg-h	intersegment direct
1111	1111	oo	101	mmm				intersegment indirect

Jump unconditional   Flags: O D I T S Z A P C

(Continued)

Operand	*Clocks*	*Transfers*	*Bytes*	*Example*
short label	15	none	2	JMP SHORT NUMB
near label	15	none	3	JMP NEAR__PLACE
far label	15	none	5	JMP FAR__WAY
memptr16	18 + EA	none	2–4	JMP TABLE[BX]
regptr16	11	none	2	JMP DX
memptr32	24 + EA	none	2–4	JMP FAR__SEG[DI]

JNC
0111 0011 disp
Jump no carry

Flags: O D I T S Z A P C

Operand	*Clocks*	*Transfers*	*Bytes*	*Example*
short label	16/4	none	2	JNC BACK

JNE/JNZ
0111 0101 disp
Jump not equal to/not 0

Flags: O D I T S Z A P C

Operand	*Clocks*	*Transfers*	*Bytes*	*Example*
short label	16/4	none	2	JNE OVER JNZ INNER

JNO
0111 0001 disp
Jump not overflow

Flags: O D I T S Z A P C

Operand	*Clocks*	*Transfers*	*Bytes*	*Example*
short label	16/4	none	2	JNO GOOD

JNS
0111 1001 disp
Jump not sign

Flags: O D I T S Z A P C

Operand	*Clocks*	*Transfers*	*Bytes*	*Example*
short label	16/4	none	2	JNS POSITIVE

(Continued)

JNP/JPO
0111 1011 disp
Jump no parity/parity odd — Flags: O D I T S Z A P C

Operand	*Clocks*	*Transfers*	*Bytes*	*Example*
short label	16/4	none	2	JNP NO JPO ODD

JO
0111 0000 disp
Jump on overflow — Flags: O D I T S Z A P C

Operand	*Clocks*	*Transfers*	*Bytes*	*Example*
short label	16/4	none	2	JO OVER__FLOW

JP/JPE
0111 1010 disp
Jump parity/parity even — Flags: O D I T S Z A P C

Operand	*Clocks*	*Transfers*	*Bytes*	*Example*
short label	16/4	none	2	JO PAIR JPE EVEN

JS
0111 1000 disp
Jump on sign — Flags: O D I T S Z A P C

Operand	*Clocks*	*Transfers*	*Bytes*	*Example*
short label	16/4	none	2	JS MINUS

LAHF
1001 1111
Load AH from flags — Flags: O D I T S Z A P C

Operand	*Clocks*	*Transfers*	*Bytes*	*Example*
none	4	none	1	LAHF

(Continued)

LDS
1100 0101 oo rrr mmm dp-l dp-h
Load pointer and DS Flags: O D I T S Z A P C

Operand	Clocks	Transfers	Bytes	Example
reg16,mem32	24 + EA	2	2–4	LDS DI,DATA

LEA
1000 1101 oo rrr mmm dp-l dp-h
Load effective address Flags: O D I T S Z A P C

Operand	Clocks	Transfers	Bytes	Example
reg16,mem16	2 + EA	none	2–4	LEA DI,[BP + SI]

LES
1100 0100 oo rrr mmm dp-l dp-h
Load pointer and ES Flags: O D I T S Z A P C

Operand	Clocks	Transfers	Bytes	Example
reg16,mem32	24 + EA	2	2–4	LES SI,MESS

LOCK
1111 0000
Lock the bus Flags: O D I T S Z A P C

Operand	Clocks	Transfers	Bytes	Example
none	2	none	1	LOCK ADD AX,BX

LODS
1010 110w
Load string Flags: O D I T S Z A P C

Operand	Clocks	Transfers	Bytes	Example
source string	12(16)	1	1	LODS WOOD
repeat	9 + 13(17)/rep	1/rep	1	REP LODS DATA

(Continued)

LOOP
1110 0010 disp
Loop while CX is not 0 Flags: O D I T S Z A P C

Operand	*Clocks*	*Transfers*	*Bytes*	*Example*
short label	17/5	none	2	LOOP BACK

LOOPE/LOOPZ
1110 0001 disp
Loop while equal to/while 0 Flags: O D I T S Z A P C

Operand	*Clocks*	*Transfers*	*Bytes*	*Example*
short label	18/6	none	2	LOOPE AGAIN

LOOPNZ/LOOPNE
1110 0000 disp
Loop while not 0/not equal to Flags: O D I T S Z A P C

Operand	*Clocks*	*Transfers*	*Bytes*	*Example*
short label	19/5	none	2	LOOPNE TOP

MOV

1000	10dw	oo	rrr	mmm	dp-l	dp-h				
1100	011w	oo	000	mmm	da-l	da-h	dp-l	dp-h		(immediate)
1011	wrrr	da-l			da-h					(reg immediate)
1010	000w	adr-l			adr-h					(mem to acc)
1010	001w	adr-l			adr-h					(acc to mem)
1000	1110	oo	0rr	mmm	dp-l	dp-h				(to segment)
1000	1100	oo	0rr	mmm	dp-l	dp-h				(from segment)

Move Flags: O D I T S Z A P C

Operand	*Clocks*	*Transfers*	*Bytes*	*Example*
mem,acc	10(14)	1	3	MOV ARRAY,AL
acc,mem	10(14)	1	3	MOV AL,DATA
reg,reg	2	none	2	MOV AX,BX
reg,mem	8(12) + EA	1	2–4	MOV DI,DATA
mem,reg	9(13) + EA	1	2–4	MOV MEM,BL
reg,imm	4	none	2–3	MOV BH,33
mem,imm	10(14) + EA	1	3–6	MOV DATA[BX + DI],1
seg,reg	2	none	2	MOV ES,AX
seg,mem	8(12) + EA	1	2–4	MOV DS,MEMORY
reg,seg	2	none	2	MOV AX,DS
mem,seg	9(13) + EA	1	2–4	MOV SAVE,DS

(Continued)

MOVS
1001 010w
Move string — Flags: O D I T S Z A P C

Operand	Clocks	Transfers	Bytes	Example
string	18(26)	2	1	MOVSB
repeat	9 + 17(25)/rep	2/rep	1	REP MOVSW

MUL
1111 011w oo 100 mmm dp-l dp-h
Multiply — Flags: O D I T S Z A P C

Operand	Clocks	Transfers	Bytes	Example
reg8	70–77	none	2	MUL CL
reg16	118–123	none	2	MUL AX
mem8	(76–83) + EA	1	2–4	MUL DATA[BP]
mem16	(128–143) + EA	1	2–4	MUL RATE

NEG
1111 011w oo 011 mmm dp-l dp-h
Negate — Flags: O D I T S Z A P C

Operand	Clocks	Transfers	Bytes	Example
reg	3	none	2	NEG AL
mem	16(24) + EA	2	2–4	NEG DATA

NOP
1001 0000
No operation — Flags: O D I T S Z A P C

Operand	Clocks	Transfers	Bytes	Example
none	3	none	1	NOP

NOT
1111 011w oo 010 mmm dp-l dp-h
Logical inversion(not) — Flags: O D I T S Z A P C

Operand	Clocks	Transfers	Bytes	Example
reg	3	none	2	NOT BX
mem	16(24) + EA	2	2–4	NOT DATA[BX]

(Continued)

OR

0000	10dw	oo	rrr	mmm	dp-l	dp-h			
1000	00sw	oo	001	mmm	da-l	da-h	dp-l	dp-h	(immediate)
0000	110w	da-l			da-h				(acc immediate)

Logical OR

Flags: O D I T S Z A P C
0 * * ? * 0

Operand	*Clocks*	*Transfers*	*Bytes*	*Example*
reg,reg	3	none	2	OR AH,AL
reg,mem	9(13) + EA	1	2–4	OR AX,DATA
mem,reg	16(24) + EA	2	2–4	OR [BX + 4],AL
acc,imm	4	none	2–3	OR AX,345
reg,imm	4	none	3–4	OR BL,12H
mem,imm	17(25) + EA	2	3–6	OR DATA,1234H

OUT

1110	011w	port	fixed
1110	111w		variable

Output a byte or word

Flags: O D I T S Z A P C

Operand	*Clocks*	*Transfers*	*Bytes*	*Example*
port,acc	10(14)	1	2	OUT 44H,AL
DX,acc	8(12)	1	1	OUT DX,AX

POP

1000	1111	oo	000	mmm	dp-l	dp-h
0101	1rrr	(register)				
000r	r111	(segment)				

Pop

Flags: O D I T S Z A P C

Operand	*Clocks*	*Transfers*	*Bytes*	*Example*
reg	12	1	1	POP CX
seg	12	1	1	POP SS
mem	25 + EA	2	2–4	POP MEMORY

POPF

1001 1100

Pop flags

Flags: O D I T S Z A P C
* * * * * * * * *

Operand	*Clocks*	*Transfers*	*Bytes*	*Example*
none	12	1	1	POPF

(Continued)

PUSH
1111 1111 oo 110 mmm dp-l dp-h
0101 0rrr (register)
000r r110 (segment)
Push Flags: O D I T S Z A P C

Operand	*Clocks*	*Transfers*	*Bytes*	*Example*
reg	15	1	1	PUSH BP
seg	14	1	1	PUSH DS
mem	24 + EA	2	2–4	PUSH DATA

PUSHF
1001 1101
Push flags Flags: O D I T S Z A P C

Operand	*Clocks*	*Transfers*	*Bytes*	*Example*
none	14	1	1	PUSHF

RCL
1101 00vw oo 010 mmm dp-l dp-h
Rotate left through carry Flags: O D I T S Z A P C
(* under O and C)

Operand	*Clocks*	*Transfers*	*Bytes*	*Example*
reg	2	none	2	RCL AX
reg,CL	8 + 4/bit	none	2	RCL BL,CL
mem	15(23) + EA	2	2–4	RCL DATA[DI]
mem,CL	20(2) + EA + 4/bit	2	2–4	RCL DATA,CL

RCR
1101 00vw oo 011 mmm dp-l dp-h
Rotate right through carry Flags: O D I T S Z A P C
(* under O and C)

Operand	*Clocks*	*Transfers*	*Bytes*	*Example*
reg	2	none	2	RCR AX
reg,CL	8 + 4/bit	none	2	RCR BL,CL
mem	15(23) + EA	2	2–4	RCR DATA[DI]
mem,CL	20(2) + EA + 4/bit	2	2–4	RCR DATA,CL

(Continued)

REP (conditional)
1111 001z
Repeat conditional

Flags: O D I T S Z A P C

Operand	*Clocks*	*Transfers*	*Bytes*	*Example*
REPE/REPZ	2	none	1	REPE CMPSB
REPNE/REPNZ	2	none	1	REPNE SCASW

RET
1100 0011 intrasegment
1100 0010 da-l da-h intrasegment with add
1100 1011 intersegment
1100 1010 da-l da-h intersegment with add
Return

Flags: O D I T S Z A P C

Operand	*Clocks*	*Transfers*	*Bytes*	*Example*
intra, no add	20	1	1	RET
intra, add	24	1	3	RET 6
inter, no add	32	2	1	RET
inter, add	31	2	3	RET 2

ROL
1101 00vw oo 000 mmm dp-l dp-h
Rotate left

Flags:	O	D	I	T	S	Z	A	P	C
	*								*

Operand	*Clocks*	*Transfers*	*Bytes*	*Example*
reg	2	none	2	ROL CL
reg,CL	8 + 4/bit	none	2	ROL BX,CL
mem	15(23) + EA	2	2–4	ROL DATA[DI]
mem,CL	20(28) + EA + 4/bit	2	2–4	ROL MEM,CL

ROR
1101 00vw oo 001 mmm dp-l dp-h
Rotate right

Flags:	O	D	I	T	S	Z	A	P	C
	*								*

Operand	*Clocks*	*Transfers*	*Bytes*	*Example*
reg	2	none	2	ROR AX
reg,CL	8 + 4/bit	none	2	ROR AX,CL
mem	15(23) + EA	2	2–4	ROR LIST
mem,CL	20(28) + EA + 4/bit	2	2–4	ROR LIST,CL

(Continued)

SAHF
1001 1110
Store AH into flags

Flags:	O	D	I	T	S	Z	A	P	C
					*	*	*	*	*

Operand	*Clocks*	*Transfers*	*Bytes*	*Example*
none	4	none	1	SAHF

SAL/SHL
1101 00vw oo 100 mmm dp-l dp-h
Shif left arithmetic/logic

Flags:	O	D	I	T	S	Z	A	P	C
	*				*	*	?	*	*

Operand	*Clocks*	*Transfers*	*Bytes*	*Example*
reg	2	none	2	SHL AX
reg,CL	8 + 4/bit	none	2	SHL AX,CL
mem	15(23) + EA	2	2–4	SHL LIST
mem,CL	20(28) + EA + 4/bit	2	2–4	SHL LIST,CL

SAR
1101 00vw oo 100 mmm dp-l dp-h
Shift right arithmetic/logic

Flags:	O	D	I	T	S	Z	A	P	C
	*				*	*	?	*	*

Operand	*Clocks*	*Transfers*	*Bytes*	*Example*
reg	2	none	2	SAR AX
reg,CL	8 + 4/bit	none	2	SAR AX,CL
mem	15(23) + EA	2	2–4	SAR LIST
mem,CL	20(28) + EA + 4/bit	2	2–4	SAR LIST,CL

SBB
0001 10dw oo rrr mmm dp-l dp-h
1000 00sw oo 011 mmm da-l da-h da-l dp-h (immediate)
0001 110w da-l da-h (acc immediate)
Subtract with borrow

Flags:	O	D	I	T	S	Z	A	P	C
	*				*	*	*	*	*

Operand	*Clocks*	*Transfers*	*Bytes*	*Example*
reg,reg	3	none	2	SBB AL,BL
reg,mem	9(13) + EA	1	2–4	SBB BX,DATA
mem,reg	16(24) + EA	2	2–4	SBB DATA,BX
acc,imm	4	none	2–3	SBB AH,12
reg,imm	4	none	3–4	SBB DI,1000H
mem,imm	17(25) + EA	2	3–6	SBB DATA,33

(Continued)

SCAS
1010 111w
Scan for byte or word

Flags:	O	D	I	T	S	Z	A	P	C
	*				*	*	*	*	*

Operand	*Clocks*	*Transfers*	*Bytes*	*Example*
string	15(19)	1	1	SCASB
repeat	9 + 15(19)/rep	1/rep	1	REPE SCASW

SHR
1101 00vw oo 101 mmm dp-l dp-h
Shift right

Flags:	O	D	I	T	S	Z	A	P	C
	*				*	*	?	*	*

Operand	*Clocks*	*Transfers*	*Bytes*	*Example*
reg	2	none	2	SHR AX
reg,CL	8 + 4/bit	none	2	SHR AX,CL
mem	15(23) + EA	2	2–4	SHR LIST
mem,CL	20(28) + EA + 4/bit	2	2–4	SHR LIST,CL

STC
1111 1001
Set carry

Flags:	O	D	I	T	S	Z	A	P	C
									1

Operand	*Clocks*	*Transfers*	*Bytes*	*Example*
none	2	none	1	STC

STD
1111 1101
Set direction

Flags:	O	D	I	T	S	Z	A	P	C
		1							

Operand	*Clocks*	*Transfers*	*Bytes*	*Example*
none	2	none	1	STD

STI
1111 1011
Set interrupt-enable flag

Flags:	O	D	I	T	S	Z	A	P	C
			1						

Operand	*Clocks*	*Transfers*	*Bytes*	*Example*
none	2	none	1	STI

(Continued)

STOS
1010 101w
Store byte or word Flags: O D I T S Z A P C

Operand	*Clocks*	*Transfers*	*Bytes*	*Example*
string	11(15)	1	1	STOSB
repeat	9 + 10(14)/rep	1/rep	1	REP STOSW

SUB
0010 10dw oo rrr mmm dp-l dp-h
1000 00sw oo 101 mmm da-l da-h dp-l dp-h (immediate)
0010 110w da-l da-h (acc immediate)

Subtract	Flags:	O	D	I	T	S	Z	A	P	C
		*				*	*	*	*	*

Operand	*Clocks*	*Transfers*	*Bytes*	*Example*
reg,reg	3	none	2	SUB AL,BL
reg,mem	9(13) + EA	1	2–4	SUB BX,DATA
mem,reg	16(24) + EA	2	2–4	SUB DATA,BX
acc,imm	4	none	2–3	SUB AH,12
reg,imm	4	none	3–4	SUB DI,1000H
mem,imm	17(25) + EA	2	3–6	SUB DATA,33

TEST
1000 01dw oo rrr mmm dp-l dp-h
1111 011w oo 000 mmm da-l da-h dp-l dp-h (immediate)
1010 100w da-l da-h (acc immediate)

Test	Flags:	O	D	I	T	S	Z	A	P	C
		0				*	*	?	*	0

Operand	*Clocks*	*Transfers*	*Bytes*	*Example*
reg,reg	3	none	2	TEST DI,BP
reg,mem	9(13) + EA	1	2–4	TEST BX,NUMB
acc,imm	4	none	2–3	TEST AX,2
reg,imm	5	none	3–4	TEST CL,10H
mem,imm	11 + EA	none	3–6	TEST MOM,01H

SN54ALS00A, SN54AS00, SN74ALS00A, SN74AS00
QUADRUPLE 2-INPUT POSITIVE-NAND GATES

D2661, APRIL 1982—REVISED MAY 1986

- **Package Options Include Plastic "Small Outline" Packages, Ceramic Chip Carriers, and Standard Plastic and Ceramic 300-mil DIPs**
- **Dependable Texas Instruments Quality and Reliability**

description

These devices contain four independent 2-input NAND gates. They perform the Boolean functions $Y = \overline{A \cdot B}$ or $Y = \overline{A} + \overline{B}$ in positive logic.

The SN54ALS00A and SN54AS00 are characterized for operation over the full military temperature range of −55°C to 125°C. The SN74ALS00A and SN74AS00 are characterized for operation from 0°C to 70°C.

FUNCTION TABLE (each gate)

INPUTS		OUTPUT
A	B	Y
H	H	L
L	X	H
X	L	H

logic symbol†

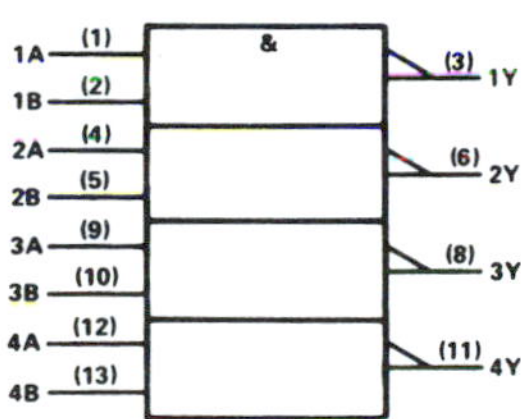

†This symbol is in accordance with ANSI/IEEE Std 91-1984 and IEC Publication 617-12.
Pin numbers shown are for D, J, and N packages.

SN54ALS00A, SN54AS00 . . . J PACKAGE
SN74ALS00A, SN74AS00 . . . D OR N PACKAGE
(TOP VIEW)

Pin		Pin	
1A	1	14	VCC
1B	2	13	4B
1Y	3	12	4A
2A	4	11	4Y
2B	5	10	3B
2Y	6	9	3A
GND	7	8	3Y

SN54ALS00A, SN54AS00 . . . FK PACKAGE
(TOP VIEW)

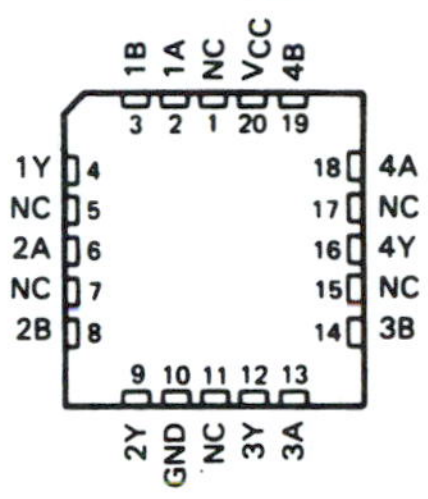

NC—No internal connection

logic diagram (positive logic)

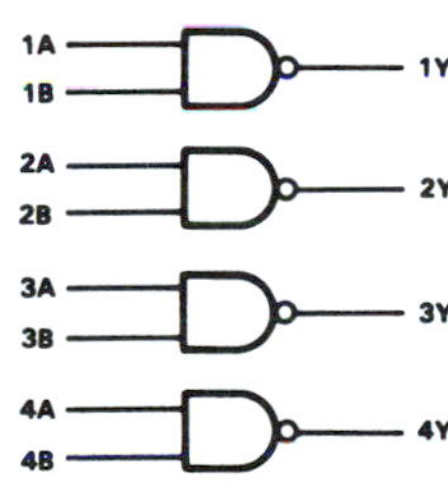

SN54ALS00A, SN74ALS00A
QUADRUPLE 2-INPUT POSITIVE-NAND GATES

absolute maximum ratings over operating free-air temperature range (unless otherwise noted)

Supply voltage, V_{CC} . . . . 7 V
Input voltage . . . . 7 V
Operating free-air temperature range: SN54ALS00A . . . . −55°C to 125°C
SN74ALS00A . . . . 0°C to 70°C
Storage temperature range . . . . −65°C to 150°C

recommended operating conditions

		SN54ALS00A			SN74ALS00A			UNIT
		MIN	NOM	MAX	MIN	NOM	MAX	
V_{CC}	Supply voltage	4.5	5	5.5	4.5	5	5.5	V
V_{IH}	High-level input voltage	2			2			V
V_{IL}	Low-level input voltage			0.7			0.8	V
I_{OH}	High-level output current			−0.4			−0.4	mA
I_{OL}	Low-level output current			4			8	mA
T_A	Operating free-air temperature	−55		125	0		70	°C

electrical characteristics over recommended operating free-air temperature range (unless otherwise noted)

PARAMETER	TEST CONDITIONS		SN54ALS00A			SN74ALS00A			UNIT
			MIN	TYP†	MAX	MIN	TYP†	MAX	
V_{IK}	V_{CC} = 4.5 V,	I_I = −18 mA			−1.5			−1.5	V
V_{OH}	V_{CC} = 4.5 V to 5.5 V,	I_{OH} = −0.4 mA	$V_{CC}-2$			$V_{CC}-2$			V
V_{OL}	V_{CC} = 4.5 V,	I_{OL} = 4 mA		0.25	0.4		0.25	0.4	V
	V_{CC} = 4.5 V,	I_{OL} = 8 mA					0.35	0.5	
I_I	V_{CC} = 5.5 V,	V_I = 7 V			0.1			0.1	mA
I_{IH}	V_{CC} = 5.5 V,	V_I = 2.7 V			20			20	μA
I_{IL}	V_{CC} = 5.5 V,	V_I = 0.4 V			−0.1			−0.1	mA
I_O‡	V_{CC} = 5.5 V,	V_O = 2.25 V	−30		−112	−30		−112	mA
I_{CCH}	V_{CC} = 5.5 V,	V_I = 0 V		0.5	0.85		0.5	0.85	mA
I_{CCL}	V_{CC} = 5.5 V,	V_I = 4.5 V		1.5	3		1.5	3	mA

†All typical values are at V_{CC} = 5 V, T_A = 25°C.
‡The output conditions have been chosen to produce a current that closely approximates one half of the true short-circuit output current, I_{OS}.

switching characteristics (see Note 1)

PARAMETER	FROM (INPUT)	TO (OUTPUT)	V_{CC} = 5 V, C_L = 50 pF, R_L = 500 Ω, T_A = 25°C	V_{CC} = 4.5 V to 5.5 V, C_L = 50 pF, R_L = 500 Ω, T_A = MIN to MAX				UNIT
			'ALS00A	SN54ALS00A		SN74ALS00A		
			TYP	MIN	MAX	MIN	MAX	
t_{PLH}	A or B	Y	7	3	16	3	11	ns
t_{PHL}	A or B	Y	5	2	13	2	8	

NOTE 1: Load circuit and voltage waveforms are shown in Section 1.

HIGH-SPEED CMOS LOGIC

TYPES SN54HC00, SN74HC00 QUADRUPLE 2-INPUT POSITIVE-NAND GATES

D2684, DECEMBER 1982—REVISED MARCH 1984

- **Package Options Include Both Plastic and Ceramic Chip Carriers in Addition to Plastic and Ceramic DIPs**
- **Dependable Texas Instruments Quality and Reliability**

SN54HC00 . . . J PACKAGE
SN74HC00 . . . J OR N PACKAGE
(TOP VIEW)

Signal	Pin	Pin	Signal
1A	1	14	V_{CC}
1B	2	13	4B
1Y	3	12	4A
2A	4	11	4Y
2B	5	10	3B
2Y	6	9	3A
GND	7	8	3Y

description

These devices contain four independent 2-input NAND gates. They perform the Boolean functions $Y = \overline{A \cdot B}$ or $Y = \overline{A} + \overline{B}$ in positive logic.

The SN54HC00 is characterized for operation over the full military temperature range of −55°C to 125°C. The SN74HC00 is characterized for operation from −40°C to 85°C.

SN54HC00 . . . FH OR FK PACKAGE
SN74HC00 . . . FH OR FN PACKAGE
(TOP VIEW)

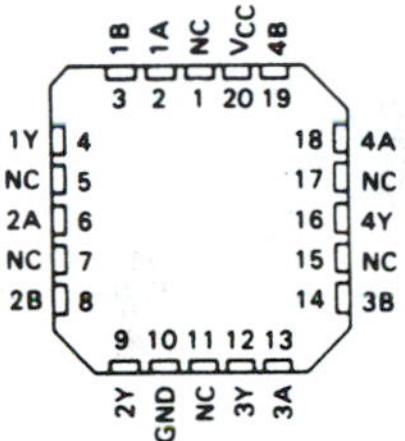

NC—No internal connection

FUNCTION TABLE (each gate)

INPUTS		OUTPUT
A	B	Y
H	H	L
L	X	H
X	L	H

logic symbol

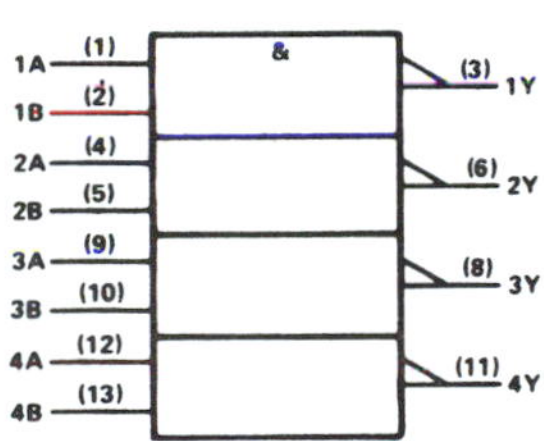

Pin numbers shown are for J and N packages.

maximum ratings, recommended operating conditions, and electrical characteristics

See Table I, page 2-4.

TYPES SN54HC00, SN74HC00
QUADRUPLE 2-INPUT POSITIVE-NAND GATES

switching characteristics over recommended operating free-air temperature range (unless otherwise noted), C_L = 50 pF (see Note 1)

PARAMETER	FROM (INPUT)	TO (OUTPUT)	V_{CC}	T_A = 25°C MIN	T_A = 25°C TYP	T_A = 25°C MAX	SN54HC00 MIN	SN54HC00 MAX	SN74HC00 MIN	SN74HC00 MAX	UNIT
t_{pd}	A or B	Y	2 V		45	90		135		115	ns
			4.5 V		9	18		27		23	
			6 V		8	15		23		20	
t_t		Y	2 V		38	75		110		95	ns
			4.5 V		8	15		22		19	
			6 V		6	13		19		16	

C_{pd}	Power dissipation capacitance per gate	No load, T_A = 25°C	20 pF typ

NOTE 1: For load circuit and voltage waveforms, see page 1-14.

3

HCMOS DEVICES

TYPES SN5476, SN54H76, SN54LS76A, SN7476, SN74H76, SN74LS76A DUAL J-K FLIP-FLOPS WITH PRESET AND CLEAR

REVISED DECEMBER 1983

- **Package Options Include Plastic and Ceramic DIPs**
- **Dependable Texas Instruments Quality and Reliability**

SN5476, SN54H76, SN54LS76A . . . J OR W PACKAGE
SN7476, SN74H76 . . . J OR N PACKAGE
SN74LS76A . . . D, J OR N PACKAGE
(TOP VIEW)

	Pin	Pin	
1CLK	1	16	1K
1 $\overline{PRE}$	2	15	1Q
1 $\overline{CLR}$	3	14	1$\overline{Q}$
1 J	4	13	GND
V_{CC}	5	12	2K
2CLK	6	11	2Q
2 $\overline{PRE}$	7	10	2$\overline{Q}$
2 $\overline{CLR}$	8	9	2J

description

The '76 and 'H76 contain two independent J-K flip-flops with individual J-K, clock, preset, and clear inputs. The '76 and 'H76 are positive-edge-triggered flip-flops. J-K input is loaded into the master while the clock is high and transferred to the slave on the high-to-low transition. For these devices the J and K inputs must be stable while the clock is high.

The 'LS76A contain two independent negative-edge-triggered flip-flops. The J and K inputs must be stable one setup time prior to the high-to-low clock transition for predictable operation. The preset and clear are asynchronous active low inputs. When low they override the clock and data inputs forcing the outputs to the steady state levels as shown in the function table.

The SN5476, SN54H76, and the SN54LS76A are characterized for operation over the full military temperature range of −55 °C to 125 °C. The SN7476, SN74H76, and the SN74LS76A are characterized for operation from 0 °C to 70 °C.

'76, 'H76
FUNCTION TABLE

INPUTS					OUTPUTS	
$\overline{PRE}$	$\overline{CLR}$	CLK	J	K	Q	$\overline{Q}$
L	H	X	X	X	H	L
H	L	X	X	X	L	H
L	L	X	X	X	H†	H†
H	H	⎍	L	L	Q_0	$\overline{Q}_0$
H	H	⎍	H	L	H	L
H	H	⎍	L	H	L	H
H	H	⎍	H	H	TOGGLE	

'LS76A
FUNCTION TABLE

INPUTS					OUTPUTS	
$\overline{PRE}$	$\overline{CLR}$	CLK	J	K	Q	$\overline{Q}$
L	H	X	X	X	H	L
H	L	X	X	X	L	H
L	L	X	X	X	H†	H†
H	H	↓	L	L	Q_0	$\overline{Q}_0$
H	H	↓	H	L	H	L
H	H	↓	L	H	L	H
H	H	↓	H	H	TOGGLE	
H	H	H	X	X	Q_0	$\overline{Q}_0$

† This configuration is nonstable; that is, it will not persist when either preset or clear returns to its inactive (high) level.

FOR CHIP CARRIER INFORMATION, CONTACT THE FACTORY

**TYPES SN5476, SN54H76,
SN7476, SN74H76
DUAL J-K FLIP-FLOPS WITH PRESET AND CLEAR**

logic diagrams

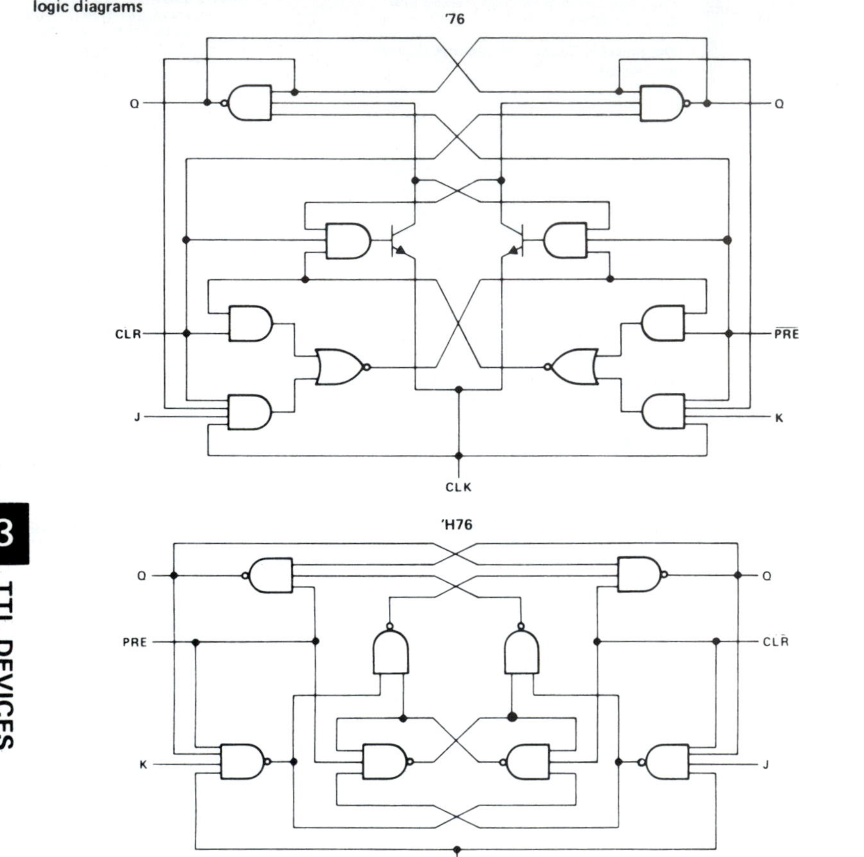

TEXAS INSTRUMENTS
POST OFFICE BOX 225012 • DALLAS, TEXAS 75265

TYPES SN5476, SN54H76, SN54LS76A, SN7476, SN74H76, SN74LS76A DUAL J-K FLIP-FLOPS WITH PRESET AND CLEAR

logic diagrams (continued)

'LS76A

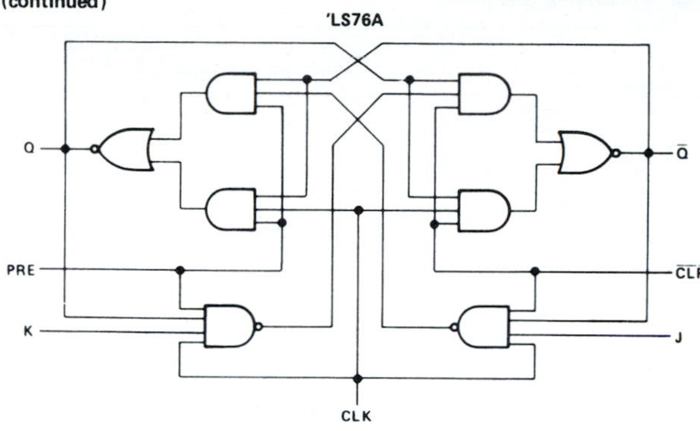

logic symbols

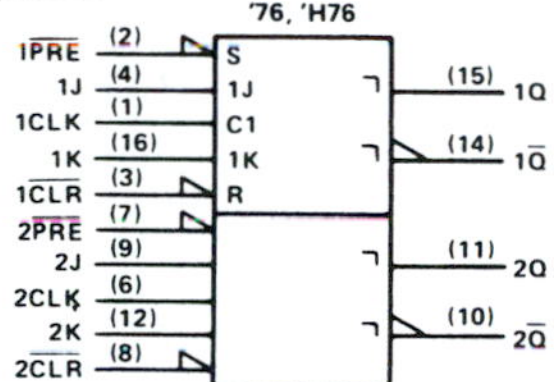

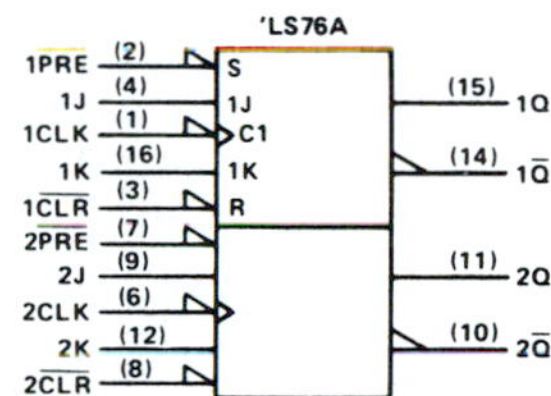

Pin numbers shown on logic notation are for D, J or N packages.

schematics of inputs and outputs

'76

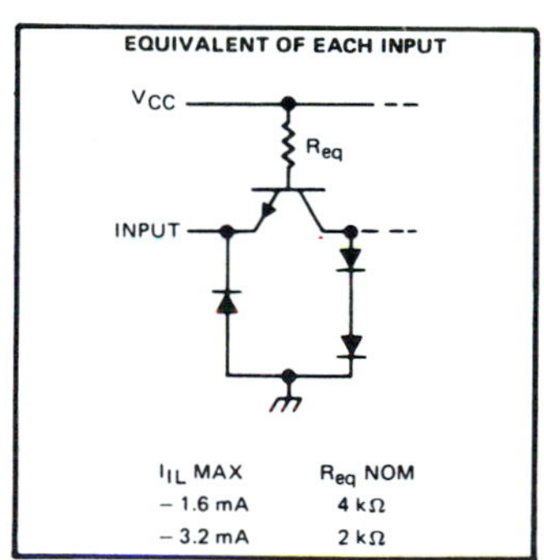

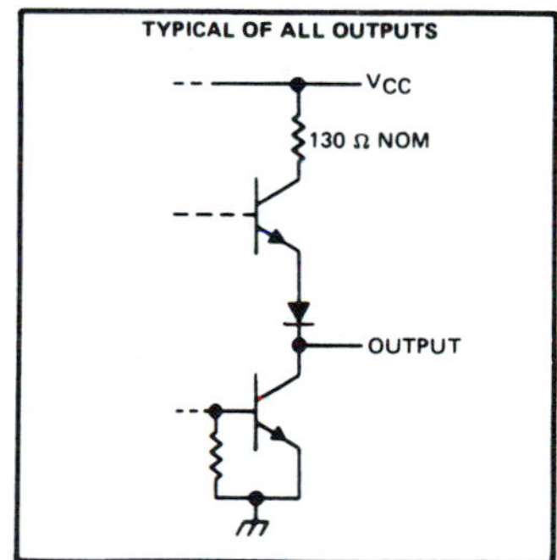

3 TTL DEVICES

TYPES SN54LS76A, SN74LS76A
DUAL J-K FLIP-FLOPS WITH PRESET AND CLEAR

recommended operating conditions

			SN54LS76A			SN74LS76A			UNIT
			MIN	NOM	MAX	MIN	NOM	MAX	
V_{CC}	Supply voltage		4.5	5	5.5	4.75	5	5.75	V
V_{IH}	High-level input voltage		2			2			V
V_{IL}	Low-level input voltage				0.7			0.8	V
I_{OH}	High-level output current				−0.4			−0.4	mA
I_{OL}	Low-level output current				4			8	mA
f_{clock}	Clock frequency		0		30	0		30	MHz
t_w	Pulse duration	CLK high	20			20			ns
		$\overline{PRE}$ or $\overline{CLR}$ low	25			25			
t_{su}	Setup time before CLK↓	data high or low	20			20			ns
		$\overline{CLR}$ inactive	20			20			
		$\overline{PRE}$ inactive	25			25			
t_h	Hold time-data after CLK↓		0			0			ns
T_A	Operating free-air temperature		−55		125	0		70	°C

electrical characteristics over recommended operating free-air temperature range (unless otherwise noted)

PARAMETER		TEST CONDITIONS†	SN54LS76A			SN74LS76A			UNIT
			MIN	TYP‡	MAX	MIN	TYP‡	MAX	
V_{IK}		V_{CC} = MIN, I_I = −18 mA			−1.5			−1.5	V
V_{OH}		V_{CC} = MIN, V_{IH} = 2 V, V_{IL} = MAX, I_{OH} = −0.4 mA	2.5	3.4		2.7	3.4		V
V_{OL}		V_{CC} = MIN, V_{IL} = MAX, V_{IH} = 2 V, I_{OL} = 4 mA		0.25	0.4		0.25	0.4	V
		V_{CC} = MIN, V_{IL} = MAX, V_{IH} = 2 V, I_{OL} = 8 mA					0.35	0.5	
I_I	J or K	V_{CC} = MAX, V_I = 7 V			0.1			0.1	mA
	$\overline{CLR}$ or $\overline{PRE}$				0.3			0.3	
	CLK				0.4			0.4	
I_{IH}	J or K	V_{CC} = MAX, V_I = 2.7 V			20			20	μA
	$\overline{CLR}$ or $\overline{PRE}$				60			60	
	CLK				80			80	
I_{IL}	J or K	V_{CC} = MAX, V_I = 0.4 V			−0.4			−0.4	mA
	All other				−0.8			−0.8	
I_{OS}§		V_{CC} = MAX, See Note 4	−20		−100	−20		−100	mA
I_{CC}		V_{CC} = MAX, See Note 2		4	6		4	6	mA

† For conditions shown as MIN or MAX, use the appropriate value specified under recommended operating conditions.
‡ All typical values are at V_{CC} = 5 V, T_A = 25°C.
§ Not more than one output should be shorted at a time, and the duration of the short circuit should not exceed one second.
NOTE 2: With all outputs open, I_{CC} is measured with the Q and $\overline{Q}$ outputs high in turn. At the time of measurement, the clock input is grounded.
NOTE 4: For certain devices where state commutation can be caused by shorting an output to ground, an equivalent test may be performed with V_O = 2.25 V and 2.125 V for the 54 family and the 74 family, respectively, with the minimum and maximum limits reduced to one half of their stated values.

switching characteristics, V_{CC} = 5 V, T_A = 25°C (see note 3)

PARAMETER	FROM (INPUT)	TO (OUTPUT)	TEST CONDITIONS	MIN	TYP	MAX	UNIT
f_{max}			R_L = 2 kΩ, C_L = 15 pF	30	45		MHz
t_{PLH}	$\overline{PRE}$, $\overline{CLR}$ or CLK	Q or $\overline{Q}$			15	20	ns
t_{PHL}					15	20	ns

NOTE 3: See General Information Section for load circuits and voltage waveforms.

3

TTL DEVICES

HIGH-SPEED CMOS LOGIC

TYPES SN54HC76, SN74HC76 DUAL J-K FLIP-FLOPS WITH CLEAR AND PRESET

D2684, DECEMBER 1982 – REVISED MARCH 1984

- **Package Options Include Both Plastic and Ceramic Chip Carriers in Addition to Plastic and Ceramic DIPs**
- **Dependable Texas Instruments Quality and Reliability**

SN54HC76 . . . J PACKAGE
SN74HC76 . . . J OR N PACKAGE
(TOP VIEW)

Signal	Pin	Pin	Signal
1CLK	1	16	1K
1$\overline{\text{PRE}}$	2	15	1Q
1$\overline{\text{CLR}}$	3	14	1$\overline{\text{Q}}$
1J	4	13	GND
V_{CC}	5	12	2K
2CLK	6	11	2Q
2$\overline{\text{PRE}}$	7	10	2$\overline{\text{Q}}$
2$\overline{\text{CLR}}$	8	9	2J

For functionally and electrically identical parts in chip carrier packages, see SN54HC112 and SN74HC112.

description

These devices contain two independent J-K negative-edge-triggered flip-flops. A low level at the Preset or Clear input sets or resets the outputs regardless of the levels of the other inputs. When Preset and Clear are inactive (high), data at the J and K inputs meeting the setup time requirements are transferred to the outputs on the negative-going edge of the clock pulse. Clock triggering occurs at a voltage level and is not directly related to the rise time of the clock pulse. Following the hold time interval, data at the J and K inputs may be changed without affecting the levels at the outputs. These versatile flip-flops can also perform as toggle flip-flops by tying J and K high.

FUNCTION TABLE
(EACH FLIP-FLOP)

INPUTS					OUTPUTS	
$\overline{\text{PRE}}$	$\overline{\text{CLR}}$	CLK	J	K	Q	$\overline{\text{Q}}$
L	H	X	X	X	H	L
H	L	X	X	X	L	H
L	L	X	X	X	H*	H*
H	H	↓	L	L	Q_0	$\overline{Q}_0$
H	H	↓	H	L	H	L
H	H	↓	L	H	L	H
H	H	↓	H	H	TOGGLE	
H	H	H	X	X	Q_0	$\overline{Q}_0$

*This configuration is nonstable; that is, it will not persist when either Preset or Clear returns to its inactive (high) level.

logic symbol

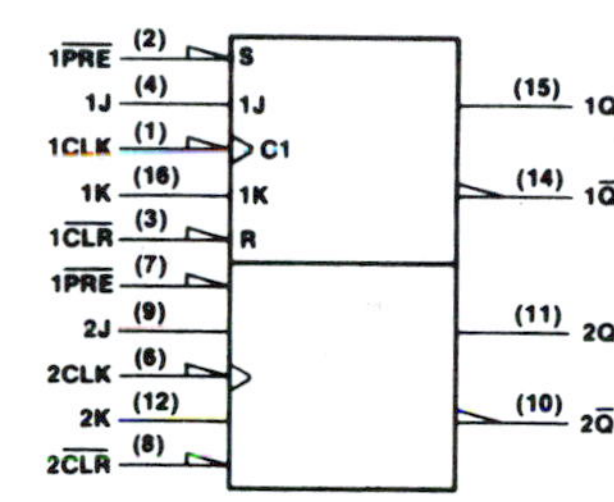

Pin numbers shown are for J and N packages.

logic diagram, each flip-flop (positive logic)

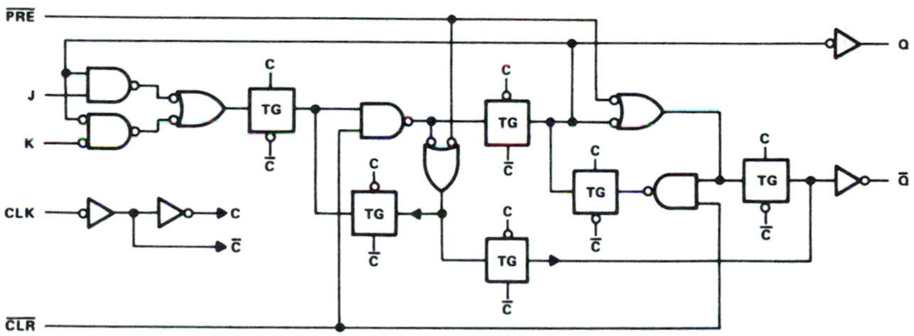

TYPES SN54HC76, SN74HC76
DUAL J-K FLIP-FLOPS WITH CLEAR AND PRESET

maximum ratings, recommended operating conditions, and electrical characteristics

See Table II, page 2-6.

timing requirements over recommended operating free-air temperature range (unless otherwise noted)

			V_{CC}	T_A = 25°C		SN54HC76		SN74HC76		UNIT
				MIN	MAX	MIN	MAX	MIN	MAX	
f_{clock}	Clock frequency		2 V	0	6	0	4.2	0	5	MHz
			4.5 V	0	31	0	21	0	25	
			6 V	0	36	0	25	0	29	
t_w	Pulse duration	$\overline{PRE}$ or $\overline{CLR}$ low	2 V	100		150		125		ns
			4.5 V	20		30		25		
			6 V	17		25		21		
		CLK high or low	2 V	80		120		100		
			4.5 V	16		24		20		
			6 V	14		20		17		
t_{su}	Setup time before CLK↓	Data	2 V	150		225		190		ns
			4.5 V	30		45		38		
			6 V	25		38		32		
		$\overline{PRE}$ or $\overline{CLR}$ inactive	2 V	100		150		125		
			4.5 V	20		30		25		
			6 V	17		25		21		
t_h	Hold time after CLK↓		2 V	0		0		0		ns
			4.5 V	0		0		0		
			6 V	0		0		0		

switching characteristics over recommended operating free-air temperture range (unless otherwise noted), C_L = 50 pF (see Note 1)

PARAMETER	FROM (INPUT)	TO (OUTPUT)	V_{CC}	T_A = 25°C			SN54HC76		SN74HC76		UNIT
				MIN	TYP	MAX	MIN	MAX	MIN	MAX	
f_{max}			2 V	6	9		4.2		5		MHz
			4.5 V	31	41		21		25		
			6 V	36	50		25		29		
t_{pd}	$\overline{PRE}$ or $\overline{CLR}$	Q or $\overline{Q}$	2 V		65	155		250		190	ns
			4.5 V		16	31		47		39	
			6 V		15	26		40		33	
t_{pd}	CLK	Q or $\overline{Q}$	2 V		70	145		220		180	ns
			4.5 V		19	29		44		36	
			6 V		16	25		37		31	
t_t		Q or $\overline{Q}$	2 V		38	75		110		95	ns
			4.5 V		8	15		22		19	
			6 V		6	13		19		16	

C_{pd}	Power dissipation capacitance per flip-flop	No load, T_A = 25°C	36 pF typ

NOTE 1: For load circuit and voltage waveforms, see page 1-14.

TEXAS INSTRUMENTS
POST OFFICE BOX 225012 • DALLAS, TEXAS 75265

00

QUADRUPLE 2-INPUT POSITIVE-NAND GATES

typical performance

TYPE	POWER	DELAY
'00	10 mW	10 ns
'ALS00A	1.25 mW	3.5 ns
'AS00	8 mW	3 ns
'H00	22 mW	6 ns
'L00	1 mW	33 ns
'LS00	2 mW	9.5 ns
'S00	19 mW	3 ns

SN5400 (J,FH)	SN7400 (J,N)
SN54ALS00A (J,FH)	SN74ALS00A (N,FN)
SN54AS00 (J,FH)	SN74AS00 (N,FN)
SN54H00 (J)	SN74H00 (J,N)
SN54L00 (J)	
SN54LS00 (J,FH)	SN74LS00 (J,N,FN)
SN54S00 (J,FH)	SN74S00 (J,N,FN)

logic symbol†

positive logic: $Y = \overline{AB}$

pin assignments

J, N PACKAGES			
1	1A	8	3Y
2	1B	9	3A
3	1Y	10	3B
4	2A	11	4Y
5	2B	12	4A
6	2Y	13	4B
7	GND	14	V_{CC}

FH, FN PACKAGES			
1	nc	11	nc
2	1A	12	3Y
3	1B	13	3A
4	1Y	14	3B
5	nc	15	nc
6	2A	16	4Y
7	nc	17	nc
8	2B	18	4A
9	2Y	19	4B
10	GND	20	V_{CC}

02

QUADRUPLE 2-INPUT POSITIVE-NOR GATES

typical performance

TYPE	POWER	DELAY
'02	14 mW	10 ns
'ALS02	1.89 mW	5.5 ns
'AS02	12 mW	3 ns
'L02	1.5 mW	33 ns
'LS02	2.75 mW	10 ns
'S02	29 mW	3.5 ns

SN5402 (J,FH)	SN7402 (J,N)
SN54ALS02 (J,FH)	SN74ALS02 (N,FN)
SN54AS02 (J,FH)	SN74AS02 (N,FN)
SN54L02 (J)	
SN54LS02 (J,FH)	SN74LS02 (J,N,FN)
SN54S02 (J,FH)	SN74S02 (J,N,FN)

logic symbol†

positive logic: $Y = \overline{A + B}$

pin assignments

J, N PACKAGES			
1	1Y	8	3A
2	1A	9	3B
3	1B	10	3Y
4	2Y	11	4A
5	2A	12	4B
6	2B	13	4Y
7	GND	14	V_{CC}

FH, FN PACKAGES			
1	nc	11	nc
2	1Y	12	3A
3	1A	13	3B
4	1B	14	3Y
5	nc	15	nc
6	2Y	16	4A
7	nc	17	nc
8	2A	18	4B
9	2B	19	4Y
10	GND	20	V_{CC}

04

HEX INVERTERS

typical performance

TYPE	POWER	DELAY
'04	10 mW	10 ns
'ALS04A	1.27 mW	3.5 ns
AS04	7.4 mW	3 ns
'H04	22 mW	6 ns
'L04	1 mW	33 ns
'LS04	2 mW	9.5 ns
'S04	19 mW	3 ns

SN5404 (J,FH)	SN7404 (J,N)
SN54ALS04A (J,FH)	SN74ALS04A (N,FN)
SN54AS04 (J,FH)	SN74AS04 (N,FN)
SN54H04 (J)	SN74H04 (J,N)
SN54L04 (J)	
SN54LS04 (J,FH)	SN74LS04 (J,N,FN)
SN54S04 (J,FH)	SN74S04 (J,N,FN)

logic symbol†

positive logic: $Y = \overline{A}$

pin assignments

J, N PACKAGES			
1	1A	8	4Y
2	1Y	9	4A
3	2A	10	5Y
4	2Y	11	5A
5	3A	12	6Y
6	3Y	13	6A
7	GND	14	V_{CC}

FH, FN PACKAGES			
1	nc	11	nc
2	1A	12	4Y
3	1Y	13	4A
4	2A	14	5Y
5	nc	15	nc
6	2Y	16	5A
7	nc	17	nc
8	3A	18	6Y
9	3Y	19	6A
10	GND	20	V_{CC}

08

QUADRUPLE 2-INPUT POSITIVE-AND GATES

TYPE	POWER	DELAY
'08	19 mW	15 ns
'ALS08	2.19 mW	6.5 ns
AS08	13 mW	4 ns
'LS08	4.25 mW	12 ns
'S08	32 mW	4.75 ns

SN5408 (J,FH)	SN7408 (J,N)
SN54ALS08 (J,FH)	SN74ALS08 (N,FN)
SN54AS08 (J,FH)	SN74AS08 (N,FN)
SN54LS08 (J,FH)	SN74LS08 (J,N,FN)
SN54S08 (J,FH)	SN74S08 (J,N,FN)

logic symbol†

positive logic: $Y = AB$

pin assignments

J, N PACKAGES			
1	1A	8	3Y
2	1B	9	3A
3	1Y	10	3B
4	2A	11	4Y
5	2B	12	4A
6	2Y	13	4B
7	GND	14	V_{CC}

FH, FN PACKAGES			
1	nc	11	nc
2	1A	12	3Y
3	1B	13	3A
4	1Y	14	3B
5	nc	15	nc
6	2A	16	4Y
7	nc	17	nc
8	2B	18	4A
9	2Y	19	4B
10	GND	20	V_{CC}

† Pin numbers shown on logic symbols are for J and N packages only.

nc — no internal connection.

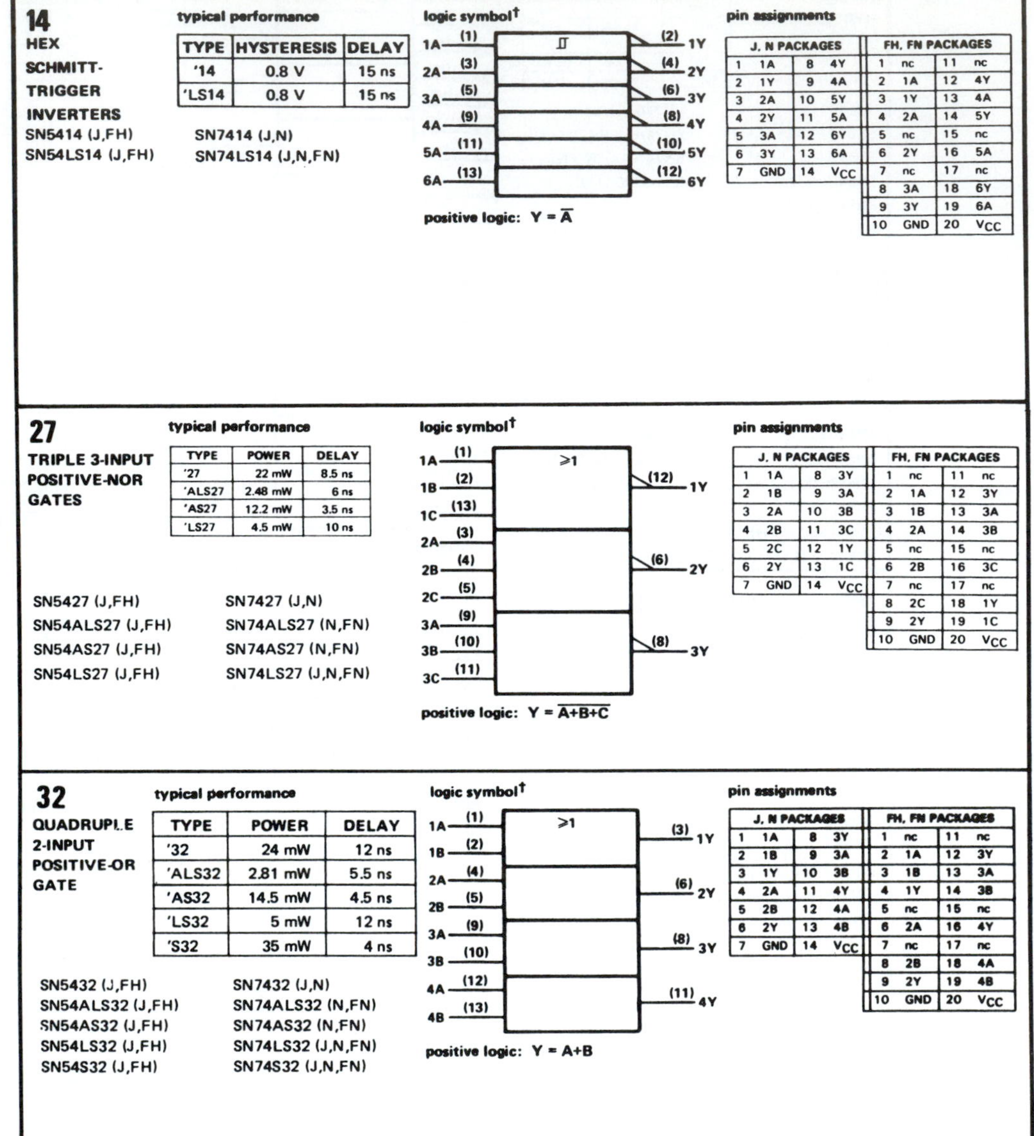

14

HEX SCHMITT-TRIGGER INVERTERS

SN5414 (J,FH) SN7414 (J,N)
SN54LS14 (J,FH) SN74LS14 (J,N,FN)

typical performance

TYPE	HYSTERESIS	DELAY
'14	0.8 V	15 ns
'LS14	0.8 V	15 ns

pin assignments

J, N PACKAGES			
1	1A	8	4Y
2	1Y	9	4A
3	2A	10	5Y
4	2Y	11	5A
5	3A	12	6Y
6	3Y	13	6A
7	GND	14	V_{CC}

FH, FN PACKAGES			
1	nc	11	nc
2	1A	12	4Y
3	1Y	13	4A
4	2A	14	5Y
5	nc	15	nc
6	2Y	16	5A
7	nc	17	nc
8	3A	18	6Y
9	3Y	19	6A
10	GND	20	V_{CC}

27

TRIPLE 3-INPUT POSITIVE-NOR GATES

typical performance

TYPE	POWER	DELAY
'27	22 mW	8.5 ns
'ALS27	2.48 mW	6 ns
'AS27	12.2 mW	3.5 ns
'LS27	4.5 mW	10 ns

SN5427 (J,FH) SN7427 (J,N)
SN54ALS27 (J,FH) SN74ALS27 (N,FN)
SN54AS27 (J,FH) SN74AS27 (N,FN)
SN54LS27 (J,FH) SN74LS27 (J,N,FN)

pin assignments

J, N PACKAGES			
1	1A	8	3Y
2	1B	9	3A
3	2A	10	3B
4	2B	11	3C
5	2C	12	1Y
6	2Y	13	1C
7	GND	14	V_{CC}

FH, FN PACKAGES			
1	nc	11	nc
2	1A	12	3Y
3	1B	13	3A
4	2A	14	3B
5	nc	15	nc
6	2B	16	3C
7	nc	17	nc
8	2C	18	1Y
9	2Y	19	1C
10	GND	20	V_{CC}

32

QUADRUPLE 2-INPUT POSITIVE-OR GATE

typical performance

TYPE	POWER	DELAY
'32	24 mW	12 ns
'ALS32	2.81 mW	5.5 ns
'AS32	14.5 mW	4.5 ns
'LS32	5 mW	12 ns
'S32	35 mW	4 ns

SN5432 (J,FH) SN7432 (J,N)
SN54ALS32 (J,FH) SN74ALS32 (N,FN)
SN54AS32 (J,FH) SN74AS32 (N,FN)
SN54LS32 (J,FH) SN74LS32 (J,N,FN)
SN54S32 (J,FH) SN74S32 (J,N,FN)

pin assignments

J, N PACKAGES			
1	1A	8	3Y
2	1B	9	3A
3	1Y	10	3B
4	2A	11	4Y
5	2B	12	4A
6	2Y	13	4B
7	GND	14	V_{CC}

FH, FN PACKAGES			
1	nc	11	nc
2	1A	12	3Y
3	1B	13	3A
4	1Y	14	3B
5	nc	15	nc
6	2A	16	4Y
7	nc	17	nc
8	2B	18	4A
9	2Y	19	4B
10	GND	20	V_{CC}

† Pin numbers shown on logic symbols are for J and N packages only.

nc — no internal connection.

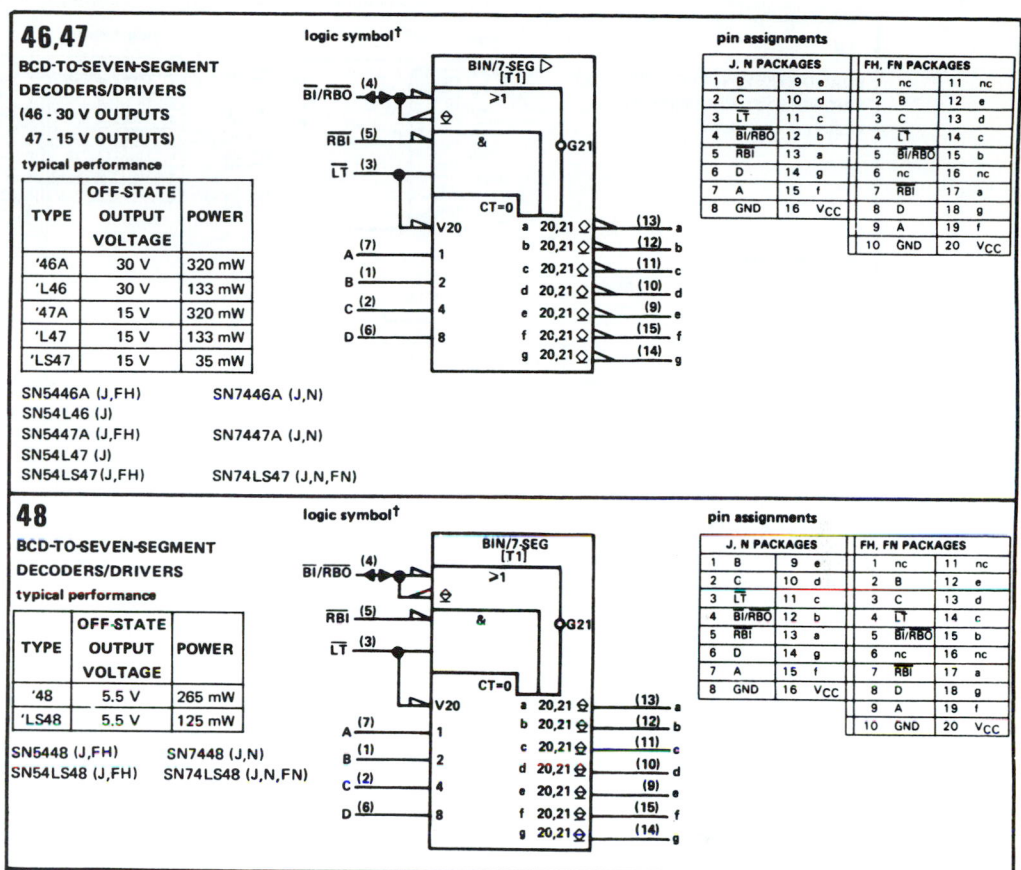

46,47

BCD-TO-SEVEN-SEGMENT DECODERS/DRIVERS

(46 - 30 V OUTPUTS
47 - 15 V OUTPUTS)

typical performance

TYPE	OFF-STATE OUTPUT VOLTAGE	POWER
'46A	30 V	320 mW
'L46	30 V	133 mW
'47A	15 V	320 mW
'L47	15 V	133 mW
'LS47	15 V	35 mW

SN5446A (J,FH) SN7446A (J,N)
SN54L46 (J)
SN5447A (J,FH) SN7447A (J,N)
SN54L47 (J)
SN54LS47(J,FH) SN74LS47 (J,N,FN)

logic symbol†

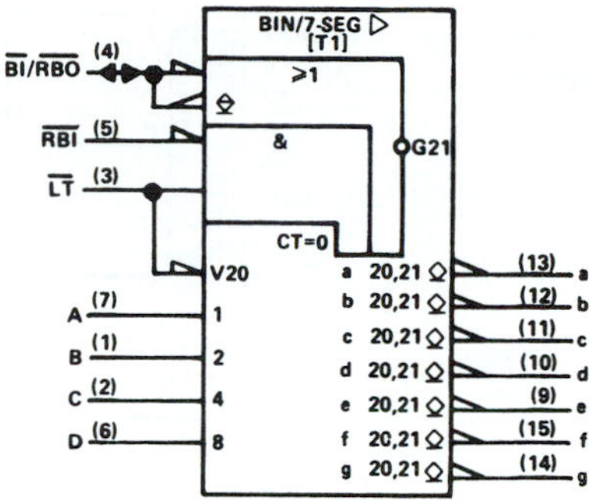

pin assignments

J, N PACKAGES				FH, FN PACKAGES			
1	B	9	e	1	nc	11	nc
2	C	10	d	2	B	12	e
3	$\overline{LT}$	11	c	3	C	13	d
4	$\overline{BI}/\overline{RBO}$	12	b	4	$\overline{LT}$	14	c
5	$\overline{RBI}$	13	a	5	$\overline{BI}/\overline{RBO}$	15	b
6	D	14	g	6	nc	16	nc
7	A	15	f	7	$\overline{RBI}$	17	a
8	GND	16	V_{CC}	8	D	18	g
				9	A	19	f
				10	GND	20	V_{CC}

48

BCD-TO-SEVEN-SEGMENT DECODERS/DRIVERS

typical performance

TYPE	OFF-STATE OUTPUT VOLTAGE	POWER
'48	5.5 V	265 mW
'LS48	5.5 V	125 mW

SN5448 (J,FH) SN7448 (J,N)
SN54LS48 (J,FH) SN74LS48 (J,N,FN)

logic symbol†

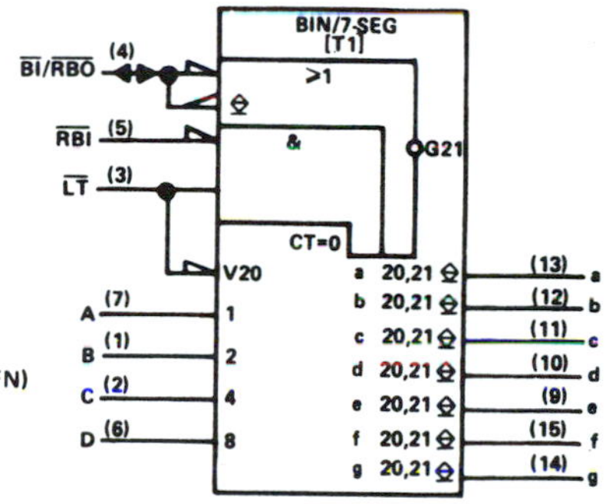

pin assignments

J, N PACKAGES				FH, FN PACKAGES			
1	B	9	e	1	nc	11	nc
2	C	10	d	2	B	12	e
3	$\overline{LT}$	11	c	3	C	13	d
4	$\overline{BI}/\overline{RBO}$	12	b	4	$\overline{LT}$	14	c
5	$\overline{RBI}$	13	a	5	$\overline{BI}/\overline{RBO}$	15	b
6	D	14	g	6	nc	16	nc
7	A	15	f	7	$\overline{RBI}$	17	a
8	GND	16	V_{CC}	8	D	18	g
				9	A	19	f
				10	GND	20	V_{CC}

† Pin numbers shown on logic symbols are for J and N packages only.
nc — no internal connection.

FONT TABLE T1 – FOR '46, '47, '48, '49

74

DUAL D-TYPE POSITIVE-EDGE-TRIGGERED FLIP-FLOPS WITH PRESET AND CLEAR

logic symbol†

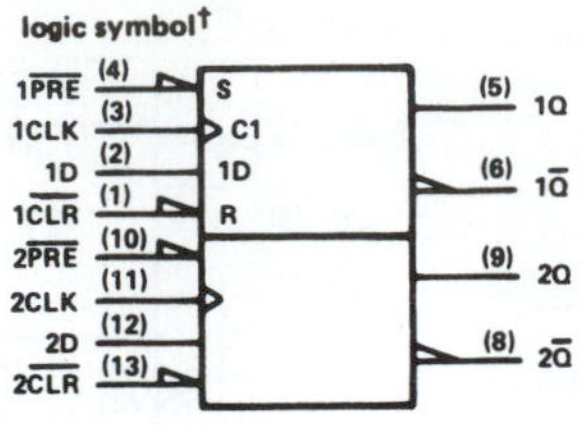

pin assignments

J, N PACKAGES			
1	$1\overline{CLR}$	8	$2\overline{Q}$
2	1D	9	2Q
3	1CLK	10	$2\overline{PRE}$
4	$1\overline{PRE}$	11	2CLK
5	1Q	12	2D
6	$1\overline{Q}$	13	$2\overline{CLR}$
7	GND	14	V_{CC}

FH, FN PACKAGES			
1	nc	11	nc
2	$1\overline{CLR}$	12	$2\overline{Q}$
3	1D	13	2Q
4	1CLK	14	$2\overline{PRE}$
5	nc	15	nc
6	$1\overline{PRE}$	16	2CLK
7	nc	17	nc
8	1Q	18	2D
9	$1\overline{Q}$	19	$2\overline{CLR}$
10	GND	20	V_{CC}

typical performance

TYPE	f_{max}	PWR/ F-F	SET-UP	HOLD
'74	25 MHz	43 mW	20 ns↑	5 ns↑
'ALS74	50 MHz	6 mW	15 ns↑	0 ns↑
'AS74	125 MHz	26 mW	4.5 ns↑	0 ns↑
'H74	43 MHz	75 mW	15 ns↑	5 ns↑
'L74	3 MHz	4 mW	50 ns↑	15 ns↑
'LS74A	33 MHz	10 mW	20 ns↑	5 ns↑
'S74	110 MHz	75 mW	3 ns↑	2 ns↑

↑ Rising edge of clock pulse.

SN5474 (J,FH)	SN7474 (J,N)
SN54ALS74 (J,FH)	SN74ALS74 (N,FN)
SN54AS74 (J,FH)	SN74AS74 (N,FN)
SN54H74 (J)	SN74H74 (J,N)
SN54L74 (J)	
SN54LS74A (J,FH)	SN74LS74A (J,N,FN)
SN54S74 (J,FH)	SN74S74 (J,N,FN)

76

DUAL J-K FLIP-FLOPS WITH PRESET AND CLEAR

typical performance

TYPE	f_{max}	PWR/ F-F	SET-UP	HOLD
'76	20 MHz	50 mW	0 ns↑	0 ns↓
'H76	30 MHz	80 mW	0 ns↑	0 ns↓
'LS76A	45 MHz	10 mW	20 ns↓	0 ns↓

↑ Rising edge of clock pulse.
↓ Falling edge of clock pulse.

SN5476 (J)	SN7476 (J,N)
SN54H76 (J)	SN74H76 (J,N)
SN54LS76A (J)	SN74LS76A (J,N)

logic symbol, '76, 'H76†

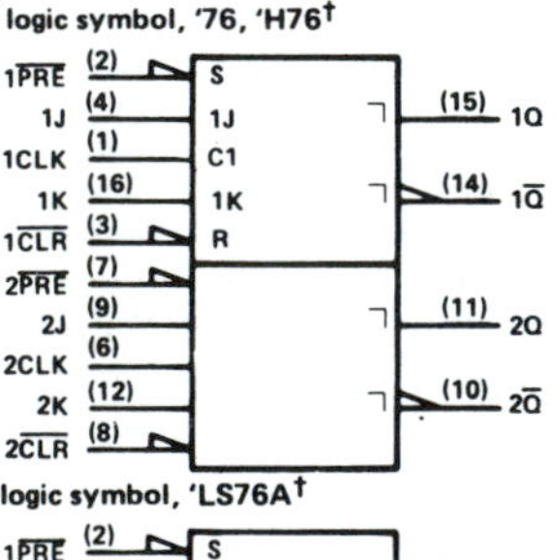

logic symbol, 'LS76A†

1PRE (2), 1J (4), 1CLK (1), 1K (16), 1CLR (3), 2PRE (7), 2J (9), 2CLK (6), 2K (12), 2CLR (8); S, 1J, C1, 1K, R; (15) 1Q, (14) $1\overline{Q}$, (11) 2Q, (10) $2\overline{Q}$

pin assignments

J, N PACKAGES			
1	1CLK	9	2J
2	$1\overline{PRE}$	10	$2\overline{Q}$
3	$1\overline{CLR}$	11	2Q
4	1J	12	2K
5	V_{CC}	13	GND
6	2CLK	14	$1\overline{Q}$
7	$2\overline{PRE}$	15	1Q
8	$2\overline{CLR}$	16	1K

For chip carrier information, contact the factory.

† Pin numbers shown on logic symbols are for J and N packages only.

nc – no internal connection.

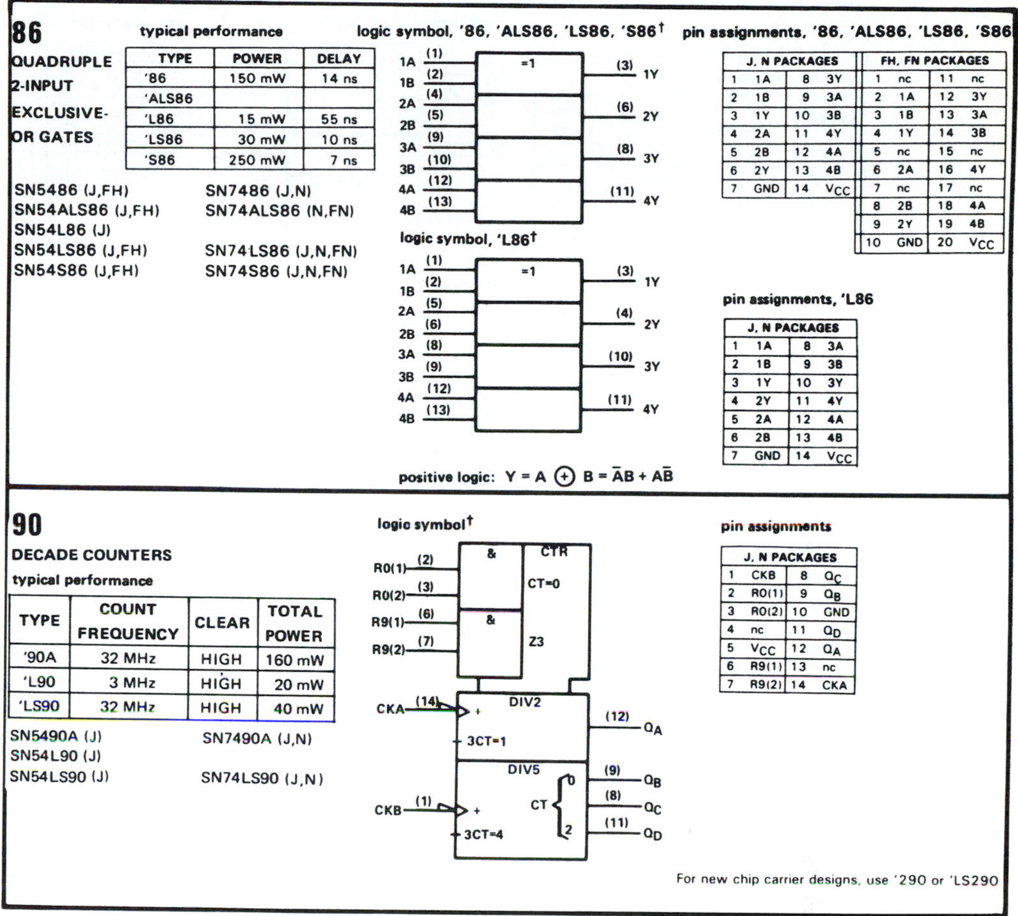

86

QUADRUPLE 2-INPUT EXCLUSIVE-OR GATES

typical performance

TYPE	POWER	DELAY
'86	150 mW	14 ns
'ALS86		
'L86	15 mW	55 ns
'LS86	30 mW	10 ns
'S86	250 mW	7 ns

SN5486 (J,FH)	SN7486 (J,N)
SN54ALS86 (J,FH)	SN74ALS86 (N,FN)
SN54L86 (J)	
SN54LS86 (J,FH)	SN74LS86 (J,N,FN)
SN54S86 (J,FH)	SN74S86 (J,N,FN)

logic symbol, '86, 'ALS86, 'LS86, 'S86†

logic symbol, 'L86†

positive logic: $Y = A \oplus B = \bar{A}B + A\bar{B}$

pin assignments, '86, 'ALS86, 'LS86, 'S86

J, N PACKAGES			
1	1A	8	3Y
2	1B	9	3A
3	1Y	10	3B
4	2A	11	4Y
5	2B	12	4A
6	2Y	13	4B
7	GND	14	V_{CC}

FH, FN PACKAGES			
1	nc	11	nc
2	1A	12	3Y
3	1B	13	3A
4	1Y	14	3B
5	nc	15	nc
6	2A	16	4Y
7	nc	17	nc
8	2B	18	4A
9	2Y	19	4B
10	GND	20	V_{CC}

pin assignments, 'L86

J, N PACKAGES			
1	1A	8	3A
2	1B	9	3B
3	1Y	10	3Y
4	2Y	11	4Y
5	2A	12	4A
6	2B	13	4B
7	GND	14	V_{CC}

90

DECADE COUNTERS

typical performance

TYPE	COUNT FREQUENCY	CLEAR	TOTAL POWER
'90A	32 MHz	HIGH	160 mW
'L90	3 MHz	HIGH	20 mW
'LS90	32 MHz	HIGH	40 mW

SN5490A (J)	SN7490A (J,N)
SN54L90 (J)	
SN54LS90 (J)	SN74LS90 (J,N)

logic symbol†

pin assignments

J, N PACKAGES			
1	CKB	8	Q_C
2	R0(1)	9	Q_B
3	R0(2)	10	GND
4	nc	11	Q_D
5	V_{CC}	12	Q_A
6	R9(1)	13	nc
7	R9(2)	14	CKA

For new chip carrier designs, use '290 or 'LS290

† Pin numbers shown on logic symbols are for J and N packages only.

nc — no internal connection.

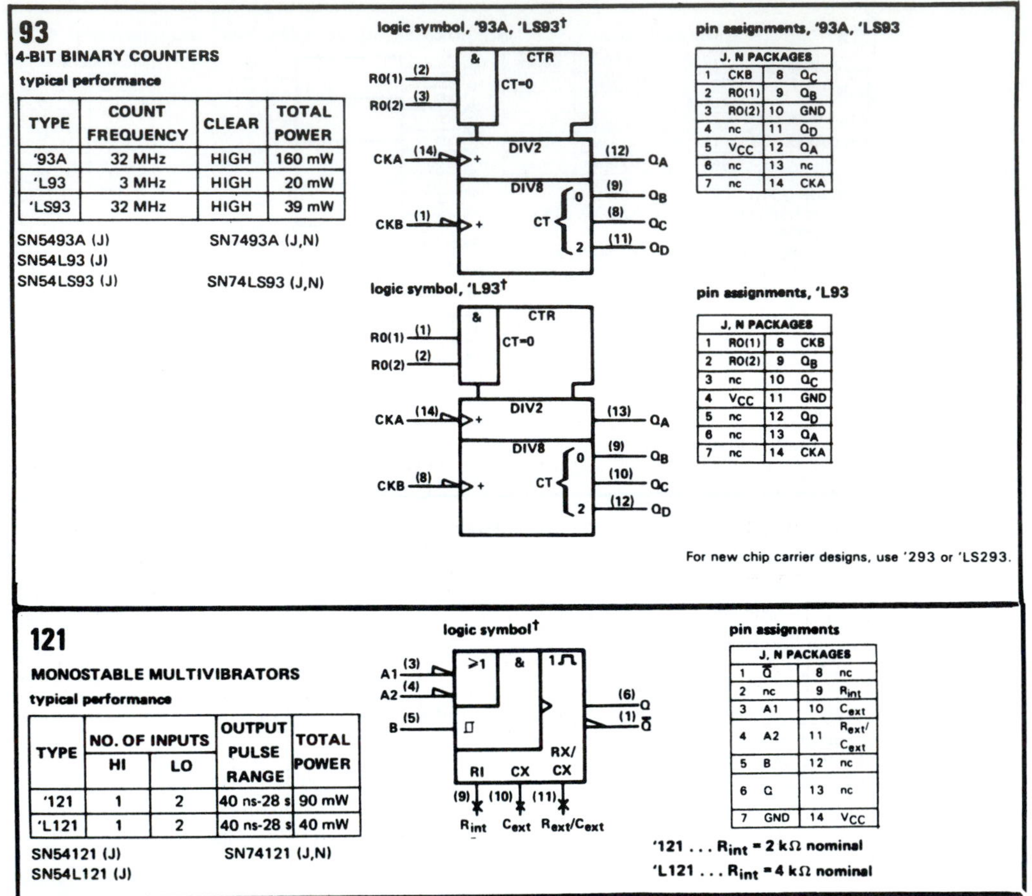

93

4-BIT BINARY COUNTERS

typical performance

TYPE	COUNT FREQUENCY	CLEAR	TOTAL POWER
'93A	32 MHz	HIGH	160 mW
'L93	3 MHz	HIGH	20 mW
'LS93	32 MHz	HIGH	39 mW

SN5493A (J) SN7493A (J,N)
SN54L93 (J)
SN54LS93 (J) SN74LS93 (J,N)

logic symbol, '93A, 'LS93†

pin assignments, '93A, 'LS93

J, N PACKAGES			
1	CKB	8	Q_C
2	R0(1)	9	Q_B
3	R0(2)	10	GND
4	nc	11	Q_D
5	V_{CC}	12	Q_A
6	nc	13	nc
7	nc	14	CKA

logic symbol, 'L93†

pin assignments, 'L93

J, N PACKAGES			
1	R0(1)	8	CKB
2	R0(2)	9	Q_B
3	nc	10	Q_C
4	V_{CC}	11	GND
5	nc	12	Q_D
6	nc	13	Q_A
7	nc	14	CKA

For new chip carrier designs, use '293 or 'LS293.

121

MONOSTABLE MULTIVIBRATORS

typical performance

TYPE	NO. OF INPUTS		OUTPUT PULSE RANGE	TOTAL POWER
	HI	LO		
'121	1	2	40 ns-28 s	90 mW
'L121	1	2	40 ns-28 s	40 mW

SN54121 (J) SN74121 (J,N)
SN54L121 (J)

logic symbol†

pin assignments

J, N PACKAGES			
1	$\bar{Q}$	8	nc
2	nc	9	R_{int}
3	A1	10	C_{ext}
4	A2	11	R_{ext}/C_{ext}
5	B	12	nc
6	Q	13	nc
7	GND	14	V_{CC}

'121 . . . R_{int} = 2 kΩ nominal
'L121 . . . R_{int} = 4 kΩ nominal

† Pin numbers shown on logic symbols are for J and N packages only.
nc — no internal connection.

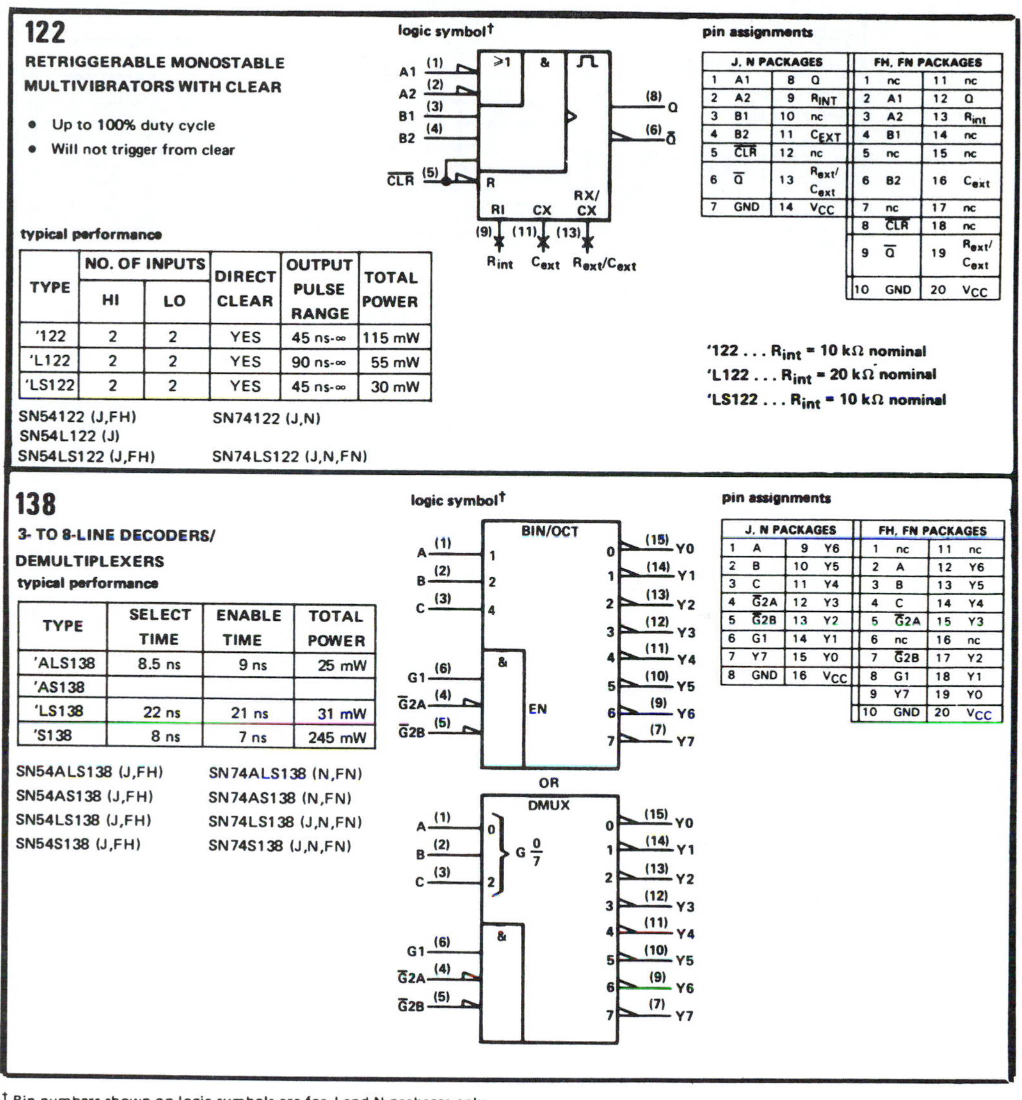

122

RETRIGGERABLE MONOSTABLE MULTIVIBRATORS WITH CLEAR

- Up to 100% duty cycle
- Will not trigger from clear

logic symbol†

typical performance

TYPE	NO. OF INPUTS		DIRECT CLEAR	OUTPUT PULSE RANGE	TOTAL POWER
	HI	LO			
'122	2	2	YES	45 ns-∞	115 mW
'L122	2	2	YES	90 ns-∞	55 mW
'LS122	2	2	YES	45 ns-∞	30 mW

SN54122 (J,FH) SN74122 (J,N)
SN54L122 (J)
SN54LS122 (J,FH) SN74LS122 (J,N,FN)

pin assignments

J, N PACKAGES			
1	A1	8	Q
2	A2	9	R_{INT}
3	B1	10	nc
4	B2	11	C_{EXT}
5	$\overline{CLR}$	12	nc
6	$\overline{Q}$	13	R_{ext}/C_{ext}
7	GND	14	V_{CC}

FH, FN PACKAGES			
1	nc	11	nc
2	A1	12	Q
3	A2	13	R_{int}
4	B1	14	nc
5	nc	15	nc
6	B2	16	C_{ext}
7	nc	17	nc
8	$\overline{CLR}$	18	nc
9	$\overline{Q}$	19	R_{ext}/C_{ext}
10	GND	20	V_{CC}

'122 . . . R_{int} = 10 kΩ nominal
'L122 . . . R_{int} = 20 kΩ nominal
'LS122 . . . R_{int} = 10 kΩ nominal

138

3- TO 8-LINE DECODERS/ DEMULTIPLEXERS

typical performance

TYPE	SELECT TIME	ENABLE TIME	TOTAL POWER
'ALS138	8.5 ns	9 ns	25 mW
'AS138			
'LS138	22 ns	21 ns	31 mW
'S138	8 ns	7 ns	245 mW

SN54ALS138 (J,FH) SN74ALS138 (N,FN)
SN54AS138 (J,FH) SN74AS138 (N,FN)
SN54LS138 (J,FH) SN74LS138 (J,N,FN)
SN54S138 (J,FH) SN74S138 (J,N,FN)

logic symbol†

pin assignments

J, N PACKAGES			
1	A	9	Y6
2	B	10	Y5
3	C	11	Y4
4	$\overline{G}2A$	12	Y3
5	$\overline{G}2B$	13	Y2
6	G1	14	Y1
7	Y7	15	Y0
8	GND	16	V_{CC}

FH, FN PACKAGES			
1	nc	11	nc
2	A	12	Y6
3	B	13	Y5
4	C	14	Y4
5	$\overline{G}2A$	15	Y3
6	nc	16	nc
7	$\overline{G}2B$	17	Y2
8	G1	18	Y1
9	Y7	19	Y0
10	GND	20	V_{CC}

† Pin numbers shown on logic symbols are for J and N packages only.
nc — no internal connection.

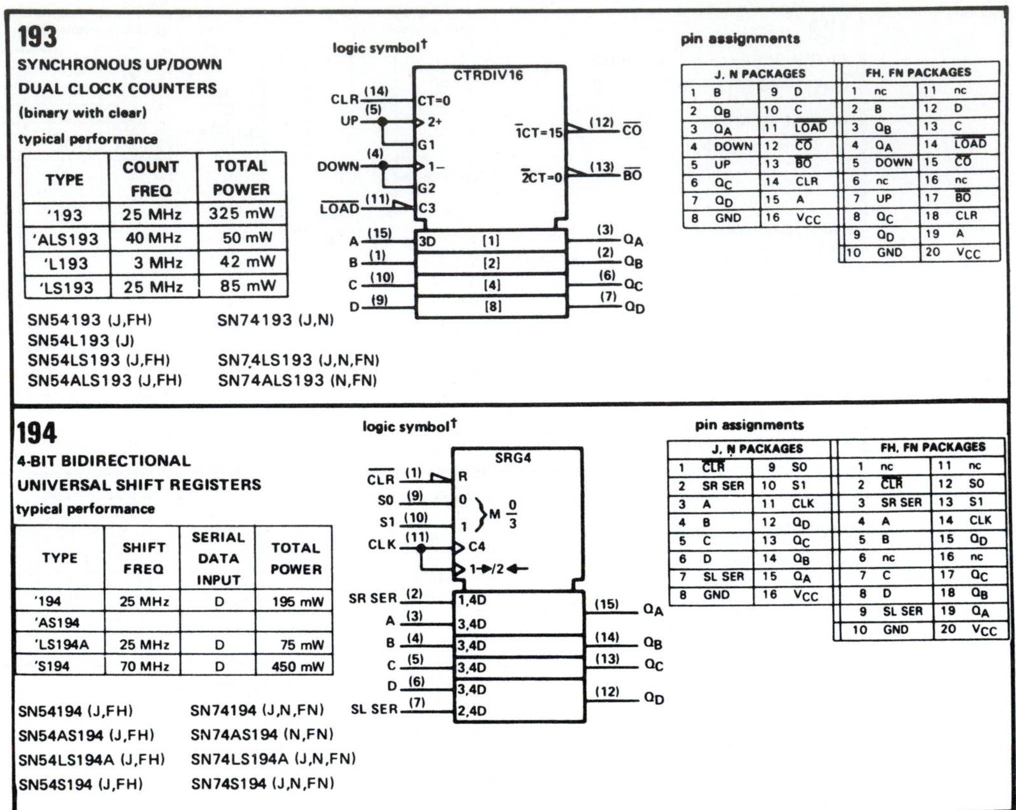

193

SYNCHRONOUS UP/DOWN DUAL CLOCK COUNTERS

(binary with clear)

typical performance

TYPE	COUNT FREQ	TOTAL POWER
'193	25 MHz	325 mW
'ALS193	40 MHz	50 mW
'L193	3 MHz	42 mW
'LS193	25 MHz	85 mW

SN54193 (J,FH) SN74193 (J,N)
SN54L193 (J)
SN54LS193 (J,FH) SN74LS193 (J,N,FN)
SN54ALS193 (J,FH) SN74ALS193 (N,FN)

logic symbol†

pin assignments

J, N PACKAGES			
1	B	9	D
2	Q_B	10	C
3	Q_A	11	$\overline{\text{LOAD}}$
4	DOWN	12	$\overline{\text{CO}}$
5	UP	13	$\overline{\text{BO}}$
6	Q_C	14	CLR
7	Q_D	15	A
8	GND	16	V_{CC}

FH, FN PACKAGES			
1	nc	11	nc
2	B	12	D
3	Q_B	13	C
4	Q_A	14	$\overline{\text{LOAD}}$
5	DOWN	15	$\overline{\text{CO}}$
6	nc	16	nc
7	UP	17	$\overline{\text{BO}}$
8	Q_C	18	CLR
9	Q_D	19	A
10	GND	20	V_{CC}

194

4-BIT BIDIRECTIONAL UNIVERSAL SHIFT REGISTERS

typical performance

TYPE	SHIFT FREQ	SERIAL DATA INPUT	TOTAL POWER
'194	25 MHz	D	195 mW
'AS194			
'LS194A	25 MHz	D	75 mW
'S194	70 MHz	D	450 mW

SN54194 (J,FH) SN74194 (J,N,FN)
SN54AS194 (J,FH) SN74AS194 (N,FN)
SN54LS194A (J,FH) SN74LS194A (J,N,FN)
SN54S194 (J,FH) SN74S194 (J,N,FN)

logic symbol†

pin assignments

J, N PACKAGES			
1	$\overline{\text{CLR}}$	9	S0
2	SR SER	10	S1
3	A	11	CLK
4	B	12	Q_D
5	C	13	Q_C
6	D	14	Q_B
7	SL SER	15	Q_A
8	GND	16	V_{CC}

FH, FN PACKAGES			
1	nc	11	nc
2	$\overline{\text{CLR}}$	12	S0
3	SR SER	13	S1
4	A	14	CLK
5	B	15	Q_D
6	nc	16	nc
7	C	17	Q_C
8	D	18	Q_B
9	SL SER	19	Q_A
10	GND	20	V_{CC}

† Pin numbers shown on logic symbols are for J and N packages only.

nc — no internal connection.

244

OCTAL BUFFERS/LINE DRIVERS/LINE RECEIVERS

(non-inverted three-state outputs)

typical performance

TYPE	DELAY	MAX SOURCE CURRENT	MAX SINK CURRENT	POWER DISSI-PATION
SN54ALS244A	7 ns	− 12 mA	12 mA	68 mW
SN74ALS244A	7 ns	− 15 mA	24 mA	
SN74ALS244A-1	7 ns	− 15 mA	48 mA	
SN54AS244	4.5 ns	− 12 mA	48 mA	235 mW
SN74AS244	4.5 ns	− 15 mA	64 mA	
SN54LS244	12 ns	− 12 mA	12 mA	127 mW
SN74LS244	12 ns	− 15 mA	24 mA	
SN54S244	6 ns	− 12 mA	48 mA	558 mW
SN74S244	6 ns	− 15 mA	64 mA	

SN54ALS244A (J,FH) SN74ALS244A (N,FN)
SNALS244A-1 (N,FN)
SN54AS244 (J,FH) SN74AS244 (N,FN)
SN54LS244 (J,FH) SN74LS244 (J,N,FN)
SN54S244 (J,FH) SN74S244 (J,N,FN)

logic symbol†

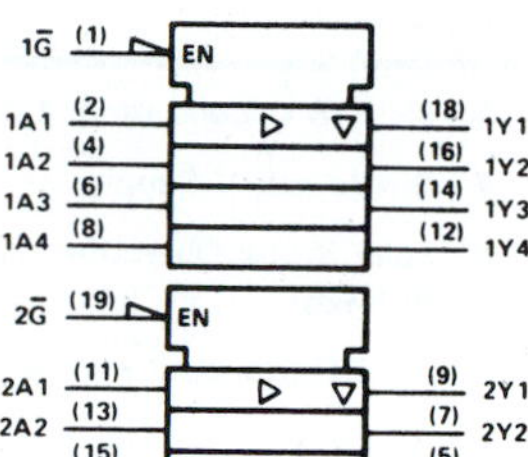

pin assignments

J, N PACKAGES			
1	1G̅	11	2A1
2	1A1	12	1Y4
3	2Y4	13	2A2
4	1A2	14	1Y3
5	2Y3	15	2A3
6	1A3	16	1Y2
7	2Y2	17	2A4
8	1A4	18	1Y1
9	2Y1	19	2G̅
10	GND	20	V_{CC}

FH, FN PACKAGES			
1	1G̅	11	2A1
2	1A1	12	1Y4
3	2Y4	13	2A2
4	1A2	14	1Y3
5	2Y3	15	2A3
6	1A3	16	1Y2
7	2Y2	17	2A4
8	1A4	18	1Y1
9	2Y1	19	2G̅
10	GND	20	V_{CC}

245

OCTAL BUS TRANSCEIVERS

(non-inverted three-state outputs)

typical performance

TYPE	DELAY	MAX SOURCE CURRENT	MAX SINK CURRENT	POWER DISSI-PATION
SN54ALS245A	6 ns	− 12 mA	12 mA	173 mW
SN74ALS245A	6 ns	− 15 mA	24 mA	
SN74ALS245A-1	6 ns	− 15 mA	48 mA	
SN54AS245	6 ns	− 12 mA	32 mA	310 mW
SN74AS245	6 ns	− 15 mA	48 mA	
SN54LS245	8 ns	− 12 mA	12 mA	290 mW
SN74LS245	8 ns	− 15 mA	24 mA	

SN54ALS245A (J,FH) SN74ALS245A (N,FN)
SN74ALS245A-1 (N,FN)
SN54AS245 (J,FH) SN74AS245 (N,FN)
SN54LS245 (J,FH) SN74LS245 (J,N,FN)

logic symbol†

pin assignments

J, N PACKAGES			
1	DIR	11	B8
2	A1	12	B7
3	A2	13	B6
4	A3	14	B5
5	A4	15	B4
6	A5	16	B3
7	A6	17	B2
8	A7	18	B1
9	A8	19	G̅
10	GND	20	V_{CC}

FH, FN PACKAGES			
1	DIR	11	B8
2	A1	12	B7
3	A2	13	B6
4	A3	14	B5
5	A4	15	B4
6	A5	16	B3
7	A6	17	B2
8	A7	18	B1
9	A8	19	G
10	GND	20	V_{CC}

† Pin numbers shown on logic symbols are for J and N packages only.

nc — no internal connection.

MOS
LSI

TMS4016
2048-WORD BY 8-BIT STATIC RAM

FEBRUARY 1981 REVISED AUGUST 1983

- 2K X 8 Organization, Common I/O
- Single +5-V Supply
- Fully Static Operation (No Clocks, No Refresh)
- JEDEC Standard Pinout
- 24-Pin 600 Mil (15.2 mm) Package Configuration
- Plug-in Compatible with 16K 5 V EPROMs
- 8-Bit Output for Use in Microprocessor-Based Systems
- 3-State Outputs with $\overline{S}$ for OR-ties
- $\overline{G}$ Eliminates Need for External Bus Buffers
- All Inputs and Outputs Fully TTL Compatible
- Fanout to Series 74, Series 74S or Series 74LS TTL Loads
- N-Channel Silicon-Gate Technology
- Power Dissipation Under 385 mW Max
- Guaranteed dc Noise Immunity of 400 mV with Standard TTL Loads
- 4 Performance Ranges:

	ACCESS TIME (MAX)
TMS4016-12	120 ns
TMS4016-15	150 ns
TMS4016-20	200 ns
TMS4016-25	250 ns

TMS4016 . . . NL PACKAGE
(TOP VIEW)

Pin	Name	Pin	Name
1	A7	24	V_{CC}
2	A6	23	A8
3	A5	22	A9
4	A4	21	$\overline{W}$
5	A3	20	$\overline{G}$
6	A2	19	A10
7	A1	18	$\overline{S}$
8	A0	17	DQ8
9	DQ1	16	DQ7
10	DQ2	15	DQ6
11	DQ3	14	DQ5
12	V_{SS}	13	DQ4

PIN NOMENCLATURE	
A0 - A10	Addresses
DQ1 - DQ8	Data In Data Out
$\overline{G}$	Output Enable
$\overline{S}$	Chip Select
V_{CC}	+5 V Supply
V_{SS}	Ground
$\overline{W}$	Write Enable

description

The TMS4016 static random-access memory is organized as 2048 words of 8 bits each. Fabricated using proven N-channel, silicon-gate MOS technology, the TMS4016 operates at high speeds and draws less power per bit than 4K static RAMs. It is fully compatible with Series 74, 74S, or 74LS TTL. Its static design means that no refresh clocking circuitry is needed and timing requirements are simplified. Access time is equal to cycle time. A chip select control is provided for controlling the flow of data-in and data-out and an output enable function is included in order to eliminate the need for external bus buffers.

Of special importance is that the TMS4016 static RAM has the same standardized pinout as TI's compatible EPROM family. This, along with other compatible features, makes the TMS4016 plug-in compatible with the TMS2516 (or other 16K 5 V EPROMs). Minimal, if any modifications are needed. This allows the microprocessor system designer complete flexibility in partitioning his memory board between read write and non-volatile storage.

The TMS4016 is offered in the plastic (NL suffix) 24-pin dual-in-line package designed for insertion in mounting hole rows on 600-mil (15.2 mm) centers. It is guaranteed for operation from 0°C to 70°C.

(Courtesy of Texas Instruments Inc.)

TMS4016
2048-WORD BY 8-BIT STATIC RAM

operation

addresses (A0 – A10)

The eleven address inputs select one of the 2048 8-bit words in the RAM. The address-inputs must be stable for the duration of a write cycle. The address inputs can be driven directly from standard Series 54/74 TTL with no external pull-up resistors.

output enable ($\overline{G}$)

The output enable terminal, which can be driven directly from standard TTL circuits, affects only the data-out terminals. When output enable is at a logic high level, the output terminals are disabled to the high-impedance state. Output enable provides greater output control flexibility, simplifying data bus design.

chip select ($\overline{S}$)

The chip-select terminal, which can be driven directly from standard TTL circuits, affects the data-in/data-out terminals. When chip select and output enable are at a logic low level, the D/Q terminals are enabled. When chip select is high, the D/Q terminals are in the floating or high-impedance state and the input is inhibited.

write enable ($\overline{W}$)

The read or write mode is selected through the write enable terminal. A logic high selects the read mode; a logic low selects the write mode. $\overline{W}$ must be high when changing addresses to prevent erroneously writing data into a memory location. The $\overline{W}$ input can be driven directly from standard TTL circuits.

data-in/data-out (DQ1 – DQ8)

Data can be written into a selected device when the write enable input is low. The D/Q terminal can be driven directly from standard TTL circuits. The three-state output buffer provides direct TTL compatibility with a fan-out of one Series 74 TTL gate, one Series 74S TTL gate, or five Series 74LS TTL gates. The D/Q terminals are in the high impedance state when chip select ($\overline{S}$) is high, output enable ($\overline{G}$) is high, or whenever a write operation is being performed. Data-out is the same polarity as data-in.

(Courtesy of Texas Instruments Inc.)

TMS4016
2048-WORD BY 8-BIT STATIC RAM

logic symbol†

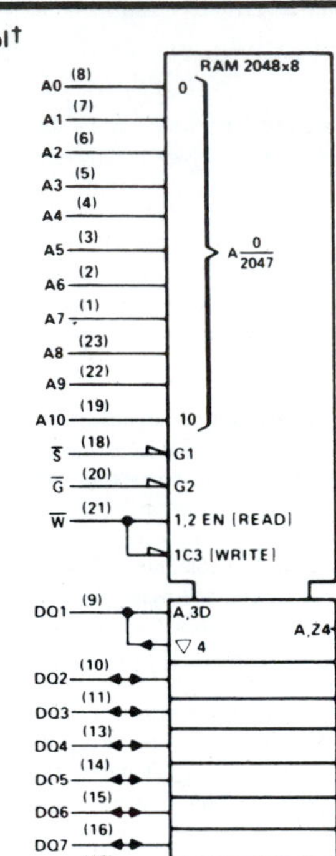

FUNCTION TABLE

$\overline{W}$	$\overline{S}$	$\overline{G}$	DQ1-DQ8	MODE
L	L	X	VALID DATA	WRITE
H	L	L	DATA OUTPUT	READ
X	H	X	HI-Z	DEVICE DISABLED
H	L	H	HI-Z	OUTPUT DISABLED

† This symbol is in accordance with IEEE Std 91/ANSI Y32.14 and recent decisions by IEEE and IEC. See explanation on page 10-1.

absolute maximum ratings over operating free-air temperature range (unless otherwise noted)†

Supply voltage, V_{CC} (see Note 1) . . . . −0.5 V to 7 V
Input voltage (any input) (see Note 1) . . . . −1 V to 7 V
Continuous power dissipation . . . . 1 W
Operating free-air temperature range . . . . 0°C to 70°C
Storage temperature range . . . . −55°C to 150°C

† Stresses beyond those listed under "Absolute Maximum Ratings" may cause permanent damage to the device. This is a stress rating only and functional operation of the device at these or any other conditions beyond those indicated in the "Recommended Operating Conditions" section of this specification is not implied. Exposure to absolute-maximum-rated conditions for extended periods may affect device reliability.

NOTE 1: Voltage values are with respect to the V_{SS} terminal.

recommended operating conditions

PARAMETER	MIN	NOM	MAX	UNIT
Supply voltage, V_{CC}	4.5	5	5.5	V
Supply voltage, V_{SS}		0		V
High-level input voltage, V_{IH}	2		5.5	V
Low-level input voltage, V_{IL} (see Note 2)	−1		0.8	V
Operating free-air temperature, T_A	0		70	°C

NOTE 2: The algebraic convention, where the more negative (less positive) limit is designated as minimum, is used in this data sheet for logic voltage levels only.

(Courtesy of Texas Instruments Inc.)

TMS4016
2048-WORD BY 8-BIT STATIC RAM

electrical characteristics over recommended operating free-air temperature range (unless otherwise noted)

PARAMETER		TEST CONDITIONS		MIN	TYP†	MAX	UNIT
V_{OH}	High level voltage	$I_{OH} = -1$ mA,	$V_{CC} = 4.5$ V	2.4			V
V_{OL}	Low level voltage	$I_{OL} = 2.1$ mA,	$V_{CC} = 4.5$ V			0.4	V
I_I	Input current	$V_I = 0$ V to 5.5 V				10	μA
I_{OZ}	Off-state output current	$\overline{S}$ or $\overline{G}$ at 2 V or $\overline{W}$ at 0.8 V, $V_O = 0$ V to 5.5 V				10	μA
I_{CC}	Supply current from V_{CC}	$I_O = 0$ mA, $T_A = 0°C$ (worst case)	$V_{CC} = 5.5$ V,		40	70	mA
C_i	Input capacitance	$V_I = 0$ V,	f = 1 MHz			8	pF
C_o	Output capacitance	$V_O = 0$ V,	f = 1 MHz			12	pF

†All typical values are at $V_{CC} = 5$ V, $T_A = 25°C$.

timing requirements over recommended supply voltage range and operating free-air temperature range

PARAMETER		TMS4016-12		TMS4016-15		TMS4016-20		TMS4016-25		UNIT
		MIN	MAX	MIN	MAX	MIN	MAX	MIN	MAX	
$t_{c(rd)}$	Read cycle time	120		150		200		250		ns
$t_{c(wr)}$	Write cycle time	120		150		200		250		ns
$t_{w(W)}$	Write pulse width	60		80		100		120		ns
$t_{su(A)}$	Address setup time	20		20		20		20		ns
$t_{su(S)}$	Chip select setup time	60		80		100		120		ns
$t_{su(D)}$	Data setup time	50		60		80		100		ns
$t_{h(A)}$	Address hold time	0		0		0		0		ns
$t_{h(D)}$	Data hold time	5		10		10		10		ns

switching characteristics over recommended voltage range, $T_A = 0°C$ to $70°C$ with output loading of Figure 1 (see notes 3 and 4)

PARAMETER		TMS4016-12		TMS4016-15		TMS4016-20		TMS4016-25		UNIT
		MIN	MAX	MIN	MAX	MIN	MAX	MIN	MAX	
$t_{a(A)}$	Access time from address		120		150		200		250	ns
$t_{a(S)}$	Access time from chip select low		60		75		100		120	ns
$t_{a(G)}$	Access time from output enable low		50		60		80		100	ns
$t_{v(A)}$	Output data valid after address change	10		15		15		15		ns
$t_{dis(S)}$	Output disable time after chip select high		40		50		60		80	ns
$t_{dis(G)}$	Output disable time after output enable high		40		50		60		80	ns
$t_{dis(W)}$	Output disable time after write enable low		50		60		60		80	ns
$t_{en(S)}$	Output enable time after chip select low	5		5		10		10		ns
$t_{en(G)}$	Output enable time after output enable low	5		5		10		10		ns
$t_{en(W)}$	Output enable time after write enable high	5		5		10		10		ns

NOTES: 3. $C_L = 100$ pF for all measurements except $t_{dis(W)}$ and $t_{en(W)}$.
$C_L = 5$ pF for $t_{dis(W)}$ and $t_{en(W)}$.
4. t_{dis} and t_{en} parameters are sampled and not 100% tested.

(Courtesy of Texas Instruments Inc.)

TMS4164
65,536-BIT DYNAMIC RANDOM-ACCESS MEMORY

MAY 1985—REVISED NOVEMBER 1985

This Data Sheet Is Applicable to All TMS4164s Symbolized with Code "A" as Described on Page 4-57.

- **65,536 X 1 Organization**
- **Single 5-V Supply (10% Tolerance)**
- **JEDEC Standardized Pinout in Dual-in-Line Package**
- **Performance Ranges:**

	ACCESS TIME ROW ADDRESS (MAX)	ACCESS TIME COLUMN ADDRESS (MAX)	READ OR WRITE CYCLE (MIN)	READ-MODIFY-WRITE CYCLE (MIN)
'4164-12	120 ns	70 ns	230 ns	255 ns
'4164-15	150 ns	85 ns	260 ns	290 ns
'4164-20	200 ns	135 ns	330 ns	345 ns

- **Upward Pin Compatible with TMS4116 (16K Dynamic RAM)**
- **First Military Version of 64K DRAM**
- **Also Available with MIL-STD-883B Processing and L(0°C to 70°C), E(−40°C to 85°C), S(−55°C to 100°C), or M(−55°C to 125°C) Temperature Ranges**
- **Operations of the TMS4164 Can Be Controlled by TI's TMS4500A and/or THCT4501 Dynamic RAM Controllers**
- **Long Refresh Period . . . 4 ms**
- **Low Refresh Overhead Time . . . As Low As 1.8% of Total Refresh Period**
- **All Inputs, Outputs, Clocks Fully TTL Compatible**
- **3-State Unlatched Output**
- **Common I/O Capability with Early Write Feature**
- **Page-Mode Operation for Faster Access**
- **Low Power Dissipation**
  - **Operating . . . 135 mW (Typ)**
  - **Standby . . . 17.5 mW (Typ)**
- **SMOS (Scaled-MOS) N-Channel Technology**

N PACKAGE (TOP VIEW)

	Pin	Pin	
NC	1	16	V_{SS}
D	2	15	$\overline{CAS}$
$\overline{W}$	3	14	Q
$\overline{RAS}$	4	13	A6
A0	5	12	A3
A2	6	11	A4
A1	7	10	A5
V_{DD}	8	9	A7

FP PACKAGE (TOP VIEW)

Top: D 2, NC 1, V_{SS} 18, $\overline{CAS}$ 17
Left: $\overline{W}$ 3, $\overline{RAS}$ 4, NC 5, A0 6, A2 7
Right: 16 Q, 15 A6, 14 NC, 13 A3, 12 A4
Bottom: A1 8, V_{DD} 9, A7 10, A5 11

PIN NOMENCLATURE	
A0-A7	Address Inputs
$\overline{CAS}$	Column-Address Strobe
D	Data In
NC	No Connection
Q	Data Out
$\overline{RAS}$	Row-Address Strobe
V_{DD}	5-V Supply
V_{SS}	Ground
$\overline{W}$	Write Enable

4 Dynamic RAMs

description

The TMS4164 is a high-speed, 65,536-bit, dynamic random-access memory, organized as 65,536 words of one bit each. It employs state-of-the-art SMOS (scaled MOS) N-channel double-level polysilicon gate technology for very high performance combined with low cost and improved reliability.

TMS4164
65,536-BIT DYNAMIC RANDOM-ACCESS MEMORY

The TMS4164 features $\overline{RAS}$ access times of 120 ns, 150 ns, and 200 ns maximum. Power dissipation is 135 mW typical operating and 17.5 mW typical standby.

Refresh period is extended to 4 milliseconds, and during this period each of the 256 rows must be strobed with $\overline{RAS}$ in order to retain data. $\overline{CAS}$ can remain high during the refresh sequence to conserve power.

All inputs and outputs, including clocks, are compatible with Series 74 TTL. All address lines and data in are latched on chip to simplify system design. Data out is unlatched to allow greater system flexibility. Pin 1 has no internal connection to allow compatibility with other 64K RAMs that use this pin for an additional function.

The TMS4164 is offered in 16-pin dual-in-line plastic (N suffix) and 18-lead plastic chip carrier (FP suffix) packages. The dual-in-line plastic package is designed for insertion in mounting-hole rows on 7,62-mm (300-mil) centers. The TMS4164 is guaranteed for operation from 0°C to 70°C.

operation

address (A0 through A7)

Sixteen address bits are required to decode 1 of 65,536 storage cell locations. Eight row-address bits are set up on pins A0 through A7 and latched onto the chip by the row-address strobe ($\overline{RAS}$). Then the eight column-address bits are set up on pins A0 through A7 and latched onto the chip by the column-address strobe ($\overline{CAS}$). All addresses must be stable on or before the falling edges of $\overline{RAS}$ and $\overline{CAS}$. $\overline{RAS}$ is similar to a chip enable in that it activates the sense amplifiers as well as the row decoder. $\overline{CAS}$ is used as a chip select activating the column decoder and the input and output buffers.

write enable ($\overline{W}$)

The read or write mode is selected through the write-enable ($\overline{W}$) input. A logic high on the $\overline{W}$ input selects the read mode and a logic low selects the write mode. The write-enable terminal can be driven from standard TTL circuits without a pull-up resistor. The data input is disabled when the read mode is selected. When $\overline{W}$ goes low prior to $\overline{CAS}$, data out will remain in the high-impedance state for the entire cycle permitting common I/O operation.

data in (D)

Data is written during a write or read-modify-write cycle. Depending on the mode of operation, the falling edge of $\overline{CAS}$ or $\overline{W}$ strobes data into the on-chip data latch. This latch can be driven from standard TTL circuits without a pull-up resistor. In an early write cycle, $\overline{W}$ is brought low prior to $\overline{CAS}$ and the data is strobed in by $\overline{CAS}$ with setup and hold times referenced to this signal. In a delayed-write or read-modify-write cycle, $\overline{CAS}$ will already be low, thus the data will be strobed in by $\overline{W}$ with setup and hold times referenced to this signal.

data out (Q)

The three-state output buffer provides direct TTL compatibility (no pull-up resistor required) with a fan out of two Series 74 TTL loads. Data out is the same polarity as data in. The output is in the high-impedance (floating) state until $\overline{CAS}$ is brought low. In a read cycle the output goes active after the access time interval $t_{a(C)}$ that begins with the negative transition of $\overline{CAS}$ as long as $t_{a(R)}$ is satisfied. The output becomes valid after the access time has elapsed and remains valid while $\overline{CAS}$ is low; $\overline{CAS}$ going high returns it to a high-impedance state. In an early write cycle, the output is always in the high-impedance state. In a delayed-write or read-modify-write cycle, the output will follow the sequence for the read cycle.

refresh

A refresh operation must be performed at least every four milliseconds to retain data. Since the ouput buffer is in the high-impedance state unless $\overline{CAS}$ is applied, The $\overline{RAS}$-only refresh sequence avoids any output during refresh. Strobing each of the 256 row addresses (A0 through A7) with $\overline{RAS}$ causes all bits in each row to be refreshed. $\overline{CAS}$ can remain high (inactive) for this refresh sequence to conserve power.

page mode

Page-mode operation allows effectively faster memory access by keeping the same row address and strobing random column addresses onto the chip. Thus, the time required to setup and strobe sequential row addresses for the same page is eliminated. To extend beyond the 256 column locations on a single RAM, the row address and $\overline{\text{RAS}}$ are applied to multiple 64K RAMs. $\overline{\text{CAS}}$ is then decoded to select the proper RAM.

power up

After power up, the power supply must remain at its steady-state value for 1 ms. In addition, $\overline{\text{RAS}}$ must remain high for 100 μs immediately prior to initialization. Initialization consists of performing eight $\overline{\text{RAS}}$ cycles before proper device operation is achieved.

logic symbol†

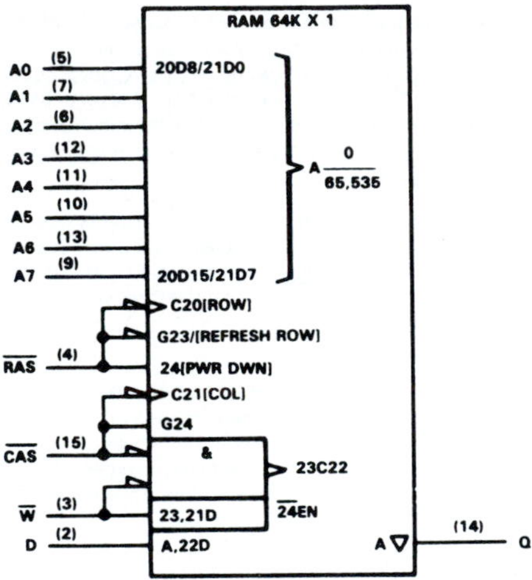

†This symbol is in accordance with ANSI/IEEE Std 91-1984 and IEC Publication 617-12.
Pin numbers shown are for the dual-in-line package.

TMS4164
65,536-BIT DYNAMIC RANDOM-ACCESS MEMORY

functional block diagram

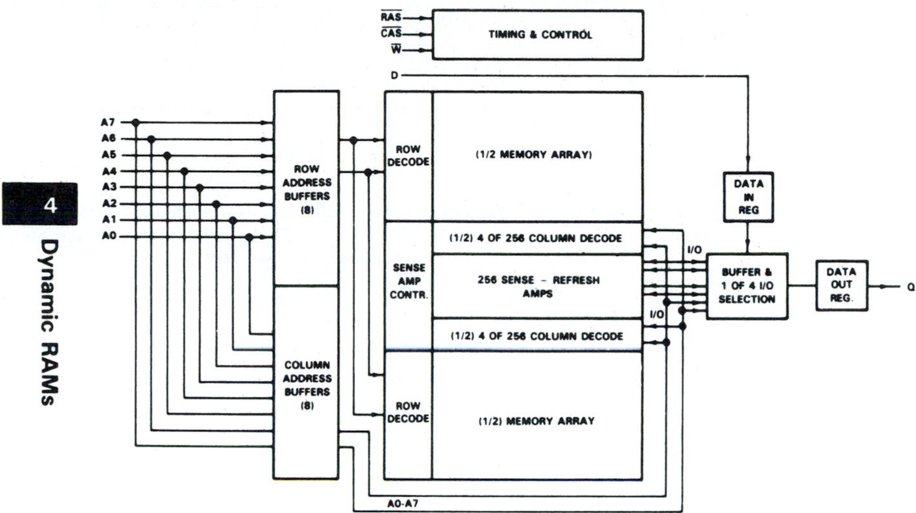

absolute maximum ratings over operating free-air temperature range (unless otherwise noted)†

Voltage on any pin except V_{DD} and data out (see Note 1) . . . . −1.5 V to 10 V
Voltage on V_{DD} supply and data out with respect to V_{SS} . . . . −1 V to 6 V
Short circuit output current . . . . 50 mA
Power dissipation . . . . 1 W
Operating free-air temperature range . . . . 0 °C to 70 °C
Storage temperature range . . . . −65 °C to 150 °C

† Stresses beyond those listed under "Absolute Maximum Ratings" may cause permanent damage to the device. This is a stress rating only and functional operation of the device at these or any other conditions beyond those indicated in the "Recommended Operating Conditions" section of this specification is not implied. Exposure to absolute-maximum-rated conditions for extended periods may affect device reliability.

NOTES: 1. All voltage values in this data sheet are with respect to V_{SS}.
2. Additional information concerning the handling of ESD sensitive devices is available in a document entitled *"Guidelines for Handling Electrostatic-Discharge-Sensitive (ESDS) Devices and Assemblies"* in Section 12.

recommended operating conditions

			MIN	NOM	MAX	UNIT
V_{DD}	Supply voltage		4.5	5	5.5	V
V_{SS}	Supply voltage			0		V
V_{IH}		V_{DD} = 4.5 V	2.4		4.8	V
		V_{DD} = 5.5 V	2.4		6	
V_{IL}	Low-level input voltage (see Notes 3 and 4)		−0.6		0.8	V
T_A	Operating free-air temperature		0		70	°C

NOTES: 3. The algebraic convention, where the more negative (less positive) limit is designated as minimum, is used in this data sheet for logic voltage levels only.
4. Due to input protection circuitry, the applied voltage may begin to clamp at −0.6 V. Test conditions must comprehend this occurrence. See application report entitled "TMS4164A and TMS4416 Input Protection Diode" on page 9-5.

electrical characteristics over full ranges of recommended operating conditions (unless otherwise noted)

PARAMETER		TEST CONDITIONS	TMS4164-12 MIN	TMS4164-12 TYP†	TMS4164-12 MAX	TMS4164-15 MIN	TMS4164-15 TYP†	TMS4164-15 MAX	UNIT
V_{OH}	High-level output voltage	I_{OH} = −5 mA	2.4			2.4			V
V_{OL}	Low-level output voltage	I_{OL} = 4.2 mA			0.4			0.4	V
I_I	Input current (leakage)	V_I=0 V to 5.8 V, V_{DD}=5 V, All other pins = 0 V			±10			±10	μA
I_O	Output current (leakage)	V_O = 0.4 to 5.5 V, V_{DD} = 5 V, $\overline{CAS}$ high			±10			±10	μA
I_{DD1}‡	Average operating current during read or write cycle	t_c = minimum cycle, All outputs open		40	48		35	45	mA
I_{DD2}§	Standby current	After 1 memory cycle, $\overline{RAS}$ and $\overline{CAS}$ high, All outputs open		3.5	5		3.5	5	mA
I_{DD3}‡	Average refresh current	t_c = minimum cycle, $\overline{CAS}$ high and $\overline{RAS}$ cycling, All outputs open		28	40		25	37	mA
I_{DD4}	Average page-mode current	$t_{c(P)}$ = minimum cycle, $\overline{RAS}$ low and $\overline{CAS}$ cycling, All outputs open		28	40		25	37	mA

† All typical values are at T_A = 25°C and nominal supply voltages.
‡ Additional information on page 4-58.
§ V_{IL} > −0.6V. See application report entitled "TMS4164A and TMS4416 Input Protection Diode" on page 9-5.

TMS4164
65,536-BIT DYNAMIC RANDOM-ACCESS MEMORY

electrical characteristics over full ranges of recommended operating conditions (unless otherwise noted)

PARAMETER		TEST CONDITIONS	TMS4164-20			UNIT
			MIN	TYP†	MAX	
V_{OH}	High-level output voltage	I_{OH} = −5 mA	2.4			V
V_{OL}	Low-level output voltage	I_{OL} = 4.2 mA			0.4	V
I_I	Input current (leakage)	V_I = 0 V to 5.8 V, V_{DD} = 5 V All other pins = 0 V			±10	μA
I_O	Output current (leakage)	V_O = 0.4 to 5.5 V, V_{DD} = 5 V, $\overline{CAS}$ high			±10	μA
I_{DD1}‡	Average operating current during read or write cycle	t_c = minimum cycle All outputs open		27	37	mA
I_{DD2}§	Standby current	After 1 memory cycle, $\overline{RAS}$ and $\overline{CAS}$ high, All outputs open		3.5	5	mA
I_{DD3}‡	Average refresh current	t_c = minimum cycle, $\overline{CAS}$ high and $\overline{RAS}$ cycling, All outputs open		20	32	mA
I_{DD4}	Average page-mode current	$t_{c(P)}$ = minimum cycle, $\overline{RAS}$ low and $\overline{CAS}$ cycling, All outputs open		20	32	mA

† All typical values are at T_A = 25°C and nominal supply voltages.
‡ Additional information on page 4-58.
§ V_{IL} > −0.6V. See application report entitled "TMS4164A and TMS4416 Input Protection Diode" on page 9-5.

capacitance over recommended supply voltage range and operating free-air temperature range, f = 1 MHz

PARAMETER		TYP†	MAX	UNIT
$C_{i(A)}$	Input capacitance, address inputs	4	5	pF
$C_{i(D)}$	Input capacitance, data input	4	5	pF
$C_{i(RC)}$	Input capacitance strobe inputs	6	8	pF
$C_{i(W)}$	Input capacitance, write enable input	6	8	pF
C_o	Output capacitance	5	6	pF

† All typical values are at T_A = 25°C and nominal supply voltages.

switching characteristics over recommended supply voltage range and operating free-air temperature range

PARAMETER		TEST CONDITIONS	ALT. SYMBOL	TMS4164-12		TMS4164-15		UNIT
				MIN	MAX	MIN	MAX	
$t_{A(C)}$	Access time from $\overline{CAS}$	C_L = 100 pF, Load = 2 Series 74 TTL gates	t_{CAC}		70		85	ns
$t_{a(R)}$	Access time from $\overline{RAS}$	C_L = 100 pF, t_{RLCL} = MAX, Load = 2 Series 74 TTL gates	t_{RAC}		120		150	ns
$t_{dis(CH)}$	Output disable time after $\overline{CAS}$ high	C_L = 100 pF, Load = 2 Series 74 TTL gates	t_{OFF}	0	40	0	40	ns

POST OFFICE BOX 1443 • HOUSTON, TEXAS 77001

switching characteristics over recommended supply voltage range and operating free-air temperature range

PARAMETER		TEST CONDITIONS	ALT. SYMBOL	TMS4164-20		UNIT
				MIN	MAX	
$t_{a(C)}$	Access time from $\overline{CAS}$	C_L = 100 pF, Load = 2 Series 74 TTL gates	t_{CAC}		135	ns
$t_{a(R)}$	Access time from $\overline{RAS}$	C_L = 100 pF, t_{RLCL} = MAX, Load = 2 Series 74 TTL gates	t_{RAC}		200	ns
$t_{dis(CH)}$	Output disable time after $\overline{CAS}$ high	C_L = 100 pF, Load = 2 Series 74 TTL gates	t_{OFF}	0	50	ns

2716*
16K (2K x 8) UV ERASABLE PROM

- **Fast Access Time**
  - **—2716-1: 350 ns Max.**
  - **—2716-2: 390 ns Max.**
  - **—2716: 450 ns Max.**
  - **—2716-5: 490 ns Max.**
  - **—2716-6: 650 ns Max.**
- **Single +5V Power Supply**
- **Low Power Dissipation**
  - **— Active Power: 525 mW Max.**
  - **— Standby Power: 132 mW Max.**
- **Pin Compatible to Intel 2732A EPROM**
- **Simple Programming Requirements**
  - **— Single Location Programming**
  - **— Programs with One 50 ms Pulse**
- **Inputs and Outputs TTL Compatible During Read and Program**
- **Completely Static**

The Intel® 2716 is a 16,384-bit ultraviolet erasable and electrically programmable read-only memory (EPROM). The 2716 operates from a single 5-volt power supply, has a static standby mode, and features fast single-address location programming. It makes designing with EPROMs faster, easier and more economical.

The 2716, with its single 5-volt supply and with an access time up to 350 ns, is ideal for use with the newer high-performance +5V microprocessors such as Intel's 8085 and 8086. A selected 2716-5 and a 2716-6 are available for slower speed applications. The 2716 is also the first EPROM with a static standby mode which reduces the power dissipation without increasing access time. The maximum active power dissipation is 525 mW while the maximum standby power dissipation is only 132 mW, a 75% savings.

The 2716 has the simplest and fastest method yet devised for programming EPROMs—single-pulse, TTL-level programming. No need for high voltage pulsing because all programming controls are handled by TTL signals. Program any location at any time—either individually, sequentially or at random, with the 2716's single-address location programming. Total programming time for all 16,384 bits is only 100 seconds.

*Part(s) also available in extended temperature range for Military and Industrial grade applications.

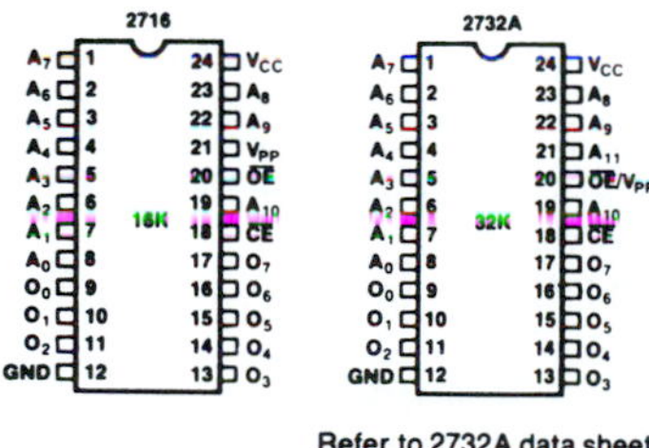

Refer to 2732A data sheet for specifications.

PIN NAMES

A_0–A_{10}	ADDRESSES
$\overline{CE}$/PGM	CHIP ENABLE/PROGRAM
$\overline{OE}$	OUTPUT ENABLE
O_0–O_7	OUTPUTS

Figure 1. Pin Configuration

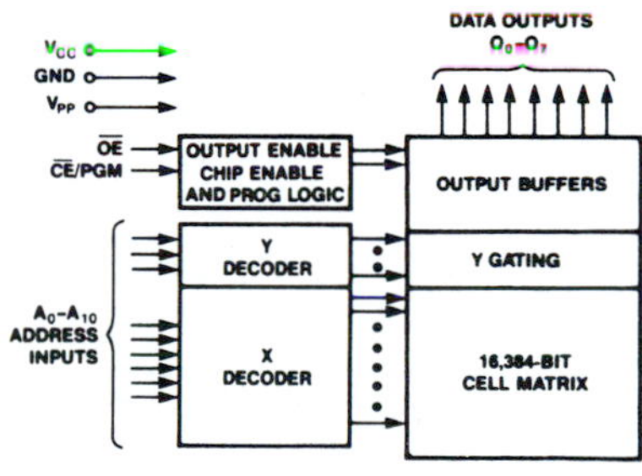

Figure 2. Block Diagram

SEPTEMBER 1981
AFN-00811B

DEVICE OPERATION

The five modes of operation of the 2716 are listed in Table 1. It should be noted that all inputs for the five modes are at TTL levels. The power supplies required are a +5V V_{CC} and a V_{PP}. The V_{PP} power supply must be at 25V during the three programming modes, and must be at 5V in the other two modes.

Read Mode

The 2716 has two control functions, both of which must be logically satisfied in order to obtain data at the outputs. Chip Enable ($\overline{CE}$) is the power control and should be used for device selection. Output Enable ($\overline{OE}$) is the output control and should be used to gate data to the output pins, independent of device selection. Assuming that addresses are stable, address access time (t_{ACC}) is equal to the delay from $\overline{CE}$ to output (t_{CE}). Data is available at the outputs t_{OE} after the falling edge of $\overline{OE}$, assuming that $\overline{CE}$ has been low and addresses have been stable for at least $t_{ACC}-t_{OE}$.

Standby Mode

The 2716 has a standby mode which reduces the active power dissipation by 75%, from 525 mW to 132 mW. The 2716 is placed in the standby mode by applying a TTL high signal to the $\overline{CE}$ input. When in standby mode, the outputs are in a high impedence state, independent of the $\overline{OE}$ input.

Output OR-Tieing

Because 2716s are usually used in larger memory arrays, Intel has provided a 2-line control function that accomodates this use of multiple memory connections. The two-line control function allows for:

a) the lowest possible memory power dissipation, and

b) complete assurance that output bus contention will not occur.

To most efficiently use these two control lines, it is recommended that $\overline{CE}$ (pin 18) be decoded and used as the primary device selecting function, while $\overline{OE}$ (pin 20) be made a common connection to all devices in the array and connected to the READ line from the system control bus. This assures that all deselected memory devices are in their low-power standby modes and that the output pins are only active when data is desired from a particular memory device.

Programming

Initially, and after each erasure, all bits of the 2716 are in the "1" state. Data is introduced by selectively programming "0's" into the desired bit locations. Although only "0's" will be programmed, both "1's" and "0's" can be presented in the data word. The only way to change a "0" to a "1" is by ultraviolet light erasure.

The 2716 is in the programming mode when the V_{PP} power supply is at 25V and $\overline{OE}$ is at V_{IH}. The data to be programmed is applied 8 bits in parallel to the data output pins. The levels required for the address and data inputs are TTL.

When the address and data are stable, a 50 msec, active-high, TTL program pulse is applied to the $\overline{CE}$/PGM input. A program pulse must be applied at each address location to be programmed. You can program any location at any time—either individually, sequentially, or at random. The program pulse

Table 1. Mode Selection

Mode \ Pins	$\overline{CE}$/PGM (18)	$\overline{OE}$ (20)	V_{PP} (21)	V_{CC} (24)	Outputs (9–11, 13–17)
Read	V_{IL}	V_{IL}	+5	+5	D_{OUT}
Standby	V_{IH}	Don't Care	+5	+5	High Z
Program	Pulsed V_{IL} to V_{IH}	V_{IH}	+25	+5	D_{IN}
Program Verify	V_{IL}	V_{IL}	+25	+5	D_{OUT}
Program Inhibit	V_{IL}	V_{IH}	+25	+5	High Z

has a maximum width of 55 msec. The 2716 must not be programmed with a DC signal applied to the $\overline{CE}$/PGM input.

Programming of multiple 2716s in parallel with the same data can be easily accomplished due to the simplicity of the programming requirements. Like inputs of the paralleled 2716s may be connected together when they are programmed with the same data. A high-level TTL pulse applied to the $\overline{CE}$/PGM input programs the paralleled 2716s.

Program Inhibit

Programming of multiple 2716s in parallel with different data is also easily accomplished. Except for $\overline{CE}$/PGM, all like inputs (including $\overline{OE}$) of the parallel 2716s may be common. A TTL-level program pulse applied to a 2716's $\overline{CE}$/PGM input with V_{PP} at 25V will program that 2716. A low-level $\overline{CE}$/PGM input inhibits the other 2716 from being programmed.

Program Verify

A verify should be performed on the programmed bits to determine that they were correctly programmed. The verify may be performed with V_{PP} at 25V. Except during programming and program verify, V_{PP} must be at 5V.

ERASURE CHARACTERISTICS

The erasure characteristics of the 2716 are such that erasure begins to occur when exposed to light with wavelengths shorter than approximately 4000 Angstroms (Å). It should be noted that sunlight and certain types of fluorescent lamps have wavelengths in the 3000–4000 Å range. Data show that constant exposure to room-level fluorescent lighting could erase the typical 2716 in approximately 3 years, while it would take approximately 1 week to cause erasure when exposed to direct sunlight. If the 2716 is to be exposed to these types of lighting conditions for extended periods of time, opaque labels are available from Intel which should be placed over the 2716 window to prevent unintentional erasure.

The recommended erasure procedure (see Data Catalog PROM/ROM Programming Instruction Section) for the 2716 is exposure to shortwave ultraviolet light which has a wavelength of 2537 Angstroms (Å). The integrated dose (i.e., UV intensity X exposure time) for erasure should be a minimum of 15 W-sec/cm^2. The erasure time with this dosage is approximately 15 to 20 minutes using an ultraviolet lamp with a 1200 $\mu W/cm^2$ power rating. The 2716 should be placed within 1 inch of the lamp tubes during erasure. Some lamps have a filter on their tubes which should be removed before erasure.

ABSOLUTE MAXIMUM RATINGS*

Temperature Under Bias .... −10°C to +80°C
Storage Temperature .... −65°C to +125°C
All Input or Output Voltages with
Respect to Ground .... +6V to −0.3V
V_{PP} Supply Voltage with Respect
to Ground During Program .... +26.5V to −0.3V

**NOTICE: Stresses above those listed under "Absolute Maximum Ratings" may cause permanent damage to the device. This is a stress rating only and functional operation of the device at these or any other conditions above those indicated in the operational sections of this specification is not implied. Exposure to absolute maximum rating conditions for extended periods may affect device reliability.*

DC AND AC OPERATING CONDITIONS DURING READ

	2716	2716-1	2716-2	2716-5	2716-6
Temperature Range	0°C–70°C	0°C–70°C	0°C–70°C	0°C–70°C	0°C–70°C
V_{CC} Power Supply[1,2]	5V ±5%	5V ±10%	5V ±5%	5V ±5%	5V ±5%
V_{PP} Power Supply[2]	V_{CC}	V_{CC}	V_{CC}	V_{CC}	V_{CC}

READ OPERATION

D.C. AND OPERATING CHARACTERISTICS

Symbol	Parameter	Limits			Units	Test Conditions
		Min.	Typ.[3]	Max.		
I_{LI}	Input Load Current			10	μA	V_{IN} = 5.25V
I_{LO}	Output Leakage Current			10	μA	V_{OUT} = 5.25V
I_{PP1}[2]	V_{PP} Current			5	mA	V_{PP} = 5.25V
I_{CC1}[2]	V_{CC} Current (Standby)		10	25	mA	$\overline{CE} = V_{IH}$, $\overline{OE} = V_{IL}$
I_{CC2}[2]	V_{CC} Current (Active)		57	100	mA	$\overline{OE} = \overline{CE} = V_{IL}$
V_{IL}	Input Low Voltage	−0.1		0.8	V	
V_{IH}	Input High Voltage	2.0		$V_{CC}+1$	V	
V_{OL}	Output Low Voltage			0.45	V	I_{OL} = 2.1 mA
V_{OH}	Output High Voltage	2.4			V	I_{OH} = −400 μA

TYPICAL CHARACTERISTICS

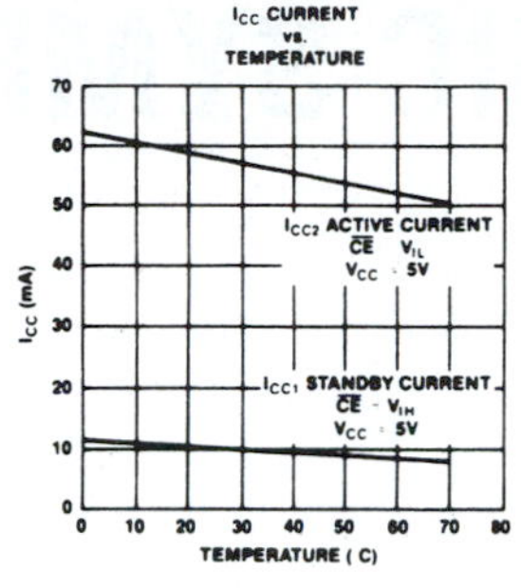

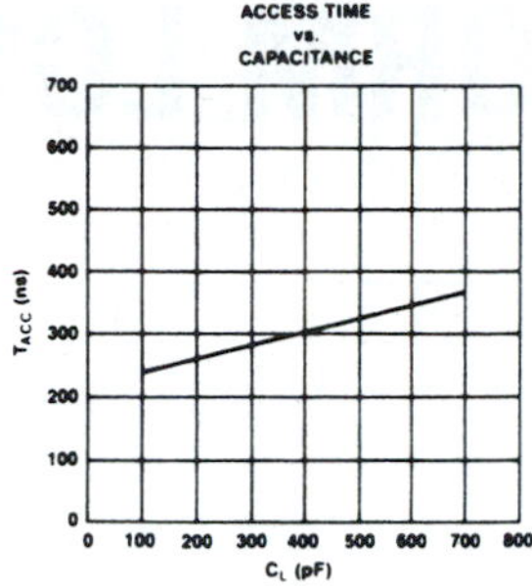

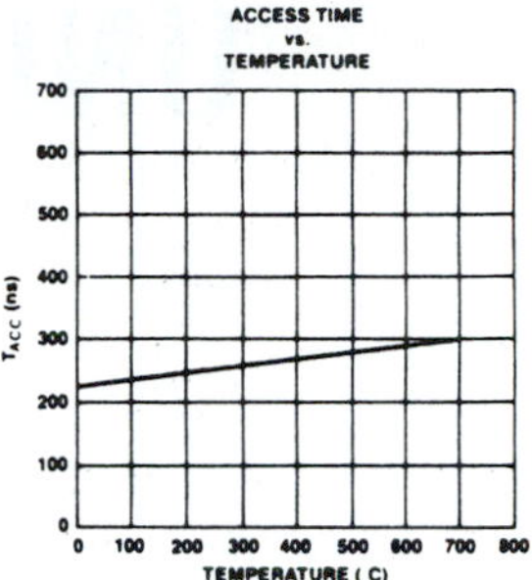

A.C. CHARACTERISTICS

Symbol	Parameter	Limits (ns)										Test Conditions
		2716		2716-1		2716-2		2716-5		2716-6		
		Min.	Max.	Min.	Max.	Min.	Max.	Min.	Max.	Min.	Max.	
t_{ACC}	Address to Output Delay		450		350		390		450		450	$\overline{CE} = \overline{OE} = V_{IL}$
t_{CE}	$\overline{CE}$ to Output Delay		450		350		390		490		650	$\overline{OE} = V_{IL}$
t_{OE} [4]	Output Enable to Output Delay		120		120		120		160		200	$\overline{CE} = V_{IL}$
t_{DF} [4]	Output Enable High to Output Float	0	100	0	100	0	100	0	100	0	100	$\overline{CE} = V_{IL}$
t_{OH}	Output Hold from Addresses, $\overline{CE}$ or $\overline{OE}$ Whichever Occurred First	0		0		0		0		0		$\overline{CE} = \overline{OE} = V_{IL}$

CAPACITANCE[4] (T_A = 25°C, f = 1 MHz)

Symbol	Parameter	Typ.	Max.	Units	Test Conditions
C_{IN}	Input Capacitance	4	6	pF	V_{IN} = 0V
C_{OUT}	Output Capacitance	8	12	pF	V_{OUT} = 0V

A.C. TEST CONDITIONS

Output Load .... 1 TTL gate and C_L = 100 pF
Input Rise and Fall Times .... ≤20 ns
Input Pulse Levels .... 0.8V to 2.2V
Timing Measurement Reference Level:
Inputs .... 1V and 2V
Outputs .... 0.8V and 2V

APPENDIX D

UNIFORM LOGIC SYMBOLS

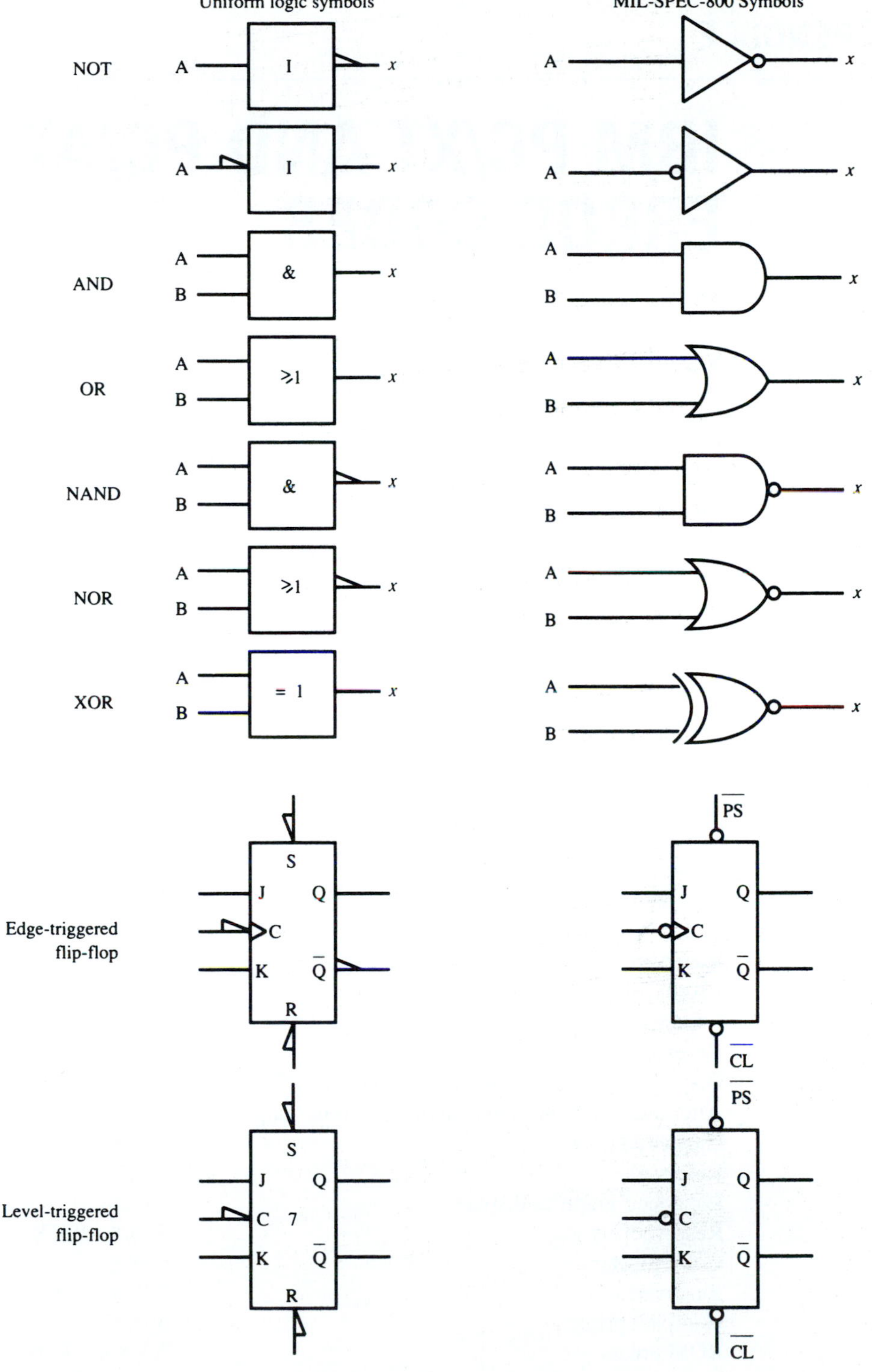
Uniform logic symbols
MIL-SPEC-800 Symbols
NOT
A
I
x
AND
B
&
OR
≥1
NAND
NOR
XOR
= 1
Edge-triggered flip-flop
S
J
Q
C
K
R
PS
CL
Level-triggered flip-flop
7

APPENDIX E

IBM PC/XT AND PC/AT ERROR CODES

Visual Error Codes

Major Subsystem	Error Code
Power	02X
System board	1XX
Memory	20X XXXX XXX20X
Keyboard	30X or XX30X
Monochrome display	4XX
Color graphics display	5XX
Diskette	6XX
Math coprocessor	7XX
Reserved	8XX
Parallel printer adapter	9XX
Reserved	10XX
Asynchronous communications adapter	11XX
Alternate asynchronous communications adapter	12XX
Game control adapter	13XX
Printer	14XX
SDLC communications adapter	15XX
Reserved	16XX
Fixed disk	17XX
Expansion unit	18XX
Reserved	19XX
Bisynchronous communications adapter	20XX
Alternate bisynchronous communications adapter	21XX
Cluster adapter	22XX
Reserved	23XX
Enhanced graphics display	24XX
Reserved	25XX–28XX
Color printer	29XX
Reserved	30XX–32XX
Compact printer	33XX
ROM errors	XXXXX ROM

Audible Error Codes

Indication	Possible Cause
No beeps or continuous tone	System board or power supply
One long and one short beep	System board
One long and two short beeps	Display
One short beep and blank display	Display or switches incorrect

Detailed Error Codes

1XX System Board Failures

101 Interrupt failure, 8259 chip and associated logic.
102 Timer failure, 8253 chip and associated logic.
103 Timer interrupt failure, 8253 or 8259 chip failure.
104 Protected mode failure (PC/AT only).
105 8042 MCU chip failure (PC/AT only).
106 Converting logic failure.
107 Hot NMI failure (NMI stuck active).
108 Timer bus test failure.
109 Direct memory access failure, 8237 chip and logic.
121 Unexpected interrupts, 8259 chip or stuck input (short).
131 Cassette interface failure, PC only.

The Following Are for the PC/AT Only

161 System option error, MC146818 chip or battery.
162 System options not set correctly, adapter board failure.
163 Time and date not set, MC146818 or battery failure.
164 Memory size error, memory failure or changed.

2XX RAM Memory Errors

201 Memory test data failure, see text.
202 Memory address test failure, check for shorts.
203 Memory address test failure, check for shorts.

3XX Keyboard Errors

301 Keyboard did not respond to software reset command correctly or stuck key detected. If a stuck key was detected, the hex keycode for the stuck key is displayed.
302 User-indicated failure or PC/AT keylock is locked.
303 Keyboard or system board error, check cable.
304 Keyboard error—CMOS (MC146818) does not match system.

4XX Monochrome Display Errors

401 Monochrome display adapter failed, adapter memory, horizontal frequency, or video failure.
408 User-indicated display attributes are incorrect.

416 User-indicated character set incorrect.
424 User 80X25 mode incorrect.
432 Parallel port failure, missing loop-back plug.

5XX Color Graphics Display Errors

501 Color graphics display adapter failed, adapter memory horizontal frequency, or video failure.
508 User-indicated display attributes are incorrect.
516 User-indicated character set is incorrect.
524 User-indicated 80X25 mode is incorrect.
532 User-indicated 40X25 mode is incorrect.
540 User-indicated 320X200 graphics mode is incorrect.
548 User-indicated 640X200 graphics mode is incorrect.

6XX Diskette Errors

601 Diskette post failed.
602 Unable to read boot record.
606 Diskette-verify function failed.
607 Write-protected diskette or sensor failure.
608 Illegal diskette command, status returned.
610 Diskette initialization failed.
611 System timeout, status returned.
612 Bad NEC, floppy disk controller chip bad.
613 Diskette DMA failure, no data transferred.
621 Seek error, status returned.
622 Read error, bad CRC.
623 Read error, record not found (could not find header).
624 Read error, bad address mark (AM) or not found.
625 Bad NEC seek, status returned.
626 Diskette data-compare error.

7XX Math Coprocessor Errors

7XX 8087 or 80287 math coprocessor chip failure.

9XX Parallel Printer Adapter Errors

901 Parallel printer adapter failed, requires test plug.

11XX Asynchronous Communications Adapter Errors

1101 Asynchronous adapter failed, requires test plug.

12XX Alternate Asynchronous Communications Adapter Errors

1201 Alternate asynchronous communications adapter failed.

13XX Game-Control Adapter Errors

1301 Game-control adapter failed.
1302 Joystick test failed.

14XX Printer Errors

1401 Printer test failed.
1404 Matrix printer failed.

15XX Synchronous Data Link (SDLC) Communications Adapter

1510 8255 PPI chip port A failure.
1511 8255 PPI chip port B failure.
1512 8255 PPI chip port C failure.
1513 8253 timer 1 did not reach terminal count.
1514 8253 timer 1 stuck on.
1515 8235 timer 0 did not reach terminal count.
1516 8253 timer 0 stuck on.
1517 8253 timer 2 did not reach terminal count.
1518 8253 timer 2 stuck on.
1519 8273 port B error.
1520 8273 port A error.
1521 8273 command/read timeout.
1522 Interrupt level 4 failure.
1523 Ring indicate stuck on.
1524 Receive clock stuck on.
1525 Transmit clock stuck on.
1526 Test indicate stuck on.
1527 Ring indicate not on.
1528 Receive clock not on.
1529 Transmit clock not on.
1530 Test indicate not on.
1531 Data set ready not on.
1532 Carrier detect not on.
1533 Clear to send not on.
1534 Data set ready stuck on.
1536 Clear to send stuck on.
1537 Level 3 interrupt failure
1538 Receive interrupt results failure.
1539 Wrap data miscompare.
1540 DMA channel 1 failure.
1541 DMA channel 1 failure.
1542 Error in 8273 error checking or status reporting.
1547 Stray interrupt level 4.
1548 Stray interrupt level 3.
1549 Interrupt presentation sequence timeout.

16XX Display Emulation Errors (327X, 5520, 525X)

No codes available at this time.

17XX Fixed Disk Errors

1701 Fixed disk POST error.
1702 Fixed disk adapter error.
1703 Fixed disk drive error.
1704 Fixed disk drive or adapter error.
1780 Fixed disk drive 0 error.
1781 Fixed disk drive 1 error.
1782 Fixed disk adapter failure.
1790 Fixed disk drive 0 error.
1791 Fixed disk drive 1 error.

18XX I/O Expansion Unit Errors

1801 I/O expansion unit POST error.
1810 Enable/disable failure.
1811 Extender card wrap test failed.
1812 High-order address lines failed.
1813 Wait state failure.
1814 Enable/disable could not be set on.
1815 Wait state failure.
1816 Extender card wrap test failed (enabled).
1817 High-order address lines failure (enabled).
1818 Disable not functioning.
1819 Wait request switch not set correctly.
1820 Receiver card wrap test failure.
1821 Receiver high-order address lines failure.

19XX 3270 PC Attachment Card Error

No codes available at this time.

20XX Binary Synchronous Communications (BSC) Adapter Errors

2010 8255 port A failure.
2011 8255 port B failure.
2012 8255 port C failure.
2013 8253 timer 1 did not reach terminal count.
2014 8253 timer 1 stuck on.
2016 8235 timer 2 did not reach terminal count or stuck on.
2017 8251 data set ready failed to come on.
2018 8251 clear to send not sensed.
2019 8251 data set ready stuck on.
2020 8251 clear to send stuck on.
2021 8251 hardware reset failed.
2022 8251 software reset failed.
2023 8251 software error reset failed.
2024 8251 transmit ready did not come on.
2025 8251 receive ready did not come on.
2026 8251 could not force overrun status to come on.

2027 Interrupt failure—on timer interrupt.
2028 Interrupt failure—transmit.
2029 Interrupt failure—transmit.
2030 Interrupt failure—receive.
2031 Interrupt failure—receive.
2033 Ring indicate stuck on.
2034 Receive clock stuck on.
2035 Transmit clock stuck on.
2036 Test indicate stuck on.
2037 Ring indicate not on.
2038 Receive clock not on.
2039 Transmit clock not on.
2040 Test indicate not on.
2041 Data set ready not on.
2042 Carrier detect not on.
2043 Clear to send not on.
2044 Data set ready stuck on.
2046 Clear to send stuck on.
2047 Unexpected transmit interrupt.
2048 Unexpected receive interrupt.
2049 Transmit data did not equal receive data.
2050 8251 detected overrun error.
2051 Lost data set ready during wrap.
2052 Receive timeout during wrap.

22XX Cluster Adapter Errors

24XX Enhanced Graphics Adapter Errors

29XX Color Matrix Printer Errors

33XX Compact Printer Errors

ANSWERS TO SELECTED ODD-NUMBERED EXERCISES

Chapter 1

1.1 Decimal, or base 10
1.3 a. 0.5 b. 0.0625 c. 0.003906 d. 0.00097656
1.5 a. 113 b. 17 c. 245 d. 47
1.7 a. 13.3125 b. 23.4375 c. 65.9375 d. 5.78125
1.9 a. 1101101 b. 10000000 c. 11111010 d. 100000001
1.11 a. 1110.0001111 b. 100111.10001111 c. 1111101.10110011
d. 1101110.0001
1.13 a. 679 b. 4095 c. 3115 d. 2842
1.15 a. 33 b. 231 c. 1656 d. 1423
1.17 a. 111 b. 110 c. 110010 d. 1001001
1.19 a. 0.001010 b. 1101.000011 c. 1110111.101011001
d. 110111101100.001011000010
1.21 a. 777 b. 142 c. 252 d. 62.72
1.23 a. 181 b. 255 c. 16 d. 41
1.25 a. 27 b. 4*E* c. 100 d. 3*E*8
1.27 a. 10100100 b. 11111101 c. 00010001 d. 000111000001
1.29 a. 333 b. *A.F* c. 968 d. 27*C*
1.31 a. 01001001 b. 10010110 c. 001000000001 d. 010100110111
1.33 a. 16 b. Not BCD c. 91 d. 273
1.35 a. 1010010 b. 1110010 c. 0111001 d. 0101010
1.37 a. 0 b. O c. Z d. A
1.39 a. Even b. Odd c. Even d. Odd
1.41

a. 1010111 0
1001111 0
1010010 0
1001011 1

b. 1010010 0
1010101 1
1001110 1

c. 0110011 1
0101011 1
0110010 0
0111101 0
0110101 1

d. 1001101 1
1000101 0
1001110 1
1010101 1
0100011 0
0110010 0

1.43 a. 011 b. 011 c. 0101 d. 0111
1.45 a. 100 b. 100100 c. 1110 d. 11001

Chapter 2

2.1 A logic gate is essentially a combination of switches.
2.3 Positive logic
2.5 5 V_{DC}; 0 V_{DC}
2.7 Duty cycle = 75 μs/250μs × 100% = 30%
2.9 See Figure 2.3.
2.11 AND function; see Figure 2.4.
OR function; see Figure 2.5.
NOT function; see Figure 2.6.
2.13 A truth table is a tabular method of expressing Boolean functions.
2.15 A 3-input AND gate with inputs *A, B, C* and output L1.
2.17 OR logic function
2.19 a. See Figure 2.4. b. See Figure 2.5. c. See Figure 2.6.
2.21 TTL and CMOS
2.23 a. 2-input AND gates (TTL) b. 3-input AND gates (TTL) c. 2-input OR gates (TTL)
d. 6 inverters (TTL) e. 2-input AND gates (CMOS) f. 4-input AND gates (CMOS)
g. 2-input OR gates (CMOS) h. 6 inverters (CMOS)
2.25 Logic 1
2.27

2.29 a. See Figure 2.21. b. See Figure 2.23, which must be modified.
2.31 See Tech Tips and Troubleshooting and Figure 2.25.
2.33

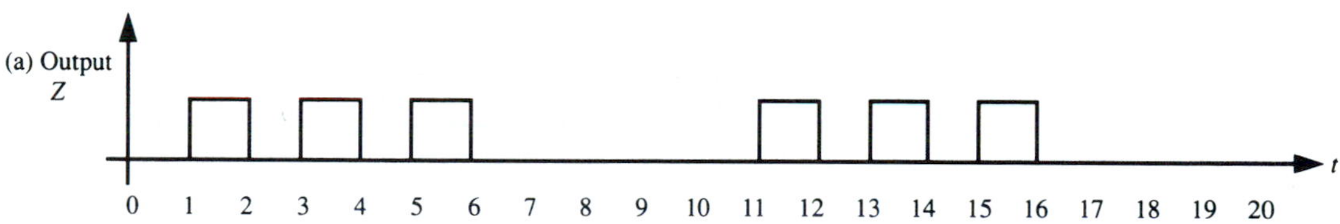

2.35

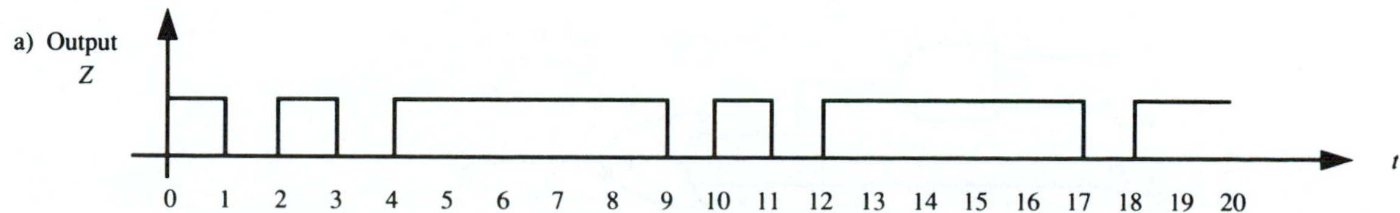

2.37

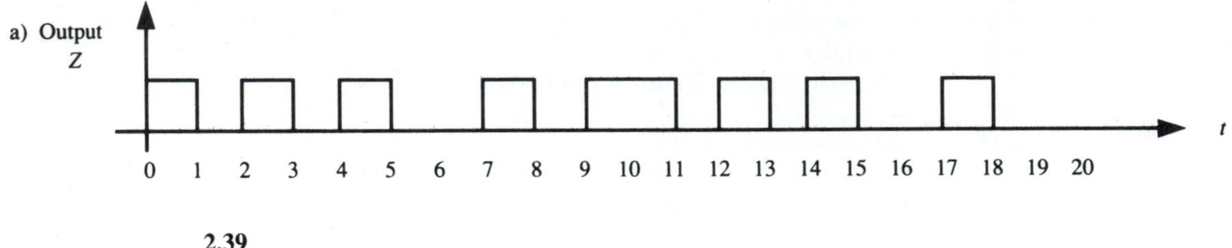

2.39

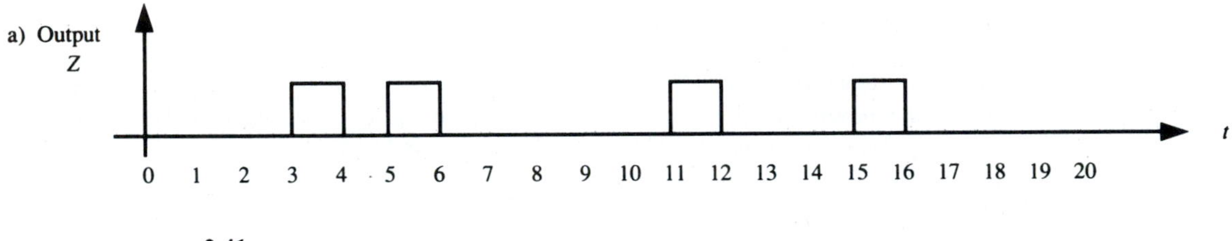

2.41

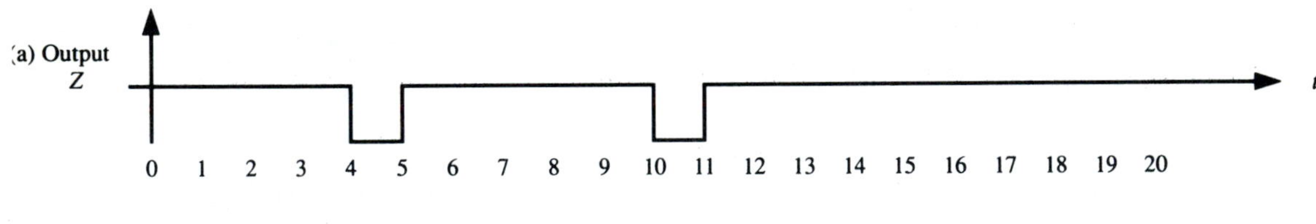

2.43

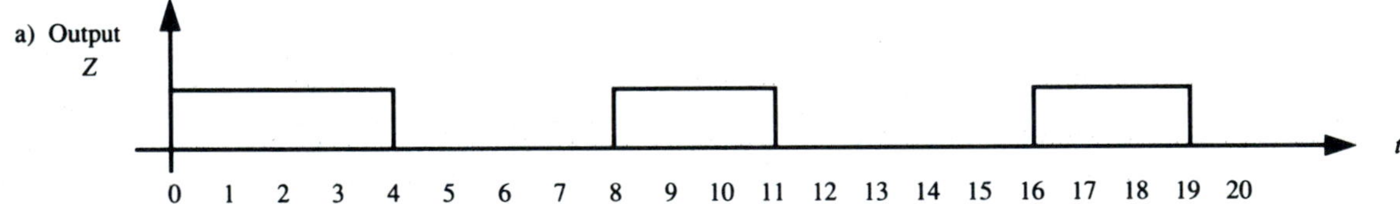

Chapter 3

3.1 See Section 3.1 and Figure 3.1.

3.3 a. $z = x + (y \cdot w)$ Theorem 7

b.

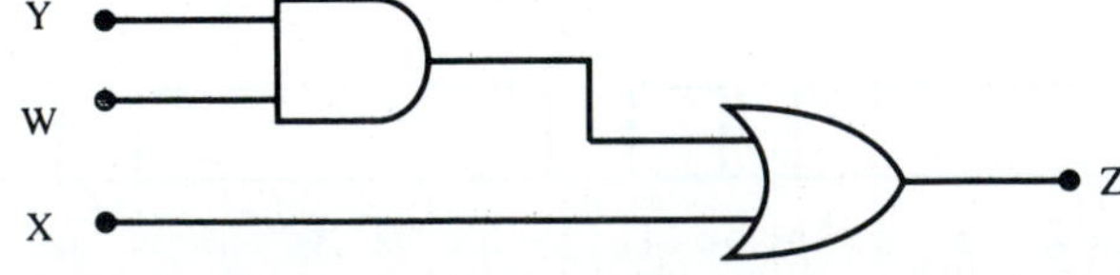

c. $z = (x + y) \cdot (x + w) = x + (y \cdot w)$

xyz	$(x + y) \cdot (x + w)$	z	$x + (y \cdot w)$	z
000	$0 \cdot 0$	0	$0 + 0$	0
001	$0 \cdot 1$	0	$0 + 0$	0
010	$1 \cdot 0$	0	$0 + 0$	0
011	$1 \cdot 1$	1	$0 + 1$	1
100	$1 \cdot 1$	1	$1 + 0$	1
101	$1 \cdot 1$	1	$1 + 0$	1
110	$1 \cdot 1$	1	$1 + 0$	1
111	$1 \cdot 1$	1	$1 + 1$	1

3.5 a. $\text{GO} = \overline{A}(\overline{B}C + B\overline{C}) + A\overline{B} = \overline{A}(B \oplus C) + A\overline{B}$ Theorem 7

b.

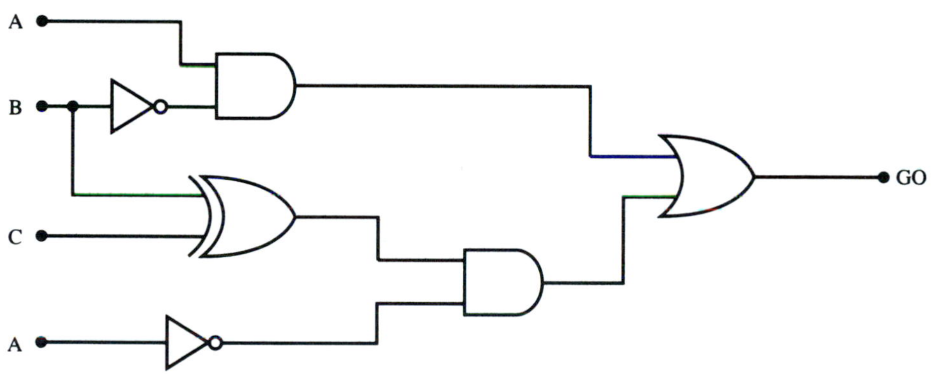

c. $\text{GO} + (\overline{A}\overline{B}C) + (A\overline{B}) + (\overline{A}B\overline{C})$

ABC	$(\overline{A}\overline{B}C) + (A\overline{B}) + (\overline{A}B\overline{C})$	GO	$\overline{A}(B \oplus C) + A\overline{B}$	GO
000	$0 + 0 + 0$	0	$(1 \cdot 0) + 0$	0
001	$1 + 0 + 0$	1	$(1 \cdot 1) + 0$	1
010	$0 + 0 + 1$	1	$(1 \cdot 1) + 0$	1
011	$0 + 0 + 0$	0	$(1 \cdot 0) + 0$	0
100	$0 + 1 + 0$	1	$(0 \cdot 0) + 1$	1
101	$0 + 1 + 0$	1	$(0 \cdot 1) + 1$	1
110	$0 + 0 + 0$	0	$(0 \cdot 1) + 0$	0
111	$0 + 0 + 0$	0	$(0 \cdot 0) + 0$	0

3.7 a. $\text{ENABLE} = ABC + \overline{AC}$ Theorem 8

b.

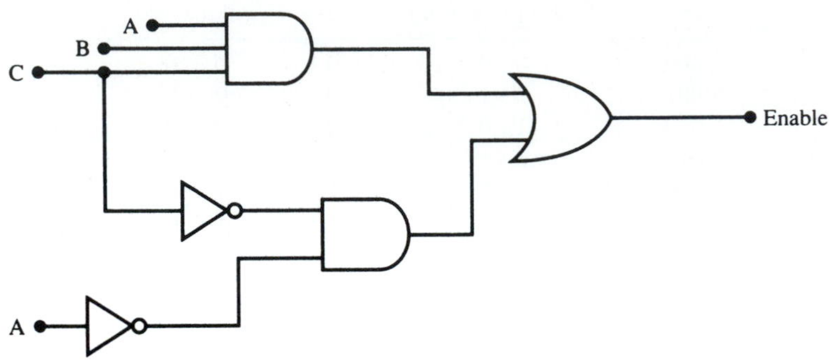

c. ENABLE $= (A + \overline{C}) \cdot (\overline{A} + ABC)$

ABC	$(A + \overline{C}) \cdot (\overline{A} + ABC)$	ENABLE	$ABC + \overline{AC}$	Enable
000	$1 \cdot 1$	1	$0 + 1$	1
001	$0 \cdot 1$	0	$0 + 0$	0
010	$1 \cdot 1$	1	$0 + 1$	1
011	$0 \cdot 1$	0	$0 + 0$	0
100	$1 \cdot 0$	0	$0 + 0$	0
101	$1 \cdot 0$	0	$0 + 0$	0
110	$1 \cdot 0$	0	$0 + 0$	0
111	$1 \cdot 1$	1	$1 + 0$	1

3.9 $(\overline{x \cdot y \cdot z}) = \overline{x} + \overline{y} + \overline{z}$

xyz	$(x \cdot y \cdot z)$	$\overline{x} + \overline{y} + \overline{z}$
000	1	$1 + 1 + 1 = 1$
001	1	$1 + 1 + 0 = 1$
010	1	$1 + 0 + 1 = 1$
011	1	$1 + 0 + 0 = 1$
100	1	$0 + 1 + 1 = 1$
101	1	$0 + 1 + 0 = 1$
110	1	$0 + 0 + 1 = 1$
111	0	$0 + 0 + 0 = 0$

3.11 BELL $= x(z + y)$

3.13 READY $= \overline{A} \cdot B \cdot \overline{C}$

3.15 $MX = J + RD$

3.17 See Section 3.6 and Figure 3.15.

3.19 $x = AB\overline{C} + \overline{ABC} + \overline{AB}C + A\overline{B}C$

$X = AB\overline{C} + \overline{A}\,\overline{B}\,\overline{C} + \overline{A}\,\overline{B}C + A\overline{B}C$

	$\overline{B}\,\overline{C}$	$\overline{B}C$	BC	$B\overline{C}$
$\overline{A}$	1	1		
A		1		1

$X = \overline{A}\,\overline{B} + \overline{B}C + AB\overline{C}$

$x = \overline{A}\,\overline{B} + \overline{B}C + AB\overline{C}$

3.21 $OFF = \overline{ABCD} + \overline{ABC}\overline{D} + CD + A\overline{B}C\overline{D}$

a.

$OFF = \overline{A}\,\overline{B}\,\overline{C}\,\overline{D} + \overline{A}\,\overline{B}C\overline{D} + CD + A\overline{B}C\overline{D}$

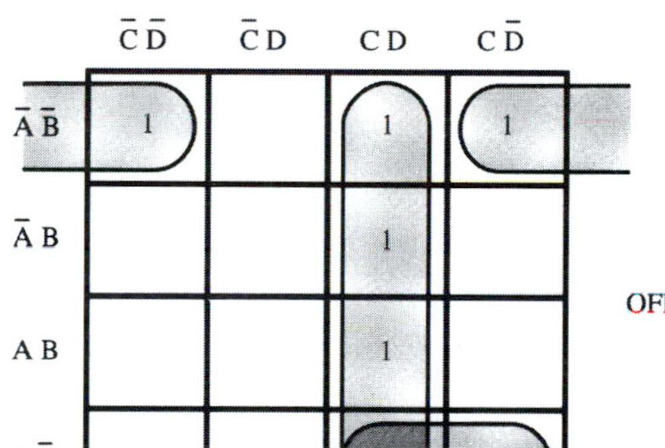

$OFF = CD + \overline{A}\,\overline{B}\,\overline{D} + A\overline{B}C$

$OFF = CD + \overline{ABD} + A\overline{B}C$

b.

$ABCD$	$\overline{ABCD} + \overline{ABC}\overline{D} + CD + A\overline{B}C\overline{D}$	OFF	$CD + \overline{ABD} + A\overline{B}C$	OFF
0000	1 + 0 + 0 + 0	1	0 + 1 + 0	1
0001	0 + 0 + 0 + 0	0	0 + 0 + 0	0
0010	0 + 1 + 0 + 0	1	0 + 1 + 0	1
0011	0 + 0 + 1 + 0	1	1 + 0 + 0	1
0100	0 + 0 + 0 + 0	0	0 + 0 + 0	0
0101	0 + 0 + 0 + 0	0	0 + 0 + 0	0
0110	0 + 0 + 0 + 0	0	0 + 0 + 0	0
0111	0 + 0 + 1 + 0	1	1 + 0 + 0	1
1000	0 + 0 + 0 + 0	0	0 + 0 + 0	0
1001	0 + 0 + 0 + 0	0	0 + 0 + 0	0

$ABCD$	$\overline{ABCD} + \overline{ABC}\overline{D} + CD + A\overline{B}C\overline{D}$	OFF	$CD + \overline{ABD} + A\overline{B}C$	OFF
1010	0 + 0 + 0 + 1	1	0 + 0 + 1	1
1011	0 + 0 + 1 + 0	1	1 + 0 + 1	1
1100	0 + 0 + 0 + 0	0	0 + 0 + 0	0
1101	0 + 0 + 0 + 0	0	0 + 0 + 0	0
1110	0 + 0 + 0 + 0	0	0 + 0 + 0	0
1111	0 + 0 + 1 + 0	1	1 + 0 + 0	1

c.

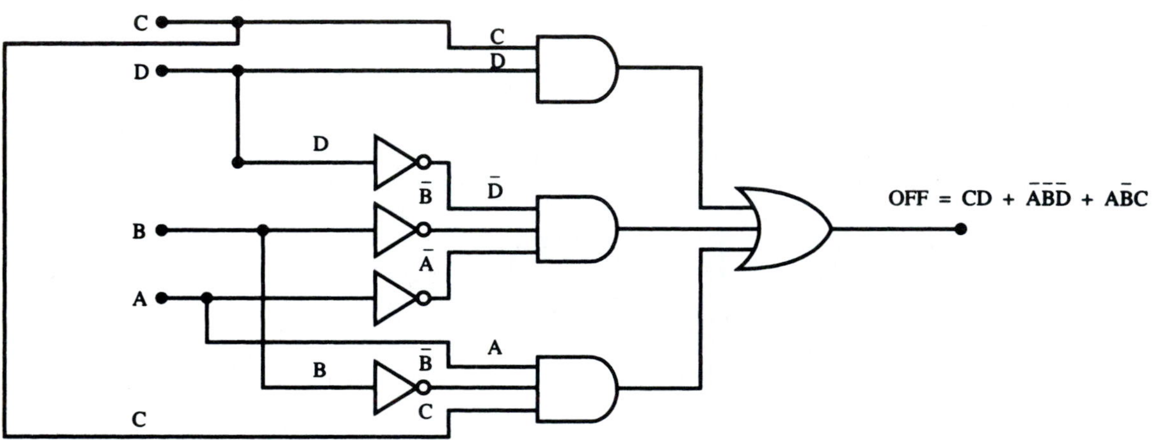

3.25

CBA	LIGHT
000	1
001	1
010	1
011	1
100	1
101	0
110	0
111	0

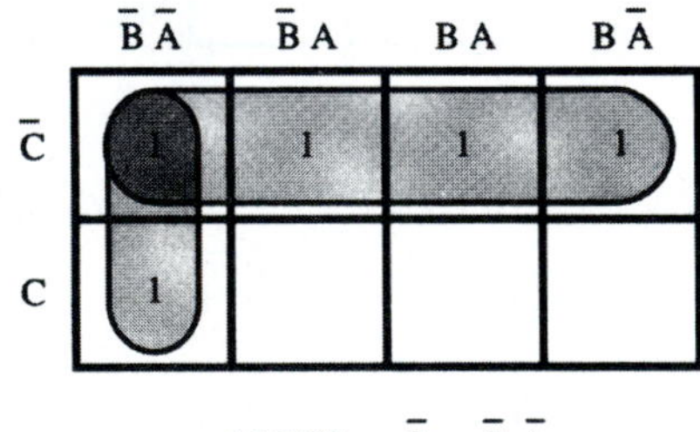

3.27 See Figure 3.27.

3.29

ABCD	OUT
0000	1
0001	1
0010	1
0011	0

ABCD	OUT
0100	1
0101	0
0110	0
0111	1
1000	1
1001	0
1010	0
1011	1
1100	0
1101	1
1110	1
1111	1

	$\bar{C}\bar{D}$	$\bar{C}D$	CD	$C\bar{D}$
$\bar{A}\bar{B}$	1	1		1
$\bar{A}B$	1		1	
AB		1	1	1
$A\bar{B}$	1		1	

$$OUT = \bar{A}\bar{B}\bar{C} + \bar{A}\bar{B}\bar{D} + A\bar{C}\bar{D} + \bar{A}C\bar{D} + ABD + ABC + BCD + ACD$$

OR

$$OUT = (\bar{A}\bar{B}\bar{C}) + (ABC) + (\bar{A}\bar{B}\bar{D}) + (ABD) + (\bar{A}\bar{C}\bar{D}) + (ACD) + (A\bar{C}\bar{D}) + (BCD)$$

Chapter 4

4.1 a. See Figure 4.4. b. See Figure 4.2.

4.3

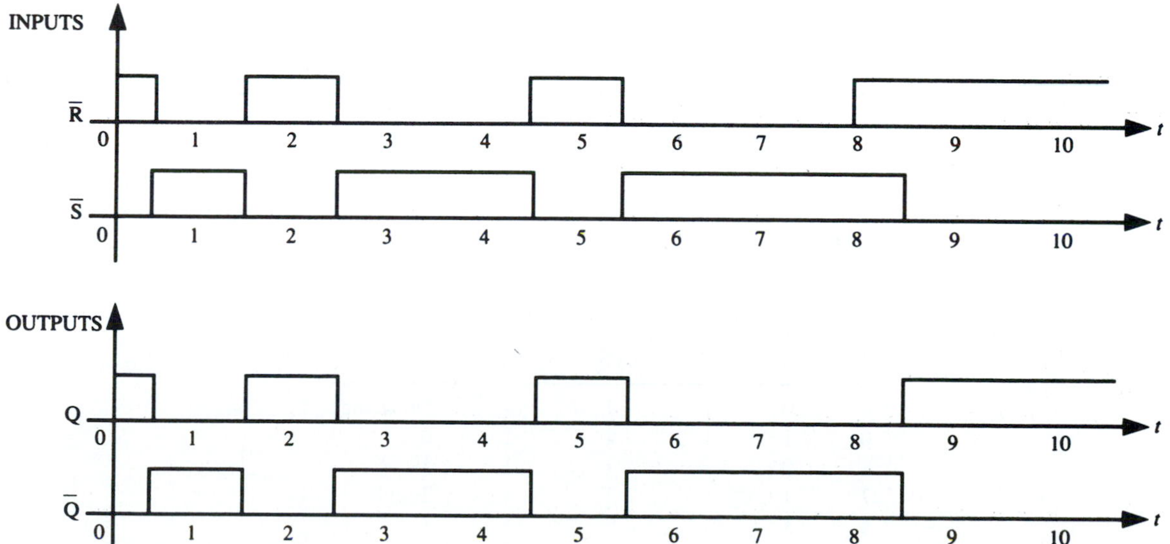

4.5

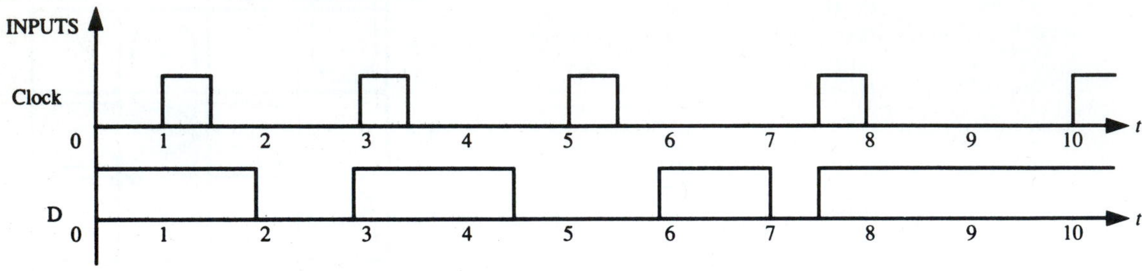

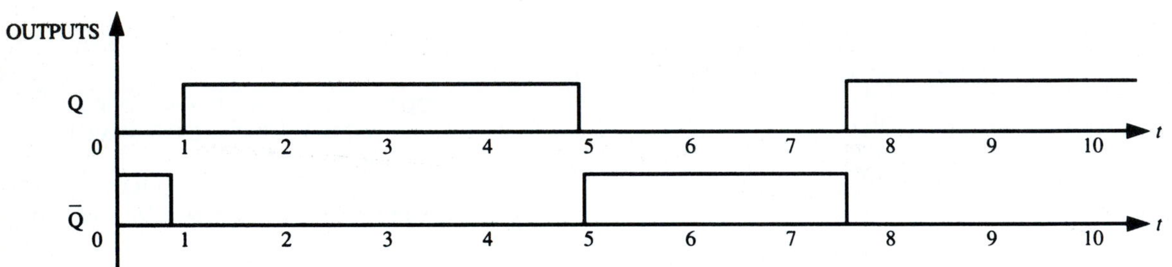

4.7 See Figure 4.11(a).

4.9

J	K	C	Q	$\overline{Q}$
0	0	↓	NC	NC
0	1	↓	0	1
1	0	↓	1	0
1	1	↓	Toggle	Toggle

4.11 5 kHz

4.13

J	K	C	$\overline{\text{PS}}$	$\overline{\text{CL}}$	Q
0	0	↓	1	1	No change
0	1	↓	1	1	0
1	0	↓	1	1	1
1	1	↓	1	1	Toggle
x	x	x	0	1	1
x	x	x	1	0	0

4.15

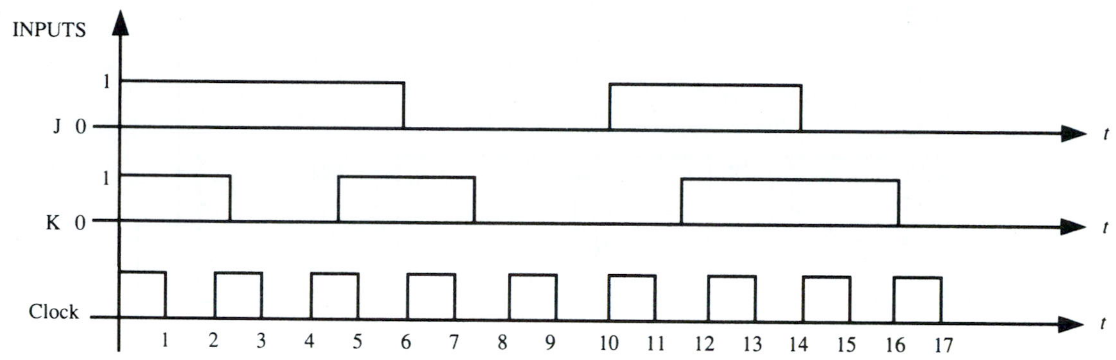

OUTPUTS
Q
t
$\overline{Q}$
t
1 2 3 4 5 6 7 8 9 10 11 12 13 14 15 16 17

4.17

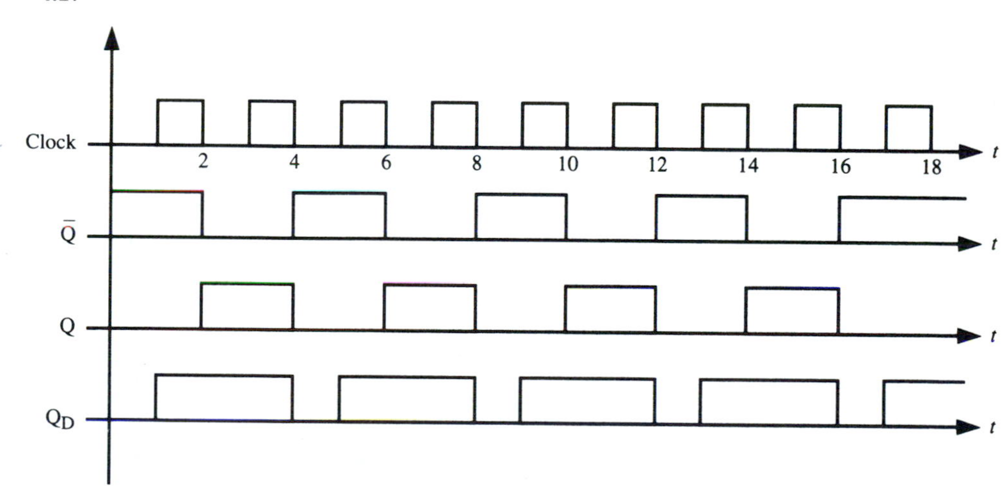

4.19

a. Setup time is the minimum time that the input data must be stable before the clock transition.

b. Propagation delay time is the time it takes for the output of a flip-flop actually to change state.

c. Asynchronous inputs are inputs that act directly on a flip-flop, regardless of the state of the clock and synchronous inputs (*JK*).

d. Race condition occurs when the input data change at the same time that the clock trigger pulse occurs.

4.21 $f_{out} = 0.391$ kHz

4.23

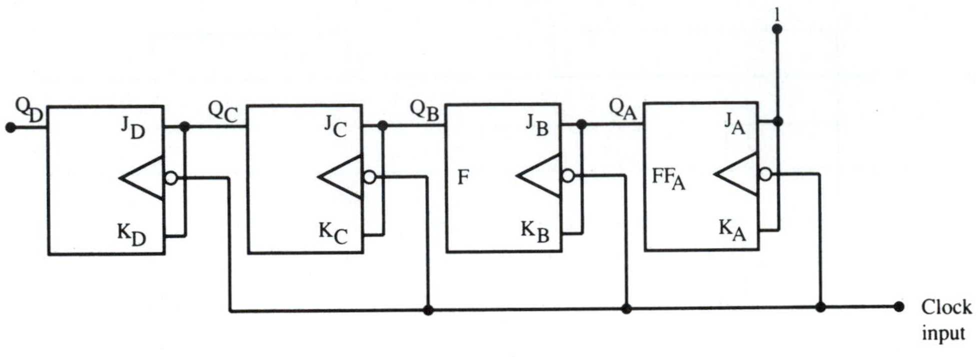

After Clock	Q_D	Q_C	Q_B	Q_A
0	0	0	0	0
1	0	0	0	1
2	0	0	1	0
3	0	0	1	1
4	0	1	0	0
5	0	1	0	1
6	0	1	1	0
7	0	1	1	1
8	1	0	0	0
9	1	0	0	1
10	1	0	1	0
11	1	0	1	1
12	1	1	0	0
13	1	1	0	1
14	1	1	1	0
15	1	1	1	1
16	0	0	0	0

4.25

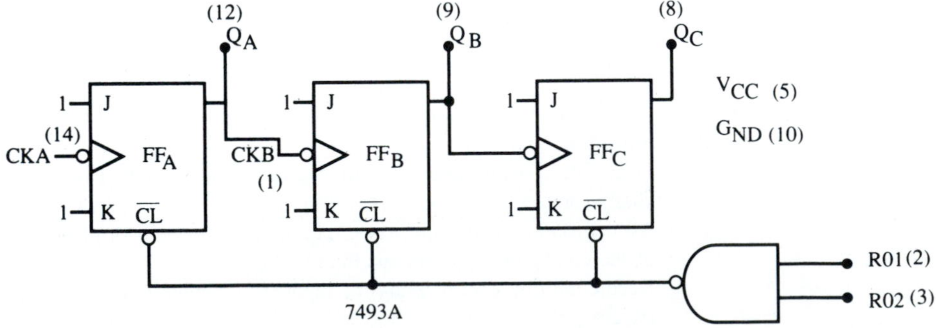

4.27 The 74122 one-shot has two positive-edge triggers instead of one as in the 74121 and a clear-line control not found in the 74121. The 74121 is a nonretriggerable, whereas the 74122 is retriggerable.

4.29

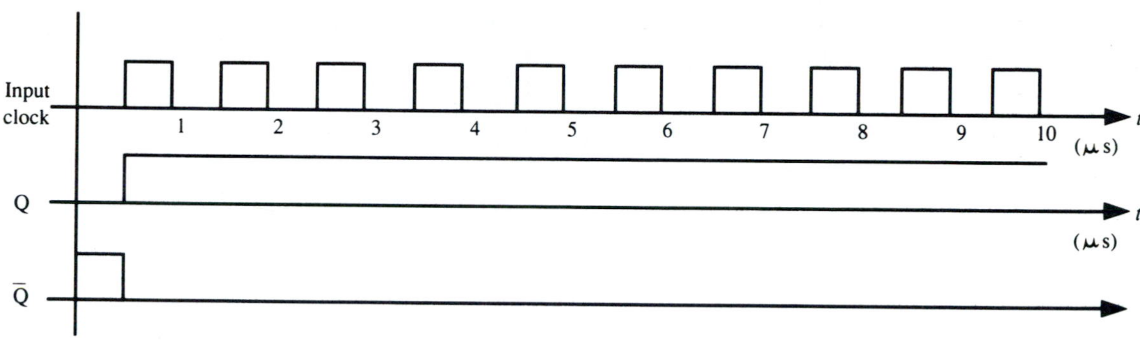

Chapter 5

5.1 See Figure 5.1.

5.3 a. See Figure 5.3, which must be modified.

b.

Data In	Clock Pulse	Q_0	Q_1	Q_2	Q_3	Q_4
0	0	0	0	0	0	0
1	1	1	0	0	0	0
0	2	0	1	0	0	0
1	3	1	0	1	0	0
0	4	0	1	0	1	0
1	5	1	0	1	0	1
0	6	0	1	0	1	0

Note: Any data input may be used.

5.5 a. See Figure 5.6, which must be modified.

b. See Figure 5.6, which must be modified.

c. Count table

Input Clock	Q_0	Q_1	Q_2
0	1	0	0
1	0	1	0
2	0	0	1
3	1	0	0

5.7 % duty cycle = $1/N \times 100\% = 1/3 \times 100\% = 33.33\%$

5.9 $f_0 = f_{in}/2n = 5\text{ MHz}/2(3) = 0.833\text{ MHz}$

5.11 a. See Figure 5.9.

b. Truth table

IN	EN	OUT
0	0	1
0	1	Hi = z
1	0	0
1	1	Hi = z

5.13 See Figure 5.11 with inverted output.

5.15

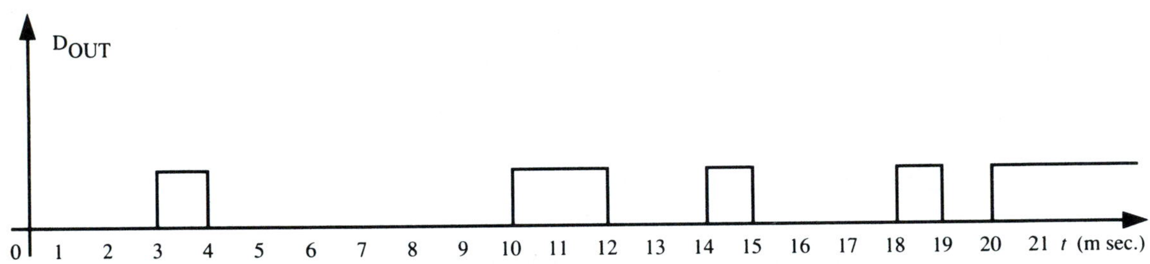

5.17 1, 1, 1, 2, 2, 4, 0, 8, 8, 0, 1, 9, 8, 2, 7, 5, 1, 2, 2, 6, 6, 6

5.19 a. Resolution = $1/2'' \times 100\% = 1/2^{10} \times 100\% = 0.0977\%$

b. Resolution = $1/2^{12} \times 100\% = 0.0244\%$

5.21

Start Bit	Data Bits	Party Bit	Stop Bits
0	00000011	1	11

Chapter 6

6.1 See Figure 6.1, two packages, 14 pin and 20 pin.

6.3 a. $V_{CC(max)} = 7\text{ V}_{DC}$ b. $V_{in(max)} = 7\text{ V}_{DC}$ c. $t_{storage(max)} = 150°\text{ C}$

6.5 a. $V_{CC(min)} = 4.75\text{ V}_{DC}$ b. $V_{IH(min)} = 2\text{ V}_{DC}$ c. $V_{OH(min)} = 2.7\text{ V}_{DC}$

6.7 a. 54LS00: $V_{NMH} = 0.5$ V; $V_{NML} = 0.3$ V

b. 74LS76: $V_{NMH} = 0.7$ V; $N_{NML} = 0.3$ V

6.9 a. 74LS00: $I_{OH} = -0.4$ ma (max)

b. 74LS76: $I_{OH} = -0.4$ ma (max)

6.11 All voltages are approximate.

	B_1	E_1	C_1	V_{out_1}	V_{in}	B_2	E_2	C_2	V_{out_2}
Logic levels	0	1	0	1	1	1	0	0	0
Voltages	0	5	0	5	5	5	0	0	0

6.13 See Figure 6.6(b), which must be modified with a third input diode.
b.

Voltage Table	Truth Table
$ABCV_{out}$	$ABCV_{out}$
0000	0000
0055	0011
0505	0101
0555	0111
5005	1001
5055	1011
5505	1101
5555	1111

6.15 The output waveform at V_{out_3} will be a straight line of approximately +5 V_{DC}.

6.17 a. See Figure 6.7(b), which must be modified with a third input diode.
b.

Voltage Table	Truth Table
$ABCV_{out}$	$ABCV_{out}$
0005	0001
0050	0010
0500	0100
0550	0110
5000	1000
5050	1010
5500	1100
5550	1110

6.19 Open-collector devices are useful when high current is necessary. The outputs of open-collector devices can be connected together to form wired-AND operations.

6.21 Increasing the supply voltage to CMOS devices will
a. Decrease propagation delay
b. Increase power dissipation

6.23 All voltages are approximate.

	V_{in}	A	B	V_{out}
Logic levels	0	1	0	1
Voltages	0	5	0	5

6.25 See Figure 6.14. The primary concerns when interfacing CMOS to TTL are the output parameters of the CMOS device and the input parameters of the TTL device. The I_{OL} of the CMOS device cannot sink enough current for the input of the TTL device when it is in a logic 0 state.

6.27 Unused inputs to CMOS devices should be connected to either V_{DD} or ground.

6.29 All CMOS devices have input circuit protection built into the device using input diodes. See Figure 6.15. TTL devices also use input diodes for protection. See Figure 6.16.

6.31 Decoupling capacitors are used to eliminate unwanted transients or spikes on power supply lines. They are installed between V_{CC} and ground as close as possible to each IC and on PC board edge connectors.

Chapter 7

7.1
a. A flip-flop that represents a single storage bit is called a memory cell.
b. A memory address is a unique location in memory. The address tells the computer where to find the memory cell.
c. The process by which a computer retrieves data from memory is called a memory-read cycle.
d. A memory-write cycle is a process in which a computer stores or writes data into memory.

7.3 Nonvolatile memory is memory that does not lose its data when the power is turned off, such as ROM memory. Volatile memory is memory that loses data whenever the power is switched off, such as RAM memory.

7.5 See equation (7.1). It would have 16 memory locations, each of which could store one word of data.

7.7 ROMs are used in applications where data is of a permanent nature, such as code converters, character generators, and data tables.

7.9 Redraw Figure 7.4(a) so that the first two bipolar cells have their fuses intact (store logic 1) and the last two bipolar cells have their fuses open (store logic 0). Thus, the data stored will be 1100.

7.11 Static RAM (SRAM) will hold its data as long as the power is applied to the device. It is essentially an array of flip-flops in which each flip-flop represents one bit or memory cell.
Dynamic RAM (DRAM) is a type of RAM in which the data must be periodically recharged or refreshed. Refreshing is accomplished by performing a repetitive read or write operation or a special refresh operation.
The added cost and complexity of DRAMs as compared to SRAMs are more than offset by the higher cell density, which lowers the per-bit cost for DRAM devices.

7.13
a. See Figure 7.7 and modify. All data inputs (pins 9 to 11 and 13 to 17) are wired to a logic 1 ($+5\ V_{DC}$). All address inputs (pins 1 to 8 and 19, 22, 23) are wired to logic 0 ($0\ V_{DC}$), and $\overline{W} = 0$ V, $\overline{S} = 0$ V, $\overline{G} = 0$ V (control lines).
b. All data pins are now used as outputs (not wired to +5 V), and $\overline{W} = +5$ V, $\overline{S} = 0$ V, $\overline{G} = 0$ V.

7.15 To save address lines and reduce the number of pins required on the IC, this device uses address multiplexing. During read or write operations the seven address lines first contain the row information and then the column information. This is controlled by the $\overline{RAS}$ and $\overline{CAS}$ control lines.

7.17 Modify Figure 7.11 by using only four devices (4116). For data D_0 to D_3 use the first four DRAMs.

7.19 U_4 must be activated to select the upper bank $(1800\text{–}1FFF)_{16}$. Therefore, $A_{11} = 1$, $A_{12} = 1$, and A_0 to A_{10} are all logic 1s. $\overline{G_4}$ must be a logic 0 in order to read U_4.

7.21 A PLD can be used to replace a large number of logic devices. The PLD is programmed to perform the same logic function as the individual logic gates.

7.23 Product line 1 would have $\overline{X}$ and Z connected.
Product line 2 would have $\overline{Y}$ and $\overline{Z}$ connected.
Product line 3 would have X, Y, and Z connected, and one of the sum lines (O_1) would be connected to the outputs of AND gates 1, 2, and 3.

7.25 Modify Figure 7.19 by changing the product lines as follows:
Product line 1: All fuses are blown, producing a logic 1.
Product line 2: $\bar{I}_1$, I_2, and $\bar{I}_3$ are left connected by the fuses.
Product line 3: All the fuses are left connected, producing a logic 0.

7.27 A 2-to-1 multiplexer uses only two input lines, I_0 and I_1, and one select line, S_0.
Out $= I_0 = (\bar{S}_0 \cdot I_0)$
Out $= I_1 = (S_0 \cdot I_1)$
Modify Figure 7.23.

7.29 In word parity or parity word, a parity bit is added to each data word (row) and also to each vertical data position (column). The parity word is calculated by determining the odd or even parity for each vertical position (column). If a single bit error occurs, it can be detected and corrected.

Chapter 8

8.1 General-purpose computing, automobile applications, calculators, digital clocks, appliances, robotics

8.3 AX (accumulator), BX (base), CX (count), DX (data)

8.5

Data Bus	Instruction Queue
8088 | 8-bit	4-byte (2-word)
8086 | 16-bit	6-byte (3-word)

8.7 Execution unit: arithmetic logic unit, flag register, general register, and pointer and index registers.
Bus interface unit: address generator, address-control unit, bus-control unit, instruction queue, instruction pointer, and segment registers.

8.9 Address and data lines 0 through 7; address and data line 15; address line 16 and status line 3; address line 17 and status line 4; address line 18 and status line 5; address line 19 and status line 6;

8.11

Address Lines	Memory Size
8	256
16	65,536
20	1,048,576

8.13 a. $\overline{\text{WR}}$ = 0, DT/$\overline{\text{R}}$ = 1 b. DT/$\overline{\text{R}}$ c. 10/$\overline{\text{M}}$ d. ALE = 1 e. $\overline{\text{INTA}}$
f. $\overline{\text{RD}}$ = O g. MN/$\overline{\text{MX}}$ = 0

8.15 The flag register is used to store information about the conditions of the operation of the ALU.

8.17 Data registers: AX (accumulator), BX (base), CX (count), and DX (data). These data registers are used for arithmetic and logic operations. They can temporarily hold the accumulated results.
Pointer and index registers: SP (stack), BP (base), SI (source), and DI (destination). These registers are used to point to or index to a specific location in memory.

8.19 CS is a code segment register.

8.21 Static RAM is memory that keeps information until power is removed. Dynamic RAM is memory that needs constant refreshing to keep information.

8.23 Modify Figure 8.14.

8.25 See Figure 8.18.

8.27 The 8088/8086 microprocessor is faster than older MPUs because it is divided into two independent fetch and execute units (BIU and EU).

8.29 *F*1200

8.31 221*F*

Chapter 9

9.1 Fortran, Pascal, BASIC, COBOL

9.3 a. Determine what the program should do.
b. Develop a flow chart.
c. Translate the flow chart into program form.
d. Enter the program and edit, assemble, and link it.
e. Run the program and verify the results.

9.5 The contents in the DX register will be moved to the CS register, so both registers will have the same contents.

9.7 BX contains the address where the content is to be moved into the AX.

9.9 MOV destination, source

9.11 LEA, PUSH, POP, MOV

9.13 ADD, SUB, MUL, DIV

9.15 AX = 0000H

9.17 OR destination, source

9.19 The loop instruction is a special form of the jump instruction with a built-in count capability.

9.21

```
MOV    AX, 0004H    ;load 4H into AX register
MOV    BX, 0003H    ;load 3H into BX register
ADD    AX, BX       ;add BX to AX, result in AX = 7H
MOV    SUM, AX      ;move result into memory address
                    ;labeled SUM
```

Chapter 10

10.1 To convert assembly language programs into machine language programs.

10.3 ASCII (ASM).

10.5 To make a file relocatable in memory.

10.7 COM.

10.9 Underscore (_).

10.11 RCX.

10.13 Ctrl-Break or Ctrl-Alt-Del

Chapter 11

11.1 a. MPU b. Input/Output c. Bus System d. Memory

11.3 Input/Output.

11.5

8088	8086
8-bit data bus	16-bit data bus
4-byte instruction queue	6-byte instruction queue
I-O/$\overline{\text{M}}$	M/$\overline{\text{I-O}}$
$\overline{\text{SSO}}$	$\overline{\text{BHE}}$/S7

11.7 The QS_0 and QS_1 provide status information about the instruction queue.

11.9 The 8284A chip is a clock generator providing the basic timing for the microprocessor.

11.11 This is used to control the bus-control signals when the system is used with the MPU in maximum mode.

11.13 DT/$\overline{\text{R}}$, $\overline{\text{DEN}}$, MCE/$\overline{\text{PDEN}}$, ALE

11.15 The 74LS245 transceiver is used to buffer and control the direction of data flow on the data bus.

11.17 Refer to Figure 11.13.

11.19 See Figure 11.16.

11.21 Idle and active cycles.

11.23 a. Load command register
b. Load internal registers
c. Enable DMA channel request

11.25 The I/O subsystem is responsible for the movement of data between the basic microprocessor system and the peripheral or external devices connected to it.

11.27 AX, AL.

11.29 Read/Write ($\overline{\text{RD}}$, $\overline{\text{WR}}$).

11.31 Instruction pointer and the affected registers.

11.33 PIC (programmable interrupt controller).

11.35 Data bus buffer, read/write logic, cascade buffer, control logic, inservice register (ISR), priority resolver (PR), interrupt request register (IRR), interrupt mask register (IMR).

11.37 See Figure 11.29.

11.39 Mode 1: programmable one-shot
Mode 2: rate generator
Mode 3: square-wave rate generator
Mode 4: software-triggered strobe
Mode 5: hardware-triggered strobe

11.41 By using the DEBUG input/output instructions for controlling devices.

11.43 The 8284A provides the basic timing requirements for the microprocessor.

11.45 *FFFF*0H

Chapter 12

12.1 Linear-pass and switching regulator types.

12.3 See Figure 12.4(a).

12.5 Basic Input/Output System—A ROM that contains a machine level program that controls system startup, the POST test, and interfaces with the rest of the system.

12.7 DMA is a fast data-transfer method that does not require the CPU.

12.9 80486—32 bits
80386—32 bits
80286—16 bits
8088—8 bits

12.11 Real, protected, virtual.
12.13 Yes.
12.15 IRQ 14.
12.17 CMOS RAM.
12.19 ISA, EISA, VL, PCI, PCMCIA.
12.21 BRDY.
12.23 The chipset controls the several different timing functions that are necessary for operation, and coordinates bus transfers.
12.25 The AGP connector provides a high-video-speed connection used exclusively for video cards.
12.27 Put on a wrist strap and discharge any static electricity in his/her body.
12.29 DMA 1.

Chapter 13

13.1 Paper tape.
13.3 Retentivity.
13.5 Iron (ferric) oxide.
13.7 Electromagnetic.
13.9 No.
13.11 Droop.
13.13 Frequency shift keying.
13.15 Twice the density of FM encoding.
13.17 False.
13.19 The best indicators of performance are area density and RPM.
13.21 Four devices (two for each controller).
13.23 Two partitions (primary and extended).
13.25 /S.
13.27 The MPC2 audio connector and cable.
13.29 8.5 gigabytes.

Chapter 14

14.1 A simple computer input device consisting of switches.
14.3 Matrix (ed).
14.5 Serial.
14.7 Cathode-ray tube.
14.9 Voltage changes between the cathode and grid 1.
14.11 Blanked.
14.13 262.5.
14.15 Filament, cathode, grids, anode.
14.17 False.
14.19 Phosphor.
14.21 True.
14.23 False.
14.25 Parabolic.
14.27 Ramp (sawtooth).
14.29 During retrace.
14.31 Phase-locked loop.

14.33 Video information.
14.35 Due to the difference in the gain of the electron guns.
14.37 See Figure 14.29(a).
14.39 Phase modulation.
14.41 Screen RAM.
14.43 Character generator (ROM).
14.45 *B*8000H.
14.47 Inked ribbon.
14.49 A matrix printer that uses the heat-transfer method.
14.51 By firing electrically charged ink droplets onto the paper.
14.53 Laser beam.
14.55 Status checking between two devices that communicate data.
14.57 Plus and minus 12 V.

Chapter 15

15.1 Broadband signaling is similar to radio and television transmission in that information is modulated onto a common carrier wave. Baseband signaling does *not* use modulation. Data levels are applied directly to the wire in the form of changing DC voltages.
15.3 Parallel.
15.5 Hub.
15.7 MAU.
15.9 Error correction.
15.11 Jam signal.
15.13 Manchester encoding.
15.15 10BASE2, 10BASE5, and 10BASET.
15.17 Cable TV wiring (Coax).
15.19 Category 5.
15.21 Dotted decimal notation.
15.23 Hosts.
15.25 Host file.
15.27 Windows Internet naming service.
15.29 Signal regeneration.
15.31 Data link layer.
15.33 Network.
15.35 Filtering, isolation.
15.37 The routing table contains all network addresses, connection options between networks, cost of data transmission (in terms of the number of "hops" required for a transmission), and host addresses (depending on the network protocol running). Using the information in its routing table, the router chooses the most efficient route, determined by the available paths and cost of transmission.
15.39 Routers provide better traffic management by not passing broadcast traffic (but can be set to pass multicast traffic), avoiding broadcast storms, not routing bad packets, providing more sophisticated filtering and isolation, only passing packets with a known network address, and listening to the network in order to determine the path that is least busy.
15.41 Gateways allow networks that are completely dissimilar to communicate with each other. They provide communication between networks with different architectures, protocols, data formatting, and languages. They provide translation of data, signal levels, and network protocols.
15.43 NDIS and ODI.

Chapter 16

16.1 Windows 95, because it will overwrite the NT boot loader.

16.3 To create a boot disk on a DOS machine, insert a blank diskette and type FORMAT A: /S at the DOS prompt. To create a boot disk on a Windows 95 machine, insert a blank formatted diskette, navigate to the floppy disk drive icon, right click on the floppy and select Format. Check the Copy System Files Only button and click Start.

16.5 Boot sequence, drive letter, bypass memory test in the POST.

16.7 Yes, when using only SCSI devices or when you need IRQ lines for another device.

16.9 System initializes motherboard, quick checks RAM, and initializes video display.

16.11 IRQ 5.

16.13 IRQ 7.

16.15 Slot 1.

16.17 34 pins.

16.19 Motherboard or controller.

16.21 40 pins.

16.23 Two.

16.25 I/O wait states and bus speed.

16.27 I/O address, DMA, and IRQ.

16.29 MSCDEX.

16.31 NTFS.

16.33 The format process will read every sector, initialize the file allocation table and root directory, and clear all sector pointers.

16.35 Hardware compatibility list, or HCL.

Chapter 17

17.1 Power-on self-test. A BIOS program that is a built-in system diagnostic.

17.3 Power supply.

17.5 The system may be unable to display the error message because of a video problem or because the error occurred before the video card was initialized.

17.7 Static or ESD.

17.9 With a soft pencil eraser or a solvent designed for the purpose.

17.11 Absolutely *not,* because you will remove the gold plating from the contacts.

17.13 Populate only one bank of memory at a time.

17.15 Partitioning a hard drive.

17.17 The video card.

17.19 The ISP.

17.21 Referring to the network diagram, use the PING command to attempt communication with the server and other workstations to narrow down the problem.

Chapter 18

18.1 Logic probe.

18.3 Logic pulser.

18.5 True.

18.7 The digital storage oscilloscope converts the incoming analog waveform into digital data and stores the data in RAM. The data can then be displayed on the screen as a waveform.

18.9 True.

GLOSSARY

access time The time delay between the instant data is requested from the memory device and the instant data is received.

accumulator An MPU register used for data operations.

accuracy The percentage of error between the expected and actual measurement.

active high The description of a signal that indicates that it is true (active) when the level of that signal is at a high voltage level (usually +5 V).

active low The description of a signal that indicates that it is true (active) when the level of that signal is at a low voltage level (usually 0 volts).

adapter A device used to expand the operation of a system (e.g., memory boards, modems, display drivers, etc.).

ADC Analog-to-digital converter.

address A value defining a memory location.

address bus The multiple lines that carry the binary coded address from the MPU throughout the system.

alphanumeric The character set that contains both letters (alpha) and numbers (numeric).

ALU Arithmetic logic unit.

analog Signals that are constantly variable over a range of levels and physical quantities (as opposed to digital).

AND gate A logic circuit that produces a 1 output only when all inputs are 1.

ASCII (American Standard Code for Information Interchange) The standard 7-bit code used for exchanging information between data processing systems, data communication systems, and any associated equipment. The ASCII character set consists of control characters, graphic characters, and alphanumeric characters.

assembler A program that converts a symbolic code (ADD) to binary machine code (1s and 0s).

astable multivibrator A circuit with no stable state. An oscillator or clock.

asynchronous Not clocked or occurring at the same time.

BASIC A high-level computer programming language.

baud A unit of measure used for serial data communication. The unit baud represents 1 bit of data per second (e.g., 1200 baud is the same as 1200 bits per second.).

BCD (Binary-coded Decimal) A 4-bit coding of decimal numbers.

Binary A numbering sequence with 2 as the base. Only 2 digits are valid, 0 or 1. This is used in digital computers where the values are similar to the *off* or *on* level of the electrical switches.

BIOS (Basic Input/Output System) The machine-level program that interfaces the computer hardware with higher-level programs (e.g., DOS, BASIC, etc.).

Bit (Binary digit) A single binary character (either a 1 or a 0).

Boolean algebra The mathematics of digital logic.

buffer A device used for temporary storage of information. This can be either a section of system memory or an integrated circuit used in combination with a data or address bus. Also used to supply additional current drive.

bus A group of signals with similar purposes such as address or data lines.

byte A binary word made up of a group of 8 bits.

CAS (Column Address Strobe) A signal used to enable a column address in a memory device.

CGA (Color Graphics Adapter) The format used for the display of low resolution color graphics. The resolution is 640 by 200 pixels with four colors or 640 by 350 pixels with four colors.

character generator A unit (usually a programmable memory device) which converts character codes into displayable characters.

check sum An error detection scheme used in ROMs.

clear An asynchronous input used to reset or clear a flip-flop. ($Q = 0, \overline{Q} = 1$)

clock An input to a digital circuit used for timing. Also an astable multivibrator or oscillator.

CMOS (Complementary Metal-Oxide Semiconductor) A type of material used in the manufacture of semiconductor devices. Devices made with a CMOS usually have lower power requirements than similar devices manufactured using standard semiconductor materials.

combinational logic A digital logic circuit containing only gates (no flip-flops).

complement The opposite, or inverse.

computer A machine designed to solve problems using a systematic program of instructions.

configuration The arrangement of a computer system that is defined by the nature of the work to be performed and the type of components used.

continguous Touching or joining at the edge or boundary; adjacent as in memory systems.

CPU (Central Processing Unit) The main processor of a computer system. Also known as the MPU in a microcomputer.

CRC (Cyclic Redundancy Check) A method used to test the accuracy of data stored using a combination of shift registers and EXCLUSIVE-OR gates.

CRT (Cathode-Ray Tube) A vacuum tube similar to a television tube used as a display device for computers.

current sinking The drawing of current by a device from a load.

current sourcing The driving current from a device to a load.

cylinder The tracks of a disk (floppy or hard) that are used for storing data.

DAC (Digital-to-Analog Converter) A device or circuit that converts digital signals to analog signals.

data Information used by a system.

debounce The method used to eliminate the make/break bouncing of keyboard switches.

decade counter A counter with ten states.

decoupling capacitor A capacitor used to provide a low-impedance path to ground to prevent interference between circuits.

DeMorgan's Theorem A Boolean algebra principle used to define the complementing of a function.
demultiplexer A switch connecting one input to one of many outputs (selector).
DIP (Dual In-Line Package) A container used on printed circuit boards for integrated circuits, resistor packages, and switches. DIPs have connection pins in two parallel rows which are generally spaced $\frac{1}{10}$ in. apart.
DIP switch A small set of switches which are mounted in a dual-in-line package.
disable To stop the operation of a circuit or device.
diskette A recording medium used for storage of information on a computer system.
diskette drive A device that magnetically stores information from the computer system onto a diskette.
display A device used with a computer to present temporary information to the operator visually.
DMA (Direct Memory Access) A method used to transfer information stored in memory to I/O devices that does not require the processor for control.
DOS (Disk Operating System) A computer program to control the operation of the disk subsystem.
dot matrix A pattern of dots used to create alphanumeric or graphic characters on a printer.
dynamic memory Random-access memory (RAM) that uses transistors and capacitors as storage elements. This type of memory requires a recharging cycle (refresh) every few milliseconds.

EEPROM (Electrically Erasable Programmable Read-Only Memory) A programmable memory device that can be erased or reprogrammed.
EMI (Electromagnetic Interference) Electrical noise.
enable To activate the operation of a circuit or device.
EOF (End of File) A label used by programs to indicate the file being used has ended.
EPROM (Erasable Programmable Read-Only Memory) A programmable memory device that can be erased and reprogrammed.
EXCLUSIVE-NOR A logic gate whose output is 1 when the inputs are the same (comparator).
EXCLUSIVE-OR A logic gate whose output is 1 when the inputs are different.

falling edge Synonym for negative-going edge of a logic signal.
fan-out The number of inputs that can be driven from a gate output.
FIFO (First In, First Out) A method used for the sequential storage and retrieval of data. As the name implies, the first information stored is the first retrieved.
fixed disk drive A unit consisting of nonremovable magnetic disks, which is used for storing and retrieving data from the disks.
flag A bit in a register that is used to indicate the occurrence of a specific condition.
flip-flop A bistable multivibrator that can store one binary digit.
floppy diskette A synonym for a diskette.
FM (Frequency Modulation) A modulation technique used to encode data.
font A family or assortment of characters that share a common style.
formatting The process of writing identification information on each track and sector of a diskette.
FSK (Frequency Shift Keying) A modulation technique which encodes digital information into two different frequencies.

gate A digital logic circuit that performs one of the basic Boolean functions.
glitch An electrical spike or pulse of a very short duration.

hard disk See *fixed disk drive.*
hardware Physical equipment used in data processing, as opposed to programs.

head The portion of the floppy or hard disk drive that reads and writes the data onto the magnetic disk.

Hertz (Hz) A unit measuring the number of cycles per second; pertaining to frequency.

hexadecimal The base sixteen number system used on personal computers pertaining to a 16-bit base, which uses the digits 0 through 9 and *A* through *F,* as follows: 0, 1, 2, 3, 4, 5, 6, 7, 8, 9, *A, B, C, D, E, F.*

high-impedance state A state when the output of a device is isolated from the circuit (e.g., tri-state buffer).

IC (Integrated Circuit) A device that incorporates many electrical components on a single semiconductor material.

index The point on a diskette at which a track begins.

inhibit To prevent the operation of a circuit or device.

initialize To set system or device variables to a known state from which to begin operation.

instruction set The set of instructions in a computer or of a programming language.

interface A circuit used to connect devices together in a compatible way.

interrupt A command that pauses the processing of the current set of instructions and calls attention to a condition which requires processor intervention.

inverter A logic circuit whose output is the opposite of its input.

I/O (Input/Output) Referring to a device or devices that perform the function of receiving or transmitting information to or from a processor or controller.

***JK* flip-flop** A type of flip-flop with *J* and *K* inputs, which can be set, reset, or toggled.

Johnson counter A type of counter in which the inverted output of the last flip-flop is fed back to the input of the first flip-flop.

k (kilo) Prefix with a value of 1000.

K Used when referring to memory or stage capacity and equal to 1024 (1024 is 2^{10}).

Karnaugh map A graphical method of representing a truth table used to minimize Boolean expressions.

LED (Light-Emitting Diode) A semiconductor that gives off visible or infrared light when activated.

LSI (Large-Scale Integration) An IC that contains thousands of electronic gates.

machine code The machine language used for entering program instructions onto the recording medium or into storage.

machine language A language used directly by a processor or computer.

main storage Addressable storage from which instructions and other data can be loaded directly into registers in the MPU for processing.

mask A pattern of characters used to control if a specific character or characters are to be saved or eliminated (ignored).

MCGA (Multicolor Graphics Array) A display format used by the models 25 and 30 of the new IBM PS/2 line. These displays output an analog signal to the monitors, which can display up to 16 colors at a resolution of 640 pixels by 200 pixels. The MCGA is also a subset of the video graphics array (VGA).

megahertz (MHz) Abbreviation for 1 million hertz.

MFM (Modified Frequency Modulation) A modulation technique used to encode data.

microcode A code that represents a machine instruction when broken down into logic ones and zeros for use by the processor hardware.

microprocessor A large-scale integrated circuit that can carry out programmed instructions. The instructions can be programmed externally or stored internally.
mnemonic A symbol used to assist the programmer in remembering instructions, such as "LD" for the command "load."
mode A method of operation; for example, binary mode, alphanumeric mode.
modulus The number of unique states in a sequential circuit or counter.
monitor A synonym for cathode-ray tube display.
monostable multivibrator A device whose output will become active for a predetermined amount of time after it is triggered. Also known as a one-shot.
MSI (Medium-scale Integration) An IC containing hundreds of electronic gates.
multiplexer A device capable of interleaving the events of two or more activities without affecting either of them (e.g., address and data).

NAND gate A logic gate whose output is a 1 whenever any input is a 0.
nibble Half of a byte; usually 4 bits.
noise immunity The difference between input and output logic voltage levels. The ability to resist electrical noise.
noise margin See *noise immunity.*
NOR gate A logic circuit that produces a 1 output only when all inputs are 0.

op-code A binary, octal, or hex number that indicates the operation a computer system is to perform.
one The high logic level.
operating system Software that controls the execution of programs.
OR gate A logic circuit that produces a 1 output whenever any input is a 1.

PAL (Programmable Array Logic) A semicustom logic device that performs Boolean logic functions.
parity Error detection technique that counts the number of 1s in a data word.
Pixel (Picture Element) The smallest displayable unit on a monitor. Controlled by the resolution of the display adapter.
polling A method of examining the electrical state of a device to determine the operational status.
port An access point for data or information entry or exit.
POS (Product of the Sums) A form of a Boolean expression that AND's together all ORed terms.
positive-going The edge of a signal that is changing in a positive direction. Synonymous with rising edge.
priority A rank assigned to a task that determines the order in which the tasks will be carried out.
processor Synonymous with microprocessor.
program A sequential set of instructions stored in a computer system's memory.
PROM (Programmable Read-Only Memory) A read-only memory that can be erased and changed by the user.
propagation delay The time necessary for a signal to travel from one point on a circuit to another; also the time delay between a signal change at an input and the corresponding change at an output of a device.
protocol The set of rules governing the operation of functional units of a communication system.

queue A line of instructions or items waiting to be serviced by a computer system.

RAM (Random-Access Memory) Memory that can be written to and read from many times.
race A logic condition where signals arrive at the same time.

radix The base of a number system.

RAS (Row Address Strobe) In memories using dynamic RAMs, the row address strobe is used to refresh the stored data.

refresh The process of rewriting data periodically into a dynamic RAM.

register A digital circuit capable of storing binary data.

ring counter A type of shift register in which the output of the last flip-flop is fed back or recirculated to the first flip-flop.

ripple counter A counter in which each flip-flop is toggled from the output of the previous flip-flop. See also *asynchronous.*

rise time The time required to go from 10% to 90% of the value of the positive-going edge of a pulse.

ROM (Read-Only Memory) A storage device whose contents cannot be modified. The memory is retained when the power is removed.

RS-232C A standard by the EIA (Electronic Industry Association) for communication between computers and external equipment over a serial data path.

Schmitt trigger A wave-shaping circuit that operates on switching threshhold voltages.

sector A part of a track or band on a magnetic storage disk that can be accessed by the read/write head.

sequential logic A logic circuit whose output depends on the present state of its input and previous memory condition. Logic circuits which involve flip flops.

shift register A register used to move data either left or right.

SOP (Sum of the Products) A form of a Boolean expression that ORs together all ANDed terms.

SSI (Small-Scale Integration) ICs containing less than a hundred electronic gates.

static memory RAM using flip-flops as storage elements. Data are retained as long as power is applied to the storage device without recharging (refreshing).

strobe A pulse used for timing and synchronization.

synchronous Occuring at the same time.

three-state logic A type of logic circuit with three unique logic states: high, low and hi-Z. Also known as tristate logic.

track A band of magnetic data created on the surface of a diskette.

trigger A pulse or signal used to initiate an action.

truth table A table that defines all possible output and input conditions for a digital circuit.

TTL (Transistor–Transistor Logic) The 7400 logic family of digital devices.

UART (Universal Asynchronous Receiver/Transmitter) A device used in digital communications systems.

unit load The loading effect of one gate's input to the output of another gate.

VDT (Video Display Terminal) The CRT terminal used to display computer data.

VGA (Video Graphics Adapter) A display format used by models 50, 60, and 80 of the IBM PS/2 computers. These adapters are capable of displaying 256 colors at a resolution of 640 by 480 pixels. The output is in an analog format.

VLSI (Very Large-Scale Integration) An IC containing tens of thousands of electronic gates.

volatile memory A type of memory that loses its data when power is removed.

Winchester drive A type of disk drive in which the read/write heads do not contact the disk surface. Also known as a hard disk drive.

XNOR See *EXCLUSIVE-NOR.*
XOR See *EXCLUSIVE-OR.*

yolk A set of coils on the neck of a CRT used to deflect the electron beam horizontally and vertically.

zero The low logic level.

INDEX

Organic Chemistry
A Short Course

EIGHTH EDITION

Organic Chemistry
A Short Course

EIGHTH EDITION

Harold Hart
Michigan State University

HOUGHTON MIFFLIN COMPANY Boston
Dallas Geneva, Illinois Palo Alto Princeton, New Jersey

Cover

"Midnight Variations," artist Kenneth Snelson's imagined atomic world. The image was created through use of a three-dimensional graphics computer.

Interior design

George McLean

Credits

Page 64, Wide World Photos; page 96, Union Carbide Agricultural Products, Inc.; page 101, Wide World Photos; 129, American Cancer Society; 149, Courtesy of the Polaroid Corporation; page 190, Courtesy of Alpha Therapeutic Corporation; page 192, Wide World Photos; page 229, Geisinger Medical Center, Joseph J. Mentrikoski and Raymond C. Carballada (photographers), Department of Medical Photography; page 232, U.S. Department of Agriculture; page 234, "Scotch" 5-minute Epoxy Adhesive by 3M; page 259, Corning Glass Works; page 296, Tennessee Valley Authority; page 345, Reprinted with permission from University Science Books; page 348, Kneeland/Phototake; page 398, Gary S. Weber/Photo Researchers, Inc.; page 458, Wide World Photos; page 476, Pfizer Inc.; page 484, Courtesy of Linus Pauling.

Printed in the U.S.A.

Library of Congress Catalog Number: 90-83279

ISBN: 0-395-43336-3

EFGHIJ–AH–9876543

The Main Functional Groups *(continued)*

Structure	*Class of Compound*	*Specific Example*	*Name*	*Use*
C. Containing nitrogen				
$-NH_2$	primary amine	$CH_3CH_2NH_2$	ethylamine	intermediate for dyes, medicinals
$-NHR$	secondary amine	$(CH_3CH_2)_2NH$	diethylamine	pharmaceuticals
$-NR_2$	tertiary amine	$(CH_3)_3N$	trimethylamine	insect attractant
$-C{\equiv}N$	nitrile (cyanide)	$CH_2{=}CHCN$	acrylonitrile	orlon manufacture
D. Containing oxygen and nitrogen				
$-\overset{+}{N}(=O)-O^-$	nitro compounds	CH_3NO_2	nitromethane	rocket fuel
$-C(=O)-NH_2$	primary amide	$NH_2C(=O)NH_2$	urea	fertilizer
E. Containing halogen				
$-X$	alkyl or aryl halide	CH_3Cl	methyl chloride	refrigerant, local anesthetic
$-C(=O)-X$	acid (acyl halide)	$CH_3C(=O)Cl$	acetyl chloride	acetylating agent
F. Containing sulfur				
$-SH$	thiol	CH_3CH_2SH	ethanethiol	odorant to detect gas leaks
$-S-$	thioether	$(CH_2{=}CHCH_2)_2S$	allyl sulfide	odor of garlic
$-S(=O)_2-OH$	sulfonic acid	$CH_3-C_6H_4-SO_3H$	*para*-toluenesulfonic acid	strong organic acid

Contents

2 *Alkanes and Cycloalkanes; Conformational and Geometric Isomerism* **41**

3 *Alkenes and Alkynes* **69**

7 *Alcohols, Phenols, and Thiols* **197**

16 *Carbohydrates* 425

17 *Amino Acids, Peptides, and Proteins* 455

Preface

Purpose

Several decades have passed since the first edition of this text was published. Although the content continues to change, my purpose in writing it remains much the same—to present as clearly as possible a brief introduction to modern organic chemistry.

This book was written for students who, for the most part, will not major in chemistry, but whose main interest—agriculture, biology, human or veterinary medicine, pharmacy, nursing, medical technology, health sciences, engineering, home economics, forestry, or whatever—requires some knowledge of organic chemistry. To maintain the interest of these students, I have made a special effort to illustrate the practical applications of organic chemistry to everyday life and to biological processes. The success of this approach is demonstrated by the widespread use of this text by hundreds of thousands of students here in the U.S., and worldwide via numerous translations.

The text is designed for a one-semester introductory course, but it is easily adapted to other formats. It is often used in either a one- or two-quarter course. In some countries (France and Japan, for example) it is an introductory text for chemistry majors, followed by a longer and more detailed full-year text. And in a number of high schools, it is used as the text for a second-year course, following the usual introductory general chemistry.

Organization

The organization is fairly classical, with some exceptions. After an introductory chapter on bonding, isomerism, and an overview of the subject (Chapter 1), the next three chapters treat in sequence saturated, unsaturated, and aromatic hydrocarbons. The concept of reaction mechanism is presented early, and examples are included in virtually all subsequent chapters. Stereoisomerism is also introduced early, briefly in Chapters 2 and 3, and then given separate attention in a full chapter (Chapter 5). Halogen compounds are used in Chapter 6 as a vehicle for introducing aliphatic substitution and elimination mechanisms and dynamic stereochemistry.

Chapters 7 through 10 take up oxygen functionality in order of increasing oxidation state of carbon (alcohols and phenols, ethers, aldehydes and ketones, acids and their derivatives). Brief mention of sulfur analogs is made in these chapters. Chapter 11 deals with amines.

Chapters 2 through 11 treat all of the main functional groups and constitute the heart of the course. Chapter 12 then takes up spectroscopy, with an emphasis on NMR and applications to structure determination. It handles the student's question—how do you know that those molecules really have the structures you say they have?

Next come two chapters on topics not always treated in introductory texts but especially important in practical organic chemistry—Chapter 13 on heterocyclic compounds and Chapter 14 on polymers. The book ends with four chapters on biologically important substances—lipids, carbohydrates, amino acids and proteins, and nucleic acids.

"A Word About" Essays

Although relevant applications of organic chemistry are stressed throughout the text, short sections under the general rubric *A Word About* emphasize applications to other branches of science and to human life. These sections, which have been a popular feature, appear at appropriate places within the text rather than as isolated essays. Numbered and printed in special type, they stand out from the text so that instructors can easily require these sections or not, as desired. There are thirty-five of these essays, ten new in this edition.

Examples and Problems

Problem solving is essential to learning organic chemistry. **Examples** (worked-out problems) appear at appropriate places within each chapter to help students develop these skills. These examples and their solutions are clearly marked. Unsolved **problems** that provide immediate learning reinforcement are included within each chapter and are supplemented with an abundance of end-of-chapter problems. The combined number of examples and problems is 956, or an average of more than 53 per chapter.

New in the Eighth Edition

The entire text was carefully revised to sharpen the writing and clarify difficult sections. In addition to many small changes, users of the previous edition will notice the following more substantial changes: (1) Chapter 5 on stereochemistry has been rewritten to present the subject from a first principles point of view rather than from a historical perspective. However, a historical account is retained in *A Word About* #9 (Pasteur's Experiments and the van't Hoff-LeBel Explanation). Reviewers of the manuscript have responded favorably to these changes. (2) Large sections of Chapter 6, those dealing with nucleophilic substitution and elimination, have been rewritten to clarify this difficult subject. (3) Former Chapter 11 on Difunctional Acids; Fats and Detergents has been eliminated, and material from that chapter has been placed either in Chapter 10 on Acids or in the new Chapter 15 on Lipids and Detergents. This change allowed the treatment of amines to be moved forward to Chapter 11. (4) Material on spectroscopy, formerly part of the final chapter, has been moved forward to Chapter 12 to immediately follow the main functional groups. (5) Chapter 13 on Heterocyclic Compounds is new. It provides a brief systematic introduction to this important topic and has received high marks from reviewers of the manuscript. (6) Chapter 14 on Synthetic Polymers is also new, though some of this material was formerly in the final chapter. Again, reviewers responded favorably to this change.

The book now contains eighteen chapters instead of the former sixteen. However, this increase is largely a consequence of reorganization. Where new material was added, some old material was deleted. I am very conscious of the need to keep the book to a manageable size for the one-semester course, and users will find that this edition is nearly identical in length to the previous one.

Although ten new *A Word About* sections have been added, six former sections of this type have been deleted; in some cases these changes involve moving material from the main body of the text to *A Word About* section or the reverse, but others are entirely new. I hope that students and teachers alike will enjoy the following timely topics: Isomers, Possible and Impossible; Graphite, Carbon Clusters, and Aromaticity; CFCs, the Ozone Layer, and Tradeoffs; NMR in Biology and Medicine; DNA and Crime; and Nucleic Acids and Viruses. Please write to me with your comments.

Ancillaries

Two accompanying books are available to help the student in this course learn organic chemistry.

The **Study Guide and Solutions Book** contains answers to all text problems, a guide on how to reason out the answers, a summary of each chapter, a summary of the new reactions in each chapter, a list of learning objectives for each chapter, a summary of important reaction mechanisms, and sample test questions.

The **Laboratory Manual** contains experiments that have been time-tested with thousands of students. In the latest edition, a substantial number of the preparative experiments contain procedures on both the **macro-** and the **microscale**, thus adding considerable flexibility for the instructor and the opportunity for both types of laboratory experience for the student. We have been careful to avoid hazardous chemicals on the OSHA list and to minimize contact with solvents, and so forth. The student and instructor are clearly warned whenever caution or special care is required, and thorough waste disposal instructions are consistently specified. The manual has tear-out, perforated report sheets convenient for student and instructor. It is also a convenient size for the nonmajor lab. Most experiments can be completed in the relatively short two- or three-hour lab period for nonmajors. The manual contains appendices giving atomic weights, other properties of common reagents, instructions for the teacher on how to make or obtain special reagents, and a list of chemicals and equipment required for each experiment that will simplify ordering and stocking the labs. Experiments are a good mix of techniques, preparations, tests, and applications.

An **Instructor's Resource Manual** and a set of fifty-two transparencies are also available.

A Personal Note

It is with special pleasure and pride that I welcome my co-authors on the ancillaries to this edition of the text. They are my oldest daughter, Dr. Leslie Elizabeth Hart Craine, and my son, Dr. David Joel Hart.

Leslie teaches organic and general chemistry at Trinity College in Hartford. She developed, with her students, the microscale procedures that appear in this edition of the laboratory manual. David, who is on the faculty at The Ohio State University in Columbus, did essentially all the revision necessary for this edition of the study guide.

Leslie and David bring with them experience in undergraduate teaching at the small college and large university levels. Both are fine writers, and I welcome them in making this project a family affair.

Acknowledgements

For their frankness and diligence in reviewing the proposed revisions and later, the completed manuscript, I would like to thank the following professors:

P. I. Abell (University of Rhode Island); R. A. Abramovitch (Clemson University); G. A. Epling (University of Connecticut); J. Grutzner (Purdue University); R. L. Jacobs (Bowling Green State University); A. P. Krapcho (University of Vermont); G. A. Kraus (Iowa State University); E. M. Nicholson (Eastern Michigan University); B. O. Mejia (California State University, Chico); R. K. Murray (University of Delaware); J. A. Roth (Northern Michigan University); J. Saltiel (The Florida State University); R. P. Thummel (University of Houston); G. H. Wahl, Jr. (North Carolina State University); F. E. Wood (University of California, Davis).

I have incorporated many of their recommendations, and the book is much improved as a consequence.

I also thank my secretary Shari Glynn for typing the entire manuscript and entering it on our computer. And of course, thanks and much more to my wife Gerry for sticking with me through yet another revision.

One pleasure of authorship is receiving letters from students who have benefited from the book, and from their teachers. I thank here all who have written to me, from all parts of the world, since the last edition; I have incorporated many of their suggestions in this revision. I will be happy to hear from users and nonusers, faculty and students, with suggestions for further improvements.

HAROLD HART
EMERITUS PROFESSOR OF CHEMISTRY
MICHIGAN STATE UNIVERSITY
EAST LANSING, MI 48824

Organic Chemistry
A Short Course

EIGHTH EDITION

INTRODUCTION

To the Student

In this introduction I will tell you briefly what organic chemistry is about and why it is important in a technological society. I will also explain how this course is organized and give you a few hints that may help you to study more effectively.

What Is Organic Chemistry About?

The term *organic* suggests that this branch of chemistry has something to do with *organisms,* or living things. Originally, organic chemistry did deal only with substances obtained from living matter. Years ago, chemists spent much of their time extracting, purifying, and analyzing substances from animals and plants. They were motivated by a natural curiosity about living matter and also by the desire to obtain from nature ingredients for medicines, dyes, and other useful products.

It gradually became clear that most compounds in plants and animals differ in several respects from those that occur in nonliving matter, such as minerals. In particular, most compounds in living matter are made up of the same few elements: **carbon, hydrogen, oxygen, nitrogen,** and sometimes sulfur, phosphorus, and a few others. Carbon is virtually always present. This fact led to our present definition: ***Organic chemistry is the chemistry of carbon compounds.*** This definition broadens the scope of the subject to include not only compounds from nature but also synthetic compounds—compounds invented by organic chemists and prepared in their laboratories.

Synthetic Organic Compounds

Scientists used to think that compounds that occurred in living matter were different from other substances and that they contained some sort of intangible **vital force** that imbued them with life. This idea discouraged chemists from trying to make organic compounds in the laboratory. But in 1828 the German chemist Friedrich Wöhler, then 28 years old, accidentally prepared **urea,** a well-known constituent of urine, by heating the inorganic (or mineral) substance ammonium cyanate. He was quite excited about this result, and in a letter to his former teacher, the Swedish chemist J. J. Berzelius, he wrote, "I can make urea without the necessity of a kidney, or even of an animal, whether man or dog." This experiment and others like it gradually discredited the vital-force theory and opened the way for modern synthetic organic chemistry.

Synthesis usually consists of piecing together small, relatively simple molecules to make larger, more complex ones. To make a molecule that contains many atoms from molecules that contain fewer atoms, one must know how to link atoms to each other—that is, how to make and break chemical bonds. Wöhler's preparation of urea was accidental, but synthesis is much more effective if it is carried out in a controlled and rational way, so that, when all the atoms are assembled, they will be connected to one another in the correct manner to give the desired product.

Chemical bonds are made or broken during chemical reactions. In this course, you will learn about quite a few reactions that can be used to make new bonds and that are therefore useful in synthesis.

Why Synthesis?

At present, the number of organic compounds that have been synthesized in research laboratories is far greater than the number isolated from nature. Why is it important to know how to synthesize molecules? There are several reasons. For one, it might be important to synthesize a natural product in the laboratory in order to make the substance more widely available at lower cost than it would be if the compound had to be extracted from its natural source. Some examples of compounds first isolated from nature but now produced synthetically for commercial use are vitamins, amino acids, the dye indigo, the moth repellent camphor, and the antibiotic penicillin. Although the term *synthetic* is sometimes frowned on as implying something artificial or unnatural, these synthetic natural products are in fact identical to the same compounds extracted from nature.

Another reason for synthesis is to create new substances that may have new and useful properties. Synthetic fibers such as nylon and Orlon, for example, have properties that make them superior for some uses to natural fibers such as silk, cotton, and hemp. Most drugs used in medicine are synthetic (including aspirin, ether, Novocain, and barbiturates). The list of synthetic products that we take for granted is long indeed—plastics, detergents, insecticides, and oral contraceptives are just a few. All of these are compounds of carbon; all are organic compounds.

Finally, organic chemists sometimes synthesize new compounds to test chemical theories, or sometimes just for the fun of it. Certain geometric structures, for example, are aesthetically pleasing, and it can be a challenge to make a molecule in which the carbon atoms are arranged in some regular way. One example is the hydrocarbon cubane, C_8H_8. First synthesized in 1964, its molecules have eight carbons at the corners of a cube, each carbon with one hydrogen and three other carbons connected to it. Cubane is more than just aesthetically pleasing. The bond angles in cubane are distorted from normal because of its geometry. Studying the chemistry of cubane therefore gives chemists information about how the distortion of carbon-carbon and carbon-hydrogen bonds affects their chemical behavior.

cubane, C_8H_8
mp 130–131°C
P. E. Eaton (U. of Chicago), 1964

Here are a few other examples of organic chemists' fascination with geometry and the challenge of synthesis.

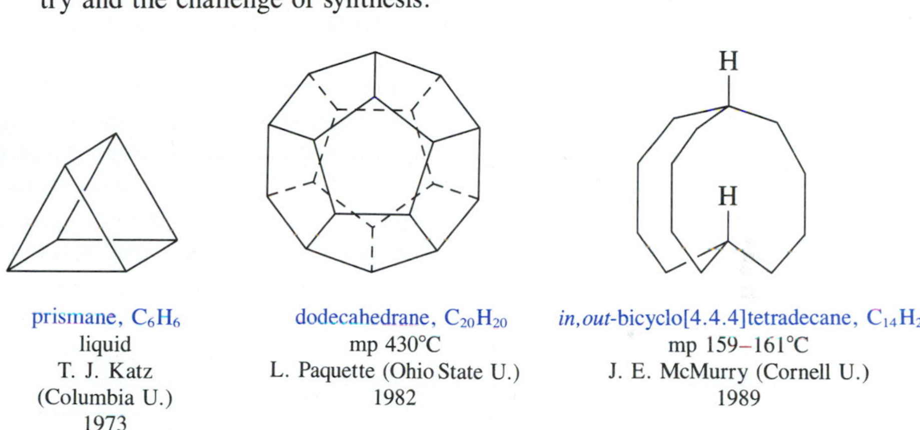

prismane, C_6H_6
liquid
T. J. Katz
(Columbia U.)
1973

dodecahedrane, $C_{20}H_{20}$
mp 430°C
L. Paquette (Ohio State U.)
1982

in,out-bicyclo[4.4.4]tetradecane, $C_{14}H_{26}$
mp 159–161°C
J. E. McMurry (Cornell U.)
1989

Organic Chemistry in Everyday Life

Organic chemistry touches our daily lives. We are made of and surrounded by organic compounds. Almost all the reactions in living matter involve organic compounds, and it is impossible to understand life, at least from the physical point of view, without knowing some organic chemistry. The major constituents of living matter—proteins, carbohydrates, lipids (fats), nucleic acids (DNA, RNA), cell membranes, enzymes, hormones—are organic, and later in the book, I will describe their chemical structures. These structures are quite complex. To understand them, we will first have to discuss simpler molecules.

Other organic substances include the gasoline, oil, and tires for our cars, the clothing we wear, the wood of our furniture and the paper of our books, the medicines we take, plastic containers, camera film, perfume, carpeting, and fabrics. Name it, and the chances are good that it is organic. Daily, in the paper or on television, we encounter references to polyethylene, epoxys, Styrofoam, nicotine, polyunsaturated fats, and cholesterol. All of these terms refer to organic substances; we will study them and many more in this book.

In short, organic chemistry is more than just a branch of science for the professional chemist or for the student preparing to become a physician, dentist, veterinarian, pharmacist, nurse, or agriculturist. It is part of our technological culture.

Organization

Organic chemistry is a vast subject. Some molecules and reactions are simple; others are quite complex. We will proceed from the simple to the complex and begin with a chapter on bonding, with special emphasis on bonds to carbon. Next, we have three chapters on organic compounds that contain only two elements, carbon and hydrogen (called hydrocarbons). These are followed by a chapter that deals with the three-dimensionality of organic compounds. Next we add other elements to the carbon and hydrogen framework, halogens in Chapter 6, oxygen and sulfur in Chapters 7 through 10, and nitrogen in Chapter 11. At that point, we will have completed an introduction to all the main classes of organic compounds.

Spectroscopy is a valuable tool in determining organic structures—that is, details of how atoms and groups are arranged in organic molecules. We take up this topic in Chapter 12. Next comes a chapter on heterocyclic compounds, many of which are important in medicine and in natural products. It is followed by a chapter on polymers, which highlights one of the most important industrial uses of organic chemistry. The last four chapters deal with the organic chemistry of four major classes of biologically important molecules: the lipids, carbohydrates, proteins, and nucleic acids. Since the structures of these molecules of nature are rather complex, we leave them for last. But with the background knowledge of simpler molecules that you will have acquired by then, these compounds and their chemistry will be more clear and understandable.

A Word About

In each chapter after the first, you will find special sections under the general rubric "A Word About." These are short, self-contained articles that expand on the main subject of the chapter. They may deal with intellectual curiosities (the first one, on impossible organic structures); industrial applications (petroleum, gasoline, and octane number in Chapter 3 or epoxy resins in Chapter 8); organic chemistry in biology or medicine (polycyclic aromatic hydrocarbons and cancer in Chapter 4 or morphine and other painkillers in Chapter 11); or just fun topics (artificial blood in Chapter 6 or sweetness and sweeteners in Chapter 16). They provide a convenient break at various points in each chapter, and I hope that you will enjoy them.

The Importance of Problem Solving

The key to success in studying organic chemistry is problem solving.

Each chapter in this book contains a large number of facts that must be digested. Also, the subject matter builds continuously, so that to understand each new topic, it is essential to have the preceding information clear in your mind and available for recall. To learn all this material, careful study of the text is necessary, but it is *not sufficient*. Practical knowledge of how to use the facts is required, and such skill can be obtained only through the solving of problems.

This book contains several types of problems. Some, called ***Examples,*** contain a ***Solution,*** so that you can see how to work such problems. Throughout a chapter examples are usually followed by similar ***Problems,*** designed to reinforce your learning immediately by allowing you to be sure that you understand the new material just presented. At the end of each chapter, ***Additional Problems*** enable you to practice your problem-solving skills. Problems that simply test your knowledge begin the list and then gradually become more challenging.

Try to work as many problems as you can. If you have trouble, seek help from your instructor or from the study guide that accompanies this text. The study guide provides answers to all the problems and explains how to solve them. Problem solving is time consuming, but it will pay off in an understanding of the subject and in a good grade.

And now let us begin.

CHAPTER 1

Bonding and Isomerism

The ways that atoms form bonds with one another to make molecules are important to understand because they help to explain the structures of molecules and why particular molecules react as they do. Perhaps you have already studied some of these ideas in a beginning chemistry course. Browse through each section of this chapter to see whether it is familiar, and try to work the problems. If you can work the problems, you can safely skip that section. But if you have difficulty with any of the problems within or at the end of this chapter, study the entire chapter carefully because we will use the ideas developed here throughout the rest of the book.

1.1 How Electrons Are Arranged in Atoms

Atoms contain a small, dense **nucleus** surrounded by **electrons.** The nucleus is positively charged and contains most of the mass of the atom. The nucleus consists of **protons,** which are positively charged, and **neutrons,** which are neutral. (The only exception is hydrogen, whose nucleus consists of only a single proton.) The positive charge of the nucleus is exactly balanced by the negative charge of the electrons that surround it. The **atomic number** of an element is equal to the number of protons in the nucleus (or the number of electrons around the nucleus; the two numbers are the same). The **atomic weight** is approximately equal to the sum of the number of protons and the number of neutrons in the nucleus; the electrons are not counted because they are very light by comparison. The periodic table on the inside back cover of this book shows all the elements with their atomic numbers and weights.

We are concerned here mainly with the atom's electrons because their number and arrangement provide the key to how a particular atom reacts with other atoms to form molecules. Also, we will deal only with electron arrangements in the lighter elements because these elements are the most important in organic molecules.

Electrons are concentrated in certain regions of space around the nucleus called **orbitals.** *Each orbital can contain a maximum of two electrons.* The orbitals, which differ in shape, are designated by the letters, *s*, *p*, and *d*. In addition, orbitals are grouped in **shells** designated by the numbers 1, 2, 3, and so on. Each shell contains different types and numbers of orbitals, corresponding to the shell number. For example, shell 1 contains only one type of orbital, designated the 1*s* orbital. Shell 2 contains two types of orbitals, 2*s* and 2*p*, and shell 3 contains three types, 3*s*, 3*p*, and 3*d*. Within a particular shell, the number of *s*, *p*, and *d* orbitals is 1, 3, and 5, respectively (Table 1.1).

These rules permit us to count how many electrons each shell will contain when it is filled (last column in Table 1.1). Table 1.2 shows how the electrons of the first 18 elements are arranged.

TABLE 1.1 Numbers of orbitals and electrons in the first three shells

	Number of orbitals of each type			
Shell number	*s*	*p*	*d*	*Total number of electrons when shell is filled*
1	1	0	0	2
2	1	3	0	8
3	1	3	5	18

TABLE 1.2 Electron arrangements of the first 18 elements

		Number of electrons in each orbital				
Atomic number	*Element*	*1s*	*2s*	*2p*	*3s*	*3p*
1	H	1				
2	He	2				
3	Li	2	1			
4	Be	2	2			
5	B	2	2	1		
6	**C**	**2**	**2**	**2**		
7	N	2	2	3		
8	O	2	2	4		
9	F	2	2	5		
10	Ne	2	2	6		
11	Na	2	2	6	1	
12	Mg	2	2	6	2	
13	Al	2	2	6	2	1
14	Si	2	2	6	2	2
15	P	2	2	6	2	3
16	S	2	2	6	2	4
17	Cl	2	2	6	2	5
18	Ar	2	2	6	2	6

TABLE 1.3 Valence electrons of the first 18 elements

Group	I	II	III	IV	V	VI	VII	VIII
	H·							He:
	Li·	Be·	·B·	·C·	·N:	·O:	:F:	:Ne:
	Na·	Mg·	·Al·	·Si·	·P:	·S:	:Cl:	:Ar:

The first shell is filled for helium (He) and all elements beyond, and the second shell is filled for neon (Ne) and all elements beyond. Filled shells play almost no role in chemical bonding. Rather, the outer shells, or **valence shells,** are mainly involved in chemical bonding, and we will focus our attention on them.

Table 1.3 shows the **valence electrons,** the electrons in the outermost shell, for the first 18 elements. The element's symbol stands for the **kernel** of the element (the nucleus plus the filled electron shells), and the dots represent the valence electrons. The elements are arranged in groups according to the periodic table, and (except for helium) these group numbers correspond to the number of valence electrons.

Armed with this information about atomic structure, we are now ready to tackle the problem of how elements combine to form chemical bonds.

1.2 Ionic and Covalent Bonding

An early, but still useful, theory of chemical bonding was proposed in 1916 by Gilbert Newton Lewis, then a professor at the University of California in Berkeley. Lewis noticed that the **inert gas** helium had only two electrons surrounding its nucleus and that the next inert gas, neon, had ten such electrons (2 + 8; see Table 1.2). He concluded that atoms of these gases must have very stable electron arrangements because these elements do not combine with other atoms. He further suggested that other atoms might react in such a way as to achieve these stable arrangements. This stability could be achieved in one of two ways: by complete transfer of electrons from one atom to another or by sharing of electrons between atoms.

1.2a Ionic Compounds ***Ionic bonds are formed by the transfer of one or more valence electrons from one atom to another.*** Since electrons are negatively charged, the atom that gives up the electron(s) becomes positively charged, a **cation.** The atom that receives the electron(s) becomes negatively charged, an **anion.** The reaction between sodium and chlorine atoms to form sodium chloride (ordinary table salt) is a typical electron-transfer reaction.

$$\underset{\text{sodium atom}}{\text{Na}\cdot} + \underset{\text{chlorine atom}}{\cdot\ddot{\underset{\cdot\cdot}{\text{Cl}}}:} \longrightarrow \underset{\text{sodium cation}}{\text{Na}^+} + \underset{\text{chloride anion}}{:\ddot{\underset{\cdot\cdot}{\text{Cl}}}:^-} \tag{1.1}$$

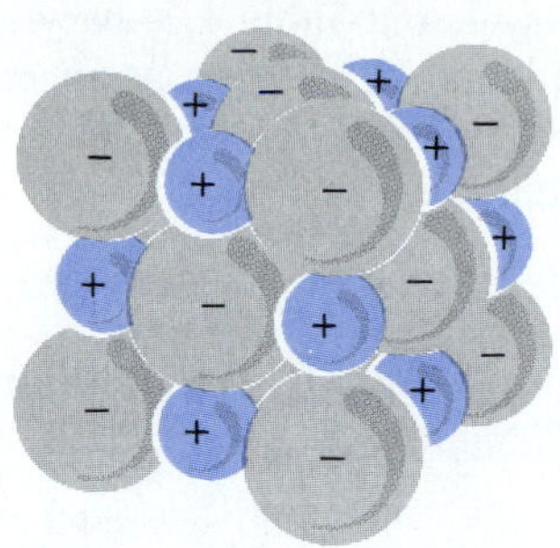

FIGURE 1.1
Sodium chloride, Na^+Cl^-, is an ionic crystal. The colored spheres represent sodium ions, Na^+, and the gray spheres are chloride ions, Cl^-. Each ion is surrounded by six oppositely charged ions, except for those ions that are at the surface of the crystal.

The sodium atom has only one valence electron (it is in the third shell; see Table 1.2). By giving up that electron it achieves the electron arrangement of neon. At the same time, it becomes positively charged, a sodium cation. The chlorine atom has seven valence electrons. By accepting an additional electron, it achieves the electron arrangement of argon and becomes negatively charged, a chloride anion. *Atoms,* such as sodium, *that tend to give up electrons are said to be* **electropositive.** *Atoms,* such as chlorine, *that tend to accept electrons are said to be* **electronegative.**

EXAMPLE 1.1 Write an equation for the reaction of magnesium (Mg) with fluorine atoms (F).

Solution

$$\dot{Mg}\cdot + :\ddot{\underset{\cdot\cdot}{F}}\cdot + \cdot\ddot{\underset{\cdot\cdot}{F}}: \longrightarrow Mg^{2+} + 2:\ddot{\underset{\cdot\cdot}{F}}:^-$$

Magnesium has two valence electrons. Since each fluorine atom can accept only one electron (from the magnesium) to complete its valence shell, two fluorine atoms are needed to react with one magnesium atom.

The product of eq. 1.1 is sodium chloride, an ionic compound made up of equal numbers of sodium and chloride ions. In general, ionic compounds form when strongly electropositive atoms and strongly electronegative atoms interact. The ions in a crystal of an ionic substance are held together by the attractive force between their opposite charges, as shown in Figure 1.1 for a sodium chloride crystal.

In a sense, the ionic bond is not really a bond at all. Being oppositely charged, the ions attract one another like the opposite poles of a magnet. In the crystal, the ions are packed in a definite arrangement, but we cannot say that any particular ion is bonded or connected to any other particular ion. And, of course, when the substance is dissolved, the ions separate and are able to move about in solution relatively freely.

EXAMPLE 1.2 What charge will a beryllium ion carry?

Solution As seen in Table 1.3, beryllium (Be) has two valence electrons. To achieve the filled-shell electron arrangement of helium, it must lose both its valence electrons. Thus, the beryllium cation will carry two positive charges and is represented by Be^{2+}.

Problem 1.1 Using Table 1.3, tell what charge the ion will carry when each of the following elements reacts to form an ionic compound: Al, Li, S, H.

EXAMPLE 1.3 Which atom is more electropositive, lithium or beryllium?

Solution To achieve the electron arrangement of helium, lithium must give up only one valence electron, whereas beryllium must give up two. Thus, lithium is the more electropositive of the two atoms. Looked at another way, the lithium nucleus is less positive (charge of +1) than the beryllium nucleus (charge of +2). It is easier, therefore, to remove one electron from lithium than it is to remove two electrons from beryllium.

In general, within a given horizontal row in the periodic table, the more electropositive elements are those farthest to the left, and the more electronegative elements are those farthest to the right. Within a given vertical column, the more electropositive elements are those toward the bottom, and the more electronegative elements are those toward the top.

Problem 1.2 Using Table 1.3, tell which is the more electropositive element: sodium or aluminum, boron or carbon, boron or aluminum.

Problem 1.3 Using Table 1.3, tell which is the more electronegative element: oxygen or fluorine, oxygen or nitrogen, fluorine or chlorine.

Problem 1.4 Judging from its position in Table 1.3, do you expect carbon to be electropositive or electronegative?

1.2b The Covalent Bond Elements that are neither strongly electronegative nor strongly electropositive tend to form bonds by sharing electron pairs instead of completely transfering electrons. ***A covalent bond involves the mutual sharing of one or more electron pairs between atoms.*** When the two atoms are identical or have equal electronegativities, the electron pairs are shared equally. The hydrogen molecule is an example.

$$\underset{\text{hydrogen atoms}}{H\cdot + H\cdot} \longrightarrow \underset{\text{hydrogen molecule}}{H{:}H} + \text{heat} \qquad (1.2)$$

Each hydrogen atom can be considered to have filled its first electron shell by the sharing process. That is, each atom is considered to "own" all the electrons it shares with the other atom, as shown by the loops in these structures.

(H:)H H(:H)

EXAMPLE 1.4 Write an equation similar to eq. 1.2 for the formation of a chlorine molecule from two chlorine atoms.

Solution $:\ddot{\underset{\cdot\cdot}{Cl}}\cdot + \cdot\ddot{\underset{\cdot\cdot}{Cl}}: \longrightarrow :\ddot{\underset{\cdot\cdot}{Cl}}:\ddot{\underset{\cdot\cdot}{Cl}}: + \text{heat}$

One electron pair is shared by the two chlorine atoms. In that way, each chlorine completes its valence shell with eight electrons (three unshared pairs and one shared pair).

When two hydrogen atoms combine to form a molecule, heat is liberated. Conversely, this same amount of heat (energy) has to be supplied to a hydrogen molecule to break it apart into atoms. To break apart 1 mol of hydrogen molecules (1 gram molecular weight, in this case 2 g) into atoms requires 104 kcal (or 435 kJ*) of heat, quite a lot of energy.

The H—H bond is a very strong bond. The main reason for this is that the shared electron pair is attracted to *both* hydrogen nuclei, whereas in a hydrogen atom, the valence electron is associated with only one nucleus. But other forces in the hydrogen molecule tend to counterbalance the attraction between the electron pair and the nuclei. These forces are the repulsion between the two like-charged nuclei and the repulsion between the two like-charged electrons. A balance is struck between the attractive and the repulsive forces. The hydrogen atoms neither fly apart nor fuse together. Instead, they remain connected, or bonded, and vibrate about some equilibrium distance, which we call the **bond length.** For a hydrogen molecule, the bond length (that is, the average distance between the two hydrogen nuclei) is 0.74 Å.**

1.3 Carbon and the Covalent Bond

Now let us look at carbon and its bonding. We represent atomic carbon by the symbol $\cdot\dot{\underset{\cdot}{C}}\cdot$ where the letter C stands for the kernel (the nucleus plus the two 1*s* electrons) and the dots represent the valence electrons.

With four valence electrons, the valence shell of carbon is half filled (or half empty). Carbon atoms have neither a strong tendency to lose all their electrons (and become C^{4+}) nor a strong tendency to gain four electrons (and become C^{4-}). Being in the middle of the periodic table, carbon is neither strongly electropositive nor strongly electronegative. Instead, it usually forms covalent bonds with other atoms by sharing electrons. For example, carbon combines with four hydrogen atoms (each of which supplies one valence electron) by sharing four electron pairs.[†] The substance formed is known as **methane.** Carbon can also share electron pairs with four chlorine atoms, forming **tetrachloromethane.**[††]

---

*Although most organic chemists use the kilocalorie as the unit of heat, the currently used international unit is the kilojoule; 1 kcal = 4.184 kJ.

** 1 Å, or angstrom unit, is 10^{-8} cm, so the H—H bond length is 0.74×10^{-8} cm.

[†] To designate electrons from different atoms, the symbols • and x are often used. But the electrons are, of course, identical.

[††] Tetrachloromethane is the systematic name, but carbon tetrachloride is the common name. We will discuss how to name organic compounds later.

```
                         H                                      Cl
  H                      |              :Cl:                    |
H:C:H        or      H — C — H     :Cl: C :Cl:     or     Cl — C — Cl
  H                      |              :Cl:                    |
                         H                                      Cl
```

methane

tetrachloromethane
(carbon tetrachloride)

By sharing electron pairs, the atoms complete their valence shells. In both examples, carbon has eight valence electrons around it. In methane, the hydrogen completes its first valence shell with two electrons, and in tetrachloromethane the chlorine atoms fill their valence shell with eight electrons. In this way, all valence shells are filled and the compounds are quite stable.

The shared electron pair is called a **covalent bond,** because it bonds or links the atoms together (by mutual attraction for their nuclei). The single bond is usually represented by a dash, or single line, as shown in the second formulas above for methane and tetrachloromethane.

■ ***EXAMPLE 1.5*** Draw the formula for chloromethane (also called methyl chloride), CH_3Cl.

Solution

```
                          H
  H                       |
H:C:Cl:        or     H — C — Cl
  H                       |
                          H
```

Problem 1.5 Draw the formulas for dichloromethane (also called methylene chloride), CH_2Cl_2, and trichloromethane (chloroform), $CHCl_3$.

1.4 Carbon–Carbon Single Bonds

The unique property of carbon atoms—that is, the property that makes it possible for chemists to construct literally millions of organic compounds—is their ability to share electrons not only with different elements but also with other carbon atoms. For example, two carbon atoms may be bonded to one another, and each of these carbon atoms may be linked to other atoms. In **ethane** and **hexachloroethane,** each carbon is connected to the other carbon *and* to three hydrogen atoms or three chlorine atoms. Although they have two carbon atoms instead of one, these compounds have chemical properties similar to those of methane and tetrachloromethane, respectively.

```
                    H   H                                      Cl  Cl
 H H                |   |          :Cl: :Cl:                   |   |
H:C:C:H    or   H — C — C — H    :Cl:C : C:Cl:    or    Cl — C — C — Cl
 H H                |   |          :Cl: :Cl:                   |   |
                    H   H                                      Cl  Cl
```

ethane

hexachloroethane

The carbon–carbon bond in ethane, like the hydrogen–hydrogen bond in a hydrogen molecule, is a purely covalent bond, with the electrons shared *equally* between the two identical carbon atoms. As with the hydrogen molecule, heat is required to break the carbon–carbon bond of ethane to give two CH_3 fragments (called **methyl radicals**). ***A radical is a molecular fragment with an odd number of unshared electrons.***

```
  H H                    H
  | |                    |
H—C:C—H   --heat-->  2 H—C·          (1.3)
  | |                    |
  H H                    H
 ethane            methyl radical
```

However, less heat is required to break the carbon–carbon bond in ethane than is required to break the hydrogen–hydrogen bond in a hydrogen molecule. The actual amount is 88 kcal (or 368 kJ) per mole of ethane. The carbon–carbon bond in ethane is longer (1.54 Å) than the hydrogen–hydrogen bond (0.74 Å) and also somewhat weaker. Breaking carbon–carbon bonds by heat, as represented in eq. 1.3, is the first step in the *cracking* of petroleum, an important process in the manufacture of gasoline (see "A Word About Petroleum" on page 101).

EXAMPLE 1.6 What do you expect the bond length of a C—H bond (as in methane or ethane) to be?

Solution It should measure somewhere between the H—H bond length in a hydrogen molecule (0.74 Å) and the C—C bond length in ethane (1.54 Å). The actual value is about 1.09 Å, close to the average of the H—H and C—C bond lengths.

Problem 1.6 The Cl—Cl bond length is 1.98 Å. Which bond will be longer, the C—C bond in ethane or the C—Cl bond in chloromethane?

There is almost no limit to the number of carbon atoms that can be linked together, and some molecules contain as many as 100 or more carbon–carbon bonds in a row. This ability of an element to form chains as a result of bonding between its own atoms is called **catenation.**

Problem 1.7 Using the structure of ethane as a guide, draw the structure for propane, C_3H_8.

1.5 Polar Covalent Bonds

As we have seen, covalent bonds can be formed not only between identical atoms (H—H, C—C) but also between different atoms (C—H, C—Cl), provided that the atoms do not differ too greatly in electronegativity. However, *if the atoms are different from one another, the electron pair may not be shared equally between them.* Such a bond is sometimes called a **polar covalent bond** because the atoms that are linked carry a partial negative and a partial positive charge.

The hydrogen chloride molecule provides an example of a polar covalent bond. Chlorine atoms are more electronegative than hydrogen atoms, but even so, the bond that they form is covalent rather than ionic. However, the shared electron pair is attracted more toward the chlorine, which therefore is slightly negative with respect to the hydrogen. This bond polarization is indicated by an arrow whose head is negative and whose tail is marked with a plus sign. Alternatively, a partial charge, written as $\delta+$ or $\delta-$ (read as "delta plus" or "delta minus"), may be shown:

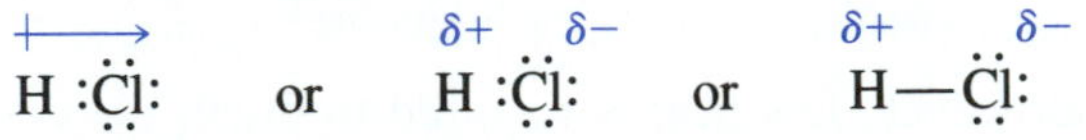

$$\overset{+\!\!\longrightarrow}{\text{H}:\ddot{\underset{\cdot\cdot}{\text{Cl}}}:} \quad \text{or} \quad \overset{\delta+}{\text{H}}\,:\overset{\delta-}{\ddot{\underset{\cdot\cdot}{\text{Cl}}}}: \quad \text{or} \quad \overset{\delta+}{\text{H}}-\overset{\delta-}{\ddot{\underset{\cdot\cdot}{\text{Cl}}}}:$$

The bonding electron pair, which is shared *unequally,* is displaced toward the chlorine.

You can usually rely on the periodic table to determine which end of a polar covalent bond is more negative and which end is more positive. As we proceed from left to right across the table within a given period, the elements become *more* electronegative, owing to increasing atomic number, or charge on the nucleus. As we proceed from the top to the bottom of the table within a given group (down a column), the elements become *less* electronegative because the valence electrons are shielded from the nucleus by an increasing number of inner-shell electrons. From these generalizations, we can safely predict that the atom on the right in each of the following bonds will be negative with respect to the atom on the left:

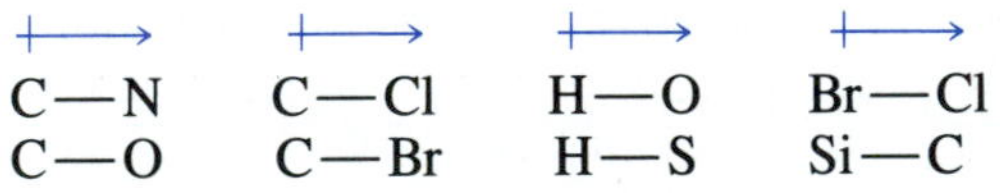

$$\begin{array}{cccc} +\!\!\longrightarrow & +\!\!\longrightarrow & +\!\!\longrightarrow & +\!\!\longrightarrow \\ \text{C—N} & \text{C—Cl} & \text{H—O} & \text{Br—Cl} \\ \text{C—O} & \text{C—Br} & \text{H—S} & \text{Si—C} \end{array}$$

The carbon–hydrogen bond, which is so common in organic compounds, requires special mention. Carbon and hydrogen have nearly identical electronegativities, so the C—H bond is almost purely covalent.

■ ***EXAMPLE 1.7*** Indicate any bond polarization in the structure of tetrachloromethane.

Solution

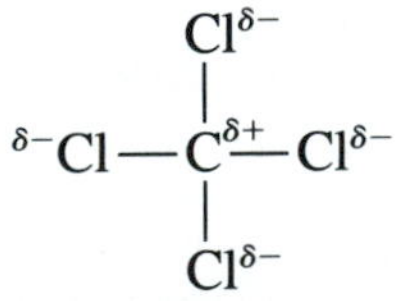

$$\begin{array}{c} \text{Cl}^{\delta-} \\ | \\ {}^{\delta-}\text{Cl}-\text{C}^{\delta+}-\text{Cl}^{\delta-} \\ | \\ \text{Cl}^{\delta-} \end{array}$$

Chlorine is more electronegative than carbon. The electrons in each C—Cl bond are therefore displaced toward the chlorine.

Problem 1.8 Predict the polarity of the N—Cl bond and of the S—O bond.

Problem 1.9 Draw the structure of the refrigerant dichlorodifluoromethane (CFC-12), and indicate the polarity of the bonds.

Problem 1.10 Draw the formula for methanol, CH_3OH, and (where appropriate) indicate bond polarity with an arrow, $\leftarrow\!\!\!+$.

1.6 Multiple Covalent Bonds

To complete their valence shells, atoms may sometimes share more than one electron pair. Carbon dioxide, CO_2, is an example. The carbon atom has four valence electrons, and each oxygen has six valence electrons. A formula that allows each atom to complete its valence shell with eight electrons is

O::C::O or O=C=O or O=C=O

A B C

In formula A, the dots represent the electrons from carbon, and the x's are the electrons from the oxygens. Formula B shows the bonds and oxygen's unshared electrons, and formula C shows only the covalent bonds. Two electron pairs are shared between carbon and oxygen. Consequently, the bond is called a **double bond.** Each oxygen atom also has two pairs of **nonbonding electrons,** or **unshared electron pairs.** The loops in the following formulas show that each atom in carbon dioxide has a complete valence shell of eight electrons.

O::C::O O::C::O O::C::O

Hydrogen cyanide, HCN, is an example of a simple compound with a **triple bond,** a bond in which three electron pairs are shared.

H:C:::N: or H—C≡N: or H—C≡N

hydrogen cyanide

Problem 1.11 Show with loops how each atom in hydrogen cyanide completes its valence shell.

■ *EXAMPLE 1.8* Tell what, if anything, is wrong with the following electron arrangement for carbon dioxide:

:O:::C::Ö:

Solution The formula contains the correct total number of valence electrons (16), and each oxygen is surrounded by 8 valence electrons, which is correct. What is wrong is that the carbon atom has 10 valence electrons, two more than is allowable.

Problem 1.12 Show what is wrong with each of the following electron arrangements for carbon dioxide.

```
a. :O:::C:::O:    b.  ..  ..  ..      c.  ..
                     :O:C:O:             :O:C:::O:
                      ..  ..  ..          ..
```

Problem 1.13 Methanal (formaldehyde) has the formula H_2CO. Draw a formula that shows how the valence electrons are arranged.

Problem 1.14 Draw an electron-dot formula for carbon monoxide, CO.

Carbon atoms can be connected to one another by double bonds or triple bonds, as well as by single bonds. Thus there are three **hydrocarbons** (compounds with just carbon and hydrogen atoms) that have two carbon atoms per molecule: ethane, ethene, and ethyne.

```
    H   H
    |   |          H         H
H — C — C — H        \     /          H — C ≡ C — H
    |   |              C = C
    H   H            /     \
                   H         H
```

ethane ethene (ethylene) ethyne (acetylene)

They differ in that the carbon–carbon bond is single, double, or triple, respectively. They also differ in number of hydrogens. As we will see later, these compounds have different chemical reactivities because of the different types of bonds between the carbon atoms.

EXAMPLE 1.9 Draw the formula for C_3H_6 having one carbon–carbon double bond.

Solution First, draw the three carbons with one double bond.

```
C = C — C
```

Then add the hydrogens in such a way that each carbon has eight electrons around it (or in such a way that each carbon has four bonds).

```
    H   H   H
    |   |   |
H — C = C — C — H
            |
            H
```

Problem 1.15 Draw three different structures that have the formula C_4H_8 and have one carbon–carbon double bond.

1.7 Valence

The valence of an element is simply the number of bonds that an atom of the element can form. The number is usually equal to the *number of electrons needed to fill the valence shell*. Table 1.4 gives the common valences of several elements.

TABLE 1.4 Valences of common elements

Element	H·	·C·	·N:	·O:	:F:	:Cl:
Valence	1	4	3	2	1	1

Notice the difference between the number of valence electrons and the valence. Oxygen, for example, has six valence electrons but a valence of only 2. The *sum* of the two numbers is equal to the number of electrons in the filled shell.

The valences in Table 1.4 apply whether the bonds are single, double, or triple. For example, carbon has four bonds in each of the formulas we have written so far: methane, tetrachloromethane, ethane, ethene, ethyne, carbon dioxide, and so on. These valences are worth remembering, because they will help you to write correct formulas.

EXAMPLE 1.10 Using dashes for bonds, draw a formula for C_3H_4 that has the proper valence of 1 for each hydrogen and 4 for each carbon.

Solution There are three possibilities:

```
     H
     |                  H         H         H           H
H—C—C≡C—H                \       /           \         /
     |                     C=C=C               C=C
     H                  H /       \ H           \ /
                                                 C
                                                / \
                                               H   H
```

A compound that corresponds to each of these three different arrangements of the atoms is known.

Problem 1.16 Use dashes for bonds, and use the valences given in Table 1.4 to write a structure for each of the following:

a. CH_5N b. CH_4O

Problem 1.17 Does C_2H_5 represent a stable molecule?

In Example 1.10, we saw that three carbon atoms and four hydrogen atoms can be connected to one another in three different ways, each of which satisfies the valences of both kinds of atoms. Let us take a closer look at this phenomenon.

1.8 *Isomerism*

The **molecular formula** of a substance tells us the numbers of different atoms present, but a **structural formula** tells us how those atoms are arranged. For example, H_2O is the molecular formula for water. It tells us that each water molecule contains two hydrogen atoms and one oxygen atom. But the structural formula H—O—H tells us more than that. It tells us that the hydrogens are connected to the oxygen (and not to each other).

It is sometimes possible to arrange the same atoms in more than one way and still satisfy their valences. Molecules that have the same kinds and numbers of atoms but different arrangements are called **isomers,** a term that comes from the Greek (*isos,* equal, and *meros,* part). **Structural (or constitutional) isomers** ***are compounds that have the same molecular formula but different structural formulas.*** Let us look at a particular pair of isomers.

Two very different chemical substances are known, each with the molecular formula C_2H_6O. One of these substances is a colorless liquid that boils at 78.5°C, whereas the other is a colorless gas at ordinary temperatures (bp −23.6°C). The only possible explanation is that the atoms must be arranged differently in the molecules of each substance and that these arrangements are somehow responsible for the fact that one substance is a liquid and the other, a gas.

For the molecular formula C_2H_6O, two (and only two) structural formulas are possible that satisfy the valence requirement of 4 for carbon, 2 for oxygen, and 1 for hydrogen. They are

```
    H  H                      H     H
    |  |                      |     |
 H—C—C—O—H       and      H—C—O—C—H
    |  |                      |     |
    H  H                      H     H
```

In one formula, the two carbons are connected to one another by a single covalent bond; in the other formula, each carbon is connected to the oxygen. When we complete the valences by adding hydrogens, each arrangement requires six hydrogens.

How can we tell which arrangement of the atoms corresponds to molecules of the liquid and which, to molecules of the gas? In this example, a simple chemical test gives the answer. The liquid C_2H_6O (called ethanol or ethyl alcohol) reacts with sodium metal to produce hydrogen gas and a new compound, C_2H_5ONa. On the other hand, the gaseous C_2H_6O (called methoxymethane or dimethyl ether) does not react with sodium metal at all. The most reasonable interpretation of this experiment is that ethanol is represented by the structural formula in which one hydrogen is different from the other five. Apparently the hydrogen on oxygen can be replaced by sodium. In methoxymethane, all six hydrogens are alike, connected to carbon, and nonreplaceable by sodium. Thus we can associate names and properties with each structural formula.

```
    H  H                  H     H
    |  |                  |     |
 H—C—C—O—H           H—C—O—C—H
    |  |                  |     |
    H  H                  H     H
```

ethanol	methoxymethane
(ethyl alcohol)	(dimethyl ether)
bp 78.5°C	bp −23.6°C

Many other kinds of experimental evidence verify these structural assignments. We leave for later (in Chapters 7 and 8) an explanation of why these arrangements of atoms produce substances that are so different from one another.

Ethanol and methoxymethane are **structural isomers.** They have the same molecular formula but different structural formulas. Ethanol and methoxymethane differ in physical and chemical properties as a consequence of their different molecular structures.

Problem 1.18 Draw structural formulas for all possible isomers of C_3H_8O (there are three).

1.9 *Writing Structural Formulas*

You will be writing structural formulas throughout this course. Perhaps a few hints about how to do so will be helpful. Let's look at another case of isomerism. Suppose we want to write out all possible structural formulas that correspond to the molecular formula C_5H_{12}. We begin by writing all five carbons in a **continuous chain.**

```
C—C—C—C—C
```

a continuous chain

This chain uses up one valence for each of the end carbons and two valences for the carbons in the middle of the chain. Each end carbon therefore has three valences left for bonds to hydrogens. Each middle carbon has only two valences for bonds to hydrogens. As a consequence, the structural formula in this case is written

```
   H  H  H  H  H
   |  |  |  |  |
H—C—C—C—C—C—H
   |  |  |  |  |
   H  H  H  H  H
```

pentane, bp 36°C

To find structural formulas for the other isomers, we must consider **branched chains.** For example, we can reduce the longest chain to only four carbons and connect the fifth carbon to one of the middle carbons, as in the following structural formula.

```
C—C—C—C
  |
  C
```

a branched chain

If we add the remaining bonds so that each carbon has a valence of 4, we see that three of the carbons have three hydrogens attached, but the other carbons have only one or two hydrogens. The molecular formula, however, is still C_5H_{12}.

```
    H   H   H   H
    |   |   |   |
H — C — C — C — C — H
    |   |   |   |
    H   |   H   H
        |
    H — C — H
        |
        H
```

2-methylbutane, bp 28°C
(isopentane)

Suppose we keep the chain of four carbons and try to connect the fifth carbon somewhere else. Consider the following chains:

```
C—C—C—C      C—C—C—C      C—C—C—C
|                  |          |
C                  C          C
```

Do we have anything new here? *No!* The first two structures have five-carbon chains, exactly as in the formula for pentane, and the third structure is identical to the branched chain we have already drawn for 2-methylbutane—a four-carbon chain with a one-carbon branch attached to the second carbon in the chain (counting now from the right instead of from the left).

But there is a third isomer of C_5H_{12}. We can find it by reducing the longest chain to only three carbons and connecting two one-carbon branches to the middle carbon.

```
    C
    |
C — C — C
    |
    C
```

If we fill in the hydrogens, we see that the middle carbon has no hydrogens attached to it.

```
            H
            |
        H — C — H
            |
    H       |       H
    |       |       |
H — C ——— C ——— C — H
    |       |       |
    H       |       H
            |
        H — C — H
            |
            H
```

2,2-dimethylpropane, bp 10°C
(neopentane)

So we can draw three (and only three) different structural formulas that correspond to the molecular formula C_5H_{12}, and in fact we find that only three different chemical substances with this formula exist. They are commonly called *n*-pentane (*n* for normal, with an unbranched carbon chain), isopentane, and neopentane.

Problem 1.19 To which isomer of C_5H_{12} does each of the following structural formulas correspond?

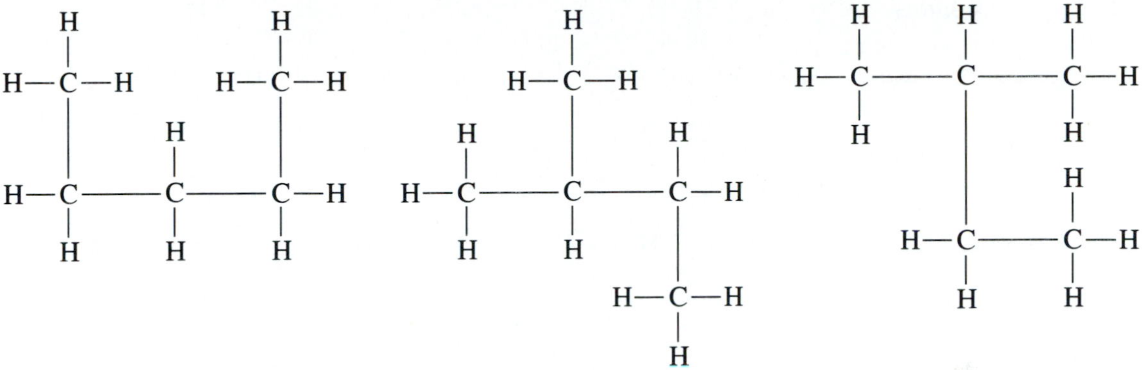

```
   H                H
   |                |
H—C—H           H—C—H
   |       H        |
   |       |        |
H—C———————C———————C—H
   |       |        |
   H       H        H
```

```
          H
          |
       H—C—H
    H     |      H
    |     |      |
 H—C—————C—————C—H
    |     |      |
    H     H      |
              H—C—H
                 |
                 H
```

```
    H        H        H
    |        |        |
 H—C————————C————————C—H
    |        |        |
    H        |        H
             |        H
             |        |
         H—C————————C—H
             |        |
             H        H
```

1.10 *Abbreviated Structural Formulas*

Structural formulas like the ones we have written so far are useful, but they are also somewhat cumbersome. They take up a lot of space and are tiresome to write out. Consequently, we often take some short cuts that still convey the meaning of structural formulas. For example, we may abbreviate the structural formula of ethanol (ethyl alcohol) from

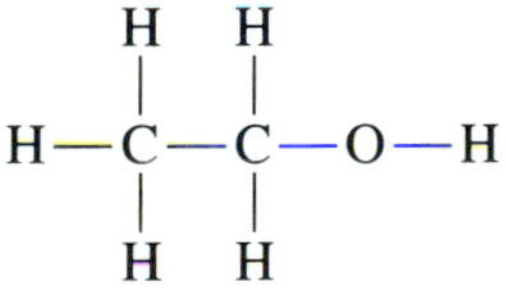

$$\begin{array}{c} \text{H}\;\;\text{H} \\ \text{H—}\overset{|}{\underset{|}{\text{C}}}\text{—}\overset{|}{\underset{|}{\text{C}}}\text{—O—H} \\ \text{H}\;\;\text{H} \end{array} \quad \text{to} \quad CH_3\text{—}CH_2\text{—}OH \quad \text{or} \quad CH_3CH_2OH$$

Each formula clearly represents ethanol rather than methoxymethane (dimethyl ether), which can be represented by any of the following structures:

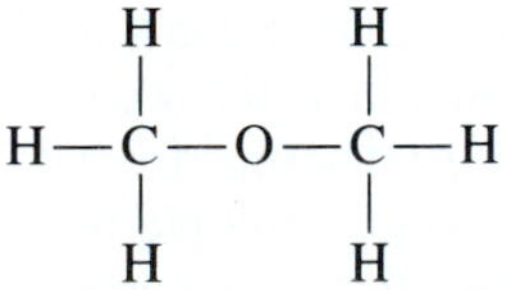

$$\begin{array}{c} \text{H}\;\;\;\;\;\text{H} \\ \text{H—}\overset{|}{\underset{|}{\text{C}}}\text{—O—}\overset{|}{\underset{|}{\text{C}}}\text{—H} \\ \text{H}\;\;\;\;\;\text{H} \end{array} \quad \text{to} \quad CH_3\text{—}O\text{—}CH_3 \quad \text{or} \quad CH_3OCH_3$$

The structural formulas for the three pentanes can be abbreviated in a similar fashion.

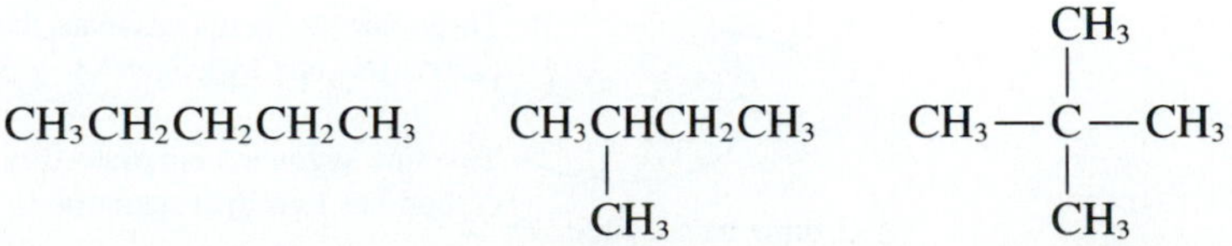

$$CH_3CH_2CH_2CH_2CH_3 \qquad \begin{array}{l} CH_3\underset{\displaystyle |\atop \displaystyle CH_3}{C}HCH_2CH_3 \end{array} \qquad \begin{array}{c} CH_3 \\ | \\ CH_3\text{—}C\text{—}CH_3 \\ | \\ CH_3 \end{array}$$

n-pentane — isopentane — neopentane

Sometimes these formulas are abbreviated even further. For example, they can be printed on a single line in the following ways:

$CH_3(CH_2)_3CH_3$ — *n*-pentane

$(CH_3)_2CHCH_2CH_3$ — isopentane

$(CH_3)_4C$ — neopentane

EXAMPLE 1.11 Write a structural formula that shows all bonds for each of the following:

a. $CH_3CCl_2CH_3$ b. $(CH_3)_2C(CH_2CH_3)_2$

Solution

```
      H  Cl  H
      |  |   |
a. H—C—C—C—H
      |  |   |
      H  Cl  H

                        H
                        |
                      H—C—H
         H   H          |          H   H
         |   |          |          |   |
b.    H—C—C————————C————————C—C—H
         |   |          |          |   |
         H   H          |          H   H
                      H—C—H
                        |
                        H
```

This is the carbon atom to which two —CH_3 and two —CH_2CH_3 groups are attached.

Problem 1.20 Write a structural formula that shows all bonds for each of the following:

a. $(CH_3)_2CHCH_2OH$ b. $CCl_2{=}CCl_2$

Perhaps the ultimate abbreviation of structures is the use of lines to represent the carbon framework:

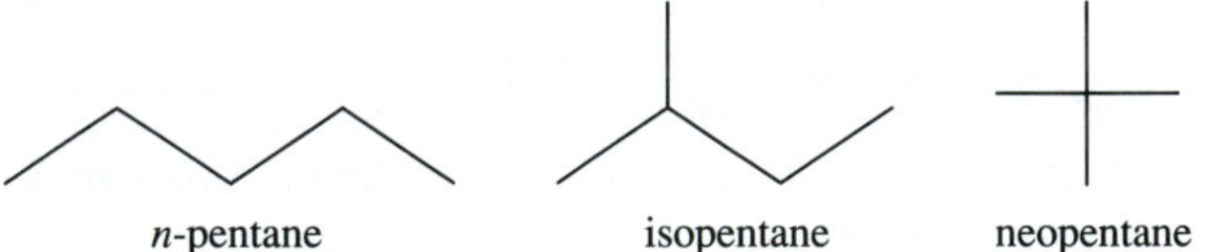

n-pentane isopentane neopentane

In these formulas, *each line segment is understood to have a carbon atom at each end*. The hydrogens are omitted, but we can quickly find the number of hydrogens on each carbon by subtracting from four the number of line segments that emanate from any point. Multiple bonds are represented by multiple line segments. For example, the hydrocarbon with a chain of five carbon atoms and a double bond between the second and third carbon atoms (that is, $CH_3CH{=}CHCH_2CH_3$) is represented as follows:

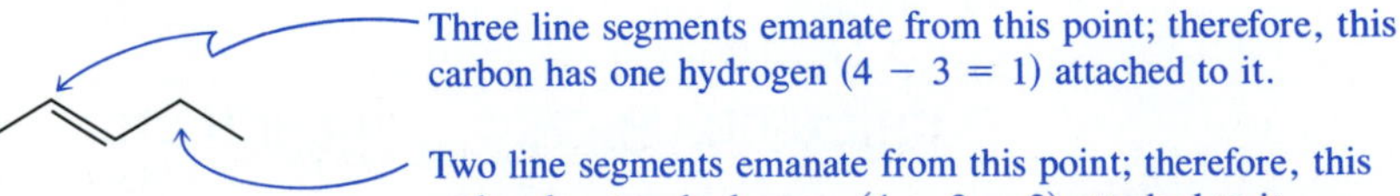

Three line segments emanate from this point; therefore, this carbon has one hydrogen (4 − 3 = 1) attached to it.
Two line segments emanate from this point; therefore, this carbon has two hydrogens (4 − 2 = 2) attached to it.

EXAMPLE 1.12 Write a more detailed structural formula for [line-segment formula]

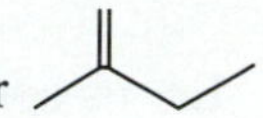

Solution

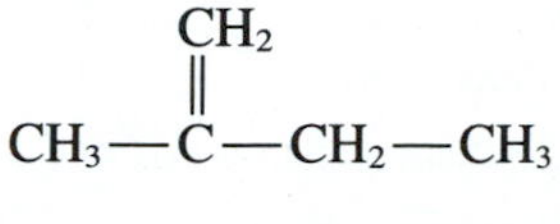

$$CH_3-\overset{\overset{\displaystyle CH_2}{\|}}{C}-CH_2-CH_3 \quad \text{or} \quad H-\underset{H}{\overset{H}{\overset{|}{C}}}-\overset{\overset{\displaystyle H\diagdown C \diagup H}{\|}}{C}-\underset{H}{\overset{H}{\overset{|}{C}}}-\underset{H}{\overset{H}{\overset{|}{C}}}-H$$

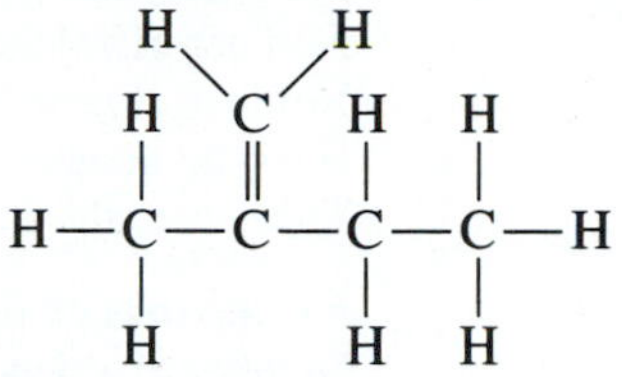

Problem 1.21 Write a line-segment formula for $(CH_3)_2CHCH(CH_3)_2$.

1.11 Formal Charge

So far we have considered only molecules whose atoms are neutral. But in some compounds one or more atoms may be charged, either positive or negative. Because such charges usually affect the chemical reactions of such molecules, it is important to know how to tell where the charge is located.

Consider the formula for hydronium ion, H_3O^+, the product of the reaction of a water molecule with a proton.

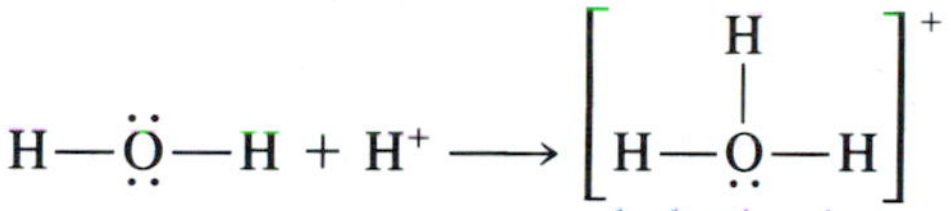

$$H-\ddot{\underset{\cdot\cdot}{O}}-H + H^+ \longrightarrow \left[H-\overset{\overset{\displaystyle H}{|}}{\underset{\cdot\cdot}{O}}-H \right]^+ \tag{1.4}$$

hydronium ion

The structure has eight electrons around the oxygen and two electrons around each hydrogen, so that all valence shells are complete. Note that there are eight valence electrons altogether. Oxygen contributes six, and each hydrogen contributes one, for a total of nine, but the ion has a single positive charge, so one electron must have been given away, leaving eight. Six of these eight electrons are used to form three O—H single bonds, leaving one unshared electron pair on the oxygen.

Although the entire hydronium ion carries a positive charge, we can ask, "Which atom, in a formal sense, bears the charge?" To determine **formal charge,** we consider that each atom "owns" *all* of its unshared electrons plus only *half* of its shared electrons. We then subtract this total from the number of valence electrons in the neutral atom to get the formal charge. This definition can be expressed in equation form as follows:

$$\text{Formal charge} = \text{number of valence electrons in the neutral atom} - \left(\text{unshared electrons} + \text{half the shared electrons}\right) \tag{1.5}$$

or, in a simplified form,

$$\text{Formal charge} = \text{number of valence electrons in the neutral atom} - (\text{dots} + \text{bonds})$$

Let us apply this definition to the hydronium ion.

For each hydrogen atom:
Number of valence electrons in the neutral atom = 1
Number of unshared electrons = 0
Half the number of the shared electrons = 1
Therefore, the formal charge = 1 − (0 + 1) = 0

For the oxygen atom:
Number of valence electrons in the neutral atom = 6
Number of unshared electrons = 2
Half the number of the shared electrons = 3
Therefore, the formal charge = 6 − (2 + 3) = +1

It is the oxygen atom that formally carries the +1 charge in the hydronium ion.

EXAMPLE 1.13 On which atom is the formal charge in the hydroxide ion, OH^-?

Solution The electron-dot formula is

[:Ö:H]⁻

Oxygen contributes six electrons, hydrogen contributes one, and there is one more for the negative charge, for a total of eight electrons. The formal charge on oxygen is 6 − (6 + 1) = −1, so the oxygen carries the negative charge. The hydrogen is neutral.

Problem 1.22 Calculate the formal charge on the nitrogen atom in ammonia, NH_3; in the ammonium ion, NH_4^+; and in the amide ion, NH_2^-.

Now let us look at a slightly more complex situation involving electron-dot formulas and formal charge.

1.12 Resonance

Sometimes an electron pair is involved with more than two atoms, in forming bonds. As an example, consider the structure of the carbonate ion, CO_3^{2-}.

The total number of valence electrons in the carbonate ion is 24 (4 from the carbon, 3 × 6 = 18 from the three oxygens, *plus* 2 more electrons that give the ion its negative charge; these 2 electrons presumably have been donated by some metal, perhaps one each from two sodium atoms). An electron-dot formula that completes the valence shell of eight electrons around the carbon and each oxygen is

[:Ö: / C::Ö / :Ö:]²⁻ or [:Ö: / C=Ö / :Ö:]²⁻

carbonate ion, CO_3^{2-}

The structure contains two carbon–oxygen *single* bonds and one carbon–oxygen *double* bond. Application of the definition for formal charge shows that the carbon is formally neutral, each singly bonded oxygen has a formal charge of −1, and the doubly bonded oxygen is formally neutral.

Problem 1.23 Show that the last sentence of the preceding paragraph is correct.

When we wrote the electron-dot formula for the carbonate ion, our choice of which oxygen atom would be doubly bonded to the carbon atom was purely arbitrary. There are in fact *three exactly equivalent* structures that we might write.

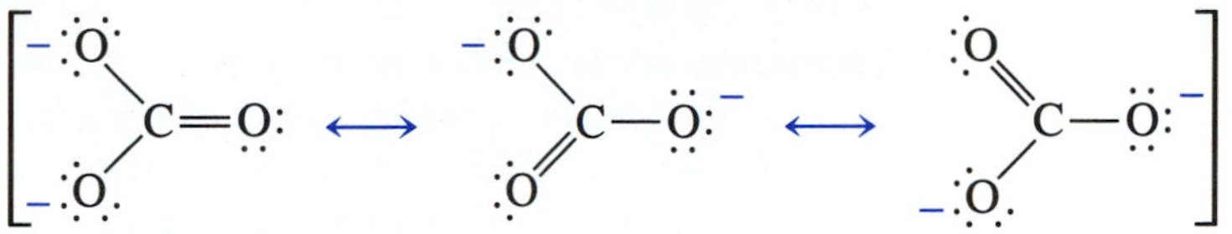

three equivalent structures for the carbonate ion

In each structure there is one C=O bond and there are two C—O bonds. These structures have the same arrangement of the atoms. They differ from one another *only* in the arrangement of the electrons.

The three structures for the carbonate ion are redrawn below, with curved arrows* to show how electron pairs can be moved to interrelate the structures:

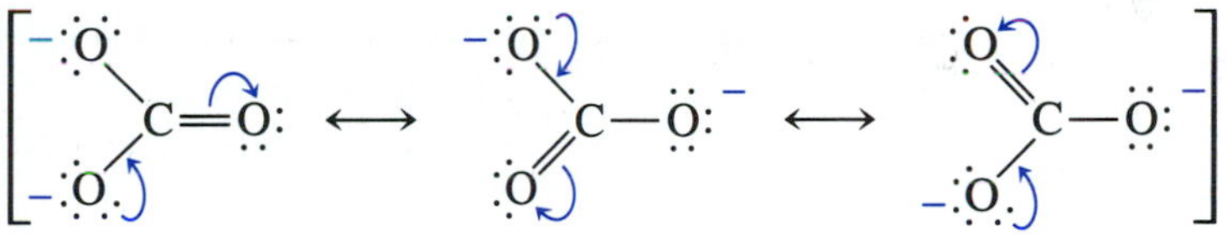

---

* Chemists use curved arrows to keep track of a change in the location of electrons. For example, the curved arrow in the structure

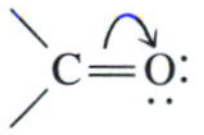

means that the two electrons in *one* of the bonds between carbon and oxygen move out onto the oxygen, to give the dipolar structure

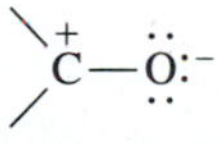

Similarly, the curved arrow in the structure

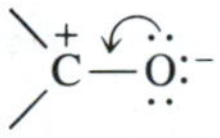

means that an unshared electron *pair* on the oxygen atom moves between the oxygen and carbon to form the C=O bond, giving

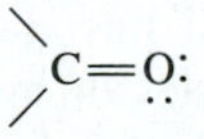

The arrows must be carefully drawn to show exactly which electron pair moves and where it moves to. We will use curved arrows throughout this text as a way of keeping track of electron movement. Several curved-arrow problems are included at the end of this chapter to help you get used to drawing them.

Physical measurements tell us that no one of the foregoing structures accurately describes the real carbonate ion. For example, although each structure shows two different types of bonds between carbon and oxygen, we find experimentally that ***all three carbon–oxygen bond lengths are identical: 1.31 Å.*** This distance is intermediate between the normal C=O (1.20 Å) and C—O (1.41 Å) bond lengths. To explain this fact, we usually say that the real carbonate ion has a structure that is a **resonance hybrid** of the three contributing resonance structures. It is as if we could take an average of the three structures. In the real carbonate ion, the two formal negative charges are spread equally over the three oxygen atoms, so that each oxygen atom carries two-thirds of a negative charge.

Resonance ***arises whenever we can write two or more structures for a molecule with different arrangements of the electrons but identical arrangements of the atoms.*** Resonance is very different from isomerism, for which the atoms themselves are arranged differently. ***When resonance is possible, the substance is said to have a structure that is a*** **resonance hybrid** ***of the various*** **contributing structures.** ***We use a double-headed arrow (↔) between contributing structures to distinguish resonance from an equilibrium, for which we use ⇌.***

Each carbon–oxygen bond in the carbonate ion is neither single nor double, but something in between—perhaps a one-and-one-third bond (any particular carbon–oxygen bond is single in two contributing structures and double in one). Sometimes we represent a resonance hybrid with one formula by writing a solid line for each full bond and a dotted line for each partial bond (in carbonate ion, the dots represent one-third of a bond).

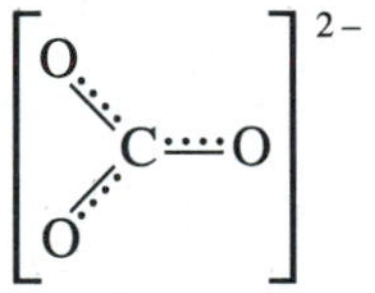

$$\left[\begin{array}{l} O \\ \quad \text{C}\cdots O \\ O \end{array} \right]^{2-}$$

carbonate ion
resonance hybrid

Problem 1.24 Draw the three equivalent contributing resonance structures for the nitrate ion, NO_3^-. What is the formal charge on the nitrogen atom and on each oxygen atom in the individual structures? What is the charge on the oxygens and on the nitrogen in the resonance hybrid structure? Show with curved arrows how the structures can be interconverted.

Although electron-dot formulas are often useful, they have some limitations. The Lewis theory of bonding itself has some limitations, especially in explaining the three-dimensional geometries of molecules. For this purpose in particular, we will discuss how another theory of bonding, involving orbitals, is more useful.

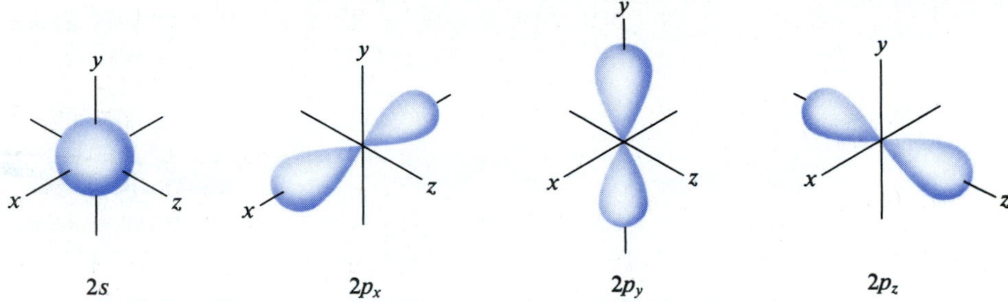

FIGURE 1.2 The shapes of the s and p orbitals used by the valence electrons of carbon. The nucleus is at the origin of the three coordinate axes.

1.13 *The Orbital View of Bonding; the Sigma Bond*

The atomic orbitals named in Sec. 1.1 have definite shapes. The *s* orbitals are spherical. The electrons that fill an *s* orbital confine their movement to a spherical region of space around the nucleus. The three *p* orbitals are dumbbell-shaped and mutually perpendicular, oriented along the three coordinate axes, *x*, *y*, and *z*. Figure 1.2 shows the shapes of these orbitals.

In the orbital view of bonding, atoms approach each other in such a way that their atomic orbitals can *overlap* to form a bond. For example, if two hydrogen atoms form a hydrogen molecule, their two spherical 1*s* orbitals combine to form a new orbital that encompasses both of the atoms (see Figure 1.3). This orbital contains both valence electrons (one from each hydrogen). Like atomic orbitals, each **molecular orbital** can contain no more than two electrons. In the hydrogen molecule these electrons mainly occupy the space between the two nuclei.

The orbital in the hydrogen molecule is cylindrically symmetric along the H—H internuclear axis. Such orbitals are called **sigma (σ) orbitals,** and the

H + H → H—H

1*s* atomic orbitals

s-*s* molecular orbital

FIGURE 1.3 The molecular orbital representation of covalent bond formation between two hydrogen atoms.

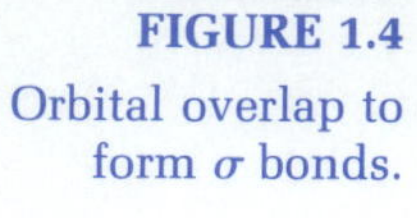

FIGURE 1.4
Orbital overlap to form σ bonds.

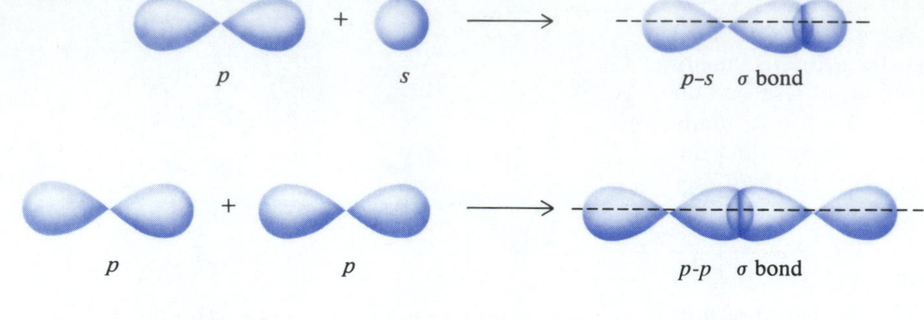

bond is referred to as a **sigma bond.** Sigma bonds may also be formed by the overlap of an s and a p orbital or of two p orbitals, as shown in Figure 1.4.*

Let us see how these ideas apply to the bonding in carbon compounds.

1.14 Carbon sp^3 Hybrid Orbitals

In a carbon atom, the six electrons are arranged as shown in Figure 1.5 (compare with carbon in Table 1.2). The 1s shell is filled, and the four valence electrons are in the 2s orbital and two different 2p orbitals. There are a few things to notice about Figure 1.5. The energy scale at the left represents the energy of electrons in the various orbitals. The farther the electron is from the nucleus, the greater its potential energy, because it takes energy to keep the electron (negatively charged) and the nucleus (positively charged) apart. The 2s orbital has a slightly lower energy than the three 2p orbitals, which have equal energies (they differ from one another only in orientation around the nucleus, as shown in Figure 1.2). The two highest-energy electrons are placed in different 2p orbitals rather than in the same orbital, because this keeps them farther apart and thus reduces the repulsion between these like-charged particles. One p orbital is vacant.

We might get a misleading idea about the bonding of carbon from Figure 1.5. For example, we might think that carbon should form only two bonds (to complete the partially filled 2p orbitals), or perhaps three bonds (if some atom donated two electrons to the empty 2p orbital). But we know from experience that this picture is wrong. Carbon usually forms *four* single bonds, and often these bonds are all equivalent, as in CH_4 or CCl_4. How can this discrepancy between theory and fact be resolved?

---

* Two properly aligned p orbitals can also overlap to form another type of bond, called a π (pi) bond. We will discuss this type of bond in Chapter 3.

p + p → π bond

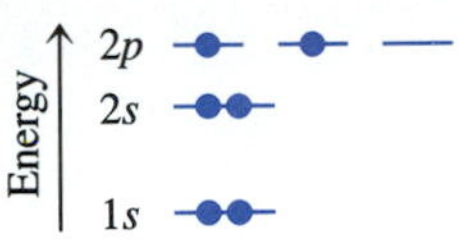
Energy
2p
2s
1s

FIGURE 1.5
Distribution of the six electrons in a carbon atom. Each dot stands for an electron.

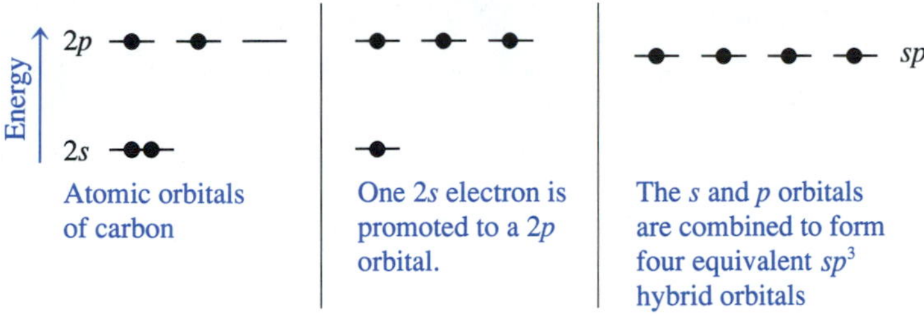
Energy
2p
2s
Atomic orbitals of carbon
One 2s electron is promoted to a 2p orbital.
sp3
The s and p orbitals are combined to form four equivalent sp3 hybrid orbitals

FIGURE 1.6
The formation of four sp^3 hybrid orbitals. The dots stand for electrons. (Only the electrons in the valence shell are shown; the electrons in the 1s orbital are omitted because they are not involved in bonding.)

One way is illustrated in Figure 1.6. First, one of the 2*s* electrons is "promoted" to the vacant 2*p* orbital. This electron arrangement is shown in the middle of the figure. Although it costs some energy to raise an electron from the 2*s* to the 2*p* level, some of this energy is recovered by reducing electron repulsion (there are no longer two electrons in any one orbital).

The picture in the middle of Figure 1.6 is still not quite satisfactory. There are four half-filled orbitals, so we can now expect four bonds to form (with, say, four hydrogen atoms). But such bonds would not have identical energies, and we know that all four C—H bonds in methane are identical. The resolution of this discrepancy is to allow the *s* and *p* orbitals to "mix" or "combine" and form four identical **sp^3 hybrid orbitals.** These hybrid orbitals are called sp^3 because each one is the result of combining one *s* orbital and three *p* orbitals. As shown in Figure 1.6, their energy is a little lower than that of the 2*p* orbitals but somewhat greater than that of the 2*s* orbital. The shape of sp^3 orbitals resembles the shape of *p* orbitals, except that the dumbbell is lopsided, and the electrons are more likely to be found in the lobe that extends out the greater distance from the nucleus, as shown in Figure 1.7. The four sp^3 hybrid orbitals of a single carbon atom are directed toward the corners of a regular tetrahedron, shown in Figure 1.7. This particular geometry puts each orbital as far from the other three orbitals as it can be and thus minimizes repulsion when the orbitals are filled with electron pairs. The angle between any two of the four bonds formed from sp^3 orbitals is approximately 109.5°, the angle made by lines drawn from the center to the corners of a regular tetrahedron.

Hybrid orbitals can form sigma bonds by overlap with other hybrid orbitals or with atomic orbitals. Figure 1.8 shows some examples.

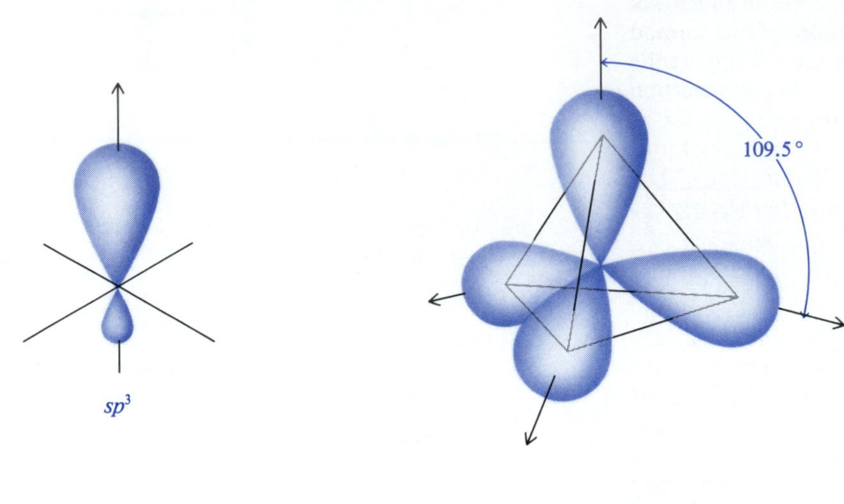

FIGURE 1.7

An sp^3 orbital extends mainly in one direction from the nucleus and forms bonds with other atoms in that direction. The four sp^3 orbitals of any particular carbon atom are directed toward the corners of a regular tetrahedron, as shown in the right-hand part of the figure (in this part of the drawing, the small "back" lobes of the orbitals have been omitted for simplification).

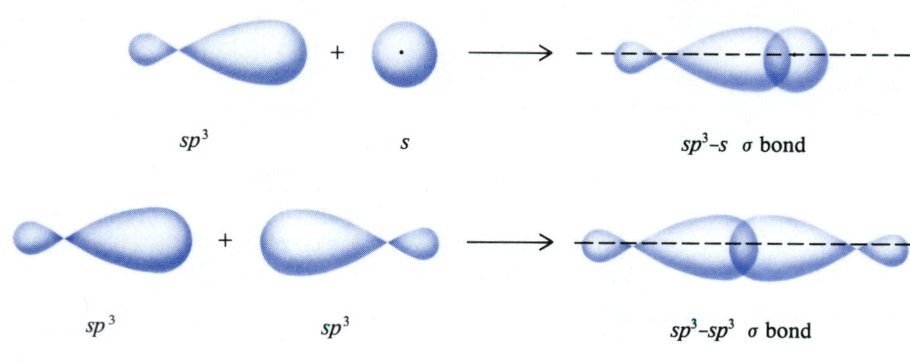

FIGURE 1.8

Examples of sigma (σ) bonds formed from sp^3 hybrid orbitals.

1.15 Tetrahedral Carbon; the Bonding in Methane

We can now describe the way a carbon atom combines with four hydrogen atoms to form methane. This process is pictured in Figure 1.9. The carbon atom is joined to each hydrogen atom by a sigma bond, which is formed by the overlap of a carbon sp^3 orbital with a hydrogen 1s orbital. The four sigma bonds are directed from the carbon nucleus to the corners of a regular tetrahedron. In this way, the electron pair in any one bond experiences minimum repulsion from the electrons in the other bonds. Each H—C—H **bond angle** is the same, 109.5°. To summarize, in methane there are four sp^3–s C—H sigma bonds, each directed from the carbon atom to one of the four corners of a regular tetrahedron.

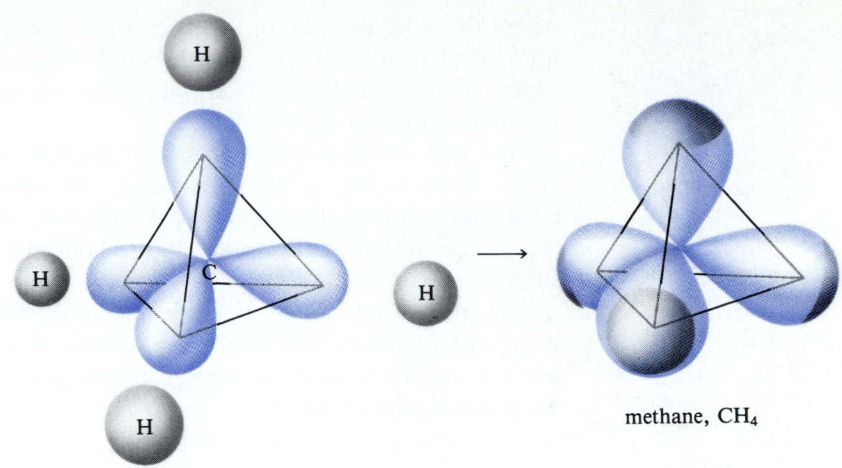

FIGURE 1.9
A molecule of methane, CH_4, formed by the overlap of the four sp^3 carbon orbitals with the $1s$ orbitals of four hydrogen atoms. The resulting molecule has the geometry of a regular tetrahedron and contains four σ bonds of the sp^3–s type.

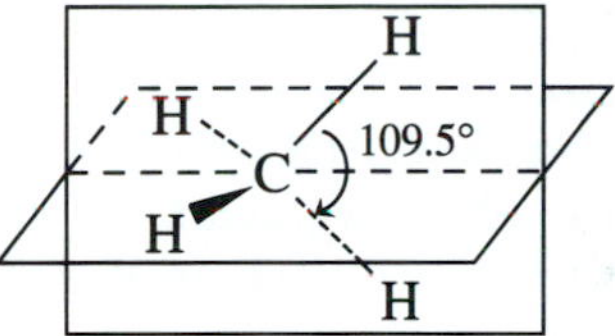

FIGURE 1.10
The carbon and two of the hydrogens in methane form a plane that perpendicularly bisects the plane formed by the carbon and the other two hydrogens.

Problem 1.25 Considering the repulsion that exists between electrons in different bonds, give a reason why a planar geometry for methane would be less stable than the tetrahedral geometry.

Because the tetrahedral geometry of carbon plays such an important role in organic chemistry, it is a good idea to become familiar with the features of a regular tetrahedron. One feature is that *the center and any two corners of a tetrahedron form a plane that is the perpendicular bisector of a similar plane formed by the center and the other two corners*. In methane, for example, any two hydrogens and the carbon form a plane that perpendicularly bisects the plane formed by the carbon and the other two hydrogens. These planes are illustrated in Figure 1.10.

The geometry of carbon with four single bonds, as in methane, may be represented as

H, C, H, H, H (109.5°) or H, C, H, H, H (109.5°) or H, C, H, H, H (109.5°)

where the solid lines lie in the plane of the page, the dashed wedge goes behind the plane of the paper, and the solid wedge extends out of the plane of the paper toward you. Formulas drawn in this way are sometimes called 3D (that is, three-dimensional) formulas.

Now that we have described single covalent bonds and their geometry, we are ready to tackle, in the next chapter, the structure and chemistry of saturated hydrocarbons. But before we do that, we will present a brief overview of organic chemistry, so that you can see how the subject will be organized for study.

Because carbon atoms can be linked to one another or to other atoms in so many different ways, the number of possible organic compounds is almost limitless. Literally millions of organic compounds have been characterized, and the number grows daily. How can we hope to study this vast subject systematically? Fortunately, organic compounds can be classified according to their structures into a relatively small number of groups. Structures can be classified both according to the carbon framework (sometimes called the carbon *skeleton*) and according to the groups that are attached to that framework.

1.16 Classification According to Molecular Framework

There are three main classes of molecular frameworks for organic structures.

1.16a Acyclic Compounds

By *acyclic* we mean *not cyclic*. Acyclic organic molecules have chains of carbon atoms but no rings. As we have seen, the chains may be unbranched or branched.

unbranched chain of eight carbon atoms

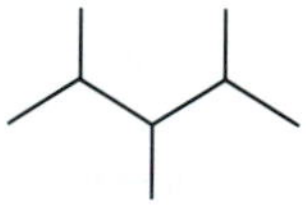

branched chain of eight carbon atoms

Pentane is an example of an acyclic compound with an unbranched carbon chain, whereas isopentane and neopentane are also acyclic but have branched carbon frameworks (Sec. 1.9). Figure 1.11 shows the structures of a few acyclic compounds that occur in nature.

FIGURE 1.11 Examples of natural acyclic compounds, their sources (in parentheses), and selected characteristics.

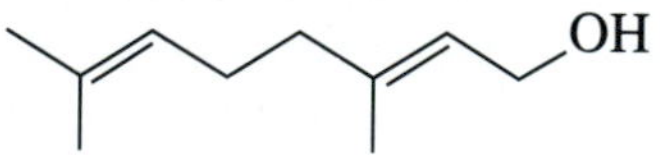

OH

geraniol
(oil of roses)
bp 229–230°C

A branched chain compound used in perfumes

$CH_3(CH_2)_5CH_3$

heptane
(petroleum)
bp 98.4°C

A hydrocarbon present in petroleum, used as a standard in testing the knock of gasoline engines

$CH_3\overset{\overset{\displaystyle O}{\|}}{C}(CH_2)_4CH_3$

2-heptanone
(oil of cloves)
bp 151.5°C

A colorless liquid with a fruity odor, in part responsible for the "peppery" odor of blue cheese

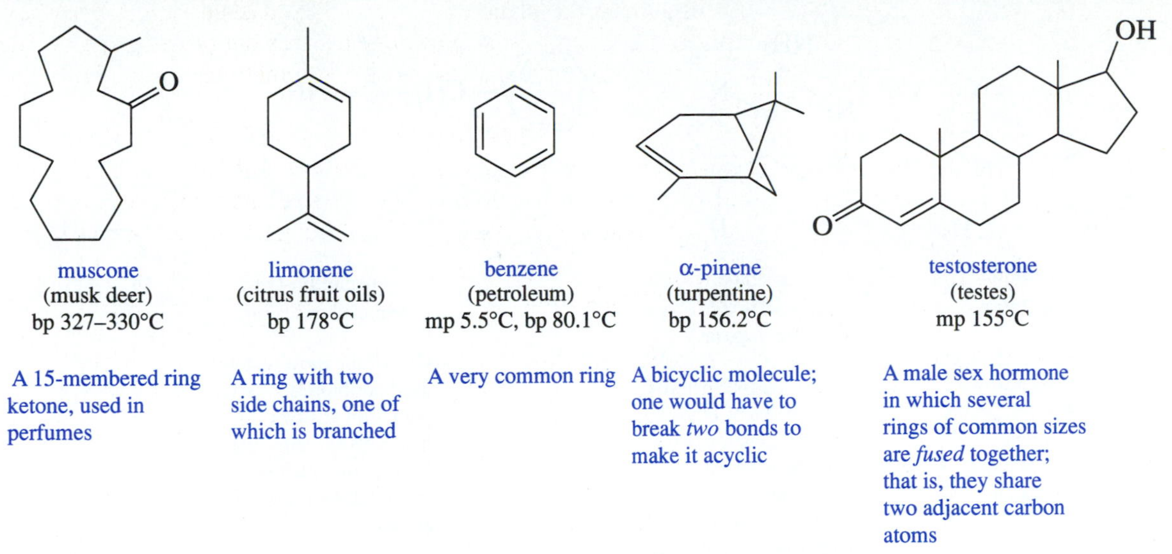
O
O
OH
muscone
(musk deer)
bp 327–330°C
A 15-membered ring ketone, used in perfumes
limonene
(citrus fruit oils)
bp 178°C
A ring with two side chains, one of which is branched
benzene
(petroleum)
mp 5.5°C, bp 80.1°C
A very common ring
α-pinene
(turpentine)
bp 156.2°C
A bicyclic molecule; one would have to break *two* bonds to make it acyclic
testosterone
(testes)
mp 155°C
A male sex hormone in which several rings of common sizes are *fused* together; that is, they share two adjacent carbon atoms

FIGURE 1.12 Examples of natural carbocyclic compounds with rings of various sizes and shapes. The source and special features of each structure are indicated below it.

1.16b Carbocyclic Compounds Carbocyclic compounds contain rings of carbon atoms. The smallest possible carbocyclic ring has three carbon atoms, but carbon rings come in many sizes and shapes. The rings may have chains of carbon atoms attached to them and may contain multiple bonds. Many compounds with more than one carbocyclic ring are known. Figure 1.12 shows the structures of a few carbocyclic compounds that occur in nature. Five- and six-membered rings are most common, but smaller and larger rings are also found.

1.16c Heterocyclic Compounds Heterocyclic compounds make up the third and largest class of molecular frameworks for organic compounds. In heterocyclic compounds, at least one atom in the ring must be a heteroatom, an atom that is *not* carbon. The most common heteroatoms are **oxygen, nitrogen,** and **sulfur,** but heterocyclics with other elements are also known. More than one heteroatom may be present, and if so, the heteroatoms may be alike or different. Heterocyclic rings come in many sizes, may contain multiple bonds, may have carbon chains or rings attached to them, and in short may exhibit a great variety of structures. Figure 1.13 shows the structures of a few natural products that contain heterocyclic rings. In these abbreviated structural formulas, the symbols for the heteroatoms are shown, but the carbons are indicated using lines only.

The formulas in Figures 1.11 through 1.13 show not only the molecular frameworks, but also various groups of atoms that may be part of or attached to the frameworks. Fortunately, these groups can also be classified in a way that helps simplify the study of organic chemistry.

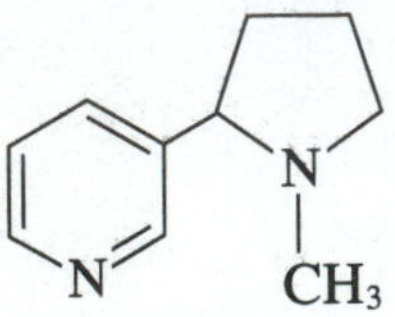

nicotine
bp 246°C

Present in tobacco, nicotine has two heterocyclic rings of different sizes, each containing one nitrogen.

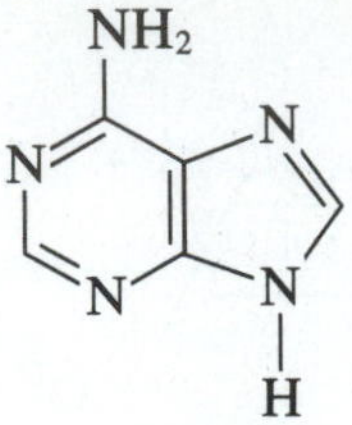

adenine
mp 360–365°C
(decomposes)

One of the four heterocyclic bases of DNA, adenine contains two fused heterocyclic rings, each of which contains two heteroatoms (nitrogen).

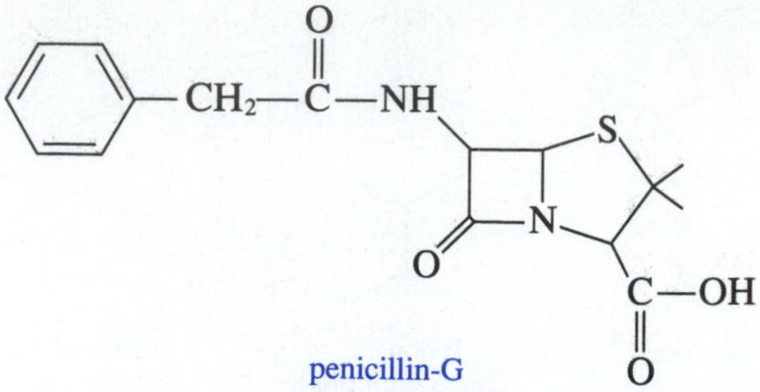

penicillin-G
(amorphous solid)

One of the most widely used antibiotics, penicillin has two heterocyclic rings, the smaller of which is crucial to biological activity.

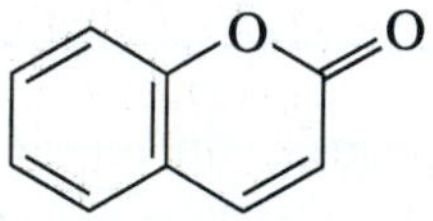

coumarin
mp 71°C

Found in clover and grasses, coumarin produces the pleasant odor of new-mown hay.

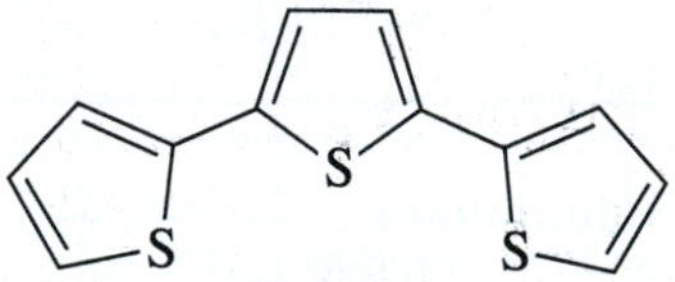

α-terthienyl
mp 92–93°C

This compound, with three linked sulfur-containing rings, is present in certain marigold species.

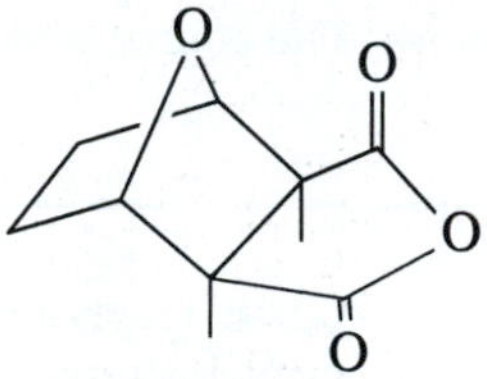

cantharidin
mp 218°C

This compound, an oxygen heterocycle, is the active principle in cantharis (also known as Spanish fly), a material isolated from certain dried beetles of the species *Cantharis vesicatoria* and incorrectly thought by some to increase sexual desire.

FIGURE 1.13 Examples of natural heterocyclic compounds having a variety of heteroatoms and ring sizes.

1.17 *Classification According to Functional Group*

Certain groups of atoms have chemical properties that depend only moderately on the molecular framework to which they are attached. These groups of atoms are called **functional groups.** The **hydroxyl group, —OH,** is an example of a functional group, and compounds with this group attached to a carbon framework are called **alcohols.** In most organic reactions, some chemical change occurs at the functional group, but the rest of the molecule keeps its original structure. For example, in the reaction of ethanol with sodium (mentioned in Sec. 1.8), the hydrogen atom of the hydroxyl group is replaced by sodium.

$$\underset{\text{ethanol}}{2\ CH_3CH_2OH} + \underset{\text{sodium}}{2\ Na} \longrightarrow \underset{\text{sodium ethoxide}}{2\ CH_3CH_2O^-Na^+} + \underset{\text{hydrogen}}{H_2} \quad (1.6)$$

The other atoms (two carbons, five hydrogens, and the oxygen) have the same arrangement in the product (sodium ethoxide) as they have in the starting material (ethanol). There are thousands of alcohols that behave similarly on treatment with sodium.

EXAMPLE 1.14 Predict the product of the reaction of isopropyl alcohol,

$$\underset{\displaystyle OH}{CH_3\overset{|}{C}HCH_3},$$

with sodium.

Solution

$$\underset{\text{isopropyl alcohol}}{2\ CH_3\underset{\displaystyle OH}{\overset{|}{C}H}CH_3} + 2\ Na \longrightarrow \underset{\text{sodium isopropoxide}}{2\ CH_3\underset{\displaystyle O^-Na^+}{\overset{|}{C}H}CH_3} + H_2$$

In the product, only the hydrogen of the hydroxyl group is replaced by sodium, just as in eq. 1.6. The oxygen in the product remains attached to the middle carbon of the three-carbon chain. Put another way, the sodium reacts only with the hydroxyl group, and the rest of the isopropyl alcohol molecule remains unchanged.

The reaction of alcohols with sodium can be written in the general form

$$2\ ROH + 2\ Na \longrightarrow 2\ RO^-Na^+ + H_2 \quad (1.7)$$

where R is some organic group.

Problem 1.26 Write the structure of the product of the reaction of sodium metal with each of the following alcohols:

a. $CH_3CH_2CH_2OH$ b. $CH_3CH(OH)CH_2CH_3$

This maintenance of most of the structural formula throughout a chemical reaction greatly simplifies our study of organic chemistry. It allows us to focus attention on the chemistry of the various functional groups. We can study classes of compounds (such as alcohols) instead of having to learn the chemistry of each individual compound.

Some of the main functional groups that we will study are listed in Table 1.5, together with a typical compound of each type. Although we will describe these classes of compounds in greater detail in later chapters, it would be a good idea for you to become familiar with their names and structures now. If a

TABLE 1.5 The main functional groups

	Structure	*Class of compound*	*Specific example*	*Common name of the specific example*
A. Functional groups that are a part of the molecular framework	$-\overset{\vert}{\underset{\vert}{C}}-\overset{\vert}{\underset{\vert}{C}}-$	alkane	$CH_3{-}CH_3$	ethane, a component of natural gas
	$>C{=}C<$	alkene	$CH_2{=}CH_2$	ethylene, used to make polyethylene
	$-C{\equiv}C-$	alkyne	$HC{\equiv}CH$	acetylene, used in welding
B. Functional groups containing oxygen *1. With carbon-oxygen single bonds*	$-\overset{\vert}{\underset{\vert}{C}}-OH$	alcohol	CH_3CH_2OH	ethyl alcohol, found in beer, wines, and liquors
	$-\overset{\vert}{\underset{\vert}{C}}-O-\overset{\vert}{\underset{\vert}{C}}-$	ether	$CH_3CH_2OCH_2CH_3$	diethyl ether, the common anesthetic
*2. With carbon-oxygen double bonds**	$-\overset{O}{\overset{\Vert}{C}}-H$	aldehyde	$CH_2{=}O$	formaldehyde, used to preserve biological specimens
	$-\overset{\vert}{\underset{\vert}{C}}-\overset{O}{\overset{\Vert}{C}}-\overset{\vert}{\underset{\vert}{C}}-$	ketone	$CH_3\overset{O}{\overset{\Vert}{C}}CH_3$	acetone, a solvent for varnish and rubber cement
3. With single and double carbon-oxygen bonds	$-\overset{O}{\overset{\Vert}{C}}-OH$	carboxylic acid	$CH_3\overset{O}{\overset{\Vert}{C}}-OH$	acetic acid, a component of vinegar
	$-\overset{O}{\overset{\Vert}{C}}-O-\overset{\vert}{\underset{\vert}{C}}-$	ester	$CH_3\overset{O}{\overset{\Vert}{C}}-OCH_2CH_3$	ethyl acetate, a solvent for nail polish and model airplane glue
*C. Functional groups containing nitrogen***	$-\overset{\vert}{\underset{\vert}{C}}-NH_2$	primary amine	$CH_3CH_2NH_2$	ethylamine, smells like ammonia
	$-C{\equiv}N$	cyanide or nitrile	$CH_2{=}CH{-}C{\equiv}N$	acrylonitrile, raw material for making Orlon

*The $>C{=}O$ group, present in several functional groups, is called a **carbonyl group.** The $-\overset{O}{\overset{\Vert}{C}}-OH$ group of acids is called a **carboxyl group** (a contraction of *carb*onyl and hydr*oxyl*).

The $-NH_2$ group is called an **amino group.

	Structure	Class of compound	Specific example	Common name of the specific example
D. Functional group with oxygen and nitrogen	$-\overset{\overset{\displaystyle O}{\parallel}}{C}-NH_2$	primary amide	$H_2N-\overset{\overset{\displaystyle O}{\parallel}}{C}-NH_2$	urea, a fertilizer and odorless component of urine
*E. Functional groups containing sulfur**	$-\overset{\vert}{\underset{\vert}{C}}-SH$	thiol (also called mercaptan)	CH_3SH	methanethiol, has the odor of rotten cabbage
	$-\overset{\vert}{\underset{\vert}{C}}-S-\overset{\vert}{\underset{\vert}{C}}-$	thioether (also called sulfide)	$(CH_2{=}CHCH_2)_2S$	diallyl sulfide, has the odor of garlic

* Thiols and thioethers are the sulfur analogs of alcohols and ethers.

particular functional group is mentioned before its chemistry is discussed in detail, and you forget what it is, you can refer to Table 1.5 or to the inside front cover of this book.

Problem 1.27 What functional groups can you find in the following natural products? (Their formulas are given in Figures 1.11 and 1.12.)

a. geraniol b. muscone c. limonene d. testosterone

Additional Problems

1.28. Show the number of valence electrons in each of the following atoms. Let the element's symbol represent its kernel, and use dots for the valence electrons.
a. carbon **b.** fluorine **c.** silicon
d. boron **e.** sulfur **f.** phosphorus

1.29. When a solution of salt (sodium chloride) in water is treated with a silver nitrate solution, a white precipitate forms immediately. When tetrachloromethane is shaken with aqueous silver nitrate, no such precipitate is produced. Explain these facts in terms of the types of bonds present in the two chlorides.

1.30. Use the relative positions of the elements in the periodic table (Table 1.3) to classify the following substances as ionic or covalent:
a. NaF **b.** F_2 **c.** $MgCl_2$ **d.** P_2S_5
e. S_2Cl_2 **f.** LiCl **g.** ClF **h.** $SiCl_4$

1.31. For each of the following elements, tell (1) how many valence electrons it has and (2) what its common valence is:
a. oxygen **b.** hydrogen **c.** chlorine
d. nitrogen **e.** sulfur **f.** carbon

1.32. Write a structural formula for each of the following compounds, using a line to represent each single bond and dots for any unshared electron pairs:
a. CH_3Cl **b.** C_3H_8 **c.** C_2H_5F
d. CH_3NH_2 **e.** CH_3CH_2OH **f.** CH_2O

1.33. Draw a structural formula for each of the following covalent molecules. Which bonds are polar? Indicate the polarity by proper placement of the symbols $\delta+$ and $\delta-$.
a. Cl_2 **b.** CH_3F **c.** CO_2 **d.** HBr
e. SF_6 **f.** CH_4 **g.** SO_2 **h.** CH_3OCH_3

1.34. Considering bond polarity, which hydrogen in acetic acid, $CH_3\overset{\overset{O}{\|}}{C}{—}OH$, do you expect to be most acidic? Write an equation for the reaction between acetic acid and metallic sodium.

1.35. Draw structural formulas for all possible isomers having the following molecular formulas:
a. C_3H_8 **b.** C_3H_7Cl **c.** $C_2H_4Cl_2$ **d.** $C_3H_6Br_2$
e. C_4H_9F **f.** $C_2H_2Cl_2$ **g.** C_3H_6 **h.** $C_4H_{10}O$

1.36. Draw structural formulas for the five isomers of C_6H_{14}. As you write them out, try to be systematic.

1.37. For each of the following abbreviated structural formulas, write a structural formula that shows all the bonds:
a. $CH_3(CH_2)_4CH_3$ **b.** $(CH_3)_3CCH_2CH_3$ **c.** $(CH_3)_2CHOH$
d. $(CH_3CH_2)_2S$ **e.** CH_2ClCH_2OH **f.** $(CH_3)_2NCH_2CH_3$

1.38. Write structural formulas that correspond to the following abbreviated structures, and show the correct number of hydrogens on each carbon:

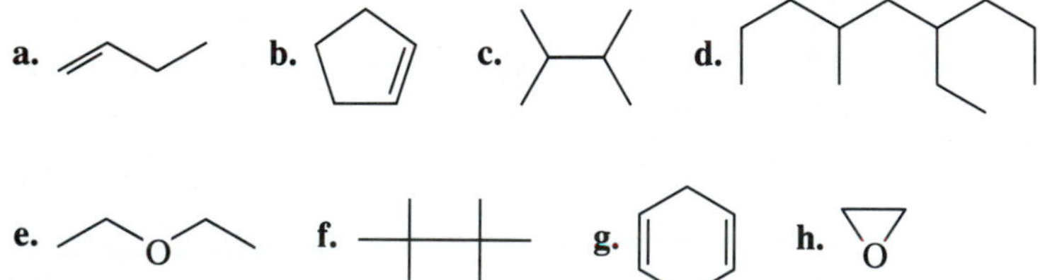

1.39. An abbreviated formula of geraniol is shown in Figure 1.11.
a. How many carbons does geraniol have?
b. What is its molecular formula?
c. Write a more detailed structural formula for it.

1.40. What is the *molecular formula* for each of the following compounds? Consult Figures 1.12 and 1.13 for the abbreviated structural formulas.
a. muscone **b.** benzene **c.** testosterone
d. nicotine **e.** limonene **f.** adenine

1.41. Write electron-dot formulas for the following species. Show where the formal charges, if any, are located.
a. nitrous acid, $HONO$ **b.** nitric acid, $HONO_2$
c. formaldehyde, H_2CO **d.** ammonium ion, NH_4^+
e. cyanide ion, CN^- **f.** carbon monoxide, CO
g. sulfate ion, SO_4^{2-} **h.** boron trifluoride, BF_3
i. hydrogen peroxide, H_2O_2 **j.** bicarbonate ion, HCO_3^-

1.42. Consider each of the following highly reactive carbon species. What is the formal charge on carbon in each species?

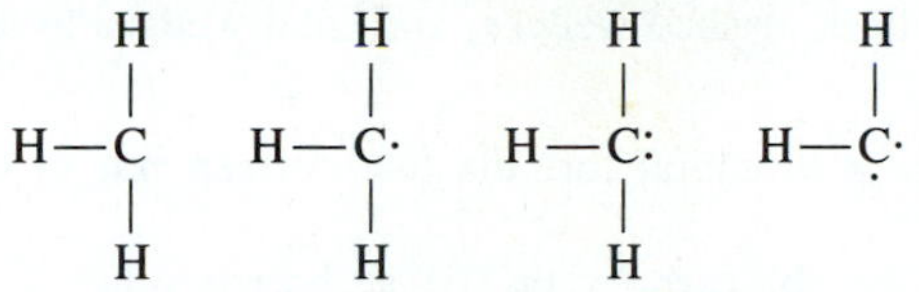

```
   H          H          H          H
   |          |          |          |
H—C       H—C·       H—C:       H—C·
   |          |          |          ·
   H          H          H
```

1.43. Draw electron-dot formulas for the two contributors to the resonance hybrid structure of the nitrite ion, NO_2^-. (Each oxygen is connected to the nitrogen.) What is the charge on each oxygen in each contributor and in the hybrid structure? Show by curved arrows how the electron pairs can relocate to interconvert the two structures.

1.44. Write the contributors to the resonance hybrid structures of

a. azide ion, a linear ion with three connected nitrogens, N_3^-

b. acetate ion, $CH_3CO_2^-$.

1.45. Write out the structure obtained when electrons move as indicated by the curved arrows in the following structure:

```
        ..
       ↷O:
        ‖
CH3—C↶—NH2
          ..
```

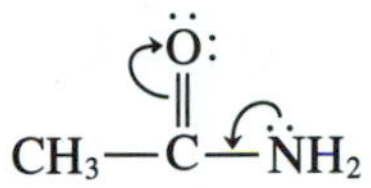

Does each atom in the resulting structure have a complete valence shell of electrons? Locate any formal charges in each structure.

1.46. Add curved arrows to the following structures to show how electron pairs must be moved to interconvert the structures, and locate any formal charges.

$$\left[C_6H_5\text{—}\ddot{O}\text{—H} \longleftrightarrow (:C_6H_5)\text{=}\ddot{O}\text{—H} \right]$$

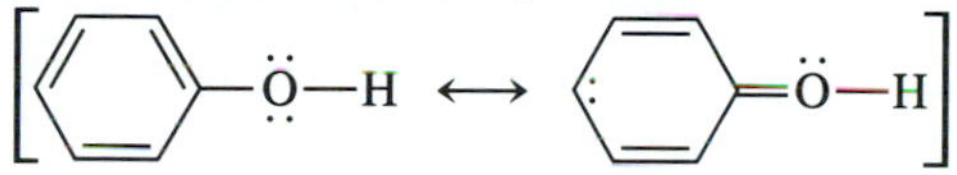

1.47. Add curved arrows to show how electrons must move to form the product from the reactants in the following equation, and locate any formal charges.

```
                        ..                      ..
                       :O                      :O:
          ..            ‖                       |
CH3—NH2 + CH3—C—OCH3   ⟶   CH3—C—OCH3
                                                |
                                             H2N—CH3
```

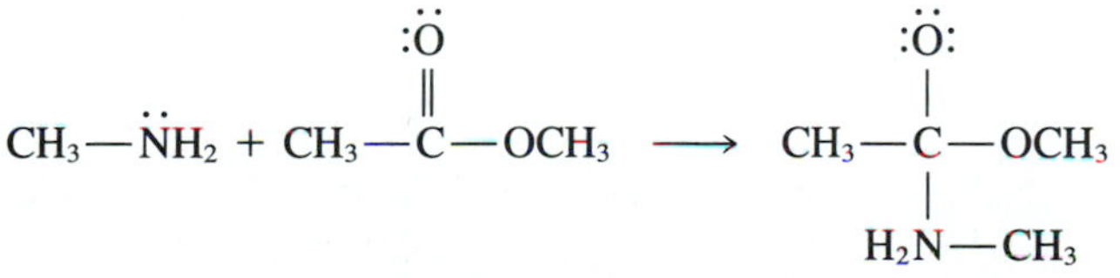

1.48. Each of the following substances contains ionic and covalent bonds. Draw their electron-dot formulas.

a. CH_3ONa **b.** NH_4Cl

1.49. Fill in any unshared electron pairs that are missing from the following formulas:

a. CH_3CH_2OH **b.** $CH_3\overset{\overset{\displaystyle O}{\|}}{C}\text{—}OH$ **c.** $(CH_3)_2NH$ **d.** $CH_3OCH_2CH_2OH$

1.50. Make a drawing (similar to the right-hand part of Figure 1.6) of the electron distribution that will be expected in nitrogen atoms if the *s* and *p* orbitals are hybridized to sp^3. Based on this model, predict the geometry of the ammonia molecule, NH_3.

1.51. The ammonium ion, NH_4^+, has a tetrahedral geometry analogous to that of methane. Explain this structure in terms of atomic and molecular orbitals.

1.52. Silicon is just below carbon in the periodic table. Predict the geometry of silane, SiH_4.

1.53. Use lines, dashed wedges, and solid wedges to show the geometry of CCl_4 and CH_3OH.

1.54. Write a structural formula that corresponds to the molecular formula $C_5H_{10}O$ and is

a. acyclic. **b.** carbocyclic. **c.** heterocyclic.

1.55. Divide the following compounds into groups that might be expected to exhibit similar chemical behavior:

a. CH_3OH **b.** CH_3OCH_3 **c.** $CH_2(OH)CH(OH)CH_2(OH)$
d. C_5H_{12} **e.** C_4H_9OH **f.** C_8H_{18}
g. C_3H_7OH **h.** $(CH_3CH_2)_2O$ **i.** $CH_3OCH_2CH_2OCH_3$

1.56. Using eq. 1.7 as a guide, write an equation for the reaction of cyclohexanol with sodium metal.

cyclohexanol

1.57. Using Table 1.5, write a structural formula for each of the following:

a. an alcohol, $C_4H_{10}O$
b. an ether, C_3H_8O
c. an aldehyde, C_3H_6O
d. a ketone, C_4H_8O
e. a carboxylic acid, $C_3H_6O_2$
f. an ester, $C_5H_{10}O_2$
g. an ester that is an isomer of the one in part f
h. an amine, C_3H_9N

CHAPTER 2

Alkanes and Cycloalkanes; Conformational and Geometric Isomerism

2.1 Introduction

The main components of petroleum and natural gas, resources that now supply most of our energy, are **hydrocarbons, compounds that contain only carbon and hydrogen.** There are three main classes of hydrocarbons, based on the types of carbon–carbon bonds present. **Saturated hydrocarbons** contain only carbon–carbon *single* bonds. **Unsaturated hydrocarbons** contain carbon–carbon *multiple* bonds—double bonds, triple bonds, or both. **Aromatic hydrocarbons** are a special class of cyclic compounds related in structure to benzene.*

Saturated hydrocarbons are known as **alkanes** if they are acyclic, or as **cycloalkanes** if they are cyclic. Let us look at the structures and properties of alkanes.

2.2 The Structures of Alkanes

The simplest alkane is methane. Its tetrahedral three-dimensional structure was described in the previous chapter (see Figure 1.9). Additional alkanes are constructed by lengthening the carbon chain and adding an appropriate number of hydrogens to complete the carbon valences (for examples, see Figure 2.1** and Table 2.1).

All alkanes fit the general molecular formula C_nH_{2n+2}, where n is the number of carbon atoms. Alkanes with carbon chains that are unbranched (Table 2.1) are called **normal alkanes.** Each member of this series differs from the

---

* Unsaturated and aromatic hydrocarbons are discussed in Chapters 3 and 4, respectively.

** Molecular models can help you visualize organic structures in three dimensions. They will be extremely useful to you throughout this course, especially when we consider various types of isomerism. Relatively inexpensive sets are usually available at stores that sell textbooks, and your instructor can suggest which kind to buy. If you cannot locate or afford a set, you can create models that are adequate for some purposes from toothpicks (for bonds) and marshmallows, gum drops, or jelly beans (for atoms).

FIGURE 2.1

Three-dimensional models of ethane, propane, and butane. The stick-and-ball models at the left show the way in which the atoms are connected and depict the correct bond angles. The space-filling models at the right are constructed to scale and give a better idea of the molecular shape.

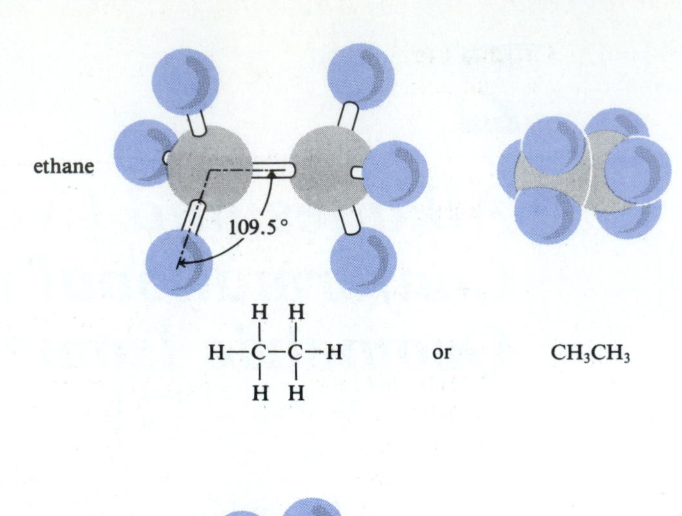

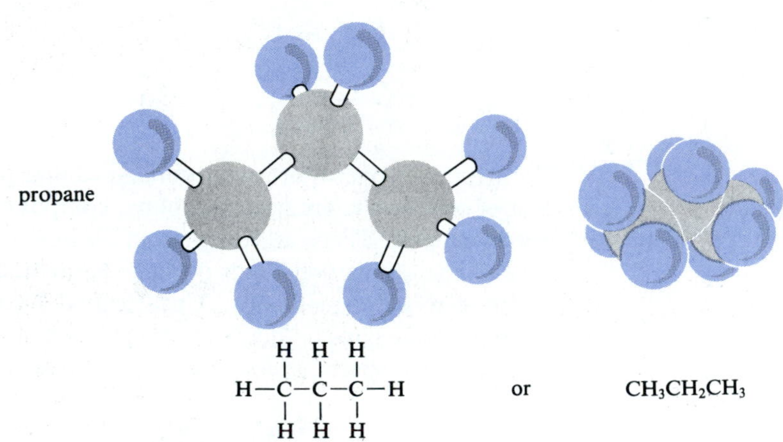

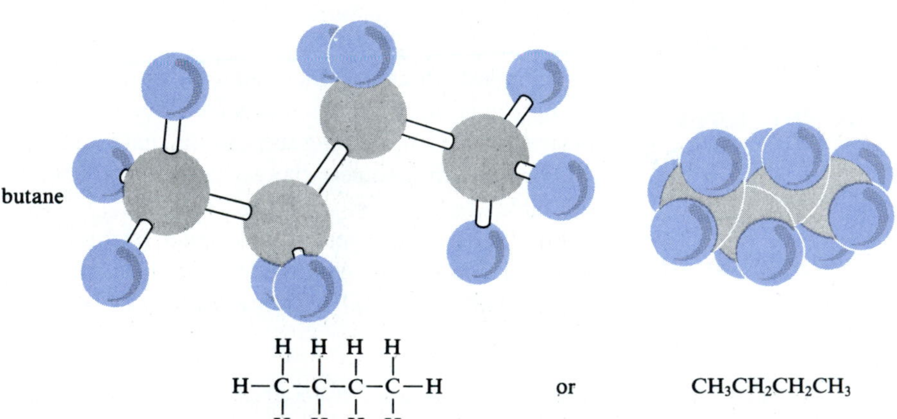

TABLE 2.1 Names and formulas of the first ten unbranched alkanes

Name	*Number of carbons*	*Molecular formula*	*Structural formula*	*Number of structural isomers*
methane	1	CH_4	CH_4	1
ethane	2	C_2H_6	CH_3CH_3	1
propane	3	C_3H_8	$CH_3CH_2CH_3$	1
butane	4	C_4H_{10}	$CH_3CH_2CH_2CH_3$	2
pentane	5	C_5H_{12}	$CH_3(CH_2)_3CH_3$	3
hexane	6	C_6H_{14}	$CH_3(CH_2)_4CH_3$	5
heptane	7	C_7H_{16}	$CH_3(CH_2)_5CH_3$	9
octane	8	C_8H_{18}	$CH_3(CH_2)_6CH_3$	18
nonane	9	C_9H_{20}	$CH_3(CH_2)_7CH_3$	35
decane	10	$C_{10}H_{22}$	$CH_3(CH_2)_8CH_3$	75

next higher and the next lower member by a —CH_2— group (called a **methylene group).** A series of compounds in which the members are built up in a regular, repetitive way is called a **homologous series.** Members of such a series have similar chemical and physical properties, which change gradually as carbon atoms are added to the chain.

■ ***EXAMPLE 2.1*** What is the molecular formula of an alkane with 6 carbon atoms?

Solution If $n = 6$, then $2n + 2 = 14$. The formula is C_6H_{14}.

Problem 2.1 What is the molecular formula of an alkane with 20 carbon atoms?

Problem 2.2 Which of the following are alkanes?

a. C_8H_{16} b. C_7H_{16} c. C_7H_{18} d. $C_{27}H_{56}$

2.3 *Nomenclature of Organic Compounds*

In the early days of organic chemistry, each new compound was given a name that was usually based on its source or use. Examples (Figures 1.12 and 1.13) include limonene (from lemons), α-pinene (from pine trees), coumarin (from the tonka bean, known to South American natives as cumaru), and penicillin (from the mold that produces it, *Pencillium notatum*). Even today, this method of naming may be used to give a short, simple name to a molecule with a complex structure. For example, cubane was named after its shape.

It became clear many years ago, however, that one could not rely only on common or trivial names and that a systematic method for naming compounds was needed. Ideally, the rules of the system should result in a unique name for each compound. Knowing the rules and seeing a structure, one should be able to write out the systematic name. Seeing the systematic name, one should able to write out the correct structure.

Eventually, an internationally recognized system of nomenclature was devised by a commission of the International Union of Pure and Applied Chemistry; it is known as the IUPAC (pronounced "eye-you-pack") system. In this book, we will use mainly IUPAC names. However, in some cases, the common name is so widely used that we will ask you to learn it (for example, formaldehyde [common] is used in preference to methanal [systematic], and cubane is much easier to remember than its systematic name pentacyclo$[4.2.0.0^{2,5}.0^{3,8}.0^{4,7}]$octane).

Let us now learn the IUPAC system for saturated hydrocarbons.

2.4 The IUPAC Rules for Naming Alkanes

1. The general name for acyclic saturated hydrocarbons is **alkanes.** The *-ane* ending is used for all saturated hydrocarbons. This is important to remember because later other endings will be used for other functional groups.
2. Alkanes without branches are named according to the *number of carbon atoms*. These names, up to 10 carbons, are given in the first column of Table 2.1.
3. For alkanes with branches, ***the root name is that of the longest continuous chain of carbon atoms.*** For example, in the structure

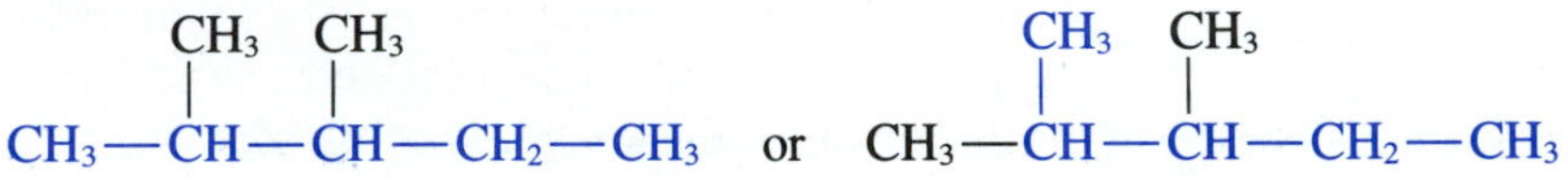

$$\begin{array}{c} \quad CH_3 \quad CH_3 \\ \quad | \qquad | \\ CH_3—CH—CH—CH_2—CH_3 \end{array} \quad \text{or} \quad \begin{array}{c} \quad CH_3 \quad CH_3 \\ \quad | \qquad | \\ CH_3—CH—CH—CH_2—CH_3 \end{array}$$

the longest continuous chain (in color) has five carbon atoms. The compound is therefore named as a substituted *pent*ane, even though there are seven carbon atoms altogether.
4. Groups attached to the main chain are called **substituents.** Saturated substituents that contain only carbon and hydrogen are called **alkyl groups.** An alkyl group is named by taking the name of the alkane with the same number of carbon atoms and changing the *-ane* ending to *-yl*.

   In the example above, each substituent has only one carbon. Derived from methane by removing one of the hydrogens, a one-carbon substituent is called a **methyl group.**

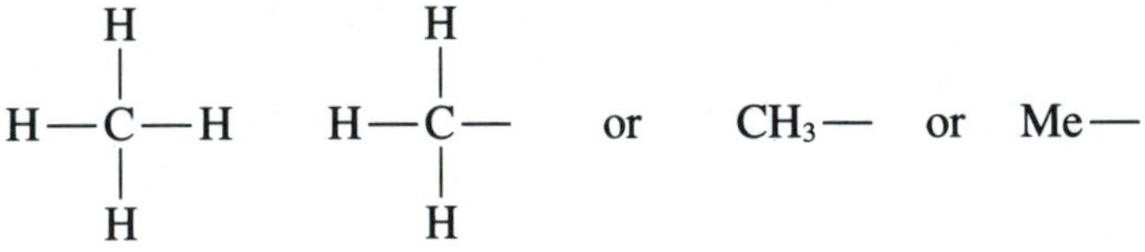

$$\underset{\text{methane}}{\begin{array}{c} H \\ | \\ H—C—H \\ | \\ H \end{array}} \qquad \begin{array}{c} H \\ | \\ H—C— \\ | \\ H \end{array} \quad \text{or} \quad CH_3— \quad \text{or} \quad Me—$$

methane　　　　methyl group

   The names of substituents with more than one carbon atom will be described in Section 2.5.
5. *The main chain is numbered in such a way that the first substituent encountered along the chain receives the lowest possible number*. Each substituent is then located by its name and by the number of the carbon atom to which it is attached. When two or more identical groups are attached to the main chain, prefixes such as *di-*, *tri-*, and *tetra-* are used. *Every substituent must be named and numbered,* even if two identical substituents are attached to

the same carbon of the main chain. The compound

$$\overset{1}{CH_3}-\overset{2}{\underset{}{CH}}(CH_3)-\overset{3}{CH}(CH_3)-\overset{4}{CH_2}-\overset{5}{CH_3}$$

is correctly named **2,3-dimethylpentane.** The name tells us that there are two methyl substituents, one attached to carbon-2 and one attached to carbon-3 of a five-carbon saturated chain.

6. If two or more different types of substituents are present, they are listed alphabetically, except that prefixes such as *di-* and *tri-* are not considered for alphabetizing.
7. *Punctuation is important in writing IUPAC names.* IUPAC names for hydrocarbons are written as one word. Numbers are separated from each other by commas and are separated from letters by hyphens. There is no space between the last named substituent and the name of the parent alkane that follows it.

To summarize and amplify these rules, we take the following steps to find the correct IUPAC name of an alkane:

1. Locate the longest continuous carbon chain. This gives the name of the parent hydrocarbon. For example,

```
        C—C                      C—C
        |                        |
  C—C—C—C        not     C—C—C—C
```

2. Number the longest chain beginning at the end nearest to the first branch point. For example,

```
          C     C                      C     C
          |     |                      |     |
  C—C—C—C—C—C        not     C—C—C—C—C—C
  6  5  4  3  2  1              1  2  3  4  5  6
```

If there are two equally long continuous chains, select the one with the most branches. For example,

```
     C                                 C
  1  2|  3  4  5  6                    |  3  4  5  6
  C—C—C—C—C—C        not      C—C—C—C—C—C
        |                              |
        C—C                            C—C
                                       2  1
  two branches                    one branch
```

If there is a branch equidistant from each end of the longest chain, begin numbering nearest to a third branch:

```
     C  C           C                  C  C           C
     |  |           |                  |  |           |
  C—C—C—C—C—C—C        not     C—C—C—C—C—C—C
  1  2  3  4  5  6  7               7  6  5  4  3  2  1
  2,3,6-trimethylheptane            2,5,6-trimethylheptane
```

If there is no third branch, begin numbering nearest the substituent whose name has alphabetic priority:

```
      C—C   C                                  C—C   C
        |   |                                    |   |
C—C—C—C—C—C—C          not          C—C—C—C—C—C—C
1  2  3  4  5  6  7                   7  6  5  4  3  2  1
```

3-ethyl-5-methylheptane — 5-ethyl-3-methylheptane

3. Write out the name as one word, placing substituents in alphabetic order and using proper punctuation.

EXAMPLE 2.2 Give the IUPAC name for $CH_3—C(CH_3)_2—CH_2CH_2CH_3$.

```
         CH3
         |
CH3—C—CH2CH2CH3
         |
         CH3
```

Solution

```
          CH3
 1      2 |   3    4    5
CH3—C—CH2CH2CH3     2,2-dimethylpentane
          |
          CH3
```

Problem 2.3 Give the IUPAC name for the following compounds:

```
                                                             CH3
                                                             |
a. CH3CHCH2CH3     b. CH3CH2CHCH3     c. CH3—C—CH3
       |                        |                            |
       CH3                      CH3                          CH3
```

2.5 Alkyl and Halogen Substituents

As illustrated for the methyl group, alkyl substituents are named by changing the *-ane* ending of alkanes to *-yl*. Thus the two-carbon alkyl group is called the **ethyl group,** from ethane.

CH_3CH_3 (ethane) $\quad CH_3CH_2—$ or $C_2H_5—$ or Et— (ethyl group)

When we come to propane, there are two possible alkyl groups, depending on which type of hydrogen is removed. If a *terminal* hydrogen is removed, the group is called a **propyl group.**

```
  H  H  H              H  H  H
  |  |  |              |  |  |
H—C—C—C—H          H—C—C—C—      or    CH3CH2CH2—    or    Pr—
  |  |  |              |  |  |
  H  H  H              H  H  H
```

propane — propyl group

But if a hydrogen is removed from the *central* carbon atom, we get a different propyl group, called the **isopropyl** (or 1-methylethyl) **group.**

$$\begin{array}{ccccccc} & H & & H & & H & \\ & | & & | & & | & \\ H- & C & - & C & - & C & -H \\ & | & & | & & | & \\ & H & & H & & H & \end{array} \qquad \begin{array}{ccccccc} & H & & H & & H & \\ & | & & | & & | & \\ H- & C & - & C & - & C & -H \\ & | & & | & & | & \\ & H & & & & H & \end{array} \quad \text{or} \quad \begin{array}{c} CH_3CHCH_3 \\ | \end{array} \quad \text{or} \quad i\text{-Pr}-$$

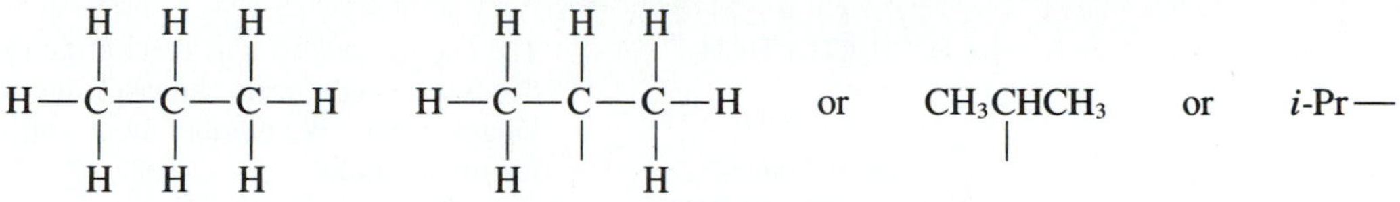

propane — isopropyl or 1-methylethyl* group

There are four different butyl groups.

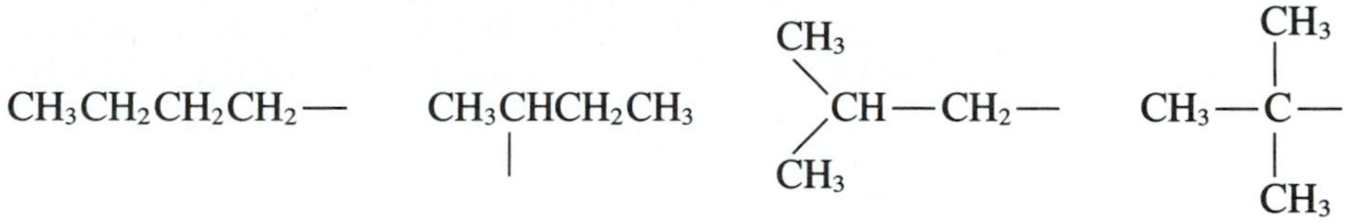

$$CH_3CH_2CH_2CH_2- \qquad \begin{array}{c} CH_3CHCH_2CH_3 \\ | \end{array} \qquad \begin{array}{l} CH_3 \\ \quad \diagdown \\ \quad\ CH-CH_2- \\ \quad \diagup \\ CH_3 \end{array} \qquad \begin{array}{c} CH_3 \\ | \\ CH_3-C- \\ | \\ CH_3 \end{array}$$

butyl	*sec*-butyl (or 1-methylpropyl)	isobutyl (or 2-methylpropyl)	*tert*-butyl (or 1,1-dimethylethyl)

These names for the alkyl groups with up to four carbon atoms are very commonly used, so you should memorize them.

The letter **R** is used as a general symbol for an alkyl group. The formula R—H therefore represents any alkane, and the formula R—Cl stands for any alkyl chloride (methyl chloride, ethyl chloride, and so on).

Halogen substituents are named by changing the *-ine* ending of the element to *-o*.

F—	Cl—	Br—	I—
fluoro-	chloro-	bromo-	iodo-

■ ***EXAMPLE 2.3*** Give the common and IUPAC names for $CH_3CH_2CH_2Br$.

Solution The common name is propyl bromide (the common name of the alkyl group is followed by the name of the halide). The IUPAC name is 1-bromopropane, the halogen being named as a substituent on the three-carbon chain.

Problem 2.4 Give the correct IUPAC name for CH_2BrCl.

Problem 2.5 Write the formula for each of the following compounds:

a. propyl iodide
b. isopropyl chloride
c. 2-chloropropane
d. *tert*-butyl iodide
e. isobutyl bromide
f. all alkyl fluorides

* The name 1-methylethyl for this group comes about by regarding it as a substituted ethyl group.

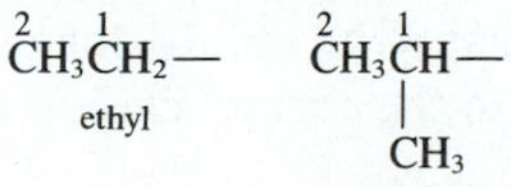

$$\overset{2}{C}H_3\overset{1}{C}H_2- \qquad \begin{array}{c} \overset{2}{C}H_3\overset{1}{C}H- \\ \quad | \\ \quad CH_3 \end{array}$$

ethyl — 1-methylethyl

TABLE 2.2 Examples of use of the IUPAC rules

Structure and name	Explanation
$\overset{5}{C}H_3\overset{4}{C}H_2\overset{3}{C}H_2\overset{2}{C}H(CH_3)\overset{1}{C}H_3$ 2-methylpentane (not 4-methylpentane)	The ending *-ane* tells us that all the carbon–carbon bonds are single; *pent-* indicates five carbons in the longest chain. We number them starting closest to the branch point.
$CH_3\overset{3}{C}H(\overset{2}{C}H_2\overset{1}{C}H_3)\overset{4}{C}H_2\overset{5}{C}H_2\overset{6}{C}H_3$ 3-methylhexane (*not* 2-ethylpentane or 4-methylhexane)	A six-carbon saturated chain with a methyl group on the third carbon. We would usually write the structure as $CH_3CH_2CH(CH_3)CH_2CH_2CH_3$
$\overset{1}{C}H_3—\overset{2}{C}(CH_3)_2—\overset{3}{C}H_2\overset{4}{C}H_3$ 2,2-dimethylbutane (*not* 2,2-methylbutane or 2-dimethylbutane)	There must be a number for each substituent, and the prefix *di-* says that there are two methyl substituents.
$\overset{1}{C}H_2(Cl)\overset{2}{C}H_2\overset{3}{C}H(Br)\overset{4}{C}H_3$ 3-bromo-1-chlorobutane (*not* 1-chloro-3-bromobutane or 2-bromo-4-chlorobutane)	First, we number the butane chain from the end closest to the first substituent. Then we name the substituents in alphabetical order, regardless of position number.

2.6 Use of the IUPAC Rules

The examples given in Table 2.2 illustrate how the IUPAC rules are applied for particular structures. Study each example to see how the correct name is obtained and how to avoid certain pitfalls.

It is important not only to be able to write the correct IUPAC name of a given structure, but also to do the converse: write the structure given the IUPAC name. In this case, first write out the longest carbon chain and number it, then add the substituents to the correct carbon atoms, and finally fill in the formula with the correct number of hydrogens at each carbon. For example, to write the formula for **2,2,4-trimethylpentane,** we go through the following steps:

C—C—C—C—C (Write down the pentane chain.) $\xrightarrow{\text{Add the numbers.}}$ C—C—C—C—C, numbered 1 2 3 4 5 from left to right

Add the three methyl substituents: two CH_3 groups on C-2 (one above, one below) and one CH_3 above C-4.

$\xleftarrow{\text{Fill in the hydrogens.}}$ CH_3—C(CH_3)(CH_3)—CH_2—CH(CH_3)—CH_3

2,2,4-trimethylpentane

Problem 2.6 Name the following compounds by the IUPAC system:

a. CH_3CHFCH_3 b. $(CH_3)_3CCH_2CHClCH_3$

Problem 2.7 Write the structure for 3,3-dimethylpentane.

Problem 2.8 Explain why 1,3-dichlorobutane is a correct IUPAC name, but 1,3-dimethylbutane is *not* a correct IUPAC name.

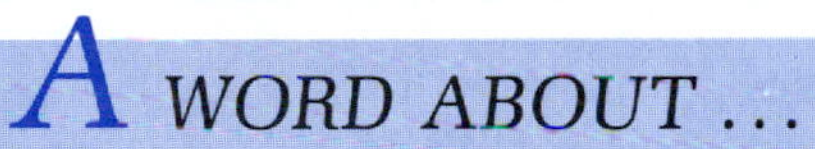

1. Isomers, Possible and Impossible

Table 2.1 shows that there are 75 structural isomers of the alkane $C_{10}H_{22}$. How many such isomers do you think there might be if we double the number of carbons ($C_{20}H_{42}$)? The answer is 366,319! And if we double the number of carbons again ($C_{40}H_{82}$)? Exactly 62,481,801,147,341. Of course, no one sits down with pencil and paper or molecular models and determines these numbers by constructing all the possibilities; it could take a lifetime. Complex mathematical formulas have been developed to compute these numbers.

Although we can write some isomers' formulas on paper, they are structurally impossible and cannot be synthesized. Consider, for example, the series of alkanes obtained by replacing the hydrogens of methane with methyl groups and then repeating that process on the product indefinitely. You can see from the drawings on the next page, even though they are only two-dimensional, that in this way we build up molecules with a central core of carbon atoms and a surface of hydrogen atoms.

$$CH_4 \longrightarrow C(CH_3)_4 \longrightarrow C[C(CH_3)_3]_4 \longrightarrow C\{C[C(CH_3)_3]_3\}_4$$

$$CH_4 \longrightarrow C_5H_{12} \longrightarrow C_{17}H_{36} \longrightarrow C_{53}H_{108}$$

In three dimensions, the molecules are nearly spherical. Of these compounds, only the first two are known (methane and 2,2-dimethylpropane). The $C_{17}H_{36}$ hydrocarbon (tetra-*t*-butylmethane or, more accurately, 3,3-di-*t*-butyl-2,2,4,4-tetramethylpentane) has not yet been synthesized, and if it ever is, it will be an exceptionally strained molecule. The reason is simply that there is not enough room for all the methyl groups on the surface of the molecule. Calculations from space-filling models show that the inner five carbons form a

$$CH_4 \longrightarrow C(CH_3)_4 \longrightarrow C(C(CH_3)_3)_4 \longrightarrow \cdots$$

methane

2,2-dimethylpropane

tetra-*t*-butylmethane ($C_{17}H_{36}$)

sphere with a surface area of about 85 Å^2 whereas the 12 methyl groups require a surface area of about 107 Å^2. So the synthesis of this $C_{17}H_{36}$ isomer may be barely possible, but if prepared, its bond angles and bond lengths are likely to be severely distorted from the normal. Synthesizing this hydrocarbon, then, presents an interesting challenge for organic chemists, one that has not yet been successfully met.

There is no possibility of synthesizing the $C_{53}H_{108}$ isomer in this series; its structure is too strained. Like these hydrocarbons, the growth of trees, sponges and other biological forms is similarly limited by the ratio of surface area to volume. (For more on this subject, see the article by R. E. Davies and P. J. Freyd, *J. Chem. Educ.* **1989,** *66,* 278–81).

2.7 Sources of Alkanes

The two most important natural sources of alkanes are **petroleum** and **natural gas.** *Petroleum* is a complex liquid mixture of organic compounds, many of which are alkanes or cycloalkanes. For more details about how petroleum is refined to obtain gasoline, fuel oil, and other useful substances, read "A Word About Petroleum, Gasoline, and Octane Number" on page 101.

Natural gas, often found associated with petroleum deposits, consists mainly of methane (about 80%) and ethane (5 to 10%), with lesser amounts of some higher alkanes. Propane is the major constituent of liquefied petroleum gas (LPG), a domestic fuel used mainly in rural areas and mobile homes. Natural gas is becoming an energy source that can compete with and possibly surpass oil. In the United States, there are over 250,000 miles of natural gas pipelines distributing this energy source to all parts of the country. Natural gas is also distributed worldwide via huge tankers. To conserve space, the gas is liquefied (−160°C), because 1 cubic meter (m^3) of liquefied gas is equivalent to about 600 m^3 at atmospheric pressure. Large tankers can carry over 100,000 m^3 of liquefied gas. In the future, less-developed countries are likely to obtain energy more cheaply by developing local resources of natural gas rather than by importing oil.

FIGURE 2.2

As shown by the curve, the boiling points of the normal alkanes rise smoothly as the length of the carbon chain increases. Note from the table, however, that chain branching causes a decrease in boiling point (each compound in the table has the same number of carbons and hydrogens, C_5H_{12}).

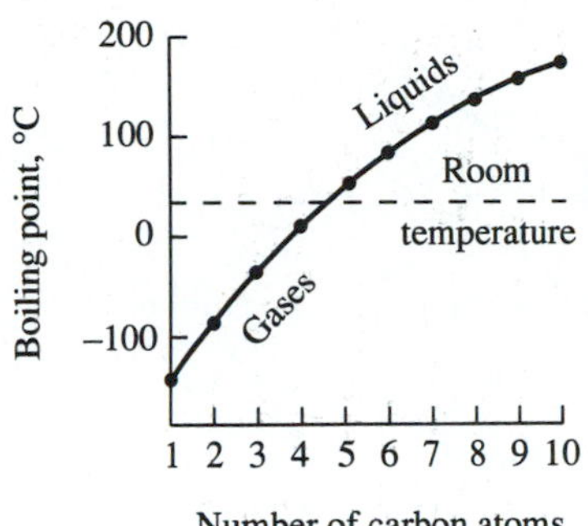

Name	Formula	Boiling point, °C
pentane	$CH_3CH_2CH_2CH_2CH_3$	36
2-methylbutane (isopentane)	$CH_3CHCH_2CH_3$ (with CH_3 on the second carbon)	28
2,2-dimethylpropane (neopentane)	CH_3 $CH_3—C—CH_3$ CH_3	10

2.8 *Physical Properties of Alkanes*

Alkanes are insoluble in water. The reason is that water molecules are polar and attract one another, whereas alkanes are nonpolar (all the C—C and C—H bonds are nearly purely covalent). To intersperse alkane and water molecules, we would have to break up the attractive force between the water molecules, which would require considerable energy.

The mutual insolubility of alkanes and water is used to advantage by many plants. Alkanes often constitute part of the protective coating on leaves and fruits. If you have ever polished an apple, you know that the skin, or cuticle, contains waxes. Among them are the normal alkanes $C_{27}H_{56}$ and $C_{29}H_{60}$. The leaf wax of cabbage and broccoli is mainly n-$C_{29}H_{60}$, and the main alkane of tobacco leaves is n-$C_{31}H_{64}$. Similar hydrocarbons are found in beeswax. The major function of plant waxes is to prevent water loss from the leaves or fruit.

Alkanes have lower boiling points for a given molecular weight than most other organic compounds. This is because the attractive forces between nonpolar molecules are weak, and the process of separating molecules from one another (which is what we do when we convert a liquid to a gas) requires relatively little energy. Figure 2.2 shows the boiling points of some alkanes. The greater the molecular surface area, the greater the attractive forces between molecules. Therefore, boiling points rise as the chain length increases and fall as the chains become branched and more nearly spherical in shape.

2.9 *Conformations of Alkanes*

The shapes of molecules often affect their properties. A simple molecule like ethane, for example, can have an infinite number of shapes as a consequence of rotating one carbon atom (and its attached hydrogens) with respect to the other carbon atom. These arrangements are called **conformations,** or **conformers.** Two possibilities for ethane are shown in Figure 2.3.

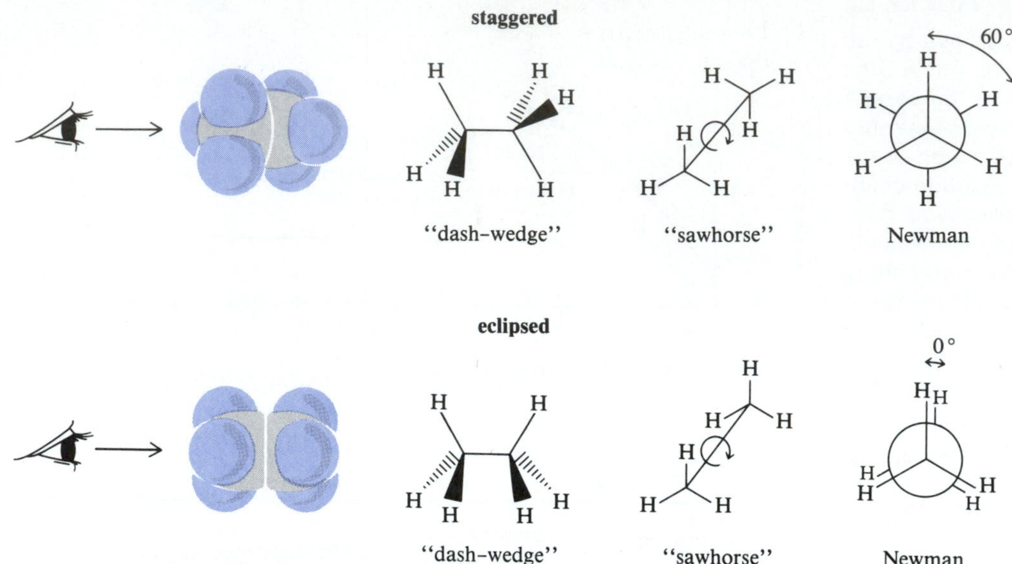

FIGURE 2.3 Two of the possible conformations of ethane: staggered and eclipsed. Interconversion is easy via a 60° rotation about the C—C bond, as shown by the curved arrows. The structures at the left are space-filling models. In each case, the next structure is a "dash-wedge" structure which, if viewed as shown by the eyes, converts to the "sawhorse" drawing, or the Newman projection at the right, an end-on view down the C—C axis. In Newman projections, bonds on the "front" carbon go to the center of the circle, and bonds on the "rear" carbon go only to the edge of the circle.

In the **staggered conformation** of ethane, each C—H bond on one carbon bisects an H—C—H angle on the other carbon. In the **eclipsed conformation,** C—H bonds on the front and back carbons are aligned. By rotating one carbon 60° with respect to the other, we can interconvert staggered and eclipsed conformations. Between these two extremes there is an infinite number of intermediate conformations of ethane.

Although they cannot be separated from one another, the staggered and eclipsed conformations of ethane can be regarded as **rotational isomers** (or **rotamers**), because each is convertible to the other by rotation about the carbon–carbon bond. Such rotation about a single bond occurs easily, because the amount of overlap of the sp^3 orbitals on the two carbon atoms is unaffected by rotation about the sigma bond (see Figure 1.8). Indeed, there is enough energy available at room temperature for the staggered and eclipsed conformers of ethane to interconvert rapidly. We know from various types of physical evidence, however, that both forms are not equally stable. The staggered conformation is much preferred, and at room temperature more than 99% of ethane molecules have the staggered arrangement. The reasons why the staggered conformation is preferred are quite complex, but one of them is probably that the bonding electrons on adjacent carbons are farthest apart and therefore experience less mutual repulsion in this arrangement.

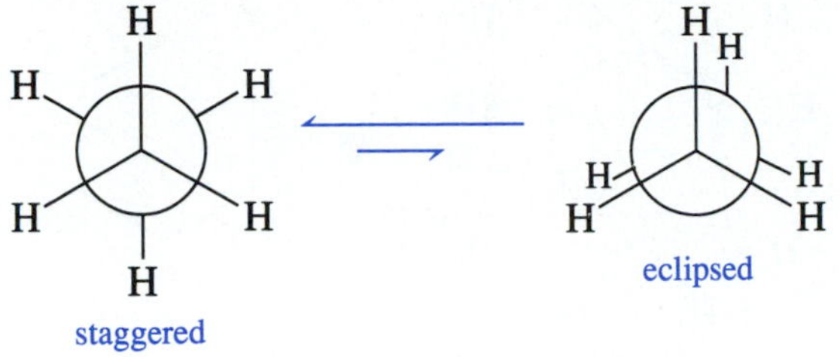

H H H H H H (staggered) ⇌ H H H H H H (eclipsed) (2.1)

staggered

eclipsed

EXAMPLE 2.4 Draw the Newman projections for the staggered and eclipsed conformations of propane.

Solution

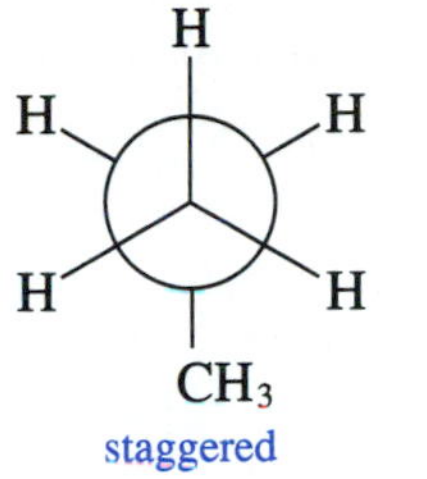

H H H H H CH_3

staggered

The projection formula is similar to that of ethane, except for the replacement of one hydrogen with methyl.

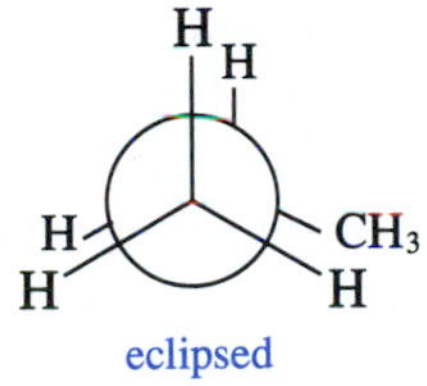

H H H H CH_3 H

eclipsed

Rotation of the "rear" carbon of the staggered conformation by 60° gives the eclipsed conformation shown.

We are looking down the C_1—C_2 bond.

Problem 2.9 Draw Newman projections for two different *staggered* conformations of butane (looking end-on at the bond between carbon-2 and carbon-3), and predict which of the two conformations is more stable.

The most important thing to remember about conformers is that they are just different forms of a single molecule that can be interconverted by rotational motions about single (sigma) bonds. More often than not, there is sufficient thermal energy for this rotation at room temperature. Consequently, at room temperature it is usually not possible to separate conformers from one another.

Now let us look at the structures of cycloalkanes and their conformations.

2.10 Cycloalkane Nomenclature and Conformation

Cycloalkanes are saturated hydrocarbons that have at least one ring of carbon atoms. A common example is cyclohexane.

Structural and abbreviated structural formulas for cyclohexane

Cycloalkanes are named by placing the prefix *cyclo-* before the alkane name that corresponds to the number of carbon atoms in the ring. The structures and names of the first six unsubstituted cycloalkanes are

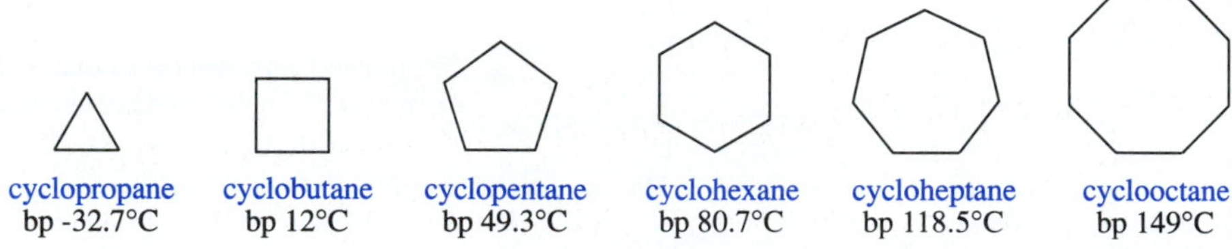

cyclopropane	cyclobutane	cyclopentane	cyclohexane	cycloheptane	cyclooctane
bp -32.7°C	bp 12°C	bp 49.3°C	bp 80.7°C	bp 118.5°C	bp 149°C

Alkyl or halogen substituents attached to the rings are named in the usual way. If only one substituent is present, no number is needed to locate it. If there are several substituents, numbers are required. One substituent is always located at ring carbon number 1, and the remaining ring carbons are then numbered consecutively in a way that gives the other substituents the lowest possible numbers. With different substituents, the one with highest alphabetic priority is located at carbon 1. The following examples illustrate the system:

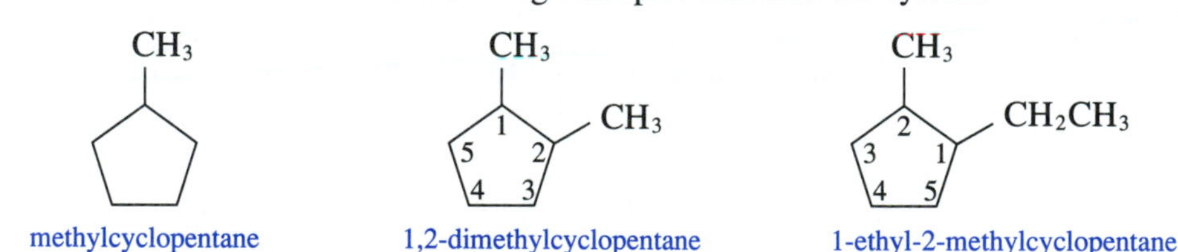

methylcyclopentane (*not* 1-methylcyclopentane)

1,2-dimethylcyclopentane (*not* 1,5-dimethylcyclopentane)

1-ethyl-2-methylcyclopentane (not 2-ethyl-1-methylcyclopentane)

Problem 2.10 Draw the structural formulas for

a. 1,3-dimethylcyclopentane.
b. 1,2,3-trichlorocyclopropane.

Problem 2.11 Give the IUPAC names for

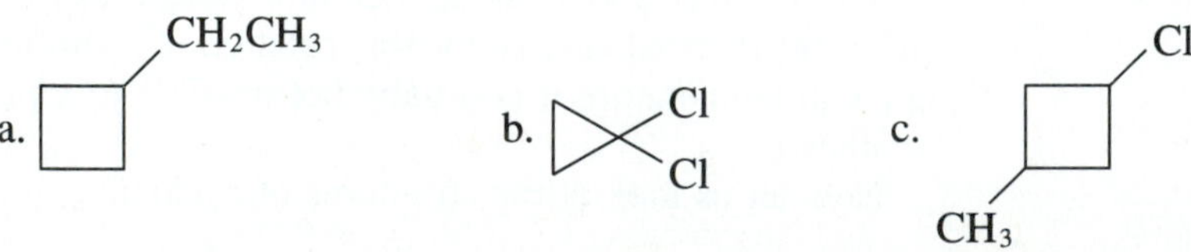

What are the conformations of cycloalkanes? Cyclopropane, with only three carbon atoms, is necessarily planar (because three points determine a plane). The C—C—C angle is only 60° (the carbons form an equilateral triangle), much less than the usual tetrahedral angle of 109.5°. The hydrogens lie above and below the carbon plane, and hydrogens on adjacent carbons are eclipsed.

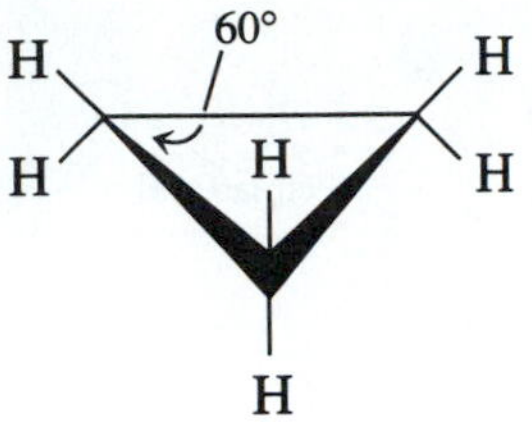

60°
H H
H H H
H
cyclopropane

EXAMPLE 2.5 Explain why the hydrogens in cyclopropane lie above and below the carbon plane.

Solution Refer to Figure 1.10. The carbons in cyclopropane have a similar geometry to that shown there, except that the C—C—C angle is "squeezed" and is smaller than tetrahedral. In compensation, the H—C—H angle is expanded and is larger than tetrahedral, approximately 120°.

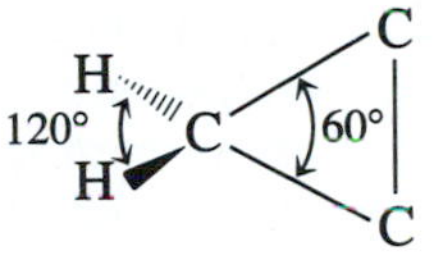

C
H
120° C 60°
H
C

The H—C—H plane perpendicularly bisects the C—C—C plane, which, as drawn here, lies in the plane of the paper.

Cycloalkanes with more than three carbon atoms are nonplanar and have "puckered" conformations. In cyclobutane and cyclopentane, puckering actually makes the C—C—C angles a little smaller than they would be if the molecules were planar, but less eclipsing of adjacent hydrogens compensates for this.

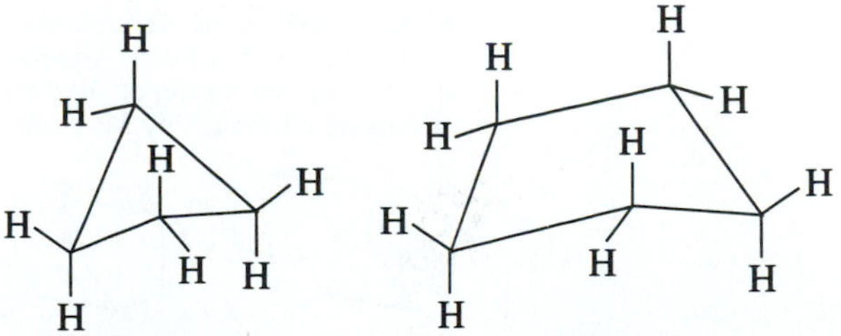

	cyclobutane	cyclopentane
∠ C—C—C planar	90°	108°
observed experimentally	88°	105°

FIGURE 2.4 The chair conformation of cyclohexane, shown in ball-and-stick and space-filling models. The axial hydrogens lie above or below the mean plane of the carbons, and the six equatorial hydrogens lie approximately in that mean plane. The origin of the chair terminology is illustrated at the right.

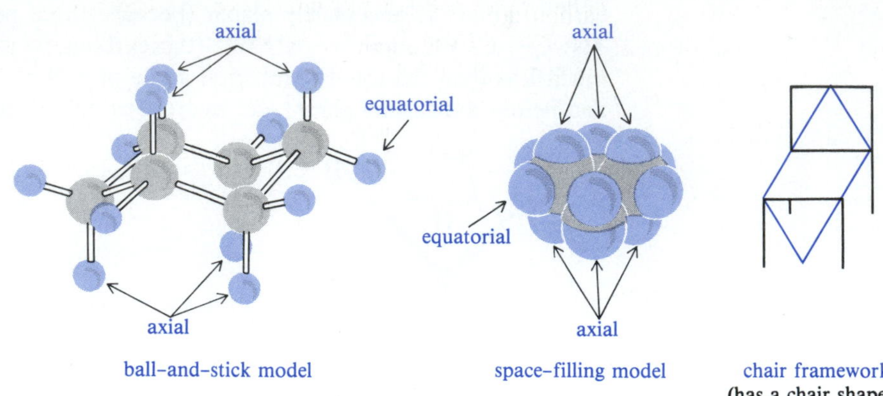

Six-membered rings are rather special and have been studied in great detail because they are so common in nature. If cyclohexane were planar, the internal C—C—C angles would be those of a regular hexagon, 120°—quite a bit larger than the normal tetrahedral angle (109.5°). The strain resulting from such angles prevents cyclohexane from being planar (flat). The most favored conformation of cyclohexane is the **chair conformation,** an arrangement in which all the C—C—C angles are 109.5° and all the hydrogens on adjacent carbon atoms are perfectly staggered. Figure 2.4 shows models of the cyclohexane chair conformation.* (If a set of molecular models is available, it would be good idea for you to construct a cyclohexane model to better visualize the concepts discussed in this and the next two sections.)

Problem 2.12 How are the H—C—H and C—C—C planes at any one carbon atom in cyclohexane related? (Refer, if necessary, to Example 2.5.)

---

* **Diamond** is one naturally occurring form of carbon. In the diamond crystal, the carbon atoms are connected to one another in a structure similar to the chair form of cyclohexane, except that all of the hydrogens are replaced by carbon atoms, resulting in a continuous network of carbon atoms. The hydrocarbons **adamantane** and **diamantane** show the beginnings of the diamond structure in their fusing of chair cyclohexanes. For a fascinating article on diamond structure, see "Diamond Cleavage" by M. F. Ansell in *Chemistry in Britain,* **1984,** 1017–21.

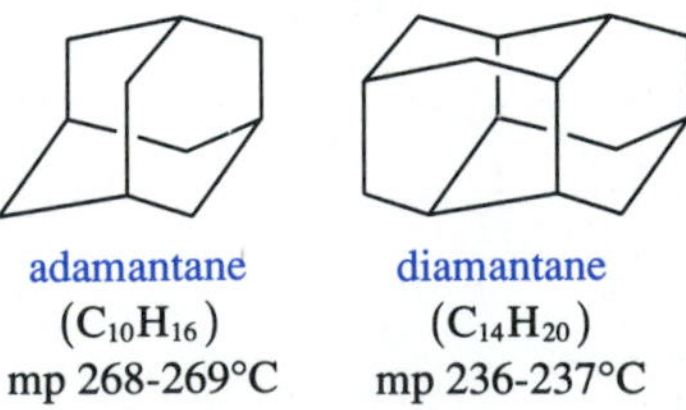

In the chair conformation, the hydrogens in cyclohexane fall into two sets, called **axial** and **equatorial.** Three axial hydrogens lie above and three lie below the average plane of the carbon atoms; the six equatorial hydrogens lie approximately in that plane. By a motion in which alternate ring carbons (say, 1, 3, and 5) move in one direction (down) and the other three ring carbons move in the opposite direction (up), one chair conformation can be converted into another chair conformation in which all axial hydrogens have become equatorial, and vice versa.

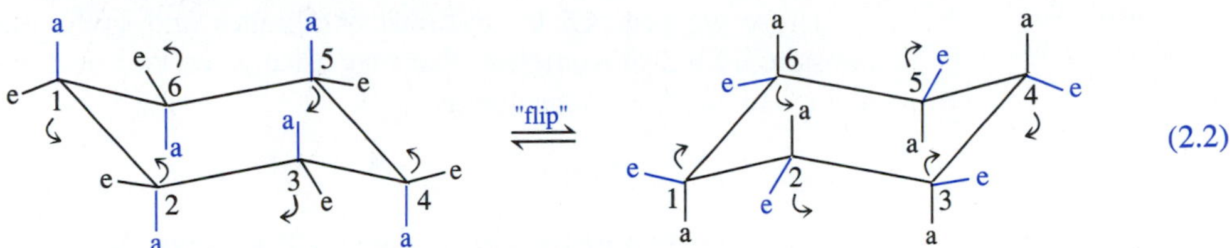

Axial bonds (in color) in the left structure become equatorial bonds (in color) in the right structure when the ring "flips."

At room temperature this flipping process is rapid, but at low temperatures (say −90°C), the flipping process slows down enough that the two different types of hydrogens can actually be detected by certain spectroscopic methods.

There is another important feature of cyclohexane conformations. If you look carefully at the space-filling model of cyclohexane (Figure 2.4), you will notice that *the three axial hydrogens on the same face of the ring nearly touch each other*. If an axial hydrogen is replaced by a larger substituent (such as a methyl group), the axial crowding is even worse. Therefore, the preferred conformation is the one in which the larger substituent, in this case the methyl group, is equatorial.

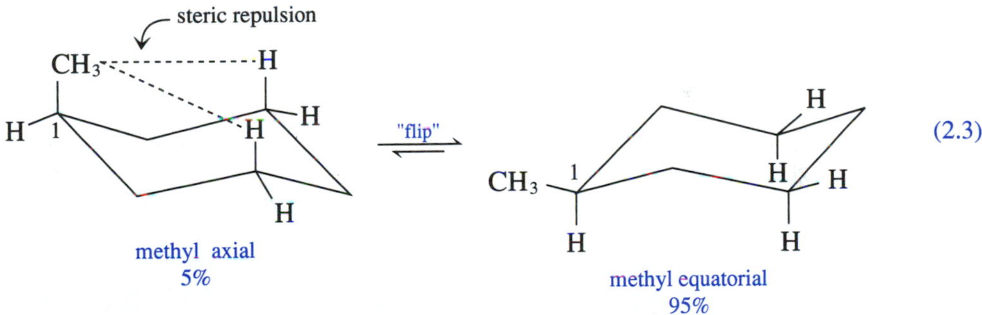

Problem 2.13 Another puckered conformation for cyclohexane, one in which all C—C—C angles are the normal 109.5°, is the boat conformation.

boat cyclohexane

Explain why this conformation is very much less stable than the chair conformation (*Hint:* Note the arrangement of hydrogens as you sight along the bond between carbon-2 and carbon-3; a molecular model will help you answer this problem).

Problem 2.14 *tert*-Butylcyclohexane molecules exist in a single conformation, with the *tert*-butyl group equatorial. Explain why this conformational preference is greater than that for methylcyclohexane.

Before we proceed to reactions of alkanes and cycloalkanes, we need to consider a kind of isomerism that may arise when two or more carbon atoms in a cycloalkane have substituents.

2.11 *Cis–trans* Isomerism in Cycloalkanes

Stereoisomerism deals with molecules that have the same order of attachment of the atoms, but different arrangements of the atoms in space. ***Cis–trans*** **isomerism** (sometimes called **geometric isomerism**) is one kind of stereoisomerism, and it is most easily understood with a specific case. Consider, for example, the possible structures of 1,2-dimethylcyclopentane. For simplicity, let us neglect the slight puckering of the ring and draw it as if it were planar. The two methyl groups may be on the same side of the ring plane or they may be on opposite sides.

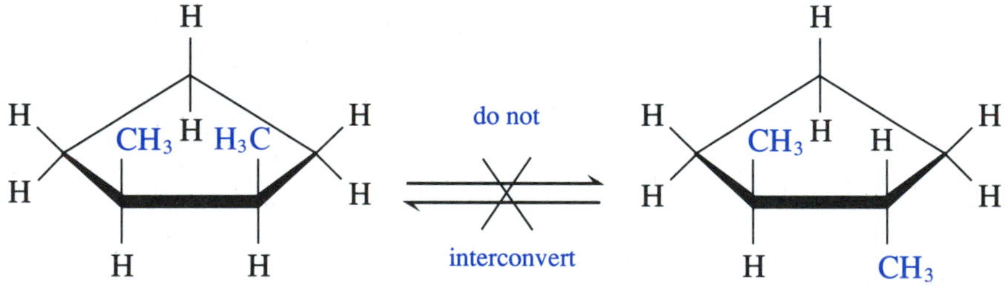

cis-1,2-dimethylcyclopentane
bp 99°C

do not interconvert

trans-1,2-dimethylcyclopentane
bp 92°C

The methyl groups are said to be ***cis*** (Latin, on the same side) or ***trans*** (Latin, across) to each other.

Cis–trans **isomers differ from one another only in the way the atoms or groups are positioned in space.** Yet this difference is sufficient to give them different physical and chemical properties (note, for example, the boiling points under the 1,2-dimethylcyclopentane structures). ***Cis–trans*** **isomers are separate and unique compounds. Unlike conformers, they cannot be interconverted by rotation around carbon-carbon bonds.** In this example, the cyclic structure prevents rotation about the ring bonds. *Cis–trans* isomers can be separated from each other and kept separate, usually without interconversion at room temperature. *Cis–trans* isomerism can be important in determining the biological properties of molecules. For example, a molecule in which two reactive groups are *cis* will interact differently with an enzyme or biological receptor site than will its isomer in which the same two groups are *trans*.

Problem 2.15 Draw the structure for the *cis* and *trans* isomers of

a. 1,2-dichlorocyclopropane.
b. 1-bromo-3-chlorocyclobutane.

2.12 Summary of Isomerism

At this point, it may be useful to summarize the relationships of the several types of isomers we have discussed so far. These relationships are outlined in Figure 2.5.

The first thing to look at in a pair of isomers is their bonding patterns (or atom connectivities). If the bonding patterns are *different,* the compounds are **structural (or constitutional) isomers.** But if the bonding patterns are the *same,* the compounds are **stereoisomers.** Examples of structural isomers are ethanol and methoxymethane (page 18) or the three isomeric pentanes (page 20). Examples of stereoisomers are the staggered and eclipsed forms of ethane (page 52) or the *cis* and *trans* isomers of 1,2-dimethylcyclopentane (page 58).

If compounds are stereoisomers, we can make a further distinction as to isomer type. If *bond rotation* easily interconverts the two stereoisomers (as with staggered and eclipsed ethane), we call them **conformational isomers, or conformers.** If the two stereoisomers can be interconverted only by breaking and remaking bonds (as with *cis-* and *trans-*1,2-dimethylcyclopentane), we call them **configurational isomers.**

Problem 2.16 Classify each of the following isomer pairs according to the scheme in Figure 2.5:

a. 1-bromopropane and 2-bromopropane
b. *cis-* and *trans-*1,2-dimethylcyclohexane
c. chair and boat forms of cyclohexane

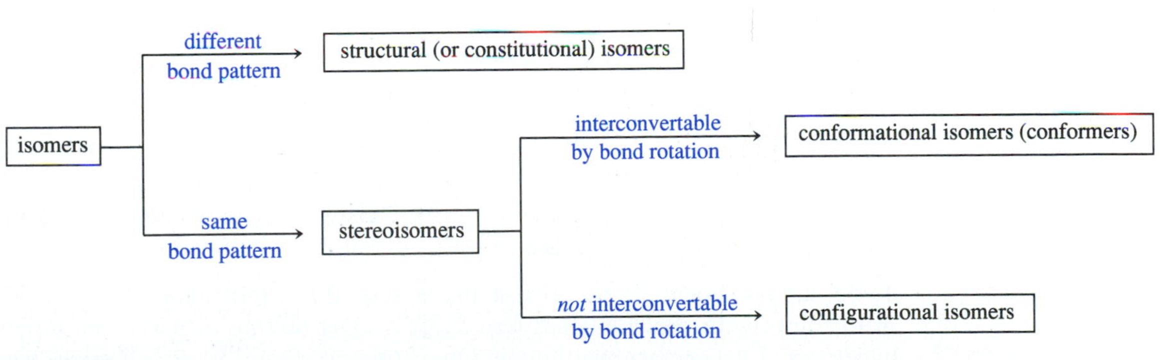

FIGURE 2.5 The relationships of the various types of isomers.

2.13 Reactions of Alkanes

All the bonds in alkanes are single, covalent, and nonpolar. Hence alkanes are relatively inert. Alkanes ordinarily do not react with most common acids, bases, or oxidizing and reducing agents. Because of this inertness, alkanes can be used as solvents for extraction or crystallization, or for carrying out chemical reactions with other substances. However, alkanes do react with some reagents, such as oxygen and the halogens. We will discuss those reactions here.

2.14 Oxidation and Combustion; Alkanes as Fuels

The most important use of alkanes is as fuels. With excess oxygen, alkanes burn to form carbon dioxide and water. Most important, the reactions evolve considerable heat (that is, the reactions are **exothermic**).

$$\underset{\text{methane}}{CH_4} + 2\,O_2 \longrightarrow CO_2 + 2\,H_2O + \text{heat (212.8 kcal/mol)} \quad (2.4)$$

$$\underset{\text{butane}}{C_4H_{10}} + \tfrac{13}{2}\,O_2 \longrightarrow 4\,CO_2 + 5\,H_2O + \text{heat (688.0 kcal/mol)} \quad (2.5)$$

These combustion reactions are the basis for the use of hydrocarbons for heat (natural gas and heating oil) and for power (gasoline). An initiation step is required—usually ignition by a spark or flame. Once initiated, the reaction proceeds spontaneously and exothermically.

In methane, all four bonds to the carbon atom are C—H bonds. In carbon dioxide, its combustion product, all four bonds to the carbon are C—O bonds. *Combustion is an oxidation reaction, the replacement of C—H bonds by C—O bonds.* In methane, carbon is in its most reduced form. In carbon dioxide, carbon is in its most oxidized form. Intermediate oxidation states of carbon are also known, in which only one, two, or three of the C—H bonds are converted to C—O bonds. It is not surprising, then, that if insufficient oxygen is available for complete combustion of a hydrocarbon, *partial* oxidation may occur, as illustrated in eqs. 2.6 through 2.9.

$$2\,CH_4 + 3\,O_2 \longrightarrow \underset{\text{carbon monoxide}}{2\,CO} + 4\,H_2O \quad (2.6)$$

$$CH_4 + O_2 \longrightarrow \underset{\text{carbon}}{C} + 2\,H_2O \quad (2.7)$$

$$CH_4 + O_2 \longrightarrow \underset{\text{formaldehyde}}{CH_2O} + H_2O \quad (2.8)$$

$$2\,C_2H_6 + 3\,O_2 \longrightarrow \underset{\text{acetic acid}}{2\,CH_3CO_2H} + 2\,H_2O \quad (2.9)$$

Toxic carbon monoxide in exhaust fumes (eq. 2.6), soot emitted copiously from trucks with diesel engines (eq. 2.7), smog resulting in part from aldehydes (eq. 2.8), and acid build-up in lubricating oils (eq. 2.9) are all prices we pay for being a motorized society. However, incomplete hydrocarbon combustion can occasionally be useful, as in the manufacture of carbon blacks (eq. 2.7), particularly lampblack, a pigment for ink.

EXAMPLE 2.6 In which compound is carbon more oxidized, formaldehyde (CH_2O) or formic acid (HCO_2H).

Solution Draw out the structures.

$$\underset{\text{formaldehyde}}{\begin{array}{c}H \\ \diagdown \\ C{=}O \\ \diagup \\ H\end{array}} \qquad \underset{\text{formic acid}}{H{-}C\begin{array}{l}{/\!\!/}O \\ \diagdown OH\end{array}}$$

formaldehyde formic acid

Formic acid is the more oxidized form (3 C—O and 1 C—H bond, compared to 2 C—O and 2 C—H bonds in formaldehyde).

Problem 2.17 Which represents the more oxidized form of carbon, (a) methanol (CH_3OH) or formaldehyde? (b) methanol or dimethyl ether (CH_3OCH_3)?

2.15 *Halogenation of Alkanes*

When a mixture of an alkane and chlorine gas is stored at low temperatures in the dark, no reaction occurs. In sunlight or at high temperatures, however, an exothermic reaction occurs. One or more hydrogen atoms of the alkane are replaced by chlorine atoms. This reaction can be represented by the general equation

$$R{-}H + Cl{-}Cl \xrightarrow[\text{heat}]{\text{light or}} R{-}Cl + H{-}Cl \tag{2.10}$$

or, specifically for methane:

$$\underset{\text{methane}}{CH_4} + Cl{-}Cl \xrightarrow[\text{or heat}]{\text{sunlight}} \underset{\substack{\text{chloromethane} \\ \text{(methyl chloride)} \\ \text{bp } -24.2°\text{C}}}{CH_3Cl} + HCl \tag{2.11}$$

The reaction is called **chlorination.** It is a **substitution reaction;** a chlorine is substituted for a hydrogen.

An analogous reaction, called **bromination,** occurs when the halogen is bromine.

$$R{-}H + Br{-}Br \xrightarrow[\text{heat}]{\text{light or}} R{-}Br + HBr \tag{2.12}$$

If excess halogen is present, the reaction can continue further to give polyhalogenated products. Thus, methane and excess chlorine can give products with two, three, or four chlorines.*

$$CH_3Cl \xrightarrow{Cl_2} \underset{\substack{\text{dichloromethane} \\ \text{(methylene chloride)} \\ \text{bp } 40°\text{C}}}{CH_2Cl_2} \xrightarrow{Cl_2} \underset{\substack{\text{trichloromethane} \\ \text{(chloroform)} \\ \text{bp } 61.7°\text{C}}}{CHCl_3} \xrightarrow{Cl_2} \underset{\substack{\text{tetrachloromethane} \\ \text{(carbon tetrachloride)} \\ \text{bp } 76.5°\text{C}}}{CCl_4} \tag{2.13}$$

* Note that we sometimes write the formula of one of the reactants (in this case Cl_2) over the arrow for convenience, as in eq. 2.13. We also sometimes omit obvious inorganic products (in this case HCl).

By controlling the reaction conditions and the ratio of chlorine to methane, we can favor formation of one or another of the possible products.

Problem 2.18 Write the names and structures of all possible products of bromination of methane.

With longer-chain alkanes, mixtures of products may be obtained even at the first step.* For example, with propane,

$$\underset{\text{propane}}{CH_3CH_2CH_3} + Cl_2 \xrightarrow[\text{or heat}]{\text{light}} \underset{\substack{\text{1-chloropropane} \\ (n\text{-propyl chloride})}}{CH_3CH_2CH_2Cl} + \underset{\substack{\text{2-chloropropane} \\ \text{(isopropyl chloride)}}}{\underset{\displaystyle Cl}{CH_3\overset{}{\underset{|}{C}}HCH_3}} + HCl \quad (2.14)$$

When larger alkanes are halogenated, the mixture of products becomes even more complex; individual isomers become difficult to separate and obtain pure, so halogenation tends not to be a useful way to synthesize specific alkyl halides. With unsubstituted *cycloalkanes,* however, where all the hydrogens are equivalent, a single pure organic product can be obtained:

$$\underset{\text{cyclopentane}}{C_5H_{10}} + Br_2 \xrightarrow{\text{light}} \underset{\substack{\text{bromocyclopentane} \\ \text{(cyclopentyl bromide)}}}{C_5H_9Br} + HBr \quad (2.15)$$

Problem 2.19 Write out the structures of all possible products of *mono*bromination of pentane. Note the complexity of the product mixture, compared to that from the corresponding reaction with cyclopentane (eq. 2.15).

Problem 2.20 How many organic products can be obtained from the monochlorination of octane? of cyclooctane?

Problem 2.21 Do you think that the chlorination of 2,2-dimethylpropane might be synthetically useful?

2.16 The Free Radical Chain Mechanism of Halogenation

One may well ask how halogenation occurs. Why is light or heat necessary? Eqs. 2.10 and 2.11 express the *overall* reaction for halogenation. They describe the structures of the reactants and the products, and they show necessary reaction conditions or catalysts over the arrow. But they do *not* tell us exactly how the products are formed from the reactants.

A reaction mechanism is a step-by-step description of the bond-breaking and bond-making processes that occur when reagents react to

---

*Note that we often do not write a balanced equation, especially when more than one product is formed from a single organic reactant. Instead, we show on the right side of the equation the structures of *all* the important organic products, as in eq. 2.14.

form products. In the case of halogenation, various experiments show that this reaction occurs in several steps, not in one. Halogenation occurs via a **free-radical chain** of reactions.

The **chain-initiating step** is the breaking of the halogen molecule into two halogen atoms.

$$\textit{initiation} \quad :\ddot{\underset{..}{Cl}}\ :\ \ddot{\underset{..}{Cl}}: \xrightarrow[\text{heat}]{\text{light or}} 2\ :\ddot{\underset{..}{Cl}}\cdot \tag{2.16}$$
*

chlorine molecule — chlorine atoms

The halogen bond is weaker than either the C—H bond or the C—C bond, and is therefore the easiest bond to break. When light is the energy source, chlorine absorbs visible light but alkanes do not, so again it is the Cl—Cl bond that breaks.

The **chain-propagating steps** are

propagation

$$R{-}H + :\ddot{\underset{..}{Cl}}\cdot \longrightarrow R\cdot + HCl \tag{2.17}$$

(R· = alkyl radical)

$$R\cdot + Cl{-}Cl \longrightarrow R{-}Cl + :\ddot{\underset{..}{Cl}}\cdot \tag{2.18}$$

(R—Cl = alkyl chloride)

Chlorine atoms are very reactive, because they have an incomplete valence shell (7 electrons instead of the required 8). They may either recombine to form chlorine molecules (the reverse of eq. 2.16) or, if they collide with an alkane molecule, they may abstract a hydrogen atom to form hydrogen chloride and an alkyl radical R·. Recall from Sec. 1.4 that a radical is a fragment with an odd number of unshared electrons. The space-filling models in Figure 2.1 show that alkanes seem to have an exposed surface of hydrogens covering the carbon skeleton. So it is most likely that, if a halogen atom collides with an alkane molecule, it will hit the hydrogen end of a C—H bond.

Like a chlorine atom, the alkyl radical formed in the first step of the chain (eq. 2.17) is very reactive (incomplete octet). If it were to collide with a chlorine molecule, it could form an alkyl chloride molecule and a chlorine atom (eq. 2.18). The chlorine atom formed in this step can then react to repeat the sequence. When you add eq. 2.17 and eq. 2.18, you get the overall equation for chlorination (eq. 2.10). *In each chain-propagating step, a radical (or atom) is consumed, but another radical (or atom) is formed and can continue the chain. Almost all of the reactants are consumed, and almost all of the products are formed in these steps.*

Were it not for **chain-terminating steps,** all of the reactants could, in principle, be consumed by initiating a single chain. However, because many chlorine molecules react to form chlorine atoms in the chain-initiating step, many chains are started simultaneously. Quite a few radicals are present as the reaction proceeds. If any two radicals combine, the chain will be terminated.

---

*Note that we use a "fishhook," or half-headed, arrow, ⌒, to show the movement of only *one* electron whereas we use a complete arrow, ⌒, to describe the movement of an electron *pair*.

Three possible chain-terminating steps are

termination

$$2\ :\ddot{\underset{..}{Cl}}\cdot \longrightarrow Cl{-}Cl \qquad (2.19)$$

$$2\ R\cdot \longrightarrow R{-}R \qquad (2.20)$$

$$R\cdot + :\ddot{\underset{..}{Cl}}\cdot \longrightarrow R{-}Cl \qquad (2.21)$$

No new radicals are formed in these reactions, so the chain is broken or, as we say, terminated.

Problem 2.22 Write equations for all the steps (initiation, propagation, termination) in the free-radical chlorination of methane to form methyl chloride.

Problem 2.23 Account for the experimental observation that small amounts of ethane and chloroethane are produced during the chlorination of methane. (*Hint:* Consider the possible chain-terminating steps).

A WORD ABOUT ...

2. Methane, Marsh Gas, and Miller's Experiment

Methane is commonly found in nature wherever bacteria decompose organic matter in the absence of oxygen, as in marshes, swamps, or the muddy sediment of lakes—hence its common name, marsh gas. In China, methane has been collected from the mud at the bottom of swamps for use in domestic cooking and lighting. Methane is similarly formed from bacteria in the digestive tracts of certain ruminant animals, such as cows.

The scale of methane production by bacteria is considerable. The earth's atmosphere contains an average of 1 part per million of methane. Because our planet is small and because methane is light compared to most other air constituents (O_2, N_2), one would expect most of the methane to escape from our atmosphere, and it has been calculated that the equilibrium concentration should be very much less than is observed. The reason for the relatively high observed concentration is that, at the same time that methane escapes from the atmosphere, it is constantly being produced by bacterial decay of plant matter.

In cities, the amount of methane in the atmosphere reaches much higher levels, up to several parts per million. The peak concentrations come in the early morning and late afternoon, a direct correlation with the peaks of automobile traffic. Fortunately, methane, which constitutes about 50% of urban atmospheric hydrocarbon pollutants, seems to have no direct harmful effect on human health.

Methane can accumulate in coal mines, where it is a hazard. Mixed with 5 to 14% of air, methane is explosive. Also, miners can be asphyxiated by it (due to lack of sufficient oxygen). Dangerous concentrations of methane are readily detected by a variety of safety devices, including canaries, which succumb to lower concentrations of methane than are harmful to humans and alert miners to the presence of a hazard.

Methane was probably one of the main components of the earth's atmosphere in its early years. Also, hydrogen is the most common element in the solar system (it constitutes about 87% of the sun's mass). It is therefore reasonable that, when the planets were formed, other elements should have been present in reduced (rather than oxidized) forms: carbon as methane, nitrogen as ammonia, and oxygen as water. Indeed, some of the larger planets (such as Saturn and Jupiter), which have very strong gravitational fields and low surface temperatures that help retain light molecules, still have atmospheres that are rich in methane and ammonia.

A now-famous experiment carried out in 1955 by Stanley L. Miller (working in the laboratory of H. C. Urey at Columbia University) supports the idea that life could have arisen in a reducing atmosphere. Miller found that when mixtures of methane, ammonia, water, and hydrogen were subjected to electric discharges to simulate lightning, some organic compounds were formed (amino acids, for example) that are important to biology and necessary for life. Similar results have since been obtained using heat or ultraviolet light in place of electric discharges (it seems likely that the earth's early atmosphere was subjected to much more ultraviolet light than it is now). When oxygen was added to these simulated primeval atmospheres, no amino acids were produced—strong evidence that the earth's original atmosphere did *not* contain free oxygen. Miller's experiment provided the model for much work in the branch of science now called **chemical evolution,** the study of chemical events that may have taken place on earth or elsewhere in the universe leading up to the appearance of the first living cell.

During the years since Miller's experiment, ideas about the chemistry of life's origin have become more precise and sophisticated as a consequence of much experimentation and of exploration in outer space. For example, it seems likely that the main carbon source in the earth's early atmosphere was carbon dioxide, *not* methane as assumed by Miller, and that hydrogen was present mainly as water rather than as hydrogen gas. (A fascinating and lucid account of our current knowledge is contained in "The Chemistry of Life's Origin," an article by James P. Ferris in *Chemical and Engineering News,* August 27, 1984, p. 22).

Additional Problems

2.24. Write structural formulas for the following compounds:

a. 3-methylpentane
b. 2,3-dimethylbutane
c. 4-ethyl-3,3-dimethylhexane
d. 2-chloro-3-methylpentane
e. 2,2,3-trimethylbutane
f. 2-bromopropane
g. 1,1-dichlorocyclopropane
h. 1,1,3,3-tetrachloropropane
i. 3-bromo-1,1-dimethylcyclopentane
j. 1,4-dichlorocyclohexane

2.25. Write expanded formulas for the following compounds, and name them using the IUPAC system:

a. $CH_3(CH_2)_3CH_3$
b. $CH_3CH(CH_3)CH_2CH_3$
c. $CH_3CH_2C(CH_3)_2CH_2CH_3$
d. $CH_3(CH_2)_2C(CH_3)_3$
e. $CH_3CH_2CHBrCH_3$
f. $CH_3CCl_2CBr_3$
g. $(CH_3CH_2)_4C$
h. CH_2ClCH_2Br
i. $CH_2BrCH(CH_3)CH(CH_3)_2$
j. $(CH_2)_5$
k. MeI
l. *i*-PrBr

2.26. Give both common and IUPAC names for the following compounds:

a. CH_3I
b. CH_3CH_2Cl
c. CH_2Cl_2
d. $CHBr_3$
e. $CH_3CH_2CH_2Cl$
f. $(CH_3)_2CHBr$
g. $CHCl_3$
h.
```
CH2—CH—Cl
|    |
CH2—CH2
```
i. $CH_3CHICH_2CH_3$
j. $(CH_3)_3CCl$

2.27. Write a structure for each of the compounds listed. Explain why the name given here is objectionable, and give a correct name in each case.
a. 1-methylpentane **b.** 2-ethylbutane
c. 2,3-dichloropropane **d.** 1,4-dimethylcyclobutane
e. 1,1,3-trimethylpropane **f.** 3-bromo-2-methylpropane

2.28. Chemicals used for communication in nature are called *pheromones*. The pheromone used by the female tiger moth to attract the male is the alkane 2-methylheptadecane. Write out its structural formula.

2.29. In 1985, the synthesis of normal alkanes with 102, 150, 198, 246, and 390 carbon atoms was reported for the first time. What is the molecular formula of each of these compounds? The C_{390} compound sets the current size record for pure alkanes. What is its molecular weight?

2.30. Write the structural formula for all the isomers (the number is indicated in parentheses) for each of the following compounds, and name each isomer by the IUPAC system.
a. C_4H_{10} (2) **b.** C_4H_9Br (4) **c.** C_6H_{14} (5)
d. $C_3H_6Br_2$ (4) **e.** $C_2H_2BrCl_3$ (3) **f.** C_3H_6BrCl (5)

2.31. The general formula for an alkane is C_nH_{2n+2}. What is the corresponding formula for a cycloalkane?

2.32. Write structural formulas and names for all possible cycloalkanes having each of the following molecular formulas. Be sure to include *cis–trans* isomers when appropriate. Name each compound by the IUPAC system.
a. C_5H_{10} **b.** C_6H_{12} (there are 16)

2.33. Without referring to tables, arrange the following five hydrocarbons in order of increasing boiling point:
a. 2-methylhexane **b.** heptane **c.** 3,3-dimethylpentane
d. hexane **e.** 2-methylpentane

2.34. In Problem 2.9 you drew two staggered conformations of butane (looking end-on down the bond between carbon-2 and carbon-3). There are also two eclipsed conformations around this bond. Draw Newman projections for them. Arrange all four conformations in order of decreasing stability.

2.35. Draw all possible staggered and eclipsed conformations of 1-bromo-2-chloroethane, using Newman projections. Underneath each, draw the corresponding "dash-wedge" and "sawhorse" structures. Rank the structures in order of decreasing stability.

2.36. Draw the formula for the preferred conformation of
a. ethylcyclohexane. **b.** *trans*-1,4-dimethylcyclohexane.
c. *cis*-1-methyl-3-(1-methylethyl)cyclohexane. **d.** 1,1-dibromocyclohexane.

2.37. Name the following *cis–trans* pairs:

a.
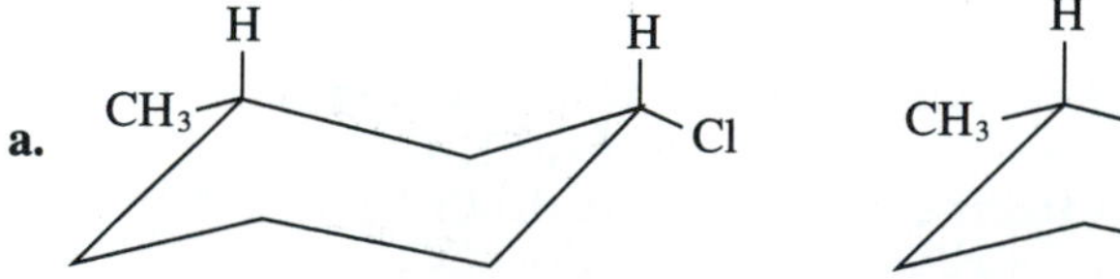

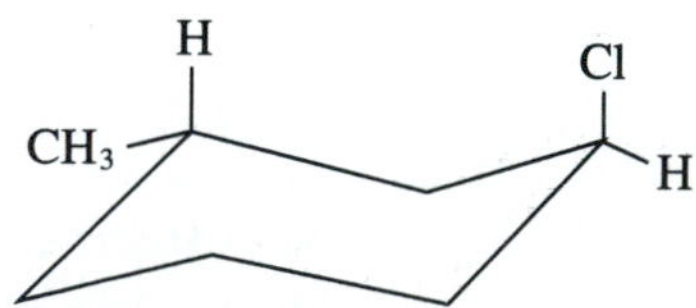

b. 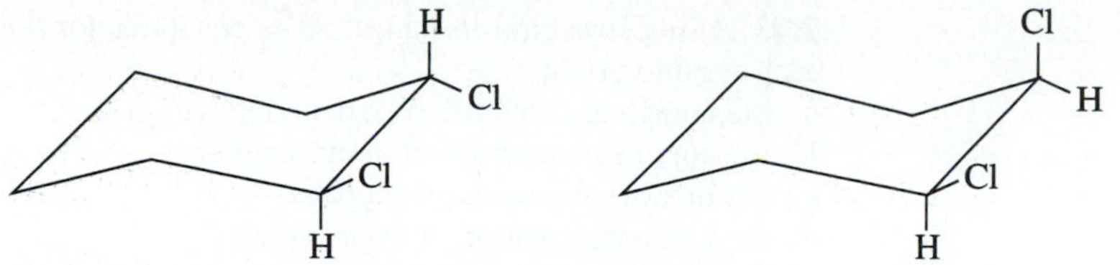

2.38. Explain with the aid of conformational structures why *cis*-1,3-dimethylcyclohexane is more stable than *trans*-1,3-dimethylcyclohexane, whereas the reverse order of stability is observed for the 1,2 and 1,4 isomers.

2.39. Which will be more stable, *cis*- or *trans*-1,4-di-*tert*-butylcyclohexane? Explain your answer by drawing conformational structures for each compound.

2.40. Classify the following pairs of structures according to the scheme in Figure 2.5:
a. the pairs of compounds in Problem 2.37

b. 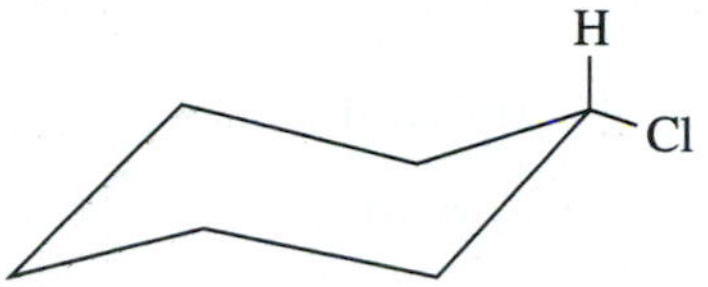 and 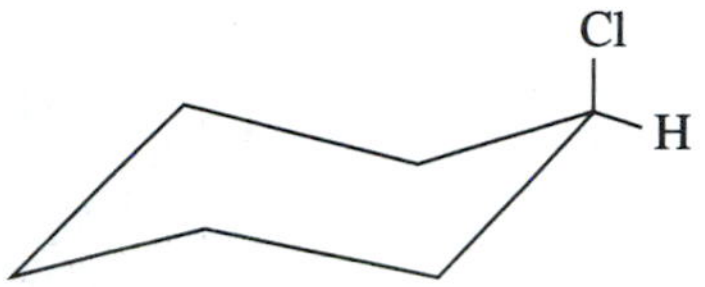

c. 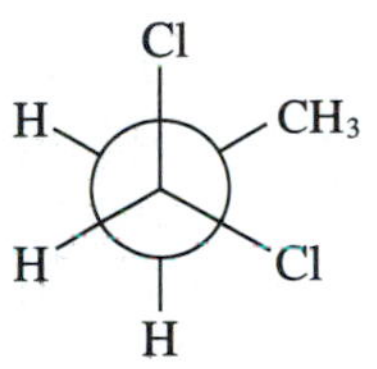 and 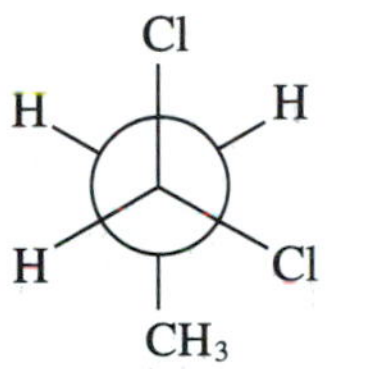

d. 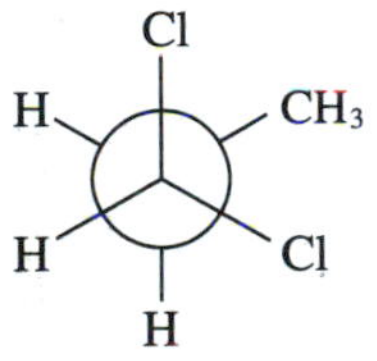 and 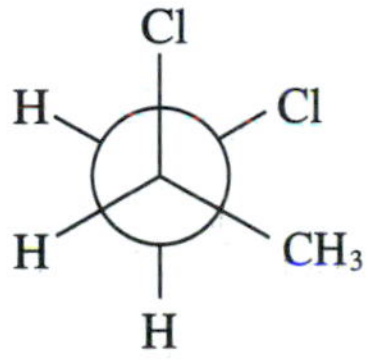

e. $CH_3CHCH_2CH_3$ and $CH_3CH_2CHCH_3$ (careful!)
CH_3 CH_3

2.41. Draw structural formulas for all possible products of the dichlorination of cyclopentane. Include *cis–trans* isomers.

2.42. How many monochlorination products can be obtained from each of the following polycyclic alkanes?

a. 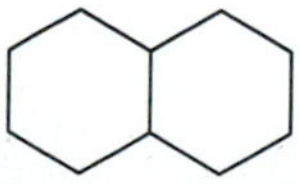

b. 

c. 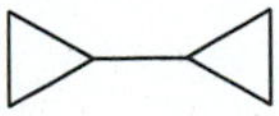

2.43. Using structural formulas, write equations for the following reactions, and name each organic product.
a. the complete combustion of pentane
b. the complete combustion of cyclopentane
c. the monobromination of propane
d. the monochlorination of cyclopentane
e. the complete chlorination of ethane

2.44. From the dichlorination of propane, four isomeric products with the formula $C_3H_6Cl_2$ were isolated and designated A, B, C, and D. Each was separated and further chlorinated to give one or more trichloropropanes, $C_3H_5Cl_3$. A and B gave three trichloro compounds, C gave one, and D gave two. Deduce the structures of C and D. One of the products from A was identical with the product from C. Deduce structures for A and B.

2.45. Write out all the steps in the free-radical chain mechanism for the monochlorination of ethane.

$$CH_3CH_3 + Cl_2 \longrightarrow CH_3CH_2Cl + HCl$$

What trace by-products would you expect to be formed as a consequence of the chain-terminating steps?

CHAPTER 3

Alkenes and Alkynes

3.1 Definition and Classification

Hydrocarbons that contain a carbon–carbon double bond are called **alkenes;** those with a carbon–carbon triple bond are **alkynes.*** Their general formulas are

C_nH_{2n} (alkenes) C_nH_{2n-2} (alkynes)

Both of these classes of hydrocarbons are **unsaturated,** because they contain fewer hydrogens per carbon than alkanes (C_nH_{2n+2}). Alkanes can be obtained from alkenes or alkynes by adding one or two moles of hydrogen.

$$\left.\begin{array}{l} RCH{=}CHR \xrightarrow[\text{catalyst}]{H_2} \\ RC{\equiv}CR \xrightarrow[\text{catalyst}]{2H_2} \end{array}\right\} \longrightarrow RCH_2CH_2R \qquad (3.1)$$

Compounds with more than one double or triple bond exist. If two double bonds are present, the compounds are called **alkadienes** or, more commonly, **dienes.** There are also trienes, tetraenes, and even polyenes (compounds with *many* double bonds, from the Greek *poly,* many). Compounds with more than one triple bond, or with double and triple bonds, are also known.

EXAMPLE 3.1 What are all the structural possibilities for the compound C_3H_4?

Solution The formula C_3H_4 corresponds to the general formula C_nH_{2n-2}. The compound could have one triple bond, two double bonds, or one ring and one double bond. For their structures, see the solution to Example 1.10 on page 17.

*An old but still used synonym for alkenes is *olefins,* which means oil-forming. This name was originally given to ethylene because it formed an oily liquid when treated with chlorine. Alkynes are also called *acetylenes* after the first member of the series.

Problem 3.1 What are all the structural possibilities for C_4H_6? (Nine compounds, four acyclic and five cyclic, are known.)

When two or more multiple bonds are present in a molecule, it is useful to classify the structure further, depending on the relative positions of the multiple bonds. Double bonds are said to be **cumulated** when they are right next to one another. When multiple bonds *alternate* with single bonds, they are called **conjugated.** When more than one single bond comes between multiple bonds, the latter are **isolated** or **nonconjugated.**

cumulated	conjugated	isolated
C=C=C	C=C—C=C	C=C—C—C=C
C=C=C=C	C=C—C≡C	C≡C—C—C—C≡C

Problem 3.2 Which of the following compounds have conjugated multiple bonds?

a. b.

c. d.

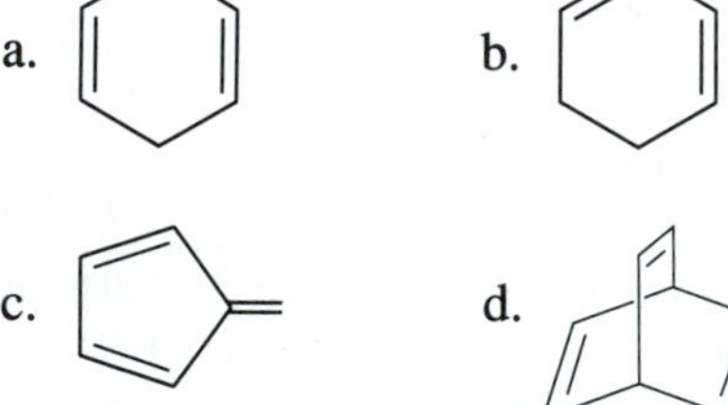

3.2 Nomenclature

The IUPAC rules for naming alkenes and alkynes are similar to those for alkanes (Sec. 2.4), but a few rules must be added for naming and locating the multiple bonds.

1. The ending *-ene* is used to designate a carbon–carbon double bond. When more than one double bond is present, the ending is *-diene, -triene,* and so on. The ending *-yne* is used for a triple bond (*-diyne* for two triple bonds, and so on). Compounds with a double *and* a triple bond are *-enynes.*

2. Select the longest chain that includes *both* carbons of the double or triple bond. For example

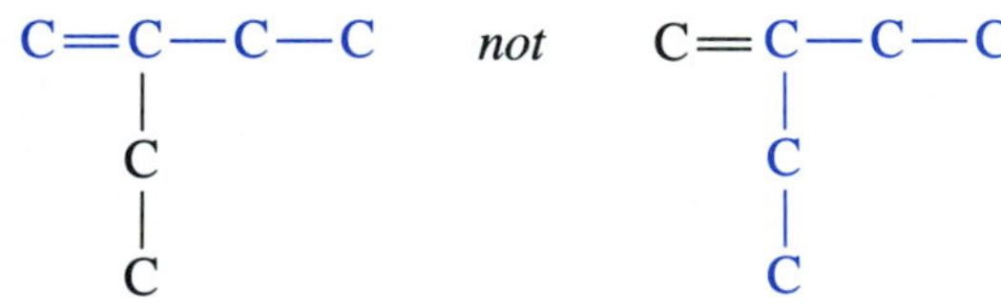

named as a butene, not as a pentene

3. Number the chain from the end nearest the multiple bond, so that the carbon atoms in that bond have the lowest possible numbers.

$\overset{1}{C}—\overset{2}{C}=\overset{3}{C}—\overset{4}{C}—\overset{5}{C}$ *not* $\overset{5}{C}—\overset{4}{C}=\overset{3}{C}—\overset{2}{C}—\overset{1}{C}$

If the multiple bond is equidistant from both ends of the chain, number from the end nearest the first branch point.

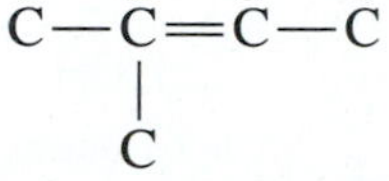

$$\overset{1}{C}-\overset{2}{\underset{|\atop C}{C}}=\overset{3}{C}-\overset{4}{C} \quad \textit{not} \quad \overset{4}{C}-\overset{3}{\underset{|\atop C}{C}}=\overset{2}{C}-\overset{1}{C}$$

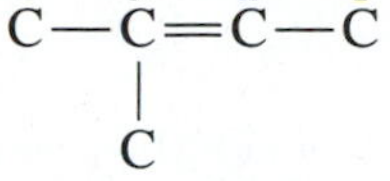

4. Indicate the position of the multiple bond using the *lowest-numbered carbon atom* of that bond. For example,

$$\overset{1}{C}H_2=\overset{2}{C}H\overset{3}{C}H_2\overset{4}{C}H_3 \quad \text{1-butene, } \textit{not} \text{ 2-butene}$$

5. If more than one multiple bond is present, number from the end nearest the first multiple bond.

$$\overset{1}{C}=\overset{2}{C}-\overset{3}{C}=\overset{4}{C}-\overset{5}{C} \quad \textit{not} \quad \overset{5}{C}=\overset{4}{C}-\overset{3}{C}=\overset{2}{C}-\overset{1}{C}$$

If a double and a triple bond are equidistant from the end of the chain, the *double* bond receives the lowest numbers. For example,

$$\overset{1}{C}=\overset{2}{C}-\overset{3}{C}\equiv\overset{4}{C} \quad \textit{not} \quad \overset{4}{C}=\overset{3}{C}-\overset{2}{C}\equiv\overset{1}{C}$$

Let us see how these rules are applied. The first two members of each series are

CH_3CH_3	$CH_2=CH_2$	$HC\equiv CH$
ethane	ethene	ethyne
$CH_3CH_2CH_3$	$CH_2=CHCH_3$	$HC\equiv CCH_3$
propane	propene	propyne

The root of the name (*eth-* or *prop-*) tells us the number of carbons, and the ending (*-ane, -ene,* or *-yne*) tells us whether the bonds are single, double, or triple. No number is necessary in these cases, because in each instance, only one structure is possible.

With four carbons, a number is necessary to locate the double or triple bond.

$\overset{1}{C}H_2=\overset{2}{C}H\overset{3}{C}H_2\overset{4}{C}H_3$	$\overset{1}{C}H_3\overset{2}{C}H=\overset{3}{C}H\overset{4}{C}H_3$	$\overset{1}{H}C\equiv\overset{2}{C}\overset{3}{C}H_2\overset{4}{C}H_3$	$\overset{1}{C}H_3\overset{2}{C}\equiv\overset{3}{C}\overset{4}{C}H_3$
1-butene	2-butene	1-butyne	2-butyne

Branches are named in the usual way.

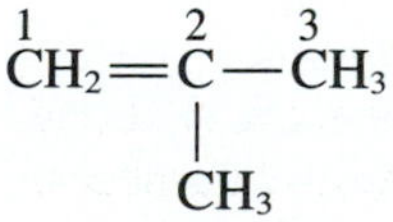

$\overset{1}{C}H_2=\overset{2}{C}(CH_3)-\overset{3}{C}H_3$	$\overset{1}{C}H_2=\overset{2}{C}(CH_3)-\overset{3}{C}H_2\overset{4}{C}H_3$	$\overset{1}{C}H_3-\overset{2}{C}(CH_3)=\overset{3}{C}H\overset{4}{C}H_3$	$\overset{1}{C}H_2=\overset{2}{C}(CH_3)-\overset{3}{C}H=\overset{4}{C}H_2$
methylpropene (isobutylene)	2-methyl-1-butene	2-methyl-2-butene	2-methyl-1,3-butadiene (isoprene)

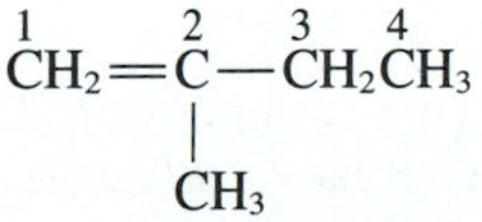

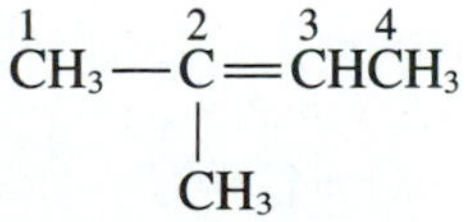

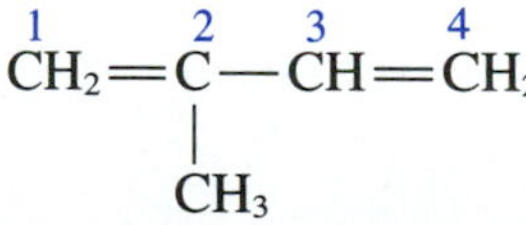

Note how the rules are applied in the following examples:

$$\overset{1}{CH_3}-\overset{2}{CH}=\overset{3}{CH}-\overset{4}{\underset{\underset{CH_3}{|}}{CH}}-\overset{5}{CH_3}$$

4-methyl-2-pentene
(Not 2-methyl-3-pentene; the chain is numbered so that the double bond gets the lowest number.)

$$\overset{1}{CH_2}=\overset{2}{\underset{\underset{CH_2CH_3}{|}}{C}}-\overset{3}{CH_2}\overset{4}{CH_3}$$

2-ethyl-1-butene
(Named this way, even though there is a five-carbon chain present, because that chain does not include both carbons of the double bond.)

$$\overset{1}{CH_2}=\overset{2}{CH}-\overset{3}{CH}=\overset{4}{CH_2}$$

1,3-butadiene
(Note the *a* inserted in the name, to help in pronunciation.)

With cyclic hydrocarbons, we start numbering the ring with the carbons of the multiple bond.

cyclopentene
(No number is necessary, because there is only one possible structure.)

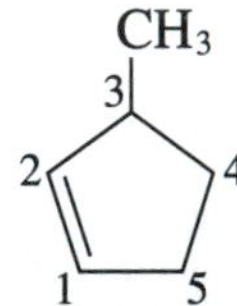

3-methylcyclopentene
(Start numbering at, and number through, the double bond; 5-methylcyclopentene and 1-methyl-2-cyclopentene are incorrect names.)

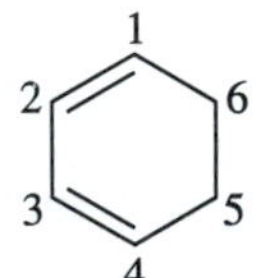

1,3-cyclohexadiene

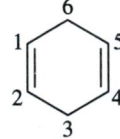

1,4-cyclohexadiene

Problem 3.3 Name each of the following structures by the IUPAC system.

a. $ClCH=CHCH_3$ b. $(CH_3)_2C=C(CH_3)_2$ c. $CH_2=C(CH_3)CH=CH_2$

d. CH_3 (1-methylcyclopentene ring structure) e. $CH_2=C(Cl)CH_3$ f. $HC\equiv C(CH_2)_3CH_3$

EXAMPLE 3.2 Write the structural formula for 3-methyl-2-pentene.

Solution To get the structural formula from the IUPAC name, first write out the longest chain or ring, number it, and locate the multiple bond. In this case, note that the chain has five carbons and that the double bond is located between carbon-2 and carbon-3:

$$\overset{1}{C}-\overset{2}{C}=\overset{3}{C}-\overset{4}{C}-\overset{5}{C}$$

Next, add the substituent:

$$\overset{1}{C}-\overset{2}{C}=\overset{3}{C}-\overset{4}{C}-\overset{5}{C}$$
$$\quad\;\; | $$
$$\quad\;\; CH_3$$

Finally, fill in the hydrogens:

$$CH_3-CH=\underset{\underset{\displaystyle CH_3}{|}}{C}-CH_2-CH_3$$

Problem 3.4 Write the structural formula for:

a. 2,4-dimethyl-2-pentene
b. 2-hexyne
c. 1,2-dibromocyclobutene
d. 2-chloro-1,3-butadiene

In addition to the IUPAC rules, it is important to learn a few common names. For example, the simplest members of the alkene and alkyne series are frequently referred to by their older common names, **ethylene, acetylene,** and **propylene.**

$CH_2=CH_2$	$HC\equiv CH$	$CH_3CH=CH_2$
ethylene	acetylene	propylene
(ethene)	(ethyne)	(propene)

Three important groups also have common names. They are the **vinyl, allyl,** and **propargyl** groups (their IUPAC names are in parentheses).

$CH_2=CH-$	$\overset{1}{C}H_2=\overset{2}{C}H-\overset{3}{C}H_2-$	$H\overset{1}{C}\equiv\overset{2}{C}-\overset{3}{C}H_2-$
vinyl	allyl	propargyl
(ethenyl)	(3-propenyl)	(3-propynyl)

These groups are used in common names.

$CH_2=CHCl$	$CH_2=CH-CH_2Cl$	$HC\equiv C-CH_2Br$
vinyl chloride	allyl chloride	propargyl bromide
(chloroethene)	(3-chloropropene)	(3-bromopropyne)

Problem 3.5 Write the structural formula for

a. vinylcyclohexane.
b. allylcyclopropane.
c. propargyl iodide.

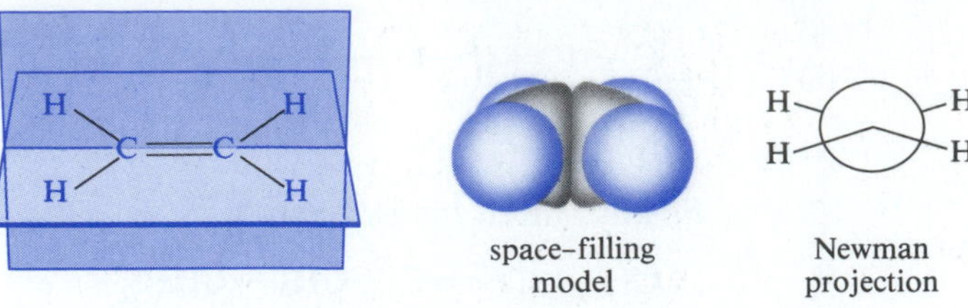

FIGURE 3.1 Three models of ethylene, each showing that the four atoms attached to a carbon–carbon double bond lie in a single plane.

3.3 Some Facts About Double Bonds

Carbon–carbon double bonds have some special features that are different from those of single bonds. For example, each carbon atom of a double bond is connected to only *three* other atoms (instead of four atoms, as with tetrahedral carbon). We speak of such a carbon as being **trigonal.** Furthermore, the two carbon atoms of a double bond and the four atoms that are attached to them lie in a single plane. This planarity is shown in Figure 3.1 for ethylene. The H—C—H and H—C═C angles in ethylene are approximately 120°. Although rotation occurs freely around single bonds, ***rotation around double bonds is restricted.*** Ethylene does not adopt any other conformation except the planar one. The doubly bonded carbons with two attached hydrogens do not rotate with respect to each other. Finally, carbon–carbon double bonds are shorter than carbon–carbon single bonds.

These differences between single and double bonds are summarized in Table 3.1. Let us see how the orbital model for bonding can explain the structure and properties of double bonds.

3.4 The Orbital Model of a Double Bond; the Pi Bond

Figure 3.2 shows what must happen with the atomic orbitals of carbon to accommodate trigonal bonding, bonding to only three other atoms. The first two parts of this figure are exactly the same as Figure 1.6. But now we combine only *three* of the orbitals, to make *three equivalent* sp^2*-hybridized orbitals* (called sp^2 because they are formed by combining one s and two p orbitals). These orbitals lie in a plane and are directed to the corners of an equilateral

TABLE 3.1 Comparison of C—C and C═C bonds

Property	*C—C*	*C═C*
1. Number of atoms attached to a carbon	4 (tetrahedral)	3 (trigonal)
2. Rotation	relatively free	restricted
3. Geometry	many conformations are possible; staggered is preferred	planar
4. Bond angle (common)	109.5°	120°
5. Bond length (common)	1.54 Å	1.34 Å

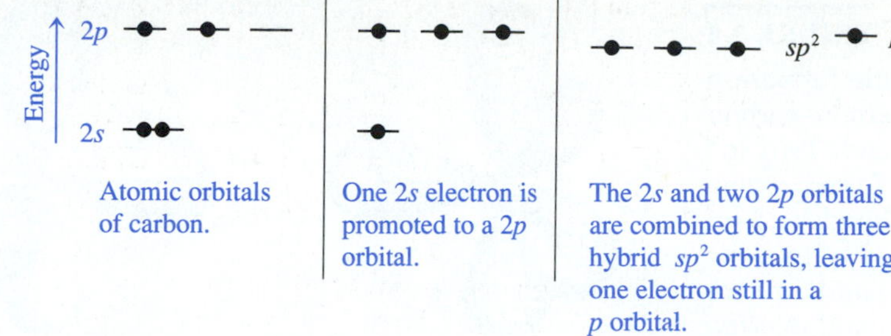
Energy
2p
2s
sp²
p
Atomic orbitals of carbon.
One 2s electron is promoted to a 2p orbital.
The 2s and two 2p orbitals are combined to form three hybrid sp² orbitals, leaving one electron still in a p orbital.

FIGURE 3.2
The formation of three sp^2 hybrid orbitals

triangle. The angle between them is 120°. This angle is preferred because repulsion between electrons in each orbital is minimized. The fourth valence electron is placed in the remaining $2p$ orbital, whose axis is perpendicular to the plane formed by the three sp^2 hybrid orbitals (see Figure 3.3).

Now let us see what happens when two sp^2-hybridized carbons are brought together to form a double bond. The process can be imagined as occurring stepwise (Figure 3.4). One of the two bonds, formed by end-on overlap of two sp^2 orbitals, is a sigma (σ) bond. The second bond of the double bond is formed differently. If the two carbons are aligned with the p orbitals on each carbon parallel, lateral overlap can occur, as shown at the bottom of Figure 3.4. The bond formed by lateral p-orbital overlap is called a **pi (π) bond.** The bonding in ethylene is summarized in Figure 3.5.

The orbital model explains the facts about double bonds listed in Table 3.1. Rotation about a double bond is restricted because, for rotation to occur, we would have to "break" the pi bond, as seen in Figure 3.6. For ethylene, it takes about 62 kcal/mol (259 kJ/mol) to break the pi bond, much more thermal energy than is available at room temperature. With the pi bond intact, the sp^2 orbitals on each carbon lie in a single plane. The 120° angle between those

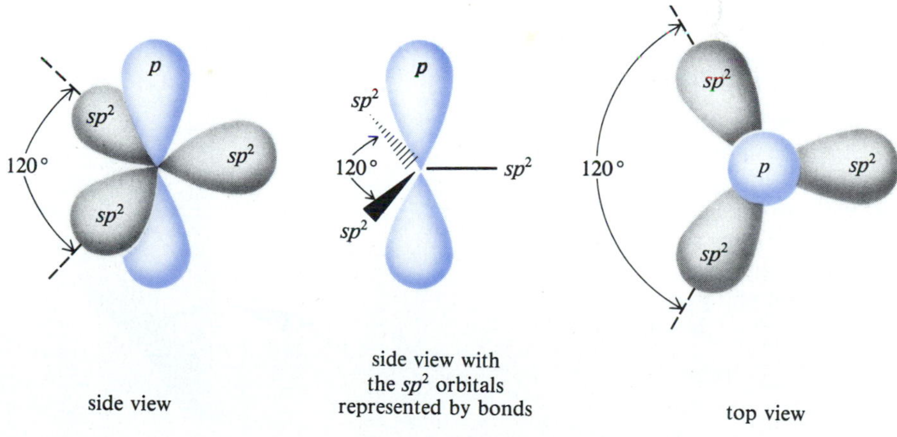
p
sp²
sp²
sp²
120°
side view
p
sp²
120°
sp²
sp²
side view with the sp² orbitals represented by bonds
sp²
120°
p
sp²
sp²
top view

FIGURE 3.3
A trigonal carbon showing three sp^2 hybrid orbitals in a plane with a 120° angle between them. The remaining p orbital is perpendicular to the sp^2 orbitals. There is a small back lobe to each sp^2 orbital, which has been omitted for ease of representation.

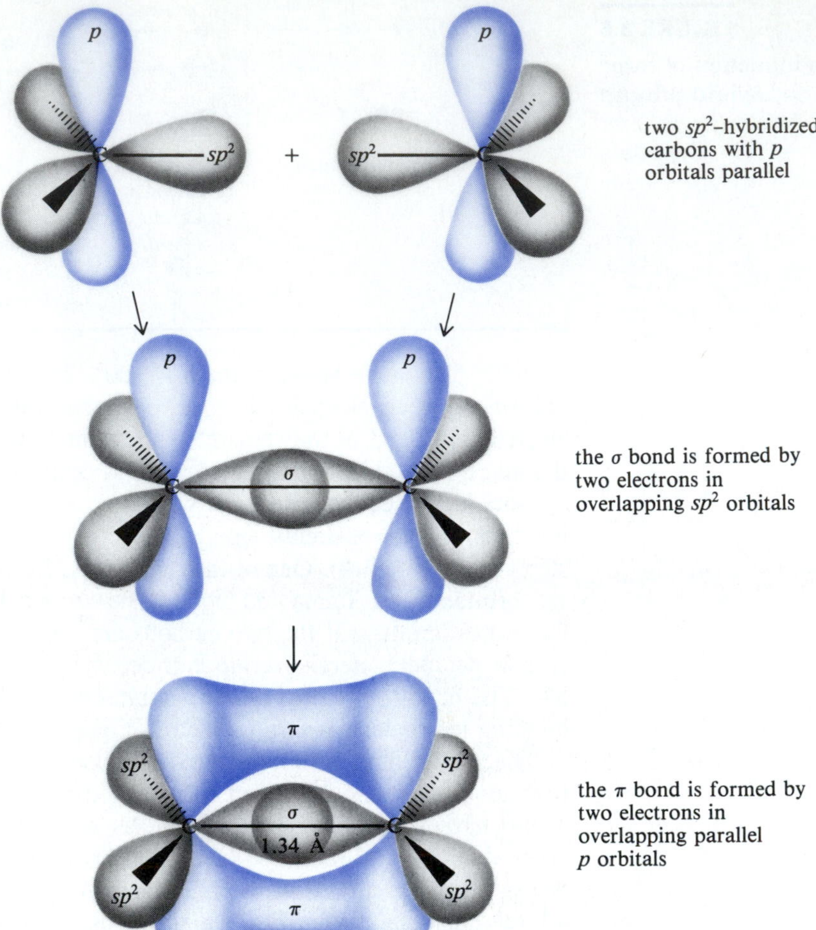

FIGURE 3.4

Schematic formation of a carbon–carbon double bond. Two sp^2 carbons form a sigma (σ) bond (end-on overlap of two sp^2 orbitals) and a pi (π) bond (lateral overlap of two properly aligned p orbitals).

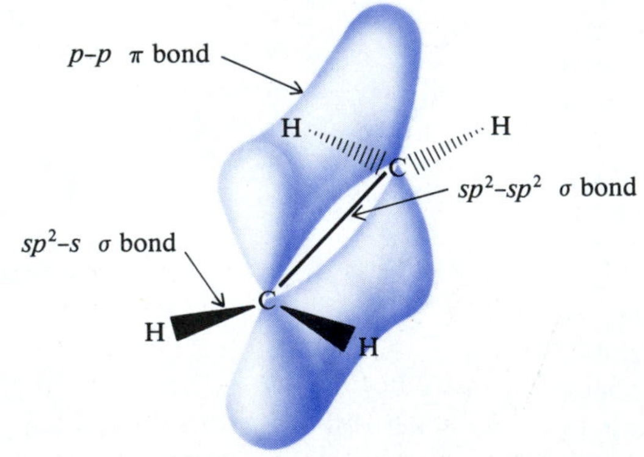

FIGURE 3.5

The bonding in ethylene consists of one sp^2–sp^2 carbon–carbon σ bond, four sp^2–s carbon–hydrogen σ bonds, and one p–p π bond.

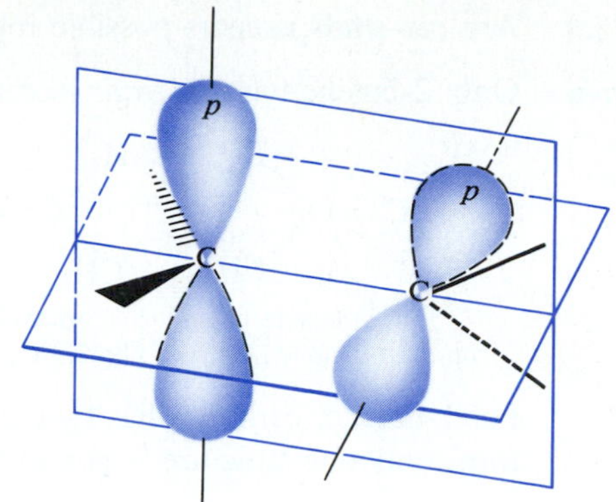

FIGURE 3.6
Rotation of one sp^2 carbon 90° with respect to another orients the p orbitals perpendicular to one another so that no overlap (and therefore no π bond) is possible.

orbitals minimizes repulsion between the electrons in them. Finally, the carbon–carbon double bond is shorter than the carbon–carbon single bond because the two shared electron pairs draw the nuclei closer together than a single pair does.

To recap, according to the orbital model, the carbon–carbon double bond consists of one sigma bond and one pi bond. The two electrons in the sigma bond lie along the internuclear axis; the two electrons in the pi bond lie in a region of space above and below the plane formed by the two carbons and the four atoms attached to them. The pi electrons are more exposed than the σ electrons and, as we will see, can be attacked by various electron-seeking reagents.

But before we consider reactions at the double bond, let us examine an important result of the restricted rotation around double bonds.

3.5 Cis–trans *Isomerism in Alkenes*

Because rotation at carbon–carbon double bonds is restricted, *cis–trans* isomerism is possible in appropriately substituted alkenes. For example, 1,2-dichloroethene exists in two different forms:

H(Cl)C=C(H)(Cl)	Cl(H)C=C(H)(Cl)
cis-1,2-dichloroethene	*trans*-1,2-dichloroethene
bp 60°C, mp −80°C	bp 47°C, mp −50°C

These stereoisomers are *not* readily interconverted by rotation around the double bond at room temperature. They are configurational stereoisomers and can be separated from one another by distillation, taking advantage of the difference in their boiling points.

EXAMPLE 3.3 Are *cis–trans* isomers possible for 1-butene and 2-butene?

Solution Only 2-butene has *cis–trans* isomers.

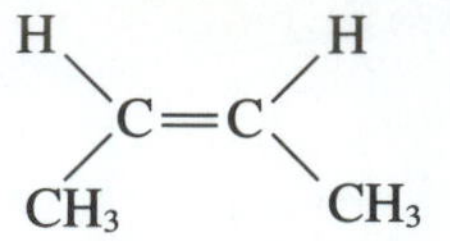

$$\underset{\text{CH}_3}{\overset{\text{H}}{\diagdown}}\text{C}{=}\text{C}\underset{\text{CH}_3}{\overset{\text{H}}{\diagup}} \qquad \underset{\text{CH}_3}{\overset{\text{H}}{\diagdown}}\text{C}{=}\text{C}\underset{\text{H}}{\overset{\text{CH}_3}{\diagup}}$$

cis-2-butene
bp 3.7°C, mp −139°C

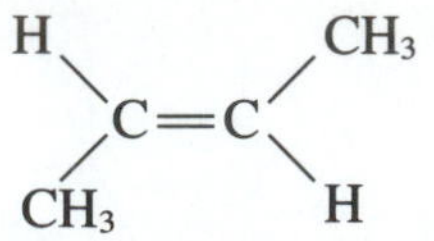

trans-2-butene
bp 0.3°C, mp −106°C

For 1-butene, carbon-1 has two identical hydrogen atoms attached to it; therefore, only one structure is possible.

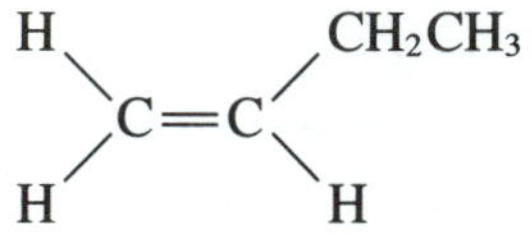

$$\underset{\text{H}}{\overset{\text{H}}{\diagdown}}\text{C}{=}\text{C}\underset{\text{H}}{\overset{\text{CH}_2\text{CH}_3}{\diagup}} \quad \text{is identical to} \quad \underset{\text{H}}{\overset{\text{H}}{\diagdown}}\text{C}{=}\text{C}\underset{\text{CH}_2\text{CH}_3}{\overset{\text{H}}{\diagup}}$$

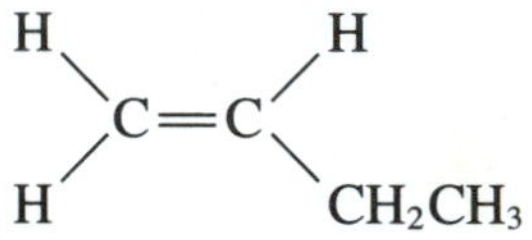

For *cis–trans* isomerism in alkenes, *each* carbon of the double bond must have two different atoms or groups attached to it.

Problem 3.6 Which of the following compounds can exist as *cis–trans* isomers? Draw their structures.

a. propene b. 3-hexene c. 2-hexene d. 2-methyl-2-butene

Geometric isomers of alkenes can be interconverted if sufficient energy is supplied to break the pi bond and allow rotation about the remaining, somewhat stronger, sigma bond (eq. 3.2). The required energy may take the form of light or heat.

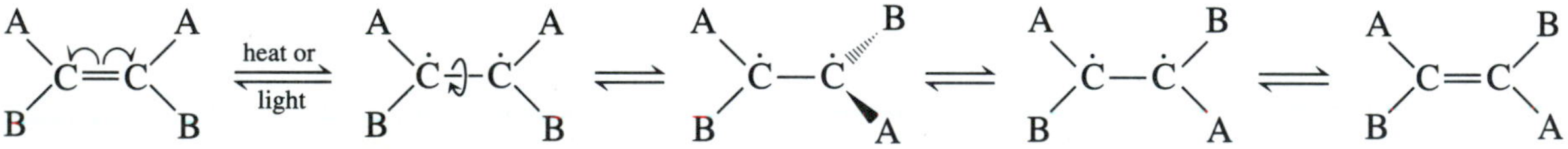

$$\underset{\textit{cis}}{\text{AB}\text{C}{=}\text{C}\text{AB}} \xrightarrow[\text{light}]{\text{heat or}} \text{AB}\dot{\text{C}}{-}\dot{\text{C}}\text{AB} \rightleftharpoons \text{AB}\dot{\text{C}}{-}\dot{\text{C}}\text{BA} \rightleftharpoons \text{AB}\dot{\text{C}}{-}\dot{\text{C}}\text{BA} \rightleftharpoons \underset{\textit{trans}}{\text{AB}\text{C}{=}\text{C}\text{BA}} \quad (3.2)$$

A WORD ABOUT ...

3. The Chemistry of Vision

Color in organic molecules is usually associated with extended conjugated systems of double bonds. A good example is ***β*-carotene,** a yellow-orange pigment found in carrots and many other plants. This $C_{40}H_{56}$ hydrocarbon has 11 carbon–carbon double bonds in conjugation. It is the biological precursor of the C_{20} unsaturated alcohol **vitamin A** (also called retinol), which in turn leads to the key substance involved in vision, **11-*cis*-retinal.** Notice in Figure 3.7 that the conversion of vitamin A to 11-*cis*-retinal involves not only oxidation of the alcohol group (—CH_2OH) to an aldehyde (—CH=O), but also *trans* → *cis* isomerism at the C_{11}–C_{12} double bond.

Cis–trans isomerism plays a key role in the process

of vision. The rod cells in the retina of the eye contain a red, light-sensitive pigment called **rhodopsin.** This pigment consists of the protein **opsin** combined at its active site with 11-*cis*-retinal. When visible light with the appropriate energy is absorbed by rhodopsin, the complexed *cis*-retinal is isomerized to the *trans isomer*. This process is fantastically fast, occurring in only picoseconds (10^{-12} seconds). As you can see from their structures, the shapes of the *cis* and *trans* isomers are very different.

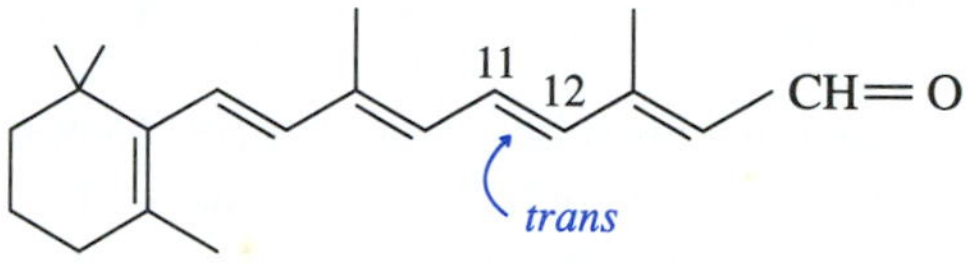

11 12 CH=O

trans

trans-retinal

The *trans*-retinal complex with opsin (called metarhodopsin-II) is less stable than the *cis*-retinal complex, and it dissociates into opsin and *trans*-retinal. This change in geometry triggers a response in the rod nerve cells that is transmitted to the brain and perceived as vision.

If this were all that happened, we would be able to see for only a few moments, because all of the 11-*cis*-retinal present in the rod cells would be quickly consumed. Fortunately, the enzyme *retinal isomerase,* in the presence of light, converts the *trans*-retinal back to the 11-*cis* isomer, so that the cycle can be repeated. Calcium ions in the cell and its membrane control how fast the visual system recovers after exposure to light. They also mediate the way in which cells adapt to various light levels. The following sequence summarizes the visual cycle.

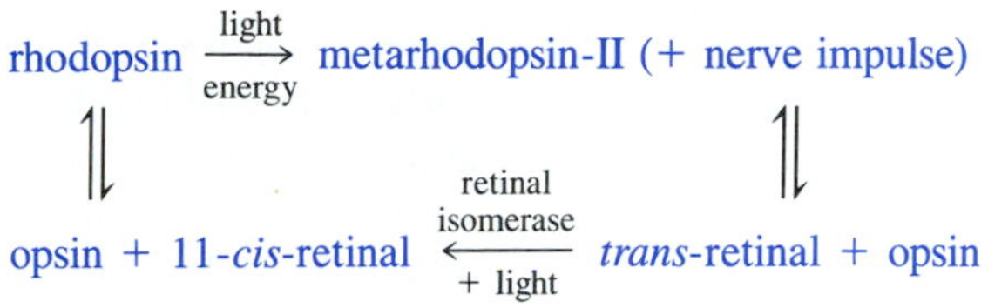

rhodopsin $\xrightarrow[\text{energy}]{\text{light}}$ metarhodopsin-II (+ nerve impulse)

$\rightleftharpoons$ $\rightleftharpoons$

opsin + 11-*cis*-retinal $\xleftarrow[\text{+ light}]{\text{retinal isomerase}}$ *trans*-retinal + opsin

This representation is simplified because there are actually several additional intermediates between rhodopsin and the fully dissociated *trans*-retinal and opsin.

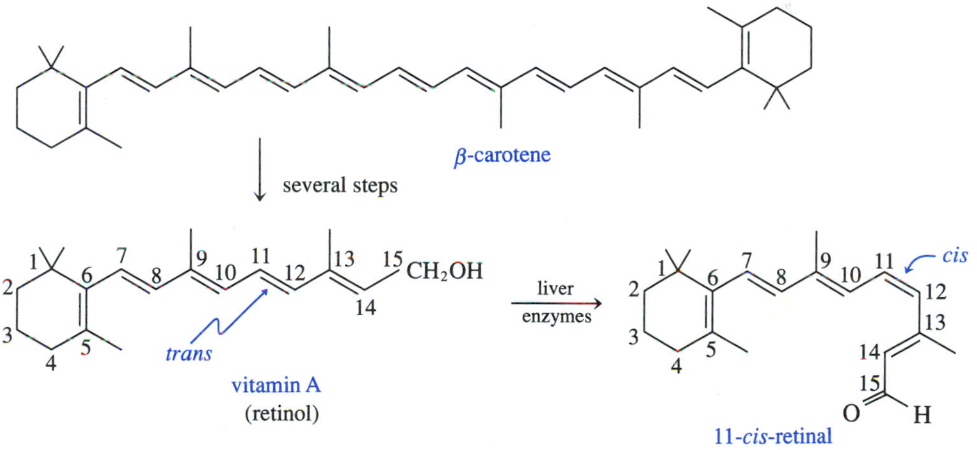

β-carotene

several steps

1 2 3 4 5 6 7 8 9 10 11 12 13 14 15 CH_2OH

trans

vitamin A
(retinol)

liver enzymes

1 2 3 4 5 6 7 8 9 10 11 12 13 14 15 *cis* O H

11-*cis*-retinal

FIGURE 3.7 In the liver, *β*-carotene is converted into vitamin A first and then into 11-cis-retinal.

3.6 *Physical Properties of Alkenes*

Unsaturated hydrocarbons have physical properties similar to those of alkanes. They are less dense than water and, being nonpolar, are not very soluble in it. As with the alkanes, compounds with four or fewer carbons are colorless gases whereas higher homologs are volatile liquids.

3.7 Addition and Substitution Reactions Compared

We saw in Chapter 2 that, aside from combustion, the most common reaction of alkanes is **substitution** (for example, halogenation). This reaction type can be expressed by general equation.

$$\text{R—H} + \text{A—B} \longrightarrow \text{R—A} + \text{H—B} \quad \text{substitution} \qquad (3.3)$$

where R—H stands for an alkane and A—B stands for the halogen molecule.

With alkenes, on the other hand, the most common reaction is **addition:**

$$\rangle\text{C}{=}\text{C}\langle + \text{A—B} \longrightarrow \text{—}\underset{\text{A}}{\overset{|}{\underset{|}{\text{C}}}}\text{—}\underset{\text{B}}{\overset{|}{\underset{|}{\text{C}}}}\text{—} \quad \text{addition} \qquad (3.4)$$

In an addition reaction, group A of the reagent A—B becomes attached to one carbon atom of the double bond, group B becomes attached to the other carbon atom, and the product has only a single bond between the two carbon atoms.

What bond changes take place in an addition reaction? The pi bond of the alkene is broken and the sigma bond of the reagent is also broken. Two new sigma bonds are formed. In other words, we break a pi and a sigma bond, and we make two sigma bonds. Because sigma bonds are usually stronger than pi bonds, the net reaction is favorable.

Problem 3.7 Why, in general, is a sigma bond between two atoms stronger than a pi bond between the same two atoms?

In the next sections, we will describe a few typical alkene addition reactions.

3.8 Addition of Hydrogen

Hydrogen adds to alkenes in the presence of an appropriate catalyst. The process is called **hydrogenation.**

$$\rangle\text{C}{=}\text{C}\langle + H_2 \xrightarrow{\text{catalyst}} \text{—}\underset{\text{H}}{\overset{|}{\underset{|}{\text{C}}}}\text{—}\underset{\text{H}}{\overset{|}{\underset{|}{\text{C}}}}\text{—} \qquad (3.5)$$

The catalyst is usually a finely divided metal, such as nickel, platinum, or palladium. These metals adsorb hydrogen gas on their surfaces and activate the hydrogen–hydrogen bond. Both hydrogen atoms usually add from the catalyst surface to the same face of the double bond. For example, 1,2-dimethylcyclopentene gives mainly *cis*-1,2-dimethylcyclopentane.

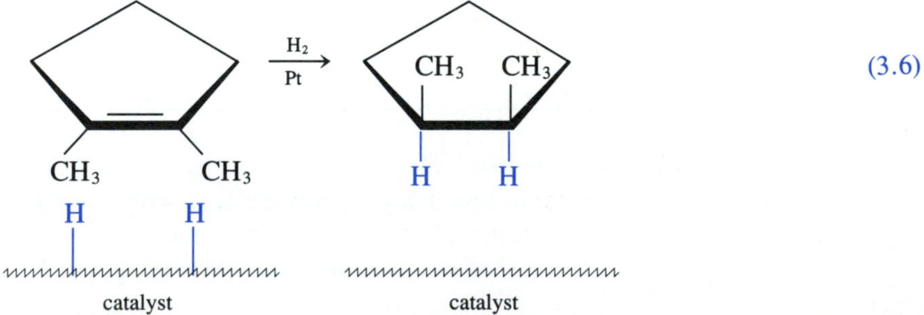

(3.6)

Catalytic hydrogenation of double bonds is used commercially to convert vegetable oils to margarine and other cooking fats (Sec. 15.3).

Problem 3.8 Write an equation for the catalytic hydrogenation of

a. methylpropene.
b. 1,2-dimethylcyclobutene.

3.9 Polar Addition Reactions

Several reagents add to double bonds by a two-step polar process. In this section and the next, we will describe examples of this reaction type, after which we will consider details of the reaction mechanism.

3.9a Addition of Halogens

Alkenes readily add chlorine or bromine.

$$\underset{\substack{\text{2-butene} \\ \text{bp 1–4°C}}}{CH_3CH{=}CHCH_3} + Cl_2 \longrightarrow \underset{\substack{\text{2,3-dichlorobutane} \\ \text{bp 117–119°C}}}{CH_3\underset{\displaystyle Cl}{\underset{|}{C}}H{-}\underset{\displaystyle Cl}{\underset{|}{C}}HCH_3} \tag{3.7}$$

$$\underset{\substack{\text{1,4-pentadiene} \\ \text{bp 26.0°C}}}{CH_2{=}CH{-}CH_2{-}CH{=}CH_2} + 2\ Br_2 \longrightarrow \underset{\substack{\text{1,2,4,5-tetrabromopentane} \\ \text{mp 85–86°C}}}{\underset{\displaystyle Br}{\underset{|}{C}}H_2{-}\underset{\displaystyle Br}{\underset{|}{C}}H{-}CH_2{-}\underset{\displaystyle Br}{\underset{|}{C}}H{-}\underset{\displaystyle Br}{\underset{|}{C}}H_2} \tag{3.8}$$

Usually the halogen is dissolved in some inert solvent such as tri- or tetrachloromethane, and then this solution is added dropwise to the alkene. Reaction is nearly instantaneous, even at room temperature or below. No light or heat is required, as in the case of substitution reactions.

Problem 3.9 Write an equation for the reaction of bromine at room temperature with

a. 1-butene. b. 2-methyl-2-butene.

The addition of bromine can be used as a **chemical test** for the presence of unsaturation in an organic compound. Bromine solutions in tetrachloromethane are dark reddish-brown, and the unsaturated compound and its bromine adduct are usually both colorless. As the bromine solution is added to the unsaturated compound, the bromine color disappears. If the compound being tested is saturated, it will not react with bromine under these conditions, and the color will persist.

3.9b Addition of Water (Hydration)

If an acid catalyst is present, water adds to alkenes. It adds as H—OH, and the products are alcohols.

$$CH_2{=}CH_2 + H{-}OH \xrightarrow{H^+} \underset{\text{ethanol}}{\underset{\displaystyle H}{\underset{|}{C}}H_2{-}\underset{\displaystyle OH}{\underset{|}{C}}H_2} \quad (\text{or } CH_3CH_2OH) \tag{3.9}$$

$$\text{cyclohexene (bp } 83.0°) + H{-}OH \xrightarrow{H^+} \text{cyclohexanol (bp } 161.1°) \tag{3.10}$$

cyclohexene
bp 83.0°

cyclohexanol
bp 161.1°

The necessary role of the acid catalyst will be explained when we discuss the mechanism of these reactions. Hydration is used industrially and occasionally in the laboratory to synthesize alcohols from alkenes.

Problem 3.10 Write an equation for the acid-catalyzed addition of water to

a. 2-butene.
b. cyclopentene.

3.9c Addition of Acids A variety of acids add to the double bond of alkenes. The hydrogen ion (or proton) adds to one carbon of the double bond, and the remainder of the acid becomes connected to the other carbon.

$$\rangle C{=}C\langle + \overset{\delta+}{H}{-}\overset{\delta-}{A} \longrightarrow -\underset{H}{\underset{|}{\overset{|}{C}}}-\underset{A}{\underset{|}{\overset{|}{C}}}- \tag{3.11}$$

Acids that add in this way are the hydrogen halides (HF, HCl, HBr, HI), sulfuric acid ($H{-}OSO_3H$), and organic carboxylic acids $\left(H{-}O\overset{O}{\overset{\|}{C}}R\right)$. Here are two typical examples:

$$CH_2{=}CH_2 + H{-}Cl \longrightarrow \underset{H}{\underset{|}{CH_2}}{-}\underset{Cl}{\underset{|}{CH_2}} \quad (\text{or } CH_3CH_2Cl) \tag{3.12}$$

ethene, hydrogen chloride, chloroethane (ethyl chloride)

$$\text{cyclopentene} + H{-}OSO_3H \longrightarrow \text{cyclopentyl hydrogen sulfate} \tag{3.13}$$

cyclopentene, sulfuric acid, cyclopentyl hydrogen sulfate

Problem 3.11 Write an equation for each of the following reactions:

a. 2-butene + HI
b. cyclopentene + HBr

Before we discuss the mechanism of addition reactions, we must introduce a complication that we have carefully avoided in all the examples given so far.

3.10 Addition of Unsymmetric Reagents to Unsymmetric Alkenes; Markovnikov's Rule

Reagents and alkenes can be classified as either symmetric or unsymmetric with respect to addition reactions. Table 3.2 illustrates what this means. If a reagent and/or an alkene is symmetric, only one addition product is possible. If you check back through all the equations and problems for addition reactions up to now, you will see that either the alkene or the reagent (or both) were symmetric. But if *both* the reagent *and* the alkene are *unsymmetric,* two products are, in principle, possible.

$$\underset{\text{unsymmetric alkene}}{\text{R(H)C}{=}\text{CH}} + \underset{\text{unsymmetric reagent}}{\text{X—Y}} \longrightarrow -\overset{\text{R}}{\underset{\text{X}}{\text{C}}}-\overset{\text{H}}{\underset{\text{Y}}{\text{C}}}- \text{ and/or } -\overset{\text{R}}{\underset{\text{Y}}{\text{C}}}-\overset{\text{H}}{\underset{\text{X}}{\text{C}}}- \quad (3.14)$$

The products of eq. 3.14 are sometimes called **regioisomers.** If a reaction of this type gives *only one* of the two possible regioisomers, it is said to be **regiospecific.** If it gives *mainly one* product, it is said to be **regioselective.**

TABLE 3.2 Classification of reagents and alkenes by symmetry with regard to addition reactions

	Symmetric	*Unsymmetric*
Reagents	Br—Br	H—Br
	Cl—Cl	H—OH
	H—H	H—OSO_3H
Alkenes	$CH_2{=}CH_2$	$CH_3CH{=}CH_2$
	$CH_3CH{=}CHCH_3$	$CH_3CH_2CH{=}CHCH_3$
	cyclopentene	1-methylcyclopentene (CH_3)

Let us consider as a specific example the acid-catalyzed addition of water to propene. In principle, two products could be formed: 1-propanol or 2-propanol.

$$\underset{\text{propene}}{\overset{3\quad 2\quad\ 1}{CH_3CH{=}CH_2}} \begin{cases} \xrightarrow[H^+]{H{-}OH} \underset{\text{2-propanol}}{CH_3\underset{OH}{\underset{|}{C}}HCH_3} \\ \xrightarrow[H^+]{H{-}OH} \underset{\text{1-propanol}}{CH_3CH_2CH_2{-}OH} \end{cases} \tag{3.15}$$

That is, the hydrogen of the water could add to C-1 and the hydroxyl group to C-2 of propene, or vice versa. When the experiment is carried out, *only one product is formed. The addition is regiospecific, and the only product is 2-propanol.*

Most addition reactions of alkenes show a similar preference for the formation of only (or mainly) one of the two possible addition products. Here are some examples.

$$CH_3CH{=}CH_2 + \overset{\delta+}{H}{-}\overset{\delta-}{Cl} \longrightarrow CH_3\underset{Cl}{\underset{|}{C}}HCH_3 \quad (\textit{not}\ CH_3CH_2CH_2Cl) \tag{3.16}$$

$$CH_3\underset{CH_3}{\underset{|}{C}}{=}CH_2 + \overset{\delta+}{H}{-}\overset{\delta-}{OH} \xrightarrow{H^+} CH_3\overset{OH}{\overset{|}{\underset{CH_3}{\underset{|}{C}}}}CH_3 \quad (\textit{not}\ CH_3\underset{CH_3}{\underset{|}{C}}HCH_2OH) \tag{3.17}$$

$$\text{1-methylcyclopentene} + \overset{\delta+}{H}{-}\overset{\delta-}{I} \longrightarrow \text{1-iodo-1-methylcyclopentane} \quad (\textit{not}\ \text{1-iodo-2-methylcyclopentane}) \tag{3.18}$$

Notice that the reagents are all polar, with a positive and a negative end. After studying a number of such addition reactions, the Russian chemist Vladimir Markovnikov formulated the following rule over 100 years ago: ***When an unsymmetric reagent adds to an unsymmetric alkene, the electropositive part of the reagent bonds to the carbon of the double bond that has the greater number of hydrogen atoms attached to it.****

Problem 3.12 Use Markovnikov's rule to predict which regioisomer predominates in each of the following reactions:

a. 1-butene + HCl b. 2-methyl-2-butene + H_2O (H^+ catalyst)

*Actually, Markovnikov stated the rule a little differently. The form given here is easier to remember and apply. For an interesting historical article on what he actually said, when he said it, and how his name is spelled, see J. Tierney, *J. Chem. Educ.* **1988,** *65,* 1053–54.

Problem 3.13 What two products are *possible* from the addition of HCl to 2-pentene? Would you expect the reaction to be regiospecific?

Let us now develop a rational explanation for Markovnikov's rule in terms of modern chemical theory.

3.11 Mechanism of Electrophilic Addition to Alkenes

The pi electrons of a double bond are more exposed to an attacking reagent than are the σ electrons. Recall that the pi bond is also weaker than the σ bond. It is the pi electrons, then, that are involved in additions to alkenes. The double bond can act as a supplier of pi electrons to an electron-seeking reagent.

Polar reactants can be classified as either **electrophiles** or **nucleophiles.** **Electrophiles (literally, electron lovers) are electron poor reagents; in reactions with some other molecule, they seek electrons.** They are often positive ions (cations) or otherwise electron-deficient species. **Nucleophiles (literally, nucleus lovers), on the other hand, are electron rich; they form bonds by donating electrons to an electrophile.**

$$E^+ + :Nu^- \longrightarrow E:Nu \qquad (3.19)$$

electrophile nucleophile

Let us now consider the mechanism of a polar addition to a carbon–carbon double bond, specifically the addition of acids to alkenes. The carbon–carbon double bond, because of its pi electrons, is a nucleophile. The proton (H^+) is the attacking electrophile. As the proton approaches the pi bond, the two pi electrons are used to form a σ bond between the proton and one of the two carbon atoms. Because this bond uses *both* pi electrons, the other carbon acquires a positive charge, producing a **carbocation.**

$$H^+ + \text{C=C} \longrightarrow \text{H–C–C}^+ \qquad (3.20)$$

carbocation

The resulting carbocations are, however, extremely reactive because there are only six electrons (instead of the usual eight) around the positive carbon. The carbocation rapidly combines with some species that can supply it with two electrons, a nucleophile.

$$\text{H–C–C}^+ + Nu:^- \longrightarrow \text{H–C–C–Nu} \qquad (3.21)$$

nucleophile product of addition of H—Nu to an alkene

Examples include the addition of HCl, $HOSO_3H$, and HOH to alkenes:

$$H^+ + \text{C}{=}\text{C} \rightleftharpoons \text{H}{-}\text{C}{-}\overset{+}{\text{C}} \xrightarrow{:\ddot{\text{Cl}}:^-} \text{H}{-}\text{C}{-}\text{C}{-}\text{Cl}; \quad \xrightarrow{^-:\ddot{\text{O}}SO_3H} \text{H}{-}\text{C}{-}\text{C}{-}\text{OSO}_3\text{H}; \quad \xrightarrow{H-\ddot{O}-H} \text{H}{-}\text{C}{-}\text{C}{-}\overset{+}{\text{O}}H_2 \rightleftharpoons \text{H}{-}\text{C}{-}\text{C}{-}\text{OH} + H^+ \qquad (3.22)$$

In these reactions, the electrophile H^+ first adds to the alkene to give a carbocation. Then the carbocation combines with a nucleophile, in these examples, a chloride ion, a bisulfate ion, or a water molecule.

With most alkenes, the first step in this process—the formation of the carbocation—is the slower of the two steps. The resulting carbocation is usually so reactive that combination with the nucleophile is extremely rapid. **Since the initiating step in these additions is attack by the electrophile, the whole process is called an electrophilic addition.**

EXAMPLE 3.4 Since carbocations are involved in electrophilic addition reactions of alkenes, it is important to understand the bonding in these chemical intermediates. Describe the bonding in carbocations in orbital terms.

Solution The carbon atom is positively charged and therefore has only three valence electrons to use in bonding. Each of these electrons is in an sp^2 orbital. The three sp^2 orbitals lie in one plane with 120° angles between them, an arrangement that minimizes repulsion between the electrons in the three bonds. The remaining p orbital is perpendicular to that plane and vacant.

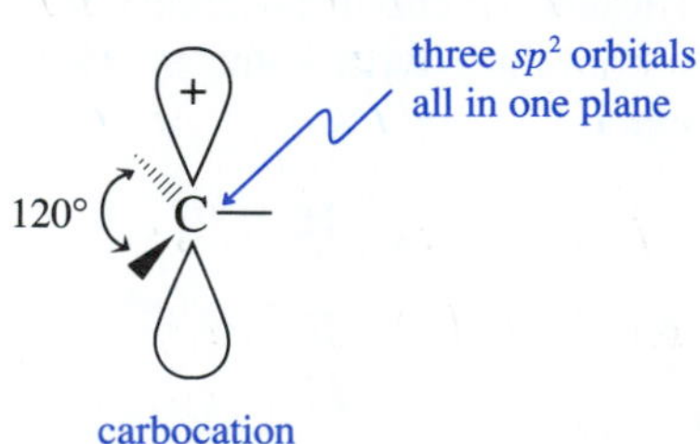

3.12 Markovnikov's Rule Explained

To explain Markovnikov's rule, let us consider a specific example, the addition of H—Cl to propene. The first step is addition of a proton to the double bond. This can occur in two ways, to give either an isopropyl cation or a propyl cation.

$$\overset{3}{CH_3}-\overset{2}{CH}=\overset{1}{CH_2} \xrightarrow{H^+} \begin{cases} \xrightarrow{\text{adds to C-1}} CH_3\overset{+}{C}HCH_3 \ \text{(isopropyl cation)} \\ \xrightarrow[\not\;]{\text{adds to C-2}} CH_3CH_2CH_2^+ \ \text{(propyl cation)} \end{cases} \quad (3.23)$$

propene

At this stage of the reaction, the structure of the product is already determined; when combining with chloride ion, the isopropyl cation can give only 2-chloropropane, and the propyl cation can give only 1-chloropropane. The only observed product is 2-chloropropane, so we must conclude that the *proton adds to C-1 to form only the isopropyl cation*. Why?

Carbocations are classified as tertiary, secondary, or primary, depending on whether the positive carbon atom has attached to it three organic groups, two groups, or only one group. From many studies, it has been established that the stability of carbocations decreases in the following order:

$$R_3C^+ > R_2\overset{+}{C}H \gg R\overset{+}{C}H_2 > \overset{+}{C}H_3$$

tertiary (3°) | secondary (2°) | primary (1°) | methyl (unique)

most stable ⟶ least stable

What is the reason for this order? One reason is the following: A carbocation will be more stable when the positive charge can be spread out, or delocalized, over several atoms in the ion, instead of being concentrated on a single carbon atom. In alkyl cations, this delocalization occurs by drift of electron density to the positive carbon from the other bonds in the ion. The more bonds, the more the charge is delocalized. If the positive carbon is surrounded by other carbon atoms (alkyl groups), instead of by hydrogen atoms, there will be more bonding electrons to help delocalize the charge. This is the main reason for the observed stability order of carbocations.

Markovnikov's rule can now be restated in modern terms: ***The electrophilic addition of an unsymmetric reagent to an unsymmetric double bond proceeds in such a way as to involve the most stable carbocation.***

Problem 3.14 Classify each of the following carbocations as primary, secondary, or tertiary:

a. $CH_3\overset{+}{C}HCH_2CH_3$ b. $(CH_3)_2CH\overset{+}{C}H_2$ c.

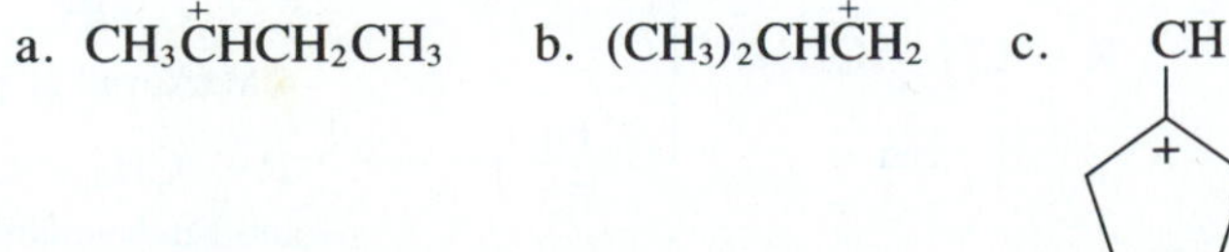

Problem 3.15 Which carbocation in Problem 3.14 is most stable? least stable?

Problem 3.16 Write out the steps in the electrophilic additions in eqs. 3.17 and 3.18, and in each case, show that reaction occurs via the most stable carbocation.

3.13 Hydroboration of Alkenes

Hydroboration was discovered by Professor Herbert C. Brown (Purdue University). This reaction is so useful in synthesis that Brown's work earned him a Nobel Prize (1979). We will describe here only one practical example of hydroboration, a two-step alcohol synthesis from alkenes.

Hydroboration involves addition of a hydrogen–boron bond to an alkene. The $H{-}B\lt$ bond is polarized with the hydrogen $\delta-$ and the boron $\delta+$. Addition occurs so that the boron (the electrophile) adds to the less substituted carbon.

$$R{-}CH{=}CH_2 + \overset{\delta-}{H}{-}\overset{\delta+}{B}\lt \longrightarrow R{-}CH_2{-}CH_2{-}B\lt \qquad (3.24)$$

Because it has three B—H bonds, one molecule of borane, BH_3, can react with three molecules of an alkene. For example, propene gives tri-*n*-propylborane.

$$\underset{\text{propene}}{3CH_3CH{=}CH_2} + \underset{\text{borane}}{BH_3} \longrightarrow \underset{\text{tri-}n\text{-propylborane}}{CH_3CH_2CH_2{-}B(CH_2CH_2CH_3)_2} \qquad (3.25)$$

The trialkylboranes made in this way are usually not isolated but are treated with some other reagent to obtain the desired final product. For example, trialkylboranes can be oxidized by hydrogen peroxide and base to give alcohols.

$$\underset{\text{tri-}n\text{-propylborane}}{(CH_3CH_2CH_2)_3B} + 3\,H_2O_2 + 3\,NaOH \longrightarrow \underset{n\text{-propyl alcohol}}{3\,CH_3CH_2CH_2OH} + \underset{\text{sodium borate}}{Na_3BO_3} + 3\,H_2O \qquad (3.26)$$

One great advantage of this hydroboration-oxidation sequence is that it provides a route to alcohols that *cannot* be obtained by the acid-catalyzed hydration of alkenes (review eq. 3.15).

$$R{-}CH{=}CH_2 \xrightarrow[H^+]{H{-}OH} \underset{\text{Markovnikov product}}{R{-}\underset{OH}{\underset{|}{CH}}{-}CH_3} \qquad (3.27)$$

$$R{-}CH{=}CH_2 \xrightarrow[2.\ H_2O_2,\ OH^-]{1.\ BH_3} \underset{\text{anti-Markovnikov product}}{R{-}CH_2{-}CH_2OH}$$

The overall result of the two-step hydroboration sequence is addition of water to the carbon–carbon double bond in the reverse of the usual Markovnikov sense.

EXAMPLE 3.5 What alcohol is obtained from this sequence?

$$CH_3-\underset{}{\overset{CH_3}{\overset{|}{C}}}=CH_2 \xrightarrow{BH_3} \xrightarrow[OH^-]{H_2O_2}$$

Solution The boron adds to the less substituted carbon; oxidation gives the corresponding alcohol. Compare this result with that of eq. 3.17.

$$CH_3-\overset{CH_3}{\overset{|}{C}}=CH_2 \xrightarrow{BH_3} (CH_3-\overset{CH_3}{\overset{|}{C}H}-CH_2)_3-B \xrightarrow[OH^-]{H_2O_2} CH_3-\overset{CH_3}{\overset{|}{C}H}-CH_2OH$$

Problem 3.17 What alcohol is obtained by applying the hydroboration-oxidation sequence to 2-methyl-2-butene?

Problem 3.18 What alkene is needed to obtain (cyclopentyl)$-CH_2CH_2OH$ via the hydroboration-oxidation sequence?

3.14 Additions to Conjugated Systems

We will describe here two ways in which the alternate double and single bonds of conjugated systems has special consequences for their addition reactions.

3.14a Electrophilic Addition to Conjugated Dienes

When 1 mole of hydrogen bromide adds to 1 mole of 1,3-butadiene, a rather surprising result is obtained. Two products are isolated.

$$\underset{\text{1,3-butadiene}}{\overset{1}{C}H_2=\overset{2}{C}H-\overset{3}{C}H=\overset{4}{C}H_2} \xrightarrow{HBr} \begin{cases} \underset{\text{3-bromo-1-butene}}{\underset{H}{\underset{|}{C}H_2}-\underset{Br}{\underset{|}{C}H}-CH=CH_2} & \text{(1,2-addition)} \\ \underset{\text{1-bromo-2-butene}}{\underset{H}{\underset{|}{C}H_2}-CH=CH-\underset{Br}{\underset{|}{C}H_2}} & \text{(1,4-addition)} \end{cases} \quad (3.28)$$

In one of these products, HBr has added to one of the two double bonds, and the other double bond is still present in its original position. We call this the product of **1,2-addition.** The other product may at first seem unexpected. The hydrogen and bromine have added to carbon-1 and carbon-4 of the original diene, and a new double bond has appeared between carbon-2 and carbon-3. This process, called **1,4-addition,** is quite a general reaction for electrophilic additions to conjugated systems. How can we explain it?

In the first step, the proton adds to the terminal carbon atom, according to Markovnikov's rule.

$$H^+ + CH_2{=}CH{-}CH{=}CH_2 \longrightarrow CH_3{-}\overset{+}{C}H{-}CH{=}CH_2 \qquad (3.29)$$

The resulting carbocation can be stabilized by resonance; in fact, it is a hybrid of two contributing resonance structures.

$$[CH_3{-}\overset{+}{C}H{-}CH{=}CH_2 \longleftrightarrow CH_3{-}CH{=}CH{-}\overset{+}{C}H_2]$$

The positive charge is delocalized over carbon-2 and carbon-4. When, in the next step, the carbocation reacts with bromide ion (the nucleophile), it can react either at carbon-2 to give the product of 1,2-addition, or at carbon-4 to give the product of 1,4-addition.

$$\left.\begin{matrix} \underset{1}{C}H_3{-}\underset{2}{\overset{+}{C}}H{-}\underset{3}{C}H{=}\underset{4}{C}H_2 \\ \updownarrow \\ \underset{1}{C}H_3{-}\underset{2}{C}H{=}\underset{3}{C}H{-}\underset{4}{\overset{+}{C}}H_2 \end{matrix}\right\} \xrightarrow{Br^-} \begin{matrix} CH_3{-}\underset{\displaystyle Br}{\underset{|}{C}H}{-}CH{=}CH_2 \\ + \\ CH_3{-}CH{=}CH{-}\underset{\displaystyle Br}{\underset{|}{C}H_2} \end{matrix} \qquad (3.30)$$

Problem 3.19 Explain why, in the first step in the addition of HBr to 1,3-butadiene, the proton adds to C-1 (eq. 3.29) and not to C-2.

The carbocation intermediate in these reactions is a single species, a resonance hybrid. ***This type of carbocation, with a carbon–carbon double bond adjacent to the positive carbon, is called an allylic cation.*** The parent allyl cation, shown below as a resonance hybrid, is a primary carbocation, but it is more stable than simple primary ions (such as propyl), because its positive charge is delocalized over the two end carbon atoms.

$$CH_2{=}CH{-}\overset{+}{C}H_2 \longleftrightarrow \overset{+}{C}H_2{-}CH{=}CH_2$$

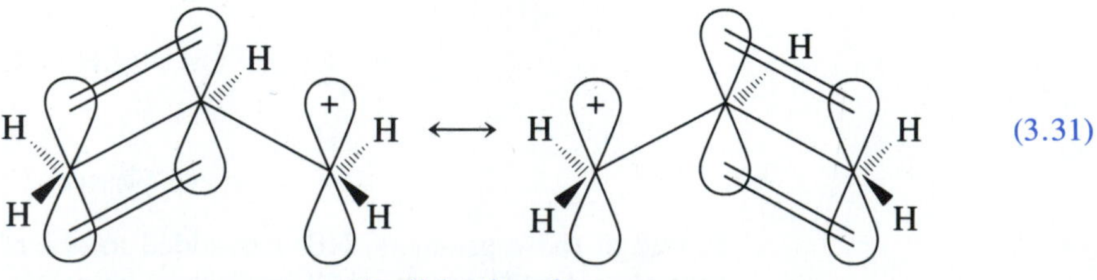

the allyl carbocation

Problem 3.20 Draw the contributors to the resonance hybrid structure of the 3-cyclopentenyl cation (cyclopentene ring with + at C-3).

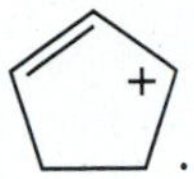

Problem 3.21 Write an equation for the expected products of 1,2-addition and 1,4-addition of bromine to 1,3-butadiene.

3.14b Cycloaddition to Conjugated Dienes; the Diels-Alder Reaction

Conjugated dienes undergo another type of 1,4-addition when they react with alkenes. The simplest example is the addition of ethylene to 1,3-butadiene to give cyclohexene.

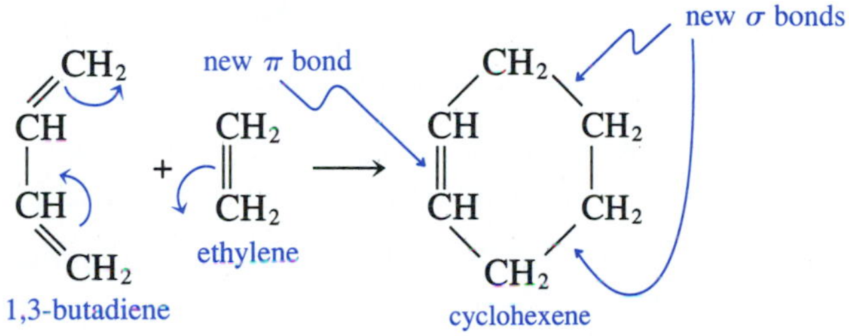

$$CH_2{=}CH{-}CH{=}CH_2 \; (\text{1,3-butadiene}) + CH_2{=}CH_2 \; (\text{ethylene}) \longrightarrow \text{cyclohexene} \qquad (3.32)$$

new π bond

new σ bonds

This reaction is an example of a **cycloaddition reaction,** an addition that results in a cyclic product. This cycloaddition, which converts three π bonds to two σ bonds and one new π bond, is called the **Diels-Alder reaction,** after its discoverers, Otto Diels and Kurt Alder. It is so useful for making cyclic compounds that it earned the 1950 Nobel Prize in Chemistry for its discoverers.

The two reactants are a **diene** and a **dienophile** (diene lover). The simple example in eq. 3.32 is not typical of most Diels-Alder reactions, in that it proceeds only under pressure and not in very good yield. However, this type of reaction gives excellent yields at moderate temperatures if the dienophile has electron-withdrawing groups attached, as in the following examples:

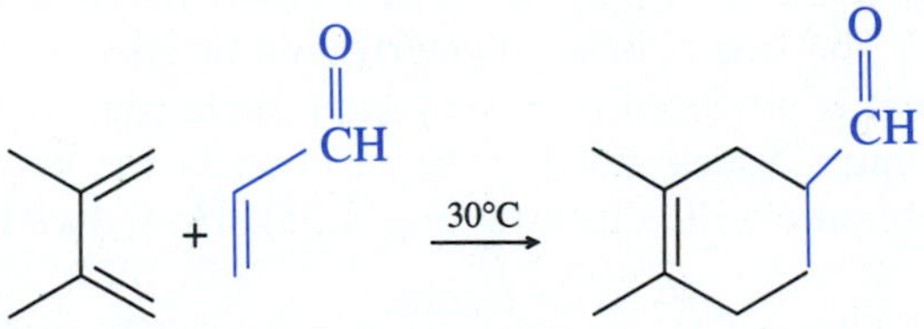

$$\text{2,3-dimethyl-1,3-butadiene} + CH_2{=}CH{-}\overset{O}{\overset{\|}{C}}H \xrightarrow{30°C} \text{3,4-dimethylcyclohex-3-ene-1-carbaldehyde} \qquad (3.33)$$

$$\text{1,3-butadiene} + (NC)_2C{=}C(CN)_2 \xrightarrow{0°C} \text{4,4,5,5-tetracyanocyclohexene} \qquad (3.34)$$

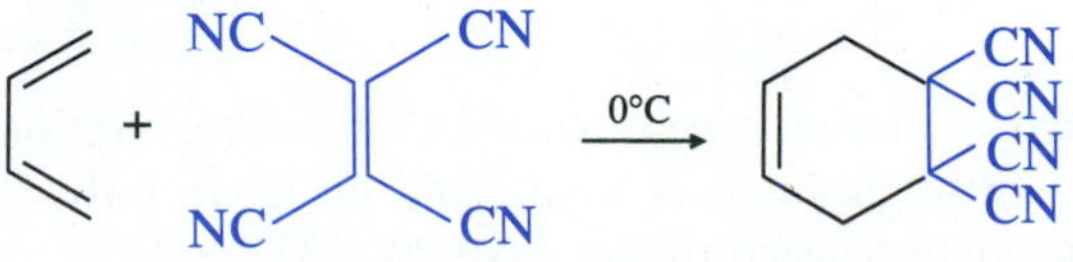

EXAMPLE 3.6 How could a Diels-Alder reaction be used to synthesize this compound:

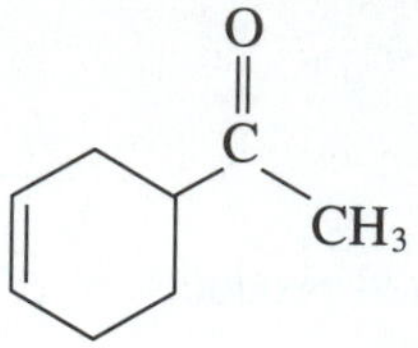

Solution Work backwards. The double bond in the product was a single bond in the starting diene. Therefore,

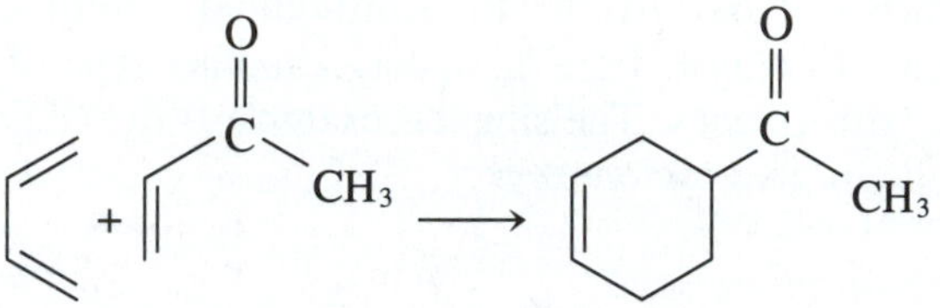

Problem 3.22 Show how limonene (Figure 1.12) could be formed by a Diels-Alder reaction of isoprene (2-methyl-1,3-butadiene) with itself.

Problem 3.23 What is the structure of the product of the following cycloaddition reactions?

a. 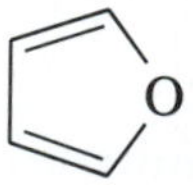 (furan) + $CH_2{=}CH{-}CN$

b. $CH_2{=}CH{-}CH{=}CH_2 + NC{-}C{\equiv}C{-}CN$

3.15 Free-Radical Additions; Polyethylene

Some reagents add to alkenes by a free-radical mechanism instead of by an ionic mechanism. From a commercial standpoint, the most important of these free-radical additions are those that lead to polymers.

A **polymer** is a large molecule, usually with a high molecular weight, built up from small repeating units. The simple molecule from which these repeating units are derived is called a **monomer,** and the process of converting a monomer to a polymer is called **polymerization.**

The free radical polymerization of ethylene gives **polyethylene,** a material that is produced on a very large scale (over 10 billion pounds annually in the United States alone). The reaction is carried out by heating ethylene under pressure with a catalyst (eq. 3.35). How does this reaction occur?

$$\underset{\text{ethylene}}{CH_2{=}CH_2} \xrightarrow[\text{1000 atm, >100°C}]{\text{ROOR}} \underset{\substack{\text{polyethylene}\\(n\,=\,\text{several thousand})}}{-\!\!\left(CH_2{-}CH_2\right)\!\!-_n} \tag{3.35}$$

One common type of catalyst for polymerization is an organic peroxide. The O—O single bond is weak, and on heating this bond breaks, with one electron going to each of the oxygens.

$$R—O—O—R \xrightarrow{\text{heat}} 2\,R—O\cdot \tag{3.36}$$

organic peroxide two radicals

A catalyst radical then adds to the carbon–carbon double bond:

$$RO\cdot \quad CH_2{=}CH_2 \longrightarrow RO—CH_2—CH_2\cdot \tag{3.37}$$

catalyst radical a carbon free radical

The result of this addition is a carbon free radical, which may add to another ethylene molecule, and another, and another, and so on.

$$ROCH_2CH_2\cdot \xrightarrow{CH_2{=}CH_2} ROCH_2CH_2CH_2CH_2\cdot \xrightarrow{CH_2{=}CH_2} ROCH_2CH_2CH_2CH_2CH_2CH_2\cdot \text{ and so on} \tag{3.38}$$

The carbon chain continues to grow in length until some chain-termination reaction occurs (perhaps combination of two radicals).

We might think that only a single long chain of carbons will be formed in this way, but this is not always the case. A "growing" polymer chain may abstract a hydrogen atom from its back, so to speak, to cause **chain branching.**

$\cdot CH_2$ H $\longrightarrow$ H— CH_2 $\xrightarrow{CH_2{=}CH_2}$ CH_3 CH_2—$CH_2\cdot$ and so on (3.39)

A giant molecule with long and short branches is thus formed:

branched polyethylene

The degree of chain branching and other features of the polymer structure can often be controlled by the choice of catalyst and reaction conditions.

A polyethylene molecule is mainly saturated despite its name (polyethyl*ene*) and consists mostly of linked CH_2 groups, but with CH groups at the branch points and CH_3 groups at the ends of the branches. It also contains an OR group from the catalyst at one end, but since the molecular weight is very large, this OR group constitutes a minor and, as far as properties go, relatively insignificant fraction of the molecule.

Polyethylene made in this way is transparent and used in packaging and film (for example, for freezer and sandwich bags).

In Chapter 14, we will describe many other polymers, some made by the process just described for polyethylene and some made by other methods.

3.16 Oxidation of Alkenes

In general, alkenes are more easily oxidized than alkanes by chemical oxidizing agents. These reagents attack the pi electrons of the double bond. The reactions may be useful as chemical tests for the presence of a double bond or for synthesis.

3.16a Oxidation with Permanganate; a Chemical Test Alkenes react with alkaline potassium permanganate to form **glycols** (compounds with two adjacent hydroxyl groups).

$$3\ \mathrm{>C{=}C<} + 2\,K^+MnO_4^- + 4\,H_2O \longrightarrow 3\ \mathrm{-\underset{OH}{\overset{|}{C}}-\underset{OH}{\overset{|}{C}}-} + 2\,MnO_2 + 2\,K^+OH^- \qquad (3.40)$$

alkene; potassium permanganate (purple); a glycol; manganese dioxide (brown-black)

As the reaction occurs, the purple color of the permanganate ion is replaced by the brown precipitate of manganese dioxide. Because of this color change, the reaction can be used as a chemical test to distinguish alkenes from alkanes, which normally do not react.

Problem 3.24 Write an equation for the reaction of 2-butene with potassium permanganate.

3.16b Ozonolysis of Alkenes Alkenes react rapidly and quantitatively with ozone, O_3. Ozone is generated by passing oxygen over a high-voltage electric discharge. The resulting gas stream is then bubbled at low temperature into a solution of the alkene in an inert solvent, such as dichloromethane. The first product, a **molozonide,** rearranges rapidly to an **ozonide.** Since these products may be explosive if isolated, they are usually treated directly with a reducing agent, commonly zinc and aqueous acid, to give carbonyl compounds as the isolated products.

$$\mathrm{>C{=}C<} \xrightarrow{O_3} \left[\text{molozonide}\right] \longrightarrow \text{ozonide} \xrightarrow[H_3O^+]{Zn} \mathrm{>C{=}O} + \mathrm{O{=}C<} \qquad (3.41)$$

alkene; a molozonide; an ozonide; two carbonyl groups

The net result of this reaction is to break the double bond of the alkene

and to form two carbon–oxygen double bonds (carbonyl groups), one at each carbon of the original double bond. The overall process is called ozonolysis.

Ozonolysis can be used to locate the position of a double bond. For example, ozonolysis of 1-butene gives two different aldehydes, whereas 2-butene gives a single aldehyde.

$$\underset{\text{1-butene}}{CH_2{=}CHCH_2CH_3} \xrightarrow[\text{2. Zn, H}^+]{\text{1. O}_3} \underset{\text{formaldehyde}}{CH_2{=}O} + \underset{\text{propanal}}{O{=}CHCH_2CH_3} \tag{3.42}$$

$$\underset{\text{2-butene}}{CH_3CH{=}CHCH_3} \xrightarrow[\text{2. Zn, H}^+]{\text{1. O}_3} 2\ \underset{\text{ethanal}}{CH_3CH{=}O} \tag{3.43}$$

Using ozonolysis, one can easily tell which butene isomer is which. By working backwards from the structures of ozonolysis products, one can deduce the structure of an unknown alkene.

EXAMPLE 3.7 Ozonolysis of an alkene produces equal amounts of acetone and formaldehyde, $(CH_3)_2C{=}O$ and $CH_2{=}O$, respectively. Deduce the alkene structure.

Solution Connect to each other by a double bond the carbons that are bound to oxygen in the ozonolysis products.

The alkene is $(CH_3)_2C{=}CH_2$.

Problem 3.25 Which alkene will give only acetone, $(CH_3)_2C{=}O$, as the ozonolysis product?

3.16c Other Alkene Oxidations Various reagents can convert alkenes to epoxides (eq. 3.44).

$$\underset{\text{alkene}}{>C{=}C<} \longrightarrow \underset{\text{epoxide}}{>C{-}C<\ (\text{bridged by O})} \tag{3.44}$$

This reaction and the chemistry of epoxides are discussed in Chapter 8.

Like alkanes (and all other hydrocarbons), alkenes can be used as fuels. Complete combustion gives carbon dioxide and water.

$$C_nH_{2n} + \tfrac{3n}{2}O_2 \longrightarrow nCO_2 + nH_2O \tag{3.45}$$

Before we turn to alkynes and their chemistry, you might want to read "A Word About" the importance of ethylene in our economy.

A WORD ABOUT ...

4. Ethylene and Acetylene

Ethylene, the simplest alkene, ranks first among organic chemicals in industrial production. Current U.S. annual production of ethylene is well over 30 billion pounds. Propene comes in second with about half that amount.

How is all this ethylene produced, and what is it used for? Most hydrocarbons can be "cracked" to give ethylene. (See "A Word About Petroleum, Gasoline, and Octane Number" on page 101.) In the United States, the major raw material for this purpose is ethane.

$$CH_3{-}CH_3 \xrightarrow{700-900°C} CH_2{=}CH_2 + H_2$$

A substantial fraction of industrial ethylene is, of course, converted to polyethylene, as described in Sec. 3.15; but ethylene is also a key raw material for the manufacture of other industrial organic chemicals, because of the reactivity of the carbon–carbon double bond. Shown in Figure 3.8 are 9 of the top 50 organic chemicals; each is produced from ethylene.

Ethylene is not only the most important industrial source of organic chemicals, it also has some biochemical properties that are crucial to agriculture. Ethylene is a **plant hormone** that can cause seeds to sprout, flowers to bloom, fruit to ripen and fall, and leaves and petals to shrivel and turn brown. It is produced naturally by plants from the amino acid *methionine* via an unusual cyclic amino acid, *1-aminocyclopropane-1-carboxylic acid (ACC),* which is then, in several steps, converted to ethylene.

$$\underset{\text{methionine}}{CH_3{-}S{-}CH_2CH_2{-}\underset{\displaystyle NH_3^+}{\overset{|}{CH}}{-}CO_2^-} \xrightarrow[\text{steps}]{\text{several}}$$

$$\underset{\text{ACC}}{\text{cyclo-}(CH_2CH_2)C(CO_2^-)(NH_3^+)} \xrightarrow[\text{steps}]{\text{several}} \underset{\text{ethylene}}{CH_2{=}CH_2} + CO_2 + HCN$$

The mode by which ethylene functions biologically is still being studied.

Chemists have prepared synthetic compounds that can release ethylene in plants in a controlled manner. One such example is 2-chloroethylphosphonic acid, $ClCH_2CH_2PO(OH)_2$. Sold by Union Carbide under the trade name *ethrel,* it is water soluble and is taken up by plants, where it breaks down to ethylene, chloride, and phosphate. It has been used commercially to induce fruits, such as pineapples and tomatoes, to ripen uniformly so that an entire field can be harvested efficiently, as shown in the photo in the preceding column. It has also been used to regulate the growth of other crops, such as wheat, apples, cherries, and cotton. Only a small amount need be used, since plants are very sensitive to ethylene and respond to concentrations lower than 0.1 part per million of the gas.

Acetylene, like ethylene, is made by pyrolysis, but from methane instead of ethane. A much higher temperature and a very short reaction time is required.

$$2CH_4 \xrightarrow[<0.1s]{1500°C} HC{\equiv}CH + 3H_2$$

Acetylene is considerably more expensive than ethylene. Most of it is used directly, in arc welding, rather than as a raw material for industrial chemicals.

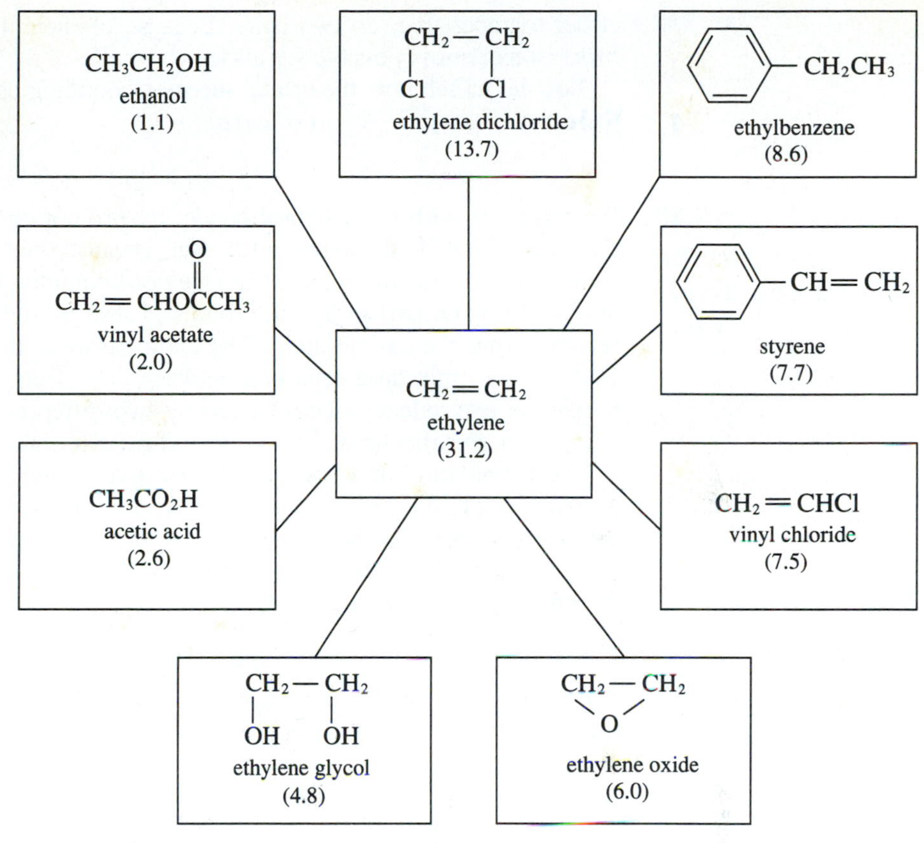

FIGURE 3.8 Ethylene is central to the manufacture of many industrial organic chemicals. The numbers in parentheses give the recent U.S. production of these chemicals in billions of pounds.

3.17 Some Facts About Triple Bonds

In the final sections of this chapter, we will describe some of the special features of triple bonds and alkynes.

A carbon atom that is part of a triple bond is directly attached to only *two* other atoms, and the bond angle is 180°. Thus, acetylene is linear, as shown in Figure 3.9. The carbon–carbon triple bond distance is about 1.21 Å, appreciably shorter than that of an ordinary double bond (1.34 Å) or single bond (1.54 Å). Apparently, three electron pairs between two carbons draw them even

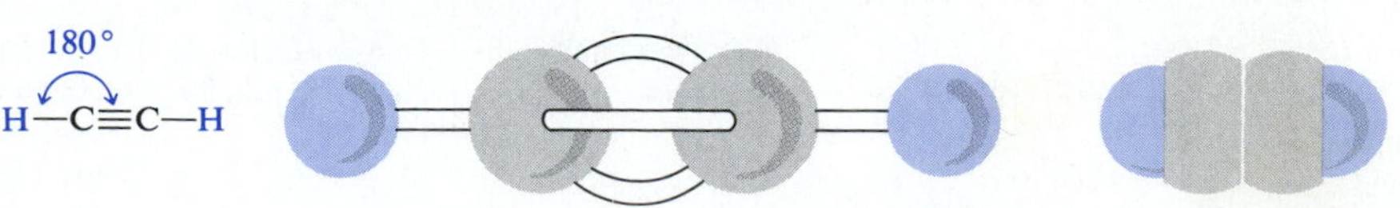

FIGURE 3.9 Models of acetylene, showing its linearity.

closer together than do two pairs. Because of their linear geometry, no *cis–trans* isomerism is possible for alkynes.

Now let us see how the orbital theory of bonding can be adapted to explain these facts.

3.18 The Orbital Model of a Triple Bond

We begin, as with other hybrid bonds, by promoting a carbon 2*s* electron to the vacant 2*p* orbital (Figure 3.10). But, because the carbon of an acetylene is connected to only *two* other atoms, we need combine the 2*s* with only one 2*p* orbital, to make two *sp* hybrid orbitals. These orbitals extend in opposite directions from the carbon atom. The angle between the two hybrid orbitals is 180° so as to minimize repulsion between any electrons placed in them. The remaining two valence electrons occupy two different *p* orbitals that are both mutually perpendicular and also perpendicular to the hybrid *sp* orbitals.

The formulation of a triple bond from two *sp*-hybridized carbons is shown in Figure 3.11. *End-on overlap of two sp orbitals forms a sigma bond between the two carbons, and lateral overlap of the properly aligned p orbitals forms two pi bonds (designated* π_1 *and* π_2 *in the figure).* This model nicely explains the linearity of acetylenes.

3.19 Addition Reactions of Alkynes

Many addition reactions described for alkenes also occur, though usually more slowly, with alkynes. For example, bromine adds as follows:

$$\underset{\text{ethyne}}{\mathrm{H{-}C{\equiv}C{-}H}} \xrightarrow{Br_2} \underset{\textit{trans}\text{-1,2-dibromoethene}}{\mathrm{(H)(Br)C{=}C(Br)(H)}} \xrightarrow{Br_2} \underset{\text{1,1,2,2-tetrabromoethane}}{\mathrm{H{-}CBr_2{-}CBr_2{-}H}} \qquad (3.46)$$

In the first step, the addition occurs mainly in a *trans* manner.

With an ordinary nickel or platinum catalyst, alkynes are hydrogenated all the way to alkanes (eq. 3.1). However, a special palladium catalyst (called

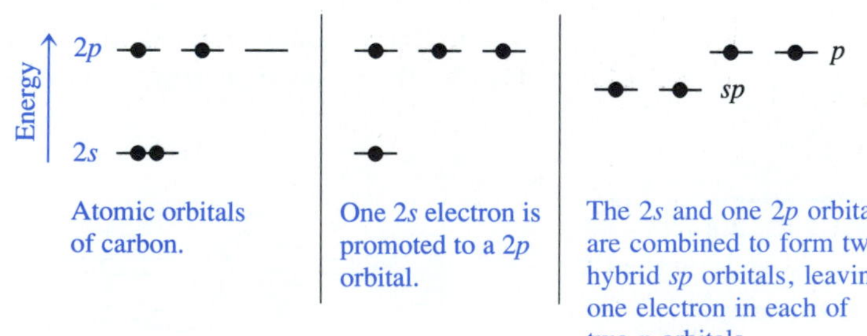
Energy
2p
2s
p
sp
Atomic orbitals of carbon.
One 2s electron is promoted to a 2p orbital.
The 2s and one 2p orbital are combined to form two hybrid sp orbitals, leaving one electron in each of two p orbitals.

FIGURE 3.10 The formation of two *sp* hybrid orbitals.

FIGURE 3.11
A triple bond consists of the end-on overlap of two *sp* hybrid orbitals to form a σ bond and the lateral overlap of two sets of parallel-oriented *p* orbitals to form two mutually perpendicular π bonds.

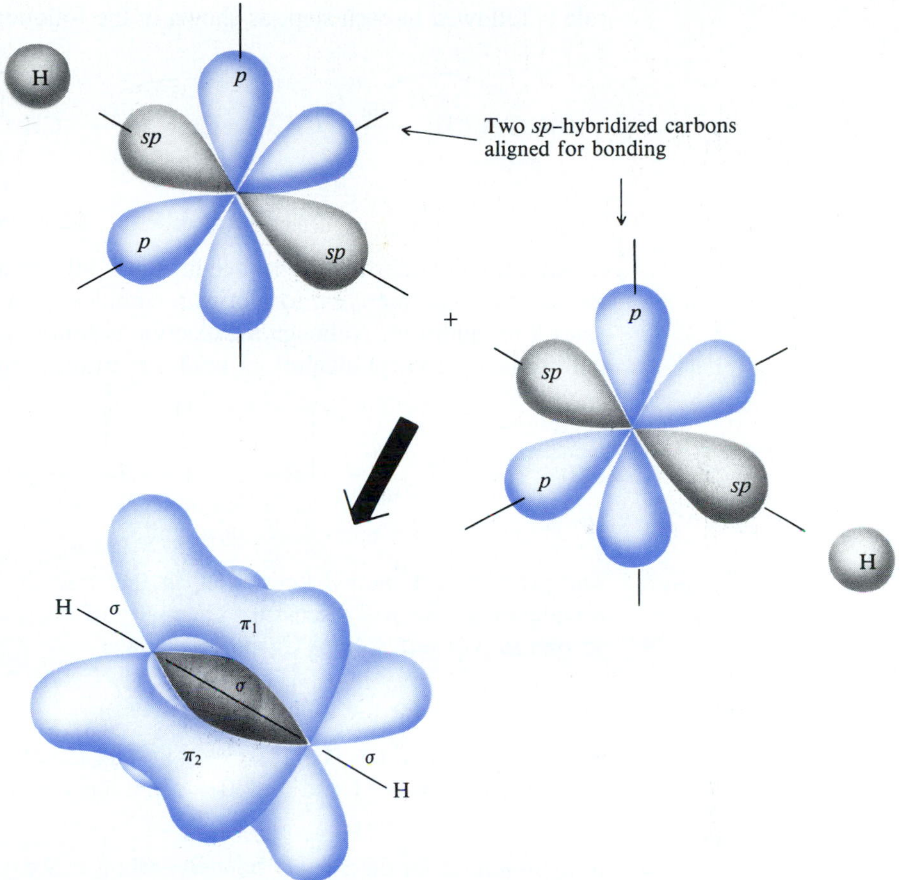

The resulting carbon–carbon triple bond, with a hydrogen atom attached to each remaining *sp* bond. (The orbitals involved in the C—H bonds are omitted for clarity.)

Lindlar's catalyst) can control hydrogen addition so that only *one* mole of hydrogen adds. In this case, the product is a *cis* alkene, because both hydrogens add to the same face of the triple bond from the catalyst surface.

$$\underset{\substack{\text{2-butyne} \\ \text{bp } 27^\circ\text{C}}}{CH_3{-}C{\equiv}C{-}CH_3} \xrightarrow[\text{Pd (Lindlar catalyst)}]{H{-}H} \underset{\substack{cis\text{-2-butene} \\ \text{bp } 3.7^\circ\text{C}}}{(CH_3)(H)C{=}C(CH_3)(H)} \qquad (3.47)$$

With unsymmetric triple bonds and unsymmetric reagents, Markovnikov's rule is followed in each step, as shown in the following example:

$$\underset{\text{propyne}}{CH_3C{\equiv}CH} \xrightarrow{HBr} \underset{\text{2-bromopropene}}{CH_3\underset{Br}{\underset{|}{C}}{=}CH_2} \xrightarrow{HBr} \underset{\text{2,2-dibromopropane}}{CH_3\overset{Br}{\overset{|}{\underset{Br}{\underset{|}{C}}}}CH_3} \tag{3.48}$$

Addition of water to alkynes requires not only an acid catalyst but mercuric ion as well. The mercuric ion forms a complex with the triple bond and activates it for addition. Although the reaction is similar to that of alkenes, the initial product—a **vinyl alcohol,** or **enol**—rearranges to a carbonyl compound.

$$R{-}C{\equiv}CH + H{-}OH \xrightarrow[HgSO_4]{H^+} \underset{\text{a vinyl alcohol, or enol}}{\left[R{-}\overset{HO}{\overset{|}{C}}{=}\overset{H}{\overset{|}{C}}{-}H\right]} \longrightarrow R{-}\overset{O}{\overset{\|}{C}}{-}CH_3 \tag{3.49}$$

The product is a methyl ketone or, in the case of acetylene itself (R=H), acetaldehyde. We will discuss the chemistry of enols and the mechanism of the second step of eq. 3.49 in Chapter 9.

Problem 3.26 Write equations for the following reactions:

a. $CH_3C{\equiv}CH + Cl_2$ (1 mol)
b. $CH_3C{\equiv}CH + Cl_2$ (2 mol)
c. 1-butyne + HBr (1 and 2 mol)
d. 1-butyne + H_2O (Hg^{2+}, H^+)

3.20 Acidity of Alkynes

A hydrogen atom on a triply bonded carbon is weakly acidic and can be removed by a very strong base. Sodium amide, for example, converts acetylenes to acetylides.

$$R{-}C{\equiv}C{-}H + \underset{\text{sodium amide}}{Na^+NH_2^-} \xrightarrow{\text{liquid } NH_3} \underset{\text{a sodium acetylide}}{R{-}C{\equiv}C{:}^-Na^+} + NH_3 \tag{3.50}$$

this hydrogen is weakly acidic

This type of reaction occurs easily with a hydrogen adjacent to a triple bond, but with increasing difficulty when the hydrogens are adjacent to a double or single bond. Why? Consider the hybridization of the carbon atom in C—H bonds.

$-CH_3$	$=CH_2$	$\equiv C{-}H$
sp^3	*sp*^{2}	*sp*
25% *s*, 75% *p*	33⅓% *s*, 66⅔% *p*	50% *s*, 50% *p*

increasing acidity →

As the hybridization at carbon becomes more s-like and less p-like, the acidity of the attached hydrogen increases. Recall that *s* orbitals are closer to the nucleus than are *p* orbitals. Consequently, the bonding electrons are closest to the carbon atom in the $\equiv$C—H bond, making it easiest for a base to remove that type of proton. Sodium amide is a sufficiently strong base for this purpose.

Problem 3.27 Write an equation for the reaction of 1-butyne with sodium amide in liquid ammonia.

Although 1-alkynes are acidic, they are much less acidic than water. Internal alkynes, in which the triple bond is not at the end of the chain, have no acidic hydrogens.

Problem 3.28 Write an equation for the reaction of a sodium acetylide with water.

Problem 3.29 Will 2-butyne react with sodium amide? Explain.

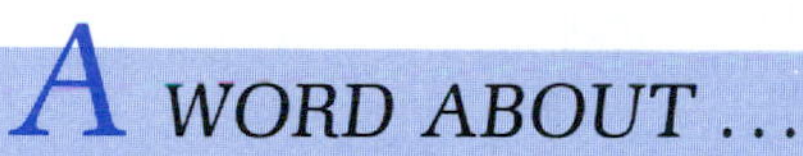

5. Petroleum, Gasoline, and Octane Number

Petroleum is at present our most important fossil fuel. The need for petroleum to keep our industrial society going sometimes seems second only to our need for food, air, water, and shelter. What is black gold, as petroleum has been called, and how do we use it?

Petroleum is a complex mixture of hydrocarbons formed over eons of time through the gradual decay of buried animal and vegetable matter. **Crude oil** is a viscous black liquid that collects in vast underground pockets in sedimentary rock (the word *petroleum* literally means rock oil, from the Latin *petra,* rock, and *oleum,* oil). It must be brought to the surface via drilling and pumping. To be most useful, the crude oil must be refined.

The first step in petroleum refining is usually **distillation.** The crude oil is heated to about 400°C, and the vapors rise through a tall fractionating column. The lower-boiling fractions rise faster and higher in the column before condensing to liquids; higher-boiling fractions do not rise so high. By drawing off liquid at various column levels, technicians separate crude oil roughly into the fractions shown in Table 3.3.

The gasoline fraction comprises only about 25% of crude oil. It is the most valuable fraction, however, both as a fuel and as a source material for the petrochemical industry, the industry that furnishes our synthetic fibers, plastics, and many other useful materials. For this reason, many processes have been developed for converting the other fractions into gasoline.

Higher-boiling fractions can be "cracked" by heat and catalysts (mainly silica and alumina), to give

TABLE 3.3 Common petroleum fractions

Boiling range, °C	*Name*	*Range of carbon atoms per molecule*	*Use*
below 20	gases	C_1 to C_4	heating, cooking, petrochemical raw material
20–200	naphtha; straight-run gasoline	C_5 to C_{12}	fuel; lighter fractions (such as petroleum ether, bp 30–60°C) also used as laboratory solvents
200–300	kerosene	C_{12} to C_{15}	fuel
300–400	fuel oil	C_{15} to C_{18}	heating homes, diesel fuel
over 400		over C_{18}	lubricating oil, greases, paraffin waxes, asphalt

products with shorter carbon chains and therefore lower boiling points. The carbon chain can break at many points.

$$\underset{\text{alkane}}{C_{10}H_{22}} \longrightarrow \begin{cases} \underset{\text{alkane}}{C_5H_{12}} + \underset{\text{alkene}}{C_5H_{10}} \\ C_8H_{18} + C_2H_4 \\ C_2H_6 + C_8H_{16} \\ C_4H_{10} + (C_4H_8 + C_2H_4) \end{cases}$$

To balance the number of hydrogens, any particular alkane must give at least one alkane and one alkene as products. Thus, catalytic cracking converts larger alkanes into a mixture of smaller alkanes and alkenes and increases the yield of gasoline from petroleum.

During cracking, large amounts of the lower gaseous hydrocarbons—ethene, propene, butanes, and butenes—are formed. Some of these, especially ethene, are used as petrochemical raw materials. In order to obtain more gasoline, scientists sought methods to convert these low-molecular-weight hydrocarbons to somewhat larger hydrocarbons that boil in the gasoline range. One such process is **alkylation,** the combination of an alkane with an alkene to form a higher-boiling alkane.

$$C_2H_6 + C_4H_8 \xrightarrow{\text{catalyst}} C_6H_{14}$$

$$C_4H_{10} + C_4H_8 \xrightarrow{\text{catalyst}} C_8H_{18}$$

These processes, which were developed in the 1930s, were important for producing aviation fuel during World War II and are still used to make high-octane gasoline.

This brings us to **octane number** and why it is important. Some hydrocarbons, especially those with highly branched structures, burn smoothly in an engine and drive the piston forward evenly. Other hydrocarbons, especially those with unbranched carbon chains, tend to explode in the cylinder and drive the piston forward violently. These undesirable explosions produce audible knocks. A scale was set up many years ago to evaluate this important knock property of gasolines. **Isooctane** (2,2,4-trimethylpentane), an excellent fuel with a highly branched structure, was arbitrarily given a rating of 100, and **heptane,** a very poor automotive fuel, was given a rating of 0. A "regular" gasoline with an octane number of 87 has the same "knock" properties as a mixture that is 87% isooctane and 13% heptane.

The addition of small amounts of **tetraethyllead,** $(CH_3CH_2)_4Pb$, to gasoline improves its octane rating but is undesirable for environmental reasons. For example, there is some evidence that, where automobile fumes are intense, young children in particular accumulate high levels of lead in their blood. However, unleaded gasoline must contain a very high percentage of hydrocarbons with a high octane rating. It became important, then, to develop methods for converting straight-chain hydrocarbons to branched-chain hydrocarbons, which have higher octane ratings, or to develop additives less toxic than tetraethyllead.

Certain catalysts can produce branched-chain alkanes from straight-chain alkanes. This process, called **isomerization,** is carried out on a large scale commercially.

$$\underset{n\text{-butane}}{CH_3CH_2CH_2CH_3} \xrightarrow[\text{alumina}]{AlCl_3,\ HCl} \underset{\text{isobutane}}{CH_3\overset{\overset{\large CH_3}{|}}{C}HCH_3}$$

Aromatic hydrocarbons, such as benzene and toluene, also have a high octane rating. A platinum catalyst used in a process called **platforming** cyclizes and dehydrogenates alkanes to cycloalkanes and to aromatic hydrocarbons. Of course, large amounts of hydrogen gas are also formed during platforming. Millions of gallons of aromatic hydrocarbons are produced daily by such processes, not only to add to unleaded gasoline to improve its octane rating, but also to supply raw materials for many other petrochemically based products, as we will see in the next chapter.

For further reading on the history of the petrochemical industry, see the book by P. H. Spitz, "*Petrochemicals: The Rise of an Industry*"; Wiley: New York, 1988.

$CH_3(CH_2)_5CH_3 \xrightarrow[\text{catalyst}]{\text{Pt}}$ methylcyclohexane $\xrightarrow[\text{catalyst}]{\text{Pt}}$ toluene (an aromatic hydrocarbon)

Additional Problems

3.30. For the following compounds, write structural formulas and IUPAC names for all possible isomers having the indicated number of multiple bonds:

a. C_4H_8(one double bond) **b.** C_5H_{10}(one double bond)
c. C_5H_8(two double bonds) **d.** C_5H_8(one triple bond)

3.31. Name the following compounds by the IUPAC system:

a. $CH_3CH_2CH{=}CHCH_3$ **b.** $(CH_3)_2C{=}CHCH_3$ **c.** 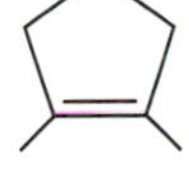
d. $CH_3C{\equiv}CCH_2CH_3$ **e.** $CH_2{=}CCl{-}CH{=}CH_2$
f. **g.**
h. $CH_3CH{=}C(CH_2CH_2CH_3)_2$ **i.** $HC{\equiv}CCH_2CH{=}CHCH_3$
j. $CH_2{=}CHCH_2C{\equiv}CH$

3.32. Write a structural formula for each of the following compounds:

a. 3-hexene **b.** cyclobutene
c. 1,3-dibromo-2-butene **d.** 3-methyl-1-pentyne
e. 1,4-hexadiene **f.** vinyl bromide
g. allyl chloride **h.** vinylcyclopentane
i. 4-methylcyclohexene **j.** 2,3-dibromo-1,3-cyclopentadiene

3.33. Explain why the following names are incorrect, and give a correct name in each case.

a. 3-butene **b.** 3-pentyne
c. 2-ethyl-1-propene **d.** 2-methylcyclopentene
e. 3-methyl-1,3-butadiene **f.** 1-methyl-2-butene
g. 3-buten-1-yne **h.** 3-butyne-1-ene

3.34.

a. What are the usual lengths for the single (sp^3–sp^3), double (sp^2–sp^2), and triple (sp–sp) carbon–carbon bonds?

b. The *single* bond in each of the following compounds has the length shown. Suggest a possible explanation for the observed shortening.

$$CH_2{=}CH\underset{\substack{\uparrow\\ 1.47\ \text{Å}}}{-}CH{=}CH_2 \qquad CH_2{=}CH\underset{\substack{\uparrow\\ 1.43\ \text{Å}}}{-}C{\equiv}CH \qquad HC{\equiv}C\underset{\substack{\uparrow\\ 1.37\ \text{Å}}}{-}C{\equiv}CH$$

3.35. Which of the following compounds can exist as *cis–trans* isomers? If such isomerism is possible, draw the structures in a way that clearly illustrates the geometry.
a. 1-pentene **b.** 2-pentene **c.** 1-chloropropene
d. 3-chloropropene **e.** 1,3,5-hexatriene **f.** 1,2-dibromocyclodecene

3.36. The mold metabolite and antibiotic *mycomycin* has the formula

$$HC{\equiv}C{-}C{\equiv}C{-}CH{=}C{=}CH{-}CH{=}CH{-}CH{=}CH{-}CH_2{-}\overset{\overset{\displaystyle O}{\|}}{C}{-}OH$$

Number the carbon chain, starting with the carbonyl carbon.
a. Which multiple bonds are conjugated?
b. Which multiple bonds are cumulative?
c. Which multiple bonds are isolated?

3.37. Write the structural formula and name of the product when each of the following reacts with 1 equivalent of bromine:
a. 2-butene **b.** vinyl bromide **c.** 1,3-cyclohexadiene
d. 1,4-cyclohexadiene **e.** 2,3-dimethyl-2-butene

3.38. Write an equation for the reaction of 1-butene with each of the following reagents:
a. chlorine **b.** hydrogen chloride
c. hydrogen (Pt catalyst) **d.** ozone, followed by Zn, H^+
e. H_2O, H^+ **f.** BH_3 followed by H_2O_2, OH^-
g. $KMnO_4$, OH^- **h.** oxygen (combustion)

3.39. What reagent will react by addition to what unsaturated hydrocarbon to form each of the following compounds?
a. $CH_3CHBrCHBrCH_3$ **b.** $(CH_3)_3COH$ **c.** $(CH_3)_2CHOSO_3H$
e. $CH_3CH{=}CHCH_2Br$ **f.** $CH_3CCl_2CCl_2CH_3$
d.

g.
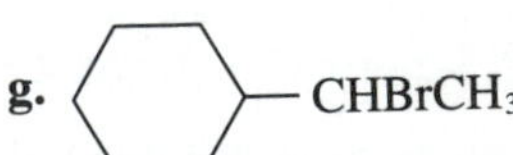

3.40. *Caryophyllene* is an unsaturated hydrocarbon mainly responsible for the odor of oil of cloves. It has the molecular formula $C_{15}H_{24}$. Hydrogenation of caryophyllene gives a saturated hydrocarbon $C_{15}H_{28}$. Does caryophyllene contain any rings? How many?

3.41. Which of the following reagents are electrophiles? Which are nucleophiles?
a. HCl **b.** Cl^- **c.** H_3O^+ **d.** $AlCl_3$ **e.** OH^-

3.42. Water can act as an electrophile or as a nucleophile. Explain.

3.43. The acid-catalyzed hydration of 1-methylcyclohexene gives 1-methylcyclohexanol.

$$\xrightarrow[H^+]{H_2O}\quad OH$$

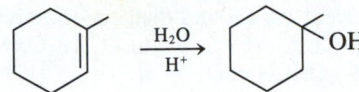

Write every step in the mechanism of this reaction.

3.44. Predict the structures of the two possible monohydration products of limonene (Figure 1.12). These alcohols are called terpineols. Predict the structure of the diol (dialcohol) obtained by hydrating both double bonds in limonene. These alcohols are used in the cough medicine "elixir of terpin hydrate."

3.45. Draw the resonance contributors to the carbocation $CH_3\overset{+}{C}HCH{=}CHCH_3$. Does the ion have a symmetric structure?

3.46. Adding 1 equivalent of hydrogen bromide to 2,4-hexadiene gives two products. Give their structures, and write all the steps in a reaction mechanism that explains how each product is formed.

3.47. Predict the product of each Diels-Alder reaction.

a. $CH_2{=}CH{-}CH{=}CH_2 + $ maleic anhydride ($CH{=}CH$ with two $C{=}O$ groups joined through O)

b. $CH_3CH{=}CH{-}CH{=}CH{-}CH_3 + (NC)_2C{=}C(CN)_2$

3.48. From what diene and dienophile could each of the following be made?

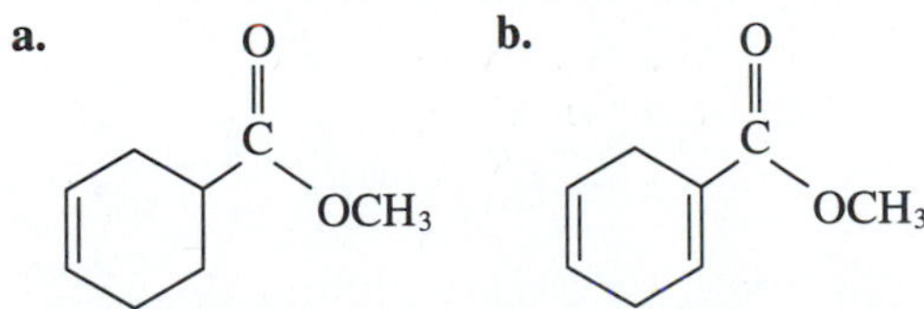

3.49. Given the information that free-radical stability follows the same order as carbocation stability (3° > 2° > 1°), predict the structure of polypropylene produced by the free-radical polymerization of propene. It should help to write out each step in the mechanism, as in eqs. 3.37 and 3.38.

3.50. Write an equation that clearly shows the structure of the alcohol obtained from the sequential hydroboration and H_2O_2/OH^- oxidation of

a. 2-methyl-1-butene. **b.** 1,2-dimethylcyclopentene.

3.51. Write equations to show how methylenecyclohexane ($=CH_2$) could be converted to

a. 1-methylcyclohexanol (CH_3, OH) **b.** cyclohexyl$-CH_2OH$

3.52. Describe two simple chemical tests that could be used to distinguish cyclohexane from cyclohexene.

3.53. Give the formulas of the alkenes that, on ozonolysis, give

a. only $CH_3CH_2CH{=}O$. **b.** $(CH_3)_2C{=}O$ and $CH_3CH{=}O$.

c. $CH_2{=}O$ and $(CH_3)_2CHCH{=}O$. **d.** $O{=}CHCH_2CH_2CH{=}O$.

3.54. Write equations for the following reactions:

a. 2-pentyne + Cl_2 (2 mol)

b. 3-hexyne + H_2 (1 mol, Lindlar's catalyst)

c. propyne + H_2O (H^+, $HgSO_4$ catalyst)

d. propyne + sodium amide in liquid ammonia

3.55. Determine what alkyne and what reagent will give

a. 2,2-dibromobutane. **b.** 2,2,3,3-tetrabromobutane.

CHAPTER 4

Aromatic Compounds

4.1 Historical Introduction

Spices and herbs have long played a romantic role in the course of history. They bring to mind frankincense and myrrh and the great explorers of past centuries—Vasco da Gama, Christopher Columbus, Ferdinand Magellan, Sir Francis Drake—whose quest for spices helped to open the Western world. Trade in spices was immensely profitable. It was natural, therefore, that spices and herbs were among the first natural products studied by organic chemists. If one could isolate from plants the pure compounds with these desirable fragrances and flavors and determine their structures, perhaps one could synthesize them in large quantity and at low cost.

It turned out that many of these aromatic substances have relatively simple structures. Many contain a six-carbon unit that passes unscathed through various chemical reactions that alter only the rest of the structure. This group, C_6H_5—, is common to many substances, including **benzaldehyde** (isolated from the oil of bitter almonds), **benzyl alcohol** (isolated from gum benzoin, a balsam resin obtained from certain Southeast Asian trees), and **toluene** (a hydrocarbon isolated from tolu balsam). When any of these three compounds is oxidized, the C_6H_5 group remains intact. The product in each case is **benzoic acid** (another constituent of gum (benzoin). The calcium salt of this acid, when heated, yields the parent hydrocarbon C_6H_6 (eq. 4.1).

$$\begin{array}{l} C_6H_5CH{=}O \text{ (benzaldehyde)} \xrightarrow{\text{oxidize}} \\ C_6H_5CH_2OH \text{ (benzyl alcohol)} \xrightarrow{\text{oxidize}} C_6H_5CO_2H \text{ (benzoic acid)} \xrightarrow[\text{2. heat}]{\text{1. CaO}} C_6H_6 \text{ (benzene)} \\ C_6H_5CH_3 \text{ (toluene)} \xrightarrow{\text{oxidize}} \end{array} \quad (4.1)$$

This same hydrocarbon, first isolated from compressed illuminating gas by Michael Faraday in 1825, is now called **benzene.** It is the parent hydrocarbon of a class of substances that we now call **aromatic compounds,** *not because of* *their aroma,* but because of their special chemical properties, in particular their stability.

4.2 Some Facts About Benzene

The carbon-to-hydrogen ratio in benzene, C_6H_6, suggests a highly unsaturated structure. Compare the number of hydrogens, for example, with that in hexane, C_6H_{14}, or in cyclohexane, C_6H_{12}, both of which also have six carbons but are saturated.

Problem 4.1 Draw at least five isomeric structures that have the molecular formula C_6H_6. Note that all are highly unsaturated or contain small, strained rings.

Despite its molecular formula, benzene for the most part does not behave as if it were unsaturated. For instance, it does not decolorize bromine solutions the way alkenes and alkynes do, nor is it easily oxidized by potassium permanganate. It does not undergo the typical addition reactions of alkenes or alkynes. Instead, *benzene reacts mainly by substitution.* For example, when treated with bromine in the presence of ferric bromide as a catalyst, benzene gives bromobenzene and hydrogen bromide.

$$\underset{\text{benzene}}{C_6H_6} + Br_2 \xrightarrow[\text{catalyst}]{FeBr_3} \underset{\text{bromobenzene}}{C_6H_5Br} + HBr \tag{4.2}$$

Chlorine, with a ferric chloride catalyst, reacts similarly.

$$\underset{\text{benzene}}{C_6H_6} + Cl_2 \xrightarrow[\text{catalyst}]{FeCl_3} \underset{\text{chlorobenzene}}{C_6H_5Cl} + HCl \tag{4.3}$$

Only *one* monobromobenzene or monochlorobenzene has ever been isolated; that is, no isomers are obtained in either of these reactions. This result implies that *all six hydrogens in benzene are chemically equivalent*. It does not matter which hydrogen is replaced by bromine; we get the same monobromobenzene. This fact has to be accounted for in any structure proposed for benzene.

When bromobenzene is treated with a second equivalent of bromine and the same type of catalyst, *three di*bromobenzenes are obtained.

$$C_6H_5Br + Br_2 \xrightarrow[\text{catalyst}]{FeBr_3} \underset{\substack{\text{dibromobenzenes} \\ \text{(three isomers)}}}{C_6H_4Br_2} + HBr \tag{4.4}$$

The isomers are not formed in equal amounts. Two of them predominate, and only a small amount of the third isomer is formed. The important point is that there are three isomers—no more and no less. Similar results are obtained when chlorobenzene is further chlorinated to give dichlorobenzenes. These facts also have to be explained by any structure proposed for benzene.

The problem of benzene's structure does not sound overwhelming, yet it took many decades to solve. Let us examine the main ideas that led to our modern view of its structure.

4.3 The Kekulé Structure of Benzene

In 1865 Kekulé proposed the first reasonable structure for benzene.* He suggested that the six carbon atoms are located at the corners of a regular hexagon, with one hydrogen atom attached to each carbon atom. To give each carbon atom a valence of 4, he suggested that single and double bonds alternate around the ring (what we now call a *conjugated* system of double bonds). But this structure is highly unsaturated. To explain benzene's negative tests for unsaturation (that is, its failure to decolorize bromine or to give a permanganate test), Kekulé suggested that the single and double bonds exchange positions around the ring *so rapidly* that the typical reactions of alkenes cannot take place.

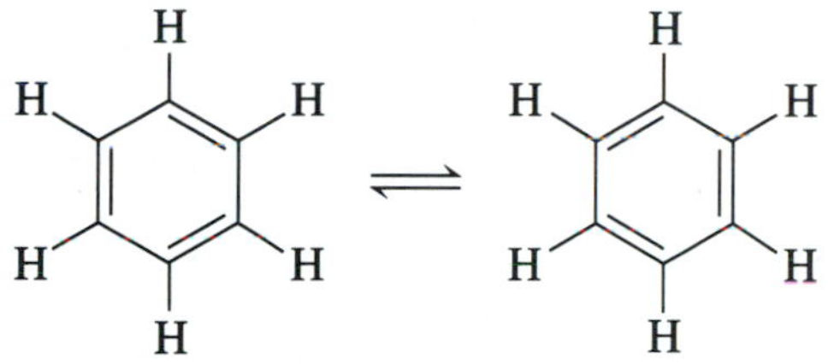

the Kekulé structures for benzene

Problem 4.2 Write out eqs. 4.2 and 4.4 using a Kekulé structure for benzene. Does this model explain the existence of only one monobromobenzene? only three dibromobenzenes?

Problem 4.3 How might Kekulé explain the fact that there is only one dibromobenzene with the bromines on adjacent carbon atoms, even though we can draw two different structures, with either a double or a single bond between the bromine-bearing carbons?

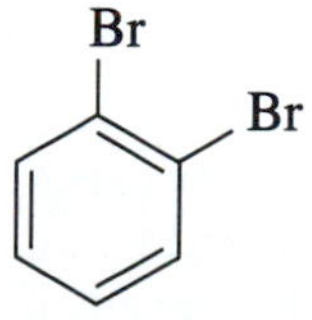

and

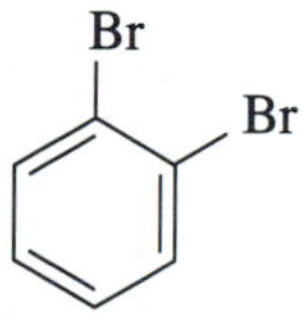

---

* Friedrich August Kekulé (1829–1896) was a pioneer in the development of structural formulas in organic chemistry. He was among the first to recognize the tetracovalence of carbon and the importance of carbon chains in organic structures. He is best known for his proposal regarding the structure of benzene and other aromatic compounds. It is interesting that Kekulé first studied architecture, and only later switched to chemistry. Judging from his contributions, he apparently viewed chemistry as molecular architecture. Kekulé is supposed to have arrived at his structure for benzene while daydreaming before a fireplace; the flames reminded him of a snake swallowing its own tail. The truth or myth of this bit of chemical folklore is discussed in an interesting article by J. H. Wotiz and S. Rudofsky in *Chemistry in Britain,* **1984,** *20,* 720–723. This controversial article stimulated a strong response from A. J. Rocke and O. B. Ramsay, *Chemistry in Britain,* **1984,** *20,* 1093. The subject, which has implications for the history and philosophy of science, is still being debated (see *Chemical and Engineering News,* Nov. 4, 1985, pp. 22–23).

FIGURE 4.1
Space-filling model of benzene.

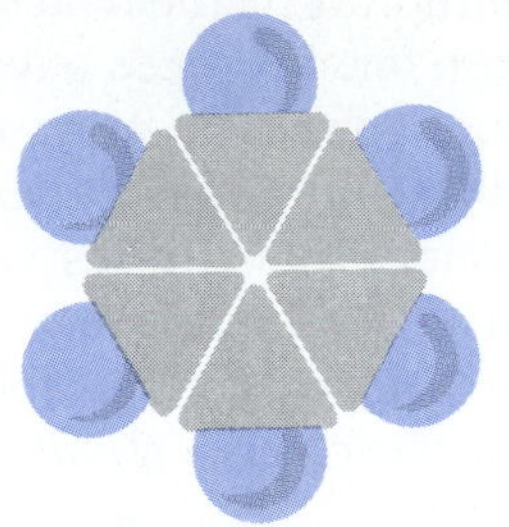

Properties
colorless liquid
bp 80 °C
mp 5.5 °C

4.4 The Resonance Model for Benzene

Kekulé's model for the structure of benzene is nearly, but not entirely, correct. *Kekulé's two structures for benzene differ only in the arrangement of the electrons;* all the atoms occupy the same positions in both structures. *This is precisely the requirement for resonance* (review Sec. 1.12). Kekulé's formulas represent two identical contributing structures to a *single* resonance hybrid structure of benzene. Instead of writing an equilibrium symbol between them, as Kekulé did, we now write the double-headed arrow used to indicate a resonance hybrid:

Benzene is a resonance hybrid of these two contributing structures.

To express this model another way, all benzene molecules are identical, and their structure is not adequately represented by either of Kekulé's contributing structures. Being a resonance hybrid, benzene is more stable than its contributing Kekulé structures. There are no single or double bonds in benzene—only one type of carbon–carbon bond, which is of some intermediate type. Consequently, it is not surprising that benzene does not react chemically exactly like alkenes.

Modern physical measurements support this model for the benzene structure. *Benzene is planar, and each carbon atom is at the corner of a regular hexagon. All the carbon–carbon bond lengths are identical,* and this bond length is 1.39 Å, intermediate between typical single (1.54 Å) and double (1.34 Å) carbon–carbon bond lengths. Figure 4.1 shows a space-filling model of the benzene molecule.*

* Notice the difference in the shapes of benzene and cyclohexane (Figure 2.4).

4.5 Orbital Model for Benzene

Orbital theory, which is so useful in rationalizing the geometries of alkanes, alkenes, and alkynes, is also useful in explaining the structure of benzene. Each carbon atom in benzene is connected to only *three* other atoms (two carbons and a hydrogen). Each carbon is therefore sp^2-hybridized, as in ethylene. Two sp^2 orbitals of each carbon atom overlap with similar orbitals of adjacent carbon atoms to form the sigma bonds of the hexagonal ring. The third sp^2 orbital of each carbon overlaps with a hydrogen 1*s* orbital to form the C—H sigma bonds. Perpendicular to the plane of the three sp^2 orbitals at each carbon is a *p* orbital containing one electron, the fourth valence electron. The *p* orbitals on all six carbon atoms can overlap laterally to form pi orbitals that create a cloud of electrons above and below the plane of the ring. The construction of a benzene ring from six sp^2-hybridized carbons is shown schematically in Figure 4.2. This model nicely explains the planarity of benzene. It also explains its hexagonal shape, with H—C—C and C—C—C angles of 120°.

4.6 Symbols for Benzene

Two symbols are used to represent benzene. One is the Kekulé structure, and the other is a hexagon with an inscribed circle, to represent the idea of a delocalized pi electron cloud.

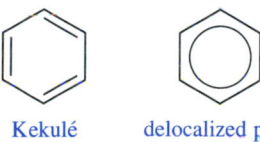

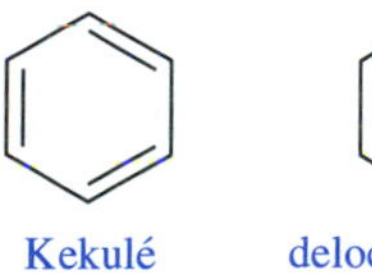

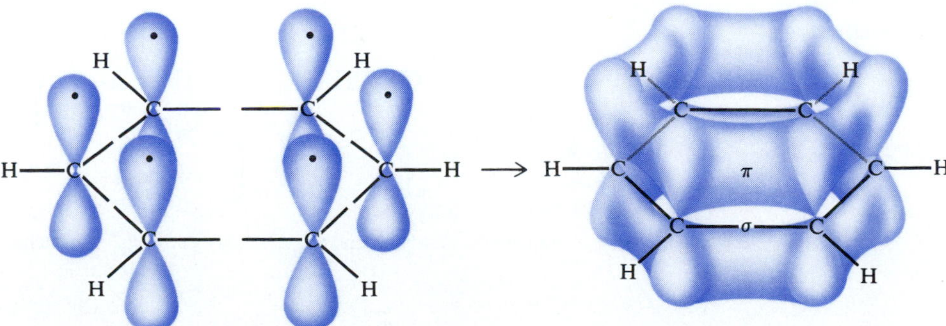

FIGURE 4.2 An orbital representation of the bonding in benzene. Sigma bonds are formed by the end-on overlap of sp^2 orbitals. In addition, each carbon contributes one electron to the pi system by lateral overlap of its *p* orbital with the *p* orbitals of its two neighbors.

Regardless of which symbol is used, the hydrogens are usually not written explicitly, but we must remember that one hydrogen atom is attached to the carbon at each corner of the hexagon.

The symbol with the inscribed circle emphasizes the fact that the electrons are distributed evenly around the ring, and in this sense, it is perhaps the more accurate of the two. The Kekulé symbol, however, reminds us very clearly that there are six pi electrons in benzene. For this reason, it is particularly useful in allowing us to keep track of the valence electrons during chemical reactions of benzene. In this book, we will use the Kekulé symbol. However, we must keep in mind that the "double bonds" are not fixed in the positions shown, nor are they really double bonds at all.

EXAMPLE 4.1 Write the structural formula for benzaldehyde (eq. 4.1).

Solution One hydrogen in the formula for benzene is replaced by the aldehyde group.

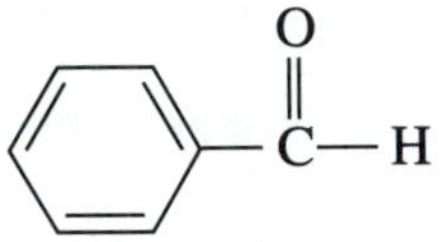

O
‖
—C—H

Problem 4.4 Write the formulas for benzyl alcohol, toluene, and benzoic acid (eq. 4.1).

4.7 Nomenclature of Aromatic Compounds

Because aromatic chemistry developed in a haphazard fashion many years before systematic methods of nomenclature were developed, common names have acquired historic respectability and are accepted by IUPAC. Examples include

	CH_3	CH_3CHCH_3	$CH{=}CH_2$	OH
benzene	toluene	cumene	styrene	phenol

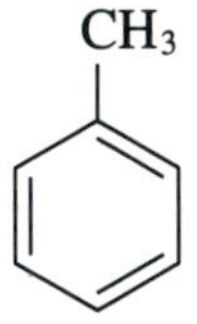

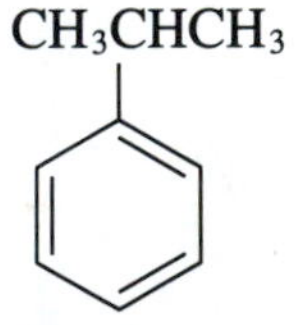

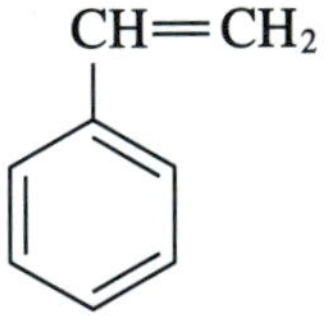

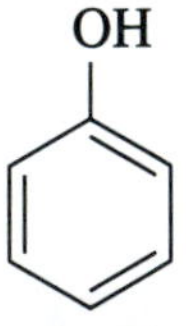

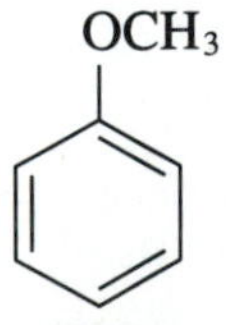

OCH_3	$CH{=}O$	$CH_3C{=}O$	CO_2H	NH_2
anisole	benzaldehyde	acetophenone	benzoic acid	aniline

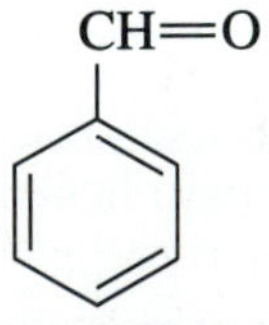

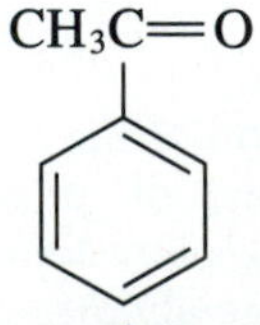

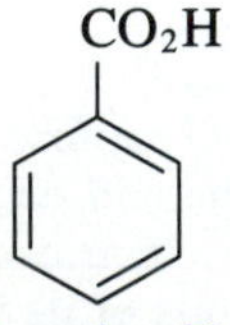

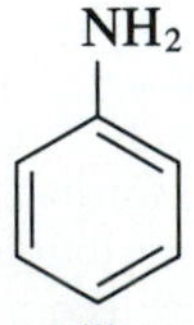

Monosubstituted benzenes are named as derivatives of benzene.

Br, Cl, NO_2, CH_2CH_3, $CH_2CH_2CH_3$

bromobenzene chlorobenzene nitrobenzene ethylbenzene propylbenzene

When two substituents are present, three isomeric structures are possible. They are designated by the prefixes *ortho-*, *meta-*, and *para-*, which are usually abbreviated as *o-*, *m-*, and *p-*, respectively. In this structure,

X; *o*, *o*; *m*, *m*; *p*

o-groups are next to X, *m*-groups are separated from X by one carbon, and *p*-groups are opposite X. Specific examples are

ortho-dichlorobenzene *meta*-dichlorobenzene *para*-dichlorobenzene *para*-xylene* *para*-chlorobenzenesulfonic acid

Problem 4.5 Draw the structures for *ortho*-xylene and *meta*-xylene.

The prefixes *ortho-*, *meta-*, and *para-* are used even when the two substituents are not identical.

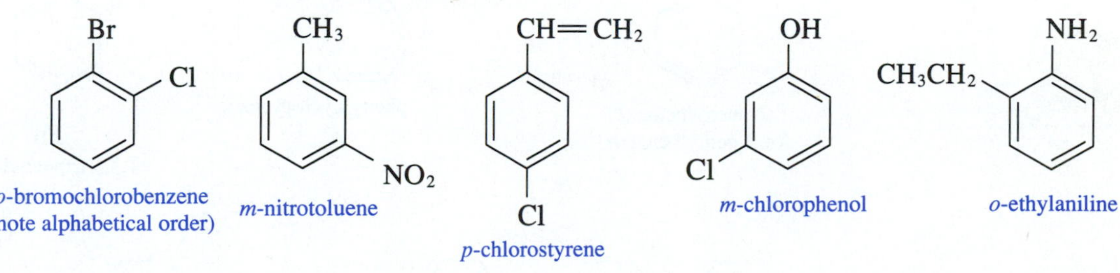

o-bromochlorobenzene (note alphabetical order) *m*-nitrotoluene *p*-chlorostyrene *m*-chlorophenol *o*-ethylaniline

* The common and IUPAC name is xylene, *not* *p*-methyltoluene.

When more than two substituents are present, their positions are designated by numbering the ring.

1,2,4-trimethylbenzene

3,5-dichlorotoluene

2,4,6-trinitrotoluene (TNT)

Problem 4.6 Draw the structure of

a. *p*-nitrotoluene. b. *o*-bromophenol.
c. *m*-dinitrobenzene. d. *p*-divinylbenzene.

Problem 4.7 Draw the structure of

a. 1,3,5-trimethylbenzene.
b. 2,6-dibromo-4-chlorotoluene.

Aromatic hydrocarbons, as a class, are called **arenes.** The symbol **Ar** is used for an **aryl group,** just as the symbol R is used for an alkyl group. The formula Ar-R would therefore represent any arylalkane.

Two groups with special names occur frequently in aromatic compounds. They are the **phenyl group** and the **benzyl group.**

$C_6H_5—$ or

phenyl group

$C_6H_5CH_2—$ or

benzyl group

The symbol Ph is sometimes used as an abbreviation for the phenyl group. The use of these group names is illustrated in the following examples:

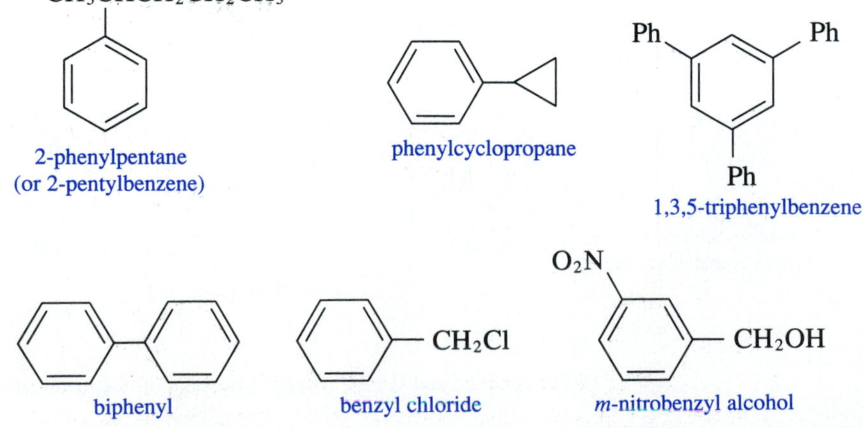

2-phenylpentane (or 2-pentylbenzene)

phenylcyclopropane

1,3,5-triphenylbenzene

biphenyl

benzyl chloride

m-nitrobenzyl alcohol

Problem 4.8 Draw the structure of

a. phenylcyclohexane.
b. benzyl alcohol.
c. *p*-phenylstyrene.
d. dibenzyl.

Problem 4.9 Name the following structures:

a. 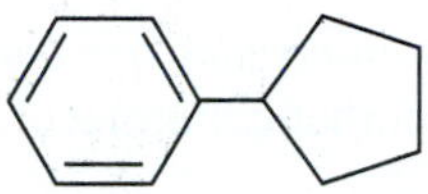 (benzene ring)–(cyclopentane ring)

b. 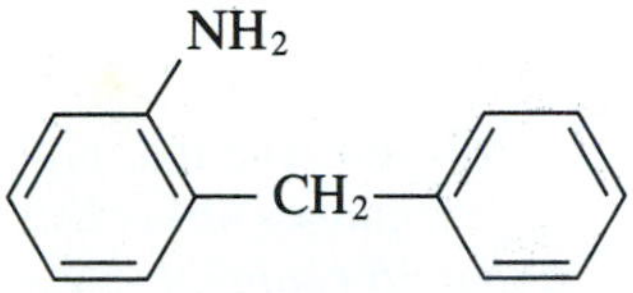 (benzene ring bearing NH_2)–CH_2–(benzene ring)

4.8 The Resonance Energy of Benzene

We have asserted that a resonance hybrid is always more stable than any of its contributing structures. Fortunately, in the case of benzene, this assertion can be proved experimentally, and we can even measure how much more stable benzene is than the hypothetical molecule 1,3,5-cyclohexatriene (the IUPAC name for one Kekulé structure).

Hydrogenation of a carbon–carbon double bond is an exothermic reaction. The amount of energy (heat) released is about 26 to 30 kcal/mol for each double bond.

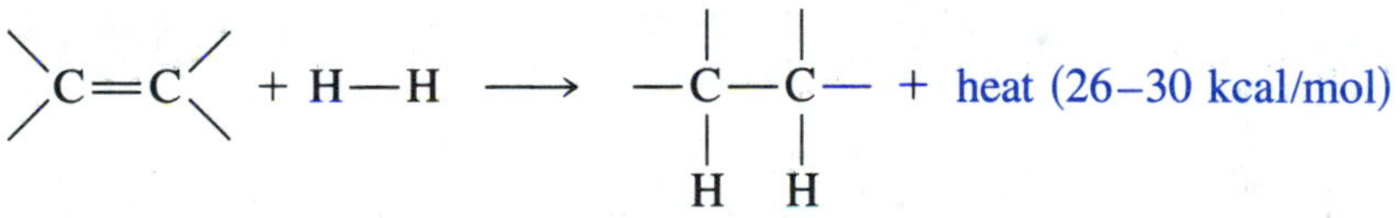

$$\text{>C=C<} + \text{H—H} \longrightarrow \text{—}\underset{\text{H}}{\overset{|}{\text{C}}}\text{—}\underset{\text{H}}{\overset{|}{\text{C}}}\text{—} + \text{heat (26–30 kcal/mol)} \qquad (4.5)$$

(The exact value depends on the substituents attached to the double bond.) When two double bonds in a molecule are hydrogenated, twice as much heat is evolved, and so on.

Hydrogenation of cyclohexene releases 28.6 kcal/mol.

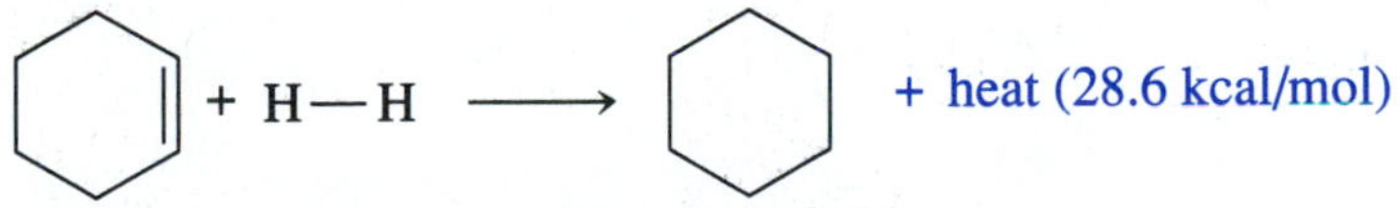

$$\text{cyclohexene} + \text{H—H} \longrightarrow \text{cyclohexane} + \text{heat (28.6 kcal/mol)} \qquad (4.6)$$

cyclohexene cyclohexane

We expect that the complete hydrogenation of 1,3-cyclohexadiene should release twice that amount of heat, or $2 \times 28.6 = 57.2$ kcal/mol; experimentally the value is close to what we expect.

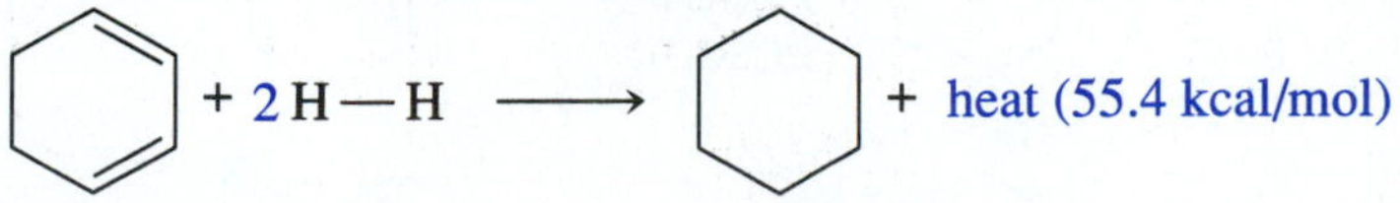

$$\text{1,3-cyclohexadiene} + 2\,\text{H—H} \longrightarrow \text{cyclohexane} + \text{heat (55.4 kcal/mol)} \qquad (4.7)$$

1,3-cyclohexadiene cyclohexane

It seems reasonable, therefore, to expect that the heat of hydrogenation of a Kekulé structure (the *hypothetical* triene 1,3,5-cyclohexatriene) should correspond to that for *three* double bonds, or about 84 to 86 kcal/mol. However, we find experimentally that benzene is more difficult to hydrogenate than simple alkenes, and the heat evolved when benzene is hydrogenated to cyclohexane is *much lower* than expected: only 49.8 kcal/mol.

$$\text{benzene} + 3\ \text{H—H} \longrightarrow \text{cyclohexane} + \text{heat (49.8 kcal/mol)} \qquad (4.8)$$

We conclude that *real benzene molecules are more stable than the contributing resonance structures* (the hypothetical molecule 1,3,5-cyclohexatriene), *by about 36 kcal/mol* (86 − 50 = 36).

We define the **stabilization energy,** or **resonance energy,** of a substance as the difference between the actual energy of the real molecule (the resonance hybrid) and the calculated energy of the most stable contributing structure. For benzene this value is about 36 kcal/mol. This is a substantial amount of energy. Consequently, as we will see, *benzene and other aromatic compounds usually react in such a way as to preserve their aromatic structure and therefore retain their resonance energy.*

4.9 Electrophilic Substitution

The most common reactions of aromatic compounds involve substitution of other atoms or groups for a ring hydrogen. Here are some typical substitution reactions of benzene.

$$\text{benzene} + Cl_2 \xrightarrow{FeCl_3} C_6H_5Cl + HCl \qquad \text{chlorination} \quad (4.9)$$

$$\text{benzene} + Br_2 \xrightarrow{FeBr_3} C_6H_5Br + HBr \qquad \text{bromination} \quad (4.10)$$

$$\text{benzene} + \underset{(HONO_2)}{HNO_3} \xrightarrow{H_2SO_4} C_6H_5NO_2 + H_2O \qquad \text{nitration} \quad (4.11)$$

$$\text{benzene} + \underset{(HOSO_3H)}{H_2SO_4} \longrightarrow C_6H_5SO_3H + H_2O \qquad \text{sulfonation} \quad (4.12)$$

$$\text{benzene} + RCl \xrightarrow{AlCl_3} C_6H_5R + HCl \qquad \text{alkylation} \quad (4.13)$$

(R = an alkyl group such as CH_3—, CH_3CH_2—)

$$C_6H_6 + CH_2{=}CH_2 \xrightarrow{H_2SO_4} C_6H_5CH_2CH_3 \qquad \text{alkylation} \tag{4.14}$$

$$C_6H_6 + R{-}\overset{\overset{\displaystyle O}{\|}}{C}Cl \xrightarrow{AlCl_3} C_6H_5{-}\overset{\overset{\displaystyle O}{\|}}{C}{-}R + HCl \qquad \text{acylation} \tag{4.15}$$

Most of these reactions are carried out at temperatures between about 0° and 50°C, but these conditions may have to be milder or more severe if other substituents are already present on the benzene ring. Also, the conditions usually can be adjusted to introduce more than one substituent, if desired.

How do these reactions take place? And why do we observe substitution instead of addition? In the next sections, we will try to answer these questions.

4.10 The Mechanism of Electrophilic Aromatic Substitution

Much evidence indicates that all of the substitution reactions listed in the previous section involve initial attack on the benzene ring by an electrophile. Consider chlorination (eq. 4.9) as a specific example. The reaction of benzene with chlorine is exceedingly slow without a catalyst, but it occurs quite briskly with one. What does the catalyst do? It acts as a Lewis acid and converts chlorine to a strong electrophile by polarizing the Cl—Cl bond.

$$\underset{\text{weak electrophile}}{:\ddot{C}l{-}\ddot{C}l:} + FeCl_3 \rightleftharpoons \underset{\text{strong electrophile}}{\overset{\delta+}{Cl}\cdots\overset{\delta-}{Cl}\cdots FeCl_3} \tag{4.16}$$

The reason why a *strong* electrophile is required will become apparent shortly.

The electrophile bonds to one carbon atom of the benzene ring, using two of the pi electrons from the pi cloud to form a sigma bond with a ring carbon atom. This carbon atom becomes sp^3-hybridized. The benzene ring acts as a pi-electron donor, or nucleophile, toward the electrophilic reagent.

$$C_6H_6 + \overset{\delta+}{Cl}{-}\overset{\delta-}{Cl}\cdots FeCl_3 \longrightarrow \underset{\substack{\text{a benzenonium ion}\\ \text{(a carbocation)}}}{C_6H_6Cl^+} + FeCl_4^- \tag{4.17}$$

This carbon is sp^3 hybridized; it is bonded to *four* other atoms, and has no double bond to it.

The resulting carbocation is a **benzenonium ion,** in which the positive charge is delocalized by resonance to the carbon atoms *ortho* and *para* to the carbon to which the chlorine atom became attached; that is, *ortho* and *para* to the sp^3 carbon atom.

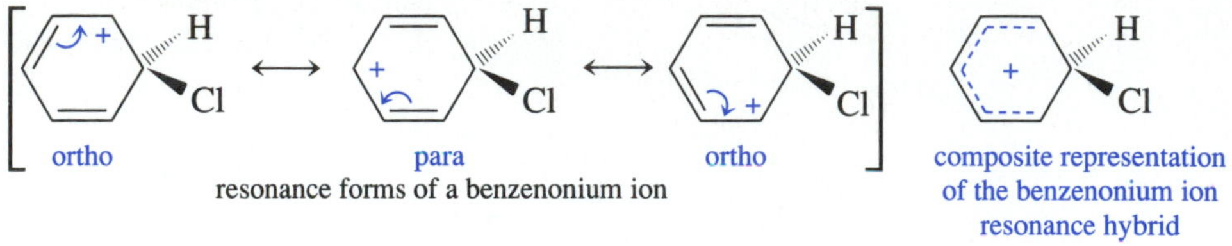

ortho para ortho
resonance forms of a benzenonium ion
composite representation of the benzenonium ion resonance hybrid

A benzenonium ion is similar to an allylic carbocation (sec. 3.14a), but the positive charge is delocalized over five carbon atoms instead of only three. Although stabilized by resonance compared with other carbocations, its resonance energy is much less than that of the starting benzene ring.

Substitution is completed by loss of a proton from the sp^3 carbon atom, the same atom to which the electrophile became attached.

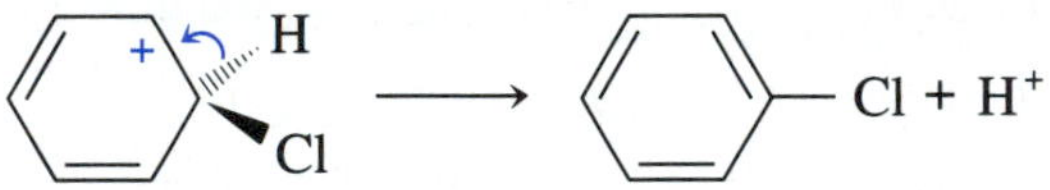

$$\longrightarrow \text{C}_6\text{H}_5\text{—Cl} + \text{H}^+ \quad (4.18)$$

We can generalize this two-step mechanism for all the electrophilic aromatic substitutions in Sec. 4.9 with the following equation:

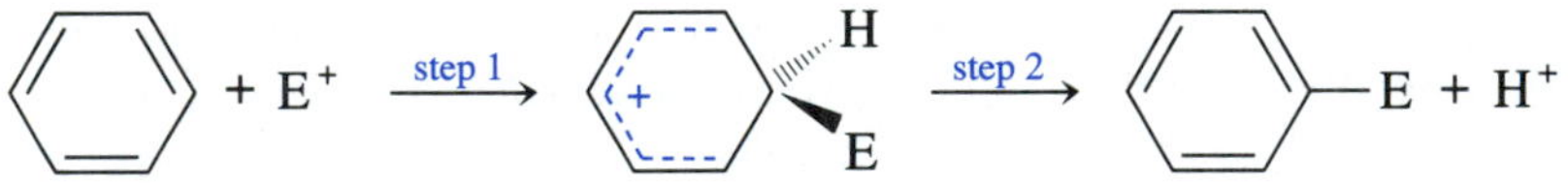

$$\text{C}_6\text{H}_6 + \text{E}^+ \xrightarrow{\text{step 1}} [\text{benzenonium ion}] \xrightarrow{\text{step 2}} \text{C}_6\text{H}_5\text{—E} + \text{H}^+ \quad (4.19)$$

The reason why a strong electrophile is important, and why we observe substitution instead of addition, now becomes clear. In step 1, the stabilization energy (resonance energy) of the aromatic ring is lost, due to disruption of the aromatic pi system. This disruption, caused by addition of the electrophile to one of the ring carbons, costs energy and requires a strong electrophile. In step 2, the aromatic resonance energy is regained by loss of a proton. This would not be the case if the intermediate carbocation added a nucleophile (as in electrophilic *additions* to double bonds, Sec. 3.11).

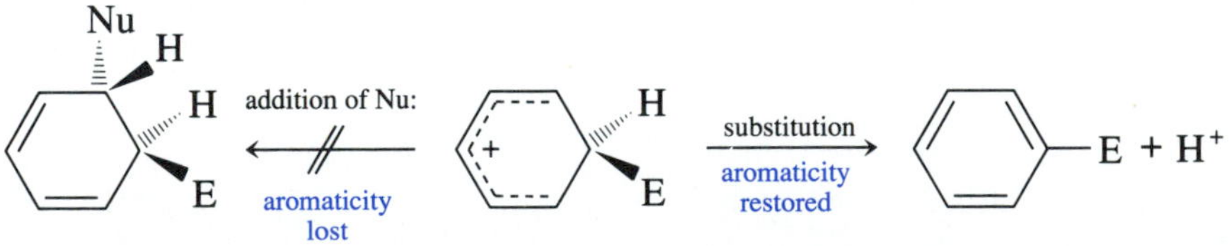

Nu H H E — addition of Nu: (aromaticity lost) — substitution (aromaticity restored) — E + H$^+$ (4.20)

The first step in eq. 4.19 is usually slow or rate-determining because it disrupts the aromatic system. The second step is usually fast because it regenerates the aromatic system.

Now let us briefly consider separately each of the various types of electrophilic aromatic substitutions listed in Sec. 4.9.

4.10a Halogenation Chlorine or bromine are readily introduced into aromatic rings by using the halogen together with the corresponding iron halide as a catalyst (that is, $Cl_2 + FeCl_3$ or $Br_2 + FeBr_3$). Usually the reaction is carried out by adding the halogen slowly to a mixture of the aromatic compound and iron filings. The iron reacts with the halogen to form the iron halide, which then catalyzes the halogenation.

Direct fluorination or iodination of aromatic rings is also possible but requires special methods.

4.10b Nitration In aromatic nitrations (eq. 4.11), the sulfuric acid catalyst protonates the nitric acid, which then loses water to generate the **nitronium ion.**

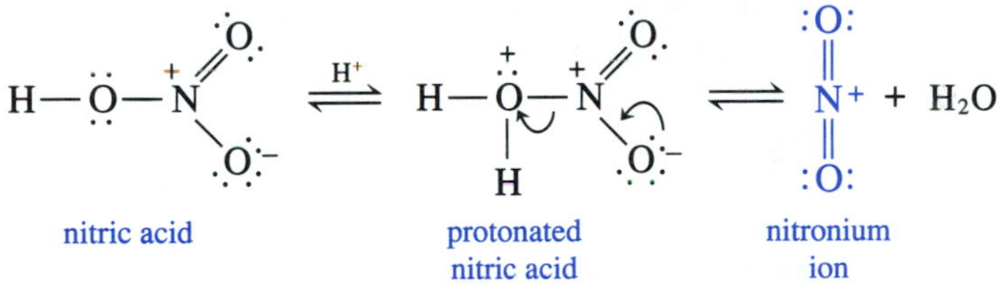

$$H{-}\ddot{O}{-}\overset{+}{N}(=O)O^- \underset{}{\overset{H^+}{\rightleftharpoons}} H{-}\overset{+}{O}(H){-}\overset{+}{N}(=O)O^- \rightleftharpoons O{=}N^+{=}O + H_2O \qquad (4.21)$$

nitric acid — protonated nitric acid — nitronium ion

This is the electrophile that then attacks the aromatic ring.

EXAMPLE 4.2 Write out the steps in the mechanism for the nitration of benzene.

Solution The first step, formation of the electrophile NO_2^+, is shown in eq. 4.21. Then

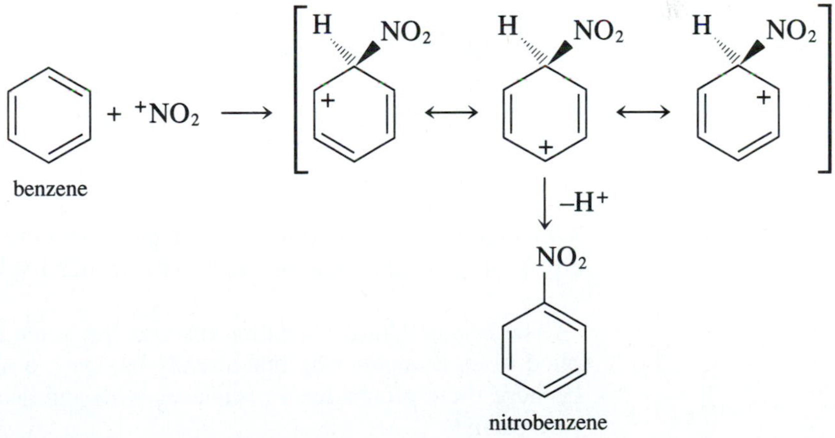

$$\text{benzene} + {}^+NO_2 \longrightarrow \left[\text{resonance structures of the } (H)(NO_2)\text{ cyclohexadienyl cation} \right] \xrightarrow{-H^+} \text{nitrobenzene}$$

benzene — nitrobenzene

4.10c Sulfonation In sulfonation (eq. 4.12), we use either concentrated or fuming sulfuric acid, and the electrophile may be sulfur trioxide or protonated sulfur trioxide, ${}^+SO_3H$. The products, sulfonic acids, are strong organic acids.

Also, they can be converted to phenols by reaction with base at high temperatures.

$$\text{benzene} \xrightarrow[H_2SO_4]{SO_3} C_6H_5SO_3H \xrightarrow[200°C]{NaOH} C_6H_5OH \tag{4.22}$$

benzenesulfonic acid phenol

Problem 4.10 Write out the steps in the mechanism for the sulfonation of benzene.

4.10d Alkylation and Acylation

Alkylation of aromatic compounds (eq. 4.13 and eq. 4.14) is referred to as the **Friedel-Crafts reaction,** after Charles Friedel (French) and James Mason Crafts (American), who first discovered the reaction in 1877. The electrophile is a carbocation, which may be formed either by removing halide ion from an alkyl halide with a Lewis acid catalyst (for example, $AlCl_3$) or by adding a proton to an alkene. For example, the synthesis of ethylbenzene may be carried out as follows:

$$AlCl_3 + ClCH_2CH_3 \rightleftharpoons AlCl_4^- + {}^+CH_2CH_3 \xleftarrow{H^+} CH_2{=}CH_2 \tag{4.23}$$

ethyl cation

$$C_6H_6 + {}^+CH_2CH_3 \longrightarrow [C_6H_6CH_2CH_3^+ \leftrightarrow \leftrightarrow] \xrightarrow{-H^+} C_6H_5CH_2CH_3 \tag{4.24}$$

benzene ethylbenzene

Problem 4.11 Which product would you expect if propene were used in place of ethene in eq. 4.14, propylbenzene or isopropylbenzene? Explain.

The Friedel-Crafts alkylation reaction has some limitations. It cannot be applied to an aromatic ring that already has on it a nitro or sulfonic acid group, because these groups form complexes with and deactivate the aluminum chloride catalyst.

Friedel-Crafts **acylations** (eq. 4.15) occur similarly. The electrophile is an acyl cation generated from an acid derivative, usually an acyl halide. The reaction provides a useful general route to aromatic ketones.

$$CH_3\overset{O}{\overset{\|}{C}}Cl + AlCl_3 \rightleftharpoons CH_3\overset{+}{C}{=}O + AlCl_4^- \tag{4.25}$$

acetyl chloride acetyl cation

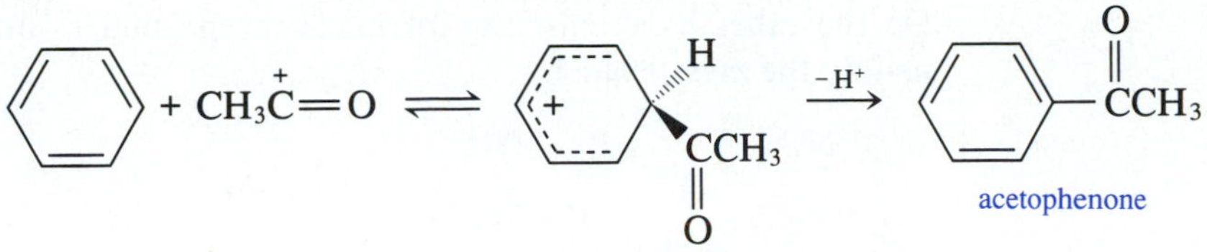

$$C_6H_6 + CH_3\overset{+}{C}{=}O \rightleftharpoons [C_6H_6(H)(C(=O)CH_3)]^{+} \xrightarrow{-H^+} C_6H_5C(=O)CH_3 \quad (4.26)$$

acetophenone

4.11 Ring-Activating and Ring-Deactivating Substituents

In this section and the next, we will present experimental evidence that supports the electrophilic aromatic substitution mechanism just described. We will do this by examining how substituents already present on an aromatic ring affect further substitution reactions.

For example, consider the relative nitration rates of the following compounds, all under the same reaction conditions:

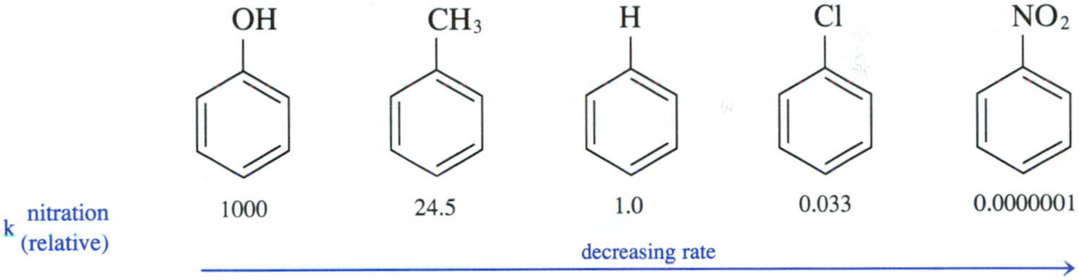

	OH	CH_3	H	Cl	NO_2
$k_{nitration}$ (relative)	1000	24.5	1.0	0.033	0.0000001

decreasing rate →

Taking benzene as the standard, we see that some substituents (for example, OH and CH_3) speed up the reaction, and other substituents (Cl and NO_2) retard the reaction. We know from other evidence that hydroxyl and methyl groups are more electron donating than hydrogen, whereas chloro and nitro groups are more electron withdrawing than hydrogen.

These observations support the electrophilic mechanism for substitution. If the reaction rate depends on electrophilic (that is, electron-seeking) attack on the aromatic ring, then substituents that donate electrons to the ring will speed up the reaction; substituents that withdraw electrons from the ring will decrease electron density in the ring and therefore slow down the reaction. This reactivity pattern is exactly what is observed, not only with nitration, but with all electrophilic aromatic substitution reactions.

4.12 Ortho, Para-Directing and Meta-Directing Groups

Substituents already present on an aromatic ring determine the position taken by a new substituent. For example, nitration of toluene gives mainly a mixture of *o*- and *p*-nitrotoluene.

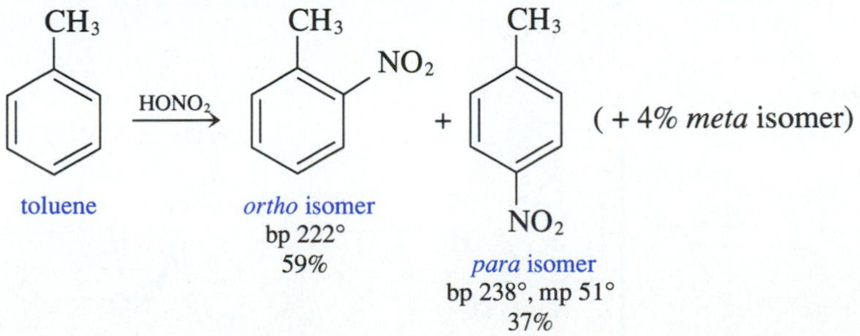

$$C_6H_5CH_3 \xrightarrow{HONO_2} o\text{-}CH_3C_6H_4NO_2 + p\text{-}CH_3C_6H_4NO_2 \quad (+\,4\%\ meta \text{ isomer}) \quad (4.27)$$

toluene

ortho isomer
bp 222°
59%

para isomer
bp 238°, mp 51°
37%

On the other hand, nitration of nitrobenzene under similar conditions gives mainly the *meta* isomer.

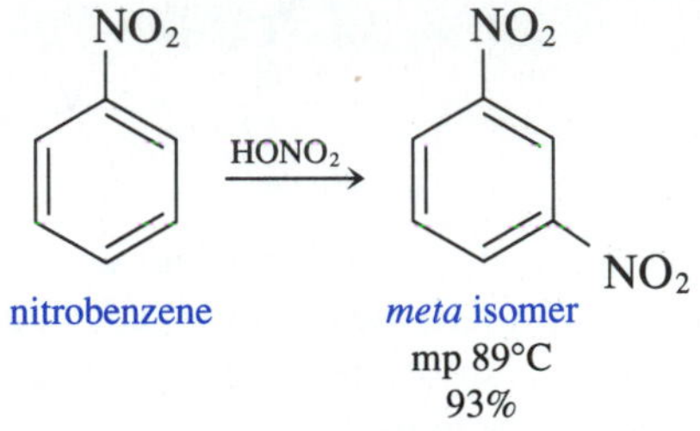

NO_2 — $\xrightarrow{HONO_2}$ — NO_2, NO_2 (+ 7% *ortho* isomer) (4.28)

nitrobenzene

meta isomer
mp 89°C
93%

This pattern is also followed for other electrophilic aromatic substitutions—chlorination, bromination, sulfonation, and so on. Toluene undergoes mainly *ortho,para* substitution, whereas nitrobenzene undergoes *meta* substitution.

In general, groups fall into one of two categories. Certain groups are ***ortho,para*-directing,** and others are ***meta*-directing.** Table 4.1 lists some of the common groups in each category. Let us see how the electrophilic substitution mechanism accounts for the behavior of these two classes of substituents.

4.12a *Ortho,Para*-Directing Groups Consider the nitration of toluene. In the first step, the nitronium ion may attack a ring carbon that is *ortho, meta,* or *para* to the methyl group.

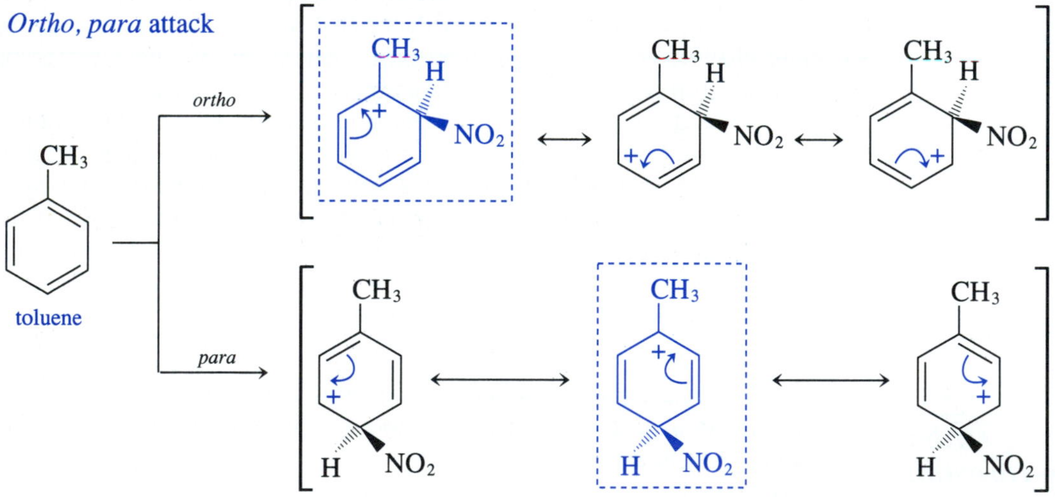

Ortho, para attack

CH_3 toluene — *ortho* → [CH_3 H NO_2 ↔ CH_3 H NO_2 ↔ CH_3 H NO_2]

para → [CH_3 H NO_2 ↔ CH_3 H NO_2 ↔ CH_3 H NO_2] (4.29)

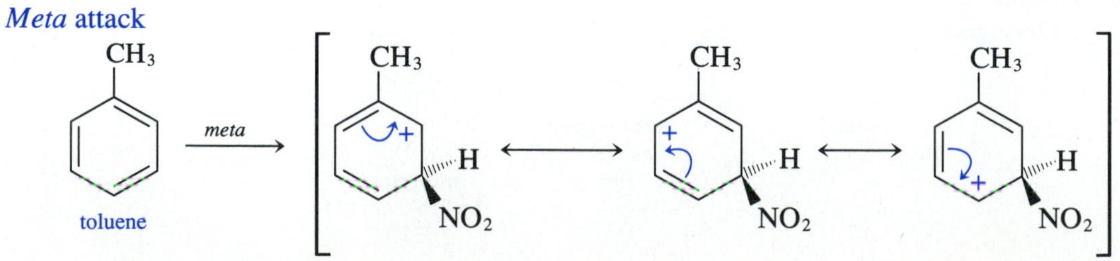

Meta attack

CH_3 toluene — *meta* → [CH_3 H NO_2 ↔ CH_3 H NO_2 ↔ CH_3 H NO_2] (4.30)

TABLE 4.1 Directing effects of common functional groups

Ortho,Para-Directing	*Meta-Directing*
$-CH_3$, $-CH_2CH_3$ (alkyl, $-R$)	$-\overset{+}{N}(=O)-O^-$, $-SO_2-OH$ ($-S(=O)(=O)-OH$)
$-F$, $-Cl$, $-Br$, $-I$	
$-OH$, $-OCH_3$, $-OR$	
$-NH_2$, $-NHR$, $-NR_2$	$-C(=O)-R$, $-C(=O)-OH$, $-C(=O)-OR$
	$-C{\equiv}N$

In one of the three resonance contributors to the benzenonium ion intermediate for *ortho* or *para* substitution (shown in dashed boxes), the positive charge is on the methyl-bearing carbon. That contributor is a *tertiary* carbocation and more stable than the other contributors, which are secondary carbocations. However, with *meta* attack, *all* the contributors are secondary carbocations; the positive charge in the intermediate benzenonium ion is never adjacent to the methyl substituent. Therefore the methyl group is *ortho,para*-directing, so that the reaction can proceed via the most stable carbocation intermediate.

Similarly, all other alkyl groups (top left line in Table 4.1) are *ortho,para*-directing.

Consider now the other *ortho,para*-directing groups listed in Table 4.1. ***In each of them, the atom attached to the aromatic ring has an unshared electron pair.***

$$-\ddot{\underset{\cdot\cdot}{F}}: \qquad -\ddot{\underset{\cdot\cdot}{O}}H \qquad -\ddot{N}H_2$$

This unshared electron pair can stabilize an adjacent positive charge. Let us consider, as an example, the bromination of phenol.

Ortho, para attack

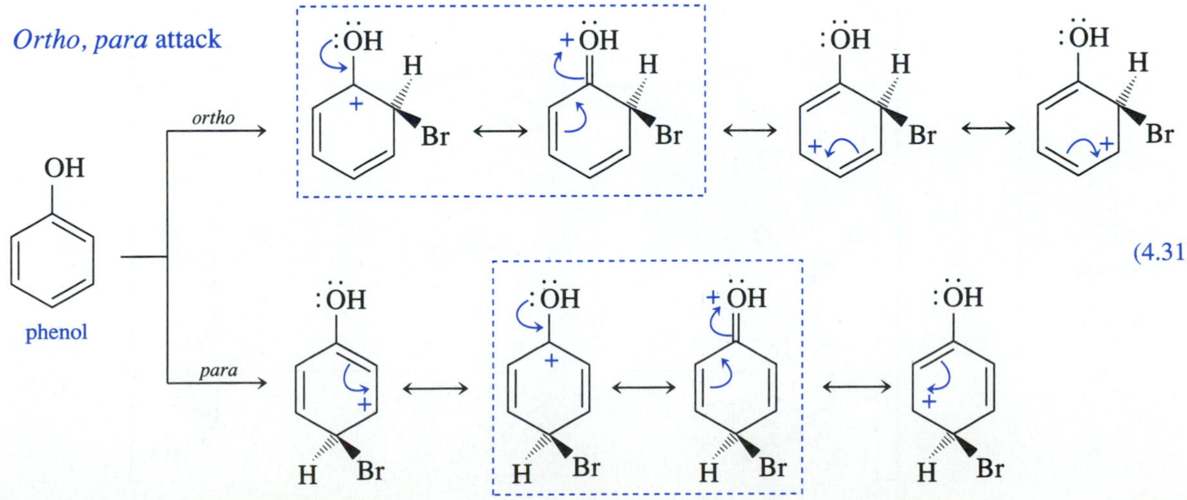

(4.31)

Meta attack

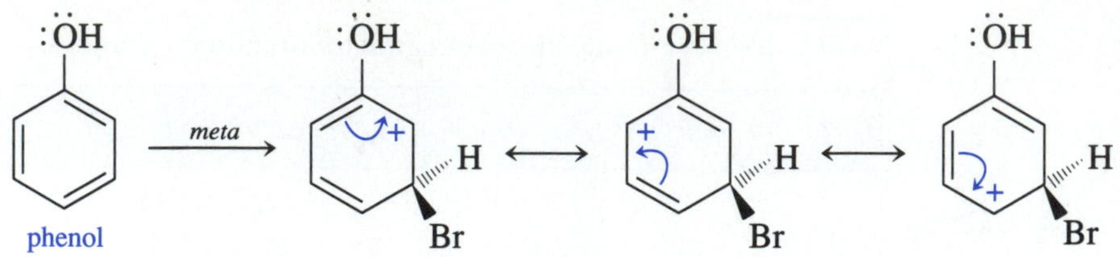

:ÖH ... phenol —*meta*→ (resonance structures with H and Br) (4.32)

In the case of *ortho* or *para* attack, one of the contributors to the intermediate benzenonium ion places the positive charge on the hydroxyl-bearing carbon. *Shift of an unshared electron pair from the oxygen to the positive carbon allows the positive charge to be delocalized onto the oxygen* (see the structures in the dashed boxes). No such structures are possible for *meta* attack. Therefore, the hydroxyl group is *ortho,para*-directing.

We can generalize this observation. ***All groups with unshared electrons on the atom attached to the ring are ortho,para-directing.***

Problem 4.12 Draw the important resonance contributors for the intermediate in the bromination of aniline, and explain why *ortho,para* substitution predominates.

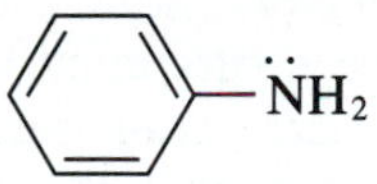

$\ddot{N}H_2$

aniline

4.12b *Meta*-Directing Groups Now let us examine the nitration of nitrobenzene in the same way, to see if we can explain the *meta*-directing effect of the nitro group. In nitrobenzene, the nitrogen has a formal charge of +1, as shown on the structures. The equations for forming the intermediate benzenonium ion are

Ortho, para attack

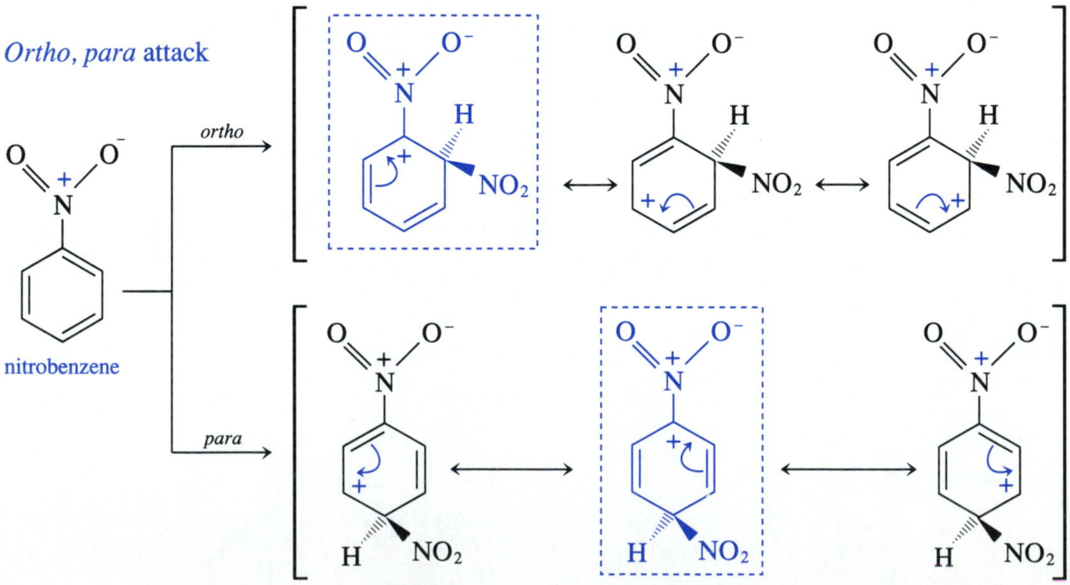

nitrobenzene —*ortho*→ [resonance structures with H and NO_2] ; —*para*→ [resonance structures with H and NO_2] (4.33)

Meta attack

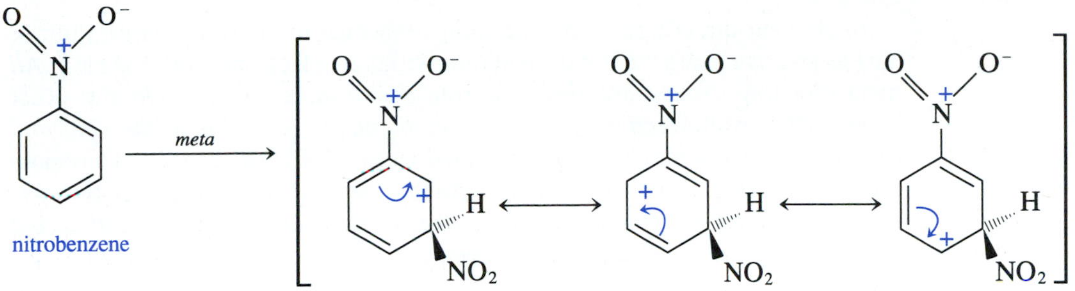

In eq. 4.33, one of the contributors to the resonance hybrid intermediate for *ortho* or *para* substitution (shown in the boxes) has *two adjacent positive charges*, a highly *undesirable* arrangement, because like charges repel each other. No such intermediate is present for *meta* substitution (eq. 4.34). For this reason, *meta* substitution is preferred.

Can we generalize this explanation to the other *meta*-directing groups in Table 4.1? Notice that each *meta*-directing group is connected to the aromatic ring by an atom that is part of a double or triple bond, at the other end of which is an atom more electronegative than carbon (for example, an oxygen or nitrogen atom). In such cases, ***the atom directly attached to the benzene ring will carry a partial positive charge*** (like the nitrogen in the nitro group). This is because of resonance contributors, such as

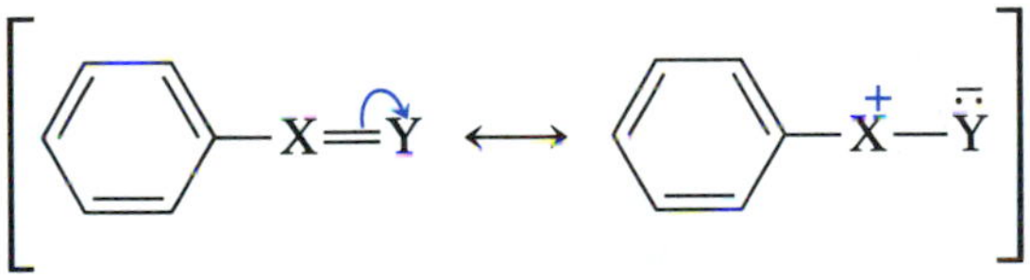

Y is an electron-withdrawing atom such as oxygen or nitrogen; atom X carries a positive charge in one of the resonance contributors.

All such groups will be *meta*-directing for the same reason that the nitro group is *meta*-directing: to avoid having two adjacent positive charges in the intermediate benzenonium ion. We can generalize. **All groups in which the atom directly attached to the aromatic ring is positively charged or is part of a multiple bond to a more electronegative element will be *meta*-directing.**

Problem 4.13 Compare the intermediate benzenonium ions for *ortho,meta* and *para* bromination of benzoic acid, and explain why the main product is *m*-bromobenzoic acid.

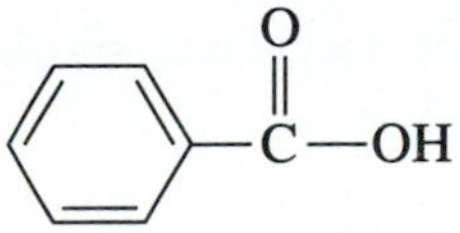

benzoic acid

4.12c Substituent Effects on Reactivity Substituents not only affect the position of substitution, they also affect the *rate* of substitution, whether it will occur slower or faster than for benzene. Is this rate effect related to the orientation effect?

In all *meta*-directing groups, the atom connected to the ring carries a full or partial positive charge and will therefore withdraw electrons from the ring. *All* meta-*directing groups are therefore ring-deactivating groups.* On the other hand, ortho,para-*directing groups in general supply electrons to the ring and are therefore ring-activating.* With the halogens (F, Cl, Br, and I), two opposing effects bring about the only important exception to these rules. *Because they are strongly electron withdrawing, the halogens are ring-deactivating; but because they have unshared electron pairs, they are* ortho,para-*directing.*

4.13 The Importance of Directing Effects in Synthesis

When designing a multistep synthesis involving electrophilic aromatic substitution, we must keep in mind the directing and activating effects of the groups involved. Consider, for example, the bromination and nitration of benzene to make bromonitrobenzene. If we brominate first and then nitrate, we will get a mixture of the *ortho* and *para* isomers.

$$\text{benzene} \xrightarrow[\text{FeBr}_3]{\text{Br}_2} \text{bromobenzene} \xrightarrow[\text{H}_2\text{SO}_4]{\text{HONO}_2} \underset{\text{ortho}}{o\text{-}BrC_6H_4NO_2} + \underset{\text{para}}{p\text{-}BrC_6H_4NO_2} \quad (4.35)$$

This is because the bromine atom in bromobenzene is *ortho,para*-directing. On the other hand, if we nitrate first and then brominate, we will get mainly the *meta isomer* because the nitro group is *meta*-directing.

$$\text{benzene} \xrightarrow[\text{H}_2\text{SO}_4]{\text{HONO}_2} \text{nitrobenzene} \xrightarrow[\text{FeBr}_3]{\text{Br}_2} \underset{\text{meta}}{m\text{-}BrC_6H_4NO_2} \quad (4.36)$$

The sequence in which we carry out the reactions of bromination and nitration is therefore very important. It determines which type of product is formed.

Problem 4.14 Devise a synthesis for each of the following, starting with benzene:

a. *m*-chlorobenzenesulfonic acid
b. *p*-nitrotoluene

Problem 4.15 Explain why it is *not* possible to prepare *m*-bromochlorobenzene or *p*-nitrobenzenesulfonic acid by carrying out two successive electrophilic aromatic substitutions.

A WORD ABOUT . . .

6. Benzene, Chemical of Commerce

When one speaks of benzene, the prototypical aromatic compound, one is speaking of big business indeed. Approximately 10 billion pounds are produced annually in the United States alone. Benzene leads the other aromatic hydrocarbons in production, though ethylbenzene, styrene, xylene, toluene, and isopropylbenzene are all among the top 50 commercial organic chemicals.

Most benzene is produced by catalytic reforming of alkanes and cycloalkanes (page 103) or by cracking certain gasoline fractions. Only about 8% of it comes from coal tar, though this was once its major source.

What is all this benzene used for? More than half is used to make styrene, by the sequence

$$\text{benzene} + CH_2{=}CH_2 \xrightarrow[\text{silica, } 300^\circ C, \text{ pressure}]{H_3PO_4} \text{ethylbenzene } (C_6H_5CH_2CH_3) \xrightarrow{\text{Fe catalyst, } 650^\circ C} \text{styrene } (C_6H_5CH{=}CH_2) + H_2$$

The first step is a Friedel-Crafts alkylation (eq. 4.14), and the second is a catalytic dehydrogenation. The styrene is converted to polymers and to synthetic rubber (Chapter 14).

Over 20% of benzene is used to make phenol and acetone. Again, the first step is a Friedel-Crafts reaction, this time with propene. Air oxidation of the resulting cumene gives a hydroperoxide, which is decomposed by acid to acetone and phenol. Both of these products have commercial uses.

$$\text{benzene} + CH_3CH{=}CH_2 \text{ (propylene)} \xrightarrow[H_3PO_4]{AlCl_3 \text{ or}} \text{cumene (isopropylbenzene)} (C_6H_5CH(CH_3)_2)$$

$$\xrightarrow{O_2} \text{cumene hydroperoxide } (C_6H_5C(CH_3)_2{-}O{-}OH) \xrightarrow[80^\circ]{\text{dil. } H_2SO_4} CH_3\overset{O}{\overset{\|}{C}}CH_3 \text{ (acetone)} + C_6H_5OH \text{ (phenol)}$$

Large quantities of benzene (over 13% of the total) are hydrogenated catalytically to cyclohexane. Cyclohexane is a major raw material for the manufacture of nylon.

$$\text{benzene} \xrightarrow[\text{catalyst}]{3\,H_2} \text{cyclohexane}$$

Together, these three commercial processes—production of styrene, phenol, and cyclohexane—consume nearly 90% of commercial benzene. The rest is used to make aniline, chlorobenzene, and a variety of other industrial chemicals. At present, all of these important materials come almost entirely from petroleum, further evidence of how dependent we are on that natural resource.

4.14 Polycyclic Aromatic Hydrocarbons

The concept of **aromaticity**—*the unusual stability of certain fully conjugated* ***cyclic*** *systems*—can be extended well beyond benzene itself or simple substituted benzenes.

Coke, required in huge quantities for the manufacture of steel, is obtained by heating coal in the absence of air. A by-product of this conversion of coal to coke is a distillate called **coal tar,** a complex mixture containing many aromatic hydrocarbons (including benzene, toluene, and xylenes). **Naphthalene,** $C_{10}H_8$, was the *first* pure compound to be obtained from the higher-boiling fractions of coal tar. It was easily isolated because it sublimes from the tar as a beautiful colorless crystalline solid, mp 80°C. *Naphthalene is a planar molecule with two fused benzene rings.* The two rings share two carbon atoms.

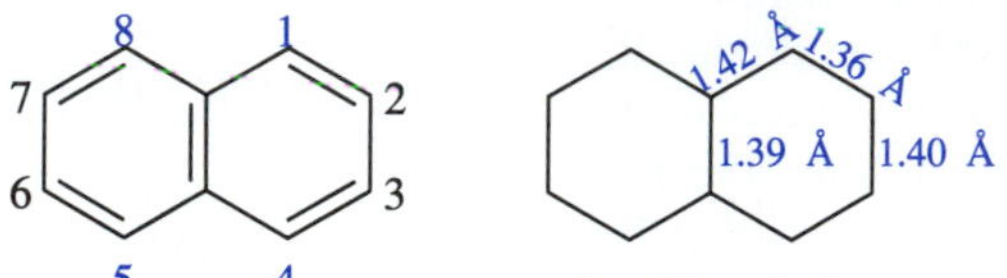

naphthalene
mp 80°C

bond lengths in naphthalene

The bond lengths in naphthalene are not all identical, but they all approximate the bond length in benzene (1.39 Å). Although it has two six-membered rings, naphthalene has a resonance energy somewhat less than twice that of benzene, about 60 kcal/mol. Because of its symmetry, naphthalene has two sets of equivalent carbon atoms: C-1, C-4, C-5, and C-8; and C-2, C-3, C-6, and C-7. Like benzene, naphthalene undergoes electrophilic substitution reactions (halogenation, nitration, and so on), usually under somewhat milder conditions than benzene. Although two monosubstitution products are possible, substitution at C-1 usually predominates.

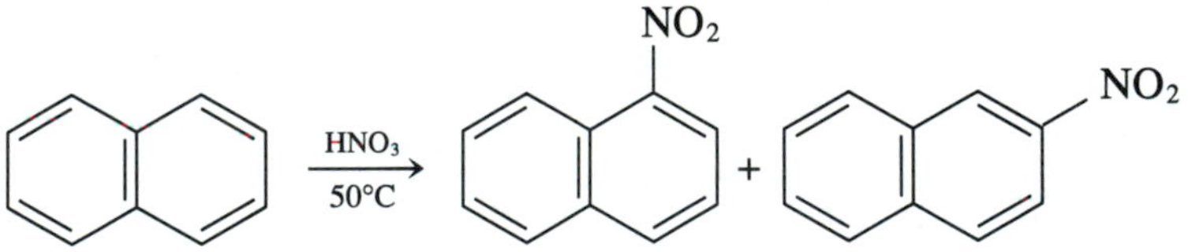

$$\text{naphthalene} \xrightarrow[50°C]{HNO_3} \text{1-nitronaphthalene} + \text{2-nitronaphthalene} \quad (4.37)$$

1-nitronaphthalene 2-nitronaphthalene
(ratio 10:1)

EXAMPLE 4.3 Draw the resonance contributors for the carbocation intermediate in nitration of naphthalene at C-1; include only structures that retain benzenoid aromaticity in the unsubstituted ring.

Solution Four such contributors are possible.

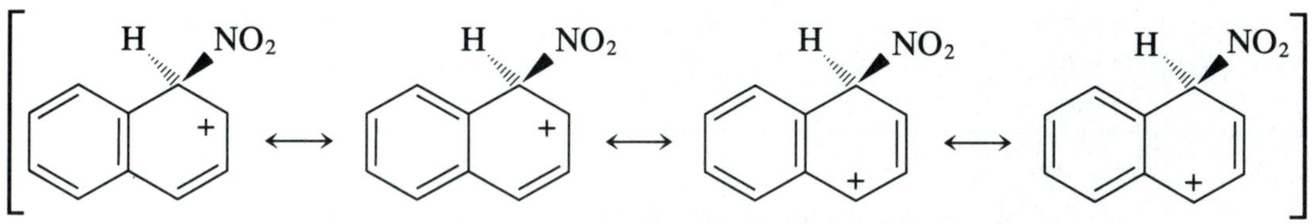

Problem 4.16 Repeat Example 4.3 for nitration at C-2. Can you suggest why substitution at C-1 is preferred?

Naphthalene is the parent compound of a series of **fused polycyclic hydrocarbons,** a few other examples of which are

anthracene
mp 217°C

phenanthrene
mp 98°C

pyrene
mp 156°C

Infinite extension of such rings leads to sheets of hexagonally arranged carbons, the structure of graphite (a form of elemental carbon).

Problem 4.17 Calculate the ratio of carbons to hydrogens in benzene, naphthalene, anthracene, and pyrene. Notice that as the number of fused rings increases, the proportion of carbon also increases, so that extrapolation to a very large number of such rings leads to graphite. (see "A Word About" on page 130).

A WORD ABOUT . . .

7. Polycyclic Aromatic Hydrocarbons and Cancer

Certain polycyclic aromatic hydrocarbons are carcinogenic (that is, they produce cancers). They can produce a tumor on mice in a short time when only trace amounts are painted on the skin. These carcinogenic hydrocarbons are present not only in coal tar but also in soot and tobacco smoke and can be formed in barbecuing meat. Their biological effect was noted as long ago as 1775 when soot was identified as the cause of the high incidence of scrotal cancer in chimney sweeps. A similar occurrence of lung and lip cancer is common in habitual smokers.

The way these carcinogens produce cancer is now fairly well understood. To eliminate hydrocarbons, the body usually oxidizes them to render them more water soluble, so that they can be excreted. The metabolic oxidation products seem to be the real culprits in causing cancer. For example, one of the most potent carcinogens of this type is benzo[a]pyrene. Enzymatic oxidation converts it to the diol-epoxide shown.

benzo[a]pyrene

enzyme →

O

HO

OH

a diol-epoxide

The diol-epoxide reacts with cellular DNA, causing mutations that eventually prevent the cells from reproducing normally.

Benzene itself is quite toxic to humans and can cause severe liver damage, but toluene is much less toxic. How can this different behavior of two very similar compounds be possible? To eliminate benzene from the body, the aromatic ring must be oxidized, and intermediates in this oxidation are damaging. However, the *methyl side chain* of toluene can be oxidized to give benzoic acid, which can be excreted. None of the intermediates in this process causes problems.

Although some chemicals may cause cancer, others can help prevent or cure it. Many substances inhibit cancer growth, and the study of cancer chemotherapy has contributed substantially to human health.

A WORD ABOUT ...

8. Graphite, Carbon Clusters, and Aromaticity

Graphite, one of the two best-known allotropic forms of carbon (the other is diamond), is the ultimate in fused aromatic rings. The carbon atoms in graphite are arranged in layers that consist of planar, hexagonal rings. Each carbon atom (except for those at the outer edges of each layer) is bonded to three other carbon atoms in the same layer by bonds that approximate the bond length in benzene (1.42 Å in graphite versus 1.39 Å in benzene). There are no covalent bonds between carbon atoms in different layers, and these layers are 3.4 Å apart. This distance, then, can be taken as the effective thickness of an aromatic ring; that is, the thickness of the ring carbons *and* the pi cloud of electrons above and below it.

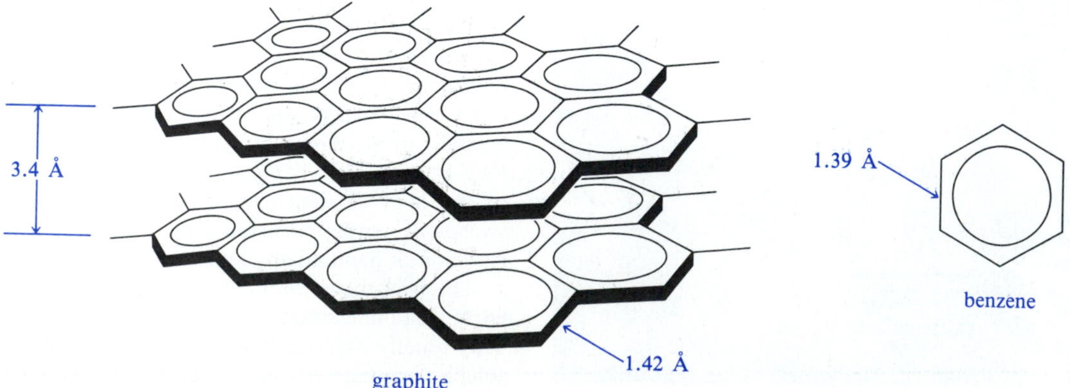

graphite

benzene

Only relatively weak forces hold the layers in graphite together. The lubricating properties of graphite (and its greasy feel) are thought to result from "slipping" of the layers with respect to each other. Graphite can also absorb or even chemically bind small molecules between its layers. For example, potassium metal can be intercalated in graphite by adding the metal to graphite that is heated above the melting point of potassium (62.3°C) in an inert atmosphere, such as argon. When the graphite-potassium ratio corresponds to C_8K at one point, the graphite rather suddenly changes from black to a lustrous golden-yellow—a beautiful and spectacular sight. This form of potassium, which is highly reactive, will spontaneously burst into flames if exposed to air.

In the mid-1980s, a remarkable discovery was made about graphite. Its laser vaporization under special conditions gave rise to even-numbered large clusters of carbon atoms, mainly C_{60} but also others. In a rather bold hypothesis, it was proposed that C_{60} has each of its carbon atoms at the vertex of a truncated icosahedron, a structure also used for soccer balls (Figure 4.3). This polygon has 60 vertices and 32 faces, 12 of which are *pentagons* and 20 of which are hexagons. The C_{60} cluster was trivially named "buckminsterfullerene" after Buckminster Fuller, who used these shapes for the construction of geodesic domes. Similar clusters with different sizes are now called fullerenes.

Each carbon atom in such a cluster is sp^2 hybridized (attached to three other carbons). At each carbon there is an electron in a p-orbital that is perpendicular to the spherical surface. Overlap of these p-orbitals results in an aromatic pi cloud inside and outside the spherical surface. Thus, the cluster is fully aromatic and exceptionally stable.

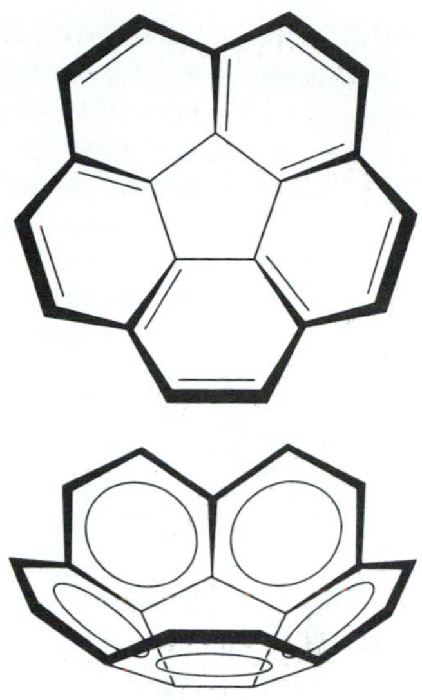

corannulene, a saucer-shaped aromatic molecule

Although a structure of regular hexagons as in graphite is necessarily flat, the incorporation of pentagons allows the structure to curve. For example, the aromatic hydrocarbon **corannulene,** with five benzene rings around a pentagon, is saucer-shaped, not flat. Incorporation of 12 pentagons allows the structure to close on itself to produce a sphere. Ejection of carbon atoms (through laser irradiation) from the flat layers of graphite leads to curvature of the layers and eventually to closed, spherelike structures. For more on fullerenes, their role in soot formation, and the presence of similar structures in cosmic space, see articles by R. F. Curl and R. E. Smalley, *Science* **1988,** *242,* 1017–22, and by H. Kroto, *Science* **1988,** *242,* 1139–45.

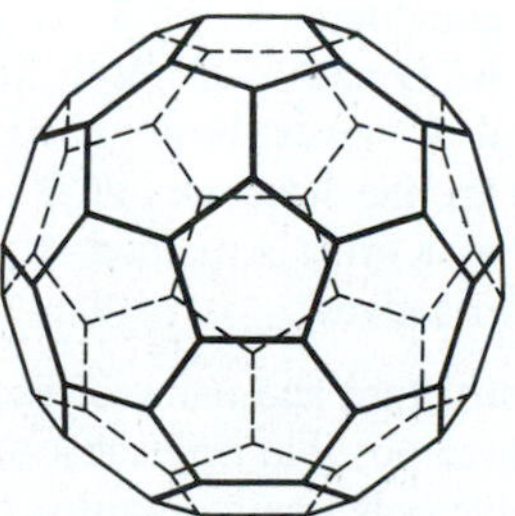

FIGURE 4.3 The structure of "buckminsterfullerene," a C_{60} aromatic carbon cluster.

Additional Problems

4.18. Write structural formulas for the following compounds:

a. 1,3,5-tribromobenzene
b. *m*-chlorotoluene
c. *o*-diethylbenzene
d. isopropylbenzene
e. benzyl bromide
f. *p*-chlorophenol
g. 2,3-diphenylbutane
h. *p*-bromostyrene
i. 2-chloro-4-ethyl-3,5-dinitrotoluene
j. *m*-chlorobenzenesulfonic acid
k. *p*-bromobenzoic acid
l. 2,4,6-trimethylaniline
m. *m*-nitroanisole
n. *p*-fluoroacetophenone
o. 3,4-dimethylbenzaldehyde
p. *m*-chlorobenzoic acid

4.19. Name the following compounds:

a. 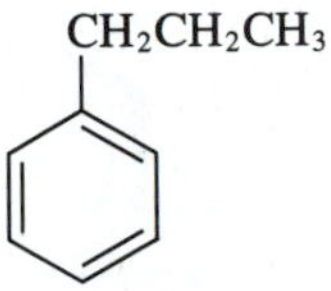 $CH_2CH_2CH_3$

b. 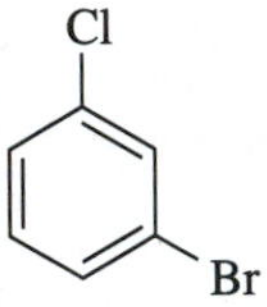 Cl, Br

c. 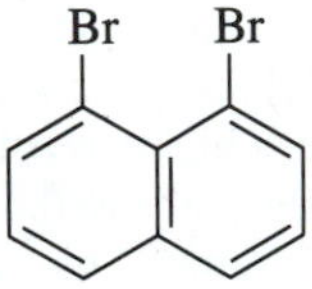 Br Br

d. 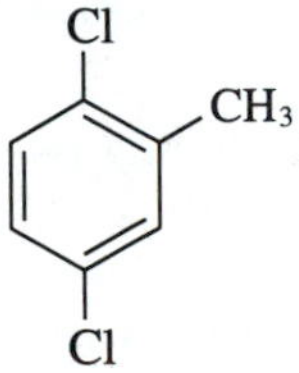 Cl, CH_3, Cl

e. 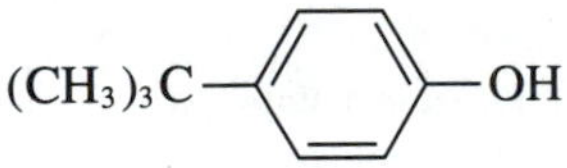 $(CH_3)_3C$—, —OH

f. 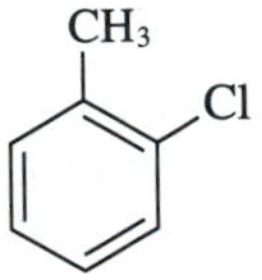 CH_3, Cl

g. 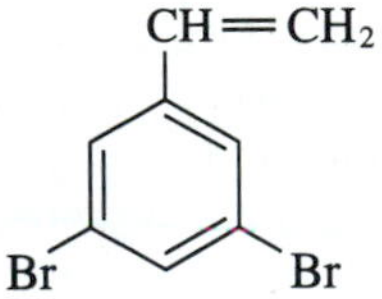 $CH{=}CH_2$, Br, Br

h. 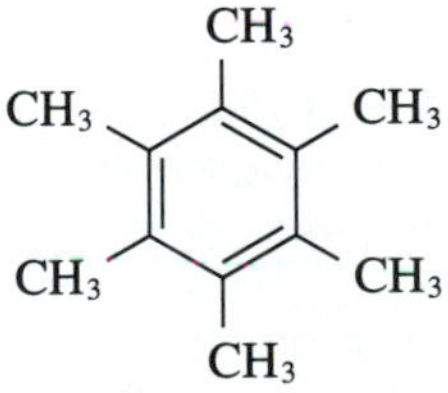 CH_3, CH_3, CH_3, CH_3, CH_3, CH_3

i. 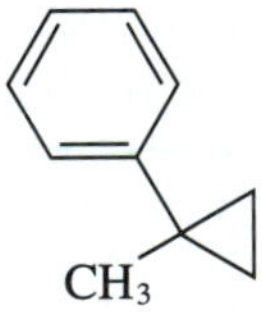 CH_3

j. Ph_3CH

4.20. Give the structures and names for all possible
a. trimethylbenzenes. **b.** dichloronitrobenzenes.

4.21. There are three dibromobenzenes (*o*-, *m*-, and *p*-). Suppose we have samples of each in separate bottles, but we don't know which is which. Let us call them A, B, and C. On nitration, compound A (mp 87°C) gives only *one* nitrodibromobenzene. What is the structure of A? B and C are both liquids. On nitration, B gives *two* nitrodibromobenzenes, and C gives *three* nitrodibromobenzenes (of course, not in equal amounts). What are the structures of B and C? of their mononitration products? (This method, known as Körner's method, was used years ago to assign structures to isomeric benzene derivatives).

4.22. Give the structure and name of each of the following aromatic hydrocarbons:
a. C_8H_{10}; has three possible ring-substituted monobromo derivatives
b. C_9H_{12}; can give only one mononitro product on nitration
c. C_9H_{12}; can give four mononitro derivatives on nitration

4.23. The observed amount of heat evolved when 1,3,5,7-cyclooctatetraene is hydrogenated is 110 kcal/mol. What does this tell you about the possible resonance energy of this compound?

4.24. The structure of the nitro group $-NO_2$ is usually shown as

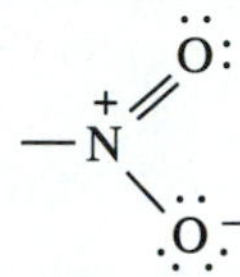

yet experiments show that the two nitrogen–oxygen bonds have the same length of 1.21 Å. This length is intermediate between 1.36 Å for the N—O single bond and 1.18 Å for the N=O double bond. Draw structural formulas that explain this observation.

4.25. Draw all reasonable electron-dot formulas for the nitronium ion, $(NO_2)^+$, the electrophile in aromatic nitrations. Show any formal charges. Which structure is favored, and why?

4.26. Write out all steps in the mechanism for the reaction of
a. *p*-xylene + nitric acid (H_2SO_4 catalyst).
b. benzene + *t*-butyl chloride + $AlCl_3$.

4.27. Draw all possible contributing structures to the carbocation intermediate in the chlorination of chlorobenzene. Explain why the major products are *o*- and *p*-dichlorobenzene. (*Note: p*-Dichlorobenzene is produced commercially this way, for use against clothes moths.)

4.28. Repeat Problem 4.27 for the chlorination of benzenesulfonic acid, and explain why the product is *m*-chlorobenzenesulfonic acid.

4.29. Indicate the main *mono*substitution products in each of the following reactions. Keep in mind that certain substituents are *meta*-directing and others are *ortho,para*-directing.
a. toluene + chlorine (Fe catalyst)
b. nitrobenzene + concentrated sulfuric acid (heat)
c. bromobenzene + chlorine (Fe catalyst)
d. chlorobenzene + bromine (Fe catalyst)
e. benzenesulfonic acid + concentrated nitric acid (heat)
f. ethylbenzene + bromine (Fe catalyst)
g. iodobenzene + bromine (Fe catalyst)
h. toluene + acetyl chloride ($AlCl_3$ catalyst)

4.30. Suggest a reason why $FeCl_3$ is used as a catalyst for aromatic chlorinations and $FeBr_3$, for brominations (that is, why the iron halide used has the same halogen as the halogenating agent).

4.31. Using benzene or toluene as the only aromatic organic starting material, devise a synthesis for each of the following:
a. *m*-bromonitrobenzene
b. *p*-toluenesulfonic acid
c. *p*-nitroethylbenzene
d. methylcyclohexane
e. 2,6-dibromo-4-nitrotoluene
f. *p*-bromonitrobenzene
g. 2-chloro-4-nitrotoluene
h. 3,5-dinitrochlorobenzene

4.32. When benzene is treated with excess D_2SO_4 at room temperature, the hydrogens on the benzene ring are gradually replaced by deuterium. Write a mechanism that explains this observation.

4.33. Predict whether the following substituents on the benzene ring are likely to be *ortho,para*-directing or *meta*-directing and whether they are likely to be ring-activating or ring-deactivating:

a. $-SCH_3$ **b.** $-\overset{+}{N}(CH_3)_3$ **c.** $-O-\overset{\overset{O}{\|}}{C}-CH_3$ **d.** $-\overset{\overset{O}{\|}}{C}-OCH_3$

4.34. The explosive TNT (2,4,6-trinitrotoluene) can be made by nitrating toluene with a mixture of nitric and sulfuric acids, but the reaction conditions must gradually be made more severe as the nitration proceeds. Explain why.

4.35. Which compound is more reactive toward electrophilic substitution (for example, nitration)?
a. anisole or benzoic acid **b.** bromobenzene or toluene

4.36. For a one-step synthesis of 3-bromo-5-nitrobenzoic acid, which is the better starting material, 3-bromobenzoic acid or 3-nitrobenzoic acid? Why?

4.37. Show how pure 3,5-dinitrobromobenzene can be prepared, starting from a disubstituted benzene.

4.38. How many possible monosubstitution products are there for each of the following?
a. anthracene **b.** phenanthrene

4.39. Nitration of anthracene gives mainly 9-nitroanthracene. Write out the steps in the mechanism of this reaction.

4.40. Draw a molecular orbital picture for the resonance hybrid benzenonium ion shown in eq. 4.19, and describe the hybridization of each ring carbon atom.

CHAPTER 5

Stereoisomerism

5.1 *Introduction*

Stereoisomers *have the same atom connectivities, or order of attachment of the atoms, but different arrangements of the atoms in space.* We have already seen that stereoisomers may be characterized according to the ease with which they can be interconverted. That is, they may be **conformational isomers,** which can be interconverted by rotation about a single bond, or they may be **configurational isomers,** which can be interconverted only by breaking and remaking covalent bonds.

Here we will consider other useful ways to categorize stereoisomers, ways that are particularly helpful in describing their properties.

5.2 *Chirality and Enantiomers*

Consider the difference between a pair of gloves and a pair of socks. A sock, like its partner, can be worn on either the left or the right foot. But a left-hand glove, unlike its partner, cannot be worn on the right hand. Like a pair of gloves, certain molecules possess this property of "handedness," which affects their chemical behavior. Let us examine the idea of molecular handedness.

A molecule (or object) is either **chiral** or **achiral.** The word *chiral,* pronounced "kai-ral" to rhyme with spiral, comes from the Greek χειρ (*cheir,* hand.) ***A chiral molecule (or object) is one that exhibits the property of handedness.*** An **achiral** molecule does not have this property.

What test can we apply to tell whether a molecule (or object) is chiral or achiral? We examine the molecule (or object) *and its mirror image*. ***The mirror image of a chiral molecule is neither identical with nor superimposable on the molecule itself. The mirror image of an achiral molecule, however, is identical with or superimposable on the molecule itself.***

Let us apply this test to some specific examples. Figure 5.1 shows one of the more obvious examples. The mirror image of a left hand is not another left hand, but a right hand. A hand and its mirror image are not superimposable. A hand is chiral. But the mirror image of a ball (sphere) is also a ball (sphere). A ball (sphere) is achiral.

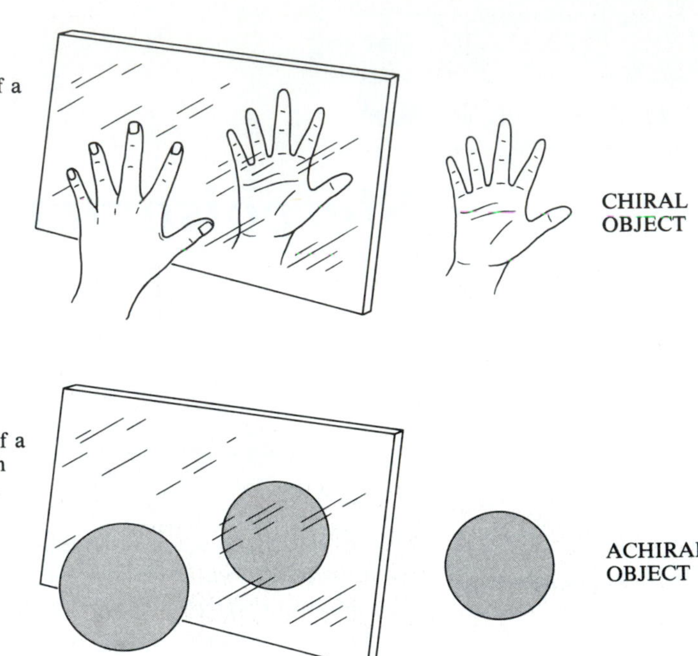
The mirror image of a left hand is not a left hand, but a right hand.
CHIRAL OBJECT
The mirror image of a ball is identical with the object itself.
ACHIRAL OBJECT

FIGURE 5.1
The mirror-image relationships of chiral and achiral objects.

Problem 5.1 Which of the following objects are chiral and which are achiral?

a. golf club b. teacup c. football d. corkscrew
e. tennis racket f. shoe g. portrait h. pencil

Now let us look at two molecules, 2-chloropropane and 2-chlorobutane, and their mirror images.

Figure 5.2 shows that 2-chloropropane is achiral. Its mirror image is superimposable on the molecule itself. Therefore 2-chloropropane has only one possible structure.

On the other hand, as Figure 5.3 shows, 2-chlorobutane has two possible structures, related to one another as nonsuperimposable mirror images. **We call a pair of molecules that are related as nonsuperimposable mirror images *enantiomers*.** Every molecule, of course, has a mirror image. Only those that are *nonsuperimposable* are called enantiomers.

5.3 Chiral Centers; the Asymmetric Carbon Atom

What is it about their structures that leads to chirality in 2-chlorobutane but not in 2-chloropropane? Notice that, in 2-chlorobutane, carbon atom 2, the one marked with an asterisk, has four different groups attached to it (Cl, H, CH_3 and CH_3CH_2). A carbon atom with four different groups attached to it is called an **asymmetric carbon atom.** This type of carbon is also called a **chiral center** because it gives rise to stereoisomers, that is, it is **stereogenic.**

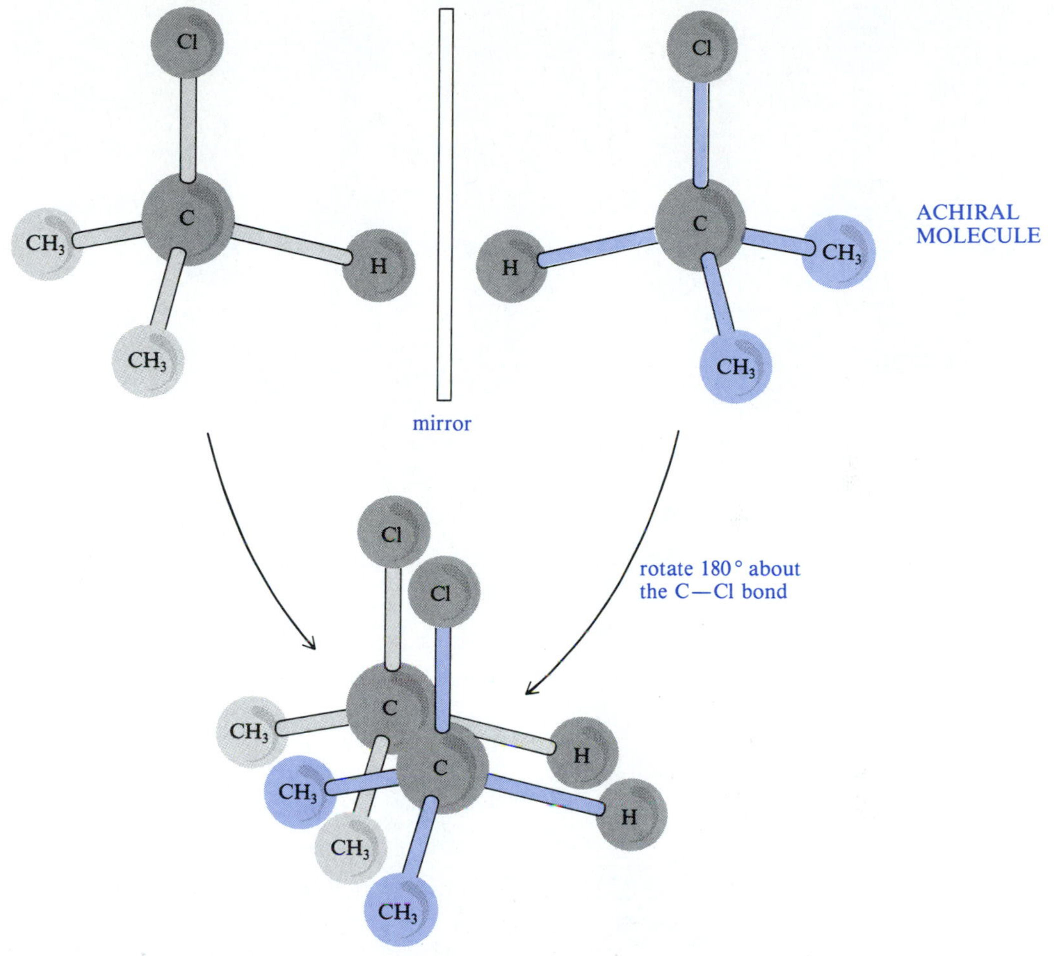

FIGURE 5.2 Model of 2-chloropropane and its mirror image. The mirror image is superimposable on the original molecule.

$$CH_3-\overset{*}{\underset{\underset{H}{|}}{\overset{\overset{Cl}{|}}{C}}}-CH_2CH_3$$

Let us examine the more general case of a carbon atom with any four different groups attached; let us call the groups A, B, D, and E. Figure 5.4 shows such a molecule and its mirror image. That the molecules on each side of the mirror in Figure 5.4 are nonsuperimposable mirror images (enantiomers) be-

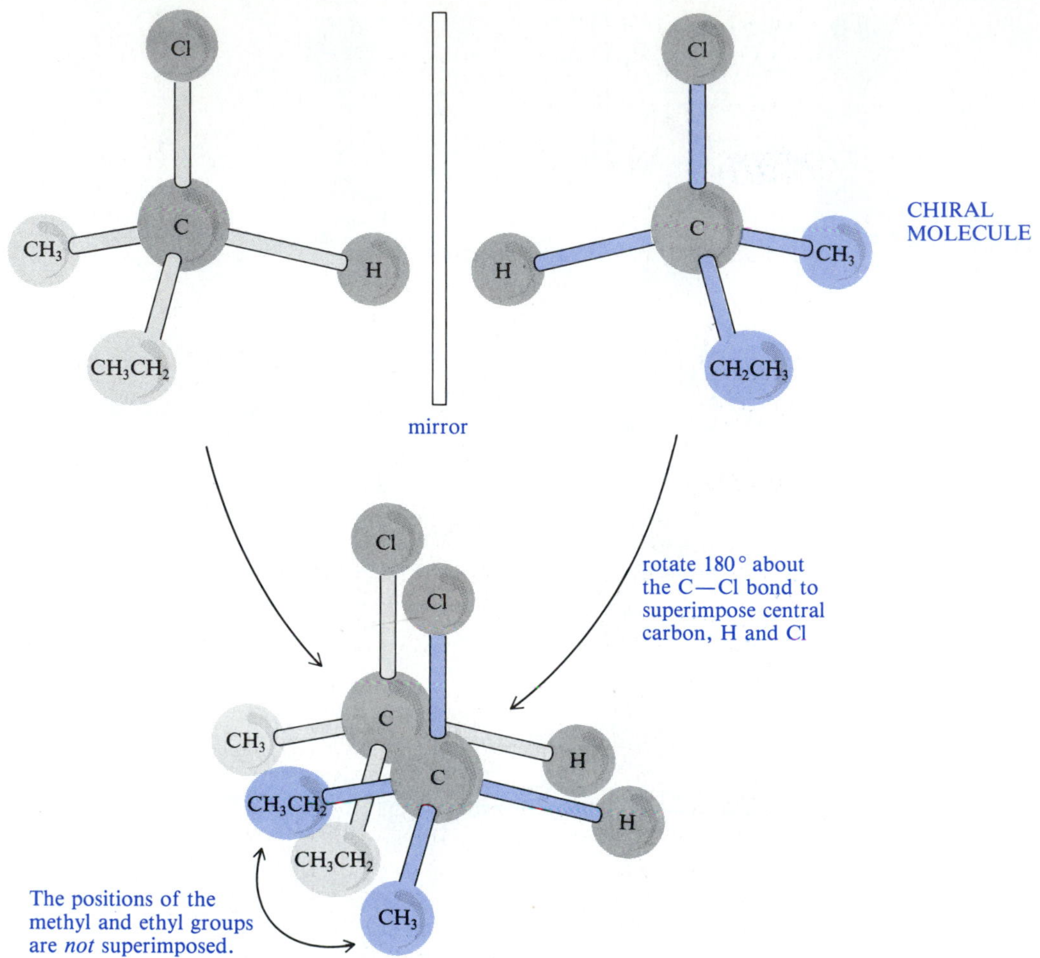

FIGURE 5.3 Model of 2-chlorobutane and its mirror image. The mirror image is *not* superimposable on the original molecule. The two forms of 2-chlorobutane are enantiomers.

comes clear by examining Figure 5.5. (We strongly urge you to use molecular models when studying this chapter. It is sometimes difficult to visualize three-dimensional structures when they are drawn on a two-dimensional surface [this page or a blackboard], though with experience, your ability to do so will improve.)

The handedness of these molecules is also illustrated in Figure 5.4, where the clockwise or counterclockwise arrangement of the groups (we might call them right- or left-handed arrangements) is apparent.

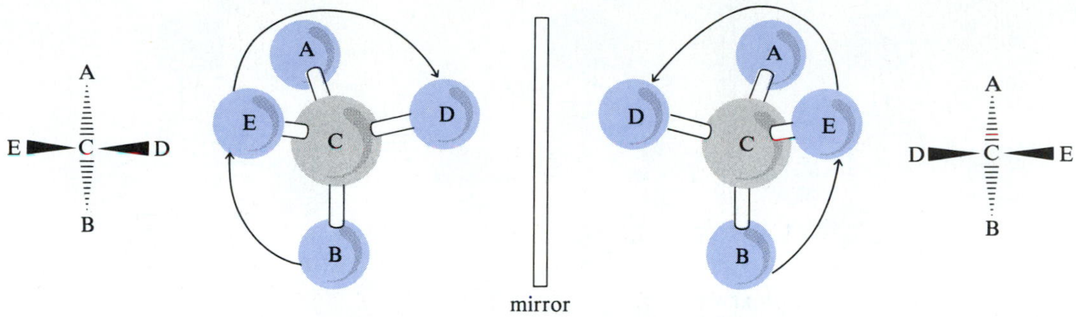

FIGURE 5.4 The chirality of enantiomers. Looking down the C—A bond, we have to read clockwise to spell BED for the model on the left, but we must read counterclockwise for its mirror image.

FIGURE 5.5
When the four different groups attached to an asymmetric carbon atom are arranged to form mirror images, the molecules are not superimposable. The models may be twisted or turned in any direction, but as long as no bonds are broken, only two of the four attached groups can be made to coincide.

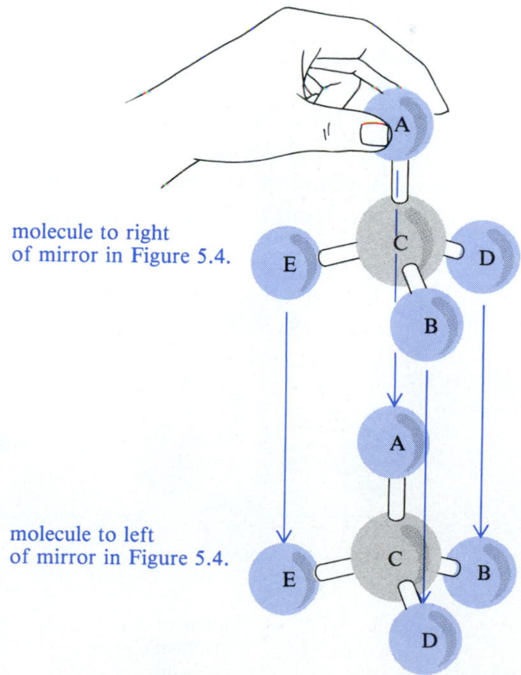

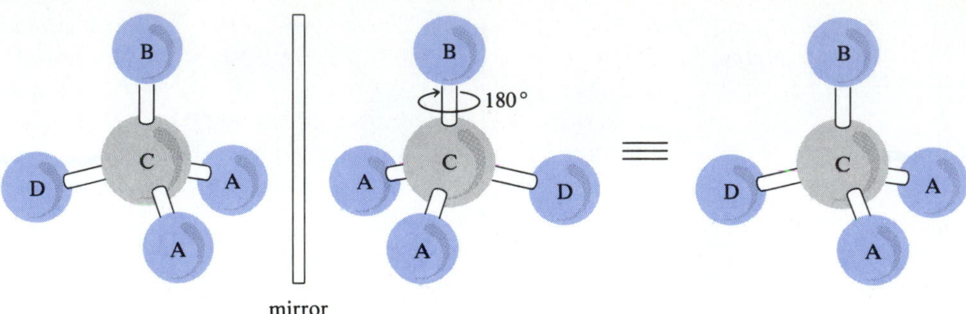

FIGURE 5.6 The tetrahedral model at the left has two corners occupied by identical groups (A). It has a plane of symmetry that passes through atoms B, C, and D and bisects angle ACA. Its mirror image is identical to itself, seen by a 180° rotation of the mirror image about the C—B bond. Hence the model is achiral.

What happens when all four of the groups attached to the central carbon atom are *not* different from one another? Suppose two of the groups are identical—say, A, A, B, and D. Figure 5.6 describes this situation. The molecule and its mirror image are now *identical,* and the molecule is achiral. This is exactly the situation with 2-chloropropane, where two of the four groups attached to carbon 2 are identical (CH_3, CH_3, H, and Cl).

Notice that the molecule in Figure 5.6 has a plane of symmetry. This plane passes through atoms B, C, and D and bisects the ACA angle. On the other hand, the molecule in Figure 5.4 does *not* have a symmetry plane.

A **plane of symmetry** (sometimes called a mirror plane) is a plane that passes through a molecule (or object) in such a way that what is on one side of

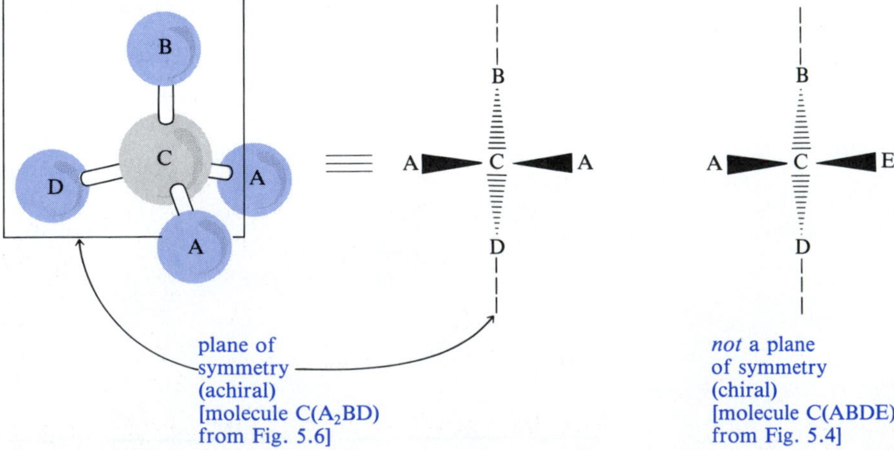

the plane is the exact reflection of what is on the other side. *Any molecule with a plane of symmetry is achiral. Chiral molecules do not have a plane of symmetry.* Seeking a plane of symmetry is usually one quick way to tell whether a molecule is chiral or achiral.

To summarize, a molecule with a chiral center (in our examples, the chiral center is a carbon atom with four different groups attached to it) can exist in two stereoisomeric forms, that is, as a pair of enantiomers. Compounds with a symmetry plane are achiral.

EXAMPLE 5.1 Locate the chiral center in 3-methylhexane.

Solution Draw the structure, and look for a carbon atom with four different groups attached.

```
1   2    3  4   5   6
CH3CH2CHCH2CH2CH3
          |
         CH3
```

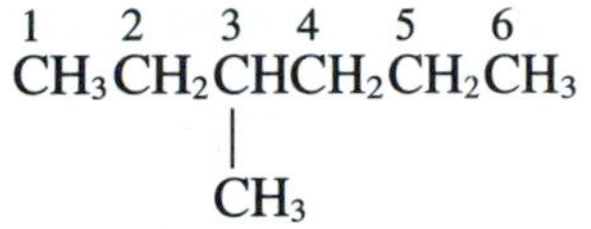

All of the carbons except carbon 3 have at least two hydrogens (two identical groups) and therefore cannot be chiral centers. But carbon 3 has four different groups attached (H, CH_3—, CH_3CH_2— and $CH_3CH_2CH_2$—) and is therefore a chiral center. By convention, we sometimes mark such centers with an asterisk.

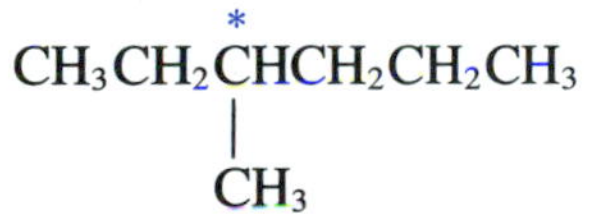

```
             *
CH3CH2CHCH2CH2CH3
          |
         CH3
```

EXAMPLE 5.2 Draw the two enantiomers of 3-methylhexane.

Solution There are many ways to do this. Here are two of them. First draw carbon 3 with four tetrahedral bonds.

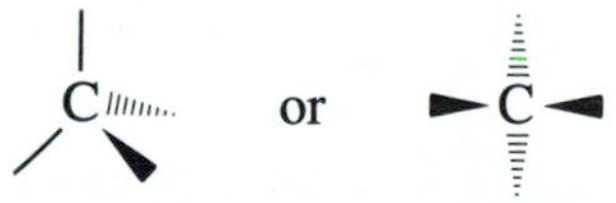

C or C

Then attach the four different groups, in any order.

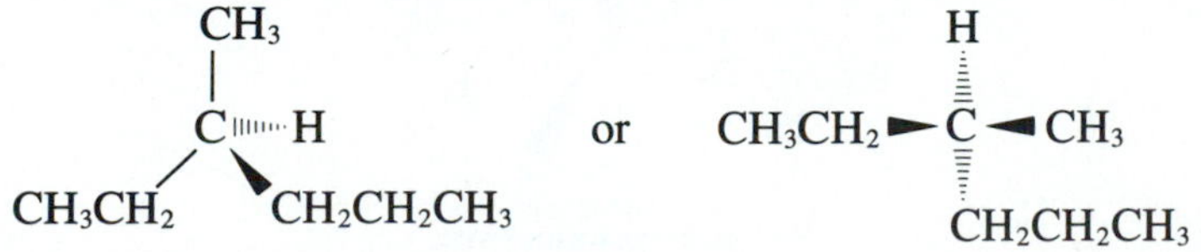

```
          CH3                                  H
          |                                    ⋮
          C····H           or     CH3CH2 ▶ C ◀ CH3
         /  \                                  ⋮
CH3CH2       CH2CH2CH3                    CH2CH2CH3
```

Now draw the mirror image, or interchange the positions of any two groups.

CH_3
H ····C
$CH_3CH_2CH_2$ CH_2CH_3

or

H
CH_3 — C — CH_2CH_3
$CH_2CH_2CH_3$

To convince yourself that the ***interchange of any two groups at a chiral center produces the enantiomer,*** work with molecular models.

Problem 5.2 Find the chiral centers in

a. 3-chlorohexane.
b. 2,3-dichlorobutane.
c. 3-methylcyclohexene.
d. 1-bromo-1-chloroethane.

Problem 5.3 Which of the following compounds is chiral?

a. 1-bromo-1-phenylethane
b. 1-bromo-2-phenylethane

Problem 5.4 Draw three-dimensional structures for the two enantiomers of the chiral compound in Problem 5.3.

Problem 5.5 Locate the planes of symmetry in the eclipsed conformation of ethane. In this conformation, is ethane chiral or achiral?

Problem 5.6 Does the staggered conformation of ethane have planes of symmetry? In this conformation, is ethane chiral or achiral? *(Careful!)*

Problem 5.7 Locate the planes of symmetry in *cis*- and *trans-1,2*-dichloroethene. Are these molecules chiral or achiral? *(Careful!)*

5.4 Configuration and the R-S Convention

Enantiomers differ in the arrangement of the groups attached to the chiral center. This arrangement of groups is called the **configuration** of the chiral center. *Enantiomers are configurational isomers; they are said to have opposite configurations.*

When referring to a particular enantiomer, we would like to be able to specify which configuration we mean without having to draw the structure. A convention for doing this is known as the *R-S* or Cahn-Ingold-Prelog (CIP)* system. Here is how it works.

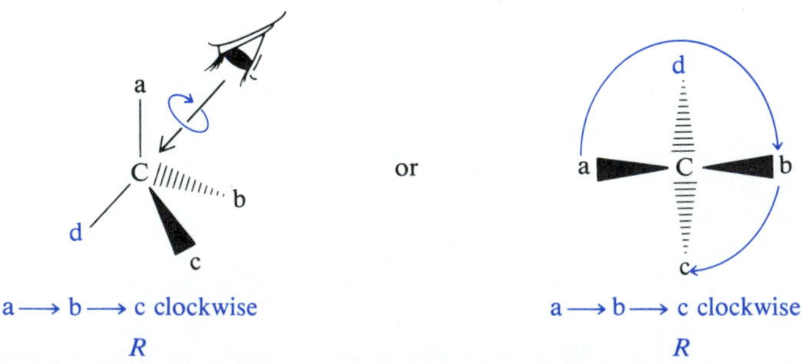

*After R. S. Cahn and C. K. Ingold, both British organic chemists, and V. Prelog, a Swiss Nobel Prize winner.

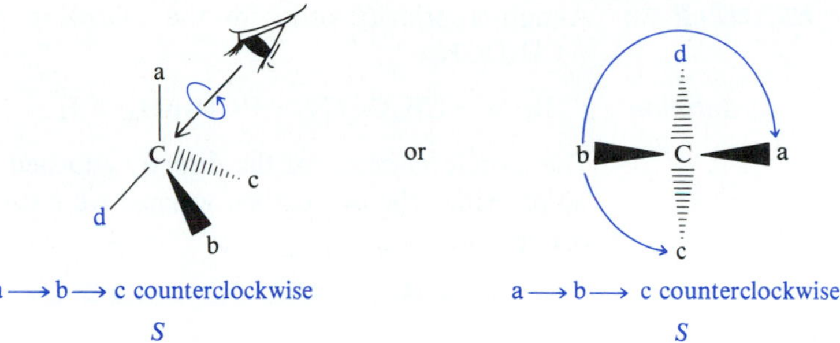

The four groups attached to the chiral center are placed in a priority order (by a system we will describe next), $a \rightarrow b \rightarrow c \rightarrow d$. The chiral center is then observed *from the side opposite the lowest-priority group, d.* If the remaining three groups ($a \rightarrow b \rightarrow c$) form a clockwise array, the configuration is designated *R* (from the Latin *rectus,* right).* If they form a *counterclockwise* array, the configuration is designated as *S* (from the Latin *sinister,* left).

The priority order of the four groups is set in the following way:

Rule 1. The atoms directly attached to the chiral center are ranked according to *atomic number:* the higher the atomic number, the higher the priority.

For example:

Cl > O > C > H

high priority ⟶ low priority

Rule 2. If a decision cannot be reached with rule 1 (that is, if two or more of the directly attached atoms are the same), work outward from the chiral center until a decision is reached. For example, the ethyl group has a higher priority than the methyl group, because at the first point of difference, working outward from the chiral center, we come to a *carbon* (higher priority) in the ethyl group and a *hydrogen* (lower priority) in the methyl group.

chiral center —CH_2—CH_3 (ethyl) > chiral center —CH_3 (methyl)

---

* More precisely, *rectus* means "right" in the sense of "correct, or proper," and not in the sense of direction (which is *dexter* = right, opposite to left). It may not be entirely coincidental that the initials of one of the inventors of this system are R. S.

EXAMPLE 5.3 Assign a priority order to the following groups: H, Br, $-CH_2CH_3$, and $-CH_2OCH_3$.

Solution $Br > -CH_2OCH_3 > -CH_2CH_3 > H$

The atomic numbers of the directly attached atoms are ordered Br > C > H. To prioritize the two carbon groups, we must go out until a point of difference is reached.

$$-CH_2OCH_3 > -CH_2CH_3 \qquad (O > C)$$

Problem 5.8 Assign a priority order to each of the following sets of groups:

a. $-CH_3$, $-C(CH_3)_3$, $-H$, $-OH$
b. $-OH$, $-OCH_3$, $-CH_3$, $-CH_2OH$
c. $-CN$, $-NHCH_3$, $-CH_2NH_2$, $-OH$
d. $-CH_3$, $-CH_2CH_3$, $-CH_2CH_2CH_3$, $-CH(CH_3)_2$

A third, somewhat more complicated, rule is required to handle double or triple bonds and aromatic rings (which are written in the Kekulé fashion).

Rule 3. Multiple bonds are treated as if they were an equal number of single bonds. For example, the vinyl group $-CH{=}CH_2$ is counted as

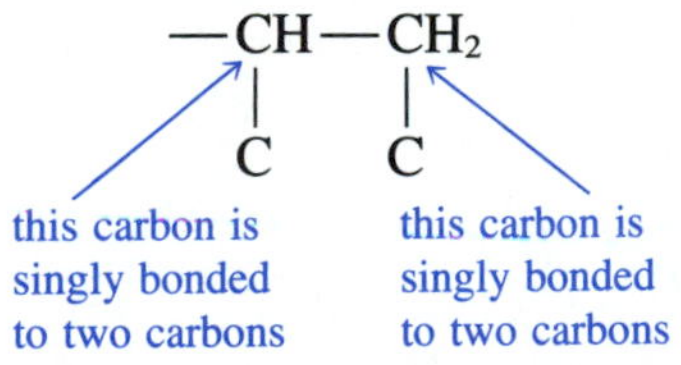

$$\begin{array}{cc} -CH- & CH_2 \\ | & | \\ C & C \end{array}$$

this carbon is singly bonded to two carbons

this carbon is singly bonded to two carbons

Similarly,

$$-C{\equiv}CH \quad \text{is treated as} \quad \begin{array}{ccccc} & C & & C & \\ & | & & | & \\ - & C & - & C & -H \\ & | & & | & \\ & C & & C & \end{array}$$

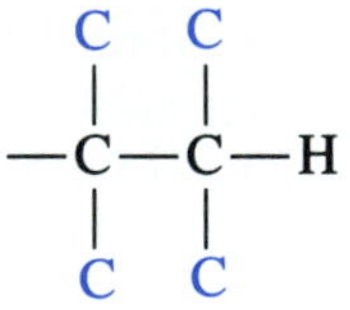

and

$$-CH{=}O \quad \text{is treated as} \quad \begin{array}{cccc} & H & & \\ & | & & \\ - & C & - & O \\ & | & & | \\ & O & & C \end{array}$$

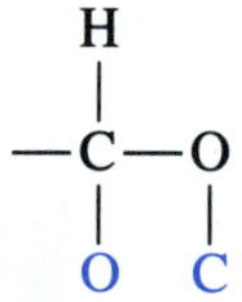

EXAMPLE 5.4 Which group has the higher priority, isopropyl or vinyl?

Solution The vinyl group has the higher priority. We go out until we reach a difference, shown in color.

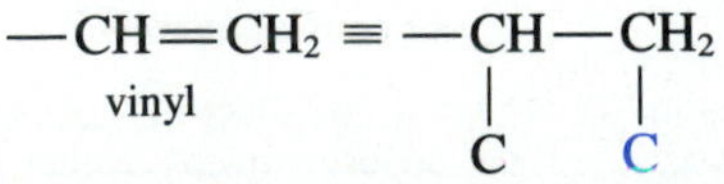

$$\underset{\text{vinyl}}{-CH{=}CH_2} \equiv \begin{array}{cc} -CH- & CH_2 \\ | & | \\ C & C \end{array}$$

$$-CH(CH_3)_2 \equiv -\underset{\large CH_3}{\underset{|}{C}H}-\underset{\large H}{\underset{|}{C}H_2}$$

isopropyl

Problem 5.9 Assign a priority order to

$-C{\equiv}CH$ and $-CH{=}CH_2$; $-CH{=}CH_2$ and —(phenyl)

$-CH{=}O$, $-CH{=}CH_2$, $-CH_2CH_3$, and $-OH$

Now let us see how these rules are applied.

EXAMPLE 5.5 Assign the configuration (*R* or *S*) to the following enantiomer of 3-methylhexane (see Example 5.2).

CH_3 (top), C with H (dashed), CH_3CH_2 (lower left), $CH_2CH_2CH_3$ (wedge, lower right)

Solution First assign the priority order to the four different groups attached to the chiral center.

$$-CH_2CH_2CH_3 > -CH_2CH_3 > -CH_3 > -H$$

Now view the molecule *from the side opposite the lowest-priority group* (—H) and determine whether the remaining three groups, from high to low priority, form a clockwise (*R*) or counterclockwise (*S*) array.

CH_3, C, H, CH_3CH_2, $CH_2CH_2CH_3$ — *R* (clockwise)

We write the name (*R*)-3-methylhexane.

If we view the other representation of this molecule shown in Example 5.2, we come to the same conclusion.

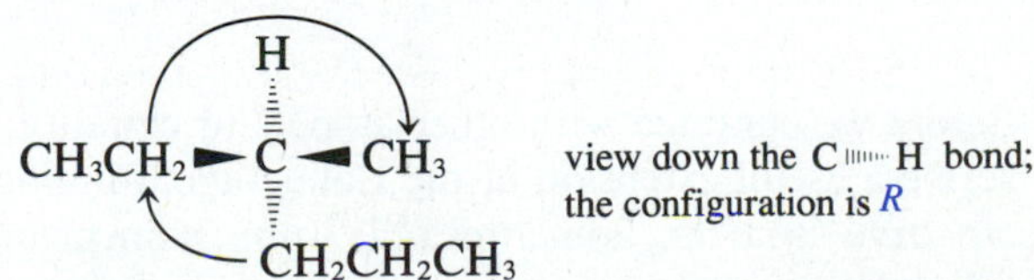

Problem 5.10 Determine the configuration (R or S) at the chiral center in

a. 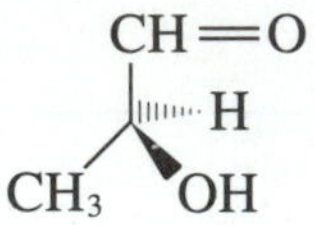 (structure: CH=O on top; C with hashed bond to H; CH$_3$ and wedge to OH)   b. 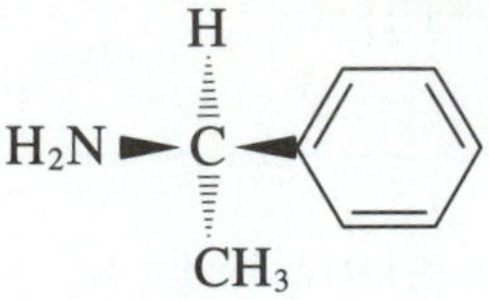 H_2N—C (H up, CH_3 down, phenyl ring)

EXAMPLE 5.6 Draw the structure of (R)-2-bromobutane.

Solution First, write out the structure and prioritize the groups attached to the chiral center.

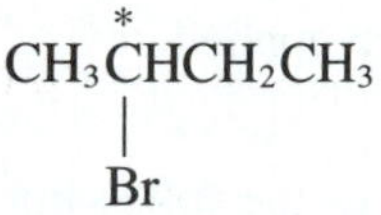

$$\overset{*}{CH_3CHCH_2CH_3} \text{ with Br on C-2}$$

$$Br > CH_3CH_2— > CH_3— > H$$

Now make the drawing with the H (lowest-priority group) "away" from you, and place the three remaining groups ($Br \rightarrow CH_3CH_2 \rightarrow CH_3$) in a clockwise ($R$) array.

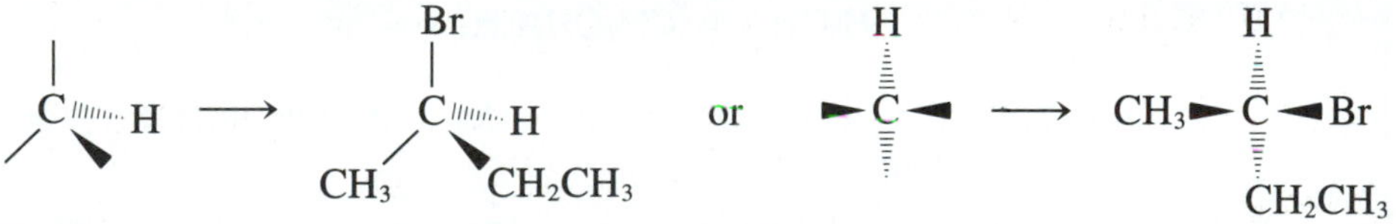

(structures: C with H back → C with Br, H, CH_3, CH_2CH_3; or → CH_3—C(H)(Br)—CH_2CH_3)

Of course, we could have started with the top-priority group at either of the other two bonds to give the following structures, which are equivalent to those above:

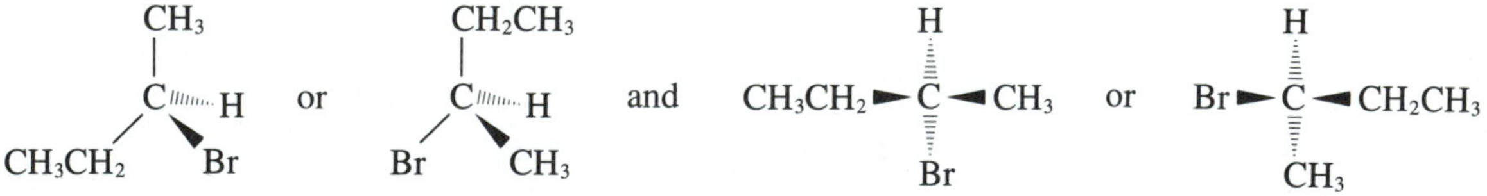

(structures: C with CH_3, H, CH_3CH_2, Br or C with CH_2CH_3, H, Br, CH_3 and CH_3CH_2—C(H)(Br)—CH_3 or Br—C(H)(CH_3)—CH_2CH_3)

Problem 5.11 Draw the structure of

a. (R)-2-phenylbutane.
b. (S)-3-methyl-1-pentene.
c. (R)-3-methylcyclopentene.

5.5 The E-Z Convention for Cis–Trans Isomers

Before we continue with other aspects of chirality, let us digress briefly to describe a useful extension of the Cahn-Ingold-Prelog system of nomenclature to *cis–trans* isomers. Sometimes *cis–trans* nomenclature is ambiguous, as in the following examples:

$$\begin{array}{c} F\diagdown \quad \diagup Br \\ C{=}C \\ Cl\diagup \quad \diagdown I \end{array} \qquad \begin{array}{c} CH_3CH_2\diagdown \quad \diagup Cl \\ C{=}C \\ CH_3\diagup \quad \diagdown Br \end{array}$$

cis or *trans?* *cis* or *trans?*

The system we have just discussed for chiral centers has been extended to double-bond isomers. We use exactly the same priority rules. *The two groups attached to each carbon of the double bond are assigned priorities*. If the two highest-priority groups are on *opposite* sides of the double bond, the prefix *E* (from the German *entgegen,* opposite) is used. If the two higher-priority groups are on the *same* side of the double bond, the prefix is *Z* (from the German *zusammen,* together). The highest-priority groups for the above examples are shown here in color, and the correct names are given below the structures.

$$\begin{array}{c} F\diagdown \quad \diagup Br \\ C{=}C \\ Cl\diagup \quad \diagdown I \end{array} \qquad \begin{array}{c} CH_3CH_2\diagdown \quad \diagup Cl \\ C{=}C \\ CH_3\diagup \quad \diagdown Br \end{array}$$

(*Z*)-1-bromo-2-chloro-2-fluoro-1-iodoethene (*E*)-1-bromo-1-chloro-2-methyl-1-butene

Problem 5.12 Name each compound by the E-Z system.

a. $$\begin{array}{c} CH_3\diagdown \quad \diagup CH_2CH_3 \\ C{=}C \\ H\diagup \quad \diagdown H \end{array}$$ b. $$\begin{array}{c} Cl\diagdown \quad \diagup H \\ C{=}C \\ Br\diagup \quad \diagdown F \end{array}$$

Problem 5.13 Write the structure for

a. (*E*)-2-pentene. b. (*Z*)-1,3-pentadiene.

5.6 Polarized Light and Optical Activity

The concept of molecular chirality follows logically from the tetrahedral geometry of carbon, as developed in Secs. 5.2 and 5.3. Historically, however, these concepts were developed in the reverse order; how this happened is one of the most elegant and logically beautiful stories in the history of science. The story began in the early eighteenth century with the discovery of polarized light and with studies on how molecules placed in the path of such a light beam affect it.

An ordinary light beam consists of waves vibrating in all possible planes perpendicular to its path. However, if this light beam is passed through certain types of substances, the transmitted beam will have all of its waves vibrating in parallel planes. Such a light beam, said to be **plane-polarized,** is illustrated in Figure 5.7. One convenient way to polarize light is to pass it through a device composed of Iceland spar (crystalline calcium carbonate) called a **Nicol prism** (invented in 1828 by the British physicist William Nicol). A more recently developed polarizing material is **Polaroid,** which was invented by E. H. Land, an American. It contains a crystalline organic compound properly oriented and

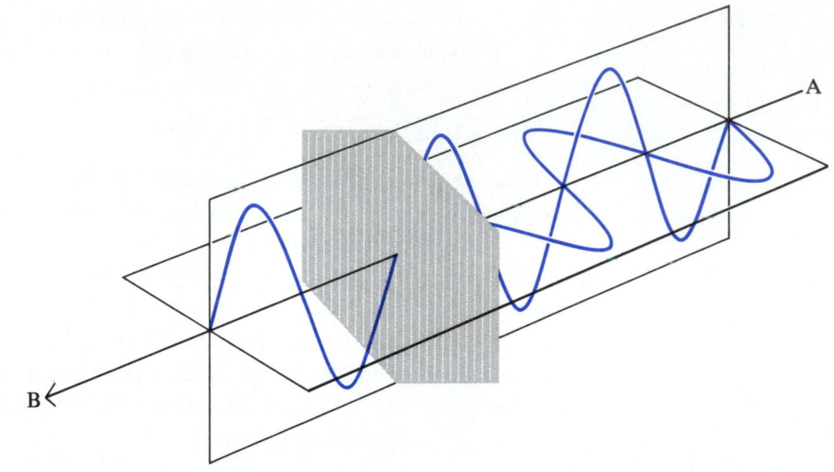

FIGURE 5.7
A beam of light, AB, initially vibrating in all directions, passes through a polarizing substance that "strains" the light so that only the vertical component emerges.

embedded in a transparent plastic. Sunglasses, for example, are often made from Polaroid.

A light beam will pass through *two* samples of polarizing material only if their polarizing axes are aligned. If the axes are perpendicular, no light will pass through. This result, illustrated in Figure 5.8, is the basis of an instrument used to study the effect of various substances on plane-polarized light.

A **polarimeter** is shown schematically in Figure 5.9. Here is how it works. With the light on and the sample tube empty, the analyzer prism is rotated so that the light beam that has been polarized by the polarizing prism is completely blocked and the field of view is dark. At this point, the prism axes of the polarizer prism and the analyzer prism are perpendicular to one another. Now the sample is placed in the sample tube. If the substance is **optically inactive,** nothing changes. The field of view remains dark. But if an **optically active** substance is placed in the tube, it rotates the plane of polarization, and some light passes through the analyzer to the observer. By turning the analyzer prism clockwise or counterclockwise, the observer can again block the light beam and restore the dark field.

The angle through which the analyzer prism must be rotated in this experiment is called α, the **observed rotation.** It is equal to the number of degrees that the optically active substance rotated the beam of plane-polarized light. If the analyzer must be rotated to the *right* (clockwise), the optically active substance is said to be **dextrorotatory** (+); if rotated to the *left* (counterclockwise), the substance is **levorotatory** (−).*

* It is not possible to tell from a single measurement whether a rotation is + or −. For example, is a reading +10° or −350°? We can distinguish between these alternatives by, for example, increasing the sample concentration by 10%. Then a +10° reading would change to +11°, and a −350° reading would change to −385° (that is, −25°).

FIGURE 5.8
The two sheets of polarizing material shown have their axes aligned perpendicularly. Although each disk alone is almost transparent, the area where they overlap is opaque. You can duplicate this effect using two pairs of Polaroid sunglasses. (Courtesy of the Polaroid Corporation.)

The observed rotation, α, of a sample of an optically active substance depends on its molecular structure and also on the number of molecules in the sample tube, as well as on certain environmental factors. All of these have to be standardized if we want to compare the optical activity of different substances. The **specific rotation** $[\alpha]$, of an optically active substance is defined as follows:

$$\text{Specific rotation} = [\alpha]_\lambda^t = \frac{\alpha}{l \times c}(\text{solvent})$$

where l is the length of the sample tube in *decimeters*, c is the concentration in *grams per milliliter*, t is the temperature of the solution, and λ is the wavelength of light. The solvent used is indicated in parentheses. Measurements are usually made at room temperature, and the most common light source is the D-

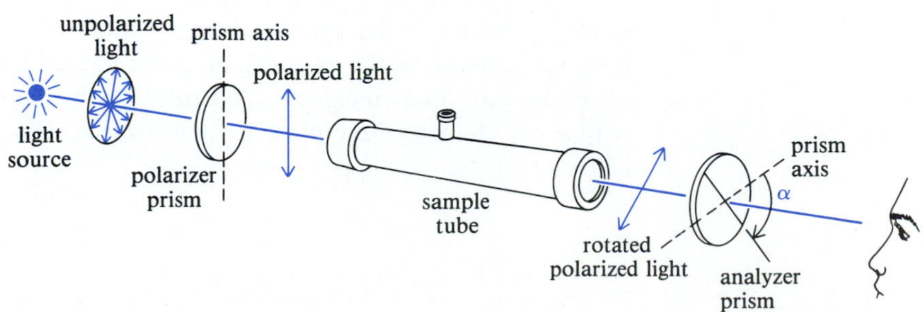

FIGURE 5.9
Diagram of a polarimeter.

line of a sodium vapor lamp (λ = 589.3 nm), although modern instruments called **spectropolarimeters** allow the light wavelength to be varied at will. The specific rotation of an optically active substance at a particular wavelength is as definite a property of the substance as its melting point, boiling point, or density.

Problem 5.14 Camphor is optically active. A camphor sample (1.5 g) dissolved in ethanol (optically inactive) to a total volume of 50 mL, placed in a 5-cm polarimeter sample tube, gives an observed rotation of $+0.66°$ at 20°C (using the sodium D-line). Calculate and express the specific rotation of camphor.

In the early nineteenth century, the French physicist Jean Baptiste Biot (1774–1862) studied the behavior of a great many substances in a polarimeter. Some, such as turpentine, lemon oil, solutions of camphor in alcohol, and solutions of cane sugar in water, were optically active. Others, such as water, alcohol, and solutions of salt in water, were optically inactive. Later, many natural products (carbohydrates, proteins, and steroids, to name just a few) were added to the list of optically active compounds. What is it about the structure of molecules that causes some to be optically active and others inactive?

When plane-polarized light passes through a single molecule, the light and the electrons in the molecule interact. This interaction causes the plane of polarization to rotate slightly.* But when we place a substance in a polarimeter, *we do not place a single molecule there, we place a large collection of molecules there* (recall that even as little as a thousandth of a mole contains 6×10^{20} molecules).

Now, if the substance is achiral, then for every single molecule in one orientation that rotates the plane of polarization in one direction, there is apt to be another molecule with the mirror-image orientation that will rotate the plane of polarization an equal amount in the opposite direction. The net result is such that the light beam passes through a sample of achiral molecules without any net change in the plane of polarization. *Achiral molecules are optically inactive.*

But for chiral molecules the situation is different. Consider a sample containing one enantiomer (say, *R*) of a chiral molecule. For any molecule with a given orientation in the sample, *there can be no mirror-image orientation* (because the mirror image gives a different molecule, the *S* enantiomer). Therefore, the rotation in the polarization plane caused by one molecule is *not* cancelled by any other molecule, and the light beam passes through the sample with a net change in the plane of polarization. *Chiral molecules are optically active.*

---

*This is because the electric and magnetic fields that result from electronic motions in the molecule affect the electric and magnetic fields of the light.

A WORD ABOUT ...

9. Pasteur's Experiments and the van't Hoff-LeBel Explanation

The great French scientist Louis Pasteur (1822–1895) was the first to recognize that optical activity is related to what we now call chirality. He realized that similar molecules that rotate plane-polarized light through equal angles but in opposite directions must be related to one another as an object and its nonsuperimposable mirror image (that is, as a pair of enantiomers). Here is how he came to that conclusion.

Working in the mid-nineteenth century in a country famous for its wine industry, Pasteur was aware of two *isomeric* acids that deposit in wine casks during fermentation. One of these, called **tartaric acid,** was optically active and dextrorotatory. The other, at the time called **racemic acid,** was optically inactive.

Pasteur prepared various salts of these acids. He noticed that *crystals* of the sodium ammonium salt of *tartaric acid* were *not* symmetric (that is, they were not identical to their mirror images). In other words, they exhibited the property of handedness (chirality). Let us say that the crystals were all left-handed.

When Pasteur next examined crystals of the same salt of *racemic acid,* he found that they, too, were chiral but that some of the crystals were left-handed and others were right-handed. The crystals were related to one another as an object and its nonsuperimposable mirror image, and they were present in equal amounts. With a magnifying lens and a pair of tweezers, Pasteur carefully separated these crystals into two piles: the left-handed ones and the right-handed ones.

Then Pasteur made a crucial observation. When he *dissolved* the two types of crystals *separately* in water and placed the solutions in a polarimeter, he found that each solution was optically active (remember, he obtained these crystals from racemic acid, which was optically inactive). One solution had a specific rotation identical to that of the sodium ammonium salt of tartaric acid! The other had an equal but *opposite* specific rotation. This meant that it must be the mirror image, or levorotatory tartaric acid. Pasteur correctly concluded that racemic acid was not really a single substance, but a 50:50 mixture of (+) and (−) tartaric acids. Racemic acid was optically inactive because it contained equal amounts of two enantiomers. *We define a **racemic mixture** as a 50 : 50 mixture of enantiomers,* and of course, such a mixture is optically inactive because the rotations of the two enantiomers cancel out.

Pasteur recognized that optical activity must be due to some property of tartaric acid molecules themselves, not to some property of the crystals, because the crystalline shape was lost when the crystals were dissolved in water in order to measure the specific rotation. However, the precise explanation in terms of molecular structure eluded Pasteur and was not to come for another 25 years.

Pasteur's experiments were performed at about the same time that Kekulé in Germany was developing his theories about organic structures. Kekulé recognized that carbon is tetravalent, and there is even a hint in some of his writings (about 1867) and also in the writings of the Russian chemist A. M. Butlerov (1862) and the Italian E. Paterno (1869) that carbon might be tetrahedral. But it was not until 1874 that the Dutch physical chemist J. H. van't Hoff (1852–1911) and the Frenchman J. A. LeBel (1874–1930) simultaneously but quite independently made a bold hypothesis about carbon that would explain the optical activity of some organic molecules and the optical inactivity of others.

These scientists knew their solid geometry. They knew that four different objects can be arranged in two different ways at the corners of a tetrahedron, and that these two ways are related to one another as an object and its nonsuperimposable mirror image. They also knew that this arrangement resulted in right- and left-handedness, as shown in the drawing.

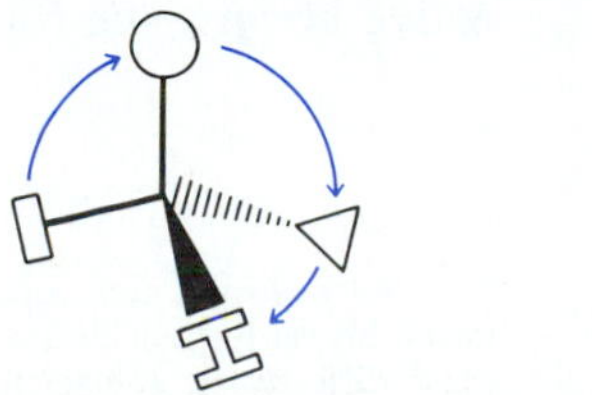

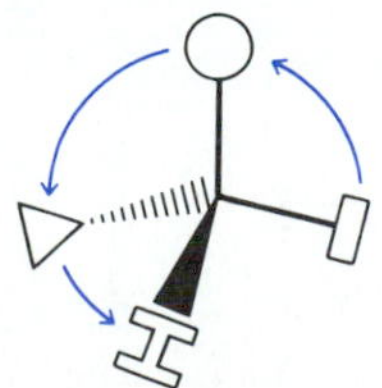

They made the bold hypothesis that the four valences of carbon were directed toward the corners of a tetrahedron, and that optically active molecules would contain at least one carbon atom with four different groups attached to it. This idea would explain why Pasteur's (+) and (−)-tartaric acids rotated plane-polarized light to equal extents but in opposite (right- and left-handed) directions. Optically inactive organic substances would either contain no asymmetric carbon atom or be a 50 : 50 mixture of enantiomers.

This, then, is how the tetrahedral geometry of carbon was first recognized.* The logic that led to the proposal is admirable, especially when one realizes that neither the electron nor the nucleus of an atom had yet been discovered and that almost nothing was then known about the physical nature of chemical bonds. At the time they made their proposal, van't Hoff and LeBel were relatively unknown chemists. Their hypothesis was ridiculed by at least one establishment chemist at the time, but it was soon generally accepted, has survived many tests, and can now be regarded as fact.

5.7 Properties of Enantiomers

In what properties do enantiomers differ from one another? Enantiomers differ only with respect to chirality. In all other respects they are identical. For this reason, they differ from one another *only in properties that are also chiral.* Let us illustrate this idea first with familiar objects.

A left-handed baseball player (chiral) can use the same ball (achiral) as can a right-handed player. But of course a left-handed player (chiral) can use only a left-handed baseball glove (chiral). A bolt with a right-handed thread (chiral) can use the same washer (achiral) as a bolt with a left-handed thread, but it can only fit into a nut (chiral) with a right-handed thread. To generalize, *the chirality of an object is most significant when the object interacts with another chiral object.*

Enantiomers have identical achiral properties, such as melting point, boiling point, density, and various types of spectra. Their solubilities in an ordinary, achiral solvent are also identical. However, *enantiomers have different chiral properties,* one of which is the *direction* in which they rotate plane-polarized light (clockwise or counterclockwise). Although enantiomers rotate plane-polarized light in opposite directions, they have specific rotations of the same magnitude (but with opposite signs), because the *number of degrees* is not a chiral property. Only the *direction* of rotation is a chiral property. Here is a specific example.

Lactic acid is an optically active hydroxyacid that is important in several biological processes. It has one chiral center. Its structure and some of its properties are shown in Figure 5.10. Note that both enantiomers have identical melting points and, *except for sign,* identical specific rotations.

A 50 : 50 mixture of enantiomers, called a **racemic mixture,** is optically inactive because the rotations of the two enantiomers cancel out.

*Actually, LeBel developed his ideas based on symmetry considerations, whereas van't Hoff based his (at least at first) on the idea of an asymmetric carbon atom. For a stimulating article on these differences, see R. B. Grossman, *J. Chem. Educ.* **1988,** *66,* 30–33.

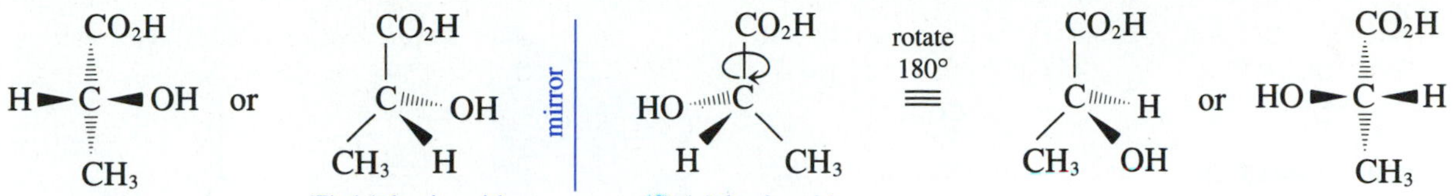

(R)-(−)-lactic acid
$[\alpha]_D^{25°C}$ −3.33(H_2O)
mp 53°C

mirror

rotate 180°

(S)-(+)-lactic acid
$[\alpha]_D^{25°C}$ +3.33(H_2O)
mp 53°C

FIGURE 5.10 The structures and properties of the lactic acid enantiomers.

There is no obvious relationship between configuration (R or S) and sign of rotation (+ or −). For example, (*R*)-lactic acid is levorotatory. When (*R*)-lactic acid is converted to its methyl ester (eq. 5.1), the configuration is unchanged because none of the bonds to the chiral carbon is involved in the reaction. Yet the sign of rotation of the product, a physical property, changes from − to +.

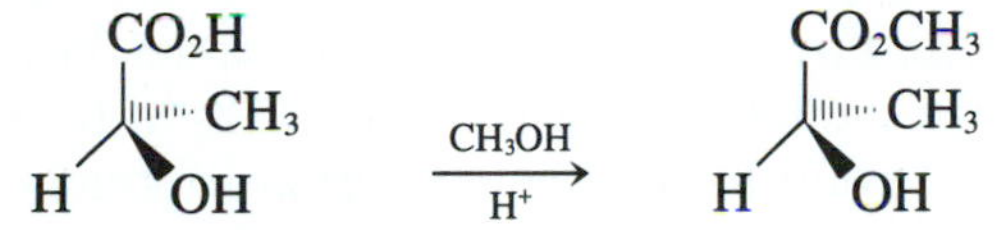

(R)-(−)-lactic acid $\xrightarrow[H^+]{CH_3OH}$ (R)-(+)-methyl lactate (5.1)

Enantiomers often have different biological properties because the biological property usually involves a reaction with another chiral molecule. For example, the enzyme *lactic acid dehydrogenase* will oxidize (+)-lactic acid to pyruvic acid, but it will *not* oxidize (−)-lactic acid.

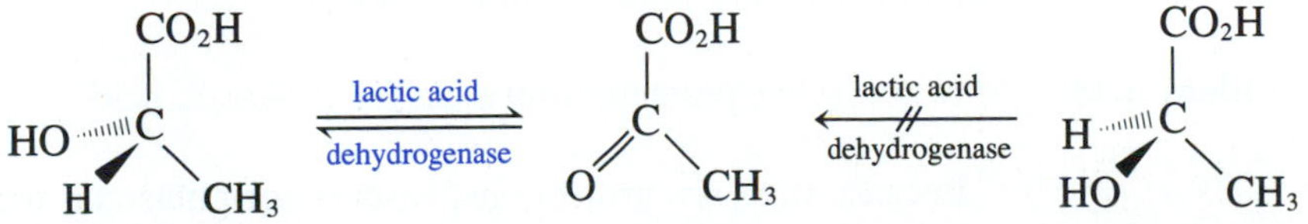

(+)-lactic acid $\xrightleftharpoons[\text{dehydrogenase}]{\text{lactic acid}}$ pyruvic acid $\xleftarrow[\text{dehydrogenase}]{\text{lactic acid}}$ (−)-lactic acid (5.2)

Why? The enzyme itself is chiral and can distinguish between right- and left-handed lactic acid molecules.

Enantiomers differ in many types of biological activity. One enantiomer may be a drug, whereas the other enantiomer may be ineffective. For example, only (−)-adrenalin is a cardiac stimulant; (+)-adrenalin is ineffective. One enantiomer may be toxic, another harmless. One may be an antibiotic, the other useless. One may be an insect sex attractant, the other without effect or actually a repellant. Chirality is of paramount importance in the biological world.

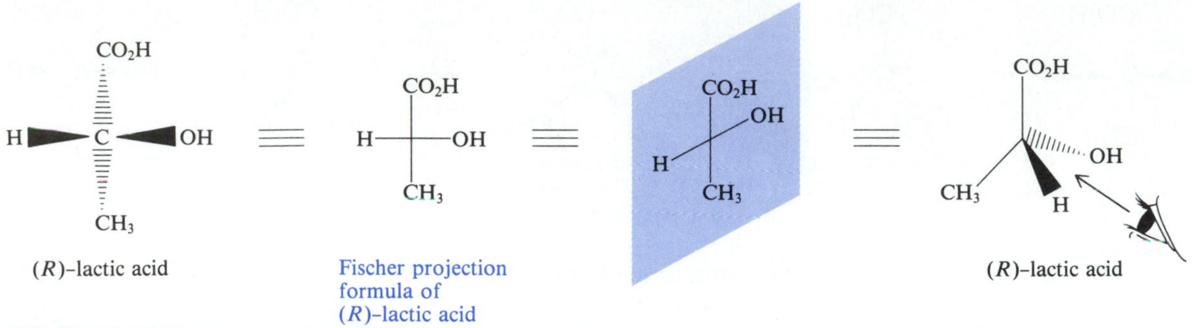

FIGURE 5.11 Projecting the model at the right onto a plane gives the Fischer projection formula.

5.8 Fischer Projection Formulas

Instead of using dashed and solid wedges to show the three-dimensional arrangements of groups in a chiral molecule, it is convenient to have a two-dimensional way of doing so. A useful way to do this was devised many years ago by Emil Fischer*; the formulas are called **Fischer projections.**

Consider the formula for (*R*)-lactic acid, to the left of the mirror in Figure 5.10. If we project that three-dimensional formula onto a plane, as illustrated in Figure 5.11, we obtain the flattened Fischer projection formula.

There are two important things to notice about Fischer projection formulas. First, the C for the asymmetric carbon atom is omitted and is represented simply as the crossing point of the horizontal and vertical lines. Second, horizontal lines connect the chiral center to groups that project *above* the plane of the page, *toward* the viewer; vertical lines lead to groups that project *below* the plane of the page, *away* from the viewer.

Problem 5.15 Draw a Fischer projection formula for (*S*)-lactic acid.

Because they are projections, Fischer formulas can only be manipulated in certain allowed ways and *still maintain their stereochemical integrity* (that is, *not* be changed to the enantiomer).

1. Fischer projections may be *turned 180° in the plane of the paper* (but *not* 90°).

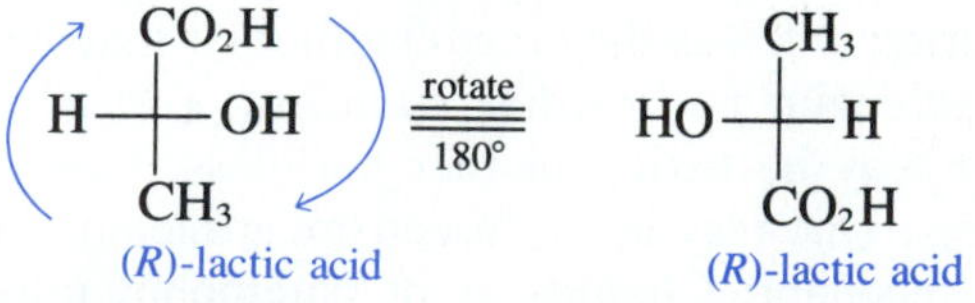

*Emil Fischer (1852–1919), who devised these formulas, was one of the early giants in the field of organic chemistry. He did much to elucidate the structures of carbohydrates, proteins, and other natural products and received the 1902 Nobel Prize in chemistry.

Rotation by 90° gives the enantiomer.

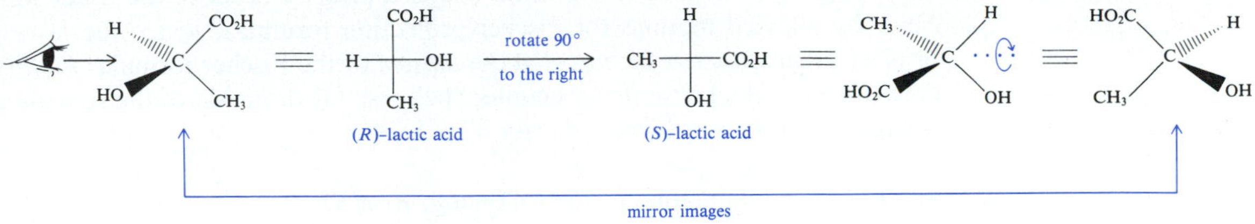

H, CO_2H, HO, CH_3 ≡ (R)-lactic acid: CO_2H / H—+—OH / CH_3 —rotate 90° to the right→ (S)-lactic acid: H / CH_3—+—CO_2H / OH ≡ CH_3, H, HO_2C, OH ≡ HO_2C, H, CH_3, OH

mirror images

Lifting a Fischer projection out of the plane of the paper and turning it over also gives the enantiomer:

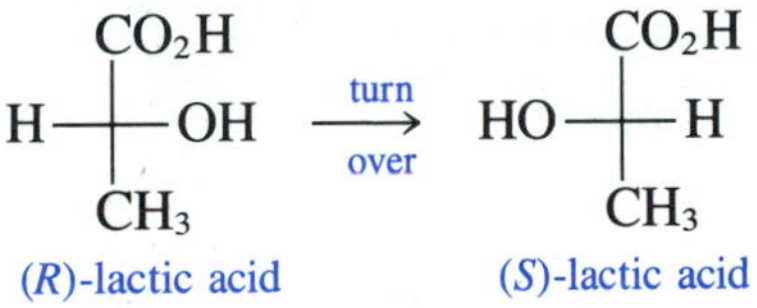

CO_2H / H—+—OH / CH_3 (R)-lactic acid —turn over→ CO_2H / HO—+—H / CH_3 (S)-lactic acid

2. While holding any one group fixed, the remaining three groups to a chiral center in a Fischer projection can be rotated either clockwise or counterclockwise without changing the configuration. For example,

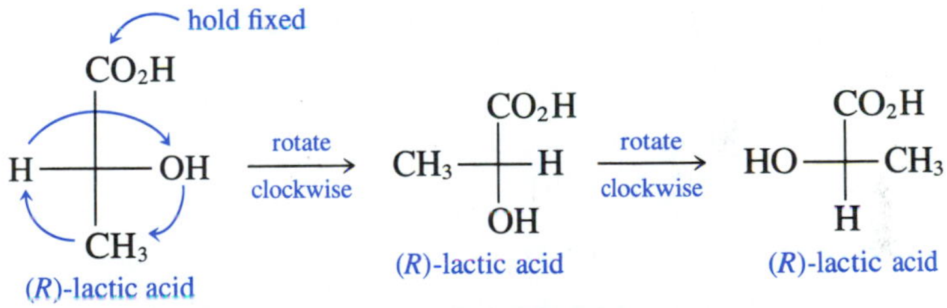

hold fixed

CO_2H / H—+—OH / CH_3 (R)-lactic acid —rotate clockwise→ CO_2H / CH_3—+—H / OH (R)-lactic acid —rotate clockwise→ CO_2H / HO—+—CH_3 / H (R)-lactic acid

This process is simply equivalent to rotating the chiral center around the C—CO_2H single bond.

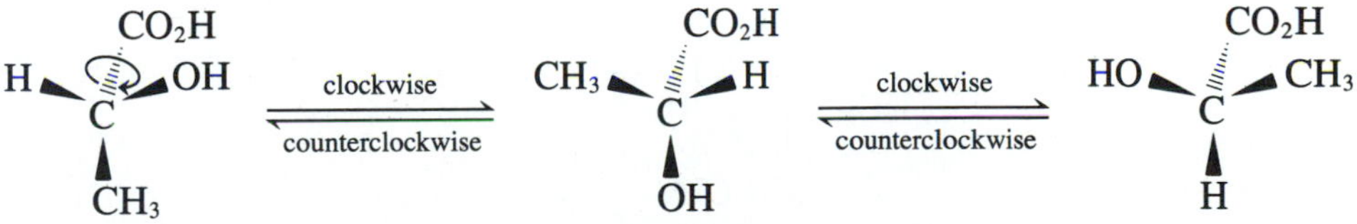

CO_2H, H, OH, C, CH_3 ⇌ (clockwise / counterclockwise) CO_2H, CH_3, H, C, OH ⇌ (clockwise / counterclockwise) CO_2H, HO, CH_3, C, H

Problem 5.16 Show that the Fischer projections

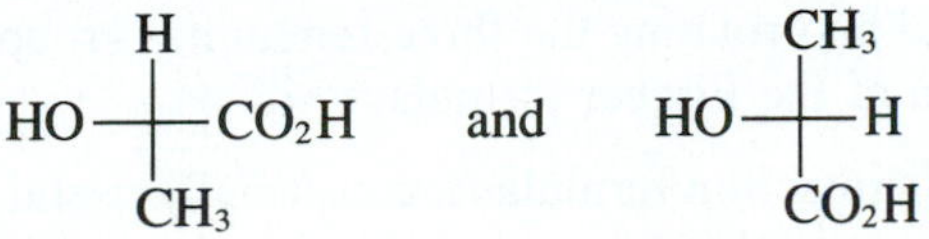

H / HO—+—CO_2H / CH_3 and CH_3 / HO—+—H / CO_2H

can be derived from CO_2H / H—+—OH / CH_3 by allowed motions and all represent (R)-lactic acid.

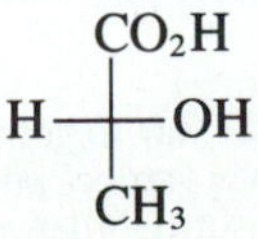

It is easy to convert from the Fischer to the *R/S* convention. *First,* assign the four groups attached to the chiral center a priority order in the usual way. *Next,* by allowed motions for Fischer projection formulas, locate the *lowest-priority* group either at the *top* or at the *bottom* of the Fischer formula. *Finally,* determine the clockwise (*R*) or counterclockwise (*S*) direction of the remaining groups as you proceed from priority $a \rightarrow b \rightarrow c$.

EXAMPLE 5.7 What is the configuration of the following, *R* or *S*?

CH_3 (top), H (left), OH (right), CH_2CH_3 (bottom)

Solution First, assign priority orders

ⓒ CH_3 (top), ⓓ H (left), ⓐ OH (right), ⓑ CH_2CH_3 (bottom)

Next, holding the ethyl group fixed, rotate the remaining three groups one turn clockwise, so as to bring the lowest-priority group (H) to the top.

ⓓ H (top), ⓐ HO (left), ⓒ CH_3 (right), ⓑ CH_2CH_3 (bottom)

Now observe that the three top-priority groups are arranged counterclockwise, so the configuration is *S*.

ⓓ H (top), ⓐ HO (left), ⓒ CH_3 (right), ⓑ CH_2CH_3 (bottom)

Problem 5.17 Show that you get the same result as in Example 5.7 by holding the methyl group fixed and rotating the three remaining groups so that the hydrogen is at the bottom of the Fischer formula.*

Fischer projection formulas are especially useful in dealing with compounds containing more than one chiral center, as we will now see.

---

*There is an even faster way to get to the R/S assignment. First assign the priorities. If the lowest-priority group is in a *vertical* position (top or bottom), then a clockwise arrangement of the remaining groups in priority order $a \rightarrow b \rightarrow c$ signifies *R* (and counterclockwise, *S*). *But,* if the lowest-priority group is in a *horizontal* position (left or right), then the reverse directions apply; a clockwise arrangement of the remaining groups $a \rightarrow b \rightarrow c$ signifies *S* (and counterclockwise, *R*).

FIGURE 5.12
The four stereoisomers of 2-bromo-3-chlorobutane, a compound with two chiral centers.

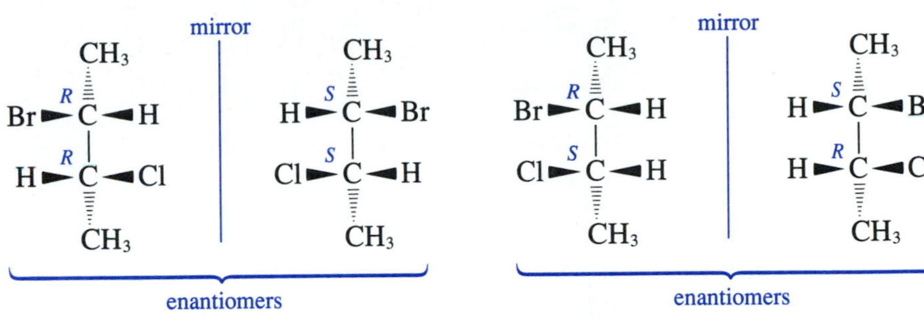

5.9 Compounds with More Than One Chiral Center; Diastereomers

Compounds may have more than one chiral center, so it is important to be able to determine how many isomers exist in such cases and how they are related to one another. Consider the molecule 2-bromo-3-chlorobutane.

$$\overset{1}{CH_3}-\overset{2*}{CH}-\overset{3*}{CH}-\overset{4}{CH_3}$$
(Br on C-2, Cl on C-3)

2-bromo-3-chlorobutane

As indicated by the asterisks, the molecule has two chiral centers. Each of these could have the configuration *R* or *S*. Thus, four isomers in all are possible: (2*R*,3*R*), (2*S*,3*S*), (2*R*,3*S*) and (2*S*,3*R*). We can draw these four isomers as shown in Figure 5.12. Note that there are two pairs of enantiomers. The (2*R*,3*R*) and (2*S*,3*S*) forms are nonsuperimposable mirror images, and the (2*R*,3*S*) and (2*S*,3*R*) forms are another such pair.

Let us see how to use Fischer projection formulas for these molecules. Consider the (2*R*,3*R*) isomer, the one at the left in Figure 5.12. The solid-dashed wedge drawing has horizontal groups projecting out of the plane of the paper toward us and vertical groups going away from us, behind the paper. These facts are expressed in the equivalent Fischer projection formula on page 158.[†]

[†] Notice that these structures are derived from an eclipsed conformation of the molecule, viewed from above so that horizontal groups project toward the viewer. The actual molecule is probably an equilibrium mixture of several staggered conformations, one of which is shown. Fischer formulas are used to represent the correct *configurations,* but not necessarily the correct *conformations* of a molecule.

eclipsed

staggered

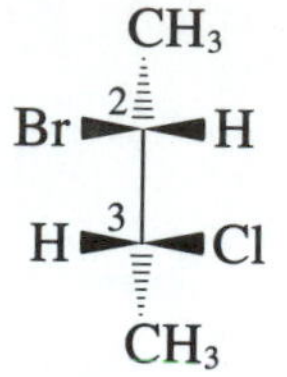

CH_3
Br 2 H
H 3 Cl
CH_3
solid-dashed wedge formula

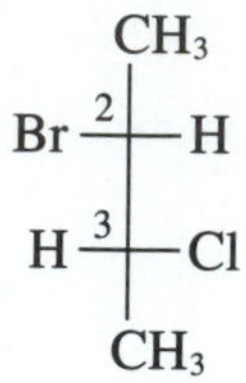

≡

CH_3
Br 2 H
H 3 Cl
CH_3
Fischer projection formula

Problem 5.18 Using the method described in Sec. 5.8, show that the above Fischer projection has the *R* configuration at carbon 2 and at carbon 3.

Problem 5.19 Draw the Fischer projection formulas for the remaining stereoisomers of 2-bromo-3-chlorobutane shown in Figure 5.12, and check the *R*/*S* assignments.

Now we come to an extremely important new idea. Consider the relationship between, for example, the (2*R*,3*R*) and (2*R*,3*S*) forms of the isomers in Figure 5.12. These forms are *not* mirror images because they have the *same* configuration at carbon 2, though they have opposite configurations at carbon 3. They are certainly stereoisomers, but they are not enantiomers. For such pairs of stereoisomers we use the term **diastereomers.** ***Diastereomers are stereoisomers that are not mirror images of one another.***

There is an important, fundamental difference between enantiomers and diastereomers. Because they are mirror images, enantiomers differ *only* in mirror-image (chiral) properties. They have the same achiral properties, such as melting point, boiling point, and solubility in ordinary solvents. Enantiomers cannot be separated from one another by methods, such as recrystallization or distillation, that depend on achiral properties. On the other hand, diastereomers are *not* mirror images. They may differ in *all* properties, whether chiral or achiral. As a consequence, diastereomers may differ in melting point, boiling point, solubility, and not only direction but also the number of degrees that they rotate plane-polarized light—in short, they behave as two different chemical substances.

Problem 5.20 How do you expect the specific rotations of the (2*R*,3*R*) and (2*S*,3*S*) forms of 2-bromo-3-chlorobutane to be related? Answer the same question for the (2*R*,3*R*) and (2*S*,3*R*) forms.

Can we generalize about the number of stereoisomers possible when a larger number of chiral centers is present? Suppose, for example, that we add a third chiral center to the compounds shown in Figure 5.12 (say, 2-bromo-3-chloro-4-iodopentane). The new chiral center added to each of the four structures can once again have either an *R* or an *S* configuration, so that with three different chiral centers, eight stereoisomers are possible. The situation is summed up in a single rule: *If a molecule has* n *different chiral centers, it may exist in a maximum of* 2^n *stereoisomeric forms. There will be a maximum of* $2^n/2$ *pairs of enantiomers.*

Problem 5.21 The Fischer projection formula for glucose (blood sugar) is

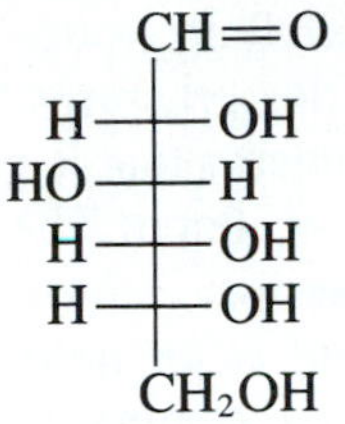

$$\begin{array}{c} CH{=}O \\ H-\!\!\!+\!\!\!-OH \\ HO-\!\!\!+\!\!\!-H \\ H-\!\!\!+\!\!\!-OH \\ H-\!\!\!+\!\!\!-OH \\ CH_2OH \end{array}$$

glucose

Altogether, how many stereoisomers of this sugar are possible?

Actually, the number of isomers predicted by this rule is the *maximum* number possible. Sometimes certain structural features reduce the actual number of isomers. In the next section, we examine a case of this type.

5.10 *Meso* Compounds; the Stereoisomers of Tartaric Acid

Consider the stereoisomers of 2,3-dichlorobutane. There are two chiral centers.

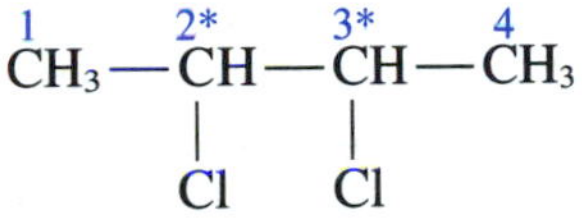

$$\overset{1}{C}H_3-\overset{2*}{C}H(Cl)-\overset{3*}{C}H(Cl)-\overset{4}{C}H_3$$

2,3-dichlorobutane

We can write out the stereoisomers just as we did in Figure 5.12; they are shown in Figure 5.13. Once again, the (*R*,*R*) and (*S*,*S*) isomers constitute a pair of nonsuperimposable mirror images, or enantiomers. *However, the other "two" structures, (R,S) and (S,R), in fact now represent a single compound.* You can easily see this by performing one of the allowed manipulations of the Fischer projection formulas at the right. A 180° rotation in the plane of the pa-

FIGURE 5.13 Fischer projections of the stereoisomers of 2,3-dichlorobutane.

$$\begin{array}{c} CH_3 \\ Cl\overset{R}{-\!\!+\!\!-}H \\ H\overset{R}{-\!\!+\!\!-}Cl \\ CH_3 \end{array} \qquad \begin{array}{c} CH_3 \\ H\overset{S}{-\!\!+\!\!-}Cl \\ Cl\overset{S}{-\!\!+\!\!-}H \\ CH_3 \end{array} \qquad\qquad \begin{array}{c} CH_3 \\ Cl\overset{R}{-\!\!+\!\!-}H \\ Cl\overset{S}{-\!\!+\!\!-}H \\ CH_3 \end{array} \equiv \begin{array}{c} CH_3 \\ H\overset{S}{-\!\!+\!\!-}Cl \\ H\overset{R}{-\!\!+\!\!-}Cl \\ CH_3 \end{array}$$

plane of symmetry

enantiomers, chiral

identical, achiral
a *meso* form

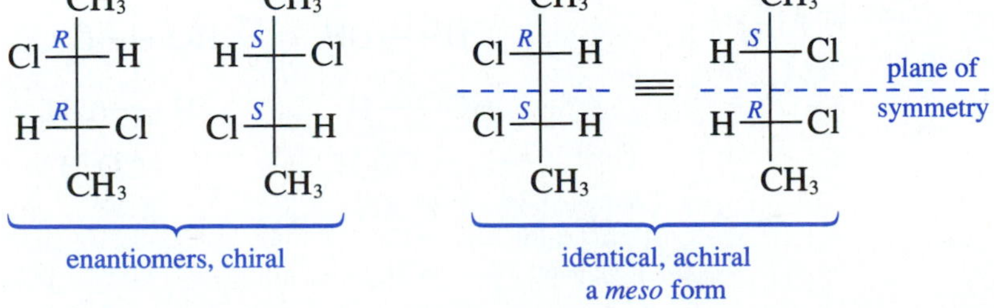

per interconverts these structures; therefore, they are identical. So altogether there are only *three* stereoisomers of 2,3-dichlorobutane, not four. The reason is that each chiral center has the *same* four groups attached.

Now let us look more closely at the structures to the right of Figure 5.13. Notice that they have a plane of symmetry that is perpendicular to the plane of the paper and bisects the central C—C bond. The structures are identical, superimposable mirror images and therefore *achiral*. We call such a structure a ***meso* compound.** ***A meso compound is an achiral diastereomer of a compound with chiral centers.*** Its chiral centers have opposite configurations. *Meso* compounds become possible whenever the chiral centers have identical groups attached. Being achiral, *meso* compounds are optically inactive.

Now let us take a look at tartaric acid, the compound whose optical activity was so carefully studied by Louis Pasteur (see "A Word About" on page 151). It has two identical chiral centers.

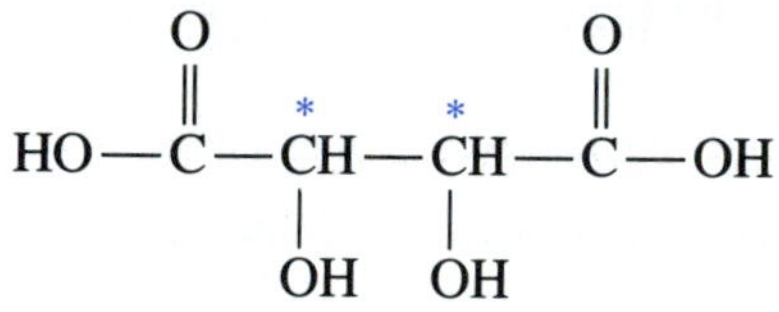

$$HO-\overset{\overset{O}{\|}}{C}-\underset{\underset{OH}{|}}{\overset{*}{C}H}-\underset{\underset{OH}{|}}{\overset{*}{C}H}-\overset{\overset{O}{\|}}{C}-OH$$

tartaric acid

The structures of these three stereoisomers and two of their properties are shown in Figure 5.14. Note that the enantiomers have identical properties except for the *sign* of the specific rotation, whereas the *meso* form, being a diastereomer of each enantiomer, differs from them in both properties.

For about 100 years after Pasteur's research, it was still not possible to determine the configuration associated with a particular enantiomer of tartaric acid. For example, it was not known whether (+)-tartaric acid had the (*R*,*R*) or the (*S*,*S*) configuration. It was known that (+)-tartaric acid had to have one of these two configurations and that (−)-tartaric acid had to have the opposite configuration, but which isomer had which?

FIGURE 5.14
The stereoisomers of tartaric acid.

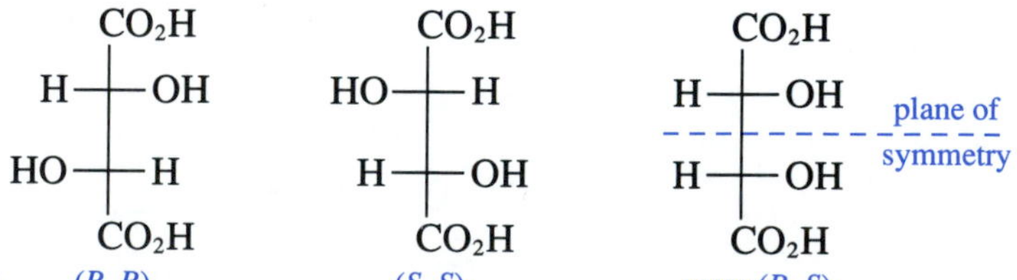

	CO_2H / H–C–OH / HO–C–H / CO_2H	CO_2H / HO–C–H / H–C–OH / CO_2H	CO_2H / H–C–OH / H–C–OH / CO_2H (plane of symmetry)
Configuration	(*R*,*R*)	(*S*,*S*)	*meso* (*R*,*S*)
$[\alpha]_D^{20^\circ}$ (H_2O)	+12	−12	0
Melting point, °C	170	170	140

In 1951, the Dutch scientist J. M. Bijvoet developed a special x-ray technique that solved the problem. Using this technique on crystals of the sodium rubidium salt of (+)-tartaric acid, Bijvoet showed that it had the (*R*,*R*) configuration. So this was the tartaric acid studied by Pasteur, and racemic acid was a 50:50 mixture of the (*R*,*R*) and (*S*,*S*) isomers. The *meso* form was not studied until later.

Since tartaric acid had been converted chemically into other chiral compounds and these in turn into still others, it became possible as a result of Bijvoet's work to assign **absolute configurations** to many pairs of enantiomers.

Problem 5.22 Show that *trans*-1,2-dimethylcyclopropane can exist in chiral, enantiomeric forms.

Problem 5.23 Is *cis*-1,2-dimethylcyclopropane chiral or achiral? What stereochemical term can we give to it?

5.11 Stereochemistry; a Recap of Definitions

We have seen here and in Sec. 2.12 that there are three different ways that *stereoisomers* can be classified. They may be either *conformational* or *configurational isomers;* they may be *chiral* or *achiral;* and they may be *enantiomers* or *diastereomers*.

A	*Conformational:*	interconvertible by rotation about single bonds
	Configurational:	not interconvertible by rotation, only by breaking and making bonds
B	*Chiral:*	mirror image not superimposable on itself
	Achiral:	molecule and mirror image are identical
C	*Enantiomers:*	mirror images; have opposite configurations at all chiral centers
	Diastereomers:	stereoisomers but not mirror images; have same configuration at one or more centers, but differ at the remaining chiral centers

Various combinations of these three sets of terms can be applied to any pair of stereoisomers. Here are a few examples:

1. *Cis*- and *trans*-2-butene

$$\begin{matrix} CH_3 & & CH_3 \\ & C{=}C & \\ H & & H \end{matrix} \quad \text{and} \quad \begin{matrix} CH_3 & & H \\ & C{=}C & \\ H & & CH_3 \end{matrix}$$

These isomers are *configurational* (not interconverted by rotation about single bonds), *achiral* (the mirror image of each is superimposable on the original), and *diastereomers* (although they are stereoisomers, they are *not* mirror images of one another; hence they must be diastereomers).

2. Staggered and eclipsed ethane

H
H H
and
H H
H

H H
H
H H
H

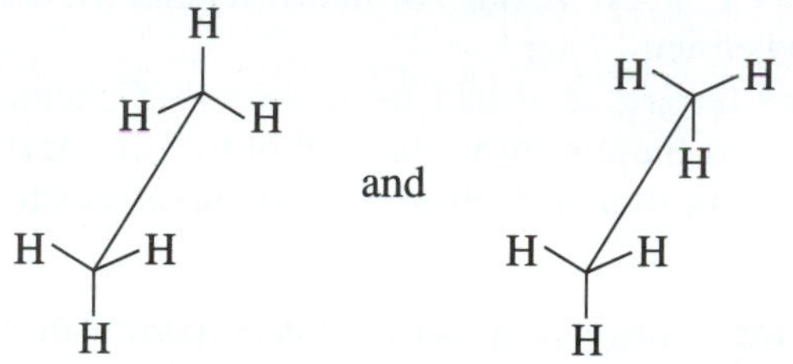

These isomers are *conformational, achiral,* and *diastereomers.*

3. (*R*)- and (*S*)-lactic acid.

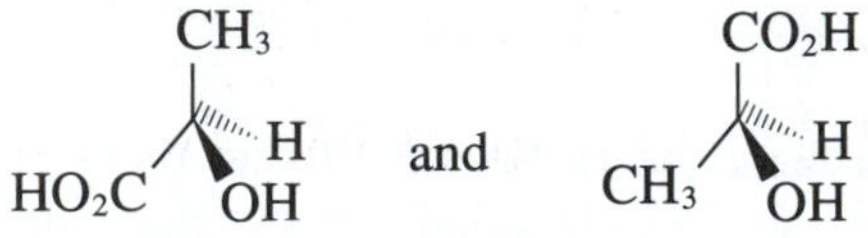

CH_3
H
HO_2C OH
and
CO_2H
H
CH_3 OH

These isomers are *configurational,* each is *chiral,* and they constitute a pair of *enantiomers.*

4. *Meso*- and (*R,R*)-tartaric acids.

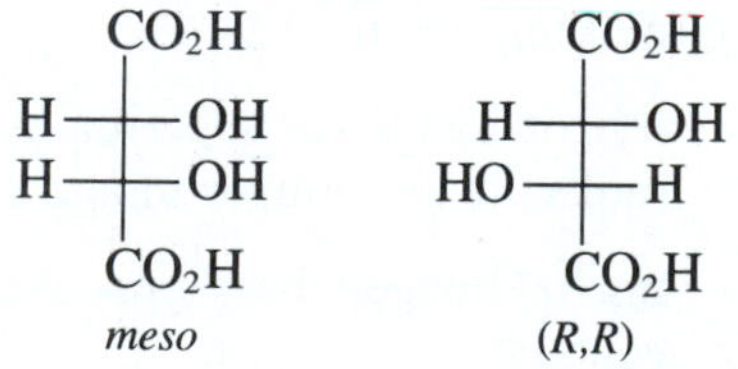

CO_2H
H—OH
H—OH
CO_2H
meso

CO_2H
H—OH
HO—H
CO_2H
(*R,R*)

These isomers are *configurational* and *diastereomers*. One is *achiral,* and the other is *chiral.*

Enantiomers, such as (*R*)- and (*S*)-lactic acid, differ only in chiral properties and therefore cannot be separated by ordinary achiral methods such as distillation or recrystallization. Diastereomers differ in all properties, chiral or achiral. *If* they are also configurational isomers (such as *cis*- and *trans*-2-butene, or *meso*- and *R,R*-tartaric acid), they can be separated by ordinary achiral methods, such as distillation or recrystallization. *If,* on the other hand, they are conformational isomers (such as staggered and eclipsed ethane), they may interconvert so readily by bond rotation as to not be separable.

Problem 5.24 Draw the two stereoisomers of 1,3-dimethylcyclobutane, and classify the pair according to the categories listed in A, B, and C above.

5.12 Stereochemistry and Chemical Reactions

How important is stereochemistry in chemical reactions? The answer depends on the nature of the reactants. First, consider the formation of a chiral product from achiral reactants; for example, the addition of hydrogen bromide to 1-butene to give 2-bromobutane in accord with Markovnikov's rule.

$$\underset{\text{1-butene}}{CH_3CH_2CH{=}CH_2} + HBr \longrightarrow \underset{\text{2-bromobutane}}{CH_3CH_2\overset{*}{C}H(Br)CH_3} \tag{5.3}$$

The product has one chiral center, marked with an asterisk, but both enantiomers are formed in exactly equal amounts. The product is a racemic mixture. Why? Although this result will be obtained *regardless* of the reaction mechanism, let us consider the generally accepted mechanism.

$$CH_3CH_2CH{=}CH_2 + H^+ \longrightarrow \underset{\text{2-butyl cation}}{CH_3CH_2\overset{+}{C}HCH_3} \xrightarrow{Br^-} CH_3CH_2CH(Br)CH_3 \tag{5.4}$$

The intermediate 2-butyl cation obtained by adding a proton to the end carbon is planar, and bromide ion can combine with it from the "top" or "bottom" side with exactly equal probability.

$$\underset{\text{2-butyl cation}}{CH_3CH_2{-}\overset{+}{C}(H)(CH_3)} \xrightarrow{:\ddot{Br}:^-} \begin{cases} \text{top} \rightarrow (S)\text{-2-bromobutane} \\ \text{bottom} \rightarrow (R)\text{-2-bromobutane} \end{cases} \tag{5.5}$$

The product is therefore a racemic mixture, an optically inactive 50 : 50 mixture of the two enantiomers.

We can generalize this result. *When chiral products are obtained from achiral reactants, both enantiomers are formed at the same rates, in equal amounts.*

Problem 5.25 Show that, if the mechanism of addition of HBr to 1-butene involved *no* intermediates, but *simultaneous one-step* addition (in the Markovnikov sense), the product would still be racemic 2-bromobutane.

Problem 5.26 Show that the chlorination of butane at carbon-2 will give a 50:50 mixture of enantiomers.

Now consider a different situation, the reaction of a *chiral* molecule with an achiral reagent to create a second chiral center. Consider, for example, the addition of HBr to 3-chloro-1-butene.

$$\underset{\text{3-chloro-1-butene}}{CH_3\overset{*}{C}H(Cl)CH{=}CH_2} + HBr \longrightarrow \underset{\text{2-bromo-3-chlorobutane}}{CH_3\overset{*}{C}H(Cl){-}\overset{*}{C}H(Br)CH_3} \tag{5.6}$$

Suppose we start with one pure enantiomer of 3-chloro-1-butene, say, the *R* isomer. What can we say about the stereochemistry of the products? One way to see the answer quickly is to draw Fischer projections.

$$\underset{(R)\text{-3-chloro-1-butene}}{\begin{array}{c} CH_3 \\ Cl\overset{R}{-\!\!\!-\!\!\!+}\!-H \\ CH \\ \| \\ CH_2 \end{array}} \xrightarrow{HBr} \underset{(2R,3R)\text{-2-bromo-3-chlorobutane}}{\begin{array}{c} CH_3 \\ Cl\overset{R}{-\!\!\!-\!\!\!+}\!-H \\ H\overset{R}{-\!\!\!-\!\!\!+}\!-Br \\ CH_3 \end{array}} + \underset{(2S,3R)\text{-2-bromo-3-chlorobutane}}{\begin{array}{c} CH_3 \\ Cl\overset{R}{-\!\!\!-\!\!\!+}\!-H \\ Br\overset{S}{-\!\!\!-\!\!\!+}\!-H \\ CH_3 \end{array}} \tag{5.7}$$

The configuration where the chloro substituent is located remains unchanged and *R*, but the new chiral center can be either *R* or *S*. Therefore, the products are *diastereomers*. Are they formed in equal amounts? No. Looking at the starting material in eq. 5.7, we can see that it has no plane of symmetry. Approach of the bromine to the double bond from the H side or from the Cl side of the chiral center should not occur with equal ease.

We can generalize this result. *Reaction of a chiral reagent with an achiral reagent, when it creates a new chiral center, leads to diastereomeric products at different rates and in unequal amounts.*

Problem 5.27 Let us say that the (2*R*,3*R*) and (2*S*,3*R*) products in eq. 5.7 are formed in a 60 : 40 ratio. What products would be formed and in what ratio by adding HBr to pure (*S*)-3-chloro-1-butene? by adding HBr to a racemic mixture of (*R*)- and (*S*)-3-chloro-1-butene?

5.13 Resolution of a Racemic Mixture

We have just seen that, when reaction between two achiral reagents leads to a chiral product, it always gives a racemic (50 : 50) mixture of enantiomers. Suppose we want to obtain each enantiomer pure and free of the other. *The process of separating a racemic mixture into its enantiomers is called* ***resolution.*** Since enantiomers have identical achiral properties, how can we resolve a racemic mixture into its components? The answer is to convert them to diastereomers, separate *them,* and then reconvert the now-separated diastereomers back to enantiomers.

To separate two enantiomers, we first let them react with a chiral reagent. The product will be a pair of *diastereomers*. These, as we have seen, differ in all types of properties and can be separated by ordinary methods. This principle is illustrated in the following equation:

$$\begin{Bmatrix} R \\ S \end{Bmatrix} + R \longrightarrow \begin{Bmatrix} R\text{—}R \\ S\text{—}R \end{Bmatrix} \qquad (5.8)$$

pair of enantiomers (not separable) — chiral reagent — diastereomeric products (separable)

After the diastereomers are separated, we then carry out reactions that regenerate the chiral reagent and the separated enantiomers.

$$R\text{—}R \longrightarrow R + R$$

and (5.9)

$$S\text{—}R \longrightarrow S + R$$

Louis Pasteur was the first to resolve a racemic mixture when he separated the sodium ammonium salts of (+)- and (−)-tartaric acid. In a sense, he was a chiral reagent, since he could distinguish between the right- and left-handed crystals. In Chapter 11, we will see a specific example of how this is done chemically.

The principle behind the resolution of racemic mixtures is the same as the principle involved in the specificity of many biological reactions. That is, a chiral reagent (in cells, usually an enzyme) can discriminate between enantiomers because the two possible products of the reaction are diastereomers.

Additional Problems

5.28. Define or describe the following terms.

a. chiral molecule **b.** enantiomers
c. polarized light **d.** specific rotation
e. chiral center **f.** plane of symmetry
g. racemic mixture **h.** diastereomers
i. *meso* form **j.** resolution

5.29. Which of the following substances can exist in optically active forms?

a. 2,2-dibromopropane **b.** 1,2-dibromopropane
c. 3-ethylhexane **d.** 2,3-dimethylhexane
e. methylcyclopentane **f.** 1-deuterioethanol (CH_3CHDOH)

5.30. Locate with an asterisk the chiral centers in the following structures.

a. $C_6H_5CH(OH)CO_2H$ **b.** $CH_2(OH)CH(OH)CH(OH)CHO$

c. 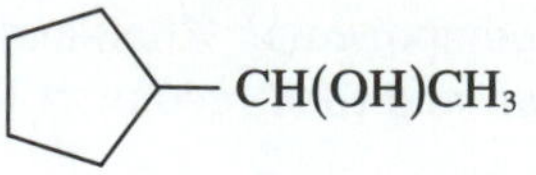 (cyclopentyl)—$CH(OH)CH_3$ **d.** $CH_3CHClCCl_3$

e.  CH_3—(cyclohexane)—CH_3 **f.** 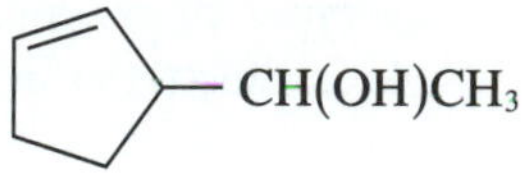 (cyclopentenyl)—$CH(OH)CH_3$

5.31. What would happen to the observed and to the *specific* rotation if, in measuring the optical activity of a solution of sugar in water, we
a. doubled the concentration of the solution?
b. doubled the length of the sample tube?

5.32. The observed rotation for 100 mL of an aqueous solution containing 1 g of sucrose (ordinary sugar), placed in a 2-decimeter sample tube, is +1.33° at 25°C (using a sodium lamp). Calculate and express the specific rotation of sucrose.

5.33. Tell whether the following structures are identical or enantiomers. (*Hint:* A quick way to do this is to see whether an odd or an even number of exchanges of groups interconverts the structures.)

a. 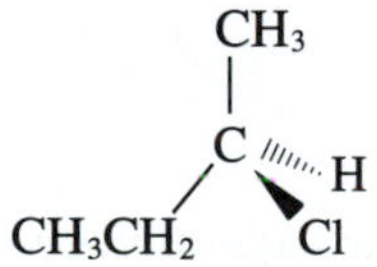 C with CH_3 (up), CH_3CH_2, H (hashed), Cl (wedged) and C with H (up), Cl, CH_3 (hashed), CH_2CH_3 (wedged)

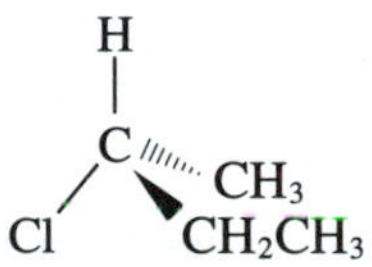

b. 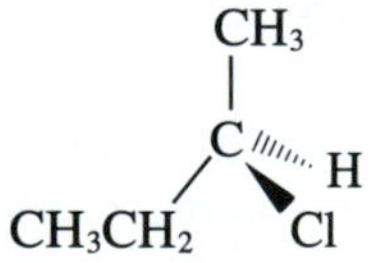 C with CH_3 (up), CH_3CH_2, H (hashed), Cl (wedged) and C with Cl (up), CH_3, CH_2CH_3 (hashed), H (wedged)

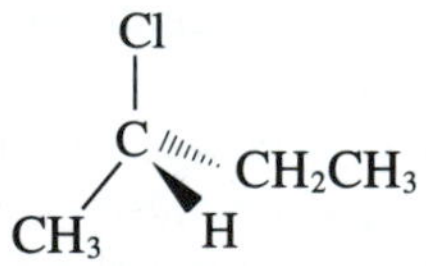

5.34. Draw a structural formula for an optically active compound with the molecular formula
a. $C_4H_{10}O$ **b.** $C_5H_{11}Br$ **c.** $C_4H_8(OH)_2$ **d.** C_6H_{12}

5.35. Draw the formula of an unsaturated chloride, C_5H_9Cl, that can show
a. neither *cis–trans* isomerism nor optical activity.
b. *cis–trans* isomerism but no optical activity.
c. no *cis–trans* isomerism but optical activity.
d. *cis–trans* isomerism and optical activity.

5.36. Place the members of the following groups in order of decreasing priority according to the *R-S* convention:
a. CH_3—, H—, HO—, CH_3CH_2—
b. H—, CH_3—, C_6H_5—, Cl—
c. CH_3—, HO—, —CH_2Cl, —CH_2OH
d. CH_3CH_2—, $CH_3CH_2CH_2$—, $CH_2{=}CH$—, —$CH{=}O$

5.37. Assume that the four groups in each part of Problem 5.36 are attached to one carbon atom. Draw a three-dimensional formula for the *R* configuration of the molecule.

5.38. Tell whether the chiral centers marked with an asterisk in the following structures have the *R* or the *S* configuration:

a. CH_3, H, H, $CH(CH_3)_2$, O

(−)-menthone
(found in peppermint)

b. CO_2H / H_2N—C^*—H / CH_2OH

(−)-serine
(an amino acid
found in proteins)

c. HO, HO, H, OH, C^*, CH_2NHCH_3

(−)-epinephrine
(also called adrenalin)

5.39. Determine the configuration, *R* or *S*, of (+)-carvone, the compound responsible for the odor of caraway seeds.

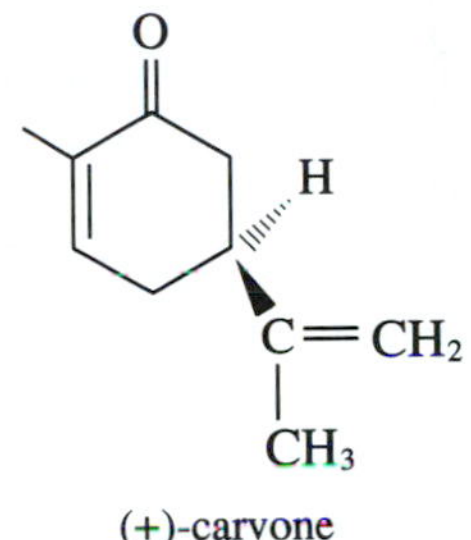

(+)-carvone

5.40. Name the following compounds, using *E-Z* notation.

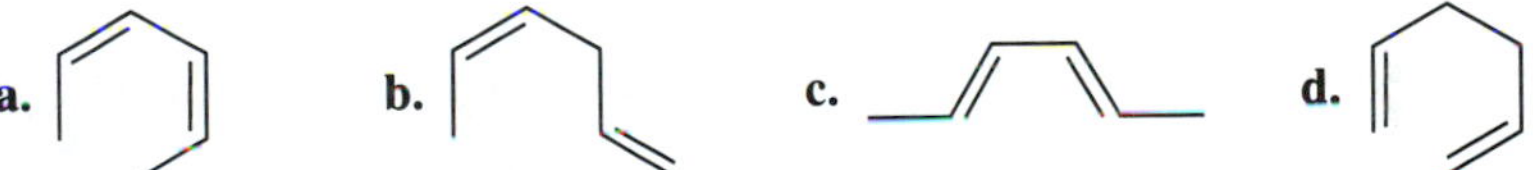

5.41. 4-Bromo-2-pentene has a double bond that can have either the *E* or the *Z* configuration and a chiral center that can have either the *R* or the *S* configuration. How many stereoisomers are possible altogether? Draw the structure of each, and group the pairs of enantiomers.

5.42. How many stereoisomers are possible for each of the following structures? Draw them, and name each by the *R-S* and *E-Z* conventions.

a. 3-methyl-1,4-pentadiene **b.** 3-methyl-1,4-hexadiene
c. 2-bromo-5-chloro-3-hexene **d.** 2,5-dibromo-3-hexene

5.43. Which of the following Fischer projection formulas have the same configuration as

CH_3 (top), HO (left), H (right), C_2H_5 (bottom) **(A)**

and which are its enantiomer?

a. OH (top), CH_3 (left), C_2H_5 (right), H (bottom)

b. H (top), C_2H_5 (left), CH_3 (right), OH (bottom)

c. CH_3 (top), H (left), OH (right), C_2H_5 (bottom)

d. C_2H_5 (top), H (left), CH_3 (right), OH (bottom)

5.44. What is the configuration, *R* or *S*, of

a. 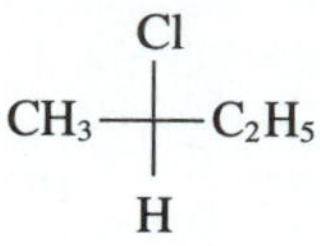 CH_3–C(Cl, H)–C_2H_5 (Cl top, H bottom) **b.** Cl–C(CH_3, C_2H_5)–$CH(CH_3)_2$ (CH_3 top, C_2H_5 bottom) **c.** CH_3–C(OH, CO_2H)–H (OH top, CO_2H bottom)

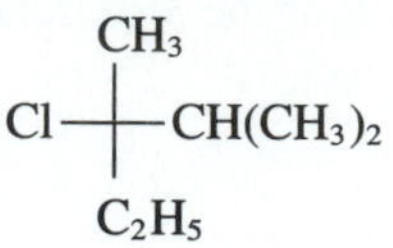

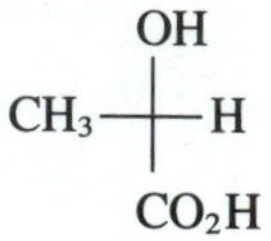

5.45. What is the configuration, *R* or *S*, at each chiral center in

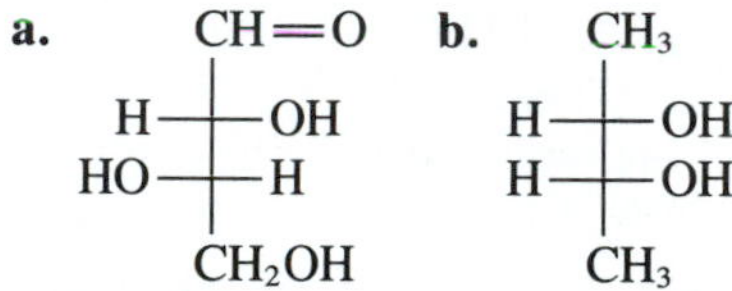

5.46. What is the configuration, *R* or *S*, at each chiral center in glucose (Problem 5.21, p. 159)?

5.47. Two possible configurations for a molecule with three different chiral centers are (*R*,*R*,*R*) and its mirror image (*S*,*S*,*S*). What are all the remaining possibilities? Repeat for a compound with four different chiral centers.

5.48. When racemic 2-chlorobutane is chlorinated, we obtain some 2,3-dichlorobutane. It consists of 71% *meso* isomer and 29% racemic isomers. Explain why the mixture need not be 50 : 50 *meso* and racemic 2,3-dichlorobutane. (*Hint:* It will help if you draw three-dimensional structures or Fischer projections.)

5.49. Below are Newman projections for the three tartaric acids (*R*,*R*), (*S*,*S*), and *meso*. Which is which?

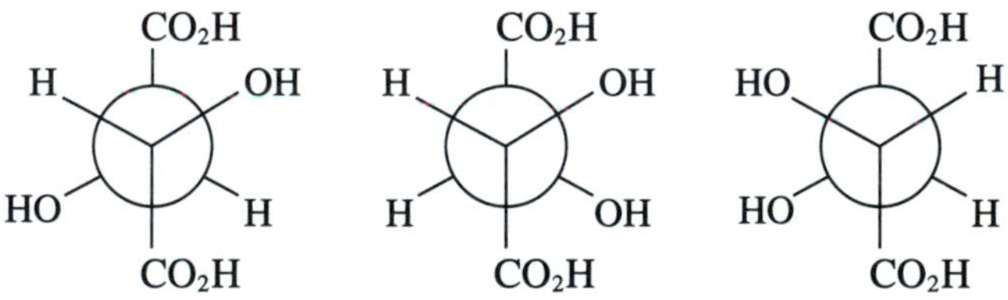

5.50. Convert the sawhorse formula below for one isomer of tartaric acid to a Fischer projection formula. Which isomer of tartaric acid is it?

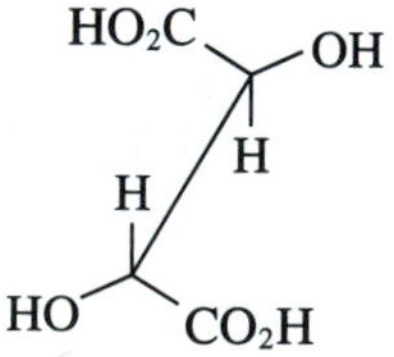

5.51. Two possible isomeric structures of 1,2-dichloroethane are

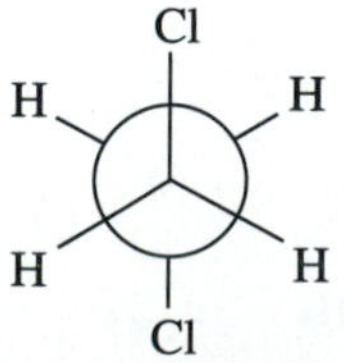

and

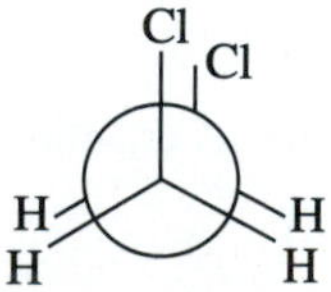

Classify them fully, according to the discussion in Sec. 5.11.

5.52. Two other possible isomeric structures of 1,2-dichloroethane are

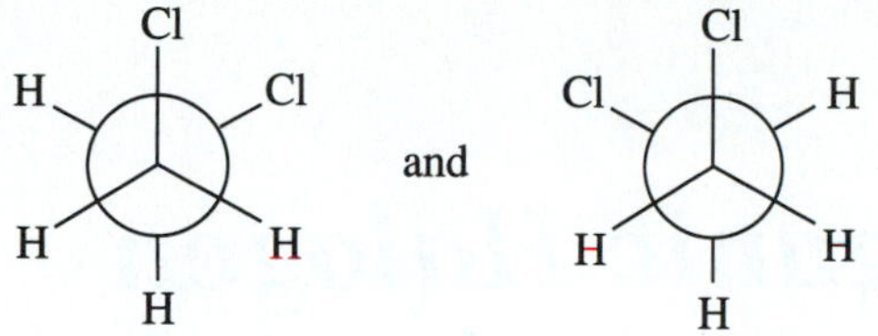

Cl, H, Cl, H, H, H and Cl, Cl, H, H, H, H

Classify them fully, as in Problem 5.51.

5.53. The formula for muscarine, a toxic constituent of poisonous mushrooms, is

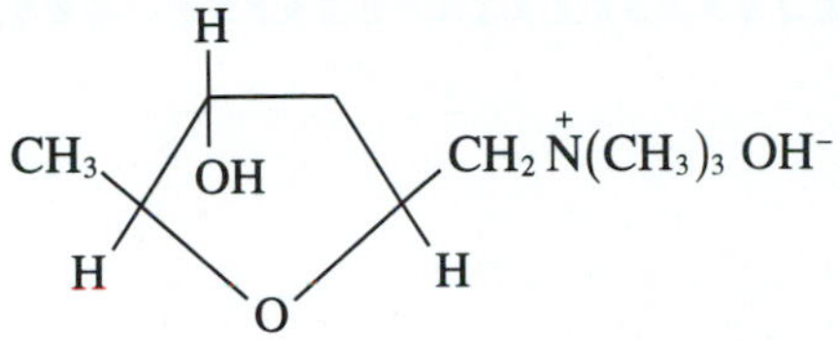

$$CH_3, H, OH, H, O, \; CH_2\overset{+}{N}(CH_3)_3\ OH^-, H$$

Is it chiral? How many stereoisomers of this structure are possible? An interesting murder mystery, which you might enjoy reading and which depends for its solution on the distinction between optically active and racemic forms of this poison, is Dorothy L. Sayers's *The Documents in the Case,* published in paperback by Avon Books. (See an article by H. Hart, "Accident, Suicide, or Murder? A Question of Stereochemistry," *J. Chem. Educ.*, **1975,** *52,* 444.)

5.54. Chloramphenicol is an antibiotic that is particularly effective against typhoid fever. Its structure is

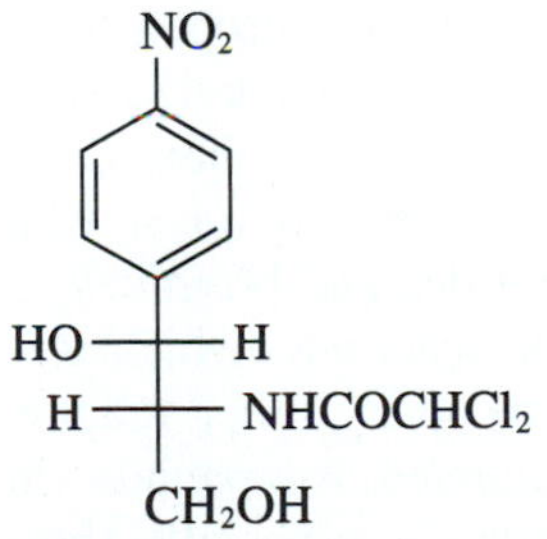

NO_2 (on the benzene ring)

HO—C—H

H—C—$NHCOCHCl_2$

CH_2OH

What is the configuration (R,S) at each chiral center?

5.55. What can you say about the stereochemistry of the products in the following reactions?

a. 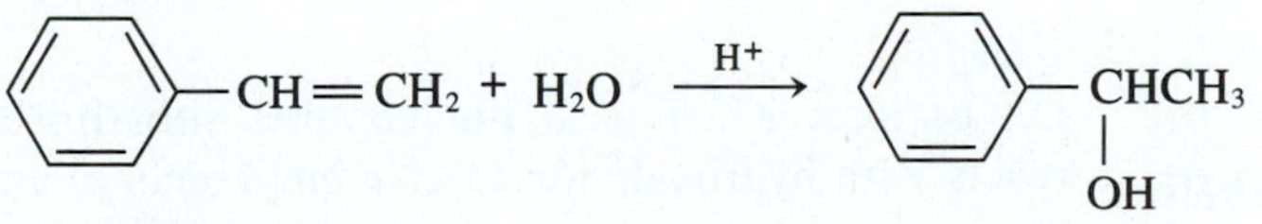 $C_6H_5-CH{=}CH_2 + H_2O \xrightarrow{H^+} C_6H_5-\underset{\displaystyle OH}{\underset{|}{C}}HCH_3$

b. $CH_3\underset{\displaystyle OH}{\underset{|}{C}}HCH{=}CH_2 + H_2O \xrightarrow{H^+} CH_3\underset{\displaystyle OH}{\underset{|}{C}}H-\underset{\displaystyle OH}{\underset{|}{C}}HCH_3$

(R-enantiomer)

5.56. (+)- and (−)-Carvone (see Problem 5.39 for the structure) are enantiomers that have very different odors, being responsible for the odors of caraway seeds and spearmint, respectively. Suggest a possible explanation.

CHAPTER 6

Organic Halogen Compounds; Substitution and Elimination Reactions

6.1 Introduction

Chlorine- and bromine-containing natural products have been isolated from various species that live in the sea—sponges, mollusks, and other ocean creatures that have adapted to their environment by metabolizing inorganic chlorides and bromides that are prevalent there. With these exceptions, most organic halogen compounds are creatures of the laboratory. We have already seen that they can be made by the direct halogenation of alkanes and aromatic compounds and by the addition of hydrogen halides to alkenes and alkynes.* And in Chapter 7, we will learn how alkyl halides can be prepared from alcohols. Why are there so many routes to these compounds?

Halogen compounds are important for several reasons. Simple alkyl and aryl halides, especially chlorides and bromides, are versatile reagents in syntheses. Through substitution reactions, which we will discuss in this chapter, halogens can be replaced by many other functional groups. Organic halides can be converted to unsaturated compounds through dehydrohalogenation. Also, some halogen compounds, especially those that contain two or more halogen atoms per molecule, have practical uses—as solvents, insecticides, herbicides, fire retardants, cleaning fluids, and refrigerants, for example. In this chapter, we will discuss all these aspects of halogen compounds.

6.2 Nucleophilic Substitution

Let us look at a typical **nucleophilic substitution reaction.** Ethyl bromide reacts with hydroxide ion to give ethyl alcohol and bromide ion.

$$\underset{}{^{-}OH} + \underset{\text{ethyl bromide}}{CH_3CH_2{-}Br} \xrightarrow{H_2O} \underset{\text{ethyl alcohol}}{CH_3CH_2{-}OH} + Br^{-} \quad (6.1)$$

* If you do not remember these reactions, review Secs. 2.15, 3.9, 3.19, and 4.9. Also review the definitions of electrophiles and nucleophiles, Sec. 3.11.

Hydroxide ion is the **nucleophile.** It reacts with the **substrate** (ethyl bromide) and displaces bromide ion. The bromide ion is called the **leaving group.**

In reactions of this type, one covalent bond is broken, and a new covalent bond is formed. In this particular example, the carbon–bromine bond is broken and the carbon–oxygen bond is formed. The leaving group (bromide) takes with it *both* of the electrons from the carbon–bromine bond, and the nucleophile (hydroxide ion) supplies *both* electrons for the new carbon–oxygen bond.

These ideas are generalized in the following equations for a nucleophilic substitution reaction:

$$\underset{\text{nucleophile (neutral)}}{\text{Nu:}} + \underset{\text{substrate}}{\text{R:L}} \longrightarrow \underset{\text{product}}{\text{R:}\overset{+}{\text{Nu}}} + \underset{\text{leaving group}}{\text{:L}^-} \quad (6.2)$$

$$\underset{\text{nucleophile (anion)}}{\text{Nu:}^-} + \underset{\text{substrate}}{\text{R:L}} \longrightarrow \underset{\text{product}}{\text{R:Nu}} + \underset{\text{leaving group}}{\text{:L}^-} \quad (6.3)$$

If the nucleophile and substrate are neutral, the product will be positively charged (eq. 6.2). If the nucleophile is a negative ion and the substrate is neutral, the product will also be neutral (eq. 6.3). In either case, an unshared electron pair on the nucleophile supplies the electrons for the new covalent bond.

In principle, of course, these reactions may be reversible because the leaving group also has an unshared electron pair that can be used to form a covalent bond. However, we can use various methods to force the reactions to go in the forward direction. For example, we can choose Nu: so it is a *stronger* nucleophile than the leaving group :L. Or we can shift the equilibrium by using a large excess of one reagent or by removing one of the products as it is formed.

6.3 Examples of Nucleophilic Substitutions

Nucleophiles can be classified according to the kind of atom that forms a new covalent bond. For example, the hydroxide ion in eq. 6.1 is an *oxygen* nucleophile. In the product, a new carbon–*oxygen* bond is formed. **The most common nucleophiles are *oxygen, nitrogen, sulfur, halogen,* and *carbon* nucleophiles.** Table 6.1 shows some examples of nucleophiles and the products that they form when they react with an alkyl halide.

Let us consider a few specific examples of these reactions, to see how they may be used in synthesis.

EXAMPLE 6.1 Use Table 6.1 to write an equation for the reaction of sodium ethoxide with bromoethane.

Solution

$$\underset{\text{sodium ethoxide}}{CH_3CH_2\ddot{\underset{..}{O}}:^-\ Na^+} + \underset{\text{bromoethane}}{CH_3CH_2Br} \longrightarrow \underset{\text{diethyl ether}}{CH_3CH_2OCH_2CH_3} + Na^+Br^-$$

Sodium ethoxide can be obtained from ethanol and sodium (eqs. 1.6 and 1.7). The product is diethyl ether, used as an anesthetic.

TABLE 6.1 Reactions of common nucleophiles with alkyl halides* (eqs. 6.2 and 6.3)

Nu		R—Nu			
Formula	*Name*	*Formula*	*Name*	*Comments*	
Oxygen nucleophiles					
1. $H\ddot{\underset{..}{O}}:^-$	hydroxide	R—OH	alcohol		
2. $R\ddot{\underset{..}{O}}:^-$	alkoxide	R—OR	ether		
3. $H\ddot{\underset{..}{O}}H$	water	$R-\overset{+}{\ddot{O}}(H)H$	alkyloxonium ion	These ions readily lose a proton.	$\xrightarrow{-H^+}$ ROH (alcohol)
4. $R\ddot{\underset{..}{O}}H$	alcohol	$R-\overset{+}{\ddot{O}}(R)H$	dialkyloxonium ion		$\xrightarrow{-H^+}$ ROR (ether)
5. $R\ddot{\underset{..}{O}}R$	ether	$R-\overset{+}{\ddot{O}}(R)R$	trialkyloxonium ion		
6. $R-C(=O)-\ddot{\underset{..}{O}}:^-$	carboxylate	$R-OC(=O)-R$	ester		
Nitrogen nucleophiles					
7. $\ddot{N}H_3$	ammonia	$R-\overset{+}{N}H_3$	alkylammonium ion	With a base, these ions readily lose a proton to give amines.	$\xrightarrow{-H^+}$ RNH_2
8. $R\ddot{N}H_2$	primary amine	$R-\overset{+}{N}H_2R$	dialkylammonium ion		$\xrightarrow{-H^+}$ R_2NH
9. $R_2\ddot{N}H$	secondary amine	$R-\overset{+}{N}HR_2$	trialkylammonium ion		$\xrightarrow{-H^+}$ R_3N
10. $R_3\ddot{N}$	tertiary amine	$R-\overset{+}{N}R_3$	tetraalkylammonium ion		
Sulfur nucleophiles					
11. $H\ddot{\underset{..}{S}}:^-$	hydrosulfide ion	R—SH	thiol		
12. $R\ddot{\underset{..}{S}}:^-$	mercaptide ion	R—SR	thioether (sulfide)		
13. $R_2\ddot{S}:$	thioether	$R-\overset{+}{\underset{..}{S}}R_2$	trialkylsulfonium ion		

*Aryl halides and vinyl halides normally do *not* undergo this type of nucleophilic substitution reaction.

Nu		R—Nu		
Formula	*Name*	*Formula*	*Name*	*Comments*
Halogen nucleophiles				
14. $:\ddot{\underset{..}{I}}:^-$	iodide	R—I	alkyl iodide	The usual solvent is acetone. Sodium iodide is soluble in acetone, but sodium bromide and sodium chloride are not.
Carbon nucleophiles				
15. $^-:C{\equiv}N:$	cyanide	R—CN	alkyl cyanide (nitrile)	Sometimes the isonitrile, RNC, is formed.
16. $^-:C{\equiv}CR$	acetylide	$R—C{\equiv}CR$	acetylene	

■ ***EXAMPLE 6.2*** Devise a synthesis for propyl cyanide using a nucleophilic substitution reaction.

Solution First, write the structure of the desired product.

$$CH_3CH_2CH_2—CN$$
propyl cyanide

If we use cyanide ion as the nucleophile (item 15 in Table 6.1), the alkyl halide must have the halogen (Cl, Br, or I) attached to a propyl group. The equation is

$$CN^- + CH_3CH_2CH_2Br \longrightarrow CH_3CH_2CH_2CN + Br^-$$

Some salt, such as sodium or potassium cyanide, can be used to supply the nucleophile.

■ ***EXAMPLE 6.3*** Show how 1-butyne could be converted to 3-hexyne using a nucleophilic substitution reaction.

Solution Compare the starting material with the product.

$$\underset{\text{1-butyne}}{CH_3CH_2C{\equiv}CH} \qquad \underset{\text{3-hexyne}}{CH_3CH_2C{\equiv}CCH_2CH_3}$$

From Table 6.1, line 16, we see that acetylides react with alkyl halides to give acetylenes. We therefore need to convert 1-butyne to an acetylide (review eq. 3.50), then treat it with a 2-carbon alkyl halide.

$$CH_3CH_2C{\equiv}CH + NaNH_2 \xrightarrow{NH_3} CH_3CH_2C{\equiv}C:^-Na^+$$

$$CH_3CH_2C{\equiv}C:^-Na^+ + CH_3CH_2Br \longrightarrow CH_3CH_2C{\equiv}CCH_2CH_3 + Na^+Br^-$$

EXAMPLE 6.4 Complete the following equation:

$$NH_3 + CH_2{=}CHCH_2Br \longrightarrow$$

Solution Ammonia is a nitrogen nucleophile (Table 6.1, line 7). Since both reactants are neutral, the product has a positive charge (the formal +1 charge is on the nitrogen—check it out!)

$$\ddot{N}H_3 + CH_2{=}CHCH_2Br \longrightarrow CH_2{=}CHCH_2\overset{+}{N}H_3 + Br^-$$

Notice that the ammonia reacts only with the C—Br bond; it does not add to the double bond. (Why not?)

Problem 6.1 Using Table 6.1, write complete equations for the following nucleophilic substitution reactions:

a. $NaOH + CH_3CH_2CH_2Br$
b. $(CH_3CH_2)_3N + CH_3CH_2Br$
c. $NaSH + C_6H_5CH_2Br$ (benzene ring—CH_2Br)

Problem 6.2 Write an equation for the preparation of each of the following compounds, using a nucleophilic substitution reaction. In each case, label the nucleophile, the substrate, and the leaving group.

a. $CH_3CH_2CH_2CH_2SH$
b. $(CH_3)_2CHCH_2OH$
c. $(CH_3CH_2CH_2)_2NH$
d. $(CH_3CH_2)_3S^+Br^-$
e. $CH_2{=}CHCH_2I$
f. $C_6H_5OCH_3$ (benzene ring—OCH_3) *(Careful!* See footnote to Table 6.1)

The substitution reactions in Table 6.1 have some serious limitations, particularly with respect to the structure of the *R* group in the alkyl halide. These limitations operate most often when the nucleophile is either an anion or a base or both. For example,

$$\underset{\text{anion}}{CN^-} + \underset{\text{primary halide*}}{CH_3CH_2CH_2CH_2Br} \longrightarrow CH_3CH_2CH_2CH_2CN + Br^- \qquad (6.4)$$

but

$$\underset{\text{anion}}{CN^-} + \underset{\text{tertiary halide*}}{CH_3-\overset{\overset{CH_3}{|}}{\underset{\underset{CH_3}{|}}{C}}-Br} \longrightarrow \underset{\text{methylpropene}}{CH_3-\overset{\overset{CH_2}{\|}}{\underset{\underset{CH_3}{|}}{C}}} + HCN + Br^- \qquad (6.5)$$

* For the definition of primary, secondary, and tertiary alkyl groups, review Sec. 3.12.

Another example is

$$H_2O + CH_3{-}\underset{CH_3}{\overset{CH_3}{\underset{|}{\overset{|}{C}}}}{-}Br \longrightarrow CH_3{-}\underset{CH_3}{\overset{CH_3}{\underset{|}{\overset{|}{C}}}}{-}\overset{+}{O}H_2 + Br^- \quad (6.6)$$

neutral, not very basic; tertiary halide; (about 80%; some methylpropene is also formed)

but

$$OH^- + CH_3{-}\underset{CH_3}{\overset{CH_3}{\underset{|}{\overset{|}{C}}}}{-}Br \longrightarrow CH_3{-}\underset{CH_3}{\overset{CH_2}{\underset{|}{\overset{\|}{C}}}} + H_2O + Br^- \quad (6.7)$$

strong base; tertiary halide; methylpropene

To understand these differences, we must consider the mechanisms by which the substitutions in Table 6.1 take place.

6.4 Nucleophilic Substitution Mechanisms

As a result of experiments that began over sixty years ago, we now understand the mechanisms of nucleophilic substitution reactions rather well. We use the plural because such ***nucleophilic substitutions occur by more than one mechanism.*** The mechanism observed in a particular case depends on the structures of the nucleophile and the alkyl halide, the solvent, the reaction temperature, and other factors.

There are two main nucleophilic substitution mechanisms. These are described by the symbols $\mathbf{S_N2}$ and $\mathbf{S_N1}$, respectively. The S_N part of each symbol stands for "substitution, nucleophilic." The meaning of the numbers 2 and 1 will become clear as we discuss each mechanism.

6.5 The S_N2 Mechanism

The S_N2 mechanism is a one-step process, represented by the following equation:

$$Nu{:} + \; C{-}L \longrightarrow \left[\overset{\delta+}{Nu} \cdots C \cdots \overset{\delta-}{L} \right] \longrightarrow \overset{+}{Nu}{-}C + {:}L^- \quad (6.8)$$

nucleophile; substrate; transition state; product

The nucleophile attacks from the *back* side of the C—L bond. At some stage (the transition state) the nucleophile *and* the leaving group are *both* partly bonded to the carbon at which substitution occurs. As the leaving group departs *with its electron pair*, the nucleophile supplies another electron pair to the carbon atom.

The number 2 is used in describing this mechanism because the reaction is *bi*molecular. That is, two molecules—the nucleophile and the substrate—are involved in the key step (the *only* step) in the reaction mechanism.

How can we recognize when a particular nucleophile and substrate react by the S_N2 mechanism? There are several tell-tale signs.

1. ***The rate of the reaction depends on both the nucleophile and the substrate concentrations.*** The reaction of hydroxide ion with ethyl bromide (eq. 6.1) is an example of an S_N2 reaction. If we double the base concentration (OH^-), the reaction goes twice as fast. The same thing happens if we double the ethyl bromide concentration. We will see shortly that this rate behavior is *not* observed in the S_N1 mechanism.
2. ***Every S_N2 displacement occurs with inversion of configuration.*** For example, if we treat (*R*)-2-bromobutane with sodium hydroxide, we obtain (*S*)-2-butanol.

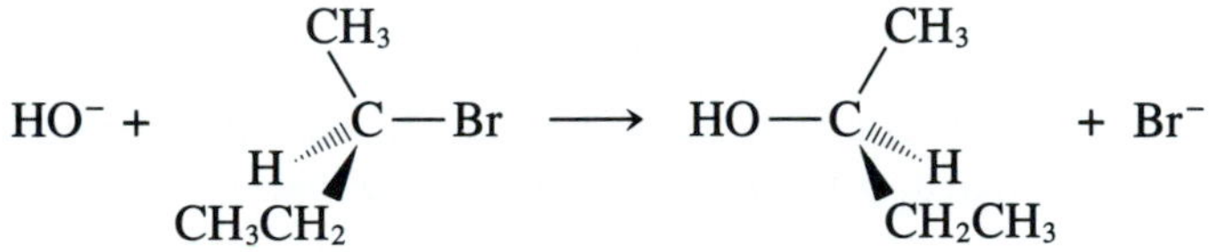

$$HO^- + \underset{(R)\text{-2-bromobutane}}{(CH_3)(H)(CH_3CH_2)C{-}Br} \longrightarrow \underset{(S)\text{-2-butanol}}{HO{-}C(CH_3)(H)(CH_2CH_3)} + Br^- \qquad (6.9)$$

This experimental result, which at first came as a surprise to chemists, meant that the OH group did *not* take the exact position occupied by the Br. If it had, the configuration would have been retained; (*R*)-bromide would have given (*R*)-alcohol. What is the only reasonable explanation?

The hydroxide ion must attack the C—Br bond from the rear. As substitution occurs, the three groups attached to the sp^3 carbon *invert*, somewhat like an umbrella caught in a strong wind.
3. ***The reaction is fastest when the alkyl group of the substrate is methyl or primary and slowest when it is tertiary.*** Secondary alkyl halides react at an intermediate rate. The reason for this reactivity order is fairly obvious if we think about the S_N2 mechanism. The rear side of the carbon, where displacement occurs, is more crowded if more alkyl groups are attached to it, thus slowing down the reaction rate.

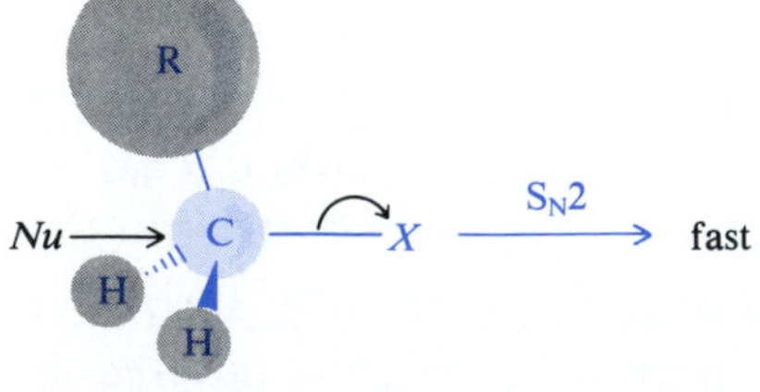

(6.10)

primary halide
(rear side not crowded)

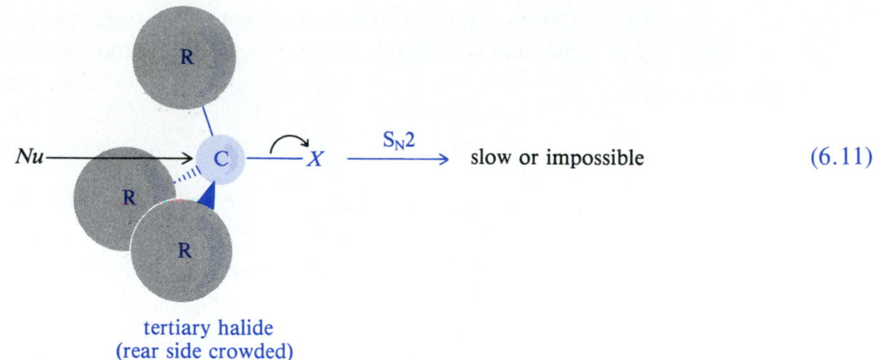

(6.11)

tertiary halide
(rear side crowded)

EXAMPLE 6.5 Predict the product from the S_N2 reaction of *cis*-1-bromo-2-methylcyclopentane with cyanide ion.

Solution

CH_3, H, Br, H (*cis*) $\xrightarrow{CN^-}$ CH_3, H, H, CN (*trans*)

CN^- attacks the C—Br bond from the rear and therefore the cyano group ends up *trans* to the methyl group.

Problem 6.3 Draw a Fischer projection formula for the product of this S_N2 reaction:

CH_3 / H—+—Br / CH_2CH_3 $\xrightarrow[\text{acetone}]{\text{NaI}}$

Problem 6.4 Arrange the following compounds in order of *decreasing* S_N2 reactivity toward sodium ethoxide:

$CH_2{=}CHCH(CH_3)Br$, $CH_2{=}C(CH_3)CH_2Br$, $CH_3CH{=}CHCH_2Br$

To summarize, the S_N2 mechanism is a one-step process favored for methyl and primary halides. It occurs more slowly with secondary halides and usually not at all with tertiary halides. An S_N2 reaction occurs with inversion of configuration, and its rate depends on the concentration of *both* the nucleophile and the substrate (the alkyl halide).

Now let us see how these features differ for the S_N1 mechanism.

6.6 The S_N1 Mechanism

The S_N1 mechanism is a two-step process. In the first step, the bond between the carbon and the leaving group breaks as the substrate dissociates (ionizes).

$$\text{C—L (substrate)} \xrightleftharpoons{\text{slow}} \text{C}^+ \text{ (carbocation)} + :\text{L}^- \tag{6.12}$$

The electrons of the C—L bond go with the leaving group, and a carbocation is formed.

In the second step, which is fast, the carbocation combines with the nucleophile to give the product.

$$\text{C}^+ \text{ (carbocation)} + :\text{Nu (nucleophile)} \xrightarrow{\text{fast}} \text{C—Nu}^+ \text{ or } {}^+\text{Nu—C} \tag{6.13}$$

The number 1 is used to designate this mechanism because the slow, or rate-determining, step involves *only one* of the two reactants: the substrate (eq. 6.12). It does *not* involve the nucleophile at all. That is, the first step is *unimolecular*.

How can we recognize when a particular nucleophile and substrate react by the S_N1 mechanism? Here are the signs:

1. ***The rate of the reaction does not depend on the concentration of the nucleophile.*** The first step is rate determining, and the nucleophile is not involved in this step. The bottleneck in the reaction rate is therefore the rate of formation of the carbocation, not its rate of reaction with the nucleophile, which is nearly instantaneous.
2. ***If the carbon bearing the leaving group is asymmetric, the reaction occurs mainly with loss of optical activity (that is, with racemization).*** In carbocations, only three groups are attached to the positive carbon. Therefore, the positive carbon is sp^2-hybridized and planar. As shown in eq. 6.13, the nucleophile can react at either "face" of the carbocation to give a 50 : 50 mixture of two enantiomers, a racemic mixture. For example, the reaction of (*S*)-3-bromo-3-methylhexane with water gives the racemic alcohol.

$$\text{(S)-3-bromo-3-methylhexane } (CH_3CH_2)(CH_3CH_2CH_2)(CH_3)C\text{–Br} \xrightarrow[\text{acetone}]{H_2O} (CH_3CH_2)(CH_3CH_2CH_2)(CH_3)C\text{–OH (50\% S)} + HO\text{–}C(CH_3)(CH_2CH_3)(CH_2CH_2CH_3) \text{ (50\% R)} \tag{6.14}$$

The intermediate carbocation is planar and achiral.

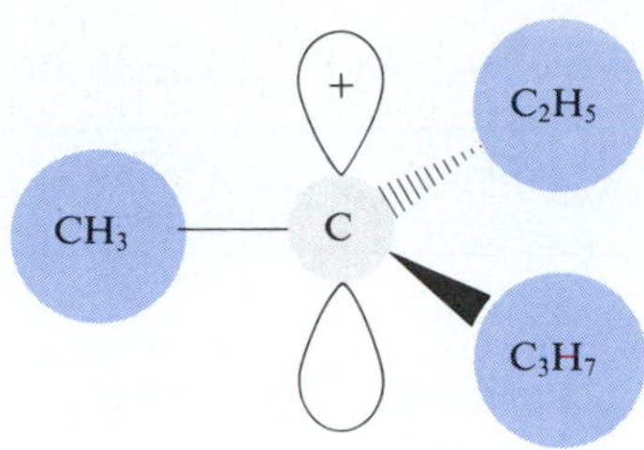

Combination with H_2O from the "top" or "bottom" is equally probable, giving the *S* and *R* alcohols, respectively, in equal amounts.

3. ***The reaction is fastest when the alkyl group of the substrate is tertiary and slowest when it is primary.*** The reason is that S_N1 reactions proceed via carbocations, so the reactivity order corresponds to that of carbocation stability (3° > 2° > 1°). That is, the easier it is to form the carbocation, the faster the reaction will proceed. For this reason, S_N1 reactivity is also favored for resonance-stabilized carbocations, such as allyl or benzyl.

Problem 6.5 Which of the following bromides will react faster with methanol (S_N1), $CH_3CH_2C(CH_3)_2Br$ or $CH_2{=}CHC(CH_3)_2Br$? What are the reaction products in each case?

To summarize, the S_N1 mechanism is a two-step process favored when the alkyl halide is tertiary. Primary halides normally do not react by this mechanism. The S_N1 process occurs with racemization, and its rate is independent of the nucleophile's concentration.

6.7 The S_N1 and S_N2 Mechanisms Compared

How can we tell whether a particular nucleophilic substitution reaction will proceed by an S_N2 or an S_N1 mechanism? And why do we care? First of all, we care for several reasons. We want to be sure that the reaction will proceed at a rate such that we obtain the product in a reasonable time. If the reaction has stereochemical consequences, we want to know in advance what that outcome will be, inversion or racemization.

Table 6.2 should be helpful in answering our first question. It summarizes what we have said so far about the two substitution mechanisms, and it compares them with respect to two other variables, solvent and nucleophile structure, which we will take up here.

Notice that primary halides almost always react by the S_N2 mechanism, whereas tertiary halides react by the S_N1 mechanism. Only with secondary halides are we likely to encounter both possibilities.

One experimental variable that we can use to help control mechanism is the solvent polarity. Water and alcohols are **polar protic solvents** (protic because of the proton-donating ability of the hydroxyl groups). How will such solvents affect S_N1 and S_N2 reactions?

The first step of the S_N1 mechanism involves the formation of ions. Since polar solvents can solvate ions, the reaction rate of S_N1 processes is enhanced

TABLE 6.2 Comparison of S_N2 and S_N1 substitutions.

Halide structure	*S_N2*	*S_N1*
Primary or CH_3	Common	Rarely*
Secondary	Sometimes	Sometimes
Tertiary	Rarely	Common
Stereochemistry	Inversion	Racemization
Solvent	Rate is retarded by polar protic solvents	Because the intermediates are ions, the rate is highly favored with polar solvents
Nucleophile	Rate depends on nucleophile concentration; mechanism is favored when the nucleophile is an anion	Rate is independent of nucleophile concentration; mechanism is more likely with neutral nucleophiles

*Allyl and benzyl substrates are the common exceptions.

by polar solvents. On the other hand, solvation of nucleophiles ties up their unshared electron pairs. Therefore, S_N2 reactions, whose rates depend on nucleophile effectiveness, are usually retarded by polar protic solvents.*

Let us apply these ideas to a *secondary halide,* which might react by either mechanism. We may choose the mechanism simply by adjusting the solvent polarity. For example, we can change the mechanism by which a secondary halide reacts with water (to form an alcohol) from S_N2 to S_N1 by changing the solvent from 95% acetone–5% water (relatively nonpolar) to 50% acetone–50% water (more polar and a better ionizing solvent).

EXAMPLE 6.6 The reaction of (*R*)-2-iodobutane with 95% acetone–5% water gives mainly (*S*)-2-butanol. With 30% acetone–70% water, the product has a much lower optical activity, namely being about 60% (*S*)-2-butanol and 40% (*R*)-2-butanol. Explain.

Solution In 95% acetone–5% water (a relatively nonpolar solvent), substitution occurs by the S_N2 mechanism with inversion.

$$H_2\ddot{O}: + \text{(R)-}CH_3CH(CH_2CH_3)I \xrightarrow[-I^-]{S_N2} H_2\overset{+}{O}{-}CH(CH_3)(CH_2CH_3) \underset{}{\overset{-H^+}{\rightleftharpoons}} HO{-}CH(CH_3)(CH_2CH_3)$$

(*R*)-2-iodobutane (*S*)-2-butanol

*Polar but *aprotic* solvents [examples are dimethyl sulfoxide, $(CH_3)_2S{=}O$, or dimethylformamide, $(CH_3)_2NCHO$] solvate cations instead of anions. These solvents *accelerate* S_N2 reactions because, by solvating the cation (say, K^+ in K^+CN^-), they leave the anion more "naked" or unsolvated and improve its nucleophilicity.

When the percentage of water in the solvent is increased, the solvent becomes more polar, and some reaction occurs by the S_N1 process.

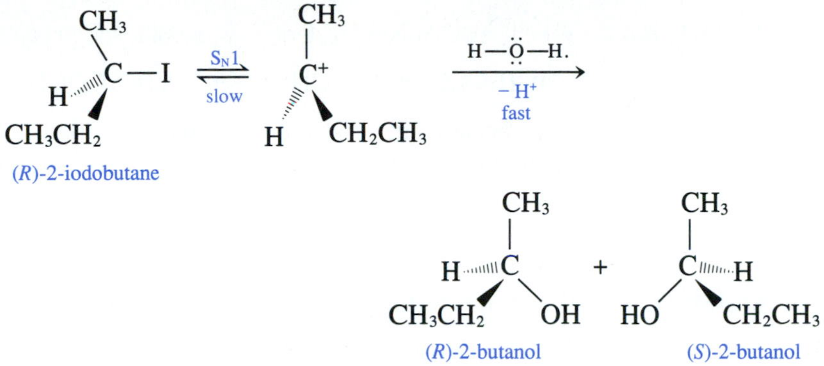

$$(R)\text{-2-iodobutane} \underset{\text{slow}}{\overset{S_N1}{\rightleftharpoons}} \text{carbocation} \xrightarrow[-H^+,\ \text{fast}]{H\ddot{O}H} (R)\text{-2-butanol} + (S)\text{-2-butanol}$$

(R)-2-iodobutane

(R)-2-butanol (S)-2-butanol

There is a slight preference for the (*S*) isomer (60 : 40) for one of two reasons. Either some of the reaction still occurs by the S_N2 mechanism, or else the departing iodide ion in the S_N1 mechanism is still close enough to the face of the carbocation from which it departed to partially block that face from nucleophilic attack.

Now let us consider the other variable in Table 6.2—the nucleophile. As we have seen, the rate of an S_N2 reaction (but *not* an S_N1 reaction) depends on the nucleophile. If a reagent is a *strong* nucleophile, the S_N2 mechanism will be favored. How can we tell whether a nucleophile is strong or weak, or whether one nucleophile is stronger than another? Here are a few useful generalizations.

1. Negative ions are more nucleophilic, or better electron suppliers, than the corresponding neutral molecules. Thus,

$$HO^- > HOH \qquad RS^- > RSH$$

$$RO^- > ROH \qquad R{-}\underset{\displaystyle O}{\overset{}{\underset{\|}{C}}}{-}O^- > R{-}\underset{\displaystyle O}{\underset{\|}{C}}{-}OH$$

2. Elements low in the periodic table tend to be more nucleophilic than elements above them in the same column. Thus,

$$HS^- > HO^- \qquad I^- > Br^- > Cl^- > F^- \quad \text{(in protic solvents)}$$

$$R\ddot{\underset{\cdot\cdot}{S}}H > R\ddot{\underset{\cdot\cdot}{O}}H \qquad (CH_3)_3P{:} > (CH_3)_3N{:}$$

3. Elements in the same row in the periodic table tend to be less nucleophilic, the more electronegative the element (that is, the more tightly it holds electrons to itself). Thus,

$$R_3C^- > R_2N^- > R{-}O^- > F^- \quad \text{and} \quad H_3N{:} > H_2\ddot{O}{:} > H\ddot{\underset{\cdot\cdot}{F}}{:}$$

Can we juggle all these factors to make some predictions about particular substitution reactions? Here are some examples.

EXAMPLE 6.7 Which mechanism, S_N1 or S_N2, would you predict for this reaction?

$$(CH_3)_3CBr + CH_3OH \longrightarrow (CH_3)_3COCH_3 + HBr$$

Solution S_N1, because the substrate is a tertiary alkyl halide. Also, the nucleophile methanol is neutral and, if used as the reaction solvent, rather polar. Thus, it favors ionization.

EXAMPLE 6.8 Which mechanism, S_N1 or S_N2, would you predict for this reaction?

$$CH_3CH_2I + NaCN \longrightarrow CH_3CH_2CN + NaI$$

Solution S_N2, because the substrate is a primary halide, and cyanide, an anion, is a rather strong nucleophile.

Problem 6.6 Which mechanism, S_N1 or S_N2, would you predict for each of the following reactions?

a. $CH_3\underset{\displaystyle Br}{\underset{|}{C}}HCH_2CH_2CH_3 + Na^+SH^- \longrightarrow CH_3\underset{\displaystyle SH}{\underset{|}{C}}HCH_2CH_2CH_3 + NaBr$

b. 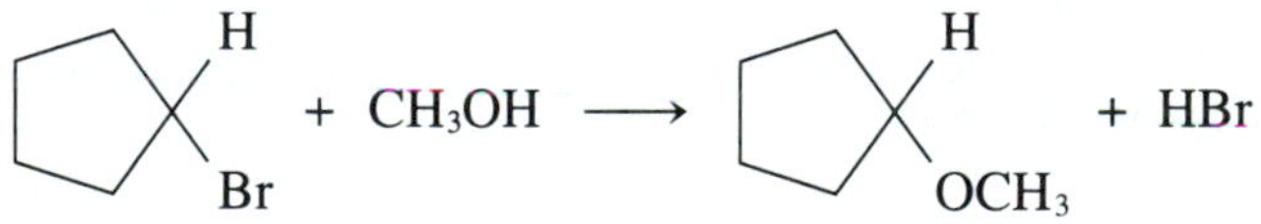 cyclopentane bearing H and Br on one carbon + $CH_3OH \longrightarrow$ cyclopentane bearing H and OCH_3 on one carbon + HBr

c. 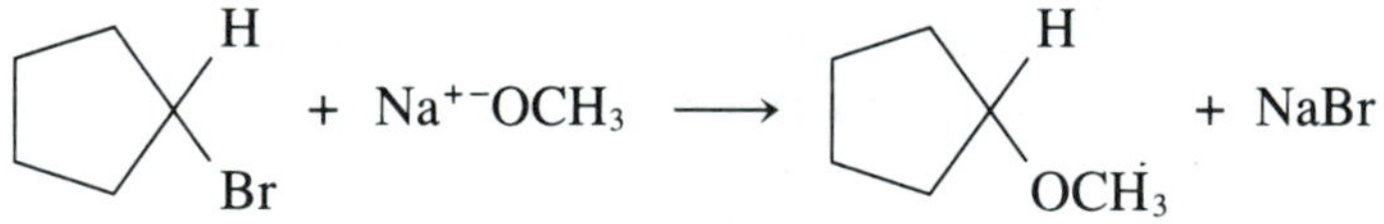 cyclopentane bearing H and Br on one carbon + $Na^{+-}OCH_3 \longrightarrow$ cyclopentane bearing H and OCH_3 on one carbon + NaBr

Problem 6.7 By what mechanism do the reactions in eqs. 6.4 and 6.6 occur? Explain what structural feature prevents the same type of reaction from occurring with tertiary butyl bromide (eqs. 6.5 and 6.7).

Now let us turn to reactions like those in eqs. 6.5 and 6.7, to try to explain why a reaction other than substitution occurred.

6.8 Dehydrohalogenation, an Elimination Reaction; the E2 and E1 Mechanisms

We have now seen several examples of reactions in which two reactants give, not a single product, but mixtures. Examples include halogenation of alkanes (eq. 2.14), addition to double bonds (eqs. 3.14 and 3.29), and electrophilic aromatic substitutions (Sec. 4.12), where more than one isomer may be formed from the same two reactants. Even in nucleophilic substitutions cases like that in eq. 6.14 exist, in which hydrolysis of a single alkyl bromide gives a mixture of two alcohols. But sometimes we find **two entirely different reaction types occurring at the same time between the same two reactants,** to give two (or more) entirely different types of products. Let us consider one example.

When an alkyl halide with a hydrogen attached to the carbon *adjacent* to the halogen-bearing carbon reacts with a nucleophile, two competing reaction paths are possible: **substitution** or **elimination.**

$$\mathrm{H{-}\overset{2}{C}{-}\overset{1}{C}{-}X} + \mathrm{Nu}: \xrightarrow{\text{substitution } (S)} \mathrm{H{-}\overset{|}{\underset{|}{C}}{-}\overset{|}{\underset{|}{C}}{-}Nu} + \mathrm{X^-} \quad (6.15)$$

$$\xrightarrow{\text{elimination } (E)} \mathrm{{>}C{=}C{<}} + \mathrm{Nu} + \mathrm{X^-} \quad (6.16)$$

In the substitution reaction, the nucleophile replaces the halogen X. In the elimination reaction, the nucleophile acts as a base and removes a proton from carbon 2, the carbon next to the one that bears the halogen X. The halogen X and the hydrogen from the *adjacent* carbon atom are *eliminated,* and a new bond (a pi bond) is formed between carbons 1 and 2. The symbol E is used to designate an elimination process. Since, in this case, a hydrogen halide is eliminated, the reaction is called **dehydrohalogenation.** Elimination reactions provide a useful way to prepare compounds with double or triple bonds.

Often substitution and elimination reactions occur simultaneously with the same set of reactants—a nucleophile and a substrate. One reaction type or the other may predominate, depending on the structure of the nucleophile, the structure of the substrate, and other reaction conditions. As with substitution reactions, **there are two main mechanisms for elimination reactions, designated E2 and E1.** To learn how to control these reactions, we must first understand their mechanisms.

Like the S_N2 mechanism, **the E2 mechanism is a one-step process.** *The nucleophile, acting as a base, removes the proton (hydrogen) on a carbon atom adjacent to the one that bears the leaving group. At the same time, the leaving group departs and a double bond is formed.* The flow of electron pairs is shown by the curved arrows:

$$\mathrm{Nu}: \; \mathrm{H{-}C{-}C{-}L} \xrightarrow{\text{E2}} \mathrm{{>}C{=}C{<}} + \mathrm{Nu{-}H} + \mathrm{L^-} \quad (6.17)$$

The preferred conformation for the substrate in an E2 reaction is shown in eq. 6.17. The H—C—C—L atoms lie in a single plane, with H and L in a *transoid* or *anti* arrangement. The reason for this preference is that the electrons in the C—H and C—L orbitals are properly aligned in this conformation to overlap and form the new pi bond. Another way to think of this is that as the nucleophile removes the proton, the electron pair that remains carries out an *intramolecular* S_N2-type displacement on the C—L bond.

The E1 mechanism has the same first step as the S_N1 mechanism, the slow and rate-determining ionization of the substrate to give a carbocation (compare with eq. 6.12).

$$\underset{\text{substrate}}{-\overset{\overset{H}{|}}{\underset{|}{C}}-\overset{|}{\underset{|}{C}}-L} \xrightleftharpoons{\text{slow}} \underset{\text{carbocation}}{-\overset{\overset{H}{|}}{\underset{|}{C}}-\overset{|}{\underset{|}{C^+}}} + L^- \qquad (6.18)$$

Two reactions are then possible for the carbocation. It may either combine with a nucleophile (the S_N1 process) or lose a proton from a carbon atom adjacent to the positive carbon, as shown by the curved arrow, to give an alkene (the E1 process).

$$\underset{\text{carbocation}}{-\overset{\overset{H}{|}}{\underset{|}{C}}-\overset{|}{\underset{|}{C^+}}} \begin{cases} \xrightarrow{Nu:} -\overset{\overset{H}{|}}{\underset{|}{C}}-\overset{|}{\underset{|}{C}}-Nu & S_N1 \\ \xrightarrow{-H^+} >C=C< \; + H^+ & E1 \end{cases} \qquad (6.19)$$

6.9 Substitution and Elimination in Competition

Now we can consider specific examples of how substitution and elimination reactions compete with one another. Let us consider the options for each class of alkyl halide.

6.9a. Tertiary Halides Substitution can only occur by the S_N1 mechanism, but elimination can occur by either mechanism, E1 or E2. With weak nucleophiles and polar solvents, S_N1 and E1 mechanisms compete. For example,

$$\underset{t\text{-butyl bromide}}{(CH_3)_3CBr} \xrightleftharpoons{H_2O} (CH_3)_3C^+ + Br^- \begin{cases} \xrightarrow{H_2O,\ S_N1} \underset{\text{(about 80\%)}}{(CH_3)_3COH} \\ \xrightarrow{E1} \underset{\text{(about 20\%)}}{(CH_3)_2C{=}CH_2} + H^+ \end{cases} \qquad (6.20)$$

Can the ratio of S_N1/E1 products be altered under these conditions? No, because both products arise from a common intermediate, the carbocation.

If we use a strong nucleophile instead of a weak one, and if we use a less polar solvent, we favor elimination by the E2 mechanism. Thus, with OH^- or CN^- as nucleophiles, only elimination occurs (eqs. 6.5 and 6.7), and the exclusive product is the alkene.

$$HO^- + \underset{t\text{-butyl bromide}}{H-CH_2-C(CH_3)_2-Br} \xrightarrow{E2} \underset{\substack{\text{methylpropene} \\ (100\%)}}{CH_2{=}C(CH_3)_2} + H_2O + Br^- \qquad (6.21)$$

Remember that the tertiary carbon is too hindered sterically for S_N2 attack, so under these conditions, substitution cannot compete with elimination.

6.9b. Primary Halides. Only the S_N2 and E2 mechanisms are possible, because ionization to a primary carbocation, the first step that would be required for the S_N1 or E1 mechanisms, does not occur.

In general, with most nucleophiles, *primary halides give mainly substitution products* (S_N2). Only with very bulky, strongly basic nucleophiles do we see the E2 process favored. For example,

$$CH_3CH_2CH_2CH_2Br \xrightarrow[\text{in ethanol}]{CH_3CH_2O^- Na^+} CH_3CH_2CH_2CH_2OCH_2CH_3 + CH_3CH_2CH{=}CH_2$$

1-bromobutane → butyl ethyl ether (S_N2; 90%) + 1-butene (E2, 10%)

$$CH_3CH_2CH_2CH_2Br \xrightarrow[\text{in } t\text{-butyl alcohol}]{(CH_3)_3C{-}O^- K^+} CH_3CH_2CH_2CH_2OC(CH_3)_3 + CH_3CH_2CH{=}CH_2 \qquad (6.22)$$

1-bromobutane → butyl *t*-butyl ether (S_N2; 15%) + 1-butene (E2, 85%)

Potassium *t*-butoxide is a bulky base. Hence substitution is retarded, and the main reaction is elimination.

6.9c. Secondary Halides. All four mechanisms, S_N2 and E2, S_N1 and E1, are possible. The product composition will be sensitive to the nucleophile (its strength as a nucleophile and as a base) and to the reaction conditions (solvent, temperature). In general, substitution is favored with good nucleophiles that are not strong bases (S_N2) or by weaker nucleophiles such as polar solvents (S_N1), but elimination is favored by strong bases (E2).

$$CH_3CH(Br)CH_3 \xrightarrow[\text{strong nucleophile}]{Na^+SH^-} CH_3CH(SH)CH_3 \; (S_N2)$$

$$CH_3CH(Br)CH_3 \xrightarrow[\text{weak nucleophile}]{CH_3CH_2OH} CH_3CH(OCH_2CH_3)CH_3 + CH_3CH{=}CH_2 \qquad (6.23)$$

2-bromopropane → (S_N1; major) + (E1; minor)

$$CH_3CH(Br)CH_3 \xrightarrow[\text{strong base}]{CH_3CH_2O^- Na^+} CH_3CH(OCH_2CH_3)CH_3 + CH_3CH{=}CH_2$$

(S_N2; minor) + (E2; major)

EXAMPLE 6.9 Predict the product of the reaction of 1-chloro-1-methylcyclopentane with

a. sodium ethoxide in ethanol.
b. refluxing ethanol.

Solution The alkyl chloride is tertiary

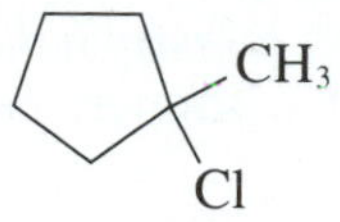

[Structure: cyclopentane ring with CH_3 and Cl on the same carbon]

a. The first set of conditions favors the E2 process, because sodium ethoxide is a strong base. Two elimination products are possible, depending on whether the base attacks a hydrogen on an adjacent CH_2 or CH_3 group.

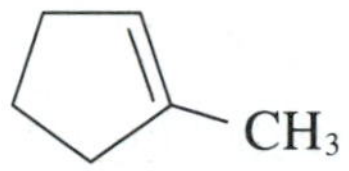

[Structures: 1-methylcyclopentene (CH_3) and methylenecyclopentane ($=CH_2$)]

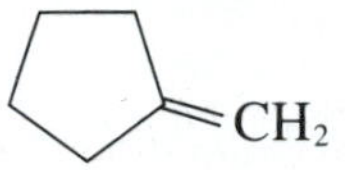

b. This set of conditions favors ionization, because the ethanol is neutral (hence a weak nucleophile) and, as a solvent, fairly polar. The S_N1 process predominates, and the main product is the ether

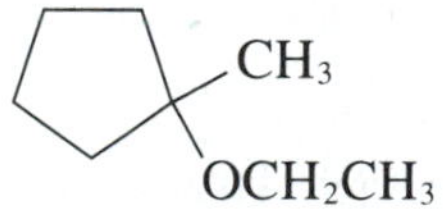

[Structure: cyclopentane ring with CH_3 and OCH_2CH_3 on the same carbon]

Some of the above alkenes will also be formed by the E1 mechanism.

EXAMPLE 6.10 Draw structures for *all* possible E2 products when 2-bromobutane reacts with concentrated sodium hydroxide.

Solution

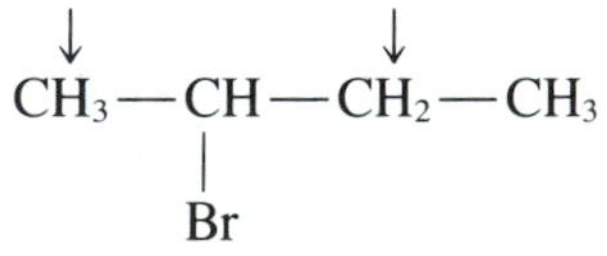

$$\overset{\downarrow}{CH_3}-\underset{\underset{Br}{|}}{CH}-\overset{\downarrow}{CH_2}-CH_3$$

The base can attack the hydrogens on the adjacent CH_3 or CH_2 group, as indicated by the arrows. Attack at the methyl group puts the double bond between carbon-1 and carbon-2.

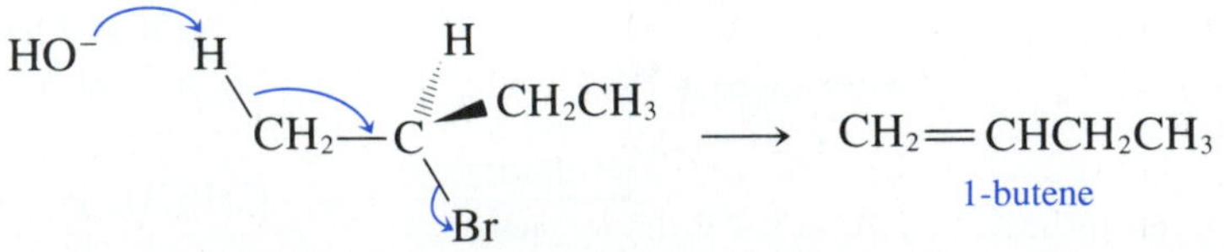

$$HO^- \quad H-CH_2-C(H)(CH_2CH_3)(Br) \longrightarrow \underset{\text{1-butene}}{CH_2{=}CHCH_2CH_3}$$

Attack at the methylene group puts the double bond between carbon-2 and carbon-3, giving 2-butene. Either *cis*- or *trans*-2-butene can be obtained, so altogether, *three* alkenes are formed.

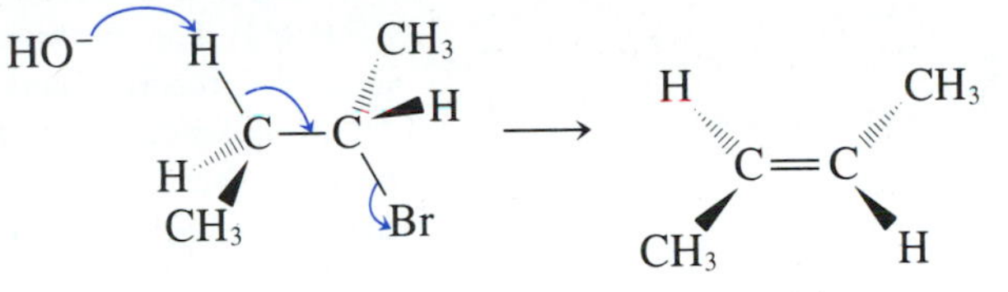

trans-2-butene

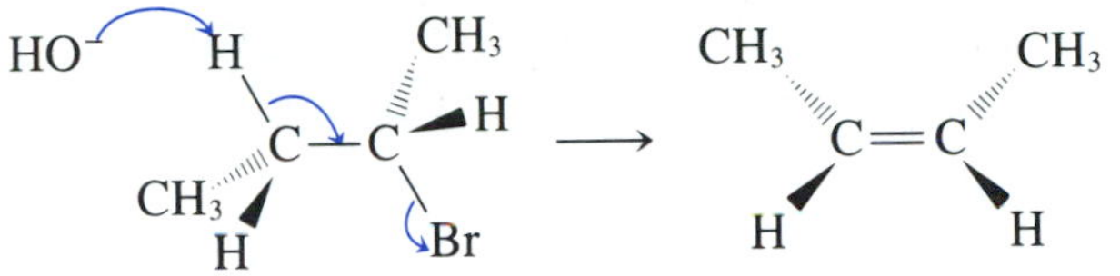

cis-2-butene

Problem 6.8 Draw structures for *all* possible elimination products obtainable from 3-bromo-3-methylhexane.

Problem 6.9 Treatment of the alkyl halide in problem 6.8 with KOH in methanol gives mainly a mixture of the alkenes whose structures you drew. But treatment with only methanol gives a different product. What is it, and by what mechanism is it formed?

A WORD ABOUT . . .

10. S_N2 Reactions in the Cell: Biochemical Methylations

Substitution and elimination reactions are so useful that it is not surprising that they occur in living matter. However, alkyl halides are not compatible with cytoplasm, being hydrocarbon-like and therefore water insoluble. In the cell, **alkyl phosphates** play the role that alkyl halides do in the laboratory. Adenosine triphosphate (ATP) is an example of a biological equivalent of an alkyl halide. We will abbreviate its structure here as Ad—O—Ⓟ—Ⓟ—Ⓟ (the full structure is given in Sec. 18.13). Ad- can be considered as a primary alkyl group, and the triphosphate group, —O—Ⓟ—Ⓟ—Ⓟ, acts as a leaving group, just like a halogen.

There are many compounds in nature with a methyl group attached to an oxygen or nitrogen atom. Examples include *mescaline* (a hallucinogen from the peyote cactus), which has three —OCH_3 groups; *morphine* (the pain-relieving drug from opium), which has an $\rangle N—CH_3$ group; and *codeine* (a close relative of morphine, used as an anticough agent), which has an $\rangle N-CH_3$ group and an $-OCH_3$ group.

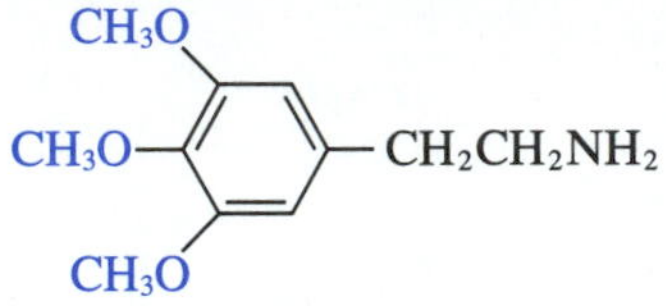

mescaline

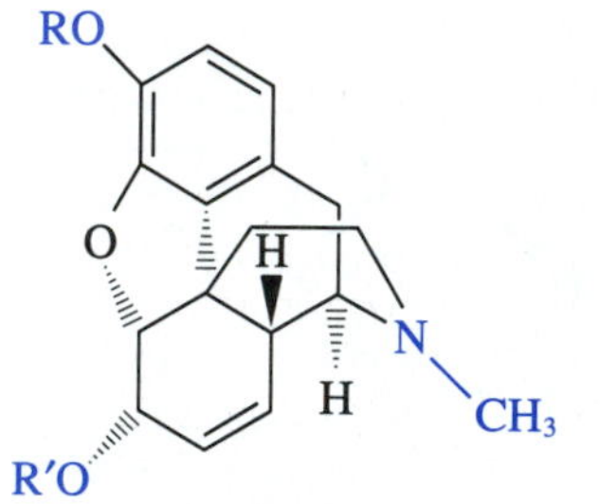

morphine ($R = R' = H$)
codeine ($R = CH_3$, $R' = H$)

How do the methyl groups get there? Two steps are involved, both of them nucleophilic substitutions.

The methyl carrier in most biochemical methylations is a sulfur-containing amino acid called *methionine*. In the first step, methionine is alkylated by ATP to form S-adenosylmethionine (shown in Figure 6.1). This reaction is a biological example of reaction 13 in Table 6.1. The methionine acts as a sulfur nucleophile in an S_N2 reaction and displaces the triphosphate ion.

In the second step, the oxygen or nitrogen atom to be methylated acts as a nucleophile. The S-adenosylmethionine acts just like a methyl halide.

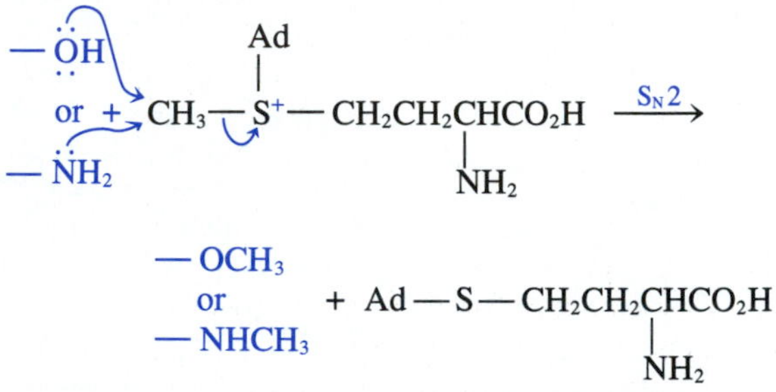

$$\begin{matrix}-\ddot{O}H \\ \text{or} \\ -\ddot{N}H_2\end{matrix} + CH_3-\overset{\overset{\displaystyle Ad}{|}}{S^+}-CH_2CH_2\underset{\underset{\displaystyle NH_2}{|}}{C}HCO_2H \xrightarrow{S_N2} \begin{matrix}-OCH_3 \\ \text{or} \\ -NHCH_3\end{matrix} + Ad-S-CH_2CH_2\underset{\underset{\displaystyle NH_2}{|}}{C}HCO_2H$$

S-adenosylhomocysteine

Eventually, the S-adenosylhomocysteine formed in the second step is converted, by enzymes, back to ATP and methionine for reuse.

Biochemical methylations are just one of many examples of nucleophilic substitutions that occur in metabolic processes.

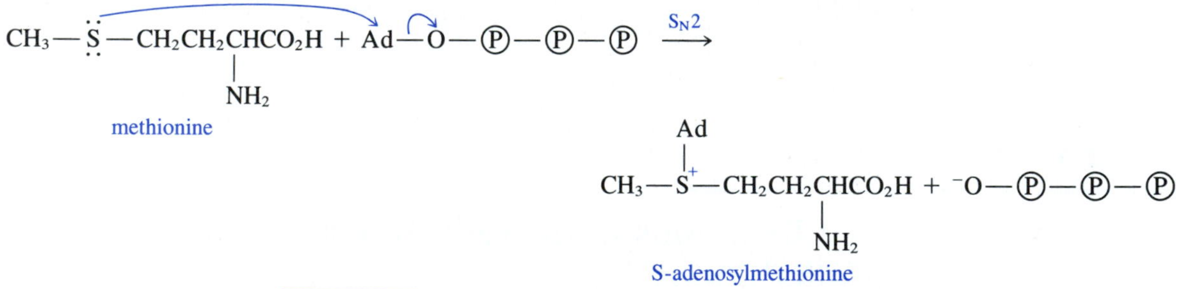

$$CH_3-\ddot{\underset{..}{S}}-CH_2CH_2\underset{\underset{\displaystyle NH_2}{|}}{C}HCO_2H + Ad-O-Ⓟ-Ⓟ-Ⓟ \xrightarrow{S_N2}$$

methionine

$$CH_3-\overset{\overset{\displaystyle Ad}{|}}{S^+}-CH_2CH_2\underset{\underset{\displaystyle NH_2}{|}}{C}HCO_2H + {}^-O-Ⓟ-Ⓟ-Ⓟ$$

S-adenosylmethionine

FIGURE 6.1 The formation of S-adenosylmethionine.

6.10 Polyhalogenated Aliphatic Compounds

Because of their useful properties, many polyhalogen compounds are produced commercially. Being industrial chemicals, they are usually given common names.

Chlorinated methanes* are made by the chlorination of methane (eqs. 2.11 and 2.13). **Carbon tetrachloride** (CCl_4, bp 77°C), **chloroform** ($CHCl_3$, bp

*The analogous F, Br, and I compounds are also known, but are much more expensive and not commercially important.

62°C), and **methylene chloride** (CH_2Cl_2, bp 40°C) are all water insoluble, but effective solvents for organic compounds. Also important for this purpose are the **tri-** and **tetrachloroethylenes,** used in dry cleaning and as degreasing agents in metal and textile processing.

$$\underset{\substack{\text{trichloroethylene}\\ \text{bp 87°C}}}{Cl_2C{=}CHCl} \qquad \underset{\substack{\text{tetrachloroethylene}\\ \text{bp 121°C}}}{Cl_2C{=}CCl_2}$$

Because some of these chlorinated compounds are suspected carcinogens, adequate ventilation is essential when they are used as solvents.

Tetrafluoroethylene is the raw material for **Teflon,** a polymer related to polyethylene (Sec. 3.15) but with all of the hydrogens replaced by fluorine atoms.

$$n\,CF_2{=}CF_2 \xrightarrow[\text{catalyst}]{\text{peroxide}} \underset{\text{Teflon}}{\left(CF_2CF_2\right)_n} \tag{6.24}$$

Teflon has exceptional properties. It is resistant to almost all chemicals and is widely used as a nonstick coating for pots, pans, and other cooking utensils. It stands up to heat and prevents food from sticking to the utensil surface, making it easy to clean. Another use of Teflon is in Gore-Tex-like fabrics, materials with as many as nine billion pores per square inch. These pores are the right size to transmit water vapor but not liquid water. Thus, perspiration vapor can pass through the fabric, but wind, rain, and snow cannot. Gore-Tex has revolutionized cold- and wet-weather gear for both military and civilian uses. It is used in skiwear, boots, sleeping bags, tents, and other rugged outdoor gear. Teflon is also used in wire insulation, as a protective coating over glass and fabric roofs, in vascular grafts and heart patches, and recently in delicate films for certain biological experiments outside the earth's atmosphere.

A WORD ABOUT ...

11. Artificial Blood

Fantastic as it may seem, it has now been proved that animals can survive after having most or all of their natural blood replaced by a suitable emulsion of perfluorochemicals—hydrocarbons, ethers, or amines in which all of the hydrogens are replaced by fluorine atoms! This discovery, which dates back to the mid-1960s, holds promise for medical use and has already saved human lives. Perfluorochemicals can dissolve as much as 60% of oxygen by volume. By contrast, whole blood dissolves only about 20% and blood plasma, about 3%. Rats totally submerged in fluorochemicals saturated with oxygen continue to "breathe" for long periods. Although perfluorochemicals are immiscible with blood, it was found that they can be emulsified (broken up into microdroplets that remain suspended in solution) by adding various surfactants (compounds related to soaps). Here is a typical example of an artificial blood "recipe": perfluorotributylamine, $(CF_3CF_2CF_2CF_2)_3N$, 11–13 mL; Pluronic F-68 (a polymeric emulsifier), 2.3–2.7 g; hydroxyethyl starch, 2.5–3.2 g; NaCl, 54 mg; KCl, 32 mg; $MgCl_2$, 7 mg; $CaCl_2$, 10 mg; Na_2HPO_4, 9.6 mg; enough Na_2CO_3 to adjust the pH to 7.44; enough water to dilute to 100 mL of artificial blood. Other perfluorochemicals, such as F-decalin ($C_{10}F_{18}$), can also be used. Several different techniques are employed to convert these mixtures into emulsions; the particle size must be carefully regulated if the blood substitute is to be effective.

Unlike ordinary blood, artifical blood is unaffected by certain poisons. "Bloodless" rats have breathed oxygen containing 10 to 50% carbon monoxide for hours and continued to live normally on return to an ordinary nitrogen-oxygen atmosphere. Of course, the artificial blood is gradually replaced by normal blood as the animal continues to live and produce blood in the usual way. This experiment proved, however, that artificial blood is fully capable of carrying out the oxygen-transport function of blood, because any red blood cells that might have been present would have been prevented by the carbon monoxide from transporting oxygen.

Perfluorochemicals that function temporarily as artificial blood are eventually excreted over time (days to weeks). Because of their chemical inertness, they apparently do no permanent harm.

What uses might artificial blood have? It might be used in emergencies and disasters when blood banks are depleted and natural blood is in short supply. No typing is required, so it can be used even with patients who have rare blood types. Artificial blood might also be used to preserve organs for transplant surgery (kidneys, hearts), to treat certain diseases (sickle-cell or cancer-associated anemia), and to administer drugs that would otherwise react with ordinary blood. Complete body washout of ordinary blood could be used to rid the body of viruses, toxins, or drug overdoses. In Japan, doctors have used artificial blood on patients threatened with death due to excessive bleeding, and artificial blood has been used to save the lives of patients who refused conventional transfusion for religious reasons.

When first prepared, perfluorochemicals were chemical curiosities. No one could have foreseen their properties or uses. They provide an outstanding example of why it is important to fund research in basic science without regard to immediate practical benefit.

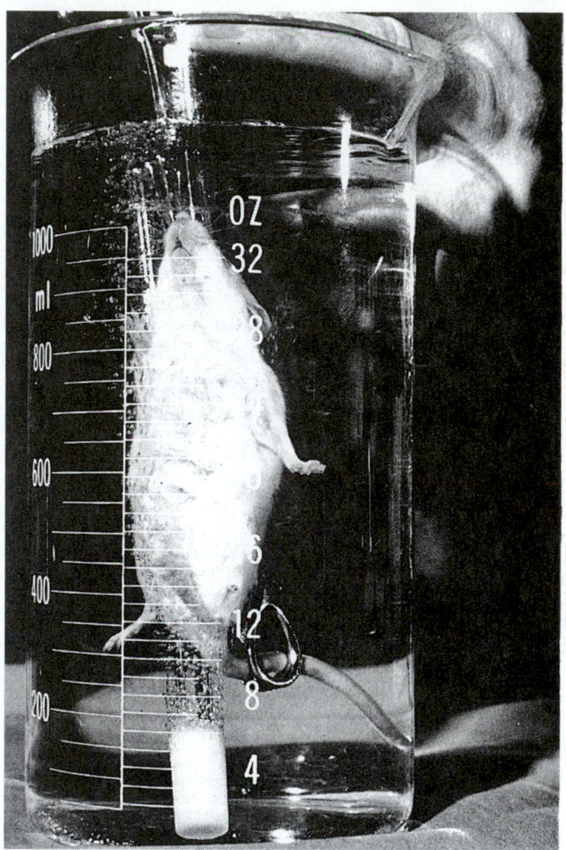

Polyhalogen compounds that contain two or three different halogens per molecule are important commercial compounds. The best known are the **chlorofluorocarbons** (**CFCs,** formerly known as **Freons**). The two that have been produced on the largest scale are CFC-11 and CFC-12, made by fluorination of carbon tetrachloride.

$$\underset{\text{(bp 77°C)}}{CCl_4} \xrightarrow[SbF_5]{HF} \underset{\substack{\text{trichlorofluoromethane} \\ \text{(CFC-11)} \\ \text{bp 24°C}}}{CCl_3F} \xrightarrow[SbF_5]{HF} \underset{\substack{\text{dichlorodifluoromethane} \\ \text{(CFC-12)} \\ \text{bp } -30°C}}{CCl_2F_2} \qquad (6.25)$$

They are used as refrigerants, as blowing agents in the manufacture of foams, as cleaning fluids, and as aerosol propellants. They are exceptionally stable. Because of this stability, they accumulate in the upper stratosphere, where they can damage the earth's ozone layer. Consequently, their use for nonessential propellant purposes is now banned in the United States, and the search is on for replacements (see "A Word About" number 12.).

Bromine-containing compounds of this type are now widely used to extinguish fires. Called **Halons,** the best known are

$CBrClF_2$	$CBrF_3$
bromochloro-difluoromethane (Halon-1211)	bromotrifluoro-methane (Halon-1301)

Halons are much more effective than carbon tetrachloride. They are very important in air safety because of their ability to douse fires within seconds.

A WORD ABOUT . . .

12. CFCs, the Ozone Layer, and Tradeoffs

The story of CFCs has many lessons for us. Fluorine's reputation as one of the most reactive of the elements was widespread. It was quite remarkable and unexpected, therefore, that fluorocarbons and chlorofluorocarbons should be so extremely *un*reactive. (Indeed, Thomas Midgley, who discovered CFCs, demonstrated their nontoxicity, nonflammability, and noncorrosiveness to the public by inhaling them and then puffing out a candle.)

Because of these and other properties, CFCs have at least *four* major commercial uses. Their low boiling points and other heat properties make them excellent *refrigerants,* far superior to ammonia, sulfur dioxide, and other rather difficult-to-handle refrigerants. They are now used in most freezers and refrigerators, as well as in all types of air conditioners. CFCs make excellent *blowing agents* for rigid foams (such as those used for ice chests, fast-food take-out boxes, and other packaging materials) and for flexible foams (like those used to make pillows and furniture cushions). Their low surface tension and low viscosity give them excellent wetting properties, which led to their use as *cleaning fluids*. They are excellent for cleaning printed computer circuits, artificial limbs, and many other products. Finally, they are used as *propellants* in aerosol sprays. CFCs are manufactured on such a scale that they constitute a multibillion-dollar industry.

But their extreme stability has led to a major world problem. CFCs are so stable that, when they are released into the atmosphere, they do not decompose in the lower atmosphere, as most other industrial chemicals do. Instead, they eventually rise to the stratosphere where, through ultraviolet radiation, the C—Cl bonds are broken and chlorine atoms are released. These chlorine atoms initiate a chain of reactions that damages the ozone layer, needed to protect life on earth from harmful ultraviolet rays.

How to solve the problem? Ban all use of CFCs? To do so could bring about a major crisis for civilization. How, without refrigeration, could we ensure safe and adequate food supplies to urban areas, or delivery of heat-sensitive medical supplies? A return to old-fashioned refrigerants is fraught with even more environmental hazards than CFCs. So the solution is not simple.

But we can cut back or ban some less essential uses, which *is* being done. For example, in the United States (but *not* worldwide), all nonessential aerosol use of CFCs is banned. In 1987, twenty-four countries signed the Montreal Protocol, an agreement that calls for cutting CFC use to half of the 1986 level by 1998. Currently, the venting of CFC-12 from automobile and truck air conditioners, when they are serviced, is the largest single source of ozone destroying CFCs in the United States. A proposal to allow auto mechanics to purify and recycle this CFC-12 could solve this problem.

Ultimately, the best solution is to seek harmless replacements for CFCs. Chemists and other scientists are working hard on this problem and are getting results. The properties of polyhalogen compounds are so uniquely useful that for the moment the best chance to find CFC substitutes is among this type of compound. For example, it was found that introducing one or more hydrogen atoms into the molecule substantially increases its decomposition rate in the lower atmosphere, where no damage to the ozone layer is possible. Compounds such as CF_3CHCl_2 (HCFC-123, bp 28°C) and CF_3CH_2F(HFC-134a, bp −26°C) are possible replacements for CFC-11 and CFC-12, respectively. These compounds are more difficult to synthesize than CFCs, however, and will be more expensive. But continued use of the cheaper CFCs and destruction of the ozone layer is not a viable alternative.

CFCs are just one of many examples of the trade-offs between beneficial and possible harmful effects of a new research product. During World War II, the use of CFCs as propellants for the insecticide DDT saved the lives of many troops in the Pacific zone who were suffering more casualties from malaria than from enemy action. But later, indiscriminate use of CFCs as propellants for all sorts of trivial purposes led to our current problems with their environmental accumulation. Chemicals (in this case, CFCs) are neither good nor evil, but we must exercise good judgment in how we use them.

A WORD ABOUT ...

13. Insecticides and Herbicides

Many polyhalogen compounds have been used as insecticides and herbicides. Perhaps the best-known is DDT (*di*chloro*di*phenyl*tri*chloroethane). It is manufactured by the acid-catalyzed reaction of chlorobenzene with trichloroacetaldehyde.

$$2\ C_6H_5\text{–}Cl + CCl_3CH{=}O \xrightarrow{H^+} Cl\text{–}C_6H_4\text{–}\underset{\displaystyle CCl_3}{\underset{|}{CH}}\text{–}C_6H_4\text{–}Cl + H_2O$$

DDT

DDT was used during World War II to control malaria by killing mosquitoes. It is also effective against flies and many agricultural pests. Unfortunately, DDT is not easily degraded biochemically, and excessive use resulted in its accumulation in the environment. It tends to accumulate in fat tissues and can cause harm, particularly to fish and birds. Currently the use of DDT is restricted, and annual world production is only about 80,000 tons.

Weeds create formidable problems for agriculture. They consume nutrients and moisture needed by crops. They also rob crops of sunlight and space, thus reducing crop yields. U.S. agricultural production is diminished by about 10% because of weeds. The annual financial loss is about \$12 billion, and an additional \$6 billion is spent annually on weed control.

One way to control weeds is through the use of herbicides. About 85% to 90% of U.S. corn, soybean, cotton, peanut, and rice acreage is sprayed with herbicides to control weeds. Some are sprayed before the crop is planted, others are applied after planting but before emergence of the crop, and still others are applied to the weed foliage itself. The rising world population makes increasing demands on food production and, ultimately, on herbicide use to meet that demand. Although most herbicides are used for this purpose, some are also employed for industrial purposes—along railway and power line rights of way, on roadsides, on rangeland, in vacant lots, and so on—as well as for lawns and gardens.

Farmers have used weed killers for many years; common salt was used for this purpose even in ancient times. Before World War II, most chemical herbicides were not very selective (they killed weeds but also damaged crops) and had to be used in large quantities per acre of land. The breakthrough came with the discovery that 2,4-dichlorophenoxyacetic acid (2,4-D) killed broadleaf weeds but allowed narrowleaf plants to grow unharmed and in greater yield. In addition, only 0.25 to 2.0 pounds were needed per acre (compared to over 200 pounds per acre for inorganic herbicides such as sodium chlorate). 2,4-D is still the most popular herbicide used on wheat fields.

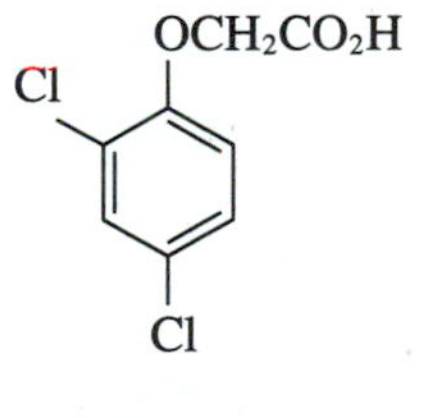

OCH_2CO_2H, Cl, Cl

2,4-dichlorophenoxyacetic acid
"2,4-D"

At least 40 herbicides are in large-scale use. Some of the most widely used are

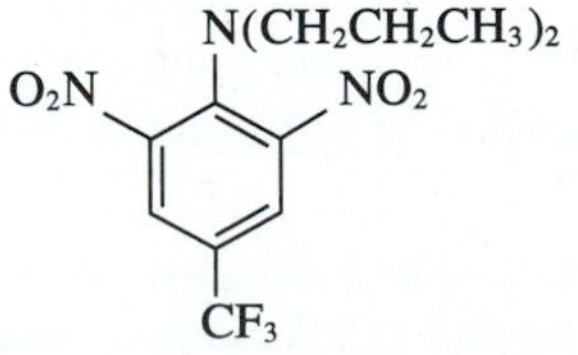

$N(CH_2CH_2CH_3)_2$, O_2N, NO_2, CF_3

trifluralin
(cotton, lima beans, cantaloupes, tomatoes, sugar beets)

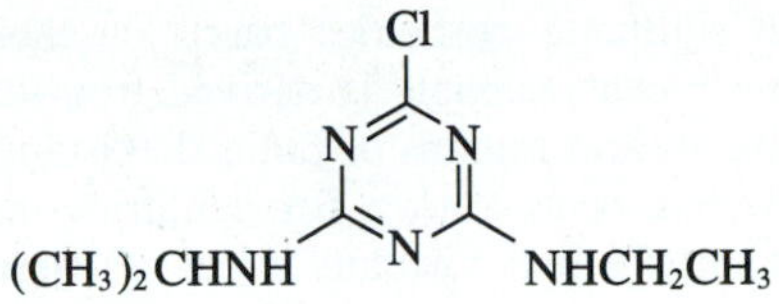

Cl, N, N, N, $(CH_3)_2CHNH$, $NHCH_2CH_3$

atrazine
(corn, sugarcane, pineapples)

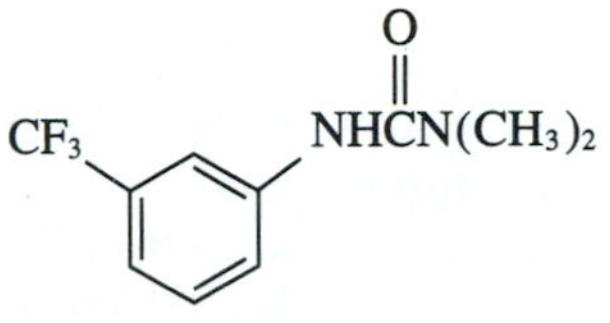

CF_3, O, $NHCN(CH_3)_2$

fluometuron
(cotton, sugarcane)

A new class of herbicides recently developed by DuPont, represents a major breakthrough. One example is Glean™.

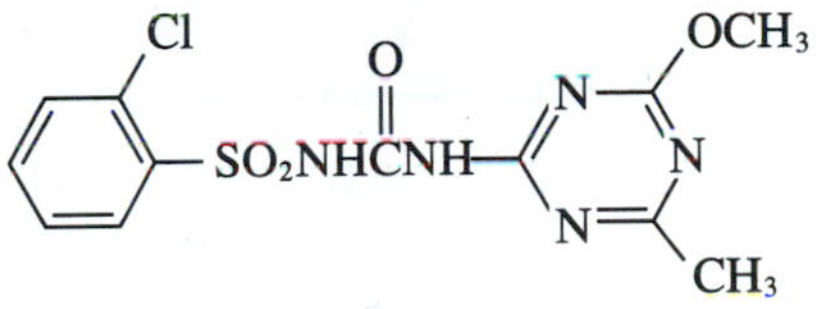

Cl, O, SO_2NHCNH, N, N, N, OCH_3, CH_3

chlorsulfuron
(Glean™)

It is effective against a broad spectrum of weeds in cereal grains (wheat, barley, oats) at the remarkably low application level of *less than one ounce per acre*. At present, at least a dozen pesticides of this type are marketed commercially, serving unique needs of farmers.

Because of the danger that, even in very low concentrations, pesticides may have harmful side effects and become environmental pollutants, and because agriculture without pesticides can no longer meet human needs, still other approaches to the important problem of pest control are being researched. For example, many plants have natural defense mechanisms against predators and pathogens; they produce substances that shield them from harm. These substances may kill harmful bacteria, repel insects or interfere with their reproductive cycles, prevent fungal spores from germinating, and in many other ways act as natural protection for a plant. One new approach to pesticides is to isolate these natural defense substances, learn how they work, and then either use these materials themselves (rotenone, the principal insecticidal constituent of derris root, is a well-known example) or design synthetic pesticides based on their structures. One recent example is derived from the neem tree, native to arid regions of India, Pakistan, and Sri Lanka. It has been known for centuries that areas where it grows are virtually free of insects, nematodes, and plant diseases. Extracts of its seeds provide effective protection against more than 100 crop pests, aphids, and boll weevils; and a pesticide based on these extracts has recently been approved for limited use.

The achievements of modern agriculture and the feeding of the ever-growing world population would not be possible without the herbicides that chemists and other scientists have developed.

Additional Problems

6.10. Using Table 6.1, write an equation for each of the following substitution reactions.

a. 1-bromobutane + sodium iodide (in acetone)
b. 2-chlorobutane + sodium ethoxide
c. *t*-butyl bromide + water
d. *p*-chlorobenzyl chloride + sodium cyanide
e. *n*-propyl iodide + sodium acetylide
f. 2-chloropropane + sodium hydrosulfide
g. allyl chloride + ammonia (2 equivalents)
h. 1,4-dibromobutane + sodium cyanide (excess)
i. 1-methyl-1-bromocyclohexane + methanol

6.11. Select an alkyl halide and a nucleophile that will give each of the following products:

a. $CH_3CH_2CH_2NH_2$
b. $CH_3CH_2SCH_2CH_3$
c. $HC{\equiv}CCH_2CH_2CH_3$
d. $(CH_3)_2CHOCH(CH_3)_2$
e. $NCCH_2-C_6H_4-CH_2CN$ (para-disubstituted benzene ring)
f. $C_6H_5-OCH_2CH_3$

6.12. Draw out each of the following equations in a way that shows clearly the stereochemistry of the reactants and products.

a. (*S*)-2-bromobutane + sodium methoxide (in methanol) $\xrightarrow{S_N2}$ 2-methoxybutane
b. (*R*)-3-bromo-3-methylhexane + methanol $\xrightarrow{S_N1}$ 3-methoxy-3-methylhexane
c. *cis*-2-bromo-1-methylcyclopentane + NaSH $\longrightarrow$ 2-methylcyclopentanethiol

6.13. Use Fischer projection formulas to show the stereochemistry of Problem 6.12, parts a and b.

6.14. Determine the order of reactivity for $(CH_3)_2CHCH_2Br$, $(CH_3)_3CBr$, and $CH_3\underset{\displaystyle Br}{\underset{|}{C}}HCH_2CH_3$ reacting with

a. sodium cyanide. **b.** 50% aqueous acetone.

6.15. When treated with sodium iodide, a solution of (*R*)-2-iodooctane in acetone gradually loses all its optical activity. Explain.

6.16. Eq. 6.20 shows that hydrolysis of *t*-butyl bromide gives about 80% $(CH_3)_3COH$ and 20% $(CH_3)_2C{=}CH_2$. The same ratio of alcohol to alkene is obtained when the starting halide is *t*-butyl chloride or *t*-butyl iodide. Explain.

6.17. Tell what product you expect, and by what mechanism it is formed, for each of

the following reactions:
a. 1-chloro-1-methylcyclohexane + ethanol
b. 1-chloro-1-methylcyclohexane + sodium ethoxide (in ethanol)

6.18. Give the structures of all possible products when 2-chloro-2-methylbutane reacts by the E1 mechanism.

6.19. Explain the different products of the following two reactions by considering the mechanism by which each reaction proceeds:

$$CH_2{=}CH{-}\underset{\displaystyle Br}{\underset{|}{CH}}{-}CH_3 + Na^+\,{}^-OCH_3 \xrightarrow{CH_3OH} CH_2{=}CH{-}\underset{\displaystyle OCH_3}{\underset{|}{CH}}{-}CH_3$$

$$CH_2{=}CH{-}\underset{\displaystyle Br}{\underset{|}{CH}}{-}CH_3 + CH_3OH \longrightarrow CH_2{=}CH{-}\underset{\displaystyle OCH_3}{\underset{|}{CH}}{-}CH_3 + \underset{\displaystyle OCH_3}{\underset{|}{CH_2}}CH{=}CHCH_3$$

6.20. Combine an electrophilic addition and a nucleophilic substitution to devise a two-step synthesis of

a. $CH_3\underset{\displaystyle OCH_3}{\underset{|}{C}H}CH_2CH_3$ from $CH_3CH{=}CHCH_3$.

b. $CH_3{-}\overset{\displaystyle CH_3}{\overset{|}{\underset{\displaystyle OCH_3}{\underset{|}{C}}}}{-}CH_2CH_3$ from $CH_3{-}\overset{\displaystyle CH_3}{\overset{|}{C}}{=}CHCH_3$.

c. 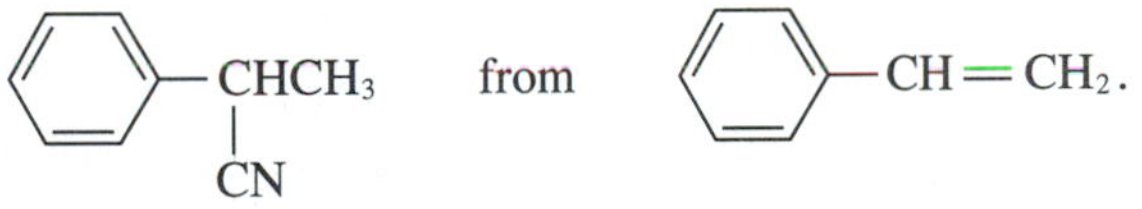 $C_6H_5{-}\underset{\displaystyle CN}{\underset{|}{CH}}CH_3$ from $C_6H_5{-}CH{=}CH_2$.

6.21. Devise a two-step synthesis of each of the following, starting with an appropriate alkene:

a. 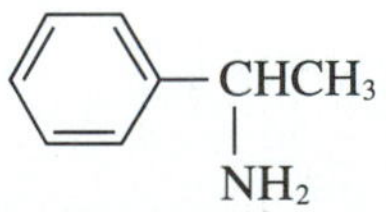 $C_6H_5{-}\underset{\displaystyle NH_2}{\underset{|}{CH}}CH_3$ **b.** $(CH_3CH_2)_2CHSH$

6.22. Combine the reaction of an alcohol with sodium and a nucleophilic substitution to devise a two-step synthesis of
a. $CH_3OCH_2CH_3$ from CH_3OH and CH_3CH_2Br.
b. $CH_3OC(CH_3)_3$ from an alcohol and an alkyl halide.

6.23. Combine a 1,4-electrophilic addition to a conjugated diene with a nucleophilic substitution to devise a two-step synthesis of
a. $CH_3CH{=}CHCH_2CN$ from $CH_2{=}CH{-}CH{=}CH_2$.
b. $NCCH_2CH{=}CHCH_2CN$ from $CH_2{=}CH{-}CH{=}CH_2$.

6.24. Combine the reaction in eq. 3.50 with a nucleophilic substitution to devise

a. a two-step synthesis of $CH_3C{\equiv}C{-}CH_2{-}C_6H_5$ from

$CH_3C{\equiv}CH$ and $C_6H_5{-}CH_2Br$.

b. a four-step synthesis of $CH_3C{\equiv}CCH_2CH_3$ from acetylene and appropriate alkyl halides.

6.25. Combine an electrophilic addition with an elimination reaction to devise a two-step synthesis of

a. $(CH_3)_2C{=}CHCH_3$ from $CH_2{=}C(CH_3)CH_2CH_3$.

b. 1-methylcyclohexene (ring with CH_3) from methylenecyclohexane-type structure bearing CH_3 (cyclohexene with CH_3 on the exocyclic position).

6.26. Combine a nucleophilic substitution with a catalytic hydrogenation to synthesize

a. $CH_3CH_2CH_2OH$ from $CH_2{=}CHCH_2Br$.

b. *cis*-2-butene from propyne and methyl iodide.

CHAPTER 7

Alcohols, Phenols, and Thiols

7.1 *Introduction*

Alcohols have the general formula **R—OH,** and are structurally similar to water but with one of the hydrogens replaced by an alkyl group. Their functional group is the **hydroxyl group,** —OH. **Phenols** have the same functional group, but it is attached to an *aromatic* ring. **Thiols** are similar to alcohols and phenols, except that the oxygen is replaced by sulfur.

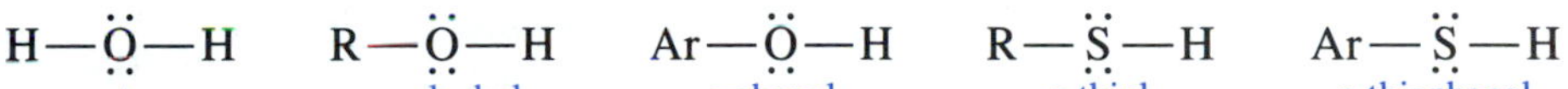

$H—\ddot{\underset{..}{O}}—H$	$R—\ddot{\underset{..}{O}}—H$	$Ar—\ddot{\underset{..}{O}}—H$	$R—\ddot{\underset{..}{S}}—H$	$Ar—\ddot{\underset{..}{S}}—H$
water	an alcohol	a phenol	a thiol	a thiophenol

Alcohols, phenols, and thiols occur in nature. In this chapter, we will discuss the physical properties and main chemical reactions of these classes of compounds. We will also describe their commercial and laboratory syntheses and give examples of their biological importance.

7.2 *Nomenclature of Alcohols*

In the IUPAC system, the hydroxyl group in alcohols is indicated by the ending **-ol.** In common names the separate word *alcohol* is placed after the name of the alkyl group. The following examples illustrate the use of IUPAC rules, with common names given in parentheses.

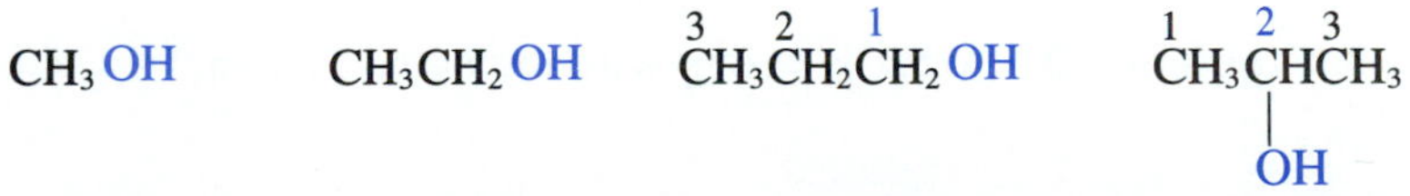

CH_3OH	CH_3CH_2OH	$\overset{3}{C}H_3\overset{2}{C}H_2\overset{1}{C}H_2OH$	$\overset{1}{C}H_3\overset{2}{C}H(OH)\overset{3}{C}H_3$
methanol (methyl alcohol)	ethanol (ethyl alcohol)	1-propanol (*n*-propyl alcohol)	2-propanol (isopropyl alcohol)

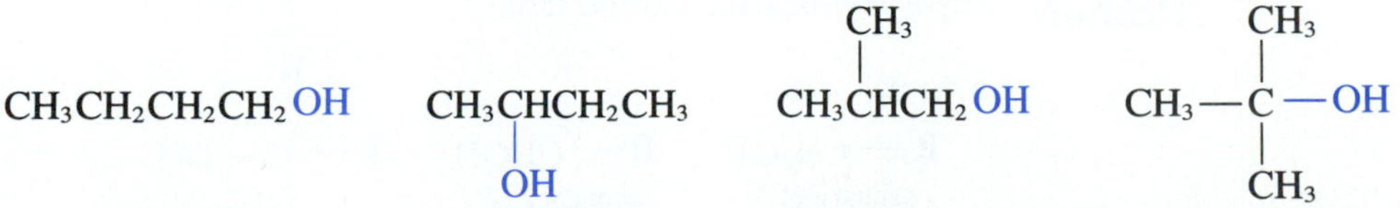

$CH_3CH_2CH_2CH_2OH$	$CH_3CH(OH)CH_2CH_3$	$CH_3CH(CH_3)CH_2OH$	$(CH_3)_3C—OH$
1-butanol (*n*-butyl alcohol)	2-butanol (*sec*-butyl alcohol)	2-methyl-1-propanol (isobutyl alcohol)	2-methyl-2-propanol (*tert*-butyl alcohol)

$CH_2{=}CHCH_2OH$	cyclohexane ring bearing H and OH	C_6H_5–CH_2OH
2-propen-1-ol	cyclohexanol	phenylmethanol
(allyl alcohol)	(cyclohexyl alcohol)	(benzyl alcohol)

With unsaturated alcohols, two endings are needed: one for the double or triple bond and one for the hydroxyl group (see the IUPAC name for allyl alcohol). In these cases, the *-ol* suffix comes last and takes precedence in numbering.

EXAMPLE 7.1 Name the following alcohols by the IUPAC system:

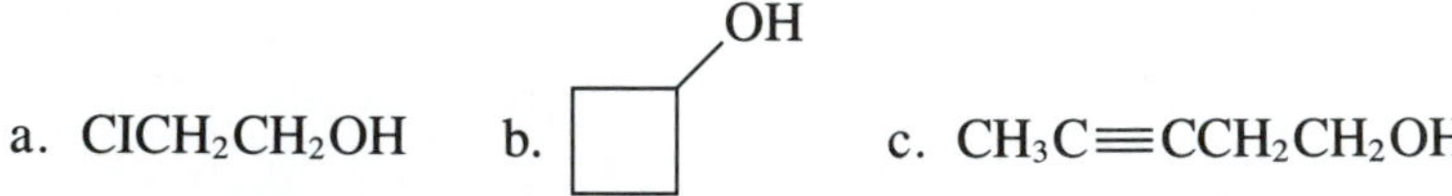

a. $ClCH_2CH_2OH$ b. (cyclobutane ring with OH) c. $CH_3C{\equiv}CCH_2CH_2OH$

Solution
a. 2-chloroethanol (number from the hydroxyl-bearing carbon)
b. cyclobutanol
c. 3-pentyne-1-ol (*not* 2-pentyne-5-ol)

EXAMPLE 7.2 What is wrong with the name *isopropanol* for $(CH_3)_2CHOH$?

Solution It mixes two naming systems; either 2-propanol or isopropyl alcohol is correct, but isopropanol follows neither the IUPAC nor the common name system and should not be used.

Problem 7.1 Name these alcohols by the IUPAC system:

a. $BrCH_2CH_2CH_2OH$ b. (cyclopentane ring bearing H and OH) c. $CH_2{=}CHCH_2CH_2OH$

Problem 7.2 Write a structural formula for

a. 2-pentanol. b. 1-phenylethanol. c. 3-penten-2-ol.

Problem 7.3 What is wrong with the name *t*-butanol for $(CH_3)_3COH$?

7.3 Classification of Alcohols

Alcohols are classified as primary (1°), secondary (2°), or tertiary (3°), depending on on whether one, two, or three organic groups are connected to the hydroxyl-bearing carbon atom.

$R{-}CH_2OH$	$R{-}CH(R)OH$	$R{-}C(R)(R){-}OH$
primary (1°)	secondary (2°)	tertiary (3°)

Methyl alcohol, which is not strictly covered by this classification, is usually grouped with the primary alcohols. This classification is similar to that for carbocations (Sec. 3.11). We will see that the chemistry of an alcohol sometimes depends on its class.

Problem 7.4 Classify as 1°, 2°, or 3° the eleven alcohols listed in Sec. 7.2.

7.4 Nomenclature of Phenols

Phenols are usually named as derivatives of the parent compounds.

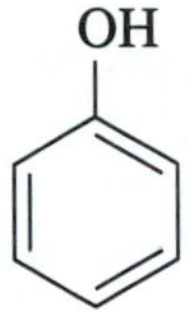

phenol *

p-chlorophenol

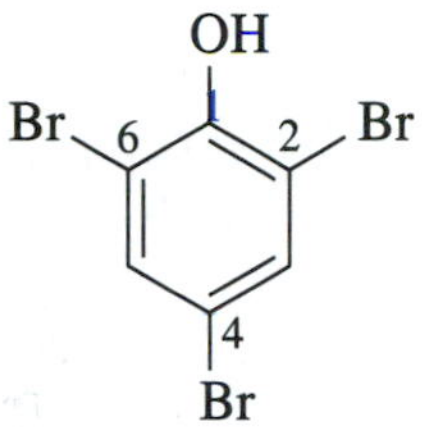

2,4,6-tribromophenol

The hydroxyl group is named as a substituent when it occurs in the same molecule with carboxylic acid, aldehyde, or ketone functionalities, which have priority in naming. Examples are

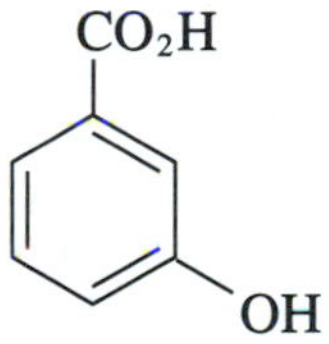

m-hydroxybenzoic acid

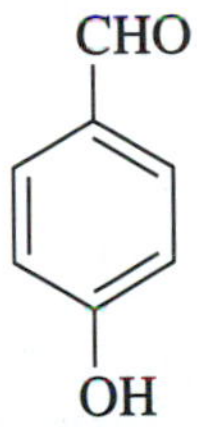

p-hydroxybenzaldehyde

but

p-nitrophenol (*not p*-hydroxynitrobenzene)

Problem 7.5 Write the structure for

a. *p*-ethylphenol.
b. pentachlorophenol (an insecticide for termite control, and a fungicide).
c. *o*-hydroxyacetophenone.

7.5 Hydrogen Bonding in Alcohols and Phenols

The boiling points of alcohols are much higher than those of ethers or hydrocarbons with similar molecular weights.

	CH_3CH_2OH	CH_3OCH_3	$CH_3CH_2CH_3$
mol wt	46	46	44
bp	+78.5°C	−24°C	−42°C

* The name *benzenol* has recently been introduced for phenol and its derivatives. Although this name is used by Chemical Abstracts Services, it is not in common use among organic chemists.

Why? Because their molecules form **hydrogen bonds** with one another. The O—H bond is polarized by the high electronegativity of the oxygen atom. This polarization places a partial positive charge on the hydrogen atom and a partial negative charge on the oxygen atom. Because of these charges and the small size of the hydrogen atom, it can link together two electronegative atoms such as oxygen.

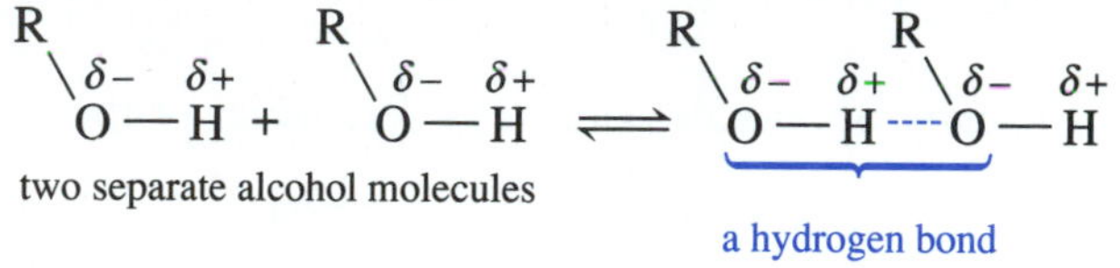

$$\underset{\text{two separate alcohol molecules}}{\overset{R}{\backslash}\overset{\delta-}{O}-\overset{\delta+}{H} + \overset{R}{\backslash}\overset{\delta-}{O}-\overset{\delta+}{H}} \rightleftharpoons \underbrace{\overset{R}{\backslash}\overset{\delta-}{O}-\overset{\delta+}{H}\cdots\overset{R}{\backslash}\overset{\delta-}{O}}_{\text{a hydrogen bond}}-\overset{\delta+}{H} \qquad (7.1)$$

Two or more alcohol molecules thus become loosely bonded to one another through hydrogen bonds.

Hydrogen bonds are weaker than ordinary covalent bonds. Nevertheless, their strength is significant, about 5 to 10 kcal/mol (20 to 40 kJ/mol). Consequently, alcohols and phenols have relatively high boiling points because we must not only supply enough heat to vaporize each molecule but also supply enough heat (energy) to break the hydrogen bonds before each molecule can be vaporized.

Water, of course, is also a hydrogen-bonded liquid. The lower-molecular-weight alcohols can readily replace water molecules in the hydrogen-bonded network.

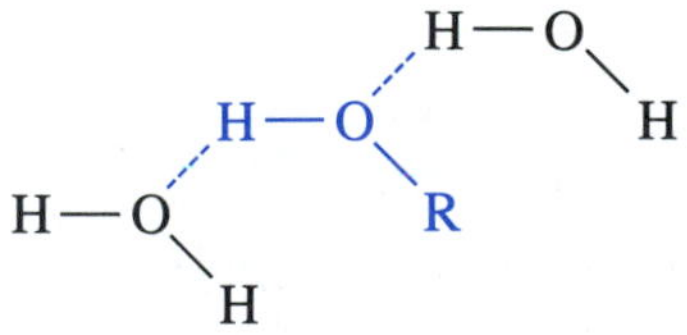

H—O(—H) ··· H—O(—R) ··· H—O—H

This accounts for the complete miscibility of the lower alcohols with water. However, as the organic chain lengthens and the alcohol becomes relatively more hydrocarbon-like, its water solubility decreases. Table 7.1 illustrates these properties.

TABLE 7.1 Boiling point and water solubility of some alcohols

Name	*Formula*	*bp, °C*	*Solubility in H_2O g/100 g at 20°C*
methanol	CH_3OH	65	completely miscible
ethanol	CH_3CH_2OH	78.5	completely miscible
1-propanol	$CH_3CH_2CH_2OH$	97	completely miscible
1-butanol	$CH_3CH_2CH_2CH_2OH$	117.7	7.9
1-pentanol	$CH_3CH_2CH_2CH_2CH_2OH$	137.9	2.7
1-hexanol	$CH_3CH_2CH_2CH_2CH_2CH_2OH$	155.8	0.59

7.6 Acidity and Basicity Reviewed

The acid–base behavior of organic compounds often helps to explain their chemistry; this is certainly true of alcohols. It is a good idea, therefore, to review the fundamental concepts of acidity and basicity.

Acids and bases are defined in two ways. According to the **Brønsted-Lowry definition,** ***an acid is a proton donor, and a base is a proton acceptor.*** For example, in eq. 7.2, which represents what occurs when hydrogen chloride dissolves in water, the water accepts a proton from the hydrogen chloride.

$$H{-}\ddot{O}{:}(H) + H{-}\ddot{Cl}{:} \rightleftharpoons H{-}\overset{+}{\ddot{O}}(H){-}H + {:}\ddot{Cl}{:}^- \qquad (7.2)$$

base; acid; conjugate acid of water; conjugate base of hydrogen chloride

Here water acts as a base or proton acceptor, and hydrogen chloride acts as an acid or proton donor. The products of this proton exchange are called the **conjugate acid** and the **conjugate base.**

The strength of an acid (or base) is measured quantitatively by its **acidity constant,** or **ionization constant, K_a,** usually calculated in reference to water. K_a is the equilibrium constant for the reaction.

$$HA + H_2O \rightleftharpoons H_3O^+ + A^- \qquad (7.3)$$

where

$$K_a = \frac{[H_3O^+][A^-]}{[HA]} \qquad (7.4)^*$$

The stronger the acid, the more this equilibrium is shifted to the right, thus increasing the concentration of H_3O^+ and the value of K_a.

For water, these expressions are

$$H_2O + H_2O \rightleftharpoons H_3O^+ + OH^- \qquad (7.5)$$

and K_a (water), usually written K_w, is

$$K_w = \frac{[H_3O^+][OH^-]}{[H_2O]} = 1.8 \times 10^{-16} \qquad (7.6)$$

Problem 7.6 Verify from eq. 7.6 and from the molarity of water (55.5 M) that the concentration of H_3O^+ or OH^- in water is 10^{-7} moles per liter.

To avoid using numbers with negative exponents, such as those we have just seen for K_w, we often express acidity as **pK_a,** ***the negative logarithm of the acidity constant.***

---

* The square brackets used in the expression for K_a indicate concentration of the enclosed species in moles per liter. The concentration of water, $[H_2O]$, is usually omitted from the denominator of this expression since it remains nearly constant at 55.5 M, very large compared to the concentrations of the other three species.

$$pK_a = -\log K_a \tag{7.7}$$

The pK_a of water is

$$-\log (1.8 \times 10^{-16}) = -\log 1.8 - \log 10^{-16} = -0.26 + 16 = +15.74$$

The mathematical relationship between the values for K_a and pK_a means that ***the smaller*** K_a ***or the larger*** pK_a***, the weaker the acid.***

It is useful to keep in mind that there is an inverse relationship between the strength of an acid and that of its conjugate base. In eq. 7.2, for example, hydrogen chloride is a *strong* acid since the equilibrium is shifted largely to the right. It follows, then, that the chloride ion must be a *weak* base since it has relatively little affinity for a proton. Similarly, since water is a *weak* acid, its conjugate base, hydroxide ion, must be a *strong* base.

Another way to define acids and bases was first proposed by G. N. Lewis. ***A Lewis acid is a substance that can accept an electron pair, and a Lewis base is a substance that can donate an electron pair.*** According to this definition, a proton is considered a Lewis acid because it can accept an electron pair from a donor (a Lewis base) to fill its 1*s* shell.

$$\underset{\text{Lewis acid}}{H^+} + \underset{\text{Lewis base}}{:\ddot{O}(H)-H} \rightleftharpoons \left[H-\ddot{O}(H)-H\right]^+ \tag{7.8}$$

Any atom with an unshared electron pair can act as a Lewis base.

Compounds with an element whose valence shell is incomplete also act as Lewis acids. For example,

$$\underset{\text{Lewis acid}}{BF_3} + \underset{\text{Lewis base}}{:\ddot{F}:^-} \rightleftharpoons \left[BF_4\right]^- \tag{7.9}$$

Similarly, when $FeCl_3$ or $AlCl_3$, acts as a catalyst for electrophilic aromatic chlorination (eqs. 4.16 and 4.17) or the Friedel-Crafts reaction (eqs. 4.23 and 4.25), they are acting as Lewis acids; the metal atom accepts an electron pair from chlorine or from an alkyl or acyl chloride to complete its valence shell of electrons.

Finally, some substances can act as either an acid or a base, depending on the other reactant. For example, in eq. 7.2, water acts as a base (a proton acceptor). However, in its reaction with ammonia, water acts as an acid (a proton donor).

$$\underset{\text{water (acid)}}{:\ddot{O}(H)-H} + \underset{\text{ammonia (base)}}{:NH_3} \rightleftharpoons \underset{\text{hydroxide ion}}{H-\ddot{O}:^-} + \underset{\text{ammonium ion}}{^+NH_4} \tag{7.10}$$

Water acts as a base toward acids that are stronger than itself (HCl) and as an acid toward bases that are stronger than itself (NH_3). Substances that can act as either an acid or a base are said to be **amphoteric.**

Problem 7.7 The K_a for ethanol is 1.0×10^{-16}. What is its pK_a?

Problem 7.8 The pK_a's of hydrogen cyanide and acetic acid are 9.2 and 4.7, respectively. Which is the stronger acid?

Problem 7.9 Which of the following are Lewis acids, and which are Lewis bases?

a. $H:^-$ b. $(CH_3)_3B$ c. Mg^{2+}
d. CH_3OCH_3 e. $(CH_3)_3C^+$ f. $(CH_3)_2C{=}\ddot{O}:$

Problem 7.10 Ammonia is a weak base (eq. 7.10). How would you characterize the ammonium ion, NH_4^+

Problem 7.11 In eq. 3.50, how is the amide ion, NH_2^-, functioning?

7.7 The Acidity of Alcohols and Phenols

Like water, alcohols and phenols are weak acids. The hydroxyl group can act as a proton donor, and dissociation occurs in a manner similar to that for water:

$$\underset{\text{alcohol}}{R\ddot{\underset{..}{O}}{-}H} \rightleftharpoons \underset{\text{alkoxide ion}}{R\ddot{\underset{..}{O}}:^-} + H^+ \tag{7.11}$$

The conjugate base of an alcohol is an **alkoxide ion** (for example, *meth*oxide ion from *meth*anol, *eth*oxide ion from *eth*anol, and so on).

Table 7.2 lists pK_a values for selected alcohols and phenols. Methanol and ethanol have nearly the same acid strength as water; bulky alcohols such as *t*-butyl alcohol are somewhat weaker because their bulk makes it difficult to solvate the corresponding alkoxide ion.

Phenol is nearly a million times stronger an acid than is ethanol. How can we explain this huge acidity difference between alcohols and phenols, since in both types of compounds the proton donor is a hydroxyl group?

Phenols are stronger acids than alcohols mainly because the corresponding phenoxide ions are stabilized by resonance. The negative charge of an alkoxide ion is concentrated on the oxygen atom, but the negative charge on a phenoxide ion can be delocalized to the *ortho* and *para* ring positions through resonance.

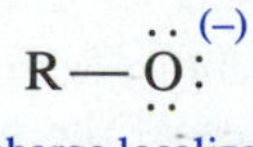

charge localized on the oxygen atom in alkoxide ions

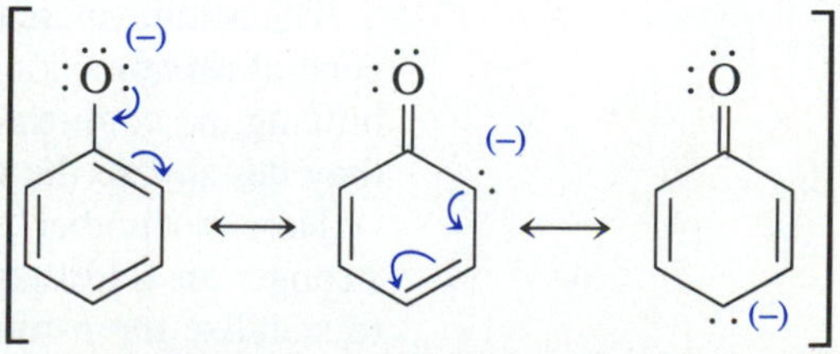

charge delocalized in phenoxide ion

TABLE 7.2 pK_a's of selected alcohols and phenols

Name	*Formula*	pK_a
water	HO—H	15.7
methanol	CH_3O—H	15.5
ethanol	CH_3CH_2O—H	15.9
t-butyl alcohol	$(CH_3)_3CO$—H	18
2,2,2-trifluoroethanol	CF_3CH_2O—H	12.4
phenol	C_6H_5—O—H	10.0
p-nitrophenol	O_2N—C_6H_4—O—H	7.2
picric acid	2,4,6-$(NO_2)_3C_6H_2$—O—H	0.25

Since phenoxide ions are stabilized in this way, the equilibrium for their formation is more favorable than that for alkoxide ions. Thus, phenols are stronger acids than alcohols.

We see in Table 7.2 that 2,2,2-trifluoroethanol is about 3000 times stronger an acid than is ethanol. How can we explain this effect of fluorine substitution? Again, think about the stabilities of the respective anions. Fluorine is a strongly electronegative element, so each C—F bond is polarized, with the fluorine partially negative and the carbon partially positive.

$CH_3CH_2O^-$

ethoxide ion

$(\delta-F)_3C^{\delta+}CH_2O^-$

2,2,2-trifluoroethoxide ion

The positive charge on the carbon is located near the negative charge on the nearby oxygen atom, where it can partially neutralize and hence stabilize it. This **inductive effect,** as it is called, is absent in ethoxide ion.

The acidity-increasing effect of fluorine seen here is not a special case, but a general phenomenon. *All electron-withdrawing groups increase acidity,* by stabilizing the conjugate base. *Electron-donating groups decrease acidity* because they destabilize the conjugate base.

Here is another example. *p*-Nitrophenol (Table 7.2) is nearly 1000 times stronger an acid than is phenol. In this case, the nitro group acts in *two* ways to stabilize the *p*-nitrophenoxide ion.

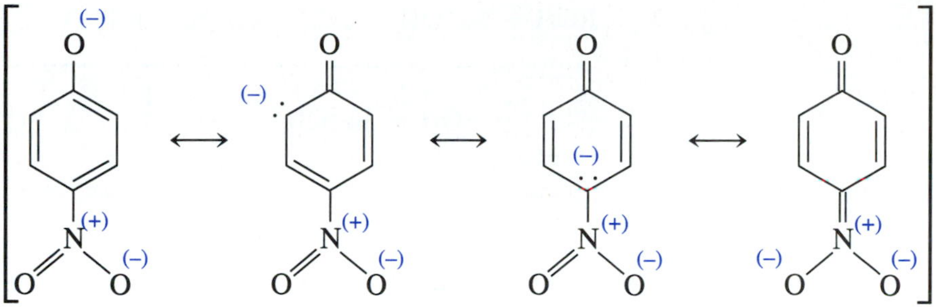

I II III IV

p-nitrophenoxide ion resonance contributors

First, the nitrogen atom has a formal positive charge and is therefore strongly electron withdrawing. It therefore increases the acidity of *p*-nitrophenol through the inductive effect. Second, the negative charge on the oxygen of the hydroxyl group can be delocalized through resonance, not only to the *ortho* and *para* ring carbons, as in phenoxide itself, *but to the oxygen atoms of the nitro group as well* (structure IV). Both the inductive and the resonance effects of the nitro group are acid-strengthening.

Additional nitro groups on the benzene ring further increase phenolic acidity. Picric acid (2,4,6-trinitrophenol) is almost 10 billion times stronger an acid than phenol.

Problem 7.12 Draw the resonance contributors for the 2,4,6-trinitrophenoxide (picrate) ion, and show that the negative charge can be delocalized to every oxygen atom.

Problem 7.13 Rank the following five compounds in order of increasing acid strength: 2-chloroethanol, *p*-chlorophenol, *p*-methylphenol, ethanol, phenol.

Alkoxide ions dissolved in alcohol, like hydroxide ion in water, are strong bases. Metal alkoxides are frequently used in organic chemistry when a strong base is called for. They may be prepared by the reaction of an alcohol with sodium or potassium metal or with a metal hydride.

$$\underset{\text{alcohol}}{2RO{-}H} + 2\,Na \longrightarrow \underset{\text{sodium alkoxide}}{2\,RO^-\,Na^+} + H_2 \tag{7.12}$$

$$RO{-}H + \underset{\text{sodium hydride}}{NaH} \longrightarrow \underset{\text{sodium alkoxide}}{RO^-\,Na^+} + H_2 \tag{7.13}$$

Problem 7.14 Write the equation for the reaction of *t*-butyl alcohol with potassium metal. Name the product.

Ordinarily, treatment of alcohols with sodium hydroxide does not convert them to their alkoxides. This is because alkoxides are stronger bases than hydroxide ion, so the reaction goes in the reverse direction. Phenols, however, can be converted to phenoxide ions in this way.

$$ROH + Na^+OH^- \not\rightleftarrows RO^-Na^+ + H_2O \qquad (7.14)$$

$$C_6H_5\text{—}OH + Na^+OH^- \longrightarrow C_6H_5\text{—}O^-Na^+ + HOH \qquad (7.15)$$

phenol — sodium phenoxide

EXAMPLE 7.3 How can we separate a mixture of 1-octanol (bp 194°C) and *o*-methylphenol (bp 191°C)?

Solution Because of their similar boiling points, separation of these two compounds by distillation would be difficult. However, we can take advantage of their different acidities.

We treat the mixture with aqueous sodium hydroxide. The *o*-methylphenol reacts to form the corresponding phenoxide, which, being ionic, dissolves in the water layer. The 1-octanol does not react. Since it is insoluble in water, it forms a separate layer.

$$o\text{-}CH_3C_6H_4OH + Na^+OH^-(aq) \longrightarrow o\text{-}CH_3C_6H_4O^-Na^+ + H_2O$$

o-methylphenol — sodium *o*-methylphenoxide

$$CH_3(CH_2)_6CH_2OH + Na^+OH^-(aq) \longrightarrow \text{no reaction}$$

1-octanol

The layers can be separated. The *o*-methylphenol can be recovered by acidifying the aqueous layer, a reaction that is essentially the reverse of that in eq. 7.15.

$$o\text{-}CH_3C_6H_4O^-Na^+ + HCl \longrightarrow o\text{-}CH_3C_6H_4OH + Na^+Cl^-$$

sodium *o*-methylphenoxide — *o*-methylphenol

Problem 7.15 Write an equation for the reaction, if any, between

a. *p*-nitrophenol and aqueous potassium hydroxide.
b. cyclohexanol and aqueous potassium hydroxide.

7.8 The Basicity of Alcohols and Phenols

Alcohols (and phenols) function not only as acids but also as bases. They have unshared electron pairs on the oxygen and are therefore Lewis bases. They can be protonated by strong acids. The product, analogous to the oxonium ion H_3O^+, is an alkyloxonium ion.

$$\underset{\text{alcohol acting as a base}}{R-\ddot{\underset{..}{O}}-H} + H^+ \rightleftharpoons \underset{\text{alkyloxonium ion}}{\left[R-\overset{\overset{\large H}{|}}{\underset{..}{O}}-H \right]^+} \tag{7.16}$$

This protonation is the first step in two important reactions of alcohols that are discussed in the following two sections: their dehydration to alkenes and their conversion to alkyl halides.

7.9 Dehydration of Alcohols to Alkenes

Alcohols can be dehydrated by heating them with a strong acid. For example, when ethanol is heated at 180°C with a small amount of concentrated sulfuric acid, a good yield of ethylene is obtained.

$$\underset{\text{ethanol}}{H-CH_2CH_2-OH} \xrightarrow{H^+,\ 180°C} \underset{\text{ethylene}}{CH_2=CH_2} + H-OH \tag{7.17}$$

This type of reaction, which can be used to prepare alkenes, is the reverse of their hydration (Sec. 3.9b). It is an *elimination reaction* and can occur by either an E1 or an E2 mechanism, depending on the class of the alcohol.

Tertiary alcohols dehydrate by the E1 mechanism. *t*-Butyl alcohol is a typical example. The first step involves reversible protonation of the hydroxyl group.

$$(CH_3)_3C-\ddot{\underset{..}{O}}H + H^+ \rightleftharpoons (CH_3)_3C-\overset{+}{\ddot{O}}-H \ (\text{with } H \text{ on } O) \tag{7.18}$$

Ionization, with water as the leaving group, occurs readily because the resulting carbocation is tertiary.

$$(CH_3)_3C-\overset{+}{\ddot{O}}-H \ (\text{with } H \text{ on } O) \rightleftharpoons \underset{t\text{-butyl cation}}{(CH_3)_3C^+} + H_2O \tag{7.19}$$

Proton loss from a carbon atom adjacent to the positive carbon completes the reaction.

$$\overset{H}{\underset{|}{}}CH_2-\overset{CH_3}{\underset{CH_3}{C^+}} \longrightarrow CH_2=C(CH_3)_2 + H^+ \tag{7.20}$$

The overall dehydration reaction is the sum of all three steps.

$$\underset{t\text{-butyl alcohol}}{\overset{\overset{\text{H}}{|}\quad\ \overset{\text{CH}_3}{|}}{\text{CH}_2\text{—}\underset{\underset{\text{CH}_3}{|}}{\text{C}}\text{—OH}}} \xrightarrow[\text{heat}]{\text{H}^+} \underset{\substack{\text{2-methylpropene}\\ \text{(isobutylene)}}}{\text{CH}_2\text{=C(CH}_3)_2} + \text{H—OH} \qquad (7.21)$$

With a primary alcohol, a primary carbocation intermediate is avoided by combining the last two steps of the mechanism. The loss of water and an adjacent proton occur simultaneously in an E2 mechanism.

$$CH_3CH_2\ddot{\underset{..}{O}}H + H^+ \rightleftharpoons CH_3CH_2\text{—}\overset{+}{\ddot{O}}\text{—}H \ (\text{with H on O}) \qquad (7.22)$$

$$\overset{\text{H}}{\underset{|}{}}\text{CH}_2\text{—CH}_2\text{—}\overset{+}{\ddot{O}}\text{—H}\ (\text{with H on O}) \longrightarrow CH_2\text{=}CH_2 + H^+ + H_2O \qquad (7.23)$$

The important things to remember about alcohol dehydrations are (1) they all begin by protonation of the hydroxyl group (that is, the alcohol acts as a base) and (2) the ease of alcohol dehydration is $3° > 2° > 1°$ (the same as the order of carbocation stability).

Sometimes a single alcohol gives two or more alkenes because the proton lost during dehydration can come from any carbon atom that is *adjacent* to the hydroxyl-bearing carbon. For example, 2-methyl-2-butanol can give two alkenes.

$$\underset{\text{2-methyl-2-butanol}}{\overset{\text{H}\qquad \text{OH}\ \text{H}}{\text{CH}_2\text{—}\underset{\underset{\text{CH}_3}{|}}{\text{C}}\text{—CH—CH}_3}} \xrightarrow[\text{heat}\ -\text{H}_2\text{O}]{\text{H}^+} \underset{\text{2-methyl-1-butene}}{\text{CH}_2\text{=}\underset{\underset{\text{CH}_3}{|}}{\text{C}}\text{—CH}_2\text{CH}_3} \quad \text{and/or} \quad \underset{\text{2-methyl-2-butene}}{\text{CH}_3\text{—}\underset{\underset{\text{CH}_3}{|}}{\text{C}}\text{=CHCH}_3} \qquad (7.24)$$

In these cases, *the alkene with the most substituted double bond usually predominates*. In the example shown, the major product is 2-methyl-2-butene.

Problem 7.16 Write the structure for all the possible dehydration products of

a. 3-methyl-3-hexanol

b. (cyclopentane ring bearing OH and CH_3 on the same carbon)

In each case, which product do you expect to predominate?

7.10 The Reaction of Alcohols with Hydrogen Halides

Alcohols react with hydrogen halides to give alkyl halides.

$$\underset{\text{alcohol}}{R\text{—}OH} + H\text{—}X \longrightarrow \underset{\text{alkyl halide}}{R\text{—}X} + H\text{—}OH \tag{7.25}$$

This substitution reaction provides a useful general route to alkyl halides. Because halide ions are good nucleophiles, we obtain mainly substitution products instead of dehydration. The reaction rate and mechanism depend on the alcohol structure (3°, 2°, or 1°); the mechanism will be either S_N1 or S_N2.

Tertiary alcohols react fastest. For example, we can convert *t*-butyl alcohol to *t*-butyl chloride simply by shaking it for a few minutes at room temperature (rt) with concentrated hydrochloric acid.

$$\underset{t\text{-butyl alcohol}}{(CH_3)_3COH} + H\text{—}Cl \xrightarrow[15\ \text{min}]{\text{rt}} \underset{t\text{-butyl chloride}}{(CH_3)_3C\text{—}Cl} + H\text{—}OH \tag{7.26}$$

On the other hand, 1-butanol, a primary alcohol, must be heated for several hours with a mixture of concentrated hydrochloric acid and a Lewis acid catalyst such as zinc chloride to accomplish the same type of reaction.

$$\underset{\text{1-butanol}}{CH_3CH_2CH_2CH_2OH} + H\text{—}Cl \xrightarrow[\text{several hours}]{\text{heat, } ZnCl_2} \underset{\text{1-chlorobutane}}{CH_3CH_2CH_2CH_2\text{—}Cl} + H\text{—}OH \tag{7.27}$$

Of course, the difference in reaction conditions is related to the difference in mechanism for a tertiary and a primary alcohol.

EXAMPLE 7.4 Write out the steps in the mechanism for eq. 7.26.

Solution In the first step, the alcohol is protonated by the acid, as in eq. 7.18. Ionization follows, as in eq. 7.19. In the final step, the *t*-butyl cation is captured by the nucleophile, chloride ion.

$$(CH_3)_3C^+ + Cl^- \xrightarrow{\text{fast}} (CH_3)_3CCl$$

EXAMPLE 7.5 Write out the steps in the mechanism for eq. 7.27, and explain the role played by the zinc chloride catalyst.

Solution In the first step, the alcohol is protonated by the acid, just as in Example 7.4.

$$CH_3CH_2CH_2CH_2\text{—}\ddot{\underset{..}{O}}H + H^+ \rightleftharpoons CH_3CH_2CH_2CH_2\text{—}\overset{+}{\ddot{O}}\text{—}H \quad (\text{with H on O})$$

In the second step, chloride ion displaces water in a typical S_N2 process.

$$Cl^- \curvearrowright \overset{CH_3CH_2CH_2}{\underset{H\ \ H}{C}}\text{—}\overset{+}{\ddot{O}}\text{—}H \ (\text{with H on O}) \longrightarrow CH_3CH_2CH_2CH_2Cl + H_2O$$

The zinc chloride can serve the same role as a proton in sharing an electron pair of the hydroxyl oxygen. It also increases the chloride ion concentration, thus speeding up the S_N2 displacement.

Problem 7.17 Explain why *t*-butyl alcohol reacts at nearly equal rates with HCl, HBr, and HI (to form, in each case, the corresponding *t*-butyl halide).

Problem 7.18 Explain why 1-butanol reacts with hydrogen halides in the rate order HI > HBr > HCl (to form, in each case, the corresponding butyl halide).

Problem 7.19 Write an equation for the reaction of 1-methylcyclopentanol with concentrated HBr.

7.11 Other Ways to Prepare Alkyl Halides from Alcohols

Since alkyl halides are extremely useful in synthesis, it is not surprising that chemists have devised several ways to prepare them from alcohols. For example, **thionyl chloride** reacts with alcohols to give alkyl chlorides.

$$ROH + \underset{\text{thionyl chloride}}{Cl-\overset{\overset{\displaystyle O}{\|}}{S}-Cl} \xrightarrow{\text{heat}} RCl + HCl\uparrow + SO_2\uparrow \qquad (7.28)$$

One advantage of this method is that two of the reaction products, hydrogen chloride and sulfur dioxide, are gases and evolve from the reaction mixture, leaving behind only the desired alkyl chloride. The method is not effective, however, for preparing low-boiling alkyl chlorides (in which R has only a few carbon atoms), because they easily boil out of the reaction mixture with the gaseous products.

Phosphorus halides also convert alcohols to alkyl halides.

$$3\,ROH + PX_3 \longrightarrow 3\,RX + H_3PO_3\ (X = Cl\text{ or }Br) \qquad (7.29)$$

In this case, the other reaction product, phosphorous acid, has a rather high boiling point. Thus, the alkyl halide is usually the lowest-boiling component of the reaction mixture and can be isolated by distillation.

Both of these methods are used mainly with primary and secondary alcohols, whose reaction with hydrogen halides is slow.

Problem 7.20 Tell how you would prepare each of the following alkyl halides from the corresponding alcohol without using HX.

a. $CH_3(CH_2)_6CH_2Cl$ b. $(CH_3)_2CHBr$

7.12 A Comparison of Alcohols and Phenols

Because they have the same functional group, alcohols and phenols have many similar properties. But whereas it is relatively easy, with acid catalysis, to break the C—OH bond of alcohols, this bond is difficult to break in phenols. Protonation of the phenolic hydroxyl group can occur, but loss of a water molecule would give a phenyl cation.

$$C_6H_5-\overset{+}{\ddot{O}}H_2 \not\rightarrow C_6H_5^+ + H_2O \quad (7.30)$$

a phenyl cation

With only two attached groups, the positive carbon in a phenyl cation should be *sp*-hybridized and linear. But this geometry is prevented by the structure of the benzene ring, so phenyl cations are exceedingly difficult to form. Consequently, phenols cannot undergo replacement of the hydroxyl group by an S_N1 mechanism. Neither can phenols undergo displacement by the S_N2 mechanism (the geometry of the ring makes the usual inversion mechanism impossible). Therefore, hydrogen halides, phosphorus halides, or thionyl halides cannot cause replacement of the hydroxyl group by halogens in phenols.

Problem 7.21 Compare the reactions of cyclohexanol and phenol with

a. HBr. b. H_2SO_4, heat. c. PCl_3.

7.13 Oxidation of Alcohols to Aldehydes and Ketones

Alcohols with at least one hydrogen attached to the hydroxyl-bearing carbon can be oxidized to carbonyl compounds. Primary alcohols give aldehydes, which may be further oxidized to acids. Secondary alcohols give ketones.

$$R-CH_2-OH \xrightarrow[\text{agent}]{\text{oxidizing}} R-\overset{H}{C}=O \xrightarrow[\text{agent}]{\text{oxidizing}} R-\overset{OH}{C}=O \quad (7.31)$$

primary alcohol — aldehyde — acid

$$R-\overset{R'}{\underset{H}{C}}-OH \xrightarrow[\text{agent}]{\text{oxidizing}} R-\overset{R'}{C}=O \quad (7.32)$$

secondary alcohol — ketone

Tertiary alcohols do not undergo this type of oxidation.

A common laboratory oxidizing agent for this purpose is chromic anhydride, CrO_3, dissolved in aqueous sulfuric acid and acetone **(Jones' reagent).** Typical examples are

$$C_6H_{11}OH \xrightarrow[\text{H}^+\text{, acetone (Jones' reagent)}]{CrO_3} C_6H_{10}=O \quad (7.33)$$

cyclohexanol — cyclohexanone

$$CH_3(CH_2)_6CH_2OH \xrightarrow[\text{reagent}]{\text{Jones'}} CH_3(CH_2)_6CO_2H \quad (7.34)$$

1-octanol — octanoic acid

With primary alcohols, oxidation can be stopped at the aldehyde stage by special reagents, such as pyridinium chlorochromate (PCC).

$$\underset{\text{1-octanol}}{CH_3(CH_2)_6CH_2OH} \xrightarrow[CH_2Cl_2,\ 25°C]{PCC} \underset{\text{octanal}}{CH_3(CH_2)_6CH{=}O} \quad (7.35)$$

PCC is prepared by dissolving CrO_3 in hydrochloric acid and then adding pyridine:

$$CrO_3 + HCl + \underset{\text{pyridine}}{C_5H_5N:} \longrightarrow \underset{\substack{\text{pyridinium chlorochromate}\\ \text{(PCC)}}}{C_5H_5N^+{-}H\ \ CrO_3Cl^-} \quad (7.36)$$

Problem 7.22 Write an equation for the oxidation of

a. 3-pentanol with Jones' reagent.
b. 4-phenylbutanol with Jones' reagent.
c. 4-phenylbutanol with PCC.

In the body, similar oxidations are accomplished by enzymes, together with a rather complex coenzyme called nicotinamide adenine dinucleotide, NAD^+ (for its structure, see Sec. 18.13). Oxidation occurs in the liver and is a key step in the body's attempt to rid itself of imbibed alcohol.

$$\underset{\text{ethanol}}{CH_3CH_2OH} + NAD^+ \underset{\text{dehydrogenase}}{\overset{\text{alcohol}}{\rightleftharpoons}} \underset{\text{acetaldehyde}}{CH_3CH{=}O} + NADH \quad (7.37)$$

The resulting acetaldehyde—also toxic—is further oxidized to acetate and eventually to carbon dioxide and water.

7.14 Alcohols with More Than One Hydroxyl Group

Compounds with two adjacent alcohol groups are called **glycols.** The most important example is **ethylene glycol.** Compounds with more than two hydroxyl groups are also known, and several, such as **glycerol** and **sorbitol,** are important commercial chemicals.

$$\underset{\substack{\text{ethylene glycol}\\ \text{(1,2-ethanediol)}\\ \text{bp 198°C}}}{\underset{OH\quad OH}{CH_2{-}CH_2}} \qquad \underset{\substack{\text{glycerol (glycerine)}\\ \text{(1,2,3-propanetriol)}\\ \text{bp 290°C (decomposes)}}}{\underset{OH\quad OH\quad OH}{CH_2{-}CH{-}CH_2}} \qquad \underset{\substack{\text{sorbitol}\\ \text{(1,2,3,4,5,6-hexanehexaol)}\\ \text{mp 110–112°C}}}{\underset{OH\quad OH\quad OH\quad OH\quad OH\quad OH}{CH_2{-}CH{-}CH{-}CH{-}CH{-}CH_2}}$$

Ethylene glycol is used as the "permanent" antifreeze in automobile radiators and as a raw material in the manufacture of Dacron. Ethylene glycol is completely miscible with water. Notice that, because of its increased capacity for hydrogen bonding, ethylene glycol has an exceptionally high boiling point for its molecular weight—much higher than that of ethanol.

Glycerol is a syrupy, colorless, water-soluble, high-boiling liquid with a distinctly sweet taste. Its soothing qualities make it useful in shaving and toilet soaps and in cough drops and syrups. It is also used as a moistening agent in tobacco.

Nitration of glycerol gives **glyceryl trinitrate** (nitroglycerine), a powerful and shock-sensitive explosive.

$$\begin{array}{l} CH_2OH \\ | \\ CHOH \\ | \\ CH_2OH \end{array} + 3\ HONO_2 \xrightarrow{H_2SO_4} \begin{array}{l} CH_2ONO_2 \\ | \\ CHONO_2 \\ | \\ CH_2ONO_2 \end{array} + 3\ H_2O \qquad (7.38)$$

glycerol — glyceryl trinitrate (nitroglycerine)

Alfred Nobel, inventor of dynamite (in 1866), found that glyceryl trinitrate could be controlled by absorbing it on an inert porous material. Dynamite contains about 15% glyceryl (and glycol) nitrate. The main explosive is ammonium nitrate (55%); the other components are sodium nitrate and wood pulp (about 15% each). Dynamite is used mainly in mining and construction.

Nitroglycerine is also used in medicine as a vasodilator, to prevent heart attacks in patients who suffer from angina.

Triesters of glycerol are fats and oils, whose chemistry is discussed in Chapter 15.

Sorbitol, with its many hydroxyl groups, is water soluble. It is almost as sweet as cane sugar and is used in candy making and as a sugar substitute for diabetics.

A WORD ABOUT ...

14. Industrial Alcohols

The lower alcohols (those with up to four carbon atoms) are manufactured on a large scale. They are used as raw materials for other valuable chemicals and also have important uses in their own right.

Methanol was at one time produced from wood by distillation and is still sometimes called wood alcohol. The word *methyl* originates from the Greek (*methy,* wine, and *yle,* wood). At present, however, methanol is manufactured from carbon monoxide and hydrogen.

$$CO + 2H_2 \xrightarrow[400°C,\ 150\ atm]{ZnO—Cr_2O_3} CH_3OH$$

The world production of methanol is approximately 10 million tons per year. Most of it is used to produce formaldehyde and other chemicals, but some is used as a solvent and an antifreeze. Recently methanol has been used as the carbon source in the commercial production of single-cell proteins. Some yeasts and bacteria (single cells) can synthesize proteins from methanol and other carbon sources in the presence of aqueous nutrient salt solutions that contain certain essential sulfur, phosphorus, and nitrogen compounds. These proteins are used as an animal food supplement and may eventually also play a part in human nutrition. Methanol itself, however, is highly toxic and can cause permanent blindness and death if taken internally.

Ethanol is prepared by the fermentation of blackstrap molasses, the residue that results from the purification of cane sugar.

$$\underset{\text{cane sugar}}{C_{12}H_{22}O_{11}} + H_2O \xrightarrow{\text{yeast}} \underset{\text{ethyl alcohol}}{4\ CH_3CH_2OH} + 4\ CO_2$$

The starch in grain, potatoes, and rice can be fermented similarly to produce ethanol, sometimes called grain alcohol.

Besides fermentation, ethanol is also manufactured by the acid-catalyzed hydration of ethylene (eq. 3.9). This method, using sulfuric acid or other acid catalysts, results in an annual world production of over 1 million tons.

Commercial alcohol is a constant-boiling mixture containing 95% ethanol and 5% water and cannot be further purified by distillation. To remove the remaining water to obtain absolute alcohol, one adds quicklime (CaO), which reacts with water to form calcium hydroxide but does not react with ethanol.

Since earlier times, ethanol has been known as an ingredient in fermented beverages (beer, wine, whiskey, and so on). The term *proof,* as used in the United States in reference to alcoholic beverages, is approximately twice the volume percentage of alcohol present. For example 100-proof whiskey contains 50% ethanol.

Ethanol is used as a solvent, as a topical antiseptic, and as a starting material for the manufacture of ether and ethyl esters. It can be used as a fuel (gasohol) and as a carbon source for single-cell proteins.

2-Propanol (isopropyl alcohol) is manufactured commercially by the acid-catalyzed hydration of propene (eq. 3.15). It is the main component of rubbing alcohol. More than half the isopropyl alcohol produced (over 1 million tons annually) is used to make acetone, by oxidation.

7.15 Aromatic Substitution in Phenols

Now we will examine some reactions that occur with phenols, but not with alcohols. Phenols undergo electrophilic aromatic substitution under very mild conditions because the hydroxyl group is strongly ring activating. For example, phenol can be nitrated with *dilute aqueous* nitric acid.

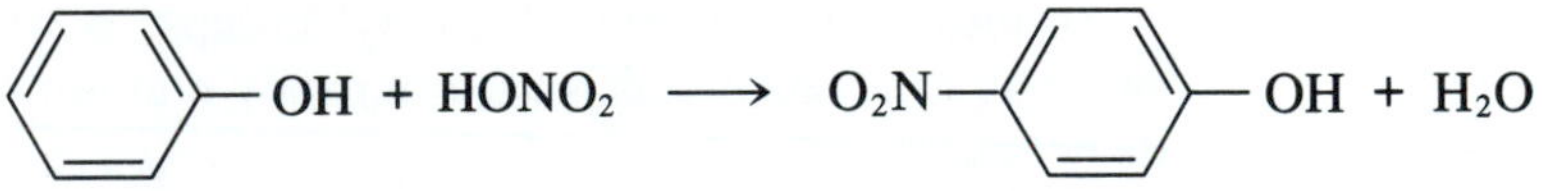

$$\underset{\text{phenol}}{C_6H_5\text{—OH}} + HONO_2 \longrightarrow \underset{p\text{-nitrophenol}}{O_2N\text{—}C_6H_4\text{—OH}} + H_2O \quad (7.39)$$

Phenol is also brominated rapidly with *bromine* water, to produce 2,4,6-tribromophenol.

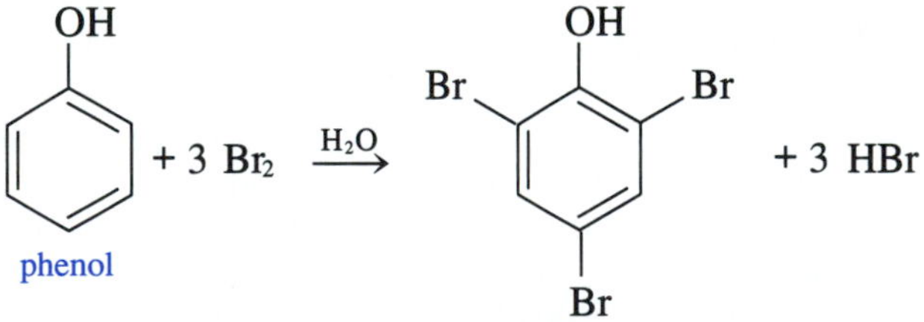

$$\underset{\text{phenol}}{C_6H_5OH} + 3\ Br_2 \xrightarrow{H_2O} \underset{\text{2,4,6-tribromophenol}}{2,4,6\text{-}Br_3C_6H_2OH} + 3\ HBr \quad (7.40)$$

EXAMPLE 7.6 Draw the intermediate in electrophilic aromatic substitution *para* to a hydroxyl group, and show how the intermediate benzenonium ion is stabilized by the hydroxyl group.

Solution

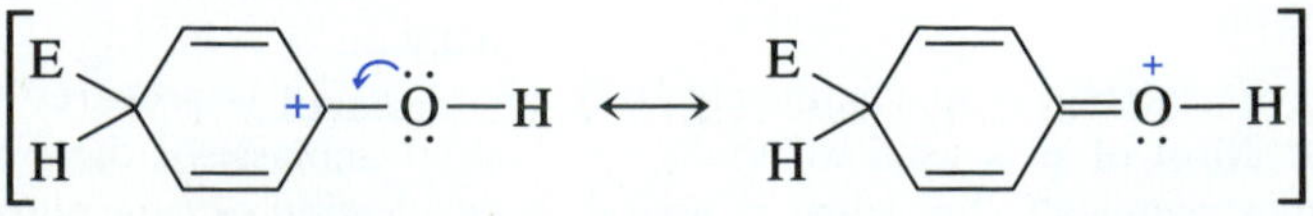

$$\left[\text{(E)(H)C}_6\text{H}_4^{+}\text{—}\ddot{\underset{..}{O}}\text{—H} \longleftrightarrow \text{(E)(H)C}_6\text{H}_4\text{=}\overset{+}{\underset{..}{O}}\text{—H} \right]$$

An unshared electron pair on the oxygen atom helps to delocalize the positive charge.

Problem 7.23 Explain why phenoxide ion undergoes electrophilic aromatic substitution even more easily than phenol does.

Problem 7.24 Write an equation for the reaction of

a. *p*-methylphenol + $HONO_2$ (1 mol).
b. *o*-chlorophenol + Br_2 (1 mol).

7.16 Oxidation of Phenols

Phenols are easily oxidized. Samples that stand exposed to air for some time often become highly colored due to the formation of oxidation products. With **hydroquinone** (1,4-dihydroxybenzene), the reaction is easily controlled to give **1,4-benzoquinone** (commonly called *quinone*).

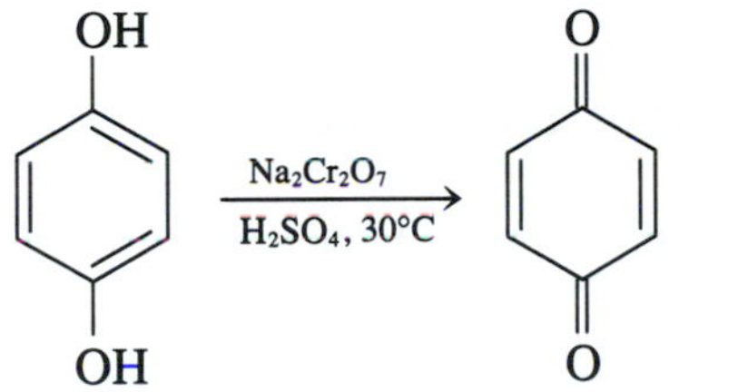

$$\text{hydroquinone} \xrightarrow[H_2SO_4,\ 30°C]{Na_2Cr_2O_7} \text{1,4-benzoquinone} \quad (7.41)$$

hydroquinone
colorless, mp 171°C

1,4-benzoquinone
yellow, mp 116°C

Hydroquinone and related compounds are used in photographic developers. They reduce silver ion that has not been exposed to light to metallic silver (and, in turn, they are oxidized to quinones). The oxidation of hydroquinones to quinones is reversible; this interconversion plays an important role in several biological oxidation–reduction reactions.

Substances that are sensitive to air oxidation, such as foods and lubricating oils, can be protected by phenolic additives. Phenols function as **antioxidants;** they are oxidized instead of the substances to which they have been added. Two commercial phenolic antioxidants are **BHA** (butylated hydroxy anisole) and **BHT** (butylated hydroxy toluene).

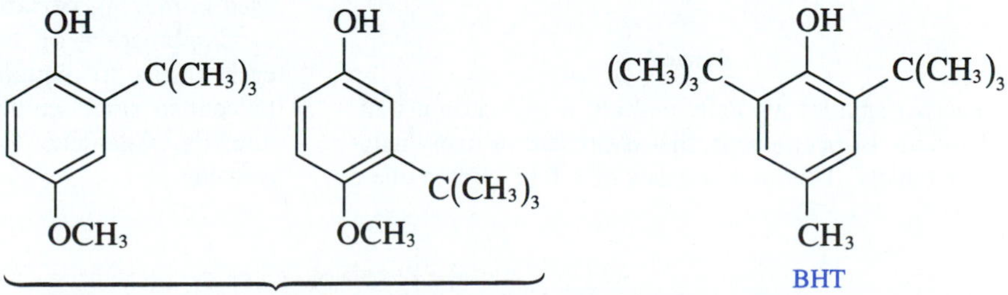

OH, $C(CH_3)_3$, OCH_3 OH, $C(CH_3)_3$, OCH_3

BHA

OH, $(CH_3)_3C$, $C(CH_3)_3$, CH_3

BHT

BHA is used as an antioxidant in foods, especially meat products. BHT is used not only in foods, animal feeds, and vegetable oils, but also in lubricating oils, synthetic rubber, and various plastics.

Vitamin E (α-tocopherol) is a common naturally occurring phenol. One of its biological functions is to act as a natural antioxidant.

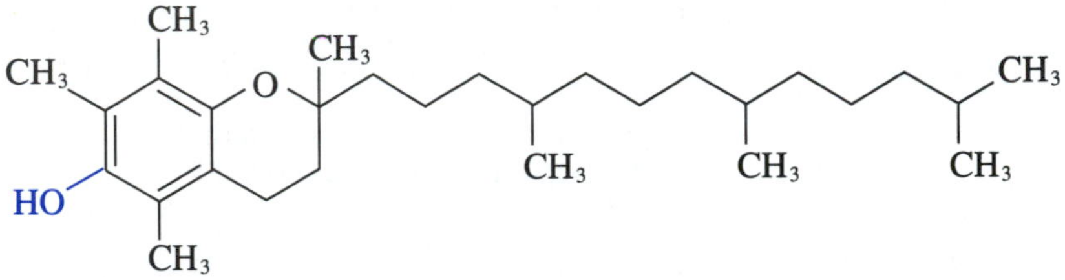

vitamin E (α-tocopherol)

A WORD ABOUT ...

15. Biologically Important Alcohols and Phenols

The hydroxyl group appears in many biologically important molecules, both as an alcohol and as a phenol.

Four metabolically important unsaturated primary alcohols are 3-methyl-2-buten-1-ol, 3-methyl-3-buten-1-ol, geraniol, and farnesol.

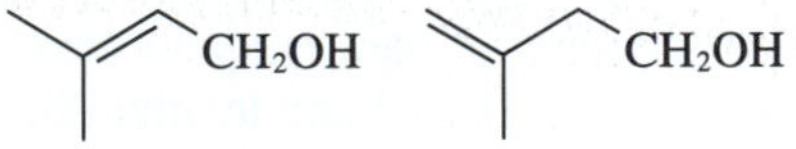

3-methyl-2-buten-1-ol 3-methyl-3-buten-1-ol

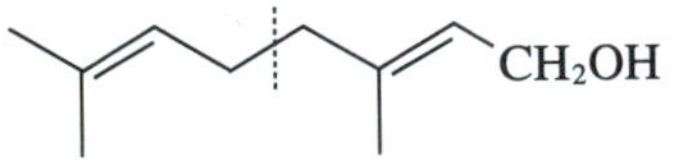

geraniol

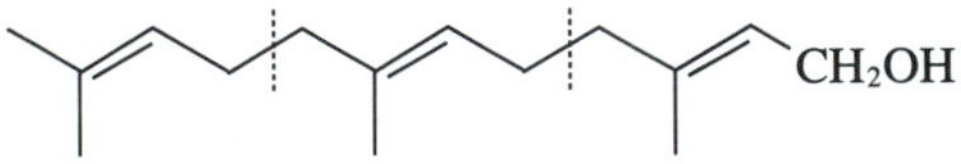

farnesol

The two smaller alcohols contain a five-carbon unit, called an **isoprene unit,** that is present in many natural products. This unit consists of a four-carbon chain with a one-carbon branch at carbon 2. These five-carbon alcohols can combine to give geraniol, which then can add yet another five-carbon unit to give farnesol. Note the isoprene units, marked off by dotted lines, in the structures of geraniol and farnesol.

Compounds of this type are called **terpenes.** Terpenes occur in the *essential oils* of many plants and flowers. They have 10, 15, 20, or more carbon atoms and are formed by linking isoprene units in various ways.

Geraniol, as its name implies, occurs in oil of geranium but also constitutes about 50% of rose oil, the extract of rose petals. Geraniol is also the biological precursor of α-pinene, a terpene that is the main component of turpentine. Farnesol, which occurs in the essential oils of rose and cyclamen, has a pleasing lily-of-the-valley odor. Both geraniol and farnesol are used in making perfumes.

Combination of two farnesol units (15 carbons each) leads to **squalene,** a 30-carbon hydrocarbon present in small amounts in the liver of most higher animals. Squalene is the biological precursor of steroids.

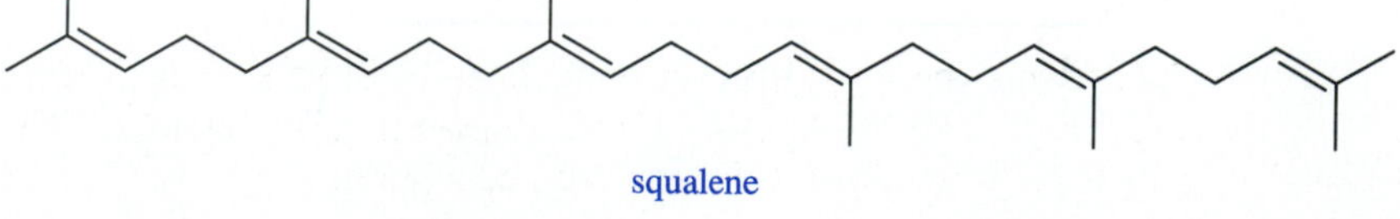

squalene

Cholesterol, a typical steroidal alcohol, has the structure

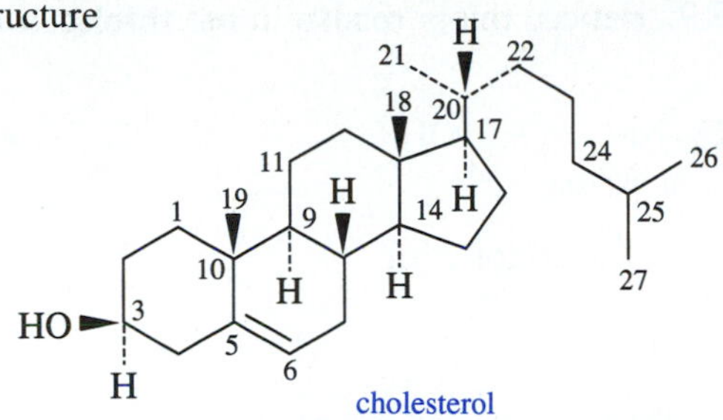

cholesterol
mp 148.5°C

Although it has 27 carbon atoms (instead of 30) and is therefore not strictly a terpene, cholesterol is synthesized in the body from the terpene squalene through a complex process that, in its final stages, involves the loss of 3 carbon atoms.

Phenols are less involved than alcohols in fundamental metabolic processes. Three phenolic alcohols do, however, form the basic building blocks of **lignins,** complex polymeric substances that, together with cellulose, form the woody parts of trees and shrubs. They have very similar structures.

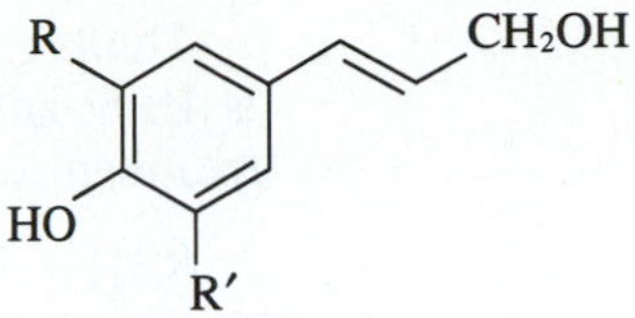

coniferyl alcohol (R = OCH_3 , R′ = H)
sinapyl alcohol (R = R′ = OCH_3)
p-coumaryl alcohol (R = R′ = H)

Some phenolic natural products to be avoided are **urushiols,** the active allergenic ingredients in poison ivy and poison oak.

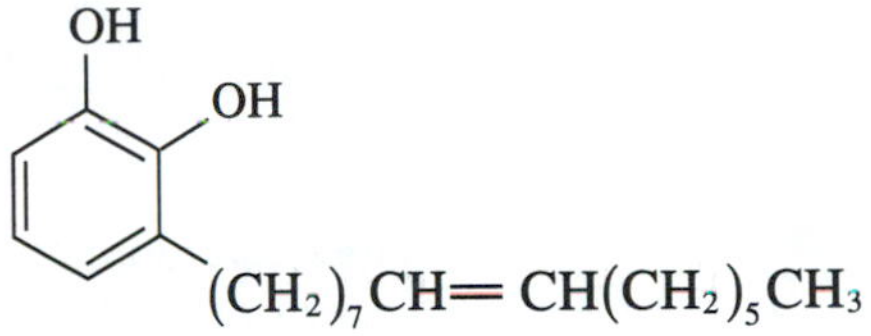

a urushiol

In other urushiols, the long side chain may be saturated, may have additional double bonds, or may have two more carbon atoms.

7.17 *Thiols, the Sulfur Analogs of Alcohols and Phenols*

Sulfur is immediately beneath oxygen in the periodic table and can often take its place in organic structures. The —SH group, called the **sulfhydryl group,** is the functional group of **thiols.** Thiols are named as follows:

CH_3SH	$CH_3CH_2CH_2CH_2SH$	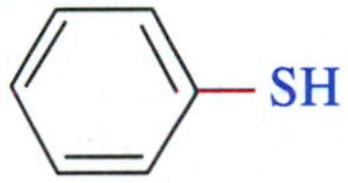
methanethiol	1-butanethiol	thiophenol
(methyl mercaptan)	(*n*-butyl mercaptan)	(phenyl mercaptan)

Thiols are sometimes called **mercaptans** because of their reaction with mercuric ion to form mercury salts called **mercaptides.**

$$2\,RSH + HgCl_2 \longrightarrow \underset{\text{a mercaptide}}{(RS)_2Hg} + 2\,HCl \qquad (7.42)$$

Problem 7.25 Draw the structure for

a. 2-butanethiol. b. isopropyl mercaptan.

Alkyl thiols can be made from alkyl halides by nucleophilic displacement with sulfhydryl ion.

$$R{-}X + {}^{-}SH \longrightarrow R{-}SH + X^{-} \qquad (7.43)$$

Perhaps the most distinctive feature of thiols is their intense and disagreeable odor. The thiols $CH_3CH{=}CHCH_2SH$ and $(CH_3)_2CHCH_2CH_2SH$, for example, are responsible for the odor of a skunk.

Thiols are more acidic than alcohols. The pK_a of ethanethiol, for example, is 10.6, and that of ethanol is 15.9. Hence, thiols readily form **thiolates** on treatment with aqueous base.

$$RSH + Na^+OH^- \longrightarrow \underset{\text{a sodium thiolate}}{RS^-Na^+} + HOH \quad (7.44)$$

Problem 7.26 Write an equation for the reaction of ethanethiol with

a. KOH. b. $HgCl_2$.

Thiols are easily oxidized to **disulfides** by mild oxidizing agents such as hydrogen peroxide or iodine.

$$\underset{\text{thiol}}{2\,RS{-}H} \underset{\text{reduction}}{\overset{\text{oxidation}}{\rightleftharpoons}} \underset{\text{disulfide}}{RS{-}SR} \quad (7.45)$$

This reaction can be reversed with a variety of reducing agents. Proteins contain disulfide links, so this reversible oxidation reduction can be used to manipulate their structures.

A WORD ABOUT ...

16. Hair, Curly or Straight

Hair consists of a fibrous protein called **keratin,** which, as proteins go, contains an unusually large percentage of the sulfur-containing amino acid **cystine.** Horse hair, for example, contains about 8% cystine:

$$\underset{\text{cystine (CyS—SCy)}}{HO_2C\underset{\substack{|\\NH_2}}{C}HCH_2S{-}SCH_2\underset{\substack{|\\NH_2}}{C}HCO_2H}$$

The disulfide link in cystine serves to cross-link the chains of amino acids that make up the protein (Figure 7.1).

The chemistry used in waving or straightening hair involves the oxidation-reduction chemistry of the disulfide bond (eq. 7.45). First, the hair is treated with a reducing agent, which breaks the S—S bonds, converting each sulfur to an —SH group. This breaks the cross-links between the long protein chains. The reduced hair can now be shaped as desired, either waved or straightened. Finally, the reduced and rearranged hair is treated with an oxidizing agent to reform the disulfide cross-links. The new disulfide bonds, no longer in their original positions, hold the hair in its new shape.

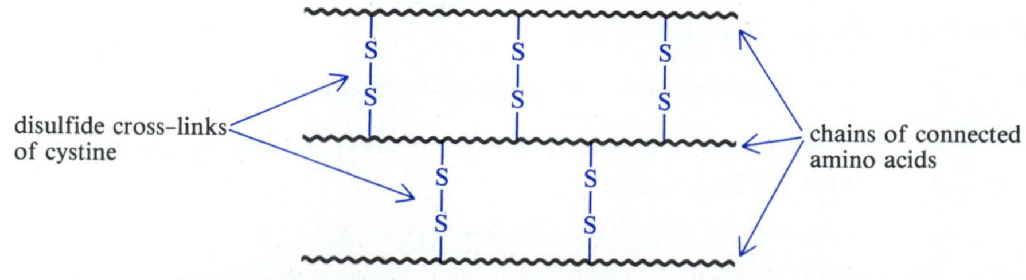

FIGURE 7.1 Schematic structure of hair.

Additional Problems

7.27. Write a structural formula for each of the following compounds:

a. 2,2-dimethyl-1-butanol
b. *o*-bromophenol
c. 2,3-pentanediol
d. 2-phenylethanol
e. ethyl nitrate
f. tricyclopropylmethanol
g. sodium ethoxide
h. 1-methylcyclopentanol
i. *trans*-2-methylcyclopentanol
j. (*R*)-2-butanol
k. 2-methyl-2-propen-1-ol
l. 2-cyclohexenol

7.28. Classify the alcohols in parts a, d, f, h, i, k, and l of Problem 7.27 as primary, secondary, or tertiary.

7.29. Name each of the following compounds:

a. $CH_3C(CH_3)_2CH(OH)CH_3$
b. $CH_3CHBrC(CH_3)_2OH$

c. OH Cl Cl
d. H OH

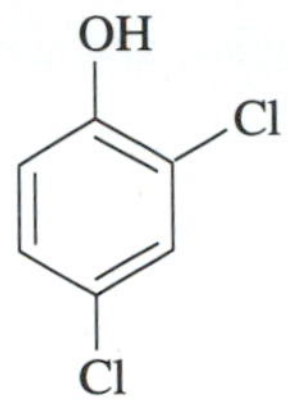

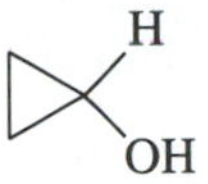

e. OH Br
f. CH OH

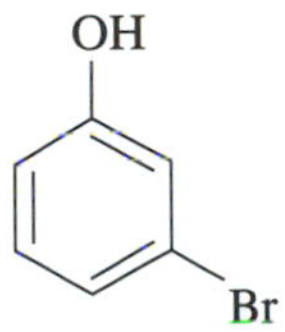

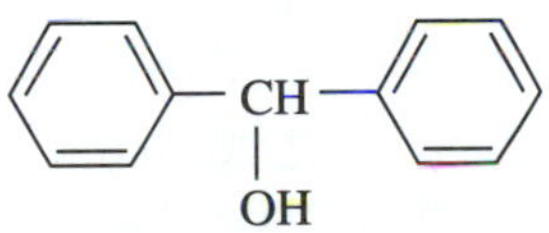

g. $CH_3CH{=}CHCH_2OH$
h. $CH_3CH(SH)CH_3$
i. $HOCH_2CH(OH)CH(OH)CH_2OH$
j. $CH_3CH_2CH_2O^-K^+$

k. CH_3 OH H H
l. SH H

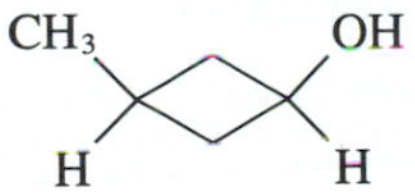

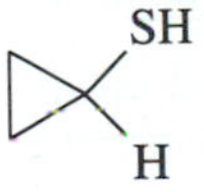

m. $-CH{=}CHCH_2OH$
n. OH CH_3

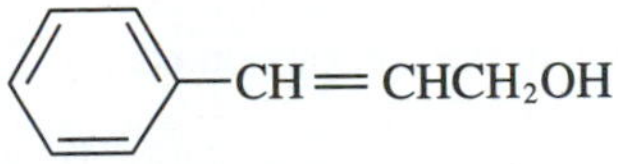

7.30. Explain why each of the following names is unsatisfactory, and give a correct name.

a. 2,2-dimethyl-3-butanol
b. 2-ethyl-1-propanol
c. 1-propene-3-ol
d. 5-chlorocyclohexanol
e. 3,6-dibromophenol
f. *sec*-butanol

7.31. Arrange the compounds in each of the following groups in order of increasing solubility in water, and briefly explain your answers.

a. ethanol; ethyl chloride; 1-hexanol
b. 1-pentanol; 1,5-pentanediol; $HOCH_2(CHOH)_3CH_2OH$

7.32. The following classes of organic compounds are Lewis bases. Write an equation that shows how each class might react with H^+.

a. ether, ROR **b.** amine, R_3N **c.** ketone, $R_2C{=}O$

7.33. Arrange the following compounds in order of increasing acidity, and explain the reasons for your choice of order: cyclohexanol, phenol, *p*-cyanophenol, 2-chlorocyclohexanol.

7.34. Which is the stronger base, potassium *t*-butoxide or potassium ethoxide? (*Hint:* Use the data in Table 7.2.)

7.35. Explain, with the aid of equations, what would happen if a solution of cyclohexanol and *p*-methylphenol in an inert solvent were successively (a) shaken with 10% aqueous sodium hydroxide, (b) separated into organic and aqueous layers, and (c) subjected to acidification of the aqueous layer.

7.36. Tell how each of the following mixtures could be separated without the use of distillation.

a. benzene and phenol
b. phenol and 1-hexanol
c. 1-propanol and 1-heptanol

7.37. Complete each of the following equations, and name the products.

a. $CH_3CH(OH)CH_2CH_3 + K \longrightarrow$
b. $(CH_3)_2CHOH + NaH \longrightarrow$

c. Cl—C_6H_4—OH + NaOH $\longrightarrow$ (Cl and OH para on a benzene ring)

d. cyclopentane ring carbon bearing H and OH + NaOH $\longrightarrow$

7.38. Show the structures of all possible acid-catalyzed dehydration products of the following. If more than one alkene is possible, predict which one will be formed in the largest amount.

a. cyclohexanol
b. 2-butanol
c. 1-methylcyclopentanol
d. 2-phenylethanol

7.39. Explain why the reaction shown in eq. 7.19 occurs much more easily than the reaction $(CH_3)_3C{-}OH \rightleftharpoons (CH_3)_3C^+ + OH^-$. (That is, why is it necessary to protonate the alcohol before ionization can occur?)

7.40. Write out all the steps in the mechanism for eq. 7.24, showing how each product is formed.

7.41. Although the reaction shown in eq. 7.26 occurs faster than that shown in eq. 7.27, the yield of product is lower. The yield of *t*-butyl chloride is only 80%, whereas the yield of *n*-butyl chloride is nearly 100%. What by-product is formed in eq. 7.26, and by what mechanism is it formed? Why is a similar by-product *not* formed in eq. 7.27?

7.42. Write an equation for each of the following reactions.

a. 2-methyl-2-butanol + HCl
b. 1-pentanol + Na
c. cyclopentanol + PBr_3
d. 1-phenylethanol + $SOCl_2$
e. 1-methylcyclopentanol + H_2SO_4, heat
f. ethylene glycol + $HONO_2$
g. 1-pentanol + aqueous NaOH
h. 1-octanol + HBr + $ZnBr_2$
i. 1-pentanol + CrO_3, H^+
j. 2-cyclohexylethanol + PCC

7.43. Treatment of 3-buten-2-ol with concentrated hydrochloric acid gives a mixture of two products, 3-chloro-1-butene and 1-chloro-2-butene. Write a reaction mechanism that explains how both products are formed.

7.44. Which four-carbon acyclic alcohols can be manufactured commercially by acid-catalyzed hydration of alkenes? (Remember Markovnikov's rule!)

7.45. Write an equation for each of the following two-step syntheses:
a. cyclohexene to cyclohexanone
b. 1-bromobutane to butanal
c. 1-butanol to 1-butanethiol

7.46. What product do you expect from the oxidation of cholesterol with CrO_3 and H^+? (See page 217 for the formula of cholesterol.)

7.47. Draw the structure of the quinone expected from the oxidation of

a. 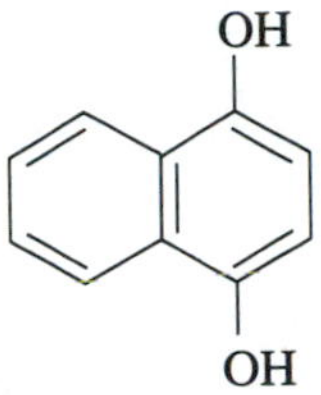 (1,4-naphthalenediol: naphthalene with OH at positions 1 and 4) **b.** 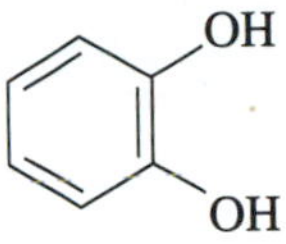 (1,2-benzenediol: benzene ring with two adjacent OH groups)

7.48. BHT (see Sec. 7.16 for the structure) is manufactured commercially from *p*-methylphenol and 2-methylpropene, using an acid catalyst. Write equations that show the steps in this Friedel-Crafts synthesis.

7.49. Mark off the isoprene units in squalene, vitamin A and β-carotene (the structures are in "A Word About" on pages 79 and 216?

7.50. Give correct IUPAC names for the two thiols that are responsible for skunk odor (Sec. 7.17).

7.51. Dimethyl disulfide, $CH_3S—SCH_3$, found in the vaginal secretions of female hamsters, acts as a sexual attractant for the male hamster. Write an equation for its synthesis from methanethiol.

7.52. The disulfide $[(CH_3)_2CHCH_2CH_2S]_2$ is a component of the odorous secretion of mink. Describe a synthesis of this disulfide, starting with 3-methyl-1-butanol.

CHAPTER **8**

Ethers and Epoxides

8.1 *Introduction*

To most people the word *ether* is synonymous with the well-known anesthetic. That particular ether, however, is but one member of a general class of organic compounds known as **ethers.** These compounds have two organic groups connected to a single oxygen atom. The general formula for ethers is R—O—R′, where R and R′ may be identical or different, and may be alkyl or aryl groups. Specifically, in the common anesthetic, both R and R′ are ethyl groups, $CH_3CH_2—O—CH_2CH_3$.

In this chapter, we will describe the physical and chemical properties of ethers. Their excellent solvent properties are applied in the preparation of Grignard reagents, organometallic compounds with a carbon-magnesium bond. We will give special attention to **epoxides,** cyclic three-membered ethers that have important industrial utility.

8.2 *Nomenclature of Ethers*

Ethers are usually named by giving the name of each alkyl or aryl group, in alphabetical order, followed by the word *ether*.

$CH_3CH_2—O—CH_3$
ethyl methyl ether

$CH_3CH_2—O—CH_2CH_3$
diethyl ether (the prefix *di-* is sometimes omitted)

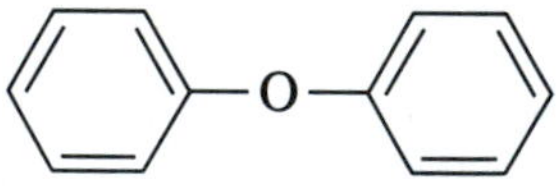

$C_6H_5—O—C_6H_5$
diphenyl ether

For ethers with more complex structures, it may be necessary to name the —**OR** group as an **alkoxy group.** In the IUPAC system, the smaller alkoxy group is named as a substituent.

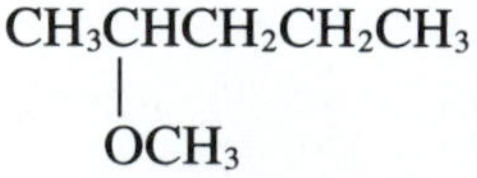

$CH_3CH(OCH_3)CH_2CH_2CH_3$
2-methoxypentane

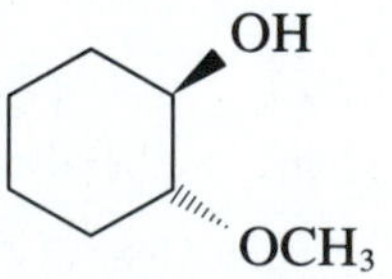

(cyclohexane ring bearing OH and OCH_3 on adjacent carbons, trans)
trans-2-methoxycyclohexanol

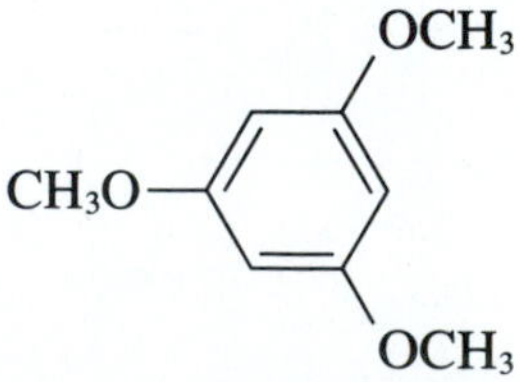

(benzene ring with three OCH_3 groups: CH_3O, OCH_3, OCH_3)
1,3,5-trimethoxybenzene

EXAMPLE 8.1 Give a correct name for $CH_3CHCH(CH_3)_2$ with OCH_2CH_3 on C-2

$$\begin{array}{l} CH_3CHCH(CH_3)_2 \\ \quad\;\;| \\ \quad\; OCH_2CH_3 \end{array}$$

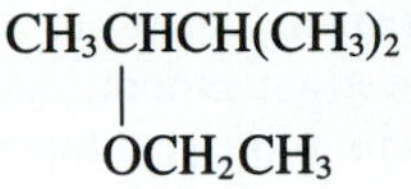

Solution

$$\begin{array}{l} \qquad\quad CH_3 \\ \quad\;\; ^{1}\;\;^{2\,3}|\;\;^{4} \\ CH_3CHCHCH_3 \\ \qquad | \\ \quad\;\; OCH_2CH_3 \end{array}$$

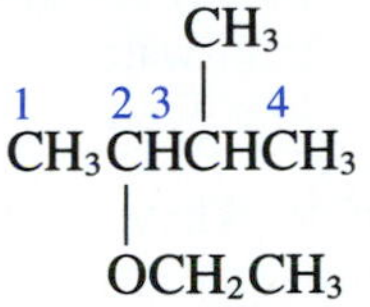

2-ethoxy-3-methylbutane

Problem 8.1 Give a correct name for

a. $(CH_3)_2CHOCH_3$ b. 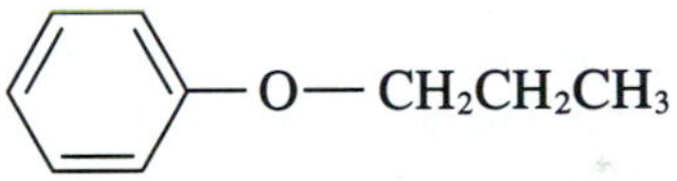 C_6H_5—O—$CH_2CH_2CH_3$ c. cyclopentane bearing CH_3 and OCH_3 on the same carbon

Problem 8.2 Write the structural formula for

a. dicyclopropyl ether. b. 2-ethoxyoctane.

8.3 Physical Properties of Ethers

Ethers are colorless compounds with characteristic, relatively pleasant odors. They have lower boiling points than alcohols with an equal number of carbon atoms. In fact, an ether has nearly the same boiling point as the corresponding hydrocarbon in which a —CH_2— group replaces the ether's oxygen. The following data illustrate these facts:

		bp	mol wt	water solubility (g/100 mL, 20°C)
1-butanol	$CH_3CH_2CH_2CH_2OH$	118°C	74	7.9
diethyl ether	CH_3CH_2—O—CH_2CH_3	35°C	74	7.5
pentane	CH_3CH_2—CH_2—CH_2CH_3	36°C	72	0.03

Because of their structures, ether molecules cannot form hydrogen bonds with one another. This is why they boil so much lower than their isomeric alcohols.

Problem 8.3 Write structures for each of the following *isomers,* and arrange them in order of decreasing boiling point: 3-methoxy-1-propanol, 1,2-dimethoxyethane, 1,4-butanediol.

Although ethers cannot form hydrogen bonds with one another, they can and do form hydrogen bonds with alcohols:

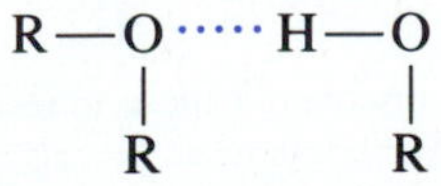

$$\begin{array}{ccc} R—O & \cdots\cdots & H—O \\ | & & | \\ R & & R \end{array}$$

For this reason, alcohols and ethers are usually mutually soluble. Low-molecular-weight ethers, such as dimethyl ether, are quite soluble in water. Likewise, the modest solubility of diethyl ether in water is similar to that of its isomer 1-butanol (see data tabulated above) because each can form a hydrogen bond to water. Ethers are less dense than water.

8.4 Ethers as Solvents

Ethers are relatively inert compounds. They do not usually react with dilute acids, with dilute bases, or with common oxidizing and reducing agents. They do not react with metallic sodium—a property that distinguishes them from alcohols.

This general inertness, coupled with the fact that most organic compounds are ether-soluble, makes ethers excellent solvents in which to carry out organic reactions.

Ethers are also used frequently to extract organic compounds from their natural sources. Diethyl ether is particularly good for this purpose. Its low boiling point makes it easy to remove from an extract and easy to recover by distillation. It is highly flammable, however, and must not be used if there are any flames in the same laboratory.

Another risk is that ethers that have been in a laboratory for a long time, exposed to air, may contain organic peroxides as a result of oxidation.

$$CH_3CH_2OCH_2CH_3 + O_2 \longrightarrow CH_3CH_2O\underset{\substack{|\\ OOH}}{C}HCH_3 \qquad (8.1)$$

an ether hydroperoxide

These peroxides are extremely explosive and must be removed before the ether can be used safely. Shaking with aqueous ferrous sulfate destroys these peroxides by reduction.

8.5 The Grignard Reagent; an Organometallic Compound

One of the most striking examples of the solvating power of ethers is in the preparation of **Grignard reagents.** These reagents, which are exceedingly useful in organic synthesis, were discovered by the French organic chemist Victor Grignard [pronounced "greenyar(d)"]. In 1912 he received a Nobel Prize for this contribution to organic synthesis.*

Grignard found that, when an anhydrous ether solution of an alkyl or aryl halide is stirred with magnesium turnings, the metal gradually dissolves. The resulting solutions contain Grignard reagents.

$$R{-}X + Mg \xrightarrow{\text{dry ether}} R{-}MgX \qquad (8.2)$$

a Grignard reagent

The magnesium inserts itself into the carbon-halogen bond.

---

* For a brief account of how Grignard discovered these reagents, see D. Hodson, *Chemistry in Britain* **1987,** 141–42.

Although the ether used as a solvent for this reaction is normally not shown as part of the Grignard reagent structure, it does play an important role. The unshared electron pairs on the ether oxygen help to stabilize the magnesium through coordination.

$$\begin{array}{c} R\quad\;\; R \\ \diagdown\ddot{}\diagup \\ \ddot{O} \\ R'\text{—}Mg\text{—}X \\ \ddot{O} \\ \diagup\ddot{}\diagdown \\ R\quad\;\; R \end{array}$$

Acting as a Lewis base, ether stabilizes a Grignard reagent.

The two ethers most commonly used in Grignard preparations are diethyl ether and the cyclic ether tetrahydrofuran, abbreviated THF (p. 235). The Grignard reagent will not form unless the ether is scrupulously dry, free of traces of water or alcohols.

Grignard reagents are named as shown in the following equations:

$$\underset{\text{methyl iodide}}{CH_3\text{—}I} + Mg \xrightarrow{\text{ether}} \underset{\text{methylmagnesium iodide}}{CH_3MgI} \tag{8.3}$$

$$\underset{\text{bromobenzene}}{C_6H_5\text{—}Br} + Mg \xrightarrow{\text{ether}} \underset{\text{phenylmagnesium bromide}}{C_6H_5\text{—}MgBr} \tag{8.4}$$

Notice that there is no space between the name of the organic group and magnesium, but that there is a space before the halide name.

Although the exact nature of Grignard solutions is still being researched, Grignard reagents usually react as if the alkyl or aryl group is negatively charged (a carbanion) and the magnesium atom is positively charged.

$$\overset{\delta-}{R}\text{—}\overset{\delta+}{MgX}$$

Carbanions are strong bases (they are the conjugate bases of hydrocarbons, which are very weak acids). It is not surprising, then, that Grignard reagents react vigorously with even such a weak acid as water, or with any other compound with an O—H, S—H, or N—H bond.

$$\overset{\delta-}{R}\text{—}MgX + \overset{\delta+}{H}\text{—}OH \longrightarrow RH + Mg^{2+}(OH)^-X^- \tag{8.5}$$

This is why the ether used as a solvent for the Grignard reagent must be scrupulously free of water or alcohol.

EXAMPLE 8.2 Is it possible to prepare a Grignard reagent from $HOCH_2CH_2CH_2Br$ and magnesium?

Solution No! Any Grignard reagent that might be formed would immediately be destroyed by protons from the OH group. Grignard and hydroxyl functionality in the same molecule are incompatible.

Problem 8.4 Is it possible to prepare a Grignard reagent from $CH_3OCH_2CH_2CH_2Br$?

The reaction of a Grignard reagent with water can be put to useful purpose. For example, if heavy water (D_2O) is used, deuterium can be substituted for a halogen.

$$CH_3-C_6H_4-Br \xrightarrow[\text{ether}]{Mg} CH_3-C_6H_4-MgBr \xrightarrow{D_2O} CH_3-C_6H_4-D \qquad (8.6)$$

p-bromotoluene; *p*-tolylmagnesium bromide; *p*-deuteriotoluene

This is a useful way to introduce an isotopic label into an organic compound.

EXAMPLE 8.3 Show how to prepare CH_3CHDCH_3 from $CH_2{=}CHCH_3$.

Solution $$CH_2{=}CHCH_3 \xrightarrow{HBr} CH_3\underset{\displaystyle Br}{\underset{|}{C}}HCH_3 \xrightarrow[\text{ether}]{Mg} CH_3\underset{\displaystyle MgBr}{\underset{|}{C}}HCH_3 \xrightarrow{D_2O} CH_3CHDCH_3$$

Problem 8.5 Show how to prepare CH_3CHDCH_3 from $(CH_3)_2CHOH$.

Grignard reagents are **organometallic compounds;** they contain a carbon-metal bond. Many other types of organometallic compounds are known (recall acetylides, eq. 3.50). Among the more useful in synthesis are **organolithium compounds,** which can be prepared in a manner similar to that for Grignard reagents.

$$R{-}X + 2\,Li \xrightarrow{\text{ether}} \underset{\text{an alkyllithium}}{R{-}Li} + Li^+X^- \qquad (8.7)$$

Problem 8.6 Write an equation for the preparation of propyllithium and for its reaction with D_2O.

Later in this chapter and elsewhere in this book, we will see examples of the synthetic utility of organometallic reagents.

8.6 Preparation of Ethers

The most important commercial ether is diethyl ether. It is prepared from ethanol and sulfuric acid.

$$\underset{\text{ethanol}}{CH_3CH_2OH + HOCH_2CH_3} \xrightarrow[140°C]{H_2SO_4} \underset{\text{diethyl ether}}{CH_3CH_2OCH_2CH_3} + H_2O \quad (8.8)$$

Note that ethanol can be dehydrated by sulfuric acid to give either ethylene (eq. 7.17) or diethyl ether (eq. 8.8). Of course, the reaction conditions are different in each case. These reactions provide a good example of how important it is to control reaction conditions and to specify them in equations.

Problem 8.7 The reaction in eq. 7.17 occurs by an E2 mechanism (review eqs. 7.22 and 7.23). By what mechanism does the reaction in eq. 8.8 occur?

Although it can be adapted to other ethers, the alcohol–sulfuric acid method is most commonly used to make symmetric ethers from primary alcohols.

Problem 8.8 Write an equation for the synthesis of dipropyl ether.

The commercial production of *t*-butyl methyl ether (and its ethyl analog) has become important in recent years. Used as an octane number enhancer in unleaded gasolines, it is prepared by the acid-catalyzed addition of methanol to 2-methylpropene.

$$\underset{\text{methanol}}{CH_3OH} + \underset{\text{2-methylpropene}}{CH_2{=}C(CH_3)_2} \xrightarrow{H^+} \underset{t\text{-butyl methyl ether}}{CH_3O{-}\overset{\overset{CH_3}{|}}{\underset{\underset{CH_3}{|}}{C}}{-}CH_3} \quad (8.9)$$

Problem 8.9 Write out the steps in the mechanism for eq. 8.9 (see eqs. 3.17 and 3.22).

Most important for the laboratory synthesis of unsymmetric ethers is the **Williamson synthesis,** named after the British chemist Alexander Williamson, who devised it. This method has two steps, both of which we have already discussed. In the first step, an alcohol is converted to its alkoxide by treatment with a reactive metal (sodium or potassium) or metal hydride (review eqs. 7.12 and 7.13). In the second step, an S_N2 displacement is carried out between the alkoxide and an alkyl halide (see Table 6.1, item 2). The Williamson synthesis is summarized by the general equations

$$2\,ROH + 2\,Na \longrightarrow 2\,RO^-Na^+ + H_2 \quad (8.10)$$

$$RO^-Na^+ + R'{-}X \longrightarrow ROR' + Na^+X^- \quad (8.11)$$

Since the second step is an S_N2 reaction, it works best if R′ in the alkyl halide is primary and not well at all if R′ is tertiary.

EXAMPLE 8.4 Write an equation for the synthesis of $CH_3OCH_2CH_2CH_3$ using the Williamson method.

Solution There are two possibilities, depending on which alcohol and which alkyl halide are used:

$$\begin{array}{ccc} CH_3OCH_2CH_2CH_3 & \text{or} & CH_3OCH_2CH_2CH_3 \\ \uparrow & & \uparrow \\ CH_3O^- \; Na^+ + XCH_2CH_2CH_3 & & CH_3X + Na^+ \; ^-OCH_2CH_2CH_3 \end{array}$$

The equations are

$$2\ CH_3OH + 2\ Na \longrightarrow 2\ CH_3O^-Na^+ + H_2$$

$$CH_3O^-\ Na^+ + CH_3CH_2CH_2X \longrightarrow CH_3OCH_2CH_2CH_3 + Na^+X^-$$

or

$$2\ CH_3CH_2CH_2OH + 2\ Na \longrightarrow 2\ CH_3CH_2CH_2O^-Na^+ + H_2$$

$$CH_3CH_2CH_2O^-\ Na^+ + CH_3X \longrightarrow CH_3CH_2CH_2OCH_3 + Na^+X^-$$

X is usually Cl, Br, or I.

Problem 8.10 Write equations for the synthesis of the following ethers by the Williamson method.

a. C_6H_5—CH_2OCH_3 b. $(CH_3)_3COCH_3$ (*Reminder*: the second step proceeds by the S_N2 mechanism).

A WORD ABOUT ...

17. Ether and Anesthesia

Prior to the 1840s, pain during surgery was relieved by various methods (asphyxiation, pressure on nerves, administration of narcotics or alcohol), but on the whole it was almost worse torture to undergo an operation than to endure the disease. Modern use of anesthesia during surgery has changed all that. Anesthesia stems from the work of several physicians in the mid-nineteenth century. The earliest experiments used nitrous oxide, ether, or chloroform. Perhaps the best known of these experiments was the removal of a tumor from the jaw of a patient anesthetized by ether, performed by Boston dentist William T. G. Morton in 1846.

Anesthetics fall into two major categories, general and local. *General anesthetics* are usually administered to accomplish three ends: insensitivity to pain (analgesia), loss of consciousness, and muscle relaxation. Gases such as nitrous oxide and cyclopropane and volatile liquids such as ether are administered by inhalation, but other general anesthetics such as barbiturates are injected intravenously.

The exact mechanism by which anesthetics affect

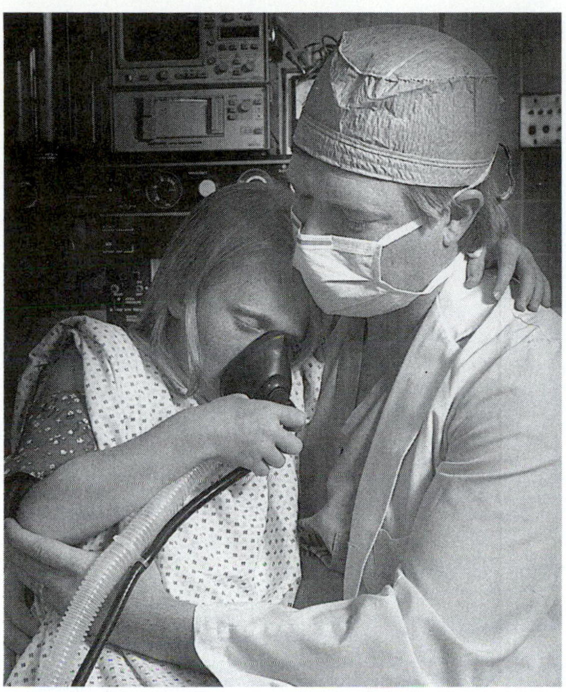

the central nervous system is not completely known. Unconsciousness may result from several factors: changes in the properties of nerve cell membranes, suppression of certain enzymatic reactions, and solubility of the anesthetic in lipid membranes.

A good inhalation anesthetic should vaporize readily and have appropriate solubility in the blood and tissues. It should also be stable, inert, nonflammable, potent, and minimally toxic. It should have an acceptable odor and cause minimal side effects such as nausea or vomiting. No anesthetic that meets *all* these specifications has yet been developed. Although *diethyl ether* is perhaps the best-known general anesthetic to the layperson, it fails on several counts (flammability, side effects of nausea or vomiting, and relatively slow action). It is quite potent, however, and produces good analgesia and muscle relaxation. The use of ether at present is rather limited, mainly because of its undesirable side effects. **Halothane,** $CF_3CHBrCl$, comes closest to an ideal inhalation anesthetic at present, but halogenated ethers such as **enflurane,** $CF_2H—O—CF_2CHClF$, are also used.

Local anesthetics are either applied to body surfaces or injected near nerves to desensitize a particular region of the body to pain. The best known of these anesthetics is procaine (Novocain), an aromatic amino-ester (see "A Word About" on page 378).

The discovery of anesthetics enabled physicians to perform surgery with deliberation and care, leading to many of the advances of modern medicine.

8.7 Cleavage of Ethers

Ethers have unshared electron pairs on the oxygen atom and are therefore Lewis bases. They react with strong proton acids and with Lewis acids such as the boron halides.

$$R-\ddot{\underset{..}{O}}-R' + H^+ \rightleftharpoons R-\underset{..}{\overset{\overset{\displaystyle H}{|}}{O}}-R' \tag{8.12}$$

$$R-\ddot{\underset{..}{O}}-R' + Br-\underset{\underset{\displaystyle Br}{|}}{B}-Br \rightleftharpoons R-\overset{..}{\underset{\underset{\displaystyle Br-\underset{\underset{\displaystyle Br}{|}}{B}-Br}{|}}{O}}-R' \tag{8.13}$$

These reactions are similar to the reaction of alcohols with strong acids (eq. 7.16). If the alkyl groups R and/or R′ are primary or secondary, the bond to oxygen can be broken by reaction with a strong nucleophile such as I^- or Br^- (by an S_N2 process). For example,

$$CH_3CH_2OCH(CH_3)_2 + HI \xrightarrow{\text{heat}} CH_3CH_2I + HOCH(CH_3)_2 \quad (8.14)$$

ethyl isopropyl ether → ethyl iodide, isopropyl alcohol

$$C_6H_5{-}OCH_3 + BBr_3 \xrightarrow[2.\ H_2O]{1.\ \text{heat}} C_6H_5{-}OH + CH_3Br \quad (8.15)$$

anisole → phenol, methyl bromide

If R or R′ is tertiary, a strong nucleophile is not required since reaction will occur by an S_N1 (or E1) mechanism.

$$C_6H_5{-}OC(CH_3)_3 \xrightarrow[H_2O]{H^+} C_6H_5{-}OH + (CH_3)_3COH \quad (8.16)$$

t-butyl phenyl ether → phenol, *t*-butyl alcohol (and $(CH_3)_2C{=}CH_2$)

The net result of these reactions is **cleavage** of the ether at one of the C—O bonds. Ether cleavage is a useful reaction for determining the structure of a complex, naturally occurring ether because it allows one to break the large molecule into more easily handled, smaller fragments.

EXAMPLE 8.5 Write out the steps in the mechanism for eq. 8.14.

Solution The ether is first protonated by the acid.

$$CH_3CH_2\ddot{\underset{\cdot\cdot}{O}}CH(CH_3)_2 \underset{}{\overset{H^+}{\rightleftharpoons}} CH_3CH_2\underset{\cdot\cdot}{\overset{H\ |+}{O}}CH(CH_3)_2$$

The resulting oxonium ion is then cleaved by S_N2 attack of iodide ion at the primary carbon (recall that 1° > 2° in S_N2 reactions).

$$I^- + CH_3CH_2{-}\overset{H\ |+}{O}CH(CH_3)_2 \longrightarrow CH_3CH_2I + HOCH(CH_3)_2$$

Problem 8.11 Write out the steps in the mechanism for eq. 8.15. Explain why the products are phenol and methyl bromide, *not* bromobenzene and methanol.

Problem 8.12 Write out the steps in the mechanism for eq. 8.16. Which C-O bond cleaves, the one to the phenyl or the one to the *t*-butyl group?

8.8 Epoxides (Oxiranes)

Epoxides (or oxiranes) are cyclic ethers with a three-membered ring containing one oxygen atom.

ethylene oxide (oxirane) bp 13.5°C

cis-2-butene oxide (*cis*-2,3-dimethyloxirane) bp 60°C

trans-2-butene oxide (*trans*-2,3-dimethyloxirane) bp 54°C

The most important commercial epoxide is ethylene oxide, produced by the silver-catalyzed air oxidation of ethylene.

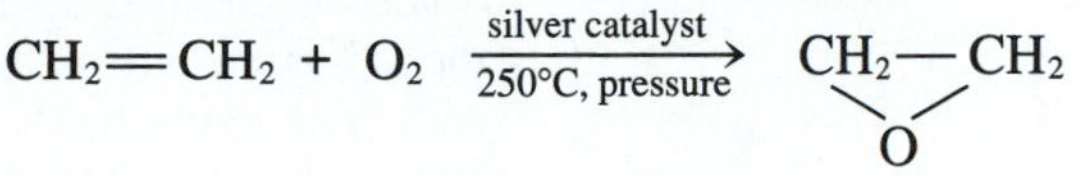

$$CH_2{=}CH_2 + O_2 \xrightarrow[250°C,\ pressure]{silver\ catalyst} \underset{ethylene\ oxide}{CH_2{-}CH_2\ (\text{bridged by } O)} \qquad (8.17)$$

Annual U.S. production of ethylene oxide exceeds 4 billion pounds. Only rather small amounts are used directly (for example, as a fumigant in grain storage). Most of the ethylene oxide constitutes a versatile raw material for the manufacture of other products, the main one being ethylene glycol.

The reaction in eq. 8.17 is suitable only for ethylene oxide. Other epoxides are usually prepared by the reaction of an alkene with an organic peroxyacid.

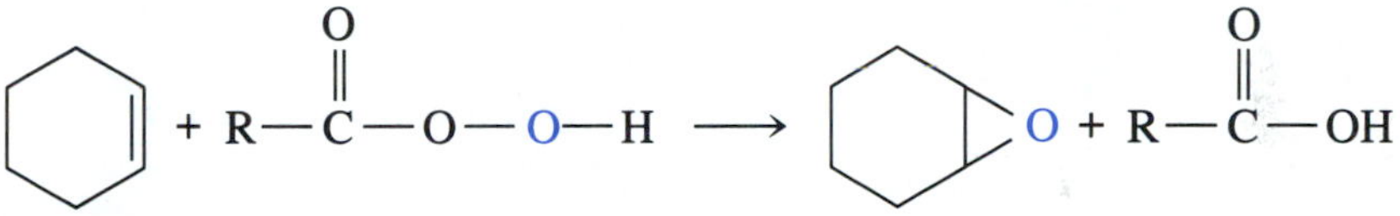

$$\underset{cyclohexene}{C_6H_{10}} + \underset{organic\ peracid}{R{-}\overset{O}{\overset{\|}{C}}{-}O{-}O{-}H} \longrightarrow \underset{cyclohexene\ oxide}{C_6H_{10}O} + \underset{organic\ acid}{R{-}\overset{O}{\overset{\|}{C}}{-}OH} \qquad (8.18)$$

Peroxyacids, like hydrogen peroxide H—O—O—H, to which they are structurally related, are good oxidizing agents. On a large scale, peroxyacetic acid ($R{=}CH_3$) is used, whereas in the laboratory the preferred reagent is *m*-chloroperoxybenzoic acid.

Problem 8.13 Write an equation for the reaction of cyclopentene with *m*-chloroperoxybenzoic acid.

A WORD ABOUT ...

18. The Gypsy Moth's Epoxide

The main mode of communication among insects is via the emission and detection of specific chemical substances. These substances are called **pheromones.** The word is from the Greek (*pherein,* to carry, and *horman,* to excite). Even though they are emitted and detected in exceedingly small amounts, pheromones have profound biological effects. One of their main effects is sexual attraction and stimulation, but they are also used as alarm substances to alert members of the same species to danger, as aggregation substances to call together both sexes of a species, and as trail substances to lead members of a species to food.

Often pheromones are chemically relatively simple compounds—alcohols, esters, aldehydes, ketones, ethers, epoxides, or even hydrocarbons. Two examples are **muscalure** and **bombykol,** the sex attractants of the common housefly and the silkworm moth, respectively.

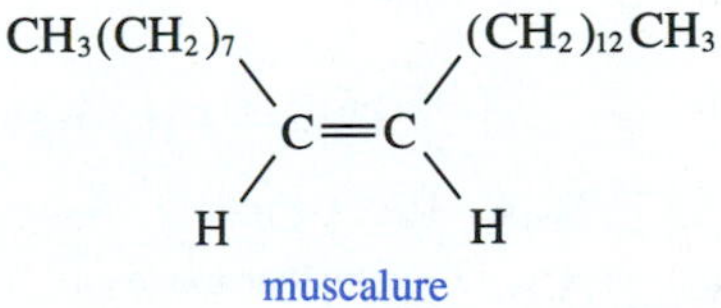

$$\underset{muscalure}{CH_3(CH_2)_7(H)C{=}C(H)(CH_2)_{12}CH_3}$$

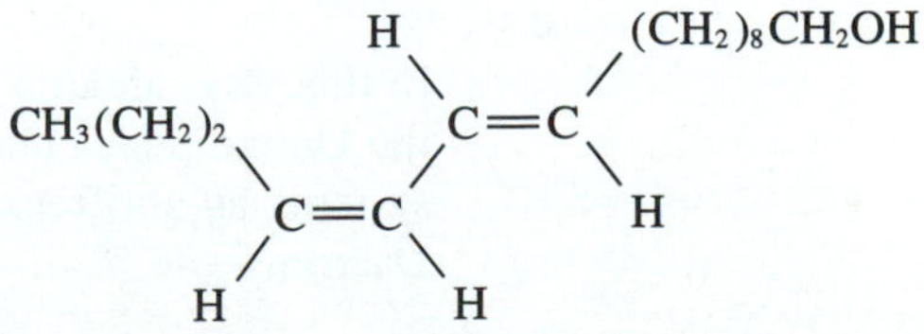

$$\underset{bombykol}{CH_3(CH_2)_2(H)C{=}C(H){-}C(H){=}C(H)(CH_2)_8CH_2OH}$$

Their molecular weights are low enough that the substances are volatile, yet not so low that they disperse too rapidly. Also, their molecular structures must be distinctive to make them species-specific; survival of the species would not be served by attracting another species. Often this specificity is attained through stereoisomerism (at double bonds and/or at chiral centers), but it can also be achieved by using specific ratios of two or more pheromones for a particular communication purpose.

Let us consider a specific pheromone, **disparlure,** the sex attractant of the gypsy moth (*Lymantria dispar*). The gypsy moth is a serious despoiler of forest and shade trees as well as fruit orchards. Gypsy moth larvae, which hatch each spring, are voracious eaters and can strip a tree bare of leaves in just a few weeks.

The abdominal tips (last two segments) of the virgin female moth contain the sex attractant. Extraction of 78,000 tips led to isolation of the main sex attractant, which was the following *cis*-epoxide:

$$(CH_3)_2CH(CH_2)_4 \cdots C^7(H)—O—C^8(H) \cdots (CH_2)_9CH_3$$

(7*R*,8*S*)-(+)-7,8-epoxy-2-methyloctadecane
(disparlure)

The active isomer has the *R* configuration at carbon 7 and the *S* configuration at carbon 8. This isomer can be detected by the male gypsy moth at a concentration as low at 10^{-10}g/mL; its enantiomer is inactive in solutions a million times more concentrated.

Disparlure has been synthesized in the laboratory. The synthetic material can be used to lure the male to traps and in that way to control the insect population. This form of insect control sometimes has advantages over spraying with insecticides.

8.9 Reactions of Epoxides

Because of the strain in the three-membered ring, epoxides are much more reactive than ordinary ethers and give products in which the ring has opened. For example, with water they undergo acid-catalyzed ring opening to give glycols.

$$\underset{\text{ethylene oxide}}{\underset{\diagdown O \diagup}{CH_2—CH_2}} + H—OH \xrightarrow{H^+} \underset{\text{ethylene glycol}}{\underset{OH \quad OH}{CH_2—CH_2}} \quad (8.19)$$

In this way, about 3 billion pounds of ethylene glycol are produced annually in the United States alone. Approximately half of it is used in automobile cooling systems as antifreeze. Most of the rest is used to prepare polyesters such as Dacron.

EXAMPLE 8.6 Write equations that show the mechanism for eq. 8.19.

Solution The first step is reversible protonation of the epoxide oxygen, as in eq. 8.12.

$$\underset{\ddot{\text{O}}}{CH_2-CH_2} + H^+ \rightleftharpoons \underset{\substack{\ddot{\text{O}}^+ \\ | \\ H}}{CH_2-CH_2}$$

The second step is a nucleophilic S_N2 displacement on the primary carbon, with water as the nucleophile. Then proton loss yields the glycol.

$$H_2\ddot{O}: + \underset{\substack{\ddot{\text{O}}^+ \\ | \\ H}}{CH_2-CH_2} \longrightarrow H-\overset{+}{\underset{\substack{| \\ H}}{\ddot{O}}}-CH_2-CH_2-OH \rightleftharpoons HO-CH_2CH_2-OH + H^+$$

Problem 8.14 Write an equation for the acid-catalyzed reaction of cyclohexene oxide with water. Predict the stereochemistry of the product.

Other nucleophiles add to epoxides in a similar way.

$$\underset{O}{CH_2-CH_2} \xrightarrow{H^+} \begin{cases} \xrightarrow{CH_3OH} HOCH_2CH_2OCH_3 \text{ (2-methoxyethanol)} \\ \xrightarrow{HOCH_2CH_2OH} HOCH_2CH_2OCH_2CH_2OH \text{ (diethylene glycol)} \end{cases} \quad (8.20)$$

2-Methoxyethanol is an additive for jet fuels, used to prevent water from freezing in fuel lines. Being both an alcohol and an ether, it is soluble in both water and organic solvents. *Diethylene glycol* is useful as a plasticizer (softener) in cork gaskets and tiles.

Grignard reagents are strong nucleophiles capable of opening the ethylene oxide ring. The initial product is a magnesium alkoxide, but after hydrolysis (as in the reverse of eq. 7.14), we obtain a primary alcohol with two more carbon atoms than the Grignard reagent.

$$\overset{\delta-}{R}-\overset{\delta+}{MgX} + \underset{O}{CH_2-CH_2} \longrightarrow \underset{\text{a magnesium alkoxide}}{RCH_2CH_2OMgX} \xrightarrow{H-OH} RCH_2CH_2OH + Mg^{2+}OH^-X^- \quad (8.21)$$

Problem 8.15 Write an equation for the preparation of 1-pentanol from a Grignard reagent and ethylene oxide.

A WORD ABOUT . . .

19. Epoxy Resins

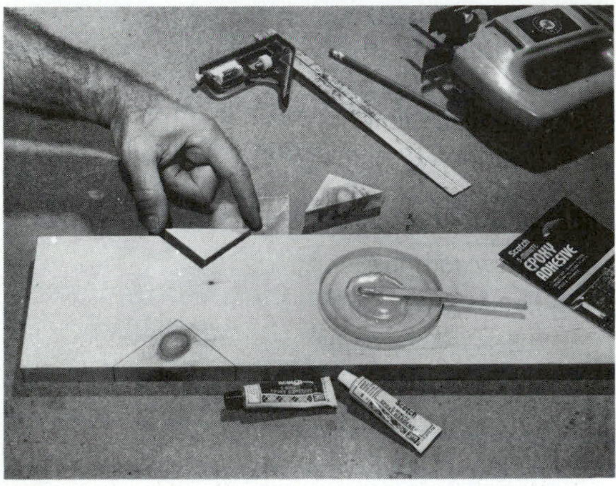

Most people hear the word *epoxy* in connection with *epoxy resins,* materials used as adhesives for bonding to metal, glass, and ceramics. Epoxy resins are also used in surface coatings (for example, paints) because of their exceptional inertness, hardness, and flexibility.

Two raw materials for the manufacture of epoxy resins are **epichlorhydrin** and **bisphenol-A.** Reaction of a mixture of these two raw materials with a base gives a "linear" epoxy resin; its structure is shown in Figure 8.1.

$$Cl-CH_2-CH-CH_2 \quad (\text{epoxide: } CH-O-CH_2)$$

epichlorhydrin

$$HO-C_6H_4-C(CH_3)_2-C_6H_4-OH$$

bisphenol-A

Commercial resins of this type range from liquids (where n is small) to viscous adhesives to solids used in surface coatings (where n may be as large as 25).

It is possible to take advantage of the remaining epoxide rings and hydroxyl groups in the "linear" polymer to form cross-links between the polymer chains, thus substantially increasing the molecular weight of the polymer. This is especially important when the end use is as a surface coating.

Epoxy resins can be varied in structure. For example, the bisphenol-A can be partially or totally replaced by other di- or polyhydroxy compounds, and epoxides other than epichlorhydrin can be used. Annual world production of epoxy resins runs to about a billion pounds.

$$CH_2-CH-CH_2- \left[-O-C_6H_4-C(CH_3)_2-C_6H_4-OCH_2-CH(OH)-CH_2- \right]_n$$

$$-O-C_6H_4-C(CH_3)_2-C_6H_4-OCH_2-CH-CH_2$$

(terminal groups are epoxide rings; C_6H_4 = p-phenylene)

FIGURE 8.1 Structure of a "linear" epoxy resin.

8.10 Cyclic Ethers

Cyclic ethers whose rings are larger than the three-membered epoxides are known. Most commonly they have five- or six-membered rings. Some examples include

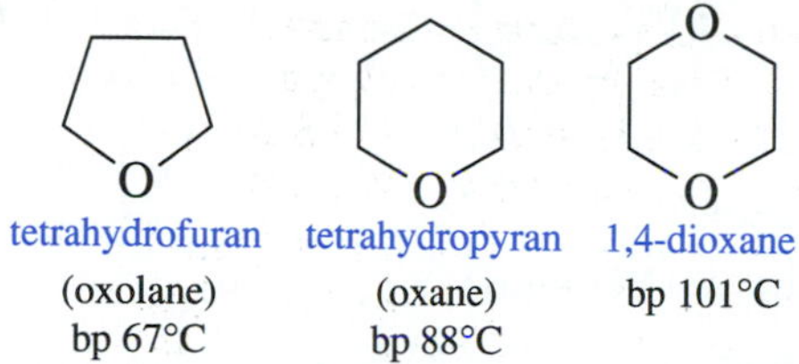

tetrahydrofuran (oxolane) bp 67°C — tetrahydropyran (oxane) bp 88°C — 1,4-dioxane bp 101°C

Tetrahydrofuran (THF), is a particularly useful solvent that not only dissolves many organic compounds but is miscible with water. THF is an excellent solvent—often superior to diethyl ether—in which to prepare Grignard reagents. Although it has the same number of carbon atoms as diethyl ether, they are "pinned back" in a ring. The oxygen in THF is therefore less hindered and better at coordinating with the magnesium in a Grignard reagent. **Tetrahydropyran** and **1,4-dioxane** are also soluble in both water and organic solvents.

The cyclic ethers most common in nature are the carbohydrates. They are usually either pyranoses (six-membered ring) or furanoses (five-membered ring). Because of their special importance, we will devote a full chapter to their chemistry (Chapter 16).

In recent years, there has been much interest in macrocylic (large-ring) polyethers. Some examples are

[18]crown-6 mp 39–40°C — dibenzo[18]crown-6 mp 164°C — [15]crown-5 (liquid) — [12]crown-4 mp 16°C

These compounds are called **crown ethers** because their molecules have a crownlike shape. The number in brackets in their common names gives the ring size, and the terminal number gives the number of oxygens. The oxygens are usually separated from one another by two carbon atoms.

Crown ethers have the unique property of forming complexes with positive ions (Na^+, K^+, and so on). The positive ions fit within the macrocyclic rings selectively, depending on the sizes of the particular ring and ion. For example, [18]crown-6 binds K^+ more tightly than it does the smaller Na^+ (too loose a fit) or the larger Cs^+ (too large to fit in the hole). Similarly, [15]crown-5 binds Na^+, and [12]crown-4 binds Li^+. The crown ethers act as hosts for their ionic guests.

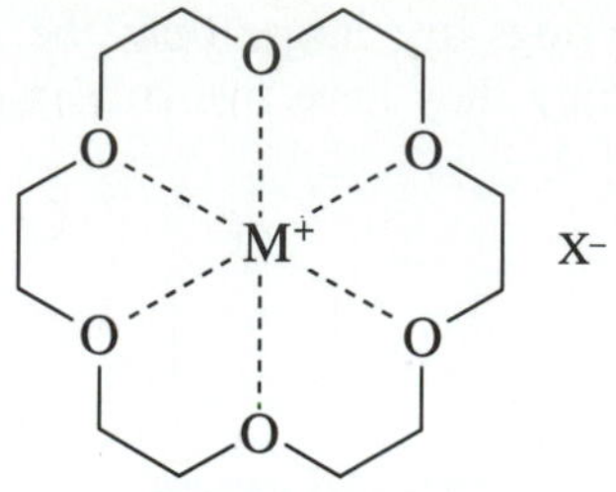

M+ complexed in [18]crown-6

Cavity diameter	2.6–3.2Å	
Ion diameter	Na^+	1.90 Å
	K^+	2.66 Å
	Cs^+	3.34 Å

only this ion achieves a snug fit

This complexing ability is so strong that ionic compounds can be dissolved in organic solvents that contain a crown ether. For example, potassium permanganate ($KMnO_4$) is soluble in water but insoluble in benzene. However, if some dicyclohexyl[18]crown-6 is dissolved in the benzene, it is possible to extract the potassium permanganate from the water into the benzene! The resulting "purple benzene," containing free, essentially unsolvated permanganate ions, is a powerful oxidizing agent.*

The selective binding of metallic ions by macrocyclic compounds is important in nature. Several antibiotics, such as **nonactin,** have large rings that contain regularly spaced oxygen atoms. Nonactin (which contains four tetrahydrofuran rings joined by four ester links) selectively binds K^+ (in the presence of Na^+) in aqueous media, thus allowing the selective transport of K^+ (but not Na^+) through cell membranes.

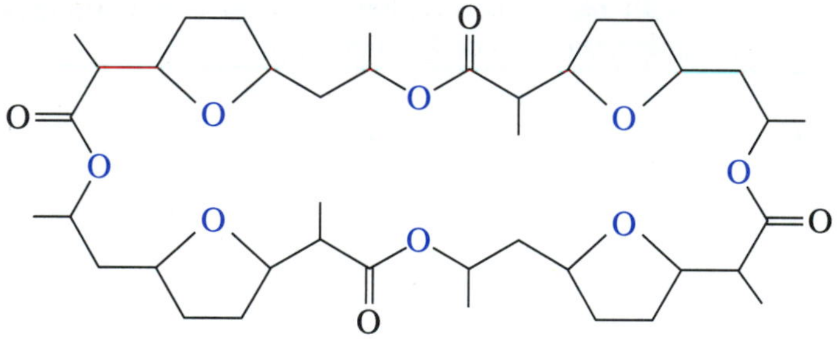

nonactin

Additional Problems

8.16. Write a structural formula for each of the following compounds:

a. dipropyl ether
b. *t*-butyl ethyl ether
c. 3-methoxyhexane
d. diallyl ether
e. *p*-bromophenyl ethyl ether
f. *cis*-2-ethoxycyclopentanol
g. ethylene glycol dimethyl ether
h. 1-methoxypropene
i. propylene oxide
j. *p*-ethoxyanisole

---

*Crown ethers were discovered by Charles J. Pedersen, working at Du Pont Company. This discovery had broad implications for a field now known as molecular recognition, or host-guest chemistry. Pedersen, Donald J. Cram (U.S.) and Jean-Marie Lehn (France) shared the 1987 Nobel Prize in chemistry for their imaginative development of this field. You might enjoy Pedersen's personal account of this discovery (*Journal of Inclusion Phenomena* **1988,** *6,* 337–50); the same journal contains the Nobel lectures by Cram and Lehn on their work.

8.17. Name each of the following compounds:

a. $(CH_3)_2CHOCH(CH_3)_2$ **b.** $(CH_3)_2CHCH_2OCH_3$

c. 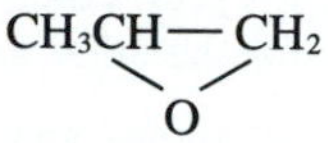 $CH_3CH—CH_2$ (with O bridging) **d.** 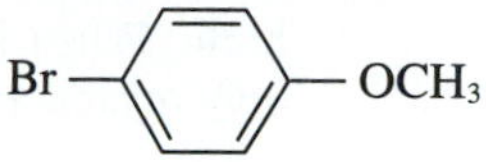 Br—⟨benzene⟩—OCH_3

e. 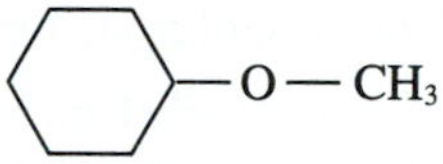 ⟨cyclohexane⟩—O—CH_3 **f.** 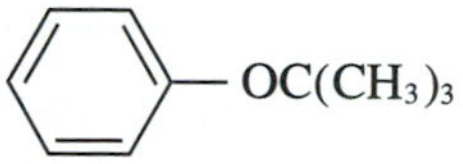 ⟨benzene⟩—$OC(CH_3)_3$

g. $CH_3CH(OCH_2CH_3)CH_2CH_2CH_3$ **h.** $CH_3OCH_2CH_2OH$

i. 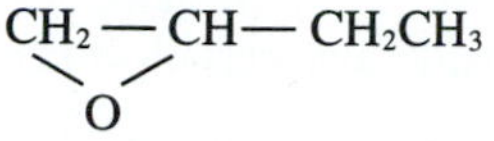 $CH_2—CH—CH_2CH_3$ (with O bridging) **j.** $CH_3OCH_2C{\equiv}CH$

8.18. Ethers and alcohols can be isomeric. Write the structures and give the names for all possible isomers with the molecular formula $C_4H_{10}O$.

8.19. Consider four compounds that have nearly the same molecular weights: 1,2-dimethoxyethane, ethyl propyl ether, hexane, and 1-pentanol. Which would you expect to have the highest boiling point? Which would be most soluble in water? Explain the reasons for your choices.

8.20. Write equations for the reaction of each of the following with (1) Mg in ether followed by (2) addition of D_2O to the resulting solution.

a. $CH_3CH_2CH_2CH_2Br$ **b.** $CH_3OCH_2CH_2CH_2Br$

8.21. The following steps can be used to convert anisole to *o*-*t*-butylanisole. Give the reagent for each step. Explain why the overall result cannot be achieved in one step by a Friedel-Crafts alkylation.

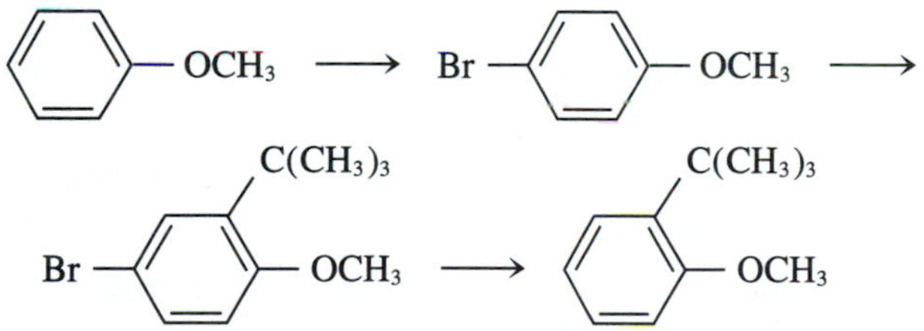

⟨benzene⟩—OCH_3 ⟶ Br—⟨benzene⟩—OCH_3 ⟶ Br—⟨benzene, $C(CH_3)_3$ ortho to OCH_3⟩—OCH_3 ⟶ ⟨benzene, $C(CH_3)_3$ ortho to OCH_3⟩—OCH_3

8.22. Write equations for the best method to prepare each of the following ethers:

a. $(CH_3CH_2CH_2CH_2)_2O$ **b.** 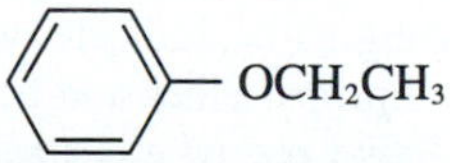 ⟨benzene⟩—OCH_2CH_3 **c.** $CH_3CH_2OC(CH_3)_3$

8.23. Explain why the Williamson synthesis cannot be used to prepare diphenyl ether.

8.24. Ethers are soluble in cold, concentrated sulfuric acid, but alkanes are not. This difference can be used as a simple chemical test to distinguish between these two classes of compounds. What chemistry (show an equation) is the basis for this difference?

8.25. Write an equation for each of the following reactions. If no reaction occurs, say so.

a. dibutyl ether + boiling aqueous NaOH ⟶
b. methyl propyl ether + excess HBr (hot) ⟶
c. dipropyl ether + Na ⟶

d. ethyl ether + cold concentrated H_2SO_4 $\longrightarrow$
e. ethyl phenyl ether + BBr_3 $\longrightarrow$

8.26. When heated with excess HBr, a cyclic ether gave 1,4-dibromobutane as the only organic product. Write a structure for the ether and an equation for the reaction.

8.27. Using the peracid epoxidation of an alkene and the ring opening of an epoxide, devise a two-step synthesis of 1,2-butanediol from 1-butene.

8.28. Write an equation for the reaction of ethylene oxide with
a. 1 mol of HBr. **b.** excess HBr. **c.** phenol + H^+.

8.29. $CH_3CH_2OCH_2CH_2OH$ (ethyl cellosolve) and $CH_3CH_2OCH_2CH_2OCH_2CH_2OH$ (ethyl carbitol) are solvents used in the formulation of lacquers. They are produced commercially from ethylene oxide and certain other reagents. Show with equations how this might be done.

8.30. 2-Phenylethanol, which has the aroma of oil of roses, is used in perfumes. Write equations to show how 2-phenylethanol can be synthesized from bromobenzene and ethylene oxide, using a Grignard reagent.

8.31. 1,1-Dimethyloxirane dissolved in excess methanol and treated with a little acid yields the product 2-methoxy-2-methyl-1-propanol (and no 1-methoxy-2-methyl-2-propanol). What reaction mechanism explains this result?

8.32. Write an equation for the reaction of ammonia with ethylene oxide. The product is a water-soluble organic base used to absorb and concentrate CO_2 in the manufacture of dry ice.

8.33. Design a synthesis of 3-pentyne-1-ol using propyne and ethylene oxide as the only sources of carbon atoms.

8.34. The first commercial method used to make ethylene oxide involved treating ethylene with hypochlorous acid (HO—Cl), followed by reaction of the product with dilute base. Write equations for these reactions, and describe the mechanism of each step.

8.35. Write a series of equations that show how epichlorhydrin, bisphenol-A, and a base might react to form the "linear" epoxy resin shown in Figure 8.1.

8.36. Write out the steps in the reaction mechanisms for the reactions given in eq. 8.20.

8.37. What chemical test will distinguish between the compounds in each of the following pairs? Indicate what is visually observed with each test.
a. dipropyl ether and hexane
b. ethyl phenyl ether and allyl phenyl ether
c. 2-butanol and methyl propyl ether
d. phenol and anisole

8.38. An organic compound with the molecular formula $C_4H_{10}O_3$ shows properties of both an alcohol and an ether. When treated with an excess of hydrogen bromide, it yields only one organic compound, 1,2-dibromoethane. Draw a structural formula for the original compound.

8.39. 1,4-Dioxane can be synthesized by slowly distilling a mixture of ethylene glycol ($HOCH_2CH_2OH$) and dilute sulfuric acid. Write a series of equations that describes the mechanism of this reaction.

CHAPTER 9

Aldehydes and Ketones

9.1 Introduction

We now come to perhaps the most important functional group in organic chemistry—the **carbonyl group,**

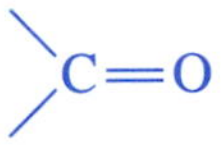

$$\rangle C{=}O$$

This group is present in aldehydes, ketones, carboxylic acids, esters, and several other classes of compounds. Many of these compounds are important commercially and in biological processes. In this chapter, we will discuss aldehydes and ketones and in the next chapter, carboxylic acids and related compounds.

Aldehydes have at least one hydrogen atom attached to the carbonyl group. The remaining group may be another hydrogen atom or any organic group.

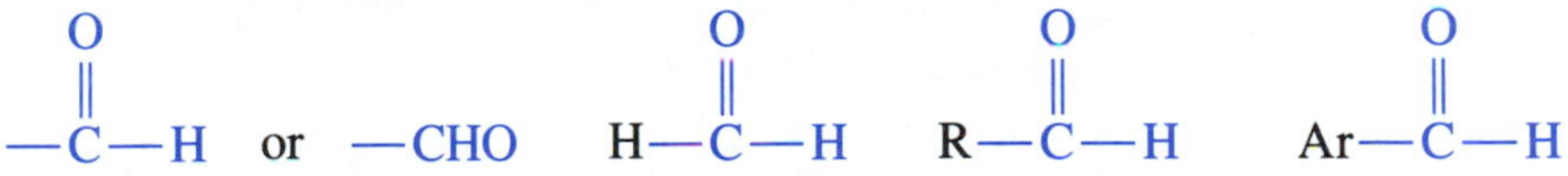

$$\underset{\text{aldehyde group}}{-\overset{\overset{\large O}{\|}}{C}-H \quad \text{or} \quad -CHO} \qquad \underset{\text{formaldehyde}}{H-\overset{\overset{\large O}{\|}}{C}-H} \qquad \underset{\text{aliphatic aldehyde}}{R-\overset{\overset{\large O}{\|}}{C}-H} \qquad \underset{\text{aromatic aldehyde}}{Ar-\overset{\overset{\large O}{\|}}{C}-H}$$

In **ketones,** the carbonyl carbon atom is connected to two other carbon atoms.

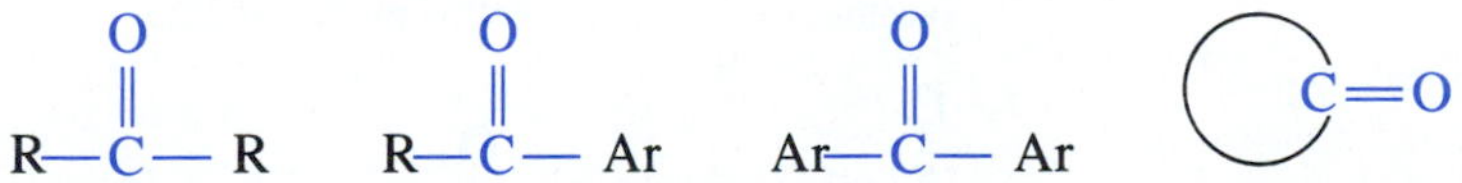

$$\underset{\text{aliphatic ketone}}{R-\overset{\overset{\large O}{\|}}{C}-R} \qquad \underset{\text{alkyl aryl ketone}}{R-\overset{\overset{\large O}{\|}}{C}-Ar} \qquad \underset{\text{aromatic ketone}}{Ar-\overset{\overset{\large O}{\|}}{C}-Ar} \qquad \underset{\text{a cyclic ketone}}{\bigcirc C{=}O}$$

9.2 Nomenclature of Aldehydes and Ketones

In the IUPAC system, the characteristic ending for aldehydes is *-al* (from the first syllable of aldehyde). The following examples illustrate the system:

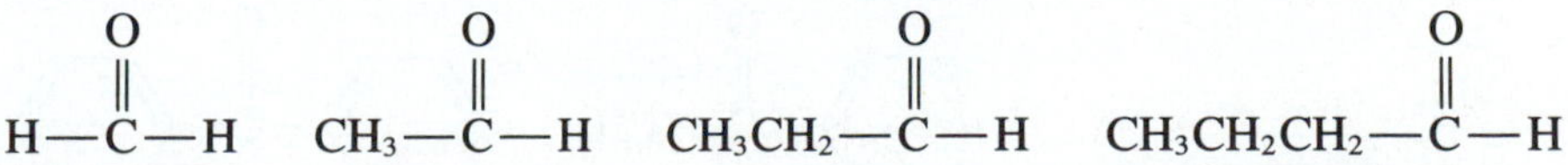

$$\underset{\substack{\text{methanal}\\ \text{(formaldehyde)}}}{H-\overset{\overset{\large O}{\|}}{C}-H} \qquad \underset{\substack{\text{ethanal}\\ \text{(acetaldehyde)}}}{CH_3-\overset{\overset{\large O}{\|}}{C}-H} \qquad \underset{\substack{\text{propanal}\\ \text{(propionaldehyde)}}}{CH_3CH_2-\overset{\overset{\large O}{\|}}{C}-H} \qquad \underset{\substack{\text{butanal}\\ (n\text{-butyraldehyde})}}{CH_3CH_2CH_2-\overset{\overset{\large O}{\|}}{C}-H}$$

The common names shown below the IUPAC names are in frequent use, so you should learn them.

For substituted aldehydes, we number the chain starting with the aldehyde carbon, as the following examples illustrate:

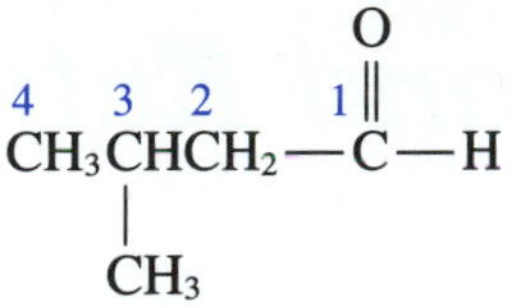

$\overset{4}{CH_3}\overset{3}{C}H(CH_3)\overset{2}{C}H_2{-}\overset{1}{C}(=O){-}H$
3-methylbutanal

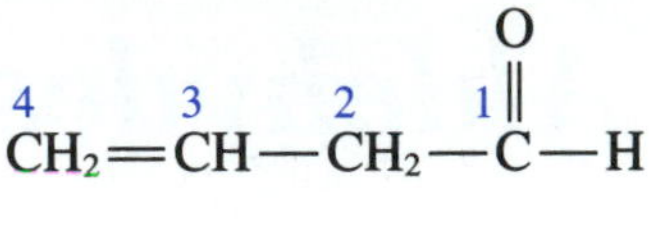

$\overset{4}{CH_2}{=}\overset{3}{C}H{-}\overset{2}{C}H_2{-}\overset{1}{C}(=O){-}H$
3-butenal

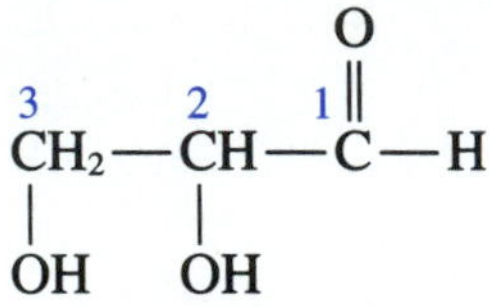

$\overset{3}{CH_2}(OH){-}\overset{2}{C}H(OH){-}\overset{1}{C}(=O){-}H$
2,3-dihydroxypropanal
(glyceraldehyde)

Notice from the last two examples than an aldehyde group has priority over a double bond or a hydroxyl group, not only in numbering, but also as the suffix. For cyclic aldehydes, the suffix *-carbaldehyde* is used. Aromatic aldehydes often have common names:

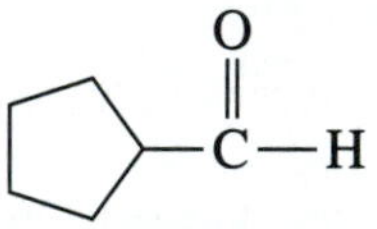

cyclopentanecarbaldehyde
(formylcyclopentane)

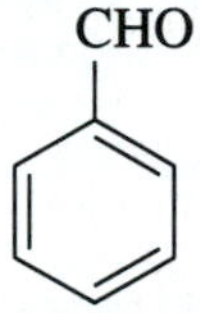

benzaldehyde
(benzenecarbaldehyde)

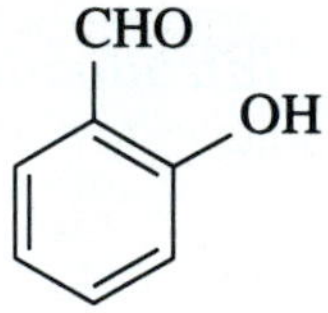

salicylaldehyde
(2-hydroxybenzenecarbaldehyde)

In the IUPAC system, the ending for ketones is *-one* (from the last syllable of ketone). The chain is numbered so that the carbonyl carbon has the lowest possible number. Common names of ketones are formed by adding the word *ketone* to the names of the alkyl or aryl groups attached to the carbonyl carbon. In still other cases, traditional names are used. The following examples illustrate these methods:

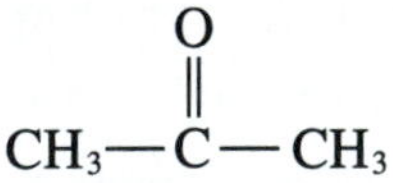

$CH_3{-}C(=O){-}CH_3$
propanone
(acetone)

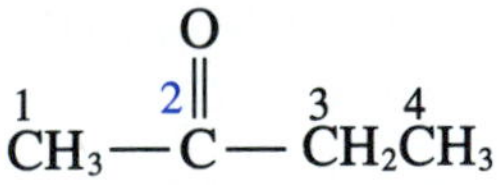

$\overset{1}{CH_3}{-}\overset{2}{C}(=O){-}\overset{3}{C}H_2\overset{4}{C}H_3$
2-butanone
(ethyl methyl ketone)

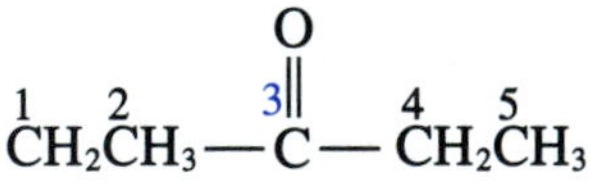

$\overset{1}{C}H_2\overset{2}{C}H_3{-}\overset{3}{C}(=O){-}\overset{4}{C}H_2\overset{5}{C}H_3$
3-pentanone
(diethyl ketone)

cyclohexanone

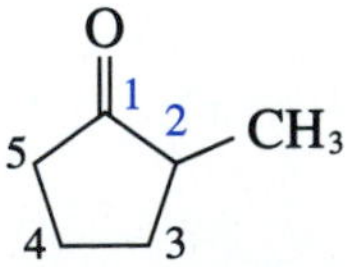

2-methylcyclopentanone

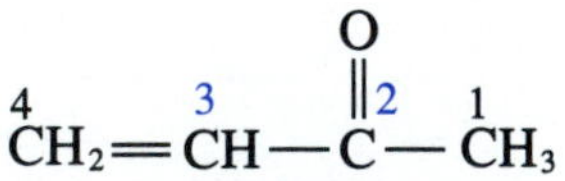

$\overset{4}{CH_2}{=}\overset{3}{C}H{-}\overset{2}{C}(=O){-}\overset{1}{C}H_3$
3-buten-2-one
(methyl vinyl ketone)

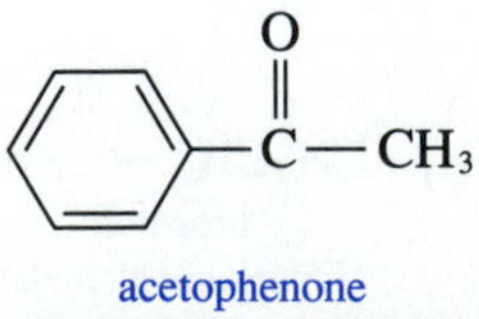

acetophenone
(methyl phenyl ketone)

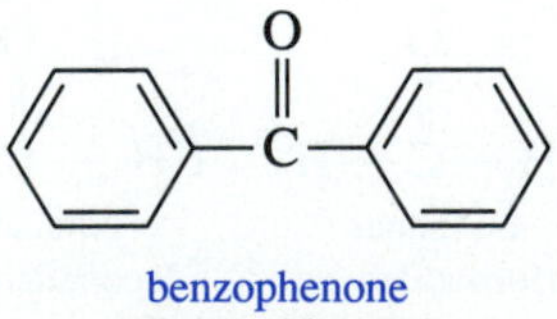

benzophenone
(diphenyl ketone)

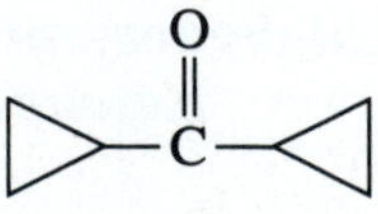

dicyclopropyl ketone

Problem 9.1 Using the examples as a guide, write a structure for

a. pentanal.
b. *p*-bromobenzaldehyde.
c. 2-pentanone.
d. *t*-butyl methyl ketone.
e. cyclohexanecarbaldehyde.
f. 3-pentyne-2-one.

Problem 9.2 Using the examples as a guide, write a correct name for

a. $(CH_3)_2CHCH_2CH{=}O$
b. $CH_3CH{=}CHCH{=}O$
c. 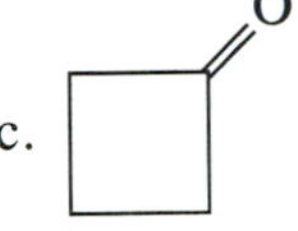

d. $(CH_3)_2CHCH_2\overset{\overset{\displaystyle O}{\|}}{C}CH_3$

9.3 Some Common Aldehydes and Ketones

Formaldehyde, the simplest aldehyde, is manufactured on a very large scale by the oxidation of methanol.

$$CH_3OH \xrightarrow[600-700°C]{\text{Ag catalyst}} \underset{\text{formaldehyde}}{CH_2{=}O} + H_2 \tag{9.1}$$

Annual world production is over 6 billion pounds. Formaldehyde is a gas (bp -21°C), but it cannot be stored in a free state because it polymerizes readily. Normally it is supplied as a 37% aqueous solution called **formalin.** In this form it is used as a disinfectant and preservative, but most formaldehyde is used in the manufacture of plastics, building insulation, particle board, and plywood.

Acetaldehyde boils close to room temperature (bp 20°C). It is manufactured mainly by the Wacker process, which involves direct selective oxidation of ethylene over a palladium-copper catalyst.

$$2CH_2{=}CH_2 + O_2 \xrightarrow[100-130°C]{\text{Pd–Cu}} 2CH_3CH{=}O \tag{9.2}$$

About half the acetaldehyde produced annually is oxidized to acetic acid. The rest is used for the production of 1-butanol and other commercial chemicals.

Acetone, the simplest ketone, is also produced on a large scale—about 4 billion pounds annually. The most common methods for its commercial synthesis are the Wacker oxidation of propene (analogous to eq. 9.2), the oxidation of isopropyl alcohol (eq. 7.32, $R{=}R'{=}CH_3$), and the oxidation of isopropylbenzene (see A Word About Benzene, Chemical of Commerce, on page 127). About 30% of the acetone is used directly, for it is not only completely miscible with water but is also an excellent solvent for many organic substances (resins, paints, dyes, and nail polish). The rest is used to manufacture other commercial chemicals, including bisphenol-A for epoxy resins (page 234).

$$\underset{\text{phenol}}{2HO{-}C_6H_4} + \underset{\text{acetone}}{CH_3\overset{\overset{\displaystyle O}{\|}}{C}CH_3} \xrightarrow[-H_2O]{H^+} \underset{\text{bisphenol-A}}{HO{-}C_6H_4{-}C(CH_3)_2{-}C_6H_4{-}OH} \tag{9.3}$$

Problem 9.3 The mechanism of eq. 9.3 involves the following steps: (a) protonation of the carbonyl oxygen to produce a tertiary carbocation, which then alkylates phenol (Friedel-Crafts electrophilic aromatic substitution), and (b) protonation of the resulting tertiary alcohol leading to another tertiary carbocation, which then alkylates a second phenol molecule. From this description, write a series of equations that shows the steps in the reaction mechanism.

9.4 Synthesis of Aldehydes and Ketones

We have already seen, in previous chapters, several ways to prepare aldehydes and ketones. One of the most useful is the oxidation of alcohols.

$$>C(H)(OH) \xrightarrow[\text{agent}]{\text{oxidizing}} >C{=}O \qquad (9.4)$$

Recall that ***primary alcohols give aldehydes, secondary alcohols give ketones,*** and CrO_3 in the form of PCC and Jones' reagent is the most common laboratory reagent for this purpose (review Sec. 7.13).

Problem 9.4 Give the structure of an alcohol that is a suitable precursor for oxidation to

a. 2-methylpropanal. b. 4-*t*-butylcyclohexanone.

Aromatic ketones can be made by Friedel-Crafts acylation of an aromatic ring (review eq. 4.15 and Sec. 4.10d). For example,

$$C_6H_6 + C_6H_5\overset{O}{\overset{\|}{C}}Cl \xrightarrow{AlCl_3} C_6H_5\overset{O}{\overset{\|}{C}}C_6H_5 + HCl \qquad (9.5)$$

benzene, benzoyl chloride, benzophenone

Problem 9.5 Complete the equation

$$CH_3{-}C_6H_5 + CH_3\overset{O}{\overset{\|}{C}}Cl \xrightarrow{AlCl_3}$$

and name the product.

Methyl ketones can be prepared by hydration of terminal alkynes, catalyzed by acid and mercuric ion (review eq. 3.48). For example,

$$\underset{\text{1-octyne}}{CH_3(CH_2)_5C{\equiv}CH} \xrightarrow[Hg^{2+}]{H^+,\ H_2O} \underset{\text{2-octanone}}{CH_3(CH_2)_5\overset{O}{\overset{\|}{C}}CH_3} \qquad (9.6)$$

Problem 9.6 What alkyne would be useful for the synthesis of 2-hexanone?

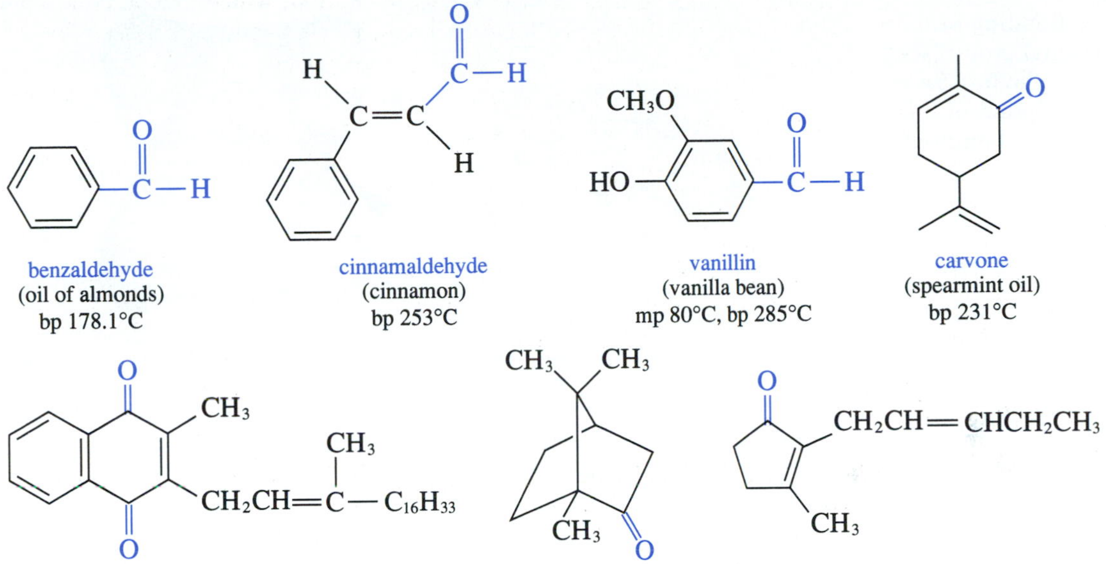

benzaldehyde
(oil of almonds)
bp 178.1°C

cinnamaldehyde
(cinnamon)
bp 253°C

vanillin
(vanilla bean)
mp 80°C, bp 285°C

carvone
(spearmint oil)
bp 231°C

vitamin K
mp −20°C

camphor
mp 179°C

jasmone
(from oil of jasmine)

FIGURE 9.1 Some naturally occurring aldehydes and ketones.

9.5 Aldehydes and Ketones in Nature

Aldehydes and ketones occur very widely in nature. Figures 1.11 and 1.12 show three examples, and Figure 9.1 gives several more. Many aldehydes and ketones have pleasant odors and flavors and are used for these properties in perfumes and other consumer products (soaps, bleaches, and air fresheners, for example). The gathering and extraction of these fragrant substances from flowers, plants, and animal glands is extremely expensive, however. Chanel No. 5, introduced to the perfume market in 1921, was the first fine fragrance to use *synthetic* organic chemicals. Today most fragrances do.*

9.6 The Carbonyl Group

To best understand the reactions of aldehydes, ketones, and other carbonyl compounds, we must first appreciate the structure and properties of the carbonyl group.

The carbon-oxygen double bond consists of a sigma bond and a pi bond (Figure 9.2). The carbon atom is sp^2-hybridized. *The three atoms attached to the carbonyl carbon lie in a plane with bond angles of 120°.* The pi bond is

* For a brief, interesting, and readable account of present-day perfumery, see C. S. Sell, *Chemistry in Britain* **1988,** 791–94.

FIGURE 9.2
Bonding in the carbonyl group (see the text for a description of this bonding).

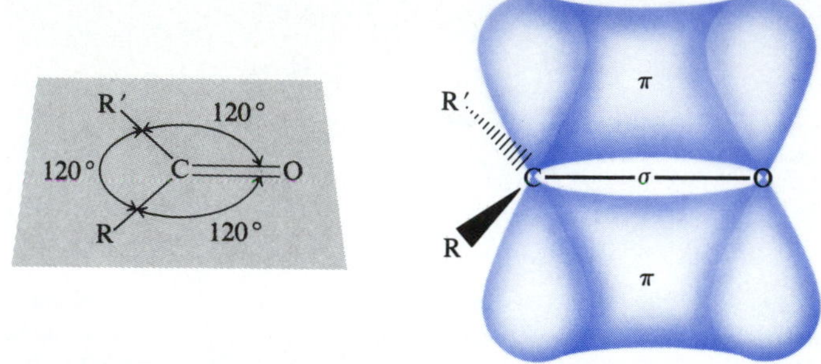

formed by overlap of a p orbital on carbon with an oxygen p orbital. There are also two unshared electron pairs on the oxygen atom. The C=O bond distance is 1.24 Å, shorter than the C—O distance in alcohols and ethers (1.43 Å).

Oxygen is much more electronegative than carbon. Therefore the electrons in the C=O bond are attracted to the oxygen, producing a highly polarized bond. This effect is especially pronounced for the pi electrons and can be expressed in the following ways:

$$\left[\ \mathrm{C{=}\ddot{O}{:}} \longleftrightarrow \mathrm{\overset{+}{C}{-}\ddot{\underset{..}{O}}{:}^{-}}\ \right] \qquad \mathrm{\overset{\delta+}{C}{=}\overset{\delta-}{O}} \qquad \mathrm{C{=}O}$$

resonance contributors to the carbonyl group polarization of the carbonyl group

As a consequence of this polarization, most carbonyl reactions involve **nucleophilic attack** at the carbonyl carbon, often accompanied by addition of a proton to the oxygen.

attack here by a nucleophile ⟶ $\mathrm{\overset{\delta+}{C}{=}\overset{\delta-}{O}}$ ⟵ may react with a proton

C=O bonds are quite different, then, from C=C bonds, where attack at carbon is usually by an electrophile (Sec. 3.11).

In addition to its effect on reactivity, polarization of the C=O bond influences the physical properties of carbonyl compounds, as illustrated in the following examples.

EXAMPLE 9.1 Explain the fact that carbonyl compounds boil at higher temperatures than hydrocarbons, but at lower temperatures than alcohols of comparable molecular weight (for example, $CH_3(CH_2)_3CH_3$, bp 36°C; $CH_3(CH_2)_2CH{=}O$, bp 75°C; $CH_3(CH_2)_2CH_2OH$, bp 118°C).

Solution Being polar, molecules of carbonyl compounds tend to associate: the positive part of one molecule is attracted to the negative part of another molecule. To overcome this attractive force, which is not significant in hydrocarbons, requires energy (heat) when the substance is converted from liquid to vapor. Having no O—H bonds, however, carbonyl compounds cannot form hydrogen bonds with one another, as alcohols can.

EXAMPLE 9.2 Explain why carbonyl compounds with low molecular weights are water soluble.

Solution Although they cannot form hydrogen bonds with themselves, carbonyl compounds readily form hydrogen bonds with O—H or N—H compounds.

$$\overset{\delta+}{C}=\overset{\delta-}{\ddot{O}}:\cdots\overset{\delta+}{H}-\overset{\delta-}{O}-H$$

Problem 9.7 Arrange benzaldehyde (mol. wt. 106), benzyl alcohol (mol. wt. 108), hydroquinone (mol. wt. 110), and *p*-xylene (mol. wt. 106) in order of

a. increasing boiling point. b. increasing water solubility.

9.7 Nucleophilic Addition to Carbonyl Groups; an Overview

Nucleophiles attack the carbon atom of a carbon–oxygen double bond because that carbon has a partial positive charge. The pi electrons of the C=O bond move to the oxygen atom, which, because of its electronegativity, can easily accommodate the negative charge that it acquires. When these reactions are carried out in a hydroxylic solvent such as alcohol or water (represented as SOH), the reaction is usually completed by addition of a proton to the negative oxygen.

$$Nu:^- + C=\ddot{O}: \rightleftharpoons Nu-C-\ddot{O}:^- \overset{SOH}{\rightleftharpoons} Nu-C-\ddot{O}H \qquad (9.7)$$

trigonal reactant — tetrahedral intermediate — tetrahedral product

The carbonyl carbon, which is trigonal and sp^2-hybridized in the starting aldehyde or ketone, becomes tetrahedral and sp^3-hybridized in the reaction product.

Because of the unshared electron pairs on the oxygen atom, carbonyl compounds are weak Lewis bases and can be protonated. ***Acids can catalyze the addition of weak nucleophiles to carbonyl compounds*** by protonating the carbonyl oxygen atom.

$$C=\ddot{O}: + H^+ \longrightarrow \left[C=\overset{+}{\ddot{O}}H \longleftrightarrow \overset{+}{C}-\ddot{O}H \right] \xrightarrow{Nu:^-} Nu-C-\ddot{O}H \qquad (9.8)$$

a resonance-stabilized carbocation

This converts the carbonyl carbon to a carbocation and enhances its susceptibility to attack by nucleophiles.

In general, ***ketones are somewhat less reactive than aldehydes toward nucleophiles.*** There are two main reasons for this reactivity difference. *The first reason is steric.* The carbonyl carbon atom is more crowded in ketones (two organic groups) than in aldehydes (one organic group and one hydrogen atom). In nucleophilic addition, we bring these attached groups closer together because the hybridization changes from sp^2 to sp^3 and the bond angles decrease from 120° to 109.5°. Less strain is involved in additions to aldehydes than in additions to ketones because one of the groups (H) is small. *The second reason is electronic.* As we have already seen in connection with carbocation stability, alkyl groups are usually electron-donating compared to hydrogen. They therefore tend to neutralize the partial positive charge on the carbonyl carbon, decreasing its reactivity toward nucleophiles. Ketones have two such alkyl groups; aldehydes have only one. If, however, the attached groups are strongly electron-withdrawing (contain halogens, for example), they can have the opposite effect and increase carbonyl reactivity toward nucleophiles.

In the following discussion, we will classify nucleophilic additions to aldehydes and ketones according to the type of new bond formed to the carbonyl carbon. We will consider oxygen, carbon, and nitrogen nucleophiles, in that sequence.

9.8 Addition of Alcohols; Formation of Hemiacetals and Acetals

The reactions discussed in this section are extremely important because they are crucial to understanding the chemistry of carbohydrates, which we will discuss later.

Alcohols are oxygen nucleophiles. They can attack the carbonyl carbon of aldehydes or ketones, resulting in addition to the C=O bond.

$$\underset{\text{alcohol}}{ROH} + \underset{\text{aldehyde}}{R'(H)C{=}O} \overset{H^+}{\rightleftharpoons} \underset{\text{hemiacetal}}{R'(H)C(OR){-}OH} \quad (9.9)$$

Because alcohols are *weak* nucleophiles, an acid catalyst is required. The product is a **hemiacetal,** which contains both alcohol and ether functional groups on the same carbon atom. The addition is reversible.

EXAMPLE 9.3 Show the steps in the mechanism for eq. 9.9.

Solution First, the carbonyl carbon is protonated by the acid catalyst, as in eq. 9.8. The alcohol oxygen then attacks the carbonyl carbon, and a proton is lost from the resulting positive oxygen. *Each step is reversible.*

R′(H)C=Ö: ⇌ (H⁺ / −H⁺) R′(H)C=ÖH⁺ ⇌ (ROH / −ROH) R′(H)C(RÖ⁺H)—ÖH ⇌ (−H⁺ / H⁺) R′(H)C(RO)—ÖH (9.10)

aldehyde — protonated aldehyde — hemiacetal

Problem 9.8 Write an equation for the formation of a hemiacetal from acetaldehyde, ethanol, and H^+. Show each step in the reaction mechanism.

In the presence of *excess alcohol,* hemiacetals react further to form **acetals.**

R′(H)C(RO)—OH + ROH ⇌ (H⁺) R′(H)C(RO)—OR + HOH (9.11)

hemiacetal — acetal

The hydroxyl group of the hemiacetal is replaced by an alkoxyl group. Acetals have *two* ether functions at the same carbon atom.

EXAMPLE 9.4 Show the steps in the mechanism for eq. 9.11.

Solution

R′(H)C(RO)—ÖH ⇌ (H⁺ / −H⁺) R′(H)C(RO)—Ö⁺H(H) ⇌ (−H₂O / +H₂O) [RÖ—C⁺(R′)(H) ⟷ RÖ⁺=C(R′)(H)]

hemiacetal — resonance-stabilized carbocation

⇅ (−ROH / ROH) (9.12)

R′(H)C(RO)—ÖR ⇌ (−H⁺ / H⁺) R′(H)C(RO)—Ö⁺R(H)

acetal

Either oxygen of the hemiacetal can be protonated. When the hydroxyl oxygen is protonated, loss of water leads to a resonance-stabilized carbocation. Reaction of this carbocation with the alcohol, *which is usually the solvent and present in large excess,* gives (after proton loss) the acetal. The mechanism is like an S_N1 reaction. *Each step is reversible.*

EXAMPLE 9.5 What would happen if the *ether* oxygen of the hemiacetal were protonated instead of the alcohol oxygen?

Solution This can happen since both oxygens have unshared electron pairs and are Lewis bases. Protonation of the ether oxygen would be the first step in the reversal of eq. 9.10. Loss of alcohol from this intermediate would continue the reversal back to aldehyde and alcohol. *Excess alcohol is present, however,* so all the equilibria in eq. 9.10 should be driven forward again to the hemiacetal. *Protonation of the ether oxygen is therefore nonproductive,* but protonation of the *alcohol* oxygen leads to the acetal.

Problem 9.9 Write an equation for the reaction of the hemiacetal

$$CH_3\underset{|}{\overset{OH}{C}}HOCH_2CH_3$$

with excess ethanol and H^+. Show each step in the mechanism.

Aldehydes that have an appropriately located hydroxyl group *in the same molecule* may exist in equilibrium with a **cyclic hemiacetal,** formed by *intramolecular* nucleophilic addition. For example, 5-hydroxypentanal exists mainly in the cyclic hemiacetal form:

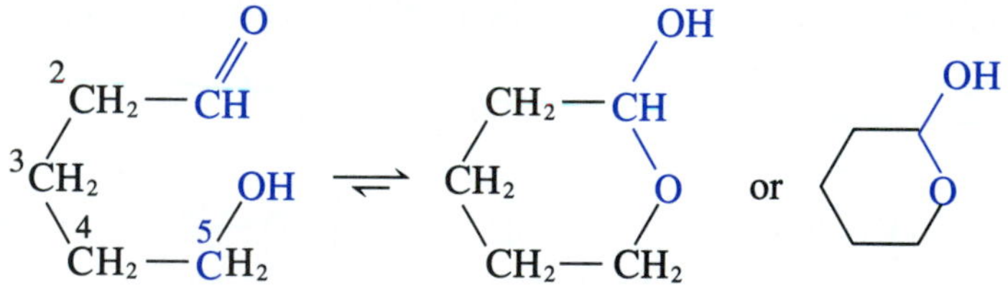

$$\text{5-hydroxypentanal} \rightleftharpoons \text{hemiacetal form of 5-hydroxypentanal} \quad (9.13)$$

5-hydroxypentanal

hemiacetal form of 5-hydroxypentanal
(also called 2-hydroxytetrahydropyran)

The hydroxyl group is favorably located to act as a nucleophile toward the carbonyl carbon, and cyclization occurs by the following mechanism:

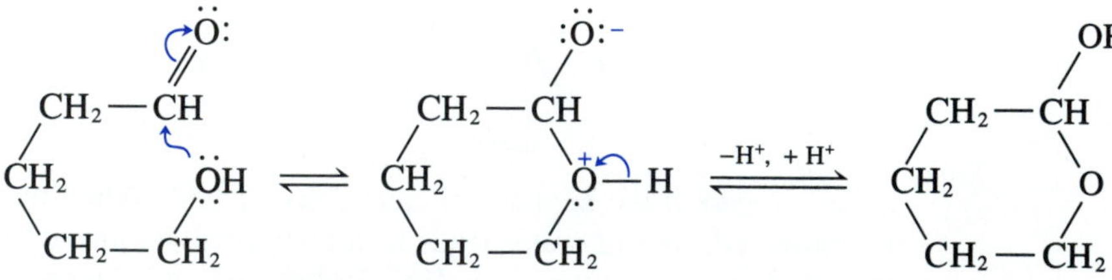

$$\rightleftharpoons \quad \xrightleftharpoons{-H^+,\ +H^+} \quad (9.14)$$

Compounds with a hydroxyl group four or five carbons from the aldehyde group tend to form cyclic acetals because the ring size (five- or six-membered) is relatively strain free. As we will see in Chapter 16, these structures are crucial to the chemistry of carbohydrates.

Ketones also form acetals. If, as in the following example, a glycol is used as the alcohol, the product will be cyclic.

$$(CH_3)_2C{=}O + HOCH_2CH_2OH \xrightleftharpoons{H^+} (CH_3)_2C\begin{matrix}O-CH_2\\ |\\ O-CH_2\end{matrix} + H_2O \qquad (9.15)$$

acetone; ethylene glycol; acetone–ethylene glycol ketal

Problem 9.10 An intermediate in eq. 9.15 is the hemiacetal

$$(CH_3)_2C(OH){-}OCH_2CH_2OH$$

With this information and Examples 9.3 and 9.4 as guides, write out the steps in the mechanism of eq. 9.15.

To summarize, aldehydes and ketones react with alcohols to form, first, hemiacetals and then, if excess alcohol is present, acetals.

$$R'{-}\overset{O}{\overset{\|}{C}}{-}H \underset{H^+}{\overset{RO-H}{\rightleftharpoons}} R'{-}\underset{H}{\overset{OH}{C}}{-}OR \underset{H^+}{\overset{RO-H}{\rightleftharpoons}} R'{-}\underset{H}{\overset{OR}{C}}{-}OR + HOH \qquad (9.16)$$

aldehyde; hemiacetal; acetal

These equilibria are driven in the forward direction by excess alcohol. On the other hand, the equilibria can be reversed. An acetal can be hydrolyzed to its component aldehyde and alcohol by treatment with *excess water* in the presence of acid. The hemiacetal intermediate in both the forward and reverse processes usually cannot be isolated when R and R′ are simple alkyl or aryl groups.

■ ***EXAMPLE 9.6*** Write an equation for the reaction of benzaldehyde dimethylacetal with aqueous acid.

Solution

$$C_6H_5{-}CH(OCH_3)_2 \xrightarrow[H^+]{H_2O} C_6H_5{-}CH{=}O + 2\ CH_3OH \qquad (9.17)$$

Problem 9.11 Show the steps in the mechanism for eq. 9.17.

The acid-catalyzed cleavage of acetals occurs much more readily than the acid-catalyzed cleavage of simple ethers (Sec. 8.7) because the intermediate carbocation is resonance-stabilized. However acetals, like ordinary ethers, are stable toward bases.

9.9 Addition of Water; Hydration of Aldehydes and Ketones

Water, like alcohols, is an oxygen nucleophile and can add reversibly to aldehydes and ketones. For example, formaldehyde in water exists mainly as its hydrate.

$$H_2C{=}O + H{-}OH \rightleftharpoons H_2C(OH)(OH) \tag{9.18}$$

formaldehyde — formaldehyde hydrate

With most other aldehydes or ketones, however, the hydrates cannot be isolated because they readily lose water to reform the carbonyl compound. An exception is trichloroacetaldehyde (chloral), which forms a stable crystalline hydrate, $CCl_3CH(OH)_2$. **Chloral hydrate** is used in medicine as a sedative and in veterinary medicine as a narcotic and anesthetic for horses, cattle, swine, and poultry. The potent drink known as a Mickey Finn is a combination of alcohol and chloral hydrate.

Problem 9.12 Hydrolysis of $CH_3CBr_2CH_3$ with sodium hydroxide does *not* give $CH_3C(OH)_2CH_3$. Instead, it gives acetone. Explain.

9.10 Addition of Grignard Reagents and Acetylides

Grignard reagents act as carbon nucleophiles toward carbonyl compounds. The R group of the Grignard reagent attacks the carbonyl carbon, forming a new carbon–carbon bond. The product is an alkoxide, which can then be hydrolyzed to an alcohol.

$$\rangle C{=}O + RMgX \xrightarrow{\text{ether}} R{-}C{-}\overset{-}{O}\overset{+}{M}gX \xrightarrow[\text{HCl}]{H_2O} R{-}C{-}OH + Mg^{2+}X^{-}Cl^{-} \tag{9.19}$$

intermediate addition product (a magnesium alkoxide) — an alcohol

The reaction is normally carried out by slowly adding an ether solution of the aldehyde or ketone to an ether solution of the Grignard reagent. After all the carbonyl compound is added and the reaction is complete, the resulting magnesium alkoxide is hydrolyzed with aqueous acid.

The reaction of a Grignard reagent with a carbonyl compound is very useful. Many alcohols can be synthesized in this way by the proper choice of reagents. The type of carbonyl compound chosen determines the class of alcohol produced. *Formaldehyde gives primary alcohols.*

$$\mathrm{R{-}MgX} + \underset{\text{formaldehyde}}{\mathrm{H{-}\overset{\overset{\displaystyle O}{\|}}{C}{-}H}} \longrightarrow \mathrm{R{-}\underset{\underset{\displaystyle H}{|}}{\overset{\overset{\displaystyle H}{|}}{C}}{-}OMgX} \xrightarrow[H^+]{H_2O} \underset{\text{a primary alcohol}}{\mathrm{R{-}\underset{\underset{\displaystyle H}{|}}{\overset{\overset{\displaystyle H}{|}}{C}}{-}OH}} \qquad (9.20)$$

Other aldehydes give secondary alcohols.

$$\mathrm{R{-}MgX} + \underset{\text{aldehyde}}{\mathrm{R'{-}\overset{\overset{\displaystyle O}{\|}}{C}{-}H}} \longrightarrow \mathrm{R{-}\underset{\underset{\displaystyle H}{|}}{\overset{\overset{\displaystyle R'}{|}}{C}}{-}OMgX} \xrightarrow[H^+]{H_2O} \underset{\text{a secondary alcohol}}{\mathrm{R{-}\underset{\underset{\displaystyle H}{|}}{\overset{\overset{\displaystyle R'}{|}}{C}}{-}OH}} \qquad (9.21)$$

Ketones give tertiary alcohols.

$$\mathrm{R{-}MgX} + \underset{\text{ketone}}{\mathrm{R'{-}\overset{\overset{\displaystyle O}{\|}}{C}{-}R''}} \longrightarrow \mathrm{R{-}\underset{\underset{\displaystyle R''}{|}}{\overset{\overset{\displaystyle R'}{|}}{C}}{-}OMgX} \xrightarrow[H^+]{H_2O} \underset{\text{a tertiary alcohol}}{\mathrm{R{-}\underset{\underset{\displaystyle R''}{|}}{\overset{\overset{\displaystyle R'}{|}}{C}}{-}OH}} \qquad (9.22)$$

Note that only *one* of the R groups (shown in black) attached to the hydroxyl-bearing carbon of the alcohol comes from the Grignard reagent. The rest of the alcohol's carbon skeleton comes from the carbonyl compound.

EXAMPLE 9.7 Show how the following alcohol can be synthesized from a Grignard reagent and a carbonyl compound.

$$\mathrm{C_6H_5{-}\overset{\overset{\displaystyle OH}{|}}{C}HCH_3}$$

Solution The alcohol is secondary, so the carbonyl compound must be an aldehyde. We can use either a methyl or a phenyl Grignard reagent.

$$\mathrm{C_6H_5{-}\overset{\overset{\displaystyle OH}{|}}{C}H{-}CH_3} \leftarrow \mathrm{CH_3MgBr} \qquad \mathrm{C_6H_5{-}\overset{\overset{\displaystyle OH}{|}}{C}H{-}CH_3} \leftarrow \mathrm{C_6H_5{-}MgBr}$$

The equations are

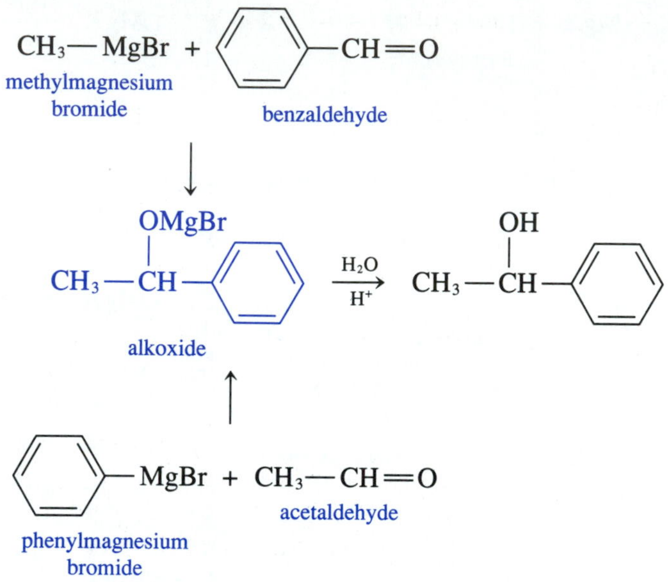

$$CH_3{-}MgBr + C_6H_5{-}CH{=}O$$

methylmagnesium bromide; benzaldehyde

↓

$$CH_3{-}\underset{}{\overset{OMgBr}{\overset{|}{C}H}}{-}C_6H_5 \xrightarrow[H^+]{H_2O} CH_3{-}\overset{OH}{\overset{|}{C}H}{-}C_6H_5 \qquad (9.23)$$

alkoxide

↑

$$C_6H_5{-}MgBr + CH_3{-}CH{=}O$$

phenylmagnesium bromide; acetaldehyde

The choice between the possible sets of reactants may be made by availability or cost, or for chemical reasons (for example, the more reactive aldehyde or ketone might be selected).

Problem 9.13 Show how each of the following alcohols can be made from a Grignard reagent and a carbonyl compound.

a. $C_6H_5{-}CH_2OH$ b. $C_6H_5{-}C(CH_3)_2OH$

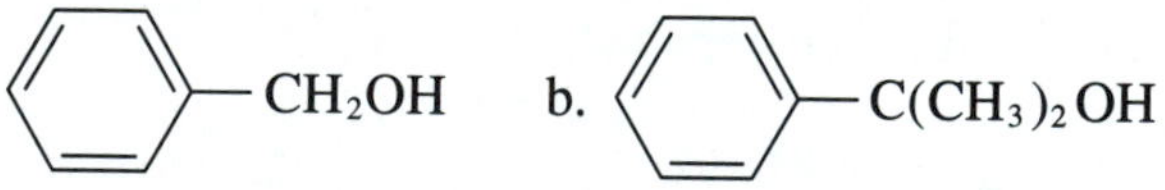

Other organometallic reagents, such as organolithium compounds and acetylides, react with carbonyl compounds similarly to Grignard reagents. For example,

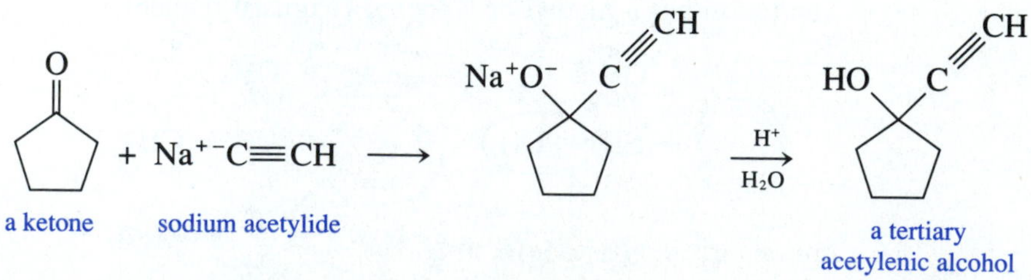

$$\text{cyclopentanone} + Na^{+}\,{}^{-}C{\equiv}CH \longrightarrow \text{1-(C}{\equiv}\text{CH)-1-}(Na^{+}O^{-})\text{cyclopentane} \xrightarrow[H_2O]{H^+} \text{1-(C}{\equiv}\text{CH)-1-(HO)cyclopentane} \qquad (9.24)$$

a ketone; sodium acetylide; a tertiary acetylenic alcohol

9.11 Addition of Hydrogen Cyanide; Cyanohydrins

Hydrogen cyanide adds to the carbonyl group of aldehydes and ketones to form **cyanohydrins,** compounds with a hydroxyl and a cyano group attached to the same carbon. A basic catalyst is required.

$$>C{=}O + HCN \xrightarrow{OH^-} (CN)C{-}OH \tag{9.25}$$

a cyanohydrin

Acetone, for example, reacts as follows:

$$CH_3{-}\overset{O}{\overset{\|}{C}}{-}CH_3 + HCN \xrightarrow{OH^-} CH_3{-}\underset{CN}{\underset{|}{\overset{OH}{\overset{|}{C}}}}{-}CH_3 \tag{9.26}$$

acetone — acetone cyanohydrin

Hydrogen cyanide has no unshared electron pair on its carbon, so it cannot function as a carbon nucleophile. The base converts some of the hydrogen cyanide to cyanide ion, however, which then acts as a carbon nucleophile.

$$>C{=}\ddot{O}: + {}^-:C{\equiv}N: \rightleftharpoons (CN)C{-}\ddot{O}:^- \overset{HCN}{\rightleftharpoons} (CN)C{-}\ddot{O}H + {}^-CN \tag{9.27}$$

cyanohydrin

Problem 9.14 Write an equation for the addition of HCN to

a. acetaldehyde. b. benzaldehyde.

9.12 Addition of Nitrogen Nucleophiles

Ammonia, amines, and certain related compounds have an unshared electron pair on the nitrogen atom and act as nitrogen nucleophiles toward the carbonyl carbon atom. For example, primary amines react as follows:

$$>C{=}O + \ddot{N}H_2{-}R \rightleftharpoons \left[(OH)C{-}NHR\right] \xrightarrow{-HOH} >C{=}NR \tag{9.28}$$

primary amine — tetrahedral addition product — imine

The tetrahedral addition product that is formed first is similar to a hemiacetal, but with an NH group in place of one of the oxygens. These addition products are normally not stable. They eliminate water to form a product with a carbon–nitrogen double bond. With primary amines, the products are called **imines.** Imines are like carbonyl compounds, except that the O is replaced by NR.

They are important intermediates in some biochemical reactions, particularly in binding carbonyl compounds to the free amino groups that are present in most enzymes.

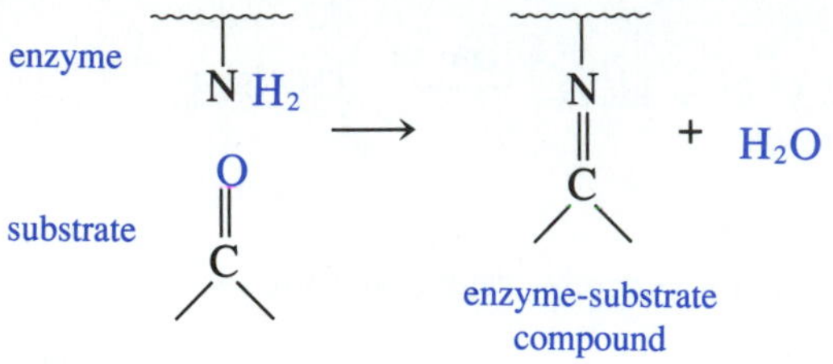

enzyme —NH_2 + substrate $C{=}O$ ⟶ enzyme—$N{=}C$ (enzyme-substrate compound) + H_2O (9.29)

For example, retinal ("A Word About" on page 79) binds to the protein opsin in this way, to form rhodopsin.

EXAMPLE 9.8 Using Example 9.3 as a guide, write the steps in the mechanism for eq. 9.28.

Solution

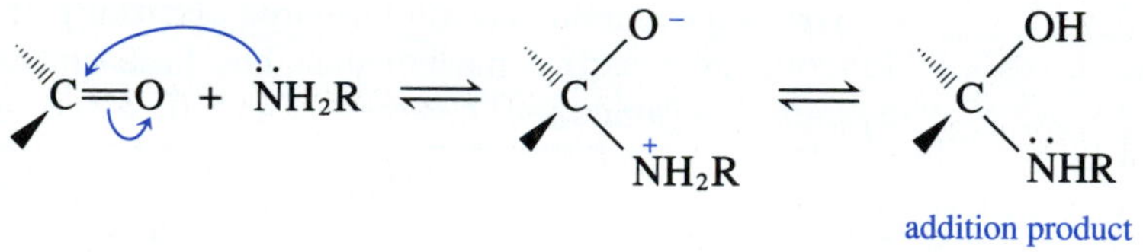

$$C{=}O + \ddot{N}H_2R \rightleftharpoons C(O^-)(\overset{+}{N}H_2R) \rightleftharpoons C(OH)(\ddot{N}HR)$$

addition product

The first product formed is a dipolar ion. The positive nitrogen (an ammonium-type ion) loses a proton, and the negative oxygen (an alkoxide ion) gains a proton, thus forming the tetrahedral addition product. The 1,2-elimination of water then gives the observed product:

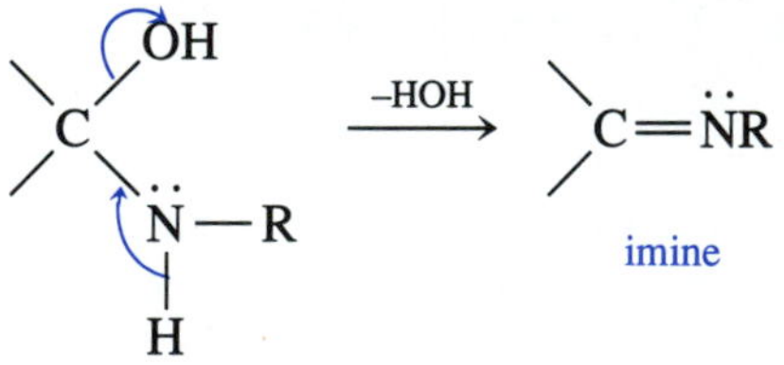

$$C(OH)(\ddot{N}H{-}R) \xrightarrow{-HOH} C{=}\ddot{N}R$$

imine

Problem 9.15 Write an equation for the reaction of benzaldehyde with aniline (the formula of which is $C_6H_5NH_2$).

Other ammonia derivatives containing an $—NH_2$ group react with carbonyl compounds similarly to primary amines. Table 9.1 lists some specific examples.

Problem 9.16 Using Table 9.1 as a guide, write an equation for the reaction of

a. propanal with hydroxylamine.
b. benzaldehyde with phenylhydrazine.

TABLE 9.1 Nitrogen derivatives of carbonyl compounds

Formula of ammonia derivative	*Name*	*Formula of carbonyl derivative*	*Name*
RNH_2 or $ArNH_2$	primary amine	$>C{=}NR$ or $>C{=}NAr$	imine
NH_2OH	hydroxylamine	$>C{=}NOH$	oxime
NH_2NH_2	hydrazine	$>C{=}NNH_2$	hydrazone
$NH_2NHC_6H_5$	phenylhydrazine	$>C{=}NNHC_6H_5$	phenylhydrazone

9.13 Reduction of Carbonyl Compounds

Aldehydes and ketones are easily reduced to primary and secondary alcohols, respectively. Reduction can be accomplished in many ways, most commonly by metal hydrides.

The most common metal hydrides used to reduce carbonyl compounds are **lithium aluminum hydride** ($LiAlH_4$) **and sodium borohydride** ($NaBH_4$). The metal-hydride bond is polarized, with the metal positive and the hydrogen negative. The reaction therefore involves nucleophilic attack of the hydride at the carbonyl carbon:

$$\overset{\delta+}{C}{=}\overset{\delta-}{\ddot{O}}\!: + \overset{\delta-}{H}{-}\overset{\delta+}{Al} \longrightarrow \underset{\text{aluminum alkoxide}}{H{-}C{-}O{-}Al} \xrightarrow[H^+]{H_2O} \underset{\text{alcohol}}{H{-}C{-}OH} \quad (9.30)$$

The initial product is an aluminum alkoxide, which is subsequently hydrolyzed by water and acid to give the alcohol. The net result is addition of hydrogen across the carbon-oxygen double bond. A specific example is

$$\underset{\text{cyclohexanone}}{C_6H_{10}{=}O} \xrightarrow[2.\ H^+,\ H_2O]{1.\ LiAlH_4} \underset{\text{cyclohexanol}}{C_6H_{10}(H)OH} \quad (9.31)$$

Since a carbon-carbon double bond is not readily attacked by nucleophiles, metal hydrides can be used to reduce a carbon-oxygen double bond to the corresponding alcohol without reducing a carbon-carbon double bond present in the same compound.

$$CH_3{-}CH{=}CH{-}\overset{\overset{\large O}{\|}}{C}H \xrightarrow{NaBH_4} CH_3CH{=}CH{-}CH_2OH \qquad (9.32)$$

2-butenal (crotonaldehyde) → 2-buten-1-ol (crotyl alcohol)

Problem 9.17 Show how (cyclohex-2-enyl)$-\overset{\overset{\large O}{\|}}{C}-CH_3$ can be reduced to (cyclohex-2-enyl)$-CH(OH)CH_3$

9.14 Oxidation of Carbonyl Compounds

Aldehydes are more easily oxidized than ketones. Oxidation of an aldehyde gives an acid with the same number of carbon atoms.

$$R{-}\overset{\overset{\large O}{\|}}{C}{-}H \xrightarrow[\text{agent}]{\text{oxidixing}} R{-}\overset{\overset{\large O}{\|}}{C}{-}OH \qquad (9.33)$$

aldehyde → acid

Since the reaction occurs easily, many oxidizing agents, such as $KMnO_4$, CrO_3, Ag_2O, and peroxyacids, will work. Specific examples are

$$CH_3(CH_2)_5CH{=}O \xrightarrow[\text{(Jones' reagent)}]{CrO_3,\ H^+} CH_3(CH_2)_5CO_2H \qquad (9.34)$$

$$\text{(cyclohex-2-enyl)}{-}CHO \xrightarrow{Ag_2O} \text{(cyclohex-2-enyl)}{-}CO_2H \qquad (9.35)$$

Silver ion as an oxidant is expensive but has the virtue that it selectively oxidizes the aldehyde group without oxidizing the double bond (eq. 9.35).

A laboratory test that distinguishes aldehydes from ketones takes advantage of their different ease of oxidation. In the **Tollens' silver mirror test,** the silver–ammonia complex ion is reduced by aldehydes (but not by ketones) to metallic silver.* The equation for the reaction may be written as follows:

---

*Silver hydroxide is insoluble in water, so the silver ion must be complexed with ammonia to keep it in solution in a basic medium.

$$\underset{\text{aldehyde}}{\overset{\overset{\displaystyle O}{\|}}{RCH}} + \underset{\substack{\text{silver-ammonia}\\\text{complex ion}\\\text{(colorless)}}}{2Ag(NH_3)_2^+} + 3OH^- \rightarrow \underset{\substack{\text{acid}\\\text{anion}}}{\overset{\overset{\displaystyle O}{\|}}{RC}-O^-} + \underset{\substack{\text{silver}\\\text{mirror}}}{2Ag\downarrow} + 4NH_3\uparrow + 2H_2O \qquad (9.36)$$

If the glass vessel in which the test is performed is thoroughly clean, the silver deposits as a mirror on the glass surface. This reaction is also employed to silver glass, using the relatively inexpensive aldehyde formaldehyde.

Problem 9.18 Write an equation for the formation of a silver mirror from formaldehyde and Tollens' reagent.

Aldehydes are so easily oxidized that stored samples usually contain some of the corresponding acid. This contamination is caused by air oxidation.

$$2RCHO + O_2 \longrightarrow 2RCO_2H \qquad (9.37)$$

Ketones can be oxidized but require special oxidizing conditions. For example, cyclohexanone is oxidized commercially to **adipic acid,** an important industrial chemical used to manufacture nylon.

one of these C—C bonds is cleaved in the oxidation

$$\underset{\text{cyclohexanone}}{\text{cyclohexanone}} + HNO_3 \xrightarrow{V_2O_5} \underset{\text{adipic acid}}{HO-\overset{\overset{\displaystyle O}{\|}}{C}-CH_2CH_2CH_2CH_2-\overset{\overset{\displaystyle O}{\|}}{C}-OH} \qquad (9.38)$$

9.15 *Keto–Enol Tautomerism*

Aldehydes and ketones may exist as an equilibrium mixture of two forms, called the **keto form** and the **enol form.** The two forms differ in the location of a proton and a double bond.

$$\underset{\text{keto form}}{-\overset{\overset{\displaystyle H}{|}}{\underset{|}{C}}-\overset{\overset{\displaystyle O}{\|}}{C}-} \rightleftharpoons \underset{\text{enol form}}{>C=C\overset{\displaystyle OH}{<}} \qquad (9.39)$$

This type of structural isomerism is called **tautomerism** (from the Greek *tauto,* the same, and *meros,* part). The two forms of the aldehyde or ketone are called **tautomers.**

EXAMPLE 9.9 Write formulas for the keto and enol forms of acetone.

Solution

$$\underset{\text{keto form}}{CH_3-\overset{\overset{\displaystyle O}{\|}}{C}-CH_3} \qquad \underset{\text{enol form}}{CH_2=\overset{\overset{\displaystyle OH}{|}}{C}-CH_3}$$

Problem 9.19 Draw the structural formula for the enol form of

a. cyclopentanone. b. acetaldehyde.

Tautomers are structural isomers, *not* contributors to a resonance hybrid. They often readily equilibrate, however, and we indicate that fact by the equilibrium symbol $\rightleftharpoons$ between their structures.

To be capable of existing in an enol form, a carbonyl compound must have a hydrogen atom attached to the carbon atom adjacent to the carbonyl group. This hydrogen is called an **α-hydrogen** and is attached to the **α-carbon atom** (from the first letter of the Greek alphabet, α, or alpha).

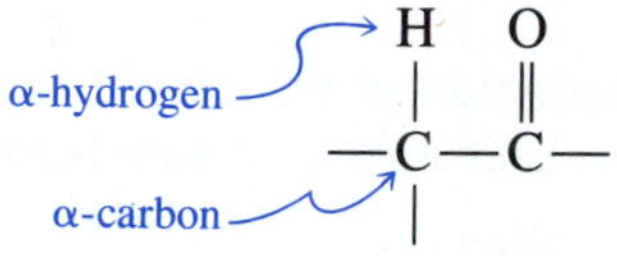

α-hydrogen

α-carbon

Most simple aldehydes and ketones exist mainly in the keto form. Acetone, for example, is 99.9997% in the keto form, with only 0.0003% of the enol present. The main reason for the greater stability of the keto form is that the C=O plus C—H bond energy present in the keto form is greater than the C=C plus O—H bond energy of the enol form. We have already encountered some molecules, however, that have mainly the enol structure—the *phenols*. In this case, the resonance stabilization of the aromatic ring is greater than the usual energy difference that favors the keto over the enol form. Aromaticity would be destroyed if the molecule existed in the keto form; therefore, the enol form is preferred.

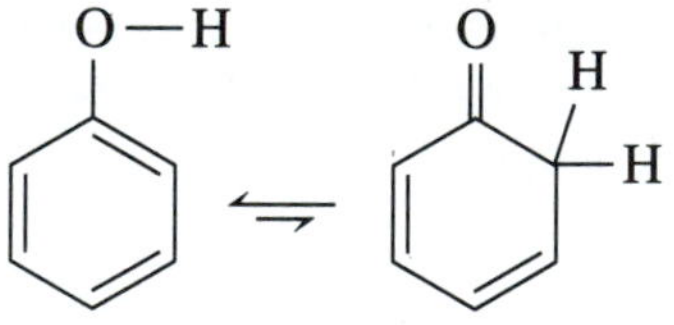

enol form of phenol $\rightleftharpoons$ keto form of phenol (9.40)

Carbonyl compounds that do not have an α-hydrogen cannot form enols and exist only in the keto form. Examples include

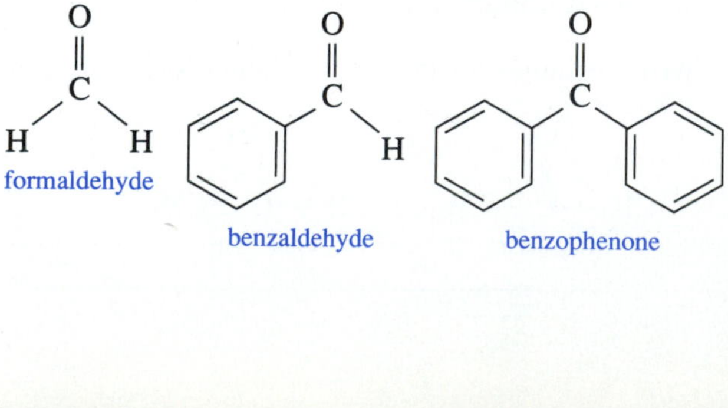

formaldehyde

benzaldehyde

benzophenone

A WORD ABOUT . . .

20. Tautomerism and Photochromism

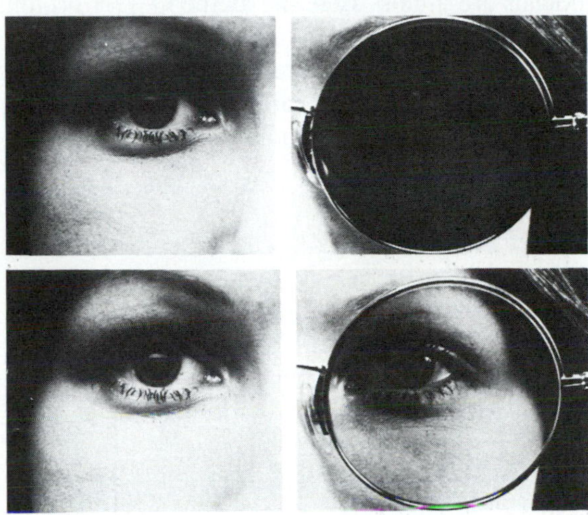

The concept of tautomerism can be expanded beyond keto and enol forms to include any pair or group of isomers that can be easily interconverted by the relocation of an atom and/or bonds. For example, imines and enamines (unsaturated amines) are tautomers whose relationship is similar to that of keto and enol forms.

an imine ⇌ an enamine

keto ⇌ enol

For some pairs of tautomers, one can be converted to the other photochemically—that is, by the absorption of light. Irradiation of the following pale yellow phenol-imine causes the hydrogen atom to shift from the oxygen to the nitrogen, with appropriate rebonding.

$\underset{\text{dark}}{\overset{h\nu}{\rightleftharpoons}}$

a phenol-imine
pale yellow
(both rings aromatic)

a keto-enamine
red
(only one ring aromatic)

If the keto-enamine product of this photochemical reaction is allowed to remain in the dark, it gradually changes back to the more stable phenol-imine. Of what use can such a cycle of reactions, with no net change, possibly be?

Note that, in this example, one tautomer is pale yellow, the other red. This phenomenon, in which two compounds undergo a reversible photochemical color change, is called **photochromism.** Photochromic substances have many practical uses. One thinks immediately of glasses that, when exposed to sunlight, become darker because their lenses are impregnated with a photochromic material. When the sunlight dims or when one goes indoors, the colored photochromic substance gradually changes back to its colorless form. Photochromic substances can be used for data storage and display (as in digital watches), for chemical switches in computers, for micro images (microfilm and microfiche), for protection against sudden light flashes, for camouflage, and in many other creative ways.

9.16 Acidity of α-Hydrogens; the Enolate Anion

The α-hydrogen in a carbonyl compound is more acidic than normal hydrogens bonded to a carbon atom. Table 9.2 shows the pK_a values for a typical aldehyde and ketone, as well as for reference compounds. The result of placing a carbonyl group adjacent to methyl protons is truly striking, an increase in their acidity of over 30 powers of 10! (Compare acetaldehyde or acetone with propane.) Indeed, these compounds are almost as acidic as the O—H protons in alcohols. Why is this?

There are two reasons. First, the carbonyl carbon carries a partial positive charge. Bonding electrons are displaced toward the carbonyl carbon and away from the α-hydrogen, making it easy for a base to remove the α-hydrogen as a proton (that is, without its bonding electrons).

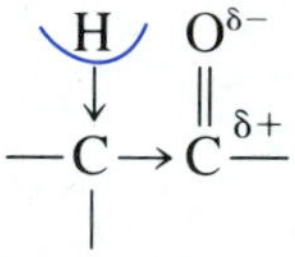

H O^δ− ; —C→C^δ+—

Second and more important, the resulting anion is stabilized by resonance.

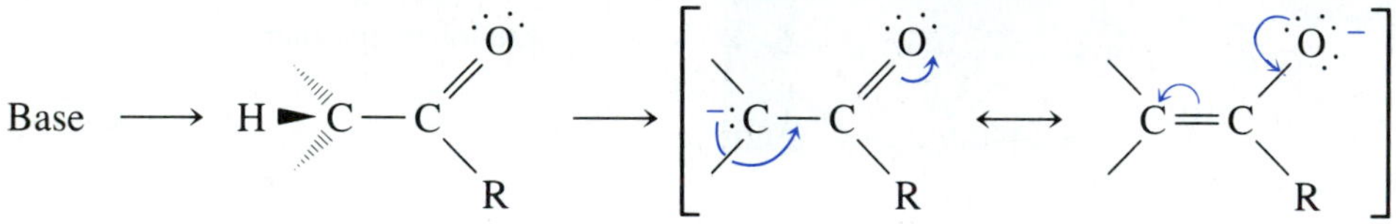

Base ⟶ H–C–C(=O)R ⟶ [⁻:C–C(=O)R ⟷ C=C(–O:⁻)R] (9.41)

enolate anion (resonance stabilized)

The anion is called an **enolate anion.** The negative charge is distributed between the α-carbon and the carbonyl oxygen atom.

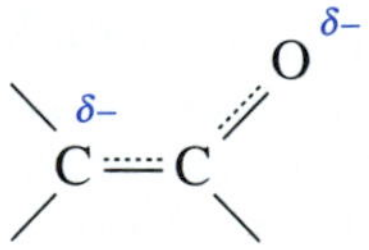

C^δ−⋯C⋯O^δ−

an enolate anion

TABLE 9.2 Acidity of α-hydrogens

Compound	*Name*	pK_a
$CH_3CH_2CH_3$	propane	~50
$CH_3C(=O)CH_3$	acetone	19
$CH_3C(=O)H$	acetaldehyde	17
CH_3CH_2OH	ethanol	15.9

EXAMPLE 9.10 Draw the formula for the enolate anion of acetone.

Solution

$$\left[\ddot{C}H_2-\overset{\overset{:\ddot{O}}{\|}}{C}-CH_3 \longleftrightarrow CH_2=\overset{\overset{:\ddot{O}:^-}{|}}{C}-CH_3\right] \quad \text{or} \quad \left[CH_2\overset{\cdots}{=}\overset{\overset{O}{\vdots\vdots}}{C}-CH_3\right]^-$$

An enolate anion is a resonance hybrid of two contributing structures that differ *only* in the arrangement of the electrons.

Problem 9.20 Draw the resonance contributors to the enolate ion of

a. cyclopentanone. b. acetaldehyde.

9.17 Deuterium Exchange in Carbonyl Compounds

Even though its concentration is very low, the presence of the enol form of ordinary aldehydes and ketones can be demonstrated experimentally. For example, the α-hydrogens can be exchanged for deuterium by placing the carbonyl compound in a solvent such as D_2O or CH_3OD that contains O—D bonds. The exchange is catalyzed by acid or base. *Only the α-hydrogens exchange*, as illustrated by the following examples:

$$\text{cyclohexanone} \xrightarrow[\text{CH}_3\text{OD (excess)}]{\text{Na}^{+\,-}\text{OCH}_3} \text{2,2,6,6-tetradeuteriocyclohexanone} \tag{9.42}$$

$$\underset{\text{butanal}}{CH_3CH_2CH_2\overset{\overset{O}{\|}}{C}H} \xrightarrow[D^+]{D_2O} \underset{\text{2,2-dideuteriobutanal}}{CH_3CH_2CD_2\overset{\overset{O}{\|}}{C}H} \tag{9.43}$$

EXAMPLE 9.11 Write a mechanism for eq. 9.42.

Solution

$$\text{cyclohexanone (}\alpha\text{-H, H)} + {}^-OCH_3 \rightleftharpoons [\text{enolate (carbanion form)} \longleftrightarrow \text{enolate (oxyanion form)}] + CH_3OH$$

$$\Big\downarrow\uparrow\ CH_3OD$$

$$\text{2-deuteriocyclohexanone (H, D)} + CH_3O^-$$

The base (methoxide ion) removes an α-proton to form the enolate anion. Reprotonation, but with CH_3OD, replaces the α-hydrogen with deuterium. With excess CH_3OD, all four α-hydrogens are eventually exchanged.

EXAMPLE 9.12 Write a mechanism for eq. 9.43.

Solution In this case, the catalyst is an acid. The keto form is first protonated and, by loss of an α-hydrogen, converted to its enol.

$$\underset{\text{keto form}}{CH_3CH_2CH_2\overset{\overset{\ddot{O}:}{\|}}{C}H} \overset{D^+}{\rightleftharpoons} CH_3CH_2\underset{\underset{H}{|}}{\overset{\overset{H}{|}}{C}}-\overset{\overset{\overset{+}{\ddot{O}}-D}{\|}}{C}H \overset{-H^+}{\rightleftharpoons} \underset{\text{enol form}}{CH_3CH_2CH=\overset{\overset{:\ddot{O}-D}{|}}{C}H}$$

In the reversal of these equilibria, the enol then adds D^+ at the α-carbon.

$$D^+ \quad CH_3CH_2CH=\overset{\overset{:\ddot{O}-D}{|}}{C}H \longrightarrow CH_3CH_2\overset{\overset{D}{|}}{C}H-\overset{\overset{:\ddot{O}}{\|}}{C}H + D^+$$

Repetition of this sequence results in exchange of the other α-hydrogen.

Problem 9.21 Identify the hydrogens that are readily exchanged for deuterium in

a. 2-methylcyclopentanone.
b. *t*-butyl methyl ketone.

Problem 9.22 Treatment of cyclohexanone with chlorine and a little acid as a catalyst gives a good yield of 2-chlorocyclohexanone. Write an equation for this reaction and the steps in its mechanism.

9.18 The Aldol Condensation

Enolate anions may act as carbon nucleophiles. They can add to the carbonyl group of another aldehyde or ketone molecule in a reaction called the **aldol condensation,** an extremely useful carbon–carbon bond-forming reaction.

The simplest example of an aldol condensation is the combination of two acetaldehyde molecules, which occurs when a solution of acetaldehyde is treated with aqueous base.

$$\underset{\text{acetaldehyde}}{CH_3\overset{\overset{O}{\|}}{C}H + CH_3\overset{\overset{O}{\|}}{C}H} \overset{OH^-}{\rightleftharpoons} \underset{\substack{\text{3-hydroxybutanal}\\ \text{(an aldol)}}}{CH_3\overset{\overset{OH}{|}}{C}H-CH_2\overset{\overset{O}{\|}}{C}H} \qquad (9.44)$$

The product is called an **aldol** (so named because the product is both an *ald*ehyde and an alcoh*ol*).

The aldol condensation of acetaldehyde occurs according to the following three-step mechanism:

$$\text{Step 1.}\quad \overset{\alpha}{CH_3}-\overset{\overset{\displaystyle \ddot{O}:}{\|}}{C}-H + OH^- \rightleftharpoons \overset{..}{\bar{C}}H_2-\overset{\overset{\displaystyle \ddot{O}:}{\|}}{C}-H + HOH \qquad (9.45)$$

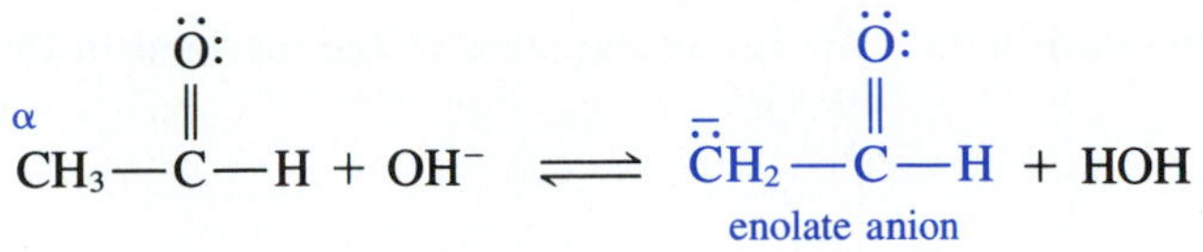

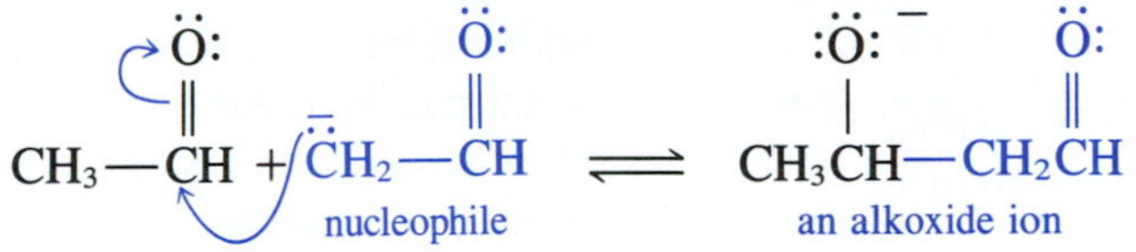

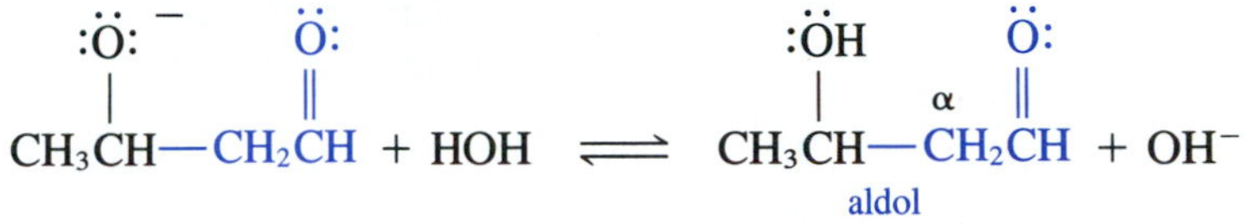

enolate anion

$$\text{Step 2.}\quad CH_3-\overset{\overset{\displaystyle \ddot{O}:}{\|}}{CH} + \overset{..}{\bar{C}}H_2-\overset{\overset{\displaystyle \ddot{O}:}{\|}}{CH} \rightleftharpoons CH_3\overset{\overset{\displaystyle :\ddot{O}:^-}{|}}{CH}-CH_2\overset{\overset{\displaystyle \ddot{O}:}{\|}}{CH} \qquad (9.46)$$

nucleophile — an alkoxide ion

$$\text{Step 3.}\quad CH_3\overset{\overset{\displaystyle :\ddot{O}:^-}{|}}{CH}-CH_2\overset{\overset{\displaystyle \ddot{O}:}{\|}}{CH} + HOH \rightleftharpoons CH_3\overset{\overset{\displaystyle :\ddot{O}H}{|}}{CH}-\overset{\alpha}{CH_2}\overset{\overset{\displaystyle \ddot{O}:}{\|}}{CH} + OH^- \qquad (9.47)$$

aldol

In Step 1, the base removes an α-hydrogen to form the enolate anion. In Step 2, this anion adds to the carbonyl carbon of another acetaldehyde molecule, forming a new carbon–carbon bond. Ordinary bases convert a small fraction of the carbonyl compound to the enolate anion, so that a substantial fraction of the aldehyde is still present in the un-ionized carbonyl form needed for this step. In Step 3, the alkoxide ion formed in Step 2 accepts a proton from the solvent, thus regenerating the hydroxide ion needed for the first step.

In the aldol condensation, the α-carbon of one aldehyde molecule becomes connected to the carbonyl carbon of another aldehyde molecule.

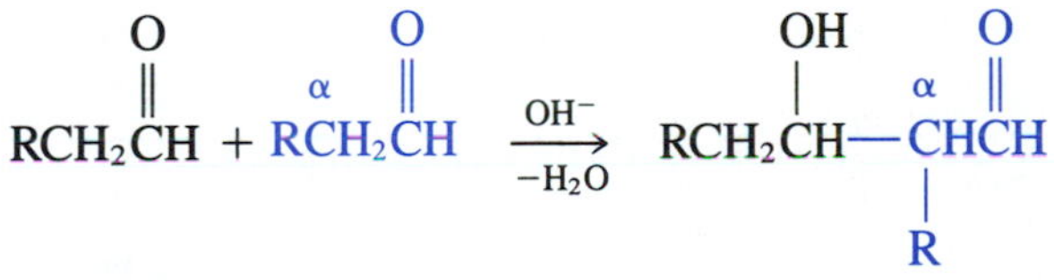

$$RCH_2\overset{\overset{\displaystyle O}{\|}}{C}H + R\overset{\alpha}{C}H_2\overset{\overset{\displaystyle O}{\|}}{C}H \xrightarrow[-H_2O]{OH^-} RCH_2\overset{\overset{\displaystyle OH}{|}}{C}H-\underset{\underset{\displaystyle R}{|}}{\overset{\alpha}{C}H}\overset{\overset{\displaystyle O}{\|}}{C}H \qquad (9.48)$$

an aldol

Aldols are therefore 3-hydroxyaldehydes. ***Since it is always the α-carbon that acts as a nucleophile, the product always has just one carbon atom between the aldehyde and alcohol carbons,*** regardless of how long the carbon chain is in the starting aldehyde.

EXAMPLE 9.13 Give the structure of the aldol that is obtained by treating propanal ($CH_3CH_2CH{=}O$) with base.

Solution Rewriting eq. 9.48 with $R{=}CH_3$, the product is

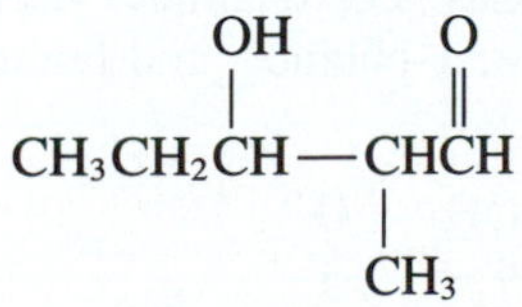

$$CH_3CH_2\overset{\overset{\displaystyle OH}{|}}{C}H-\underset{\underset{\displaystyle CH_3}{|}}{C}H\overset{\overset{\displaystyle O}{\|}}{C}H$$

Problem 9.23 Write out the steps in the mechanism for formation of the product in Example 9.13.

9.19 The Mixed Aldol Condensation

The aldol condensation is very versatile, in that the enolate anion of *one* carbonyl compound can be made to add to the carbonyl carbon of *another*, provided that the reaction partners are carefully selected. Consider, for example, the reaction between acetaldehyde and benzaldehyde, when treated with base. Only acetaldehyde can form an enolate anion (benzaldehyde has no α hydrogen). If the enolate ion of acetaldehyde adds to the benzaldehyde carbonyl group, a mixed aldol condensation occurs.

$$C_6H_5\overset{O}{\overset{\|}{C}}H + \overset{\alpha}{C}H_3\overset{O}{\overset{\|}{C}}H \underset{}{\overset{OH^-}{\rightleftharpoons}} C_6H_5\overset{OH}{\overset{|}{C}}H-CH_2\overset{O}{\overset{\|}{C}}H \xrightarrow[-H_2O]{heat} C_6H_5CH=CH\overset{O}{\overset{\|}{C}}H \quad (9.49)$$

a mixed aldol — cinnamaldehyde

In this particular example, the resulting mixed aldol eliminates water on heating to give **cinnamaldehyde** (the flavor constituent of cinnamon).

EXAMPLE 9.14 Write the structure of the mixed aldol obtained from acetone and formaldehyde.

Solution Of the two reactants, only acetone has α-hydrogens.

$$H-\overset{O}{\overset{\|}{C}}-H + CH_3\overset{O}{\overset{\|}{C}}CH_3 \xrightarrow{base} H-\underset{H}{\underset{|}{\overset{OH}{\overset{|}{C}}}}-CH_2\overset{O}{\overset{\|}{C}}CH_3$$

Problem 9.24 Using eqs. 9.45 through 9.47 as a guide, write out the steps in the mechanism for eq. 9.49.

Problem 9.25 Write the structure of the mixed aldol obtained from propanal and benzaldehyde. What structure is obtained from dehydration of this mixed aldol?

9.20 Commercial Syntheses via the Aldol Condensation

Aldols are useful in synthesis. For example, acetaldehyde is converted commercially to crotonaldehyde, 1-butanol, and butanal using the aldol condensation.

$$\underset{\text{acetaldehyde}}{2\ CH_3\overset{O}{\overset{\|}{C}}H} \xrightarrow{OH^-} \underset{\text{aldol}}{CH_3\overset{OH}{\overset{|}{C}}HCH_2\overset{O}{\overset{\|}{C}}H} \xrightarrow[-H_2O]{H^+} \underset{\text{crotonaldehyde}}{CH_3CH=CH\overset{O}{\overset{\|}{C}}H} \xrightarrow[catalyst]{H_2}$$

$$\underset{\text{butanal}}{CH_3CH_2CH_2\overset{\overset{\displaystyle O}{\|}}{C}H} \quad \text{or} \quad \underset{\text{1-butanol}}{CH_3CH_2CH_2CH_2OH} \tag{9.50}$$

The particular product obtained in the hydrogenation step depends on the catalyst and reaction conditions.

Butanal is the starting material for the synthesis of the mosquito repellent "6-12" (2-ethylhexane-1,3-diol). The first step is an aldol condensation, and the second step is reduction of the aldehyde group to a primary alcohol.

$$\underset{\text{butanal}}{2\ CH_3CH_2CH_2\overset{\overset{\displaystyle O}{\|}}{C}H} \xrightarrow{OH^-} \underset{\text{butanal aldol}}{CH_3CH_2CH_2\overset{\overset{\displaystyle OH}{|}}{C}H\underset{\underset{\displaystyle CH_3CH_2}{|}}{C}H\overset{\overset{\displaystyle O}{\|}}{C}H} \xrightarrow[Ni]{H_2} \underset{\text{2-ethylhexane-1,3-diol ("6-12")}}{CH_3CH_2CH_2\overset{\overset{\displaystyle OH}{|}}{C}H\underset{\underset{\displaystyle CH_3CH_2}{|}}{C}HCH_2OH} \tag{9.51}$$

The aldol condensation is also used in nature to build up (and, in the case of *reverse* aldol condensations, to break down) carbon chains.

Problem 9.26 2-Ethylhexanol, used commercially in the manufacture of plasticizers and synthetic lubricants, is synthesized from butanal via its aldol. Devise a route to it.

A WORD ABOUT ...

21. Quinones, Dyes, and Electron Transfer

Quinones constitute a unique class of carbonyl compounds. They are cyclic, conjugated diketones. The simplest example is 1,4-benzoquinone. All quinones are colored. Many are naturally occurring plant pigments, which often exhibit special biological activity.

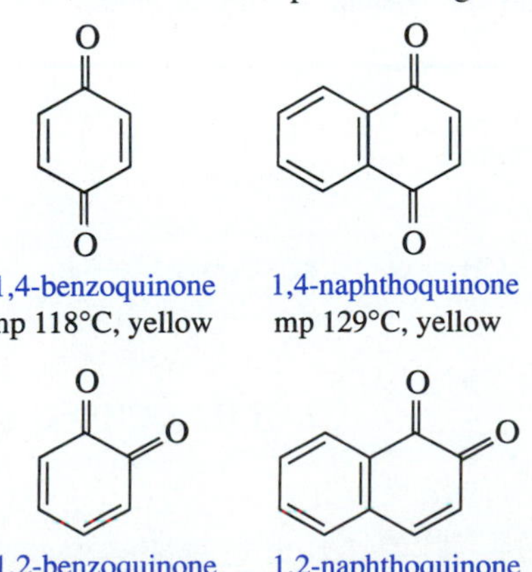

1,4-benzoquinone mp 118°C, yellow

1,4-naphthoquinone mp 129°C, yellow

1,2-benzoquinone mp 70°C, red

1,2-naphthoquinone mp 146°C, yellow-red

Alizarin is a quinone dye that was known in ancient Egypt, Persia, and India. It was extracted from the madder root (an herb) and fixed to cloth by various metal ions (for example, aluminum). The red coats used by the British army during the American Revolution were dyed with alizarin. In the mid-nineteenth century, madder was a substantial agricultural crop in Europe, with as many as 70,000 tons in annual production. Later, organic chemists determined alizarin's structure and were able to work out a commercial synthesis that dramatically reduced the price. Many alizarin-type pigments are still used today.

alizarin
mp 290°C, orange-red

Lawsone is a quinone pigment extracted from the tropical henna shrub (*Lawsonia inermis*). It dyes wool and silk orange and can tint hair red. The Islamic prophet Muhammad is said to have dyed his beard with henna.

Juglone was first isolated from walnut shells (*Juglans regia*), where it occurs as the colorless hydroquinone (1,4,5-trihydroxynaphthalene). In air, it is oxidized to the quinone. Juglone, which also occurs in pecans, stains skin brown.

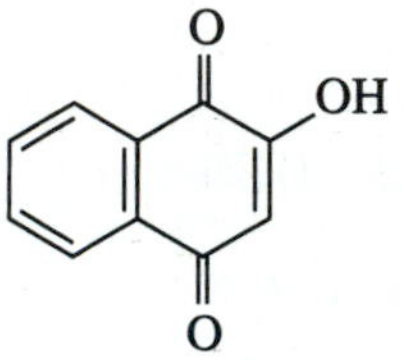

lawsone
mp 192°C, red-brown

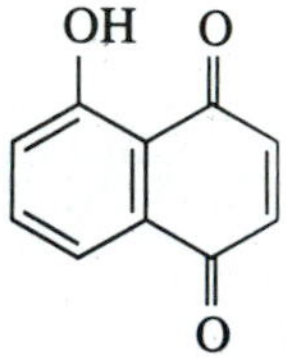

juglone
mp 155°C, yellow

Perhaps the most important property of quinones is their *reversible reduction* to hydroquinones.

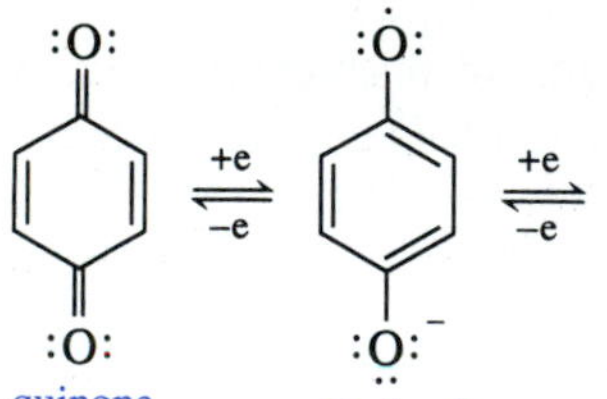

quinone $\underset{-e}{\overset{+e}{\rightleftharpoons}}$ radical anion $\underset{-e}{\overset{+e}{\rightleftharpoons}}$

dianion $\underset{-2H^+}{\overset{+2H^+}{\rightleftharpoons}}$ hydroquinone

Virtually all quinones undergo this reaction. The reduction involves stepwise addition of two electrons to give first a radical anion and then a dianion. It is this property that permits quinones to play an important role in reversible biochemical oxidation-reduction (electron transport) reactions. A group of enzymes called **coenzymes Q** (also known as *ubiquinones* because of their ubiquitous, or widespread, occurrence in animal and plant cells) participate in electron transport in mitochondria, the granular bodies in the cell that are involved in the metabolism of lipids, carbohydrates, and proteins.

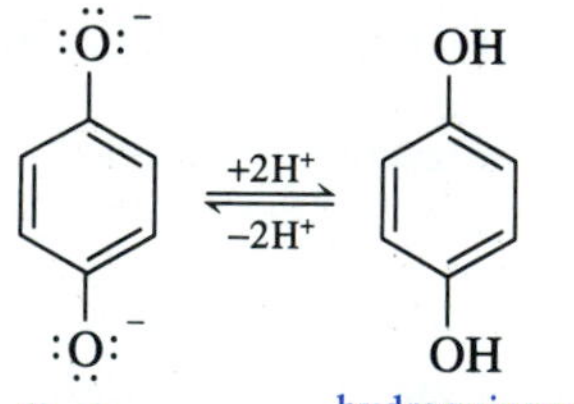

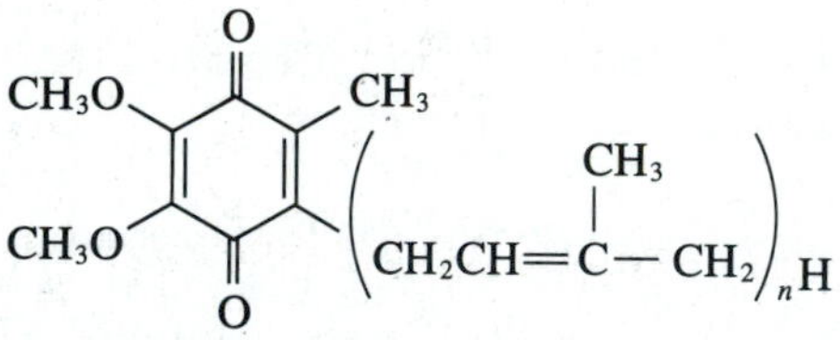

CH_3O, CH_3O, CH_3, $(CH_2CH{=}C(CH_3){-}CH_2)_nH$

coenzymes Q
(n = 10 is common)

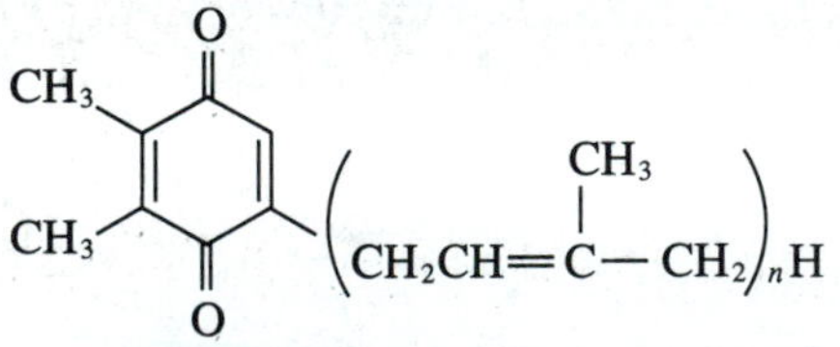

CH_3, CH_3, $(CH_2CH{=}C(CH_3){-}CH_2)_nH$

plastoquinones
(n = 9 is common)

In plant tissues the **plastoquinones** perform a similar function in photosynthesis. The long isoprenoid carbon chain in these quinones is no doubt necessary to promote fat solubility of these coenzymes.

Vitamin K (Figure 9.1) is a quinone that is required for the normal clotting of blood.

Additional Problems

9.27. Name each of the following compounds.

a. $CH_3CH_2\overset{O}{\overset{\|}{C}}CH_2CH_3$ **b.** $CH_3(CH_2)_4CH{=}O$ **c.** $(C_6H_5)_2C{=}O$

d. 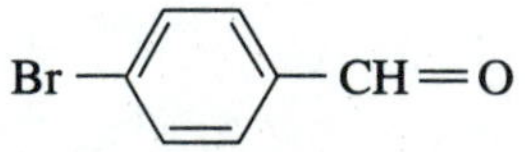 Br–C$_6$H$_4$–CH=O **e.** 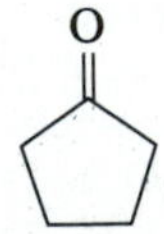 cyclopentanone **f.** $(CH_3)_3CCH{=}O$

g. 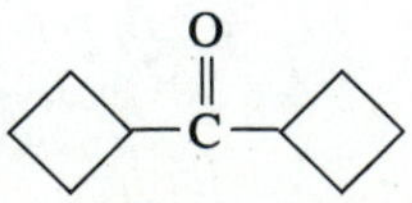 cyclobutyl–$\overset{O}{\overset{\|}{C}}$–cyclobutyl **h.** $CH_3CH{=}CH\overset{O}{\overset{\|}{C}}CH_3$ **i.** $CH_2Br\overset{O}{\overset{\|}{C}}CH_3$

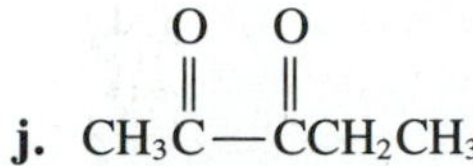

9.28. Write a structural formula for each of the following:

a. 2-octanone
b. 4-methylpentanal
c. *m*-chlorobenzaldehyde
d. 3-methylcyclohexanone
e. 2-butenal
f. benzyl phenyl ketone
g. *p*-tolualdehyde
h. *p*-benzoquinone
i. 2,2-dibromohexanal
j. 1-phenyl-2-butanone

9.29. Give an example of each of the following:

a. acetal
b. hemiacetal
c. cyanohydrin
d. imine
e. oxime
f. phenylhydrazone
g. enol
h. aldehyde with no α-hydrogen

9.30. Write an equation for the synthesis of 2-pentanone by

a. oxidation of an alcohol.
b. hydration of an alkyne.

9.31. Write an equation, using the Friedel-Crafts reaction, for the preparation of methyl 1-naphthyl ketone.

9.32. Write an equation for the reaction, if any, of *p*-bromobenzaldehyde with each of the following reagents, and name the organic product.

a. Tollens' reagent
b. hydroxylamine
c. CrO_3, H^+
d. ethylmagnesium bromide, then H_3O^+
e. phenylhydrazine
f. aniline (C_6H_5—NH_2)
g. cyanide ion
h. excess methanol, dry HCl
i. ethylene glycol, H^+
j. lithium aluminum hydride

9.33. What simple chemical test can distinguish between the members of the following pairs of compounds?

a. hexanal and 2-hexanone
b. benzyl alcohol and benzaldehyde
c. cyclopentanone and 2-cyclopentenone

9.34. Use the structures shown in Figure 9.1 to write equations for the following reactions of natural products:

a. cinnamaldehyde + Tollens' reagent
b. vanillin + hydroxylamine
c. carvone + sodium borohydride
d. camphor + (1) methylmagnesium bromide and (2) H_3O^+

9.35. The boiling points of the isomeric carbonyl compounds heptanal, 4-heptanone, and 2,4-dimethyl-3-pentanone are 155°C, 144°C, and 124°C, respectively. Suggest a possible explanation for the observed order of boiling points.

9.36. Complete each of the following equations.

a. butanal + excess ethanol, H^+ $\longrightarrow$
b. $CH_3CH(OCH_3)_2$ + H_2O, H^+ $\longrightarrow$
c. (2-methoxytetrahydropyran: six-membered ring with ring O, OCH_3 on the adjacent carbon) + H_2O, H^+ $\longrightarrow$

d. 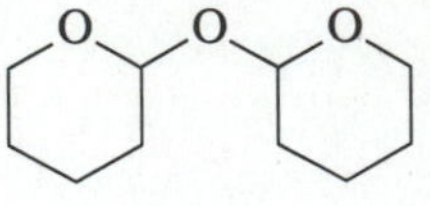 [structure: tetrahydropyranyl–O–tetrahydropyranyl ether] + $H_2O, H^+ \longrightarrow$

e. 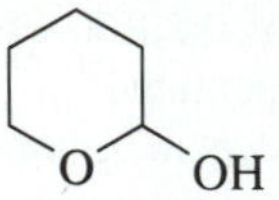 [structure: 2-hydroxytetrahydropyran, ring O, OH] + excess $CH_3OH, H^+ \longrightarrow$

9.37. Write an equation for the reaction of each of the following with methylmagnesium bromide, followed by hydrolysis with aqueous acid.

a. acetaldehyde **b.** acetophenone **c.** formaldehyde **d.** cyclohexanone

9.38. Using a Grignard reagent and the appropriate aldehyde or ketone, show how each of the following can be prepared:

a. 1-pentanol **b.** 3-pentanol
c. 2-methyl-2-butanol **d.** 1-cyclopentylcyclopentanol
e. 1-phenyl-1-propanol **f.** 3-butene-2-ol

9.39. Complete the equation for the reaction of

a. cyclohexanone + $Na^{+}\,{}^{-}C{\equiv}CH \longrightarrow \xrightarrow[H^+]{H_2O}$

b. cyclopentanone + HCN $\longrightarrow$

c. 2-butanone + $NH_2OH \xrightarrow{H^+}$

d. *p*-tolualdehyde + benzylamine $\longrightarrow$

e propanal + phenylhydrazine $\longrightarrow$

9.40. Write out each step in the mechanism for

a. 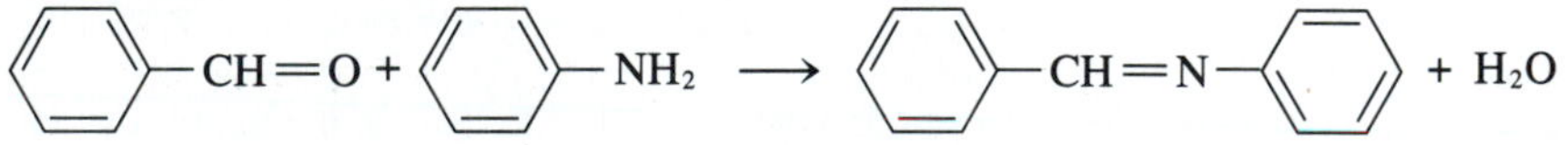 $C_6H_5-CH{=}O + C_6H_5-NH_2 \longrightarrow C_6H_5-CH{=}N-C_6H_5 + H_2O$

b. $(CH_3)_2C{=}O + NH_2OH \longrightarrow (CH_3)_2C{=}NOH + H_2O$

9.41. Give the structure of each product.

a. 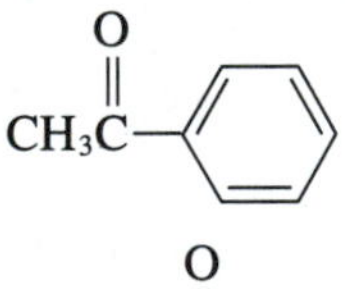 $CH_3\overset{O}{\overset{\|}{C}}-C_6H_5 \xrightarrow[2.\ H_2O,\ H^+]{1.\ LiAlH_4}$

b. 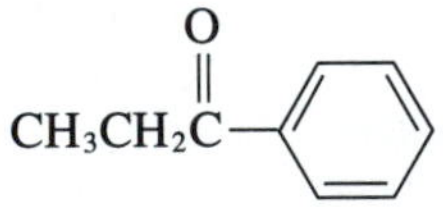 $CH_3CH_2\overset{O}{\overset{\|}{C}}-C_6H_5 \xrightarrow[CH_3OD\ (excess)]{CH_3O^-Na^+}$

c. 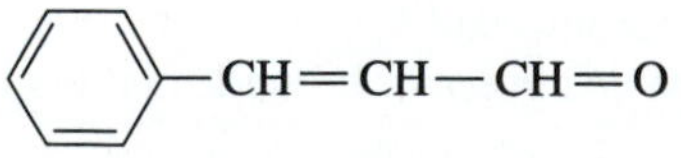 $C_6H_5-CH{=}CH-CH{=}O \xrightarrow[Ni,\ heat]{excess\ H_2}$

d. 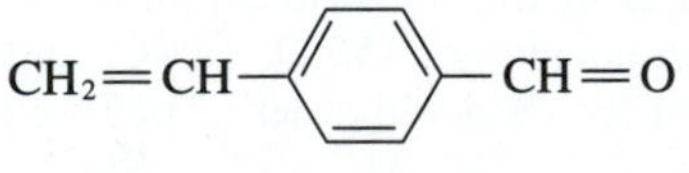 $CH_2{=}CH-C_6H_4-CH{=}O \xrightarrow[2.\ H_2O,\ H^+]{1.\ NaBH_4}$

e.  $C_6H_5-CH_2CH_2CH{=}O \xrightarrow[reagent]{Jones'}$

f. $CH_3CH{=}CHCHO \xrightarrow{Ag_2O}$

9.42. Write the structural formulas for all possible enols of
a. 2-butanone. **b.** phenylacetaldehyde. **c.** 2,4-pentanedione.

9.43. How many hydrogens are replaced by deuterium when each of the following compounds is treated with NaOD in D_2O?
a. 3-methylcyclopentanone **b.** 2-methylbutanal

9.44. Write out the steps in the mechanism for the aldol condensation of butanal (the first step in the synthesis of the mosquito repellent "6-12," eq. 9.51).

9.45. Treatment of the unsaturated diketone

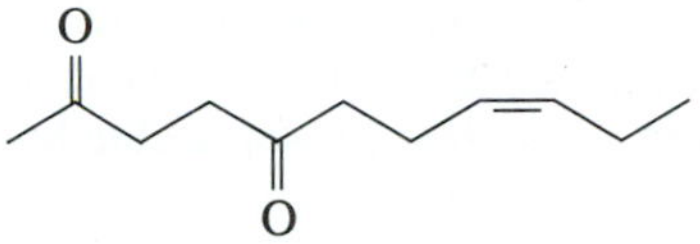

with sodium hydroxide in ethanol gives *jasmone,* the fragrant component of the jasmine flower, used in perfumes.

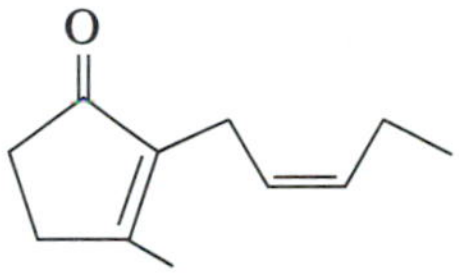

jasmone

Explain how this reaction takes place.

9.46. Excess benzaldehyde reacts with acetone and base to give a yellow crystalline product, $C_{17}H_{14}O$. Deduce its structure and explain how it is formed.

9.47. The final steps in the synthesis of two oral contraceptives, Enovid and Norlutin, are shown. For each step, supply the missing reagent, and tell what general type of reaction is involved.

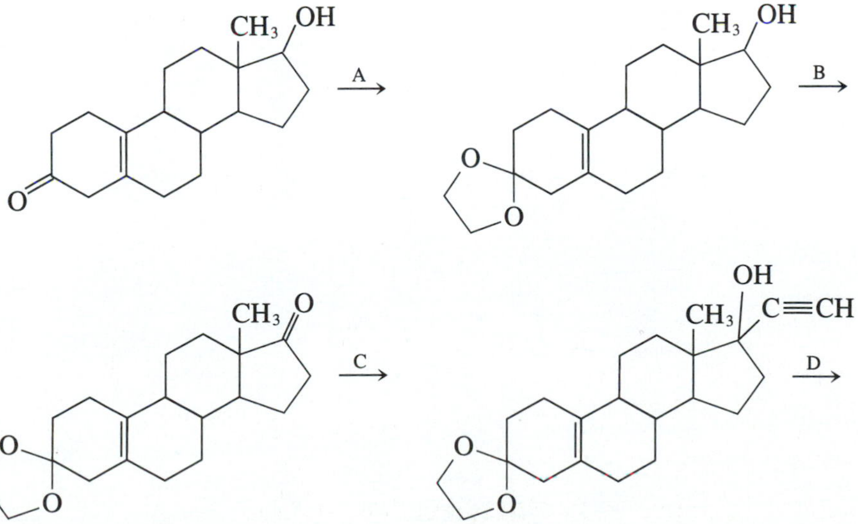

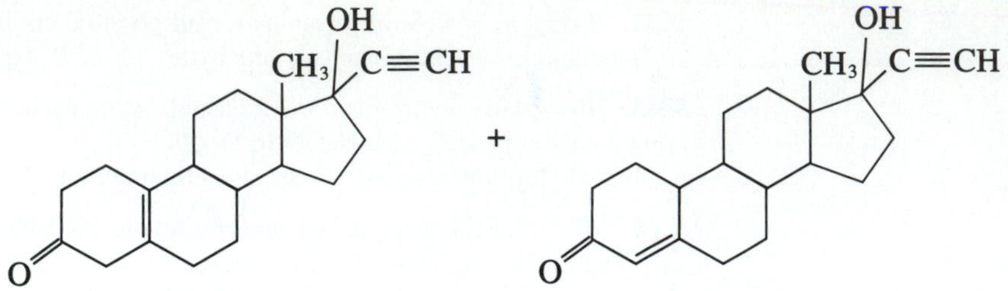

Enovid + Norlutin

Explain why the carbonyl group in the starting material was converted to an acetal before the rest of the synthesis proceeded, since in the final products, that carbonyl group is to be there. Explain how the carbon-carbon double bond ends up where it does in Norlutin.

9.48. Lily aldehyde, used in perfumes, can be made starting with a mixed aldol condensation. Show how its carbon skeleton could be assembled.

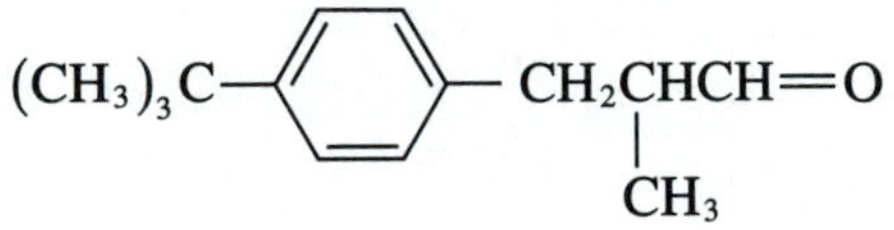

$(CH_3)_3C-C_6H_4-CH_2CH(CH_3)CH=O$

lily aldehyde

CHAPTER 10

Carboxylic Acids and Their Derivatives

10.1 Introduction

Carboxylic acids are the most important organic acids; their functional group is the **carboxyl group.** This name is a contraction of the parts: the *carb*onyl and hydr*oxyl* groups. The general formula for a carboxylic acid can be written in expanded or abbreviated forms.

$$-\mathrm{C}\overset{\displaystyle /\!\!/ \mathrm{O}}{\underset{\displaystyle \backslash \mathrm{OH}}{}} \qquad \mathrm{R}-\mathrm{C}\overset{\displaystyle /\!\!/ \mathrm{O}}{\underset{\displaystyle \backslash \mathrm{OH}}{}} \quad \text{or} \quad \mathrm{RCOOH} \quad \text{or} \quad \mathrm{RCO_2H}$$

carboxyl group — three ways to write a carboxylic acid

In this chapter, we will describe the structure, acidity, preparation, and reactions of carboxylic acids. We will also discuss some common related classes of compounds called **acid derivatives,** in which the —OH group of an acid is replaced by other functions (—OR, halogen, or others).

10.2 Nomenclature of Acids

Because of their abundance in nature, carboxylic acids were among the earliest classes of compounds studied by organic chemists. It is not surprising, then, that many of them have common names. These names usually come from some Latin or Greek word that indicates the original source of the acid. Table 10.1 lists the first ten unbranched carboxylic acids, with their common and IUPAC names.

To obtain the IUPAC name of a carboxylic acid, we replace the final *e* in the name of the corresponding alkane with the suffix *-oic* and add the word *acid*. Substituted acids are named in two ways. In the IUPAC system, *the chain is numbered beginning with the carboxyl carbon atom,* and substituents are located in the usual way. If the common name of the acid is used, substituents are located with Greek letters, beginning with the α-carbon atom.

TABLE 10.1 Aliphatic carboxylic acids

Carbon atoms	*Formula*	*Source*	*Common name*	*IUPAC name*
1	HCOOH	ants (Latin, *formica*)	formic acid	methanoic acid
2	CH_3COOH	vinegar (Latin, *acetum*)	acetic acid	ethanoic acid
3	CH_3CH_2COOH	milk (Greek, *protos pion,* first fat)	propionic acid	propanoic acid
4	$CH_3(CH_2)_2COOH$	butter (Latin, *butyrum*)	butyric acid	butanoic acid
5	$CH_3(CH_2)_3COOH$	valerian root (Latin, *valere,* to be strong)	valeric acid	pentanoic acid
6	$CH_3(CH_2)_4COOH$	goats (Latin, *caper*)	caproic acid	hexanoic acid
7	$CH_3(CH_2)_5COOH$	vine blossom (Greek, *oenanthe*)	enanthic acid	heptanoic acid
8	$CH_3(CH_2)_6COOH$	goats (Latin, *caper*)	caprylic acid	octanoic acid
9	$CH_3(CH_2)_7COOH$	pelargonium (an herb with stork-shaped seed capsules; Greek, *pelargos,* stork)	pelargonic acid	nonanoic acid
10	$CH_3(CH_2)_8COOH$	goats (Latin, *caper*)	capric acid	decanoic acid

IUPAC and common naming systems should not be mixed.

$$\overset{\beta}{\overset{3}{CH_3}}-\overset{\alpha}{\overset{2}{CH}}(Br)-\overset{1}{CO_2H}$$
2-bromopropanoic acid
(α-bromopropionic acid)

$$\overset{3}{CH_2}=\overset{2}{C}H\overset{1}{C}O_2H$$
propenoic acid
(acrylic acid)

$$\overset{\gamma}{\overset{4}{CH_3}}\overset{\beta}{\overset{3}{CH}}(OH)\overset{\alpha}{\overset{2}{CH_2}}\overset{1}{CO_2H}$$
3-hydroxybutanoic acid
(β-hydroxybutyric acid)

The carboxyl group has priority over alcohol, aldehyde, or ketone functionality in naming. In the latter cases, the prefix *oxo-* is used to locate the carbonyl group of the aldehyde or ketone, as in these examples:

$$H\overset{O}{\overset{\|}{\underset{3}{C}}}-\overset{2}{CH_2}\overset{1}{CO_2H}$$
3-oxopropanoic acid

$$\overset{5}{CH_3}\overset{O}{\overset{\|}{\underset{4}{C}}}\overset{3}{CH_2}\overset{2}{CH}(Br)\overset{1}{CO_2H}$$
2-bromo-4-oxopentanoic acid

Problem 10.1 Write the structure for:

a. 3-chlorobutanoic acid
b. 2-hydroxy-2-methylpropanoic acid
c. 2-butynoic acid
d. 5-methyl-6-oxohexanoic acid

Problem 10.2 Give the IUPAC name for

a. 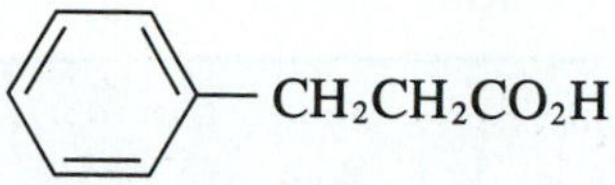 C_6H_5—$CH_2CH_2CO_2H$ b. CCl_3CO_2H

c. $HOCH_2CH{=}CHCO_2H$ d. $(CH_3)_3CCO_2H$

When the carboxyl group is attached to a ring, the ending *-carboxylic acid* is added to the name of the parent cycloalkane.

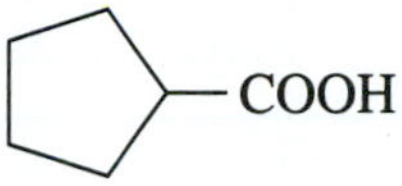

cyclopentanecarboxylic acid

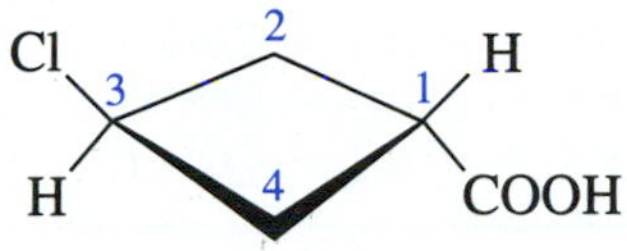

trans-3-chlorocyclobutanecarboxylic acid

Aromatic acids are named by attaching the suffix *-oic acid* or *-ic acid* to an appropriate prefix derived from the aromatic hydrocarbon.

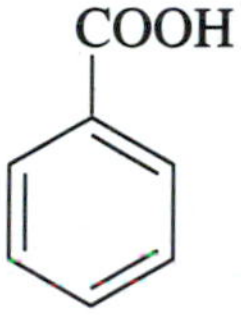

benzoic acid (benzenecarboxylic acid)

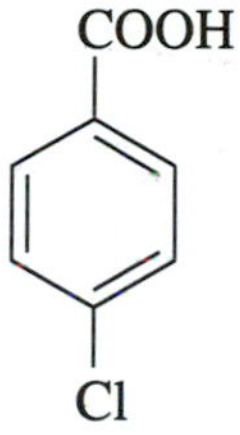

p-chlorobenzoic acid (4-chlorobenzenecarboxylic acid)

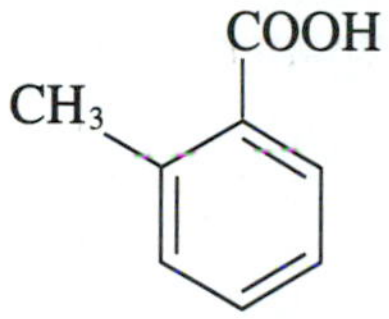

o-toluic acid (2-methylbenzenecarboxylic acid)

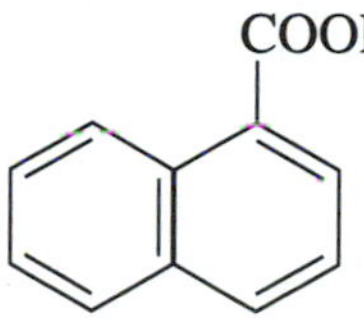

1-naphthoic acid (1-naphthalenecarboxylic acid)

Problem 10.3 Write the structure for

a. 4,4-dimethylcyclohexanecarboxylic acid.
b. *m*-nitrobenzoic acid.

Problem 10.4 Give the correct name for:

a. 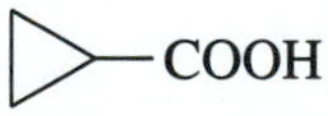 cyclopropyl—COOH b. 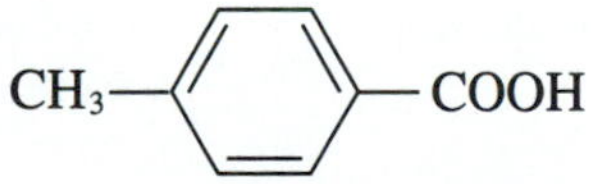 CH_3—C_6H_4—COOH

Aliphatic dicarboxylic acids are given the suffix *-dioic acid* in the IUPAC system. For example,

$$\overset{1}{HO_2C}—\overset{2}{CH_2}\overset{3}{CH_2}—\overset{4}{CO_2H}$$
butanedioic acid

$$HO_2C—C{\equiv}C—CO_2H$$
butynedioic acid

Many dicarboxylic acids occur in nature and go by their common names, which are based on their source. Table 10.2 lists some common aliphatic

TABLE 10.2 Aliphatic dicarboxylic acids

Formula	*Common name*	*Source*	*IUPAC name*
HOOC—COOH	oxalic acid	plants of the *oxalic* family (for example, sorrel)	ethanedioic acid
HOOC—CH_2—COOH	malonic	apple (Gk. *malon*)	propanedioic
HOOC—$(CH_2)_2$—COOH	succinic	amber (L. *succinum*)	butanedioic
HOOC—$(CH_2)_3$—COOH	glutaric	gluten	pentanedioic
HOOC—$(CH_2)_4$—COOH	adipic	fat (L. *adeps*)	hexanedioic
HOOC—$(CH_2)_5$—COOH	pimelic	fat (Gk. *pimele*)	heptanedioic

diacids.* The most important commercial compound in this group is adipic acid, used to manufacture nylon.

The two butenedioic acids played an important role in the discovery of *cis–trans* isomerism and are usually known by their common names **maleic**** and **fumaric† acid.**

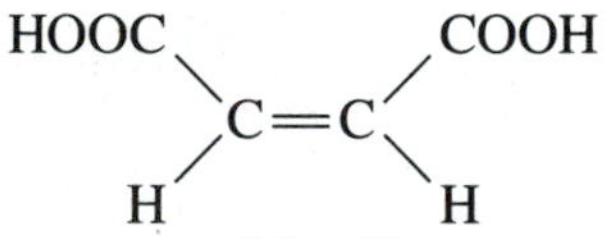

HOOC(H)C=C(H)COOH (cis) and HOOC(H)C=C(COOH)H (trans)

maleic acid
(*cis*-2-butenedioic acid)

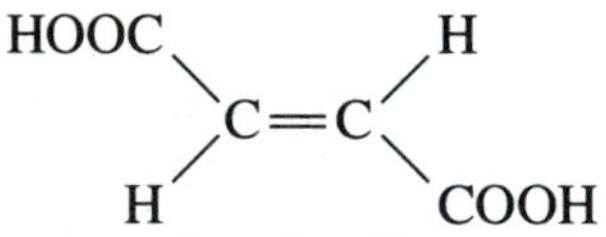

fumaric acid
(*trans*-2-butenedioic acid)

The three benzenedicarboxylic acids are generally known by their common names.

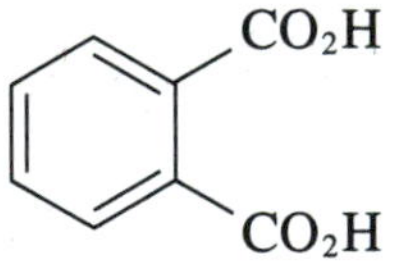

CO_2H CO_2H — phthalic acid

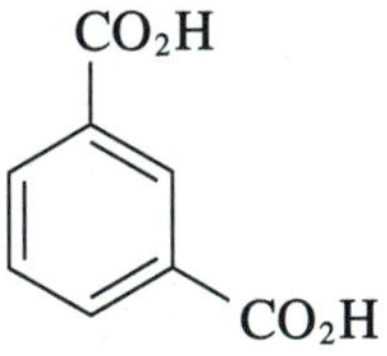

CO_2H CO_2H — isophthalic acid

CO_2H CO_2H — terephthalic acid

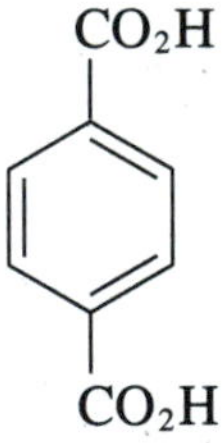

All three are important commercial chemicals, used to make polymers and other useful materials.

Finally, it is useful to have a name for an **acyl group.** Particular acyl groups are named from the corresponding acid by changing the *-ic* ending to *-yl.*

---

* The first letter of each word in the sentence "Oh my, such good apple pie" gives, in order, the first letters of the common names of these acids and can help you to remember them.

** From the Latin *malum* (apple). Malic acid (2-hydroxybutanedioic acid), found in apples, can be dehydrated on heating to give maleic acid.

† Found in fumitory, an herb of the genus *fumaria.*

$R-\overset{O}{\overset{\|}{C}}-$ an acyl group $H-\overset{O}{\overset{\|}{C}}-$ formyl (methanoyl) $CH_3-\overset{O}{\overset{\|}{C}}-$ acetyl (ethanoyl) $CH_3CH_2\overset{O}{\overset{\|}{C}}-$ propanoyl $C_6H_5-\overset{O}{\overset{\|}{C}}-$ benzoyl

Problem 10.5 Write the formula for

a. 4-acetylbenzoic acid.
b. benzoyl bromide.
c. butanoyl chloride.
d. formylcyclopentane.

10.3 Physical Properties of Acids

The first members of the carboxylic acid series are colorless liquids with sharp or unpleasant odors. Acetic acid, which constitutes about 4% to 5% of vinegar, gives it its characteristic odor and flavor. Butyric acid gives rancid butter its disagreeable odor, and the goat acids (caproic, caprylic, and capric in Table 10.1) smell like goats. Table 10.3 lists some physical properties of selected carboxylic acids.

Carboxylic acids are polar. Like alcohols, they form hydrogen bonds with themselves or with other molecules. They therefore have rather high boiling points for their molecular weights—higher even than those of comparable alcohols. For example, acetic acid and propyl alcohol, which have the same formula weights (60), boil at 118°C and 97°C respectively. Carboxylic acids form dimers, with the units firmly held together by *two* hydrogen bonds.

$$R-C\begin{matrix}\nearrow O\cdots H-O\searrow \\ \searrow O-H\cdots O\nearrow\end{matrix}C-R$$

Hydrogen bonding also explains the water-solubility of the lower-molecular-weight carboxylic acids.

TABLE 10.3 Physical properties of some carboxylic acids

Name	*bp, °C*	*mp, °C*	*Solubility, g/100 g H_2O at 25°C*
formic	101	8	miscible (∞)
acetic	118	17	miscible (∞)
propanoic	141	−22	miscible (∞)
butanoic	164	−8	miscible (∞)
hexanoic	205	−1.5	1.0
octanoic	240	17	0.06
decanoic	270	31	0.01
benzoic	249	122	0.4 (but 6.8 at 95°C)

10.4 Acidity and Acidity Constants

Carboxylic acids dissociate in water, yielding a **carboxylate anion** and an oxonium ion.

$$R-C(=O)OH + H\ddot{O}H \rightleftharpoons \underset{\text{carboxylate ion}}{R-C(=O)O^-} + \underset{\text{oxonium ion (hydroniuum ion)}}{H-\ddot{O}^+(H)-H} \qquad (10.1)$$

Their **acidity constant** K_a is given by the expression

$$K_a = \frac{[RCO_2^-][H_3O^+]}{[RCO_2H]} \qquad (10.2)$$

(Before proceeding further, it would be a good idea for you to review Secs. 7.6 and 7.7.)

Table 10.4 lists the acidity constants for some carboxylic and other acids. In comparing data in this table, remember that the larger the value of K_a or the smaller the value of pK_a, the stronger the acid.

EXAMPLE 10.1 Which is the stronger acid, formic or acetic, and by how much?

Solution Formic acid is stronger; it has the larger K_a. The ratio of acidities is

$$\frac{2.1 \times 10^{-4}}{1.8 \times 10^{-5}} = 1.16 \times 10^{1} = 11.6$$

This means that formic acid is 11.6 times stronger than acetic acid.

TABLE 10.4 The ionization constants of some acids

Name	*Formula*	K_a	pK_a
formic	$HCOOH$	2.1×10^{-4}	3.68
acetic	CH_3COOH	1.8×10^{-5}	4.74
propanoic	CH_3CH_2COOH	1.4×10^{-5}	4.85
butanoic	$CH_3CH_2CH_2COOH$	1.6×10^{-5}	4.80
chloroacetic	$ClCH_2COOH$	1.5×10^{-3}	2.82
dichloroacetic	$Cl_2CHCOOH$	5.0×10^{-2}	1.30
trichloroacetic	CCl_3COOH	2.0×10^{-1}	0.70
2-chlorobutanoic	$CH_3CH_2CHClCOOH$	1.4×10^{-3}	2.85
3-chlorobutanoic	$CH_3CHClCH_2COOH$	8.9×10^{-5}	4.05
benzoic acid	C_6H_5COOH	6.6×10^{-5}	4.18
o-chlorobenzoic	*o*-$Cl-C_6H_4COOH$	12.5×10^{-4}	2.90
m-chlorobenzoic	*m*-$Cl-C_6H_4COOH$	1.6×10^{-4}	3.80
p-chlorobenzoic	*p*-$Cl-C_6H_4COOH$	1.0×10^{-4}	4.00
p-nitrobenzoic	*p*-$NO_2-C_6H_4COOH$	4.0×10^{-4}	3.40
phenol	C_6H_5OH	1.0×10^{-10}	10.00
ethanol	CH_3CH_2OH	1.0×10^{-16}	16.00
water	HOH	1.8×10^{-16}	15.74

Problem 10.6 Using the data given in Table 10.4, determine which is the stronger acid, acetic or chloroacetic, and by how much.

Before we can explain the acidity differences in Table 10.4, we must examine the structural features that make carboxylic acids acidic.

10.5 *Resonance in the Carboxylate Ion*

You might wonder why carboxylic acids are so much more acidic than alcohols or phenols, since all three classes ionize by losing H^+ from a hydroxyl group. One answer lies in a comparison of the possibilities for charge delocalization in the resulting anions. Let us use a specific example.

From Table 10.4 we see that acetic acid is approximately 10^{11}, or one hundred thousand million, times stronger an acid than ethanol.

$$CH_3CH_2OH \rightleftharpoons \underset{\text{ethoxide ion}}{CH_3CH_2O^-} + H^+ \qquad K_a = 10^{-16} \tag{10.3}$$

$$CH_3\overset{\overset{\large O}{\|}}{C}{-}OH \rightleftharpoons \underset{\text{acetate ion}}{CH_3\overset{\overset{\large O}{\|}}{C}{-}O^-} + H^+ \qquad K_a = 10^{-5} \tag{10.4}$$

In ethoxide ion, *the negative charge is localized on a single oxygen atom.* In acetate ion, on the other hand, *the negative charge can be localized through resonance.*

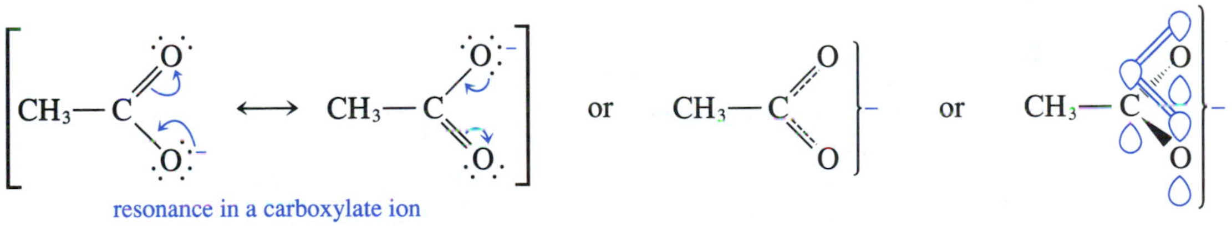

$$\left[CH_3{-}C\begin{matrix}\diagup\!\!\diagup \ddot{O}: \\ \diagdown \ddot{O}:^-\end{matrix} \longleftrightarrow CH_3{-}C\begin{matrix}\diagup \ddot{O}:^- \\ \diagdown\!\!\diagdown \ddot{O}:\end{matrix}\right] \text{ or } \left[CH_3{-}C\begin{matrix}O \\ O\end{matrix}\right]^- \text{ or } \left[CH_3{-}C\begin{matrix}O \\ O\end{matrix}\right]^-$$

resonance in a carboxylate ion

The negative charge is spread *equally* over the two oxygens, so that each oxygen in the carboxylate ion carries only half the negative charge. The acetate ion is stabilized by resonance compared to the ethoxide ion. This stabilization drives the equilibrium more to the right in eq. 10.4 than that in eq. 10.3. Consequently, more H^+ is formed from acetic acid than from ethanol. That is, acetic acid is a stronger acid than ethanol.

EXAMPLE 10.2 Phenoxide ions are also stabilized by resonance (Sec. 7.7). Why aren't phenols as strong acids as carboxylic acids?

Solution Charge delocalization is not as great in phenoxide ions as in carboxylate ions because the contributors to the resonance hybrid are not equivalent. Some of them put the negative charge on carbon instead of on oxygen and disrupt aromaticity. In the carboxylate ion, both contributors are identical and have the negative charge on oxygen, a more electronegative atom than carbon.

Physical data support the importance of resonance in carboxylate ions. In formic acid molecules, the two carbon-oxygen bonds have different lengths. But in sodium formate, both carbon-oxygen bonds of the formate ion are identical, and their length is between those of normal double and single carbon-oxygen bonds.

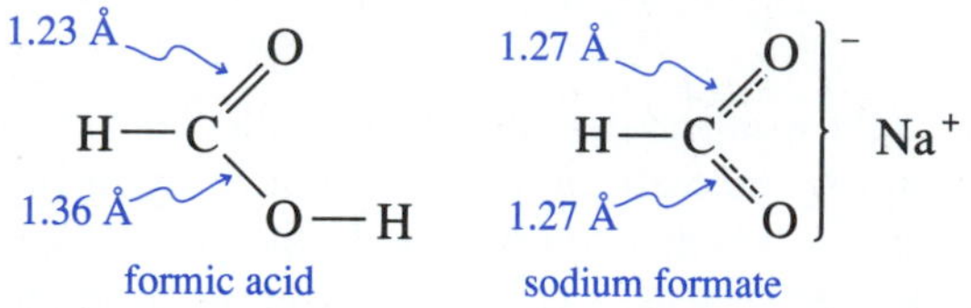

1.23 Å (C=O), 1.36 Å (C–O)
formic acid

1.27 Å, 1.27 Å Na^+
sodium formate

10.6 Effect of Structure on Acidity; the Inductive Effect Revisited

The data in Table 10.4 show that even among carboxylic acids (where the ionizing functional group is kept constant), acidities can vary depending on what other groups are attached to the molecule. Compare, for example, the K_a of acetic acid with those of mono-, di-, and trichloroacetic acids, and note that the acidity varies by a factor of 10,000.

The most important factor operating here is the **inductive effect** of the groups close to the carboxyl group. This effect relays charge through bonds, by displacing bonding electrons toward electronegative atoms, or away from electropositive atoms. Recall that *electron-withdrawing groups enhance acidity, and electron-releasing groups reduce acidity.*

Let us examine the carboxylate ions formed when acetic acid and its chloro derivatives ionize:

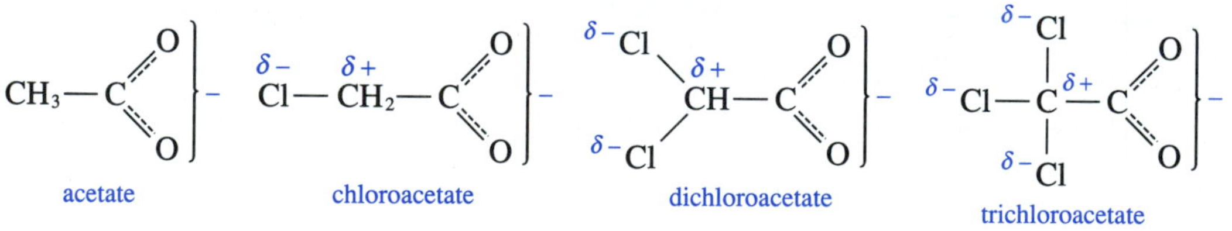

$CH_3-CO_2^-$ acetate

$\overset{\delta-}{Cl}-\overset{\delta+}{CH_2}-CO_2^-$ chloroacetate

$(\overset{\delta-}{Cl})_2\overset{\delta+}{CH}-CO_2^-$ dichloroacetate

$(\overset{\delta-}{Cl})_3\overset{\delta+}{C}-CO_2^-$ trichloroacetate

Because chlorine is more electronegative than carbon, the C—Cl bond is polarized with the chlorine partially negative and the carbon partially positive. Thus, electrons are pulled away from the carboxylate end of the ion toward the chlorine. The effect tends to spread the negative charge over more atoms than in acetate ion itself and thus stabilizes the ion. The more chlorines, the greater the effect and the greater the strength of the acid.

EXAMPLE 10.3 Explain the acidity order in Table 10.4 for butanoic acid and its 2- and 3-chloro derivatives.

Solution The 2-chloro substituent increases the acidity of butanoic acid substantially, due to its inductive effect. In fact, the effect is about the same as for chloroacetic and acetic acids. The 3-chloro substituent exerts a similar *but much smaller* effect, because the C—Cl bond is now farther away from the carboxylate group. *Inductive effects fall off rapidly with distance.*

Problem 10.7 Account for the relative acidities of benzoic acid and its *ortho, meta,* and *para* chloro derivatives (Table 10.4).

We saw in Example 10.1 that formic acid is substantially stronger than acetic acid. This suggests that the methyl group is more electron-releasing (hence acidity-reducing) than hydrogen. This observation is consistent with what we have already learned about carbocation stabilities—that alkyl groups are more effective than hydrogen atoms at releasing electrons to, and therefore stabilizing, a positive carbon atom.

10.7 Conversion of Acids to Salts

Carboxylic acids, when treated with a strong base, form salts. For example,

$$\underset{\text{carboxylic acid}}{R{-}C({=}O){-}OH} + \underset{\text{strong base}}{Na^+\,OH^-} \longrightarrow \underset{\text{a sodium salt}}{R{-}C({=}O){-}O^-Na^+} + HOH \qquad (10.5)$$

The salt can be isolated by evaporating the water.

Salts are named as shown in the following examples:

$$\underset{\substack{\text{sodium acetate}\\ \text{(sodium ethanoate)}}}{CH_3{-}C({=}O){-}O^-Na^+} \qquad \underset{\text{potassium benzoate}}{C_6H_5{-}C({=}O){-}O^-K^+} \qquad \underset{\text{calcium propanoate}}{\left(CH_3CH_2C({=}O){-}O^-\right)_2\;Ca^{2+}}$$

The cation is named first, followed by the name of the carboxylate ion, which is obtained by changing the *-ic* ending of the acid to *-ate*.

EXAMPLE 10.4 Name the following salt:

$$CH_3CH_2CH_2C({=}O){-}O^-NH_4^+$$

Solution The salt is ammonium butanoate (IUPAC) or ammonium butyrate (common).

Problem 10.8 Write an equation, analogous to eq. 10.5, for the preparation of potassium 3-bromopropanoate from the corresponding acid.

As we will see in Chapter 15, salts of certain acids are useful as soaps and detergents.

10.8 Preparation of Acids

Organic acids can be prepared in many ways, four of which will be described here: (1) oxidation of primary alcohols or aldehydes, (2) oxidation of alkyl side chains on aromatic rings, (3) reaction of Grignard reagents with carbon dioxide, and (4) hydrolysis of alkyl cyanides (nitriles).

10.8a Oxidation of Primary Alcohols or Aldehydes The oxidation of primary alcohols (eq. 7.31) and aldehydes (eq. 9.33) to carboxylic acids has already been mentioned. It is easy to see that these are oxidation reactions because going from an alcohol to an aldehyde to an acid requires replacement of C—H bonds by C—O bonds.

$$R{-}CH_2{-}OH \longrightarrow R(H)C{=}O \longrightarrow R{-}C({=}O){-}OH \qquad (10.6)$$

alcohol (one C—O bond) aldehyde (two C—O bonds) acid (three C—O bonds)

The most commonly used oxidizing agents for these purposes are potassium permanganate ($KMnO_4$), chromic acid (CrO_3), nitric acid, and with aldehydes only, silver oxide (Ag_2O). For specific examples, see eqs. 7.34, 9.34, 9.35 and 9.38.

10.8b Oxidation of Aromatic Side Chains Aromatic acids can be prepared by oxidizing an alkyl side chain on an aromatic ring.

$$C_6H_5{-}CH_3 \xrightarrow[\text{heat}]{KMnO_4} C_6H_5{-}C({=}O){-}OH \qquad (10.7)$$

toluene benzoic acid

This reaction illustrates the striking stability of aromatic rings; it is the alkane-like methyl group, not the aromatic ring, that is oxidized. The reaction involves attack of the oxidant at a C—H bond adjacent to the benzene ring. Longer side chains are also oxidized to a carboxyl group.

$$C_6H_5{-}CH_2CH_2CH_3 \xrightarrow[\text{heat}]{KMnO_4} C_6H_5{-}CO_2H \qquad (10.8)$$

If no C—H bond is in the benzylic position, however, the aromatic ring is oxidized.

$$C_6H_5{-}C(CH_3)_3 \xrightarrow{KMnO_4} (CH_3)_3CCO_2H \qquad (10.9)$$

With oxidants other than potassium permanganate, this reaction is commercially important. For example, **terephthalic acid,** one of the two raw materials needed to manufacture Dacron, is produced in this way, using a cobalt catalyst for the air oxidation.

$$CH_3{-}C_6H_4{-}CH_3 \xrightarrow[CH_3CO_2H]{O_2,\,Co(III)} HOOC{-}C_6H_4{-}COOH \qquad (10.10)$$

p-xylene terephthalic acid

Phthalic acid, used for making plasticizers, resins, and dyestuffs, is manufactured by similar oxidations, starting with *o*-xylene.

$$\text{o-xylene} \xrightarrow[\text{CH}_3\text{CO}_2\text{H}]{\text{O}_2,\text{Co(III)}} \text{phthalic acid} \qquad (10.11)$$

o-xylene phthalic acid

10.8c Reaction of Grignard Reagents with Carbon Dioxide As we saw previously, Grignard reagents add to the carbonyl groups of aldehydes or ketones to give alcohols. In a similar manner, they add to the carbonyl group of carbon dioxide to give acids.

$$R{-}MgX + O{=}C{=}O \longrightarrow R{-}\overset{\overset{\large O}{\|}}{C}{-}OMgX \xrightarrow{HX} R{-}\overset{\overset{\large O}{\|}}{C}{-}OH + Mg^{2+}X_2^- \qquad (10.12)$$

This reaction gives good yields and is an excellent laboratory method for preparing both aliphatic and aromatic acids. Note that the acid obtained has one more carbon atom than the alkyl or aryl halide from which the Grignard reagent is prepared, so the reaction provides a way to increase the length of a carbon chain.

EXAMPLE 10.5 Show how $(CH_3)_3CBr$ can be converted to $(CH_3)_3CCO_2H$.

Solution

$$(CH_3)_3CBr \xrightarrow[\text{ether}]{Mg} (CH_3)_3CMgBr \xrightarrow[2.\ H_3O^+]{1.\ CO_2} (CH_3)_3CCO_2H$$

Problem 10.9 Devise a synthesis of butanoic acid from 1-propanol.

10.8d Hydrolysis of Cyanides (Nitriles) The carbon-nitrogen triple bond of organic cyanides can be hydrolyzed to a carboxyl group. The overall reaction is as follows:

$$\underset{\text{a cyanide, or nitrile}}{R{-}C{\equiv}N} + 2\,H_2O \xrightarrow[OH^-]{H^+ \text{ or}} \underset{\text{an acid}}{R{-}C({=}O)OH} + NH_3 \qquad (10.13)$$

The reaction requires either an acid or a base as a catalyst, and the nitrogen atom of the cyanide is converted to ammonia. Alkyl cyanides are generally made from the corresponding alkyl halide (usually primary) and sodium cyanide by an S_N2 displacement. For example,

$$\underset{\substack{\text{propyl bromide}\\ \text{(1-bromopropane)}}}{CH_3CH_2CH_2Br} \xrightarrow{NaCN} \underset{\substack{\text{butyronitrile}\\ \text{(butanenitrile)}}}{CH_3CH_2CH_2CN} \xrightarrow[H^+]{H_2O} \underset{\substack{\text{butyric acid}\\ \text{(butanoic acid)}}}{CH_3CH_2CH_2CO_2H} + NH_4^+ \qquad (10.14)$$

Problem 10.10 Why is it *not* possible to convert bromobenzene to benzoic acid by the nitrile method? How could this conversion be accomplished?

Organic cyanides are commonly named after the corresponding acid, by changing the *-ic* or *-oic* suffix to *-onitrile* (hence butyronitrile in eq. 10.14). In the IUPAC system, the suffix *-nitrile* is added to the name of the hydrocarbon with the same number of carbon atoms (hence butanenitrile in eq. 10.14). Sometimes these cyanides are named as alkyl cyanides.

EXAMPLE 10.6 Name CH_3CN in three ways.

Solution Acetonitrile (it gives acetic acid on hydroylsis), ethanenitrile (IUPAC), or methyl cyanide.

Note that with the hydrolysis of nitriles, as with the Grignard method, the acid obtained has one more carbon atom than the alkyl halide from which the cyanide is prepared. Consequently, both methods provide ways of increasing the length of a carbon chain.

Problem 10.11 Write equations for synthesizing phenylacetic acid from benzyl bromide by two routes.

10.9 Carboxylic Acid Derivatives

Carboxylic acid derivatives are compounds in which the hydroxyl part of the carboxyl group is replaced by various other groups. All acid derivatives can be hydrolyzed to the corresponding acid. In the remainder of this chapter, we will consider the preparation and reactions of the more important of these acid derivatives. Their general formulas are as follows:

$$\underset{\text{ester}}{R-\overset{\overset{\displaystyle O}{\|}}{C}-OR'} \qquad \underset{\text{acyl halide}}{R-\overset{\overset{\displaystyle O}{\|}}{C}-X} \quad \begin{pmatrix}\text{X is usually}\\ \text{Cl or Br}\end{pmatrix} \qquad \underset{\text{acid anhydride}}{R-\overset{\overset{\displaystyle O}{\|}}{C}-O-\overset{\overset{\displaystyle O}{\|}}{C}-R} \qquad \underset{\text{primary amide}}{R-\overset{\overset{\displaystyle O}{\|}}{C}-NH_2}$$

Esters and amides occur widely in nature. Anhydrides, however, are uncommon in nature, and acyl halides are strictly creatures of the laboratory.

10.10 Esters

Esters are derived from acids by replacing the —OH group by an —OR group. They are named in a manner analogous to that for salts. The R part of the —OR group is named first, followed by the name of the acid, with the *-ic* ending changed to *-ate*.

$$\underset{\substack{\text{methyl acetate}\\ \text{(methyl ethanoate)}\\ \text{bp 57°C}}}{CH_3\overset{\overset{\displaystyle O}{\|}}{C}-OCH_3} \qquad \underset{\substack{\text{ethyl acetate}\\ \text{(ethyl ethanoate)}\\ \text{bp 77°C}}}{CH_3\overset{\overset{\displaystyle O}{\|}}{C}-OCH_2CH_3} \qquad \underset{\substack{\text{methyl butanoate}\\ \text{bp 102.3°C}}}{CH_3CH_2CH_2\overset{\overset{\displaystyle O}{\|}}{C}-OCH_3}$$

Notice the different names of the following pair of isomeric esters, where the R and R′ groups are interchanged.

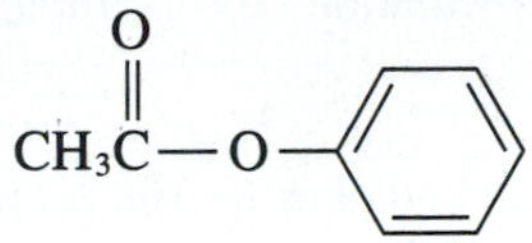

$$CH_3\overset{\overset{O}{\|}}{C}-O-C_6H_5 \qquad C_6H_5-\overset{\overset{O}{\|}}{C}-OCH_3$$

phenyl acetate
bp 195.7°C

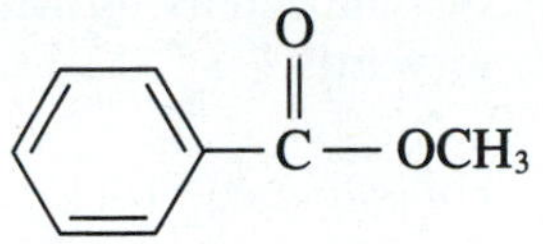

methyl benzoate
bp 196.6°C

EXAMPLE 10.7 Name $CH_3CH_2CO_2CH(CH_3)_2$.

Solution The related acid is $CH_3CH_2CO_2H$, so the last part of the name is *propanoate* (change the *-ic* of propanoic to *-ate*). The alkyl group that replaces the hydrogen is *isopropyl,* or *2-propyl,* so the correct name is *isopropyl propanoate*, or *2-propyl propanoate*.

Problem 10.12 Write the IUPAC name for

a. $H-\overset{\overset{O}{\|}}{C}-OCH_3$ **b.** $CH_3CH_2\overset{\overset{O}{\|}}{C}-OCH_2CH_2CH_3$

Problem 10.13 Write the structure of

a. 2-pentyl acetate.
b. ethyl 2-methylpropanoate.

Many esters are rather pleasant-smelling substances, which are responsible for the flavor and fragrance of fruits and flowers. Among the more common are pentyl acetate (bananas), octyl acetate (oranges), ethyl butanoate (pineapples), and pentyl butanoate (apricots). Natural flavors can be exceedingly complex. For example, no fewer than 53 esters have been identified among the volatile constituents of Bartlett pears! Mixtures of esters are used as perfumes and artificial flavors.

10.11 Preparation of Esters; Fischer Esterification

When a carboxylic acid and an alcohol are heated in the presence of an acid catalyst (usually HCl or H_2SO_4), an equilibrium is established with the ester and water.

$$\underset{\text{acid}}{R-\overset{\overset{O}{\|}}{C}-OH} + \underset{\text{alcohol}}{HO-R'} \overset{H^+}{\rightleftharpoons} \underset{\text{ester}}{R-\overset{\overset{O}{\|}}{C}-OR'} + H_2O \qquad (10.15)$$

The process is called **Fischer esterification,** after Emil Fischer (page 154), who developed the method. Although the reaction is an equilibrium, it can be used to make esters in high yield by shifting the equilibrium to the right. This can be accomplished in several ways. If either the alcohol or the acid is inexpensive, a large excess can be used. Alternatively, the ester and/or water may be removed as formed (by distillation, for example), thus driving the reaction forward.

Problem 10.14 Following eq. 10.15, write an equation for the preparation of propyl butanoate from the correct acid and alcohol.

10.12 *The Mechanism of Acid-catalyzed Esterification; Nucleophilic Acyl Substitution*

We can ask the following simple mechanistic question about Fischer esterification: Is the water molecule formed from the hydroxyl group of the acid and the hydrogen of the alcohol (as shown in color in eq. 10.15), or from the hydrogen of the acid and the hydroxyl group of the alcohol? This question may seem rather trivial, but the answer provides a key to understanding much of the chemistry of acids, esters, and their derivatives.

This question was resolved using isotopic labeling. For example, Fischer esterification of benzoic acid with methanol that had been enriched with the ^{18}O isotope of oxygen gave labeled methyl benzoate.

$$C_6H_5-\overset{O}{\overset{\|}{C}}-OH + H^{18}OCH_3 \xrightleftharpoons{H^+} C_6H_5-\overset{O}{\overset{\|}{C}}-{}^{18}OCH_3 + HOH \qquad (10.16)$$

methyl benzoate

None of the ^{18}O appeared in the water. Thus it is clear that ***the water was formed using the hydroxyl group of the acid and the hydrogen of the alcohol.*** In other words, in Fischer esterification, the —OR group of the alcohol replaces the —OH group of the acid.

How can we explain this experimental fact? A mechanism consistent with this result is as follows (the oxygen atom of the alcohol is shown in color so that its path can be traced):

$$R-\overset{O}{\overset{\|}{C}}-OH \underset{①}{\xrightleftharpoons{H^+}} R-\overset{^+OH}{\overset{\|}{C}}-OH \;(+\; R'OH) \underset{}{\xrightleftharpoons{②}} R-\overset{OH}{\underset{R'\overset{+}{O}H}{\overset{|}{\underset{|}{C}}}}-OH \underset{③}{\xrightleftharpoons{-H^+}} \boxed{R-\overset{OH}{\underset{R'O}{\overset{|}{\underset{|}{C}}}}-OH}$$

$$\boxed{R-\overset{OH}{\underset{R'O}{\overset{|}{\underset{|}{C}}}}-OH} \underset{④}{\xrightleftharpoons{H^+}} R-\overset{OH}{\underset{R'O}{\overset{|}{\underset{|}{C}}}}-\overset{+}{O}H_2 \underset{⑤}{\xrightleftharpoons{-H_2O}} R-\overset{^+OH}{\underset{R'O}{\overset{\|}{\underset{|}{C}}}} \underset{⑥}{\xrightleftharpoons{-H^+}} R-\overset{O}{\overset{\|}{C}}-OR' \qquad (10.17)$$

nucleophilic addition (step ②); elimination step (step ⑤)

Let us go through this mechanism, which looks more complicated than it really is, one step at a time.

Step 1. The carbonyl group of the acid is reversibly protonated. This step explains how the acid catalyst works. Protonation increases the positive charge on the carboxyl carbon and enhances its reactivity toward nucleophiles (recall the similar effect of acid catalysts with aldehydes and ketones, eq. 9.8).

Step 2. *This is the crucial step.* The alcohol, as a nucleophile, attacks the carbonyl carbon of the protonated acid. This is the step in which the new C—O bond (the ester bond) is formed.

Steps 3 and 4. These steps are equilibria in which oxygens lose or gain a proton. Such acid-base equilibria are reversible and rapid and go on constantly in any acidic solution of an oxygen-containing compound. In step 4, it doesn't matter which —OH group is protonated since these groups are equivalent.

Step 5. This is the step in which water, one product of the overall reaction, is eliminated. For this step to occur, an —OH group must be protonated to improve its leaving-group capacity. (This step is similar to the reverse of step 2.)

Step 6. This deprotonation step gives the ester and regenerates the acid catalyst. (This step is similar to the reverse of step 1.)

Some other features of the mechanism in eq. 10.17 are worth examining. The reaction begins with an acid, in which the carboxyl carbon is trigonal and sp^2-hybridized. The end product is an ester; the ester carbon is also trigonal and sp^2-hybridized. However, *the reaction proceeds through a neutral tetrahedral intermediate* (shown in a box in eq. 10.17 and in color in eq. 10.18), in which the carbon atom has four groups attached to it and is thus sp^3-hybridized. If we omit all of the proton-transfer steps in eq. 10.17, we can focus on this feature of the reaction:

$$\underset{sp^2}{R-\overset{O}{\overset{\|}{C}}-OH} + R'OH \rightleftharpoons \underset{sp^3}{R-\overset{OH}{\overset{|}{\underset{\underset{R'O}{|}}{C}}}-OH} \rightleftharpoons \underset{sp^2}{R-\overset{O}{\overset{\|}{C}}-OR'} + HOH \tag{10.18}$$

The net result of this process is substitution of the —OR′ group of the alcohol for the —OH group of the acid. Hence the reaction is referred to as **nucleophilic acyl substitution.** But the reaction is not a direct substitution. Instead, it occurs in two steps: (1) nucleophilic addition, followed by (2) elimination. We will see in the next and subsequent sections of this chapter that this is a general mechanism for nucleophilic substitutions at the carbonyl carbon atoms of acid derivatives.

Problem 10.15 Following eq. 10.17, write out the steps in the mechanism for the acid-catalyzed preparation of ethyl acetate from ethanol and acetic acid. In the

United States, this method is used commercially to produce over 100 million pounds of ethyl acetate annually, mainly for use as a solvent in the paint industry, but also as a solvent for nail polish and various glues.

10.13 Lactones

Hydroxy acids contain both functional groups required for ester formation. If these groups can come in contact through bending of the chain, they may react with one another to form **cyclic esters** called **lactones.** For example,

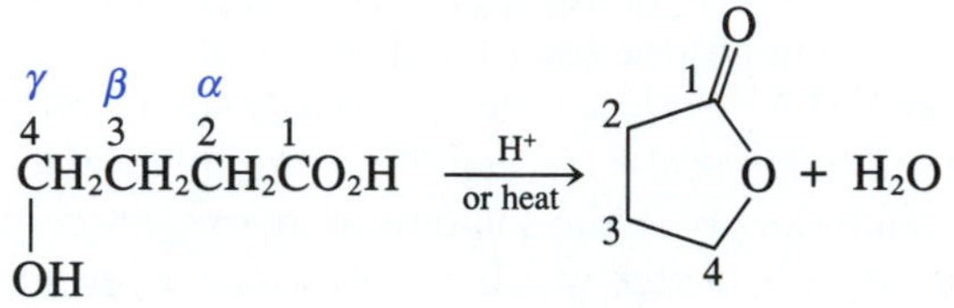

$$\underset{\text{OH}}{\overset{\gamma\ \ \beta\ \ \alpha}{\overset{4\ \ 3\ \ 2\ \ 1}{CH_2CH_2CH_2CO_2H}}} \xrightarrow[\text{or heat}]{H^+} \gamma\text{-butyrolactone} + H_2O \qquad (10.19)$$

γ-butyrolactone

Most common lactones have five- or six-membered rings, although lactones with smaller or larger rings are known. Two examples of six-membered lactones from nature are **coumarin,** which is responsible for the pleasant odor of newly mown hay, and **nepatalactone,** the compound in catnip that excites cats.

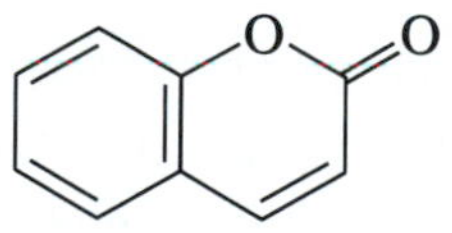

coumarin

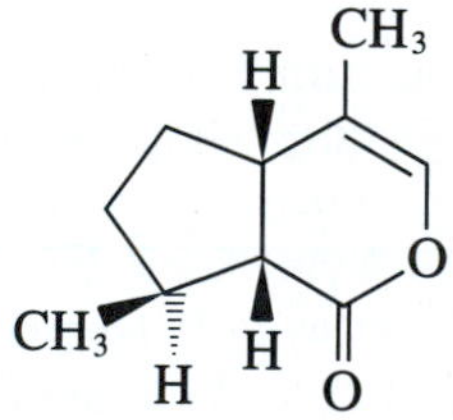

nepatalactone

Problem 10.16 Write out the steps in the mechanism for the acid-catalyzed reaction in eq. 10.19.

10.14 Saponification of Esters

Esters are commonly hydrolyzed with base. The reaction is called **saponification** (from the Latin *sapon,* soap) because this type of reaction is used to make soaps from fats (Chapter 15). The general reaction is as follows:

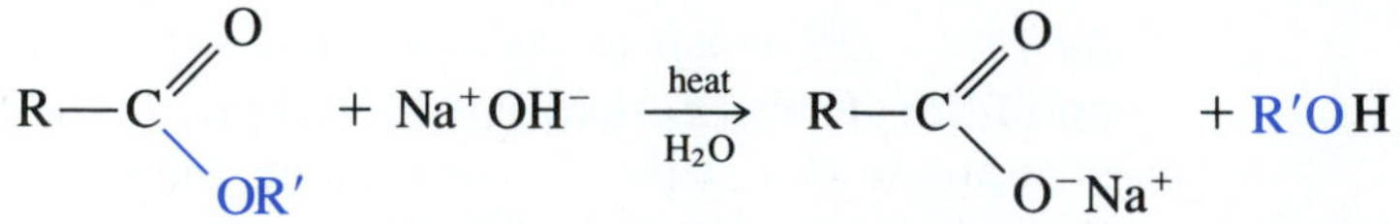

$$\underset{\text{ester}}{R{-}C({=}O){-}OR'} + \underset{\text{base}}{Na^+OH^-} \xrightarrow[H_2O]{\text{heat}} \underset{\text{salt of an acid}}{R{-}C({=}O){-}O^-Na^+} + \underset{\text{alcohol}}{R'OH} \qquad (10.20)$$

The mechanism is another example of a nucleophilic acyl substitution. It involves nucleophilic attack by hydroxide ion, a strong nucleophile, on the carbonyl carbon of the ester.

$$HO^- + R{-}\overset{\overset{O}{\|}}{C}{-}OR' \rightleftharpoons R{-}\underset{\underset{OH}{|}}{\overset{\overset{O^-}{|}}{C}}{-}OR'$$

$$\downarrow\uparrow$$

$$R{-}\overset{\overset{O}{\|}}{C}{-}OH + {}^-{:}OR' \longrightarrow R{-}\overset{\overset{O}{\|}}{C}{-}O^- + R'OH \tag{10.21}$$

strong base — weak base

The key step is nucleophilic addition to the carbonyl group. The reaction proceeds via a tetrahedral intermediate, but the reactant and the product are trigonal. *Saponification is not reversible;* in the final step, the strongly basic alkoxide ion removes a proton from the acid to form a carboxylate ion and an alcohol molecule—a step that proceeds completely in the forward direction.

Saponification is especially useful for breaking down an unknown ester, perhaps isolated from a natural source, into its component acid and alcohol for structure determination.

Problem 10.17 Following eq. 10.20, write an equation for the saponification of methyl benzoate.

10.15 *Ammonolysis of Esters*

Ammonia converts esters to amides.

$$\underset{\text{ester}}{R{-}C({=}O){-}OR'} + NH_3 \longrightarrow \underset{\text{amide}}{R{-}C({=}O){-}NH_2} + R'OH \tag{10.22}$$

For example,

$$\underset{\text{methyl benzoate}}{C_6H_5{-}C({=}O){-}OCH_3} + NH_3 \xrightarrow{\text{ether}} \underset{\text{benzamide}}{C_6H_5{-}C({=}O){-}NH_2} + CH_3OH \tag{10.23}$$

The reaction mechanism is very much like that of saponification. The unshared electron pair on the ammonia nitrogen initiates nucleophilic attack on the ester carbonyl group.

Problem 10.18 Using eq. 10.21 as a guide, write out the steps in the mechanism for eq. 10.23.

10.16 Reaction of Esters with Grignard Reagents

Esters react with two equivalents of a Grignard reagent to give tertiary alcohols. The reaction proceeds by nucleophilic attack of the Grignard reagent on the ester carbonyl group. The initial product, a ketone, reacts further in the usual way to give the tertiary alcohol.

$$\underset{\text{ester}}{R-\overset{O}{\overset{\|}{C}}-OR'} + 2\,R''MgBr \xrightarrow{\text{overall}} R-\overset{OMgBr}{\underset{R''}{C}}-R'' \xrightarrow[H^+]{H_2O} \underset{\text{tertiary alcohol}}{R-\overset{OH}{\underset{R''}{C}}-R''} \qquad (10.24)$$

$$\text{ester} \xrightarrow{R''MgBr} R-\overset{OMgBr}{\underset{R''}{C}}-OR' \xrightarrow{-R'OMgBr} \underset{\text{ketone}}{R-\overset{O}{\overset{\|}{C}}-R''} \xrightarrow{R''MgBr} R-\overset{OMgBr}{\underset{R''}{C}}-R''$$

This method is useful for making tertiary alcohols in which at least two of three alkyl groups attached to the hydroxyl-bearing carbon atom are identical.

Problem 10.19 Using eq. 10.24 as a guide, write the structure of the tertiary alcohol that is obtained from

$$\text{(cyclopropyl)}-\overset{O}{\overset{\|}{C}}-OCH_3 + \text{excess } C_6H_5-MgBr$$

10.17 Reduction of Esters

Esters can be reduced to alcohols by lithium aluminum hydride.

$$\underset{\text{ester}}{R-\overset{O}{\overset{\|}{C}}-OR'} \xrightarrow[\text{ether}]{LiAlH_4} \underset{\text{primary alcohol}}{RCH_2OH} + R'OH \qquad (10.25)$$

The mechanism is similar to the hydride reduction of aldehydes and ketones (eq. 9.30).

$$\underset{\text{ester}}{R-\overset{O}{\overset{\|}{C}}-OR'} \xrightarrow{\overset{\delta-\ \delta+}{H-Al\lessgtr}} R-\overset{O-Al\lessgtr}{\underset{H}{C}}-OR' \xrightarrow{-R'OAl\lessgtr}$$

$$\underset{\text{aldehyde}}{R-\overset{O}{\overset{\|}{C}}-H} \xrightarrow{H-Al\lessgtr} R-\overset{O-Al\lessgtr}{\underset{H}{C}}-H \xrightarrow[H^+]{H_2O} \underset{1^\circ\text{ alcohol}}{RCH_2OH} \qquad (10.26)$$

The intermediate aldehyde is not usually isolable and reacts rapidly with additional hydride to produce the alcohol.

It is possible to reduce an ester carbonyl group without reducing a C=C bond in the same molecule. For example,

$$\underset{\text{ethyl 2-butenoate}}{CH_3CH{=}CH\overset{\overset{\displaystyle O}{\|}}{C}{-}OCH_2CH_3} \xrightarrow[\text{2. } H_2O,\ H^+]{\text{1. } LiAlH_4} \underset{\text{2-buten-1-ol}}{CH_3CH{=}CHCH_2OH} + CH_3CH_2OH \tag{10.27}$$

10.18 The Need for Activated Acyl Compounds

As we have seen, most reactions of carboxylic acids, esters, and related compounds involve, as the first step, nucleophilic attack on the carbonyl carbon atom. Examples are Fischer esterification, saponification and ammonolysis of esters and the first stage of the reaction of esters with Grignard reagents or lithium aluminum hydride. All of these reactions can be summarized by a single mechanistic equation:

$$\underset{sp^2}{\overset{R}{\underset{L}{>}}C{=}\ddot{O}:} + :Nu \overset{①}{\rightleftharpoons} \underset{\text{tetrahedral intermediate}}{R{-}\overset{\overset{\displaystyle :\ddot{O}:^-}{|}}{\underset{\underset{\displaystyle L}{|}}{C}}{-}Nu} \overset{②}{\rightleftharpoons} \underset{sp^2}{\overset{R}{\underset{Nu}{>}}C{=}\ddot{O}:} + :L \tag{10.28}$$

The carbonyl carbon, initially trigonal, is attacked by a nucleophile, Nu:, to form a **tetrahedral intermediate** (step 1). Loss of a leaving group L (step 2) then regenerates the carbonyl group with its trigonal carbon atom. The net result is the replacement of L by Nu.

Biochemists look at eq. 10.28 in a slightly different way. They refer to the overall reaction as **acyl transfer.** The acyl group is transferred from L in the starting material to Nu in the product.

Regardless of how we consider the reaction, one important feature that can affect the rate of both steps is the nature of the leaving group. ***The rates of both steps in a nucleophilic acyl substitution reaction are enhanced by increasing the electron-withdrawing properties of the leaving group.*** Step 1 is favored because the more electronegative L is, the more positive the carbonyl carbon becomes, and therefore the more susceptible it is to nucleophilic attack. Step 2 is also facilitated because the more electronegative L is, the better leaving group it becomes.

In general, *esters are less reactive toward nucleophiles than are aldehydes or ketones* because the positive charge on the carbonyl carbon in esters can be delocalized to the oxygen atom.

the carbonyl carbon has a partial positive charge

$$\left[\begin{matrix} R \\ \quad \diagdown \\ \quad C{=}\ddot{O}\!: \\ \quad \diagup \\ R' \end{matrix} \longleftrightarrow \begin{matrix} R \\ \quad \diagdown \\ \quad C^{+}{-}\ddot{O}\!:^{-} \\ \quad \diagup \\ R' \end{matrix} \right]$$

resonance in aldehydes and ketones

the positive charge can be delocalized to the oxygen

$$\left[\begin{matrix} R \\ \quad \diagdown \\ \quad C{=}\ddot{O}\!: \\ \quad \diagup \\ R'\ddot{O}\!: \end{matrix} \longleftrightarrow \begin{matrix} R \\ \quad \diagdown \\ \quad C^{+}{-}\ddot{O}\!:^{-} \\ \quad \diagup \\ R'\ddot{O}\!: \end{matrix} \longleftrightarrow \begin{matrix} R \\ \quad \diagdown \\ \quad C{-}\ddot{O}\!:^{-} \\ \quad /\!/ \\ R'\overset{+}{O} \end{matrix} \right]$$

resonance in esters

Consequently, the carbonyl carbon is less positive in esters than it is in aldehydes or ketones and therefore less susceptible to nucleophilic attack.

Now let us examine some of the ways in which the carboxyl group can be modified to *increase* its reactivity toward nucleophiles.

10.19 Acyl Halides

Acyl halides are among the most reactive of carboxylic acid derivatives. *Acyl chlorides* are more common and less expensive than bromides or iodides. They are usually prepared from acids by reaction with thionyl chloride or phosphorus pentachloride (compare with Sec. 7.11).

$$R{-}C(=O){-}OH + \underset{\text{thionyl chloride}}{SOCl_2} \longrightarrow R{-}C(=O){-}Cl + HCl\uparrow + SO_2\uparrow \qquad (10.29)$$

$$R{-}C(=O){-}OH + \underset{\text{phosphorus pentachloride}}{PCl_5} \longrightarrow R{-}C(=O){-}Cl + HCl + \underset{\text{phosphorus oxychloride}}{POCl_3} \qquad (10.30)$$

Problem 10.20 Rewrite eq. 10.29 to show the preparation of benzoyl chloride.

Acyl halides react rapidly with most nucleophiles. For example, they are rapidly hydrolyzed by water.

$$\underset{\text{acetyl chloride}}{CH_3{-}\overset{\overset{\displaystyle O}{\|}}{C}{-}Cl} + HOH \xrightarrow{\text{rapid}} \underset{\text{acetic acid}}{CH_3{-}\overset{\overset{\displaystyle O}{\|}}{C}{-}OH} + \underset{\text{(fumes)}}{HCl} \qquad (10.31)$$

For this reason, acyl halides have irritating odors. Benzoyl chloride, for example, is a lachrymator (tear gas).

Problem 10.21 Explain why acyl halides may be irritating to the nose.

Acyl halides react rapidly with alcohols to form esters.

$$C_6H_5-\overset{\displaystyle O}{\overset{\|}{C}}-Cl + CH_3OH \xrightarrow[\text{temp.}]{\text{room}} C_6H_5-\overset{\displaystyle O}{\overset{\|}{C}}-OCH_3 + HCl \qquad (10.32)$$

benzoyl chloride — methyl benzoate

Indeed, the most common way to prepare an ester *in the laboratory* is to convert an acid to its acid chloride, then react the latter with an alcohol. Even though two steps are necessary (compared with one step for Fischer esterification), the method may be preferable, especially if either the acid or the alcohol is expensive. (Recall that Fischer esterification is an equilibrium reaction and must often be carried out with a large excess of one of the reactants.)

Acyl halides react rapidly with ammonia to form amides.

$$CH_3\overset{\displaystyle O}{\overset{\|}{C}}-Cl + 2NH_3 \longrightarrow CH_3\overset{\displaystyle O}{\overset{\|}{C}}-NH_2 + NH_4^+Cl^- \qquad (10.33)$$

acetyl chloride — acetamide

The reaction is much more rapid than the ammonolysis of esters. Two equivalents of ammonia are required, however—one to form the amide and one to neutralize the hydrogen chloride.

EXAMPLE 10.8 Explain why acyl chlorides are more reactive than esters toward nucleophiles.

Solution The electronegativity order is Cl > OR. Therefore, the carbonyl carbon is more positive in acyl halides than in esters and more reactive toward nucleophiles. Also, Cl^- is a better leaving group (weaker nucleophile) than OR^-.

Acyl halides are used to synthesize aromatic ketones, through Friedel-Crafts acylation of aromatic rings (review Sec. 4.10d).

Problem 10.22 Devise a synthesis of 4-methylphenyl propyl ketone from toluene and butanoic acid as starting materials.

10.20 *Acid Anhydrides*

Acid anhydrides are derived from acids by removing water from two carboxyl groups and connecting the fragments.

$$R-\overset{\displaystyle O}{\overset{\|}{C}}-OH \quad HO-\overset{\displaystyle O}{\overset{\|}{C}}-R \qquad\qquad R-\overset{\displaystyle O}{\overset{\|}{C}}-O-\overset{\displaystyle O}{\overset{\|}{C}}-R$$

two acid molecules — an acid anhydride

The most important aliphatic anhydride commercially is **acetic anhydride ($R{=}CH_3$).** About 1 million tons are manufactured annually, mainly to react with alcohols to form acetates. The two most common uses are in making cellulose acetate (rayon) and aspirin.

Problem 10.23 Write the structural formula for

a. butanoic anhydride.
b. benzoic anhydride.

Dicarboxylic acids with appropriately spaced carboxyl groups lose water on heating to form cyclic anhydrides with five- or six-membered rings. For example,

$$\underset{\text{maleic acid}}{\text{HC(COOH)=CH(COOH)}} \xrightarrow{135°\text{C}} \underset{\text{maleic anhydride}}{\text{maleic anhydride}} + H_2O \qquad (10.34)$$

Problem 10.24 Predict and name the product of the following reaction:

$$C_6H_4(COOH)_2 \xrightarrow{\text{heat}}$$

Problem 10.25 Do you expect fumaric acid to form a cyclic anhydride on heating? Explain.

Acid anhydrides are more reactive than esters, but less reactive than acyl halides, toward nucleophiles. Some typical reactions of acetic anhydride are the following:

$$\underset{\substack{\text{acetic anhydride}\\\text{bp 139.5°C}}}{CH_3-\overset{O}{\overset{\|}{C}}-O-\overset{O}{\overset{\|}{C}}-CH_3} \begin{cases} \xrightarrow{HO-H} \underset{\text{acid}}{CH_3\overset{O}{\overset{\|}{C}}-OH} + CH_3\overset{O}{\overset{\|}{C}}-OH \\ \xrightarrow{RO-H} \underset{\text{ester}}{CH_3\overset{O}{\overset{\|}{C}}-OR} + CH_3\overset{O}{\overset{\|}{C}}-OH \\ \xrightarrow{NH_2-H} \underset{\text{amide}}{CH_3\overset{O}{\overset{\|}{C}}-NH_2} + CH_3\overset{O}{\overset{\|}{C}}-OH \end{cases} \qquad (10.35)$$

Water hydrolyzes an anhydride to the corresponding acid. Alcohols give esters, and ammonia gives amides. In each case, one equivalent of acid is also produced.

Problem 10.26 Write an equation for the reaction of acetic anhydride with 1-butanol.

The reaction of acetic anhydride with **salicylic acid** (*o*-hydroxybenzoic acid) is used to synthesize **aspirin.** In this reaction, the phenolic hydroxyl group is **acetylated** (converted to its acetate ester).

Annual aspirin production in the United States is more than 50 million pounds, enough to produce over 50 billion standard 5-grain tablets. Aspirin is widely used, either by itself or mixed with other drugs, as an analgesic and antipyretic. It is not without dangers, however. Repeated use may cause gastrointestinal bleeding, and a large single dose (10 to 20g) can cause death.

$$\underset{\text{salicylic acid}}{o\text{-}HOC_6H_4CO_2H} + \underset{\text{acetic anhydride}}{CH_3\overset{O}{\overset{\|}{C}}-O-\overset{O}{\overset{\|}{C}}CH_3} \longrightarrow \underset{\substack{\text{acetylsalicylic acid}\\ \text{(aspirin)}}}{o\text{-}CH_3\overset{O}{\overset{\|}{C}}OC_6H_4CO_2H} + CH_3CO_2H \qquad (10.36)$$

Problem 10.27 Methyl salicylate is the chief component of oil of wintergreen. It is used to flavor gum and candy and in rubbing liniments, where its mild irritating action on skin provides a counterirritant for sore muscles. Write an equation to show how methyl salicylate can be prepared from salicylic acid and methanol.

A WORD ABOUT ...

22. Thioesters, Nature's Acyl-Activating Groups

Acyl transfer plays an important role in many biochemical processes. However, acyl halides and anhydrides are far too corrosive to be cell constituents—they are hydrolyzed quite rapidly by water and are therefore incompatible with cellular fluid. Most ordinary esters, on the other hand, react too slowly with nucleophiles for acyl transfer to be carried out efficiently at body temperatures. Consequently, other groups have evolved to activate acyl groups in the cell. The most important of these is **coenzyme A** (the A stands for acetylation, one of the functions of this enzyme). Coenzyme A is a complex *thiol* (Figure 10.1). It is usually abbreviated by the symbol **CoA—SH.** Though its structure is made up of three parts—adenosine diphosphate (ADP), pantothenic acid (a vitamin), and 2-aminoethanethiol—it is the thiol group that gives coenzyme A its most important functions.

Coenzyme A can be converted to **thioesters,** which are the active acyl-transfer agents in the cell. Of the thioesters that coenzyme A forms, the acetyl ester, called **acetyl-coenzyme A** and abbreviated as

$$CH_3\overset{O}{\overset{\|}{C}}-S-CoA$$

is the most important. Acetyl-CoA reacts with many nucleophiles to transfer the acetyl group.

$$\underset{\text{acetyl-CoA}}{CH_3\overset{O}{\overset{\|}{C}}-S-CoA} + Nu{:} \xrightarrow[\text{enzyme}]{H_2O} CH_3\overset{O}{\overset{\|}{C}}-Nu + CoA-SH$$

The reactions are usually enzyme-mediated and occur rapidly at ordinary cell temperatures.

Why are thioesters superior to ordinary esters as acyl-transfer agents? Part of the answer lies in the acidity difference between alcohols and thiols (Sec.7.17). Since thiols are approximately 1 million times stronger acids than are alcohols, their conjugate bases, ^{-}SR, are a million times weaker bases than ^{-}OR. Thus the —SR group of thioesters is a much better leaving group, in nucleophilic substitutions, than is the —OR group of ordinary esters. Thioesters are not so reactive that they hydrolyze in cellular fluid, but they are appreciably more reactive than simple esters. Nature makes use of this fact.

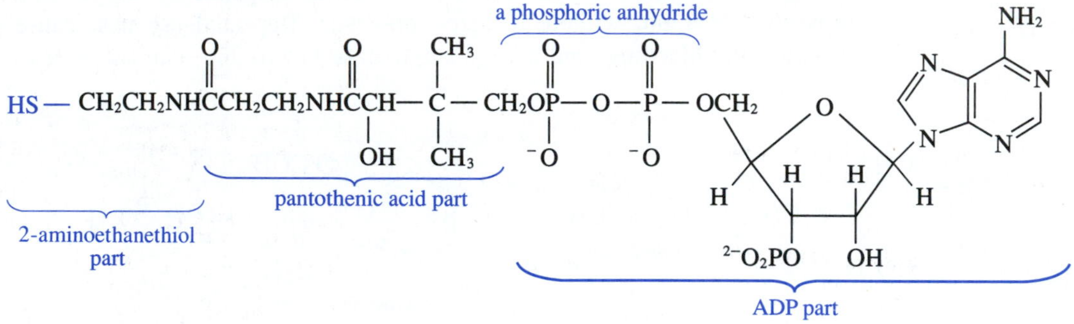

FIGURE 10.1 Coenzyme A.

10.21 Amides

Amides are the least reactive of the common carboxylic acid derivatives. They occur widely in nature. The most important amides are the proteins, whose chemistry we will discuss later. Here we will concentrate on just a few properties of simple amides.

Primary amides have the general formula $RCONH_2$. They can be prepared by the reaction of ammonia with esters (eq. 10.22), with acyl halides (eq. 10.33), or with acid anhydrides (eq. 10.35). Amides can also be prepared by heating the ammonium salts of acids.

$$R{-}\overset{O}{\overset{\|}{C}}{-}OH + NH_3 \longrightarrow \underset{\text{ammonium salt}}{R{-}\overset{O}{\overset{\|}{C}}{-}O^{-}NH_4^{+}} \xrightarrow{\text{heat}} \underset{\text{amide}}{R{-}\overset{O}{\overset{\|}{C}}{-}NH_2} + H_2O \qquad (10.37)$$

Amides are named by replacing the *-ic* or *-oic* ending of the acid name, either the common or the IUPAC name, with the *-amide* ending.

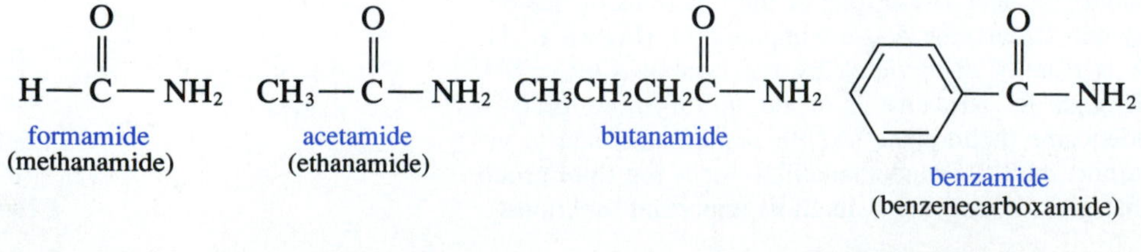

$$\underset{\substack{\text{formamide}\\\text{(methanamide)}}}{H{-}\overset{O}{\overset{\|}{C}}{-}NH_2} \quad \underset{\substack{\text{acetamide}\\\text{(ethanamide)}}}{CH_3{-}\overset{O}{\overset{\|}{C}}{-}NH_2} \quad \underset{\text{butanamide}}{CH_3CH_2CH_2\overset{O}{\overset{\|}{C}}{-}NH_2} \quad \underset{\substack{\text{benzamide}\\\text{(benzenecarboxamide)}}}{C_6H_5{-}\overset{O}{\overset{\|}{C}}{-}NH_2}$$

Problem 10.28 a. Name $(CH_3)_2CHCONH_2$.
b. Write the structure of 1-methylcyclopropancarboxamide.

The above examples are all primary amides. Secondary and tertiary amides, in which one or both of the hydrogens on the nitrogen atom are replaced by organic groups, are described in the next chapter.

Amides have a planar geometry. Even though the carbon–nitrogen bond is normally written as a single bond, rotation around that bond is restricted, because resonance is very important in amides.

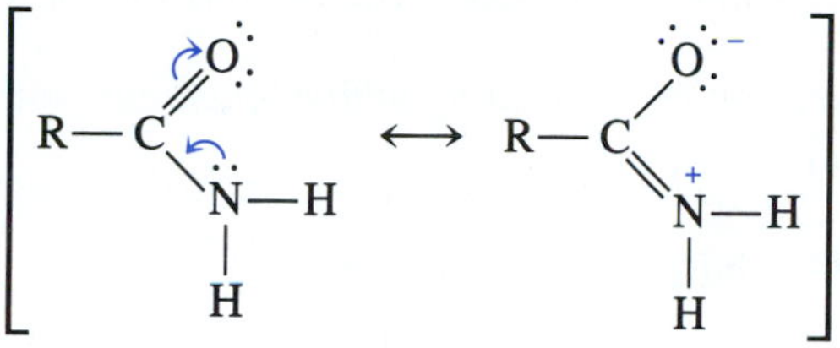

$$\left[R-C(=\ddot{O}:)-\ddot{N}H_2 \longleftrightarrow R-C(-\ddot{O}:^-)=\overset{+}{N}H_2 \right]$$

amide resonance

The dipolar contributor is so important that the carbon-nitrogen bond behaves much like a double bond. Consequently, the nitrogen and the carbonyl carbon, and the two atoms attached to each of them, lie in the same plane, and rotation at the C—N bond is restricted. Indeed, the C—N bond in amides is only 1.32 Å long—much shorter than the usual carbon-nitrogen single bond (which is about 1.47 Å).

As the dipolar resonance contributor suggests, amides are highly polar and form strong hydrogen bonds.

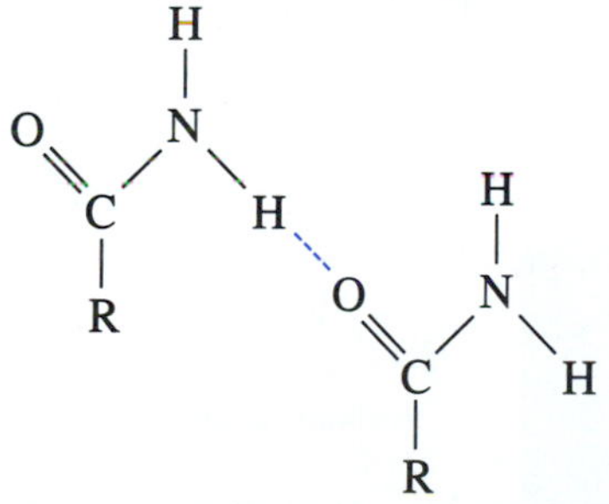

$$R-C(=O)-N(H)-H\cdots O=C(R)-NH_2 \quad \text{or} \quad \text{cyclic dimer: } R-C(=O\cdots H-N(H)-)\ldots(-N(H)-H\cdots O=)C-R$$

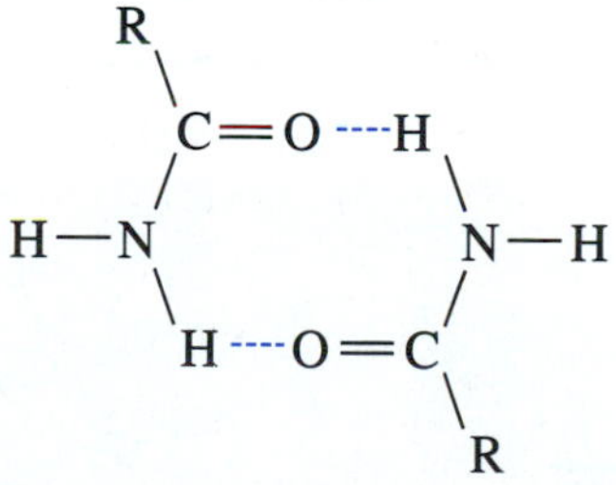

They have exceptionally high boiling points for their molecular weights, although alkyl substitution on the nitrogen lowers the boiling and melting points by decreasing the hydrogen-bonding possibilities.

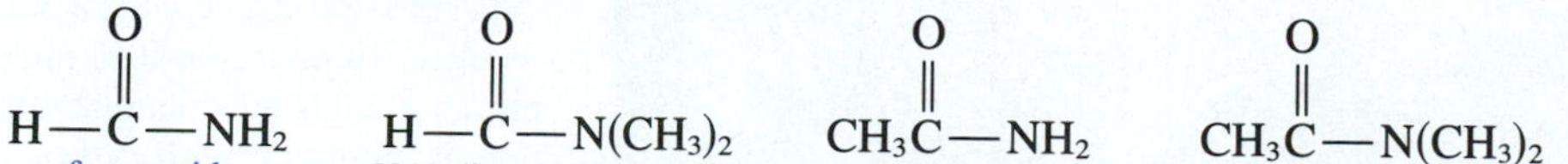

	$H-\overset{O}{\overset{\|}{C}}-NH_2$ formamide	$H-\overset{O}{\overset{\|}{C}}-N(CH_3)_2$ *N,N*-dimethylformamide	$CH_3\overset{O}{\overset{\|}{C}}-NH_2$ acetamide	$CH_3\overset{O}{\overset{\|}{C}}-N(CH_3)_2$ *N,N*-dimethylacetamide
bp	210°C	153°C	222°C	165°C
mp	2.5°C	−60.5°C	81°C	−20°C

Problem 10.29 Show that hydrogen bonding is possible for acetamide, but not for *N,N*-dimethylacetamide.

Like other acid derivatives, amides react with nucleophiles. For example, they can be hydrolyzed by water.

$$\underset{\text{amide}}{R{-}\overset{\overset{\displaystyle O}{\|}}{C}{-}NH_2} + H{-}OH \xrightarrow[OH^-]{H^+ \text{ or}} \underset{\text{acid}}{R{-}\overset{\overset{\displaystyle O}{\|}}{C}{-}OH} + NH_3 \tag{10.38}$$

The reactions are slow, however. Prolonged heating and acid or base catalysis are usually necessary.

Problem 10.30 Using eq. 10.38 as a model, write an equation for the hydrolysis of acetamide.

Amides can be reduced by lithium aluminum hydride to give amines.

$$\underset{\text{amide}}{R{-}\overset{\overset{\displaystyle O}{\|}}{C}{-}NH_2} \xrightarrow[\text{ether}]{LiAlH_4} \underset{\text{amine}}{RCH_2NH_2} \tag{10.39}$$

This is an excellent way to make primary amines, whose chemistry is discussed in the next chapter.

Problem 10.31 Outline steps for the synthesis of benzylamine

$$C_6H_5{-}CH_2NH_2$$

starting with benzoic acid.

A WORD ABOUT ...

23. Urea

Urea is a special amide, the diamide of carbonic acid.

$$\underset{\text{carbonic acid}}{HO{-}\overset{\overset{\displaystyle O}{\|}}{C}{-}OH} \qquad \underset{\substack{\text{urea}\\ \text{mp } 133°C}}{H_2N{-}\overset{\overset{\displaystyle O}{\|}}{C}{-}NH_2}$$

A colorless, water-soluble, crystalline solid, urea is the normal end product of protein metabolism. An average adult excretes approximately 30 g of urea in the urine daily.

Urea is produced commercially, mainly for use as a fertilizer (it contains 40% nitrogen by weight). Over 2 billion pounds are manufactured annually by the reaction of ammonia with carbon dioxide.

$$CO_2 + 2\ NH_3 \xrightarrow[\text{pressure}]{150\text{–}200°C} H_2N{-}\overset{\overset{\displaystyle O}{\|}}{C}{-}NH_2 + H_2O$$

Urea is also used as a raw material in the manufacture of certain drugs and plastics.

10.22 A Summary of Carboxylic Acid Derivatives

We have studied a rather large number of reactions in this chapter. However, most of them can be summarized in a single chart, shown in Table 10.5.

The four types of acid derivatives are listed at the left of the chart in order of decreasing reactivity toward nucleophiles. Three common nucleophiles are listed across the top. Note that the main organic product in each column is the same, regardless of which type of acid derivative we start with. For example, **hydrolysis** gives the corresponding organic acid, whether we start with an acyl halide, acid anhydride, ester, or amide. Similarly, **alcoholysis** gives an ester, and **ammonolysis** gives an amide. Note also that the *other* reaction product is generally the same from a given acid derivative (horizontally across the table), regardless of the nucleophile. For example, starting with an ester, RCO_2R', we obtain as the second product the alcohol R′OH, regardless of whether the reaction type is hydrolysis, alcoholysis, or ammonolysis.

All of the reactions in Table 10.5 take place via attack of the nucleophile on the carbonyl carbon of the acid derivative, as described in eq. 10.28. Indeed, most of the reactions from Secs. 10.11 through 10.20 occur by that same mechanism. We can sometimes use this idea to predict new reactions.

TABLE 10.5 Reactions of acid derivatives with certain nucleophiles

Acid derivative (↓ decreasing reactivity)	Nucleophile: HOH (hydrolysis)	R′OH (alcoholysis)	NH_3 (ammonolysis)
$R-C(=O)-Cl$ acyl halide	$R-C(=O)-OH + HCl$	$R-C(=O)-OR' + HCl$	$R-C(=O)-NH_2 + NH_4^+Cl^-$
$R-C(=O)-O-C(=O)-R$ acid anhydride	$2\,R-C(=O)-OH$	$R-C(=O)-OR' + RCO_2H$	$R-C(=O)-NH_2 + RCO_2H$
$R-C(=O)-O-R''$ ester	$R-C(=O)-OH + R''OH$	$R-C(=O)-OR' + R''OH$ (ester interchange)	$R-C(=O)-NH_2 + R''OH$
$R-C(=O)-NH_2$ amide	$R-C(=O)-OH + NH_3$	$R-C(=O)-OR' + NH_3$	no reaction
Main organic product	acid	ester	amide

EXAMPLE 10.9 The reaction of esters with Grignard reagents (Sec. 10.16) involves nucleophilic attack of the Grignard reagent on the ester carbonyl group. Keeping in mind that all acid derivatives are susceptible to nucleophilic attack, predict the reaction of an acyl halide with a Grignard reagent.

Solution The first step is as follows:

$$R-\overset{\overset{O}{\|}}{C}-Cl + R'MgX \longrightarrow R-\underset{\underset{R'}{|}}{\overset{\overset{O^-\overset{+}{MgX}}{|}}{C}}-Cl \longrightarrow R-\overset{\overset{O}{\|}}{C}-R' + MgXCl$$

The ketone can sometimes be isolated, but usually it reacts with a second mole of Grignard reagent to give a tertiary alcohol.

$$R-\overset{\overset{O}{\|}}{C}-R' + R'MgX \longrightarrow R-\underset{\underset{R'}{|}}{\overset{\overset{O^-\overset{+}{MgX}}{|}}{C}}-R' \xrightarrow{H_3O^+} R-\underset{\underset{R'}{|}}{\overset{\overset{OH}{|}}{C}}-R'$$

This reaction is analogous to eq. 10.24 for esters.

Problem 10.32 Modified metal hydrides, such as lithium tri-*t*-butoxyaluminum hydride $Li^{+-}Al[OC(CH_3)_3]_3H$, react with acyl chlorides to give aldehydes. Write an equation for the reaction of *p*-nitrobenzoyl chloride with this reagent, and suggest a possible reaction mechanism.

10.23 The α-Hydrogen of Esters; the Claisen Condensation

In this final section, we describe an important reaction of esters that resembles the aldol condensation of aldehydes and ketones (Sec. 9.18). It makes use of the **α-hydrogen** of an ester.

Being adjacent to a carbonyl group, **the α-hydrogens of an ester can be removed by a strong base. The product is an ester enolate.**

$$-\underset{|}{\overset{\overset{H}{|}\alpha}{C}}-C\overset{\diagup\!\!\diagup O}{\underset{\diagdown OR}{}} \underset{}{\overset{\text{base}}{\rightleftharpoons}} \left[-\underset{|}{\overset{\ominus\,\ddot{}}{C}}-C\overset{\diagup\!\!\diagup \ddot{O}:}{\underset{\diagdown OR}{}} \longleftrightarrow -\underset{|}{C}=C\overset{\diagup :\ddot{O}:^{\ominus}}{\underset{\diagdown OR}{}} \right] \quad (10.40)$$

resonance contributors to an ester enolate

Common bases used for this purpose are sodium alkoxides or sodium hydride. The ester enolate, once formed, can act as a carbon nucleophile and add to the carbonyl group of another ester molecule. This reaction is called the **Claisen condensation.** It is a way of making **β-keto esters.** We can use ethyl acetate as an example to see how the reaction works.

Treatment of ethyl acetate with sodium ethoxide in ethanol produces the β-keto ester **ethyl acetoacetate:**

$$\underset{\text{ethyl acetate}}{CH_3\overset{O}{\overset{\|}{C}}—OCH_2CH_3} + \underset{\text{ethyl acetate}}{H—\overset{\alpha}{CH_2}—\overset{O}{\overset{\|}{C}}—OCH_2CH_3} \xrightarrow[\text{in ethanol}]{NaOCH_2CH_3}$$

$$\underset{\substack{\text{ethyl acetoacetate}\\ \text{(ethyl 3-oxobutanoate)}}}{CH_3\overset{O}{\overset{\|}{C}}—CH_2—\overset{O}{\overset{\|}{C}}—OCH_2CH_3} + CH_3CH_2OH \quad (10.41)$$

This reaction takes place in three steps.

Step 1.

$$CH_3\overset{O}{\overset{\|}{C}}—OCH_2CH_3 + \underset{\text{sodium ethoxide}}{Na^{+}\,{}^{-}OCH_2CH_3} \rightleftharpoons$$

$$\underset{\text{ester enolate}}{Na^{+}\,{}^{-}CH_2\overset{O}{\overset{\|}{C}}OCH_2CH_3} + CH_3CH_2OH \quad (10.42)$$

Step 2.

$$CH_3\overset{O}{\overset{\|}{C}}—OCH_2CH_3 + {}^{-}CH_2\overset{O}{\overset{\|}{C}}OCH_2CH_3 \rightleftharpoons$$

$$CH_3\overset{O^{-}}{\overset{|}{\underset{\underset{\overset{\|}{O}}{CH_2C—OCH_2CH_3}}{\underset{|}{C}}}}—OCH_2CH_3 \rightleftharpoons CH_3\overset{O}{\overset{\|}{C}}CH_2\overset{O}{\overset{\|}{C}}OCH_2CH_3 + {}^{-}OCH_2CH_3 \quad (10.43)$$

Step 3.

$$CH_3\overset{O}{\overset{\|}{C}}CH_2\overset{O}{\overset{\|}{C}}OCH_2CH_3 + {}^{-}OCH_2CH_3 \longrightarrow$$

$$\underset{\text{enolate ion of a }\beta\text{-keto ester}}{CH_3\overset{O}{\overset{\|}{C}}\cdots\overset{\ominus}{CH}\cdots\overset{O}{\overset{\|}{C}}OCH_2CH_3} + CH_3CH_2OH \quad (10.44)$$

In Step 1, the base (sodium ethoxide) removes an α-hydrogen from the ester to form an ester enolate. In Step 2, this ester enolate, acting as a nucleophile, adds to the carbonyl group of a second ester molecule, displacing ethoxide ion. This step follows the mechanism in eq. 10.28. These first two steps of the reaction are completely reversible.

Step 3 drives the equilibria forward. In this step, the β-keto ester is converted to its enolate anion. The methylene (CH_2) hydrogens in ethyl acetoacetate are *α to two carbonyl groups* and hence are appreciably more acidic than ordinary α-hydrogens. They are easily removed by the base (ethoxide ion) to form a resonance-stabilized β-keto enolate ion, *with the negative charge delocalized to both carbonyl oxygen atoms.*

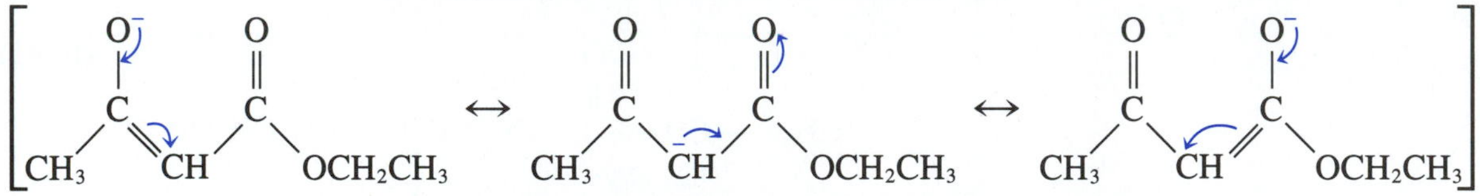

$$\left[CH_3\overset{O^-}{\overset{|}{C}}{=}CH\overset{O}{\overset{\|}{C}}OCH_2CH_3 \longleftrightarrow CH_3\overset{O}{\overset{\|}{C}}\overset{-}{C}H\overset{O}{\overset{\|}{C}}OCH_2CH_3 \longleftrightarrow CH_3\overset{O}{\overset{\|}{C}}CH{=}\overset{O^-}{\overset{|}{C}}OCH_2CH_3 \right]$$

resonance contributors to ethyl acetoacetate enolate anion

To complete the Claisen condensation, the solution is acidified, to regenerate the β-keto ester from its enolate anion.

EXAMPLE 10.10 Identify the product of the Claisen condensation of ethyl propanoate:

$$CH_3CH_2\overset{O}{\overset{\|}{C}}{-}OCH_2CH_3$$

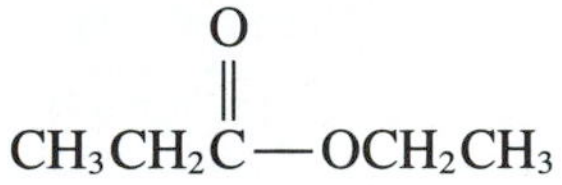

Solution The product is

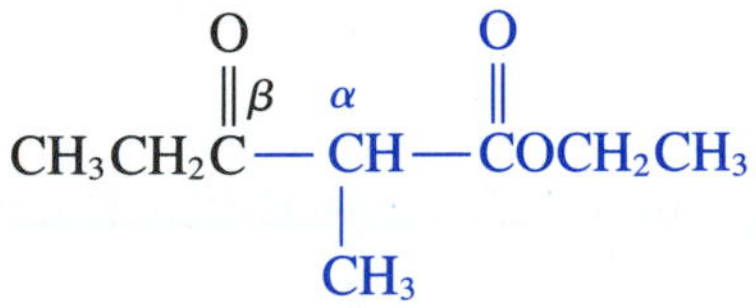

$$CH_3CH_2\overset{O}{\overset{\|\beta}{C}}{-}\overset{\alpha}{C}H{-}\overset{O}{\overset{\|}{C}}OCH_2CH_3 \quad (\text{with } CH_3 \text{ on the } \alpha\text{-carbon})$$

The α-carbon of one ester molecule displaces the —OR group and becomes joined to the carbonyl carbon of the other ester. The product is always a β-keto ester.

Problem 10.33 Using eqs. 10.42 through 10.44 as a model, write out the steps in the mechanism for the Claisen condensation of ethyl propanoate.

The Claisen condensation, like the aldol condensation, is useful for making new carbon-carbon bonds. The resulting β-keto esters can be converted to a variety of useful products. For example, ethyl acetate can be converted to ethyl butanoate by the following sequence.

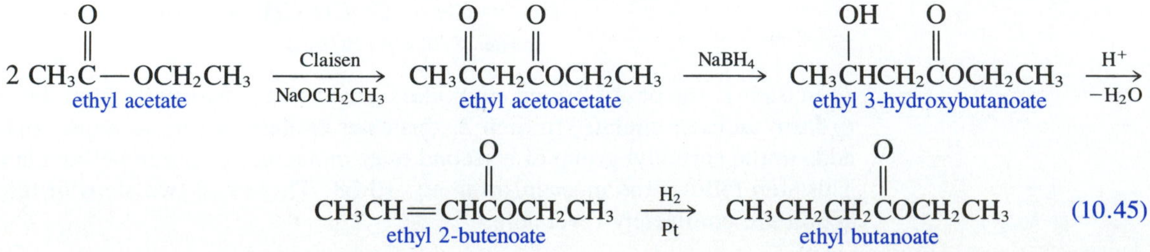

$$\underset{\text{ethyl acetate}}{2\,CH_3\overset{O}{\overset{\|}{C}}{-}OCH_2CH_3} \xrightarrow[NaOCH_2CH_3]{\text{Claisen}} \underset{\text{ethyl acetoacetate}}{CH_3\overset{O}{\overset{\|}{C}}CH_2\overset{O}{\overset{\|}{C}}OCH_2CH_3} \xrightarrow{NaBH_4} \underset{\text{ethyl 3-hydroxybutanoate}}{CH_3\overset{OH}{\overset{|}{C}}HCH_2\overset{O}{\overset{\|}{C}}OCH_2CH_3} \xrightarrow[-H_2O]{H^+}$$

$$\underset{\text{ethyl 2-butenoate}}{CH_3CH{=}CH\overset{O}{\overset{\|}{C}}OCH_2CH_3} \xrightarrow[Pt]{H_2} \underset{\text{ethyl butanoate}}{CH_3CH_2CH_2\overset{O}{\overset{\|}{C}}OCH_2CH_3} \qquad (10.45)$$

In this way, the acetate chain is lengthened by two carbon atoms. Nature makes use of a similar process, catalyzed by various enzymes, to construct the long-chain carboxylic acids that are components of fats and oils (Chapter 15).

Additional Problems

10.34. Write a structural formula for each of the following acids:

a. 3-methylpentanoic acid
b. 2,2-dichlorobutanoic acid
c. 4-hydroxyhexanoic acid
d. *p*-toluic acid
e. cyclobutanecarboxylic acid
f. 2-propanoylbenzoic acid
g. phenylacetic acid
h. 2-naphthoic acid
i. 2,3-dimethyl-3-butenoic acid
j. 3-oxobutanoic acid
k. 2,2-dimethylbutanedioic acid
l. α-methyl-γ-butyrolactone

10.35. Name each of the following acids:

a. $(CH_3)_2CHCH_2CH_2COOH$
b. $CH_3CHBrCH(CH_3)COOH$
c. 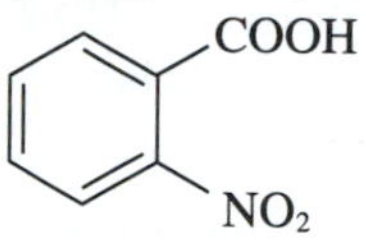 [benzene ring with COOH and ortho NO_2]
d. $CH_3CH(C_6H_5)COOH$
e. $CH_2{=}CHCOOH$
f. 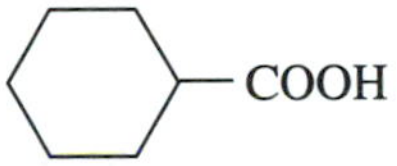 [cyclohexane ring]—COOH
g. CH_3CF_2COOH
h. 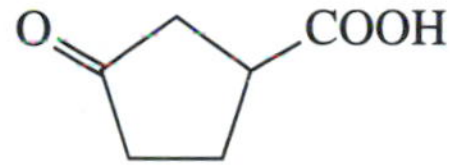 [cyclopentanone ring with O= and COOH]
i. 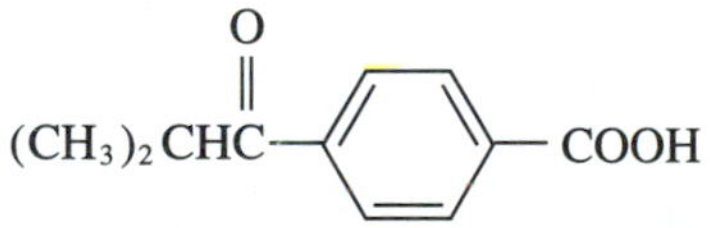 $(CH_3)_2CHC(=O)$—[benzene ring]—COOH
j. $HC{\equiv}CCH_2CO_2H$

10.36. *Ibuprofen,* an anti-inflammatory agent used to treat rheumatoid arthritis and other diseases, is chemically called 2(*p*-isobutylphenyl)propanoic acid. Draw its structure.

10.37. Which will have the higher boiling point? Explain your reasoning.

a. CH_3CH_2COOH or $CH_3CH_2CH_2CH_2OH$
b. $CH_3CH_2CH_2CH_2COOH$ or $(CH_3)_3CCOOH$

10.38. In each of the following pairs of acids, which would be expected to be the stronger, and why?

a. CH_2ClCO_2H and CH_2BrCO_2H
b. *o*-$BrC_6H_4CO_2H$ and *m*-$BrC_6H_4CO_2H$
c. CCl_3CO_2H and CF_3CO_2H
d. $C_6H_5CO_2H$ and *p*-$CH_3OC_6H_4CO_2H$
e. $ClCH_2CH_2CO_2H$ and $CH_3CHClCO_2H$

10.39. Write a balanced equation for the reaction of

a. chloroacetic acid with potassium hydroxide.
b. decanoic acid with calcium hydroxide.

10.40. Give equations for the synthesis of

a. $CH_3CH_2CH_2CO_2H$ from $CH_3CH_2CH_2CH_2OH$.
b. $CH_3CH_2CH_2CO_2H$ from $CH_3CH_2CH_2OH$ (two ways).
c. 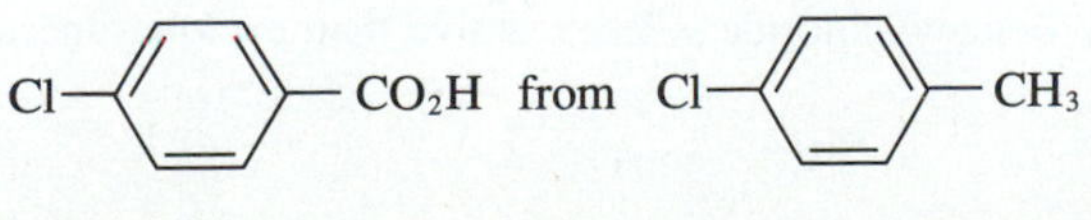 Cl—[benzene ring]—CO_2H from Cl—[benzene ring]—CH_3

d. cyclopentyl–CO_2H from cyclopentane

e. $CH_3OCH_2CO_2H$ from $CH_2—CH_2$ (epoxide, bridged by O) (two steps).

f. phenyl–CO_2H from phenyl–Br

10.41. The acid-catalyzed hydrolysis of an alkyl cyanide (eq. 10.13) involves, as the first step, nucleophilic attack of water on the protonated cyanide. Write out the steps of a plausible mechanism for the reaction.

10.42. The Grignard route for the synthesis of $(CH_3)_3CCO_2H$ from $(CH_3)_3CBr$ (Example 10.5) is far superior to the nitrile route. Explain why.

10.43. Write a structure for each of the following compounds:

a. sodium 2-chloropropanoate
b. calcium acetate
c. isopropyl acetate
d. ethyl formate
e. phenyl benzoate
f. benzonitrile
g. propanoic anhydride
h. *m*-toluamide
i. 4-chlorobutanoyl chloride
j. 3-formylcyclopentanecarboxylic acid

10.44. Name each of the following compounds:

a. Cl–C_6H_4–$COO^-NH_4^+$ (para-substituted benzene ring)

b. $[CH_3(CH_2)_2CO_2^-]_2Ca^{2+}$

c. $(CH_3)_2CHCOOC_6H_5$

d. $CF_3CO_2CH_3$

e. CH_3COSH

f. CH_3COSCH_3

g. $HCONH_2$

h. cyclopropyl–C(=O)–O–C(=O)–cyclopropyl

10.45. Write out each step in the Fischer esterification of benzoic acid with methanol (you may wish to use eq. 10.17 as a model).

10.46. Write an equation for the reaction of ethyl benzoate with

a. hot aqueous sodium hydroxide.
b. ammonia (heat).
c. propylmagnesium bromide (two equivalents), then H_3O^+.
d. lithium aluminum hydride (two equivalents), then H_3O^+.

10.47. Write out all the steps in the mechanism for

a. saponification of $CH_3CH_2CO_2CH_3$.
b. ammonolysis of $CH_3CH_2CO_2CH_3$.

10.48. Explain each difference in reactivity toward nucleophiles.

a. Esters are less reactive than ketones.
b. Benzoyl chloride is less reactive than cyclohexanecarbonyl chloride.

10.49. Identify the Grignard reagent and the ester that would be used to prepare

a. $CH_3CH_2CH_2{-}\overset{\displaystyle OH}{\underset{\displaystyle C_6H_5}{\overset{|}{\underset{|}{C}}}}{-}CH_2CH_2CH_3$ **b.** $CH_3CH_2CH_2C(C_6H_5)_2OH$

10.50. Write an equation for

a. hydrolysis of acetyl chloride.
b. reaction of benzoyl chloride with methanol.
c. esterification of 1-pentanol with acetic anhydride.
d. ammonolysis of 4-bromobutanoyl bromide.
e. Fischer esterification of pentanoic acid with ethanol.
f. 2-methylpropanoyl chloride + ethylbenzene + $AlCl_3$.
g. succinic acid + heat (235°C).
h. phthalic anhydride + methanol (1 equiv.) + H^+.
i. phthalic anhydride + methanol (excess) + H^+.
j. oxalyl chloride + ammonia (excess).

10.51. Complete the equation for each of the following reactions:

a. $CH_3CH_2CH_2CO_2H + PCl_5 \rightarrow$
b. $CH_3(CH_2)_8CO_2H + SOCl_2 \rightarrow$
c. 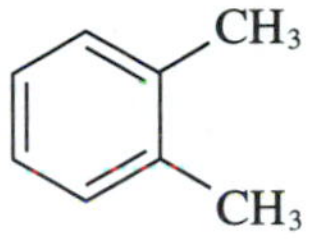 (benzene ring with two adjacent CH_3 groups) $+ KMnO_4 \longrightarrow$
d. 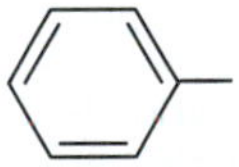 (benzene ring)$-CO_2^-NH_4^+$ + heat $\longrightarrow$
e. $CH_3(CH_2)_5CONH_2 + LiAlH_4 \rightarrow$
f. 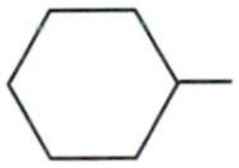 (cyclohexane ring)$-CO_2CH_2CH_3 + LiAlH_4 \longrightarrow$

10.52. Write the important resonance contributors to the structure of propanamide, and tell which atoms will lie in a single plane.

10.53. Adapt the explanation (given in "A Word About Thioesters", page 293) for the enhanced reactivity of thioesters over ordinary esters to account for the fact that thioesters are *less* reactive toward nucleophiles than are acid anhydrides or acyl chlorides.

10.54. Considering the relative reactivities of ketones and esters toward nucleophiles, which of the following products seems the more likely?

$$CH_3\overset{\overset{\displaystyle O}{\|}}{C}CH_2CH_2CO_2CH_3 \xrightarrow{NaBH_4} CH_3\overset{\overset{\displaystyle O}{\|}}{C}CH_2CH_2CH_2OH \quad \text{or} \quad CH_3\overset{\overset{\displaystyle OH}{|}}{C}HCH_2CH_2CO_2CH_3$$

10.55. Mandelic acid, which has the formula $C_6H_5CH(OH)COOH$, can be isolated from bitter almonds (called *Mandel* in German). It is sometimes used in medicine to treat urinary infections. Devise a two-step synthesis of mandelic acid from benzaldehyde, using the latter's cyanohydrin as an intermediate.

10.56. Consider the structure of the catnip ingredient nepatalactone (page 286).
a. Show with dotted lines that the structure is composed of two isoprene units.
b. Circle the chiral centers and determine their configurations (*R* or *S*).

10.57. A phenolic acid, $C_9H_8O_3$, exists in two isomeric forms. Both rapidly decolorize permanganate and, on moderate oxidation, yield salicylic and oxalic acids as the only organic products. One isomer easily loses water, when heated, to yield $C_9H_6O_2$. The other fails to dehydrate under the same conditions. Suggest structural formulas for the isomers.

10.58. When maleic acid is heated at reflux with a little concentrated hydrochloric acid, it is gradually converted to fumaric acid. Explain how this isomerization might occur.

10.59. Write the structure of the Claisen condensation product of ethyl phenylacetate, and show the steps in its formation.

10.60. Diethyl adipate, when heated with sodium ethoxide, gives the product shown, by an *intra*molecular Claisen condensation.

$$CH_3CH_2O\overset{\overset{\large O}{\|}}{C}-(CH_2)_4-\overset{\overset{\large O}{\|}}{C}OCH_2CH_3 \xrightarrow{NaOCH_2CH_3} \text{ethyl 2-oxocyclopentanecarboxylate}$$

diethyl adipate

ethyl 2-oxocyclopentanecarboxylate

Write out the steps in a plausible mechanism for the reaction.

10.61. Analogous to the mixed aldol condensation (Sec. 9.19), mixed Claisen condensations are possible. Predict the structure of the product obtained when a mixture of ethyl benzoate and ethyl acetate is heated with sodium ethoxide in ethanol. How is it formed?

CHAPTER 11

Amines and Related Nitrogen Compounds

11.1 Introduction

In this chapter, we discuss the last of the major, simple functionalities—the **amines.**

Amines are organic relatives of ammonia, derived by replacing one, two, or all three hydrogens of ammonia with organic groups. Like ammonia, amines are bases. In fact, *amines are the most important type of organic base that occurs in nature.* In this chapter, we will first describe the structure, preparation, chemical properties, and uses of some simple amines. Later in the chapter, we will discuss a few natural and synthetic amines with important biological activity.

11.2 Classification and Structure of Amines

The relation between ammonia and amines is illustrated by the following structures:

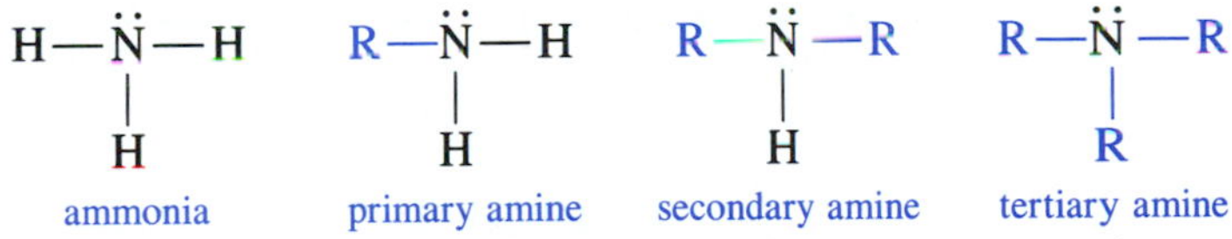

For convenience, amines are classified as **primary, secondary,** or **tertiary,** depending on whether one, two, or three organic groups are attached to the nitrogen. The *R* groups in these structures may be alkyl or aryl, and when two or more *R* groups are present, they may be identical to or different from one another. In some secondary and tertiary amines, the nitrogen may be part of a ring.

Problem 11.1 Classify each of the following amines as primary, secondary, or tertiary:

a. $(CH_3)_3CNH_2$

b. (five-membered ring containing N–H)

c. CH_3–(benzene ring)–NH_2

d. $(CH_3)_2N$–(benzene ring)

FIGURE 11.1

(a) An orbital view of the pyramidal bonding in trimethylamine. (b) Top view of a space-filling model of trimethylamine. The center ball represents the orbital with the unshared electron pair.

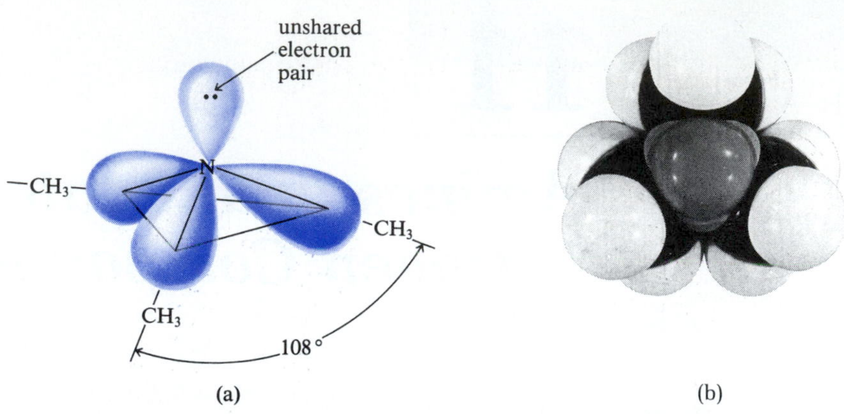

The nitrogen atom in amines is trivalent. In addition, the nitrogen carries an unshared electron pair. The nitrogen orbitals are therefore sp^3-hybridized, and the overall geometry is pyramidal (nearly tetrahedral), as shown for trimethylamine in Figure 11.1. From this geometry, one might think that an amine with three different groups attached to the nitrogen would be chiral, with the unshared electron pair acting as the fourth group. This is true in principle, but in practice the two enantiomers usually interconvert rapidly through inversion, via an "umbrella-in-the-wind" type of process, and are not resolvable.

$$N(R_1)(R_2)(R_3) \rightleftharpoons \left[R_1\text{—}N(R_2)(R_3) \right] \rightleftharpoons N(R_1)(R_2)(R_3) \quad (11.1)$$

planar transition state

11.3 Nomenclature of Amines

Amines can be named in several different ways. Commonly, simple amines are named by specifying the alkyl groups attached to the nitrogen and adding the suffix *-amine*.

$CH_3CH_2NH_2$	$(CH_3CH_2)_2NH$	$(CH_3CH_2)_3N$
ethylamine	diethylamine	triethylamine
(primary)	(secondary)	(tertiary)

In the IUPAC system, the **amino group, —NH_2,** is named as a substituent, as in the following examples:

$CH_3CH_2NH_2$
aminoethane

$\overset{1}{C}H_3\overset{2}{C}H(NH_2)\overset{3}{C}H_2\overset{4}{C}H_2\overset{5}{C}H_3$
2-aminopentane

H_2N, NH_2, H, H (cyclobutane ring)
cis-1,3-diaminocyclobutane

In this system, secondary or tertiary amines are named by using a prefix that includes all but the longest carbon chain, as in

$CH_3NH\overset{1}{C}H_2\overset{2}{C}H_2\overset{3}{C}H_3$
1-methylaminopropane

$CH_3N(CH_2CH_3)—\overset{1}{C}H_2\overset{2}{C}H_2\overset{3}{C}H_3$
1-(ethylmethylamino)propane

$(CH_3)_2N$–cyclohexane
dimethylaminocyclohexane

Recently, *Chemical Abstracts* (CA) introduced a system for naming amines that is rational and easy to use. In this system, amines are named as **alkanamines.** For example,

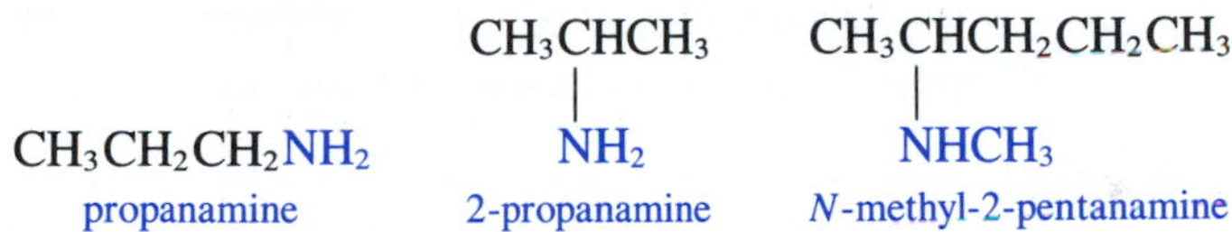

propanamine 2-propanamine *N*-methyl-2-pentanamine

EXAMPLE 11.1 Name (cyclohexyl)—$N(CH_3)_2$ by the *Chemical Abstracts* system.

Solution The largest alkyl group attached to nitrogen is used as the root of the name. The compound is *N,N*-dimethylcyclohexanamine.

Problem 11.2 Name $CH_3CH_2CHCH_2CH_3$ (with $N(CH_3)_2$ on the central carbon) by the CA system.

When other functional groups are present, the amino group is named as a substituent:

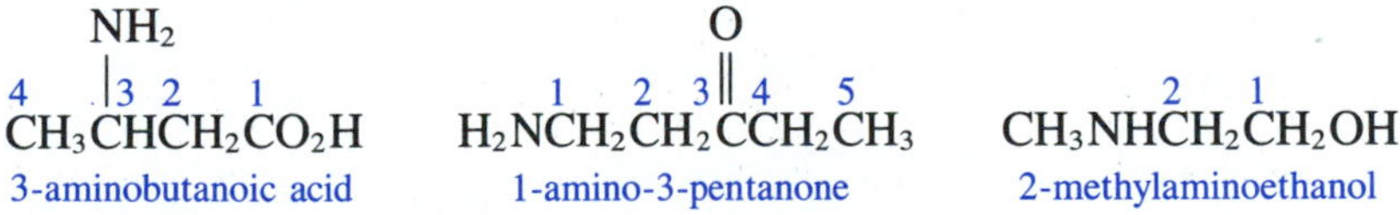

3-aminobutanoic acid 1-amino-3-pentanone 2-methylaminoethanol

Aromatic amines are named as derivatives of aniline. In the CA system, aniline is called benzenamine; these CA names are shown in parentheses.

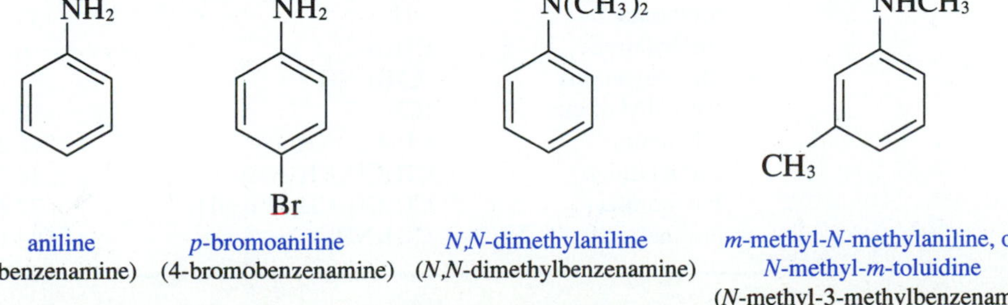

aniline (benzenamine) *p*-bromoaniline (4-bromobenzenamine) *N,N*-dimethylaniline (*N,N*-dimethylbenzenamine) *m*-methyl-*N*-methylaniline, or *N*-methyl-*m*-toluidine (*N*-methyl-3-methylbenzenamine)

EXAMPLE 11.2 Give an acceptable name for the following compounds:

a. $(CH_3)_2CHCH_2NH_2$ b. $CH_3NHCH_2CH_3$

c. (3,5-dibromo-substituted benzene ring with NH_2; substituents: Br, Br, NH_2)

d. (cyclopentane ring bearing OH (wedge) and NH_2 (hashed))

Solution a. isobutylamine (common); 1-amino-2-methylpropane (IUPAC); 2-methylpropanamine (CA).

b. ethylmethylamine (common); methylaminoethane (IUPAC); *N*-methylethanamine (CA).

c. 3,5-dibromoaniline (common, IUPAC); 3,5-dibromobenzenamine (CA).

d. *trans*-2-aminocyclopentanol (only name).

Problem 11.3 Give an acceptable name for the following compounds:

a. $(CH_3)_3CNH_2$ b. $H_2NCH_2CH_2OH$

c. O_2N—(benzene ring)—NH_2

Problem 11.4 Write the structure for

a. dipropylamine. b. 3-aminohexane.
c. pentamethylaniline. d. *N,N*-dimethyl-2-butanamine.

11.4 Physical Properties of Amines

Table 11.1 lists the boiling points of some common amines. Methylamine and ethylamine are gases, but primary amines with three or more carbons are liquids. Primary amines boil well above alkanes with comparable molecular weights, but below comparable alcohols, as shown in Table 11.2. Intermolecu-

TABLE 11.1 The boiling points of some simple amines

Name	*Formula*	*bp, °C*
ammonia	NH_3	−33.4
methylamine	CH_3NH_2	−6.3
dimethylamine	$(CH_3)_2NH$	7.4
trimethylamine	$(CH_3)_3N$	2.9
ethylamine	$CH_3CH_2NH_2$	16.6
propylamine	$CH_3CH_2CH_2NH_2$	48.7
butylamine	$CH_3CH_2CH_2CH_2NH_2$	77.8
aniline	$C_6H_5NH_2$	184.0

TABLE 11.2 A comparison of alkane, amine, and alcohol boiling points*

alkane	CH_3CH_3 (30) bp −88.6°C	$CH_3CH_2CH_3$ (44) bp −42.1°C
amine	CH_3NH_2 (31) bp −6.3°C	$CH_3CH_2NH_2$ (45) bp +16.6°C
alcohol	CH_3OH (32) bp +65.0°C	CH_3CH_2OH (46) bp +78.5°C

* Molecular weights are given in parentheses.

lar N—H···N hydrogen bonds are important and raise the boiling points of primary and secondary amines but are not as strong as the O—H···O bonds of alcohols. The reason for this is that nitrogen is not as electronegative as oxygen.

Problem 11.5 Explain why the tertiary amine $(CH_3)_3N$ boils so much lower than its primary isomer $CH_3CH_2CH_2NH_2$.

All three classes of amines can form hydrogen bonds with the —OH group of water (that is, O—H···N). Thus, most simple amines with up to five or six carbon atoms are either completely or appreciably soluble in water.

Now we will describe some ways that amines can be prepared.

11.5 Preparation of Amines; Alkylation of Ammonia and Amines

Ammonia reacts with alkyl halides to give amines via a two-step process. The first step is a nucleophilic substitution reaction (S_N2).

$$\underset{\text{ammonia}}{H_3N\!:} + R\text{—}X \longrightarrow \underset{\text{alkylammonium halide}}{R\text{—}\overset{+}{N}H_3 + X^-} \tag{11.2}$$

The free amine can then be obtained from its salt by treatment with a strong base.

$$R\text{—}\overset{+}{N}H_3\ X^- + NaOH \longrightarrow \underset{\text{primary amine}}{RNH_2} + H_2O + Na^+X^- \tag{11.3}$$

Primary, secondary and tertiary amines can be similarly alkylated.

$$\underset{\text{primary amine}}{R\ddot{N}H_2} + R\text{—}X \longrightarrow R_2\overset{+}{N}H_2\ X^- \xrightarrow{NaOH} \underset{\text{secondary amine}}{R_2NH} \tag{11.4}$$

$$\underset{\text{secondary amine}}{R_2\ddot{N}H} + R\text{—}X \longrightarrow R_3\overset{+}{N}H\ X^- \xrightarrow{NaOH} \underset{\text{tertiary amine}}{R_3N} \tag{11.5}$$

$$R_3\ddot{N} + R{-}X \longrightarrow R_4N^+X^- \qquad (11.6)$$

tertiary amine → quaternary ammoniun salt

Unfortunately, mixtures of products are often obtained in these reactions because the starting ammonia or amine and the alkylammonium ion formed in the S_N2 step can equilibrate, as in the following equation:

$$NH_3 + R\overset{+}{N}H_3\,X^- \rightleftharpoons NH_4^+\,X^- + RNH_2 \qquad (11.7)$$

So, in the reaction of ammonia with an alkyl halide (eq. 11.2), some primary amine is formed (eq. 11.7), and it may be further alkylated (eq. 11.4) to give a secondary amine, and so on. By adjusting the ratio of the reactants, however, a good yield of one desired amine may be obtained. For example, with a large excess of ammonia, the primary amine is the major product.

Aromatic amines can often be alkylated selectively.

$$C_6H_5NH_2 \xrightarrow{CH_3I} C_6H_5NHCH_3 \xrightarrow{CH_3I} C_6H_5N(CH_3)_2 \qquad (11.8)$$

aniline, *N*-methylaniline, *N,N*-dimethylaniline

The alkylation can be intramolecular, as in the following final step in a laboratory synthesis of nicotine:

$$\text{(3-pyridyl)CH(Br)}{-}CH_2{-}CH_2{-}CH_2{-}\ddot{N}H{-}CH_3 \xrightarrow{-HBr} \text{nicotine} \qquad (11.9)$$

nicotine

EXAMPLE 11.3 Write an equation for the synthesis of benzylamine, $C_6H_5{-}CH_2NH_2$

Solution

$$C_6H_5{-}CH_2X + 2\,\ddot{N}H_3 \longrightarrow C_6H_5{-}CH_2\ddot{N}H_2 + NH_4^+X^-$$

(X = Cl, Br, or I)

Use of excess ammonia helps prevent further substitution.

Problem 11.6 Complete equations for the following reactions.

a. $CH_3CH_2CH_2Br + 2\ NH_3 \longrightarrow$
b. $CH_3CH_2I + 2(CH_3CH_2)_2NH \longrightarrow$
c. $(CH_3)_3N + CH_3I \longrightarrow$
d. $CH_3CH_2CH_2NH_2 + C_6H_5{-}CH_2Br \longrightarrow$

Problem 11.7 Give a synthesis of $C_6H_5{-}NHCH_2CH_3$ from aniline.

11.6 Preparation of Amines; Reduction of Nitrogen Compounds

All bonds to the nitrogen atom in amines are either N—H or N—C bonds. Nitrogen in ammonia or amines is therefore in a reduced form. It is not surprising, then, that organic compounds in which a nitrogen atom is present in a more oxidized form can be reduced to amines by appropriate reducing agents. Several examples of this useful synthetic approach to amines are described here.

The best route to aromatic primary amines is by reduction of the corresponding nitro compounds, which are in turn prepared by electrophilic aromatic nitration. The nitro group is easily reduced, either catalytically with hydrogen or by chemical reducing agents.

$$CH_3{-}C_6H_4{-}NO_2 \xrightarrow[SnCl_2,\ HCl]{3H_2,\ \text{Ni catalyst} \text{ or}} CH_3{-}C_6H_4{-}NH_2 + 2H_2O \qquad (11.10)$$

p-nitrotoluene → *p*-toluidine

EXAMPLE 11.4 Devise a synthesis of *p*-chloroaniline, $Cl{-}C_6H_4{-}NH_2$, from chlorobenzene.

Solution Chlorobenzene is first nitrated; —Cl is an *o,p*-directing group, so the major product is *p*-chloronitrobenzene. This product is then reduced.

$$C_6H_5Cl \xrightarrow[H_2SO_4]{HONO_2} Cl{-}C_6H_4{-}NO_2 \xrightarrow[Ni]{H_2} Cl{-}C_6H_4{-}NH_2$$

Problem 11.8 Give a synthesis for $H_2N-C_6H_3(NH_2)-CH_3$ (2,4-diaminotoluene, with NH_2 ortho to CH_3) from toluene.

As described in the previous chapter (eq. 10.36), *amides can be reduced to amines with lithium aluminum hydride.*

$$R-\overset{\overset{\displaystyle O}{\|}}{C}-N\begin{matrix} R' \\ R'' \end{matrix} \xrightarrow{LiAlH_4} RCH_2N\begin{matrix} R' \\ R'' \end{matrix} \quad \text{(R' and R'' may be H or organic groups)} \tag{11.11}$$

Depending on the structures of R′ and R″, we can obtain primary, secondary or tertiary amines in this way.

■ ***EXAMPLE 11.5*** Complete the equation $CH_3\overset{\overset{\displaystyle O}{\|}}{C}NHCH_2CH_3 \xrightarrow{LiAlH_4}$

Solution The C=O group is reduced to CH_2. The product is the secondary amine $CH_3CH_2NHCH_2CH_3$.

Problem 11.9. Show how $CH_3CH_2N(CH_3)_2$ can be synthesized from an amide.

Reduction of nitriles (cyanides) gives primary amines.

$$R-C\equiv N \xrightarrow[\text{or } H_2,\ Ni]{LiAlH_4} RCH_2NH_2 \tag{11.12}$$

■ ***EXAMPLE 11.6*** Complete the equation $NCCH_2CH_2CH_2CH_2CN \xrightarrow[\text{Ni catalyst}]{\text{excess } H_2}$

Solution Both CN groups are reduced. The product $H_2N-(CH_2)_6-NH_2$, or 1,6-diaminohexane, is one of two raw materials for the manufacture of nylon (p. 382).

Problem 11.10 Devise a synthesis of $C_6H_5-CH_2CH_2NH_2$ from $C_6H_5-CH_2Br$

Aldehydes and ketones undergo ***reductive amination*** when treated with ammonia, primary, or secondary amines, to give primary, secondary, or tertiary amines, respectively. The most commonly used laboratory reducing agent for this purpose is the metal hydride sodium cyanoborohydride, $NaBH_3(CN)$.

$$\gt C{=}\ddot{O}: + R\ddot{N}H_2 \underset{}{\overset{-H_2O}{\rightleftharpoons}} \left[\gt C{=}NR \right] \xrightarrow{NaBH_3CN} \gt CHNHR \qquad (11.13)$$

aldehyde or ketone — primary amine — imine — secondary amine

The reaction involves nucleophilic attack on the carbonyl group, leading to an imine (in the case of ammonia or primary amines; compare with eq. 9.28) or an iminium ion with secondary amines. The reducing agent then reduces the C=N bond.

Problem 11.11 Using eq. 11.13 as a guide, devise a synthesis of 3-aminopentane from 3-pentanone.

Now that we know several ways to make amines, let us examine some of their properties.

11.7 *The Basicity of Amines*

The unshared pair of electrons on the nitrogen atom dominates the chemistry of amines. Because of this electron pair, amines are both basic and nucleophilic.

Aqueous solutions of amines are basic because of the following equilibrium:

$$\gt N: + H{-}\ddot{O}H \rightleftharpoons \gt N^{+}{-}H + {}^{-}:\ddot{O}H \qquad (11.14)$$

amine — ammonium ion — hydroxide ion

EXAMPLE 11.7 Write an equation that shows why aqueous solutions of ethylamine are basic.

Solution

$$CH_3CH_2\ddot{N}H_2 + H_2O \rightleftharpoons CH_3CH_2\overset{+}{N}H_3 + OH^-$$

ethylamine — ethylammonium ion

Amines are more basic than water. They accept a proton from water, producing hydroxide ion, so their solutions are basic.

Problem 11.12 Write an equation representing the equilibrium in an aqueous solution of trimethylamine.

An amine and its ammonium ion (eq. 11.14) *are related as a* base *and its* conjugate acid. For example, RNH_3^+ is the conjugate acid of the primary amine RNH_2. It is convenient, when comparing basicities of different amines, to compare instead the acidity constants (pKa's) of their conjugate acids. Eq. 11.15 expresses this acidity for a primary alkylammonium ion.

$$\underset{\text{conjugate acid}}{R\overset{+}{N}H_3} + H_2O \rightleftharpoons \underset{\text{base}}{RNH_2} + H_3O^+ \tag{11.15}$$

$$K_a = \frac{[RNH_2][H_3O^+]}{[RNH_3^+]}$$

The larger the K_a (or the smaller the pK_a) the stronger $R\overset{+}{N}H_3$ is as an acid, or the weaker RNH_2 is as a base.

EXAMPLE 11.8 The p*Ka*'s of NH_4^+ and $CH_3\overset{+}{N}H_3$ are 9.30 and 10.64, respectively. Which is the stronger base, NH_3 or CH_3NH_2?

Solution NH_4^+ is the stronger acid (lower pK_a). Therefore, NH_3 is the *weaker* base, and CH_3NH_2 is the *stronger*.

Table 11.3 lists some amine basicities. Alkylamines are approximately 10 times as basic as ammonia. Recall that alkyl groups are electron-donating relative to hydrogen $R \rightarrow \ddot{N}\begin{smallmatrix} H \\ \\ H \end{smallmatrix}$. This electron-donating effect stabilizes the am-

TABLE 11.3 Basicities of some common amines, expressed as pK_a of the corresponding ammonium ions.

	Formula		
Name	*Amine*	*Ammonium Ion*	*pK_a of the ammonium ions*
ammonia	$\ddot{N}H_3$	$\overset{+}{N}H_4$	9.30
methylamine	$CH_3\ddot{N}H_2$	$CH_3\overset{+}{N}H_3$	10.64
dimethylamine	$(CH_3)_2\ddot{N}H$	$(CH_3)_2\overset{+}{N}H_2$	10.71
trimethylamine	$(CH_3)_3\ddot{N}$	$(CH_3)_3\overset{+}{N}H$	9.77
ethylamine	$CH_3CH_2\ddot{N}H_2$	$CH_3CH_2\overset{+}{N}H_3$	10.67
propylamine	$CH_3CH_2CH_2\ddot{N}H_2$	$CH_3CH_2CH_2\overset{+}{N}H_3$	10.58
aniline	$C_6H_5\ddot{N}H_2$	$C_6H_5\overset{+}{N}H_3$	4.62
N-methylaniline	$C_6H_5\ddot{N}HCH_3$	$C_6H_5\overset{+}{N}H_2(CH_3)$	4.85
N,N-dimethylaniline	$C_6H_5\ddot{N}(CH_3)_2$	$C_6H_5\overset{+}{N}H(CH_3)_2$	5.04
p-chloroaniline	$p\text{-}ClC_6H_4\ddot{N}H_2$	$p\text{-}ClC_6H_4\overset{+}{N}H_3$	3.98

monium ion (positive charge) relative to the free amine (eq. 11.14). Hence it decreases the acidity of the ammonium ion, or increases the basicity of the amine. In general, *electron-donating groups increase the basicity of amines, and electron-withdrawing groups decrease their basicity.*

Problem 11.13 Do you expect $ClCH_2CH_2NH_2$ to be a stronger or weaker base than $CH_3CH_2NH_2$? Explain.

Aromatic amines are much weaker bases than aliphatic amines or ammonia. For example, aniline is less basic than cyclohexylamine by nearly a million times.

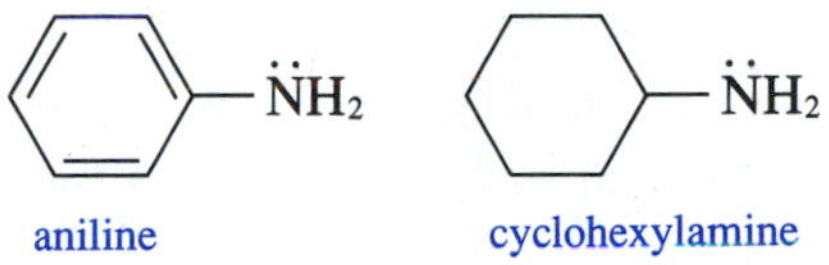

aniline

cyclohexylamine

The reason for this huge difference is the resonance delocalization of the unshared electron pair that is possible in aniline but not in cyclohexylamine.

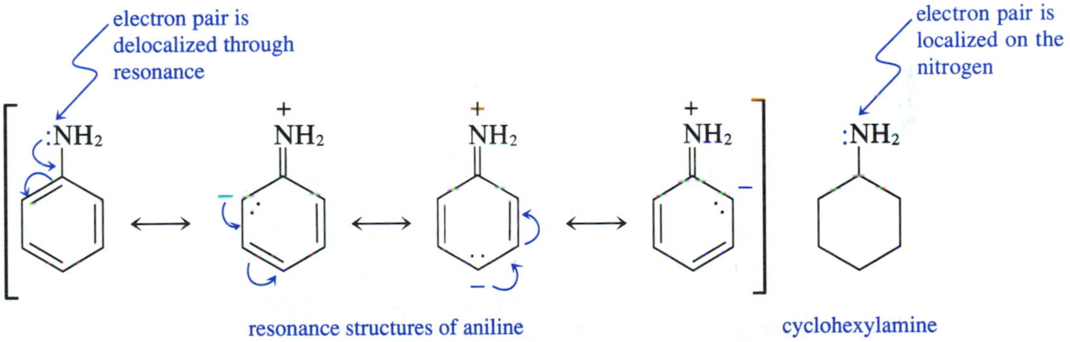

electron pair is delocalized through resonance

electron pair is localized on the nitrogen

resonance structures of aniline

cyclohexylamine

Resonance stabilizes the unprotonated form of aniline. This shifts the equilibrium in eq. 11.15 to the right, increasing the acidity of the anilinium ion or decreasing the basicity of aniline. Another way to describe the situation is to say that the unshared electron pair in aniline is delocalized and therefore less available for donation to a proton than is the electron pair in cyclohexylamine.

Problem 11.14 Compare the basicities of the last four amines in Table 11.3, and explain the reasons for the observed basicity order.

Problem 11.15 Place the following amines in order of increasing basicity: aniline, *p*-nitroaniline, *p*-toluidine.

11.8 Comparison of the Basicity of Amines and Amides

Both amines and amides have nitrogens with an unshared electron pair. There is a huge difference, however, in their basicities. Aqueous solutions of amines are basic; aqueous solutions of amides are essentially neutral. Why this striking difference?

The answer lies in their structures, as illustrated in the following comparison of a primary amine with a primary amide:

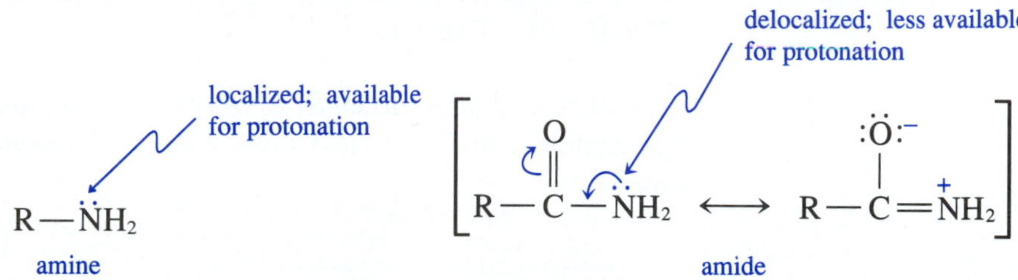

In the amine, the electron pair is mainly localized on the nitrogen. In the amide, the electron pair is delocalized to the carbonyl oxygen. The effect of this delocalization is seen in the low pK_a values for amides, compared with those for amines, for example:

$CH_3CH_2NH_2$	$CH_3C(=O)NH_2$
ethylamine	acetamide
pK_a = 10.67	pK_a = −0.6

Primary and secondary amines and amides have N—H bonds, and one might expect that they would on occasion behave as acids (proton donors).

$$R{-}\ddot{N}H_2 \rightleftharpoons R{-}\ddot{N}H^- + H^+ \qquad K_a \cong 10^{-40} \tag{11.16}$$

Primary amines are exceedingly weak acids, much weaker than alcohols. Their pK_a is about 40, compared with about 16 for alcohols. The main reason for the difference is that nitrogen is much less electronegative than oxygen and thus cannot stabilize a negative charge nearly as well.

Amides, on the other hand, are *much* stronger acids than amines; in fact, their pK_a (about 15) is comparable to that of alcohols:

$$R{-}\overset{O}{\overset{\|}{C}}{-}\ddot{N}H_2 \rightleftharpoons \left[R{-}\overset{\ddot{O}:}{\overset{\|}{C}}{-}\ddot{N}H^- \longleftrightarrow R{-}\overset{:\ddot{O}:^-}{\overset{|}{C}}{=}\ddot{N}H \right] + H^+ \qquad K_a \cong 10^{-15} \tag{11.17}$$

amidate anion

One reason is that the negative charge of the **amidate anion** can be delocalized through resonance. Another reason is that the nitrogen in an amide carries a partial positive charge, making it easy to lose the attached proton, which is also positive.

It is important to understand these differences between amines and amides, not only because they involve important chemical principles, but also because they help us understand the chemistry of certain natural products, such as peptides and proteins.

Problem 11.16 Place the following compounds in order of increasing basicity; in order of increasing acidity.

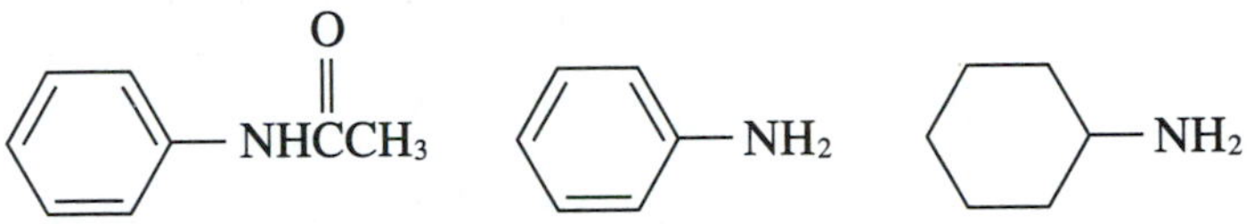

C_6H_5—$NHCCH_3$ (C=O) acetanilide; C_6H_5—NH_2 aniline; C_6H_{11}—NH_2 cyclohexylamine

11.9 Reaction of Amines with Strong Acids; Amine Salts

Amines react with strong acids to form **alkylammonium salts.** An example of this reaction for a primary amine and HCl is as follows:

$$\underset{\text{primary amine}}{R\text{—}\ddot{N}H_2} + HCl \longrightarrow \underset{\text{an alkylammonium chloride}}{R\overset{+}{N}H_3\ Cl^-} \qquad (11.18)$$

■ *EXAMPLE 11.9* Complete the following acid-base reactions, and name the products.

a. $CH_3CH_2NH_2 + HBr \longrightarrow$ b. $(CH_3)_3N + HCl \longrightarrow$

Solution

a. $$\underset{\text{ethylamine}}{CH_3CH_2\ddot{N}H_2} + HBr \longrightarrow \underset{\text{ethylammonium bromide}}{CH_3CH_2N^+H_3\ Br^-}$$

b. $$\underset{\text{trimethylamine}}{(CH_3)_2\ddot{N}\text{—}CH_3} + HCl \longrightarrow \underset{\text{trimethylammonium chloride}}{(CH_3)_3N^+\text{—}H\ Cl^-}$$

Problem 11.17 Complete the following equation, and name the product.

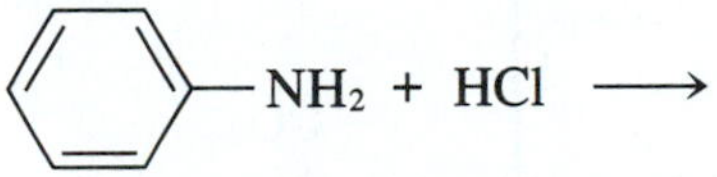

$$C_6H_5\text{—}NH_2 + HCl \longrightarrow$$

This type of reaction is used to separate or extract amines from neutral or acidic water-insoluble substances. Consider, for example, a mixture of *p*-toluidine and *p*-nitrotoluene, which might arise from a preparation of the amine that for some reason does not go to completion (eq. 11.10). The amine can be separated from the unreduced nitro compound by the following scheme:

p-toluidine (bp 200°C) + p-nitrotoluene (bp 238°C) —shake with aqueous HCl and ether→ ether layer: p-nitrotoluene (evaporate the ether); water layer: amine salt ($\overset{+}{N}H_3\ Cl^-$) —$NaOH$, H_2O→ free amine + $Na^+Cl^- + H_2O$ (11.19)

The mixture, neither component of which is water soluble, is dissolved in an inert, low-boiling solvent such as ether and is shaken with aqueous hydrochloric acid. The amine reacts to form a salt, which is ionic and dissolves in the water layer. The nitro compound does not react and remains in the ether layer. The two layers are then separated. The nitro compound can be recovered by evaporating the ether. The amine can be recovered from its salt by making the aqueous layer alkaline with a strong base such as NaOH.

11.10 Chiral Amines as Resolving Agents

Amines also form salts with organic acids. This reaction is used to resolve enantiomeric acids (Sec. 5.13). For example, (*R*)- and (*S*)-lactic acids can be resolved by reaction with a chiral amine such as (*S*)-1-phenylethylamine:

[(*R*)-lactic acid + (*S*)-lactic acid] + (*S*)-1-phenylethylamine ⟶ [(*R,S*) salt + (*S,S*) salt] (11.20)

The salts are diastereomers, not enantiomers, and can be separated by ordinary methods, such as fractional crystallization. Once separated, each salt can be treated with a strong acid, such as HCl, to liberate one enantiomer of lactic acid. For example,

$$(R,S)\text{ salt} + \text{HCl} \longrightarrow \underset{(R)\text{-lactic acid}}{\text{H}(\text{HO})\text{C}(\text{CO}_2\text{H})\text{CH}_3} + \underset{(S)\text{-1-phenylethylammonium chloride}}{\text{CH}_3(\text{H})\text{C}(\overset{+}{\text{N}}\text{H}_3\ \text{Cl}^-)\text{C}_6\text{H}_5} \qquad (11.21)$$

The chiral amine can be recovered for reuse by treating its salt with sodium hydroxide (as in the last step of eq. 11.19).

Numerous chiral amines are available from natural products and can be used to resolve acids. Conversely, some chiral acids are available to resolve amine enantiomers.

So far we have considered reactions in which amines act as bases. Now we will examine some reactions in which they act as nucleophiles.

11.11 Acylation of Amines with Acid Derivatives

Amines are nitrogen nucleophiles. They react with the carbonyl group of acid derivatives (acyl halides, anhydrides and esters) by nucleophilic acyl substitution (Sec. 10.12).

Looked at from the viewpoint of the amine, we can say that the N—H bond in primary and secondary amines can be **acylated** by acid derivatives. For example, primary and secondary amines react with acyl halides to form amides (compare with eq. 10.32).

$$\underset{\text{acyl halide}}{\text{R—}\overset{\text{O}}{\overset{\|}{\text{C}}}\text{—Cl}} + \underset{\text{primary amine}}{\text{H}_2\ddot{\text{N}}\text{—R}'} \longrightarrow \underset{\text{secondary amide}}{\text{R—}\overset{\text{O}}{\overset{\|}{\text{C}}}\text{—NHR}'} + \text{HCl} \qquad (11.22)$$

$$\underset{\text{acyl halide}}{\text{R—}\overset{\text{O}}{\overset{\|}{\text{C}}}\text{—Cl}} + \underset{\text{secondary amine}}{\text{H}\ddot{\text{N}}(\text{R}')(\text{R}'')} \longrightarrow \underset{\text{tertiary amide}}{\text{R—}\overset{\text{O}}{\overset{\|}{\text{C}}}\text{—N}(\text{R}')(\text{R}'')} + \text{HCl} \qquad (11.23)$$

If the amine is inexpensive, two equivalents are used—one to form the amide and the second to neutralize the HCl. Alternatively, an inexpensive base may be added for the latter purpose. This can be sodium hydroxide (especially if R is *aromatic*), or a tertiary amine; having no N—H bonds, tertiary amines cannot be acylated, but they can neutralize the HCl.

■ ***EXAMPLE 11.10*** Using eq. 10.27 as a guide, write out the steps in the mechanism for eq. 11.22.

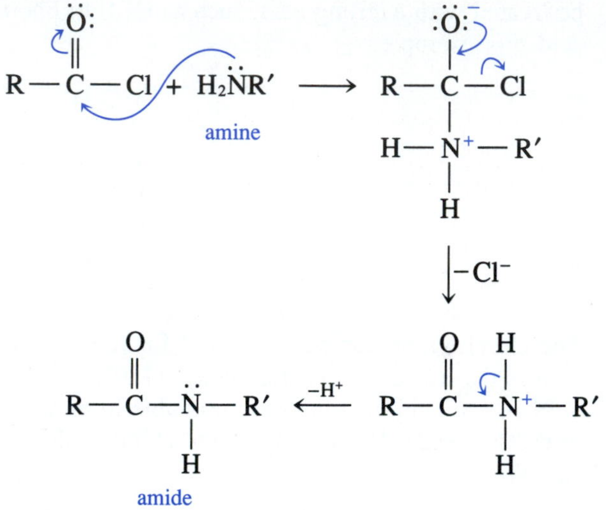

$$R{-}\overset{\overset{\displaystyle \ddot{O}:}{\|}}{C}{-}Cl + H_2\ddot{N}R' \longrightarrow R{-}\overset{\overset{\displaystyle :\ddot{O}:^-}{|}}{\underset{\underset{\displaystyle H{-}N^+{-}R'}{|}}{C}}{-}Cl$$

amine

$$\downarrow -Cl^-$$

$$\underset{\text{amide}}{R{-}\overset{\overset{\displaystyle O}{\|}}{C}{-}\underset{\underset{\displaystyle H}{|}}{\ddot{N}}{-}R'} \xleftarrow{-H^+} R{-}\overset{\overset{\displaystyle O}{\|}}{C}{-}\overset{\overset{\displaystyle H}{|}}{\underset{\underset{\displaystyle H}{|}}{N^+}}{-}R'$$

The first step involves nucleophilic addition to the carbonyl group. Elimination of HCl completes the substitution reaction.

Acylation of amines is put to practical use. For example, the insect repellent Off is the amide formed in the reaction of *m*-toluyl chloride and diethylamine.

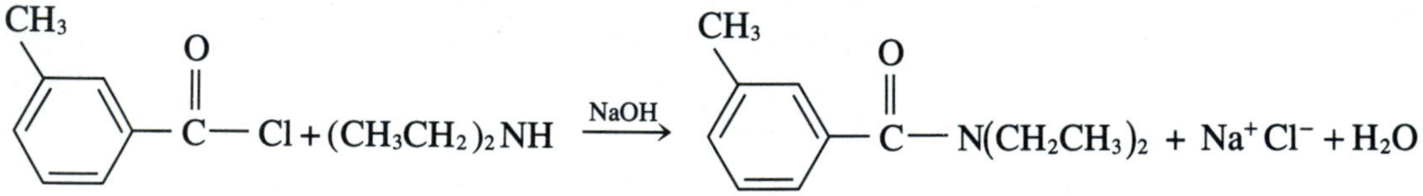

$$m\text{-}CH_3C_6H_4{-}\overset{\overset{\displaystyle O}{\|}}{C}{-}Cl + (CH_3CH_2)_2NH \xrightarrow{NaOH} m\text{-}CH_3C_6H_4{-}\overset{\overset{\displaystyle O}{\|}}{C}{-}N(CH_2CH_3)_2 + Na^+Cl^- + H_2O \quad (11.24)$$

m-toluyl chloride — diethylamine — *N,N*-diethyl-*m*-toluamide (the insect repellent Off)

Problem 11.18 Write out the steps in the mechanism for the synthesis of Off (eq. 11.24).

The antipyretic (fever-reducing substance) acetanilide is an amide made from aniline and acetic anhydride.

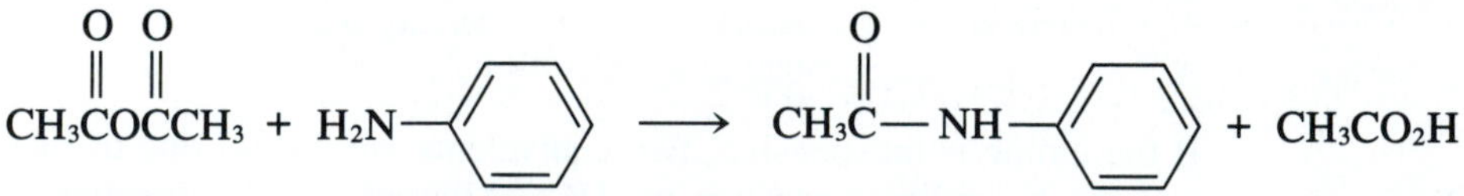

$$CH_3\overset{\overset{\displaystyle O}{\|}}{C}O\overset{\overset{\displaystyle O}{\|}}{C}CH_3 + H_2N{-}C_6H_5 \longrightarrow CH_3\overset{\overset{\displaystyle O}{\|}}{C}{-}NH{-}C_6H_5 + CH_3CO_2H \quad (11.25)$$

acetic anhydride — aniline — acetanilide

Problem 11.19 Write out the steps in the mechanism of the preparation of acetanilide from aniline and acetic anhydride (eq. 11.25).

Problem 11.20 Complete the following equation:

$$CH_3\overset{\overset{\displaystyle O}{\|}}{C}-O-\overset{\overset{\displaystyle O}{\|}}{C}CH_3 + HN\langle\text{piperidine ring}\rangle \longrightarrow$$

A classic laboratory test for distinguishing among primary, secondary, and tertiary amines, the **Hinsberg test,** is based on their reaction with **benzenesulfonyl chloride,** the acid chloride of benzenesulfonic acid.

$$C_6H_5-\overset{\overset{\displaystyle O}{\|}}{\underset{\underset{\displaystyle O}{\|}}{S}}-Cl + R\ddot{N}H_2 \longrightarrow C_6H_5-\overset{\overset{\displaystyle O}{\|}}{\underset{\underset{\displaystyle O}{\|}}{S}}-NHR \;(+\ HCl) \qquad (11.26)$$

benzenesulfonyl chloride; primary amine; a sulfonamide; this H is acidic; this type of sulfonamide is soluble in base

$$C_6H_5-\overset{\overset{\displaystyle O}{\|}}{\underset{\underset{\displaystyle O}{\|}}{S}}-Cl + R_2\ddot{N}H \longrightarrow C_6H_5-\overset{\overset{\displaystyle O}{\|}}{\underset{\underset{\displaystyle O}{\|}}{S}}-NR_2 \;(+\ HCl) \qquad (11.27)$$

secondary amine; a sulfonamide; no acidic H; this type of sulfonamide is insoluble in base

Primary amines form sulfonamides that have an N—H bond. This proton is acidic, and such sulfonamides are soluble in aqueous sodium hydroxide. Sulfonamides derived from secondary amines, have no N—H bond, are insoluble in aqueous base. Tertiary amines cannot eliminate HCl with sulfonyl chlorides (they have no N—H bond that can be acylated) and show no net reaction.

Problem 11.21 How could you distinguish between this pair of isomers?

$$CH_3-C_6H_4-NH_2 \quad \text{and} \quad C_6H_5-NHCH_3$$

A WORD ABOUT . . .

24. Sulfanilamide and Sulfa Drugs

Sulfa drugs were among the first antibiotics. They were developed about the time of World War II and are credited with saving thousands of lives by preventing infection in the wounded. Even today they still see use, although in many instances they have been replaced by more effective and safer antibiotics.

Sulfanilamide is the parent compound of all the sulfa drugs. Its synthesis, which begins with **acetanilide** (eq. 11.25), illustrates several important reaction types that we have already studied. The first step is an electrophilic aromatic substitution, very much like sulfonation (Sec. 4.10c), but using chlorosulfonic acid (the acid chloride of sulfuric acid) as the sulfonating agent.

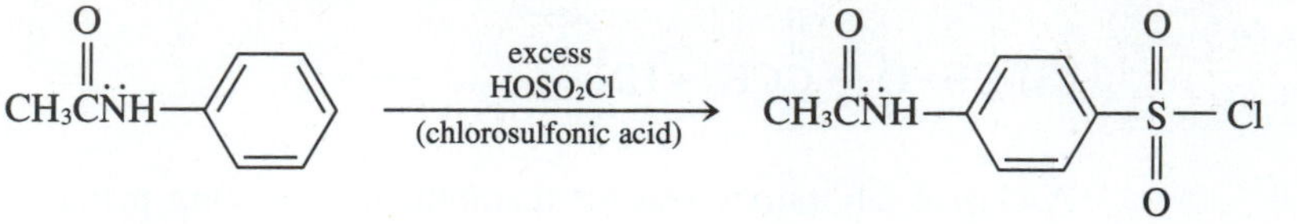

$$CH_3C(=O)\ddot{N}H-C_6H_5 \xrightarrow[\text{(chlorosulfonic acid)}]{\text{excess } HOSO_2Cl} CH_3C(=O)\ddot{N}H-C_6H_4-SO_2-Cl$$

acetanilide → 4-acetamidobenzenesulfonyl chloride

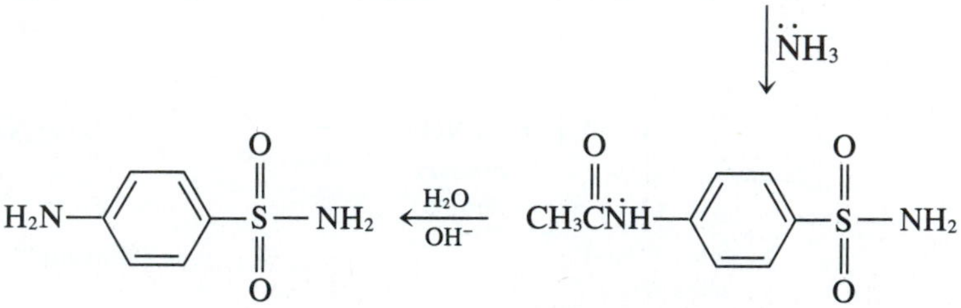

$$\downarrow \ddot{N}H_3$$

$$H_2N-C_6H_4-SO_2-NH_2 \xleftarrow[OH^-]{H_2O} CH_3C(=O)\ddot{N}H-C_6H_4-SO_2-NH_2$$

sulfanilamide ← 4-acetamidobenzenesulfonamide

The product is an aromatic sulfonyl chloride. Like other acid chlorides, it reacts with ammonia or amines to form amides—in this case, sulfonamides. The product contains two types of amide substituents, a carboxamide and a sulfonamide. In the third and final step of the synthesis, the carboxamide group is *selectively* hydrolyzed; the sulfonamide group is unaffected.

Sulfanilamide acts by binding to an enzyme site that is normally occupied by ***p*-aminobenzoic acid (PABA),** a compound with a similar structure and shape. PABA is used to synthesize folic acid, a compound essential for normal bacterial growth. By binding to the enzyme site, sulfanilamide blocks folic acid synthesis, and thus inhibits growth of the infecting bacteria.

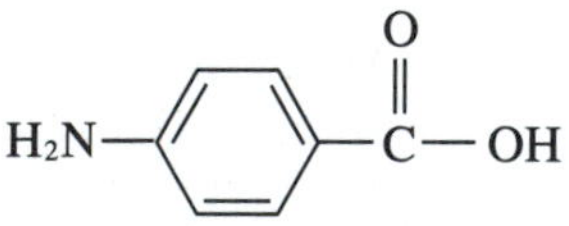

$$H_2N-C_6H_4-C(=O)-OH$$

p-aminobenzoic acid
(PABA)

One problem with sulfanilamide is its low solubility. It can crystallize out in the kidneys or urine, causing discomfort and tissue damage. Within a short time of its discovery, over 5000 analogs of sulfanilamide were synthesized, to correct the deficiencies of the parent drug. The structure was modified mainly by using various amines in place of ammonia, in the second step of the synthesis. **Sulfadiazine,** perhaps the most common of these analogs, is used to treat meningitis, dysentery and urinary infections.

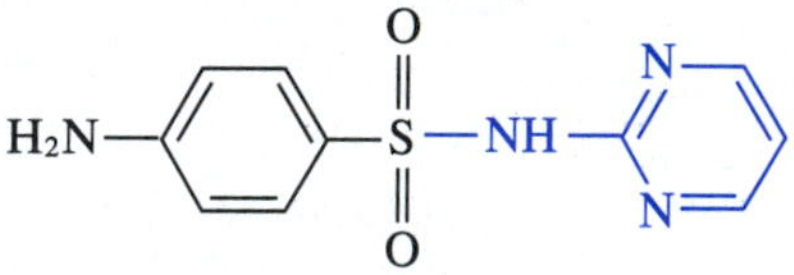

$$H_2N-C_6H_4-SO_2-NH-C_4H_3N_2$$

sulfadiazine

Sulfa drugs were discovered by accident. It was found that a dye called **prontosil,** used to stain bacteria, was an effective antibacterial agent.

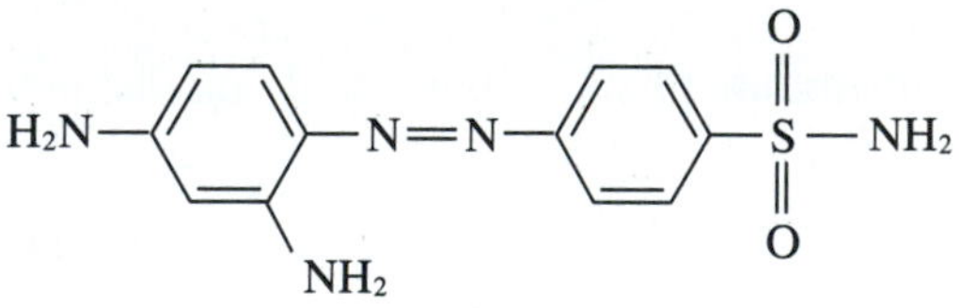

$$H_2N-C_6H_3(NH_2)-N{=}N-C_6H_4-SO_2-NH_2$$

prontosil

It was later discovered that the active agent was not prontosil itself, but sulfanilamide, formed from it in the cell. This discovery was then exploited rapidly and systematically, through the preparation of many variations on the basic structure in order to improve the pharmacological properties. Some of these variations led to new drugs with entirely new activity. This pattern of serendipitous discovery, followed by scientific development that engenders another new discovery, is not unusual in drug research.

11.12 Quaternary Ammonium Compounds

Tertiary amines react with primary or secondary alkyl halides by an S_N2 mechanism (eq. 11.6). The products are **quaternary ammonium salts,** in which all four hydrogens of ammonium ion are replaced by organic groups. For example,

$$(CH_3CH_2)_3\ddot{N} + C_6H_5CH_2{-}Cl \longrightarrow (CH_3CH_2)_3\overset{+}{N}CH_2{-}C_6H_5 + Cl^- \tag{11.28}$$

triethylamine; benzyl chloride; benzyltriethylammonium chloride

Quaternary ammonium compounds are important in biological processes. One of the most common natural quaternary ammonium ions is **choline,** which is present in phospholipids (Sec. 15.7).

$$(CH_3)_3\overset{+}{N}{-}CH_2CH_2OH\ OH^- \qquad (CH_3)_3\overset{+}{N}{-}CH_2CH_2{-}O{-}\overset{O}{\overset{\|}{C}}{-}CH_3\ OH^-$$

choline; acetylcholine

Choline not only is involved in various metabolic processes, but is also the precursor of **acetylcholine,** a compound that plays a key role in the transmission of nerve impulses.

11.13 Aromatic Diazonium Compounds

Primary aromatic amines react with nitrous acid at 0°C to yield ***aryldiazonium ions.*** *The process is called* ***diazotization.***

$$C_6H_5{-}NH_2 + HONO + H^+Cl^- \xrightarrow[\text{aqueous solution}]{0\text{–}5^\circ C} C_6H_5{-}N_2^+Cl^- + 2\,H_2O \tag{11.29}$$

aniline; nitrous acid; benzenediazonium chloride

Diazonium compounds are extremely useful synthetic intermediates. Before we describe their chemistry, let us try to understand the steps in eq. 11.29. First we need to examine the structure of nitrous acid.

Nitrous acid decomposes rather rapidly at room temperature. It is therefore prepared as needed by treating an aqueous solution of sodium nitrite with a strong acid at ice temperature. At that temperature, nitrous acid solutions are reasonably stable.

$$\underset{\text{sodium nitrite}}{Na^+NO_2^-} + H^+Cl^- \xrightarrow{0\text{–}5°C} \underset{\text{nitrous acid}}{H{-}O{-}\ddot{N}{=}\ddot{O}\!:} + Na^+Cl^- \quad (11.30)$$

The reactive species in reactions of nitrous acid is the **nitrosonium ion** NO^+. It is formed by protonation of the nitrous acid, followed by loss of water (compare with eq. 4.21):

$$H\ddot{O}{-}\ddot{N}{=}\ddot{O}\!: + H^+ \rightleftharpoons H\overset{\oplus}{\ddot{O}}(H){-}\ddot{N}{=}\ddot{O}\!: \rightleftharpoons H_2O + \underset{\text{nitrosonium ion}}{:\overset{\oplus}{N}{=}\ddot{O}\!:} \quad (11.31)$$

How do the two nitrogens, one from the amine and one from the nitrous acid, become bonded to one another, as they appear in diazonium ions? This happens in the first step of diazotization (eq. 11.29), which involves nucleophilic attack of the primary amine on the nitrosonium ion, followed by proton loss.

$$Ar\ddot{N}H_2 + :\overset{+}{N}{=}\ddot{O}\!: \longrightarrow ArN^+H_2{-}\ddot{N}{=}\ddot{O}\!: \rightleftharpoons \underset{\text{a primary nitrosamine}}{Ar\ddot{N}(H){-}\ddot{N}{=}\ddot{O}\!:} + H^+ \quad (11.32)$$

Protonation of the oxygen in the resulting nitrosamine, followed by elimination of water then gives the aromatic diazonium ion.

$$Ar\ddot{N}(H){-}\ddot{N}{=}\ddot{O}\!: + H^+ \longrightarrow Ar\overset{+}{N}(H){=}\ddot{N}{-}\ddot{O}H \xrightarrow{-H_2O} \underset{\text{aryldiazonium ion}}{Ar\overset{+}{N}{\equiv}N\!:} \quad (11.33)$$

Notice that in the final product there are no N—H bonds; both hydrogens of the amino group are lost, the first in eq. 11.32 and the second in eq. 11.33. Therefore, *only primary amines can be diazotized.* (Secondary and tertiary amines do react with nitrous acid, but their reactions are less important in synthesis.)

Solutions of aryldiazonium ions are moderately stable and can be kept at 0°C for several hours. They are useful in synthesis because the **diazonio group** (—N_2^+) can be replaced by nucleophiles; the other product is nitrogen gas.

$$Ar{-}\overset{+}{N}{\equiv}N\!: + Nu\!:^- \longrightarrow Ar{-}Nu + N_2 \quad (11.34)$$

Specific useful examples are shown in eq. 11.35. The nucleophile always takes the position on the benzene ring that was occupied by the diazonio group.

CN
Cl
H
KCN
$Cu_2(CN)_2$
$\overset{+}{N}_2$
HCl
Cu_2Cl_2
OH
H_3PO_2
H_2O, H^+
(11.35)
I
KI
HBr
Cu_2Br_2
Br
HBF_4,
heat
F

Conversion of diazonium compounds to aryl chlorides, bromides, or cyanides is usually accomplished using cuprous salts, and is known as the **Sandmeyer reaction.** Since a CN group is easily converted to a CO_2H group (eq. 10.13), this provides another route to aromatic carboxylic acids. The reaction with KI gives aryl iodides, usually not easily accessible by direct electrophilic iodination. Similarly, direct aromatic fluorination is difficult, but aromatic fluorides can be prepared from diazonium compounds and tetrafluoroboric acid, HBF_4.

Phenols can be prepared by adding diazonium compounds to hot aqueous acid. This reaction is important because there are not many ways to introduce an —OH group directly on an aromatic ring.

Finally, we sometimes use the orienting effect of a nitro or amino group and afterwards remove this substituent from the aromatic ring. This can be done by diazotization followed by reduction. A common reducing agent for this purpose is **hypophosphorous acid,** H_3PO_2.

Here are some examples of ways that diazonium compounds can be used in synthesis:

EXAMPLE 11.11 How can *m*-dibromobenzene be prepared?

Solution It *cannot* be prepared by direct electrophilic bromination of bromobenzene, because the Br group is *o,p*-directing (Sec. 4.12). But we can take advantage of the *m*-directing effect of a nitro group and then convert the nitro group to a bromine atom, as follows:

NO_2 (benzene) $\xrightarrow[FeBr_3]{Br_2}$ NO_2, Br (meta) $\xrightarrow[HCl]{SnCl_2}$ NH_2, Br $\xrightarrow[0°]{HONO}$ N_2^+, Br $\xrightarrow[Cu_2Br_2]{HBr}$ Br, Br

EXAMPLE 11.12 How can *o*-toluic (*o*-methylbenzoic) acid be prepared from *o*-toluidine (*o*-methylaniline)?

Solution

NH_2, CH_3 $\xrightarrow[2.\ Cu_2(CN)_2,\ KCN]{1.\ HONO}$ CN, CH_3 $\xrightarrow{H_3O^+}$ CO_2H, CH_3

o-toluidine — *o*-toluic acid

EXAMPLE 11.13 Design a route to 1,3,5-tribromobenzene from aniline.

Solution First brominate; the amino group is *o,p*-directing and ring-activating. Then remove the amino group by diazotization and reduction.

NH_2 $\xrightarrow{3Br_2}$ NH_2, Br, Br, Br $\xrightarrow[2.\ H_3PO_2]{1.\ HONO}$ Br, Br, Br

Problem 11.22 Design a synthesis of each of the following compounds, using a diazonium ion intermediate.

a. *m*-bromochlorobenzene from benzene
b. *m*-nitrophenol from *m*-nitroaniline
c. 2,4-difluorotoluene from toluene
d. 3,5-dibromotoluene from *p*-toluidine

11.14 Diazo Coupling; Azo Dyes

Being positively charged, aryldiazonium ions are electrophiles. They are *weak* electrophiles, however, because the positive charge can be delocalized through resonance.

EXAMPLE 11.14 Write the resonance contributors for the benzenediazonium ion that show how the nitrogen furthest from the benzene ring can become electrophilic.

Solution

$$\left[C_6H_5{-}\overset{+}{N}{\equiv}N: \longleftrightarrow C_6H_5{-}\ddot{N}{=}\overset{+}{N}: \right]$$

In the second contributor, the nitrogen at the right has only six electrons; it can react as an electrophile.

Problem 11.23 Draw resonance contributors which show that the positive charge in benzenediazonium ion can also be delocalized to the ortho and para carbons of the benzene ring. (CAREFUL! These contributors have two positive charges and one negative charge.)

Aryldiazonium ions react with strongly activated aromatic rings (phenols and aromatic amines) to give **azo compounds.** For example,

$$C_6H_5{-}\overset{+}{N}{\equiv}N: + C_6H_5{-}OH \xrightarrow{OH^-} C_6H_5{-}N{=}N{-}C_6H_4{-}OH + H_2O \quad (11.36)$$

benzenediazonium ion — phenol — *p*-hydroxyazobenzene, yellow leaflets, mp 155–157°C

The nitrogen atoms are retained in the product. This electrophilic aromatic substitution reaction is called **diazo coupling,** because in the product, two aromatic rings are coupled by the azo, or —N=N—, group. *Para* coupling is preferred, as in eq. 11.36, but if the *para* position is blocked by another substituent, *ortho* coupling can occur. *All azo compounds are colored, and many are used commercially as dyes for cloth and in color photography.*

EXAMPLE 11.15 Write out the steps in the mechanism for eq. 11.36.

Solution The reaction is an electrophilic aromatic substitution. The phenoxide ion, formed by dissociation of phenol, is readily attacked, even though the diazonium ion is a weak electrophile.

$$C_6H_5{-}\overset{+}{N}{\equiv}N: + C_6H_5{-}\ddot{O}:^- \longrightarrow \left[C_6H_5{-}\ddot{N}{=}\ddot{N}{-}C_6H_5(H){=}\ddot{O}: \right] \xrightarrow{-H^+,+H^+} C_6H_5{-}\ddot{N}{=}\ddot{N}{-}C_6H_4{-}\ddot{O}H$$

Problem 11.24 Methyl orange is an azo dye used as an indicator in acid-base titrations. (It is yellow-orange above pH 4.5 and red below pH 3.) Show how it can be synthesized from *p*-aminobenzenesulfonic acid (sulfanilic acid) and *N,N*-dimethylaniline.

$(CH_3)_2N-C_6H_4-N{=}N-C_6H_4-SO_3^-Na^+$

methyl orange

At this point, we have completed a survey of the main functional groups in organic chemistry. By now, all of the structures in the table inside the front cover of this book should seem familiar to you. In the next chapter, we will describe some modern techniques that permit us, rather quickly, to assign a structure to a particular molecule. After that, we will conclude with a series of chapters on important commercial and biological applications of organic chemistry.

Additional Problems

11.25. Give an example of each of the following:

a. a primary amine
b. a cyclic secondary amine
c. a tertiary aromatic amine
d. a quaternary ammonium salt
e. an aryldiazonium salt
f. an azo compound
g. a primary amide
h. a sulfonamide

11.26. Write a structural formula for each of the following compounds:

a. *m*-chloroaniline
b. *sec*-butylamine
c. 2-aminohexane
d. dimethylpropylamine
e. benzylamine
f. 1,2-diaminopropane
g. *N,N*-dimethylaminocyclohexane
h. tetraethylammonium bromide
i. triphenylamine
j. *o*-toluidine
k. 2-methyl-2-propanamine
l. *N,N*-dimethyl-3-pentanamine

11.27. Write a correct name for each of the following compounds:

a. $Br-C_6H_4-NH_2$ (para)
b. $CH_3NHCH_2CH_2CH_3$
c. $(CH_3CH_2)_2NCH_3$
d. $(CH_3)_4N^+Cl^-$
e. $CH_3CH(OH)CH_2CH_2NH_2$
f. $H_2N-C_6H_{10}{=}O$ (4-aminocyclohexanone ring)
g. $Br-C_6H_4-N_2^+Cl^-$ (para)
h. $CH_3-C_6H_4-NHCH_3$ (para)
i. $C_5H_9-NH_2$ (cyclopentyl)
j. $H_2N(CH_2)_6NH_2$

11.28. Draw the structures for, name, and classify as primary, secondary, or tertiary the eight isomeric amines with the molecular formula $C_4H_{11}N$.

11.29. Explain why the difference in the boiling points of isobutane (2-methylpropane; bp −10.2°C) and trimethylamine (bp 2.9°C) is much smaller than the difference in the boiling points of butane (bp −0.5°C) and propylamine (bp 48.7°C). All four compounds have nearly identical formula weights.

11.30. The formula weights of propylamine and 1,2-diaminoethane are nearly identical, yet the diamine boils 60°C *higher* than the monoamine. Explain.

11.31. Place the following substances, which have nearly identical formula weights, in order of increasing boiling point: 1-aminobutane, 1-butanol, methyl propyl ether, pentane.

11.32. Give equations for the preparation of the following amines from the indicated precursor.

a. *N,N*-diethylaniline from aniline **b.** *m*-chloroaniline from benzene

c. *p*-chloroaniline from benzene **d.** 1-aminopentane from 1-bromobutane

11.33. Complete the following equations:

a. (cyclohexyl)—NH_2 + $CH_2{=}CHCH_2Br$ $\xrightarrow{\text{heat}}$

b. $CH_3\overset{O}{\overset{\|}{C}}Cl + H_2NCH_2CH(CH_3)_2 \longrightarrow A \xrightarrow{LiAlH_4} B$

c. $CH_3O\overset{O}{\overset{\|}{C}}$—(phenyl) $\xrightarrow[H^+]{HONO_2}$ C $\xrightarrow[\text{excess}]{LiAlH_4}$ D

d. (phenyl)—CH_2Br $\xrightarrow{NaCN}$ E $\xrightarrow{LiAlH_4}$ F $\xrightarrow{(CH_3\overset{O}{\overset{\|}{C}})_2O}$ G

e. (cyclopentanone)=O + $CH_3CH_2CH_2NH_2$ $\xrightarrow{NaBH_3CN}$ H

11.34. Adamantadine hydrochloride is used (under the name SYMMETREL®) to treat influenza, a viral respiratory tract illness. Suggest how it might by synthesized from the hydrocarbon adamantane.

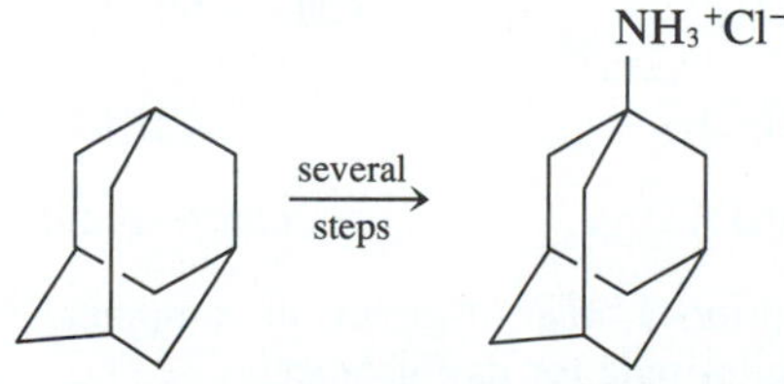

adamantane adamantadine hydrochloride

11.35. Tell which is the stronger base and why.

a. aniline or *p*-cyanoaniline **b.** aniline or diphenylamine

11.36. Write out a scheme similar to eq. 11.19 to show how you could separate a mixture of *p*-toluidine, *p*-methylphenol, and *p*-xylene.

CH_3—(C₆H₄)—NH_2 CH_3—(C₆H₄)—OH CH_3—(C₆H₄)—CH_3

p-toluidine *p*-methylphenol *p*-xylene

11.37. Draw the important contributors to the resonance hydrid structure of *p*-nitroaniline.

11.38. When an amide is dissolved in concentrated sulfuric acid, it is protonated on the oxygen, not on the nitrogen.

$$R\text{—}\overset{\overset{\displaystyle O}{\|}}{C}\text{—}NH_2 \xrightarrow{H^+} R\text{—}\overset{\overset{\displaystyle {}^{+}OH}{\|}}{C}\text{—}NH_2$$

$$R\text{—}\overset{\overset{\displaystyle O}{\|}}{C}\text{—}NH_2 \;\not\xrightarrow{H^+}\; R\text{—}\overset{\overset{\displaystyle O}{\|}}{C}\text{—}\overset{+}{N}H_3$$

Consider the resonance contributors to the various species involved, and explain this observation.

11.39. Write an equation for the reaction of
a. aniline with hydrochloric acid.
b. triethylamine with sulfuric acid.
c. diethylammonium chloride with sodium hydroxide.
d. *N,N*-dimethylaniline with methyl iodide.
e. cyclohexylamine with acetic anhydride.

11.40. Write out the steps in the mechanism for the following reaction:

$$CH_3CH_2NH_2 + CH_3\overset{\overset{\displaystyle O}{\|}}{C}O\overset{\overset{\displaystyle O}{\|}}{C}CH_3 \longrightarrow CH_3CH_2NH\overset{\overset{\displaystyle O}{\|}}{C}CH_3 + CH_3COOH.$$

Explain why only one of the hydrogens of the amine is replaced by an acetyl group, even if a large excess of acetic anhydride is used.

11.41. Explain why compound A can be separated into its *R*- and *S*-enantiomers, but compound B cannot.

$$CH_3\text{—}\overset{+}{N}(CH_2CH_3)(CH_2CH_2CH_3)\text{—}CH_2C_6H_5\;\;Cl^-$$

compound A

$$CH_3\text{—}\ddot{N}(CH_2CH_3)(CH_2CH_2CH_3)$$

compound B

11.42. Give the priority order of groups in compound A (problem 11.41), and draw a Fischer projection formula for its *R*-isomer.

11.43. Write equations that show the different behavior of aniline, *N*-methylaniline, and *N,N*-dimethylaniline in the Hinsberg test (Sec. 11.11).

11.44. Answer the following questions with regard to the sulfanilamide synthesis (page 322).
a. Why can we not use aniline in place of acetanilide in the first step (reaction with chlorosulfonic acid)?
b. What would the structure of the final product be if we were to use $H_2\ddot{N}\text{—}$(2-pyrimidinyl) in place of ammonia in the second step?
c. Write equations for the mechanism of the last step.

11.45. Choline (Sec. 11.12) can be prepared by the reaction of trimethylamine with ethylene oxide. Write an equation for the reaction, and show its mechanism.

11.46. Acetylcholine (Sec. 11.12) is synthesized in the cell body of neurons. The enzyme choline acetyltransferase catalyzes its synthesis from acetyl-CoA (page 293) and choline. Write an equation for the reaction, using the formula $CH_3\overset{\overset{\large O}{\|}}{C}-S-CoA$ for acetyl-CoA.

11.47. Primary aliphatic amines RNH_2 react with nitrous acid in the same way that primary arylamines $ArNH_2$ do, to form diazonium ions. But alkyldiazonium ions RN_2^+ are much less stable than aryldiazonium ions ArN_2^+ and readily lose nitrogen even at 0°C. Explain the difference.

11.48. Write an equation for the reaction of $CH_3-C_6H_4-N_2^+\,HSO_4^-$ with

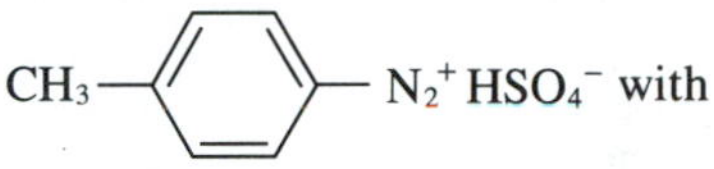

a. KCN and cuprous cyanide.
b. aqueous acid, heat.
c. HCl and cuprous chloride.
d. potassium iodide.
e. *p*-methylphenol and OH^-.
f. *N,N*-dimethylaniline and base.
g. hypophosphorous acid.
h. fluoroboric acid, then heat.

11.49. Show how diazonium ions could be used to synthesize
a. *p*-bromobenzoic acid from *p*-bromoaniline.
b. *m*-iodobromobenzene from benzene.
c. *m*-iodoacetophenone from benzene.
d. 3-cyano-4-methylbenzenesulfonic acid from toluene.

11.50. Congo red is used as a direct dye for cotton. Write equations to show how it can be synthesized from benzidine and 1-aminonaphthalene-4-sulfonic acid.

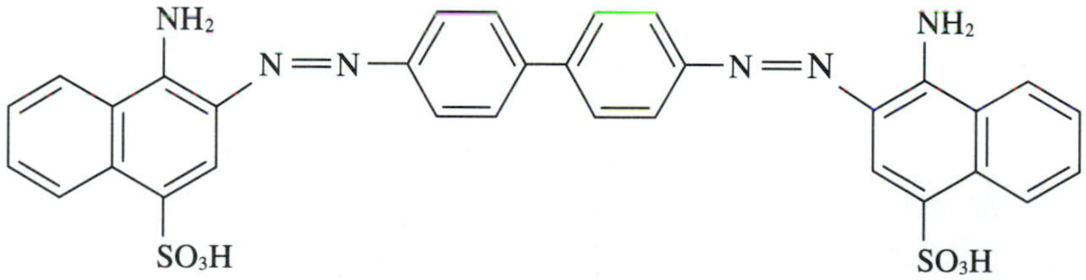

NH_2 N=N N=N NH_2

SO_3H SO_3H

Congo red

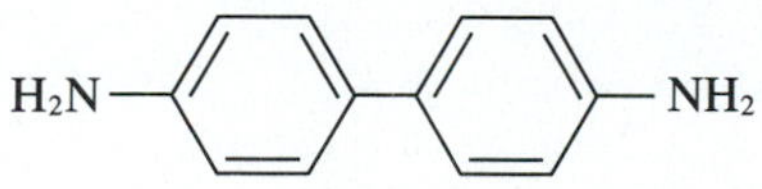

H_2N NH_2

benzidine

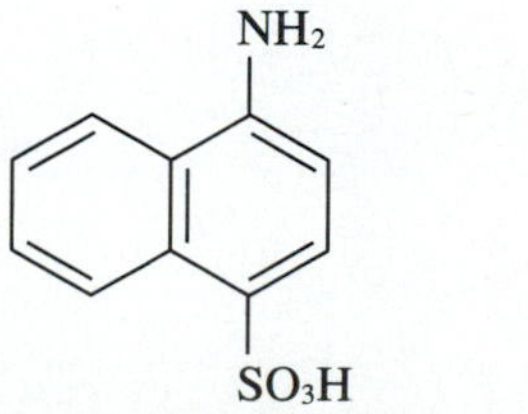

NH_2

SO_3H

1-aminonaphthalene-4-sulfonic acid

11.51. Show how prontosil (page 322) could be synthesized from *p*-aminobenzenesulfonic acid and *m*-dinitrobenzene.

11.52. Sunset yellow is a food dye that can be used to color Easter eggs. Write an equation for an azo coupling reaction that will give this dye.

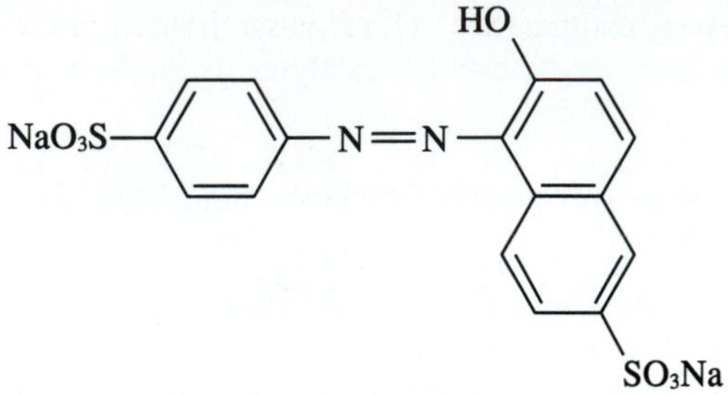

sunset yellow

CHAPTER 12

Spectroscopy and Structure Determination

12.1 Introduction

In the early years of organic chemistry, determining the structure of a new compound was often a formidable task. The first step, of course, was an elemental analysis. Knowing the percentage of each element present allowed the empirical formula to be calculated; the molecular formula was then either the same as or a multiple of that formula. Elemental analysis is also an important criterion of the *purity* of a compound.

But how are the atoms arranged? What functional groups are present? And what about the carbon skeleton? Is it acyclic or cyclic, are there branches and where are they located, are benzene rings present, and so on? All these questions had to be answered by chemical means. Reactions such as ozonolysis or saponification could be used to convert complex molecules to simpler ones whose structures were easier to determine. To identify functional groups, various chemical tests could be applied (such as the bromine or permanganate tests for unsaturation, or the Tollens' silver mirror test for an aldehyde group, or the Hinsberg test for the class of amine, and so on).

Once the functionality was known, reactions whose chemistry was well understood could be used to convert the unknown compound in one or more steps to a compound whose structure was already known. For example, if the compound was an aldehyde suspected to have the same R- group as a known acid, it could be oxidized. If the physical properties (bp, mp, specific rotation if chiral, and so on) and chemical reactions of the acid obtained from the aldehyde

$$RCH{=}O \xrightarrow{KMnO_4} RCO_2H \tag{12.1}$$

agreed with those of the known acid, it could safely be concluded that the two R- groups *were* the same, and the structure of the aldehyde also became known. If they did *not* agree, one had to do some rethinking about the suspected structure. Ultimate structure proof came through synthesis of the unknown from compounds whose structures were already known, by reactions whose outcome was unambiguous. Gradually, over the years, a vast network of compounds with known structures was built up and catalogued in reference books.

These methods—which often required weeks, months, even years—are still used in appropriate situations. But since the 1940s, various types of spectroscopy have simplified and speeded up the process of structure determination greatly. Automated instruments have been developed that permit us to determine and record spectroscopic properties often with little more effort than pushing a button. And these spectra, if properly interpreted, yield a great deal of structural information.

Spectroscopic methods have many advantages. Usually only a very small sample of material is required, and it can often be recovered if necessary. The methods are rapid, sometimes requiring only a few minutes. And frequently we can obtain more detailed structural information from spectra than from ordinary laboratory methods.

In this chapter, we will describe some of the more important spectroscopic techniques used today and how they can be applied to structural problems. But first, let us examine some principles that form the basis of most of these techniques.

12.2 Principles of Spectroscopy

Equation 12.2 describes the relationship between the energy of light, (or any other form of radiation) ***E***, and its **frequency,** ν (Greek nu, pronounced "new").

$$E = h\nu \tag{12.2}$$

The equation says that there is a direct relationship between the frequency of light and its energy: the higher the frequency, the higher the energy. The proportionality constant between the two is known as **Planck's constant, *h*.** Because the frequency of light and its wavelength are *inversely* proportional, the equation can also be written

$$E = hc/\lambda\text{, because } \nu = c/\lambda \tag{12.3}$$

where λ is the **wavelength** of light and c is the speed of light. In this form, the equation tells us that the shorter the wavelength of light, the higher its energy.

Molecules can exist at various energy levels. For example, the bonds in a given molecule may stretch, bend, or rotate; electrons may move from one orbital to another; and so on. These processes are quantized; that is, bonds may

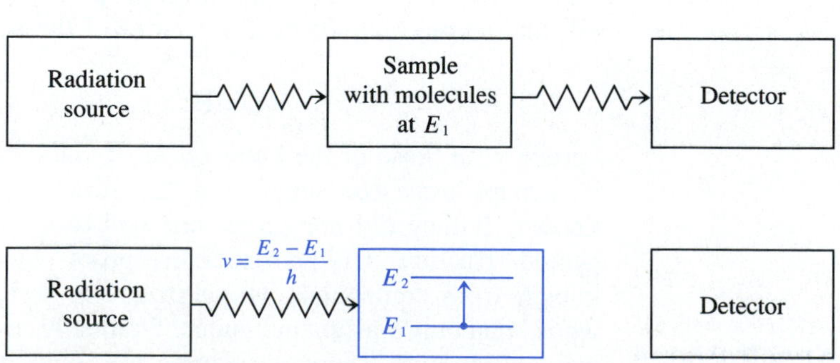

FIGURE 12.1 Radiation passes through the sample unchanged, except when its frequency corresponds to the energy difference between two energy states of the molecule.

TABLE 12.1 Types of spectroscopy and the electromagnetic spectrum

		Region of the spectrum			
Type of spectroscopy	*Radiation source*	*Frequency (hertz)*	*Wavelength (meters)*	*Energy (kcal/mol)*	*Type of transition*
nuclear magnetic resonance	radio waves	$60–600 \times 10^6$ (depends on magnet strength of the instrument)	5–0.5	$6–60 \times 10^{-6}$	nuclear spin
infrared	infrared light	$0.2–1.2 \times 10^{14}$	$15.0–2.5 \times 10^{-6}$	2–12	molecule vibrations
visible-ultraviolet (electronic)	visible or ultraviolet light	$0.375–1.5 \times 10^{15}$	$8–2 \times 10^{-7}$	37–150	electronic states

stretch, bend, or rotate only with certain frequencies (or energies; the two are proportional), and electrons may only jump between orbitals with well-defined energy differences. It is these energy (or frequency) differences that we measure by various types of spectra.

The idea behind most forms of spectroscopy is very simple and is expressed schematically in Figure 12.1. A molecule at some energy level, say E_1, is exposed to radiation. The radiation passes through the molecule to a detector. As long as the molecule does not absorb the radiation, the amount of radiation detected will be equal to the amount of radiation emitted by the source (top part of Figure 12.1). At a frequency that corresponds to some molecular energy transition, say from E_1 to E_2, the radiation will be absorbed by the molecule and will *not* appear at the detector (bottom part of Figure 12.1). The spectrum, then, consists of a record or plot of the amount of energy (radiation) received by the detector as the input energy is gradually varied.

Some transitions require more energy than others, so we must use radiation of the appropriate frequency to determine them. In this chapter, we will discuss three types of spectroscopy that depend on such transitions. They are nuclear magnetic resonance (NMR), infrared (IR), and ultraviolet-visible (UV-vis) spectroscopy. Table 12.1 summarizes the regions of the electromagnetic spectrum in which transitions for these three types of spectroscopy can be observed. We will begin with NMR spectroscopy and nuclear spin transitions, which require exceedingly small amounts of energy.

12.3 Nuclear Magnetic Resonance Spectroscopy (NMR)

The kind of spectroscopy that has had by far the greatest impact on the determination of organic structures is nuclear magnetic resonance (NMR) spectroscopy. Commercial instruments became available in the late 1950s, and since then, NMR spectroscopy has become an indispensable tool for the organic chemist. Let us look briefly at the theory and then see what practical information we can obtain from an NMR spectrum.

FIGURE 12.2 Orientation of nuclei in an applied field, and excitation of nuclei from the lower to the higher energy spin state.

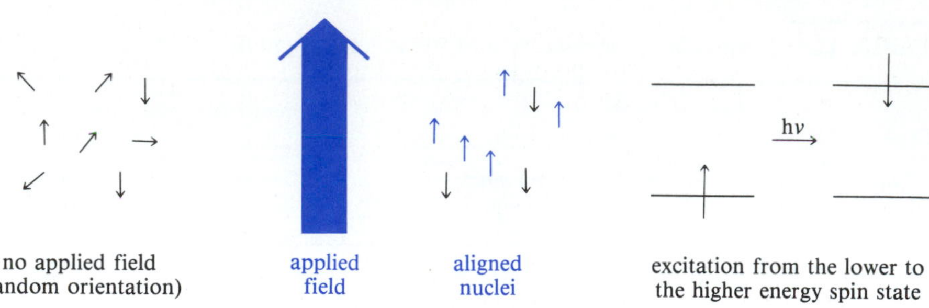

Certain nuclei behave as though they are spinning. Because nuclei are charged and a spinning charge creates a magnetic field, these spinning nuclei behave like tiny magnets. The most important nuclei for organic structure determination are ^1H (ordinary hydrogen) and ^{13}C, a stable, nonradioactive isotope of ordinary carbon. Although ^{12}C and ^{16}O are present in most organic compounds, they do not possess a spin and do not give NMR spectra.

When nuclei with spin are placed between the poles of a powerful magnet, they align themselves *with* or *against* the field of the magnet. Nuclei aligned with the field have a slightly lower energy than those aligned against the field (Figure 12.2). By applying energy in the radiofrequency range, it is possible to excite nuclei in the lower energy state to the higher energy spin state (we sometimes say that the spins "flip").

The energy gap between the two spin states depends on the strength of the applied magnetic field; the stronger the field, the larger the energy gap. Instruments currently in use have magnetic fields that range from about 14,000 to 140,000 gauss (by comparison, the earth's magnetic field is only about 0.5 gauss). At these field strengths, the energy gap corresponds to a radiofrequency of 60 to 600 MHz (megahertz; 1 MHz = 10^6 Hz or 10^6 cycles per second). Translated to energy units to which chemists are more accustomed, the energy gap between the spin states is only $6–60 \times 10^{-6}$ kcal/mol. Even though this gap is exceedingly small, modern technology permits its detection with great accuracy.

12.3a. Measuring an NMR Spectrum A proton* NMR spectrum is usually obtained in the following way. A sample of the compound being studied (usually only a few milligrams) is dissolved in some inert solvent that does not contain protons. Examples of such solvents are CCl_4, or solvents with the hydrogens replaced by deuterium, such as $CDCl_3$ (deuteriochloroform) and CD_3COCD_3 (hexadeuterioacetone). A small amount of a reference compound is also added (we will say more about this in the next section). The solution, in

* The term *proton* is often used interchangeably with *hydrogen* in discussing NMR spectra, even though the hydrogens are covalently bound (and not H^+). This is an incorrect, but common usage.

a thin glass tube, is placed in the center of a radiofrequency (rf) coil, between the pole faces of a powerful magnet. The nuclei align themselves with or against the field. Continuously increasing amounts of energy can then be applied to the nuclei by the rf coil. When this energy corresponds exactly to the energy gap between the lower and higher energy spin states, it is absorbed by the nuclei. At this point, the nuclei are said to be in resonance with the applied frequency—hence the term **nuclear magnetic resonance.** A plot of the energy absorbed by the sample against the applied frequency of the rf coil gives an NMR spectrum.

In practice, it is usually easier to apply a *constant* rf frequency and slightly vary the strength of the applied magnetic field. One then measures exactly the strength of the magnetic field that corresponds to the applied radiofrequency. The spectra given in this book were obtained in this way. The **applied magnetic field** increases as we go from left to right in the recorded spectra.

12.3b Chemical Shifts and Peak Areas All protons do not flip their spins at precisely the same radiofrequency because they may differ in chemical (and, more particularly, electronic) environment. We will return to this point, but first let us examine some spectra.

Figure 12.3 shows the proton NMR spectrum of *p*-xylene. The spectrum is very simple and consists of two peaks. The positions of the peaks are mea-

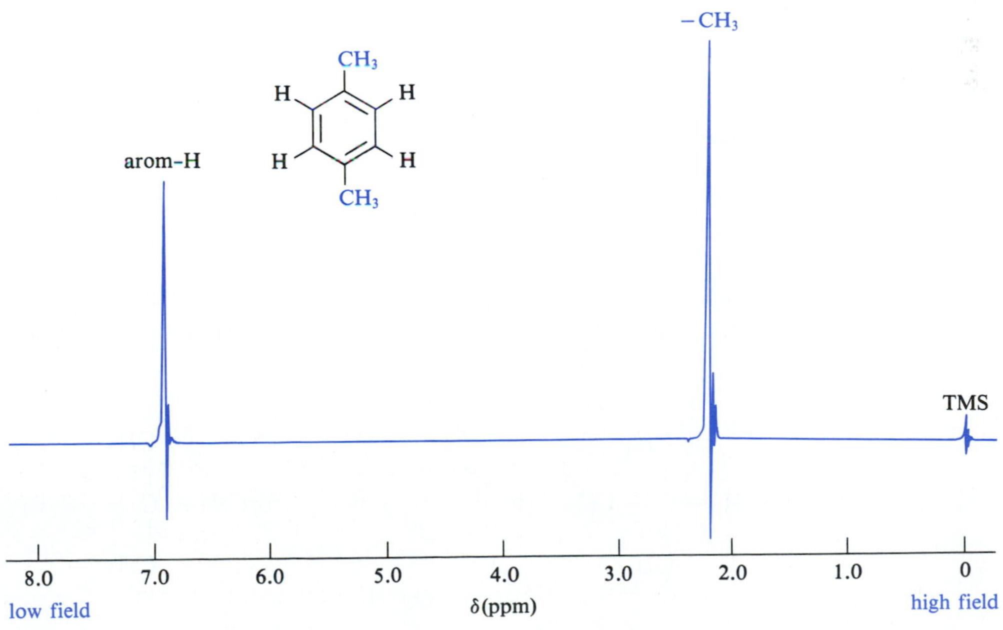

FIGURE 12.3 NMR spectrum of p-xylene.

sured in δ (delta) units from the peak of a reference compound, which is **tetramethylsilane (TMS),** $(CH_3)_4Si$. The reasons for selecting TMS as a reference compound are (1) all 12 of its protons are equivalent, so it shows only one sharp NMR signal, which serves as a reference point; (2) its protons appear at higher field than do most protons in organic compounds, thus making it easy to identify the TMS peak; and (3) TMS is inert, so it does not react with most organic compounds, and it is low-boiling and can be removed easily at the end of a measurement.

Most organic compounds have peaks *downfield* from TMS and are given positive δ values. A δ value of 1.00 means that a peak appears 1 part per million (ppm) downfield from the TMS peak. If the spectrum is measured at 60 MHz (60×10^6Hz), then 1 ppm is 60 Hz (one-millionth of 60 MHz) downfield from TMS. If the spectrum is run at 100 MHz, a δ value of 1 ppm is 100 Hz downfield from TMS, and so on. *The **chemical shift** of a particular kind of proton is its δ value with respect to TMS.* It is called a *chemical* shift because it depends on the chemical environment of the protons. The chemical shift is independent of the instrument on which it is measured.

$$\text{Chemical shift} = \delta = \frac{\text{distance of peak from TMS, in Hz}}{\text{spectrometer frequency in MHz}}\ \text{ppm} \qquad (12.4)$$

In the spectrum of *p*-xylene, we see a peak at δ 2.20 and another at δ 6.95. It seems reasonable that these peaks are caused by the two different "kinds" of protons in the molecule: the methyl protons and the aromatic ring protons. How can we tell which is which?

One way is to integrate the area under each peak. *The **peak area** is directly proportional to the number of protons responsible for the particular peak.* All commercial NMR spectrometers are equipped with electronic integrators that print out these areas. Thus, we find that the areas of the peaks at δ 2.20 and δ 6.95 in the *p*-xylene spectrum give a ratio of 3:2 (or 6:4). These areas allow us to assign the peak at δ 2.20 to the six methyl protons and the peak at δ 6.95 to the four aromatic ring protons.

EXAMPLE 12.1 How many peaks do you expect to see in the NMR spectrum of each of the following compounds? If you expect several peaks, what will their relative areas be?

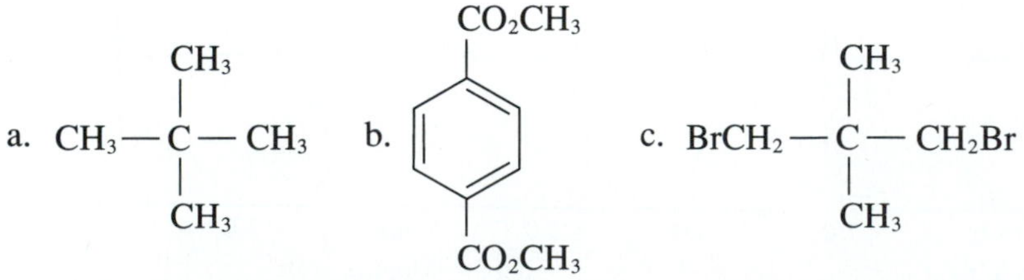

a. CH_3—C(CH_3)(CH_3)—CH_3

b. CO_2CH_3 / benzene ring / CO_2CH_3 (para)

c. $BrCH_2$—C(CH_3)(CH_3)—CH_2Br

Solution a. All twelve protons are equivalent and appear as a single peak.

b. The four aromatic protons are equivalent, and the six methyl protons on the ester functions are equivalent. There will be two peaks in the spectrum, with an area ratio of 4:6 (or 2:3).

c. There are two kinds of protons, $CH_3—C$ and $CH_2—Br$. There will be two peaks, with the area ratio 6 : 4 (or 3 : 2).

Problem 12.1 Which of the following compounds show only a single peak in their NMR spectrum?

a. CH_3OCH_3 b. $CH_3CH_2OCH_2CH_3$ c. (cyclopentane)

Problem 12.2 Each of the following compounds shows more than one peak in its NMR spectrum. What will the area ratios be?

a. CH_3OH b. $CH_3\overset{\overset{\displaystyle O}{\|}}{C}OCH_3$ c. $CH_3CH_2\overset{\overset{\displaystyle O}{\|}}{C}CH_2CH_3$

Problem 12.3 How could ^{1}H NMR spectroscopy be used to distinguish 1,1-dichloroethane from 1,2-dichloroethane?

A more general way to assign peaks is to compare chemical shifts with those of similar protons in a known reference compound. For example, benzene has six equivalent protons and shows a single peak in its NMR spectrum, at δ 7.24. Other aromatic compounds also show a peak in this region. We can conclude that most aromatic ring protons will have chemical shifts at about δ 7. Similarly, most $CH_3—Ar$ protons appear at δ 2.2–2.5 (see Figure 12.3).

The chemical shifts of protons in various chemical environments have been determined by measuring the NMR spectra of a large number of compounds with known, relatively simple structures. Table 12.2 gives the chemical shifts for several common types of protons.

EXAMPLE 12.2 Using the data in Table 12.2, describe the expected NMR spectrum of

a. $CH_3\overset{\overset{\displaystyle O}{\|}}{C}—OCH_3$ b. $Cl_2CH—\underset{\underset{\displaystyle CH_3}{|}}{\overset{\overset{\displaystyle CH_3}{|}}{C}}—CH_2Cl$

Solution a. The spectrum will consist of two peaks, equal in area, at about δ 2.3 (for the $CH_3\overset{\overset{\displaystyle O}{\|}}{C}—$ protons) and δ 3.6 (for the $—OCH_3$ protons).

b. The spectrum will consist of three peaks, with relative areas 6 : 2 : 1 at δ 0.9 (the two methyls), δ 3.5 (the $—CH_2—Cl$ protons), and δ 5.8 (the $—CHCl_2$ proton).

TABLE 12.2 Typical proton chemical shifts (relative to tetramethylsilane)

Type of proton	δ *(ppm)*	*Type of proton*	δ *(ppm)*
C—CH_3	0.85–0.95	—CH_2—F	4.3–4.4
C—CH_2—C	1.20–1.35	—CH_2—Br	3.4–3.6
C—CH(—C)—C	1.40–1.65	—CH_2—I	3.1–3.3
		CH_2=C	4.6–5.0
CH_3—C=C	1.6–1.9	—CH=C	5.2–5.7
CH_3—Ar	2.2–2.5	Ar—H	6.6–8.0
CH_3—C(—)=O	2.1–2.6	—C≡C—H	2.4–2.7
CH_3—N<	2.1–3.0	—C(=O)—H	9.5–9.7
CH_3—O—	3.5–3.8	—C(=O)—OH	10–13
—CH_2—Cl	3.6–3.8	R—OH	0.5–5.5
—$CHCl_2$	5.8–5.9	Ar—OH	4–8

Problem 12.4 Describe the expected NMR spectrum of

a. $CH_3C(=O)OH$ b. $(CH_3)_2C{=}CH_2$

Problem 12.5 An ester is suspected of being either $(CH_3)_3CC(=O)OCH_3$ or $CH_3C(=O)$—$OC(CH_3)_3$. Its NMR spectrum consists of two peaks at δ 0.9 and δ 3.6 (relative areas 3 : 1). Which compound is it? Describe the spectrum that would be expected if it had been the other ester.

Now let us return to the point mentioned at the beginning of this section, the factors that influence chemical shifts. One important factor is the **electronegativity of groups in the immediate environment of the protons. Electron-withdrawing groups generally cause a downfield chemical shift.** Compare, for example, the following chemical shifts from Table 12.2:

—CH_3	—CH_2Cl	—$CHCl_2$
~ 0.9	~ 3.7	~ 5.8

Electrons in motion near a proton create a small magnetic field in its microenvironment that tends to shield the proton from the externally applied magnetic field. Chlorine is an electron-withdrawing group. Withdrawal of electron density by the chlorines therefore "deshields" the proton, allowing it to flip its spin at a lower applied external field. The more chlorines, the larger the effect.

EXAMPLE 12.3 Predict the order of chemical shifts of the various protons in 1-bromopropane.

Solution

$$\overset{3}{C}H_3\overset{2}{C}H_2\overset{1}{C}H_2Br$$

The protons at C1 will be at lowest field because they are closest to the electron-withdrawing Br atom. The methyl protons will be at highest field because they are farthest from the Br, and the peak for the C2 protons will appear between the other two. The inductive effect falls off rapidly with distance, as seen by the actual chemical shift values.

$$\delta\ \underset{CH_3}{1.06}\!-\!\underset{CH_2}{1.81}\!-\!\underset{CH_2}{3.47}\!-\!Br$$

Problem 12.6 Explain the following chemical shifts:

δ	0.23	3.05	2.68	2.16
	CH_4	CH_3Cl	CH_3Br	CH_3I

A second factor that influences chemical shifts is the presence of pi electrons. Protons attached to a carbon that is part of a multiple bond or aromatic ring usually appear downfield from protons attached to saturated carbons. Compare these values from Table 12.2:

$C-CH_2-C$	$CH_2{=}C$	$-CH{=}C$	H–(benzene ring)
δ 1.2–1.35	4.6–5.0	5.2–5.7	6.6–8.0

The reasons for this effect are complex, but it is useful in assigning structures.

Problem 12.7 Describe the ^{1}H NMR spectrum of *trans*-2,2,5,5-tetramethyl-3-hexene.

12.3c Spin–Spin Splitting Many compounds give spectra that show more complex peaks than just single peaks **(singlets)** for each type of proton. Let us examine some spectra of this type to see what additional structural information they convey.

Figure 12.4 shows the NMR spectrum of diethyl ether, $CH_3CH_2OCH_2CH_3$. From the information given in Table 12.2, we might have expected the NMR spectrum of diethyl ether to consist of two lines: one in the region of δ 0.9 for the six equivalent CH_3 protons and one at about δ 3.5 for the four equivalent CH_2 protons adjacent to the oxygen atom, with relative areas 6 : 4. Indeed, in

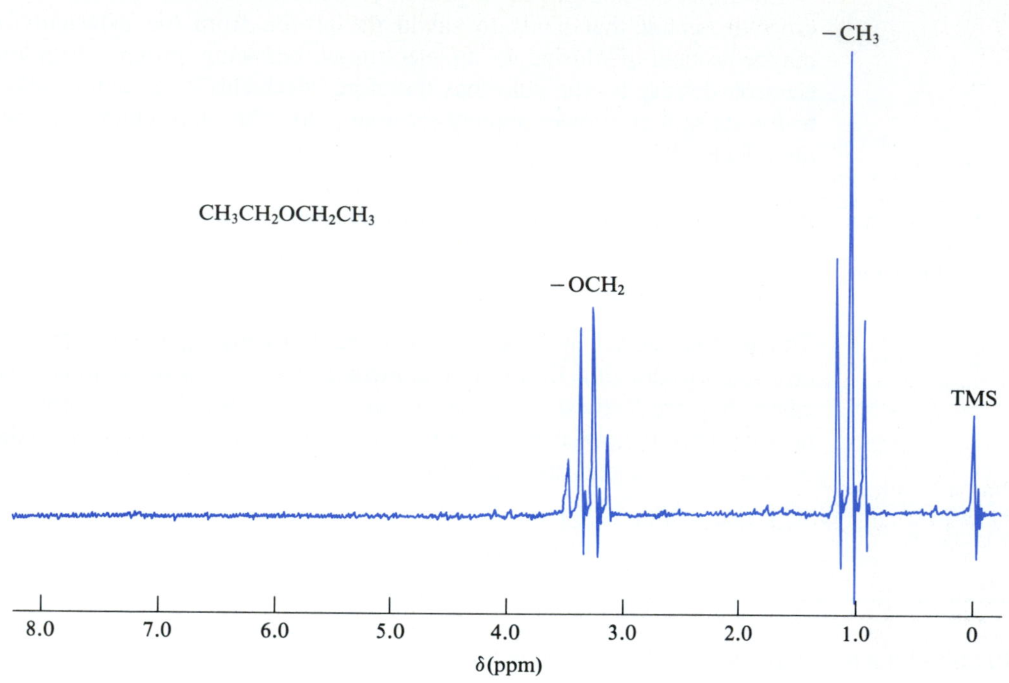

FIGURE 12.4 NMR spectrum of diethyl ether, showing spin–spin splitting.

Figure 12.4 we see absorptions in each of these regions, with the expected total area ratio. But we do not see singlets! Instead, the methyl signal is split into three peaks, a **triplet,** with relative areas 1 : 2 : 1; and the methylene signal is split into four peaks, a **quartet,** with relative areas 1 : 3 : 3 : 1. These **spin–spin splittings,** as they are called, tell us quite a bit about molecular structure, and they arise in the following way.

We know that each proton in the molecule acts as a tiny magnet. When we run an NMR spectrum, each proton "feels" not only the very large applied magnetic field but also a tiny field due to its neighboring protons. During the time that we sweep through one signal, the protons on neighboring carbons can be in either the lower or the higher energy spin state, with nearly equal probabilities (nearly equal because, as we have said, the energy difference between the two states is exceedingly small). So the magnetic field of the protons whose peak we sweep through is perturbed slightly by the tiny fields of its neighboring protons.

We can predict the splitting pattern by **the $n + 1$ rule: if a proton or a set of equivalent protons has n proton neighbors with a substantially different chemical shift, its NMR signal will be split into $n + 1$ peaks.** In diethyl ether, each CH_3 proton has *two* proton neighbors (on the CH_2 group). Therefore, the CH_3 signal is split into 2 + 1 = 3 peaks. At the same time, each CH_2 proton has *three* proton neighbors (on the CH_3 group). The CH_2 signal is therefore split into 3 + 1 = 4 peaks. Let us see why this rule works and why the split peaks have the area ratios they do.

Consider first the system

```
   |    |
 —C—C—
   |    |
  Ha   Hb
```

Proton H_a has *one* nonequivalent proton neighbor, H_b. At the time we pass through the H_a signal, H_b can be in either the lower or the higher energy spin state. Because these two possibilities are nearly equal, the H_a signal will be split into two *equal* peaks: a **doublet.** The same is true for the peak due to H_b.

Now consider the system

```
        Hb
   |    |
 —C—C—
   |    |
  Ha   Hb
```

Proton H_a has *two* proton neighbors, H_b. At the time we pass through the H_a signal, there are three possibilities for the two H_b protons:

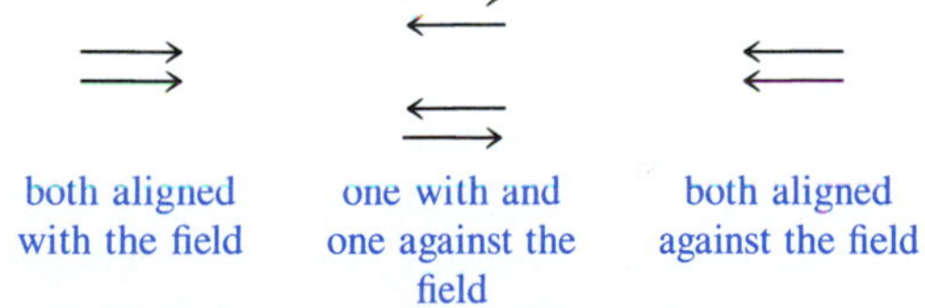

Both can be in the lower energy state, both can be in the higher energy state, or one can be in each state, with this latter arrangement being possible in two ways. Hence the H_a signal will be a *triplet* with relative areas 1 : 2 : 1. The H_b signal, on the other hand, will be a *doublet* because of the two possible spin states for H_a.

EXAMPLE 12.4 Explain why the signal of H_a with *three* neighboring protons H_b is a *quartet* with relative areas 1 : 3 : 3 : 1.

Solution The system is

```
        Hb
   |    |
 —C—C—Hb      or      >CH—CH3
   |    |
  Ha   Hb
```

The possibilities for the spin states of the three H_b protons are

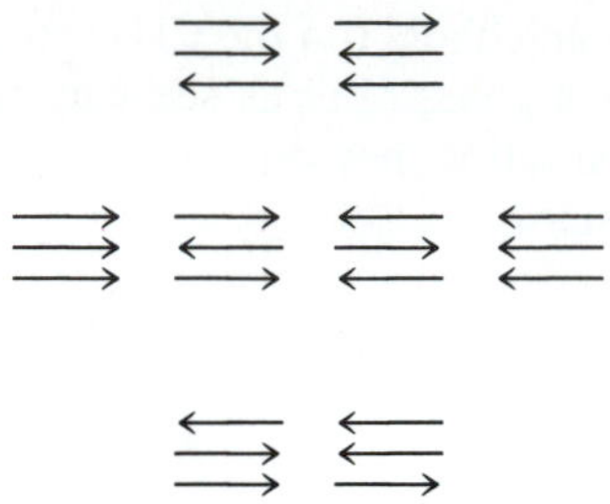

$$\begin{array}{cc} \longrightarrow & \longrightarrow \\ \longrightarrow & \longleftarrow \\ \longleftarrow & \longleftarrow \end{array}$$

$$\begin{array}{cccc} \longrightarrow & \longrightarrow & \longleftarrow & \longleftarrow \\ \longrightarrow & \longleftarrow & \longrightarrow & \longleftarrow \\ \longrightarrow & \longrightarrow & \longleftarrow & \longleftarrow \end{array}$$

$$\begin{array}{cc} \longleftarrow & \longleftarrow \\ \longrightarrow & \longleftarrow \\ \longrightarrow & \longrightarrow \end{array}$$

Hence the H_a signal will appear as four peaks (a quartet), with the area ratio 1 : 3 : 3 : 1. The H_b peak will appear as a doublet, because the possible alignments of H_a are $\longrightarrow$ or $\longleftarrow$.

Problem 12.8 Use the data in Table 12.1 to predict the NMR spectrum of CH_3CHCl_2. Give the approximate chemical shifts *and* the splitting patterns of the various peaks.

Protons that split one another's signals are said to be **coupled.** The extent of the coupling, or the number of hertz by which the signals are split, is called the **coupling constant** (abbreviated J). A few typical coupling constants are shown in Table 12.3. ***Spin–spin splitting falls off rapidly with distance.*** Whereas protons on adjacent carbons may show appreciable splitting (J = 6–8 Hz), protons farther apart hardly "feel" each other's presence (J = 0–1 Hz).

TABLE 12.3 Some typical coupling constants

Group	*J (Hz)*	*Group*	*J (Hz)*
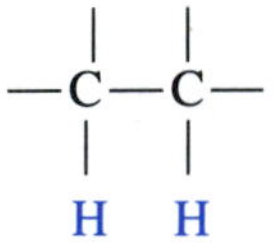 H–C–C–H (—C—C—, each C bearing H)	6–8	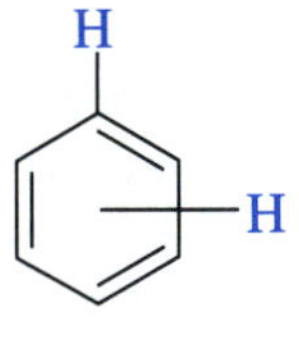 benzene ring with H and H	*ortho:* 6–10 *meta:* 1–3 *para:* 0–1
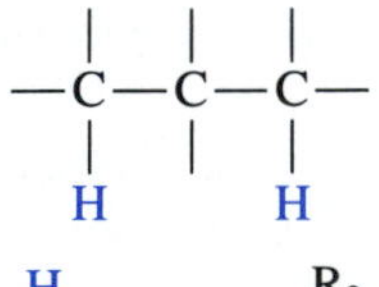 H–C–C–C–H (—C—C—C—, H on terminal carbons)	0–1	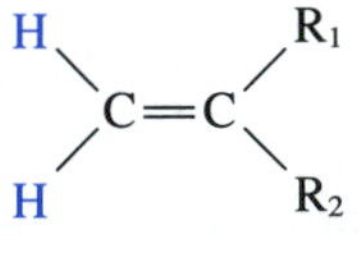 (H)(H)C=C(R_1)(R_2)	0–3
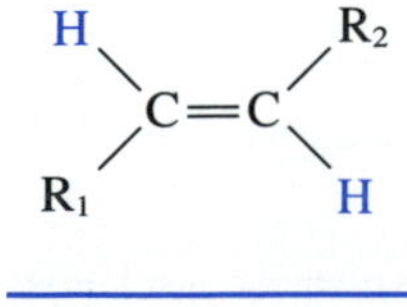 (H)(R_1)C=C(R_2)(H) (trans H)	12–18	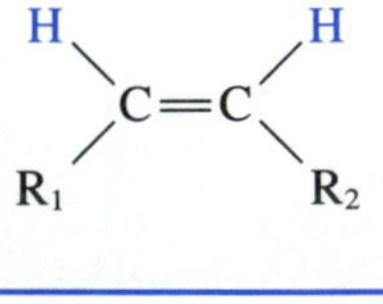 (H)(R_1)C=C(H)(R_2) (cis H)	6–12

As seen in Table 12.3, coupling constants can even be used at times to distinguish between *cis–trans* isomers or between positions of substituents on a benzene ring.

Chemically equivalent protons do not split each other. For example, $BrCH_2CH_2Br$ shows only a sharp singlet in its NMR spectrum for all four protons. Even though they are on adjacent carbons, the protons do not split each other because they have identical chemical shifts.

Problem 12.9 Describe the NMR spectrum of

a. $BrCH_2CH_2Cl$ b. $ClCH_2CH_2Cl$

Not all NMR spectra are simple; they may sometimes be quite complex. This complexity can arise when adjacent protons have nearly the same, but not identical, chemical shifts. An example is phenol (see Figure 12.5). We can easily distinguish the aromatic protons (δ 6.6–7.4) from the hydroxyl proton (δ 5.85), but the splitting pattern of the complex **multiplet** seen for the aromatic protons cannot be analyzed with the simple $n + 1$ rule. Such spectra can, however, be thoroughly analyzed by specially designed computer programs.

In summary, then, proton NMR spectroscopy can give us the following kinds of structural information:

1. **The number of signals and their chemical shifts can be used to identify the kinds of chemically different protons in the molecule.**

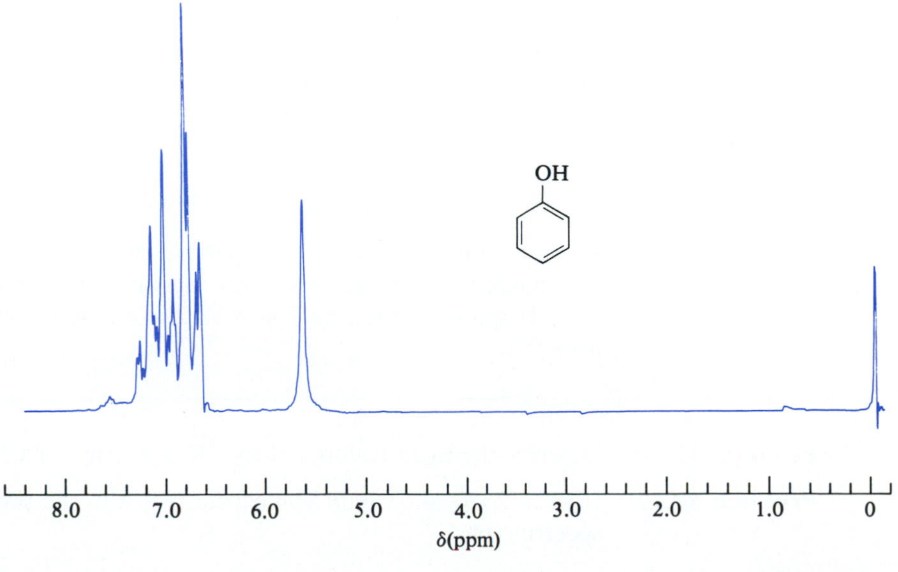

FIGURE 12.5 The proton NMR spectrum of phenol. Note the complexity in the aromatic proton region (δ 6.6–7.4). Reprinted with permission from University Science Books.

2. The peak areas tell us how many protons of each kind are present.
3. The spin–spin splitting pattern gives us information about the number of nearest proton neighbors that a particular kind of proton may have.

12.4 ^{13}C NMR Spectroscopy

Whereas proton NMR spectroscopy gives information about the arrangement of protons in a molecule, ^{13}C NMR spectroscopy gives information about the carbon skeleton. The ordinary isotope of carbon, carbon-12, does not have a nuclear spin, but carbon-13 does. ^{13}C constitutes only 1.1% of naturally occurring carbon atoms. Also, the energy gap between the higher and lower spin states of ^{13}C is very small. For these two reasons, ^{13}C NMR spectrometers must be exceedingly sensitive. Nevertheless, the use of such instruments has become fairly routine in recent years.

^{13}C spectra differ from proton spectra in several ways. ^{13}C chemical shifts occur over a wider range than those of protons. They are measured against the same reference compound, TMS, whose methyl carbons are all equivalent and give a sharp signal. Chemical shifts for ^{13}C are reported in δ units, but the usual range is about 0 to 200 ppm downfield from TMS (instead of the smaller range of 0 to 10 ppm observed for protons). This wide range of chemical shifts tends to simplify ^{13}C spectra relative to ^{1}H spectra.

Because of the low natural abundance of ^{13}C, the chance of finding two adjacent ^{13}C atoms in the same molecule is small. Hence $^{13}C{-}^{13}C$ spin–spin splitting is ordinarily not seen. This feature simplifies ^{13}C spectra. However, $^{13}C{-}^{1}H$ spin–spin splitting can occur. A spectrum can be run in such a way as to show this splitting or not, as desired. Figure 12.6 shows the ^{13}C NMR spectrum of 2-butanol measured with and without $^{13}C{-}^{1}H$ splitting. The spectrum without proton splitting (called a **proton-decoupled spectrum**) shows four sharp singlets, one for each type of carbon atom. The hydroxyl-bearing carbon occurs at the lowest field (δ 69.3), and the two methyl carbons are well separated (δ 10.8 and 22.9). In the spectrum *with* $^{13}C{-}^{1}H$ splitting, the $n + 1$ rule applies. Both CH_3 signals are quartets (three hydrogens, therefore $n + 1 = 4$), the CH_2 carbon is a triplet, and the CH carbon is a doublet.

EXAMPLE 12.5 Describe the ^{13}C spectrum of CH_3CH_2OH.

Solution The spectrum without $^{13}C{-}^{1}H$ splitting consists of two lines because there are two nonequivalent carbons (their signals come at δ 18.2 and 57.8). With $^{13}C{-}^{1}H$ splitting, the signal at δ 18.2 is a quartet, and the one at δ 57.8 is a triplet.

Problem 12.10 Describe the main features of the ^{13}C spectrum of $CH_3CH_2CH_2OH$.

Problem 12.11 How many peaks would you expect to see in the proton-decoupled ^{13}C NMR spectrum of

a. cyclohexane?
b. 2-methyl-2-propanol?
c. 2-methyl-1-propanol?
d. *cis*-1,3-dimethylcyclopentane?

FIGURE 12.6
The ^{13}C NMR spectrum of 2-butanol without (bottom) and with (top) ^{13}C–^{1}H coupling. δ values are shown in the lower spectrum.

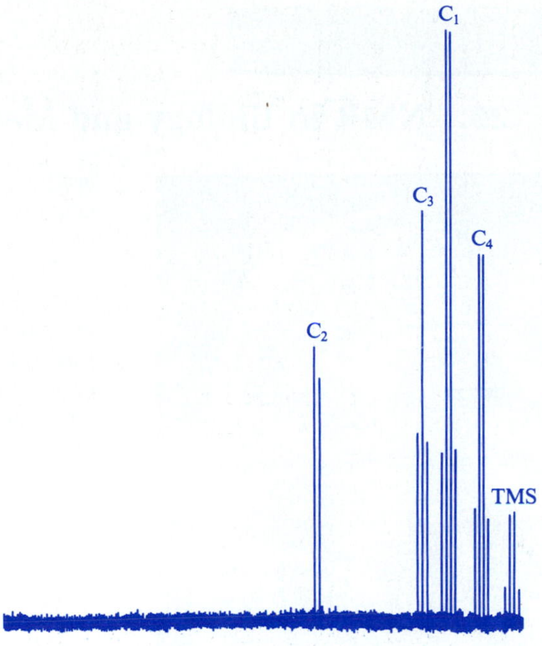

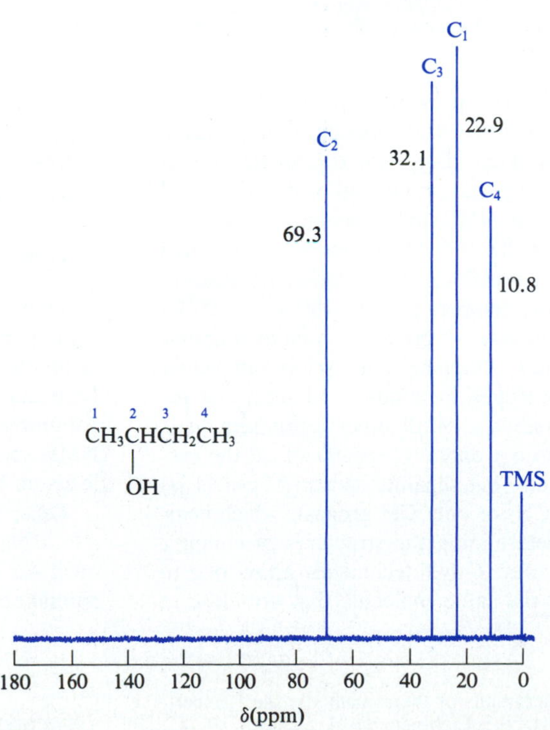

A WORD ABOUT ...

25. NMR in Biology and Medicine

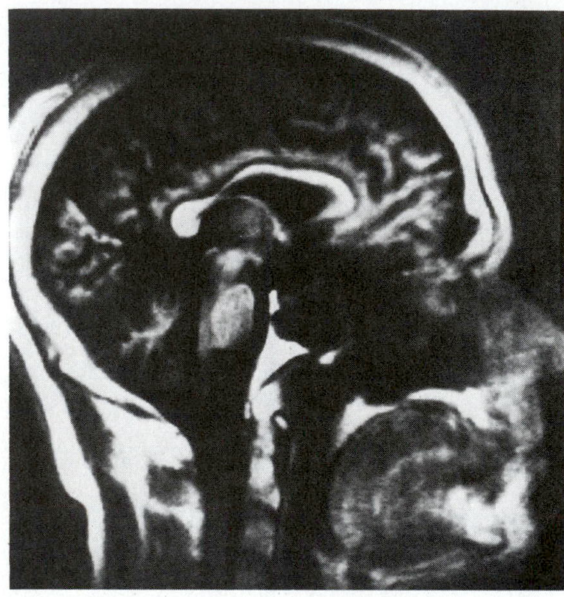

In this chapter, we have presented only the bare bones of NMR spectroscopy, with examples of organic structure determination. But with the application of computer technology, particularly minicomputers and microprocessors, to NMR instrumentation, the field has developed rapidly and added many sophisticated techniques.* It is possible, for example, to measure spectra at variable temperatures (−180 to +200°C) and to measure the rates of processes such as rotations about single bonds, "flipping" of cyclohexane chair conformations, nitrogen inversion in amines, and reactions of free radicals. With other techniques, it is possible to obtain not only ^{13}C spectra of *all* the carbons in a molecule, but separate spectra of *only* CH_3 groups (or only CH_2 or only CH groups), which helps enormously in determining the structures of complex biological molecules. Other techniques allow one to locate groups in the same molecule that are close in space even though they may seem to be far apart if we count by bonds, a real help in studying conformations of complex molecules.

The potential of NMR spectroscopy for solving biological and medical problems is now being realized. For example, instruments are now sufficiently sensitive that one can study intact bodily fluids such as urine, blood plasma, seminal fluid, cerebrospinal fluid, and eye fluid. **Creatinine,** for example, is a normal organic protein metabolite excreted in the urine, where it is present at millimolar levels. The singlets due to its CH_3 (δ 3.1) and CH_2 (δ 4.2) groups are easily observed at these low concentrations with a high resolution (500 MHz) spectrometer, even in the presence of all the other constituents of urine. Its concentration, which can be determined in just a few minutes, provides a useful indicator of kidney function. NMR spectroscopy has been used to detect inherited metabolic diseases by studying the urine of newborn infants.

CH_3
N NH
CH_2
NH
O

creatinine

Other bodily fluids also provide useful medical information. Monitoring low-density and high-density lipoproteins in blood plasma can help in understanding heart disease. Studies of seminal fluid could be of use for problems of infertility. Neurologists are examining NMR spectra of cerebrospinal fluid in infants for new leads on brain disorders.**

Other biologically important nuclei with spin are ^{31}P, ^{23}Na, and ^{19}F. Instrumentation has been developed for studying NMR spectra of intact human and animal body parts. For example, ^{31}P spectra of human

* For excellent descriptions of these methods, see Lambert, J. B.; Shurvell, H. F.; Lightner, D. A.; Cooks, R. G. "Introduction to Organic Spectroscopy," Macmillan: New York, 1987.

** For further information, see Bell, J. D.; Brown, J. C. C.; and Sadler, P. J., "NMR Spectroscopy of Body Fluids" *Chemistry in Britain* **1988,** 1021–24.

forearm muscle (the forearm is placed directly in the magnetic field) taken before, during, and after exercise has allowed the monitoring of several phosphorus-containing components of muscle tissue—for example, adenosine triphosphate (ATP), phosphocreatine, and inorganic phosphate. By comparing concentration changes of these components in normal people with those in patients with muscle disorders, the nature of the disorders and methods of treatment can be devised.*

In **topical NMR,** the NMR magnet is brought to the object being studied, instead of vice versa. A magnetic probe of some sort placed on the surface induces resonances in molecules close to the surface, and useful ^{1}H, ^{13}C or ^{31}P spectra of molecules within living bodies can be achieved in this way. The method has been used, for example, to monitor effects of various drugs on metabolic processes.

Magnetic resonance imaging (MRI) is a new technique that has been used clinically in hospitals only since the mid-1980s. The method allows one to obtain internal images of whole-body parts and has several advantages over x-rays. For one, it is far less hazardous, since it does not cause radiation damage. For another, it gives good images of soft tissues, which are more difficult to obtain with x-rays. How does it work?

A large percentage of the human body is water. Most MRI uses the proton spectra of water in various tissues (although ^{31}P has also been used) to create images. The magnet must be very large, with about a 25-cm-diameter gap to accommodate a human head, and an even larger one for whole-body work. Fortunately, the magnetic field need not be as uniform as in high-resolution NMR for structure determination, so that construction of such large magnets becomes feasible. A gradient field, instead of a uniform field, is essential for imaging. The field strength can also be appreciably lower than for laboratory NMR work. This is fortunate as well because, as you can imagine, the large magnets required for imaging are quite expensive.

The time required for MRI varies from about thirty seconds to ten minutes, with about two minutes being fairly typical. Imaging can be done at preselected planar sections through the body or in three dimensions. The picture at the beginning of this "A Word About" is a cross-section of a human head obtained by MRI.

The principal use of MRI is, of course, in *diagnostic medicine,* but it can also be used in *food science* to assess correct conditions for picking, storing, or marketing food; in *agriculture* to study seed germination; in the *building industry* to study water concentration and distribution in timbers; and in many other areas.**

Nuclear magnetic resonance was discovered by physicists and developed by chemists and is now being applied by biologists and others. The benefits to humanity could not have been foreseen in the beginning—another example of the wisdom of investing in fundamental research without immediate regard for practical applications.

*For more details, see Radda, G. K., "The Use of NMR Spectroscopy for the Understanding of Disease," *Science* **1986,** *233,* 640–45.

**For more about MRI see the book by Morris, P. G. "Nuclear Magnetic Resonance in Medicine and Biology"; Clarendon Press: Oxford, England, 1986.

12.5 Infrared Spectroscopy

Even though NMR spectroscopy is a powerful tool for deducing structures, it is usually supplemented by other spectroscopic methods that provide additional structural information. One of the more important of these is **infrared spectroscopy.**

Infrared frequency is usually expressed in units of **wavenumber,** defined as *the number of waves per centimeter*. Ordinary instruments scan the range of about 700 to 5000 cm^{-1}. This frequency range corresponds to energies of about 2 to 12 kcal/mol (Table 12.1). This amount of energy is sufficient to affect bond vibrations (motions such as *bond stretching* or *bond bending*) but is appreciably less than would be needed to break bonds. These motions are exemplified for a CH_2 group in Figure 12.7.

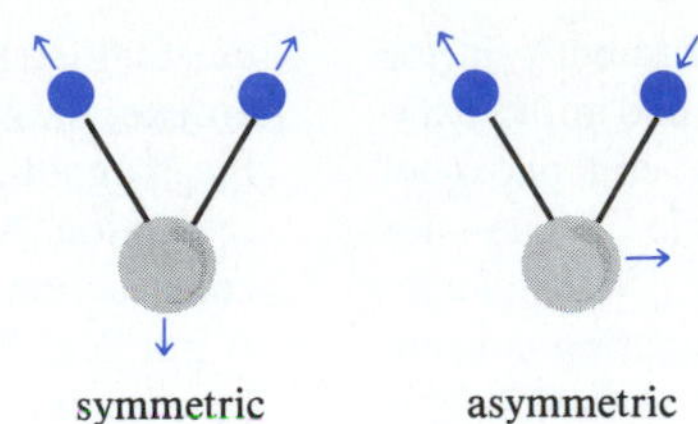

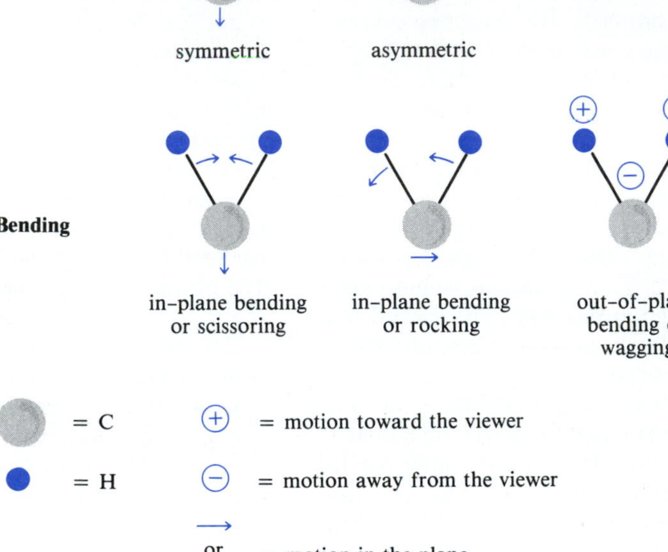

FIGURE 12.7 Stretching and bending motions of a CH_2 group that require energies in the infrared region.

TABLE 12.4 Infrared stretching frequencies of some typical bonds

Bond type	*Group*	*Class of compound*	*Frequency range (cm^{-1})*
single bonds to hydrogen	C—H	alkanes	2850–3000
	=C—H	alkenes and aromatic compounds	3030–3140
	≡C—H	alkynes	3300
	O—H	alcohols and phenols	3500–3700 (free) 3200–3500 (hydrogen-bonded)
		carboxylic acids	2500–3000
	N—H	amines	3200–3600
	S—H	thiols	2550–2600
double bonds	C=C	alkenes	1600–1680
	C=N	imines, oximes	1500–1650
	C=O	aldehydes, ketones, esters, acids	1650–1780
triple bonds	C≡C	alkynes	2100–2260
	C≡N	nitriles	2200–2400

Particular types of bonds usually stretch within certain rather narrow frequency ranges. ***Infrared spectroscopy is particularly useful for determining the types of bonds that are present in a molecule.*** Table 12.4 gives the ranges of stretching frequencies for some bonds commonly found in organic molecules.

The infrared spectrum of a compound can easily be obtained in a few minutes. A small sample of the compound is placed in an instrument with an infrared radiation source. The spectrometer automatically scans the amount of radiation that passes through the sample over a given frequency range and records on a chart the percentage of radiation that is transmitted. Radiation absorbed by the molecule appears as a band in the spectrum.

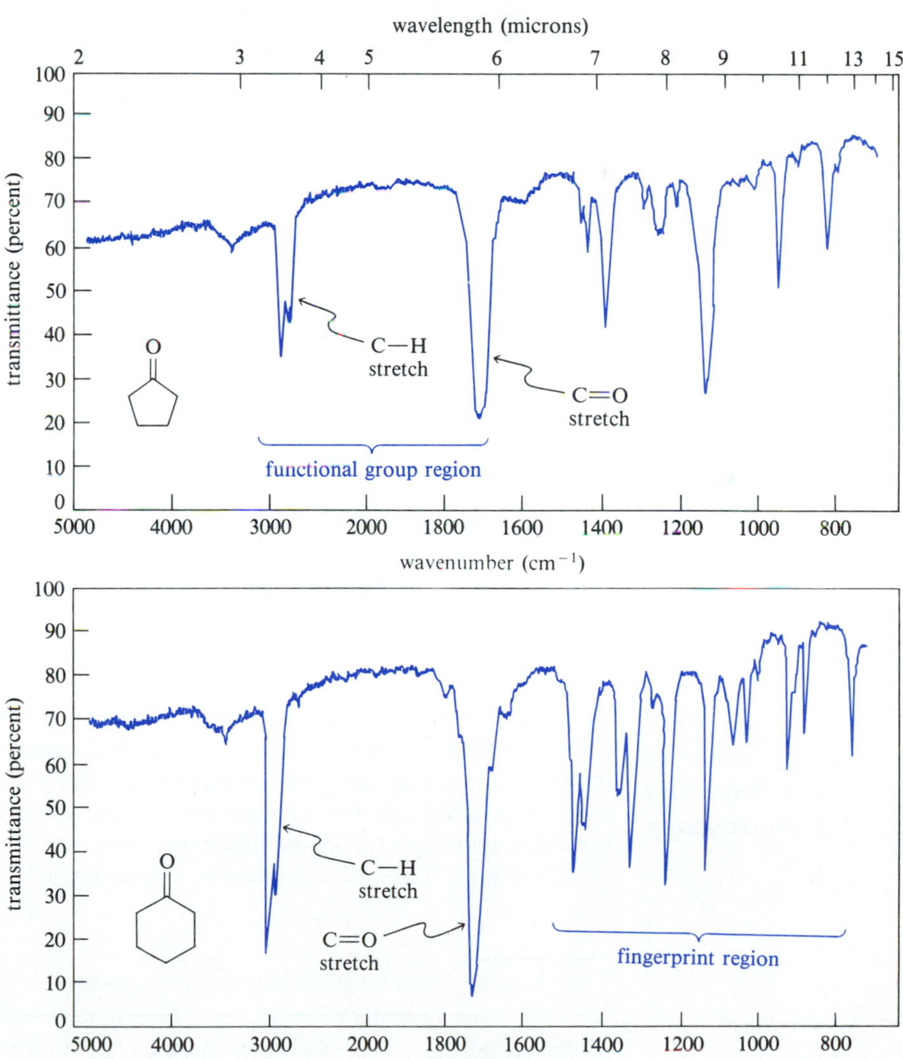

FIGURE 12.8
The infrared spectra of two similar ketones. The spectra have similar functional group bands but differ in the fingerprint region.

Figure 12.8 shows two typical infrared spectra. Both spectra show C—H stretching bands near 3000 cm^{-1} and a C=O stretching band near 1700 cm^{-1}. **Functional group bands** will appear in the same range regardless of the details of the molecular structure. Both cyclopentanone and cyclohexanone have C—H and C=O bonds and therefore have similar spectra in the functional group region of the spectrum (1500 to 4000 cm^{-1}).

The spectra of cyclopentanone and cyclohexanone differ in the low frequency or **fingerprint region,** from 700 to 1500 cm^{-1}. Bands in this region result from combined bending and stretching motions of the atoms and *are unique for each particular compound.*

EXAMPLE 12.6 Using infrared spectroscopy, how could you quickly distinguish between the structural isomers benzyl alcohol and anisole?

C_6H_5—CH_2OH C_6H_5—OCH_3

benzyl alcohol   anisole

Solution Benzyl alcohol's infrared spectrum will show a band in the O—H stretching frequency region (3200 to 3700 cm^{-1}); the spectrum of anisole will not have a band in that frequency region.

Problem 12.12 How could you use infrared spectroscopy to distinguish between the isomers 1-hexyne and 1,3-hexadiene?

To summarize, infrared spectra can be used to tell what types of bonds may be present in a molecule (by using the functional group region) and to tell whether two substances are identical or different (by using the fingerprint region).

12.6 Visible and Ultraviolet Spectroscopy

The visible region of the spectrum (visible to the human eye, that is) corresponds to light with wavelengths of 400 to 800 **nanometers** (nm; 1 nm = 10^{-9} meters). Ultraviolet light has a shorter wavelength, about 200 to 400 nm, whereas infrared light has a longer wavelength, above 2500 nm. Older units for reporting these spectra are **millimicrons** ($m\mu$ = 1 nm) or **angstrom units** (Å; 10Å = 1 nm). Table 12.5 summarizes these units.

TABLE 12.5 Units for visible-ultraviolet spectra

Visible (vis)	400–800 nm (or $m\mu$)	4000–8000 Å
Ultraviolet (uv)	200–400 nm (or $m\mu$)	2000–4000 Å

The amounts of energy associated with light are 37 to 75 kcal/mol for the visible region and 75 to 150 kcal/mol for the ultraviolet region (Table 12.1). These energies are much larger than those involved in infrared spectroscopy (2 to 12 kcal/mol). They correspond to the amounts of energy needed to cause an electron to jump from a filled molecular orbital to a higher-energy, vacant molecular orbital. Such electron jumps are called **electronic transitions.**

Figure 12.9 shows a typical ultraviolet absorption spectrum. Unlike infrared spectra, visible-ultraviolet spectra are quite broad and generally show only a small number of peaks. The peaks are reported as the wavelengths where maxima occur. The conjugated, unsaturated ketone whose spectrum is shown in Figure 12.9 has an intense absorption at λ_{max} = 232 nm and a much weaker absorption at λ_{max} = 330 nm. The band at shorter wavelength corresponds to a pi electron transition, whereas the longer-wavelength, weaker-intensity band corresponds to a transition of the nonbonding electrons on the carbonyl oxygen atom.

The intensity of an absorption band can be expressed quantitatively. Band intensity depends on the particular molecular structure and also on the number of absorbing molecules in the light path. **Absorbance,** which is the log of the ratio of light intensities entering and leaving the sample, is given by the equation

$$A = \epsilon c l \text{ (Beer's law)} \qquad (12.5)$$

where ϵ is the **molar absorptivity** (sometimes called the **extinction coefficient),** c is the concentration of the solution in moles per liter, and l is the length in centimeters of the sample through which the light passes. The value of ϵ for any peak in the spectrum of a compound is a constant characteristic of that particular molecular structure. For example, the values of ϵ for the peaks in the spectrum of the unsaturated ketone shown in Figure 12.9 are λ_{max} = 232 nm (ϵ = 12,600) and λ_{max} = 330 nm (ϵ = 78).

FIGURE 12.9
The absorption spectrum of 4-methyl-3-penten-2-one.

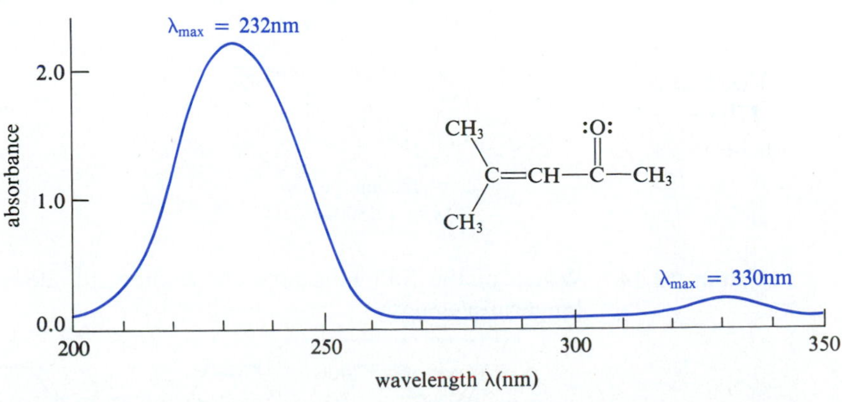

EXAMPLE 12.7 What is the effect of doubling the concentration of a particular absorbing sample on A? on ϵ?

Solution The observed absorbance A will be doubled since A is directly proportional to c. The value of ϵ, however, is a function of molecular structure and is a constant, independent of the concentration.

Problem 12.13 A particular solution of $(CH_3)_2C{=}CH{-}\overset{\overset{\displaystyle O}{\|}}{C}{-}CH_3$, the ketone whose spectrum is shown in Figure 12.9, placed in a 1-cm absorption cell shows a peak at λ_{max} = 232 nm with an observed absorbance A = 2.2. Calculate the concentration of the solution, using the value of ϵ given in the text.

Visible-ultraviolet spectra are most commonly used to detect conjugation. In general, molecules with no double bonds or with only one double bond do not absorb in the visible-ultraviolet region (200 to 800 nm). Conjugated systems do absorb there, however, and the greater the conjugation, the longer the wavelength of maximum absorption, as seen in the following examples:

$CH_2{=}CH{-}CH{=}CH_2$
λ_{max} = 220 nm
(ϵ = 20,900)

$CH_2{=}CH{-}CH{=}CH{-}CH{=}CH_2$
λ_{max} = 257 nm
(ϵ = 35,000)

$CH_2{=}CH{-}CH{=}CH{-}CH{=}CH{-}CH{=}CH_2$
λ_{max} = 287 nm
(ϵ = 52,000)

λ_{max} = 255 nm
(ϵ = 215)

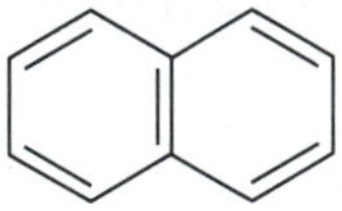

λ_{max} = 314 nm
(ϵ = 289)

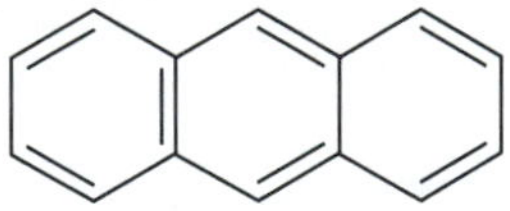

λ_{max} = 380 nm
(ϵ = 9000)

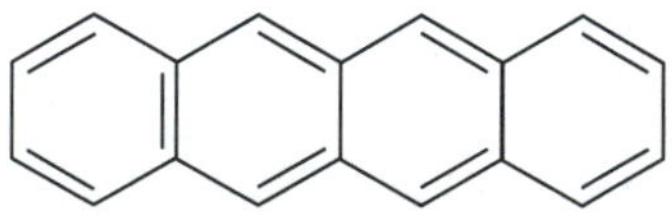

λ_{max} = 480 nm: yellow
(ϵ = 12,500)

Problem 12.14 Which of the following aromatic compounds do you expect to absorb at the longer wavelength?

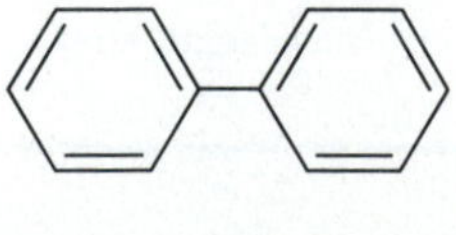

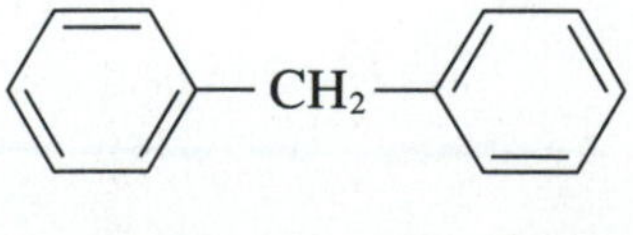

CH_2

Problem 12.15 Naphthalene is colorless, but its isomer azulene is blue. Which compound has the lower-energy pi electronic transition?

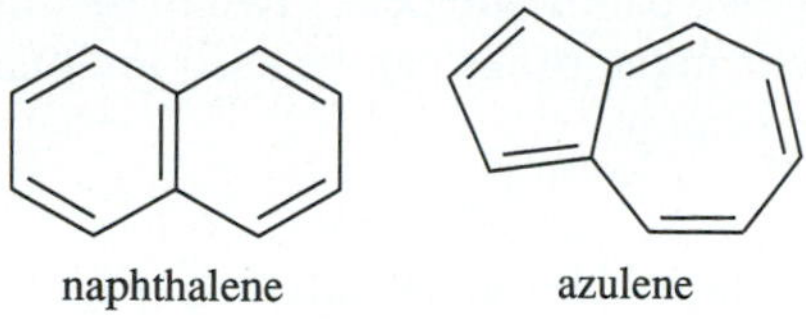

12.7 Mass Spectrometry

Mass spectrometry differs from the other types of spectroscopy discussed in this chapter, in that it does not depend on transitions between energy states. Instead, a mass spectrometer converts molecules to ions, sorts them according to their mass-to-charge ratio (m/e), and determines the relative amounts of each ion present. A small sample of the substance is introduced into a high-vacuum chamber, where it is vaporized and bombarded with high-energy electrons. These bombarding electrons eject an electron from the molecule M, to give a **cation radical** called the **molecular ion** $M^{+\cdot}$ (sometimes referred to as the **parent ion**).

$$M + e^- \longrightarrow \underset{\text{molecular ion}}{M^{+\cdot}} + 2\,e^- \tag{12.6}$$

Methanol, for example, forms a molecular ion in the following way:

$$e^- + CH_3\ddot{\underset{\cdot\cdot}{O}}H \longrightarrow \underset{\text{methanol molecular ion } (m/e = 32)}{[CH_3\dot{\underset{\cdot\cdot}{O}}H]^+} + 2\,e^- \tag{12.7}$$

The beam of these parent ions then passes between the poles of a powerful magnet, which deflects the beam. The extent of the deflection depends on the mass of the ion. Since $M^{+\cdot}$ has a mass that is essentially identical to the mass of the molecule M (the mass of the ejected electron is trivial compared to the mass of the rest of the molecule), **mass spectrometers can be used to determine molecular weights.**

Frequently, mass spectra show a peak one or two mass units *higher* than the molecular weights. How can this be? Recall that the isotope ^{13}C (one mass unit higher than ordinary ^{12}C) has a natural abundance of about 1.1%. This gives rise to an $(M + 1)^{+\cdot}$ peak in carbon compounds. The intensity of this peak relative to the $M^{+\cdot}$ peak is approximately 1.1% times the number of carbons in the compound (because the chance of finding a ^{13}C atom in a compound is proportional to the number of carbon atoms present).

Problem 12.16 An alkane shows an $M^{+\cdot}$ peak at m/e 142. What is its molecular formula? What will be the relative intensities of the 143/142 peaks?

Other isotopic peaks can also be useful. For example, chlorine consists of a mixture of ^{35}Cl (75%) and ^{37}Cl (25%), and bromine consists of a 50 : 50 mixture of ^{79}Br and ^{81}Br. A monochloroalkane will therefore show two parent ion peaks, two mass units apart and in an intensity ratio of 3 : 1. Monobromoalkanes also show two parent ion peaks two mass units apart, but in a 1 : 1 intensity ratio. These isotopic peaks can be used to obtain structural information, as in the following example:

EXAMPLE 12.8 A bromoalkane shows two equal-intensity parent ion peaks at m/e 136 and 138. Deduce its molecular formula.

Solution Only one Br can be present (the molecular weight is not high enough for two bromines). Subtract 136–79 (or 138–81) to get 57 for the mass of the carbons and hydrogens. Dividing 57 by 12 (the mass of carbon), we get 4, with 9 mass units left over for the hydrogens. The formula is C_4H_9Br.

Problem 12.17 A compound containing only C, H, and Cl shows parent ion peaks at m/e 74 and 76 in a ratio of 3 : 1. Suggest possible structures for the compound.

If bombarding electrons have enough energy, they produce not only parent ions but also fragments called **daughter ions.** That is, the original molecular ion breaks into smaller fragments, some of which are ionized and get sorted on an m/e basis by the spectrometer. A prominent peak in the mass spectrum of methanol, for example, is the $M^+ - 1$ peak at $m/e = 31$. This peak arises through loss of a hydrogen atom from the molecular ion.

$$\mathrm{H{-}\overset{H}{\underset{H}{\dot{\dot{C}}}}{-}\overset{\cdot+}{\underset{\cdot\cdot}{O}}{-}H} \longrightarrow \mathrm{H{-}\underset{H}{C}{=}\overset{+}{\underset{\cdot\cdot}{O}}{-}H + H\cdot} \qquad (12.8)$$

$m/e = 32$ $\qquad$ $m/e = 31$

You will recognize this daughter ion as protonated formaldehyde, a resonance-stabilized carbocation.

A mass spectrum consists, then, of a series of signals of varying intensities at different m/e ratios. In practice, most of the ions are singly charged ($e = 1$), so that we can readily obtain their masses, m. Figure 12.10 shows a mass spectrum printed as the computer output of a mass spectrometer. It is the mass spectrum of a typical ketone, 4-octanone. The peak at $m/e = 128$ is the most intense high mass peak in the spectrum and corresponds to the molecular weight of the ketone. In addition, we see certain prominent daughter ion peaks. For example, the peaks at $m/e = 85$ and $m/e = 71$ correspond in mass to $C_4H_9CO^+$ and $C_3H_7CO^+$, respectively. This suggests that one easy fragmentation path for the parent ion is to break the carbon-carbon bond adjacent to the carbonyl group. Ion fragmentation paths depend on ion structure, and the interpretation of mass spectral fragmentation patterns can give significant information about molecular structure.

FIGURE 12.10 The mass spectrum of 4-octanone.

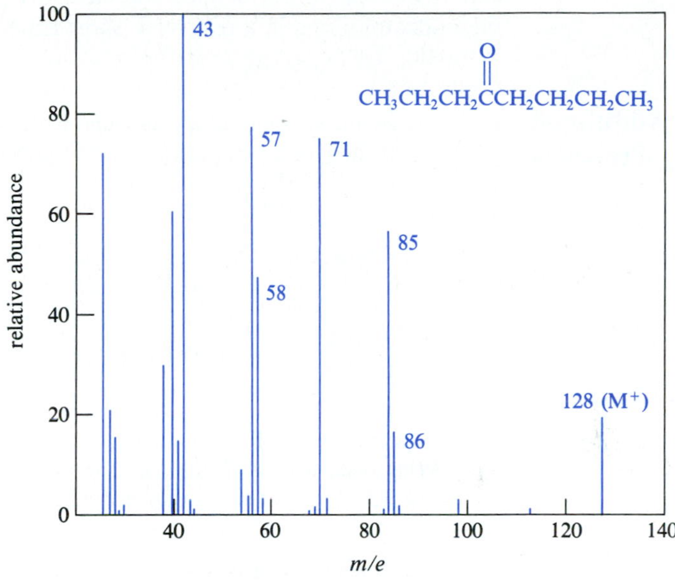

EXAMPLE 12.9 The most intense peak (called the *base peak*) in Figure 12.10 occurs at m/e = 43. Suggest how it might arise.

Solution This peak corresponds to the m/e for $C_3H_7^+$, suggesting that the daughter ion $C_3H_7CO^+$ loses carbon monoxide to give $C_3H_7^+$. This explanation becomes more plausible when we consider that the spectrum also contains an intense peak at m/e = 57, corresponding to the analogous process for $C_4H_9CO^+$. We can summarize these conclusions in the following "family tree" of ions:

$$C_8H_{16}O^{+\cdot}\ M^+\ (128) \xrightarrow{-C_3H_7^{\cdot}} \underset{(85)}{C_4H_9CO^+} \xrightarrow{-CO} \underset{(57)}{C_4H_9^+}$$

$$C_8H_{16}O^{+\cdot}\ M^+\ (128) \xrightarrow{-C_4H_9^{\cdot}} \underset{(71)}{C_3H_7CO^+} \xrightarrow{-CO} \underset{(43)}{C_3H_7^+}$$

Problem 12.18 In what ways will the mass spectrum of 4-heptanone be similar to and in what ways will it differ from the mass spectrum in Figure 12.10?

The types of spectroscopy that we have described here are routinely used in research laboratories. With modern instrumentation, each type of spectrum can be obtained in a few minutes to an hour, including time to prepare the sample. Interpretation of the spectra may take longer, but investigators with experience can often deduce the structure of even complex molecules from their spectra alone in a relatively short time.

Additional Problems

12.19. Draw the structure of a compound with each of the following molecular formulas that will show only one peak in its proton NMR spectrum.
a. C_6H_{12} **b.** $C_3H_6Cl_2$ **c.** C_4H_6
d. $C_{12}H_{18}$ **e.** C_2H_6O **f.** C_5H_{12}

12.20. Tell how many chemically different types of protons are present in
a. $(CH_3)_2CHCH_2CH_3$ **b.** $(CH_3)_2NCH_2CH_3$
c. 1,2-dimethylcyclohexene (ring structure with two CH_3 groups on the double-bonded carbons) **d.** CH_3CH_2OH

12.21. The proton NMR spectrum of a compound, C_4H_9Br, consists of a single sharp peak. What is its structure? The spectrum of an isomer of this compound consists of a doublet at δ 3.2, a complex pattern at δ 1.9, and a doublet at δ 0.9, with relative areas 2 : 1 : 6. What is its structure?

12.22. The chemical shifts of the protons in 2,2-dimethylpropane and TMS are δ 0.95 and δ 0.0, respectively. From these data, what can you deduce about the relative electronegativities of carbon and silicon?

12.23. How could you distinguish between the following pairs of isomers by proton NMR spectroscopy?
a. CH_3CCl_3 and $CH_2ClCHCl_2$ **b.** $CH_3CH_2CH_2OH$ and $(CH_3)_2CHOH$
c. $CH_3CH_2C(=O)—OCH_3$ and $CH_3C(=O)—OCH_2CH_3$
d. $C_6H_5—CH_2—CH=O$ and $C_6H_5—C(=O)—CH_3$

12.24. The proton NMR spectrum of cyclohexane, measured at room temperature, consists of a single peak. At very low temperatures, however, the spectrum consists of two sets of peaks that are equal in area. Suggest an explanation.

12.25. The NMR spectrum of methyl *p*-toluate consists of a singlet at δ 2.35, a singlet at δ 3.82, and two doublets, at δ 7.15 and δ 7.87; relative areas are 3 : 3 : 2 : 2. Draw the structure and determine which protons are responsible for each peak, as far as you can tell (use Table 12.2).

12.26. Using the information in Table 12.2, sketch the proton NMR spectrum of each of the following compounds. Be sure to show all splitting patterns.
a. CH_3CHO **b.** $(CH_3)_2CHOCH(CH_3)_2$
c. $Cl_2C=CH(CH_3)$ **d.** 1,3,5-trimethylbenzene (benzene ring with three CH_3 groups)

12.27. The proton NMR spectrum of a compound, $C_3H_3Cl_5$, consists of a triplet at δ 4.5 and a doublet at δ 6.0 ($J = 7$ Hz), with relative areas 1 : 2. What is its structure?

12.28. A compound is known to be the methyl ester of a toluic acid, but the orientation of the two substituents ($—CH_3$ and $—CO_2CH_3$) on the aromatic ring is not known. The ^{13}C NMR spectrum shows seven peaks. Which isomer is it? What will the proton NMR spectrum look like?

12.29. Do you expect the ^1H and ^{13}C NMR spectra of (*R*)-2-butanol and (*S*)-2-butanol to differ? Explain.

12.30. Do you expect the ^1H and ^{13}C NMR spectra of meso and (2*R*,3*R*)-2,3-butanediol to differ? Explain.

12.31. A compound, C_3H_6O, has no bands in the infrared region around 3500 or 1720 cm^{-1}. What structures can be eliminated by these data? Suggest a possible structure, and tell how you could determine whether it is correct.

12.32. From Table 12.4, deduce the order of ease with which C—H bonds in alkanes, alkenes, and alkynes are stretched. Suggest an explanation for the observed order.

12.33. A very dilute solution of ethanol in carbon tetrachloride shows a sharp infrared band at 3580 cm^{-1}. As the solution is made more concentrated, a new, rather broad band appears at 3250 to 3350 cm^{-1}. Eventually the sharp band disappears and is replaced entirely by the broad band. Explain.

12.34. For the following pairs of compounds, give at least one major peak in the infrared spectrum of one member of the pair that will enable you to distinguish it from the other member.

a. $CH_3CH_2\overset{\overset{O}{\|}}{C}CH_2CH_3$ and $CH_3CH_2OCH_2CH_3$

b. 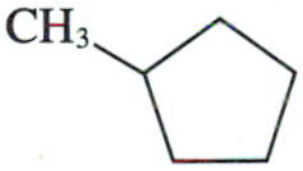 CH_3–(cyclopentane) and CH_3–(cyclopentene)

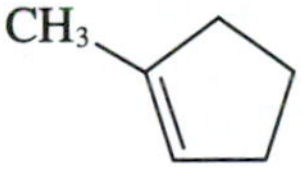

c. $CH_3CH_2OCH_2CH_3$ and $CH_3CH_2CH_2CH_2OH$

d. $CH_3CH_2\overset{\overset{O}{\|}}{C}H$ and $CH_3CH_2\overset{\overset{O}{\|}}{C}—OH$

12.35. What features will be similar in the infrared spectra of the following compounds, and how will their infrared spectra differ?

$CH_3\overset{\overset{O}{\|}}{C}CH_2CH_2CH_2OH$ and $CH_3CH_2\overset{\overset{O}{\|}}{C}CH_2CH_2OH$?

12.36. How could you use infrared spectroscopy to distinguish between the following pairs of isomers?

a. $CH_3\overset{\overset{O}{\|}}{C}CH_2CH_3$ and $CH_3CH(OH)CH{=}CH_2$

b. 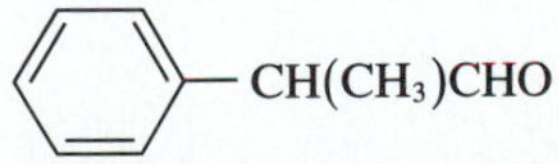 $C_6H_5—CH(CH_3)CHO$ and $C_6H_5—CH{=}CHOCH_3$

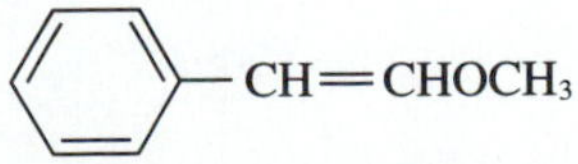

c. $(CH_3CH_2)_3N$ and $(CH_3CH_2CH_2)_2NH$

12.37. A compound, $C_5H_{10}O$, has an intense infrared band at 1725 cm^{-1}. Its proton NMR spectrum consists of a quartet at δ 2.7 and a triplet at δ 0.9, with relative areas 2:3. What is its structure?

12.38. A compound, $C_5H_{10}O_3$, has a strong infrared band at 1745 cm^{-1}. Its proton NMR spectrum consists of a quartet at δ 4.15 and a triplet at δ 1.20; relative areas are 2:3. What is the correct structure?

12.39. You are oxidizing cyclohexanol with CrO_3 to obtain cyclohexanone (eq. 7.33). How could you use infrared spectroscopy to tell that the reaction was complete and that the product was free of starting material?

12.40. Which of the following compounds are not likely to absorb ultraviolet radiation in the range 200 to 400 nm?

a. CH_3CH_2OH **b.**  (benzene) **c.**  (cyclohexane)

d. 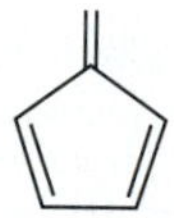 (methylenecyclopentadiene) **e.** $CH_3CH_2OCH_2CH_3$ **f.** $CH_2{=}CHCH_2CH_2CH{=}CH_2$

12.41. The unsaturated aldehydes $CH_3(CH{=}CH)_nCH{=}O$ have ultraviolet absorption spectra that depend on the value of n; the λ_{max} values are 220, 270, 312, and 343 nm as n changes from 1 to 4. Explain.

12.42. The λ_{max} for *cis*-1,2-diphenylethene is at shorter wavelength (280 nm) than for *trans*-1,2-diphenylethene (295 nm). Suggest an explanation.

12.43. A sample of cyclohexane is suspected of being contaminated with benzene, from which it had been prepared by hydrogenation. At $\lambda = 255$ nm, benzene has the molar absorptivity $\epsilon = 215$, whereas cyclohexane does not absorb at that wavelength ($\epsilon = 0$). An ultraviolet spectrum of the contaminated cyclohexane (obtained in a 1.0-cm cell) shows an absorbance $A = 0.43$ at 255 nm. Calculate the concentration of benzene in the cyclohexane.

12.44. Write a formula for the molecular ion of ethanol.

12.45. The mass spectrum of 1-pentanol shows an intense daughter ion peak at $m/e = 31$. Explain how this peak might arise.

12.46. An alcohol, $C_5H_{12}O$, shows a daughter ion peak at $m/e = 59$ in its mass spectrum. An isomeric alcohol shows no daughter ion at $m/e = 59$ but does have a peak at $m/e = 45$. Suggest possible structures for each isomer. How could you confirm these structures by proton NMR spectroscopy? by ^{13}C NMR spectroscopy?

12.47. A hydrocarbon shows a parent ion peak in its mass spectrum at $m/e = 102$. Its proton NMR spectrum shows peaks at δ 2.7 and δ 7.4, with relative areas 1:5. What is the correct structure?

12.48. A compound containing only C, H, and Br shows three parent ion peaks: at m/e 198, 200, and 202 with relative intensities 1:2:1. Deduce its molecular formula.

CHAPTER 13

Heterocyclic Compounds

13.1 Introduction

From an organic chemist's viewpoint, heteroatoms are atoms other than carbon or hydrogen that may be present in organic compounds. The most common heteroatoms are oxygen, nitrogen, and sulfur. In heterocyclic compounds, one or more of these heteroatoms replaces carbon in a ring.

Heterocycles form the largest class of organic compounds. In fact, most natural products and drugs contain heterocyclic rings; indeed, well over half of all organic chemical publications deal in one way or another with heterocycles.

Heterocycles can be divided into two subgroups: **nonaromatic** and **aromatic.** We have already encountered a few nonaromatic heterocycles—ethylene oxide and other cyclic ethers (Chapter 8), cyclic acetals (Chapter 9), and cyclic esters (lactones, Chapter 10). In general, these nonaromatic heterocycles behave a great deal like their acyclic analogs with the same functional groups and do not require special discussion.

Much more important are the **aromatic heterocycles;** in this chapter, we will focus our attention on them. We begin with an important six-membered aromatic nitrogen heterocycle, **pyridine.**

13.2 Pyridine: Bonding and Basicity

Pyridine has a structure similar to that of benzene, except that **one CH unit is replaced by a nitrogen atom.**

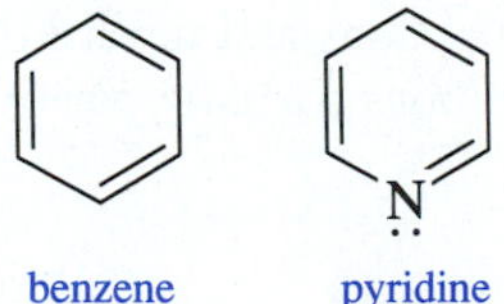

benzene (bp 80°C) pyridine (bp 115°C)

The orbital pictures for benzene and pyridine are similar (review Sec. 4.5 and Figure 4.2). The nitrogen atom, like the carbons, is sp^2 hybridized, with one electron in a p orbital perpendicular to the ring plane.

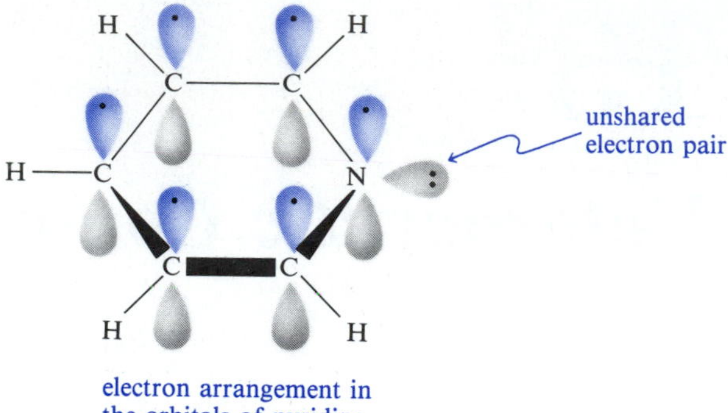

electron arrangement in the orbitals of pyridine

Thus, the nitrogen contributes *one* electron to the six electrons that form the aromatic pi cloud above and below the ring plane. On the other hand, the unshared electron pair on nitrogen lies in the ring plane (just like the C—H bonds) in an sp^2 orbital.

Pyridine is a resonance hybrid of Kekulé-type structures.

Because of the similarities in bonding, pyridine resembles benzene in shape. It is planar, with nearly perfect hexagonal geometry. It is aromatic, and tends to undergo substitution rather than addition reactions.

But the substitution of nitrogen for carbon changes many of the properties. Like benzene, pyridine is miscible with most organic solvents, but unlike benzene, pyridine is also completely miscible with water! One explanation lies in its hydrogen-bonding capability.

N:---H—O—H

Another reason is that pyridine is much more polar than benzene. The nitrogen atom is electron-withdrawing compared with carbon; hence there is a shift of electrons away from the ring carbons and toward the nitrogen, making it partially negative and the ring carbons partially positive.

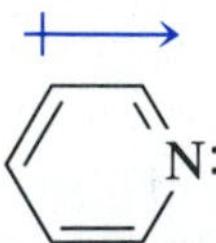

This polarity enhances the solubility of pyridine in polar solvents like water.

Pyridine is a weakly basic tertiary amine, with $pK_a = 5.29$. It is a much weaker base than aliphatic amines ($pK_a \cong 10$; see Table 11.3), mainly because of the different hybridization of the nitrogen (sp^2 in pyridine, sp^3 in aliphatic amines). The greater *s*-character ($\frac{1}{3}s$ in pyridine, $\frac{1}{4}s$ in aliphatic amines) means that the unshared electron pair is held closer to the nitrogen nucleus in pyridine, decreasing its basicity.

Pyridine reacts with strong acids to form **pyridinium salts**.

pyridinium chloride

(13.1)

For this reason, pyridine is often used as a scavenger in acid-producing reactions (for example, in the reaction of thionyl chloride with alcohols, eq. 7.28).

Problem 13.1 Write an equation for the reaction of pyridine with

a. cold sulfuric acid.
b. methyl iodide (see Sec. 11.12).

13.3 *Substitution in Pyridine*

Though aromatic, *pyridine is very resistant to electrophilic aromatic substitution* and undergoes reaction only under drastic conditions. For example, nitration or bromination requires high temperatures and forcing strong acid catalysis.

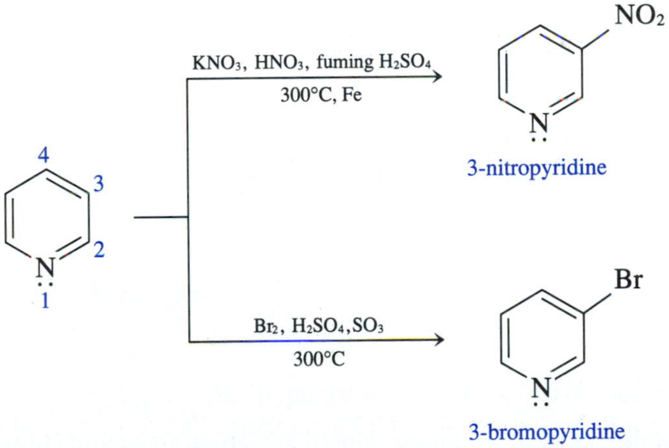

(13.2)

One reason for this sluggishness is that electron withdrawal by the nitrogen makes the ring partially positive and therefore not receptive to attack by electrophiles, which are also positive. A second reason is that, under the acidic conditions for these reactions, most of the pyridine is protonated and present as the positive pyridinium ion, which is even more unlikely to be attacked by electrophiles than neutral pyridine.

When substitution does occur, electrophiles attack pyridine mainly at C3. The cationic intermediate (review Sec. 4.10) is *least unfavorable* in this case, because it does not put a positive charge on the electron-deficient nitrogen (especially bad if the nitrogen is protonated).

EXAMPLE 13.1 Draw all contributors to the resonance hybrid for electrophilic attack at C3 of pyridine.

Solution

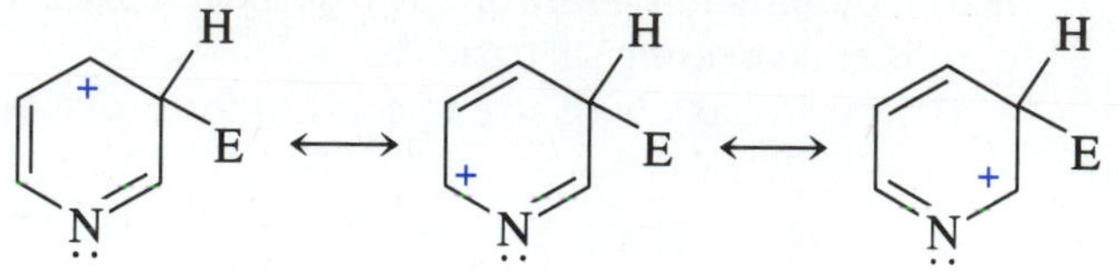

For substitution at C3, the "pyridinonium ion" positive charge is delocalized to C2, C4, and C6, but not to the nitrogen.

Problem 13.2 Repeat Example 13.1, but for electrophilic substitution at C2 or C4 of pyridine. Explain why substitution at C3 (eq. 13.2) is preferred.

Although resistant to electrophilic substitution, *pyridine undergoes nucleophilic substitution.* The pyridine ring is partially positive (due to electron withdrawal by the nitrogen) and is therefore susceptible to attack by nucleophiles. Here are two examples:

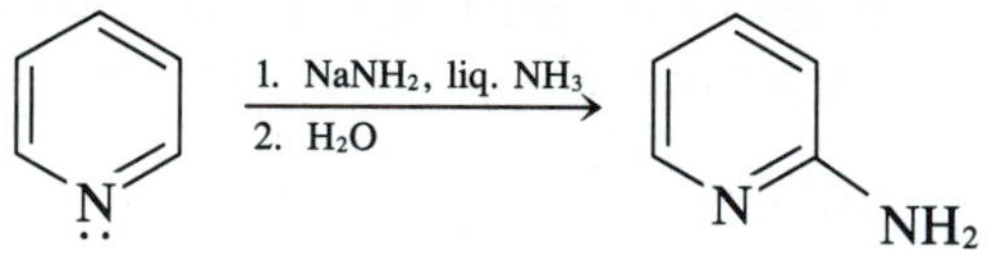

1. $NaNH_2$, liq. NH_3
2. H_2O

2-aminopyridine (13.3)

$NaOCH_3, CH_3OH$

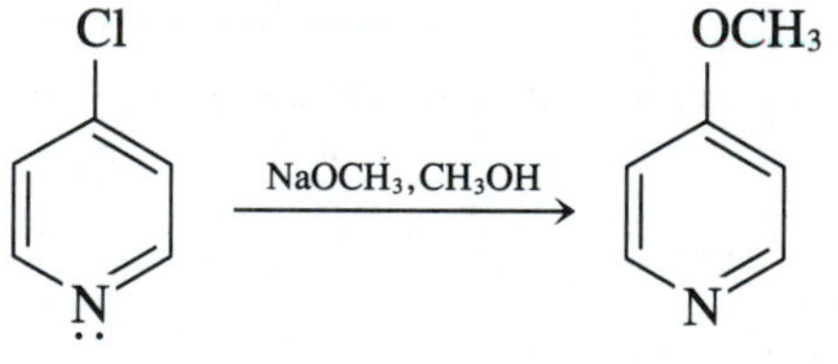

4-chloropyridine 4-methoxypyridine (13.4)

EXAMPLE 13.2 Write a mechanism for eq. 13.3.

Solution Attack of amide ion at C2 gives an anionic intermediate with the negative charge mainly on nitrogen.

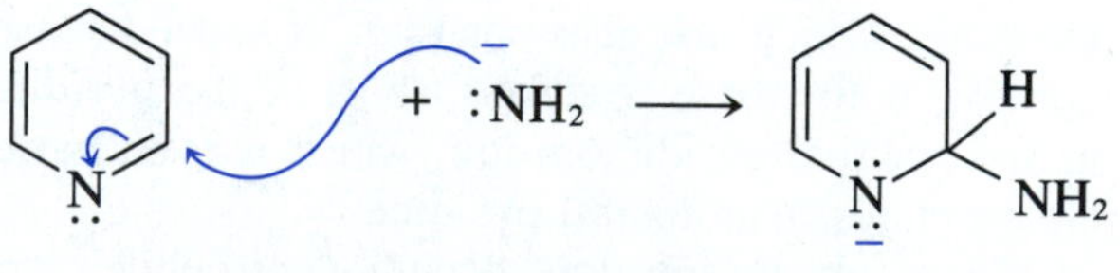

You can think of this as nucleophilic addition to a C=N bond, analogous to addition of nucleophiles to a C=O bond (Sec. 9.7).

To restore aromaticity, hydride ion is displaced. The hydride then attacks the amino group to give hydrogen gas and an amide-type anion (cf. eq. 11.17).

$$\text{(2-amino-1,2-dihydropyridinide)} \longrightarrow \text{2-aminopyridine} + H^- \longrightarrow \text{2-pyridylamide ion } (\bar{N}H) + H_2$$

In the final step, this ion is protonated by water.

$$\text{2-pyridylamide ion } (\bar{N}H) + H_2O \longrightarrow \underset{\text{2-aminopyridine}}{\text{(pyridine-2-}NH_2)} + OH^-$$

Problem 13.3 Write a mechanism for eq. 13.4.

Pyridine and alkylpyridines are found in coal tar. The monomethyl pyridines (called **picolines**) undergo side-chain oxidation to carboxylic acids (review Sec. 10.8b). For example, 3-picoline gives **nicotinic acid** (or **niacin**), a vitamin essential in the human diet, to prevent the disease *pellagra*.

$$\underset{\text{3-picoline}}{(3\text{-}CH_3\text{-pyridine})} \xrightarrow{KMnO_4} \underset{\text{nicotinic acid}}{(3\text{-}CO_2H\text{-pyridine})} \tag{13.5}$$

Pyridine can be reduced to the fully saturated secondary amine **piperidine.**

$$\underset{\text{pyridine}}{\text{}} \xrightarrow[Pt]{3\,H_2} \underset{\text{piperidine}}{(\text{N–H})} \tag{13.6}$$

(13.6)

The pyridine and piperidine rings are found in many natural products. Examples include **nicotine** (the major alkaloid in tobacco, used as an agricultural insecticide and highly toxic to humans), **pyridoxine** (vitamin B_6, a coenzyme), and **coniine** (the toxic principle of poison hemlock, taken by Socrates).

nicotine (N–CH_3) pyridoxine (CH_2OH, CH_2OH, HO, CH_3) (+)-coniine (H, $CH_2CH_2CH_3$, N–H)

Problem 13.4 Naturally occurring coniine is the (+)-isomer shown. What is its configuration, *R* or *S*?

Problem 13.5 Naturally occurring nicotine is the *(S)-(−)* isomer. Locate the chiral center, and draw the three-dimensional structure.

Problem 13.6 Nicotine contains two nitrogens, one in a pyridine ring and one in a pyrrolidine ring. It reacts with *one* equivalent of HCl to form a crystalline salt, $C_{10}H_{15}N_2Cl$. Draw its structure. Nicotine also reacts with *two* equivalents of HCl to form another crystalline salt, $C_{10}H_{16}N_2Cl_2$. Draw its structure.

13.4 Other Six-Membered Heterocycles

The pyridine ring can be fused with benzene rings to produce polycyclic aromatic heterocycles. The most important examples are **quinoline** and **isoquinoline,** analogs of naphthalene (Sec. 4.14) but with N in place of CH at C1 or C2.

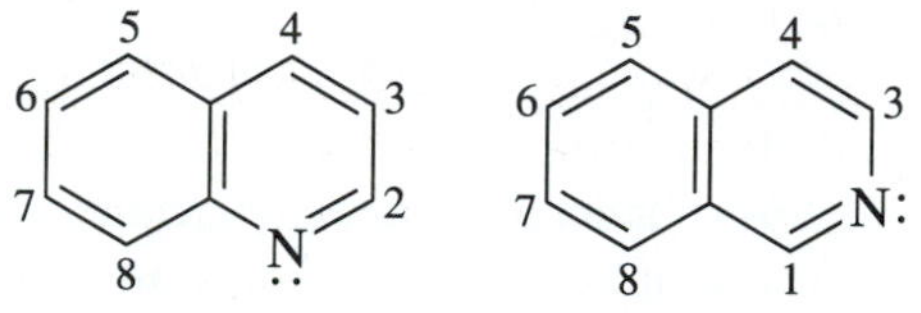

quinoline
bp 237°C

isoquinoline
bp 243°C, mp 26.5°C

Electrophilic substitution in these amines occurs in the carbocyclic ring, illustrating the inactivity towards electrophiles of the pyridinoid vis-à-vis the benzenoid rings.

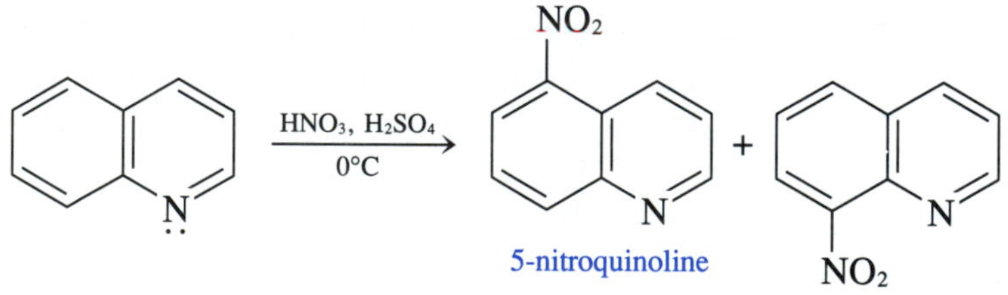

$$\text{quinoline} \xrightarrow[0°C]{HNO_3,\ H_2SO_4} \text{5-nitroquinoline} + \text{8-nitroquinoline} \quad (13.7)$$

The stability of the pyridine ring is also illustrated by its resistance to oxidation. Thus, when quinoline is treated with potassium permanganate, the benzenoid ring is oxidized.

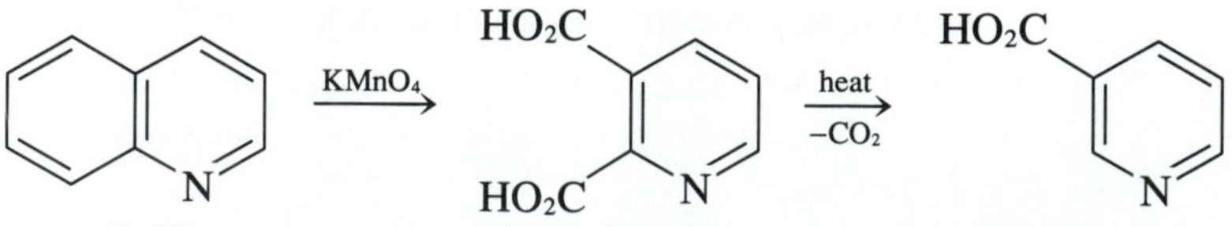

$$\text{quinoline} \xrightarrow{KMnO_4} \text{quinolinic acid} \xrightarrow[-CO_2]{\text{heat}} \text{nicotinic acid} \quad (13.8)$$

The quinoline and isoquinoline rings occur in many natural products. Good examples are **quinine** (which occurs in cinchona bark and is used to treat malaria) and **papaverine** (present in opium and used as a muscle relaxant).

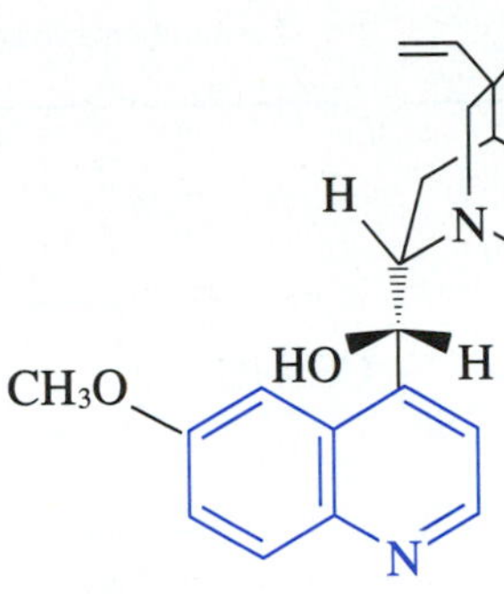

quinine

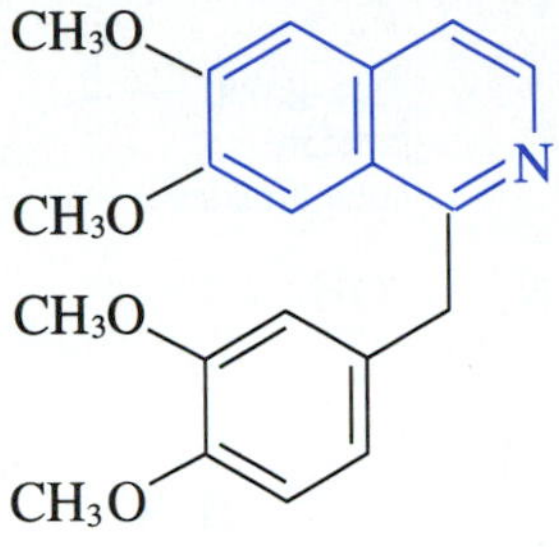

papaverine

It is logical to ask: If we can replace one CH with N in a benzene ring to create pyridine, can we replace more than one of the benzene CH groups with nitrogens? The answer is yes. For example, there are *three* **diazines.**

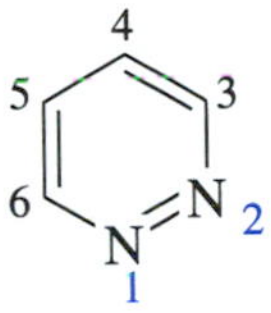

pyridazine
bp 208°C

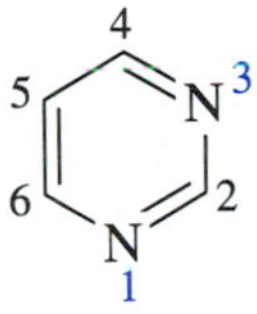

pyrimidine
bp 134°C

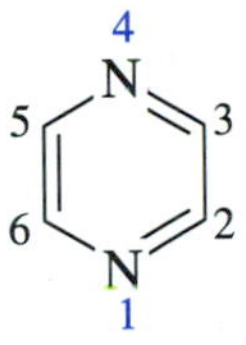

pyrazine
bp 118°C

Of these, the most important are the **pyrimidines,** derivatives of which (**cytosine, thymine,** and **uracil**) are important bases in nucleic acids (DNA and RNA; see Chapter 18).

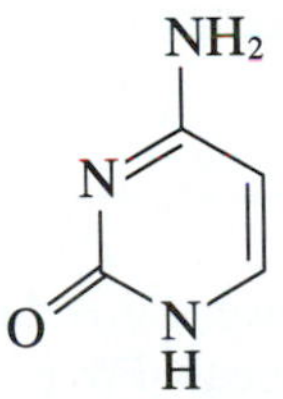

cytosine

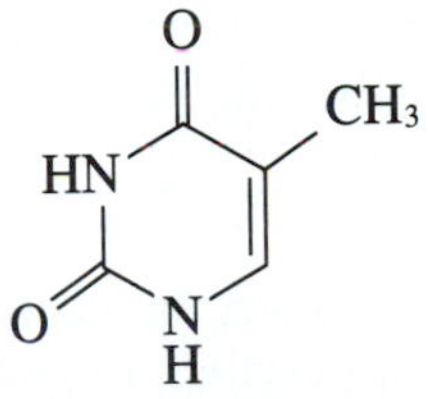

thymine

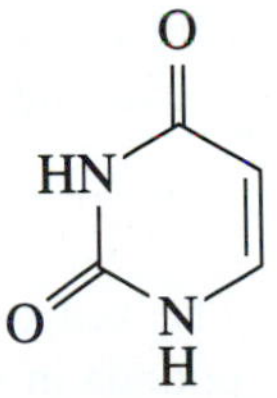

uracil

Triazines and tetrazines are also known, but neither pentazine nor hexazine (which would really not be a heterocycle, since it would contain only one element and be an allotrope of nitrogen) are known.

Similar analogs of naphthalene with more than one nitrogen are also known. The **pteridine** ring, with nitrogens replacing C1, C3, C5, and C8 in naphthalene, is present in many natural products, such as the butterfly wing pigment **xanthopterin** and the blood-forming vitamin **folic acid.** The analog of folic acid, but with an NH_2 group in place of the OH group on the pteridine ring and a methyl group on the first nitrogen in the side chain, is useful in cancer chemotherapy (it is called **methotrexate**).

pteridine part | 4-aminobenzoic acid part | glutamic acid part

NH_2 group here in methotrexate

methyl group here in methotrexate

xanthopterin

folic acid

If we replace a benzene CH group with oxygen (instead of nitrogen), we obtain an aromatic cation called a **pyrylium ion.**

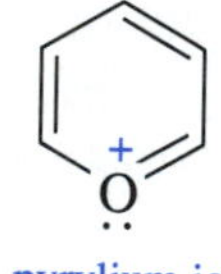

pyrylium ion

EXAMPLE 13.3 Describe the bonding in a pyrylium ion.

Solution The oxygen is sp^2 hybridized, with only five electrons (hence the +1 formal charge). Two of these, the unshared electron pair, are in an sp^2 orbital that lies in the ring plane. Two others are in sp^2 orbitals that form the sigma bonds with the adjacent carbon atoms, also sp^2 hybridized. The fifth electron is in a p orbital perpendicular to the ring plane; it overlaps with similar orbitals on each ring carbon atom (as in benzene) to form the aromatic six-electron pi cloud above and below the ring plane.

The red and blue colors of many flowers are due to **anthocyanines,** compounds in which a pyrylium ring is present. For example, the pigment responsible for the color of red roses is

red rose pigment
(Gl = Glucose)

The glucose units solubilize the pigment in the aqueous cellular fluid. The color of blue cornflowers is due to the same pigment, but complexed with metallic ions, such as Fe^{3+} or Al^{3+}. Other flower pigments have the same basic structures, but with fewer, more, or differently located hydroxyl groups.

To summarize, six-membered heterocycles with nitrogen (one or more) in place of the CH groups in benzenoid aromatics are also aromatic. Each nitrogen atom contributes one electron to the 6π aromatic system. The nitrogens are also basic, because of unshared electron pairs located in the ring plane. The nitrogen, being electron-withdrawing, deactivates the ring toward electrophilic aromatic substitution but activates it toward nucleophilic aromatic substitution. These heterocyclic aromatic rings are present in many natural products.

Now let us examine rather different types of heteroaromatic compounds: those with five-membered rings.

13.5 Five-Membered Heterocycles: Furan, Pyrrole, and Thiophene

Furan, pyrrole, and thiophene are important five-membered ring heterocycles with one heteroatom.

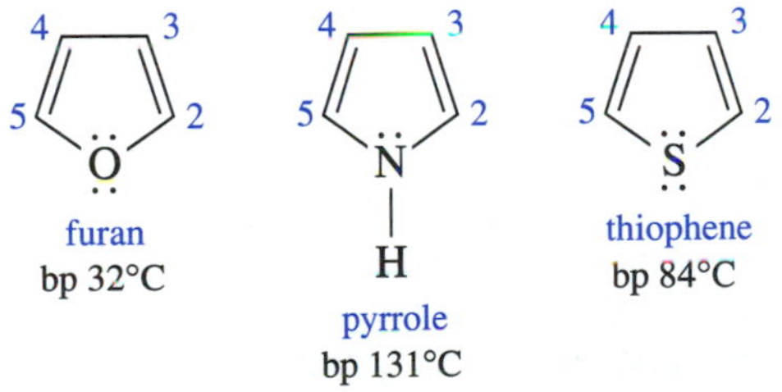

furan
bp 32°C

pyrrole
bp 131°C

thiophene
bp 84°C

Numbering begins with the heteroatom and proceeds around the ring.

The most important commercial source of furans is **furfural** (2-furaldehyde), obtained by heating oat hulls, corn cobs, or straw with strong acid. These naturally occurring materials are polymers of a five-carbon sugar, which is dehydrated by the acid to furfural.

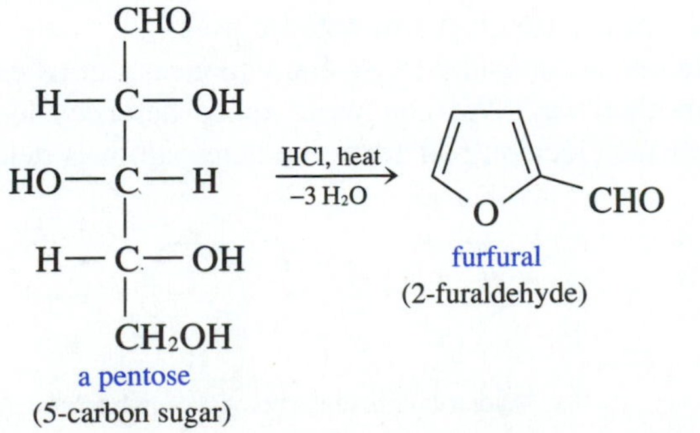

$$\text{CHO–CH(OH)–CH(OH)–CH(OH)–CH}_2\text{OH} \xrightarrow[-3\,H_2O]{\text{HCl, heat}} \text{furfural} \qquad (13.9)$$

a pentose
(5-carbon sugar)

furfural
(2-furaldehyde)

Pyrrole is obtained commercially by distillation of coal tar or from furan, ammonia, and a catalyst. Thiophene is obtained by heating a mixture of butanes and butenes with sulfur.

As drawn, the structures of these heterocycles look as if they ought to be dienes, but in fact, these ring systems are aromatic; they behave like benzene in many ways, particularly in their tendency to undergo electrophilic aromatic substitution. The reasons for this behavior will become clearer if we examine the bonding in these molecules.

Furan has a planar, pentagonal structure in which each ring atom is sp^2 hybridized.

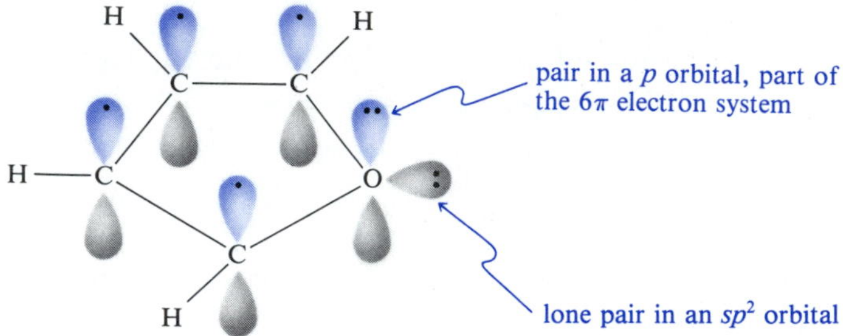

orbital structure of furan

Each ring atom uses two of these orbitals to form sigma bonds with its neighbors. Each carbon also uses one sp^2 orbital to form a sigma bond in the ring plane with a hydrogen atom and has one electron in a p orbital perpendicular to the ring plane. Now look at the oxygen. It has an unshared electron pair in an sp^2 orbital in the ring plane and **two electrons in a *p* orbital perpendicular to the ring plane.** These two electrons overlap with those in the p orbitals on the carbons to form a 6π electron cloud above and below the ring plane, just as in benzene. The bonding in pyrrole and thiophene is similar to that in furan.

The important difference between five- and six-membered aromatic heterocycles is that, **in the five-membered heterocycles, the heteroatom contributes *two* electrons to the aromatic 6π systems, whereas in six-membered heterocycles, the heteroatom contributes only one electron to that system.** This difference has important consequences for the chemical behavior of the two types of heterocycles.

Before we consider those consequences, let us examine the bonding in furan in another way. We can write these heterocycles as a resonance hybrid in which the electron pair from the heteroatom is delocalized to all ring atoms.

contributors to the resonance hybrid structure of furan

Notice that four of these structures are dipolar and place a *negative* charge on the ring carbons. As we will see, this enhances their susceptibility to attack by electrophiles.

EXAMPLE 13.4 Although pyrrole is an amine, it is an exceedingly weak base. It has a pK_a of -4.4, nearly 10^{10} times weaker than pyridine ($pK_a = 5.29$). Suggest an explanation.

Solution In pyrrole, the unshared electron pair on nitrogen is part of the aromatic 6π electron system.

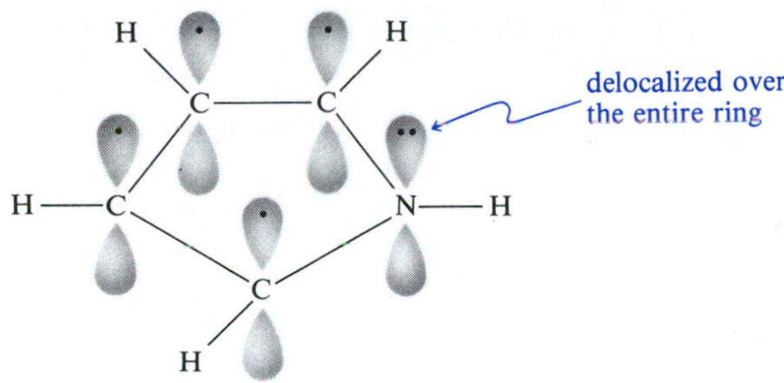

Protonation of the nitrogen would destroy the aromatic system, thus losing its resonance energy. Hence pyrrole is a very weak base; in very strong acids, *it is protonated on carbon rather than on the nitrogen.*

pyrrole (N–H) + H^+ ⇌ 2-H-pyrrolium ion ($\overset{+}{N}$–H, CH_2)

$pK_a = -4.4$

In pyridine, the unshared electron pair is *not* part of the aromatic pi system and is available for adding a proton (eq. 13.1).

Problem 13.7 Pyrrole is insoluble in water, but its saturated analog, pyrrolidine, is completely miscible with water. Suggest an explanation.

CH_2—CH_2 / CH_2 CH_2 / $\ddot{N}$ / H

pyrrolidine

13.6 Electrophilic Substitution in Furan, Pyrrole, and Thiophene

Furan, pyrrole, and thiophene are all much more reactive than benzene toward electrophilic substitution. Each reacts predominantly at the 2-position (and if that position is already substituted, at the 5-position). Here are typical examples:

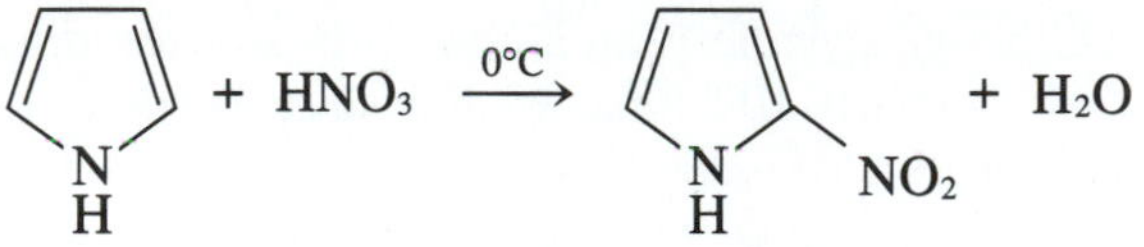

pyrrole + HNO_3 $\xrightarrow{0°C}$ 2-nitropyrrole + H_2O (13.10)

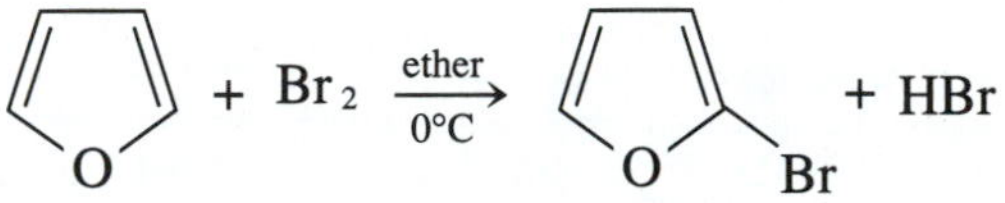

2-nitropyrrole

furan + Br_2 $\xrightarrow[0°C]{ether}$ 2-bromofuran + HBr (13.11)

2-bromofuran

2-methylthiophene + CH_3CCl (O) $\xrightarrow{SnCl_4}$ 2-acetyl-5-methylthiophene + HCl (13.12)

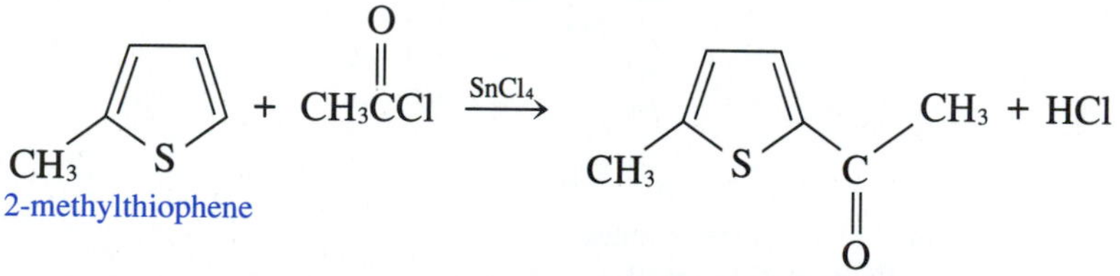

2-methylthiophene

2-acetyl-5-methylthiophene

The reason for predominant attack at C2 (instead of the other possibility, C3) becomes clear if we examine the carbocation intermediate in each case:

Attack of electrophile at C2 (X = NH, O, or S)

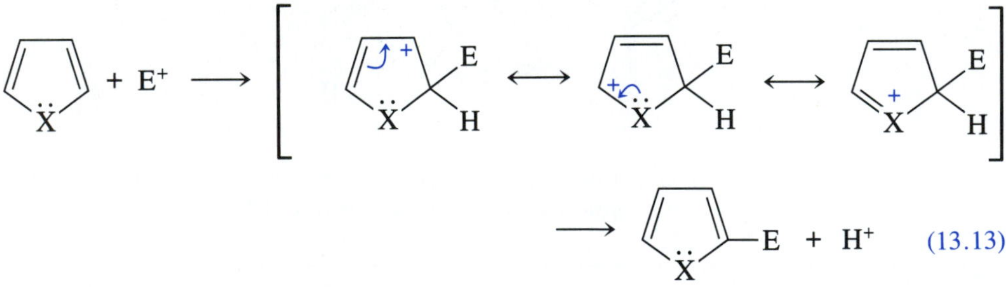

$$+ E^+ \longrightarrow [\ldots \longleftrightarrow \ldots \longleftrightarrow \ldots] \longrightarrow \text{2-E product} + H^+ \qquad (13.13)$$

Attack of electrophile at C3

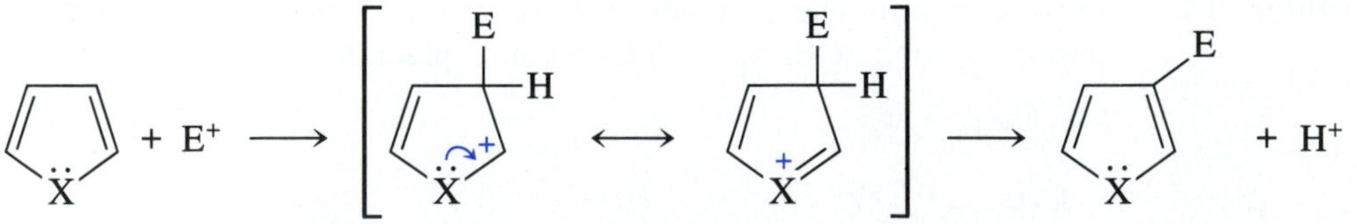

$$+ E^+ \longrightarrow [\ldots \longleftrightarrow \ldots] \longrightarrow \text{3-E product} + H^+ \qquad (13.14)$$

Attack at C2 is preferred because, in the carbocation intermediate, the positive charge can be delocalized over *three* atoms, whereas attack at C3 allows delocalization of the charge over only two positions.

Problem 13.8 Write out the steps in the mechanism for bromination of furan (eq. 13.11).

A WORD ABOUT . . .

26. Porphyrins: What Makes Blood Red and Grass Green?

Pyrrole rings form the building blocks of several biologically important pigments. The **porphyrins** are macrocyclic compounds that contain four pyrrole rings linked by one-carbon bridges. The molecules are flat and have a conjugated system of 18π electrons, shown in color in the parent molecule, **porphine.**

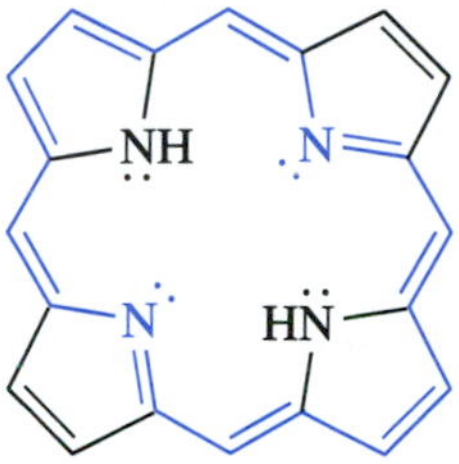

porphine
red crystals

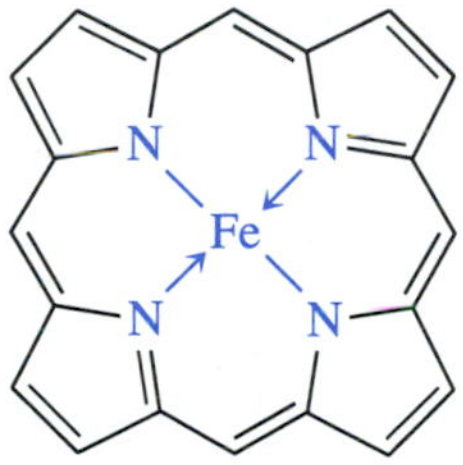

the Fe^{2-} porphine complex
brown cubic crystals

Porphyrins are exceptionally stable and highly colored. They form complexes with metallic ions. In these complexes, the hydrogens bonded to two of the nitrogens are absent, and each of the four nitrogens donates an electron pair to the metal, which sits in the middle of the structure.

Porphine itself does not occur in nature, but analogous compounds with various side chains on the pyrrole rings are some of the most important life-sustaining compounds of nature. One example is **heme,** the iron-porphyrin complex responsible for the red color of arterial blood.

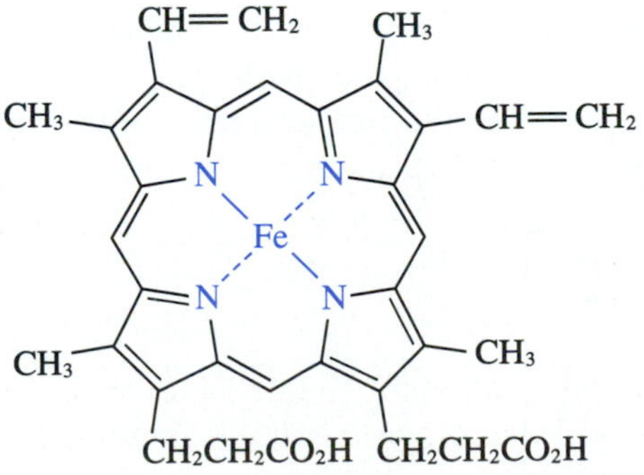

heme
brown needles with violet sheen

Heme is present in red blood cells as a complex with a protein called **globin.** The complex is called **hemoglobin.** This complex is responsible for binding molecular oxygen and transporting it to sites where it is needed. The iron atom is complexed with the four porphyrin nitrogens but also has two additional coordination sites, above and below the plane of the porphyrin ring. One of these sites is occupied by an imidazole ring from a histidine unit in the protein, but the second site is available for **reversible binding to oxygen.** Carbon monoxide is toxic because it, too, can bind to this site, preventing oxygen transport to the lungs and elsewhere and ultimately leading to death by suffocation. Fortunately, this CO binding can be reversed if oxygen is administered quickly to a patient poisoned by the gas.

Heme is also associated with **myoglobin,** the oxygen-transport protein of muscle, where it performs a similar function.

The green color of plants is due to **chlorophyll,** a magnesium complex of a modified porphyrin. This pigment is present in chloroplasts, one or more of which can be present in cells, depending on the plant. Chlorophyll-*a* is an ester of the long-chain alcohol **phytol,** which helps solubilize the pigment in the chloroplasts.

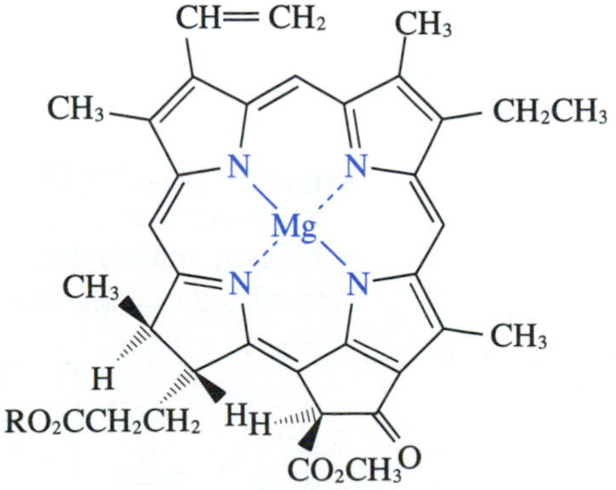

chlorophyll-*a*
blue-black crystals, mp 117-120°C

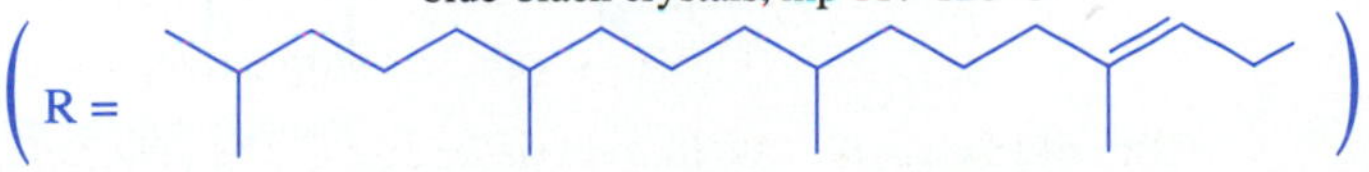

Chlorophyll-*b*, another plant pigment, differs from chlorophyll-*a* only in the replacement of one ring methyl group by an aldehyde group. The mechanism of photosynthesis (the conversion of CO_2 and H_2O by plants to carbohydrates) is quite complex and involves many steps. The first of these is the absorption of sunlight by chlorophyll and its conversion into chemical energy. Thus, chlorophyll is, in a sense, a solar energy converter.

13.7 Other Five-Membered Heterocycles: Azoles

It is possible to introduce a second heteroatom (and even a third and fourth) into five-membered heterocycles. The most important of these are the **azoles,** in which the second heteroatom is nitrogen.

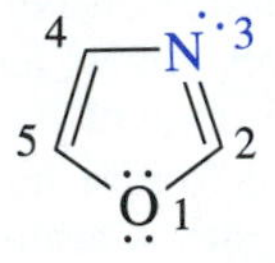

oxazole 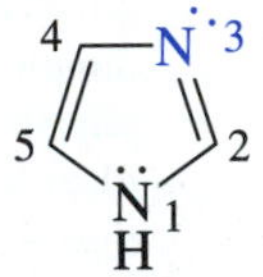 imidazole 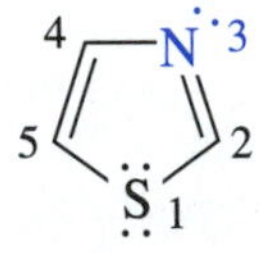 thiazole

Analogs in which the two heteroatoms are adjacent are also known.

As with pyridine, these heterocycles can be thought of as derived by replacing an aromatic CH group by N. *The unshared electron pair on this nitrogen (at position 3) is therefore* not *part of the 6π aromatic system,* as seen in the following orbital picture for imidazole:

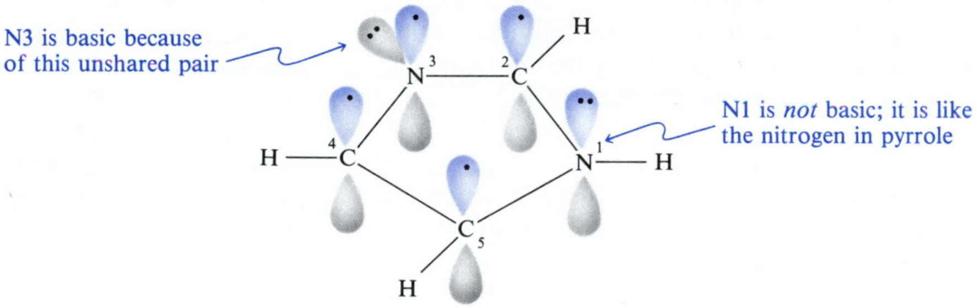

N3 is basic because of this unshared pair

N1 is *not* basic; it is like the nitrogen in pyrrole

bonding in imidazole

Consequently, the N3 nitrogen is basic and can be protonated. Imidazole is even more basic than pyridine ($pK_a = 7.0$; compare with pK_a 5.2 for pyridine), mainly because the positive charge in the imidazolium ion can be delocalized equally over both nitrogens.

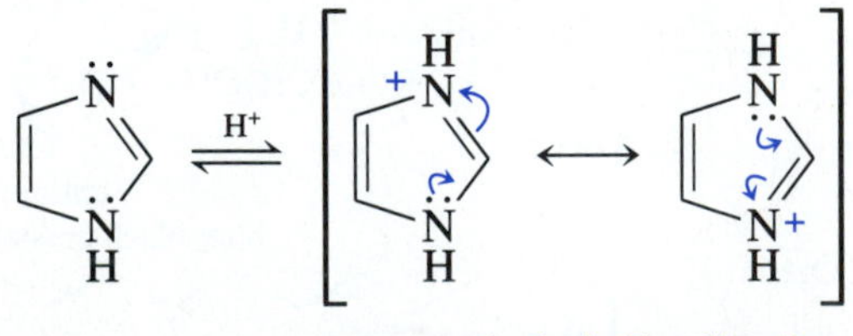

$$\xrightleftharpoons{H^+} \qquad (13.15)$$

resonance in the imidazolium ion

These ring systems occur in nature. For example, the imidazole skeleton is present in the amino acid **histidine,** where it plays an important role in the reactions of many enzymes. Decarboxylation of histidine gives **histamine,** a toxic substance present in combination with proteins in body tissues. It is released as a consequence of allergic hypersensitivity or inflammation (for example, in hay fever sufferers). Many **antihistamines,** compounds that counteract the effects of histamine, have been developed. One of the better known of these is the drug **benadryl** (diphenylhydramine).

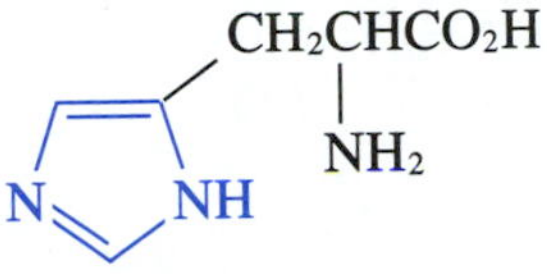

CH_2CHCO_2H / NH_2 — histidine

$CH_2CH_2NH_2$ — histamine

$(C_6H_5)_2CHOCH_2CH_2N(CH_3)_2$

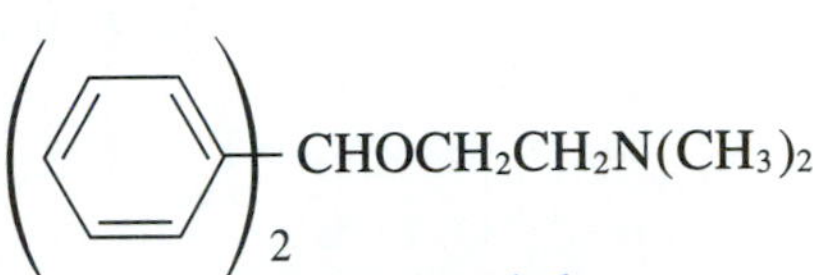

benadryl
(an antihistamine)

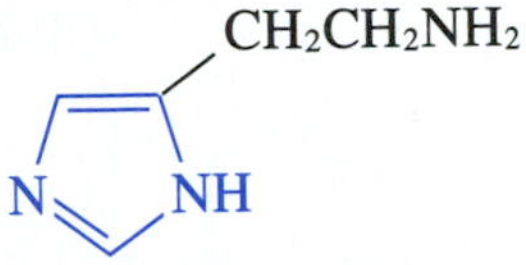

The thiazole ring occurs in **thiamin** (vitamin B_1), a coenzyme required for certain metabolic processes and hence essential to life (thiamin also contains a pyrimidine ring). In its reduced form, the tetrahydrothiazole ring appears in penicillins, important antibiotics.

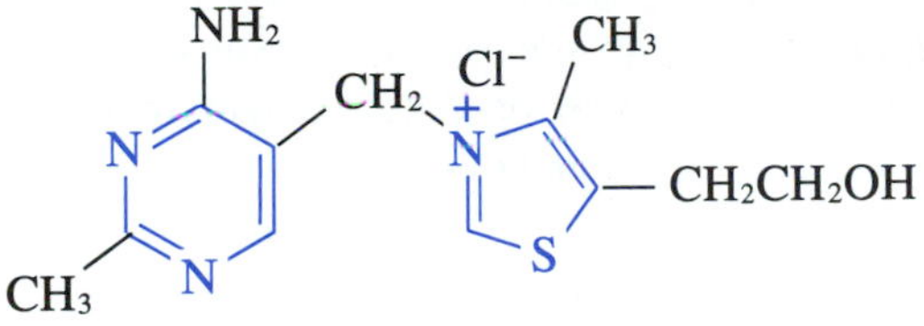

thiamin
(vitamin B_1)

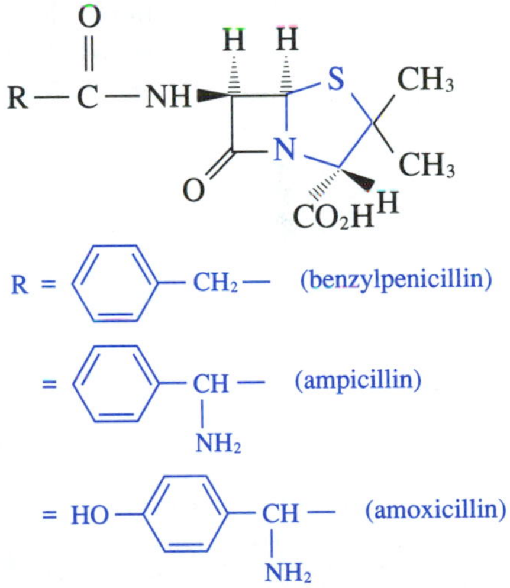

R = C_6H_5—CH_2— (benzylpenicillin)

= C_6H_5—CH(NH_2)— (ampicillin)

= HO—C_6H_4—CH(NH_2)— (amoxicillin)

13.8 Fused-Ring Five-Membered Heterocycles: Indoles and Purines

Another aromatic or heteroaromatic ring can be fused to the double bonds of five-membered heterocycles. For example, **indole** has a benzene ring fused to the C2–C3 bond of pyrrole.

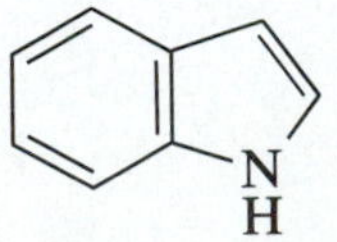

indole

The indole ring system, which is prevalent in a number of important natural products, is usually biosynthesized from the amino acid **tryptophan,** one of the building blocks of proteins. Indole itself and its 3-methyl derivative, **skatole,** are formed during protein decay.

Decarboxylation of tryptophan gives **tryptamine.** Several compounds with this skeleton have a profound effect on the brain and nervous system. One example is **serotonin** (5-hydroxytryptamine), a neurotransmitter and vasoconstrictor active in the central nervous system.

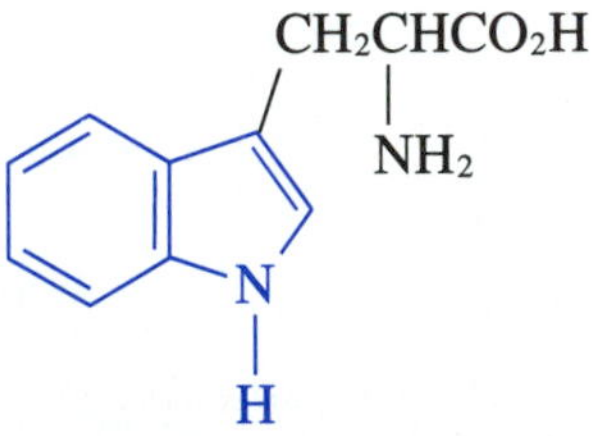

CH_2CHCO_2H / NH_2 / N / H

tryptophan

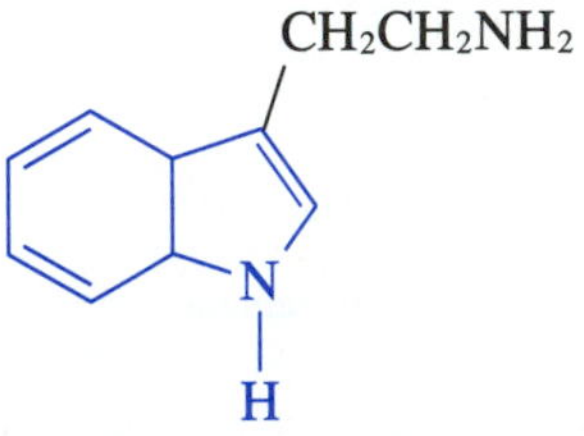

$CH_2CH_2NH_2$ / N / H

tryptamine

HO / $CH_2CH_2NH_2$ / N / H

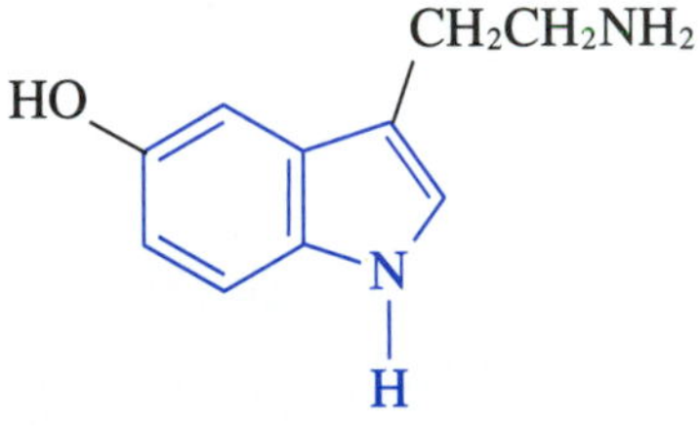

serotonin

The tryptamine skeleton is disguised but present (shown in color) in more complex molecules. **Reserpine**, present in Indian snake root (*Rauwolfia serpintina*), which grows wild on the foothills of the Himalayas, has been used medically for centuries. It lowers blood pressure and is used to calm schizophrenics and improve their accessibility to psychiatric treatment. **Lysergic acid** is present in the fungus *ergot,* which grows on rye and other grains. Conversion of the carboxyl group to its diethylamide gives the extremely potent hallucinogen LSD.

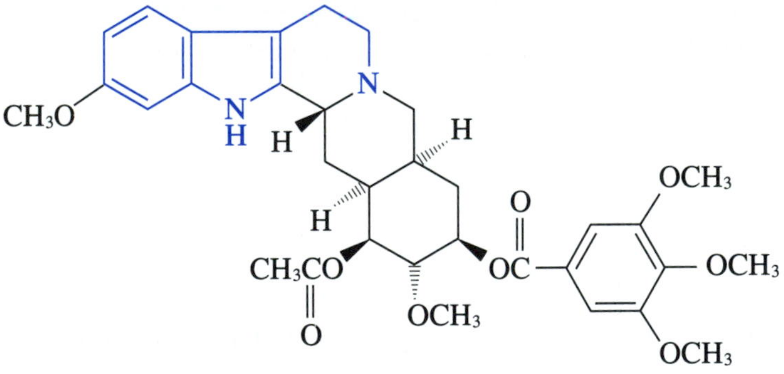

CH_3O / N / H / N / H / H / H / CH_3CO / O / OCH_3 / O / OC / OCH_3 / OCH_3 / OCH_3

reserpine

H / HO_2C / NCH_3 / H / N / H

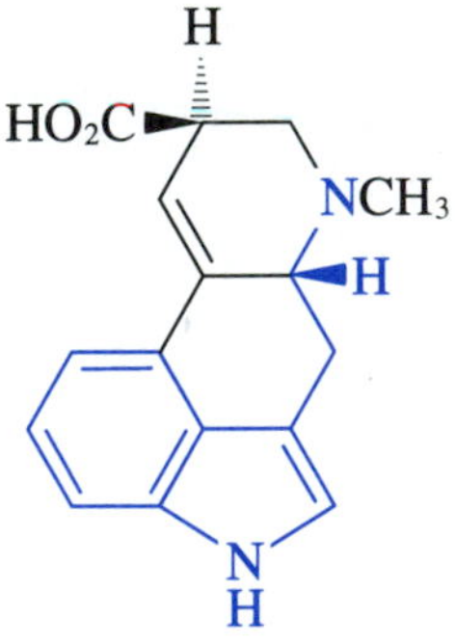

lysergic acid

The **purines** are another biologically important class of fused-ring heterocycles. They contain a pyrimidine ring fused to an imidazole ring.

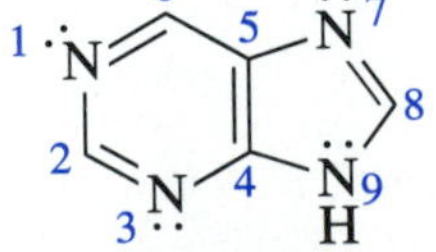

1 N, 2, 3 N, 4, 5, 6, N 7, 8, N 9, H

purine
mp 217°C

Uric acid is present in the urine of all carnivores and is the main product of nitrogen metabolism in the excrement of birds and reptiles. The disease gout results from deposition of sodium urate (the salt of uric acid) in joints and tendons. **Caffeine**, present in coffee, tea, and cola beverages, and **theobromine** (in cocoa) are also purines.

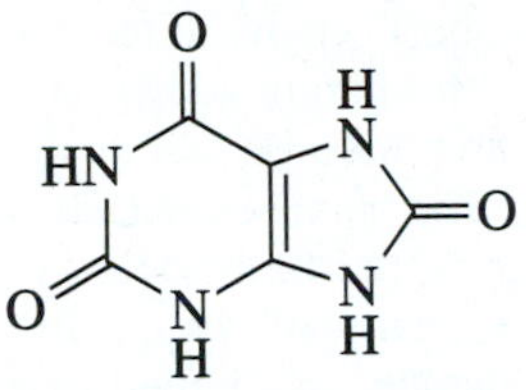

uric acid

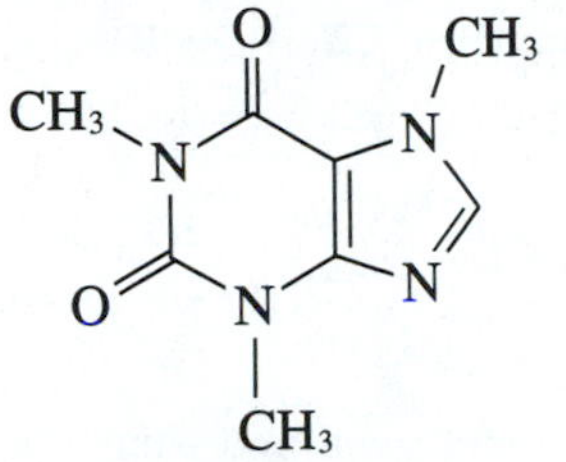

caffeine

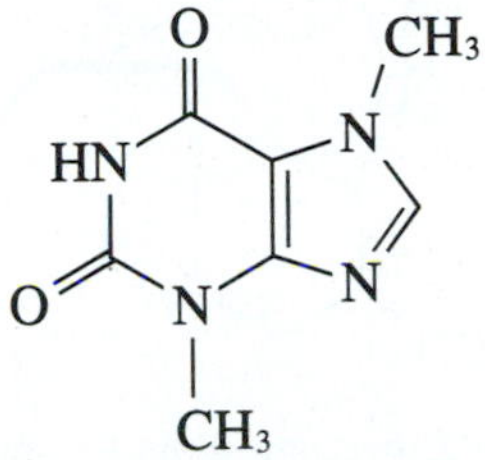

theobromine

Perhaps the most important purines in nature are **adenine** and **guanine,** two of the nitrogen bases that are present in nucleic acids (DNA and RNA; for further details, see Chapter 18).

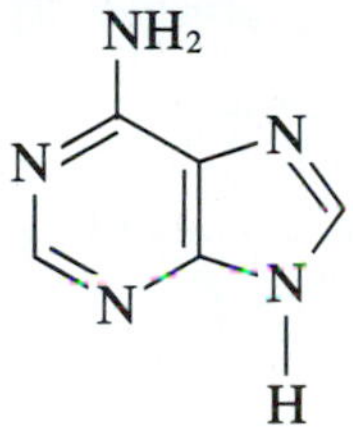

adenine

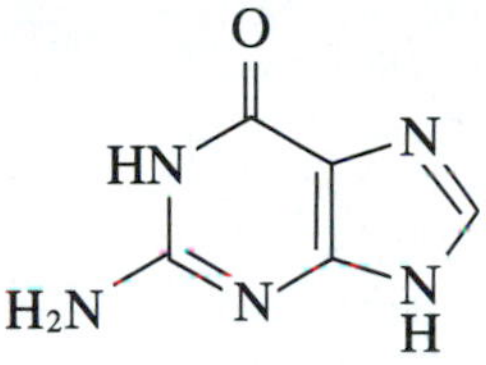

guanine

Many nitrogen heterocycles play a role in medicine. One leading actor in this field is morphine.

A WORD ABOUT ...

27. Morphine and Other Nitrogen-Containing Drugs

Morphine (named after Morpheus, the Greek god of dreams) is the major alkaloid present in opium. An **alkaloid** is any basic, nitrogen-containing plant product, often with a complex structure and significant pharmacological properties. Quinine, papaverine, and caffeine are examples of alkaloids already mentioned in this chapter.

Opium is the dried sap of the unripe seed capsule of the poppy *Papaver somniferum;* its medical properties have been known since ancient times. Morphine was not isolated in pure form until 1805; its correct structure was not established until 1925, and it was not synthesized in the laboratory until 1952.

Pain is a major problem in medicine, and relief of pain has long been a medical goal. Morphine is an **analgesic,** a substance that relieves pain without causing unconsciousness. It was used for the large-scale relief of pain from battle wounds during the American Civil War (largely as a consequence of the invention, at about that time, of the hypodermic syringe). But morphine has serious side effects. It is addictive and also can cause nausea, a decrease in blood pressure, and a depressed breathing rate that can be fatal to the very young or the severely debilitated.

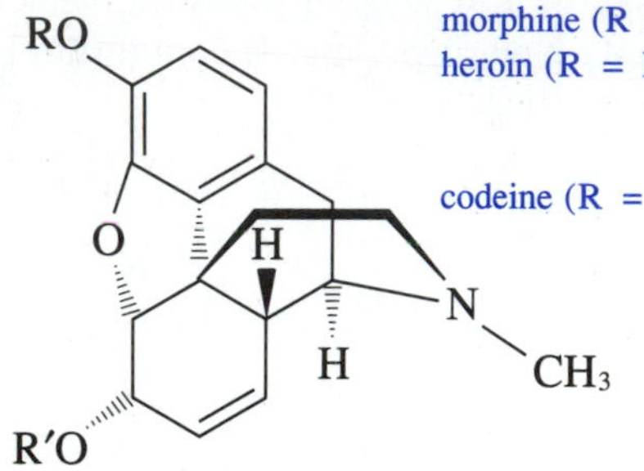

morphine (R = R′ = H)
heroin (R = R′ = $-\overset{\displaystyle O}{\overset{\|}{C}}CH_3$)
codeine (R = CH_3, R′ = H)

The first attempts to find a substance with morphine's benefits but without its side effects involved minor modification of its structure. Acetylation with acetic anhydride gave its diacetyl derivative **heroin,** which is a good analgesic with less of a respiratory depressant effect than morphine. But heroin is severely addictive, and its abuse has become a serious problem. Partial methylation of morphine gave **codeine,** which is useful as an antitussive (anticough) agent. Unfortunately, it is less than one-tenth as effective as morphine as an analgesic.

Many compounds similar to various parts of the morphine structure have been synthesized and tested for their analgesic properties. Two of these are shown here. Their structural similarity to morphine is evident in the colored parts of the structures.

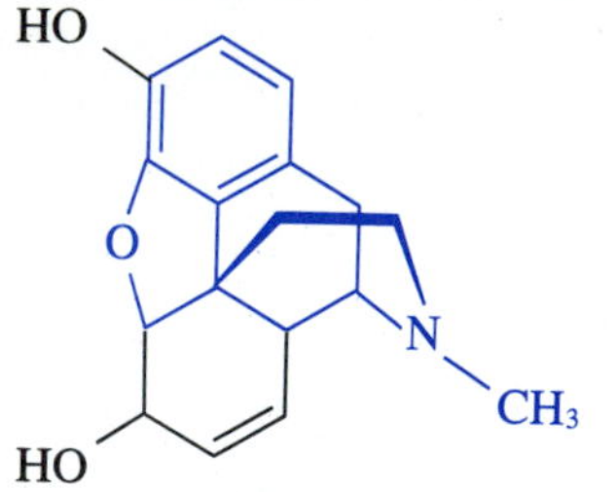

morphine

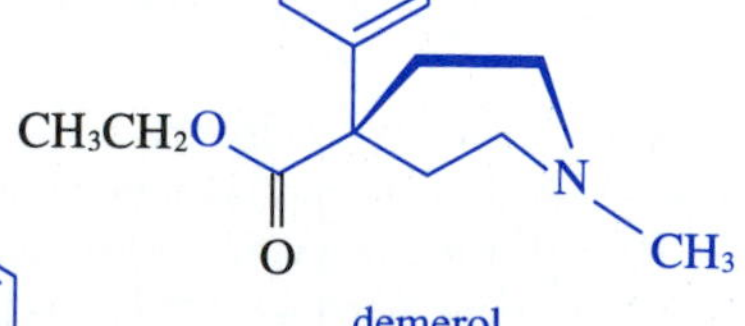

demerol
(meperidine)

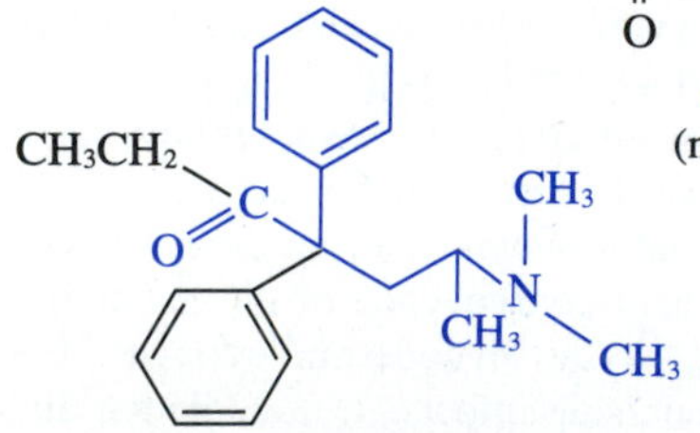

methadone

Demerol is an effective analgesic with a relatively simple structure compared to that of morphine. Notice that it still retains the piperidine ring present in morphine. **Methadone,** which retains the nitrogen of morphine but is no longer heterocyclic, was synthesized and used as an analgesic by the Germans during World War II, when natural sources of morphine were in short supply. Later it was used in substitution therapy for heroin addiction, but it, too, is addictive. The search for a perfect analgesic still goes on.

The pain associated with surgery or injury can sometimes be treated with local or regional anesthetics, many of which are nitrogen-containing drugs. **Cocaine,** an alkaloid found in the plant *Erythroxylum coca,* was one of the first anesthetics of this type. It constricts blood vessels, thus producing a bloodless surgical area. But cocaine is addictive and has other undesirable properties. Its medical use (it is also used illegally for nonmedical purposes) has been supplanted largely by **procaine hydrochloride** (Novocain).

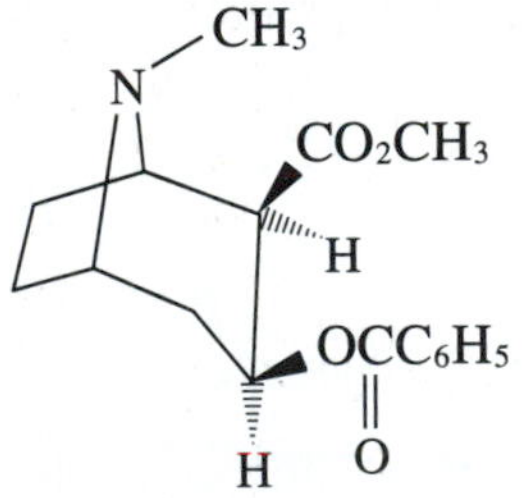

cocaine
mp 98°C

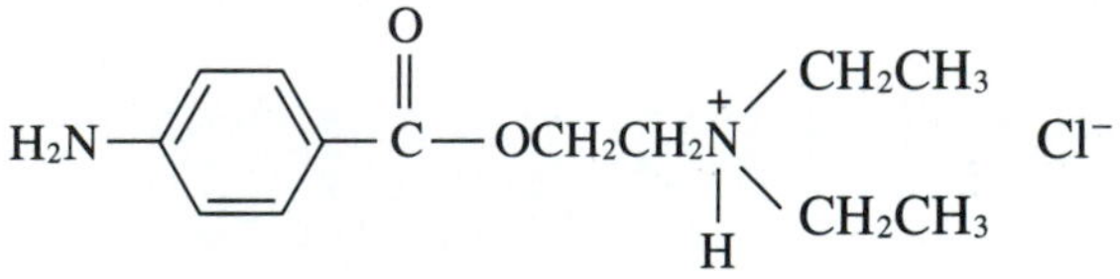

procaine hydrochloride
mp 153–156°C

Procaine is less toxic, is easier to synthesize and sterilize, and has a desirably shorter period of action than cocaine. It is usually injected into a nerve to anesthetize a small region of the body. It acts by inhibiting nerve impulse transmission by acetylcholine. Procaine hydrochloride is a widely used drug, sold to dentists and to doctors of human and veterinary medicine under at least 27 different trade names.

Benzocaine (ethyl *p*-aminobenzoate), which has a very simple structure, is used as a mild topical anesthetic in ointments for burns, insect bites, and open wounds. Note that procaine and benzocaine are different esters of the same acid, *p*-aminobenzoic acid.

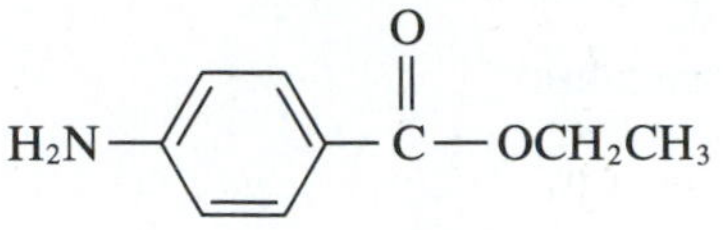

$$H_2N-C_6H_4-\overset{O}{\overset{\|}{C}}-OCH_2CH_3$$

benzocaine
mp 88–90°C

Fentanyl, a very short-acting drug about 100 times as potent as morphine, is extremely important in medicine today. It is used as an anesthetic in perhaps 70% of all surgical procedures in the United States.

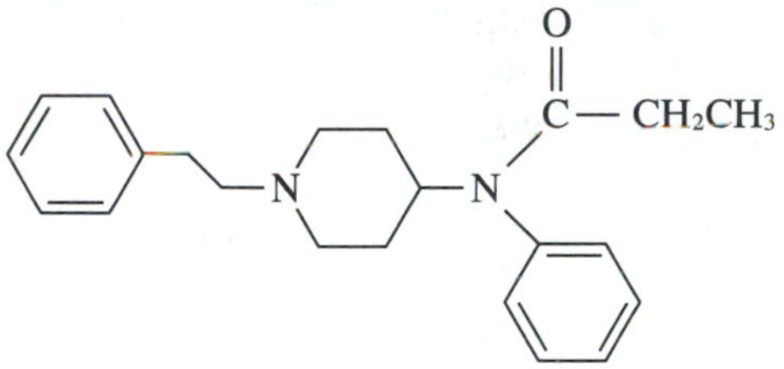

fentanyl

There are, of course, many types of pain. Sometimes a mild tranquilizer can be medically useful. Two of the most commonly prescribed contain seven-membered heterocyclic rings; they are the well-known twins of modern psychiatry: **librium** and **valium.**

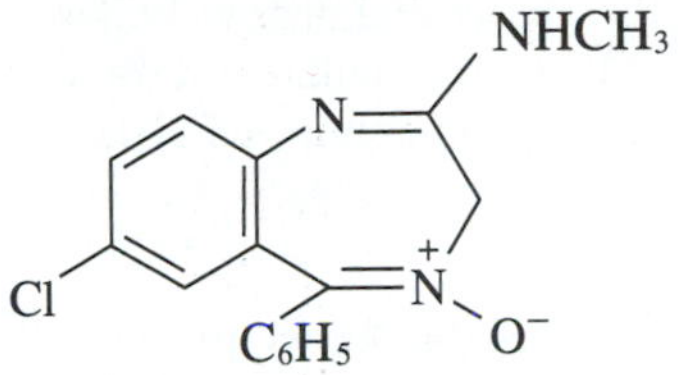

librium

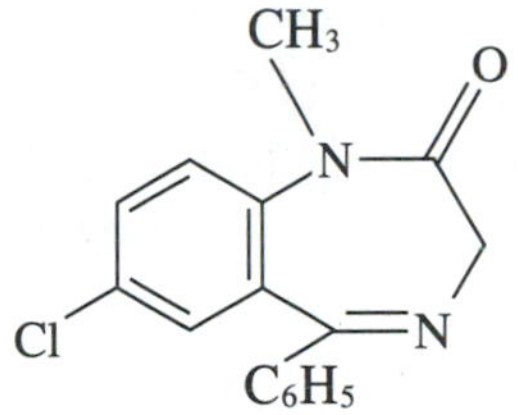

valium

Additional Problems

13.9. In addition to the Kekulé-type contributors to the pyridine resonance hybrid shown on page 362, there are three minor dipolar contributors. Draw their structures. Do they suggest a reason why pyridine is deactivated (relative to benzene) toward reaction with electrophiles and a reason why substitution, when it does occur, takes place at the 3-position?

13.10. Although nitration of pyridine requires a temperature of 300°C (eq. 13.2), 2,6-dimethylpyridine is readily nitrated at 100°C. Write an equation for the reaction, and explain why milder conditions suffice.

13.11. Pyridine reacts with phenyllithium to give a good yield of 2-phenylpyridine. Write an equation and a mechanism for the reaction.

13.12. Oxidation of nicotine with $KMnO_4$ gives nicotinic acid. Write an equation for the reaction.

13.13. Write equations for the reaction of coniine (2-propylpiperidine) with

a. hydrochloric acid. **b.** methyl iodide (1 equivalent).
c. methyl iodide (2 equivalents). **d.** acetic anhydride.
e. benzenesulfonyl chloride.

13.14. Explain why nitration of quinoline (eq. 13.7) occurs mainly at C5 and C8.

13.15. Write an equation for each of the following reactions:

a. quinoline + HCl **b.** nitration of isoquinoline
c. quinoline + $NaNH_2$ **d.** isoquinoline + phenyllithium

13.16. What is the configuration (*R* or *S*) of the hydroxyl-bearing carbon atom in quinine?

13.17. In contrast with phenol, which exists almost entirely in the enol form, 2-hydroxypyridine exists mainly in the keto form. Draw the structures, and suggest a reason for the difference.

13.18. Write an equation for each of the following reactions:
a. piperidine + acetic anhydride
b. pyrrolidine + HCl
c. pyrimidine + HBr
d. quinoline + CH_3I
e. thiophene + HNO_3
f. furan + acetyl chloride + $AlCl_3$

13.19. Write equations to show how furfural (2-furaldehyde) could be converted to
a. 2-furylmethanol. **b.** 2-furoic acid.

13.20. Although electrophilic substitution occurs at C2 in pyrrole, it occurs predominantly at C3 in indole. Suggest an explanation.

13.21. Draw a molecular orbital picture of the bonding in oxazole (Sec. 13.7), using the bonding in imidazole as an example. Do you expect oxazole to be more or less basic than pyrrole? Explain.

13.22. Using structures for the parent compounds given in the text, write the formula for
a. 3-bromothiophene.
b. 2,5-dimethylpyrrole.
c. 5-hydroxyindole.
d. 4-hydroxyisoquinoline.
e. 8-hydroxyquinoline.
f. 2-methylimidazole.
g. 4-pyridinecarboxylic acid (isonicotinic acid).
h. 3-bromopyridinium bromide.
i. 4-methylpyrimidine.
j. tetramethylfuran.

13.23. Benadryl can be synthesized in two steps from diphenylmethane, according to the following sequence:

$$C_6H_5-CH_2-C_6H_5 \xrightarrow[\text{light}]{Br_2} A \xrightarrow{(CH_3)_2NCH_2CH_2OH} \text{benadryl}$$

a. What is the structure of A?
b. How can the aminoalcohol needed for the second step be synthesized from ethylene oxide?
c. Write a mechanism for each step.

13.24. The water solubility of morphine is only 0.2 g/L, but morphine hydrochloride has a water solubility of 57 g/L. Write equations to show how morphine could be extracted and isolated from opium.

13.25. Identify the chiral centers in benzylpenicillin, and assign an absolute configuration (*R* or *S*) to each.

13.26. Uric acid has four hydrogens. Which do you expect to be the most acidic, and why?

CHAPTER 14

Synthetic Polymers

14.1 Introduction

Polymers, or **macromolecules,** as they are sometimes called, are large molecules that are built up by repetitive linking of many smaller units called **monomers.** Polymers may be **natural** or **synthetic.** The most important natural polymers are carbohydrates (starch, cellulose), proteins, and nucleic acids (DNA, RNA). We will study these **biopolymers** in the last three chapters of this book. In this chapter, we will focus on some of the most important *synthetic* polymers.

Synthetic polymers whose names may already be familiar to you include polyethylene (Sec. 3.15), Teflon, Styrofoam, nylon, Dacron, saran, and polyurethanes. There are many others. In the United States alone, annual synthetic polymer production exceeds 50 billion pounds. Synthetic polymers touch our daily lives perhaps to a greater extent than any other synthetic organic chemicals. They make up a substantial percentage of our clothing, appliances, vehicles, homes, packaging, toys, paints, plywood and fiberboard, and tires. It is virtually impossible to live in the modern world without, dozens of times each day, making use of these materials—*all of which were totally unknown less than a century ago*. Our standard of living could not possibly approach what it is today without products of the synthetic organic chemical industry.

14.2 Classification of Polymers

Synthetic polymers can be classified into two main types, depending on how they are made. **Chain-growth polymers** (also called *addition polymers)* are made by the addition of one monomer unit to another in a repetitive manner. A catalyst or initiator is required, and it adds to a carbon–carbon double bond to form a reactive intermediate. This intermediate then adds to the double bond of a second monomer unit to yield a new intermediate. The process continues until the polymer chain is built up. Eventually the process is terminated in some way. *Chain-growth polymers retain all the atoms of the monomer units in the polymer*. A good example of a chain-growth polymer is polyethylene (Sec. 3.15).

$$\underset{\text{ethylene}}{n\,CH_2{=}CH_2} \xrightarrow[\text{initiator}]{\text{polymerization}} \underset{\text{polyethylene}}{\left(CH_2{-}CH_2 \right)_n} \qquad (14.1)$$

Step-growth polymers (also called *condensation polymers*) are usually formed by a reaction between two different functional groups, with the loss of some small molecule, such as water, between them. Thus, *a step-growth polymer does not contain all the atoms initially present in the monomers;* some atoms are lost in the small molecule that is eliminated. The monomer units are usually di- or polyfunctional, and the monomers usually appear in alternating order in the polymer chain. Perhaps the best-known example of a step-growth polymer is the polyamide **nylon,** prepared from 1,6-diaminohexane (hexamethylenediamine) and hexanedioic acid (adipic acid).

$$\underset{\substack{\text{1,6-diaminohexane}\\ \text{(hexamethylenediamine)}}}{H_2N(CH_2)_6NH_2} + \underset{\substack{\text{hexanedioic acid}\\ \text{(adipic acid)}}}{HO\overset{\overset{O}{\|}}{C}(CH_2)_4\overset{\overset{O}{\|}}{C}OH} \xrightarrow{200\text{-}300°C} \underset{\text{nylon, a polyamide}}{\left[NH(CH_2)_6NH\overset{\overset{O}{\|}}{C}(CH_2)_4\overset{\overset{O}{\|}}{C} \right]_n} + 2nH_2O \qquad (14.2)$$

Let us consider each of these ways of making polymers in greater detail.

14.3 Free-Radical Chain-Growth Polymerization

The free-radical chain mechanism described earlier for polyethylene (Sec. 3.15) is typical of chain-growth polymers. The overall reaction is

$$\underset{\text{vinyl monomer}}{\underset{\underset{L}{|}}{CH_2{=}CH}} \xrightarrow[\text{initiator}]{\text{radical}} \underset{\text{vinyl (or chain-growth) polymer}}{—CH_2\underset{\underset{L}{|}}{CH}\left(CH_2\underset{\underset{L}{|}}{CH}\right)_n CH_2\underset{\underset{L}{|}}{CH}—} \qquad (14.3)$$

where L is some substituent. Table 14.1 lists some common commercial chain-growth polymers and their uses.

Free-radical chain-growth polymerization requires a radical initiator, of which **benzoyl peroxide** is an example. It decomposes at about 80°C to give benzoyloxy radicals. These radicals may initiate chains or may lose carbon dioxide to give phenyl radicals that can also initiate chains.

$$\underset{\text{benzoyl peroxide}}{C_6H_5\overset{\overset{O}{\|}}{C}{-}O\underset{\text{weak bond}}{-}O{-}\overset{\overset{O}{\|}}{C}C_6H_5} \xrightarrow{80°C} 2\underset{\text{benzoyloxy radicals}}{C_6H_5\overset{\overset{O}{\|}}{C}{-}O\cdot} \xrightarrow{-CO_2} 2\underset{\text{phenyl radicals}}{C_6H_5\cdot} \qquad (14.4)$$

For simplicity, we can represent the initiator radicals by the symbol In·

TABLE 14.1 Some commercial chain-growth (vinyl) polymers prepared by free-radical polymerization

Monomer name	*Formula*	*Polymer*	*Uses*
ethylene (ethene)	$CH_2{=}CH_2$	polyethylene	sheets and films, blow-molded bottles, injection-molded toys and housewares, wire and cable coverings, shipping containers
propylene (propene)	$CH_2{=}CHCH_3$	polypropylene	fiber products such as indoor-outdoor carpeting, car and truck parts, packaging, toys, housewares
styrene	$CH_2{=}CH{-}C_6H_5$ (phenyl ring drawn)	polystyrene	packaging and containers (Styrofoam), toys, recreational equipment, appliance parts, disposable food containers and utensils, insulation
acrylonitrile (propenenitrile)	$CH_2{=}CHCN$	polyacrylonitrile (Orlon, Acrilan)	sweaters and other clothing
vinyl acetate (ethenyl ethanoate)	$CH_2{=}CH{-}O\overset{\overset{\displaystyle O}{\|}}{C}CH_3$	polyvinyl acetate	adhesives, latex paints
methyl methacrylate (methyl 2-methyl-propenoate)	$CH_2{=}C(CH_3){-}\overset{\overset{\displaystyle O}{\|}}{C}OCH_3$	polymethyl methacrylate (Plexiglas, Lucite)	objects that must be clear, transparent, and tough
vinyl chloride (chloroethene)	$CH_2{=}CHCl$	polyvinyl chloride (PVC)	plastic pipe and pipe fittings, films and sheets, floor tile, records, coatings
tetrafluoroethylene (tetrafluoroethene)	$CF_2{=}CF_2$	polytetrafluoroethylene (Teflon)	coatings for utensils, electric insulators

Initiator radicals add to the carbon–carbon double bond of the vinyl monomer to produce a carbon radical.

Initiation step

$$\text{In}\cdot + \underset{\text{monomer}}{\text{CH}_2\!=\!\underset{\displaystyle \text{L}}{\overset{}{\text{CH}}}} \longrightarrow \underset{\text{carbon radical}}{\text{In}-\text{CH}_2-\underset{\displaystyle \text{L}}{\dot{\text{C}}\text{H}}} \qquad (14.5)$$

Experience shows that the initiator usually adds to the *least* substituted carbon of the monomer, that is, to the CH_2 group. This gives a carbon radical adjacent to the substituent. There are two reasons for this preference: first, the terminal vinylic carbon is less hindered and therefore more easily attacked, and second, the substituent L usually can stabilize an adjacent radical by resonance.

The carbon radical formed in the initiation step then adds to another monomer molecule, and the adduct adds to another, and so on.

Propagation steps

$$\text{In}\,\text{CH}_2\underset{\displaystyle \text{L}}{\dot{\text{C}}\text{H}} \xrightarrow{\text{CH}_2=\text{CHL}} \text{In}\,\text{CH}_2\underset{\displaystyle \text{L}}{\text{CH}}\text{CH}_2\underset{\displaystyle \text{L}}{\dot{\text{C}}\text{H}} \xrightarrow{\text{CH}_2=\text{CHL}}$$

$$\underset{\text{growing polymer chain}}{\text{In}\,\text{CH}_2\underset{\displaystyle \text{L}}{\text{CH}}\text{CH}_2\underset{\displaystyle \text{L}}{\text{CH}}\text{CH}_2\underset{\displaystyle \text{L}}{\dot{\text{C}}\text{H}}} \longrightarrow \text{and so on} \qquad (14.6)$$

Chain propagation (eq. 14.6) occurs in the same sense as initiation (eq. 14.5), so that the monomer units are linked in a head-to-tail manner, with the substituent on alternate carbon atoms.

Problem 14.1 In polystyrene, the chain grows in a strictly head-to-tail arrangement because the intermediate radical is stabilized through resonance. Draw the intermediate radical, and show how it is stabilized through resonance.

Chain propagation may continue until anywhere from a few hundred to several thousand monomer units are linked. The extent of reaction depends on several factors, some of which are the reaction conditions (temperature, pressure, solvent, concentration of monomer and catalyst, for example); the nature of the monomer, especially the substituent L; and the rates of competing reactions, which may terminate the chain. Two common chain-terminating reactions are **radical coupling** and **radical disproportionation.**

Termination steps

$$2 \sim\text{CH}_2\underset{\displaystyle \text{L}}{\dot{\text{C}}\text{H}} \xrightarrow[\text{coupling}]{\text{radical}} \sim\text{CH}_2\overbrace{\underset{\displaystyle \text{L}}{\text{CH}}-\underset{\displaystyle \text{L}}{\text{CH}}}^{\text{head-to-head}}\text{CH}_2\sim \qquad (14.7)$$

$$2 \;\sim\!CH_2\dot{C}H(L) \xrightarrow[\text{disproportionation}]{\text{radical}} \underset{\text{alkane}}{\sim\!CH_2CH_2(L)} + \underset{\text{alkene}}{\sim\!CH{=}CHL} \quad (14.8)$$

Radical coupling (eq. 14.7) gives rise to a head-to-head arrangement of two monomers. In radical disproportionation (eq. 14.8), one radical abstracts a hydrogen atom from the carbon adjacent to another radical site, producing a saturated and an unsaturated polymer.

EXAMPLE 14.1 What feature distinguishes propagation steps from termination steps?

Solution In propagation steps, one radical is destroyed, but another radical is created. In other words, the number of radicals on the left and on the right side of the equation for a propagation step is always the same. In termination steps, however, radicals are destroyed, and no new radicals are generated. Therefore, chain growth terminates.

If eqs. 14.5 through 14.8 told the whole story, all polymers formed by free-radical chain growth would be linear. But we know from physical measurements that many such polymers have branched chains, so something is missing from the scheme.

What is missing is a **chain-transfer reaction.** A growing polymer radical may abstract a hydrogen atom from another polymer chain.

Chain-transfer step

$$\sim\!CH_2\dot{C}H(L) + \sim\!CH_2CH(L)\!\sim \longrightarrow \sim\!CH_2CH_2(L) + \sim\!CH_2\dot{C}(L)\!\sim \quad (14.9)$$

This step terminates one chain but initiates another chain somewhere along the length of the polymer, not at its end, so that, when the polymerization continues, a branch is produced.

$$\underset{\text{mid-chain radical from eq. 14.9}}{\sim\!CH_2\dot{C}(L)\!\sim} \xrightarrow{CH_2=CHL} \sim\!CH_2C(L)(CH_2\dot{C}HL)\!\sim \quad (14.10)$$

(the substituted carbon is labeled "branch point")

Chain-transfer and radical disproportionation are similar in that both involve a hydrogen abstraction reaction.

The extent of chain branching in a particular polymer depends on the relative rates of the chain-propagating and chain-transfer steps. If radical addition is very fast compared with hydrogen abstraction, the polymer will be mainly linear. On the other hand, if the chain-transfer rate were, for example, 1/10 that of the addition rate, we might expect an average of 1 branch for every 10 linearly linked monomers.

Chain-transfer reactions can be used to control the molecular weight of a polymer. Certain reagents, such as thiols, have a hydrogen that is easily abstracted. The resulting RS· radical is not reactive enough to add to double bonds. Instead, it dimerizes to a disulfide.

$$\sim CH_2\dot{C}H(L) + \underset{\text{thiol}}{RSH} \xrightarrow[\text{abstraction}]{\text{hydrogen}} \sim CH_2CH_2(L) + RS\cdot \longrightarrow RSSR \qquad (14.11)$$

does not add to double bonds; dimerizes instead

Thiols are therefore chain terminators, and when added to a polymerization reaction mixture in small amounts, they limit the polymer chain length.

Free-radical chain-growth polymerization is a very fast reaction. A chain may grow to 1000 monomer units or more in less than a second! The polymer chains contain one or two groups derived from the initiator radical, but those groups make up only a small fraction of the polymer molecule, so that the polymer properties are determined largely by the particular monomer used.

Now let us consider two typical free-radical growth polymers: polystyrene and polyvinyl chloride.

Styrene is easily polymerized by benzoyl peroxide, and the product **polystyrene** has a molecular weight in the range of 1 to 3 million.

$$\underset{\text{styrene}}{CH_2{=}CH{-}C_6H_5} \xrightarrow{\text{benzoyl peroxide}} \underset{\text{polystyrene}}{\left(CH_2{-}CH(C_6H_5) \right)_n} \qquad (14.12)$$

Polystyrene is an amorphous, thermoplastic polymer. By **amorphous,** we mean that the polymer chains are irregularly arranged in a random manner; they are not regularly aligned, as in a crystal. By **thermoplastic,** we mean that the polymer melts or softens on heating and hardens again on cooling. Polystyrene can be molded or extruded to produce parts for housewares, toys, radio and television chassis, and bottles, jars, and containers of all kinds. **Styrofoam** is produced by including a low-boiling hydrocarbon such as pentane in the processing. When the polymer is heated, the pentane volatilizes, producing bubbles that expand the polymer into a foam. These foams are used for insulation, packaging, cups for hot drinks, egg cartons, and many other purposes. Annual U.S. production of polystyrene is over 5 billion pounds.

Polystyrene can be modified in various ways. For example, it can be rigidified through **cross-linking,** by including small amounts of *p*-divinylbenzene with the monomer.

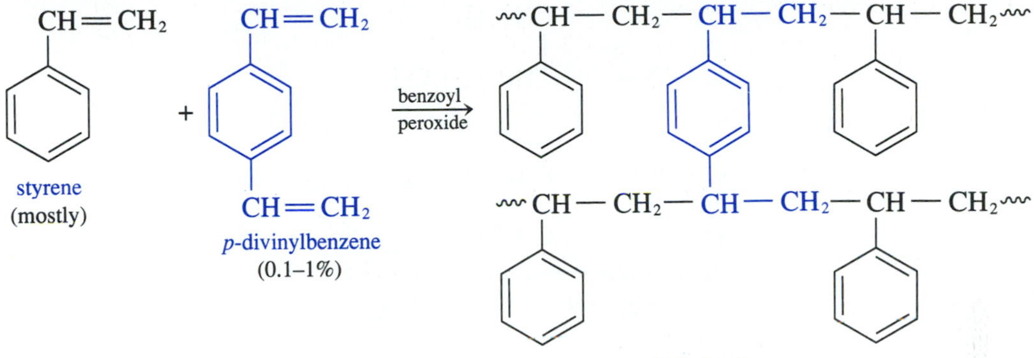

$CH{=}CH_2$ (styrene (mostly)) + $CH{=}CH_2$–C_6H_4–$CH{=}CH_2$ (*p*-divinylbenzene (0.1–1%)) $\xrightarrow[\text{peroxide}]{\text{benzoyl}}$ cross-linked polystyrene (14.13)

styrene
(mostly)

p-divinylbenzene
(0.1–1%)

cross-linked polystyrene

The resulting polymer is more rigid and less soluble in organic solvents than ordinary polystyrene. By sulfonation, this cross-linked polymer can be converted to an ion-exchange resin used in water softeners. As hard water percolates through the resin, Ca^{2+} and Mg^{2+} ions are exchanged for Na^+ ions.

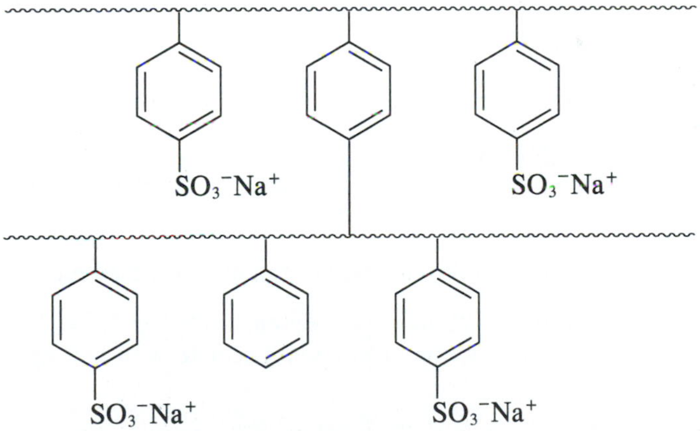

$SO_3^-Na^+$ $SO_3^-Na^+$

$SO_3^-Na^+$ $SO_3^-Na^+$

schematic structure of an ion-exchange resin

Problem 14.2 Explain why sulfonation of polystyrene occurs in the positions shown in the ion-exchange resin formula.

Polyvinyl chloride (PVC) can be represented by the general formula

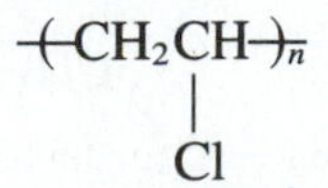

$$\left(CH_2\underset{\displaystyle Cl}{\underset{|}{C}}H\right)_n$$

polyvinyl chloride (PVC)

It has a head-to-tail structure. PVC is a hard polymer but can be softened by adding **plasticizers,** usually low-molecular-weight esters that act as lubricants between the polymer chains. A good example is bis-2-ethylhexyl phthalate.

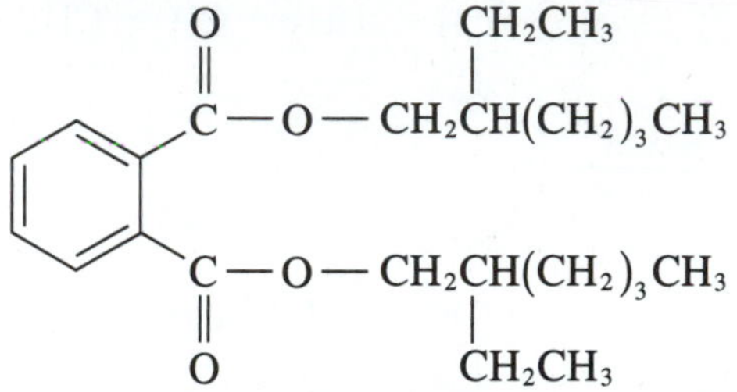

$$C_6H_4[C(=O)-O-CH_2CH(CH_2CH_3)(CH_2)_3CH_3]_2$$

bis-2-ethylhexyl phthalate
(a plasticizer)

Problem 14.3 Bis-2-ethylhexyl phthalate can be prepared from 2-ethylhexanol (review Problem 9.26) and phthalic anhydride (review Problem 10.24). Write an equation for the reaction.

PVC is used to make floor tiles, vinyl upholstery (imitation leather), plastic pipes, plastic squeeze bottles, and so on. Annual U.S. production exceeds 6 billion pounds.

Problem 14.4 Write the structural formula for a three-monomer segment of

a. polypropylene.
b. polyvinyl acetate.
c. poly(methyl methacrylate).
d. polyacrylonitrile.

Problem 14.5 Using a three-monomer segment, write an equation for

a. the reaction of polystyrene with Cl_2 + $FeCl_3$.
b. the reaction of polyvinyl acetate with hot aqueous sodium hydroxide.

14.4 *Cationic Chain-Growth Polymerization*

Certain vinyl compounds are best polymerized via cationic rather than free-radical intermediates. The most common commercial example is isobutylene (2-methylpropene), which can be polymerized with Friedel-Crafts catalysts via tertiary carbocation intermediates.

Initiation

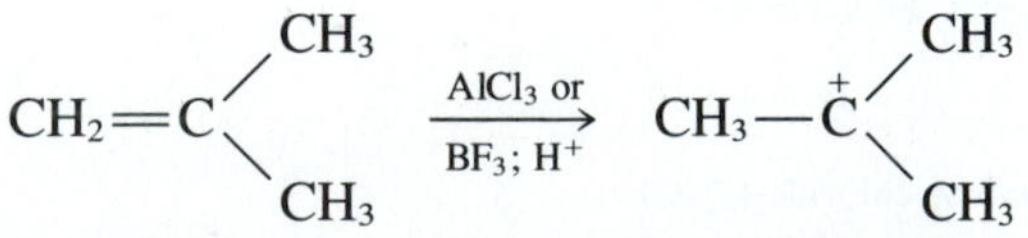

$$\underset{\text{isobutylene}}{CH_2{=}C(CH_3)_2} \xrightarrow[BF_3;\ H^+]{AlCl_3 \text{ or}} \underset{\text{tertiary carbocation}}{CH_3-\overset{+}{C}(CH_3)_2} \qquad (14.14)$$

Propagation

$$\mathrm{CH_3{-}\overset{\displaystyle CH_3}{\underset{\displaystyle CH_3}{\overset{|}{\underset{|}{C^+}}}} + CH_2{=}C\begin{matrix}\diagup CH_3\\ \diagdown CH_3\end{matrix} \longrightarrow CH_3{-}\overset{\displaystyle CH_3}{\underset{\displaystyle CH_3}{\overset{|}{\underset{|}{C}}}}{-}CH_2{-}\overset{\displaystyle CH_3}{\underset{\displaystyle CH_3}{\overset{|}{\underset{|}{C^+}}}} \longrightarrow}$$

$$\mathrm{CH_3{-}\overset{\displaystyle CH_3}{\underset{\displaystyle CH_3}{\overset{|}{\underset{|}{C}}}}{-}\left[CH_2{-}\overset{\displaystyle CH_3}{\underset{\displaystyle CH_3}{\overset{|}{\underset{|}{C}}}}\right]_n{-}CH_2{-}\overset{\displaystyle CH_3}{\underset{\displaystyle CH_3}{\overset{|}{\underset{|}{C^+}}}}} \quad (14.15)$$

Termination

$$\mathrm{(CH_3)_3C{-}\left[CH_2{-}\overset{\displaystyle CH_3}{\underset{\displaystyle CH_3}{\overset{|}{\underset{|}{C}}}}\right]_n{-}CH_2{-}\overset{\displaystyle CH_3}{\underset{\displaystyle CH_3}{\overset{|}{\underset{|}{C^+}}}} \xrightarrow{-H^+}}$$

$$\mathrm{(CH_3)_3C{-}\left[CH_2{-}\overset{\displaystyle CH_3}{\underset{\displaystyle CH_3}{\overset{|}{\underset{|}{C}}}}\right]_n{-}CH_2{-}C\begin{matrix}\diagup\!\!\diagup CH_2\\ \diagdown CH_3\end{matrix}} \quad (14.16)$$

polyisobutylene

Initiation gives the *tert*-butyl cation (eq. 14.14) which, in the propagation step, adds to the CH_2 carbon of the double bond in a Markovnikov manner to produce another tertiary carbocation, and so on (eq. 14.15). The chain is terminated by loss of a proton from a carbon atom adjacent to the positive carbon (eq. 14.16).

Polyisobutylenes prepared this way (n = about 50) are used as additives in lubricating oil and as adhesives in pressure-sensitive tape and removable paper labels. Higher-molecular-weight polymers are used in the manufacture of inner tubes for truck and bicycle tires.

14.5 Anionic Chain-Growth Polymerization

Alkenes with electron-withdrawing substituents can be polymerized via carbanionic intermediates. The catalyst may be an organometallic compound, for example, an alkyllithium.

Initiation

$$\mathrm{CH_2{=}CHL + RLi \longrightarrow RCH_2\underset{\displaystyle L}{\underset{|}{\ddot{C}H}}\ \ \overset{+}{Li}} \quad (14.17)$$

Propagation

$$\mathrm{RCH_2\underset{\displaystyle L}{\underset{|}{CH^-}}\ \ Li^+ \xrightarrow{CH_2=CHL} RCH_2\underset{\displaystyle L}{\underset{|}{C}}HCH_2\underset{\displaystyle L}{\underset{|}{\bar{C}}}HLi^+,\ \text{and so on}} \quad (14.18)$$

Addition of the catalyst to the double bond gives a carbanion intermediate (eq. 14.17) in which the substituent L usually delocalizes the negative charge through resonance. Common L groups of this type are cyano, carbomethoxy, phenyl, and vinyl.

EXAMPLE 14.2 Draw the carbanion intermediate for anionic polymerization of acrylonitrile ($CH_2{=}CHCN$), and show how it is stabilized by resonance.

Solution

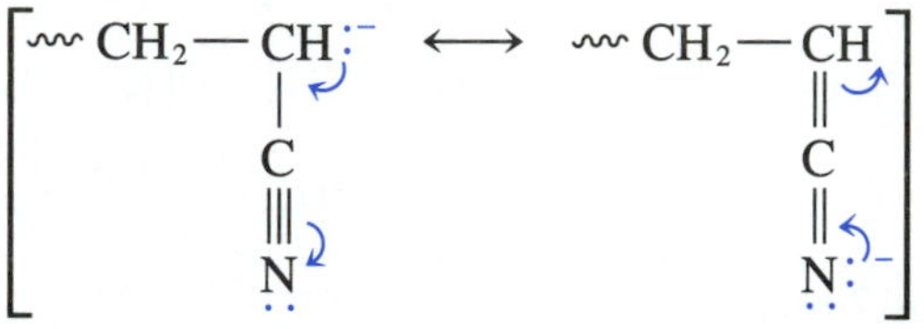

$$\left[\sim CH_2-\overset{..^-}{C}H-C\equiv\ddot{N} \longleftrightarrow \sim CH_2-CH=C=\ddot{N}:^- \right]$$

Problem 14.6 Methyl methacrylate (Table 14.1) can be polymerized by *n*-butyllithium at −78°C. Using eqs. 14.17 and 14.18 as a model, write a mechanism for the reaction. Show how the intermediate carbanion is resonance-stabilized.

Anionic polymerizations are terminated by quenching the reaction mixture with a proton source (water or alcohol).

Problem 14.7 Ethylene oxide can be polymerized by base to give carbowax, a water-soluble wax.

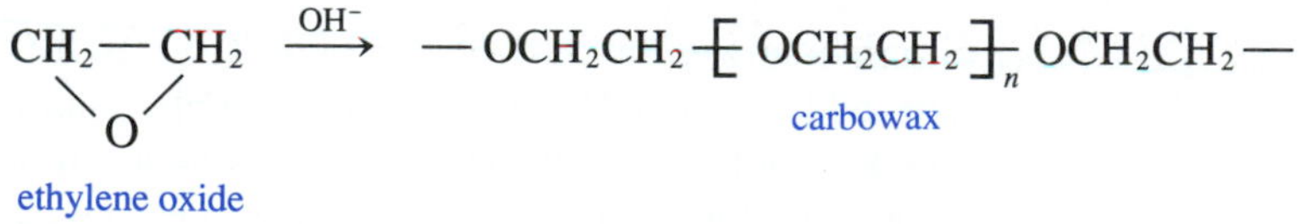

$$\underset{\text{ethylene oxide}}{CH_2-CH_2\ (\text{bridged by O})} \xrightarrow{OH^-} \underset{\text{carbowax}}{-OCH_2CH_2-\!\!\left[OCH_2CH_2\right]_n\!\!-OCH_2CH_2-}$$

Suggest a mechanism for the reaction.

14.6 Stereoregular Polymers; Ziegler–Natta Polymerization

When a monosubstituted vinyl compound is polymerized, every other carbon atom in the chain becomes a chiral center:

$$\underset{\textstyle L}{CH_2{=}\underset{|}{CH}} \longrightarrow -CH_2-\underset{L}{\overset{*}{C}H}-CH_2-\underset{L}{\overset{*}{C}H}-CH_2-\underset{L}{\overset{*}{C}H}- \qquad (14.19)$$

The carbons marked with an asterisk have four different groups attached and are therefore chiral centers. Three classes of such polymers are recognized:

atactic: stereocenters have random configurations
isotactic: all stereocenters have the same configuration
syndiotactic: stereocenters alternate in configuration

An atactic polymer is **stereorandom,** but an isotactic or syndiotactic polymer is **stereoregular.** These three classes of polymers, *even if derived from the same monomer,* will have different physical properties.

EXAMPLE 14.3 Draw a chain segment of isotactic polypropylene.

Solution For polypropylene, the group L in eq. 14.19 is $—CH_3$. With the chain extended in zigzag fashion, all methyl substituents occupy identical positions.

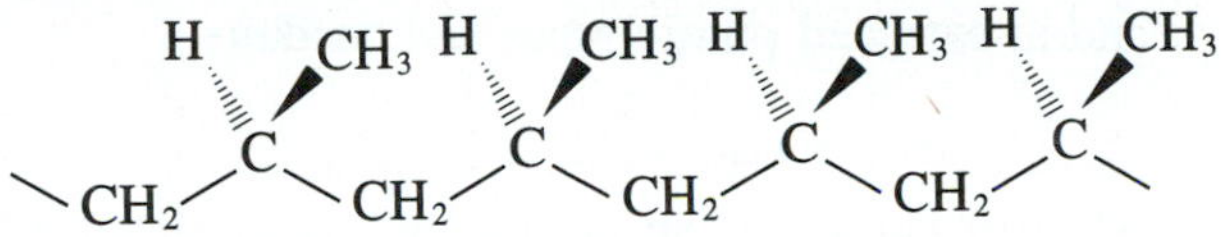

$$\sim CH_2-\overset{H\ \ CH_3}{C}-CH_2-\overset{H\ \ CH_3}{C}-CH_2-\overset{H\ \ CH_3}{C}-CH_2-\overset{H\ \ CH_3}{C}\sim$$

Problem 14.8 Using the definitions above, draw a chain segment of

a. syndiotactic polypropylene.
b. atactic polypropylene.

Stereoregularity imparts certain favorable properties to polymers. Since free-radical polymerization usually results in an atactic polymer, the discovery in the 1950s by Ziegler and Natta* of mixed organometallic catalysts that produce stereoregular polymers was a landmark in polymer chemistry. One such catalyst system is a mixture of triethylaluminum (or other trialkylaluminums) and titanium tetrachloride. With this catalyst, for example, propylene gives a polymer that is more than 98% isotactic.

The mechanism of Ziegler–Natta catalysis is quite complex. A key step in the chain growth involves an alkyl–titanium bond and coordination of the monomer to a vacant site on the metal. The coordinated monomer then inserts into the carbon-titanium bond, and the process continues.

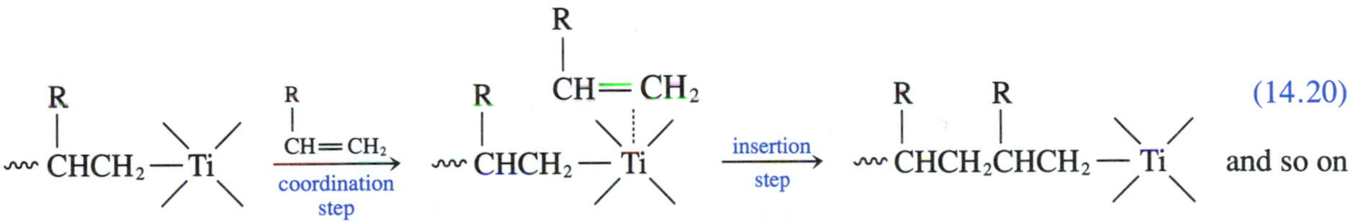

$$\sim\overset{R}{\overset{|}{C}}HCH_2-Ti \xrightarrow[\text{coordination step}]{\overset{R}{\overset{|}{C}}H=CH_2} \sim\overset{R}{\overset{|}{C}}HCH_2-Ti\cdots\overset{R}{\overset{|}{C}}H=CH_2 \xrightarrow[\text{step}]{\text{insertion}} \sim\overset{R}{\overset{|}{C}}HCH_2\overset{R}{\overset{|}{C}}HCH_2-Ti \text{ and so on} \quad (14.20)$$

Because of the various ligands attached to the titanium atom, coordination and insertion occur in a stereoregular manner and can be controlled to give either an isotactic or a syndiotactic polymer.

Commercial production of polypropylene is performed exclusively with Ziegler–Natta catalysts. A stereoregular, isotactic polymer which is highly crystalline is obtained. It is used for interior trim and battery cases in automobiles, for packaging (for example, containers for nested potato chips), and for furniture (such as plastic stacking chairs). It is also spun into fibers for ropes that float (an advantage for sailors and dockers), synthetic grass, carpet backings, and related materials.

Polyethylene obtained through Ziegler–Natta catalysis is linear, in contrast with the highly branched polyethylene obtained through free-radical processes.

---

* Karl Ziegler (Germany) and Giulio Natta (Italy) shared the 1963 Nobel Prize in chemistry for this discovery.

Linear polyethylene has a more crystalline structure, a higher density, and great tensile strength and hardness than the branched polymer. It is used in thin-wall containers like those used to hold laundry bleach and detergents; in molded housewares such as mixing bowls, refrigerator containers, and toys; and in extruded plastic pipes and conduits.

14.7 Diene Polymers: Natural and Synthetic Rubber

Natural rubber is an unsaturated hydrocarbon polymer. It is obtained commercially from the milky sap (latex) of the rubber tree. Its chemical structure was deduced in part from the observation that, when latex is heated *in the absence of air,* it breaks down to give mainly a single unsaturated hydrocarbon product, **isoprene.**

$$\text{natural rubber} \xrightarrow{\text{heat}} CH_2{=}\underset{\underset{CH_3}{|}}{C}{-}CH{=}CH_2 \qquad (14.21)$$

isoprene
(2-methyl-1,3-butadiene)

It is possible to synthesize a material that is nearly identical to natural rubber by treating isoprene with a Ziegler–Natta catalyst, such as a mixture of triethylaluminum, $(CH_3CH_2)_3Al$, and titanium trichloride, $TiCl_3$. The isoprene molecules add to one another by a **head-to-tail 1,4-addition**.

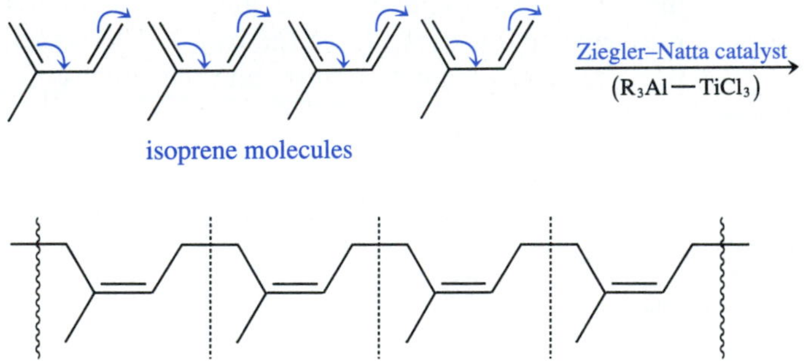

Ziegler–Natta catalyst ($R_3Al{-}TiCl_3$)

isoprene molecules

(14.22)

natural rubber segment (all *cis*)

The double bonds in natural rubber are *isolated;* that is, they are separated from one another by three single bonds. They have a *cis* geometry. By this we mean that, to proceed down the length of the carbon chain, we must enter and leave each double bond from the same "side."

Problem 14.9 Gutta-percha is a less common form of natural rubber. It is also a 1,4-polymer of isoprene, but with *trans* double bonds. Draw the structural formula for a three-monomer segment of gutta-percha.

Most rubber has a molecular weight in excess of 1,000,000, though the value varies with the source and method of processing. This corresponds to about 15,000 isoprene monomers per rubber molecule. Crude plantation rubber contains, in addition to polyisoprene, about 2.5 to 3.5% protein, 2.5 to 3.2% fats, 0.1 to 1.2% water, and traces of inorganic matter.

Although natural rubber has many useful properties, it also has some undesirable ones. Early manufactured rubber goods were often sticky and smelly, and they softened in warm weather and hardened in cold. Some of these weaknesses were overcome when Charles Goodyear invented **vulcanization,** a process of cross-linking polymer chains by heating rubber with sulfur. The cross-links add strength to the rubber and act as a kind of "memory" that helps the polymer recover its original shape after stretching.

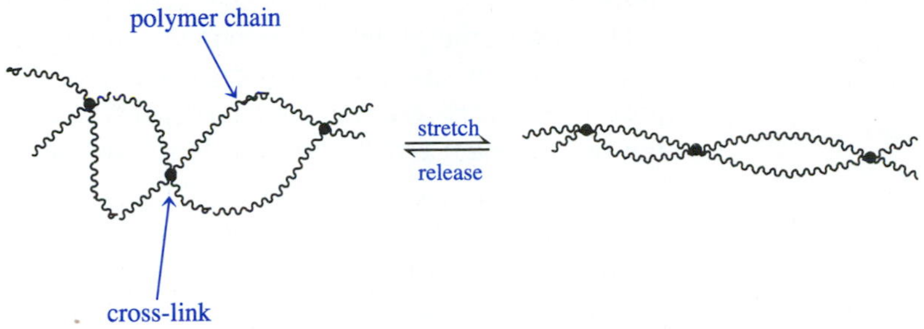

polymer chain

stretch / release

cross-link

In spite of such improvements, there were still problems. For example, it was not uncommon years ago to have to check the air pressure in tires almost every time one purchased gasoline, because the rubber inner tubes were somewhat porous. Therefore, there was a need to develop **synthetic rubber,** a name given to polymers with properties similar to those of natural rubber but superior to and somewhat different chemically from it.

Many monomers or mixtures of monomers form **elastomers** (rubberlike substances) when they are polymerized. The largest-scale commercial synthetic rubber is a **copolymer** of 25% styrene and 75% 1,3-butadiene, called SBR (styrene-butadiene rubber).

$$\underset{\text{styrene}}{n\mathrm{CH_2{=}CHC_6H_5}} + \underset{\text{butadiene}}{3n\mathrm{CH_2{=}CH{-}CH{=}CH_2}} \xrightarrow[\text{initiator}]{\text{free radical}} \qquad (14.23)$$

C_6H_5

SBR

The structure is approximately as shown, although about 20% of the butadiene adds 1,2- instead of 1,4-. Unlike those in natural rubber, the double bonds in this polymer have a *trans* geometry. The dashed lines in the structure show the units from which the polymer is constructed. About two-thirds of SBR goes into tires. Its annual production exceeds that of natural rubber by a factor of two.

Problem 14.10 Draw the structural formula for a three-monomer segment of poly(1,3-butadiene) in which

a. addition is 1,4 and double bonds are *cis*.
b. addition is 1,4 and double bonds are *trans*.
c. addition is 1,2 for the middle unit and 1,4 for the outer units, with double bonds *cis*.

14.8 Copolymers

Most of the polymers we have described so far are **homopolymers,** polymers made from a single monomer. But the variety and utility of chain-growth polymerization can sometimes be enhanced (as we have just seen with SBR synthetic rubber) by using mixtures of monomers, to give **copolymers.** Figure 14.1 summarizes some of the ways that monomers can be arranged in homo- and copolymers. The copolymers depicted are limited to two different monomers (A and B), though in principle, of course, the possibilities are unlimited.

The exact arrangement of monomers along a copolymer chain will depend on a number of factors. One of these is the relative reactivity of the two monomers. Let us assume that we polymerize a 1 : 1 mixture of A and B by a free-radical chain-growth process. Here are some of the possibilities:

1. A· reacts rapidly with B but slowly with A, and B· reacts rapidly with A but slowly with B. The polymer will then be alternating: —ABABAB—. Many copolymers tend toward this arrangement, though not always perfectly.
2. A and B are equally reactive toward radicals, and each reacts readily with A· or B·. The polymer will then be random: —AABABBA—.
3. A is much more reactive than B toward all radicals. In this case, A will be consumed first, followed more slowly by B. We will obtain a mixture of two homopolymers, $—(A)_n—$ and $—(B)_m—$.

Problem 14.11 1,1-Dichloroethene and vinyl chloride form a copolymer called saran, used in food packaging. The monomer units tend to alternate in the chain. Draw the structural formula for a 4-monomer segment of the chain.

FIGURE 14.1 Arrangements of monomers in polymers.

Homopolymers

```
                          AA—
                          |
 —AAAAA—             —AAAAA—             —AAAAA—
  linear               |                     |
                       AA—                —AAAAA—
                     branched            cross-linked
```

Copolymers

```
—ABABAB—      —AABABBA—      —AAAAABBBB—      —AAAAAAA—
alternating     random          block            |    |
                                              —BBB   BBB—
                                                  graft
```

Block and graft copolymers are made by special methods. If we first initiate polymerization of monomer A, then add some B, then add A again, and so on, we can obtain a block polymer with alternating segments of blocks of A units, then B units, and so on. This is particularly easy with anionic polymerizations, where there are no significant termination steps.

Graft polymers are made by taking advantage of functionality present in a homopolymer. For example, if a polymer contains double bonds (as in poly-1,3-butadiene), addition of a free-radical initiator R· and second monomer (such as styrene) will "graft" polystyrene chains onto the polybutadiene backbone.

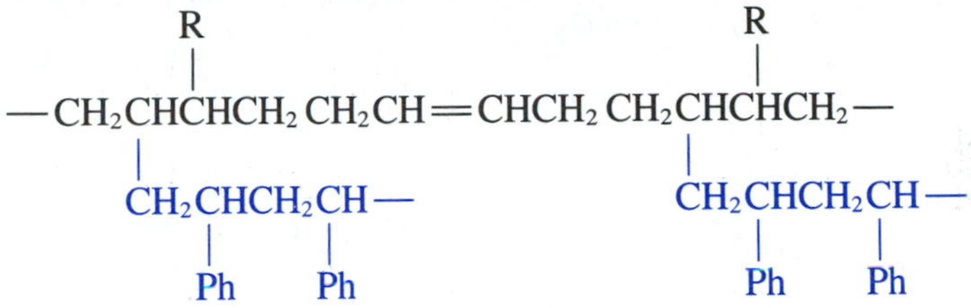

$$
\begin{array}{l}
\qquad\ \ \mathrm{R} \qquad\qquad\qquad\qquad\qquad\qquad\qquad \mathrm{R} \\
\qquad\ \ | \qquad\qquad\qquad\qquad\qquad\qquad\qquad\ \ | \\
-\mathrm{CH_2CHCHCH_2\,CH_2CH{=}CHCH_2\,CH_2CHCHCH_2}- \\
\qquad | \qquad\qquad\qquad\qquad\qquad\qquad\quad | \\
\qquad \mathrm{CH_2CHCH_2CH}- \qquad\qquad\qquad \mathrm{CH_2CHCH_2CH}- \\
\qquad\qquad\ | \qquad\quad | \qquad\qquad\qquad\qquad\quad | \qquad\quad | \\
\qquad\qquad \mathrm{Ph} \quad\ \ \mathrm{Ph} \qquad\qquad\qquad\qquad \mathrm{Ph} \quad\ \ \mathrm{Ph}
\end{array}
$$

poly-1,3-butadiene with polystyrene grafts

This particular graft polymer is used to make rubber soles for shoes.

14.9 Step-Growth Polymerization: Dacron and Nylon

Step-growth polymers are usually produced by a reaction between two monomers, each of which is at least difunctional. Many of them can be represented by overall equation 14.24,

$$
\mathrm{A{\sim}A} + \mathrm{B{\sim}B} \longrightarrow -\mathrm{A{\sim}A}-\mathrm{B{\sim}B}-\mathrm{A{\sim}A}-\mathrm{B{\sim}B}- \tag{14.24}
$$

where A~A and B~B are difunctional molecules with groups A and B that can react with one another. For example A might be an OH group, and B might be a CO_2H group, in which case A~A would be a diol, B~B would be a dicarboxylic acid and ~A—B~ would be an ester. The polymer would be a polyester.

Unlike chain-growth polymers, which grow by one monomer unit at a time, step-growth polymers are formed in steps (or leaps), often by reaction of one polymer molecule with another. The way this works is best illustrated with a specific example.

Consider the formation of a polyester from a diol and a diacid. In the first step, the product will be an ester, with an alcohol group at one end and an acid at the other (eq. 14.25).

$$
\underset{\text{diol}}{\mathrm{HO{\sim}OH}} + \underset{\text{diacid}}{\mathrm{HO_2C{\sim}CO_2H}} \xrightarrow{-\mathrm{H_2O}} \underset{\text{alcohol}}{\mathrm{HO{\sim}}}\underset{\text{ester}}{\mathrm{O{-}\overset{\overset{\large O}{\|}}{C}}}\underset{\text{acid}}{\mathrm{{\sim}CO_2H}} \tag{14.25}
$$

At the next stage, the alcohol-ester-acid can react with another diol, with another diacid, *or with another trifunctional molecule like itself*.

$$\underset{\text{alcohol-ester-acid}}{HO\!\sim\!O\!-\!\overset{\overset{\displaystyle O}{\|}}{C}\!\sim\!CO_2H} \xrightarrow[-H_2O]{HO\sim OH} \underset{\text{diester-diol}}{HO\!\sim\!O\!-\!\overset{\overset{\displaystyle O}{\|}}{C}\!\sim\!\overset{\overset{\displaystyle O}{\|}}{C}\!-\!O\!\sim\!OH}$$

$$\xrightarrow[-H_2O]{HO_2C\sim CO_2H} \underset{\text{diester-diacid}}{HO_2C\!\sim\!\overset{\overset{\displaystyle O}{\|}}{C}\!-\!O\!\sim\!O\!-\!\overset{\overset{\displaystyle O}{\|}}{C}\!\sim\!CO_2H} \qquad (14.26)$$

$$\xrightarrow[-H_2O]{HO\sim O-\overset{\overset{O}{\|}}{C}\sim CO_2H} \underset{\text{alcohol-triester-acid}}{HO\!\sim\!O\!-\!\overset{\overset{\displaystyle O}{\|}}{C}\!\sim\!\overset{\overset{\displaystyle O}{\|}}{C}\!-\!O\!\sim\!O\!-\!\overset{\overset{\displaystyle O}{\|}}{C}\!\sim\!CO_2H}$$

The consequences of these alternatives are different; each of the first two products contains three monomer units, but in the third alternative, we go directly from two-monomer fragments to a product with four monomer units. Since the reactivity of the —OH group or of the —CO_2H group in all these reactants is quite similar, there is no particular preference among them, and the rates of the various reactions will depend mainly on the concentrations of the particular reactants.

Problem 14.12 How many monomer units will be present in the next product if the diester-diol and diester-diacid in eq. 14.26 react? Draw the structure of the product.

If we start with exactly one equivalent each of a diol and a diacid, we should, in principle, be able to form one giant polyester molecule. In practice, this does not happen. In fact, to form a polymer with an average of 100 monomer units or more, the reaction must go to at least 99% completion. Consequently, the starting materials for this type of polymerization must be exceedingly pure, their mole ratio must be controlled precisely, and the reaction must be forced to completion, usually by distilling or otherwise removing, as it is formed, the small molecule that is eliminated.

Problem 14.13 What product will mainly be formed if a diacid is treated with a *large excess* of a diol? if a diol is treated with a *large excess* of a diacid? These reactions represent two extremes of what can happen as the ratio of reactant concentrations deviates from 1 : 1 in a step-growth polymerization.

Although many polyesters are known, the most common example is **Dacron, the polyester of terephthalic acid and ethylene glycol.**

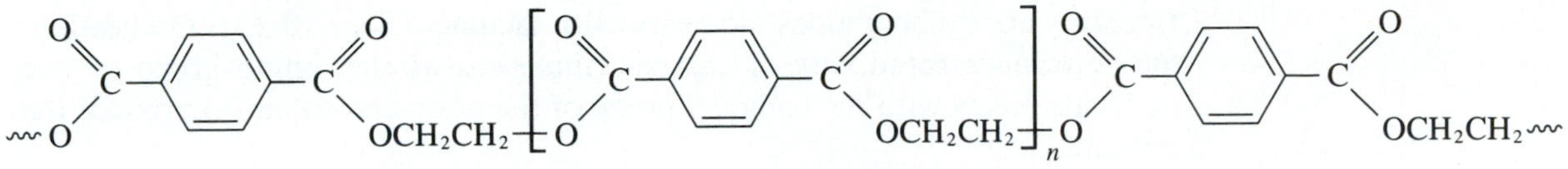

$$\sim O\text{-}C(=O)\text{-}C_6H_4\text{-}C(=O)OCH_2CH_2\left[O\text{-}C(=O)\text{-}C_6H_4\text{-}C(=O)OCH_2CH_2\right]_n O\text{-}C(=O)\text{-}C_6H_4\text{-}C(=O)OCH_2CH_2\sim$$

the polyester Dacron,
poly(ethylene terephthalate)

The value of n is about 100 ± 20. The crude polyester can be spun into fibers for use in textiles. The fibers are highly resistant to wrinkling.

The same polyester can also be fabricated into a particularly strong film called **Mylar.** Mylar polyester film is used for the long-term protection of artwork and historical documents because of its transparency, strength, and inertness. Because of their extraordinary resistance to tear, polyesters are used to make magnetic tapes for the recording industry. In the United States, production of polyester fibers exceeds 4 billion pounds per year.

Problem 14.14 *Kodel* is a polyester with the following structure:

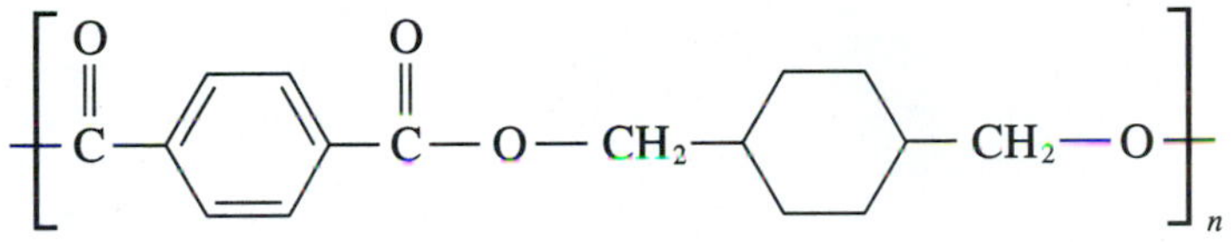

$$\left[C(=O)\text{-}C_6H_4\text{-}C(=O)\text{-}O\text{-}CH_2\text{-}C_6H_{10}\text{-}CH_2\text{-}O\right]_n$$

From what two monomers is it made?

Nylons are polyamide step-growth polymers. The formula for **nylon-6,6,** so called because each monomer (diamine and diacid) has six carbon atoms, is shown in eq. 14.2 (page 382). This polymer was first made by W. H. Carothers at the Du Pont Company in 1933 and was commercialized five years later.* When mixed, the two monomers form a polysalt which, on heating, loses water to form a polyamide. The molten polymer can be molded or spun into fibers.

The second most important polyamide is **nylon-6,** made from **caprolactam.**

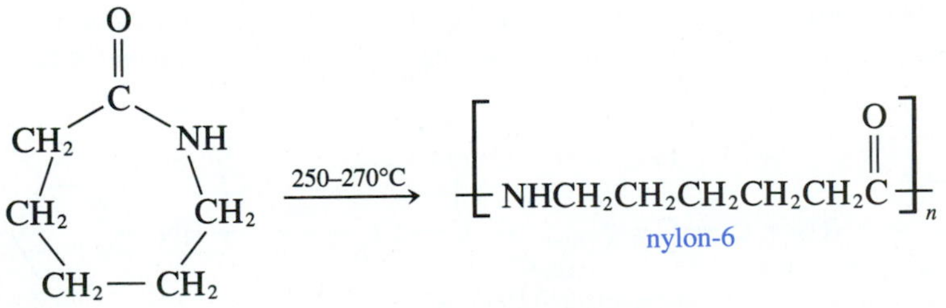

$$\text{caprolactam (cyclic } C(=O)\text{-}NH\text{-}CH_2CH_2CH_2CH_2CH_2\text{)} \xrightarrow{250\text{–}270°C} \left[NHCH_2CH_2CH_2CH_2CH_2C(=O)\right]_n \quad (14.27)$$

caprolactam — nylon-6

* For an account of its discovery, the strange origin of its name, and a description of its properties and many uses, see the article by G. B. Kauffman in *J. Chem. Education* **1988,** *65,* 803–8.

Lactams are cyclic amides (compare with lactones, Sec. 10.13). On heating, the seven-membered ring of caprolactam opens as the amino group of one molecule reacts with the carbonyl group of the next, and so on, to produce the polyamide.

Nylons are extremely versatile polymers that can be processed to give materials as delicate as sheer fabrics, as long-wearing as carpets, as tough as molded automobile parts, or as useful as Velcro fasteners.

A WORD ABOUT ...

28. Aramids, the Latest in Polyamides

Aromatic polyamides (called aramids) are being produced at a rapidly growing rate because of their special properties of heat resistance, low flammability, and exceptional strength. The best known is **Kevlar** (see Figure 14.2). Because of the aromatic rings, this type of polyamide has a much stiffer structure than do the nylons. It is being used in place of steel to make tire cord for radial tires, since a Kevlar fiber is five times stronger than an equal-weight steel fiber. Kevlar is used in lightweight personal body armor (such as bullet-resistant vests), which offers protection against handgun fire, shotgun pellets, and knife slashes.

Nomex has a structure similar to Kevlar's but uses *meta-* instead of *para*-oriented monomers. It is used in flame-resistant clothing because its fibers char rather than melt when exposed to flame. It has wide applications, ranging from firefighters' coats to racing drivers' uniforms. It has also been used in flame-resistant building materials. Honeycomb core constructions of Nomex are used in many internal and external parts of military and commercial jets, helicopters, and space vehicles. For example, close to 25,000 square feet of Boeing 747's exterior structures include Nomex construction. The combination of strength and light weight has also made both Nomex and Kevlar popular in boat construction. For an account of the Kevlar story, see Tanner, D.; Fitzgerald, J. A.; and Phillips, B. R. *Angewandte Chemie International Edition in English, Advanced Materials* **1989,** *28,* 649–54.

$$H_2N-C_6H_4-NH_2 + Cl-\overset{O}{\overset{\|}{C}}-C_6H_4-\overset{O}{\overset{\|}{C}}-Cl \xrightarrow{\text{base}} \left(HN-C_6H_4-NH-\overset{O}{\overset{\|}{C}}-C_6H_4-\overset{O}{\overset{\|}{C}} \right)_n$$

p-phenylenediamine — terephthaloyl chloride — Kevlar

FIGURE 14.2

Reaction used to synthesize Kevlar.

14.10 Polyurethanes and Other Step-Growth Polymers

A **urethane** (also called a **carbamate**) is a functional group that is simultaneously an ester and an amide, $RNH\overset{O}{\overset{\|}{C}}OR'$. Urethanes are commonly prepared from isocyanates and alcohols.

$$\underset{\text{isocyanate}}{R{-}N{=}C{=}O} + \underset{\text{alcohol}}{R'OH} \longrightarrow \underset{\text{urethane}}{RNH\overset{O}{\overset{\|}{C}}OR'} \qquad (14.28)$$

EXAMPLE 14.4 Write a mechanism for eq. 14.28.

Solution The reaction is an example of a nucleophilic addition to the carbonyl group (compare with Secs. 9.7 and 10.12).

$$R{-}\ddot{N}{=}C{=}\ddot{O}\!: + R'\ddot{O}H \longrightarrow R{-}\ddot{N}{=}\overset{:\ddot{O}:^-}{\overset{|}{C}}{-}\underset{H}{\overset{..+}{O}}{-}R' \xrightarrow{\;+H^+,\;-H^+\;} R{-}\underset{H}{\ddot{N}}{-}\overset{O}{\overset{\|}{C}}{-}OR'$$

Problem 14.15 The highly effective, biodegradable insecticide Sevin is a urethane called 1-naphthyl-*N*-methylcarbamate. It is made from methyl isocyanate and 1-naphthol. Using eq. 14.28 as a guide, write an equation for its preparation.

Polyurethanes are made from *di*isocyanates and *di*ols. The most important commercial example is prepared from 2,4-tolylenediisocyanate (TDI) and ethylene glycol.

$$\underset{\text{tolylene diisocyanate (TDI)}}{CH_3C_6H_3(N{=}C{=}O)_2} + \underset{\text{ethylene glycol}}{HOCH_2CH_2OH} \longrightarrow \underset{\text{a polyurethane}}{\sim\!\left[{-}C_6H_3(CH_3){-}\underbrace{NH{-}\overset{O}{\overset{\|}{C}}{-}}_{\text{urethane group}}OCH_2CH_2O{-}\overset{O}{\overset{\|}{C}}{-}NH{-}\right]\!\sim} \qquad (14.29)$$

This reaction is a little different from most step-growth polymerizations in that no small molecule is eliminated. Like other step-growth polymerizations, however, it does involve a reaction of two different difunctional monomers.

Problem 14.16 Rewrite the partial polyurethane structure shown in eq. 14.29, but with one more monomer unit attached to each wavy bond.

The reaction in eq. 14.29 produces a polyurethane but does not produce a foam. To obtain a foam as the product, the polymerization is carried out with a little water present. The water reacts with isocyanate groups in the starting material or in the growing polymer to produce a **carbamic acid.** This acid spontaneously loses carbon dioxide, which creates bubbles as the polymer forms.

$$\sim N{=}C{=}O + HOH \longrightarrow \sim NH\overset{\overset{\displaystyle O}{\|}}{C}OH \longrightarrow \sim NH_2 + CO_2\uparrow \qquad (14.30)$$

isocyanate, water, carbamic acid, amine

The amount of carbon dioxide formed, which determines the density of the foam, can be controlled by the amount of water used. The resulting amine can also react with isocyanate groups to form a urea, which may serve as a cross-link between polymer chains.

$$\sim N{=}C{=}O + H_2N\sim \longrightarrow \sim NH\overset{\overset{\displaystyle O}{\|}}{C}NH\sim \qquad (14.31)$$

isocyanate, amine, urea link

Polyurethanes have a wide range of applications. Polymers with relatively little cross-linking produce stretchable fibers (Spandex, Lycra) used for bathing suits. Polyurethane foams are used in furniture, mattresses, and car seats and as insulation in portable ice chests. Cross-linked polyurethanes form very tough surface coatings in paints and varnishes. The major component of the Jarvik-7 artificial heart is also a polyurethane.

Several commercially important step-growth polymers are based on reactions of formaldehyde. **Bakelite,** the oldest totally synthetic polymer, was invented by Leo Baekeland in 1907. It is prepared from phenol and formaldehyde. The polymer is highly cross-linked, with methylene groups *ortho* and/or *para* to the phenolic hydroxyl group.

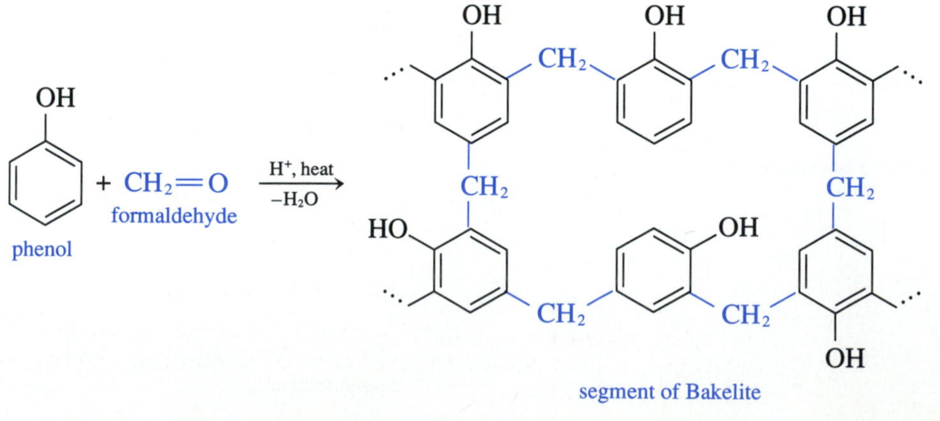

$$\text{phenol} + CH_2{=}O \text{ (formaldehyde)} \xrightarrow[-H_2O]{H^+,\ \text{heat}} \text{segment of Bakelite} \qquad (14.32)$$

segment of Bakelite

EXAMPLE 14.5 Write a mechanism for the formation of Bakelite.

Solution The first step in this acid-catalyzed polymerization involves electrophilic aromatic substitution.

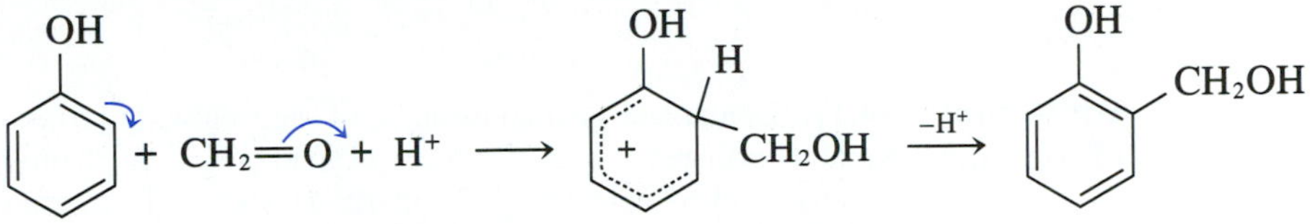

The next step involves protonation of the alcohol, formation of a benzyl cation, and another electrophilic aromatic substitution. In this step, a molecule of water is eliminated.

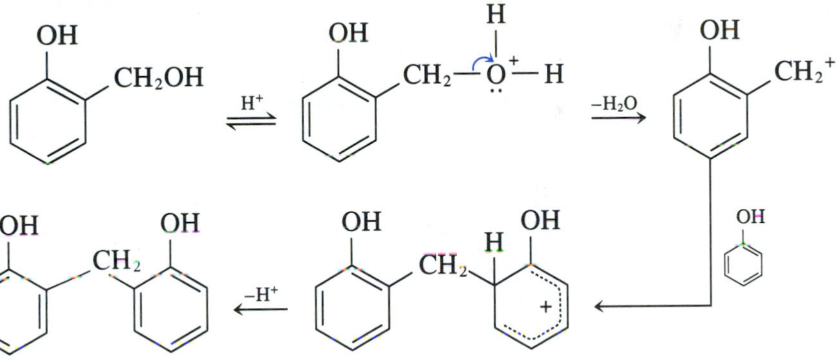

Repetition at *ortho* and *para* positions gives the polymer.

Bakelite is a **thermosetting polymer.** Heating leads to further cross-linking, producing a hard, infusible material. This process cannot be reversed, and once setting has occurred, the polymer cannot be melted. Bakelite is used for molded plastic parts, such as appliance handles, and for applications requiring light materials that can withstand high temperatures, such as missile nose cones.

Urea and formaldehyde also form an important commercial polymer.

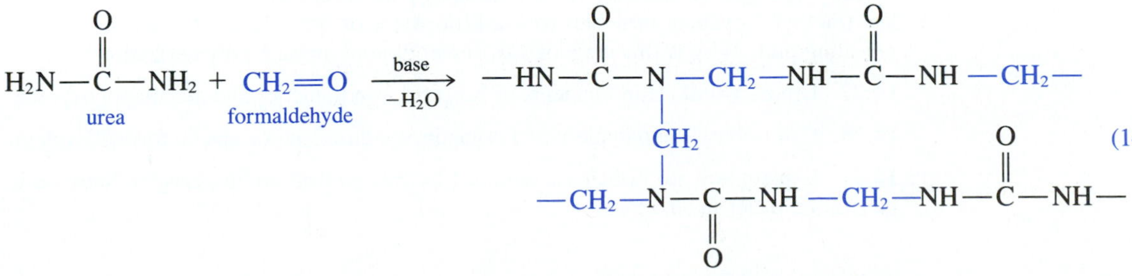

$$\underset{\text{urea}}{H_2N-\overset{\overset{\displaystyle O}{\|}}{C}-NH_2} + \underset{\text{formaldehyde}}{CH_2{=}O} \xrightarrow[-H_2O]{\text{base}} \text{urea-formaldehyde polymer} \quad (14.33)$$

urea-formaldehyde polymer

This kind of polymer is used in molded materials (electrical fittings and kitchenware), in laminates such as Formica, in plywood and particle board as adhesive, and in foams.

We have barely touched on polymer chemistry. The field is vast and constantly developing. No doubt you will see many new types of polymers reach the marketplace during your lifetime.*

Additional Problems

14.17. Define and give an example of the following terms:

a. homopolymer **b.** copolymer **c.** chain-growth polymerization
d. cross-linked polymer **e.** thermoplastic **f.** thermosetting
g. isotactic **h.** atactic **i.** chain transfer

14.18. Write out all the steps in the free-radical chain-growth polymerization of vinyl chloride.

14.19. Draw the structure of a chain segment of polyvinyl alcohol. This polymer cannot be made by polymerizing its monomer. Why not? It is usually made from polyvinyl acetate. (see Problem 14.6b).

14.20. Although propylene can be polymerized with a free-radical initiator, the molecular weight of the polymer is never very high by this method because chain transfer from the methyl group of propylene keeps the chains short. Explain why this chain transfer occurs so easily.

14.21. Consider a polyethylene molecule containing 1000 monomer units initiated by one benzoyloxy radical (from benzoyl peroxide). What percentage of the molecular weight is due to the catalyst radical? Repeat the calculation for a similar polystyrene molecule.

14.22. Although radical combination is a chain-terminating step in free-radical chain-growth polymerization, no comparable step is possible in ionic chain-growth polymerizations. Why not? How, then, might chains be terminated in such reactions?

14.23. The polymerization of methyl methacrylate to Plexiglas or Lucite is catalyzed by butyllithium and carried out at $-78°C$. What other reaction might occur if this reaction were performed at room temperature?

14.24. Draw the expected structure of "propylene tetramer," made by the acid-catalyzed polymerization of propene.

14.25. Propylene oxide can be converted to a polyether by anionic chain-growth polymerization. What is the structure of the polymer, and how is it formed?

14.26. Superglue is made by the on-the-spot polymerization of methyl α-cyanoacrylate (methyl 2-cyanopropenoate) with a little water or base. Draw the structure of the repeating unit. Why is this monomer so susceptible to anionic polymerization?

14.27. Draw six-unit chain segments of isotactic, syndiotactic, and atactic polystyrene.

14.28. Can isobutylene polymerize in isotactic, syndiotactic, or atactic forms? Explain.

14.29. Explain how polyethylenes obtained by free-radical and by Ziegler-Natta polymerization differ in structure.

---

* For further reading, try "Practical Macromolecular Organic Chemistry," 3rd ed., by D. Braun, H. Cherdon, and W. Kern (Harwood Academic Publishers, New York, 1984, translated by K. J. Ivin) or "Industrial Organic Chemicals in Perspective" by H. A. Wittcoff and B. G. Reuben (John Wiley, New York, 1980, Vols. 1 and 2).

14.30. The ozonolysis of natural rubber gives levulinic aldehyde, $CH_3\overset{\displaystyle O}{\overset{\|}{C}}CH_2CH_2CH{=}O$

Explain how this result is consistent with natural rubber's formula (eq. 14.22).

14.31. Write out the steps in a mechanism that explains the free-radical copolymerization of 1,3-butadiene and styrene to give the synthetic rubber SBR (eq. 14.23).

14.32. **Neoprene** is a synthetic rubber invented more than 50 years ago by the Du Pont Company. It is used to manufacture industrial hoses, drive belts, window gaskets, shoe soles, and packaging materials. Neoprene is a polymer of 2-chloro-1,3-butadiene. Assuming mainly 1,4-addition, draw the structure of the repeating unit in neoprene.

14.33. Draw the structural formula for a segment of alternating copolymer made from styrene and methyl methacrylate.

14.34. Draw the repeating unit in the polymer that results from each of the following step-growth polymerizations:

a. $Cl{-}\overset{\displaystyle O}{\overset{\|}{C}}(CH_2)_6\overset{\displaystyle O}{\overset{\|}{C}}{-}Cl + H_2N(CH_2)_6NH_2$

b. $O{=}C{=}N{-}C_6H_4{-}CH_2{-}C_6H_4{-}N{=}C{=}O + HOCH_2CH_2OH \longrightarrow$

c. $CH_3O\overset{\displaystyle O}{\overset{\|}{C}}(CH_2)_4\overset{\displaystyle O}{\overset{\|}{C}}OCH_3 + HOCH_2CH_2OH \xrightarrow{H^+}$

14.35. **Lexan** is a tough polycarbonate used to make molded articles. It is made from diphenyl carbonate and bisphenol A. Draw the structure of its repeating unit.

$C_6H_5{-}O{-}\overset{\displaystyle O}{\overset{\|}{C}}{-}O{-}C_6H_5$

diphenyl carbonate

$HO{-}C_6H_4{-}C(CH_3)_2{-}C_6H_4{-}OH$

bisphenol A

14.36. Formaldehyde polymerizes in aqueous solution to give paraformaldehyde, $HO{-}(CH_2O)_n{-}H$. Although high molecular weights are achieved, the polymer "unzips" readily. However, if the polymer is treated with acetic anhydride, the resulting material (an important commercial polymer called **Delrin**) no longer "unzips." Explain the chemistry described.

14.37. Reaction of phthalic anhydride with glycerol gives a cross-linked polyester called a *glyptal resin*. Draw a partial structure, showing the cross-linking.

14.38. Using eq. 14.32 as a model, draw a segment of the polymer that will be formed from formaldehyde and *p*-methylphenol. Will the polymer be cross-linked? Explain.

14.39. To make polymers commercially, it is necessary to make the monomers inexpensively and on a large scale. One commercial method for making hexamethylenediamine (for nylon-6,6) starts with the 1,4-addition of chlorine to 1,3-butadiene. Suggest a possibility for the remaining steps.

14.40. Methyl methacrylate (methyl 2-methylpropenoate), the monomer for Plexiglas and Lucite, is made by reaction of acetone cyanohydrin (Sec. 9.11) with methanol and sulfuric acid. The reaction involves a methanolysis and a dehydration. Write an equation for the reaction and mechanistic equations for the two processes.

CHAPTER 15

Lipids and Detergents

15.1 Introduction

In this chapter, we take up the first of four major classes of biologically important substances, the **lipids.** And because they are related to one type of lipid, we will also discuss here the chemistry of **detergents.**

Lipids (from the Greek *lipos,* fat) are constituents of plants or animals that are characterized by their solubility properties. In particular, *lipids are insoluble in water but soluble in nonpolar organic solvents,* such as ether. Lipids can be extracted from cells and tissues by organic solvents. This solubility property distinguishes lipids from three other major classes of natural products, carbohydrates, proteins and nucleic acids, which in general are *not* soluble in organic solvents.

Lipids may vary considerably in chemical structure, even though they have similar solubility properties. Some are esters, some are hydrocarbons; some are acyclic and others are cyclic, even polycyclic. We will take up each structural type separately.

15.2 Fats and Oils; Triesters of Glycerol

Fats and oils are familiar parts of daily life. Common fats include butter, lard, and the fatty portions of meat. Oils come mainly from plants and include corn, cottonseed, olive, peanut, and soybean oils. Although fats are solids and oils are liquids, they have the same basic organic structure. ***Fats and oils are triesters of glycerol and are called triglycerides.*** When we boil a fat or oil with alkali and acidify the resulting solution, we obtain glycerol and a mixture of **fatty acids.** The reaction is called **saponification** (Sec. 10.14).

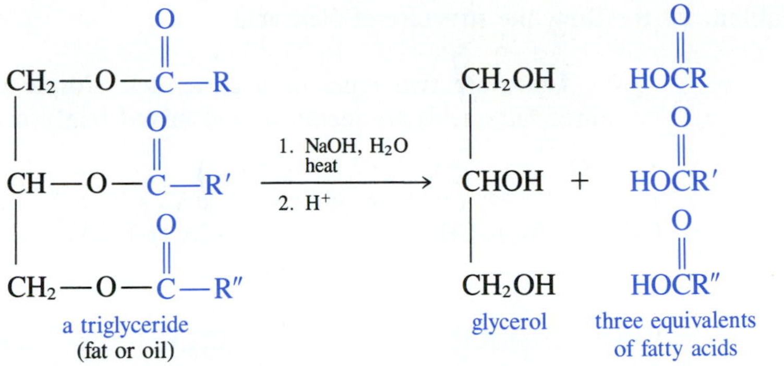

$$\begin{array}{l} CH_2-O-\overset{\overset{\Large O}{\|}}{C}-R \\ | \\ CH-O-\overset{\overset{\Large O}{\|}}{C}-R' \\ | \\ CH_2-O-\overset{\overset{\Large O}{\|}}{C}-R'' \end{array} \xrightarrow[2.\ H^+]{1.\ NaOH,\ H_2O \text{ heat}} \begin{array}{l} CH_2OH \\ | \\ CHOH \\ | \\ CH_2OH \end{array} + \begin{array}{l} HO\overset{\overset{\Large O}{\|}}{C}R \\ HO\overset{\overset{\Large O}{\|}}{C}R' \\ HO\overset{\overset{\Large O}{\|}}{C}R'' \end{array} \quad (15.1)$$

a triglyceride (fat or oil) — glycerol — three equivalents of fatty acids

The most common saturated and unsaturated fatty acids obtained in this way are listed in Table 15.1. Although exceptions are known, *most fatty acids are unbranched and contain an even number of carbon atoms.* If double bonds are present, they usually have the *cis* (or *Z*) configuration and are not conjugated.

EXAMPLE 15.1 Draw the structure of linoleic acid, showing the geometry at each double bond.

Solution Both double bonds have the *Z* configuration.

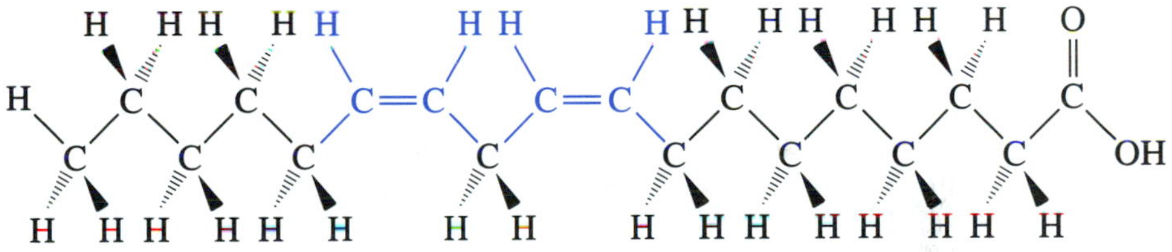

$CH_3(CH_2)_4CH{=}CHCH_2CH{=}CH(CH_2)_7COOH$ (fully extended zigzag drawing with wedge/dash H atoms; both C=C *Z*)

The preferred conformation is fully extended, with a staggered arrangement at each C—C single bond.

TABLE 15.1 Common acids obtained from fats

	Common name	*Number of carbons*	*Structural formula*	*mp, °C*
Saturated	lauric	12	$CH_3(CH_2)_{10}COOH$	44
	myristic	14	$CH_3(CH_2)_{12}COOH$	58
	palmitic	16	$CH_3(CH_2)_{14}COOH$	63
	stearic	18	$CH_3(CH_2)_{16}COOH$	70
	arachidic	20	$CH_3(CH_2)_{18}COOH$	77
Unsaturated	oleic	18	$CH_3(CH_2)_7CH{=}CH(CH_2)_7COOH$ (*cis*)	13
	linoleic	18	$CH_3(CH_2)_4CH{=}CHCH_2CH{=}CH(CH_2)_7COOH$ (*cis*, *cis*)	−5
	linolenic	18	$CH_3CH_2CH{=}CHCH_2CH{=}CHCH_2CH{=}CH(CH_2)_7COOH$ (all *cis*)	−11

Problem 15.1 Draw the structure of oleic acid.

There are two types of triglycerides: **simple triglycerides,** in which all three fatty acids are identical, and **mixed triglycerides.**

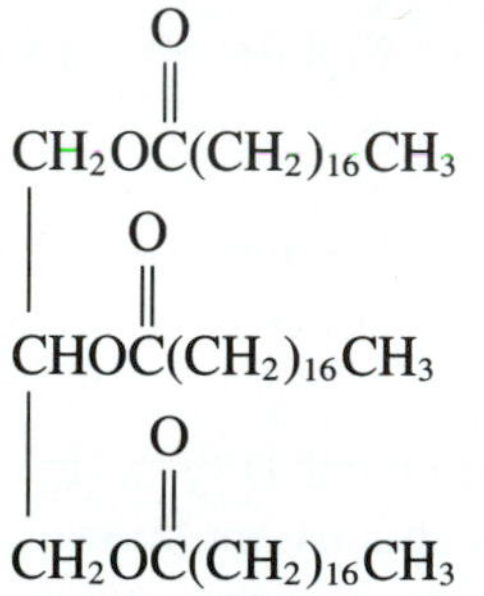

$$\begin{array}{l} CH_2O\overset{O}{\overset{\|}{C}}(CH_2)_{16}CH_3 \\ | \\ CHO\overset{O}{\overset{\|}{C}}(CH_2)_{16}CH_3 \\ | \\ CH_2O\overset{O}{\overset{\|}{C}}(CH_2)_{16}CH_3 \end{array}$$

a simple triglyceride
(glyceryl tristearate or tristearin)

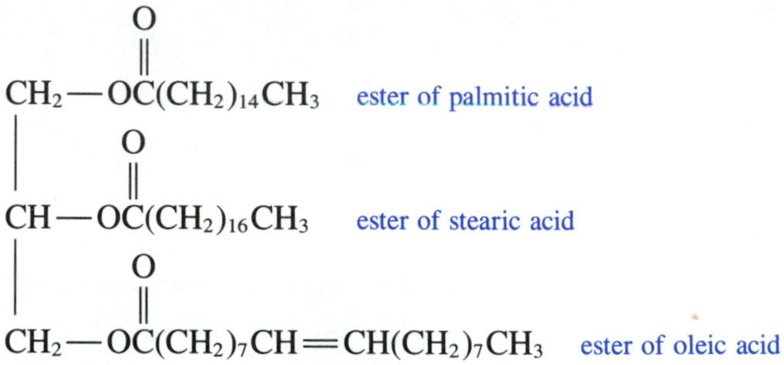

$$\begin{array}{ll} CH_2{-}O\overset{O}{\overset{\|}{C}}(CH_2)_{14}CH_3 & \text{ester of palmitic acid} \\ | & \\ CH{-}O\overset{O}{\overset{\|}{C}}(CH_2)_{16}CH_3 & \text{ester of stearic acid} \\ | & \\ CH_2{-}O\overset{O}{\overset{\|}{C}}(CH_2)_7CH{=}CH(CH_2)_7CH_3 & \text{ester of oleic acid} \end{array}$$

a mixed triglyceride
(glyceryl palmitostearoöleate)

EXAMPLE 15.2 Draw the structure of glyceryl stearopalmitoöleate, an isomer of the mixed triglyceride shown above.

Solution

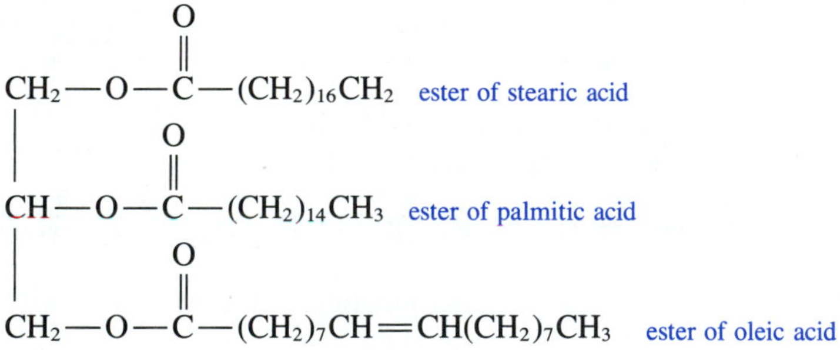

$$\begin{array}{ll} CH_2{-}O{-}\overset{O}{\overset{\|}{C}}{-}(CH_2)_{16}CH_2 & \text{ester of stearic acid} \\ | & \\ CH{-}O{-}\overset{O}{\overset{\|}{C}}{-}(CH_2)_{14}CH_3 & \text{ester of palmitic acid} \\ | & \\ CH_2{-}O{-}\overset{O}{\overset{\|}{C}}{-}(CH_2)_7CH{=}CH(CH_2)_7CH_3 & \text{ester of oleic acid} \end{array}$$

Notice that each of these triglycerides would give the same saponification products.

Problem 15.2 Draw the structure for

a. glyceryl tripalmitate.
b. glyceryl palmitoöleostearate.

Problem 15.3 What saponification products would be obtained from each of the triglycerides in Problem 15.2?

In general, a particular fat or oil consists, not of a single triglyceride, but of a complex mixture of triglycerides. For this reason, the composition of a fat or oil is usually expressed in terms of the percentages of the various acids obtained from it by saponification (Table 15.2). Some fats and oils give mainly one or two acids, with only minor amounts of other acids. Olive oil, for example, gives 83% oleic acid. Palm oil gives 43% palmitic acid and 43% oleic

TABLE 15.2 Fatty acid composition of some fats and oils (approximate)

	Saturated acids (%)					*Unsaturated acids (%)*	
Source	C_{10} *and less*	C_{12} *lauric*	C_{14} *myristic*	C_{16} *palmitic*	C_{18} *stearic*	C_{18} *oleic*	C_{18} *linoleic*
Animal Fats							
Butter	12	3	12	28	10	26	2
Lard	—	—	1	28	14	46	5
Beef tallow	—	0.2	3	28	24	40	2
Human	—	1	3	25	8	46	10
Vegetable Oils							
Olive	—	—	1	5	2	83	7
Palm	—	—	2	43	2	43	8
Corn	—	—	1	10	2	40	40
Peanut	—	—	—	8	4	60	25

acid, with lesser amounts of stearic and linoleic acids. Butterfat, on the other hand, gives at least 14 different acids on hydrolysis and is somewhat exceptional in that about 9% of these acids have fewer than 10 carbon atoms.

Problem 15.4 From the data in Table 15.2, what can you say in general about the ratio of saturated to unsaturated acids in fats and oils?

What is it that makes some triglycerides solids (fats) and others liquids (oils)? The distinction is clear from their composition. *Oils contain a much higher percentage of unsaturated fatty acids than do fats.* For example, most vegetable oils (such as corn oil or soybean oil) give about 80% unsaturated acids on hydrolysis. For fats (such as beef tallow) the figure is much lower, just a little over 50%.

Table 15.1 shows that the melting points of unsaturated fatty acids are appreciably lower than those of saturated acids. Compare, for example, the melting points of stearic and oleic acids, which differ structurally by only one double bond. The same difference applies to triglycerides: *the more double bonds in the fatty acid portion of the triester, the lower its melting point.*

The reason for the effect of saturation or unsaturation on the melting point becomes apparent when we examine space-filling models. Figure 15.1 shows a model of a fully saturated triglyceride. The long, saturated chains have fully extended, staggered conformations. They can therefore pack together fairly regularly, as in a crystal. Consequently, saturated triglycerides are usually solids at room temperature.

The result of introducing just one *cis* double bond in one of the chains is shown in Figure 15.2. Clearly, the chains in this kind of molecule (and the molecules themselves, when many are close to one another) cannot align nicely in a crystalline array. The substance therefore remains a liquid. The more double bonds, the more disorderly the structure and the lower the melting point.

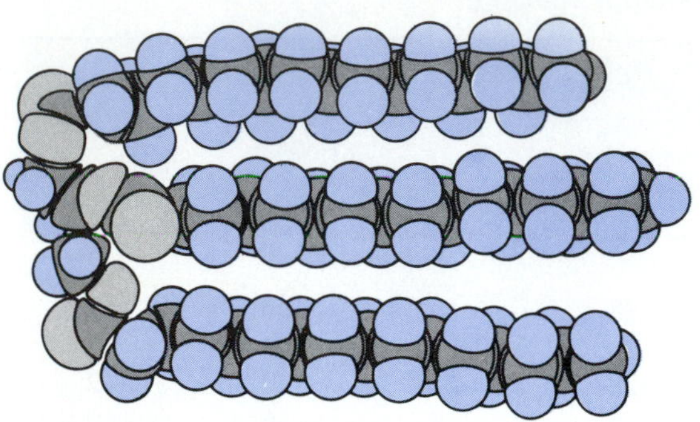

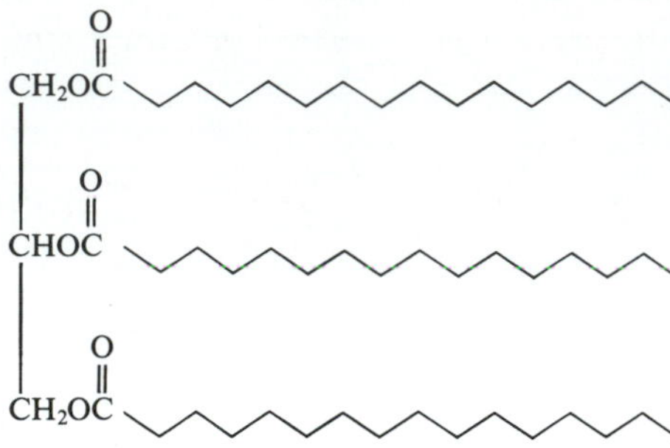

FIGURE 15.1
Space-filling and schematic models of glyceryl tripalmitate.

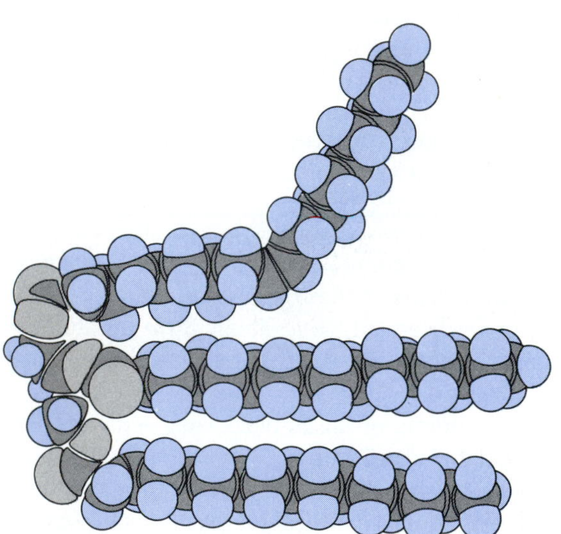

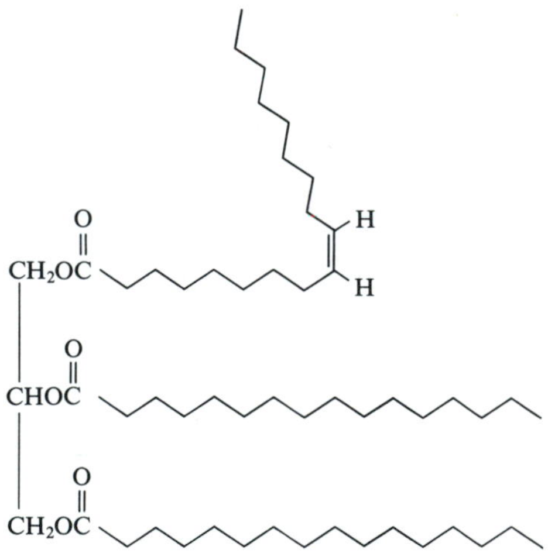

FIGURE 15.2
Space-filling and schematic models of glyceryl dipalmitoöleate.

15.3 *Hydrogenation of Vegetable Oils*

Vegetable oils, which are highly unsaturated, are converted into solid vegetable fats, such as Crisco, by catalytically hydrogenating some or all of the double bonds. This process, called **hardening,** is illustrated by the hydrogenation of glyceryl trioleate to glyceryl tristearate.

$$\begin{array}{l} \overset{\displaystyle O}{\|} \\ CH_2OC(CH_2)_7CH{=}CH(CH_2)_7CH_3 \\ |\quad \overset{\displaystyle O}{\|} \\ CHOC(CH_2)_7CH{=}CH(CH_2)_7CH_3 \\ |\quad \overset{\displaystyle O}{\|} \\ CH_2OC(CH_2)_7CH{=}CH(CH_2)_7CH_3 \end{array} \xrightarrow[\text{Ni catalyst, heat}]{3\,H_2} \begin{array}{l} \overset{\displaystyle O}{\|} \\ CH_2OC(CH_2)_{16}CH_3 \\ |\quad \overset{\displaystyle O}{\|} \\ CHOC(CH_2)_{16}CH_3 \\ |\quad \overset{\displaystyle O}{\|} \\ CH_2OC(CH_2)_{16}CH_3 \end{array} \qquad (15.2)$$

glyceryl trioleate mp −17°C → glyceryl tristearate mp 55°C

Margarine is made by hydrogenating cottonseed, soybean, peanut, or corn oil until the desired butterlike consistency is obtained. The product may be churned with milk and artifically colored to mimic butter's flavor and appearance.

15.4 *Saponification of Fats and Oils; Soap*

When a fat or oil is heated with alkali, the ester is converted to glycerol and the salts of fatty acids. The reaction is illustrated here with the saponification of glyceryl tripalmitate.

$$\begin{array}{l} \overset{\displaystyle O}{\|} \\ CH_2OC(CH_2)_{14}CH_3 \\ |\quad \overset{\displaystyle O}{\|} \\ CHOC(CH_2)_{14}CH_3 \\ |\quad \overset{\displaystyle O}{\|} \\ CH_2OC(CH_2)_{14}CH_3 \end{array} + 3\,Na^+OH^- \xrightarrow{\text{heat}} \begin{array}{l} CH_2OH \\ | \\ CHOH \\ | \\ CH_2OH \end{array} + 3\,CH_3(CH_2)_{14}CO_2^-Na^+ \qquad (15.3)$$

glyceryl tripalmitate (from palm oil); glycerol; sodium palmitate (a soap)

The salts (usually sodium) of long-chain fatty acids are **soaps.**

The conversion of animal fats (for example, goat tallow) into soap by heating with wood ashes (which are alkaline) is one of the oldest of chemical processes. Soap has been produced for at least 2300 years, having been known to the ancient Celts and Romans. Yet, as recently as the sixteenth and seventeenth centuries, soap was still a rather rare substance, used mainly in medicine. But by the nineteenth century, soap had come into such widespread use that the German organic chemist Justus von Liebig was led to remark that the quantity of soap consumed by a nation was an accurate measure of its wealth and civilization. At present, annual world production of ordinary soaps (not including synthetic detergents) is well over 6 million tons.

Soaps are made by either a batch process or a continuous process. In the batch process, the fat or oil is heated with a slight excess of alkali (NaOH) in an open kettle. When saponification is complete, salt is added to precipitate the soap as thick curds. The water layer, which contains salt, glycerol, and excess alkali, is drawn off, and the glycerol is recovered by distillation. The crude soap curds, which contain some salt, alkali, and glycerol as impurities, are purified by boiling with water and reprecipitating with salt several times. Finally, the curds are boiled with enough water to form a smooth mixture that, on standing, gives a homogeneous upper layer of soap. This soap may be sold without further processing, as a cheap industrial soap. Various fillers, such as sand or pumice, may be added, to make scouring soaps. Other treatments transform the crude soap to toilet soaps, powdered or flaked soaps, medicated or perfumed soaps, laundry soaps, liquid soaps, or (by blowing air in) floating soaps.

In the continuous process, which is more common today, the fat or oil is hydrolyzed by water at high temperatures and pressures in the presence of a catalyst, usually a zinc soap. The fat or oil and the water are introduced continuously into opposite ends of a large reactor, and the fatty acids and glycerol are removed as formed, by distillation. The acids are then carefully neutralized with an appropriate amount of alkali to make soap.

15.5 How Do Soaps Work?

Most dirt on clothing or skin adheres to a thin film of oil. If the oil film can be removed, the dirt particles can be washed away. A soap molecule consists of a long, hydrocarbon-like chain of carbon atoms with a highly polar or ionic group at one end (Figure 15.3). The carbon chain is **lipophilic** (attracted to or soluble in fats and oils), and the polar end is **hydrophilic** (attracted to or soluble in water). In a sense, soap molecules are "schizophrenic," having two different "personalities." Let us see what happens when we add soap to water.

When soap is shaken with water, it forms a colloidal dispersion—not a true solution. These soap solutions contain aggregates of soap molecules called **micelles.** The nonpolar, or lipophilic, carbon chains are directed toward the center of the micelle. The polar, or hydrophilic, ends of the molecule form the "surface" of the micelle that is presented to the water (Figure 15.4). In ordinary soaps, the outer part of each micelle is negatively charged, and the positive sodium ions congregate near the periphery of each micelle.

In acting to remove dirt, soap molecules surround and emulsify the droplets of oil or grease. The lipophilic "tails" of the soap molecules dissolve in the oil.

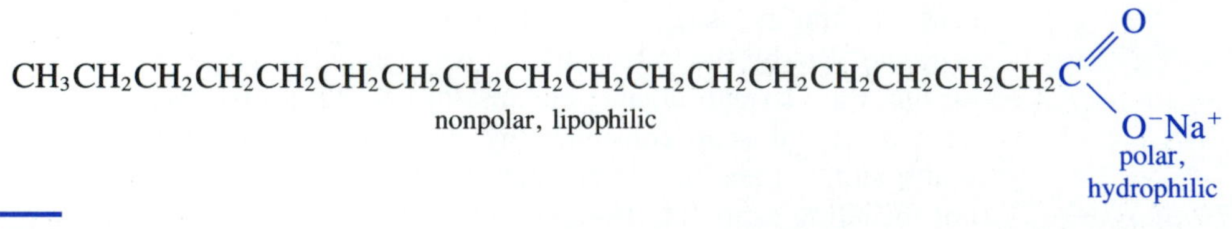

$$\underbrace{CH_3CH_2CH_2CH_2CH_2CH_2CH_2CH_2CH_2CH_2CH_2CH_2CH_2CH_2CH_2CH_2CH_2}_{\text{nonpolar, lipophilic}}C\underbrace{\overset{\displaystyle O}{\parallel}\!-O^-Na^+}_{\text{polar, hydrophilic}}$$

FIGURE 15.3
Sodium stearate, an ordinary soap.

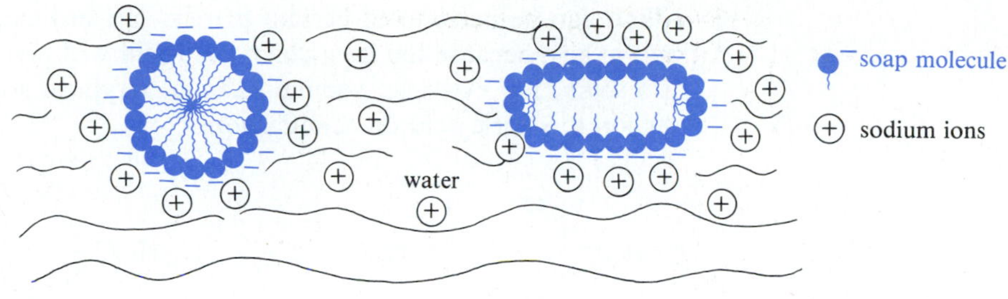

FIGURE 15.4
Soap molecules form micelles when "dissolved" in water.

The hydrophilic ends extend out of the oil droplet toward the water. In this way, the oil droplets are stabilized in the water solution because the negative surface charge of the droplets prevents their coalescence (as shown in Figure 15.5).

Another striking property of soap solutions is their unusually low surface tension, which gives a soap solution more "wetting" power than plain water has. As a consequence, soaps belong to a class of substances called **surfactants.** A combination of the emulsifying power and the surface action of soap solutions enables them to detach dirt, grease, and oil particles from the surface being cleaned and to emulsify them so that they can be washed away. These same principles of cleansing action apply to synthetic detergents.

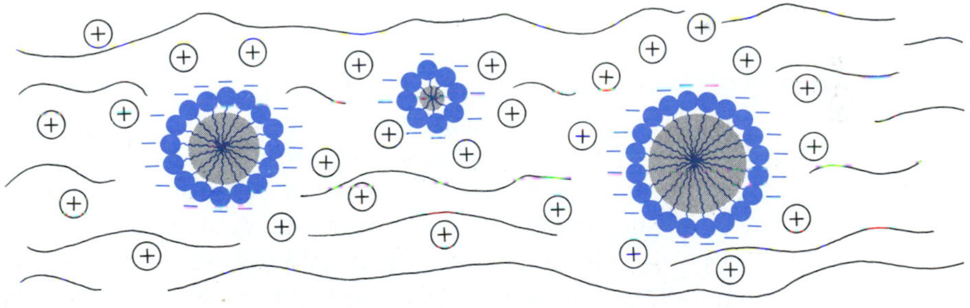

FIGURE 15.5
Oil droplets become emulsified by soap molecules.

15.6 *Synthetic Detergents (Syndets)*

Annual world production of synthetic detergents (sometimes called **syndets**) now exceeds that of ordinary soaps. Syndets evolved in response to two problems with the use of ordinary, unimproved soaps. First, being salts of weak acids, *soaps give rather alkaline solutions in water*. This is due to partial hydrolysis of the sodium salts.

$$\underset{\text{soap}}{R{-}\overset{\overset{\displaystyle O}{\|}}{C}{-}O^-Na^+} + H{-}OH \rightleftharpoons R{-}\overset{\overset{\displaystyle O}{\|}}{C}{-}OH + \underset{\text{alkali}}{Na^+OH^-} \qquad (15.4)$$

Alkali can be harmful to certain fabrics. Yet ordinary soaps cannot function well in acid because the long-chain fatty acid will precipitate from the solution as a scum. For example, sodium stearate, a typical soap, is destroyed by conversion to stearic acid on acidification.

$$\underset{\substack{\text{sodium stearate}\\ \text{(soluble)}}}{C_{17}H_{35}C(=O)O^{-}Na^{+}} + H^{+}Cl^{-} \longrightarrow \underset{\substack{\text{stearic acid}\\ \text{(insoluble)}}}{C_{17}H_{35}C(=O)OH} \downarrow + Na^{+}Cl^{-} \qquad (15.5)$$

The second problem with ordinary soaps is that *they form insoluble salts with the calcium, magnesium, or ferric ions that may be present in hard water*.

$$2\ \underset{\substack{\text{sodium stearate}\\ \text{(soluble)}}}{C_{17}H_{35}C(=O)O^{-}Na^{+}} + Ca^{2+} \longrightarrow \underset{\substack{\text{calcium stearate}\\ \text{(insoluble)}}}{(C_{17}H_{35}COO^{-})_2Ca^{2+}} \downarrow + 2\ Na^{+} \qquad (15.6)$$

The insoluble salts are responsible for the "rings" around bathtubs or collars and for films that dull the look of clothing and hair.

These problems with ordinary soaps have been solved or diminished in several ways. For example, water can be "softened," either municipally or in individual households, to remove the offending calcium or magnesium ions. In softened water, these ions are replaced by sodium ions. If this water is also used for drinking, however, it may cause health problems, especially for people who have to limit their sodium ion intake.

Phosphates can also be added to soaps. Phosphates form soluble complexes with metal ions, thus keeping these ions from forming insoluble salts with the soap. But widespread use of phosphates in the past has created environmental problems. Because of their use in detergents, tremendous quantities of phosphates eventually found their way into lakes, rivers, and streams. Since they are fertilizers, these phosphates stimulated plant growth to such an extent that the plants exhausted the dissolved oxygen in the water, in turn causing fish to die. Phosphates are still used in detergents, but their use is now limited by law to levels that are unlikely to be harmful.

Another way to eliminate the problems associated with ordinary soaps is to design more effective detergents. These syndets must have several features. Like ordinary soaps, they must have a long lipophilic chain and a polar, or ionic, hydrophilic end. However, the polar end should not form insoluble salts with metal ions present in water and should not affect the acidity of water.

The first syndets were sodium salts of alkyl hydrogen sulfates. A long-chain alcohol was prepared by the **hydrogenolysis** of a fat or oil. For example, glyceryl trilaurate can be reduced to a mixture of 1-dodecanol (from the acid part of the fat) and glycerol.

$$\begin{array}{l} CH_3(CH_2)_{10}\overset{\overset{\displaystyle O}{\|}}{C}-OCH_2 \\ \qquad\qquad\qquad\quad | \\ CH_3(CH_2)_{10}\overset{\overset{\displaystyle O}{\|}}{C}-OCH \\ \qquad\qquad\qquad\quad | \\ CH_3(CH_2)_{10}\overset{\overset{\displaystyle O}{\|}}{C}-OCH_2 \end{array} + 6\,H_2 \xrightarrow[\text{heat, pressure}]{\text{copper chromite}} 3\,CH_3(CH_2)_{10}CH_2OH + \begin{array}{c} HOCH_2 \\ | \\ HOCH \\ | \\ HOCH_2 \end{array} \tag{15.7}$$

glyceryl trilaurate; 1-dodecanol (lauryl alcohol); glycerol

Since glycerol is water-soluble, whereas the long-chain alcohol is not, the two hydrogenolysis products can be easily separated. The long-chain alcohol was next treated with sulfuric acid to make the alkyl hydrogen sulfate, which was then neutralized with base.

$$\underset{\text{lauryl alcohol}}{CH_3(CH_2)_{10}CH_2OH} + \underset{\text{sulfuric acid}}{HOSO_2OH} \longrightarrow \underset{\text{lauryl hydrogen sulfate}}{CH_3(CH_2)_{10}CH_2OSO_2OH} + H_2O$$

$$\downarrow NaOH \tag{15.8}$$

$$\overbrace{CH_3CH_2CH_2CH_2CH_2CH_2CH_2CH_2CH_2CH_2CH_2CH_2}^{\text{lipophilic chain}}\underbrace{-O-\overset{\overset{\displaystyle O}{\|}}{\underset{\underset{\displaystyle O}{\|}}{S}}-O^-Na^+}_{\text{polar, hydrophilic end}} + H_2O$$

sodium lauryl sulfate

Sodium lauryl sulfate is an excellent detergent. Because it is a salt of a *strong* acid, its solutions are nearly neutral. Its calcium and magnesium salts do not precipitate from solution, so it is effective in hard as well as soft water. Unfortunately, its supply is too limited to meet the demand, so the need for other syndets persisted.

At present, *the most widely used syndets are straight-chain alkylbenzenesulfonates*. They are made in three steps, shown in eq. 15.9. A straight-chain alkene with 10 to 14 carbons is treated with benzene and a Friedel-Crafts catalyst ($AlCl_3$ or HF) to form an alkylbenzene. Sulfonation and neutralization of the sulfonic acid with base complete the process.

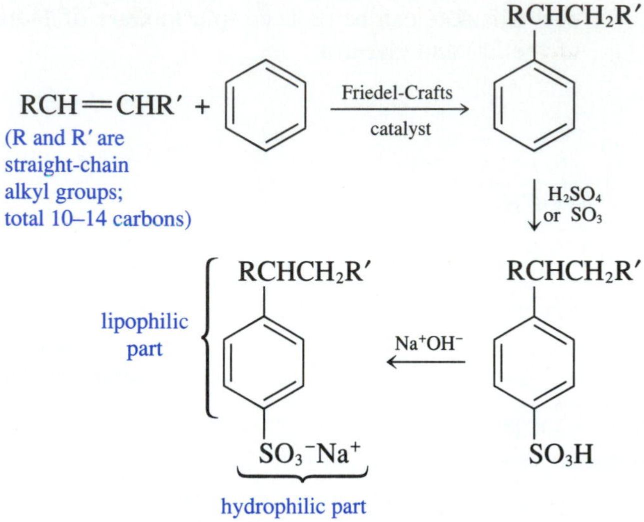

$$RCH{=}CHR' + C_6H_6 \xrightarrow[\text{catalyst}]{\text{Friedel-Crafts}} RCH(CH_2R')C_6H_5 \xrightarrow{H_2SO_4 \text{ or } SO_3} RCH(CH_2R')C_6H_4SO_3H \xrightarrow{Na^+OH^-} RCH(CH_2R')C_6H_4SO_3^-Na^+ \quad (15.9)$$

(R and R′ are straight-chain alkyl groups; total 10–14 carbons)

lipophilic part

hydrophilic part

a sodium alkylbenzenesulfonate

It is important that the alkyl chain in these detergents have no branches. The first alkylbenzenesulfonates had branched side chains and were found to be nonbiodegradable. They created severe pollution problems in the 1950s, causing foaming in sewage treatment plants, lakes, and rivers. But after about 1965, alkylbenzenesulfonates with unbranched side chains became available through further research. They are fully biodegradable and do not accumulate in the environment.

The soaps and syndets we have mentioned so far are *anionic* detergents; they have a lipophilic chain with a *negatively charged* polar end. But there are also *cationic, neutral,* and even *amphoteric* detergents, in which the polar portion of the molecule is *positive, neutral,* or *dipolar,* respectively. Here are some examples:

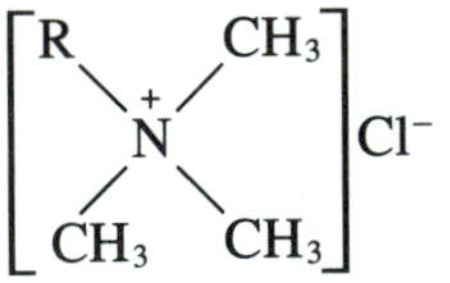

$$[R(CH_3)_3N^+]Cl^-$$
cationic detergent
($R = C_{16-18}$)

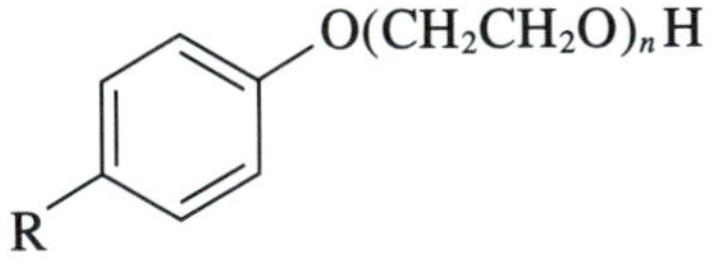

$$R{-}C_6H_4{-}O(CH_2CH_2O)_nH$$
neutral detergent
($R = C_{8-12}; n = 5-10$)

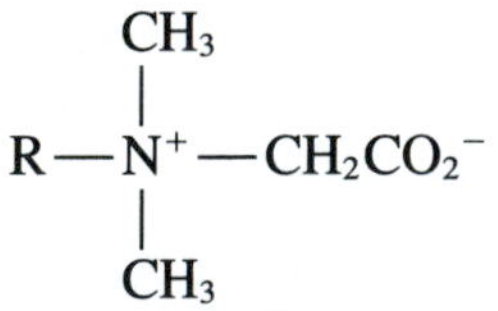

$$R{-}N^+(CH_3)_2{-}CH_2CO_2^-$$
amphoteric detergent
($R = C_{12-18}$)

The essential features of all these detergents are a lipophilic portion with a hydrocarbon chain of appropriate length to dissolve in oil or grease droplets and a polar portion to create a micelle surface that is attractive to water.

EXAMPLE 15.3 Design a synthesis of the cationic detergent shown above ($R = C_{16}$).

Solution

$$CH_3(CH_2)_{14}CH_2Cl + (CH_3)_3N{:} \longrightarrow \left[CH_3(CH_2)_{14}CH_2{-}N^+(CH_3)_2{-}CH_3 \quad Cl^-\right]$$

The reaction is an S_N2 displacement (see Table 6.1, entry 10, and eqs. 11.6 and 11.28).

Problem 15.5 Design a synthesis of the neutral detergent shown above ($R = C_8$, $n = 5$), starting with *p*-octylphenol and ethylene oxide (review Sec. 8.9).

Problem 15.6 Design a synthesis of the amphoteric detergent shown above, using an S_N2 displacement with $CH_3(CH_2)_{10}CH_2\ddot{N}(CH_3)_2$ as the nucleophile and an appropriate halide.

29. Commercial Detergents

The design and manufacture of a successful commercial detergent is a highly sophisticated process because, more and more, detergents are developed for a specific purpose and market. No one formulation fits every use. An inventory of the cleaning materials in a typical household might disclose half a dozen or more products designed to be most suitable for a specific job—to clean clothes, dishes, floors, automobiles, the human body (special hand and bath soaps and hair shampoo), and so on.

In Secs. 15.4 through 15.6 we described the structures of common surface active agents (*surfactants*). As we saw, these are organic molecules with a lipophilic portion and a polar portion, whose function is to emulsify and disperse oil and grease and to lower the surface tension of water to aid in wetting clothes, dishes, or other surfaces. Surfactants form an important part of all commercial detergents, but depending on the end use, these detergents will also contain a variety of other components, such as builders, bleaches, fabric softeners, enzymes, antiredeposition agents, and optical brighteners.

Builders are perhaps the second most important component of commercial detergents. They are added to remove calcium and magnesium ions ("hardness") from wash water. They may do this by chelation (forming a complex) or by exchanging these ions for sodium ions. Builders also raise pH to aid oil emulsification and buffer against pH changes. The most common builder is **sodium tripolyphosphate** ($5Na^+ \ P_3O_{10}^{5-}$), but because waste phosphates can be environmental pollutants, the amount that may be used is restricted by law; recently, sodium citrate, sodium carbonate, and sodium silicate have begun to replace it as builders. Zeolites (sodium aluminosilicates) are used as ion exchangers, especially for calcium ions.

Bleaches that contain chlorine (hypochlorites) have been used for years, sometimes separately and sometimes incorporated into the detergent. The chlorine acts as an oxidizing agent. Problems with residual chlorine and its odor have led more recently to the use of peroxides, especially sodium perborate ($NaBO_3$). Hydrolysis of the perborate produces hydrogen peroxide, which does the bleaching.

Fabric softeners are cationic surfactants that give clothes a soft feel. They may be incorporated in a laundry detergent or added to the wash separately.

Enzymes are added to detergents to deal with specific types of clothing stains. *Proteases* hydrolyze protein-based stains, and *amylases* convert starch-based stains to soluble materials that are easily washed away.

Antiredeposition agents are compounds added to laundry detergents to prevent the redeposit of soil on clothes. The most common examples are cellulose ethers or esters.

Optical brighteners are usually organic dyes that absorb ultraviolet light and fluoresce blue. In this way, a yellowed appearance of some "white" clothing can be avoided. In the mid-nineteenth century, "bluing" (a blue dye, such as ultramarine) used to be used for this purpose and is sometimes still used separately from the detergent. Brighteners that are incorporated in the detergent formulation are usually aromatic or heteroaromatic amines.

In addition to all these components, a modern commercial detergent may contain *antistatic agents* (cationic surfactants added to reduce static cling); *hydrotropes* (compounds added to liquid detergents to hold less-soluble surfactants and other compounds in solution); and, of course, *fragrances* and *perfumes* and inert fillers and formulation aids that keep powdered detergents free flowing.

Next time you do the laundry, bathe, or wash your hair, think about the complexity and diversity of the product you are using. It isn't just soap.

15.7 Phospholipids

Phospholipids constitute about 40% of cell membranes, the remaining 60% being proteins. Phospholipids are related structurally to fats and oils, except that one of the three ester groups is replaced by a phosphatidylamine.

$$\underbrace{\begin{array}{l} CH_3CH_2CH_2CH_2CH_2CH_2CH_2CH_2CH_2CH_2CH_2CH_2CH_2CH_2CH_2\overset{\overset{\displaystyle O}{\|}}{C}-O-CH_2 \\ CH_3CH_2CH_2CH_2CH_2CH_2CH_2CH_2CH_2CH_2CH_2CH_2CH_2CH_2CH_2\overset{\overset{\displaystyle O}{\|}}{C}-O-CH \end{array}}_{\text{nonpolar tail}}$$

$$CH_2-O-\underbrace{\overset{\overset{\displaystyle O^-}{|}}{\underset{\underset{\displaystyle O}{\|}}{P}}-OCH_2CH_2\overset{+}{N}H_3}_{\text{polar head}}$$

a phospholipid

The fatty acid portions are usually palmityl, stearyl, or oleyl. The structure shown is a **cephalin;** the three protons on the nitrogen are replaced by methyl groups in the **lecithins.** Both types of phospholipids are widely distributed in the body, especially in the brain and nerve tissues.

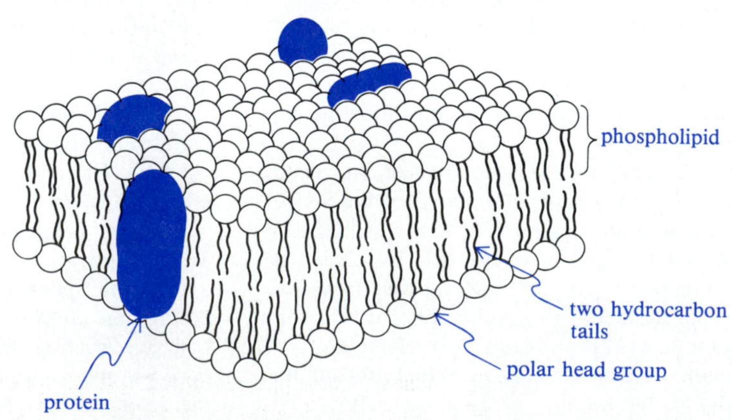

FIGURE 15.6 Schematic diagram of a cell membrane.

Phospholipids arrange themselves in **bilayers** in membranes, with the two hydrocarbon "tails" pointing in and the polar phosphatidylamine ends constituting the membrane surface, as shown in Figure 15.6. Membranes play a key role in biology, controlling diffusion of substances into and out of cells.

15.8 *Prostaglandins, Leukotrienes, and Lipoxins*

Prostaglandins are a group of compounds related to the unsaturated fatty acids. They were discovered in the 1930s, when it was found that human semen contained substances that could stimulate smooth muscle tissue, such as uterine muscle, to contract. On the assumption that these substances came from the prostate gland, they were named prostaglandins. We now know that prostaglandins are widely distributed in almost all human tissues, that they are biologically active in minute concentration, and that they have various effects on fat metabolism, heart rate, and blood pressure.

Prostaglandins have 20 carbon atoms. They are synthesized in the body by oxidation and cyclization of the 20-carbon unsaturated fatty acid **arachidonic acid.** Carbon-8 through carbon-12 of the chain are looped to form a cyclopentane ring, and an oxygen function (carbonyl or hydroxyl group) is always present at carbon-9. Various numbers of double bonds or hydroxyl groups may also be present elsewhere in the structure.

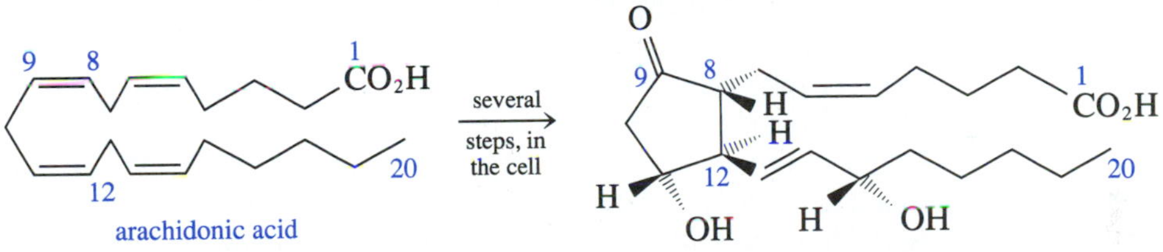

arachidonic acid —several steps, in the cell→ prostaglandin E_2 (PGE_2) (15.10)

Prostaglandins have excited much interest in the medical community, where they may find use in the treatment of inflammatory diseases, such as asthma and rheumatoid arthritis; treatment of peptic ulcers; control of hypertension; regulation of blood pressure and metabolism; and inducing of labor and therapeutic abortions.

Enzymatic oxidation of arachidonic acid also leads to two important classes of *acyclic* products, **leukotrienes** and **lipoxins,** that arise from oxidation at C5 and/or C15.

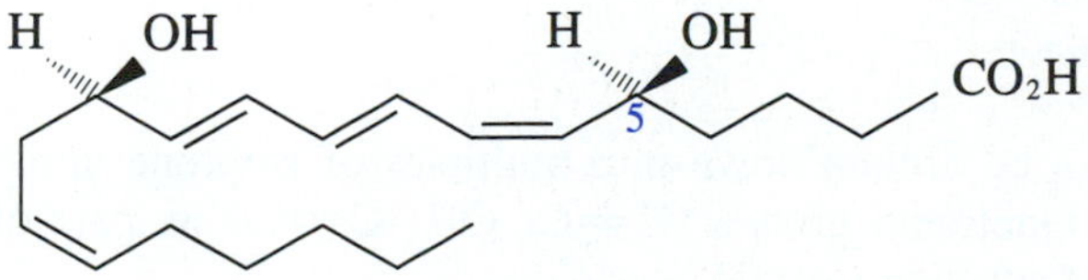

leucotriene B_4

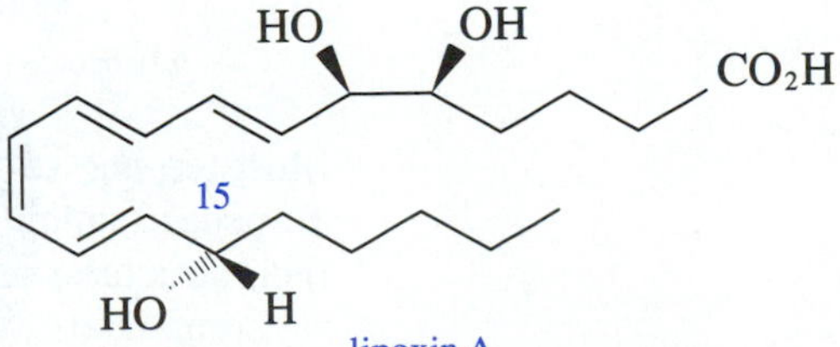

lipoxin A

These classes of compounds can regulate specific cellular responses important in inflammation and immune reactions and are the subject of intense current research.

15.9 Waxes

Waxes differ from fats and oils in that they are simple monoesters. The acid and alcohol portions of a wax molecule both have long saturated carbon chains.

$$CH_3(CH_2)_{13}CH_2\overset{\overset{\displaystyle O}{\|}}{C}{-}O(CH_2)_{15}CH_3$$

cetyl palmitate
(component of spermaceti, a wax in sperm whale oil)

$$C_{25\text{-}27}H_{51\text{-}55}\overset{\overset{\displaystyle O}{\|}}{C}{-}OC_{30\text{-}32}H_{61\text{-}65}$$

components of beeswax

Some plant waxes are simply long-chain saturated hydrocarbons (Sec. 2.8).

Waxes are more brittle, harder, and less greasy than fats. They are used to make polishes, cosmetics, ointments, and other pharmaceutical preparations, as well as candles and phonograph records. In nature, waxes coat the leaves and stems of plants that grow in arid regions, thus reducing evaporation. Similarly, insects with a high surface-area-to-volume ratio are often coated with a natural protective wax.

15.10 Terpenes and Steroids

Essential oils of many plants and flowers are obtained by distilling the plant with water. The water-insoluble oil that separates usually has an odor characteristic of the particular plant (rose oil, geranium oil, and others). Compounds isolated from these oils contain multiples of five carbon atoms (that is, 5, 10, 15, and so on) and are called **terpenes** (some compounds of this type were described in "A Word About Biologically Important Alcohols and Phenols", page 216). They are synthesized in the plant from acetate by way of an important biochemical intermediate, **isopentenyl pyrophosphate.** The five-carbon unit with a four-carbon chain and a one-carbon branch at C2 is called an **isoprene unit.**

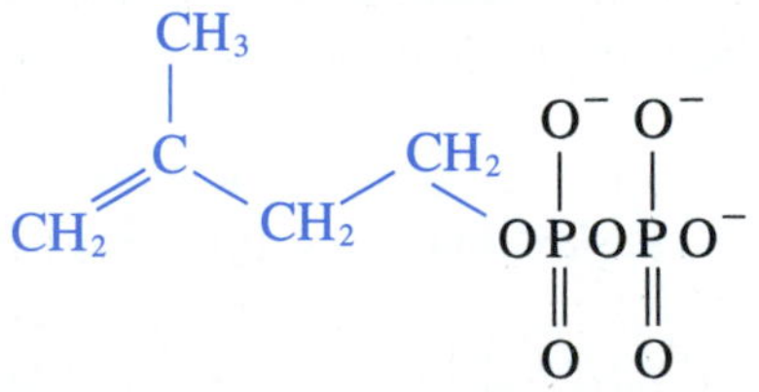

$$CH_2{=}C(CH_3){-}CH_2{-}CH_2{-}O\overset{\overset{\displaystyle O^-}{|}}{\underset{\underset{\displaystyle O}{\|}}{P}}O\overset{\overset{\displaystyle O^-}{|}}{\underset{\underset{\displaystyle O}{\|}}{P}}O^-$$

isopentenyl pyrophosphate

$$C{-}\overset{\overset{\displaystyle C}{|}}{C}{-}C{-}C$$

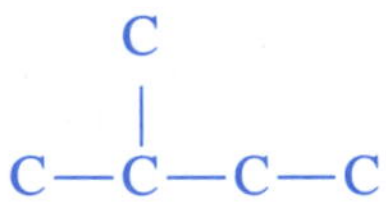

isoprene unit

Most terpene structures can be broken down into multiples of isoprene units. Terpenes contain various functional groups (C=C, OH, C=O) as part of their structures and may be acyclic or cyclic.

Compounds with a single isoprene unit (C_5) are relatively rare in nature, but compounds with two such units (C_{10}), called **monoterpenes,** are common. Examples include geraniol (page 216), **citronellal,** and **myrcene** (all acyclic) and **menthol** and **β-pinene** (both cyclic).

citronellal
(lemon oil)

myrcene
(bay leaves)

menthol
(peppermint oil)

β-pinene
(turpentine)

EXAMPLE 15.4 Mark off the isoprene units in citronellal and menthol.

Solution

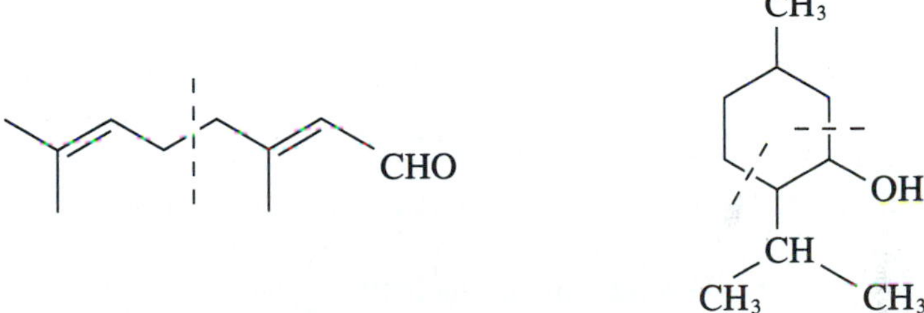

The dashed lines divide the molecules into two isoprene units.

Problem 15.7 There is another way to divide menthol into two isoprene units. Can you find it? Notice that both ways of dividing the structure join the isoprene units in a head-to-tail manner.

We have already seen examples of a **sesquiterpene** (C_{15}, farnesol, page 216), a **diterpene** (C_{20}, retinal, page 79), a **triterpene** (C_{30}, squalene, page 216), and even a **tetraterpene** (C_{40}, β-carotene, page 79).

Problem 15.8 Draw the structures of farnesol, retinal, squalene, and β-carotene, and mark off the component isoprene units.

Steroids constitute a major class of lipids. They are related to terpenes in the sense that they are biosynthesized by a similar route. Through a truly remarkable reaction sequence, the acyclic triterpene squalene is converted *stereospecifically* to the tetracyclic steroid **lanosterol,** from which other steroids are subsequently synthesized.

1. O_2, enzyme
2. H^+, enzyme

squalene (C_{30})

CH_3 H C CH_3 H CH_3 CH_3 HO H CH_3 H CH_3

lanosterol (C_{30})

(15.11)

Problem 15.9 How many chiral centers are present in squalene? In lanosterol?

The common structural feature of steroids is a system of four fused rings. The A, B, and C rings are six-membered, and the D ring is five-membered, usually all fused in a *trans* manner.

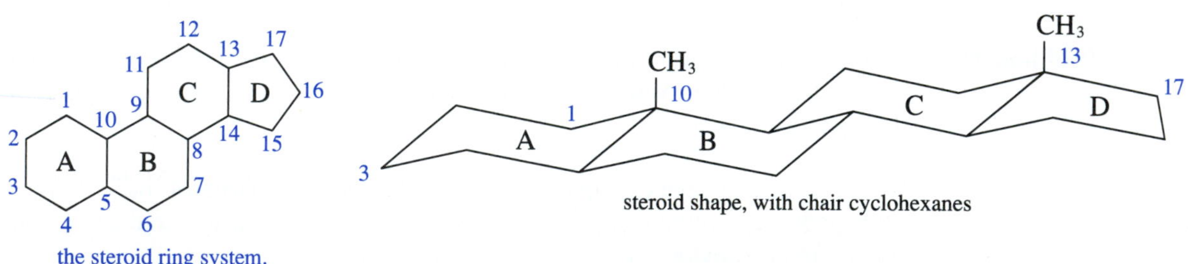

the steroid ring system. showing the numbering

steroid shape, with chair cyclohexanes

In most steroids, the six-membered rings are not aromatic, although there are exceptions. Usually there are methyl substituents attached to C-10 and C-13 (called "angular" methyl groups) and some sort of side chain attached to C-17.

Perhaps the best-known steroid is **cholesterol.** With 27 carbons, it is biosynthesized from lanosterol by a sequence of reactions that includes removing three carbon atoms.

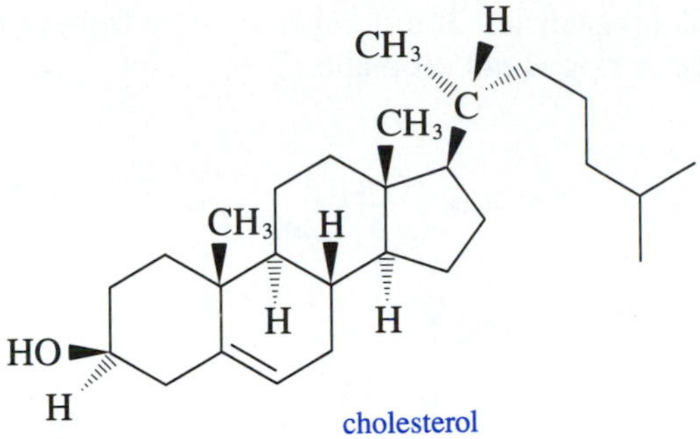

cholesterol

Cholesterol is present in all animal cells but is mainly concentrated in the brain and spinal cord. It is also the chief constituent of gallstones. The total amount of cholesterol present in an average human is about half a pound! There appears to be some connection between the concentration of blood serum cholesterol and coronary heart disease. Levels below 200 mg/dL seem desirable, whereas levels above 280 mg/dL may constitute a high risk.

Other steroids are common in animal tissues and play important biological roles. **Cholic acid,** for example, occurs in the bile duct, where it is present mainly in the form of various amide salts. These salts have a polar region (hydrophilic) and a largely hydrocarbon region (lipophilic) and function as emulsifying agents to facilitate the absorption of fats in the intestinal tract. They are, in a sense, biological soaps.

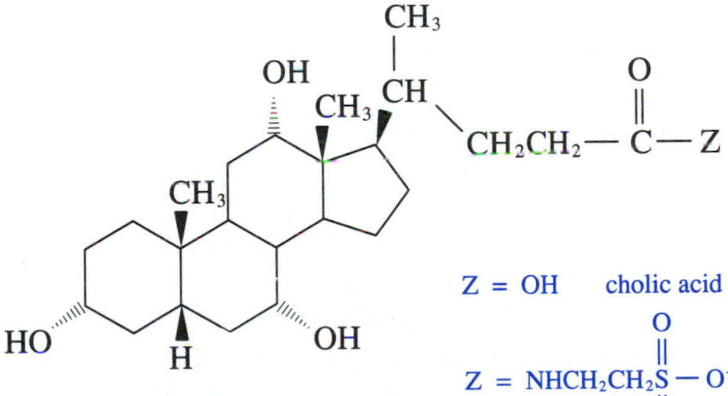

Z = OH cholic acid

$Z = NHCH_2CH_2S(=O)_2-O^-Na^+$ a bile salt

The **sex hormones** are compounds, produced in the ovaries and testes, that control reproductive physiology and secondary sex characteristics. Those sex hormones that predominate in females are of two types. The **estrogens,** of which the most plentiful is **estradiol,** are essential for initiating changes during the menstrual cycle and for the development of female secondary sex characteristics. **Progesterone,** which prepares the uterus for implantation of the fertilized egg, also maintains a pregnancy and prevents further ovulation during that time. Progesterone is administered clinically to prevent abortion in difficult pregnancies. It differs structurally from estrogens, such as estradiol, in that the A ring is not aromatic.

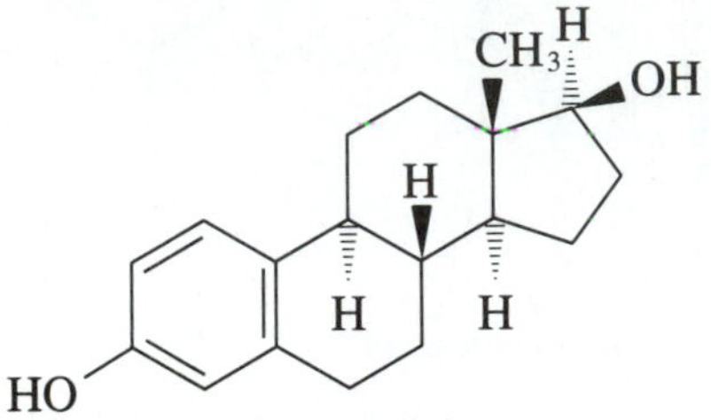

estradiol

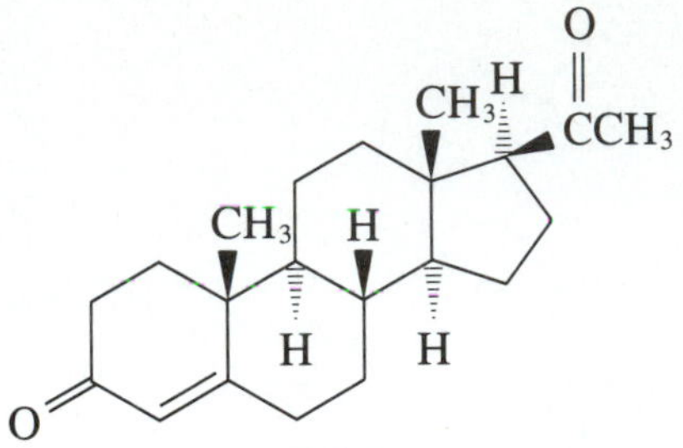

progesterone

Oral contraceptives, sometimes called "the pill," have structures similar to that of progesterone. An example is the acetylenic alcohol **norethindrone,** which prevents conception. On the other hand, **RU 486,** which also resembles progesterone, is a contragestive. It interferes with gestation of a fertilized ovum and, if taken in conjuction with prostaglandins, terminates a pregnancy within the first nine weeks of gestation more effectively and safely than surgical methods. It was discovered and is mainly available in France.

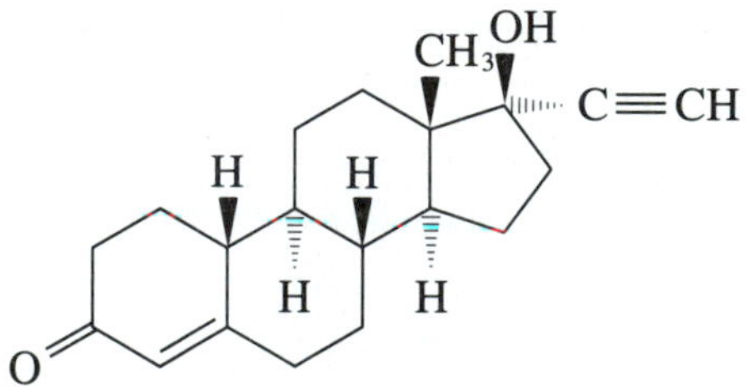

norethindrone (Norlutin)

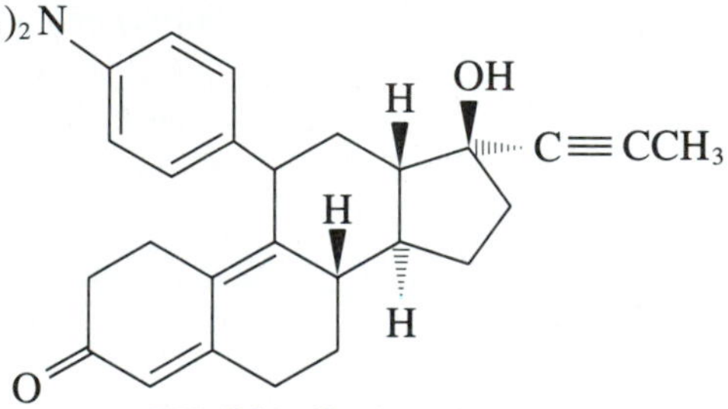

RU 486 (mifepristone)

Other steroid combinations show promise as "morning after" contraceptives, a current area of intense research.

Sex hormones that predominate in males are called **androgens.** They regulate the development of male reproductive organs and secondary sex characteristics, such as facial and body hair, deep voice, and male musculature. Two important androgens are **testosterone** and **androsterone.**

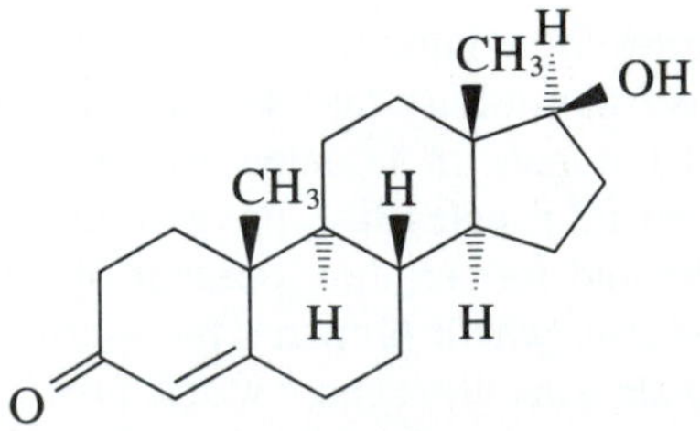

testosterone

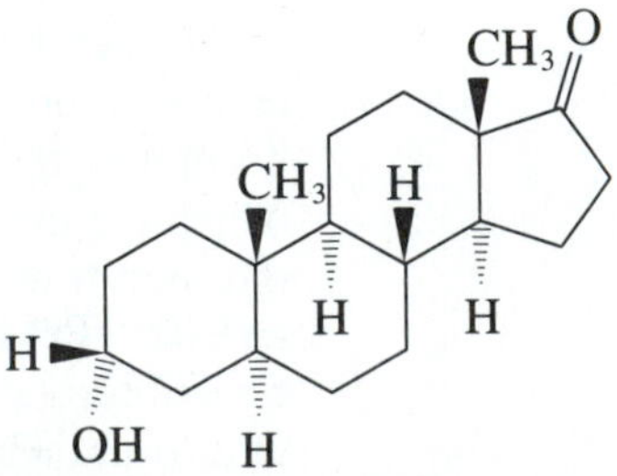

androsterone

Testosterone is an **anabolic** (muscle-building) steroid. Drugs based on its structure are sometimes administered to prevent withering of muscle in people recovering from surgery, starvation, or similar trauma. These same drugs, however, are sometimes illegally administered to healthy athletes and race horses to increase muscle mass and endurance. If taken in high doses, they can have serious side effects, including sexual malfunctions and liver tumors.

The only structural difference between testosterone and progesterone is the replacement of a hydroxyl group by an acetyl group at C-17 in the D ring. This great change in bioactivity, due to a seemingly minor change in structure, exemplifies the extreme specificity of biochemical reactions.

Additional Problems

15.10. Using Table 15.1 as a guide, write the structural formula for

a. potassium palmitate. **b.** magnesium oleate.
c. glyceryl trilaurate. **d.** glyceryl butyropalmitoöleate.
e. myristyl linoleate. **f.** ethyl arachidate.

15.11. Write equations for the (a) saponification, (b) hydrogenation, and (c) hydrogenolysis of glyceryl trilinolenate.

15.12. Saponification of castor oil gives glycerol and mainly (80 to 90%) **ricinoleic acid,** also called 12-hydroxyoleic acid. Draw the structure of the main component of castor oil.

15.13. Complete the equation for each of the following reactions:

a. $C_{15}H_{31}\overset{\overset{O}{\|}}{C}O^-Na^+ + HCl \longrightarrow$ **b.** $C_{15}H_{31}\overset{\overset{O}{\|}}{C}O^-Na^+ + Mg^{2+} \longrightarrow$

15.14. Using eq. 15.9 as a model, write equations for the preparation of an alkylbenzenesulfonate synthetic detergent, starting with 1-decene and benzene.

15.15. A synthetic detergent widely used in dishwashing liquids has the structure $CH_3(CH_2)_{11}(OCH_2CH_2)_3OSO_3^-Na^+$. Write a series of equations showing how this detergent can be synthesized from $CH_3(CH_2)_{10}CH_2OH$ and ethylene oxide.

15.16. List important characteristics of a good syndet.

15.17. What difficulty might there be with a commercial detergent that includes nearly equal amounts of cationic and anionic surfactants?

15.18. Write the general structure for

a. a fat. **b.** a vegetable oil. **c.** a wax.
d. an ordinary soap. **e.** a synthetic detergent. **f.** a steroid.
g. a phospholipid. **h.** a terpene. **i.** an isoprene unit.

15.19. The central carbon of the glycerol unit in a phospholipid is chiral and, in nature, has the *R* configuration. Draw the structure of a lecithin that has this arrangement.

15.20. What is the configuration (*Z* or *E*) of each double bond in arachidonic acid? For the structure, see eq. 15.10.

15.21. Consider the structure of prostaglandin E_2 (page 417).
a. How many chiral centers are present?
b. What is the configuration (*R* or *S*) of each?
c. What is the configuration of the double bonds (*Z* or *E*) in the two side chains?
d. Are the two side chains *cis* or *trans* to one another?

15.22. Write an equation for the saponification of cetyl palmitate, the main component of spermaceti.

15.23. When boiled with concentrated aqueous alkali, fats and oils dissolve, but waxes do not. Explain the difference.

15.24. Divide the structures of (a) myrcene and (b) β-pinene into their component isoprene units.

15.25. Notice (from your answer to Problem 15.8) that all the isoprene units in farnesol and retinal are joined in a head-to-tail manner, but that they are not all arranged that way in squalene and β-carotene. Where does the head-to-tail arrangement break down in these two terpenes? What does this suggest about the way these two terpenes might be biosynthesized?

15.26. Predict the products of the following reactions. (Consult the text for the structures of the starting materials).
a. estradiol + acetic anhydride **b.** progesterone + $LiAlH_4$
c. testosterone + peracetic acid **d.** androsterone + chromic acid

15.27. Cortisone is a drug used in the treatment of arthritis.

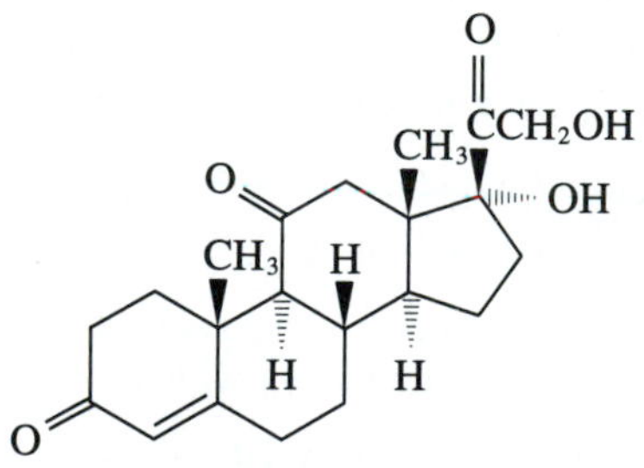

cortisone

a. Number all the carbon atoms according to the steroidal system.
b. Which carbons are chiral centers, and what is the configuration (*R* or *S*) of each?

CHAPTER **16**

Carbohydrates

16.1 *Introduction*

Carbohydrates occur in all plants and animals and are essential to life. Through photosynthesis, plants convert carbon dioxide to carbohydrates, mainly **cellulose, starch,** and **sugars.** Cellulose is the building block of rigid cell walls and woody tissue in plants, whereas starch is the chief storage form of carbohydrates for later use as a food or energy source. Some plants (cane and sugar beets) produce sucrose, ordinary table sugar. Another sugar, glucose, is an essential component of blood. Two other sugars, ribose and 2-deoxyribose, are components of the genetic materials RNA and DNA. Other carbohydrates are important components of coenzymes, antibiotics, cartilage, the shells of crustaceans, and bacterial cell walls.

In this chapter, we will describe the structures and a few reactions of the more important carbohydrates.

16.2 *Definitions and Classification*

The word ***carbohydrate*** arose because molecular formulas of these compounds can be expressed as *hydrates* of *carbon*. Glucose, for example, has the molecular formula $C_6H_{12}O_6$, which might be written as $C_6(H_2O)_6$. Although this type of formula is useless in studying the chemistry of carbohydrates, the old name persists.

We can now define carbohydrates more precisely in terms of their organic structures. *Carbohydrates are polyhydroxyaldehydes, polyhydroxyketones, or substances that give such compounds on hydrolysis.* The chemistry of carbohydrates is mainly the combined chemistry of two functional groups: the hydroxyl group and the carbonyl group.

Carbohydrates are usually classified according to their structure as **monosaccharides, oligosaccharides,** or **polysaccharides.** The term *saccharide* comes from Latin (*saccharum,* sugar) and refers to the sweet taste of some simple carbohydrates. The three classes of carbohydrates are related to each other through hydrolysis.

$$\text{Polysaccharide} \xrightarrow[H^+]{H_2O} \text{oligosaccharides} \xrightarrow[H^+]{H_2O} \text{monosaccharides} \quad (16.1)$$

For example, hydrolysis of starch, a polysaccharide, gives first maltose and then glucose.

$$\underset{\substack{\text{starch} \\ \text{(a polysaccharide)}}}{[C_{12}H_{20}O_{10}]_n} \xrightarrow[H^+]{n\,H_2O} \underset{\substack{\text{maltose} \\ \text{(a disaccharide)}}}{n\,C_{12}H_{22}O_{11}} \xrightarrow[H^+]{n\,H_2O} \underset{\substack{\text{glucose} \\ \text{(a monosaccharide)}}}{2n\,C_6H_{12}O_6} \quad (16.2)$$

Monosaccharides (or *simple sugars,* as they are sometimes called) *are carbohydrates that cannot be hydrolyzed to simpler compounds.* **Polysaccharides** contain many monosaccharide units—sometimes hundreds or even thousands. Usually, but not always, the units are identical. Two of the most important polysaccharides, starch and cellulose, contain linked units of the same monosaccharide, glucose. **Oligosaccharides** (from the Greek *oligos,* few) contain at least two and generally no more than a few linked monosaccharide units. They may be called **disaccharides, trisaccharides,** and so on, depending on the number of units, which may be the same or different. Maltose, for example, is a disaccharide made of two glucose units, but sucrose, another disaccharide, is made of two different monosaccharide units: glucose and fructose.

In the next section, we will describe the structures of monosaccharides. Later, we will see how these units are linked together to form oligosaccharides and polysaccharides.

16.3 Monosaccharides

Monosaccharides are classified according to the number of carbon atoms present (**triose, tetrose, pentose, hexose,** and so on) and according to whether the carbonyl group is present as an aldehyde (**aldose**) or as a ketone (**ketose**).

There are only two trioses: **glyceraldehyde** and **dihydroxyacetone.** Each has two hydroxyl groups, on separate carbon atoms, and one carbonyl group.

$$\underset{\substack{\text{glyceraldehyde} \\ \text{(an aldose)}}}{\begin{array}{l} {}^{1}CH{=}O \\ \;| \\ {}^{2}CHOH \\ \;| \\ {}^{3}CH_2OH \end{array}} \qquad \underset{\substack{\text{dihydroxyacetone} \\ \text{(a ketose)}}}{\begin{array}{l} {}^{1}CH_2OH \\ \;| \\ {}^{2}C{=}O \\ \;| \\ {}^{3}CH_2OH \end{array}} \qquad \underset{\text{glycerol}}{\begin{array}{l} CH_2OH \\ \;| \\ CHOH \\ \;| \\ CH_2OH \end{array}}$$

Glyceraldehyde is the simplest aldose, and dihydroxyacetone is the simplest ketose. Each is related to glycerol in having a carbonyl group in place of one of the hydroxyl groups.

Other aldoses or ketoses can be derived from glyceraldehyde or dihydroxyacetone by adding carbon atoms, each with a hydroxyl group. In aldoses, the chain is numbered from the aldehyde carbon. In most ketoses, the carbonyl group is located at C-2.

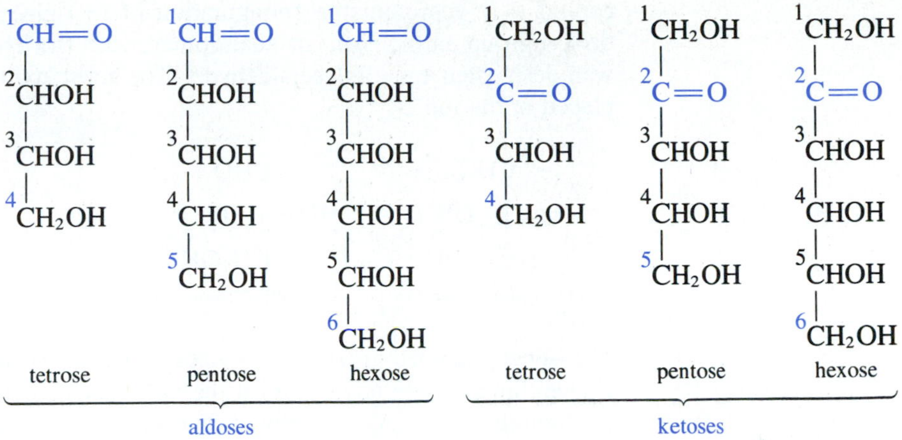

aldoses: tetrose	aldoses: pentose	aldoses: hexose	ketoses: tetrose	ketoses: pentose	ketoses: hexose
1 CH=O	1 CH=O	1 CH=O	1 CH_2OH	1 CH_2OH	1 CH_2OH
2 CHOH	2 CHOH	2 CHOH	2 C=O	2 C=O	2 C=O
3 CHOH	3 CHOH	3 CHOH	3 CHOH	3 CHOH	3 CHOH
4 CH_2OH	4 CHOH	4 CHOH	4 CH_2OH	4 CHOH	4 CHOH
	5 CH_2OH	5 CHOH		5 CH_2OH	5 CHOH
		6 CH_2OH			6 CH_2OH

16.4 Chirality in Monosaccharides; Fischer Projection Formulas and D,L-Sugars

You will notice that glyceraldehyde, the simplest aldose, has one chiral carbon atom (C-2) and hence can exist in two enantiomeric forms.

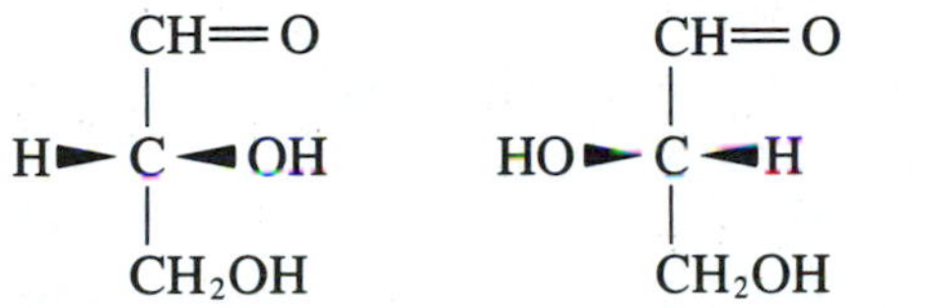

CH=O / H▸C◂OH / CH_2OH

R-(+)-glyceraldehyde
$[\alpha]_D^{25}$ +8.7(c = 2, H_2O)

CH=O / HO▸C◂H / CH_2OH

S-(−)-glyceraldehyde
$[\alpha]_D^{25}$ −8.7(c = 2, H_2O)

The dextrorotatory form has the *R* configuration.

It was in connection with his studies on carbohydrate stereochemistry that Emil Fischer invented his system of projection formulas. Since we will be using these formulas here, it might be wise for you to review Secs. 5.8 through 5.10. Recall that, in a Fischer projection formula, *horizontal* lines show groups that project *above* the plane of the paper *toward* the viewer; *vertical* lines show groups that project *below* the plane of the paper *away* from the viewer. Thus, *R*-(+)-glyceraldehyde can be represented as

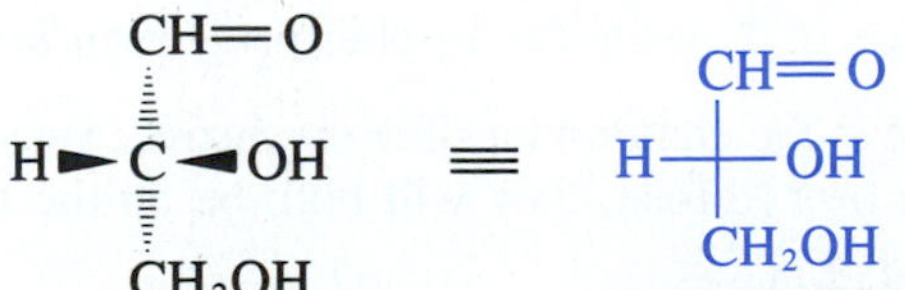

CH=O / H▸C◂OH / CH_2OH

R-(+)-glyceraldehyde

≡

CH=O / H—+—OH / CH_2OH

Fischer projection formula for *R*-(+)-glyceraldehyde

with the chiral center represented by the intersection of two crossed lines.

Fischer also introduced a stereochemical nomenclature that preceded the *R,S* system and is still in common use for sugars and amino acids. He used a small-capital D to represent the configuration of (+)-glyceraldehyde, with the hydroxyl group on the *right;* its enantiomer, with the hydroxyl group on the *left*, was designated L-(−)-glyceraldehyde. The most oxidized carbon (CHO) was placed at the top.

```
     CHO              CHO
H —+— OH         HO —+— H
   CH2OH            CH2OH
```

D-(+)-glyceraldehyde L-(−)-glyceraldehyde

Fischer extended his system to other monosaccharides in the following way. If the chiral carbon *farthest* from the aldehyde or ketone group had the same configuration as D-glyceraldehyde (hydroxyl on the right), the compound was called a D-sugar. If the configuration at the remote carbon had the same configuration as L-glyceraldehyde (hydroxyl on the left), the compound was an L-sugar.

```
                                         CH2OH            CH2OH
                                           |                |
   CH=O             CH=O                  C=O              C=O
    |                |                     |                |
 (CHOH)n          (CHOH)n               (CHOH)n          (CHOH)n
H —+— OH        HO —+— H              H —+— OH         HO —+— H
  CH2OH            CH2OH                 CH2OH            CH2OH
```

a D-aldose an L-aldose a D-ketose an L-ketose

Figure 16.1 shows the Fischer projection formulas for all the D-aldoses through the hexoses. Starting with D-glyceraldehyde, one CHOH unit at a time is inserted in the chain. This carbon, which adds a new chiral center to the structure, is shown in black. In each case, the new chiral center can have the hydroxyl group at the right or at the left in the Fischer projection formula (R or S absolute configuration).

■ ***EXAMPLE 16.1*** Using Figure 16.1, write the Fischer projection formula for L-erythrose.

Solution L-Erythrose is the enantiomer of D-erythrose. Since both —OH groups are on the right in D-erythrose, they will both be on the left in its mirror image.

```
     CH=O
HO —+— H
HO —+— H
    CH2OH
```

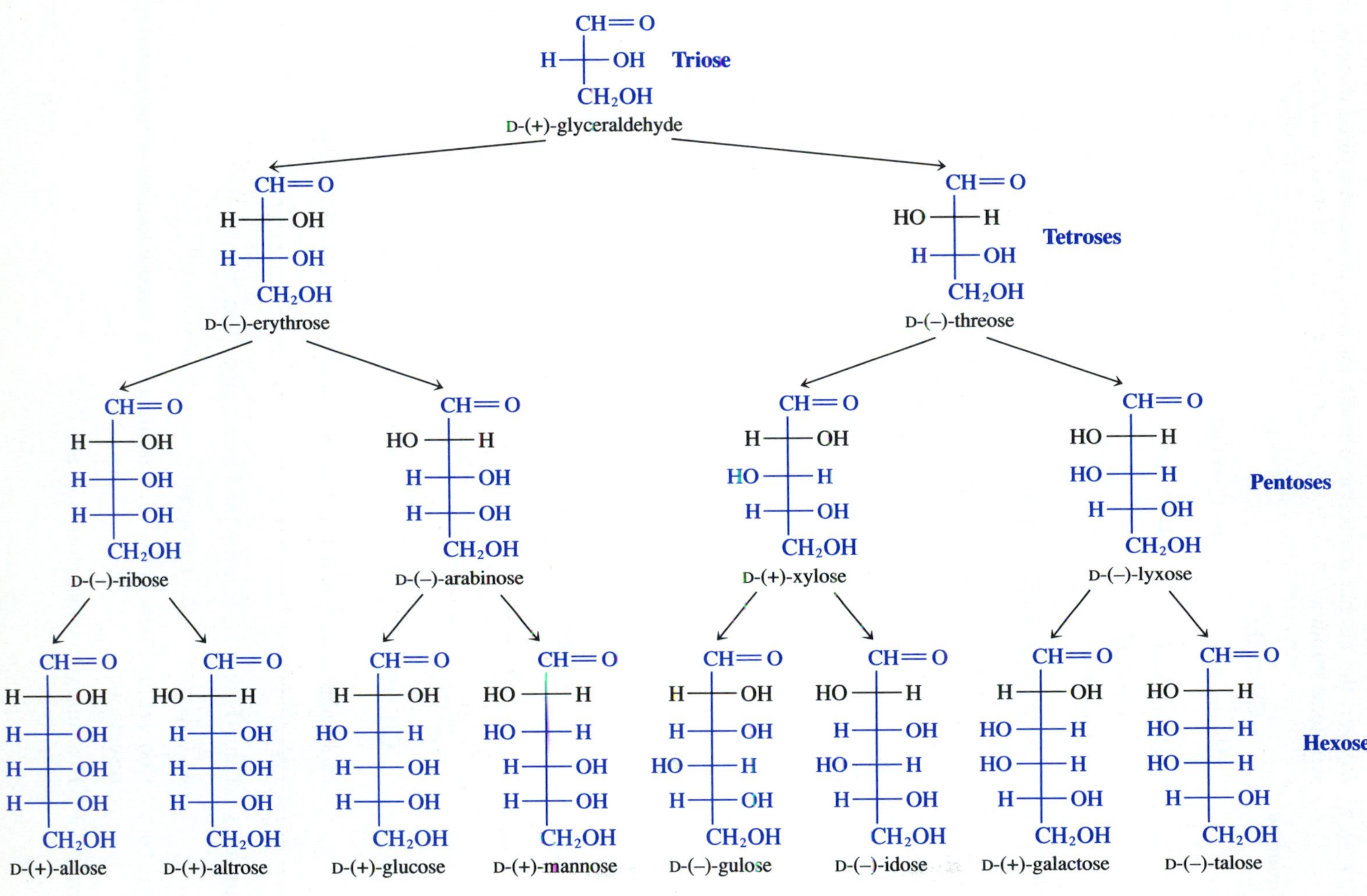

FIGURE 16.1 Fischer projection formulas and genealogy chart for the D-aldoses with up to six carbon atoms.

EXAMPLE 16.2 Convert the Fischer projection formula for D-erythrose to a three-dimensional structural formula.

Solution

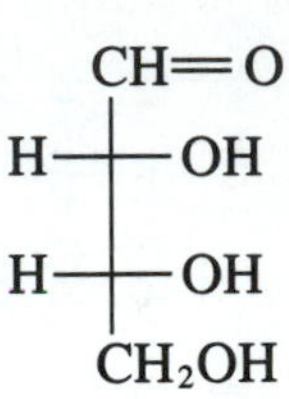

CH=O
H—OH
H—OH
CH_2OH
D-erythrose

≡

CH=O
H—C—OH
H—C—OH
CH_2OH

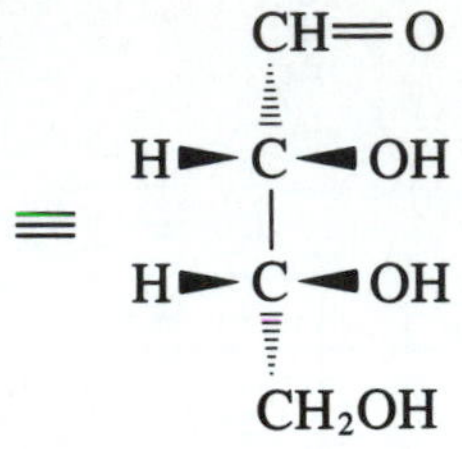

We can also write the structure as

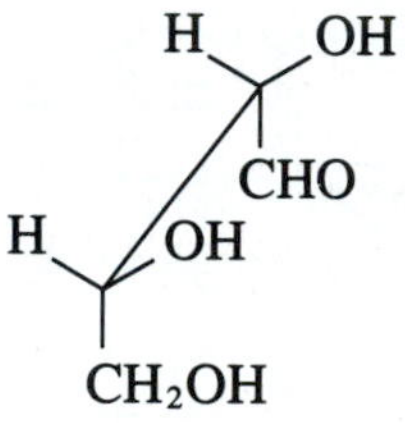

sawhorse

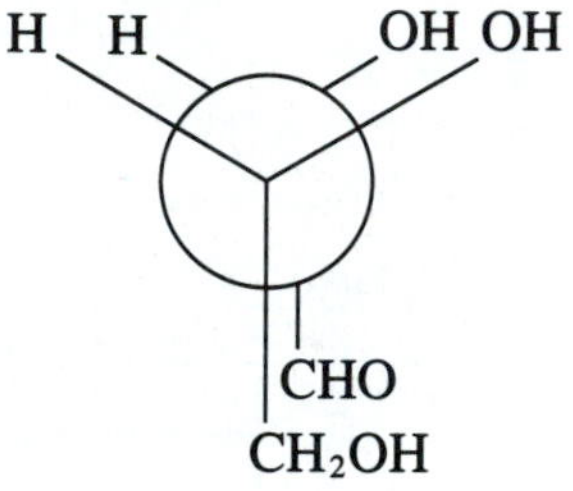

Newman

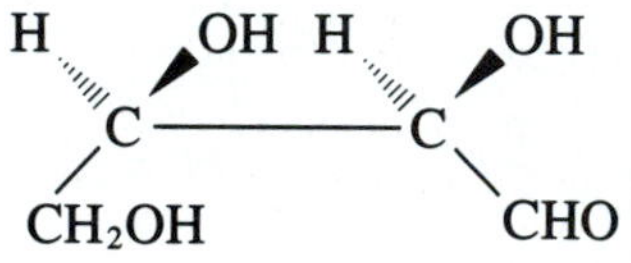

dash-wedge

and we can then rotate around the central C—C bond to more favorable staggered (instead of eclipsed) conformations, such as

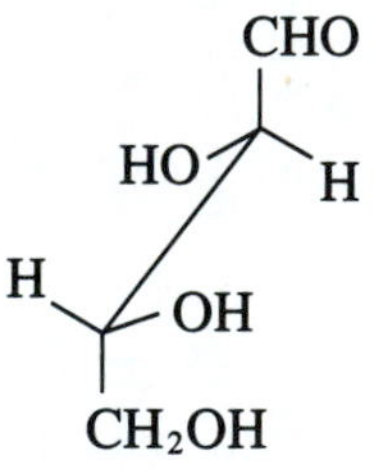

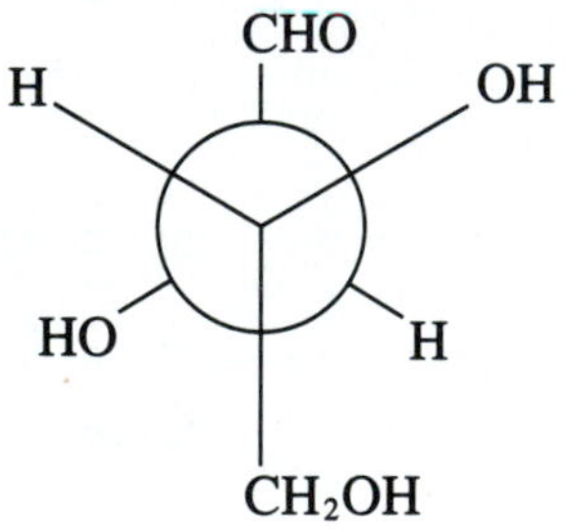

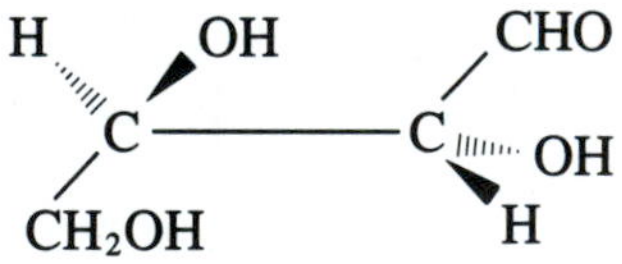

Molecular models may help you to follow these interconversions.

Problem 16.1 Using Figure 16.1, write the Fischer projection formula for

a. L-threose. b. L-glucose.

Problem 16.2 Convert the Fischer projection formula for D-threose to three-dimensional representations.

Problem 16.3 How many D-aldoheptoses are possible?

How are the sugars with identical numbers of carbon atoms, shown horizontally across Figure 16.1, related to one another? Compare, for example, D-(−)-erythrose and D-(−)-threose. They have the same configuration at C-3 (D, with the OH on the right), but opposite configurations at C-2. These sugars are stereoisomers, but *not* mirror images (*not* enantiomers). In other words, they are **diastereomers** (review Sec. 5.9). Similarly, there are four diastereomeric D-pentoses and eight diastereomeric D-hexoses.

A special name is given to diastereomers that differ in configuration *at only one chiral center;* they are called **epimers.** D-(−)-erythrose and D-(−)-threose are not only diastereomers, they are epimers. Similarly, D-glucose and D-mannose are epimers (at C-2), and D-glucose and D-galactose are epimers (at C-4). Each pair has the same configurations at all chiral centers except one.

Problem 16.4 What pairs of D-pentoses are epimeric at C-3?

Notice that there is no direct relationship between configuration and the sign of optical rotation. Although all the sugars in Figure 16.1 are D-sugars, some are dextrorotatory (+) and others are levorotatory (−).

16.5 The Cyclic Hemiacetal Structures of Monosaccharides

The monosaccharide structures described so far are consistent with much of the known chemistry of these compounds, but they are oversimplified. Let us now examine the true structures of these compounds.

We learned earlier that alcohols undergo rapid and reversible addition to the carbonyl group of aldehydes and ketones, to form hemiacetals (review Sec. 9.8). This can happen *intramolecularly* when the hydroxyl and carbonyl groups are properly located in the same molecule (eqs. 9.13 and 9.14), just the situation in many monosaccharides. *Monosaccharides exist mainly in cyclic, hemiacetal forms* and not in the acyclic aldo- or keto-forms we have depicted so far.

As an example, consider D-glucose. First, let us rewrite its Fischer projection formula in a way that brings the OH group at C-5 within bonding distance of the carbonyl group (as in eq. 9.13). This is shown in Figure 16.2. The Fischer projection is first converted to its three-dimensional (dash-wedge) structure, which is then turned on its side and bent around so that C-1 and C-6 are close to one another. Finally, rotation about the C-4–C-5 bond brings the hydroxyl oxygen at C-5 close enough for nucleophilic addition to the carbonyl carbon (C-1). Reaction then leads to the cyclic, hemiacetal structure shown at the bottom left of the figure.

The British carbohydrate chemist W. N. Haworth (Nobel Prize, 1937) introduced a useful way of representing the cyclic forms of sugars. In a **Haworth projection,** the ring is represented as if it were planar and viewed edge on, with the oxygen at the upper right. **The carbons are arranged clockwise numerically, with C-1 at the right.** Substituents attached to the ring lie above or

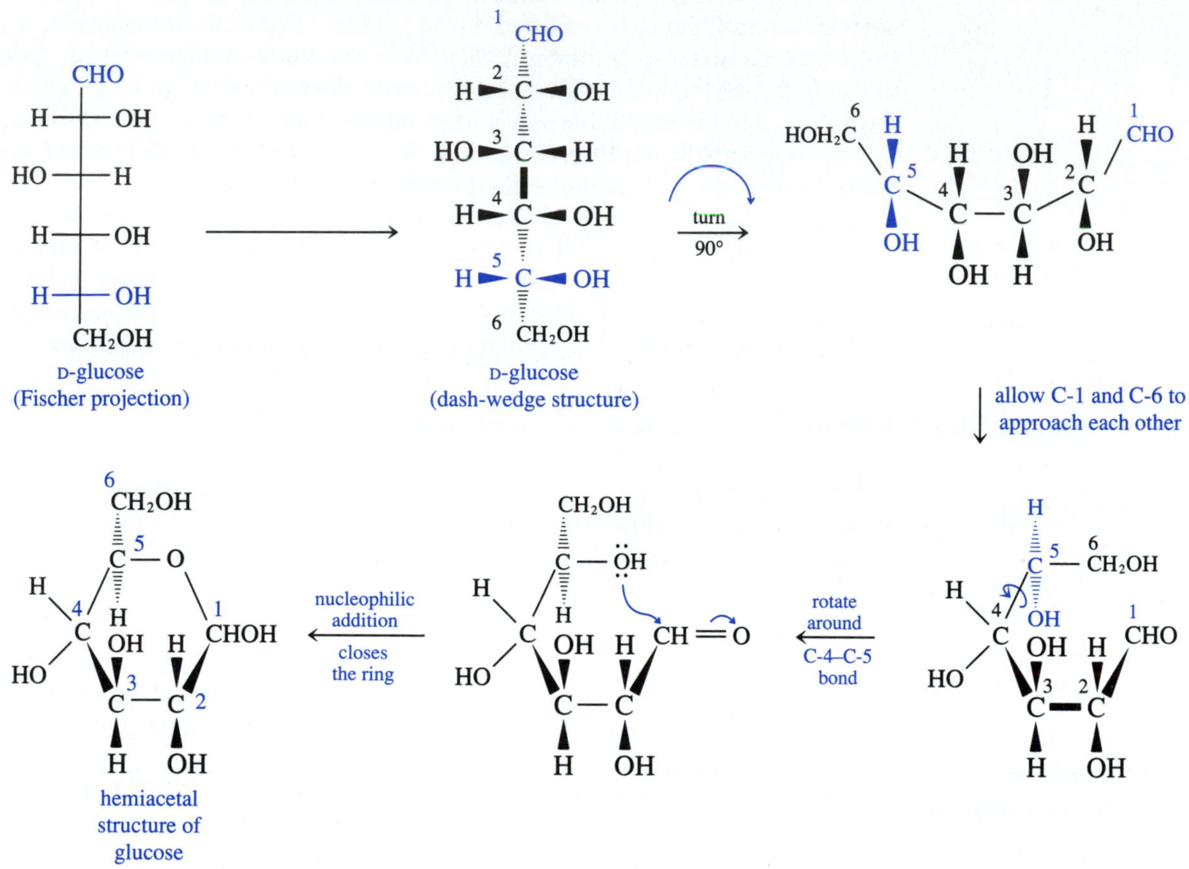

FIGURE 16.2 Manipulation of the Fischer projection formula of D-glucose to bring the C-5 hydroxyl group in position for cyclization to the hemiacetal form.

below the plane. For example, the Haworth formula for D-glucose (Figure 16.2) is written

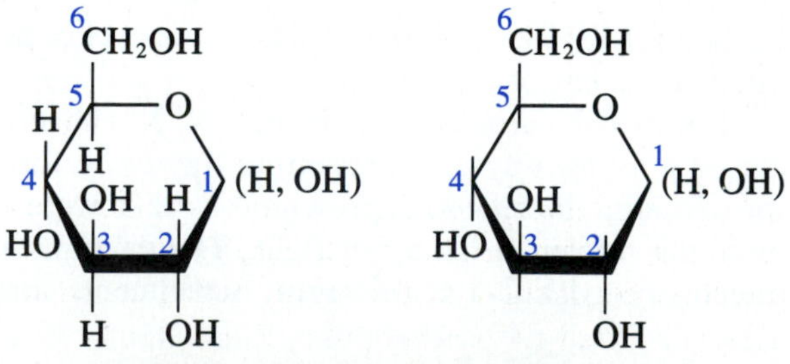

Haworth projection formulas for D-glucose

Sometimes, as in the structure at the right, the ring hydrogens are omitted for clarity.

In converting from one type of projection formula to another, notice that hydroxyl groups on the *right* in the Fischer projection are *down* in the Haworth projection (and conversely, hydroxyl groups on the *left* in the Fischer projection are *up* in the Haworth projection). For D-sugars, the terminal —CH_2OH group is *up* in the Haworth projection; for L-sugars, it is down.

EXAMPLE 16.3 Draw the Haworth projection for the six-membered cyclic structure of D-mannose.

Solution Notice from Figure 16.1 that D-mannose differs from D-glucose *only* in the configuration at C-2. In the Fischer projection formula, the C-2 hydroxyl is on the *left;* therefore, it will be *up* in the Haworth projection. Otherwise, the structure is identical to that of D-glucose.

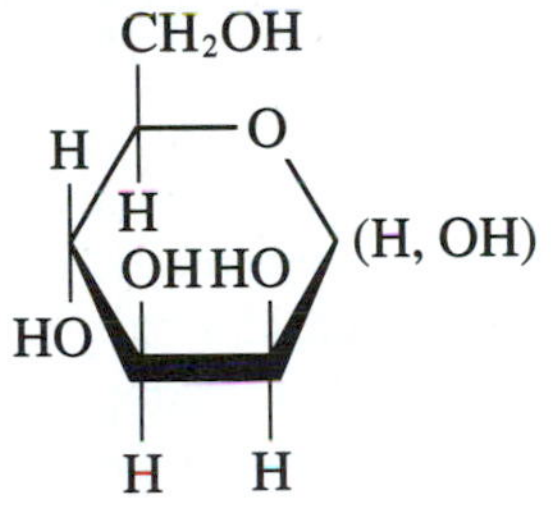

CH_2OH
H O
H
OH HO (H, OH)
HO
H H

D-mannose

Problem 16.5 Draw the Haworth projection formula for the six-membered cyclic structure of D-galactose.

Now notice three important features of the hemiacetal structure of D-glucose. First, *the ring is heterocyclic,* with five carbons and an oxygen. Carbons 1 through 5 are part of the ring structure, but carbon 6 (the -CH_2OH group) is a substituent on the ring. Next, C-1 is special. *C-1 is the hemiacetal carbon,* simultaneously an alcohol and an ether carbon (it carries a hydroxyl group, and it is also connected to C-5 by an ether link). In contrast, all the other carbons are monofunctional. C-2, C-3, and C-4 are secondary alcohol carbons; C-6 is a primary alcohol carbon; and C-5 is an ether carbon. These differences show up in the different chemical reactions of D-glucose. Finally, *C-1 in the cyclic, hemiacetal structure is chiral.* It has four different groups attached to it (H, OH, OC5, and C2) and can therefore exist in two configurations, *R* or *S*. Let us consider this last feature in greater detail.

16.6 *Anomeric Carbons; Mutarotation*

In the acyclic, aldehyde form of glucose, C-1 is achiral, but in the cyclic structures, this carbon becomes chiral. Consequently, *two* hemiacetal structures are possible, depending on the configuration at the new chiral center. The hemiacetal carbon, the carbon that forms the new chiral center, is called the **anomeric carbon.** Two monosaccharides that differ only in configuration at the anomeric center are **anomers** (a special kind of epimers). Anomers are called α or β, depending on the position of the hydroxyl group. For monosaccharides in the D-series, the hydroxyl group is "down" in the α anomer and "up" in the β anomer, when the structure is written in the usual way (eq. 16.3).

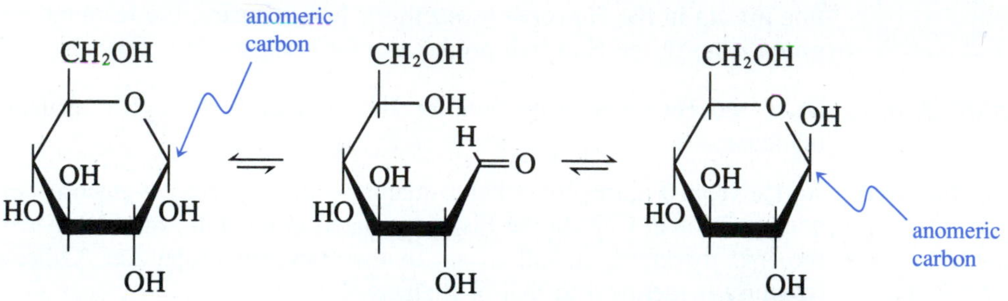

anomeric carbon

anomeric carbon

(16.3)

α-D-glucose (36%)
mp 146°C
[α] +112°

D-glucose
(aldehyde form)

β-D-glucose (64%)
mp 150°C
[α] +19°

The α and β forms of D-glucose have identical configurations at every chiral center *except at C-1, the anomeric carbon.*

How do we know that monosaccharides exist mainly as cyclic hemiacetals? There is direct physical evidence. For example, if D-glucose is crystallized from methanol, the pure α form is obtained. On the other hand, crystallization of glucose from acetic acid gives the β form. The α and β forms of D-glucose are *diastereomers*. Being diastereomers, they have different physical properties, as shown under their structures in eq. 16.3, where we see that they have different melting points and different specific optical rotations.

The α and β forms of D-glucose interconvert in aqueous solution. For example, if crystalline α-D-glucose is dissolved in water, the specific rotation drops gradually from an initial value of +112° to an equilibrium value of +52°. Starting with the pure crystalline β form results in a gradual rise in specific rotation from an initial +19° to the *same equilibrium value* of +52°. These changes in optical rotation are called **mutarotation.** They can be explained by the equilibria shown in eq. 16.3. Recall that hemiacetal formation is a reversible equilibrium process (Sec. 9.8). Starting with either pure hemiacetal form, the ring can open to the acyclic aldehyde, which can then recyclize to give either the α or the β form. Eventually, an equilibrium mixture is obtained.

At equilibrium, an aqueous solution of D-glucose contains 35.5% of the α form and 64.5% of the β form. There is only about 0.003% of the open-chain aldehyde form present.

■ *EXAMPLE 16.4* Show that the percentages of α- and β-D-glucose in aqueous solution at equilibrium can be calculated from the specific rotations of the pure α and β forms and the specific rotation of the solution at equilibrium.

Solution The equilibrium rotation is +52°, and the rotations of pure α and β forms are +112° and +19°, respectively. Assuming that no other forms are present, we can express these values graphically as follows:

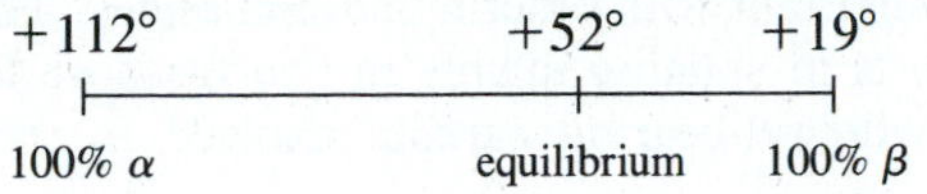

+112°	+52°	+19°
100% α	equilibrium	100% β

The percentage of the β form at equilibrium is then

$$\frac{112 - 52}{112 - 19} \times 100 = \frac{60}{93} \times 100 = 64.5\%$$

The percentage of the α form at equilibrium is $100 - 64.5 = 35.5\%$.

16.7 Pyranose and Furanose Structures

The six-membered cyclic form of most monosaccharides is the preferred structure. These structures are called **pyranose** forms after the six-membered oxygen heterocycle **pyran.** The formula at the extreme left of eq. 16.3 is more completely named **α-D-glucopyranose,** the last part of the name showing the ring size.

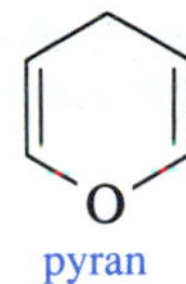

pyran

Pyranoses are formed by reaction of the hydroxyl group at C-5, with the carbonyl group. With some sugars, however, the hydroxyl group at C-4 reacts instead. In these cases, the cyclic hemiacetal that is formed has a *five*-membered ring. This type of cyclic monosaccharide is called a **furanose,** after the parent five-membered oxygen heterocycle **furan.**

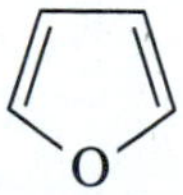

furan

For example, D-glucose could, in principle, exist in two furanose forms (α and β at C-1) through attack of the C-4 hydroxyl on the aldehyde carbon.

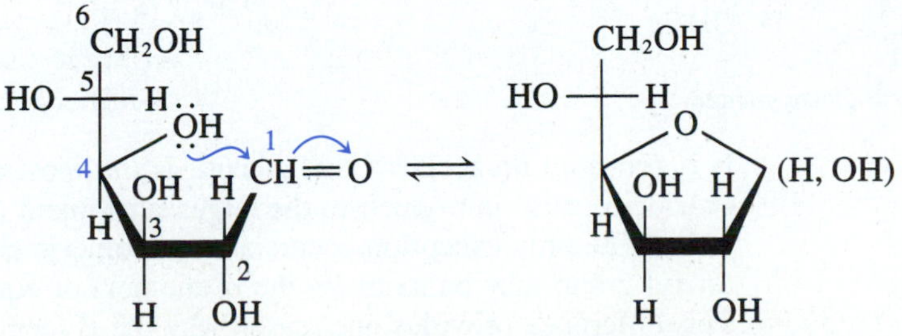

D-glucose ⇌ α- and β-D-glucofuranose (16.4)

In practice, these forms are present to less than 1% in glucose solutions, but they are important with other monosaccharides. The ketose **D-fructose,** for example, exists in solution mainly in two furanose forms. The carbonyl carbon and the hydroxyl-bearing carbon involved in the cyclization are in a 1,4-relationship.

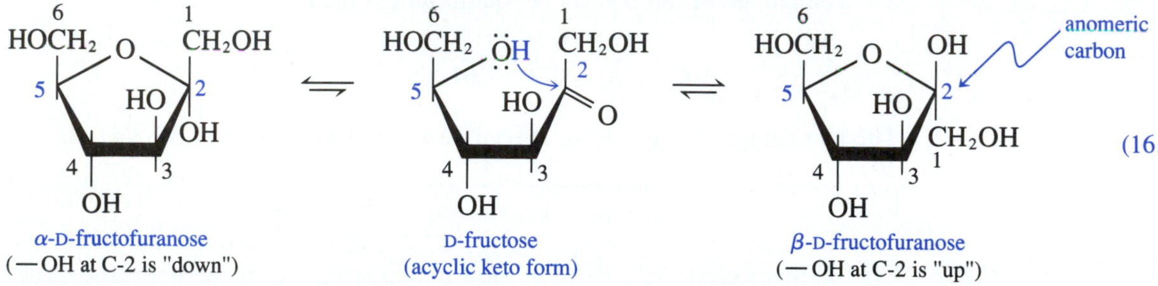

α-D-fructofuranose (—OH at C-2 is "down") ⇌ D-fructose (acyclic keto form) ⇌ β-D-fructofuranose (—OH at C-2 is "up") (16.5)

Problem 16.6 D-Erythrose cannot exist in pyranose forms, but furanose cyclic forms are possible. Explain. Draw the structure for α-D-erythrofuranose.

16.8 Conformations of Pyranoses

Haworth projections depict pyranose rings as planar. However, as with cyclohexane, the rings generally prefer a chair conformation. Consequently, we can rewrite eq. 16.3 more accurately as eq. 16.6.

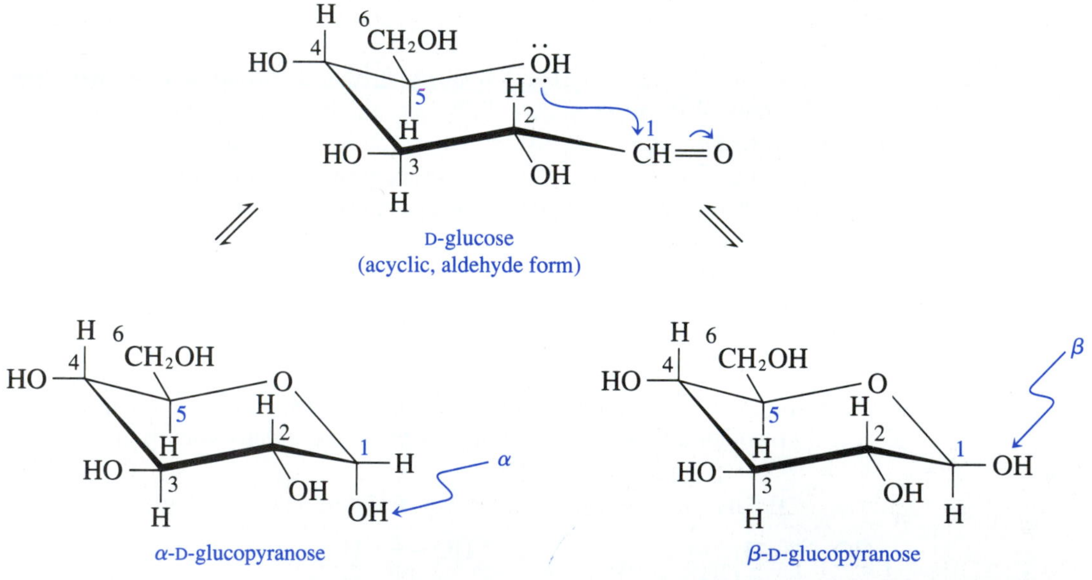

D-glucose (acyclic, aldehyde form)

(16.6)

α-D-glucopyranose β-D-glucopyranose

It is probably no accident that glucose is the most abundant natural monosaccharide because in D-glucose, the larger substituent at each ring carbon is equatorial. The only exception occurs at the anomeric carbon (C-1), where the hydroxyl group may be axial (in the α anomer) or equatorial (in the β anomer). This difference provides one reason why the β form is preferred at equilibrium (eq. 16.3).

EXAMPLE 16.5 Draw the conformation of α-D-mannopyranose.

Solution Recall from Example 16.3 that D-mannose differs from D-glucose only at C-2. Using the cyclic structure at the left of eq. 16.6 as a guide, we can write

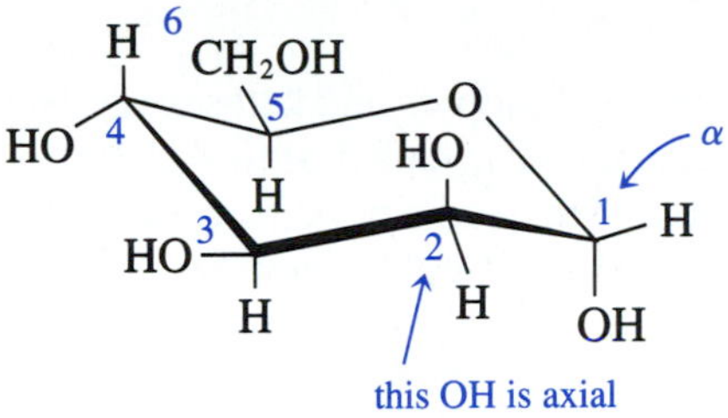

this OH is axial

α-D-mannopyranose

Problem 16.7 D-Galactose differs from D-glucose only in the configuration at C-4. Draw the chair conformation of β-D-galactopyranose.

Now that we have described the structures of monosaccharides, let us examine some of their common reactions.

16.9 Esters and Ethers from Monosaccharides

Monosaccharides contain hydroxyl groups. It is not surprising, then, that they undergo reactions typical of alcohols. For example, they can be converted to esters by reaction with acid halides or anhydrides. The conversion of β-D-glucose to its pentaacetate by reaction with excess acetic anhydride is typical; all five hydroxyl groups, including the hydroxyl at anomeric C-1, are esterified. (To clarify the structure, the ring H's are omitted).

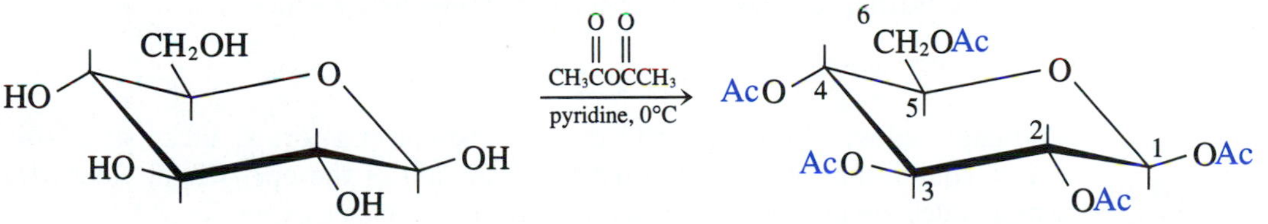

β-D-glucopyranose $\xrightarrow[\text{pyridine, 0°C}]{CH_3COCCH_3}$ β-D-glucopyranose pentaacetate (16.7)

Ac = $CH_3C(=O)$—

The hydroxyl groups can also be converted to ethers by treatment with an alkyl halide and a base (the Williamson synthesis, Sec. 8.6). Because sugars are sensitive to strong bases, the mild base silver oxide is preferred.

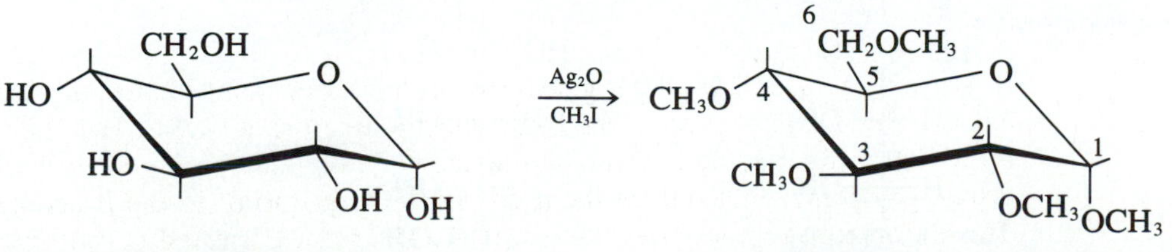

α-D-glucopyranose $\xrightarrow[CH_3I]{Ag_2O}$ α-D-glucopyranose pentamethyl ether (16.8)

Whereas sugars tend to be soluble in water and insoluble in organic solvents, the reverse is true for their esters and ethers. This often facilitates their purification and manipulation with organic reagents.

16.10 Reduction of Monosaccharides

The carbonyl group of aldoses and ketoses can be reduced by various reagents. The products are **polyols,** called **alditols.** For example, catalytic hydrogenation or reduction with sodium borohydride ($NaBH_4$) converts D-glucose to D-glucitol (also called sorbitol; review Sec. 9.13).

D-glucose (cyclic): CH_2OH, O, HO, HO, OH, OH ⇌ D-glucose (acyclic): $CH{=}O$; H–C–OH; HO–C–H; H–C–OH; H–C–OH; CH_2OH $\xrightarrow[\text{or } NaBH_4]{H_2,\text{ catalyst}}$ D-glucitol (sorbitol): CH_2OH; H–C–OH; HO–C–H; H–C–OH; H–C–OH; CH_2OH (16.9)

Reaction occurs by reduction of the small amount of aldehyde in equilibrium with the cyclic hemiacetal. As that aldehyde is reduced, the equilibrium shifts to the right, so that eventually all of the sugar is converted. Sorbitol is used commercially as a sweetener and sugar substitute.

Problem 16.8 D-Mannitol, which occurs naturally in olives, onions, and mushrooms, can be made by $NaBH_4$ reduction of D-mannose. Draw its structure.

16.11 Oxidation of Monosaccharides

Although aldoses exist primarily in cyclic hemiacetal forms, these structures are in equilibrium with a small but finite amount of the open-chain aldehyde. These aldehyde groups can be easily oxidized to acids (review Sec. 9.14). The products are called **aldonic acids.** For example, D-glucose is easily oxidized to D-gluconic acid.

D-glucose: $CH{=}O$; H–C–OH; HO–C–H; H–C–OH; H–C–OH; CH_2OH $\xrightarrow[\text{or } Ag^+ \text{ or } Cu^{2+}]{Br_2,\ H_2O}$ D-gluconic acid: COOH; H–C–OH; HO–C–H; H–C–OH; H–C–OH; CH_2OH (16.10)

The oxidation of aldoses is so easy that they react with such mild oxidizing agents as Tollens' reagent (Ag^+ in aqueous ammonia), Fehling's reagent (Cu^{2+} complexed with tartrate ion), or Benedict's reagent (Cu^{2+} complexed with citrate ion). With Tollens' reagent, they give a silver mirror test (Sec. 9.14), and with the copper reagents, the blue solution gives a red precipitate of cuprous oxide, Cu_2O.

$$\underset{}{RCH{=}O} + \underset{\text{blue solution}}{2\,Cu^{2+}} + 5\,OH^- \longrightarrow RC(=O)O^- + \underset{\text{red precipitate}}{Cu_2O} + 3\,H_2O \tag{16.11}$$

Problem 16.9 Write an equation for the reaction of D-mannose with Fehling's reagent (Cu^{2+}) to give D-mannonic acid.

Stronger oxidizing agents, such as aqueous nitric acid, attack the aldehyde group *and* the primary alcohol group, producing dicarboxylic acids called **aldaric acids.** For example, D-glucose gives D-glucaric acid.

$$\begin{array}{c} CH{=}O \\ H{-}C{-}OH \\ HO{-}C{-}H \\ H{-}C{-}OH \\ H{-}C{-}OH \\ CH_2OH \\ \text{D-glucose} \end{array} \xrightarrow{HNO_3} \begin{array}{c} COOH \\ H{-}C{-}OH \\ HO{-}C{-}H \\ H{-}C{-}OH \\ H{-}C{-}OH \\ COOH \\ \text{D-glucaric acid} \end{array} \tag{16.12}$$

Problem 16.10 Write the structure of D-mannaric acid.

16.12 *Formation of Glycosides from Monosaccharides*

Because monosaccharides exist as cyclic hemiacetals, they can react with one equivalent of an alcohol to form acetals. An example is the reaction of β-D-glucose with methanol.

$$\underset{\beta\text{-D-glucopyranose}}{\text{(CH}_2\text{OH, OH, HO, OH ring)}} + CH_3OH \xrightarrow{H^+} \underset{\substack{\text{methyl } \beta\text{-D-glucopyranoside} \\ \text{mp 115–116°C}}}{\text{(CH}_2\text{OH, OCH}_3\text{, OH, HO, OH ring)}} + H_2O \tag{16.13}$$

Note that *only the —OH on the anomeric carbon is replaced by an —OR group.* Such acetals are called **glycosides,** and the bond from the anomeric carbon to the OR group is called the **glycosidic bond.** Glycosides are named from the corresponding monosaccharide by changing the *-e* ending to *-ide*. Thus, glucose gives glucosides, mannose gives mannosides, and so on.

EXAMPLE 16.6 Write a Haworth formula for ethyl α-D-mannoside.

Solution

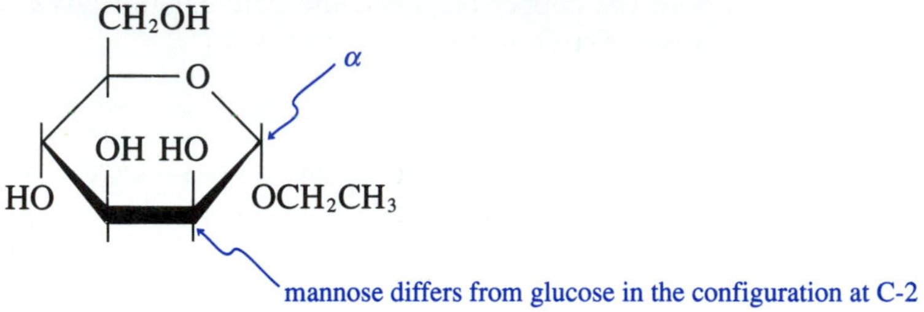

Problem 16.11 Write an equation for the acid-catalyzed reaction of β-D-galactose with methanol.

EXAMPLE 16.7 Write out a mechanism that explains why only the —OH group at C-1 is replaced by $—OCH_3$ in eq. 16.13.

Solution The acid catalyst can, of course, protonate any of the six oxygen atoms, since each has unshared electron pairs and is basic. However, *only protonation of the hydroxyl oxygen at C-1 leads, after water loss, to a resonance-stabilized carbocation.*

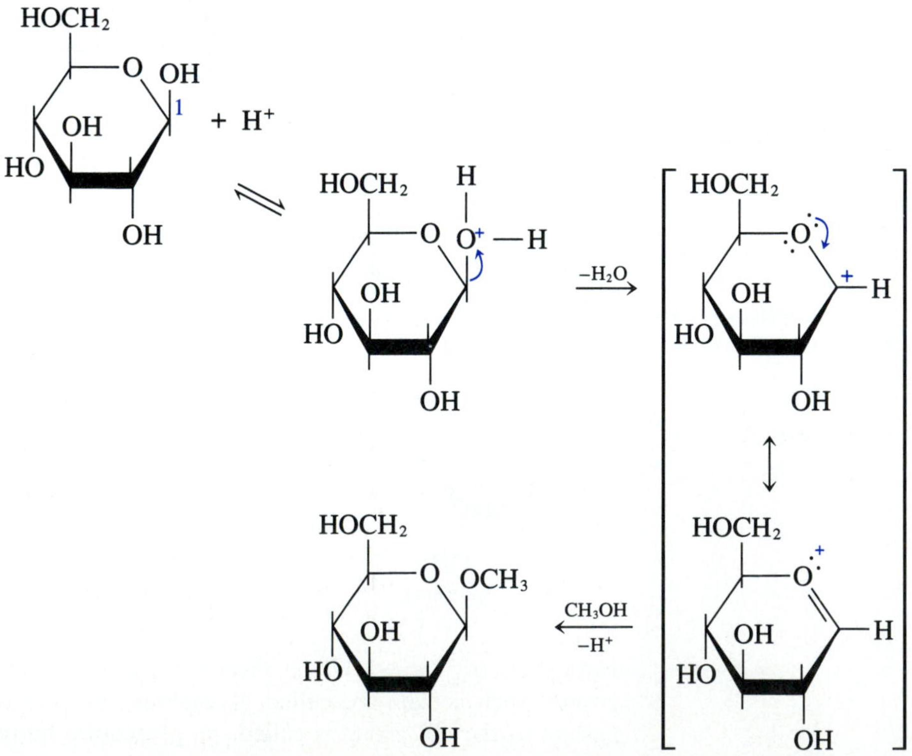

The reaction mechanism is exactly the same as was described in Example 9.4 in Sec. 9.8. Hemiacetal hydroxyl groups are converted to ethers (acetals) under much milder conditions than are the hydroxyl groups of primary or secondary alcohols.

In the final step, methanol can attack from either "face" of the six-membered ring, to give either the β-glucoside as shown, or the α-glucoside.

Problem 16.12 Protonate each of the other hydroxyl groups in β-D-glucose (at C-2, C-3, C-4, or C-6), and show that the carbocation obtained after water loss is *not* stabilized by resonance and is just an ordinary primary or secondary carbocation.

Naturally occurring alcohols or phenols often occur in cells combined as a glycoside with some sugar—most commonly, glucose. In this way, the many hydroxyl groups of the sugar portion of the glycoside solubilize compounds that would otherwise be incompatible with cellular protoplasm. An example is the bitter-tasting glucoside **salicin,** which occurs in willow bark and whose fever-reducing power was known to the ancients.

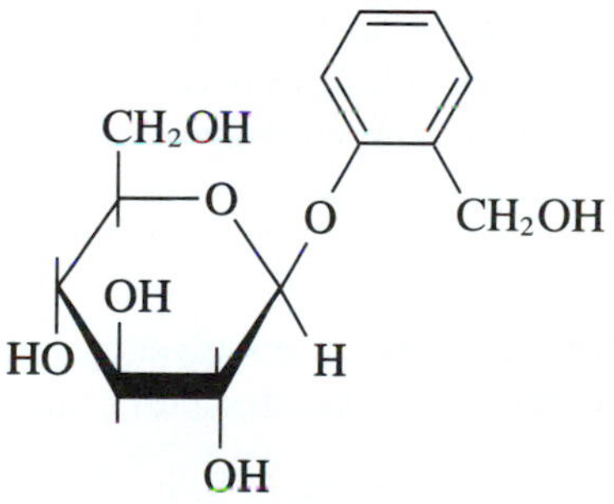

CH_2OH O O CH_2OH OH HO H OH

salicin
(the β-D-glucoside of salicyl alcohol)

The glycosidic bond is the key to understanding the structure of oligosaccharides and polysaccharides, as we will see in the following sections.

16.13 Disaccharides

The most common oligosaccharides are **disaccharides**. *In a disaccharide, two monosaccharides are linked by a glycosidic bond between the anomeric carbon of one monosaccharide unit and a hydroxyl group on the other unit.* In this section, we will describe the structure and properties of four important disaccharides.

16.13a Maltose **Maltose** is the disaccharide obtained by the partial hydrolysis of starch. Further hydrolysis of maltose gives only D-glucose (eq. 16.2). Maltose must, therefore, consist of two linked glucose units. It turns out that the anomeric carbon of the left unit is linked to the C-4 hydroxyl group of the unit at the right as an acetal (glycoside). The configuration at the anomeric carbon of the left unit is α. In the crystalline form, the anomeric carbon of the right unit has the β configuration. Both units are pyranoses, and the right-hand unit fills the same role as the methanol in eq. 16.13.

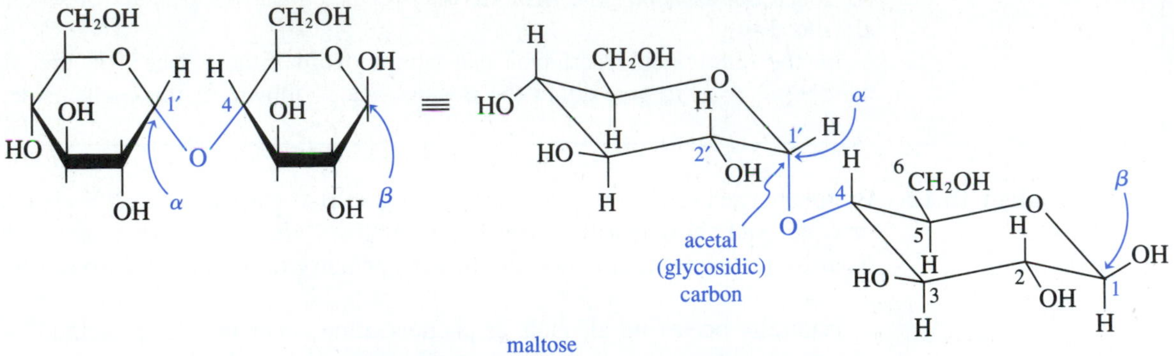

maltose
4-O-(α-D-glucopyranosyl)-β-D-glucopyranose

The systematic name for maltose, shown beneath the common name, describes the structure fully, including the name of each unit (D-glucose), the ring sizes (pyranose), the configuration at each anomeric carbon (α or β), and the location of the hydroxyl group involved in the glycosidic link (4-O).

The anomeric carbon of the right glucose unit in maltose is a hemiacetal. Naturally, when maltose is in solution, this hemiacetal function will be in equilibrium with the open-chain aldehyde form. Maltose therefore gives a positive Tollens' test and other reactions similar to those of the anomeric carbon in glucose.

Problem 16.13 When crystalline maltose is dissolved in water, the initial specific rotation changes and gradually reaches an equilibrium value. Explain.

16.13b Cellobiose **Cellobiose** is the disaccharide obtained by the partial hydrolysis of cellulose. Further hydrolysis of cellobiose gives only D-glucose. Cellobiose must therefore be an isomer of maltose. In fact, *cellobiose differs from maltose* only *in having the β configuration at C-1 of the left glucose unit.* Otherwise, all other structural features are identical, including a link from C-1 of the left unit to the hydroxyl group at C-4 in the right unit.

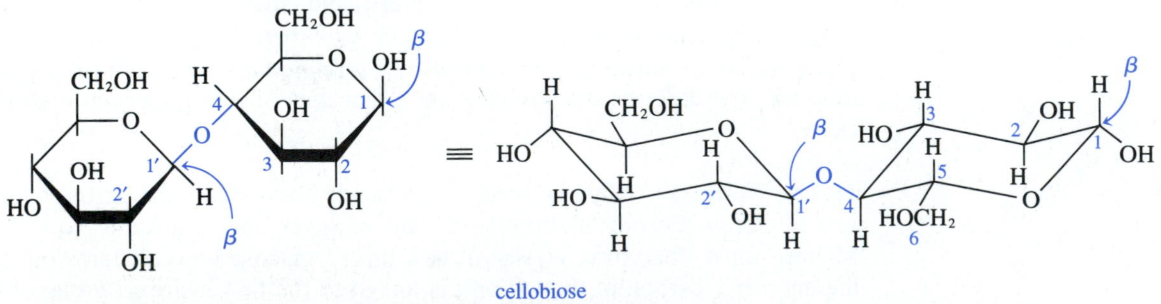

cellobiose
4-O-(β-D-glucopyranosyl)-β-D-glucopyranose

Note that, in the conformational formula for cellobiose, one ring-oxygen is drawn to the "rear" and one to the "front" of the molecule. This is the way the rings exist in the cellulose chain.

16.13c Lactose **Lactose** is the major sugar in human and cow's milk (4 to 8% lactose). Hydrolysis of lactose gives equimolar amounts of D-galactose and D-glucose. The anomeric carbon of the galactose unit has the β configuration at C-1 and is linked to the hydroxyl group at C-4 of the glucose unit. The crystalline anomer that has the α configuration at the glucose unit is made commercially from cheese whey.

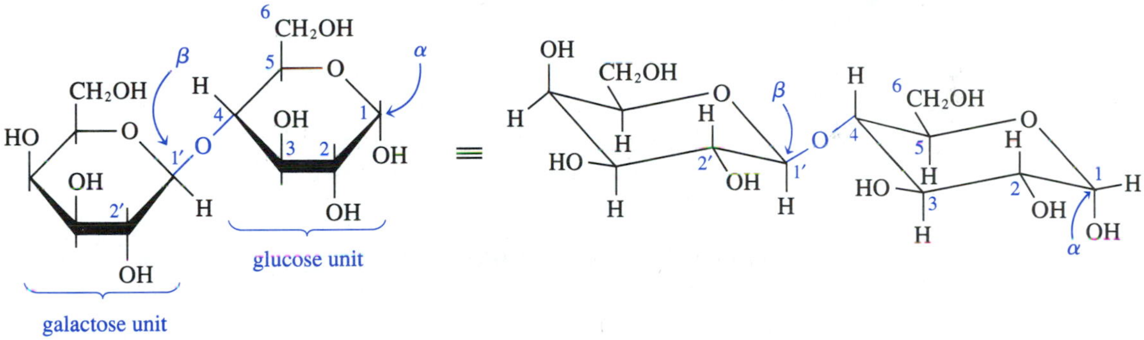

galactose unit

glucose unit

lactose
4-O-(β-D-galactopyranosyl)-α-D-glucopyranose

Problem 16.14 Will lactose give a positive Fehling's test? Will it mutarotate?

Some human infants are born with a disease called *galactosemia*. They lack the enzyme that isomerizes galactose to glucose and cannot digest milk. If milk is excluded from such infants' diets, the disease symptoms caused by accumulation of galactose can be avoided.

16.13d Sucrose The most important commercial disaccharide is **sucrose,** ordinary table sugar. Over 100 million tons are produced annually worldwide. Sucrose occurs in all photosynthetic plants, where it functions as an energy source. It is obtained commercially from sugar cane and sugar beets, in which it constitutes 14 to 20% of the plant juices.

Hydrolysis of sucrose gives equimolar amounts of D-glucose and the ketose D-fructose. *Sucrose differs from the other disaccharides we have discussed in that the anomeric carbons of* ***both*** *units are involved in the glycosidic link.* That is, C-1 of the glucose unit is linked, via oxygen, to C-2 of the fructose unit. A further difference is that the fructose unit is in the furanose form.

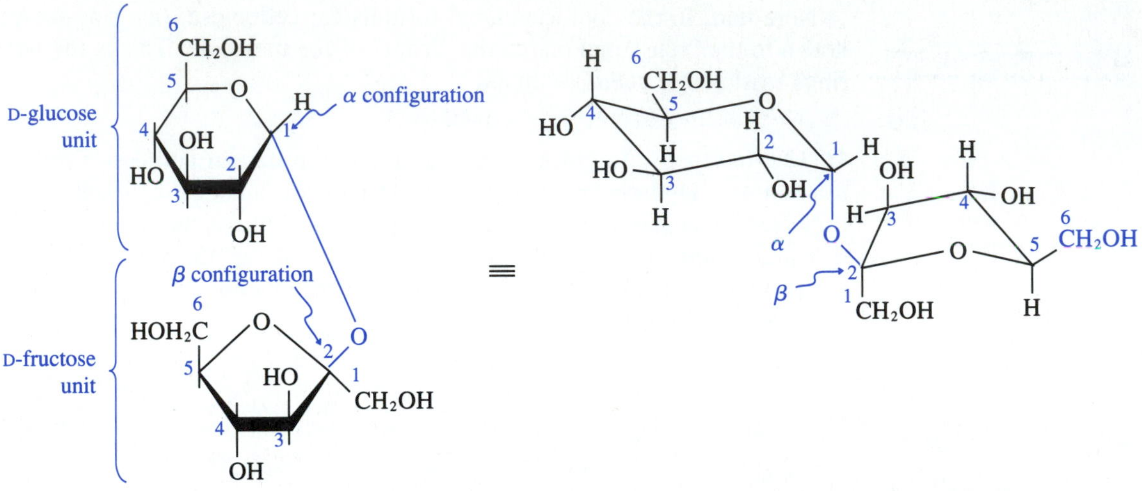

D-glucose unit

α configuration

D-fructose unit

β configuration

sucrose
α-D-glucopyranosyl-β-D-fructofuranoside
(or β-D-fructofuranosyl- α-D-glucopyranoside)

Since both anomeric carbons are linked in the glycosidic bond, neither monosaccharide unit has a hemiacetal group. Therefore, neither unit is in equilibrium with an acyclic form. Sucrose cannot mutarotate. And, because there is no free or potentially free aldehyde group, sucrose cannot reduce Tollens', Fehling's, or Benedict's reagent. Sucrose is therefore referred to as a *nonreducing sugar,* in contrast with the other disaccharides and monosaccharides we have discussed, which are reducing sugars.

Problem 16.15 Although β-D-glucose is a reducing sugar, methyl β-D-glucopyranoside (eq. 16.13) is not. Explain.

Sucrose has an optical rotation of $[\alpha] = +66°$. When sucrose is hydrolyzed to an equimolar mixture of D-glucose and D-fructose, the optical rotation changes value and sign and becomes $[\alpha] = -20°$. This is because the equilibrium mixture of D-glucose anomers (α and β) has a rotation of $+52°$, but the mixture of fructose anomers has a strong negative rotation, $[\alpha] = -92°$. In the early days of carbohydrate chemistry, glucose was called **dextrose** (because it was dextrorotatory), and fructose was called **levulose** (because it was levorotatory). Because hydrolysis of sucrose inverts the sign of optical rotation (from + to −), enzymes that bring about sucrose hydrolysis are called **invertases,** and the resulting equimolar mixture of glucose and fructose is called **invert sugar.** A number of insects, including the honeybee, possess invertases. *Honey* is largely a mixture of D-glucose, D-fructose, and some unhydrolyzed sucrose. It also contains flavors from the particular flowers whose nectars are collected.

A WORD ABOUT ...

30. Sweetness and Sweeteners

Sweetness is literally a matter of taste. Although individuals vary greatly in their sensory perceptions, it is possible to make some quantitative comparisons of sweetness. For example, we can take some standard sugar solution (say 10% sucrose in water) and compare its sweetness with that of solutions containing other sugars or sweetening agents. If a 1% solution of some compound tastes as sweet as the 10% sucrose solution, we can say that the compound is 10 times sweeter than sucrose.

D-Fructose is the sweetest of the simple sugars—almost twice as sweet as sucrose. D-Glucose is almost as sweet as sucrose. On the other hand, sugars like lactose and galactose have less than 1% of the sweetness of sucrose.

Many synthetic sweeteners are known, perhaps the most familiar being **saccharin.** It was discovered in 1879 in the laboratory of Professor Ira Remsen at the Johns Hopkins University. Its structure has no relation whatever to that of the saccharides, but saccharin is about 300 times sweeter than sucrose. For most tastes, 0.5 grain (0.03 g) of saccharin is equivalent in sweetness to a heaping teaspoon (10 g) of sucrose. Saccharin is made commercially from toluene, as shown in Figure 16.3.

Saccharin is very sweet yet has virtually no caloric content. It is useful as a sugar substitute for those who must restrict their sugar intake and also for those who wish to control their weight but still have a desire for sweets.

In 1981, **aspartame** became the first new sweetener to be approved by the U.S. Food and Drug Administration (FDA) in nearly 25 years. It is about 160 times sweeter than sucrose. Structurally, aspartame is the methyl ester of a dipeptide of two amino acids that occur naturally in proteins—aspartic acid and phenylalanine—and is sold under the trade name NutraSweet®.

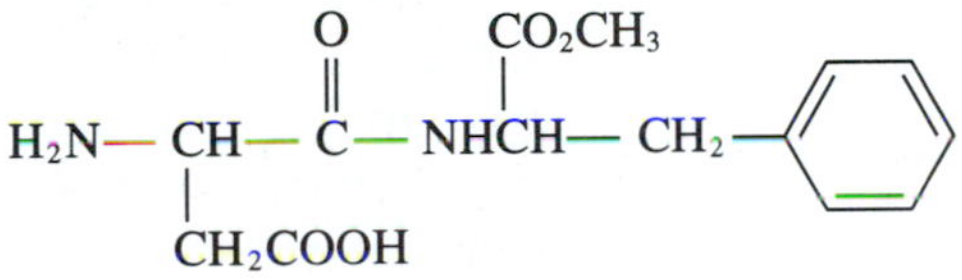

$$H_2N-\underset{\displaystyle CH_2COOH}{CH}-\overset{\displaystyle O}{\overset{\|}{C}}-NH\overset{\displaystyle CO_2CH_3}{CH}-CH_2-C_6H_5$$

the methyl ester of
N-L-α-aspartyl-L-phenylalanine
(aspartame)

Aspartame has about the same caloric content as sucrose, but because of its intense sweetness, much less is used, so its energy value becomes insignificant.

FIGURE 16.3
Synthesis of saccharin.

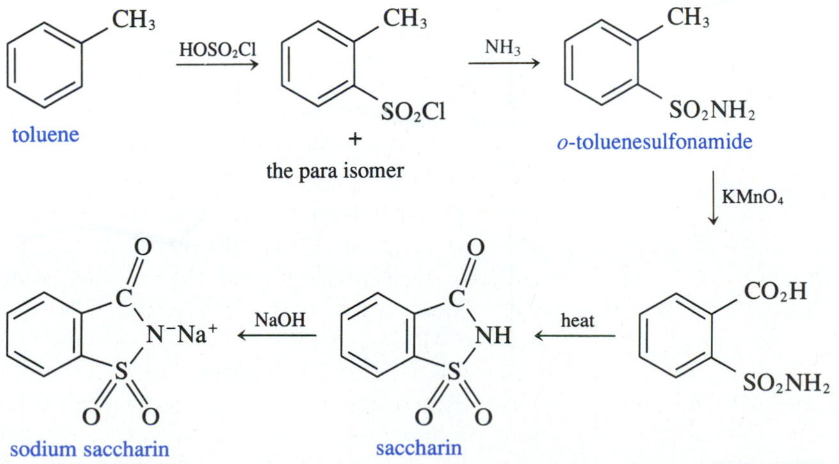

16.14 Polysaccharides

Polysaccharides contain many linked monosaccharides and vary in chain length and molecular weight. Most polysaccharides give a single monosaccharide on complete hydrolysis. The monosaccharide units may be linked linearly, or the chains may be branched. In this section, we will describe a few of the more important polysaccharides.

16.14a Starch and Glycogen **Starch** is the energy-storing carbohydrate of plants. It is a major component of cereals, potatoes, corn, and rice. It is the form in which glucose is stored by plants for later use.

Starch is made up of glucose units joined mainly by 1,4-α-glycosidic bonds, although the chains may have a number of branches attached through 1,6-α-glycosidic bonds. Partial hydrolysis of starch gives maltose, and complete hydrolysis gives only D-glucose.

Starch can be separated by various solution and precipitation techniques into two fractions: amylose and amylopectin. In **amylose,** which constitutes about 20% of starch, the glucose units (50 to 300) are in a continuous chain, with 1,4 linkages (Figure 16.4).

Amylopectin (Figure 16.5) is highly branched. Although each molecule may contain 300 to 5000 glucose units, chains with consecutive 1,4 links average only 25 to 30 units in length. These chains are connected at branch points by 1,6 linkages. Because of this highly branched structure, starch granules swell and eventually form colloidal solutions in water.

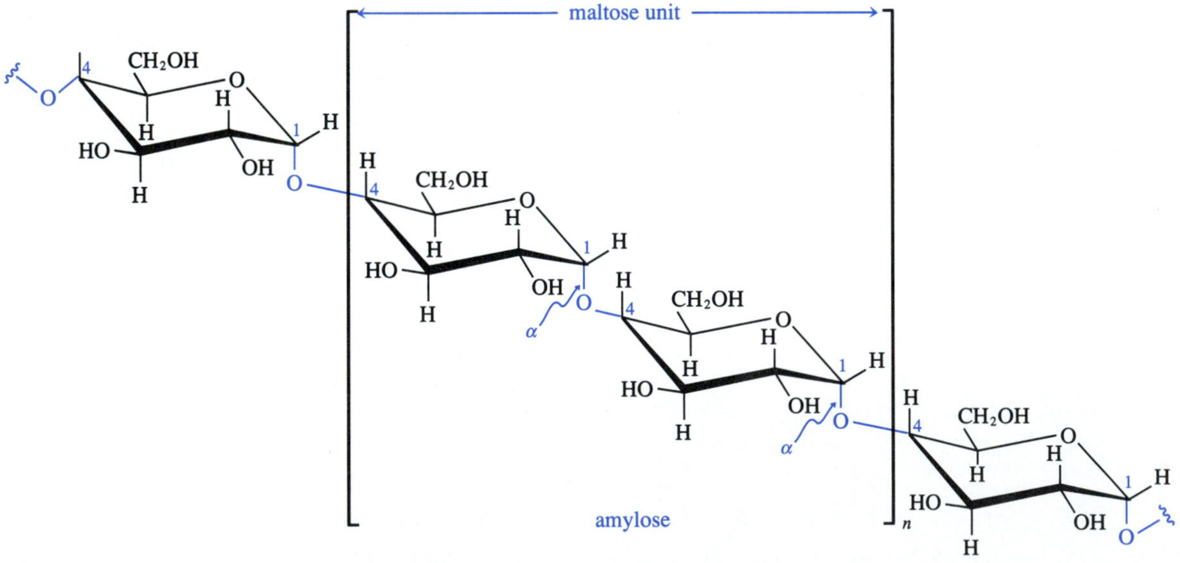

FIGURE 16.4 Structure of the amylose fraction of starch.

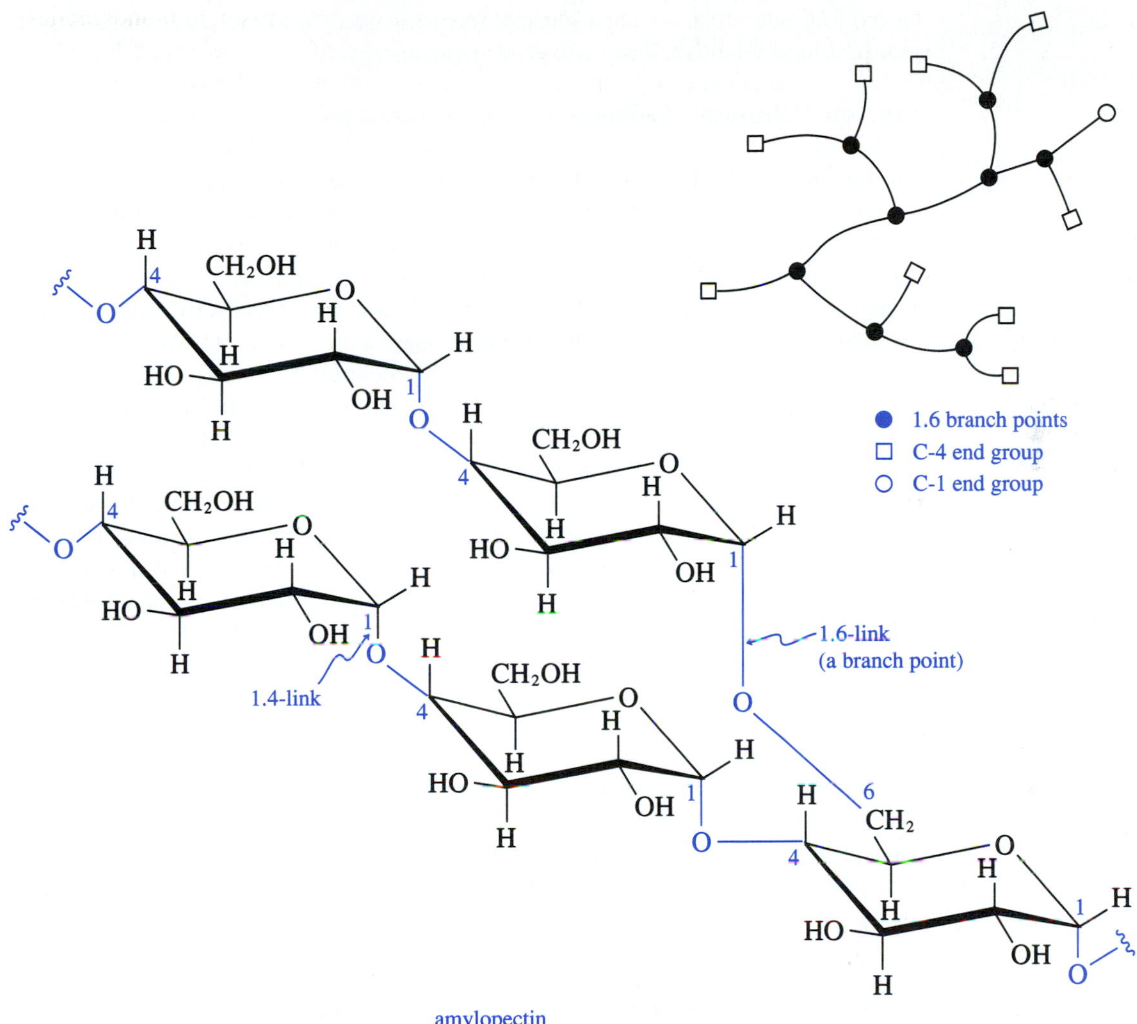

FIGURE 16.5 Structure of the amylopectin fraction of starch. (Adapted from Ferrier, R. J., and Collins, P. M. "Monosaccharide Chemistry"; Penguin Books, Ltd.; England. Used by permission.)

Glycogen is the energy-storing carbohydrate of animals. Like starch, it is made of 1,4- and 1,6-linked glucose units. Glycogen has a higher molecular weight than starch (perhaps 100,000 glucose units). Its structure is even more branched than that of amylopectin, with a branch every 8 to 12 glucose units. Glycogen is produced from glucose that is absorbed from the intestines into the blood; transported to the liver, muscles, and elsewhere; and then polymerized enzymatically. Glycogen helps maintain the glucose balance in the body, by removing and storing excess glucose from ingested food and later supplying it to the blood when various cells need it for energy.

16.14b Cellulose **Cellulose** is an *unbranched* polymer of glucose joined by 1,4-β-glycosidic bonds. X-ray examination of cellulose shows that it consists of linear chains of cellobiose units, in which the ring oxygens alternate in "forward" and "backward" positions (Figure 16.6). These linear molecules, containing an average of 5000 glucose units, aggregate to give fibrils bound together by hydrogen bonds between hydroxyls on adjacent chains. Cellulose fibers having considerable physical strength are built up from these fibrils, wound spirally in opposite directions around a central axis. Wood, cotton, hemp, linen, straw, and corncobs are mainly cellulose.

Although humans and other animals can digest starch and glycogen, they cannot digest cellulose. This is a truly striking example of the specificity of biochemical reactions. *The only chemical difference between starch and cellulose is the stereochemistry of the glucosidic link*—more precisely, the stereochemistry at C-1 of each glucose unit. The human digestive system contains enzymes that can catalyze the hydrolysis of α-glucosidic bonds, but it lacks the enzymes necessary to hydrolyze β-glucosidic bonds. Many bacteria, however, do contain β-glucosidases and can hydrolyze cellulose. Termites, for example, have such bacteria in their intestines and thrive on wood (cellulose) as their main food. Ruminants (cud-chewing animals such as cows) can digest grasses and other forms of cellulose because they harbor the necessary microorganisms in their rumen.

Cellulose is the raw material for several commercially important derivatives. Each glucose unit in cellulose contains three hydroxyl groups. These hydroxyl groups can be modified by the usual reagents that react with alcohols. For example, cellulose reacts with acetic anhydride to give **cellulose acetate.**

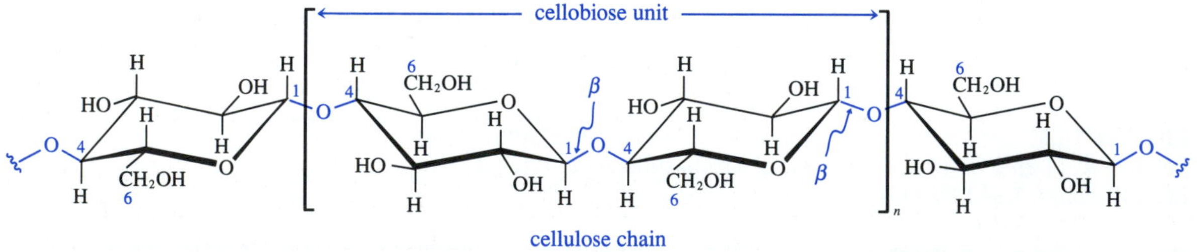

cellobiose unit

cellulose chain

FIGURE 16.6 Partial structure of a cellulose molecule showing the β linkages of each glucose unit.

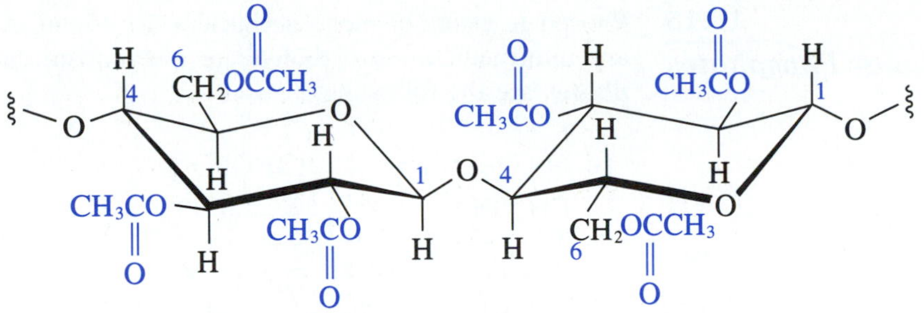

segment of a cellulose acetate molecule

Cellulose with about 97% of the hydroxyl groups acetylated is used to make acetate rayon.

Cellulose nitrate is another useful cellulose derivative. Like glycerol, cellulose can be converted with nitric acid to a nitrate ester (compare eq. 7.37). The number of hydroxyl groups nitrated per glucose unit determines the properties of the product. **Guncotton,** a highly nitrated cellulose, is an efficient explosive used in smokeless powders.

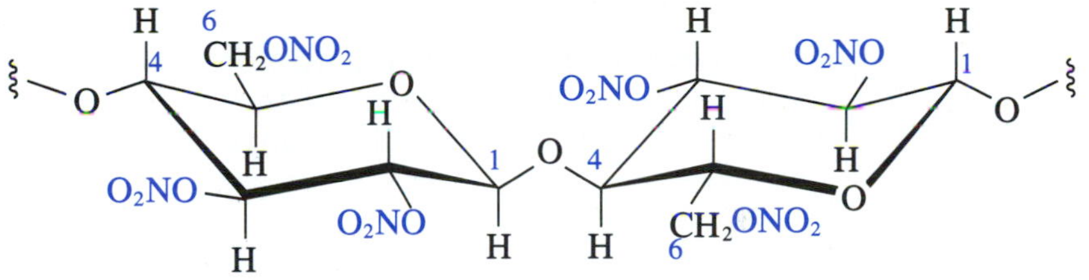

segment of a cellulose nitrate molecule

16.14c Other Polysaccharides **Chitin** is a nitrogen-containing polysaccharide that forms the shells of crustaceans and the exoskeletons of insects. It is similar to cellulose, except that the hydroxyl group at C-2 of each glucose unit is replaced by an acetylamino group, CH_3CONH-.

Pectins, which are obtained from fruits and berries, are polysaccharides used in making jellies. They are linear polymers of D-galacturonic acid, linked with 1,4-α-glycosidic bonds. D-Galacturonic acid has the same structure as D-galactose, except that the C-6 primary alcohol group is replaced by a carboxyl group.

Numerous other polysaccharides are known, such as gum arabic and other gums and mucilages, chondroitin sulfate (found in cartilage), the blood anticoagulant heparin (found in the liver and heart), and the dextrans (used as blood plasma substitutes).

Some saccharides have structures that differ somewhat from the usual polyhydroxyaldehyde or polyhydroxyketone pattern. In the final sections of this chapter, we will describe a few such modified saccharides that are important in nature.

16.15 Sugar Phosphates

Phosphate esters of monosaccharides are found in all living cells, where they are intermediates in carbohydrate metabolism. Some common **sugar phosphates** are the following:

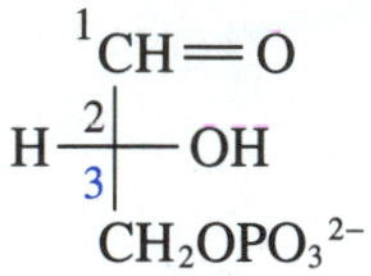

D-glyceraldehyde-3-phosphate

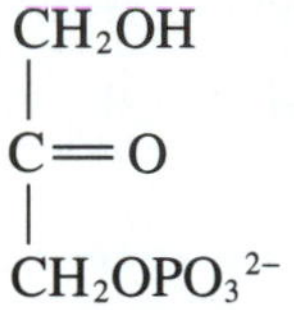

dihydroxyacetone-phosphate

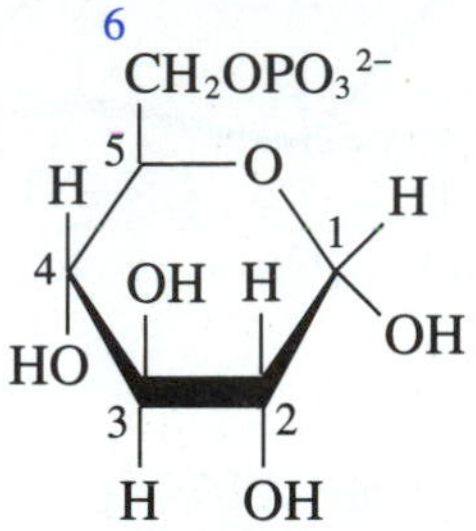

α-D-glucose-6-phosphate

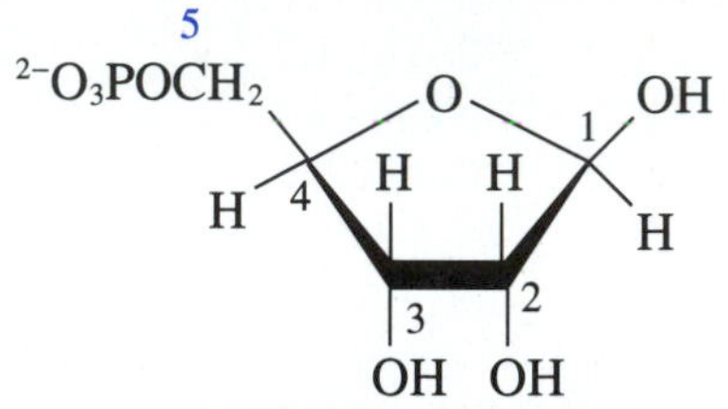

β-D-ribose-5-phosphate

Phosphates of the five-carbon sugar ribose and its 2-deoxy analog are important in nucleic acid structures (DNA, RNA) and in other key biological compounds (Sec. 18.14).

16.16 Deoxy Sugars

In **deoxy sugars,** one or more of the hydroxyl groups is replaced by a hydrogen atom. The most important example is **2-deoxyribose,** the sugar component of DNA. It lacks the hydroxyl group at C-2 and occurs in DNA in the furanose form.

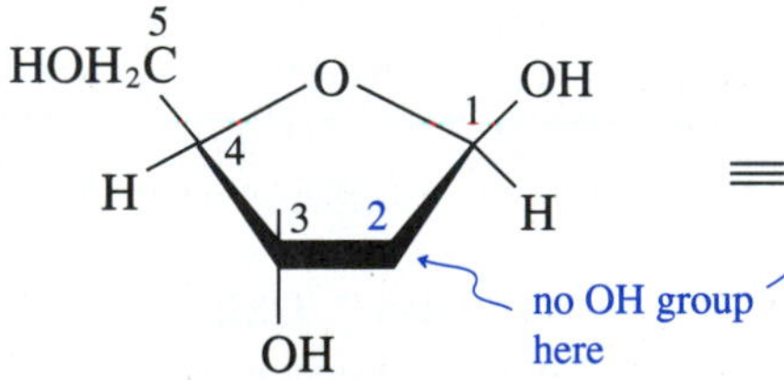

no OH group here

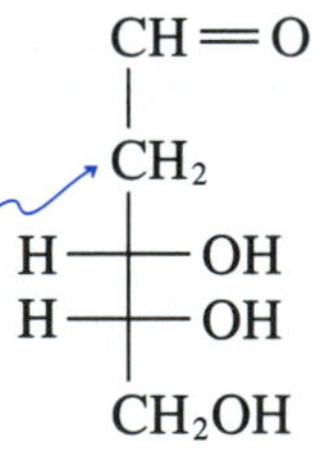

β-D-deoxyribofuranose
(the sugar of DNA)

16.17 Amino Sugars

In **amino sugars,** one of the sugar hydroxyl groups is replaced by an amino group. Usually the —NH_2 group is also acetylated. **D-Glucosamine** is one of the more abundant amino sugars.

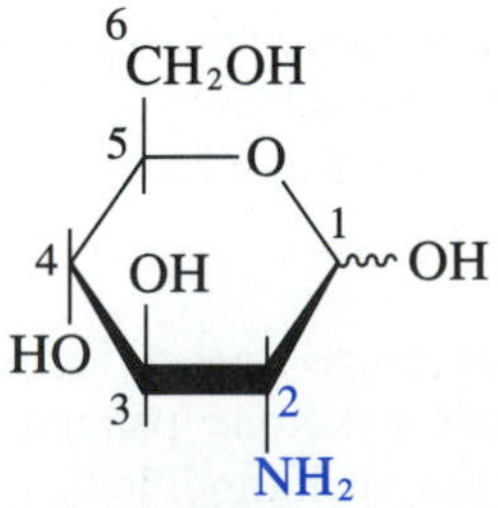

D-glucosamine
α, mp 88°C
β, mp 110°C (decomposes)

In its *N*-acetyl form, β-D-glucosamine is the monosaccharide unit of chitin, which forms the shells of lobsters, crabs, shrimp, and other shellfish.

16.18 *Ascorbic Acid (Vitamin C)*

L-Ascorbic acid (vitamin C) resembles a monosaccharide, but its structure has several unusual features. The compound has a five-membered unsaturated lactone ring with two hydroxyl groups attached to the doubly bonded carbons. This **enediol** structure is relatively uncommon.

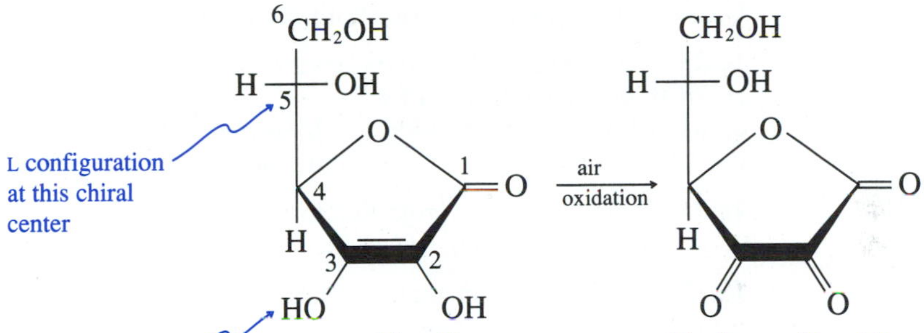

L configuration at this chiral center

acidic proton

L-ascorbic acid (vitamin C)
mp 192°C (decomposes)
pleasant, sharp-acid taste

$\xrightarrow[\text{oxidation}]{\text{air}}$

dehydroascorbic acid

(16.14)

As a consequence of this structural feature, ascorbic acid is easily oxidized to dehydroascorbic acid. Both forms are biologically effective as a vitamin.

There is no carboxyl group in ascorbic acid, but it is nevertheless an acid with a pK_a of 4.17. The proton of the hydroxyl group at C-3 is acidic, because the anion resulting from its loss is resonance-stabilized and similar to a carboxylate anion.

Humans, monkeys, guinea pigs, and a few other vertebrates lack an enzyme that is essential for the biosynthesis of ascorbic acid from D-glucose. Hence ascorbic acid must be included in the diet of humans and these other species. Ascorbic acid is abundant in citrus fruits and tomatoes. Its lack in the diet causes scurvy, a disease that results in weak blood vessels, hemorrhaging, loosening of teeth, lack of ability to heal wounds, and eventual death. Ascorbic acid is needed for collagen synthesis (collagen is the structural protein of skin, connective tissue, tendon, cartilage, and bone). In the eighteenth century, British sailors were required to eat fresh limes (a vitamin C source) to prevent outbreaks of the dreaded scurvy; hence the nickname "limeys."

Additional Problems

16.16. Define each of the following, and give the structural formula of one example.

a. aldohexose **b.** ketopentose **c.** monosaccharide
d. disaccharide **e.** polysaccharide **f.** furanose
g. pyranose **h.** glycoside **i.** anomeric carbon

16.17. Explain, using formulas, the difference between a D-sugar and an L-sugar.

16.18. Three of the four hydroxyl groups at the chiral centers in the Fischer projection of D-talose (Figure 16.1) are on the left, yet it is called a D-sugar. Explain.

16.19. What is the absolute configuration (*R* or *S*) of the chiral centers at C-2 and C-3 of D-erythrose? (Use Figure 16.1 for this and other problems if you have not yet memorized the structures of the monosaccharides.)

16.20. What is the absolute configuration (*R* or *S*) at each chiral center in the acyclic form of D-glucose? at the new chiral center in β-D-glucose?

16.21. What term would you use to describe the stereochemical relationship between D-glucose and D-idose?

16.22. Construct an array analogous to Figure 16.1 for the D-ketoses as high as the hexoses. Dihydroxyacetone should be at the head, in place of glyceraldehyde.

16.23. Using Figure 16.1 if necessary, write a Fischer projection formula and a Haworth projection formula for

a. methyl α-D-glucopyranoside. **b.** α-D-gulopyranose.
c. β-D-arabinofuranose. **d.** methyl α-L-glucopyranoside.

16.24. Draw the Fischer projection formula for

a. L-(−)-mannose. **b.** L-(+)-fructose.

16.25. At equilibrium in aqueous solution, D-ribose exists as a mixture containing 20% α-pyranose, 56% β-pyranose, 6% α-furanose, and 18% β-furanose forms. Draw Haworth formulas for each of these forms.

16.26. Write Fischer, Haworth, and conformational structures for β-D-allose.

16.27. The solubilities of α- and β-D-glucose in water at 25°C are 82 and 178 g/100 mL, respectively. Why are their solubilities not identical?

16.28. The specific rotations of pure α- and β-D-fructofuranose are +21° and −133°, respectively. Solutions of each isomer mutarotate to an equilibrium specific rotation of −92°. Assuming that no other forms are present, calculate the equilibrium concentrations of the two forms.

16.29. D-Threose can exist in a furanose form but *not* in a pyranose form. Explain. Draw the β-furanose structure.

16.30. Draw the Fischer and Newman projection formulas for L-erythrose.

16.31. Starting with β-D-glucose and using acid (H^+) as a catalyst, write out all the steps in the mechanism for the mutarotation process. Use Haworth projections for the cyclic structures.

16.32. Oxidation of either D-erythrose or D-threose with nitric acid gives tartaric acid. In one case, the tartaric acid is optically active; in the other, it is optically inactive. How can these facts be used to assign stereochemical structures to erythrose and threose?

16.33. Write a structure for

a. D-galactonic acid **b.** D-galactaric acid.

16.34. Using complete structures, write out the reaction of D-mannose with

a. bromine water. **b.** nitric acid.
c. sodium borohydride. **d.** acetic anhydride.

16.35. Reduction of D-fructose with $NaBH_4$ gives a mixture of D-glucitol and D-mannitol. What does this result prove about the configurations of D-fructose, D-mannose, and D-glucose?

16.36. Although D-galactose contains five chiral centers, its oxidation with nitric acid gives an optically inactive dicarboxylic acid (called galactaric or mucic acid). What is the structure of this acid, and why is it optically inactive?

16.37. Write equations that clearly show the mechanism for the acid-catalyzed hydrolysis of

a. maltose to glucose.
b. lactose to galactose and glucose.
c. sucrose to fructose and glucose.

16.38. Write equations for the reaction of maltose with

a. methanol and H^+. **b.** Tollens' reagent.
c. bromine water. **d.** acetic anhydride.

16.39. Trehalose is a disaccharide that is the main carbohydrate component in the blood of insects. Its structure is

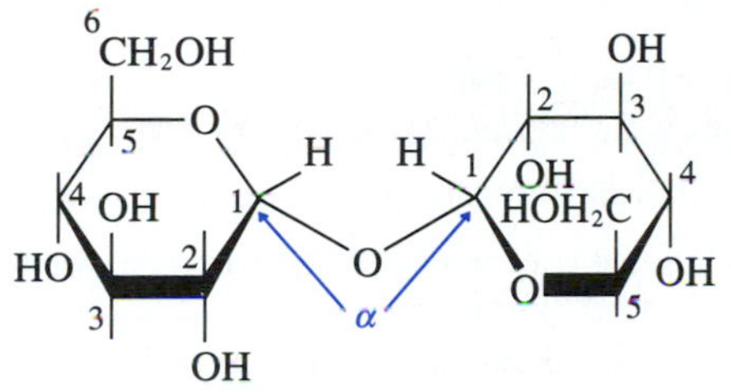

trehalose

a. What are its hydrolysis products?
b. Will trehalose give a positive or a negative test with Fehling's reagent? Explain.

16.40. Write an equation for the acid-catalyzed hydrolysis of salicin (Sec. 16.12). Notice that one of the products is structurally similar to aspirin (eq. 10.36), perhaps accounting for salicin's fever-reducing property.

16.41. Lactose exists in α and β forms, with specific rotations of +92.6° and +34°, respectively.

a. Draw their structures.
b. Solutions of each isomer mutarotate to an equilibrium value of +52°. What is the percentage of each isomer at equilibrium?

16.42. Write a balanced equation for the reaction of D-(+)-glucose (use either an acyclic or a cyclic structure, whichever seems most appropriate) with each of the following:

a. acetic anhydride (excess) **b.** bromine water
c. hydrogen, catalyst **d.** hydroxylamine (to form an oxime)
e. methanol, H^+ **f.** hydrogen cyanide (to form a cyanohydrin)
g. Fehling's reagent

16.43. Explain why sucrose is a nonreducing sugar but maltose is a reducing sugar.

16.44. Explain what type of reaction takes place in each of the five steps in the synthesis of sodium saccharin (Figure 16.3). Write a mechanism for steps 1, 2, and 4.

16.45. The last step in Figure 16.3 shows that saccharin is acidic and readily forms a sodium salt. How do you account for its acidity?

16.46. Using the descriptions in Sec. 16.14c, write a formula for
a. chitin. **b.** pectin.

16.47. L-Fucose is a component of bacterial cell walls. It is also called 6-deoxy-L-galactose. Write its Fischer projection formula.

16.48. Daunosamine is an amino sugar that forms part of the structure of doxorubicin (adriamycin), a tetracycline anticancer agent. From its Haworth projection shown below, draw its conformational (chair) structure.

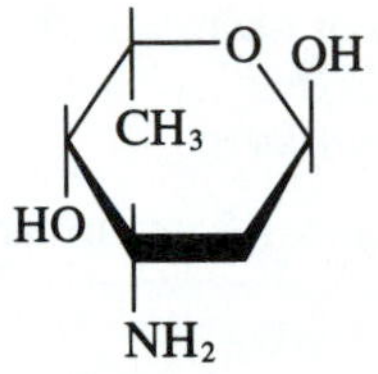

daunosamine

Is daunosamine a D or an L sugar?

16.49. Write the main contributors to the resonance hybrid anion formed when ascorbic acid acts as an acid (loses the proton from the —OH group on C-3).

16.50. Hemicelluloses are noncellulose materials produced by plants and found in straw, wood, and other fibrous tissues. Xylans are the most abundant hemicelluloses. They consist of 1,4-β-linked D-xylopyranoses. Draw the structure for the repeating unit in xylans.

16.51. Inositols are hexahydroxycyclohexanes, with one hydroxyl group on each carbon atom of the ring. Although not strictly carbohydrates, they are obviously similar to pyranose sugars and do occur in nature. There are nine possible stereoisomers. Draw Haworth formulas for all possibilities (all are known), and tell which are chiral.

CHAPTER 17

Amino Acids, Peptides, and Proteins

17.1 Introduction

Proteins are natural polymers composed of **amino acid** units joined one to another by amide (or peptide) bonds. Proteins are essential to the structure, function, and reproduction of living matter. In this chapter, we will first discuss the structure and properties of amino acids. We will next describe the properties of **peptides,** which consist of just a few amino acids linked together, and, finally, the structures of proteins.

17.2 Naturally Occurring Amino Acids

The amino acids obtained from protein hydrolysis are α-amino acids. That is, the amino group is on the α-carbon atom, the one adjacent to the carboxyl group.

$$\mathrm{R{-}\overset{\alpha}{C}H(NH_2){-}C({=}O)OH}$$

an α-amino acid

With the exception of glycine, where R = H, α-amino acids have a chiral center at the α-carbon. All except glycine are therefore optically active. They have the L configuration relative to glyceraldehyde (Figure 17.1). Note that the Fischer convention, used with carbohydrates, is also applied to amino acids.

Table 17.1 lists the 20 α-amino acids commonly found in proteins. The amino acids are known by common names. Each also has a three-letter abbreviation based on this name, which is used when writing the formulas of peptides and proteins. The amino acids in Table 17.1 are grouped to emphasize structural similarities. Of the 20 amino acids listed in the table, 12 can be synthesized in the body from other foods. The other 8, those with names shown in color and referred to as **essential amino acids,** cannot be synthesized by adult humans and therefore must be included in the diet in the form of proteins.

TABLE 17.1 Names and formulas of the common amino acids

Name	*Abbreviation (Isoelectric point)*	*Formula*	*R*
A. One amino group and one carboxyl group			
1. glycine	Gly (6.0)	$H{-}CH{-}CO_2H$ | NH_2	
2. alanine	Ala (6.0)	$CH_3{-}CH{-}CO_2H$ | NH_2	
3. valine	Val (6.0)	$CH_3CH{-}CH{-}CO_2H$ | | CH_3 NH_2	R is hydrogen or an alkyl group.
4. leucine	Leu (6.0)	$CH_3CHCH_2{-}CH{-}CO_2H$ | | CH_3 NH_2	
5. isoleucine	Ile (6.0)	$CH_3CH_2CH{-}CH{-}CO_2H$ | | CH_3 NH_2	
6. serine	Ser (5.7)	$CH_2{-}CH{-}CO_2H$ | | OH NH_2	
7. threonine	Thr (5.6)	$CH_3CH{-}CH{-}CO_2H$ | | OH NH_2	R contains an alcohol function.
8. cysteine	Cys (5.0)	$CH_2{-}CH{-}CO_2H$ | | SH NH_2	
9. methionine	Met (5.7)	$CH_3S{-}CH_2CH_2{-}CH{-}CO_2H$ | NH_2	R contains sulfur.
10. proline	Pro (6.3)	$CH_2{-}CH{-}CO_2H$ | | CH_2 NH \ / CH_2	The amino group is secondary and part of a ring.

TABLE 17.1 *(continued)*

Name	*Abbreviation (Isoelectric point)*	*Formula*	*R*
11. phenylalanine	Phe (5.5)	$C_6H_5—CH_2—CH(NH_2)—CO_2H$	One hydrogen in alanine is replaced by an aromatic or heteroaromatic (indole) ring.
12. tyrosine	Tyr (5.7)	$HO—C_6H_4—CH_2—CH(NH_2)—CO_2H$	
13. tryptophan	Trp (5.9)	(indol-3-yl)$—CH_2—CH(NH_2)—CO_2H$	
B. One amino group and two carboxyl groups			
14. aspartic acid	Asp (3.0)	$HOOC—CH_2—CH(NH_2)—CO_2H$	
15. glutamic acid	Glu (3.2)	$HOOC—CH_2CH_2—CH(NH_2)—CO_2H$	
16. asparagine	Asn (5.4)	$H_2N—C(=O)—CH_2—CH(NH_2)—CO_2H$	These are the primary amides of aspartic and glutamic acids.
17. glutamine	Gln (5.7)	$H_2N—C(=O)—CH_2CH_2—CH(NH_2)—COOH$	
C. One carboxyl group and two basic groups			
18. lysine	Lys (9.7)	$CH_2(NH_2)CH_2CH_2CH_2—CH(NH_2)—CO_2H$	The second basic group is a primary amine, a guanidine, or an imidazole.
19. arginine	Arg (10.8)	$(NH_2)(HN=)C—NH—CH_2CH_2CH_2—CH(NH_2)—CO_2H$	
20. histidine	His (7.6)	$CH=C(—NH—CH=N—)—CH_2—CH(NH_2)—CO_2H$ (imidazole ring)	

FIGURE 17.1
Naturally occurring α-amino acids have the L configuration.

CHO / HO—C—H / CH_2OH
L-(−)-glyceraldehyde

CO_2H / NH_2—C—H / R
a naturally occurring L-amino acid

CO_2H / NH_2—+—H / R
Fischer projection formula of an L-amino acid

CO_2H / NH_2—+—H / CH_3
L-(+)-alanine

A WORD ABOUT . . .

31. Amino Acid Dating

The question of age is one of the first that archaeologists seek to answer when they find artifacts or skeletons at a "dig." Are the bones or the pot shards ancient or modern? If they are ancient, *how* ancient? Knowing the age helps to answer other questions, such as how the people lived, with what other groups they traded or had contact, who they followed and who followed them, and so on.

Chemists have helped archaeologists with this problem. One of the best-known methods is carbon-14, or radioactive, dating, first proposed in 1947 by Willard F. Libby (Nobel Prize winner in 1960). The isotope ^{14}C decays with a half-life of 5730 years, which is sufficiently long for a steady-state equilibrium concentration to be established in the biosphere. A tiny but constant fraction (about $1.2 \times 10^{-10}\%$) of the carbon in live plants and animals is ^{14}C. After death, when ^{14}C is no longer taken in from the environment (as food, carbon dioxide, and so on), its concentration decreases. Knowing the decay rate of ^{14}C and comparing the ^{14}C content of the ancient material with that of modern allows the age of the ancient material to be calculated. The practical limit of the method is about 10 half-lives of ^{14}C, or about 50,000 years.

Amino acids can be found in fossil bones, shells, and teeth. In living systems, amino acids have the L configuration and are optically pure. Once death occurs, however, the biochemical reactions that prevent equilibration of the L and D forms are terminated, and gradual thermal equilibration of the two forms begins. This reaction can be used for dating because the amount of racemization is a function of the material's age.

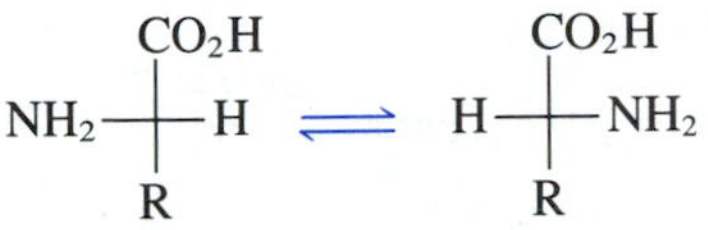

$$\begin{array}{c} CO_2H \\ NH_2-\!\!\!\!+\!\!\!-H \\ R \end{array} \rightleftharpoons \begin{array}{c} CO_2H \\ H-\!\!\!\!+\!\!\!-NH_2 \\ R \end{array}$$

L-form D-form

Racemization rates differ for different amino acids. For example, the half-life at 25°C and pH 7 for aspartic acid is about 3000 years; for alanine, it is about 12,000 years. The racemization rate also depends on temperature. For example, the half-life at 0°C for aspartic acid increases to about 430,000 years. So it is necessary, for accurate dating, to know the temperature at which the material was stored. Fortunately, the temperature below ground at a given depth and climate often remains constant over long periods and can be estimated fairly accurately.

Accuracy can be improved by using a combination of dating methods or by calibrating one method with another. One advantage of amino acid dating is that the sample size can be very much smaller than is needed for carbon-14 dating. Also, a range of time spans can be covered by using different amino acids. Amino acid dating can be extended back in time well beyond the limits of ^{14}C dating, even to ice ages 100 to 400 thousand years ago!

17.3 The Acid-Base Properties of Amino Acids

The carboxylic acid and amine functional groups are *simultaneously* present in amino acids, and we might ask whether they are mutually compatible, since one group is acidic and the other is basic. Although we have represented the amino acids in Table 17.1 as having amino and carboxyl groups, these structures are oversimplified.

Amino acids with one amino group and one carboxyl group are better represented by a **dipolar ion structure.***

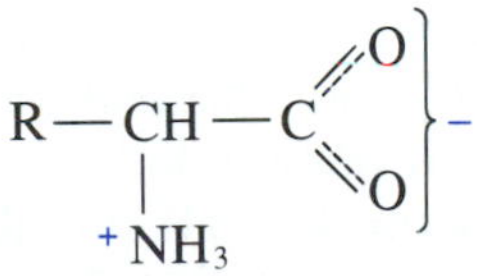

$$R-\underset{\underset{^{+}NH_3}{|}}{CH}-\left\{C\begin{matrix} \nearrow O \\ \searrow O \end{matrix}\right\}^{-}$$

dipolar structure of an α-amino acid

The amino group is protonated and present as an ammonium ion, whereas the carboxyl group has lost its proton and is present as a carboxylate anion. This dipolar structure is consistent with the saltlike properties of amino acids, which have rather high melting points (even the simplest, glycine, melts at 233°C) and relatively low solubilities in organic solvents.

Amino acids are amphoteric. They can behave as acids and donate a proton to a strong base, or they can behave as bases and accept a proton from a

*Such structures are sometimes called *zwitterions* (from a German word for hybrid ions).

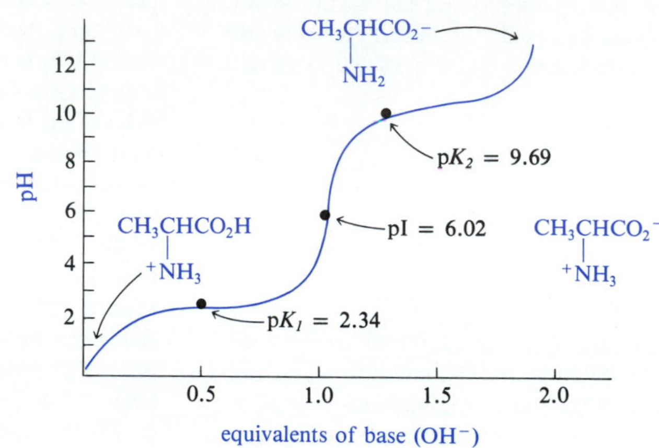

FIGURE 17.2 Titration curve for alanine, showing how its structure varies with pH.

strong acid. These behaviors are expressed in the following equilibria for an amino acid with one amino and one carboxyl group:

$$\underset{\substack{\text{amino acid}\\\text{at low pH}\\\text{(acid)}}}{\underset{\substack{|\\ {}^{+}NH_3}}{RCHCO_2H}} \underset{H^+}{\overset{OH^-}{\rightleftharpoons}} \underset{\substack{\text{dipolar ion}\\\text{form}\\\text{(neutral)}}}{\underset{\substack{|\\ {}^{+}NH_3}}{RCHCO_2^-}} \underset{H^+}{\overset{OH^-}{\rightleftharpoons}} \underset{\substack{\text{amino acid}\\\text{at high pH}\\\text{(base)}}}{\underset{\substack{|\\ NH_2}}{RCHCO_2^-}} \tag{17.1}$$

Figure 17.2 shows a titration curve for alanine, a typical amino acid of this kind. At low pH (acidic solution), the amino acid is in the form of a substituted ammonium ion. At high pH (basic solution), it is present as a substituted carboxylate ion. At some intermediate pH (for alanine, pH 6.02), the amino acid is present as the dipolar ion. A simple rule to remember for any acidic site is that *if the pH of the solution is* less *than the* pK_a, *the proton is on; if the pH of the solution is* greater *than the* pK_a, *the proton is off.*

EXAMPLE 17.1 Starting with alanine hydrochloride (its structure at low pH in hydrochloric acid; see lower left corner of the curve in Figure 17.2), write equations for its reaction with one equivalent of sodium hydroxide and then with a second equivalent of sodium hydroxide.

Solution

$$\underset{\text{ammonium salt}}{\underset{\substack{|\\ {}^{+}NH_3\ Cl^-}}{CH_3CHCO_2H}} + Na^+OH^- \longrightarrow \underset{\text{dipolar ion}}{\underset{\substack{|\\ {}^{+}NH_3}}{CH_3CHCO_2^-}} + Na^+Cl^- + H_2O \tag{17.2}$$

$$CH_3CH(^+NH_3)CO_2^- + Na^+OH^- \longrightarrow CH_3CH(NH_2)CO_2^-Na^+ + H_2O \quad (17.3)$$

dipolar ion → carboxylate salt

The first equivalent of base removes a proton from the carboxyl group to give the dipolar ion, and the second equivalent of base removes a proton from the ammonium ion to give the sodium carboxylate.

Problem 17.1 Starting with the sodium carboxylate salt of alanine, write equations for its reaction with one equivalent of hydrochloric acid and then with a second equivalent, and explain what each equivalent of acid does.

Problem 17.2 Which group in the ammonium salt form of alanine is more acidic, the $-\overset{+}{N}H_3$ group or the $-CO_2H$ group?

Problem 17.3 Which group in the carboxylate salt form of alanine is more basic, the $-NH_2$ group or the $-CO_2^-$ group?

Note from Figure 17.2 and from eq. 17.1 that the charge on an amino acid changes as the pH changes. At low pH, for example, the sign on alanine is positive, at high pH it is negative, and near neutrality the ion is dipolar. If placed in an electric field, the amino acid will therefore migrate toward the cathode (negative electrode) at low pH and toward the anode (positive electrode) at high pH (Figure 17.3). At some intermediate pH, called the **isoelectric point (pI),** the amino acid will be dipolar and have a net charge of zero. It will be unable to move toward either electrode. The isoelectric points of the various amino acids are listed in Table 17.1.

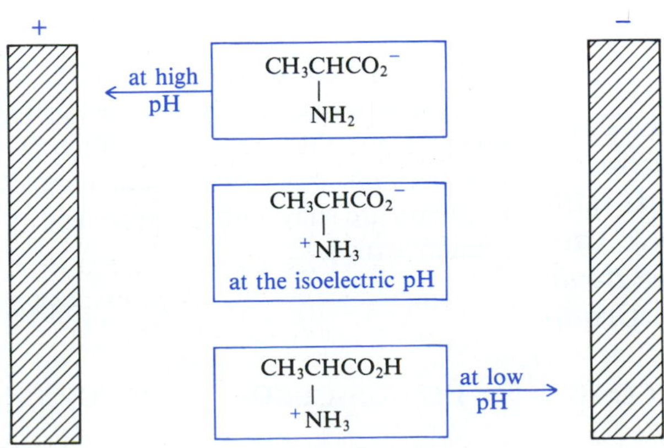

FIGURE 17.3 The migration of an amino acid (such as alanine) in an electric field depends on pH.

EXAMPLE 17.2 Write the structure of valine

a. at the pI. b. at high pH. c. at low pH.

Solution a. $(CH_3)CHCHCO_2^-$ with $^+NH_3$ on the α-carbon — dipolar and neutral

b. $(CH_3)_2CHCHCO_2^-$ with NH_2 on the α-carbon — negative

c. $(CH_3)_2CHCHCO_2H$ with $^+NH_3$ on the α-carbon — positive

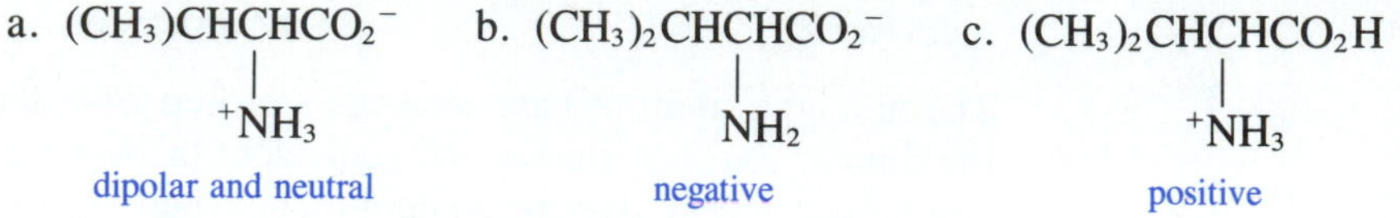

Problem 17.4 Write the structure for the predominant form of each of the following amino acids at the indicated pH. If placed in an electric field, toward which electrode (+ or −) will each amino acid migrate?

a. phenylalanine at its pI b. methionine at low pH
c. serine at high pH

In general, amino acids with one amino group and one carboxyl group, and no other acidic or basic groups in their structure, have two pK_a values: one around 2 to 3 for proton loss from the carboxyl group and the other around 9 to 10 for proton loss from the ammonium ion. The isoelectric point is about halfway between the two pK_a values, near pH 6.

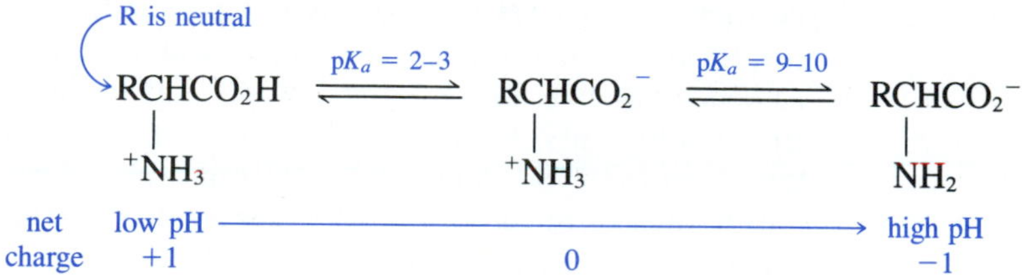

R is neutral

$$\underset{\substack{|\\ ^+NH_3}}{RCHCO_2H} \underset{}{\overset{pK_a = 2\text{–}3}{\rightleftharpoons}} \underset{\substack{|\\ ^+NH_3}}{RCHCO_2^-} \overset{pK_a = 9\text{–}10}{\rightleftharpoons} \underset{\substack{|\\ NH_2}}{RCHCO_2^-} \quad (17.4)$$

net charge: low pH +1 → 0 → high pH −1

The situation is more complex with amino acids containing two acidic or two basic groups.

17.4 *The Acid-Base Properties of Amino Acids with More Than One Acidic or Basic Group*

Aspartic and glutamic acids (numbers 14 and 15 in Table 17.1) have two carboxyl groups and one amino group. In strong acid (low pH) all three of these groups are in their acidic form (protonated). As the pH is raised and the solution becomes more basic, each group in succession gives up a proton. The equilibria are shown for **aspartic acid,** with the three pK_a values over the equilibrium arrows:

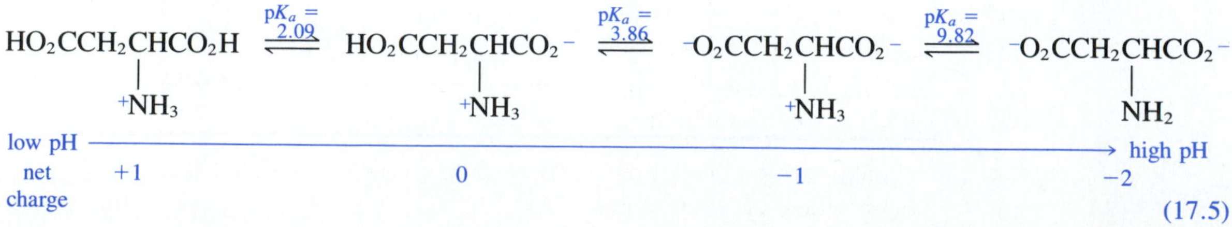

$$\underset{\substack{|\\ ^+NH_3}}{HO_2CCH_2CHCO_2H} \overset{pK_a = 2.09}{\rightleftharpoons} \underset{\substack{|\\ ^+NH_3}}{HO_2CCH_2CHCO_2^-} \overset{pK_a = 3.86}{\rightleftharpoons} \underset{\substack{|\\ ^+NH_3}}{^-O_2CCH_2CHCO_2^-} \overset{pK_a = 9.82}{\rightleftharpoons} \underset{\substack{|\\ NH_2}}{^-O_2CCH_2CHCO_2^-}$$

low pH → high pH
net charge: +1, 0, −1, −2

(17.5)

The isoelectric point for aspartic acid, the pH at which it is mainly in the neutral dipolar form, is 2.87 (in general, the pI is close to the average of the two pK_a's on either side of the neutral, dipolar species).

EXAMPLE 17.3 Which carboxyl group is the stronger acid in the most acidic form of aspartic acid?

Solution As shown at the extreme left of eq. 17.5, the first proton to be removed from the most acidic form of aspartic acid is the proton on the carboxyl group nearest the $^+NH_3$ substituent. The $^+NH_3$ group is electron-withdrawing due to its positive charge and enhances the acidity of the carboxyl group closest to it. Also, the resulting dipolar ion has the opposite charges closer to each other than would be the case if the other carboxyl group had given up its proton.

Problem 17.5 Use eq. 17.5 to tell which is the least acidic group in aspartic acid and why.

The situation differs for amino acids with two basic groups and only one carboxyl group (numbers 18, 19, and 20 in Table 17.1). With **lysine,** for example, the equilibria are

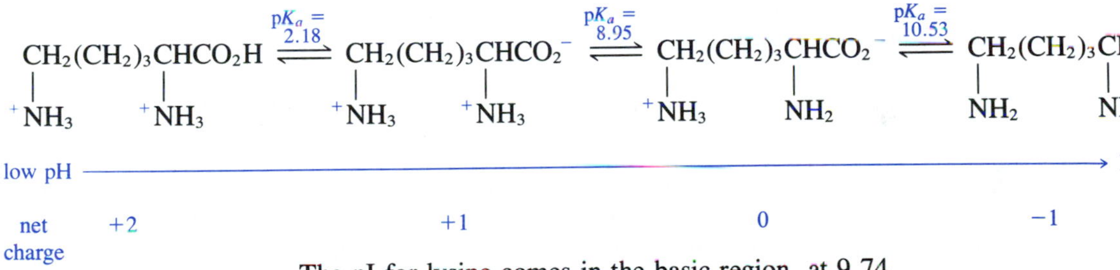

$$\underset{^+NH_3\qquad\qquad ^+NH_3}{CH_2(CH_2)_3CHCO_2H} \underset{}{\overset{pK_a = 2.18}{\rightleftharpoons}} \underset{^+NH_3\qquad\qquad ^+NH_3}{CH_2(CH_2)_3CHCO_2^-} \overset{pK_a = 8.95}{\rightleftharpoons} \underset{^+NH_3\qquad\qquad NH_2}{CH_2(CH_2)_3CHCO_2^-} \overset{pK_a = 10.53}{\rightleftharpoons} \underset{NH_2\qquad\qquad NH_2}{CH_2(CH_2)_3CHCO_2^-}$$

low pH ⟶ high pH

net charge: +2, +1, 0, −1 (17.6)

The pI for lysine comes in the basic region, at 9.74.

The second basic groups in arginine and histidine are not simple amino groups. They are a **guanidine** group and an **imidazole** ring, respectively, shown in color. The most protonated forms of these two amino acids are

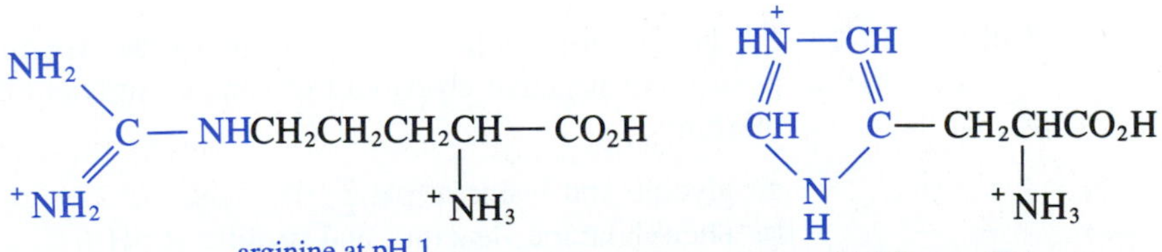

$$(NH_2)(^+NH_2{=})C{-}NHCH_2CH_2CH_2\underset{^+NH_3}{CH}{-}CO_2H$$

arginine at pH 1

$$\text{(imidazole ring: } \overset{+}{H}N{-}CH{=}C{-}NH{-}CH{=}\text{)}{-}CH_2\underset{^+NH_3}{CH}CO_2H$$

histidine at pH 1

Problem 17.6 Arginine shows three pK_a's: at 2.17 (the —COOH group), at 9.04 (the $-\overset{+}{N}H_3$ group), and at 12.48 (the guanidinium ion). Write equilibria (similar to eq. 17.6) for its dissociation. At approximately what pH will the isoelectric point come, and what is the structure of the dipolar ion?

Table 17.2 summarizes the approximate pK_a values and isoelectric points for the three types of amino acids.

TABLE 17.2 Approximate acidity constants and isoelectric points (pI) for the three types of amino acids

Type		pK_a 1	2	3	p*I*
1 acidic and 1 basic group		2.3	9.4	—	6.0
2 acidic and 1 basic group		2.2	4.1	9.8	3.0
1 acidic and 2 basic groups	(Lys, Arg)	2.2	9.0	11.5	10.0
	(His)	1.8	6.0	9.2	7.6

17.5 Electrophoresis

As seen in eqs. 17.4 through 17.6, the charge on an amino acid depends on the pH of the solution. **Electrophoresis,** an important method for separating amino acids and proteins, takes advantage of these charge differences. It is based on the differential rates and directions of migration of amino acids or proteins in an electric field at a controlled pH.

EXAMPLE 17.4 Predict the direction of migration (toward the positive or negative electrode) of alanine in an electrophoresis apparatus at pH 5. Do the same for aspartic acid.

Solution A pH of 5 is *less* than the pI of alanine (~6). Therefore, the dipolar ions will be protonated (positive) and migrate toward the negative electrode. But pH 5 is *greater* than the pI of aspartic acid (~3). Therefore aspartic acid will exist mainly as the −1 ion (eq. 17.5) and migrate toward the positive electrode. A mixture of the two amino acids could therefore easily be separated in this way.

Problem 17.7 Predict the direction of migration in an electrophoresis apparatus (toward the positive or negative electrode) of each component of the following amino acid mixtures:

a. glycine and lysine at pH 7
b. phenylalanine, leucine, and proline at pH 6

17.6 Reactions of Amino Acids

In addition to their acidic and basic behavior, amino acids undergo other reactions typical of carboxylic acids or amines. For example, the carboxyl group can be esterified.

$$\underset{\displaystyle {}^{+}NH_3}{R-\underset{|}{C}H}-CO_2^- + R'OH + H^+ \xrightarrow{\text{heat}} \underset{\displaystyle {}^{+}NH_3}{R-\underset{|}{C}H}-CO_2R' + H_2O \qquad (17.7)$$

The amino group can be acylated to an amide.

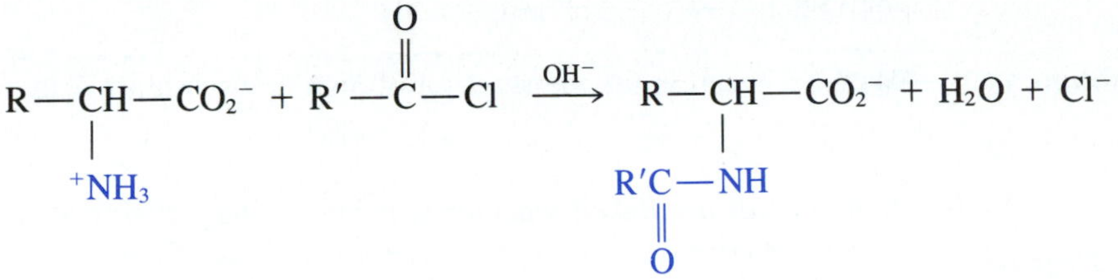

$$R{-}\underset{\underset{^{+}NH_3}{|}}{CH}{-}CO_2^- + R'{-}\overset{\overset{O}{\|}}{C}{-}Cl \xrightarrow{OH^-} R{-}\underset{\underset{R'\underset{\underset{O}{\|}}{C}{-}NH}{|}}{CH}{-}CO_2^- + H_2O + Cl^- \tag{17.8}$$

These types of reactions are useful in temporarily modifying or protecting either of the two functional groups, especially during the controlled linking of amino acids to form peptides or proteins.

Problem 17.8 Using eqs. 17.7 and 17.8 as models, write equations for the following reactions:

a. phenylalanine + CH_3OH + HCl $\longrightarrow$

b. valine + benzoyl chloride + NaOH $\longrightarrow$

c. glycine + acetic anhydride $\xrightarrow{\text{heat}}$

17.7 The Ninhydrin Reaction

Ninhydrin is a useful reagent for detecting amino acids and determining the concentrations of their solutions. It is the hydrate of a cyclic triketone, and when it reacts with an amino acid, a violet dye is produced. The overall reaction, whose mechanism is complex and need not concern us in detail here, is as follows:

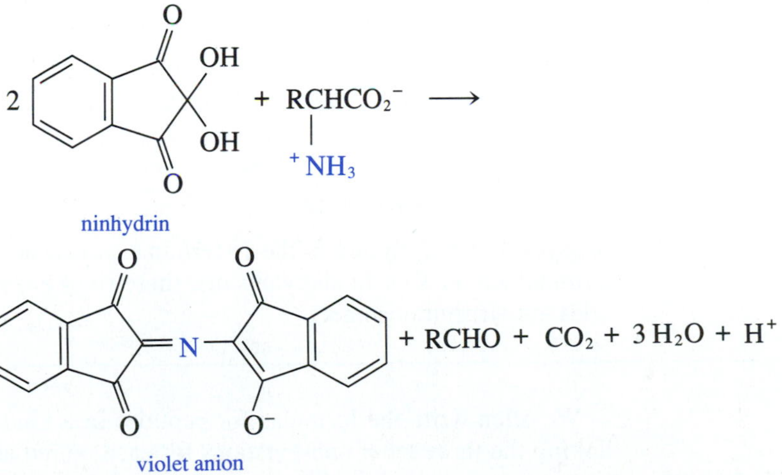

$$2\ \text{ninhydrin} + R\underset{\underset{^{+}NH_3}{|}}{CH}CO_2^- \longrightarrow \text{violet anion} + RCHO + CO_2 + 3\,H_2O + H^+ \tag{17.9}$$

ninhydrin

violet anion

Only the nitrogen atom of the violet dye comes from the amino acid. The rest of the amino acid is converted to an aldehyde and carbon dioxide. Therefore, *the same violet dye is produced from all α-amino acids with a primary amino group,* and the intensity of its color is directly proportional to the concentration of the amino acid present. Only proline, which has a secondary amino group, reacts differently to give a yellow dye, but this, too, can be used for analysis.

Problem 17.9 Write an equation for the reaction of alanine with ninhydrin.

17.8 Peptides

Amino acids are linked together in peptides and proteins by an amide bond between the carboxyl group of one amino acid and the α-amino group of another amino acid. Emil Fischer, who first proposed this structure, called this amide bond a **peptide bond.** A molecule containing only *two* amino acids (the shorthand aa is used for amino acid) joined in this way is called a **dipeptide:**

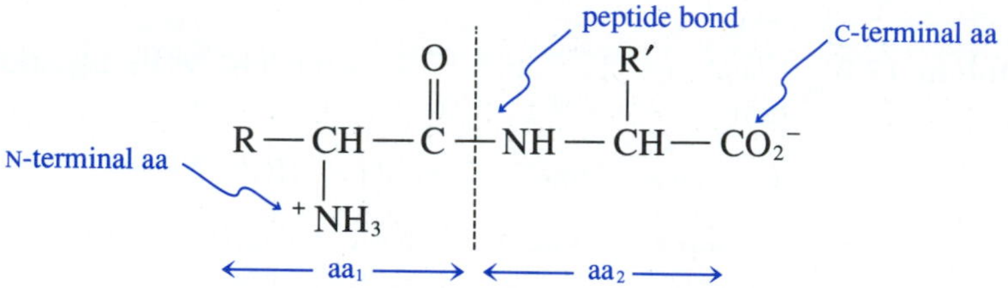

$$\underset{\text{N-terminal aa}\;\to\;{}^{+}NH_3}{R-\underset{|}{CH}}-\overset{O}{\overset{\|}{C}}\;\vdots\;NH-\overset{R'}{\overset{|}{CH}}-CO_2^-$$

peptide bond; C-terminal aa; ← aa₁ → ← aa₂ →

By convention, the peptide bond is written with the amino acid having a free $^{+}NH_3$ group at the left and the amino acid with a free CO_2^- group at the right. These amino acids are called, respectively, the **N-terminal amino acid** and the **C-terminal amino acid.**

EXAMPLE 17.5 Write the dipeptide structures that can be made by linking alanine and glycine with a peptide bond.

Solution There are two possibilities:

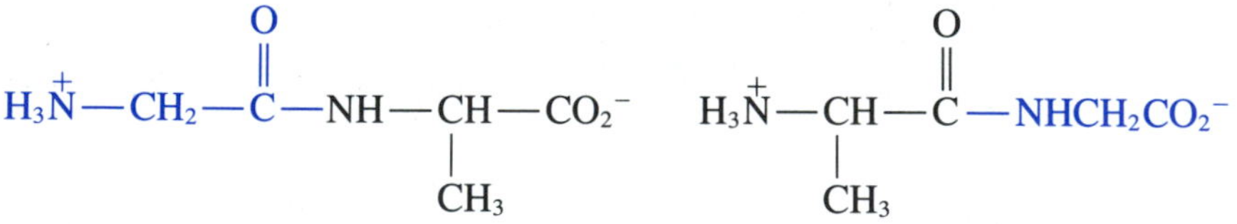

$$H_3\overset{+}{N}-CH_2-\overset{O}{\overset{\|}{C}}-NH-\underset{CH_3}{\underset{|}{CH}}-CO_2^- \qquad H_3\overset{+}{N}-\underset{CH_3}{\underset{|}{CH}}-\overset{O}{\overset{\|}{C}}-NHCH_2CO_2^-$$

glycylalanine — alanylglycine

In glycylalanine, glycine is the N-terminal amino acid, and alanine is the C-terminal amino acid. In alanylglycine, these roles are switched. The two dipeptides are structural isomers.

We often write the formulas for peptides in a kind of shorthand by simply linking the three-letter abbreviations for each amino acid, starting with the N-terminal one at the left. For example, glycylalanine is Gly—Ala, and alanylglycine is Ala—Gly.

Problem 17.10 In example 17.5 the formulas for Gly—Ala and Ala—Gly are written in their dipolar forms. At what pH do you expect these structures to predominate? Draw the expected structure of Gly—Ala in solution at pH 3; at pH 9.

Problem 17.11 Write the dipolar structural formula for

a. valylalanine. b. alanylvaline.

EXAMPLE 17.6 Consider the abbreviated formula Gly—Ala—Ser for a tripeptide. Which is the N-terminal amino acid, and which is the C-terminal amino acid?

Solution Such formulas always read from the N-terminal amino acid at the left to the C-terminal amino acid at the right. Glycine is the N-terminal amino acid, and serine is the C-terminal amino acid. Both the amino group *and* the carboxyl group of the middle amino acid, alanine, are tied up in peptide bonds.

Problem 17.12 Write out the complete structural formula for Gly—Ala—Ser.

Problem 17.13 Write out the *abbreviated* formulas for all possible tripeptide isomers of Gly—Ala—Ser.

The complexity that is possible in peptide and protein structures is truly astounding. For example, Problem 17.13 shows that there are 6 possible arrangements of 3 different amino acids in a tripeptide. For a tetrapeptide this number jumps to 24, and for an octapeptide (constructed from 8 different amino acids) there are 40,320 possible arrangements!

Now we must introduce one small additional complication before we consider the structures of particular peptides and proteins.

17.9 The Disulfide Bond

Aside from the peptide bond, the only other type of covalent bond between amino acids in peptides and proteins is the **disulfide bond.** It links two **cysteine** units. Recall that thiols are easily oxidized to disulfides (eq. 7.45). Two cysteine units can be linked by a disulfide bond.

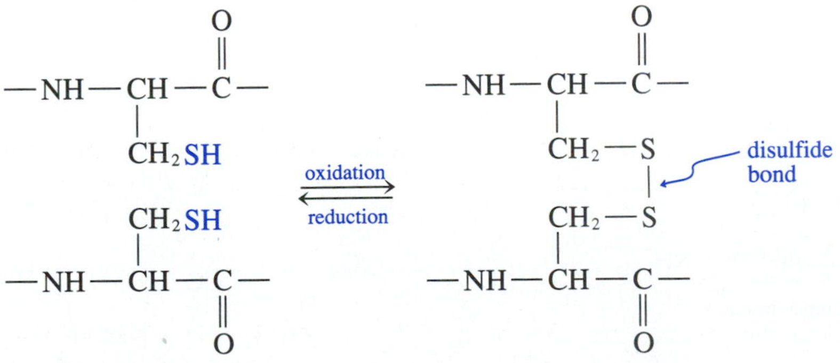

$$\begin{matrix} & O & \\ & \| & \\ —NH—CH—C— & & \\ | & & \\ CH_2SH & & \\ CH_2SH & & \\ | & & \\ —NH—CH—C— & & \\ & \| & \\ & O & \end{matrix} \underset{\text{reduction}}{\overset{\text{oxidation}}{\rightleftharpoons}} \begin{matrix} & O \\ & \| \\ —NH—CH—C— & \\ | & \\ CH_2—S & \\ \ \ \ \ \ \ \ \ | & \leftarrow \text{disulfide bond} \\ CH_2—S & \\ | & \\ —NH—CH—C— & \\ & \| \\ & O \end{matrix} \qquad (17.10)$$

two cysteine units — Cys—S—S—Cys —

If the two cysteine units are in different parts of the *same* chain of a peptide or protein, a disulfide bond between them will form a "loop," or large ring. If the two units are on different chains, the disulfide bond will cross-link the two chains. We will see examples of both arrangements. Disulfide bonds can easily be broken by mild reducing agents (see "A Word About Hair" on page 218).

A WORD ABOUT ...

32. Some Naturally Occurring Peptides

Peptides with just a few linked amino acids per molecule have been isolated from living matter, where they often perform important roles in biology. Here are a few examples.

Bradykinin is a nonapeptide present in blood plasma and involved in regulating blood pressure. Several peptides found in the brain act as chemical transmitters of nerve impulses. One of these is the decapeptide **substance P,** thought to be a transmitter of pain impulses.

Arg—Pro—Pro—Gly—Phe
|
Ser—Pro—Phe—Arg

bradykinin

Arg—Pro—Lys—Pro—Gln
|
Phe
|
Phe—Gly—Leu—Met

substance P

Oxytocin (Figure 17.4) and **vasopressin** are two *cyclic* nonapeptide hormones produced by the posterior pituitary gland. Oxytocin regulates uterine contraction and lactation and may be administered when it is necessary to induce labor at childbirth. Note that its structure includes two cysteine units joined by a disulfide bond. In addition, the C-terminal amino acid, glycine, is present as the amide. This end group is quite common in peptide chains.

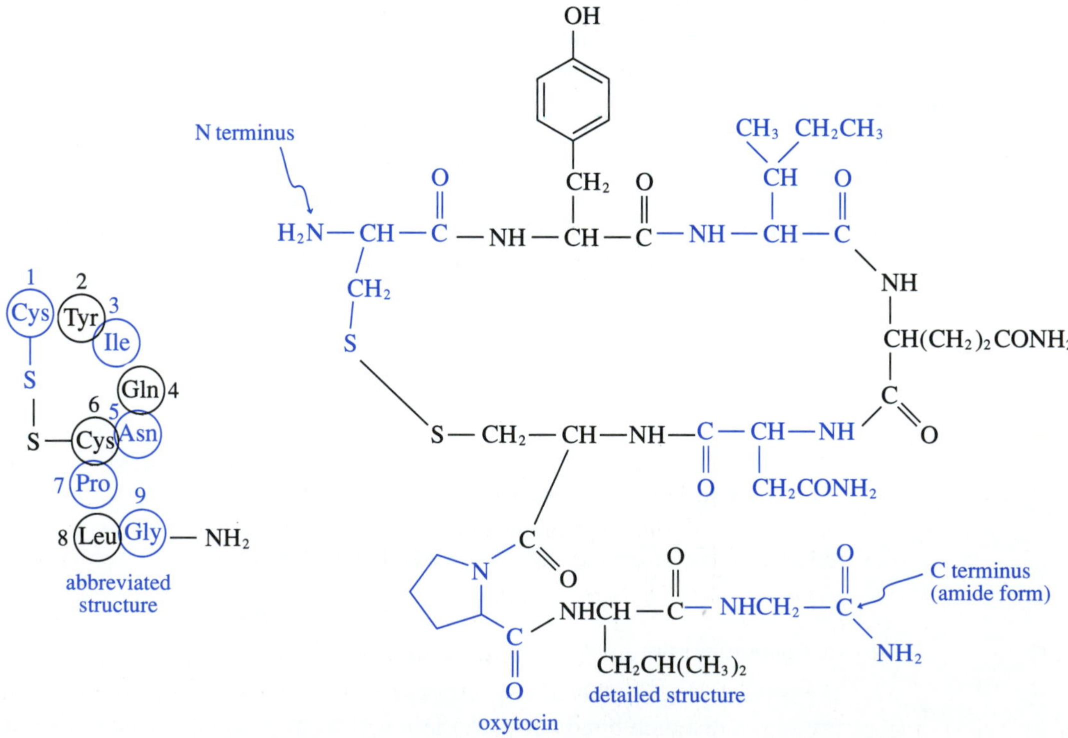

FIGURE 17.4 The nonapeptide hormone oxytocin.

Vasopressin differs from oxytocin only in the substitution of Phe for Ile and Arg for Leu. Vasopressin regulates the excretion of water by the kidneys and also affects blood pressure. The disease *diabetes insipidus,* in which too much urine is excreted, is a consequence of vasopressin deficiency and can be treated by administering this hormone.

One important area of current research is modifying the structures of natural, biologically important peptides (by replacing one amino acid by another or by altering side-chain structures) with the hope of developing new, useful drugs.

In the remainder of this chapter, we will describe the main features of peptide and protein structure. This description can be given at several levels of detail. We can say which amino acids are present and how many of each per peptide or protein molecule. Or we can give their sequence in the chain. Or we can describe more gross aspects of their structure, such as molecular shape. Are the molecules helical, spherical, or sheetlike? Do the molecules aggregate?

Four levels of description, called the primary, secondary, tertiary, and quaternary structures, are usually used. We begin with the primary structure.

17.10 The Primary Structure of Proteins

The backbone of proteins is a repeating sequence of one nitrogen and two carbon atoms.

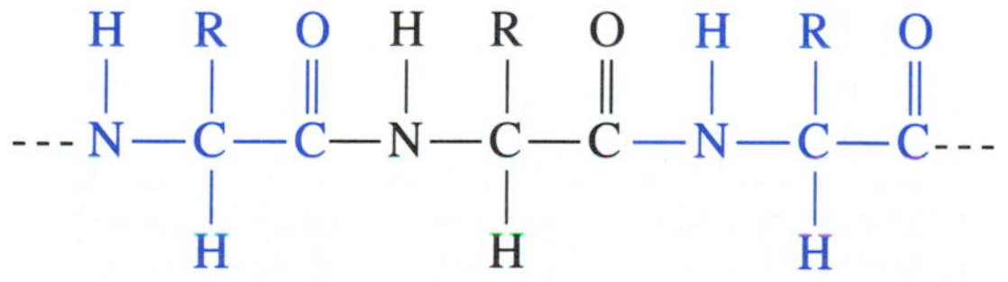

```
       H   R   O   H   R   O   H   R   O
       |   |   ‖   |   |   ‖   |   |   ‖
  --- N — C — C — N — C — C — N — C — C ---
           |           |           |
           H           H           H
```

protein chain, showing amino acids linked by amide groups

Things we must know about a peptide or protein, if we are to write down its structure, are (1) which amino acids are present and how many of each there are and (2) the sequence of the amino acids in the chain. In this section, we will briefly describe ways to obtain this kind of information.

17.10a Amino Acid Analysis Complete hydrolysis of a peptide or protein gives a mixture of amino acids. This hydrolysis is typically accomplished by heating the peptide or protein with 6*N* HCl at 110°C for 24 hours. Analysis of the resulting amino acid mixture requires a procedure that separates the amino acids from one another, identifies each amino acid present, and determines its amount.

An instrument called an **amino acid analyzer** performs these tasks automatically in the following way. The amino acid mixture from the complete hydrolysis of a few milligrams of the peptide or protein is placed at the top of a column packed with material that selectively absorbs amino acids. The packing is an insoluble resin that contains strongly acidic groups. These groups protonate the amino acids. Next, a buffer solution of known pH is pumped through the column. The amino acids pass through the column at different rates, depending on their structure and basicity, and are thus separated.

The column effluent is met by a stream of ninhydrin reagent. Therefore, the effluent is alternately violet or colorless, depending on whether or not an amino acid is being eluted from the column. The intensity of the color is automatically recorded as a function of the volume of effluent. Calibration with known amino acid mixtures allows each amino acid to be identified by the appearance time of its peak. Furthermore, the intensity of each peak gives a quantitative measure of the amount of each amino acid that is present. Figure 17.5 shows a typical plot obtained from an automatic amino acid analyzer.

17.10b Sequence Determination Frederick Sanger* devised a method for sequencing peptides based on the observation that the N-terminal amino acid differs from all others in the chain by having a free amino group. If that amino group were to react with some reagent *prior* to hydrolysis, then after hydrolysis, that amino acid would be labeled and could be identified. **Sanger's**

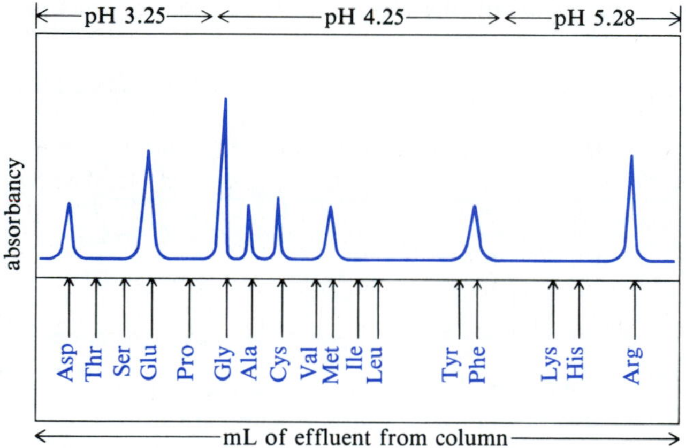

FIGURE 17.5

Different amino acids in a peptide hydrolyzate are separated on an ion-exchange resin. Buffers with different pH's elute the amino acids from the column. Each amino acid is identified by comparing the peaks with the standard elution profile shown near the bottom on the figure. The amount of each amino acid is proportional to the area under its peak.

* Frederick Sanger (Cambridge University, England) received *two* Nobel Prizes, the first in 1958 for his landmark work in amino acid sequencing and the second in 1980 for methodology in the base sequencing of RNA and DNA.

reagent is 2,4-dinitrofluorobenzene, which reacts with the NH_2 group of amino acids and peptides to give yellow 2,4-dinitrophenyl (DNP) derivatives.

$$\underset{\text{2,4-dinitrofluorobenzene}}{2,4\text{-}(O_2N)_2C_6H_3\text{—F}} + \underset{\text{N-terminal amino acid}}{H_2N\text{—}\overset{R_1}{\overset{|}{C}H}\text{—}\overset{O}{\overset{\|}{C}}\text{----}} \xrightarrow[\text{base}]{\text{mild}}$$

$$\underset{\text{DNP-peptide, labeled at the N terminus}}{2,4\text{-}(O_2N)_2C_6H_3\text{—NH—}\overset{R_1}{\overset{|}{C}H}\text{—}\overset{O}{\overset{\|}{C}}\text{----}} + F^- \qquad (17.11)$$

Hydrolysis of a peptide treated this way (eq. 17.11) would give the DNP derivative of the N-terminal amino acid; other amino acids in the chain would be unlabeled. In this way, the N-terminal amino acid could be identified.

EXAMPLE 17.7 How might alanylglycine be distinguished from glycylalanine?

Solution Both dipeptides will give one equivalent each of alanine and glycine on hydrolysis. Therefore, we cannot distinguish between them without applying a sequencing method.

Treat the dipeptide with 2,4-dinitrofluorobenzene and *then* hydrolyze. If the dipeptide is alanylglycine, we will obtain DNP-alanine and glycine; if the dipeptide is glycylalanine, we will get DNP-glycine and alanine.

Problem 17.14 Write out equations for the reactions described in Example 17.7.

Sanger used his method with great ingenuity to deduce the complete sequence of insulin, a protein hormone with 51 amino acid units. But the method suffers in that it identifies only the N-terminal amino acid.

An ideal method for sequencing a peptide or protein would have a reagent that clips off just one amino acid at a time from the end of the chain, and identifies it. Just such a method was devised by Pehr Edman (Professor at the University of Lund in Sweden), and it is now widely used.

Edman's reagent is phenyl isothiocyanate, $C_6H_5N{=}C{=}S$. The steps in selectively labeling and releasing the N-terminal amino acid are shown in Figure 17.6. In the first step, the N-terminal amino acid acts as a nucleophile toward the C=S bond of the reagent to form a thiourea derivative. In the second step, the N-terminal amino acid is removed in the form of a heterocyclic compound, a phenylthiohydantoin. The specific phenylthiohydantoin that is

FIGURE 17.6 The Edman degradation of peptides.

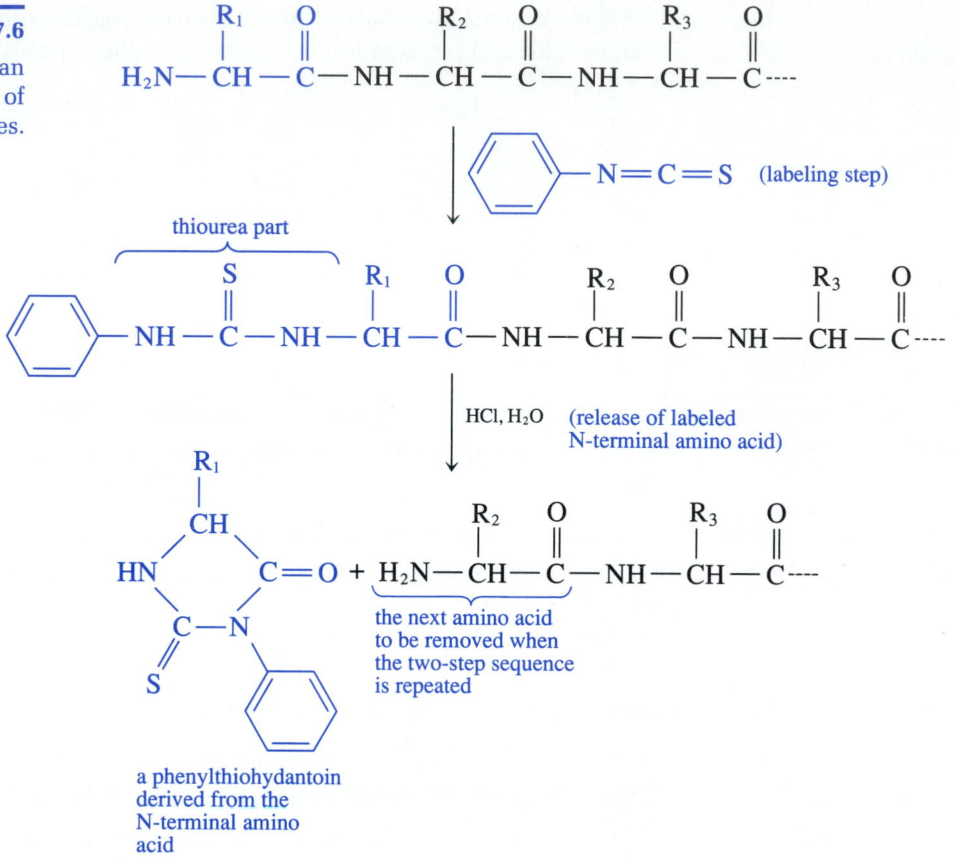

$H_2N-CH(R_1)-C(=O)-NH-CH(R_2)-C(=O)-NH-CH(R_3)-C(=O)$----

$C_6H_5-N{=}C{=}S$ (labeling step)

thiourea part

$C_6H_5-NH-C(=S)-NH-CH(R_1)-C(=O)-NH-CH(R_2)-C(=O)-NH-CH(R_3)-C(=O)$----

HCl, H_2O (release of labeled N-terminal amino acid)

a phenylthiohydantoin derived from the N-terminal amino acid + $H_2N-CH(R_2)-C(=O)-NH-CH(R_3)-C(=O)$----

the next amino acid to be removed when the two-step sequence is repeated

formed can be identified by comparison with reference compounds prepared from the known amino acids. Then the two steps are repeated, to identify the next amino acid, and so on. The method has been automated, so currently amino acid "sequenators" can easily determine, in a day, the sequence of the first 20 or so amino acids in a peptide, starting at the N-terminal end. But the Edman method cannot be used indefinitely, due to the gradual buildup of impurities. It is most effective with peptides containing up to 20 to 25 amino acid units.

17.10c Cleavage of Selected Peptide Bonds If a protein contains several hundred amino acid units, it is best to first partially hydrolyze the chain to smaller fragments that can be separated and subsequently sequenced by the Edman method. Certain chemicals or enzymes are used to cleave proteins at *particular* peptide bonds. For example, the enzyme *trypsin* (an intestinal digestive enzyme) specifically hydrolyzes polypeptides only at the carboxy end of arginine and lysine. A few of the many reagents of this type are listed in Table 17.3.

TABLE 17.3 Reagents for specific cleavage of polypeptides

Reagent	*Cleavage site*
trypsin	carboxyl side of Lys, Arg
chymotrypsin	carboxyl side of Phe, Tyr, Trp
cyanogen bromide (CNBr)	carboxyl side of Met
carboxypeptidase	a C-terminal amino acid

EXAMPLE 17.8 Consider the following peptide:

Ala—Gly—Tyr—Trp—Ser—Lys—Gly—Leu—Met—Gly

Determine what fragments will be obtained when this peptide is hydrolyzed with

a. trypsin. b. chymotrypsin. c. cyanogen bromide.

Solution a. The enzyme trypsin will split the peptide on the carboxyl side of lysine, to give

Ala—Gly—Tyr—Trp—Ser—Lys and Gly—Leu—Met—Gly

b. The enzyme chymotrypsin will split the peptide on the carboxyl sides of tyrosine and tryptophan, to give

Ala—Gly—Tyr and Trp and Ser—Lys—Gly—Leu—Met—Gly

c. Cyanogen bromide will split the peptide on the carboxyl side of methionine, thus splitting off the C-terminal glycine and leaving the rest of the peptide untouched. (Carboxypeptidase would do the same thing, confirming that the C-terminal amino acid is glycine.)

Problem 17.15 Determine what fragments will be obtained if bradykinin (its abbreviated formula is given on page 468) is hydrolyzed enzymatically with

a. trypsin. b. chymotrypsin.

During the past 15 years, the methods discussed here have been improved and expanded; the separation and sequencing of peptides and proteins can now be accomplished even if only minute amounts are available.

17.11 The Logic of Sequence Determination

A specific example will illustrate the reasoning that is used to fully determine the amino acid sequence in a particular peptide with 30 amino acid units.

First we hydrolyze the peptide completely, subject it to amino acid analysis, and find that it has the formula

$Ala_2ArgAsnCys_2GlnGlu_2Gly_3His_2Leu_4LysPhe_3ProSerThrTyr_2Val_3$

Using the Sanger method, we find that the N-terminal amino acid is Phe.

Since the chain is too long to degrade completely by the Edman method, we decide to simplify the problem by digesting the peptide with chymotrypsin. (We select chymotrypsin because the peptide contains three Phe's and two Tyr's and will undoubtedly be cleaved by that reagent.) When we carry out this cleavage, we get three fragment peptides. In addition, we get two equivalents of Phe and one of Tyr. After separation, we subject the three fragment peptides to Edman degradation and obtain their structures.

A. Leu—Val—Cys—Gly—Glu—Arg—Gly—Phe

B. Val—Asn—Gln—His—Leu—Cys—Gly—Ser—His—Leu—Val—Glu—Ala—Leu—Tyr

27 28 29 30
C. Thr—Pro—Lys—Ala

We still cannot write a unique structure for the intact peptide, but we can say that the C-terminal amino acid must be Ala and that the last four amino acids must be in the sequence shown for fragment C. We deduce this because we know that Ala is *not* cleaved at its carboxyl end by chymotrypsin, yet it appears at the C-terminal end of one of the fragments. (Note that the C-terminal amino acids in fragments A and B are Phe and Tyr, both cleaved at the carboxyl ends by chymotrypsin.) That the C-terminal amino acid is Ala can be confirmed using carboxypeptidase. We can number the amino acids in fragment C as 27 through 30 in the chain.

What to do next? Cyanogen bromide is no help, because the peptide does not contain Met. But the peptide does contain Lys and Arg, so we go back to the beginning and digest the intact peptide with trypsin, which cleaves peptides on the carboxy side of these amino acid units. We obtain (not surprisingly) some Ala (the C-terminal amino acid) because it comes right after a Lys. We also obtain two peptides. One of them is relatively short, so we determine its sequence by the Edman method and find it to be

23 24 25 26 27 28 29
D. Gly—Phe—Phe—Tyr—Thr—Pro—Lys

Because the last three amino acids in fragment D *overlap* with 27, 28, and 29 of fragment C, we can number the rest of the chain, back to 23. We now note that amino acids 23 and 24 appear at the end of fragment A, so originally A must have been connected to C. The only place left for fragment B is in front of A. This leaves only one of the Phe's unaccounted for, and it must occupy the N-terminal position (recall the Sanger result). We can now write out the complete sequence!

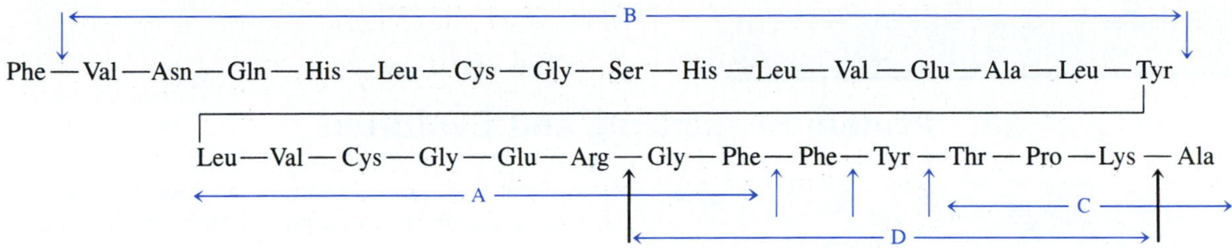

The colored vertical arrows show the cleavage points with chymotrypsin, and the black ones show the cleavage points with trypsin.

The peptide just used for illustration is the B chain of the protein hormone **insulin,** whose structure was first determined by Sanger and is shown schematically in Figure 17.7. Insulin consists of an A chain with 21 amino acid units and a B chain with 30 amino acid units. The two chains are joined by two disulfide bonds, and the A chain also contains a small disulfide loop.

Problem 17.16 How could the A and B chains of insulin be separated chemically?

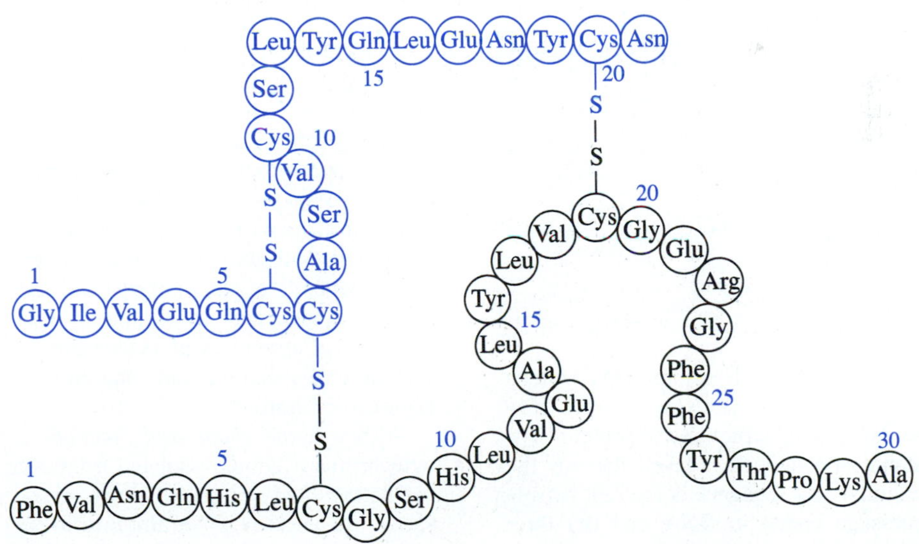

FIGURE 17.7 Primary structure of beef insulin. The A chain is in color, and the B chain, whose structure determination is described in the text, is shown in black.

A WORD ABOUT ...

33. Protein Sequencing and Evolution

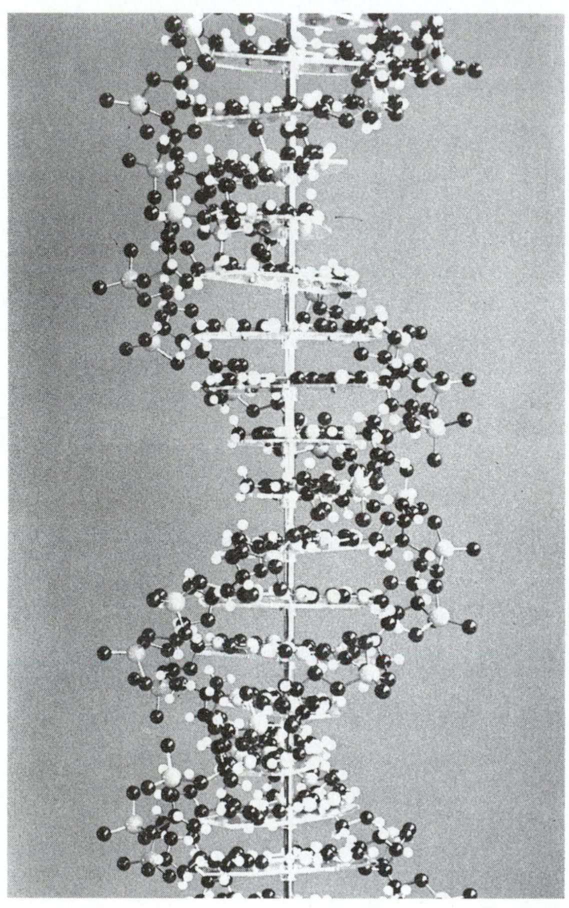

There are many reasons why it is important to determine the sequences of amino acids in proteins. First, we must know the detailed structures of proteins if we are to understand, at a molecular level, the way they function. The amino acid sequence is the link between the genetic message coded in DNA and the three-dimensional protein structure that forms the basis for biological function.

There are medical benefits to knowing amino acid sequences. Certain genetic diseases, such as sickle-cell anemia, can result from the change of a single amino acid unit in a protein. In this case, it is replacement of glutamic acid at position 6 in the β chain of hemoglobin by a valine unit. Thus, sequence determination is an important part of medical pathology. One future possible application of genetic engineering is devising ways to correct these amino acid sequence errors.

Another important reason for determining amino acid sequences is that they provide a chemical tool for studying our evolutionary history. Proteins resemble one another in amino acid sequence if they have a common evolutionary ancestry. Let's look at a specific example.

Cytochrome *c*, an enzyme important in the respiration of most plants and animals, is an electron-transport protein with 104 amino acid residues. It is involved in oxidation-reduction processes. In these reactions, cytochrome *c* must react with and transfer an electron from one enzyme complex (cytochrome reductase) to another (cytochrome oxidase).

Cytochrome *c* probably evolved more than 1.5 billion years ago, even before the evolutionary divergence of plants and animals. The function of this protein has been preserved all that time! We know this because the cytochrome *c* isolated from any eucaryotic microorganism (one that contains a cell nucleus) reacts in vitro with the cytochrome oxidase of every other species tested so far. For example, wheat germ cytochrome *c*, a plant-derived enzyme, reacts with human cytochrome *c* oxidase. Also, the three-dimensional structures of cytochrome *c* isolated from such diverse species as tuna and photosynthetic bacteria are very similar.

Although *the shape and functions are similar* for cytochrome *c* samples isolated from different sources, *the amino acid sequence varies somewhat* from one species to another. The amino acid sequence of cytochrome *c* isolated from humans differs from that of monkeys in *just 1* (out of 104) amino acid residues! On the other hand, cytochrome *c* from dogs, a more distant evolutionary relative of humans, differs from the human protein by 11 amino acid residues.

17.2 Peptide Synthesis

Once we know the amino acid sequence in a peptide or protein, we are in a position to synthesize it from its amino acid components. Why would we want to do this? There are several reasons. For example, we might wish to verify a particular peptide structure by comparing the properties of the synthetic and natural substances. Or we might wish to study the effect of substituting one amino acid for another on the biological properties of a peptide or protein. Such modified proteins could be valuable in treating disease or in understanding how the protein functions.

Many methods have been developed to link amino acids in a controlled manner. They require careful strategy. Amino acids are bifunctional. To link the carboxyl group of one amino acid to the amino group of a second amino acid, we must first prepare each compound by protecting the amino group of the first and the carboxyl group of the second.

$$\underset{aa_1}{H_2N-\overset{\overset{R_1}{|}}{C}H-CO_2H} \xrightarrow[\text{amino group}]{\text{protect the}} \boxed{P_1}-NH-\overset{\overset{R_1}{|}}{C}H-CO_2H \tag{17.12}$$

$$\underset{aa_2}{H_2N-\overset{\overset{R_2}{|}}{C}H-CO_2H} \xrightarrow[\text{carboxyl group}]{\text{protect the}} H_2N-\overset{\overset{R_2}{|}}{C}H-\overset{\overset{O}{\|}}{C}-\boxed{P_2} \tag{17.13}$$

In this way, we can control the linking of the two amino acids so that the carboxyl group of aa_1 combines with the amino group of aa_2.

$$\boxed{P_1}-NH\overset{\overset{R_1}{|}}{C}HCO_2H + H_2N-\overset{\overset{R_2}{|}}{C}H-\overset{\overset{O}{\|}}{C}-\boxed{P_2} \xrightarrow{-H_2O}$$

$$\underset{\text{doubly protected dipeptide}}{\boxed{P_1}-NH\overset{\overset{R_1}{|}}{C}H-\underbrace{\boxed{\overset{\overset{O}{\|}}{C}-NH}}_{\text{peptide bond}}-\overset{\overset{R_2}{|}}{C}H-\overset{\overset{O}{\|}}{C}-\boxed{P_2}} \tag{17.14}$$

Later we will give specific examples of protecting groups and of a reagent that can be used to form the peptide bond.

EXAMPLE 17.9 What would happen if we tried to combine aa_1 with aa_2 without using protecting groups?

Solution Since each amino acid could react either as an amine or as a carboxylic acid, we could get not only $aa_1—aa_2$ but also $aa_2—aa_1$, $aa_1—aa_1$, and $aa_2—aa_2$. Furthermore, since the resulting dipeptides would still have a free amino and a free carboxyl group, we could also get trimers, tetramers, and so on. In other words, a mess.

After the peptide bond is formed, we must be able to *remove the protecting groups under conditions that do not hydrolyze the peptide bond*. Or, if more amino acids are to be added to the chain, we must be able to *selectively* remove *one* of the two protecting groups from the doubly protected dipeptide before joining the next amino acid to it. All of this can be quite a tricky and tedious process. Yet these methods were used by Vincent du Vigneaud* and his colleagues to synthesize oxytocin and vasopressin (page 468), the first naturally occurring polypeptides to be synthesized in the laboratory.

In 1965, R. Bruce Merrifield** developed a technique that revolutionized peptide synthesis. This **solid-phase technique** avoids many of the tedious aspects of previous methods and is now universally used. The principle is to *assemble the peptide chain while one end of it is chemically anchored to an insoluble inert solid.* In this way, excess reagents and by-products can be removed simply by washing and filtering the solid. The growing peptide chain does not need to be purified at any intermediate stage. When the peptide is fully constructed, it is cleaved chemically from the solid support.

Typically, the solid phase is a cross-linked polystyrene in which some (usually 1 to 10%) of the aromatic rings contain chloromethyl ($ClCH_2—$) groups.

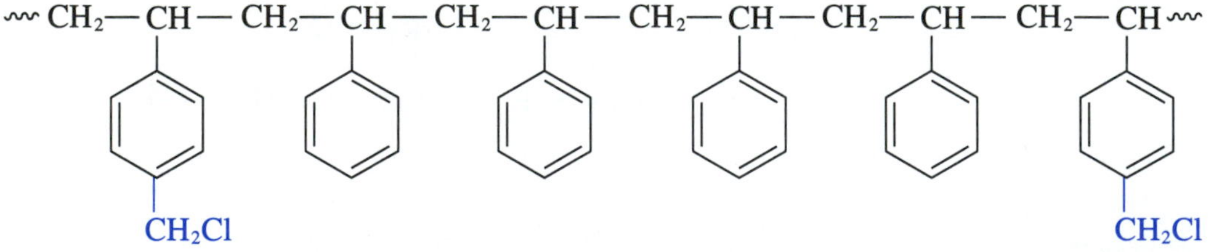

~ CH_2—CH—CH_2—CH—CH_2—CH—CH_2—CH—CH_2—CH—CH_2—CH ~

CH_2Cl CH_2Cl

The polymer behaves chemically like benzyl chloride, an alkyl halide that is quite reactive in nucleophilic substitution reactions (S_N2).

* Vincent du Vigneaud (Cornell University) was awarded the 1955 Nobel Prize in chemistry for this achievement.

R. Bruce Merrifield (Rockefeller University) received the 1984 Nobel Prize in chemistry for his contribution, which not only revolutionized peptide synthesis but also affected many other areas of chemistry through the use of polymer-bound reagents. For an account of the history of this discovery, see the article by Merrifield in *Science* **1986, *232,* 341, and for a personal history, see *Chemistry in Britain* **1987,** 816.

The steps in a Merrifield synthesis are summarized for a dipeptide in Figure 17.8. In step 1, the polymer is first treated with an N-protected amino acid. The carboxylate ion acts as an oxygen nucleophile and displaces the chloride ion from the polymer, thus forming an ester link. *The first amino acid attached to the polymer will eventually become the C-terminal amino acid of the synthetic peptide.*

Many protecting groups are known, but in solid-phase peptide synthesis, the most frequently used N-protecting group is the ***t*-butoxycarbonyl (Boc)** group. The amino acid is protected by reaction with di-*t*-butyl dicarbonate.

$$[(CH_3)_3CO{-}C({=}O)]_2O + H_3\overset{+}{N}{-}CH(R){-}CO_2^- \xrightarrow{\text{base}} \underbrace{(CH_3)_3CO{-}\overset{O}{\overset{\|}{C}}}_{\boxed{P}}{-}NH{-}CH(R){-}CO_2H \qquad (17.15)$$

di-*t*-butyl dicarbonate

Problem 17.17 Write out the steps in the mechanism of eq. 17.15 (a nucleophilic acyl substitution reaction). Why is the base necessary?

After the amino acid is attached to the polymer, the protecting group is removed. This is accomplished (step 2) by reaction with acid under mild conditions.

$$CH_3{-}C(CH_3)(CH_2{-}H){-}O{-}\overset{O}{\overset{\|}{C}}{-}NHCH(R){-}\overset{O}{\overset{\|}{C}}{-}O{-}CH_2{-}C_6H_4{-}\text{polymer} \xrightarrow[CH_2Cl_2]{CF_3CO_2H}$$

$$CH_3{-}C(=CH_2){-}CH_3 + CO_2 + H_2N{-}CH(R_1){-}\overset{O}{\overset{\|}{C}}{-}O{-}CH_2{-}C_6H_4{-}\text{polymer} \qquad (17.16)$$

2-methylpropene; deprotected amino group; ester group still intact

Two products of deprotection are gaseous (2-methylpropene and carbon dioxide) and are thus easily removed from the reaction mixture. The ester group that links the first amino acid to the polymer is *not* hydrolyzed under these conditions.

In step 3, the next N-protected amino acid is linked to the first one. This is accomplished with the aid of **dicyclohexylcarbodiimide (DCC).** DCC is able

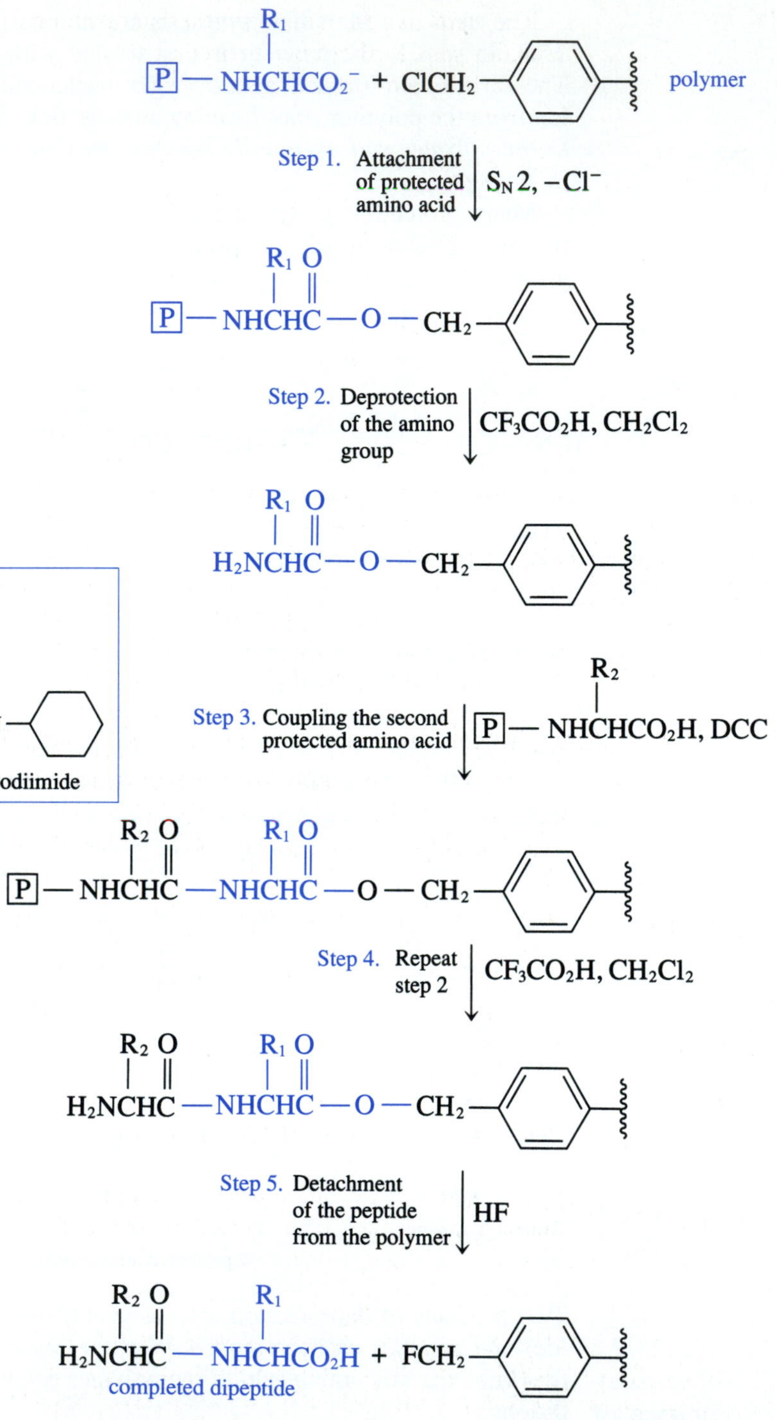

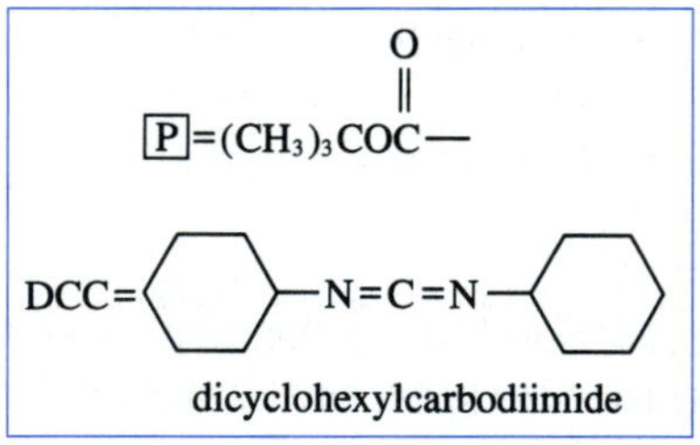

FIGURE 17.8 Solid-phase synthesis of a dipeptide.

to link carboxyl and amino groups in a peptide bond; in the process, the DCC is hydrolyzed to dicyclohexylurea.

$$\sim\overset{\overset{\displaystyle O}{\|}}{C}-OH + H_2N\sim + C_6H_{11}-N{=}C{=}N-C_6H_{11} \longrightarrow$$

dicyclohexylcarbodiimide (DCC)

$$\sim\overset{\overset{\displaystyle O}{\|}}{C}-NH\sim + C_6H_{11}-NH-\overset{\overset{\displaystyle O}{\|}}{C}-NH-C_6H_{11} \quad (17.17)$$

peptide bond — dicyclohexylurea

Steps 2 and 3 may be repeated to add a third amino acid, a fourth, and so on. Finally, when the desired amino acids have been connected in the proper sequence and the N-terminal amino group has been deprotected (step 4 in Figure 17.8), the complete polypeptide chain is detached from the polymer. This can be accomplished by treatment with anhydrous hydrogen fluoride, which cleaves the benzyl ester without hydrolyzing the amide bonds in the polypeptide (step 5 in Figure 17.8).

All operations in solid-phase peptide synthesis have been automated. The reactions occur in a single reaction vessel, with reagents and wash solvents automatically added from reservoirs by mechanical pumps. Working around the clock, the programmer can incorporate eight or more amino acids into a polypeptide in a day. Merrifield synthesized the nonapeptide bradykinin (page 468) in just 27 hours using this technique. And, in 1969, he used the automated synthesizer to prepare the enzyme ribonuclease (124 amino acid residues), the first enzyme to be prepared synthetically from its amino acid components. The synthesis, which required 369 chemical reactions and 11,391 steps, was completed in only six weeks. Automated peptide synthesis, though still not without occasional problems, is now a fairly routine matter.

Problem 17.18 Write all the equations for the synthesis of Gly—Ala—Phe using the Merrifield solid-phase technology.

We have seen how the primary structure of peptides and proteins can be determined and how peptides can be synthesized in the laboratory. Now let us examine some further details of protein structure.

17.13 *Secondary Structure of Proteins*

Because proteins consist of long chains of amino acids strung together, one might think that their shapes are rather amorphous, or "floppy" and ill-defined. This is incorrect. Many proteins have been isolated in pure crystalline form and are polymers with very well defined shapes. Indeed, even in solution, the shapes seem to be quite regular. Let us examine some of the structural features of peptide chains that are responsible for their definite shapes.

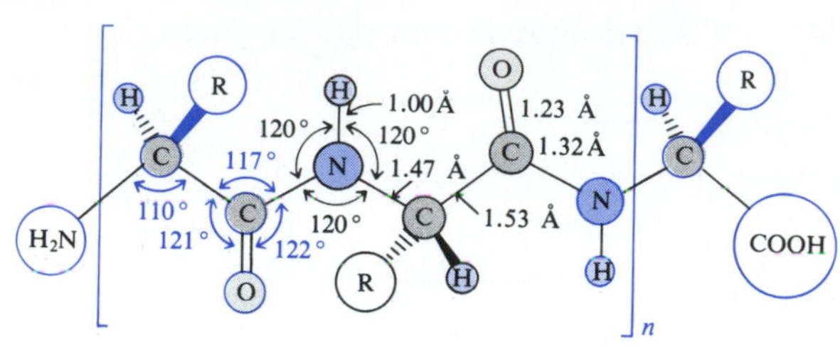

FIGURE 17.9 The characteristic bond angles and bond lengths in peptide bonds.

17.13a Geometry of the Peptide Bond We pointed out earlier that simple amides have a planar geometry, that the amide C—N bond is shorter than usual, and that rotation around that bond is restricted (Sec. 10.21). Bond planarity and restricted rotation, which are consequences of resonance, are also important in peptide bonds.

X-ray studies of crystalline peptides by Linus Pauling and his colleagues determined the precise geometry of peptide bonds. The characteristic dimensions, which are common to all peptides and proteins, are shown in Figure 17.9.

Things to notice about peptide geometry are as follows: (1) the amide group is flat; the carbonyl carbon, the nitrogen, and the four atoms connected to them all lie in a single plane. (2) The short amide C—N distance (1.32 Å, compared with 1.47 Å for the other C—N bond) and the 120° bond angles around that nitrogen show that it is essentially sp^2 hybridized and that the bond between it and the carbonyl carbon is like a double bond. (3) Although each amide group is planar, two adjacent amide groups need not be coplanar because of rotation about the other single bonds in the chain; that is, rotation can occur around the two single bonds to the —CHR— group.

The rather rigid geometry and restricted rotation of the peptide bond help to impart a definite shape to proteins.

17.13b Hydrogen Bonding We pointed out earlier that amides readily form intermolecular hydrogen bonds between the carbonyl group and the N—H group, bonds of the type C=O ··· H—N. Such bonds are present and important in peptide chains. The chain may coil in such a way that the N—H of one peptide bond can hydrogen-bond with a carbonyl group of another peptide bond farther down the *same* chain, thus rigidifying the coiled structure. Alternatively, carbonyl groups and N—H groups on *different* peptide chains may hydrogen-bond, linking the two chains. Although a single hydrogen bond is relatively weak (perhaps only 5 kcal/mol of energy), the possibility of forming multiple intrachain or interchain hydrogen bonds makes this a very important factor in protein structure, as we will now see.

17.13c The α Helix and the Pleated Sheet X-ray studies of α-keratin, a structural protein present in hair, wool, horns, and nails, showed that some

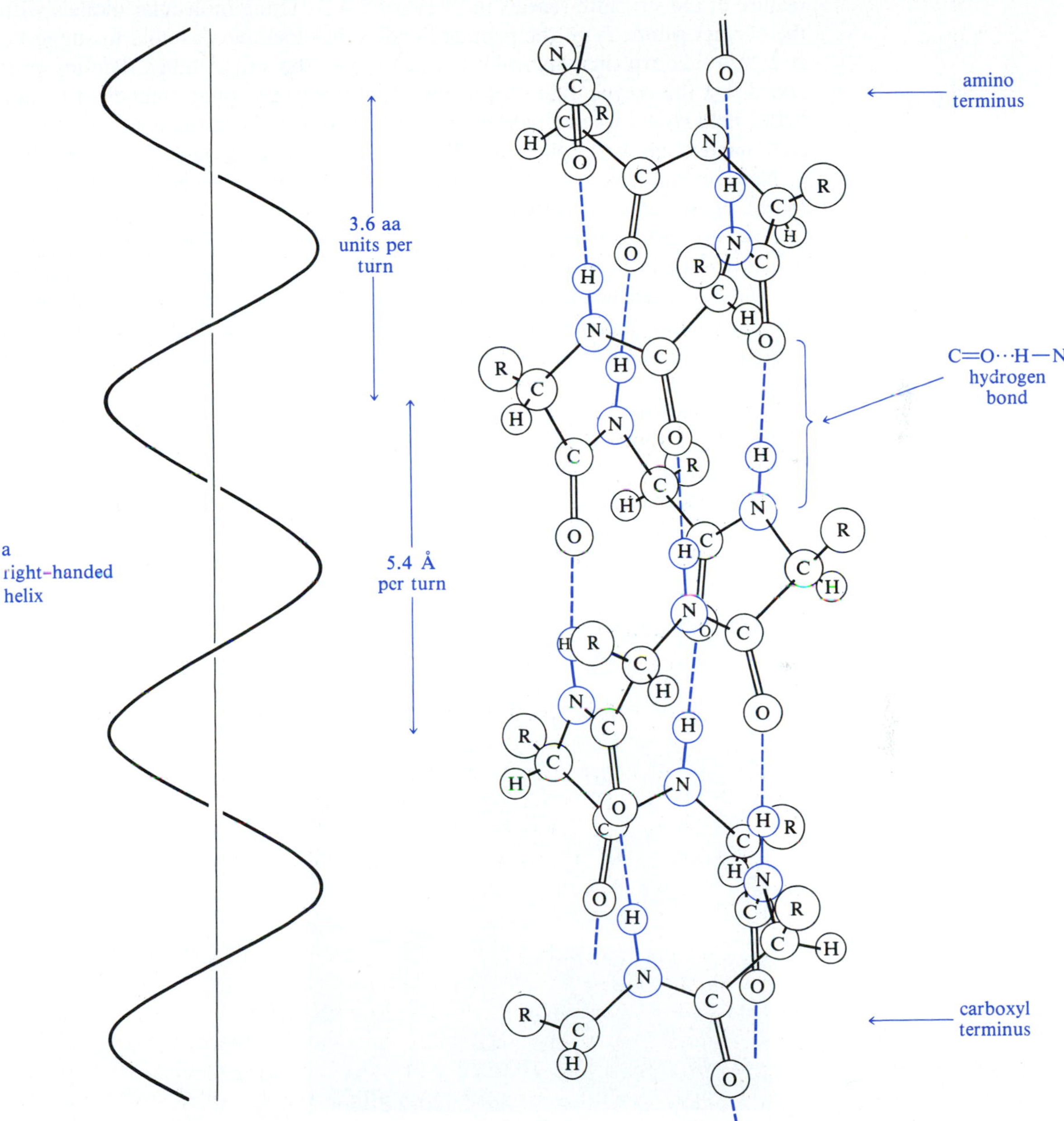

FIGURE 17.10 Segment of an α helix, showing three turns of the helix, with 3.6 amino acid units per turn. Hydrogen bonds are shown as dashed colored lines.

feature of the structure repeats itself every 5.4 Å. Using molecular models with the correct geometry of the peptide bond, Linus Pauling was able to suggest a structure that explains this and other features of the x-ray studies. Pauling proposed that the polypeptide chain coils about itself in a spiral manner to form a helix, held rigid by intrachain hydrogen bonds. The **α helix,** as it is called, is right-handed and has a pitch of 5.4 Å, or 3.6 amino acid units (Figure 17.10).

Note several features of the α helix. Proceeding from the N terminus (at the top of the structure as drawn in the figure), each carbonyl group points ahead or down toward the C terminus and is hydrogen-bonded to an N—H bond farther down the chain. The N—H bonds all point back toward the N terminus. All the hydrogen bonds are roughly aligned with the long axis of the helix. The very large number of hydrogen bonds (one for each amino acid unit) strengthen the helical structure. The R groups of the individual amino acid units are directed *outward* and do not disrupt the central core of the helix. It turns out that the α helix is a natural pattern into which many proteins fold. Figure 17.11 shows Professor Pauling seated by a scale model of the α helix.

The structural protein β-keratin, obtained from silk fibroin, shows a different repeating pattern (7 Å) in its x-ray structure. To explain the data, Pauling suggested a **pleated-sheet** arrangement of the peptide chain (Figure 17.12). In the pleated sheet, peptide chains lie side by side and are held together by *inter-*

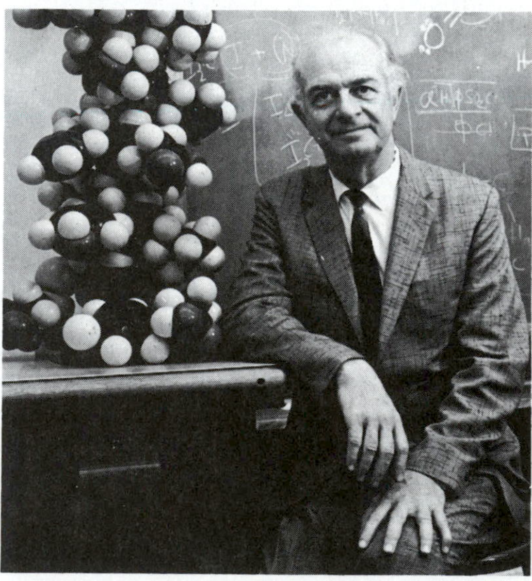

FIGURE 17.11 Linus Pauling (California Institute of Technology and Stanford University), who has made many contributions to our knowledge of organic structures. He did fundamental work on the theory of resonance, on the measurement of bond lengths and energies, and on the structure of proteins and the mechanism of antibody action. A truly exceptional person with wide-ranging interests, he received the Nobel Prize in chemistry in 1954 and the Nobel Peace Prize in 1962.

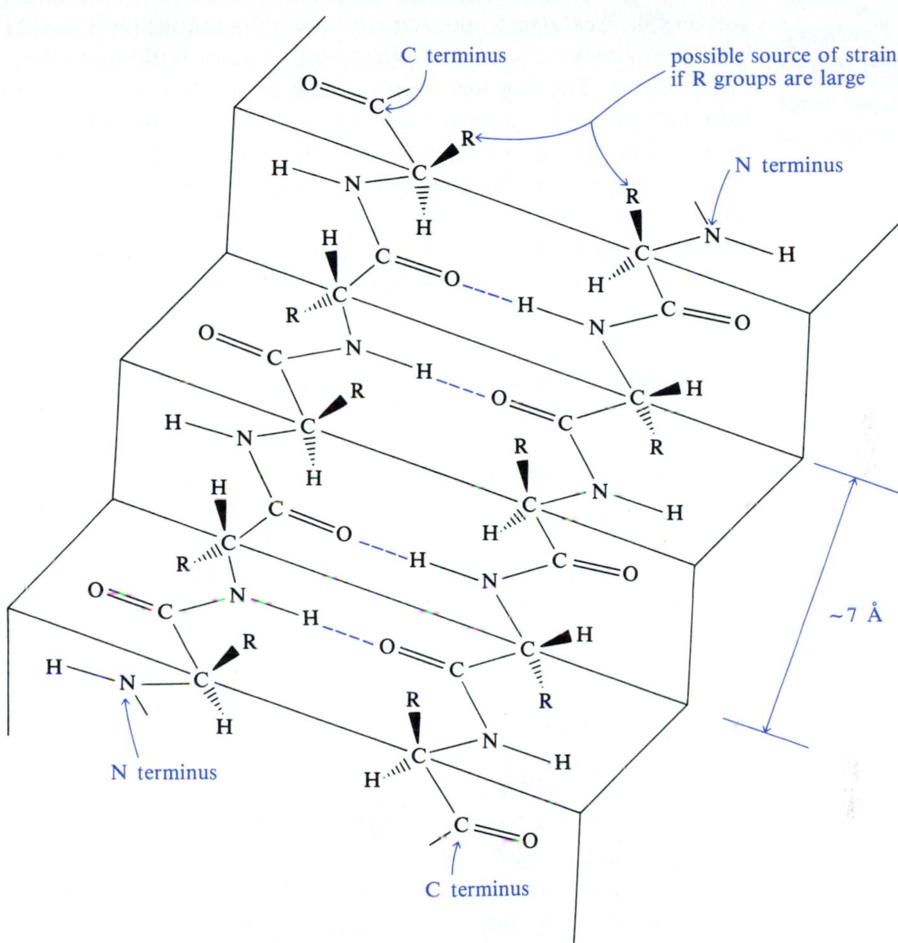

FIGURE 17.12 A segment of the pleated sheet structure of β-keratin. Adjacent chains run in opposite directions and are held together by hydrogen bonds (shown in color). R groups project above or below the mean plane of the sheet.

chain hydrogen bonds. Adjacent chains run in opposite directions. The repeating unit in each chain, which is stretched out compared with the α helix, is about 7 Å. In the pleated-sheet structure, the R groups of amino acid units in any one chain alternate above and below the mean plane of the sheet. If the R groups are large, there will be appreciable steric repulsion between them on adjacent chains. For this reason, the pleated-sheet structure is important *only* in proteins that have a high percentage of amino acid units with *small* R groups. In the β-keratin of silk fibroin, for example, 36% of the amino acid units are glycine (R = H) and another 22% are alanine (R = CH_3). Because this type of repulsion between R groups is not encountered in the α helix, the α helix is by far the more common structure of the two.

17.14 Tertiary Structure: Fibrous and Globular Proteins

We may well ask how materials as rigid as horses' hoofs, as springy as hair, as soft as silk, as slippery and shapeless as egg white, as inert as cartilage, and as reactive as enzymes can all be made of the same building blocks: amino acids and proteins. The key lies mainly in the amino acid makeup itself. So far we have focused on the protein backbone and its shape. But what about the diverse R groups of the various amino acids? How do they affect protein structure?

Some amino acids have nonpolar R groups, simple alkyl or aromatic groups. Others have highly polar R groups, with carboxylate or ammonium ions and hydroxyl or other polar groups. Still others have flat, rigid aromatic rings that may interact in specific ways. *Different R groups affect the gross properties of a protein.*

Problem 17.19 Which amino acids in Table 17.1 have nonpolar R groups? highly polar groups? relatively flat R groups?

Proteins generally fall into one of two main classes: **fibrous** or **globular.** *Fibrous proteins* are animal structural materials and hence are water-insoluble. They fall into three general categories: the **keratins,** which make up protective tissue, such as skin, hair, feathers, claws, and nails; the **collagens,** which form connective tissue, such as cartilage, tendons, and blood vessels; and the **silks,** such as the fibroin of spider webs and cocoons.

Keratins and collagens have helical structures, whereas silks have pleated-sheet structures. A large fraction of the R groups attached to these frameworks are nonpolar, accounting for the insolubility of these proteins in water. In hair, three α helices are braided to form a rope, the helices being held together by disulfide cross-links. The ropes are further packed side by side in bundles that ultimately form the hair fiber. The α-keratin of more rigid structures, such as nails and claws, is similar to that of hair, except that there is a higher percentage of cysteine amino acid units in the polypeptide chain. Therefore, there are more disulfide cross-links, giving a firmer, less flexible overall structure.

To summarize, nonpolar R groups and disulfide cross-links, together with helical or sheetlike backbones, tend to give fibrous proteins their rather rigid, insoluble structures.

Globular proteins are very different from fibrous proteins. They tend to be water-soluble and have roughly spherical shapes, as their name suggests. Instead of being structural, globular proteins perform various other biological functions. They may be **enzymes** (biological catalysts), **hormones** (chemical messengers that regulate biological processes), **transport proteins** (carriers of small molecules from one part of the body to another, such as hemoglobin, which transports oxygen in the blood), or **storage proteins** (which act as food stores; ovalbumin of egg white is an example).

Globular proteins have more amino acids with polar or ionic side chains than the water-insoluble fibrous proteins. An enzyme or other globular protein that carries out its function mainly in the aqueous medium of the cell will adopt a structure in which the nonpolar, hydrophobic R groups point in toward the center and the polar or ionic R groups point out toward the water.

Globular proteins are mainly helical, but they have folds that permit the overall shape to be globular. One of the 20 amino acids, proline, has a secondary amino group. Wherever a proline unit occurs in the primary peptide structure, there will be no N—H group available for intrachain hydrogen bonding.

---- NH—CH(R)—C(=O)—N(proline ring)—CH—C(=O)—NH—CH(R)—C(=O) ----

no H for hydrogen bonding

proline unit

Proline units tend therefore to disrupt an α helix, and we frequently find proline units at "turns" in a protein structure.

Myoglobin, the oxygen-transport protein of muscle, is a good example of a globular protein (Figure 17.13). It contains 153 amino acid units, yet is extremely compact, with very little empty space in its interior. Approximately 75% of the amino acid units in myoglobin are part of eight right-handed α-helical sections. There are four proline units, and each occurs at or near "turns" in the structure. There are also three other "turns" caused by structural features of other R groups. The interior of myoglobin consists almost entirely of nonpolar R groups, such as those of leucine, valine, phenylalanine, and methionine. The only interior polar groups are two histidines. These perform a

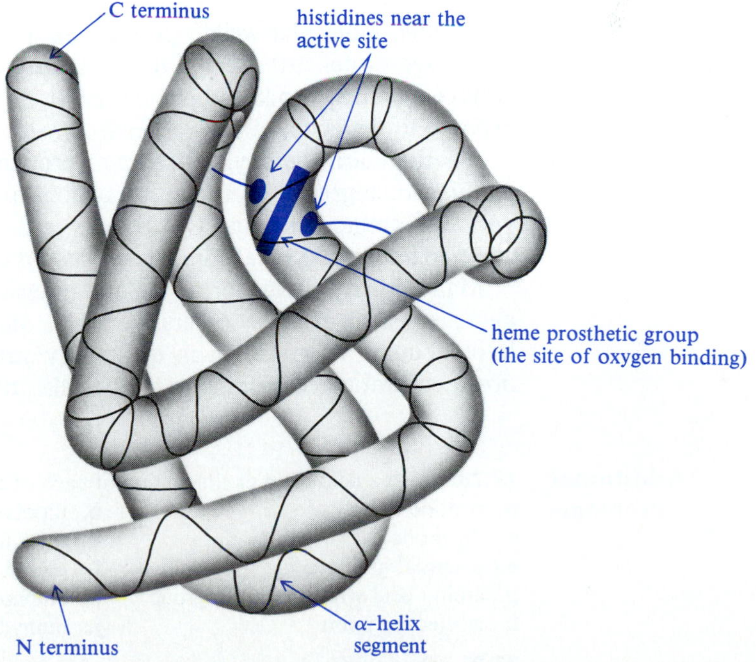

FIGURE 17.13

Schematic drawing of myoglobin. Each of the tubular sections is a segment of α helix, but the overall shape is globular.

FIGURE 17.14
Schematic drawing of the four hemoglobin subunits.

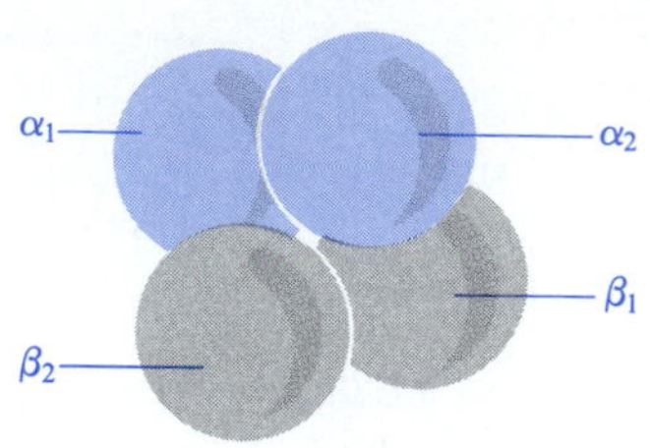

necessary function at the *active site* of the protein, where the nonprotein portion, a molecule of the porphyrin *heme* (page 373) binds the oxygen. The outer surface of the protein includes many highly polar amino acids residues (lysine, arginine, glutamic acid, and so on).

To summarize this section, we see that the particular amino acid content of a peptide or protein influences its shape. These interactions are mainly a consequence of disulfide bonds and of the polarity or nonpolarity of the R groups, their shape, and their ability to form hydrogen bonds. When we refer to the **tertiary structure** of a protein, we refer to all the contributions of these factors to its three-dimensional structure.

17.15 *Quaternary Protein Structure*

Some high-molecular-weight proteins exist as aggregates of several subunits. These aggregates are referred to as the **quaternary structure** of the protein. Aggregation helps to keep nonpolar portions of the protein surface from being exposed to the aqueous cellular environment.

Hemoglobin, the oxygen-transport protein of red cells, provides an example of such aggregation. It consists of four almost spherical units, two α units with 141 amino acids and two β units with 146 amino acids. The four units come together in a tetrahedral array, shown in Figure 17.14.

Many other proteins form similar aggregates. Some are active only in their aggregate state, whereas others are active only when the aggregate dissociates into subunits. Aggregation in quaternary structures, then, provides an additional control mechanism over biological activity.

Additional Problems

17.20. Give a definition or illustration of each of the following terms:

a. peptide bond
b. dipolar ion
c. dipeptide
d. L configuration of amino acids
e. essential amino acid
f. amino acid with a nonpolar R group
g. amino acid with a polar R group
h. amphoteric compound
i. isoelectric point
j. ninhydrin

17.21. Draw a Fischer projection for L-leucine. What is the priority order of the groups attached to the chiral center? What is the absolute configuration, *R* or *S*?

17.22. Write Fischer projection formulas for
a. L-phenylalanine. **b.** L-proline.

17.23. Illustrate the amphoteric nature of amino acids by writing an equation for the reaction of alanine in its dipolar ion form with one equivalent of
a. hydrochloric acid. **b.** sodium hydroxide.

17.24. Write the formula for each of the following in its dipolar ion form:
a. valine **b.** serine **c.** proline **d.** tyrosine

17.25. Locate the most acidic proton in each of the following species, and draw the structure of the product formed by reaction with one equivalent of base (OH^-).

a. $HOOC—CH_2CH_2\underset{\overset{|}{^+NH_3}}{CH}CO_2H$ **b.** $HOCH_2—\underset{\overset{|}{^+NH_3}}{CH}CO_2^-$

c. $(CH_3)_2CH\underset{\overset{|}{^+NH_3}}{CH}CO_2H$ **d.** 

17.26. What species is obtained by adding a proton to each of the following?

a. $CH_3\underset{\overset{|}{OH}}{CH}—\underset{\overset{|}{^+NH_3}}{CH}CO_2^-$ **b.** $^-O_2CCH_2\underset{\overset{|}{^+NH_3}}{CH}—CO_2^-$

17.27. Protonated alanine, $CH_3CH(\overset{+}{N}H_3)CO_2H$, has a p$K_a$ of 2.34, whereas propanoic acid, $CH_3CH_2CO_2H$, has a pK_a of 4.85. Explain the increase in acidity due to replacing an α hydrogen with an $—\overset{+}{N}H_3$ substituent.

17.28. The pK_a's of glutamic acid are 2.19 (the α carboxyl group), 4.25 (the other carboxyl group), and 9.67 (the α ammonium ion). Write equations for the sequence of reactions that occurs when base is added to a strongly acidic (pH = 1) solution of glutamic acid.

17.29. Tyrosine shows three pK_a's, at 2.20, 9.11 and 10.07. Write an equation (similar to eq. 17.5) to show the equilibria between the various forms of this amino acid.

17.30. The pK_a's of arginine are 2.17 for the carboxyl group, 9.04 for the ammonium ion, and 12.48 for the guanidinium ion. Write equations for the sequence of reactions that occurs when acid is gradually added to a strongly alkaline solution of arginine.

17.31. Draw the structure of histidine at pH 1, and show how the positive charge in the second basic group (the imidazole ring) can be delocalized.

17.32. Predict the direction of migration in an electrophoresis apparatus of each component in a mixture of asparagine, histidine, and aspartic acid at pH 6.

17.33. Write equations for the reaction of alanine with
a. CH_3CH_2OH + HCl. **b.** C_6H_5COCl + base. **c.** acetic anhydride.

17.34. Write equations for the following reactions:
a. serine + excess acetic anhydride →
b. tyrosine + bromine →
c. threonine + excess benzoyl chloride →
d. glutamic acid + excess methanol + HCl →
e. glutamine + aqueous NaOH + heat →

17.35. Write the equations that describe what occurs when phenylalanine is treated with ninhydrin.

17.36. Write structural formulas for the following peptides:
a. alanylalanine **b.** valyltryptophan
c. tryptophanylvaline **d.** glycylalanylglycine
e. serylleucylarginine **f.** histidylglycylglycylglutamic acid

17.37. Write an equation for the hydrolysis of
a. leucylserine. **b.** serylleucine. **c.** valyltyrosylmethionine.

17.38. Write an equation for the acid-catalyzed hydrolysis of the artificial sweetener aspartame (page 445).

17.39. Write formulas that show how the structure of alanylglycine changes as the pH of the solution changes from 1 to 10. Estimate the pI (isoelectric point) of this dipeptide.

17.40. Use the three-letter abbreviations to write out all possible tetrapeptides containing one unit each of glycine, alanine, valine, and leucine. How many structures are possible?

17.41. Write the structure of the product expected from the reaction of glycylcysteine with a mild oxidizing agent, such as hydrogen peroxide.

17.42. Write equations for the following reactions of Sanger's reagent:
a. 2,4-dinitrofluorobenzene + glycine →
b. excess 2,4-dinitrofluorobenzene + lysine →

17.43. A pentapeptide was converted to its 2,4-dinitrophenyl (DNP) derivative, then completely hydrolyzed and analyzed quantitatively. It gave DNP-methionine, 2 moles of methionine, and 1 mole each of serine and glycine. The peptide was then partially hydrolyzed, the fragments were converted to their DNP derivatives, and each of them was hydrolyzed and analyzed quantitatively. Two tripeptides and two dipeptides isolated in this way gave the following products:
Tripeptide A: DNP-methionine and 1 mole each of methionine and glycine
Tripeptide B: DNP-methionine and 1 mole each of methionine and serine
Dipeptide C: DNP-methionine and 1 mole of methionine
Dipeptide D: DNP-serine and 1 mole of methionine
Deduce the structure of the original pentapeptide, and explain your reasoning.

17.44. Write out the mechanism for the first step in an Edman degradation (Figure 17.6), the reaction of an N-terminal amino acid with phenyl isothiocyanate.

17.45. Write the equations for the removal of one amino acid from the peptide alanylglycylvaline by the Edman method. What is the name of the remaining dipeptide?

17.46. The second step in an Edman degradation is mechanistically trickier than it appears. It begins with nucleophilic attack by the sulfur on the carbonyl carbon of the adjacent peptide bond to form a thiazolinone, which rearranges in aqueous acid to the N-phenylthiohydantoin. Write equations that show these steps.

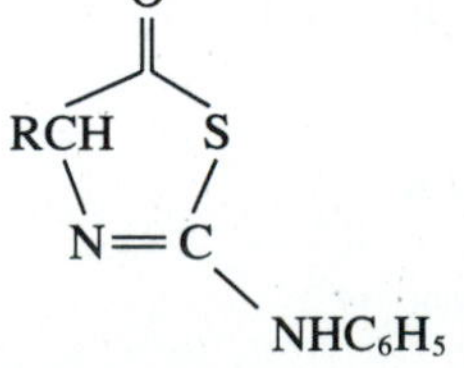

```
      O
      ‖
   ／   ＼
RCH      S
  \     /
   N═C
      \
       NHC6H5
```

a thiazolinone

17.47. Insulin (Figure 17.7), when subjected to the Edman degradation, gives *two* phenylthiohydantoins. From which amino acids are they derived? Draw their structures.

17.48. The following compounds are isolated as hydrolysis products of a peptide: Ala—Gly, Tyr—Cys—Phe, Phe—Leu—Try, Cys—Phe—Leu, Val—Tyr—Cys, Gly—Val, and Gly—Val—Tyr. Complete hydrolysis of the peptide shows that it contains one unit of each amino acid. What is the structure of the peptide, and what are its N- and C-terminal amino acids?

17.49. Simple pentapeptides called *enkephalins* are abundant in certain nerve terminals. They have opiatelike activity and are probably involved in organizing sensory information pertaining to pain. An example is *methionine enkephalin,* Try—Gly—Gly—Phe—Met. Write out its complete structure, including all the side chains.

17.50. Angiotensin II is an octapeptide with vasoconstrictor activity. Complete hydrolysis gives one equivalent each of Arg, Asp, His, Ile, Phe, Pro, Tyr, and Val. Reaction with Sanger's reagent gives, after hydrolysis,

$$2,4\text{-}(O_2N)_2C_6H_3\text{—}NHCH(CH_2CO_2H)CO_2H$$

and seven amino acids. Treatment with carboxypeptidase gives Phe as the first released amino acid. Treatment with trypsin gives a dipeptide and a hexapeptide, whereas with chymotrypsin, two tetrapeptides are formed. One of these tetrapeptides, by Edman degradation, had the sequence Ile—His—Pro—Phe. From these data, deduce the complete sequence of angiotensin II.

17.51. *Endorphins* were isolated from the pituitary gland in 1976. They are potent pain relievers. β-Endorphin is a polypeptide containing 32 amino acid residues. Digestion of β-endorphin with trypsin gave the following fragments:

Lys
Gly—Gln
Asn—Ala—His—Lys
Asn—Ala—Ile—Val—Lys
Tyr—Gly—Gly—Phe—Leu—Met—Thr—Ser—Glu—Lys
Ser—Gln—Thr—Pro—Leu—Val—Thr—Leu—Phe—Lys

From these data only, what is the C-terminal amino acid of β-endorphin?
Treatment with cyanogen bromide gave the hexapeptide
Tyr—Gly—Gly—Phe—Leu—Met
and a 26-amino acid fragment. From these data only, what is the N-terminal amino acid of β-endorphin? Digestion of β-endorphin with chymotrypsin gave, among other fragments, a 15-unit fragment identified as
Leu—Met—Thr—Ser—Glu—Lys—Ser—Gln—Thr—Pro—Leu—Val—Thr—Leu—Phe

You should now be able to locate 22 of the 32 amino acid units. Write out as much as you can of the sequence. What further information do you need to complete the sequence?

17.52. The attachment of the N-protected C-terminal amino acid to the polymer in solid-phase peptide synthesis (Figure 17.8) is an S_N2 displacement reaction. What is the nucleophile? What is the leaving group? Write an equation that clearly shows the reaction mechanism.

17.53. The detachment of the peptide chain from the polymer in solid phase peptide synthesis (Figure 17.8) occurs by an acid-catalyzed S_N2 mechanism. Write an equation to show this mechanism.

17.54. Write out all the steps in a Merrified solid-phase synthesis of Leu—Pro.

17.55. Write the structure for glycylglycine, and show the resonance contributors to the peptide bond. At which bond is rotation restricted?

17.56. *Glucagon* is a polypeptide hormone secreted by the pancreas when the blood sugar level is low. It increases the blood sugar level by stimulating the breakdown of glycogen in the liver. The primary structure of glucagon is

His—Ser—Glu—Gly—Thr—Phe—Thr—Ser—Asp—Tyr—Ser—Lys—Tyr—Leu—
Asp—Ser—Arg—Arg—Ala—Gln—Asp—Phe—Val—Gln—Trp—Leu—
Met—Asn—Thr

What fragments would you expect to obtain from digestion of glucagon with:

a. trypsin **b.** chymotrypsin

17.57. In a globular protein, which of the following amino acid's side chains are likely to point toward the center of the structure? Which will point toward the surface when the protein is dissolved in water?

a. arginine **b.** phenylalanine **c.** isoleucine
d. glutamic acid **e.** asparagine **f.** tyrosine

CHAPTER 18

Nucleotides and Nucleic Acids

18.1 Introduction

DNA, the double helix, and the genetic code—through the media's popularization of science, these have become household words. And they represent one of the greatest triumphs ever for chemistry and biology.

In this chapter, we will describe the structure of the nucleic acids, DNA and RNA. We will first look at their building blocks, the nucleosides and nucleotides, and then describe how these building blocks are linked together to form giant nucleic acid molecules. Later we will consider the three-dimensional structures of these vital biopolymers and how the information they contain (the genetic code) was unraveled.

18.2 The General Structure of Nucleic Acids

Nucleic acids are linear, chainlike macromolecules that were first isolated from cell nuclei. **Hydrolysis of nucleic acids gives nucleotides,** which are the building blocks of nucleic acids, just as amino acids are the building blocks of proteins. A complete description of the primary structure of a nucleic acid requires knowledge of its nucleotide sequence, comparable to knowing the amino acid sequence in a protein.

Hydrolysis of a nucleotide gives 1 mole each of phosphoric acid and a nucleoside. The nucleoside can be hydrolyzed further, to 1 equivalent each of a sugar and a heterocyclic base.

$$\text{nucleic acid} \xrightarrow[\text{enzyme}]{H_2O} \underset{\text{(phosphate-sugar-heterocyclic base)}}{\text{nucleotide}}$$

$$\downarrow H_2O,\ OH^- \qquad (18.1)$$

$$\underset{\text{base}}{\text{heterocyclic}} + \text{sugar} \xleftarrow[H^+]{H_2O} \underset{\text{(sugar-base)}}{\text{nucleoside}} + H_3PO_4$$

The overall structure of the nucleic acid itself, then, is a macromolecule with a backbone of sugar molecules connected by phosphate links and with a base attached to each sugar unit.

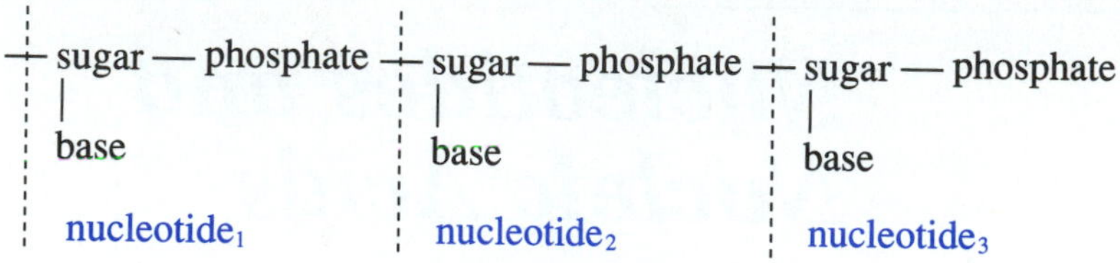

schematic structure of a nucleic acid

18.3 Components of Deoxyribonucleic Acid (DNA)

Complete hydrolysis of DNA gives phosphoric acid, a single sugar, and a mixture of four heterocyclic bases. The sugar is **2-deoxy-D-ribose.**

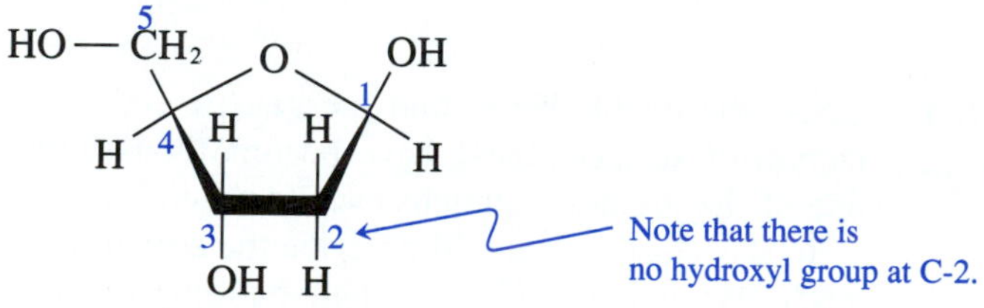

2-deoxy-D-ribose

The heterocyclic bases (Figure 18.1) fall into two categories, the pyrimidines (**cytosine** and **thymine;** review Sec. 13.4) and the purines (**adenine** and **guanine;** review Sec. 13.8). When we refer to these bases later, especially in connection with the genetic code, we will use the first letters of their names (capitalized) as abbreviations for their structures.

Now let us see how the sugar and bases are linked.

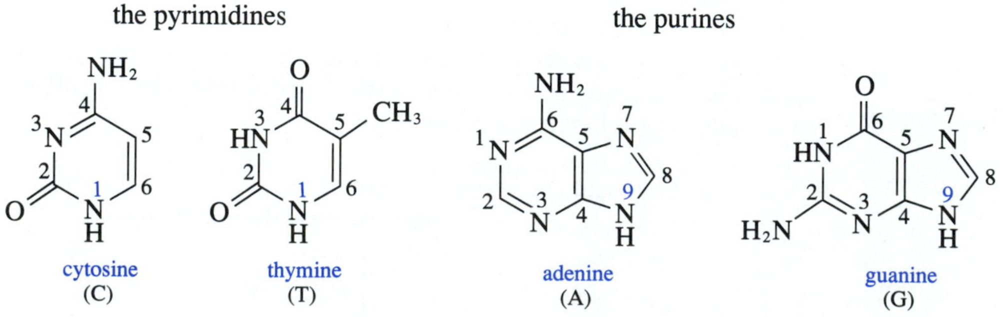

cytosine (C)

thymine (T)

adenine (A)

guanine (G)

FIGURE 18.1 The DNA bases.

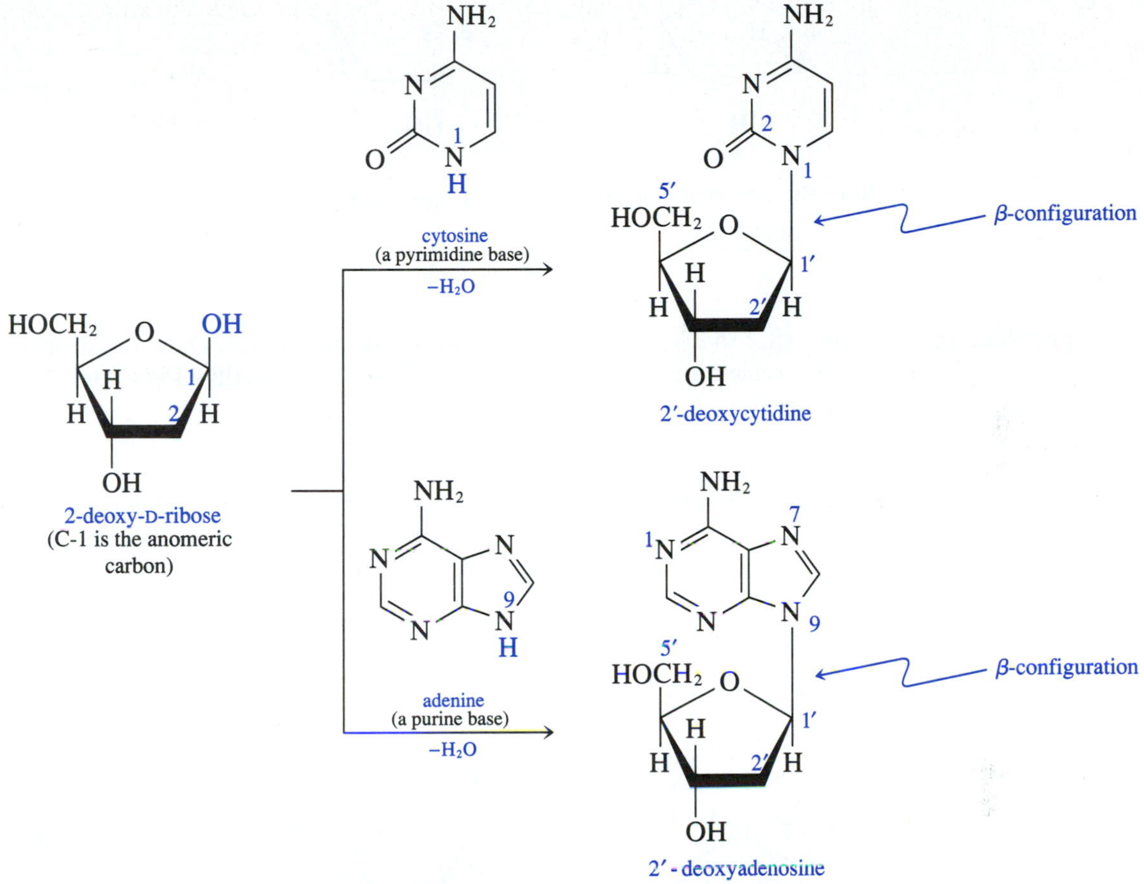

NH$_2$ N O N H 1 cytosine (a pyrimidine base) $-H_2O$

NH$_2$ N 2 O N 1 5′ HOCH$_2$ O H 1′ H 2′ H OH ← β-configuration

2′-deoxycytidine

HOCH$_2$ O OH H 1 H 2 H OH

2-deoxy-D-ribose (C-1 is the anomeric carbon)

NH$_2$ N N 9 N N H adenine (a purine base) $-H_2O$

NH$_2$ 7 N 1 N N N 9 5′ HOCH$_2$ O H 1′ H 2′ H OH ← β-configuration

2′-deoxyadenosine

FIGURE 18.2 Schematic formation of nucleosides.

18.4 *Nucleosides*

A nucleoside is an *N-glycoside*. The pyrimidine or purine base is connected to the anomeric carbon (C-1) of the sugar. The pyrimidines are connected at N-1 and the purines at N-9 (Figure 18.2). Nucleoside structures are numbered in the same way as their component bases and sugars, except that primes are added to the numbers for the sugar part.

N-glycosides have structures similar to those of O-glycosides (Sec. 16.12). In O-glycosides, the —OH group on the anomeric carbon is replaced by —OR; in N-glycosides, that group is replaced by —NR$_2$.

EXAMPLE 18.1 Draw the structure of

a. the β-O-glycoside of 2-deoxy-D-ribose and methanol.
b. the β-N-glycoside of 2-deoxy-D-ribose and dimethylamine.

Solution a. 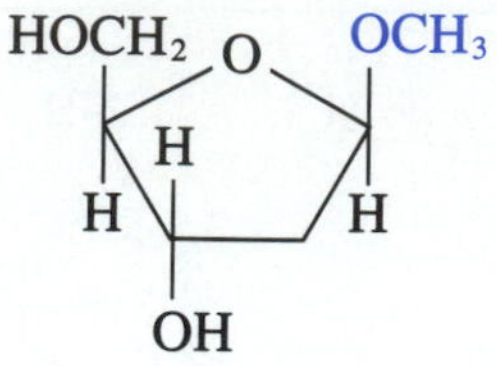 HOCH$_2$ O OCH$_3$ (H, H, H, OH) b. 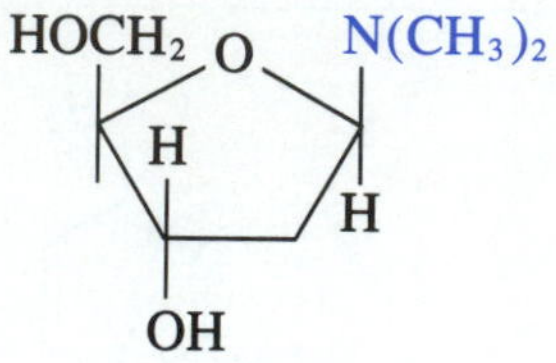 HOCH$_2$ O N(CH$_3$)$_2$ (H, H, H, OH)

Note the similarity between N- and O-glycosides.

Problem 18.1 Figure 18.2 shows the structures of two DNA nucleosides. Draw the structures for the remaining two nucleosides of DNA: 2′-deoxythymidine and 2′-deoxyguanosine.

Because of their many polar groups, nucleosides are water-soluble. Like other glycosides, they can be hydrolyzed readily by aqueous acid (or by enzymes) to the sugar and the heterocyclic base. For example,

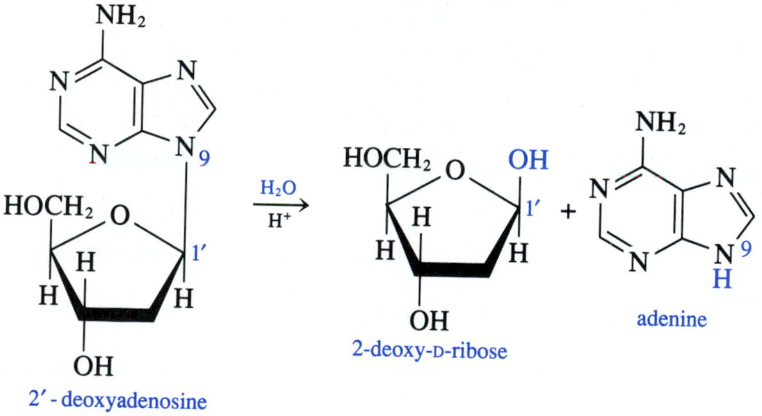

$$\text{2'-deoxyadenosine} \xrightarrow[H^+]{H_2O} \text{2-deoxy-D-ribose} + \text{adenine} \qquad (18.2)$$

2′-deoxyadenosine

2-deoxy-D-ribose

adenine

Problem 18.2 The mechanisms for hydrolysis of O- and N-glycosides are similar. Write out the mechanistic steps for eq. 18.2. Begin by adding a proton (from the acid catalyst) to N-9 of the adenine part. Keep track of the charges.

18.5 *Nucleotides*

Nucleotides are phosphate esters of nucleosides. A hydroxyl group in the sugar part of a nucleoside is esterified with phosphoric acid. In DNA nucleotides, either the 5′ or the 3′ hydroxyl group of 2-deoxy-D-ribose is esterified.

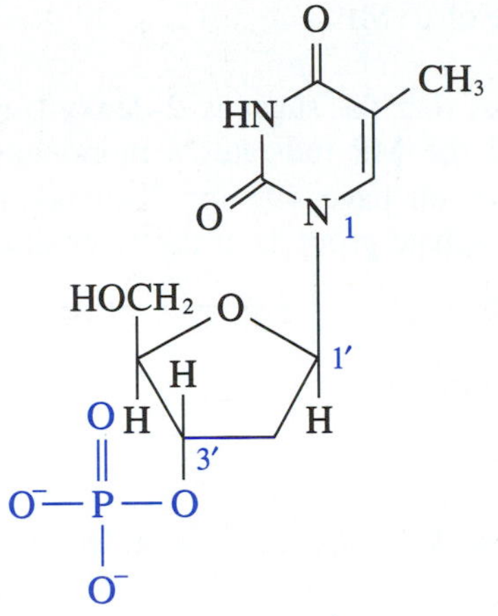

2′-deoxythymidine-3′-monophosphate

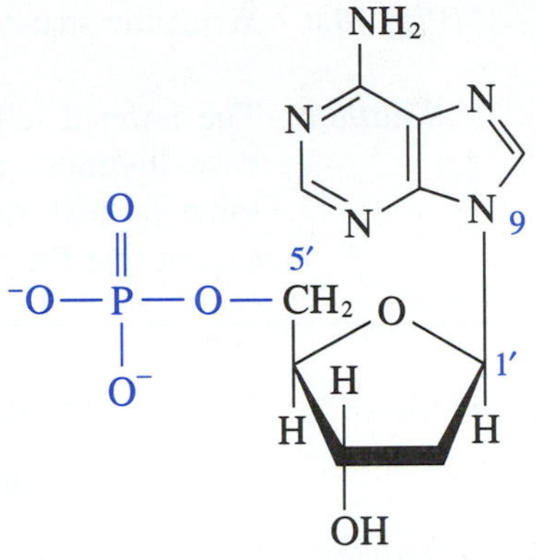

2′-deoxyadenosine-5′-monophosphate

Nucleotides are named as the 3′- or 5′-monophosphate esters of a nucleoside, as shown above. These names are frequently abbreviated as shown in Table 18.1. In these abbreviations, the d stands for 2-deoxy-D-ribose, the next letter refers to the base, and MP stands for monophosphate. (Later we will see that some nucleotides are diphosphates, abbreviated DP, or triphosphates, TP.) Unless otherwise stated, the abbreviations usually refer to the 5′-phosphates.

TABLE 18.1 The common 2-deoxyribonucleotides:

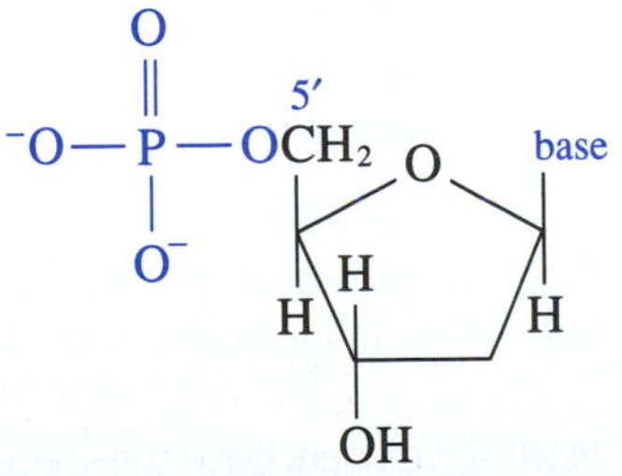

Base	*Monophosphate name*	*Abbreviation*
cytosine (C)	2′-deoxycytidine monophosphate	dCMP
thymine (T)	2′-deoxythymidine monophosphate	dTMP
adenine (A)	2′-deoxyadenosine monophosphate	dAMP
guanine (G)	2′-deoxyguanosine monophosphate	dGMP

EXAMPLE 18.2 Write the structure of dTMP.

Solution The letter d tells us that the sugar is 2-deoxy-D-ribose. The T stands for the base thymine, and the MP indicates a monophosphate. The structure is the same as that shown on page 497 for 2′-deoxythymidine-3′-monophosphate, except that the phosphate group is at the 5′ position.

Problem 18.3 Write out the structure for

a. dCMP. b. dGMP.

The phosphoric acid groups of nucleotides are acidic, and at pH 7, these groups exist mainly as dianions, as shown in the structures.

Nucleotides can be hydrolyzed by aqueous base (or by enzymes) to nucleosides and phosphoric acid. Phosphoric acid is sometimes abbreviated P_i, meaning inorganic phosphate.

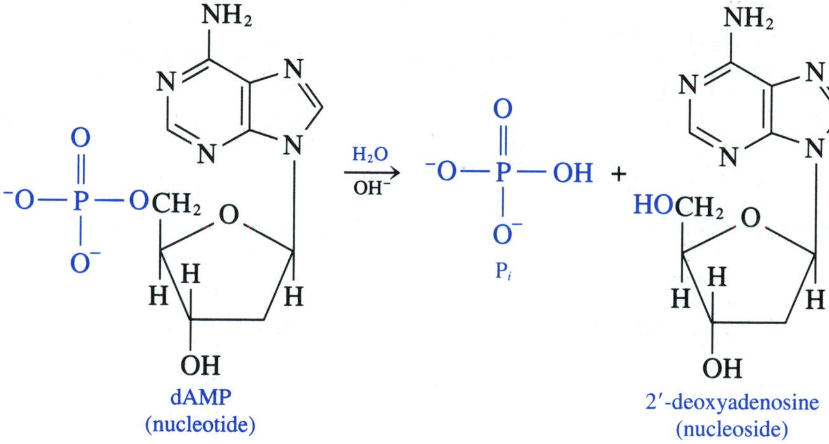

$$\text{dAMP (nucleotide)} \xrightarrow[\text{OH}^-]{\text{H}_2\text{O}} {}^-\text{O}-\text{P(=O)(O}^-)-\text{OH} \; (\text{P}_i) + \text{2}'\text{-deoxyadenosine (nucleoside)} \qquad (18.3)$$

dAMP
(nucleotide)

P_i

2′-deoxyadenosine
(nucleoside)

Problem 18.4 Write a sequence of two equations showing the stepwise hydrolysis of dTMP first to its nucleoside, then to the sugar and free base.

Now let us see how the nucleotides are linked to one another in DNA.

18.6 *The Primary Structure of DNA*

In *deoxyribonucleic acid (DNA)*, 2-deoxy-D-ribose and phosphate units alternate in the backbone. The 3′ hydroxyl of one ribose unit is linked to the 5′ hydroxyl of the next ribose unit by a phosphodiester bond. The heterocyclic base is connected to the anomeric carbon of each deoxyribose unit by a β-N-glycosidic bond. Figure 18.3 shows a schematic drawing of a DNA segment.

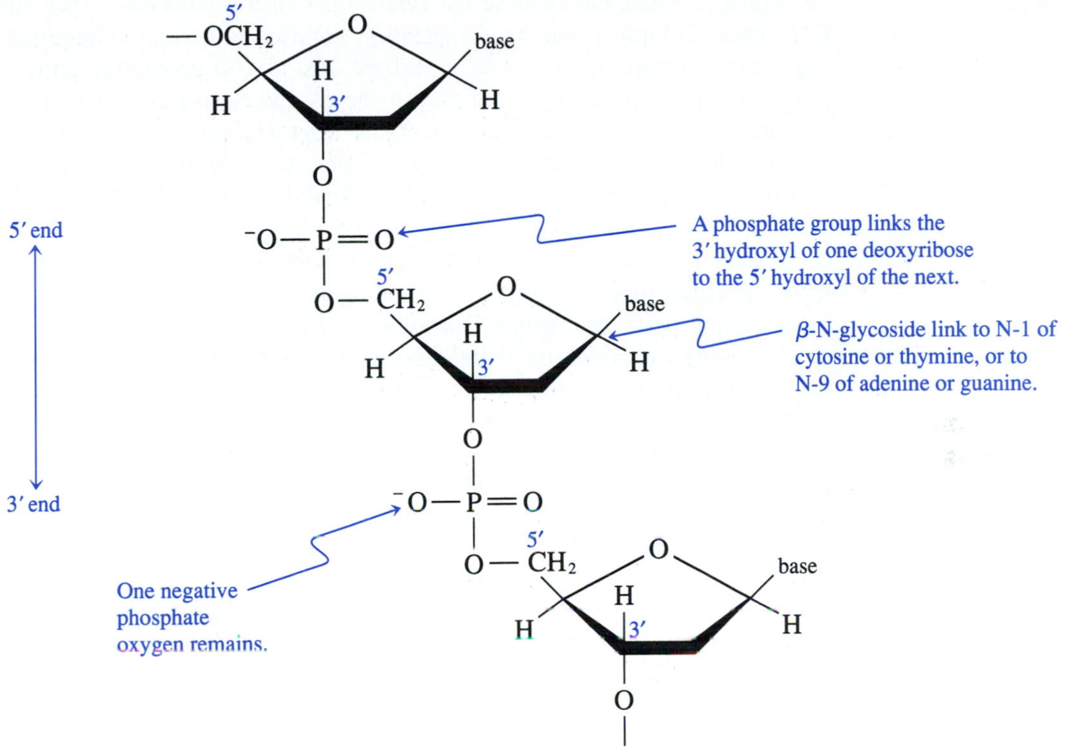

FIGURE 18.3 A segment of a DNA chain.

In DNA, there are no remaining hydroxyl groups on any deoxyribose unit. Each phosphate, however, still has one acidic proton that is usually ionized at pH 7, leaving a negatively charged oxygen, as shown in Figure 18.3. If this proton were present, the substance would be an acid; hence the name *nucleic acid*. A complete description of any particular DNA molecule, which may contain thousands or even millions of nucleotide units, would have to include the exact sequence of heterocyclic bases (A, C, G, and T) along the chain.

18.7 Sequencing Nucleic Acids

The problem of sequencing nucleic acids is in principle similar to that of sequencing proteins. At first, the job might appear to be easier because there are only 4 bases compared to 20 common amino acids. In fact, it is much more difficult. Even the smallest DNA molecule contains at least 5000 nucleotide units, and some DNA molecules may contain 1 million or more nucleotide units. To determine the exact base sequence in such a molecule is a task of considerable magnitude.

Without trying to discuss nucleic acid sequencing in detail, we can describe the strategy. First, enzymes called **restriction endonucleases,** which split the DNA chain at known four-base sequences, are used to break the huge molecule down into smaller fragments with perhaps 100 to 150 nucleotide units. These fragments are separated and labeled at the 5′ end with radioactive phosphate (containing ^{32}P). Next, these fragments are further degraded using four different, carefully controlled reaction conditions and specific reagents, each of which selectively splits the chains at a particular base, A, G, C, or T. The result is a group of pieces containing one, two, three, and more nucleotide units, each labeled at the 5′ end with ^{32}P. The products of these four parallel experiments are then subjected to gel electrophoresis (a technique similar to that used in separating peptides), which separates them depending on the number of nucleotide units they contain. If the dinucleotide were split after A, it would appear in that column; if the trinucleotide were split after G, it would appear in that column; and so on. The DNA sequence can be read directly from the positions of the radioactive spots on the electrophoresis plate.

Progress in nucleic acid sequencing has been spectacular. In 1978, the longest known nucleic acid sequences (in RNA chains, which are shorter than DNA chains) were of about 200 nucleotide units. Later, the base sequence of DNA in a virus chromosome with 5375 nucleotide units was worked out by F. Sanger, who earned his second Nobel Prize in chemistry in 1980 for this achievement. In 1977, the Maxam-Gilbert* method for sequencing was introduced; and by 1985, base sequences of more than 170,000 nucleotide units became known. With present instrumentation, several thousand nucleotide bases can be sequenced in a day. Indeed, DNA sequencing, together with a knowledge of the genetic code (Sec. 18.12), is now sometimes used to sequence very large proteins.

A program of enormous magnitude—sequencing the entire human genome—now seems possible and is being undertaken on an international scale, with the hope of achieving the goal by the year 2000. This will require coding perhaps 3 billion base pairs, an impossible dream just a decade ago, but potential reality now.

34. DNA and Crime

DNA profiling, sometimes called **DNA fingerprinting,** is one of the most powerful new techniques of forensic science. Here's how it works.

A small quantity of DNA is obtained from some source associated with the crime—semen, blood, or hair roots, perhaps associated with a rape, murder, or other violent crime. The DNA is purified, cut with established restriction enzymes, and sequenced as described in Sec. 18.7. Functional genes—those that code for enzymes, hormones, and other peptides com-

* Walter Gilbert (Harvard University) shared a Nobel Prize in 1980 with F. Sanger and P. Berg.

mon to all humans—do not vary from person to person, but these genes account for only about 5% of human DNA. The remaining DNA varies enormously from person to person, and in fact (except for identical twins) is characteristic of the individual. *The DNA in these genes can identify a single person.* And one tremendous advantage of the method is that the DNA sample can be very small, as little as a few micrograms.

There are two main uses of DNA profiling in dealing with a crime. One is to compare a suspect's profile with a sample from the scene of the crime. A few years ago, in one of the first applications of the DNA technique, this kind of pairwise comparison cleared one individual suspected of two murders and led to the conviction of another.

But DNA profiling also has the potential for use in crime investigation, through the accumulation of databases similar to those used for fingerprinting. Once that is done, search of the database should answer such questions as: Does the assailant in this crime match up with one from a previous crime? Is there a match with an individual already on file?

Despite the tremendous potential of DNA profiling, it isn't without problems; a New York murder case that came to trial in 1989 identified some of them. As with all analytical methods, DNA typing must be done carefully and with proper controls; otherwise, errors are possible, especially with band shifting, when one lane in the gel electrophoresis step runs faster or slower than another. When comparing patterns from two DNA samples, it is essential to show that the bands match in order to conclude that the samples come from the same individual. When the job is done properly and a match is obtained, the odds for identity are very long indeed—usually better than one in a hundred million. But in their zeal to apply the method, some firms have been less careful than necessary about controls, and as a result of the New York trial, the method may be "on hold" for a few years. But it is hoped that guidelines soon will be set up by independent agencies (for example in the U.S., the National Academy of Sciences) for appropriate practice of DNA profiling. Once that is done, the method will be a great boon in solving crimes.

18.8 *Laboratory Synthesis of Nucleic Acids*

It is important to be able to synthesize specific DNA segments in the laboratory, just as it is important to synthesize specific peptide segments. Short nucleotide chains of known sequence were needed, as we will see, to solve the genetic code. Longer chains of known sequence could, through genetic engineering techniques, be used to induce microorganisms to synthesize useful proteins—insulin, for example. Synthetic **oligonucleotides** of known sequence can be inserted into the DNA of microorganisms, where they can serve as templates for biological DNA synthesis.

The problems of laboratory DNA synthesis are similar to, but even greater than, those of peptide synthesis, mainly because nucleotides have more complicated structures than do amino acids. Each nucleotide has several functional groups that must be protected, and later deprotected, during synthesis. Despite these difficult chemical problems, methods for oligonucleotide synthesis have been developed in various laboratories. Using these methods, Khorana* and coworkers were able to synthesize a gene for the first time, by a combination of chemical and enzymatic means.

---

* Har Gobind Khorana (Massachusetts Institute of Technology) won the Nobel Prize in medicine, 1968.

More recently, automated gene synthesizers have been developed which operate on principles similar to the Merrifield solid-phase technique for peptides. A protected nucleotide is covalently bonded to a polymer. Other protected nucleotides are then added sequentially to the chain, using a coupling reagent. Eventually, the protecting groups are removed, and the synthetic oligonucleotide is then detached from the solid support. Fairly rapid and reliable synthesis of oligonucleotides is now possible, and further progress in this area is likely to be rapid.

Let us now proceed from the primary DNA structure, the specific base sequences, to the secondary DNA structure: the double helix and the genetic code.

18.9 Secondary DNA Structure; the Double Helix

It has been known since 1938 that DNA molecules have a discrete shape, because x-ray studies of DNA threads showed a regular stacking pattern with some periodicity. A key observation by E. Chargaff (Columbia University) in 1950 provided an important clue to the structure. Chargaff analyzed the base content of DNA from many different organisms and found that the amounts of A and T are always equal and the amounts of G and C are also equal. For example, human DNA contains about 30% each of A and T and 20% each of G and C. Other DNA sources give different percentages, but the ratios of A to T and of G to C are always unity.

The meaning of these equivalences was not evident until 1953, when Watson and Crick,* working together in Cambridge, England, proposed the double helix model for DNA. They received simultaneous supporting x-ray data for their proposal from Rosalind Franklin and Maurice Wilkins in London. The important features of their model are:

1. DNA consists of two helical polynucleotide chains coiled around a common axis.
2. The helixes are right-handed, and the two strands run in opposite directions with regard to their 3′ and 5′ ends.
3. The purine and pyrimidine bases lie *inside* the helix, in planes perpendicular to the helical axis; the deoxyribose and phosphate groups form the outside of the helix.
4. The two chains are held together by purine-pyrimidine base pairs connected by hydrogen bonds. ***Adenine is always paired with thymine, and guanine is always paired with cytosine.***
5. The diameter of the helix is 20 Å. Adjacent base pairs are separated by 3.4 Å and oriented through a helical rotation of 36°. There are therefore 10

---

*James D. Watson (Harvard University) and Francis H. Crick (Cambridge University) won the Nobel Prize in medicine in 1960.

base pairs for every turn of the helix (360°), and the structure repeats every 34 Å.

6. There is no restriction on the sequence of bases along a polynucleotide chain. The exact sequence carries the genetic information.

Figure 18.4 shows schematic models of the double helix. The key feature of the structure is the complementarity of the base pairing: **A—T** and **G—C.** Only purine–pyrimidine base pairs fit into the helical structure. There is not enough room for two purines and too much room for two pyrimidines, which would be too far apart to form hydrogen bonds. Of the purine–pyrimidine pairs, the hydrogen-bonding possibilities are best for A—T and G—C pairing.

FIGURE 18.4

Model and schematic representations of the DNA double helix. The space-filling model at the left clearly shows the base pairs in the helix interior, in planes perpendicular to the main helical axis. The center drawing shows the structure more schematically, including the dimensions of the double helix. At the right is a schematic method for showing base pairing in the two strands.

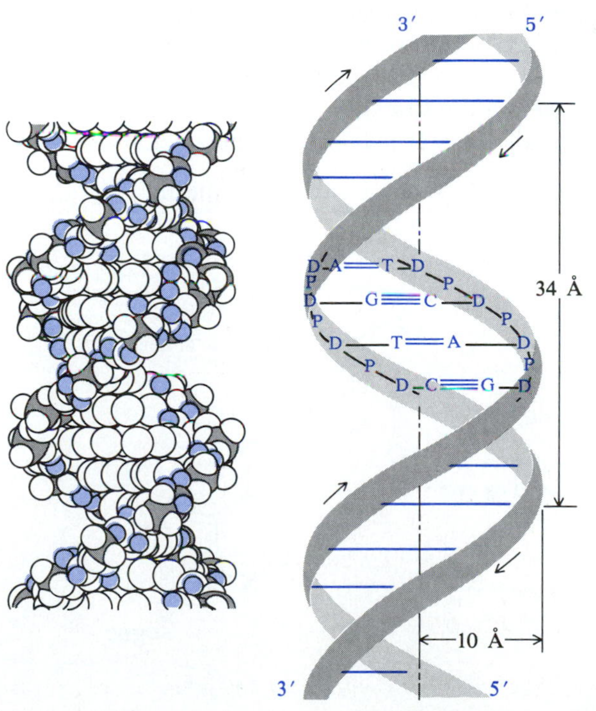

5′ ↓ 3′

A=T
G≡C
C≡G
C≡G
A=T
T=A
G≡C
T=A

3′ ↑ 5′

A=T pairs have two hydrogen bonds

G≡C pairs have three hydrogen bonds

D = deoxyribose
P = phosphate
A = adenine
T = thymine
G = guanine
C = cytosine

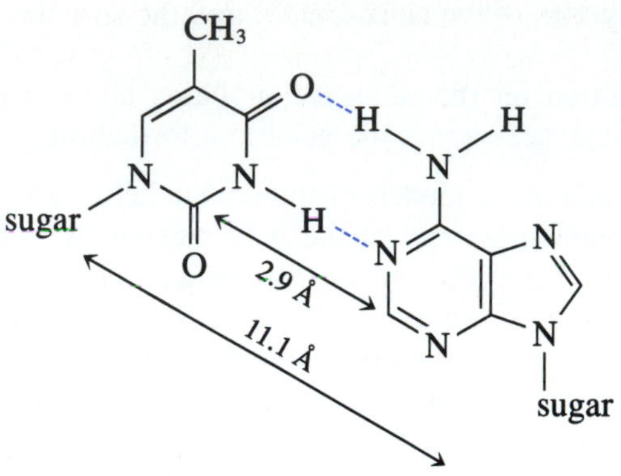
2.9 Å
11.1 Å
sugar
sugar

T — A base pair (2 H bonds)

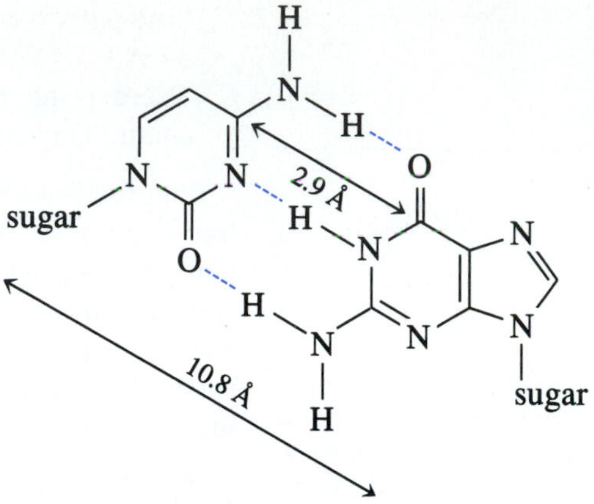
2.9 Å
10.8 Å
sugar
sugar

C — G base pair (3 H bonds)

The A — T pair is joined by two hydrogen bonds and the G — C pair by three. The geometries of the two pairs are nearly identical.

■ *EXAMPLE 18.3* Draw out the structure for the A — C pair, and explain why it is less favorable than the A — T and G — C pairs.

Solution Two possibilities for an A — C pair are

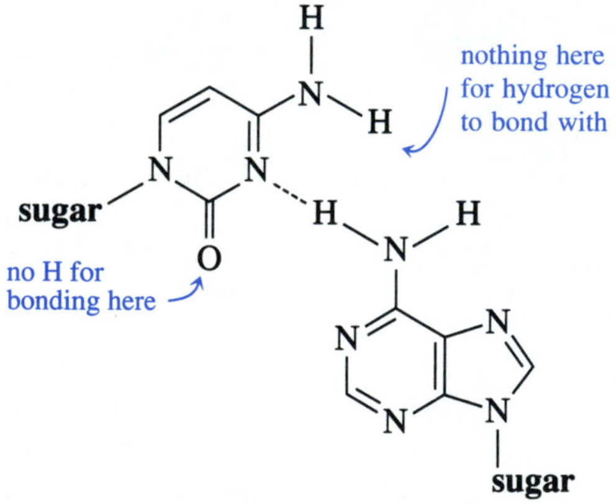
nothing here for hydrogen to bond with
sugar
no H for bonding here
sugar

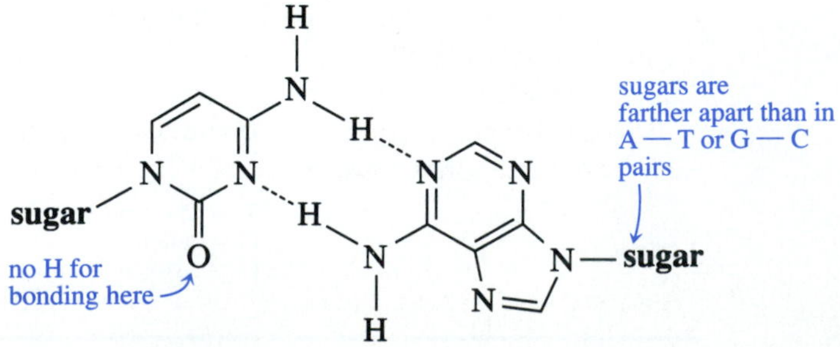
sugars are farther apart than in A — T or G — C pairs
sugar
no H for bonding here
sugar

The first structure has only one hydrogen bond. The second has two hydrogen bonds but substantially increases the sugar-to-sugar distance and would distort the helix.

Problem 18.5 Repeat Example 18.3 for the G—T pair.

Problem 18.6 Consider the following sequence of bases from one strand of DNA: —AGCCATGT— (written from 5′ to 3′). What will the sequence of bases on the other strand be?

We now know that, although the Watson–Crick model for the double helix is essentially correct, it is oversimplified. Helical conformations of DNA can now be classified into three general families, called the A-, B- and Z-forms. B-DNA, the predominant form, is the regular right-handed helix of Watson and Crick, with the base pairs essentially perpendicular to the helix axis. In the A-form, base pairs may be tilted by as much as 20° to the helix axis, and the sugar rings are puckered differently from the way they are in the B-form. And in the Z-form we see a 180° rotation of some of the bases about the C—N glycosidic bond, resulting in a *left*-handed helix.

The particular overall conformation adopted by a DNA molecule depends in part on the actual base sequence. For example, synthetic DNAs made of alternating purine-pyrimidine units have different conformations from DNAs made of blocks of purine bases followed by blocks of pyrimidine bases. Also, A—T and G—C base-pairing with different H-bonds from that originally proposed by Watson and Crick has been observed.

These variations in details of DNA structures lead, instead of the rigid helical column shown in Figure 18.4, to DNA molecules with bends, hairpin loops, supercoils, single-stranded loops, and even cruciforms in which single intrastrand H-bonded loops are extruded from the double helix. These structural changes add flexibility to the way DNA molecules are able to recognize and interact with other cellular components to perform their functions.*

18.10 DNA Replication

The beauty of the DNA double helix model was that it immediately suggested a molecular basis for transmitting information from one generation to the next: **DNA replication.** In 1954, Watson and Crick proposed that, as the two strands of a double helix separate, a new complementary strand is synthesized from nucleotides in the cell, using one strand as a template for the other. Figure 18.5 schematically depicts the process.

Though simple in principle, replication is quite a complex process in practice. The nucleotides must be present as triphosphates (not monophosphates),

* For an excellent, readable article on this subject, see J. K. Barton, *Chem. and Eng. News* **1988,** Sept. 26, pp. 30–42.

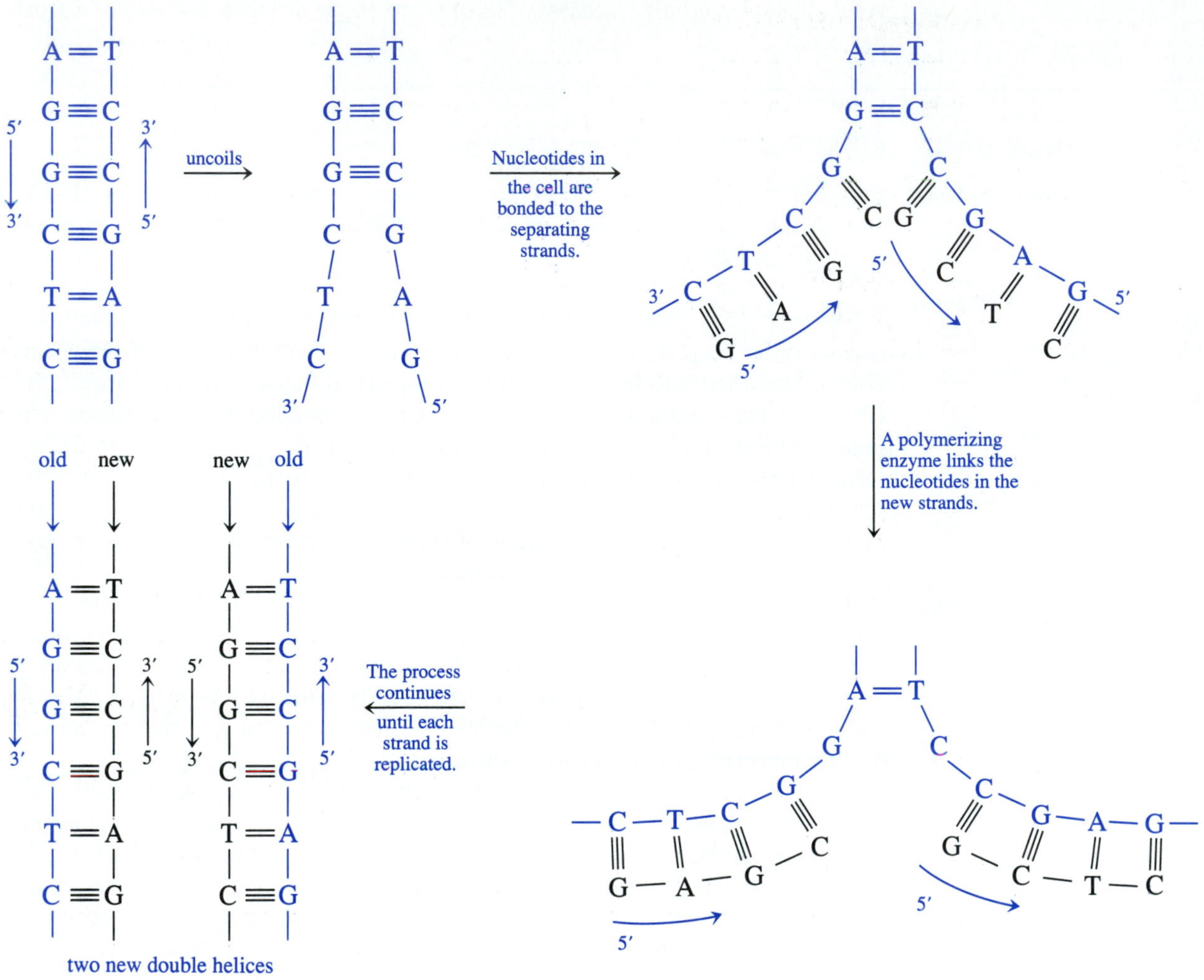

uncoils

Nucleotides in the cell are bonded to the separating strands.

A polymerizing enzyme links the nucleotides in the new strands.

The process continues until each strand is replicated.

old new new old

two new double helices

FIGURE 18.5 Schematic representation of DNA replication. As the double helix uncoils, nucleotides in the cell bond to the separate strands, following the base-pairing rules. A polymerizing enzyme links the nucleotides in the new strands to one another. Both new strands are assembled from the 5′ to the 3′ end.

an enzyme (DNA-polymerase) adds the nucleotides to a primer chain, other enzymes link DNA chains (DNA-ligase), there are specific places at which replication starts and stops, and so on. Our knowledge of the details of this process has increased considerably since the DNA double helix was proposed over three decades ago, and as in all science, it seems that the more we know, the more we need to learn.

Before we turn to the role of DNA in protein synthesis, we must first consider another type of nucleic acid, RNA, which plays a key role in this process.

18.11 Ribonucleic Acids; RNA

Ribonucleic acids (RNA) differ from DNA in three important ways: (1) the sugar is D-ribose; (2) uracil replaces thymine as one of the four heterocyclic bases; and (3) most RNA molecules are single-stranded, although helical regions may be present by looping of the chain back on itself.

The RNA sugar D-ribose differs from the DNA sugar in that it has a hydroxyl group at C-2. Otherwise, the nucleosides and nucleotides of RNA have structures similar to those of DNA.

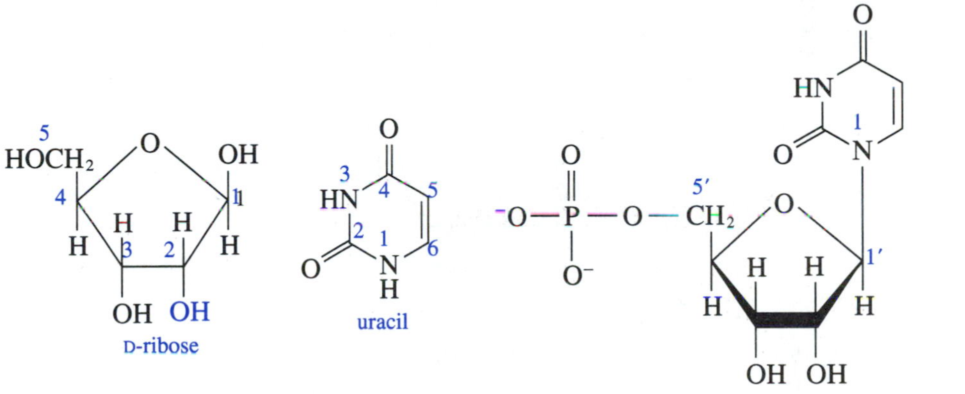

D-ribose

uracil

uridine-5′-monophosphate (UMP)

Uracil differs from thymine only in lacking the C-5 methyl group. Like thymine, it forms nucleotides at N-1, and their names are similar to those in Table 18.1. In the abbreviations of these names, the d (which stands for deoxyribose) is omitted because the sugar is ribose.

Problem 18.7 Draw the full structure of

a. AMP.
b. the RNA trinucleotide UCG (the 5′ and 3′ hydroxyl groups are again used in linking the mononucleoside units).

Cells contain three major types of RNA. ***Messenger RNA*** *(mRNA) is involved in* ***transcription*** *of the genetic code and is the template for protein synthesis.* There is a specific *m*RNA for every protein synthesized by the cell. The base sequence of *m*RNA is complementary to the base sequence in a single strand of DNA, with U the complement of A.

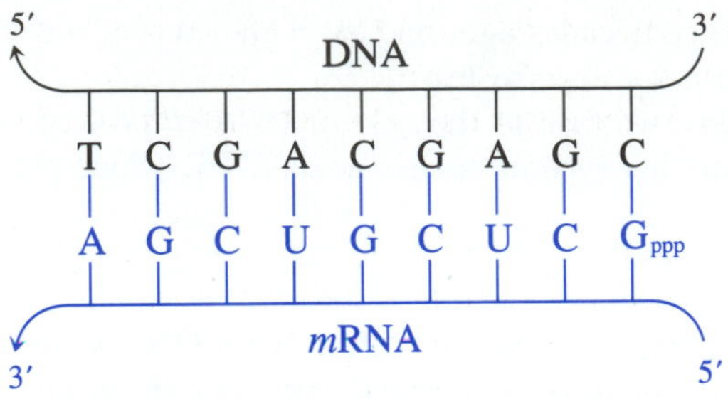

Transcription proceeds in the 3′-to-5′ direction along the DNA template. That is, the *m*RNA chain grows from its own 5′ end. The 5′ terminal nucleotide in *m*RNA is usually present as a triphosphate, not a monophosphate, and is commonly pppG or pppA. An enzyme called RNA-polymerase is essential for transcription. Usually only one strand of DNA is transcribed. It contains base sequences, called *promoter sites,* which initiate transcription. It also contains certain termination sequences, which signal the completion of transcription.

At the 3′ end of *m*RNA, there is usually a special sequence of about 200 successive nucleotide units of the same base, adenine. This sequence plays a role in transporting the *m*RNA from the cell nucleus to the *ribosomes,* the cellular structures where proteins are synthesized.

Transfer RNA *(tRNA) carries amino acids in an activated form to the ribosome for peptide bond formation,* in a sequence determined by the *m*RNA template. There is at least one *t*RNA for each of the 20 amino acids. Transfer RNA molecules are relatively small as nucleic acids go, with about 70 to 90 nucleotide units. Each *t*RNA has a three-base sequence, C—C—A, at the 3′ hydroxyl end, where the amino acid is attached as an ester. Each *t*RNA also has an **anticodon loop** quite remote from the amino acid attachment site. This loop contains seven nucleotides, the middle three of which are complementary to the three-base code word on the *m*RNA for that particular amino acid.

The third type of RNA is **ribosomal RNA** (*r*RNA). It comprises about 80% of the total cellular RNA (*t*RNA = 15%, *m*RNA = 5%) and is the main component of the ribosomes. Its molecular weight is large, and each molecule may contain several thousand nucleotide units.

Until recently, it was thought that all enzymes are proteins. But this dogma of biochemistry was recently overthrown by the discovery that some types of RNA can function as biocatalysts. They can cut, splice and assemble themselves without outside help of conventional enzymes. This discovery* of **ribozymes,** as they are called, had a major impact on theories of the origin of life. The question was: Which came first in the primordial soup from which life began, the proteins or the nucleic acids? Proteins could be enzymes and catalyze the reactions needed for life, but they could not store genetic information. The reverse was thought to be true for nucleic acids. But, with the discovery of catalytic activity in certain types of RNA, it now seems almost certain that the

---

*Sidney Altman (Yale) and Thomas R. Cech (University of Colorado) received the 1989 Nobel Prize in chemistry for this discovery.

earth of 4 billion years ago was an RNA world, in which RNA molecules carried out all the processes of life without the help of proteins or DNA—even though it is the latter that now contains the genetic code.

18.12 The Genetic Code and Protein Biosynthesis

It is beyond the scope of this book to give a detailed account of the genetic code, how it was unraveled, and how over a hundred types of macromolecules must interact to translate that code into the synthesis of a protein. But we can present a few of the main concepts.

The **genetic code** is the relationship between the base sequence in DNA, or its RNA transcript, and the amino acid sequence in a protein. A three-base sequence, called a **codon,** corresponds to *one* amino acid. Because there are 4 bases in RNA (A, G, C, and U), there are $4 \times 4 \times 4 = 64$ possible codons. However, there are only 20 common amino acids in proteins. Each codon corresponds to only one amino acid, but the code is *degenerate;* that is, *several different codons may correspond to the same amino acid*. Of the 64 codons, 3 are codes for "stop" (UAA, UAG, and UGA). Each signals that the particular protein synthesis is complete. One codon, AUG, serves double duty. It is the initiator codon, but if it appears again after a chain has been initiated, it codes for the amino acid methionine. The entire code is summarized in Table 18.2.

TABLE 18.2 The genetic code; translation of the codons into amino acids

First base (5′ end)	*Second base*	*Third base (3′ end)* U	C	A	G
U	U	Phe	Phe	Leu	Leu
	C	Ser	Ser	Ser	Ser
	A	Tyr	Tyr	Stop	Stop
	G	Cys	Cys	Stop	Trp
C	U	Leu	Leu	Leu	Leu
	C	Pro	Pro	Pro	Pro
	A	His	His	Gln	Gln
	G	Arg	Arg	Arg	Arg
A	U	Ile	Ile	Ile	Met (start)
	C	Thr	Thr	Thr	Thr
	A	Asn	Asn	Lys	Lys
	G	Ser	Ser	Arg	Arg
G	U	Val	Val	Val	Val
	C	Ala	Ala	Ala	Ala
	A	Asp	Asp	Glu	Glu
	G	Gly	Gly	Gly	Gly

How was the genetic code solved? The first successful experiment was done by Marshall Nirenberg (United States 1961; Nobel Prize, 1968). Nirenberg added a synthetic RNA, polyuridine (an RNA in which all the bases were uracil, U) to a cell-free protein-synthesizing system containing all the amino acids. He found a tremendous increase in the incorporation of phenylalanine in the resulting polypeptides. Since UUU is the only codon present in polyuridine, it must be a codon for phenylalanine. Similarly, polyadenosine led to the synthesis of polylysine, and polycytidine led to polyproline. Thus AAA ≡ Lys, and CCC ≡ Pro. Later, other synthetic polyribonucleotides with known repeating sequences were found to give other polypeptides with repeating amino acid sequences, and in this way, the complete code was unraveled.

EXAMPLE 18.4 A polyribonucleotide was prepared from the tetranucleotide UAUC. When it was subjected to peptide-synthesizing conditions, the polypeptide (Tyr—Leu—Ser—Ile)$_n$ was obtained. What are the codons for these four amino acids?

Solution The polyribonucleotide must have the sequence

$$\underline{\text{U A U C}}\,\underline{\text{U A U C}}\,\underline{\text{U A U C}}\cdots$$

If we divide the chain into codons, we get

$$\left.\begin{array}{l}\text{UAU—CUA—UCU—AUC}\cdots\\ \text{Tyr — Leu — Ser — Ile}\ \ \cdots\end{array}\right\}\ \text{this sequence repeats}$$

In this way, the meaning of four codons is disclosed.

Problem 18.8 A polynucleotide made from the dinucleotide UA turned out to be (Tyr—Ile)$_n$. How does this outcome confirm the results in Example 18.4? What new information does this experiment give?

The genetic code is universal for all organisms on earth and has remained invariant through all the years of evolution. Consider what would happen if the "meaning" of a codon were changed. The result would be a change in the amino acid sequence of most proteins synthesized by that organism. Many of these changes would undoubtedly be disadvantageous. Hence there is a strong natural selection *against* changing the code. It was recently demonstrated, from a statistical analysis of *t*RNAs, that the genetic code cannot be older than 3.8 (±0.6) billion years and thus is not older than, but almost as old as, our planet.*

*Manfred Eigen (Nobel Prize, 1967) and coworkers, *Science* **1989,** *244,* 673–679.

Problem 18.9 Mutations (caused by radiation, cancer-producing agents, or other means) may replace one base with another or may add or delete a base. What would happen to the protein produced if the sequence UUU were mutated to UCU? If UCU were mutated to UCC? What advantage is there to the genetic code in having redundant codons?

Proteins are biosynthesized using *m*RNA as a template. Amino acids, each attached to its own unique *t*RNA, are brought up to the *m*RNA, where the anticodon on the *t*RNA matches up, through hydrogen bonding, with the codon on the *m*RNA. Enzymes then link the amino acids together, detach them from their *t*RNAs, and detach the *t*RNAs from the *m*RNA so that the process can be repeated.

Coordinated reactions involving many types of molecules are required for protein biosynthesis. The molecules include *m*RNA, *t*RNA, scores of enzymes, the amino acids, phosphate, and many others. In spite of these requirements, this complex process happens with remarkable speed. It is estimated that a protein containing as many as 150 amino acid units can be assembled biosynthetically in less than a minute! When we compare this with solid-phase peptide synthesis (Sec. 17.12), efficient as it is for a laboratory procedure, we see that as chemists we still have a long way to go if we are to emulate nature.

Reprinted by permission of NEA, Inc.

18.13 Other Biologically Important Nucleotides

The nucleotide structure is a part not only of nucleic acids, but also of several other biologically active substances. Some of the more important of these species are described here.

Adenosine exists in several different phosphate forms. The 5′-*mono*phosphate, *di*phosphate, and *tri*phosphate, as well as the 3′,5′-cyclic monophosphate, are key intermediates in many biological processes.

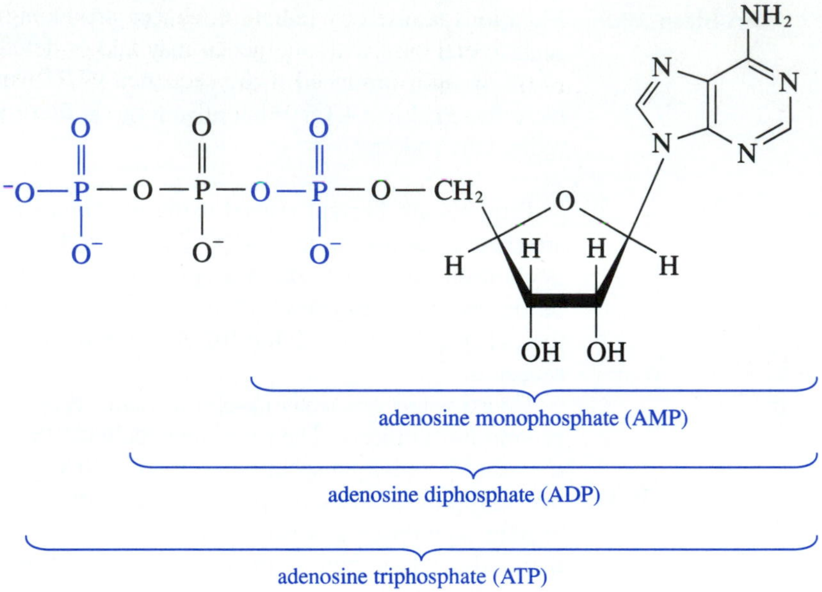

adenosine monophosphate (AMP)

adenosine diphosphate (ADP)

adenosine triphosphate (ATP)

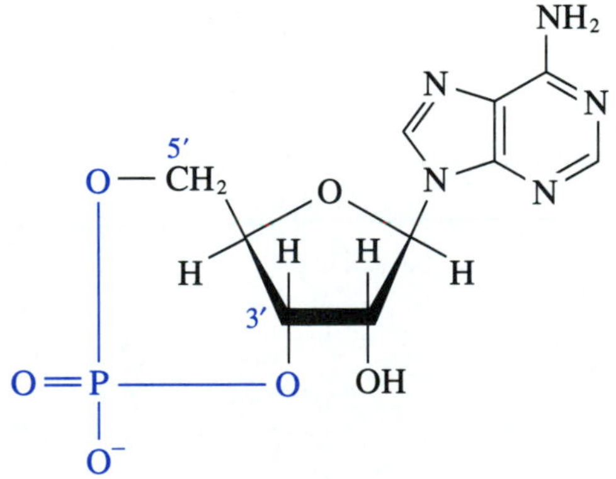

adenosine 3′,5′-cyclic monophosphate
(cyclic AMP; cAMP)

ATP contains two phosphoric anhydride bonds, and considerable energy is released when ATP is hydrolyzed to ADP and further to AMP. These reactions often provide energy for other biological reactions.

Cyclic AMP is a mediator of certain hormonal activity. When a hormone outside a cell interacts with a receptor site on the cell membrane, it may stimulate cAMP synthesis inside the cell. The cAMP in turn acts *within* the cell to regulate some biochemical process. In this way, a hormone need not penetrate a cell to exert its effect.

Four important coenzymes contain nucleotides as part of their structures. We have already mentioned **coenzyme A** (for its structure, see Figure 10.1), which contains ADP as part of its structure. It is a biological acyl-transfer agent and plays a key role in fat metabolism. **Nicotinamide adenine dinucleotide (NAD)** is a coenzyme that dehydrogenates alcohols to aldehydes or ketones, or the reverse process: reduces carbonyl groups to alcohols. It consists of two nucleotides linked by the 5′ hydroxyl group of each ribose unit.

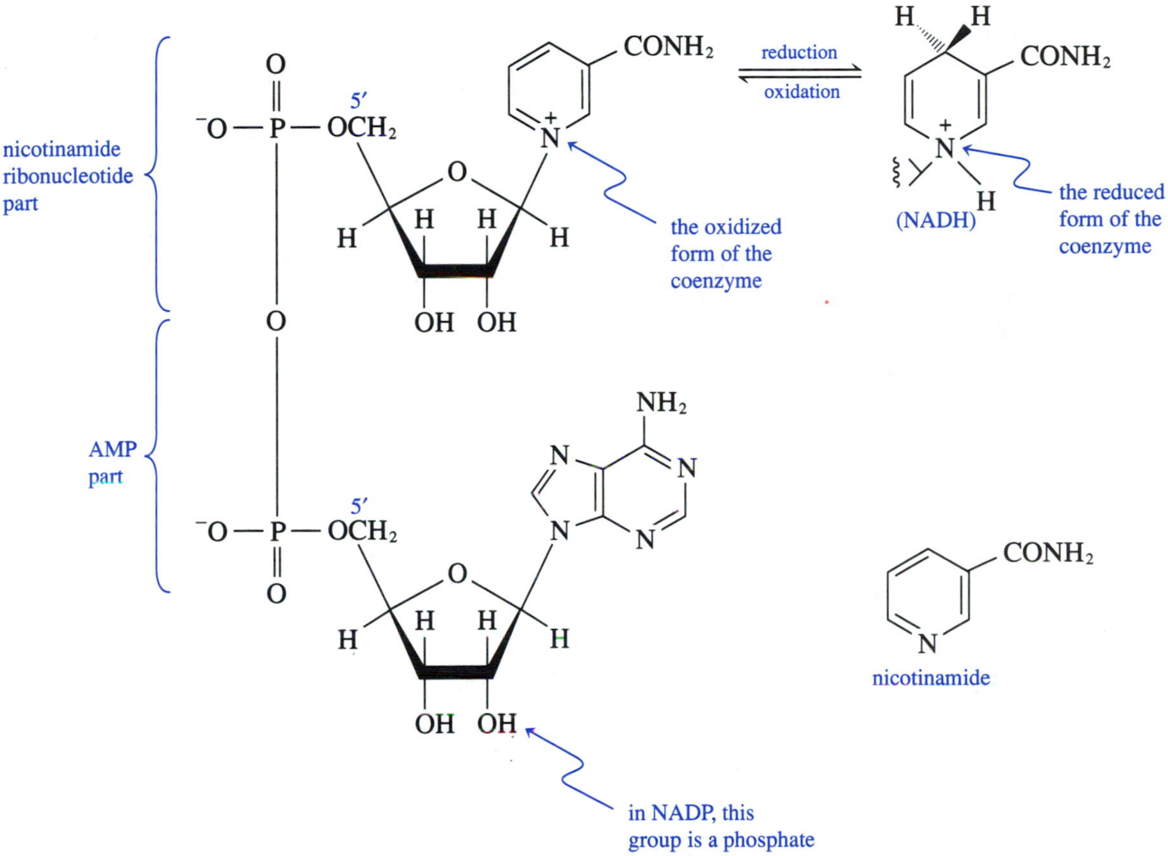

nicotinamide ribonucleotide part

AMP part

reduction

oxidation

the oxidized form of the coenzyme

(NADH)

the reduced form of the coenzyme

nicotinamide

in NADP, this group is a phosphate

nicotinamide adenine dinucleotide (NAD)

When NADP (a phosphate ester of NAD) oxidizes an alcohol to a carbonyl compound, the pyridine ring in the nicotinamide part of the coenzyme is reduced to a dihydropyridine, giving NADPH. The reverse process occurs when NADPH reduces a carbonyl compound to an alcohol. Nicotinic acid is a B vitamin needed for synthesis of this coenzyme. Its deficiency causes the chronic disease pellagra.

Flavin adenine dinucleotide (FAD) is a yellow coenzyme involved in many biological oxidation-reduction reactions. It consists of a riboflavin part (vitamin B_2) connected to ADP. The reduced form has two hydrogens attached to the riboflavin part.

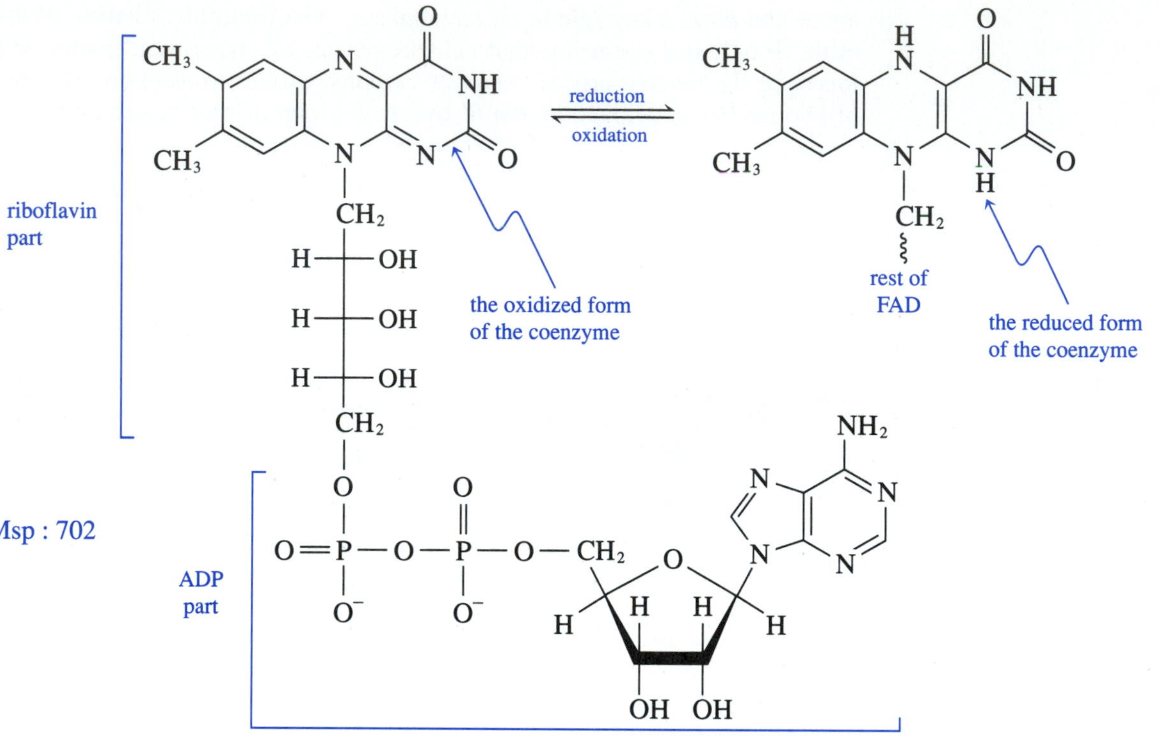

riboflavin part

reduction

oxidation

the oxidized form of the coenzyme

rest of FAD

the reduced form of the coenzyme

Msp : 702

ADP part

flavin adenine dinucleotide (FAD)

Vitamin B_{12} (cobalamine), which is essential for the maturation and development of red blood cells, is an incredibly complex molecule that includes a nucleotide as part of its structure. The related **coenzyme B_{12}** contains a second nucleotide unit. Both of these molecules have a central cobalt atom surrounded by a macrocyclic molecule containing four nitrogens, similar to a porphyrin (see "A Word About Porphyrins," p. 373). But the cobalt has two additional ligands attached to it, above and below the mean plane of the nitrogen-containing rings. One of these ligands is a ribonucleotide of the unusual base, **5,6-dimethylbenzimidazole.** The other ligand is a cyanide group in the vitamin and a 5-deoxyadenosyl group in the coenzyme. In each case, there is a direct carbon-cobalt bond. The reactions catalyzed by coenzyme B_{12} usually involve replacement of the Co—R group by a Co—H group.

Vitamin B_{12}, which is produced by certain microorganisms, cannot be synthesized by humans and must be ingested. Only minute amounts are required, but pernicious anemia can result from its deficiency.

Vitamin B_{12}, with its remarkable array of functionality and chirality, is one of the most complex molecules ever to have been created in an organic laboratory. Its synthesis was completed in 1973 by R. B. Woodward* and A. Eschenmoser and their students.

---

*Robert Burns Woodward (Harvard University) received the Nobel Prize in chemistry in 1965 for his many contributions to the "art of organic synthesis." He is regarded by many organic chemists as perhaps the greatest practitioner of this art.

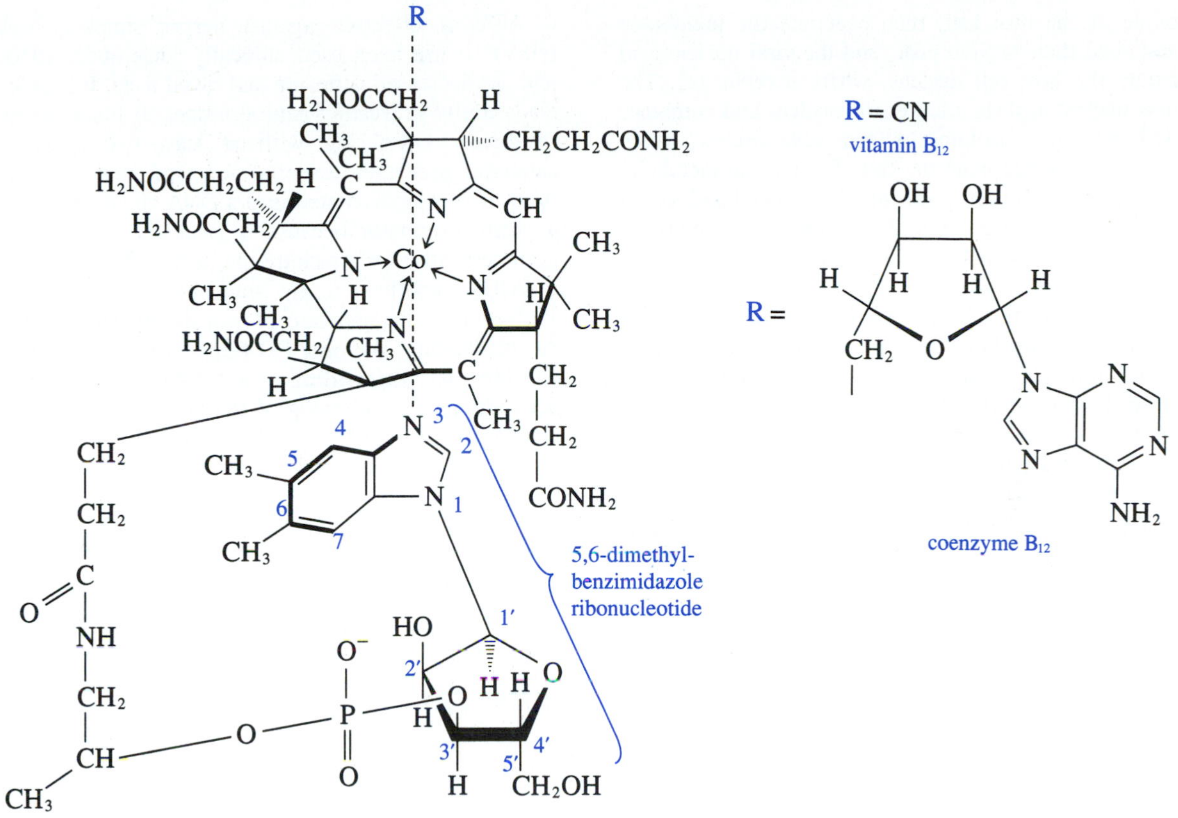

$R = CN$
vitamin B_{12}

5,6-dimethylbenzimidazole ribonucleotide

coenzyme B_{12}

A WORD ABOUT ...

35. Nucleic Acids and Viruses

Viral infections—from influenza and the common cold to the more serious herpes infections and AIDS (acquired immune deficiency syndrome)—account for about 60% of illnesses, contrasted with about 15% for bacterial infections. Since the 1940s, chemists have developed all sorts of highly effective antibiotics (sulfa drugs, penicillins, tetracyclines, and others) that are effective against bacterial infections, but progress with antiviral agents has been slower and more difficult. Why is this so, and what hope do we have for combating viral infections?

The problem is one of selectivity. Any drug must *selectively* kill pathogens in the presence of other living cells. Fortunately, there are sufficient biochemical differences between the metabolisms of bacterial and of mammalian cells to allow selectivity; thus, safe antibiotics have been developed. Viruses present a more difficult problem because, during their replicative cycle, they become physically and functionally incorporated into host cells, and one must find biochemical features that selectively attack the virus without damaging the host.

Viruses are exceedingly small; they consist of a protein coat surrounding an inner core of nucleic acid. The core carries all the genetic information for their reproduction. Broadly, there are two classes of viruses, in which the nucleic acid is either DNA (herpes viruses) or RNA (flu virus and HIV, the human immunodeficiency virus responsible for AIDS). When viruses infect a host cell, they first attach to the membrane of the host cell, then penetrate the membrane and shed their protein coat, and the viral nucleic acid enters the host cell nucleus, where it replicates. The new nucleic acid then leaves the nucleus and combines with structural proteins to form new viruses, which are then expelled from the cell. The normal metabolic machinery of the cell is essentially hijacked and forced into manufacturing the viral components instead of its own. The host cell becomes, in a sense, a virus factory.

Any one of the these steps (from attachment to the membrane to final ejection) are targets for interference by an antiviral agent. One trick, for example, might be to design a molecule very much like the DNA nucleosides, but sufficiently different for it to put a roadblock in the DNA synthesis scheme. One example of a successful antiviral agent based on this idea is **acyclovir** [9-(2-hydroxyethoxymethyl)guanine **(ACV)**, developed by Gertrude B. Elion* and coworkers at the Wellcome Research Laboratories (North Carolina). Notice that ACV mimics one of the natural nucleosides, 2′-deoxyguanosine, differing only (but importantly) in lacking part of the sugar portion of the molecule (shown in color).

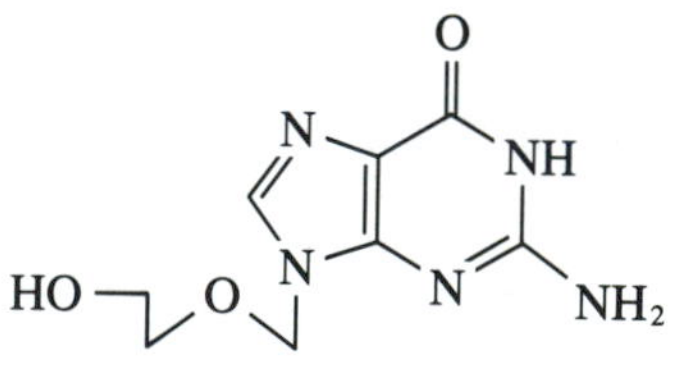

ACV

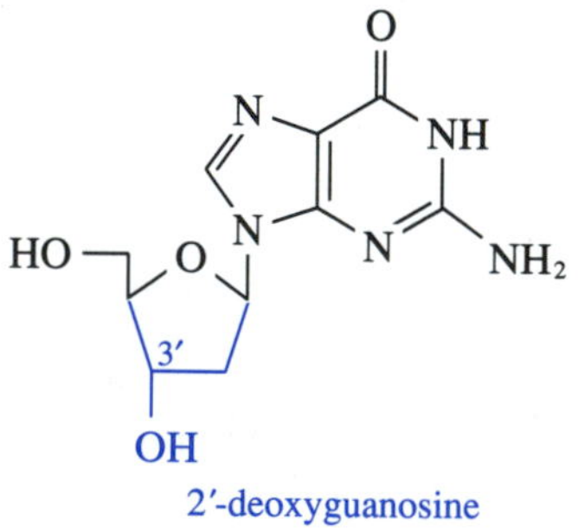

2′-deoxyguanosine

ACV is effective against herpes simplex virus (HSV). It has been used clinically since about 1980s and has decreased suffering and saved lives. It is especially useful in treating genital herpes. It functions by selectively interfering with at least two enzyme-catalyzed processes essential for viral reproduction. Also, if it is incorporated into a DNA chain, it acts as a chain terminator because it lacks the 3′-hydroxyl group essential for attaching the next nucleotide unit.

AIDS, another virally transmitted disease, has evolved into an epidemic of worldwide proportions. At this writing, the only FDA-approved drug for this purpose is **zidovudine** (or **AZT,** 3′-azido-3′-deoxythymine), an analog of the DNA nucleoside 2′-deoxythymidine.

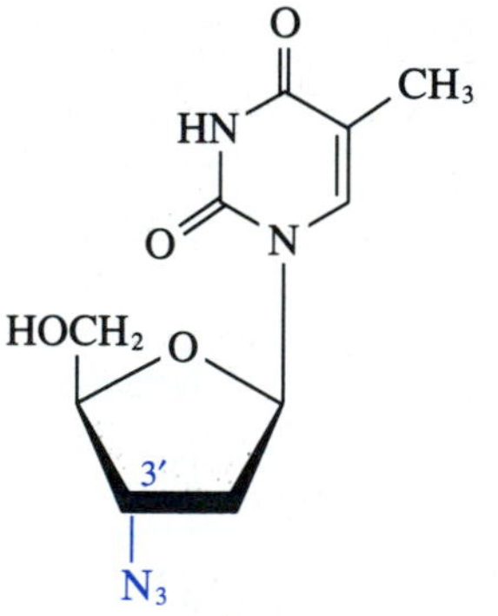

zidovudine (AZT)

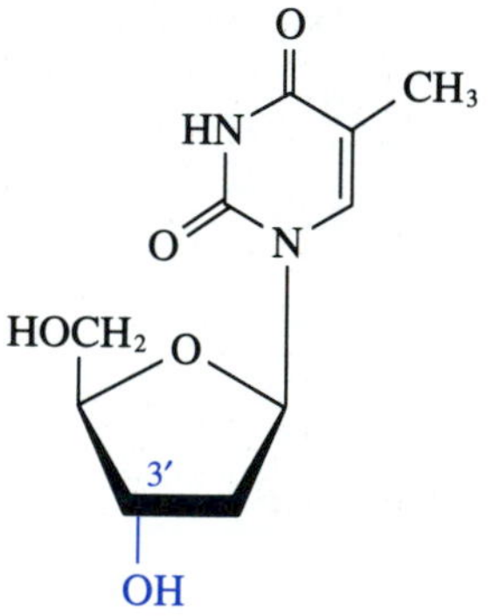

2′-deoxythymidine

AZT interferes with reverse transcriptase, the enzyme HIV uses to transcribe its RNA genome into a DNA copy. The enzyme is fooled into incorporating the drug instead of the natural nucleotide into the growing DNA chain, but the growth then stops because there is no OH group (an azido group instead) at the 3′-position. But AZT is hardly an ideal drug, due to its toxicity to bone marrow (where blood cells are produced). Also, after 12 to 18 months it tends to lose its effectiveness, perhaps because the AIDS virus becomes resistant to it. So other drugs are needed; several, such as 2′,3′-dideoxycytidine (DDC) and 2′,3′-dideoxyinosine (DDI), show considerable promise.

Many other approaches to the control of the HIV virus are also being researched. For example, CD4 is the protein on cell surfaces to which HIV binds when it infects a cell; molecules that combine CD4 with

*Gertrude B. Elion, George Hitchings, and James Black shared the 1988 Nobel Prize in physiology and medicine.

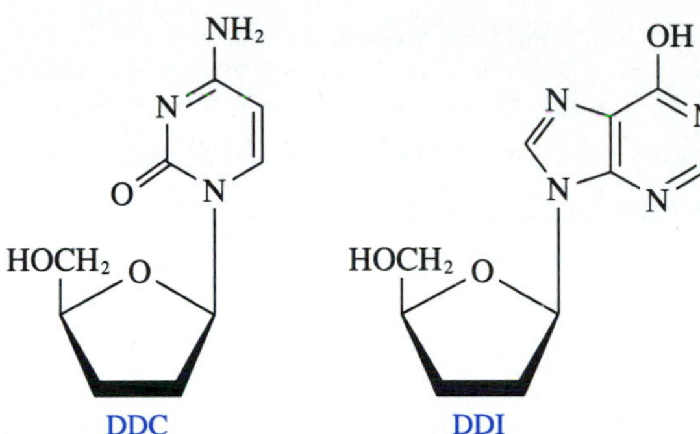

protein toxins might inhibit viral entry into cells. And vaccines are also being worked on.

One day, perhaps through application of methods like those outlined here or new ones yet undiscovered, effective antiviral agents will be developed. It is certainly one of the major medical challenges of our times.

Additional Problems

18.10. Write the structural formula for an example of each of the following:
a. a pyrimidine base **b.** a purine base **c.** a nucleoside **d.** a nucleotide

18.11. The DNA bases in Figure 18.1 can exist in other tautomeric forms. Draw all possible tautomers of cytosine.

18.12. Examine the structures of adenine and guanine (Figure 18.1). Do you expect their rings to be planar or puckered? Explain. What about the pyrimidine bases, cytosine and thymine?

18.13. Draw the structure of each of the following nucleosides:
a. cytidine (from β-D-ribose and cytosine)
b. deoxyadenosine (from β-2-deoxy-D-ribose and adenine)
c. uridine (from β-D-ribose and uracil)
d. deoxyguanosine (from β-2-deoxy-D-ribose and guanine)

18.14. Write an equation for the complete hydrolysis of adenosine-5′-monophosphate (AMP) to its component parts.

18.15. Using Table 18.1 as a guide, write the structures of the following nucleotides:
a. guanosine 5′-monophosphate **b.** 2′-deoxythymidine 5′-monophosphate

18.16. Write the steps in the mechanism for the base-catalyzed hydrolysis of AMP (eq. 18.3). What type of reaction mechanism is involved?

18.17. Draw the structures of the following DNA-derived dinucleotides:
a. A—T **b.** G—T **c.** A—C

18.18. Draw the structures of the following RNA-derived dinucleotides:
a. A—U **b.** G—U **c.** A—C

18.19. Consider the DNA-derived tetranucleotide A—A—T—C. What products will be obtained when this tetranucleotide is hydrolyzed by each of the following?
a. base **b.** base, followed by acid

18.20. Draw the structures of the following RNA components:
a. UUU **b.** UAA **c.** GCA

18.21. Draw a structure showing the hydrogen bonding between uracil and adenine, and compare it with that for thymine and adenine (page 504).

18.22. A segment of DNA contains the following base sequence:

5′ A—A—G—C—T—G—T—A—C 3′

Draw the sequence of the complementary segment, and label its 3′ and 5′ ends.

18.23. For the DNA segment in Problem 18.22, write the *m*RNA complement, and label its 3′ and 5′ ends.

18.24. Consider the following *m*RNA sequence:

5′ A—G—C—U—G—C—U—C—A 3′

Draw the segment of DNA double helix from which this sequence was derived, using the schematic method at the right of Figure 18.4. Be sure to show the 5′ and 3′ ends of each strand.

18.25. Explain how the double-helical structure of DNA is consistent with Chargaff's analyses for the purine and pyrimidine content of DNA samples from various sources.

18.26. The codon CAU corresponds to the amino acid histidine (His). How will this codon appear on the DNA strand from which it was transcribed? on the complement of that strand? Be sure to label the 5′ and 3′ directions.

18.27. Consider Table 18.2. Will any changes occur in the resultant biosynthesized protein by a purine → purine mutation in the third base of a codon? in a pyrimidine → pyrimidine mutation in the third base of a codon? If so, describe the change.

18.28. From Table 18.2, are mutations in the first or second base of a codon more or less serious than mutations at the third base?

18.29. A *m*RNA strand has the sequence

$-{}^{5'}$CCAUGCAGCAUGCCAAACUAAUUAACUAGC$^{3'}-$

What peptide would be produced? (Don't forget the start and stop codons!)

18.30. What would happen if the first U in the sequence in Problem 18.29 were deleted?

18.31. What peptide would be synthesized from the following DNA sequence?

$^{5'}$TTACCGTCTGCTGCCCCCCAT$^{3'}$

18.32. What products would you expect to obtain from the complete hydrolysis of nicotine adenine dinucleotide (NAD)? See page 513 for its structure.

18.33. Using the following formula

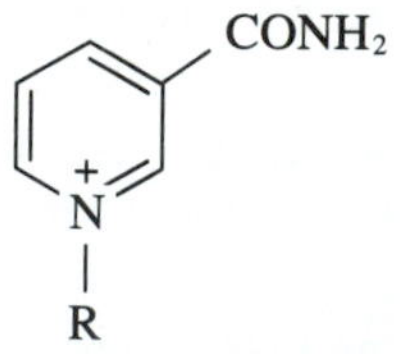

$CONH_2$

$\overset{+}{N}$

|

R

to represent NADP, write an equation for its reaction with ethanol.

18.34. UDP-glucose is an activated form of glucose involved in the synthesis of glycogen. It is a nucleotide in which α-D-glucose is esterified at C-1 by the terminal phosphate of uridine diphosphate (UDP). From this description, draw the structure of UDP-glucose.

18.35. Caffeine, the alkaloid stimulant in coffee and tea, is a purine with the following formula:

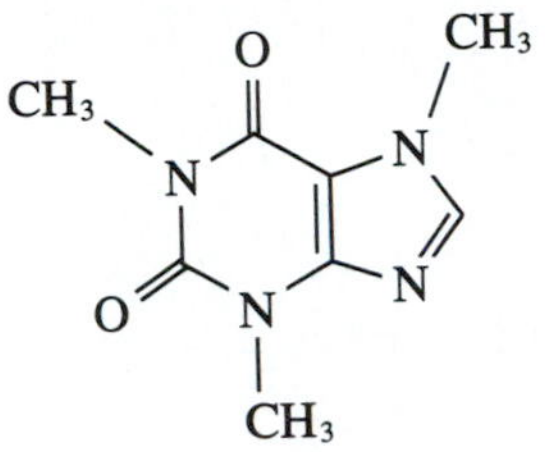

caffeine

Compare its formula with those of adenine and guanine. Do you expect caffeine to form N-glycosides with sugars such as 2-deoxy-D-ribose?

18.36. *5-Fluorouracil-2-deoxyriboside* (FUdR) is used in medicine as an antiviral and antitumor agent. From its name, draw its structure.

18.37. *Psicofuranine* is a nucleoside used in medicine as an antibiotic and antitumor agent. Its structure differs from that of adenosine only in having $—CH_2OH$ attached with α geometry at C-1′. Draw its structure.

Index